Aquatic Monocotyledons of North America

Ecology, Life History, and Systematics

Donald H. Les

CRC Press
Taylor & Francis Group
Boca Raton London New York

CRC Press is an imprint of the
Taylor & Francis Group, an **informa** business

CRC Press
Taylor & Francis Group
6000 Broken Sound Parkway NW, Suite 300
Boca Raton, FL 33487-2742

First issued in paperback 2021

ISBN 13: 978-1-138-05493-6 (hbk)
ISBN 13: 978-1-03-223587-5 (pbk)

DOI: 10.1201/b22197

Visit the Taylor & Francis Web site at
http://www.taylorandfrancis.com

and the CRC Press Web site at
http://www.crcpress.com

Publisher's Note
The publisher has gone to great lengths to ensure the quality of this reprint but points out that some imperfections in the original copies may

This book is dedicated to Elias Landolt (1926–2013), whose incredible contributions to the study of the world's smallest monocotyledons (duckweeds) represent some of the most remarkable taxonomic insights of the past century.

Contents

The Monocotyledons

Acknowledgments

My sincere thanks go out to all of the people who have influenced my study of aquatic plants. I especially recognize R.O. Belcher, E.G. Voss, J. Hiltunen, R.L. Stuckey, R.P. Wunderlin, and D.J. Crawford, who provided inestimable encouragement, inspiration, and mentorship during the earliest stages of my career. So many others, including numerous colleagues and graduate students, contributed invaluable expertise and camaraderie during the years leading to the completion of this manuscript. Finally, I thank my wife Jane, who has faithfully endured my monomaniacal pursuit of this book's completion.

Introduction

This book is intended to serve as a companion volume to *Aquatic Dicotyledons of North America* (Les, 2017) and follows a similar format by providing a comprehensive overview of the life history, ecology, and systematics of those monocotyledon species designated as obligate wetland indicators (abbreviated throughout the text as OBL) in North America. Following the previous treatment (Les, 2017), aquatic plants are defined as: "indicative taxa capable of perpetuating their life cycles and continuing their existence in still or flowing standing water or upon inundated or non-inundated hydric soils." Additional clarification should be sought in the more detailed discussion provided in the previous volume (Les, 2017).

The primary focus of this book is to provide information on the indicative taxa, which include those flowering plant species most readily recognized as being adapted to aquatic habitats. As in the companion volume, the species selected for inclusion here meet the criterion of "indicative" by their recognition as OBL wetland indicators in at least some portion of their North American range (Lichvar et al., 2016; USACE, 2016). Consequently, only those species most likely to occur in aquatic and wetland habitats are included. It is important to emphasize that the OBL category can include species that require water to complete an essential part of their life history, even though they might often appear to be growing on dryland following the subsidence of the water table.

This volume follows version 3.3 of the National wetland plant list (http://rsgisias.crrel.usace.army.mil/NWPL/), which designates more than 800 monocotyledon species as OBL North American indicators. Of these, 305 species in 97 genera, 34 families, and 11 orders have been included here in exclusion of species belonging to three graminoid families: Cyperaceae, Juncaceae, and Poaceae (see explanation below). Previous versions of the wetland plant list defined the indicator categories quantitatively, using numerical percentages of occurrence in wetlands as delimiters. However, subsequent versions of the list have abandoned that practice in favor of a qualitative assessment, which delimits the various indicator categories as follows:

1. Obligate (OBL): almost always occur in wetlands
2. Facultative wetland (FACW): usually occur in wetlands, but may occur in non-wetlands
3. Facultative (FAC): occur in wetlands and non-wetlands
4. Facultative upland (FACU): usually occur in non-wetlands, but may occur in wetlands
5. Upland (UPL): almost never occur in wetlands

The life history information summarized here for these OBL taxa is intended to facilitate their evaluation as wetland indicators with respect to the categories defined above.

Although this book focuses strictly on aquatic monocotyledonous flowering plants, it is not a taxonomic guide for aquatic plant species identification but should be regarded as a source of information on their life history, ecology, and systematics. Necessarily, one must first identify a species in order to apply this information effectively, using one of the several available manuals to facilitate species identifications (e.g., those referenced in Les, 2017). The geographical coverage of this book is restricted to North America, which is defined here as comprising the lower 48 United States, Alaska, and Canada. Because many North American aquatics also are found throughout the world, this reference also should be useful elsewhere. Except for the taxonomic outline, the following remarks essentially reiterate those of the companion volume (Les, 2017), but are repeated here for convenience.

The entries in this book are first organized as follows:

Monocotyledons

    Monocotyledons I: Early diverging monocotyledons

        A group of uncertain phylogenetic position (Acorales)
        Alismatid monocotyledons (Alismatidae)
        Aroids

    Monocotyledons II: Lilioid monocotyledons ("Liliidae")
    Monocotyledons III: Commelinoid monocots (Commelinidae)

The entries then are grouped secondarily by their respective orders (arranged alphabetically), which except for Alismatales and Arales, closely follow the APG IV classification (APG, 2016). A bracketed number follows each ordinal name to indicate the total number of families recognized. A systematic overview (with text citations) follows each order and indicates specifically those families that contain OBL indicator taxa. An overview of each family containing OBL taxa follows sequentially in alphabetical order. Family names are followed by a bracketed estimate of the total number of genera contained. An overview of each family follows (again with cited references) to summarize information on the systematics and economic importance. Each family treatment ends with an alphabetical list of genera that contain OBL species, listing the taxonomic authority for each genus.

Phylogenetic trees describing ordinal and familial relationships are included whenever available. However, it is important to emphasize that the putative relationships indicated by the organization of this book or by the phylogenetic trees included, *all represent hypotheses rather than fact* and that no assurance can be given regarding their accuracy beyond the conclusions advocated by the relevant studies cited.

The main part of the text focuses on genera and species. Each genus name is arranged alphabetically within a family and is followed (in parentheses) by a common name or names in English and French (when available), with the latter separated by a semicolon. Generic treatments follow the same format, listing first the etymology of the name, major synonyms, and the global and North American distribution and species numbers. Compilation of synonyms was facilitated by consulting *The Plant List* (2013) (www.theplantlist.org/). "In part" is used for synonyms that do not apply exclusively to the taxon. Asterisks designate nonindigenous distributional regions. Next is a list of any species designated as OBL along with any alternative regional rankings (i.e., FACW, FAC, FACU, or UPL) applied to these species. Generic and species names primarily are those accepted by ITIS (www.itis.gov/); otherwise, nomenclatural deviations are explained in the relevant "*Systematics*" section.

Following that listing is a description of habitat, which indicates salinity (freshwater to marine) and general community type. The latter includes four categories: lacustrine, palustrine, riverine, and oceanic. Lacustrine refers to those species found primarily in the deeper waters of lakes, riverine to those species preferentially occupying watercourses, and palustrine to "wetland" communities, often including those that occur along the shorelines of rivers and lakes. The pH value given next represents the complete range of values (or a qualitative assessment) reported among the individual species or is listed as "unknown" if reports were not found. The entry for "depth" also summarizes the full range of water depth associated with all the OBL species in the genus.

Life-forms are summarized for all the OBL species discussed. Herbs are herbaceous species where the leaves are scattered along an elongate stem (vittate) or extend from a common basal point (rosulate). For herbs, "emergent" refers to those species where substantial vegetative growth (shoot and leaf production) occurs above the water surface; "submersed" plants grow with their shoots and leaves underwater (excluding flowering portions in some cases); "suspended" plants are those whose shoots and leaves remain buoyant just beneath the water surface without any attachment to the substrate; "floating-leaved" plants produce leaves on the surface of water bodies and are rooted in the substrate; all structures of "free-floating" plants float on the water surface and do not attach to the substrate unless stranded in very shallow water or along shorelines. Woody plants grow only as emergents and include trees (strong ascending trunks exceeding 6 m in height) and shrubs (multiply branched, usually less than 6 m in height).

The section on "key morphology" is somewhat nontraditional in that it describes the genus with respect to the salient characteristics summarized only for the OBL species. Consequently, a measurement such as "to 50 cm" indicates that at least one OBL species in the genus is characterized by this value, even though others might be substantially smaller in stature; or potentially, some of the non-OBL members could be larger. Where several discrete ranges occur among the OBL species, they are separated by a comma, e.g. "to 5–10, 50–60 mm." The values provided indicate the longest axis of a structure, which is not necessarily its length. For monotypic genera or where a single OBL species exists, the description provided is of that species.

The life history section summarizes information for all OBL species in the genus. For duration, the relevant structure is indicated in parentheses; for fruits, an indication of frequency is indicated parenthetically as well as the presumed vectors associated with the structures involved in local and long-distance dispersal (additional details are provided within the included species treatments).

The imperilment status is summarized for each species using the data reported by NatureServe (www.natureserve.org/). The global [G] rank and regional ranks (where applicable) are listed for each OBL species using standard abbreviations for the US states and Canadian provinces; the latter are underlined for distinction. Taxa that are secure globally will include only the global [G] ranking. Because infraspecific designations are merged with a species name, the same state or province may appear in different categories when subspecies or varieties are given a different imperilment status. The NatureServe website should be consulted to clarify such discrepancies. It is important to realize that the rankings listed reflect the status when each treatment was prepared and always should be verified because they are updated continually to incorporate various taxonomic and ranking changes.

A general ecological overview of each genus is provided, which indicates the frequency of OBL taxa, and summarizes basic common features such as reproductive modes, pollinators, and seed ecology. When a genus is monotypic, information is excluded from this section and is simply summarized under the species treatment.

Each OBL species within the genus is treated successively in alphabetical order. Species names are highlighted in bold and include the names of taxonomic authorities. Habitat types reported in literature accounts or from specimen label descriptions are listed alphabetically. Some designations (e.g., "dunes") might appear unusual for aquatic plants, but refer to wet areas (pools, swales, etc.) occurring within these landforms. For additional clarification, adjectives for many of the habitat designations are provided in parentheses. For example, an entry of "prairies (mesic, wet)" would indicate that some records simply have described the habitat as "prairies," whereas others more specifically have cited "mesic prairies" or "wet prairies." Elevation data have been determined from literature reports or from specimen databases and represent the highest reliably documented values. Thus, a value of "to 3489 m" does not mean that the plants cannot occur above this specific elevation, but only that the designation represents the highest credible value found in the literature and database searches. Evidently anomalous values (due apparently to conversion errors, etc.) have been excluded at the discretion of the author. All available life history information that could be reasonably extracted from literature accounts and specimen database records is summarized. A conscientious effort has been made to locate information pertaining to reproductive ecology, pollination biology, seed germination, dispersal, and other important life history information for each species. When such information is incomplete or essentially absent for a species, a comment usually is made regarding the need for additional research, hopefully as an inspiration to others.

Where specific organisms are mentioned by their common name (e.g., "honeybees" as pollinators), additional taxonomic information (e.g., "Insecta: Hymenoptera: Apidae: *Apis*) is provided parenthetically to enable the reader to more specifically identify the intended taxon. Although it might seem obvious and/or redundant to constantly repeat groups like Aves (birds) and Insecta (insects), this convention has been followed for two reasons. First, a designation simply of "insects (Insecta)" indicates that no additional taxonomic information (e.g., order, family, etc.) was available for that particular entry at the time of consultation. Second, the use of this book by non-English speaking readers hopefully will be facilitated by including the scientific names along with the English common names, with which they might be unfamiliar.

To conserve space, none of the information summarized in the species treatments contains any literature citations; however, any literature from which information has been obtained is compiled at the end of each generic treatment under "References." Admittedly, although more space efficient, this convention does make it somewhat difficult to track down the original source of a specific entry, especially where lengthy reference lists occur.

Comprehensive lists of "Reported associates" are provided for each species, relative to the information available. The names presented here are those accepted in ITIS (www.itis.gov), except for a relatively few instances of disagreement at the author's discretion. These lists reflect the reported ecological association of any plant genus or species (including non-angiosperms) with the subject species in a variety of resources including published books and papers as well as various data repositories (see Les, 2017 for additional clarification).

Because these lists summarize data throughout the range of a species, they may or may not provide an accurate account of the likely or possible associations at a particular locality, especially for widespread species. Also, there are many factors that influence the inclusion of species in these lists. Records of imperiled taxa often exclude associated species, even when reported on database records, if those records have been "masked" to ensure site confidentiality (which is a widespread convention). Herbarium records for very weedy species often do not mention many associated species because such plants frequently grow in dense monocultures where associated species may not be evident to a collector. Sometimes the associated reports reflect a broader community that includes the habitat of a species, and it often is difficult to distinguish such reports from those where species might be growing closely together in one portion of a habitat. Consequently, these lists should be viewed as identifying those plants most likely to be found in habitats similar to that of the subject species. At the discretion of the author, highly improbable associations have been omitted, but these occurrences represented relatively few of the records evaluated. Unlike the preceding volume, the genus names of associates have not been abbreviated where multiple species occur. Although less space efficient, this convention made it easier to alphabetize the lists and more readily accommodate any necessary nomenclatural changes during the preparation of the text. Listing the full binomial also makes it easier for users of electronic text versions to search for a particular species.

Because infraspecific taxa are not reported, the accuracy of species identities depends entirely on the source from which the data were obtained. In some cases there will be inaccuracies. For instance, if a source cited only "*Xyris caroliniana*" as an associated species, it is impossible to tell whether the author of the record was referring to *Xyris caroliniana* Walter, rather than to an infraspecific taxon such as *X. caroliniana* var. *olneyi* Alph. Wood, which in this case would represent an entirely different species (i.e., *X. smalliana* Nash). Such errors are unavoidable but hopefully represent only a small fraction of those records reported.

A key rationale for including the listing of ecological associates is to provide an alternative means for evaluating the overall wetland indicator status of a particular species. By enumerating and comparing the wetland designations of all the associated species, one can potentially obtain a more accurate perspective of the typical wetland affinity of a single target species. If the associates of a particular OBL species also are typically designated as OBL, then that species itself would serve as a reliable wetland indicator; however, if the associates commonly include species ranked as facultative or upland, then that species would be a poor wetland indicator overall and a re-evaluation of its OBL status might be warranted.

Following the species accounts is a section on their use by wildlife. The species are discussed alphabetically as consistently as possible. Scientific names (at various ranks determined to provide adequate categorization) are provided for all associated organisms mentioned. In addition to published literature accounts, this information also relied on several websites, which are enumerated in the companion volume (Les, 2017).

The economic importance of species in each genus was summarized primarily from literature accounts and information provided by the Native American Ethnobotany website (http://naeb.brit.org/). **CAUTION:** *The information regarding food and medicinal uses of plant species provided in this book does not represent an endorsement of their edibility, medical efficacy, or toxicity; but simply reports information that exists in the literature.* Consequently, no wild plant ever should be used for food or medicinal purposes based solely on the information reported here.

The RHS horticultural database (http://apps.rhs.org.uk/horticulturaldatabase/) served as a primary source for information on species that are maintained regularly under cultivation. To provide additional information on weeds, the database on herbicide resistant weeds (www.weedscience.org/summary/home.aspx) was consulted.

The systematics section was compiled with the objective of evaluating recent literature on each included genus to provide reasonable syntheses of appropriate classifications, overviews of phylogenetic relationships, cytological information, hybridization, and other systematically pertinent topics. Although infraspecific taxa are not included in the individual OBL species accounts, they are discussed in this section whenever deemed relevant. Phylogenetic trees have been included whenever available and are provided to convey current (or at least recent) hypotheses regarding evolutionary relationships among the pertinent groups shown. These trees have been simplified without any indication of branch lengths, internal branch support, or other statistics, which usually can be found in the original source that is referenced in the figure caption. Because the simplified trees depicted here may differ in some respects from those presented in the original citation, the original source always should be consulted for a definitive interpretation.

A convention followed throughout the book is to indicate any taxon (at any rank) that contains at least one OBL species to be set in bold type when appearing on a tree diagram. Non-OBL taxa appearing on diagrams typically appear in lighter-shaded (50% gray) type and also display their wetland indicator status where applicable. In many cases, this representation provides a fair estimate of how many independent origins of the OBL habit have occurred in the group shown (depending on the accuracy of the given phylogenetic tree, of course). It also is important to realize that taxa occurring only outside of North America never will be designated as OBL (because they have not been ranked), even though they may very well represent true aquatic plants.

The "Distribution" section primarily summarizes the geographical distributions of each species in North America and elsewhere, occasionally adding additional important information not covered in other sections.

Lastly, each genus treatment ends with a list of the primary references from which any specific information conveyed in the preceding account was obtained. A number of "general" references (e.g., Sculthorpe, 1967) were consulted throughout the book and are not listed repetitively in each reference section.

The Monocotyledons

The term "monocotyledons" (or "monocots") refers anatomically to those flowering plants (angiosperms) having a shoot apex that is lateral to a single embryonic seed leaf, which otherwise is known as the cotyledon (Fahn, 1982). The primary root is short-lived in most of the species, resulting in a proliferation of adventitious roots, which often take on a highly fibrous appearance. All of the species lack secondary growth from a vascular cambium, but some (e.g., *Smilax laurifolia*) possess shoots that appear to be "woody" due to a cork cambium and its associated secondary tissues. Most of the species are small herbs, but some (e.g., *Acoelorraphe wrightii*) can reach arborescent stature by means of diffuse secondary growth or a secondary thickening meristem (Esau, 1977; Fahn, 1982). The shoot vasculature of monocots varies, but is organized commonly as a series of scattered bundles, which also occur in some early diverging dicotyledon groups.

A trimerous floral perianth was once thought to reliably distinguish between monocots and dicots; however, many dicot families (e.g., Annonaceae, Aristolochiaceae, Berberidaceae, Cabombaceae, Elatinaceae, Haloragaceae, Lauraceae, Lythraceae, Magnoliaceae, Myristicaceae, Papaveraceae, and Polygonaceae) also possess trimerous flowers. Monocots uniformly possess monosulcate pollen, which occurs also in some dicotyledons.

Monocotyledon leaf venation usually is described as "parallel" because of its superficial appearance as prominent, parallel vascular strands. Technically the veins all connect ultimately through smaller lateral veins, rather than ending blindly within the mesophyll as in dicot leaves (Fahn, 1982). Sheathed leaves are common in the group. Although toothed leaf margins are generally uncommon in monocots, a number of aquatic species (e.g., many Hydrocharitaceae) do possess serrate foliage. Compound leaves occur rarely (except for the families Araceae and Arecaceae), and are found in only one OBL monocot (Arecaceae: *Acoelorraphe wrightii*). Many monocots possess linear or elongate "grass-like" leaves, which are described as "graminoid" (see below).

Monocotyledons have long been regarded as a monophyletic group and numerous phylogenetic analyses have substantiated that conclusion (Judd et al., 2016). However, their conceptual systematic relationship to dicotyledons has changed considerably over the past decades as a consequence of those same analyses. The long-standing concept of monocots and dicots representing two distinct sister clades was first seriously questioned as the result of a phylogenetic analysis of plastid (*rbcL*) gene sequence data, which resolved the clade as embedded within the monosulcate dicotyledons (Chase et al., 1993). Although most molecular studies have shown similar results, they have relied on highly divergent sequences (e.g., gymnosperms or *Amborella*) as outgroups, the consequences of which are not fully understood.

One problem is that despite its association with various monosulcate dicot families, the monocot clade resolves in different positions among them. In many cases, the topology of the angiosperm clade is such that simply moving the root actually can resolve the monocots and dicots as sister clades. These factors stress a precarious reliance on divergent outgroup sequences to place the angiosperm root properly in phylogenetic analyses. In other cases, the topology rendered obviously is problematic. Notable examples include profile alignments of 18s+26s rDNA data, which resolve the anomalous dicotyledon family Ceratophyllaceae within the monocots as the sister group of Tofieldiaceae (Maia et al., 2014), and *PHYC* data or cpDNA genome sequences, which place Acoraceae among the dicotyledons (Goremykin et al., 2005; Hertweck et al., 2015). As yet, it seems best to accept the monocots as a clade, but one for which a precise phylogenetic placement within the angiosperms remains equivocal.

Plants with grass-like foliage are found throughout the monocots but occur predominately in three families (Cyperaceae, Juncaceae, and Poaceae), which are denoted here as the "graminoid monocots." Although these families are rich in wetland species (comprising about 65% of all OBL monocots), more than half of their OBL species belong to just four genera (*Carex*, *Eleocharis*, *Juncus*, and *Rhynchospora*), making it difficult to treat them while avoiding excessively redundant information. Moreover, because of their highly modified structure, the species in these families are notoriously difficult to identify, and therefore, less useful as wetland indicators to those who are not taxonomic specialists in these groups. Because their inclusion in a book of this nature (which relies on the prior identification of a species) would be of questionable utility, the three graminoid families have been excluded.

1 Monocotyledons I
Early Diverging Monocotyledons

A GROUP OF UNCERTAIN PHYLOGENETIC POSITION (ACORALES)

ORDER 1: ACORALES [1]

1.1. **Acoraceae** Martinov

Family 1.1. Acoraceae [1]

Because the Acorales contain only the single extant family Acoraceae, the two groups are discussed here simultaneously, with the characteristics of the order being those of the family. Acorales/Acoraceae (as *Acorus*) had been assigned formerly to the family Araceae (Arales). However, morphological and molecular evidence (e.g., Grayum, 1987; Duvall, 1993a) convincingly indicated that *Acorus* should be assigned to its own family (i.e., Acoraceae), a conclusion that later was substantiated by several phylogenetic analyses, which allied the group with other early diverging monocots but in various topological positions (Figure 1.1). Although many authors accept without reservation that Acorales represent the sister group of all monocotyledons, the fundamentally different placement of this clade in different phylogenetic analyses (e.g., Petersen et al., 2016) indicates that the question of its relationship to other monocots is still far from being settled.

Acoraceae contain fewer than six species, all within the genus *Acorus*. The family is distinctive and undeniably monophyletic, although no more than three of the presumed taxa (Ryuk et al., 2014) have yet been analyzed simultaneously in comparative genetic studies. All of the species are aromatic, rhizomatous perennial herbs with equitant, elongate, gladiate leaves. The leaf apex is acute and bent asymmetrically as in the genus *Iris* (Iridaceae), with which it can be confused when found in vegetative condition. The flowers of all species are bisexual, trimerous (six tepals), and arranged densely in a cylindrical spadix, which is associated with a sympodial, spathe-like leaf. The flowers are pollinated either by insects (Insecta) or by wind. The berrylike fruits have thin but leathery pericarps, which are dispersed abiotically by water.

The family is distributed widely throughout the temperate Northern Hemisphere and extends into the higher elevations of tropical Asia; some species have been introduced to the Southern Hemisphere (Thompson, 2000a).

Acorus plants have been used in traditional medicinal practices throughout many parts of the world. Numerous cultivars exist, which are planted as marginal water garden ornamentals.

The OBL North American indicators occur within one genus:

1.1.1. *Acorus* L.

1.1.1. Acorus

Ratroot, sweetflag; acore, belle-angélique
Etymology: from the Latin *ăcor*, meaning "sour taste"
Synonyms: *Calamus* Garsault [nom. illeg.]
Distribution: global: Africa; Asia; North America; **North America:** widespread
Diversity: global: 3–6 species; **North America:** 2 species
Indicators (USA): obligate wetland (OBL): *Acorus americanus, A. calamus*
Habitat: freshwater; palustrine; **pH:** 5.5–8.6; **depth:** <1 m; **life-form(s):** emergent herb
Key morphology: rhizomes (to 25 cm) branched, aromatic, pinkish, thick; leaves basal, aromatic, equitant, sessile, the blades (to 175 cm) gladiate, flattened medially, the apex acute, bent asymmetrically; inflorescence a dense, cylindrical spadix (to 9 cm), projecting laterally from a sympodial, leaflike bract (to 184 cm); flowers (to 4 mm) numerous (to 652), radial; tepals (to 4 mm) 6; stamens 6; ovary 1, 3-locular; berries (to 6 mm) obpyramidal, the pericarp thin, leathery; seeds (to 4 mm) 1–14, apical, embedded in mucilage
Life history: duration: perennial (rhizomes); **asexual reproduction:** rhizomes; **pollination:** insect or wind (uncertain); **sexual condition:** hermaphroditic; **fruit:** leathery berries (absent or common); **local dispersal:** water (rhizome fragments, seeds); **long-distance dispersal:** water (seeds)
Imperilment: 1. *Acorus americanus* [G5]; SH (DE, MT); S1 (GA, NF, NJ, PA); S2 (BC, NE, OH); S3 (AB, IL, ND, PE); **2.** *A. calamus* [G4]; S1 (CO, MS); S2 (NC); S3 (IA)
Ecology: general: All *Acorus* species are obligate wetland plants, which occur only where high substrate moisture levels are maintained. They grow in open meadow-like exposures along with a variety of other graminoid wetland plants. The pollination biology of *Acorus* is uncertain but is assumed to be mediated by insects or wind. Although the triploid taxa are sterile, fairly high fertility can occur in diploids or tetraploids. The spadices of some diploid species emit a large volume of volatile floral substances (e.g., tetrahydrofurans), which are believed to attract insects. In fertile plants, the seeds are morphophysiologically dormant and require a period of cold stratification to induce germination. They are dispersed by water and will lose viability if stored dry for extended periods. The plants reproduce vegetatively by the extension of thick rhizomes, which can fragment; the fragments also are dispersed by water, particularly along watercourses. Due to taxonomic confusion between the following two species (see *Systematics*), it is possible that some of the subsequent

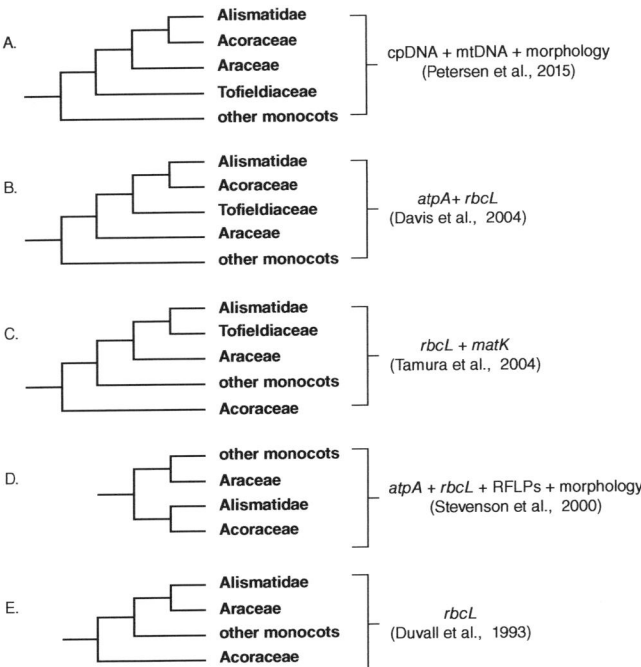

FIGURE 1.1 The groups of early diverging monocotyledons (i.e., Acoraceae, Alismatidae, Araceae, Tofieldiaceae) represent distinct clades; however, their interrelationships remain tenuous as indicated by the varied results obtained in different phylogenetic analyses (A–E). Although many authors have redefined the order "Alismatales" to include Alismatidae, Araceae, and Tofieldiaceae, such a disposition is inconsistent with several of these analyses (e.g., A, B, D). In contrast, the present treatment recognizes Alismatales as one of the two major clades resolved consistently within the alismatid monocots (Alismatidae).

information was assigned incorrectly in the literature consulted; however, care was taken to minimize that possibility.

1.1.1.1. *Acorus americanus* (Raf.) Raf. inhabits bogs, bottomlands, coves, depressions, ditches, fens, floodplains, levees, marshes, meadows, mudflats, prairies, seeps, swales, swamps (in openings), and the margins of lakes, ponds, reservoirs, rivers, streams, and swamps at elevations to 915 m. The plants can tolerate shallow standing water and grow primarily in exposures that receive full sunlight. The best growth occurs in freshwaters, with declining growth rates observed at 25–100 mM NaCl concentrations. Conductivity levels are low to moderate (10–240 µmhos cm⁻¹) with low alkalinity (5–115 mg l⁻¹ CaCO₃). Reported substrates span a range of acidity (pH: 5.5–8.6) and include clay, loam, muck, and mud. Experimentally, the plants grow optimally at pH 7.0 but decline substantially at pH levels >8. Flowering occurs from May to September, with fruiting extending into October. Although the precise pollination mechanism is unknown, each spadix can produce from 492 to 652 berries yielding a total of 1,700–2,700 seeds. The seeds are dispersed by water. They can be stored in water (at 1°C–3°C) in which viability will be retained for at least 7 months. The seeds are dormant and require a period of cold stratification and light in order to germinate. Satisfactory germination has been reported for seeds dried at 20°C–25°C, then stratified at 2°C–4°C for 4 weeks, and then incubated under a 25°C/15°C day/night

temperature regime. Under those conditions, the seedlings emerge after 10–30 days. The plants also can be cultivated readily using rhizome cuttings. The plants lack drought tolerance adaptations, but develop drought avoidance characteristics such as fewer stomata having smaller pores, and produce less biomass when growing under water-deprived conditions.

Reported associates: *Acer saccharinum, Agalinis purpurea, Alisma, Alopecurus aequalis, Amorpha fruticosa, Apocynum cannabinum, Arnoglossum plantagineum, Asclepias incarnata, Betula nigra, Bidens laevis, Boehmeria cylindrica, Bolboschoenus fluviatilis, Calamagrostis canadensis, Calla palustris, Carex aquatilis, Carex cristatella, Carex haydenii, Carex hystericina, Carex lacustris, Carex pellita, Carex stricta, Carex utriculata, Carex vesicaria, Carex vulpinoidea, Cephalanthus occidentalis, Chelone glabra, Cicuta bulbifera, Clinopodium glabrum, Comarum palustre, Cornus amomum, Cornus racemosa, Crataegus, Cyperus strigosus, Dasiphora floribunda, Deschampsia cespitosa, Dulichium arundinaceum, Echinochloa crus-galli, Eleocharis palustris, Equisetum fluviatile, Eupatorium perfoliatum, Eutrochium maculatum, Filipendula rubra, Galium obtusum, Galium triflorum,* Gentiana andrewsii, *Glyceria grandis, Glyceria striata, Gratiola aurea, Helenium autumnale, Helianthus grosseserratus, Hippuris vulgaris, Howellia aquatilis, Hypericum kalmianum, Hypericum sphaerocarpum, Impatiens capensis, Iris pseudacorus, Iris virginica, Juncus dudleyi, Juncus nodosus, Juncus torreyi, Leersia oryzoides, Liparis loeselii, Lobelia spicata, Lycopus americanus, Lycopus uniflorus, Lysimachia ciliata, Lysimachia quadriflora, Lysimachia thyrsiflora, Lythrum alatum, Lythrum salicaria, Mimulus ringens, Muhlenbergia mexicana, Myosotis scorpioides, Osmunda regalis, Parnassia glauca, Pedicularis lanceolata, Peltandra virginica, Persicaria amphibia, Persicaria coccinea, Persicaria hydropiperoides, Persicaria maculosa, Persicaria pensylvanica, Persicaria punctata, Phalaris arundinacea, Phleum, Phlox glaberrima, Physostegia virginiana, Platanthera flava,* Platanthera *leucophaea, Populus deltoides, Prenanthes racemosa, Pycnanthemum virginianum, Rumex britannica, Sagittaria cuneata, Sagittaria latifolia, Salix exigua, Salix interior, Salix nigra, Schoenoplectus acutus, Schoenoplectus tabernaemontani, Scolochloa festucacea, Scutellaria galericulata, Sium suave, Solidago gigantea, Solidago ohioensis, Solidago riddellii, Sparganium emersum, Sparganium eurycarpum, Spartina pectinata, Sphagnum, Stachys pilosa, Stellaria longifolia, Symphyotrichum ericoides, Symphyotrichum firmum, Symphyotrichum laeve, Symphyotrichum novae-angliae, Thelypteris palustris, Triadenum virginicum, Typha angustifolia, Typha latifolia, Verbena hastata, Veronicastrum virginicum, Viburnum lentago, Zizania aquatica.*

1.1.1.2. *Acorus calamus* L. is a nonindigenous species, which grows in bogs, brooks, canals, depressions, ditches, floodplains, marshes, meadows, pools, prairies, rivulets, roadsides, seeps, sloughs, springs, swales, swamps (openings), and along the margins of lakes, ponds, reservoirs, rivers, and streams at elevations to 1060 m. The plants will tolerate shallow standing water (5–45 cm deep) and primarily occupy exposures of full sunlight. The substrates, primarily silt (~49%) and sand (~37%), tend to be acidic (pH: <6) but can span a broader range (pH: 5.6–7.2). The nitrogen content of sediments tends to be low (e.g., 3.2 µg l⁻¹), but the organic matter (1–2%) and clay (e.g., 14%) content are relatively high. The substrates are described as alluvium, clay, gravel, humus, loam, muck, mud, and sand. Flowering occurs from May to October (fruiting is not observed in North America). Although sexually sterile, the plants can disperse effectively by vegetative propagules derived from rhizome fragments. These propagules are dispersed by water, wherein they can remain buoyant for at least 6 months and retain an adequate capacity (e.g., 58%) to resprout. Rhizome length has been found to correlate positively with aluminum, calcium, and magnesium levels in the sediment. The rhizomes are tolerant of flooding or drawdown conditions, light or moderate shade, and nitrogen limitation.

Reported associates (North America): *Acer saccharinum, Alisma subcordatum, Amphicarpa bracteata, Angelica atropurpurea, Asclepias incarnata, Asparagus, Boehmeria cylindrica, Calamagrostis canadensis, Caltha palustris, Carex conjuncta, Carex crus-corvi, Carex frankii, Carex grayi, Carex jamesii, Carex leavenworthii, Carex lupulina, Carex muskingumensis, Carex typhina, Celtis occidentalis, Cephalanthus occidentalis, Chasmanthium latifolium, Cladium mariscoides, Convolvulus sepium, Cyperus, Dulichium arundinaceum, Equisetum arvense, Festuca subverticillata, Fraxinus pennsylvanica, Fraxinus profunda, Glyceria striata, Gratiola aurea, Hypericum mutilum, Hypericum prolificum, Impatiens capensis, Iodanthus pinnatifidus, Iris pseudacorus, Juglans, Laportea canadensis, Leersia oryzoides, Leersia virginica, Lemna minor, Lolium perenne, Lycopus virginicus, Lysimachia ciliata, Lysimachia thyrsiflora, Mentha, Parthenocissus, Pastinaca sativa, Persicaria amphibia, Persicaria punctata, Phalaris arundinacea, Phalaris arundinacea, Phleum pratense, Pilea pumila, Poa nemoralis, Prunella vulgaris, Rhus, Robinia pseudoacacia, Rosa, Rubus occidentalis, Rumex, Sagittaria latifolia, Salix interior, Salix nigra, Sambucus, Sanicula canadensis, Scirpus, Scutellaria lateriflora, Solidago, Sparganium eurycarpum, Spiraea, Symphyotrichum lanceolatum, Symphyotrichum lateriflorum, Symphyotrichum ontarionis, Symplocarpus foetidus, Toxicodendron radicans, Typha latifolia, Ulmus americana, Utricularia macrorhiza, Verbena hastata, Viola sororia, Vitis cinerea, Vitis riparia.*

Use by wildlife: *Acorus americanus* and *A. calamus* rhizomes are eaten by muskrats (Mammalia: Cricetidae: *Ondatra zibethicus*) and other mammals. Neither species is particularly valued as a waterfowl food but the seeds of *A. calamus* are eaten by redhead ducks (Aves: Anatidae: *Aythya americana*). The foliage is used as cover by waterfowl. The plants are rated as highly resistant to grazing by deer (Mammalia: Cervidae: *Odocoileus*). The leaves of *A. americanus* are oviposition sites for damselflies (Insecta: Odonata: Aeshnidae: *Aeshna constricta*). *Acorus calamus* is a host to several Fungi (Ascomycota: Glomerellaceae: *Colletotrichum*; Dermateaceae: *Cylindrosporium acori*; Mycosphaerellaceae: *Ramularia aromatica, Septocylindrium, Sphaerulina acori*;

Phaeosphaeriaceae: *Phaeosphaeria acori*; Basidiomycota: Ceratobasidiaceae: *Rhizoctonia solani*; Doassansiaceae: *Nannfeldtiomyces sparganii*; Pucciniaceae: *Nigredo pyriformis, Uromyces pyriformis*; *U. sparganii*; Oomycota: Pythiaceae: *Pythium*).

Economic importance: food: Consumption of *Acorus* can be dangerous given its potentially toxic properties, and its use as food is not recommended. Deleterious effects can include constipation, diarrhea, digestive disorders, gastroenteritis, and internal bleeding. Some constituents (e.g., β-asarone) are known to induce tumors in laboratory animals. However, the young flower stems and leaf stalks, which emerge in early spring, reportedly are edible in raw form and the candied rhizomes have been made into a confection. Because *A. americanus* lacks the carcinogenic β-asarone and its phenylpropane derivatives, presumably it would be less dangerous to consume; **medicinal:** The medicinal and pharmacological literature on *Acorus* is voluminous. The essential oil of *Acorus* contains over 240 volatile components, including β-asarone, β-farnesene, cis-methylisoeugenol, epishyobunone, geranylacetate, isoshyobunone, methyleugenol, and shyobunone. Concentrations of these substances can differ widely among the taxa; e.g., β-asarone, which occurs at low levels (5%) in *A. calamus* but is absent in *A. americanus*. The plants have been used for centuries to treat numerous ailments worldwide. The rhizomes of *A. americanus* contain α-asarone (similar chemically to the alkaloid mescaline), which is a stimulant at low concentrations but hallucinogenic if ingested in "higher" amounts; β-asarone also is hallucinogenic. Both the essential oil and the asarones are smooth muscle relaxants, having effects similar to those of papaverine (a constituent of opium). The oils also have a wide range of antibacterial and fungicidal properties. *Acorus* species produce lectins, which are mitogenic to human lymphocytes and inhibitory to some cancer cell lines. Many indigenous North American tribes regarded *Acorus* as a medicinal panacea. The plants were particularly important medicinally to the Aboriginal inhabitants of Alberta, Canada. The Chippewa, Hocąk, Ojibwa, and Potawatomi people used small amounts of the plants to treat colds. The Meskwaki treated burns and sores from boiled rhizome extracts. In other parts of North America the plants have been used as a stimulant and also to treat diabetes, diarrhea, fever, flatulence, heartburn, hyperlipidity, indigestion, liver disorders, pain, rheumatism, stomach upset, and many other disorders; **cultivation:** *Acorus* was first cultivated in Europe during the 16th century and is grown today as an ornamental marginal plant for water gardens. Cultivars of *A. calamus* include 'Argenteostriatus', 'Purpureus', and 'Variegatus'; **misc. products:** *Acorus* has long been used for various ceremonial purposes, with remains of the plants recovered from the tomb of King Tutankhamen (dating to the 14th century BC). In modern times, *A. americanus* extracts have been sprayed around camping tents to repel spiders (Arthropoda: Arachnida: Araneae) and snakes (Reptilia: Squamata: Serpentes). Essential oils from the rhizome of *A. calamus* have been used as a household deodorant, as a flavoring for alcoholic beverages (e.g., beer, bitters, gin, vermouth, and wine), as a perfume essence, as a pest repellant, for ritual oils, and to tan leather. The matted foliage was once used as a floor covering. Exposure to the volatile oils of *A. calamus* causes sterility in male domestic house flies (Insecta: Diptera: Muscidae: *Musca domestica*); **weeds:** *Acorus calamus* can compete with more useful wildlife plants; **nonindigenous species:** The introduction of *Acorus calamus* to North America probably occurred during the mid-18th century.

Systematics: Although the systematic position of *Acorus* remains uncertain, even after the phylogenetic evaluation of a large amount of molecular data (e.g., Figure 1.1), the genus clearly is monophyletic. However, the taxonomy of the genus itself is problematic, with no consensus on the number of species that should be recognized. Molecular phylogenetic analyses have distinguished at least three distinct taxa: *A. calamus*, *A. gramineus* (much smaller in stature), and *A. angustatus* (= *A. tatarinowii*, etc.). The latter possibly is conspecific with *A. americanus* (not included in simultaneous phylogenetic analyses), but this issue has not yet been resolved. The chloroplast genome sequences of *A. americanus* and *A. calamus* in GenBank are virtually identical (>99.9% sequence identity), which indicates that the triploid (or hexaploid) *A. calamus* likely is derived from an ancestor similar to the diploid (or tetraploid) *A. americanus* (or *A. angustatus* if conspecific with the latter). Some of the taxonomic problems are associated with inconsistent cytological observations. The basic chromosome number of *Acorus* has been difficult to determine due to numerous conflicting accounts. It is widely regarded as $x=12$, with the fertile *A. americanus* ($2n=24$) interpreted as diploid and the sterile *A. calamus* ($2n=36$) as a triploid. Tetraploid ($2n=48$) and hexaploid ($2n=72$) races of *A. calamus* also have been reported outside of North America. However, other researchers have found $x=7$ (and an aneuploid derivative $x=9$) as the base numbers in numerous Asian accessions. Several other "anomalous" counts reported (e.g., $2n=54$, 66, 72) could be explained readily by a base number of $x=6$ (rather than 12), which also would accommodate the more widely accepted counts. In that case, *A. americanus* would represent a tetraploid and *A. calamus* a hexaploid. A more definitive cytological assessment of *Acorus* seems necessary in order to reconcile these inconsistencies.

Distribution: *Acorus americanus* is distributed across northern North America; *A. calamus* is widespread throughout the eastern United States and occurs sporadically in the western United States and eastern Canada.

References: Angier, 1974; Azuma & Toyota, 2012; Bains et al., 2005; Baskin & Baskin, 1998; Beal, 1977; Bogner & Mayo, 1998; Calvo-Polanco et al., 2014; Cottam, 1939; Crothers, 2012; Davis et al., 2004; Dolan, 2014; Dozier, 1945; Duvall et al., 1993a; Elmore et al., 2016; Fernald & Kinsey, 1943; Geiger & Banker, 2012; Haines, 2000a; Kindscher & Hurlburt, 1998; Kumar & Singh, 2015; Les & Mehrhoff, 1999; Marles, 2001; Mathur & Saxena, 1975; Morgan, 1980; Morton, 1963; Motley, 1994; Muenscher, 1936a; Nichols, 1999; Packer & Ringius, 1984; Pai & McCarthy, 2005; 2010; Perdomo et al., 2004; Petersen et al., 2016; Romanello et al., 2008; Rudgley, 1999; Sarneel, 2013; Smreciu et al., 2014; 2015; Stevenson et al., 2000; Strong, 2000; Tamura et al., 2004; Thompson, 2000a; Wang et al., 2001; Wein, 1939; 1940a; 1940b; Wieffering, 1972.

ALISMATID MONOCOTYLEDONS (ALISMATIDAE)

The alismatid monocotyledons consist entirely of aquatic and wetland species, which represent the greatest hydrophyte radiation within the angiosperms. All of the North American members are designated as OBL indicators throughout their ranges. The 11–12 "core" families (a distinction between Cymodoceaceae and Ruppiaceae remains unsettled) comprise a distinct, monophyletic group, which has been recognized at various taxonomic levels (Les & Tippery, 2013; Ross et al., 2016). Although once included in the group, the families Petrosaviaceae and Triuridaceae now are regarded as unrelated (Les & Tippery, 2013).

A long-standing tradition has been to treat this aquatic clade as a subclass (Alismatidae) containing two to four orders (e.g., Takhtajan, 1969; Cronquist, 1981). Older literature refers to the same taxon by the illegitimate ordinal name Helobiae. However, many recent authors (e.g., APG, 2016) have adopted a broader classification circumscribing one large order (Alismatales sensu lato), which contains the alismatid clade along with the former members of Arales (Araceae, Lemnaceae), and Tofieldiaceae. Although the broader concept of Alismatales has been widely followed, several phylogenetic studies (e.g., Figure 1.1A, B, D) do not support that assemblage as monophyletic. Until more consistent and compelling phylogenetic information is available, it seems prudent to restrict the circumscription of alismatids to include only the core families while excluding Araceae, Lemnaceae, and Tofieldiaceae (Les & Tippery, 2013). Full plastid genome sequence analyses (Ross et al., 2016) also indicate that the core alismatids are quite distinct genetically from both Arales and Tofieldiaceae, which also would argue for the distinction of these groups as smaller, ordinal subdivisions. Consequently, the treatment followed here is to recognize the alismatids as a subclass (Alismatidae) containing the twelve core aquatic families (Figure 1.2).

The Alismatidae can be subdivided effectively into two orders. Various molecular phylogenetic analyses (e.g., Les et al., 1993; 1997c; Les & Haynes, 1995; Les & Tippery, 2013; Ross et al., 2016) consistently have resolved two major clades within the group. These clades correspond with distinct (i.e., petaloid vs. tepaloid) floral morphologies (Posluszny et al., 2000) and have been recognized respectively as the orders Alismatales (sensu stricto) and Potamogetonales (Les & Tippery, 2013), which is the convention followed here (Figure 1.2). In this interpretation, the Alismatidae contain 12 families, 57 genera, and about 480 species.

Alismatids are highly diverse, making it difficult to identify consistent synapomorphic traits. Although the group is regarded as representing unspecialized or "primitive" monocots by its inclusion of plants having a combination of polypetalous, hermaphroditic flowers with radial symmetry, apocarpous gynoecia, and numerous stamens and carpels, it also contains many species having derived traits such as

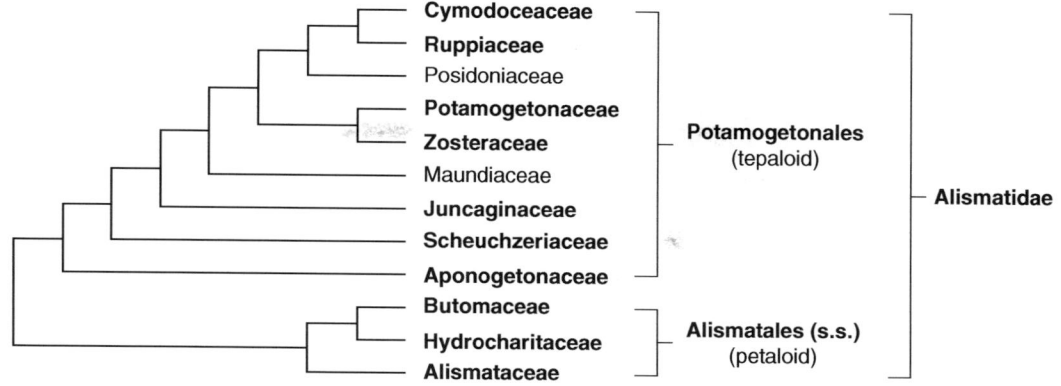

FIGURE 1.2 Phylogenetic relationships of alismatid families as indicated by a likelihood analysis of 83 plastid gene sequences (adapted from Ross et al., 2016). The North American families containing OBL indicators are highlighted in bold (Ruppiaceae are merged with Cymodoceaceae in the present treatment).

unisexuality, syncarpy, epigyny, and the only known flowering plants adapted to life in marine habitats. The subclass also exhibits a broad spectrum of pollination mechanisms incorporating biotic (Insecta) and abiotic (water, wind) vectors as well as self-pollination (Les, 1988; Les et al., 1997a). Representatives of the subclass are found throughout the world.

ORDER 2: ALISMATALES [3]

In molecular phylogenetic analyses, Alismatales consistently comprise a well-defined clade which is sister to the Potamogetonales (Figure 1.2). Referred to commonly as the "petaloid clade," these species are distinguished from other Alismatidae by having well-differentiated perianths consisting of both a calyx and corolla (Posluszny & Charlton, 1993; Posluszny et al., 2000). Most of the flowers are relatively showy and entomophilous. In some species they are unisexual and arranged in dioecious or monoecious sexual conditions. Until recently, the order segregated *Butomopsis*, *Hydrocleys*, and *Limnocharis* as a fourth family, Limnocharitaceae. Although these three genera associate as a clade, phylogenetic analyses resolve them as being nested in Alismataceae, within which they have been subsumed (Les & Tippery, 2013; Ross et al., 2016).

Alismatales are cosmopolitan in distribution. The order contains three families, all having OBL North American representatives:

2.1. **Alismataceae** Ventenat
2.2. **Butomaceae** Mirbel
2.3. **Hydrocharitaceae** Jussieu

Family 2.1. Alismataceae [17]

The water-plantains (Alismataceae) are herbaceous annuals (rarely) or rhizomatous perennials, which grow in standing water or in wetlands. Their foliage commonly contains laticifers (except *Sagittaria*) and abundant airspace lacunae. The leaves typically are basal and are differentiated into blade and petiole, except when growing

under submersed conditions when they can assume an elongate, graminoid shape (Sculthorpe, 1967; Cronquist, 1981; Judd et al., 2016). The hypogynous flowers are perfect or unisexual, often in whorls, and consist of a well-differentiated perianth of three sepals and three white, pink, or yellowish petals, six–numerous stamens, and an apocarpous gynoecium of three–numerous pistils; the fruits are aggregates of achenes or follicles (Cronquist, 1981; Haynes & Holm-Nielsen, 1994; Haynes et al., 1998a; Judd et al., 2016). Pollination of the nectariferous flowers is facilitated by insects (Insecta: Diptera; Hymenoptera). The fruits are dispersed by water or transported endozoically by waterfowl (Aves: Anatidae).

The family includes three genera (*Butomopsis*, *Hydrocleys*, *Limnocharis*), which formerly were assigned to Butomaceae or were segregated as the family Limnocharitaceae. However, most molecular phylogenetic analyses (e.g., Figure 1.3) resolve these genera as embedded within the clade containing the traditional genera of Alismataceae, where they have been included here. However, conflicting phylogenetic results have been obtained by analyses incorporating morphological characters, which resolve the "Limnocharit" genera as a basal grade of Alismataceae (Lehtonen, 2009).

Members of some genera (e.g., *Echinodorus*, *Hydrocleys*, *Limnocharis*, *Sagittaria*) are grown as ornamental water garden specimens. The underground organs of some species have been used as a source of food or medicine.

Alismataceae have a nearly cosmopolitan distribution. OBL indicators occur in six North American genera:

2.1.1. ***Alisma*** L.
2.1.2. ***Damasonium*** Mill.
2.1.3. ***Echinodorus*** Rich. & Engelm. ex A. Gray
2.1.4. ***Helanthium*** (Benth. & Hook.f.) Engelm. ex J.G.Sm.
2.1.5. ***Hydrocleys*** Rich.
2.1.6. ***Sagittaria*** L.

2.1.1. Alisma

Water plantain; plantain d'eau

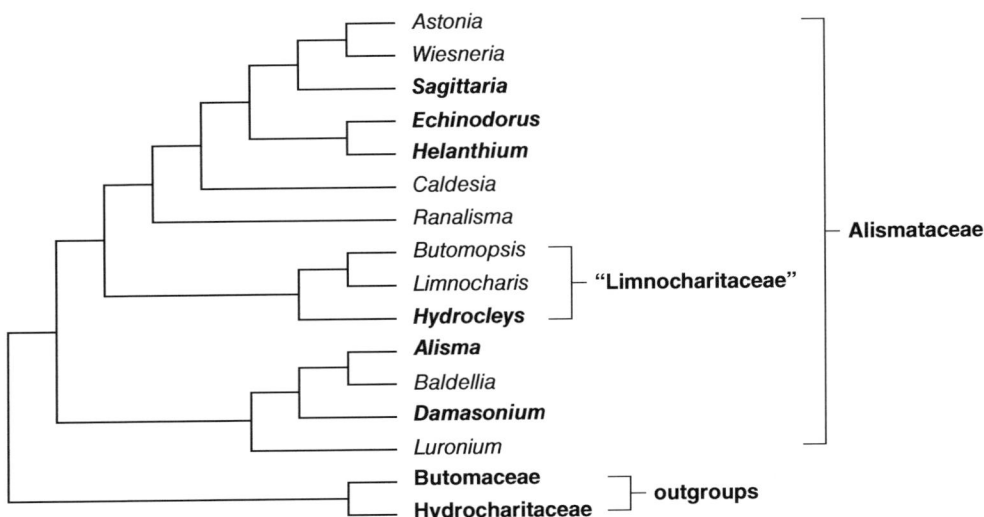

FIGURE 1.3 Intergeneric relationships in Alismataceae as indicated by a likelihood analysis of 83 plastid gene sequences (adapted from Ross et al., 2016). The North American groups containing OBL indicators are highlighted in bold.

Etymology: an ancient name of unknown meaning but possibly derived from alcea, another Greek name for these plants

Synonyms: none

Distribution: global: nearly cosmopolitan; **North America:** widespread

Diversity: global: 9 species; **North America:** 4 species

Indicators (USA): obligate wetland (OBL): *Alisma gramineum, A. lanceolatum, A. subcordatum, A. triviale*

Habitat: brackish to freshwater; lacustrine, palustrine, riverine; **pH:** 4.5–9.6; **depth:** <1 m; **life-form(s):** emergent herb, floating-leaved, submersed (rosulate)

Key morphology: plants (to 1 m) arising from a corm-like rhizome (to 3.5 cm) with non-septate roots; leaves basal, the blades (to 35 cm) linear-lanceolate to elliptic or ovate and petiolate (to 55 cm) when emersed or floating, linear (to 100 cm) and sessile when submersed, the margins entire; panicles (to 1 m) spreading, with whorls (to 10) of verticillate cymes containing pedicellate (to 47 mm) flowers; flowers bisexual, radial, cleistogamous (submersed) or chasmogamous (emergent); perianth 3-merous, differentiated into calyx and corolla, petals (to 6.5 mm) pink, purplish, or white; stamens 6–8; gynoecium apocarpous, the pistils (to 28) arranged in a flat ring (to 8 mm); achenes (to 3.1 mm) compressed laterally

Life history: duration: perennial (rhizomes); **asexual reproduction:** rhizomes (lateral); **pollination:** insect, self; **sexual condition:** dioecious, hermaphroditic, monoecious; **fruit:** aggregates of achenes or follicles (common); **local dispersal:** seeds (water); **long-distance dispersal:** seeds (birds)

Imperilment: 1. *Alisma gramineum* [G5]; S1 (CA, <u>MB</u>); S2 (NY, WY); S3 (<u>AB</u>, <u>BC</u>, <u>QC</u>, <u>SK</u>); **2.** *A. lanceolatum* [G4]; **3.** *A. subcordatum* [G5]; S1 (MS, <u>NB</u>, WY); S3 (NC, <u>QC</u>); **4.** *A. triviale* [G5]; S1 (<u>NF</u>, NJ, PA, <u>YT</u>); S2 (IL); S3 (OH, WY)

Ecology: general: *Alisma* is entirely aquatic and all of the North American species are designated as OBL indicators throughout their ranges. The genus is perennial, with some species behaving on occasion as annuals. The flowers contain septal nectaries and are produced in profusion, numbering up to 1,000 in some plants. They are self-compatible and lack dichogamy, but can be self-pollinating or outcrossed by small insects such as flies (e.g., Insecta: Diptera: Syrphidae) and also by the wind. Submersed, self-pollinating, cleistogamous flowers are produced in some species. The average seed set typically is high, in the range of 88–96%. The seeds are buoyant, dispersed by water, and can remain afloat for extended periods of up to several months. Some seeds reportedly can germinate without treatment but nearly all become physiologically dormant and require a period of cold stratification (e.g., 30–210 days at 2°C) for optimal germination (at 20°C or 25°C/15°C). Germination rates typically are low, usually only 5–10%. Slightly higher rates (to 15%) occur under maximum moisture conditions. The seeds of some species are known to remain viable for more than 10 years and seed banks can develop. Vegetative reproduction can occur by means of lateral offshoots arising from the bulbous rhizome and is more prevalent in plants growing under shaded conditions.

Because of historical taxonomic misunderstandings (see *Systematics*), information can be difficult to associate unambiguously with occurrences of *Alisma subcordatum* and *A. triviale*, which at one time were recognized in North America as subspecies or varieties of *A. plantago-aquatica* (i.e., *A. plantago-aquatica* var. *parviflorum* and *A. plantago-aquatica* var. *americanum*, respectively). It is now quite evident that both *A. subcordatum* and *A. triviale* are distinct from *A. plantago-aquatica* (see *Systematics*) and that the latter does not actually occur in the New World, despite numerous conflicting reports. Consequently, although designated as an OBL North American indicator, *A. plantago-aquatica* has been excluded here.

2.1.1.1. ***Alisma gramineum*** **Lej.** grows in shallow to moderately deep (to 1.6 m), fresh to brackish, still to flowing waters of backwashes, beaches, bottoms, canals (dry), channels, depressions, ditches, flats, floodplains, marshes (tidal), mudflats, pools, puddles, roadsides, shores, slopes (to 1%), sloughs, springs, streams, vernal pools, washes, and along the margins of canals, channels, lakes, playas, ponds, reservoirs, and rivers at elevations to 3180 m. The plants can grow completely submersed (then with ribbon-like foliage and cleistogamous flowers) but also occur often on drying or exposed substrates where ephemeral waters have receded. Their optimal depth has been reported as 50–70 cm. Exposures typically receive full sunlight. The substrates are described as alkaline (e.g., pH: 7.5–9.6; total alkalinity: 48–108 ppm) and include clay, clay loam, clay muck, clay mud, clay silt, gravel, limestone, muck, mucky clay, mud, muddy clay, rock, sand, sandstone, sandy gravel, sandy muck, sandy silt, shales, silty clay, and stones. The flowers of emergent plants are chasmogamous and normally insect pollinated, whereas those of submersed plants are cleistogamous and self-pollinated. Flowering and fruiting occur from June to September. The flowers remain open from 6:30 am to 10:00 pm. Seed set in open-pollinated plants has been estimated at 88%. The achenes are buoyant and dispersed by water, but under continuous agitation, they sink within 128 h. A persistent seed bank develops. Dried seeds (at 15% relative humidity) that are frozen (at −20°C) have remained viable for at least a year. Scarification of the pericarp or stratification is necessary to promote germination. A 60% germination rate has been reported for dried, stored seeds that have been re-exposed to high humidity for 1 day at 21°C, then physically scarified and incubated at 26°C under a 12/12-h photoperiod. High germination rates (56%) have been obtained for seeds that have been cold stratified (e.g., 10 months at 6°C) and then incubated in shallow water under constant (25°C) or fluctuating (25°C/10°C) temperatures. Germination rates are reduced nearly by half if the seeds are stratified for a comparable time at 20°C. The best germination rates (94–100%) have been obtained for cold, wet-stored seeds that are pre-treated with bleach and then incubated at 20°C–30°C. The seedlings have a relatively high vulnerability to winter conditions, and do not fare well unless they initially establish soundly on exposed substrates during the late summer and fall. Because new seedlings are killed by high water conditions, much of the annual reproduction occurs by seed. In subsequent years, the plants must overwinter under water to avoid being killed by frost. Overall, the plants are poorly adapted to changing habitat conditions, although they tolerate constant submergence well. Established stands are known to reduce wave energies, resulting in higher local sedimentation rates. Denser stands can produce 85 culms m^{-2} and achieve a mean biomass of 125 g ash-free dry mass [DM] m^{-2}. **Reported associates:** *Acorus calamus, Alisma triviale, Allium geyeri, Alopecurus aequalis, Alopecurus saccatus, Artemisia nova, Aulosira, Bacopa monnieri, Beckmannia syzigachne, Bidens beckii, Bidens laevis, Bidens polylepis, Bolboschoenus fluviatilis, Bolboschoenus maritimus, Butomus umbellatus, Carex praegracilis, Ceratophyllum demersum, Chara vulgaris, Chenopodium glaucum, Chenopodium rubrum, Deschampsia danthonioides, Distichlis spicata, Downingia elegans, Elaeagnus angustifolia, Elatine, Eleocharis acicularis, Eleocharis macrostachya, Eleocharis palustris, Elodea canadensis, Elymus, Epilobium, Eryngium petiolatum, Glyceria leptostachya, Gnaphalium palustre,*

Gratiola ebracteata, Grindelia squarrosa, Heliotropium, Heteranthera dubia, Hippuris, Hippuris vulgaris, Honckenya peploides, Hordeum jubatum, Hydrocharis morsus-ranae, Isoetes howellii, Iva axillaris, Juncus articulatus, Juncus balticus, Juncus effusus, Juncus tenuis, Lemna minor, Lemna trisulca, Leptochloa fusca, Limosella aquatica, Myosotis laxa, Myosurus minimus, Myriophyllum sibiricum, Najas flexilis, Navarretia intertexta, Navarretia squarrosa, Nuphar, Oryza sativa, Oscillatoria, Panicum capillare, Paspalum distichum, Persicaria amphibia, Persicaria coccinea, Persicaria lapathifolia, Persicaria punctata, Phalaris arundinacea, Plagiobothrys leptocladus, Plantago maritima, Pluchea odorata, Polygonum ramosissimum, Polypogon monspeliensis, Potamogeton amplifolius, Potamogeton berchtoldii, Potamogeton friesii, Potamogeton pusillus, Potamogeton richardsonii, Potamogeton zosteriformis, Potentilla anserina, Puccinellia nuttalliana, Ranunculus aquatilis, Ranunculus subrigidus, Rorippa curvisiliqua, Rorippa sinuata, Rorippa teres, Rumex maritimus, Rumex stenophyllus, Rumex triangulivalvis, Ruppia maritima, Sagittaria cuneata, Salicornia depressa, Salix caroliniana, Salix exigua, Sarcobatus vermiculatus, Schoenoplectus acutus, Schoenoplectus pungens, Schoenoplectus tabernaemontani, Sium suave, Sparganium eurycarpum, Spergularia, Spirodela, Stuckenia filiformis, Stuckenia pectinata, Stuckenia vaginata, Triglochin scilloides, Typha angustifolia, Typha latifolia, Vallisneria americana, Zannichellia palustris.

2.1.1.2. *Alisma lanceolatum* With. is a nonindigenous species, which grows in shallow standing waters or on exposed substrates in channels (bottoms), depressions, ditches, floodplains, gravel pits, marshes, meadows, mudflats, pools, seeps, streams (intermittent), swamps (drying), washes, and along the margins of lakes, ponds, rivers, and streams at elevations to 1267 m. The plants originate from regions having relatively warmer climate conditions. Exposures range from open sunlight to partial shade. The substrates are circumneutral (pH: 6.5–8.5), tend to be nutrient-rich, and include clay, gravel, mud, sand, and silt. Flowering and fruiting occur from April to August in North America. Fluctuating water conditions appear to promote the flowering response. The flowers remain open from 7:30 am to 7:00 pm and are pollinated by insects (Insecta). Typical seed set in open-pollinated plants has been estimated at 89%. The flat achenes are buoyant and are dispersed by water; however, under continuous agitation, they sink within 96 h. They also are known to be spread by attachment to agricultural machinery. The seeds are dormant when shed and require a period of cold stratification to induce germination. Germination rates often are above 30%. Optimal germination rates (62–75%) have been obtained for seeds that have been cold stratified (e.g., 3–10 months at 6°C) and then incubated in shallow water under constant (15°C, 25°C) or fluctuating (25°C /10°C) temperatures. Stratification at higher temperatures (e.g., 20°C) greatly decreases their germination rates. The plants adapt only moderately to changing habitat conditions. However, when conditions are favorable, seed germination is rapid and synchronous, which facilitates colonization and establishment. Seedlings establish more successfully when they remain emergent or develop under fluctuating water conditions. Higher seedling survival rates occur when growing under submersed conditions (e.g., 40–80 cm depth). The plants have been described both as mycorrhizal (arbuscular) and non-mycorrhizal. **Reported associates (North America):** *Adiantum, Alopecurus aequalis, Artemisia douglasiana, Athyrium, Bidens, Convolvulus arvensis, Cotula coronopifolia, Downingia, Festuca perennis, Gratiola, Isolepis carinata, Lemna, Melilotus indicus, Oryza sativa, Rubus, Rubus ulmifolius, Rumex, Rumex conglomeratus, Salix, Salix laevigata, Tribulus terrestris, Typha latifolia, Urtica dioica, Veronica.*

2.1.1.3. *Alisma subcordatum* **Raf.** is found in shallow (e.g., 10–30 cm) standing water or on exposed substrates in alluvial fans, beaches, bogs, bottomlands, bottoms (stream), canals, channels (river), depressions, ditches, flats, floodplains, gravel bars, marshes (ephemeral), meadows, mudflats, oxbows, ponds (intermittent), pools (artificial; ephemeral), prairies (mesic; remnant), puddles, right-of-ways (powerline; railroad), roadsides (springy), sandbars, seeps, shores (marshy), sloughs, springs (outflow), streams (shallow), swamps, thickets (seepy), swales, woodlands, and along the margins of lagoons, lakes, ponds, reservoirs, rivers, streams, swamps, and woodlands at elevations to 1448 m. The plants are heterophyllous, producing floating leaves when grown in standing water (e.g., 25 cm deep) and emergent foliage when growing under shallow conditions (e.g., 2 cm water). Exposures range from fully open conditions to partial shade. The substrates are circumneutral (pH: 6.2–8.4; total alkalinity: 11–290 mg l⁻¹) and are described as alluvium, clay loam muck, cobble, gravel, Hepler silt loam, loamy clay, loamy muck, muck, mud, peat, peaty muck, sand, sandy clay, sandy loam, sandy muck (organic), shale, and silt. Flowering and fruiting occur from May to September. Warming global temperatures have been implicated in a 2 month earlier flowering initiation of some populations during the past two decades. The flowers are self-compatible, but remain open from 11:30 am to 7:30 pm, which enables cross-pollination by insects (see *Use by Wildlife*). Typical seed set in open-pollinated plants has been estimated to approach 90%. Each flower produces from 6 to 18 fruits. The achenes float and are dispersed locally by water, but will sink within 128 hr if the surface is agitated periodically. The seeds are physiologically dormant and require a period of cold stratification (e.g., 210 days at 2°C–5°C) to induce germination (under a 25°C/15°C day/night temperature regime). Successful germination also has been reported for seeds stratified at 5°C for 60 days, and then placed on sand under water. Reported germination rates vary from 0.4% to 35%. The seeds germinate optimally under flooded conditions (up to 62%) but remain dormant under unflooded conditions or when buried. A persistent seed bank can develop and refrigerated achenes have retained 50% viability after 9 years of storage. Adult plants more typically occur in non-flooded sites. Exposed substrates are necessary for the effective establishment of seedlings. The roots reportedly have weak arbuscular mycorrhizal development. **Reported associates:** *Acalypha rhomboidea, Acer negundo, Acer saccharinum, Acorus calamus, Agrimonia parviflora, Agrostis gigantea, Agrostis hyemalis, Agrostis perennans, Agrostis stolonifera, Alisma triviale, Ambrosia artemisiifolia, Ammannia coccinea, Ammannia robusta, Amorpha fruticosa, Apios americana, Asclepias incarnata, Asclepias syriaca, Bacopa rotundifolia, Betula nigra, Bidens cernuus, Bidens connatus, Bidens polylepis, Bidens tripartitus, Boehmeria cylindrica, Bolboschoenus fluviatilis, Boltonia asteroides, Calamagrostis canadensis, Campsis radicans, Cardamine bulbosa, Carex caroliniana, Carex conjuncta, Carex corrugata, Carex cristatella, Carex frankii, Carex granularis, Carex grayi, Carex haydenii, Carex hyalinolepis, Carex lacustris, Carex lupulina, Carex lurida, Carex muskingumensis, Carex shortiana, Carex stipata, Carex stricta, Carex tribuloides, Carex typhina, Cephalanthus occidentalis, Chelone glabra, Chelone obliqua, Chenopodium simplex, Cicuta maculata, Cladium jamaicense, Coleataenia longifolia, Conoclinium coelestinum, Cornus amomum, Cuscuta gronovii, Cyperus acuminatus, Cyperus bipartitus, Cyperus esculentus, Cyperus odoratus, Cyperus strigosus, Dichanthelium clandestinum, Dipsacus fullonum, Echinochloa crus-galli, Echinodorus berteroi, Eclipta prostrata, Eleocharis acicularis, Eleocharis engelmannii, Eleocharis erythropoda, Eleocharis macrostachya, Eleocharis obtusa, Eleocharis ovata, Eleocharis palustris, Eleocharis quadrangulata, Epilobium coloratum, Equisetum arvense, Equisetum fluviatile, Erechtites hieraciifolius, Eupatorium perfoliatum, Eupatorium serotinum, Eutrochium maculatum, Fallopia scandens, Fimbristylis autumnalis, Fraxinus pennsylvanica, Fraxinus profunda, Galium obtusum, Galium tinctorium, Gentiana andrewsii, Geum laciniatum, Glyceria striata, Gratiola aurea, Gratiola neglecta, Helenium autumnale, Helianthus grosseserratus, Helianthus mollis, Hibiscus moscheutos, Holcus lanatus, Humbertacalia, Hypericum canadense, Impatiens capensis, Ipomoea lacunosa, Iris virginica, Juncus acuminatus, Juncus effusus, Juncus marginatus, Juncus tenuis, Juncus torreyi, Laportea canadensis, Leersia lenticularis, Leersia oryzoides, Leersia virginica, Lemna minor, Leucospora multifida, Lindernia dubia, Lobelia siphilitica, Ludwigia alternifolia, Ludwigia palustris, Ludwigia polycarpa, Lycopus americanus, Lycopus virginicus, Lysimachia ciliata, Lysimachia nummularia, Lysimachia quadriflora, Lysimachia terrestris, Lythrum alatum, Lythrum salicaria, Mentha arvensis, Mentha ×piperita, Mimulus ringens, Myriophyllum aquaticum, Nymphoides peltata, Nyssa sylvatica, Oenothera pilosella, Onoclea sensibilis, Panicum dichotomiflorum, Panicum virgatum, Panicum virgatum, Parthenocissus quinquefolia, Pedicularis lanceolata, Peltandra virginica, Penthorum sedoides, Persicaria amphibia, Persicaria careyi, Persicaria hydropiperoides, Persicaria lapathifolia, Persicaria pensylvanica, Persicaria punctata, Persicaria setacea, Phalaris arundinacea, Phragmites australis, Phyla lanceolata, Pilea pumila, Platanthera peramoena, Platanus occidentalis, Poa nemoralis, Polygonum ramosissimum, Pontederia cordata, Populus deltoides, Potamogeton nodosus, Proserpinaca, Prunella vulgaris, Pycnanthemum virginianum, Quercus bicolor, Quercus macrocarpa, Quercus* palustris, *Ranunculus hispidus, Ranunculus pensylvanicus, Rhamnus frangula, Rhynchospora capitellata, Rosa multiflora, Rubus setosus, Rudbeckia subtomentosa, Rumex obtusifolius, Rumex verticillatus, Sabatia angularis, Saccharum, Sagittaria australis, Sagittaria latifolia, Sagittaria montevidensis, Salix discolor, Salix interior, Salix myricoides, Salix nigra, Samolus valerandi, Saururus cernuus, Schoenoplectus hallii, Schoenoplectus mucronatus, Schoenoplectus pungens, Schoenoplectus tabernaemontani, Scirpus atrovirens, Scirpus cyperinus, Scutellaria galericulata, Scutellaria lateriflora, Sium suave, Smilax, Solanum dulcamara, Solidago riddellii, Sparganium americanum, Sparganium androcladum, Sparganium emersum, Sparganium eurycarpum, Spartina pectinata, Sphagnum, Spiraea tomentosa, Spirodela polyrhiza, Stachys palustris, Stellaria longifolia, Symphyotrichum lanceolatum, Symphyotrichum novi-belgii, Symphyotrichum ontarionis, Symphyotrichum praealtum, Teucrium canadense, Thalictrum clavatum, Thelypteris palustris, Toxicodendron radicans, Tridens flavus, Typha angustifolia, Typha latifolia, Verbena hastata, Vernonia fasciculata, Veronicastrum virginicum, Viola sororia, Vitis cinerea, Vitis riparia, Xanthium strumarium.*

2.1.1.4. ***Alisma triviale* Pursh** occurs in shallow (0.05–1.0 m) fresh to somewhat saline standing waters, or on exposed substrates in backwaters, bays (shallow), bogs, canals, channels (intermittent), ditches (irrigation; roadside), flats (river), floodplains, glades, impoundments, lakes, marshes, meadows, mudflats, oxbows (swampy), ponds (beaver; intermittent; stock), pools (dry; seasonal; vernal), potholes, prairies, puddles, reservoirs (dry), rice fields, roadsides, shores, slopes (to 6%), sloughs, springs, stream beds, swales (drying), swamps, tanks, and along the margins of lagoons, lakes, ponds, pools, reservoirs, rivers (estuaries), and streams at elevations to 2830 m. In experimental manipulations, the optimal water depth ranged from 15 to 20 cm. Exposures can vary from full to partial sunlight. The substrates have been described as acidic or alkaline (pH: 4.5–8.0; total alkalinity: 13.5–47.5 mg l^{-1}) and include adobe, bentonite, clay (Gumbo; Pierre), gravel, loam, loamy clay, logs (decayed), muck (organic), mud, peat, peaty mud, sand, sandy clay loam, sandy loam, silt, silty mud, and stones. Flowering and fruiting extend from June to November. The flowers remain open from 8:00 am to 6:30 pm. Average seed set in open-pollinated plants has been estimated at 91%. The achenes are dispersed by water, and under intermittent agitation, will sink within 96 h. Specific conditions for germination have not been reported, but germination rates have been observed to increase proportionally with water depth. The mean density of germinating seeds can reach 128.2 seeds m^{-2}. The greatest seedling growth occurs at water depths from 2 to 7 cm. The roots reportedly are colonized by arbuscular mycorrhizae. The plants are known to occur commonly along hedgerows that are adjacent to agricultural fields.

Reported associates: *Acer glabrum, Acer negundo, Acer saccharinum, Achillea millefolium, Acmispon wrightii, Acorus calamus, Agrostis exarata, Agrostis scabra, Agrostis stolonifera, Alisma gramineum, Alisma subcordatum, Alnus, Alopecurus aequalis, Alopecurus geniculatus, Amaranthus powellii, Amaranthus tuberculatus, Ambrosia tomentosa, Ammannia, Amorpha fruticosa, Apocynum cannabinum, Arctium minus, Aristida oligantha, Asclepias incarnata, Astragalus humistratus, Azolla cristata, Baccharis douglasii, Baccharis salicifolia, Beckmannia syzigachne, Bidens aristosus, Bidens beckii, Bidens cernuus, Bidens connatus, Bidens frondosus, Bidens trichospermus, Bistorta bistortoides, Bolboschoenus fluviatilis, Bolboschoenus maritimus, Bolboschoenus robustus, Boltonia asteroides, Bouteloua curtipendula, Brassica, Brickellia californica, Bromus inermis, Bromus tectorum, Calamagrostis canadensis, Callitriche heterophylla, Callitriche palustris, Capsella bursa-pastoris, Carex annectens, Carex aquatilis, Carex athrostachya, Carex bebbii, Carex brevior, Carex buxbaumii, Carex canescens, Carex comosa, Carex cristatella, Carex densa, Carex frankii, Carex lasiocarpa, Carex leptalea, Carex lupuliformis, Carex lyngbyei, Carex meadii, Carex muskingumensis, Carex normalis, Carex occidentalis, Carex pellita, Carex praegracilis, Carex retrorsa, Carex scoparia, Carex siccata, Carex simulata, Carex stipata, Carex utriculata, Carex vesicaria, Carex ×cayouettei, Castilleja angustifolia, Ceanothus fendleri, Ceratophyllum demersum, Chara, Chelone glabra, Chenopodium, Cicuta bulbifera, Cicuta maculata, Cirsium arvense, Cirsium undulatum, Cirsium vulgare, Cirsium wheeleri, Clintonia, Conium maculatum, Convolvulus, Coreopsis, Cornus amomum, Cornus drummondii, Cornus sericea, Crypsis vaginiflora, Cuscuta glomerata, Cyperus difformis, Cyperus eragrostis, Cyperus strigosus, Damasonium californicum, Decodon verticillatus, Deschampsia elongata, Dichanthelium oligosanthes, Distichlis spicata, Downingia laeta, Dracocephalum parviflorum, Echinochloa crus-galli, Echinodorus berteroi, Echinodorus cordifolius, Eclipta prostrata, Elatine californica, Elatine chilensis, Elatine rubella, Eleocharis acicularis, Eleocharis engelmannii, Eleocharis erythropoda, Eleocharis macrostachya, Eleocharis montana, Eleocharis ovata, Eleocharis palustris, Eleocharis parishii, Eleocharis rostellata, Eleocharis tenuis, Eleocharis wolfii, Elodea canadensis, Elymus elymoides, Epilobium campestre, Epilobium ciliatum, Epilobium torreyi, Equisetum arvense, Equisetum fluviatile, Equisetum laevigatum, Eragrostis pectinacea, Erigeron divergens, Erigeron speciosus, Erigeron strigosus, Eriogonum racemosum, Eryngium articulatum, Eupatorium perfoliatum, Euphorbia marginata, Eustoma exaltatum, Euthamia occidentalis, Eutrochium maculatum, Fallopia convolvulus, Fallugia paradoxa, Festuca arizonica, Fimbristylis spadicea, Flaveria campestris, Forestiera pubescens, Fraxinus pennsylvanica, Galium aparine, Galium trifidum, Geranium caespitosum, Geranium richardsonii, Geum laciniatum, Geum macrophyllum, Glyceria borealis, Glyceria elata, Glyceria grandis, Glyceria septentrionalis, Glyceria striata, Glycyrrhiza lepidota, Gnaphalium exilifolium, Gnaphalium uliginosum, Gratiola neglecta, Grindelia squarrosa, Hackelia floribunda, Helianthus annuus, Helianthus grosseserratus, Helianthus maximiliani, Heteranthera, Hippuris vulgaris, Hordeum jubatum, Hymenoxys hoopesii, Hymenoxys subintegra, Hypericum* *anagalloides, Impatiens capensis, Impatiens pallida, Iris missouriensis, Isoetes, Juncus acuminatus, Juncus articulatus, Juncus balticus, Juncus bufonius, Juncus effusus, Juncus ensifolius, Juncus interior, Juncus oxymeris, Juncus patens, Juncus torreyi, Juncus xiphioides, Juniperus monosperma, Juniperus osteosperma, Juniperus scopulorum, Lactuca serriola, Landoltia punctata, Lappula occidentalis, Leersia lenticularis, Leersia oryzoides, Leersia virginica, Lemna minor, Liatris lancifolia, Lilaeopsis, Limosella acaulis, Lindernia dubia, Lobelia cardinalis, Lobelia siphilitica, Ludwigia palustris, Ludwigia peploides, Lupinus argenteus, Lupinus kingii, Lycopus americanus, Lycopus asper, Lycopus uniflorus, Lysimachia terrestris, Lythrum alatum, Lythrum hyssopifolia, Lythrum salicaria, Madia, Marsilea mollis, Marsilea vestita, Medicago lupulina, Medicago sativa, Melilotus albus, Melilotus officinalis, Mentha arvensis, Mentha ×piperita, Mikania scandens, Mimulus guttatus, Mimulus ringens, Muhlenbergia frondosa, Muhlenbergia wrightii, Myosotis laxa, Myosurus, Myriophyllum pinnatum, Myriophyllum sibiricum, Najas flexilis, Nasturtium, Navarretia intertexta, Nuphar polysepala, Oenothera rhombipetala, Onoclea sensibilis, Orthocarpus luteus, Ottelia alismoides, Panicum virgatum, Parthenocissus quinquefolia, Pascopyrum smithii, Paspalum distichum, Peltandra virginica, Pennisetum glaucum, Penstemon barbatus, Penstemon linarioides, Penthorum sedoides, Persicaria amphibia, Persicaria coccinea, Persicaria coccinea, Persicaria hydropiper, Persicaria hydropiperoides, Persicaria lapathifolia, Persicaria maculosa, Persicaria pensylvanica, Persicaria punctata, Phalaris arundinacea, Phleum pratense, Phragmites australis, Phyla lanceolata, Phyla nodiflora, Picea mariana, Pinus ponderosa, Plagiobothrys, Plantago, Platanus occidentalis, Poa arida, Poa nemoralis, Poa pratensis, Polygonum aviculare, Polygonum douglasii, Polypogon monspeliensis, Pontederia cordata, Populus angustifolia, Populus deltoides, Populus tremuloides, Porterella carnosula, Portulaca oleracea, Potamogeton amplifolius, Potamogeton berchtoldii, Potamogeton diversifolius, Potamogeton foliosus, Potamogeton gramineus, Potamogeton natans, Potamogeton nodosus, Potamogeton pusillus, Potamogeton richardsonii, Potamogeton zosteriformis, Potentilla crinita, Potentilla gracilis, Potentilla hippiana, Potentilla norvegica, Prunella, Prunus virginiana, Pseudocymopterus montanus, Pseudognaphalium luteoalbum, Quercus bicolor, Quercus gambelii, Quercus grisea, Ranunculus aquatilis, Ranunculus cymbalaria, Ranunculus trichophyllus, Rhus aromatica, Ribes cereum, Robinia neomexicana, Rorippa curvipes, Rorippa islandica, Rorippa sphaerocarpa, Rosa arkansana, Rosa woodsii, Rotala, Rumex crispus, Rumex maritimus, Rumex salicifolius, Rumex triangulivalvis, Sagittaria cuneata, Sagittaria latifolia, Sagittaria montevidensis, Salix amygdaloides, Salix exigua, Salix gooddingii, Salix nigra, Salix interior, Sambucus nigra, Schizachyrium scoparium, Schoenoplectus acutus, Schoenoplectus pungens, Schoenoplectus tabernaemontani, Schoenoplectus triqueter, Scirpus atrovirens, Scirpus cyperinus, Scirpus lineatus, Scirpus microcarpus, Scrophularia parviflora, Scutellaria lateriflora, Senecio spartioides, Senecio wootonii, Setaria parviflora, Setaria viridis, Sidalcea, Sisyrinchium angustifolium, Sium suave, Solanum dulcamara, Solidago missouriensis, Sparganium angustifolium, Sparganium emersum, Spartina pectinata, Sphenoclea zeylanica, Sphenopholis obtusata, Spiraea, Spiranthes cernua, Spiranthes diluvialis, Spirodela polyrhiza, Sporobolus compositus, Stachys palustris, Stachys pilosa, Stuckenia pectinata, Suaeda calceoliformis, Symphoricarpos, Symphyotrichum ericoides, Symphyotrichum lanceolatum, Symphyotrichum lateriflorum, Symphyotrichum subulatum, Taraxacum officinale, Thalictrum fendleri, Thermopsis, Torreyochloa pallida, Toxicodendron radicans, Tragopogon, Triglochin scilloides, Typha angustifolia, Typha latifolia, Urtica dioica, Utricularia, Verbascum thapsus, Verbena bracteata, Verbena hastata, Verbena urticifolia, Vernonia fasciculata, Veronica anagallis-aquatica, Veronica peregrina, Veronica serpyllifolia, Vicia cracca, Viola sororia, Vitis riparia, Wolffia brasiliensis, Wolffia columbiana, Xanthium strumarium, Zannichellia palustris.*

Use by wildlife: The achenes of *Alisma gramineum* are eaten by various ducks (Aves: Anatidae: *Anas*), including baldpates (*A. americana*), blue-winged teal (*A. discors*), gadwells (*A. strepera*), and shovelers (*A. clypeata*). *Alisma subcordatum* is grazed by Canada geese (Aves: Anatidae: *Branta canadensis*) and European ambersnails (Gastropoda: Succineidae: *Succinea putris*). Its flowers are visited (and likely pollinated) by flies (Insecta: Diptera) and occasionally by skippers (Insecta: Lepidoptera: Hesperiidae: *Ancyloxypha numitor*). *Alisma triviale* is a host to moth larvae (Lepidoptera: Noctuidae: *Hypena madefactalis*) and spittlebugs (Insecta: Hemiptera: Cercopidae). The roots are host to slime molds (Fungi: Myxomycota: Plasmodiophoraceae: *Ligniera junci*). The plants serve as an oviposition substrate for aster leafhoppers (Insecta: Hemiptera: Cicadellidae: *Macrosteles fascifrons*). The achenes are eaten by blue-winged teal (Aves: Anatidae: *Anas discors*) and mallard ducks (*Anas platyrhynchos*). In North America, *Alisma* plants (species uncertain) are host to several Fungi (Ascomycota: Mycosphaerellaceae: *Cercospora alismatis*, *C. callae*; Basidiomycota: Doassansiaceae: *Doassansia alismatis*; Blastocladiomycota: *Cladochytrium alismatis*).

Economic importance: food: After washing and drying, the bulblike leaf bases of *Alisma* plants were eaten by the Calmucks; however, some sources suggest that the foliage (especially when bruised) can be poisonous or cause skin irritation; **medicinal:** Some *Alisma* species contain protostane triterpenoids, which exhibit antiplasmodial activity against *Plasmodium falciparum*, a causative agent of malaria. *Alisma gramineum* has been used to treat gastrointestinal disorders, skin diseases, and to treat abrasions and cuts in parts of Asia. *Alisma subcordatum* was used by the Cherokee to treat gastrointestinal disorders and skin ailments. The Iroquois used *A. triviale* to treat kidney ailments and tuberculosis; **cultivation:** Various *Alisma* species are planted occasionally as ornamental plants in water gardens or along river gardens; **misc. products:** *Alisma gramineum* has been used as a natural insect repellant. Seeds of *A. subcordatum* have been included in mixes used for wetland restoration programs. *Alisma triviale* has been planted in free water surface constructed wetlands; **weeds:** *Alisma lanceolatum* and *A. triviale* have been reported as rice field weeds in western North America. Strains resistant to bensulfuron-methyl herbicides have been reported in *A. lanceolatum*; **nonindigenous species:** *Alisma lanceolatum* likely was introduced to California and Oregon from Eurasia by means of contaminated rice stock. *Alisma plantago-aquatica* has been characterized as nonindigenous to North America; however, no verifiable records of this species are known to occur in the region.

Systematics: Various analyses of DNA sequence data (e.g., Figures 1.3 and 1.4) confirm that *Alisma* is monophyletic and indicate that it is related most closely to *Baldellia*. Analysis of nrITS sequence data (Figure 1.4) effectively distinguishes *A. gramineum* and *A. lanceolatum* from the majority of species but fails to resolve relationships among *A. plantago-aquatica*, *A. subcordatum*, and *A. triviale*, which indicates their relatively recent divergence. However, all three taxa are clearly separable using RAPDs markers. Although North American specimens of *A. subcordatum* and *A. triviale* were long treated taxonomically as variants of the Old World diploid species *A. plantago-aquatica*, many morphological, chromosomal, and genetic isolating distinctions eventually became evident. Even though these taxa are similar morphologically, no crosses have succeeded despite numerous reciprocal attempts involving *A. plantago-aquatica* × *A. subcordatum*, *A. plantago-aquatica* or *A. subcordatum* × *A. triviale*, which indicates that all three species are well isolated genetically. In contrast, some successful crosses (i.e., those resulting in seed set) have been obtained involving either *A. lanceolatum* or *A. gramineum* × *A. plantago-aquatica*, and *A. lanceolatum* × *A. gramineum*, which exhibit more distant interrelationships. The basic chromosome number of *Alisma* is *x* = 7. *Alisma*

gramineum and *A. subcordatum* (2*n* = 14) are diploids; *A. lanceolatum* (2*n* = 26, 28) and *A. triviale* (2*n* = 28) are tetraploids. The two cytotypes of *A. lanceolatum* are well isolated genetically. No spontaneous interspecific *Alisma* hybrids have been reported in North America. There have been no successful intergeneric hybrids resulting from crossing attempts involving *Alisma*, *Baldellia*, and *Luronium*.

Distribution: *Alisma gramineum* and *A. triviale* occur throughout central North America and *A. subcordatum* is widespread in the eastern United States. The nonindigenous *A. lanceolatum* is restricted to California and Oregon.

References: Adams et al., 2011; Andersson et al., 2000; Baldwin, Jr. & Speese, 1955; Barrett & Seaman, 1980; Baskin & Baskin, 1998; Beas et al., 2013; Bergeron & Pellerin, 2014; Björkqvist, 1967; 1968; Boutin et al., 2002; Burgess, 1970; Calha et al., 1998; Conover & Pelikan, 2010; Countryman, 1968; Duke, 2000; Fernald, 1946; Fernald & Kinsey, 1943; Grau & Leonard, 1978; Haynes & Holm-Nielsen, 1994; Haynes & Hellquist, 2000a; Haynes et al., 1998a; Hellquist & Crow, 1981; Hendricks, 1957; Hopkins, 1969; Hroudová et al., 2004; Hudon, 1997; Johnson, 1999; Kaul, 1978; 1985; Keith, 1961; Kellogg et al., 2003; Knight, 1965; Kratzer, 2014; Les & Tippery, 2013; Lieneman, 1929; Lovell, 1899; McIntyre & Newnham, 1988; Mollik et al., 2010; Moravcová et al., 2001; Parker et al., 1969; Pogan, 1963; Putman, 1953; Ransom & Oelke, 1983; Rhoades, 1962; Rogers, 1983; Ryser et al., 2011; Scott, 2014; Seabloom et al., 1998; Smith, 1953; Swanson & Bartonek, 1970; Todorova et al., 2013; Tucker et al., 2015; Ungar, 1964; Vermaat et al., 2000; Vymaza, 2013; Wagner & Oplinger, 2017; Wang & Qiu, 2006; Way, 1982; Weiher et al., 2003; Yatskievych & Raveill, 2001; Zákravsky & Hroudová, 1998.

2.1.2. Damasonium

Fringed water plantain

Etymology: an ancient name of unknown derivation, but possibly commemorating Damasos, a Trojan soldier of the Iliad

Synonyms: *Actinocarpus*; *Machaerocarpus*

Distribution: global: Australia; Europe; North America; **North America:** western

Diversity: global: 5 species; **North America:** 1 species

Indicators (USA): obligate wetland (OBL): *Damasonium californicum*

Habitat: freshwater; lacustrine, palustrine, riverine; **pH:** alkaline; **depth:** <1 m; **life-form(s):** emergent or floating-leaved herb

Key morphology: foliage arising from a corm; the leaves erect, blades undeveloped or lanceolate to ovate (to 9 cm), the margins entire, petioles (to 15 cm) triangular in section; inflorescence an erect raceme (rarely a panicle), the flowers (to 22 mm) bisexual, trimerous, apocarpous, pedicellate (to 65 mm), in whorls (to 9); sepals (to 5 mm) persistent, somewhat hoodlike; petals (to 10 mm) white with basal yellow spot, rarely pink, their apex erose; stamens 6; pistils (to 15) in a ring, spreading or radiating in a star-like pattern; follicles (to 5.5 mm) laterally compressed, beaked (to 6 mm), dehiscing basally, sometimes indehiscent and achene-like

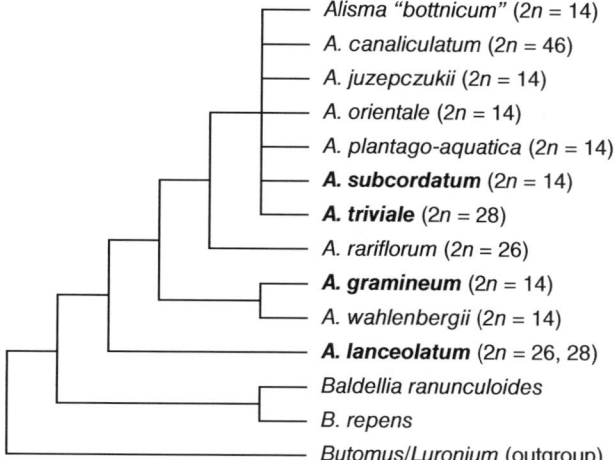

FIGURE 1.4 Phylogenetic relationships of *Alisma* derived from analysis of nrITS sequence data (adapted from Jacobson & Hedrén, 2007). The North American species categorized as OBL indicators are highlighted in bold. Confirmed ploidy levels for the *Alisma* species are shown in parentheses.

Life history: duration: annual (seeds) or perennial (corms); **asexual reproduction:** corm buds; **pollination:** insect or self; **sexual condition:** hermaphroditic; **fruit:** aggregates (follicles) (common); **local dispersal:** mud, water; **long-distance dispersal:** mud, water

Imperilment: 1. *Damasonium californicum* [G4]; S1 (WA); S2 (ID)

Ecology: general: All *Damasonium* species occur in wetlands. The plants are perennial or annual, but all North American representatives are perennial. All of the species are self-compatible, with most being highly autogamous or even cleistogamous. The North American plants are outcrossing, which is associated with protandry, higher pollen quantities (i.e., greater pollen:ovule ratios), and fewer but heavier seeds than the inbreeding Australian or European species, which are homogamous, produce less pollen, and have lighter seeds. The seeds are dispersed in mud or by water. Vegetative reproduction occurs in two species by the production of corm buds. **2.1.2.1.** *Damasonium californicum* **Torr.** occurs in shallow waters (e.g., 15–61 cm deep) or on exposed substrates associated with bottoms (pool; stream), canals (drying), depressions, ditches (roadside), flats, floodplains, marshes, meadows, mudflats (drying), playas, ponds, pools, prairies, rice fields, roadsides, seeps, shores, sloughs (roadside; vernal), streams (flowing), swales, vernal pools, and along the margins of lakes, ponds, reservoirs, and streams (intermittent) at elevations to 2057 m. Although typically emergent, the plants will develop floating leaves when growing in deeper water. The habitats occur in open exposures that receive full sunlight. The substrates are alkaline and include adobe, alluvium, clay, loamy clay, muck, mud, rock (basaltic), sand, sandy silt, serpentine, silty clay, silty mud, stony Tuscan loam, and Tuscan mudflow. Flowering and fruiting occur from May to September, often after standing waters recede. Plants grown in cultivation have flowered from April to June. The variance in flower production is substantial, but averages about 47 flowers/plant. The flowers are strongly protandrous, have the highest pollen:ovule ratios (1,000–1,200:1) in the genus, and are assumed to be highly outcrossed, presumably by insects (Insecta). The male phase occurs from 9 to 14 hr after the onset of flowering, with pollen often accumulating near the base of the petals. The female phase initiates on the second day (27–32 hr after the onset of flowering), as the styles spread radially to expose the stigmas. The petals remain open throughout anthesis. Although the flowers are highly self-compatible, their protandry prevents spontaneous self-pollination. The seeds (averaging 1.42 mg) are dispersed by water or in mud that becomes attached to animal vectors. **Reported associates:** *Alisma, Allium, Arnica chamissonis, Artemisia cana, Artemisia tridentata, Arthrocnemum subterminale, Callitriche hermaphroditica, Carex praegracilis, Comarum palustre, Crypsis schoenoides, Cynodon dactylon, Distichlis spicata, Downingia bacigalupii, Downingia concolor, Downingia insignis, Downingia ornatissima, Downingia yina, Eleocharis macrostachya, Eleocharis palustris, Epilobium campestre, Epilobium cleistogamum, Eryngium alismifolium, Eryngium aristulatum, Eryngium castrense, Frankenia salina, Hippuris vulgaris,* *Hydrocotyle ranunculoides, Isoetes howellii, Juncus arcticus, Juncus balticus, Juncus nevadensis, Lasthenia conjugens, Lasthenia glaberrima, Limosella, Lythrum hyssopifolia, Marsilea vestita, Muhlenbergia richardsonis, Myriophyllum sibiricum, Navarretia leucocephala, Nuphar polysepala, Orcuttia tenuis, Pascopyrum smithii, Phyla, Plagiobothrys leptocladus, Plagiobothrys mollis, Plagiobothrys scouleri, Plagiobothrys stipitatus, Plantago elongata, Pleuropogon, Porterella carnosula, Potamogeton amplifolius, Potamogeton gramineus, Psilocarphus brevissimus, Ranunculus aquatilis, Rosa, Rumex, Sagittaria cuneata, Sparganium, Triglochin scilloides, Typha latifolia, Utricularia macrorhiza, Xanthium strumarium, Zannichellia palustris.*

Use by wildlife: The seeds of *Damasonium californicum* are eaten by waterfowl (Aves: Anatidae). The plants are a host of nematodes (Nematoda: Hoplolaimidae: *Pratylenchus morettoi*).

Economic importance: food: no reported uses; **medicinal:** no reported uses; **cultivation:** not in cultivation; **misc. products:** none; **weeds:** none; **nonindigenous species:** none.

Systematics: In molecular phylogenetic analyses (e.g., Figure 1.3), *Damasonium* resolves as the sister group to a clade containing *Alisma* and *Baldellia*. Although the genus appears to be monophyletic, a comprehensive phylogenetic survey has yet to be conducted for the group. The basic chromosome number of *Damasonium* is $x = 7$. Counts are unavailable for *D. californicum*.

Distribution: *Damasonium californicum* occurs in the western United States.

References: Cook, 1996; Haynes & Hellquist, 2000a; Kaul, 1976; Les & Tippery, 2013; Martin & Uhler, 1951; Michel et al., 1986; Ramaley, 1919; Ross et al., 2016; Turner et al., 2012; Vuille, 1987.

2.1.3. **Echinodorus**

Burhead, swordplant; plante-épée

Etymology: from the Greek *echinos doro* ("spiny spear") in reference to the spiny fruiting heads and sword-like leaves

Synonyms: *Alisma* (in part); *Sagittaria* (in part)

Distribution: global: New World; **North America:** southern

Diversity: global: 28 species; **North America:** 3 species

Indicators (USA): obligate wetland (OBL): *Echinodorus berteroi, E. cordifolius, E. grandiflorus*

Habitat: freshwater; lacustrine, palustrine, riverine; **pH:** 4.8–7.4; **depth:** 0–2 m; **life-form(s):** emergent herb, floating-leaved, submersed (rosulate)

Key morphology: shoots herbaceous, rosulate, rhizomatous; leaves all basal, the blades (to 32 cm) emersed or submersed, lanceolate, ovate, or elliptic, and marked by translucent lines or dots, the base attenuate, cordate, or truncate, the margins entire, often undulating, the petioles (to 115 cm) triangular (rarely roundish) or ridged; inflorescence (to 131 cm) an erect or decumbent (then arching and proliferating) raceme or erect panicle, the flowers (1.1–4 cm) in 1–9 whorls, bisexual, pedicellate (to 7.5 cm), trimerous; sepals (to 21-veined) erect, recurved, or spreading; petals (to 15 mm) white, clawed or sessile; stamens numerous (to 22), the anthers versatile; pistils numerous (to 250), arranged spirally in a head; achenes (to 2.5 mm) oblanceolate, 3–5-ribbed, beaked (to 1.3 mm), sometimes flattened

Life history: duration: annual (fruits/seeds); perennial (rhizomes); **asexual reproduction:** rhizomes; **pollination:** insect, self; **sexual condition:** hermaphroditic; **fruit:** aggregates (achenes) (common); **local dispersal:** rhizomes, achenes (water, wind); **long-distance dispersal:** achenes (waterfowl)

Imperilment: 1. *Echinodorus berteroi* [G5]; SX (IN); S1 (AZ, TN, UT); S2 (AR, KY, OH); **2.** *E. cordifolius* [G5]; SX (DC); S1 (IN, MD); S2 (KS); S3 (IL, NC, VA); **3.** *E. grandiflorus* [GNR]

Ecology: general: All *Echinodorus* species characteristically are aquatic, growing either in standing waters or on moist substrates at sites where the waters have receded. The three North American species are designated as OBL indicators throughout their distributional ranges. Although fairly showy, the flowers of most species appear to be strongly self-compatible, which allows for self-pollination to occur. Otherwise they are believed to be pollinated by insects (Insecta). The plants most often are perennial but reportedly can behave as annuals in more temperate areas. The seeds generally germinate well at high temperatures (25°C–30°C) and any dormancy usually can be broken by incubating them in a 50°C water bath for 10 min, followed by warm incubation at 30°C–35°C until germination occurs. They are buoyant and are dispersed locally by water or wind, or to greater distances via epizoic transport by waterfowl (Aves: Anatidae) or other animal vectors. Most of the species are rhizomatous. When submersed, some taxa develop proliferous inflorescences, which produce vegetative propagules (bulbils) or adventitious plantlets. Because of their popularity as aquarium specimens, many of the species present a risk of becoming nonindigenous introductions via their escape from cultivation.

2.1.3.1. ***Echinodorus berteroi* (Spreng.) Fassett** is an amphibious annual or perennial, which grows in standing waters (e.g., 7–180 cm deep) or on exposed substrates of beaches, canals, depressions, dikes, ditches (roadside), flats, floodplains, lake beds (dried), marshes, meadows, mudflats, playas, ponds (stock), pools, rice fields, river bottoms, riverbeds (drying), roadsides, sandbars, seeps, shores (drying), silt bars, sloughs (vernal), stream beds (dry), swamps, tire ruts, washes, and along the margins of borrow pits, gravel pits, lakes (drying), ponds, reservoirs, and streams at elevations to 2080 m. The plants occur typically in small, temporary water bodies. In deeper waters (and under short photoperiods) the plants remain sterile and produce elongate, ribbon-like leaves. Water temperatures from 18°C to 27°C are considered optimal. Habitat exposures range from full sun to partial shade. The substrates are characterized as alkaline or saline and include alluvium, clay, clay loam, gravel, humus, loamy clay, muck, mud, Riverwash, sand (coarse), sandstone, sandy gravel, sandy loam, sandy muck, silt, silty clay, and stones. Flowering and fruiting extend from April to October. The flowers are self-compatible and have high seed set, presumably as a consequence of self-pollination. The achenes often remain attached to the receptacle when mature, even after the inflorescences

senesce. When shed, they float for prolonged periods and also can be dispersed by their attachment to fur, feathers, or in mud that clings to a potential animal vector. The achenes require no pretreatment and germinate readily at 25°C–30°C. In North America, the plants usually perennate and disperse locally by means of rhizomes; however, they reportedly behave as annuals when growing at more temperate latitudes. **Reported associates:** *Acer negundo, Acer saccharinum, Alisma subcordatum, Alisma triviale, Amaranthus tuberculatus, Ambrosia psilostachya, Ammannia coccinea, Ammannia robusta, Artemisia californica, Atriplex serenana, Azolla filiculoides, Baccharis salicifolia, Baccharis sarothroides, Baccharis sergiloides, Bassia hyssopifolia, Bergia texana, Bidens connatus, Bidens frondosus, Bidens tripartitus, Bolboschoenus fluviatilis, Brassica nigra, Brickellia californica, Celtis laevigata, Cenchrus setaceus, Cephalanthus occidentalis, Ceratophyllum demersum, Chaenactis glabriuscula, Chara, Cirsium vulgare, Coleataenia longifolia, Conoclinium coelestinum, Conyza canadensis, Cressa truxillensis, Crypsis schoenoides, Crypsis vaginiflora, Cylindropuntia prolifera, Cynodon dactylon, Cyperus acuminatus, Cyperus eragrostis, Cyperus erythrorhizos, Cyperus esculentus, Cyperus squarrosus, Cyperus strigosus, Datura wrightii, Digitaria sanguinalis, Echinochloa crus-galli, Echinochloa muricata, Eclipta prostrata, Eleocharis acicularis, Eleocharis macrostachya, Eleocharis obtusa, Eleocharis palustris, Epilobium ciliatum, Epilobium densiflorum, Epilobium torreyi, Eragrostis hypnoides, Eriogonum fasciculatum, Ferocactus viridescens, Frankenia salina, Fraxinus, Gratiola neglecta, Helianthus annuus, Heliotropium, Heteranthera limosa, Hordeum murinum, Ibicella lutea, Iva axillaris, Juglans, Lasthenia glaberrima, Leersia oryzoides, Lemna gibba, Leptochloa fusca, Lindernia dubia, Ludwigia palustris, Ludwigia peploides, Ludwigia polycarpa, Lythrum hyssopifolia, Malvella leprosa, Marsilea vestita, Mimulus cardinalis, Najas, Nasturtium officinale, Navarretia fossalis, Oryza sativa, Panicum capillare, Panicum dichotomiflorum, Paspalum dilatatum, Paspalum distichum, Paspalum pubiflorum, Penthorum sedoides, Persicaria coccinea, Persicaria lapathifolia, Phalaris arundinacea, Phyla lanceolata, Phyla nodiflora, Plagiobothrys leptocladus, Plagiobothrys undulatus, Plantago major, Platanus racemosa, Pluchea odorata, Polypogon monspeliensis, Pontederia cordata, Populus fremontii, Potamogeton nodosus, Pseudognaphalium luteoalbum, Quercus berberidifolia, Raphanus sativus, Rotala ramosior, Rumex persicarioides, Rumex pulcher, Sagittaria graminea, Sagittaria latifolia, Sagittaria montevidensis, Salix gooddingii, Salix laevigata, Schinus molle, Schoenoplectus acutus, Schoenoplectus californicus, Schoenoplectus hallii, Schoenoplectus mucronatus, Schoenoplectus tabernaemontani, Sphenoclea zeylanica, Stuckenia pectinata, Tamarisk, Trichocoronis wrightii, Typha angustifolia, Typha domingensis, Typha latifolia, Veronica anagallis-aquatica, Veronica peregrina, Xanthium strumarium, Zannichellia palustris.*

2.1.3.2. *Echinodorus cordifolius* **(L.) Griseb.** is a perennial, which grows in shallow (e.g., 1–20 cm depth), tidal or non-tidal waters or on exposed substrates of backwaters, bayous, bottomlands, depressions, ditches (roadside), flood-plains, marshes, meadows, mudflats, oxbows, ponds (depression), pools, prairies, rice fields (abandoned), roadsides, shores (seepy), sloughs, springs, streams, swales, swamps, and along the margins of canals, Carolina bays, channels, lakes, ponds, reservoirs, and rivers at elevations to 500 m. The plants frequently occur as emergents in drawdown areas but will develop submersed leaves when growing under photoperiods of less than 12 h. All the leaves will die back when temperatures fall below 10°C. Site exposures range from full sunlight to partial shade. The substrates (pH: 5.8–7.4) are described as calcareous and include alluvium (clayey), clay, clay mud, humus, loam, muck, mucky Dowling clay, mud, sand, sandy loam (Cecil), sandy silt, silt, silty loam, and silty sand. Flowering occurs under long-day conditions from May to October; fruiting has been observed from July to October. Low seed set typically has been reported, which might indicate a greater requirement for pollinator visitation (and increased outcrossing). A seed bank can develop with recovery densities of germinating seeds ranging from 0.7 (non-flooded sites) to 1.4 (flooded sites) seeds m^{-2}. The arching inflorescences proliferate by producing adventitious plant-lets, which can function as vegetative propagules. Vegetative reproduction also occurs by means of rhizomes. The plants have been micropropagated successfully (from rhizomes) using a combination of 24.6 µM *N*-isopentenyladenine and 2.68 µM naphthalene acetic acid. **Reported associates:** *Acer saccharinum, Ammannia coccinea, Ampelopsis arborea, Asclepias perennis, Bacopa, Bergia texana, Boehmeria cylindrica, Bolboschoenus fluviatalis, Campsis radicans, Carex comosa, Carex crus-corvi, Carex longii, Carex lupulina, Carpinus caroliniana, Celtis laevigata, Cephalanthus occidentalis, Chamaecrista fasciculata, Chasmanthium latifolium, Cornus foemina, Cynosciadium digitatum, Cyperus pseudovegetus, Cyperus squarrosus, Diospyros virginiana, Echinochloa colona, Elaeagnus umbellata, Elatine californica, Eragrostis hypnoides, Eupatorium, Forestiera acuminata, Hibiscus, Hydrolea uniflora, Ipomoea hederacea, Iris tridentata, Juncus effusus, Leersia oryzoides, Lemna, Leptochloa fusca, Leptochloa panicoides, Leucospora, Lindernia dubia, Liquidambar styraciflua, Lobelia cardinalis, Ludwigia alternifolia, Ludwigia peploides, Ludwigia repens, Marsilea vestita, Micranthemum umbrosum, Mimulus alatus, Najas flexilis, Nelumbo lutea, Nuphar, Nyssa biflora, Persicaria hydropiperoides, Phyla lanceolata, Phyla nodiflora, Pilea, Pinus taeda, Planera aquatica, Pluchea camphorata, Populus heterophylla, Quercus lyrata, Quercus nigra, Quercus palustris, Quercus shumardii, Rhynchospora corniculata, Rotala ramosior, Sabal minor, Sabatia calycina, Sagittaria latifolia, Sagittaria platyphylla, Salix caroliniana, Salix nigra, Sarracenia flava, Saururus cernuus, Schoenoplectus tabernaemontani, Scirpus cyperinus, Scleria, Senna marilandica, Sium, Sparganium americanum, Taxodium distichum, Torilis, Toxicodendron radicans, Triadica sebifera, Tridens flavus, Typha latifolia, Ulmus alata, Ulmus rubra, Utricularia gibba, Vitis, Xanthium strumarium, Zannichellia palustris.*

2.1.3.3. ***Echinodorus grandiflorus*** **(Cham. & Schltdl.) Micheli** is a nonindigenous perennial, which occurs in swamps at elevations to 10 m. The plants can tolerate a fairly wide range of acidity (pH: 4.8–7.4) and in North America, occur on sand substrates. Flowering (North America) occurs from summer through the fall. In its native range, flowering occurs in the mornings of the rainy season and lasts about 8 hr in duration. The plants are self-compatible but attract various pollinating bees (Insecta: Hymenoptera: Andrenidae; Anthophoridae; Apidae; Colletidae; Halictidae). Seed set can range from 0.2% to 73% but typically is low; it is substantially higher in open-pollinated plants than in selfed individuals. The unscarified seeds germinate moderately (18–35%) when incubated under continuous illumination at 25°C–30°C, but not under continuous darkness. Scarified seeds (the seed coat slit near the site of embryo emergence) germinate well (98–99%) under constant illumination or in darkness. The inflorescences are proliferous and produce numerous vegetative plantlets. Vegetative reproduction also occurs by means of rhizomes. **Reported associates (North America):** *Acer rubrum, Colocasia esculenta, Juncus effusus, Magnolia virginiana, Mikania scandens, Myrica cerifera, Nyssa biflora, Osmunda regalis, Osmundastrum cinnamomeum, Persea palustris, Persicaria punctata, Pinus elliottii, Rhynchospora miliacea, Sabal minor, Saururus cernuus, Smilax walteri, Taxodium distichum, Thelypteris palustris, Woodwardia areolata.*

Use by wildlife: The achenes of several *Echinodorus* species are eaten occasionally by waterfowl (e.g., Aves: Anatidae: *Anas platyrhynchos*), which have been suggested as potential long-distance dispersal agents. The achenes of *E. berteroi* are eaten by cinnamon teal (Aves: Anatidae: *Anas cyanoptera*). *Echinodorus berteroi* is used frequently as a nesting site of coots (Aves: Rallidae: *Fulica americana, F. caribaea*). *Echinodorus berteroi* and *E. cordifolius* are the hosts of sac Fungi (Ascomycota: Mycosphaerellaceae: *Cercospora echinodori*). *Echinodorus cordifolius* is a larval host plant of moths (Insecta: Lepidoptera: Crambidae: *Synclita occidentalis*); it also is fed upon by weevils (Insecta: Coleoptera: Curculionidae: *Listronotus echinodori*).

Economic importance: food: *Echinodorus cordifolius* plants can contain up to 81.6% digestible dry matter and 17.2% protein; however, they have not been used as a human food; **medicinal:** Root decoctions of *E. berteroi* have been used as a folk remedy for epilepsy and have been found to exhibit neuroleptic and antiepileptic activity. Aqueous extracts of *E. grandiflorus* have been shown to exhibit potent vasodilator activity and also may be a potential therapeutic agent for asthma; ethanolic extracts from the plants are antihypertensive. Foliar alcoholic extracts of *E. grandiflorus* are rich in diterpenes and flavonoids, which have been linked to their effectiveness in treating inflammatory conditions; **cultivation:** Several *Echinodorus* species are cultivated widely as ornamental aquarium plants, including *E. berteroi, E. cordifolius,* and *E. grandiflorus.* Cultivars of *E. cordifolius* include 'Marble Queen', 'Oriental', and 'Ozelot Green'; **misc. products:** *Echinodorus cordifolius* has been recommended as a potentially effective phytoremedial agent for the removal of azo dyes, ethylene glycol, and phosphorous from contaminated wastewater; **weeds:** *Echinodorus berteroi* is a weed of California rice fields; **nonindigenous species:** *Echinodorus grandiflorus* was introduced to Florida as an escape from cultivation, sometime before 1981.

Systematics: *Echinodorus* resolves as a clade after the removal of species assigned previously to subgenus *Helanthium*, which have been transferred to the genera *Albidella* and *Helanthium*. Its sister group is not certain, but has included combinations of various genera (*Alisma, Astonia, Baldellia, Helanthium, Sagittaria,* and *Wiesneria*) depending on the analysis and data sets analyzed. Originally named as a distinct species, *E. floridanus* does not appear to differ from *E. grandiflorus* (Figure 1.5), and it has been treated here as a synonym of the latter (along with the transfer of its OBL indicator status to *E. grandiflorus*). Phylogenetic analyses have indicated that the formerly proposed sectional divisions of *Echinodorus* are not entirely monophyletic. However, the assignment of *E. cordifolius* and *E. grandiflorus* to the same section (sect. *Cordifolii*) and *E. berteroi* to a different section (sect. *Berteroii*) is consistent with the results of phylogenetic analyses, which resolve *E. berteroi* as a distinct group and unite *E. cordifolius* and *E. grandiflorus* within the same clade (Figure 1.5). The basic chromosome number of *Echinodorus* is $x=11$. All counted species in the genus, including *E. berteroi, E. cordifolius,* and *E. grandiflorus*, are uniformly diploid ($2n=22$). Hybridization is known to occur commonly among species in cultivation; however, no accounts of natural hybridization involving *Echinodorus* have been reported in North America. Putative intergeneric hybrids involving *Alisma* and *Echinodorus* have been called ×*Alismodorus* H.R.Wehrh.

Distribution: *Echinodorus berteroi* occurs across southern North America, with *E. cordifolius* restricted to the southeastern United States; both species extend into South America. The nonindigenous *E. grandiflorus* currently occurs only in Florida, but is native to South America.

References: Beal, 1977; Boyd & McGinty, 1981; Braun et al., 2014; Brugiolo et al., 2011; Buznego & Pérez-Saad, 2006; Cook, 1996; Correll & Correll, 1975; de Faria Garcia et al., 2010; Dissanayake et al., 2007; DiTomaso & Healy, 2003; Fassett, 1955; 1957; Gordón, 1997; Harms & Grodowitz, 2009; Haynes & Burkhalter, 1998; Haynes & Hellquist, 2000a; Haynes & Holm-Nielsen, 1994; Hohman & Ankney, 1994; Kasselmann, 2001; Knobloch, 1972; Lehtonen, 2009a; Lehtonen & Myllys, 2008; Lessa et al., 2008; McAtee, 1918; McNair & Cramer-Burke, 2006; Middleton, 2009; Mühlberg, 1982; Munz & Johnston, 1922; Noonpui & Thiravetyan, 2011; Rataj, 1975; 2004; Reese & Lubinski, 1983; Teamkao & Thiravetyan, 2010; Tibiriçá et al., 2007; Torit et al., 2012; Tucker et al., 2015; Usinger, 1956; Vieira & de Souza Lima, 1997.

2.1.4. Helanthium

Marsh flower, mudbabies

Etymology: from the Greek *helos anthos* ("marsh flower") in reference to the habitat

Synonyms: *Alisma* (in part); *Echinodorus* (in part); *Helanthium*

Distribution: global: New World; **North America:** eastern United States

Diversity: global: 3 species; **North America:** 1 species

Indicators (USA): obligate wetland (OBL): *Helanthium tenellum*

Habitat: freshwater; freshwater (tidal); palustrine; **pH:** 6.1–7.6; **depth:** <1 m; **life-form(s):** emergent or submersed (rosulate) herbs

Key morphology: foliage (to 6 cm) arising from rosettes, connected by fine pseudostolons; submersed leaves (to 10 cm) sessile and linear, emersed leaves with ridged petioles (to 9.5 cm), the blades (to 7.4 cm) narrowly lanceolate to ovate; umbels or racemes (to 8 cm) pedunculate (to 4 cm), with 1–2 whorls of 4–16 flowers; flowers (to 10 mm) pedicellate (to 3 cm), trimerous, radial; petals (to 4 mm) clawed, white; stamens 9; gynoecium apocarpous (to 20 pistils); fruit a headlike aggregate of flattened achenes (to 1.5 mm) with lateral beaks (to 0.2 mm)

Life history: duration: annual (fruits/seeds); **asexual reproduction:** pseudostolons; **pollination:** unknown; **sexual condition:** hermaphroditic; **fruit:** aggregates (achenes) (common); **local dispersal:** achenes (water?), pseudostolons; **long-distance dispersal:** achenes (animals)

Imperilment: 1. *Helanthium tenellum* [G5]; SX (DE, NY); SH (LA, MA, NC); S1 (CT, IL, KS, MI, MS, VA); S2 (SC)

Ecology: general: All *Helanthium* species occur in wetlands and are amphibious plants capable of growing under emersed or submersed conditions. Their reproductive biology and seed ecology have not been described in any detail and require further study.

2.1.4.1. *Helanthium tenellum* **Britt.** is an annual, which inhabits shallow waters (up to at least 50 cm depth) or colonizes exposed substrates in borrow pits, canals, depressions, ditches, flats, meadows, pools (gum; swamp), prairies, sinkholes, swales, and along the margins of small lakes (depression), ponds (depression; drying; limesink depression; sink hole; temporary; swale), rivers, and streams at elevations to 591 m. Although most often found on wet, exposed substrates, the plants also grow well when fully submersed. Exposures typically receive full sunlight, but also include partially shaded conditions. Occurrences span a pH range from 6.1 to 7.6, but the plants thrive in weakly acidic waters of soft to medium hardness at temperatures from 18°C to 28°C (but up to 33°C). The substrates typically are acidic and include clay, Cowarts (typic hapludults), loam, muck (organic), mud (exposed), peat, peaty sand, Plummer (grossarenic paleaquults), Rutlege (typic humaquepts), sand, sandy peat, and silty loam. Flowering occurs only when the inflorescences are emersed. Flowering and fruiting extend from March to September. The pollination biology of this species has not been described, but seed set tends to be high as in most annuals. In its South American range, this species can dominate the seed bank, comprising more than 24% of the total germinating propagules. The mechanism of seed dispersal also has not been elucidated but likely involves water over short distances

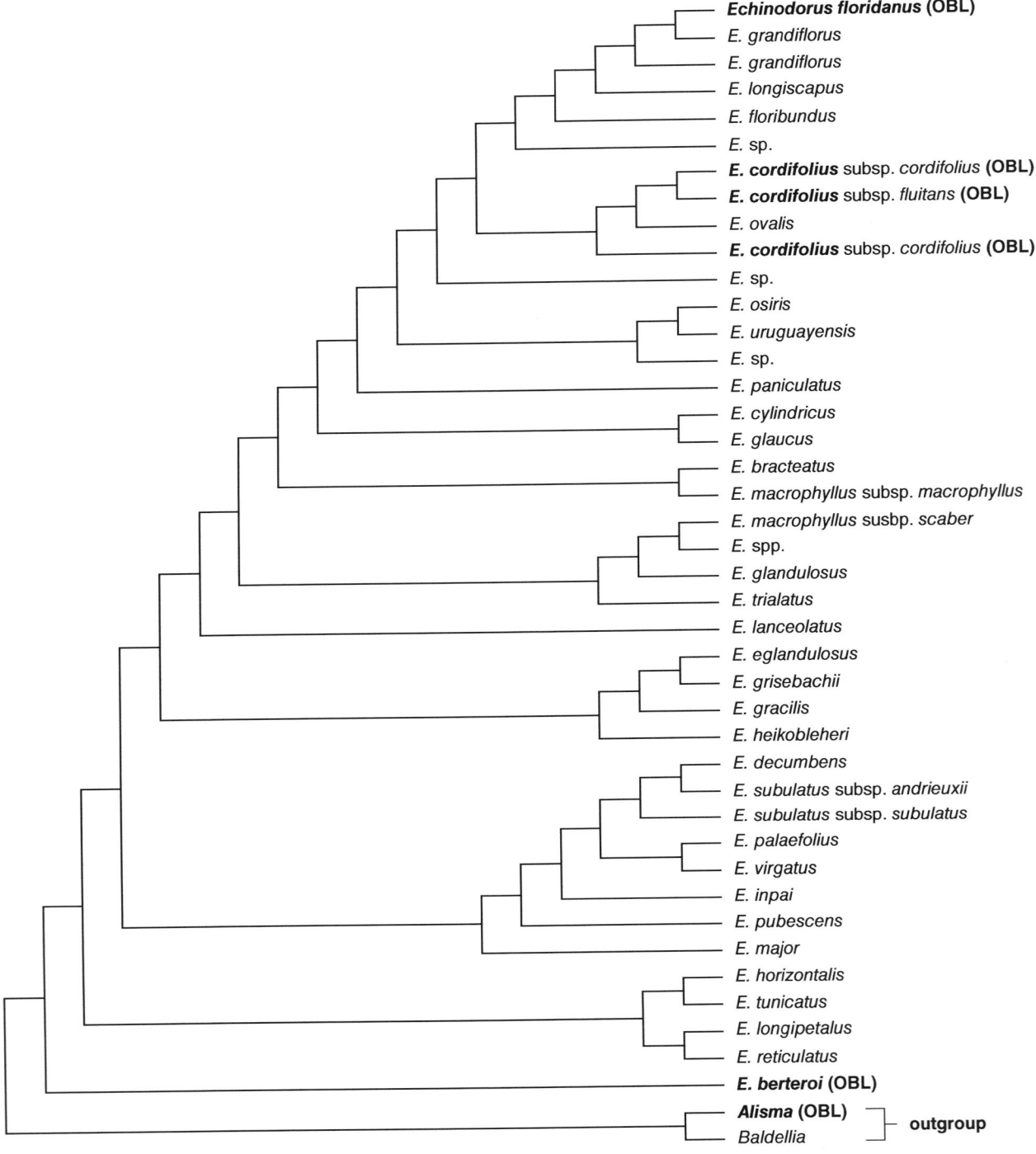

FIGURE 1.5 Interspecific relationships in *Echinodorus* as indicated by phylogenetic analysis of combined morphological and DNA sequence data (adapted from Lehtonen & Myllys, 2008). The OBL North American wetland indicators (in bold) represent diverse origins in the genus. These results do not distinguish *E. floridanus* and *E. grandiflorus* (the latter name has been retained in the present treatment); the currently defined subspecific divisions of *E. cordifolius* also are untenable.

and attachment to various animal vectors over greater distances. The plants can be connected by delicate "pseudostolons," i.e., sterilized inflorescences from which vegetative buds can develop. Although unusual for an annual species, this means of vegetative reproduction can quickly result in the development of large, dense "cushions" of plants during the growing season. **Reported associates:** *Alisma triviale, Bidens cernuus, Bidens connatus, Bidens frondosus, Bidens*

vulgatus, Boehmeria cylindrica, Bolboschoenus fluviatilis, Callitriche heterophylla, Cardamine pensylvanica, Carex barrattii, Carex lasiocarpa, Carex lurida, Cephalanthus occidentalis, Chelone glabra, Coleataenia longifolia, Cyperus dentatus, Cyperus diandrus, Cyperus esculentus, Cyperus haspan, Cyperus strigosus, Echinochloa crus-galli, Eleocharis acicularis, Eleocharis melanocarpa, Eleocharis obtusa, Eleocharis ovata, Eleocharis palustris, Fimbristylis vahlii, Gratiola aurea, Helenium virginicum, Heteranthera limosa, Hypericum mutilum, Hyptis alata, Isoetes virginica, Juncus acuminatus, Leersia oryzoides, Lindernia dubia, Lipocarpha maculata, Lipocarpha micrantha, Ludwigia palustris, Lycopus uniflorus, Lysimachia terrestris, Lythrum salicaria, Mayaca fluviatilis, Mentha arvensis, Nelumbo lutea, Nuphar variegata, Nymphaea odorata, Oldenlandia boscii, Oldenlandia uniflora, Panicum hemitomon, Panicum verrucosum, Penthorum sedoides, Persicaria hydropiperoides, Persicaria lapathifolia, Persicaria longiseta, Persicaria maculosa, Persicaria pensylvanica, Phalaris arundinacea, Physostegia leptophylla, Pilea pumila, Polypremum procumbens, Pontederia cordata, Proserpinaca palustris, Rhexia virginica, Rorippa palustris, Rotala ramosior, Rotala ramosior, Sagittaria graminea, Sagittaria latifolia, Schoenoplectus erectus, Schoenoplectus hallii, Schoenoplectus smithii, Scirpus cyperinus, Scutellaria lateriflora, Serenoa repens, Solanum dulcamara, Sparganium americanum, Sphagnum, Stachys hyssopifolia, Symphyotrichum lateriflorum, Taxodium, Typha latifolia, Xanthium strumarium, Xyris jupicai.

Use by wildlife: none reported.

Economic importance: food: not reported as edible; **medicinal:** *Helanthium tenellum* has been used in South America in preparing treatments for headaches, rheumatism, and syphilis; **cultivation:** *Helanthium tenellum* is grown as an aquarium plant; **misc. products:** none; **weeds:** none; **nonindigenous species:** none.

Systematics: Once treated as a section or subgenus of *Echinodorus*, *Helanthium* has more recently been recognized as a distinct genus in accordance with results of phylogenetic analyses. The species assigned to *Helanthium* group consistently as a clade, which resolves as the sister group of *Echinodorus* in several studies, including analyses of complete cpDNA gene sequences (Figure 1.3); however, other studies indicate *Ranalisma* as a possible sister group (Figure 1.6). Phylogenetic analyses of the three consistently recognized species using combined molecular and morphological data resolve *H. tenellum* as the sister to the remainder of the genus (Figure 1.6). Some authors have recognized as many as 10 species in the genus, indicating that a more comprehensive taxonomic evaluation is necessary. The base

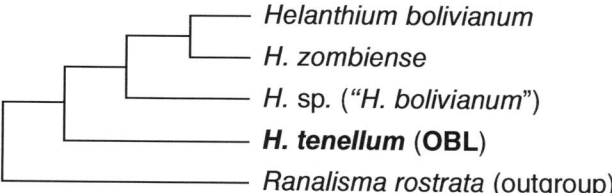

FIGURE 1.6 Phylogenetic relationships in *Helanthium* as indicated by analysis of combined molecular and morphological data (adapted from Lehtonen & Myllys, 2008). These results resolve the OBL *H. tenellum* (in bold) as the sister to the remainder of the genus. They also indicate the possibility of a fourth species, which had been identified provisionally as *H. bolivianum*, but does not associate with that species.

chromosome number of *Helanthium* is *x*=11. *Helanthium tenellum* has diploid and triploid cytotypes (2*n*=22, 33). The triploids have meiotic abnormalities and reduced pollen stainability (~50%) with respect to the diploids (normal meiosis and 100% pollen stainability). Chromosomal (C-banding) analyses indicate that the triploids are of autopolyploid origin.

Distribution: *Helanthium tenellum* occurs sporadically throughout the eastern United States.

References: Bercu, 2015; Brooks & Wardrop, 2013; Charlton, 1979; Chester & Palmer-Ball, 2011; Chester & Souza, 1984; Costa, 2004; Haynes & Hellquist, 2000a; Haynes & Holm-Nielsen, 1994; Kasselmann, 2003; Lehtonen, 2007; 2009; Lehtonen & Myllys, 2008; Les & Tippery, 2013; Pagotto et al., 2011; Rataj, 2004.

2.1.5. Hydrocleys

Waterpoppy

Etymology: probably from the Greek *hydro keleuthos* ("leaving a water path"), in reference to the stepping-stone-like pattern displayed on the water by the floating leaves

Synonyms: *Hydrocleis*; *Ostenia*; *Vespuccia*

Distribution: global: New World; **North America:** southern United States

Diversity: global: 5 species; **North America:** 1 species

Indicators (USA): obligate wetland (OBL): *Hydrocleys nymphoides*

Habitat: freshwater; lacustrine, palustrine; **pH:** 5.5–6.0; **depth:** 0–2 m; **life-form(s):** floating-leaved

Key morphology: plants (to 50 cm) stoloniferous (to 45 cm); leaves submersed (sessile and phyllodial) or floating, then the petioles long (to 40 cm), septate, sheathing (to 8.5 cm), the blades (to 11.9 cm) ovate to orbiculate, the base rounded or cordate; inflorescence (to 6 flowers) umbellate, proliferous, the peduncle (to 30 cm) septate; flowers (to 6.5 cm) pedicellate (to 17.5 cm); petals (to 4.1 cm) spreading, yellow to white, yellow at base; stamens (to 25) and outer staminodes (to 20+) numerous; gynoecium apocarpous, pistils (5–8; to 10 mm) attenuate to an inwardly curved style; follicles (to 14.5 mm) beaked (to 5.5 mm), dehiscing adaxially; seeds (to 50+) small (~1 mm), with glandular trichomes (to 0.15 mm)

Life history: duration: perennial (stolons); **asexual reproduction:** stolons; **pollination:** insect; **sexual condition:** hermaphroditic; **fruit:** aggregates of follicles (common); **local dispersal:** seeds (water), stolons, vegetative plantlets (water); **long-distance dispersal:** seeds (water, waterfowl?)

Imperilment: 1. *Hydrocleys nymphoides* [G5]

Ecology: general: All *Hydrocleys* species are floating-leaved hydrophytes (also producing submersed foliage), which occupy relatively shallow lentic or standing waters. Although normally rooted in the muddy substrate, the plants can dislodge and assume a free-floating habit. They produce floating inflorescences with showy flowers that are pollinated primarily by bees (Insecta: Hymenoptera: Apidae; Colletidae) or flies (Insecta: Diptera), which are attracted to their bright perianths by volatile methoxylated aromatics (e.g., ρ-methylanisole). Unlike many monocots, the perianth is not shed, but becomes extremely soft when water-soaked, eventually withering. Flowering can extend year-round in the native South American range. The sterility of some nonindigenous populations indicates the possibility of self-compatibility in the genus, but that prospect has not been investigated. The seeds are dispersed by water and possibly to greater distances by attachment to the plumage of waterfowl (Aves: Anatidae). There are no known specific germination requirements. The inflorescences of some species proliferate and produce vegetative buds, which produce adventitious roots and can develop into new plantlets as the peduncles arch over and contact the substrate.

2.1.5.1. *Hydrocleys nymphoides* (Humb. & Bonpl. ex Willd.) Buch.

is a nonindigenous perennial, which grows in ditches, pools (shallow), and along the margins of lakes at elevations to 70 m. It can tolerate depths to 2 m, but thrives in shallower waters from 20 to 45 cm. Exposures can range from fully open to strongly shaded sites. Waters in the native South American habitats are acidic (pH: 5.5–6.0), soft, and can range from 6°C to 24°C; the native substrates include loam and sandy gravel. Flowering (North America) has been observed in July, but the pollination biology has not been studied in this nonindigenous region. In their native South American range, the flowers are visited by flies (Insecta: Diptera) and bees (Insecta: Hymenoptera: Apidae: *Geotrigona argentina*) and the pollen is found in various honey samples, which indicates that the plants normally are entomophilous. Plants introduced to Australia completely fail to set seed. The inflorescence often proliferates into leaves or stolons, which are capable of vegetative reproduction (the primary reproductive mode at least in nonindigenous populations). Detached, floating vegetative plantlets can facilitate late seasonal dispersal. Unlike many indigenous species, the photosynthetic surface area (and biomass accumulation) increases proportionally with nutrient levels, which facilitates their invasion into new territories. The plants are non-mycorrhizal. **Reported associates (North America):** *Acmella repens, Alternanthera philoxeroides, Cyperus flavicomus, Diodia virginiana, Eleocharis obtusa, Hydrocotyle verticillata, Kyllinga brevifolia, Sagittaria platyphylla*.

Use by wildlife: In the Pantanal of Brazil, *Hydrocleys nymphoides* is consumed frequently by marsh deer (Mammalia: Cervidae: *Blastocerus dichotomus*). The plants are known hosts of powdery mildew (Fungi: Ascomaycota: Erysiphaceae: *Erysiphe*).

Economic importance: food: *Hydrocleys nymphoides* supposedly is edible, but no specific uses have been reported; **medicinal:** Extracts of *Hydrocleys nymphoides* have exhibited antimicrobial activity against pathogenic bovine mastitis microbes; **cultivation:** *Hydrocleys nymphoides* is a popular ornamental plant for aquariums and garden ponds; **misc. products:** none; **weeds:** *Hydrocleys nymphoides* is a "vigorously controlled" weed in New Zealand; **nonindigenous species:** *Hydrocleys nymphoides* is nonindigenous to North America, having escaped from cultivation sometime during the 20th century. The plants were introduced to Hawaii before 1938. They also have become naturalized in Australia, Japan, Korea, New Zealand, and Taiwan.

Systematics: *Hydrocleys* formerly was placed in the Butomaceae or Limnocharitaceae until several phylogenetic analyses consistently indicated that the group was nested within Alismataceae. However, the sister group of *Hydrocleys* remains somewhat uncertain, with either *Butomopsis* or *Butomopsis + Limnocharis* resolving in that position depending on the specific data set analyzed. A closer relationship between *Butomopsis* and *Hydrocleys* is indicated by their highly similar chromosome morphology and symmetry; chromosomally, they are both interpreted as relatively primitive elements compared to *Limnocharis*. Interspecific relationships in *Hydrocleys* have been investigated to some degree using morphological phylogenetic analyses, which indicate that the group is monophyletic. However, those analyses are incongruent with respect to the relationships depicted when using different outgroups (i.e., *Butomopsis* vs. *Limnocharis*), which is problematic given the uncertain sister-group relationships (see above). *Hydrocleys nymphoides* resolves either as the sister species of *H. martii* or *H. mattogrossensis* in the morphological cladograms. A phylogenetic evaluation of the genus using molecular data potentially would be helpful in resolving the interspecific relationships and also might help to clarify the sister group of *Hydrocleys*. The chromosomal base number is $x=7$ or $x=8$. *Hydrocleys nymphoides* ($2n=16$; $x=8$) is a diploid, with a triploid cytotype ($2n=24$) also reported. The diploid *H. modesta* ($2n=14$; $x=7$) differs by having a pair of telocentric chromosomes and a pair of acrocentric chromosomes replaced by a pair of metacentric chromosomes as a result of Robertsonian fission or fusion. No natural hybrids involving *H. nymphoides* have been reported.

Distribution: The nonindigenous *Hydrocleys nymphoides* reportedly is naturalized in Florida, Louisiana, and Texas.

References: Aston & Jacobs, 1980; Boxell, 2014; Carvalho et al., 2014; Clayton, 1996; Cook et al., 1997; Edgerton, 2014; Fagúndez & Caccavari, 2006; Forni-Martins & Calligaris, 2002; Haynes, 2004; Haynes & Holm-Nielsen, 1992; Haynes et al., 1998b; Kadono, 2004; Kasselmann, 2003; Kenton, 1981; Les & Tippery, 2013; MacRoberts & MacRoberts, 2010; McKenzie & Lovell, 1992; Nesom, 2009; Ross et al., 2016; Rossi et al., 2011; Shin et al., 2012; Tomas & Salis, 2000; Vossler et al., 2010; Wang & Qiu, 2006; Wester, 1992; Wu et al., 2010.

2.1.6. *Sagittaria*

Arrowhead, wapato; flèche d'eau, sagittaire
Etymology: derived from the Latin *sagitta* ("arrow"), in reference to the arrow-like leaves of some species
Synonyms: *Alisma* (in part); *Echinodorus* (in part); *Lophiocarpus* (later homonym); *Lophotocarpus*
Distribution: global: Eurasia; Western hemisphere; **North America:** widespread
Diversity: global: 40 species; **North America:** 26 species
Indicators (USA): obligate wetland (OBL): *Sagittaria ambigua, S. australis, S. brevirostra, S. calycina, S. cristata, S. cuneata, S. demersa, S. engelmanniana, S. fasciculata, S. filiformis, S. graminea, S. guayanensis, S. isoetiformis, S. kurziana, S. lancifolia, S. latifolia, S. longiloba, S. macrocarpa,*

S. montevidensis, S. papillosa, S. platyphylla, S. rigida, S. sanfordii, S. secundifolia, S. subulata, S. teres
Habitat: brackish, freshwater; freshwater (tidal); lacustrine, palustrine, riverine; **pH:** 4.4–9.2; **depth:** 0–2 m; **life-form(s):** emergent, floating-leaved, or submersed (rosulate) herbs
Key morphology: rosettes emersed, floating-leaved, or submersed, with stolons or rhizomes (sometimes producing tubers), the roots septate; leaves (when submersed) phyllodial (to 7–70 cm; to 250 cm in some species), leaves (when emersed or floating) with blades cordate, elliptic, hastate, lance-elliptic, lanceolate, linear, linear-lanceolate, linear-oblanceolate, linear-ovate, ovate, sagittate, spatulate (to 2.5–35 cm), margins entire, petioles (to 4–100 cm) circular or triangular; inflorescences (to 4–72 cm) erect, floating, or emergent racemes or panicles, the flowers (to 0.8–4.0 cm) sessile (rarely) or stalked (to 1.0–6.5 cm), in whorls (to 4–17); flowers bisexual (rarely) or unisexual and in andromonoecious (rarely; normally pistillate flowers with a ring of stamens), dioecious (rarely), or monoecious (distally male) arrangements, perianth trimerous; sepals recurved or erect; petals white (rarely with a purplish spot); stamens numerous (to 30); gynoecium apocarpous, pistils numerous (to 1,500+), spirally arranged; fruit an aggregate head (to 0.5–3.0 cm) of achenes (to 1.5–4.0 mm); achenes compressed, abaxial keel and/or abaxial/lateral wings absent or present, with an erect or lateral beak (to 0.2–2.1 mm; to 17 mm in one species)
Life history: duration: annual (fruits/seeds); perennial (corms, rhizomes, stolons); **asexual reproduction:** rhizomes, stolons, tubers; **pollination:** insect; **sexual condition:** dioecious, monoecious; **fruit:** aggregates (achenes) (common); **local dispersal:** fish, water (achenes); **long-distance dispersal:** birds (achenes)
Imperilment: 1. *Sagittaria ambigua* [G2]; S1 (AR, MO); S2 (KS); **2.** *S. australis* [G5]; S1 (DC, IL, NJ); S2 (IN); **3.** *S. brevirostra* [G5]; SH (MN, VA); S1 (TN); S2 (CO); S3 (IL); **4.** *S. calycina* [G5]; SH (DC, NH); S1 (CO, DE, MA, NC, PA, QC, VA, WI, WV); S2 (MD, NB, NJ, NY, OR, WV); S3 (CT, DE, IL, ME, NJ); **5.** *S. cristata* [G4]; S3 (ON); **6.** *S. cuneata* [G5]; S1 (CT, NH, NJ, OH, OK, PA); S2 (MA, YT); S3 (IA, WY); **7.** *S. demersa* [G2/G4]; **8.** *S. engelmanniana* [G5]; SH (VA); S1 (WV); S2 (DE, MD, NC); **9.** *S. fasciculata* [G2]; S2 (NC, SC); **10.** *S. filiformis* [G4/G5]; SX (PA); SH (NC, NJ); S2 (ME); **11.** *S. graminea* [G5]; S1 (AZ, CO, KY, NC, OH, PE, SC, TN); S2 (DE, KS, NC, NE, RI); S3 (GA, IA, IL, NC, NF, VA); **12.** *S. guayanensis* [G5]; **13.** *S. isoetiformis* [G4]; S1 (MS); S2 (AL, NC); S3 (SC); **14.** *S. kurziana* [G4]; S1 (MS); **15.** *S. lancifolia* [G5]; S1 (DE); S3 (NC); **16.** *S. latifolia* [G5]; S1 (AZ, NF, WY); S2 (AB, NC); S3 (DE, OH); **17.** *S. longiloba* [G5]; S1 (AZ, NE); **18.** *S. macrocarpa* [G5T2]; S2 (NC); **19.** *S. montevidensis* [G5]; **20.** *S. papillosa* [G5]; **21.** *S. platyphylla* [G5]; S1 (GA, KS, KY, MO, NC, OH, WV); S2 (TN); **22.** *S. rigida* [G5]; SX (DC); SH (AR, DE, NH, PE); S1 (ID, KS, KY, MD, VA, TN); S2 (ME, MB, MA, OH); S3 (NE, QC, VT); **23.** *S. sanfordii* [G3]; S3 (CA); **24.** *S. secundifolia* [G1]; S1 (AL, GA); **25.** *S. subulata* [G4]; SH (RI); S1 (MA); S2 (DE, NC, NJ); S3 (CT, NY, PA, VA); **26.** *S. teres* [G3]; S1 (NH, NJ, NY, RI); S3 (MA)

Ecology: general: *Sagittaria* is entirely aquatic, with all 26 North American species designated as OBL indicators throughout their ranges. Although not recognized in the National list, *S. calycina* and *S. macrocarpa* have been reinstated here as distinct species and are included as OBL indicators (see *Systematics*). Three species (*S. chapmanii, S. spathulata, S. weatherbiana*) retain OBL rankings on the National Wetland list, but are excluded here on taxonomic grounds (see *Systematics*). Information attributed to these taxa has been included with the species to which they currently are assigned. Most of the species are perennials (from coarser rhizomes or finer stolons) but a few behave as annuals. The stolons often are terminated by tubers, which some refer to as corms. The plants grow as submersed, floating-leaved, and emergent life-forms, with the various species producing leaves that range from narrow, submersed phyllodes to emergent broad-bladed structures. The flowers almost always are unisexual (sometimes the pistillate appearing to be bisexual by the presence of sterile stamens), rarely perfect, and are arranged in andromonoecious (having some perfect flowers), dioecious, or monoecious sexual conditions; sexual conditions sometimes are mixed within a species. Monoecious individuals have the staminate flowers situated above the pistillate flowers. They are self-compatible (i.e., capable of geitonogamy) but most of the flowers produce nectar and display a showy corolla of prominent white petals, which serves to attract various bees, beetles, butterflies, flies, and wasps (Insecta: Coleoptera; Diptera; Hymenoptera; Lepidoptera) as potential pollinators. Seed production generally tends to be high in both monoecious and dioecious populations. Seed germination requirements are poorly understood for many species and can be complex. Varied reports indicate germination in water without any pretreatment, germination following a five- to seven-month cold treatment, germination requiring warm stratification, and germination under constant or fluctuating incubation temperatures. Physical scarification (by puncture of the integument) is known to increase germination rates in some species. Seeds originating from dioecious plants can produce germination inhibitors, which require stratification and/or scarification to overcome. Usually, the seeds are consumed only occasionally by waterfowl (Aves: Anatidae); however, locally they can be important components of their meals, with several hundred achenes being consumed during a feeding period. The mature achenes will float (for up to 2 months) and are dispersed by water. In some species they are eaten by fish (e.g., Osteichthyes: Teleostei: *Cyprinus carpio*) and are excreted in viable state, which facilitates their transport within water bodies. They also are transported over longer distances by their external attachment to or consumption and subsequent egestion by birds (e.g., Aves: Anatidae; Charadriidae). Vegetative reproduction can occur in most species by the proliferation of stolons (the thicker structures categorized as rhizomes) and associated tubers. This is a taxonomically difficult group and a good number of literature accounts (even those authored by experts) often involve misidentifications or improper species assignments.

2.1.6.1. *Sagittaria ambigua* **J.G. Sm.** is a perennial, which inhabits shallow waters or wet exposed sites in depressions, ditches, gravel pits, mudflats, ponds (shallow), pools, prairies, roadsides, sloughs, swamps, and along the margins or shores of lakes, ponds, and streams at elevations to 230 m. The plants are common along shores where water levels fluctuate and have been characterized as pioneer species of shallow pond bottoms. The substrates often are described as mud. Flowering occurs from April to September. There is no information on the reproductive biology or seed ecology of this species. Seed dispersal likely occurs primarily by water. **Reported associates:** *Alisma subcordatum, Ammannia auriculata, Callitriche heterophylla, Cephalanthus occidentalis, Echinochloa crus-galli, Eclipta prostrata, Eleocharis engelmannii, Eleocharis macrostachya, Eleocharis obtusa, Fimbristylis autumnalis, Gratiola virginiana, Helenium flexuosum, Hydrolea ovata, Juncus brachycarpus, Juncus torreyi, Ludwigia alternifolia, Ludwigia glandulosa, Ludwigia palustris, Nelumbo lutea, Persicaria pensylvanica, Rhexia mariana, Rotala ramosior, Sagittaria brevirostra, Sagittaria calycina, Sagittaria latifolia, Samolus ebracteatus, Schoenoplectus pungens, Schoenoplectus tabernaemontani, Typha latifolia, Verbena hastata.*

2.1.6.2. *Sagittaria australis* **(J.G. Sm.) Small** is a perennial, inhabits exposed sediments or shallow waters (e.g., 15 cm depth) within bottomlands (alluvial; dry; creek), channels (river), depressions, ditches (banks), floating mats, floodplains, marshes, meadows (alluvial), oxbows, ponds (spring-fed; vernal), slopes (to 40%), sloughs, streams (lentic), swales, swamps, and along the margins of lakes, ponds, reservoirs, streams, and swamps at elevations to 488 m. The plants can tolerate exposures ranging from fully open to shaded conditions. The substrates are consistently circumneutral (pH: 6.0–7.6) and include Cecil clay loam, Cecil sandy loam, Congaree silt loam, dredge material, gravel, mud, and sand. Flowering and fruiting in this monoecious species extend from July to October. The flowers are pollinated by generalist bees (see *Use by wildlife*). Seeds collected from late August through early October, and stored moist or in water, have germinated well on saturated media (23% to 52%) within 10–11 weeks after planting. The breeding system is not well correlated with subsequent seed germination, with higher, lower, and equal rates reported for inbred vs. outcrossed seeds. **Reported associates:** *Acer floridanum, Acer negundo, Acer rubrum, Acorus calamus, Alisma subcordatum, Arundinaria tecta, Bacopa, Betula nigra, Bidens frondosus, Bidens tripartitus, Boltonia, Calamagrostis canadensis, Carex alata, Carex lurida, Carex vulpinoidea, Cicuta maculata, Commelina virginica, Conoclinium coelestinum, Cyperus flavescens, Cyperus squarrosus, Cyperus strigosus, Eleocharis ovata, Epilobium coloratum, Fraxinus pennsylvanica, Geum canadense, Glyceria striata, Hydrocotyle verticillata, Hypericum mutilum, Ipomoea lacunosa, Isoetes engelmannii, Juncus effusus, Kosteletzkya, Leersia oryzoides, Leersia virginica, Leptochloa, Liquidambar styraciflua, Liriodendron tulipifera, Lobelia, Ludwigia alternifolia, Lycopus virginicus, Microstegium vimineum, Mimulus alatus, Murdannia keisak,* *Nitella, Nuphar advena, Nymphaea odorata, Nyssa, Oxypolis rigidior, Panicum, Peltandra virginica, Persicaria amphibia, Persicaria longiseta, Persicaria setacea, Potamogeton, Ptilimnium costatum, Quercus, Rhexia virginica, Rosa palustris, Rumex conglomeratus, Sagittaria calycina, Sagittaria latifolia, Sagittaria platyphylla, Salix caroliniana, Salix nigra, Samolus valerandi, Saururus cernuus, Scirpus cyperinus, Sesbania, Sparganium americanum, Sphenoclea zeylanica, Taxodium distichum, Typha latifolia.*

2.1.6.3. *Sagittaria brevirostra* **Mackenzie & Bush** is a perennial, which occurs on exposed substrates or in shallow waters (to 60 cm deep) of alluvial fans, bottomlands, canals, channels (ravine), ditches (dredged; roadside), floodplains, gravel pits, lakebeds, marshes, meadows, mud bars, mudflats, oxbows, ponds (dessicated; sand), prairies, ravines (dried), seeps, shores, slopes (to 5%), sloughs, springs, swamps (dried), and along the margins of lakes, ponds, rivers, streams, and swamps at elevations to 1539 m. Exposures from open sunlight to shade are tolerated. The substrates are described as clay loam, clay mud, loamy clay, loam muck, mud, rock, sand, sandy loam (Bolivar), and silt. The plants exhibit a highly plastic morphology. Smaller leaves develop when growing in dry and deep-water conditions than those when growing in shallow water. In deep waters, the leaves are quite narrow. These monoecious plants flower and fruit from June to October. Plants growing in deep waters are sterile. The individuals are self-compatible, allowing for either geitonogamous or xenogamous pollination to occur. The pollinators consist of numerous types of small insects (Insecta). Inflorescence size varies but does not influence seed set, which typically is high in all female flowers. Inflorescences are larger and produce more flowers under wet vs. dry conditions. Male flowers outnumber females in either dry or wet conditions, but sex ratios are more highly male biased in plants growing under dry conditions. Details on seed germination are unavailable. The achenes are dispersed by water. **Reported associates:** *Acalypha rhomboidea, Acer negundo, Acer saccharinum, Agalinis purpurea, Alisma subcordatum, Ambrosia, Ammannia coccinea, Ammannia robusta, Asclepias incarnata, Azolla, Bacopa rotundifolia, Betula nigra, Boehmeria cylindrica, Bolboschoenus fluviatilis, Calamagrostis canadensis, Callitriche heterophylla, Campanula americana, Campanula aparinoides, Campsis radicans, Carex caroliniana, Carex cristatella, Carex frankii, Carex grayi, Carex muskingumensis, Carex stipata, Carex stricta, Carex tribuloides, Carex typhina, Carex vesicaria, Carya laciniosa, Cephalanthus occidentalis, Cicuta, Coleataenia longifolia, Cornus amomum, Cornus drummondii, Cornus racemosa, Doellingeria umbellata, Echinochloa crus-galli, Eleocharis erythropoda, Eleocharis macrostachya, Eleocharis obtusa, Eleocharis ovata, Epilobium leptophyllum, Equisetum, Eupatorium perfoliatum, Fraxinus pennsylvanica, Galium tinctorium, Gleditsia aquatica, Glyceria septentrionalis, Gratiola virginiana, Helenium autumnale, Helianthus hirsutus, Heteranthera limosa, Heteranthera rotundifolia, Hibiscus laevis, Hypericum mutilum, Ilex decidua, Ilex vomitoria, Impatiens capensis, Ipomoea lacunosa, Iris virginica, Juncus canadensis,*

Juncus dudleyi, Juncus effusus, Juncus torreyi, Justicia americana, Laportea canadensis, Leersia oryzoides, Lemna minor, Lindernia dubia, Lobelia cardinalis, Ludwigia alternifolia, Ludwigia palustris, Ludwigia peploides, Ludwigia polycarpa, Lycopus americanus, Lycopus rubellus, Lycopus virginicus, Lysimachia ciliata, Lysimachia quadriflora, Lythrum, Marsilea vestita, Mentha arvensis, Mimulus ringens, Myrica cerifera, Nymphaea odorata, Onoclea sensibilis, Panicum virgatum, Peltandra virginica, Penthorum sedoides, Persicaria amphibia, Persicaria bicornis, Persicaria coccinea, Persicaria hydropiperoides, Persicaria pensylvanica, Persicaria punctata, Phalaris arundinacea, Phragmites australis, Phyla, Pinus taeda, Platanus occidentalis, Polygonum ramosissimum, Populus deltoides, Potamogeton nodosus, Potamogeton pusillus, Proserpinaca pectinata, Pycnanthemum virginianum, Quercus macrocarpa, Quercus palustris, Ranunculus flabellaris, Rhynchospora, Rorippa aquatica, Rorippa palustris, Rumex altissimus, Rumex crispus, Rumex stenophyllus, Sagittaria ambigua, Sagittaria calycina, Sagittaria graminea, Sagittaria latifolia, Sagittaria rigida, Salix amygdaloides, Salix exigua, Salix interior, Salix myricoides, Salix nigra, Schoenoplectus heterochaetus, Scirpus atrovirens, Scirpus cyperinus, Scutellaria lateriflora, Sium suave, Solidago nemoralis, Solidago ulmifolia, Spiraea tomentosa, Stachys palustris, Symphyotrichum ontarionis, Symphyotrichum patens, Teucrium canadense, Thelypteris palustris, Triadenum fraseri, Tridens flavus, Typha latifolia, Verbena hastata, Vitis cinerea, Wolffia brasiliensis.

2.1.6.4. *Sagittaria calycina* Engelm. is an annual, which inhabits shallow (<100 cm), tidal or non-tidal, brackish to freshwater sites including canals, depressions, ditches (irrigation), estuaries, floodplains, gravel bars, marshes, meadows, mudflats (lake; river), ponds (artificial; catchment; sand), pools, ravines (dried), rice fields, roadsides, saltmarshes, sandbars, seeps, shores, sinkholes, sloughs, swales, swamps, and the margins of lakes (artificial; small), ponds, reservoirs, rivers, and stock tanks at elevations to 1829 m. Although the plants can grow either emersed or submersed, they regularly appear as emergents on mudflats and at other sites with open exposures where standing waters have receded. The substrates are described as alkaline (e.g., pH: 8.5) or calcareous, and include alluvium, ballast (railroad), clay, clay loam (over limestone), loam, muck, mud, sand, sandy gravel, silt, silty loam (Humphrey's Cherty), and silty sand. Flowering occurs from June to October, with developing fruits observed by late June. Individuals are described as being andromonoecious, i.e., bearing perfect and staminate flowers. However, the perfect flowers are described as possessing either functional stamens or staminodes, depending on the source consulted. Whether these flowers retain any capacity for autogamy should be clarified, especially given the annual habit of the plants. The pollination biology of this species has not been described in any detail; however, the fruits are known to mature within a few weeks following anthesis. Mature plants can reach densities of 20–30 plants m^{-2} and are known to produce up to 2,000 achenes per inflorescence with up to 32 inflorescences per plant, yielding an average of 20,000 achenes per plant. Other estimates indicate achene yields of 73 cubic cm m^{-2}, or by mass, an average of 400 kg ha^{-1}. As the fruiting peduncles become prostrate along the substrate, the pedicels recurve to push the fruiting heads into the mud. Consequently, many seeds are planted near the maternal plant. However, the achenes also are buoyant and can remain afloat for up to 2 weeks. They can be dispersed locally by water, and potentially are transported over greater distances by waterfowl (Aves: Anatidae). The achenes remain viable in the seed bank where they retain 54% viability after 1 year and 45% viability after 3 years. The seeds will germinate at temperatures below 10°C, but much higher rates are observed above 11°C (optimal germination occurs at 12°C–24°C). **Reported associates:** *Abutilon theophrasti, Acalypha gracilens, Acalypha rhomboidea, Acer negundo, Acer saccharinum, Agrimonia parviflora, Alisma subcordatum, Amaranthus, Ammannia coccinea, Andropogon virginicus, Apocynum cannabinum, Bergia texana, Betula nigra, Bidens aristosus, Bidens bidentoides, Bidens connatus, Bidens frondosus, Bidens tripartitus, Boehmeria cylindrica, Bolboschoenus novae-angliae, Carex hyalinolepis, Celtis, Cephalanthus occidentalis, Chamaecrista fasciculata, Chasmanthium latifolium, Coleataenia longifolia, Conoclinium coelestinum, Conyza canadensis, Cyperus acuminatus, Cyperus bipartitus, Cyperus erythrorhizos, Cyperus esculentus, Cyperus flavescens, Cyperus odoratus, Cyperus squarrosus, Cyperus strigosus, Dichanthelium clandestinum, Dichanthelium dichotomum, Digitaria ischaemum, Diodella teres, Diodia virginiana, Dioscorea, Dysphania ambrosioides, Echinochloa muricata, Echinodorus berteroi, Eclipta prostrata, Eleocharis acicularis, Eleocharis ambigens, Eleocharis diandra, Eleocharis erythropoda, Eleocharis macrostachya, Eleocharis obtusa, Eleocharis ovata, Eleocharis parvula, Eleusine indica, Eragrostis hypnoides, Eragrostis pectinacea, Erigeron annuus, Erigeron philadelphicus, Erigeron strigosus, Eriocaulon parkeri, Eupatorium perfoliatum, Eupatorium serotinum, Euphorbia maculata, Euphorbia nutans, Fimbristylis autumnalis, Fragaria virginiana, Galium triflorum, Gamochaeta purpurea, Hedeoma hispida, Helenium flexuosum, Heteranthera limosa, Heteranthera reniformis, Hypericum mutilum, Hypericum punctatum, Impatiens capensis, Isoetes riparia, Juncus acuminatus, Juncus diffusissimus, Juncus effusus, Juncus tenuis, Justicia americana, Kummerowia stipulacea, Kummerowia striata, Kyllinga brevifolia, Kyllinga pumila, Leersia oryzoides, Leersia virginica, Leptochloa fusca, Lespedeza procumbens, Lespedeza thunbergii, Leucanthemum vulgare, Lilaeopsis chinensis, Lindernia dubia, Lobelia cardinalis, Lobelia inflata, Lobelia siphilitica, Ludwigia palustris, Ludwigia peploides, Lycopus americanus, Lycopus rubellus, Microstegium vimineum, Mimulus alatus, Mimulus ringens, Mollugo verticillata, Nuphar advena, Panicum dichotomiflorum, Panicum philadelphicum, Paronychia fastigiata, Paspalum, Peltandra virginica, Penthorum sedoides, Persicaria hydropiper, Persicaria hydropiperoides, Persicaria lapathifolia, Persicaria maculosa, Persicaria pensylvanica, Persicaria posumbu, Persicaria punctata, Persicaria sagittata, Persicaria virginiana, Phlox bifida,*

Phytolacca americana, Plantago lanceolata, Plantago rugelii, Pluchea camphorata, Pluchea odorata, Pontederia cordata, Potentilla norvegica, Potentilla simplex, Prunella vulgaris, Ptelea, Quercus bicolor, Ranunculus, Rhexia mariana, Rorippa palustris, Rorippa sylvestris, Rotala ramosior, Rumex crispus, Rumex obtusifolius, Sagittaria ambigua, Sagittaria australis, Sagittaria brevirostra, Sagittaria graminea, Sagittaria latifolia, Sagittaria longiloba, Sagittaria platyphylla, Sagittaria subulata, Salix, Samolus valerandi, Schoenoplectus hallii, Schoenoplectus pungens, Schoenoplectus smithii, Scirpus polyphyllus, Scutellaria lateriflora, Senna marilandica, Setaria viridis, Sida spinosa, Solanum carolinense, Solidago canadensis, Sonchus asper, Sparganium americanum, Spartina alterniflora, Spartina cynosuroides, Symphyotrichum lanceolatum, Symphyotrichum ontarionis, Symphyotrichum pilosum, Symphyotrichum racemosum, Tradescantia bracteata, Trifolium repens, Tussilago farfara, Typha angustifolia, Typha latifolia, Urochloa platyphylla, Verbascum thapsus, Verbena hastata, Verbena urticifolia, Verbesina alternifolia, Vernonia gigantea, Vitis, Xanthium strumarium, Zizania aquatica.

2.1.6.5. ***Sagittaria cristata* Engelm.** is an annual or a perennial, which inhabits shallow to moderately deep waters (to 2 m) in lakes, rivers, sloughs, and along the margins of lakes, ponds, and swamps at elevations to 1000 m. The habitats are categorized as hard water (pH: 7.0–8.6; total alkalinity >40 ppm; e.g., 54.0 mg l^{-1}; sulfates <50 ppm). The sediments (means: organic matter: 1.5%; pH: 5.5; Ca: 10/50 μg g^{-1}; K: 5/7.3 μg g^{-1}; Mg: 5/14.8 μg g^{-1}; P: 5/8.4 μg g^{-1}) are described as muck and sand. This is one of the more aquatic species in the genus, found normally in deeper standing water and less often on exposed shorelines. The foliage has relative low ratios of C:P (400), C:N (16), and N:P (20). The plants are monoecious, with flowering and fruiting observed from July to September. Additional details on the reproductive biology or seed ecology of this species are lacking. **Reported associates:** *Bidens beckii, Brasenia schreberi, Ceratophyllum demersum, Chara, Drepanocladus, Dulichium arundinaceum, Elatine minima, Eleocharis acicularis, Elodea canadensis, Eriocaulon aquaticum, Heteranthera dubia, Isoetes echinospora, Juncus pelocarpus, Lemna minor, Lemna trisulca, Lobelia dortmanna, Myriophyllum sibiricum, Myriophyllum tenellum, Najas flexilis, Najas guadalupensis, Nitella, Nuphar advena, Nuphar variegata, Nymphaea odorata, Persicaria amphibia, Potamogeton amplifolius, Potamogeton foliosus, Potamogeton gramineus, Potamogeton illinoensis, Potamogeton natans, Potamogeton praelongus, Potamogeton pusillus, Potamogeton richardsonii, Potamogeton robbinsii, Potamogeton strictifolius, Potamogeton zosteriformis, Ranunculus flammula, Ruppia maritima, Schoenoplectus acutus, Schoenoplectus subterminalis, Sparganium americanum, Sparganium angustifolium, Sparganium eurycarpum, Sparganium fluctuans, Stuckenia filiformis, Stuckenia pectinata, Utricularia cornuta, Utricularia macrorhiza, Utricularia minor, Utricularia resupinata, Vallisneria americana.*

2.1.6.6. ***Sagittaria cuneata* Sheldon** is a perennial, which inhabits still or slow-moving standing waters (to 1.2 m deep) or exposed substrates within backwaters, beaches, bogs, borrow pits, canals (irrigation), depressions, ditches (roadside), draws, fens, flats, lake beds (dry), marshes, meadows, mudflats, ponds (intermittent; montane; retention; spring-fed; stock), pools (culvert; dried; roadside), potholes, riverbeds, rivers, roadsides (disturbed; gravelly), slopes (to 6%), sloughs, streambeds, swales, swamps, and along the margins or receding shores of channels, lakes, ponds, potholes, reservoirs, rivers, and streams at elevations to 3048 m. The plants are highly polymorphic, producing linear or sagittate emergent foliage when growing in shallow sites but developing phyllodial submersed or cordate/sagittate floating leaves in deeper water. Exposures can range from full sun to shade. The surrounding waters can vary widely in terms of alkalinity (from 5 to 628 mg l^{-1} CaCO$_3$) but tend to be moderately alkaline to calcareous (e.g., pH: 6.8–8.9; $\bar{x}$ = 7.4; total hardness: 184 mg l^{-1}; Ca: 2.7 mg l^{-1}; Mg 0.8 mg l^{-1}). The substrates are described as acidic (e.g., pH: 5.5–7.0) or brackish and include alluvium, clay (black; Pierre), clay loam (Shingle-Samday; Worthenton), clay silt, gravel, gravelly sand, gumbo, loam (black; organic), muck (alkaline; black), mud (alluvial; desiccated), peat, rock, sand, sandy clay, sandy loam, sandy ooze, Shingle-Haverdad series, silt, silty clay, silty loam, silty muck, silty sand, and Worfka-Shingle-Samday series. Flowering occurs from April to September, with fruiting extending into October. Individuals are monoecious but self-compatible, theoretically enabling geitonogamous pollination to occur; however, the reproductive ecology of this species has not been characterized in any detail. Outcrossing presumably is carried out by bees (Insecta: Hymenoptera) and flies (Insecta: Diptera). A plant typically produces six flowers, which can yield about 5,150 seeds, each weighing about 0.235 mg. A seed bank develops. The highest experimental germination rates occurred for seeds buried at less than 1 cm (3-week stratification at 10°C; they were then placed under ambient greenhouse conditions: 15°C–30°C; 14–16 hr light). The seeds are dispersed by water and perhaps to greater distances by their adherence to various animals. **Reported associates:** *Acer negundo, Acer saccharinum, Agrostis gigantea, Alisma subcordatum, Alisma triviale, Alopecurus aequalis, Amaranthus, Anagallis minima, Angelica atropurpurea, Arnica chamissonis, Artemisia frigida, Artemisia nova, Artemisia tridentata, Asclepias incarnata, Bacopa rotundifolia, Beckmannia syzigachne, Berula erecta, Bidens cernuus, Boehmeria cylindrica, Bouteloua gracilis, Calla, Callitriche hermaphroditica, Callitriche palustris, Carex aquatilis, Carex arcta, Carex athrostachya, Carex comosa, Carex cristatella, Carex emoryi, Carex lasiocarpa, Carex molesta, Carex nebrascensis, Carex rostrata, Carex stipata, Carex utriculata, Carex utriculata, Carex vesicaria, Carex vulpinoidea, Cephalanthus occidentalis, Ceratophyllum demersum, Ceratophyllum echinatum, Chara, Chenopodium glaucum, Cicuta bulbifera, Comarum palustre, Convolvulus arvensis, Cornus amomum, Cyperus bipartitus, Cyperus odoratus, Damasonium californicum, Deschampsia cespitosa, Downingia bella, Downingia* *yina, Dulichium arundinaceum, Echinochloa crus-galli, Elaeagnus angustifolia, Elatine heterandra, Eleocharis acicularis, Eleocharis erythropoda, Eleocharis macrostachya, Eleocharis palustris, Eleocharis quinqueflora, Elodea canadensis, Equisetum fluviatile, Eupatorium perfoliatum, Fimbristylis autumnalis, Glyceria borealis, Glyceria grandis, Glyceria septentrionalis, Glyceria striata, Gnaphalium palustre, Gratiola heterosepala, Gratiola neglecta, Gutierrezia sarothrae, Helenium, Hippuris vulgaris, Hordeum jubatum, Iris virginica, Isoetes bolanderi, Juncus alpinoarticulatus, Juncus balticus, Juncus effusus, Juncus nodosus, Juncus torreyi, Leersia oryzoides, Lemna minor, Lepidium virginicum, Limosella, Lindernia dubia, Ludwigia palustris, Lycopus americanus, Lycopus asper, Lycopus uniflorus, Lysimachia nummularia, Lythrum salicaria, Madia glomerata, Marsilea vestita, Melilotus officinalis, Mentha arvensis, Mimulus primuloides, Mimulus ringens, Myriophyllum sibiricum, Myriophyllum spicatum, Najas flexilis, Najas marina, Nuphar polysepala, Nuphar variegata, Peltandra virginica, Persicaria amphibia, Persicaria coccinea, Persicaria hydropiper, Persicaria hydropiperoides, Persicaria maculosa, Persicaria punctata, Phalaris arundinacea, Phleum pratense, Phragmites australis, Physostegia virginiana, Pilea pumila, Poa nemoralis, Poa pratensis, Polypogon monspeliensis, Populus angustifolia, Populus balsamifera, Populus deltoides, Porterella carnosula, Potamogeton berchtoldii, Potamogeton diversifolius, Potamogeton epihydrus, Potamogeton foliosus, Potamogeton gramineus, Potamogeton illinoensis, Potamogeton natans, Potamogeton pusillus, Potamogeton richardsonii, Potamogeton zosteriformis, Potentilla gracilis, Ranunculus aquatilis, Ranunculus pensylvanicus, Ranunculus sceleratus, Rhamnus cathartica, Rhaponticum repens, Rorippa islandica, Rorippa palustris, Rosa multiflora, Rosa setigera, Rotala ramosior, Rudbeckia laciniata, Rumex altissimus, Rumex salicifolius, Rumex verticillatus, Sagittaria latifolia, Salix amygdaloides, Salix exigua, Salix interior, Salix lasiandra, Salix nigra, Sambucus nigra, Schedonorus arundinaceus, Schoenoplectus acutus, Schoenoplectus pungens, Schoenoplectus tabernaemontani, Scirpus atrovirens, Scutellaria lateriflora, Sium suave, Solanum dulcamara, Solidago altissima, Solidago gigantea, Sparganium angustifolium, Sparganium emersum, Sparganium eurycarpum, Spartina pectinata, Sphagnum, Spirodela polyrhiza, Stuckenia filiformis, Stuckenia pectinata, Tamarix ramosissima, Thalictrum dasycarpum, Trifolium, Triglochin scilloides, Typha angustifolia, Typha latifolia, Typha ×glauca, Ulmus americana, Urtica dioica, Vallisneria americana, Verbesina alternifolia, Vitis riparia, Zannichellia palustris.*

2.1.6.7. *Sagittaria demersa* J.G. Sm. is an annual, which occurs in standing waters (0.2–1.2 m), or (more rarely) is stranded on the exposed substrates of lakes, pools, and streams at elevations from 1500 to 2082 m. The plants produce phyllodial leaves, which are submersed or are floating. In Mexico, this species occurs in montane, clean, medium-hard to soft waters (pH: 6.0–7.2; CaCO$_3$: 48.0–88.0 mg l^{-1}). The sediments contain average levels of nitrogen, total phosphorus,

and organic matter, and are described as clay, clayey sand, and sandy sandstone. Further life history information is unavailable for this species, particularly for the few North American occurrences. The plants apparently are intolerant of pollution and readily succumb to eutrophication and other human disturbances. **Reported associates:** none.

2.1.6.8. *Sagittaria engelmanniana* J.G. Sm. is a perennial, which inhabits shallow waters (e.g., 30 cm deep) of alluvial fans, backwaters, beaches, bogs (peat), bottoms (exposed; stream), Carolina bays, channels (drainage), depressions, ditches, fens, floodplains, lakes, marshes, oxbows, pocosins (streamside), ponds (dried), reservoirs, shores, sloughs, swales, swamps, woodlands, and the margins of lakes, ponds, rivers, and streams at elevations to 127 m. The sites typically occur in open exposures with waters that characteristically are acidic (pH: 5.6–6.3). The substrates are described as muck, mucky peat, mud, peat, sand, and silty sand. The plants are monoecious and are in flower from June to September with fruiting extending into October. Individuals are self-compatible, which allows for geitonogamous pollination to occur (but to what extent has not been determined). The usual pollinators presumably are insects (Insecta). Other life history details are lacking for this species. **Reported associates:** *Acer rubrum, Agrostis hyemalis, Alnus maritima, Alnus serrulata, Andropogon glomeratus, Andropogon virginicus, Apios americana, Aronia arbutifolia, Bartonia paniculata, Bartonia virginica, Bidens frondosus, Bidens mitis, Bidens trichospermus, Burmannia capitata, Calamagrostis canadensis, Callitriche heterophylla, Carex albolutescens, Carex atlantica, Carex bullata, Carex canescens, Carex exilis, Carex exilis, Carex lasiocarpa, Carex striata, Cephalanthus occidentalis, Chamaecyparis thyoides, Chamaedaphne calyculata, Cladium mariscoides, Clethra alnifolia, Coleataenia longifolia, Cuscuta, Cyperus dentatus, Cyperus squarrosus, Cyrilla racemiflora, Dichanthelium acuminatum, Dichanthelium dichotomum, Dichanthelium ensifolium, Dichanthelium scabriusculum, Drosera intermedia, Drosera rotundifolia, Dulichium arundinaceum, Eleocharis acicularis, Eleocharis equisetoides, Eleocharis flavescens, Eleocharis robbinsii, Eleocharis tenuis, Eleocharis tuberculosa, Eriocaulon aquaticum, Eriocaulon compressum, Eriocaulon decangulare, Eriophorum virginicum, Eubotrys racemosa, Eupatorium perfoliatum, Eupatorium pilosum, Eupatorium resinosum, Eupatorium ×pinnatifidum, Euthamia graminifolia, Eutrochium dubium, Fuirena squarrosa, Galium tinctorium, Gaylussacia dumosa, Glyceria obtusa, Gratiola aurea, Hydrocotyle verticillata, Hypericum canadense, Hypericum densiflorum, Hypericum denticulatum, Hypericum mutilum, Ilex glabra, Iris versicolor, Itea virginica, Juncus canadensis, Juncus effusus, Juncus militaris, Juncus pelocarpus, Juncus trigonocarpus, Kalmia angustifolia, Kalmia buxifolia, Lachnanthes caroliniana, Leersia oryzoides, Lindernia dubia, Lobelia canbyi, Lobelia elongata, Lobelia nuttallii, Ludwigia palustris, Ludwigia sphaerocarpa, Lycopus amplectens, Lycopus uniflorus, Lycopus virginicus, Lyonia ligustrina, Lyonia mariana, Lysimachia terrestris, Magnolia virginiana, Micranthemum* *umbrosum, Microstegium vimineum, Mikania scandens, Muhlenbergia uniflora, Murdannia keisak, Nuphar variegata, Nymphaea odorata, Nymphoides aquatica, Nyssa sylvatica, Oclemena nemoralis, Orontium aquaticum, Osmunda regalis, Osmundastrum cinnamomeum, Oxypolis rigidior, Panicum hemitomon, Panicum verrucosum, Panicum virgatum, Peltandra virginica, Persicaria hydropiperoides, Pinus rigida, Pogonia ophioglossoides, Polygala cruciata, Polygala ramosa, Pontederia cordata, Potamogeton confervoides, Potamogeton oakesianus, Rhexia mariana, Rhexia nashii, Rhexia virginica, Rhododendron viscosum, Rhynchospora alba, Rhynchospora capitellata, Rhynchospora chalarocephala, Rhynchospora fusca, Rhynchospora microcephala, Rubus hispidus, Sabatia difformis, Sagittaria latifolia, Sagittaria macrocarpa, Sarracenia purpurea, Sassafras albidum, Schizachyrium scoparium, Schoenoplectus subterminalis, Scirpus cyperinus, Smilax glauca, Smilax herbacea, Smilax laurifolia, Smilax rotundifolia, Smilax walteri, Sparganium americanum, Sphagnum, Spiraea tomentosa, Spiranthes cernua, Symphyotrichum dumosum, Symphyotrichum novi-belgii, Taxodium, Toxicodendron radicans, Triadenum virginicum, Triadenum walteri, Utricularia cornuta, Utricularia gibba, Utricularia juncea, Utricularia striata, Vaccinium corymbosum, Vaccinium macrocarpon, Viburnum nudum, Viola lanceolata, Woodwardia areolata, Woodwardia virginica, Xyris difformis, Xyris fimbriata, Zizania aquatica.*

2.1.6.9. *Sagittaria fasciculata* E.O. Beal is a short-lived perennial (usually <2 years), which grows in shallow (<5 cm deep), freshwater alluvial floodplains, seeps, or streams, at elevations to 1000 m. The plants require cool, clean waters with continuous flow, moderate levels of dissolved oxygen (3–7 mg l^{-1}), and low conductivities (20–50 µS). They are intolerant of droughts, floods, and other hydrological fluctuations. The exposures become well shaded by late summer. The substrates are acidic (e.g., pH: 4.8–6.8) and have been described as hydrated muck–sand suspension, sandbars, or sandy mucks with a high organic matter content (6.5–10%) and high Mg content (108.6 µg g^{-1}). Ecological details for this rare species are difficult to compile, due in part to the redaction of information from electronic database records. Flowering occurs from May to July, with fruiting extending into August. The reproductive ecology of this species has not been elucidated. Genetic analyses using ISSR markers have indicated some genetic differentiation among watersheds but little differentiation within watersheds. The plants reproduce vegetatively by stolons (also referred to as rhizomes), which are produced in greater prevalence when flow rates or shading conditions are altered. The plants are of conservation concern, with their status interpreted as "declining." Their threats include alterations of hydrological conditions, shading by competitive vegetation, and trampling. **Reported associates:** *Carex, Decumaria barbara, Glyceria striata, Gratiola virginiana, Murdannia keisak.*

2.1.6.10. *Sagittaria filiformis* J.G. Sm. is a perennial, which inhabits still or flowing, shallow (0.01–1.8 m deep), tidal or non-tidal waters in canals (drainage), depressions, ditches, lakes, marshes, pits (borrow; phosphate), ponds (artificial; karst; seasonal), pools, sloughs, streams (blackwater), swamps, and the shores of lakes at elevations to 100 m. The shoots are heterophyllous, developing linear submersed leaves and ovate floating leaves. Exposures range from open to shaded conditions. The plants supposedly inhabit more lotic sites in the north of their distribution and more lentic waters in the southern portion of their range. The waters are somewhat alkaline (e.g., pH: 7.1; alkalinity: 35.0 mg l^{-1}). The sediments have been described as muck, mud, and sand. Flowering and fruiting occur from January to October. Some of the lower scape verticels are proliferous, capable of producing roots, phyllodial leaves, and flowering pedicels from the bract axils. The pollination biology is unknown. The seed germination requirements also have not been determined; however, a persistent seed bank does develop. Vegetative reproduction occurs by the expansion of stolons (also referred to as rhizomes). **Reported associates:** *Acer rubrum, Alternanthera philoxeroides, Ampelopsis arborea, Azolla filiculoides, Bacopa caroliniana, Bidens discoideus, Boltonia diffusa, Brasenia schreberi, Carya cordiformis, Cephalanthus occidentalis, Ceratophyllum muricatum, Clethra alnifolia, Crotalaria spectabilis, Cynodon dactylon, Cyperus erythrorhizos, Echinodorus, Eclipta prostrata, Eleocharis baldwinii, Eubotrys racemosa, Eupatorium capillifolium, Fimbristylis autumnalis, Fraxinus, Hibiscus moscheutos, Hydrilla verticillata, Hydrocotyle ranunculoides, Hydrocotyle verticillata, Ilex myrtifolia, Juncus diffusissimus, Juncus effusus, Landoltia punctata, Leersia hexandra, Lemna valdiviana, Limnobium spongia, Lindernia dubia, Liquidambar styraciflua, Ludwigia, Lycopus, Lyonia lucida, Myrica cerifera, Myriophyllum aquaticum, Myriophyllum pinnatum, Nelumbo lutea, Nitella translucens, Nymphaea odorata, Nymphoides cordata, Nyssa biflora, Nyssa sylvatica, Panicum dichotomiflorum, Panicum hemitomon, Paspalum urvillei, Peltandra virginica, Persea palustris, Persicaria hydropiperoides, Persicaria punctata, Pinus taeda, Pistia stratiotes, Pontederia cordata, Proserpinaca palustris, Quercus phellos, Rhynchospora glomerata, Rhynchospora macrostachya, Rhynchospora microcephala, Rotala ramosior, Rumex verticillatus, Saccharum, Sagittaria graminea, Sagittaria latifolia, Saururus cernuus, Scirpus cyperinus, Smilax rotundifolia, Spirodela, Taxodium distichum, Thalia, Typha domingensis, Typha latifolia, Utricularia gibba, Utricularia purpurea.*

2.1.6.11. *Sagittaria graminea* Michx. is a perennial, which grows in shallow to moderately deep waters (30–180 cm; average: 60 cm depth) in bogs, borrow pits, Carolina bays, cypress domes, depressions, ditches (drainage), flats, flatwoods, floodplains, gravel pits, hammocks, lakebeds, marshes, meadows, mudflats, pinelands, pocosins, pools (savannah), ponds (railroad; roadside; sand; sinkhole; stock), prairies, rice fields, roadsides, seeps, sloughs (dried), swales, swamps (blackwater; dried; river), and along the margins or shores of canals, lakes, ponds, rivers, and streams at elevations to 789 m. The waters are broadly circumneutral (pH: 5.7–8.4; $\bar{x}=7.0$) with low conductivity (<200 µmhos cm^{-1}) and low total alkalinity (<112 mg l^{-1} CaCO$_3$; $\bar{x}=18.7$ mg l^{-1}).

The substrates are more acidic (pH: 4.4–7.7). Exposures can range from open (full sunlight) to deep shade. The substrates have been characterized as clay, clay loam, gravelly alluvium, loam, loamy sand (Ocilla), muck, mucky loam, mud (over dolomite), peat, peaty mud, Placid (typic humaquepts), psammaquents (Basinger; St. Johns), sand (organic), sandy clay loam, sandy muck, sandy peat, silt, silty loam, silty muck, and silty sandy loam. The leaf morphology varies substantially from phyllodial to broad-bladed; however, experimental transplant studies indicate only minor plasticity due to water level variation, and that plants having different leaf types are genetically distinct and adapted differentially to the specific habitats in which they are found. Flowering and fruiting occur from March to December. Submersed plants remain sterile (unless under very high irradiance and shallow conditions), with flowering generally requiring the emergence of non-phyllodial leaves from the water. Flowering (usually initiating from 5:00 to 6:00 am) can occur under daily photoperiods ranging from 7 to 15 hr of light. The plants are monoecious. A greater proportion of male flowers is produced under high phosphorus and low nitrogen conditions; under opposite conditions (low phosphorus, high nitrogen), more female flowers, and subsequently more seeds, are produced. All flowers (and seeds) increase in number when both nutrients are elevated. Individuals are self-compatible (but protogynous within an inflorescence) and require insects (Insecta) in order to achieve successful pollination and fruit set. Female and male floral phases overlap less in smaller populations than in larger populations. Once deposited, pollen must remain undisturbed for a period of 3 hr in order to achieve fertilization. Seed set is highly reduced (by nearly 90%) in plants subjected to tidal activity, which is attributed to inviability or washing away of the pollen when exposed to water. Higher fruit set occurs in populations with male-biased floral sex ratios. The fruits ripen and disperse within 50 days. Seeds have germinated following 5 days of cold stratification, when subjected to a 30°C/15°C day/night temperature regime. Isozyme analyses (13 loci) of the typical subspecies indicated that a single multilocus genotype characterized 97% of the accessions surveyed, which indicates extremely low levels of genetic variation in the species (see *Systematics*). Vegetative reproduction is well developed. High phosphorous levels yield larger corms, whereas high nitrogen levels yield smaller corms but more numerous and longer stolons (also characterized as coarse rhizomes), which produce tubers. In late fall, tuber biomass of mudflat plants (124 g m^{-2}) can greatly exceed that of pond dwelling plants (44 g m^{-2}). Dispersal occurs by means of the achenes or tubers, which typically are transported by water, especially during flood conditions. **Reported associates:** *Acer rubrum, Acer saccharinum, Aletris lutea, Alisma subcordatum, Alternanthera philoxeroides, Amaranthus, Amsonia rigida, Andropogon, Asclepias incarnata, Asclepias perennis, Azolla cristata, Bacopa rotundifolia, Bidens cernuus, Bidens frondosus, Boehmeria cylindrica, Bolboschoenus fluviatilis, Brasenia schreberi, Calamagrostis canadensis, Callitriche heterophylla, Carex gigantea, Carex haydenii, Carex scoparia, Carex striata, Carex tribuloides,*

Carex vesicaria, Carex vulpinoidea, Centella asiatica, Cephalanthus occidentalis, Ceratophyllum demersum, Chara, Cladium jamaicense, Commelina, Cuscuta gronovii, Cyperus haspan, Cyperus squarrosus, Cyrilla racemiflora, Didiplis diandra, Digitaria serotina, Diodella teres, Diospyros virginiana, Drosera capillaris, Dulichium arundinaceum, Elatine minima, Eleocharis acicularis, Eleocharis baldwinii, Eleocharis equisetoides, Eleocharis microcarpa, Eleocharis palustris, Eleocharis quadrangulata, Elodea canadensis, Equisetum fluviatile, Eriocaulon compressum, Eupatorium perfoliatum, Eupatorium serotinum, Fimbristylis autumnalis, Fuirena, Glossostigma cleistanthum, Glyceria septentrionalis, Glyceria striata, Gratiola aurea, Gratiola brevifolia, Gratiola virginiana, Helanthium tenellum, Helenium flexuosum, Helenium pinnatifidum, Heteranthera limosa, Heteranthera reniformis, Heteranthera rotundifolia, Hordeum jubatum, Hydrocotyle, Hymenocallis crassifolia, Hypericum fasciculatum, Hypericum mutilum, Ilex, Iris virginica, Isoetes lacustris, Juncus acuminatus, Juncus effusus, Juncus megacephalus, Juncus repens, Lachnanthes caroliniana, Larix laricina, Lemna minor, Lindernia grandiflora, Lindernia monticola, Liquidambar styraciflua, Ludwigia lanceolata, Ludwigia polycarpa, Lycopodiella appressa, Lysimachia nummularia, Magnolia grandiflora, Marsilea vestita, Mayaca fluviatilis, Myrica cerifera, Myriophyllum pinnatum, Myriophyllum spicatum, Myriophyllum tenellum, Najas flexilis, Najas guadalupensis, Nitella, Nuphar microphylla, Nuphar polysepala, Nuphar variegata, Nymphaea odorata, Nyssa aquatica, Nyssa biflora, Nyssa ogeche, Osmunda regalis, Panicum hemitomon, Paspalum praecox, Persea borbonia, Persicaria amphibia, Persicaria bicornis, Persicaria coccinea, Persicaria hirsuta, Persicaria hydropiperoides, Persicaria lapathifolia, Persicaria pensylvanica, Persicaria punctata, Phalaris arundinacea, Phyla lanceolata, Picea mariana, Pinus echinata, Pinus elliottii, Pinus palustris, Pinus taeda, Pluchea baccharis, Pluchea foetida, Polygala lutea, Polypremum procumbens, Pontederia cordata, Potamogeton amplifolius, Potamogeton foliosus, Potamogeton gramineus, Potamogeton nodosus, Potamogeton pusillus, Potamogeton richardsonii, Potamogeton robbinsii, Potamogeton zosteriformis, Proserpinaca pectinata, Quercus laurifolia, Quercus nigra, Quercus virginiana, Ranunculus flabellaris, Ranunculus laxicaulis, Ranunculus longirostris, Rhexia aristosa, Rhexia nashii, Rhynchospora chalarocephala, Rhynchospora inundata, Rhynchospora macra, Rhynchospora macrostachya, Rhynchospora tracyi, Rorippa aquatica, Rorippa palustris, Rubus, Rumex altissimus, Rumex crispus, Rumex stenophyllus, Sabal minor, Sabal palmetto, Sabatia, Sagittaria brevirostra, Sagittaria calycina, Sagittaria filiformis, Sagittaria lancifolia, Sagittaria latifolia, Sagittaria papillosa, Sagittaria rigida, Sagittaria subulata, Sagittaria teres, Salix amygdaloides, Salix nigra, Samolus valerandi, Saururus cernuus, Schoenoplectus acutus, Schoenoplectus heterochaetus, Schoenoplectus pungens, Schoenoplectus tabernaemontani, Scirpus cyperinus, Scirpus pendulus, Scutellaria lateriflora, Serenoa reprens, Sium suave, Smilax, Solanum carolinense, Solidago gigantea, Sparganium eurycarpum, Spartina cynosuroides, Sphagnum, Spirodela polyrhiza, Symphyotrichum novae-angliae, Taxodium distichum, Thalia geniculata, Tridens ambiguus, Tridens strictus, Typha angustifolia, Typha domingensis, Typha latifolia, Ulmus americana, Utricularia gibba, Utricularia macrorhiza, Vaccinium elliottii, Vallisneria americana, Verbena hastata, Viola, Wisteria frutescens, Wolffia brasiliensis, Woodwardia virginica, Xanthium strumarium, Xyris difformis.

2.1.6.12. ***Sagittaria guayanensis* Kunth** is a nonindigenous annual, which grows in ditches, pools (ephemeral), rice fields, and streams at elevations to 100 m. The plants typically produce broad floating leaves, which superficially resemble lily pads; only the inflorescence extends above the water surface. The substrates are described as mud. Flowering occurs from summer to fall. Unlike most of their congeners, these plants are andromonoecious, with individuals bearing staminate and perfect flowers. Here, the condition is believed to have been derived secondarily from a monoecious ancestry in order to facilitate self-pollination. The flowers are preanthesis cleistogamous. Consequently, they typically attract few pollinator visits, rarely outcross, and primarily are self-pollinating. In its native range, the seeds are well represented in the seed bank (up to 37 seeds m^{-2}) and germinate well under ambient greenhouse temperatures. The plants have no well-developed means of asexual reproduction. **Reported associates (North America):** *Chara, Bacopa repens, Ammannia auriculata, Dopatrium junceum, Eriocaulon cinereum, Heteranthera limosa, Rotala ramosior, Sphenoclea zeylanica.*

2.1.6.13. ***Sagittaria isoetiformis* J.G. Sm.** is a perennial, which grows in shallow waters (2–70 cm; typically 20–40 cm) in bottoms (pond), clay pits, depressions, ditches, ponds (karst; sinkhole), pools, and along the shores of lakes, limesinks, ponds (drying), and rivers at elevations to 73 m. Exposures tend to receive full sunlight. The substrates are acidic (e.g., pH: 5.4) and are described as Basinger (spodic psammaquents), loamy fine sand (Albany), loamy sand, muck, mucky sand, peat, peaty sand, sand, sandy muck, sandy peat, and Troup (grossarenic paleudults). They contain relatively low levels of Ca, K, and Mg. Flowering and fruiting have been observed from June to October. These monoecious plants are highly sexual, with most female flowers producing numerous (up to 200) achenes. Although individuals are self-compatible, they primarily are outcrossed (the male and female flowers rarely open simultaneously) by insects (Insecta). Populations maintain relatively high levels of genetic variation (94% polymorphic loci; $H_{ES}=0.399$); however, they are fairly differentiated ($G_{ST}=0.399$). Although possible, geitonogamous pollen transfer probably occurs rarely, being limited by the temporally exclusive sex expression and by the close proximity of different phase genets, which simultaneously offer flowers of both sexes to potential pollinators. Though capable of clonal growth, the populations are not structured genetically as being highly clonal, but exhibit patterns of diverse genetic variation, which has been attributed to the recruitment of sexual propagules from the seed bank. The seeds can germinate in wet soil

or in several centimeters of water. Seed dispersal (primarily by water) likely occurs over small distances due to the lack of any obvious adaptations for long-distance dispersal. Vegetative reproduction occurs by stolons (also referred to as rhizomes), which produce small corms (yielding as many as 20 ramets annually). The corms persist for only a few weeks. Stolon connections do not persist between ramets. The plants differ from most of their congeners by their ability to implement a CAM photosynthetic pathway. **Reported associates:** *Andropogon glomeratus, Andropogon gyrans, Brasenia schreberi, Cabomba, Carex striata, Centella asiatica, Cephalanthus occidentalis, Coleataenia tenera, Cyperus, Dichanthelium acuminatum, Drosera, Echinodorus, Eleocharis melanocarpa, Eleocharis obtusa, Eleocharis robbinsii, Eriocaulon compressum, Eubotrys racemosa, Eupatorium leptophyllum, Eupatorium leucolepis, Eupatorium mohrii, Fimbristylis perpusilla, Fuirena, Gratiola ramosa, Helanthium tenellum, Hydrocotyle umbellata, Hypericum denticulatum, Juncus repens, Lachnanthes caroliniana, Lachnocaulon anceps, Lechea, Leersia hexandra, Lobelia boykinii, Ludwigia linearis, Ludwigia spathulata, Lysimachia loomisii, Nymphaea odorata, Nymphoides cordata, Nyssa biflora, Panicum hemitomon, Panicum verrucosum, Panicum virgatum, Polygala cymosa, Pontederia, Proserpinaca palustris, Rhynchospora perplexa, Rhynchospora pusilla, Riccia, Scleria reticularis, Sclerolepis uniflora, Serenoa repens, Smilax walteri, Spartina, Syngonanthus flavidulus, Taxodium distichum, Typha, Utricularia cornuta, Utricularia gibba, Utricularia resupinata, Verbena, Viola lanceolata, Xyris.*

2.1.6.14. ***Sagittaria kurziana* Glück** is a perennial, which grows in tidal or non-tidal canals, ditches (flowing), rivers (spring-fed), springs, and streams at elevations below 100 m. It inhabits cool, clear waters that are fresh to brackish (specific conductance: 36–878 µS cm^{-1}; chloride: 4–236 mg l^{-1}), clear, fast flowing (e.g., 300 cm s^{-1}), and moderately deep (to 1.8 m). They are circumneutral (pH: 4.4–8.7; $\bar{x}$: 6.9); and usually hard ($\bar{x}$ total CaCO$_3$ alkalinity: 36.2 mg l^{-1}) with a high total nitrogen ($\bar{x}$: 860 µg l^{-1}) and sulfur content ($\bar{x}$ sulfate: 21.8 mg l^{-1}). The substrates are alkaline ($\bar{x}$: pH: 8.3), low in organic matter, and high in Ca ($\bar{x}$: 1830 µg g^{-1}) and are described as limestone-shell sand or sandy. Unlike most of its congeners, this is an entirely submersed species, which is restricted to lotic riverine habitats, where it can grow densely to become the dominant component (by biomass). Transplant studies have failed to induce floating or emergent leaves on the plants. Flowering occurs from March to May. The flowers emerge from the otherwise inundated inflorescence during anthesis but sometimes open while underwater. The plants are monoecious, with the lower inflorescence whorl bearing three female flowers (occasionally two females and one male), and the next whorl is entirely male (rarely with one male and two female flowers or two male and one female flower); the remaining flowers are male. Further studies are needed on the pollination biology and seed ecology of this species. The plants reproduce vegetatively by the production of coarse rhizomes. **Reported associates:** *Cabomba caroliniana, Ceratophyllum demersum, Chara, Cladophora,*

Cryptocoryne beckettii, Egeria densa, Eichhornia crassipes, Hydrilla verticillata, Hymenocallis rotata, Lemna, Lobelia cardinalis, Myriophyllum aquaticum, Myriophyllum heterophyllum, Myriophyllum laxum, Myriophyllum spicatum, Najas guadalupensis, Nuphar, Pistia stratiotes, Pontederia cordata, Potamogeton, Vallisneria americana, Zizania aquatica.

2.1.6.15. *Sagittaria lancifolia* L. is a perennial, which grows in shallow (to 1.2 m), sluggish to still, tidal or non-tidal, brackish (oligohaline) to freshwaters on beaches or in baygalls, bogs, borrow pits, canals (drainage), depressions, ditches, dunes, flats, flatwoods, floodplains, lakebeds (dry), marshes (depression; floating; freshwater; salt), meadows, ponds (spring-fed), pools, rice fields, roadsides, savannahs, scrub, sinkholes, slopes (to 5%), sloughs, swales, swamps (dome; strand), and along the margins of bayous, lagoons, lakes, ponds, rivers, and streams (estuarine) at elevations to 31 m. The plants consistently occur in open exposures, where they receive full sunlight, and less often in partial shade. They can tolerate some brackish conditions but are killed at salinity levels of 15% (especially if grazed by herbivores). The leaves elongate proportionally with increased water depth. The waters are circumneutral (pH: 5.3–8.7; $\bar{x}$: 7.2), alkaline ($\bar{x}$ total alkalinity: 32.9 mg l^{-1}; $\bar{x}$ Ca: 23.9 mg l^{-1}; $\bar{x}$ chloride: 30.8 mg l^{-1}), and high in nitrogen ($\bar{x}$ total N: 1020 µg l^{-1}). The substrates often are brackish, but somewhat acidic ($\bar{x}$ = pH: 6.1), and have been characterized as clay (Schriever), clayey alluvium, dredge spoil, gravel, Lafitte-Clovelly association, loamy sand, marl, muck (Allemands; Kenner), Newhan-Corolla, peat, peaty muck, rock, sand (Duckston; Leon; Waveland), sandy loam (fine; Harleston), silt, silty clay loam (Cancienne), and silty loam. Although described as monoecious, the plants actually are subandrodioecious (mixtures of male and cosexual plants). The flowers are pollinated by insects (Insecta). Flowering and fruiting have been observed throughout the year (from January to December). The achenes float and are dispersed by water; however, the presence of resinous glands confers adhesiveness, which is believed to facilitate their attachment to animal vectors for epizoochorous dispersal. A seed bank (having up to 167 seeds m^{-2}) develops; however, many of the seeds are not intact and are inviable. The achenes require light for germination, which occurs readily (100%) at 25°C. Warm stratification can promote germination. Germination rates are highest under inundated conditions and are reduced under anaerobic conditions; higher rates (67%) occur in saturated sites than in flooded sites (33%). Seed germination is inhibited completely at salinities of 2‰ or higher. Seedling densities of 20 m^{-2} have been observed at several sites. The adult plants contain a milky latex sap and reproduce vegetatively by means of a thick rhizome. Genetic analyses (using allozymes) indicate that populations are characterized by only low levels of clonality. The leaves and inflorescences are larger on plants growing under high phosphorous conditions, but more biomass is allocated to roots and rhizomes under low phosphorous conditions. The plants can attain mean plot biomass values of 254 g DM m^{-2}. The roots are colonized by arbuscular mycorrhizae and dark septate endophytic Fungi. **Reported**

associates: *Acer rubrum, Acmella repens, Aeschynomene, Alternanthera philoxeroides, Ambrosia artemisiifolia, Ambrosia trifida, Ammannia latifolia, Ampelaster carolinianus, Ampelopsis arborea, Amphicarpum, Aristida palustris, Asclepias perennis, Asclepias rubra, Baccharis halimifolia, Bacopa monnieri, Berchemia scandens, Boehmeria cylindrica, Bolboschoenus robustus, Calopogon tuberosus, Celtis laevigata, Cephalanthus occidentalis, Cicuta maculata, Cirsium horridulum, Cirsium nuttallii, Cladium jamaicense, Cyperus articulatus,, Cyperus odoratus, Cyperus strigosus, Cyperus virens, Decodon verticillatus, Desmanthus illinoensis, Diospyros virginiana, Echinochloa walteri, Eichhornia crassipes, Eleocharis cellulosa, Eleocharis fallax, Eleocharis montevidensis, Elephantopus carolinianus, Eragrostis elliottii, Eryngium aquaticum, Eupatorium, Fimbristylis spadicea, Fraxinus, Fuirena scirpoidea, Galium tinctorium, Gonolobus, Hibiscus laevis, Hydrocotyle verticillata, Hydrolea corymbosa, Hymenocallis rotata, Hypericum cistifolium, Hypericum fasciculatum, Ilex cassine, Ilex glabra, Ilex opaca, Ilex vomitoria, Ipomoea sagittata, Iris hexagona, Itea virginica, Iva annua, Juncus acuminatus, Juncus effusus, Juncus roemerianus, Juniperus virginiana, Lachnanthes caroliniana, Leersia hexandra, Leersia oryzoides, Leptochloa fusca, Liquidambar styraciflua, Ludwigia glandulosa, Ludwigia octovalvis, Ludwigia peploides, Ludwigia peruviana, Lythrum alatum, Lythrum lineare, Mikania scandens, Morus rubra, Myrica cerifera, Myriophyllum aquaticum, Nymphaea odorata, Nymphoides aquatica, Nyssa aquatica, Orontium aquaticum, Osmunda regalis, Packera glabella, Panicum amarum, Panicum dichotomiflorum, Panicum hemitomon, Panicum repens, Panicum virgatum, Parthenocissus, Paspalum distichum, Paspalum urvillei, Peltandra virginica, Persea, Persicaria glabra, Persicaria hydropiperoides, Persicaria punctata, Phragmites australis, Phyla lanceolata, Phyla nodiflora, Physostegia leptophylla, Pinus elliottii, Platanthera nivea, Polygala rugelii, Pontederia cordata, Ptilimnium capillaceum, Quercus, Ranunculus sardous, Rhapidophyllum hystrix, Rhynchospora caduca, Rhynchospora elliottii, Rhynchospora fascicularis, Rhynchospora inundata, Rubus argutus, Rumex conglomeratus, Rumex crispus, Rumex obovatus, Sabal minor, Sabal palmetto, Sabatia calycina, Sacciolepis striata, Sagittaria graminea, Sagittaria latifolia, Sagittaria platyphylla, Salix caroliniana, Salix nigra, Sambucus, Samolus valerandi, Saururus cernuus, Schoenoplectus californicus, Schoenoplectus tabernaemontani, Scleria lacustris, Serenoa, Smilax smallii, Spartina alterniflora, Spartina bakeri, Spartina cynosuroides, Spartina patens, Sphagnum, Symphyotrichum elliotii, Symphyotrichum novi-belgii, Symphyotrichum subulatum, Taxodium distichum, Thelypteris palustris, Tillandsia, Toxicodendron radicans, Triadica sebifera, Typha domingensis, Typha latifolia, Utricularia purpurea, Vigna luteola, Vitis, Woodwardia areolata, Xyris difformis, Zizania aquatica, Zizaniopsis miliacea.*

2.1.6.16. *Sagittaria latifolia* Willd. is a perennial, which inhabits shallow (average depth of 10 cm, but up to 1.2 m), intertidal, non-tidal, or tidal freshwaters of backwaters, baygalls, beaches, bogs (quaking; sandy), bottomlands, canals (drainage), channels, depressions, dikes, ditches (roadside; verbal), fens (calcareous; poor; prairie; rich), flats (sand), floodplains, gravel pits, lagoons (dessicated), lakes (playa), marshes, meadows, mud flats, oxbows, pans (dune), ponds (sand), pools (ephermeral; roadside), prairies, rice fields, roadsides, seeps, shores, silt bars, slopes (to 1%), sloughs, springs, streams (intermittent), swales, swamps, vernal pools, woodlands, and the margins of lagoons, lakes, ponds, potholes, reservoirs, rivers, sloughs, and streams at elevations to 2625 m. The waters span a broad range of conditions (pH: 4.7–9.2; $\bar{x}$: 6.9–7.3; conductivity: 20–510 µmhos cm^{-1}; total alkalinity: 1.5–290 mg l^{-1} CaCO$_3$; $\bar{x} = 26.4$ mg l^{-1}; total N: 380–2470 µg l^{-1}). Exposures range from open sites in full sunlight to areas of partial shade. The substrates include alluvium, clay (heavy), clayey sand, granite, gravel, loam, muck, mucky sand, mud (organic; soft), peaty muck, peaty mud, rocky serpentine, sand, sandy mud, sandy silty alluvium, silt (river bottom), silty clay, silty gravel, and silty sand. Flowering occurs from May to November, with fruiting observed from June to October. The flowers are pollinated by generalist insects (Insecta), including bees and wasps (Hymenoptera) or flies (Diptera). Individuals can be monoecious or dioecious. Sex expression is controlled by two Mendelian loci (male and female sterility), where dioecious males are heterozygous at both loci, and monoecious individuals are homozygous for alleles dominant to male sterility in females and recessive to female sterility in males. Allozyme data indicate that individuals of each type remain genetically distinct and that both suffer substantially from inbreeding depression. Dioecious populations contain only a subset of the haplotype diversity found in monoecious populations, indicating the derivation of the former from the latter. Reproduction primarily is sexual in monoecious populations but vegetative in dioecious populations. Selfing rates are higher in monoecious populations. In dioecious individuals, the males have larger flowers but smaller floral displays than the female flowers, which remain open for a shorter time period than the males. The larger male flowers experience higher pollinator visitation rates, but larger male floral displays result in lower pollination success. Sexual expression is highly plastic in monoecious populations with higher numbers of female flowers produced at higher nutrient levels. In monoecious individuals, the female flowers increase proportionally with plant size, while the males do not. The flat seeds are dispersed by water. A persistent seed bank develops. Germination requires light and flooded conditions. Seeds from monoecious plants germinate freely, whereas those of dioecious plants require a period of stratification. The seeds can be recalcitrant but have been stored successfully in water at 3°C. They have germinated following 150 days of cold stratification under a 21°C/16°C day/night temperature regime. Vastly higher germination rates have been obtained for seeds whose seed coats are first "nipped" using scissors. Densities of up to 20 seedlings m^{-2} have been observed at field sites. Vegetative propagation occurs by the formation of rhizomes, which produce large tubers. The rhizomes normally extend from 4 to 6 cm below the surface, but can occur as deep as 10 cm. An essentially 1:1 trade-off exists between sexual reproduction (primarily the production of female flowers) and the extent of vegetative reproduction. The roots can harbor large colonies of methanotrophic bacteria. The plants can tolerate short-term intrusions of salinity but suffer from decreased seed germination, and seedling emergence, survival, and growth rates when exposed to saline conditions.

Reported associates: *Abutilon theophrasti, Acalypha, Acer negundo, Acer rubrum, Acer saccharinum, Acmella, Agalinis, Agrostis gigantea, Agrostis stolonifera, Alisma subcordatum, Alisma triviale, Alnus serrulata, Alternanthera philoxeroides, Amaranthus tuberculatus, Ambrosia artemisiifolia, Ammannia coccinea, Ammannia robusta, Amorpha fruticosa, Andropogon glomeratus, Anemone canadensis, Angelica atropurpurea, Apios americana, Apocynum cannabinum, Aristida purpurascens, Asclepias incarnata, Asclepias syriaca, Azolla, Baccharis halimifolia, Baccharis salicifolia, Betula, Bidens aristosus, Bidens beckii, Bidens cernuus, Bidens frondosus, Bidens laevis, Bistorta bistortoides, Boehmeria cylindrica, Bolboschoenus fluviatilis, Boltonia asteroides, Brasenia schreberi, Butomus umbellatus, Calamagrostis canadensis, Calla palustris, Caltha palustris, Campanula aparinoides, Carex bebbii, Carex comosa, Carex cristatella, Carex exilis, Carex haydenii, Carex hystericina, Carex lasiocarpa, Carex limosa, Carex livida, Carex lupuliformis, Carex lurida, Carex molesta, Carex muskingumensis, Carex pellita, Carex praegracilis, Carex rostrata, Carex stipata, Carex stricta, Carex utriculata, Carex vesicaria, Carex vulpinoidea, Cephalanthus occidentalis, Ceratophyllum demersum, Chamaedaphne calyculata, Chara, Chelone glabra, Cicuta bulbifera, Comarum palustre, Conium maculatum, Conoclinium, Coreopsis gladiata, Cornus amomum, Cornus sericea, Crataegus douglasii, Cyperus erythrorhizos, Cyperus esculentus, Cyperus odoratus, Cyperus squarrosus, Cyperus strigosus, Decodon verticillatus, Distichlis spicata, Doellingeria umbellata, Drosera anglica, Dryopteris, Dulichium arundinaceum, Echinochloa crus-galli, Echinochloa walteri, Eclipta prostrata, Egeria densa, Elaeagnus angustifolia, Eleocharis acicularis, Eleocharis bella, Eleocharis erythropoda, Eleocharis obtusa, Eleocharis ovata, Eleocharis palustris, Eleocharis parishii, Eleocharis quadrangulata, Eleocharis robbinsii, Elodea canadensis, Elymus canadensis, Equisetum arvense, Equisetum fluviatile, Eragrostis hypnoides, Eriocaulon aquaticum, Eupatorium perfoliatum, Eutrochium maculatum, Fallopia scandens, Festuca, Fimbristylis vahlii, Forestiera acuminata, Fraxinus nigra, Fraxinus pennsylvanica, Galium obtusum, Galium trifidum, Galium triflorum, Gentiana andrewsii, Glyceria borealis, Glyceria grandis, Glyceria striata, Gnaphalium palustre, Hedeoma hispida, Helenium autumnale, Helianthus tuberosus, Heliopsis helianthoides, Hippuris vulgaris, Hordeum jubatum, Humulus lupulus, Hydrocotyle verticillata, Hypericum canadense, Hypericum sphaerocarpum, Ilex cassine, Impatiens capensis, Iris pseudacorus, Iris virginica, Isoetes lacustris, Juncus acuminatus, Juncus effusus, Juncus nodosus, Laportea canadensis, Lathyrus palustris, Leersia oryzoides, Lemna minor, Lemna trisulca, Lepidium* *latifolium*, *Leptochloa panicea*, *Lilaeopsis*, *Lilium michiganense*, *Lindernia dubia*, *Lipocarpha micrantha*, *Ludwigia palustris*, *Ludwigia peploides*, *Ludwigia polycarpa*, *Lycopus americanus*, *Lycopus rubellus*, *Lycopus uniflorus*, *Lycopus virginicus*, *Lysimachia ciliata*, *Lysimachia nummularia*, *Lysimachia vulgaris*, *Lythrum salicaria*, *Mazus pumilus*, *Melilotus albus*, *Mentha arvensis*, *Menyanthes trifoliata*, *Mimulus ringens*, *Murdannia*, *Myriophyllum sibiricum*, *Navarretia*, *Nuphar polysepala*, *Nuphar variegata*, *Nymphaea odorata*, *Nyssa biflora*, *Oenothera biennis*, *Panicum dichotomiflorum*, *Panicum hemitomon*, *Panicum virgatum*, *Paspalum dilatatum*, *Peltandra virginica*, *Penthorum sedoides*, *Persicaria amphibia*, *Persicaria arifolia*, *Persicaria coccinea*, *Persicaria hydropiperoides*, *Persicaria lapathifolia*, *Persicaria pensylvanica*, *Persicaria punctata*, *Persicaria sagittata*, *Phalaris arundinacea*, *Phanopyrum gymnocarpon*, *Phragmites australis*, *Phyla lanceolata*, *Physostegia virginiana*, *Poa nemoralis*, *Poa pratensis*, *Polytrichum*, *Pontederia cordata*, *Populus angustifolia*, *Potamogeton amplifolius*, *Potamogeton berchtoldii*, *Potamogeton crispus*, *Potamogeton foliosus*, *Potamogeton gramineus*, *Potamogeton illinoensis*, *Potamogeton natans*, *Potamogeton natans*, *Potamogeton pusillus*, *Potamogeton richardsonii*, *Potamogeton zosteriformis*, *Potentilla norvegica*, *Proserpinaca palustris*, *Prunus virginiana*, *Puccinellia*, *Pycnanthemum virginianum*, *Ranunculus aquatilis*, *Ranunculus flammula*, *Ranunculus pensylvanicus*, *Ranunculus recurvatus*, *Rhexia mariana*, *Rhododendron canescens*, *Rhus glabra*, *Rhynchospora inundata*, *Ribes americanum*, *Rorippa islandica*, *Rorippa palustris*, *Rosa setigera*, *Rubus argutus*, *Rudbeckia laciniata*, *Rumex altissimus*, *Rumex verticillatus*, *Sagittaria ambigua*, *Sagittaria australis*, *Sagittaria calycina*, *Sagittaria cuneata*, *Sagittaria engelmanniana*, *Sagittaria filiformis*, *Sagittaria graminea*, *Sagittaria lancifolia*, *Sagittaria montevidensis*, *Sagittaria platyphylla*, *Sagittaria rigida*, *Sagittaria sanfordii*, *Sagittaria teres*, *Salix amygdaloides*, *Salix exigua*, *Salix gooddingii*, *Salix interior*, *Salix lasiandra*, *Salix nigra*, *Sambucus nigra*, *Saururus cernuus*, *Scheuchzeria palustris*, *Schoenoplectus acutus*, *Schoenoplectus americanus*, *Schoenoplectus hallii*, *Schoenoplectus lacustris*, *Schoenoplectus tabernaemontani*, *Schoenoplectus triqueter*, *Scirpus atrovirens*, *Scirpus cyperinus*, *Scutellaria lateriflora*, *Sesbania drummondii*, *Sicyos angulatus*, *Silene nivea*, *Silphium perfoliatum*, *Sium suave*, *Solidago gigantea*, *Sonchus arvensis*, *Sparganium androcladum*, *Sparganium emersum*, *Sparganium eurycarpum*, *Sparganium natans*, *Sphagnum angustifolium*, *Sphagnum centrale*, *Sphagnum teres*, *Spiraea douglasii*, *Spirodela polyrhiza*, *Stuckenia pectinata*, *Stuckenia vaginata*, *Stylisma aquatica*, *Symphyotrichum lanceolatum*, *Symphyotrichum pilosum*, *Symphyotrichum puniceum*, *Symphyotrichum subulatum*, *Symplocarpus foetidus*, *Taxodium distichum*, *Thalia geniculata*, *Thalictrum dasycarpum*, *Toxicodendron radicans*, *Toxicodendron vernix*, *Tradescantia virginiana*, *Triadenum virginicum*, *Typha angustifolia*, *Typha domingensis*, *Typha latifolia*, *Urtica dioica*, *Utricularia macrorhiza*, *Utricularia purpurea*, *Vallisneria americana*, *Verbascum thapsus*, *Verbena bonariensis*, *Verbena hastata*, *Vernonia fasciculata*, *Veronica anagallis-aquatica*, *Veronica catenata*, *Veronica peregrina*, *Viburnum lentago*, *Viola lanceolata*, *Vitis riparia*, *Wolffia brasiliensis*, *Wolffia columbiana*, *Xanthium strumarium*, *Zannichellia palustris*, *Zizania palustris*.

2.1.6.17. **Sagittaria longiloba Engelm. ex J.G. Sm.** is a perennial, which grows in shallow waters (to 45 cm) in canals (irrigation), cienegas, ditches (drainage; roadside), marshes, mudflats, playas, ponds, pools (alkaline; ephemeral), rice fields, and along the margins of lakes, ponds, rivers, and streams at elevations to 1676 m. This is a common species of playa lakes and similar habitats. The exposures typically are open and receive full sunlight. The substrates are described as alkaline and include clay, clay loam, mud, and sand. Flowering occurs from May to October, with fruiting observed from July to October. The plants are monoecious and probably are pollinated by insects (Insecta). Viable achenes are known to be passed through the digestive tracts of birds (Aves: Anatidae: *Anas platyrhynchos*; Charadriidae: *Charadrius vociferus*) after retention times ranging from 8 to 16 h, which implicates birds as agents of their long-distance dispersal. The achenes germinate in a few centimeters of water under ambient greenhouse conditions. The seedlings typically establish in sites that have been shallowly flooded for more than 2 weeks. In appropriately flooded sites, germination is continuous throughout the growing season. Seedling densities from 25 to 75 m^{-2} have been observed in the field. Vegetative reproduction occurs by the production of tubers. **Reported associates:** *Agropyron*, *Alisma triviale*, *Ambrosia grayi*, *Ammannia coccinea*, *Bacopa rotundifolia*, *Callitriche heterophylla*, *Chara*, *Cyperus digitatus*, *Digitaria sanguinalis*, *Echinochloa crus-galli*, *Echinodorus berteroi*, *Eleocharis macrostachya*, *Eleocharis palustris*, *Helianthus ciliaris*, *Heteranthera dubia*, *Heteranthera limosa*, *Heteranthera mexicana*, *Juncus*, *Juniperus*, *Lemna*, *Leptochloa fusca*, *Limosella aquatica*, *Ludwigia glandulosa*, *Ludwigia palustris*, *Marsilea vestita*, *Myriophyllum pinnatum*, *Paspalum distichum*, *Persicaria bicornis*, *Persicaria glabra*, *Persicaria hydropiperoides*, *Persicaria lapathifolia*, *Pontederia cordata*, *Populus fremontii*, *Potamogeton natans*, *Potamogeton nodosus*, *Rumex chrysocarpus*, *Rumex crispus*, *Sagittaria calycina*, *Sagittaria latifolia*, *Schoenoplectus californicus*, *Spirodela polyrhiza*, *Stuckenia pectinata*, *Symphyotrichum subulatum*, *Typha domingensis*, *Xanthium*.

2.1.6.18. **Sagittaria macrocarpa J.G. Sm.** is a perennial (or sometimes annual?), which is narrowly distributed and endemic to the inner coastal plain Sandhills Region of the southeastern United States. The plants occur in very shallow waters of ponds (artificial; beaver), pools, seepage bogs, streamheads (blackwater), and lakeshores at low elevations. The waters are acidic, darkly stained (blackwaters), and low in nutrients. The substrates have been described as clay, clayey sand, mud, peaty sand, and sand. The plants are monoecious and flower during summer (their specific flowering period has not been clarified). The plants reproduce vegetatively by the production of slender stolons. Much life history information is needed for this species, which until recently had been regarded as a synonym of *S. graminea* (see *Systematics*). **Reported associates:** *Dichanthelium, Eleocharis flavescens, Eleocharis robbinsii, Juncus debilis, Lachnanthes caroliniana, Mayaca fluviatilis, Nymphaea odorata, Nymphoides, Nyssa biflora, Panicum hemitomon, Potamogeton diversifolius, Sagittaria engelmanniana, Sagittaria graminea, Schoenoplectus etuberculatus, Sparganium americanum, Sphagnum.*

2.1.6.19. *Sagittaria montevidensis* Cham. & Schlecht. is a perennial, which colonizes shallow waters or exposed substrates in ditches (landscaped), mudflats (riverine), sloughs (roadside), and along the margins of swamps (roadside) at elevations to 10 m. The substrates are described as muddy clay. The plants are monoecious, with the female flowers lacking the ring of stamens found in *S. calycina*. The petals possess a bright crimson spot, which presumably serves as an attractant to potential pollinators. Flowering occurs from July to October. Aside from these few details, most of the North American information for this taxon has been attributed incorrectly and refers instead to *S. calycina*, a distinct, native species (see *Systematics*). Reliable information pertaining correctly to the North American occurrences of this nonindigenous species is scarce and should be compiled. **Reported associates (North America):** *Nymphaea capensis, Sagittaria latifolia, Sagittaria platyphylla.*

2.1.6.20. *Sagittaria papillosa* Buch. is a perennial, which grows in shallow (e.g., 6.5–20 cm depth), brackish to freshwaters of bogs, borrow pits, ditches (roadside), flats, flatwoods (pine), marshes, meadows, pools (roadside), prairies (coastal; saline), savannahs, swamps, terraces (river), wastelands, woodlands (sparse), and along the margins of lakes and ponds at elevations to 67 m. Exposures are described as open. The substrates are acidic (mean values: pH: 4.9–5.2; organic matter: <1–3.5%; Ca (as CaO): 118–308 ppm; K (as K_2O): 34–78 ppm; Mg (as MgO): 30–104 ppm; P (as P_2O_5): 7–39.5 ppm). Flowering and fruiting occur from April to September. The plants normally are racemose and monoecious, but occasionally produce branched inflorescences (panicles) having staminate and bisexual flowers on the lowest branches (i.e., they are andromonoecious); the stamens of the bisexual flowers are functional. Individuals are self-compatible, which allows for self-pollination to occur. The natural pollinators of open-pollinated plants have not been enumerated, but likely are insects (Insecta). Only racemose, monoecious plants have been observed to develop under greenhouse conditions (and from F_2 crosses), indicating that the anomalous floral features are environmentally plastic and not strictly determined genetically. Additional information on the pollination biology or seed ecology of this species is unavailable. The plants reproduce vegetatively by the production of coarse rhizomes, from which new vegetative plantlets develop. The connections with the maternal plant sever by the onset of anthesis, which yields clonal ramets in near proximity. **Reported associates:** *Acer negundo, Baccharis halimifolia, Bacopa caroliniana, Boehmeria cylindrica, Carex amphibola, Carex arkansana, Celtis laevigata, Centella asiatica, Cephalanthus occidentalis, Cornus foemina, Crataegus mollis, Crataegus viridis,* *Cyrilla racemiflora, Eleocharis macrostachya, Eleocharis quadrangulata, Eleocharis wolfii, Eupatorium rotundifolium, Euthamia leptocephala, Helenium flexuosum, Hydrolea ovata, Hymenocallis liriosme, Ilex, Iris brevicaulis, Iris hexagona, Isoetes melanopoda, Juncus acuminatus, Juncus brachycarpus, Juncus dichotomus, Juncus effusus, Juncus marginatus, Juncus nodatus, Limnosciadium pinnatum, Ludwigia, Ludwigia palustris, Mikania scandens, Myrica, Oplismenus hirtellus, Panicum hemitomon, Paspalidium geminatum, Paspalum wrightii, Persicaria hydropiperoides, Pinus palustris, Pluchea, Pontederia cordata, Proserpinaca palustris, Quercus, Ranunculus hispidus, Rhynchospora corniculata, Sabal minor, Sagittaria graminea, Salix nigra, Sambucus nigra, Samolus valerandi, Sarracenia, Schoenolirion croceum, Scutellaria integrifolia, Sesbania drummondii, Steinchisma hians, Thalia dealbata, Tridens strictus, Typha latifolia.*

2.1.6.21. *Sagittaria platyphylla* (Engelm.) J.G. Sm. is a perennial, which grows in shallow (e.g., 8–50 cm deep), tidal or non-tidal, brackish or freshwaters of bayous, bogs (hillside), bottoms (lake), canals (drainage), cobble bars, ditches (drainage; roadside), flats, flatwoods, gravel pits, lagoons, levees (wooded), mudflats, lakes, marshes, mudflats, ponds (farm; retention; seasonal), pools, rice fields, roadsides, sand pits, spillways, streams, swales, swamps (roadside), woodlands, and along the margins or shores of lakes, ponds, pools, and reservoirs at elevations to 316 m. Although the plants occur occasionally in brackish sites, their biomass declines at salinities of 3 ppt; salinity levels of 6‰ are lethal. In cultivation, the plants thrive in water temperatures from 20°C to 24°C. Exposures typically are open sites with full sunlight, and uncommonly are characterized by shade. The substrates are acidic (mean pH: 5.8), high in organic matter (mean: 23.8%), and include clay, clayey alluvium, gravel, muck (tidal), mud, sand (fine; Mascotte), sandy clay, sandy silt, silt, silty clay, and silty clay loam (Schriever). Flowering and fruiting extend from April to December. The plants are monoecious and likely pollinated by insects (Insecta). The fruits are buoyant and dispersed by water currents. Although no persistent seed banks have been reported, the seeds are known to remain viable for several years when stored in uncontaminated water. In one study, all *S. platyphylla* accessions surveyed genetically possessed a single multilocus isozyme genotype (13 loci), indicating extremely low levels of genetic variation in the species. Those results contrast with a study of invasive populations in Australia, where each surveyed individual possessed a unique (AFLP) genotype, which indicated that seeds are the primary propagules involved in dispersal. Vegetative reproduction occurs by means of rhizomes, from which large tubers are produced. Detached tubers sink initially, then float to the surface once the shoots develop, but eventually sink again when the roots develop. They also are dispersed by water. Compared to other aquatic species, *S. platyphylla* has a relative slow exponential leaf decay rate (0.053 day^{-1}). **Reported associates:** *Acmella repens, Alternanthera philoxeroides, Ammannia coccinea, Ampelopsis arborea, Asclepias, Azolla filiculoides, Bacopa monnieri, Briza minor,*

Bromus catharticus, Caperonia, Cephalanthus occidentalis, Ceratophyllum demersum, Colocasia esculenta, Commelina diffusa, Corchorus, Cyperus difformis, Cyperus flavicomus, Cyperus haspan, Cyperus pseudovegetus, Cyperus strigosus, Cyperus virens, Dichanthelium, Diodia virginiana, Diospyros virginiana, Echinochloa crus-galli, Echinodorus berteroi, Echinodorus cordifolius, Egeria densa, Eleocharis ambigens, Eleocharis baldwinii, Eleocharis montana, Eleocharis obtusa, Elymus virginicus, Eupatorium capillifolium, Euphorbia, Fagus grandifolia, Fimbristylis dichotoma, Gleditsia aquatica, Hydrilla verticillata, Hydrocleys nymphoides, Hydrocotyle umbellata, Hydrocotyle verticillata, Hydrolea ovata, Hydrolea uniflora, Ilex glabra, Juncus effusus, Juncus megacephalus, Juncus repens, Justicia ovata, Kyllinga brevifolia, Leersia oryzoides, Lepidium virginicum, Leptochloa, Liquidambar styraciflua, Lobelia, Lonicera japonica, Ludwigia alternifolia, Ludwigia octovalvis, Ludwigia palustris, Ludwigia peploides, Ludwigia repens, Marsilea vestita, Micranthemum umbrosum, Mikania scandens, Myrica cerifera, Najas guadalupensis, Najas minor, Nyssa biflora, Paspalum urvillei, Persicaria punctata, Phyla nodiflora, Pinus taeda, Plantago patagonica, Potamogeton crispus, Potamogeton diversifolius, Potamogeton illinoensis, Potamogeton nodosus, Quercus nigra, Rhynchospora colorata, Rhynchospora corniculata, Rotala ramosior, Rubus trivialis, Rumex conglomeratus, Rumex crispus, Sabal palmetto, Sagittaria australis, Sagittaria calycina, Sagittaria lancifolia, Sagittaria latifolia, Sagittaria montevidensis, Salix, Schoenoplectus americanus, Sesbania, Solidago altissima, Sphenoclea zeylanica, Taxodium distichum, Triadica sebifera, Typha domingensis, Typha latifolia, Vallisneria americana, Verbena brasiliensis, Zizania texana.

2.1.6.22. Sagittaria rigida Pursh is a perennial, which grows in shallow to deep (e.g., 0.2–1.8 m), still or flowing, brackish to fresh, tidal or non-tidal waters of bays, bogs, canals, channels, depressions, ditches, meadows, mudflats, oxbows, ponds (artificial; dried; farm; natural; shallow; small; stock; upland), sinkholes (depression), swales, thickets (swamp), and along the margins or shores (receding) of lakes, ponds, reservoirs, rivers, streams, and swamps at elevations to 1385 m. The plants are tolerant of turbidity and fairly eutrophic conditions. In standing waters, they often occur at greater depths (>1.0 m). Exposures typically receive full sunlight. The waters (pH: 5.4–8.9; $\bar{x}$ =7.4; specific conductance: 319–340 μS cm^{-1}; total alkalinity as $CaCO_3$: 10–140 mg l^{-1}) have been characterized as brackish, calcareous, or hard. The substrates (pH: 5.9; 2.7% organic matter) include muck (marly), mucky sand, mud (alluvial), rock, sand, and sandy muck. Flowering and fruiting extend from May to September. The plants are monoecious. The pollination system has not been described but probably is mediated by small insects (Insecta). A persistent seed bank develops, with a mean density of up to 119 seeds m^{-2} reported from shallowly flooded sites. Seeds that are freshly collected in the spring have germinated well following several days of cold (4°C), wet storage. Seedling establishment occurs optimally during drawdown conditions. The plants reproduce vegetatively by rhizomes, from which tubers can develop.

Reported associates: *Alisma triviale, Bacopa rotundifolia, Bolboschoenus fluviatilis, Brasenia schreberi, Butomus umbellatus, Carex gynandra, Carex interior, Carex lupulina, Cephalanthus occidentalis, Ceratophyllum demersum, Ceratophyllum echinatum, Cyperus bipartitus, Dulichium arundinaceum, Eleocharis erythropoda, Eleocharis obtusa, Eleocharis palustris, Elodea canadensis, Eriophorum angustifolium, Heteranthera limosa, Heteranthera multiflora, Heteranthera reniformis, Heteranthera rotundifolia, Hibiscus, Hypericum mutilum, Iris pseudacorus, Isoetes riparia, Juncus acuminatus, Justicia americana, Kyllinga gracillima, Leersia oryzoides, Lemna minor, Lindernia dubia, Ludwigia palustris, Lythrum salicaria, Marsilea vestita, Menyanthes trifoliata, Myriophyllum sibiricum, Nelumbo lutea, Nuphar polysepala, Nymphaea odorata, Orontium aquaticum, Peltandra virginica, Persicaria amphibia, Persicaria bicornis, Persicaria coccinea, Persicaria hydropiperoides, Persicaria punctata, Phalaris arundinacea, Phragmites australis, Phyla lanceolata, Pontederia cordata, Potamogeton foliosus, Potamogeton nodosus, Potamogeton richardsonii, Potamogeton vaseyi, Ranunculus ambigens, Rumex altissimus, Rumex crispus, Rumex stenophyllus, Sagittaria brevirostra, Sagittaria graminea, Sagittaria latifolia, Sagittaria sanfordii, Saururus cernuus, Schoenoplectus acutus, Schoenoplectus heterochaetus, Schoenoplectus pungens, Schoenoplectus smithii, Schoenoplectus tabernaemontani, Scirpus cyperinus, Sparganium americanum, Sparganium emersum, Sphagnum magellanicum, Sphagnum palustre, Sphagnum subsecundum, Thelypteris palustris, Typha angustifolia, Typha latifolia, Utricularia macrorhiza, Wolffia columbiana, Zizania aquatica, Zizania palustris.*

2.1.6.23. Sagittaria sanfordii Greene is a perennial, which inhabits waters (to 1 m deep) in canals, ditches, floodplains, lakes (vernal), marshes, mudflats, pools (alkaline; vernal), rice fields, river bottoms (drying), sloughs, streams and the margins of canals, lakes, ponds (beaver; stock), rivers, and streams at elevations to 609 m. The habitats are categorized as alkaline, with substrates described as alluvium (rocky), clay, loam (gummy), mud (drying), and Tuscan mudflow. Flowering occurs from May to October. The plants are monoecious, but occasionally produce perfect (but normally pistillate) flowers with functional stamens. The plants reproduce vegetatively by rhizomes, which produce spherical tubers. Little additional information exists on the reproductive or seed ecology. This is a relatively rare species, which has become extirpated in many areas due to habitat loss. Its life history deserves a comprehensive evaluation. **Reported associates:** *Alisma, Artemisia douglasiana, Callitriche, Carex, Cyperus eragrostis, Eleocharis macrostachya, Eleocharis quadrangulata, Iris pseudacorus, Isoetes, Juncus effusus, Legenere valdiviana, Ludwigia peploides, Marsilea, Polygonum, Potamogeton, Ranunculus, Sagittaria calycina, Sagittaria latifolia, Sagittaria rigida, Schoenoplectus acutus, Typha.*

2.1.6.24. Sagittaria secundifolia Kral is a perennial, which inhabits clear, slow to swift, riverine waters associated with pools (shallow), rapids, riffles (shallow), and shoals at elevations to 466 m. The plants can grow in exposed shallows or at depths of up to 75 cm. They are intolerant of pollution and epiphytic growths of filamentous algae. Exposures typically receive full sunlight, with populations declining where a wooded overstory is present. The substrates are characterized as boulders (small), cobble, rocky (sandstone) crevices, sand, and slabs (flat; jointed; sandstone). The plants do not occur in highly turbulent water, deep pools, or on substrates consisting of large boulders. Plants growing in swift, shallow water are short in stature, sterile, and reproduce vegetatively by thick, elongate rhizomes (to 10 cm), which anchor amidst the cracks and fissures in the rocky substrate. In quieter waters, the plants produce longer leaves, shorter rhizomes, and are fertile. Flowering occurs from May to August, with fruiting from July to September. The flowers (which appear to float on the water surface) develop only under full sunlight and on plants growing near the water surface, which are conditions associated with low water flow periods. Most of the plants are devoid of flowers and reproduce asexually; however, preliminary studies indicate that populations retain a fair level of genetic variability. The perianth is essentially absent on pistillate flowers, but of normal proportions in the staminate flowers. Pollination presumably is facilitated by bees (Insecta: Hymenoptera), but the reproductive biology and seed ecology of this species remain virtually unknown. **Reported associates:** *Justicia americana, Lindernia, Ludwigia, Myriophyllum, Persicaria, Potamogeton, Najas, Podostemum ceratophyllum.*

2.1.6.25. ***Sagittaria subulata*** **(L.) Buch.** is a perennial, which inhabits shallow (to 61 cm), coastal, brackish (salinity: 0.5–18 ppt), tidal or non-tidal waters in bays, canals, depressions, hammocks, marshes (tidal), mudflats, streams, and along the margins of ponds and rivers at elevations <100 m. The plants occur commonly in areas that become exposed at low tide. The waters are circumneutral (pH: 4.8–8.3; $\bar{x}$: 6.6–6.7), low in total alkalinity (0.4–56.4 mg l^{-1}; $\bar{x}$: 18.5–20.3 mg l^{-1}), with a relatively low total nitrogen content ($\bar{x}$: 700 µg l^{-1}). In cultivation, the plants thrive in water temperatures from 18°C to 28°C. The substrates include mud, muddy peat, sand, sandy gravel, silty clay, and silty mud. Flowering can extend from March to November. The individuals are monoecious, with the flowers pollinated by small flies (Insecta: Diptera). In deeper waters, the upper (male) flowers open only partially, with the lower (pistillate) flowers rarely becoming emersed above the water surface. Few seeds are produced and nothing is known regarding their germination requirements. The plants primarily reproduce vegetatively by the production of stolons, from which tubers can develop. The leaves lack Krantz anatomy but exhibit CAM-like concentrations of malate, which indicates a similar but modified C$_4$ photosynthetic pathway. **Reported associates:** *Acer, Alternanthera philoxeroides, Azolla filiculoides, Bacopa rotundifolia, Bidens bidentoides, Bidens eatonii, Bidens hyperboreus, Callitriche heterophylla, Cardamine longii, Carex glaucescens, Carya cordiformis, Ceratophyllum demersum, Cornus foemina, Crassula aquatica, Crotalaria spectabilis, Cyperus virens, Echinodorus berteroi, Egeria densa, Eichhornia crassipes, Eleocharis acicularis, Eleocharis diandra, Eleocharis engelmannii, Eleocharis flavescens, Eleocharis obtusa, Eleocharis*

parvula, Eriocaulon parkeri, Festuca filiformis, Fimbristylis autumnalis, Fraxinus, Heteranthera multiflora, Heteranthera reniformis, Hydrilla verticillata, Hydrocotyle umbellata, Hydrolea quadrivalvis, Isoetes mattaponi, Lemna valdiviana, Lilaeopsis chinensis, Limosella australis, Lindernia dubia, Ludwigia palustris, Luziola fluitans, Micranthemum micranthemoides, Najas guadalupensis, Nuphar advena, Nymphaea, Nyssa biflora, Orontium aquaticum, Persicaria arifolia, Persicaria hydropiperoides, Persicaria punctata, Pistia stratiotes, Planera aquatica, Pontederia cordata, Potamogeton diversifolius, Potamogeton illinoensis, Potamogeton pusillus, Proserpinaca palustris, Ptilimnium capillaceum, Rhynchospora glomerata, Rhynchospora microcephala, Ruppia maritima, Sagittaria calycina, Sagittaria graminea, Salvinia, Schoenoplectus smithii, Spergularia salina, Taxodium distichum, Typha domingensis, Utricularia gibba, Utricularia inflata, Vallisneria americana, Zannichellia palustris.

2.1.6.26. ***Sagittaria teres*** **S. Wats.** is a perennial, which grows in shallow (to 100 cm) waters of borrow pits, swamps, and along the margins or shores of lakes and ponds at elevations to 64 m. Exposures are open and receive full sunlight. The waters are acidic (pH: 4.5–6.7; $\bar{x}$ = 5.2) and very soft (total alkalinity 4–7 mg l^{-1}; $\bar{x}$ = 5.5 mg l^{-1}). The substrates can include cobble, gravel, or sand (mucky; pure; silty; white). Flowering and fruiting extend from July to September. The plants are monoecious. The fruits normally develop and ripen underwater and can differ morphologically if ripened above water. Only scarce information exists on the reproductive biology and seed ecology. Allozyme analyses indicate high levels of genetic variation in populations, which implicates recruitment from the seed bank as a principal means of population persistence (although seed bank data are not available). Vegetative reproduction occurs by the production of stolons, from which minute tubers can develop. However, genetic analyses also indicate that populations are not highly clonally structured, despite their capacity for asexual reproduction. **Reported associates:** *Carex longii, Coreopsis rosea, Cyperus squarrosus, Drosera filiformis, Drosera intermedia, Drosera rotundifolia, Dulichium arundinaceum, Eleocharis robbinsii, Eleocharis tricostata, Eleocharis tuberculosa, Eriocaulon aquaticum, Eupatorium leucolepis, Eupatorium perfoliatum, Fimbristylis autumnalis, Fuirena pumila, Gratiola aurea, Hypericum adpressum, Hypericum mutilum, Isoetes echinospora, Juncus militaris, Juncus pelocarpus, Lipocarpha micrantha, Lobelia dortmanna, Najas flexilis, Nymphaea odorata, Nymphoides cordata, Panicum verrucosum, Pontederia cordata, Potamogeton illinoensis, Potamogeton perfoliatus, Proserpinaca pectinata, Rhexia virginica, Rhynchospora inundata, Rhynchospora macrostachya, Rhynchospora scirpoides, Rotala ramosior, Sabatia kennedyana, Sagittaria graminea, Sagittaria latifolia, Schoenoplectus etuberculatus, Scleria reticularis, Sparganium angustifolium, Utricularia cornuta, Utricularia geminiscapa, Utricularia gibba, Utricularia resupinata, Vallisneria americana, Viola lanceolata, Xyris difformis, Xyris smalliana.*

Use by wildlife: The starchy tubers of various *Sagittaria* species are eaten by beavers (Mammalia: Castoridae: *Castor canadensis*), muskrats (Mammalia: Cricetidae: *Ondatra zibethicus*), and waterfowl (Aves: Anatidae). The plants (species unspecified) host many insects (Insecta), including beetles and weevils (Coleoptera: Chrysomelidae: *Donacia caerulea, Galerucella nymphaea, Plateumaris rufa*; Curculionidae: *Brachybamus electus, Listronotus turbatus, Onychylis angustus, O. nigrirostris*), flies (Diptera: Ephydridae *Hydrellia deceptor, H. griseola*), and larval moths (Lepidoptera: Noctuidae: *Acronicta oblinita, Argyrogramma verruca, Bellura obliqua, Homophoberia cristata*). The seeds of *Sagittaria ambigua* reportedly are eaten by birds (Aves). The flowers of *S. australis* are visited (and pollinated primarily by generalist bees (Insecta: Hymenoptera: Apidae: *Ceratina*; Halictidae: *Augochlora, Lasioglossum*). The tubers are eaten by weevils (Insecta: Coleoptera: Curculionidae: *Listronotus caudatus, L. echinodori*); the flowers are attacked by *L. appendiculatus*. The plants provide cover and shade for small fish. The nutlets of *S. calycina* are eaten by canvasbacks (Aves: Anatidae: *Aythya valisineria*) and by other ducks (Aves: Anatidae: *Anas*) in low to locally high quantities. *Sagittaria cristata* is colonized by various freshwater sponges (Bryozoa: Cristatellidae: *Cristatella mucedo*; Fredericellidae: *Fredericella sultana*; Paludicellidae: *Paludicella articulata*; Pectinatellidae: *Pectinatella magnifica*; Plumatellidae: *Hyalinella punctata, Plumatella emarginata, P. repens*). It is a minor food item for ducks (Aves: Anatidae). The foliage of *S. cuneata* contains 21.8% crude protein, 17.3% crude fiber and is high in K, Fe, and Mo. The plants are grazed by muskrats (Mammalia: Cricetidae: *Ondatra zibethicus*) and domestic livestock (Mammalia: Bovidae) including cattle (*Bos taurus*) and sheep (*Ovis aries*). The achenes and tubers are eaten by various waterfowl (Aves: Anatidae) including Canada geese (*Branta canadensis*), mallards (*Anas platyrhynchos*), and wood ducks (*Aix sponsa*). The plants also are host to weevils (Insecta: Coleoptera: Curculionidae: *Listronotus rubtzoffi*). They have been found to harbor more than 39 million nitrifying bacteria. The flower heads, leaves, and roots of *S. engelmanniana* are fed upon by weevils (Insecta: Coleoptera: Curculionidae: *Listronotus appendiculatus, L. neocallosus, L. sordidus*). Various weevils (Insecta: Coleoptera: Curculionidae) also are hosted by *S. graminea* (*Listronotus neocallosus*), *S. filiformis* (*L. neocallosus*) and *S. lancifolia* (*Listronotus cryptops*). The foliage of *S. graminea* is eaten by grass carp (Vertebrata: Osteichthyes: Cyprinidae: *Ctenopharyngodon idella*) and nutria (Mammalia: Myocastoridae: *Myocastor coypus*); its tubers (15.9% crude protein; 16.1% crude fiber) are consumed by canvasbacks (Aves: Anatidae: *Aythya valisineria*) and wood ducks (Aves: Anatidae: *Aix sponsa*) during winter months. The plants are categorized overall as a good food for ducks (Aves: Anatidae). In Asia, *S. guayanensis* is fed to pigs (Mammalia: Suidae: *Sus scrofa*) as fodder. The leaves of *S. kurziana* support large populations of larval flies (Insecta: Diptera: Chironomidae) and provide habitat for the Ichetucknee siltsnail (Mollusca: Gastropoda: Hydrobiidae: *Floridobia mica*). *Sagittaria lancifolia* is a host of smut (Fungi: Basidiomycota: Doassansiaceae: *Doassansia sagittariae*; *Doassansiopsis deformans*). The foliage is used for oviposition by Florida applesnails (Mollusca: Gastropoda: Ampullariidae: *Pomacea paludosa*), and is eaten by grasshoppers (Insecta: Orthoptera: Romaleidae: *Romalea microptera*) and white-tailed deer (Mammalia: Cervidae: *Odocoileus virginianus seminolus*); the flower buds (primarily the males) are eaten by weevils (Insecta: Coleoptera: Curculionidae: *Listronotus appendiculatus*). *Sagittaria latifolia* is eaten by moose (Mammalia: Cervidae: *Alces americanus*) and white-tailed deer (Mammalia: Cervidae: *Odocoileus virginianus*); its tubers are consumed by pintail ducks (Aves: Anatidae: *Aythya valisineria*) and other waterfowl. The seeds and foliage are eaten by American wigeons (Aves: Anatidae: *Anas americana*) and occasionally by bullfrogs (Amphibia: Ranidae: *Rana catesbeiana*). The plants host numerous insects (Insecta), including aphids (Homoptera: Aphididae: *Rhopalosiphum nymphaeae*), beetles and weevils (Coleoptera: Chrysomelidae: *Donacia caerulea, D. subtilis*; Coccinellidae: *Coleomegilla maculata*; Curculionidae: *Hyperodes solutus, Listronotus appendiculatus, L. delumbis, L. echinodori, L. squamiger, Lixellus lutulentus*), caddis flies (Trichoptera: Lemnephilidae: *Pycnopsyche*; Leptoceridae: *Leptocerus americanus, Nectopsyche, Triaenodes abus, T. ignitus, T. injustus, T. marginatus*; Polycentropodidae: *Neureclipsis crepuscularis, Polycentropus*), flies (Diptera: Itonididae), grasshoppers (Orthoptera: Acrididae: *Paroxya atlantica, P. clavuliger, Romalea microptera*), leafhoppers (Homoptera: Cicadellidae: *Draeculacephala*), moths (Lepidoptera: Crambidae: *Munroessa icciusalis, Paraponyx obscuralis*; Noctuidae: *Cryptocala acadiensis*), and stink bugs (Hemiptera: Pentatomidae: *Amaurochrous cinctipes*). The flowers are visited by a variety of potentially pollinating insects (Insecta), including bees (Hymenoptera: Apidae: *Apis mellifera, Bombus, Ceratina*; Halictidae: *Augochlora, Halictus, Lasioglossum*), beetles (Coleoptera: Cantharidae: *Chauliognathus pennsylvanicus*; Chrysomelidae: *Diabrotica undecimpunctata*; Coccinellidae: *Coleomegilla maculata*), butterflies (Lepidoptera: Hesperiidae: *Ancyloxypha numitor*), flies (Diptera: Syrphidae: *Allograpta, Criorhina, Eristalis, Heliophilus*; Tachinidae: *Linnaemya*), and wasps (Hymenoptera: Vespidae: *Euodynerus*). The foliage is colonized by freshwater sponges (Paludicellidae: *Paludicella articulata*; Plumatellidae: *Plumatella repens*). *Sagittaria longiloba* is eaten by sandhill cranes (Aves: Gruidae: *Grus canadensis*) and waterfowl (Aves: Anatidae). The seeds are eaten by ducks (Aves: Anatidae: *Anas platyrhynchos*) and killdeer (Charadriidae: *Charadrius vociferus*). The plants are fed on by weevils (Insecta: Coleoptera: Curculionidae: *Listronotus manifestus, L. scapularis*). *Sagittaria platyphylla* is eaten by feral swine (Mammalia: Suidae: *Sus scrofa*), nutria (Mammalia: Myocastoridae: *Myocastor coypus*), and white-tailed deer (Mammalia: Cervidae: *Odocoileus virginianus*). The leaf axils provide refuge for larval dragonflies (Insecta: Odonata: Libellulidae: *Pachydiplax longipennis*), and emergent portions of the plant are used as mating sites by adult dragonflies (Insecta: Odonata: Coenagrionidae *Telebasis salva*). The foliage provides an oviposition site for fountain darters (Osteichthyes: Percidae: *Etheostoma fonticola*), cover and habitat for fish (Osteichthyes: Cyprinidae: *Notropis chalybaeus*), and serves as an attachment substrate for ciliophorans (Protozoa: Peritrichida: Operculariidae: *Opercularia wallgreni*). The large (e.g., 2.5 cm) tubers (19.9% crude protein; 56.1% total digestible nutrients) provide an important food for turtles (Reptilia: Emydidae: *Pseudemys texana*) and various waterfowl (Aves: Antidae), including canvasbacks (*Aythya valisineria*), gadwalls (*Anas strepera*), mallards (*Anas platyrhynchos*), pintails (*Anas acuta*), and ring-necked ducks (*Aythya collaris*). The foliage of *S. rigida* (14.78% crude protein; 23.69% crude fiber; Fe: 2083 ppm) is eaten by muskrats (Mammalia: Cricetidae: *Ondatra zibethicus*). The plants are inhabited by beetles (Insecta: Coleoptera: Chrysomelidae: *Donacia caerulea*). The tubers (11.6% crude protein; 2.0% crude fiber; 0.6% crude fat) are consumed by canvasback ducks (Aves: Anatidae: *Aythya valisineria*) and tundra swans (Aves: Anatidae: *Cygnus columbianus*). The seeds and tubers of *S. subulata* are consumed occasionally by waterfowl (Aves: Anatidae). *Sagittaria teres* is eaten by pintail ducks (Aves: Anatidae: *Anas acuta*).

Economic importance: food: The tubers of *Sagittaria* species are rich in starch and were eaten by Native Americans and European settlers, who boiled them to temper the bitter taste imparted by their milky sap. Tubers of *S. cuneata* were eaten by the Klamath and Menominee, Montana, Ojibwa, and Paiute tribes. *Sagittaria guayanensis* is eaten as a rainy season vegetable in Laos. *Sagittaria latifolia* contains 13.7% crude protein. Its tubers were eaten by the Chippewa, Cocopa, Dakota, Klamath, Lakota, Meskwaki, Omaha, Pawnee, Pomo, Potawatomi, Thompson, and Winnebago tribes, who often stored them as a winter food; **medicinal:** The Cheyenne and Chippewa administered *S. cuneata* medicinally for various ailments. The plants were regarded as an analgesic by the Navajo, and the corms were eaten by the Ojibway to treat indigestion. Decoctions prepared from *Sagittaria guayanensis* are used in parts of Asia as a body wash ('Banarbhega') to control fever; the plant juices (mixed with honey) are drunk to facilitate recovery following childbirth. *Sagittaria lancifolia* was used by the Florida Seminoles to treat alligator bites. Extracts from the plant have been found to inhibit several Bacteria (Pseudomonadaceae: *Pseudomonas aeruginosa*; Staphylococcaceae: *Staphylococcus aureus*). Infusions of *S. latifolia* were used by the Cherokee to treat fevers (leaves) and by the Chippewa as a remedy for gastrointestinal disorders (roots). The Iroquois used infusions from the plant for treating rheumatism and fabricated various compound decoctions as remedies for boils, constipation, and itching. The species also was used medicinally by the Lakota and the Potawatomi, the latter applying a poultice from the corms to heal wounds; **cultivation:** Several *Sagittaria* species are grown as ornamental water garden plants. 'Benni' is a patented cultivar of *S. australis* (US PP12,198 P2); 'Crushed Ice' is a cultivar of *S. graminea*. *Sagittaria lancifolia* and *S. montevidensis* are used as water garden ornamentals. *Sagittaria*

platyphylla is distributed occasionally as an aquarium plant, whereas *S. subulata* is a popular decorative aquarium species; **misc. products:** *Sagittaria lancifolia* and *S. latifolia* have been used in wetland restoration projects. The Iroquois used a root decoction of *S. latifolia* as a medicine to improve corn yields; **weeds:** *Sagittaria guayanensis* is a rice-field weed in Louisiana and also in Brazil and Peru. It has developed resistance to ALS inhibitor herbicides (bensulfuron-methyl) in Malaysia; bensulfuron-methyl resistant strains of *S. montevidensis* occur in California. *Sagittaria latifolia* has been reported as a rice-field weed in California; *S. longiloba* is a common rice-field weed in the western United States. *Sagittaria montevidensis* is a nuisance rice-field weed in South America; herbicide resistant strains have evolved (various ALS inhibitors) in Australia, Brazil, China, and the United States (California). *Sagittaria sanfordii* is regarded as a rice-field weed in California; **nonindigenous species:** *Sagittaria graminea* has been introduced to Washington state. It is invasive in China and has been introduced to Australia and France, where it also is problematic. *Sagittaria guayanensis* was introduced to Louisiana from tropical America. *Sagittaria latifolia* has been introduced widely throughout Europe. It is possible that *S. longiloba* was introduced to California, where nearly all of the records are associated with rice fields. *Sagittaria montevidensis* has been introduced to Australia, China, and the southeastern United States, where it has been distributed commercially as an ornamental; it also has been introduced to California, reportedly as an inadvertent consequence of disposed shipping ballast. *Sagittaria platyphylla* is nonindigenous to Australia and European Russia. *Sagittaria rigida* has been introduced to California, Idaho, and Washington in the United States, and to Ireland and the United Kingdom in Europe. *Sagittaria subulata* has been introduced to the British Isles.

Systematics: *Sagittaria* remains in need of taxonomic revision, with even basic species boundaries yet to be determined reliably. There is still some debate whether to subdivide the genus by excluding *Lophotocarpus*, which contains those annual species having perfect flowers with hypogynous stamens. Because the *Lophotocarpus* group resolves phylogenetically as the sister clade to the remainder of *Sagittaria* (Figure 1.7), either taxonomic disposition could be reconciled. In either case, *Sagittaria* appears to be monophyletic, with *Astonia*, *Limnophyton*, and *Wiesneria* among its most closely related genera (e.g., Figures 1.3 and 1.7). Several species (which retain OBL designations) are now regarded as various synonyms. *Sagittaria chapmanii* and *S. weatherbiana* are treated as subspecies of *S. graminea*. Although each taxon appears to be distinct phylogenetically (Figure 1.7), the nuclear data used to construct that tree do not exclude the possibility that *S. chapmanii* and *S. weatherbiana* could be of hybrid origin, with both having *S. graminea* as one parent. This hypothesis is consistent with results from crossing studies, which have shown *S. graminea* to be highly interfertile with *S. isoetiformis* (with which *S. chapmanii* resolves), and at least partially interfertile with *S. fasciculata* (with which *S. weatherbiana* resolves; Figure 1.7). Although *S. chapmanii*

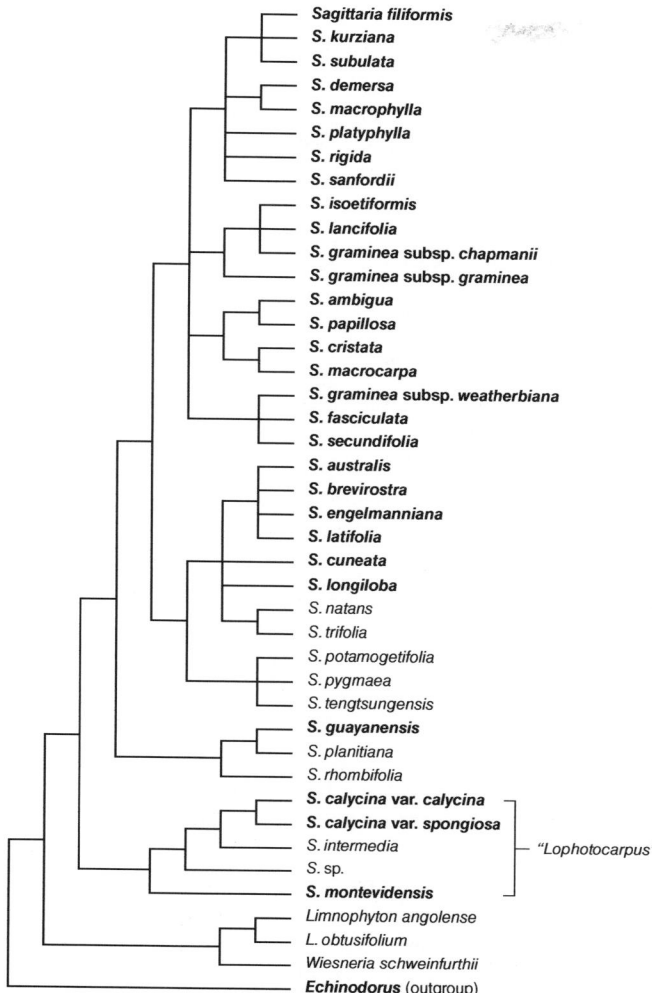

FIGURE 1.7 Interspecific relationships in *Sagittaria* as indicated by analysis of 5S-NTS data (modified from Keener, 2005). North American taxa designated as OBL indicators are highlighted in bold. The species recognized by some authors within the segregate genus *Lophotocarpus* resolve within a clade as indicated. The disjunct placement of *S. graminea* subspecies possibly reflects their hybrid origin (see text).

(= *S. graminea* subsp. *chapmanii*) is interfertile with *S. graminea*, the allozymically determined genetic identity of these two taxa is remarkably low (0.473). Also, in contrast with the typical subspecies, *S. graminea* subsp. *chapmanii* is quite variable genetically. Both factors could be a consequence of hybridization involving more distant species. *Sagittaria graminea* is well isolated reproductively from *S. cristata* and *S. teres*. *Sagittaria spathulata* is treated as a synonym of *S. calycina*. Although regarded by some authors as a synonym of *S. montevidensis*, *S. calycina* is retained here as a distinct species in accordance with its unique morphological features and genetic distinctness from the former (e.g., Figure 1.7). It seems clear that *S. montevidensis* and *S. calycina* should not be treated as synonyms. However, *S. calycina* includes a variety (*S. calycina* var. *spongiosa*), which some authors have distinguished as *S. spathulata*. In any case, both are closely related

sister taxa (Figure 1.7). *Sagittaria macrocarpa* has been recognized as a synonym of *S. fasciculata*; however, morphological and genetic evidence (e.g., Figure 1.7) indicate its status as a distinct species, which has been accepted here (but does not preclude the possibility that *S. fasciculata* might be of hybrid origin). DNA sequence data resolve *S. macrocarpa* as the sister species of *S. cristata* (Figure 1.7). Though *S. cuneata* and *S. latifolia* bear a superficial morphological similarity, they are distinct genetically (e.g., by both isozyme and DNA markers). Furthermore, crossing studies have demonstrated that strong infertility barriers exist among *S. australis*, *S. brevirostra*, *S. cuneata*, *S. engelmanniana*, and *S. latifolia*, which all are well isolated reproductively. Allozyme data indicate that *S. isoetiformis* and *S. teres* are a closely related progenitor-derivative species pair. Specimens of *S. filiformis*, *S. kurziana*, and *S. subulata* often show a history of confused identifications, which has been attributed to extensive phenotypic plasticity relating to water depth. Observations of extensive variation in leaf morphology have raised doubts regarding the distinctness of these three taxa, which are separated by relatively minor fruit features and share a close phylogenetic interrelationship (Figure 1.7). Furthermore, because artificial hybridizations between *S. filiformis* and *S. kurziana* (Figure 1.7) have produced fertile F_1 offspring, the recognition of these three taxa at the species level should be re-evaluated by additional genetic studies.

Distribution: *Sagittaria cuneata* is widespread throughout northern North America; *S. graminea* occurs in eastern North America and is disjunct in Washington State; *S. latifolia* is widespread in eastern North America and sporadic in the west. *Sagittaria australis*, *S. brevirostra*, *S. calycina*, and *S. platyphylla* occur primarily in the eastern United States; *S. rigida* occurs in the eastern United States and southeastern Canada, with nonindigenous stations in Idaho and Washington; *S. cristata* occurs throughout the upper Great Lakes region. *Sagittaria engelmanniana*, *S. filiformis*, *S. isoetiformis*, and *S. lancifolia* occur along the southeastern Coastal Plain of the United States. *Sagittaria subulata* occurs along the eastern and southeastern coastal areas of the United States; *S. teres* is restricted to the coastal plain of the northeastern United States. *Sagittaria longiloba* occurs in the southwestern United States, and extends into Mexico and Central America. Species with more restricted distributions include *S. ambigua* (Illinois, Indiana, Kansas, and Missouri), *S. demersa* (New Mexico and elsewhere restricted to a small region of central Mexico), *S. fasciculata* (North and South Carolina), *S. kurziana* (Florida), *S. papillosa* (Arkansas, Louisiana, Oklahoma, and Texas), *S. macrocarpa* (endemic to the Sandhills Region of the Carolinas), *S. sanfordii* (endemic to California), and *S. secundifolia* (Alabama and Georgia). *Sagittaria guayanensis* has been introduced to Louisiana from the New World tropics; *S. montevidensis* was introduced to the southeastern United States and to California.

References: Abbitt & Scott, 2001; Adair et al., 2012; Adams & Godfrey, 1961; Ainsworth, 2006; Alves Pagotto et al., 2011; Atwood, 1950; Badzinski et al., 2006; Baldwin & Mendelssohn, 1998; Baskin & Baskin, 1998; 2015; Baxter, 2007; Beal, 1977;

Beal et al., 1982; Bowes et al., 2002; Bowker, 1991; Bowyer et al., 2005; Boyd & McGinty, 1981; Broadhurst & Chong, 2011; Brewis, 1975; Brown, 1942; Bushnell, Jr., 1966; Calhoun & King, 1998; Chafin, 2007; Colle et al., 1978; Collon & Velasquez, 1989; Cooperrider, 1955; Coulter, 1955; Craven & Hunt, 1984; Crocker, 1907; Crow, 1969; Cruden, 1988; Das et al., 2006; Davis, 1993; Delesalle & Blum, 1994; De Oliveira et al., 2015; DeVlaming & Proctor, 1968; Diamond, 2016; Dirrigl, Jr. & Mohlenbrock, 2012; DiTomaso & Healy, 2003; Dorken & Barrett, 2003; 2004a; 2004b; 2004c; Dorken et al., 2002; Edwards & Sharitz, 2000; 2003; Emerson, 1921; Enser & Caljouw, 1989; Everitt et al., 1999; Fassett, 1957; Ferren, Jr., 1973; Ferren, Jr. & Schuyler, 1980; Fields et al., 2003; Fraser et al., 1980; Frost & Hicks, 2012; Glaettli & Barrett, 2008; Gleason et al., 2003; Gluck, 1927; Godfrey & Adams, 1964; Gonzalez et al., 1983; Grabowski, 2001; Grace & Ford, 1996; Graves, 1984; Grier, 1916; Hamel & Parsons, 2001; Harms & Grodowitz, 2009; Harrison & Knapp, 2010; Harvey & Haines, 2003; Hauber & Lege, 1999; Haukos & Smith, 1997; 2001; Haynes & Hellquist, 2000a; Haynes & Holm-Nielsen, 1994; Haynes & Les, 2004; Heilman & Carlton, 2001; Hellquist & Crow, 1981; Hendricks & Goodwin, Jr., 1952; Hestand & Carter, 1974; Hoagland & Buthod, 2007; Hoagland et al., 2001; Hohman et al., 1990; Hooper, 1951; Howard & Mendelssohn, 1995; Hoyer et al., 1996; Huang, 2003; Huang et al., 2000; Hussner, 2012; Jacono, 2001; 2002; Johnson & Rohwer, 2000; Kandalepas et al., 2010; Kasselmann, 2003; Kaul, 1976; 1979; 1991; Kenow & Lyon, 2009; Kissoon et al., 2013; 2015; Knapton & Pauls, 1994; Koch, 1970; Korschgen & Moyle, 1955; Kral, 1982; Labisky et al., 2003; Lagueux et al., 1995; Laidig & Zampella, 1996; Landers et al., 1977; Lavoie et al., 2003; Leck & Graveline, 1979; Lehman et al., 2009; Lehtonen, 2009; Les & Tippery, 2013; Linn et al., 1975; Little, Jr., 1938; Lot et al., 2002; Low & Bellrose, Jr., 1944; Luken & Thieret, 2001; MacGillivray, 1903; MacRoberts & MacRoberts, 2010; Manandhar, 1985; Mancera et al., 2005; Manolis, 2016; Marburger, 1993; Martin & Shaffer, 2005; Mason, 1957; Mattson et al., 1995; Matulewich & Finstein, 1978; McAtee, 1939; McGaha, 1952; McKee & Mendelssohn, 1989; McPherson & Paskewitz, 1984; Middleton, 1989; Mitchell, 1926; Mohr, 1901; Morrone, 2013; Moyle, 1945; Muenchow, 1998; Muenchow & Delesalle, 1992; 1994; Mulhouse, 2004; Nakamura & Nelson, 2001; Newberry, 1991; Norquist, 1990; O'Brien, 1997; Padgett et al., 2004; Pates & Madsen, 1955; Perkin et al., 2012; Perry & Dorken, 2011; Pezeshki et al., 2000; Phillips, 1911; Phillips et al., 2011; Pollux, 2011; Power, 1996; Profous & Loeb, 1984; Quinlan et al., 2008; Reid et al., 2010; Rhodes, 1978; Richards & Ivey, 2004; Robinson & Frye, 1986; Rosen, 2007; Rubtsov, 1975; Sarkissian et al., 2001; Sasser et al., 1995a; Sawyer et al., 2005; Schlickeisen et al., 2003; Schuyler & Gordon, 2002; Shaffer et al., 1992; Schaffner, 1924; Self et al., 1975; Sharitz et al., 2010; Skinner & Telfer, 1974; Sletten & Larson, 1984; Smith, 1895; 1900; Smith & Capinera, 2005; Sneddon & Lamont, 2010; Snyder, 1988; Sorrie et al., 2006; 2007; Stevens, 1957; Strong & Kelloff, 1994; Sugden & Driver, 1980; Sullivan, 1981; Takos, 1947; Terrell et al., 1978; Thieret, 1969; 1970; Thompson &

McKinney, 2006; Tom & Karr, 2013; Turner, 1996; Turner et al., 2012; USFWS, 1983; 2014; Vamosi et al., 2006; Van der Valk & Rosburg, 1997; Van der Valk et al., 2009; Van Drunen & Dorken, 2012; Vidal, 1960; Wallis, 2007; Wang et al., 2006; Wang et al., 2012a; Wellborn & Robinson, 1987; Weller, 1988; Wells & Alexander, 2014; West, 1945; Whetstone et al., 1987; Whyte & Cain, 1981; Wilson, 1937; Wooten, 1970; 1971; 1973a; 1973b; 1978; 1986; Wooten & Brown, 1983; Yanovsky, 1936; Yatskievych & Jenkins, 1981; Zaremba & Lamont, 1993; Zebryk, 2004; Zhang et al., 2010a; 2013a; Zhang et al., 2014a.

Family 2.2. Butomaceae [1]

Although once recognized as a somewhat broader taxon that included members of the former Limnocharitaceae (now merged with Alismataceae), Butomaceae currently are circumscribed as a monotypic family comprising only *Butomus* and the single species *B. umbellatus* (Cronquist, 1981). Phylogenetic analysis of sequence data from the *rbcL* gene provided supportive evidence that Butomaceae were not only distinct from Limnocharitaceae but resolved as the sister group to Hydrocharitaceae rather than Alismataceae, where the limnocharit genera were included (Les et al., 1997a). The same sister-group association of Butomaceae and Hydrocharitaceae has been recovered in many subsequent analyses (see Les & Tippery, 2013), including those incorporating sequence data from complete cpDNA genomes (Ross et al., 2016; Figure 1.2).

This is a fairly distinctive group of aquatics bearing linear, distichous, sheathing leaves that arise from an elongate rhizome, and perfect, hypogynous, trimerous flowers with an apocarpous gynoecium of nectariferous, basally connate, distally unsealed carpels, which mature into follicles; placentation is laminar (Cronquist, 1981). The flowers are arranged in large, showy, determinate umbels, which terminate an elongate, naked scape. Other aspects of the life history, economic uses, etc. are discussed below under the generic treatment.

Butomaceae are indigenous to temperate Eurasia and North Africa, but introduced to North America. One North American genus contains OBL indicators:

2.2.1. *Butomus* L.

2.2.1. Butomus

Flowering rush; butome à ombelle

Etymology: after the Greek *bous tomos* ("ox cutting") in reference to the unsuitability of the plant for fodder

Synonyms: none

Distribution: global: Africa (northern); Eurasia; North America; **North America:** north central

Diversity: global: 1 species; **North America:** 1 species

Indicators (USA): obligate wetland (OBL): *Butomus umbellatus*

Habitat: freshwater; lacustrine, palustrine, riverine; **pH:** 7.3–7.6; **depth:** 0–2.5 m; **life-form(s):** emergent, floating-leaved, or submersed (rosulate) herb

Key morphology: plants stemless, rhizomatous (to 40 cm), emersed, floating-leaved, or submersed; leaves (to 2.7 m)

basal, sessile, sheathed, linear, triangular in section, flattened distally; inflorescence an erect, scapose, umbel (to 25-flowered), the scape (to 150 cm) naked, triangular in section; flowers (to 2.5 cm) bisexual, pedicellate (to 10 cm), hypogynous, trimerous; tepals (to 11.5 mm) six, pinkish purple, the veins darker; stamens (to 5.5 mm) nine, in two cycles; gynoecium apocarpous (but coherent basally), pistils six, placentation laminar, ovules ~50; follicles (to 1 cm) leathery, beaked

Life history: duration: perennial (bulbils, rhizomes); **asexual reproduction:** bulbils, rhizomes; **pollination:** self; insect; **sexual condition:** hermaphroditic; **fruit:** follicles (infrequent); **local dispersal:** bulbils (water), rhizome fragments (water), seeds (gravity, wind); **long-distance dispersal:** bulbils (water), seeds (birds, water)

Imperilment: 1. *Butomus umbellatus* [G5]

Ecology: general: Monotypic (see next).

2.2.1.1 *Butomus umbellatus* L. is a nonindigenous perennial, which inhabits exposed substrates to relatively deeper waters (typically 0.9–1.3 m but up to 4.5 m) in backwaters (diked), bays, bogs (floating), channels (dried), ditches, flats (river), floodplains, gravel pits, lakes, marshes, meadows, mudflats, ponds (retention), rivers, sand/mud bars, shoals, shores (receding), sloughs, streams, swales, and along the margins of canals, lakes, ponds, reservoirs, rivers, and streams at elevations to 1372 m. Site exposures are open and receive full sunlight. The sites typically are alkaline (e.g., pH: 7.3–7.6; total alkalinity: 55–153 mg l^{-1}). The plants grow as emergent, floating-leaved, or completely submersed life-forms and occur often in swift currents. The biomass peaks at 1.3 m depth and decreases in deeper water. The plants have a strong affinity for coarse substrates but occur on clay, clayey sand, gravel, muck, mucky cobble, mud, rock, sand, and silt. Flowering occurs from June to October with fruiting extending from July to October. The flowers of diploids are self-compatible (the triploids being sterile). However, they are protandrous (with a short, initial male phase) and require pollinators to achieve seed set. At the umbel level, the plants are synchronously dichogamous, with nearly three quarters of the female flowers being receptive while the male phase is absent. Pollen is shed sequentially from 3 to 11 hr after anthesis and is completely removed from the flowers by 12 hr; the stigmas sequentially become receptive at about 15 hr from the onset of anthesis, continuing until about 21 hr after the initial floral opening. Pollen retains high viability for 7 hr, but becomes inviable after 24 hr; wet pollen rapidly loses viability. During the male phase, the dehiscence of anthers is "reversible"; i.e., being able to reclose in wet conditions and reopen subsequently (within 13–20 min) under dryer conditions. The insect (Insecta) pollinators include bees (Hymenoptera: Apidae) and flies (Diptera: Syrphidae). The highest floral visitation rates occur on sunny days when the flowers are dry. Polymorphic gynoecium color (pink or white) has been observed in China, where pink morphs (the common type) produce more pollen and ovules (and exhibit correspondingly higher seed production) than the white morphs. A single plant can produce 35,000 small (0.16 mg) seeds on average, with larger plants yielding an estimated 500 fruits and 258,000 seeds. Typically, the seeds fall directly on the

ground beneath the maternal plant or are dispersed locally by wind; however, they are buoyant and can remain afloat for several days, which facilitates longer distance dispersal by water. The seeds also are believed to be dispersed by becoming entangled within the feathers of birds (Aves). The seeds reportedly germinate adequately (20–50%) after being cold stratified for several months in water and then subjected to an 8/16-hr light/dark regime at 25°C/10°C. They require light for germination, which also has been achieved successfully for stratified seeds incubated at 20°C–30°C under long-day conditions. The seeds are known to retain up to 68% viability after 5 years of storage in cold water. The seedlings grow well in cultivation under a 14/10 hr, 30°C/20°C, light/temperature regime (70% relative humidity). The seedlings grow slowly, are capable of floatation, and require at least 2 months of unshaded conditions in order to establish. Vegetative bulbils can replace flowers in the inflorescence and also develop from the rhizomes; either can function as asexual propagules. There are somewhat conflicting reports regarding the production of vegetative structures relative to different cytotypes. In native European populations, triploid plants reproduce primarily by vegetative reproduction involving prolific, detachable, rhizome buds and the development of inflorescence bulbils, which both are capable of water dispersal; the diploids produce fewer rhizome buds and lack floral bulbils. Individual plants consistently can produce upward of 200 vegetative bulbils annually. However, in the introduced North American plants, diploid plants were found to invest highly in sexual structures as well as in clonal bulbils, whereas the triploids were sterile and did not produce vegetative bulbils. Plants in introduced regions also produce a greater number of sexual inflorescences and asexual bulbils and invest more biomass in each reproductive mode than those in native populations, which may account for their successful invasion ability. The bulbils appear in mid-August when water temperatures approach 22.5°C and can be produced by plants less than 25 cm in length. They reach their highest density (averaging 200–300 buds m^{-2}) at depths from 0.3 to 1.6 m. The plants spread rapidly during periods of low water, which results in more substrate for colonization by sexual and vegetative propagules. The average ramet density can approach 100 m^{-2}. Genetic studies have confirmed that reduced variability and fewer genets characterize the introduced North American populations, which presumably underwent a severe bottleneck at their time of introduction. North American plants emerge from May to June and reach peak DM biomass by late July, which can exceed 108 g m^{-2}.

Reported associates (North America): *Alisma subcordatum, Asclepias incarnata, Bidens cernuus, Bidens trichospermus, Bolboschoenus fluviatilis, Carex cristatella, Ceratophyllum demersum, Chara, Drepanocladus, Echinochloa crusgalli, Eleocharis erythropoda, Eleocharis obtusa, Elodea canadensis, Glyceria grandis, Glyceria striata, Heteranthera dubia, Hippuris vulgaris, Hydrocharis morsus-ranae, Iris pseudacorus, Iris virginica, Leersia oryzoides, Lemna minor, Lindernia dubia, Ludwigia palustris, Ludwigia polycarpa, Lycopus virginicus, Lythrum salicaria, Mentha arvensis, Myriophyllum hippuroides, Myriophyllum sibiricum,*

Myriophyllum spicatum, Najas flexilis, Nitella, Nuphar advena, Nuphar variegata, Nymphaea odorata, Nymphoides peltata, Peltandra virginica, Persicaria amphibia, Persicaria lapathifolia, Phalaris arundinacea, Phragmites australis, Poa palustris, Pontederia cordata, Populus angustifolia, Populus deltoides, Potamogeton amplifolius, Potamogeton crispus, Potamogeton foliosus, Potamogeton gramineus, Potamogeton illinoensis, Potamogeton nodosus, Potamogeton praelongus, Potamogeton pusillus, Potamogeton richardsonii, Potamogeton robbinsii, Potamogeton zosteriformis, Potentilla anserina, Ranunculus aquatilis, Ranunculus flabellaris, Ricciocarpus, Rorippa aquatica, Rosa woodsii, Rumex altissimus, Rumex crispus, Rumex verticillatus, Sagittaria cuneata, Sagittaria latifolia, Salix exigua, Schoenoplectus acutus, Schoenoplectus pungens, Schoenoplectus tabernaemontani, Scirpus atrovirens, Sparganium androcladum, Sparganium eurycarpum, Stuckenia pectinata, Trapa natans, Typha angustifolia, Typha latifolia, Utricularia macrorhiza, Vallisneria americana, Xanthium strumarium, Zannichellia palustris, Zizania aquatica.

Use by wildlife: The leaves of *Butomus umbellatus* are eaten by muskrats (Mammalia: Cricetidae: *Ondatra zibethicus*), and the bulbils are consumed by blue-winged teal (Aves: Anatidae: *Anas discors*). North American plants also are host to chytrids (Fungi: Blastocladiomycota: Physodermataceae: *Physoderma butomi*), Fungi (Ascomycota: Nectriaceae: *Fusarium oxysporum*; Pestalotiopsidaceae: *Pestalotiopsis guepinii*; Xylariaceae: *Dicyma ovalispora, Virgaria nigra*; Basidiomycota: Agaricaceae: *Phoma*), larval caddisflies (Insecta: Trichoptera: Leptoceridae: *Ylodes*; *Ceraclea, Oecetis, Triaenodes*), larval flies (Insecta: Diptera: Chironomidae: *Chironomus, Rheotanytarsus*), and snails (Mollusca: Gastropoda: Lymnaeidae: *Lymnaea stagnalis*; Physidae: *Physa*).

Economic importance: food: The rhizomes and bulbils of *Butomus umbellatus* are high in starch and have been eaten (when dried or roasted) or ground into flour for bread making; **medicinal:** Acetone extracts of *Butomus umbellatus* are inhibitory to some microbes (Bacteria: Bacillaceae: *Bacillus subtilis*); **cultivation:** *Butomus umbellatus* is grown as an ornamental water garden plant and has been distributed under the cultivar names 'Rosenrot' and 'Schneeweisschen'; **misc. products:** The Iroquois made a decoction using whole plants of *Butomus*, which they added to livestock feed as a treatment for parasitic worms; **weeds:** *Butomus umbellatus* is a weed of North American wetlands, where it has been described as being more invasive than purple loosestrife (Lythraceae: *Lythrum salicaria*); **nonindigenous species:** *Butomus umbellatus* is nonindigenous to the New World, introduced to North America sometime prior to 1897. Morphological and genetic data indicate that there have been several separate introductions of the species from different regions of Eurasia. The first introductions presumably resulted from the seed-contaminated ballast and packing materials disposed by ships along the St. Lawrence River. Introductions have continued, primarily via the escape of horticultural specimens.

Systematics: Molecular phylogenetic analyses consistently resolve the monotypic *Butomus* as the sister group of Hydrocharitaceae (e.g., Figures 1.2 and 1.3). Being monotypic, the genus also is monophyletic. Although two species have been proposed by some researchers, most evidence (including transplant studies) indicates that there is a single but cytotypically variable species. The base chromosome number of *Butomus* is $x = 13$. *Butomus umbellatus* has diploid ($2n = 26$) and triploid ($2n = 39$) cytotypes, which both occur in New World populations. Triploid populations are believed to have multiple origins. All coding regions for the chloroplast and mitochondrial genomes of *B. umbellatus* have been sequenced, providing ample data for phylogenetic studies. *Butomus* is not known to hybridize.

Comments: *Butomus umbellatus* occurs throughout central North America, north of the glacial maxima.

References: Anderson et al., 1974; Barrett et al., 2008; Bhardwaj & Eckert, 2001; Brown & Eckert, 2005; Core, 1941; Countryman, 1970; Cuenca et al., 2013; Delisle et al., 2003; Eckert et al., 2000; Fernald & Kinsey, 1943; Fernando & Cass, 1996; Gunderson et al., 2016; Harms & Shearer, 2015; Haynes, 2000a; Hroudová & Zákravský, 1993; 2003; Hroudová et al., 2004; Huang & Tang, 2008; Les & Mehrhoff, 1999; Les & Tippery, 2013; Li et al., 2012; Lohammar, 1954; Lui et al., 2005; Madsen et al., 2016; Maki & Galatowitsch, 2004; Muenscher, 1936; Özbay & Alim, 2009; Ross et al., 2016; Salisbury, 1976; Sparrow, 1974; Stuckey, 1968; Turnage et al., 2012.

Family 2.3. Hydrocharitaceae [17]

With approximately 127 entirely aquatic species, Hydrocharitaceae are among the most diverse hydrophyte families. Every member of the group is designated as an OBL indicator throughout North America. Although not large in terms of species, the family contains an impressive assortment of growth forms (emergent, floating-leaved, free-floating, submersed), habits (annual, perennial, rosulate, vitatte), and pollination systems (anemophily, entomophily, epihydrophily, hypohydrophily, self-pollination). It is the only known angiosperm family to contain an unusual water-mediated pollination system, which involves detached, free-floating, male flowers. Moreover, it is one of only three angiosperm families to contain "seagrasses" (i.e., angiosperms adapted marine habitats) as well as freshwater species. Leaf and floral morphology are equally diverse.

Various phylogenetic studies (e.g., Les & Haynes, 1995; Les & Tippery, 2013; Les et al., 1993; 1997c; 2006a; Petersen et al., 2006; Ross et al., 2016; Shaffer-Fehre, 1991) have clarified the circumscription of the family to include the former Najadaceae, whose highly reduced morphology has made it difficult to place among other alismatid monocots. All of these studies, which encompass morphological as well as DNA sequence data (from mitochondrial, nuclear, and plastid genomes), have consistently supported the merger of Najadaceae and Hydrocharitaceae as a single clade. Similar lines of evidence (e.g., Figures 1.2 and 1.3) support Butomaceae as the sister group.

Economically, the family is important as a source of ornamental water garden and aquarium plants, which are represented by a large number of genera. For decades, introductory botany courses have used *Egeria densa* as a "classic example" for illustrating the phenomenon of cytoplasmic streaming. Many species provide food and cover for a variety of wildlife. The three marine genera (*Enhalus*, *Halophila*, and *Thalassia*) are of particular importance to the ecology of coastal marine habitats. However, a number of genera (e.g., *Egeria*, *Hydrilla*, *Hydrocharis*, *Lagarosiphon*, *Najas*) also include noxious, weedy species, which have seriously disrupted the ecology of many freshwater systems. In North America, all such cases represent nonindigenous introductions. Some of the resulting infestations have had significant economic consequences and are responsible for millions of dollars in annual expenditures directed at their control.

Eleven genera occur in North America and all contain OBL indicators:

2.3.1. **Blyxa** Noronha ex Thouars
2.3.2. **Egeria** Planch.
2.3.3. **Elodea** Michx.
2.3.4. **Halophila** Thouars
2.3.5. **Hydrilla** Rich.
2.3.6. **Hydrocharis** L.
2.3.7. **Limnobium** Rich.
2.3.8. **Najas** L.
2.3.9. **Ottelia** Pers.
2.3.10. **Thalassia** Banks & Sol. ex K.D. Koenig
2.3.11. **Vallisneria** L.

2.3.1. Blyxa

Bamboo plant, blyxa
Etymology: thought to be derived from the Greek *blyzo* ("to flow"), in reference to the habitat
Synonyms: *Diplosiphon*; *Enhydrias*; *Hydrilla* (in part); *Hydrolirion*; *Hydrotrophus*; *Saivala*; *Vallisneria* (in part)
Distribution: global: Africa; Asia; Australia; North America*; **North America:** southern United States
Diversity: global: 12 species; **North America:** 1 species
Indicators (USA): obligate wetland (OBL): *Blyxa aubertii*
Habitat: brackish, freshwater; lacustrine; **pH:** 6.0–7.4; **depth:** <1 m; **life-form(s):** submersed (rosulate)
Key morphology: stems (to 3 cm) cormose; leaves (to 60 cm) basal, submersed, linear, sessile, the margins finely serrate; inflorescence solitary, peduncled (to 50 cm) or sessile (rarely), with flattened spathe; flowers (to 15 cm) bisexual, trimerous, emersed (then chasmogamous) or submersed (then cleistogamous); sepals (to 10 mm) green; petals (to 17 mm) linear, white or reddish; stamens (to 5.8 mm) 3; capsules (to 80 mm) terete, cylindrical, 1-locular, style single, stigmas (to 15 mm) linear; seeds (to 1.8 mm) numerous, the surface smooth or with up to 12 warty ridges
Life history: duration: annual (fruits/seeds); **asexual reproduction:** none; **pollination:** autogamy; **sexual condition:** hermaphroditic; **fruit:** capsules (common); **local dispersal:** seeds (water); **long-distance dispersal:** seeds (birds)

Imperilment: 1. *Blyxa aubertii* [GNR]
Ecology: general: All *Blyxa* species are submersed, obligately aquatic plants, with only their flowers emerging from the water during sexual reproduction. The species include rosulate or vittate annuals and perennials, with the latter developing stolons. The flowers can be bisexual or unisexual (then arranged dioeciously). The bisexual flowers are either emergent and chasmogamous or submersed and cleistogamous (when growing in deeper water), with both types being highly autogamous. The unisexual flowers are emergent and are pollinated by insects (Insecta: Diptera; Odonata). The seeds are smooth or spiny (sometimes variable within a species), with the latter capable of tangling together and then being dispersed as a larger unit. Local dispersal occurs by water and probably over greater distances by animal vectors (presumably by seeds attached in adhering mud). The seeds are known to be eaten by birds (Aves), but it has not determined whether they remain viable after passage through their digestive tract. Large seed banks can develop in some species. The highest seed germination occurs on the sediment surface.
2.3.1.1. *Blyxa aubertii* **Rich.** is an annual which grows in borrow pits, lakes (artificial), pools, and rice fields at elevations to 70 m. In North America, the plants occur primarily in turbid waters (to 91 cm depth), which are associated with anthropogenically disturbed habitats. In portions of the native range, the waters are circumneutral (pH: 6.0–7.4) and often are characterized by iron precipitation. The plants have been cultivated successfully in subalkaline waters of medium hardness at 20°C–28°C. In North America, the substrates are nutrient poor and have been described as clay, clay mud, and mud. Flowering in North America has been observed from July to December. The plants can allocate up to 38.1% of their DM biomass to sexual reproduction. The flowers are bisexual, self-compatible, and chasmogamous when emersed but cleistogamous when submersed (which occurs when excessive water depth precludes floral emergence). Because the anthers dehisce before the perianth and the stigmas expand beyond the calyx (pollen is collected as their expansion ensues), both types are highly autogamous and have comparable levels of seed set. During anthesis (lasting just 2 days), little pollen is exposed to potential pollinators, and no biotic pollinators have been observed. Studies of populations in Japan found there to be from 10 to 71 flowers/plant, 45.3 seeds/fruit, and 92.9% seed set on average. Individual plants are capable of producing from 340 to 4224 seeds, with larger plants producing higher seed numbers. Mechanisms of seed dispersal in this species are uncertain, but presumably include water (locally) and adherence to mud on the feet of migrating waterfowl (Aves: Anatidae) over greater distances. Seeds also have been dispersed inadvertently by transport within contaminated rice (Poaceae: *Oryza*) seed. **Reported associates (North America):** *Ceratophyllum demersum*, *Egeria densa*, *Juncus repens*, *Limnophila indica*, *Limnophila* ×*ludoviciana*, *Najas minor*, *Nymphaea odorata*, *Potamogeton diversifolius*, *Sagittaria platyphylla*.
Use by wildlife: *Blyxa* seeds are eaten by black-winged stilts (Recurvirostridae: *Himantopus himantopus*). The foliage is eaten by snails (Mollusca: Gastropoda).

Economic importance: food: *Blyxa aubertii* is eaten as a green leaf vegetable in Vietnam; **medicinal:** *Blyxa* is mentioned in Ayurveda medical texts as a universal remedy; **cultivation:** Due to its fragile leaves (and annual habit), *Blyxa aubertii* is cultivated only occasionally as an aquarium plant; **misc. products:** none; **weeds:** *Blyxa aubertii* is a rice-field weed in Asia and North America. Korean populations of *B. aubertii* have become resistant to ALS inhibitor type herbicides; **nonindigenous species:** *Blyxa aubertii* was introduced to Louisiana rice fields around the 1960s.

Systematics: Molecular phylogenetic analyses consistently resolve *Blyxa* and *Ottelia* as sister genera; however, further taxon sampling will be necessary in order to determine whether *Ottelia* is monophyletic and distinct from *Blyxa* (e.g., Figure 1.8). *Blyxa aubertii* possesses a seed amino acid profile similar to that of *B. echinosperma*, which some authors treat as a variety of the former. The two taxa undoubtedly are closely related if not conspecific. The basic chromosome number of *Blyxa* probably is $x=6$. Several anomalous counts ($2n=16$, 24, and 40) have been reported for *B. aubertii* and the cytology of this species requires further clarification. No *Blyxa* hybrids have been reported.

Distribution: In North America, *Blyxa aubertii* currently is known only from Louisiana and Mississippi.

References: Bhunia & Mondal, 2014; Cook, 1996a; 1996b; Cook & Lüönd, 1983; Feitoza et al., 2009; Haynes, 2000b; Jiang & Kadono, 2001; Kasselmann, 2003; McNair & Alford, 2014; Ogle et al., 2001; Quattrocchi, 2012; Sullivan, 1981; Thieret et al., 1969.

2.3.2.   Egeria

Brazilian elodea, Brazilian waterweed, South American waterweed

Etymology: after the Latin *Aegeria*, a legendary Roman nymph of springs

Synonyms: *Anacharis* (in part); *Elodea* (in part); *Philotria* (in part); *Udora* (in part)

Distribution: global: New World; **North America:** United States, British Columbia (Canada)

Diversity: global: 3 species; **North America:** 2 species

Indicators (USA): obligate wetland (OBL): *Egeria densa*, *E. naias*

Habitat: brackish (coastal), freshwater; lacustrine, riverine; **pH:** 5.2–8.9; **depth:** to 3 m; **life-form(s):** submersed (vittate)

Key morphology: stems (to 3 m+) completely submersed or floating, single or branched, rooted in the substrate (or water); leaves cauline, in whorls (of 4–5+), dense (internodes to 2.4 cm), sessile, the blades (to 40 mm) linear, flat or recurved, the margins finely to strongly serrate, the apex tipped by a single spine; flowers unisexual (plants dioecious); inflorescence (♂) 2–5-flowered, enclosed in a sessile spathe (to 12 mm), the flowers raised above the surface by an elongate peduncle (to 80 mm); sepals 3, green (to 4.4 mm); petals (to 10.5 mm) 3, white, shiny; stamens 9, the filaments (to 4.5 mm) clavate, papillose, yellow, contracted below anthers (to 1.8 mm); pistillate flowers (♀) either absent or smaller than the male flowers; when present, the spathes (to 8 mm) 1–2-flowered, the flowers

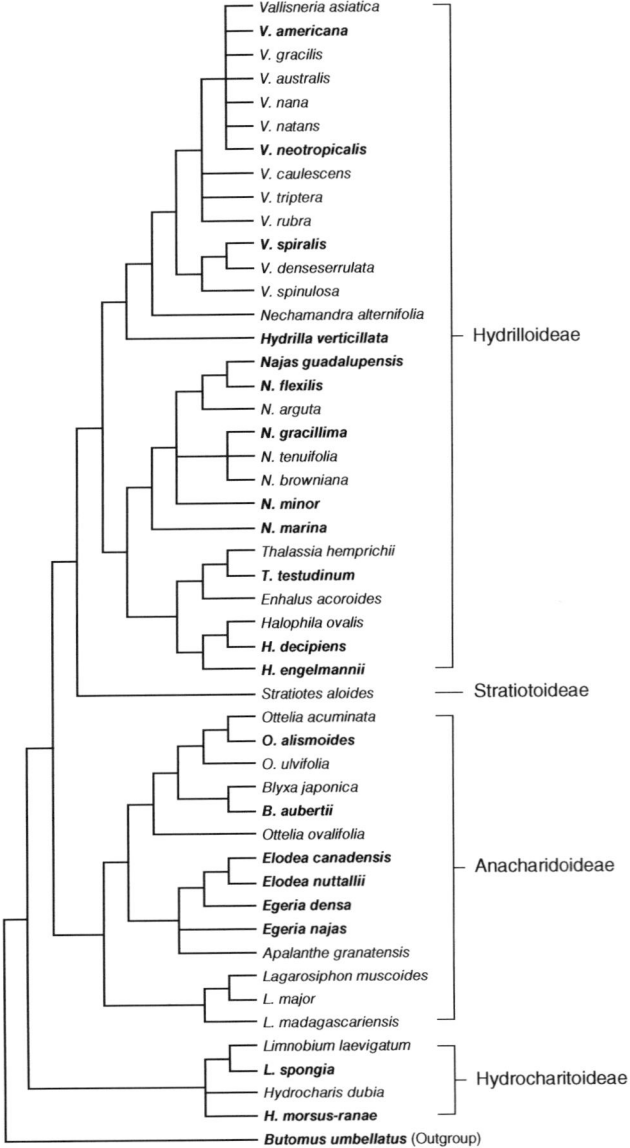

FIGURE 1.8 Phylogenetic relationships in Hydrocharitaceae as indicated by a 167-taxon analysis of *rbcL* sequence data (adapted from Les & Tippery, 2013). The result is similar to an analysis of full cpDNA data (Ross et al., 2016) by resolving four distinct clades corresponding to subfamilies Anacharidoideae, Hydrilloideae, Hydrocharitoideae, and Stratiotoideae; however, the interrelationships among these clades are resolved somewhat differently in the two studies. Although the entire family is aquatic, the OBL North American taxa (highlighted in bold) are scattered throughout the phylogeny.

raised above the water surface by their elongate hypanthium (to 48+ mm); the petals 3 (to 8.5 mm), white, conspicuous, with 3 orangish or reddish, papillose staminodes (to 2.4 mm); nectaries (3) present; ovary (to 3 mm) inferior, styles (to 3.8 mm) 3, each with 2–4 lobed stigmas, ovules (to 9) erect; capsules (to 14.5 mm) irregularly dehiscent; seeds (to 7.2 mm) beaked (to 3.7 mm), ellipsoidal

Life history: duration: perennial (whole plants); **asexual reproduction:** shoot fragments; **pollination:** insect; **sexual condition:** dioecious; **fruit:** capsules (absent); **local**

dispersal: shoot fragments (water); **long-distance dispersal:** shoot fragments (humans, water)

Imperilment: 1. *Egeria densa* [G5]; *E. naias* [unranked]

Ecology: general: The genus *Egeria* is entirely aquatic, with all species occurring as submersed growth forms where only the flowers emerge above the water surface. The flowers are unisexual and are arranged in a dioecious condition. They are produced almost continuously when water temperatures are within 15°C–25°C. The flowers are nectariferous, scented, and appear to be pollinated by small insects such as flies (Insecta: Diptera). However, in at least one species, the non-indigenous North American material is comprised entirely of male plants, thereby lacking any ability for sexual reproduction. Little is known regarding the seed production or germination requirements of the plants in their indigenous range. In North America, the plants are vegetatively vigorous and exclusively reproduce asexually by the production and spread of vegetative shoot fragments.

2.3.2.1. *Egeria densa* Planch. inhabits still or flowing (e.g., 30 cm s^{-1}), tidal or non-tidal, brackish to freshwaters (to 3.0 m; to 7.0 m in its native range) in bayous, canals, channels (drainage), ditches (irrigation), floodplains, lagoons (gravel pit), lakes, ponds (irrigation), pools, puddles, reservoirs, rivers, sloughs, springs, and streams at elevations to 2165 m. The plants can tolerate exposures ranging from full sunlight to deep shade; however, light reduction (e.g., to 25%) has been observed to reduce relative growth rates. The experimentally determined optimal light intensity is only about 108 lumens and the plants do not acclimate well to higher light intensities. The waters span a broad range of acidity (pH: 6.0–8.9; $\bar{x}$: 7.4) and are relatively high in alkalinity ($\bar{x}$: 38.2 mg l^{-1} CaCO$_3$) and specific conductance ($\bar{x}$: 184 µS cm^{-1}), but relatively low in inorganic nitrogen ($\bar{x}$: 100 µg l^{-1}) and phosphorous ($\bar{x}$: 68 µg l^{-1}). The plants are particularly susceptible to iron deficiency. Optimal water temperatures range from 16°C to 28°C, but temperatures up to 35°C are tolerated. Fertilization experiments indicate that the plants do not respond positively to increased sediment nutrient levels as do potential competitors such as *Hydrilla* (Hydrocharitaceae). Growth rates and biomass production generally correlate negatively with pH, which is due to the higher concentrations of free CO$_2$ in more acidic waters. However, the plants are adapted to the lower inorganic carbon levels in waters of higher pH by their ability to use bicarbonate (HCO$_3^-$) as an alternative carbon source. They also possess a high-affinity nitrate transporter (NRT$_2$), which facilitates nitrate uptake, especially by the shoots. Under stressful conditions (high light and temperature), the shoots induce a C4-like photosynthetic mechanism by increasing expression levels of PEP carboxylase and malic enzyme. Studies in the western United States indicate that the lowest total nonstructural carbohydrate reserves (35–51% starch) occur from March to June. The substrates include boulders, clay, clay loam, logs (submerged), mud (organic), sand (granite), sandy shell, sandy silt, silt, and silty loam (Congaree). In North America the plants flower from March to November; however, only male plants of this dioecious species have been introduced to the United States, making it reliant entirely on asexual reproduction. The plants essentially are evergreen and overwinter on the bottom (even under ice cover) as intact, short, dormant shoot fragments or rootstocks (sometimes more than 113 m^{-2}), with their vegetative growth resuming when water temperatures warm to 10°C–15°C. The starch concentration is highest in leaves (25.4%) and stems (22.6%) during the winter months. Optimal growth (in cultivation) occurs at 20.7°C and the plants are strong competitors in warm water environments (20°C–30°C). Bimodal growth peaks (late spring and fall) have been observed in some areas. Dispersal occurs primarily by the generation of shoot fragments, which float on the water surface and are transported by currents. The fragments average 7.3 cm in length and 88% retain full regeneration capacity after 4 weeks. They also survive desiccation well (e.g., during drawdown), especially if mixed among mounds of other stranded vegetation. Fragments on the exposed sediment surface die within 22 days, but 32% of shoot apices on fragments at the bottom of exposed vegetation mounds have remained viable for more than 34 days. The fragments are tolerant of chlorates (ClO$_3^-$) and significant exposure (to 1,000 mg l^{-1}) does not influence their extent of desiccation. Human-mediated transport of the fragments on boating and fishing equipment also facilitates their spread. Surveys of nonindigenous populations using RAPDs markers have shown very limited levels of variation, which indicates that genetic bottlenecks have occurred at the times of introduction, and/or that introduced plants have spread widely as clonally derived ramets. **Reported associates (North America):** *Bidens beckii, Brasenia schreberi, Cabomba caroliniana, Callitriche, Ceratophyllum demersum, Ceratophyllum echinatum, Eichhornia crassipes, Eleocharis acicularis, Eleocharis palustris, Elodea canadensis, Heteranthera, Hydrilla verticillata, Isoetes, Lemna minor, Limnobium spongia, Ludwigia, Myriophyllum heterophyllum, Myriophyllum hippuroides, Myriophyllum spicatum, Najas flexilis, Najas guadalupensis, Nelumbo lutea, Nitella, Nuphar polysepala, Nymphaea capensis, Nymphaea odorata, Peltandra virginica, Pistia stratiotes, Potamogeton crispus, Potamogeton diversifolius, Potamogeton foliosus, Potamogeton richardsonii, Potamogeton robbinsii, Ranunculus aquatilis, Sagittaria montevidensis, Spirodela polyrhiza, Utricularia macrorhiza, Vallisneria americana, Vallisneria neotropicalis, Wolffia, Zizaniopsis miliacea.*

2.3.2.2. *Egeria naias* Planch. is a nonindigenous, submersed perennial, which grows in lakes to depths of 1 m. It occurs preferentially in warmer (optimum temperature: 15°C–26°C), clear to tannin stained, somewhat acidic to alkaline (pH: 5.2–7.6) waters. In its native range, the plants are more tolerant to low light (compensation point: 6–22 µM m^{-2} s^{-1} PAR) and turbid water conditions than are those of the congeneric *E. densa* (see above). Although they are bicarbonate (HCO$_3^-$) users, they thrive less in higher bicarbonate/conductivity waters than *E. densa*, which exhibits a higher carbon uptake efficiency under low CO$_2$ conditions. The plants rely on sediment nutrients (N and P) for optimal growth, but neither nutrient appears to be a growth limiting factor. In the indigenous range, the total number of male flowers ordinarily outnumbers the female flowers by ratios of 3–4:1. Levels of sexual reproduction appear to be reasonably high in native populations, which contain a fair level of genetic variability as indicated by allozyme and RAPD markers. Although the plants can tolerate turbid, eutrophic conditions, the regeneration of vegetative fragments occurs most rapidly in less-turbid, oligotrophic, and sand-dominated sites. Higher growth rates occur in vegetative fragments that retain the apical meristems. In their native range, the plants have exhibited low tolerance to drawdown conditions, which substantially impact the populations. **Reported associates (North America):** none. This species has only recently been found in Florida (USA) where it is known from a single locality.

Use by wildlife: *Egeria densa* contains 75.1% digestible dry matter and 17.4% crude protein. North American plants are eaten by apple snails (Mollusca: Gastropoda: Ampullariidae: *Pomacea*), crayfish (Crustacea: Decopoda: Cambaridae: *Procambarus acutus*, *P. spiculifer*), grass carp (Vertebrata: Teleostei: Cyprinidae: *Ctenopharyngodon idella*), larval leaf-cutter moths (Insecta: Lepidoptera: Crambidae: *Parapoynx diminutalis*), redbelly tilapia (Vertebrata: Teleostei: Cichlidae: *Tilapia zilli*), and Suwannee River cooters (Reptilia: Emydidae: *Pseudemys concinna suwanniensis*). The plants also provide rearing habitat for largemouth bass (Vertebrata: Teleostei: Centrarchidae: *Micropterus salmoides*). Several species of birds (Aves) intensely forage in *Egeria densa* beds, including common moorhens (Rallidae: *Gallinula chloropus*), little blue herons (Ardeidae: *Egretta caerulea*), and tricolored herons (Ardeidae: *Egretta tricolor*); they are foraged to a lesser extent by blue-winged teals (Anatidae: *Anas discors*), great blue herons (Ardeidae: *Ardea herodias*), pied-billed grebes (Podicipedidae: *Podilymbus podiceps*), snowy egrets (Ardeidae: *Egretta thula*), and wood ducks (Anatidae: *Aix sponsa*).

Economic importance: food: Neither *Egeria densa* nor *E. naias* is reported to be edible; **medicinal:** no medicinal uses for *Egeria densa* or *E. naias* are known; **cultivation:** *Egeria densa* is one of the most common plants sold commercially for freshwater aquariums (see also next). *E. naias* also is widely grown as an aquarium plant but only recently has become popular in the United States; **misc. products:** *Egeria densa* has been used widely as a botanical subject for various physiological experiments pertaining to cyclosis and photosynthesis. Meal made from dried *E. densa* has been used as a feed for chickens (Aves: Phasianidae: *Gallus gallus*). *Egeria densa* is under investigation as a potential renewal energy source (biomethane) via the process of continuous anaerobic digestion. Dead *E. densa* plants have shown to be effective biosorptives for the removal of copper ions (Cu_2^+) from contaminated water; **weeds:** *Egeria densa* and *E. naias* are weedy in their native range and seriously invasive throughout their nonindigenous range; **nonindigenous species:** *Egeria densa* was introduced to the United States sometime before 1893 and also has been introduced (with variable persistence) to Africa, Australia, Chile, Denmark, England, France, Germany, Iceland (geothermal springs), Italy, Japan, The Netherlands, New Zealand, Russia, and Spain. The introductions have occurred by the careless disposal of plants used as botanical

laboratory specimens or as aquarium ornamentals. The presence of plants in nonindigenous sites has been detected successfully from environmental DNA (eDNA) samples. *Egeria naias* was first detected in 2017 from a site in west-central Florida, which appears to have resulted from the careless disposal of aquarium plants.

Systematics: The monophyly of *Egeria* is uncertain, with phylogenetic analyses placing *E. densa* closer to *Elodea* than to its congener *Egeria najas* (Figure 1.8). Without further sampling of *Egeria* (*E. heterostemon*) and *Elodea* species, it is unclear whether *Egeria* should best be retained as a distinct genus or merged with *Elodea* (which often has been done in the past). The basic chromosome number of *Egeria* also is uncertain. Several different cytotypes ($2n = 16$, 24, 46) have been attributed to plants identified putatively as *Egeria densa*, which raises doubts on the identification of the source material, especially because some of the cytotypes ($2n = 16$, 24) also are quite divergent genetically. No natural hybrids involving *E. densa* or *E. najas* have been reported. Experimental reciprocal crosses between these two species also have failed to set seed, which indicates the presence of an effective genetic isolating mechanism. Intergeneric crosses involving *Elodea*, *Hydrilla*, and *Lagarosiphon* also have been unsuccessful.

Distribution: *Egeria densa* occurs sporadically throughout the warmer parts of the United States; *E. naias* currently is known only in west-central Florida.

References: Baker et al., 2010; Barnes et al., 2013; Bartodziej & Weymouth, 1995; Bini & Thomaz, 2005; Blackburn et al., 1961; Boschilia et al., 2012; Boyd & McGinty, 1981; Carter & Sytsma, 2001; Catling & Wojtas, 1986; Catling & Mitrow, 2001; Conrad et al., 2016; Cook & Urmi-König, 1984a; Crutchfield, Jr. et al., 1992; De Abreu Pietrobelli et al., 2009; DiTomaso & Healy, 2003; Dugdale et al., 2012; Ernst-Schwarzenbach, 1945; 1953; Fujiwara et al., 2016; Haramoto & Ikusima, 1988; Haynes, 2000b; Hoyer et al., 1996; Hussner, 2012; Hussner et al., 2010a; 2016; Kasselmann, 2003; Kadono, 2004; Kadono et al., 1997; Koehler & Bove, 2001; Lagueux et al., 1995; Lara et al., 2002; Les & Tippery, 2013; Maurice et al., 1984; Mony et al., 2007; Mori et al., 1999; 2012; Mühlberg, 1982; Palma et al., 2013; Parker & Hay, 2005; Pennington & Sytsma, 2009; Pierinia & Thomaz, 2004; Pine & Anderson, 1991; Pulgar & Izco, 2005; Redekop et al., 2016; Riis et al., 2012; Rixon et al., 2005; Santos et al., 2011; Silveira et al., 2009; Sultana et al., 2013; Takahashi & Asaeda, 2014; Takayanagi et al., 2012; Tanaka et al., 2004; Tavechio & Thomaz, 2003; Thomaz et al., 2007; Umetsu et al., 2012; Wasowicz et al., 2014; Yarrow et al., 2009; Yeo, 1966; Zhen et al., 2016.

2.3.3. Elodea

Waterweed; Elodée

Etymology: after the Greek *helodes* ("wetland") in reference to the habitat

Synonyms: *Anacharis*; *Diplandra*; *Helodea*; *Hydora*; *Philotria*; *Serpicula*; *Udora*

Distribution: global: Asia*; Australia*; Europe*; New World; New Zealand*; **North America:** widespread

Diversity: global: 5 species; **North America:** 3 species
Indicators (USA): obligate wetland (OBL): *Elodea bifoliata, E. canadensis, E. nuttallii*
Habitat: freshwater; lacustrine, riverine; **pH:** 5.6–10.2; **depth:** 0.1–12 m; **life-form(s):** submersed (vittate)
Key morphology: stems (to 4+ m) submersed, branching, flexuous, sometimes with turion-like buds and adventitious roots; foliage opposite or in whorls of 3- and 6-ranked, the leaves (to 15.5–24.8 mm) sessile, linear to ovate, often clustered apically; plants dioecious; spathes (♂ to 8.2–42.0 mm; ♀ to 17.6–67.0 mm) sessile, with 1–2 flowers; ♂ flowers: essentially sessile (to 0.5 mm) and detaching or long-pedicelled (to 15+ cm) and attached, floating and opening at water surface; sepals (to 3.4–6.1 mm) 3, green; petals (to 2.6–6.2 mm) 3, reduced, whitish, translucent; stamens 9; ♀ flowers: sepals (to 3.5 mm) 3, green; petals (to 3 mm) 3, reduced, whitish, translucent; staminodia (to 1.4 mm) 3, hypanthium elongate (to 32 cm); ovary syncarpous (3 carpels), inferior, styles (to 4 mm) 3, recurved, entire or bifid; capsules (to 10 mm) irregularly dehiscent, several-seeded (to 10), beaked (to 6.5–20.0 mm); seeds (to 5.7 mm) fusiform, smooth or covered by hairs (to 1 mm)
Life history: duration: perennial (dormant apices, rhizomes, winter buds); **asexual reproduction:** rhizomes, shoot fragments, winter buds; **pollination:** epihydrophily; **sexual condition:** dioecious; **fruit:** capsules (rare to common); **local dispersal:** seeds, shoot fragments (water); **long-distance dispersal:** shoot fragments (birds, water)
Imperilment: 1. *Elodea bifoliata* [G4/G5]; S1 (KS, NE, UT); S2 (AB, MT, SK, WY); **2.** *E. canadensis* [G5]; S1 (AL, NC, UT); S2 (SK, WY); S3 (IL, KY, NS); **3.** *E. nuttallii* [G5]; S1 (KS, MB, NS, PE, WY); S2 (KY, MT, NB, NC, TN); S3 (BC, IL, ON, QC, VT, WV)
Ecology: general: All *Elodea* species are submersed aquatic plants, which extend only their flowers above the water surface. All three North American species are ranked as OBL indicators throughout their ranges. Although retained as OBL indicators in the National list, *Elodea callitrichoides* and *E. schweinitzii* have been excluded here. The former was erroneously reported for North America and the latter is not accepted taxonomically as a distinct species (see *Systematics*). When water clarity is high, the species can colonize depths up to 12 m, which is the greatest maximum depth recorded for any freshwater aquatic angiosperm. All of the species are dioecious perennials, which possess an epihydrophilous water pollination system. The male flowers either detach from the submersed male plants and rise to the surface where they float and open or they remain attached to the male plants by long pedicels and open once they reach the surface. In either case, the pollen is shed from the male flowers onto the water surface, where it is transported by currents to the female plants. The female flowers also float, but remain attached to the maternal plants by an elongate hypanthium; once reaching the surface, they open and extend their stigmas into the water, where the floating pollen grains are encountered. Despite this rather intricate pollination system, there typically are few seeds produced (except in one species) and most of the reproduction is

asexual. Vegetative reproduction occurs by the fragmentation of the shoots, which can remain viable having only a few nodes intact. The plants (or at least large fragments) also are known to overwinter readily, even in waters having thick ice cover. Requirements for seed germination have not been elucidated. The seeds and vegetative fragments are dispersed locally by water, and over greater distances by birds (Aves) as well as by human-mediated transport. Phenotypic plasticity is extensive and facilitates adaptation of the plants to variable local conditions. It is strongly advised that the taxonomic issues involving *E. canadensis* and *E. nuttallii* (see *Systematics*) be taken into account before attempting to interpret the ecological data provided for these species.

2.3.3.1. *Elodea bifoliata* **H. St. John** inhabits still or slow-moving (e.g., 0.57 m^{-3} s^{-1}) waters (to 7.0 m deep) of canals, lakes, ponds, pools, potholes, reservoirs, rivers, sloughs, and streams at elevations to 3088 m. The plants occur in open exposures to partial shade. They often are found floating on the surface of the water, or are washed ashore from sites established in deeper water. The waters are described as alkaline. The substrates have been characterized as mud, rock, sand, and sandy clay. Flowering occurs from July to August with fruiting extending through September. The plants are highly fertile and there has been some suspicion (but no definitive evidence) that they might be monoecious. Further life history information is lacking for this species.
Reported associates: *Alisma subcordatum, Azolla filiculoides, Carex microptera, Ceratophyllum demersum, Chara, Crypsis schoenoides, Elatine brachysperma, Elatine californica, Elatine rubella, Eleocharis acicularis, Eleocharis palustris, Eleocharis parishii, Glyceria, Gratiola, Lemna minor, Limosella, Marsilea mollis, Myriophyllum sibiricum, Myriophyllum spicatum, Najas guadalupensis, Najas marina, Persicaria amphibia, Persicaria coccinea, Potamogeton diversifolius, Potamogeton foliosus, Potamogeton gramineus, Potamogeton natans, Potamogeton nodosus, Potamogeton pusillus, Potamogeton richardsonii, Potamogeton richardsonii, Ranunculus aquatilis, Ranunculus longirostris, Schoenoplectus acutus, Sparganium, Stuckenia pectinata, Utriculata, Veronica, Wolffia columbiana, Zannichellia palustris.*

2.3.3.2. *Elodea canadensis* **Michx.** inhabits standing freshwaters (to 12.0 m deep) in backwashes, backwaters, bays, beaches, bottomlands (river), canals, channels, depressions, ditches (drainage; irrigation; roadside), floodplains, flumes, gravel pits, lagoons, lakes, mudflats, ponds, pools, potholes, reservoirs, rivers, sloughs, springs, and streams at elevations to 3326 m. The plants occur occasionally in slightly brackish water but can tolerate only low salinity levels (<3.5 ppt). Habitats can span a broad range of acidity (pH: 6.5–10.2), but typically are mesotrophic to eutrophic, alkaline waters (pH: $\bar{x}$ = 7.6–7.8) of higher alkalinity (to 300 mg l^{-1} CaCO$_3$) and conductivity (to 650 µmhos cm^{-1}). Optimum water temperatures range from 15°C to 20°C. Exposures range from open sites to partial shade. The plants tolerate relatively high light levels with an optimum near 16,000 lumens m^{-2}. Net photosynthesis is saturated at light intensities from 18,000 to 38,000 lumens m⁻² (15–30% of full sunlight). However, they also tolerate deep shade, suffering less than a 20% decrease in biomass compared to unshaded conditions. Shoot fragments can sustain growth at light levels as low as 90% shade, at which point they can still survive but without additional growth. The plants also are fairly tolerant of disturbances due to cutting and competition. The shoots take up nutrients and minerals directly from the water and are capable of removing large amounts of nitrogen (e.g., 39.8 kg ha⁻¹) and phosphorous (12.1 kg ha⁻¹). The foliage can concentrate heavy metals (e.g., As: 307 mg kg⁻¹; Cd: 32.33 µg g⁻¹; Pb: 160.9 µg g⁻¹) at up to several thousand times their ambient concentrations. The plants are capable of using bicarbonate (HCO_3^-) when concentrations of free CO_2 are low in the ambient waters. Bicarbonate use is reduced proportionally with increasing CO_2 concentrations. The majority of sediment-derived phosphorous (54%) is translocated from the roots to shoots. The plants are limited by sediment nitrogen (rather than phosphorous) levels. The substrates include alluvium, clay, cobbly sandy loam, gravel, humic soil over clay, loam, marl, muck, mud, ooze, peat, rocks (basalt), sand, sandstone, sandy gravel, sandy muck, silt, silty mud, and stones. Flowering has been observed from June to October; the fruits are seldom encountered but have been reported in August. Most reproduction is asexual and involves the production and dispersal of vegetative shoot fragments, which are produced readily by the relatively brittle stems. The apical foliage often becomes tightly clustered, developing into well-insulated winter buds, which enable the plants to persist throughout the cold season. The plants die back during autumn, releasing manganese and other ions into the water. They decompose rapidly, losing 50% of their biomass (DM) within a week and 95% of their biomass within 47–57 days. Spring regrowth occurs from the dormant stem fragments or winter buds. The fragments rapidly regenerate roots and quickly establish, even when in deep (e.g., 94%) shade. The fragments have high survival rates (>80%) whether rooted or not. Their ability to establish on sediments is higher in the autumn than in the spring. Fragments have been observed to be draped around the necks and backs of waterfowl (Aves: Anatidae), which can disperse them over fairly long distances. The fragments are fairly tolerant to desiccation and can remain viable in the open air for at least 23 hr. On a dry mass basis, the plants consume 0.2–2.8 mg O_2 g⁻¹ hr⁻¹ while fixing 1.16–2.97 mg C g⁻¹ hr⁻¹ during the growing season. Biomass production (DM) can reach levels to 11.7 kg ha⁻¹ day⁻¹ for a seasonal total of up to 2320 kg ha⁻¹. Genetic analyses (microsatellite markers) of introduced populations (where reproduction is entirely asexual) have shown an unexpectedly high level of genetic variation (e.g. 2–5 alleles/locus; 0.19–0.37 expected heterozygosity), which indicates an elevated level of somatic mutation, given a low probability of multiple introductions. The roots are not colonized by mycorrhizal Fungi. **Reported associates:** *Acorus calamus, Alisma subcordatum, Beckmannia syzigachne, Bidens beckii, Brasenia schreberi, Cabomba caroliniana, Callitriche palustris, Carex gynandra, Carex pellita, Cephalanthus occidentalis, Ceratophyllum demersum, Chara, Eleocharis acicularis, Eleocharis palustris, Elodea nuttallii, Equisetum fluviatile, Fontinalis, Glyceria borealis, Gnaphalium, Gratiola, Heteranthera dubia, Hippuris vulgaris, Hydrocharis morsus-ranae, Hydrocotyle vulgaris, Iris pseudacorus, Isoetes bolanderi, Juncus balticus, Juncus marginatus, Justicia americana, Lemna minor, Lemna obscura, Lemna trisulca, Lilaeopsis, Marsilea mollis, Marsilea vestita, Myosotis laxa, Myriophyllum heterophyllum, Myriophyllum sibiricum, Myriophyllum spicatum, Najas flexilis, Najas guadalupensis, Nasturtium officinale, Nitella, Nuphar advena, Nuphar polysepala, Nuphar variegata, Nymphaea odorata, Nyssa, Persicaria amphibia, Persicaria coccinea, Persicaria hydropiper, Phalaris arundinacea, Potamogeton alpinus, Potamogeton amplifolius, Potamogeton berchtoldii, Potamogeton crispus, Potamogeton diversifolius, Potamogeton foliosus, Potamogeton friesii, Potamogeton gramineus, Potamogeton illinoensis, Potamogeton natans, Potamogeton nodosus, Potamogeton perfoliatus, Potamogeton praelongus, Potamogeton pusillus, Potamogeton richardsonii, Potamogeton robbinsii, Potamogeton zosteriformis, Ranunculus aquatilis, Ranunculus flabellaris, Ranunculus flammula, Ranunculus longirostris, Ranunculus subrigidus, Sagittaria cuneata, Sagittaria latifolia, Sagittaria rigida, Schoenoplectus acutus, Schoenoplectus hallii, Schoenoplectus triqueter, Scirpus microcarpus, Sium suave, Sparganium angustifolium, Sparganium emersum, Spiraea douglasii, Spirodela polyrhiza, Stuckenia filiformis, Stuckenia pectinata, Stuckenia vaginata, Typha latifolia, Utricularia gibba, Utricularia macrorhiza, Utricularia minor, Vallisneria americana, Veronica, Zannichellia palustris.*

2.3.3.3. *Elodea nuttallii* (Planch.) **H. St. John** inhabits flowing or still freshwater, tidal or non tidal waters (to 6.1 m deep) in backwaters, bayous, bottoms (lake; tank), brooks, canals, ditches (channelized; drainage; roadside), flats (mud; river), gravel pits, lakes, marshes, oxbows (drying), ponds (hatchery; sinkhole; strip mine), pools (channel), potholes, reservoirs, rivers, sloughs, streams, and swamps at elevations to 3201 m. The waters (to 3 m deep) are described as calcareous (pH: 5.6–9.2; $\bar{x}$: 6.9–7.0; alkalinity to 160 mg l⁻¹ $CaCO_3$; chloride: to 1.6 ppm; conductivity to 310 µmhos cm⁻¹). The plants are fairly resistant to currents; however, turbulence can decrease their total chlorophyll content by as much as 40%. They typically are found in sunny exposures but are fairly shade tolerant; they are less tolerant of anoxic or hypoxic conditions. The plants are very resistant to desiccation and have been observed to rapidly recolonize a wetland, which had been drawn down for 10 weeks during summer. Increased sediment PO_4^{3-} concentration has been observed to reduce growth, while increasing growth in shoots that have been trimmed; an opposite effect occurs when PO_4^{3-} and NH_4^+ levels are increased simultaneously. Like the preceding species, *E. nuttallii* similarly possesses the ability to use bicarbonate ions (HCO_3^-) during conditions of low free CO_2 availability (e.g., at higher pH levels). The ability to uptake and accumulate heavy metals is similar to that of the preceding species as well. Experiments have indicated a higher overall growth rate in *Elodea nuttallii* than *E. canadensis*, but with foliar (vs. root) uptake being greater in the latter. The former also is more shade tolerant than the latter. The substrates include cobble (river), gravel, marl, mucky silt, mud, rock, sand, sandy mud, sandy silt, silt, and silty gravel. Flowering has been observed from June to August, with fruiting in August. The species is believed to form a transient soil seed bank, with the seeds persisting for <1 year. However, fruiting is rare and most reproduction is asexual, occurring by means of stem fragmentation. During the growing season, the shoots usually do not fragment unless they are subjected to current velocities above 2.4 m s^{-1}; however, fragmentation becomes widespread at the end of the season. The fragments (with as few as four nodes) have high survival rates (>70%) whether rooted or not. They are known to establish effectively under virtually any light regime including 94% shade. The fragments survive cold water conditions well and can resume growth when mean daily water temperatures surpass 4°C. Their ability to establish on sediments is higher in the spring than in the autumn. The plants produce a number of allelochemicals (α-linolenic acid, β-ionone, dihydroactinidiolide, linoleic acid, pentadecanoic acid, and stearic acid), which inhibit the growth of phytoplankton. **Reported associates:** *Azolla, Brasenia schreberi, Cabomba caroliniana, Callitriche hermaphroditica, Carex diandra, Carex stipata, Carex vesicaria, Cephalanthus occidentalis, Ceratophyllum demersum, Ceratophyllum echinatum, Chara, Eichhornia crassipes, Elatine chilensis, Elatine minima, Elatine triandra, Eleocharis acicularis, Elodea canadensis, Eriocaulon aquaticum, Glossostigma cleistanthum, Gratiola aurea, Heteranthera dubia, Hippuris vulgaris, Isoetes echinospora, Isoetes ×eatoni, Juncus balticus, Juncus bulbosus, Juncus nevadensis, Leersia oryzoides, Lemna minor, Lemna trisulca, Lemna turionifera, Ludwigia palustris, Lythrum salicaria, Menyanthes trifoliata, Myriophyllum heterophyllum, Myriophyllum heterophyllum × Myriophyllum laxum, Myriophyllum humile, Myriophyllum pinnatum, Myriophyllum sibiricum, Myriophyllum spicatum, Myriophyllum tenellum, Myriophyllum verticillatum, Najas canadensis, Najas flexilis, Najas gracillima, Najas guadalupensis, Najas marina, Najas minor, Nasturtium officinale, Nelumbo lutea, Nuphar polysepala, Nuphar variegata, Nymphaea odorata, Persicaria amphibia, Persicaria glabra, Potamogeton amplifolius, Potamogeton berchtoldii, Potamogeton bicupulatus, Potamogeton crispus, Potamogeton epihydrus, Potamogeton foliosus, Potamogeton gramineus, Potamogeton natans, Potamogeton nodosus, Potamogeton perfoliatus, Potamogeton praelongus, Potamogeton pulcher, Potamogeton pusillus, Potamogeton richardsonii, Potamogeton robbinsii, Potamogeton spirillus, Potamogeton zosteriformis, Proserpinaca palustris, Ranunculus aquatilis, Ranunculus longirostris, Riccia fluitans, Ricciocarpus natans, Sagittaria cuneata, Schoenoplectus acutus, Sparganium eurycarpum, Spirodela polyrhiza, Stuckenia filiformis, Stuckenia pectinata, Typha latifolia, Utricularia gibba, Utricularia macrorhiza, Utricularia minor, Utricularia purpurea, Utricularia radiata, Vallisneria americana, Wolffia brasiliensis, Wolffia columbiana, Zannichellia palustris.*

Use by wildlife: *Elodea canadensis* comprises 18–27% protein and is consumed by a number of herbivores. The plants are eaten by rusty crayfish (Crustacea: Decapoda: Cambaridae: *Orconectes rusticus*), muskrats (Mammalia: Cricetidae: *Ondatra zibethicus*), and mute swans (Aves: Anatidae: *Cygnus olor*). Insect (Insecta) herbivores include larval caddisflies (Trichoptera: Leptoceridae: *Leptocerus americanus, Nectopsyche*), flies (Diptera: Ephydridae: *Hydrellia discursa, H. harti, H. trichaeta*), and moths (Lepidoptera: Crambidae: *Acentria ephemerella, A. nivea*). The plants also are host to several Fungi (Ascomycota: Helotiaceae: *Varicosporium elodeae*; Chytridiomycota: Chytridiaceae: *Chytridium elodeae*; Cladochytriaceae: *Megachytrium westonii*). The presence of *E. canadensis* can reduce the level of larval mosquitoes (Insecta: Diptera: Culicidae) by as much as 34%. Dense growths of the plants also interfere with foraging activities of juvenile bluegills (Osteichthyes: Centrarchidae: *Lepomis macrochirus*) and other fish.

Economic importance: food: *Elodea* shoots are eaten as a snack in Estonia. However, given the ability of the shoots to concentrate heavy metals, consumption of the plants is inadvisable; **medicinal:** The Iroquois prepared an infusion from *Elodea canadensis*, which was used as an emetic. *Elodea nuttallii* contains β-sitosterol, which is used to treat enlarged prostates; **cultivation:** *Elodea canadensis* and *E. nuttallii* are widely grown as cold water aquarium plants; **misc. products:** *Elodea canadensis* and *E. nuttallii* are used routinely in various plant physiological experiments. *Elodea nuttallii* has been used as a co-substrate (with maize silage) for biogas generation. The plants also have been considered for wastewater treatment, where they are capable of achieving high annual removal rates of ammonia (75%), 5-day biochemical oxygen demand (90%), total nitrogen (47%), and total phosphorus (38%) from primary wastewater; **weeds:** *Elodea canadensis* and *E. nuttallii* are regarded as weeds throughout their nonindigenous range (and often also within their indigenous range); **nonindigenous species:** *Elodea* (*E. canadensis* and/ or *E. nuttallii*) has been introduced throughout the world, including Africa, Australia (~1931), Europe [Croatia (~2006), England (~1847), France (~1850), Germany (~1860), Hungary (~1870), Ireland (~1836), Italy (~1892), Switzerland (~1869)], New Zealand (~1872), and Russia (~1970)].

Systematics: Phylogenetic analyses indicate that *Elodea* is most closely related to *Egeria* (Figure 1.9) and somewhat more distantly to *Apalanthe*. Although some authors have advocated the merger of *Elodea* and *Egeria*, crosses between these genera (also between *Elodea* and *Hydrilla* or *Lagarosiphon*) are inviable, which indicates that the genera are well isolated genetically. Accessions of three *Elodea* species included in phylogenetic analyses resolve as a clade, which substantiates the monophyly of the genus as far as it has been sampled (Figure 1.9). *Elodea* is extremely difficult taxonomically. Due to their frequent sterility, specimens of *E. canadensis* and *E. nuttallii* typically are identified by their leaf morphology (shorter, ovate leaves in the former; elongate, linear leaves in the latter). However, genetic analyses have indicated that many of the specimens observed in the field are *E. canadensis* × *E. nuttallii* hybrids, and that "*E. canadensis*" phenotypes commonly associate with "*E. nuttallii*" genotypes

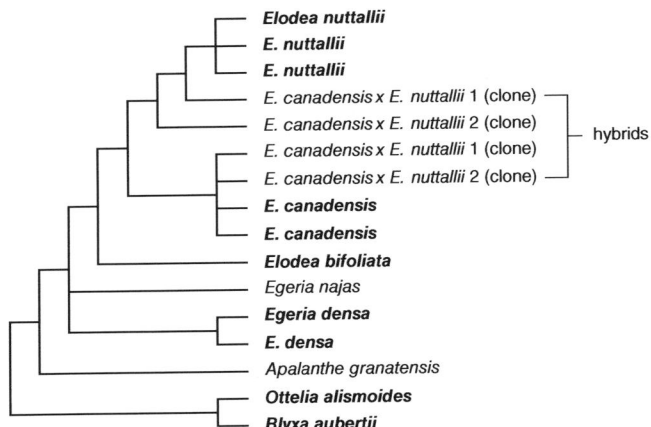

FIGURE 1.9 Phylogenetic relationships of *Elodea* and related genera as indicated by analysis of nrITS sequence data (adapted from Les & Tippery, 2013). As sampled, *Elodea* is monophyletic and closely related to *Egeria* (the placement of *Egeria najas* notwithstanding). Many *E. canadensis* × *E. nuttallii* hybrids have been detected by the recovery of cloned parental alleles originating from both species. The hybrids can resemble either parental species, at least vegetatively. Taxa designated as OBL North American indicators are highlighted in bold.

and vice versa. The author also has observed phenotypically plastic specimens clearly having "*E. canadensis*" foliage on one portion of the shoot and distinctly "*E. nuttallii*" foliage on another part of the same shoot. Consequently, it is virtually impossible to identify material of either taxon morphologically, at least when in vegetative condition. Yet, the two taxa are distinct genetically (Figure 1.9), despite their bewildering mosaic of phenotypic variability. *Elodea* also is problematic cytologically. The basic chromosome number of *Elodea* is uncertain but probably is $x=8$, given a $2n=16$ count reported for the South American *Elodea callitrichoides*. However, that base number would indicate that *E. bifoliata* ($2n=24$) is triploid, which would be unusual given its relatively high fertility. The counts reported for *E. canadensis* and *E. nuttallii* are so variable ($2n=24, 32, 40, 48, 56, 64, 72, 96$) that their ploidy levels are indeterminable; yet they consistently are based on $x=8$. Extensive somatically derived aneuploidy ($2n=32, 42, 43, 44, 48$) reportedly has arisen within a single strain of *E. nuttallii* that was cultivated over a 73-year period. Experimental hybridizations involving *E. callitrichoides* × *E. canadensis* are inviable; however, reciprocal crosses between *E. canadensis* and *E. nuttallii* are fully interfertile. Genetically determined hybrids between *E. canadensis* and *E. nuttallii* have been collected throughout their range; however, it is not understood whether hybrid plants occur frequently or if their widespread distribution is due to the broad dispersal of fewer ramets. The complete chloroplast genome of *E. canadensis* has been sequenced.

Distribution: *Elodea bifoliata* occurs in western North America. *Elodea canadensis* and *E. nuttallii* are essentially sympatric throughout much of North America, except for the extreme northern and southern regions where neither is common.

References: Abernethy et al., 1996; Atapaththu & Asaeda, 2015; Bailey et al., 2008; Barrat-Segretain, 2004; Barrat-Segretain et al., 2002; Barrat-Segretain & Cellot, 2007; Beck-Nielsen & Madsen, 2001; Bishop & Eighmy, 1989; Catling & Wojtas, 1986; Ciutti et al., 2011; Cook & Urmi-König, 1985; Crow & Hellquist, 1982; Eighmy et al., 1991; Ernst-Schwarzenbach, 1945; 1953; Eugelink, 1998; Fulmer & Robinson, 2008; Green, 2016; Greulich & Tremolieres, 2006; Gross et al., 2001; Harms & Grodowitz, 2009; Harrel & Dibble, 2001; Haynes, 2000b; Hoffmann et al., 2014a; Huotari & Korpelainen, 2012; Huotari et al., 2011; Hussner et al., 2010b; Kalle & Sõukand, 2013; Kasselmann, 2003; Király et al., 2007; Kočić et al., 2014; Kravtsova et al., 2010; Kunii, 1981; Kuntz et al., 2014; Les & Philbrick, 1993; Mayes et al., 1977; Mielecki & Pieczynska, 2005; Löve, 1982; Peters et al., 2008; Preston & Croft, 1997; Riis et al., 2010; Robinson et al., 2007; Sand-Jensen & Gordon, 1986; Sheldon & Boylen, 1977; Spencer & Ksander, 2003; Thiébaut et al., 2010; Vincent & Bertola, 2012; Wang et al., 2014a; Zaman & Asaeda, 2013; Zefferman, 2014.

2.3.4. Halophila

Seagrass, star grass

Etymology: from the Greek *hals phílos* ("salt loving") in reference to the marine habitat

Synonyms: *Barkania*; *Caulinia* (in part); *Kernera* (in part); *Lemnopsis*; *Serpicula* (in part); *Thalassia* (in part); *Zostera* (in part)

Distribution: global: cosmopolitan; **North America:** southern (Gulf and Atlantic coasts)

Diversity: global: 10–13 species; **North America:** 3 species

Indicators (USA): obligate wetland (OBL): *Halophila decipiens*, *H. engelmannii*, *H. ovalis*

Habitat: marine; oceanic; **pH:** 7.9–8.3; **depth:** to 90 m; **life-form(s):** submersed (vittate)

Key morphology: foliage arising directly from rhizomes or in pseudowhorls (to 8 leaves) from short, erect, unbranched stems, the rhizomes thin or thick (to 2 mm), with 2 scales per node; leaves sessile or petiolate (to 12 cm), the blades (to 30 mm) linear-oblong, linear-lanceolate, oblong, or oblong-elliptic, the margins entire or serrulate; flowers unisexual (the plants dioecious or monoecious), enclosed in a spathe; ♀ flowers (to 9 mm) essentially sessile, styles (to 30 mm) 3; ♂ flowers (to 4 mm), pedicellate (to 2 cm), the pollen released in chains from linear anthers (to 2 mm); capsules (to 6 mm) ovoid, beaked (to 6 mm); seeds (to 0.3 mm) few-30, reticulate

Life history: duration: perennial (rhizomes); **asexual reproduction:** rhizomes; **pollination:** water (hypohydrophily); **sexual condition:** dioecious, monoecious; **fruit:** capsules (common); **local dispersal:** fruits (gravity), shoot fragments (water); **long-distance dispersal:** fruits (water), seeds (birds, water)

Imperilment: 1. *Halophila decipiens* [G4/G5]; **2.** *H. engelmannii* [G3/G5]; S1 (LA, MS); **3.** *H. ovalis* [G2]; S2 (FL)

Ecology: general: All *Halophila* species are submersed, perennial, marine, hydrophilous plants, and all North American taxa are OBL wetland indicators. The genus occurs within one of three major clades of "seagrasses," which represent the relatively few angiosperms adapted to life in the ocean. By their ability to survive in saltwater, the species typically encounter salinities of 34–35‰ and an alkaline pH of 8.1. Both features can vary due to freshwater intrusion from rainfall, proximity to freshwater sources, temperature, and other factors. *Halophila* species can occur in fairly deep waters and have been recorded at depths of up to 90 m. Although these are the smallest of the seagrasses physically, they also exhibit the highest rates of sexual reproduction. All of the species have unisexual flowers (the plants dioecious or monoecious) and are water pollinated. The pollen grains are ellipsoid and form linear moniliform strands, which typically are dispersed while submersed but occasionally float to the water surface. The grains are captured by the elongate, papillate stigmas of the female flowers. The fruits contain from 7 to 60 seeds, which are not particularly buoyant, are released at or below the sediment surface, and typically are dispersed within distances of only a few centimeters. Although the fruits are not particularly vagile, their concentration of stored seeds facilitates the rapid colonization of newly exposed sites. In some cases, the fruits are buoyant and able to float for some distance. Recent studies have shown that the seeds can retain viability after passage through the digestive tract of waterfowl (Aves: Anatidae), which implicates birds as potential vectors for long-distance dispersal. Transient or persistent seed banks can develop at densities of up to 70,000 seeds m^{-2}. The seeds are not dormant but do require light for germination. Rates of natural seed germination range from 12% to 63%. Like all seagrasses, *Halophila* species also reproduce asexually by means of rhizomes, and occasionally by stem fragmentation. Although "*Halophila johnsonii*" is designated as an OBL indicator, that taxon has been merged here with *H. ovalis* in accordance with the results of various systematic investigations (see *Systematics*). However, the ecological information compiled for this taxon is based primarily on information from reports specifically referencing "*H. johnsonii*."

2.3.4.1. *Halophila decipiens* Ostenf. inhabits coastal marine communities in waters ranging from 24.3‰ to 38.0‰ salinity and extends from subtidal zones to depths as great as 85 m (in clear waters). The plants are ruderal and able to survive in reduced or fluctuating light; however, deeper water populations are intolerant of high irradiance levels (>1000 µE m^{-2} s^{-1}) and reduced salinity (<35 ppt). Typical water depths are 10–30 m with temperatures ranging from 21.0°C to 36.0°C. Plant cover diminishes noticeably when illumination falls below 10% of surface incident light. The plants thrive in sites where disturbance has eliminated competition by ordinarily more persistent species. The substrates can include coral, limestone, muddy coral sand, muddy sand, sand, shell hash, and silt. The plants are monoecious and the flowers are protandrous. Flower density can vary from 50 to 1925 m^{-2}. Pollination is hypohydrophilous, occurring entirely underwater. In North America, flowering has been reported from July to September with fruiting from July to August. The flowers typically are open during a 9-hr period extending from first light until mid-afternoon. On average the fruits contain 36.8

seeds, which readily sink into the substrate. Large seed banks (to 13,500 seeds m^{-2}) have been observed within the upper 3 cm of sediments; more typical values range from 134 to 3,414 seeds m^{-2}. Seed germination has been observed from October to January. Buried seeds germinate well (to 86%) within 2–9 days after exposure to light, but will remain dormant in the dark at 24°C. Seed germination occurs at salinities from 25% to 34% but is inhibited at 42%. Although the seeds are deposited locally, their movement *en masse* within large quantities of sediment dislodged by disturbance (e.g. hurricanes) can result in far greater dispersal distances. Seedling densities are highest during the winter. The biomass attained by the plants varies widely, ranging from 0.02 to 12 g m^{-2}; it is lowest during fall and winter. Production rates (e.g., 0.023 g C m^{-2} day^{-1}) can reach a total of 4.56 × 10^8 g C day^{-1} during the peak growing season. The ratio of aboveground to belowground biomass is higher in shallow sites, with more storage occurring in underground organs at deeper sites. The plants can occupy extensive ranges of 7,500 km^2 or more. The foliage decomposes rapidly, losing more than 50% of its original mass within 3 days. The plants have been cultured successfully *in vitro*. **Reported associates:** *Halodule wrightii, Halophila engelmannii, Halophila ovalis, Ruppia maritima, Syringodium filiforme, Thalassia testudinum.*

2.3.4.2. *Halophila engelmannii* Asch. inhabits coasts (oceanic), lagoons, and tidal flats at depths to 90 m. The plants can tolerate salinity to 60‰ and occur where water temperatures vary from 13.3°C during winter to 31.6°C during summer. They decline during periods of increased precipitation, which favors species with lower salinity tolerances. The plants are intolerant of ultraviolet-A and UV-B radiation and typically are shaded by epiphytes or occur in the shade of other seagrasses. The substrates include limestone, mud, muddy clay, muddy sand, sand (gray), and sandy mud. The plants are dioecious. Flowering has been observed from March to June; however, under laboratory conditions, flowers have been produced continuously from January to September. The flowers open during a 9-hr period extending from first light until mid-afternoon. Although typically hypohydrophilous, some pollen has been observed to entrap gas bubbles, which enables them to rise to the water surface, where epihydrophily (surface pollination) conceivably could occur. An extensive seed bank can develop (100–800 m^{-2}; $\bar{x}$: 74 m^{-2}). Seeds collected from mid-May to mid-June are not dormant and have germinated within 4 weeks. Germination requires light and at 24°C–27°C, initiates within 7–10 days (if collected from November to May) and in 27–33 days if collected in October. Seed collected in June (and stored for 18 weeks in the dark at 24°C–27°C) germinated in the light within 27–38 days; seed stored for 30 weeks in the dark eventually germinated after 9–12 days. Many seeds most likely remain dormant in the sediment until they are exposed to light by disturbance. Seeds stored in the dark can remain viable for at least 2 years. Seedlings that have germinated under laboratory conditions represented a 64:46 ratio of male and female flowers; however, a 42:58 ratio of staminate to pistillate shoots has been observed in a natural population. Biomass has been observed to range from 0.25 to 6.0 g m^{-2}. The plants have been cultured successfully *in vitro*. **Reported associates:** *Acetabularia, Caulerpa, Codium, Halimeda, Halodule wrightii, Halophila decipiens, Halophila ovalis, Penicillus, Ruppia maritima, Syringodium filiforme, Thalassia testudinum, Udotea*.

2.3.4.3. ***Halophila ovalis* (R. Br.) Hook. f.** inhabits the intertidal and subtidal zones (to 3 m depth) of coastal marine and riverine environments, including channel bottoms, lagoons, sand bars, and sand flats. These physiologically plastic plants are tolerant of higher irradiances, turbidity, fluctuations in salinity and temperature, and contain intracellular UV-absorbing pigments (numerous flavonoids), all factors that enable them to colonize shallower water depths in exclusion of deeper dwelling seagrass taxa. They acclimate rapidly to high UV-B and PAR levels. Their maximum growth occurs at salinities of 30 ppt. The plants will experience complete mortality if rapidly exposed to 10‰ salinity for 10 days; however, tolerance to 10‰ can be maintained if the reduction in salinity occurs gradually or in brief pulses. The substrates have been characterized as mud and sand (coarse). The plants are dioecious, with flowering (in this case ♀) observed from April to October; no seeds are produced. The North American populations probably are derived from the founding of a single female clone, which is incapable of seed set due to the lack of male plants and associated pollination. The plants can be dispersed vegetatively by fragments; however, these are relatively short-lived and remain viable only for 4–8 days. Radioisotope tracers have shown that the plants store few carbon reserves. Carbon is allocated more to younger portions of a genet and to ramets in proportion to their proximity, but is not allocated preferentially to ramets experiencing stress. These factors make it difficult for the plants to persist vegetatively over unsuitable substrates, which compromise their ability to withstand disturbances resulting in habitat loss. Biomass has been measured at 43.0 (aboveground) and 53.5 (belowground) g DM m^{-2}. **Reported associates (North America):** *Caulerpa prolifera* (Chlorophyta: Caulerpaceae), *Halodule wrightii, Halophila decipiens, Halophila engelmannii, Ruppia maritima, Syringodium filiforme, Thalassia testudinum*.

Use by wildlife: The queen conch (Mollusca: Gastropoda: Strombidae: *Stombus gigas*) is known to graze on epiphytes that clothe the foliage of *Halophila decipiens*. The foliage of *H. decipiens* is low in total protein (2.5–6.2% by DM) and soluble carbohydrate (4.3–8.2%) but is eaten by green sea turtles (Reptilia: Testudines: Cheloniidae: *Chelonia mydas*); *H. engelmannii* also is eaten by the turtles. Various organisms graze on the epiphytic algae found on the foliage of *H. ovalis*.

Economic importance: food: not edible; **medicinal:** Crude extracts of *Halophila decipiens* are inhibitory to some Fungi (Ascomycota: Lulworthiaceae: *Lindra thalassiae*; Nectriaceae: *Fusarium*); **cultivation:** not cultivated; **misc. products:** none; **weeds:** none; **nonindigenous species:** The occurrence of *Halophila ovalis* along a small portion of the Atlantic coast is enigmatic, given the existence there only of genetically similar female plants, which have spread as asexual ramets. Although these plants have been treated as an

endemic, federally (USA) threatened taxon ("*H. johnsonii*"), it is almost certain that a single female clone was introduced to the region either by natural dispersal vectors or as a consequence of the intense international shipping in that area. Morphologically and genetically similar plants of *H. ovalis* (both with narrower leaves and fewer ovules than typical populations) have been documented in Antigua, indicating at least that the plants are not endemic to Florida.

Systematics: Various phylogenetic analyses have confirmed that *Halophila* is monophyletic. However, there remains some discrepancy on the number of species in the genus, even though most taxa have now been sampled and included in phylogenetic analyses. The most contentious issue involving North American material is the taxonomic status of "*Halophila johnsonii*," which was discovered relatively recently (1980) along the Atlantic coast of Florida. These plants consist entirely of genetically similar, female clones and reproduce only asexually. Given its rarity and precarious reproductive issues, *H. johnsonii* was recognized as a threatened endemic species in the United States, following numerous studies which demonstrated its morphological, ecological, and genetic distinctness from the more common *H. decipiens*. However, it was later shown that the Florida plants actually were not only unrelated to *H. decipiens*, but instead were closely related to *H. ovalis* and resolved within that clade in several phylogenetic studies (e.g., Figure 1.10). Although the Florida populations have narrower leaves and fewer ovules than other *H. ovalis* populations surveyed, they are similar genetically (identical nrITS sequences) and morphologically to *H. ovalis* populations from Antigua. Together, these results indicate that "*H. johnsonii*" should not be regarded as a distinct endemic species, but at best as a thin-leaved variant of the more widely ranging *H. ovalis*. The basic chromosome number of *Halophila* is $x=9$. *Halophila ovalis* ($2n=18$) is diploid; counts for *H. decipiens* and *H. engelmannii* are unavailable. No hybrids involving any *Halophila* species have been reported.

Distribution: *Halophila decipiens* occurs along the Gulf Coast of Florida and *H. ovalis* occurs along the Atlantic coast of Florida; both are widespread in tropical and subtropical waters worldwide. *Halophila engelmannii* occurs on the Atlantic coast of Florida and along the Gulf of Mexico coast.

References: Ackerman, 2006; Ballantine & Humm, 1975; Bell et al., 2008; Bird & Jewett-Smith, 1994; Bird et al., 1998; Bortone, 2000; Dawes et al., 1989; Dean & Durako, 2007; Duarte, 1991; Duarte & Chiscano, 1999; Durako et al., 2003; Eiseman & McMillan, 1980; Fourqurean et al., 2001; Gabriel et al., 2015; Gavin & Durako, 2011; 2012; 2014; Green & Short, 2003; Griffin & Durako, 2012; Hall et al., 2006; Hammerstrom et al., 2006; Hartog, 1970; Haynes, 2000b; Herbert, 1986; Howell, 2012; Humm, 1956; Inglis, 1999; Jewett-Smith & McMillan, 1990; Jewett-Smith et al., 1997; Josselyn, 1986; Kahn & Durako, 2008; 2009; Kahn et al., 2013; Kenworthy, 1999; Kunzelman et al., 2005; Larkum, 1995; Les, 1988; Les & Tippery, 2013; Les et al., 1997a; McDermid et al., 2007; McMillan, 1976; 1987; 1988; 1988a; 1988b; 1989; 1990; 1991; McMillan & Jewett-Smith, 1988; Meng et al., 2008; MNFS, 2002; Onuf, 1996;

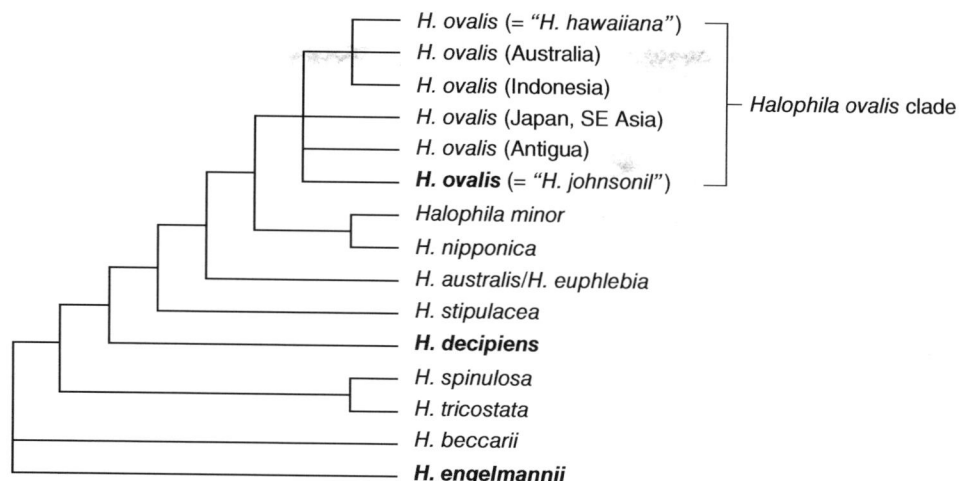

FIGURE 1.10 Phylogenetic tree constructed for *Halophila* using nrITS sequence data (adapted from Short et al., 2010). These results (and those of Uchimura et al., 2008) show "*Halophila johnsonii*" to be embedded within a large *H. ovalis* clade. The nrITS sequence of "*H. johnsonii*" is identical to material from Antigua, which also is similar morphologically. The three OBL North American *Halophila* species (in bold) originate from various parts of the tree, indicating a broad phylogenetic diversity in the region.

Orth et al., 2000; 2006; Pettitt & Jermy, 1974; Phillips, 1960; Phillips & Meñez, 1988; Ross et al., 2008; Short & Wyllie-Echeverria, 1996; Short et al., 2010; Torquemada et al., 2005; Trocine et al., 1981; 1982; Uchimura et al., 2008; Waycott et al., 2002; 2006; Wu et al., 2016; Xu et al., 2011; York et al., 2008.

2.3.5. Hydrilla

Florida elodea, hydrilla, water thyme; hydrille

Etymology: derived from the Greek *Hydra*, a mythological lake-dwelling serpent

Synonyms: *Elodea* (in part); *Epigynanthus*; *Hydora* (in part); *Hydrospondylus*; *Ixia* (in part); *Serpicula* (in part); *Udora* (in part); *Vallisneria* (in part)

Distribution: global: Africa, Australia, Eurasia, North America*, South America (northern)*; **North America:** eastern, southern, and western United States

Diversity: global: 1 species; **North America:** 1 species

Indicators (USA): obligate wetland (OBL): *Hydrilla verticillata*

Habitat: brackish (coastal), freshwater, freshwater (tidal); lacustrine, palustrine, riverine; **pH:** 5.4–8.6; **depth:** to 15 m; **life-form(s):** submersed (vittate)

Key morphology: subterranean shoots horizontal, rhizomatous or stoloniferous, sometimes producing white to brownish turion-like tubers at the stolon termini, roots unbranched; leafy shoots (to 8.5 m) submersed, branched or unbranched, erect in water column, sometimes bearing hard, olive-green, conelike turions; leaves cauline, in whorls of 3–8, the blades (to 20 mm) linear, sessile, single-nerved, margins finely serrulate, abaxial surface with midvein prickles, base tapering to stem; inflorescences single-flowered, sessile to subsessile (pedicels to 0.5 mm), enclosed by a spathe of 2 connate bracts; flowers unisexual (plants dioecious or monoecious), perianth trimerous, translucent or reddish to whitish; ♂ flowers released under water in bud, rising to surface, the enclosed staminal filaments distinct, springing apart explosively to release pollen from anthers as the free-floating flowers open at water surface; ♀ flowers floating on perianth, attached by an elongate, translucent to purplish floral hypanthium (to 50 mm); staminodes (3) minute; ovary tricarpellate, 1-locular, styles (to 0.75 mm) 3; fruits indehiscent, cylindrical, smooth or spiny; seeds (to 5) fusiform, smooth

Life history: duration: perennial (rhizomes, stolons, turions, whole plants); **asexual reproduction:** rhizomes, shoot fragments, stolons, tubers, turions (stem, stolons); **pollination:** abiotic (ballistic); **sexual condition:** dioecious, monoecious; **fruit:** capsular [indehiscent] (rare); **local dispersal:** shoot fragments, tubers, turions (water, waterfowl); **long-distance dispersal:** shoot fragments, tubers, turions (humans, water, waterfowl)

Imperilment: 1. *Hydrilla verticillata* [GNR]

Ecology: general: Although viewed almost universally as monotypic, *Hydrilla* represents a complex of cryptic, genetically variable taxa, which exhibit a number of cytological, ecological, and morphological differences. Nevertheless, all of the ecological information assembled for the genus over the past century has been attributed to one single, variable, wide-ranging species and is summarized below in that respect.

2.3.5.1. ***Hydrilla verticillata* (L. f.) Royle** is a nonindigenous species, which inhabits still or flowing, tidal, intertidal, or nontidal, brackish to fresh (0–11‰ salinity) waters in backwaters, bayous, canals, coves, ditches (drainage, irrigation, roadside), gravel pits, lagoons (river), lakes (freshwater), marshes (floating, tidal), ponds (artificial, farm, fire, fish, floodplain, irrigation, koi, retention, sinkhole, stagnant), pools (riverbed, shallow), potholes (drying), reservoirs, spillways, springs, and the margins of channels (bayou, boat), embayments, lakes, rivers, and streams (spring-fed) at elevations (North America) to 860 m. Exposures can range from full sunlight to shade (e.g., along banks). The plants can colonize a variety of still to flowing waters from 0.7 to 15.0 m deep, with slow flow rates <0.2 m s^{-1} apparently being optimal. They can grow in clear to turbid waters (e.g., Secchi depth: 0.6 m) at light levels as low as 10–12 µE m^{-2} s^{-1}. The plants are known to grow where percent transmittance of light is only 0.46 of that at the surface; they exhibit a decreasing chlorophyll A:B ratio with depth, which is attributed as an adaptation to reduced light levels. Shoot growth has been observed to continue for up to 10 weeks when the plants are kept in total darkness. The substrates (North America) are diverse and include alluvium (clayey, loamy), granite, gravel (limestone), limestone, marl, mud, rocky silt, sand, sandy silt, silt, silty clay, and silty clay loam (Cancienne); however, a high sediment organic content (e.g., 13% dw^{-1}) is deleterious to growth. The plants occur most often in freshwater (salinity <0.5‰) with their growth seriously impacted at salinities >13‰. Otherwise, broad spectrum of water conditions is tolerated (water temperature: 18.2°C–18.8°C; specific conductance: 27.0–400.8 µS cm^{-1}; $\bar{x} = 183$ µS cm^{-1}; pH: 5.4–8.6; $\bar{x} = 7.1$; total alkalinity: 1.7–164.0; $\bar{x} = 29.9$ mg l^{-1} CaCO$_3$; total P: 10–106 ($\bar{x} = 49$) µg l^{-1}; total N: 310–1,820 ($\bar{x} = 900$) µg l^{-1}). Although exhibiting C$_3$ photosynthesis, hydrilla also is capable of fixing CO$_2$ into C$_4$ acids for subsequent decarboxylation through an inducible C$_4$-like mechanism. Flowering (North America) occurs nearly year-round (January–November). Two major genetic strains (commonly referred to as "biotypes") occur in North America, and are represented either by monoecious individuals or as pistillate clones of a dioecious strain (entirely staminate plants remain unknown in North America). The abiotic pollination system of *Hydrilla* is unique and difficult to characterize. The staminate flowers are released in bud from either male or monoecious shoots, and float individually to the water surface. The stamens remain reflexed under tension until the flower buds reach the surface before opening as distinct, free-floating structures. Concurrently, the pistillate flowers arise from female or monoecious shoots on which they remain attached by an elongate floral hypanthium, which extends them to the water surface. The pistillate flowers open at the water surface, where the stigmatic surfaces are exposed to the air (but are kept dry) within the floating flowers. A sudden release of the stamen filaments from their tensed orientation occurs when the staminate flowers open at the water surface, which results in an explosive, ballistic discharge of pollen in the form of an aerial shower. The fortuitous landing of some grains upon a nearby, receptive stigma is necessary in order to achieve pollination. Both pollen and stigma remain dry during the process. From this description it is easy to see why seed set often is low to non-existent, and seeds rarely are found in (monoecious) North American plants. However, high seed production has occurred in Asian plants and experimentally manipulated crosses using various strains of international provenance also have produced numerous seeds, with a success rate of 71% between dioecious and monoecious strains. Such crosses have produced highly viable (90%) seeds characterized by high seedling survival. Detached shoots can remain buoyant for 1–6 days, during which time they can transport attached sexual or vegetative propagules across the water surface. Seeds in Asian

populations have germinated within 1 week when incubated in light at 23°C–28°C. Those stored in darkness (in water or dry) for up to 1 year have germinated readily when brought under light. *Hydrilla* plants are perennials, which possess an incredible capacity for rapid, vegetative growth and reproduction by means of stem fragmentation, turions, stolons, and tubers (a.k.a. subterranean turions). They are known to reach expansive surface coverages ranging from 8.5 to 405 ha. Biomass estimates up to 2.275 kg m^{-2} (FW) have been reported. The doubling time of shoots is about 19.8 days, but occurs much more rapidly (2.5–11 days) for tubers. Stem fragments comprising a single node are capable of regrowth, with 68%+ regrowth observed for fragments having 3–5 nodes. Tubers and turions can develop from October to April. Individual tuber production can reach 375 month^{-1} resulting in sediment densities of 6046 m^{-2} and several million ha^{-1}. Although the tubers can survive in the sediments for more than 4 years, the turions are thought to live for only a single year. Natural germination rates under drawdown conditions have reached 80%. The tubers germinate best (88–100%) in the light at temperatures from 22°C to 30°C; however, low temperatures (15°C) have resulted in reduced tuber germination in the monoecious (35–68%) and dioecious (~3%) strains. Both tubers and turions have survived desiccation by maintaining physiological activity after drying (e.g., 16 hr @ 30°C, 40% RH). However, tolerance to drying increases with the number of leaf whorls present and single node fragments can succumb to drying in as short as 2 hr. The tubers survive longer, with 16.7% viability observed after drying for 64 hr; whereas turion viability declines sharply as drying time approaches 8 hr. Turion production is enhanced under short photoperiods (<12 hr light) and in free-floating (non-rooted) plants but declines as plant density increases. The subterranean turions of monoecious plants are much more resistant to cold temperatures than intact dioecious plants. When kept in cold (4°C) water for 105 days, mortality increased to 98% for dioecious plants but only to 48% for monoecious turions. **Reported associates (North America):** *Acorus calamus, Alternanthera philoxeroides, Bacopa monnieri, Brasenia schreberi, Cabomba caroliniana, Callitriche, Ceratophyllum demersum, Ceratophyllum echinatum, Chara contraria, Cynodon dactylon, Cyperus strigosus, Echinodorus cordifolius, Egeria densa, Eichhornia crassipes, Elodea nuttallii, Heteranthera dubia, Heteranthera multiflora, Hydrocotyle umbellata, Hygrophila polysperma, Juncus acuminatus, Lemna minor, Lemna perpusilla, Limnobium spongia, Ludwigia grandiflora, Ludwigia octovalvis, Luziola fluitans, Myriophyllum aquaticum, Myriophyllum laxum, Myriophyllum spicatum, Najas flexilis, Najas gracillima, Najas guadalupensis, Najas minor, Nelumbo lutea, Nitella, Nuphar, Nymphaea odorata, Nymphoides aquatica, Nyssa aquatica, Ottelia alismoides, Pistia stratiotes, Pontederia cordata, Potamogeton amplifolius, Potamogeton crispus, Potamogeton diversifolius, Potamogeton foliosus, Potamogeton nodosus, Potamogeton perfoliatus, Potamogeton pusillus, Potamogeton spirillus, Potamogeton zosteriformis, Sagittaria kurziana, Sagittaria latifolia, Salvinia minima, Spirodela polyrrhiza, Stuckenia*

pectinata, Taxodium distichum, Trapa natans, Utricularia foliosa, Utricularia gibba, Vallisneria americana, Xyris.
Use by wildlife: *Hydrilla verticillata* is grazed by Suwannee cooters (Reptilia: Testudines: Emydidae: *Pseudemys concinna suwanniensis*), which on average consume 850 g (WM) [68 g DM] turtle⁻¹ day⁻¹. *Hydrilla* collections from the southern United States have documented as many as 131 associated insect (Insecta) species. Of these, the plants are damaged or grazed primarily by aphids (Hemiptera: Aphididae: *Rhopalosiphum nymphaeae*), caddisflies (Trichoptera: Leptoceridae: *Leptocerus americanus, Nectopsyche tavara, Oecetis cinerascens, Triaenodes*), midges (Diptera: Chironomidae: *Ablabesmyia rhamphe, Apedilum elachistus, Cricotopus bicinctus, C. tricinctus, C. politus, C. sylvestris, Dicrotendipes, Endochironomus, Glyptotendipes, Larsia decolorata, Nanocladius alternantherae, Parachironomus hazelriggi, Pentaneura inconspicua, Pseudochironomus richardsoni, Tanytarsus buckleyi*), moths (Lepidoptera: Crambidae: *Oxyelophila callista, Paraponyx allionealis, P. diminulatis, P. obscuralis, Synclita obliteralis*), and shore flies (Diptera: Ephydridae: *Hydrellia bilobifera*). The plants also are eaten by grass carp (Vertebrata: Teleostei: Cyprinidae: *Ctenopharyngodon idella*). Although hydrilla typically is viewed as a noxious weed, some studies have found enhanced benthic productivity where the plants have been established in previously unvegetated zones. The plants also have been found to increase foraging access in several bird (Aves) species, including American coots (Rallidae: *Fulica americana*), ducks (Anatidae), and pied-billed grebes (Podicipedidae: *Podilymbus podiceps*). *Hydrilla* is used for nest construction by the invasive brown hoplo catfish (Vertebrata: Teleostei: Callichthyidae: *Hoplosternum littorale*). *Hydrilla* plants also host epiphytic Cyanobacteria (Notocales: Hapalosiphonaceae: *Aetokthonos hydrillicola*), which produce a toxin known to cause avian vacuolar myelinopathy in the federally endangered Florida snail kite (Aves: Accipitridae: *Rostrhamus sociabilis*).
Economic importance: food: *Hydrilla* foliage contains about 19–25% (dw) crude protein and from 0.1% to 2% of 16 amino acids (nine essential); however, it has not been used as a human food; **medicinal:** *Hydrilla* extracts contain natural antioxidants (free radical scavenging activity: 29.6% inhibition at 100 µg ml⁻¹) and exhibit antimicrobial activity (primarily against Gram-positive bacteria); **cultivation:** *Hydrilla verticillata* has been popular as an aquarium plant, but due to its aggressive growth characteristics, its cultivation is emphatically discouraged anywhere beyond its indigenous range; **misc. products:** *Hydrilla* plants have been used for the phytoremediation of aqueous solutions contaminated by dyes and heavy metals and for domestic wastewater treatment. They also have been evaluated for compost production, biofuel production, and in the phytofabrication of silver nanoparticles. Feed for domestic laying hens (Aves: Phasianidae: *Gallus gallus domesticus*) has been supplemented by including up to 7.5% hydrilla without inducing any deleterious dietary effects. *Hydrilla* plants also have been considered as a viable feed for domestic cattle

(Mammalia: Bovidae: *Bos taurus*); **weeds:** *Hydrilla verticillata* is listed as a Federal noxious weed in the United States and is regarded as a notorious weed throughout its nonindigenous range. Some Florida populations exhibit varying levels of resistance to carotenoid biosynthesis inhibiting herbicides (e.g., Fluridone), which arise from somatic mutations occurring at a single codon in the phytoene desaturase (*pds*) gene; **nonindigenous species:** *Hydrilla* is nonindigenous throughout the New World. It was introduced to North America in the mid-20th century (~1959) through the careless disposal of aquarium plants. The nonindigenous New World range of *Hydrilla* has expanded southward into Central America and more recently (~2005) to South America (Brazil). Molecular data indicate that the US "monoecious biotype" originated from China or South Korea, and the US "dioecious biotype" from India. Recently, a third genetic strain has been documented from the United States (Connecticut River, CT), which is similar genetically to plants occurring in Ireland, Japan, Korea, and Latvia. Adventive *Hydrilla* populations also occur in Ireland, New Zealand, and South Africa. Ecological niche models have predicted a future increase of about 17% in potential North American hydrilla habitat due to changing climatic conditions.
Systematics: Phylogenetic analyses typically place the monotypic *Hydrilla* within subfamily Hydrilloideae of Hydrocharitaceae, either in proximity of *Nechamandra* and *Vallisneria* (e.g., Figure 1.8), or closer to *Najas*, depending on the data source and method of analysis implemented. A relationship to *Najas* has been suspected based on the shared occurrence of unusual abaxial midvein teeth in both genera, a morphological feature that occurs in no other alismatid genus. The genetic structure of *Hydrilla verticillata* has been studied quite extensively by analyses of various markers (e.g., isozymes, ISSRs, RAPDs) as well as by nuclear and plastid DNA sequence data. Isozyme data provided early genetic evidence for two independent origins of hydrilla plants in North America, which since have been characterized as dioecious and monoecious "biotypes" with respect to their corresponding sexual conditions. These two, distinct, nonindigenous North American strains have been confirmed subsequently by various genetic data; however, recent studies also have discovered a third strain (introduced to New England, USA) for which the sexual condition has not yet been determined (Figure 1.11). Although several *Hydrilla* haplotypes (5–9) have been described on the basis of plastid sequence data, analyses of nuclear sequence data have disclosed that substantially greater genetic diversity exists in the species due to extensive hybridization, which has occurred among all the strains throughout their Old World range. Even though the US dioecious and monoecious biotypes (Figure 1.11) represent specific haplotypes (i.e., H1/H2 and H5, respectively), both were derived initially through hybridization involving other Old World haplotypes. However, field studies and genetic analyses conducted at one North American reservoir indicate that hybridization between the dioecious and monoecious strains did not occur there, despite their sympatric occurrence. The basic chromosome

H1/H2 (USA dioecious)
H5 (USA monoecious)
H3/H4
H6/H7
H8/H9 (USA Connecticut River)

FIGURE 1.11 Genetic substructure in *Hydrilla verticillata* as illustrated by plastid *trnL-F* haplotype sequences (adapted from Tippery et al., 2020). Geographical affinities of coded haplotypes are as follows: H1/H2 (Brazil, Burundi, China, India, Nepal, Pakistan, the United States [dioecious strain], Vietnam); H3/H4 (Australia, China, New Zealand); H5 (China, Korea, the United States [monoecious strain]); H6/H7 (Australia, China, Indonesia, Malaysia, South Africa, Taiwan, Thailand, Vietnam); H8/H9 (China, Ireland, Japan, Korea, Latvia, Poland, the United States [Connecticut River]). The three nonindigenous US strains are highlighted in bold. Extensive hybridization has occurred among these basic strains throughout various portions of the Old World (see text).

number of *Hydrilla* is $x = 8$. *Hydrilla verticillata* contains diploid, triploid, and tetraploid cytotypes ($2n = 16, 24, 32$), with North American populations commonly being triploid. Chromosome variation in *Hydrilla* originates at least in part by endopolyploidy, which has been indicated by documentation of triploid plants at sites where only diploids had occurred previously, and moreover, by the observation of different polyploid cells ($2n = 16, 24, 32$) found within single, clonal populations or even within individual root tips. *Hydrilla* occurrences recently have been detected successfully from environmental DNA (eDNA) samples.

Distribution: *Hydrilla verticillata* occurs in the eastern and western United States, with a southward extension into South America (Brazil).

References: Abdul-Alghaffar & Al-Dhamin, 2016; Adler et al., 2018; Balciunas & Minno, 1985; Baniszewski et al., 2016; Basiouny et al., 1978; Benoit, 2011; Benoit & Les, 2013; Benoit et al., 2019; Bianchini, Jr. et al., 2010; Bowes et al., 1977; 1979; Boyd, 1969; Canfield et al., 1985; Cook, 1982; 1996a; Cook & Lüönd, 1982a; Dodd et al., 2016; Easley & Shirley, 1974; Esler, 1990; Evans & Wilkie, 2010; Godfrey & Wooten, 1979; Grajczyk et al., 2009; Haller et al., 1976; Haug et al., 2019; Haynes, 2000b; Holaday & Bowes, 1980; Hoyer et al., 1996; Kanabkaew & Udomphon, 2004; King & Les, 2016; Lal & Gopal, 1993; Langeland, 1989; Langeland & Sutton, 1980; Langeland et al., 1992; Les et al., 1997c; Low et al., 1994; Madeira et al., 2000; Maki & Galatowitsch, 2008; Matsuhashi et al., 2019; McDowell et al., 1990; Michel et al., 2004; Miller et al., 1993; Ming et al., 1994; Netherland, 1997; Nico & Muench, 2004; Osborne & Sassic, 1981; Owens et al., 2001; Pandi & Rajkumar, 2015; Pieterse, 1981; Posey et al., 1993; Promdee et al., 2018; Reiskind et al., 1997; Sable et al., 2012; Scott & Osborne, 1981; Simberloff & Leppanen, 2019; Sousa, 2011; Steward, 1991; 1993; Steward & Van, 1987; Stratman et al., 2013; Sutton et al., 1992; Tippery et al., 2020; True-Meadows et al., 2016; Van & Steward, 1990; Van et al., 1977; Verkleij et al., 1983; Williams et al., 2017; Wood & Netherland, 2017; Zhu et al., 2017.

2.3.6. Hydrocharis

Frog-bit; grenouillette, hydrocharide

Etymology: from the Greek *hydro chaeri*, meaning "water delight"

Synonyms: *Bootia* (in part); *Hydrochaeris*; *Monochoria* (in part); *Ottelia* (in part); *Pontederia* (in part)

Distribution: global: Africa, Australia, Eurasia, North America*; **North America:** northeastern

Diversity: global: 3 species; **North America:** 1 species

Indicators (USA): obligate wetland (OBL): *Hydrocharis morsus-ranae*

Habitat: freshwater; lacustrine, palustrine, riverine; **pH:** 6.5–7.8; **depth:** <1 m; **life-form(s):** floating-leaved or free-floating

Key morphology: plants floating freely or rooted in sediment; shoots contracted and rosulate or stoloniferous (to 20 cm), bearing simple or branched adventitious roots (to 50 cm), and terminating in summer buds (to 15 mm) or turions (to 9 mm); leaves aerenchymatous on underside, the blades (to 6.3 cm) orbicular, cordate at base, petioled (to 14 cm), with paired, free stipules (to 2.5 cm); flowers trimerous, unisexual, the plants dioecious or (rarely) monoecious; ♂ flowers (1–5) pedicelled (to 4 cm), arising from a peduncled (to 5.5 cm) spathe (to 1.2 cm); sepals (to 5.5 mm) greenish or white; petals (to 1.9 cm) white, obovate to orbicular; stamens (9–12) in 4 whorls; ♀ flowers solitary, pedicelled (to 9 cm); sepals (to 5 mm) greenish; petals (to 1.5 cm) white, obovate to orbicular, 3 staminodes present, styles (to 5 mm) 6, bifid (to ⅔ of length), ovary spherical, 6-carpelled; capsules (to 1 cm) berrylike, with numerous, bluntly papillate seeds (to 1.3 mm)

Life history: duration: perennial (stolons, turions); **asexual reproduction:** stolons, tubers, turions (stolons); **pollination:** insect; **sexual condition:** dioecious, monoecious; **fruit:** berrylike capsule (infrequent); **local dispersal:** seeds, turions (water); **long-distance dispersal:** whole plants or fragments (birds, humans [boating and fishing equipment]); seeds, turions (water, waterfowl)

Imperilment: 1. *Hydrocharis morsus-ranae* [GNR]

Ecology: general: All *Hydrocharis* species are perennial, floating-leaved aquatics. The plants grow either as free-floating rosettes on the water surface or can become rooted in the sediment when in shallower water. The flowers are unisexual and are arranged in a dioecious or monoecious condition; single rosettes mostly bear flowers of the same sex. The flowers of both sexes possess showy perianths, produce nectar (♀ only), and are pollinated primarily by insects. Vegetative reproduction can occur by the extension of stolons and also by the production of tubers during the summer, and by turions, which detach and overwinter. The characteristics of all three species are quite similar and are exemplified well by the following species.

2.3.6.1. *Hydrocharis morsus-ranae* L. is a nonindigenous species, which occupies shallow, mesotrophic North American waters (e.g., 0.5–0.6 m) associated with canals, channels, ditches, floodplains, impoundments, lagoons, lakes, marshes, ponds, rivers, streams, and swamps at elevations to 378 m (to 1500 m elsewhere). The plants have been observed in open and shaded sites (usually sheltered) and in clear to turbid waters, which often are rich in organic matter. Shading has little influence on growth unless levels exceeding 50% are maintained. The plants have been described as favoring both calcium-poor waters and calcareous waters; North American sites represent circumneutral conditions (pH: 6.5–7.8). The plants do not appear to be well adapted to acidic, nutrient-poor (oligotrophic) conditions. They have been reported on substrates described as clay, detritus, mud, and rocky sandy silt. Flowering occurs from mid-June through August (North America). Floral sex expression is quite labile. Typically, the rosettes are unisexual (the plants being dioecious); however, a plant infrequently can produce both male and female rosettes (the plant then monoecious). Plants of a known sex also have been observed to produce plants of the opposite sex, which were derived vegetatively from their turions. The flowers open by early morning, become receptive by mid-day, and are pollinated by insects (see *Use by wildlife*). Despite the unisexual flowers and prevalent dioecy, genetic data (allozyme markers) have indicated low levels of variation and a relatively high degree of inbreeding. When formed, the maturing fruits are pushed beneath the water surface by a curving peduncle. Seed production often is uncommon (due in part to low pollinator visitation rates), but reportedly initiates in mid-August. It can be locally abundant (over 250 m^{-2}), reaching 2,000–3,000 seeds m^{-2} in some cases; however, few seedlings survive, even in such cases. An average of 26–42 seeds (to a maximum of 74 seeds) is produced per flower. Individual seed production generally is greater (e.g., 55–60 seeds) in high-density mats than in low density mats (e.g., 15–20 seeds). The seeds are enclosed in mucilage and are released when the capsules swell and burst after about 5 weeks following peduncle curvature. The seeds (averaging 0.17 g) float upon their release and are dispersed primarily by water currents, although the mucilaginous coating also is suspected to facilitate their attachment to the feet of waterfowl (Aves: Anatidae). The seeds germinate as the water temperatures rise above 15°C. A fairly high experimental seed germination rate (69%) has been achieved after 16 months of cold storage and subsequent exposure to high light levels. The seeds are known to retain viability after 2 months of storage at −20°C. The principal means of reproduction is vegetative through the production of stolons and turions. Individual plants can produce up to 150 turions. Between 15°C and 25°C, the turions are initiated by photoperiod, with higher temperatures requiring shorter photoperiods. No turions form below 10°C, but they develop readily at temperatures above 25°C regardless of photoperiod. Turions can remain dormant for at least 2 years, with greater dormancy characteristic of those that initiate at higher temperatures. Mature turions detach and sink to the bottom, where they remain for about 7 months; they cannot withstand prolonged periods of freezing (i.e., >10 days to several weeks). Turions require light for germination unless they have been thoroughly vernalized (i.e., cold stratified). The turions germinate from April to May and float to the surface. Individual turions can develop into fairly large plants (to 1 m^2) within a single growing season. Great blue herons (Aves: Ardeidae: *Ardea herodias*) have been seen with whole, entangled plants attached for significant distances (e.g., 2 km) and turions (of uncertain viability) have been recovered from the digestive tracts of American black ducks (Anatidae: *Anas rubripes*). **Reported associates (North America):** *Butomus umbellatus, Carex, Cephalanthus occidentalis, Equisetum fluviatile, Hottonia inflata, Lemna minor, Lythrum salicaria, Myriophyllum sibiricum, Myriophyllum spicatum, Nuphar variegata, Nymphaea odorata, Phalaris arundinacea, Phragmites australis, Pontederia cordata, Potamogeton crispus, Potamogeton epihydrus, Potamogeton pusillus, Potamogeton vaseyi, Sagittaria graminea, Sagittaria rigida, Salix, Scirpus, Sparganium, Spirodela polyrhiza, Trapa natans, Typha angustifolia, Typha latifolia, Utricularia macrorhiza.*

Use by wildlife: *Hydrocharis morsus-ranae* is eaten by various ducks (Aves: Anatidae), grass carp (Teleostei: Cyprinidae: *Ctenopharyngodon idella*), mice (Mammalia: Rodentia: Muridae), and snails (Mollusca: Gastropoda: Lymnaeidae: *Lymnaea*; Physidae: *Physa*; Planorbidae: *Helisoma*). The leaves also are eaten by larval moths (Insecta: Lepidoptera: Crambidae: *Elophila icciusalis, Synclita occidentalis*). In North America, the flowers are visited by numerous insects (Insecta) including aphids (Hemiptera: Aphididae) and flies (Diptera: Ephydridae: *Hydrellia, Notiphila*), although the principal pollinators appear to be fairly specialized hover flies (Diptera: Syrphidae: *Toxomerus marginatus*) and solitary bees (Hymenoptera: Halictidae: *Lasioglossum*).

Economic importance: food: *Hydrocharis* plants contain from 22% to 24% crude protein, but have been described as "almost tasteless" and are of little culinary interest. They impart a reddish coloration to cooking water; **medicinal:** no uses reported; **cultivation:** *Hydrocharis morsus-ranae* is grown as an ornamental aquarium and water garden plant; **misc. products:** *Hydrocharis morsus-ranae* is easily cultivated and has been used as an experimental plant for physiological and developmental studies; **weeds:** *Hydrocharis morsus-ranae* can develop into dense floating mats and has been reported as a nuisance in Canada (Ontario and Quebec) and the United States (New York); **nonindigenous species:** In North America, *Hydrocharis morsus-ranae* was originally cultivated by an arboretum in Ottawa, Canada from material sent from Zürich, Switzerland in 1932; it was first observed as an escape in the region in 1939. The subsequent spread of plants through North America (at a rate of roughly 17 km year^{-1}) has been facilitated by their dispersal on boats and boat trailers. Plants also have been observed for sale at some local retail shops in the United States, and likely also have escaped from cultivation in water gardens.

Systematics: Analyses of DNA sequence data for two of the three *Hydrocharis* species indicate that the genus is related most closely to (but not necessarily distinct from) *Limnobium* (Figure 1.8). However, a simultaneous phylogenetic analysis of all three *Hydrocharis* species and both *Limnobium* species will be necessary to determine whether these two genera actually resolve as distinct sister clades, which appears to be the case when fewer exemplars are evaluated (Figure 1.12). Such an analysis also is necessary to determine the interspecific

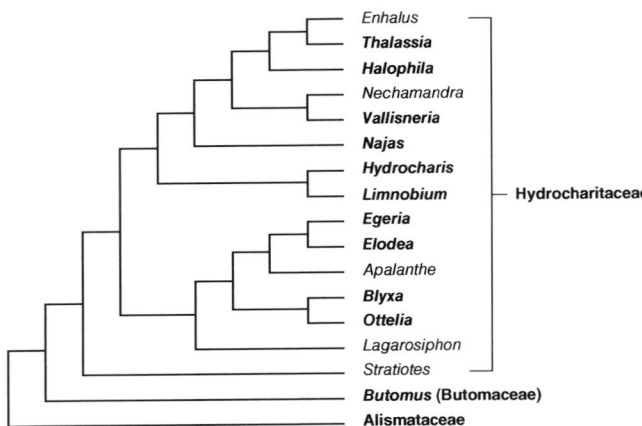

FIGURE 1.12 Intergeneric relationships in Hydrocharitaceae based on analysis of complete cpDNA gene sequence data (adapted from Ross et al., 2016). Although only those taxa designated as OBL in North America are indicated in bold, the entire group shown comprises aquatic species.

relationships in *Hydrocharis*. DNA sequence data indicate that *Hydrocharis* and *Limnobium* have diverged approximately 11.7 million years before present (MYBP). The base chromosome number of *Hydrocharis* is $x = 7$ or 8. *Hydrocharis morsus-ranae* ($2n = 28$) presumably is tetraploid. No hybrids involving any member of the genus have been reported, due at least in part to the highly allopatric distributions of the three species.

Distribution: *Hydrocharis morsus-ranae* occurs in northeastern North America.

References: Casper & Krausch, 1980; Catling & Dore, 1982; Catling & Porebski, 1995; Catling et al., 1988; 2003; Cook & Lüönd, 1982b; Dore, 1954; 1968; Haynes, 2000b; Lamoureux, 1987; Les & Tippery, 2013; Les et al., 2003; Mühlberg, 1982; Scribailo & Posluszny, 1983; 1984; 1985; Scribailo et al., 1984; Zhu, 2014; Zhu et al., 2014.

2.3.7. Limnobium

American frog-bit, American spongeplant

Etymology: from the Greek *limno bios* ("lake living") in reference to the typical habitat

Synonyms: *Hydrocharella*; *Hydrocharis* (in part); *Hydromystria*; *Jalambicea*; *Rhizakenia*; *Trianea*

Distribution: global: New World (Americas); **North America:** southeastern United States

Diversity: global: 2 species; **North America:** 1 species

Indicators (USA): obligate wetland (OBL): *Limnobium spongia*

Habitat: freshwater; lacustrine, palustrine, riverine; **pH:** 5.6–8.6; **depth:** <1 m; **life-form(s):** emergent, floating-leaved, free-floating

Key morphology: plants floating freely or rooted in sediment; shoots dimorphic: contracted and rosulate, bearing branched adventitious roots (to 60 cm), or stoloniferous (to 15+ cm), terminating in new rosettes or compact winter buds; foliage dimorphic: aerial leaves with flat, leathery blades, and long, erect petioles (to 40 cm); floating leaves with thick spongy

pad of tissue on lower surface, and shorter petioles (0.8–12 cm); blades (to 10 cm), broadly ovate, apex acute, the base often cordate, stipules median, conspicuous (to 8 cm), longer on erect leaves; flowers trimerous, unisexual, the rosettes unisexual or monoecious; ♂ flowers (to 25), pedicelled (to 9 cm), arising from a peduncled (to 3.5 cm) spathe (to 6.3 cm); sepals (to 9 mm) greenish or yellowish, becoming reflexed; petals (to 9.4 mm) whitish, linear to lanceolate, stamens (3–16) inserted on a column, in 5–6 whorls, the latter three usually staminodal; ♀ flowers (to 6) pedicelled (to 12 cm, but elongating further in fruit), arising from a peduncled (to 1.5 cm) spathe (to 5.5 cm); sepals (to 9.3 mm) elliptical, spreading; petals (to 14 mm) spreading, greenish-white; staminodes (to 3.3 mm) 2–6; ovary (to 8.5 mm) unilocular, styles (to 20 mm) 6–9, divided to ¼ their length; capsule (to 18 mm) oblong, berry-like; seeds (to 2.3 mm) numerous (to 200), covered by spike-like trichomes (to 0.4 mm)

Life history: duration: perennial (stolons, turions, whole plants); **asexual reproduction:** stolons, turions; **pollination:** wind; **sexual condition:** dioecious, monoecious; **fruit:** capsule (berrylike) (common); **local dispersal:** water (seeds, turions); **long-distance dispersal:** waterfowl (seeds, seedlings)

Imperilment: 1. *Limnobium spongia* [G4]; S1 (IL, IN, MD, NJ, WV); S2 (DE, KY, MO, NC); S3 (GA, VA)

Ecology: general: Both *Limnobium* species are perennial, emergent or floating-leaved aquatics. As in *Hydrocharis*, the plants can float freely as rosettes on the water surface (and grow in mats of other floating vegetation), or become rooted in the sediment when growing in sufficiently shallow water or on exposed substrates. Floating leaves develop on plants growing in more than 10 cm of water; the roots can anchor in the substratum when the water depth is below 60 cm. Vegetative plasticity (e.g., see above description) can be fairly extensive and represents an adaptive response to the various habitats (e.g., exposed sites vs. inundated sites) in which the plants may be found. The root caps are easily removed without damage, and are thought to facilitate the liberation of rooted plants from the sediment as water levels rise. The flowers are unisexual and although primarily of mixed sexes (the plants then monoecious), a single rosette also can be unisexual (individual plants, or at least separate ramets, then potentially dioecious). Unlike *Hydrocharis*, the flowers are not showy, lack nectar, and are pollinated abiotically by the wind.

2.3.7.1. ***Limnobium spongia* (Bosc) Rich. ex Steud.** grows in shallow (e.g., 0.5–100 cm), still or slowly moving waters, or on exposed, saturated substrates associated with baygalls, bayous, borrow pits, bottomlands, canals, channels (roadside), ditches (roadside), flats, floodplains, gravel bars, lakes, marshes (depression), meadows (flooded), mudflats, ponds, prairies, reservoirs, rivers, roadsides, shell pits, sinkholes, sloughs, springs (edges), streams, swales, and swamps at elevations to 204 m. Site exposures can range from full sun to shaded conditions. The substrates have been characterized as alluvium, clay, entisols (loamy), gravel, muck (Barbary, organic, Shenks), mud, Nittau (typic argiaquolls), peat, sand (coarse), silt, and silty clay. The plants are variable phenotypically, producing long-stalked, erect, coriaceous leaves when emergent, and short-stalked, floating leaves underlain by a broad lacunal pad of air-space tissue when growing in deeper sites. The plants occur primarily in alkaline (pH: 5.6–8.6; $\bar{x} = 7.1$), hard water (total alkalinity: $\bar{x} = 29.9$ mg l⁻¹ [as Ca CO₃]; Ca: $\bar{x} = 29.7$ mg l⁻¹), nutrient-rich (total P: $\bar{x} = 56$ µg l⁻¹; total N: $\bar{x} = 990$ µg l⁻¹) water bodies. Sexual reproduction occurs primarily on emergent plants that are rooted in the mud. Flowering has been observed from February to October with fruiting reported from June to December. Most rosettes are monoecious. Pollination is abiotic and mediated by the wind. Dehisced, dry pollen collects in the perianth, but is heavy and can fall onto lower occurring pistillate flowers to effect self (geitonogamous)-pollination. Although some reports indicate the possibility of water pollination, that method seems unlikely and never has been demonstrated. The peduncle recurves after pollination, enabling the developing fruit to mature below the water surface or become pressed into the soft muddy substrate. The seeds are released in a gelatinous mass and germinate while they remain underwater; however, the seedlings float to the surface, where they are dispersed by water currents. A seed bank of uncertain persistence develops. The seeds have germinated successfully under ambient greenhouse conditions after 3 months of cold stratification (7°C). Seeds and seedlings conceivably could be transported via exozoic (external) transport, e.g., while in mud attached to the feet of waterfowl (Aves: Anatidae); however, the dispersal mechanisms for this species have not been elucidated adequately. Although the seeds do not appear to be well adapted for endozoic transport, they have been recovered from the stomachs and feces of waterfowl (Aves: Anatidae) and crocodilians (Reptilia: Alligatoridae: *Alligator mississippiensis*) and possibly are dispersed by this means. **Reported associates:** *Acer rubrum, Alnus serrulata, Alternanthera philoxeroides, Azolla cristata, Azolla filiculoides, Bacopa caroliniana, Berchemia scandens, Betula nigra, Bidens cernuus, Bidens laevis, Brasenia schreberi, Cabomba caroliniana, Carex hyalinolepis, Carya, Cephalanthus occidentalis, Ceratophyllum demersum, Ceratophyllum echinatum, Ceratopteris pteridoides, Chasmanthium laxum, Cladium jamaicense, Crataegus viridis, Cyperus odoratus, Decodon verticillatus, Egeria densa, Eichhornia crassipes, Erechtites hieraciifolius, Eupatorium capillifolium, Eupatorium serotinum, Hydrilla verticillata, Hydrocotyle bonariensis, Hydrocotyle ranunculoides, Juncus coriaceus, Juncus diffusissimus, Juncus effusus, Juncus validus, Kosteletzkya, Leersia oryzoides, Leersia virginica, Lemna minor, Lemna perpusilla, Lemna trisulca, Lemna valdiviana, Liquidambar styraciflua, Lonicera japonica, Ludwigia brevipes, Ludwigia hexapetala, Ludwigia peploides, Mikania scandens, Myriophyllum aquaticum, Najas flexilis, Najas guadalupensis, Nelumbo lutea, Nuphar, Nymphaea mexicana, Nymphaea odorata, Nymphoides aquatica, Nyssa biflora, Onoclea sensibilis, Osmunda regalis, Osmundastrum cinnamomeum, Oxycaryum cubense, Panicum hemitomon, Paspalum repens, Persicaria hydropiperoides, Persicaria punctata, Persicaria sagittata, Pistia stratiotes, Pontederia cordata, Potamogeton diversifolius, Quercus nigra, Rhynchospora, Riccia fluitans,* *Sabal palmetto, Sacciolepis striata, Sagittaria lancifolia, Sagittaria platyphylla, Salix nigra, Salvinia minima, Saururus cernuus, Schoenoplectus pungens, Sesbania drummondii, Setaria magna, Smilax rotundifolia, Sparganium americanum, Spirodela polyrhiza, Taxodium distichum, Typha domingensis, Typha latifolia, Utricularia gibba, Viburnum nudum, Wolffia brasiliensis, Wolffia columbiana, Wolffiella gladiata, Woodwardia virginica, Zizaniopsis miliacea.*

Use by wildlife: The fruits and seeds of *Limnobium spongia* are eaten by alligators (Reptilia: Alligatoridae: *Alligator mississippiensis*) and various waterfowl (Aves: Anatidae), including goldeneye (*Bucephala clangula*), green-winged teal (*Anas crecca*), mallard (*Anas platyrhynchos*), oldsquaw (*Clangula hyemalis*), pintail (*Anas acuta*), and ring-necked (*Aythya collaris*) ducks. Although it has not been determined whether the seeds remain viable after consumption by these agents, such feeding activities should at least increase the chance of adhesion and exozoic dispersal. The foliage is readily consumed by snails (Mollusca: Gastropoda: Ampullariidae: *Pomacea insularum*) and by Amazonian manatees (Mammalia: Sirenia: Trichechidae: *Trichechus inunguis*) [the latter account probably actually referring to *L. laevigatum*]. The leaves, petioles, and stolons are damaged by several insects (Insecta), including flies (Diptera: Ephydridae: *Hydrellia limnobii*), grasshoppers (Orthoptera: Acrididae: *Paroxya clavuliger, Romalea microptera*), moths (Lepidoptera: Crambidae: *Munroessa gyralis*), and weevils (Coleoptera: Curculionidae: *Bagous lunatoides, M. icciusalis*). The plants also are host to plant pathogenic Fungi (Ascomycota: Mycosphaerellaceae: *Cercospora limnobii*). The foliage provides resting sites for some aquatic insects (e.g., Insecta: Hempitera: Gerridae: *Limnoporus canaliculatus*). The roots are known to harbor resting stages of various freshwater zooplankton.

Economic importance: food: not reported as edible; **medicinal:** no uses reported; **cultivation:** *Limnobium spongia* is easily cultivated and is grown in artificial garden ponds, often as a shade plant to suppress the growth of submersed species; **misc. products:** none; **weeds:** Populations of *Limnobium spongia* are often characterized as weedy, even within its indigenous range; **nonindigenous species:** *Limnobium spongia* is suspected of being introduced to portions of the United States outside of its indigenous range. *Limnobium laevigatum* has reportedly escaped from a few ornamental ponds in California; however, it does not appear to be established in that area.

Systematics: Molecular data (e.g., Figure 1.8) resolve *Limnobium* as a clade, which is closely associated with *Hydrocharis*. Although some authors have advocated to merger of the two *Limnobium* species and their recognition as subspecies, the taxa appear to be clearly differentiated genetically. Less certain is whether *Hydrocharis* and *Limnobium* should be merged or retained as distinct genera; the groups differ primarily by a suite of adaptations related to a shift from insect to wind pollination and often are extremely difficult to differentiate when in vegetative state. The basic chromosome number of *Limnobium* is $x = 6$ or 7; *L. spongia* ($2n = 24$) presumably is a tetraploid. Numerous aneuploid counts

($2n = 26$–29) have been observed in the genus, which appear to reflect actual differences rather than technical errors. The two species are distinctly allopatric and no hybrids have been reported.

Distribution: *Limnobium spongia* occurs primarily in the southeastern United States, but recently has been extending northward.

References: Baker et al., 2010; Battauz et al., 2017; Bernardello & Moscone, 1986; Bowles, 2013; Cook & Urmi-König, 1983; Dammer, 1888; DiTomaso & Healy, 2003; Guterres-Pazin et al., 2014; Harms & Grodowitz, 2009; Haynes, 2000b; Herring, 1950; Howard & Wells, 2009; Hoyer et al., 1996; Les & Capers, 1999; Lowden, 1992; Mohlenbrock, 1959a; Moscone & Bernardello, 1985; Mühlberg, 1982; Platt et al., 2013; Smith & Capinera, 2005; Smock et al., 1981; Wilder, 1974.

2.3.8. Najas

Naiad, water nymph, water sprite; naïade

Etymology: after the naiads, the freshwater nymphs of Greek mythology

Synonyms: *Caulinia* (in part); *Fluvialis*; *Ittnera*

Distribution: global: cosmopolitan; **North America:** widespread

Diversity: global: 40 species; **North America:** 9 species

Indicators (USA): obligate wetland (OBL): *Najas canadensis, N. filifolia, N. flexilis, N. gracillima, N. graminea, N. guadalupensis, N. marina, N. minor, N. wrightiana*

Habitat: brackish (coastal, inland); freshwater; freshwater (tidal); lacustrine, riverine; **pH:** 5.6–10.2; **depth:** to 14 m; **lifeform(s):** submersed (vittate)

Key morphology: shoots (to 2.1+ m) limp to somewhat rigid, branching distally, internodes smooth or prickly; leaves pseudowhorled but appearing opposite, the blades (to 3.9 cm) sessile, linear, flat or recurved, flexuous or stiff, the margins entire or strongly toothed, coarsely or weakly denticulate, abaxial midvein smooth or with spines, leaf sheaths auriculate, tapered, or truncate; flowers (1–3) axillary, unisexual (the plants dioecious or monoecious), minute (to 4 mm), naked; ♂ flowers in membranous spathe(s) (1–2), consisting of a single stamen, the anthers 1–4 locular; ♀ flowers sessile, occasionally within a membranous spathe, gynoecium unilocular, styles branched (3–4 parted); achenes (to 4.5 mm) ovoid to fusiform, straight or recurved, the surface smooth and shiny or areolate

Life history: duration: annual (fruits/seeds); perennial (dormant apices, whole plants); **asexual reproduction:** shoot fragments; **pollination:** water (hypohydrophily); **sexual condition:** dioecious, monoecious; **fruit:** achenes (common); **local dispersal:** fish (seeds), water (shoot fragments, seeds); **long-distance dispersal:** boating equipment (plant fragments, seeds), waterfowl (seeds)

Imperilment: 1. *Najas canadensis* [unranked] S1 (NY); **2.** *N. filifolia* [G1]; S1 (FL, GA); **3.** *N. flexilis* [G5]; SH (MO); S1 (AK, KY, PE, SC, WY, YT); S2 (AB, NF, SK, WV); S3 (IL, MD); **4.** *N. gracillima* [G5]; S1 (AL, DE, IN, NS, OH, VT); S2 (IL, KY, MO, NB, NC, ON, WV); S3 (ME, MN, NJ); **5.**

N. graminea [GNR]; **6.** *N. guadalupensis* [G5]; SH (DE); S1 (ME, MT, ND, NH, QC, UT, WV, WY); S2 (ON, VT); S3 (IL, MD, NC, NJ, ON); **7.** *N. marina* [G5]; S1 (IL, NE, ND, NY, ON); S2 (UT); S3 (IN, MN); **8.** *N. minor* [GNR]; **9.** *N. wrightiana* [GNR];

Ecology: general: All *Najas* species are obligately submersed aquatics, which complete virtually every aspect of their life cycle (except dispersal) beneath the surface of standing waters. This is a taxonomically problematic genus, making it difficult at times to associate published research reports with the correct taxon. An additional species (*N. canadensis*) has been added to the list of OBL indicators, given its fairly recent disclosure as a cryptic (but distinct), allopolyploid derivative of *N. flexilis* (see *Systematics*). Many North American (and all European) accounts of "*N. flexilis*" actually refer to *N. canadensis*. Whenever practical, data (e.g., associated species) have been based on accessions verified using DNA sequence data. There are so many misidentified *Najas* specimens that it often is difficult to rely on the reported data; every practical effort (e.g., by examining accompanying specimen images) has been made to minimize such errors. Most of the North American species are annuals, but the most widespread taxon (*N. guadalupensis*) is a perennial, which overwinters as whole plants. Thickened perennating shoot apices (interpreted as turions) have been observed in an Old World population of *Najas marina*, but have not been reported anywhere in its New World range. Turion-like structures also have been observed in *Najas guadalupensis*, but their function has not yet been evaluated adequately. All *Najas* species possess unisexual flowers, which are in a monoecious arrangement in all but one case (*N. marina*), which is dioecious. Pollination in all *Najas* species takes place entirely beneath the water surface (hypohydrophily); however, the proximity of male and female flowers often results in within-plant pollen transfer and inbreeding (i.e., geitonogamy). Although the extent of outcrossing appears to be relatively low, several interspecific hybrids have been documented genetically (see *Systematics*), which provides clear evidence of xenogamy. Even though most species are annuals, they are able to reproduce vegetatively during the growing season by fragmentation of the often brittle shoots (many containing seeds), which at least facilitates within-lake dispersal. Seed production in the annual species often is prolific, and extensive seed banks can develop. The seeds can be stored successfully in water kept at 3°C. They are relished by numerous species of waterfowl (Aves: Anatidae), which disperse them over wide distances through endozoic (unlikely as the fragile seeds normally are destroyed during digestion) or exozoic transport. However, endozoic dispersal can occur when the seeds are ingested by various fish (Vertebrata: Teleosti: Cichlidae, Cyprinidae). Vegetative fragments are able to facilitate among-lake dispersal, when they become attached inadvertently to boating and other water-recreational equipment.

2.3.8.1. ***Najas canadensis*** Michx. is a common annual, which inhabits shallow to moderately deep (e.g., 0.1–2.0 m) standing waters in canals, lakes, ponds, rivers (slow-moving), and streams at elevations to 702 m. Due to its relatively recent disclosure as distinct from *N. flexilis*, few limnological data are available; however, the two species are broadly sympatric and occur commonly within the same lakes and habitats, which indicates their similar ecological tolerances (see *Najas flexilis* below). The substrates have been characterized as cobble, muck, mucky sand, mucky silt, sand, sandy gravel, sandy muck, silt, and silty sand. The plants are monoecious, with flowering and fruiting observed from June to September. Fragments of plants (often with seeds attached) are observed commonly during the growing season and undoubtedly facilitate their dispersal within water bodies, at least via the translocation of seeds. The seeds are intolerant of desiccation, completely losing their viability after 2 hr of drying at 15°C. **Reported associates:** *Bidens beckii, Brasenia schreberi, Callitriche, Ceratophyllum demersum, Chara, Egeria densa, Elatine minima, Eleocharis acicularis, Elodea canadensis, Elodea nuttallii, Equisetum, Eriocaulon aquaticum, Glossostigma cleistanthum, Gratiola aurea, Heteranthera dubia, Hydrodictyon reticulatum, Isoetes, Lemna, Lobelia dortmanna, Ludwigia palustris, Myriophyllum heterophyllum, Myriophyllum sibiricum, Myriophyllum spicatum, Myriophyllum tenellum, Myriophyllum verticillatum, Najas flexilis, Najas gracillima, Najas guadalupensis, Najas marina, Najas minor, Nitella, Nuphar polysepala, Nuphar variegata, Nymphaea odorata, Potamogeton amplifolius, Potamogeton berchtoldii, Potamogeton bicupulatus, Potamogeton crispus, Potamogeton diversifolius, Potamogeton foliosus, Potamogeton illinoensis, Potamogeton nodosus, Potamogeton praelongus, Potamogeton pusillus, Potamogeton richardsonii, Potamogeton robbinsii, Potamogeton spirillus, Potamogeton vaseyi, Ranunculus, Sagittaria, Stuckenia pectinata, Utricularia macrorhiza, Utricularia radiata, Vallisneria americana, Wolffia.*

2.3.8.2. *Najas filifolia* R.R. Haynes is a rare annual, which usually occurs in shallow waters (but to 3.0 m) of lakes, ponds (spring-fed), rivers (still bays), and streams (tidal) at elevations to 45 m. The waters often are tannin stained (darkwater), but the plants also thrive in crystal-clear waters. The substrates include clay and sand. The plants are monoecious, with flowering observed in June and fruiting occurring from August to October. Further details on the life history of this species are scarce. **Reported associates:** *Brasenia schreberi, Cabomba, Didiplis diandra, Eleocharis baldwinii, Ludwigia arcuata, Mayaca fluviatilis, Myriophyllum heterophyllum, Nitella, Nuphar advena, Nymphaea odorata, Nymphoides aquatica, Typha, Utricularia gibba.*

2.3.8.3. *Najas flexilis* (Willd.) Rostk. & W.L.E. Schmidt inhabits shallow to deep waters (0.1–14.0 m; $\bar{x} = 1.2$ m) within lakes (marl), ponds, and rivers at elevations to 1153 m. This plant is regarded as a typical "hard water" species, but one that also can occur in soft waters. The waters characteristically are clear and alkaline (pH: 6.0–10.2; $\bar{x} = 7.4$–8.9; median: 8.3; total alkalinity: 5–308 mg l^{-1} as $CaCO_3$; $\bar{x} = 36.5$ mg l^{-1}; median 122.5 mg l^{-1}; conductivity: 10–686 μmhos cm^{-1}; $\bar{x} = 461$ μmhos cm^{-1}; SO_4 to 308 ppm). The plants are intolerant of turbidity (disappearing from sites with declining water clarity) and usually require at least 3.1% of surface illumination to

persist, although they have been found to occur at even lower light values (0.5–1.0%). They often grow at relatively shallow depths (e.g., 1.2 m) but also have been found in waters up to 12–14 m deep. The chloroplast genome has lost the entire NDH gene complex, contains a unique insert in a photosystem gene spacer region, and has a highly modified RNA polymerase, all features which have been interpreted as physiological adaptations to the light regime of aquatic habitats. The substrates are described as gravel, gravelly sand, gravelly silt, marl, muck, sand (marly), sandy gravel, sandy gravel silt, sandy muck, sandy silt (marly), sandy silty muck, sand, sandy gravel, silty muck, and silty sand. The shoots have been shown to obtain more than 99% of their phosphorous from the sediments. The plants are monoecious and water pollinated (hypohydrophilous), with the male and female flowers borne in close proximity. Although within-plant pollen transfer (geitonogamy) probably occurs frequently, the extent of hybridization in this species (see *Systematics*) also provides evidence of xenogamy. A description of an embryo-like structure that developed prior to pollination raises some suspicion that the plants might be at least facultative apomicts. Flowering commences in early summer with fruiting observed by late summer through August. Seed densities in the field as high as 11 seeds m^{-2} have been reported. A persistent seed bank (of undetermined duration) occurs, with densities of up to 17.9 seedlings m^{-2} obtained from some wetland cores. The seeds are somewhat desiccation sensitive but have been stored successfully (optimally at 11–15% moisture content) at $-20°C$ (3 months) and in liquid nitrogen (1 week). They have germinated successfully (to 20%) in lake water in the dark at 20°C–25°C after 2 months of cold stratification at 4°C. Light (e.g., 600 lux) inhibits germination but a brief exposure to heat (35°C–42°C for 24–72 hr) enhances germination rates by nearly twofold levels. Dispersal within water bodies occurs from the movement of seeds by water currents, or from the reestablishment of vegetative fragments produced during the growing season. Seeds are the primary propagules for long-distance dispersal and are transported (at least exozoically) by the numerous waterfowl (Aves: Anatidae), which feed upon them. Because of their annual habit, the plants are regarded as a pioneer species capable of invading disturbed areas. They are particularly resistant to extended winter drawdown conditions, are known to rapidly colonize formerly acidic lakes that have been treated with lime, and have become dominant in areas of lakes that have been dredged. Field densities (standing crops) can reach 2,500 plants m^{-2}. Experimental studies have shown that the plants attain 60 times greater biomass (>200 g m^{-2}) in fish-free enclosures than in those stocked with fish (<5 g m^{-2}), which indicates a high susceptibility to herbivory. **Reported associates:** *Bidens beckii, Butomus umbellatus, Cabomba caroliniana, Callitriche, Ceratophyllum demersum, Chara braunii, Chara virgata, Chara vulgaris, Eleocharis acicularis, Elodea canadensis, Elodea nuttallii, Eriocaulon aquaticum, Gratiola aurea, Heteranthera dubia, Hippuris, Isoetes, Juncus pelocarpus, Lemna trisulca, Myriophyllum alterniflorum, Myriophyllum heterophyllum, Myriophyllum sibiricum, Myriophyllum spicatum, Myriophyllum tenellum,*

Myriophyllum verticillatum, Najas canadensis, Najas gracillima, Najas guadalupensis, Najas marina, Najas minor, Nitella, Nuphar polysepala, Nymphaea odorata, Nymphaea tetragona, Potamogeton alpinus, Potamogeton amplifolius, Potamogeton epihydrus, Potamogeton foliosus, Potamogeton friesii, Potamogeton gramineus, Potamogeton illinoensis, Potamogeton natans, Potamogeton obtusifolius, Potamogeton praelongus, Potamogeton pusillus, Potamogeton richardsonii, Potamogeton robbinsii, Potamogeton strictifolius, Potamogeton zosteriformis, Ranunculus flammula, Ranunculus longirostris, Ranunculus trichophyllus, Sagittaria graminea, Schoenoplectus subterminalis, Sparganium, Stuckenia pectinata, Utricularia gibba, Utricularia inflata, Utricularia intermedia, Utricularia macrorhiza, Utricularia purpurea, Vallisneria americana, Zannichellia palustris, Zizania aquatica.

2.3.8.4. *Najas gracillima* (A. Braun ex Engelm.) Magnus is an uncommon annual, which includes both indigenous and nonindigenous populations that occur in bays, canals, channels (boggy), lagoons, lakes, ponds (man-made), rice fields, shores, sinkholes, and along the margins of bogs at elevations to 538 m. The plants thrive across a broad depth gradient (0.1–3.6 m; $\bar{x} = 1.5$ m) and can be found as often growing directly along the shore as occurring in deeper waters. The waters typically are clear, cool, oligotrophic (conductivity <90 µmhos cm^{-1}), and soft (pH: <7.5; total alkalinity: <40 mg l^{-1} as CaCO$_3$; $\bar{x} = 4.6$ mg l^{-1}); however, they span a broadly circumneutral range of acidity (pH: 5.6–9.1; $\bar{x} = 6.7$–7.2). Some populations south of the glacial maximum occur in quite warm, turbid, and more eutrophic conditions; such populations also are divergent genetically. The substrates have been described as muck, mucky sand, mud (black), sand, sandy gravelly silt, sandy muck, sandy peat, silt, and silty sand. The plants are monoecious with flowering and fruiting extending from July to September. Fruit production usually is fairly high. The seeds are not particularly durable but are known to be transported endozoically by various waterfowl (Aves: Anatidae). **Reported associates (North America):** *Acorus calamus, Agalinis purpurea, Arethusa bulbosa, Asclepias incarnata, Bidens beckii, Bidens cernuus, Brasenia schreberi, Callitriche palustris, Campanula aparinoides, Carex comosa, Carex hystericina, Carex lacustris, Carex lasiocarpa, Carex scoparia, Carex stricta, Ceratophyllum demersum, Ceratophyllum echinatum, Chamaedaphne calyculata, Chara, Cicuta bulbifera, Cladium mariscoides, Dulichium arundinaceum, Elatine minima, Eleocharis acicularis, Eleocharis palustris, Eleocharis robbinsii, Elodea canadensis, Elodea nuttallii, Equisetum fluviatile, Eriocaulon aquaticum, Eriocaulon aquaticum, Eupatorium perfoliatum, Glyceria borealis, Glyceria canadensis, Glyceria grandis, Glyceria striata, Heteranthera dubia, Heteranthera rotundifolia, Hydrilla verticillata, Iris versicolor, Isoetes echinospora, Isoetes lacustris, Isoetes tuckermanii, Juncus militaris, Juncus pelocarpus, Lemna minor, Lemna trisulca, Lemna turionifera, Littorella uniflora, Lobelia dortmanna, Lysimachia terrestris, Lythrum salicaria, Myriophyllum farwellii, Myriophyllum humile, Myriophyllum sibiricum, Myriophyllum spicatum,*

Myriophyllum tenellum, Myriophyllum verticillatum, Najas canadensis, Najas flexilis, Najas guadalupensis, Najas minor, Nitella, Nuphar variegata, Nymphaea odorata, Nymphoides cordata, Persicaria amphibia, Persicaria coccinea, Phalaris arundinacea, Polygonum achoreum, Pontederia cordata, Potamogeton amplifolius, Potamogeton berchtoldii, Potamogeton bicupulatus, Potamogeton confervoides, Potamogeton diversifolius, Potamogeton epihydrus, Potamogeton foliosus, Potamogeton friesii, Potamogeton gramineus, Potamogeton natans, Potamogeton nodosus, Potamogeton oakesianus, Potamogeton obtusifolius, Potamogeton praelongus, Potamogeton pusillus, Potamogeton robbinsii, Potamogeton spirillus, Potamogeton vaseyi, Potamogeton zosteriformis, Ranunculus flammula, Ranunculus trichophyllus, Rumex britannica, Sagittaria cristata, Sagittaria graminea, Sagittaria latifolia, Sagittaria rigida, Schoenoplectus pungens, Schoenoplectus subterminalis, Schoenoplectus tabernaemontani, Schoenoplectus torreyi, Scirpus microcarpus, Sparganium angustifolium, Sparganium emersum, Sparganium fluctuans, Sphagnum, Stuckenia pectinata, Triadenum fraseri, Utricularia geminiscapa, Utricularia gibba, Utricularia intermedia, Utricularia macrorhiza, Utricularia minor, Utricularia purpurea, Utricularia radiata, Utricularia resupinata, Vallisneria americana, Zizania palustris.

2.3.8.5. *Najas graminea* Delile is a nonindigenous species, which occurs in shallow waters (e.g., 0.5 m) of ditches (irrigation), ponds, reservoirs, and rice fields at elevations to 511 m. In the Old World, the habitats are described as clear, mesoeutrophic waters and the species as being intolerant of hypereutrophic conditions. The plants are, however, quite tolerant of salt (to 255 mM NaCl). Fruiting (North America) has been observed in August. Although pollen:ovule ratios are low (e.g., 400), seed production can be prolific, with as many as 70,045 seeds m^{-2} recovered from some Old World localities (rice fields). In the Old World, the fruits are consumed by several species of waterfowl that are also common to the New World (see *Use by wildlife*). However, it has not been established whether the seeds can survive their digestion. Fluctuating diurnal temperatures reportedly provide cues that trigger underwater seed germination. Germinable seeds also are known to persist in Old World seed banks. Within lakes, the seeds and plant fragments (often containing seeds) are dispersed by water currents. **Reported associates (North America):** none.

2.3.8.6. *Najas guadalupensis* (Spreng.) Magnus is a submersed perennial, which inhabits brackish (chloride: 7.3–1,201.2 mg l^{-1}; Na: 3.9–45.1 mg l^{-1}; salinity: to 10.9‰) or fresh (salinity: <0.5‰), warm (19°C–30°C), still or flowing, shallow to deep waters (0.08–4.5 m) in backwaters (shallow), bayous, borrow pits, canals (irrigation, roadside), channels (spring-fed), coves, ditches (drainage, irrigation, roadside,), estuaries (margins), flats (tidal), floodplains, gravel pits, lakes (marl, spring-fed), marshes (roadside), millponds, ponds (artificial, brackish, farm, irrigation, sinkhole, stagnant), pools (alkaline), quarries (limestone, stone), reservoirs, rice fields, rivers (brackish, swiftly flowing), sand pits (brackish), sloughs, springs (calcareous, intermittent), stock tanks, streams (intermittent, tidal), swamps (shallow), and vernal pools at elevations to 3950 m. These ecologically versatile plants can be found rooted to the bottom or suspended in masses (occasionally quite thick) just beneath the water surface. They occur in sunny and shaded exposures; however, shading of the foliage due to the growth of periphyton has been shown to accelerate their seasonal senescence. The habitats can comprise clear, marly, or brown-stained waters. Typically they are alkaline, hard water sites (pH: 5.9–10.0; $\bar{x} = 7.2$–7.9; total alkalinity: 19–240 mg l^{-1} as $CaCO_3$; conductivity: 80–660 µmhos cm^{-1}; Ca: $\bar{x} = 28$ mg l^{-1}; Mg: $\bar{x} = 12.8$ mg l^{-1}). The substrates have been described as adobe, alluvium (clayey, loamy), clay (black, Schriever, sticky), clay loam, gravel, limestone (Edwards), muck (glacial, light), mucky sand, mucky silt, mud, muddy gravel, muddy ooze, muddy silty clay, rock, sand (coarse, glacial), sandstone, sandy gravel, sandy muck, sandy rock, rocky sandy cobble, sandy muck, sandy silt, silt, silty sand, silty clay loam (Cancienne), and silty sandy gravel. Flowering and fruiting have been observed from January to November, but sexual reproduction is observed infrequently in many collections. When sexual, the plants are monoecious and are pollinated underwater (hypohydrophilous). Sexual reproduction is virtually absent in northern populations (which then must rely exclusively on clonal reproduction), but can be locally common in some southern and western portions of the range (e.g., 78,500 seeds reportedly were produced during one season in a small culture of California plants). The seeds (when present) are probably dispersed by waterfowl (Aves: Anatidae) but usually are too scarce to constitute any significant seed bank. Although often mistakenly characterized as annuals, there have been numerous intact, healthy plants collected during winter (e.g., January–February) from northern ponds (in Ohio, Ontario, and Wisconsin) covered in some cases by 30 cm of ice and 13 cm of snow. Fully grown plants also have been observed in early spring (e.g., April), which further indicates their perennial habit. Some plants produce unusual, thickened axillary buds, which likely function as hibernacula and propagules (and if so, would properly be designated as turions). Fragmentation of the shoots results in prolific local dispersal within water bodies but is difficult to rationalize as a significant means of long-distance dispersal (unless facilitated by human activities). Yet, the enigmatic vagility of this species is evidenced by one record, which reported a collection of plants from within the water inside of a wooden barrel. Populations across North America are diverse genetically, relative to all other *Najas* taxa present (also see *Systematics* below). This species has been expanding northward in recent years (especially in artificial lakes and man-made ponds), and there have been increasing numbers of accounts reporting the development of aggressive populations in that region. **Reported associates:** *Alternanthera philoxeroides, Bacopa monnieri, Brasenia schreberi, Cabomba caroliniana, Callitriche, Cephalanthus occidentalis, Ceratophyllum demersum, Chara globularis, Chara vulgaris, Echinodorus berteroi, Echinodorus cordifolius, Egeria densa, Eichhornia crassipes, Elatine minima, Eleocharis acicularis, Eleocharis quadrangulata, Elodea canadensis, Elodea nuttallii, Glossostigma cleistanthum, Glyceria borealis, Gratiola aurea, Heteranthera dubia, Heteranthera limosa, Hydrilla verticillata, Hydrocotyle verticillata, Hymenoxys odorata, Justicia americana, Leersia hexandra, Lemna minor, Lemna perpusilla, Lemna trisulca, Lemna turionifera, Lemna valdiviana, Lilaeopsis chinensis, Limnobium spongia, Lindernia dubia, Ludwigia octovalvis, Ludwigia palustris, Ludwigia peploides, Ludwigia repens, Lysimachia terrestris, Mayaca fluviatilis, Myriophyllum heterophyllum, Myriophyllum humile, Myriophyllum sibiricum, Myriophyllum spicatum, Myriophyllum verticillatum, Najas canadensis, Najas filifolia, Najas flexilis, Najas gracillima, Najas marina, Najas minor, Nitella, Nelumbo lutea, Nuphar variegata, Nymphaea mexicana, Nymphaea odorata, Nymphoides cristata, Nyssa aquatica, Ottelia alismoides, Paspalum denticulatum, Pistia stratiotes, Potamogeton amplifolius, Potamogeton berchtoldii, Potamogeton bicupulatus, Potamogeton crispus, Potamogeton diversifolius, Potamogeton epihydrus, Potamogeton foliosus, Potamogeton friesii, Potamogeton illinoensis, Potamogeton nodosus, Potamogeton obtusifolius, Potamogeton perfoliatus, Potamogeton praelongus, Potamogeton pusillus, Potamogeton richardsonii, Potamogeton robbinsii, Potamogeton spirillus, Potamogeton vaseyi, Potamogeton zosteriformis, Ranunculus longirostris, Rhynchospora indianolensis, Rumex maritimus, Ruppia maritima, Sagittaria cuneata, Sagittaria graminea, Sagittaria kurziana, Sagittaria longiloba, Sagittaria rigida, Sagittaria subulata, Salvinia minima, Schoenoplectus acutus, Schoenoplectus hallii, Schoenoplectus triqueter, Sparganium emersum, Sparganium fluctuans, Spirodela polyrhiza, Stuckenia pectinata, Taxodium distichum, Typha domingensis, Utricularia foliosa, Utricularia geminiscapa, Utricularia gibba, Utricularia macrorhiza, Utricularia purpurea, Utricularia radiata, Vachellia farnesiana, Vallisneria americana, Vallisneria neotropicalis, Veronica anagallis-aquatica, Wolffia, Zannichellia palustris.*

2.3.8.7. *Najas marina* **L.** is a submersed annual, which occurs in brackish to fresh, shallow to deep waters (0.4–4.5 m) in bays, canals, channels, coves, fens, gravel pits, lagoons, lakes (marl), pits (dolomite mine, rock crusher), ponds (artificial, kettle, marl), pools (channel, hot springs, tidal), quarries (marl), reservoirs, rivers, sloughs, springs (warm), and streams at elevations to 2048 m. Although tolerant of freshwaters, the plants grow optimally at a salt concentration of 37–55 µM NaCl (chloride: >10 mg l^{-1}; conductivity: 280–780 µmhos cm^{-1}; SO_4: 50–1297 mg l^{-1}). Experimentally, the plants reach their light compensation point at 5 µE m^{-2} s^{-1}, which indicates a typical maximum depth of about 1.5 m; in the field, their typical depth range is from 0.4 to 3.0 m ($\bar{x} = 1.6$ m). Field studies also indicate that the plants colonize deeper waters (e.g., 2.3 m) earlier in the season but occupy shallower depths later in the season. Exposures include fully open sites to partial shade. The waters range from clear (often blue-tinged) to somewhat turbid conditions and can reach extremely warm temperatures (to 40°C). Characteristically, they are hard and alkaline or calcareous (pH: 7.8–10.2; $\bar{x} = 8.6$; total alkalinity [as $CaCO_3$]: 89–376 mg l^{-1}). Experimentally the plants have been shown to tolerate more acidic conditions (to pH 6.2) at least temporarily but decline rapidly at pH levels below 4.7. The sediments are described as alkaline, soft or flocculent, and include alluvium (sandy loamy), clay, clay loam (Cecil), gravel, marl, muck (hard), mud (black, soft), peat (fibrous, pulpy), sand (coarse, marly), sandy clay, sandy loam, sandy marl, sandy muck, sandy silt, silt, silty sand, and silty rock. The plants are dioecious, though many collections are described as sterile. Sex ratios observed in the field usually are close to unity (e.g., 1.2:1 females to males), but some skewed ratios (e.g., 10:1) also have been observed. Male plants can grow from 20% to 40% faster than female plants under certain environmental conditions. They also can flower earlier and senesce earlier than female plants, which raises some questions regarding how they successfully achieve pollination. Conversely, in some populations where male flowers had never been observed, a more thorough survey indicated their appearance much later in the season than the female flowers. In the latter case, widespread maturation of fruits occurred much earlier than the appearance of the male flowers, which indicated that some plants might be at least facultative apomicts, having an early apomictic phase and later sexual phase. The temporal expression of sexes in this species has been postulated as a means of reducing intergender competition. In North America, flowering can occur as early as January and as late as October, with fruiting reported from June to October. The mechanism of underwater pollination has been studied in some detail. At anthesis the male flower pedicels elongate rapidly and curve to situate the anthers opposite the branches for greater pollen dispersal efficiency. Subsequently, the anthers dehisce to release large clouds consisting of numerous pollen grains, from which the pollen tubes emerge precociously (elongating to 2 mm) as a means of enhancing their capture on the stigmas of the female flowers. Pollen production per flower ranges from 22,800 to 39,800 grains. Elongation of the polar axis of the pollen also occurs. The pollen is released for 5–60 min and remains viable for up to 30 hr. High pollen:ovule ratios (24,200–31,700) have been calculated. Pollen germination is very low (<2.5%) unless the grains come in close proximity or contact with the stigma, whereupon germination then occurs in about 30 min. In Europe, natural seed germination has been observed from April to June in more northern latitudes. Seed germination occurs reliably from 20°C to 25°C (optimally at 24°C) at low redox potential (−300 to −440 mV) and is enhanced under dark conditions. Under long-day conditions, no germination has been observed at temperatures below 16°C. Germination of untreated seeds is low (e.g., 11%), but increases substantially (to 62%) once the seed coat has been scarified or if the seeds have been cold stratified (e.g., 84–112 days @ 3.8°C–4°C). Germination rates also are higher for seeds collected later in the season (e.g., late October–November), which are darker in color and arguably more mature. The seeds (stored wet or dry) can remain viable for more than 4 years and will survive within the digestive tracts of fish (Vertebrata: Teleosti: Cichlidae, Cyprinidae) for up to 82 hr. Although most seeds (e.g., 70%) are digested when eaten by waterfowl (Aves: Anatidae), the rest exhibit enhanced

germination when egested. They can be retained in waterfowl digestive tracts for more than 10 hr, which results in significant long-distance dispersal potential. Several types of seeds are produced, which vary in their hardness, durability, and relative survival attributes. The seedlings establish best when situated on softly cohesive sediments because their roots do not penetrate firm substrates well. **Reported associates:** *Alisma, Arthrocnemum subterminale, Berula erecta, Bolboschoenus robustus, Ceratophyllum demersum, Chara contraria, Cyperus erythrorhizos, Cyperus odoratus, Distichlis spicata, Eleocharis montevidensis, Elodea canadensis, Elodea nuttallii, Heteranthera dubia, Lemna trisulca, Ludwigia peploides, Myriophyllum aquaticum, Myriophyllum sibiricum, Myriophyllum spicatum, Myriophyllum verticillatum, Najas canadensis, Najas flexilis, Najas guadalupensis, Najas minor, Nitella tenuissima, Nuphar variegata, Nymphaea odorata, Pluchea odorata, Potamogeton amplifolius, Potamogeton crispus, Potamogeton foliosus, Potamogeton friesii, Potamogeton gramineus, Potamogeton illinoensis, Potamogeton natans, Potamogeton nodosus, Potamogeton pusillus, Potamogeton richardsonii, Potamogeton zosteriformis, Proserpinaca palustris, Ruppia maritima, Sagittaria cuneata, Sagittaria latifolia, Sarcocornia pacifica, Schoenoplectus acutus, Stuckenia pectinata, Typha angustifolia, Utricularia macrorhiza, Utricularia minor, Vallisneria americana, Wolffia, Zannichellia palustris.*

2.3.8.8. *Najas minor* **All.** is a nonindigenous annual, which inhabits fresh to brackish waters (salinity: 0–15 ppt) on mud/sand flats or in non-tidal or tidal borrow pits, coves, deltas, ditches (roadside), gravel pits, lakes (eutrophic, marl, mesotrophic, outlet, oxbow), ponds (artificial, beaver, calcareous, farm, mine, muddy, quarry), pools, quarries (sand), reservoirs, rivers, sloughs, and streams at elevations to 662 m. The plants occur primarily in shallower waters (0.08–2.0 m depth) and in sites (North America) having higher specific conductivity, with a high probability of occurrence (>50%) in waters where conductance values exceed >180 µS cm^{-1}. Water clarity varies from high to turbid with flow rates ranging from still waters to slow currents. In North America the waters are typically alkaline (total alkalinity $\bar{x} = 101$ mg l^{-1}; pH: 7.3–8.2; $\bar{x} = 7.8$) and often marly. The substrates have been characterized as muck (oozy), mucky sand, mucky silt (soft), mud, rocky sand, sand (marly), sandy gravel, sandy muck, sandy mud, silt, and silty clay. Flowering and fruiting (North America) extend from May to October. The plants are monoecious and water pollinated (hypohydrophilous), with a pollen:ovule ratio estimated at 1:500. Seed production usually is high. The seeds can float for an undetermined period of time. Within water bodies, they are dispersed by the water and have been recovered from the water surface and from along the shoreline drift. The seeds also are eaten by waterfowl (Aves: Anatidae), which likely disperse them between water bodies. The seeds are not particularly durable and do not appear to survive through the complete digestive tract of waterfowl; therefore, it is more likely that they are transported exozoically in mud, feathers, or by other means. The seeds are quite resistant to desiccation and a persistent seed bank (of undetermined duration) can develop, with an average of 26.6 seedlings germinating from 15 cm sediment cores taken in one Old World locality. The highest germination success requires shallow inundation (e.g., 3.0 cm), and has been achieved in some samples (up to 100%) after several months of cold stratification (at 4°C). Although annuals, the plants can propagate vegetatively during the growing season by shoot fragmentation. Experimentally, planted fragments have exhibited >70% viability. The plants are particularly resilient to lake drawdown conditions, likely because of their persistent seed bank. This species exhibits a dramatic niche shift between its native habitats and those in nonindigenous regions. Genetic studies have identified at least two distinct genotypes introduced into North America, which appear to differ in their ecological preferences. The maximum monthly biomass productivity can be as high as 224.4 g DM m^{-2} (July). **Reported associates (North America):** *Alternanthera, Bacopa, Brasenia schreberi, Cabomba, Callitriche, Ceratophyllum demersum, Ceratophyllum echinatum, Chara, Echinodorus, Egeria densa, Elatine minima, Eleocharis acicularis, Elodea canadensis, Elodea nuttallii, Glossostigma cleistanthum, Gratiola aurea, Heteranthera dubia, Hydrilla verticillata, Hydrodictyon reticulatum, Justicia americana, Lemna minor, Lemna trisulca, Ludwigia palustris, Luziola fluitans, Lysimachia terrestris, Lythrum salicaria, Murdannia, Myriophyllum sibiricum, Myriophyllum spicatum, Najas canadensis, Najas flexilis, Najas gracillima, Najas guadalupensis, Najas marina, Nelumbo lutea, Nitella flexilis, Nuphar, Nymphaea odorata, Peltandra virginica, Persicaria amphibia, Persicaria coccinea, Potamogeton amplifolius, Potamogeton berchtoldii, Potamogeton bicupulatus, Potamogeton crispus, Potamogeton diversifolius, Potamogeton epihydrus, Potamogeton foliosus, Potamogeton foliosus, Potamogeton gramineus, Potamogeton illinoensis, Potamogeton natans, Potamogeton nodosus, Potamogeton pusillus, Potamogeton richardsonii, Potamogeton robbinsii, Potamogeton spirillus, Sagittaria montevidensis, Sagittaria platyphylla, Spirodela polyrhiza, Stuckenia pectinata, Utricularia foliosa, Utricularia gibba, Utricularia macrorhiza, Utricularia radiata, Vallisneria americana, Wolffia brasiliensis, Zannichellia palustris.*

2.3.8.9. *Najas wrightiana* A. Braun is an annual, which occurs in canals, marshes (basin), ponds, ruts, and swamps at elevations to at least 9 m. North American localities tend to be still, shallow, temporary waters (to 0.5 m deep). However, in Belize the plants reportedly inhabit inundated rocks in fast-flowing rivers, along with members of Podostemaceae. A fair number of specimens of other species are misidentified as this taxon in collections. In Cuba the plants remain in flower and fruit throughout the year. Several North American sites are characterized by extreme water level fluctuations. There the plants can appear to be completely absent for 1 year, only to reappear in numbers during the next growing season. These observations indicate that a persistent seed bank develops and is essential for long-term survival. Additional life history information is wanting. **Reported associates:** *Taxodium.*

Use by wildlife: *Najas* species (especially *N. flexilis* and *N. guadalupensis*) rank highly as food plants for various waterfowl and shorebirds (e.g., Aves: Anatidae: *Anas acuta, A. clypeata, A. americana, A. strepera, Aythya affinis, A. americana, A. marila, A. valisineria, Bucephala albeola, B. clangula, Oxyura jamaicensis*; Rallidae: *Fulica americana*), which readily consume the fruits, seeds, and vegetative tissues, especially in areas following disturbances (e.g., hurricanes). Some individuals of these birds have been found to contain upward of 22,000 *Najas* seeds in their guts. The plants are consumed by several turtles (Reptilia: Testudines), including Blanding's (Emydidae: *Emydoidea blandingii*), musk (Kinosternidae: *Sternotherus odoratus*), painted (Emydidae: *Chrysemys picta*), and snapping (Chelydridae: *Chelydra serpentina*) turtles. *Najas* plants are used in Africa as fodder for sheep (Mammalia: Bovidae: *Ovis aries*) and goats (Mammalia: Bovidae: *Capra hircus*). *Najas flexilis* also is eaten by crayfish (Crustacea: Cambaridae: *Orconectes rusticus, O. virilis*), moose (Mammalia: Cervidae: *Alces alces*), muskrats (Mammalia: Cricetidae: *Ondatra zibethicus*), mute swans (Aves: Anatidae: *Cygnus olor*), and snails (Mollusca: Physidae: *Physa gyrina*). The plants are favored by the herbivorous (and nonindigenous) rudd (Vertebrata: Teleostei: Cyprinidae: *Scardinius erythrophthalmus*). Their foliage has been observed to support up to 36 different invertebrate species and also has been considered for use as livestock fodder. Assayed *N. flexilis* plants range from 14.2% to 14.6% dry matter, with 12.7–14.4% crude protein and 15.2–22.7% crude fiber. Remains of mastodons (Mammalia: Mammutidae: *Mammut americanum*) from 11,440 to 11,630 years before present (BP) have been found associated with *N. flexilis* seeds. In Asia, the seeds of *N. graminea* are eaten by several species of waterfowl (Aves: Anatidae) that also occur in North America, including gadwall (*Anas strepera*), pintail (*Anas acuta*), and shoveler (*Spatula clypeata*) ducks. The plants are eaten in the Old World by turtles (Reptilia: Testudines). The crude DM protein content of fresh *N. guadalupensis* plants has been shown to peak in April and July and can reach 46%. The digestible dry matter in *N. guadalupensis* can reach 98%, with the total digestibility of plants fed to fish (Teleostei: Cichlidae: *Tilapia zillii*) estimated at 29.3%. The plants are readily eaten by young hybrid grass carp (Teleostei: Cyprinidae: *Ctenopharyngodon idella* × *Hypophthalmichthys nobilis*) and are fed upon during summer by beavers (Mammalia: Castoridae: *Castor canadensis*). They also are eaten by manatees (Vertebrata: Trichechidae: *Trichechus manatus*) and to a slight degree by white-tailed deer (Mammalia: Cervidae: *Odocoileus virginianus*). The foliage of *N. guadalupensis* is colonized by periphyton (e.g., Chromista: Bacillariophyceae; Cyanobacteria: Cyanophyceae) and is eaten by several insects (Insecta), including caddisflies (Trichoptera: Leptoceridae: *Nectopsyche tavara*) and flies (Diptera: Ephydridae: *Hydrellia bilobifera, H. najadis*). It also is consumed by snails (Gastropoda: Ampullariidae: *Pomacea maculata*) and turtles (Reptilia: Emydidae: *Pseudemys alabamensis, Trachemys scripta elegans*). The shoots of *N. guadalupensis* provide oviposition sites for damselflies (Insecta: Odonata: Coenagrionidae: *Neoerythromma gladiolatum*) and a substrate for developing larval midges (Insecta: Diptera:

Chironomidae: *Dicrotendipes hulberti*). The fruits and seeds of *N. marina* are eaten by fish (Vertebrata: Teleosti: Cichlidae: *Oreochromis*; Cyprinidae: *Ctenopharyngodon idella*) and waterfowl (Aves: Anatidae), including blue-winged teal (*Anas discors*), mallard (*Anas platyrhynchos*), and redhead (*Aythya americana*) ducks. *Najas minor* is eaten by grass carp (Vertebrata: Teleosti: Cyprinidae: *Ctenopharyngodon idella*) and by a variety of dabbling ducks (Aves: Anatidae), including gadwalls (*Anas strepera*), lesser scaups (*Aythya affinis*), mallards (*Anas platyrhynchos*), pintails (*Anas acuta*), and wigeons (*Anas americana*). The plants provide cover for the northern pike (Vertebrata: Teleostei: *Esox lucius*). Ethyl acetate extracts of *N. minor* are inhibitory to some algae (Cyanobacteria: Chroococcaceae: *Microcystis aeruginosa*) and Gram-positive bacteria (Bacteria: Bacillaceae: *Bacillus subtilis*; Staphylococcaceae: *Staphylococcus aureus*); some suppression of Gram-negative biofilm bacteria (Bacteria: Enterobacteriaceae: *Escherichia coli*) also occurs.

Economic importance: food: *Najas* species are not reported to be edible; **medicinal:** A paste made from the leaves of *N. graminea* has been used in Asia to treat boils and goiters. Exposure to the spiny leaves of *N. minor* has caused dermatitis in some rice-field workers; **cultivation:** Because of their annual habit and brittle shoots, most *Najas* species are unsuitable as aquarium plants. *Najas guadalupensis* (a perennial) is cultivated in aquariums, particularly as a cover plant for fish breeding; **misc. products:** *Najas graminea* has been shown to effectively remove heavy metals from wastewater. In India, *N. minor* is used as packing material for fish baskets; **weeds:** In some parts of the world, *Najas graminea* has been described as the most abundant submersed species in rice agroecosystems. *Najas guadalupensis* has been characterized as aggressive and weedy in California (rice fields), Connecticut, Massachusetts, Missouri, and Wisconsin. *Najas marina* occasionally is reported as a weed in golf course ponds and recreational lakes in the western United States; **nonindigenous species:** Although once uncertain, it is now clear that Alaskan populations of *N. flexilis* are indigenous to that region due to their possession of a unique genotype found nowhere else in North America. Genetic and morphological data have confirmed that nonindigenous populations of *N. gracillima* were introduced to northern California rice fields from Asia sometime prior to 1966; *N. graminea* was introduced to California rice fields sometime before 1946 (specimens confirmed by DNA analysis). Since 1933, it also has been introduced to several European countries, including Albania, Bulgaria, Croatia, Greece, Romania, and Turkey. Although some authors have questioned whether *N. guadalupensis* is indigenous in eastern North America, that conclusion is supported by genetic data. In the Old World, *N. guadalupensis* has been introduced to Bangladesh, Hungary, and Israel. Despite the existence of fossils more than 12,000 years old, which document that *Najas marina* is indigenous to North America, some localities (e.g., the western Great Lakes) are thought to have been recolonized within the past century or so. Genetic data indicate that *N. minor* was introduced at least twice to eastern North America in the early 1930s. Its remote disjunction in northern

California (genetically similar to populations in the eastern United States) is enigmatic. It is unclear why some authors have categorized *N. wrightiana* as nonindigenous rather than as a natural range extension into North America. There are Florida specimens dating back to 1913 and there seems to be no evidence to indicate that it is nonindigenous to the region. Its status as an indigenous species must be reconsidered and is accepted here.

Systematics: Formerly, *Najas* had been assigned to a distinct family (Najadaceae), which was thought to be related closely to *Potamogeton* and *Zannichellia* (Potamogetonales; Potamogetonaceae). However, the results of various phylogenetic studies (e.g., Figure 1.8) have now clarified its placement within Hydrocharitaceae (subfamily Hydrilloideae) in the order Alismatales (refer to earlier discussion of Hydrocharitaceae). All relevant evidence examined thus far indicates that *Najas* is monophyletic. Several classifications have distinguished two subgenera: *Caulinia* (recognized at one time as a distinct genus) and *Najas*, which are supported by their resolution as distinct clades in phylogenetic analyses (e.g., Figure 1.13). Subgenus *Caulinia* also has been subdivided into as many as four sections (*Americanae*, *Euvaginatae*, *Nudae*, and *Spathaceae*), although only two (*Americanae* and *Euvaginatae*) are defensible on the basis of results from phylogenetic analyses. Most indigenous North American *Najas* occur either within subgenus *Najas* (*N. marina*), or within subgenus *Caulinia*, section *Americanae* (*N. canadensis*, *N. filifolia*, *N. flexilis*, *N. guadalupensis*, and *N. wrightiana*); *N. gracillima*, along with the nonindigenous *N. graminea* and *N. minor*, would be placed in section *Euvaginatae* (Figure 1.13). Two distinct diploid ($2n = 12$) "cytodemes" or karyotypes ("A" and "B") have been identified in *N. marina* and represent at least five rearrangements resulting from spontaneous chromosome breakage in heterochromatic regions. These genetically divergent karyotypes now are recognized as delimiting separate, cryptic species, which have been distinguished nomenclaturally as *N. major* and *N. marina*, respectively. All New World plants surveyed genetically associate with the "B" karyotype, which retains the name *N. marina*. *Najas canadensis* recently has been distinguished as a cryptic allotetraploid derivative of *N. flexilis* and *N. guadalupensis*. Although some authors have merged *N. filifolia* with the South American *N. conferta*, material of the North American *N. filifolia* appears to be quite distinct and it is retained here as a separate species. Currently regarded as a North American endemic, it is possible that *N. filifolia* (as "*N. conferta*") extends into the Caribbean islands as well, a possibility that deserves further evaluation. It is related more closely to the South American rather than North American members of section *Americanae*. Populations of *N. gracillima* from the south-central portion of its range are ecologically and genetically distinct. The nonindigenous strains of *N. minor* also appear to differ ecologically and apparently have hybridized in some zones of overlap. *Najas guadalupensis* is a complex of genetically divergent and polyploid segregates, which are inconsistent taxonomically with formerly recognized subspecies. Extensive hybridization among the genetic races, as well as interspecific

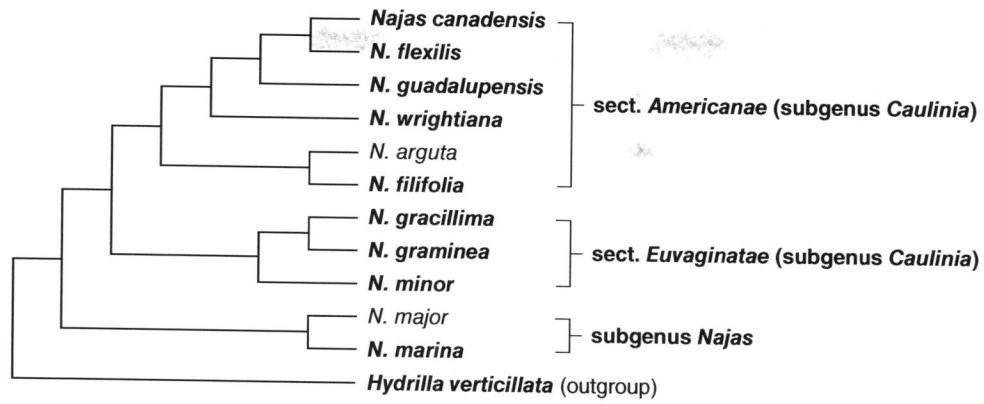

FIGURE 1.13 Interspecific relationships in North American *Najas* (modified and simplified extensively from Ito et al., 2017). Numerous Old World taxa have been omitted from the original representation due to many evidently erroneous identifications. Although the results of that study are based on analyses with extensive amounts of missing data, they are fairly consistent with those of several other phylogenetic studies of the genus (e.g., Les et al., 2010; 2015a; King et al., 2017). The entire genus is aquatic, but only those species categorized as OBL indicators in North America (also including *N. canadensis* – see text) are depicted in bold.

crosses involving *N. flexilis* and *N. canadensis*, has made it difficult to delimit discrete taxonomic entities within this species. It is likely that sufficient evidence eventually will materialize to warrant the recognition of "*N. guadalupensis* subsp. *floridana*" (2*n*=48) as a distinct species. The taxon known formerly as "*N. olivacea*" (2*n*=36) represents a hybrid of multiple origins involving *N. flexilis* and *N. guadalupensis*; that known as "*N. muenscheri*" is indistinguishable genetically from *N. canadensis*. The "rough-seeded" forms (often associated with "*N. muenscheri*") and "smooth-seeded" forms (typical of *N. canadensis*) co-occur on specimens collected in close proximity from several lakes. In contrast to *N. filifolia*, *N. wrightiana* is related more closely to the North American representatives of section *Americanae*. The basic chromosome number in *Najas* is *x*=6. Consistent diploids (2*n*=12) include *N. flexilis* and *N. marina*, although autotetraploids of *N. marina* (2*n*=24) also have been observed and can be differentiated from diploids by their allozyme profile. *Najas canadensis* (an allopolyploid), *N. gracillima*, and *N. minor* are tetraploids (2*n*=24); *N. minor* has an estimated genome size (1C) of 3560 Mbp (mega base pairs). Multiple counts have been reported for *N. graminea* (2*n*=12, 24, 36) as well as for *N. guadalupensis* (2*n*=12, 36, 42, 48, 54, 60). Counts have not yet been reported for *N. filifolia* or *N. wrightiana*. Various interspecific hybrids have involved *N. canadensis*, *N. flexilis*, and *N. guadalupensis*. *Najas major* and *N. marina* are known to hybridize in the Old World. Seed macrofossils attributed to *Najas flexilis* can number in the thousands, but they usually are not illustrated or vouchered, which precludes any distinction from *N. canadensis*.

Distribution: Fairly widespread species include *Najas canadensis* and *N. flexilis* (sympatric throughout much of northern North America), *N. guadalupensis* (the United States and southern Canada), and *N. minor* (the eastern United States and adjacent Ontario, Canada; disjunct in California). *Najas marina* occurs in the western United States, Great Lakes region, and intermittently across the southern United States. *Najas gracillima* (indigenous populations) inhabits

eastern North America; nonindigenous (Asian) populations are established in northern California. Restricted distributions characterize *N. graminea* (nonindigenous in northern California but possibly extirpated), *N. filifolia* (a few sites in Florida and Georgia), and *N. wrightiana* (rare in central and south Florida).

References: Adair et al., 1994; Agami et al., 1980; 1984; 1984a; 1986; Agami & Waisel, 1984; 1986; 1988; Alderson & Rawlins, 1925; Bailey et al., 2008; Baker et al., 2010; Bartonek & Hickey, 1969; Baskin & Baskin, 1998; Beal, 1977; Begum et al., 2006; Bellrose, 1941; Bengtson, 1983; Bhatia, 1970; Birks, 2002; Black et al., 2000; Boyd & Blackburn, 1970; Boyd & McGinty, 1981; Bräuchler, 2015; Brochet et al., 2009; 2012; Buddington, 1979; Burks, 1995; Campbell, 1897; Cassani, 1981; Catling et al., 1986; Chase, 1947; Clausen, 1936; Cottam, 1939; Crowder et al., 1977a; Cruz et al., 1998; Dale & Gillespie, 1977; De la Vega et al., 1993; DeMarco et al., 2016; DiTomaso & Healy, 2003; Epler, 2016; Fernald, 1923a; Figuerola & Green, 2002; Fraser et al., 1980; Ge et al., 2013; González-Soriano et al., 2004; Gortner, 1934; Gupta & Pandey, 2014; Hagley et al., 1996; Hamerstrom, Jr. & Blake, 1939; Handley & Davy, 2002; 2005; Harms & Grodowitz, 2009; Havera, 1986; Hay & Muir, 2000; Haynes, 1979; 1985a; 2000c; 2000e; Haynes et al., 1998c; Hellquist, 1997; Hellquist & Crow, 1980; Hidalgo et al., 2015; Hoffmann, 2014b; Hogsden et al., 2007; Hopfensperger & Baldwin, 2009; Hossain et al., 2010; Hoyer et al., 1996; Huang et al., 2001; Hussner, 2012; Ismail & Phaik-Hong, 2004; Ismail Sahid & Ho, 1995; Ito et al., 2017; Izzati, 2016; Jha & Chaudhary, 2011; Jones & Drobney, 1986; June-Wells et al., 2013; Juraimi et al., 2012; Kadono, 1982; Kaisar et al., 2016; Kapuscinski et al., 2014; Kasselmann, 2003; Kimble & Ensminger, 1959; King et al., 2017; Knapton & Pauls, 1994; Knapton & Petrie, 1999; Krull, 1970; Lagler, 1943; Lansdown et al., 2016; Lee et al., 1999; Les & Mehrhoff, 1999; Les et al., 1993; 2012a; 2012b; 2013; 2015a; 2015b; Léveillé-Bourret et al., 2017; Lodge et al., 1994; Lowden, 1986; Lyon & Eastman, 2006; Mallik et al., 2011; Martin & Sauerborn, 2000; Martin & Uhler, 1951;

Martin et al., 1992; Mauermann, 1995; McFarland & Rogers, 1998; Mendall, 1949; Meriläinen, 1968; Miller, 1987; Moeller, 1984; Moeller et al., 1988; Morton & Hogg, 1989; Moyle, 1944; 1945; Muenscher, 1936; Near & Belcher, 1974; Nichols, 1975; 1984; 1999; O'Kennon & McLemore, 2004; Padgett & Crow, 1994a; Patterson, 1982; Paulus, 1982a; Peredo et al., 2012; 2013; Pip & Simmons, 1986; Pollux, 2011; Radomski & Perleberg, 2012; Reynolds et al., 2015; Roberts & Arner, 1984; Rosendahl & Butters, 1935; Rothe, 2011; Rüegg et al., 2017; Rybicki et al., 2001; Schloesser & Manny, 1986; Seaman et al., 1968; Sekercioglu et al., 2016; Self et al., 1975; Shaffer-Fehre, 1991a; 1991b; Sheldon & Boylen, 1977; Shields et al., 2012; Siver et al., 1986; Sletten & Larson, 1984; Stuckey, 1985; Stuckey et al., 1978; Sullivan, 1981; Thacker et al., 1993; Tietje & Teer, 1996; Titus & Grisé, 2009; Topuzovic et al., 2015; Toriyama, 2005; Triest, 1988; 1989; Triest et al., 1989; Turner & Nelson, 2001; Urban et al., 2006; Van Vierssen, 1982a; Viinikka, 1973; 1976; 1977; Viinikka & Kotimäki, 1979; Viinikka et al., 1978; 1987; Wang et al., 2010; Wang et al., 2017; Wentz & Stuckey, 1971; Westcott et al., 1997; Wetzel & McGregor, 1968; White & James, 1978; Winge, 1927; Witztum & Chaouat, 1991; Wood, 1950; Xiao et al., 2010; Yacoub, 2009; Yeo, 1966; You et al., 1985.

2.3.9. Ottelia

Duck lettuce

Etymology: believed to be derived from *ottel-ambel*, the ancient Malabar name for the plants

Synonyms: *Abildgaardia*; *Beneditaea*; *Bootia*; *Damasonium* (in part); *Hydrocharis* (in part); *Oligolobos*; *Stratiotes* (in part); *Xystrolobos*

Distribution: global: Africa, Asia, Australia, Europe*, North America*, South America; **North America:** southern and western United States (sporadic)

Diversity: global: 21 species; **North America:** 1 species

Indicators (USA): obligate wetland (OBL): *Ottelia alismoides*

Habitat: freshwater; lacustrine, palustrine, riverine; **pH:** 7.5–8.2; **depth:** <2 m; **life-form(s):** floating-leaved, submersed (rosulate)

Key morphology: plants (to 75 cm) arising from an erect, cormose stem; leaves rosulate, submersed, petioled (to 50+ cm), the blades (to 40 cm) thin, translucent, narrowly ovoid (juvenile foliage) to broadly ovate (adult foliage), the margins denticulate or entire, the bases cuneate to cordate; inflorescence enclosed by a "spathe" of two bracts (to 50 mm) having up to 12 crisped, longitudinal wings or ridges and containing bisexual or unisexual, chasmogamous (emergent) or cleistogamous (submersed), trimerous flowers (solitary and sessile if bisexual or ♂; pedicelled and numerous if ♂), borne on an elongate peduncle (to 60 cm); sepals (to 24 mm) green with brown stripes; petals (to 30 mm) clawed, bluish or purple to pink or white, yellow toward base; stamens 3–12, filaments elongate (to 6 mm); ovary (to 50 mm) inferior, consisting of 3–9 carpels and, respectively, 3–9 deeply bifid styles; fruit (to 50 mm) a fleshy capsule persistent calyx; seeds (to 1.8 mm) numerous (to 2,000+ per fruit), fusiform, purple to black when mature [the unisexual flowers differ primarily by the missing sexual organs; ♂ flowers also contain staminodes and pistillodes]

Life history: duration: annual (fruits/seeds); perennial (corm) **asexual reproduction:** none; **pollination:** insect; **sexual condition:** dioecious, hermaphroditic; **fruit:** capsules (common); **local dispersal:** water (fruit, seeds); **long-distance dispersal:** water birds (seeds)

Imperilment: 1. *Ottelia alismoides* [GNR]

Ecology: general: All *Ottelia* species are obligate, freshwater, aquatic plants with entirely submersed or floating-leaved foliage. The species primarily are monocarpic annuals, but in some cases can live for several years by means of a persistent corm if they do not flower. One species (*Ottelia acuminata* – not in North America) produces vegetative bulbils in the inflorescence, which are capable of vegetative reproduction. In another species (*O. muricata* – not in North America), the stems are creeping, rhizomatous, and irregularly forked. Most *Ottelia* species reproduce sexually by the production of emergent, showy, nectariferous flowers, which are fragrant and apparently adapted for entomophily. Cleistogamous flowers (autogamous) are produced in some species under high water or fast current conditions.

2.3.9.1. *Ottelia alismoides* (L.) Pers. is a nonindigenous, submersed annual or perennial, which grows in bayous, canals (drainage), coves, deltas, ditches (access, agricultural, roadside), impoundments, lakes (artificial), marshes, ponds, pools, reservoirs, rice fields, rivers (slow-moving), and swamps at elevations to 90 m. The plants inhabit still to slow-moving waters but prefer sites that experience moderately fluctuating water levels. They thrive under more eutrophic conditions if the water clarity is not severely impacted; however, they have been found in waters ranging from clear to murky conditions. Dense stands of plants often are observed in highly illuminated sites. In North America, the plants have been reported from water depths up to 2.0 m, but 50 cm is the optimal depth for material grown under laboratory conditions. Optimal water temperatures range from 22°C to 26°C. This species is effective at carbon uptake by combing the ability to use bicarbonate (HCO_3^-) with the possession of modified types of C_4 photosynthesis, including NAD-ME decarboxylation and CAM-like metabolism. The few acidity measures available indicate alkaline conditions (pH: 7.5–8.2). The substrates (North America) include clay, humus, mud, sand, and silt. More than 30% of the total plant biomass can be allocated to sexual reproduction. The developing flowers are enclosed within a protective spathe, which herbivorous snails (Mollusca: Gastropoda) appear to avoid. Flowering and fruiting (North America) have been observed from May to December. Usually, the flowers are emergent and polygamous (bisexual or unisexual; hermaphroditic or dioecious); open flowers sometimes become submersed. The petals wither after 1 day. In deeper waters (>80 cm), where flowers cannot reach the surface, they remain cleistogamous and are autogamous. The emergent flowers are fragrant and can attract various insects (e.g., Insecta: Coleoptera; Diptera: Hymenoptera); however, they also appear to be strongly autogamous, relying minimally on insects for pollination. In these flowers, the pollen often germinates precociously (even within the anther) and reaches the stigmas before the flowers open. Although the bisexual flowers of some greenhouse cultivated plants reportedly did not spontaneously self, other studies using bagged flowers (to exclude pollinators) have clearly demonstrated that the plants are highly autogamous. Comparably high levels of fruit set (to 100%) and seed set (averaging 72%) have been obtained for both bagged and open-pollinated flowers. A prevalent inbreeding system has been corroborated by genetic analyses of native Asian populations (using ISSR markers), which indicate low within-population variability attributed to high levels of autogamy. The fruits ripen within 2 weeks. They are winged and buoyant (dispersed on the surface if detached) but usually mature underwater (being pushed downward by the peduncle), where the seeds (up to 2,000 capsule^{-1}) are released within a pulpy mass. This aggregation of seed remains afloat only for distances up to 40 m before sinking (dislodged seeds will sink immediately). The small seeds (to 1.6 mm) are dormant when shed. They are resistant to dessication, with viability retained in some seeds after 4 years of dry storage. A persistent seed bank (of uncertain duration) develops, with an average density of 311 seeds m^{-2} observed in some native Asian populations. The seeds require light and a period of cold stratification (e.g., 3–5 months @ 4°C) in order to germinate (at 25°C–30°C; 12-hr photoperiod). Germination is inhibited in buried seeds. The germination rates increase substantially (e.g., from 19% to 72%) when seed densities are high (e.g., clusters of 50 seeds). Physical scarification makes it possible to germinate (up to 53%) unstratified seeds within 2 weeks when they are incubated at 20°C. Germinating seeds initially become affixed to the bottom by the development of hypocotylar hairs, and subsequently by the extension of the primary and adventitious roots. The biomass of standing crops can reach yields of 1.2 kg m^{-2}. **Reported associates (North America):** *Alisma triviale, Alternanthera philoxeroides, Azolla cristata, Cabomba, Ceratophyllum demersum, Echinodorus cordifolius, Eichhornia crassipes, Eleocharis, Heteranthera, Hydrilla verticillata, Ludwigia peploides, Luziola, Myriophyllum laxum, Myriophyllum pinnatum, Nelumbo lutea, Nuphar advena, Nymphaea, Pistia stratiotes, Pontederia cordata, Potamogeton crispus, Potamogeton foliosus, Sagittaria calycina, Salvinia minima, Sphenoclea zeylanica, Utricularia foliosa, Vallisneria americana, Zizania aquatica, Zizaniopsis miliacea.*

Use by wildlife: The seed masses of *Ottelia alismoides* are eaten by fish (Vertebrata: Gnathostomata); however, it is unknown whether they remain viable afterward or are dispersed by them.

Economic importance: food: The young leaves, petioles, and fruits of *O. alismoides* comprise about 2.3% nitrogen and are eaten raw or cooked in Asia; the raw fruits have a somewhat salty flavor; **medicinal:** *Ottelia alismoides* is credited with having anti-microbial activity against the agent for human tuberculosis (Bacteria: Mycobacteriaceae: *Mycobacterium tuberculosis*). The foliage contains two diastereomeric 4-methylene-2-cyclohexenones (otteliones A and

B), which are cytotoxic against several types of cancer cells. The plants are used as a diuretic in some parts of India; **cultivation:** *Ottelia alismoides* is sold as an aquarium plant; **misc. products:** none; **weeds:** *Ottelia alismoides* is considered to be a weed in southern European rice fields; **nonindigenous species:** *Ottelia alismoides* was introduced to North America (Louisiana) sometime before 1939, most likely as a contaminant of planted rice. It spread to California's rice growing region by 1977.

Systematics: *Ottelia* consistently allies with *Blyxa* in various molecular phylogenetic analyses, although neither of these large genera has been sampled well in such studies. The most inclusive analysis (e.g., Figure 1.8) evaluates only 6 of the 30 species and is based entirely on *rbcL* sequence data. Because those results do not clearly resolve the genera as distinct, additional sampling of taxa and genetic loci is needed in order to more thoroughly assess their monophyly. In any case, a very close relationship between the groups is apparent. The basic chromosome number of *Ottelia* is $x = 11$. *Ottelia alismoides* is variable cytologically ($2n = 22$, 44, 66, and 132), with all of the polyploid cytotypes forming bivalents and apparently fertile. Numerous aneuploid counts also have been reported. Partially fertile hybrids have been synthesized successfully using *O. alismoides* as the pollen donor and *O. cordata* as the female parent.

Distribution: *Ottelia alismoides* occurs sporadically throughout the southern and western United States (Arkansas, California, Florida, Louisiana, Missouri, and Texas).

References: Anderson, 2007; Ayyad et al., 1998; Barrett & Seaman, 1980; Brahma et al., 2013; Kasselmann, 2003; Chen et al., 2008; Collectanea, 1976; Cook & Urmi-König, 1984b; Cook et al., 1984; Cunningham, 1887; Dewanji et al., 1997; Edwards, 1980; Gledhill, 2002; Holmes, 1978; Jiang & Kadono, 2001a; 2001b; Kaul, 1969; 1970; 1978; Les & Tippery, 2013; Les et al., 1997a; 2006a; Liu et al., 2006; Misra, 1974; Mühlberg, 1982; Saini et al., 2011; Sarkar et al., 2008; Shao et al., 2017; Thieret, 1970; Turner, 1980; Yatskievych & Raveill, 2001; Yin et al., 2009; 2013; 2017; Yu & Yu, 2009; Zhang et al., 2014b.

2.3.10. Thalassia

Seagrass, turtlegrass

Etymology: After *Thalassa*, the primordial sea goddess of Greek mythology

Synonyms: *Schizotheca; Thalassiophila*

Distribution: global: Indian and western Pacific Oceans; **North America:** southern (Caribbean and Gulf of Mexico)

Diversity: global: 2 species; **North America:** 1 species

Indicators (USA): obligate wetland (OBL): *Thalassia testudinum*

Habitat: brackish (coastal); marine; oceanic; **pH:** 7.7–8.4; **depth:** to 30 m; **life-form(s):** submersed (rosulate)

Key morphology: rhizomes (to 6 mm thick) terete, with a single, unbranched adventitious root and leaf-like scale at each node; leaves (to 6) distichous, arising in tufts from short lateral shoots, the blades (to 60 cm) ribbon-like, entire, basally sheathing (to 10 cm), the apex rounded, finely serrate; flowers unisexual, each solitary and enveloped by a peduncled (to 7.5 cm) spathe (to 3 cm); ♂ flowers pedicelled (to 2.5 cm); perianth of 3 strap-like segments (to 12 mm), bland; stamens 9, the anthers linear (to 9 mm); pollen grains numerous, released in a gelatinous matrix, forming chains, germinating precociously; ♀ flowers nearly sessile; perianth of 3 strap-like segments (to 12 mm), bland, arising from a hypanthium (to 2 cm); ovary (to 1 cm) inferior, with several (to 8) styles (to 2.5 mm), stigmas (to 13 mm) dehiscent; capsules (to 2.5 cm) elliptic to globose, beaked (to 7 mm), the surface prickly or warty, dehiscing into valves (to 8); seeds (1–6) pear-shaped (to 10 mm)

Life history: duration: perennial (rhizomes); **asexual reproduction:** rhizomes (fragments and proliferation); **pollination:** water (hypohydrophilous; zoobenthophilous); **sexual condition:** dioecious; **fruit:** capsules (common); **local dispersal:** vector water (fruits, seeds); **long-distance dispersal:** water (fruits, seeds)

Imperilment: 1. *Thalassia testudinum* [G4/G5]; S1 (MS); S2 (LA)

Ecology: general: Both *Thalassia* species are perennial, oceanic seagrasses, which grow entirely submersed in coastal marine habitats. They possess unisexual flowers in a dioecious arrangement. The flowers normally are water pollinated (hypohydrophilous); their pollination is assisted sometimes by invertebrates (zoobenthophilous), which actively collect and transfer the pollen in calm waters. The pollen grains are shed in a gelatinous matrix, but often organize into moniliform chains. Like many hydrophiles, the pollen exine is rudimentary. The fruits are buoyant and are dispersed on the water surface, sometimes for considerable distances. There is no seed dormancy. The plants reproduce vegetatively by the extension of coarse rhizomes, which also can disperse and establish as fragments.

2.3.10.1. ***Thalassia testudinum*** **Banks & Sol. ex K.D. Koenig** inhabits warm (23°C–31°C), brackish to saline (salinity: 12–40‰) waters of bays, flats (alluvial), oceans, salt marshes, shores (lagoon), and swamps (mangrove) at elevations from 0 to −30 m. Although the optimal salinity range (for both adult plants and seedlings) is 30–40‰, substantially lower salinities (e.g., 6–16‰) are tolerated, but result in reduced growth. Populations often occur under somewhat "suboptimal" conditions (e.g., salinity: 24.0–26.7‰). Seedlings are more sensitive to salinity than adult plants and are killed at extremely low (0–10‰) or high (50–70‰) salt concentrations; they are most sensitive to hypersaline conditions. Population impacts and slow recovery have been observed at some sites following prolonged, heavy rainfall, which results in decreased salinity. The plants exhibit intermediate tolerance to cold water temperatures (northern populations generally are more tolerant than southern populations), with most populations unable to withstand prolonged exposure (e.g., >12 hr) to water temperatures of 2°C. Although deeper occurrences (e.g., 10–30 m) have been reported, most coastal North American records are from waters less than 2.5 m deep. The maximum depth occurs at 22.5% of the subsurface irradiance; however, higher or lower loads of epiphytes (which increase shading) can alter that

limit accordingly. The average ocean pH is 8.1 (historically 8.2), which would generally characterize most *Thalassia* habitats. Variation in pH is slight, with ranges of 7.68–8.15 (diurnal), 7.95–8.24 (four-day period), and 8.17–8.39 (three-day period) observed in different *Thalassia* field sites. The substrates include mud, muddy sand, and sand. The growth of plants increases proportionally with increasing sediment nitrogen (NH_4^+) concentrations; however, higher tissue nitrogen contents also result in enhanced levels of herbivory. Inflorescence growth (North America) initiates in April and peaks during June. Flowering plants have been observed from May to June with fruiting in July; however, flowering periods vary substantially over space and seasons and can extend from March to September in more southern (e.g., Mexican) waters. Flowering rates can vary from 1% to 21%, with values generally increasing inversely with latitude. Low porewater nitrogen levels have been implicated in the initiation of sexual reproduction. The importance of sexual reproduction is indicated by genetic data (RAPD and microsatellite markers), which have revealed fairly high levels of genetic variation and fairly high heterozygosity ($H_O = 0.68$) within populations coupled with low fixation indices ($F_{ST} = 0.10$). Despite their clonal growth, biparental inbreeding has not been observed in populations. Pollination takes place entirely underwater (hypohydrophily) in this dioecious species. The floral sex ratio has been estimated as a male bias of 12:1 but can vary considerably from year to year due primarily to the unpredictable lability of male flower production. In subtidal zones, the pollen is released from the male plants within an hour of dusk, arguably to limit capture by grazing parrotfish (Vertebrata: Teleostei: Scaridae: *Sparisoma cretense*) and/or when the activity of potential pollinating invertebrates is high. The pollen is released throughout the night (for 1–2 hr per plant) within a neutrally buoyant mass of mucilage and is transported through the water passively by currents or actively by invertebrates (see *Use by wildlife*), which visit the male flowers to forage for the pollen and then transfer the pollen-containing mass to the female flower stigmas (zoobenthophilous pollination). The female flowers remain receptive for 72 hr. The observed range of pollen dispersal is from 0.3 to 1.6 m, with a maximum dispersal distance of 4.8 m. The fruits are buoyant and disperse their seed mass as they float on the water surface for up to 10 days to distances up to 360 km (their average transport speed is estimated at 1.5 km day^{-1}). There also is corresponding genetic evidence for long-distance dispersal between populations. However, seed dispersal in some populations also can be quite limited (e.g., only to 1.8 m). Densities of up to 180 fruits m^{-2} of substrate have been reported. If released at the surface, the seeds remain buoyant for less than 1 day. Seed densities on the sediment average 66 seeds m^{-2}. The photosynthetic epidermis of the seeds facilitates their colonization of sites having low light intensity. There is no seed dormancy. Densities up to 14.8 seedlings m^{-2} have been observed in the field. A survival rate of 11% has been reported for seedlings in the field after 1 year of growth. Maximum aboveground biomass (June–July) has been estimated at 300 g dwt·m^{-2}. The extensive rhizomes (e.g., 900 g dwt·m^{-2}) occur beneath 5–25 cm of sediment and can constitute more than 85% of the total plant biomass and enable the plants to be highly tolerant of erosion. However, genetic data indicate that populations are not monoclonal but contain fairly diverse arrays of genets even over small spatial scales (<0.25 m). The maximum observed extension of genets is 19.2 m. Accordingly, the detection of identical ramets (indicated by high-resolution genetic analysis) over distances of 230 m indicates their ability to propagate vegetatively by fragmentation. Rapid die-offs of plants have been attributed to reduced water oxygen content, which promotes sulfide intrusion into the rhizomes and concomitant root anoxia. **Reported associates:** *Avrainvillea, Caulerpa Cladophora fascicularis, Halimeda incrassata, H. opuntia, Halodule wrightii, Halophila decipiens, Halophila engelmannii, Halophila ovalis, Halophila stipulacea, Penicillus pyriformis, Ruppia maritima, Syringodium filiforme, Udotea flabellum.*

Use by wildlife: The foliage of *Thalassia testudinum* is colonized by numerous algal epiphytes (Chlorophyta: Cladophoraceae: *Cladophora brasilian*; Ulvaceae: *Enteromorpha flexuosa*; Chromista: Bacillariophyceae: Bacillariaceae: *Nitzschia liebetruthii*; Brachysiraceae: *Brachysira aponina*; Fragillariaceae: *Reimerothrix floridensis*; Phaeophyta: Chordariaceae: *Cladosiphon occidentalis, C. zosterae*; Myriotrichiaceae: *Myriotrichia subcorymbosa*; Sphacelariaceae: *Sphacelaria rigidula*; Striariaceae: *Hummia onusta*; Rhodophyta: Ceramiaceae: *Centroceras clavulatum, Ceramium gracillimum*; Corallinaceae: *Epilithon membranaceum, Fosliella farinosa, Melobesia membranacea, Pneophyllum fragile*; Delesseriaceae: *Hypoglossum involvens*; Rhodomelaceae: *Herposiphonia tenella, Polysiphonia macrocarpa*; Stylonemataceae: *Stylonema alsidii*; Wrangeliaceae: *Griffithsia*). Other epiphytes include foraminiferans (Rhizaria: Retaria), hydroids (Cnidaria: Hydrozoa), and worms (Annelida: Polycvhaeta: Serpulidae: *Spirobis*). The plants also harbor large populations of zooplankton (up to 10,728 individuals m^{-2}) such as copepods (Crustacea: Copepoda: Harpacticidae: *Harpacticus*), as well as crabs (Crustacea: Diogenidae: *Calcinus tibicen, Clibanarius antillensis, Paguristes limonensis*; Mithracidae: *Mithraculus forceps*; Panopeidae: *Panopeus occidentalis*; Porcellanidae: *Megalobrachium poeyi, Petrolisthes armatus, P. galathinus*), molluscs (Gastropoda: Cerithiidae: *Cerithium eburneum*; Fasciolariidae: *Fasciolaria tulipa*; Modulidae: *Modulus modulus*), nematodes (Nematoda), ostracods (Crustacea: Ostracoda), polychaetes (Annelida: Polychaeta), shrimp (Crustacea: Alpheidae: *Alpheus armillatus, A. viridari*; Hippolytidae: *Thor manningi*; Paguridae: *Pagurus bonairensis*), and turbellarians (Platyhelminthes). The bases of the plants are colonized by sponges (Porifera: Halichondriidae: *Halichondria melanadocia*). The foliage, seeds, and seedlings are eaten by crabs (Crustacea: Decapoda: Majoidea; Portunoidea: Portunidae: *Callinectes sapidus*), green sea turtles (Reptilia: Cheloniidae: *Chelonia mydas*), manatees (Mammalia: Trichechidae: *Trichechus manatus*), parrotfish (Vertebrata: Teleostei: Scaridae: *Nicholsina usta,*

Scarus, Sparisoma radians), pinfish (Vertebrata: Teleostei: Sparidae: *Lagodon rhomboides*), planehead filefish (Vertebrata: Teleostei: Monacanthidae: *Stephanolepis hispidus*), sea urchins (Echinodermata: Diadematidae: *Diadema antillarum*; Toxopneustidae: *Lytechinus variegatus, Tripneustes esculentus, T.* ventricosus), and surgeonfishes (Vertebrata: Teleostei: Acanthuridae: *Acanthurus*). The leaves host numerous fungal epiphytes (Fungi: Ascomycota: Cephalothecaceae: *Cephalosporium*; Davidiellaceae: *Hormodendrum*; Pleosp[oraceae: *Scolecobasidium arenarium*; Myxomycota: Labyrinthulaceae: *Labyrinthula*) and endophytes (Fungi: Ascomycota: Chaetomiaceae: *Humicola alopallonella*; Cladosporiaceae: *Cladosporium*; Herpotrichiellaceae: *Phialophora*; Hypocreaceae: *Trichoderma*; Leotiaceae: *Zalerion varia*; Lulworthiaceae: *Lindra thalassiae*; Nectriaceae: *Fusarium*; Orbiliaceae: *Arthrobotrys*; Pestalotiopsidaceae: *Pestalotiopsis*; Pleosporaceae: *Exserohilum rostratum, Scolecobasidium arenarium*; Pleosporales: *Phoma*; Trichocomaceae: *Aspergillus, Penicillium*). The plants provide important habitat for the imperiled spotted seahorse (Vertebrata: Teleostei: Syngnathidae: *Hippocampus erectus*), as well as other fish (Vertebrata: Teleostei: Gerreidae: *Gerres cinereus*; Gobiidae: *Gobiosoma robustum*) and shrimp (Crustacea: Malacostraca: Penaeidae: *Farfantepenaeus duorarum*). The flowers of *T. testudinum* are visited by an array of invertebrate species, which forage for the pollen and can effect pollination when water currents are minimal. These include amphipods (Crustacea: Malacostraca: Peracarida: Amphipoda: Pontogeneiidae: *Tethygeneia*), copepods (Crustacea: Maxillopoda: *Acarthia*), isopods (Crustacea: Malacostraca: Peracarida: Isopoda: Bopyridae), leptocheliids (Crustacea: Malacostraca: Peracarida: Tanaidacea: Leptochellidae: *Chondrochelia dubia*), mysid shrimp (Crustacea: Malacostraca: Mysidacea), nannastacids (Crustacea: Malacostraca: Peracarida: Cumacea: Nannastacidae: *Cumella*), ostracods (Ostracoda: Cypridinidae: *Kornickeria, Skogsbergia*), polychaetes (Annelida: Polychaeta: Alciopidae: *Rhynchonerella petersi*; Nereididae: *Rullierinereis mexicana*; Syllidae: *Salvatoria*), stomatopod shrimps (Crustacea: Stomatopoda), and several Zoea larvae (Crustacea: Decapoda: Caridea; Thalassinidea; Brachyura; Majidae).

Economic importance: food: *Thalassia testudinum* is not eaten by humans; **medicinal:** *Thalassia testudinum* contains thalassiolin B and other flavonoids/phenolics, which reportedly exhibit antioxidant, anti-inflammatory, and neuroprotective effects in rodents (Mammalia: Muridae: *Meriones unguiculatus, Mus, Rattus*). Other studies using rodent systems have shown that *T. testudinum* extracts can induce CYP1A1/2 enzymes, which are known to produce DNA protective as well as highly carcinogenic effects. The aqueous ethanolic extracts also are potential moderators of multidrug resistance effects. The plants contain an unknown (but non-phenolic) substance, which is active against pathogenic protists (Myxomycota: Labyrinthulaceae: *Labyrinthula*), the causitive agents of "wasting disease" in segrasses; **cultivation:**

Thalassia testudinum is not cultivated; **misc. products:** none; **weeds:** none; **nonindigenous species:** none.

Systematics: Several phylogenetic analyses have corroborated the monophyly of *Thalassia*, which includes only two species. Although some authors have advocated the segregation of *Thalassia* within its own family (Thalassiaceae) or subfamily (Hydrocharitaceae subfamily Thalassioideae), phylogenetic analyses consistently nest the genus deeply within Hydrocharitaceae, where it resolves in a clade with *Enhalus* and *Halophila*, the other seagrass genera in that family (e.g., Figure 1.8). Most analyses (including evaluation of full cpDNA genome data) depict *Enhalus* as the sister genus of *Thalassia*. The seagrass clade resolves within a larger group, which has been recognized as subfamily Hydrilloideae. The basic chromosome number of *Thalassia* is $x=9$; both *T. hemprichii* and *T. testudinum* are diploid ($2n=18$). Hybridization involving *T. testudinum* has not been reported and is precluded naturally by geographical distance, given that *T. hemprichii* (its sister species and only conceivable prospect for hybridization) occurs exclusively in the Old World.

Distribution: *Thalassia testudinum* occurs along oceanic coastal areas of Florida, Louisiana, Mississippi, and Texas.

References: Allen et al., 2017; Archer et al., 2015; Barry et al., 2013; Baskin & Baskin, 1998; Borum et al., 2005; Brook, 1978; Cabaço et al., 2008; Campanella et al., 2015; Celdran, 2017; Cho et al., 2002; Chollett et al., 2007; Darnell & Dunton, 2015; 2016; 2017; Davis et al., 1999; Dawes et al., 1995; Dias et al., 2002; Dixon & Leverone, 1995; Doering & Chamberlain, 2000; Durako & Howarth, 2017; Durako & Moffler, 1985; 1987; Frankovich et al., 2006; García et al., 2017; Goecker et al., 2005; Green & Short, 2003; Guerra et al., 2014; Hall & Bell, 1993; Hartog, 1970; Hartog et al., 1979; Heck, Jr. & Wetstone, 1977; Heck, Jr. et al., 2015; Herzka & Dunton, 1998; Humm, 1956; Kahn & Durako, 2006; Kaldy & Dunton, 1999; Kendrick et al., 2012; Kirsten et al., 1998; Lee & Dunton, 1999; Les & Tippery, 2013; Les et al., 1997a; Lewis & Hollingworth, 1982; Lewis & Phillips, 1980; Low-Décarie et al., 2014; Mata & Cebrián, 2013; McMillan, 1979; Miguel et al., 2015; Moran & Bjorndal, 2005; Nelsen, Jr. & Ginsburg, 1986; Orpurt et al., 1964; Philbrick & Les, 1996; Phillips & Meñez, 1988; Ross et al., 2016; Thorhaug et al., 2006; Tomlinson, 1969; Trevathan-Tackett et al., 2015; Van Dijk & Van Tussenbroek, 2010; Vanitha et al., 2016; Van Tussenbroek & Muhlia-Montero, 2013; Van Tussenbroek et al., 2016a; 2016b; Waycott et al., 2006; Witz & Dawes, 1995; Yates et al., 2007; Zieman et al., 1984.

2.3.11. Vallisneria

Tape grass, water celery, wild celery; célerie d'eau, herbe à la barbotte

Etymology: after Antonio Vallisneri (1661–1730)

Synonyms: *Physkium*

Distribution: global: cosmopolitan; **North America:** eastern and western (disjunct)

Diversity: global: 12–15 species; **North America:** 3 species

Indicators (USA): obligate wetland (OBL): *Vallisneria americana, V. neotropicalis, V. spiralis*

Habitat: brackish (coastal), freshwater; freshwater (tidal); lacustrine, riverine; **pH:** 5.6–9.9; **depth:** to 7.0 m; **life-form(s):** submersed (rosulate)

Key morphology: shoots compressed (to 1 cm), rosulate, with simple adventitious roots; stolons elongate (to 19+ cm), terminated by indurate tubers or small plantlets; leaves (to 10+) basal, the blades (to 2.0 m) ribbon-like, sessile, opaque with numerous reddish streaks, or uniformly green with a prominent translucent medial lacunal band, the margins smooth or strongly denticulate; flowers unisexual (the plants ordinarily dioecious; occasionally monoecious with up to 60 stalked flowers of intergrading sexuality produced in umbels); ♂ flowers short-stalked, initially clustered (released at pollination) within a peduncled (to 160 mm) spathe (to 13 mm), the released ♂ flowers minute (to 1.5 mm), floating, sepals (to 5.5 mm) 3, unequal; petal 1, rudimentary; stamens 3, 2 fertile (filaments fused), 1 staminodal, pollen grains few (~68 per flower); ♀ flowers (to 20 mm) solitary, arising from a spathe, long-peduncled (to 100+ cm); sepals (to 6.5 mm) ovoid or elliptical, terminating an elongate hypanthium; petals (to 1 mm) rudimentary, linear; ovary unilocular, stigmas (to 6.5 mm) 3, shallow to deeply divided; capsules (to 15 cm) berry-like, many-seeded (to 500), dehiscing longitudinally; seeds (to 2.6 mm) numerous, enclosed within a gelatinous matrix

Life history: duration: perennial (stolons, tubers, whole plants); **asexual reproduction:** stolons, tubers; **pollination:** water-mediated surface contact (type III-B); **sexual condition:** dioecious; **fruit:** capsules (common); **local dispersal:** fruit/seeds (water), stolons, tubers (water); **long-distance dispersal:** fruit/seeds (waterfowl)

Imperilment: 1. *Vallisneria americana* [G5]; S1 (AR, ID, NE, PE, SC); S2 (KY, NS); S3 (DE, IA, IL, NC); **2.** *V. neotropicalis* [unranked]; **3.** *V. spiralis* [unranked]

Ecology: general: All *Vallisneria* species are obligately submersed aquatics. They include annuals (but not in New World) or perennials, with the latter condition being prevalent. Most species (and all North American taxa) occur as a rosulate growth form having long, ribbon-like leaves, although several Old World species are shorter-leaved, vittate plants. All the species share a similar but highly unusual pollination mechanism, which occurs also in the genera *Appertiella*, *Enhalus*, and *Lagarosiphon*. This method of pollination defies classification and has been categorized unceremoniously as "Group III, Subgroup B," a type that mimics water pollination but differs by keeping the pollen and stigmatic surfaces dry throughout the process. Some authors incorrectly have categorized the system as "hypohydrophilous," a case where both of these reproductive structures function when wet (e.g., *Elodea*). The process in *Vallisneria* occurs as follows: While remaining anchored to the substrate, the male plants release numerous spherical flower buds, which float individually to the surface before opening, keeping their contents dry as they rise in the water column. The buds open on the surface to extend the pollen-bearing anthers, which remain elevated above the minute, floating boat-like flowers. Moving by surface currents either singly or in aggregations, the floating male flowers eventually must come in physical contact with a solitary pistillate flower, where their open perianths are oriented (in a minute meniscus) at the water surface. The stigmatic surfaces are exposed but remain dry within the cup-like depression. The male flowers either tumble into the female flower, depositing their pollen on a stigma; or, the elevated pollen-bearing anthers will directly contact an extended stigma as the floating male flowers brush alongside the female flower. Pollen in any male flowers that capsize becomes wet and inviable. The female flower remains attached at all times to the maternal plant (which also is anchored to the bottom) by an elongate hypanthium and peduncle. Once the female flower has been fertilized, the peduncle begins to coil, which pulls the developing fruit beneath the water surface where it ripens. The fruits are many-seeded, berrylike capsules, which float and detach from the maternal plant while containing masses of non-buoyant seeds within a gelatinous matrix. The seeds are released as the fruits dehisce or deteriorate. Depending on the species, the seeds can be dormant or non-dormant. Seed dispersal (while retained by the fruit) occurs locally by water currents and over longer distances via the transport of individual seeds by water birds (Aves: Anatidae). The surrounding gelatinous matrix possibly assists in the adhesion of seeds to potential vector surfaces. However, it is uncertain whether the seeds are dispersed primarily exozoically (e.g., on the muddy feet or feathers of birds) or endozoically (by passage of viable seeds through the avian digestive tract). The latter has not been demonstrated conclusively. Because the birds feed primarily on the vegetative structures (ingesting seeds infrequently), some form of exozoic seed transport seems to be more plausible, at least as a primary means of long-distance dispersal. Vegetative reproduction (in the perennials) occurs by means of stolon proliferation. In North America, the stolons produce terminal, indurate tubers (also characterized as turions or winter buds) or plantlets, which serve as overwintering structures and/or as vegetative propagules. Fragmentation of the stolons and their attached tubers or plantlets results in numerous ramets, which are capable of dispersal along the substrate by water currents. At least in warmer climates, whole plants of some species are capable of overwintering by dormant apices.

2.3.11.1. ***Vallisneria americana*** **Michx.** grows in non-tidal or tidal backwaters, borrow pits, canals, channels, ditches (roadside), flowages, impoundments, lakes, mill races, oxbows, ponds (beaver), pools, reservoirs, rivers, sloughs, and streams at elevations to 701 m. The waters generally are alkaline but span a wide range of conditions (pH: 5.6–9.9; conductivity: 5–600 µmhos cm^{-1}; total alkalinity [as $CaCO_3$]: 5–290 mg l^{-1}). Experimentally, the plants can endure even higher acidity (pH: 5.0) for at least 60 days but then exhibit substantially reduced growth, which has been attributed to the toxic effects of elevated tissue aluminum and iron levels. Nutrients are taken up by the foliage and roots. The plants are able to use aqueous bicarbonate as a carbon source but can assimilate as much as 75% of their CO_2 by uptake from the sediment when CO_2 concentrations in the water column are low (e.g., 16 µmol l^{-1}). The uptake of phosphate (but not ammonium nitrogen) is facilitated by aquatic vesicular-arbuscular mycorrhizal fungi. The salinity tolerance of this species is uncertain

due to the uncertain identity of experimental plants, which originated from the southern portion of the range where *V. neotropicalis* (see below) also occurs and often is misidentified. Exposures range from open conditions to partial shade (e.g., under docks). The plants are well adapted to low light conditions (light extinction coefficient of 0.013–0.019 m^2·g^{-1}) but are most abundant at sites receiving at least 38% of the surface illumination and exhibit increasingly poorer survival and reduced rosette and leaf numbers when subjected to incrementally lower light levels (from 53% to 7% total incident light). Water depths also vary (0.3–7.0 m) but typically average about 1.3 m. The optimum water temperature (for maximum photosynthesis) has been determined as 32.6°C. Water velocities range from still to quickly flowing conditions; the net photosynthetic rate can increase by an order of magnitude in weak currents (0.066 m s^{-1}) relative to still water. The substrates have been characterized as clay (Schriever), clayey alluvium, clayey mud, cobble, gravel, gravelly sandy clay, loamy sand, marl, muck, mucky sand, mud, muddy gravel, rock, rocky mud, sand (Duckston), sandy muck, sandy mud, sandy rock, silt, and silty clay. The maximum biomass (August) varies considerably (50–217 g dry wt·m^{-2}) and can be reduced by as much as 50% if plants are exposed to small (0.15 m) but periodic waves. The maximum productivity has been recorded at 3.2 g m^{-2} day^{-1}, with leaves accounting for up to 70% of the summer biomass. Biomass allocation to sexual reproduction (flower production) is relatively low (e.g., 3–13%). Flowering has been observed from July to September with fruiting from August to October. Flowering rates can vary from 5% to 60% seasonally and also are higher at more southerly latitudes. Sex ratios vary considerably, with more male-biased ratios (e.g., 10:1) associated with lower flowering frequencies (e.g., 27%) and less male bias (e.g., 3:1) occurring when flowering frequencies increase (e.g., 42%). Male-biased ratios also are higher in shallow (e.g., 0.5 m) waters and decline with depth (e.g., 2.5 m), where female-biased ratios can occur. Each male inflorescence contains numerous (e.g., 817–1,418) flowers. Each male flower produces only about 68 pollen grains on average, and a significant number (up to half) can be entirely sterile (lack pollen). The average fruit set varies from 0% to 97% in natural populations, but varies inversely with respect to water currents (which carry pollen-bearing flowers away), reaching zero at current velocities above 0.30 m·s^{-1}. Understandably, field studies have indicated that populations often are pollen limited as evidenced by the lower seed set per fruit in comparison to controlled experimental pollinations. However, in other cases, seed production appears to be tied more closely to variation in the induction of female flowers. Female plants generally are larger than the males. Because the female flowers must reach the surface at anthesis, the peduncles elongate rapidly (to 2 cm h^{-1}) for up to 2 days. Deeper water prevents the female flowers from reaching the surface and strong surface currents or wind can carry male flowers away from female plants. Each female flower contains numerous (e.g., 167–288) ovules, resulting in a large number of viable seeds (e.g., 156–282 per fruit). The entire fruit (which contains the seeds in a gelatinous matrix) is buoyant and capable of transport on the surface by wind and water currents. Any seeds that become dislodged from the fruit sink quickly. Seed viability is high (e.g., 93–98%), with the greatest proportion of viable seeds located closest to the stylar end of the ovary. The seeds remain viable when stored in water at 3°C–5°C (at least 3 months), at salinities up to 10‰, and when desiccated to a moisture content of 4%. Densities as high as 16,000 seeds m^{-2} have been observed, but only 1–13% of deposited seeds are retained in a persistent seed bank (of unknown maximum duration). The seeds are non-dormant and have germinated well under a day/night temperature regime of 19°C/15°C; however, stratified seeds have a higher percent germination. Scarification of the seed coat can enhance germination substantially (to 90% after 60 days). Seed germination occurs independent of light, but is somewhat higher at those levels above 9% of surface illumination. Optimal germination conditions include high oxygen levels (e.g., 8.00 mg l^{-1}), warm temperatures (>22°C), low salinities (<1‰), and sandy sediments (>40%) with low organic matter content (≤3%). Germination occurs uniformly across burial depths from 0.2 to 10.0 cm. Although many seeds can be produced, recruitment via seedlings occurs rarely, an outcome due at least in part by the earlier emergence and growth of plants originating from vegetative propagules. Vegetative reproduction is stimulated at temperatures above 25°C and results in a well-developed system of elongating stolons (up to 20 per plant each season), which produce durable, almost turion-like tubers (referred to by some as "winter buds"). Roughly 13–27% of the total biomass is allocated to stolon production with up to 11% of the total biomass allocated to tuber formation. The tubers (each up to 0.18 g) typically occur as sediment depths from 10 to 20 cm in silty clay and slightly less (5–15 cm) in sand, where their survival is lower; they do not survive when buried by more than 15 cm of sediment. Tuber size and numbers decline if plants are subjected to 2 weeks of darkness. The density of ramets can reach 677 m^{-2}. Complexes of stolon fragments bearing tubers or newly emerging plantlets can be transported by water currents as vegetative propagules. Although the plants reproduce both sexually and asexually, genetic markers (allozymes) have shown that populations can retain fairly high levels of variation overall (e.g., $H_T = 0.3403$), yet because that variation is partitioned at a small scale, they still can be dominated by only a few clonal genotypes. Intra- and interpopulational crosses have detected no significant effects of inbreeding or outbreeding depression, which indicates that differences in seed production and germination primarily reflect local adaptations. Genetic studies (using microsatellite markers) have demonstrated that restored *Vallisneria* populations can contain levels of genetic variation comparable to those of natural populations. **Reported associates:** *Alternanthera philoxeroides, Brasenia schreberi, Cabomba caroliniana, Callitriche, Ceratophyllum demersum, Ceratophyllum echinatum, Chara vulgaris, Egeria densa, Eleocharis acicularis, Eleocharis acicularis, Eleocharis palustris, Eleocharis quadrangulata, Elodea canadensis, Elodea nuttallii, Gratiola aurea, Hydrodictyon reticulatum, Isoetes echinospora, Lemna minor, Lemna trisulca, Myriophyllum sibiricum, Myriophyllum spicatum, Myriophyllum heterophyllum × Myriophyllum laxum,*

Myriophyllum sibiricum × Myriophyllum spicatum, Najas flexilis, Najas gracillima, Najas guadalupensis, Najas marina, Najas minor, Nuphar variegata, Nymphaea odorata, Pontederia cordata, Potamogeton amplifolius, Potamogeton bicupulatus, Potamogeton crispus, Potamogeton epihydrus, Potamogeton foliosus, Potamogeton friesii, Potamogeton gramineus, Potamogeton illinoensis, Potamogeton natans, Potamogeton nodosus, Potamogeton perfoliatus, Potamogeton praelongus, Potamogeton pusillus, Potamogeton richardsonii, Potamogeton robbinsii, Potamogeton spirillus, Potamogeton zosteriformis, Ranunculus longirostris, Rhizoclonium, Ruppia maritima, Sagittaria cuneata, Sagittaria latifolia, Schoenoplectus pungens, Sparganium, Stuckenia pectinata, Utricularia macrorhiza, Utricularia purpurea.

2.3.11.2. ***Vallisneria neotropicalis* Vict.** inhabits non-tidal or tidal canals, lakes, reservoirs, rivers, springs (mineralized), and streams at elevations to at least 3 m. The known habitats are characterized by clear, alkaline (pH: 7.25) waters ranging in depth from 1.5 to 3 m. The chlorophyll A:B ratio remains stable across a depth gradient from 0 to 1.5 m. These are predominantly freshwater plants that can tolerate somewhat brackish (0–9‰) tidal conditions, but they do not extend into more saline sites. The substrates include gravel and mud. Flowering and fruiting have been observed from March to October. The plants exhibit low to moderate tolerance to sites contaminated by arsenic and mercury. Their total biomass can reach 1939 g m^{-2} (23.7% belowground). Life history information for this species is difficult to elucidate from the literature due to its confused history of misidentification as *V. americana*. A thorough re-evaluation needs to be conducted based on properly identified plants. **Reported associates:** *Azolla, Ceratophyllum demersum, Chara, Egeria densa, Eleocharis, Fissidens fontanus, Heteranthera dubia, Hydrilla verticillata, Najas guadalupensis, Nymphaea ampla, Potamogeton illinoensis, Potamogeton nodosus, Riccia fluitans, Ruppia maritima, Sagittaria, Salvinia natans.*

2.3.11.3. ***Vallisneria spiralis* L.** is a nonindigenous species, which in North America is known only from one freshwater river at an elevation of 172 m. Old World populations commonly grow on muddy substrates to depths of 3 m in stagnant or slow-moving, nutrient-rich waters. Optimal culturing conditions indicate a preference for alkaline, medium-hard to hard waters high in carbonate and a sandy substrate. Aquarium-grown plants thrive at water temperatures from 15°C to 30°C. Information derived from virtually all North American accounts pertaining to "*V. spiralis*" is inapplicable, being based on the improper synonymy of that species with *V. americana*. The ecology and life history of the nonindigenous North American populations of *V. spiralis* require extensive investigation, as nearly no reliable information currently exists. **Reported associates (Texas):** *Ceratophyllum demersum, Heteranthera dubia, Hydrilla verticillata, Hydrocotyle, Hygrophila polysperma, Potamogeton illinoensis, Sagittaria platyphylla, Zizania texana.*

Use by wildlife: *Vallisneria* plants are a rich food source for various wildlife and are fairly high in calcium (1.8% of DM). Their maximum calorific content (at peak biomass) can approach 3200 kJ m^{-2}, with a crude protein content of

15% and crude fiber content of 27%. The crude protein content of the flowers or fruits can average as much as 21.8% of the dry mass. *Vallisneria americana* plants harbor numerous invertebrates including amphipods (Crustacea: Gammaridae: *Gammarus tigrinus*), bivalves (Mollusca: Bivalvia: Dreissenidae: *Dreissena polymorpha*; Pisidiidae: *Pisidium*), cladocerans (Crustacea: Sididae: *Sida crystallina*), hydras (Cnidaria: Hydridae: *Hydra*), insects (Insecta: Diptera: Chironomidae: *Cricotopus bicinctus*, *Harnischia curtilamellata*, *Phaenopsectra*, *Polypedilum*, *Procladius*, *Rheotanytarsus*, *Tanytarsus*; Coleoptera: Chrysomelidae: *Galerucella nymphaeae*; Lepidoptera: Crambidae: *Parapoynx obscuralis*), nematodes (Nematoda: Actinolaimidae: *Actinolaimus*; Aporcelaimidae: *Idiodorylaimus novaezealandiae*; Dorylaimidae: *Dorylaimus stagnalis*; Mermithidae), and oligochaete worms (Annelida: Naididae: *Arcteonais lomondi*, *Aulodrilus americanus*, *Limnodrilus hoffmeisteri*, *L. udekemianus*, *Nais variabilis*, *Potamothrix moldaviensis*, *Pristina leidy*, *Stylaria lacustris*). The foliage and tubers (and occasionally the fruits) of *V. americana* are eaten by at least 19 species of waterfowl (Aves: Anatidae) mainly by baldpates (*Anas americana*), black ducks (*Anas rubripes*), blue-winged teal (*Anas discors*), buffleheads (*Bucephala albeola*), canvasbacks (*Aythya valisineria*), gadwalls (*Anas strepera*), goldeneyes (*Bucephala clangula*), greater scaups (*Aythya marila*), green-winged teal (*Anas crecca*), lesser scaups (*Aythya affinis*), mallards (*Anas platyrhynchos*), redhead ducks (*Aythya americana*), and swans (*Cygnus columbianus*). *Vallisneria neotropicalis* is eaten by West Indian manatees (Mammalia: Trichechidae: *Trichechus manatus*).

Economic importance: food: In Asia, young leaves of *Vallisneria* (*V. natans*) are eaten in salads. However, due to their ability to concentrate heavy metals and pesticides, these plants probably should be avoided as a food, at least if collected from the wild; **medicinal:** *Vallisneria* plant extracts have been used in Asia to treat leucorrhoea, stomach-ache, and as a demulcent; **cultivation:** *Vallisneria* is one of the earliest genera to be cultivated in aquaria and persists in its popularity to this day. *Vallisneria americana* has been associated with the cultivar 'Rubra'; cultivars of *V. spiralis* (the most widely distributed species commercially) include 'Contortionist', 'Portugalensis', 'Torta', and 'Tortifolia'; **misc. products:** *Vallisneria americana* has been evaluated as a biomonitor of organochlorine and septic contamination; **weeds:** *Vallisneria* generally is not regarded as a weed in North America; **nonindigenous species:** *Vallisneria spiralis* was documented as an introduction to the San Marcos River, Texas by DNA sequence analysis of a specimen collected in 2003. It is believed to have been introduced there by the careless disposal of cultivated aquarium plants.

Systematics: Phylogenetic analyses including nearly all recognized *Vallisneria* species have demonstrated the monophyly of the genus, once the vittate (i.e., leafy-stemmed) genus *Maidenia* is included; all three vittate species nest within the group, with *Nechamandra* and *Vallisneria* consistently resolved as sister genera (Figures 1.8 and 1.14). The incorporation of DNA sequence analysis has expanded previous taxonomic concepts, which recognized *Vallisneria* as containing only two species, to one that recognizes the genus as being far more diverse with 12–15 species worldwide. Early taxonomic

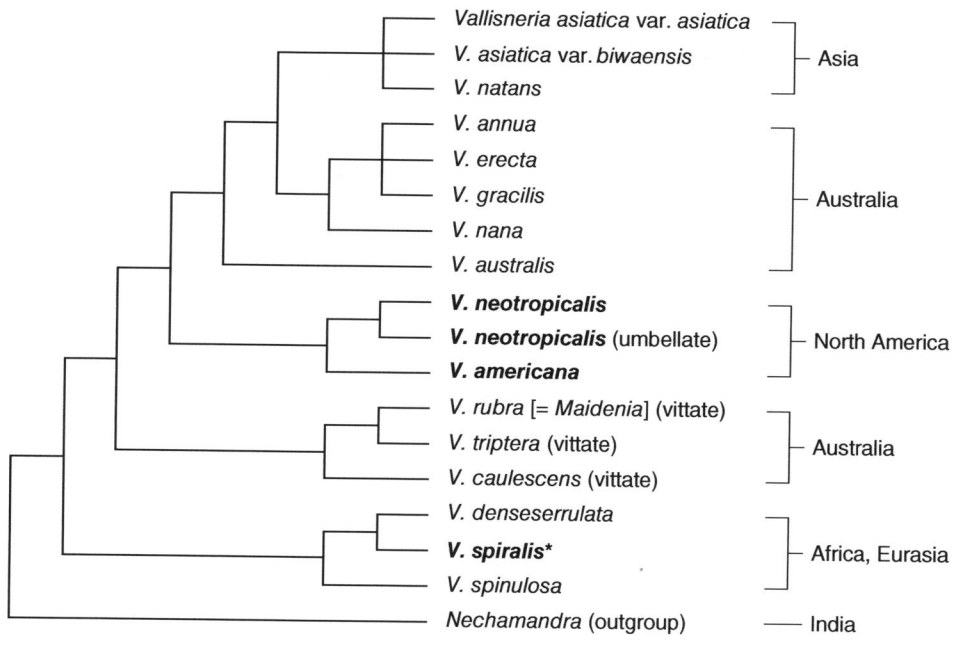

FIGURE 1.14 Phylogenetic relationships in *Vallisneria* as indicated by an analysis of combined molecular and morphological data (adapted from Les et al., 2008). These results argue strongly for the recognition of at least a dozen distinct *Vallisneria* species in contrast to earlier taxonomic treatments (e.g., Lowden, 1982), which recognized only two. The vittate Australian clade (which includes the former genus *Maidenia*) is embedded within the rosulate species. All *Vallisneria* species (and their sister genus *Nechamandra*) are submersed aquatic plants. The OBL North American indicators (in bold) include both of the indigenous species as well as the introduced *V. spiralis* (asterisked), which once was treated as synonymous with the distantly related *V. americana*.

treatments initially regarded the North American material (all rosulate) as being conspecific with the European *V. spiralis*, but eventually its distinctness (as *V. americana*) was realized. That conclusion was substantiated by DNA sequence analysis, which indicated that only a remote relationship existed between *V. americana* and *V. spiralis* (Figure 1.14). However, genetic analyses also confirmed that "true" *V. spiralis* actually did occur in North America, in at least one site where it had been introduced. Although all North American *Vallisneria* material has long been treated as a single species (*V. americana*), phylogenetic analyses have demonstrated the distinctness of *V. neotropicalis* as another indigenous component of the flora. The basic chromosome number of *Vallisneria* is $x = 10$. *Vallisneria americana* ($2n = 20$) is a diploid; whereas numerous cytotypes at various ploidy levels have been reported for *V. spiralis* ($2n = 16, 20, 22, 24, 28, 30, 33, 40$), perhaps due at least in part to taxonomic misidentifications of the material examined. Counts are unavailable for *V. neotropicalis*. Certainly, the cytology of the genus should be re-evaluated in the context of the revised taxonomic framework. No natural hybrids have been reported in the genus.

Distribution: *Vallisneria americana* is widespread in the eastern United States and adjacent Canada, and is disjunct in the northwestern United States and adjacent Canada. *Vallisneria neotropicalis* occurs near southern coastal regions of Alabama, Florida, Louisiana, Mississippi, and Texas; *V. spiralis* currently is known only from the upper San Marcos River in Texas.

References: Badzinski et al., 2006; Beal, 1977; Benson et al., 2008; Brunel, 2009; Campbell & Irvine, 1977; Capers, 2003a; Catling et al., 1994; Cook, 1982; Cowell & Resico, Jr., 1975; Custer & Custer, 1996; Donnermeyer & Smart, 1985; Doyle, 2001; Doyle & Smart, 2001; Doyle et al., 2014; Ferasol et al., 1995; Grisé et al., 1986; Jarvis & Moore, 2008; Jones & Drobney, 1986; Kasselmann, 2003; Kauth & Biber, 2015; Kimber et al., 1995; Korschgen et al., 1988; 1997; Kreiling et al., 2007; Lafabrie et al., 2011; Lamoureux, 1987; Lauer et al., 2011; Les & Tippery, 2013; Les et al., 1997a; 2006a; 2008; Linn et al., 1975; Lloyd et al., 2009; 2012; Lokker et al., 1994; 1997; Lovett Doust & Laporte, 1991; Lovett Doust et al., 1994; Lowden, 1982; Mabbott, 1920; Marie-Victorin, 1943; Marsden et al., 2013; Martin & Uhler, 1951; McAtee, 1939; McFarland, 2006; Miller et al., 2011; Mühlberg, 1982; Nishihara & Ackerman, 2006; Owens et al., 2010; Partridge et al., 2012; Prameela & Venkaiah, 2016; Ross et al., 2016; Rybicki & Carter, 1986; Schloesser & Manny, 2007; Spencer et al., 2000; Strayer et al., 2003; Sullivan & Titus, 1996; Svedelius, 1932; Titus & Adams, 1979; Titus & Stephens, 1983; Titus & Stone, 1982; Van et al., 1977; Wigand, 1997; Winkel & Borum, 2009; Wylie, 1917; Zhongqiang et al., 2005.

ORDER 3: POTAMOGETONALES [8]

Molecular phylogenetic analyses consistently resolve Potamogetonales (= Zosterales Nakai) as a discrete clade, which is the sister group of the Alismatales (Figure 1.2). Also referred to as the "tepaloid" clade, these species differ from members of the Alismatales by possessing simplified flowers that lack a differentiated calyx and corolla (Posluszny & Charlton, 1993; Posluszny et al., 2000). The flowers of this group are interpreted as developmentally decoupled unisexual modules, characterized by a superposition of stamens and tepals (Posluszny et al., 2000). In this way they represent a highly primitive floral structure reminiscent of the very earliest angiosperm ancestors where the modern floral axis presumably was derived by condensation of an inflorescence bearing unisexual flowers. Not surprisingly, unisexual flowers (in dioecious or monoecious arrangement) are widespread in the group. As such, the floral development of this group deserves more intensive study. The flowers then yield several synapomorphies for the order including stamen/tepal superposition and the lack of a differentiated or petaloid perianth. The entire order comprises aquatic or wetland plants, including both freshwater and marine ("seagrass") species. Unlike Alismatales, most Potamogetonales are pollinated abiotically via wind (anemophily) or by water (hydrophily). Somewhat showy flowers (essentially by coloration of the tepals and/or sexual organs) occur occasionally within the relatively unspecialized Aponogetonaceae, which is sister to the remainder of the order (Figure 1.2). These species also are the only Potamogetonales that are insect pollinated (entomophilous).

Recent phylogenetic analyses have recommended several taxonomic modifications in the order including the abandonment of Lilaeaceae (by merging *Lilaea* with *Triglochin* in Juncaginaceae), the transfer of *Maundia* from Juncaginaceae to a monotypic family (i.e., Maundiaceae), inclusion of Ruppiaceae within Cymodoceaceae, and the merger of Zannichelliaceae with Potamogetonaceae (Les & Tippery, 2013). As a result, the taxonomic concept of Potamogetonales followed here recognizes six families in North America (Aponogetonaceae, Cymodoceaceae, Juncaginaceae, Potamogetonaceae, Scheuchzeriaceae, and Zosteraceae) in addition to Maundiaceae and Posidoniaceae (Old World). The circumscription of families within the "Cymodoceaeae complex" (currently Cymodoceaceae and Posidoniaceae) remains contentious (Les & Tippery, 2013).

The economic importance of the order is confined mainly to Aponogetonaceae (*Aponogeton*), in which several aquarium and water garden specimens are cultivated.

Potamogetonales are cosmopolitan. All six of the North American families include OBL indicators:

3.1. **Aponogetonaceae** Planch.
3.2. **Cymodoceaceae** Vines
3.3. **Juncaginaceae** Rich.
3.4. **Potamogetonaceae** Bercht. & J. Presl
3.5. **Scheuchzeriaceae** F. Rudolphi
3.6. **Zosteraceae** Dumort.

Family 3.1. Aponogetonaceae [1]

Aponogetonaceae are a small but well-defined family consisting entirely of aquatic plants having floating or submersed foliage. Phylogenetically, the group represents an early diverging lineage of the "tepaloid" alismatids, which possess simplified, modular flowers (Figure 1.2). The flowers of some species are notably pseudanthial (Cronquist, 1981). Additional features are discussed in the generic treatment below.

The family is indigenous to the Old World but is introduced widely elsewhere. It includes a single genus, which also contains OBL North American indicators:

3.1.1. *Aponogeton* L. f.

3.1.1. Aponogeton

Cape pondweed, water hawthorn

Etymology: after the *Aponi fons* (a spring) and *geitōn*, which is Greek for "resident"

Synonyms: *Amogeton*; *Apogeton*; *Hydrogeton* (in part); *Ouvirandra*; *Spathium*; *Uvirandra*

Distribution: global: Africa, Asia, Australia, Europe*, New World*; **North America:** western (California)

Diversity: global: 59 species; **North America:** 1 species

Indicators (USA): obligate wetland (OBL): *Aponogeton distachyos*

Habitat: freshwater; lacustrine, palustrine, riverine; **pH:** 4.5–8.0; **depth:** <2 m; **life-form(s):** floating-leaved

Key morphology: stems cormose (to 6 cm) the foliage natant, rosulate; floating leaf blades (to 23 cm) ovate to lanceolate, rounded or tapered at base, the margins entire, long-petioled (to 100 cm); peduncle (to 80 cm) arising from a caducous spathe (to 3 cm), terminating in a pair of spikes (to 4.5 cm); flowers fragrant, secund, arranged in two rows, bisexual; perianth of a single white (or pinkish) tepal (to 15 mm); stamens (to 4.5 mm) 8–16, the filaments basally dilated, the anthers purple; gynoecium apocarpous, of 2–6 pistils (to 3 mm), ovules few (~4); follicles (to 22 mm) with an apical beak (to 5 mm); seeds (to 17 mm) with one seed coat

Life history: duration: perennial (corms); **asexual reproduction:** corms; **pollination:** insect; **sexual condition:** hermaphroditic; **fruit:** follicles (common); **local dispersal:** water (seeds); **long-distance dispersal:** unknown (seeds)

Imperilment: 1. *Aponogeton distachyos* [GNR]

Ecology: general: All *Aponogeton* species are floating-leaved or submersed aquatic perennials, which occur in freshwater (occasionally in brackish sites). The flowers are bisexual or (less commonly) unisexual and then dioecious or monoecious; they extend above the water surface where they are insect pollinated or self-pollinated. In some cases, they possibly are agamospermous apomicts. The seeds are dispersed locally by water. Vegetative propagation occurs by means of corms or rhizomes.

3.1.1.1. *Aponogeton distachyos* L. f. is a nonindigenous perennial, which inhabits ditches (roadside), ponds, pools (roadside), streams (slow-moving), and vernal pools at elevations to 135 m. The plants can grow in 1–2 m of water but normally are found in shallower sites (<1 m). This is a cool climate species that does not thrive during periods of summer heat; however, the corms are not hardy and do not survive periods of winter freezing. The plants can survive on exposed substrates following drawdown conditions but do not thrive under those conditions. They can occur in exposures from light shade to full sunlight. The substrates (North America)

have been described as clay. The plants can tolerate a wide range of acidity (pH: 4.5–8.0) but grow optimally under circumneutral conditions (pH: 6.0–7.5). Flowering (California) has been observed from February to September. The flowers are fragrant and attract bees (Insecta: Hymenoptera: Apidae) in their indigenous range (Africa). The seeds are produced in abundance, even without artificial pollination. **Reported associates (North America):** *Carex*, *Zannichellia palustris*. **Use by wildlife:** The foliage is relished by snails (Mollusca: Gastropoda).

Economic importance: food: *Aponogeton distachyos* is eaten in South Africa where it has become a part of the "traditional Cape cuisine" served by some restaurants. The roasted corms are considered by some to be a delicacy. The flowers are pickled, prepared as a vegetable, or added to stews as a substitute for green beans; **medicinal:** The juicy foliage of *A. distachyos* has been used to treat abrasions, burns, and sunburn; **cultivation:** *Aponogeton distachyos* has been cultivated in botanic gardens for more than two centuries. It is distributed commercially as an ornamental plant for garden ponds. In South Africa, *A. distachyos* has been in cultivation as a food plant (known as Waterblommetjie) since the 1970s. Even during this short time period, the size and mass of cultivated plant flowers have increased significantly through artificial selection; **misc. products:** none; **weeds:** *Aponogeton distachyos* sometimes is regarded as an invasive weed; **nonindigenous species:** *Aponogeton distachyos* became naturalized in California sometime before 1947 as an escape from cultivation. It also has been introduced to Argentina, Australia, Belgium, England, France (around 1830), Ireland, New Zealand, Peru, Scotland, The Netherlands, and Wales.

Systematics: Early analyses of protein sequence data placed the monotypic family Aponogetonaceae (i.e., *Aponogeton*) at the base of those monocots surveyed, which indicated (as presumed from non-molecular data) its relatively basal position among monocotyledons. Subsequent studies (using DNA sequence data) resolved the group as the sister clade to the remainder of Potamogetonales within subclass Alismatidae (e.g., Figure 1.2). Several phylogenetic investigations of interspecific relationships (which included from 21 to 42 *Aponogeton* species) have consistently demonstrated the monophyly of the genus. These studies also have resolved the Australian *A. hexatepalus* as the sister group to the remainder of the species. The most comprehensive phylogenetic analysis to date (Figure 1.15) situates *A. distachyos* in a relatively isolated position among other African species. However, its placement is not too distant from a clade containing *A. junceus*, which an earlier taxonomic treatment regarded as being closely related. The rare *A. viridis*, another proposed relative, has yet to be included in any phylogenetic analysis. *Aponogeton distachyos* currently is placed within section *Pleuranthus* subsection *Pleuranthus*. However, the existing infrageneric classification of *Aponogeton* (sections and subsections) is not entirely compatible with recent phylogenetic results and needs to be revised accordingly. The basic chromosome number of *Aponogeton* is $x = 8$. *Aponogeton distachyos* ($2n = 16, 24, 32$) exhibits a range of polyploid cytotypes.

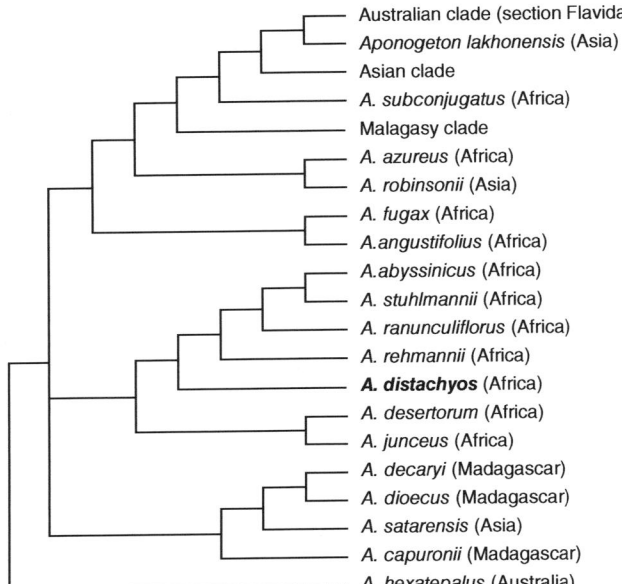

Australian clade (section Flavida)
Aponogeton lakhonensis (Asia)
Asian clade
A. subconjugatus (Africa)
Malagasy clade
A. azureus (Africa)
A. robinsonii (Asia)
A. fugax (Africa)
A. angustifolius (Africa)
A. abyssinicus (Africa)
A. stuhlmannii (Africa)
A. ranunculiflorus (Africa)
A. rehmannii (Africa)
A. distachyos (Africa)
A. desertorum (Africa)
A. junceus (Africa)
A. decaryi (Madagascar)
A. dioecus (Madagascar)
A. satarensis (Asia)
A. capuronii (Madagascar)
A. hexatepalus (Australia)

FIGURE 1.15 Interspecific relationships in *Aponogeton* as indicated by analysis of combined DNA sequence data (modified and simplified from Chen et al., 2015). This analysis places *Aponogeton distachyos*, the only OBL North American indicator (in bold), in an isolated position among other African species. The relationships depicted are consistent with those found in previous studies (Les et al., 2005; Les & Tippery, 2013), which included fewer species.

Although hybridization occurs commonly in *Aponogeton*, no hybrids involving *A. distachyos* have been reported.

Distribution: In North America, *Aponogeton distachyos* is known to occur only in California, where it was introduced.

References: Chen et al., 2015a; Cook, 1985; 1996a; De Vynck et al., 2016; Hellquist & Haynes, 2000; Hussner, 2012; Les & Tippery, 2013; Les et al., 2005; Martin & Dowd, 1986; Mühlberg, 1982; Pemberton, 1998; 2000; Quattrocchi, 2012; Van Bruggen, 1973; 1985; 1990; Van Wyk, 2011.

Family 3.2. Cymodoceaceae [6]

Hartog (1970) recognized five seagrass genera (*Amphibolis, Cymodocea, Halodule, Syringodium* and *Thalassodendron*) as a subfamily (Cymodoceoideae) of Potamogetonaceae. That unconventional scheme was never widely accepted, with most classifications instead placing these genera within a distinct family, Cymodoceaceae. In a broad, but single gene (*rbcL*) phylogenetic study of alismatid monocots, Les et al. (1997) introduced the "Cymodoceaceae complex," a group of three families (Cymodoceaceae, Posidoniaceae, and Ruppiaceae), which previously were not thought to be closely related. However, the interrelationships among those three families were not well established in that study, with Posidoniaceae resolving as the sister group to a polytomy comprising members of Cymodoceaceae and Ruppiaceae. Nevertheless, the *rbcL* data demonstrated the monophyly of Cymodoceaceae and its distinctness from both Potamogetonaceae and Zosteraceae, the latter being another family merged with Potamogetonaceae by Hartog (1970). Several subsequent studies using combined DNA sequence data (Petersen et al., 2006;

Ross et al., 2016) also recovered the "Cymodoceaceae complex" as a clade, but with Ruppiaceae resolving either within or outside of Cymodoceaceae proper (Les & Tippery, 2013). Analysis of full cpDNA genome sequences (Ross et al., 2016) retained Ruppiaceae as the sister group of Cymodoceaceae. Taking all of the phylogenetic results into account, reasonable arguments could be made for either retaining Ruppiaceae or merging it with Cymodoceaceae. The latter approach has been taken here, as recommended by Les & Tippery (2013).

In comprehensive analyses using cpDNA or mtDNA sequences (Les et al., 1997a; Petersen et al., 2006; Ross et al., 2016), Cymodoceaceae, together with Posidoniaceae, consistently comprise the sister group to a clade containing Potamogetonaceae and Zosteraceae (e.g., Figures 1.2 and 1.16). Currently, six genera (*Amphibolis, Cymodocea, Halodule, Ruppia, Syringodium* and *Thalassodendron*) are assigned to the family. However, "splitting" at various taxonomic levels seems to pervade the different seagrass groups, where much of the nomenclature has been authored by researchers trained primarily in fields other than systematics. The low level of molecular divergence observed among *Amphibolis, Cymodocea, Syringodium*, and *Thalassodendron*, which all contain species once recognized within *Cymodocea*, warrants a renewed evaluation of generic limits in the group, especially with respect to *Cymodocea*, which does not appear to be monophyletic (Les & Tippery, 2013; Petersen et al., 2014). Hence, the "Cymodoceaceae complex" remains complex taxonomically.

As treated here, Cymodoceaceae represent a clade of obligately aquatic, submersed, water-pollinated (hypohydrophilous) marine plants, which have simplified bisexual or unisexual flowers devoid of a perianth and consisting of two stamens and an apocarpous gynoecium. Cymodoceaceae are distributed across temperate and tropical oceans worldwide. The group is of no particular commercial importance. Three genera include OBL North American indicators:

3.2.1. ***Halodule*** Endl.
3.2.2. ***Ruppia*** L.
3.2.3. ***Syringodium*** Kütz.

3.2.1. Halodule

Shoal-grass

Etymology: from the Greek *háls doulé* ("servant of the sea") referring to the marine habit of the plants

Synonyms: *Diplanthera*

Distribution: global: Atlantic, Indian, and Pacific Oceans; **North America:** southern (Atlantic, Caribbean, and Gulf of Mexico)

Diversity: global: 2–3 species; **North America:** 1 species

Indicators (USA): obligate wetland (OBL): *Halodule wrightii*

Habitat: brackish (coastal), marine; oceanic; **pH:** 7.9–8.2; **depth:** to 2.0 [20.4] m; **life-form(s):** submersed (rosulate/vittate)

Key morphology: rhizomes with short internodes (to 4 cm), bearing scales (to 1 cm) and adventitious roots (2–5 at a node); foliage leaves alternate, but arising from short, rosette-like clusters, the blades (to 32 cm) flat, linear, sheathed (to 6 cm) at base, the apex variably toothed or emarginate; flowers solitary, unisexual, naked, subterranean, borne within the leaf sheaths; ♂ flower emerging at anthesis from an elongating stalk (to 25 mm) to expose a pair of stamens, their anthers (to 5 mm) fused; pollen grains elongate, numerous; ♀ flower with a pair of compressed, ovoid, ovaries (to 2mm), each elongating style (to 28 mm) emerging at anthesis to expose the stigmatic surface; achenes (to 2 mm) with a hard exocarp

Life history: duration: perennial (rhizomes); **asexual reproduction:** rhizomes, shoot fragments; **pollination:** water (hypohydrophily); **sexual condition:** dioecious; **fruit:** achenes (infrequent); **local dispersal:** sediments, water (rhizomes, fruits); **long-distance dispersal:** animals (uncertain), water (fruits, shoot fragments)

Imperilment: 1. *Halodule wrightii* [G5]; S1 (LA, MS)

Ecology: general: *Halodule* species are obligately submersed, marine, perennial aquatics ("seagrasses"), which propagate from an elongate rhizome. All of the species are dioecious, bearing highly reduced, unisexual flowers that are water pollinated (hypohydrophilous). Given that the genus may contain as few as two distinct species (see *Systematics* next), the following treatment should also provide an adequate general ecological overview for the group.

3.2.1.1. *Halodule wrightii* Asch. inhabits brackish to saline (e.g., salinity: 14–31‰; mean: 24.7‰) bays, canals, coasts, coves, ditches, flats (tidal), marshes (brackish), oceans, ponds (tidal), reefs (limestone), rivers (coastal), salt marshes, and sounds at elevations from 0.03 to –20.4 m. This species usually occupies shallow (0.07–2.0 m), alkaline (pH: 7.9–8.2), protected waters but some stands have been found at depths up to 20.4 m. It can tolerate steadily increasing salinities from 1‰ to 65‰ for at least 30 days; however, prolonged (e.g. week-long) high salinity pulses (>50‰) are detrimental to growth. Experimentally, a salinity of 10‰ has yielded the highest biomass production, with a linear decline observed at values increasing up to 40‰. The plants are fairly heat-tolerant (to 31.0°C), especially those having narrower leaves. The plants also are fairly cold-tolerant and have survived well after a 36-hr exposure to 2°C water temperature. *Halodule wrightii* has a relatively higher light demand than other seagrasses (18–37% surface illumination) and tends to occur in clear waters (turbidity: <30 NTU). Photosynthetic rates up to 47 mg $Cm^{-2}h^{-1}$ have been recorded. The reduction of light due to algal blooms (e.g., Cyanobacteria: Oscillatoriaceae: *Lyngbya majuscula*) results in leaf elongation and reduced biomass. The substrates have been described as clay, limestone, mud, muddy sand, sand, sandy shells, shells (terraces), and silt. Experimental manipulations have observed optimal growth rates on sediments having a moderate level of H_2S (0.5 mM) and low soluble Fe (0.5–2.0 µg ml^{-1}). The plants are dioecious and hypohydrophilous. The male flowers (only one per shoot) are protected by the leaf sheaths and remain beneath the sediment until 4–6 hr before anthesis, when they emerge from the sheaths as they elongate. They release their copious pollen (averaging 0.46×10^6 pollen grain per anther) during a roughly 4-hr period, which can occur at any time during the day or twilight. The pollen is released near the substrate where it is confined to the plant canopy. The solitary female flowers also develop beneath the sediment, extending their stigmas toward the water surface during anthesis by elongation of the style. Consequently, the often clonal plants are more highly inbred despite their sexual condition. This inference is corroborated by genetic data (AFLP markers), which indicate that the plants contain levels of genetic variation (e.g., heterozygosity) that are 2–3 times lower than would be expected in a widespread, dioecious species. However, preliminary surveys using microsatellite markers have detected levels of observed heterozygosity (H_o) ranging from 0.080 to 0.88 in some populations. Many collections of this frequently clonal species are sterile (e.g., less than 8% flowering) due perhaps to logistic factors (e.g., consumption of reproductive structures by herbivores); however, flowering and fruiting can be locally common and have been observed from March to October. Both processes have been related to a warm water temperature requirement. Flowering has been induced under constant illumination at temperatures from 20°C to 26°C. Fruiting also occurs at the sediment level, resulting in the immediate burial of the seeds, which can reach densities of 20,000 m^{-2} in the sediment (the highest densities occurring in sediment furrows). The achenes are not buoyant and are believed to be fairly dispersal limited (over only a few centimeters from the maternal plant). However, the hard coat of the achenes also is suspected to facilitate endozoic dispersal by various animal agents. The seeds have an extended longevity (up to 46 months) and have germinated under laboratory conditions at 24°C–27°C over a 3-year period. The seed longevity results in a long-term seed bank. Natural germination rates tend to be low (~2%); however, the seeds can germinate over a range of salinities from 5‰ to 50‰. Seedling survivorship also is normally quite low (<2%). The ability to maintain high productivity under a wide range of environmental conditions enables the plants to function as colonizers and as early successional species during secondary seagrass community succession. The plants have a relatively high nutrient requirement and thrive in warm water temperatures on organic-rich sediments with high sulfate reduction activity. Growth rates of up to 0.85 cm·day^{-1} have been observed for the leaves. Productivity during June–August can reach 160 g dwt·m^{-2} (to 614 g dwt·m^{-2} in more southern latitudes), with most of the biomass (>66%) being allocated to belowground structures (e.g., aboveground: 50 g dwt·m^{-2}; belowground: 140 g dwt·m^{-2}). The rhizomes can extend at rates as high as 365 cm year^{-1}. Genetic analyses using microsatellite markers have detected identical, widely dispersed genotypes, which indicates that rhizome fragments are dispersed in water and can establish afterward. They can remain viable for up to 4 weeks, especially when shed during the spring seasonal period. Plants growing in the vicinity of human disturbance (e.g., marinas) exhibit lower chlorophyll content and density. **Reported associates:** *Caulerpa prolifera, Halophila decipiens, Halophila engelmannii, Halophila ovalis, Hypnea, Ruppia maritima, Sargassum, Spartina alterniflora, Syringodium filiforme, Thalassia testudinum, Zostera marina.*

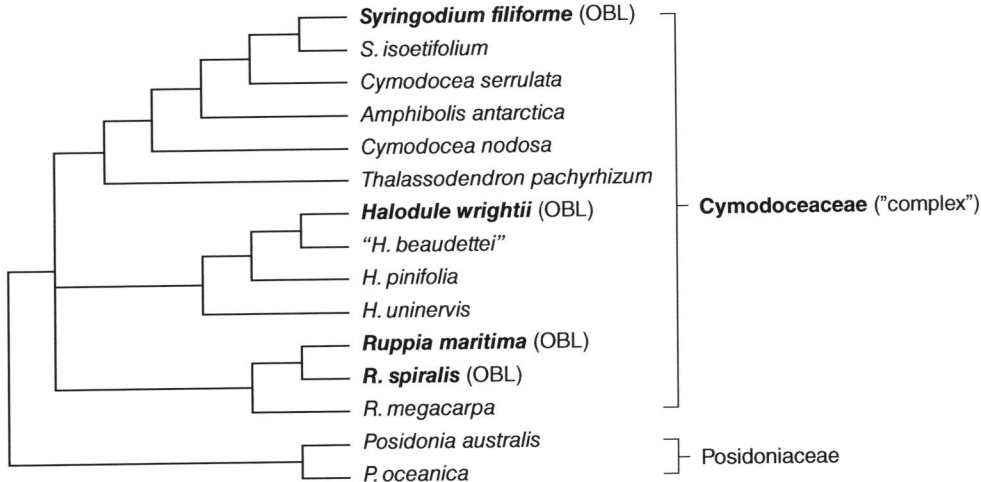

FIGURE 1.16 Phylogenetic relationships in the "Cymodoceaceae complex" as indicated by analysis of *rbcL* sequence data (adapted from Les & Tippery, 2013). Although the placement of *Ruppia* is not resolved here, an analysis of full cpDNA sequence data (but for fewer taxa) places it as the sister group of Cymodoceaceae (Ross et al., 2016). All of the species shown are obligate, submersed aquatics, with the four North American OBL indicators highlighted in bold.

Use by wildlife: *Halodule wrightii* is important ecologically as a cover plant and is preferred habitat for young spotted weakfish (Vertebrata: Teleostei: Sciaenidae: *Cynoscion nebulosus*). Numerous organisms inhabit *H. wrightii* meadows with one study documenting 58 species and abundances (April) as high as 104,338 organisms m^{-2}. The foliage is colonized by protozoans (Protozoa: Plasmodiophoromycota: Plasmodiophoraceae: *Plasmodiophora diplantherae*) and other epiphytes. Cover by epiphytes has been shown to reduce the severity of photosynthetic inhibition caused by excessive UV-β exposure; however, dense coverage can result in reduced photosynthetic rates and tissue necrosis. The leaf epiphytes are grazed by amphipods (Crustacea: Ampithoidae: *Cymadusa compta*; Aoridae: *Grandidierella bonnieroides*; Gammaridae: *Gammarus mucronatus*; Melitidae: *Melita nitida*), buttonsnails (Gastropoda: Modulidae: *Modulus modulus*), grass shrimp (Crustacea: Decapoda: Palaemonidae: *Palaemonetes pugio*), and parrot fish (Vertebrata: Teleostei: Scaridae: *Sparisoma radians*). The plants are eaten by redhead ducks (Aves: Anatidae: *Aythya americana*). The flowers are eaten by various fish (Vertebrata: Teleostei) including grunts (Haemulidae: *Haemulon sciurus*), ocean surgeon (Acanthuridae: *Acanthurus bahianus*), parrot fish (Scaridae: *Sparisoma viride*), puffer fish (Tetraodontidae: *Canthigaster rostrata*, *Sphoeroides spengleri*) and wrasse (Labridae: *Halichoeres bivittatus*). The foliage contains roughly 18% crude protein, 0.8% crude fat, and 34% non-fiber carbohydrate. It is grazed by Florida manatees (Mammalia: Sirenia: Trichechidae *Trichechus manatus latirostris*), green sea urchins (Echinodermata: Toxopneustidae: *Lytechinus variegatus*), green sea turtles (Reptilia: Testudines: Cheloniidae: *Chelonia mydas*), redhead ducks (Aves: Anatidae: *Aythya americana*), and various fish (Vertebrata: Teleostei) including emerald parrot fish (Scaridae: *Nicholsina usta*), filefish (Monacanthidae: *Stephanolepis hispidus*), ocean sunfish (Molidae: *Mola mola*), and pinfish (Sparidae: *Lagodon rhomboides*).

Economic importance: food: *Halodule wrightii* is not eaten by humans; **medicinal:** *Halodule wrightii* contains sulfated polysaccharides, which exhibit high antioxidant and anticoagulant activity; **cultivation:** *Halodule wrightiana* is not cultivated; **misc. products:** *Halodule wrightiana* has been used in some seagrass restoration projects; **weeds:** *Halodule wrightiana* is not regarded as weedy; **nonindigenous species:** none.

Systematics: Several phylogenetic studies have confirmed that *Halodule* is monophyletic. However, its placement within Cymodoceacee is somewhat uncertain with respect to *Ruppia* and other members of the family (e.g., Figure 1.16). Analysis of full cpDNA sequence data (but including only one *Halodule* representative) resolves the genus in a position intermediate to *Ruppia* and the other Cymodoceaceae. Although as many as six *Halodule* species have been recognized, a profusion of names has resulted from an inappropriate focus on leaf blade dimensions and leaf-tip morphology, an approach that has not held up to scrutiny when several of the putative taxa were evaluated genetically, in common culture, or across a broad geographical range. Both features are known to vary considerably depending on their degree of tidal exposure and other environmental influences. Estimates of diversity resulting from independent phylogenetic analyses (e.g., Figure 1.17) indicate no more than three species, with two of these (*H. pinifolia* and *H. uninervis*) being very closely related (if at all distinct) and able to hybridize. There has been no compelling evidence (genetic or otherwise) to distinguish *H. beaudettei*, *H. bermudensis*, or *H. ciliata* from *H. wrightii* and they are regarded here as synonyms of the latter. Older records of *H. wrightiana* in North America often refer mistakenly to *H. beaudettei*. There have been several attempts to develop DNA "barcodes" to distinguish seagrass species, which, in the case of *Halodule*, is irrational given that the taxa still remain so poorly circumscribed. The basic chromosome number of Cymodoceaceae is uncertain ($x = 5, 6, 7,$

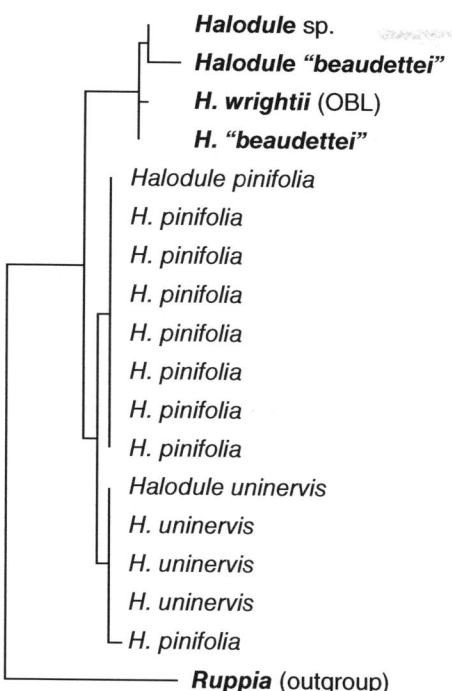

FIGURE 1.17 Phylogenetic associations of various *Halodule* taxa as indicated by analysis of *rbc*L data (modified from Ito & Tanaka, 2011), showing no clear distinction between those of "*H. beaudettei*" and *H. wrightiana* as well as overlap between accessions of *H. pinifolia* and *H. uninervis* (hybrids between *H. pinifolia* and *H. uninervis* also have been documented genetically). These results essentially agree with earlier analyses by Waycott et al. (2006), which recommend the recognition of only 2–3 *Halodule* species worldwide. Taxa with OBL North American indicators are highlighted in bold.

or 8), with that of *Halodule* also vague. Counts of ($2n=44$) for *H. wrightii* initially indicated that it was a polyploid, a conclusion later refuted by the recovery of only a single copy of the nuclear *phy*B gene. Subsequent research demonstrated that the chromosome number of *H. wrightii* within some populations exhibits considerable aneuploid variation ($2n=24–39$), which results from cytomixis (a process involving the formation of cytoplasmic bridges between adjacent pollen mother cells) rather than polyploidy. Consequently, despite some attempts to associate different species with specific chromosome numbers, chromosome number *per se* is of little taxonomic value in the genus. No hybrids involving *H. wrightii* have been reported.

Distribution: *Halodule wrightii* occurs along the southern and southeastern coastal regions of the United States. It extends southward along the Atlantic and Pacific coasts of Mexico and Central America, to northern and eastern Atlantic coastal South America, and also to coastal Africa, India, and Oman.

References: Bakenhaster & Knight-Gray, 2016; Biber & Cho, 2017; Bragg & McMillan, 1986; Cho & May, 2008; Darnell & Dunton, 2016; Da Silva et al., 2017; Dawes et al., 1995; Dunton, 1994; 1996; Ferguson et al., 1993; Fourqurean et al., 1995; Gallegos et al., 1994; Gama et al., 2016; Garrote-Moreno et al., 2015; Green & Short, 2003; Hall et al., 2006; Hartog, 1970; Haynes, 2000f; Herrera-Silveira, 1994; Ito & Tanaka, 2011; Kennedy, 2016; Kenworthy & Fonseca, 1996; Koch et al., 2007; Kowalski & DeYoe, 2016; Kuo, 2013; Larkin et al., 2012; 2017; Lefebvre et al., 2017; Lobel & Ogden, 1981; Lucas et al., 2012; McGovern & Blankenhorn, 2007; McMillan, 1976; 1979; 1981; 1983; 1985; 1991; McMillan et al., 1981; Mitchell et al., 1994; Mueller, 2004; Nguyen et al., 2015; Orth et al., 2006; Phillips & Meñez, 1988; Prado & Heck, Jr., 2011; Pulich, Jr., 1982; 1985; Sheridan & Livingston, 1983; Siegal-Willott et al., 2010; Silva et al., 2012; Stafford & Bell, 2006; Tiling & Proffitt, 2017; Touchette, 2007; Travis & Sheridan, 2006; Uhler, 1982; Vanitha et al., 2016; Van Montfrans et al., 1984; Van Tussenbroek & Muhlia-Montero, 2013; Waycott et al., 2006.

3.2.2. Ruppia
Ditch-grass, widgeon-grass; ruppelle
Etymology: after Heinrich Bernhard Ruppius (1689–1719)
Synonyms: *Bucafer*; *Buccaferrea*; *Dzieduszyckia*
Distribution: global: cosmopolitan; **North America:** widespread
Diversity: global: 5 species; **North America:** 2 species
Indicators (USA): obligate wetland (OBL): *Ruppia maritima*, *R. spiralis*
Habitat: brackish (coastal), freshwater; marine; saline (inland); lacustrine, oceanic, palustrine, riverine; **pH:** 6.7–10.1; **depth:** to 2.5 m; **life-form(s):** submersed (vittate), suspended
Key morphology: rhizomes (to 80 cm) slender, stoloniferous, with flexuous, slender upright stems (to 55 cm); leaves alternate to sub-opposite, the blades (to 45.1 cm) filiform, single-nerved, basally sheathed (adnate to 2 cm), the sheath open, sometimes inflated; inflorescence a spike of a few (usually 2) naked, hermaphroditic flowers (i.e., reduced floral-like inflorescences) borne at the end of a short (<25 mm) or long (>30 cm) flexuous or strongly coiled peduncle; stamens 2, subsessile; gynoecium apocarpous (2–16 pistils), each pistil borne on a slender stipe (to 3.5 cm) in fruit, as an umbelliform cluster; fruits (to 2 mm) single-seeded, achene- or drupe-like, terminally beaked (to 1 mm)
Life history: duration: annual (fruits/seeds); perennial (rhizomes, turions, whole plants); **asexual reproduction:** rhizomes, shoot fragments, turions; **pollination:** water; **sexual condition:** hermaphroditic; **fruit:** achenes (common); **local dispersal:** water (fruits, turions); **long-distance dispersal:** fish, waterfowl (fruits)
Imperilment: 1. *Ruppia maritima* [G5]; S1 (MI, MO); S2 (ON, SK); S3 (GA, NC, NF, NY; QC); **2.** *R. spiralis* [G5]; S1 (AB, IA, KS, YT); S2 (WY); S3 (MB, MN, SK)
Ecology: general: All *Ruppia* species are obligately submersed aquatics, which normally are associated with brackish sites or strongly alkaline freshwater sites. They comprise a basal trophic level in salt marsh ecosystems. The plants also can persist as a suspended growth form if dislodged from the sediment. They usually are rhizomatous perennials when growing under inundated conditions but can assume an

annual habit when subjected to temporary water conditions. The flowers (probably representing compacted inflorescences) are reduced and devoid of a perianth. Stomata in the upper carpel surfaces produce gas bubbles that facilitate floatation of the inflorescence. Accordingly, the flowers likely are unisexual, but are functionally hermaphroditic with the stamens situated adjacent to the pistils. The genus is categorized as hydrophilous (epihydrophilous) even though technically it is not, due to its non-wettable reproductive structures, which are protected by air bubbles. However, the pollen is transported more or less by the water in some cases. In some species the boomerang-like (V-shaped) pollen grains are released below the water surface within bubbles (which also enclose the stigmas) resulting in self-pollination (hydroautogamy). In others, the pollen is transported to the surface within detached, floating anthers, which dehisce to shed the grains in chains or clumps. The unusual shape of the pollen grains facilitates their aggregation. The floating pollen eventually contacts the funnel-like stigmas, which facilitate their capture (essentially epihydrophilous). Once pollinated, the flowers mature underwater. If not already submersed on short peduncles, they are drawn under the surface by the rapid coiling of an elongate peduncle. Genetic studies indicate that the hydroautogamous species are well adapted to self-pollination, which then results in high seed production; the epihydrophilous taxa are more highly outcrossed. The seeds are dispersed within the water column. Although the fruits are not buoyant, they also can disperse by remaining attached to floating vegetative fragments, which drift across the water surface. Genetic evidence has confirmed that ingested seeds are dispersed successfully by fish (Vertebrata: Teleostei) and/or various waterfowl (Aves: Anatidae). Seed germination is stimulated by receding water conditions. Perennial plants can produce turions, which serve as vegetative propagules. *Ruppia* has been so confused taxonomically that it is difficult to determine whether information reported in the literature has been attributed correctly to a species. Where the information reported below was derived from a database, the consultation of accompanying images helped to confirm species identities in some but not all cases. Inland (non-coastal) reports of *R. maritima* are likely to be erroneous, referring instead to *R. spiralis*; however, coastal records could be of either species.

3.2.2.1. ***Ruppia maritima*** **L.** is an annual or perennial, which inhabits coastal backchannels (brackish), borrow pits, canals (drainage), channels (estuary, seepage), depressions, ditches, estuaries, flats (saline, tidal), inlets, lagoons (brackish, saline), meadows (halophilic), mudflats (estuary, intertidal, tidal), pannes, ponds (brackish, roadside, salt marsh, tidal), pools (intertidal, saline, tidal), puddles, salt marshes, sloughs, sounds, springs (limestone), and streams (brackish) at elevations from −1 to 65 m. In coastal habitats, the plants generally occur from +20 to −100 cm elevations. The plants usually occur in shallow water <60 cm but have been reported in sites up to 2.5 m deep. Their substrates include clayey silty ooze (organic), muck (Dirego, organic), mud (organic), peat (decomposing, intertidal), sand (Duckston, St. Lucie), sandy clay, sandy gravel, silt, and silty mud. More aerobic sediments with

low H_2S levels (e.g., average fall concentration of 50 μM in the upper 10 cm of substrate) are preferred. The plants grow in brackish or saline waters (mesohaline to hyperhaline) of high alkalinity, conductivity, and pH (e.g., pH: 6.7–10.1; conductivity: 260–680 μmhos cm^{-1}; alkalinity: 130–225 mg l^{-1} [as $CaCO_3$]; chloride >10 mg l^{-1}). The typical pH averages about 8.1. Experimentally, *R. maritima* can tolerate salinities from 0 to 70‰ (mean: 24.7‰); however, photosynthesis is optimal (highest quantum yields) at levels between 10‰ and 20‰. The plants appear to succumb more to pulsed salinity increases. Although the plants can tolerate a wide range in water temperature (12.2°C–31.0°C), some studies indicate that the most vigorous growth occurs under cool spring temperatures (e.g., 12°C–18°C) rather than at low water salinities. The mean water temperature has been reported as 21.5°C. The photosynthetic rate increases steadily with water temperatures up to about 23°C. Although generally found in low nutrient sites, the plants respond positively to enhanced nitrate nitrogen levels in the water column. They uptake nutrients (i.e., ammonium nitrogen and phosphate) through their roots and shoots, but leaf uptake is reduced when an adequate nutrient supply to the roots exists. The source of root nitrogen uptake primarily is ammonium nitrogen, despite the observation that the roots release a fair amount of oxygen [average rate of 2–3 μg O_2 (mg dry wt)$^{-1}$ h^{-1}] leading to the development of a distinct oxygenated zone (0.75–1.25 mm radius) in the sediments surrounding the roots, which would promote nitrification. The shoots are capable of direct H^+ driven bicarbonate (HCO_3^-) uptake. A broad range of turbidity is tolerated (1.4–30 NTU) but higher turbidity often is limiting to growth. Although individual leaves are fine filaments, the foliage can "fan out" on the water surface to mimic a larger floating leaf, as a means of withstanding more highly turbid conditions (a similar growth form and strategy occurs in *Stuckenia* – see later). Flowering and fruiting extend from March to September, with mature inflorescence density reaching 40 inflorescences m^{-2} and representing up to 27% of the shoot mass at 14.3 g DM m^{-2}. Optimal flower production and pollination occur at temperatures from 24°C to 30°C. The plants are self-pollinated as their pollen is transferred within flowers on the surface of air bubbles. Genetic studies using microsatellite markers have indicated high levels of inbreeding (e.g., H_O: 0–0.182) in Europe, but lower levels in Asia (e.g., H_O: 0.240–0.701). Seed production and maturation occur from 26°C to 32°C and can reach densities of up to 20,390 seeds m^{-2}. The seeds are dispersed endozoically by ducks (Aves: Anatidae: *Anas*) and fish (Vertebrata: Teleostei). However, studies have failed to detect significant differences in germination rates between seeds whether ingested or not by ducks. Some fish species like grass carp (Cyprinidae: *Ctenopharyngodon idella*) and tilapia (Cichlidae: *Tilapia*) can retain viable seeds for up to 65 hr. Although those seeds that are purged earlier retain higher germination rates, their maximum dispersal range (by ducks) is estimated at 40–280 km. Bird dispersal of the seeds results in scattered, isolated, and unique genotypes. Buried seeds can suffer over-winter mortality rates as high as 50%. Seed banks are scarce in ecotones, but can reach significant numbers in reproductive meadows (e.g., 25,398 total seeds m^{-2}; 3,556 viable seeds m^{-2}). Seed germination is enhanced by aerobic conditions, cold stratification (2 weeks at 4°C–10°C), and low salinity levels (<10‰); their viability is reduced after prolonged drying. Germination in the field occurs from 10°C to 18°C and is inhibited below 10°C. Some ecotypic differentiation in germination has been detected between northern and southern populations with respect to salinity but not temperature. Seed germination normally occurs only at lower salinities (<25‰), especially when conditions shift from high to low salinity. Germination is highest on the sediment surface and decreases sharply at burial depths to 3 cm, where less than 6% germination has been recorded. Seeds from brackish estuarine populations lack dormancy and exhibit reduced (but earlier) germination and lower viability when dry stratified. *In vitro*, the seeds have germinated successfully after a surface washing with 96% ethyl alcohol (pH: 2), 2–3 months of cold stratification (7°C), and subsequent incubation at 25°C. Seeds must be stored in highly saline water (to avoid precocious germination), but their germination rates increase when they are transferred subsequently to freshwater. Seedling emergence is greater in finer sediments than in coarse sand and their growth is most vigorous from 18°C to 25°C. Seedlings develop best at low salinity (<10‰) but can tolerate concentrations to 50‰ if the levels increase slowly. They are susceptible to salinity fluctuations, especially if occurring rapidly (e.g., 2.5–20‰ in less than a 24-hr period). Average total biomass can approach 250 g DM m^{-2}; peak biomass depends on the latitude and has reached 620 g DM m^{-2} in some tropical localities. Depending on depth, belowground biomass (averaging 30–45% of total) can vary from 2% in deep sites to 76% in shallow sites. Essentially all the belowground components occur within 5–10 cm of the substrate surface. During winter, the rhizomes and shoot nodes of some plants produce turions (to 1.2 cm), which are dispersed by water currents and function as vegetative propagules. Turion production in these perennial plants can represent up to 37% of the total shoot biomass and can reach densities as high as 452 turions m^{-2}. Ecologically, these plants often are opportunistic and readily colonize bare areas. They can complete their entire growth cycle within 4 months. Populations have been observed to overwinter (perennating by rhizomes) at lower nutrient and higher salinity sites but remain annual at higher latitudes or at high nutrient, low salinity sites. Perennial plants occupy only deeper, permanent waters that do not experience rapid salinity fluctuations. The coverage of water surfaces by stands of filamentous algae is believed to reduce desiccation of underlying *Ruppia* seeds upon recession of the shallow waters. **Reported associates:** *Alternanthera philoxeroides, Ampelopsis arborea, Atriplex watsonii, Baccharis halimifolia, Bacopa monnieri, Bolboschoenus maritimus, Bolboschoenus robustus, Carex comosa, Carex lyngbyei, Caulerpa taxifolia* (Chlorophyta: Caulerpaceae), *Centella asiatica, Ceratophyllum demersum, Chara zeylanica, Cladium jamaicense, Convolvulus sepium, Cressa truxillensis, Cyperus odoratus, Deschampsia cespitosa, Distichlis littoralis, Distichlis spicata, Eleocharis*

palustris, Eleocharis parvula, Festuca, Frankenia salina, Halodule wrightii, Hordeum brachyantherum, Ilex vomitoria, Juncus roemerianus, Kosteletzkya pentacarpos, Lathyrus japonicus, Ludwigia peploides, Lysimachia maritima, Myriophyllum pinnatum, Najas guadalupensis, Najas marina, Nymphaea, Panicum repens, Panicum virgatum, Persicaria hydropiperoides, Persicaria punctata, Phragmites australis, Potamogeton diversifolius, Potamogeton foliosus, Potamogeton pusillus, Potentilla anserina, Ranunculus orthorhynchus, Rubus trivialis, Rumex verticillatus, Salicornia depressa, Schoenoplectus americanus, Schoenoplectus tabernaemontani, Seutera angustifolia, Spartina alterniflora, Spartina bakeri, Spartina patens, Stuckenia pectinata, Syringodium filiforme, Thalassia testudinum, Triglochin maritima, Typha angustifolia, Typha domingensis, Typha latifolia, Utricularia foliosa, Vallisneria, Vigna luteola, Vitis rotundifolia, Zannichellia palustris, Zostera japonica, Zostera marina.

3.2.2.2. *Ruppia spiralis* L. ex Dumort. (= *R. cirrhosa*) is an annual or perennial, which grows in still or flowing, saline to fresh (salinity: 0–35‰) waters in canals (irrigation, saltwater), channels (drainage, tidal), ditches (irrigation, roadside), estuaries, fens, gravel pits, impoundments, lagoons (coastal), lakes (borax, soda), ponds (borax, gravel), pools (alkaline, estuarine, mineralized, saline, thermal), reservoirs (margins), rivers, salt marshes, sinks (alkali), springs (hot), and streams at elevations from −76 to 3640 m. The waters are alkaline (e.g., pH: 8.5) and the plants are capable of bicarbonate HCO$_3^-$ uptake when growing across a range of alkaline conditions (pH: 7.5–9.5). They can tolerate salinities from 0‰ to 75‰ but occur normally where values are less than 30‰. Most reports are from shallow waters (e.g., <50 cm) near the lower end of the depth range (5–150 cm). Generally, the plants favor sites with relatively higher pH, total dissolved solids (500–3000 mg·l^{-1}), conductivity, and dissolved oxygen levels. The exposures occur in full sunlight where very warm water temperatures (to 45°C) can occur. Many reports (perhaps tracing back to the same original reference) indicate that this species occurs mainly in inland, freshwater sites; however, it also is reported widely from brackish coastal regions in western North America as well as in numerous Old World localities. The substrates are described as basalt, clay, clay loam, colluvium, cobble, granite, mud, muddy gravel, rock, sand, silt, and silty loam. Flowering and fruiting occur from March to November. At anthesis, the anthers are released and float to the surface, whereupon they dehisce to release the pollen, which floats in bubbles on the water surface, forming chains or clumps that eventually encounter the stigmas. Due to the released anthers, there is a higher potential for cross-pollination than in *R. maritima*. Although usually perennial (when inundated), an annual habit can develop when plants occur in temporarily flooded sites. Annual plants produce much less biomass, but more flowers, which results in higher fruit production (up to 590 m^{-2}). Large seed banks (to 2852 seeds m^{-2}) have been reported from some African localities. The seeds do not germinate well and none will germinate at salinity levels above 35‰. The maximum net productivity

has been reported as 361.1 g C m^{-2} year^{-1}. A maximum biomass of 495–532 g DM m^{-2} results, of which 47–91% is produced aboveground. The plants can tolerate some desiccation (63% desiccated after 30 min of exposure) but are killed if they are exposed to daily (5 hr) or extended (1 week) desiccation periods. Much of the ecological information for this species is derived from studies of Old World populations but should also apply to New World plants. **Reported associates:** *Allenrolfea, Anemopsis californica, Atriplex prostrata, Baccharis sergiloides, Bolboschoenus maritimus, Bolboschoenus robustus, Carex alma, Ceratophyllum demersum, Chara evoluta, Chenopodium glaucum, Chenopodium rubrum, Cressa truxillensis, Distichlis spicata,Eleocharis acicularis, Eleocharis palustris, Eleocharis parishii, Elodea, Frankenia, Galium trifidum, Hippuris, Lemna gibba, Lemna minor, Marsilea vestita, Mimulus guttatus, Myriophyllum sibiricum, Najas marina, Potamogeton alpinus, Potamogeton diversifolius, Potamogeton foliosus, Potamogeton friesii, Potamogeton illinoensis, Potamogeton pusillus, Potamogeton richardsonii, Potamogeton zosteriformis, Ranunculus aquatilis, Schoenoplectus americanus, Sparganium angustifolium, Spergularia salina, Stuckenia pectinata, Suaeda calceoliformis, Symphyotrichum frondosum, Triglochin maritima, Typha angustifolia, Utricularia macrorhiza, Zannichellia palustris.*
Use by wildlife: *Ruppia maritima* is an extremely important wildlife food. It is eaten by coots (Aves: Rallidae: *Fulica americana*) and by many ducks (Aves: Anatidae) including the American scoter (*Melanitta nigra americana*), American wigeon (*Anas americana*), Barrow's goldeneye (*Bucephala islandica*), black duck (*Anas rubripes*), blue-winged teal (*Anas discors*), bufflehead (*Bucephala albeola*), canvasback (*Aythya valisineria*), common goldeneye (*Bucephala clangula*), Florida duck (*Anas fulvigula*), gadwall (*Anas strepera*), greater scaup (*Aythya marila*), green-winged teal (*Anas crecca*), king eider (*Somateria spectabilis*), lesser scaup (*Aythya affinis*), mallard (*Anas platyrhynchos*), northern shoveler (*Anas clypeata*), oldsquaw (*Clangula hyemalis*), pintail (*Anas acuta*), redhead (*Aythya americana*), ring-necked duck (*Aythya collaris*), ruddy duck (*Oxyura jamaicensis*), surf scoter (*Melanitta perspicillata*), and wood duck (*Aix sponsa*), as well as by brant geese (*Branta bernicla*), mute swans (*Cygnus olor*), and whistling swans (*Cygnus columbianus*). The plants can make up to 38% of the total biomass consumed by some species. They also provide food for various fish (Vertebrata: Teleostei) including the Amargosa pupfish (Cyprinodontidae: *Cyprinodon nevadensis*), oyster toadfish (Batrachoididae: *Opsanus tau*), pinfish (Sparidae: *Lagodon rhomboides*), and sheepshead (Sparidae: *Archosargus probatocephalus*). They are grazed by the Antillean manatee (Mammalia: Trichechidae: *Trichechus manatus manatus*), domestic cattle (Mammalia: Bovidae: *Bos taurus*), nutira (Mammalia: Myocastoridae: *Myocastor coypus*), and white-tailed deer (Mammalia: Cervidae: *Odocoileus virginianus*). *Ruppia maritima* also has been used as feed for rearing rabbits (Mammalia: Leporidae) and young lambs (Mammalia: Bovidae: *Ovis aries*). The seeds are consumed by fish (Vertebrata: Teleostei: Cichlidae: *Tilapia*; Cyprinidae: *Ctenopharyngodon idella*) and flamingos (Aves: Phoenicopteridae: *Phoenicopterus roseus*). The foliage supports epiphytic diatoms (e.g., Chromista: Bacillariophyceae: Naviculaceae: *Navicula pavillardi*) and blue-green algae (Bacteria: Cyanobacteria). In turn, the epiphytes provide food for grass shrimp (Crustacea: Decapoda: Palaemonidae: *Palaemonetes*), whose grazing reduces die-back of the plants. *Ruppia spiralis* is grazed by coots (Aves: Rallidae: *Fulica atra*) and mallard ducks (Aves: Anatidae: *Anas platyrhynchos*).
Economic importance: food: not reported as edible; **medicinal:** *Ruppia maritima* contains polyhydroxylated sterols (ergosta-8,22-diene-3β,6β,7α-triol and ergosta-8(14),22-diene-3β,6β,7α-triol), which have been investigated for their cancer cell cytotoxicity. The powdered tissues of *Ruppia spiralis* reputedly exhibit antimicrobial and antifungal properties. Both *R. maritima* and *R. spiralis* contain several polyphenolics (chicoric acid and flavonoids), which exhibit antioxidant and DPPH radical scavenging activity; **cultivation:** *In vitro* propagation of *Ruppia maritima* has been investigated for use in seagrass meadow restoration; **misc. products:** *Ruppia maritima* contains at least seven *ent*-labdane diterpenes, which are inhibitory to the growth of some algae (e.g., Chlorophyta: Selenastraceae: *Raphidocelis subcapitata*). The plants have been used to treat highly saline municipal wastewater in Iran. They also have been used as a fertilizer for cultivated tomatoes (Solanaceae: *Solanum lycopersicum*). *Ruppia spiralis* contains more than 2% lipids and has been evaluated as a source of biofuel; **weeds:** not usually reported as weedy; **nonindigenous species:** none in North America; *Ruppia maritima* has been introduced recently to western Africa.
Systematics: For many years *Ruppia* was recognized in a distinct family (Ruppiaceae), which was believed to be allied with Potamogetonaceae and sometimes placed within it. However, early analyses of DNA sequence (i.e., *rbcL*) data indicated to the contrary a close relationship of *Ruppia* with members of the "Cymodoceaceae complex," which contained various seagrasses from the families Cymodoceaceae and Posidoniaceae. That association was upheld by subsequent analyses of additional cpDNA data and mtDNA data, although without a consistent distinction between Cymodoceaceae and Ruppiaceae (Figure 1.16). These circumstances led to the more recent inclusion of *Ruppia* within Cymodoceaceae and abandonment of Ruppiaceae, although some studies have retained Ruppiaceae as the sister group of Cymodoceaceae (Figure 1.2). In addition to the phylogenetic affinities indicated by molecular data, palynological studies also have shown that *Ruppia* exhibits a pattern of pollen development that is similar to members of Cymodoceaceae. *Ruppia* itself also has undergone a history of taxonomic turmoil. Until recently, species circumscriptions were extremely vague with estimates ranging from 2 to 10 worldwide. That uncertainty is responsible for much nomenclatural confusion in the literature including the misapplication of the synonym name *Ruppia cirrhosa* to *Ruppia spiralis*, which has nomenclatural priority. Molecular phylogenetic studies now support the recognition of at least five distinct species; however, several morphologically similar (but cytologically heterogeneous) entities occur within the

cosmopolitan *Ruppia maritima* "complex," which is undergoing further taxonomic evaluation (Figure 1.18). Both *R. maritima* and *R. spiralis* occur within this complex, which is relatively derived within the genus (Figure 1.18). The basic chromosome number in *Ruppia* is *x* = 10. *Ruppia maritima* (2*n* = 20) is diploid, whereas *R. spiralis* (2*n* = 40) is tetraploid. **Distribution:** *Ruppia maritima* occurs along the Atlantic and Pacific coastal regions of North America; *R. spiralis* is widespread across North America except for the southeastern United States.

References: Adams & Bate, 1994a; 1994b; Agami & Waisel, 1988; Ahmadi et al., 2017; Ailstock et al., 2010a; 2010b; Allen et al., 2017b; Bartonek & Hickey, 1969; Beal, 1977; Beer et al., 2002; Bigley, 1981; Bird et al., 1994; Botero & Rusch, 1994; Burkholder et al., 1994; Chamberlain, 1959; Charalambidou et al., 2003; Cho & Poirrier, 2005; Cho & Sanders, 2009; Cook, 1996; Cottam, 1939; Cottam et al., 1944; Cronquist, 1981; Darnell, 1958; DellaGreca et al., 2000; 2001a; Dunton, 1990; El-Hady et al., 2007; Evans et al., 1986; Fernald & Wiegand, 1914; Figuerola & Green, 2004; Garner, 1963; Gesti et al., 2005; Graves, 1908; Green et al., 2016; Halloran, 1943; Harrison, 1982; Hartke et al., 2009; Haynes, 2000d; Heilman & Carlton, 2001; Husband & Hickman, 1989; Ito et al., 2010; 2013; 2015; 2017b; Kahn & Durako, 2005; Kantrud, 1991; Kaul, 1993; Koch & Dawes, 1991; Ktita et al., 2010; 2014; Lacroix & Kemp, 1997; Lafferty et al., 2006; Landers et al., 1976; La Peyre & Rowe, 2003; Les & Tippery, 2013; Les et al., 1993; 1997c; Marco-Méndez et al., 2015; Martínez-Garrido et al., 2016; 2017a; 2017b; McAtee, 1922; McCall & Rakocinski, 2007; McMahan, 1970; McMillan, 1985; Mendall, 1949; Menéndez, 2002; Menéndez & Comín, 1989; Miller & Hoven, 2007; Murphy et al., 2003; Orth & Moore, 1988; Perry & Uhler, 1988; Preston & Croft, 1997; Pulich, Jr., 1985; Renzi et al., 2013; Richardson, 1980; Riddin & Adams, 2009; Rogers & Korschgen, 1966; Ross et al., 2016; Schwartz & Dutcher, 1963; Seeliger et al., 1984; Self et al., 1975; Serie & Swanson, 1976; Setchell, 1924; Stewart & Manning, 1958; Stieglitz, 1972; Strazisar et al., 2013a; 2013b; 2015; Sullivan, 1977; Talavera et al., 1993; Taylor et al., 2018; Thursby, 1984; Thursby & Harlin, 1984; Triest & Sierens, 2014; 2015; Van Vierssen et al., 1981; 1982; Verhoeven, 1979; Vest & Conover, 2011; Walsh & Grow, 1972; Williams & Grosholz, 2002; Willner et al., 1979; Yu et al., 2009.

3.2.3. Syringodium
Manatee-grass
Etymology: a Latinized form of the Greek *syringodes* ("pipe-like") for the tubular leaves
Synonyms: *Cymodocea* (in part); *Phucagrostis*; *Phycoschoenus*
Distribution: global: Atlantic, Indian, and Pacific Oceans; **North America:** southern (Atlantic, Caribbean, and Gulf of Mexico)
Diversity: global: 2 species; **North America:** 1 species
Indicators (USA): obligate wetland (OBL): *Syringodium filiforme*
Habitat: brackish (coastal); marine; oceanic; **pH:** 7.85–8.09; **depth:** to 25 m; **life-form(s):** submersed (rosulate/vittate)
Key morphology: rhizomes coarse, scaly (to 6 mm), the internodes to 5 cm, rooted at nodes; shoots (short) erect, arising from nodes, each bearing 2–3 leaves; leaves (to 30 cm) filiform, sheathed (to 7 cm); plants dioecious, inflorescences cymose; ♂ flowers axillary in reduced, leaf-like bracts (to 10 mm) comprised mainly of inflated sheaths (to 9 mm); perianth absent; anthers (to 5 mm) stalked (to 10 mm); ♀ flowers sessile; perianth absent; ovary (to 1.5 mm) bicarpellate, ellipsoid, style (to 4mm) terminating in 2 linear stigmas (to 27 mm); fruit (to 9 mm) elliptic, compressed, with 3 blunt dorsal ridges, and short (to 3 mm), terminally bifid beak
Life history: duration: perennial (rhizomes); **asexual reproduction:** rhizomes, rhizome fragments; **pollination:** water (hypohydrophily); **sexual condition:** dioecious; **fruit:** drupe-like achenes (common); **local dispersal:** seeds, vegetative fragments (water); **long-distance dispersal:** seeds, vegetative fragments (water)
Imperilment: 1. *Syringodium filiforme* [G4]; S1 (MS)
Ecology: general: Both *Syringodium* species are obligately submersed, perennial seagrasses, which occur only in brackish to saline (normally oceanic) waters, where they occupy regions from the upper sublittoral zone to greater ocean depths (to 20+ m). The brittle, linear, and grass-like foliage can form a canopy up to 45 cm but is not particularly dessication-resistant, which limits occurrences in shallow tidal areas. Their foliage arises from durable, elongating rhizomes, from which the plants perennate. These are C_3 plants but they have acquired the ability to use bicarbonate (HCO_3^-) as a carbon source. Sexual reproduction occurs routinely but not to a

FIGURE 1.18 Phylogenetic overview of *Ruppia* summarized from analyses of cpDNA and nuclear *phyB* sequence data (modified from Ito et al., 2015). The diploid *R. maritima* and tetraploid *R. spiralis* occur within the cosmopolitan *Ruppia maritima* "complex," which represents a highly specialized clade in the genus relative to the Australian and African species. All *Ruppia* taxa are obligately aquatic; however only the two North American taxa (in bold) are designated as OBL indicators.

prolific extent. The combination of dioecy and hypohydrophilous pollination typically results in a low level of seed set and populations can be fairly clonal in nature. *Syringodium* plants grow intermixed with other seagrasses or in monospecific meadows, and provide food, shelter, and substrates for numerous epiphytes and various pelagic marine animals. They also are important ecologically for their ability to attenuate wave action (up to 40% under certain conditions).

3.2.3.1. *Syringodium filiforme* **Kütz.** occurs in brackish to saline bays, beds (seagrass), canals, channels, flats, gulfs, lagoons, rivers (tidal), and along beaches (strand), coasts (oceanic), and shores (oceanic) at elevations to 1.9 m. The plants have been reported at a wide range of depths (from 0.1 to 25 m), but occur typically at depths nearer to 1.2 m (e.g., 0.75–1.5 m). The most luxuriant stands occur in shallow waters (0.5–0.7 m deep). The pH is relatively stable at values ranging from 7.85 to 8.09. Habitats can vary in salinity from 5.5‰ to 52‰ with the plants responding variously. There is some indication that they prefer stable salinity conditions. Some studies have reported their restriction to high salinity environments (38–52‰), although they reportedly are unable to withstand hypersalinity (50–70‰) for extended periods (e.g., in lagoons experiencing prolonged drought conditions). The substrates have been described as mud, muddy sand, sand, sandy mud, shells, and silt. Although the plants are nitrogen limited when growing on terrigenous sediments, they can become phosphorous limited when growing on carbonate sediments. An annual temperature range of 13.7°C–34°C is tolerated; however, maximum growth has been observed to occur at temperatures from 23°C to 29°C. The plants are not particularly heat-tolerant. Elevated temperatures will increase winter productivity but will reduce summer productivity once the optimal range has been surpassed. The foliage also is highly susceptible to chilling, with 100% of leaf tissue likely to be lost following a 24-hr exposure to 2°C water. Chilling for 48 hr has more serious consequences, ranging from severe leaf damage to complete plant (adult or seedling) mortality. Reduced light (attenuation at depths below 1.5 m) and salinity (≤20‰) resulting from freshwater intrusion due to hurricanes also are known to impact plant cover substantially. Minimal light requirements have been estimated to fall between 24% and 37% of the surface illumination. The plants are highly susceptible to shading caused by the presence of boat docks. Photosynthesis is of the typical C_3 type. In North America, flowering and fruiting in this dioecious species reportedly occurs from January to November. The extent of sexual reproduction is regulated by temperature (initiated at 20°C–24°C) under day lengths ranging from 12 hr to continuous illumination. Thus, flowering initiates as water temperatures and day lengths increase. About 6.4% of the shoots flower during the season with flowering shoot density capable of reaching 1,000 shoots m^{-2}. Pollination is hypohydrophilous, whereby pollen from the anthers of the male plants is released underwater in sinking clumps or strands and must be captured by the stigmas of a pistillate plant in order to initiate fertilization. Because the pollen is not dispersed widely, seed set is inversely proportional to the distance between male and female plants. Populations can contain anywhere from 0% to 100% female plants, but the

highest seed set occurs in stands where a mixture of the sexes is present. The non-buoyant seeds are shed above the sediment level at densities that can reach 10,000 m^{-2}. They (along with vegetative fragments) are dispersed in the water by currents, a process facilitated by hurricane activity. The seeds have short-term dormancy, do not develop into a long-term seed bank, and can germinate across a broad salinity range from 20‰ to 50‰. Seeds stored in saltwater have remained viable for more than 4 years. This is a pioneer species whose high growth rates (e.g., photosynthetic rates up to 1.72 mg C g^{-1} DM h^{-1}) result in a relatively high nutrient demand. Root to shoot biomass ratios change seasonally and can vary from 0.44 to 2.56. Each shoot lives for about 182 days. The shoots can achieve a mean density of 7139 m^{-2} and a biomass of 446.9 g DM m^{-2}. Vertical shoot growth averages 3.36 cm $year^{-1}$. From 3 to 4 leaves are produced within the first 45 days on newly developing shoots, with a new leaf developing subsequently every 40 days or so. Individual leaves can grow at rates as high as 3.11 cm day^{-1}. Annual leaf production is around 2234 g DM m^{-2} $year^{-1}$ (summer: up to 4.0 g DM m^{-2} day^{-1}); about 47% of the leaves are shed on average during the growing season. Rhizome production has been estimated at 466 g DM m^{-2} $year^{-1}$ with each rhizome elongating apically at an average rate of 51 cm $year^{-1}$. The rhizomes can grow to an average density of 7,468 internodes m^{-2} and achieve a biomass of 233 g DM m^{-2}. The rhizomes also can produce "aerial" (i.e., above the sediment surface) "runners," which enable the plants to spread vegetatively by layering subsequent to initial colonization. The rhizomes expand much more quickly than those of the climax species *Thalassia testudinum* (Hydrocharitaceae), which enables the plants to quickly colonize bare substrates; however, the plants do not compete well in the long term with *T. testudinum*, which better exploits light and sediment nutrients when the two species coexist. Soluble carbohydrates in the rhizomes are highest in the fall and lowest in the spring and act as a nutrient reserve to sustain the plants during the winter. Clonally derived rhizome fragments (ramets) can survive and grow well if not too young of age; the survival rate of older ramets is far higher than that of younger offshoots that become detached prematurely. **Reported associates:** *Halodule wrightii, Halophila decipiens, Halophila engelmannii, Halophila ovalis, Halophila stipulacea, Thalassia testudinum, Ruppia maritima.*

Use by wildlife: The leaves of *Syringodium filiforme* contain high levels of soluble carbohydrate (16%), have a lipid content of 1.5%, and provide a mean caloric value of 2.9 kcal g dry wt^{-1}. They are grazed by several fish (Vertebrata: Teleostei) including the bucktooth parrotfish (Scaridae: *Sparisoma radians*), emerald parrotfish (Scaridae: *Nicholsina usta*), filefish (Monacanthidae: *Stephanolepis hispidus*), and pinfish (Sparidae: *Lagodon rhomboides*), as well as by green sea turtles (Reptilia: Testudines: Cheloniidae: *Chelonia mydas*), green sea urchins (Echinodermata: Toxopneustidae: *Lytechinus variegatus*), and manatees (Mammalia: Trichechidae: *Trichechus manatus*). The foliage can be colonized by numerous (up to 8 mg) epiphytes including various foraminifera (Protozoa: Foraminifera: Acervulinidae: *Planogypsina squamiformis*; Discorbidae: *Discorbis rosea*; Planorbulinidae: *Planorbulina* *acervalis, P. mediterranensis*; Rosalinidae: *Tretomphalus bulloides*; Soritidae: *Archaias angulatus, Sorites orbiculus*), brown algae (Phaeophyta: Acinetosporaceae: *Hincksia mitchelliae*; Sphacelariaceae:*Sphacelaria rigidula*), green algae (Chlorophyta: Ulvaceae: *Enteromorpha flexuosa*), and red algae (Rhodophyta: Acrochaetiaceae: *Acrochaetium seriatum*; Ceramiaceae: *Centroceras clavulatum, C. gasparrinii, Ceramium, Gayliella, Griffithsia, Spyridia filamentosa*; Champiaceae: *Champia parvula*; Cystocloniaceae: *Hypnea musciformis, H. spinella*; Rhodomelaceae: *Chondria collinsiana, Herposiphonia tenella, Polysiphonia flaccidissima*; Stylonemataceae: *Stylonema alsidii*). The foliar epiphytes (and incidentally ingested leaf material) provide food for shrimp (Crustacea: Decapoda: Hippolytidae: *Hippolyte zostericola*; Palaemonidae: *Palaemonetes intermedius*) and other grazing organisms.

Economic importance: food: not reportedly edible; **medicinal:** *Syringodium filiforme* contains chicoric acid, which has been attributed with immunostimulative, anti-hyaluronidase, and antioxidant properties, is inhibitory to HIV-1 integrase and replication, and protects against free radical-induced collagen degradation. The plant extracts are rich in alkaloids, anthocyanins, flavonoids, phenols, reducing sugars, and terpenes, which exhibit significant free radical scavenging properties; **cultivation:** *Syringodium filiforme* has been cultured primarily for use in habitat restoration projects; **misc. products:** Sods of *Syringodium filiforme* cut from healthy "donor" populations, have been successfully used as transplants to restore degraded seagrass meadows; **weeds:** not weedy; **nonindigenous species:** none.

Systematics: *Syringodium* consists of only two morphologically similar species, which are broadly disjunct between the New World (*S. filiforme*) and Old World (*S. isoetifolium*). Analyses of cpDNA and mtDNA sequences (e.g., Figure 1.16) indicate that although *Syringodium* is monophyletic, the closely related *Cymodocea* is not, and both genera probably should be merged, perhaps along with *Amphibolis* and *Thalassodendron*. Both species of *Syringodium* had once been recognized as species of *Cymodocea* as well as *Amphibolis* and *Thalassodendron*, which were classified previously within *Cymodocea* as section *Amphibolis*. A similar taxonomic disposition might be better justified given the phylogenetic relationships indicated. All four genera are members of the "Cymodoceaceae complex," which has been recalcitrant to any resolution corresponding with the smaller generic circumscriptions that have been accepted since the 1970s. The base chromosome number of *Syringodium* probably is $x = 10$, with *Syringodium filiforme* ($2n = 20$) being diploid. No hybrids involving either *Syringodium* species have been reported.

Distribution: *Syringodium filiforme* occurs in marine habitats from the eastern Atlantic Florida coast to the Texas gulf coast southward to northern South America.

References: Barber & Behrens, 1985; Beer & Wetzel, 1982a; Bjorndal, 1996; Bortone, 2000; Buzzelli et al., 2012; Cox et al., 1990; Dawes, 1986; Dawes & Lawrence, 1980; Dawes et al., 1995; Fonseca & Cahalan, 1992; Fonseca et al., 1984; Fry, 1983; Fry & Virnstein, 1988; Fujita & Hallock, 1999; Gallegos et al., 1994; García et al., 2011; 2017; Green & Short, 2003; Gutierrez et al., 2010; Hartog, 1970; Hartog et al.,

1979; Haynes, 2000f; Heffernan & Gibson, 1983; Johansson et al., 2009; Kendall et al., 2004; Kenworthy & Fonseca, 1996; Kenworthy & Schwarzschild, 1998; Les & Haynes, 1995; Les & Tippery, 2013; Les et al., 1993; 1997c; Lobel & Ogden, 1981; McMillan, 1979; 1980; 1981; 1982; 1983; 1984; 1991; Nuissier et al., 2010; Orth et al., 2000; 2006; Petersen et al., 2014; Phillips & Meñez, 1988; Prado & Heck, Jr., 2011; Provancha & Hall, 1991; Ross et al., 2016; Schwarzschild & Zieman, 2008; Short, 1985; Short et al., 1990; 1993; Waycott & Les, 1996; Waycott et al., 2006; Williams, 1987; Willette & Ambrose, 2012; Wilson, 1998; Wilson & Dunton, 2012; Wilson & Dunton, 2017; Won et al., 2010; Zupo & Nelson, 1999.

Family 3.3. Juncaginaceae [3]

The Juncaginaceae (arrow grasses) are a relatively small group of wetland plants, which are assigned to three genera: *Cycnogeton* (2 spp.), *Tetroncium* (1 sp.), and *Triglochin* with 25–35 species (Koecke et al., 2010; von Mering & Kadereit, 2010; Les & Tippery, 2013). Although *Maundia* had once been included in the family, several phylogenetic analyses (e.g., von Mering & Kadereit, 2010; Les et al., 2013; Les & Tippery, 2013; Ross et al., 2016) have resolved *Maundia* in an isolated position outside of Juncaginaceae as the sister clade to an assemblage comprising the Cymodoceaceae complex, Posidoniaceae, Potamogetonaceae, and Zosteraceae (Figure 1.2). Given this consistent phylogenetic placement, *Maundia* currently is recognized within its own family Maundiaceae. Similarly, *Scheuchzeria* also has been included in Juncaginaceae by some authors, but represents yet another isolated clade (Figure 1.2) worthy of recognition as a distinct family (Les et al., 1997a; von Mering & Kadereit, 2010; Les & Tippery, 2013; Ross et al., 2016). Another taxonomic modification of Juncaginaceae has been the merger of the monotypic genus *Lilaea* with *Triglochin*, another outcome of phylogenetic analyses (e.g., von Mering & Kadereit, 2010; 2015; Les & Tippery, 2013), which show *Lilaea* to resolve within *Triglochin* (Figures 1.19 and 1.20). Arrow grasses represent another lineage of "tepaloid" alismatids

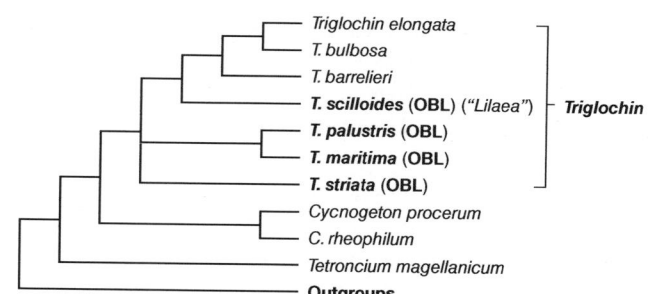

FIGURE 1.19 Inter- and intra-generic relationships summarized for *Triglochin* and other taxa of Juncaginaceae based on analysis of *rbcL* sequence data (modified from Les & Tippery, 2013). This and similar analyses (e.g., von Mering and Kadereit, 2010; 2015) show the monotypic "*Lilaea scilloides*" to be nested phylogenetically within the genus *Triglochin*, to which it has been transferred (as *T. scilloides*). Four of the five included OBL North American indicators (in bold) resolve as closely related species. *Cycnogeton* resolves consistently as the sister group of *Triglochin*, and some authors have merged the two genera.

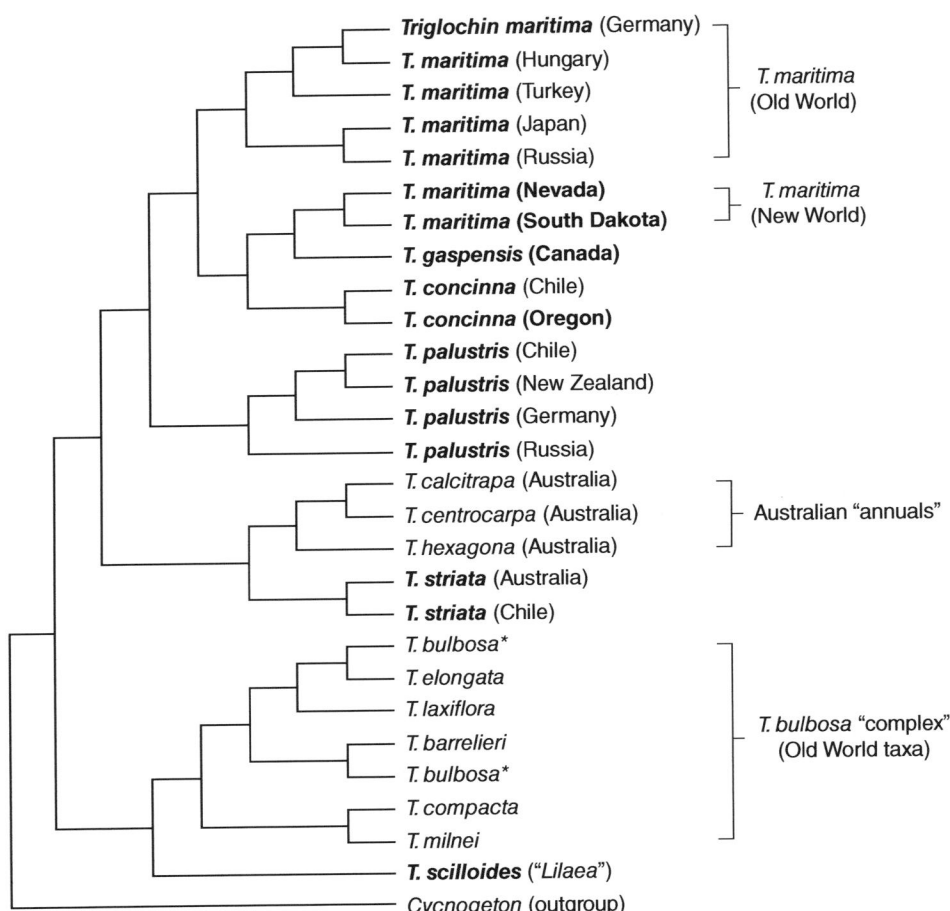

FIGURE 1.20 Interspecific relationships in *Triglochin* as indicated by phylogenetic analysis of combined DNA sequence data (modified from von Mering & Kadereit, 2015). The species currently recognized as "*T. maritima*" resolves in two distinct clades representing New World vs. Old World accessions. The treatment of *T. concinna* and *T. gaspensis* as distinct from *T. maritima* would be warranted only if the Old and New World clades of *T. maritima* are assigned to different species. The widespread *T. striata* resolves as the sister group to a clade comprising annual Australian species. The annual New World *T. scilloides* (long recognized as *Lilaea scilloides*) is sister to a clade of perennial Old World species, which comprise the *T. bulbosa* "complex"; *T. bulbosa* itself (asterisked accessions) resolves as polyphyletic in this analysis. Taxa recognized as OBL indicators in North America are highlighted in bold.

(Figure 1.2), which are distinguished by their small, simplified, bisexual or unisexual, wind-pollinated flowers and linear, grass-like leaves. The fruits are schizocarps, which dehisce as mericarps or achenes. They are dispersed primarily by water (Cook, 1996; Haynes & Hellquist, 2000b).

In some species the cyanogenic foliage (producing HCN when bruised) has been implicated in cases of livestock poisoning (Cronquist, 1981). The plants are not cultivated and are of no major economic value except for some minor tribal use as food. Despite its small size, the family is widespread and distributed in cooler temperate regions throughout both hemispheres of the New and Old Worlds. In North America, OBL indicators occur within a single genus:

3.3.1. *Triglochin* L.

3.3.1. Triglochin

Arrow grass; troscart

Etymology: from the Greek *treis glochis* (three-pointed) in reference to the 3-parted fruits of some species

Synonyms: *Abbotia*; *Cycnogeton* (in part); *Lilaea*; *Phalangium*

Distribution: global: cosmopolitan; **North America:** widespread (except south-central USA)

Diversity: global: 30 species; **North America:** 6 species

Indicators (USA): obligate wetland (OBL): *Triglochin concinna, T. gaspensis, T. maritima, T. palustris, T. scilloides, T. striata*

Habitat: brackish (coastal), freshwater; freshwater (tidal); saline (inland); palustrine; **pH:** 5.1–9.6; **depth:** <1 m; **lifeform(s):** emergent herb

Key morphology: rhizomes absent (annuals) or present (to 10 cm) and stout (to 1 cm thick), arising from a thickened caudex; leaves (to 80 cm) basal, erect, filiform, with basal, broadened, ligulate (to 5 mm) sheaths (to 15 cm); scapes (to 90 cm) terminating in a spike-like raceme (to 50 cm), bearing numerous whorls of 3–4 sessile or pedicelled (to 7 mm) bisexual (most species) flowers (mixtures of bisexual, staminate, and pistillate flowers in *T. scilloides*); perianth (most species) of 6 concave tepals (to 2 mm), tepals fewer (0–1) in *T. scilloides*;

stamens (most species) 6, only 1 in *T. scilloides* (bisexual or ♂ flowers); pistils (most species) of 6 carpels (to 1.5 mm; 3 or 6 being fertile), a single carpel in *T. scilloides* (bisexual or ♀ flowers), the styles short (most species, but elongate (to 30 cm) and threadlike in the paired, basal ♀ flowers of *T. scilloides*); fruit (most species) a schizocarp, the mature mericarps (to 7 mm) winged or ribbed, dehiscing completely or remaining attached apically (achenes [to 10 mm] in *T. scilloides*)

Life history: duration: annual (fruits/seeds); perennial (rhizomes, stolons, tubers); **asexual reproduction:** rhizomes, stolons, tubers; **pollination:** wind; **sexual condition:** hermaphroditic, polygamomonoecious (trimonoecious); **fruit:** achenes or schizocarps (common); **local dispersal:** rhizomes, seeds (water); **long-distance dispersal:** seeds (water, waterfowl [endozoic], wrack)

Imperilment: 1. *Triglochin concinna* [G5]; S2 (BC); 2. *T. gaspensis* [G4/G5]; S1 (ME, NS); S2 (NF, PE); S3 (NB, QC); 3. *T. maritima* [G5]; SX (PA); S1 (KS, IL, NJ, VT); S2 (AZ, IA, OH); S3 (LB, WI); 4. *T. palustris* [G5]; SX (PA); SH (RI); S1 (IL, NE); S2 (IA, IN, NY, OH, PE); S3 (CA, LB, WI, WY); 5. *T. scilloides* [G5]; SH (MT); S1 (AB, SK); S2 (BC); S3 (OR); 6. *T. striata* [G5]; SH (DE); S1 (LA, MD); S3 (NC, VA)

Ecology: general: All *Triglochin* species are obligate inhabitants of highly alkaline freshwater or saline wetlands. They share an emergent rosulate growth habit with wiry (sometimes succulent) filiform leaves. The species include annuals or perennials, with rhizomes or stolons produced by the latter. The perennial species overwinter by buds that are protected by the persistent leaf bases. Their caudex-like stems occur at a depth of about 10 cm under the soil surface, with roots that can extend to 60 cm deep. Their branching internodes can produce up to 20 tillering shoots, which are able to detach and survive as ramets. Individual perennial plants are known to live from 15 to 20 years. The flowers are similar among most of the species, which share inconspicuous tepaloid perianths (absent or highly reduced in some species). Depending on the taxon they can be bisexual or unisexual (or mixed). Pollination is anemophilous, with selfing often precluded by strongly protandrous dichogamy. One species (*T. scilloides*) also develops a basal pair of unisexual (female) flowers, which produce filiform styles up to 30 cm in length. If the plants become inundated during anthesis, the elongate styles enable the stigmas to float on the water surface as a means of capturing wind-borne pollen. There is no evidence whatsoever of hydrophily as some have suggested. Most of the species produce schizocarps, which dehisce as floating, water-dispersed mericarps; the fruit of *T. scilloides* is an achene. Dispersal by animals (endozoic or epizoic) also has been implicated. The abraded foliage of all of the species probably contains cyanide to some degree, with the cyanogenic glycoside triglochinin routinely found in those surveyed (also known from Araceae and Scheuchzeriaceae). Some of the plants are known to cause livestock poisoning. For widespread species (e.g., *T. maritima*, and *T. palustris*), the ecological information was compiled as much as possible from studies of North American populations.

3.3.1.1. **Triglochin concinna Burtt Davy** is a perennial, which grows in or on beaches, bottomland, cienega (alkaline), ditches (roadside), estuaries, flats (alkali, mud, tidal), marshes, meadows (alkaline, drying, saline, wet), playas, pools (brackish), roadsides, salt marshes (coastal), seeps (alkaline, saline), sinks, slopes, spits (sand), springs (warm), and along the margins of lakes, reservoirs, sloughs, and streams (alkaline) at elevations from −1 to 2409 m. The two recognized varieties (see *Systematics*) occupy quite different ecological niches. In general, the salt marsh variety inhabits brackish to saline sites; whereas the more inland variety occurs more under freshwater conditions. The substrates are alkaline (e.g., pH: 9.6) or saline and include alluvium, caliche, clay, granite, gravel, limestone (Tertiary Claron), muck, mud (saline), rock, sand (saline), sandstone, sandy gravel (crusty), shale (Hillard, Mancos), silt (alkaline), silty clay (alkaline), silty loam, silty pumice (alkaline), silty sand, and travertine. Flowering and fruiting occur from April to October. Flowering of the salt marsh variety occurs in the early spring whereafter the leaves senesce and decay during summer, releasing nitrogen to the sediments. The schizocarps dehisce completely into six distinct mericarps when mature. They are dispersed directly by water or are distributed within the drifting shoreline wrack. Hand-cleaned seeds (i.e., mericarps) collected during July–October have germinated well (75%) following a 17-day period of cold stratification at 4°C. The seeds have germinated most profusely at an average salinity of 25.9‰ (range: 11–32‰). The mericarps persist for an undetermined time in the seed bank (at densities to 3,320.6 m^{-2}), and the resulting seedlings have been observed to constitute up to 70.8% of the emerging seedling pool at some localities. The plants reproduce vegetatively by their slender, creeping rhizomes. They become tufted or mat-forming but typically are low in stature (<6 cm) and percent cover (<20%) relative to other coexisting species. Growth occurs primarily from late winter to early spring, when nitrogen levels are released by associated species such as *Salicornia bigelovii* (Amaranthaceae). Productivity estimates (from individual potted greenhouse plants) have reached 3.3 g (aboveground biomass) and 1 g (belowground biomass) when grown under high water and high nitrogen (15 g m^{-2}) conditions; the observed productivity of specimens in the field is less. The plants can become infected at low rates by the parasitic saltmarsh dodder (Convolvulaceae: *Cuscuta salina*), which facilitates their survival by its far higher infection rates on competing species. Consequently, the elimination of the parasite has been shown to cause up to a 7.5% decline in *T. concinna* abundance within populations. *Triglochin concinna* also is regarded as a principal host of the hemiparasitic *Chloropyron maritimum* (Orobanchaceae). The plants can suffer high winter mortality due to coverage by sediments and algal mats. **Reported associates:** *Agoseris glauca, Agrostis, Alisma gramineum, Anemopsis californica, Arthrocnemum subterminale, Atriplex parryi, Atriplex prostrata, Atriplex watsonii, Avena, Baccharis douglasii, Baccharis salicina, Batis maritima, Bolboschoenus maritimus, Bromus diandrus, Calochortus striatus, Carex praegracilis, Carex scirpoidea, Carex simulata, Castilleja ambigua, Centaurea, Chloropyron maritimum, Chloropyron tecopense, Cirsium mohavense, Cleomella brevipes, Cleomella parviflora, Crepis runcinata,*

Cressa truxillensis, Cuscuta salina, Deschampsia cespitosa, Distichlis littoralis, Distichlis spicata, Eleocharis rostellata, Ericameria albida, Ericameria nauseosa, Erigeron formosissimus, Extriplex californica, Frankenia salina, Gentianella amarella, Gentianopsis thermalis, Grindelia hirsutula, Hainardia cylindrica, Halogeton glomeratus, Honckenya peploides, Isocoma menziesii, Isolepis cernua, Jaumea carnosa, Juncus balticus, Juncus bufonius, Juncus cooperi, Lasthenia glabrata, Leymus cinereus, Limonium californicum, Lysimachia maritima, Muhlenbergia, Nitrophila occidentalis, Oxytenia acerosa, Parnassia palustris, Peritoma lutea, Plantago maritima, Poa secunda, Polemonium occidentale, Polygonum marinense, Potentilla anserina, Potentilla gracilis, Primula pauciflora, Puccinellia lemmonii, Puccinellia nutkaensis, Pyrrocoma lanceolata, Pyrrocoma racemosa, Rumex occidentalis, Salicornia bigelovii, Salicornia depressa, Sarcobatus vermiculatus, Sarcocornia pacifica, Sarcocornia perennis, Schoenoplectus americanus, Schoenoplectus pungens, Senecio hydrophilus, Sisyrinchium bellum, Solidago multiradiata, Spartina densiflora, Spartina foliosa, Spergularia canadensis, Spergularia macrotheca, Spergularia rubra, Sporobolus airoides, Sporobolus airoides, Suaeda calceoliformis, Suaeda esteroa, Suaeda taxifolia, Symphyotrichum subulatum, Tamarix, Triglochin maritima, Triglochin palustris, Typha, Valeriana edulis.

3.3.1.2. **Triglochin gaspensis Lieth & D. Löve** is a perennial species of beaches, coves, estuaries, flats, marshes (brackish), meadows (brackish, rivershore), salt marshes (intertidal, subtidal, tidal), shores, and along the margins of ponds (barrier, tidal) at elevations near sea level (~0 m). The plants are restricted to the tidal zone below the high-water mark (low marsh), where they undergo daily cycles of inundation during hide tide intervals. The sediments are described as organic peat and sand. Flowering and fruiting occur from July to September. The mature schizocarps dehisce completely into six separate mericarps, which are dispersed by water. The low-growing plants can form expansive lawns by their dense, rhizomatous growth. Additional life history information for this rare endemic would be desirable. **Reported associates:** *Carex mackenziei, Carex recta, Eleocharis uniglumis, Festuca rubra, Juncus gerardii, Lilaeopsis chinensis, Lysimachia maritima, Plantago maritima, Potentilla anserina, Ruppia maritima, Salicornia maritima, Samolus valerandi, Solidago sempervirens, Spartina alterniflora, Spartina patens, Spergularia canadensis, Suaeda calceoliformis, Teucrium canadense.*

3.3.1.3. **Triglochin maritima L.** is a perennial, which occurs in or on beaches, bogs (quaking, *Sphagnum*), bottoms (coulee, pothole), channels (estuarine), coasts (rocky), depressions (calcareous), ditches (irrigation, roadside), draws (alkaline), fens (calcareous, moderately rich to rich), flats (alkaline, calcareous, dried, grass, marly, saline, salt, sedge, tidal), floodplains, fountainheads, gravel bars, lakebeds (dry), marshes (alkaline, brackish, coastal, inland saline, sandy, tidal), meadows (alkaline, estuarine), mudflats (alkali, intertidal, saline), muskeg, pannes (saline, salt), patterned bogs (flarks, strings), playas (alkaline, salt), pools (brackish), roadsides, salt marshes

(coastal), sand spits, seeps (alkaline, mineral), shores (brackish, gravelly, marly, rocky), sinks (marshy), slopes (alkaline, gravel; to 10%), sloughs (brackish, dried, tidal), sloughlets, springs (hot, mineral), swales, swamps (alkali), tinajas, tundra, vernal pools (dry), and along the margins of bays, cienegas, estuaries, lagoons (brackish), lakes (marly), ponds, reservoirs, rivers, seeps, sloughs, and streams (tidal) at elevations to 5200 m. The plants occur in open sites (full sunlight) in up to 15 cm of standing water. Some shade can be tolerated, but experimental shading of more than 50% has been shown to substantially reduce the biomass. The plants occur in strongly alkaline, brackish, or saline (even hypersaline) sites, which span a wide range of ecological conditions. When growing in salt marshes, they primarily occupy the upper (high) and middle marsh (~1.4–2.5 m elevation). The water (pH: 5.1–8.5) typically is at a salinity level near 10‰ but the plants can tolerate salinities up to 31‰. The substrates range from 5% to 80% organic matter and are alkaline (pH: 6.8–8.9 [–10]), often brackish or saline (e.g., to 1,031 mM NaCl; conductivity: to 34 mmhos cm^{-1}) and are characterized by a relatively high SO_4^{2-} content (to 7,600 ppm). They are high in total soluble salts (to 25,152 ppm) and various nutrients (e.g., N: 43 ppm; P: 22 ppm; K: >600 ppm; Ca: 71,450 ppm; Cl: 463 ppm; Mg: 891 ppm; Na: 12,408 ppm). The reported substrates include alluvium (quaternary, sandy, silty), boulders (basalt), clay (alkaline, heavy, mineralized), clay humus (alkaline), clay loam, clay silt, cobble, colluvium (tertiary claron), gravel, hardpan (silty), humus (acidic, salty), igneous detritus, limestone (Claron), loam (alkaline, black, heavy, humic, saline), muck (saline), mud (alkaline), muddy pebbles, peat (marly), rocks, sand, sandy clay, sandy loam, sandy marly peat, sawdust, schist (gravelly), serpentine, shale (calcareous), silt (alluvial, organic), silty gravel, silty hardpan, silty loam (salty), talus (limestone), and travertine (boggy). Flowering occurs from March to October, with fruiting taking place from May to October (mainly September–October). Flowers first initiate in plants that are 2 years old but will not develop in sites of high salinity or with high stem density. The wind-pollinated flowers are strongly protogynous and do not shed their pollen until they have been fertilized (a period from 5 to 8 days), which promotes complete outcrossing. Natural populations exhibit fairly high levels of genetic variation (e.g., 2–12 alleles per locus; average of 6) based on microsatellite data. Each plant produces an average of six flowering scapes. A single inflorescence averages 250 flowers (but up to 320) and can produce as many as 1,200 seeds, with a maximum total yield estimated at nearly 9,000 seeds per plant (although much lower averages, e.g., only 12 seeds/flowering stem, have been observed at some field sites). The schizocarps detach completely into six distinct mericarps when mature. About 90% of the seeds produced are viable. The intact schizocarps are buoyant, dispersed by water, and can remain afloat for more than 6 months (50% remaining afloat after 1 month); however, individual mericarps sink within 5 days. The floating seeds remain highly viable for more than 5 months. Presumably, the seeds also are dispersed endozoically by waterfowl (see *Use by Wildlife*). Genetic analyses (using microsatellite markers) indicate that overland dispersal of seeds (which likely involves a biotic vector) contributes to the genetic structure of populations in addition to coastal dispersal of seeds by water. A persistent seed bank of at least 1 year in duration develops with densities of up to 10,000 seeds m^{-2}; however, much lower densities (e.g., 2.5 seeds m^{-2}) have been observed at some field sites. The seeds are physiologically dormant and require 10–30 days of cold stratification (at 1°C–3°C) followed by incubation at 20°C–30°C (or a 25°C/5°C day/night temperature regime) in the light to induce germination. No germination occurs in the dark. High germination rates (e.g., 40%) occur in freshwater and decline progressively as salinity increases. Germination declines markedly at a 400 mM concentration of NaCl and is completely inhibited at salinities above 24‰. The rates in freshwater can be increased by treatment of seeds with dormancy alleviating compounds such as ethephon, fusicoccin, kinetin, proline, and thiourea. Ethephon, kinetin, nitrate, and thiourea also can counteract to some degree the inhibition of germination at higher levels of salinity. However, immersion in seawater can help to break dormancy and seeds kept in seawater at 3°C for 60–80 days have germinated when brought to 25°C under dark or light conditions. Manipulated seeds that have been cold stratified (2°C–3°C) for 4 months in 8–16 mM PEG (polyethylene glycol) have germinated to 100% at 20°C–25°C. In the field, seed germination usually commences in the spring as the new foliage is produced by plants emerging from dormancy. The plants (living 15–20 years or more) are highly clonal and produce densely packed rhizomes that exclude other species. They have been described as "ecosystem engineers" because their growth form develops into circular "rings" of densely packed shoots, which by elevating the plants enhances biomass production (by up to 250%), retains shed seeds, and reduces physiological stress when the areas become waterlogged. Sediments trapped by the plants reduce erosion and facilitate salt marsh expansion. Burial of shoots by up to 15 cm of sediment does not seriously affect their density. Salt marsh plants grow taller toward the lower marsh than in the higher marsh. Estimates for Old World plants indicate that an annual biomass production of up to 783 g m^{-2} year^{-1} can be attained. From 10% to 20% of the shoot dry mass can comprise the amino acid proline, which is believed to be responsible in part for salt tolerance by acting as an intracellular solute to facilitate osmotic regulation. An increase in fall proline levels also has been implicated as a protective response associated with frost tolerance. There are conflicting accounts of whether the plants are colonized by vesicular-arbuscular mycorrhizal Fungi. **Reported associates:** *Achillea millefolium, Agalinis calycina, Agalinis maritima, Agoseris glauca, Agropyron, Agrostis gigantea, Agrostis stolonifera, Allenrolfea occidentalis, Allium schoenoprasum, Almutaster pauciflorus, Amphiscirpus nevadensis, Andromeda polifolia, Andropogon gerardii, Antennaria microphylla, Antennaria pulcherrima, Anthoxanthum nitens, Apocynum cannabinum, Arnoglossum plantagineum, Artemisia cana, Astragalus bodinii, Atriplex argentea, Atriplex calotheca, Atriplex dioica, Atriplex hymenelytra, Atriplex patula, Aulacomnium palustre, Baccharis, Bassia hyssopifolia, Batis maritima, Bromus japonicus, Berula erecta, Betula glandulosa, Betula occidentalis, Betula pumila, Bistorta bistortoides, Blysmopsis rufa, Bolboschoenus maritimus, Brachythecium mildeanum, Brachythecium turgidum, Bryum pseudotriquetrum, Cakile, Calamagrostis stricta, Calliergon giganteum, Calopogon, Caltha palustris, Camassia, Campylium stellatum, Capsella bursa-pastoris, Cardamine pratensis, Carex aquatilis, Carex bigelowii, Carex buxbaumii, Carex capillaris, Carex chordorrhiza, Carex crawei, Carex cusickii, Carex diandra, Carex douglasii, Carex exilis, Carex gynocrates, Carex hystericina, Carex interior, Carex lasiocarpa, Carex leptalea, Carex limosa, Carex livida, Carex lyngbyei, Carex mackenziei, Carex magellanica, Carex maritima, Carex membranacea, Carex microglochin, Carex nebrascensis, Carex obnupta, Carex paleacea, Carex parryana, Carex pellita, Carex pluriflora, Carex praegracilis, Carex ramenskii, Carex rariflora, Carex retrorsa, Carex rostrata, Carex rotundata, Carex salina, Carex sartwellii, Carex saxatilis, Carex scirpoidea, Carex simulata, Carex sterilis, Carex subspathacea, Carex tetanica, Carex utriculata, Carex vaginata, Carex viridula, Castilleja ambigua, Cercocarpus montanus, Chamaedaphne calyculata, Chenopodium glaucum, Chenopodium rubrum, Chloropyron maritimum, Chloropyron tecopense, Chrysanthemum arcticum, Cicuta douglasii, Cicuta maculata, Cirsium arvense, Cirsium hydrophilum, Cirsium wrightii, Cladium mariscoides, Cleome, Cochlearia groenlandica, Comarum palustre, Coreopsis, Cornus sericea, Crepis runcinata, Cuscuta salina, Cyperus squarrosus, Danthonia intermedia, Dasiphora floribunda, Deschampsia cespitosa, Distichlis spicata, Drepanocladus aduncus, Drepanocladus sendtneri, Drosera anglica, Drosera intermedia, Drosera rotundifolia, Dryas integrifolia, Elaeagnus angustifolia, Eleocharis macrostachya, Eleocharis palustris, Eleocharis quinqueflora, Eleocharis rostellata, Eleocharis uniglumis, Elodea, Elymus lanceolatus, Epilobium ciliatum, Equisetum arvense, Equisetum fluviatile, Equisetum hyemale, Equisetum laevigatum, Equisetum palustre, Equisetum variegatum, Ericameria albida, Ericameria nauseosa, Erigeron gracilis, Erigeron lonchophyllus, Eriophorum angustifolium, Eriophorum crinigerum, Eriophorum gracile, Eriophorum russeolum, Eupatorium perfoliatum, Euthamia graminifolia, Euthamia occidentalis, Eutrochium maculatum, Fimbristylis puberula, Frankenia salina, Galium boreale, Galium labradoricum, Galium trifidum, Gentianella amarella, Gentianopsis crinita, Gentianopsis detonsa, Glyceria grandis, Grindelia hirsutula, Hamatocaulis vernicosus, Helenium autumnale, Helianthus grosseserratus, Helianthus nuttallii, Helianthus paradoxus, Hippuris montana, Hippuris tetraphylla, Hippuris vulgaris, Holcus lanatus, Honckenya peploides, Hordeum brachyantherum, Hordeum jubatum, Horkelia, Impatiens capensis, Iris missouriensis, Iris versicolor, Iva frutescens, Jaumea carnosa, Juncus alpinoarticulatus, Juncus arcticus, Juncus balticus, Juncus biglumis, Juncus bufonius, Juncus ensifolius, Juncus gerardii, Juncus interior, Juncus longistylis, Juncus nodosus, Juncus tenuis, Juncus torreyi, Juncus triglumis, Kalmia* *microphylla, Kochia, Lactuca serriola, Larix laricina, Lathyrus japonicus, Lepidium virginicum, Leymus arenarius, Leymus cinereus, Liatris spicata, Lilaeopsis occidentalis, Lilium philadelphicum, Limonium californicum, Limonium carolinianum, Limprichtia revolvens, Lobelia kalmii, Lomatogonium rotatum, Lycopus asper, Lycopus uniflorus, Lysimachia maritima, Lysimachia quadriflora, Lythrum californicum, Lythrum salicaria, Maianthemum stellatum, Medicago polymorpha, Meesia triquetra, Mentha arvensis, Menyanthes trifoliata, Mimulus, Muhlenbergia asperifolia, Muhlenbergia filiformis, Muhlenbergia glomerata, Muhlenbergia racemosa, Muhlenbergia richardsonis, Myrica gale, Myriophyllum quitense, Nasturtium officinale, Oryzopsis, Oxytropis, Packera debilis, Packera paupercula, Packera pseudaurea, Parapholis incurva, Parnassia glauca, Parnassia palustris, Pascopyrum smithii, Pedicularis crenulata, Pedicularis groenlandica, Pedicularis lanceolata, Pedicularis sudetica, Penstemon rydbergii, Phalaris arundinacea, Phlox kelseyi, Phragmites australis, Picea glauca, Picea mariana, Pinguicula villosa, Pinguicula vulgaris, Plantago eriopoda, Plantago lanceolata, Plantago maritima, Poa arida, Poa nemoralis, Poa nervosa, Poa palustris, Poa pratensis, Poa secunda, Pogonia ophioglossoides, Polypogon monspeliensis, Populus angustifolia, Populus deltoides, Potamogeton amplifolius, Potamogeton natans, Potentilla anserina, Potentilla gracilis, Primula egaliksensis, Primula incana, Primula pauciflora, Prunella vulgaris, Pseudotsuga menziesii, Puccinellia distans, Puccinellia maritima, Puccinellia nutkaensis, Puccinellia nuttalliana, Puccinellia phryganodes, Puccinellia pumila, Pyrola asarifolia, Pyrrocoma apargioides, Pyrrocoma racemosa, Ranunculus cymbalaria, Ranunculus gmelinii, Rhamnus frangula, Rhododendron groenlandicum, Rhododendron lapponicum, Rhododendron tomentosum, Rhynchospora alba, Rhynchospora capillacea, Rhynchospora fusca, Ribes aureum, Rosa, Rubus arcticus, Rumex britannica, Rubus chamaemorus, Rudbeckia fulgida, Rumex crispus, Ruppia maritima, Salicornia depressa, Salicornia maritima, Salicornia rubra, Salix arbusculoides, Salix arctica, Salix bebbiana, Salix brachycarpa, Salix candida, Salix exigua, Salix fuscescens, Salix geyeriana, Salix lasiandra, Salix myrtillifolia, Salix pedicellaris, Salix planifolia, Salix serissima, Sarcobatus vermiculatus, Sarcocornia perennis, Sarracenia purpurea, Schedonorus pratensis, Scheuchzeria palustris, Schoenoplectus acutus, Schoenoplectus americanus, Schoenoplectus pungens, Schoenoplectus tabernaemontani, Scirpus microcarpus, Scorpidium scorpioides, Scutellaria, Senecio crassulus, Senecio hydrophilus, Senecio sphaerocephalus, Shepherdia argentea, Sidalcea oregana, Sisyrinchium idahoense, Sisyrinchium pallidum, Solidago ohioensis, Sonchus arvensis, Spartina alterniflora, Spartina foliosa, Spartina gracilis, Spartina patens, Spartina pectinata, Spergularia canadensis, Spergularia macrotheca, Spergularia media, Spergularia salina, Sphagnum magellanicum, Sphagnum subsecundum, Sporobolus airoides, Stipa, Stuckenia filiformis, Suaeda calceoliformis, Suaeda nigra, Symphyotrichum boreale, Symphyotrichum ciliatum, Symphyotrichum firmum, Symphyotrichum foliaceum, Symphyotrichum laeve, Symphyotrichum lentum, Symphyotrichum puniceum, Symphyotrichum spathulatum, Tamarix ramosissima, Taraxacum officinale, Thalictrum alpinum, Thalictrum dasycarpum, Thelypodium crispum, Thelypteris palustris, Thermopsis montana, Tofieldia pusilla, Tomenthypnum nitens, Torreyochloa pallida, Toxicodendron vernix, Triantha glutinosa, Trichophorum alpinum, Trichophorum cespitosum, Trichophorum pumilum, Triglochin concinna, Triglochin palustris, Triglochin striata, Typha angustifolia, Typha latifolia, Utricularia intermedia, Utricularia macrorhiza, Utricularia ochroleuca, Vaccinium oxycoccos, Vaccinium uliginosum, Valeriana edulis, Valeriana sitchensis, Veronica arvensis, Viola adunca, Viola sororia, Zannichellia palustris, Zeltnera exaltata, Zizia, Zostera marina.*

3.3.1.4. ***Triglochin palustris*** **L.** is a perennial, which occurs in or on beaches (back, backwater, shingle), bogs (calcareous, Travertine), bottomland (alkaline), channels (loamy, river), depressions (calcareous, marly, sandy), ditches (bottom, drainage, ephemeral, gravel), draws (alkaline, muddy), fens (calcareous), flats (gravel, river, salt, silt), floodplains, gullies, hummocks, levees (slough), marshes (brackish, coastal, inland, peaty), meadows (alkaline, brackish tidal, estuarine, halophilic, saline, sedge, streamside), mudflats, muskeg, pannes, patterned bogs (flarks, strings), ponds (beaver, dried, mud-pan, upper estuary), prairies, saltmarshes, seeps (calcareous), shores (clay), silt pits, slicks (alkali), slopes (fellfield, tundra; to 10%), springs (hot, mineral), streambeds (dry), swales, swamps (alkali, marly seepage), swards, tundra, wallows (elk), and along the margins of channels (spring-fed), lakes (alkali, calcareous), ponds (alkaline, brackish), pools (roadside, saline, tundra), rivers, sloughs, and streams at elevations to 4500 m. The plants grow in open exposures and at water depths to 30 cm. They have moderate salinity tolerance (e.g., conductivity from 0 to 20 mS cm^{-1}) and occur in the upper intertidal zone when growing in salt marshes. The substrates are alkaline or calcareous (pH: 7.4–8.3; Ca^{2+} 50 mg kg^{-1}) and include alluvium (glacial), clay (alkaline, salty), clay loam, colluvium, gravel (coarse), gravelly loam, limestone (Mississippian), loam (mollic), marl (calcareous), muck, mud (hard-packed), muddy sand, peat, peaty marl, sand (alkaline, brackish, Rupert), sandstone (red), sandy clay, sandy clay loam, sandy loam, sandy silt, shale (Belden), silt (alluvial), silty clay loam, silty loam, silty sand, and silty shale. Flowering occurs from June to October with fruiting extending from July to October. Individual plants produce about 100 flowers (averaging three scapes and 45 flowers per scape) and up to 400 seeds per plant. The flowers are wind pollinated. When the schizocarps mature, the three mericarps do not separate completely but adhere distally to the fruit axis. A short-term persistent seed bank (1–5 years) develops at densities of up to 14.3 seeds m^{-1}. The mericarps (averaging 1.19 mg) are buoyant and are dispersed locally by water. They also have been shown to adhere (for up to 3 hr) to the fur of grazing animals like deer (Mammalia: Cervidae), which likely facilitate their overland dispersal at distances up to 1242 m. Seed germination (in light under a 25°C/15°C; day/night temperature regime) is substantially enhanced by a dry after-ripening period of 28 days at room temperature. Scarcely any germination occurs in the dark. The plants can reach a density of 14 stems m^{-2} and reproduce vegetatively by producing slender stolons that terminate in small tubers. The plants are not resistant to burial and can exhibit a 76–100% reduction in shoot density when covered by 5–15 cm of sediment. The plants often reproduce by vegetative means. Although the plants are more abundant in grazed versus ungrazed plots, the fertilization of sediments from defecation by grazing herbivore can affect populations negatively by enhancing interspecific competition for light. Herbivory also consumes inflorescences and reduces plant size, which results in decreased flowering. However, herbivory on competing species decreases survivorship of plants by making them more visible to herbivores. The plants have been described either as non-mycorrhizal or as possessing arbuscular mycorrhizae.

Reported associates: *Abies lasiocarpa, Achillea millefolium, Agalinis purpurea, Agrostis scabra, Agrostis stolonifera, Allium schoenoprasum, Alnus tenuifolia, Alnus viridis, Alopecurus, Andromeda polifolia, Andropogon gerardii, Antennaria pulcherrima, Anthoxanthum nitens, Anthoxanthum odoratum, Anticlea elegans, Arctagrostis latifolia, Arctophila fulva, Arctopoa eminens, Artemisia frigida, Artemisia tilesii, Atriplex dioica, Atriplex gmelinii, Betula glandulosa, Betula nana, Betula neoalaskana, Betula pumila, Bidens trichospermus, Bistorta vivipara, Bolboschoenus maritimus, Bouteloua gracilis, Bromus marginatus, Cakile edentula, Calamagrostis canadensis, Calamagrostis deschampsioides, Calamagrostis stricta, Calamovilfa longifolia, Calliergon trifarium, Campanula parryi, Campylium stellatum, Carex aquatilis, Carex atrofusca, Carex aurea, Carex bigelowii, Carex buxbaumii, Carex canescens, Carex disperma, Carex douglasii, Carex exilis, Carex glareosa, Carex gynocrates, Carex hallii, Carex hirta, Carex hystericina, Carex illota, Carex interior, Carex krausei, Carex lasiocarpa, Carex limosa, Carex livida, Carex lyngbyei, Carex mackenziei, Carex magellanica, Carex microglochin, Carex nebrascensis, Carex nigricans, Carex nova, Carex paleacea, Carex parryana, Carex pellita, Carex petasata, Carex pluriflora, Carex prairea, Carex ramenskii, Carex rariflora, Carex rostrata, Carex rotundata, Carex saxatilis, Carex scirpoidea, Carex simulata, Carex sterilis, Carex stricta, Carex subspathacea, Carex ursina, Carex utriculata, Carex viridula, Castilleja cusickii, Castilleja sulphurea, Castilleja unalaschcensis, Chamerion latifolium, Chara contraria, Chelone glabra, Chloropyron maritimum, Chrysanthemum arcticum, Chrysosplenium tetrandrum, Chrysothamnus, Cinclidium stygium, Cladium mariscoides, Claytonia sibirica, Comarum palustre, Crepis runcinata, Cycloloma atriplicifolium, Cynoglossum officinale, Cyperus bipartitus, Danthonia intermedia, Dasiphora floribunda, Deschampsia cespitosa, Distichlis spicata, Drosera anglica, Drosera intermedia, Drosera linearis, Drosera rotundifolia, Dryas, Dupontia fisheri, Eleocharis acicularis, Eleocharis compressa, Eleocharis elliptica, Eleocharis erythropoda, Eleocharis palustris, Eleocharis quinqueflora, Eleocharis rostellata, Elymus canadensis, Empetrum nigrum, Epilobium palustre, Equisetum arvense, Equisetum scirpoides, Equisetum variegatum, Equisetum variegatum, Erigeron gracilis, Eriophorum angustifolium, Eriophorum scheuchzeri, Eriophorum viridicarinatum, Eupatorium perfoliatum, Euphrasia frigida, Eutrochium maculatum, Festuca baffinensis, Festuca rubra, Filipendula rubra, Fissidens adianthoides, Gaillardia aristata, Galium boreale, Galium triflorum, Gaultheria hispidula, Gentianopsis crinita, Gentianopsis detonsa, Gentianopsis thermalis, Geranium erianthum, Glyceria grandis, Glyceria striata, Hedysarum alpinum, Heracleum sphondylium, Hesperostipa comata, Hippuris tetraphylla, Hippuris vulgaris, Honckenya peploides, Hordeum jubatum, Iris missouriensis, Juncus alpinoarticulatus, Juncus articulatus, Juncus balticus, Juncus brachycephalus, Juncus castaneus, Juncus ensifolius, Juncus longistylis, Juncus mertensianus, Juncus nodosus, Juncus tenuis, Juncus torreyi, Juncus triglumis, Kalmia microphylla, Kobresia myosuroides, Kobresia simpliciuscula, Kochia scoparia, Larix laricina, Leontodon autumnalis, Lepidium campestre, Leymus arenarius, Liatris ligulistylis, Ligusticum scoticum, Lilium philadelphicum, Limprichtia revolvens, Linum lewisii, Lobelia kalmii, Lomatogonium rotatum, Luzula, Lysimachia maritima, Lysimachia quadriflora, Malus pumila, Mimulus guttatus, Mentha arvensis, Menyanthes trifoliata, Mimulus primuloides, Moerckia hibernica, Monarda fistulosa, Monolepis nuttalliana, Muhlenbergia filiculmis, Muhlenbergia filiformis, Muhlenbergia glomerata, Muhlenbergia mexicana, Muhlenbergia richardsonis, Nephrophyllidium crista-galli, Oenothera coronopifolia, Oxypolis rigidior, Packera debilis, Panicum virgatum, Parnassia glauca, Parnassia kotzebuei, Parnassia palustris, Parnassia parviflora, Pedicularis groenlandica, Pedicularis lanceolata, Penstemon angustifolius, Phlox kelseyi, Phragmites australis, Picea engelmannii, Picea glauca, Picea mariana, Pilea fontana, Pinguicula vulgaris, Plantago maritima, Platanthera aquilonis, Platanthera hyperborea, Poa alpina, Poa arctica, Poa nervosa, Poa pratensis, Populus tremuloides, Populus trichocarpa, Potentilla anserina, Potentilla pulchella, Primula egaliksensis, Primula incana, Primula pauciflora, Prunella vulgaris, Puccinellia nutkaensis, Puccinellia nuttalliana, Puccinellia phryganodes, Puccinellia vaginata, Pycnanthemum virginianum, Pyrola asarifolia, Pyrrocoma apargioides, Pyrrocoma integrifolia, Ranunculus cymbalaria, Ranunculus flammula, Ranunculus gmelinii, Ranunculus pensylvanicus, Ranunculus uncinatus, Rhinanthus minor, Rhododendron columbianum, Rhododendron groenlandicum, Rhynchospora alba, Rhynchospora capillacea, Rosa acicularis, Rosa woodsii, Rubus, Rudbeckia hirta, Rumex britannica, Rumex transitorius, Salicornia rubra, Salix brachycarpa, Salix candida, Salix cordata, Salix exigua, Salix interior, Salix longifolia, Salix myrtillifolia, Salix petiolaris, Salix planifolia, Salix serissima, Sarcobatus vermiculatus, Sarcocornia utahensis, Sarracenia purpurea, Saussurea angustifolia, Saxifraga aizoides, Schizachyrium scoparium, Schoenoplectus acutus, Schoenoplectus pungens, Schoenoplectus tabernaemontani, Scirpus microcarpus, Scleria verticillata, Scorpidium scorpioides, Selaginella apoda, Senecio sphaerocephalus, Senecio* *triangularis, Sisyrinchium idahoense, Solidago ohioensis, Solidago patula, Solidago uliginosa, Sorghastrum nutans, Spergularia salina, Sphagnum, Spiraea stevenii, Splachnum, Sporobolus airoides, Stellaria calycantha, Stellaria crassifolia, Stellaria humifusa, Stellaria longipes, Suaeda calceoliformis, Symphyotrichum firmum, Symphyotrichum puniceum, Tephroseris palustris, Thalictrum alpinum, Thalictrum sparsiflorum, Thelypodium sagittatum, Thelypteris palustris, Tomenthypnum nitens, Torreyochloa pallida, Triantha glutinosa, Trichophorum alpinum, Trichophorum cespitosum, Trichophorum pumilum, Trientalis europaea, Trifolium repens, Triglochin concinna, Triglochin maritima, Tussilago farfara, Typha latifolia, Utricularia cornuta, Utricularia intermedia, Utricularia ochroleuca, Vaccinium oxycoccos, Vaccinium uliginosum, Verbena hastata, Viola, Wilhelmsia physodes, Zizia aurea.*

3.3.1.5. *Triglochin scilloides* (Poir.) Mering & Kadereit

is an annual, which inhabits channels, depressions, ditches (roadside), flats (alkali, dry, seasonal, tidal), marshes (high, intertidal, tidal), meadows (bog, rush), mudflats (tidal), playas, ponds (temporary), potholes, puddles, seeps (dry), sloughs (estuarine), swales, vernal pools, wallows, washes, and the margins of lakes, reservoirs, rivers, and streams (dried) at elevations to 1709 m. These plants are specialists of vernal pools and other temporary water sites. However, they can withstand extended periods of inundation (to 65 days) in shallow standing waters (to 25 cm deep). The sites are characterized by open exposures (full sunlight to partial shade). The substrates have been described as adobe (gravelly), clay (muddy), clay loam (mucky), hardpan (Tuscan), loamy sand, muck, mud (granitic), sand, schist (Julian), and silt. Flowering and fruiting occur from June to September. The plants are wind pollinated and polygamous (polygamomonoecious), bearing bisexual, pistillate, and staminate flowers (with or without a perianth member) occurring on the same individual. The bisexual and "normal" unisexual flowers are produced in the upper inflorescence, whereas naked, unisexual pistillate flowers with extremely long styles (to 30 cm) occur at the inflorescence base. Although sexual reproduction normally takes place during drying conditions after waters recede, the elongate styles of the lower flowers allow them to maintain sexual reproduction if standing waters persist during the flowering period by enabling their stigmas to float on the water surface. A seed bank of more than 1,000 seeds m^{-2} can develop at depths from 0 to 5 cm. The seeds are physiologically dormant and will not germinate unless they remain inundated. Germination occurs primarily in February (at temperatures <20°C) but extends until April. Genetic studies of Mexican plants (using allozyme data) indicate low levels of intrapopulational genetic variation and a high proportion of identical multilocus genotypes. Although the latter potentially had been explained as indicative of clonal reproduction, such an explanation is untenable given the annual habit of the plants and lack of vegetative propagules. The genetic data more likely indicate a high level of inbreeding. **Reported associates:** *Achyrachaena mollis, Agrostis hooveri, Agrostis lacuna-vernalis, Alisma*

triviale, Alopecurus saccatus, Amsinckia intermedia, Anagallis minima, Angelica arguta, Artemisia cana, Avena barbata, Baccharis salicifolia, Bidens cernuus, Briza minor, Brodiaea, Bromus hordeaceus, Callitriche heterophylla, Callitriche longipedunculata, Callitriche marginata, Caltha palustris, Carex lyngbyei, Centromadia pungens, Cirsium scariosum, Cotula coronopifolia, Crassula aquatica, Crassula connata, Croton setiger, Cyperus squarrosus, Damasonium, Delphinium hansenii, Deschampsia cespitosa, Deschampsia danthonioides, Dianthus armeria, Dichelostemma capitatum, Downingia bicornuta, Downingia cuspidata, Downingia pusilla, Elatine brachysperma, Elatine californica, Elatine chilensis, Eleocharis acicularis, Eleocharis macrostachya, Eleocharis palustris, Eleocharis parishii, Epilobium ciliatum, Epilobium cleistogamum, Equisetum fluviatile, Equisetum palustre, Equisetum telmateia, Erodium, Eryngium aristulatum, Eryngium vaseyi, Galium trifidum, Gayophytum ramosissimum, Glyceria ×occidentalis, Gnaphalium palustre, Gratiola ebracteata, Gratiola virginiana, Holcus lanatus, Hordeum brachyantherum, Hordeum depressum, Hordeum marinum, Hordeum murinum, Hydrocotyle ranunculoides, Hydrocotyle umbellata, Hypochaeris glabra, Impatiens capensis, Iris missouriensis, Isoetes howellii, Isoetes nuttallii, Isoetes orcuttii, Juncus articulatus, Juncus bufonius, Juncus oxymeris, Juncus phaeocephalus, Juncus sphaerocarpus, Juncus xiphioides, Lasthenia californica, Lasthenia fremontii, Lasthenia glaberrima, Lasthenia platycarpha, Lathyrus palustris, Leersia oryzoides, Legenere valdiviana, Lemna gibba, Lepidium, Lilaeopsis occidentalis, Limnanthes alba, Limnanthes douglasii, Limosella acaulis, Limosella australis, Logfia gallica, Lomatium bicolor, Lythrum hyssopifolia, Lythrum salicaria, Marsilea vestita, Mentha arvensis, Mentha aquatica, Menyanthes trifoliata, Mimulus guttatus, Mimulus tricolor, Montia fontana, Muhlenbergia rigens, Myosotis scorpioides, Myosurus, Nasturtium officinale, Navarretia leucocephala, Navarretia prostrata, Pascopyrum smithii, Phalaris arundinacea, Phalaris caroliniana, Phalaris lemmonii, Pilularia americana, Plagiobothrys stipitatus, Plagiobothrys trachycarpus, Plagiobothrys undulatus, Plantago bigelovii, Platanthera dilatata, Platystemon californicus, Pleuropogon californicus, Poa annua, Poa nemoralis, Poa palustris, Poa trivialis, Pogogyne abramsii, Pogogyne clareana, Potamogeton foliosus, Potentilla anserina, Psilocarphus brevissimus, Psilocarphus tenellus, Ranunculus aquatilis, Ranunculus bonariensis, Ranunculus californicus, Ranunculus lobbii, Ranunculus sceleratus, Rorippa curvisiliqua, Rumex conglomeratus, Rumex crispus, Sagittaria cuneata, Sagittaria latifolia, Salix lasiandra, Salix lasiolepis, Schedonorus arundinaceus, Schoenoplectus acutus, Schoenoplectus tabernaemontani, Schoenoplectus triqueter, Sidalcea hendersonii, Sium suave, Sparganium emersum, Spergula, Symphyotrichum eatonii, Symphyotrichum subspicatum, Tamarix ramosissima, Taraxacum officinale, Trifolium depauperatum, Trifolium variegatum, Trifolium wormskioldii, Triteleia hyacinthina, Typha latifolia, Veronica anagallis-aquatica, Veronica peregrina, Vulpia bromoides, Vulpia myuros, Zannichellia palustris.

3.3.1.6. *Triglochin striata* Ruiz & Pav. is a perennial, which inhabits coastal environments including beaches, canals (drainage), channels (river), depressions (interdunal), dikes (salt marsh), ditches (brackish, drainage, oligohaline, tidal), estuaries (tidal), fens (seepage), flats (alkaline, mud, sand, tidal), hammocks (coastal), marshes (brackish, coastal alkaline, freshwater, interdunal, tidal), prairies (brackish), puddles (mud), salt marshes, seeps, shores (sandy), sloughs (tidal), springs (inland), strands (coastal), streambeds (dried), swales, swamps (interdunal), and the margins of lagoons (salt), lakes, ponds, rivers, streams (tidal), and swamps (mangrove) at elevations to 24 m. The plants are found in shallow water (to at least 5 cm) and grow in exposures of full sunlight, partial shade, and even in deep shade. They grow on alkaline or saline sites and can tolerate salinities up to 46‰. It is believed that osmoregulation is facilitated by the accumulation of proline, which can increase by 200% (shoots) and 500% (roots) at 400 mol m^{-3} NaCl. The sediments (pH: 5.4–7.2) include clay (heavy), Felda (arenic ochraqualfs), gravel, loamy mud, loamy sand, muck, mud (limey), peat, sand, sandy muck, spoil (quartzipsamment). Flowering occurs from May to October. The flowers are wind pollinated. The schizocarps dehisce as three separate mericarps when mature. The seeds germinate best in freshwater (<250 mmol l^{-1} NaCl) and light under a 30°C/20°C (12/12 hr) temperature regime; the rates decrease with increasing salinity. Studies in the Old World indicate that the seeds might be dispersed endozoically by waterfowl (see *Use by wildlife*). Vegetative reproduction occurs by means of a stoloniferous rhizome. Other life history details are lacking for North American populations and deserve further inquiry. **Reported associates:** *Acrostichum danaeifolium, Agalinis maritima, Agrostis, Alternanthera philoxeroides, Amaranthus australis, Ammannia coccinea, Ammannia latifolia, Ammophila arenaria, Asclepias lanceolata, Baccharis angustifolia, Baccharis halimifolia, Bacopa monnieri, Batis maritima, Blutaparon vermiculare, Bolboschoenus maritimus, Bolboschoenus robustus, Borrichia frutescens, Carex albolutescens, Casuarina, Centella asiatica, Cicuta maculata, Cladium jamaicense, Convolvulus sepium, Corethrogyne filaginifolia, Cotula coronopifolia, Cryptantha leiocarpa, Cuscuta, Cyperus filiculmis, Cyperus flavescens, Cyperus haspan, Cyperus odoratus, Cyperus strigosus, Distichlis littoralis, Distichlis spicata, Echinochloa walteri, Eleocharis albida, Eleocharis cellulosa, Eleocharis flavescens, Eleocharis palustris, Eleocharis parvula, Equisetum arvense, Euthamia graminifolia, Fimbristylis spadicea, Fuirena pumila, Fuirena squarrosa, Grindelia, Hibiscus moscheutos, Hydrocotyle bonariensis, Hydrocotyle umbellata, Isolepis cernua, Jaumea carnosa, Juncus biflorus, Juncus bufonius, Juncus dichotomus, Juncus scirpoides, Juniperus virginiana, Lilaeopsis carolinensis, Limonium carolinianum, Ludwigia, Lysimachia maritima, Lythrum lineare, Myrica cerifera, Nuttallanthus canadensi, Oenothera fruticosa, Paspalum, Persicaria, Phacelia argentea, Phyla lanceolata, Phyla nodiflora, Pluchea camphorata, Pontederia cordata, Potentilla anserina, Pteridium aquilinum, Ptilimnium capillaceum, Rhynchospora colorata, Saccharum giganteum, Sacciolepis striata, Sagittaria lancifolia, Sagittaria latifolia, Salicornia depressa, Salix hookeriana, Salix sitchensis, Samolus valerandi, Sarcocornia pacifica, Saururus cernuus, Schinus terebinthifolius, Schoenoplectus pungens, Sesuvium portulacastrum, Seutera angustifolia, Spartina bakeri, Spergularia salina, Spiranthes praecox, Sporobolus virginicus, Tanacetum bipinnatum, Teucrium canadense, Trifolium wormskioldii, Triglochin maritima, Typha angustifolia, Typha domingensis, Typha latifolia, Vaccinium, Zizania aquatica.*

Use by wildlife: Although the forage value of *Triglochin palustris* (i.e., C:N; C:P; N:P ratios) is somewhat higher than that of *T. maritima*, the toxicity of *Triglochin* plants (see *medicinal*) limits their use by wildlife. The plants were long recognized as a livestock poison by the Blackfoot Indians. They are toxic to grazing livestock (Mammalia: Bovidae), with the lethal amount of *T. maritima* ingested by cattle (*Bos taurus*) ranging from 1% to 7% of body mass and for sheep (*Ovis aries*) from 0.5% to 2% of body mass. The highest toxicity is found in fresh, green foliage, which can contain up to 0.035% hydrogen cyanide (HCN); the levels generally are lower in dried foliage, but can become elevated in foliage consumed from drought-stricken plants. Intraperitoneal injections (1 g sodium nitrite, 2 g sodium thiosulfate; 20% aqueous solution) have been administered successfully as an antidote for sheep poisoning, unless several lethal doses have been consumed. In addition to toxins, *T. maritima* has a relatively high phenolic content, which also lessens its palatability to some herbivores. Otherwise, *Triglochin concinna* is a host of Basidiomycota Fungi (Uropyxidaceae: *Aecidium triglochinis*). *Triglochin maritima* is used for cover and habitat by rails (Aves: Rallidae), short-eared owls (Aves: Strigidae: *Asio flammeus*), and various shorebirds. The plants are grazed frequently, and the fruits (high in soluble carbohydrates; 25% crude protein; 16% crude fiber; 6% lipids) are eaten by various waterfowl (Aves: Anatidae) including the American wigeon (*Anas americana*), brent goose (*Branta bernicla*), Canada goose (*Branta canadensis*), green-winged teal (*Anas crecca*), lesser snow goose (*Chen caerulescens caerulescens*), mallard (*Anas platyrhynchos*), pintail (*Anas acuta*), and wood duck (*Aix sponsa*). *Triglochin maritima* also is consumed by snow buntings (Aves: Calcariidae: *Plectrophenax nivalis*). In the Old World, the rhizomes are known to be gnawed by rats (Mammalia: Muridae: *Rattus norvegicus*). The stems are used by aphids (Insecta: Hemiptera: Aphididae) and are bored by larval plume moths (Insecta: Lepidoptera: Pterophoridae). Fungi hosted by *T. maritima* include Ascomycota (Erysiphaceae: *Leveillula taurica*; Pleosporaceae: *Pleospora herbarum, Wettsteinina operculata*) and Basidiomycota (Pucciniaceae: *Puccinia aristidae*). The bulb-like bases of *Triglochin palustris* contain 13% crude protein, 6% crude fiber, 1% lipid and are eaten by waterfowl (Aves: Anatidae) including black brant geese (*Branta bernicla nigricans*), Canada geese (*Branta canadensis*), and lesser snow geese (*Chen caerulescens caerulescens*). The leaves (to 4.7% total DM N) are the preferred food of emperor goose (*Chen canagica*) goslings. The plants are host to some Fungi (Ascomycota: Pleosporaceae: *Pleospora herbarum*). In the Old World, *T. striata* is eaten by various waterfowl (Aves: Anatidae). The recovery of viable seeds from their gizzards indicates the potential for endozoic dispersal of the propagules.

Economic importance: food: Because of their potential cyanide content, the foliage of *Triglochin* species never should be eaten. Yet, despite their toxicity, plants of *Triglochin maritima* were eaten as a vegetable by the Klamath and Salish tribes. The seeds (which contain little cyanide) were used as food by the Gosiute and Montana tribes and were roasted (which volatilizes the toxins) by the Klamath people as a coffee substitute; **medicinal:** The mature foliage and seedlings of *Triglochin maritima* contain the toxic cyanogenic glucosides taxiphyllin and triglochinin. The substances are synthesized by two P450 enzymes (CYP79E1 and CYP79E2), which encode multifunctional *N*-hydroxylases that catalyze the conversion of tyrosine to *p*-hydroxyphenylacetonitrile, the preferred substrate for cyanide production in the biosynthetic pathway. Annually, the highest concentration of triglochinin (up to 4% of the DM biomass) occurs in the new spring foliage and inflorescences. Lower triglochinin levels occur in saline compared to non-saline sites. Concentrations of the toxic glycosides also can become elevated during prolonged periods of moisture deficit. *Triglochin palustris* and *T. scilloides* also contain triglochinin; **cultivation:** *Triglochin* species are not distributed in the horticultural trade. *Triglochin concinna* has been planted in salt marsh restoration projects in California; **misc. products:** Seeds of *Triglochin maritimum* have been used in wetland restoration projects; **weeds:** none reported in North America; **nonindigenous species:** *Triglochin scilloides* is naturalized in Australia, Portugal, and Spain; *T. striata* has been introduced to the Iberian Peninsula and elsewhere in western Europe.

Systematics: Molecular phylogenetic studies resolve *Triglochin* as a monophyletic sister genus to *Cycnogeton*, but only if the monotypic genus *Lilaea* (i.e., *L. scilloides*) is included (Figures 1.19 and 1.20). Several issues remain to be reconciled within *Triglochin*, which has long been problematic taxonomically. *Triglochin concinna* has been merged with *T. maritima* (which is regarded as a taxonomic species "complex") but is distinct in phylogenetic analyses (Figure 1.20) if the Old World populations of *T. maritima* are distinguished taxonomically from New World populations. Previously, some polyploid ($2n = 144$) eastern North American plants have been segregated from the Old World plants as *T. elata* Nutt. Despite the difficulty of distinguishing taxa within the *T. maritima* complex morphologically (e.g., by multivariate analyses), it remains possible that the name *T. elata* eventually might be applied to this chromosomally distinct North American segregate of *T. maritima* given the discrete phylogenetic relationships (Figure 1.20). Until the taxonomy of the *T. maritima* complex has been resolved more satisfactorily, it seems prudent to retain the recognition of *T. concinna* and *T. gaspensis*. *Triglochin scilloides* (formerly recognized as the monotypic *Lilaea scilloides*) presents another interesting anomaly. Despite is distinct morphology and annual habit, "*Lilaea*"

nests within other *Triglochin* species, as the sister to a large Old World clade of perennials (Figure 1.20). The variable and wide-ranging perennial *T. striata* resolves as the sister group to a clade of Australian annuals (Figure 1.20). The basic chromosome number of *Triglochin* is $x = 6$. *Triglochin scilloides* ($2n = 12$) is a diploid, *T. palustris* and *T. striata* (both $2n = 24$) are tetraploids. *Triglochin concinna* has been subdivided into two varieties: *T. concinna* var. *concinna* (a $2n = 48$ octoploid of coastal salt marshes) and *T. concinna* var. *debilis* (a $2n = 96$ sedecaploid of freshwater, higher elevation sites); *T. gaspensis* ($2n = 96$) also is a sedecaploid. Different populations of *T. maritima* can occur at various ploidy levels ($2n = 12$, 24, 36, 48, 96, 120, 144, 156) and likely represent distinct taxa and/or misidentifications of material, in at least some cases. Alleged natural hybrids (*T. gaspensis* × *T. maritima*) have been reported but have not been verified experimentally.

Distribution: *Triglochin maritima* and *T. palustris* are fairly widespread in North America except for much of the south-central and southeastern United States. More restricted species include *T. concinna* (western North America but extending into South America), *T. striata* (southeastern and western coastal regions of the United States; extending into South America; also Africa, Australasia, Madagascar, West Europe), *T. scilloides* (sporadic in western North America, but extending to Mexico and South America), and *T. gaspensis* (confined to a small portion of northern Maine and eastern Canada).

References: Aston, 1967; Barbour et al., 2005; Baskin & Baskin, 1998; Beal, 1977; Benedict, 1983; Björk & Dunwiddie, 2004; Blaney et al., 2014; Bliss & Zedler, 1998; Boestfleisch & Papenbrock, 2017; Bouchard et al., 1991; Bowles, 1991; Bradfield & Porter, 1982; Buchsbaum et al., 1984; Buffington et al., 2018; Burchill & Kenkel, 1991; Burk, 1962; Buzgo et al., 2006; Chang et al., 2001; Chee & Vitt, 1989; Clawson & Moran, 1937; Colwell & Oring, 1988; Cook, 1988; Cooke, 1997; Cooper, 1996; Cooper & MacDonald, 2000; Cutler et al., 1981; Davy & Bishop, 1991; Dodd & Coupland, 1966; Doherty et al., 2011; Drummond, 1960; Eleuterius & Caldwell, 1984; Erfanzadeh et al., 2010; Eyjólfsson, 1970; Fernald & Kinsey, 1943; Fernández-Pascual et al., 2013; Ford & Ball, 1988; Gates, 1910; Glaser et al., 1981; 1990; Glawe et al., 2005; Gorham et al., 1987; Grewell, 2008; Grewell et al., 2007; Griese et al., 1980; Griffen, 1975; Handa et al., 2002; Hanson, 1951; Harris, 1990; Haynes & Hellquist, 2000b; Hegnauer & Ruijgrok, 1971; Heimbinder, 2001; Herzog & Sedinger, 2004; Hildebrand, 2012; Huffman & Tucker, 1984; Huiskes et al., 1995; Hutchings & Russell, 1989; Janousek & Folger, 2013; Jensen, 2004; Johnson & Steingraeber, 2003; Johnson-Green et al., 1995; Johnston, 1956a; 1956b; Jorgenson & Ely, 2001; Keer & Zedler, 2002; Keil, 2012; Kelly & Fletcher, 1994; Kemp et al., 2013; Khan & Ungar, 1997; 1999; 2001; Khan et al., 2001; Kiviniemi, 1996; Kiviniemi & Telenius, 1998; Kopecko & Lathrop, 1975; Laing & Raveling, 1993; Les, 2017; Les & Tippery, 2013; Lesica, 1986; Liehrmann et al., 2018; Lindig-Cisneros & Zedler, 2002; Looman, 1976; Löve & Lieth, 1961; Magallán et al., 2013; Majak et al., 1980; Mazerolle et al., 2014; McAtee, 1939; Moran et al., 1940; Morzaria-Luna & Zedler, 2007; 2014; Mouissie et al., 2005;

Mulder & Ruess, 1998a; 1998b; Mulder et al., 1996; Nahrsted et al., 1979; Naidoo, 1994; Naidoo & Naicker, 1992; Nava et al., 2000; Ņečajeva & Ievinsh, 2007; Ngai & Jefferies, 2004; Nicholson & Vitt, 1994; Nielsen & Møller, 1999; 2000; Ott-Conn et al., 2014; Peck, 1925; Peterson et al., 2011; Pinkava, 1963; Posluszny et al., 1986; Prevett et al., 1979; 1985; Raulings et al., 2011; Ritchie, 1957; Roberts & Robertson, 1986; Rochefort & Vitt, 1988; Rouger & Jump, 2013; 2014; Rozema et al., 1986; Ruch et al., 2008; Saarela et al., 2017; Sedinger & Raveling, 1984; Seigler, 1976; Skougard & Brotherson, 1979; Stewart & Lee, 1974; Thomas & Prevett, 1980; Thorne & Lathrop, 1969; 1970; Timoney, 2001; Tiner, 1993; Ungar, 1970; 1974; Van der Valk, 1975; Van der Valk et al., 1983; Vasey et al., 2012; Viereck et al., 1992; Vinson & Bushman, 2005; von Mering & Kadereit, 2010; 2015; Wang & Qiu, 2006; Wilson et al., 1996; Wolters et al., 2005; Zedler et al., 1999; 2003; Zepeda et al., 2014.

Family 3.4. Potamogetonaceae [6]

Potamogetonaceae ("pondweeds") comprise a clade of obligate aquatic plants represented by six genera: *Groenlandia* Fourr., *Potamogeton* L., and *Stuckenia* Börner (Potamogetonaceae sensu stricto), and *Althenia* F. Petit [includes *Lepilaena* Harv.], *Pseudalthenia* (Graebn.) Nakai, and *Zannichellia* L., which have been included following the recommended subsummation of Zannichelliaceae (Les & Tippery, 2013). Recent phylogenetic evidence (Ito et al., 2016a) has recommended the merger of *Lepilaena* with *Althenia*. Species assigned to *Stuckenia* (= *Coleogeton*) once had been distinguished within *Potamogeton* as subgenus *Coleogeton* (Les & Haynes, 1996; Haynes et al., 1998e); however, their high degree of morphological and genetic divergence has warranted recognition as a distinct genus. This diverse family includes annuals and perennials, floating-leaved or submersed life-forms, wind and water pollination systems, and even marine species (Waycott et al., 2006).

The reduced and simplified floral structure of all the species has made it difficult to elucidate their interrelationships; however, molecular phylogenetic analyses (see Les & Tippery, 2013; Ross et al., 2016) consistently have resolved the former segregate genera of Zannichelliaceae near or among *Groenlandia*, *Potamogeton*, and *Stuckenia* (Figure 1.21) while failing to demonstrate any close relationship of the group to the formerly included *Ruppia* (see Cymodoceaceae discussion above). The family Zosteraceae resolves consistently as the sister family of Potamogetonaceae (Figure 1.21).

Pondweeds essentially are cosmopolitan but are of little economic importance. A few species of *Potamogeton* are noxious weeds and several (e.g., *P. gayi*, *P. schweinfurthii*, *P. wrightii*) are grown occasionally as ornamental aquarium or pond plants (Kasselmann, 2003). However, the group is of paramount importance to wildlife, with numerous animal species depending on the plants as a source of food and shelter.

Three genera contain OBL indicators in North America:

    3.4.1. ***Potamogeton*** L.
    3.4.2. ***Stuckenia*** Börner
    3.4.3. ***Zannichellia*** L.

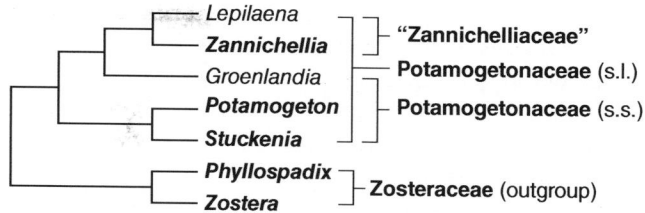

FIGURE 1.21 Intergeneric relationships of Potamogetonaceae *sensu lato* and related groups as indicated by the phylogenetic analysis of full chloroplast genome sequences (adapted from Ross et al., 2016). These results mirror other studies by showing no clear phylogenetic distinction between genera assigned previously to Zannichelliaceae and those of Potamogetonaceae sensu stricto (s.s.). Consequently, Potamogetonaceae sensu lato (s.l) have been redefined to include the former segregate genera of Zannichelliaceae (see Les & Tippery, 2013). The placement of Zosteraceae as the sister group of Potamogetonaceae is a consistent result of numerous molecular phylogenetic analyses. Although the members of all depicted taxa are obligate aquatics, those that include OBL North American indicators are highlighted in bold.

3.4.1. Potamogeton

Pondweed; potamot

Etymology: from the Greek *potamós geítōn* ("river neighbor") indicating an affinity for riverine habitats

Synonyms: *Hydrogeton* (in part); *Peltopsis*; *Spirillus*

Distribution: global: cosmopolitan; **North America:** widespread

Diversity: global: 72 species; **North America:** 33 species

Indicators (USA): obligate wetland (OBL): *Potamogeton alpinus, P. amplifolius, P. berchtoldii, P. bicupulatus, P. clystocarpus, P. confervoides, P. crispus, P. diversifolius, P. epihydrus, P. floridanus, P. foliosus, P. friesii, P. gramineus, P. hillii, P. illinoensis, P. natans, P. nodosus, P. oakesianus, P. obtusifolius, P. ogdenii, P. perfoliatus, P. polygonifolius, P. praelongus, P. pulcher, P. pusillus, P. richardsonii, P. robbinsii, P. sibiricus, P. spirillus, P. strictifolius, P. tennesseensis, P. vaseyi, P. zosteriformis*

Habitat: brackish (coastal), freshwater, freshwater (tidal); lacustrine, palustrine, riverine; **pH:** 3.1–10.6; **depth:** 0.05–14.0 m; **life-form(s):** floating-leaved, submersed (vittate)

Key morphology: rhizomes absent or, if present (to 48+ cm), with or without terminal turions; upright shoots (to 250 cm) round to strongly flattened, frequently with axillary or terminal turions (to 9.5 cm), heterophyllous or homophyllous, nodal oil glands (to 1 mm) present or absent; leaves alternate (submersed) to opposite (near water surface), arranged spirally or strongly 2-ranked, stipules convolute or tubular with fused margins, free or fused partly to leaf base; floating leaves (heterophyllous species) opaque, leathery, sometimes nearly sessile but mainly long-petioled (to 29 cm), the stipules (to 8 cm) free, the blades (to 19 cm) elliptic to ovate, with up to 49 fine veins, the apices acute to obtuse, the bases cordate, rounded, or tapering to petiole, the margins entire; submersed leaves translucent, sessile or petioled (to 13 cm), the stipules (to 11.7 cm) free or adnate to leaf base (for up to ½ their length), the blades (to 25 cm) phyllodial to narrowly linear to roundish, flat or folded medially or falcate, with up to 49 fine veins, with or without a medial lacunal band, the apices hooded (cucullate) or flat and obtuse to awn-like, the bases acute to nearly perfoliate, margins straight to undulate, smooth to denticulate or serrate; inflorescences monomorphic (emersed or submersed) or dimorphic (emersed and submersed), the spikes simple or branched, moniliform to densely flowered, those submersed being axillary, capitate (to 7 mm), and short-stalked (to 10 mm), those emersed being terminal and capitate or cylindric (to 7.5 cm), and pedunculate (to 53 cm); achenes (to 6.7 mm) rounded or 1- to 3-keeled, beaked (to 3 mm) or beakless, the embryo coiled

Life history: duration: annual (vegetative); perennial (dormant apices, rhizomes, stolons, turions/winter buds, whole plants); **asexual reproduction:** rhizomes, shoot fragments, stolons, tubers, turions (stem), turions/winter buds (rhizome); **pollination:** self, water, wind; **sexual condition:** hermaphroditic; **fruit:** achenes (common); **local dispersal:** water, fish (fruits, rhizome/shoot fragments, tubers, turions, winter buds); **long-distance dispersal:** animals (fruits)

Imperilment: 1. *Potamogeton alpinus* [G5]; SH (MA); S1 (NH, NJ, PE); S2 (NY, WY); S3 (AB, LB, NF, NU); **2.** *P. amplifolius* [G5]; SH (DC); S1 (GA, IA, IL, KY, MD, NC, NE, SD, SK, TN, VA, WY); S2 (KS, MB, ND, NF); **3.** *P. berchtoldii* [unranked]; **4.** *P. bicupulatus* [G4]; S1 (IN, MN, VA); S2 (MI, VT, WI); S3 (ME, ON); **5.** *P. clystocarpus* [G1]; S1 (TX); **6.** *P. confervoides* [G4]; SH (RI); S1 (CT, LB, SC); S2 (MA, NC, NJ, ON, PA, QC, VT, WI); S3 (ME, MI, NF NH, NY); **7.** *P. crispus* [G5]; **8.** *P. diversifolius* [G5]; S1 (CO, ID, MA, MN, NY, OR, WY); S2 (ND, NE, WI); S3 (IA, IL); **9.** *P. epihydrus* [G5]; SH (KY, LA, MO); S1 (IL, IN, MS, TN, UT); S2 (CA, PE, SK, WY); S3 (NC); **10.** *P. floridanus* [G1]; S1 (AL, FL); **11.** *P. foliosus* [G5]; SX (DC); SH (NH, OR); S1 (CA, MD, NF, UT); S2 (AB, GA, NC, YT); S3 (IL, MB, PE, QC, WY); **12.** *P. friesii* [G5]; SH (VA); S1 (CT, IL, IN, MA, ME, NB, NF, OH, PA, UT, WY); S2 (NE, NS); S3 (IA, NU, QC, VT); **13.** *P. gramineus* [G5]; SH (KS); S1 (IL, OH, PA); S3 (IA, NU, WY); **14.** *P. hillii* [G3]; S1 (CT, OH, PA, VA, WI); S2 (MI, NY, ON); S3 (MA, VT); **15.** *P. illinoensis* [G5]; SH (DC); S1 (KS, MD, NC, NJ, UT, WY); S2 (IL, KY, MB, QC, WV); S3 (BC, VA); **16.** *P. natans* [G5]; SH (NC); S1 (DE, KS, OK, UT, WV); S2 (ND, WY, YT); S3 (AB, IL, NJ, OH); **17.** *P. nodosus* [G5]; SH (NC); S1 (AB, BC, NB, NH, NS, WY); S3 (IL, KY, NJ, QC); **18.** *P. oakesianus* [G5]; SH (LB, WV); S1 (IN, PA, PE); S2 (BC, NJ, VA); S3 (NB, QC); **19.** *P. obtusifolius* [G5]; SH (NH); S1 (MA, NF, NJ, PA, WY, YT); S2 (AB, AK, KS, NS, NU, PE, SK, WA); S3 (MB, MT, NB, QC, VT); **20.** *P. ogdenii* [G1/G2]; SH (ON); S1 (CT, MA, NY, VT); **21.** *P. perfoliatus* [G5]; SH (LA, OH); S1 (WI); S2 (BC, LB, MD); S3 (NC, PA, VA); **22.** *P. polygonifolius* [G5]; S1 (NS); S3 (NF); **23.** *P. praelongus* [G5]; SH (NH); S1 (IA, IL, IN, NE, ND, NJ, OH, PA, WY); S2 (CA, NB, NF); S3 (AB, NS, NU); **24.** *P. pulcher* [G5]; SH (ON); S1 (IL, IN, KY, ME, NS, PA, WI, WV); S2 (MI, MO, NY, OH); S3 (AR, NJ); **25.** *P. pusillus* [G5]; SH (DC, NH); S1 (CT, DE, GA, MD, MO, QC, RI, WV); S2 (IN, KS, MB, SK, WV); S3 (IA, IL, MA, MS, NF, QC, SK, WV, WY); **26.** *P. richardsonii* [G5]; SH (MD); S1 (IL, NF); S2 (IN, NB, NS); S3 (NU, OH, PA, WY); **27.** *P. robbinsii* [G5]; SX (DE); SH (DC, MD, VA); S1 (AB, AK, AL, IL, NF, OH, UT, WY); S2 (IN, MB, NJ, NU, SK); S3 (CA); **28.** *P. sibiricus* [G3/G4]; SH (ON); S2 (YT); S3 (AK, NU, QC); **29.** *P. spirillus* [G5]; SH (IA, OH); S1 (KY, MD, VA); S2 (MB, NF, WV); S3 (NJ); **30.** *P. strictifolius* [G5]; SH (IA, IL, OH, PA); S1 (CT, IN, MA, ME, NB, NF, ND, NE, NY, UT, VA, WY, YT); S2 (AB, BC, QC, SK, VT); S3 (MB); **31.** *P. tennesseensis* [G2/G3]; S1 (NC, PA, VA); S2 (OH, TN, WV); **32.** *P. vaseyi* [G4]; SX (IL); SH (MI, NJ, OH); S1 (CT, IA, IN, MA, NH, PA); S2 (ME, NB, QC, VT); S3 (MN, WI); **33.** *P. zosteriformis* [G5]; SH (KS, WV); S1 (MD, NF, NH, NJ, UT, VA, WY); S2 (NS, PA); S3 (AB, CA, IA, IL, OH, QC)

Ecology: general: All *Potamogeton* species are OBL aquatic indicators. *Potamogeton* is difficult taxonomically and widespread misidentification of material makes it extremely difficult to compile ecological data from databases and literature sources with confidence. The prevalence of hybridization in the genus adds an additional layer of complexity in attempts to attribute published information to the correct taxon. These difficulties were alleviated somewhat by seeking information associated with images (from which identifications could be checked) as well as by obtaining data derived from specimens that were annotated or collected by recognized experts. In fewer cases, information was obtained from material identified conclusively using molecular (DNA) markers. These approaches were useful for reconciling problems associated with several similar and commonly misidentified species such as *P. richardsonii* and *P. perfoliatus*. Some issues were extremely difficult to address, such as the correct parsing of information for *P. berchtoldii* and *P. pusillus*, which North American botanists had regarded as synonymous for decades. In that case, information was sought primarily from those references that at least distinguished the taxa at an infraspecific (e.g., varietal) level. Pondweeds include heterophyllous plants with waxy, coriaceous floating leaves and membranous, translucent broad to linear submersed leaves as well as homophyllous species where the broad to linear foliage is strictly submersed. The production of floating leaves is regulated to some degree by levels of abscisic acid (ABA). The floating and submersed leaves of heterophyllous species are not only distinct morphologically but also differ chemically with the former possessing a far greater assortment of secondary compounds (e.g., flavonoids), which serve to filter out UV radiation (submersed foliage is protected from UV damage by dissolved organic material in the water). Although the foliage is entirely floating and/or submersed, most of the flowers are produced on emergent, aerial spikes, which extend above the water surface. They have relatively high pollen:ovule ratios (i.e., $\bar{x} > 9,000:1$), and are thought to be primarily wind pollinated (anemophilous). The flowers of most pondweeds are strongly to at least partially self-incompatible (gametophytic) with a few (e.g., *P. alpinus*) being fully self-compatible. As each flower opens, the structure of the perianth and androecium enables the pollen to be distributed on the stigmas of the same flower, shed to other flowers on the same axis (geitonogamy), or released into the air or water to facilitate xenogamy. Many of the species have been characterized as being capable of both anemophily (wind pollination) and "anemo-epihydrophily," whereby their hydrophobic pollen is moved across the water surface by the wind to be captured by the stigmas of flowers lying along the surface. Although the transport of the pollen on the water surface has been clearly documented by several authors, there is still some question whether the water-transported pollen remains viable, or if seed set in such instances actually has taken place by self- or geitonogamous pollinations. Due to their ability to set seed, the underwater flowers produced by some species (e.g., *P. pusillus*) have been categorized as water pollinated (hydrophilous), a system usually associated with xenogamous pollen transfer. Despite their stigmas and pollen wet being when functional, these flowers have relatively low pollen:ovule ratios and have been shown more specifically to be "hydroautogamous;" i.e., a type of autogamy, whereby the pollen grains are transferred within underwater flowers on the surface of air bubbles. Even here free pollen grains have been observed as being shed within the water; however, their reproductive fate in such cases has not been determined. Apomictic fruit production has been documented to occur in *P. obtusifolius* and is suspected to occur (but is yet to be confirmed) in *P. bicupulatus, P. confervoides, P. foliosus, P. hillii,* and *P. spirillus.* The fruit walls often are hardened and impermeable to water, which facilitates their dormancy. Some seeds can remain dormant for 3–4 years before germinating. The thick achene walls deter seed germination unless they become physically abraded. Higher germination rates have been observed in scarified seeds, including those of several species that germinated readily after being removed from the stomachs of waterfowl (Aves: Anatidae). To overcome physiological dormancy, many of the species require a period of cold stratification (e.g., 60–180+ days @ 1°C–3°C) to induce germination (at room temperature). The germination rate of dried seeds is miniscule (<1%). The fruits of several species are buoyant and are known to remain afloat for short (e.g., 12 hr) to extended times (>12 months), which enables their dispersal by surface water currents. Some fish (Vertebrata: Teleostei) can ingest and disperse seeds locally within water bodies. The seeds are transported over longer distances by waterfowl (Aves: Anatidae), which consume them readily and convey them endozoically, or carry them unwittingly when trapped in their foliage or in mud that adheres to their feet. Several vegetative structures also can serve as effective propagules. Field studies indicate that seedling recruitment is of minimal importance in the persistence of populations, which exhibit prolific vegetative reproduction by means of rhizomes, shoot fragments, tubers, turions, and winter buds. Some perennial species die back almost entirely each season and apparently overwinter primarily by dispersed vegetative propagules rather than by seeds. Although these plants creatively have been categorized as "vegetative annuals," in some cases they also behave as true annuals, by overwintering entirely as seeds. Vegetative reproduction can occur by the production of rhizomes, on which develop dense, hardened structures inconsistently designated as turions or "winter buds." Turions also can develop on the stems, which frequently is the case in species that are not rhizomatous. In either case, the turions are highly effective vegetative propagules once detached. Some pondweed species are capable of overwintering as whole plants beneath the ice, especially when growing at greater depths (e.g., 3+ m). So many of the species have such broad ranges and ecological tolerances that at one time or another they can be found growing in quite different habitats as their individual accounts will demonstrate. Unavoidably, the comprehensive information provided in the individual species treatments also tends to obscure the most typical features of the habitat associated with a particular species. Some associations of this type have been demonstrated statistically. These include several species (*P. amplifolius, P. epihydrus, P. friesii, P. gramineus, P. praelongus, P. richardsonii, P. robbinsii,* and *P. zosteriformis*) that are most likely to occur in lakes and another (*P. obtusifolius*) that occurs most often in flowing water. A few species (*P. amplifolius, P. friesii, P. praelongus,* and *P. richardsonii*) are noticeably more common on sandy substrates, sandy organic substrates (*P. foliosus*), or on those derived from granitic bedrock (*P. amplifolius* and *P. epihydrus*). In general, sand supports the greatest diversity of pondweed species and silt the least. Five species (*P. amplifolius, P. epihydrus, P. obtusifolius, P. robbinsii,* and *P. spirillus*) inhabit lower alkalinity sites ($\bar{x}$: <55 mg l^{-1}); whereas several others (*P. richardsonii, P. foliosus, P. friesii, P. natans,* and *P. pusillus*) typically occur where mean alkalinity values exceed 100 mg l^{-1}. *Potamogeton epihydrus* and *P. obtusifolius* are characteristic of sites having very low levels of dissolved inorganic matter. Most of the North American pondweeds occur in alkaline waters (mean pH >7.0) except for six species (*P. bicupulatus, P. confervoides, P. diversifolius, P. polygonifolius, P. pulcher,* and *P. tennesseensis*), which all favor more acidic waters (mean pH: 6.0–7.0).

3.4.1.1. *Potamogeton alpinus* Balb. is a perennial, which occurs in backwaters, bogs, brooks, canals, channels (loamy, river), coves, ditches, fens (organic), flats (sandy, swampy), gravel pits, holes (bog, meadow), impoundments (beaver dam), lakes (kettle), marshes, oxbows, ponds (beaver, meadow, muskeg, seepage, slough, thermokarst, tundra), pools (alpine, bog, mudflat), potholes, sloughs (muddy, river meander), streams (bog, lentic), and along the margins of rivers at elevations to 3537 m. These are northerly distributed plants that occur most often in cold (but from 7.7°C to 22.2°C), shallow (usually <1.5 m but up to 4.0 m) spring-fed waters that span a broad range of acidity (pH: 4.9–9.2; $\bar{x}$: 7.3–8.0) and are characterized by moderate conductivity (10–490 μmhos cm^{-1}) and intermediate total alkalinity (2.4–225.9 mg HCO$_3^-$ l^{-1}; 4–275; $\bar{x}$: 40.5–81.0 mg l^{-1} CaCO$_3$). They can tolerate some shade and are positively associated with sites having high levels of dissolved organic matter. The substrates have been described as alluvium (sandy, stony), boulders, clay, clayey silt, gravel, marl, muck (organic, silty), mucky peat, mud, muddy gravel, ooze, peat (fibrous), rocks, sand, sandy muck, silt (brown, fine, mineral), silty clay loam, silty sand, and volcanic. Although a heterophyllous species, the floating leaves often are not produced (the plants then entirely submersed) or they develop in a somewhat rudimentary fashion. Flowering and fruiting have been reported from June to September. The flowers are wind pollinated and self-compatible. Despite being protogynous (the female phase and male phase each lasting for 2–3 days), the flowers are believed to be almost obligately self-pollinating (facilitated by the wind). Seed set on the emergent spikes is high, typically exceeding 50%. Observations that some flowers are produced underwater (with no explanation of pollination) suggest that further inquiry into the reproductive biology is necessary for clarification. The seeds of European plants have germinated within 2–4 weeks at 20°C (under a 16/8 hr light/dark regime) after 5 months of cold storage in water at 4°C. Germination of the seeds is enhanced after passage through the gut of carp (Vertebrata: Teleostei: Cyprinidae: *Cyprinus carpio*), which implicates endozoic transport by fish as a possible dispersal mechanism. The plants are rhizomatous and produce stolon-like branches, which often terminate as dense overwintering buds that have been characterized as turions. There also are reports that the plants overwinter entirely by these buds, which can serve as effective vegetative propagules. **Reported associates:** *Alisma, Arctophila fulva, Bidens beckii, Callitriche hermaphroditica, Callitriche palustris, Cardamine pensylvanica, Carex aquatilis, Carex lasiocarpa, Carex rostrata, Carex utriculata, Carex vesicaria, Ceratophyllum demersum, Chara vulgaris, Cicuta bulbifera, Crassula aquatica, Elatine minima, Eleocharis acicularis, Eleocharis palustris, Elodea canadensis, Elodea nuttallii, Equisetum fluviatile, Eriocaulon aquaticum, Fontinalis, Geum rivale, Glyceria borealis, Gratiola aurea, Heteranthera dubia, Hippuris vulgaris, Isoetes bolanderi, Isoetes echinospora, Isoetes lacustris, Isoetes occidentalis, Juncus pelocarpus, Lemna minor, Lemna trisulca, Limosella aquatica, Lobelia dortmanna, Lysimachia thyrsiflora, Menyanthes trifoliata, Myriophyllum alterniflorum, Myriophyllum sibiricum, Myriophyllum spicatum, Myriophyllum tenellum, Najas flexilis, Nitella, Nuphar polysepala, Nuphar variegata, Nymphaea odorata, Persicaria amphibia, Potamogeton amplifolius, Potamogeton berchtoldii, Potamogeton epihydrus, Potamogeton foliosus, Potamogeton friesii, Potamogeton gramineus, Potamogeton natans, Potamogeton nodosus, Potamogeton obtusifolius, Potamogeton perfoliatus, Potamogeton praelongus, Potamogeton pusillus, Potamogeton richardsonii, Potamogeton robbinsii, Potamogeton sibiricus, Potamogeton spirillus, Potamogeton strictifolius, Potamogeton zosteriformis, Ranunculus aquatilis, Ranunculus flammula, Ranunculus gmelinii, Ranunculus longirostris, Ranunculus trichophyllus, Ricciocarpus natans, Ruppia, Sagittaria cuneata, Salix geyeriana, Schoenoplectus acutus, Sparganium angustifolium, Sparganium emersum, Sparganium natans, Stuckenia filiformis, Stuckenia pectinata, Stuckenia vaginata, Torreyochloa pallida, Typha latifolia, Utricularia gibba, Utricularia macrorhiza, Utricularia minor, Vallisneria americana, Veronica americana, Zannichellia palustris, Zizania palustris.*

3.4.1.2. ***Potamogeton amplifolius*** **Tuck.** is a perennial, which inhabits backwaters, bays, canals (drainage), channels (rivulet), ditches, lakes (brown water), oxbows, ponds, pools

(meander), rivers, sloughs, springs, and streams at elevations to 2895 m. The plants occur primarily in sunny, open exposures at water depths from 20 cm to 5+ m. Growth is optimal at a depth of 2.0–2.5 m (the stems can grow more than 2 m in length). Although the plants occur down to 5.5 m, their growth declines rapidly from 3.5 to 4.0 m and deeper. The plants are intolerant of eutrophic conditions and favor high light habitats that are low in total phosphorous (which can promote algal blooms that intensify turbidity). They are negatively associated with sites having high amounts of dissolved organic matter. Growth increases by up to 154% under higher light conditions and by up to 255% under higher nutrient conditions, with an increased root/shoot biomass ratio under high light but a reduced root/shoot ratio under high nutrients. Leaf/root nitrogen increases by up to 53%/50% under low light and by up to 40%/77% under high nutrient conditions resulting in a decreased C:N ratio (by up to 20–40%). Shallow water plants are often found with fruit, but then die back during winter. Although fruiting does not occur at depths beyond 3 m, deeper populations remain stable throughout the year, with the plants able to survive the winter (in 2°C water) in a physiologically active (but at ~5.5% of the summer photosynthetic rate) vegetative state; leafy, green plants have been recovered from below the ice during February. The waters typically are alkaline (pH: 5.6–10.0; $\bar{x}$: 7.5–8.1) and span an intermediate range of alkalinity (5–260 [$\bar{x}$: 37.2–60.0] mg l^{-1} CaCO$_3$) and conductivity (20–540 [$\bar{x}$: 125] μmhos cm^{-1}). The substrates have been characterized as clay, granite, gravel, marl, marly muck, marly peat, muck, mucky marl, mud (organic), muddy gravel, peat (fibrous, pulpy), rock, rubble, sand, sandy cobble, sandy gravel, sandy marl, sandy muck, sandy mud, shale, silt (flocculent), silty muck, and silty sand. Although this is a heterophyllous species, it often grows completely submersed without floating leaves; however, floating leaves are present during anthesis and facilitate the elevation of the floral spike above the water surface. The floral spikes (inflorescences) are known to release lacunal gases, which include CH$_4$ and CO$_2$. Flowering and fruiting extend from June to October. The flowers are borne on emergent spikes, which are supported by floating leaves. They have relatively high pollen:ovule ratios (mean: 27,000) and presumably are wind pollinated; however, there is little detailed information on the reproductive biology or pollination ecology of this species. The fruits are eaten by waterfowl and potentially are dispersed (endozoically) by them. Moderate seed germination (31–47%) has been obtained under ambient greenhouse conditions after 2–6 months of dark storage in cold (1°C–3°C) water. None survived 1-year of cold water storage. In any case, the plants rarely establish by seedling recruitment as evidenced by repeated observations of rhizomatous tissue attached to newly emerging spring shoots. By late summer, the leaves can become encrusted with up to 105 mg Ca g^{-1} leaf (as "calcite" or "marl" deposits), which co-precipitates with P as the result of photosynthetic activity. Productivity estimates up to 21.12 g DM biomass m^{-2} have been reported. The plants reproduce luxuriantly by the vegetative proliferation of rhizomes and vertical vegetative shoots, which function as winter buds. In one case, small 4-leaved cuttings (devoid of rhizome tissue) grew into plants having 52 leaves and a 47.5 cm rhizome system within a single growing season. They compete well with Eurasian water milfoil (Haloragaceae: *Myriophyllum spicatum*) unless sunlight is occluded by overtopping plants. **Reported associates:** *Alisma, Bidens beckii, Brasenia schreberi, Butomus umbellatus, Cabomba caroliniana, Calamagrostis canadensis, Callitriche, Carex rostrata, Carex stricta, Carex vesicaria, Cephalanthus occidentalis, Ceratophyllum demersum, Chara contraria, Chara globularis, Chara vulgaris, Drepanocladus, Dulichium arundinaceum, Elatine minima, Eleocharis acicularis, Eleocharis palustris, Elodea canadensis, Elodea nuttallii, Equisetum fluviatile, Eriocaulon aquaticum, Glyceria borealis, Gratiola aurea, Heteranthera dubia, Hippuris vulgaris, Isoetes echinospora, Isoetes lacustris, Juncus pelocarpus, Lemna minor, Lemna trisulca, Lobelia dortmanna, Lobelia kalmii, Ludwigia, Lychnothamnus barbatus, Menyanthes trifoliata, Myriophyllum alterniflorum, Myriophyllum heterophyllum, Myriophyllum sibiricum, Myriophyllum spicatum, Myriophyllum tenellum, Myriophyllum verticillatum, Najas canadensis, Najas flexilis, Najas gracillima, Najas minor, Nitella flexilis, Nitella opaca, Nuphar advena, Nuphar microphylla, Nuphar polysepala, Nuphar variegata, Nymphaea odorata, Phalaris arundinacea, Pontederia cordata, Potamogeton alpinus, Potamogeton berchtoldii, Potamogeton bicupulatus, Potamogeton crispus, Potamogeton diversifolius, Potamogeton epihydrus, Potamogeton foliosus, Potamogeton friesii, Potamogeton gramineus, Potamogeton hillii, Potamogeton illinoensis, Potamogeton natans, Potamogeton nodosus, Potamogeton oakesianus, Potamogeton obtusifolius, Potamogeton perfoliatus, Potamogeton praelongus, Potamogeton pulcher, Potamogeton pusillus, Potamogeton richardsonii, Potamogeton robbinsii, Potamogeton spirillus, Potamogeton strictifolius, Potamogeton vaseyi, Potamogeton zosteriformis, Ranunculus flammula, Ranunculus longirostris, Rumex britannica, Ruppia spiralis, Sagittaria graminea, Sagittaria latifolia, Sagittaria rigida, Schoenoplectus acutus, Schoenoplectus tabernaemontani, Sparganium emersum, Spiraea douglasii, Spirodela polyrhiza, Stuckenia pectinata, Stuckenia filiformis, Thelypteris palustris, Typha angustifolia, Typha latifolia, Utricularia gibba, Utricularia macrorhiza, Utricularia purpurea, Vallisneria americana, Stuckenia pectinata, Wolffia brasiliensis, Zannichellia palustris, Zizania palustris.*

3.4.1.3. ***Potamogeton berchtoldii* Fieber** is an annual (vegetative) or perennial, which grows in backwaters (river), bogs, borrow pits, brooks, channels (slough), ditches (roadside), flats (drawdown, tidal), gravel pits, lakes (bog, kettle), marshes, ponds (artificial, beaver, floodplain, kettle, roadside, strip mine, subalpine), pools (bog, canyon, stagnant), potholes, reservoirs, sloughs (margins), springs, streams, swamps (backwater), and along the margins of rivers at elevations to 3049 m. The substrates have been characterized as clay, cobble, gravel, marl, muck (organic), mucky loam, mucky sand, mud, muddy rock, muddy silt, ooze (organic flocculent), peat (fibrous, pulpy), rock (Beckmantown, marly), sand, sandy

clay, sandy gravel, sandy sod, and silt. The habitats (depths to 2.6 m) can span a fairly wide range of water conditions (pH: 4.8–10.2; $\bar{x} = 7.1$–7.6; total alkalinity [as HCO_3^-]: 2.5–300.6 mg l^{-1}; $\bar{x} = 32.4$–63.3 mg l^{-1}; conductivity: 10–390 µmhos cm^{-1}; $\bar{x} = 95$ µmhos cm^{-1}). This is a submersed, homophyllous, linear-leaved aquatic, which is confined to standing waters. However, emergent wetland species occasionally are reported as associates (see below) for occurrences in very shallow water (sometimes as little as 1 cm depth) and/or near wetland margins. Flowering occurs from June to August with fruiting observed from July to September. The flowers are protogynous, self-incompatible (at least in North America), and allegedly outcrossed; however, those of European plants are self-compatible and predominantly self-pollinating. They develop in monomorphic inflorescences that are aerial (selfed, geitonogamous, or wind pollinated) or become submersed, with the latter releasing the pollen on bubbles that contact the submersed stigmas of the same flowers, resulting in successful self-pollination. Underwater pollen released in this way also drifts away from the flowers, but its potential role in reproduction (i.e., by possible hypohydrophily) remains uncertain. Pollen that is deposited on and transferred on the water surface (i.e., via epihydrophily) also apparently results in successful pollination. Genetic (allozyme) data obtained for European populations have indicated low within-population levels of variation and high among-population differentiation, which most likely reflects frequent vegetative reproduction (by winter buds) and limited seedling recruitment. The achenes are dispersed locally by water currents (individually or while attached to dislodged stems) and possibly over longer distances by waterfowl (Aves: Anatidae). The seeds have retained fair viability (59%) after 5 years of storage in water at 1°C–3°C. Although the plants fruit commonly, most of their reproduction is attributed to asexual means and dispersal occurs primarily by water transport of vegetative propagules. Specialized shoot-borne, dormant winter buds (referred to as "turions" by some) usually far outnumber the seeds in sediment propagule banks. The winter buds are dispersed by water currents either while still attached to dislodged and floating stems (which commonly are collected) or as individual propagules. When formed early (e.g., August) the structures can germinate within 2 weeks if exposed to light, but become dormant as water temperatures fall <10°C and the photoperiod decreases to less than 11 hr of daylight. Germination of dormant winter buds is regulated by temperature, with a subsequent cold period (e.g., 2°C for 60 days; 75 days at 5°C–8°C) then necessary to break dormancy. The winter buds are killed if they remain frozen (−2°C) for more than 30 days. Biomass estimates as high as 86.3 g m^{-2} have been reported for populations residing in deeper water sloughs. **Reported associates:** *Agrostis gigantea, Alisma triviale, Alopecurus aequalis, Beckmannia syzigachne, Brasenia schreberi, Callitriche, Caltha palustris, Carex aquatilis, Carex brevior, Carex cusickii, Carex microptera, Carex pellita, Carex scoparia, Carex utriculata, Carex vesicaria, Cephalanthus occidentalis, Ceratophyllum demersum, Chamerion angustifolium, Chara vulgaris, Chiloscyphus* *polyanthos, Comarum palustre, Danthonia parryi, Decodon verticillatus, Egeria densa, Elatine minima, Eleocharis acicularis, Eleocharis obtusa, Eleocharis palustris, Eleocharis robbinsii, Elodea canadensis, Elodea nuttallii, Eriocaulon aquaticum, Festuca arizonica, Fontinalis antipyretica, Glyceria borealis, Heteranthera dubia, Hippuris vulgaris, Iris missouriensis, Isoetes echinospora, Isoetes occidentalis, Juncus balticus, Juncus effusus, Juncus militaris, Juncus pelocarpus, Juncus supiniformis, Lemna minor, Lemna trisulca, Limosella aquatica, Lobelia dortmanna, Lycopus, Myosotis laxa, Myriophyllum alterniflorum, Myriophyllum farwellii, Myriophyllum heterophyllum, Myriophyllum sibiricum, Myriophyllum spicatum, Myriophyllum tenellum, Myriophyllum verticillatum, Najas flexilis, Nitella, Nuphar polysepala, Nuphar variegata, Nymphaea odorata, Nymphoides cordata, Nyssa, Persicaria amphibia, Pontederia cordata, Potamogeton alpinus, Potamogeton amplifolius, Potamogeton bicupulatus, Potamogeton clystocarpus, Potamogeton confervoides, Potamogeton crispus, Potamogeton diversifolius, Potamogeton epihydrus, Potamogeton foliosus, Potamogeton friesii, Potamogeton gramineus, Potamogeton natans, Potamogeton nodosus, Potamogeton oakesianus, Potamogeton obtusifolius, Potamogeton praelongus, Potamogeton pulcher, Potamogeton richardsonii, Potamogeton robbinsii, Potamogeton strictifolius, Potamogeton zosteriformis, Potentilla anserina, Ranunculus aquatilis, Ranunculus flammula, Ranunculus longirostris, Rorippa sphaerocarpa, Ruppia spiralis, Sagittaria cuneata, Schoenoplectus acutus, Schoenoplectus subterminalis, Schoenoplectus tabernaemontani, Sium suave, Sparganium angustifolium, Sparganium emersum, Sphagnum, Spirodela, Stuckenia filiformis, Stuckenia pectinata, Typha latifolia, Utricularia gibba, Utricularia intermedia, Utricularia macrorhiza, Utricularia purpurea, Utricularia resupinata, Vallisneria americana, Veronica americana, Zannichellia palustris.*

3.4.1.4. ***Potamogeton bicupulatus* Fernald** is a perennial, which inhabits still or sluggish waters in depressions (peat), ditches, flowages, lakes (bog), ponds, pools, rivers, and streams at elevations to 266 m. The substrates have been described as gravel, muck, mud, peat, rock, sand, sandy muck, and silt. The plants grow at depths to 2 m (usually shallower) in soft, acidic (pH: 5.4–7.3; $\bar{x} = 6.4$–6.9; total alkalinity [as $CaCO_3^-$] 1.5–23.0 mg l^{-1}; $\bar{x} = 6.4$–12.2 mg l^{-1}), often oligotrophic waters having high clarity. Although heterophyllous, entirely submersed individuals lacking floating leaves can be found when growing in deeper waters. Flowering and fruiting extend from May to October. The inflorescences are dimorphic, bearing flowers in emergent spikes and in submersed, head-like clusters. Although the reproductive ecology of this species has not been investigated independently, it likely resembles that of the related *P. spirillus* (see below) in having protogynous but self-compatible flowers that are facultatively (emergent) or obligately (submersed) autogamous. In any case, the plants produce numerous seeds, which are dispersed locally by water currents, and perhaps to greater distances by adhesion to waterfowl (Aves: Anatidae). The

seed ecology also remains unstudied. The plants overwinter (and reproduce asexually) by rhizomes. **Reported associates:** *Brasenia schreberi, Cabomba caroliniana, Callitriche heterophylla, Carex limosa, Cephalanthus occidentalis, Ceratophyllum demersum, Ceratophyllum echinatum, Chara, Elatine minima, Eleocharis acicularis, Eleocharis robbinsii, Elodea canadensis, Elodea nuttallii, Eriocaulon aquaticum, Fontinalis antipyretica, Glossostigma cleistanthum, Gratiola aurea, Heteranthera dubia, Hydrilla verticillata, Isoetes echinospora, Isoetes engelmannii, Isoetes lacustris, Juncus militaris, Juncus pelocarpus, Lemna minor, Littorella uniflora, Lobelia dortmanna, Ludwigia palustris, Lycopus, Myriophyllum farwellii, Myriophyllum heterophyllum, Myriophyllum heterophyllum* × *Myriophyllum laxum, Myriophyllum humile, Myriophyllum spicatum, Myriophyllum tenellum, Najas flexilis, Najas gracillima, Najas guadalupensis, Najas minor, Najas canadensis, Nitella, Nuphar variegata, Nymphaea odorata, Nymphoides cordata, Persicaria amphibia, Pontederia cordata, Potamogeton amplifolius, Potamogeton berchtoldii, Potamogeton confervoides, Potamogeton crispus, Potamogeton diversifolius, Potamogeton epihydrus, Potamogeton foliosus, Potamogeton gramineus, Potamogeton natans, Potamogeton oakesianus, Potamogeton perfoliatus, Potamogeton pulcher, Potamogeton pusillus, Potamogeton robbinsii, Potamogeton spirillus, Potamogeton vaseyi, Riccia, Sagittaria graminea, Sagittaria latifolia, Schoenoplectus subterminalis, Schoenoplectus tabernaemontani, Schoenoplectus torreyi, Sparganium americanum, Sparganium angustifolium, Sphagnum, Stuckenia pectinata, Utricularia gibba, Utricularia intermedia, Utricularia macrorhiza, Utricularia minor, Utricularia purpurea, Utricularia radiata, Utricularia resupinata, Vallisneria americana.*

3.4.1.5. ***Potamogeton clystocarpus* Fernald** is a perennial, which is known only from several shallow rock pools along an ephemeral stream at elevations from 1524 to 1615 m. The stream habitat is subject to intense annual flash flooding and periods of severe drought. The site receives an average annual rainfall of 37.18 cm with the majority (13 cm) occurring from July to August. The substrate has been characterized as igneous-derived alluvium. Flowering and fruiting reportedly occur from May to October. There are conflicting reports whether the plants lack rhizomes and it is uncertain whether they produce turions or winter buds (they are described as perennial). Reproduction is believed to occur by sexual means (seeds) or by the vegetative dispersal of rooting stem fragments. A comprehensive life history study is sorely needed for this imperiled taxon, which is questionably distinct from *P. berchtoldii* (see *Sysytematics* below). **Reported associates:** *Elatine, Heteranthera limosa, Ludwigia palustris, Marsilea vestita, Najas guadalupensis, Potamogeton berchtoldii, Potamogeton diversifolius, Potamogeton foliosus, Potamogeton nodosus, Stuckenia pectinata, Zannichellia palustris.*

3.4.1.6. ***Potamogeton confervoides* Rchb.** is a perennial, which grows in bogs, channels, coves, lakes, ponds, pools (bog), and streams (cold water, sluggish, spring-fed), at elevations to 866

m. The plants occur in exposures of full sunlight at water depths to 4+ m. This is a characteristic species of acidic habitats. The sediments are acidic (e.g., pH: 4.6–5.8), relatively low in nutrients (e.g., total P: 1.13 mg g dry wt^{-1}), and have been described as gravel, muck, mud, peat (fibrous, soupy), peaty muck, rock, rocky clay loam, sand, or sandy mud. The waters are acidic (pH: 3.7–8.4; $\bar{x} = 6.3$), clear (Secchi depths to 18+ m), oligotrophic (e.g., total P: <1 µg l^{-1}), and soft (total alkalinity [as CaCO$_3^-$]: 0–165 mg l^{-1}; $\bar{x} = 4.2$–45 mg l^{-1}; conductivity: 15–310 µmhos cm^{-1}). Flowering and fruiting occur from June to September. Sexual reproduction is common, but the floral biology and pollination ecology remain uninvestigated. The seed ecology also is not well understood. The seeds exhibit dormancy and have retained high viability (86% germination) after 5–6 months of storage in cold (1°C–3°C) water, with 71% germination occurring after 1 year of storage. The plants reproduce vegetatively by an intertwining rhizome system and by the production of turions, which rise from the axils of shed leaves on deteriorating shoots. Because the plants often are found as floating fragments, shoot fragments also likely function as important propagules, which facilitate local dispersal. **Reported associates:** *Brasenia schreberi, Calamagrostis canadensis, Callitriche, Carex lasiocarpa, Carex oligosperma, Cephalozia, Chamaedaphne calyculata, Dulichium arundinaceum, Elatine minima, Eleocharis acicularis, Eleocharis microcarpa, Eleocharis palustris, Eleocharis robbinsii, Equisetum fluviatile, Eriocaulon aquaticum, Fontinalis novae-angliae, Glyceria borealis, Glyceria canadensis, Gratiola aurea, Hypericum mutilum, Isoetes acadiensis, Isoetes lacustris, Isoetes tuckermanii, Juncus militaris, Juncus pelocarpus, Lobelia dortmanna, Lycopus, Lysimachia terrestris, Myriophyllum farwellii, Myriophyllum tenellum, Myriophyllum verticillatum, Najas flexilis, Najas gracillima, Nuphar variegata, Nymphaea odorata, Nymphoides cordata, Orontium aquaticum, Pallavicinia lyellii, Pontederia cordata, Potamogeton berchtoldii, Potamogeton bicupulatus, Potamogeton epihydrus, Potamogeton gramineus, Potamogeton natans, Potamogeton oakesianus, Potamogeton pusillus, Potamogeton robbinsii, Potamogeton spirillus, Ranunculus flammula, Sagittaria engelmanniana, Sagittaria graminea, Sagittaria latifolia, Schoenoplectus subterminalis, Sparganium americanum, Sparganium angustifolium, Sparganium fluctuans, Sphagnum macrophyllum, Sphagnum pylaesii, Sphagnum torreyanum, Stuckenia filiformis, Utricularia geminiscapa, Utricularia gibba, Utricularia intermedia, Utricularia macrorhiza, Utricularia minor, Utricularia purpurea, Utricularia radiata, Utricularia resupinata, Vallisneria americana.*

3.4.1.7. *Potamogeton crispus* **L.** is a nonindigenous perennial, which inhabits brackish (to 5.3‰) to freshwater bottoms (river), brooks, canals, catch basins, channels (oxbow, river), clay pits, depressions, ditches (drainage, irrigation, roadside, tidal), estuaries, floodbars, floodways, lagoons (river), lakes (artificial), marshes (freshwater, intertidal), mudflats, ponds (artificial, spring-fed), pools (bedrock, ephemeral, seasonal, still), reservoirs (backwaters), rivers (cold, swift), sandbars, shores (lake, river), sloughs, streams, and washes at elevations

to 2529 m. The plants occur in still or flowing (currents to at least 0.3 m s^{-1}) waters under exposures that range from full sun to partial shade. They thrive at depths of 1–3 m but can become dominant at greater depths (5 m) and can persist at depths of 7 m under conditions of high water clarity. The substrates typically have an organic matter content of 10–25% and have been described as clay, cobble, granite, gravel, loam, loamy mud (organic), marl, muck, mud, muddy sand, rocky mud, sand (fine), sandy gravel, sandy loam, sandy marl, sandy muck, silt, silty mud, and stony loam (Ephrata). Although it can occur in more pristine waters, this species (at least in its North American range) is regarded as an indicator of eutrophic conditions because of its high tolerance of turbid (<1% of surface irradiance), eutrophic conditions (e.g., phosphate to 7.5 mg l^{-1}; total nitrogen to 75 mg l^{-1}). The plants are intolerant of acidic waters and occur nearly exclusively in alkaline waters (pH: 5.6–9.9; $\bar{x} = 7.6$–8.2; total alkalinity [as CaCO$_3$]:19.4–265.0 mg l^{-1}; $\bar{x} = 66.2$–140.0 mg l^{-1}; conductivity to 750 µmhos cm^{-1}). Although perennial, the plants exhibit an unusual life cycle like that of winter annuals. Throughout much of the summer season, many of the plants persist entirely as dormant, vegetative turions (note: despite this widespread perception, it also is not unusual to find fully intact plants persisting throughout the summer). Turions that germinate during the fall (September–October) develop into finer-leaved plants that will overwinter under the ice by slender rhizomes until the subsequent spring when sexual and vegetative reproduction resume. The plants can persist under a fairly thick ice cover unless there is an extended period of winter snow cover (which prevents light penetration). As day length and water temperatures increase (e.g., 12–16 hr, 13°C–35°C) coarser, dentate foliage and new turions begin to develop, first appearing at the end of April and maturing between mid-May and October. A single plant can produce as many as 23,520 turions in one growing season. The turions are dormant initially but remain metabolically active for a 6-week period. The turion-bearing plants begin to senesce and deteriorate from mid-May. It is uncertain whether plant senescence is related to maximum growing time (i.e., 5 months) or to increasing water temperatures. The turions sink (usually) or float if detached but also are dispersed by floating shoots if they remain attached. They also are transported by water currents, with reports of as many as 30 routinely entering and leaving sites each day due to tidal activity. Averaging 5–13 per shoot, the turions can attain extremely high sediment densities approaching 9600 m^{-2} year^{-1}. Fall turion germination (as high as 80%) is associated with reduced day length (11 hr) and cooler water temperatures (<20°C). Higher overall germination rates occur under light conditions but turions germinate readily in the light or dark once temperatures fall <12°C. In one study, 100% of shoots were inviable (unable to produce turions) after 3 hr of desiccation. Unsprouted turions acquire a deeper level of dormancy similar to those that already have overwintered. Flowering typically extends from mid-April to June (through August in the west) with peak flower production during May. The emergent inflorescences bear several flowers (averaging 4.7/spike). The spikes are retracted underwater subsequent to pollination, which occurs primarily by wind. Seed set occurs within a week after the inflorescences first emerge above the water. Fruiting (North America) occurs from June to October (fruits in European populations reportedly mature primarily in early July). Fruiting occurs in shallow water and is induced when water levels fall <30 cm. North American fruit production varies (averaging 3.5 fruits/inflorescence) but can be fairly prolific (e.g., occurring on 25% of plants in early June; up to 1,394 fruits m⁻²; up to 5.6 million fruits acre⁻¹). Each plant is capable of producing upward of 960 single-seeded fruits during the growing season. The fruits are dispersed locally by water currents (remaining afloat for less than a week) and by shifting sediments but are believed to be transported over longer distances by waterfowl (possibly endozoically). The specific conditions necessary for fruit germination remain unclear; however, the fruit wall is not particularly thick and is quite permeable, which facilitates water imbibition relative to other pondweed species. In North America, no germination has been observed for fruits stored dry from 2 to 12 months or from fruits collected at field sites; however, fruits from some European populations have germinated well (30–50%) after 1–2 years of water storage. Seedlings have not been observed in any North American population, even following directed searches. Yet, a seed bank apparently does develop as indicated by the emergence of seedlings within 1 week from field-collected (Wisconsin) lake sediments that were incubated in the lab at 25°C (constant) under a 14-hr photoperiod at 350 µE m⁻² s⁻¹ PAR (photosynthetically active radiation). In that case, the seedling densities were estimated at 21.0 m⁻². In any event, the dispersal of North American plants has been attributed primarily to water (or occasionally waterfowl) transport of turions and to human-mediated movement of plants among water bodies, with seeds playing a minor (if any) role. Large plant fragments also have been observed attached externally to flying waterfowl (Aves: Anatidae), which then transport them exozoically. Due to the apparently low rates of seed germination and seedling establishment, most reproduction presumably is vegetative; however, genetic surveys of nonindigenous (New Zealand) populations using allozyme data indicated that while most individual populations were homogeneous, they were differentiated genetically from other populations. The plants take up nutrients from both their roots (especially phosphorous) and shoots (especially nitrogen and potassium) and can use bicarbonate (HCO₃⁻) as a dissolved inorganic carbon source for photosynthesis. Vegetative growth is minimal at temperatures <5°C but resumes rapidly at 10°C–15°C, even under low light conditions (e.g., 7 µE m⁻² s⁻¹). Although a cold-tolerant species, the maximum photosynthetic rate has been recorded at 30°C. Growth rates can increase by as much as 33% in turbulent waters; however, smaller turions are produced in lotic environments. Biomass production can be extremely high (e.g., 190–6410 g DM m⁻²). Depending on the season, turions can represent 23–50% of the total biomass. The rhizomes comprise a much higher proportion of biomass in spring (e.g., 34%) than in fall (e.g., 1.7%). **Reported associates (North America):** *Bidens beckii, Brasenia schreberi, Cabomba*

caroliniana, Callitriche fassettii, Carex stricta, Ceratophyllum demersum, Ceratophyllum echinatum, Chara, Egeria densa, Eleocharis acicularis, Eleocharis erythropoda, Eleocharis palustris, Elodea canadensis, Elodea nuttallii, Equisetum fluviatile, Heteranthera dubia, Hydrilla verticillata, Lemna minor, Lemna minuta, Lemna trisulca, Lilaeopsis, Limosella acaulis, Ludwigia palustris, Mimulus guttatus, Myriophyllum sibiricum, Myriophyllum spicatum, Najas canadensis, Najas flexilis, Najas gracillima, Najas guadalupensis, Najas minor, Nelumbo lutea, Nuphar variegata, Ottelia alismoides, Persicaria amphibia, Persicaria coccinea, Persicaria hydropiperoides, Phalaris arundinacea, Potamogeton amplifolius, Potamogeton berchtoldii, Potamogeton bicupulatus, Potamogeton diversifolius, Potamogeton epihydrus, Potamogeton foliosus, Potamogeton friesii, Potamogeton gramineus, Potamogeton hillii, Potamogeton illinoensis, Potamogeton natans, Potamogeton nodosus, Potamogeton obtusifolius, Potamogeton perfoliatus, Potamogeton praelongus, Potamogeton pusillus, Potamogeton richardsonii, Potamogeton robbinsii, Potamogeton spirillus, Potamogeton strictifolius, Potamogeton vaseyi, Potamogeton zosteriformis, Ranunculus longirostris, Riccia fluitans, Sagittaria cristata, Sagittaria latifolia, Sagittaria platyphylla, Schoenoplectus acutus, Schoenoplectus triqueter, Sparganium androcladum, Sparganium emersum, Spirodela polyrhiza, Stuckenia filiformis, Stuckenia pectinata, Typha latifolia, Utricularia gibba, Utricularia macrorhiza, Vallisneria americana, Wolffia columbiana, Zannichellia palustris, Zizania aquatica, Zizania palustris.

3.4.1.8. *Potamogeton diversifolius* **Raf.** is a perennial, which grows in brackish (rarely) to freshwater bayous, bays, bogs, borrow pits, canals, depressions, ditches (irrigation, roadside), flatwoods, gravel pits, lakes (playa), marshes (floodplain, shallow), mires, ponds (artificial, beaver, farm, mudflat, natural, oxbow, roadside, shallow, spring-fed, stock, temporary, vernal), pools (shallow, vernal), reservoirs (drying), rivers (slow-moving), ruts (shallow), seeps, sinkholes, sloughs (prairie), springs, streams (blackwater, shallow), swales (peaty, seasonal), and swamps (cypress, roadside) at elevations to 2744 m. Although the plants ordinarily are heterophyllous with well-developed floating leaves, they grow occasionally as an entirely submersed form. Most occurrences are in shallow, still, or flowing waters in exposures ranging from full sunlight to partial shade. The plants occupy the floating-leaved zone comprised of other natant macrophytes at depths ranging from 0.1 to 1.9 m. The numerous reports of associated wetland species (see below) in addition to more hydrophytic species indicate that the plants can occur at times in extremely shallow wetland waters in addition to deeper water habitats such as ponds (frequently) or lakes. In general, this species tolerates a broad range of habitats with respect to acidity and alkalinity (pH: 4.9–9.2; $\bar{x}=6.3–7.0$; total alkalinity [as CaCO₃]: 1.3–43.5 mg l⁻¹; $\bar{x}=2.0–11.6$ mg l⁻¹). Typically, it grows in sites having higher than average levels of Ca (2.1–39.5 mg l⁻¹; $\bar{x}=11.8$ mg l⁻¹), conductivity (to 200 µmhos cm⁻¹; 18–182 µS cm⁻¹ @ 25°C; $\bar{x}=96$ µS cm⁻¹ @ 25°C), total N (275–899 µg l⁻¹; $\bar{x}=542$ µg l⁻¹), total P (9–40 µg l⁻¹; $\bar{x}=21$ µg l^{-1}), and water clarity (Secchi depth: 0.8–3.9 m; $\overline{x}$ = 2.3 m) but lower than average levels of chloride (2.9–28.0 mg l^{-1}; $\overline{x}$ = 15.7 mg l^{-1}). The substrates are characterized as clay, clay loam, clayey sand, cobble, gravelly loam (Corning), limestone, muck, mucky clay, mud, peat, rock, sand (Carrizo), sandy clay, sandy loam, sandy mud, silt, silty loam, and typic fragiaquult. Flowering and fruiting extend from April to October. Like the related *P. bicupulatus*, the flowers occur in both emergent spikes (supported by floating leaves) and submersed (capitate) inflorescences. The reproductive ecology has not been investigated specifically, but likely is similar to that of the related *P. spirillus* (see below) in having protogynous but self-compatible flowers that are facultatively (emergent) or obligately (submersed) autogamous. Most reports indicate that the plants reproduce primarily by seed, which are produced in quantity. A seed bank (with a potential seedling density of 2.3 m^{-2}) of uncertain duration persists, based on estimates obtained from seeds that germinated from sediment cores kept under flooded or saturated/flooded conditions. The plants also reproduce vegetatively by the production of rhizomes and can attain a standing crop biomass of more than 510 kg DM acre^{-1}. The decomposing plants release various ions (Ca, Fe, K, Mg, Mn, Na, Zn) into the water column, at first by physical leaching, and after 10 days by microbial degradation. There is some indication that strong grazing pressure on periphyton by fish (Vertebrata: Teleostei) is detrimental to the growth of the plants. **Reported associates:** *Alisma triviale, Alopecurus geniculatus, Alternanthera philoxeroides, Ammannia coccinea, Aneilema, Azolla cristata, Azolla filiculoides, Bacopa rotundifolia, Barbarea orthoceras, Batrachospermum, Beckmannia syzigachne, Bidens beckii, Boehmeria cylindrica, Brasenia schreberi, Butomus umbellatus, Cabomba caroliniana, Callicarpa americana, Callitriche hermaphroditica, Callitriche heterophylla, Callitriche palustris, Carex frankii, Carex glaucescens, Carex utriculata, Cephalanthus occidentalis, Ceratophyllum demersum, Chara, Chloracantha spinosa, Coreopsis, Cynodon dactylon, Cyperus acuminatus, Cyperus articulatus, Cyperus entrerianus, Cyperus esculentus, Cyperus ochraceus, Cyperus odoratus, Cyperus polystachyos, Cyperus virens, Cyrilla racemiflora, Damasonium californicum, Didiplis diandra, Downingia bacigalupii, Downingia ornatissima, Drepanocladus, Drosera brevifolia, Dulichium arundinaceum, Echinochloa colona, Echinochloa crus-galli, Echinodorus cordifolius, Eclipta prostrata, Eichhornia crassipes, Elatine californica, Elatine minima, Elatine rubella, Eleocharis acicularis, Eleocharis baldwinii, Eleocharis cellulosa, Eleocharis macrostachya, Eleocharis montana, Eleocharis montevidensis, Eleocharis obtusa, Eleocharis palustris, Eleocharis quadrangulata, Eleocharis vivipara, Elodea canadensis, Elodea nuttallii, Eriocaulon compressum, Eriophorum chamissonis, Eryngium castrense, Eryngium nasturtiifolium, Glyceria borealis, Glyceria septentrionalis, Gratiola, Helanthium tenellum, Heteranthera dubia, Heteranthera limosa, Hibiscus laevis, Hippuris vulgaris, Hordeum, Hydrilla verticillata, Hydrocotyle ranunculoides, Hydrocotyle umbellata, Hydrodictyon, Hydrolea quadrivalvis, Hypericum majus, Iris tridentata, Iris virginica, Isoetes hyemalis, Isoetes lacustris, Iva annua, Juncus acuminatus, Juncus balticus, Juncus debilis, Juncus effusus, Juncus interior, Juncus pelocarpus, Juncus repens, Juncus tenuis, Leersia oryzoides, Lemna minor, Lemna perpusilla, Lemna trisulca, Lemna turionifera, Lemna valdiviana, Leptochloa fusca, Limnobium spongia, Limosella, Lindernia dubia, Lobelia, Ludwigia glandulosa, Ludwigia peploides, Ludwigia sphaerocarpa, Luziola fluitans, Magnolia virginiana, Marsilea macropoda, Marsilea vestita, Mecardonia procumbens, Myosurus, Myriophyllum aquaticum, Myriophyllum heterophyllum, Myriophyllum hippuroides, Myriophyllum laxum, Myriophyllum pinnatum, Myriophyllum sibiricum, Myriophyllum spicatum, Myriophyllum verticillatum, Najas flexilis, Najas gracillima, Najas guadalupensis, Najas marina, Najas minor, Navarretia intertexta, Nelumbo lutea, Nitella flexilis, Nuphar advena, Nymphaea odorata, Nyssa biflora, Orontium aquaticum, Packera tampicana, Panicum hemitomon, Panicum repens, Paspalum distichum, Paspalum denticulatum, Paspalum pubiflorum, Peltandra virginica, Persicaria amphibia, Persicaria coccinea, Persicaria hydropiperoides, Persicaria pensylvanica, Phalaris arundinacea, Phalaris caroliniana, Phragmites australis, Pilularia americana, Pinus palustris, Pinus taeda, Pithophora, Plagiobothrys hispidulus, Plagiobothrys stipitatus, Pontederia cordata, Porterella carnosula, Potamogeton amplifolius, Potamogeton berchtoldii, Potamogeton clystocarpus, Potamogeton crispus, Potamogeton epihydrus, Potamogeton floridanus, Potamogeton foliosus, Potamogeton friesii, Potamogeton gramineus, Potamogeton hillii, Potamogeton illinoensis, Potamogeton natans, Potamogeton nodosus, Potamogeton oakesianus, Potamogeton obtusifolius, Potamogeton praelongus, Potamogeton pulcher, Potamogeton pusillus, Potamogeton richardsonii, Potamogeton robbinsii, Potamogeton strictifolius, Potamogeton zosteriformis, Psilocarphus, Quercus incana, Quercus laurifolia, Ranunculus aquatilis, Ranunculus flammula, Ranunculus longirostris, Ranunculus pusillus, Ricciocarpus natans, Rumex crispus, Ruppia spiralis, Saccharum giganteum, Sacciolepis striata, Sagittaria brevirostra, Sagittaria cuneata, Sagittaria graminea, Sagittaria lancifolia, Sagittaria latifolia, Sagittaria longiloba, Sagittaria macrocarpa, Sagittaria montevidensis, Sagittaria platyphylla, Sagittaria subulata, Sarracenia flava, Saururus cernuus, Schoenoplectus acutus, Schoenoplectus californicus, Schoenoplectus etuberculatus, Scirpus cyperinus, Sisyrinchium, Sparganium americanum, Sparganium angustifolium, Sphagnum, Spirodela polyrhiza, Steinchisma hians, Stuckenia pectinata, Stuckenia vaginata, Taxodium distichum, Triadenum walteri, Tridens albescens, Triglochin scilloides, Typha angustifolia, Typha domingensis, Typha latifolia, Utricularia gibba, Utricularia inflata, Utricularia macrorhiza, Utricularia radiata, Vallisneria americana, Veronica peregrina, Wolffia brasiliensis, Wolffia columbiana, Wolffiella gladiata, Xanthium, Zannichellia palustris, Zizaniopsis miliacea.*

3.4.1.9. **Potamogeton epihydrus Raf.** is a perennial, which occurs in still or flowing, non-tidal or tidal waters of bogs, brooks (floodplain), canals (slow flowing), channels (boat), depressions, ditches (irrigation), floodplains, gravel pits (roadside), lakes (seepage), marshes, meadows, ponds (beaver, bog, muskeg, oxbow), pools (brook, spill), reservoirs, rivers, sloughs (peaty), and streams at elevations to 3061 m. Exposures generally are in full sunlight. This is a very common submersed aquatic plant owing much to its ecological breadth. The plants can be found from near sea level to elevations over 3 km, and at depths from only 0.05 m up to 2.9 m. They are found in diverse habitats but perhaps are most common in waters of lower alkalinity and conductivity (pH: 3.1–10.1; $\bar{x} = 6.8$–7.6; total alkalinity [as $CaCO_3$]: 0.5–245.0 mg l^{-1}; $\bar{x} = 19.6$–62.5 mg l^{-1}; conductivity <200 μmhos cm^{-1}; nitrate N: 0–1.4 mg l^{-1}; $\bar{x} = 0.8$ mg l^{-1}; Mg usually <10 mg l^{-1}). In one study, the plants rapidly became dominant in a formerly acidic lake following its neutralization with calcite to more circumneutral conditions. The reported substrates include clay, cobble, gravel, gravelly mud, loam, muck (organic, soft), mucky sand, ooze, peat (*Sphagnum*), rock (granite), sand, silt, silty clay loam, and silty sand. These are heterophyllous plants, which are capable of taking up free atmospheric CO_2 by their floating leaves, but also can use dissolved aqueous bicarbonate (HCO_3^-) as a carbon source. Flowering and fruiting occur from June to October. The monomorphic inflorescences are elevated above the water by the floating leaves, with each spike producing from 12 to 30 emergent aerial flowers. The flowers are protogynous with an intermediate pollen:ovule ratio ($\bar{x} = 9{,}000{:}1$) but are self-compatible and capable of 100% selfing if not outcrossed during the initial receptive stigmatic phase, which persists through subsequent anther dehiscence. Surveys of British populations have indicated a complete lack of detectable genetic (i.e., allozyme) variation, which is consistent with a high level of clonal reproduction and/or inbreeding. The seeds are dormant but will germinate at room temperature following 150 days of cold (1°C–3°C) stratification; dried seeds lose their viability. Seeds kept in cold (1°C–3°C) water have remained 78–82% viable after 6–12 months of storage. The seeds are dispersed locally by water currents and shifting sediments and probably over greater distances by waterfowl (Aves: Anatidae). Vegetative reproduction occurs by means of rhizomes (late May biomass averaging 11 g m^{-2}), which achieve their highest belowground:aboveground biomass ratio ($\bar{x} = 0.41$–0.43:1) in the autumn or when exposed to high levels of dissolved inorganic carbon. Maximum plant biomass ($\bar{x} = 94$ g m^{-2}) can be attained within 85 days, with most ($\bar{x} = 81$ g m^{-2}) being allocated to aboveground structures. The plants are impacted by, but are fairly resistant to, episodic drawdown conditions. **Reported associates:** *Aldrovanda vesiculosa, Alnus viridis, Azolla cristata, Batrachospermum moniliforme, Bidens beckii, Brasenia schreberi, Cabomba caroliniana, Calamagrostis canadensis, Calamagrostis stricta, Callitriche hermaphroditica, Callitriche heterophylla, Callitriche stagnalis, Caltha natans, Carex aquatilis, Carex canescens, Carex lacustris, Carex pellita, Carex rostrata, Carex scoparia, Carex stylosa, Carex utriculata, Ceratophyllum demersum, Ceratophyllum echinatum, Chara vulgaris, Cladopodiella fluitans, Comarum palustre,* *Cornus sericea, Drosera rotundifolia, Dulichium arundinaceum, Egeria densa, Elatine minima, Eleocharis acicularis, Eleocharis flavescens, Eleocharis obtusa, Eleocharis ovata, Eleocharis palustris, Eleocharis robbinsii, Elodea canadensis, Elodea nuttallii, Equisetum fluviatile, Eriocaulon aquaticum, Euthamia occidentalis, Fontinalis antipyretica, Fontinalis novae-angliae, Fontinalis sullivantii, Glyceria borealis, Glyceria canadensis, Glyceria striata, Gratiola aurea, Heteranthera dubia, Hydrocotyle ranunculoides, Isoetes acadiensis, Isoetes bolanderi, Isoetes echinospora, Isoetes lacustris, Isoetes occidentalis, Isoetes tuckermanii, Juncus effusus, Juncus militaris, Juncus pelocarpus, Juncus supiniformis, Juncus tenuis, Leersia oryzoides, Lemna minor, Lemna trisulca, Lindernia dubia, Lobelia dortmanna, Ludwigia palustris, Lycopus uniflorus, Myriophyllum alterniflorum, Myriophyllum farwellii, Myriophyllum heterophyllum, Myriophyllum heterophyllum × Myriophyllum laxum, Myriophyllum hippuroides, Myriophyllum humile, Myriophyllum sibiricum, Myriophyllum spicatum, Myriophyllum tenellum, Myriophyllum verticillatum, Najas canadensis, Najas flexilis, Najas gracillima, Najas guadalupensis, Najas minor, Nasturtium officinale, Nitella flexilis, Nuphar advena, Nuphar microphylla, Nuphar polysepala, Nuphar variegata, Nymphaea odorata, Nymphoides cordata, Oenanthe sarmentosa, Pallavicinia lyellii, Persicaria amphibia, Persicaria coccinea, Persicaria hydropiperoides, Persicaria lapathifolia, Phalaris arundinacea, Phragmites australis, Podostemum ceratophyllum, Pontederia cordata, Potamogeton alpinus, Potamogeton amplifolius, Potamogeton berchtoldii, Potamogeton bicupulatus, Potamogeton confervoides, Potamogeton crispus, Potamogeton diversifolius, Potamogeton foliosus, Potamogeton friesii, Potamogeton gramineus, Potamogeton natans, Potamogeton nodosus, Potamogeton oakesianus, Potamogeton oblongus, Potamogeton obtusifolius, Potamogeton perfoliatus, Potamogeton praelongus, Potamogeton pulcher, Potamogeton pusillus, Potamogeton richardsonii, Potamogeton robbinsii, Potamogeton sibiricus, Potamogeton spirillus, Potamogeton strictifolius, Potamogeton tennesseensis, Potamogeton vaseyi, Potamogeton zosteriformis, Ranunculus flabellaris, Ranunculus flammula, Ranunculus longirostris, Ranunculus subrigidus, Ranunculus trichophyllus, Rhynchospora macrostachya, Riccia fluitans, Rosa woodsii, Sagittaria cuneata, Sagittaria graminea, Sagittaria latifolia, Sagittaria rigida, Sagittaria teres, Salix bebbiana, Salix exigua, Schoenoplectus acutus, Schoenoplectus pungens, Schoenoplectus smithii, Schoenoplectus subterminalis, Schoenoplectus tabernaemontani, Scirpus atrovirens, Scirpus cyperinus, Scirpus microcarpus, Scutellaria galericulata, Sparganium americanum, Sparganium angustifolium, Sparganium emersum, Sparganium eurycarpum, Sparganium fluctuans, Sparganium hyperboreum, Sparganium natans, Sphagnum pylaesii, Sphagnum subsecundum, Spiraea douglasii, Spirodela polyrhiza, Stuckenia filiformis, Stuckenia pectinata, Stuckenia vaginata, Torreyochloa pallida, Utricularia geminiscapa, Utricularia gibba, Utricularia intermedia, Utricularia macrorhiza, Utricularia minor, Utricularia* *purpurea, Utricularia radiata, Utricularia resupinata, Vallisneria americana, Veronica catenata, Warnstorfia exannulata, Wolffia brasiliensis, Zannichellia palustris, Zizania.*

3.4.1.10. *Potamogeton floridanus* Small is a narrowly restricted perennial, which inhabits flowing streams and tidal channels (bayou) at elevations to 10 m. The plants occur at depths from 0.6 to 1.2 m. Flowering reportedly occurs during the early spring but fruiting never has been observed in these plants, which recently have been determined to represent hybrids (see *Systematics* below). The plants reproduce vegetatively by means of rhizomes. There is no other life history information for this species. **Reported associates:** *Cladium jamaicense, Eleocharis, Hypericum fasciculatum, Mayaca fluviatilis, Myriophyllum laxum, Nymphoides aquatica, Pluchea foetida, Pontederia cordata, Potamogeton diversifolius, Utricularia gibba, Zizaniopsis miliacea.*

3.4.1.11. *Potamogeton foliosus* Raf. is an annual or a perennial, which grows in calm to lotic, tidal or non-tidal brooks (muddy), canals (drainage, irrigation), channels (river), depressions, ditches (dredged, irrigation, roadside), floodplains, gravel bars, gravel pits, lagoons (spring-fed), lakes, marshes (vernal), mudflats, oxbows, ponds (beaver, fish hatchery, hydrothermal, livestock, ornamental, prairie, roadside, siltation, spring-fed, strip mine), pools (dune, intermittent, isolated, spring-fed, vernal), potholes, reservoirs, rice fields, rivers, sand pits, seeps, sinkholes, sloughs, sluiceways, springs (hot [e.g., 30°C]), streams (intermittent, shallow, sluggish, spring-fed, swift), swales, troughs (water), and washes (rocky, sandy) at elevations to 3472 m. The plants are found primarily in freshwaters; at least some of the reports from brackish sites evidently are based on misidentifications of *Ruppia maritima* (Cymodoceaceae). Exposures range from full sunlight to partial shade. Although capable of growing in much shallower waters (e.g., 2.5 cm), the plants are known to occur at maximum depths of 4.3–14.0 m where light levels are only 0.5–1.0% of the surface illumination. They have disappeared in some areas that experienced excessive turbidity; however, such absences also may have been due to grazing by carp (Vertebrata: Teleostei: Cyprinidae: *Cyprinus carpio*). The substrates have been described as clay, clay loam, clayey muck, gravel (coarse), limestone, loam, muck, mud, organic, rock, sand, sandy stony alluvium, shale, silt, silty clay, and stony mud. A broad range of ecological conditions is tolerated, although the habitats more typically are alkaline (pH: 6.5–9.8; $\bar{x} = 7.8$–7.9; total alkalinity [as $CaCO_3$] 4.6–335.0 mg l^{-1}; $\bar{x} = 65.1$–106.0 mg l^{-1}; conductivity to 610 µmhos cm^{-1} [@ 25°C]). The plants have appeared rapidly in formerly acidic lakes (pH: 3.0–4.3) after the pH was increased to 6.0 or more by the addition of lime. The waters often are fairly high in nutrients (e.g., total N: 800 µg l^{-1}, total P: 42 µg l^{-1}) and occurrences are associated positively with total phosphate levels. The plants also are common in sites with fluctuating water levels. Flowering and fruiting have been reported from April to September. The flowers are emersed in capitate clusters, which also can float on the water surface. There is no detailed information on the pollination biology of this species, but its prolific seed production (as high as 69,400 plant^{-1} season^{-1})

and relatively low pollen:ovule ratio (~2,000) would implicate self-pollination (possibly hydroautogamy) as the prevalent reproductive mode. However, the existence of interspecific hybrids (see *Systematics* below) also indicates that xenogamous pollen transfer must occur in some fashion, at least occasionally. The fruits are dispersed locally by water, with a sampling of irrigation water finding 0.27 fruits 254 kl^{-1}, which maintained a 42% germination rate. The fruits also have remained viable after passing through the digestive track of mallard ducks (Aves: Anatidae: *Anas platyrhynchos*) following a 12-hr retention time. They undoubtedly are dispersed over considerable distances via this means of endozooic transport. A seed bank (an average of 32.9 seedlings m^{-2} germinating from lake sediment samples) of uncertain duration develops. Seeds that have been stored in water at room temperature retained 60% viability after 3 months; those stored in water at 1°C–3°C retained more than 81% viability after 210 days of storage when incubated under an approximately 19°C/15°C day/night temperature regime. A rhizome system does not develop. Winter buds are produced sporadically (up to 720/plant in one season); however, the plants grow frequently as annuals. The plants are submersed, rooted to the bottom, and homophyllous (linear-leaved) but their foliage often forms a floating mat on the water surface. The foliar nitrogen content (1.7–2.5%) is directly proportional to the amount of organic matter in the sediments. Standing crops up to 272 kg (dw) acre^{-1} have been reported, which can represent up to 11.8% of the total lake macrophyte standing crop biomass. Despite a fairly diverse list of associates (see below), it is common to find this species growing as a monoculture or at least as a dominant. The presence of plants possibly has an allelopathic effect on the distribution of calanoid copepods (Crustacea: Copepoda: Diaptomidae: *Diaptomus clavipes*). **Reported associates:** *Alisma triviale, Alopecurus carolinianus, Baccharis salicifolia, Baccharis sergiloides, Bidens beckii, Bidens frondosus, Boehmeria cylindrica, Brasenia schreberi, Butomus umbellatus, Calamagrostis canadensis, Callitriche heterophylla, Callitriche longipedunculata, Caltha natans, Carex brevior, Carex rostrata, Carex simulata, Carex utriculata, Ceratophyllum demersum, Chara globularis, Chara vulgaris, Conium maculatum, Crassula aquatica, Cyperus eragrostis, Datura wrightii, Downingia cuspidata, Drepanocladus, Echinochloa crus-galli, Echinochloa walteri, Echinodorus berteroi, Egeria densa, Elatine californica, Elatine chilensis, Eleocharis acicularis, Eleocharis equisetoides, Eleocharis macrostachya, Eleocharis montana, Eleocharis montevidensis, Eleocharis ovata, Eleocharis palustris, Eleocharis quinqueflora, Eleocharis rostellata, Elodea canadensis, Elodea nuttallii, Elymus repens, Epilobium ciliatum, Fissidens fontanus, Fontinalis antipyretica, Fontinalis neomexicana, Glossostigma cleistanthum, Glyceria borealis, Glyceria striata, Heteranthera dubia, Hippuris vulgaris, Holcus lanatus, Hydrocotyle ranunculoides, Hypericum scouleri, Impatiens capensis, Isoetes howellii, Isoetes orcuttii, Juncus balticus, Juncus bufonius, Juncus dudleyi, Juncus pelocarpus, Juncus phaeocephalus, Leersia oryzoides, Lemna gibba, Lemna* *minor, Lemna trisulca, Lilaeopsis, Lobelia cardinalis, Marsilea vestita, Mentha arvensis, Mimulus guttatus, Myriophyllum quitense, Myriophyllum sibiricum, Myriophyllum spicatum, Myriophyllum verticillatum, Najas flexilis, Najas guadalupensis, Najas marina, Najas minor, Nasturtium officinale, Navarretia intertexta, Nitella flexilis, Nuphar advena, Nuphar polysepala, Nuphar variegata, Nymphaea odorata, Paspalum distichum, Persicaria amphibia, Persicaria coccinea, Persicaria lapathifolia, Persicaria punctata, Phalaris arundinacea, Pilea pumila, Pontederia cordata, Potamogeton alpinus, Potamogeton amplifolius, Potamogeton berchtoldii, Potamogeton bicupulatus, Potamogeton clystocarpus, Potamogeton crispus, Potamogeton diversifolius, Potamogeton epihydrus, Potamogeton friesii, Potamogeton gramineus, Potamogeton hillii, Potamogeton illinoensis, Potamogeton natans, Potamogeton nodosus, Potamogeton oakesianus, Potamogeton obtusifolius, Potamogeton perfoliatus, Potamogeton praelongus, Potamogeton pulcher, Potamogeton pusillus, Potamogeton richardsonii, Potamogeton robbinsii, Potamogeton strictifolius, Potamogeton vaseyi, Potamogeton zosteriformis, Ranunculus abortivus, Ranunculus aquatilis, Ranunculus cymbalaria, Ranunculus flammula, Ranunculus gmelinii, Ranunculus longirostris, Ranunculus sceleratus, Ranunculus trichophyllus, Rhizoclonium crassipellitum, Rhizoclonium hieroglyphicum, Riccia fluitans, Rorippa palustris, Ruppia maritima, Ruppia spiralis, Sagittaria cuneata, Sagittaria graminea, Sagittaria latifolia, Salix exigua, Salix geyeriana, Salix gooddingii, Salix laevigata, Salix nigra, Schoenoplectus acutus, Schoenoplectus americanus, Schoenoplectus subterminalis, Scirpus cyperinus, Scirpus microcarpus, Scutellaria lateriflora, Sparganium emersum, Sparganium eurycarpum, Spirodela polyrhiza, Stuckenia filiformis, Stuckenia pectinata, Stuckenia vaginata, Triglochin scilloides, Typha domingensis, Typha latifolia, Utricularia macrorhiza, Vahlodea atropurpurea, Vallisneria americana, Vaucheria, Veronica anagallis-aquatica, Veronica peregrina, Wolffia brasiliensis, Wolffia columbiana, Zannichellia palustris.*

3.4.1.12. **Potamogeton friesii** Rupr. is an annual (vegetative) or perennial, which inhabits depressions, ditches (mucky), lakes (prairie), mudflats, ponds (beaver, mill, pothole), pools, rivers (backwaters), sloughs, and streams at elevations to 2707 m. The plants occur in open to shaded exposures and colonize still or flowing waters typically at depths from 1.8 to 3.0 m, but can grow in waters as shallow as 10 cm and as deep as 6.0 m. The substrates have been described as alluvium (silty peaty), clay, clayey silty muck (organic), cobble, detritus, gravel, marl, muck, mud (organic), muddy gravel, muddy sand, peat (fibrous, flocculent, marly, pulpy), rock, rubble, sand, sandy clay, sandy gravel, sandy muck, sandy silt (marly), and silt (fine). The surrounding waters are mesotrophic to eutrophic, brackish (rarely) or fresh and almost always alkaline (pH: 5.6–9.8; $\bar{x}$ = 7.7–8.6; total alkalinity [as CaCO$_3$]: 20.6–376.1 mg l^{-1}; $\bar{x}$ = 71.2–111.0 mg l^{-1}; conductivity [@ 25°C] to 550 µmhos cm^{-1}). Flowering and fruiting occur from June to September, with the flowers produced on

monomorphic, emergent spikes. The flowers are self-incompatible and protogynous, with the female and male phases each lasting for about 2 days. Pollination reportedly occurs both by wind (anemophily) and anemo-epihydrophily (wind-mediated water surface pollination). There is little agreement on the extent of fruiting in this species with some reports suggesting that it occurs "rather infrequently" or "very rarely" and others suggesting that the plants are "heavily fruiting". The fruits reportedly will remain viable if stored in cool water. In any case, most accounts agree that reproduction occurs primarily by vegetative means (in this case via turions), given that seedlings have not been encountered in the field even after extensive searches. The ecology of this species apparently varies substantially among different localities. The plants are reported both as being susceptible to and as tolerant to eutrophication and disturbance. Although they have disappeared from several urbanized sites over time, they also persist in turbid, eutrophic lakes (e.g., Secchi disk depth: 0.71) as well as in mesotrophic waters that are quite clear (e.g., Secchi disk depth: 4.0 m). The foliage is homophyllous (linear-leaved) and no rhizome ordinarily is produced. Vegetative reproduction by the distinctive 4-ranked turions appears to be the primary means of dispersal (by water), establishment, and perennation. The means of long-distance dispersal is uncertain, but likely involves the transport of seeds in some way. The mean annual biomass production has been estimated at 16.1–27.6 g (dw) m^{-2}. **Reported associates:** *Acorus calamus, Alisma gramineum, Azolla cristata, Bidens beckii, Brasenia schreberi, Calamagrostis canadensis, Calamagrostis stricta, Calla palustris, Callitriche hermaphroditica, Callitriche palustris, Carex alopecoidea, Carex comosa, Carex diandra, Carex hystericina, Carex interior, Carex lasiocarpa, Carex pseudocyperus, Carex rostrata, Carex scoparia, Carex viridula, Ceratophyllum demersum, Chara contraria, Chara globularis, Chara vulgaris, Drepanocladus aduncus, Dulichium arundinaceum, Eleocharis acicularis, Eleocharis erythropoda, Eleocharis palustris, Elodea canadensis, Elodea nuttallii, Equisetum fluviatile, Glyceria borealis, Heteranthera dubia, Hippuris vulgaris, Hydrocotyle ranunculoides, Isoetes echinospora, Juncus effusus, Lemna minor, Lemna trisulca, Lyngbya, Lysimachia, Myriophyllum aquaticum, Myriophyllum heterophyllum, Myriophyllum sibiricum, Myriophyllum spicatum, Myriophyllum tenellum, Myriophyllum verticillatum, Najas flexilis, Najas guadalupensis, Najas marina, Nitella flexilis, Nitellopsis obtusa, Nuphar advena, Nuphar polysepala, Nuphar variegata, Nymphaea odorata, Osmunda regalis, Persicaria amphibia, Phalaris arundinacea, Pontederia cordata, Potamogeton alpinus, Potamogeton amplifolius, Potamogeton berchtoldii, Potamogeton crispus, Potamogeton diversifolius, Potamogeton epihydrus, Potamogeton foliosus, Potamogeton gramineus, Potamogeton hillii, Potamogeton illinoensis, Potamogeton natans, Potamogeton nodosus, Potamogeton obtusifolius, Potamogeton perfoliatus, Potamogeton praelongus, Potamogeton pusillus, Potamogeton richardsonii, Potamogeton robbinsii, Potamogeton sibiricus, Potamogeton spirillus, Potamogeton* *strictifolius, Potamogeton vaseyi, Potamogeton zosteriformis, Ranunculus flammula, Ranunculus longirostris, Ranunculus subrigidus, Ranunculus trichophyllus, Ricciocarpus, Ruppia spiralis, Sagittaria cuneata, Sagittaria latifolia, Sagittaria rigida, Salix candida, Salix exigua, Salix petiolaris, Schoenoplectus acutus, Schoenoplectus tabernaemontani, Sium suave, Sparganium angustifolium, Sparganium emersum, Sparganium eurycarpum, Sparganium fluctuans, Sparganium natans, Spirodela polyrhiza, Stuckenia filiformis, Stuckenia pectinata, Stuckenia vaginata, Tolypella intricata, Typha angustifolia, Typha latifolia, Utricularia gibba, Utricularia intermedia, Utricularia macrorhiza, Vallisneria americana, Wolffia brasiliensis, Wolffia columbiana, Zannichellia palustris.*

3.4.1.13. *Potamogeton gramineus* L.

is a perennial, which grows in backwaters, bays (shallow), canals (irrigation), channels, coves (shallow), ditches (roadside), fens, flats (tidal), floodplains, lagoons, lakes (bog, kettle, oxbow), pannes (calcareous, interdunal), ponds (beaver, boggy, drying, glacial, kettle, sag, stock), pools (brook, meadow, muskeg, river, roadside), potholes, rivers, sloughs, streams (outlet), and tanks (cattle) at elevations to 3238 m. Exposures typically occur in full sun. The substrates include cobble, gravel (coarse), gravelly loam, marl, muck, mucky gravel, mud, muddy clay, ooze, peat (fibrous, sedge), sand (coarse), sandstone (Coconino), sandy gravel, sandy muck, sandy silt (marly), silt (igneous), silty mud, and silty sand. In flowing systems, the plants occur primarily where the interstitial sediment water temperatures (at 10 cm) remain fairly high ($\bar{x} = 19.5°C$) as a result of warm surface water infiltration. Although normally heterophyllous, submersed homophyllous (broad-leaved) forms also are common. The floating leaves take up free CO_2 directly from the air, whereas the submersed leaves are capable of bicarbonate (HCO_3^-) uptake. Because submersed leaves have less hydraulic resistance to water currents, the plants can tolerate currents (e.g., 46 cm s^{-1}) by means of adopting a submersed habit (and development of a strong anchoring rhizome system). They occupy a broad depth profile (range: 0.15–6.0 m) but usually occur in shallow waters (median depth: 1.1 m). The habitats represent a diversity of ecological conditions (pH: 3.7–10.0; $\bar{x} = 7.1$–8.1; total alkalinity [as $CaCO_3$]: 0.5–476.8 mg l^{-1}; $\bar{x} = 1.8$–109.0 mg l^{-1}; conductivity to 450 μmhos cm^{-1}) but most occurrences are in northern (>44°N latitude) soft water sites characterized by a short growing season, low sediment nutrient levels (total P <20 μmol g^{-1} dw), and relatively clear water (e.g., Secchi depth: 3.2–4.1 m). The plants are strongly competitive when light levels are high. In some regions (e.g., Europe) the plants are considered to be reliable indicators of nutrient-poor conditions. They have disappeared from a number of North American lakes that became eutrophic as a consequence of agriculture and urban development. However, the plants are categorized as "shade tolerant" and have a relatively low light compensation point, which enables them to tolerate low light levels. Consequently, various records document their occurrence in waters that are quite turbid (e.g., Secchi depth: 0.60–0.71 m). They are fairly intolerant of salinity and occur in fresh or only slightly brackish waters. In sites frequented by anadromous fish (e.g., Vertebrata: Teleostei: Salmonidae: *Oncorhynchus kisutch*), the plants can become enriched with nitrogen that originates from marine sources. Flowering and fruiting extend from June to October, with the monomorphic inflorescences (spikes) held above the water by the floating leaves. The flowers are self-incompatible and protogynous, with the female phase lasting from 1 to 3 days and successive male phase lasting for 1–2 days. Pollination occurs by wind (anemophily) or by anemo-epihydrophily (wind-mediated water surface pollination). Although fruiting has been observed on plants growing as deeply as 3 m, sexual reproduction presumably is relatively unimportant in the persistence and dispersal of the plants due to the apparent lack of a seed bank. However, seed production is likely to become more important during periods of dessication. Drops in water levels lead to the development of stranded plants (stunted "terrestrial" growth forms), which can reach densities from 50 to 746 m^{-2}. This is one of the few *Potamogeton* species whose fruits can withstand dry conditions for up to a year, while retaining at least some (albeit low) viability (e.g., 0.5% germination). Seeds stored for 2–3 months in water kept at room temperature also exhibit low germination rates (~1%) but at much lower levels compared to those stored for a year in water at 1°C–3°C (49%). By all accounts, most reproduction occurs vegetatively by means of rhizomatous buds (prevalent under low disturbance), rhizomes (prevalent under high disturbance), and plant fragments. The plants do not produce turions but do develop subterranean rhizomatous buds (called "winter buds" by some), which function as vegetative propagules. They develop under shorter photoperiods (10 hr) but not under longer day conditions (14 hr). The propagules occur in clumped distributions in patch sizes ranging from 0.07 to 1.7 m in diameter and can reach an average maximum "bud bank" density of 761 m^{-2}. The buds are not innately dormant but germinate across a fairly wide range of temperatures from 10°C to 25°C, consuming the majority of propagule carbon reserves within 16–30 days after sprouting. They sprout more rapidly when buried at greater sediment depths (e.g., 25 cm) than at shallow depths (e.g., 5–15 cm). The stoloniferous rhizomes have a relatively shallow rooting depth (4–14 cm; $\bar{x} = 8.7$ cm). Shoot fragmentation represents a more minor reproductive mode, but large numbers of plant fragments have been found attached to boat trailers, which likely facilitate their dispersal. The roots hairs are colonized by VA-mycorrhizal fungi. In calcareous waters, the leaves can become encrusted with marl, which is deposited more densely ($\bar{x} = 0.7$ g g^{-1} plant) in deeper waters (e.g., 30 cm) than at shallower depths (0.63 g g^{-1} plant at 10 cm). The total plant carbohydrate reserves are highest during late night (~11 pm) and lowest in early morning (~5 am). Standing crop biomass has been estimated at 0.12–0.87 g m^{-2}. **Reported associates:** *Agalinis purpurea, Alisma, Alopecurus aequalis, Arethusa bulbosa, Betula cordifolia, Bidens beckii, Bolboschoenus fluviatilis, Brasenia schreberi, Bulbochaete, Cabomba caroliniana, Callitriche hermaphroditica, Callitriche palustris, Caltha palustris, Carex chordorrhiza, Carex lacustris, Carex lasiocarpa, Carex lyngbyei, Carex rostrata, Carex scoparia,*

Carex senta, Carex stipata, Carex stricta, Carex utriculata, Carex vesicaria, Cephalanthus occidentalis, Ceratium, Ceratophyllum demersum, Chamaedaphne calyculata, Chara globularis, Cicuta bulbifera, Coelosphaerium, Comarum palustre, Damasonium californicum, Decodon verticillatus, Downingia, Dulichium arundinaceum, Elatine chilensis, Elatine minima, Eleocharis acicularis, Eleocharis erythropoda, Eleocharis macrostachya, Eleocharis palustris, Elodea canadensis, Elodea nuttallii, Equisetum fluviatile, Eriocaulon aquaticum, Fontinalis, Gloeocystis, Gloeotrichia, Glyceria borealis, Glyceria elata, Gratiola aurea, Heteranthera dubia, Hippuris vulgaris, Howellia aquatilis, Hydrocotyle umbellata, Impatiens capensis, Isoetes bolanderi, Isoetes echinospora, Isoetes howellii, Isoetes lacustris, Juncus balticus, Juncus brevicaudatus, Juncus effusus, Juncus mertensianus, Juncus pelocarpus, Lemna minor, Lemna trisulca, Limosella aquatica, Lobelia dortmanna, Lychnothamnus barbatus, Lycopus uniflorus, Lythrum salicaria, Marsilea vestita, Menyanthes trifoliata, Microcystis, Mimulus primuloides, Mougeotia, Myriophyllum alterniflorum, Myriophyllum farwellii, Myriophyllum hippuroides, Myriophyllum sibiricum, Myriophyllum spicatum, Myriophyllum tenellum, Myriophyllum verticillatum, Najas flexilis, Najas gracillima, Najas guadalupensis, Nitella flexilis, Nuphar advena, Nuphar microphylla, Nuphar polysepala, Nuphar variegata, Nymphaea odorata, Nymphaea tetragona, Oedogonium, Persicaria amphibia, Persicaria coccinea, Persicaria punctata, Poa pratensis, Pontederia cordata, Potamogeton alpinus, Potamogeton amplifolius, Potamogeton berchtoldii, Potamogeton bicupulatus, Potamogeton confervoides, Potamogeton crispus, Potamogeton diversifolius, Potamogeton epihydrus, Potamogeton foliosus, Potamogeton friesii, Potamogeton hillii, Potamogeton illinoensis, Potamogeton natans, Potamogeton nodosus, Potamogeton oakesianus, Potamogeton obtusifolius, Potamogeton perfoliatus, Potamogeton praelongus, Potamogeton pulcher, Potamogeton pusillus, Potamogeton richardsonii, Potamogeton robbinsii, Potamogeton sibiricus, Potamogeton spirillus, Potamogeton strictifolius, Potamogeton vaseyi, Potamogeton zosteriformis, Potentilla anserina, Ranunculus flabellaris, Ranunculus flammula, Ranunculus gmelinii, Ranunculus longirostris, Ranunculus sceleratus, Ranunculus trichophyllus, Rhizoclonium, Ricciocarpus, Sagittaria cristata, Sagittaria cuneata, Sagittaria graminea, Sagittaria latifolia, Sagittaria rigida, Salix geyeriana, Schoenoplectus acutus, Schoenoplectus subterminalis, Schoenoplectus tabernaemontani, Scirpus cyperinus, Sium suave, Sparganium angustifolium, Sparganium emersum, Sparganium fluctuans, Sparganium hyperboreum, Sparganium natans, Spirodela polyrhiza, Staurastrum paradoxum, Stuckenia filiformis, Stuckenia pectinata, Subularia aquatica, Triadenum virginicum, Triglochin maritima, Typha angustifolia, Typha latifolia, Utricularia cornuta, Utricularia gibba, Utricularia intermedia, Utricularia macrorhiza, Utricularia minor, Utricularia purpurea, Utricularia radiata, Vallisneria americana, Veronica scutellata, Wolffia brasiliensis, Wolffia columbiana, Zannichellia palustris, Zizania palustris.

3.4.1.14. ***Potamogeton hillii* Morong** is an annual (vegetative) or perennial, which inhabits still to slow-moving, calcareous waters associated with brooks (muddy), ditches (partially dry, roadside), lakes, marshes, oxbows (shallow), ponds (beaver, farm, floodplain, marshy, mill, oxbow, shallow, small), pools (stream-fed), rivulets, rivers, and streams (meadow, roadside, small, slow flowing) at elevations to 400 m. The plants occur in water depths up to 2.0 m. Their habitats typically are characterized by waters that are clear, cold, and highly alkaline (pH: 7.1–8.2; $\bar{x}$ = 7.5; total alkalinity [as $CaCO_3$]: 53–317 mg l^{-1}; $\bar{x}$ = 141.5 mg l^{-1}). The substrates are calcareous muck or mud derived primarily from dolomitic limestone. The flowers are borne in clusters on short, axillary peduncles, which lie at or just beneath the water surface. Flowering initiates in late July, with fruits developing from August to September. Nearly nothing is known about the floral biology or the means of pollination. The fruits are produced commonly in both shallow and deeper waters and have been recovered from sediments collected at water depths ranging from 0.4 to 4.1 m; however, their viability and germination requirements have yet to be determined. The plants lack rhizomes but reproduce vegetatively by the production of winter buds, which develop during the fall, and by fragmentation of the decumbent stems that are provisioned with nodal roots. Local dispersal occurs by the water transport of fruits and/or winter buds; long-distance dispersal must be facilitated in some way by waterfowl (Aves: Anatidae) or other animals. In one case, the plants were suspected to have been dispersed inadvertently during an interlake transport of beavers (Mammalia: Castoridae: *Castor canadensis*), either along with plant materials or as seeds carried within the fur. This is categorized as an early successional species that is intolerant of turbidity or pollution. It can develop into fairly thick growths (sometimes described as "weedy") under optimal conditions, with some field density estimates approaching 2,000 plants m^{-2}. **Reported associates:** *Ceratophyllum demersum, Chara, Dulichium arundinaceum, Eleocharis acicularis, Elodea canadensis, Equisetum fluviatile, Heteranthera dubia, Hydrocharis morsus-ranae, Myriophyllum sibiricum, Myriophyllum spicatum, Najas flexilis, Najas marina, Nuphar advena, Phalaris arundinacea, Pontederia cordata, Potamogeton amplifolius, Potamogeton crispus, Potamogeton diversifolius, Potamogeton foliosus, Potamogeton friesii, Potamogeton friesii, Potamogeton gramineus, Potamogeton illinoensis, Potamogeton natans, Potamogeton nodosus, Potamogeton pusillus, Potamogeton strictifolius, Potamogeton zosteriformis, Spirodela polyrhiza, Stuckenia pectinata, Utricularia macrorhiza, Vallisneria americana.*

3.4.1.15. ***Potamogeton illinoensis* Morong** is a perennial, which occurs in brooks, canals (drainage), channels (backwater), ditches, estuaries, fens (graminoid, prairie), lagoons, lakes (marl), marl pits, marshes (pothole), pannes (calcareous, interdunal), ponds (kettle, pineland, roadside), pools (marshy, quarry, woodland), reservoirs, rivers, sand pits, sloughs, springs (warm), and streams (blackwater) at elevations to 3088 m. The plants grow in sunny to partially shaded exposures at depths from 0.15 to 3.4 m. The waters are of varying clarity (Secchi depth: 0.5–3.6 m) and velocity, ranging from lentic sites to rapids and other swift currents of 1.3 cm s^{-1} or more. The sediments have been characterized as basalt, boulders, clay, clay alluvium, cobble, detritus (over sand), gravel, gravelly sandy silt, marl, marly muck, marly sand, muck, mud, organic matter, peat (fibrous, pulpy), rocks (limestone), rubbly mucky gravel, sand, sandy gravel, sandy marl, sandy muck, sandy shells, sandy silt, shale pebbles, silt, silty clay, and silty muck. Although this species has a broad ecological range of tolerance, it favors alkaline, hard water sites (pH: 5.6–10.6; $\bar{x}=7.8$–8.3; Ca: $\bar{x}=38.7$ mg l^{-1}; total alkalinity [as CaCO$_3$]: 10.0–307.5 mg l^{-1}; $\bar{x}=49.4$–135.0 mg l^{-1}; specific conductance [@ 25°C]: 15–600 µmhos cm^{-1}; $\bar{x}=238$ µmhos cm^{-1}), with high light levels (Secchi depth: $\bar{x}=1.7$ m), that are relatively low in total P ($\bar{x}=30$ µg l^{-1}) but high in total N ($\bar{x}=1010$ µg l^{-1}) and higher than average values for Cl$^-$ ($\bar{x}=35.3$ mg l^{-1}), K ($\bar{x}=3.5$ mg l^{-1}), Mg ($\bar{x}=17.5$ mg l^{-1}), Na ($\bar{x}=20.1$ mg l^{-1}), and SO$_4{}^{2-}$ ($\bar{x}=21.0$ mg l^{-1}). The nutrient status can range from mesotrophic to eutrophic; however, some intolerance of eutrophic conditions is indicated by the disappearance of the species from formerly less-fertile lakes that have become nutrient-rich. Although heterophyllous, the plants occur commonly as homophyllous (broad-leaved) forms, especially when in deeper waters. The foliage often is coated by marl deposits. Flowering and fruiting have been observed from June to October, yet few details exist on the floral biology and reproductive ecology. The flowers extend from the water surface on monomorphic spikes and probably are wind pollinated. The overall viability of the seeds is uncertain; however, there is some indication that a small, persistent seed bank can develop. Seeds of unknown age (collected from sediment cores) have germinated in the laboratory, in water at 20°C (14/10 hr day/night light regime) following 20 weeks of cold (4°C) stratification. Most often the plants reproduce vegetatively by prolific rhizome production. The plants die back each winter and overwinter by means of the persistent rhizomes. Vegetative reproduction also can occur by fragmentation of the shoot system, which (in flowing waters) can grow up to 6 m in length. The maximum per plant total biomass (determined experimentally) is allocated mostly to the leaves (27.21 g dw), less to the rhizomes (17.40 g dw), and least to the stems (11.87 g dw). Maximum peak (October) total biomass in the field has been estimated at 54 g m^{-2}. Invertebrate herbivory has been shown to reduce plant biomass by 63%.

Reported associates: *Alisma triviale, Alopecurus aequalis, Arundo donax, Azolla filiculoides, Baccharis salicifolia, Batrachospermum, Bidens beckii, Brasenia schreberi, Cabomba caroliniana, Callitriche heterophylla, Callitriche palustris, Cardamine bulbosa, Carex scoparia, Carex stricta, Carex vesicaria, Carex viridula, Ceratophyllum demersum, Ceratopteris thalictroides, Chara globularis, Chara vulgaris, Egeria densa, Eichhornia crassipes, Elatine chilensis, Elatine minima, Eleocharis acicularis, Eleocharis cellulosa, Eleocharis palustris, Elodea canadensis, Glyceria borealis, Glyceria elata, Heteranthera dubia, Hippuris vulgaris, Hydrilla verticillata, Hydrocotyle verticillata, Hygrophila lacustris, Isoetes echinospora, Juncus, Justicia americana,* *Lemna minor, Leptodictyum riparium, Limnophila sessiliflora, Ludwigia palustris, Ludwigia repens, Lythrum hyssopifolia, Myriophyllum aquaticum, Myriophyllum heterophyllum, Myriophyllum sibiricum, Myriophyllum spicatum, Myriophyllum tenellum, Najas flexilis, Najas guadalupensis, Najas marina, Nasturtium officinale, Nitella flexilis, Nostoc, Nuphar polysepala, Nymphaea odorata, Nymphoides humboldtiana, Persicaria amphibia, Persicaria coccinea, Persicaria hydropiperoides, Pistia stratiotes, Poa annua, Pontederia cordata, Potamogeton amplifolius, Potamogeton crispus, Potamogeton diversifolius, Potamogeton foliosus, Potamogeton friesii, Potamogeton gramineus, Potamogeton hillii, Potamogeton natans, Potamogeton nodosus, Potamogeton praelongus, Potamogeton pusillus, Potamogeton richardsonii, Potamogeton robbinsii, Potamogeton strictifolius, Potamogeton vaseyi, Potamogeton zosteriformis, Ranunculus aquatilis, Ranunculus subrigidus, Ranunculus trichophyllus, Riccia fluitans, Rumex obtusifolius, Sagittaria cuneata, Sagittaria platyphylla, Sagittaria rigida, Schoenoplectus acutus, Schoenoplectus californicus, Schoenoplectus lacustris, Schoenoplectus subterminalis, Schoenoplectus tabernaemontani, Sium suave, Sparganium americanum, Sparganium emersum, Sparganium eurycarpum, Spirodela polyrhiza, Stuckenia filiformis, Stuckenia pectinata, Typha latifolia, Utricularia intermedia, Utricularia macrorhiza, Utricularia gibba, Vallisneria americana, Vallisneria neotropicalis, Veronica catenata, Veronica peregrina, Wolffia brasiliensis, Zannichellia palustris, Zizania palustris, Zizania texana.*

3.4.1.16. *Potamogeton natans* L. is a perennial, which grows in bayous, bogs (roadside), borrow pits, canals (irrigation), channels (drainage, slow-moving), coves, deltas, ditches (irrigation, roadside), fens, gravel pits, lagoons (interdunal), lakes (oxbow), marshes (intertidal, spring-fed), ponds (artificial, beaver, cow, muskeg, pothole, stock), pools (shallow, stagnant), potholes, reservoirs, seeps, sloughs, streams, and streamlets at elevations to 3110 m. The plants can occur at depths from 0.2 to 2.9 m but usually are found in waters <1.5 m deep. They tolerate full sun to partial shade, in oligotrophic to eutrophic conditions, and slow-moving currents. The substrates include basalt, clay, gravel, limestone (Kaibab), marl, muck, mud, muddy sand, peat (fibrous), rock, sand, silt, silty loam (granitic), silty muck, and tephra. The habitats tend to be low in chloride levels (0–251 mg l^{-1}; $\bar{x}=5.0$ mg l^{-1}) but otherwise represent a broad ecological spectrum (pH: 3.7–10.0; $\bar{x}=7.1$–7.9; conductivity to 660 µmhos cm^{-1} [@ 25°C]; total alkalinity (as CaCO$_3$): 3.9–476.8 mg l^{-1}; $\bar{x}=36.7$–112.0 mg l^{-1}). Flowering extends from May to September with fruiting from June to October. Results of experimental manipulations have predicted that a 3°C global temperature rise would result in accelerated flowering. The spikes produce from 12 to 23 alternating, trimerous (usually) whorls of flowers, which are wind pollinated (or anemo-epihydrophilous), strongly self-incompatible, and protogynous, with the female phase and successive male phase each lasting for 2–3 days. The fruit-laden peduncles collapse into the water as the fruits mature; an individual plant can produce upward of 1,000 seeds in a season. The fruits have a fleshy mericarp and can remain afloat for over a year. Germination has been achieved for seeds that were stratified in cold (4°C) water for 180 days and then incubated at 20°C. Physical scarification of the fruits by rupture of the pericarp greatly enhances germination rates (e.g., 51–91%) within 10 days compared to unscarified fruits (e.g., 2–7%). Local dispersal (endozooic) is possibly facilitated by seeds that remain viable (positive tetrazolium test) after passage through the gut of mallard ducks (Aves: Anatidae: *Anas platyrhynchos*) or herbivorous fish (e.g., Vertebrata: Teleostei: Cyprinidae: *Cyprinus carpio*); however, there is some evidence that ingestion by fish can delay germination of the fruits. The fruits also remain viable after passing through the digestive tract of ducks (Aves: Anatidae: *Anas*), which likely disperse them endozoically to greater distances. The failure to capture the fruits in propagule trapping studies has led to the conclusion that most dispersal involves the water transport of rhizome fragments, which have been recovered in quantity from trapping studies over a 2-month period, peaking in November. Yet, although such studies report that a propagule bank does not develop, others have recovered large numbers of fruits (to 13,442 m⁻²) from sites or have demonstrated the persistence of viable fruits that were removed from excavated "ghost ponds" (former pond sites filled in conversion to agricultural land) as much as 150 years of age. Therefore, the role of fruits vs. fragments in dispersal most likely varies depending on specific site conditions. The plants are heterophyllous but rarely are found without floating leaves. The resilient petiole apex of the floating leaves flexes elastically to restore their contact with the water if they are lifted (e.g., by wind, waves) from the surface. The phyllodial submersed leaves appear to be derived from highly modified petioles. The plants are unable to use aqueous bicarbonate to any extent but uptake free CO_2 from the atmosphere through their floating leaves. Although their floating leaves enable the plants to persist in extremely turbid or tannin-stained waters, the species has disappeared from sites that became increasingly eutrophic. Dense growths of the plants can lead to substantially reduced dissolved oxygen levels (<2 mg l⁻¹) in the waters beneath their canopy. The rhizomes reportedly contain water-soluble allelochemicals, which are of uncertain ecological significance. The rhizome system enables the plants to withstand high disturbance (catastrophic or erratic)/low stress, or high stress/low disturbance conditions. The seasonal biomass has been estimated at 87 g DM m⁻² (aboveground) and up to 418 g DM m⁻² (total). **Reported associates:** *Acorus calamus, Agalinis purpurea, Alisma triviale, Alopecurus aequalis, Beckmannia syzigachne, Bidens beckii, Boltonia asteroides, Brasenia schreberi, Cabomba caroliniana, Calamagrostis stricta, Callitriche hermaphroditica, Callitriche heterophylla, Callitriche palustris, Callitriche stagnalis, Caltha, Carex alopecoidea, Carex aquatilis, Carex athrostachya, Carex buxbaumii, Carex canescens, Carex comosa, Carex lasiocarpa, Carex lenticularis, Carex limosa, Carex nebrascensis, Carex scopulorum, Carex subfusca, Carex tenuiflora, Carex utriculata, Carex vesicaria, Cephalanthus occidentalis, Ceratophyllum demersum, Ceratophyllum echinatum, Chara globularis, Chara vulgaris, Cicuta douglasii, Cicuta virosa, Cirsium arvense, Comarum palustre, Dasiphora floribunda, Deschampsia cespitosa, Dulichium arundinaceum, Egeria densa, Elatine minima, Eleocharis acicularis, Eleocharis macrostachya, Eleocharis obtusa, Eleocharis palustris, Elodea canadensis, Elodea nuttallii, Equisetum fluviatile, Eriocaulon aquaticum, Eupatorium perfoliatum, Eurhynchium pulchellum, Glyceria borealis, Glyceria grandis, Heteranthera dubia, Hippuris vulgaris, Hydrocotyle ranunculoides, Hydrocotyle umbellata, Iris missouriensis, Iris versicolor, Isoetes bolanderi, Isoetes occidentalis, Juncus articulatus, Juncus balticus, Juncus coriaceus, Juncus effusus, Juncus ensifolius, Juncus nevadensis, Juncus supiniformis, Lemna gibba, Lemna minor, Lemna minuta, Lemna trisulca, Lemna turionifera, Lipocarpha micrantha, Ludwigia palustris, Lycopus americanus, Lycopus uniflorus, Lythrum salicaria, Mentha, Menyanthes trifoliata, Mimulus guttatus, Myosotis laxa, Myosotis scorpioides, Myriophyllum alterniflorum, Myriophyllum aquaticum, Myriophyllum farwellii, Myriophyllum heterophyllum, Myriophyllum hippuroides, Myriophyllum sibiricum, Myriophyllum spicatum, Myriophyllum tenellum, Myriophyllum verticillatum, Najas flexilis, Najas gracillima, Najas minor, Nasturtium officinale, Nelumbo lutea, Nuphar advena, Nuphar polysepala, Nuphar variegata, Nuphar ×rubrodisca, Nymphaea leibergii, Nymphaea loriana, Nymphaea odorata, Nymphaea tetragona, Oenanthe sarmentosa, Peltandra virginica, Persicaria amphibia, Persicaria coccinea, Persicaria hydropiper, Persicaria hydropiperoides, Persicaria punctata, Phalaris arundinacea, Poa pratensis, Polygonum aviculare, Pontederia cordata, Potamogeton alpinus, Potamogeton amplifolius, Potamogeton berchtoldii, Potamogeton bicupulatus, Potamogeton confervoides, Potamogeton crispus, Potamogeton diversifolius, Potamogeton epihydrus, Potamogeton foliosus, Potamogeton friesii, Potamogeton gramineus, Potamogeton hillii, Potamogeton illinoensis, Potamogeton nodosus, Potamogeton oakesianus, Potamogeton obtusifolius, Potamogeton perfoliatus, Potamogeton praelongus, Potamogeton pulcher, Potamogeton pusillus, Potamogeton richardsonii, Potamogeton robbinsii, Potamogeton spirillus, Potamogeton strictifolius, Potamogeton vaseyi, Potamogeton zosteriformis, Ranunculus aquatilis, Ranunculus flammula, Ranunculus longirostris, Ranunculus subrigidus, Ranunculus trichophyllus, Rhododendron columbianum, Ribes aureum, Ricciocarpus natans, Rorippa palustris, Rorippa sphaerocarpa, Rorippa sylvestris, Sagittaria cuneata, Sagittaria latifolia, Sagittaria rigida, Salix exigua, Salix geyeriana, Salix scouleriana, Scheuchzeria palustris, Schoenoplectus acutus, Schoenoplectus subterminalis, Schoenoplectus tabernaemontani, Scirpus cyperinus, Scirpus microcarpus, Scutellaria galericulata, Senecio hydrophiloides, Sium suave, Sparganium angustifolium, Sparganium emersum, Sparganium eurycarpum, Sparganium fluctuans, Sparganium natans, Sphagnum, Spiraea douglasii, Spirodela polyrhiza, Stuckenia filiformis, Stuckenia pectinata, Stuckenia* *vaginata, Torreyochloa pallida, Trapa natans, Triglochin maritima, Triglochin scilloides, Typha angustifolia, Typha latifolia, Utricularia gibba, Utricularia intermedia, Utricularia macrorhiza, Utricularia minor, Utricularia purpurea, Vaccinium uliginosum, Vallisneria americana, Veronica americana, Veronica anagallis-aquatica, Veronica catenata, Veronica scutellata, Wolffia brasiliensis, Zannichellia palustris, Zizania aquatica, Zizania palustris.*

3.4.1.17. *Potamogeton nodosus* Poir. is a perennial, which inhabits intertidal, non-tidal, or subtidal, fresh to brackish waters (salinity to 1.3‰) in bayous, borrow pits (marshy), canals (irrigation), channels (backwater), coves, depressions, ditches (drainage, irrigation), estuaries, floodplains, gravel pits, impoundments, lagoons, lakes (eutrophic, playa), marshes, mudflats, oxbows, ponds (artificial, ephemeral, marsh, mill, quarry, sand, sinkhole, stock watering, vernal), pools (alkaline, plunge, rock basin), reservoirs, rivers, sandpits, sloughs, spillways, streams, and swamps (floodplain) at elevations to 3078 m. The plants grow in clear to turbid, alkaline (pH: 6.3–9.5; $\bar{x}$ = 7.8; total alkalinity [as $CaCO_3$]: 5.0–312.0 mg l^{-1}; $\bar{x}$ = 77.8–160.0 mg l^{-1}; conductivity 55–660 µmhos cm^{-1} [@ 25°C]) waters (to 30°C) at depths from 0.10 to 2.1 m. They tolerate slow to swift currents and often are observed in lotic waters. However, they also are commonly found as diminutive "mudflat forms," when stranded on exposed or desiccating sediments during drawdown conditions (which accounts for many of the more helophytic associates reported). Exposures can range from full sunlight to partial shade. The sediments have an organic matter content averaging about 5.6% and include alluvium (clayey), clay, clay loam muck, gravel (fine, glacial), limestone, loamy clay, loamy muck, marl, muck, mucky sand, mud, muddy sandy gravel, rock, rocky clay loam, rocky silt, sand (coarse, glacial, granite), sandstone, sandy clay, sandy silt (fine), silt (alkaline), and silty sand. Flowering and fruiting occur from May to October. The flowers are borne in monomorphic spikes, which are raised above the water surface on an erect peduncle. A relatively high pollen:ovule ratio (mean: 26,000:1) indicates that the plants (presumably wind pollinated) are probably outcrossed. Seed production can be high, with as many as 11,800 fruits (1985 m^{-2}) produced in one season by plants growing within a 5.9 m^2 plot. The inner fruit wall (mesocarp) is parenchymatous and contains air spaces, which confer some degree of buoyancy and facilitate water dispersal. Once dispersed, the seeds are physically dormant and require a period of cold stratification (dark storage in cold water [4°C] for 12 weeks) followed by incubation in the light (at 22°C) to induce germination. Seeds buried by more than 1 cm of substrate do not germinate. The seedlings (and adult plants) thrive on nutrient-rich clay and organic substrates and do not persist on poor mineral substrates. Seeds ingested by Eurasian teal (Aves: Anatidae: *Anas crecca*) remained highly viable ($\bar{x}$ = 77%) and germinated well ($\bar{x}$ = 83%) after their excretion (retention time: $\bar{x}$ = 4.8 hr); both viability and germination rates were higher than uningested control seeds. Consequently, it is evident that endozoic transport by waterfowl and other birds is a likely means of long-distance dispersal in this species.

Although there are discrepancies in the literature concerning whether the plants reproduce sexually or vegetatively most often, either means can become prevalent depending on the specific environmental conditions that the plants experience in any given year. Under favorable conditions, sexual reproduction and seed set can be prolific; vegetative reproduction dominates otherwise. Biomass allocation to the rhizome system increases proportionally with sediment fertility. However, the addition of more than 5% organic matter to sediments has been shown to greatly suppress plant growth. The plants develop axillary shoot turions ("winter buds" to some), rhizome tubers and which function as vegetative propagules and provide a means of withstanding high environmental stress and continuous or cyclic disturbances. As many as 2,130 tubers (358 m^{-2}) and 64 axillary turions have been produced in one season by plants growing in a 5.9 m^2 plot. Under shorter day lengths (<12 hr), tuber production increases (for roughly 8 weeks) to a maximum of about five per plant, which can account for 27% of the total plant dry mass. They are strongly dormant when mature. The propagules germinate from late winter to spring (February–April; 298 degree-days to 50% sprouting) as the water temperatures rise between 10°C and 20°C (optimally from 15°C to 20°C); germination rates often are 100%. Their germination is inhibited in total darkness or by high levels (>2 × 10^{-5} M) of ABA (abscisic acid) but can be enhanced by exposure to cold temperatures followed by high temperatures (e.g., 3 days @ 32°C), or by exposure to 1000 ppm IAA (indole-3-acetic acid) for 18 hr. They are killed upon exposure to low concentrations (2–5% by volume) of acetic acid. The vegetative propagules are transported locally by water currents, with an average of six moving in and out of a given tidal site each day. Although the plants are heterophyllous, they often shed their submersed foliage and persist entirely as a floating-leaved form. The different foliage types are not only morphologically heterophyllous but differ physiologically as well. Flavonoid compounds (which protect against UV damage) are common in the floating leaves but absent in the submersed foliage. A mitochondrial electron transport system provides the floating leaves with a higher photosynthetic efficiency and a better ability to withstand photoinhibition than the submerged leaves. The lower catalase activity in submersed leaves leads to substantially higher (almost twofold) concentrations of H_2O_2, which is thought to function as a deterrent to herbivory and pathogens. The floating leaves can develop into a dense surface canopy, which heavily shades the waters beneath. Experimentally, the plants have attained a substantially higher biomass when growing in mixed cultures (5 g m^{-2}) than in monocultures (2.9 g m^{-2}). The highest shoot biomass occurs under high light (1500 µE m^{-2} s^{-1}) and high temperature (32°C) conditions, whereas the highest rhizome biomass occurs under high light (1500 µE m^{-2} s^{-1}) and more moderate temperatures (20°C). In the field, total plot biomass can reach considerably higher levels ($\bar{x}$: to 200 g m^{-2}), with about ⅔ of it allocated to underground tissues. Invertebrate herbivory can reduce plant biomass by as much as 40%. **Reported associates:** *Acorus calamus, Alisma triviale, Alopecurus aequalis, Alternanthera philoxeroides,*

Amorpha fruticosa, Andropogon glomeratus, Azolla filiculoides, Beckmannia syzigachne, Bidens beckii, Bolboschoenus fluviatilis, Brasenia schreberi, Cabomba caroliniana, Callitriche heterophylla, Callitriche palustris, Carex athrostachya, Carex buxbaumii, Carex comosa, Carex diandra, Carex emoryi, Carex hystericina, Carex stipata, Carex utriculata, Carex vesicaria, Cephalanthus occidentalis, Ceratophyllum demersum, Ceratophyllum echinatum, Ceratopteris thalictroides, Chara contraria, Chara vulgaris, Cicuta bulbifera, Conyza canadensis, Crassula aquatica, Cyperus bipartitus, Dulichium arundinaceum, Echinodorus berteroi, Eclipta prostrata, Egeria densa, Eichhornia crassipes, Elatine minima, Eleocharis acicularis, Eleocharis equisetoides, Eleocharis macrostachya, Eleocharis montevidensis, Eleocharis ovata, Eleocharis palustris, Eleocharis rostellata, Elodea canadensis, Elodea nuttallii, Epilobium ciliatum, Eriocaulon aquaticum, Eupatorium perfoliatum, Fuirena simplex, Glyceria borealis, Glyceria grandis, Helenium elegans, Helenium virginicum, Heteranthera dubia, Hippuris vulgaris, Hydrilla verticillata, Hydrodictyon, Hygrophila lacustris, Hypericum mutilum, Isoetes melanopoda, Iva annua, Juncus effusus, Juncus saximontanus, Juncus torreyi, Justicia americana, Leersia oryzoides, Lemna minor, Lemna trisulca, Leptodictyum riparium, Leucospora multifida, Limnophila sessiliflora, Lobelia cardinalis, Lobelia dortmanna, Ludwigia palustris, Ludwigia peploides, Ludwigia repens, Lycopus uniflorus, Lysimachia thyrsiflora, Lythrum salicaria, Marsilea vestita, Mentha arvensis, Mimulus guttatus, Myosotis laxa, Myriophyllum aquaticum, Myriophyllum heterophyllum, Myriophyllum sibiricum, Myriophyllum spicatum, Myriophyllum verticillatum, Najas flexilis, Najas guadalupensis, Nelumbo lutea, Nitella flexilis, Nuphar advena, Nuphar variegata, Nymphaea odorata, Nymphoides cordata, Panicum virgatum, Paspalum distichum, Persicaria amphibia, Persicaria coccinea, Persicaria hydropiperoides, Persicaria lapathifolia, Phalaris arundinacea, Phyla lanceolata, Pistia stratiotes, Polypogon monspeliensis, Pontederia cordata, Potamogeton alpinus, Potamogeton amplifolius, Potamogeton berchtoldii, Potamogeton clystocarpus, Potamogeton crispus, Potamogeton diversifolius, Potamogeton epihydrus, Potamogeton foliosus, Potamogeton friesii, Potamogeton gramineus, Potamogeton hillii, Potamogeton illinoensis, Potamogeton natans, Potamogeton perfoliatus, Potamogeton praelongus, Potamogeton pusillus, Potamogeton richardsonii, Potamogeton robbinsii, Potamogeton spirillus, Potamogeton strictifolius, Potamogeton tennesseensis, Potamogeton vaseyi, Potamogeton zosteriformis, Ranunculus aquatilis, Ranunculus sceleratus, Ranunculus trichophyllus, Riccia fluitans, Rorippa palustris, Rumex crispus, Rumex salicifolius, Sagittaria cuneata, Sagittaria graminea, Sagittaria kurziana, Sagittaria latifolia, Sagittaria platyphylla, Sagittaria rigida, Salix exigua, Salix interior, Salix nigra, Salvinia, Schoenoplectus acutus, Schoenoplectus lacustris, Schoenoplectus subterminalis, Schoenoplectus tabernaemontani, Scirpus cyperinus, Scutellaria lateriflora, Sesbania, Sium suave, Sparganium americanum, Sparganium angustifolium, Sparganium emersum, Sparganium eurycarpum, Sparganium fluctuans, Sphagnum, Spirodela polyrhiza, Stuckenia filiformis, Stuckenia pectinata, Symphyotrichum praealtum, Symphyotrichum subulatum, Typha angustifolia, Typha domingensis, Typha latifolia, Utricularia gibba, Utricularia macrorhiza, Vallisneria americana, Wolffia brasiliensis, Xanthium strumarium, Zannichellia palustris, Zizania aquatica, Zizania palustris, Zizania texana.

3.4.1.18. ***Potamogeton oakesianus* J.W. Robbins** is a perennial, which occurs in quiescent waters and along the sheltered shorelines of bogs, ditches (roadside), lakes (marshy), marshes, mudflats, ponds (beaver, bog, pothole, sinkhole), pools (beach, bog), springs, streams, and swamps at elevations to 555 m. The plants grow primarily in acidic, soft water habitats (pH: 4.3–8.9; $\bar{x} = 6.5$; total alkalinity [as $CaCO_3$]: 2.0–220.0 mg l^{-1}; $\bar{x} = 8.1$–45 mg l^{-1}; conductivity to 215 µmhos cm^{-1} [@ 25°C]) in shallow waters at depths ranging from 0.01 to 1.4 m (average about 0.6 m). The substrates have been described as boulders, gravel, muck, mud, peat (fibrous), rock (peaty), and sand (fine). There is only scarce life history information available for this species. Flowering and fruiting occur from June to August, but otherwise there are no details available on the floral or seed ecology. The plants are heterophyllous and are found always with the floating leaves present. They assimilate by direct CO_2 uptake (from the floating leaves) but are believed to have at least some minor ability to use HCO_3^- as a dissolved inorganic carbon source. They propagate vegetatively by means of rhizomes. The belowground to total biomass ratio is approximately 0.55–0.56. Some errors in reported ecological data surely exist due to frequent misidentifications involving *Potamogeton epihydrus*, *P. natans*, *P. nodosus*, and other pondweed species. **Reported associates:** *Bidens beckii, Brasenia schreberi, Calamagrostis canadensis, Calla palustris, Callitriche heterophylla, Carex aquatilis, Cephalanthus occidentalis, Ceratophyllum echinatum, Comarum palustre, Decodon verticillatus, Dulichium arundinaceum, Elatine minima, Eleocharis acicularis, Eleocharis palustris, Eleocharis robbinsii, Equisetum fluviatile, Eriocaulon aquaticum, Fissidens fontanus, Fontinalis antipyretica, Fontinalis hypnoides, Fontinalis novae-angliae, Glyceria borealis, Glyceria canadensis, Hottonia inflata, Hypericum anagalloides, Isoetes acadiensis, Isoetes tuckermanii, Juncus militaris, Juncus pelocarpus, Lemna turionifera, Lobelia dortmanna, Ludwigia palustris, Lycopus, Lysimachia terrestris, Myriophyllum farwellii, Myriophyllum heterophyllum, Myriophyllum humile, Myriophyllum sibiricum, Myriophyllum tenellum, Najas flexilis, Najas gracillima, Nasturtium officinale, Nitella flexilis, Nitella tenuissima, Nuphar polysepala, Nuphar variegata, Nymphaea odorata, Nymphoides cordata, Persicaria amphibia, Pontederia cordata, Potamogeton amplifolius, Potamogeton berchtoldii, Potamogeton bicupulatus, Potamogeton confervoides, Potamogeton diversifolius, Potamogeton epihydrus, Potamogeton foliosus, Potamogeton gramineus, Potamogeton natans, Potamogeton pusillus, Potamogeton robbinsii, Potamogeton spirillus, Potamogeton strictifolius, Rhododendron groenlandicum, Sagittaria cristata, Sagittaria* *graminea, Sagittaria latifolia, Schoenoplectus subterminalis, Schoenoplectus tabernaemontani, Sparganium americanum, Sparganium angustifolium, Sparganium fluctuans, Sparganium natans, Sphagnum, Sphagnum cuspidatum, Sphagnum subsecundum, Spiraea douglasii, Spirodela polyrhiza, Stuckenia pectinata, Torreyochloa pallida, Utricularia geminiscapa, Utricularia gibba, Utricularia intermedia, Utricularia macrorhiza, Utricularia purpurea, Utricularia radiata, Utricularia resupinata, Vallisneria americana, Warnstorfia exannulata, Zizania palustris.*

3.4.1.19. ***Potamogeton obtusifolius*** **Mert. & W.D.J. Koch** is an annual (vegetative) or perennial, which grows in bays, bogs, brooks, lakes (brownwater), pits (clay, gravel), ponds (beaver, glacial, ice, log, roadside), pools, potholes, rivers (backwaters, slow-moving), sloughs (backwater), springs, and streams (boggy, shallow, sluggish) at elevations to 2456 m. Although often associated with cold water, oligotrophic conditions, the plants have quite broad ecological tolerances, being tolerant of at least low salinity levels and also warmer, eutrophic sites. The substrates are characterized as clay (marine), granitic, gravel, muck (sandy silty), mud (loose), organic, peat (pulpy), rock, sand (fine, rocky), sandy silt, silt (over sandstone bedrock), and silty sand. The habitats include shallow (0.03–1.8 m depth), still to slow-moving waters (but with a significantly higher occurrence in lotic habitats) of moderate alkalinity (typically <67.1 mg l^{-1}), often with fairly high levels of dissolved nutrients and organic matter (pH: 6.5–10.0; $\bar{x}$ =7.3–7.5; total alkalinity [as CaCO$_3$]: 13.5–165 mg l^{-1}; $\bar{x}$ =40.0–58.3 mg l^{-1}; conductivity to 315 µmhos cm^{-1} [@ 25°C]). Exposures tend to receive full sunlight in waters of high to moderate clarity (e.g., Secchi depth: 1.8 m); however, experiments using European plants have shown them to possess a high degree of shade tolerance. Flowering and fruiting occur from July to September. The flowers arise from an erect spike, which normally is emergent above the water surface. The reproductive biology remains enigmatic. There typically is high seed set in flowers on plants that remain submersed in deep water, which reportedly is a consequence of diplospory, a type of gametophytic apomixis. It has not been determined whether the aerial flowers undergo normal sexual reproduction. In any case, seed production is prolific. The fruits are somewhat buoyant but remain afloat for less than a week. Water is the primary vector for local seed dispersal. The seeds are dormant and require a period of cold stratification to induce germination. Experimentally, about 16% of seeds germinated after 2 months of storage in water at room temperature, whereas much higher germination (>75%) was obtained for seeds stored for 2–3 months in water at 1°C–3°C. Natural seed germination has been observed in the field during late November on the bottom mud beneath an ice-covered pond. The fruits also have been found in the guts of several species of waterfowl (Aves: Anatidae) and have remained viable (positive tetrazolium test) after passage through the digestive tracts of carp (Vertebrata: Teleostei: Cyprinidiae: *Cyprinus carpio*), coots (Aves: Rallidae: *Fulica atra*), and mallard ducks (Aves: Anatidae: *Anas platyrhynchos*), which potentially could transport and disperse them over longer distances via endozoic transport. Yet, despite their high seed production, most of the plants appear to arise annually from turions, which are produced in abundance from July to November. Experiments conducted in England indicate that the plants do not reproduce sexually at depths greater than 3.5 m, where they establish by means of the turions. Because the plants lack rhizomes, their shoots are particularly susceptible to anchorage dislodgement when subjected to hydraulic forces due to waves and water currents. **Reported associates:** *Bidens beckii, Bidens cernuus, Brasenia schreberi, Callitriche palustris, Carex aquatilis, Carex maritima, Ceratophyllum demersum, Chara vulgaris, Drepanocladus, Echinochloa muricata, Elatine minima, Eleocharis acicularis, Eleocharis erythropoda, Elodea canadensis, Elodea nuttallii, Epilobium coloratum, Equisetum fluviatile, Eriocaulon aquaticum, Fontinalis antipyretica, Heteranthera dubia, Hippuris vulgaris, Hydrodictyon reticulatum, Isoetes lacustris, Juncus militaris, Juncus pelocarpus, Lemna minor, Lemna trisulca, Lobelia dortmanna, Lycopus, Lyngbya aestuarii, Menyanthes trifoliata, Mimulus ringens, Myriophyllum alterniflorum, Myriophyllum farwellii, Myriophyllum heterophyllum, Myriophyllum sibiricum, Myriophyllum spicatum, Myriophyllum tenellum, Najas flexilis, Nasturtium officinale, Nitella tenuissima, Nuphar polysepala, Nuphar variegata, Nymphaea odorata, Nymphaea tetragona, Nymphoides cordata, Persicaria amphibia, Pontederia cordata, Potamogeton alpinus, Potamogeton amplifolius, Potamogeton berchtoldii, Potamogeton diversifolius, Potamogeton epihydrus, Potamogeton foliosus, Potamogeton friesii, Potamogeton gramineus, Potamogeton natans, Potamogeton perfoliatus, Potamogeton praelongus, Potamogeton pusillus, Potamogeton richardsonii, Potamogeton robbinsii, Potamogeton spirillus, Potamogeton strictifolius, Potamogeton vaseyi, Potamogeton zosteriformis, Ranunculus aquatilis, Ranunculus flammula, Ranunculus longirostris, Ranunculus trichophyllus, Rhamnus alnifolia, Sagittaria cuneata, Sagittaria graminea, Sagittaria latifolia, Schoenoplectus subterminalis, Schoenoplectus tabernaemontani, Sium suave, Sparganium angustifolium, Sparganium emersum, Sparganium fluctuans, Sparganium natans, Spirodela, Stuckenia filiformis, Torreyochloa pallida, Utricularia gibba, Utricularia intermedia, Utricularia macrorhiza, Utricularia purpurea, Utricularia resupinata, Vallisneria americana, Wolffia columbiana, Zizania palustris.*

3.4.1.20. ***Potamogeton ogdenii*** **Hellq. & R.L. Hilton** is not recognized in this treatment (see *Systematics* below) and is given no further consideration here.

3.4.1.21. ***Potamogeton perfoliatus*** **L.** is a perennial, which inhabits fresh to mesohaline (salinity: 0–12‰) backwaters, bays, ditches, estuaries, lakes, reservoirs, rivers, and streams at elevations to 468 m. Although most North American records are from relatively shallow waters (e.g., 0.3–1.2 m depth), European plants have been found at depths up to 6.5 m. Water clarity can range from clear to turbid (Secchi depth: 0.3–2.2 m) but the minimum light level necessary to sustain survival has been estimated to be above 11% of ambient (i.e., surface) conditions. The plants respond to lower light levels by increasing their chlorophyll *a* content and overall photosynthetic efficiency; however, they are not shade tolerant and require a relatively high summer (July) irradiance level (487.5 µE m^{-2} s^{-1}) to saturate photosynthesis. The plants also can withstand warm water temperatures up to 35°C but falter at temperatures above 30°C, likely due to the non-induction of *HSFA2a2*, a heat shock protein gene expressed by more thermotolerant pondweed species; water temperatures of 45°C are lethal to the plants. The substrates have been described as loamy sand, mud, sand, and silty clay. The waters typically are fairly alkaline (pH: 6.7–9.8; $\bar{x}$ =7.7; total alkalinity [as CaCO$_3$]: 6.1–167.8 mg l^{-1}; $\bar{x}$ =36.3–47.2 mg l^{-1}). Individuals are homophyllous (broad-leaved) and lack floating leaves; however, there have been rare mudflat forms reported, which develop coriaceous floating-type foliage when stranded on sandy sediments as the waters recede. Normally, the flowering spikes emerge above the water surface from the terminus of the submersed shoots. Flowering and fruiting occur from June to November, with the incidence of flowering decreasing as salinity levels increase. Fruit production can be prolific. The plants average 2.4 inflorescences and 115 seeds per stem, with potential yields of up to 10,100 seeds kg^{-1}. The fruits are light (averaging 1.4 mg) and are easily dispersed locally by slight currents (5–10 cm s^{-1}) or by wave action. The frequent adherence of sand grains to the pericarp and seedling rootlets is believed to help reduce their post-germination displacement. Although some reports indicate that the fruits can float for up to 6 months, they have a settling velocity of 3 cm s^{-1}, which increases (to 3.5–6.5 cm s^{-1}) in currents above 8 cm s^{-1}. Fruits released at 50 cm above the substrate surface (in a 20 cm s^{-1} current) had an average dispersal distance of only 2.2 m. Substantial subsurface germination (e.g., 37.6%) can occur at planting depths to 3 cm but the seeds must lie on the sediment surface before they can establish successfully. Fewer than 2% of seedlings emerge from seeds buried at depths of 2 cm or more. Successful germination has been achieved for seeds dark-stored for 6–8 months in aerated, deionized water at 4°C–5°C and then incubated in deionized water at 21°C–23.8°C for 4–9 days under 12–24 hr fluorescent illumination (70–100 µmol m^{-2} s^{-1}). The fruits have been stored successfully for prolonged periods when kept under cold (4°C), aerated, conditions. Germination is reduced under conditions of elevated salinity (e.g., 15‰), even when they have been stored at the same salinity. Longer distance dispersal also can occur via exozoic transport of entire plants (up to 1 m in length), which can become draped around waterfowl such as mallard ducks (Aves: Anatidae: *Anas platyrhynchos*). Clonal growth occurs by means of an extensive rhizome system, which can expand by as much as 63 cm year^{-1}, reaching combined plant lengths up to 11 m m^{-2} annually. The rhizomes proliferate as an integrated clonal network of ramets, which share resources. When young ramets are strongly shaded, they direct additional resources to subsequently developing ramets, which encourages stronger growth and expansion of the plants beyond the limits of heavily shaded clumps. Vegetative reproduction occurs effectively by shoot and rhizome fragmentation. From 72%

to 100% of shoot and rhizome fragments regenerate, resulting in highly successful colonization rates (65–92%). Even at sites where the plants are abundant, seedlings typically are rare or absent with most propagation apparently involving vegetative means. Attempted restoration projects using seeds also have failed where transplants using stem cuttings have succeeded. The plants oxygenate the rhizosphere by releasing photosynthetically derived oxygen (O$_2$) from the rhizomes at a calculated rate of 120 µmol g^{-1} DM h^{-1}. The enhanced sediment oxygen levels enhance the growth of denitrifying bacterial communities, which reduce nitrate (NO$_3^-$) to ammonium (NH$_4^+$), the preferred form of nitrogen taken up by the plants. Nitrogen uptake is highest via the root system in spring but shifts more toward shoot uptake as the plants grow larger throughout summer. Productivity estimates differ markedly among sites of different fertility but have averaged more than 235 mg DM day^{-1} in some cases; from 64% to 80% of the total biomass is allocated to shoot production. The root hairs are colonized by VA-mycorrhizal Fungi. These plants are categorized as a productive, competitive canopy-forming species. High CO$_2$/low pH conditions that arise from environmental acidification decrease the production of protective phenolic compounds in the plants, thereby subjecting them to increased rates of herbivory. Numerous specimens of *P. richardsonii* have been misidentified as *P. perfoliatus*; consequently, some of the listed associates may be misapplied. The following compilation is brief because the plants frequently occur as monocultures with few associates, especially in the more brackish sites. **Reported associates:** *Bolboschoenus fluviatilis, Cabomba caroliniana, Callitriche, Ceratophyllum demersum, Elatine minima, Eleocharis, Elodea canadensis, Eriocaulon aquaticum, Gratiola aurea, Heteranthera dubia, Juncus pelocarpus, Myriophyllum spicatum, Najas flexilis, Najas guadalupensis, Najas minor, Potamogeton alpinus, Potamogeton amplifolius, Potamogeton crispus, Potamogeton epihydrus, Potamogeton foliosus, Potamogeton friesii, Potamogeton gramineus, Potamogeton natans, Potamogeton obtusifolius, Potamogeton praelongus, Potamogeton pusillus, Potamogeton robbinsii, Potamogeton spirillus, Potamogeton zosteriformis, Ruppia maritima, Sagittaria, Stuckenia pectinata, Typha angustifolia, Utricularia macrorhiza, Vallisneria americana, Zannichellia palustris.*

3.4.1.22. *Potamogeton polygonifolius* **Pourr.** (= *P. oblongus*) is a perennial, which grows in brooks, coves, ponds, pools (ephemeral), and streams at elevations to 100 m. The plants inhabit relatively shallow, oligotrophic–mesotrophic, acidic waters (pH: 5.8–6.2; $\bar{x}$ =6.0), which occur over muddy substrates. They are heterophyllous, producing large, well-developed floating leaves (and lesser developed submersed foliage) and are not very shade tolerant. The leaf morphology is variable in form depending on the water depth, and short-statured plants can develop on shores when the water levels recede. Flowering and fruiting occur from July to September. The flowers are wind pollinated and protogynous, with the anthers being enclosed behind the perianth at the start of anthesis, and subsequently exposed by the spreading structures during the male phase. Vegetative reproduction occurs by means of a well-developed rhizome network, from which overwintering buds are produced. The plants reportedly are non-mycorrhizal. There is little additional life history information available for the North American populations, with most of the literature pertaining to European sites. The New World occurrences are believed to have arisen via the transport of viable seeds from migrating European birds. **Reported associates:** *Hippuris vulgaris, Persicaria hydropiperoides, Potamogeton epihydrus, Schoenoplectus pungens.*

3.4.1.23. ***Potamogeton praelongus*** **Wulfen** is a perennial, which occurs in bogs, gravel pits, lakes (oxbow), ponds, reservoirs, rivers, sloughs, and streams (slow-moving) at elevations to 3231 m. The exposures usually are described as receiving full sunlight. This species generally is regarded as occurring in deeper water (optimal depth around 2.5 m) but its depth range can extend from 0.1 to 7.0 m. Ecological models have predicted that its zero equilibrium biomass would not be reached until depths from 7 to 8 m. The plants are categorized as shade tolerant; however, they rarely are light limited because of their ability to produce very long shoots (2–4 m in length by mid-June). The substrates most often are soft, with growth and survival being much higher on mud than on sand. However, the plants have been found on a variety of substrate types including gravel, loam, marl, marly sand, muck (organic, peaty), mucky gravel, mucky sand, mucky silt, mud, muddy gravel, peat (fibrous, pulpy), sand, sandy cobbles, sandy gravel, sandy loam, sandy muck, sandy silt, silt, and silty muck. This species occurs in oligotrophic to mesotrophic waters and is known to decline or even disappear from sites that have become eutrophic. However, the reasons for its susceptibility to eutrophication are not obvious. This is a "northern" species geographically and is well adapted to survive in cold water conditions. The plants die back in shallower lakes as fall approaches but can overwinter as whole plants beneath the ice at depths exceeding 3 m, while maintaining a photosynthetic rate (at 2°C) of 0.25 mg C g^{-1} DM hr^{-1}. Yet, despite their cold tolerance, they also have been found to thrive in the warmer waters (to 27°C) of cooling ponds. The plants are described as turbidity intolerant, but Secchi depths recorded in their habitats span a fairly wide range (e.g., 0.7–5.9 m) of depths. The affinity for northern localities has led to its perception as having narrow ecological tolerances; yet, habitat data indicate that this species actually occurs under quite a broad spectrum of ecological conditions (pH: 5.6–10.0; $\bar{x}$ =7.7–8.5; total alkalinity [as mg l^{-1} $CaCO_3$]: 2.0–307.7; $\bar{x}$ =47.2–101.0; specific conductivity [@ 25°C]: 40–480 µmhos cm^{-1}). Although the leaves have an anatomy similar to that (i.e., "Kranz anatomy") of C4 species, metabolic studies indicate that this is a C3 species photosynthetically. Flowering occurs from June to August with fruiting from July to November. In deeper waters, the plants will not flower unless the shoots are in close proximity to the water surface. Pollination probably occurs by wind, but further investigations of the floral biology are necessary to evaluate the possibility of autogamy or apomixis. The spikes are monomorphic, emergent, and each capable of producing up to 108 fruits. Ripening seeds acquire deep physiological dormancy and must undergo a period of

cold stratification (or scarification) to induce their germination. The most successful experimental rates (18–32.7% germination) have been obtained by subjecting the fruits to 2–12 months of cold storage (1°C–8°C) in water followed by 14 days incubation at 21°C–28°C (optimal). Physical scarification has been found to reduce rather than promote germination. A period of dry storage (at 21°C) preceding the wet, cold scarification also enhances germination rates. Freezing is harmful to the seeds and reduces their germination considerably. Although the plants can flower and fruit prolifically, most reproduction occurs by vegetative means involving the extensive rhizome system and its tuberous buds (sometimes referred to as "turions"). Studies of European populations have shown very low levels of genetic (i.e., AFLP) variation and no recruitment via seeds, with reproduction being entirely clonal. The fruits probably are involved primarily in long-distance dispersal and initial colonization of sites. The starch content is highest in early June (e.g., 147.0 mg g^{-1}) and declines (to 90.1 mg g^{-1}) as growth proceeds. The plants have achieved biomass values up to 235 g DM m^{-2} (above ground); their biomass is proportional to the degree of sediment fertility. The root hairs are colonized by VA-mycorrhizal Fungi. Unlike some pondweeds, the wounded stems heal by the production of thickened, suberized cells. **Reported associates:** *Acorus calamus, Asclepias incarnata, Bidens beckii, Brasenia schreberi, Butomus umbellatus, Cabomba caroliniana, Callitriche, Ceratophyllum demersum, Ceratophyllum echinatum, Chara contraria, Chara globularis, Chara vulgaris, Elatine minima, Eleocharis acicularis, Eleocharis palustris, Elodea canadensis, Elodea nuttallii, Eriocaulon aquaticum, Fontinalis antipyretica, Fontinalis novae-angliae, Gratiola aurea, Heteranthera dubia, Hippuris vulgaris, Isoetes echinospora, Isoetes lacustris, Juncus pelocarpus, Lemna minor, Lemna trisulca, Lobelia dortmanna, Lychnothamnus barbatus, Menyanthes trifoliata, Myriophyllum alterniflorum, Myriophyllum hippuroides, Myriophyllum sibiricum, Myriophyllum spicatum, Myriophyllum tenellum, Myriophyllum verticillatum, Najas canadensis, Najas flexilis, Najas guadalupensis, Najas marina, Nitella flexilis, Nitella opaca, Nitellopsis obtusa, Nuphar polysepala, Nuphar variegata, Nymphaea odorata, Persicaria amphibia, Pontederia cordata, Potamogeton alpinus, Potamogeton amplifolius, Potamogeton berchtoldii, Potamogeton crispus, Potamogeton diversifolius, Potamogeton epihydrus, Potamogeton foliosus, Potamogeton friesii, Potamogeton gramineus, Potamogeton illinoensis, Potamogeton natans, Potamogeton nodosus, Potamogeton obtusifolius, Potamogeton perfoliatus, Potamogeton pusillus, Potamogeton richardsonii, Potamogeton robbinsii, Potamogeton spirillus, Potamogeton strictifolius, Potamogeton vaseyi, Potamogeton zosteriformis, Ranunculus aquatilis, Ranunculus flammula, Ranunculus longirostris, Ranunculus subrigidus, Ranunculus trichophyllus, Ricciocarpus, Sagittaria cuneata, Sagittaria graminea, Sagittaria rigida, Schoenoplectus acutus, Schoenoplectus subterminalis, Schoenoplectus tabernaemontani, Scirpus cyperinus, Sparganium angustifolium, Sparganium eurycarpum, Sparganium fluctuans, Sparganium natans,*

Spirodela polyrhiza, Stuckenia filiformis, Stuckenia pectinata, Subularia aquatica, Typha angustifolia, Typha latifolia, Utricularia geminiscapa, Utricularia gibba, Utricularia intermedia, Utricularia macrorhiza, Utricularia minor, Utricularia purpurea, Vallisneria americana, Wolffia brasiliensis, Wolffia columbiana, Zannichellia palustris, Zizania palustris.

3.4.1.24. ***Potamogeton pulcher* Tuck.** is a perennial, which inhabits bottomlands (inundated), Carolina bays, channels (humic stained), ditches, impoundments (blackwater), lakes, marshes (beaver), ponds (artificial, beaver, coastal, depression, floodplain, sinkhole), pools (floodplain, swamp), potholes, rivers, sandbars, sloughs, streams (blackwater), and swamps (pasture) at elevations to 893 m. The plants occur in full sunlight to shaded exposures and most often are reported from still or slow-moving, shallow (e.g., 0.15–0.50 m) waters but occasionally are found in faster currents at depths up to 1.8 m. The substrates are described as gravel, limestone, muck, mud, peat (soft), and sand (Sparta). This species is found primarily in soft, acidic waters of low alkalinity (pH: 5.6–7.5; $\bar{x} = 6.6$–6.9; total alkalinity [as mg l^{-1} CaCO$_3$]: 3.0–46.4; $\bar{x} = 10.3$–15.3) but sometimes in more alkaline sites in the western portion of its range. Flowering and fruiting occur from March to September. The flowers are emergent on monomorphic spikes and can be numerous. There are no detailed accounts of the reproductive biology, but pollination occurs presumably by wind. The mature seeds are physiologically dormant and are slow to break dormancy. The highest germination success (15%) has been achieved by stratification of seeds in cold (0°C–5°C) tap water for 13.5 months, followed by incubation (at room temperature) for 30–38 months under a 16/8 hr day/night fluorescent light regime. Seeds that were cold stratified for 20.5 months began to germinate (2.4%) after 1 month of incubation as described above. In cold water storage (0°C–5°C), the seeds have remained viable for a period somewhere between 20 and 40 months. A seed bank of unknown duration is produced at densities of up to 144 seeds m^{-2}; natural germination is highest while under flooded conditions. Other than local dispersal by water, the mechanisms of long-distance dispersal have not been elucidated; however, their clarification might help to explain the sporadic occurrences of the plants throughout their range. The plants are heterophyllous (broad-leaved), a feature which enabled them to survive in areas of higher turbidity. Vegetative reproduction occurs by means of slender rhizomes. The plants can achieve a total biomass of at least 57.8 g DM m^{-2}. Because of its proclivity for shallow waters, the following list of associated species includes a broad diversity of hydrophytic as well as helophytic (i.e., wetland) species. **Reported associates:** *Acer rubrum, Alisma subcordatum, Apios americana, Arthraxon hispidus, Asclepias incarnata, Azolla cristata, Azolla filiculoides, Batrachospermum, Bidens, Boehmeria cylindrica, Brasenia schreberi, Cabomba caroliniana, Callitriche palustris, Cardamine longii, Carex aquatilis, Carex atlantica, Carex comosa, Carex lupulina, Carex lurida, Carex stricta, Carex utriculata, Cephalanthus occidentalis, Ceratophyllum demersum, Ceratophyllum echinatum, Chara vulgaris,*

Conyza canadensis, Cuscuta gronovii, Cyperus pseudovegetus, Cyperus refractus, Decodon verticillatus, Dulichium arundinaceum, Echinochloa muricata, Eleocharis acicularis, Eleocharis flavescens, Eleocharis obtusa, Eleocharis quadrangulata, Eleocharis robbinsii, Eleocharis tuberculosa, Elodea canadensis, Elodea nuttallii, Eupatorium, Fontinalis, Fraxinus pennsylvanica, Galium obtusum, Galium tinctorium, Galium trifidum, Glyceria canadensis, Glyceria obtusa, Glyceria striata, Heteranthera reniformis, Hibiscus moscheutos, Hydrolea quadrivalvis, Hypericum mutilum, Iris pseudacorus, Iris versicolor, Itea virginica, Juncus canadensis, Juncus coriaceus, Juncus effusus, Juncus marginatus, Juncus repens, Leersia oryzoides, Lemna minor, Liquidambar styraciflua, Ludwigia alternifolia, Ludwigia palustris, Ludwigia peploides, Ludwigia sphaerocarpa, Mayaca fluviatilis, Microstegium vimineum, Mikania scandens, Murdannia keisak, Myriophyllum aquaticum, Myriophyllum heterophyllum, Myriophyllum heterophyllum × Myriophyllum laxum, Myriophyllum humile, Myriophyllum laxum, Myriophyllum pinnatum, Myriophyllum spicatum, Myriophyllum verticillatum, Najas flexilis, Najas gracillima, Nitella, Nuphar advena, Nuphar variegata, Nymphaea odorata, Nyssa aquatica, Nyssa biflora, Onoclea sensibilis, Orontium aquaticum, Osmundastrum cinnamomeum, Peltandra virginica, Persicaria arifolia, Persicaria hydropiperoides, Persicaria lapathifolia, Persicaria pensylvanica, Persicaria punctata, Persicaria sagittata, Pilea pumila, Pontederia cordata, Potamogeton amplifolius, Potamogeton berchtoldii, Potamogeton bicupulatus, Potamogeton diversifolius, Potamogeton epihydrus, Potamogeton foliosus, Potamogeton gramineus, Potamogeton natans, Potamogeton richardsonii, Potamogeton robbinsii, Potamogeton spirillus, Potamogeton zosteriformis, Proserpinaca palustris, Ranunculus, Rhexia virginica, Rhynchospora macrostachya, Riccia fluitans, Rosa multiflora, Sagittaria latifolia, Salix nigra, Saururus cernuus, Schoenoplectus tabernaemontani, Schoenoplectus torreyi, Scirpus ancistrochaetus, Scirpus cyperinus, Scutellaria lateriflora, Sium suave, Sparganium americanum, Sparganium fluctuans, Spirodela polyrhiza, Stuckenia pectinata, Symphyotrichum novi-belgii, Taxodium ascendens, Taxodium distichum, Triadenum virginicum, Typha angustifolia, Typha domingensis, Typha latifolia, Utricularia gibba, Utricularia intermedia, Utricularia macrorhiza, Utricularia minor, Utricularia purpurea, Utricularia radiata, Vallisneria americana, Verbena hastata, Wolffia brasiliensis, Wolffia columbiana, Wolffiella gladiata, Xyris difformis.

3.4.1.25. ***Potamogeton pusillus* L.** is an annual (vegetative) or a perennial, which grows in fresh to brackish (salinity levels to 3.5–5.3‰), still to flowing, tidal or non-tidal backwaters (shallow), bayous, bays (small), borrow pits, canals, channels (boat, river), ditches (irrigation, roadside), flowages, gravel bars, lakes (boggy, eutrophic, glacial, prairie), marshes (freshwater, oligohaline), ponds (beaver, glacial, mesotrophic, seepage), pools (swampy, tidal), reservoirs (stockpond), river bottoms, rivers (mouths, shores), sloughs, streams (alkaline), and vernal pools at elevations to 3300 m. The plants occur in full sun and occupy a broad depth gradient (0.15–4.0 m depth). They tend to occur on softer substrates but are reported from alluvium (clayey, loamy), clay, gravel, muck (organic, soft), mud, muddy silt, ooze (flocculent), sand (organic), sandy clay, sandy gravel, silt, silty clay (fine), silty muck, and silty sand. The habitats span a wide range of ecological conditions but usually represent more alkaline sites (pH: 5.3–10.2; $\bar{x} = 8.0$–8.1; total alkalinity [as mg l^{-1} $CaCO_3$]: 5.0–220.0; $\bar{x} = 64.1$–140.0; specific conductivity [20°C]: 20–110 μS cm^{-1}). The plants apparently have at least some ability to use bicarbonate (HCO_3^-) as a carbon source for photosynthesis. Despite their occurrence across a spectrum of environmental conditions, they do not appear to be particularly tolerant of eutrophication. Although there are numerous reports of the plants from eutrophic waters, they also are known to have disappeared from culturally eutrophied sites where they later reappeared after decades of absence once the water clarity had been restored. Flowering and fruiting occur from May to September. The inflorescences develop as monomorphic spikes, which normally extend above the water. Each spike produces several strongly protogynous but self-compatible flowers. Pollination can occur in several different ways. The pollen is shed within and among flowers on a spike, which promotes self-pollination or geitonogamy. Some outcrossing (or at least xenogamy) via wind pollination (anemophily) also is possible. When shed onto the water, the pollen is hydrophobic, floats, and can be captured by the stigmas of other flowers (i.e., via epihydrophily). Pollen that is released in underwater flowers is transported to the stigmas of the same flowers on the surface of air bubbles, a process referred to as "hydroautogamy." Pollen escaping from underwater flowers also potentially can be transferred underwater (i.e., by hypohydrophily); successful fruit set has been reported for all but the latter method. Genetic (allozyme) data for European plants are consistent with the floral observations. Those data indicate low levels of within-population variation along with high between-population differentiation and some populations with fixed heterozygosity, all of which can be attributed to self-fertilization (with low seedling recruitment) and a predominantly asexual means of reproduction (see below). Seed set generally is fairly high in any case, but the seeds are believed to lie dormant until appropriate environmental cues (e.g., receding, warming waters) trigger their germination. Dried seeds have failed to germinate, but those stratified for 3–12 months in cold (1°C–3°C) tap water under dark conditions have germinated reasonably well (21–57% germination) at ambient greenhouse temperatures. High germination has also been obtained for seeds stratified for 12 weeks in mineral water at 4°C in the dark followed by incubation at a mean temperature of 22.3°C. Due to its hard pericarp, the best germination (100% after 4 days) has been obtained for fruits pretreated with concentrated sulfuric acid (H_2SO_4) for 50 min. The plants lack rhizomes but reproduce vegetatively by the formation of axillary, reduced, bud-like shoots modified as turions, which begin to germinate in early May. Even where seed production is prolific, seedling recruitment is rare or non-existent, with most new plants apparently arising from the turions, which are regarded as the primary means of reproduction. However, because seedlings can take on the appearance of germinated turions (and can be mistaken for them), their importance in annual recruitment possibly has been underestimated. Otherwise, turions are the only means of overwintering after the shoots die back during autumn. In this way the plants behave as "vegetative annuals". The turions and vegetative fragments all are dispersed by water locally, and by waterfowl (Aves: Anatidae), at least over short distances. The seeds remain highly viable (40–70%) after being eaten and egested by herbivorous fish such as the common carp (Vertebrata: Teleostei: Cyprinidae: *Cyprinus carpio*), which also could potentially disperse them (endozoically) within a water body. Long-distance dispersal most likely occurs by the endozooic transport of the small seeds ($\bar{x}$: 0.67 mg) by waterfowl (Aves: Anatidae); the fruits have a mean retention time of 5.2 hr and have remained highly viable (78%) after passage through the gut of teal (*Anas crecca*). The role of shoot fragments as a means of dispersal is difficult to evaluate. Some field studies indicate that many shoot fragments are unable to regenerate or to establish, suffering more than 70% mortality after 10 weeks, with no indication of rooting or growth; some fragments eventually develop turions during autumn. Other studies have demonstrated that both the turions and shoot fragments remain viable in the vegetative propagule bank, which can range from non-existent to densities approaching 767 propagules m^{-2} depending on the site. Overall productivity estimates from plots tend to be relatively low (e.g., mean maximum biomass: 4.6–16.1 g m^{-2}); however, whole lake crops can reach fairly high levels (e.g., 33.66 kg). It is unusual in that the biomass is not directly proportional to the plant surface area, which precludes the estimation of biomass based on the latter. The plants have been characterized as being disturbance tolerant. **Reported associates:** *Acorus, Alisma gramineum, Alisma triviale, Alternanthera philoxeroides, Azolla cristata, Beckmannia syzigachne, Bidens beckii, Brasenia schreberi, Cabomba caroliniana, Calla palustris, Callitriche hermaphroditica, Callitriche heterophylla, Callitriche stagnalis, Carex lasiocarpa, Carex pellita, Carex scoparia, Carex utriculata, Carex vesicaria, Carex vulpinoidea, Ceratophyllum demersum, Ceratophyllum echinatum, Chara globularis, Chara vulgaris, Cladium jamaicense, Drepanocladus, Dulichium arundinaceum, Echinochloa, Egeria densa, Eichhornia crassipes, Elatine californica, Elatine minima, Elatine rubella, Eleocharis acicularis, Eleocharis macrostachya, Eleocharis montana, Eleocharis palustris, Eleocharis robbinsii, Elodea canadensis, Elodea nuttallii, Enteromorpha, Equisetum fluviatile, Eriocaulon aquaticum, Glyceria borealis, Glyceria canadensis, Gratiola aurea, Heteranthera dubia, Hippuris vulgaris, Hydrilla verticillata, Hydrocotyle ranunculoides, Iris pseudacorus, Isoetes acadiensis, Isoetes bolanderi, Isoetes echinospora, Isoetes lacustris, Isoetes occidentalis, Isoetes tuckermanii, Juncus effusus, Juncus laccatus, Juncus militaris, Juncus pelocarpus, Juncus supiniformis, Lemna minor, Lemna trisulca, Lemna turionifera, Lilaeopsis occidentalis, Limosella aquatica, Lobelia dortmanna, Ludwigia palustris, Marsilea vestita, Myriophyllum alterniflorum, Myriophyllum* *aquaticum, Myriophyllum farwellii, Myriophyllum hetero-phyllum, Myriophyllum hippuroides, Myriophyllum sibiri-cum, Myriophyllum spicatum, Myriophyllum tenellum, Myriophyllum verticillatum, Najas flexilis, Najas gracillima, Najas guadalupensis, Najas marina, Najas minor, Navarretia intertexta, Nelumbo lutea, Nitella, Nuphar advena, Nuphar microphylla, Nuphar polysepala, Nuphar variegata, Nymphaea mexicana, Nymphaea odorata, Nymphoides cor-data, Nyssa aquatica, Paspalum distichum, Persicaria amphibia, Persicaria coccinea, Phalaris arundinacea, Phragmites australis, Poa pratensis, Pontederia cordata, Potamogeton alpinus, Potamogeton amplifolius, Potamogeton bicupulatus, Potamogeton confervoides, Potamogeton cris-pus, Potamogeton diversifolius, Potamogeton epihydrus, Potamogeton foliosus, Potamogeton friesii, Potamogeton gramineus, Potamogeton hillii, Potamogeton illinoensis, Potamogeton natans, Potamogeton nodosus, Potamogeton oakesianus, Potamogeton obtusifolius, Potamogeton perfoli-atus, Potamogeton praelongus, Potamogeton richardsonii, Potamogeton robbinsii, Potamogeton spirillus, Potamogeton strictifolius, Potamogeton vaseyi, Potamogeton zosterifor-mis, Proserpinaca, Ranunculus aquatilis, Ranunculus flam-mula, Ranunculus longirostris, Ranunculus subrigidus, Ranunculus trichophyllus, Ricciocarpus natans, Rorippa, Rumex, Ruppia maritima, Ruppia spiralis, Sagittaria cune-ata, Sagittaria graminea, Sagittaria lancifolia, Sagittaria latifolia, Sagittaria rigida, Salvinia minima, Salvinia molesta, Schoenoplectus acutus, Schoenoplectus pungens, Schoenoplectus subterminalis, Schoenoplectus tabernae-montani, Sparganium americanum, Sparganium angustifo-lium, Sparganium emersum, Sparganium eurycarpum, Spartina cynosuroides, Spartina patens, Sphagnum, Spiraea douglasii, Spirodela polyrhiza, Stuckenia filiformis, Stuckenia pectinata, Stuckenia vaginata, Taxodium distichum, Typha angustifolia, Typha domingensis, Typha latifolia, Utricularia geminiscapa, Utricularia intermedia, Utricularia macro-rhiza, Utricularia minor, Utricularia purpurea, Utricularia radiata, Vallisneria americana, Veronica catenata, Wolffia borealis, Wolffia brasiliensis, Zannichellia palustris, Zizania aquatica.*

3.4.1.26. *Potamogeton richardsonii* (A. Benn.) Rydb. is a perennial, which inhabits fresh to slightly brackish bays, bor-row pits, bottoms (river), canals (drainage), channels (backwa-ter, muddy), ditches (irrigation), flats (tidal), floodplains, gravel pits, lagoons (freshwater), lakes (boggy, oligotrophic), marshes (intertidal, tidal), mudflats, oxbows, ponds (beaver, irrigation, quarry, roadside, stock, tailing, tundra), pools (perennial), potholes, reservoirs, rivers, sloughs (river), springs (swampy), streams, and swales at elevations to 3169 m. Exposures range from full sunlight to partial shade. The plants grow in still waters or in fairly swift currents at depths ranging from 0.03 to 4.1 m (optimally at 2.0–2.5 m) and prefer sites where water level fluctuations occur. The substrates include clay, clayey sand, gravel, gravelly mud, loam, marl, muck, mud (organic), rock, rocky gravel, sand (fine), sandy gravel, sandy muck, sandy mud, sandy silt, silt, silty muck, silty mud, and silty sand. The plants are tolerant of turbidity (e.g., Secchi depth:

0.71–1.8 m) and persist at sites that have become enriched with nutrients. The waters primarily are alkaline (pH: 5.6–10.2; $\bar{x} = 7.6$–8.3; total alkalinity [as mg l^{-1} CaCO$_3$]: 5.9–448.0; $\bar{x} = 46.0$–117.0; specific conductivity [@ 25°C]: 20–510 μmhos cm^{-2}). In addition to free CO$_2$, the plants are able to use bicarbonate (HCO$_3^-$) as a carbon source for photosynthe-sis. By late summer, the leaves can become encrusted with up to 180 mg Ca g^{-1} leaf (as "calcite" or "marl" deposits [CaCO$_3$]), which co-precipitates with P as the result of photosynthetic activity. The precipitation of marl on the upper leaf surfaces provides a substrate for colonizing epiphytes. Flowering and fruiting occur from June to October. The inflorescences are emergent spikes, which can produce up to 27 flowers arranged in alternating whorls of three. Experimental manipulations have shown that clipping of apical portions (mimicking her-bivory) results in a significant increase in branches, flowers, and leaves, which consequently can increase reproductive potential. Fruiting is reported commonly, and large numbers of seeds often are produced. The seeds are dispersed locally by water currents and are known to be transported in irriga-tion waters. The seeds are eaten by waterfowl (Aves: Anatidae), but it is uncertain what role they play in long-distance disper-sal. Seed bank densities up to 19.7 m^{-2} have been reported. In one case, seeds planted in a 1.5 m^2 pool yielded plants bearing a total of up to 29,200 seeds and 2,370 vegetative propagules, with single plants producing up to 40 tubers by the end of one growing season. Consequently, it might seem likely that both seeds and vegetative propagules would be important in the yearly reestablishment of this species; however, very few seedlings have been observed in the field, indicating that veg-etative propagation remains most important in this regard. The plants senesce from late August through October. They overwinter as rhizomes, which remain buried in the sediment until their rapid development in early spring (June) when water temperatures rise above 9°C. The rhizomes produce elongate tuberous shoot outgrowths (called turions by some), which also function as vegetative propagules. Shoot fragmen-tation is observed commonly throughout the growing season and undoubtedly accounts for much local dispersal. The foli-age is slow to desiccate (mean rate <2 g hr^{-1} g^{-1}) and more than 10% of shoot fragments remain viable after 3 hr of desic-cation. It is presumed that longer-distance (among-lake) dis-persal occurs by the exozoic transport of shoot fragments, which is facilitated by a favorable microclimate in the plum-age of waterfowl (Aves: Anatidae). Anthropogenic dispersal also occurs by the inadvertent attachment of shoot fragments to trailers and other boating equipment. The foliage is broad-leaved and homophyllous but variable. At least some of the vegetative plasticity has been explained as a response to vary-ing light, nutrients, and temperatures. Experiments indicate that under full sunlight the plants primarily are nutrient lim-ited, but at 12% daylight they are light limited. At comparable (0.5 m) depths, the leaf area index is an order of magnitude lower for plants growing on silty sand (0.4) than for those growing on silty loam (4.0). Internodes are longest in plants growing in cool, deep waters (e.g., 1.8 m; 9°C–15°C), and shorten progressively as water depth decreases. The plants have an average specific leaf area of 766 cm^2 g^{-1}, are fast-growing (mean primary growth rate near 11% day^{-1}), and can attain productivity levels from 2.4 to 14.4 g DM biomass m^{-2} at rates up to 9.9 mg C m^{-2} day^{-1}. The below:total biomass ratio can vary from 0.30 to 0.47 depending on the availability of dissolved inorganic carbon. **Reported associates:** *Alisma gramineum, Asclepias incarnata, Bidens beckii, Bolboschoenus fluviatilis, Brasenia schreberi, Butomus umbellatus, Cabomba caroliniana, Callitriche hermaphroditica, Callitriche heterophylla, Callitriche palustris, Catabrosa aquatica, Cephalanthus occidentalis, Ceratophyllum demersum, Ceratophyllum echinatum, Chara globularis, Chara vulgaris, Cicuta bulbifera, Crassula aquatica, Drepanocladus, Elatine californica, Elatine chilensis, Elatine minima, Eleocharis acicularis, Eleocharis palustris, Elodea canadensis, Elodea nuttallii, Equisetum fluviatile, Eriocaulon aquaticum, Fontinalis antipyretica, Glyceria borealis, Gratiola aurea, Heteranthera dubia, Hippuris vulgaris, Isoetes bolanderi, Isoetes echinospora, Isoetes lacustris, Isoetes occidentalis, Juncus bufonius, Juncus effusus, Juncus militaris, Juncus pelocarpus, Lemna minor, Lemna trisulca, Lilaeopsis, Limosella aquatica, Lobelia dortmanna, Lycopus, Lythrum salicaria, Marsilea vestita, Menyanthes trifoliata, Myosotis laxa, Myriophyllum alterniflorum, Myriophyllum farwellii, Myriophyllum heterophyllum, Myriophyllum hippuroides, Myriophyllum quitense, Myriophyllum sibiricum, Myriophyllum spicatum × Myriophyllum sibiricum, Myriophyllum spicatum, Myriophyllum tenellum, Myriophyllum verticillatum, Najas canadensis, Najas flexilis, Najas marina, Nitella allenii, Nitella tenuissima, Nitellopsis obtusa, Nuphar polysepala, Nuphar variegata, Nuphar ×rubrodisca, Nymphaea odorata, Nymphaea tetragona, Nymphoides cordata, Persicaria amphibia, Phalaris arundinacea, Pontederia cordata, Potamogeton alpinus, Potamogeton amplifolius, Potamogeton berchtoldii, Potamogeton crispus, Potamogeton diversifolius, Potamogeton epihydrus, Potamogeton foliosus, Potamogeton friesii, Potamogeton gramineus, Potamogeton illinoensis, Potamogeton natans, Potamogeton nodosus, Potamogeton obtusifolius, Potamogeton praelongus, Potamogeton pulcher, Potamogeton pusillus, Potamogeton robbinsii, Potamogeton sibiricus, Potamogeton spirillus, Potamogeton strictifolius, Potamogeton vaseyi, Potamogeton zosteriformis, Ranunculus flammula, Ranunculus gmelinii, Ranunculus longirostris, Ranunculus subrigidus, Ranunculus trichophyllus, Ricciocarpus, Sagittaria cuneata, Sagittaria latifolia, Sagittaria rigida, Schoenoplectus acutus, Schoenoplectus tabernaemontani, Schoenoplectus triqueter, Sium suave, Sparganium angustifolium, Sparganium emersum, Sparganium eurycarpum, Sparganium fluctuans, Sparganium hyperboreum, Sparganium natans, Spirodela polyrhiza, Spirogyra, Stuckenia filiformis, Stuckenia pectinata, Typha angustifolia, Typha latifolia, Utricularia geminiscapa, Utricularia gibba, Utricularia intermedia, Utricularia macrorhiza, Utricularia minor, Utricularia purpurea, Vallisneria americana, Wolffia brasiliensis, Wolffia columbiana, Zannichellia palustris, Zizania palustris.*

3.4.1.27. ***Potamogeton robbinsii* Oakes** is a perennial, which grows in bays, channels, coves, lakes (alpine, boggy), ponds, rivers, sloughs (backwater), and streams at elevations to 3293 m. The plants can be found across an incredible depth gradient ranging from 0.2 to 14 m (median depth near 1.8 m) in still to slow-moving, clear (e.g., Secchi depth: 4.1 m) waters. Experimentally, they achieve their highest biomass under shallow water (1.5 m), high nutrient conditions. They are fairly intolerant of winter drawdown conditions. The substrates are characterized as marl, muck (organic), mucky sand, mud, muddy ooze, peat (fibrous, pulpy), rocky rubble, sand, silt, silty muck, and silty sand. The habitats also span a broad range of ecological conditions, but typically are alkaline (pH: 5.0–10.0; $\bar{x} = 7.2$–8.9; total alkalinity [as mg l^{-1} CaCO$_3$]: 3.0–144.0; $\bar{x} = 26.1$–45.0; conductivity [@ 25°C] to 290 μmhos cm^{-1}). The plants are homophyllous and grow completely submersed. They can use both free CO$_2$ and bicarbonate (HCO$_3^-$) as carbon sources for photosynthesis. Flowering and fruiting occur from July to October as the upper portions of the inflorescences extend above the water surface. Little is known specifically about the floral and pollination biology. Fruiting is uncommon but can produce a maximum of 6 fruits on each spike. The rare seed production probably is mainly a consequence of the deeper waters typically inhabited by this species, combined with the short stature of its shoots (e.g., <30 cm), which prevent them from reaching the water surface. Numerous deformed and sterile pollen grains also have been observed; however, at least some viable seeds have been recovered from seed bank core studies. These have germinated under a 14:10 hr light:dark regime at temperatures between 21°C and 26°C. Vegetative reproduction occurs by rhizomes and by shoot fragments (commonly observed in the drift), which root readily on muddy substrates. Entire plants also are known to survive throughout the winter (in 2°C water) in a physiologically active vegetative state (<12.5% of the summer photosynthetic rate). The plants have a relatively low mean primary growth rate (1.32 × 10^{-2} g g^{-1} day^{-1}). In the field their maximum aboveground biomass values (<153 g DM m^{-2}; $\bar{x} = 13.1$) is not particularly impressive. Yet, despite being short in stature, the plants can densely "carpet" the bottom (up to 900 plants m^{-2} at depths from 5 to 7 m) to become one of the most abundant plant species in a lake, occurring as often as 36–43% in surveys and representing as much as 46% of total lake macrophyte biomass. Enrichment of dissolved inorganic carbon significantly increases the underground:total biomass ratio. Although resistant to decomposition, the foliage eventually decays, which increases the sediment organic content and lowers the pH, making nutrients more available to other species (e.g., *Zizania aquatica*), which grow within the clumps. However, the presence of this species has been shown to correlate with a general decrease in richness of other macrophytes. **Reported associates:** *Asclepias incarnata, Bidens beckii, Brasenia schreberi, Butomus umbellatus, Cabomba caroliniana, Callitriche hermaphroditica, Ceratophyllum demersum, Chara globularis, Cladophora, Decodon verticillatus, Drepanocladus, Egeria densa, Elatine minima, Eleocharis acicularis, Eleocharis* *palustris, Elodea canadensis, Elodea nuttallii, Eriocaulon aquaticum, Fontinalis antipyretica, Fontinalis hypnoides, Gratiola aurea, Heteranthera dubia, Hippuris vulgaris, Isoetes echinospora, Isoetes lacustris, Isoetes occidentalis, Isoetes tuckermanii, Juncus militaris, Juncus pelocarpus, Juncus supiniformis, Lemna minor, Lemna trisulca, Lilaeopsis occidentalis, Lobelia dortmanna, Myriophyllum alterniflorum, Myriophyllum farwellii, Myriophyllum heterophyllum, Myriophyllum hippuroides, Myriophyllum sibiricum, Myriophyllum spicatum, Myriophyllum tenellum, Myriophyllum verticillatum, Najas flexilis, Najas guadalupensis, Najas minor, Nitella furcata, Nitella tenuissima, Nuphar advena, Nuphar polysepala, Nymphaea odorata, Nymphaea tetragona, Persicaria amphibia, Pontederia cordata, Potamogeton alpinus, Potamogeton amplifolius, Potamogeton berchtoldii, Potamogeton bicupulatus, Potamogeton confervoides, Potamogeton crispus, Potamogeton diversifolius, Potamogeton epihydrus, Potamogeton foliosus, Potamogeton friesii, Potamogeton gramineus, Potamogeton illinoensis, Potamogeton natans, Potamogeton nodosus, Potamogeton oakesianus, Potamogeton obtusifolius, Potamogeton perfoliatus, Potamogeton praelongus, Potamogeton pulcher, Potamogeton pusillus, Potamogeton richardsonii, Potamogeton spirillus, Potamogeton strictifolius, Potamogeton vaseyi, Potamogeton zosteriformis, Ranunculus flammula, Ranunculus longirostris, Ranunculus subrigidus, Ranunculus trichophyllus, Sagittaria cuneata, Sagittaria graminea, Sagittaria latifolia, Sagittaria rigida, Schoenoplectus acutus, Schoenoplectus subterminalis, Sparganium angustifolium, Sparganium emersum, Sparganium fluctuans, Spirogyra, Stuckenia filiformis, Stuckenia pectinata, Typha latifolia, Utricularia gibba, Utricularia intermedia, Utricularia macrorhiza, Utricularia purpurea, Utricularia resupinata, Vallisneria americana, Veronica anagallis-aquatica, Wolffia brasiliensis, Zizania aquatica, Zannichellia palustris.*

3.4.1.28. ***Potamogeton sibiricus* A. Benn.** is a perennial, which grows in brackish to fresh, shallow (to 1.5 m deep) waters of deltas, ditches (drainage), floodplains, lagoons (brackish), lakes (shallow), ponds (beaver, muskeg, stagnant, tundra), pools (rock, tundra), and streams (sluggish) at elevations to 914 m. The ecology of this submersed, homophyllous (linear-leaved) species is not well-known. The reported substrates include muck (organic), mud, sandy mud, and stones. Flowering and fruiting occur from early to late summer (August). The flowers are borne in 3–4 whorls on cylindrical spikes, which extend above the water on a peduncle (to 50 mm). The reproductive biology and seed ecology of this species are poorly understood. The plants have been described as having filiform rhizomes and also as lacking rhizomes. The production of lateral turions also has been reported in some (but not all) accounts. **Reported associates:** *Callitriche hermaphroditica, Hippuris vulgaris, Menyanthes trifoliata, Myriophyllum sibiricum, Nuphar polysepala, Persicaria amphibia, Potamogeton alpinus, Potamogeton epihydrus, Potamogeton friesii, Potamogeton gramineus, Potamogeton richardsonii, Potamogeton zosteriformis, Sparganium*

angustifolium, Sparganium hyperboreum, Stuckenia filiformis, Stuckenia pectinata, Stuckenia vaginata, Triglochin maritima, Utricularia macrorhiza.

3.4.1.29. ***Potamogeton spirillus* Tuck.** is an annual or a perennial, which grows in bays (shallow, sheltered), bogs, channels, coves, flats (tidal), lakes, marshes, ponds, pools (meadow, shallow, shoreline), reservoirs, rivers, sloughs, and streams at elevations to 424 m. These heterophyllous plants occur in still or flowing, shallow water (mainly <1 m depths) to 4 m in exposures of full sunlight. They often are collected without floating leaves, especially at deeper sites. The substrates include boulders, clay, clayey mud, granite, gravel, muck, mucky sand, mud (soft), rock, rocky sand, rocky silt, sand, sandy cobble, sandy muck, silt, and silty sand. The waters are circumneutral and usually of low alkalinity and conductivity (pH: 5.9–10.2; $\bar{x} = 7.0$–7.6; total alkalinity [as mg l^{-1} CaCO$_3$]: 2.5–85.0; $\bar{x} = 16.2$–37.0; conductivity [@ 25°C]: <150 µmhos cm^{-1}). Flowering and fruiting occur from June to October. Dimorphic inflorescences (aerial and submersed) are produced. The aerial flowers occur in emergent spikes and are protogynous but self-compatible. Despite having higher pollen:ovule ratios ($\bar{x} = 3,800$:1) relative to the submersed flowers ($\bar{x} = 900$:1), they are facultatively autogamous and quite capable of selfing if outcrossed pollen is not first deposited during the receptive stigmatic phase. The anthers within the heads of submersed flowers are small and dehisce prior to the opening of the flowers, which results in obligate self-pollination. Close to 100% seed set occurs in both the aerial and submersed flowers, even on plants growing at depths up to 3 m. Despite the frequent production of fruits, which are important in the annual recruitment and reestablishment of plants, there is limited information on the seed ecology. The fruits do not survive dry storage. Fruits kept in cold water (1°C–3°C) retain fair viability (18% germination) after 5 months of storage but have not remained viable after 1 year of storage. Consequently, it is unlikely that a seed bank persists for any prolonged length of time. The plants ordinarily are perennials, overwintering by means of their rhizomes; however, they can behave as annuals when growing in cold, spring-fed waters. This species has been observed to increase in relative frequency following periods of lake drawdown. **Reported associates:** *Bidens beckii, Brasenia schreberi, Cabomba caroliniana, Callitriche heterophylla, Callitriche palustris, Campanula aparinoides, Ceratophyllum demersum, Chara braunii, Chara vulgaris, Cladium mariscoides, Drepanocladus, Dulichium arundinaceum, Elatine minima, Eleocharis acicularis, Eleocharis erythropoda, Eleocharis obtusa, Eleocharis palustris, Elodea canadensis, Elodea nuttallii, Equisetum fluviatile, Eriocaulon aquaticum, Fontinalis antipyretica, Glyceria borealis, Glyceria canadensis, Heteranthera dubia, Isoetes echinospora, Isoetes lacustris, Juncus militaris, Juncus pelocarpus, Lemna gibba, Lemna trisulca, Lobelia dortmanna, Ludwigia palustris, Lythrum salicaria, Myriophyllum alterniflorum, Myriophyllum farwellii, Myriophyllum spicatum, Myriophyllum tenellum, Najas flexilis, Najas gracillima, Najas minor, Nitella tenuissima, Nuphar variegata, Nymphaea odorata, Nymphoides cordata,*

Polygonum achoreum, Pontederia cordata, Potamogeton alpinus, Potamogeton amplifolius, Potamogeton bicupulatus, Potamogeton confervoides, Potamogeton crispus, Potamogeton epihydrus, Potamogeton friesii, Potamogeton gramineus, Potamogeton natans, Potamogeton nodosus, Potamogeton oakesianus, Potamogeton obtusifolius, Potamogeton perfoliatus, Potamogeton praelongus, Potamogeton pulcher, Potamogeton pusillus, Potamogeton richardsonii, Potamogeton robbinsii, Potamogeton strictifolius, Potamogeton vaseyi, Potamogeton zosteriformis, Ranunculus flammula, Ranunculus longirostris, Ranunculus trichophyllus, Sagittaria cuneata, Sagittaria graminea, Sagittaria latifolia, Sagittaria rigida, Schoenoplectus acutus, Schoenoplectus subterminalis, Schoenoplectus torreyi, Scirpus cyperinus, Sparganium americanum, Sparganium angustifolium, Sparganium fluctuans, Sphagnum, Stuckenia pectinata, Subularia aquatica, Utricularia geminiscapa, Utricularia gibba, Utricularia intermedia, Utricularia macrorhiza, Utricularia minor, Utricularia purpurea, Utricularia radiata, Utricularia resupinata, Vallisneria americana, Zannichellia palustris, Zizania palustris.

3.4.1.30. *Potamogeton strictifolius* **A. Benn.** is a perennial, which inhabits clear (e.g., Secchi depth: 4.5 m), usually shallow (but to 2.8 m) waters of bays (lake, river), channels, coves, ditches, flowages, inlets (drainage), lakes (seepage, swampy), ponds, pools, rivers, shores (boggy), and streams at elevations to 2439 m. These are submersed, homophyllous (linear-leaved) plants capable of using bicarbonate ions as a photosynthetic carbon source. Their foliage typically becomes encrusted with marl due to the characteristically calcareous conditions. The substrates often are characterized as hard bottoms but can include marl, muck (marly), mucky sand, mud, organic matter (e.g., 1.7%), sand, and sandy gravel. The plants are associated with waters of relatively high alkalinity, pH, and specific conductance (pH: 4.8–9.3; $\bar{x} = 7.7$–8.1; total alkalinity [as mg l^{-1} $CaCO_3$]: 31.8–262.5; $\bar{x} = 42.2$–84.3; specific conductance: to 305 μmhos cm^{-1} [@ 25°C]). Some authors indicate an affinity of this species to clear, unpolluted waters; however, others have found the plants growing under a wider range of ecological conditions including more degraded and nutrient-rich sites. Flowering and fruiting occur from June to August. From 3 to 4 whorls of flowers are produced on cylindrical spikes, which are the only part of the plant to emerge from the water. Pollination presumably occurs by wind or water but is not well documented. Fruit production is common, making this an important food for waterfowl, which readily consume the seeds (see *Use by wildlife*). Vegetative propagation occurs by the production of winter buds (also referred to as turions), which can number up to 20 per plant. Although numerous seeds can be produced, the winter buds are the most important means of propagation. In fall, the plants die back completely leaving only the seeds and/or winter buds to persist through winter. Although barren plants sometimes are regarded as annuals, they actually would represent "vegetative annuals," which technically are perennials (see *Ecology* above). Plants arising from seed can undergo several seasons of vegetative reproduction before becoming fertile. **Reported associates:**

Acorus calamus, Agalinis purpurea, Alisma, Bidens beckii, Brasenia schreberi, Carex lasiocarpa, Ceratophyllum demersum, Chara, Cicuta bulbifera, Dulichium arundinaceum, Eleocharis palustris, Elodea canadensis, Elodea nuttallii, Equisetum fluviatile, Heteranthera dubia, Hippuris vulgaris, Iris pseudacorus, Isoetes, Lemna minor, Lemna trisulca, Lemna turionifera, Lycopus uniflorus, Myriophyllum heterophyllum, Myriophyllum sibiricum, Myriophyllum spicatum, Myriophyllum verticillatum, Najas flexilis, Najas guadalupensis, Nitella, Nuphar polysepala, Nuphar variegata, Nymphaea odorata, Persicaria amphibia, Potamogeton alpinus, Potamogeton amplifolius, Potamogeton berchtoldii, Potamogeton crispus, Potamogeton diversifolius, Potamogeton epihydrus, Potamogeton foliosus, Potamogeton friesii, Potamogeton gramineus, Potamogeton hillii, Potamogeton illinoensis, Potamogeton natans, Potamogeton nodosus, Potamogeton oakesianus, Potamogeton obtusifolius, Potamogeton praelongus, Potamogeton pusillus, Potamogeton richardsonii, Potamogeton robbinsii, Potamogeton spirillus, Potamogeton zosteriformis, Ranunculus longirostris, Rorippa aquatica, Rumex maritimus, Sagittaria cuneata, Sagittaria latifolia, Sagittaria rigida, Schoenoplectus acutus, Schoenoplectus subterminalis, Schoenoplectus tabernaemontani, Sparganium emersum, Sparganium eurycarpum, Spirodela polyrhiza, Stuckenia pectinata, Trapa natans, Typha latifolia, Utricularia gibba, Utricularia intermedia, Utricularia macrorhiza, Vallisneria americana, Zizania palustris.

3.4.1.31. *Potamogeton tennesseensis* **Fernald** is a rare perennial, which occurs in still to rapidly flowing, shallow waters of lagoons (river), ponds (sinkhole), rivers, and streams at elevations to 938 m. This is a heterophyllous plant having linear submersed leaves; however, the floating leaves are not always present. Ecological information is difficult to obtain for this imperiled species because most of the database records unfortunately mask (i.e., exclude) not only specific locality data but virtually all ancillary information. Exposures of partial sunlight are tolerated. The plants occur in soft, acidic waters (pH: 6.4–7.1; $\bar{x} = 6.8$; total alkalinity [as mg l^{-1} $CaCO_3$]: 2–32; $\bar{x} = 12$). Flowering and fruiting extend at least from June to July. The flowers are produced on emergent spikes, but their reproductive biology and seed ecology remains unstudied. Perennation occurs by means of rhizomes; no turions are produced. Little additional life history information is available for this species. **Reported associates:** *Boldia erythrosiphon, Callitriche heterophylla, Fontinalis, Isoetes lacustris, Nitella, Podostemum ceratophyllum, Potamogeton epihydrus, Potamogeton nodosus.*

3.4.1.32. *Potamogeton vaseyi* **J.W. Robbins** is a perennial, which inhabits bays, channels (river), coves, flowages, lakes (boggy, marshy), marshes (shallow), ponds (beaver), pools, rivers (outlet), shores (boggy, lake), and streams (mouths) at elevations to 500 m. The plants are heterophyllous (with linear submersed leaves) at depths to about 1.8 m; however, individuals in deeper waters (to 4.6 m) are sterile and lack floating leaves. They grow primarily in soft waters (pH: 5.4–8.5; $\bar{x} = 7.1$; total alkalinity [as $CaCO_3$] to 60 mg l^{-1}; specific conductivity <200 μmhos cm^{-1}). The substrates include clay (blue-gray), marl, muck, mucky silt (organic), mud (loose), rocky sand, sand, sandy clay (soft), and silty sand. Flowering and fruiting extend from May to September. The flowers are produced on spikes, which extend above the water surface, in association with the floating leaves. Fruiting occurs only on plants that produce floating leaves. When fertile, the plants can fruit heavily. Few other details are known regarding the floral biology or seed ecology. No rhizomes develop, but vegetative reproduction occurs by means of axillary turions (also called winter buds), which are common on both fertile and sterile plants. The latter exemplify a "vegetative annual" life history (see *Ecology* above). **Reported associates:** *Acorus, Arethusa bulbosa, Asclepias incarnata, Betula pumila, Bidens beckii, Brasenia schreberi, Brasenia schreberi, Calla palustris, Callitriche palustris, Campanula aparinoides, Carex crinita, Carex lasiocarpa, Carex utriculata, Ceratophyllum demersum, Ceratophyllum echinatum, Chara, Cicuta bulbifera, Comarum palustre, Dulichium arundinaceum, Elatine minima, Eleocharis acicularis, Eleocharis erythropoda, Eleocharis palustris, Elodea canadensis, Elodea nuttallii, Equisetum fluviatile, Eriocaulon aquaticum, Fontinalis novae-angliae, Glyceria borealis, Glyceria grandis, Heteranthera dubia, Hydrocharis morsus-ranae, Iris, Isoetes echinospora, Isoetes lacustris, Juncus militaris, Juncus pelocarpus, Lemna minor, Lemna trisulca, Lobelia dortmanna, Ludwigia palustris, Lysimachia terrestris, Mimulus ringens, Myriophyllum alterniflorum, Myriophyllum farwellii, Myriophyllum heterophyllum, Myriophyllum sibiricum, Myriophyllum spicatum, Myriophyllum tenellum, Myriophyllum verticillatum, Najas flexilis, Najas gracillima, Najas guadalupensis, Nitella, Nuphar microphylla, Nuphar variegata, Nuphar ×rubrodisca, Nymphaea odorata, Persicaria coccinea, Persicaria punctata, Polygonum achoreum, Pontederia cordata, Potamogeton amplifolius, Potamogeton bicupulatus, Potamogeton crispus, Potamogeton epihydrus, Potamogeton foliosus, Potamogeton friesii, Potamogeton gramineus, Potamogeton illinoensis, Potamogeton natans, Potamogeton nodosus, Potamogeton obtusifolius, Potamogeton perfoliatus, Potamogeton praelongus, Potamogeton pusillus, Potamogeton richardsonii, Potamogeton robbinsii, Potamogeton spirillus, Potamogeton zosteriformis, Ranunculus flammula, Ranunculus longirostris, Sagittaria cristata, Sagittaria graminea, Sagittaria latifolia, Sagittaria rigida, Salix alba, Salix nigra, Salix ×rubens, Schoenoplectus acutus, Schoenoplectus subterminalis, Schoenoplectus tabernaemontani, Schoenoplectus torreyi. Scirpus cyperinus, Sium suave, Sparganium americanum, Sparganium angustifolium, Sparganium emersum, Sparganium eurycarpum, Sparganium fluctuans, Spartina pectinata, Stuckenia pectinata, Torreyochloa pallida, Typha latifolia, Utricularia gibba, Utricularia intermedia, Utricularia macrorhiza, Utricularia minor, Utricularia purpurea, Utricularia radiata, Vallisneria americana, Zizania palustris.*

3.4.1.33. **Potamogeton zosteriformis Fernald** is a perennial, which occurs in backwaters (floodplain), bays (river), bottoms

(river), channels (calcareous), depressions (meadow, organic), gravel pits, impoundments, inlets (shallow), lagoons, lakes (boggy, fen, glacial), marshes (deepwater, meadow, open, outwash), oxbows, ponds (beaver, stock), pools, potholes (permanent, sand prairie), rivers, shores (lake), sloughs, streams (slow), and washes at elevations to 2713 m. These plants are found in still to fast-moving, clear to brown waters at depths usually less than 4.0 m; however, they have been collected in waters up to 14 m deep. They are not tolerant of turbidity and typically occur in exposures that receive full sunlight. The substrates (usually soft) include clay, gravel (organic), marl, muck (loose, organic), mucky loam, mucky marl, mucky sand, mud (loose, soft), muddy cobble, muddy sand, peat (fibrous, pulpy), sand, sandy gravel, sandy muck, shale pebbles, silt, silty detritus, and silty sand. The habitats span a broad ecological spectrum, but with more of a tendency toward harder alkaline, calcareous waters (pH: 5.6–10.0; $\bar{x}=7.9$; total alkalinity [as mg l^{-1} CaCO$_3$]: 10.0–300.6; $\bar{x}=81–91$; specific conductivity [@ 25°C] to 610 μmhos cm^{-1}). Flowering and fruiting have been recorded from June to November. As many as 20 flowers can be produced on the spikes, which emerge from and are held above the water surface. The floral biology has not been investigated in any detail. Although seed production occurs fairly commonly, the extent of sexual recruitment (i.e., by seed) is uncertain. The seeds have germinated well under greenhouse conditions (14:10 hr light:dark at ambient temperatures from 21°C to 26°C). Rhizomes do not develop, but vegetative reproduction occurs by means of turions, which commonly develop in early summer, and persist throughout the winter. Because many young plants observed in early spring appear to have arisen from turions rather than seeds, it is likely that the plants behave primarily as "vegetative annuals" (see *Ecology* above) for much of their existence. **Reported associates:** *Acorus calamus, Alisma, Alopecurus aequalis, Azolla, Beckmannia syzigachne, Bidens beckii, Brasenia schreberi, Cabomba caroliniana, Calamagrostis stricta, Calla palustris, Callitriche hermaphroditica, Carex alopecoidea, Carex comosa, Carex diandra, Carex hystericina,Carex interior, Carex scoparia, Carex stricta, Carex vesicaria, Carex viridula, Cephalanthus occidentalis, Ceratophyllum demersum, Ceratophyllum echinatum, Chara globularis, Chara vulgaris, Egeria densa, Eleocharis acicularis, Eleocharis palustris, Elodea canadensis, Elodea nuttallii, Equisetum fluviatile, Glyceria borealis, Heteranthera dubia, Hippuris vulgaris, Hydrocotyle, Isoetes echinospora, Juncus effusus, Juncus marginatus, Lemna minor, Lemna trisulca, Lychnothamnus barbatus, Lythrum salicaria, Menyanthes trifoliata, Myosotis laxa, Myriophyllum heterophyllum, Myriophyllum sibiricum, Myriophyllum spicatum, Myriophyllum verticillatum, Najas canadensis, Najas flexilis, Najas guadalupensis, Najas marina, Nitella, Nuphar polysepala, Nuphar variegata, Nymphaea odorata, Persicaria hydropiper, Phalaris arundinacea, Potamogeton alpinus, Potamogeton amplifolius, Potamogeton berchtoldii, Potamogeton crispus, Potamogeton diversifolius, Potamogeton epihydrus, Potamogeton foliosus, Potamogeton friesii, Potamogeton gramineus, Potamogeton hillii, Potamogeton illinoensis, Potamogeton* natans, *Potamogeton nodosus*, *Potamogeton obtusifolius*, *Potamogeton perfoliatus*, *Potamogeton praelongus*, *Potamogeton pulcher*, *Potamogeton pusillus*, *Potamogeton richardsonii*, *Potamogeton robbinsii*, *Potamogeton sibiricus*, *Potamogeton spirillus*, *Potamogeton strictifolius*, *Potamogeton vaseyi*, *Ranunculus longirostris*, *Ranunculus subrigidus*, *Ranunculus trichophyllus*, *Sagittaria cuneata*, *Sagittaria rigida*, *Salix candida*, *Salix exigua*, *Schoenoplectus acutus*, *Schoenoplectus hallii*, *Schoenoplectus subterminalis*, *Schoenoplectus tabernaemontani*, *Sium suave*, *Sparganium eurycarpum*, *Sparganium fluctuans*, *Spirodela polyrhiza*, *Stuckenia filiformis*, *Stuckenia pectinata*, *Stuckenia vaginata*, *Typha angustifolia*, *Typha latifolia*, *Typha ×glauca*, *Utricularia geminiscapa*, *Utricularia gibba*, *Utricularia intermedia*, *Utricularia macrorhiza*, *Utricularia minor*, *Vallisneria americana*, *Wolffia brasiliensis*, *Wolffia columbiana*, *Zannichellia palustris*, *Zizania aquatica*, *Zizania palustris*.

Use by wildlife: The genus *Potamogeton* is regarded universally as a premier wildlife food resource owing to its production of nutritious fruits, which provide nourishment to numerous species of waterfowl (Aves: Anatidae). Some examples (but not an exhaustive list) include the American wigeon (*Anas americana*), which consumes the achenes of *P. foliosus*, *P. nodosus*, *P. pusillus*, and *P. richardsonii*; black ducks (*Anas rubripes*), which eat those of *P. epihydrus*, *P. gramineus*, *P. natans*, *P. nodosus*, *P. obtusifolius*, *P. perfoliatus*, *P. pusillus*, and *P. spirillus*; blue-winged teal (*Anas discors*), which eats those of *P. epihydrus*, *P. nodosus*, and *P. pusillus*; buffleheads (*Bucephala albeola*), which eat those of *P. pusillus*; canvasbacks (*Aythya valisineria*), which eat those of *P. foliosus*, *P. nodosus*, *P. perfoliatus* (especially in Chesapeake Bay), *P. pusillus*, and *P. richardsonii*; common goldeneye (*Bucephala clangula*), which eats those of *P. nodosus*; gadwall (*Anas strepera*), which eats those of *P. foliosus*, *P. nodosus*, and *P. pusillus*; greater scaups (*Aythya marila*), which feed on *P. natans* and *P. pusillus*; green-winged teal (*Anas crecca*), which eats those of *P. crispus*, *P. epihydrus*, *P. foliosus*, *P. nodosus*, and *P. pusillus*; lesser scaups (*Aythya affinis*), which eat those of *P. foliosus*, *P. nodosus*, *P. praelongus*, and *P. pusillus*; mallards (*Anas platyrhynchos*), which eat those of *P. amplifolius*, *P. foliosus*, *P. natans*, *P. nodosus*, *P. pusillus*, *P. perfoliatus*, *P. praelongus*, *P. richardsonii*, and *P. strictifolius*; northern shovelers (*Anas clypeata*), which eat those of *P. nodosus*; pintails (*Anas acuta*), which eat those of *P. foliosus*, *P. nodosus*, and *P. pusillus*; redhead ducks (*Aythya americana*), which eat those of *P. foliosus*, *P. nodosus*, *P. perfoliatus*, and *P. pusillus*; ring-necked ducks (*Aythya collaris*), which eat those of *P. epihydrus*, *P. foliosus*, *P. gramineus*, *P. natans*, *P. nodosus*, *P. obtusifolius*, *P. perfoliatus*, *P. praelongus*, and *P. pusillus*; ruddy ducks (*Oxyura jamaicensis*), which eat those of *P. foliosus*, *P. nodosus*, and *P. pusillus*; trumpeter swans (*Cygnus buccinator*), which eat those of *P. richardsonii*, and wood ducks (*Aix sponsa*), which eat those of *P. epihydrus*, *P. natans*, and *P. nodosus*. The achenes of *P. epihydrus* also are consumed in quantity by spotted rails (Aves: Rallidae: *Pardirallus maculatus*) and those of *P. foliosus* by American

coots (Aves: Rallidae: *Fulica americana*). Alkaloids have been detected in several *Potamogeton* species (*P. amplifolius*, *P. crispus*, *P. epihydrus*, *P. foliosus*, *P. natans*, *P. praelongus*, *P. richardsonii*, and *P. robbinsii*) at concentrations believed to be potentially deterrent to herbivores (e.g., 0.16–0.56 mg g^{-1} dry wt). At least several North American *Potamogeton* species (*P. alpinus*, *P. berchtoldii*, *P. gramineus*, *P. natans*, and *P. perfoliatus*) also contain ecdysteroids (20-hydroxyecdysone and ecdysone), which are known to regulate arthropod development and metamorphosis and presumably act as deterrents to arthropod herbivory. Yet, *Potamogeton* plants host numerous insects (Insecta) including aphids (Hemiptera: Aphididae: *Rhopalosiphum nymphaeae* on *P. natans* and *P. nodosus*); beetles (Insecta: Coleoptera: Dytiscidae: *Acilius*, *Dytiscus*, *Hydroporus*; Haliplidae: *Haliplus*, *Peltodytes*) on *P. zosteriformis*; biting midges (Diptera: Ceratopogonidae) on *P. zosteriformis*; caddisflies (Trichoptera: Hydroptilidae: *Agraylea* on *P. zosteriformis*; Leptoceridae: *Ceraclea* on *P. zosteriformis*; *Leptocerus americanus* on *P. nodosus*; *Nectopsyche albida* on *P. amplifolius*, *P. gramineus*, *P. praelongus*, and *P. robbinsii*; *Triaenodes abus* on *P. amplifolius*; *Triaenodes injustus* on *P. amplifolius*, *P. friesii*, *P. richardsonii*, and *P. robbinsii*; *Triaenodes marginatus* on *P. alpinus*; *Triaenodes tardus* on *P. amplifolius*; Phryganeidae: *Ptilostomis* on *P. amplifolius*); damselflies (Odonata: Zygoptera: on *P. nodosus*; Zygoptera: Lestidae: *Lestes* on *P. zosteriformis*) and dragonflies (Odonata: Anisoptera: Aeshnidae; Libellulidae) on *P. nodosus*; dung flies (Diptera: Scathophagidae: *Hydromyza confluens* on *P. alpinus*); giant water bugs (Hemiptera: Belostomatidae) on *P. nodosus*; leaf beetles (Coleoptera: Chrysomelidae: *Neohaemonia nigricornis* on *P. illinoensis*, *P. natans*, and *P. richardsonii*; *Donacia cincticornis* on *P. alpinus*, *P. amplifolius*, *P. gramineus*, *P. natans*, and *P. richardsonii*; *Donacia hirticollis* on *P. alpinus*, *P. amplifolius*, *P. epihydrus*, *P. natans*, and *P. richardsonii*); mayflies (Ephemeroptera) on *P. nodosus* [Baetidae: *Baetis*] and on *P. zosteriformis* [Baetidae: *Baetis*; Caenidae: *Caenis*]; midges (Insecta: Diptera: Chironomidae) on *P. zosteriformis* [Chironominae: Chironomini; Orthocladiinae; Tanypodinae: Tanytarsini] and on *P. nodosus* [*Chironomus*]; *Cricotopus elegans* and *Cricotopus trifasciatus* on *P. amplifolius*, *P. natans*, and *P. nodosus*; *Cricotopus flavipes* on *P. amplifolius*, *P. epihydrus*, *P. illinoensis*, *P. praelongus*, *P. richardsonii*, and *P. robbinsii*; *Cricotopus myriophylli* on *P. natans* (isolated feeding on dead tissue); *Cricotopus trifasciatus* on *P. amplifolius*, *P. natans*, and *P. nodosus*; *Endochironomus nigricans* on *P. amplifolius*, *P. gramineus*, *P. illinoensis*, and *P. richardsonii*; *Glyptotendipes dreisbachi* on *P. amplifolius*, *P. gramineus*, *P. natans*, *P. nodosus*, *P. praelongus*, *P. richardsonii*, and *P. robbinsii*; *Glyptotendipes lobiferus* on *P. amplifolius*, *P. gramineus*, and *P. robbinsii*; *Polypedilum illinoense* and *Polypedilum ophioides* on *P. natans*; *Polypedilum sordens* on *P. amplifolius*, *P. gramineus*, *P. natans*, *P. richardsonii*, and *P. robbinsii*; moths (Lepidoptera: Crambidae: *Langessa nomophilalis* and *Munroessa gyralis* on *P. diversifolius* and *P. pulcher*; *Munroessa icciusalis* on *P. diversifolius*, *P. natans*, and *P. pulcher*; *Parapoynx allionealis* on *P.*

natans and *P. diversifolius*; *Parapoynx obscuralis* on *P. natans*; *Parapoynx badiusalis* on *P. amplifolius*, *P. natans*, *P. praelongus*, *P. richardsonii*, and *P. zosteriformis*; *Parapoynx obscuralis* on *P. amplifolius*, *P. diversifolius*, *P. natans*, *P. praelongus*, *P. pulcher*, and *P. richardsonii*; *Parapoynx seminealis* on *P. diversifolius*); parasitic wasps (Hymenoptera: Braconidae: numerous on other insect larvae; Ichneumonidae: *Apsilops hirtifrons* on *P. natans*); planthoppers (Hemiptera: Delphacidae: *Gamelus davisi*) on *P. natans*; shore flies (Diptera: Ephydridae: *Hydrellia ascita*) on *P. alpinus*, *P. amplifolius*, *P. epihydrus*, *P. foliosus*, *P. illinoensis*, *P. oakesianus*, and *P. richardsonii*; *Hydrellia bergi* on *P. natans*, *P. richardsonii*, and *P. zosteriformis*; *Hydrellia caliginosa* on *P. praelongus*; *Hydrellia cruralis* on *P. alpinus*, *P. amplifolius*, *P. epihydrus*, *P. foliosus*, *P. gramineus*, *P. illinoensis*, *P. natans*, *P. nodosus*, *P. praelongus*, *P. richardsonii*, and *P. zosteriformis*; *Hydrellia luctuosa* on *P. alpinus*, *P. amplifolius*, *P. natans*, *P. richardsonii*, and *P. zosteriformis*; *Hydrellia pulla* on *P. amplifolius*, *P. gramineus*, and *P. richardsonii*; *Notiphila loewi* on *P. alpinus* and *P. richardsonii*; and water bugs (Insecta: Hemiptera: Corixidae; Notonectidae) on *P. zosteriformis*. In addition to the preceding examples, additional wildlife uses are summarized as follows: Complete plants of many *Potamogeton* species are consumed by deer (Mammalia: Cervidae: *Odocoileus*). *Potamogeton alpinus* foliage (cellulose: 28.1% dw; crude protein: 18.1% dw; Na: 0.84% dw) yields an energy equivalent of 3832 cal g⁻¹ DM and is eaten by moose (Mammalia: Cervidae: *Alces americanus*). *Potamogeton amplifolius* is colonized by freshwater bryozoans (Bryozoa: Fredericellidae: *Fredericella sultana*; Paludicellidae: *Paludicella articulata*; Plumatellidae: *Hyalinella orbisperma*, *Plumatella emarginata*, *P. repens*) and provides cover for various fish such as bluegills (Vertebrata: Teleostei: Centrarchidae: *Lepomis macrochirus*). Its leaves contain (dw) from 16.5% to 23.0% crude fiber, 10.4–13.7% crude protein, 1.3–6.6% Ca, 2.25–3.30% K, 0.32–0.66% Na, have an energy equivalent 4416 cal g⁻¹, and are eaten by moose (Mammalia: Cervidae: *Alces americanus*). The foliage, rhizomes, and seeds are of some importance as a food for waterfowl (Aves: Anatidae) such as black ducks (*Anas rubripes*) and mallard ducks (*Anas platyrhynchos*). The plants are colonized by several cladocerans (Crustacea: Branchiopoda: Chydoridae: *Alona affinis*, *A. barbulata*, *A. guttatus*, *A. costata*, *A. rustica*, *A. setulosa*, *Alonella exigua*, *Chydorus faviformis*, *C. sphaericus*, *Graptoleberis testudinaria*, *Paralona pigra*, *Pleuroxus procurvus*, *Pseudochydorus globosus*; Daphniidae: *Scapholeberis kingi*; Ilyocryptidae: *Ilyocryptus spinifer*; Sididae: *Pseudosida bidentata*, *Sida crystallina*) and oligochaete worms (Annelida: Naididae: *Stylaria lacustris*). They are grazed by snails (Mollusca: Gastropoda: Physidae: *Physa gyrina*) and rusty crayfish (Crustacea: Decapoda: Cambaridae: *Orconectes rusticus*). The leaves can harbor large densities (to 1.4×10^8 diatoms m⁻²) of diatoms (Bacillariophyceae). *Potamogeton berchtoldii* is eaten by deer (Mammalia: Cervidae: *Odocoileus virginianus*). The plants are hosts to shore flies (Diptera: Ephydridae: *Hydrellia bilobifera*). *Potamogeton confervoides* is colonized

by various planktonic Crustacea including copepods (Copepoda: Cyclopoida) and water fleas (Cladocera). The foliage, fruits, and turions of *P. crispus* are eaten by coots (Aves: Rallidae: *Fulica atra*) and by various dabbling and diving ducks (Aves: Anatidae). The plants (1.25–3.87% Ca; 1.55–3.28% Cl; 1.70–3.30% K; 2.81–3.27% N; 0.64–1.37% Na; 3774 cal g⁻¹ dw) are eaten (but not highly preferred) by grass carp (Vertebrata: Teleostei: Cyprinidae: *Ctenopharyngodon idella*) and its F_1 hybrid (*C. idella* × *Hypophthalmichthys nobilis*). The plants also support large populations of invertebrates (to 374 per m stem), especially early in the season. These include insects (Insecta) such as beetles (Coleoptera), caddisflies (Trichoptera: Hydroptilidae; Leptoceridae), damselflies (Odonata: Zygoptera), dragonflies (Odonata: Anisoptera), mayflies (Ephemeroptera: Ephemeridae), midges (Diptera: Chironomidae), as well as amphipods (Crustacea: Amphipoda: Hyalellidae: *Hyalella*), bryozoans (Bryozoa: Plumatellidae: *Plumatella*), flatworms (Platyhelminthes: Planariidae: *Planaria*; Stenostomidae: *Stenostomum*), hydras (Cnidaria: Hydrozoa: Hydridae: *Hydra*), nematodes (Nematoda), oligochaete worms (Annelida: Naididae: *Chaetogaster*, *Dero*, *Naias*, *Stylaria*), rotifers (Rotifera: Flosculariidae: *Beauchampia*), and snails (Mollusca: Gastropoda: Ancylidae: *Ancylus*; Physidae: *Physella*; Planorbidae: *Helisoma*). The leaves are host to several Fungi (Ascomycota: Dipodascaceae: *Geotrichum*; Nectriaceae: *Fusarium cerealis*; Sordariomycetes: *Papulaspora aspera*; Oomycota: Pythiaceae: *Pythium carolinianum*) and are colonized by freshwater bryozoans (Bryozoa: Cristatellidae: *Cristatella mucedo*; Fredericellidae: *Fredericella sultana*; Pectinatellidae: *Pectinatella magnifica*; Paludicellidae: *Paludicella articulata*; Plumatellidae: *Plumatella emarginata*, *P. repens*). The foliage of *P. diversifolius* (0.65% Ca; 1.35% K; 3.47% N; 0.45% Na) is colonized by snails (Mollusca: Gastropoda: Physidae: *Physella virgata*). The plants also host several Fungi (Oomycota: Saprolegniaceae: *Achlya dubia*, *A. hypogyna*). They are consumed readily by grass carp (Vertebrata: Teleostei: Cyprinidae: *Ctenopharyngodon idella*). The foliage of *P. epihydrus* (4429 cal g⁻¹ dw; 16% DM crude protein; 0.80% Ca; 3.30% K; 3.16% N; 0.24% Na) is fairly high in sodium and other minerals (by % dw: Ca: 1.16–1.55; Fe: 0.27; K: 3.12; Mg: 0.40; Na: 0.57–0.72; SO₄: 1.70), and is eaten by moose (Mammalia: Cervidae: *Alces americanus*), muskrats (Mammalia: Cricetidae: *Ondatra zibethicus*), and white-tailed deer (Mammalia: Cervidae: *Odocoileus virginianus*). The plants also harbor oligochate worms (Annelida: Naididae: *Ripistes parasita*), rotifers (Rotifera), and other macroinvertebrates (averaging 1.9 [2.2 mg] individuals g⁻¹ dw). The leaves provide sites for the deposition of egg masses by caddis flies (Insecta: Trichoptera: Sericostomatidae) and also host several Fungi (Basidiomycota: Doassansiaceae: *Doassansia martianoffiana*, *Doassansia occulta*). The foliage of *P. foliosus* (3010–4536 cal g⁻¹ dw; crude protein: 20.30% dw; Na: 0.83% dw) is eaten by common carp (Vertebrata: Teleostei: Cyprinidae: *Cyprinus carpio*), grass carp (Vertebrata: Teleostei: Cyprinidae: *Ctenopharyngodon idella*), moose (Mammalia: Cervidae: *Alces americanus*), and muskrats (Mammalia: Cricetidae: *Ondatra zibethicus*). It can comprise up to 75% of the latter's stomach contents and also is used as a house building material by the animals. It is a substrate for epiphytic red algae (Rhodophyta: Batrachospermaceae: *Batrachospermum*) and is colonized by freshwater bryozoans (Bryozoa: Lophopodidae: *Lophopodella carteri*; Plumatellidae: *Plumatella emarginata, Stolella indica*). The plants also provide important cover for larval red-spotted newts (Vertebrata: Salamandridae: *Notophthalmus viridescens*) and painted turtles (Reptilia: (Emydidae: *Chrysemys picta marginata*). Huge numbers of invertebrates can be found on the plants, with some estimates (means) ranging up to 3,876.4 m^{-2} (184,052.5 kg^{-1} plant). The foliage and turions of *P. friesii* (total lipids: 7.9 mg g^{-1} [wet wt]) are eaten by several ducks (Aves: Anatidae) including the American wigeon (*Anas americana*) and gadwall (*Anas strepera*). The leaves of *P. gramineus* (4221–4363 cal g^{-1} dw; crude protein: 14.4–15.5% dw; Ca: 1.71% dw; Na: 0.61–0.66% dw) are eaten by caribou (Mammalia: Cervidae: *Rangifer tarandus*), muskrats (Mammalia: Cricetidae: *Ondatra zibethicus*), and moose (Mammalia: Cervidae: *Alces americanus*). They also are host to a fungus (Basidiomycota: Doassansiaceae: *Doassansia martianoffiana*). The "winter buds" (4198 cal g^{-1} dw) have an average nitrogen content of 2.25%. The mean caloric content of the rhizomes is 3957 cal g^{-1} dw. The foliage harbors small numbers (e.g., 2 100 cm^{-2}) of amphipods (Crustacea: Amphipoda: *Hyalella azteca*) and other invertebrates such as oligochaete worms (Annelida: Naididae: *Arcteonais lomondi*). The plants also host a fungus (Basidiomycota: Doassansiaceae: *Doassansia occulta*) and are colonized by freshwater bryozoans (Bryozoa: Cristatellidae: *Cristatella mucedo*; Fredericellidae: *Fredericella sultana*; Paludicellidae: *Paludicella articulata*; Pectinatellidae: *Pectinatella magnifica*; Plumatellidae: *Plumatella emarginata, P. fruticosa, P. repens*). *Potamogeton illinoensis* provides habitat for the imperiled fountain darter (Teleostei: Percidae: *Etheostoma fonticola*), especially in more lotic sites. The plants harbor roughly 4.2 invertebrate individuals g^{-1} (fw), which, in addition to insects (see above), include amphipods (Crustacea: Amphipoda: *Hyalella*), flatworms (Platyhelminthes: Planariidae: *Planaria*), water mites (Arachnida: Acari: Hydrachnidae), and oligochaete worms (Annelida: Clitellata). They also are colonized by freshwater bryozoans (Bryozoa: Paludicellidae: *Paludicella articulata*; Plumatellidae: *Plumatella repens*) and are used by broad-winged damselflies (Insecta: Odonata: Calopterygidae: *Calopteryx aequabilis*) and snails (Mollusca: Gastropoda) as a substrate for the attachment of egg masses. The foliage (crude protein: 12%; nutrients [means]: Ca: to 208 mg g^{-1} dw; N: to 16.5 mg g^{-1} dw; Na: to 3419 µg g^{-1} dw) is readily eaten by giant rams-horn snails (Mollusca: Gastropoda: Ampullariidae: *Marisa cornuarietis*). The plants are regarded as a minor food of ducks (Aves: Anatidae: *Anas*) in some parts of the United States. The foliage of *P. natans* (3069–4491 cal g^{-1} dw; crude protein: 14.55% dw; 1.09–4.93% K: 1.0–1.5%; Ca: 0.95–20.0%; N content; Na: 0.24–0.42%) is eaten by caribou (Mammalia: Cervidae: *Rangifer tarandus*), muskrats

(Mammalia: Cricetidae: *Ondatra zibethicus*), and moose (Mammalia: Cervidae: *Alces americanus*). The rhizomes ($\bar{x} = 24.7$ g DM m^{-2}; starch content: $\bar{x} = 28.5\%$; crude fiber: 5.9–6.9%; digestibility: $\bar{x} = 68.4\%$) are eaten by grizzly bears (Mammalia: Ursidae: *Ursus arctos*). The plants host several Fungi (Basidiomycota: Doassansiaceae: *Doassansia martianoffiana, Doassansia occulta*) and are colonized by numerous cladocerans (Crustacea: Branchiopoda: Chydoridae: *Acroperus harpae, Alona barbulata, A. costata, A. guttata, A. rustica, A. setulosa, Chydorus sphaericus, Graptoleberis testudinaria, Picripleuroxus denticulatus, Pleuroxus procurvus*; Daphniidae: *Scapholeberis kingi*; Ilyocryptidae: *Ilyocryptus spinifer*; Sididae: *Sida crystallina*) and freshwater bryozoans (Bryozoa: Cristatellidae: *Cristatella mucedo*; Paludicellidae: *Paludicella articulata*; Plumatellidae: *Plumatella emarginata, P. fruticosa, P. repens*). They also support large populations (up to 4.2 individuals g^{-1} fw) of amphipods (Crustacea: Amphipoda: Hyalellidae: *Hyalella*), flatworms (Platyhelminthes: Planariidae: *Planaria*), insects (see above), snails (Mollusca: Gastropoda), water mites (Arachnida: Acari: Hydrachnidae), and oligochaete worms (Annelida: Clitellata). The region between the stipule and stem of the plants is used as an oviposition site by female leaf beetles (Insecta: Coleoptera: Chrysomelidae: *Donacia hirticollis, Neohaemonia nigricornis*). The fruits of *P. nodosus* are highly rated as a food for waterfowl (see above). The foliage (4292 cal g^{-1} dw) has an average nitrogen content of 3.40%; the "winter buds" (4077 cal g^{-1} dw) have an average nitrogen content of 2.24%. The plants are eaten by several fish (Vertebrata: Teleostei) including silver dollar fish (Characidae: *Metynnis lippincottianus, Mylossoma duriventris*) and triploid grass carp (Cyprinidae: *Ctenopharyngodon idella*). The rhizomes ($\bar{x} = 24.7$ g DM m^{-2}; starch content: $\bar{x} = 28.5\%$; crude fiber: 5.9–6.9%; digestibility: $\bar{x} = 68.4\%$) are eaten by grizzly bears (Mammalia: Ursidae: *Ursus arctos*). *Potamogeton nodosus* also supports a rich invertebrate fauna (5 g^{-1}; up to 555 per plant), which in addition to insects (see above), includes amphipods (Crustacea: Amphipoda: Hyalellidae: *Hyalella*), flatworms (Platyhelminthes: Planariidae: *Planaria*), nematodes (Nematoda), snails (Mollusca: Gastropoda: Hydrobiidae: *Amnicola limosus*; Planorbidae: *Gyraulus parvus*), oligochaete worms (Annelida: Clitellata: Naididae), and water mites (Arachnida: Acari: Hydrachnidae). Typical plant beds can harbor more than 30 million invertebrates during the summer months (e.g., June, August), supporting densities ($\bar{x} = 17,170$ m^{-2}) as much as seven times higher than that of adjacent unvegetated zones. In turn, these animals provide food for foraging fish like largemouth bass (Vertebrata: Teleostei: Centrarchidae: *Micropterus salmoides*). The leaves also host several Fungi (Ascomycota: Mycosphaerellaceae: *Entylomella aquatilis, Ramulispora zonata*; Nectriaceae: *Cylindrocarpon, Fusarium cerealis, Heliscus lugdunensis*; Pleosporales: *Mycocentrospora acerina*; Sclerotiniaceae: *Botrytis cinerea*; Sordariomycetes: *Papulaspora aspera*; Basidiomycota: Ceratobasidiaceae: *Rhizoctonia*; Doassansiaceae: *Doassansia martianoffiana, Doassansia occulta*). In tidal habitats, the plants provide important habitat for numerous fish (e.g., Vertebrata: Teleostei: Cyprinodontidae: *Cyprinodon variegatus*; Fundulidae: *Lucania parva*) and crustaceans (Crustacea: Decapoda) such as grass shrimp (Palaemonidae: *Palaemonetes paludosus*) and blue crabs (Portunidae: *Callinectes sapidus*). Invasive red imported fire ants (Insecta: Hymenoptera: Formicidae: *Solenopsis invicta*) use the floating leaves of *P. nodosus* as platforms, upon which to establish foraging trails. Leaves of *P. obtusifolius* are eaten by snails (Mollusca: Gastropoda), but are not particularly palatable to them, with a maximum consumption rate of about 1.3 mg day^{-1}. *Potamogeton perfoliatus* (total lipids: 5.9–100 mg g^{-1} wet wt) is eaten by moose (Mammalia: Cervidae: *Alces americanus*) and all parts of the plant including the seeds (see above) are eaten by waterfowl (Aves: Anatidae) such as Canada geese (*Branta canadensis*), tundra swans (*Cygnus columbianus*), and the duck species mentioned previously. The plants are host to a fungus (Basidiomycota: Doassansiaceae: *Doassansia occulta*). *Potamogeton polygonifolius* is eaten by snails (Mollusca: Gastropoda: Lymnaeidae: *Lymnaea stagnalis*), which can consume 5–10 mg DM day^{-1}. *Potamogeton praelongus* is colonized by freshwater bryozoans (Bryozoa: Cristatellidae: *Cristatella mucedo*; Fredericellidae: *Fredericella sultana*; Pectinatellidae: *Pectinatella magnifica*; Plumatellidae: *Plumatella repens*) and its foliage is inhabited by cladocerans (Crustacea: Branchiopoda: Chydoridae: *Pleuroxus procurvus*; Sididae: *Sida crystallina*). The leaves (4581 cal g^{-1} dw; crude protein: 15.8–17.8% dw; Na: 0.26–0.42% dw) are eaten by moose (Mammalia: Cervidae: *Alces americanus*). *Potamogeton pulcher* contains 1.21% crude protein (submersed leaves) and relatively low levels of Cl (0.74%), N (0.83–2.88%), and Na (0.2%), but fairly high levels of K (1.15–2.77%). It is a host plant for several larval moths (Insecta: Lepidoptera: Crambidae: *Munroessa gyralis*, *Parapoynx obscuralis*). The foliage of *P. pusillus* (Ca: 0.55% dw; total lipids: 12.3 mg g^{-1} wet w; Cl: 1.7%; K: 3.25%; N: 3.25%; Na: 0.21%) is eaten by Canada geese (Aves: Anatidae: *Branta canadensis*), grass carp (Cyprinidae: *Ctenopharyngodon idella*), moose (Mammalia: Cervidae: *Alces americanus*), muskrats (Mammalia: Cricetidae: *Ondatra zibethicus*), and waterfowl (Aves: Anatidae). The plants are host to a fungus (Basidiomycota: Doassansiaceae: *Doassansia occulta*). *Potamogeton richardsonii* is eaten by carp (Vertebrata: Teleostei: Cyprinidae: *Cyprinus carpio*), moose (Mammalia: Cervidae: *Alces americanus*), and rusty crayfish (Crustacea: Decapoda: Cambaridae: *Orconectes rusticus*). The shoots are consumed by mute swans (Aves: Anatidae: *Cygnus olor*), whereas the rhizomes (7.4% crude protein; 74.4% non-structural carbohydrates) are relatively low in fiber (13.8%) and are eaten preferentially by Canada geese (Aves: Anatidae: *Branta canadensis*). The plants have been observed to harbor as many as 19 different genera and a total of 3,660 invertebrate organisms m^{-2} (7,881 animals kg^{-1} plant material [wet mass]) at a particular site. These include amphipods (Crustacea: Amphipoda: Gammaridae: *Gammarus*), caddisflies (Insecta: Trichoptera: Hydroptilidae: *Agraylea*; Limnephilidae: *Limnephilus*; Polycentropodidae: *Neureclipsis bimaculata*),

damselflies (Insecta: Odonata: Coenagrionidae: *Enallagma*, *Ischnura*), flies (Insecta: Diptera: Ceratopogonidae: *Bezzia*; Chironomidae: *Cricotopus*, *Glypotendipes*, *Eukiefferiella*, *Psectrocladius*), freshwater bryozoans (Bryozoa: Cristatellidae: *Cristatella mucedo*; Fredericellidae: *Fredericella sultana*; Paludicellidae: *Paludicella articulata*; Plumatellidae: *Hyalinella punctata*, *Plumatella repens*), cladocerans (Crustacea: Cladocera: Chydoridae: *Alona affinis*, *Chydorus sphaericus*, *Eurycercus lamellatus*; Sididae: *Sida crystallina*), flatworms (Platyhelminthes: Dugesiidae: *Dugesia*), gastropods (Mollusca: Gastropoda: Hydrobiidae: *Amnicola*; Planorbidae: *Gyraulus*; Physidae: *Physella*; Valvatidae: *Valvata*), hydras (Cnidaria: Hydridae: *Hydra*), mayflies (Insecta: Ephemeroptera: Baetidae: *Baetis*; Caenidae: *Caenis*), moths (Insecta: Lepidoptera: Crambidae: *Paraponyx*), nematodes (Nematoda), water mites (Arthropoda: Acari: Hydrachnidae), and worms (Annelida: Naididae: *Chaetogaster diaphanus*, *Stylaria lacustris*). The plants are susceptible to non-consumptive mortality by California crayfishes (Crustacea: Decapoda: Astacidae: *Pacifastacus leniusculus*). *Potamogeton robbinsii* (4511 cal g^{-1} dw; crude protein: 15.50% dw; Na: 0.52% dw) is eaten by moose (Vertebrata: Cervidae: *Alces americanus*). The leaves are colonized by several cladocerans (Crustacea: Branchiopoda: Chydoridae: *Alonella excisa*, *A. exiqua*; Sididae: *Sida crystallina*) and freshwater bryozoans (Bryozoa: Fredericellidae: *Fredericella sultana*; Paludicellidae: *Paludicella articulata*; Plumatellidae: *Plumatella repens*). They can be covered by numerous epiphytic diatoms (Chromista: Bacillariophyceae); e.g., 133 taxa representing more than 13 ×10^5 cells cm^{-2}, which aggregate near the leaf margins. The plants provide habitat for red-spotted newts (Amphibia: Caudata: Salamandridae: *Notophthalmus viridescens*), especially in deeper waters (>5 m). *Potamogeton spirillus* is eaten by moose (Vertebrata: Cervidae: *Alces americanus*). The fruits are host to water molds (Fungi: Oomycota: Pythiaceae: *Cornumyces muenscheri*; *Lagenidium*). The foliage of *P. vaseyi* hosts a fungus (Basidiomycota: Doassansiaceae: *Doassansia occulta*). The foliage of *P. zosteriformis* (4436 cal g^{-1} dw; crude protein: 18.80% dw; Na: 0.41% dw; 3.37–3.70% total N; 0.35–0.59% total P) is eaten by moose (Mammalia: Cervidae: *Alces americanus*). It is colonized by cladocerans (Crustacea: Branchiopoda: Chydoridae: *Alona rustica*; Sididae: *Sida crystallina*) and by freshwater bryozoans (Bryozoa: Paludicellidae: *Paludicella articulata*; Plumatellidae: *Plumatella repens*). In addition to those listed above, the plants host a wide variety of invertebrates including amphipods (Crustacea: Amphipoda: Hyalellidae: *Hyalella azteca*); branchiopods (Crustacea: Branchiopoda: Laevicaudata); copepods (Crustacea: Copepoda: Calanoida: Diaptomidae: *Diaptomus nudus*; Cyclopoida; Harpacticoida); flatworms (Platyhelminthes: Typhloplanidae: *Mesostoma*); hydras (Cnidaria: Hydridae: *Hydra oligactis*); oligochaete worms (Annelida: Naididae: *Chaetogaster*, *Stylaria lacustris*); ostracods (Ostracoda); rotifers (Rotifera); snails (Mollusca: Gastropoda); water fleas (Crustacea: Diplostraca: Cladocera: Bosminidae: *Bosmina longirostris*; Chydoridae: *Alona*,

Alonella excisa, *Camptocercus*, *Chydorus*, *Eurycercus longirostris*, *Graptoleberis testudinaria*, *Picripleuroxus denticulatus*, *Pleuroxus aduncus*, *Pleuroxus procurvus*, *Pseudochydorus globosus*; Daphniidae: *Ceriodaphnia dubia*, *Daphnia magna*, *D. pulex*, *D. rosea*, *Scapholeberis kingi*, *Simocephalus vetulus*; Sididae: *Diaphanosoma birgei*); and water mites (Acari: Prostigmata: Hydrachnoidea: Hydrachnidae: *Hydrachna*). Mercury-resistant Bacteria (e.g., Pseudomonadaceae: *Pseudomonas chlororaphis*; Shewanellaceae: *Shewanella putrefaciens*), which are implicated in macrophyte mercury resistance, have been isolated from the foliage of *Potamogeton crispus*, *P. diversifolius*, *P. pusillus*, and *P. richardsonii*.

Economic importance: food: The site of collection must always be considered before attempting to eat any plants taken from their natural habitats; e.g., *P. epihydrus* can bio-concentrate levels of mercury (to 25.4 ng g^{-1}) and methyl mercury (to 9.7 ng g^{-1}), which although low, represent up to a 640-fold increase relative to water concentrations. That being said, a nutritional analysis of *P. diversifolius* indicated that the plants contain about 17.3% crude protein, 2.9% crude fat, 30.9% cellulose, and have a caloric content of 3.4 kcal g^{-1}. The starchy rhizomes of *Potamogeton natans* reportedly are edible (but see next); **medicinal:** *Potamogeton crispus* contains the flavonoid luteolin-3'-*O*-β-D-glucopyranoside, which has exhibited anti-metastatic properties against human ovarian cancer (ES-2) cells. *Potamogeton natans* contains potamogetonan PN-300, a pectin that exhibits an antiinflammatory effect when administered orally to mice (Mammalia: Rodentia: Muridae). *Potamogeton nodosus* has been used as a folk medicine to treat various ailments including acne, coughs, diarrhea, dysentery, jaundice, and tuberculosis. Ethanolic extracts of the plants have antioxidant activity. A furanoid diterpene [15,16-epoxy-12-oxo-8(17),13(16),14-labdatrien-20,19-olide] and 2-hydroxyheptane-3,5-dione isolated from *P. nodosus* are inhibitory against Gram-positive and Gram-negative bacteria. The rhizomes contain several lectins, which exhibit cytotoxicity in a standard assay using brine shrimp (Crustacea: Artemiidae: *Artemia salina*). Methanolic extracts of *P. perfoliatus* were inhibitory against all Gram-positive bacteria tested and some Gram-negative bacteria. Methanolic extracts of *P. pusillus* have antioxidant activity and inhibit the growth of some Gram-negative bacteria; **cultivation:** *Potamogeton crispus* is not cultivated but occasionally is inadvertently mixed with shipments of other horticultural specimens; **misc. products:** *Potamogeton amplifolius* has properties similar to those of alfalfa and has been considered for use as silage. The plants have established successfully in lake revegetation programs. The high nitrogen content of *P. crispus* foliage has led to its consideration for use as compost and as an animal food. Ducks (Aves: Anatidae) feeding on *P. crispus* lay naturally red-yolked eggs due to the plant's carotenoid content. Plants of *P. crispus* have been used to remove a dye (Reactive Red 198) from aqueous solutions by adsorption. They also have been considered for the removal of low-level liquid radioactive wastes such as cerium, cesium, and cobalt contamination. The Kawaiisu people used the dried stem fibers of *P. diversifolius* as a durable cordage to construct fishing and hunting

nets. *Potamogeton epihydrus* has been considered for use in the phytoremediation of waters contaminated with aluminum, with foliar uptake rates averaging 1.68 mg Al g^{-1} day^{-1}. A decoction made from *P. natans* was taken by the Navajo and Ramah tribes as a ceremonial emetic. *Potamogeton natans* contains potamogetonin and at least six other furano-diterpenes, which exhibit anti-algal properties. The plants also contain several lactone diterpenes, which have similar properties. *Potamogeton natans* also has been considered for use in the phytofiltration of storm waters and the removal of waterborne heavy metals such as zinc. Dense stands of *P. nodosus* have been found to suppress the growth of a problematic mat-forming cyanobacterium (Cyanobacteria: Oscillatoriaceae: *Lyngbya wollei*). Chloroplasts/thylakoids isolated from *P. nodosus* have been used to generate Au nanoparticles, which are used in a variety of medical, scientific, and technical applications. *Potamogeton nodosus* has been evaluated as an excellent candidate for wetland restoration projects, with transplanted rhizomes exhibiting high (80–100%) survival; **weeds:** *Potamogeton crispus* is regarded as an aggressive weed throughout North America. Although indigenous, *P. foliosus* often is categorized as a weed. It has encroached spawning sites of Chinook salmon (Vertebrata: Teleostei: Salmonidae: *Oncorhynchus tshawytscha*). *Potamogeton nodosus* is regarded as a weed in the western United States, where it can proliferate in irrigation canals. It often is characterized as weedy due to its ability to spread rapidly on the surface of water; **nonindigenous species:** *Potamogeton crispus* was introduced to North America near Philadelphia, Pennsylvania (USA) sometime before 1841 and reached Ontario, Canada by 1891. It is believed to have spread across North America as a contaminant in hatchery stocks of coldwater fish. The plants also can be spread inadvertently in contaminated shipments of aquatic horticultural specimens. *Potamogeton epihydrus* has been introduced to England, Scotland, and Wales. *Potamogeton foliosus* and *P. nodosus* are believed to have been accidentally introduced to the Hawaiian Islands in the early 19th century (<1825). *Potamogeton nodosus* also has been introduced to Luxembourg. Although some accounts mistakenly designate *P. polygonifolius* as nonindigenous to North America, it is an amphi-Atlantic species, with North American records dating back into the early 19th century.

Systematics: Phylogenetic analyses consistently have resolved *Potamogeton* as a monophyletic sister genus to *Stuckenia* (e.g., Figure 1.21), which had long been subsumed within the former. However, the delimitation of pondweed species and their interrelationships has not been as forthcoming, with countless discrepancies arising among various studies. *Potamogeton* species exhibit broad phenotypic plasticity, which has confounded their systematic study using morphological data and has promoted the use of molecular (DNA) markers instead. However, the molecular data also are problematic and have yielded numerous discordant results. Some of the uncertainty can be attributed to concerted evolution of the nuclear ribosomal repeat region, which has been used routinely as a source of DNA markers. As one result, hybridization can result in progeny possessing the DNA sequence of just one parent (either maternal or paternal) rather than maintaining both parental copies as a polymorphism (which also can occur). Thus, if a taxon of hybrid origin has undergone concerted evolution, its DNA profile can be quite misleading. Less is known regarding the influence of concerted evolution between the genomes of polyploid individuals, which occur throughout the genus. Genetic analyses have indicated extensive inter- and intraspecific variation in the rDNA loci and rapid divergence of 45S rDNA (i.e., 18S, 5.8S, and 26S rRNA genes) as a consequence of hybridization, indicating that chromosomal repatterning could result in the gain or loss of the rDNA loci altogether. How these factors might influence the results of molecular phylogenetic analyses has not been given an appropriate evaluation. Also, preliminary studies have indicated that some chloroplast genes (e.g., *rbcL*) have undergone positive selection in *Potamogeton*, a factor not yet considered (or accounted for) in the molecular phylogenetic studies conducted. Moreover, other anomalies occur between different analyses analyzing similar DNA markers. Although more difficult to explain, such outcomes likely involve material that has been misidentified, even where noted experts have made the determinations. Several of the more notable incongruencies include the following: *P. confervoides* and *P. epihydrus* resolve as sister groups to a clade primarily of broad-leaved species in one analysis but are embedded within a clade primarily of linear-leaved taxa in another. *Potamogeton bicupulatus* and *P. spirillus* resolve in one analysis as sister to the remainder of the genus but are embedded within a clade primarily of linear-leaved taxa in another. In another evaluation *P. clystocarpus* and *P. foliosus* resolve as sister species that are distant from *P. pusillus*, while a differnt study resolves *P. clystocarpus* with *P. berchtoldii* accessions in a position distant from a clade comprising *P. pusillus* and *P. foliosus*, which resolve as sister species. Furthermore (e.g., Figure 1.22), different accessions of *P. gramineus*, *P. berchtoldii*, *P. pusillus*, and *P. groenlandicus* resolve not together but in different places, with only the latter demonstrated as the result of hybridization. Even more perplexing is the consistent placement of some species in highly unlikely associations. A good example here is the position of *P. praelongus* as a close relative of the morphologically disparate *P. natans* (e.g., Figure 1.21) in a clade distant to morphologically similar species (i.e., *P. perfoliatus* and *P. richardsonii*). Even in this case where other data (e.g., flavonoid compounds) show differences between *P. praelongus* and *P. perfoliatus* or *P. richardsonii*, it is difficult to envision the derivation of *P. natans* and *P. praelongus* from a recent common ancestor, at least on morphological and anatomical grounds. It is evident that despite having much phylogenetic information currently available for *Potamogeton*, further analyses will be essential before the interspecific relationships can be ascertained with confidence. At least many of the results depicted in various phylogenetic analyses (e.g., the close relationship of *P. bicupulatus* and *P. spirillus*) appear to be reasonable. Some of the most confounding taxonomic issues in North America have involved basic species delimitations. North American botanists have accepted the synonmy of *P. berchtoldii* and *P. pusillus* (the

former often segregated as *P. pusillus* subsp. *tenuissimus*) since the last major revision of the group was published in 1974. However, several morphological and genetic (multiple allozyme and DNA sequence analyses) studies have clearly shown that these taxa not only are distinct but are not even

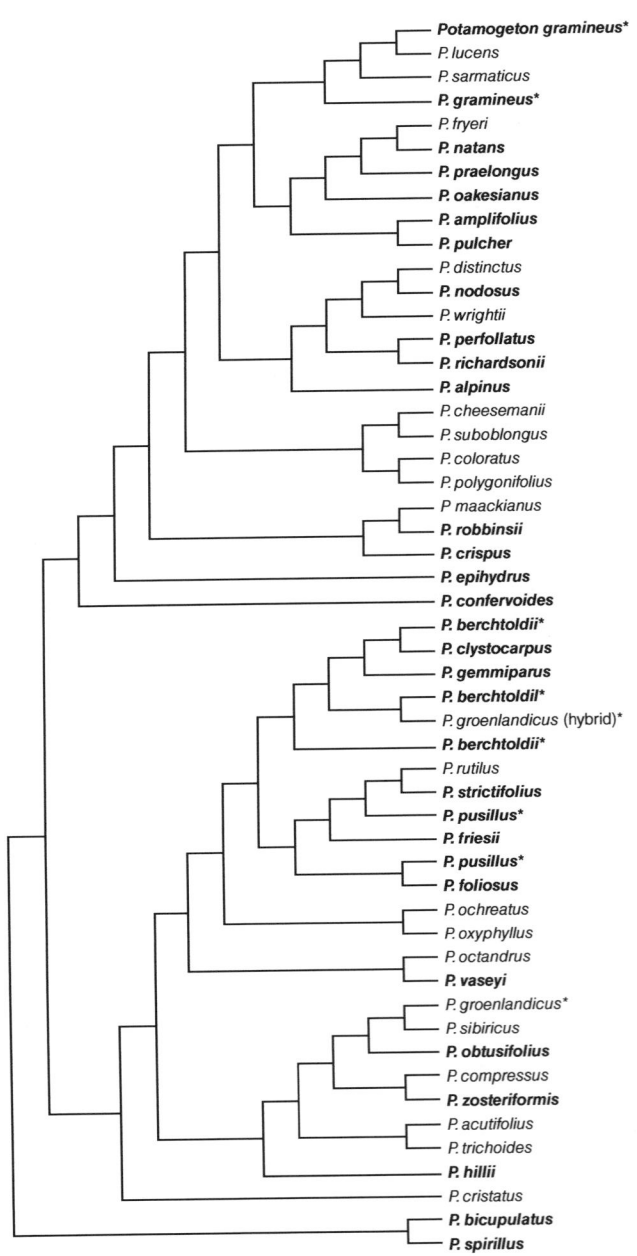

FIGURE 1.22 Species relationships in *Potamogeton* as indicated by analysis of 5S-NTS sequence data (modified from Kaplan et al., 2013). These molecular data segregate pondweed species into two principal groups comprising primarily linear-leaved species (lower clade) and broad-leaved species (upper clade). Two heterophyllous species (*P. bicupulatus* and *P. spirillus*) form a sister group to the remainder of the genus. Although many of the relationships depicted here are reasonable, some of them (e.g., the close association of *P. natans* and *P. praelongus*) are unusual and warrant further evaluation. Also, the accessions of several species (indicated by asterisks) do not resolve as clades, which indicates problems due to hybridization or other complicating factors (see text). The OBL North American indicators (shown in bold) have originated throughout the genus.

particularly closely related. The putative narrow endemic known as *P. clystocarpus* is particularly problematic, especially given its current imperilment status as federally (USA) listed endangered species. It is only slightly (if at all) distinct morphologically from *P. berchtoldii* (differing essentially by minor fruit features that appear only after the material has been dried) and has been regarded as a difficult-to-identify "cryptic" species on the basis of AFLP and DNA sequence data analyses. It did resolve as a distinct species (closely related to *P. foliosus*) in one fairly comprehensive phylogenetic study of the genus; however, the results of that study disagreed with several subsequent studies, which demonstrated that the DNA sequences published for *P. clystocarpus* are quite distant from *P. foliosus* but instead are similar to those of *P. berchtoldii* (see Figure 1.22), with which it hybridizes. In any case, it is evident that at least some of the "*P. clystocarpus*" material provided for genetic analyses has been misidentified (even by experts), and a more critical evaluation of this taxon is necessary. Although dubiously distinct, *P. clystocarpus* has been retained here as a species pending further study. Severe floods of the type locality in 1932, 1991, and 1992 had "washed the canyon clean of aquatic vegetation," which makes it uncertain how [or if?] the species has recolonized the site subsequently. Several attempts to relocate the species subsequent to those flood years were unsuccessful and it is conceivable that *bona fide P. clystocarpus* no longer exists but has been mistaken for other similar species in more recent collections. Another putative narrow endemic is *P. floridanus*, which also was regarded skeptically as a distinct species by many authors, who suggested it to be of hybrid origin (fruits have yet to be observed on the plants). Intial analyses using DNA sequence data (5S ribosomal RNA nontranscribed spacer sequences) showed the taxon to be distinct and closely allied with *P. oakesianus*, from which it differs essentially by minor variation in the shape of the floating leaf blade. Yet, 10 of the 11 5S nucleotide sites which differed between the taxa were polymorphic, which also indicated the hybrid origin of the plants. Subsequent analyses eventually confirmed that *P. floridanus* was indeed a hybrid (now recognized as *P. ×floridanus*) involving *P. oakesianus* and *P. pulcher*, with the latter as the maternal parent. It has been included here only because of the recent nature of that disclosure. "*Potamogeton ogdenii*" should not be recognized as a distinct species. The flavonoid evidence originally used to support its status was unconvincing because the compounds were identified only by *rf* values and not by spectral or other more precise means of determination. Moreover, the biochemical profile of the putative hybrid species was remarkably similar to that of *P. hillii*, with the few possibly unique compounds not being additive between the suspected parental species. Along with its morphological intermediacy, this taxon at best seemed to represent a slightly fertile F_1 hybrid derivative of *P. hillii*. As suspected, a more intensive evaluation using DNA sequence data verified that "*P. ogdenii*" is not a distinct hybrid species, but a recent hybrid derivative of *P. hillii* ($2n=26$) and *P. zosteriformis* ($2n=28$). The plants also possess a chromosome number ($2n=27$), which is intermediate between the parental species. Several

nomenclatural issues have been clarified in recent studies. *Potamogeton sibiricus* is the correct name for the taxon long known under the synonyms *P. porsildiorum* and *P. subsibiricus*. The name "*Potamogeton oblongus*," which also has been used in the wetland indicator list, should be abandoned for the North American plants (as done here), given that morphological and DNA sequence data clearly indicate no differences from *P. polygonifolius*. Some authors have summarily dismissed the possibility of intraspecific chromosome number variation despite a fair amount of rather persuasive evidence to the contrary. Nonetheless, the counts associated with the majority of pondweed species fall within two chromosomal series based on $x=14$ and $x=13$ and its polyploid derivatives. These are as follows: $x=14$: $2n=28$ (*P. polygonifolius, P. spirillus, P. zosteriformis*); $x=13$: $2n=26$ (*P. berchtoldii, P. clystocarpus, P. epihydrus, P. foliosus, P. friesii, P. gramineus, P. hillii, P. obtusifolius, P. pusillus, P. strictifolius*); $2n=52$ (*P. alpinus, P. amplifolius, P. crispus, P. natans, P. nodosus, P. perfoliatus, P. praelongus, P. richardsonii, P. robbinsii*); $2n=104$ (*P. illinoensis*). Several attempts to deduce the ancestral base of the genus using phylogenetic trees have been made; however, topological variation among the placement of key clades with respect to chromosomal series remains unsettled making it difficult to settle that question with certainty. Reported *Potamogeton* hybrids abound, which primarily appear to represent F_1 progeny. Although inadvisable from a systematic standpoint, many of these spontaneous F_1 plants have been given binomial names (along with a multiplication sign), which should be applied only to distinct species of known hybrid origin. The list of potential North American hybrids (binomial names in parentheses) includes: *P. alpinus* × *P. crispus* (= *P. ×olivaceus*); *P. alpinus* × *P. gramineus* (= *P. ×nericius*); *P. alpinus* × *P. nodosus* (= *P. ×subobtusus*); *P. alpinus* × *P. perfoliatus* (= *P. ×prussicus*); *P. alpinus* × *P. praelongus* (= *P. ×griffithii*); *P. berchtoldii* × *P. clystocarpus*; *P. berchtoldii* × *P. gemmiparus*; *P. berchtoldii* × *P. vaseyi*; *P. bicupulatus* × *P. epihydrus* (= *P. ×aemulans*); *P. crispus* × *P. friesii* (= *P. ×lintonii*); *P. crispus* × *P. obtusifolius*; *P. crispus* × *P. perfoliatus* (= *P. ×cooperi*); *P. crispus* × *P. praelongus* (= *P. ×undulatus*); *P. epihydrus* × *P. perfoliatus* (= *P. ×versicolor*); *P. foliosus* × *P. pusillus*; *P. friesii* × *P. obtusifolius* (= *P. ×semifructus*); *P. friesii* × *P. pusillus* (= *P. ×pusilliformis*); *P. gemmiparus* × *P. vaseyi*; *P. gramineus* × *P. illinoensis* (= *P. ×deminutus*); *P. gramineus* × *P. natans* (= *P. ×sparganiifolius*); *P. gramineus* × *P. oakesianus* (= *P. ×mirabilis*); *P. gramineus* × *P. perfoliatus* (= *P. ×nitens*); *P. gramineus* × *P. nodosus* (= *P. ×faxonii*); *P. hillii* × *P. zosteriformis* (*P. ×ogdenii*); *P. natans* × *P. nodosus* (= *P. ×schreberi*); *P. natans* × *P. pusillus* (= *P. ×variifolius*); *P. natans* × *P. praelongus* (= *P. ×vepsicus*), *P. perfoliatus* × *P. praelongus* (= *P. ×cognatus*); *P. perfoliatus* × *P. pusillus* (= *P. ×mysticus*); *P. perfoliatus* × *P. richardsonii* (= *P. ×absconditus*); *P. strictifolius* × *P. zosteriformis* (= *P. ×haynesii*).

Distribution: *Potamogeton* species can be found throughout North America. *Potamogeton alpinus* (circumboreal) and *P. zosteriformis* occur across Canada and the northern United States; *P. amplifolius* grows throughout central North

America; *P. berchtoldii* and *P. pusillus* (also Africa, Eurasia, South America) are widespread throughout North America; *P. bicupulatus* occurs primarily along the coastal plain of New England but is disjunct in the upper Midwest; *P. clystocarpus* is a narrow endemic known only from Little Aguja Creek, Texas; *P. confervoides, P. oakesianus, P. spirillus,* and *P. vaseyi* grow in northeastern North America (*P. oakesianus* also is disjunct in British Columbia); *P. crispus* (also Africa, Australia, Eurasia, South America) occurs throughout the United States and South Canada; *P. diversifolius* is mainly found in the eastern United States but is sporadic in the western United States and extends into Mexico; *P. epihydrus* occurs in eastern and western North America and is disjunct in the Hebrides of Europe; *P. floridanus* (recently shown to be a hybrid – see above) is a narrow endemic known only from a few streams in the Florida panhandle; *P. foliosus* is widespread throughout North America and extends into Central America; *P. friesii* (circumpolar) is found in the glaciated parts of the northern United States and in Canada; *P. gramineus* (circumpolar) is widespread throughout central and northern North America; *P. hillii* is restricted to the northeastern United States and adjacent portions of Canada; *P. illinoensis* is fairly widespread throughout central and southern North America; *P. natans* (circumpolar) is widespread throughout all but southernmost North America and also occurs in much of Eurasia; *P. nodosus* is widespread, mainly south of the glacial maximum, and extends into South America and Eurasia; *P. obtusifolius* (circumpolar) and *P. strictifolius* occur throughout northern North America; *P. perfoliatus* (also Africa, Australia, Eurasia) occurs in eastern North America, but is disjunct in the southern United States; other disjunct localities have been reported in Alaska and British Columbia. It is widely distributed globally except for South America; *P. polygonifolius* occurs in far eastern Canada but extends throughout Europe and northern Africa; *P. praelongus* (circumboreal) and *P. sibiricus* (also in Eurasia) occur across northern North America; *P. pulcher* occurs in eastern North America; *P. richardsonii* and *P. robbinsii* extend across Alaska, Canada, and more northern portions of the continental United States; *P. tennesseensis* is restricted to the eastern United States.

References: Aalto, 1970; 1974; Ailstock & Shafer, 2004; 2006; Ailstock et al., 2010a; 2010b; Alam et al., 1999; Alderton et al., 2017; Alexander & Phillips, 2012; Alix & Scribailo, 2006; Amano et al., 2011; Anderson, 1959; Anderson, 1969; Anderson, 1978; 1982; 2003; Arata, 1959; Armellina et al., 1996; Armstrong et al., 2003; Arnold et al., 2012; Arts et al., 1990; Austin & Cooper, 2016; Bailey et al., 2008; Baker, 1975; Baldridge & Lodge, 2014; Barko & Smart, 1983; Barko et al., 1982; Barnes et al., 2013; Baron & Ostrofsky, 2010; Barrat-Segretain, 1996; Barrat-Segretain & Bornette, 2000; Barret-Segretain et al., 1998; Barrett & Seaman, 1980; Bartgis, 1992; Beck & Alahuhta, 2017; Beckett et al., 1992a; 1992b; Beck-Nielsen & Madsen, 2001; Beer & Wetzel, 1982b; Bennett, 1901; Bennett & DuPont, 1993; Berg, 1949; 1950a; 1950b; Berger et al., 1995; Bergerud, 1972; Bergman & Bump, 2014; Bernhardt & Duniway, 1985; Block & Rhoads, 2011; Bodner, 1994; Boedeltje, 2005; Boedeltje et al., 2002; 2003a; 2015; 2016; Bolda & Anderson, 2011; Bolduan et al., 1994; Borman et al., 1997; Bowles & Dodd, 2015; Boyd, 1968; 1975; Boyd & McGinty, 1981; Boylen & Sheldon, 1976; Brewer, 1995; Brochet et al., 2009; 2010a; 2010b; Brouwer et al., 2002; Brunton et al., 1994; Brux et al., 1987; Burnham, 1917; Burns, Jr., et al., 1995; Bushmann & Ailstock, 2006; Bushnell, Jr., 1966; Caffrey & Kemp, 1990; 1991; 1992; Cangiano et al., 2001; Cannings, 2003; Capers, 2003a; Capers et al., 2010; Carpenter & Titus, 1984; Caslake et al., 2006; Castellanos & Rozas, 2001; Catling & Dobson, 1985; Catling et al., 1984; 1986; Cattaneo & Kalff, 1978; Cellot et al., 1998; Ceska & Warrington, 1976; Ceska et al., 1986; Chabreck et al., 1974; Chadin et al., 2003; Chambers & Kalff, 1987; Chandler, 1937; Charalambidou & Santamaría, 2002; Charlton & Posluszny, 1991; Cheruvelil et al., 2001; Cherry & Gough, 2006; Cho et al., 2012; Christy & Garvey, 2015; Clark, 1956; Cody et al., 2003; Coleman & Boag, 1987; Collins et al., 1987; Colt & Hellquist, 1974; Colt et al., 1971; Combroux et al., 2001; Cook, 1988; Cooper & Andrus, 1994; Coops et al., 1994; COSEWIC, 2005; Coughlan et al., 2015; 2017a; Coulter, 1955; Craven & Hunt, 1984; Crocker, 1907; Cronan, Jr., 1957; Cronin & Lodge, 2003; Cronin et al., 2002; Cronk et al., 2015; Crow, 1969; Crow & Hellquist, 2000; Crowder et al., 1977a; Cutter, Jr., 1943; Davidson et al., 2008; DellaGreca et al., 2001b; Dennis et al., 1979; DeVlaming & Proctor, 1968; De Vos, 1958; Dibble & Harrel, 1997; Dieffenbacher-Krall & Halteman, 2000; Dindorf & Kjelland, 2006; Downing, 1981; Doyle & Smart, 1998; Drouet, 1933; Du et al., 2015; Egertson et al., 2004; Eichler, 2009; 2011; Eichler & Boylen, 2013; Eichler et al., 2006; Eleuterius & Caldwell, 1984; Eleuterius & McDaniel, 1978; Elger & Willby, 2003; Elser et al., 1994; Enders, 1932; Engel & Nichols, 1994; Engelhardt & Ritchie, 2001; 2002; Fernald, 1932; Fernald & Kinsey, 1943; Feuchtmayr et al., 2009; Fleming & Van Alstine, 1999; Fleming et al., 2011; Flint & Madsen, 1995; Fortner & White, 1988; Frank, 1966; Fraser & Hristienko, 1983; Fraser et al., 1980; 1982; 1984; Frodge et al., 1990; Frost, 1980; Frost-Christensen & Sand-Jensen, 1995; Fryer et al., 1915; Garon-Labrecque et al., 2016; Gehrs, 1974; George et al., 1977; Gerloff & Krombholz, 1966; Gilman et al., 2008; Goldsborough & Kemp, 1988; Gorham & Pearsall, 1956; Gortner, 1934; Goulet et al., 2005; Grabowski, 1973; Grainger, 1947; Green et al., 2016; Gulnaz et al., 2011; Guy, 1988; Haag, 1983; Hafez et al., 1992; Hall & Penfound, 1943; Hall & Werner, 1977; Hamerstrom, Jr. & Blake, 1939; Hann, 1995; Harley & Findlay, 1994; Harman, 1974; Harms & Grodowitz, 2009; Harper & Daniel, 1934; Harris & Gutzmer, 1996; Havera, 1986; Haynes, 1974; 1985b; 1988; Haynes & Hellquist, 2000c; Haynes et al., 1998d; 1998e; Heckscher, 1984; Heilman & Carlton, 2001; Heisey & Damman, 1982; Hellquist, 1971; 1972; 1977; 1980; 1984; Hellquist & Crow, 1980; Hellquist & Hilton, 1983; Hellquist & Pike, 2004; Hellquist et al., 1988; 2014; Hendricks & Goodwin, Jr., 1952; Herb & Stefan, 2003; Heuschele & Gleason, 2014; Heslop-Harrison, 1952; Hicks et al., 2005; Hiebert et al., 1986; Hill, 1898; Hitchin et al., 1984; Hodgdon et al., 1952; Hoffman, 1940; Hofstra et al., 1995; Hollingsworth et al., 1998a; 1998b; Hotchkiss & Stewart, 1947; Hoyer et al., 1996; Hudon, 2004; Hunter, Jr. et al., 1986; Hussner, 2012; Hutton & Clarkson, 1961; Iida et al., 2004; 2009; Jennings, 1919; Johnson & Ostrofsky, 2004; Johnston et al., 2007; Judd, 1953; 1964; Kadono, 1980; 1984; Kaplan, 2002; Kaplan & Reveal, 2013; Kaplan & Štěpánek, 2003; Kaplan et al., 2009; 2013; 2018; Kapoor & Vijayaraghavan, 1991; Karol et al., 2017; Kaskey & Tindall, 1979; Kautsky, 1988; 1990; Keddy, 1983; Keith, 1961; Kelley & Bruns, 1975; Kennedy, 1994a; 1994b; Kilgen & Smitherman, 1971; Kirby & Ringler, 2015; Kirschner & Kaplan, 2002; Kissoon et al., 2013; Kitner et al., 2013; Knapton & Pauls, 1994; Koch, 1970; Koch et al., 2010; Korschgen et al., 1988; Koryak, 1978; Krecker, 1939; Krull, 1969; 1976; Kuehne et al., 2016; Kujawski & Thompson, 2000; Kula & Zolnerowich, 2008; Lamont et al., 2013; Lee, 1987; Lemke, 1989; LeResche & Davis, 1973; Les, 1983; Les & Haynes, 1996; Les & Sheridan, 1990; Les et al., 2009; 2015b; Lillie, 1986; Lind & Cottam, 1969; Lindqvist et al., 2006; Linn et al., 1975; Lock et al., 2009; Lombardo, 1997; Lougheed et al., 2001; Love, 1975; Lupoae et al., 2015; Lynch et al., 1947; Maass, 1967; Maberly, 1993; MacRae et al., 1990; Madsen, 1991; Madsen & Wersal, 2009; Madsen et al., 1991; 2012; Majure et al., 2011; Maki & Galatowitsch, 2004; Mandissian & McIntosh, 1960; Marsden & Ladago, 2017; Martin et al., 1992; Martine et al., 2015; Massoud, 2012; Mattson et al., 2005; McDonald, 1991; McFarland & Rogers, 1998; McFarland et al., 1992; McMillan, 1953; McShane & Mehigan, 2012; Meeker et al., 2018; Mendall, 1949; & Mendall & Gashwiler, 1940; Merz et al., 2008; Middleton, 1995; Millsap, 2009; Minckley & Tindall, 1963; Moen, 1953; Mohlenbrock, 1959a; 1959b; Mohlenbrock et al., 1961; Moore, 1913; 2000; Morong, 1893; Morris, 2014; Mrachek, 1966; Muenscher, 1936a; 1936b; 1943; Munro, 1936; Murray, 2000; Myers, 1942; Nachtrieb et al., 2011; Nault & Mikulyuk, 2009; Neck & Schramm, 1992; Nichols, 1975; 1988; 1990; Nichols & Buchan, 1997; Nichols & Shaw, 1986; Nichols et al., 2000; Nielson, 2016; Oberhauser & McAtee, 1920; Olson & Doherty, 2014; Ophel & Fraser, 1970; Ostrofsky & Miller, 2017; Ostrofsky & Zettler, 1986; Owens et al., 2001; Padgett & Crow, 1994a; Pagano & Titus, 2007; Parkinson et al., 2016; Parkes et al., 1978; Paterson, 1993; Patrock, 2007; Patterson, 1982; Paulus, 1982b; 1984; PCA, 2012; Penfound, 1953; Penfound et al., 1945; Perleberg, 2008; Perleberg & Loso, 2009a; 2009b; Perry & Uhler, 1988; Pervin et al., 2006; Peters et al., 2008; Pfauth & Sytsma, 2004; 2005; Philbrick, 1984; 1988; Philbrick & Anderson, 1987; Philbrick & Les, 1996; Phillips, 2008; Pierce & Jensen, 2002; Pine & Anderson, 1991; Pip, 1979; 1987; Pip & Simmons, 1986; Pípalová, 2002; Poirrier et al., 2010; Pollux, 2011; Poole et al., 2007; Popov et al., 2007; Posluszny, 1981; Posluszny & Sattler, 1974; Poster et al., 2013; Prasad, 2007; Prausová et al., 2013; 2014; 2015; Preston, 1995; Pritchard, 1935; Qais et al., 1998; Quade, 1969; Quayyum et al., 1999; Ren & Zhang, 2008; Reznicek, 1994; Reznicek & Bobbette, 1976; Rich et al., 1971; Richardson & Clifford, 1983; Richardson et al., 2002; Riemer, 1975; Riemer & Toth, 1969; Riis et al., 2009; Roberts & Haynes, 1986; Roberts et al., 1985; Robinson, 1956; Robles

et al., 2008; Robson et al., 2016; Rodger, 1933; Rogers, 1993; Rooney et al., 2013; Rosenthal et al., 2006; Rosine, 1955; Rosso, 1977; Rothlisberger et al., 2010; Rowell, Jr., 1983; Rozentsvet et al., 1995a; 2002; Ryan et al., 1972; Rydberg, 1914; Sabbatini & Murphy, 1996; Sass et al., 2010; Sastroutomo, 1980; 1981a; 1981b; Sastroutomo et al., 1979; Saunders, 2005; Schincariol et al., 2004; Schloesser et al., 1986; Schmid, 1965; Scholtens, 1996; Schults et al., 1976; Schuyler et al., 1993; Seaman & Porterfield, 1964; Self et al., 1975; Sexton, 1959; Shabnam & Pardha-Saradhi, 2013; 2016; Shabnam et al., 2015; Sheldon, 1986; 1987; Sheldon & Boylen, 1975; Schutten & Davy, 2000; Schutten et al., 2005; Singer, 1983; Singleton, 1951; Siver, 1978; 1980; Sjöberg & Danell, 1982; Smith, 2014; Smits et al., 1988; 1989; Søndergaard et al., 2010; Sorrie et al., 2007; Southwick & Pine, 1975; Spence & Chrystal, 1970; Spence & Dale, 1978; Spencer & Anderson, 1987; Spencer & Ksander, 1992; 1995; 1996; 1997; 2001; 2011; Spencer & Rejmánek, 2010; Spencer et al., 1994; 1997; 2003; Spindler & Hall, 1991; Sprenkle et al., 2004; Squires & Anderson, 1995; Srivastava et al., 1995; Stewart & Kantrud, 1972; Steyermark & Moore, 1933; Stoops et al., 1998; Storch et al., 1986; Stoudt, 1944; Stuckey, 1971; 1979; Stuckey & Moore, 1995; Stuckey et al., 1978; Swindale & Curtis, 1957; Sytsma & Pfauth, 2006; Takos, 1947; Taylor & Helwig, 1995; Taylor et al., 2008; Teltscherová & Hejný, 1973; Tennessen, 1993; Teryokhin et al., 2002; Thompson et al., 2009; Thorp et al., 1997; Titcomb, 1923; Titus, 1983; Tobiessen & Snow, 1984; Tracy et al., 2003; Transeau, 1903; Tsuchiya, 1991; Turnage & Shoemaker, 2018; Turner et al., 2005; Twilley & Barko, 1990; Uhler, 1982; Ungar, 1964; Ungar et al., 1969; Vannatta, 2016; Van Onsem et al., 2018; Vári, 2013; Volker & Smith, 1965; Wakeman & Les, 1994a; 1994b; Wallace, 1942; Wan et al., 2012; Wang & Min, 2007; Wang & Qiu, 2006; Warrington, 1986; Weatherbee & Crow, 1992; Weber, 1940; Wehrmeister, 1978; Weiher et al., 2011; Wersal et al., 2005; 2010; Wester, 1992; Wetzel, 1960; Whittall et al., 2004; Wiegleb & Brux, 1991; Wiegleb & Kaplan, 1998; Wilcox, 2012; Wilcox & Meeker, 1991; Wiley et al., 1986; Wilhelm & Mohlenbrock, 1986; Wilkinson & Beckett, 2002; Wilson, 1935; 1937; Wohler et al., 1975; Wolfer & Straile, 2004; 2012; Wood et al., 1982; Woodruffe-Peacock, 1917; Woolf & Madsen, 2003; Xiao & Desser, 1998; Xie et al., 2013; Yan et al., 1985; Yeo, 1966; 1967; Yin et al., 2017; Zalewska-Gałosz & Ronikier, 2012; Zebryk, 2004; Zhu et al., 2006; Zika, 1996; Zimba et al., 1993; Zundel, 1920.

3.4.2. Stuckenia

Pondweed; potamot

Etymology: after Wilhelm Adolf Stucken (1852–1901)

Synonyms: *Buccaferrea* (in part), *Coleogeton*, *Spirillus* (in part)

Distribution: global: Africa, Australia, Eurasia, New World; **North America:** widespread

Diversity: global: 7 species; **North America:** 4 species

Indicators (USA): obligate wetland (OBL): *Stuckenia fili-formis*, *S. pectinata*, *S. striata*, *S. vaginata*

Habitat: brackish (coastal), freshwater; freshwater (tidal); saline (inland); lacustrine, palustrine, riverine; **pH:** 4.6–10.7; **depth:** to 10 m; **life-form(s):** "floating-leaved" (see *S. pectinata* below); submersed (vittate)

Key morphology: rhizomes bearing tubers (singly or in groups up to 5) or with dense, elongate, modified branches; shoots (to 2 m) submersed, herbaceous, flexuous, branching distally; leaves (to 21 cm) alternate, linear, rigid, sessile, 1–5-veined, the apex round to apiculate, the stipules (to 9.5 cm) adnate to the blade for more than 2/3 their length, the sheathing portion (to 7.3 cm) connate or involute at base, inflated or not, the free apex forming a ligule (0.2–20 mm); inflorescence a cylindric to moniliform spike (to 8 cm), floating on water surface, peduncles (to 35 cm) axillary, flexuous, flowers inconspicuous, in 3–12 whorls; fruits (to 4.2 mm) rounded on bottom, beaked (to 1.1 mm) or essentially beakless, embryo coiled less than 360 degrees

Life history: duration: annual (fruits/seeds); perennial (dormant apices, rhizomes, tubers, whole plants); **asexual reproduction:** rhizomes, shoot fragments, tubers; **pollination:** water (epihydrophily); **sexual condition:** hermaphroditic; **fruit:** achenes (common); **local dispersal:** water (achenes, rhizome fragments, tubers); **long-distance dispersal:** waterfowl (achenes)

Imperilment: 1. *Stuckenia filiformis* [G5]; SH (NH, OH, PA); S1 (ME, NB, NE, NY, PE, UT, VT, WA); S2 (ME, NB, NS, WY); S3 (CA, LB, ND, NF, OR, QC, WY); **2.** *S. pectinata* [G5]; S1 (NH); S2 (NC, NF); S3 (IL, WV, WY); **3.** *S. striata* [G5/G4Q]; S1 (OR, TX); **4.** *S. vaginata* [G5]; SH (OR, SD); S1 (MN, UT, WI); S2 (WY); S3 (BC, ND, NU)

Ecology: general: All *Stuckenia* species are OBL aquatic indicators. Extensive morphological variability coupled with a high degree of phenotypic plasticity has made it difficult to identify *Stuckenia* species, which can maintain strikingly different phenotypes even when transplanted to common gardens. For exampe, *Stuckenia* plants often produce wider leaves when in running water compared to standing water. Unfortunately, there is no good way of determining whether information in the published literature was attributed to a species correctly by the original authors, so that possibility should be considered as the following treatments are evaluated. In most cases, such discrepancies would involve the misidentification of atypical *S. pectinata* specimens as one of the less common species (e.g., *S. filiformis*, *S. striata*, or *S. vaginata*), which could be particularly critical where the latter are managed as imperiled taxa. All *Stuckenia* species are submersed, herbaceous, rhizomatous perennials, although some species (referred to as "pseudo-annuals") sometimes persist entirely by seed or vegetative propagules (tubers). Most of the species are versatile ecologically and capable of colonizing brackish as well as freshwater sites. Despite their submersed habit, the plants often dispose their foliage across the water surface to maximize exposure to light. Self-compatibility and protogyny characterize those flowers evaluated. Pollination is hydrophilous (epihydrophily), with released pollen becoming buoyant and eventually contacting stigmas of flowers situated across the water surface by the floating inflorescences. Flowers that

remain underwater are pollinated less efficiently by sinking pollen. Autogamy also occurs. Vegetative reproduction occurs by means of elongating rhizomes. Those of several species produce numerous tubers, which are efficient propagules. In some cases, dense, elongate branches develop on the rhizome, which also function as propagules when becoming detached. Local dispersal of fruits and vegetative propagules is mediated by water; whereas long-distance dispersal occurs primarily by the transport of the fruits (usually endozoically) by migrating waterfowl (Aves: Anatidae).

3.4.2.1. *Stuckenia filiformis* **(Pers.) Börner** is a perennial, which inhabits still to flowing (e.g., 0.17–0.51 m s^{-1}) fresh to brackish (salinity: 0–8‰) waters of alluvial fans, bays, channels (overflow, slough), deltas, ditches (irrigation, roadside), fens (calcareous, rich, spring-fed), flats (marl, mud, saltwater, tidal), floodplains, gravel pits, lagoons, lakes (alkaline, alpine, brackish), marshes (estuarine, shallow), oxbows, ponds (artificial, beaver, brackish, calcareous, dune, interdunal, marl, plunge, rich, sag, seasonal, tundra), pools (beach, brackish, woodland), potholes, reservoirs (drying), rivers (margins), shores (marl), sloughs (tidal), springs (marl, subsaline), and streams (spring-fed) at elevations to 3279 m [much higher in Asia]. Occurrences are reported mostly in open exposures. The substrates have been described as alluvium (glacial), clay, gravel (limestone), gravelly silt, limestone, marl, muck, mud (subalkaline), muddy rock, muddy silt, organic, peat, rubble (limestone), sand, sandy clay, sandy gravel, silt, silty clay, silty ooze, stones, and volcanic. The best growth reportedly occurs on muddy substrates, but higher fruit production occurs when growing on sandy substrates. This species is categorized as a calciphile, which can use bicarbonate (HCO_3^-) as a carbon source. The plants occur in cold (but up to 29°C), shallow (depth usually <1 m but up to 2.7 m), alkaline, mesotrophic to eutrophic waters (pH: 6.2–9.6; $\bar{x} = 7.6–8.2$; alkalinity [as $CaCO_3$]: 5–180 mg l^{-1}; $\bar{x} = 66–93$ mg l^{-1}; conductivity: to 500 μmhos cm^{-1} [@ 25°C]; $\bar{x} = 100$ μmhos cm^{-1}; total P: to 0.78 mg l^{-1}; Secchi depth: 0.7–4.0 m). The shoots have a relatively high tensile strength (33.8 mN m^{-2}), which enables them to resist tearing by current forces in flowing waters. Enhanced growth also can occur on clay substrates, which contain higher phosphorous levels. Flowering and fruiting can extend from June to October. The flowers are self-compatible but protogynous, with the female phase lasting from 2 to 3 days, followed by a male phase from 1 to 2 days. Although pollination potentially is hypohydrophilous (as evidenced by confirmed interspecific hybrids), sexual reproduction most often involves autogamy, whereby the air bubbles that cause anther dehiscence will usually transport pollen from the anthers directly to the closely situated stigmas (a type of "hydroautogamy"). Plants capable of producing normal fruits when growing in still waters can become sterile if grown in flowing waters. In studies conducted in the Old World, growth of the plants was limited mainly by wave action and grazing by waterfowl (Aves: Anatidae). European plants have achieved shoot densities as high as 3216 m^{-2}, production rates of 816 mg C m^{-2}day^{-1}, and a biomass up to 243 g DM m^{-2}. Vegetative reproduction occurs by the production of rhizomes from which axial tubers arise. Occasionally, tubers can develop from the shoot portion of the plants as well. Despite a fairly high fruit set, the tubers represent the predominant means of annual perennation and local dispersal. However, dispersal among populations occurs by means of seed. Tuber densities as high as 2800 m^{-2} have been observed. Genetic analyses using isozymes have confirmed the highly clonal nature of populations with genetic variation partitioned mainly between rather than within populations. **Reported associates:** *Agrostis stolonifera, Alisma gramineum, Alisma gramineum, Beckmannia syzigachne, Brasenia schreberi, Calamagrostis canadensis, Callitriche hermaphroditica, Callitriche heterophylla, Callitriche palustris, Carex aquatilis, Carex lyngbyei, Carex mackenziei, Carex membranacea, Carex nebrascensis, Carex utriculata, Ceratophyllum demersum, Chara vulgaris, Comarum palustre, Eleocharis acicularis, Eleocharis equisetoides, Eleocharis erythropoda, Eleocharis palustris, Elodea canadensis, Elodea nuttallii, Equisetum fluviatile, Fontinalis, Glyceria borealis, Glyceria grandis, Heteranthera dubia, Hippuris tetraphylla, Hippuris vulgaris, Isoetes bolanderi, Juncus nodosus, Lemna minor, Lemna trisulca, Limosella aquatica, Ludwigia palustris, Lythrum salicaria, Menyanthes trifoliata, Mimulus glabratus, Mimulus guttatus, Myriophyllum sibiricum, Myriophyllum spicatum, Myriophyllum verticillatum, Najas flexilis, Najas guadalupensis, Nasturtium officinale, Nitella flexilis, Nuphar polysepala, Nuphar variegata, Nymphaea odorata, Persicaria amphibia, Persicaria coccinea, Pontederia cordata, Potamogeton alpinus, Potamogeton amplifolius, Potamogeton berchtoldii, Potamogeton crispus, Potamogeton foliosus, Potamogeton friesii, Potamogeton gramineus, Potamogeton illinoensis, Potamogeton natans, Potamogeton nodosus, Potamogeton obtusifolius, Potamogeton perfoliatus, Potamogeton praelongus, Potamogeton pusillus, Potamogeton richardsonii, Potamogeton robbinsii, Potamogeton sibiricus, Potamogeton strictifolius, Potamogeton zosteriformis, Ranunculus flammula, Ranunculus gmelinii, Ranunculus longirostris, Ranunculus subrigidus, Ranunculus trichophyllus, Ricciocarpus natans, Sagittaria cuneata, Sagittaria latifolia, Sagittaria rigida, Schoenoplectus subterminalis, Schoenoplectus tabernaemontani, Sparganium angustifolium, Sparganium emersum, Sparganium eurycarpum, Sparganium hyperboreum, Sparganium natans, Sphagnum contortum, Stuckenia pectinata, Stuckenia vaginata, Subularia aquatica, Torreyochloa pallida, Triglochin maritima, Triglochin palustris, Typha, Utricularia intermedia, Utricularia macrorhiza, Vallisneria americana, Veronica anagallis-aquatica, Veronica catenata, Zannichellia palustris.*

3.4.2.2. *Stuckenia pectinata* (L.) **Börner** is a perennial, which occurs in fresh to hypersaline (salinity: 5.5–104‰), still to flowing (to 2 m s^{-1}) waters of backwaters, bayous, bogs, canals (irrigation), channels (drainage, irrigation, river, tidal), coves (stagnant), ditches (irrigation, roadside), estuaries, gravel pits, inlets, lagoons, lakes (alkaline, artificial, oxbow, prairie, strip mine, subalkaline), marshes (brackish, coastal, freshwater, tidal), meadows (saline), mudflats, ponds (artificial, beaver, cattle, coastal, evaporating, floodplain, golf course, irrigation, mill, roadside, seepage, spring-fed, stock, stormwater, thaw, thermokarst), pools (marly, river, shallow, stagnant, vernal, warm), potholes (glacial), reservoirs (cement, margins), rivers, sloughs (shallow, tidal), springs (warm), streams (warm), swamps, and water tanks at elevations to 2988 m. The plants occur in open exposures under full sunlight. The substrates often have a low clay content and high organic matter content but include alluvium (sandy, sandy loam), basalt, clay (organic), clay loam, cobble river), gravel, gravelly rubble, limestone (gravelly), loam, loamy sand (fine), marl, muck, mucky marl, mud (clay, organic), muddy marl, muddy sand, peat, rocks, sand, sandy clay, sandy gravel, sandy loam, sandy mud, sandy silt, sandy silt loam, silt, silty clay, silty clay loam, silty loam, silty loess, silty mud, silty sand, and sludge. Although the distribution of the plants does not seem to be dependent on the substrate composition, it is influenced more by wave action, which secondarily affects turbidity and substrate texture. This species is ecologically versatile but shows a preference for waters that are clear (Secchi depth: >0.2 m; $\bar{x}$ = 1.7 m), shallow (depth: 0.2–10.0 m but usually <1.5 m), and characterized by low exposure to wave action. It is particularly common in alkaline, calcareous, brackish, or even hypersaline sites that are high in chloride and sulfate (pH: 5.7–10.7; $\bar{x}$ = 7.0–8.5; alkalinity [as CaCO$_3$]: 2.3–315 mg l^{-1}; $\bar{x}$ = 91.7–140 mg l^{-1}; chloride: 3.3–4095 mg l^{-1}; $\bar{x}$ = 408.2; conductivity: 68–789; $\bar{x}$ = 238 µS cm^{-1}@ 25°C; salinity: 0–104‰; SO$_4$: 0–3403 mg l^{-1}; $\bar{x}$ = 76 mg l^{-1}). The plants are adapted locally to brackish or freshwaters, with individuals from brackish waters producing more shoots and greater total biomass at low salinities, compared to those originating from freshwaters, which exhibit their optimal growth in freshwater. Higher reproductive resource allocation (e.g., 4%) has been observed in exposed habitats compared to sheltered sites (e.g., 1%), where more resources are sequestered in overwintering vegetative structures. However, flower production reportedly is highest (e.g., 945 m^{-2}) in sheltered habitats and declines (e.g., 672 m^{-2}) as exposure increases or when salinity is high (>45‰). In North America, flowering occurs primarily from mid-May through mid-July, but can extend for up to 5 months. The flowers are borne at the end of an elongating peduncle (to 30 cm), which extends across the water surface. Pollen is liberated from the anthers (sometimes forcefully) once the floral tepals retract. The hydrophobic pollen floats to facilitate surface water pollination of the flowers (epihydrophily), which attain a seed set between 6.5% and 40.0%; ($\bar{x}$ = 25.35%). Less efficient transport of pollen reduces seed set considerably in underwater flowers (2.5–7.5%; $\bar{x}$ = 3.95%). The fruits develop at the water surface from mid-June (roughly 3 weeks after flowering) remaining attached to the peduncle. Mature fruits (7–20 per spike) appear from June through November. Seed mass ($\bar{x}$ = 0.24 g 100 seeds^{-1}) varies due to developmental temperature fluctuations. As many as 63,300 fruits have been produced within a year by a single plant grown from one seed cultured within a 23.6 m^2 enclosure. One oligosaline site yielded densities of 3,707 achenes m^{-2}. The plants can produce a persistent seed bank comprising upward of 16,000 fruits m^{-2} (representing a biomass of more than 60 g DM m^{-2}) when growing under optimal conditions; sexual reproduction ceases in flowing waters. The fleshy mesocarps contain large air spaces, which enable fruits to remain buoyant for a brief time and facilitate their local dispersal by water currents. The fruits also are consumed and dispersed by various birds (Aves). They remain viable after passing through their digestive tracts with gut retention times lasting as long as 76 hr for killdeer (Charadriidae: *Charadrius vociferus*) and 73 hr for mallard ducks (Anatidae: *Anas platyrhynchos*). Consequently, endozooic transport could potentially achieve dispersal distances of more than 4100 km. Although seedlings rarely are observed in the field, the seeds readily germinate from sediment samples that have been incubated during laboratory seed bank studies. Seed banks do not develop in some sites. Fruit germination typically is rather low (<50%) but can be enhanced by physically cutting or rupturing the exocarp. It occurs from late March through summer in natural populations. In laboratory studies, germination has been optimized at 25°C (12/12 hr light/dark; 70 μE m^{-2} s^{-1}) for fruits stratified for 6 months @ 4°C or for those desiccated in sediments for 3 months. It is highest in freshwater with reduced rates (but still occurring) at salinities up to 6‰. Air-dried fruits can remain viable for more than 9 months, although their germination is reduced (to <20%). Most annual recruitment is accomplished by vegetative propagules, with the fruits functioning mainly as agents of long-distance dispersal and long-term survival. Vegetative reproduction occurs by rhizomes (up to 22 per plant at depths to 15 cm), which can fragment, and by tubers (also called turions by some authors), which occur singly or in chainlike groups of up to five. The tubers initiate under long-day photoperiods (10–12 hr) but their production does not appear to be governed by average daily light levels; however, short-day conditions will favor tuber production over new shoot growth. There is evidence of local adaptation to latitude, with one study finding tuber production to increase from 2% to 87% in direct proportion to the latitude of plant origin. Plants in freshwaters produce tubers earlier in their growth phase than do plants growing in brackish sites. Tubers occur at sediment depths up to 47 cm and can approach yields of up to 450 g m^{-2}. Their individual mass (up to 1 g) increases with increased depth and can represent up to 38% of the total plant biomass. Substrate heterogeneity has resulted in local adaptation in tuber size, with higher fitness in sandy substrates for clones producing larger tubers compared to higher fitness in clay-rich substrates for clones producing smaller tubers. Tuber production (averaging 12 per plant) typically is higher in exposed sites (e.g., to 115 m^{-2}) and decreases (e.g., to 45 m^{-2}) with decreasing exposure. The subterranean tubers can reach densities as high as 4909 m^{-2} and 4.1 100 cm^{-3}; some also can develop on the plant shoot from axillary side branches. Single tubers predominate in brackish sites (98% of plants), whereas up to 70% of plants produce double tubers when growing in freshwater. An astounding 36,000 tubers have been recovered within a year from a single plant cultured from a tuber in a 23.6 m^2 enclosure. Higher growth rates characterize plants derived from more shallowly buried tubers (e.g., 10 cm) than

those buried at greater depths (e.g., 20 cm). There is evidence that the plants can "adjust" their tuber burial depth (i.e., to increased depth) when subjected to grazing pressure by herbivorous birds. The tubers can be transported by water currents at distances of up to at least 10 km in riverine systems. In North America the tubers germinate from March through June (once water temperatures exceed 10°C) and develop into mature flowering plants from May to July. Their germination is governed primarily by temperature with the highest rates occurring at 20°C–25°C. Smaller tubers germinate much faster (e.g., 10 days) than do large tubers (e.g., 110 days). The tubers also germinate more rapidly when incubated at higher temperatures (e.g., within 4 days @ 16°C vs. 9 days @ 10°C) and grow optimally at temperatures from 23°C to 30°C. Bulk translocation of carbohydrates from tubers to shoots occurs within 3 weeks of germination. Prolonged cold stratification can result in germination at much lower temperatures (e.g., 5.5°C), but without significant subsequent growth until the water temperatures increase. Tubers that are subject to temperatures above 13°C–15°C directly after winter stratification can develop secondary dormancy, which arguably facilitates their survival in sites that experience summer drought or enables them to remain dormant over two growing seasons. Tubers do not tolerate desiccation well, with 60% failing to germinate after 2 weeks exposure to sediment moisture levels below 23%. Peak total plant biomass occurs from August to September, with growth typically lasting for about 113 days in temperate North American localities. Belowground biomass can vary from 4% to 78% of the total biomass. European plants have achieved shoot densities as high as 640 m^{-2}, production rates of 1400 mg C m^{-2} day^{-1}, net photosynthetic rates of 1.51 mg C g^{-1} ash-free DM h^{-1} (at 25°C–30°C), resulting in biomass levels up to 1312.5 g DM m^{-2} (the latter in protected brackish sites). The highest photosynthetic rates occur at a pH <7.0. Biomass is considerably lower in exposed sites and also can be reduced by as much as 17% by grazing waterfowl (Aves: Anatidae). Biomass has been found to increase proportionally with sediment organic matter content (at least to 26 mg C g^{-1}) due to enhanced nitrogen mineralization. Reduced biomass occurs at sites with higher levels of organic matter (because of the associated anoxic conditions) and also where high ionic interstitial water concentrations of Fe^{2+} and S^{2-} exist. Potassium, nitrogen, and phosphorous are taken up by the roots and shoots. Although free CO$_2$ is the major source of carbon, bicarbonate ions (HCO$_3^-$) also can be taken up, but only by the shoot. PEP carboxylase activity is very low, indicating a lack of C$_4$ metabolism. The plants tolerate differences in irradiance and photoperiod by their high degree of phenotypic plasticity. Senescence occurs late August through October and most of the plants will die back completely (to their vegetative perennating organs) by early November. Despite their highly clonal growth, genetic surveys (using allozyme, microsatellite, RAPD, and RFLP markers) have disclosed high clonal diversity in many populations, which is attributed to high rates of colonization via sexual propagules. These genetic studies indicate that pond populations have higher clonal diversity and limited seedling recruitment, which occurs primarily at distances less than 5 m. River populations comprise few, large, unrelated clones whose diversity decreases along the downstream gradient. Different clones display latitudinal variation with those originating from northern localities having a shortened life cycle with rapid and high tuber biomass investment, whereas those from lower latitudes have a longer life cycle that delays reproduction, resulting in higher total biomass. The foliage frequently becomes encrusted with marl ($CaCO_3$), which can represent up to 12% of the total biomass. Leaf surface area is relatively low, representing about 500 cm^{-2} g^{-1} DM. Although the plants occur commonly in turbid waters, their leaves are not shade tolerant and biomass is reduced when the plants are shaded. Rather, in still waters, the leaves often cluster into dense, flat, fan-like sprays, which spread out across the water surface. In this way, the plants mimic a "floating-leaved" habit, despite their finely dissected foliage. The aggregation of leaf biomass at the surface enables the plants to inhabit highly turbid waters, which otherwise might preclude their survival as a submersed life-form where low light levels prevail. This strategy also provides a competitive advantage over plants having a strictly submersed life-form. Not surprisingly, higher light levels (e.g., 416 μmol m^{-2} s^{-1}) result in higher biomass, which can decline by as much as 80% at lower light levels (e.g., 36 μmol m^{-2} s^{-1}). Some populations are evergreen, with their foliage surviving throughout the winter. In contrast, some plants die back completely in winter, surviving entirely by seed; such populations have been referred to as "pseudo annuals." Surface winds greater than 29.6 km hr^{-1} will result in the uprooting of plants. The herbage contains several *ent*-labdane diterpenes, which have exhibited potent algicidal activity against some algae (Chlorophyceae: Selenastraceae: *Raphidocelis subcapitata*). Numerous fungal endophytes have been cultured from the roots and foliage. **Reported associates:** *Acorus calamus, Agrostis stolonifera, Alisma triviale, Alopecurus aequalis, Alopecurus pratensis, Ambrosia psilostachya, Anemopsis californica, Arctophila, Asclepias incarnata, Azolla filiculoides, Baccharis salicifolia, Beckmannia syzigachne, Berula erecta, Bidens beckii, Bidens laevis, Bolboschoenus maritimus, Bothriochloa laguroides, Brasenia schreberi, Bromus japonicus, Bromus rubens, Calibrachoa parviflora, Callitriche hermaphroditica, Callitriche palustris, Carex emoryi, Carex hallii, Carex lasiocarpa, Carex lyngbyei, Carex nebrascensis, Carex subfusca, Ceratophyllum demersum, Chara globularis, Chara vulgaris, Cirsium arvense, Cladophora glomerata, Conium maculatum, Conyza canadensis, Cotula coronopifolia, Croton setiger, Cynoglossum officinale, Cyperus squarrosus, Descurainia pinnata, Distichlis spicata, Echinodorus berteroi, Egeria densa, Elatine, Eleocharis acicularis, Eleocharis macrostachya, Eleocharis ovata, Eleocharis palustris, Eleocharis parishii, Eleocharis quadrangulata, Elodea canadensis, Elodea nuttallii, Epilobium ciliatum, Epilobium glaberrimum, Equisetum fluviatile, Equisetum laevigatum, Eriogonum latifolium, Eriogonum polycladon, Eriophorum, Glyceria grandis, Gnaphalium palustre, Helianthus annuus, Heteranthera dubia, Heterotheca,* *Heterotheca subaxillaris, Hippuris tetraphylla, Hippuris vulgaris, Hordeum jubatum, Horkelia, Hydrocotyle ranunculoides, Hymenothrix loomisii, Impatiens capensis, Juncus articulatus, Juncus balticus, Juncus bufonius, Juncus drummondii, Juncus dudleyi, Juncus effusus, Juncus nevadensis, Juncus phaeocephalus, Juncus torreyi, Juncus torreyi, Kickxia elatine, Lathyrus jepsonii, Leersia oryzoides, Lemna minor, Lemna minuta, Lemna trisulca, Lemna turionifera, Limosella aquatica, Lindernia dubia, Ludwigia peploides, Lychnothamnus barbatus, Lycopus, Lythrum hyssopifolia, Lythrum salicaria, Melilotus albus, Mentha arvensis, Mimulus guttatus, Myosotis laxa, Myriophyllum sibiricum, Myriophyllum spicatum, Najas flexilis, Najas guadalupensis, Najas marina, Najas minor, Nasturtium officinale, Nitella subglomerata, Nitellopsis obtusa, Nuphar polysepala, Nuphar variegata, Nymphaea odorata, Panicum capillare, Paspalum distichum, Persicaria amphibia, Persicaria lapathifolia, Phalaris arundinacea, Phragmites australis, Pluchea odorata, Poa annua, Poa pratensis, Polypogon monspeliensis, Pontederia cordata, Populus fremontii, Potamogeton alpinus, Potamogeton amplifolius, Potamogeton berchtoldii, Potamogeton crispus, Potamogeton diversifolius, Potamogeton epihydrus, Potamogeton foliosus, Potamogeton friesii, Potamogeton gramineus, Potamogeton illinoensis, Potamogeton natans, Potamogeton nodosus, Potamogeton praelongus, Potamogeton pusillus, Potamogeton richardsonii, Potamogeton robbinsii, Potamogeton zosteriformis, Potentilla anserina, Pseudognaphalium luteoalbum, Pseudognaphalium stramineum, Ranunculus aquatilis, Ranunculus cymbalaria, Ranunculus flammula, Ranunculus longirostris, Ranunculus sceleratus, Ranunculus subrigidus, Ranunculus trichophyllus, Rorippa palustris, Rosa woodsii, Rumex conglomeratus, Rumex crispus, Rumex salicifolius, Ruppia spiralis, Sagittaria cuneata, Sagittaria latifolia, Sagittaria platyphylla, Sagittaria subulata, Salix exigua, Salix gooddingii, Salix lasiolepis, Salvia mellifera, Schoenoplectus acutus, Schoenoplectus californicus, Schoenoplectus pungens, Schoenoplectus tabernaemontani, Scirpus microcarpus, Sium suave, Sparganium angustifolium, Sparganium emersum, Spartina pectinata, Sphenopholis obtusata, Spirodela polyrhiza, Stuckenia filiformis, Stuckenia pectinata, Stuckenia striata, Symphyotrichum lentum, Tamarix chinensis, Thalictrum dasycarpum, Triglochin palustris, Typha angustifolia, Typha domingensis, Typha latifolia, Urtica dioica, Utricularia gibba, Utricularia macrorhiza, Utricularia minor, Vallisneria americana, Verbena bracteata, Verbena hastata, Verbena lasiostachys, Vernonia fasciculata, Veronica anagallis-aquatica, Veronica peregrina, Wolffia columbiana, Xanthium spinosum, Xanthium strumarium, Zannichellia palustris, Zostera.*

3.4.2.3. ***Stuckenia striata* (Ruiz & Pav.) Holub** is a perennial, which grows in canals (irrigation), channels (drainage), ditches (irrigation), floodplains, lakes, marshes, ponds (alkaline), pools, reservoirs, rivers, sloughs, springs (hot), and streams (artesian) at elevations to 2164 m. Although a fair amount of literature exists for this species from sites in Mexico southward through South America, few ecological accounts exist for North America populations and additional investigations are encouraged. The substrates are described as gravel, muck, mud, and sand. North American plants have been associated with clear, cool, shallow (e.g., 0.6–2.4 m deep), brackish or fresh, alkaline (alkalinity [methyl orange]: 48.0–102.0 mg l^{-1}; $\bar{x} = 84.0$ mg l^{-1}), still to slow-moving waters; however, many of the Latin American sites are characterized as eutrophic. Flowering and fruiting extend from May to September in North America. Although presumably similar to its congeners, the sexual reproductive biology of this species remains uninvestigated and warrants study. Vegetative reproduction occurs by production of creeping rhizomes, which apparently are devoid of tubers. **Reported associates:** *Azolla cristata, Ceratophyllum demersum, Egeria densa, Lemna, Myriophyllum sibiricum, Najas marina, Nasturtium officinale, Persicaria amphibia, Potamogeton foliosus, Potamogeton illinoensis, Potamogeton nodosus, Potamogeton pusillus, Potamogeton richardsonii, Ranunculus aquatilis, Ruppia maritima, Stuckenia pectinata, Typha domingensis, Typha latifolia, Veronica anagallis-aquatica, Zannichellia palustris.*

3.4.2.4. ***Stuckenia vaginata* (Turcz.) Holub** is a perennial, which inhabits bogs, bottoms (river), canals, channels (stream), ditches, lakes (marshy), marshes (coastal), ponds (beaver, brackish, thaw, tundra), puddles, reservoirs, rivers, sloughs (saline), and streams at elevations to 3291 m. The exposures are open and receive full sunlight. The plants often occur on fine sediments (in areas of low exposure), but more widely on clay (fine), cobble, gravel (coarse), muck, mud (loose), peat, sand, silt (alluvial), silty mud, and silty sand. The waters range from still to flowing. Although reported at sites subjected to currents of 2.6 m s^{-1}, plant biomass was found to decrease linearly with increasing current speed over a range of 0–1 m s^{-1}. This species has been characterized as preferring deep (e.g., 1.5–2.0 m), hard, alkaline or brackish (oligohaline), cold, spring-fed waters, but it certainly is found at much shallower water depths (e.g., 10–50 cm) and in a wider variety of conditions overall (pH: 5.9–9.6; alkalinity: to 175 mg l^{-1} [as CaCO$_3$]; conductivity to 1083 μS cm^{-1} @ 25°C; salinity: 0.25–6‰). In Siberia, the waters characteristically are of a calcium–magnesium hydrocarbonate type (oligotrophic–eutrophic; pH: 6.55–8.55; salinity: 0.160–2.136 g dm^{-3}). Fruiting occurs later than its congeners, from July to October. The plants reportedly produce seed only rarely. Compared to a single cultured plant of the related *S. pectinata* (see above), which yielded 2,600 seeds m^{-2}, a single *S. vaginata* plant (grown under similar conditions) yielded only 60 seeds m^{-2}. The somewhat buoyant fruits (which contain large air spaces in their fleshy mesocarps) are dispersed locally within waterbodies by currents and probably are transported over greater distances endozoically by migrating waterfowl (Aves: Anatidae). The extent of seed bank formation is unclear. In one study, no seedlings germinated from sediments collected at sites of known plant occurrence after an initial incubation under natural light (120 days @ 19°C) followed by a secondary cold stratification (120 days @ 4°C) and final 90-day incubation under ambient greenhouse conditions. Plant biomass

declined significantly at sites where aqueous total phosphorous levels were reduced by 80% and even more so at sites where aqueous total ammonia nitrogen levels were reduced by 53%. The maximum biomass attained in several Siberian lakes has been recorded as 347.31 ± 65.47 g m^{-2}. The plants reproduce vegetatively by rhizomes and their modified shoots (called tubers by some authors), which have yielded densities up to 108 m^{-2}. **Reported associates:** *Alisma gramineum, Bidens beckii, Callitriche hermaphroditica, Callitriche stagnalis, Carex subspathacea, Ceratophyllum demersum, Chara vulgaris, Elodea canadensis, Elodea nuttallii, Fontinalis, Heteranthera dubia, Hippuris tetraphylla, Hippuris vulgaris, Hydrocharis morsus-ranae, Hypericum mutilum, Hypericum ellipticum, Lemna minor, Lemna trisulca, Myriophyllum heterophyllum, Myriophyllum sibiricum, Myriophyllum spicatum, Myriophyllum verticillatum, Najas flexilis, Najas guadalupensis, Nitella flexilis, Nitellopsis obtusa, Nuphar polysepala, Nuphar variegata, Nymphaea odorata, Persicaria amphibia, Persicaria coccinea, Pithophora, Potamogeton amplifolius, Potamogeton crispus, Potamogeton diversifolius, Potamogeton foliosus, Potamogeton friesii, Potamogeton gramineus, Potamogeton illinoensis, Potamogeton natans, Potamogeton praelongus, Potamogeton pusillus, Potamogeton richardsonii, Potamogeton sibiricus, Potamogeton zosteriformis, Ranunculus longirostris, Ranunculus subrigidus, Ranunculus trichophyllus, Ruppia maritima, Sagittaria latifolia, Schoenoplectus tabernaemontani, Sparganium angustifolium, Spirodela polyrhiza, Stuckenia filiformis, Stuckenia pectinata, Typha latifolia, Utricularia macrorhiza, Utricularia minor, Vallisneria americana, Wolffia columbiana, Zannichellia palustris.*

Use by wildlife: The total lipid content of *Stuckenia filiformis* foliage (48.8 mg g^{-1} dry wt) comprises neutral lipids (34.6%), glycolipids (47.4 %), and phospholipids (18.0%) such as diphosphatidylglycerol: 17.6%; phosphatidylglycerol: 28.2%; phosphatidylethanolamine: 15.6%; phosphatidylcholine: 25.6%; phosphatidylethanolamine: 5.2%; phosphatidylserine: 5.5%; and phosphafidic acid: 2.3%. The plants are relished by moose (Mammalia: Cervidae: *Alces alces*) and also are eaten by waterfowl (Aves: Anatidae) including Canada geese (*Branta canadensis*) and trumpeter swans (*Cygnus buccinator*). The tubers and fruits are consumed by various "dabbling ducks" (Aves: Anatidae) such as blue-winged teal (*Anas discors*), buffleheads (*Bucephala albeola*; as many as 332 fruits in a single individual), green-winged teal (*Anas crecca*), mallards (*Anas platyrhynchos*), pintails (*Anas acuta*), ring-necked ducks (*Aythya collaris*), and shovelers (*Anas clypeata*). Worldwide, *Stuckenia pectinata* may well represent the most important food of waterfowl, which can consume up to 9% of the annual standing crop in some regions. The foliage (2.0–4.7 kcal g^{-1} ash-free dry wt) is relatively high in minerals (e.g., Ca, Fe, K, Mg, Na) and contains (% of dry wt) up to 17.1% protein, 40.3% crude fiber, 2.7% crude fat, and 57.8% soluble carbohydrates. Higher levels of fat (to 6.9% dw) are contained in the achenes and lesser amounts (1.1% dw) in the tubers. Despite their high ecdysteroid content (170–250 μg g^{-1} dry wt), which can deter feeding by arthropods, the leaves of *S. pectinata* are grazed by amphipods (Crustacea: Ampithoidae: *Ampithoe valida*; Gammaridae: *Gammarus daiberi*), fish (Vertebrata; Teleostei: Cichlidae: *Tilapia zillii*; Cyprinidae: *Scardinius erythrophthalmus*), isopods (Crustacea: Isopoda: Munnidae: *Uromunna ubiquita*; Sphaeromatidae: *Gnorimosphaeroma oregonensis*), muskrats (Mammalia: Rodentia: Cricetidae: *Ondatra zibethicus*), and snails (Mollusca: Gastropoda: Physidae: *Physa gyrina*). The plants harbor numerous animals (about 1.6 cm^{-1}; up to 18,890 kg^{-1}) such as amphipods (Crustacea: Hyalellidae: *Hyalella*), bryozoans (Bryozoa: Plumatellidae: *Plumatella*), flatworms (Platyhelminthes: Planariidae: *Planaria*; Stenostomidae: *Stenostomum*), hydras (Cnidaria: Hydridae: *Hydra*), seed shrimp (Crustacea: Ostracoda), snails (Mollusca: Gastropoda: Physidae: *Physa*; Planorbidae: *Helisoma*), oligochaete worms (Annelida: Naididae: *Chaetogaster, Naias, Stylaria*), water mites (Arthropoda: Arachnida: Hydrachnidia), and various insects (Insecta) including brineflies (Diptera: Ephydridae: *Hydrellia*), caddisflies (Trichoptera: Leptoceridae: *Mystacides longicornis, Oecetis*), damselflies (Odonata: Coenagrionidae: *Enallagma, Ischnura*), mayflies (Ephemeroptera: Caenidae: *Caenis*; Ephemeridae), micro-caddisflies (Trichoptera: Hydroptilidae), midges (Diptera: Chironomidae: *Chironomus*), mosquitoes (Diptera: Culicidae: *Anopheles*), and parasitic wasps (Hymenoptera: Braconidae: *Ademon*). The foliage and fruits are eaten by various birds (Aves) including killdeer (Charadriidae: *Charadrius vociferus*) and numerous waterfowl (Anatidae) such as baldpates (*Anas americana*), Barrow's goldeneye (*Bucephala islandica*; as many as 200 fruits in a single individual), black ducks (*Anas rubripes*), buffleheads (*Bucephala albeola*; as many as 210 fruits in a single individual), canvasbacks (*Aythya valisineria*), cinnamon teal (*Anas cyanoptera*), common goldeneye (*Bucephala clangula*), gadwalls (*Anas strepera*), greater scaup (*Aythya marila*), green-winged teal (*Anas carolinensis*), lesser scaup (*Aythya affinis*), mallards (*Anas platyrhynchos*), pintails (*Anas acuta*), redhead ducks (*Aythya americana*), ring-necked ducks (*Aythya collaris*; as many as 250 fruits in a single individual), ruddy ducks (*Oxyura jamaicensis*), spectacled eiders (*Somateria fischeri*), surf scoters (*Melanitta perspicillata*), white-winged scoters (*Melanitta fusca*), and wood ducks (*Aix sponsa*). The rhizomes and tubers also are eaten by numerous birds (Aves) including Hudsonian and Marbled godwits (Scolopacidae: *Limosa haemastica*; *L. fedoa*) and trumpeter swans (Anatidae: *Cygnus buccinator*). At wastewater sites in Wyoming, the selenium content of *Stuckenia vaginata* plants (to 104 μg g^{-1}) greatly exceeded the safe dietary threshold (3 μg g^{-1}) established for water birds (i.e., levels known to be lethal or to cause deformed embryos). The foliage also contained elevated levels (by dry wt) of Al (to 5236 μg g^{-1}), As (to 45 μg g^{-1}), B (to 820 μg g^{-1}), Ba (to 117 μg g^{-1}), Cr (to 6.32 μg g^{-1}), Cu (to 8.3 μg g^{-1}), Fe (to 5559 μg g^{-1}), Mg (to 11,183 μg g^{-1}), Mn (to 819 μg g^{-1}), Ni (to 7.22 μg g^{-1}), Pb (to 5.74 μg g^{-1}), Sr (to 480 μg g^{-1}), V (to 12.37 μg g^{-1}), and Zn (to 51.9 μg g^{-1}). Concentrations of selenium and other elements often are higher at sites where substantial water evaporation occurs. Yet, *S. vaginata* is eaten by several waterfowl (Aves:

Anatidae) including brant geese (*Branta bernicla*), canvasback ducks (*Aythya valisineria*), and tundra swans (*Cygnus columbianus columbianus*). It provides vital nesting habitat for the latter. Although not documented specifically, it is likely that many other waterfowl consume this species, which seldom is distinguished from *S. pectinata* in feeding studies. Floating masses of *S. vaginata* plants are used as nesting platforms by eared grebes (Aves: Podicipedidae: *Podiceps nigricollis*). The foliage is colonized by rotifers (Rotifera: Notommatidae: *Cephalodella montana*).

Economic importance: food: *Stuckenia* has not been used as a human food. Because the plants effectively concentrate heavy metals and other toxic substances (see *misc. products* below; *Use by wildlife* above), it is best to avoid their consumption; **medicinal:** *Stuckenia pectinata* has a high phenolic content and decoctions have been used to treat liver ailments; **cultivation:** not cultivated; **misc. products:** *Stuckenia pectinata* has been evaluated for wastewater removal of heavy metals. Experimental manipulations have shown that the leaves can concentrate high levels of Cd (to 596 mg kg^{-1} dw), Pb (318 mg kg^{-1} dw), Cu (62.4 mg kg^{-1} dw), Zn (6590 mg kg^{-1} dw), and Mn (16,000 mg kg^{-1} dw), which makes them highly suitable for use in remediation of contaminated waters. Experimentally, the plants have removed up to 85% of Cu, Fe, Pb, and Zn from wastewater. The plant extracts have been used to synthesize gold and silver nanoparticles as well as silver-reduced graphene oxide nanocomposites; **weeds:** *Stuckenia pectinata* can become a nuisance by obstructing flow in irrigation canals; **nonindigenous species:** none.

Systematics: Formerly recognized as an infrageneric taxon within *Potamogeton*, *Stuckenia* is now regarded as a morphologically and genetically distinct genus. Phylogenetic analyses consistently resolve the species as a clade, which is situated among the other recognized genera of Potamogetonaceae in a position that usually is close to (but not always the sister group of) *Potamogeton* (Figure 1.23). From the four species surveyed, the widespread *S. pectinata* is most divergent genetically; whereas the genetic distances among the remaining three species are slight (not evident in Figure 1.23) and do not resolve several conspecific accessions as clades. In any case, several molecular data analyses have indicated a very close relationship between *S. filiformis* and *S. vaginata*. The high genetic similarity among these species raises the question of whether such genetically similar (yet morphologically distinct) taxa would better be recognized as infraspecific entities rather than as separate species. Misidentifications have created havoc in this genus and the incorporation of molecular data has proven to be invaluable in sorting out numerous taxonomic issues. The application of a DNA "barcoding" approach (*rbcL* & *matK* sequence data) has been used to identify several *Stuckenia* taxa. Methods also have been developed, which allow for the successful detection of *S. pectinata*, *S. filiformis*, and *S. vaginata* from environmental DNA (eDNA) samples. Because numerous *Stuckenia* specimens certainly have been misidentified (even by experts), the evaluation of material using genetic data is of paramount importance, especially with respect to conservation programs. The base chromosome

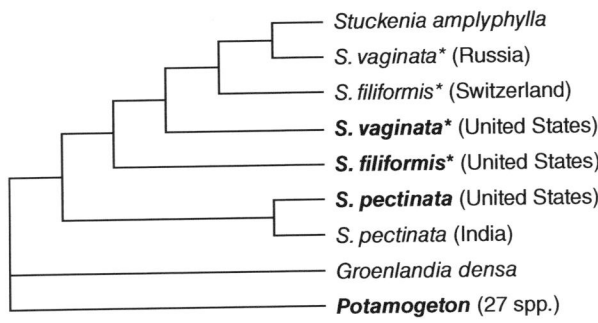

FIGURE 1.23 Phylogenetic relationships in *Stuckenia* as indicated by analysis of nrITS sequence data (adapted from Kaplan et al., 2013). Although *Stuckenia* is monophyletic, several species accessions (asterisked) from geographically distant localities do not resolve as clades (North American OBL indicators are highlighted in bold). Genetically, *S. pectinata* is the most divergent species among those surveyed, whereas genetic distances among the other *Stuckenia* species are extremely low. The ITS data also do not resolve the *Stuckenia* clade any closer to *Potamogeton* than *Groenlandia* (see also Les & Tippery, 2013), which is consistent with its recognition as a distinct genus.

number of *Stuckenia* is $x = 13$. All four of the North American species are hexaploid ($2n = 78$), although some plants appear to have higher chromosome numbers, which has been attributed to the difficulty in distinguishing *B*-chromosomes. The binomial *Stuckenia* ×*suecica* has been applied to F₁ hybrids between *S. filiformis* and *S. pectinata*, which have arisen on multiple occasions.

Distribution: *Stuckenia pectinata* is widespread throughout North America and essentially cosmopolitan; *S. filiformis* is found in northern and western North America and also in Eurasia; *S. vaginata* ranges from northern North America into the central Rocky Mountains and extends into Eurasia; *S. striata* occurs sporadically in the southwestern United States but is more common southward into South America.

References: Aalto, 1970; Abdel-Hamid et al., 2013; Alexander et al., 1996; Anderson, 1978; Bance, 1946; Bergersen, 1969; Bergey et al., 1992; Bergman, 1973; Brewer, 1995; Brewer & Parker, 1990; Brotherson, 1981; Cappers, 1993; Chadin et al., 2003; Chambers et al., 1991a; Cottam, 1939; Cottam et al., 1944; Craven & Hunt, 1984; del Moral & Watson, 1978; DeVelice et al., 1999; DeVlaming & Proctor, 1968; Earnst & Rothe, 2004; Engel & Nichols, 1994; Fehr, 1989; Feijoó & Lombardo, 2007; Fortner & White, 1988; Fraser & Hristienko, 1983; Ganie et al., 2016; Grant et al., 1994; Guo & Cook, 1989; Haag, 1983; Hammer & Heseltine, 1988; Hangelbroek et al., 2002; 2003; Haynes & Hellquist, 2000c; Hellquist, 1975; 1980; Hellquist & Crow, 1980; Hellquist et al., 2014; Hidding et al., 2009; Hollingsworth et al., 1996a; 1996b; Johnson, 1941; Juday, 1992; Jupp & Spence, 1977; Kalkman & Van Wijk, 1984; Kantrud, 1990; Kaplan, 2008; Kaplan et al., 2013; Kapuscinski et al., 2014; Karol et al., 2017; Kautsky, 1987; 1988; 1990; Keith, 1961; Kipriyanova et al., 2017; Kopec, 2015; Krecker, 1939; Kuzmina et al., 2018; LaMontagne et al., 2003; Legner & Fisher, 1980; Li et al., 2015; Lindqvist et al., 2006; Maberly & Spence, 1983; Mabbott, 1920; Mader et al.,

1998; Madsen, 1991; Madsen & Adams, 1988; 1989; Madsen et al., 1992; Marsden & Ladago, 2017; Martin & Uhler, 1951; Mason, 1957; McAtee, 1918; 1939; McMullan et al., 2011; Monda et al., 1994; Moyle, 1945; Moyle & Hotchkiss, 1945; Mrachek, 1966; Muenscher, 1948; Munro et al., 2014; Myers, 1942; Nedeau et al., 2003; Nichols, 1990; 1999; Nies & Reusch, 2004; Ohanjanian & Carli, 2010; Patten, 2016; Peng et al., 2008; Pfauth & Sytsma, 2005; Pilon & Santamaria, 2002; Pilon et al., 2003; Pip, 1979; Pip & Stewart, 1976; Preston, 1995; Preston & Croft, 1997; Quattrocchi, 2012; Radomski & Perleber, 2012; Ramirez, Jr. & Dickerson, 1997; 1999; Rozentsvet et al., 1995b; Saarela et al., 2013; Sandberg et al., 2014; Santamaría & Llano García, 2004; Santamaría et al., 2002; Sedki et al., 2015; See et al., 1992; Serie & Swanson, 1976; Shay, 1999; Sher-Kaul et al., 1995; Singh et al., 2014; Sosiak, 2002; Spencer, 1986; 1987; Spencer & Anderson, 1987; Squires & Lesack, 2003; Srivastava & Jefferies, 2002; Stewart & Kantrud, 1972; Teryokhin, 1996; Teryokhin et al., 2002; Thieret, 1963; Thorne, 1982; Tracy et al., 2003; Van den Berg et al., 1998; Van der Bijl et al., 1989; Van Wijck et al., 1992; Van Wijk, 1988; 1989a; 1989b; 1989c; Van Wijk et al., 1988; Vermaat & Hootsmans, 1994; Viereck et al., 1992; Waridel et al., 2003; Wersal et al., 2005; Winter, 1978; Yeo, 1966; Yin et al., 2017.

3.4.3. Zannichellia

Horned pondweed; alguette

Etymology: after Giovanni Gerolamo Zannichelli (1662–1729)

Synonyms: *Aponogeton* Hill non L.; *Pelta*

Distribution: global: Africa, Australia, Eurasia, New World; **North America:** widespread

Diversity: global: 5–6 species; **North America:** 1 species

Indicators (USA): obligate wetland (OBL): *Zannichellia palustris*

Habitat: brackish, freshwater; freshwater (tidal); saline (coastal, inland); lacustrine, palustrine, riverine; **pH:** 6.5–8.8; **depth:** to 3.5 m; **life-form(s):** submersed (vittate)

Key morphology: shoots (to 1 m) slender, arising from a fine, creeping rhizome; leaves (to 10 cm) opposite or pseudo-whorled, linear, filiform, 1-nerved, sessile, enveloped basally by membranous stipules; flowers axillary, minute, unisexual (appearing bisexual by close proximity at node), the ♂ flower consisting of a single, naked stamen, the ♀ flowers (2–6) surrounded by a membranous envelope, each comprising 2 carpels terminated by peltate, funnel-like stigmas; fruits (to 4 mm) short-stipitate (to 4 mm), drupaceous, banana-like, beaked (to 2 mm), warty-glandular, often with a crenulate-dentate, winged keel

Life history: duration: annual (fruits/seeds); perennial (rhizomes); **asexual reproduction:** rhizomes; **pollination:** water (autogamy; hypohydrophily); **sexual condition:** monoecious; **fruit:** achenes (common); **local dispersal:** water (seeds); **long-distance dispersal:** birds (seeds)

Imperilment: 1. *Zannichellia palustris* [G5]; S1 (NH, VT, WV); S2 (AR, DE, IN, ME, NC, NF, NU, PE, YT); S3 (AB, AK, IL, KY, MB, NB, QC, WY)

Ecology: general: *Zannichellia* is monotypic in North America (see next).

3.4.3.1. *Zannichellia palustris* L. is an annual or perennial, which inhabits still to flowing, fresh to mesohaline (salinity: 0–13‰) backwaters (stream), bayous, bays (brackish), bogs (drying), brooks, canals (brackish, shallow), channels (concrete, muddy, river, tidal), cienega, coves (intertidal, protected), culverts, depressions (tidal), ditches (drainage, irrigation, roadside), draws, estuaries (tidal), fens (graminoid), flats (clay, salt), floodplains, gravel pits, lagoons (tidal), lakes (saline), livestock troughs, marshes (brackish, drying, river, salt, tidal), meadows (boggy, wet), mudflats (tidal), ponds (alkali, artificial, beaver, coastal, dry, farm, holding, mudflat, oxbow, roadside, sag, stock, strip mine, tundra), pools (alkali, backwater, brackish, drying, marl, river, rocky, shallow, tidal, vernal), ravines, reservoirs (margins), rice fields, river bottoms (dry), potholes, rivers, seepages, shores (lake, rocky), sloughs (tidal), springs (thermal, warm), streams (artesian, boggy, intermittent, saline, spring-fed, tidal), swamps (lakeshore), tanks (stock, water), and waterholes at elevations from −55 to 3100 m. The plants occur in exposures that receive full sunlight. They are found most often on harder substrates and are averse to accumulations of peat or to layers of muck more than 1 mm thick. However, they have been reported from a variety of substrates including basalt, clay, clay loam (Parmleed–Worfka association), cobbles, gravel, limestone, loam (fine, granitic), loamy sand, muck (fetid shoe-sucking), mucky peat, mud, rock, sand, sandy loam, sandy silt, silt, silty loam, silty mud, and silty sand. Experimental studies have indicated that sediment nitrogen rather than phosphorous limits the growth of the plants. This species can tolerate turbidity and generally inhabits more eutrophic sites. It is found most often in shallow waters (<1 m), but can occur at depths from 0.05 to 3.5 m. The plants grow most often in highly alkaline (inland) or saline (coastal) waters of high pH and conductivity (pH: 6.5–8.8; $\bar{x}$ =6.9–8.1; alkalinity [as $CaCO_3$]: 7.5–277.5 mg l^{-1}; $\bar{x}$ =13.5–145 mg l^{-1}; chloride: 185.0–1201.6 mg l^{-1}; $\bar{x}$ =693.6 mg l^{-1}; conductivity to 580 µmhos cm^{-1} [@ 25°C]; $SO_4{}^{2-}$: up to 1290 mg l^{-1}). Flowering takes place nearly year-round (March–December), with fruiting extending from April to December. At anthesis the somewhat adhesive pollen is released in small clouds, which soon dissociate to release the individual, sinking grains. Contact of the pollen grains with the stigma stimulates their germination. Most pollination is autogamous (i.e., geitonogamous) due to the proximity of male and female reproductive organs. High levels of inbreeding have been indicated by studies using a variety of genetic markers. However, some xenogamous pollen transfer (i.e., to different individuals) also occurs. Overall, the percent seed set can range from 55.6% to 91%. Emasculated plants do not set seed, which indicates that they are not agamospermous. The seeds are not buoyant and sink when shed but are dispersed by water currents (tidal or otherwise). Seed trap data demonstrate that about 353 seeds on average can enter a site during rising tidal activity. Fewer seeds ($\bar{x}$ =48) are carried out during falling tides, which leads to a net accumulation in the seed bank. Though ingested by many birds, studies with captive killdeer (Aves: Charadriidae: *Charadrius vociferus*) showed that viable seeds to have a mean retention time of only 5 hr (much lower than other aquatics evaluated). The low retention time would suggest that endozoic dispersal, while possible, might be effective only over relatively short distances. Seed production can be prolific, capable of achieving an astounding density of 799,630 seeds m^{-2} from cultured material. Mean densities of 54 seeds 100 cm^{-3} have been found in some sediments, with greater numbers from non-channeled sites. Seedlings have been recorded at densities of up to 90 m^{-2} from some North American seed bank studies. Although the plants often occur at low frequencies in natural sites (e.g., 4.2%), they can become locally dense and can represent even higher frequencies of sediment seedling emergence (e.g., 133–266 plants m^{-2}). In many cases, they dominate the seed banks of permanently inundated sediments and represent the only submersed plant species exhibiting any significant germination therein. Natural seed bank densities of up to 5,920 seeds m^{-2} ($\bar{x}$ = 1,219 m^{-2}) have been reported; however, the burial of seeds by more than 2 cm of sediment inhibits their germination and seedling emergence. In one study, seeds collected from natural sediments began to germinate within 5–7 days, and continued to sprout for about 30 days, while under ambient greenhouse conditions at temperatures from 18°C to 25°C. Fresh seeds have germinated following 30–60 days of cold stratification (@ 4°C) when incubated at 20°C–24°C. The seeds are capable of germinating at temperatures ranging from 4.2°C to 34°C but maximum rates (total germination) have been observed at 24.4°C (69±5%) and 29.5°C (73±3%), temperatures that also have yielded the fastest median germination time (9 days). Germination rates and times are highly reduced under dark conditions. Seedlings have been observed at relatively high densities in saline salt flats; however, studies of Australian plants indicated that germination ceases at NaCl concentrations above 6‰. Seeds of European plants have germinated at salinities up to 4‰. Dry-stored seeds have remained viable (11% germination) after 60 months, whereas wet-stored seeds do not remain viable after 9 months. The ability of the seeds to retain their viability when dried represents an important drought-tolerance mechanism. Although densities of up to 2,530 plants m^{-2} have been observed in some (e.g., tidal freshwater) sites, biomass (<0.1–5.5 g m^{-2}) and percent cover (0.1–3.3%) often are low. Characteristically, the plants reach their peak biomass early in the growing season (spring) but decline rapidly in early summer as coarser, competing species become more prominent. Technically, the plants are perennials due to their production of a slender rhizome during the growing season. However, it is uncertain how persistent the rhizomes are afterward. In most cases, the plants appear to behave more like annuals, by dying back completely during the summer, and persisting through the next growing season entirely by seed. Yet, further studies of the extent of rhizomatous growth would be useful. Despite their extremely fine texture, the shoots have a surprisingly high mean breaking strength (8.22 ± 0.87 (MN m^{-2}). Even though annual recruitment occurs predominantly by seed, some populations can consist of a single multilocus genotype (determined using microsatellite markers), which reflects either strong clonal growth or high levels of inbreeding. The plants are among the early colonizing species of newly available sites such as beaver ponds, gravel pits, and created wetlands. **Reported associates:** *Acorus calamus, Agrostis stolonifera, Alisma gramineum, Alisma triviale, Alopecurus aequalis, Anemopsis californica, Atriplex argentea, Azolla cristata, Azolla filiculoides, Baccharis salicifolia, Beckmannia syzigachne, Berula erecta, Betula occidentalis, Bidens beckii, Bidens cernuus, Bidens laevis, Bolboschoenus maritimus, Bromus japonicus, Cabomba caroliniana, Callitriche fassettii, Callitriche hermaphroditica, Callitriche heterophylla, Callitriche palustris, Callitriche stagnalis, Carex aquatilis, Carex athrostachya, Carex lyngbyei, Carex nebrascensis, Carex pellita, Carex praegracilis, Carex simulata, Carex vulpinoidea, Castilleja minor, Ceratophyllum demersum, Ceratophyllum echinatum, Ceratopteris thalictroides, Chara braunii, Chara globularis, Chara vulgaris, Chara zeylanica, Cicuta, Cirsium vulgare, Cladophora, Cotula coronopifolia, Crassula aquatica, Cynodon dactylon, Cynoglossum officinale, Cyperus bipartitus, Cyperus elegans, Datisca glomerata, Distichlis spicata, Echinochloa, Egeria densa, Eichhornia crassipes, Elatine californica, Eleocharis acicularis, Eleocharis intermedia, Eleocharis kamtschatica, Eleocharis palustris, Eleocharis parishii, Eleocharis parvula, Eleocharis quinqueflora, Elodea canadensis, Elodea nuttallii, Epilobium ciliatum, Equisetum variegatum, Erigeron divergens, Eryngium sparganophyllum, Euthamia occidentalis, Fontinalis antipyretica, Fraxinus velutina, Glyceria borealis, Glyceria grandis, Heteranthera dubia, Hippuris tetraphylla, Hippuris vulgaris, Hydrilla verticillata, Hydrocotyle ranunculoides, Hygrophila lacustris, Iris missouriensis, Isoetes engelmannii, Juncus balticus, Juncus bufonius, Juncus compressus, Juncus effusus, Juncus saximontanus, Kochia scoparia, Leersia oryzoides, Lemna gibba, Lemna minor, Lemna trisulca, Leptochloa fusca, Leptodictyum riparium, Lilaeopsis, Limnanthes douglasii, Limnophila sessiliflora, Limosella acaulis, Limosella aquatica, Ludwigia palustris, Ludwigia repens, Lycopus americanus, Lythrum hyssopifolia, Marsilea vestita, Melilotus albus, Mentha arvensis, Menyanthes trifoliata, Mimulus glabratus, Mimulus guttatus, Muhlenbergia asperifolia, Myosotis laxa, Myosurus minimus, Myriophyllum aquaticum, Myriophyllum heterophyllum, Myriophyllum sibiricum, Myriophyllum spicatum, Myriophyllum verticillatum, Najas flexilis, Najas gracillima, Najas guadalupensis, Najas marina, Najas minor, Nasturtium officinale, Nitella, Nuphar polysepala, Nuphar variegata, Oenothera curtiflora, Paspalum distichum, Persicaria amphibia, Persicaria lapathifolia, Phalaris arundinacea, Pistia stratiotes, Pluchea sericea, Poa bigelovii, Poa nemoralis, Polygonum aviculare, Polypogon monspeliensis, Polypogon viridis, Populus fremontii, Potamogeton amplifolius, Potamogeton berchtoldii, Potamogeton crispus, Potamogeton epihydrus, Potamogeton foliosus, Potamogeton friesii, Potamogeton gramineus, Potamogeton illinoensis, Potamogeton natans, Potamogeton nodosus, Potamogeton perfoliatus, Potamogeton praelongus, Potamogeton pusillus, Potamogeton richardsonii, Potamogeton zosteriformis, Potentilla anserina, Prosopis juliflora, Ranunculus aquatilis, Ranunculus cymbalaria, Ranunculus flabellaris, Ranunculus flammula, Ranunculus gmelinii, Ranunculus longirostris, Ranunculus subrigidus, Ranunculus trichophyllus, Riccia fluitans, Ricciocarpus natans, Rorippa curvisiliqua, Rumex crispus, Rumex salicifolius, Ruppia maritima, Ruppia spiralis, Sagittaria latifolia, Sagittaria platyphylla, Salix exigua, Salix gooddingii, Salix lutea, Salvinia molesta, Schedonorus arundinaceus, Schoenoplectus acutus, Schoenoplectus americanus, Schoenoplectus pungens, Schoenoplectus tabernaemontani, Scirpus microcarpus, Sisyrinchium demissum, Sium suave, Sorghum halepense, Sparganium emersum, Sparganium emersum, Spirodela polyrhiza, Sporobolus airoides, Stuckenia filiformis, Stuckenia pectinata, Stuckenia vaginata, Symphyotrichum eatonii, Symphyotrichum subulatum, Tamarix ramosissima, Trapa natans, Triglochin maritima, Triglochin palustris, Triglochin scilloides, Typha angustifolia, Typha domingensis, Typha latifolia, Utricularia gibba, Utricularia macrorhiza, Vallisneria americana, Veronica americana, Veronica anagallis-aquatica, Veronica catenata, Wolffia brasiliensis, Wolffia columbiana, Zeltnera exaltata, Zizania texana, Zostera japonica, Zostera marina.*

Use by wildlife: *Zannichellia palustris* can be an excellent wildlife food where locally abundant. Plants from some western North American sites have contained somewhat elevated levels of toxic substances such as arsenic (2.8 μg g^{-1}, dry w) and boron (230 μg g^{-1}, dry w), but both values are below levels believed harmful to aquatic wildlife (19 and 300 μg g^{-1}, respectively). The plants harbor scores of invertebrates, which are eaten by trout (Vertebrata: Teleostei: Salmonidae) and other fish. Their foliage contains from 19% to 45% C ($\bar{x}$ = 38.15%), 2–6% N ($\bar{x}$ = 3.74%), and 26–242 μM g^{-1} phenolic acids and is grazed by several species of waterfowl (Aves: Anatidae) including Canada geese (*Branta canadensis occidentalis*), mallards (*Anas platyrhynchos*), pintails (*Anas acuta*), and shovelers (*Anas clypeata*). The fruits are eaten by baldpates (*Anas americana*), Barrow's goldeneye (*Bucephala islandica*; >100 seeds eaten by a single bird), blue-winged teal (*Anas discors*), canvasbacks (*Aythya valisineria*), cinnamon teal (*Anas cyanoptera*), common goldeneye (*Bucephala clangula*), gadwall (*Anas strepera*), greater scaup (*Aythya marila*), green-winged teal (*Anas crecca*), lesser scaup (*Aythya affinis*; >10,000 seeds eaten by a single bird), mallards (*Anas platyrhynchos*), oldsquaw (*Clangula hyemalis*), pintails (*Anas acuta*), redhead ducks (*Aythya americana*), ring-necked ducks (*Aythya collaris*), ruddy ducks (*Oxyura jamaicensis*), spectacled eiders (*Somateria fischeri*), surf scoters (*Melanitta perspicillata*), and white-winged scoters (*Melanitta fusca*).

Economic importance: food: not edible; **medicinal:** no uses reported; **cultivation:** not in cultivation; **misc. products:** *Zannichellia palustris* has been used in baseline ecological studies because of its rapid response to environmental perturbation; **weeds:** *Zannichellia palustris* is regarded as a weed of cultivated rice (Poaceae: *Oryza sativa*); **nonindigenous species:** none.

Systematics: The former genera of "Zannichelliaceae" (now transferred to Potamogetonaceae) resolve phylogenetically as two major clades, which depict *Althenia* (including *Lepilaena*) as the sister group of *Pseudalthenia* + *Zannichellia* (Figure 1.24). The precise number of *Zannichellia* species worldwide remains debateable. Though regarded previously as a monotypic genus, molecular data (Figure 1.24) have helped to clarify that at least four distinct *Zannichellia* species should be recognized (a lack of sequence divergence between *Z. major* and *Z. palustris* accessions from Asia, Europe, and North America argues against their continued segregation as distinct species). Molecular data (Figure 1.24) also indicate that *Z. palustris* is related most closely to *Z. pedunculata*. The base chromosome number of *Zannichellia* is $x=6$ (or 7?); *Z. palustris* ($2n=24$ [or 28?]) is a tetraploid. No hybrids involving *Z. palustris* occur in North America given that it is the only species of the genus found in this region.

Distribution: *Zannichellia palustris* is widespread throughout all but the northernmost and southeastern portions of North America.

References: Aronson, 1989; Bartonek & Hickey, 1969; Baskin & Baskin, 1998; Bornette & Puijalon, 2011; Bytnerowicz & Carruthers, 2014; Capers, 2003a, 2003b; Cappers, 1993; Carter & Rybicki, 1986; Chase, 2007; Comes et al., 1978; Cottam, 1939; Crow, 1979; Crowder et al., 1977b; Davis, 1985; DeVlaming & Proctor, 1968; Fassett, 1957; Greenwood & DuBowy, 2005; Guo et al., 1990; Haag, 1983; Hothem & Ohlendorf, 1989; Hughes & Young, 1982; Ito et al., 2016a; Keith & Stanislawski, 1960; Kadlec & Smith, 1984; Krapu, 1974; Lemke, 1989; Mabbott, 1920; Madsen, 1991; McAtee, 1925; Montz, 1978; Moyle, 1945; Orth et al., 1990; Poiani &

Johnson, 1989; Poirrier et al., 2010; Posluszny & Sattler, 1976; Ribicki et al., 2001; Rooney et al., 2013; Roze, 1887; Schuyler et al., 1993; Serie & Swanson, 1976; Schutten et al., 2005; Smith, Jr., 1983; Smith & Kadlec, 1983; 1985; Spencer & Ksander, 1999; 2002; 2003; Stevenson et al., 1993; Stuckey & Moore, 1995; Triest & Vanhecke, 1991; Triest et al., 2007; 2010; Van Vierssen, 1982b; Westcott et al., 1997.

Family 3.5. Scheuchzeriaceae [1]

The monotypic Scheuchzeriaceae were once merged with Juncaginaceae, to which the group is closely related (Figure 1.2). Both families contain triglochinin, a cyanogenic glycoside that otherwise is rare in angiosperms (Ruijgrok, 1974). However, Scheuchzeriaceae differ from Juncaginaceae by several traits including leaves with an apical pore (vs. pore lacking), infravaginal hairs (vs. scales), pollen in dyads (vs. monads), and follicles (vs. achenes) (Haynes et al., 1998f). Early molecular phylogenetic studies confirmed the distinctness of *Scheuchzeria palustris* (the family's only representative) from Juncaginaceae (Les & Haynes, 1995), a result that was consistently reiterated (e.g., Les et al., 1997a; Petersen et al., 2006; Les & Tippery, 2013; Ross et al., 2016). Consequently, nearly all contemporary authors accept its recognition as a distinct family.

The family is of no known economic importance.

Scheuchzeriaceae have a circumboreal distribution that extends into temperate regions of the Northern Hemisphere. North American OBL indicators are represented by a single genus:

3.5.1. *Scheuchzeria* L.

3.5.1. Scheuchzeria

Pod-grass, Rannoch-rush; scheuchzérie des marais
Etymology: after the brothers Johann Gaspar Scheuchzer (1684–1738) and Johann Jakob Scheuchzer (1672–1733)
Synonyms: none
Distribution: global: Eurasia, North America; **North America:** northern
Diversity: global: 1 species; **North America:** 1 species
Indicators (USA): obligate wetland (OBL): *Scheuchzeria palustris*
Habitat: freshwater; palustrine; **pH:** 3.8–7.6; **depth:** <1 m; **life-form(s):** emergent herb
Key morphology: rhizomes sympodial, creeping, freely branching, bearing fibrous leaf remains, aerial stems (to 40 cm) erect, unbranched; leaves alternate, striate, the blades (to 41 cm) linear, terminating in an apical pore, sessile, their bases sheathing (to 10 cm), sheath open, dilated, hyaline-margined, with a terminal pair of membranous auricles (to 12 mm); racemes (to 10 cm) 3–12–flowered, with reduced, leaf-like bracts; flowers bisexual, hypogynous, trimerous, pedicelled (to 25 mm); tepals (to 3 mm) 6 (3+3), white to yellow-green, free, lanceolate-elliptic; stamens 6, fused to tepals; pistils (to 7 mm) 3–6, free or slightly connate basally, stigmas ventral, decurrent, 1–3 ovules per locule; follicles (to 10 mm) 1–4, leathery, spreading, inflated, ovoid, greenish, brown, or purplish, beaked (to 1 mm), dehiscent adaxially; seeds (to 5 mm) 1–3, ovoid, smooth, brown to black

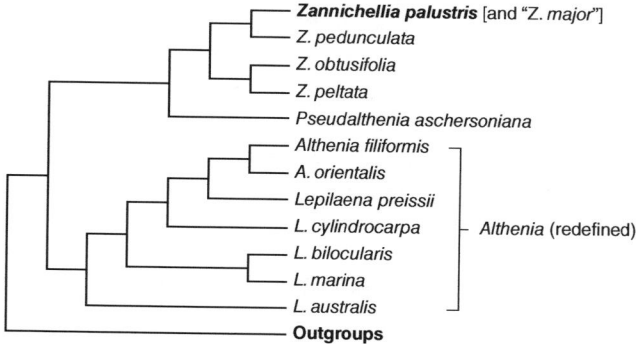

FIGURE 1.24 Relationships within the "Zannichelliaceae" clade of Potamogetonaceae as indicated by phylogenetic analysis of combined nuclear and plastid DNA sequence data (adapted from Ito et al., 2016). Two main groups of genera are evident: an upper clade, which comprises species of *Althenia* and *Lepilaena*, and a lower clade, which includes species of *Pseudalthenia* and *Zannichellia*. This nested placement of *Althenia* within *Lepilaena* has led to the persuasive recommendation that *Lepilaena* be merged with *Althenia* (which has nomenclatural priority). The five putative *Zannichellia* species surveyed resolve only as four distinct taxa (there is no sequence divergence between "*Z. major*" and *Z. palustris* for the multiple regions surveyed). All genera shown are obligate aquatic plants; *Z. palustris* (in bold) is the only North American OBL indicator.

Life history: duration: perennial (rhizomes); **asexual reproduction:** rhizomes; **pollination:** wind; **sexual condition:** hermaphroditic; **fruit:** aggregates (follicles) (common); **local dispersal:** rhizomes, fruits (gravity, water); **long-distance dispersal:** fruits (birds)

Imperilment: 1. *Scheuchzeria palustris* [G5]; SX (IL); SH (IA, WV); S1 (CA, CT, MA, ND, OH, PA, RI, WY); S2 (NU, VT); S3 (AK, MB, MT, NF, NY, YT)

Ecology: general: *Scheuchzeria* is monotypic (see next).

3.5.1.1. *Scheuchzeria palustris* L. grows in bogs (lakeshore, roadside, *Sphagnum*), depressions (boggy, muddy), fens (poor, rich), flarks, flats (bog, swampy), floodplains, marshes, meadows, muskegs, ponds (beaver), shores (peaty), string mires, swamps (*Sphagnum*), and along the margins of lakes (kettle), mats (floating, *Sphagnum*), and pools at elevations to 2073 m. Most occurrences are in open exposures, which receive full sunlight. Slight shade is tolerated, but the plants do not compete well when shaded by shrubby vegetation. The substrates are organic and acidic (e.g., pH: 5.5) and include humus, muck (quaking), ooze, and peat; they can be underlain by basaltic bedrock or clay (marine). The habitats are characterized by permanent, acidic, shallow (to 0.6 m deep), soft waters (pH: 3.8–7.6; alkalinity: 4.0–48 mg l^{-1}conductivity: 16–97 mmhos, cm^{-1}; hardness [CaCO$_3$]: 9.2–42.8 mg l^{-1}; Ca: 2.0–6.4 mg l^{-1}; K: 0.8–0.9 mg l^{-1}; Mg: 0.5–2.4 mg l^{-1}). Flowering and fruiting occur from April to September. The protogynous flowers are pollinated by wind (anemophily) and potentially are outcrossed; however, the reproductive biology of this species has not been studied in any detail. Each shoot produces from 7.17 to 13.43 seeds on average, each weighing from 2.3 to 6.3 mg. Studies of natural populations have found seed viability to be about 59% with a field germination rate of 22%. Seed germination rates generally are higher on saturated substrates (exposed peat or cover of *Sphagnum*; e.g., *S. cuspidatum*) that are devoid of standing water. Germination increases incrementally (9%, 15%, 20%) when the seeds occur on beds of *Sphagnum magellanicum*, *Cladopodiella fluitans*, or *Sphagnum cuspidatum*, respectively. These rate differences have been attributed to the different microclimatic conditions provided by each species. Mean laboratory survivorship of plants is relatively high (72–87%). The buoyant follicles can be dispersed locally by gravity or transported within water bodies on the surface. The documentation of seeds in the digestive tracts of birds (Aves) indicates that they potentially are dispersed over longer distances by means of endozoic transport. Seedling establishment is poor and requires suitable microhabitat. The plants have failed to reestablish at some available sites even after a decade of observation. Vegetative reproduction occurs by rhizomes, which produce diffuse ramets. Clonal growth occurs frequently, and plants often are devoid of flowers or fruit. The average density of plants can vary from 342 to 708 shoots m^{-2}. Their net aboveground biomass (determined experimentally) remained stable (43–58 g m^{-2}) over varying water table conditions but increased by 63% under high infrared loading conditions (a proxy for climatic warming). The plants have no drought tolerance and require sites that remain permanently saturated throughout the year. They are early colonizers of

saturated soils during the thermokarst bog stage but decline as peat accumulation and decreasing substrate moisture favors the colonization of ericaceous shrubs. This species has a high mean coefficient of conservatism (0.87–0.98), which indicates its inability to withstand any dramatic changes to or losses of habitat. **Reported associates:** *Agrostis scabra*, *Alnus rugosa*, *Andromeda polifolia*, *Arethusa bulbosa*, *Betula glandulosa*, *Betula nana*, *Betula pumila*, *Calamagrostis canadensis*, *Calopogon tuberosus*, *Carex aquatilis*, *Carex buxbaumii*, *Carex chordorrhiza*, *Carex crinita*, *Carex cusickii*, *Carex echinata*, *Carex interior*, *Carex lasiocarpa*, *Carex lenticularis*, *Carex limosa*, *Carex livida*, *Carex muricata*, *Carex oligosperma*, *Carex pauciflora*, *Carex praeceptorum*, *Carex rostrata*, *Carex utriculata*, *Chamaedaphne calyculata*, *Cicuta bulbifera*, *Cicuta douglasii*, *Cladium mariscoides*, *Comarum palustre*, *Cypripedium acaule*, *Decodon verticillatus*, *Deschampsia cespitosa*, *Drosera anglica*, *Drosera rotundifolia*, *Dulichium arundinaceum*, *Eleocharis*, *Eriophorum angustifolium*, *Eriophorum chamissonis*, *Eriophorum gracile*, *Eriophorum russeolum*, *Eriophorum tenellum*, *Eriophorum vaginatum*, *Eriophorum virginicum*, *Gaultheria hispidula*, *Gentiana sceptrum*, *Hypericum anagalloides*, *Iris versicolor*, *Juncus balticus*, *Juncus effusus*, *Kalmia latifolia*, *Kalmia microphylla*, *Kalmia polifolia*, *Larix laricina*, *Lycopodiella inundata*, *Lycopus uniflorus*, *Maianthemum trifolium*, *Menyanthes trifoliata*, *Myrica gale*, *Nuphar polysepala*, *Nuphar variegata*, *Pedicularis macrodonta*, *Phragmites australis*, *Picea engelmannii*, *Picea mariana*, *Pinus contorta*, *Pinus strobus*, *Platanthera dilatata*, *Pogonia ophioglossoides*, *Potamogeton gramineus*, *Rhododendron columbianum*, *Rhynchospora alba*, *Rhynchospora fusca*, *Rubus chamaemorus*, *Salix*, *Sanguisorba*, *Sarracenia purpurea*, *Schoenoplectus acutus*, *Sparganium*, *Sphagnum angustifolium*, *Sphagnum balticum*, *Sphagnum capillifolium*, *Sphagnum centrale*, *Sphagnum cuspidatum*, *Sphagnum lindbergii*, *Sphagnum papillosum*, *Sphagnum pulchrum*, *Sphagnum riparium*, *Sphagnum subsecundum*, *Sphagnum warnstorfii*, *Spiraea douglasii*, *Spiranthes romanzoffiana*, *Thelypteris palustris*, *Thuja plicata*, *Triadenum virginicum*, *Triantha glutinosa*, *Trichophorum alpinum*, *Trichophorum cespitosum*, *Trientalis europaea*, *Triglochin*, *Tsuga heterophylla*, *Typha latifolia*, *Utricularia macrorhiza*, *Vaccinium macrocarpon*, *Vaccinium oxycoccos*, *Vaccinium uliginosum*, *Viola*.

Use by wildlife: *Scheuchzeria palustris* provides habitat for spreadwing damselflies (Insecta: Odonata: Lestidae: *Lestes disjunctus*, *L. forcipatus*). The seeds are eaten by black ducks (*Anas rubripes*).

Economic importance: food: Because the foliar cyanogenic glycosides can liberate cyanide upon ingestion, *Scheuchzeria palustris* should be regarded as poisonous and never eaten; **medicinal:** The plants contain triglochinin, a cyanide-producing glycoside; **cultivation:** not cultivated; **misc. products:** Organic matter extracted from *Scheuchzeria* peat deposits has been used as an additive in methods for the hydrophobization of mineral binders and associated building materials; **weeds:** none; **nonindigenous species:** Viable rhizome fragments of

Scheuchzeria palustris have been found to be intermixed in commercial shipments of peat moss (*Sphagnum*); however, there have been no reported nonindigenous populations introduced by this means.

Systematics: The monotypic status of *Scheuchzeria* leaves little phylogenetic uncertainty regarding infrageneric relationships (Figure 1.2). Some authors have treated Old and New World populations as distinct varieties due to slight floral and fruit features, which appear to be quite variable and of questionable taxonomic merit. The basic chromosome number of *Scheuchzeria* is $x=11$; *S. palustris* ($2n=22$) is diploid. No hybrids involving *S. palustris* have been reported and are unlikely due to the isolated phylogenetic position of this species.

Distribution: *Scheuchzeria palustris* is circumboreal, occuring from the Arctic to Temperate regions of the Northern Hemisphere.

References: Cannings & Simaica, 2005; Crow & Hellquist, 1982; DuBois et al., 2009; Fernald, 1923b; Flensburg & Sparling, 1973; Gibson et al., 2018; Green et al., 2016; Haynes et al., 1998f; Knuth, 1909; Laberge et al., 2015; Les & Tippery, 2013; Mendall, 1949; Minayeva, 2010; Misnikov, 2006; 2018; Nienaber, 2000; Posluszny, 1983; Poulin et al., 2012; Ridley, 1930; Ruijgrok, 1974; Seigler, 1976; Sledge, 1948; Transeau, 1903; Valentine, 1976; Volkova et al., 2016; Weltzin et al., 2000.

Family 3.6. Zosteraceae [2]

The "eelgrass" family (Zosteraceae) has experienced a tumultuous taxonomic history, during which *Phyllospadix* and *Zostera* were misplaced among peculiar, heterogenous assemblages of aquatic taxa whose phylogenetic affinities also were poorly known. Such were Najadaceae sensu Morong (1893), which included the two genera along with *Najas, Potamogeton, Ruppia, Scheuchzeria, Triglochin*, and *Zannichellia*; and Potamogetonaceae sensu Hartog (1970), which included them along with *Amphibolis, Cymodocea, Halodule, Posidonia, Potamogeton, Ruppia, Syringodium*, and *Thalassodendron*. Due much in part to subsequent phylogenetic studies (especially those incorporating molecular

data), the circumscription of Zosteraceae has been clarified to represent 13 species in only two genera (*Phyllospadix* and *Zostera*), which together resolve as the sister group of Potamogetonaceae sensu lato (Figures 1.21 and 1.25). Both *Phyllospadix* and *Zostera* are monophyletic as demonstrated by the phylogenetic analysis of morphological (Les et al., 2002) as well as molecular data sets (Les et al., 2002; Tanaka et al., 2003; Coyer et al., 2013).

Phylogenetically, *Zostera* comprises three subclades (Figure 1.25), which are recognized at the rank of subgenera (Les et al., 2002). One subclade includes the former, monotypic genus *Heterozostera* (Setch.) Hartog, which was merged with *Zostera* (as subgenus *Heterozostera*) after phylogenetic analyses found it to nest within the latter (Les et al., 1997a; 2002; Tanaka et al., 2003). However, some authors (e.g., Tomlinson & Posluzny, 2001) have advocated retention of *Heterozostera* by similarly elevating the few species in subgenus *Zosterella* to generic rank as "*Nanozostera*". Although doing so would be compatible with the phylogenetic results (i.e., the same three subclades would be recognized), the resulting propagation of taxa and new names would unnecessarily clutter and complicate the taxonomy of an already problematic group. There are many arguments for retaining these three taxa as subgenera rather than elevating them to genera (Les et al., 2002). These include trivial and/or poorly defined synapomorphies, few total species (eight), and a lack of appreciable morphological or DNA sequence divergence among the proposed segregate taxa (Soros-Pottruff & Posluszny, 1995; Les et al., 2002; Coyer et al., 2013). Moreover, members of all three segregate taxa together comprise a homogenous group of quite similar species, which by all accounts, reflects a natural genus. With no compelling reason to subdivide this already species-depauperate group into additional, poorly defined genera, the recognition of "*Nanozostera*" has not been accepted here or by most North American taxonomists.

Zosteraceae represent one of three major seagrass radiations in the flowering plants (Les et al., 1997a). All the species are submersed, marine, rhizomatous perennials with grass-like or filiform foliage and unisexual (dioecious or monoecious), water-pollinated (hydrophilous), reduced, naked

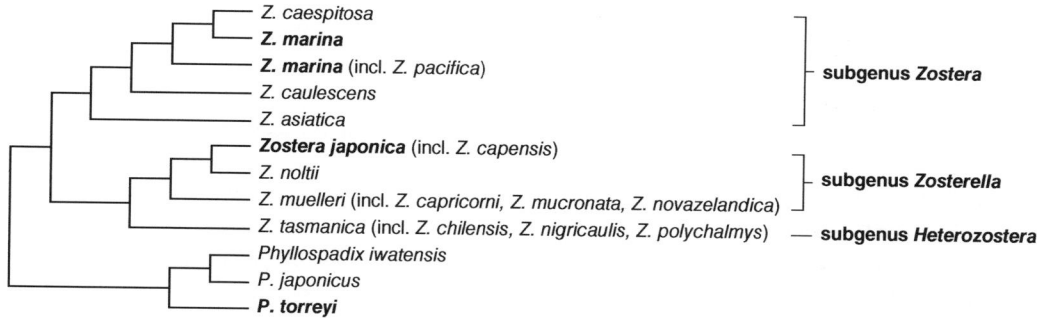

FIGURE 1.25 Infrafamilial relationships in Zosteraceae as indicated by the phylogenetic analysis of nrITS sequence data (adapted and simplified from Coyer et al., 2013). The *Zostera* species fall within three subclades, which are recognized here as subgenera (elevated unnecessarily to genera by some authors). The nesting of *Z. caespitosa* among *Z. marina* accessions is problematic and warrants further evaluation. Although all members of the family are obligate aquatics, only the three OBL North American indicators are highlighted in bold (*Phyllospadix serrulatus* and *P. scouleri* were not included in the analysis).

flowers (Cronquist, 1981; Les, 1988; Les et al., 1997a). In North America they occur exclusively along the Atlantic and Pacific coasts and are absent from inland localities (Green & Short, 2003).

Zostera is an important food source for aquatic birds (Aves) and provides vital habitat for countless marine animals. *Phyllospadix* species protect marine substrates from erosion, are important in nutrient cycling, and provide food and habitat for various fish, invertebrates, and shore birds (Soros-Pottruff & Posluszny, 1994). Species in both genera have been used as a source of human food. Eelgrass (*Zostera*) and surfgrass (*Phyllospadix*) foliage has minor economic importance as a packing material, mattress stuffing, and thatch.

Although Zosteraceae are most prominent in the temperate oceans, the family also extends into the tropical waters of coastal Africa and Australia. Two genera contain OBL indicators in North America:

3.6.1. ***Phyllospadix*** Hook.
3.6.2. ***Zostera*** L.

3.6.1. Phyllospadix

Surfgrass; phyllospadix
Etymology: from the Greek *phyllon spadix* ("palm leaf") referring to the palm-like inflorescence
Synonyms: none
Distribution: global: Asia (Pacific); North America (Pacific); **North America:** Western Pacific coast
Diversity: global: 5 species; **North America:** 3 species
Indicators (USA): obligate wetland (OBL): *Phyllospadix scouleri, P. serrulatus, P. torreyi*
Habitat: brackish (coastal), marine; oceanic; **pH:** 7.3–8.6; **depth:** to 15 m; **life-form(s):** submersed (vittate)
Key morphology: rhizomes with internodal roots or root clusters and fibrous leaf-sheath remains (to 5 cm); leaves (to 2.0 m) filiform, grass-like, or linear, coriaceous, 3–7-veined, margins entire, basally sheathing (to 55 cm); plants dioecious; reproductive shoots (to 60 cm) branched (bearing up to 20 spadices) or with a single spadix, peduncled (to 6.0 cm); spathal sheath (to 7.0 cm) subtending the linear-lanceolate spadix; ♂ spadix usually 20-flowered; subtended by ligulate retinacula (to 5.5 mm), ♀ spadix 8–26-flowered, with elliptic, linear, ligulate, or spatulate retinacula (to 8.0 mm); flowers naked, reduced (♂ a bilocular stamen; ♀ a simple pistil with 2 styles); fruits (to 5.5 mm) drupe-like, crescent-shaped, single-seeded, developing a bristly clamp-like structure; seeds (to 3 mm) ovoid, brown
Life history: duration: perennial (rhizomes); **asexual reproduction:** rhizomes; **pollination:** water (epihydrophilous, hypohydrophilous); **sexual condition:** dioecious, **fruit:** drupe-like (common); **local dispersal:** seeds (water); **long-distance dispersal:** seeds (water)
Imperilment: 1. *Phyllospadix scouleri*, [G5]; **2.** *P. serrulatus* [G4]; S1 (OR); S2 (AK); **2.** *P. torreyi* [G4/G5]; S2 (WA)
Ecology: general: All *Phyllospadix* species are seagrasses, i.e., marine flowering plants. Their common name of "surf grass" reflects their frequent occupation of the highly

energetic surf zone, where they occur on hard, rocky substrates. The three North American species occupy different depth niches, but they all are long-lived perennials, form dominant, persistent beds (>80% cover), preclude invading algal species, and recover slowly from disturbances. These characteristics have led to their ecological categorization as late-successional species. Most uptake occurs via the leaves, given that the rhizomes attach primarily to rocky substrates, which are devoid of available nutrients. All the species are dioecious and bear highly reduced flowers (the ♂ comprising only a bilocular stamen; the ♀ consisting entirely of one single-seeded carpel terminating in a pair of filiform, laciniate stigmas). The flowers are pollinated abiotically by water (epihydrophily and hypohydrophily), which is particularly ineffective in the dioecious *Phyllospadix* where populations often are characterized by female-biased sex ratios (e.g., <24% male plants, >90% female bias). These discrepancies have been shown to result primarily from male-biased mortality rather than differential reproductive costs. Because pollen limitation is proportional to the distance from the nearest male plants (which can occur either more shoreward or deeper than the females), and because the pollen also is quickly scattered and/or destroyed by the tumultuous surf, fewer than 1% of ovules mature into seeds at some sites. The flowers (both sexes) are arranged in two adjacent rows along the spadix and are enclosed behind flap-like scales called retinacula or retinacules. Seed set data indicate that the position of ♀ flowers within the inflorescence affects the likelihood of successful pollination more in the submersed inflorescences than in the surface-borne inflorescences. At anthesis (♂ flowers), the retinacula reflex to expose the anthers incrementally along the spadix axis. As the anthers desiccate during ebbing tides, the filiform pollen grains (to 1 mm in length), initially enveloped by air, are released in tetrad bundles and rise immediately to the surface where they repel one another to form a web-like surface film approximately 1 cm in diameter. These floating films achieve pollination (i.e., epihydrophily) when contacting the exposed stigmata of the female plants, which extend beyond, but remain covered by the non-reflexing retinacula. Some pollen does not rise to the surface but remains within the water column where it is dispersed by currents and effects pollination when contacting a submersed stigma (i.e., hypohydrophily). The pollen grains are known to remain viable for at least 3 days. The seeds lack dormancy (or have minimal dormancy requirements) and germinate during the ensuing growing season; no seed bank develops. The plants reproduce vegetatively by durable rhizomes. The rhizomes are thick, have substantial hypodermal fiber and root hair development, and possess smaller lacunae, which are believed to represent anatomical adaptations enabling their attachment to rocky substrates. The plants are strongly clonal and, due to dioecy, often persist as unisexual populations. They can develop into dense tussocks from 0.25 to 0.5 m in diameter. The fruits are the only functional propagules for dispersal because rhizome fragments are unable to re-establish.

3.6.1.1. ***Phyllospadix scouleri*** Hook. occurs throughout the lower intertidal and shallow subtidal zones (from −0.85

to −15 m) of the coastal Pacific shoreline, in coves, on rocky reefs, and in tidepools (rocky). The substrates are boulders, cobble, gravel, rock, or sand-inundated rock. The plants occupy surf-beaten, extremely high-energy coastal waters at temperatures near 15°C; however, they can survive across a much broader temperature range (1.5°C–25°C). Maximum cover (e.g., 37.7%) has been observed between +30 cm and mean lower low water (MLLW), but can reach 100% in tidepools at +60 cm. A maximum percent cover of ~16% has been reported from sites at 40–50 cm below mean low water level. Although the average oceanic pH is 8.1, local metabolic processes of *P. scouleri* (i.e., photosynthesis and respiration) can induce diurnal habitat variation (especially in tide pools) ranging from pH 7.3 to 8.6 (lower values during the night). Flowering initiates in April with fruiting extending from June to October. The basic reproductive biology is typical for the genus (see details in *Ecology: general:* above). Species-specific observations emphasize the occurrence of female-biased populations (e.g., male spadix density: 0.48–0.72 m^{-2}; patch-based sex ratio [m:f]: 0.01–0.03) and low seed:ovule ratios (e.g., 0.02–0.70), which are highly correlated. The seeds are dispersed within the water from September to March by currents. They possess bristled comb- or clamp-like structures, which facilitate their initial attachment to benthic algae. The seeds can reach densities up to 1000 m^{-2}, are non-dormant, and germinate readily from April to July. Germination rates typically are high (>90%). Subsequent leaf and root growth result in a firm attachment of the seedlings to the rocky algal substrate. However, seedling survival is low (e.g., 93% mortality within 7 months), which ultimately results in low recruitment rates (e.g., 0.4–13.3 m^{-2}; <1% cover after 3 years). High hydrogen sulfide (H$_2$S) levels have been found to decrease photosynthetic rates and are believed to represent one factor that limits seedling recruitment (LD$_{50}$: 430 μM @ 48 hr; 86 μM after 7 days). Studies using NaH^{14}CO$_3$ have shown carbon uptake to occur primarily via the foliage rather than rhizomes. Gross photosynthesis has been estimated at 13 mg CO$_2$ g dry wt^{-1} hr^{-1}. The plants continue to grow throughout the year. North American populations have exhibited mean production rates of 0.7 mg C fixed hr^{-1} and average maximum rates of 13.7 g dw m^{-2} day^{-1} (shoot) and 10.5 g dw m^{-2} day^{-1} (belowground). The average maximum biomass for *P. scouleri* has been estimated at 615.2 g dw m^{-2} (shoot) and 418.1 g dw m^{-2} (belowground). Exceptionally high annual productivity levels (>8000 g dw m^{-2} year^{-1}) have been measured in Mexican coastal waters. The plants are somewhat tolerant to desiccation but net photosynthesis and production decline with increasing exposure to air. Vegetative reproduction occurs by means of rhizomes, which elongate at rates from 0.24 to 0.58 mm day^{-1} and achieve an annual growth rate from 6.0 to 8.6 cm year^{-1}. Clones can cover areas up to 0.3 m^2. The mean maximized population growth rate has been estimated as $\lambda = 0.985$. **Reported associates:** *Acrosiphonia, Alaria marginata, Bossiella, Calliarthron tuberculosum, Callithamnion pikeanum, Corallina officinalis, Corallina pinnatifolia, Corallina vancouveriensis, Cystoseiria, Fucus distichus, Gelidium pusillum, Gigartina, Haliptylon gracile, Halosaccion glandiforme, Hedophyllum*

sessile, Hildenbrandia, Iridaea, Lithothrix aspergillum, Macrocystis, Melobesia, Microcladia, Neorhodomela larix, Odonthalia floccosa, Petrocelis middendorffii, Phyllospadix serrulatus, Phyllospadix torreyi, Plocamium pacificum, Polysiphonia, Porphyra, Pterocladia capillacea, Ralfsia pacifica, Rhodomela larix, Sargassum muticum, Ulva californica.

3.6.1.2. *Phyllospadix serrulatus* **Rupr. ex Aschers.** inhabits the middle to upper intertidal zone of the coastal Pacific shoreline and adjacent tidepools at elevations from +1.5 m to MLLW (rarely to −6.0 m). The plants are less resistant to wave energy than their North American congeners and occupy more sheltered sites than either. Their higher intertidal occurrence indicates that they are better adapted for photosynthesis when emersed. Nevertheless, aerial exposure of plants for even short duration results in a significantly lower biomass. The plants also exhibit shorter leaves, denser shoots, and lower seedling recruitment when growing in the shallower zones. Generally, they occur along surf-pounded coasts on rock substrates (rarely on deep clay). Flowering initiates in April with fruiting beginning in June. The basic reproductive biology is typical for the genus (see details in *Ecology: general:* above). Species-specific observations reiterate the predominance of female-biased sex ratios (e.g., male spadix density: 0–0.63 m^{-2}; patch-based sex ratio [m:f]: 0–0.23) and poor seed:ovule ratios (e.g., 0.004–0.37), which are highly correlated. The non-dormant seeds germinate from April to July. Seeds produced later during the growing season are dispersed in September. The rate of seedling recruitment is extremely low (0–1.3 seedlings m^{-2}) with seedlings found only rarely during site surveys. The plants continue to grow throughout the year in most (except subtidal) localities. The mean maximized population growth rate has been estimated as $\lambda = 0.982$. The plants are less influenced by the male-biased sex ratios and exhibit lower rates of sexual reproduction than their North American congeners. Compared to *P. scouleri*, they produce fewer rhizomes and attain a lower biomass, but facilitate higher sediment accretion. The clones can cover areas up to 0.5 m^2. **Reported associates:** *Alaria, Callithamnion, Fucus, Macrocystis pyrifera, Phyllospadix scouleri, Phyllospadix torreyi, Ulva lactuca, Zostera marina.*

3.6.1.3. *Phyllospadix torreyi* **S. Watson** inhabits oceanic coastlines and adjacent tidal pools in the lower intertidal zone (0 to −0.6 m) and shallow subtidal zones (0 to −15 m; typically −4 to −5 m). Vegetative shoot density decreases linearly as the water depth drops below 1 m. The substrates include rock and sand. The plants are broadly tolerant to variations in light, salinity, and temperature, but grow best at higher salinity (29‰), moderate light intensity (1075 lumens m^{-2}; 12-hr day), and moderate water temperatures (12°C–14°C). Less growth occurs under lower salinity or light, or at higher temperatures. Despite the average oceanic pH of 8.1, a wider diurnal range (pH: 7.3–8.6) can occur in *P. torreyi* habitats (especially tide pools) due to photosynthetic and respiratory processes, which result in lower values during the night. Flowering occurs year-round but is most prolific from May to September. Observations of intermingled male and female plants indicate that sex expression is under genetic rather than environmental control. On average, 24% of the shoots bear flowers during the peak flowering period (June). Flowering shoot density (May–August) can range from 20 to 854 shoots m^{-2} (maximum at 1.5 m depth) and correlates with the highest water temperatures and light availability. Populations are characterized by strongly female-biased sex ratios (51–90%; $\bar{x}$: 75% females) due primarily to male-biased mortality, which is not attributable to differential reproductive costs. The biased ratios have been explained by the greater ramification of female clones. Furthermore, female plants are more tenacious and require a greater wave force to dislodge them from the substrate (78.4 kg m^{-2} s^{-1}) than do male plants (57.4 kg m^{-2} s^{-1}). On the other hand, the more readily dislodged male plants (at least for flowering individuals) are believed to facilitate a broader distribution of pollen, which has been estimated to surpass 50 m in some cases. The male plants occur rarely at depths shallower than 3 m. From 16 to 19 anthers are produced per spadix on average. Male plants (up to 13%) can incur substantial damage to their spadix from grazing. Female spadix predation can vary from 7% to 24%. The average densities of male flowers (5568 m^{-2}), female flowers (9756 m^{-2}), and receptive female flowers (2027 m^{-2}) all reach their peak in waters 1.5 m deep. The mean flowering shoot mass of both sexes can decline by up to 60% in waters 6.1 m deep compared to those at a depth of 1.5 m. The average mass of female flowering shoots is 25% greater than that of males. A female spadix contains 10 flowers on average. The near complete fertilization of ovules can occur despite the highly skewed sex ratios, which appear to be offset by high mean populational pollen:ovule ratios (e.g., 18,950–58.375:1), which lead to high mean pollen densities (e.g., 184,879,872 grains m^{-2} at 1.5 m depth). The pollen tetrads develop as filiform bundles, which increases their capture rate by enabling them to rotate within the water column. Pollen limitation in some populations is indicated by a strong correlation between the density of males and seed production. Most seed maturation begins in July and peaks in August, whereas earlier developing flowers (e.g., January–February) produce smaller fruits. The highest female fitness (seed set) occurs at shallow depths at maximum light availability. Mean seed production can vary from 334 m^{-2} (June; 7.6 m depth) to 11,718 m^{-2} (July; 1.5 m depth). From 20% to 40% of shoots and numerous seeds (roughly 400 seeds m^{-2}) fail to mature but instead decompose during August and September. Surviving seeds exhibit high germination rates. Peak seed dispersal has been observed from September to November. The seeds sink when detached, and drift along the bottom at distances (up to 50 m or more) until becoming attached to a host alga by their bristly clamp-like appendages. Germination rates of cultured seeds are high (70+% within 4 weeks) when incubated under simulated natural environmental conditions (16°C, 40–50 µE m^{-2} s^{-1}; 14:10 hr light:dark photoperiod, mild agitation). The seeds remain dormant for 7–14 days, but for up to 83 days if kept under cold, dark (4°C) conditions. Germination occurs rapidly (95% within a week) for the latter when transferred to the simulated natural environmental conditions. However, seedling recruitment/survival typically is extremely low (e.g.,

roughly 1 seedling^{-1} 60 m^{-2} of seed-bearing sediment) and their survival rate is low (e.g., 30%). From 7 to 15% of seeds can be eaten by predators prior to dispersal with 50% or more consumed post-dispersal, factors which have been implicated in the low seedling numbers observed in field studies. The most successful attachment of seeds to algal substrates occurs under a moderate current velocity (max. 85 cm s^{-1}) and is substantially less under lower (max. 45 cm s^{-1}) or higher (max. 180 cm s^{-1}) flow rates. The most frequent algal substrates are red algae (Rhodophyta: Corallinaceae: *Lithothrix aspergillum*; Gigartinaceae: Chondracanthus canaliculatus). Seedlings transplanted intentionally during attempted restoration projects also suffered poor survivorship, which was lower at subtidal sites (0.9%) than at intertidal depths (2.3%). Plant cover often is continuous (100%) throughout the lower intertidal zone (80–90 cm below MLWL). The plants possess a mechanism for bicarbonate (HCO$_3^-$) uptake. Nitrogen (and presumably most nutrients) is taken up primarily by the foliage, with the rhizomes contributing little. North American populations have achieved average maximum production rates of 14.2 g dw m^{-2} day^{-1} (shoot) and 11.3 g dw m^{-2} day^{-1} (belowground). The average maximum biomass for *P. torreyi* has been estimated at 586.4 g dw m^{-2} (shoot) and 485.9 g dw m^{-2} (belowground). Although male and female plants are indistinguishable vegetatively, significantly higher biomass is allocated to female flowering shoots than to males across different water depths. The plants have low desiccation tolerance, with net photosynthesis and production declining considerably upon increasing exposure to air. Vegetative reproduction occurs by means of rhizomes, which elongate at rates from 0.24 to 0.58 mm day^{-1} and achieve an annual growth rate near 8.6 cm year^{-1}. The rhizomes are tolerant to burial and have survived well under simulated burial depths of 3, 9, and 15 cm; however, shoot numbers have been observed to decrease significantly when buried at a depth of 25 cm. Shoot growth (mm day^{-1}) generally decreases linearly with burial depth. The survivorship of transplanted plugs consisting of intertwined rhizomes and shoots has been excellent (often 100%) in some restoration projects. **Reported associates:** *Bossiella orbigniana*, Chondracanthus canaliculatus, *Corallina chilensis*, *Corallina officinalis*, *Corallina pinnatifolia*, *Corallina vancouveriensis*, *Endocladia muricata*, *Gelidium purpurascens*, *Gelidium pusillum*, *Haliptylon gracile*, *Lithothrix aspergillum*, *Macrocystis pyrifera*, *Melobesia mediocris*, *Phyllospadix scouleri*, *Phyllospadix serrulatus*, *Sargassum muticum*, *Smithora naiadum*, *Ulva californica*.

Use by wildlife: *Phyllospadix scouleri* is eaten by several marine fishes (Vertebrata: Teleostei) including monkeyface and rock pricklebacks (Stichaeidae: *Cebidichthys violaceus*, *Xiphister mucosus*). The leaves (8–12% protein, 0.8–1.5% lipids, and 40–45% carbohydrates by DM) are grazed by black abalone (Mollusca: Gastropoda: Haliotidae: *Haliotis cracherodii*), isopods (Crustacea: Isopoda: Idoteidae: *Pentidotea montereyensis*; Ligiidae: *Ligia pallasii*), sea urchins (Echinodermata: Echinoidea: Strongylocentrotidae: *Strongylocentrotus purpuratus*), snails (Mollusca: Gastropoda: Littorinidae: *Lacuna*), and surfgrass limpets

(Mollusca: Gastropoda: Acmaeidae: *Notoacmea paleacea*), which can reduce winter surfgrass biomass substantially. *Phyllospadix scouleri* is one of few obligate hosts of an epiphytic red marine alga (Rhodophyta: Erythrotrichiaceae: *Smithora naiadum*). Its foliage also harbors amphipods (Crustacea: Amphipoda: Isaeidae: *Photis conchicola*), gastropods (Mollusca: Gastropoda: Haminoeidae: *Haminoea callidegenita*), and proboscis worms (Nemertea: Tetrastemmatidae: *Tetrastemma phyllospadicola*); its leaf bases and rhizomes are colonized by hydroids (Cnidaria: Hydrozoa: Corynidae: *Coryne eximia*; Phialellidae: *Phialella*; Rhysiidae: *Rhysia fletcheri*). In some regions, *P. scouleri* provides critical foraging habitat for endangered green sea turtles (Reptilia: Testudines: Cheloniidae: *Chelonia mydas*). The rhizomes have been found to contain up to 27 different taxa of bacterial and fungal endophytes. The plants can host a pathogenic slime mold (Myxomycota: Labyrinthulaceae: *Labyrinthula*), which is the causative agent of "wasting disease." Several crabs (Crustacea; Decapoda: Pisidae: *Loxorhynchus crispatus*, *Loxorhynchus grandis* "decorate" their exoskeletons using foliage from *P. scouleri*. Plant extracts can inhibit detoxification systems in certain mussels (Mollusca: Mytilidae: *Mytilus californianus*). *Phyllospadix serrulatus* contains relatively high levels of polyunsaturated fatty acids (especially 18:3n–3, 20:0, and 22:0), which provide dietary essentials for amphipods (Crustacea: Amphipoda) and other grazing animals. *Phyllospadix torrey*i is eaten by abalone (Mollusca: Gastropoda: Haliotidae: *Haliotis corrugata*, *Haliotis fulgens*), amphipods (Crustacea: Amphipoda: Talitridae: *Megalorchestia corniculata*), black surfperch (Vertebrata: Teleostei: Embiotocidae: *Embiotoca jacksoni*), green sea turtles (Reptilia: Testudines: Cheloniidae: *Chelonia mydas*), and snails (Mollusca: Gastropoda: Turbinidae: *Megastraea undosa*). The male spadices are grazed by opaleye fish (Vertebrata: Teleostei: Kyphosidae: *Girella nigricans*). The seeds are eaten by crabs (Crustacea: Decapoda: Epialtidae: *Pugettia producta*; Grapsidae: *Pachygrapsus crassipes*), which can deplete up to 50% of the available crop, and isopods (Crustacea: Isopoda: Idoteidae: *Pentidotea resecata*). The plants are hosts to several epiphytic algae (Rhodophyta: Corallinaceae: *Melobesia mediocris*; Erythrotrichiaceae: *Smithora naiadum*) and limpets (Mollusca: Gastropoda: Acmaeidae: *Notoacmea paleacea*). Detritus derived from *P. torreyi* provides a food source for other benthic invertebrates. Leaves of *P. torreyi* are used as nesting materials by the Brandt's cormorant (Aves: Phalacrocoracidae: *Phalacrocorax penicillatus*). Stands of both *P. scouleri* and *P. torreyi* are known to nurture larval California spiny lobsters (Crustacea: Decapoda: Palinuridae: *Panulirus interruptus*). Rhizome mats formed by these species provide habitat for numerous benthic polychaetes (Annelida: Polychaeta). Massive numbers of invertebrates ($n = 18,329–20,152$) have been collected from beds of *P. serrulatus* and *P. scouleri*.

Economic importance: food: The cooked leaves of *Phyllospadix scouleri* were eaten on occasion by members of the Hesquiat tribe. The raw rhizomes of *P. scouleri*, *P. serrulatus*, and *P. torreyi* were consumed by the Makah people.

The Chumash ate the plants to ward off faintness; **medicinal:** Methanolic extracts of *P. scouleri* are moderately inhibitory to Potato virus X (PVX); **cultivation:** not in cultivation; **misc. products:** Extracts of *P. scouleri* can inhibit *Taq* polymerase in PCR reactions. The Hesquiat, Makah, and Nitinaht people fashioned baskets out of sun-bleached *P. torreyi* leaves. The Chumash used the foliage of *P. torreyi* to stuff their mattresses, to make knee pads used for canoeing, and as a covering when drying fish; **weeds:** none; **nonindigenous species:** none.

Systematics: A simultaneous phylogenetic analysis of all five *Phyllospadix* species has not yet been undertaken; however, independent analyses of subsets of the species (*P. scouleri* + *P. torreyi*; *P. iwatensis* + *P. japonica* + *P. torreyi*) consistently indicate that the genus is monophyletic and the sister clade of *Zostera* (e.g., Figure 1.25). A close relationship between *P. scouleri* and *P. torreyi* (as well as possible hybrids) and their more distant relationship to *P. serrulatus* have been indicated by isozyme data, which is in agreement with morphological evaluations. Efforts to develop DNA barcodes for seagrasses have recommended the use of nuclear *phyB* or plastid *matK* sequences to differentiate the various *Phyllospadix* (and other seagrass) species. The basic chromosome number of *Phyllospadix* is uncertain ($x = 4$ or 5), with the occurrence of sex chromosomes ($2n = 16$ ♂, 20 ♀) reported in *P. iwatensis*, the only species for which counts are available. *Phyllospadix scouleri* and *P. torreyi* reputedly hybridize; however, this possibility has not yet been demonstrated conclusively and warrants further investigations

Distribution: *Phyllospadix scouleri* extends southward along the Pacific coastline from the southern tip of Alaska; *P. torreyi* extends southward from the north side of Vancouver Island, Canada; both species reach their southern limit in Baja California, Mexico at 24°31 N latitude. *Phyllospadix serrulatus* ranges along the Pacific coast from San Francisco Bay, California northward to the southwestern Alaskan coast. All three species are endemic to the northeast Pacific region.

References: Ackerman, 1995; 2006; Barbour & Radosevich, 1979; Blanchette et al., 1999; 2008; Brinckmann-Voss, 1996; Brinckmann-Voss et al., 1993; Buckel et al., 2012; Bull et al., 2004; Campbell & Fourqurean, 2013; Campbell & Stirling, 1968; Carefoot, 1973; Carter, 1982; Cooper & McRoy, 1988a; 1988b; Cox & Murray, 2006; Cox et al., 1992a; Coyer et al., 2013; Craig et al., 2008; Crouch, 1991; Daru & Yessoufou, 2016; Dayton, 1975; Dethier, 1984; Dethier & Duggins, 1988; Dooley et al., 2015; Drysdale & Barbour, 1975; Duarte & Chiscano, 1999; Dudley, 1893; 1894; Fox et al., 2014; 2018; Garcias-Bonet et al., 2011; Gibson & Chia, 1989; Green & Short, 2003; Guzmán del Próo et al., 2003; Harlin, 1973; Holbrook et al., 2000; 2002; Horn, 1983b; Jin et al., 1997; Kendrick et al., 2017; Koch et al., 2013; Lastra et al., 2008; Lee, 1966; Leighton & Boolootian, 1963; Les, 1988; Littler & Littler, 1981; Littler et al., 1983; Lüning & Freshwater, 1988; McDermott, 1988; McMillan & Phillips, 1981; Moulton & Hacker, 2011; Orth et al., 2000; Pardee et al., 2004; Phillips & Meñez, 1988; Soros-Pottruff & Posluszny, 1994; Quattrocchi, 2012; Ramírez-García et al., 1998; 2002; Reed et al., 1998;

2009; Schmitt & Holbrook, 1984; Shelton, 2008; 2010a; 2010b; Shoemaker & Wyllie-Echeverria, 2013; Sonnenholzner et al., 2011; Stewart & Rüdenberg, 1980; Stricker, 1985; Terrados & Williams, 1997; Tharaldson, 2018; Timbrook, 1990; Turner, 1983; 1985; Turner & Lucas, 1985; Walker et al., 2001; Wicksten, 1979; Willcocks, 1982; Williams, 1995; 2007; Zidorn, 2016.

3.6.2. Zostera

Eelgrass, seawrack; zostère

Etymology: from the Greek *zoster* ("belt") referring to the strap-like leaves

Synonyms: *Heterozostera*; *Nanozostera*

Distribution: global: Africa, Australia, Eurasia, North America; **North America:** Atlantic and Pacific oceanic coasts

Diversity: global: 7–8 species; **North America:** 2 species

Indicators (USA): obligate wetland (OBL): *Zostera japonica*, *Z. marina*

Habitat: brackish (coastal); marine; oceanic; **pH:** 7.7–8.2; **depth:** to 15 m; **life-form(s):** submersed (vittate)

Key morphology: rhizomes with distinct internodes (to 35 mm) and 2–20 nodal roots; leaves arising along rhizome (1 per node), ribbonlike (to 2 m), 3–11-veined, apex emarginate, obtuse, or mucronate, the sheaths (to 20 cm) open or closed, wider than blade; reproductive shoots (to 1.5 m) lateral or terminal, simple to highly branched, each branch bearing 1–5 spadices; spadix linear, pedunculate (to 130 mm), with sheathing spathe (to 85 mm), monoecious, bearing numerous, naked, alternating ♀ and ♂ flowers (4–20 each), a bract (retinacula) present or absent on all ♂ flowers, or sometimes subtending only the proximal ♂ flowers, in ♀ flowers the styles (to 2.5 mm) bifurcating into linear stigmas (to 3.5 mm); fruits (to 5 mm) achene-like, elliptic to ovoid, brownish, usually beaked (to 1.5 mm); seeds (to 4 mm) smooth or up to 25-ribbed

Life history: duration: annual (fruits/seeds); perennial (rhizomes, whole plants); **asexual reproduction:** rhizomes, shoot fragments; **pollination:** water; **sexual condition:** monoecious; **fruit:** achenes (common); **local dispersal:** seeds, plantlets, vegetative fragments (animals, water); **long-distance dispersal:** seed-bearing fragments (water, watercraft), seeds (animals, water)

Imperilment: 1. *Z. japonica* [GNR]; **2.** *Z. marina* [G5]; S1 (MB); S2 (LB, ON); S3 (NF, NU, NY, QC)

Ecology: general: All *Zostera* species are obligately aquatic, submersed, perennial "seagrasses" (marine flowering plants), which inhabit coastal oceanic environments and inland seas. Under certain conditions, the plants can behave as annuals, which persist from season to season entirely by means of seed. Unlike the related *Phyllospadix* (see above), they occur mainly on loose substrates (mud and sand) rather than on rock. Although primarily temperate plants, the different species can tolerate a wide range of salinity and temperature. All the species that have been evaluated are bicarbonate (HCO_3^-) users, which obtain their nutrients primarily from the sediments. *Zostera* species are monoecious, with "self-compatible" (i.e., within inflorescences) but protogynous, water-pollinated (hydrophilous), unisexual flowers. Further

details of their reproductive biology are discussed in the ensuing species treatments. Despite their ability to grow in dense clones, genetic evidence indicates that the plants primarily are outcrossed. Seed production ranges from rare to abundant, and functions mainly in dispersal and colonization; whereas annual recruitment occurs primarily by means of persistent, rhizomatous growth in most of the species. The seeds can be dormant or non-dormant but must become buried within the upper sediment layer to maximize their viability (which typically is short-lived). Most dispersal occurs abiotically by water. Currents move seeds along the bottom sediments, or transport seed-laden vegetative fragments through the water column. Viable seeds also can be passed through the digestive tracts of waterfowl (Aves: Anatidae) and large marine herbivores such as dugongs (Mammalia: Sirenia: Dugongidae: *Dugong dugon*) and green sea turtles (Vertebrata: Reptilia: Cheloniidae: *Chelonia mydas*), which are capable of dispersing them at distances of up to 650 km. The plants are important ecologically by stabilizing sediments, by providing habitat for countless marine animals and epiphytes, and as a food source for waterfowl and various oceanic animals.

3.6.2.1. ***Zostera japonica* Asch. & Graebn.** is a nonindigenous annual or perennial, which grows along oceanic beaches, coastlines, mudflats (intertidal, tidal), saltmarshes, sandbars, shallow pools, shores, and streams (salt) within the intertidal zone (0.1–3.0 m above mean lower low water [MLLW]). The plants occur in exposures that receive full sunlight. The substrates consist primarily of sand (71.3–91.2%), with lower proportions of silt (6.7–22.4%) and clay (2.1–6.3%). They also have been characterized as gravel (marine), muck, mud (intertidal, sandy, tidal), and silt. These plants are categorized as "*r*-strategists"; i.e., they grow rapidly and devote a large proportion of their resources to sexual reproduction. Although oceanic, they colonize relatively harsher (i.e., intertidal) sites, which experience greater extremes in temperature (2°C–23°C), salinity (0–37‰), and desiccation stress. Estimated optimal temperatures (20°C–30°C) are higher than those of *Zostera marina*. Salinities near 20‰ appear to be optimal for photosynthesis, but the plants can maintain comparable photosynthetic rates and thereby endure long-term exposures to salinities ranging from 5 to 35‰. When grown under high temperature stress (35°C), the plants survive best at higher salinities (20–35‰) and poorly at low salinity (5‰). They have a higher growth rate when fully submerged than when exposed during low tides. Those growing in shallower sites also are shorter (e.g., 20 cm) and occur at higher densities (e.g., 3800 m^{-2}). Although often described as annuals (mostly more northern populations), some plants perennate through winter (in sheltered sites) with their shoots and rhizomes dying in the spring (short-lived perennials). In more southerly sites the plants usually persist throughout the year as perennials. In North America, sexual reproduction can occur anytime from May to December but reaches a maximum level (e.g., up to 70% flowering shoots; 340–461 flowering shoots m^{-2}) from August to October. Flowering is correlated with temperature and begins and ends earlier at higher elevations, than in deeper water plants. The extent of flowering can be reduced substantially under unusually cool, wet, and overcast conditions. Otherwise (unlike its congener *Z. marina*), there is little specific information on the floral biology or pollination of this species; presumably it is quite similar to that of *Z. marina*. Annual recruitment occurs primarily by seed, with as much as 25% of the total plant biomass allocated to sexual reproduction. It takes about 7 months for newly established seedlings to mature, flower, and set seed. Annual establishment from seed is relatively high (12–57%) compared to the primarily perennial *Z. marina* (1–12%). The seeds are not buoyant and are dispersed locally within a few meters of their maternal source. The young plantlets also are thought to function locally as water-dispersed propagules. Longer-distance dispersal occurs (6–100 km) when seed-bearing vegetative fragments ("wrack") are transported by tidal activity or by watercraft currents. Even greater dispersal distances presumably can be achieved by the endozoic transport of seeds by migratory waterfowl. The dormant seeds require a period of cold stratification (e.g., 7°C) followed by fluctuating water temperatures (e.g., 20°C/10°C) for maximum germination (54%) to occur. Seed germination rates increase at higher temperatures (15°C–20°C) and are highest at 0‰ salinity. The seeds germinate in spring in the mid to low intertidal zones and the resulting plants produce numerous seeds before becoming dislodged by late Autumn storm activity (due to their shallow rhizome system). The plants growing in shallower sites also become dormant earlier in the fall. The seeds have remained viable after periods of up to 302 days of cold storage (4°C) and germinate well (to 80%) when incubated subsequently for 21 days at 23°C, at an illumination of 3000 lux (16/8hr light/dark photoperiod). Seeds stored for longer periods (26 months) exhibited a higher germination rate (21.6%) than those stored for just 6 weeks (17.1%). The prolonged viability of stored seeds indicates that a persistent seed bank (at least short-term) potentially can develop. The rate of seedling survival through their first year is low. In North America, the mean percent cover ranges from 0.7–6.0% (midwinter) to 6.0–64.0% in early autumn. Shoot density consistently is high, varying from 3,803 to 11,000 shoots m^{-2}. Biomass ranges from 40 to 160 g DM m^{-2} (aboveground) and 70–170 g DM m^{-2} (belowground), with leaf production rates of 0.1–1.9 g DM m^{-2} day^{-1}. The mean growth rate varies from 0.19 to 0.52 mg DM $shoot^{-1}$ day^{-1}. The mean annual productivity in North America has been estimated at 314 g DM m^{-2} $year^{-1}$. Although *Z. japonica* suffers greater biomass reduction (96%) when competing with *Z. marina* (44%), its fitness and productivity increase when the plants are exposed to disturbance. Vegetative growth under submersed, simulated summer conditions (18°C, 14:10 hr light:dark) is half that of *Z. marina*; however, this disadvantage is offset by the lower desiccation tolerance of the latter, which limits its extension into the intertidal zone where *Z. japonica* thrives. The plants aid in sediment stabilization because they preferentially colonize barren sediments that have been denuded by storms and wave action during the previous winter. Consequently, they can alter substrate conditions substantially and become nutrient sinks for NO_3, NH_4, and dissolved reactive phosphorous.

Asexual reproduction occurs by branching of the rhizomes and possibly by establishment of vegetative fragments along uncolonized shorelines. Southern populations have lower levels of sexual reproduction along with higher clonality. Microsatellite markers have shown that populations can vary widely in their number of alleles/locus (2–13) and level of observed heterozygosity (0.00–1.00). **Reported associates (North America):** *Atriplex dioica, Enteromorpha linza, Fucus gardneri, Gigartina, Juncus gerardii, Melobesia, Ruppia maritima, Salicornia depressa, Smithora naiadum, Ulva lactuca, Zostera marina.*

3.6.2.2. *Zostera marina* L. is an indigenous annual (rarely) or perennial, which grows in brackish to saline waters associated with beaches (sea), canals, channels (edge), coves, depressions (intertidal), estuaries (marine), flats (tidal), lagoons (tidal), lakes (saltwater), mud flats, ponds (brackish), pools (salt), river mouths (brackish), salt marshes (coastal), shoals, shorelines (oceanic, rocky), sloughs, sounds, streams (saltwater), tidal drainages, tidal flats, and tide pools, within the low intertidal to subtidal zones (+2 to −5.5 m mean sea level) out to oceanic depths of 15 m. Growth of plants at greater depths is limited by light with their depth distribution governed primarily by daily light periods. The substrates can consist of cobbles, gravel (intertidal, marine), mud (subtidal), muddy sand, rock, rocky sand, sand, sandy gravel, sandy mud, sandy silt, silt, or silty sand. Generally, this species is adapted to colder waters at optimum temperatures ranging from 6°C to 13°C. Growth is reduced at lower (5°C) or higher (15°C) temperatures and very high temperatures (25°C–30°C) result in increased mortality. Yet some populations tolerate temperatures well above that optimal growth range, including prolonged exposures to temperatures above 15°C, with no apparent ill-effect. Although net photosynthesis increases with temperature up to 25°C–30°C, the dark respiration increases as well, resulting in the maximum photosynthetic:respiration rate occurring at 5°C (and declining substantially at higher or lower temperatures). Total chlorophyll content declines at temperatures above 19°C. The plants are not resistant to desiccation, which can be a limiting factor for intertidal establishment and persistence. They exhibit a higher rate of leaf elongation when submerged than when exposed at low tide. Average leaf longevity also is greater (48 days) in low intertidal sites than in high intertidal sites (36 days). The plants can tolerate currents of up to 120–150 cm s^{-1} and occur across a very broad range of salinity (5–36‰). Their abundance correlates positively with salinity with 10–25‰ being the optimal range determined by laboratory studies. Higher photosynthetic quantum efficiency and production have been found to occur at salinities of 22‰ than at 32‰. However, low salinity (e.g., 2.5–5.0‰) increases their mortality substantially. Plants have been reported from habitats ranging in acidity from more brackish sites (e.g., pH: 7.7, total alkalinity: 30.0 mg l^{-1}, chlorides: 3822 mg l^{-1}) to average oceanic conditions (pH: 8.1–8.2). The plants are capable of flowering as early as their second year of growth. In North America, flowering often is described as uncommon, but anthesis varies substantially between oceans (Atlantic vs.

Pacific) and across latitudes within them. In both Atlantic and Pacific populations, the onset of flowering correlates directly with latitude; i.e., flowering begins incrementally later (differing by as much as 100 Julian days) as latitude increases northward. Atlantic populations initiate flowers and mature their fruits later in the season than those in the Pacific Ocean. Both factors result from the earlier onset of warmer water temperatures and narrower temperature range in Pacific Ocean habitats. Developing (but not necessarily mature) flowers have been observed as early as January (water temperatures: 15°C–18°C) and February (water temperature: 3°C–25°C); reported floral observations extend into September. Fruiting extends from April to January, with large numbers of seeds produced in some localities (see below). Some of the phenological variability is illustrated by the following examples. In a Pacific (Puget Sound) population, anthesis initiated at water temperatures of 7°C–9°C with seeds maturing as temperatures warmed to 11°C–14°C. In a Chesapeake Bay Atlantic population, pollen was shed in mid-April (water temperature: 14.3°C), with loss of stigmatic receptivity initiated in late April (average water temperature >16°C), and completion of pollination by 19 May; seeds were released by early June. The floral cycle of a population further north (New York) was similar but began about a month later. There, anthesis extended from April to June, with pollination taking place from mid-May to mid-June when water temperatures were between 17°C and 20°C; seeds were released from June 17 to July 9. Even further north (Nova Scotia), flowering occurred at water temperatures of only 8°C–9°C. The flowers are "self-compatible" (i.e., compatible male and female flowers within an inflorescence) but are strongly protogynous, which minimizes (but does not preclude) the possibility of geitonogamous selfing within an inflorescence. Genetic analyses indicate that the plants are primarily outcrossed, have large genetic neighborhoods, and incur high fitness costs (e.g., reduced seed set) associated with geitonogamy. Genetic diversity (ascertained using microsatellite markers) declines with increasing latitude; e.g., values of average heterozygosity and allelic richness both are higher for populations at 25°N (0.6 and 4.0, respectively) than at 60°N (0.3 and 2.5, respectively). More northern populations also flower less frequently and have a higher degree of clonality than southern populations. As the female flowers initiate the reproductive cycle, their styles exert from the spadix, awaiting arrival of the pollen. In one study (New York) an average of 48 pistillate flowers (achieving 72% seed set) was produced on each flowering shoot. In the male flowers, the pollen is released from the anthers under water or at the water surface and is transported in tandem groups, but pollination primarily is epihydrophilous. The emergence of the inflorescence alters the fluid shear stress properties of the water surface, which, in turn, affects the release, transport, and capture of the pollen grains. As the male flowers dehisce at the water surface the pollen is drawn out of the anther thecae by surface tension, which then spreads the grains into a floating network and prevents the grains from readily becoming submerged. It is more difficult for the pollen to be transported underwater, which predictably occurs only

under higher-energy (wave) conditions. Pollen movement is precluded in stagnant water or by high current velocities but occurs maximally at the flow rates typical of the diurnal tidal cycles (>0–0.30 m s⁻¹). The filiform shape of the pollen allows the grains to rotate within the water, thereby facilitating their capture by the stigmas, which reflex after making contact. Water currents also orient the inflorescences optimally to maximize pollen capture. The pollen remains viable from 7 to 48 hr and most (>90%) is dispersed within a 15 m radius. Seed production can vary widely by site, with densities ranging from 700 to 78,224 m⁻² reported among different North American populations. Once shed from the maternal plant, the seeds sink at a rate of 5.96 cm s⁻¹ and are dispersed locally at distances from 10 to 50 m (83% of seeds are deposited within a 5 m radius). Subtidal populations reportedly produce heavier seeds (which sink more quickly) than intertidal sites. The fruits are dispersed primarily while remaining attached to detached, waterborne vegetative material. Although vegetative fragmentation has been cited as a dispersal mechanism in *Z. marina*, the fragments survive poorly in the water (60% survivorship after 6 weeks), and themselves do not appear to be effective dispersal or recruitment propagules. However, detached reproductive shoots bearing mature seeds can remain buoyant for up to 2 weeks and can retain the seeds for up to 3 weeks before their release. Water transport of seed-bearing fragments has been estimated to be capable of achieving dispersal distances from 10 to 23 km. The seeds also have been found to remain viable after passing through the digestive tracts of fish (Vertebrata: Teleosei: Fundulidae: *Fundulus heteroclitus*; Tetraodontidae: *Sphoeroides maculatus*; Sparidae: *Lagodon rhomboides*), terrapins (Vertebrata: Reptilia: Emydidae: *Malaclemys terrapin*), and waterfowl (Aves: Anatidae: *Aythya affinis*), which potentially disperse them endozoically over longer distances. To remain viable, dispersed seeds eventually must become buried in the sediments post-dispersal, a process that can be facilitated by wave action (causing substrate drift) or by burrowing benthic invertebrates (e.g., Annelida: Polychaeta: Maldanidae: *Clymenella torquata*). In one North American study (from Chesapeake Bay), seeds buried in the sediment began to germinate in October, when water temperatures fell to 15°C (salinity 16–19‰), with nearly complete germination occurring by December. There was no effect of burial depth (5–25 mm) on germination rates. In contrast, seeds held in the water column (ambient seawater conditions) remained dormant until mid-January. However, the germination pattern for seeds placed in the light and in deoxygenated water (devoid of sediment) was the same for those residing in the sediments, implicating anoxia and water temperature rather than light as cues for germination. Although laboratory studies of European plants found maximum seed germination to occur at 30°C and 1.0‰ salinity (an unrealistic value for natural habitats), they also found that stratification did enhance germination rates at higher salinities (>20‰), which are more realistic values for natural conditions. Other studies have concluded that seeds kept in diluted seawater are conditionally dormant and require physical scarification or a period of cold stratification (incubation at 5°C–15°C), whereas those in undiluted seawater appear to be non-dormant. Those observations are consistent with findings that seeds stored at 20°C in seawater (22‰) do not readily germinate until they are exposed to lower temperatures. Thus, it would appear that high temperatures are mainly responsible for inducing seed dormancy. Germination rates and subsequent successful seedling establishment are not density dependent. Field germination rates can be quite high, with 76–93% of seeds collected in July at one New York locality germinating within 3–4 months. Even though the plants almost always develop only a short-lived (annual) seed bank, viable seeds have been stored in sediments for as long as 22 months without germinating when kept in sterilized seawater at 20°C–22°C. High seed viability also has been reported for seeds stored at –1°C. The extent of variation in seed viability observed among different field sites is difficult to elucidate. Although viable seed density has been found to correlate with patch size, the density of non-viable seeds does not. Abundant seedlings have been noted to occur as early as February, with seedling reports extending through July. The seedlings grow and develop during the autumn and spring, but not during winter, when progressive losses begin. A complete loss of seedlings (from the previous autumn's cohort) can occur by May of the subsequent spring. The survival rate of seedlings reportedly is highest at 10°C and at 10.0–20.0‰ salinity. Seedling patches require irradiance levels greater than 7.9 E m^{-2} day^{-1} in order to thrive. Their survivorship can decline by as much as 74% when incident light is decreased by 10%. *Zostera marina* is categorized ecologically as a "*K*-strategist"; i.e., the plants devote much of their resources toward vegetative (rhizome) growth as a means of maximizing their persistence at a site. Most of their biomass is allocated to vegetative growth and reproduction; whole plants can overwinter in more sheltered sites within the low intertidal to subtidal zones. However, "annual" forms having extraordinarily high seed production (i.e., 4,889 seeds 625 cm^{-2}) also have been reported for individuals growing in ephemeral sites such as shallow tidal depressions. Vegetative reproduction and a dense, perennial habit dominate in subtidal sites, which experience more stable salinity in contrast to intertidal sites. The latter are characterized by higher rates of flowering and enhanced seed germination associated with low, seasonal salinity values. Clones can persist for 20–50 years. In one analysis 20% of clones sampled represented unique genotypes. At normal seawater pH (8.2), most carbon is taken up in the form of bicarbonate (HCO$_3^-$) with free CO$_2$ contributing less than 20% to photosynthesis. The plants cannot use carbonate (CO$_3^{2-}$) as a carbon source. The plants can take up nitrogen (NH$_4^+$) from both the water (by leaves) and sediments (by roots) but most is obtained from the sediments due to their higher nitrogen levels. The levels of interstitial sediment ammonium nitrogen correlate positively with leaf production, but negatively with shoot density and flower production. Shoot density (339–2,576 shoot m^{-2}) as well as maximum biomass (80.3–463 g DM m^{-2}) can vary substantially, particularly with respect to latitude and depth. In Atlantic waters, shoot density generally declines with

increasing latitude, while leaf biomass and overall plant size increase along that same gradient. In one study, plants from the high intertidal zone produced 143 g DM m^{-2} compared to those from the low intertidal zone, which produced 463 g DM m^{-2}. In another study, the biomass remained consistent (near 245 g DM m^{-2}) regardless of the shoot density, which varied from 150 to 2,500 shoots m^{-2}. Average productivity has been estimated at 0.88 mg C g^{-1} h^{-1}, with average annual productivity at 0.9 g C m^{-2} day^{-1}, and annual production rates of 84–480 g C m^{-2} year^{-1}. More biomass is allocated to aboveground organs (e.g., 161–336 g DM m^{-2}) than to belowground structures (61–155 g DM m^{-2}). Genomic studies of *Z. marina* have documented the losses of stomatal genes, genes involved in terpenoid synthesis and ethylene signaling, ultraviolet protective genes, and far-red sensing phytochromes, all of which arguably are superfluous in submersed marine plants. The plants also have acquired adaptive cell wall polyanionic, low-methylated pectins and sulfated galactans, which facilitate ion homoeostasis, nutrient uptake, and leaf epidermal gas exchange. Massive losses of plants (~90%) occurred along the North American and European Atlantic coasts in the 1930s as a consequence of "wasting disease" (see *Use by wildlife* next). They can become dislodged and killed by the activities of cownose rays (Vertebrata: Chondrichthyes: Rhinopteridae: *Rhinoptera bonasus*), which dig for food items in the beds. The plants also are intolerant of shading due to algal blooms (e.g., Chromista: Ochrophyta: Pelagomonadaceae: *Aureococcus anophagefferens*), dock construction, or similar processes. **Reported associates:** *Acrosorium, Atriplex, Blidingia minima, Carex lyngbyei, Ceramium cimbricum, Desmarestia ligulata, Ectocarpus, Enteromorpha flexuosa, Enteromorpha intestinalis, Enteromorpha linza, Fucus, Gracilariopsis lemaneiformis, Halodule wrightii, Melobesia, Monostroma leptodermum, Pachydictyon, Phyllospadix, Ruppia maritima, Sarcocornia perennis, Sargassum muticum, Smithora naiadum, Ulva fenestrata, Ulva lactuca, Zostera japonica.*

Use by wildlife: *Zostera* plants serve as a nutritious food source for many animals. The mean foliar energy content of *Z. japonica* (18.145 kJ g^{-1}) and *Z. marina* (16.817 kJ g^{-1}) is higher than that of their rhizomes (15.387 kJ g^{-1} and 15.713 kJ g^{-1}, respectively); the mean energy content of *Z. japonica* seeds (18.999 kJ g^{-1}) is higher still. Both species are grazed by isopods (Crustacea: Isopoda), which can remove up to 8 g DM of their biomass m^{-2} day^{-1}. Generally, the introduction of *Z. japonica* has affected large infaunal invertebrates negatively but has benefitted small infaunal invertebrate populations positively. In North America, this nonindigenous species is eaten (sometimes preferentially over the native *Z. marina*) by many waterfowl (Aves: Anatidae) including American wigeon (*Anas americana*), brant goose (*Branta bernicla*), green-winged teal (*Anas crecca*), mallard (*Anas platyrhynchos*), and northern pintail (*Anas acuta*). The plants also are grazed by amphipods (Crustacea: Amphipoda: Aoroidae: *Aoroides columbiae*; Caprellidae: *Caprella laeviuscula*; Isaeidae: *Photis brevipes*), gastropods (Mollusca: Gastropoda: Littorinidae: *Lacuna variegata*), and isopods (Crustacea: Isopoda: Idoteidae:

Pentidotea resecata). Patches of the plants support numerous invertebrates including amphipods (Crustacea: Amphipoda: Ampithoidae: *Ampithoe valida*; Corophiidae: *Corophium, Monocorophium acherusicum*; Phoxocephalidae: *Eobrolgus spinosus*), cumaceans (Crustacea: Cumacea: Nannastacidae: *Cumella vulgaris*), flies (Insecta: Diptera), oligochaetes (Annelida: Clitellata), polychaetes (Annelida: Polychaeta: Capitellidae: *Capitella*; Phyllodocidae: *Eteone californica*; Spionidae: *Boccardia proboscidea, Pseudopolydora kempi, Pygospio elegans, Streblospio benedicti*), and tanaidaceans (Crustacea: Tanaidacea: Leptocheliidae: *Leptochelia dubia*). As many as 31 endophyte taxa have been found to inhabit the rhizomes. The foliage and rhizomes of *Z. marina* are eaten by many ducks and geese (Aves: Anatidae) including the American scoter (*Melanitta americana*), American wigeon (*Anas americana*), black brant goose (*Branta bernicla nigricans*), black duck (*Anas rubripes*), blue-winged teal (*Anas discors*), bufflehead (*Bucephala albeola*), Canada goose (*Branta canadensis*), canvasback (*Aythya valisineria*), common and Pacific eider (*Somateria mollissima*), Eurasian wigeon (*Anas penelope*), gadwall (*Anas strepera*), goldeneye (*Bucephala clangula*), greater scaup (*Aythya marila*), green-winged teal (*Anas crecca*), king eider (*Somateria spectabilis*), lesser scaup (*Aythya affinis*), mallard (*Anas platyrhynchos*), northern pintail (*Anas acuta*), oldsquaw (*Clangula hyemalis*), redhead (*Aythya americana*), ruddy duck (*Oxyura jamaicensis*), Steller's eider (*Polysticta stelleri*), surf scoter (*Melanitta perspicillata*), and white-winged scoter (*Melanitta fusca*). The foliage is grazed by amphipods (Crustacea: Amphipoda: Aoroidae: *Aoroides columbiae*; Caprellidae: *Caprella laeviuscula*; Corophiidae: *Monocorophium acherusicum*; Isaeidae: *Photis brevipes*), gastropods (Mollusca: Gastropoda: Littorinidae: *Lacuna variegata*), and isopods (Crustacea: Isopoda: Idoteidae: *Pentidotea resecata*). The plants also have been fed to cattle (Mammalia: Bovidae: *Bos taurus*). *Zostera marina* provides essential habitat for fish (Vertebrata: Teleostei), migratory waterfowl (Aves: Anatidae), and wading birds (Aves). Beds of the plants have been found to accommodate numerous fish (Vertebrata: Teleostei), including anchovies (Engraulidae: *Anchoa mitchilli*), Atlantic cod (Gadidae: *Gadus morhua*), Atlantic silversides (Atherinopsidae: *Menidia menidia*), black sea bass (Serranidae: *Centropristis striata*), blennies (Blenniidae: *Chasmodes bosquianus, Hypsoblennius hentz*), pipefish (Syngnathidae: *Syngnathus floridae, S. fuscus*), rainbow smelt (Osmeridaae: *Osmerus mordax*), silver perch (Sciaenidae: *Bairdiella chrysoura*), spot (Sciaenidae: *Leiostomus xanthurus*), striped bass (Moronidae: *Morone saxatilis*), summer flounder (Paralichthyidae: *Paralichthys dentatus*), tautog (Labridae: *Tautoga onitis*), and winter flounder (Pleuronectidae: *Pseudopleuronectes americanus*). They also are used by bay scallops (Mollusca: Bivalvia: Pectinidae: *Argopecten irradians*), blue mussels (Mollusca: Bivalvia: Mytilidae: *Mytilus edulis*), eelgrass limpets (Mollusca: Gastropoda: Lottiidae: *Lottia alveus*) [extinct because of wasting disease], many different crustaceans (Crustacea) including American lobsters (Decapoda: Nephropidae: *Homarus americanus*), American prawns

(Decapoda: Palaemonidae: *Palaemonetes vulgaris*), blue crabs (Decapoda: Portunidae: *Callinectes sapidus*), brown shrimp (Decapoda: Penaeidae: *Farfantepenaeus aztecus*), grass shrimp (Decapoda: Palaemonidae: *Palaemonetes intermedius, P. pugio*), isopods (Isopoda: Idoteidae: *Idotea balthica*), opossum shrimp (Decapoda: Mysidae: *Mysis stenolepis*), sand shrimp (Decapoda: Crangonidae: *Crangon septemspinosa*), and periwinkles (Mollusca: Gastropoda: Littorinidae: *Littorina neglecta, L. obtusata*). Seed removal by predators such as blue crabs (Crustacea: Decapoda: Portunidae: *Callinectes sapidus*) and Atlantic croakers (Vertebrata: Teleostei: Sciaenidae: *Micropogonias undulatus*) can account for up to 65% of annual seed losses. The leaves are used as a substrate for roe deposition by Pacific herrings (Vertebrata: Teleostei: Clupeidae: *Clupea pallasii*). Nearly 24% of the total plant biomass can be due to algal epiphytes and this is one of few obligate hosts of an epiphytic red marine alga (Rhodophyta: Erythrotrichiaceae: *Smithora naiadum*). The grass cerith (Mollusca: Gastropoda: Cerithiidae: *Bittium varium*) removes periphyton from the leaves by grazing. More than 115 infaunal macroinvertebrate taxa have been identified from the plant beds, which contain large numbers of mollusks (Mollusca: Bivalvia: Mactridae: *Spisula solidissima*; Myidae: *Mya arenaria*; Pharidae: *Ensis directus*; Pholadidae: *Cyrtopleura costata*; Tellinidae: *Macoma, Tellina agilis*; Gastropoda: Nassariidae: *Nassarius obsoletus, N. trivittatus*), oligochaetes (Annelida: Clitellata), polychaetes (Annelida: Polychaeta: Flabelligeridae: *Pherusa affinis*; Glyceridae: *Glycera*; Lumbrineridae: *Lumbrineris*; Maldanidae: *Clymenella torquata*; Nephtyidae: *Nephtys*; Nereididae: *Neanthes virens, Nereis*; Oenonidae: *Drilonereis*; Pectinariidae: *Pectinaria gouldii*), and sipunculids (Sipuncula: Sipunculidea: Sipunculidae). The rhizomes have been found to harbor up to 37 endophyte taxa. Both *Z. marina* and *Z. japonica* are used as a spawning substrate by the Pacific herring (Vertebrata: Teleostei: Clupeidae: *Clupea pallasii*). Both species are hosts for parasitic slime molds (Fungi: Myxomycota: Labyrinthulaceae: *Labyrinthula zosterae*), which are responsible for "wasting disease" in seagrasses. In addition to those already mentioned, the two species provide habitat for numerous other invertebrates including barnacles (Crustacea: Sessilia: Balanidae: *Balanus*), caridean shrimp (Crustacea: Decapoda: Caridea), clams (Mollusca: Bivalvia: Heterodonta), copepods (Crustacea: Copepoda), limpets (Mollusca: Gastropoda: Lottiidae: *Lottia pelta*), mites (Arachnida: Acari: Halacaridae), mussels (Mollusca: Mytilidae: *Mytilus trossulus*), polychaetes (Annelida: Polychaeta: *Platynereis bicanaliculata*), sea snails (Mollusca: Gastropoda: Littorinidae: *Littorina*), and tanaidaceans (Crustacea: Tanaidacea: Leptocheliidae: *Leptochelia*).

Economic importance: food: Raw *Zostera marina* plants are mixed with grease from the eulachon (Vertebrata: Teleostei: Osmeridae: *Thaleichthys pacificus*) before being eaten (Bellabella tribe). The Cowichan and Saanich people added the roots and leaf bases to various meats as a spice. The clean, washed rhizomes were eaten raw by the Hesquiat and Nitinaht tribes. The Kwakiutl people dipped the stems and rhizomes in oil to eat during feasts. The Oweekeno people selected leaves covered in herring spawn (Vertebrata: Teleostei: Clupeidae) as a delicacy. The Seri people of northern Mexico harvest eelgrass seeds in the spring to use as a grain; **medicinal:** Crude extracts from *Z. japonica* have been found to possess anti-inflammatory properties, and to exhibit inhibitory effects on the growth of AGS, HT-1080, and MCF-7 human cancer cells. Ethanolic extracts from *Z. marina* inhibit production of LPS-induced pro-inflammatory mediators, which indicates its potential use as an anti-inflammatory agent; **cultivation:** *Zostera marina* was once considered for use as a cultivar in coastal desert sites; **misc. products:** Hesquiat and Nitinaht fishermen used the leaves of *Z. marina* to collect herring (Vertebrata: Teleostei: Clupeidae) spawn. The plants also have been used as a green mulch. Fibers derived from *Z. marina* are stiff and of low density, which has led to their consideration as a reinforcement for composite materials and for use in fabricating bio-degradable materials. The plants have been considered as environmental biomonitors for detecting and quantifying levels of tributyltin, an anti-fouling substance used widely in marine paints. Historically, *Z. marina* was used to construct many of the dikes in the Netherlands; **weeds:** Cover of *Zostera japonica* within an introduced British Columbia population increased by more than 17-fold from 1970 to 1991. In Oregon, a 400% increase in cover was recorded within a 9-year period; **nonindigenous species:** *Zostera japonica* was introduced to the western North American coast from Japan as a seed contaminant during oyster enhancement programs in the early 20th century (prior to 1950s).

Systematics: Phylogenetic analyses have consistently supported the monophyly of *Zostera* along with its sister-group relationship to *Phyllospadix* (e.g., Figures 1.21 and 1.25). Although many taxonomic studies of *Zostera* have been carried out, the appropriate delimitation of taxa remains contentious. While treated here as a single genus of three subgenera and eight species, some authors have recognized as many as three different genera (*Heterozostera*, *Nanozostera*, and *Zostera*) and 20+ species in the group (see preceding discussion under Zosteraceae). Phylogenetic analyses using various molecular data (e.g., Figure 1.25) have failed to distinguish several Australian taxa (*Z. capricorni*, *Z. mucronata*, *Z. novazelandica*) from *Z. muelleri*. These taxa, which were delimited primarily by their leaf-tip morphology, subsequently have been merged with *Z. muelleri*. Cryptic introductions likely have led to the description of "new" species in North America (i.e., *Z. americana*) and Africa (*Z. capensis*), which are indistinguishable from *Z. japonica* and *Z. tasmanica*, respectively, by both morphological and molecular data. The distinctness of *Z. caespitosa* also is not evident in molecular analyses (e.g., Figure 1.25) and deserves further scrutiny. Broader-leaved individuals from the Pacific North American coast have been interpreted variously as introduced plants of the Asian *Z. asiatica*, or as an endemic North American species (i.e., "*Z. latifolia*" or "*Z. pacifica*"). Sequence data from putatively species-differentiating "DNA barcode" regions (nrITS region, *matK*) not only demonstrate conclusively that the broad-leaved plants are not conspecific with

Z. asiatica, but also fail to distinguish them from *Z. marina*. Representatives of both leaf types were included in common garden experiments among several other ecotypes, which disclosed a wide range of phenotypic plasticity in *Z. marina* in addition to discrete genetic differences (which accounted for 14% of the total variation observed). Although the broad and narrow-leaved plants can be differentiated using high-resolution (i.e., microsatellite) markers, the same markers are recognized as being of limited utility in delimiting species, and also have provided evidence of hybridization between the ecotypes. Genetic studies of *Z. marina* further indicate that pairwise relatedness estimates based on neutral molecular markers such as microsatellites perform poorly for differentiating traits among different genotypes. Given their single differentiating character (leaf width), high degree of genetic similarity, and ability to hybridize, the recognition of broad-leaved and narrow-leaved individuals of *Z. marina* as distinct species is not compelling and has not been followed here. It is noteworthy that other distinctly broader-leaved or more narrowly leaved plants occur in Europe and the northeastern Pacific, respectively, and similarly are indistinguishable from *Z. marina* by DNA sequence data. To date, *Zostera marina* has the smallest mitochondrial genome (191,481 bp) of any autotrophic angiosperm sequenced. More than 20% of the *Zostera* mitogenome (40 kb) consists of foreign (plastid) DNA. The phylogenetic significance of these abnormalities warrants further consideration. The base chromosome number of *Zostera* is $x=6$. *Zostera asiatica*, *Z. japonica*, and *Z. marina* all are diploid ($2n=12$). Intraspecific hybridization has been reported between narrow-leaved and broad-leaved ecotypes of *Z. marina* (the latter recognized by some authors as *Z. pacifica*).

Distribution: *Zostera marina* is widespread along both the temperate Atlantic and Pacific coasts, extending along the latter to Baja California, Mexico. The introduced *Zostera japonica* extends from coastal northern California to British Columbia, Canada.

References: Abbott et al., 2018; Ackerman, 1997a; 1997b; Backman, 1991; Baldwin & Lovvorn, 1994; Bando, 2006; Baskin & Baskin, 1998; Beer & Rehnberg, 1997; Bigley & Barreca, 1982; Bigley & Harrison, 1986; Bintz & Nixon, 2001; Boese et al., 2003; 2005; Bohlmann et al., 2018; Churchill, 1983; Churchill & Riner, 1978; Cottam, 1939; Coyer et al., 2008; 2013; Cox et al., 1992b; Cullain et al., 2018; Davies et al., 2007; De Cock, 1980; Dennison & Alberte, 1985; Dennison et al., 1989; Evans et al., 1986; Ewanchuk & Williams, 1996; Felger & Moser, 1973; Fishman & Orth, 1996; Fonseca et al., 1983; Francois et al., 1989; Ganter, 2000; Green & Short, 2003; Harlin, 1973; Harrison, 1979; 1982a, 1982b; Harrison & Bigley, 1982; Hartog, 1970; 1989; Harwell & Orth, 2002; Haynes, 2000g; Heck, Jr. & Orth, 1980; Hootsmans et al., 1987; Hua et al., 2006; Jacobs & Les, 2009; Jarvis & Moore, 2010; Jiang et al., 2011; Jung et al., 2012; Kaldy, 2006; Kaldy & Shafer, 2013; Kaldy et al., 2015a; 2015b; Keddy, 1987; Kentula & DeWitt, 2003; Kentula & McIntire, 1986; Kim et al., 2015; Kishima et al., 2011; Knight et al., 2015; Larned, 2003; Les et al., 2002; Lovvorn & Baldwin, 1996; Luckenbach & Orth,

1999; Mabbott, 1920; Mach et al., 2014; Marsh, Jr. et al., 1986;
Mattila et al., 1999; McAtee, 1918; Morita et al., 2011; Moore
et al., 1993; Muehlstein et al., 1988; 1991; Nejrup & Pedersen,
2008; Olsen et al., 2014; 2016; Orth, 1973; 1975; Orth & Heck,
Jr., 1980; Orth & Moore, 1986; 1988; Orth et al., 2000; 2003;
Penhale, 1977; Petersen et al., 2017; Phillips & Echeverria,
1990; Phillips & Meñez, 1988; Phillips et al., 1983a; 1983b;
Posey, 1988; Renn, 1935; Reusch, 2001; Robertson & Mann,
1982; Ruckelshaus, 1995; 1996; Ruesink et al., 2010; Shafer
et al., 2007; 2008; 2011; 2014; Shanks et al., 2003; Shoemaker
& Wyllie-Echeverria, 2013; Short, 1983; Short & McRoy,
1984; Silberhorn et al., 1983; Soros-Pottruff & Posluszny,
1995; Stubler et al., 2017; Sumoski & Orth, 2012; Talbot et al.,
2006; 2016; Thom et al., 1995; Tol et al., 2017; Tomlinson &
Posluzny, 2001; Van Montfrans et al., 1982.

AROIDS

For many years, the "aroids" represented those species classified within four orders (Arales, Arecales, Cyclanthales, and Pandanales) of subclass Arecidae (Cronquist, 1981). Subsequent molecular phylogenetic studies provided a substantially different perspective among these groups, including evidence of a much closer relationship of the type order (Arecales) to several "commelinid" monocots than to any of the other three orders (Judd et al., 2016). Consequently, the transfer of Arecales to subclass Commelinidae, together with Commelinales, Poales, and Zingiberales (Soltis et al., 2018), has been followed here. Evidence for the inclusion of Cyclanthales within Pandanales and their distant relationship to both Arales and Arecales (Judd et al, 2016; Soltis et al., 2018) also is compelling, but of little consequence here given that neither contains OBL indicators. However, the taxonomic disposition of the remaining former order Arales is more contentious and warrants further discussion.

Throughout much of the 20th century, Arales had been defined as including just two families: Araceae and Lemnaceae (Cronquist, 1981). Both contain OBL indicators. One major change has involved the proposed merger of Lemnaceae with Araceae, under the impression that phylogenetic studies (Cabrera et al., 2008; Cusimano et al., 2011; Nauheimer et al., 2012; Henriquez et al., 2014) have conclusively shown the former to be "embedded" within the latter (Soltis et al., 2018). The unfortunate trend to merge the Lemnaceae with Araceae dates to an earlier phylogenetic era, when poor sampling of taxa or loci produced inaccurate trees depicting Lemnaceae as nesting deeply within Araceae (e.g., Duvall et al., 1993b; Chase et al., 1993; 1995; French et al., 1995; Mayo et al., 1998a). However, as taxon density and gene sampling improved, the duckweed clade eventually resolved well outside of the "core Araceae" (e.g., Figure 1.26).

An alternative interpretation (e.g., Figure 1.26) must be considered before simply dismantling a taxonomic convention that has spanned several centuries. Lemnaceae clearly are differentiated from nearly all Araceae (sometimes designated as the "true Araceae") in exclusion of four genera assigned alternatively to a third clade, which has been designated "Proto-Araceae" by many authors (e.g., French et al., 1995; Cabrera et al., 2008; Cusimano et al., 2011; Nauheimer et al., 2012; Henriquez et al., 2014; Sree et al., 2016; Choi et al., 2017; Tippery & Les, 2020). Both clades also exhibit substantial molecular divergence from the remainder of Araceae, at a level comparable to (or exceeding) that distinguishing other angiosperm families. Consequently, in this case, it seems more reasonable to retain the exceptionally distinctive Lemnaceae as a family (as has been done here). To maintain phylogenetic integrity, the retention of Lemnaceae also necessitates recognition of the "Proto-Araceae" clade as another family distinct from Araceae, in this case Orontiaceae (see further discussions in Sree et al., 2016; Tippery & Les, 2020). Orontiaceae are distinctive morphologically in having tepalate flowers (i.e., with 4 or 6 tepals) and an equal number of stamens (Lehmann & Sattler, 1992).

A more serious consequence of the widely accepted merger of Lemnaceae and Araceae has been the taxonomic mayhem resulting from ordinal revisions, which were made to accommodate the diminished number of families. With no subclass rank remaining (Arecidae being subsumed within Commelinidae), a revised concept of "Alismatales" emerged (APG, 2016) apparently to provide some higher-level taxonomic organization of the group. In the "APG Alismatales," the expanded Araceae were paired with the putatively most closely related clades (Tofieldiaceae and the entire subclass Alismatidae), despite major inconsistencies among their higher-level phylogenetic interrelationships (Les & Tippery, 2013; Figure 1.1). In doing so, the subclass Alismatidae was discarded and the nomenclatural recognition of two clearly differentiated clades within it (recognized previously as distinct orders) was abandoned. The resulting concept placed all former Arales, all former Alismatales, all former Potamogetonales, and Tofieldiaceae within a single, highly heterogeneous order redefined as "Alismatales" (APG, 2016), a group not yet assigned to any subclass (Soltis et al., 2018).

Without reiterating arguments already made above (see section on Alismatidae), the classification implemented here retains the more traditionally defined Alismatales (see preceding discussion), along with Arales (Araceae + Lemnaceae + Orontiaceae + Tofieldiaceae), and Potamogetonales, all recognized at the ordinal level. Tofieldiaceae have been included in Arales tentatively, given their uncertain higher-level phylogenetic placement (Soltis et al., 2018; Figure 1.1).

Perhaps a compromise to these competing classifications could be achieved by redefining subclass Alismatidae to contain the orders just mentioned (making it essentially equivalent to the sensu lato "APG Alismatales"). Although Alismatidae then would no longer represent a strictly aquatic clade, the resulting ordinal classification would be far more consistent with years of previous taxonomic usage and all the groups would be assigned to a subclass, which seems necessary to be consistent with the classification of other angiosperm groups. Yet, given the uncertain higher-level phylogenetic interrelationships of these groups (e.g., Figure 1.1) even such a "compromise" is not very persuasive as being one that would guarantee the accurate portrayal of higher relationships.

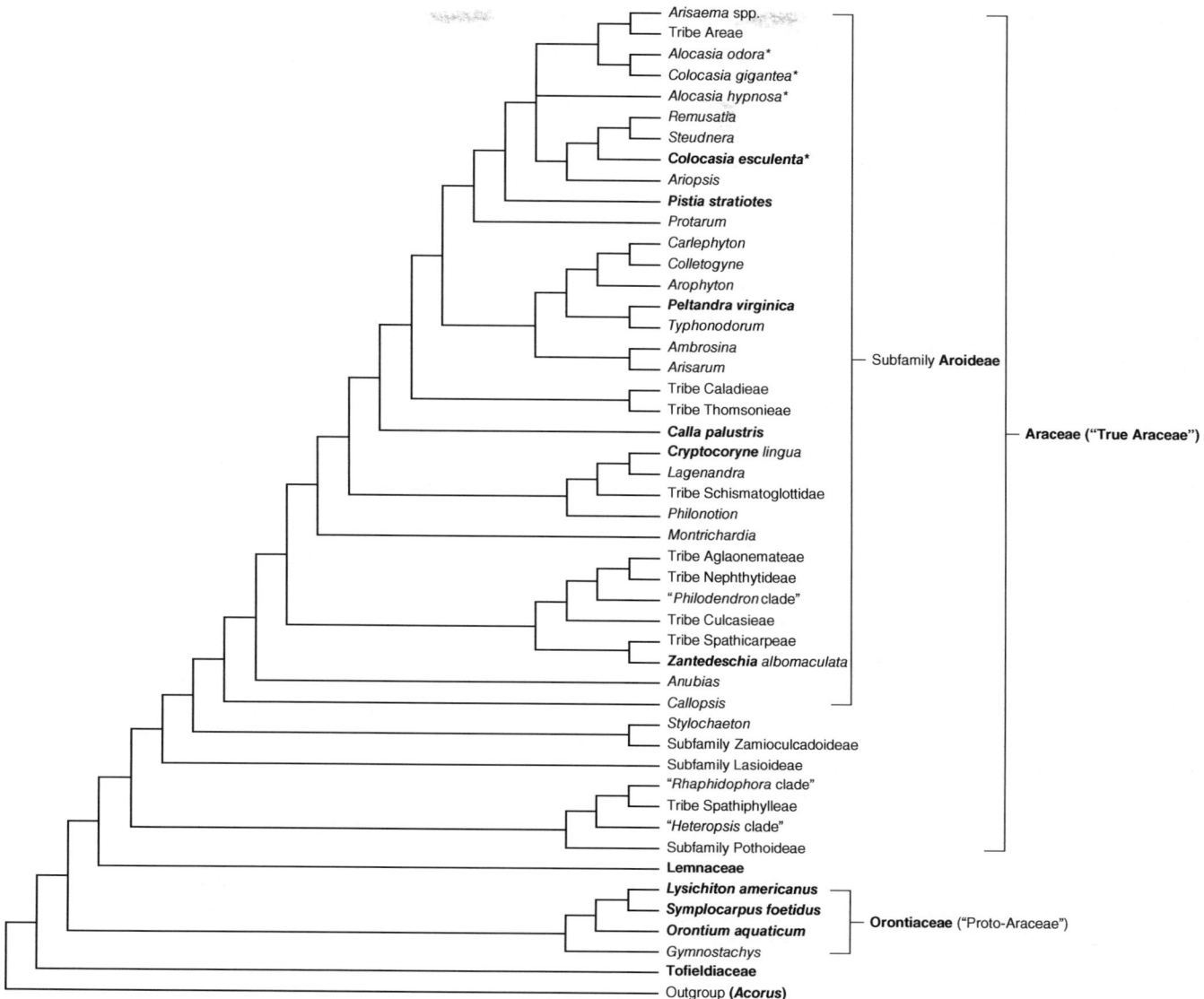

FIGURE 1.26 Phylogenetic relationships in Arales ("aroids") as indicated by the analysis of 4.3+ kb of plastid DNA sequence data (adapted and modified from Nauheimer et al., 2012). In this interpretation (see Tippery & Les, 2020), Lemnaceae represents the sister family of Araceae sensu stricto (referred to by some as "True Araceae"). Several outlier genera comprise a clade known as "Proto-Araceae," which are recognized here as the family Orontiaceae. Although included tentatively in Arales, the precise phylogenetic placement of Tolfieldiaceae remains uncertain (see Figure 1.1). Several genera (asterisked) are not monophyletic, including *Colocasia*, which contains OBL North American indicators (bold type). The OBL North American species of *Cryptocoryne* and *Zantedeschia* were not surveyed, so the placement of the OBL indicators must assume that the genera are monophyletic, which is probable but not necessarily the case. The OBL North American indicators have originated throughout the Arales; however, a concentration of aquatic species occurs in both Lemnaceae (entirely aquatic) and Orontiaceae.

ORDER 4: ARALES [4]

Phylogenetic analyses consistently recover four clades within the Arales, which are treated here as separate families (Figure 1.26). Given that the circumscription of Arales with respect to the placement of Tofieldiaceae remains uncertain, it would be premature to attempt any meaningful description of the order or to enumerate potential synapomorphies (other than molecular data). Although the APG classification (APG, 2016) has persuaded many that Arales/Tofieldiaceae are the sister group to alismatid monocots, that relationship is only one of many possibilities depicted in various studies

that have evaluated different DNA sequence data (Figure 1.1). Consequently, the higher-level placement of Arales also remains unresolved, at least to the author's satisfaction.

Arales are cosmopolitan in distribution. The order includes four families, which all have OBL North American representatives:

4.1. **Araceae** Juss.
4.2. **Lemnaceae** Gray
4.3. **Orontiaceae** Bartl.
4.4. **Tofieldiaceae** (Kunth) Takht.

Family 4.1. Araceae [100]

As viewed here, the "arum" family (Araceae) represents a well-defined clade (the "true Araceae" of some authors) that is distinct from both Lemnaceae and Orontiaceae (Figure 1.26). In this sense the family contains about 100 genera and 3,300 species (Mayo et al., 1998a). Most of the species are terrestrial climbing plants or epiphytes, which inhabit tropical rainforests (Ivancic & Lebot, 2000). Yet, six of the seven North American genera (all but *Arisaema*) are entirely aquatic. Diagnostic features include the presence of grooved calcium oxalate raphide crystals, alternate, simple leaves with a well-developed blade, and an indeterminate inflorescence consisting of a leaf-like, often colorful bract (spathe) enveloping a fleshy spike (spadix) of small, mostly unisexual flowers (naked or with small or reduced perianths), arranged in a monoecious sexual condition (Cronquist, 1981; Judd et al., 2016). Pollination occurs primarily by insects (Insecta) such as beetles (Coleoptera), bees (Hymenoptera), and flies (Diptera), which are attracted to the inflorescence by various odors and/or by metabolically generated heat (Cronquist, 1981; Judd et al., 2016). The fruits are green or brightly-colored (e.g., red or orange) berries dispersed by birds (Aves) or other animals. The fruits of some species are dispersed by ants (Insecta: Hymenoptera) or by water.

Arums are of great significance economically, with numerous cultivated garden specimens and houseplants (over 50 million sold each year) found in genera such as *Aglaonema*, *Alocasia*, *Amorphophallus*, *Anthurium*, *Arisaema*, *Arum*, *Biarum*, *Caladium*, *Colocasia*, *Dieffenbachia*, *Monstera*, *Philodendron*, *Spathiphyllum*, and *Zantedeschia* (Bown, 2000). Commercially, the "swiss-cheese plant" (*Monstera deliciosa*) is the most widely distributed member of the family (Bown, 2000). *Anubias* and *Cryptocoryne* are important sources of cultivated aquarium plants (Kasselmann, 2003) and *Pistia* is widely distributed as an ornamental water garden specimen. The jack-in-the-pulpit (*Arisaema triphyllum*) is among the most iconic wildflower species of North America. Although many arums (e.g., *Dieffenbachia seguine*) are poisonous, taro (*Colocasia esculenta*) provides food for hundreds of millions of people and is the principal ingredient of "poi," a familiar novelty sampled by countless visitors to the Hawaiian Islands (Bown, 2000). Other edible arums are found in *Alocasia*, *Amorphophallus*, *Cyrtosperma*, and *Xanthosoma* (Bown, 2000). Many arums have been used medicinally as contraceptives, expectorants, hallucinogens, insecticides, sedatives, stimulants, and as anti-cancer remedies (Bown, 2000).

Araceae are cosmopolitan but the family is distributed primarily in tropical to subtropical regions. OBL North American indicators occur within six genera, which are dispersed throughout the family (Figure 1.26):

4.1.1. *Calla* L.
4.1.2. *Colocasia* Schott
4.1.3. *Cryptocoryne* Fisch. ex Wydler
4.1.4. *Peltandra* Raf.
4.1.5. *Pistia* L.
4.1.6. *Zantedeschia* Spreng.

4.1.1. Calla

Bog arum, water arum, water dragon, wild calla; Calla des marais

Etymology: derived from *calsa* an old Latin form of *calyx*, in reference to the hood-like spathe
Synonyms: *Callaion*; *Dracunculus* (in part); *Provenzalia*
Distribution: global: circumboreal; **North America:** northern
Diversity: global: 1 species; **North America:** 1 species
Indicators (USA): obligate wetland (OBL): *Calla palustris*
Habitat: freshwater; palustrine; **pH:** 4.8–8.4; **depth:** <1 m; **life-form(s):** emergent herb
Key morphology: rhizomes (to 3 cm in diameter) horizontal, branching, rooting at nodes; the leaves arising singly along the rhizome, petioled (to 40 cm), the blades (to 14 cm) simple, ovate to circular, the base cordate, the apex with an extended finger-like tip; inflorescence a cylindric, andromonoecious spadix borne on a peduncle (to 40 cm) and enveloped by a leaflike spathe; the spathe (to 8 cm) green (pre-anthesis) to bright white, the apex long-apiculate (to 10 mm); the spadix with 40–60 naked flowers, the uppermost (4–6) staminate, the lower (to 56) bisexual; stamens 6–12 (to 691 per inflorescence), dimorphic, the outer filaments broad, the inner filaments narrow; ovaries (to 56 per inflorescence) unilocular, with 1–19 ovules; infructescence (to 5 cm) of 4–11 red, pear-shaped berries (to 12 mm); seeds (to 5 mm) brown, cylindric, embedded in mucilage
Life history: duration: perennial (rhizomes, whole plants); **asexual reproduction:** rhizome fragments; **pollination:** insect; **sexual condition:** andromonoecious; **fruit:** an infructescence of berries (common); **local dispersal:** rhizomes, berries, seeds (water); **long-distance dispersal:** berries, seeds (birds)
Imperilment: 1. [G5]; S1 (IL, IN, MD); S2 (LB, NF, ND, RI); S3 (NJ, OH)
Ecology: general: *Calla* is monotypic (see next).
4.1.1.1. *Calla palustris* L. inhabits bays (marshy), beaches (marshy), bogs (cold, floating, mucky, quaking, *Sphagnum*, streamside), depressions, ditches (muddy, roadside), fens, flats (alluvial, mud), hollows, marshes (mesotrophic), mats (bog, *Sphagnum*), meadows (sedge), oxbows, peat pits, potholes (drying), roadsides (disturbed), sloughs, swales (mucky), swamps (cedar, mud, muskeg, *Sphagnum*), springs, and the margins of flowages, lakes (boggy), ponds (beaver, boggy, dune, roadside), pools (backwater), rivers, and streams at elevations to 1580 m. The plants occur under exposures of full sun to partial shade. Their substrates have been described as gravel, gravelly sand, humus (charcoally), muck (rotted), mud, ooze, peat (*Sphagnum*), peaty muck, sand, and sandy muck. The habitats are characterized by shallow water, which usually is <0.25 m deep, but some occurrences (perhaps with surface mats of plants) have been reported at depths to 1 m. Although this a species often associated with bogs, the plants typically occur in circumneutral to acidic waters (pH: 4.8–8.4; $\bar{x} = 7.0$) of low alkalinity (<105 mg l^{-1} CaCO$_3$; $\bar{x} = 60$ mg l^{-1}) and conductivity (<200 μmhos cm^{-1} @ 25°C; $\bar{x} = 60$ μmhos cm^{-1}). Flowering in North America occurs from early

May to September (cultivated greenhouse plants have been induced to flower during January) with fruiting extending from June to October. The plants are andromonoecious. Prior to anthesis the enveloping green spathe opens and turns white in color to provide pollinator access. Some accounts report a fetid odor emanating from the inflorescence, while others assert that no detectable floral scent is produced. The spadix produces from 40 to 60 flowers with the uppermost 4–8 flowers being staminate and the remainder (lower flowers) all being bisexual. The inflorescences and flowers are pre-formed during the summer (early August) prior to the season when sexual reproduction occurs. The flowers are protogynous with anthesis (total inflorescence) lasting about 10 days. Stigma receptivity initiates from the lowermost flowers and proceeds upward to the spadix tip. Anther dehiscence follows a similar pattern but initiates near the end of, or after completion of, the female phase. The outer stamens have broader filaments and dehisce before the inner ones. Pollinator exclusion experiments have estimated that about 20% of total fruit production occurs by selfing (or apomixis). The breeding system apparently is mixed, with an undetermined proportion of outcrossing and selfing. Cross-pollination is achieved primarily by flies (Insecta: Diptera) and beetles (Insecta: Coleoptera), having evolved arguably as a "rewarding mutualism" (evidenced by correlations between these insect groups and specific floral traits). Pollination by slugs and snails (Mollusca: Gastropoda) or by wind also has been proposed. Syrphid flies (Insecta: Diptera: Syrphidae: *Sphegina, Toxomerus geminatus*) are documented as pollinators of North American plants. The berries are bright red when mature and are eaten by several species of ducks (Aves: Anatidae), which likely transport them endozoically. The seeds are buoyant by virtue of airspaces that develop in the seed coat. They can remain afloat for at least a week, which facilitates local dispersal by water. They do not remain viable when immersed in saltwater. The seeds are dormant when shed and poor germination has been reported for unstratified or dried seeds. Exposure to light is required for germination. Seeds have germinated well (81–86%) and relatively quickly ($\bar{x} = 7.2$ days) following 60–90 days of cold stratification (5°C) and incubation at 22°C (also under a 20°C/15°C day/night temperature regime). The seeds are well represented in some (European) seed bank studies ($\bar{x} = 1,177$ seedlings m^{-2}), but seedling survival also has been shown to be poor (~1%) when the seeds were planted on exposed shorelines. The plants are strongly perennial with the rhizomes (sometimes foliated by reduced leaves) persisting throughout the winter. The rhizomes branch alternately with each meristem remaining active for two seasons. Several leaves can be produced during the first year, but only two leaves are produced the second year followed by the production of an inflorescence, which arrests any further growth of a branch. The branches detach readily (~5 cm pieces) producing vegetative propagules. The propagules are buoyant (100% have remained afloat after 187 days) and retain high resprouting capacity (72%). *Calla* plants have been factored into a number of climatological studies because they can emit relatively large quantities of methane gas (an estimated 280 μmol

g^{-1} day^{-1} on average). When growing from the bases of trees or on detached substrates, the plants can initiate the development of floating substrates which expand over the water surface.

Reported associates: *Abies balsamea, Acer rubrum, Acer saccharum, Acorus americanus, Acorus calamus, Alisma, Alnus glutinosa, Alnus rugosa, Alnus tenuifolia, Andromeda polifolia, Aralia nudicaulis, Arctophila fulva, Aronia melanocarpa, Aronia ✕prunifolia, Aulacomnium palustre, Betula alleghaniensis, Betula glandulosa, Betula populifolia, Betula pumila, Bidens cernuus, Bolboschoenus fluviatilis, Calamagrostis canadensis, Calliergon, Caltha natans, Caltha palustris, Carex aquatilis, Carex brunnescens, Carex canescens, Carex chordorrhiza, Carex comosa, Carex crawfordii, Carex crinita, Carex diandra, Carex hystericina, Carex lacustris, Carex lasiocarpa, Carex limosa, Carex lurida, Carex magellanica, Carex oligosperma, Carex pseudocyperus, Carex rostrata, Carex stricta, Carex trisperma, Carex utriculata, Ceratophyllum demersum, Chamaedaphne calyculata, Cicuta bulbifera, Cicuta virosa, Cladina rangiferina, Cladina stellaris, Clethra alnifolia, Clintonia borealis, Comarum palustre, Coptis trifolia, Cornus amomum, Cornus canadensis, Cystopteris, Decodon verticillatus, Dicranum polysetum, Drepanocladus, Drepanocladus aduncus, Drosera intermedia, Drosera rotundifolia, Dryopteris carthusiana, Dulichium arundinaceum. Eleocharis erythropoda, Epilobium ciliatum, Epilobium palustre, Equisetum fluviatile, Eriophorum angustifolium, Eriophorum vaginatum, Eriophorum virginicum, Eubotrys racemosa, Eutrochium maculatum, Fagus grandifolia, Fraxinus nigra, Galium labradoricum, Galium obtusum, Galium trifidum, Gaultheria procumbens, Gaylussacia baccata, Gaylussacia frondosa, Glyceria canadensis, Glyceria grandis, Glyceria melicaria, Gymnocarpium dryopteris, Hippuris vulgaris, Hypericum canadense, Hypericum ellipticum, Hypericum majus, Hypericum mutilum, Ilex laevigata, Ilex mucronata, Ilex verticillata, Iris versicolor, Juncus canadensis, Kalmia angustifolia, Kalmia polifolia, Larix laricina, Larix occidentalis, Leersia oryzoides, Lemna minor, Lemna minuta, Lycopodium annotinum, Lycopus uniflorus, Lyonia ligustrina, Lysimachia terrestris, Lysimachia thyrsiflora, Maianthemum canadense, Maianthemum trifolium, Menyanthes trifoliata, Myrica gale, Myriophyllum sibiricum, Nuphar variegata, Nymphaea odorata, Osmunda regalis, Osmundastrum cinnamomeum, Peltandra virginica, Persicaria amphibia, Phalaris arundinacea, Phragmites australis, Picea engelmannii, Picea glauca, Picea mariana, Pinus resinosa, Pinus rigida, Pinus strobus, Platanthera clavellata, Pleurozium schreberi, Poa palustris, Pohlia nutans, Polytrichum commune, Polytrichum strictum, Pontederia cordata, Populus tremuloides, Potamogeton berchtoldii, Potamogeton foliosus, Ranunculus gmelinii, Rhododendron groenlandicum, Rhododendron maximum, Rhododendron viscosum, Rhynchospora alba, Rubus hispidus, Rumex occidentalis, Sagittaria cuneata, Sagittaria latifolia, Sagittaria rigida, Salix discolor, Salix pyrifolia, Sarracenia purpurea, Scheuchzeria palustris, Schoenoplectus tabernaemontani, Scirpus atrocinctus, Scirpus cyperinus,*

Scutellaria galericulata, Sium suave, Sparganium angustifolium, Sparganium emersum, Sparganium eurycarpum, Sparganium glomeratum, Sparganium natans, Sphagnum angustifolium, Sphagnum capillifolium, Sphagnum compactum, Sphagnum fallax, Sphagnum magellanicum, Sphagnum riparium, Sphagnum rubellum, Sphagnum squarrosum, Spiraea alba, Spiraea tomentosa, Spirodela polyrhiza, Stellaria longifolia, Stuckenia filiformis, Tephroseris palustris, Thelypteris palustris, Thuja occidentalis, Tomenthypnum nitens, Triadenum fraseri, Triadenum virginicum, Trientalis borealis, Tsuga canadensis, Typha angustifolia, Typha latifolia, Utricularia intermedia, Utricularia macrorhiza, Vaccinium angustifolium, Vaccinium corymbosum, Vaccinium macrocarpon, Vaccinium myrtilloides, Vaccinium oxycoccos, Viburnum edule, Viola, Warnstorfia trichophylla, Wolffia columbiana, Woodwardia virginica, Xyris montana.

Use by wildlife: *Calla palustris* is a preferred food of beavers (Mammalia: Castoridae: *Castor canadensis*) and also is eaten routinely by black bears (Mammalia: Ursidae: *Ursus americanus*). The berries contain an average of 17.91% protein, 7.48% fat, and 15.14% crude fiber. They have been found in the digestive tracts of various species of European ducks (Aves: Anatidae), several of which also occur in North America. The rhizomes are eaten by Eurasian wild boars (Mammalia: Suidae: *Sus scrofa*) and muskrats (Mammalia: Cricetidae: *Ondatra zibethicus*). The foliage hosts a plant parasitic fungus (Ascomycota: Mycosphaerellaceae: *Cercospora callae*). The plants provide cover and effective escape habitat for wood frogs (Vertebrata: Amphibia: Ranidae: *Rana sylvatica*) and nesting cover for lesser scaups (Aves: Anatidae: *Aythya affinis*).

Economic importance: food: Because all parts of the *Calla* plant contain acicular raphides crystals and other toxic substances (e.g., cyanogenic glycosides, proteolytic enzymes), it is probably best to completely avoid eating them. If eaten, care must be taken to prepare the rhizomes properly to remove the various toxins and the needle-like crystals, which otherwise can cause serious irritation. *Calla* rhizomes collected in early spring (before the leaves appear) are bitter and poisonous when raw, but when dried, chopped, and boiled they have been made into a meal for adding to flour. Such flour was used to prepare a famine food (known as "Missen bread") in Europe. The seeds also can be dried and ground into flour, which reduces their bitterness; **medicinal:** *Calla* herbage releases histamine when eaten and potentially is fatal if ingested by dogs, cats, and other domestic pets. Members of the Gitksan tribe prepared a root decoction for treating influenza and hemorrhages. A poultice made from the mashed roots was used by the Potawatomi to alleviate swelling. The Iroquois made a root and stem decoction into a poultice for treating snake bites; **cultivation:** *Calla palustris* is distributed commercially as an ornamental water garden plant; **misc. products:** none; **weeds:** none; **nonindigenous species:** none.

Systematics: Although *Calla* is monotypic, its systematic position within Araceae has been problematic. The genus occupies a relatively isolated position in the family. Different molecular and morphological data analyses place it within subfamily Aroideae (e.g., Figure 1.26), but in quite different topological association with other genera. In addition to various morphological inconsistencies, its nested placement within subfamily Aroideae (ranging from shallow to deep) also contradicts its distinctive pollen type, which is quite similar to that of subfamily Zamioculcadoideae but differs substantially from that in subfamily Aroideae. Such inconsistencies raise questions about the ability of molecular data to effectively reconcile the placement of divergent taxa like *Calla*. The basic chromosome number of *Calla* is $x = 6$ or 9. The North American populations reportedly are diploid ($2n = 36$); however, some Eurasian populations have yielded tetraploid ($2n = 72$) or aneuploid ($2n = 63, 69, 70$) counts.

Distribution: The circumboreal *Calla palustris* is distributed throughout Canada and northern portions of the United States.

References: Adams, 1927; Airaksinen et al., 1986; Allen et al., 2002a; Baltisberger et al., 2010; Baskin & Baskin, 1998; Beasley, 1999; Bellis, 1962; Bratton, 1974;Cabrera et al., 2008; Camill, 1999; Chartier et al., 2014; Cusimano et al., 2011; Davis, 1903; Dudley, 1937; Fernald & Kinsey, 1943; Fornier & Hines, 2001; Garon-Labrecque et al., 2016; Genaust, 1999; Green et al., 2016; Guppy, 1906; Hamerstrom, Jr. & Blake, 1939; Henriquez et al., 2014; Johri, 1984; Jutila, 2002; Karlin & Lynn, 1988; Lehmann & Sattler, 1992; Lieffers, 1984; Lynn, 1984; Marchant, 1970; Marion et al., 2017; Mitchell & Niering, 1993; Morton, 1963; Noyce et al., 1997; Pellerin et al., 2009; Perry, 1961; Persson & Shacklette, 1959; Pfauth & Sytsma, 2005; Pindel, 2001; Pip, 1979; Racine et al., 1998; Redmond et al., 1993; Sarma et al., 2007; Sarneel, 2013; Sarneel & Soons, 2012; Seigler, 1976; Shannon et al., 1996; Spinner & Bishop, 1950; Thompson, 2000b; Ulrich et al., 2013; Walton, 1995; Wetmore, 2001.

4.1.2. Colocasia

Coco yam, dasheen, elephant's ear, wild taro; taro ou colocase

Etymology: derived from *kolokasion*, an ancient Greek name for the edible rootstock

Synonyms: *Alocasia* (in part); *Aron* (in part); *Arum* (in part); *Caladium* (in part); *Calla* (in part); *Leucocasia*; *Steudnera* (in part); *Zantedeschia* (in part)

Distribution: global: Africa*, Asia, Australia*, New World*; **North America:** southern

Diversity: global: 7 species; **North America:** 1 species

Indicators (USA): obligate wetland (OBL); facultative wetland (FACW): *Colocasia esculenta*

Habitat: freshwater, freshwater (tidal); palustrine; **pH:** unknown; **depth:** <1 m; **life-form(s):** emergent herb

Key morphology: corms subterranean, producing elongate, stolon-like rhizomes (or rhizomatous stolons) with nodes near the substrate surface, from which smaller corms arise; leaves simple, clustered at apex, the blades (to 70 cm) sagittate or sagittate-cordate, glaucous above, margins entire, apex with a projecting finger-like tip, peltate (for up to 7 cm), the petioles (to 180 cm) spongy with airspaces, reddish to purple at point of attachment; spadix (to 15 cm) slender, protracted into a sterile apical appendage, surrounded by an elongate, orange spathe (to 35 cm); flowers naked, unisexual (spadix monoecious; ♂ and ♀ separated by sterile flowers), the ♂ flowers apical, with 3–6 connate stamens; the ♀ flowers basal, their ovary unilocular, containing numerous (to 67) ovules; fruits (absent in North America) orange, with numerous (to 35) seeds (to 1.5 mm) when fertile

Life history: duration: perennial (corms, rhizomes/stolons); **asexual reproduction:** corms, rhizomes/stolons; **pollination:** insect; **sexual condition:** monoecious; **fruit:** berries (rare); **local dispersal:** rhizomes/stolons; **long-distance dispersal:** corms, stolon fragments (water)

Imperilment: 1. *Colocasia esculenta* [GU]

Ecology: general: *Colocasia* is a small genus of herbaceous perennials, which are associated with damp to wet habitats including humid forests, rain forests, moist thickets, streamsides, and humid forest margins as well as wetlands. Because the monophyly of the genus is doubtful (see *Systematics* below; Figure 1.26), it is premature to provide an ecological appraisal for the genus at present. However, some general characteristics can be ascertained from the following account, which summarizes information on the only naturalized species currently occurring in North America.

4.1.2.1. *Colocasia esculenta* **(L.) Schott** is a nonindigenous perennial, which grows in or on beaches (freshwater), bottomlands (hardwood), canals, depressions (roadside, sandbar), ditches (roadside, stormwater), floating mats, floodplains, glades (marshy), hammocks (floating, hydric, mesic hardwood), marshes (muddy), meadows, ponds, sandbars, seeps, sloughs, springs, swamps (freshwater tidal), woodlands, and along the margins of bayous, canals, ponds, reservoirs, rivers, and streams at elevations to 302 m. The plants can occur in exposures ranging from dense shade to full sun. The substrates have been described as alluvium (clayey), clay (mineral, Schreiver), clay loam, loam, muck, mud, sand, shell middens, silt, and silty clay loam. This is an ecologically versatile species wherein three main genotype groups have been delineated by their habitat affinities: upland (dry conditions), intermediate (moderately wet conditions), and wetland (wet conditions). The corms of the wetland group are elongate. Within the wetland group are six subgroups: wet soil genotypes (adapted to high water tables, irrigation, or frequent rain), paddy genotypes (adapted to permanent circulating waters), riverine genotypes (adapted to river bank conditions), swamp genotypes (adapted to stagnant permanent waters), wetland genotypes (adapted to quickly circulating waters), and "floating" genotypes (adapted to floating organic mat growing conditions). A wetland strain (*C. esculenta* var. *aquatilis*) has been identified as being the most common type naturalized in continental North America, with *C. esculenta* var. *antiquorum* being found less often, and *C. esculenta* var. *nymphaeifolia* the least common; *C. esculenta* var. *esculenta* is the variety grown most widely as an ornamental. The wetland plants exhibit poor drought tolerance, whereas *C. esculenta* var. *antiquorum* is somewhat more drought resistant. Because these plants have been cultivated asexually for centuries, it was widely believed that they had lost their ability to reproduce sexually. However, taro barrenness is due in large

part to its cytotype, with the triploid plants being sterile (see *Systematics* below). Most cultivars actually are fertile diploids, with many of these capable of producing viable seeds (sterility can occur in some diploid cultivars for other reasons). The inflorescences of sexual plants are highly fragrant (a sweet, fruity scent) and attract flies (Insecta: Diptera) as pollinators. At anthesis, the spathe opens at the base and reflexes apically to provide pollinator access to the spadix. The spadix is thermogenic (reaching 6.8°C above ambient air temperature) for two successive nights. The plants are self-compatible (within an inflorescence) but are protogynous, with thermogenesis facilitating the dispersion of odor during the female phase but ending 1–1.5 hr after pollen is released during the male phase. The stigmas remain receptive (♀ phase) from 6 days prior to pollen dispersal until up to 4 days afterward, which provides opportunities for both cross- and self-pollination to occur. Self-pollination normally does not occur but can be achieved by hand pollinations. Numerous genetic surveys of flowering diploids have indicated that the plants routinely undergo natural pollination, primarily are outcrossed, and are highly heterozygous. The fruits are fully mature within 4–8 weeks, coincident with the withering of the peduncle. The seeds, which can average up to 6,576 per inflorescence, have germinated without pretreatment (60–70% within a week) when sowed in "water pots," or when placed between moist layers of filter paper at 25°C. Refrigerated seeds have been stored successfully for up to 2 years when kept within a desiccator. Continental North American plants reputedly are sterile, but the precise causative factors have not been explicated. They possibly represent triploids, which are sterile, and have been associated with those individuals identified as *C. esculenta* var. *antiquorum*. Seed set has not been observed anywhere in continental North America and Florida plants have failed to set seed even after being hand-pollinated. Despite their sterility, the plants are capable of flowering, which can occur from late June to October. North American plants propagate clonally by proliferation of the rhizomes/stolons (a distinction between these two organs is difficult here), especially once the main corm has been removed. Vegetative reproduction is highly efficient. A laboratory study of Texas plants reported a 2.6–5.6-fold increase in total biomass (mostly corms) and development of 20 asexual progeny arising from the rhizomes/stolons within a 5-month period. North American plants are dispersed entirely by vegetative fragments (corms and rhizomes/stolons), which are transported by water currents. A single plant can produce up to 10 rhizomes/stolons, each capable of producing a new plant offshoot at each node. **Reported associates (North America):** *Acer negundo, Acer rubrum, Acrostichum aureum, Alstroemeria pulchella, Alternanthera philoxeroides, Ambrosia trifida, Ampelopsis arborea, Arthraxon hispidus, Atriplex lentiformis, Bidens bipinnatus, Bidens laevis, Boehmeria cylindrica, Bolboschoenus maritimus, Briza minor, Bromus catharticus, Campsis radicans, Carex, Celtis laevigata, Cephalanthus occidentalis, Cicuta maculata, Cladium jamaicense, Clematis catesbyana, Clematis drummondii, Clematis terniflora, Commelina communis, Commelina diffusa, Cynodon dactylon, Cyperus* *difformis, Cyperus erythrorhizos, Cyperus ochraceus, Cyperus odoratus, Decodon verticillatus, Dichanthelium dichotomum, Diodella teres, Dioscorea bulbifera, Echinodorus grandiflorus, Eichhornia crassipes, Eleocharis baldwinii, Elymus virginicus, Eragrostis, Fallopia scandens, Fraxinus pennsylvanica, Galium aparine, Geranium carolinianum, Hydrocotyle umbellata, Ipomoea cairica, Iva annua, Juglans nigra, Juncus effusus, Kosteletzkya pentacarpos, Kyllinga brevifolia, Leersia oryzoides, Lepidium virginicum, Ligustrum japonicum, Liquidambar styraciflua, Ludwigia grandiflora, Ludwigia peruviana, Lycium fremontii, Lycopus rubellus, Magnolia virginiana, Malvaviscus arboreus, Melothria pendula, Merremia dissecta, Mikania scandens, Modiola caroliniana, Morus alba, Myrica cerifera, Nasturtium officinale, Nymphaea odorata, Nyssa biflora, Osmunda regalis, Osmundastrum cinnamomeum, Pancratium maritimum, Panicum dichotomiflorum, Panicum hemitomon, Parkinsonia aculeata, Parthenocissus quinquefolia, Paspalum denticulatum, Paspalum urvillei, Persea palustris, Persicaria punctata, Phanopyrum gymnocarpon, Phytolacca americana, Pinus elliottii, Pinus taeda, Pistia stratiotes, Populus deltoides, Prosopis, Quercus virginiana, Ranunculus platensis, Rhynchospora corniculata, Rhynchospora miliacea, Ruellia simplex, Rumex verticillatus, Sabal minor, Sabal palmetto, Sagittaria lancifolia, Sagittaria latifolia, Sagittaria platyphylla, Salix interior, Salix nigra, Sambucus nigra, Saururus cernuus, Schoenoplectus deltarum, Schoenoplectus tabernaemontani, Scirpus cyperinus, Sisyrinchium, Smilax walteri, Solidago, Sorghum halepense, Stenotaphrum secundatum, Suaeda nigra, Symphyotrichum, Taxodium distichum, Teucrium canadense, Thelypteris palustris, Toxicodendron, Triadenum virginicum, Triadica sebifera, Typha, Veronica persica, Vitis, Woodwardia areolata, Xanthosoma sagittifolium, Zephyranthes atamasco, Zizaniopsis miliacea.*

Use by wildlife (North America): *Colocasia* plants are eaten by apple snails (Mollusca: Gatropoda: Ampullariidae: *Pomacea canaliculata*), crayfish (Crustacea: Decapoda: Cambaridae: *Procambarus acutus, P. spiculifer*), grass carp (Vertebrata: Teleostei: Cyprinidae: *Ctenopharyngodon idella*), and muskrats (Mammalia: Cricetidae: *Ondatra zibethicus*). They are eaten sparingly by nutria (Mammalia: Rodentia: Myocastoridae: *Myocastor coypus*). *Colocasia esculenta* is a larval (caterpillar) host for various moths (Insecta: Lepidoptera: Cosmopterigidae: *Pyroderces rileyi*; Crambidae: *Spoladea recurvalis*; Hesperiidae: *Calpodes ethlius*; Noctuidae *Bellura densa*). The plants also host several Fungi including several plant pathogenic blights, leafspots, and rots (Ascomycota: Botryosphaeriaceae: *Diplodia*; Ceratocystidaceae: *Ceratocystis colocasiae*; Nectriaceae: *Fusarium solani*; Pleosporaceae: *Macrosporium epiphyllum*; Basidiomycota: Typhulaceae: *Sclerotium rolfsii*; Oomycota: Pytiaceae: *Pythium debaryanum*).

Economic importance: food: The corms of *Colocasia esculenta* contain 25–29% starch, 1.2% protein, 1.4% sucrose, and small amounts of vitamin C (13 mg 100 g^{-1}). The digestibility of taro starch is very high (97–99%) and taro flour provides a gluten-free alternative to persons suffering from celiac disease. The corms have been used as a food for more than 28,000 years and have been in cultivation for more than 10,000 years. Today they are consumed by nearly a half billion people and are regarded as the world's second most importance source of plant calories (after coconut, *Cocos nucifera*). The foliage (7 mg 100 g^{-1} carotene, 52 g 100 g^{-1} vitamin C) also is eaten, as are the cooked flowers, which are regarded as a delicacy. Like most aroids, the plants are bitter and inedible when raw and must be boiled to render them palatable. The well-known Hawaiian "poi" is a paste fermented from the cooked, mashed, corms. The plants were among the foods used by the Florida Seminole people; **medicinal:** The juice of *C. esculenta* reputedly stops bleeding. The plants can cause dermatitis in some people; **cultivation:** *Colocasia esculenta* has been given the "Award of Garden Merit" by the Royal Horticultural Society. In addition to more than 150 varieties developed for food, the numerous garden cultivars include: 'Black Beauty', 'Black Coral', 'Black Magic', 'Black Marble', 'Black Ruffles', 'Black Runner', 'Blue Hawaii', 'Bun-long', 'Chicago Harlequin', 'Diamond Head', 'Elepaio Keiki', 'Emerald', 'Euchlora', 'Fontanesii', 'Hawaiian Eye', 'Hawaiian Punch', 'Hilo Bay', 'Hilo High Colour', 'Illustris', 'Jack's Giant', 'Japanese Cranberry', 'Kona Coffee', 'Little Black Magic', 'Mammoth', 'Maui Magic', 'Mojito', 'Nancy's Revenge', 'Nigrescens', 'Palau Keiki', 'Pineapple Princess', 'Pink China', 'Purple Stem', 'Ruffles', 'Sangria', 'Silver Splash', 'Tea Cup', 'Ulaula Kumu-oha' and 'Xcintho'. In 2004, sales of cultivated taro in Florida alone were estimated to yield more than 2 million dollars and to provide 20 jobs annually; **misc. products:** none; **weeds:** *Colocasia esculenta* is an invasive species of wetlands in the southern United States. It is designated as a "category I" invasive species in Florida and an "F1" invasive in Texas (both categories representing the most serious level); **nonindigenous species:** Genetic data (microsatellite markers) indicate that diploid *Colocasia esculenta* was introduced to the Caribbean islands of the New World not from Africa (as widely had been assumed) but from Pacific plants probably brought there as a foodstuff. However, the geographical origin of plants introduced to continental North America has not yet been determined and it is noteworthy that Costa Rican and South African plants (other potential source populations for the continental plants) have been confirmed as triploid cytotypes, which are sterile like the American plants. The U.S. Department of Agriculture deliberately introduced several *Colocasia* cultivars to the southern United States early in the 20th century as potential vegetable crops.

Systematics: As evidenced by the other aroid genera that appear in the list of synonyms, the taxonomic delimitation of *Colocasia* has been problematic. Not surprisingly, preliminary molecular phylogenetic analyses (e.g., Figure 1.26) have indicated that the genus is not monophyletic as currently circumscribed and a comprehensive study of all the *Colocasia* species and putatively allied aroid genera is needed to establish generic limits and the correct application of generic names. Various analyses of DNA sequence data resolve *Colocasia esculenta* close to *Remusatia* and *Steudnera* but show *Colocasia gigantea* to be more closely allied with *Alocasia* (e.g., Figure 1.26). *Colocasia esculenta* itself has been recognized as two separate species, with the name "*C. antiquorum*" applied to those genotypes having a longer spadix appendage and smaller main corm with well-developed lateral corms. More often these taxa have been distinguished at the varietal level (i.e., *C. esculenta* var. *esculenta* vs. *C. esculenta* var. *antiquorum*) along with several others (*C. esculenta* var. *aquatilis*, *C. esculenta* var. *nymphaeifolia*, etc.). The basic chromosome number of *Colocasia* is $x = 7$ or 14, with the latter favored by genetic studies. *Colocasia esculenta* is known to occur either as diploid cytotypes ($2n = 28$) or as triploids ($2n = 3x = 42$). Most of the cultivated plants are fertile diploids, while the triploids are sterile. Hybridization between and among the numerous cultivars is extensive, but mainly is performed artificially with cultivated material.

Distribution: *Colocasia esculenta* occurs in tropical and subtropical regions of Africa, Asia, and Australia. In the New World (where introduced), it extends from the southern United States into South America.

References: Akridge & Fonteyn, 1981; Bowles & Bowles, 2015; Bown, 2000; Cabrera et al., 2008; Chaïr et al., 2016; Cusimano et al., 2011; Godfrey & Wooten, 1979; Haynes & Burkhalter, 1998; Henriquez et al., 2014; Howells et al., 2006; Ivancic & Lebot, 2000; Ivancic et al., 2004; Kikuta et al., 1938; Lampe, 1986; Li & Boyce, 2010; Moran & Yang, 2012; Morgan & Overholt, 2005; Nauheimer et al., 2012; Nesom, 2009; Nolfo-Clements, 2006; Owens et al., 2001; Parker & Hay, 2005; Potgieter, 1940; Prakash & Nayar, 2000; Prescott-Allen & Prescott-Allen, 1990; Serviss et al., 2000; Stutzenbaker, 1999; Tarver et al., 1978; Youngken, 1919; Wirth et al., 2004.

4.1.3. Cryptocoryne

Crypt, water trumpet; crypto

Etymology: from the Greek *kryptos koryne* ("hidden club") in reference to the concealed nature of the clavate spadix within the spathe

Synonyms: *Ambrosina*; *Arum* (in part); *Lagenandra* (in part)

Distribution: global: Asia; **North America:** southern

Diversity: global: ~50 species; **North America:** 1 species (see *Systematics* below)

Indicators (USA): obligate wetland (OBL): *Cryptocoryne walkeri*

Habitat: freshwater; palustrine, riverine; **pH:** 7.6–8.3; **depth:** to 1.5 m; **life-form(s):** emergent herb, submersed (rosulate)

Key morphology: rhizomes branching or stoloniferous, horizontal with contractile adventitious roots, producing bulbils (to 20); leaves rosulate, spiral, vernation convolute, the blades (to 13 cm) ovate, petiolate (to 20 cm), sheathed (inconspicuously) at base, apex apiculate, base obtuse to cordate, margins entire to undulate; spathes (to 28 cm) short-stalked, prolonged into a greenish, yellowish, or brown, recurved or twisted limb (to 6 cm), the spadix (to 1 cm) enclosed within; flowers naked, unisexual (the spadix monoecious), the upper ♂ flowers (to 90) whorled, separated from the lower ♀ flowers (to 8) by a sterile region (to 15 mm); fruit a spherical syncarp; i.e., an aggregate of coalesced berries; seeds numerous

Life history: duration: perennial (rhizomes, stolons); **asexual reproduction:** bulbils, rhizomes (fragments), stolons (fragments); **pollination:** insect; **sexual condition:** monoecious; **fruit:** an infructescence of coalesced berries (rare); **local dispersal:** bulbils, rhizomes, seeds, stolons (water); **long-distance dispersal:** seeds (water)

Imperilment: 1. *Cryptocoryne walkeri* [GNR]

Ecology: general: *Cryptocoryne* comprises various freshwater aquatic and wetland plants, which can grow either as submersed or as emergent life-forms. One species occurs in brackish tidal sites. Many of the species possess tapering leaves, which offer little resistance to the water currents that are typical of their habitats. Most of the species occur in warm waters (24°C–29°C; minimum of 20°C) but are intolerant of excessive turbidity or nutrient levels. All the species that have been surveyed are capable of using bicarbonate (HCO_3^-) as a carbon source when growing in alkaline waters. Unlike most aquatics, "crypts" occur mainly under shaded conditions rather than in bright exposures. Sexual reproduction is not common and occurs most often when the plants inhabit shallower, marginal sites. Inflorescences can develop on submersed plants but are non-functional unless they become emersed to provide pollinator access. The plants are obligately outcrossed. On the first day of anthesis, small fruit flies (Insecta: Diptera: Drosophilidae), the primary pollinators, are attracted to the spadix by the colorful spathe and a carrion-like odor, which emanates from scent glands located above the pistillate flowers in the lower region. The flies can enter the watertight spadix tube only through the open throat at the base of the limb, which is secured internally by a moveable, valve-like membranous flap of tissue. The pollinators crawl (or fall) down the open tube of the spathe and into the expanded "kettle," where they encounter unisexual flowers that are isolated physically by an elongate sterile region widely separating the upper (♂) and lower (♀) flowers. Because of the strongly protogynous floral maturation, the flies bypass the undehisced stamens and move into the lower portion of the "kettle" where they deposit any foreign pollen clinging to their bodies onto the receptive stigmas. The flap then moves to block access to the tube, thereby trapping the insects within the kettle for a period of 1 day. The slippery kettle walls initially prevent the flies from climbing upward toward the male flowers. During this time the female flowers are no longer receptive, the kettle wall surfaces lose their slippery texture, and the staminate flowers shed their viscous pollen, which readily adheres to the flies as they climb upward. On the third day, the flap deteriorates, enabling the pollen-laden insects to escape from the spathe. Together, the herkogamous floral separation and protogyny effectively preclude any self-pollination within an inflorescence. The fruits (coalesced aggregates of berries) can take up to 9 months to mature. At maturity the fruits rupture apically and release their seeds within hours. The waxy seeds are buoyant and float to the water surface where they are dispersed by currents. They lose all viability if becoming dried. The seeds are not dormant and germinate within days at 28°C. All the species perennate from horizontal rhizomes, which can become stoloniferous and produce readily detaching bulbil-like propagules.

4.1.3.1. ***Cryptocoryne walkeri* Schott** is a nonindigenous perennial, which grows in rivers (pools, riffles) and springs at elevations to 170 m. The plants occur in densely to partly shaded exposures. The substrates (North America) are described as silt (organic), pebbles (limestone), or rocks (limestone). North American habitats are characterized by flowing (e.g., 0.6–0.8 m s^{-1}) alkaline (pH: 7.6–8.3; hardness [as CaCO$_3$]: 58 mg l^{-1}) waters of about a meter in depth. Flowering is rare, but submersed inflorescences have been observed in Florida from April to June. The plants have flowered from December to July when grown under greenhouse conditions. The sexual reproductive cycle follows that already summarized above (see *Ecology: general*). Seeds (when produced) are short-lived but are buoyant and dispersed by water currents. In North America, the plants propagate almost entirely by vegetative means. They reproduce asexually by the development of a horizontal rhizome, which can branch into stoloniferous sections that can extend to water depths of 1.5 m. The stolons can develop into bulbil-like propagules, which easily fragment and become dispersed by water currents. In Florida the plants have reached densities up to 1880 m^{-2} and a total biomass of 207 g DM m^{-2}, most of which (89%) is allocated to the rhizomes. **Reported associates (North America):** *Chara, Fontinalis, Heteranthera dubia, Hydrilla verticillata, Hydrocotyle umbellata, Hygrophila polysperma, Justicia americana, Leersia, Ludwigia repens, Myriophyllum heterophyllum, Nasturtium floridanum, Nuphar, Rhynchospora colorata, Sagittaria kurziana, Sagittaria lancifolia, Sagittaria platyphylla, Utricularia, Vallisneria americana, Zizania aquatica, Zizania texana.*

Use by wildlife (North America): none reported.
Economic importance: food: no reports of use as human food; **medicinal:** Extracts from some *Cryptocoryne* species exhibit antimicrobial activity and have been used as various folk medicines in Asia; **cultivation:** *Cryptocoryne walkeri* (including *C. beckettii, C. undulata,* and *C. wendtii*) is widely distributed commercially as an ornamental aquarium plant; **misc. products:** none; **weeds:** *Cryptocoryne walkeri* is highly invasive. At one Texas locality, introduced plants of *C. walkeri* increased from 11 to 63 colonies (171 to 646 m^{-2}) within a 2-year period, averaging an 80% annual increase. Plant densities of up to 1880 m^{-2} have been observed in Florida; **nonindigenous species:** *Cryptocoryne walkeri* (a native of Sri Lanka) was introduced to Florida before 1998 and to Texas by the year 2000 likely by the careless disposal of aquarium plants.
Systematics: Most molecular phylogenetic studies resolve *Cryptocoryne* as the sister group of the morphologically similar *Lagenandra* (e.g., Figure 1.26), although at least one study did not clearly differentiate the genera. The most comprehensive survey to date (including nearly half the *Cryptocoryne* species but only two of the 14 *Lagenandra* species) resolved the *Cryptocoryne* accessions as a clade, which supports the monophyly of the genus. The "*Cryptocoryne beckettii* complex" includes four closely related and interfertile endemic Sri Lankan taxa (*C. beckettii, C. undulata, C. walkeri,* and *C. wendtii*), which often have proven nearly impossible to

differentiate morphologically given the malleability of their key distinguishing features (e.g., via slightly different spathe coloration, which is known to vary considerably). Preliminary (but unpublished) molecular analyses of 25 species (incorporating various cpDNA RFLP data as well as nuclear and plastid DNA sequence data and even RAPD markers) consistently indicated that the "*C. beckettii* complex" taxa are not well differentiated genetically. The low observed level of molecular divergence separating the taxa (e.g., only 1–3 unique RAPD markers out of 30 polymorphic loci) suggests either that they are of very recent origin or simply represent minor variants within a single species. Given their dubious morphological distinction and interfertility, they are considered here as conspecifics under the name of *C. walkeri* Schott (which has priority of publication). Consequently, all previous reports of *C. beckettii, C. undulata,* and *C. wendtii* in North America are attributed here to *C. walkeri*. The basic chromosome number of *Cryptocoryne* is $x=7$ or 14. *Cryptocoryne walkeri* has diploid ($2n=28$) and triploid ($2n=3x=42$) cytotypes.
Distribution: *Cryptocoryne walkeri* has been introduced to Florida and Texas.
References: Bown, 2000; Cook, 1996; de Wit, 1964; Doyle, 2000; In't Veen, 1981; Kasselmann, 2003; Jacobsen, 1982; Jacono, 2002; Mühlberg, 1982; Othman, 1997; Petch, 1928; Rosen, 2002; Schöpfel, 1975; Wadkar et al., 2017; Wong et al., 2010.

4.1.4. Peltandra
Arrow arum, hog wampee, spoonflower, tuckahoe, wampee; peltandre
Etymology: from the Greek *pelte andros* ("male shield") referring to the peltate synandria
Synonyms: *Alocasia* (in part); *Arum* (in part); *Caladium* (in part); *Calla* (in part); *Houttinia; Lecontia; Rensselaeria; Xanthosoma* (in part)
Distribution: global: North America; **North America:** eastern
Diversity: global: 2 species; **North America:** 2 species
Indicators (USA): obligate wetland (OBL): *Peltandra sagittifolia, P. virginica*
Habitat: brackish (coastal), freshwater; freshwater (tidal); palustrine; **pH:** 5.6–9.5; **depth:** <1.1 m; **life-form(s):** emergent herb
Key morphology: rhizome erect, branching, bearing up to 9 expanded leaves; leaf blades (to 57 cm) variable, ovoid to hastate or sagittate, pinnately veined, petiolate (to 98 cm), sheathed basally (to 65 cm), the apex rolled, digitate; inflorescences (up to 6) on elongate peduncles (to 58 cm); the spathe (to 25 cm) leaflike, its tube (to 22 cm) greenish, closed, its limb (to 5 cm) green, yellowish-green, or white, narrowly to widely open; flowers unisexual, naked, borne on a cylindric, monoecious spadix (to 24 cm), the ♂ and ♀ flowers contiguous or separated by a sterile or transitional region, the distal end sometimes sterile; upper portion of spadix with many flattened, peltate synandria, each resulting from the fused anthers (4–8) of a ♂ flower, the lower portion consisting of numerous (to 65) unilocular ♀ flowers, each surrounded by a synandrodium

of 1–5 fused staminodes; the fruits (1–3-seeded), berry-like, green, purple, or red, enclosed by the persistent, leathery base of the spathe, the peduncle recurved in fruit; seeds (to 17 mm) embedded in mucilage

Life history: duration: perennial (rhizomes); **asexual reproduction:** rhizome fragments; **pollination:** insect; **sexual condition:** monoecious; **fruit:** aggregates (berries) (common); **local dispersal:** fruits, inflorescences (water); **long-distance dispersal:** fruits (animals, water)

Imperilment: 1. *Peltandra sagittifolia* [G3/G4]; S2 (AL, GA, NC, SC); S3 (FL, MS); **2.** *P. virginica* [G5]; S1 (IA, OK); S2 (QC, VT, WV); S3 (IL, ME, ON)

Ecology: general: Both *Peltandra* species are OBL wetland indicators. Because they can develop into dense, marginal stands, the plants often create a shoreline buffer by deflecting wave energy and also serve as an excellent cover for waterfowl (Aves: Anatidae). They possess unisexual, protogynous flowers that are cross-pollinated by insects (Insecta). The seeds are not strongly dormant, but their germination can be enhanced by a period of cold stratification or physical scarification. They can germinate without pretreatment, under water, and in substrates that are nearly devoid of oxygen. The fruits (but not the seeds) are buoyant and are dispersed locally by water or over longer distances by frugivorous animals. Annual perennation occurs by means of a short, subterranean rhizome. As the rhizome decays, separation of the branches results in local asexual reproduction and the development of small clonal clusters of plants. More specific biological and ecological details are discussed in the respective treatments that follow.

4.1.4.1. *Peltandra sagittifolia* **(Michx.) Morong** inhabits baygalls, bayheads, bogs (*Sphagnum*), bottomlands (stream), brooks, depressions, ditches (railroad), floating mats (peat), floodplain wetlands, hammocks, meadows (boggy, mowed, open, swampy), pinelands (peaty), pocosins (streamhead), seeps, spillways (wooded), swales, swamps (hardwood, roadside, shrub, springy), woodlands, and the margins of lakes, ponds (roadside, swampy), and streams (seepage) at elevations to 60 m. The plants usually occur under shaded exposures. The substrates (or ground layers) have been described as muck, mud, mulch (oak leaf, pine needle), organics (fibrous-brown), peat (*Sphagnum*), pine litter, Plummer (grossarenic paleaquults), and sand (boggy). The plants are found in moist soils or in shallow standing water. The habitats consistently are acidic (pH: 4.0–6.9; $\bar{x} = 4.8$). Flowering occurs from May to August, with fruiting extending from June to September. Few details are known about the reproductive biology or seed ecology of this species, especially in comparison to the next. The spathe opens widely at anthesis, which would allow access to the inner flowers by larger insects (Insecta). The flowers produce smooth pollen and reportedly are pollinated by bees (Insecta: Hymenoptera). The red fruits probably are dispersed by water or by animals. Annual perennation occurs by means of a short rhizome. **Reported associates:** *Acer rubrum, Apios americana, Arisaema triphyllum, Aronia arbutifolia, Arundinaria gigantea, Baccharis halimifolia, Balduina uniflora, Bidens laevis, Blechnum serrulatum,* *Boehmeria cylindrica, Carex atlantica, Carex comosa, Carex elliottii, Carex glaucescens, Carex intumescens, Carex vexans, Cephalanthus occidentalis, Chamaecyparis thyoides, Chasmanthium laxum, Cliftonia monophylla, Coleataenia longifolia, Cornus foemina, Cyperus tetragonus, Cyperus virens, Dichanthelium dichotomum, Drosera capillaris, Drosera intermedia, Drosera tracyi, Dryopteris ludoviciana, Dulichium arundinaceum, Echinochloa muricata, Eleocharis baldwinii, Eleocharis elongata, Encyclia tampensis* (epiphyte), *Eriocaulon decangulare, Eupatorium capillifolium, Gaylussacia mosieri, Gordonia lasianthus, Habenaria floribunda, Hydrocotyle umbellata, Hydrocotyle verticillata, Hypericum brachyphyllum, Hypericum fasciculatum, Hypericum hypericoides, Hypericum mutilum, Ilex cassine, Ilex coriacea, Ilex vomitoria, Itea virginica, Juncus canadensis, Juncus marginatus, Juncus polycephalus, Juncus tenuis, Juncus trigonocarpus, Lachnanthes caroliniana, Lachnocaulon beyrichianum, Lilium iridollae, Lindera subcoriacea, Liquidambar styraciflua, Lophiola aurea, Ludwigia maritima, Lycopodiella alopecuroides, Lyonia lucida, Magnolia virginiana, Megathyrsus maximus, Mikania scandens, Myrica cerifera, Nephrolepis cordifolia, Nephrolepis exaltata, Nuphar advena, Nymphaea odorata, Nyssa sylvatica, Oplismenus hirtellus, Orontium aquaticum, Osmunda regalis, Osmundastrum cinnamomeum, Panicum hemitomon, Peltandra virginica, Persea palustris, Persicaria hydropiperoides, Persicaria punctata, Pieris phillyreifolia, Pinguicula primuliflora, Pinus elliottii, Platanthera blephariglottis, Platanthera cristata, Polygala brevifolia, Polygala hookeri, Quercus laurifolia, Quercus nigra, Rhapidophyllum hystrix, Rhexia petiolata, Rhododendron viscosum, Rhynchospora capitellata, Rhynchospora cephalantha, Rhynchospora chalarocephala, Rhynchospora corniculata, Rhynchospora curtissii, Rhynchospora fascicularis, Rhynchospora filifolia, Rhynchospora glomerata, Rhynchospora gracilenta, Rhynchospora macra, Rhynchospora miliacea, Rhynchospora rariflora, Rhynchospora stenophylla, Sabal palmetto, Sacciolepis striata, Sambucus nigra, Sarracenia leucophylla, Sarracenia minor, Sarracenia psittacina, Sarracenia purpurea, Sarracenia rubra, Saururus cernuus, Schoenoplectus etuberculatus, Scirpus cyperinus, Scirpus cyperinus, Scleria muehlenbergii, Scleria triglomerata, Setaria magna, Smilax laurifolia, Smilax smallii, Sparganium americanum, Sphagnum, Strophostyles helvola, Taxodium ascendens, Thelypteris interrupta, Thelypteris kunthii, Tillandsia setacea* (epiphyte), *Tillandsia simulata* (epiphyte), *Tillandsia usneoides* (epiphyte), *Tillandsia utriculata* (epiphyte), *Tripsacum dactyloides, Ulmus americana, Utricularia cornuta, Utricularia gibba, Utricularia subulata, Vaccinium corymbosum, Viburnum nudum, Vitis rotundifolia, Websteria confervoides, Woodwardia areolata, Woodwardia virginica, Xyris fimbriata, Xyris smalliana.*

4.1.4.2. *Peltandra virginica* **(L.) Schott** occurs in fresh to brackish waters (salinity: 0–2‰) associated with "batteries" (floating peat), bayheads, bayous, beaches, bogs, bottomlands (alluvial, swampy), bottoms (river, stream), canals (drainage), channels (stream), depressions (swampy), ditches (drainage, roadside), fens, floating mats, floodplains, hammocks (hydric), impoundments, levees (bayou), marshes (brackish, freshwater, intertidal, roadside, tidal), mats (*Sphagnum*), meadows, oxbows (spring-fed), pocosins, pools (floodplain, marshy), ponds (beaver), prairies (wet), rice fields (abandoned), seeps (hillside), sloughs (roadside), springheads, swales, swamps (basin, beaver, cypress, deciduous, dome, floodplain, sinkhole), woodlands (floodplain, pine), and the margins or shores of bay-galls, bayous, lakes, ponds (artificial, beaver, tidal), rivers, and streams (blackwater, flowing, shallow), at elevations to 1168 m. Exposures can range from full sunlight to deep shade. The substrates include clay (heavy), clay loam, clayey sand, gravel, Iredell/Elbert series, loamy sand (Plummer), muck (alluvial, putrid), mud, peat, peaty muck, sand, silt, silty loam (Carecay), silty mud, and silty sand. The habitats are characterized by saturated or shallow standing waters (0.1–1.1 m deep), which span a broad range of conditions (e.g.: pH: 5.0–9.6; $\bar{x} = 6.5$–7.1; total alkalinity [as $CaCO_3$]: 1.3–110.8; $\bar{x} = 34.6$ mg l^{-1}; conductivity [@ 25°C]: 35–347 μS cm^{-1}; $\bar{x} = 153$ μS cm^{-1}; Ca: 0.9–73.3 mg l^{-1}; $\bar{x} = 26.3$ mg l^{-1}; Cl: 6–37 mg l^{-1}; $\bar{x} = 22.2$ mg l^{-1}; Fe: 0.1–0.5 mg l^{-1}; $\bar{x} = 0.2$ mg l^{-1}; K: 0.2–5.1 mg l^{-1}; $\bar{x} = 2.2$ mg l^{-1}; Mg: 0.8–32. mg l^{-1}; $\bar{x} = 12.0$ mg l^{-1}; Na: 3.4–20.4 mg l^{-1}; $\bar{x} = 12.5$ mg l^{-1}; Si: 0.3–5.0 mg l^{-1}; $\bar{x} = 2.3$ mg l^{-1}; SO_4^{2-}: 3–26.4 mg l^{-1}; $\bar{x} = 13.0$ mg l^{-1}; total N: 300–2010 μg l^{-1}; $\bar{x} = 1050$ μg l^{-1}; total P: 10–105 μg l^{-1}; $\bar{x} = 41$ μg l^{-1}). Although often found growing in brackish conditions, the belowground biomass production ceases and aboveground biomass declines at salinity levels of 2–4‰ or higher; most sites are freshwater. The complex reproductive biology has been studied in detail. Floral buds are initiated approximately 20 months before they become reproductively functional. Flowering occurs from April to August with the fruits ripening from June to November. The unisexual flowers are spatially separated (i.e., herkogamous) and are strongly protogynous within the inflorescence. Although successful hand pollinations have demonstrated that within-spadix geitonogamy is possible, unmanipulated inflorescences are incapable of self-pollination. Thus, sexual reproduction relies on insect pollinators, which must reach the receptive pistillate flowers while they are completely enclosed by the spathe. Pollination is oligolectic. The primary pollinators are tiny chloropid flies (Insecta: Diptera: Chloropidae: *Elachiptera formosa*), which are attracted to the inflorescence by a floral odor that emanates from the spathe at the onset of anthesis. The fragrance is composed mainly (70%) of a novel nonane compound (1,3,6-trimethyl-2,5-dioxabicyclo [3.2.1.] nonane), which arguably provides the primary cue for pollination. These flies have a similar distribution to *P. virginica* and are not known to associate with any other plant species. Initially, the spathe limb opens slightly from the base as the undulate, overlapping margins spread apart. The small opening prevents all but very small insects from entering the interior where the floral organs reside. Inside, the flowers are in the pistillate phase as the chloropid flies enter the floral chamber on the first day of anthesis. This phase is rather short and ends by the second day as the stigmas wither prior to anther dehiscence and subsequent pollen release. Meanwhile, in the

process of mating, the flies effect outcrossing by depositing any attached conspecific pollen onto the receptive stigmas. Fruit flies (Insecta: Diptera: Drosophilidae: *Drosophila subpalustris*) and syrphid flies (Insecta: Diptera: Syrphidae: *Helophilus*) also have been found in the inflorescences and are thought to be infrequent, incidental pollinators. During the staminate phase of the second day, the pollen is shed in confluent masses, with most falling into the lower part of the spathe. The flies mating in that region acquire a fresh load of pollen (eating some in the process), which they ultimately will transport to a different genet. They also oviposit within the enclosed floral chamber. As the larvae hatch, they eat some of the pollen for nourishment and eventually provide a continuous source of adult flies, thereby ensuring the perpetuation of the reproductive process. Plant fecundity is uncompromised by the resident pollinators because an excess of pollen is produced and neither the larvae nor the adult flies feed on the ovules. Each inflorescence produces 45 fruits on average. The staminate portion of the spadix and distal part of the spathe deteriorate once pollination occurs. As the fruits develop, the peduncles reflex to submerse the developing infructescences in the surrounding water where they mature. Despite this intricate reproductive mechanism, average field seed production can be relatively low (~25 m^{-2}). The plants disperse almost entirely by their buoyant fruits, which are transported on the water surface either while still within the adhering spathe or individually. The seeds are not buoyant, sink quickly if released from the fleshy, mucilaginous pericarp, and have not been recovered from drift samples taken within the water column. A pervasive seed bank exists in many sites, where the propagules can represent more than 7% of the total seed population; however, the seeds do not persist beyond 1 year. Germination occurs in the field as temperatures approach 10°C with the highest percent seed germination (53–63%) observed during late April; however, some seeds germinate in late summer while still attached to the parental plant upon which they were produced. The seeds are not strongly dormant but generally will not germinate at temperatures at or below 5°C; however, occasional germination has occurred among seeds stored at 5°C–7°C for 6 months. The highest germination rates (95%) have been achieved using an incubation temperature of 24°C. Seed germination can be enhanced by cold stratification (e.g., 80–100% germination at room temperature when stored at 5°C for 1–4 weeks). Seed germination also is enhanced by removal of the pericarp and the enveloping mucilage (83.5% after 15 days), with even higher rates (96.7%) achieved by also removing the seed coat from the radicle end. Seeds of intact fruits stratified at 5°C for 2–4 weeks have exhibited 90–100% germination. Seed germination is reduced substantially at burial depths >2 cm. Seeds within intact fruits can withstand a short period of drying (e.g., 70% germination after 14 days), whereas liberated seeds do not remain viable beyond 7 days of desiccation (<5 days if their mucilage coating also is removed). Seedling establishment is influenced negatively by water currents and is associated positively with light. However, one study found that the highest densities of established plants (e.g., 5.3 m^{-2}) occurred within pond-like settings, while maximum mean seed densities (e.g., 22.5 m^{-2}) occurred in higher marsh localities. The seedlings grow optimally at 29°C. Seedling densities were found to be much lower along stream banks or in cattail (*Typha*) stands (<10 m^{-2}) than in stands of mixed vegetation (up to 290 m^{-2}). In greenhouse studies, up to 300 seedlings m^{-2} have emerged under flooded conditions in substrate samples taken from mixed vegetation sites. However, the seeds and seedlings comparably tolerate flooded and non-flooded conditions. There is no efficient means of vegetative reproduction beyond the local dismemberment of the rhizome branches that occurs by decay of the main rhizome axis. A significantly higher proportion of the total biomass is allocated to belowground structures compared to other wetland perennials, especially when the plants are exposed to stress. Yet, aboveground biomass is high (e.g., up to 1547 g m^{-2}; $\bar{x} = 423.40$ g m^{-2}) and can represent more than half (e.g., 55%) of the total community productivity. The plants facilitate the release of methane from the substrate (to 670 mg CH$_4$ m^{-2} day^{-1}). The leaves decompose rapidly, with less than 25% of their biomass (ash-free DM) remaining after 40 days.

Reported associates: *Acer rubrum, Acer saccharinum, Acorus calamus, Aeschynomene virginica, Aldrovanda vesiculosa, Alisma subcordatum, Alnus, Alnus serrulata, Alternanthera philoxeroides, Amaranthus cannabinus, Amorpha fruticosa, Apios americana, Arisaema triphyllum, Arisaema triphyllum, Aristida, Arundinaria gigantea, Asclepias incarnata, Asclepias perennis, Athyrium, Athyrium filix-femina, Baccharis halimifolia, Betula alleghaniensis, Betula populifolia, Bidens cernuus, Bidens laevis, Bidens mitis, Bidens tripartitus, Bidens vulgatus, Bignonia capreolata, Blechnum serrulatum, Boehmeria cylindrica, Bolboschoenus fluviatilis, Brunnichia ovata, Calamagrostis, Calla palustris, Calliergonella, Calopogon, Caltha palustris, Calycanthus floridus, Campylium stellatum, Cardamine bulbosa, Carex alata, Carex albolutescens, Carex comosa, Carex conjuncta, Carex corrugata, Carex crinita, Carex crus-corvi, Carex debilis, Carex decomposita, Carex glaucescens, Carex granularis, Carex grayi, Carex hyalinolepis, Carex intumescens, Carex lasiocarpa, Carex lonchocarpa, Carex louisianica, Carex lupuliformis, Carex lupulina, Carex lurida, Carex muskingumensis, Carex radiata, Carex seorsa, Carex tribuloides, Carex typhina, Carex utriculata, Carex vexans, Carpinus caroliniana, Carya laciniosa, Celtis occidentalis, Cephalanthus occidentalis, Chamaecyparis thyoides, Chamaedaphne calyculata, Chasmanthium laxum, Chelone glabra, Cicuta bulbifera, Cinna arundinacea, Cladium mariscoides, Claytonia, Clematis virginiana, Clethra alnifolia, Coleataenia longifolia, Cornus drummondii, Cornus florida, Cornus foemina, Cornus foemina, Cornus sericea, Cuscuta gronovii, Cyperus tetragonus, Cyperus virens, Cyrilla racemiflora, Dasiphora floribunda, Decodon verticillatus, Dichanthelium boscii, Dichanthelium dichotomum, Diospyros virginiana, Drosera intermedia, Drosera rotundifolia, Dryopteris ludoviciana, Echinochloa muricata, Eleocharis baldwinii, Eleocharis elongata, Eleocharis equisetoides, Eleocharis palustris, Encyclia tampensis (epiphyte), Equisetum fluviatile, Eriocaulon compressum, Eriophorum virginicum, Eryngium aquaticum, Eubotrys racemosa, Euonymus americanus, Eupatorium capillifolium, Eupatorium perfoliatum, Eupatorium serotinum, Eutrochium fistulosum, Fraxinus caroliniana, Fraxinus nigra, Fraxinus pennsylvanica, Fraxinus profunda, Fuirena, Galium obtusum, Galium trifidum, Gaylussacia frondosa, Glyceria, Glyceria septentrionalis, Gordonia lasianthus, Habenaria floribunda, Hackelia virginiana, Helianthus tuberosus, Hibiscus laevis, Hibiscus moscheutos, Humulus lupulus, Hydrocotyle umbellata, Hydrocotyle verticillata, Hymenocallis, Hypericum hypericoides, Hypericum mutilum, Ilex cassine, Ilex decidua, Ilex laevigata, Ilex mucronata, Impatiens capensis, Iris versicolor, Iris virginica, Itea virginica, Juncus acuminatus, Juncus effusus, Juniperus virginiana, Kalmia angustifolia, Kalmia polifolia, Larix laricina, Leersia lenticularis, Leersia oryzoides, Leersia virginica, Lemna trisulca, Leucothoe axillaris, Limnobium spongia, Liquidambar styraciflua, Lobelia cardinalis, Ludwigia linearis, Ludwigia palustris, Lycopus americanus, Lycopus rubellus, Lycopus uniflorus, Lyonia ligustrina, Lysimachia ciliata, Lysimachia nummularia, Lythrum salicaria, Magnolia grandiflora, Magnolia virginiana, Megathyrsus maximus, Microstegium vimineum, Mikania scandens, Mimulus glabratus, Mitchella repens, Morus rubra, Murdannia, Myrica cerifera, Myrica gale, Nephrolepis cordifolia, Nephrolepis exaltata, Nuphar advena, Nymphaea odorata, Nyssa aquatica, Nyssa biflora, Nyssa sylvatica, Onoclea sensibilis, Oplismenus hirtellus, Orontium aquaticum, Osmunda regalis, Osmundastrum cinnamomeum, Oxypolis rigidior, Packera glabella, Panicum hemitomon, Passiflora, Peltandra sagittifolia, Persea palustris, Persicaria arifolia, Persicaria hydropiperoides, Persicaria punctata, Persicaria sagittata, Persicaria virginiana, Phalaris arundinacea, Phragmites australis, Physostegia leptophylla, Picea mariana, Pilea pumila, Pinus elliottii, Pinus strobus, Platanus occidentalis, Pluchea odorata, Poa nemoralis, Polygala, Pontederia cordata, Populus heterophylla, Proserpinaca palustris, Pteridium aquilinum, Quercus alba, Quercus bicolor, Quercus laurifolia, Quercus lyrata, Quercus nigra, Quercus rubra, Ranunculus flabellaris, Ranunculus hispidus, Ranunculus pensylvanicus, Rhapidophyllum hystrix, Rhododendron maximum, Rhynchospora alba, Rhynchospora chalarocephala, Rhynchospora corniculata, Rhynchospora fascicularis, Rhynchospora inundata, Rhynchospora miliacea, Ribes americanum, Riccia fluitans, Robinia pseudoacacia, Rosa palustris, Rosa setigera, Rubus, Rudbeckia laciniata, Ruellia, Rumex britannica, Rumex verticillatus, Rumex verticillatus, Sabal palmetto, Sabatia calycina, Sacciolepis striata, Sagittaria brevirostra, Sagittaria graminea, Sagittaria lancifolia, Sagittaria latifolia, Salix caroliniana, Salix nigra, Sambucus nigra, Samolus valerandi, Sarracenia purpurea, Sassafras albidum, Sassafras albidum, Saururus cernuus, Schoenoplectus americanus, Schoenoplectus californicus, Schoenoplectus tabernaemontani, Scirpus cyperinus, Scleria muehlenbergii, Scleria triglomerata, Scutellaria lateriflora, Setaria magna, Silphium perfoliatum, Smilax,*

Solidago patula, Sparganium americanum, Sparganium eurycarpum, Spartina alterniflora, Spartina cynosuroides, Sphagnum, Spiraea alba, Spiranthes, Spirodela, Stachys tenuifolia, Symphyotrichum lanceolatum, Symphyotrichum ontarionis, Symphyotrichum subspicatum, Symplocarpus foetidus, Taxodium distichum, Thalictrum dasycarpum, Thelypteris interrupta, Thelypteris kunthii, Thelypteris palustris, Tillandsia setacea (epiphyte),*Tillandsia simulata* (epiphyte), *Tillandsia usneoides* (epiphyte), *Tillandsia utriculata* (epiphyte), *Toxicodendron radicans, Toxicodendron vernix, Toxicodendron vernix, Triadenum virginicum, Triadenum walteri, Tripsacum dactyloides, Tsuga canadensis, Typha angustifolia, Typha latifolia, Ulmus americana, Urtica dioica, Utricularia macrorhiza, Utricularia purpurea, Vaccinium corymbosum, Vaccinium macrocarpon, Vaccinium oxycoccos, Veronica scutellata, Viburnum lentago, Viburnum nudum, Viola sororia, Vitis rotundifolia, Wolffia, Woodwardia areolata, Woodwardia virginica, Xyris smalliana, Zizania aquatica.*

Use by wildlife: *Peltandra sagittifolia* hosts several plant pathogenic fungi (Fungi) including leaf spots (Ascomycota: Glomerellaceae: *Colletotrichum*; Mycosphaerellaceae: *Cercospora callae*) and rusts (Basidiomycota: Pucciniaceae: *Uromyces caladii*). Fresh leaves of *Peltandra virginica* have a relatively high nitrogen content (25 mg of nitrogen g^{-1} of ash-free DM; 2–3% ash-free DM). The foliage and/or fruits are eaten by black bears (Mammalia: Ursidae: *Ursus americanus*), black ducks (Aves: Anatidae: *Anas rubripes*), isopods (Crustacea: Isopoda: Asellidae: *Caecidotea forbesi*), rails (Aves: Rallidae), muskrats (Mammalia: Cricetidae: *Ondatra zibethicus*), and wood ducks (Aves: Anatidae: *Aix sponsa*). The rhizomes are eaten by feral pigs (Mammalia: Suidae: *Sus scrofa*). Eastern mud turtles (Reptilia: Testudines: Kinosternidae: *Kinosternon subrubrum*) maintain their standing against extreme tidal currents by lodging themselves among stems of *P. virginica* plants that they grasp with their forearms. The plants are visited by a diverse assortment of insects (Insecta), including ants (Hymenoptera: Formicidae), bees (Hymenoptera: Apidae), beetles (Coleoptera: Chrysomelidae; Curculionidae; Coccinellidae; Erotylidae; Histeridae; Lampyridae), bugs (Hemiptera: Anthocoridae; Aphididae; Cicadellidae; Membracidae; Miridae), flies (Diptera: Chironomidae; Chloropidae: *Elachiptera formosa*; Culicidae; Dolichopodidae; Drosophilidae: *Drosophila subpalustris*; Muscidae; Otitidae; Sciaridae; Sciomyzidae; Simuliidae; Syrphidae: *Helophilus*; Tabanidae; Tachinidae; Tipulidae), and grasshoppers (Orthoptera: Acrididae). They have been implicated as providing breeding habitat for a known mosquito vector (Insecta: Diptera: Culicidae: *Coquillettidia perturbans*) of Eastern equine encephalomyelitis and West Nile virus. The foliage harbors snails (Mollusca: Gastropoda: Ancylidae: *Laevapex fuscus*) and hosts various plant pathogenic Fungi including leaf spots (Ascomycota: Amphisphaeriaceae: *Pestalotia aquatica*; Dermateaceae: *Gloeosporium paludosum*; Mycosphaerellaceae: *Cercospora callae, Ramularia*; Basidiomycota: Pucciniales: *Aecidium importatum*; Typhulaceae: *Sclerotium caladii*) and rusts (Basidiomycota: Pucciniaceae: *Uromyces ari-triphylli, U. caladii*).

Economic importance: food: The rootstocks of *Peltandra sagittifolia* were used by the Seminoles and other Native American tribes in the southeastern states, who roasted the tubers and ate them like a potato; the boiled spadix and berries were consumed as well. The Nanticoke tribe added the grated roots to milk given to their babies. The thick (to 12 cm in diameter), starchy rhizomes of *Peltandra virginica* were used for subsistence by Native tribes of the Eastern Woodlands. Their bitter properties can be eliminated by cooking them at 110°C for 9–12 hr; **medicinal:** *Peltandra virginica* contains relatively high levels of melatonin (585 ng g^{-1}). It is an ingredient of the herbal products distributed as *"Extrim plus," "Imelda perfect Slim," "Japan lingzhi 24 hr diet," "Perfect slim," "Perfect slim 5x," "Proslim plus," "Royal slimming formula," "2 Day diet," "3X Slimming power,"* and *"5X Imelda perfect slim,"* all of which have been banned by the United States Food and Drug Administration for their potentially dangerous side effects; **cultivation:** *Peltandra virginica* is distributed commercially as an ornamental pond and water garden specimen under the species name or the cultivar names 'Snow Splash' and 'Yellow Veins'; **misc. products:** Aqueous whole plant extracts of *Peltandra sagittifola* and aqueous rootstock extracts of *P. virginica* are toxic to American cockroaches (Insecta: Blattodea: Blattidae: *Periplaneta americana*). *Peltandra virginica* plants are effective at removing nitrogen and phosphorus from dairy wastewaters and other surface waters; **weeds:** Neither *Peltandra* species has been categorized as a weed; **nonindigenous species:** *Peltandra virginica* has been introduced to California and Oregon. Nonnative genotypes also have been distributed throughout its range as part of wetland restoration and mitigation programs.

Systematics: Within Araceae *Peltandra* is assigned to the small tribe *Peltandrae* along with the monotypic Malagasy genus *Typhonodorum* (*T. lindleyanum*), which also inhabits wetlands. As unusual an alliance this might seem geographically, the two genera associate as sister groups in phylogenetic analyses (e.g., Figure 1.26). As few as one and as many as eight species have been recognized in the endemic North American *Peltandra*, described often from coloration differences of the inflorescence. Several forms have been named based on the extensive degree of phenotypic plasticity exhibited by the foliage, which can readily be seen among plants growing at different depths. The two species currently accepted are distinct ecologically and morphologically but further insight regarding their taxonomic standing would benefit by a rigorous genetic evaluation. Although both *Peltandra* species have yet to be analyzed simultaneously in a phylogenetic study, a BLAST comparison in GenBank indicates their highly similar sequences for *rbcL* and *matK* regions (>99.6%). Surprisingly, each *Peltandra* species is comparably similar to *Typhonodorum lindleyanum* for the same regions (>99%). Better insight into the relationship between *P. sagittifolia* and *P. virginica* might be gained by conducting additional reproductive studies to determine whether the taxa are interfertile (no hybrid reports have been found), and whether *P.*

sagittifolia relies on the same oligolectic pollinator as *P. virginica*. The basic chromosome number of *Peltandra* is *x*=7; *P. virginica* (2*n*=112) is highly polyploid (a hexadecaploid).

Distribution: *Peltandra sagittifolia* is narrowly restricted to the coastal plain of the southeastern United States; *P. virginica* is widespread in the eastern United States and adjacent southern Canada with scattered, introduced populations on the West Coast.

References: Beal, 1977; Blackwell, Jr. & Blackwell, 1974; Blake, 1912; Boland & Burk, 1992; Braun et al., 2014; Bown, 2000; Brown et al., 2018; Chanton et al., 1992; Cordero & Swarth, 2010; Corogin & Judd, 2009; Cypert, 1972a; DeBusk et al., 1995; Doumlele, 1981; Duncan & DeGiusti, 1976; Edwards, 1933; 1934; Eleuterius & Caldwell, 1984; Gates, 1911; Godfrey & Wooten, 1979; Goldberg, 1941; Heal et al., 1950; Holt et al., 2011; Homoya & Hedge, 1982; Hopfensperger & Baldwin, 2009; Hoyer et al., 1996; Koca Çalişkan et al., 2017; Landers et al., 1979; Leck & Graveline, 1979; Leck & Simpson, 1987; 1995; Lynn, 1984; Mabbott, 1920; Mayo et al., 1998a; McAtee, 1913; Messner & Schindler, 2010; Morris et al., 1990; Motzkin, 1994; Odum & Heywood, 1978; Odum et al., 1979; Padgett & Crow, 1994b; Patt et al., 1992; 1995; Peterson & Baldwin, 2004a; Plowman, 1969; Rejmánek & Randall, 1994; Simpson et al., 1979; Singh et al., 2012; Smock & Harlowe, 1983; Sorrie, 2000; Surdick & Jenkins, 2010; Sutter et al., 2014; Thompson, 2000b; Thompson, 2010; Tiner, 1993; Toner et al., 1995; West & Whigham, 1975; Whigham & Simpson, 1978; Whigham et al., 1979.

4.1.5. Pistia

Water lettuce; laitue d'eau

Etymology: from the Greek *pisti* ("liquid") referring to the aquatic habitat

Synonyms: *Apiospermum*; *Limnonesis*; *Zala*

Distribution: global: Africa; Eurasia*; Australia*; Central America; Mexico; North America*; South America; **North America:** southern

Diversity: global: 1 species; **North America:** 1 species

Indicators (USA): obligate wetland (OBL): *Pistia stratiotes*

Habitat: freshwater, freshwater (tidal); lacustrine, palustrine, riverine; **pH:** 5.4–8.8; **depth:** 0 m; **life-form(s):** free-floating

Key morphology: roots (to 50 cm) highly branching, feathery, root-cap prominent, removeable; shoots floating, compressed, producing elongate stolons from which terminal plantlets develop; leaves aerial or floating, arranged in a dense, circular, "lettuce-like" rosette, the blades (to 20 cm) broadly cuneate, succulent, spongy, densely pubescent, tapering to base, the veins (to 15) in prominent furrows, the apex nearly truncate, sometimes notched medially; spathe (to 4 cm) convolute, constricted medially, greenish to white or translucent, the margins densely ciliate, fused to base of spadix; spadix somewhat clavate, monoecious, the flowers naked (the perianths modified as perigonial structures), ♂ above, ♀ below; ♂ flowers (2–8) arranged in a single verticel of paired stamens, subtended by a circular or crenulate perigonium; the ♀ flower solitary, its ovary fused to base of spadix, the style short (to 3 mm), terminated by a papillate stigma, the stigma shielded

apically by a membranous perigonial flap; ovules numerous (to 30+); fruits greenish to brown, many-seeded (4–30); seeds (~1 mm) truncate

Life history: duration: perennial (stolons); **asexual reproduction:** stolons, plantlets; **pollination:** insect or insect-mediated geitonogamy; **sexual condition:** monoecious; **fruit:** berries (uncommon); **local dispersal:** plantlets, stolons (water); **long-distance dispersal:** whole plants (humans), seeds (birds)

Imperilment: 1. *Pistia stratiotes* [G5]

Ecology: general: *Pistia* is monotypic (see next).

4.1.5.1. ***Pistia stratiotes* L.** is a nonindigenous species, which grows on the water surfaces of bayous, bays (lagoon), canals (oil-field, roadside), channels (bayou, river), coves (reservoir), ditches (drainage, irrigation, roadside), floodplains, lagoons (shaded), lakes (oxbow), marshes, ponds (fish production, lily, retention), pools, reservoirs, rivers, sloughs (roadside), springs, streams (blackwater, spring-fed), and swamps (hardwood, roadside), at elevations to 2377 m. The plants occur in exposures of full sunlight to partial shade. They are most common in warm, alkaline, hard water, eutrophic lakes. Despite their free-floating habit, the plants can survive for up to 4 months when stranded on wet substrates (e.g., drying mud). The floating habit also provides a competitive advantage over submersed hydrophytes because the plants are not restricted by water depth or clarity (e.g., can grow at Secchi depth of 0.4). Consequently, they can tolerate a wide range of conditions (pH: 5.4–8.8; $\bar{x}$=7.3; alkalinity [as $CaCO_3$]: 2.2–98.4 mg l^{-1}; $\bar{x}$=33.8 mg l^{-1}; conductivity [@ 25°C]: 58–519 µS cm^{-1}; $\bar{x}$=214 µS cm^{-1}; total N: 280–1960 µg l^{-1}; $\bar{x}$=1030 µg l^{-1}; total P: 10–129 µg l^{-1}; $\bar{x}$=68 µg l^{-1}; Ca: 3.6–90.6 mg l^{-1}; $\bar{x}$=30.4 mg l^{-1}; Cl: 8.4–129.0 mg l^{-1}; $\bar{x}$=38.0 mg l^{-1}; Fe: 0–0.4 mg l^{-1}; $\bar{x}$=0.2 mg l^{-1}; K: 0.2–8.8 mg l^{-1}; $\bar{x}$=3.3 mg l^{-1}; Mg: 2.2–40.2 mg l^{-1}; $\bar{x}$=15.0 mg l^{-1}; Na: 5.0–71.5 mg l^{-1}; $\bar{x}$=21.6 mg l^{-1}; Si: 0.2–4.8 mg l^{-1}; $\bar{x}$=2.1 mg l^{-1}; SO_4^{2-}: 3.5–50.6 mg l^{-1}; $\bar{x}$=20.3 mg l^{-1}). Adult plants are not frost-tolerant but can maintain growth at temperatures from 15°C to 35°C; they grow optimally from 22°C to 30°C. Growth occurs only at pH >4.0 and is optimal near pH 7.0. The plants are intolerant of salinity and succumb to salt concentrations above 1.7‰. Although North American plants can flower continuously throughout the year, their mechanism of seed production remains a mystery. The floral sex ratio within the inflorescences is male biased (4:1 ♂:♀). Sexual reproduction occurs commonly in their native range with the plants producing from 4 to 30 seeds per fruit. Yet, to date, no published study has elucidated the pollination method of *Pistia*, which is generally believed to be carried out by insects (Insecta). Because flowers within an inflorescence are cross-compatible, it has been suspected that "self-pollination" (i.e., geitonogamy) also might occur. If so, then it probably is insect-mediated as a means of delivering pollen to the shielded stigma (see *Key morphology* above). The discovery of a new ceratopogonid fly species (Insecta: Diptera: Ceratopogonidae Forcipomyiinae: *Forcipomyia dolichopodida*) on *Pistia* plants is intriguing in this respect, given the known role of this genus as a pollinator of several other plants. However, there is no evidence of a similar role with *Pistia*. Initially, it was thought that *Pistia* could not produce seeds in North America because an appropriate pollinator was lacking. This widely held supposition was contradicted when a Florida population was found to produce prodigious quantities of highly viable seed ($\bar{x} = 726$ seeds m^{-2}; viability = 80%). The seeds can float where they are dispersed on the water surface, or sink, where they become lodged within the fibrous root system or settle to the bottom. Sediments from the fertile Florida locality also contained dense seed deposits (4,196 seeds m^{-2}). Viable seed production had already been documented in an adventive population in The Netherlands and was reported subsequently in nonindigenous populations from Australia, Germany, Japan, and Slovenia. Fecundity was high in the nonindigenous German population, where the seeds retained good viability (36–40% germination) after being stratified at 5°C for 2–10 weeks. Submerged seeds from introduced Slovenian plants germinated within a month when incubated at 20°C under an 8 hr/16 hr light/dark regime. Seeds collected from the sediment in Japan also were viable. Together, these studies indicate: The seeds of temperate plants remain viable after 2 months of storage at 4°C, or when stored in ice for several weeks. Submersed seeds require warm stratification; they remain dormant when dry, or at temperatures below 20°C, but germinate optimally in the light (red light or white light) at 20°C–25°C after a 6-week after-ripening period. Dark conditions (e.g., deep water) or far red light will suspend or reduce germination and increase germination times. From these reports, it is apparent that nonindigenous temperate populations not only are capable of seed production, but that the seeds also are adapted for surviving through winter in deeper bottom waters, where temperatures are maintained near 4°C. Several studies have confirmed the existence of viable seed banks in temperate regions. However, there are yet to be confirmed reports of seedling establishment in any nonindigenous temperate locality. The plants reproduce vegetatively by prolific production of stolons, which develop terminally into new plantlets. As the stolons dissociate, the ramets (plantlets) are spread along the water surface by wind and water currents. Large numbers (1,177–1,826) of these vegetative propagules have been recovered in studies of nonindigenous German populations. Because of the functional size of the vegetative propagules (whole plants), transport between water bodies relies either on the physical movement of plants (primarily by humans via boating equipment, etc.) or on the dispersal of the much smaller fruits or seeds. Details regarding the efficacy of the latter means for long-distance dispersal are not adequately understood, but the fruits and seeds are thought to be transported by their attachment to water birds (Aves). The maintenance of clonal integration enables the smaller lateral plantlets to compete more effectively with larger floating species such as the water hyacinth (*Eichhornia crassipes*). Total biomass in tropical populations can reach 1088 g m^{-2} and annual production rates up to 108.8 t DM ha^{-1}. Cultured plants have attained a similar maximum DM biomass of 1006 g m^{-2}. Resource allocation patterns can vary widely among sites (e.g., 20–69% shoot:root biomass) depending on water

fertility and other associated factors. The plants have been shown to be capable of storing up to 237.75 g m^{-2} of carbon. However, they can decompose rapidly with a mean biomass loss of 1,023.4 g DM m^{-2} recorded after 185 days of decomposition. **Reported associates (North America):** *Acer rubrum, Acrostichum danaeifolium, Alternanthera philoxeroides, Annona glabra, Azolla cristata, Azolla filiculoides, Baccharis halimifolia, Bacopa, Bidens, Boehmeria cylindrica, Cabomba caroliniana, Cenchrus echinatus, Cephalanthus occidentalis, Ceratophyllum demersum, Ceratopteris thalictroides, Chloracantha spinosa, Cladium jamaicense, Colocasia esculenta, Cornus foemina, Cynodon dactylon, Cyperus ligularis, Cyperus ochraceus, Cyperus polystachyos, Cyperus virens, Diospyros virginiana, Echinochloa walteri, Echinodorus cordifolius, Egeria densa, Eichhornia crassipes, Eupatorium serotinum, Forestiera acuminata, Fraxinus pennsylvanica, Gleditsia aquatica, Habenaria repens, Heteranthera dubia, Hibiscus laevis, Hydrilla verticillata, Hydrocotyle ranunculoides, Hygrophila lacustris, Hygrophila polysperma, Hymenocallis, Ilex cassine, Ipomoea alba, Lantana camara, Lemna minor, Limnobium spongia, Ludwigia grandiflora, Ludwigia leptocarpa, Ludwigia peruviana, Ludwigia repens, Megathyrsus maximus, Mikania scandens, Myrica cerifera, Myriophyllum aquaticum, Myriophyllum spicatum, Najas, Nelumbo lutea, Nuphar, Nymphaea odorata, Oenothera simulans, Oxycaryum cubense, Panicum hemitomon, Panicum repens, Paspalum distichum, Paspalum notatum, Persea palustris, Persicaria hydropiperoides, Persicaria pensylvanica, Persicaria punctata, Phalaris arundinacea, Phanopyrum gymnocarpon, Phragmites australis, Phyla nodiflora, Pontederia cordata, Potamogeton illinoensis, Potamogeton nodosus, Sagittaria platyphylla, Salix nigra, Salvinia minima, Sesbania drummondii, Sesbania herbacea, Setaria magna, Spirodela polyrhiza, Symphyotrichum, Taxodium distichum, Triadica sebifera, Typha angustifolia, Typha domingensis, Urena lobata, Utricularia macrorhiza, Vallisneria americana, Vigna luteola, Wolffia, Zizania texana, Zizaniopsis miliacea.*

Use by wildlife (North America): Surveys in Florida have collected numerous invertebrates from *Pistia* stands, including more than 300 species of insects (e.g., Insecta: Diptera: Chironomidae, Culicidae: *Anopheles, Mansonia* [vectors of various diseases]; Coleoptera; Hemiptera; Hymenoptera: Braconidae; Odonata). The *Mansonia* larvae pierce the aerenchymatous roots and foliage to obtain air. Several phytophagous insect (Insecta) species use the plants as a food source including aphids (Hemiptera: Aphididae: *Rhopalosiphum nymphaeae*), leafhoppers (Hemiptera: Cicadellidae: *Draeculacephala inscripta*), moths (Lepidoptera: Crambidae: *Petrophila drumalis, Samea multiplicalis, Synclita obliteralis*), and weevils (Coleoptera: Erirhinidae: *Tanysphyrus lemnae*). The plants release allelopathic chemicals, which inhibit the growth of some Cyanobacteria (Chroococcaceae: *Microcystis aeruginosa*).

Economic importance: food: Cooked *Pistia* plants have been eaten in India as a famine food; **medicinal:** Worldwide, *Pistia stratiotes* has been used to treat numerous ailments.

It is regarded as an effective antioxidant, bronchodilator, diuretic, and emollient, and as having antidiabetic, antimicrobial, antifungal, antiprotease, antitumor, and diuretic properties. Various leaf extracts are thought to lessen cellular injury by reducing levels of harmful superoxides, nitric oxide radicals, and free radicals. Ethanolic extracts prevent uric acid formation by inhibiting xanthine oxidase (a strategy used to treat gout). The extracts also are antipyretic and can alleviate fevers. The foliage is used as a disinfectant and in various remedies to treat dysentery, eczema, inflammation, leprosy, parasites, piles, syphilis, tuberculosis, and ulcers. Ash made from the leaves is used as a cure for ringworm; **cultivation:** *Pistia stratiotes* is one of the most widely distributed ornamental pond and water garden plants worldwide. The typical species and a variegated form are distributed commercially; **misc. products:** *Pistia* plants have been used to synthesize gold nanoparticles, to generate biogas, and as a bio-sorbent for removal of crude oil from saltwater and metals in industrial applications. They also have been evaluated as a means of removing heavy metals, pharmaceuticals, personal care product residues, and radioisotopes from natural and wastewaters; **weeds:** *Pistia stratiotes* is regarded as a noxious weed in North America and throughout its nonindigenous (and even native) range elsewhere. The state of Florida alone expends several million dollars annually on control measures. The recent discovery of *Pistia* in more northern localities (e.g., Great Lakes) has raised question whether the plants might be able to persist at higher latitudes as the global climate continues to warm; however, such seems unlikely for these surface plants unless abnormally warm, ice-free waters become available throughout the winter; **nonindigenous species:** *Pistia stratiotes* presumably is nonindigenous to North America although impassioned arguments have been made for its indigenous status, based primarily on fossil evidence. In any case, aggressive Florida populations already were present in the St. John's River and Suwanee River before 1765. The species has been disseminated widely across the globe due to the continuing sales and careless disposal of cultivated water garden specimens.

Systematics: Although several species of *Pistia* have been proposed, the genus currently is recognized as monotypic throughout its global distributional range. From phylogenetic studies, the genus has become the namesake of the "pistia clade," which contains some of the most familiar Araceae (including Tribe Areae). The name reflects the position of the genus as an isolated sister group to the remainder of the clade in many (but not all – see Figure 1.26) molecular phylogenetic studies. The basic chromosome number of *Pistia* is $x = 7$. *Pistia stratiotes* ($2n = 28$) is a diploid. Given its isolated, monotypic status, there are no reported hybrids that involve *Pistia*.

Distribution: The pantropical *Pistia stratiotes* has become established in the southern portions of the United States and is adventive northward, where it remains uncertain whether those populations can persist in the colder climate.

References: Adebayo et al., 2011; Anuradha et al., 2015; Baskin & Baskin, 1998; Bogner, 2009; Chan & Linley, 1989; Cook, 1996; 2004; Dray, Jr. & Center, 1989; Dray, Jr. et al., 1988; 1993; Eid et al., 2016; Escher & Lounibos, 1993; Evans, 2013; Gibemau, 2011; Godfrey & Wooten, 1979; Hall & Okali, 1974; Haller et al., 1974; Harley, 1990; Heidbüchel et al., 2016; Henry-Silva et al., 2008; Humaida et al., 2016; Hussner et al., 2014; Kan & Song, 2008; Khan et al., 2014; Klotzsch, 1852; Kurugundla, 2014; Lin & Li, 2016; Mayo et al., 1998a; Neuenschwander et al., 2009; Owens et al., 2001; Pieterse et al., 1981; Šajna et al., 2007; Sánchez-Galván et al., 2013; Sharma, 1984; Stuckey & Les, 1984; Sutherland, 1986; Tamada et al., 2015; Thompson, 2000b; Wang et al., 2016; Wu et al., 2015; Zhou et al., 2018.

4.1.6. Zantedeschia

African lily, arum-lily, calla-lily, Egyptian lily, lily of the Nile, trumpet lily; arum d'Éthiopie, calla d'Éthiopie, lis du Nil

Etymology: after Giovanni Zantedeschi (1773–1846)

Synonyms: *Arodes*; *Calla* (in part); *Otosma*; *Pseudohomalomena*; *Richardia* (in part)

Distribution: global: South Africa; **North America:** western United States

Diversity: global: 6–8 species; **North America:** 1 species

Indicators (USA): obligate wetland (OBL): *Zantedeschia aethiopica*

Habitat: freshwater; palustrine; **pH:** alkaline; **depth:** <1 m; **life-form(s):** emergent herb

Key morphology: plants to 2.5 m (usually <60 cm) rhizomatous, the rhizome fleshy, producing up to 55 offsets; leaves evergreen, the blades (to 20 cm) hastate to broadly ovate-cordate, the petioles sheathing at base; peduncle (to 60 cm) erect, triangular; spathe (to 23 cm) ivory-white or creamy inside, flaring apically into a funnelform limb, the apex with a recurved tip (to 2 cm); spadix sessile, the upper portion (to 7 cm) with numerous ♂ flowers, each comprising 1–3 sessile anthers (to 2 mm) (to 1,495 total anthers); lower spathe (to 1.8 cm) with 48–97 ♀ flowers, each subtended by 3 spatulate staminodes, ovary (to 4 mm) 3-locular with 3–12 ovules, style short (to 1.5 mm); fruits (to 1.2 cm) numerous, berry-like, turning orange when ripe; seeds (to 12) leathery, roundish

Life history: duration: perennial (rhizomes, whole plants); **asexual reproduction:** rhizomes; **pollination:** insect; **sexual condition:** monoecious; **fruit:** aggregates of berries (common); **local dispersal:** fruits (gravity, water); **long-distance dispersal:** fruits/seeds (animals)

Imperilment: 1. *Zantedeschia aethiopica* [GNR]

Ecology: general: *Zantedeschia* species occupy a variety of habitats ranging from mountainous rocky, grassy hillsides, and semi-shaded forest margins to shady streamsides and damp depressions. All the species perennate from a rhizome. One species is evergreen. Within an inflorescence, the unisexual flowers are either self-incompatible or self-compatible, with the latter capable of geitonogamous self-pollination (at least if facilitated artificially). All the species primarily are outcrossed by insects. The fruits are dispersed locally by gravity or water, or more broadly by animal vectors.

4.1.6.1. *Zantedeschia aethiopica* **(L.) Spreng.** occurs in depressions, floodplains, marshes, roadsides, swales, thickets, woodlands (pine), and along the margins of ditches, rivers, and streams at elevations to 184 m. In North America the plants occur most commonly in shaded exposures on wet substrates described as sandy gravel and sandy loam. The plants are not salt-tolerant and high salinity significantly reduces their growth as well as reducing and delaying seed germination. They are not frost-hardy but can survive brief exposures to temperatures as low as 1°C. The plants are in flower (California) from March to June. Prior to anthesis the spathe is wrapped tightly around the spadix. The ovary:anther ratio of the inflorescence can vary from 1:6 to 1:24. The flowers are self-incompatible (within an inflorescence) and strongly protogynous. Even hand-manipulated geitonogamous pollinations fail to set seed, resulting in an obligately outcrossing breeding system. As the spathe unfurls, the stigmas of the pistillate flowers become receptive while the anthers remain immature. A mild scent is emitted in the vicinity of the pistillate flowers, most likely from osmophores in the staminodes. The staminodes also contain tannins, which are thought to dissuade herbivores from feeding on the maturing ovaries. After 6 days, the stigmas degenerate and all anthers in the inflorescence dehisce simultaneously to shed long, fine viscous strands of psilate pollen, which can adhere to the bodies of visiting insects. The principal native pollinators are scarab beetles (Insecta: Coleoptera: Scarabaeidae), which inadvertently collect the sticky pollen as they mate in the lower portion of the inflorescence. Incidental floral visitors like arachnids (Arthropoda: Arachnida), bees (Insecta: Hymenoptera), and flies (Insecta: Diptera) potentially function as accidental pollinators as they seek refuge or prey within the inflorescence. Pollinators have not been reported in North America. Developing fruits mature in about 30 days while the peduncle remains upright. During this time the apical portion of the spathe is shed (to expose the fruits), while the lower part is retained and becomes photosynthetic. The berries turn orange and become mucilaginous when ripe. In their indigenous habitat they are dispersed locally by water and more widely by birds (Aves) or mammals (Mammalia). Seed viability declines rapidly during dry storage and a persistent seed bank does not develop. The low optimum germination temperature range (14°C–17°C) induces seed germination during fall and winter. Germination rates are unaffected by light. In the laboratory, fresh, ripe seeds have germinated well (82%) at 20°C (which results in lower but more rapid germination) under a 16-hr photoperiod. In milder climates, the adult plants are evergreen and do not die back during winter. They propagate vegetatively by means of a rhizome from which numerous offshoots can arise. Healthy plantlets have been propagated artificially through tissue culture. **Reported associates (North America):** *Alisma, Alnus rubra, Artemisia californica, Baccharis pilularis, Conium maculatum, Cortaderia selloana, Eleocharis, Eriogonum fasciculatum, Geranium, Heteromeles arbutifolia, Juncus, Ludwigia peploides, Nasturtium officinale, Platanus racemosa, Quercus agrifolia, Rhus integrifolia, Rubus bifrons,* *Salix lasiandra, Sambucus nigra, Toxicodendron, Triglochin scilloides, Typha latifolia, Veronica anagallis-aquatica.*

Use by wildlife (North America): *Zantedeschia aethiopica* is a larval host of tiger moths (Insecta: Lepiodptera: Erebidae: *Pyrrharctia isabella, Spilosoma virginica*). Plants cultivated in the United States have been found to host several plant pathogenic Fungi including various blights (Basidiomycota: Typhulaceae: *Sclerotium rolfsii*; Oomycota: Pithiaceae: *Phytophthora cryptogea, P. erythroseptica*), gray mold (Ascomycota: Sclerotiniaceae: *Botrytis cinerea*), leaf rots (Oomycota: Pythiaceae: *Pythium*), leaf spots (Ascomycota: Corynesporascaceae: *Corynespora*; Dermateaceae: *Gloeosporium callae*; Didymellaceae: *Didymella glomerate, Phoma glomerata*; Glomerellaceae: *Colletotrichum gloeosporioides, Glomerella cingulata*; Mycosphaerellaceae: *Cercospora callae, C. richardiicola*; Pleosporaceae: *Alternaria*), and root rots (Basidiomycota: Physalacriaceae: *Armillaria mellea*; Oomycota: Pithiaceae: *Phytophthora richardiae*). The plants are poisonous to rabbits (Mammalia: Leporidae) and to many other small pets.

Economic importance: food: *Zantedeschia aethiopica* is poisonous and never should be eaten. The foliage and inflorescence contain various oxalate compounds and proteolytic enzymes, which can cause serious illness if ingested (see next); **medicinal:** The flowers and foliage of *Z. aethiopica* are toxic if eaten and can cause acute gastritis, diarrhea, and swelling of the mouth, which can lead to exhaustion and shock. Tannins in the leaf extracts of *Z. aethiopica* exhibit anticoagulant properties and inhibit thrombin-induced clotting. Some people have developed a contact dermatitis allergy or eczema from handling the plants. Herbal remedies made from the plant roots are used to treat infertility in South Africa. There the leaves also are applied directly to the skin as a remedy for healing boils, sores, and wounds; **cultivation:** *Zantedeschia* species are grown commercially in California and Florida to provide ornamental garden stocks. The plants have been used as water garden ornamentals for more than a century. *Zantedeschia aethiopica* is one of the most widely cultivated aroids and there are many commercially distributed cultivars including 'Angel White One', 'Apple Court Babe', 'Black-eyed Beauty', 'Caerwent', 'Childsiana', 'Compacta', 'Crowborough' (RHS Award of Garden Merit), 'Flamingo' (PBR), 'Flamingo', 'Gigantea', 'Glencoe', 'Glow', 'Goldstar', 'Green Goddess' (RHS Award of Garden Merit), 'Green Gold', 'Hercules', 'Hong Gan', 'Kiwi Blush', 'Little Gem', 'Little Suzy', 'Luzon Lovely', 'Marshmallow', 'Mr Martin', 'Mr Sam', 'Pershore Fantasia', 'Pink Mist', 'Red Desire', 'Royal Princess', 'Snow White', 'Spotted Giant', 'Tiny Tim', 'Whipped Cream', 'White Cutie', 'White Dream', 'White Gnome', 'White Mischief', 'White Sail', and numerous others; **misc. products:** *Zantedeschia aethiopica* has been evaluated for use in constructed wastewater treatment wetlands. The plants are moderately iron tolerant and have been considered for phytoremediation of metal-contaminated wetlands. They have been used to make a greenish dye in parts of Africa; **weeds:** *Zantedeschia aethiopica* is regarded as a serious weed in Australia, Indonesia, and Mexico; **nonindigenous species:**

Zantedeschia aethiopica was introduced to California and Oregon as an escape from cultivation. (The California naturalization was predicted in 2002 based on the characteristics of nonindigenous Australian plants). Naturalized plants (due to garden escapes) also occur in Australia, Brazil, Canary Islands, Central America, France, Greece, Italy, Juan Fernández Archipelago (Chile), Kashmir Himalaya, Mexico, New Zealand, Portugal, Tunisia, and Zimbabwe. Nonindigenous plants also have appeared in Indonesian forest restoration sites.

Systematics: *Zantedeschia* is a South African genus of six to eight species. Although some authors recognize a "*Zantedeschia* clade" (including the "*Philodendron* clade" along with members of tribes Aglaonemateae and Spathicarpeae), the genus resolves differently in different phylogenetic analyses (e.g., Figure 1.26) depending on the species included (i.e., *Z. aethiopica* or *Z. albomaculata*). Because *Z. aethiopica* differs the most from the other species (it has been treated as a distinct, monotypic section), additional systematic investigations of the genus are warranted. In particular, the monophyly of the group has not yet been corroborated by simultaneous phylogenetic analysis of all potentially related taxa, including species assigned currently to other genera. The basic chromosome number of *Zantedeschia* is $x = 16$; *Z. aethiopica* ($2n = 32$) is a diploid. The 2C nuclear DNA content has been determined as 3.72 ± 0.10 pg (picograms) (= 3,638.16 Mbp). Artificial tetraploids ($2n = 4x = 64$) have been induced using colchicine. Various chromosomal rearrangements isolate *Z. aethiopica* genetically from other species in the genus, precluding its parentage of viable interspecific hybrids. Microsatellite markers to facilitate genetic research have been developed from expressed sequence tags (EST-SSR markers).

Distribution: *Zantedeschia aethiopica* currently is known to be naturalized in California and Oregon.

References: Bown, 2000; Casierra-Posada et al., 2014; Chester, 1930; Cortinovis & Caloni, 2013; das Neves & Pais, 1980; de Almeida & Freitas, 2006; El-Shamy et al., 2009; Espinosa-Garcia et al., 2006; Expósito et al., 2018; Ghimire et al., 2012; Greimler et al., 2017; Hakim & Miyakawa, 2015; Henriquez et al., 2014; Jeanmonod et al., 2011; Johnson & Johnson, 2006; Kee et al., 2008; Khuroo et al., 2007; Lastrucci et al., 2012; Letty, 1973; Mabona et al., 2013; Mason, 1957; Minciullo et al., 2007; Mokni & Aouni, 2012; Nauheimer et al., 2012; Ngamau, 2001; 2006; Panetta, 1988; Perry, 1961; Randall & Lloyd, 2002; Rolim et al., 2015; Scott, 2012; Scott & Neser, 1996; Singh et al., 1996; Slaughter et al., 2012; Steenkamp, 2003; Tjia, 1985; Wei et al., 2012; Williams, 1981; Wu et al., 2008; Yannitsaros, 1998; Yao et al., 1994; Zika et al., 2000; Zurita et al., 2008.

Family 4.2. Lemnaceae [5]

Recognition of Lemnaceae as a separate family (see preceding "*Aroids*" section; Tippery & Les, 2020) is appropriate given that duckweeds represent one of the most distinctive groups not only of monocotyledons, but of all angiosperms. These are minute aquatic plants, which exhibit such extreme reduction that the terms "frond" or "thallus" have been used to describe their flattened, floating shoot system because a distinction between stem and leaf is impossible. *Wolffia* contains the smallest known individuals of any flowering plant, with some mature fronds only 0.5 mm in size (Landolt, 1994). The capacity for vegetative reproduction is astounding. It has been estimated that a 0.4 ha water surface can contain several hundred times more *Wolffia* individuals than the entire human population of the earth (Hicks, 1937). These astronomical numbers also raise questions on the evolutionary role of somatic mutations, which could reach very high levels under such circumstances. Some of the species lack roots and the flowers of all the species are simplified to the reproductive minimum; i.e., naked, unisexual flowers consisting of just one or two stamens (♂) and one pistil (♀) (Landolt et al., 1998). Most species are protogynous, with stigma receptivity indicated by the exudation of a sucrose droplet. The flowers are pollinated by small invertebrate vectors such as insects (Insecta: Aphididae, Diptera) or mites and spiders (Arachnida); they can self-pollinate if not protogynous. Contrary to some authors, they are not water pollinated (Landolt, 1986; Landolt et al., 1998). The minute seeds are dispersed abiotically by water (Sculthorpe, 1967) or are transported over greater distances by water birds (Aves) (Landolt, 1981). Because seed set is negligible in most duckweed species (Landolt et al., 1998), their dispersal is accomplished more often by the transport of the tiny plants themselves as they adhere to the plumage or muddy feet of water birds, or cling to the fur or skin of various amphibians (Amphibia), mammals (Mammalia), or reptiles (Reptilia) (Sculthorpe, 1967; Landolt, 1986; 1998). Years ago, the author barely escaped being eaten by an alligator (Vertebrata: Reptilia: Alligatoridae: *Alligator mississippiensis*), which he mistook for a duckweed-covered log! Such a dispersal vector could easily transport thousands of duckweed plants, at least over short distances (although the seeds are resistant to desiccation, the fronds are not; Landolt, 1998). The tiny plants also are dispersed abiotically by strong winds; in one instance, viable *Wolffia* fronds were recovered from melted hailstones (Landolt, 1986). Despite their proclivity for clonal growth, it is common to find mixed stands of duckweed species, where up to 10 different species have been observed growing together.

Duckweeds are eaten by humans as a salad or vegetable plant and are used as a food for many domestically raised animals (Landolt et al., 1998). They contain 20–35% protein, 4–7% fat, 4–10% starch (% DM), and all the essential mammalian amino acids (close to World Health Organization recommendations), yielding an energy content from 9000 to 8000 J g^{-1} DM (Landolt et al., 1998; Appenroth et al., 2017). As their name implies, duckweeds are relished and are eaten by many wild ducks (Aves: Anatidae) (McAtee, 1918; Mabbott, 1920; Oberholser & McAtee, 1920; Martin & Uhler, 1951). The plants have been used for the removal of heavy metals, nutrients, and pesticides from natural and wastewaters. Duckweeds also are important as model laboratory organisms because of their simple culture, small size, and rapid growth. Their ability to grow faster than any other higher plant provides them with the potential to surpass the productivity of all land-based agricultural crops at an estimated 4 metric tons fresh mass ha⁻¹ day⁻¹ (Skillicorn et al., 1993; Ziegler et al., 2015). The fronds can double by vegetative reproduction within 24 hr. Some duckweeds (*Lemna*) have been cultivated.

Lemnaceae are cosmopolitan. As treated here, the family includes 37 species distributed among five genera (Tippery & Les, 2020). The former genera *Pseudowolffia* and *Wolffiopsis*

(Hartog & Plas, 1970) are not recognized given that they would render *Wolffiella* as polyphyletic (Les et al., 2002); however, the genus *Landoltia* is distinguished because of its genetic distinctness and isolated phylogenetic position (Crawford & Landolt, 1993; Les & Crawford, 1999; Martirosyan et al., 2009; Bog et al., 2015; Ding et al., 2017a; Tippery & Les, 2020; Figure 1.27). Two subfamilies have been recognized (Lemnoideae: *Landoltia*, *Lemna*, *Spirodela*; Wolffioideae:

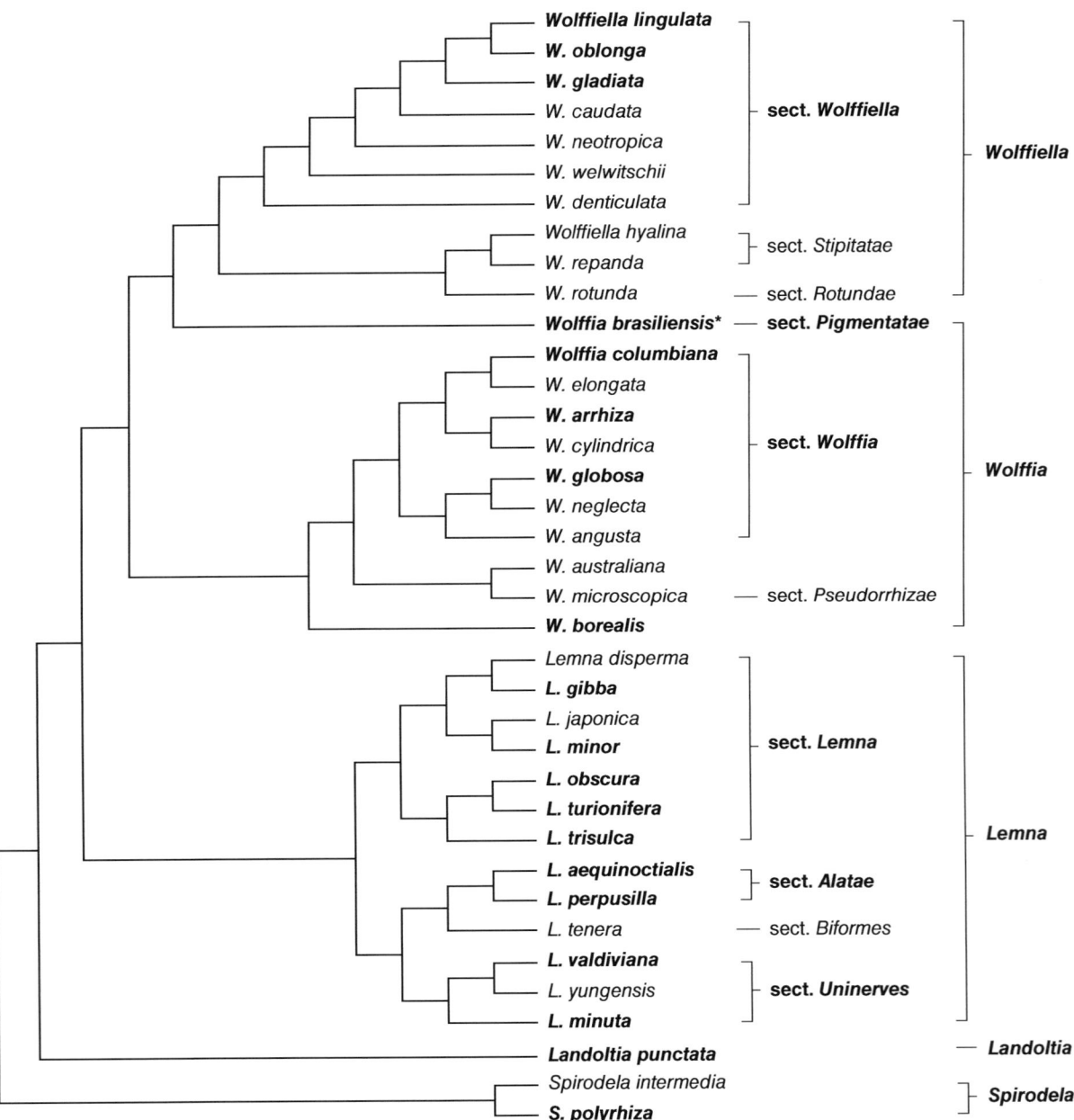

FIGURE 1.27 Interspecific relationships in Lemnaceae (duckweeds) based on phylogenetic analysis of combined nuclear and plastid DNA sequence data (adapted from Tippery et al., 2015; Tippery & Les, 2020). All recognized duckweed species appear in the analysis including the 19 OBL North American indicators (bold), which have originated throughout the family. Nearly all species and sectional relationships have been resolved satisfactorily except for those *Wolffia* species outside of section *Wolffia*. Notably, *Wolffia brasiliensis* (asterisked) associates weakly with *Wolffiella* in this analysis, which likely is an artifact given its resolution among other *Wolffia* species by plastid DNA sequence data alone (Tippery et al., 2015), or by various combinations of allozyme, micromolecular, morphological, and plastid data (Les et al., 1997b; 2002). Sections have not been assigned to two species (*Wolffia australiana* and *W. borealis*) pending further clarification of their relationships.

Wolffia, *Wolffiella*) but the former is paraphyletic (Les et al., 2002). The 19 OBL North American indicators are distributed among all five duckweed genera:

4.2.1. ***Landoltia*** Les & D. J. Crawford
4.2.2. ***Lemna*** L.
4.2.3. ***Spirodela*** Schleid.
4.2.4. ***Wolffia*** Horkel ex Schleid.
4.2.5. ***Wolffiella*** Hegelm.

4.2.1. Landoltia

Dotted duckmeat, dotted water flax-seed
Etymology: after Elias Landolt (1926–2013)
Synonyms: *Lemna* (in part); *Spirodela* (in part)
Distribution: global: Africa; Asia (southeast); Australia; Europe*; North America*; South America; **North America:** southern United States
Diversity: global: 1 species; **North America:** 1 species
Indicators (USA): obligate wetland (OBL): *Landoltia punctata*
Habitat: freshwater; lacustrine, palustrine, riverine; **pH:** 3.0–10.5; **depth:** 0 m; **life-form(s):** free-floating
Key morphology: roots (to 7 cm) 2–7 [rarely 1–12], adventitious, perforating the scale-like prophyllum on lower frond surface; fronds (to 8 mm) floating, oblong-obovate to lanceolate, asymmetric, upper surface shiny, green, finely nerved (3–7), with medial row of distinct papules, lower surface reddish, apex bluntly pointed, turions absent; flowers bisexual, enclosed by a utricular membranous scale (open along one side), perianth absent; stamens (to 1 mm) 2, 4-loculed, the external locules higher than internal locules, pollen spinulose; ovary (to 0.5 mm) 1, bottle-shaped, style short (to 0.2 mm), stigma funnelform, ovules 1–2; fruits (to 1.2 mm) circular follicles, winged (to 0.1 mm) laterally to apex, single-seeded; seeds (to 1.0 mm) with 10–15 distinct ribs
Life history: duration: perennial (whole plants); **asexual reproduction:** frond division; **pollination:** arthropod; **sexual condition:** hermaphroditic; **fruit:** utricular follicle (rare); **local dispersal:** fronds, seeds (water); **long-distance dispersal:** fronds, seeds (animals)
Imperilment: 1. *Landoltia punctata* [G5]; S2 (IL)
Ecology: general: *Landoltia* is monotypic (see next).

4.2.1.1. ***Landoltia punctata*** Les & D. J. Crawford is a nonindigenous perennial, which inhabits the water surfaces of bayous, borrow pits, bottomlands, canals, channels (bayou, stream), depressions (ponded), ditches (roadside), impoundments, floodplains, lakes, marshes, outflows (artesian), oxbows, ponds (beaver, dredge, floodplain, roadside, sewage treatment, woodland), pools (roadside, stagnant), reservoirs, rivers, sinks (ravine), sloughs (blackwater, river), spillways, springs (hot), streams (blackwater), swamps (beaver, cypress, gum, roadside, strand), and water tanks (cattle), and along the margins of hammocks at elevations to 2042 m. The plants favor warm temperate to tropical climates having mean winter and summer temperatures ranging from 1°C to 28°C. Annually, they require 200 days of mean temperatures above 10°C. The fronds grow in open exposures under full sunlight or in partial shade. Normally, they float on the surface of still or slow-moving standing waters but can become stranded on the surface of leaves, logs, stumps, and other debris as water levels recede. The reported substrates include clayey silt, mud, sand, sandy loam (Congaree), silt and silty clay loam (Cancienne). Although the plants appear to be relatively cold-sensitive, different ecotypes have exhibited a range of responses to cold, heat, and UV-B stresses. Optimum growth occurs at temperatures near 23°C. The fronds and roots can take up nitrogen in the form of NH_4^+ (preferentially) or NO_3^-. In North America, *Landoltia* grows across a broad range of environmental conditions (conductivity: 41–331 μS cm^{-1}; Ca: 3.0–41 mg l^{-1}; K: 0.6–34 mg l^{-1}; Mg: 0.8–35 mg l^{-1}; N: 0.4–1.7 mg l^{-1}; Na: 2.9–66 mg l^{-1}; P: 0.003–0.06 mg l^{-1}). Experimentally the plants survive at pH values from 3.0 to 10.5, but grow optimally at a pH of 6.0–7.0, a range consistent with the few pH values reported from established North American populations (pH: 6.1–7.3). The lower tolerance range is from pH 3.0–4.0, with growth also decreasing at pH values above 7.0. Flowering and fruiting occur rarely in North America (5% and 3% of plants, respectively), but potentially occurs from early summer through early fall (July–October). Salicylic acid and EDDHA (ethylenediamine-di-hydroxyphenylacetic acid) have induced profuse flowering in some strains. The flowers are protogynous and normally pollinated by various arthropods (Arthropoda). Seeds are uncommon in the region, and probably of little consequence in annual persistence or dispersal. There is no reliable information regarding the requirements (if any) for seed germination. The plants reproduce vegetatively by the proliferation of clonally derived plantlets. Most populations are highly clonal with some found to lack detectable genetic variation entirely. Total genetic diversity is low (H_T=0.072) with only 14 multi-locus genotypes detected among 43 different clones sampled worldwide. The values of mean allele number/locus and proportion of polymorphic loci also are low and more comparable to selfing or asexual species than to outcrossed, wide-ranging species. Some caution should be taken when evaluating data from neutral genetic markers given the likelihood that at least some of the detected variation results from somatic mutations. Dispersal in North America relies principally on the transport of the minute plants by animals, boating equipment, or other means of human-mediated transfer. The plants compete poorly in the presence of the larger, rapidly growing *Salvinia minima*.
Reported associates (North America): *Azolla filiculoides, Bacopa monnieri, Bolboschoenus robustus, Cabomba caroliniana, Ceratophyllum demersum, Chlorella, Cinna arundinacea, Cyperus, Decodon verticillatus, Eichhornia crassipes, Fimbristylis, Habenaria repens, Hydrilla verticillata, Hydrocotyle ranunculoides, Juncus, Leersia hexandra, Lemna gibba, Lemna minor, Lemna minuta, Lemna obscura, Lemna turionifera, Lemna valdiviana, Ludwigia brevipes, Marsilea vestita, Myrica cerifera, Myriophyllum aquaticum, Najas guadalupensis, Nelumbo lutea, Nuphar, Nymphaea, Nymphoides, Nyssa aquatica, Panicum repens, Persicaria sagittata, Potamogeton diversifolius, Proserpinaca, Riccia fluitans, Sacciolepis striata, Sagittaria, Schinus terebinthifolius,*

Schoenoplectus, *Spirodela polyrhiza*, *Spirogyra*, *Taxodium distichum*, *Typha latifolia*, *Verbena brasiliensis*, *Wolffia brasiliensis*, *Wolffia columbiana*, *Wolffiella gladiata*, *Xyris*.

Use by wildlife: *Landoltia* can contain up to 40.9% crude protein (when cultured in anaerobic waste lagoons) and has been used to feed livestock. Dense mats of the plants are detrimental to larval mosquito survival (Insecta: Diptera: Culicidae) by preventing their access to the surface where they must obtain oxygen.

Economic importance: food: *Landoltia punctata* has not been used as a human food; **medicinal:** no uses reported; **cultivation:** *Landoltia punctata* is not cultivated but can be distributed inadvertently with other cultivated aquatic plants; **misc. products:** *Landoltia punctata* has been used as a fermentation substrate for Bacteria (Actinobacteria: Corynebacteriaceae: *Corynebacterium crenatum*) and yeast (Fungi: Ascomycota: Saccharomycetaceae: *Saccharomyces cerevisiae*) in the production of bioethanol. It also has been evaluated as a pet food additive for domestic dogs (Mammalia: Canidae: *Canis familiaris*). The plants are well suited for use in removal of cadmium (Cd) and other heavy metals from aqueous solutions. Dried plants have been used as an adsorbent for removal of aqueous lead (Pb^{2+}); **weeds:** Some Florida populations of *Landoltia* have developed resistance to bipyridylium herbicides (diquat and paraquat). Due to their lack of turions and cold intolerance, some authors believe that the plants will not establish effectively in northern portions of North America (although they already have been reported in Massachusetts); **nonindigenous species:** *Landoltia punctata* initially was observed in the United States (Missouri) in 1930, possibly introduced inadvertently along with a shipment of goldfish (Vertebrata: Teleostei: Cyprinidae: *Carassius auratus*) purchased for an ornamental pond. The plants spread to California by 1957, to Florida, Illinois, and Louisiana by the early 1960s, and to Pennsylvania by 1965. In the United States, their dispersal over 1400 km within 70 years indicates a rate of about 20 km year^{-1}. *Landoltia* also has been introduced to France, Italy, Japan, Maltese islands, the Netherlands, Portugal, and Sweden.

Systematics: The distinctness of the monotypic *Landoltia* (formerly placed in *Spirodela*) was first demonstrated genetically by an allozyme survey (16 loci), which found there to be no alleles in common with either *Spirodela* species. That result was confirmed subsequently by numerous phylogenetic studies, which have resolved *L. punctata* outside of the *Spirodela* clade, albeit in somewhat different positions. Most data sets place *Landoltia* between *Spirodela* and the remainder of the family (e.g., Figure 1.27); yet others (e.g., full plastid genomes, *atpF-atpH*+*psbK-psbI* intergenic spacers, or *trnK* intron data) resolve *Landoltia* as the sister group of *Lemna*, or (*matK* data) as the sister group of the Wolffioideae. In any case, it is clear that *Landoltia* is distinct from *Spirodela* and should be maintained as a separate genus (nomenclatural arguments to the contrary have since been discredited). The genome size of *L. punctata* is estimated at 372–397 Mb; various genomic and transcriptomic studies are ongoing with the species. Although anomalous counts have been recorded (e.g.,

$2n = 46$), the basic chromosome number of *Landoltia* appears to be $x = 10$, with accessions of *L. punctata* being tetraploid ($2n = 40$) or pentaploid ($2n = 50$). There has been no evidence of hybridization between *Landoltia* and any other duckweed species.

Distribution: The cosmopolitan *Landoltia punctata* occurs primarily throughout the warmer, southerly portions of the United States.

References: Appenroth et al., 2015; 2017; Beal, 1977; Bog et al., 2015; Brown et al., 2013; Chen et al., 2012; Crawford & Landolt, 1993; Daubs, 1962; 1965; Ding et al., 2017a; Edwards, 1980; Fang et al., 2007; Furlow & Hays, 1972; Grippo et al., 2017; Hussner, 2012; Jansen et al., 1999; Jordan et al., 1996; Koschnick et al., 2006; Landolt, 1981; 1986; 1998; 2000; Landolt & Wildi, 1977; Les & Crawford, 1999; Les et al., 1997b; Martirosyan et al, 2009; Mason, 1957; McLay, 1976; Miki, 1934; Pieterse, 2013; Ryman & Anderberg, 1999; Soreng et al., 1984; Tang et al., 2017; Thompson, 1898; Saeger, 1934; Sree et al., 2016; Su et al., 2014; Urbanska-Worytkiewicz, 1980; Van Valkenburg & Pot, 2008; Wang et al., 2018a; Ward, 2011; Wiersema, 2015; Wohler et al., 1965; Xu et al., 2018a; Ziegler et al., 2015.

4.2.2. Lemna

Duckweed; lenticule

Etymology: an ancient Greek name, possibly derived from the Greek *limna* (lake or marsh)

Synonyms: *Hydrophace*; *Lenticula*; *Lenticularia*; *Staurogeton*; *Telmatophace*

Distribution: global: Cosmopolitan; **North America:** widespread

Diversity: global: 13 species; **North America:** 9 species

Indicators (USA): obligate wetland (OBL): *Lemna aequinoctialis*, *L. gibba*, *L. minor*, *L. minuta*, *L. obscura*, *L. perpusilla*, *L. trisulca*, *L. turionifera*, *L. valdiviana*

Habitat: brackish (coastal), freshwater; freshwater (tidal); saline (inland); lacustrine, palustrine, riverine; **pH:** 4.5–10.4; **depth:** 0–14 m; **life-form(s):** free-floating, submersed (thalloid), suspended

Key morphology: plants minute, floating, submersed, or suspended, thalloid, the shoot system a flattened frond, the fronds solitary or connected by fine green or white stipes (to 20 mm) into coherent groups or chains (of 2–50); the roots (to 15 cm) adventitious (rarely absent), one per frond, their sheath winged or wingless; individual fronds (to 8–15 mm) green or reddish, lanceolate, obovate, ovate, ovate-lanceolate, or ovate-obovate, flat or gibbous, 1–7-veined, the margins entire or minutely denticulate, the upper surface smooth or with 1-several minute raised papules, turions (to 1.6 mm) absent or present; flowers microscopic, naked, 1–2 per frond, subtended by a utricular scale, the scale opening apically or laterally on one side; style minute (0.05–0.2 mm), ovary with 1–5 ovules; fruits (to 1.35 mm) follicular, utricular, wingless or winged laterally toward apex; seeds (0.4–2.0 mm) with 8–29 distinct ribs, or 30–70 indistinct ribs, shed or retained in frond when ripe

Life history: duration: annual (fruits/seeds); perennial (resting fronds, turions, whole plants); **asexual reproduction:**

frond division, turions; **pollination:** insect, self; **sexual condition:** hermaphroditic; **fruit:** utricular follicle (rare to common); **local dispersal:** fronds, seeds (water, wind); **long-distance dispersal:** fronds, seeds (animals, water)

Imperilment: 1. *Lemna aequinoctialis* [G5]; S2 (KY); **2.** *L. gibba* [G4/G5]; S1 (UT); S2 (IL, NE); **3.** *L. minor* [G5]; S3 (WY); **4.** *L. minuta* [G4]; S1 (KS, NE, WV); S2 (IL); S3 (MO, WV); **5.** *L. obscura* [G5]; SX (PA); S2 (UT); S3 (IL, KY); **6.** *L. perpusilla* [G5]; SX (IN); SH (VT, WV); S1 (NJ, NY); S3 (IL, KY); **7.** *L. trisulca* [G5]; S1 (KS, MD, NH, UT, VA); S2 (MO, NJ, <u>NU</u>, WY); S3 (IL, <u>NB</u>, NV); **8.** *L. turionifera* [G5]; SH (MO, VT); S2 (KY, NY); S3 (KS, WY); **9.** *L. valdiviana* [G5]; SX (MI); SH (NH, PA); S1 (IN, KS, NJ, NY, UT, WY); S3 (IL, NC, WV)

Ecology: general: All *Lemna* species are obligately aquatic. They usually are perennials with frost-resistant or frost-sensitive fronds, or (rarely) annuals, which persist by seeds. The fronds typically float on the upper water surface or are suspended slightly beneath the surface. Some species also grow in an entirely submersed phase for at least part of the year. Perennation can occur by survival of the whole frond (in sufficiently mild climates) or by the production of "resting fronds." These are dormant, starch-laden buds that develop on senescing fronds, which sink to the bottom. In some cases, entire plants (rather than just their buds) can develop as resting fronds. A few species produce morphologically distinctive turions, which are shed from the surface fronds, sink, and are well adapted for survival through winter. Some studies indicate that the roots are less important for nutrient uptake than the lower frond surface, whereas others show comparable uptake from both structures. In any case, the primary function of the roots is for stability, which also is enhanced by the connection of the fronds into surface groups. In one suspended/submersed species (*L. trisulca*), most nutrient uptake occurs by the frond surface and the roots often are undeveloped or shed. Nitrogen fixing Cyanobacteria occur in the fronds of several species. At least some *Lemna* species can be grown "heterotrophically" in darkness, on media provided with sucrose. Duckweeds exhibit an incredible tolerance to water conditions. Their surface habit enables them to avoid light-related stress factors (e.g., turbidity) which can severely limit growth in potentially competitive submersed species. They also are broadly adapted physiologically. Although able to tolerate greater extremes at times, the different species (or strains) generally grow well at temperatures ranging from 4°C to 32°C and at pH from 4.5 to 7.5. Most duckweeds reproduce clonally by vegetative multiplication of the fronds or by turions. Except for a few annual species (section *Alatae*), flowering is rarely observed. The flowers of the annual species also lack dichogamy, are self-compatible, and frequently self-pollinate. Flowers of the perennial species are protogynous and either self-incompatible or self-compatible. The perennial species are thought to be outcrossed primarily by tiny arthropods (Arthropoda). The seeds produced by most duckweed species are non-dormant and are able to germinate immediately after ripening. Seeds of some annual species (e.g. *L. perpusilla*) require an after-ripening treatment (e.g., cold stratification) to

induce germination. Most local dispersal occurs abiotically, facilitated by water currents, which transport entire fronds (or groups of fronds). Long-distance dispersal primarily is biotic, occurring as the small fronds become entangled within the plumage of migrating waterfowl (Aves: Anatidae). Although the minute seeds are produced rarely by most species, they undoubtedly are involved in dispersal by adhering to muddy waterfowl feet (or comparable means) when present. Because duckweed identification is so difficult, it should be anticipated that at least some information reported in the literature could be inaccurate due to misidentifications. Although every effort has been made here to attribute information correctly, some errors undoubtedly will have occurred. In particular, it is extremely difficult to evaluate reports for widespread species like *L. minor*, which can represent an amalgam of erroneous information pertaining to virtually any North American *Lemna* species (see further discussion under *Systematics*).

4.2.2.1. *Lemna aequinoctialis* **Welw.** is a nonindigenous annual or perennial, which grows on the water surfaces of backwaters (slough, still), canals (irrigation), cattle troughs, channels (river), depressions, ditches (drainage, overflow), flats (salt), lagoons, lakes, marshes (depression), ponds (artificial, cattle, cypress, drying, farm, gum, meadow, permanent, receding, sewage, sink hole, stock), pools (artificial, marshy, seep, shaded, shallow, swamp), puddles, resacas, reservoirs, rivers, sand pits, sink holes (seasonally flooded), stock tanks, streams (intermittent, marshy, tidal), and swamps at elevations to 2630 m. The plants grow in open exposures under full sunlight or in the partial shade of taller emergent vegetation. However, in one field study, frond production by shaded plants was only 20% of that for plants growing in sunny exposures. Although normally floating, the fronds can become stranded on exposed muddy substrates after the waters recede. Once water levels are restored, the stranded plants are released quickly (90% within 25 hr) from substrates that remain soft, their buoyancy facilitated by photosynthetically derived gases. From hardened mud, a comparable release of plants may take more than ten times longer (up to 280 hr). The few substrates described include clay and clay loam. The plants occur in still or slow-moving, mesotrophic to eutrophic waters (pH: 4.5–9.4; conductivity: 41–2100 µS cm^{-1}; Ca: 2.4–64 mg l^{-1}; K: 0.6–36 mg l^{-1}; Mg: 0.8–75 mg l^{-1}; N: 0.2–3.7 mg l^{-1}; Na: 2.9–500 mg l^{-1}; P: 0.003–0.10 mg l^{-1}). The plants occur only sparsely in waters of higher pH (e.g., pH: 8.2–8.5). Their fronds are frost-sensitive and have no specialized vegetative overwintering structures. Although they can tolerate cold temperatures of −20°C, their growth is maintained only at temperatures of 13°C–16°C or higher (the approximate northern limit of these plants coincides with the 8°C mean winter temperature isotherm). They normally are perennial (the entire plants persisting year-round) but can behave as annuals (persisting entirely by seed) at desiccated sites where water levels have receded (see below). Floral development in these short-day plants (6–11-hr photoperiod; flowering inhibited >15 hr) begins when the fronds are only 0.08 mm in size [see comments pertaining to strain 6746 under *Systematics*]. The flowers are self-compatible and lack dichogamy, with the pistillate and staminate phases overlapping in most strains. Consequently, self-pollination is prevalent as the pollen simply drops onto the stigma. An autogamous breeding system has been indicated by allozyme data, which revealed populations containing several multilocus genotypes but a deficiency or lack of heterozygotes. Clones sampled throughout much of the geographical range possess the same alleles as those detected in local populations. Under warm conditions, the plants flower and fruit often (19% and 16% of plants, respectively) and nearly year-long, but most frequently from spring through fall (May–November); however, neither flowers nor fruits develop in colder climates. Flowering induction occurs when there is a high abscisic acid (ABA) content and a low gibberellic acid (GA_3) content and also is promoted when the fronds grow in dense mats. This is one of few duckweed species capable of persisting in sites characterized by periodical drying. In warmer regions, the non-dormant seeds are released readily from the ripe fruits and germinate quickly. They cannot facilitate overwintering unless the ripe seeds become stranded and desiccate in the drying sediments. In that case, the seeds acquire dormancy, which will last throughout the winter season. Experimentally, they have remained viable after 6 months of desiccation at 20°C. Frond density can approach 48 cm^{-2} in thick stands. A higher proportion of fronds detach to form ramets in smaller than in larger plants. The growth rate of the plants also declines as their density increases. **Reported associates (North America):** *Acer, Azolla cristata, Azolla filiculoides, Eichhornia crassipes, Hydrilla verticillata, Lemna gibba, Lemna minor, Lemna obscura, Lemna perpusilla, Lemna turionifera, Lemna valdiviana, Limnobium spongia, Lindera, Lindernia grandiflora, Ludwigia, Micranthemum umbrosum, Myriophyllum, Najas graminea, Nyssa sylvatica, Persicaria coccinea, Quercus, Rhizoclonium, Riccia, Sagittaria, Salvinia minima, Schoenoplectus californicus, Spirodela polyrhiza, Stuckenia pectinata, Taxodium, Typha, Utricularia, Wolffia columbiana, Wolffiella gladiata, Wolffiella oblonga.*

4.2.2.2. **Lemna gibba L.** is an indigenous annual or perennial, which floats on the water surfaces of brackish or freshwater arroyos, brooks, canals, channels (perennial, quiet, river, spring), cienegas, culverts, ditches (drainage, irrigation, roadside), drains (marsh), estuaries, floodplains, lagoons, lakes, marshes (dune, salt, spring-fed), ponds (beaver, kettle, salt marsh, stagnant, stock, wastewater), pools (dry, man-made, meadow, overflow, riverbed, spring-fed, still), reservoirs (small), rivers, seepages (marshy), sloughs, springs (cold, stagnant), stock tanks, streams (runoff), swales, swamps, tar pits (La Brea), and vernal pools at elevations to 2286 m. Preferred exposures are open, sunny waters but the plants also occur in sites that are deeply shaded by the marginal vegetation. The specific epithet refers to the gibbous (i.e., "humped") fronds, which result from prominent air-space tissue, and are more evident in plants growing under sunny exposures. The gibbous fronds also develop when cultivated on medium containing EDDHA. The extent of gibbosity is regulated by ethylene, with the fronds becoming flat at concentrations below 16 nl l^{-1} air. The plants typically are found floating on the water

surface but can become stranded on the surface of mud in areas where waters have receded. The few substrate descriptions include adobe, clay (heavy), mud, and sand. This is an ecologically versatile species, which has been observed in habitats ranging from various freshwater communities to salt marshes, and even the oily waters of California's La Brea tar pits. The plants thrive in moderate climates characterized by a mean winter isotherm of −1°C and a mean summer isotherm of 26°C. Their optimal growing temperature is relatively high and can approach 30°C in some strains. They are not found in areas of excessive precipitation (>90 mm $year^{-1}$) or where no dry period occurs during the growing season. They are tolerant of salinity and experimental cultures have maintained their growth at salinity levels as high as 250 mol m^{-3}. The plants require high levels of Ca, Mg, and other nutrients, and occur most often in warm, alkaline, more eutrophic sites (pH: 5.5–9.8; conductivity: 200–2600 µS cm^{-1}; Ca: 6.5–115 mg l^{-1}; K: 4.7–30 mg l^{-1}; Mg: 7.5–145 mg l^{-1}; N: 0.2–10.6 mg l^{-1}; Na: 28–850 mg l^{-1}; P: 0.002–0.99 mg l^{-1}). They grow well (e.g., 55 kg DM ha^{-1}) in wastewater at pH of 6.8–7.8 but cannot tolerate pH levels above 9.8 due to toxic levels of unionized ammonia (>8 mg NH_3-N l^{-1}). These are long-day plants (critical daylength of 10–14 hr) whose floral development begins when the fronds are only 0.05–0.07 mm in size. A minimum of one long day induces flower primordia formation, but six or more long days are needed for early flower development before mature flowers will be produced. Formation of daughter fronds also inhibits the floral meristem. Flowers and fruits are produced often (14% and 9% of plants, respectively) and have been observed on specimens collected from March to September. Some clones have been found to be male sterile. Ordinarily, the protogynous flowers are pollinated by small arthropods (Arthropoda). The pollen tubes reach the micropyle within 24 hr after the pollen germinates. Enhanced seed set has been obtained in some strains by artificial cross-pollination. The seeds are shed from the fruits when ripe. They are not dormant initially and germinate well (to 100%) after a short dark period and exposure to warm water temperatures. High germination rates (87–93%) also have been observed for cold stratified seeds (kept at 0°C). Although the seeds can germinate at low water temperatures (14°C), their maximum germination occurs more rapidly (within 6–8 days) at water temperatures of 25°C or 33°C than at 17°C (12–26 days). Germination will not occur in the dark. The seeds enable the plants to survive dry conditions associated with water drawdown. They can remain viable in a dried state for almost 5 years, although their germination rate declines markedly (to 3–7.6%) after the first year (17 months). Because the seeds become inviable within 2 years, only a short-term seed bank can develop. The plants multiply vegetatively by the proliferation of lateral plantlets, which can persist throughout the year under favorable climatic conditions; otherwise, they can perennate by means of resting fronds (which sink and overwinter on the bottom) or survive (as annuals) by seeds. This species is believed to compete poorly with its congeners (*L. obscura, L. turionifera*). **Reported associates:** *Anemopsis californica, Arthrocnemum subterminale, Azolla cristata,*

Azolla filiculoides, Baccharis salicifolia, Berula erecta, Bidens laevis, Bolboschoenus maritimus, Callitriche heterophylla, Carex hystericina, Carex pellita, Carex praegracilis, Ceratophyllum demersum, Chilopsis linearis, Conium maculatum, Conyza canadensis, Cynodon dactylon, Cyperus elegans, Cyperus eragrostis, Cyperus esculentus, Calibrachoa parviflora, Boerhavia coccinea, Cyperus odoratus, Dieteria canescens, Distichlis spicata, Echinochloa crus-galli, Echinodorus berteroi, Eleocharis engelmannii, Eleocharis parishii, Elodea canadensis, Epilobium ciliatum, Equisetum, Equisetum laevigatum, Eragrostis intermedia, Euphorbia vermiculata, Frankenia salina, Fraxinus velutina, Geranium richardsonii, Hydrocotyle ranunculoides, Juncus balticus, Juncus ensifolius, Juncus saximontanus, Juncus tenuis, Juncus torreyi, Juncus xiphioides, Landoltia punctata, Leersia oryzoides, Lemna aequinoctialis, Lemna minor, Lemna minuta, Lemna obscura, Lemna trisulca, Lemna valdiviana, Leptochloa fusca, Limosella acaulis, Ludwigia palustris, Marrubium vulgare, Mentha spicata, Mimulus glabratus, Mimulus guttatus, Mimulus ringens, Najas marina, Nasturtium officinale, Nuphar advena, Persicaria amphibia, Persicaria coccinea, Persicaria lapathifolia, Persicaria maculosa, Pluchea odorata, Polanisia dodecandra, Polypogon monspeliensis, Pontederia cordata, Populus fremontii, Populus ×acuminata, Potamogeton foliosus, Pseudognaphalium luteoalbum, Ranunculus trichophyllus, Ribes aureum, Rumex crispus, Rumex maritimus, Ruppia spiralis, Salicornia depressa, Salix bonplandiana, Salix exigua, Salix gooddingii, Sambucus nigra, Samolus valerandi, Schedonorus pratensis, Schoenoplectus acutus, Schoenoplectus americanus, Schoenoplectus californicus, Schoenoplectus tabernaemontani, Scirpus microcarpus, Sidalcea neomexicana, Sorghum halepense, Sparganium eurycarpum, Sphenopholis intermedia, Spirodela polyrhiza, Spirogyra, Sporobolus wrightii, Stuckenia pectinata, Symphyotrichum subulatum, Tamarix chinensis, Thinopyrum elongatum, Triglochin scilloides, Typha domingensis, Typha latifolia, Veronica anagallis-aquatica, Wolffia borealis, Wolffia brasiliensis, Wolffia columbiana, Wolffiella, Xanthium strumarium, Zannichellia palustris.

4.2.2.3. ***Lemna minor* L.** is an indigenous perennial, which colonizes fresh to brackish water surfaces of bays, bogs (roadside, *Sphagnum*), bottomlands, brooks, canals (irrigation), channels (river), depressions, ditches (drainage, irrigation, roadside, seepage), estuaries (freshwater), fens, floodplains, gullies, impoundments (river), lagoons (sewage), lakes, marshes (freshwater, roadside), meadows (sedge, spring-fed), oxbows, playas (alkali), ponds (backwater, beaver, dredge pit, fertilizer, floodplain, freshwater, marshy, mill, reservoir, sinkhole, spring-fed, stagnant, stock, temporary, vernal), pools (ephemeral, spring, stagnant), potholes (prairie), reservoirs, rivers, seepages (alkaline, hillside), sloughs (roadside), springs (cold), stock tanks, streams (drying, perennial), swamps, vernal pools, watering holes, and woodlands (alluvial) at elevations to 3201 m. Exposures range from full sunlight to shade. Although primarily a floating plant (pleustophyte), the fronds commonly become stranded on wet mud, on logs, or on other

debris at sites where the standing waters have receded. The substrates have been characterized as clay, clay loam, gravel (granitic), gravel loam, humus, loamy sand, muck, mucky rock, mud, sand, silty loam, and silty ooze. The plants do not occur in dry regions but are distributed widely throughout cool temperate regions located between the $-15°C$ mean winter isotherm and the $24°C$ mean summer isotherm. Experimental studies have indicated an optimal temperature of $29°C$. The plants are salinity tolerant and have been grown experimentally over a range of salinities from 8.3 to 33.3‰; the fronds are killed at salinities >166.5‰. Uptake of NH_4^+ and NO_3^- occurs comparably through both the roots and fronds, with greater root uptake (to 73%) occurring under N depleted conditions. Studies of natural Ohio populations found the plants at sites ranging in pH from 5.0 to 7.5, with most occurrences in lakes having pH values between 6 and 7. Those observations are consistent with pH values associated with laboratory grown plants, which tolerate pH values from 4 to 10, but grow optimally at pH 6.2. The sites are devoid of any appreciable currents. Otherwise, this species is adapted to a wide range of environmental conditions, with a tendency toward more eutrophic waters (pH: 5.0–9.5; $\bar{x} = 7.2$; alkalinity [as $CaCO_3$]: 2.5–315 mg l^{-1}; $\bar{x} = 36.2$ mg l^{-1}; $\bar{x} = 115$ mg l^{-1}; conductivity: 55–1657 μS cm^{-1}; 5–690 $\mu mhos$ cm^{-1}; Ca: 2.5–168 mg l^{-1}; K: 1–53 mg l^{-1}; Mg: 2.6–85 mg l^{-1}; N: 0.1–2.8 mg l^{-1}; Na: 5.4–115 mg l^{-1}; P: 0.003–0.27 mg l^{-1}). Flowering and fruiting can occur from May to October. However, the self-incompatible and protogynous flowers seldom are produced (only 5% of individuals), with fruit set occurring even more rarely (0.6% overall; <20% of fertile individuals). Flowering *in vitro* can be induced by addition of the chelate EDDHA. Hand pollinations have resulted in fruit set when the cross-pollinations were made between geographically different clones. The seeds are believed to be intolerant of drying, but it remains unknown whether they might require an after-ripening treatment such as cold stratification to induce germination. Because of the poor seed set, most reproduction occurs by vegetative means. On average, each frond lives for 31.3 days and produces 14 daughter fronds (0.45 fronds day^{-1}). The growth rate calculated in one river population (0.066 day^{-1}) was logarithmic (log_{10}) over time and yielded an estimated doubling time of 4.5 days. Extensive clonal reproduction is evidenced by genetic (allozyme) data, which have demonstrated an extensive propagation of multilocus genotypes and excess of homozygotes in populations rather than evidence of genetic recombination. Populations are highly differentiated genetically, which indicates extremely low levels of gene flow. Yet, high genotypic diversity can be maintained within populations (e.g., $D = 0.973$) as a result of somatic mutations, multiple introductions of clones, or by rare episodes of sexual reproduction. The fronds are highly resistant to desiccation (remaining viable for up to 22 hr after being removed from the water) and grow rapidly when water is available. Although they are transported frequently in the plumage of ducks (Aves: Anatidae), they are thought to be dispersed over only a few kilometers in this way, due to the more rapid desiccation that occurs during flight. However, if the fronds are exposed to rain or are buried deeply within the plumage, it is estimated that the fronds could survive for up to 6 hr, which would extend their potential dispersal distance to a 100 km or more, and even up to 1000 km under optimal flight conditions. The growth rate declines as plant density on the water surface increases, with the maximum relative rate (0.3 day^{-1}) found to occur at a density of 9 g DM m^{-2}. During winter, the fronds either remain alive on the water surface (able to continue their growth at temperatures low as 4°C) or die back and sink, while retaining meristematic activity in some buds (i.e., resting fronds); in the latter case, the fronds overwinter on the bottom of water bodies. Competitively, *L. minor* dominates in shaded eutrophic sites, where its high densities inhibit *L. trisulca*; however, *L. trisulca* can dominate in less eutrophic conditions. The relative growth rate of *L. minuta* is higher than *L. minor* in high nutrient conditions, whereas that of *L. minor* is higher under low nutrient conditions. *Lemna minor* has been observed to decrease in the aftermath of hurricanes as it is washed out of communities by the rising waters. Mats from 8 to 16 mm thick have been shown to reduce the rate of atmospheric/water oxygen transport by as much as 53–96%. [*Note:* The author has observed extremely thick mats in southern Florida that were more than 30 cm thick.] **Reported associates:** *Acalypha rhomboidea, Acer rubrum, Acer saccharinum, Acorus calamus, Agrimonia parviflora, Agrostis gigantea, Alisma subcordatum, Alnus rugosa, Alternanthera philoxeroides, Amaranthus tuberculatus, Ambrosia artemisiifolia, Ambrosia psilostachya, Ambrosia trifida, Ammannia robusta, Amphicarpaea bracteata, Anagallis arvensis, Anemopsis californica, Apios americana, Artemisia annua, Artemisia tridentata, Arundo donax, Asclepias incarnata, Azolla cristata, Azolla filiculoides, Baccharis salicifolia, Bacopa monnieri, Berula erecta, Bidens cernuus, Bidens frondosus, Bidens laevis, Boehmeria cylindrica, Bolboschoenus fluviatilis, Brasenia schreberi, Brunnichia ovata, Calamagrostis canadensis, Callitriche heterophylla, Caltha palustris, Camassia, Campanula aparinoides, Carex aquatilis, Carex blanda, Carex cristatella, Carex diandra, Carex emoryi, Carex frankii, Carex grayi, Carex grisea, Carex haydenii, Carex hystericina, Carex lacustris, Carex lasiocarpa, Carex lurida, Carex microptera, Carex nebrascensis, Carex praegracilis, Carex rostrata, Carex sartwellii, Carex scoparia, Carex shortiana, Carex simulata, Carex stipata, Carex stricta, Carex tribuloides, Carex utriculata, Carex vesicaria, Cephalanthus occidentalis, Ceratophyllum demersum, Ceratophyllum echinatum, Chamaedaphne calyculata, Chara, Cicuta bulbifera, Cicuta virosa, Cirsium arvense, Claytonia, Comarum palustre, Convolvulus sepium, Conyza canadensis, Cornus, Cyperus eragrostis, Cyperus esculentus, Cyperus niger, Cyperus odoratus, Cyperus squarrosus, Decodon verticillatus, Dichanthelium clandestinum, Distichlis spicata, Dulichium arundinaceum, Echinochloa crus-galli, Eclipta prostrata, Egeria densa, Eichhornia crassipes, Eleocharis acicularis, Eleocharis obtusa, Eleocharis palustris, Eleocharis parishii, Eleocharis parvula, Elodea canadensis, Elodea nuttallii, Elymus, Epilobium ciliatum, Epilobium coloratum, Epilobium leptophyllum, Epilobium oreganum, Epilobium palustre, Epilobium palustre, Equisetum arvense, Equisetum fluviatile, Equisetum hyemale, Eragrostis hypnoides, Eragrostis pectinacea, Ericameria nauseosa, Erigeron philadelphicus, Eupatorium serotinum, Euthamia occidentalis, Fallopia scandens, Fraxinus profunda, Fraxinus nigra, Fraxinus pennsylvanica, Fraxinus velutina, Galium tinctorium, Gleditsia aquatica, Glyceria leptostachya, Glyceria septentrionalis, Glyceria striata, Gratiola neglecta, Gratiola virginiana, Helenium puberulum, Helianthus grosseserratus, Hippuris vulgaris, Hopia obtusa, Hordeum jubatum, Hydrocotyle ranunculoides, Hymenoxys hoopesii, Hypericum mutilum, Hypericum scouleri, Impatiens capensis, Ipomoea costellata, Iris virginica, Juncus acuminatus, Juncus articulatus, Juncus balticus, Juncus bufonius, Juncus effusus, Juncus laccatus, Juncus nodatus, Juncus tenuis, Juncus xiphioides, Justicia americana, Landoltia punctata, Laportea canadensis, Leersia oryzoides, Lemna aequinoctialis, Lemna gibba, Lemna minuta, Lemna obscura, Lemna perpusilla, Lemna trisulca, Lemna turionifera, Lemna valdiviana, Lilaeopsis, Limnanthes douglasii, Lonicera involucrata, Ludwigia palustris, Ludwigia peploides, Lycopus americanus, Lycopus rubellus, Lycopus uniflorus, Lysimachia nummularia, Lythrum californicum, Madia, Marrubium vulgare, Medicago lupulina, Melilotus albus, Mentha arvensis, Menyanthes trifoliata, Mimulus cardinalis, Mimulus guttatus, Muhlenbergia asperifolia, Muhlenbergia frondosa, Muhlenbergia rigens, Myosotis scorpioides, Myriophyllum aquaticum, Myriophyllum sibiricum, Myriophyllum spicatum, Myriophyllum verticillatum, Najas flexilis, Najas guadalupensis, Najas minor, Nasturtium officinale, Nicotiana glauca, Nitella, Nuphar polysepala, Nuphar variegata, Nymphaea odorata, Nyssa, Onoclea sensibilis, Oxalis albicans, Oxalis stricta, Oxypolis rigidior, Panicum capillare, Paspalum distichum, Peltandra virginica, Persicaria amphibia, Persicaria arifolia, Persicaria coccinea, Persicaria hydropiperoides, Persicaria lapathifolia, Persicaria maculosa, Persicaria pensylvanica, Persicaria punctata, Persicaria sagittata, Phalaris arundinacea, Philadelphus, Phragmites australis, Physostegia virginiana, Picea mariana, Pilea pumila, Platanthera sparsiflora, Platanus occidentalis, Pluchea odorata, Poa nemoralis, Poa palustris, Poa sylvestris, Polygonum aviculare, Polypogon interruptus, Polypogon monspeliensis, Populus balsamifera, Populus deltoides, Populus fremontii, Potamogeton alpinus, Potamogeton amplifolius, Potamogeton berchtoldii, Potamogeton crispus, Potamogeton diversifolius, Potamogeton foliosus, Potamogeton gramineus, Potamogeton illinoensis, Potamogeton natans, Potamogeton nodosus, Potamogeton pusillus, Potamogeton zosteriformis, Potentilla norvegica, Quercus bicolor, Quercus macrocarpa, Quercus palustris, Ranunculus cymbalaria, Ranunculus hyperboreus, Ranunculus longirostris, Ranunculus repens, Ranunculus sceleratus, Rhamnus frangula, Rhododendron, Riccia fluitans, Ricciocarpus natans, Rorippa palustris, Rorippa sylvestris, Rudbeckia laciniata, Rumex acetosella, Rumex britannica, Rumex maritimus, Rumex verticillatus, Ruppia* maritima, *Sagittaria cuneata, Sagittaria graminea, Sagittaria latifolia, Sagittaria rigida, Salix amygdaloides, Salix bebbiana, Salix discolor, Salix exigua, Salix gooddingii, Salix interior, Salix lasiandra, Salix lasiolepis, Salix nigra, Sambucus nigra, Schedonorus arundinaceus, Schoenoplectus acutus, Schoenoplectus californicus, Schoenoplectus hallii, Schoenoplectus heterochaetus, Schoenoplectus pungens, Schoenoplectus tabernaemontani, Scirpus atrovirens, Scirpus cyperinus, Scirpus microcarpus, Scirpus pendulus, Scutellaria lateriflora, Selaginella eclipes, Setaria faberi, Sicyos angulatus, Sidalcea, Smilax tamnoides, Solanum ptychanthum, Solidago confinis, Sonchus arvensis, Sorghum halepense, Sparganium americanum, Sparganium emersum, Sparganium eurycarpum, Sparganium natans, Spartina pectinata, Sphagnum, Spiraea alba, Spirodela polyrhiza, Sporobolus wrightii, Sporobolus wrightii, Stachys palustris, Stuckenia filiformis, Stuckenia pectinata, Stuckenia vaginata, Symphyotrichum lateriflorum, Symphyotrichum puniceum, Symplocarpus foetidus, Tamarix chinensis, Taxodium distichum, Thelypteris palustris, Trifolium variegatum, Triglochin, Typha domingensis, Typha latifolia, Typha ×glauca, Urtica dioica, Utricularia gibba, Utricularia macrorhiza, Verbena hastata, Verbena urticifolia, Veronica americana, Veronica anagallis-aquatica, Veronica catenata, Viola sororia, Vitis riparia, Wolffia borealis, Wolffia brasiliensis, Wolffia columbiana, Wolffiella gladiata, Wolffiella lingulata, Xanthium strumarium, Zannichellia palustris.*

4.2.2.4. ***Lemna minuta* Kunth** is an indigenous freshwater perennial, which grows on the water surfaces of alluvial benches, alluvial washes, bogs (peat), brooks, canals, channels (freshwater, river), cienega, depressions (spring-fed), ditches (diversion, drainage, irrigation, railroad. roadside), gullies, lagoons, lakes, marshes (freshwater, perennial), meadows (marshy), oxbows, ponds (artificial, beaver, irrigation, perennial, shady, sinkhole, spring-fed), pools (roadside, shaded, stagnant), ravines (woodland), potholes, reservoirs, rice fields, river margins, seepages, sloughs, springs (cold, perennial, warm), stock ponds, stock tanks, streams, swamps, and water holes at elevations to 2877 m [up to 4000 m in South America]. Exposures can vary from full sun to shade. The plants occur where mean winter and summer temperatures range from 1°C to 26°C. Although normally a floating species, the fronds also can attach to wet substrates after the standing waters have receded. The substrates have been described as clay (heavy, organic), clay loam, gravel, humus, muck, rocks (wet), sand, sandy loam, silt, and silty loam. The plants generally occur in dry to moderately humid, tropical to subtropical climates where mean winter temperatures remain above −1°C and mean summer temperatures fall between 16°C and 26°C. They are found primarily in calm, mesotrophic to eutrophic waters (pH: 6.6–9.1; conductivity: 55–2100 µS cm^{-1}; Ca: 1.0–79 mg l^{-1}; K: 1.6–22 mg l^{-1}; Mg: 0.7–110 mg l^{-1}; N: 0.2–2.8 mg l^{-1}; Na: 8.5–500 mg l^{-1}; P: 0.00–0.13 mg l^{-1}). Sexual reproduction is uncommon with only 5% of collected specimens possessing flowers and only 2% bearing fruits. When fertile, flowering and fruiting can extend from April to November. The flowers presumably are protogynous and pollinated by arthropods (Arthropoda), but specific studies of the floral biology are lacking. The plants perennate by means of resting fronds. The normal fronds have limited desiccation tolerance and survive poorly under low relative humidity (RH) (~1 hr at 18% RH), especially at elevated temperatures; however, they can survive from 4 to 5 hr at 60% RH. They are dispersed over longer distances by becoming enveloped within the plumage of mallards (Aves: Anatidae: *Anas platyrhynchos*) and other ducks. Plants cultured in Hoagland's E+ medium have exhibited a high frond production rate (0.16 fronds day^{-1}), long mean life span (13.8 d), and low mortality rate (0.11 fronds day^{-1}), with an estimated mean doubling time of 4.61 days. In competition experiments with *L. minor* and *L. minuta* exhibited a higher relative growth rate under medium to high light intensities and in high nutrient conditions, but a lower rate under low nutrient conditions. **Reported associates:** *Alnus rubra, Anemopsis californica, Apium graveolens, Azolla cristata, Azolla filiculoides, Baccharis pilularis, Baccharis salicifolia, Baccharis sarothroides, Bacopa monnieri, Barbarea orthoceras, Berula erecta, Bidens discoideus, Brickellia californica, Calla palustris, Callitriche, Carex aquatilis, Carex nebrascensis, Carex praegracilis, Carex utriculata, Castilleja linariifolia, Ceratophyllum demersum, Chara, Cicuta douglasii, Cornus amomum, Cyperus eragrostis, Cyperus erythrorhizos, Datura wrightii, Deschampsia cespitosa, Echinochloa, Eichhornia crassipes, Eleocharis erythropoda, Eleocharis parishii, Elodea canadensis, Epilobium ciliatum, Equisetum, Erigeron divergens, Fontinalis, Fragaria, Glyceria striata, Habenaria repens, Helenium bigelovii, Helminthotheca echioides Holcus lanatus, Horkelia rydbergii, Hydrilla verticillata, Hydrocotyle ranunculoides, Isocoma menziesii, Juncus balticus, Juncus balticus, Juncus bufonius, Juncus dudleyi, Landoltia punctata, Leersia oryzoides, Lemna gibba, Lemna minor, Lemna trisulca, Lemna turionifera, Lemna valdiviana, Leymus triticoides, Ludwigia peploides, Lycopus americanus, Malosma laurina, Medicago lupulina, Mentha arvensis, Mimulus cardinalis, Mimulus guttatus, Myriophyllum aquaticum, Myriophyllum spicatum, Nasturtium officinale, Navarretia, Nuphar polysepala, Nyssa, Paspalum distichum, Perideridia parishii, Persicaria lapathifolia Phalaris arundinacea, Phragmites australis, Plantago major, Platanus racemosa, Pluchea odorata, Poa pratensis, Polypogon monspeliensis, Pontederia cordata Populus fremontii, Potamogeton foliosus, Potamogeton nodosus, Pseudognaphalium luteoalbum, Quercus agrifolia, Ranunculus hydrocharoides, Rhamnus frangula, Riccia fluitans, Rudbeckia laciniata, Rumex conglomeratus, Salix exigua, Salix gooddingii, Salix laevigata, Salix lasiolepis, Sambucus nigra, Schoenoplectus acutus, Schoenoplectus americanus, Schoenoplectus californicus, Schoenoplectus lacustris, Schoenoplectus tabernaemontani, Scirpus microcarpus, Sidalcea, Solidago spectabilis, Sparganium eurycarpum, Spirodela polyrhiza, Stellaria longipes, Stuckenia pectinata, Symphyotrichum lanceolatum, Tamarix, Taxodium distichum, Toxicodendron diversilobum, Trifolium, Typha domingensis, Typha latifolia, Typha ×glauca, Urtica dioica, Utricularia, Veronica anagallis-aquatica, Vitis riparia,*

Wolffia arrhiza, Wolffia borealis, Wolffia brasiliensis, Wolffia columbiana, Wolffia globosa, Wolffiella gladiata, Wolffiella lingulata, Wolffiella oblonga.

4.2.2.5. *Lemna obscura* **(Austin) Daubs** is an indigenous perennial, which grows on the brackish to freshwater surfaces of alligator holes, backwaters, bayous, canals (drainage), channels (stream), cypress brakes, depressions, ditches (drainage, drying, irrigation, mosquito control, roadside, shaded), floodplains, hammocks (hydric), lagoons (sewage), lakes (spring-fed), marshes (brackish, fresh, open, salt), meadows, ponds (fertilizer, sinkhole), pools (freshwater, marsh, river), rivers, sand pits, seeps, sinks (lime), sloughs (marshy), spillways, springs, and swamps at elevations to 2710 m. The plants occur in warm temperate zones with mean winter and summer temperatures ranging from 1°C to 28°C. Exposures can vary from full sunlight to shade. The sediments have been described as clay (alluvial), muck (Kenner), mud, peaty loam, sand (alluvial, Corolla-Duckston), sandy loam, silty clay, and silty clay loam (Cancienne). These plants are found in warm temperate regions of North America where the mean winter temperatures are above 1°C and the mean summer temperatures occur between 18°C and 28°C. They occur primarily in calm, mesotrophic to eutrophic waters (pH: 6.4–7.9; conductivity: 99–1060 µS cm^{-1}; Ca: 4.8–68 mg l^{-1}; K: 1.7–35 mg l^{-1}; Mg: 2.7–100 mg l^{-1}; N: 0.2–11.2 mg l^{-1}; Na: 90–800 mg l^{-1}; P: 0.004–0.06 mg l^{-1}). Sexual reproduction is uncommon with only 2% of collected specimens possessing flowers and just 1.3% having fruits. Flowering (which is day-neutral) has been induced in cultured plants by addition of salicylic acid to the growth media. When fertile, flowering and fruiting can occur from February to October. No turions develop; the plants overwinter intact or by resting fronds. Because seed production is rare, dispersal occurs primarily by the transport of whole plants via water, waterfowl (Aves: Anatidae), humans, or other means. Mean daily production rates as high as 82.4 g m^{-2} have been recorded for plants grown in agricultural wastewater. On average the plants can double their biomass or number of fronds every 4.54 days. **Reported associates:** *Acer rubrum, Aesculus, Alternanthera philoxeroides, Azolla, Bacopa monnieri, Bidens cernuus, Caltha palustris, Carex haydenii, Carex stipata, Carex stricta, Carex trichocarpa, Carpinus, Cephalanthus occidentalis, Ceratophyllum demersum, Chara zeylanica, Cladium, Conocarpus erectus, Crinum americanum, Cyperus odoratus, Decumaria barbara, Echinochloa walteri, Eichhornia crassipes, Eleocharis, Eutrochium maculatum, Galium tinctorium, Hibiscus coccineus, Hydrilla verticillata, Hydrocotyle bonariensis, Hydrocotyle ranunculoides, Hydrocotyle umbellata, Ilex cassine, Ilex vomitoria, Impatiens capensis, Iris hexagona, Itea virginica, Landoltia punctata, Leersia oryzoides, Leitneria floridana, Lemna aequinoctialis, Lemna gibba, Lemna minor, Lemna trisulca, Lemna valdiviana, Liquidambar styraciflua, Liriodendron, Magnolia grandiflora, Magnolia virginiana, Marsilea vestita, Mikania scandens, Myrica cerifera, Myrica cerifera, Najas guadalupensis, Nymphaea, Nyssa aquatica, Nyssa biflora, Oxalis dillenii, Panicum, Paspalum repens, Persicaria punctata, Phalaris arundinacea, Phragmites australis, Pinus*

elliottii, Pistia stratiotes, Plantago virginica, Potamogeton diversifolius Quercus virginiana, Rhapidophyllum hystrix, Riccia fluitans, Rudbeckia laciniata, Sabal palmetto, Salix caroliniana, Salix caroliniana, Salix discolor, Salix exigua, Salvinia minima, Samolus, Schoenoplectus, Scirpus cyperinus, Serenoa repens, Sparganium, Spermolepis echinata, Spirodela polyrhiza, Taxodium distichum, Typha domingensis, Typha latifolia, Utricularia gibba, Utricularia macrorhiza, Utricularia purpurea, Vallisneria americana, Wolffia brasiliensis, Wolffia columbiana, Wolffiella gladiata, Wolffiella lingulata, Wolffiella oblonga.

4.2.2.6. *Lemna perpusilla* **Torr.** is an indigenous annual, which occurs on the water surfaces of bogs, ditches, fens, lakes, marshes (spring-fed), meadows (swampy), ponds (depression, drying, farm, roadside, sinkhole), pools (river, shallow), ravines, rivers (quiet), ruts, sloughs, springs, streams, swales, and swamps at elevations to 512 m. Exposures range from full sunlight to shade. Although normally floating, the fronds can become stranded on exposed muddy substrates after the waters recede. The few reported substrates include mud, sand, and sandy mud. These plants have frost-sensitive fronds and occur in temperate regions of North America having milder winters and where the mean low winter temperatures and mean high summer temperatures fall between −8°C and 24°C. They inhabit tidal, intertidal or non-tidal, still, mesotrophic to eutrophic waters (pH: 5.8–8.6; $\bar{x} = 7.7$; alkalinity [as CaCO$_3$]: 14–175 mg l^{-1}; $\bar{x} = 115$ mg l^{-1}; conductivity: 116–199 µS cm^{-1}; 55–370 µmhos cm^{-1}; $\bar{x} = 200$ µmhos cm^{-1}; Ca: 3–27 mg l^{-1}; Cl$^-$: 21.4–61.8 mg l^{-1}; K: 1.1–6 mg l^{-1}; Mg: 2.4–7.4 mg l^{-1}; N: 0.7–2.9 mg l^{-1}; Na: 7.5–35 mg l^{-1}; P: 0.005–0.12 mg l^{-1}). The pH range among North Carolina populations (5.8–6.6) was somewhat lower. Nitrogen is taken up preferentially in the form of ammonium (NH$_4^+$). These are highly sexual short-day plants, which flower and fruit from January to November. The non-dichogamous flowers allow for self-pollination to occur during the overlapping sexual phases. Autogamy is prevalent and ensures the production of seeds, which are necessary for annual reestablishment. Flower and fruit production are extremely high, with 48% of collections containing flowers and 45% having fruits. The seeds are retained within the ripe fruits and eventually sink to the bottom along with the decomposing fronds. They are resistant to cold temperatures and desiccation. Unlike most *Lemna* species, the seeds remain dormant through winter until undergoing a period of cold stratification (e.g., 1 month at 4°C–6°C). The plants survive primarily by overwintering seeds (i.e., they are annuals), which is unusual for duckweeds. The plants are dispersed either by whole plants or seeds, facilitated by water or animal vectors, most likely by waterfowl (Aves: Anatidae). **Reported associates:** *Acorus americanus, Alternanthera philoxeroides, Angelica atropurpurea, Azolla filiculoides, Baccharis, Bacopa monnieri, Bolboschoenus robustus, Campanula aparinoides, Carex lacustris, Centella asiatica, Cephalanthus occidentalis, Ceratophyllum demersum, Chara, Cicuta bulbifera, Cladium jamaicense, Dulichium arundinaceum, Eleocharis, Elodea nuttallii, Heteranthera dubia, Hottonia inflata, Hydrilla verticillata, Hydrocotyle umbellata, Juncus* *coriaceus, Juncus effusus, Juncus polycephalus, Juncus roemerianus, Lemna aequinoctialis, Lemna minor, Lemna trisulca, Lemna turionifera, Ludwigia leptocarpa, Myrica cerifera, Myriophyllum spicatum, Najas guadalupensis, Najas minor, Nymphaea odorata, Persicaria, Phragmites australis, Pontederia cordata, Potamogeton crispus, Sagittaria, Salix discolor, Spartina alterniflora, Spirodela polyrhiza, Symphyotrichum firmum, Trapa natans, Typha angustifolia, Typha latifolia, Vallisneria americana, Wolffia brasiliensis, Wolffia columbiana, Wolffiella gladiata.*

4.2.2.7. *Lemna trisulca* L. is an indigenous perennial, which grows submersed or suspended just beneath the surface of backwaters, bayous, bogs (*Sphagnum*, drained, tamarack), brooks, canals (abandoned), depressions, ditches (drainage, roadside), gravel pits (abandoned), lagoons, lakes (alpine, montane), marshes (seasonal), meadows (boggy), moats, oxbows, ponds (beaver, brackish, intermittent, kettle, spring-fed, stock), pools (quiet, roadside, shaded, streamside), potholes, reservoirs, rivers, seeps (mineral, swampy), sloughs (muddy, river), springs (artesian, cold), streams (perennial, swift), swales, and swamps at elevations to 3006 m. These plants can occur in full sunlight but are highly shade tolerant and often grow in well-shaded sites or under mats of other floating vegetation. They have been collected at rather extraordinary depths of 12–14 m. Photochemical reactions are suppressed when the plants grow under reduced light, which results in the enhanced synthesis and accumulation of C_4-acids (aspartate, malate); however, there is no evidence of a C_4 photosynthetic pathway. The associated substrates most often are gravel or sand but also can be clay, cobble, muck, mud, sandy clay, silt, or organic substrates and occasionally on rock (granite, limestone). Generally, *L. trisulca* occurs in cooler temperate climates and can withstand surface temperatures low as −40°C. It requires 50 days where mean temperatures exceed 10°C but with a maximum mean temperature of 22°C. Although not highly tolerant of salinity, the plants are collected occasionally from brackish water and reportedly are somewhat more tolerant of higher conductivity than *L. minor*. They can tolerate fairly brisk currents (often becoming entangled on submersed branches or other debris) and have an affinity for waters containing higher levels of dissolved organic matter (optical density @ 275 nm: 0.03–1.68; $\bar{x}=0.38$–0.42). Most occurrences are found within stable sites characterized by alkaline, calcareous, mesotrophic, freshwaters (pH: 4.7–9.0; $\bar{x}=7.3$; $\bar{x}=7.7$; alkalinity [as $CaCO_3$]: 4–558 mg l^{-1}; $\bar{x}=60.4$–134; $\bar{x}=115$ mg l^{-1}; conductivity: 286–352 µS cm^{-1}; 20–560 µmhos cm^{-1}; $\bar{x}=240$ µmhos cm^{-1}; Ca: 45–70 mg l^{-1}; Cl$^-$: 0–520 mg l^{-1}; $\bar{x}=22$ mg l^{-1}; K: 5.2–12 mg l^{-1}; Mg: 10–93 mg l^{-1}; N: 0.2–0.5 mg l^{-1}; Na: 7.2–25 mg l^{-1}; P: 0.001–0.005 mg l^{-1}; SO_4^{2-}: 0–334 mg l^{-1}; $\bar{x}=26$ mg l^{-1}). In a survey of Ohio populations, most plants occurred in slightly acidic waters (pH: 6.1–6.7); in Indiana they also occurred primarily in acidic lakes (pH: 5.1–6.9). Laboratory grown plants grow well from pH 5.6 to 7.0 but succumb at a pH <5.0 or >8.0. Although fertile plants can occur from May to October, the level of sexual reproduction is extremely low, with only 1.5% of collected specimens having flowers and just 0.4% bearing

fruits. Various treatments have successfully induced flowers in laboratory cultures. No seed bank is known to develop. The whole plants perennate. Where the winters remain mild, the plants persist throughout the season on the water surface, becoming sparser as ice begins to form. Otherwise, their diminished air-space tissue and high starch reserves cause the fronds to sink to the bottom in the autumn. There they overwinter as intact green plants in lotic waters or in deeper lentic waters, which become no colder than 4°C. They can form a bottom layer up to 5 cm thick and are able to persist through the winter beneath a surface cover of ice as much as 3 m thick. Although nutrients are taken up primarily by the fronds, the roots often are lacking; however, the structures facilitate attachment of the plants to the substrate, especially when currents are present. The fronds are dispersed locally by water currents and wave action, which is evidenced by the large numbers recovered from drift samples. Because seed production virtually is non-existent, the principal natural means of long-distance dispersal must be by the carriage of plants (or viable portions thereof) by waterfowl (Aves: Anatidae). Presumably the fronds become tangled within their plumage or adhere to their muddy feet. Their dispersal limits are restricted by their poor drought tolerance with only 38% of fronds surviving after an hour of desiccation. The plants have a relatively slow growth rate, which has been estimated at only 0.024 new fronds day^{-1}. The fronds are kept clear of epiphytic algae by producing 12-oxoPDA, an algicidal compound. The plants can form dense mats and dominate under optimal conditions; e.g., in one study they comprised more than 85% of the total macrophyte biomass (2 m depth). However, the plants are sensitive to waterborne nutrients and have disappeared entirely following whole lake treatments with slaked lime [$Ca(OH)_2$] or calcite [$CaCO_3$]. Due to its superior ability to acquire waterborne nutrients, *L. trisulca* competes more effectively with *L. minor* in mesotrophic sites, but less effectively in eutrophic sites, where it eventually becomes shaded by dense surface growths of the latter.

Reported associates: *Acer rubrum, Alisma subcordatum, Alisma triviale, Alnus rubra, Alopecurus geniculatus, Amblystegium, Azolla cristata, Batrachospermum, Berula erecta, Bidens beckii, Bolboschoenus fluviatilis, Brasenia schreberi, Butomus umbellatus, Calamagrostis, Calla palustris, Callitriche hermaphroditica, Callitriche heterophylla, Cardamine bulbosa, Carex atherodes, Carex hystericina, Carex lacustris, Carex lupuliformis, Carex lupulina, Carex lurida, Carex rostrata, Carex stipata, Carex viridula, Cephalanthus occidentalis, Ceratophyllum demersum, Chara globularis, Chara vulgaris, Cicuta bulbifera, Cirsium parryi, Comarum palustre, Cornus sericea, Drepanocladus, Dulichium arundinaceum, Egeria densa, Eleocharis acicularis, Eleocharis palustris, Elodea canadensis, Equisetum fluviatile, Fontinalis, Fraxinus profunda, Galium obtusum, Geum rivale, Gleditsia aquatica, Glyceria striata, Heteranthera dubia, Hippuris vulgaris, Hydrocharis morsus-ranae, Iris virginica, Isoetes echinospora, Juncus balticus, Juncus nodosus, Justicia americana, Larix laricina, Leersia oryzoides, Lemna gibba, Lemna minor, Lemna minuta,*

Lemna obscura, Lemna perpusilla, Lemna turionifera, Limnobium spongia, Lobelia cardinalis, Ludwigia palustris, Lychnothamnus barbatus, Mentha aquatica, Menyanthes trifoliata Mimulus guttatus, Myriophyllum heterophyllum, Myriophyllum sibiricum, Myriophyllum spicatum, Myriophyllum verticillatum, Najas flexilis, Najas guadalupensis, Najas marina, Najas minor, Nasturtium officinale, Nelumbo lutea, Nitella flexilis, Nostoc, Nuphar advena, Nuphar polysepala, Nuphar variegata, Nymphaea odorata, Nymphaea tetragona, Persicaria amphibia, Persicaria coccinea, Persicaria hydropiperoides, Phalaris arundinacea, Phragmites australis, Physostegia virginiana, Poa annua, Pontederia cordata, Populus heterophylla, Potamogeton amplifolius, Potamogeton berchtoldii, Potamogeton crispus, Potamogeton diversifolius, Potamogeton epihydrus, Potamogeton foliosus, Potamogeton friesii, Potamogeton gramineus, Potamogeton illinoensis, Potamogeton natans, Potamogeton nodosus, Potamogeton obtusifolius, Potamogeton perfoliatus, Potamogeton praelongus, Potamogeton pusillus, Potamogeton richardsonii, Potamogeton robbinsii, Potamogeton zosteriformis, Quercus palustris, Ranunculus flabellaris, Ranunculus longirostris, Ranunculus trichophyllus, Rhizoclonium hieroglyphicum, Riccia fluitans, Ricciocarpus natans, Rudbeckia laciniata, Rumex obtusifolius, Ruppia spiralis, Sagittaria cuneata, Sagittaria latifolia Salix geyeriana, Schedonorus arundinaceus, Schoenoplectus acutus, Schoenoplectus americanus, Schoenoplectus tabernaemontani, Scirpus cyperinus, Scolochloa festucacea, Senecio hydrophilus, Sidalcea neomexicana, Sium suave, Solidago spectabilis, Sparganium americanum, Sparganium androcladum, Sparganium angustifolium, Sparganium eurycarpum, Sparganium natans, Sphagnum, Spirodela polyrhiza, Stuckenia filiformis, Stuckenia pectinata, Triglochin maritima, Typha angustifolia, Typha latifolia, Typha ×glauca, Utricularia geminiscapa, Utricularia intermedia, Utricularia macrorhiza, Utricularia minor, Vallisneria americana, Veronica americana, Veronica catenata, Wolffia brasiliensis, Wolffia columbiana, Wolffiella gladiata, Zannichellia palustris, Zizania aquatica.

4.2.2.8. **Lemna turionifera Landolt** is an indigenous perennial, which grows on the water surfaces of alluvial fans, backwaters (river), borrow pits, canals (backwater), cattle wallows, channels (stream), depressions, ditches (drainage, irrigation, roadside), fens (calcareous), flowages, impoundments, lagoons, lakes, marshes (alkali, dune, freshwater, peaty, seasonal), meadows (marshy), mudflats, oxbows, ponds (artificial, beaver, boggy, stagnant), pools (calcareous, stagnant, swamp, woodland), potholes, reservoirs, rivers, rivulets, seepages, sloughs, streams, swales, and swamps at elevations to 3008 m. The exposures range from full sunlight to partial shade. The substrates include alluvium (glacial), clay, loam (alluvial), muck (humic), mucky sand, mud, sand (marly), sandy peat, silt, and silty loam. These plants occur in cooler temperate regions characterized by an annual mean air temperature range of −40°C to 26°C but require a minimum of 80 days when the mean temperature is above 10°C. They grow in stagnant conditions or in sites with moderate currents, in

alkaline, mesotrophic to eutrophic waters (pH: 3.5–10.4; conductivity: 85–2230 μS cm^{-1}; alkalinity: e.g., 177.5 mg l^{-1}; Ca: 4.6–73 mg l^{-1}; K: 3.3–18 mg l^{-1}; Mg: 3.9–95 mg l^{-1}; N: 0.2–2.9 mg l^{-1}; Na: 7.0–290 mg l^{-1}; P: 0.00–0.04 mg l^{-1}). In nature, this species is found more often in more alkaline conditions than other duckweeds but can be grown in chelated cultures at a pH as low as 3.2–3.5. The reproductive biology has not been studied in detail, but the flowers are thought to be self-compatible and protogynous. In any case, sexual reproduction is uncommon with only 4% of collected specimens bearing flowers and just 1.2% with fruits. Fertile plants can flower and fruit from April to October. The seeds are retained within the fruits when ripe; their dormancy requirements (if any) have not been determined. As the specific epithet implies, the plants perennate vegetatively by the formation of compact, morphologically distinct turions, which detach from the fronds and sink to the bottom where they overwinter. They have a high anthocyanin content, which imparts an olive-brown color. Turion germination requires water temperatures of at least 10°C. If removed from the water, they desiccate rapidly and die within a few hours. Due to the low fertility rate, most reproduction occurs vegetatively by the proliferation of new, daughter fronds. Presumably, the populations are highly clonal; however, the only corroborating genetic evidence is their lack of appreciable variation at one allozyme locus (MDH). Despite among-strain variation in some features (e.g., cumulative fecundity, frond shape), the main demographic characteristics of the plants (e.g., fecundity trajectories, lifespan, mortality, survivorship) remain fairly stable across accessions. All populations are subject to increased age-related mortality and reduced fecundity. However, parental frond age was found to have no effect on the shape (symmetry) of the fronds, despite their successive decline in size. Cadmium and salt tolerance of the plants has been linked to the expression of the *NHX1* gene. They are thought to be inferior competitors with either *L. minor* or *L. obscura*. **Reported associates:** *Acorus, Alisma subcordatum, Alnus rubra, Azolla, Bassia hyssopifolia, Beckmannia syzigachne, Berula erecta, Bidens beckii Bidens cernuus, Bidens frondosus, Brasenia schreberi, Callitriche hermaphroditica, Callitriche heterophylla, Callitriche palustris, Callitriche stagnalis, Carex atherodes, Carex fuliginosa, Carex hystericina, Carex lasiocarpa, Carex obnupta, Cephalanthus occidentalis, Ceratophyllum demersum, Ceratophyllum echinatum, Chara, Conyza canadensis, Cornus racemosa, Damasonium californicum, Dulichium arundinaceum, Eleocharis acicularis, Eleocharis obtusa, Elodea canadensis, Elodea nuttallii, Epilobium coloratum, Equisetum arvense, Equisetum fluviatile, Glyceria, Heteranthera dubia, Hippuris vulgaris, Hosackia oblongifolia, Hydrocotyle ranunculoides, Impatiens capensis, Iris pseudacorus, Juncus arcticus, Juncus effusus, Landoltia punctata, Lemna aequinoctialis, Lemna minor, Lemna minuta, Lemna perpusilla, Lemna trisulca, Lemna valdiviana, Lobelia siphilitica, Lonicera maackii, Ludwigia peploides, Marsilea vestita, Menyanthes trifoliata, Mimulus guttatus, Muhlenbergia richardsonis, Myriophyllum aquaticum, Myriophyllum sibiricum, Myriophyllum spicatum,*

Myriophyllum verticillatum, Najas canadensis, Najas flexilis, Nasturtium officinale, Nuphar variegata, Nuphar variegata, Nymphaea odorata, Peltandra virginica, Persicaria amphibia, Phalaris arundinacea, Phragmites australis, Platanthera, Potamogeton alpinus, Potamogeton berchtoldii, Potamogeton crispus, Potamogeton diversifolius, Potamogeton epihydrus, Potamogeton foliosus, Potamogeton friesii, Potamogeton gramineus, Potamogeton illinoensis, Potamogeton natans, Potamogeton nodosus, Potamogeton praelongus, Potamogeton pusillus, Potamogeton richardsonii, Potamogeton strictifolius, Potamogeton zosteriformis, Ranunculus longirostris, Rhamnus cathartica, Riccia fluitans, Ricciocarpus natans, Sagittaria cuneata, Sagittaria latifolia, Salix, Saururus cernuus, Schoenoplectus acutus, Schoenoplectus americanus, Scirpus atrovirens, Scirpus microcarpus, Scolochloa festucacea, Sidalcea neomexicana, Solanum dulcamara, Solidago patula, Sparganium emersum, Sparganium eurycarpum, Spiraea douglasii, Spirodela polyrhiza, Spirogyra, Stachys ajugoides, Stuckenia filiformis, Stuckenia pectinata, Symplocarpus foetidus, Typha angustifolia, Typha latifolia, Urtica dioica, Utricularia macrorhiza, Vallisneria americana, Veronica anagallisaquatica, Veronica scutellata, Wolffia borealis, Wolffia brasiliensis, Wolffia columbiana, Wolffiella lingulata, Zannichellia palustris.

4.2.2.9. *Lemna valdiviana* Phil.

is an indigenous perennial, which floats on the water surfaces of backwaters, bayheads, bayous, borrow pits, brooks, bottoms (river), canals (abandoned), channels (spring), clay pits, depressions, ditches (overflow, pocosin, roadside), flatwoods, floodplains, hammocks (hydric, maritime), impoundments, lagoons (saline), lakes, marshes (freshwater, roadside, saline, seasonal, tidal), meadows (marshy, montane), mudflats, oxbows, ponds (artificial, beaver, dune, marshy, spring, stock, warm, woodland), pools (backwater, dune, lime-sink, quarry, roadside, spring, stagnant, swamp), puddles, rivers, sinks (lime), sloughs, springs (boggy, montane, roadside, seepage), streams (blackwater, meadow, stagnant), swamps (basin, calcareous, floodplain, riverine), tar pits, and terraces (river) at elevations to 2591 m [to 3800 m in Peru]. Exposures range from full sun to shade, with a greater tendency toward shaded sites. The reported substrates include alluvium (clayey, loamy), clay loam, gravel, loam, muck (black), rock (vertical), sand, sandy gravelly loam, sandy loam, silty clay loam (Cancienne), and silty loam. These are plants of warmer temperate to tropical climates, where the mean annual temperatures occur from 1°C to 30°C; they require at least 170 days when the mean temperatures remain above 10°C. The habitats are still to slow flowing, mesotrophic waters (pH: 5.2–7.6; $\bar{x}$ =6.5; conductivity: 41–199 μS cm^{-1}; Ca: 4.6–31 mg l^{-1}; Cl: 1–2.9 mg l^{-1}; K: 0.6–13.5 mg l^{-1}; Mg: 0.8–17 mg l^{-1}; N: 0.4–3.6 mg l^{-1}; Na: 2.9–47 mg l^{-1}; P: 0.003–0.09 mg l^{-1}). The plants are collected with flowers rarely (3% of collected specimens) and fruit production is rare (1.6% of collections). No detailed study has been conducted on the reproductive biology of this species; however, the flowers presumably are self-compatible but protogynous. Fertile plants can occur from March to November.

Although normally floating, the plants can withstand periods of submergence as resting fronds. These are characterized by reduced air-space tissue and high starch content, which enables them to sink to the bottom until more favorable conditions for growth return. The plants are more tolerant of high sulfite (SO_3^{2-}) concentrations than other duckweeds evaluated, which is attributed to their rapid frond replication rate. They require adequate levels of manganese (Mn; ~1 mg l^{-1}), and exhibit growth deficiency symptoms within 3 weeks when reared in Mn-free cultures. **Reported associates:** *Acer rubrum, Acrostichum danaeifolium, Alternanthera philoxeroides, Amaranthus australis, Ambrosia acanthicarpa, Ambrosia psilostachya, Amorpha fruticosa, Ampelopsis arborea, Amsonia rigida, Aristolochia serpentaria, Artemisia dracunculus, Asclepias perennis, Axonopus furcatus, Azolla cristata, Azolla filiculoides, Baccharis pilularis, Baccharis salicifolia, Bacopa monnieri, Bacopa rotundifolia, Bartonia paniculata, Begonia cucullata, Berchemia scandens, Berula erecta, Bidens cernuus, Bidens laevis, Boehmeria cylindrica, Bolboschoenus robustus, Brasenia schreberi, Cabomba caroliniana, Calibrachoa parviflora, Callitriche heterophylla, Caltha palustris, Campsis radicans, Cardamine hirsuta, Carex atlantica, Carex glaucescens, Carex joorii, Carex lasiocarpa, Carex louisianica, Carex lupulina, Carex rostrata, Carex striata, Carpinus caroliniana, Carya aquatica, Centaurea solstitialis, Centella asiatica, Cephalanthus occidentalis, Ceratophyllum australe, Ceratophyllum demersum, Ceratophyllum echinatum, Chara zeylanica, Cicuta douglasii, Cladium jamaicense, Claytonia, Clematis crispa, Clethra alnifolia, Coleataenia anceps, Coleataenia longifolia, Conoclinium coelestinum, Conyza canadensis, Coreopsis leavenworthii, Cornus foemina, Cotula coronopifolia, Crataegus viridis, Cynosurus echinatus, Cyperus eragrostis, Cyperus haspan, Cyperus retrorsus, Cyperus strigosus, Cyperus virens, Cyrilla racemiflora, Datisca glomerata, Desmodium, Dichanthelium commutatum, Dichanthelium dichotomum, Dichanthelium laxiflorum, Dieteria canescens, Diodia virginiana, Diospyros virginiana, Ditrysinia fruticosa, Dyschoriste humistrata, Echinodorus berteroi, Echinodorus cordifolius, Eichhornia crassipes, Eleocharis acicularis, Eleocharis baldwinii, Eleocharis dulcis, Eleocharis elongata, Eleocharis interstincta, Eleocharis montevidensis, Eleocharis obtusa, Eleocharis parishii, Eleocharis vivipara, Elymus glaucus, Elytraria caroliniensis, Elytraria caroliniensis, Equisetum laevigatum, Eragrostis, Erechtites hieracifolius, Eupatorium compositifolium, Forestiera acuminata, Fraxinus caroliniana, Funastrum clausum, Galium tinctorium, Gleditsia aquatica, Gonolobus suberosus, Hoita orbicularis, Hosackia oblongifolia, Hottonia inflata, Hydrilla verticillata, Hydrocotyle ranunculoides, Hydrocotyle umbellata, Hydrocotyle verticillata, Hydrolea quadrivalvis, Hydrolea uniflora, Hymenocallis duvalensis, Hymenocallis rotata, Hypericum galioides, Hypoxis curtissii, Ilex decidua, Impatiens capensis, Ipomoea cordatotriloba, Ipomoea sagittata, Itea virginica, Juncus nevadensis, Juncus xiphioides, Justicia americana, Justicia ovata, Landoltia punctata, Leersia virginica, Lemna* aequinoctialis, Lemna gibba, Lemna minor, Lemna minuta, Lemna obscura, Lemna turionifera, Lilium parryi, Limnobium spongia, Liquidambar styraciflua, Lobelia cardinalis, Ludwigia palustris, Ludwigia peploides, Ludwigia repens, Luziola fluitans, Lycopus rubellus, Melilotus albus, Melothria pendula, Mentha pulegium, Menyanthes trifoliata, Micranthemum umbrosum, Mikania scandens, Mimulus guttatus, Mitchella repens, Mitreola petiolata, Myrica cerifera, Myriophyllum heterophyllum, Myriophyllum pinnatum, Myriophyllum verticillatum, Najas flexilis, Najas guadalupensis, Nasturtium officinale, Nelumbo lutea, Nuphar advena, Nymphaea odorata, Nyssa aquatica, Nyssa biflora, Nyssa sylvatica, Oldenlandia uniflora, Oplismenus hirtellus, Osmunda regalis, Oxalis corniculata, Packera glabella, Panicum virgatum, Paspalum repens, Peltandra virginica, Persea borbonia, Persicaria arifolia, Persicaria glabra, Persicaria hydropiperoides, Persicaria punctata, Persicaria setacea, Phanopyrum gymnocarpon, Physostegia leptophylla, Pilea pumila, Pistia stratiotes, Planera aquatica, Planera aquatica, Polypogon monspeliensis, Polypogon viridis, Pontederia cordata, Potamogeton clystocarpus, Potamogeton diversifolius, Potamogeton friesii, Potamogeton nodosus, Potamogeton pulcher, Potentilla anserina, Proserpinaca palustris, Pteridium aquilinum, Ptilimnium capillaceum, Quercus phellos, Ranunculus laxicaulis, Rhynchospora caduca, Rhynchospora corniculata, Rhynchospora miliacea, Ribes aureum, Rosa palustris, Rudbeckia laciniata, Sabal minor, Sabal palmetto, Saccharum baldwinii, Sagittaria brevirostra, Sagittaria calycina, Sagittaria lancifolia, Sagittaria latifolia, Sagittaria platyphylla, Sagittaria subulata, Salix caroliniana, Salix nigra, Salvinia minima, Samolus valerandi, Saururus cernuus, Schedonorus arundinaceus, Schinus terebinthifolius, Schoenoplectus californicus, Schoenoplectus pungens, Schoenoplectus tabernaemontani, Scirpus cyperinus, Scirpus microcarpus, Scleria triglomerata, Senna marilandica, Serenoa repens, Sideroxylon reclinatum, Sisyrinchium atlanticum, Smilax bona-nox, Smilax tamnoides. Sparganium eurycarpum, Spiranthes cernua, Spirodela polyrhiza, Steinchisma hians, Stuckenia pectinata, Symphyotrichum dumosum, Symphyotrichum lateriflorum, Symphyotrichum subulatum, Tamarix, Taxodium distichum, Teucrium canadense, Teucrium canadense, Toxicodendron radicans, Triadenum walteri, Trifolium, Typha domingensis, Typha latifolia, Ulmus alata, Ulmus americana, Urtica dioica, Utricularia foliosa, Utricularia gibba, Utricularia inflata, Utricularia intermedia, Utricularia purpurea, Utricularia subulata, Veratrum californicum, Veronica anagallis-aquatica, Veronica catenata Wolffia brasiliensis, Vitis aestivalis, Vitis cinerea, Vitis rotundifolia, Wolffia columbiana, Wolffiella gladiata, Wolffiella oblonga, Woodwardia virginica, Zizania aquatica, Zizaniopsis miliacea.

Use by wildlife: Many *Lemna* species are consumed by waterfowl (Aves: Anatidae); however, because processed food contents are difficult to identify, the food items often are attributed simply to "duckweeds" rather than to a particular species (or even genus). *Lemna aequinoctialis* is eaten by various ducks (Aves: Anatidae) including blue-winged teal (*Anas discors*), canvasback (*Aythya valisineria*), cinnamon teal (*Anas cyanoptera*), goldeneye (*Bucephala*), mallard (*Anas platyrhynchos*), pintail (*Anas acuta*), scaup (*Aythya*) shovelers (*Anas clypeata*), and wood ducks (*Aix sponsa*). Its fronds have been found to harbor unique bacterial endophytes (Bacteria: Paenibacillaceae: *Paenibacillus lemnae*; Rhizobiaceae: *Rhizobium lemnae*). *Lemna minor* is a common food of ducks (Aves: Anatidae) such as the European wigeon (*Anas penelope*), green-winged teal (*Anas crecca*), mallard (*Anas platyrhynchos*), and wood duck (*Aix sponsa*). It hosts a rich invertebrate fauna (which also serves as duck food) including flatworms (Platyhelminthes: Planariidae: *Planaria maculata*), hydras (Cnidaria: Hydraenidae: *Hydraena vulgaris*; Hydridae: *Chlorohydra viridissima*), springtails (Collembola: Poduridae: *Podura aquatica*; Sminthuridae: *Sminthurus aquaticus*), and numerous insects (Insecta) such as aphids (Hemiptera: Aphididae: *Rhopalosiphum nymphaeae*), beetles (Coleoptera: Dytiscidae: *Hydroporus, Laccophilus*; Erirhinidae: *Tanysphyrus lemnae*; Haliplidae: *Haliplus*; Scirtidae: *Scirtes tibialis*; Hydrophilidae: *Tropisternus lateralis*), bugs (Hemiptera: Corixidae: *Corixa*; Gerridae: *Gerris, Trepobates pictus*; Mesoveliidae: *Mesovelia mulsanti*; Nepidae: *Ranatra*; Notonectidae: *Notonecta*; Pleidae: *Neoplea striola*), caddis flies (Trichoptera: Limnephilidae: *Limnephilus rhombicus*), fairy flies (Hymenoptera: Mymaridae: *Patasson pullicrura, Polynema*), flies (Diptera: Ceratopogonidae: *Dasyhelea traverae*; Chironomidae: *Chironomus, Corynoneura scutellata*; Culicidae: *Anopheles, Culex*; Ephydridae: *Lemnaphila scotlandae*; Sciomyzidae: *Tetanocera*; Stratiomyidae: *Odontomyia*; Syrphidae: *Eristalis tenax*), moths (Lepidoptera: Crambidae: *Nymphula obliteralis*), parasitic wasps (Hymenoptera: Braconidae: *Opius lemnaphilae*; Ichneumonidae: *Trichopria angustipennis, T. paludism*; Scelionidae: *Tiphodytes gerriphagus*), and thrips (Thysanoptera: Thripidae: *Limothrips cerealium*). The aforementioned caddis fly and moth species construct their larval cases from the fronds. The plants also host pathogenic protists (Chromista: Hyphochytridiomycota: Rhizidiomycetaceae: *Reessia amoeboidea*). They are colonized by numerous epiphytes in numbers far greater than those on inert substrates. *Lemna minuta* can contain up to 26.4% protein if grown under highly eutrophic conditions. *Lemna trisulca* fronds (and seeds when present) are eaten by virile crayfish (Crustacea: Decapoda: Cambaridae: *Orconectes virilis*), American coots (Aves: Rallidae: *Fulica americana*), and by various ducks (Aves: Anatidae) including the European wigeon (*Anas penelope*), green-winged teal (*Anas crecca*), lesser scaup (*Aythya affinis*), mallard (*Anas platyrhynchos*), ruddy duck (*Oxyura jamaicensis*), and wood duck (*Aix sponsa*). In addition to being eaten directly by waterfowl, the fronds can support a substantial biomass of invertebrates (also relished by ducks) of up to 152 animals at 2059 mg 100 g^{-1} plant mass. These include amphipods (Crustacea: Amphipoda: Hyalellidae: *Hyalella azteca*), and insects such as damselflies (Odonata: Coenagrionidae: *Ischnura*), giant water bugs (Hemiptera: Belostomatidae: *Belostoma*), midges (Diptera: Chironomidae), snails (Mollusca: Gastropoda: Physidae: *Physa*; Planorbidae: *Gyraulus*), water beetles (Coleoptera: Haliplidae: *Haliplus*), and water boatman (Hemiptera: Corixidae: *Trichocorixa*). *Lemna trisulca* is host to an endophytic alga (Chlorophyta: Chlorochytriaceae: *Chlorochytrium lemnae*). *Lemna turionifera* provides cover and food (associated invertebrates) for fish (Vertebrata: Teleostei). *Lemna valdiviana* is eaten by wood ducks (Aves: Anatidae: *Aix sponsa*) and pond slider turtles (Vertebrata: Reptilia: Emydidae: *Trachemys scripta*). It is a host to a leaf-mining fly (Insecta: Diptera: Ephydridae: *Lemnaphila scotlandae*). Consumption of duckweed fronds (*L. minor, L. trisulca*) is known to reduce the severity of lead poisoning in waterfowl (Aves: Anatidae), which can inadvertently consume stray lead hunting shot when feeding. Numerous fronds of *Lemna minor* and *L. trisulca* have been found in the stomach contents of mink frogs (Amphibia: Ranidae: *Rana septentrionalis*) and leopard frogs (Amphibia: Ranidae: *Rana pipiens*); but they are believed to be ingested accidentally as the frogs capture other prey at the water surface.

Economic importance: food: Because of their high protein contents, *Lemna gibba* (to 16.9%) and *L. minor* (to 13.4%) have been potentially evaluated as an aquatic crop. The growth of *Lemna aequinoctialis* is not affected by low gravity and the plants have been considered as a source of food for astronaut crews during long-duration space missions. *Lemna gibba* is grown as a salad vegetable in Israel; *L. obscura* is eaten by tribal native Americans in northeastern Oklahoma; **medicinal:** The Iroquois added moist fronds of *Lemna trisulca* to a poultice used to reduce swelling; **cultivation:** *Lemna minor* includes the following cultivars: 'Henry Allegro', 'Henry Blanke', 'Henry Dacapo', 'Henry Forte', 'Henry Josef', 'Henry Legato', 'Henry Maria', and 'Henry Vitesse', which have been patented for biotechnological applications. It also is the most common accidental inclusion among orders of other species distributed as water garden ornamentals; **misc. products:** Duckweed plants are widely used in bioassays to screen for various potentially toxic substances. The "*Lemna* bioassay" has been categorized as "the most standardized plant bioassay for aquatic toxicology." *Lemna aequinoctialis* (misidentified as "*Lemna perpusilla* strain 6746") has been the subject of countless experimental studies on flowering and photoperiodism. Powdered plants of *Lemna aequinoctialis* have been used as a biosorbent material to remove cadmium (Cd^{2+}) from aqueous solutions. *Lemna gibba* has been used to remove nutrients (N, P) from domestic wastewaters and heavy metals (Cd, Cu, Zn) from aqueous solutions. It has been the subject of countless physiological studies, with many devoted to the mechanisms involved in flowering or nutrient uptake. The plants have been used as an additive (up to 15%) for poultry food, and as a substrate to produce bioethanol fuels. *Lemna minor* has been used for removal of heavy metals and nutrients from wastewaters, as a fertilizer additive for rice (*Oryza*) crops, and as a model organism in many toxicity bioassays including those for evaluating human microbial pathogenesis. It has been evaluated as a feed additive (up to 20%) for commercially reared fish (Vertebrata: Teleostei:

Cichlidae: *Oreochromis niloticus*; Cyprinidae: *Cyprinus carpio*) and for the production of ethanol biofuels. Mats of *L. minor* sometimes are referred to as "Jenny Greenteeth," a fictitious literary bogey who drags unwary children beneath the water. *Lemna minuta* has been evaluated as a means of processing agricultural animal wastes for nutrient removal and for decoloring waters stained by industrial textile manufacturing wastes. *Lemna obscura* has a high capacity for the removal of heavy metals from aqueous wastes; e.g., it can accumulate up to 4.95% selenium (Se) mg^{-1} of WM biomass and remove up to 97% of lead (Pb) within a week of exposure. The plants also can remove waterborne phosphorous at a rate of 20 mg P m^{-2} day^{-1}. They have been added as meal to fish food for commercially grown tilapia (Vertebrata: Teleostei: Cichlidae: *Orechromis*) and have been used to generate bioethanol (producing up to 0.41 g g^{-1} glucose day^{-1}). *Lemna perpusilla* has been used as fish food for commercially grown Nile tilapia (Vertebrata: Teleostei: Cichlidae: *Oreochromis niloticus*). Dried, pulverized *L. perpusilla* fronds are adsorptive and effectively remove lead ions (Pb^{2+}) from aqueous solutions. The ashed plants have been used as a catalyst in biofuel production. The plants are used to remove nitrogen (and other nutrients) from aquaculture (shrimp farm) wastewaters. *Lemna trisulca* has provided a study system for investigating chloroplast movement in plants. The plants also have successfully degraded the neurotoxin anatoxin-a (ANTX-a) in phytoremediation trials. Fronds of *L. trisulca* produce an algicide known as 12-oxoPDA, which has been synthesized for commercial use. The plants have been evaluated as having high potential for use in the phytoremediation of waters contaminated by metals (Cd, Hg, Pb, and Zn). *Lemna valdiviana* has been used for wastewater and agricultural sewage treatments and as a suitable food additive for commercially raised fish (Vertebrata: Teleostei); **weeds:** *Lemna minor* is considered to be a weed of botanical gardens, aquatic plant nurseries, and irrigation canals. *Lemna minuta* is seriously invasive in Europe, where it was first recorded from Portugal in 1941. The nonindigenous plants can seriously compete with the indigenous *L. minor*; **nonindigenous species:** *Lemna aequinoctialis* was introduced to North America and to northern China, northern Japan, and southern Europe (France, Germany, Greece, and Italy) primarily through rice (*Oryza*) culture. Other introduced duckweeds include *L. minuta* (eastern Asia and Europe), *L. obscura* (Hawaii), and *L. turionifera* (Europe), which appear to have been introduced primarily as human-mediated escapes relating to fish, aquarium, or water garden cultivation. *Lemna perpusilla* reportedly has been introduced to India, Malaysia, and other parts of Asia.

Systematics: Identification of duckweed species is quite problematic due to their highly reduced and convergent morphology, which makes it difficult even to distinguish distantly related species assigned to different sections of the genus (e.g., *L. minor* and *L. minuta*). Their identification has been facilitated by the development of high-resolution genetic markers (e.g., AFLP, ISSR) and attempts to develop simple DNA barcoding approaches. However, the observation that high-resolution markers (e.g., microsatellites) generate greater intraspecific than interspecific genetic distances in duck-weeds clearly demonstrates their unsuitability for evaluating interspecific relationships in the family. Similarly, AFLP markers can effectively distinguish different duckweed species, but are less successful in reconstructing relationships among species. Consequently, estimates of higher-level (e.g., interspecific) relationships have best been resolved using DNA sequence data derived from less rapidly diverging loci. Such phylogenetic analyses (e.g., Figure 1.27) have included all recognized duckweed species and have incorporated data from both plastid and nuclear loci. These analyses have confirmed the monophyly of *Lemna* as well as the validity of four sections: *Alatae, Biformes, Lemna,* and *Uninerves.* The genus itself typically resolves as the sister group of the *Wolffia/Wolfiella* clade (e.g., Figure 1.27), which some authors recognize as subfamily *Wolffioideae*; however, in some analyses (e.g., *atpF-atpH, psbK-psbI* intergenic spacer data) *Lemna* has associated more closely with *Landoltia.* In any case, an intermediate phylogenetic position between *Spirodela* and *Wolffia/Wolffiella* is indicated and coincides with its intermediate degree of morphological specialization compared to those genera. Previously, *L. perpusilla* had been placed in synonymy with *L. aequinoctialis.* As a result, much confusion arises in the literature; e.g., with "*Lemna perpusilla* 6746," so widely used in physiological experiments, which actually is a strain of *L. aequinoctialis.* The integrity of section *Alatae,* its sister-group relationship with section *Biformes* (i.e., *L. tenera*), and recognition of *L. aequinoctialis* and *L. perpusilla* as closely related, but genetically distinct species, all are in complete accord with allozyme and DNA sequence data (e.g., Figure 1.27). The distinctness of all three species also is supported by AFLP marker data. Divergence of *L. aequinoctialis* and *L. perpusilla* is estimated to have occurred 2.4 million years ago, likely in association with ecological adaptations for cold tolerance. The largest section (sect. *Lemna*) contains seven species. It includes the similar *L. disperma* and *L. gibba,* which once had been regarded as conspecific but were found to have highly divergent allozyme loci and a genetic identity of only 0.404. They are estimated to have diverged via an allopatric mode approximately 5.5 MYBP. The two taxa resolve consistently as distinct sister species in phylogenetic analyses of DNA sequence data, with *L. japonica* and *L. minor* comprising their immediate sister group (e.g., Figure 1.27). Phylogenetically, *L. obscura* and *L. turionifera* associate as a subclade within section *Lemna* (Figure 1.27). "*Lemna ecuadoriensis*" had been recognized as a separate species closely related to *L. obscura,* but eventually was placed in synonymy with the latter by its original author (E. Landolt). Although the distinctive *L. trisulca* was once recognized as a different genus (*Staurogeton*), a relatively close relationship of *L. trisulca* and *L. obscura* was first indicated by their similar flavonoid profiles. That hypothesis eventually was corroborated by phylogenetic analyses of DNA sequence data (e.g., Figure 1.27), which have shown *L. trisulca* to be an integral member of *Lemna* section *Lemna,* where it resolves as the sister species of the *L. obscura + L. turionifera* subclade. DNA sequence data support section *Uninerves* as a clade containing

L. minuta, L. valdiviana, and *L. yungensis* (Figure 1.27). The accessions of these three species also cluster together by AFLP data. The distinctness of the highly similar *L. minuta* (formerly "*L. minuscula*") and *L. valdiviana* was confirmed initially by micromolecular (flavonoid) and allozyme data, which indicated that they were well differentiated genetically if not morphologically. Allozyme data also indicated their relatively recent origin in the family by yielding a genetic identity higher than that estimated for other duckweed species ($I = 0.70$). Until the discovery of *L. yungensis* (an unusual Bolivian duckweed found to inhabit wet, vertical rocks), *L. minuta* and *L. valdiviana* were regarded as sister species. Instead, phylogenetic analyses resolved *L. yungensis* as the sister species of *L. valdiviana* (Figure 1.27), a result that had been predicted from morphological characteristics. The two represent an example of a progenitor-derivative species pair. Divergence times estimated from DNA sequence data are 1.2 MYBP (*L. yungensis* and *L. valdiviana*) and 0.93 MYBP (*L. minuta* vs. *L. yungensis + L. valdiviana*). All three species are believed to have diverged ecologically by the acquisition of different physiological adaptations. Genome size has been estimated (flow-cytometrically) for *L. aequinoctialis* (424–760 Mbp), *L. gibba* (440–486 Mbp), *L. minor* (356–604 Mbp), *L. obscura* (487 Mbp), *L. trisulca* (446–709 Mbp), and *L. valdiviana* (323 Mbp). Complete cpDNA genome sequences are available for *L. minor.* A wide extent of chromosome number variation occurs in *Lemna* ($2n = 20, 30, 36, 40, 42, 50, 60, 70, 80$) and can be manifest at the individual, populational, or broader levels. The basic chromosome number is $x = 5$ or 10. The following counts have (both aneuploid and euploid) been reported: *Lemna aequinoctialis* ($2n = 40, 42, 50, 60, 80, 84$); *L. gibba* ($2n = 40, 42, 44, 50$); *L. minuta* ($2n = 36, 40, 42$); *L. minor*: ($2n = 20, 30, 40, 50$); *L. obscura* ($2n = 40, 42, 50$); *L. perpusilla* ($2n = 40, 42$); *L. trisulca* ($2n = 40, 42, 44, 60, 63, 80$; [$2n = 20$ in Australia]); *L. turionifera* ($2n = 40, 42, 50, 80$); *L. valdiviana* ($2n = 40, 42$). Although *Lemna* species often occur in mixed arrays, their predominantly asexual mode of reproduction likely is a strong deterrent to hybridization. To date, no genetic evidence supports the hypothesized origin of *L. perpusilla* as a hybrid involving *L. aequinoctialis* × *L. turionifera.*

Distribution: *Lemna minor* and *L. trisulca* are widespread throughout North America; *L. turionifera* is widespread throughout North America except in the southeastern United States. *Lemna minuta* extends from western North America to the southern United States; *L. perpusilla* is restricted to eastern North America. *Lemna valdiviana* is found throughout much of the United States with *L. obscura* limited to the eastern United States. The pantropical *L. aequinoctialis* occurs in the central and southern United States with *L. gibba* distributed in the central and southwestern United States.

References: Adamec, 2018; Afton et al., 1991; Ahmad et al., 1990; Aliferis et al., 2009; Ankutowicz & Laird, 2018; Armstrong et al., 1989; Auclair et al., 1973; Barks et al., 2018; Bart, 1989; Baskin & Baskin, 1998; Bergmann et al., 2000; Bhalla & Sabharwal, 1972; Boedeltje et al., 2003a; 2003b; Bog et al., 2010; Bokhari et al., 2016; Borisjuk et al., 2015;

Bouchard & Bjorndal, 2005; Bowles & Dodd, 2015; Brain et al., 2004; Buckingham, 1989; Calicioglu & Brennan, 2018; Carvalho & Martin, 2001; Cedergreen & Madsen, 2002; Cellot et al., 1998; Ceschin et al., 2016; 2018; Chabreck & Palmisano, 1973; Chambers et al., 1991b; Chen et al., 2015b; Chouhan & Sarma, 2013; Cleland & Briggs, 1967; Cleland et al., 1982; Cole & Voskuil, 1996; Coler & Gunner, 1969; Collins, 1905; Coughlan et al., 2015a; 2015b; 2017b; 2018; Crawford, 2010; Crawford et al., 1996; 2001; 2005; 2006; Crombie, 1999; Crowder et al., 1977a; Darst et al., 2002; Daubs, 1965; DeBusk et al., 1995; de Lange & Westinga, 1979; Dickinson & Miller, 1998; Dister, 2017; DiTomaso & Healy, 2003; Driever et al., 2005; Drobney & Fredrickson, 1979; Dudley, 1987; Duong & Tiedje, 1985; Easley & Judd, 1993; Elam et al., 2009; El-Shafai et al., 2004; Elzenga et al., 1980; Fassett, 1957; Foerste, 1883; França et al., 2009; Frick, 1994; Fu et al., 2017; Galatowitsch & van der Valk, 1996; Gallardo-Williams et al., 2002; Garvin et al., 2018; Giles, 1977; Gusain & Suthar, 2017; Haag & Gorham, 1977; Halder & Venu, 2012; Haller et al., 1974; Hanson et al., 2002; Hardy & Raymond, 1991; Hartman, 1985; Harvey & Fox, 1973; Hassan & Edwards, 1992; Haustein et al., 1994; Hedeen, 1972; Hegelmaier, 1868; Hendricks & Goodwin, Jr., 1952; Hicks, 1932a; 1932b; 1937; Hillman, 1961; Hossell & Baker, 1979; Hussner, 2012; Hussner et al., 2010b; Jovet & Jovet-Ast, 1967; Kaminski et al., 2014; Karol et al., 2017; Keddy, 1976; Kittiwongwattana & Thawai, 2014; 2015; Kline & McCune, 1987; Kling, 1986; Körner & Vermaat, 1998; Körner et al., 2001; Krull, 1970; 1976; Landesman et al., 2005; Landolt, 1975; 1981; 1986; 1997; 1998; 2000; Landolt & Wildi, 1977; Landolt et al., 1998; Lansdown, 2008; Lapolli et al., 2008; Larson & Searcy, 2007; Lemon et al., 2001; Les et al., 1997b; Les et al., 2002; Löve & Löve, 1954; Madsen et al., 2006; 2012; Maki & Galatowitsch, 2004; Mallin et al., 2002; Mansor, 1996; Mardanov et al., 2008; McAllister et al., 2017; McAtee, 1925; McClure & Alston, 1966; McIlraith et al., 1989; McLay, 1974; 1976; Megateli et al., 2009; Mifsud, 2010; Miretzky et al., 2004; Mkandawire et al., 2014; Mohlenbrock, 1959a; 1959b; Morris & Barker, 1977; Mühlberg, 1982; Nekrasova et al., 2003; Newton, 2008; Nichols, 1999; Njambuya et al., 2011; Oda, 1962; Olney, 1960; Paolacci et al., 2018; Perleberg & Loso, 2009b; Peters et al., 2009; Pfauth & Sytsma, 2005; Pip, 1979; 1988; Pip & Simmons, 1986; Porath et al., 1979; Prepas et al., 2001; Quattrocchi, 2012; Ramirez-Babativa & Ramírez, 2018; Rejmankova, 1976; Reveal, 1990; Reynolds et al., 2015; Rizwana et al., 2014; Ruenglertpanyakul et al., 2004; Ryman & Anderberg, 1999; Saeger, 1933; Scharfetter et al., 1987; Scotland, 1934; 1940; Siegfried, 1973; Sivakumar et al., 2015; Soreng et al., 1984; Stalter, 1985; Stehle, 1920; Stollberg, 1949; 1950; Strong & Kelloff, 1994; Sugden & Driver, 1980; Takemoto & Noble, 1986; Takimoto & Tanaka, 1973; Tan & Judd, 1995; Tang et al., 2013; Tang et al., 2014; Tavares et al., 2010; Thompson, 1898; Tippery & Les, 2020; Tippery et al., 2015; Tsao et al., 1986; Urbanska-Worytkiewicz, 1975; Vaithiyanathan & Richardson, 1999; Van Dine, 1922; Vasseur et al., 1993; Velichkova et al., 2017; Vickery, 1983; Volker & Smith, 1965; Walker & Wehrhahn, 1971; Wang, 1990; Wang et al., 2011a; Witztum, 1977; 1986; Xue et al., 2012; Yang et al., 2017; Yilmaz, 2007; Yilmaz et al., 2004; Yuan & Xu, 2017; Zhang et al., 2010b; Zurzycki, 2015.

4.2.3. Spirodela

Duckmeat, greater duckweed; spirodèle

Etymology: from the Greek *speiro dēlos* ("visible coil") in reference to the conspicuous spiral tracheids

Synonyms: *Lemna* (in part); *Lenticula* (in part); *Telmatophace* (in part)

Distribution: global: cosmopolitan; **North America:** widespread except far north

Diversity: global: 2 species; **North America:** 1 species

Indicators (USA): obligate wetland (OBL): *Spirodela polyrhiza*

Habitat: freshwater; freshwater (tidal); lacustrine, palustrine, riverine; **pH:** 5.5–9.2; **depth:** 0 m; **life-form(s):** free-floating

Key morphology: roots (to 4 cm) 7–21, adventitious, 1 [rarely 2] perforating the scale-like prophyllum on lower frond surface; fronds (to 10 mm) floating, roundish, finely nerved (7–21), the upper surface shiny, green, lacking papules, often with a red "nodal" spot, the lower surface reddish, the apex pointed or round, turions (to 3 mm) present, brownish or olive, rootless; flowers bisexual, enclosed by a utricular membranous scale (opening narrowly at top), perianth absent; stamens (to 1 mm) 2, 4-loculed, the external locules not higher than internal locules, pollen grains long spinulose; ovary 1, bottle-shaped, style short (to 0.3 mm), stigma funnelform, ovules 1–2; fruits (to 1.5 mm) circular follicles, winged (to 0.15 mm) laterally to apex, single-seeded; seeds (to 1.0 mm) with 12–20 distinct ribs

Life history: duration: perennial (turions, whole plants); **asexual reproduction:** frond division, turions; **pollination:** insect; **sexual condition:** hermaphroditic; **fruit:** utricular follicle (rare); **local dispersal:** fronds, turions (water); **long-distance dispersal:** fronds, turions (animals)

Imperilment: 1. *Spirodela polyrhiza* [G5]; S1 (UT, WY); S2 (NC); S3 (AB, MB, NB, PE, WV)

Ecology: general: Both *Spirodela* species are obligately aquatic floating plants (pleustophytes) with similar life histories. Among other Lemnaceae, the genus is characterized by the relatively large size of the fronds and by each frond possessing multiple roots, which are thought to function primarily as stabilization organs rather than for nutrient assimilation. Aqueous nutrient uptake occurs primarily by the fronds, which also can assimilate carbon both from the air and water. The reproductive biology of both species is similar, in that neither produces many flowers or fruits with most reproduction resulting from vegetative multiplication of the fronds. Floral and fruit production are somewhat higher in *S. intermedia*; however, turion production (a major means of vegetative reproduction and annual re-establishment) occurs only in *S. polyrhiza*, a feature that enables it to extend further into cooler temperate regions. The mean relative growth rate also is somewhat higher in *S. intermedia* (0.327 day^{-1}) than for *S. polyrhiza* (0.294 day^{-1}). Otherwise, more specific ecological and life history details are summarized below for *S.*

polyrhiza, which is the only North American representative of the genus.

4.2.3.1. *Spirodela polyrhiza* (L.) Schleid. inhabits the water surfaces of backwaters, bayous, bays (sheltered), bogs (Sphagnum), borrow pits, brooks, canals, channels (still), coves (sheltered), ditches (drainage, dredged, irrigation, roadside, seepage), floodplains, lakes (spring-fed), marshes (brackish, floodplain, grassy, intertidal, roadside), meadows (marshy), mill races, oxbows, ponds (artificial, beaver, farm, interdunal, meadow, mill, overflow, seasonal, shallow, sink hole, spring-fed, stagnant, warm), pools (drying, hydrothermal, vernal), potholes (permanent), puddles, quarries (abandoned, limestone), reservoirs, rivers (spring-fed), seepages, shores (lake), sink holes, sloughs (bog, bottomland), springs, streams (meadow, slow-moving), and swamps (bottomland, cypress) at elevations to 2792 m. Exposures most often receive full sunlight; shaded sites are occupied occasionally, but usually only for a short duration. The frond cells possess a DNA repair mechanism for removing UV-induced thymine dimers from their DNA. The substrates typically are gravel or sand, with fewer occurrences of clay, loam, muck, mucky clay, mud (gravelly, peaty), rock, silt, silty mud, or organic material. The plants inhabit tropical to temperate regions, which sustain more than 110 days with mean temperatures above 10°C. By overwintering (as turions) on the water bottom, they can withstand surface (air) temperatures ranging from −40° to >30°C. The adult plants deteriorate after they have been exposed to temperatures of 7°C or below. Turion formation ceases at 32°C and temperatures of 38°C or higher are lethal. Growth in the laboratory is optimal from 25°C to 28°C and declines rapidly at higher temperatures; minimum growth occurs at 10°C. The plants acclimate to different temperatures regardless of their genetic background and geographical source of origin. Although typically floating, the fronds often become stranded on muddy sediments after the standing surface waters have receded. There they can persist for 9 months or more (even producing turions), as long as the substrates remain moist. The fronds are intolerant of salinity and exhibit substantially reduced growth at salinities higher than 0.3‰ NaCl. Turion production is far more sensitive and slows when salinity values rise above 0.01‰ NaCl. The habitats are quiet, permanent, alkaline, hard, eutrophic waters (pH: 5.5–9.2; $\bar{x}$ = 6.7–7.9; $\bar{x}$ = 7.5; alkalinity [as $CaCO_3$]: 2.3–300 mg l^{-1}; $\bar{x}$ = 39.7–71.0; $\bar{x}$ = 90 mg l^{-1}; conductivity: 26–34 µS cm^{-1}; $\bar{x}$ = 237 µS cm^{-1}; 40–600 µmhos cm^{-1} [@ 25°C]; $\bar{x}$ = 180 µmhos cm^{-1}; Ca: 2.5–94.2 mg l^{-1}; $\bar{x}$ = 38.6 mg l^{-1}; Cl$^-$: 3.3–171.1 mg l^{-1}; $\bar{x}$ = 43.7 mg l^{-1}; K: 0.2–5.5 mg l^{-1}; $\bar{x}$ = 2.3 mg l^{-1}; Mg: 2.1–46.5 mg l^{-1}; $\bar{x}$ = 15.6 mg l^{-1}; total N: 430–1760 µg l^{-1}; $\bar{x}$ = 1050 µg l^{-1}; Na: 2.1–89.6 mg l^{-1}; $\bar{x}$ = 25.3 mg l^{-1}; NO_2^-: 0.4–2.6 mg l^{-1}; $\bar{x}$ = 1.0 mg l^{-1}; total P: 12–234 µg l^{-1}; $\bar{x}$ = 80 µg l^{-1}; SO_4^{2-}: 1–619 mg l^{-1}; $\bar{x}$ = 20.9 mg l^{-1}). A comparable range of acidity (pH: 5.9–7.9) was observed among natural Ohio populations with the greatest abundance of occurrences at pH 6.3–7.5 (a range corresponding well with the overall mean and modal values reported). Different optimal pH values have been claimed for cultured plants (e.g., pH: 4.7–4.8; 6.2–6.8; 6.6–7.3), but typically are near pH 7.0. In any case, an absolute minimum pH of 3.9–5.0 and maximum pH of 10.5 are indicated. The preferred form of nitrogen taken up by the plants is NH_4^+, with optimal growth occurring at concentrations from 3.5 to 20 mg l^{-1}; NO_3^- also is assimilated but at a lower level. Fertile plants are almost never observed in the field, with flowers present on only 0.3% of collected specimens, and fruits on just 0.1%. The few fertile specimens have been seen from July to September. Flowering has been induced in laboratory cultures where casein hydrolysate, gibberellic acid, salicylate, and sucrose were added to the growth medium. When produced, the flowers are strongly protogynous, with the stigmas becoming receptive 2 days prior to the shed of pollen and wilting at that time. Presumably, they are pollinated by small arthropods (Arthropoda). Even when produced, the seeds are non-dormant, and in some cases, germinate while still attached to the parental frond. Therefore, they probably are of little consequence in annual re-establishment or in dispersal over any significant distance. As a result, nearly all reproduction is vegetative, either by proliferation (and subsequent separation) of the fronds or by turions, which are produced commonly. The highly clonal nature of populations is evidenced by allozyme data, which detected only four multilocus genotypes among 67 clones examined. Unless conditions are mild enough for whole plants to overwinter on the water surface, perennation occurs exclusively by the turions, which (regardless of climate) can be produced at temperatures ranging from 10°C to 35°C. Turion production is regulated by phytochrome and is promoted by P_{fr}, which increases as the daily photoperiod is lengthened. Turion production in natural populations is profuse during July (initiated by phosphorous depletion from plant growth) and continues until the plants succumb to frost (e.g., October). Lower daytime (e.g., 18°C) or night temperatures (e.g., 15°C) enhance turion production. Unlike normal fronds, turions abscise readily from the parental frond and sink immediately to the bottom where they overwinter. Mean annual temperature strongly influences turion survival rates, which decrease substantially at lower temperatures. They can withstand prolonged freezing within ice (at least 3 months at −4°C), and have been known to survive for more than 21 months at 6°C; however, they are killed within 48 hr when kept at −8° to 12°C. Facilitated by currents, flooding, and wind, the fronds and turions are dispersed abiotically by water throughout contiguous water bodies. Like other duckweeds, the fronds also can become entangled within the plumage of ducks and other waterfowl (Aves: Anatidae), which by virtue of its insulative properties, maintains a humid environment more conducive to extended propagule viability. The turions of *S. polyrhiza* have been recovered from clumps of mud adhering to the feet of ducks (Aves: Anatidae), which evidently disperse them in this manner. However, the turions cannot be dispersed endozoically by water birds because they are fully digested when eaten and are not viable when excreted. The fronds also are known to cling to the pelts of mammals such as muskrats (Mammalia: Cricetidae: *Ondatra zibethicus*), which can easily transport them over short distances between water bodies. The same is surely true for a variety of other aquatic animals. The long-range dispersal of

fronds or turions is limited by desiccation; when exposed to air, they will remain viable for only up to 2.5 hr (often much shorter) but can survive for up to 2 months if embedded in clay or peat. Although no seed bank is produced, the number of viable vegetative propagules (turions or fronds) in the sediments can approach densities as high as 1747 m^{-2}. Most turions (except in some plants from warmer climates) are dormant when shed and remain so indefinitely when kept at low temperatures (0°C–14°C). Dormancy is broken after 2 weeks of cold stratification (–4° to 10°C) followed by incubation at water temperatures in excess of 15°C. Generally, those turions formed at lower temperatures (10°C–15°C) germinate rapidly when exposed to higher temperatures (20°C–30°C). Turion germination is regulated by concentrations of ambient nitrate (an essential requirement) and the presence of readily metabolized carbohydrates. Once the stratified turions are exposed to light, a photosynthetically derived gas bubble develops, which causes them to rise to the water surface where they eventually develop into normal fronds during early spring (e.g., May). Dark germination can be induced by culturing turions under supplemental sucrose levels (10–50 mM sucrose). Turion germination is enhanced either by mild desiccation in air or by exposure to polyethylene glycol (PEG). It also has been enhanced in laboratory cultures that were supplemented with gibberellic acid, which also obviates the need for a prior cold treatment. Turion germination can be inhibited in overcrowded conditions due to anaerobiosis resulting from turion respiration. Potential annual yields of 36.9 t DM ha^{-1} year^{-1} have been estimated from production (total biomass) studies. On average, the fronds have a relatively short lifespan (12.1 days), producing only 1.1 daughter fronds at a rate of only 0.08 day^{-1}. *Spirodela polyrhiza* competes effectively with the much larger floating *Salvinia minima* due to its more efficient nutrient uptake and lower susceptibility to herbivory. **Reported associates:** *Acer saccharinum, Agrimonia parviflora, Alisma subcordatum, Alnus rubra, Alternanthera philoxeroides, Ammannia coccinea, Azolla cristata, Azolla filiculoides, Bacopa egensis, Bacopa rotundifolia, Betula nigra, Bidens aristosus, Bidens cernuus, Bidens frondosus, Brasenia schreberi, Brunnichia ovata, Cabomba caroliniana, Callitriche, Carex aquatilis, Carex cusickii, Carex hystericina, Carex lasiocarpa, Carex tribuloides, Carex utriculata, Cephalanthus occidentalis, Ceratophyllum demersum, Chara, Cinna arundinacea, Cornus drummondii, Crataegus douglasii, Cuscuta, Cyperus erythrorhizos, Cyperus flavescens, Cyperus iria, Decodon verticillatus, Diodella teres, Diospyros virginiana, Echinochloa crus-galli, Echinodorus berteroi, Echinodorus cordifolius, Eclipta prostrata, Eclipta prostrata, Eichhornia crassipes, Eleocharis acicularis, Eleocharis intermedia, Eleocharis microcarpa, Eleocharis obtusa, Eleocharis obtusa, Eleocharis ovata, Elodea canadensis, Elodea nuttallii, Epilobium coloratum, Eupatorium serotinum, Euploca procumbens, Fimbristylis autumnalis, Fimbristylis littoralis, Forestiera acuminata, Fraxinus pennsylvanica, Fraxinus pennsylvanica, Fraxinus pennsylvanica, Gleditsia aquatica, Glyceria, Gratiola neglecta, Gratiola virginiana, Habenaria repens, Heliotropium curassavicum, Heteranthera reniformis, Hibiscus laevis, Hydrocotyle ranunculoides, Hydrocotyle umbellata, Hydrolea quadrivalvis, Hygrophila lacustris, Hypericum mutilum, Ipomoea, Iris virginica, Juncus acuminatus, Juncus effusus, Juncus nodatus, Landoltia punctata, Leersia oryzoides, Leersia virginica, Lemna aequinoctialis, Lemna gibba, Lemna minor, Lemna minuta, Lemna perpusilla, Lemna trisulca, Lemna turionifera, Lemna valdiviana, Leptochloa fusca, Limnobium spongia, Lindera benzoin, Lindernia dubia, Liquidambar styraciflua, Lonicera maackii, Ludwigia decurrens, Ludwigia grandiflora, Ludwigia leptocarpa, Ludwigia palustris, Ludwigia peploides, Ludwigia sphaerocarpa, Lycopus americanus, Marsilea vestita, Mentha arvensis, Menyanthes trifoliata, Mikania scandens, Mimosa strigillosa, Mimulus guttatus, Myriophyllum aquaticum, Myriophyllum pinnatum, Myriophyllum sibiricum, Myriophyllum spicatum, Myriophyllum ussuriense, Myriophyllum verticillatum, Najas flexilis, Najas guadalupensis, Nasturtium officinale, Nelumbo lutea, Nuphar polysepala, Nuphar variegata, Nymphaea odorata, Nyssa aquatica, Nyssa biflora, Oenanthe, Oxycaryum cubense, Packera aurea, Panicum, Paspalum vaginatum, Peltandra virginica, Persicaria amphibia, Persicaria coccinea, Persicaria hydropiperoides, Phalaris arundinacea, Phanopyrum gymnocarpon, Pistia stratiotes, Planera aquatica, Pluchea foetida, Poa annua, Poa autumnalis, Poa nemoralis, Pontederia cordata, Populus heterophylla, Potamogeton amplifolius, Potamogeton diversifolius, Potamogeton gramineus, Potamogeton natans, Potamogeton praelongus, Potamogeton pulcher, Potamogeton pusillus, Potamogeton richardsonii, Potamogeton zosteriformis, Potentilla anserina, Proserpinaca, Ranunculus cymbalaria, Ranunculus flabellaris, Ranunculus longirostris, Ranunculus sceleratus, Rhynchospora corniculata, Riccia fluitans, Ricciocarpus natans, Sacciolepis striata, Sagittaria graminea, Sagittaria latifolia, Salix interior, Salix nigra, Salvinia minima, Salvinia molesta, Samolus valerandi, Saururus cernuus, Schoenoplectus acutus, Schoenoplectus tabernaemontani, Scirpus cyperinus, Scirpus microcarpus, Sesbania herbacea, Sparganium americanum, Sparganium angustifolium, Sparganium emersum, Sparganium eurycarpum, Sparganium natans, Sphagnum teres, Sphenoclea zeylanica, Stuckenia pectinata, Symphyotrichum lateriflorum, Taxodium distichum, Toxicodendron radicans, Triadenum walteri, Typha angustifolia, Typha latifolia, Ulmus crassifolia, Urtica dioica, Utricularia gibba, Utricularia inflata, Utricularia macrorhiza, Utricularia purpurea, Salix discolor, Verbena hastata, Vitis rotundifolia, Wolffia brasiliensis, Wolffia columbiana, Wolffiella gladiata, Xanthium strumarium, Zannichellia palustris.*

Use by wildlife: *Spirodela polyrhiza* fronds contain 10.6% to 31.5% crude protein, 8.8–19.3% crude fiber, 4.5% fat, and all amino acids essential to vertebrate diets. Of the latter, all but methionine occur at levels (g 100 g^{-1} protein) recommended by the FAO. The starch content of plants can reach 46%; individual grains have an amylose content of about 21%. The fronds are eaten by turtles (Reptilia: Emydidae) such as the common slider (*Trachemys scripta*) and Florida red-bellied cooter (*Pseudemys nelsoni*), by musk-rats (Mammalia: Cricetidae: *Ondatra zibethicus*), by nutria (Mammalia: Rodentia: Myocastoridae: *Myocastor coypus*), and by many birds (Aves), including waterfowl (Anatidae) such as the blue-winged teal (*Anas discors*), canvasback (*Aythya valisineria*), mallard (*Anas platyrhynchos*), ruddy duck (*Oxyura jamaicensis*), and wood duck (*Aix sponsa*), as well as pied-billed grebes (Podicipedidae: *Podilymbus podiceps*), coots, gallinules, and soras (Rallidae: *Fulica, Gallinula chloropus*; *Porzana carolina*). They also are eaten by several fish (Vertebrata: Teleostei) such as common carp (Cyprinidae: *Cyprinus carpio*) and grass carp (Cyprinidae: *Ctenopharyngodon idella*). *Spirodela polyrhiza* fronds have been found in the stomach contents of mink frogs (Amphibia: Ranidae: *Rana septentrionalis*) and leopard frogs (Amphibia: Ranidae: *Rana pipiens*) but are believed to be ingested accidentally as the frogs capture other prey at the water surface.

Economic importance: food: *Spirodela polyrhiza* has been evaluated as a potentially suitable food for humans. However, given its affinity for eutrophic sites and ability to concentrate heavy metals and other toxins, any such application would require stringently controlled, sanitary culture conditions. The plants also have been evaluated as being suitable for life-support systems to use in space missions; **medicinal:** *Spirodela polyrhiza* is a medicinal herb in Asia, where it is used as a general diuretic and to treat acute nephritis, hives, inflammation, influenza, measles, and various skin diseases. Flavonoids extracted from the plants are anti-inflammatory and potently antiadipogenic (i.e., they prohibit fatty tissue genesis). Clinical tests have shown that topical applications of dried frond extracts can reduce atopic dermatitis. The fronds contain a fibrinolytic protease, which exhibits anticoagulant activity by hydrolyzing fibrin and fibrinogen; **cultivation:** *Spirodela polyrhiza* is grown as an ornamental aquarium, pond, and water garden plant. One cultivar is 'Henry Big Mama'; **misc. products:** *Spirodela polyrhiza* has been the subject of innumerable physiological experiments relating to nutrient uptake and waterborne toxin removal. It is tolerant to a number of metals (e.g., Cr, Mn) and has been considered for use as a practical phytoremediation system for aquatic environments and for wastewater treatment. The plants have been evaluated as a starch source for production of biofuels. Dried plants have been used as an adsorbent for removal of aqueous lead (Pb^{2+}). The fronds have been included as a dietary food supplement (up to 30%) for the commercial production of Nile tilapia (Vertebrata: Teleostei: Cichlidae: *Oreochromis niloticus*). The properties of *S. polyrhiza* protein make it potentially suitable for use in improving the water resistance of protein-based adhesives; **weeds:** Although bearing the common name of "greater duckweed," *Spirodela polyrhiza* seldom is reported as problematic; **nonindigenous species:** *Spirodela polyrhiza* occurs throughout the world, a distribution believed to have been obtained by natural dispersal events.

Systematics: The monophyly of *Spirodela* and its distinctness from *Landoltia* were first indicated by their flavonoid chemistry, and later by allozyme data, which yielded a low genetic identity ($I=0.404$) between *S. intermedia* and *S. polyrhiza*

but showed a complete lack of shared alleles ($I=0$) between either species and *L. punctata*. These observations were supported later by phylogenetic analyses of DNA sequence data (e.g., Figure 1.27), which consistently resolved the two currently recognized *Spirodela* species as a distinct, sister clade to the remainder of Lemnaceae. AFLP data distinguish both *Spirodela* species and *Landoltia* from one another but are inadequate for evaluating their interrelationships. All three genomes of *S. polyrhiza* have been sequenced. The nuclear genome size of *Spirodela polyrhiza* has been estimated at 160 (150–165) Mbp. Whole-genome sequencing has revealed a relatively low number of protein-coding genes (19,623), contracted gene families, evidence of two ancient genome duplications estimated to have occurred 95 MYBP, and extremely low levels of methylation (9%). Despite the relatively large size of *Spirodela* plants, their genome size is smallest when compared to other Lemnaceae. The genome sequence has been mapped to the respective chromosomes, which has provided a reference for facilitating the study of karyotype evolution in the genus. Such investigations already have indicated that although 10 chromosome pairs are conserved between *S. polyrhiza* and *S. intermedia*, the remainder are characterized by several translocations, inversion, and fissions. Various gene functions currently are being investigated by transcriptome analyses. The chloroplast genome is 168,788 bp in size and contains numerous RNA-edited sites. The mitochondrial genome is extremely compact (228,493 bp) with a total of only 57 genes. Nineteen polymorphic microsatellite markers also have been developed for use in population genetic studies. The basic chromosome number of *Spirodela* is $x=10$. Numerous cytotypes (both euploid and aneuploid) have been found in *S. polyrhiza* ($2n=30, 38, 40, 50, 80$).

Distribution: *Spirodela polyrhiza* is widespread throughout North America except for the far north and higher Rocky Mountain regions.

References: Appenroth, 2002; Appenroth & Nickel, 2010; Appenroth et al., 1989; 1990a; 1990b; 1992; 2013a; 2015; 2017; Aresco, 2010; Babayemi et al., 2006; Bjorndal & Bolten, 1992; Bog et al., 2015; Caicedo et al, 2000; Cao et al., 2016; Chang et al., 2012; Cheng, 2011; Cheng & Stomp, 2009; Cho & Choi, 2003; Cottam, 1939; Crawford & Landolt, 1993; Cui et al., 2011; Das & Gopal, 1969; Davidson & Simon, 1981; Degani et al., 1980; Eshel & Beer, 1986; Hall & Penfound, 1943; Hedeen, 1972; Hicks, 1932a; 1932b; Hoang & Schubert, 2017; Howard & Wells, 2009; Jacobs, 1947; Kasselmann, 2003; Khurana & Maheshwari, 1980; Kim et al., 2010; Kuehdorf & Appenroth, 2012; Kuehdorf et al., 2014; Lacor, 1968; 1969; 1970; Landers et al., 1977; Landolt, 1981; 1986; 1997; 2000; Landolt & Wildi, 1977; Lee et al., 2016; Lemon et al., 2001; Les & Crawford, 1999; Les et al., 1997b; 2002; Lien et al., 2012; Liu et al., 2017; Maheshwari, 1958; Malek, 1981; McClure & Alston, 1966; McLay, 1976; Mejbel & Simons, 2018; Michael et al., 2017; Muhonen et al., 1983; Newton et al., 1978; Nichols, 1999; Oláh et al., 2016; Oron et al., 1986; Pankey et al., 1965; Perry, 1968; Pípalová, 2003; Rusoff et al., 1980; Satake & Shimura, 1983; Seo et al., 2012; Tang et al., 2017; Tippery & Les, 2020; Tipping et al., 2009;

Vaithiyanathan & Richardson, 1999; Van Zuidam et al., 2012; Wang, 1990; Wang & Messing, 2011; 2015; Wang et al., 2007; Wang et al., 2011a; 2012b; 2014b; 2014c; 2015; Wilsey et al., 1991; Xu et al., 2012; Xu et al., 2018b; Yu et al., 2011; Ziegler et al., 2015.

4.2.4. Wolffia

Water meal; wolffie

Etymology: after Johann Friedrich Wolff (1778–1806)

Synonyms: *Bruniera*; *Grantia* (in part); *Horkelia* (in part); *Lemna* (in part); *Lenticula* (in part); *Telmatophace* (in part)

Distribution: global: Africa, Americas, Australia, Eurasia; **North America:** widespread except far north and high elevations

Diversity: global: 10 species; **North America:** 5 species

Indicators (USA): obligate wetland (OBL): *Wolffia arrhiza*, *W. borealis*, *W. brasiliensis*, *W. columbiana*, *W. globosa*

Habitat: freshwater; lacustrine, palustrine, riverine; **pH:** 3.0–10.5; **depth:** 0 m; **life-form(s):** free-floating

Key morphology: roots absent; fronds (to 1.6 mm) minute, free-floating, single or coherent in pairs, boat-shaped, globular, or ovoid, veinless, producing globular turions (to 0.7 mm), the upper surface smooth or with a prominent medial papule, deep to translucently green; flowers solitary, bisexual, protogynous; stamen 1; pistil 1, uniovulate; fruits wingless; seed 1, smooth

Life history: duration: perennial (turions, resting fronds, whole plants); **asexual reproduction:** frond division, turions; **pollination:** insect, self; **sexual condition:** hermaphroditic; **fruit:** utricular follicle (rare); **local dispersal:** fronds, turions (water); **long-distance dispersal:** fronds, turions (birds [exozoic])

Imperilment: 1. *Wolffia arrhiza* [G5]; **2.** *W. borealis* [G5]; S1 (OR, UT); S2 (KS, <u>QC</u>); S3 (<u>AB</u>, <u>BC</u>, KY, VA); **3.** *W. brasiliensis* [G5]; S1 (CA, KS, MI, NE); S2 (NC, <u>ON</u>); S3 (GA, IL, MD, MT, NJ, WV); **4.** *W. columbiana* [G5]; SH (DC); S1 (<u>MB</u>, <u>NB</u>, OR, <u>SK</u>, VA, WV); S2 (<u>AB</u>, ME, MT, NC, ND); S3 (<u>BC</u>, GA, <u>QC</u>); **5.** *W. globosa* [G5]

Ecology: general: All North American *Wolffia* species are minute (<1.6 mm), extraordinarily reduced, obligately aquatic, rootless, floating perennials (pleustophytes), which overwinter by producing small, spherical turions only 0.2–0.7 mm in size. Because the floating plants die back at the end of the season and persist entirely by means of these asexually derived structures (which sink to the bottom), this unusual life history has been described as a "vegetative annual" type. The turions germinate and rise again to the surface in the spring. There, the fronds multiply rapidly, typically developing into dense mats of plants on the water surface, which can substantially reduce or even eliminate light penetration to the waters below. However, it is unusual to find pure stands of a given species; mixtures of congeners, other Lemnaceae, and other free-floating plants are encountered more often. Fertility is low in all the North American representatives, which rarely flower or fruit. Thus, the seeds apparently are of minor significance in the overall life history of these plants. Although pleustophytes (where depth might seem inconsequential),

most *Wolffia* species inhabit fairly shallow waters, presumably to facilitate their annual cycle of re-establishment from turions, which reside winter-long at the water bottom. Local dispersal occurs by the transport of fronds or turions by water currents. Most long-distance dispersal is facilitated by the adherence of the small plants to the fur or plumage of various animals. The tiny, rootless fronds assimilate all aqueous nutrients from their lower surface and atmospheric gases (O_2, CO_2) from their upper surface, which can contain up to 100 stomata. Because of their rapid growth rates and ease of culture, the plants have been used in countless physiological experiments and in bioassays for waterborne pollutants and toxins. At least some *Wolffia* species can be grown in darkness "heterotrophically," when their growth medium is supplemented with sucrose.

4.2.4.1. ***Wolffia arrhiza* (L.) Horkel ex Wimm.** is a nonindigenous perennial, which currently is known only from the water surfaces of a freshwater California pond at an elevation of 76 m. Worldwide, the plants inhabit temperate to subtropical regions where the lowest mean winter temperatures remain above −8°C and the mean summer temperatures range from 18°C to 28°C. Their floating fronds can withstand temperatures of −1°C for up to 18 days or −2°C for 10 days. The floating fronds have a high water content (95–97%), which enables them to tolerate 5 hr of desiccation, losing up to 80% of their fresh mass before they are killed; whereas turions have a lower moisture content (61.8–78.9) and exhibit a steady decline in viability following more than an hour of desiccation. However, fronds and turions embedded within finely grained sediments (clay or peat) have been shown to survive for up to 2 months when exposed to dry air. The habitats in other regions are mesotrophic to eutrophic waters. Because of the single known North American locality (which has not been analyzed limnologically), there are few environmental data available. Cultured plants can tolerate a broad pH range (pH: 4.5–10), with their optimal growth occurring at pH 5.0. Laboratory cultured plants have grown well at a salinity of 5‰, with their growth declining substantially at salinities from 10‰ to 40‰. These are termed long-short-day plants, which flower in response to a dual daylength cycle; i.e., they require a sequence of long days followed by short days to induce flowering. The flowers are distinctly protogynous and self-compatible, but do not self-pollinate. Sexual reproduction occurs rarely in natural populations (worldwide), where only 3% of collections have contained flowers and just 0.8% fruits. Unless the climate is mild enough for the adult fronds to survive annually, perennation is achieved mainly by production of resting fronds (relatively unmodified fronds that sink) or turions, which develop in response to diminished nutrient levels, low temperatures, or poor light conditions. When ripe, the turions drop from the floating fronds, coming to rest on the bottom. They are non-dormant when shed and are capable of germinating immediately once conditions are favorable. The fronds and/or turions are dispersed locally by water currents and over longer distances by their adherence to various water birds (Aves). The plants have a high relative growth rate (0.243–0.426), but their biomass production rate is lower than for many other Lemnaceae. **Reported associates (North America):** *Azolla filiculoides, Berula erecta, Cyperus erythrorhizos, Lemna minuta, Ludwigia peploides, Pluchea odorata, Typha latifolia.*

4.2.4.2. ***Wolffia borealis* (Engelm.) Landolt & Wildi ex Gandhi, Wiersema & Brouillet** is a perennial, which inhabits backwaters (river), ditches, lakes, marshes, ponds (artificial, beaver, dredge, irrigation, seepage, sinkhole, stagnant), pools (stagnant), rivers, sloughs (river), and swamps at elevations to 1350 m. The plants can occur in full sunlight to shaded exposures. They are distributed in temperate regions, which have a minimum of 140 days with a mean temperature above 10°C, mean summer temperatures from 18°C to 22°C, and minimum mean winter temperatures at or above −8°C. The habitats are alkaline waters with negligible currents (pH: 6.1–7.9; conductivity: 182–538 μS cm^{-1}; Ca: 43–124 mg l^{-1}; K: 1–11 mg l^{-1}; Mg: 0.7–13 mg l^{-1}; total N: 200–2300 μg l^{-1}; Na: 13–63 mg l^{-1}; total P: 1–94 μg l^{-1}). Sexual reproduction is extremely rare with fewer than 0.2% flowering and <0.2% fruiting specimens collected. Although rare, fertile plants have been observed from July to September. The fronds have a very fast growth rate (0.62 fronds day^{-1}) and an average life span of 15.8 days, during which time they each produce an average of 9.8 daughter fronds. The plants overwinter entirely by the production of turions, thus behaving as "vegetative annuals" (see *Ecology* above). **Reported associates:** *Alnus rubra, Azolla cristata, Azolla filiculoides, Butomus umbellatus, Ceratophyllum demersum, Chara, Cyperus erythrorhizos, Cyperus niger, Decodon verticillatus, Echinodorus berteroi, Eclipta prostrata, Elodea canadensis, Juncus effusus, Leersia oryzoides, Lemna gibba, Lemna minor, Lemna trisulca, Lemna turionifera, Lemna valdiviana, Ludwigia grandiflora, Marsilea, Myriophyllum aquaticum, Nuphar polysepala, Nuphar variegata, Nymphaea odorata, Paspalum distichum, Peltandra virginica, Phalaris arundinacea, Phyla lanceolata, Pluchea odorata, Potamogeton crispus, Riccia fluitans, Ricciocarpus natans, Rosa nutkana, Salix lasiandra, Samolus valerandi, Schoenoplectus acutus, Spirodela polyrhiza, Stuckenia pectinata, Taxodium, Vallisneria americana, Veronica serpyllifolia, Wolffia brasiliensis, Wolffia columbiana.*

4.2.4.3. ***Wolffia brasiliensis* Wedd.** floats on the water surfaces of backwaters, bogs, canals (abandoned), ditches (drainage), impoundments (river), lagoons, lakes (spring-fed), marshes (roadside), oxbows, ponds (artificial, backwater, beaver, dune, mill, roadside, sand, shallow, sink hole), pools, reservoirs, rivers, sloughs (spring-fed), streams, and swamps (cypress, roadside) at elevations to 460 m. The plants grow in sunny to partially shaded exposures on waters underlain by muck, mud, or silt. They occur in warmer temperate to subtropical regions, which have a minimum of 170 days with a mean temperature above 10°C, mean summer temperatures from 22°C to 28°C, and minimum mean winter temperatures at or above −1°C. The habitats are quiet, mesotrophic to eutrophic waters (pH: 6.1–7.9; conductivity: 41–740 μS cm^{-1}; 105–438 μmhos cm^{-1}; Ca: 4.5–110 mg l^{-1}; K: 1–45 mg l^{-1}; Mg: 2.1–30 mg l^{-1}; total N: 200–3700 μg l^{-1}; Na: 3.8–135 mg l^{-1}; total P: 2–30

μg l^{-1}). The pH range observed among Ohio populations was similar (pH: 6.0–7.7), with the intermediate values (pH: 6.4–7.4) associated with better growth. The pH ranged from 5.8 to 6.6 among North Carolina populations. Montana populations were found to restricted to those waters having relatively lower conductivity (<438 μmhos cm^{-1}). The protogynous flowers are self-compatible but do not self-pollinate. Sexual reproduction in these short-day plants is rare with only 2% of collected specimens bearing flowers and 0.8% with fruits. Fertile plants are uncommon but have been observed from May to November. The fronds overwinter exclusively by producing turions, which develop in response to low temperatures (e.g., 18°C) and low nutrient conditions (e.g., 0.5 mg N l^{-1}, 0.083 mg P l^{-1}). The turions and fronds are dispersed locally by water; however, the minute fronds are known to be transported over longer distances on or within the plumage of waterbirds (e.g., Aves: Anhimidae: *Anhima cornuta*). **Reported associates:** *Agrimonia parviflora, Alisma triviale, Alnus rubra, Alternanthera philoxeroides, Asclepias incarnata, Azolla cristata, Azolla filiculoides, Bidens aristosus, Bidens discoideus, Bidens frondosus, Bolboschoenus fluviatilis, Boltonia asteroides, Brasenia schreberi, Cabomba caroliniana, Callitriche heterophylla, Carex comosa, Carex tribuloides, Carex vesicaria, Cephalanthus occidentalis, Ceratophyllum demersum, Ceratophyllum echinatum, Ceratopteris thalictroides, Chara vulgaris, Cornus amomum, Eclipta prostrata, Egeria densa, Eichhornia crassipes, Eleocharis obtusa, Eleocharis ovata, Elodea canadensis, Eupatorium perfoliatum, Eupatorium serotinum, Glyceria septentrionalis, Gratiola neglecta, Gratiola virginiana, Heteranthera dubia, Hibiscus moscheutos, Hydrilla verticillata, Hydrocotyle ranunculoides, Hygrophila lacustris, Hypericum mutilum, Juncus acuminatus, Juncus effusus, Juncus nodatus, Landoltia punctata, Lemna aequinoctialis, Lemna minor, Lemna minuta, Lemna obscura, Lemna valdiviana, Leptodictyum riparium, Limnobium spongia, Limnophila sessiliflora, Ludwigia decurrens, Ludwigia repens, Lycopus americanus, Mimulus ringens, Myriophyllum aquaticum Myriophyllum heterophyllum, Myriophyllum spicatum, Najas guadalupensis, Najas minor, Nelumbo lutea, Nyssa aquatica, Oedogonium, Panicum hemitomon, Panicum philadelphicum, Penthorum sedoides, Persicaria amphibia, Persicaria coccinea, Persicaria hydropiperoides, Phalaris arundinacea, Pistia stratiotes, Poa nemoralis, Pontederia cordata, Pontederia cordata, Potamogeton crispus, Potamogeton illinoensis, Potamogeton nodosus, Potamogeton pusillus, Riccia fluitans, Rorippa palustris, Sagittaria graminea, Sagittaria latifolia, Sagittaria platyphylla, Salix interior, Schoenoplectus californicus, Schoenoplectus heterochaetus, Scirpus cyperinus, Sparganium americanum, Sparganium natans, Spirodela polyrhiza, Spirogyra, Stuckenia pectinata, Symphyotrichum lanceolatum, Taxodium distichum, Typha angustifolia, Typha latifolia, Utricularia gibba, Utricularia macrorhiza, Utricularia minor, Utricularia purpurea, Vallisneria americana, Verbena hastata, Wolffia columbiana, Wolffiella gladiata, Xanthium strumarium, Zannichellia palustris, Zizania texana.*

4.2.4.4. ***Wolffia columbiana*** **H. Karst.** occurs on the water surfaces of backwaters (river, stagnant), baygalls, bayous (stagnant), bays, bogs (tamarack), borrow pits, canals (abandoned), depressions (pond), ditches (drainage, irrigation, roadside, seepage), estuaries, flats (cypress), floodplains, gravel pits, impoundments (river), lagoons, lakes (dune, spring-fed), marshes (depression, floodplain), ponds (artificial, beaver, bog, cattle, farm, fish, golf course, marsh, mill, pothole, reclamation, reservoir, river, settling, sinkhole, stagnant, stock, woodland), pools (sewage disposal), potholes, quarries, reservoirs, rivers, sinkholes, sloughs (artificial, spring-fed), spillways, springs, streams (inlet, sluggish, tidal), and swamps (bottomland, cypress, tamarack) at elevation to 1645 m. The plants occur typically under full sunlight exposures, but also are shaded often by overstory shrubs and trees. Their substrates have been described as limestone, marl, muck (Allemands), mud, humus, sand, and silt. This is a species of warmer temperate to subtropical regions, which have a minimum of 140 days with a mean temperature above 10°C, mean summer temperatures from 18°C to 28°C, and minimum mean winter temperatures at or above −8°C. The typical habitats are calm, fresh, alkaline, mesotrophic to eutrophic waters (pH: 5.0–9.8; $\bar{x} = 7.9$; alkalinity [as $CaCO_3$]: 15–235 mg l^{-1}; $\bar{x} = 135$ mg l^{-1}; conductivity: 124–1300 µS cm^{-1}; 60–480 µmhos cm^{-1} [@ 25°C]; $\bar{x} = 290$ µmhos cm^{-1}; Ca: 5.3–124 mg l^{-1}; K: 1–45 mg l^{-1}; Mg: 3.1–95 mg l^{-1}; total N: 100–2300 µg l^{-1}; Na: 8.9–225 mg l^{-1}; total P: 1–94 µg l^{-1}). The pH range in surveyed Ohio populations (pH: 5.9–7.8) and Indiana populations (pH: 5.8–7.9) were similar, with optimal growth in both cases associated with pH values from 6.4 to 7.4; a similar pH range of 6.7–7.3 was observed among North Carolina populations. Montana populations were found to be restricted to those waters having relatively lower conductivity (<438 µmhos cm^{-1}). The protogynous flowers are self-compatible, but do not self-pollinate. Sexual reproduction is extremely rare. Out of nearly 1,200 specimens examined, only 18 of them (1.5%) bore flowers and just 3 (<0.3%) had fruits. The occasional fertile fronds have been observed from June to October and tend to occur in waters of higher pH (8.1–8.4) and conductivity (204–1664 µS cm^{-1}). With virtually no seed production, the plants overwinter entirely by the production of turions, which detach and settle to the bottom. Their sinking is due to accumulation of starch. As the turions germinate, photosynthetically derived oxygen accumulates initially in parenchymatous intercellular spaces, which provides buoyancy and causes them to rise. Specialized air-filled substomatal chambers then provide buoyancy and floatation as the turions develop into surface fronds. The buoyancy of the fronds is facilitated by the waxy, cuticularized walls of the substomatal chambers. The mean relative growth rate (0.249) is low compared to other duckweeds, and yields a frond doubling time of 2.82 days. By their dense growth habit, the ratio of evapotranspiration to evaporation for the tiny fronds is low ($E/E_0 = 0.89$). **Reported associates:** *Abutilon theophrasti, Acer rubrum, Acer saccharinum, Agrimonia parviflora, Alisma triviale, Alnus rubra, Alternanthera philoxeroides, Amaranthus tuberculatus, Ampelopsis arborea, Apios americana, Asclepias incarnata, Asclepias perennis, Azolla cristata, Azolla filiculoides, Baccharis sarothroides, Bacopa rotundifolia, Bidens aristosus, Bidens cernuus, Bidens frondosus, Boehmeria cylindrica, Bolboschoenus fluviatilis, Boltonia asteroides, Brasenia schreberi, Brickellia californica, Brunnichia ovata, Cabomba caroliniana, Caltha palustris, Campsis radicans, Carex comosa, Carex cusickii, Carex hystericina, Carex tribuloides, Cephalanthus occidentalis, Ceratophyllum demersum, Ceratophyllum echinatum, Chamaedaphne calyculata, Chara braunii, Chara contraria, Chara globularis, Chara inconnexa, Chara zeylanica, Cornus amomum, Cyperus erythrorhizos, Cyperus esculentus, Cyperus niger, Cyperus odoratus, Cyperus strigosus, Datura wrightii, Decodon verticillatus, Echinochloa colona, Echinochloa muricata, Echinochloa walteri, Echinodorus berteroi, Eclipta prostrata, Eichhornia crassipes, Elaeagnus angustifolia, Eleocharis obtusa, Eleocharis palustris, Elodea canadensis, Elodea nuttallii, Equisetum arvense, Eupatorium perfoliatum, Eupatorium serotinum, Fraxinus americana, Fraxinus profunda, Gleditsia aquatica, Gratiola neglecta, Gratiola virginiana, Heteranthera dubia, Hibiscus moscheutos, Hydrocotyle bonariensis, Hydrocotyle verticillata, Hypericum mutilum, Iris pseudacorus, Iris versicolor, Itea virginica, Juncus acuminatus, Juncus effusus, Juncus nodatus, Landoltia punctata, Larix laricina, Leersia oryzoides, Lemna gibba, Lemna minor, Lemna minuta, Lemna obscura, Lemna perpusilla, Lemna trisulca, Lemna turionifera, Lemna valdiviana, Leptodictyum riparium, Limnobium spongia, Ludwigia grandiflora, Ludwigia palustris, Ludwigia peploides, Ludwigia repens, Lycopus americanus, Lycopus europaeus, Lycopus rubellus, Lythrum salicaria, Melilotus albus, Mikania scandens, Mimulus cardinalis, Mimulus ringens, Myrica cerifera, Myriophyllum aquaticum, Myriophyllum pinnatum, Myriophyllum sibiricum, Myriophyllum spicatum, Myriophyllum verticillatum, Najas flexilis, Najas gracillima, Najas guadalupensis, Najas minor, Nelumbo lutea, Nitella flexilis, Nuphar advena, Nuphar microphylla, Nuphar polysepala, Nuphar variegata, Nymphaea odorata, Nyssa aquatica, Onoclea sensibilis, Panicum virgatum, Paspalum distichum, Peltandra virginica, Penthorum sedoides, Persicaria coccinea, Persicaria hydropiperoides, Persicaria lapathifolia, Persicaria pensylvanica, Phalaris arundinacea, Phragmites australis, Phyla lanceolata, Phytolacca americana, Pistia stratiotes, Platanus racemosa, Pluchea odorata, Poa nemoralis, Pontederia cordata, Populus deltoides, Populus heterophylla Potamogeton amplifolius, Potamogeton crispus, Potamogeton diversifolius, Potamogeton epihydrus, Potamogeton foliosus, Potamogeton friesii, Potamogeton natans, Potamogeton nodosus, Potamogeton praelongus, Potamogeton pusillus, Potamogeton richardsonii, Potamogeton zosteriformis, Potentilla anserina, Ranunculus longirostris, Riccia fluitans, Ricciocarpus natans, Rosa nutkana, Rosa palustris, Rumex verticillatus, Sagittaria australis, Sagittaria latifolia, Salix amygdaloides, Salix exigua, Salix caroliniana, Salix interior, Salix lasiandra, Salix nigra, Samolus valerandi, Schoenoplectus acutus, Schoenoplectus californicus, Schoenoplectus tabernaemontani, Sparganium* *eurycarpum, Sphagnum, Spirodela polyrhiza, Stuckenia pectinata, Symphyotrichum lanceolatum, Symplocarpus foetidus, Taxodium distichum, Triadenum walteri, Typha angustifolia, Typha latifolia, Utricularia macrorhiza, Vallisneria americana, Verbena hastata, Wolffia borealis, Wolffia brasiliensis, Wolffiella gladiata, Wolffiella lingulata, Xanthium strumarium, Zizania aquatica, Zizania palustris, Zizaniopsis miliacea.*

4.2.4.5. *Wolffia globosa* (Roxb.) Hartog & Plas is a nonindigenous perennial, which floats atop the water surfaces of backwaters, ditches (irrigation), lakes, marshes (floodplain), ponds (beaver, river, wastewater treatment), and rivers at elevations to 104 m. The normal range of this species extends from warm temperate to tropical regions, which have a minimum of 200 days with a mean temperature above 10°C, mean summer temperatures from 22°C to 28°C, and minimum mean winter temperatures at or above 1°C. The habitats are characterized by calm, permanent, mesotrophic to eutrophic waters (pH: 5.0–7.2; conductivity: 55–250 μS cm^{-1}; Ca: 3.0–6.5 mg l^{-1}; K: 2.2–16 mg l^{-1}; Mg: 2.6–11 mg l^{-1}; total N: 800–2700 μg l^{-1}; Na: 13–34 mg l^{-1}; total P: 62–740 μg l^{-1}). These are short-day plants with protogynous, but likely self-compatible, flowers. Although fertility is low, it is much higher than many other *Wolffia* species, with 6% of collected specimens having flowers and 3% bearing fruits. It is likely (but not yet confirmed) that self-pollination occurs. Perennation and vegetative reproduction occur by the production of turions. Clones originating from different regions vary widely in their mean relative growth rates, which can range from 0.155 to 0.559 day^{-1}; the respective doubling times range from 4.47 days to 29.8 hr. **Reported associates (North America):** *Azolla filiculoides, Ceratophyllum demersum, Echinodorus berteroi, Hydrocotyle bonariensis, Hydrocotyle verticillata, Landoltia punctata, Lemna gibba, Lemna minuta, Lemna obscura, Lemna turionifera, Lemna valdiviana, Ludwigia peploides, Myrica cerifera, Myriophyllum pinnatum, Peltandra virginica, Salix caroliniana, Schoenoplectus californicus, Spirodela polyrhiza, Typha latifolia, Wolffia columbiana, Wolffiella gladiata, Wolffiella lingulata.*

Use by wildlife: *Wolffia* species commonly are consumed by waterfowl (Aves: Anatidae). *Wolffia arrhiza* and *W. columbiana* are eaten by several species of herbivorous fish (Vertebrata: Teleostei). *Wolffia brasiliensis* and *W. columbiana* are eaten by muskrats (Mammalia: Cricetidae: *Ondatra zibethicus*) and ducks (e.g., Aves: Anatidae: *Anas platyrhynchos*). The plants provide a symbiotic benefit to various epiphytic diatoms (Bacillariophyta: Bacillariophyceae), which utilize the seasonal life cycle of submergence and emergence as a means of optimizing their own light availability.

Economic importance: food: Unlike other duckweed genera, the oxalate content of *Wolffia* occurs in soluble rather than in crystalline form (which can cause digestive problems), making the plants more suitable for human consumption. *Wolffia arrhiza* comprises 19–40% protein, 11–44% carbohydrate, and 5.5% lipids. It is eaten as a vegetable (known as "khai-nam") in southeast Asia. The plants have been tested as a life-support system to accommodate outer space exploration missions. Sun-dried plants have been used as a pork substitute in Asian "pork ball" products. The mean fatty acid content of *W. borealis* (14.2% DM) is higher than for any other duckweed surveyed. *Wolffia columbiana* has a high water content (95%) but comprises 36.5–44.7% crude protein and 6.6% fat (% DM), and most essential amino acids required in vertebrate diets at levels recommended by the FAO. *Wolffia globosa* is eaten in parts of Asia. Because its growth rate increases under microgravity conditions, it has been considered for use in space mission life-support systems. It is a cultural plant in Thailand known as "pum"; **medicinal:** no medicinal uses reported; **cultivation:** *Wolffia arrhiza* is cultivated as an ornamental aquarium plant; **misc. products:** *Wolffia arrhiza* has been the subject in many physiological investigations pertaining to phytoremediation of aqueous systems (e.g., metal, nutrient, and toxin uptake). Cultivated plants are estimated as capable of removing 126 mg N m^{-2} day^{-1} and 38 mg P m^{-2} day^{-1} from enriched waters, while producing 6 g starch m^{-2} day^{-1}. The composition of commercial feeds for Nile tilapia (Vertebrata: Teleostei: Cichlidae: *Oreochromis niloticus*) have been supplemented successfully by up to 15% with *W. arrhiza* meal as a replacement for soybeans (Fabaceae: *Glycine max*). *Agrobacterium*-mediated methods have successfully generated transgenic lines of *W. arrhiza* for producing recombinant proteins. *Wolffia columbiana* is highly digestible (80% *in vitro* organic matter digestion) and has been used as an additive (with *Lemna gibba*) in feed used for growing carp (Vertebrata: Teleostei: Cyprinidae; *Ctenopharyngodon Idella, Hypophthalmichthys nobilis*) and other agricultural animals. The fronds have been transformed genetically using a gold nanoparticle biolistic approach as well as by *Agrobacterium*-mediated methods. The plants also have been used for bioethanol production, achieving rates of 0.50 g ethanol g^{-1} glucose. *Wolffia globosa* has been evaluated extensively for use in the phytoremediation of waters contaminated by heavy metals such as cadmium and chromium. It also can tolerate high levels of arsenic, which it detoxifies by phytochelatins. The plants have been used to produce biofuels and succinate (at yields up to 0.58 and 0.68 kg m^{-2} year^{-1}, respectively). Processed fronds have been used as a substitute for soybean (Fabaceae: *Glycine max*) meal in feed used to raise laying hens (Aves: Phasianidae: *Gallus domesticus*) and other domestic birds. The plants have been transformed genetically using *Agrobacterium tumefaciens*–mediated methods; **weeds:** The various *Wolffia* species are sometimes regarded as weeds due to their capacity for dense mat-like growth. Invasive growth of *W. columbiana* has been suppressed by applications of detergent (0.1% Tween20), which, by disrupting their buoyancy, exposes the plants to fungal infection. *Wolffia globosa* is a rice field weed in Asia; **nonindigenous species:** *Wolffia arrhiza* has been introduced to North America, being first reported from California in 1989. *Wolffia columbiana* recently has been introduced to Europe, likely through the careless disposal of aquarium plants. *Wolffia globosa* has been introduced to California and Florida and has become naturalized in Japan.

Systematics: Given the highly simplified structure and miniscule size of all the *Wolffia* species, their relationships have been difficult to ascertain using morphological and micromolecular data. Allozymes have been only slightly more successful due to a lack of shared alleles detected between most species ($I=0$), and extremely low genetic identities ($I=0.14–0.40$) in the cases having shared alleles. A genetic identity of 0.34 was reported between *W. globosa* and *W. cylindracea*; however, these taxa later were synonymized. Low allozyme identities also occur between *W. arrhiza* and *W. columbiana* ($I=0.27$) or *W. globosa* ($I=0.19$), which all are assigned to section *Wolffia* (Figure 1.27). *Wolffia* appears to be monophyletic, despite the anomalous placement of *W. brasiliensis* as the sister group of *Wolffiella* in some analyses (e.g., Figure 1.27), which are characterized by excessive branch lengths. Because that unusual association conflicts with morphological and anatomical evidence, as well as the fact that the species resolves with *Wolffia* in other molecular phylogenetic analyses, the monophyly of *Wolffia* is accepted here until any data eventually might prove otherwise. In any case, appropriate phylogenetic analyses consistently resolve all *Wolffia* and *Wolffiella* taxa within a clade that has been recognized as subfamily Wolffioideae. Due to their minute size and simplified morphology, species identifications in *Wolffia* have relied increasingly more on the use of various DNA "bar-coding" strategies, which have included various types of fragment analysis (AFLP, ISSR, and RAPD markers) as well as direct sequencing approaches. As is typical, the high-resolution fragment-based methods perform well for delimiting taxa but are less effective in reconstructing interspecific relationships. The same is true for some DNA sequence data such as those derived from intergenic spacers. Combined nuclear and plastid sequence data have consistently recovered a clade corresponding to the taxonomic section *Wolffia* (Figure 1.27). Within section *Wolffia*, *W. columbiana* and *W. elongata* associate as sister species within a relatively specialized subclade, with *W. arrhiza* and *W. cylindrica* as their sister subclade; a subclade comprising the very closely related *W. globosa*, *W. neglecta*, and *W. angusta* resolves as the sister group to the other four species (e.g., Figure 1.27). The highly infertile *W. columbiana* recently (~2.1 MYBP) is thought to have diverged ecologically from is more fertile sister species *W. elongata* (Figure 1.27), where higher seed production in the latter enabled it to colonize more seasonal habitats. Population genetic studies in China have detected two distinct cpDNA haplotypes (*glo* and *un*), which correlate with AFLP markers in *W. globosa*. The *rps16* sequences of *W. globosa* have a highly elevated transversion substitution rate. The placement of *W. borealis* and *W. brasiliensis* has been problematic. Once assigned to section *Pigmentatae* (by virtue of their shared pigment cells), both species are characterized by very long branch lengths in phylogenetic evaluations, and do not resolve consistently in analyses using different data sets. It remains unclear why the same data that consistently place other Lemnaceae in a topology are inconsistent in the placement of *W. borealis* and *W. brasiliensis* as well as two other *Wolffia* species (*W. australiana*, *W. microscopica*), which also are characterized by long branch lengths. Consequently, any relationships depicted for these four *Wolffia* species (e.g. Figure 1.27) should be regarded as dubious, other than their placement outside of section *Wolffia*. Genome sizes in *Wolffia* are on average larger than other Lemnaceae and have been estimated for *Wolffia arrhiza* (1,881 Mbp [the largest in Lemnaceae]), *W. borealis* (889 Mbp), *W. brasiliensis* (776 Mbp), *W. columbiana* (874 Mbp), and *W. globosa* (1,295 Mbp). The basic chromosome number of *Wolffia* is $x=10$; extensive cytotypic variation (aneuploid and euploid) occurs in the genus: *W. arrhiza*: $2n=30, 40, 42, 44–46, 50, 60, 62$; *W. borealis*: $2n=20, 22, 30, 40$; *W. brasiliensis*: $2n=20, 40, 42, 50, 60, 80$; *W. columbiana*: $2n=30, 40, 42, 50, 70, 80$; *W. globosa*: $2n=30, 60$.

Distribution: *Wolffia borealis*, *W. brasiliensis*, and *W. columbiana* occur mainly in the eastern United States but with sporadic localities in southern Canada and the western United States; the latter two species extend to South America. Two *Wolffia* species are nonindigenous: *W. arrhiza* (Old World), which has been introduced to California, and *W. globosa* (Old World), which has been introduced to California and Florida.

References: Appenroth et al., 2013b; Ardenghi et al., 2017; Armstrong, 2012; Armstrong & Thorne, 1984; Armstrong et al., 1989; Bhanthumnavin & McGarry, 1971; Beal, 1977; Blüm & Paris, 2001; Blüm et al., 1995; Board et al., 1993; Boehm et al., 2001; Bog et al., 2013; Boonyapookana et al., 2002; Borisjuk et al., 2015; Boyd, 1987; Calicioglu & Brennan, 2018; Cassani et al., 1982; Chantiratikul et al., 2010; Chareontesprasit & Jiwyam, 2001; Chase, 2003; Cicero et al., 2011; Crawford & Landolt, 1995; Crawford et al., 2006; Daubs, 1965; Fassett, 1957; Frick, 1994; Fujita et al., 1999; Gandhi et al., 2012; Godziemba-Czyż, 1970; Green et al., 2016; Harvey & Fox, 1973; Heenatigala et al., 2018; Hicks, 1932; 1937; Hillman, 1961; Huffman & Lonard, 1983; Hunt et al., 1995; Jäger, 1964; Jervis, 1969; Kadono, 2004; Kandeler, 1984; Kasselmann, 2003; Kuoh et al., 2000; Khvatkov et al., 2015; 2018; Kline & McCune, 1987; Krajnčić & Devidé, 1982; Kruse et al., 2002; Landolt, 1981; 1986; 1997; 2000; Landolt & Wildi, 1977; Lemke, 1989; Lemon et al., 2001; Les et al., 1997; 2002; 2003; Looman, 1983; Makkay et al., 2008; Mandal et al., 2010; McAtee, 1918; McCann, 2016; McLay, 1976; Mohlenbrock, 1959a; Montz, 1978; Pholhiamhan et al., 2017; Profous & Loeb, 1984; Ruekaewma et al., 2015; Rusoff et al., 1980; Schmitz et al., 2014; Sherman et al., 1996; Siripahanakul et al., 2013; Soda et al., 2015; Sree et al., 2015; Subudhi et al., 2015; Thomson, 2005; Tippery & Les, 2020; Van Dine, 1922; Vogel, 1977; Wang et al., 2011a; 2011b; White & Wise, 1998; White et al., 2010; Witty, 2009; 2011; Wooten & Leonard, 1984; Wunderlin & Wunderlin, 1968; Xie et al., 2013; Xu et al., 2015; Yan et al., 2013; York, 1905; Yuan & Xu, 2017; Yuan et al., 2011; Zhang et al., 2012; Ziegler et al., 2015.

4.2.5. Wolffiella

Bogmat, mudmidget

Etymology: a diminutive name commemorating Johann Friedrich Wolff (1778–1806)

Synonyms: *Lemna* (in part); *Pseudowolffia*; *Wolffia* (in part); *Wolffiopsis*

Distribution: global: Africa, North America, South America; **North America:** southern and southeastern United States
Diversity: global: 10 species; **North America:** 3 species
Indicators (USA): obligate wetland (OBL): *Wolffiella gladiata, W. lingulata, W. oblonga*
Habitat: freshwater; lacustrine, palustrine, riverine; **pH:** 5.0–7.7; **depth:** <0.02 m; **life-form(s):** free-floating, suspended
Key morphology: roots absent; fronds (to 9 mm) single or cohering in groups of 2–20, suspended just below the water surface, their distal end free-floating, ligulate, linear, ovate or ribbon-like, flat, veinless, pigment cells present; flowers (1–2), bisexual, protogynous; stamen 1, the filament (to 0.76 mm) stout; pistil (to 0.5 mm) 1, flask-like, uniovulate, the stigma circular, concave; fruit (to 0.44 mm) with persistent style; seed (to 0.44 mm) 1, smooth
Life history: duration: perennial (resting fronds/turions, whole plants); **asexual reproduction:** frond division; **pollination:** insect, self; **sexual condition:** hermaphroditic; **fruit:** indehiscent utricle (rare); **local dispersal:** fronds, resting fronds/turions (water); **long-distance dispersal:** fronds (animals)
Imperilment: **1.** *Wolffiella gladiata* [G]; **2.** *W. lingulata* [G]; **3.** *W. oblonga* [G]
Ecology: general: All *Wolffiella* species are obligate aquatic perennials, which float at the water surface or are suspended slightly beneath it. Most of the species grow having much of their frond submersed but with a distal portion immersed. None of the species possesses roots, but the fronds of some (but not in North America) are modified as root-like stabilizing organs. The tissues are non-lignified. Flowering in the North American species, as well as fruiting, occurs rarely. The surface fronds multiply vegetatively by continual division. Perennation (and further vegetative reproduction) occurs by the persistence of whole plants (where climate permits), or by development of slightly modified fronds (referred to as resting fronds or turions), which overcome inclement seasonal periods by sinking (as air spaces decline and starch contents increase) and resting on the bottom until favorable conditions return. The fronds are dispersed locally by water currents, whereas long-distance dispersal presumably is facilitated by the transport of fronds that adhere to the fur, plumage, or skin of various animals.

4.2.5.1. ***Wolffiella gladiata* (Hegelm.) Hegelm.** floats, or is suspended just beneath (1 cm) the water surfaces of backwaters, bayous, borrow pits, canals, depressions, ditches (roadside), hammocks (hydric), inlets (river), lakes (oxbow, shallow), marshes (basin, canal, outwash plain), oxbows, ponds (artificial, beaver, blackwater, dune, gravel pit, mill, roadside, semi-permanent), pools (marsh, stagnant), reservoirs, sinkholes, sloughs, stream margins, and swamps (cypress, floodplain, hardwood, river, roadside) at elevations to 400 m. The exposures commonly are characterized by partial shade. Associated substrates rarely have been described but include several types of muck (Allemands, alluvial). The plants occupy warm temperate regions, which have a minimum of 170 days with a mean temperature above 10°C, mean summer temperatures from 18°C to 28°C, and minimum mean

winter temperatures at or above 1°C. The habitats are characterized by shallow (0.2–1.6 m), acidic, mesotrophic, freshwaters (pH: 5.0–7.4; conductivity: 41–199 µS cm^{-1}; Ca: 4.8–26 mg l^{-1}; K: 0.6–7 mg l^{-1}; Mg: 0.8–11 mg l^{-1}; total N: 300–3700 µg l^{-1}; Na: 2.9–26 mg l^{-1}; total P: 3–20 µg l^{-1}). A similar pH range (pH: 5.0–6.5) was observed among different sites in Ohio, with optimum growth occurring within the pH range from 5.2 to 6.2. In one study, the plants occurred more often in sites that were highly enriched with phosphate (~68 mg l^{-1}). These are categorized as long-day plants. As in most duckweeds, their flowers are protogynous. They are thought to be self-compatible but are known not to self-pollinate. Fertility is low, with only 4% of plants observed in flower and 0.8% in fruit. Reproductive specimens have been collected from February to May and in November. Despite pervasive vegetative reproduction, allozyme data have indicated intermediate levels of genetic variability in the species. To overcome harsh seasonal conditions, the surface fronds can develop into slightly modified resting fronds, to which some authors refer as turions. Resting frond induction involves gibberellin production and has been induced in plants cultured in modified Hutner's medium containing 3% sucrose. The normal fronds are characterized by the slowest relative growth rate (0.14) observed among any of the tested Lemnaceae. **Reported associates:** *Acer rubrum, Acorus calamus, Alternanthera philoxeroides, Apios americana, Azolla cristata, Azolla filiculoides, Betula nigra, Bidens cernuus, Bidens frondosus, Boehmeria cylindrica, Brasenia schreberi, Cabomba caroliniana, Carex alata, Carex joorii, Cephalanthus occidentalis, Ceratophyllum demersum, Ceratophyllum echinatum, Chara zeylanica, Cornus foemina, Cyperus odoratus, Cyperus retrorsus, Decodon verticillatus, Dichanthelium dichotomum, Dulichium arundinaceum, Dyschoriste humistrata, Echinodorus cordifolius, Eichhornia crassipes, Forestiera acuminata, Fraxinus caroliniana, Fraxinus pennsylvanica, Fraxinus profunda, Gleditsia aquatica, Habenaria repens, Heteranthera multiflora, Hibiscus, Hydrocotyle bonariensis, Hydrocotyle ranunculoides, Hydrocotyle umbellata, Hydrocotyle verticillata, Hydrolea quadrivalvis, Hymenachne amplexicaulis, Itea virginica, Justicia americana, Landoltia punctata, Leitneria floridana, Lemna aequinoctialis, Lemna minor, Lemna minuta, Lemna obscura, Lemna perpusilla, Lemna valdiviana, Limnobium spongia, Ludwigia octovalvis, Ludwigia peploides, Lycopus rubellus, Magnolia virginiana, Mikania scandens, Myrica cerifera, Myriophyllum aquaticum, Myriophyllum pinnatum, Najas flexilis, Najas guadalupensis, Nelumbo lutea, Nuphar advena, Nymphaea mexicana, Nymphaea odorata, Nymphoides, Nyssa aquatica, Osmunda regalis, Peltandra virginica, Persicaria amphibia. Persicaria glabra, Persicaria hydropiperoides, Persicaria punctata, Phalaris arundinacea, Pistia stratiotes, Planera aquatica, Pontederia cordata, Populus heterophylla, Potamogeton diversifolius, Potamogeton pulcher, Quercus palustris, Riccia fluitans, Ricciocarpus natans, Rosa palustris, Sagittaria latifolia, Salix caroliniana, Salix nigra, Salvinia minima, Solanum dulcamara, Sparganium americanum, Spirodela polyrhiza, Spirogyra, Styrax* *americanus, Taxodium ascendens, Taxodium distichum, Teucrium canadense, Triadenum walteri, Typha angustifolia, Typha domingensis, Typha latifolia, Utricularia foliosa, Utricularia gibba, Utricularia subulata, Wolffia brasiliensis, Wolffia columbiana, Zizaniopsis miliacea.*

4.2.5.2. ***Wolffiella lingulata*** **(Hegelm.) Hegelm.** floats, or grows suspended shallowly beneath the water surfaces of bogs, canals, ditches (irrigation), lakes, marshes (river), ponds (beaver, dune), pools (lagoon, roadside, shallow), reservoirs, shorelines (lake), sloughs (river), streams, swales (dune), and swamps (cypress, willow) at elevations to 364 m. The plants occur primarily in tropical to sub-tropical climates with at least 260 days where mean temperatures exceed 10°C, mean winter temperatures are no lower than 8°C, and mean summer temperatures are maintained from 24°C to 30°C. The fronds grow in open to deeply shaded exposures but are more vigorous in the latter. Few details are reported for associated substrates, which include sandy humus. The typical habitats are calm, shallow, mesotrophic waters (pH: 5.2–7.7; conductivity: 345–1150 µS cm^{-1}; Ca: 5.3–24 mg l^{-1}; K: 4.9–23 mg l^{-1}; Mg: 12–59 mg l^{-1}; total N: 200–600 µg l^{-1}; Na: 47–224 mg l^{-1}; total P: 0–50 µg l^{-1}). The fertility rate is relatively high for duckweeds, with 13% of collected specimens in flower, and 2.6% bearing fruit. Sexual reproduction occurs nearly year-round, extending from January to November. The protogynous flowers exude a liquid droplet from the stigma, which facilitates the capture of pollen from wind or insect vectors. Because the stamen does not develop until after the droplet disappears, the plants do not self-pollinate. The fruits mature while still attached to the senescing fronds, which remain buoyant. Local dispersal by water surface currents can occur at this time. The non-dormant seeds germinate soon afterward and develop into new floating fronds. Longer dispersal distances rely on the attachment of fronds to the fur, plumage, or skin of animals. Despite the somewhat elevated rate of sexual reproduction compared to its congeners, the level of genetic variation detected by allozymes is lower than in most other widespread plant species. In any case, asexual reproduction by frond division is prevalent. The surface fronds withstand unfavorable conditions by developing into slightly modified resting fronds, which settle to the bottom in await of more favorable conditions. Because the resting fronds only tolerate low temperatures to 10°C, they do not enable the plants to overwinter in colder climates. The normal fronds have an estimated relative growth rate of 0.260. **Reported associates:** *Azolla filiculoides, Ceratophyllum demersum, Hydrocotyle ranunculoides, Lemna minor, Lemna minuta, Lemna obscura, Lemna valdiviana, Ludwigia peploides, Myriophyllum hippuroides, Oenanthe sarmentosa, Riccia fluitans, Salix, Schoenoplectus acutus, Sparganium eurycarpum, Spirodela polyrhiza, Taxodium distichum, Typha angustifolia, Typha latifolia, Wolffia brasiliensis, Wolffia columbiana, Wolffiella oblonga.*

4.2.5.3. ***Wolffiella oblonga*** **(Phil.) Hegelm.** floats, or grows suspended shallowly beneath the water surfaces of bogs, borrow pits, canals, ditches, marshes (roadside), ponds (artificial), pools, streams (slow), and swamps at elevations to 60

m. Exposures range from open to shaded sites. This species occurs in warm temperate to subtropical regions with at least 260 days where the mean temperature exceeds 10°C, mean winter temperature is no lower than 8°C, and the mean summer temperatures are maintained from 12°C to 28°C. This species is similar ecologically to *W. lingulata*, except by having a slightly greater frost tolerance and lower tolerance for warmer climates. The habitats are characterized by mesotrophic, freshwaters (pH: 6.6–7.4; conductivity: 120–425 µS cm^{-1}; Ca: 4.8–68 mg l^{-1}; K: 1.7–20 mg l^{-1}; Mg: 4.8–23 mg l^{-1}; total N: 300–600 µg l^{-1}; Na: 9–95 mg l^{-1}; total P: 11–50 µg l^{-1}). The protogynous flowers are thought to be self-incompatible. Fertile plants are uncommon (4% of records flowering, 0.4% fruiting) but have been observed sporadically from February to October. Data on the floral and seed biology are scarce, but presumably are similar to those of *W. lingulata* (see above). Unfavorable conditions are endured by the development of slightly modified resting fronds, which sink to the bottom until more favorable conditions are restored. However, the plants are unable to overwinter in colder climates because the resting fronds cannot tolerate temperatures below 10°C. Most reproduction occurs vegetatively through frond division of the surface plants, which results in widespread clonal growth. Lower than expected levels of genetic (i.e., allozyme) variation have been detected compared to widespread species in other plant families. Dispersal vectors are the same as those described above for *W. lingulata*. **Reported associates:** *Acer, Azolla filiculoides, Hydrocotyle ranunculoides, Landoltia punctata, Lemna gibba, Lemna minuta, Lemna obscura, Lemna turionifera, Lemna valdiviana, Ludwigia leptocarpa, Panicum hemitomon, Salvinia minima, Sparganium eurycarpum, Typha latifolia, Utricularia, Wolffia globosa, Wolffiella lingulata.*

Use by wildlife: *Wolffiella gladiata* is eaten by ring-necked ducks (Aves: Anatidae: *Aythya collaris*) and other waterfowl. **Economic importance: food:** *Wolffiella oblonga* has a relatively low protein content (8.9%) compared to other duckweeds and is not eaten; **medicinal:** no uses reported; **cultivation:** Only the non-North American *Wolffiella welwitschii* has been cultivated to any extent, for use as an aquarium plant; **misc. products:** *Wolffiella gladiata* has been studied for its ability to uptake cadmium from aqueous environments; **weeds:** no weedy infestations reported; **nonindigenous species:** *Wolffiella gladiata* has been introduced to Washington state. **Systematics:** Phylogenetic analyses consistently resolve *Wolffiella* as a clade, although *Wolffia brasiliensis* associates as its sister species in some evaluations (e.g., Figure 1.27). The occasional misplacement of *W. brasiliensis* with *Wolffiella* is not well supported and has been interpreted as an analytical artifact resulting from its high degree of divergence from other duckweed species. Biogeographical analyses have indicated that *Wolffiella* originated in Africa and dispersed to the New World, where it underwent further speciation. A close relationship among the three American species (*W. gladiata, W. lingulata,* and *W. oblonga*), which had been presumed by their morphological similarity, was first substantiated by their comparable flavonoid chemistry. The flavonoid data later were corroborated by allozymes, which recovered the highest genetic identity in the genus (I=0.940) between the morphologically similar *W. lingulata* and *W. oblonga*, and the next highest (I=0.816–0.845) between that species pair and *W. gladiata*. The same pattern of relationship among these three taxa was confirmed eventually by analyses of various DNA sequence data, which resolved them as comprising the most highly derived clade in the genus (e.g., Figure 1.27). The allozyme data initially raised questions regarding the distinctness of *W. lingulata* and *W. oblonga*, as well as their possible hybridization. More recently, their lack of discrete genetic divergence and likelihood of hybridization have been confirmed by analyses of AFLP and DNA sequence data. Consequently, the continued recognition of *W. lingulata* and *W. oblonga* as distinct species warrants serious reconsideration. However, it also is possible that these taxa represent incipient species, which have begun to diverge ecologically quite recently (at an estimated 0.87 MYBP). The integrity of the three sectional divisions of the genus (*Rotundae*, *Stipitatae*, and *Wolffiella*) has been maintained in various phylogenetic analyses (e.g., Figure 1.27). All the New World taxa occur within section *Wolffiella*. The relative genome size of *Wolffiella* is intermediate between that of *Spirodela* and *Wolffia* and shows little inter- or intraspecific variation (*W. gladiata*: 623 Mbp; *W. lingulata*: 629–655 Mbp). The complete chloroplast genome of *W. lingulata* has been sequenced. Probable hybrids between *W. lingulata* and *W. oblonga* have been indicated by allozyme, AFLP, and cpDNA sequence data.

Distribution: *Wolffiella gladiata* occurs in the eastern half of the United States and extends into Mexico; it is introduced in Washington state. *Wolffiella lingulata* and *W. oblonga* grow in scattered localities in the southeastern and western United States; both extend to South America.

References: Basinger et al., 1997; Bergmann et al., 2000; Blazey & McClure, 1968; Bog et al., 2018; Cottam, 1939; Crawford et al., 1997; 2006; Darst et al., 2002; Elam et al., 2009; Giardelli, 1935; Hall & Penfound, 1943; Hartman & English, 1959; Hicks, 1932a; 1932b; Huffman & Lonard, 1983; Kasselmann, 2003; Kimball et al., 2003; Kurz & Crowson, 1948; Landolt, 1981; 1986; 2000; Landolt & Wildi, 1977; Les et al., 1997; 2002; Mason, 1938; McClure & Alston, 1966; Mohlenbrock, 1959a; Patton & Judd, 1986; Pieterse et al., 1970; 1971; Schreinemakers, 1986; Scribailo & Alix, 2002; Sundue, 2007; Tippery & Les, 2020; Tippery et al., 2015; Uphof, 1922; Vaithiyanathan & Richardson, 1999; Wang & Messing, 2011; Wang et al., 2011a; Ziegler et al., 2015.

Family 4.3. Orontiaceae [4]

The reestablishment of Orontiaceae Bartl. as a distinct family is reconciled in the preceding discussions on Araceae and Lemnaceae. Although different attributions exist in the literature, the family name is credited correctly to Friedrich Bartling (Nicolson, 1984; Reveal, 2012). Bartling's concept of Orontiaceae predated the establishment of *Lysichiton* but included *Gymnostachys*, *Orontium*, and *Symplocarpus* (Bartling, 1830). Although these genera were transferred subsequently to Araceae, their distinctness from most of the

species was indicated taxonomically by their recognition as subfamilies: Gymnostachydoideae (*Gymnostachys*) and Orontioideae (*Lysichiton*, *Orontium*, *Symplocarpus*) (Mayo et al., 1998a; 1998b). That distinctness is exemplified well by phylogenetic analyses incorporating as many as 77 protein-coding sequences (Choi et al., 2017). With phylogenetic analyses resolving all four genera as a clade (e.g., Figures 1.26 and 1.28), these subfamilies simply have been elevated here to family rank in accordance with Bartling's original circumscription but could be retained at the same rank within Orontiaceae rather than Araceae.

In this sense, Orontiaceae comprise species of aquatic (*Orontium*), wetland (*Lysichiton*, *Symplocarpus*), or moist forest (*Gymnostachys*) habitats. The plants have vesselless, condensed, non-cormlike, thickened underground stems with contractile roots, stomata lacking subsidiary cells, bisexual flowers with a perigone of fornicate tepals, unilocular ovaries, locules with 1–2 anatropous or hemianatropous ovules, copious, sparse or absent endosperm, and pollen with a sporopollenin ektexine (Bown, 2000; Mayo et al., 1998a; 1998b; Hesse, 2006; Carlquist & Schneider, 2014).

The flowers are pollinated by various insects (Insecta) including bees (Hymenoptera: Apidae: *Apis*), beetles (Coleoptera: Staphylinidae), flies (Diptera: Chirnomidae; Mycetophilidae), spring stoneflies (Nemouridae), and thrips (Thysanoptera) (Gibernau, 2016). The fruits are dispersed abiotically by water or biotically by various animals.

Several genera (*Lysichiton*, *Orontium*, *Symplocarpus*) provide ornamental horticultural specimens for water gardens (Perry, 1961). *Symplocarpus* was once used pharmaceutically as a drug known as "dracontium." The starchy rhizomes and seeds of *Orontium* have been eaten as food (Bown, 2000).

Orontiaceae are distributed disjunctly among temperate regions of eastern Asia, eastern North America, western North America, and Australia. Three genera contain OBL North American indicators (Figures 1.26 and 1.28):

4.3.1. ***Lysichiton*** Schott
4.2.2. ***Orontium*** L.
4.3.3. ***Symplocarpus*** Salisb. ex W.P.C.Barton

4.3.1.   Lysichiton

American skunk-cabbage, swamp lantern; lysichiton d'Amérique

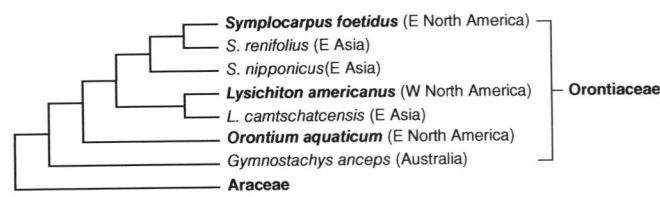

FIGURE 1.28 Phylogenetic relationships in Orontiaceae based on analysis of combined *trn*L-F and *ndh*F sequence data (adapted from Nie et al., 2006). The OBL North American indicators are highlighted in bold. Diversification of the group has involved several divergence events between Asia and North America.

Etymology: from the Greek *lysis chiton* ("dissolving tunic") in reference to the withering post-anthesis spathe

Synonyms: *Arctiodracon*; *Dracontium* (in part); *Pothos* (in part); *Symplocarpus* (in part)

Distribution: global: Asia, North America; **North America:** western

Diversity: global: 2 species; **North America:** 1 species

Indicators (USA): obligate wetland (OBL): *Lysichiton americanus*

Habitat: freshwater; palustrine; **pH:** 5.5–7.3; **depth:** <1 m; **life-form(s):** emergent herb

Key morphology: rhizome (to 30+ cm) stout (to 5 cm broad), vertical, with contractile roots; leaves appearing after flowering, clustered at apex, the blades large (to 135 cm), shiny, elliptic to oblanceolate, veins pinnate, the midvein thickened, apex obtuse to acute, base cuneate or nearly truncate, petiole (to 40 cm) stout; inflorescence a thick, cylindrical, pulpy spadix (to 14 cm) of numerous, densely embedded flowers, spadix surrounded by a yellow, boat-shaped, indole-scented spathe, the upper hood-like portion (to 25 cm) tapering to a sheathing base, stipe (to 40 cm) arising from rhizome; flowers bisexual, protogynous, perigone present; tepals (to 4 mm) 4, yellow-green, fleshy, concave; stamens 4, anthers exserted, pollen copious; ovary ovoid, conical, constricted basally, 2-locular, ovules 1–2 in each locule, style distinct, attenuate; berries fleshy, white and green-tipped at maturity, congealed in an infructescence (to 15 cm), becoming gelatinous as spadix decomposes; seeds (to 11 mm) ellipsoid to ovoid, gray to reddish brown

Life history: duration: perennial (persistent crown, rhizomes); **asexual reproduction:** rhizomes; **pollination:** insect; **sexual condition:** hermaphroditic; **fruit:** multiple [infructescence] (berries) (common); **local dispersal:** fruits (water); **long-distance dispersal:** fruits, seeds (animals [exozoic])

Imperilment: 1. *Lysichiton americanus* [G5]; SH (WY)

Ecology: general: Both *Lysichiton* species inhabit wet ground or shallow waters of swamps or other wetlands. The Asian *L. camtschatcense* is very similar ecologically, differing primarily by its white spathe and smaller, odorless flowers (also protogynous), which are pollinated primarily by flies (Insecta: Diptera) instead of beetles (Insecta: Coleoptera).

4.3.1.1. *Lysichiton americanus* **Hultén & H. St. John** inhabits bogs (peat, snow), bottoms (stream), depressions, ditches (roadside, stagnant), fens, flats (marshy), floodplains, gullies (swamp), marshes, meadows, mudflat, muskeg, ponds (beaver), ravines, seepages, slopes (to 10%), springs (roadside), streams (cold), swales, swamps (alluvial, coastal, roadside), thickets, woodlands, and the margins of rivers and streams at elevations to 1524 m. The plants occur in open to shaded exposures, but most often in the latter. The substrates include alluvium, clay (heavy), loam (humus, organic), loamy silt, muck (organic), mucky loam, mud, peat, sand, serpentine (gravelly), silt, and silty loam. The plants grow on wet substrates or in shallow (to 21 cm) standing waters (e.g., pH: 5.5–7.3; total alkalinity: 48.0 mg l^{-1}; conductivity: 26.8–35.8 μS cm^{-1}; Br$^-$: 0.019–036 mg l^{-1}; Ca^{2+}: 2.071–4.071 mg l^{-1}; Cl$^-$: 0.063–0.441 mg l^{-1}; Fl$^-$: 0.004–0.010 mg l^{-1}; K$^+$: 0.708–1.665

mg l^{-1}; Mg^{2+}: 1.198–2.483 mg l^{-1}; Na$^+$: 0.408–0.681 mg l^{-1}; NH$_4$: 0.036–0.404 mg l^{-1}; NO$_3^-$: 0.000–0.010 mg l^{-1}; PO$_4^{3-}$: 0.000–0.019 mg l^{-1}; SO$_4^{2-}$: 0.002–0.023 mg l^{-1}). Flowering is precocious (occurring before the leaves appear) and commences during early spring (March–April) with the fruits maturing from April to September. Flowering within an inflorescence essentially is synchronous. The individual period of anthesis is long (10–14 days) compared to most species evaluated in the related Araceae (<5 days). Flowering plants are more prevalent in sunny than in shaded sites and flowering within a stand can persist for up to 5 weeks. The flowers are self-compatible but strongly protogynous. The initial female phase lasts from 1 to 8 days, with the stigmas remaining receptive for 1–6 days. The subsequent male phase lasts from 4 to 18 days and overlaps slightly with the end of the receptive female phase. The brief period of overlap between the female and male phases can allow for self-pollination. The estimated pollen:ovule ratio is high (76,912:1) but wind pollination does not occur as once was suspected. Pollination is achieved primarily by rove beetles (Insecta: Coleoptera: Staphylinidae: *Pelecomalius testaceum*), which are attracted to the yellow color and odor of the spathes; the inflorescences are not thermogenic. The attractive scents have been found to contain three specific odorants [(E)-4 nonene, (E)-5-undecene, and indole], with the latter having the most pronounced effect. As they mate, the beetles feed on the pollen released during the male phase. Pollinator frequencies are associated with floral sex expression and do not appear to be influenced by differing surrounding forest use or management. Seed set is high (often near 100%). Inflorescences senesce and collapse by late summer (August) and in doing so, provide the seeds with a gelatinous coating by hydration of enveloping "mealy" material. The coating appears to reduce the incidence of some fungal infections and minimally reduces germination rates. Many seeds fall and germinate near the maternal plant. Seed germination is highest from July to August. Otherwise, the seeds are buoyant (remaining afloat for at least 4 days) and are dispersed locally by water. Their adherence to the muddy feet of various birds (see *Use by wildlife*) is thought to be responsible for longer dispersal distances. **Reported associates:** *Abies amabilis, Abies grandis, Acer circinatum, Acer glabrum, Acer macrophyllum, Acmispon, Adiantum pedatum, Agrostis, Agrostis stolonifera, Alnus rubra, Alnus rugosa, Alnus viridis, Amelanchier, Angelica arguta, Angelica genuflexa, Antennaria argentea, Asarum caudatum, Athyrium filix-femina, Berberis aquifolium, Betula glandulosa, Betula nana, Betula papyrifera, Bistorta bistortoides, Blechnum spicant, Boykinia major, Calamagrostis canadensis, Calamagrostis stricta, Callitriche heterophylla, Calocedrus decurrens, Caltha leptosepala, Cardamine, Carex amplifolia, Carex aquatilis, Carex arcta, Carex cusickii, Carex deweyana, Carex diandra, Carex echinata, Carex laeviculmis, Carex lasiocarpa, Carex leptalea, Carex leptopoda, Carex lyngbyei, Carex nigricans, Carex obnupta, Carex pellita, Carex scopulorum, Carex simulata, Carex stipata, Carex utriculata, Chamaecyparis nootkatensis, Cicuta bulbifera, Cicuta douglasii, Cinna latifolia, Circaea alpina, Cirsium* *arvense, Claytonia cordifolia, Claytonia sibirica, Clintonia uniflora, Comarum palustre, Coptis aspleniifolia, Cornus canadensis, Cornus sericea, Darlingtonia californica, Deschampsia cespitosa, Dicentra formosa, Dichanthelium acuminatum, Dicranum, Drosera anglica, Drosera rotundifolia, Dryopteris expansa, Dulichium arundinaceum, Eleocharis palustris, Eleocharis quinqueflora, Elymus glaucus, Empetrum nigrum, Epilobium ciliatum, Epilobium glaberrimum, Equisetum arvense, Equisetum fluviatile, Equisetum telmateia, Eriophorum chamissonis, Fraxinus latifolia, Galium aparine, Galium triflorum, Gaultheria shallon, Gentiana sceptrum, Geum macrophyllum, Glyceria striata, Gymnocarpium dryopteris, Heracleum sphondylium, Holcus lanatus, Hypericum anagalloides, Impatiens capensis, Juncus balticus, Juncus effusus, Juncus ensifolius, Juncus lesueurii, Juncus nevadensis, Kalmia microphylla, Larix, Lemna minor, Lemna turionifera, Linnaea borealis, Lonicera involucrata, Ludwigia palustris, Lycopus, Lycopus uniflorus, Maianthemum dilatatum, Maianthemum dilatatum, Malus fusca, Menyanthes trifoliata, Menziesia ferruginea, Mimulus guttatus, Mimulus primuloides, Mnium, Muhlenbergia filiformis, Myosotis laxa, Myrica gale, Nasturtium officinale, Neottia cordata, Nephrophyllidium crista-galli, Nuphar polysepala, Oemleria cerasiformis, Oenanthe sarmentosa, Oplopanax horridus, Oxalis oregana, Pedicularis, Persicaria hydropiper, Phalaris arundinacea, Phragmites australis, Physocarpus malvaceus, Picea engelmannii, Picea mariana, Picea sitchensis, Pinus contorta, Pinus monticola, Platanthera dilatata, Pleurozium schreberi, Poa palustris, Poa trivialis, Podagrostis thurberiana, Polypodium glycyrrhiza, Polystichum munitum, Populus trichocarpa, Potamogeton gramineus, Potentilla anserina, Primula jeffreyi, Prunella vulgaris, Pseudotsuga menziesii, Ranunculus repens, Rhamnus purshiana, Rhododendron columbianum, Ribes bracteosum, Rosa pisocarpa, Rubus laciniatus, Rubus parviflorus, Rubus pedatus, Rubus spectabilis, Rubus ursinus, Salix alaxensis, Salix commutata, Salix geyeriana, Salix hookeriana, Salix lasiandra, Salix myrtillifolia, Salix scouleriana, Salix sitchensis, Sanguisorba officinalis, Scheuchzeria palustris, Scirpus microcarpus, Scutellaria lateriflora, Senecio triangularis, Sequoia sempervirens, Solanum dulcamara, Sparganium angustifolium, Sphagnum mendocinum, Sphagnum pacificum, Sphagnum papillosum, Spiraea douglasii, Spiranthes romanzoffiana, Spiranthes romanzoffiana, Stachys chamissonis, Stachys rigida, Stellaria calycantha, Symphoricarpos albus, Symphyotrichum foliaceum, Thuja plicata, Tiarella trifoliata, Tolmiea menziesii, Torreyochloa pallida, Trautvetteria caroliniensis, Triantha occidentalis, Triglochin maritima, Trillium ovatum, Tsuga heterophylla, Tsuga mertensiana, Typha latifolia, Urtica dioica, Utricularia macrorhiza, Vaccinium ovalifolium, Vaccinium ovatum, Vaccinium parvifolium, Vaccinium uliginosum, Veratrum californicum, Veratrum fimbriatum, Veratrum viride, Veronica americana, Veronica scutellata, Veronica serpyllifolia, Vicia nigricans, Viola glabella, Viola palustris.*

Use by wildlife: The plants are eaten by black bears (Mammalia: Ursidae: *Ursus americanus*), grizzly bears (Mammalia: Ursidae: *Ursus arctos*), elk (Mammalia: Cervidae: *Cervus elaphus*), muskrats (Mammalia: Cricetidae: *Ondatra zibethicus*), raccoons (Mammalia: Procyonidae: *Procyon lotor*), and Sitka black-tailed deer (Mammalia: Cervidae: *Odocoileus hemionus sitkensis*). The seeds are eaten by frugivorous rodents (Mammalia: Rodentia) and by several birds (Aves) including chestnut-backed chickadees (Paridae: *Poecile rufescens*) and Steller's jays (Corvidae: *Cyanocitta stelleri*). The plants host several Fungi (Ascomycota: Helotiaceae: *Ombrophila lysichitonis*; Mycosphaerellaceae: *Septoria*; Pleosporaceae: *Stemphylium*) and are a monoecious holocyclic host to a specialist aphid (Insects: Hemiptera: Aphididae: *Macrosiphum oregonense*). The pollen is eaten by rove beetles (Insecta: Coleoptera: Staphylinidae: *Eusphalerum pothos*, *Pelecomalius testaceum*).

Economic importance: food: *Lysichiton americanus* should be avoided as a food because it contains calcium oxalate crystals, which can be harmful when ingested. Yet, the plants were eaten by numerous indigenous western North American tribes. The Quinault ate the roasted petioles, the Cowlitz ate the steamed inflorescences, the Skokomish and Twana ate the young leaves, and the Lower Chinook, Pemberton Lillooet, Quileute, Tolowa, and Yurok people ate the rhizomes (which reputedly have a ginger-like taste). The Kwakwaka'wakw used the dried, powdered leaves to thicken boiled currants when making berry cakes; the Hoh and Quileute people wrapped leaves around various cooked fruits to preserve them when buried. The leaves often were used to steam salmon and also mixed with seaweed (*Porphyra*) and eaten; **medicinal:** The leaves of *Lysichiton americanus* have antifungal properties and the roots antiviral properties. The extracts are cytotoxic to certain drug-resistant cancer cell lines. The plants have been used medicinally by many First Nations people. Their inflorescences were employed as an antirheumatic (Cowlitz, Gitksan). Various leaf preparations were used as an analgesic (Makah, Quileute, Shuswap, Skokomish), stimulant (Kwakiutl), and as a treatment for boils (Kwakiutl), burns (Hesquiat, Nitinaht), carbuncles (Kwakiutl), cuts (Quileute, Skokomish), fever (Quileute, Skokomish), scrofula (Klallam), stomach problems, sores (Kwakiutl, Shuswap), swellings (Quileute, Skokomish), and (when heated) to remove splinters (Kwakiutl). The rhizomes were prepared as an abortive (Makah), anti-hemorrhagic (Gitksan), antirheumatic (Gitksan, Makah, Tolowa, Yurok), bladder cleanser (Quinault), blood purifier (Makah), cathartic (Skokomish), emetic, hair tonic (Kwakiutl), sedative (Gitksan), to ease childbirth (Quileute), and as a treatment for boils (Gitksan), burns, blood poisoning (Gitksan), bloody urine (Haisla, Hanaksiala), burns (Haisla, Hanaksiala), carbuncles (Klallam), influenza (Gitksan), sores (Clallam), stomach ailments (Bella Coola, Makah), stroke (Tolowa, Yurok), swellings (Kwakiutl), and when charred, for assorted bites, infections, and wounds (Thompson). The acrid sap was used to treat ringworm; **cultivation:** *Lysichiton americanus* is grown as an ornamental water garden and pond plants. The cultivars ×*hortensis*, ×*hortensis* 'Billy', and ×*hortensis* 'Devonshire Cream' are derived from interspecific hybrids between *L. americanus* and *L. camtschatcensis*;

misc. products: The leaves of *L. americanus* were used to dry berries (Kwakiutl, Makah), as berry basket covers (Kwakiutl, Makah, Tsimshian) and as a type of wax paper for lining berry baskets (Bella Coola, Nitinaht, Oweekeno, Poliklah), as mats for berry drying racks (Haisla, Hanaksiala, Nitinaht, Quileute, Tsimshian), and as a wrap to add flavor when baking, boiling, or steaming miscellaneous foods (Bella Coola, Haisla, Hanaksiala, Hoh, Kwakiutl, Makah, Okanagan-Colville, Quileute, Tolowa, Tsimshian, Yurok). The folded leaves were used as drinking cups or water dippers (Bella Coola, Coast, Oweekeno, Nitinaht Salish, Samish, and Swinomish tribes). Kitasoo children competed by seeing how far the spadices could be thrown; **weeds:** *Lysichiton americanus* is regarded as invasive in Europe; **nonindigenous species:** *Lysichiton americanus* has been introduced to various parts of Europe as an ornamental escape from cultivation since the early 20th century.

Systematics: Both *Lysichiton* species resolve as a clade that is sister to *Symplocarpus* (Figure 1.28). The two genera are estimated to have diverged approximately 31–40 MYBP, with their divergence from *Orontium* taking place about 72–75 MYBP. Some older North American literature refers to the American species as *L. camtschatcense*, but the taxa are distinct morphologically and genetically. The basic chromosome number of *Lysichiton* is *x* = 14. *Lysichiton americanus* (2*n* = 28) presumably is diploid (as well as *L. camtschatcense*). Although their broad disjunction precludes any opportunity for natural hybridization, synthetic hybrids have been made between *L. americanus* and *L. camtschatcensis*.

Distribution: *Lysichiton americanus* occurs in western North America.

References: Armitage & Phillips, 2011; Barriault et al., 2009; Brodie et al., 2018; Brousil et al., 2015; Christy, 2004; Cooke, 1997; Doering & Coxson, 2010; Doyle & Duckett, 1985; Hanley et al., 2014; Howie & Van Meerveld, 2016; Jensen, 2000; Jones et al., 2016; Karadeniz et al., 2015; Kowarik, 2003; Mason, 1957; Mohagheghzadeh et al., 2006; Morgan & Sytsma, 2009; Mowat et al., 2017; Nie et al., 2006; Peeters-Van der Meijde & Rotteveel, 2006; Pellmyr & Patt, 1986; Ronse, 2011; Tanaka, 2004; Thompson, 2000b; Turner, 1995; 2003; Turesson, 1916; Willson & Hennon, 1997; Zanetti, 2014.

4.3.2. Orontium

Golden club, neverwet

Etymology: after the Neo-Latin *auranti* ("orange-colored") in reference to the distinctive spadix

Synonyms: *Amidena* (in part); *Aronia* (in part); *Pothos* (in part)

Distribution: global: North America; **North America:** eastern United States

Diversity: global: 1 species; **North America:** 1 species

Indicators (USA): obligate wetland (OBL): *Orontium aquaticum*

Habitat: freshwater; freshwater (tidal); lacustrine, palustrine, riverine; **pH:** 4.7–7.3; **depth:** to 1.5 m; **life-form(s):** emergent herb, floating-leaved, submersed (rosulate)

Key morphology: rhizome vertical, stout (to 3 cm in diameter), with contractile roots; leaves (to 45 cm) precocious, clustered, emergent, floating, or submersed, oblong-elliptic, the upper surface velvety, bluish-green, waxy (hydrophobic), lower surface pale green, veins parallel, margins entire, petiole (to 60 cm) dark green or reddish; spadix (to 10 cm) elongate-conical, swollen, dense with numerous, embedded, bisexual flowers (those apical sometimes ♂), bright yellow to orange in anthesis, terminating a fleshy, reddish stipe (to 75+ cm), spathe green, membranous, inconspicuous, arising at base of stipe, fugacious at anthesis; perigone present; tepals 2–6, arching atop ovary; stamens 2–6, 1–2 staminodes often present; ovary superior, unilocular but pseudomonomerous (2–3 carpels), with 1 basal ovule; infructescence of multiple berries; individual berries (to 2 cm) green, single-seeded, pericarp thin; seeds (to 1.5 cm) mucilaginous

Life history: duration: perennial (rhizome); **asexual reproduction:** none; **pollination:** insect; **sexual condition:** hermaphroditic [or ♂ distally]; **fruit:** multiple [infructescence] (berries) (common); **local dispersal:** fruits (water); **long-distance dispersal:** fruits, seeds (animals, water)

Imperilment: 1. *Orontium aquaticum* [G5]; S1 (MA, RI); S2 (KY, NY); S3 (CT, WV)

Ecology: general: *Orontium* is monotypic (see next).

4.3.2.1. ***Orontium aquaticum*** L. grows in backwaters, baygalls, bogs (quaking), bottoms (boggy, river), depressions, floating peat "batteries," ditches, floodplains, gravel bars, hammocks (hydric), lagoons, lakes, marshes (freshwater, tidal), meadows (floodplain), mudflats, ponds (beaver, cold, spring-fed, sediment control), pools (streamside), rice fields (abandoned), prairies (wet), rivers, seeps, shores, streams (artesian, blackwater, outflow, roadside, seepage), swamps (blackwater, floodplain, hardwood, riverine, roadside, *Sphagnum*), and woodlands (swampy) at elevations to 1219 m. Exposures can range from open sites to fairly dense shade. The substrates have been categorized as Basinger (spodic psammaquents), clay, Dorovan-Johnston series, humus, marl (calcareous), muck, mud, peat, peaty sand, and sand. This is a versatile species, which thrives in a variety of lentic or lotic habitats by its capability to grow as an emergent, floating-leaved form, or even as a submersed rosette. The high degree of plasticity enables the plants to occupy dynamic habitats characterized by tidally influenced or seasonally fluctuating waters. The plants occur most often in tidal or non-tidal coastal, acidic freshwaters (pH: 4.7–7.3; $\bar{x}$ = 6.0) to a maximum depth of about 1.5 m. Unlike many aquatic plants, *Orontium* undergoes frequent sexual reproduction, with the seeds filling a major reproductive role. Flowering occurs from late February through June with the fruits maturing from June to August. The bisexual flowers open acropetally, although the uppermost occasionally are staminate. Pollination is thought to be facilitated by insects (Insecta: Coleoptera; Diptera) but that assumption is documented only circumstantially by pollen characteristics. It has not been determined whether the high fertility might be due to self-pollination or apomixis, which should be evaluated. Most other aspects of the floral biology remain unstudied as well. The change in spadix (tepal) color from green to orange (during anthesis) is a potential pollinator cue and results from a decrease in chlorophyll and not an increase in the carotenoid content. After fertilization, the scape arches to reposition the spadix underwater where the fruits develop. When mature, the fruits abscise and the spadix dissociates. The fruits are buoyant and can remain afloat for about 1 week while they are dispersed by surface water currents. They sink as their pericarp becomes waterlogged, ultimately releasing their seeds (which are not buoyant) about a week after settling on the bottom. The seeds lack dormancy and can germinate immediately when ripe, especially if incubated under a 19°C/15°C temperature regime. They sometimes germinate viviparously, while still attached to the flowering spadix. However, germination can be delayed if the seeds are placed in cold water (3°C–5°C), where they have remained highly viable (92%) after 5–7 months of storage. They are intolerant of desiccation and will shrink severely and perish if dried. They are not persistent and have not been recovered in seed bank studies. Long-distance dispersal of fruits by animals has been postulated, but not documented. The common name of "neverwet" alludes to the effective water repellency of the leaves, which is attributable to their papillose epidermal cells and epicuticular wax platelets. Photosynthetic rates increased by 54–71% when the plants were grown under elevated CO_2. In some experiments, no significant difference in biomass was observed but methane (CH_4) release was enhanced. In others, elevated CO_2 levels resulted in a 17–55% increase in biomass. Higher biomass (by 29–57%) also has been observed in flooded treatments. The plants overwinter by their dormant, vertical rhizome, which can extend for 30+ cm into the substrate. Unless it would turn out that the seeds develop apomictically (which never has been evaluated), there is no specialized means of vegetative reproduction. **Reported associates:** *Acer rubrum, Acorus calamus, Agrostis perennans, Alisma triviale, Alnus serrulata, Alternanthera philoxeroides, Amaranthus cannabinus, Ampelaster carolinianus, Andropogon virginicus, Asclepias perennis, Batrachospermum vagum, Bidens connatus, Bidens eatonii, Bidens frondosus, Bidens laevis, Bidens mitis, Bolboschoenus fluviatilis, Boltonia asteroides, Brasenia schreberi, Calamagrostis canadensis, Callitriche heterophylla, Canna flaccida, Cardamine pensylvanica, Carex atlantica, Carex collinsii, Carex decomposita, Carex exilis, Carex folliculata, Carex frankii, Carex glaucescens, Carex hyalinolepis, Carex lurida, Carex scoparia, Carex striata, Carex stricta, Carex trisperma, Carpinus caroliniana, Carya aquatica, Cephalanthus occidentalis, Ceratophyllum demersum, Chamaecrista fasciculata, Chamaecyparis, Chenopodium album, Cicuta maculata, Cicuta maculata, Cladium jamaicense, Clethra alnifolia, Cliftonia monophylla, Conoclinium coelestinum, Coptis trifolia, Cornus amomum, Crassula aquatica, Crinum americanum, Cuscuta compacta, Drosera capillaris, Drosera intermedia, Drosera rotundifolia, Dulichium arundinaceum, Egeria densa, Elatine triandra, Eleocharis elongata, Eleocharis palustris, Eleocharis parvula, Eleocharis quadrangulata, Eleocharis robbinsii, Elodea canadensis, Elodea nuttallii, Equisetum arvense, Eriocaulon aquaticum, Eriocaulon compressum, Eriocaulon decangulare, Eriocaulon parkeri, Eriochloa michauxii, Eriophorum virginicum, Eryngium aquaticum, Eupatorium album, Eutrochium fistulosum, Fontinalis, Galium, Gaultheria procumbens, Glyceria canadensis, Glyceria obtusa, Glyceria septentrionalis, Gratiola aurea, Gratiola virginiana, Harperella nodosa, Helenium autumnale, Helenium flexuosum, Heteranthera reniformis, Hibiscus moscheutos, Hydrocotyle umbellata, Hygrophila lacustris, Hymenocallis crassifolia, Hymenocallis franklinensis, Hymenocallis rotata, Ilex vomitoria, Impatiens capensis, Iris hexagona, Iris versicolor, Iris virginica, Isoetes caroliniana, Isoetes engelmannii, Isoetes flaccida, Isoetes mattaponica, Isoetes riparia, Itea virginica, Juncus acuminatus, Juncus biflorus, Juncus coriaceus, Juncus effusus, Juncus elliottii, Juncus militaris, Juncus pelocarpus, Juncus roemerianus, Justicia americana, Lachnanthes caroliniana, Leersia lenticularis, Leersia oryzoides, Lilaeopsis chinensis, Limosella australis, Liquidambar styraciflua, Lobelia cardinalis, Lobelia dortmanna, Ludwigia palustris, Ludwigia palustris, Lyonia, Lysimachia terrestris, Lythrum lineare, Magnolia virginiana, Mayaca fluviatilis, Mikania scandens, Mitchella repens, Myosotis laxa, Myrica pensylvanica, Myriophyllum spicatum, Najas flexilis, Nuphar advena, Nuphar sagittifolia, Nuphar variegata, Nymphaea odorata, Nymphoides aquatica, Nyssa aquatica, Nyssa biflora, Onoclea sensibilis, Osmunda regalis, Osmundastrum cinnamomeum, Oxalis stricta, Packera glabella, Panicum hemitomon, Peltandra virginica, Persicaria arifolia, Persicaria glabra, Persicaria hydropiper, Persicaria hydropiperoides, Persicaria punctata, Persicaria sagittata, Physostegia leptophylla, Pilea pumila, Pinus palustris, Plantago cordata, Platanthera ciliaris, Platanthera clavellata, Pluchea camphorata, Podostemum ceratophyllum, Pogonia ophioglossoides, Pontederia cordata, Potamogeton berchtoldii, Potamogeton confervoides, Potamogeton epihydrus, Potamogeton nodosus, Potamogeton perfoliatus, Potamogeton spirillus, Proserpinaca palustris, Ptilimnium capillaceum, Quercus laurifolia, Ranunculus ambigens, Ranunculus hispidus, Rhapidophyllum hystrix, Rhododendron viscosum, Rhynchospora alba, Rhynchospora fusca, Rhynchospora glomerata, Rhynchospora inundata, Rhynchospora macrostachya, Rhynchospora miliacea, Rosa palustris, Rubus hispidus, Rumex verticillatus, Sabal, Sabatia calycina, Sabatia kennedyana, Sabatia stellaris, Sacciolepis striata, Sagittaria engelmanniana, Sagittaria graminea, Sagittaria lancifolia, Sagittaria latifolia, Sagittaria rigida, Sagittaria subulata, Salix caroliniana, Salix sericea, Samolus valerandi, Sanguinaria canadensis, Sarracenia flava, Sarracenia purpurea, Sarracenia rubra, Saururus cernuus, Schoenoplectus acutus, Schoenoplectus pungens, Schoenoplectus subterminalis, Schoenoplectus tabernaemontani, Scutellaria elliptica, Sium suave, Solidago sempervirens, Sparganium americanum, Sparganium eurycarpum, Spartina cynosuroides, Sphagnum, Styrax, Symphyotrichum novi-belgii, Symphyotrichum subulatum, Symphyotrichum tenuifolium, Symplocarpus foetidus, Taxodium ascendens,*

Taxodium distichum, Thalictrum pubescens, Thelypteris serrata, Torreyochloa pallida, Toxicodendron radicans, Triadenum virginicum, Typha angustifolia, Typha latifolia, Ulmus americana, Utricularia geminiscapa, Utricularia gibba, Vaccinium corymbosum, Vallisneria americana, Verbena scabra, Verbesina occidentalis, Viola, Woodwardia areolata, Woodwardia virginica, Xyris smalliana, Zizania aquatica, Zizaniopsis miliacea.

Use by wildlife: *Orontium aquaticum* is eaten by various mammals (Vertebrata: Mammalia), including beavers (Castoridae: *Castor canadensis*), cattle (Bovidae: *Bos taurus*), deer (Cervidae), marsh rabbits (Leporidae: *Sylvilagus palustris paludicola*), muskrats (Cricetidae: *Ondatra zibethicus*), and swine (Suidae: *Sus scrofa*). The plants host larval moths (Insecta: Lepidoptera: Crambidae: *Munroessa icciusalis, Paraponyx maculalis, P. obscuralis, P. seminealis*) and several Fungi (Ascomycota: Botryosphaeriaceae: *Phyllosticta orontii*; Hyponectriaceae: *Physalospora orontii*; Mycosphaerellaceae: *Mycosphaerella, Ramularia orontii, Sphaerella orontii*; Nectriaceae: *Fusarium, Volutella diaphana*; Sclerotiniaceae: *Polyactis streptothrix, Streptobotrys streptothrix*; incertae sedis: *Stilbella aciculosa*; Basidiomycota: Ceratobasidiaceae: *Rhizoctonia*; Chionosphaeraceae: *Stilbum aciculosum*).

Economic importance: food: As a potential food, *Orontium* contains 19.8–21.2% protein (DM), 7.9% crude fat (DM), and has a caloric content of 3.7 kcal g^{-1} (DM). The foliage is rich in boron (10.7 ppm DM). However, the plants are regarded as poisonous because all the tissues contain proteolytic enzymes and styloid calcium oxalate crystals, which are harmful if ingested without proper preparation (e.g., extensive drying and/or cooking). The plants were eaten by Native Americans, who cooked them for up to half a day to remove the acrid and harmful properties. They thoroughly dried the starchy rhizomes to grind into flour by roasting them for as long as 2 days. The large seeds were boiled (for 45+ min) through several water changes before eating; **medicinal:** vaguely referenced as having general medical use; **cultivation:** *Orontium aquaticum* is grown as an ornamental pond and water garden marginal, especially in the southern United States; **misc. products:** Root exudates from *Orontium aquaticum* have been used to detoxify metal-contaminated waters; **weeds:** *Orontium* has been reported as an invasive weed in Europe.; **nonindigenous species:** *Orontium aquaticum* has been introduced to Denmark, Finland, Norway, and Sweden.

Systematics: Phylogenetic analyses consistently resolve the monotypic *Orontium aquaticum* as the sister group to the *Lysichiton/Symplocarpus* clade (e.g., Figure 1.28). The genome size of *Orontium* has been estimated at 30 pg 2C^{-1}. The base chromosome number of *Orontium* is $x=13$; *O. aquaticum* ($2n=26$) is a diploid. No hybrids involving *Orontium* have been reported or are likely, due to its isolated phylogenetic position.

Distribution: *Orontium aquaticum* is distributed primarily along the coastal plain of the eastern United States.

References: Aulbach-Smith & de Kozlowski, 1996; Barabe & Labrecque, 1985; Bartgis, 1997; Beal, 1977; Beaven & Oosting, 1939; Bitton et al., 2005; Blair, 1936; Bleuel, 1923;

Bliss & Suzuki, 2012; Bown, 2000; Boyd, 1968; Boyd & Goodyear, 1971; Boyd & Walley, 1972; Capers & Les, 2005; Carr, 1940; Carter, Jr. & Jones, Jr., 1968; Casadoro et al., 1982; Chabreck, 1958; Cook, 1996; Core, 1967; Crocker, 1938; Cypert, 1961; 1972a; Ellis & Everhart, 1885; Fernald & Kinsey, 1943; Fleming & Van Alstine, 1999; Gerritsen & Greening, 1989; Gibernau, 2016; Gifford, 1893; Grear, 1966; Grear, Jr., 1973; Hann, 1986; Harms & Grodowitz, 2009; Holm, 1891; Hotchkiss & Stewart, 1947; Hussner, 2012; Klotz, 1992; Lantz, 1910; Layne & Johns, 1965; Martin, 1886; Megonigal & Schlesinger, 1997; Megonigal et al., 2005; Muenscher, 1936a; Musselman et al., 1995; 2001; Nie et al., 2006; Perry, 1961; Peterson, 1977; Sawyer et al., 2005; Schmalzer et al., 1985; Smith et al., 2001; Stalter & Baden, 1994; Stoops et al., 1998; Stuckey & Gould, 2000; Thompson, 2000; Tiner, 1987; 1993; Yanovsky, 1936; Zampella et al., 2001.

4.3.3. Symplocarpus

Skunk-cabbage, skota; chou puant

Etymology: from the Greek *symplokos karpos* ("interwoven fruit") in reference to the infructescence

Synonyms: *Dracontium* (in part); *Ictodes*; *Pothos* (in part); *Spathyema*

Distribution: global: Asia (eastern); North America; **North America:** east-central

Diversity: global: 3–4 species; **North America:** 1 species

Indicators (USA): obligate wetland (OBL): *Symplocarpus foetidus*

Habitat: freshwater; palustrine; **pH:** 3.4–6.7; **depth:** <1 m; **life-form(s):** emergent herb

Key morphology: rhizome thick (to 30 cm), vertical, dense with fleshy, contractile roots; leaves clustered, erect, the blades (to 60 cm) oblong to ovate, the base cordate to truncate, the petiole (to 57 cm) sheathed basally; flowering precocious; inflorescence thermogenic, situated at ground level from short, subterranean, peduncle; spadix (to 3 cm) ovoid to globose, with up to 73 bisexual flowers, becoming spongy in fruit, enclosed by a fleshy, coriaceous spathe (to 18 cm), the spathe yellow-green to dark purple, variously spotted and/or striped, cowl-like, opening only apically, apex indurate, twisted, apiculate; perigone present, the tepals 4, yellowish to purple; stamens 4; ovary unilocular, uniovulate; fruits brown to greenish purple, embedded in the fleshy spadix to form an infructescence (to 10 cm); seeds (to 15 mm) brown, globose

Life history: duration: perennial (rhizomes); **asexual reproduction:** none; **pollination:** insect; **sexual condition:** hermaphroditic; **fruit:** multiple [infructescence] (berries) (common); **local dispersal:** fruits (water); **long-distance dispersal:** fruits (animals, water)

Imperilment: 1. *Symplocarpus foetidus* [G5]; S1 (TN); S2 (NB); S3 (IA, IL, NC, NS)

Ecology: general: *Symplocarpus* species are similar ecologically, with all residing in low elevation marshes, swamps, or other wetlands. All the species are perennials with bisexual, entomophilous flowers. Depending on the species, the foliage expands precociously or subsequent to flower production. The multiple fruits are borne as a dense infructescence of berries.

The fruits of one Asian species (*S. nipponicus*) do not mature until the following year. The individual berries are dispersed by water or over greater distances by small frugivorous animals, which often bury and cache them undamaged.

4.3.3.1. *Symplocarpus foetidus* (L.) Salisb. ex W.P.C. Barton.

grows in bogs (cedar, seepage), bottoms (stream), fens (prairie), flats (alluvial), floodplains (stream), glades, marshes (hillside, seep), meadows (marshy, wet), oxbows, pools (ephemeral), ravines (boggy), seeps (calcareous, floodplain, forested, hillside), springs, swales (floodplain), swamps (bottomland, hardwood, roadside, seepage, tamarack), terraces (stream), thickets (seepy), woodlands (alluvial, lowland), and along the margins of ponds and streams (spring-fed) at elevations to 777 m. Exposures in full sun are uncommon, with most occurrences in partially to densely shaded sites. The reported substrates include alluvium, clay (black), gravel, humus, loam, muck (Adrian, terric medisaprist), mud, peat, sand, sandy loam, and silt. The plants are most common at sites with seasonal surface waters associated with acidic and low conductivity environments (e.g., pH: 3.4–6.7; specific conductivity: 66–78 µs cm^{-1}). However, they have been found to occur more broadly including richer sites such as calcareous fens. Flowering occurs during winter and early spring (January–March). Emergence of the inflorescence from snow covered ground is facilitated by the durable, hornlike spathe and by thermogenic properties, which elevate internal spathe temperatures to 15°C–35°C above ambient. The heat is essential in melting ice and snow, which otherwise adhere tightly to the spathe surface. The aerodynamic properties of the spathe distribute the air temperature evenly around the spadix to prevent freezing damage to any portion, even under windy conditions. Thermogenesis in the spadix has been linked to the expression of two, cold-inducible mitochondrial uncoupling proteins known as *SfUCPa* and *SfUCPb*. The flowers are protogynous, with the female phase lasting an average of 6.8 days, followed by a short overlapping bisexual phase (2 days) and the final male stage (16.7 days). The heated inflorescence also more effectively dissipates volatile pollinator attractants, especially during the female phase. A malodorous oligosulfide (dimethyl disulfide) predominates the floral odor composition of both phases and is similar to attractants known for other sapromyophilous species (i.e., those pollinated by vectors attracted to carrion or dung). However, female phase plants apparently emit unique hydrocarbons and indole compounds, which are thought to provide specific pollinator cues. Heat also might play a role in mimicry of mammal carrion or dung to entice certain pollinating vectors. The insect (Insecta) pollinators of the North American plants include flies (Diptera) and honeybees (Hymenoptera) (see *Use by wildlife*). Wind pollination might occur on occasion. Because potential pollinators often are rare during the early season of anthesis, some investigators have contemplated that the thermogenic properties may be relictual evolutionarily. The spathe decays and withers soon after pollination (usually by mid-May), as the spadix softens and swells to enclose the individual fruits within a papery sheath. The fruits harden as they mature, eventually bursting through the sheath and falling to the ground. The seeds remain dormant until the subsequent spring, often germinating close to the parental plant. Seeds collected in late July and stratified at 10°C for 3 months germinated well (37%) the next year and >90% by the subsequent year. Accelerated germination (up to 40%) has been achieved by stratifying seeds at 10°C for 30 days and then sowing them aseptically on a Murashige and Skoog (MS) growth medium. Short range dispersal of the fruits can occur by water during flooded conditions. Longer distance dispersal likely results from the transport of the fruits by frugivorous mammals. In some habitats, the roots are colonized by arbuscular mycorrhizal Fungi and dark septate endophytic Fungi. **Reported associates:** *Abies balsamea, Acer negundo, Acer rubrum, Acer saccharum, Agrimonia parviflora, Agrostis gigantea, Alliaria petiolata, Allium canadense, Allium vineale, Alnus rugosa, Andromeda polifolia, Apios americana, Apocynum cannabinum, Arisaema triphyllum, Aronia melanocarpa, Arundinaria, Asclepias incarnata, Asplenium montanum, Betula alleghaniensis, Betula lenta, Betula pumila, Boehmeria cylindrica, Bromus ciliatus, Calla palustris, Calopogon tuberosus, Caltha palustris, Cardamine bulbosa, Cardamine pensylvanica, Cardamine pratensis, Cardamine rotundifolia, Carex aggregata, Carex atlantica, Carex blanda, Carex bromoides, Carex comosa, Carex crinita, Carex disperma, Carex exilis, Carex folliculata, Carex granularis, Carex hyalinolepis, Carex hystericina, Carex interior, Carex lacustris, Carex leptalea, Carex sterilis, Carex stipata, Carex stricta, Carex tetanica, Carex utriculata, Carex vesicaria, Carex vulpinoidea, Carpinus caroliniana, Carya cordiformis, Carya ovata, Celtis occidentalis, Chaerophyllum procumbens, Chamaedaphne calyculata, Chelone glabra, Chrysosplenium americanum, Cicuta maculata, Cinna arundinacea, Cirsium muticum, Clethra alnifolia, Clintonia borealis, Collinsonia canadensis, Comandra umbellata, Coptis trifolia, Cornus amomum, Cornus canadensis, Cornus sericea, Corydalis flavula, Corylus americana, Cryptotaenia canadensis, Cypripedium calceolus, Cypripedium reginae, Dasiphora floribunda, Dennstaedtia punctilobula, Deparia acrostichoides, Desmodium canadense, Dichanthelium, Doellingeria umbellata, Drosera rotundifolia, Dryopteris carthusiana, Dryopteris cristata, Dryopteris intermedia, Dryopteris ×triploidea, Duchesnea indica, Eleocharis elliptica, Elymus virginicus, Epilobium ciliatum, Epilobium coloratum, Epilobium hirsutum, Equisetum arvense, Equisetum fluviatile, Equisetum palustre, Equisetum sylvaticum, Eriophorum virginicum, Euonymus atropurpureus, Euonymus obovatus, Euphorbia purpurea, Eupatorium perfoliatum, Eutrochium fistulosum, Eutrochium maculatum, Fagus grandifolia, Festuca subverticillata, Floerkea proserpinacoides, Fraxinus nigra, Fraxinus pennsylvanica, Galium aparine, Galium asprellum, Galium labradoricum, Galium triflorum, Gaylussacia baccata, Gaylussacia dumosa, Geranium maculatum, Geum rivale, Glechoma hederacea, Glyceria striata, Hackelia virginiana, Hamamelis virginiana, Helonias bullata, Hydrocotyle americana, Hypericum densiflorum, Hypoxis hirsuta, Ilex mucronata, Ilex verticillata, Impatiens pallida, Impatiens capensis, Iris virginica,*

Juglans cinerea, Juncus brachycephalus, Juncus dudleyi, Juniperus communis, Kalmia angustifolia, Kalmia latifolia, Kalmia polifolia, Laportea canadensis, Larix laricina, Leersia virginica, Lilium michiganense, Lindera benzoin, Linnaea borealis, Liparis loeselii, Liquidambar styraciflua, Liriodendron tulipifera, Lithospermum canescens, Lobelia inflata, Lobelia siphilitica, Lonicera dioica, Lonicera japonica, Lycopus uniflorus, Lyonia ligustrina, Lysimachia ciliata, Lysimachia nummularia, Lysimachia thyrsiflora, Lythrum salicaria, Magnolia grandiflora, Magnolia virginiana, Maianthemum canadense, Maianthemum racemosum, Maianthemum stellatum, Maianthemum trifolium, Mentha arvensis, Micranthes pensylvanica, Microstegium vimineum, Microstegium vimineum, Milium effusum, Mimulus ringens, Mitchella repens, Mitella diphylla, Mitella nuda, Monarda fistulosa, Muhlenbergia glomerata, Muhlenbergia mexicana, Myosotis, Nasturtium officinale, Nyssa sylvatica, Oclemena nemoralis, Onoclea sensibilis, Osmorhiza claytonii, Osmunda claytoniana, Osmunda regalis, Osmundastrum cinnamomeum, Oxypolis rigidior, Packera aurea, Panax trifolius, Parnassia glauca, Parthenocissus quinquefolia, Parthenocissus quinquefolia, Persicaria arifolia, Persicaria arifolia, Persicaria longiseta, Persicaria perfoliata, Persicaria sagittata, Persicaria virginiana, Phalaris arundinacea, Phlox maculata, Physocarpus opulifolius, Picea mariana, Picea rubens, Pilea pumila, Pinus rigida, Pinus strobus, Platanthera hyperborea, Platanus occidentalis, Poa alsodes, Poa compressa, Poa nemoralis, Poa paludigena, Poa trivialis, Podophyllum peltatum, Pogonia ophioglossoides, Polemonium reptans, Polygonatum biflorum, Polygonatum pubescens, Polystichum acrostichoides, Populus tremuloides, Prunus nigra, Prunus serotina, Prunus virginiana, Pycnanthemum virginianum, Pyrola asarifolia, Quercus alba, Quercus macrocarpa, Quercus palustris, Quercus rubra, Ranunculus abortivus, Ranunculus hispidus, Ranunculus recurvatus, Ranunculus sceleratus, Rhamnus alnifolia, Rhamnus frangula, Rhododendron periclymenoides, Rhus aromatica, Rhynchospora alba, Ribes americanum, Ribes hirtellum, Robinia pseudoacacia, Rosa palustris, Rubus hispidus, Rubus pensilvanicus, Rubus phoenicolasius, Rubus pubescens, Rubus repens, Rudbeckia fulgida, Rudbeckia laciniata, Rumex britannica, Rumex obtusifolius, Sagittaria latifolia, Salix candida, Salix nigra, Salix pedicellaris, Sambucus nigra, Sanguisorba canadensis, Sanicula, Sarracenia purpurea, Sassafras albidum, Schoenoplectus tabernaemontani, Scirpus expansus, Scirpus pendulus, Senecio suaveolens, Senecio vulgaris, Silene nivea, Smilax, Solidago patula, Solidago riddellii, Solidago rugosa, Solidago uliginosa, Solidago uliginosa, Sorbus decora, Sphagnum flavicomans, Sphagnum papillosum, Sphagnum pulchrum, Spiraea, Spiranthes lucida, Stachys tenuifolia, Symphyotrichum boreale, Symphyotrichum dumosum, Symphyotrichum firmum, Symphyotrichum puniceum, Taxus canadensis, Thalictrum pubescens, Thalictrum thalictroides, Thelypteris noveboracensis, Thelypteris palustris, Thuja occidentalis, Tiarella cordifolia, Tilia americana, Toxicodendron radicans, Toxicodendron vernix, Trientalis *borealis, Trillium erectum, Trillium grandiflorum, Trillium recurvatum, Trollius laxus, Tsuga canadensis, Typha angustifolia, Typha latifolia, Ulmus americana, Ulmus rubra, Vaccinium corymbosum, Vaccinium macrocarpon, Vaccinium oxycoccos, Veratrum viride, Verbena hastata, Viburnum dentatum, Viburnum lentago, Viburnum nudum, Viburnum recognitum, Viola blanda, Viola cucullata, Viola labradorica, Viola macloskeyi, Viola rotundifolia, Viola sororia, Viola striata, Vitis riparia, Zizia aurea.*

Use by wildlife: *Symplocarpus foetidus* is eaten by black bears (Mammalia: Ursidae: *Ursus americanus*), snapping turtles (Vertebrata: Reptilia: Chelydridae: *Chelydra serpentina*), and wood ducks (Aves: Anatidae: *Aix sponsa*). The foliage is grazed by slugs (Mollusca: Gastropoda: Arionidae: *Arion circumscriptus*; Limacidae: *Deroceras reticulatum, D. laeve*). The oily seeds are highly nutritious (23.3% DM protein; 37.7% DM fat), persist long into winter, and are eaten by several birds (Aves: Phasianidae) including grouse (*Bonasa umbellus*), pheasants (*Phasianus colchicus*), and quails (*Colinus virginianus*). The seeds can provide indirect nourishment when eaten by prey that later are ingested by predators such as snowy owls (Aves: *Bubo scandiacus*). Various pollen-coated arthropods (Insecta; Arachnida) have been observed within the inflorescences during early anthesis, including bees (Hymenoptera: Apidae: *Apis mellifera*), bugs (Hemiptera: Reduviidae: *Zelus exsanguis*), flies (Diptera: Trichoceridae: *Trichocera garretti*; Phoridae: *Triphleba subfusca*), and spiders (Sclerosomatidae: *Leiobunum*; Theridiidae: *Enoplognatha marmorata*). Although beetles (Coleoptera) have been suggested as potential pollinators, they appear to play at most a minor role in the process. The plants also are inhabited by mites (Arachnida: Tetranychidae *Tetranychus schoenei, T. urticae*) and slugs (e.g., Mollusca: Gastropoda), which do not transfer pollen. The decomposing plants are fed on by numerous fly larvae (Insecta: Diptera: Cecidomyiidae: *Planetella davisi*; Ceratopogonidae: *Dasyhelea oppressa*; Chironomidae: *Bryophaenocladius*; Chloropidae: *Elachiptera costata, Tricimba lineella*; Drosophilidae: *Chymomyza amoena, Drosophila busckii, D. falleni, D. palustris, D. putrida, D. quinaria, D. recens, Scaptomyza graminum*; Ephydridae: *Athyroglossa granulosa*; Psychodidae: *Psychoda alternata, P. satchelli*). The foliage hosts larval moths (Insecta: Lepidoptera: Erebidae: *Lymantria dispar, Phragmatobia fuliginosa*; Noctuidae: *Bellura obliqua*) and various Fungi (Ascomycota: Glomerellaceae: *Colletotrichum lineola*; Mycosphaerellaceae: *Cercospora symplocarpi, Septoria spiculosa*; Sclerotiniaceae: *Botrytis streptothrix*; incertae sedis: *Ascochyta, Clarireedia bennettii, Pseudodidymaria symplocarpi*).

Economic importance: food: *Symplocarpus foetidus* is edible but contains calcium oxalate crystals and other potentially harmful inclusions that must first be neutralized by thorough drying before eating. Some have mistaken the highly poisonous *Veratrum viride* for *S. foetidus*, so extreme caution should be taken if contemplating the latter plant as food. Raw portions of the plants never should be consumed. Boiling thrice in a solution of baking soda and salt is said to remove the unpleasant properties. When prepared properly, the young foliage has been eaten as a salad vegetable or made into soups or stews. The desiccated rhizomes (best gathered in early spring or late autumn) can be ground into flour, which allegedly has a cocoa-like flavor. The Iroquois cooked and ate the rootstocks and seasoned young leaves and shoots as vegetables; **medicinal:** *Symplocarpus* was the primary ingredient of "dracontium," a standard 19th-century pharmaceutical used to treat nervous, respiratory, and other disorders. Native North Americans treated numerous ailments using *Symplocarpus* as a medicinal. These treatments included its use as a general drug (Malecite, Meskwaki) or as an analgesic (Delaware, Menominee, Micmac), anticonvulsive (Delaware, Menominee, Mohegan), anthelmintic (Iroquois), antirheumatic (Abnaki, Iroquois), cardiac medicine (Menominee), cold remedy (Nanticoke), cough medicine (Chippewa, Delaware), gynecological aid (Iroquois), and for treatment of hemorrhages (Menominee), swelling (Meskwaki), toothaches (Meskwaki), tuberculosis (Iroquois), and miscellaneous wounds (Iroquois, Menominee); **cultivation:** *Symplocarpus foetidus* sometimes is grown as a curiosity in water gardens; **misc. products:** The Iroquois made a body deodorant from powdered root infusions of *Symplocarpus foetidus*; **weeds:** none; **nonindigenous species:** none.

Systematics: The three *Symplocarpus* species that have been included in phylogenetic analyses resolve as a sister clade to *Lysichiton* (Figure 1.28), which diverged approximately 31–40 MYBP. All but the endemic North American *S. foetidus* are Asian. Two additional Asian species have been proposed (*S. egorovii* and *S. nabekuraensis*), but a further evaluation of their status awaits more critical study. *Symplocarpus renifolius* (another precocious flowering species) is most closely related to *S. foetidus* (Figure 1.28), with an estimated divergence time of 3.4–6.9 MYBP. Intergenic spacer data (cpDNA) have shown higher haplotype diversity (0.7425) to characterize populations residing in unglaciated portions of the range than those in glaciated regions (0.6099) and indicate that glacial refugia were sought primarily in the northeastern portion of the range. The basic chromosome number of *Symplocarpus* is $x = 15$; *S. foetidus* ($2n = 60$) is a tetraploid.

Distribution: *Symplocarpus foetidus* occurs in east-central North America.

References: Aldrich et al., 1985; Almquist & Calhoun, 2003; Blair, 1975; Block et al., 2013; Bown, 2000; Camazine & Niklas, 1984; Cheplick, 2006; Core, 1967; Dalby, 2000; Fernald & Kinsey, 1943; Getz, 1959; 1961; Gibernau, 2003; Goslee et al., 1997; Grimaldi & Jaenike, 1983; Gross, 1944; Hartley, 1960; Hussey, 1974; Ito, 1999; Iverson et al., 2001; Judd, 1961; Kevan, 1989; Kim et al., 2018; Kitamura et al., 2008; Knollenberg et al., 1985; Knutson, 1972; Kozen, 2013; Lagler, 1943; Loeffler & Wegner, 2000; Lynn & Karlin, 1985; Mabbott, 1920; Miyazawa et al., 2015; Moodie, 1976; Nie et al., 2006; Ogle, 1989; Perry, 1961; Peterson, 1977; Petty & Petty, 2005; Picking & Veneman, 2004; Prasad, 2017; Ramachandran & Nosonovsky, 2014; Ruch et al., 2008; 2009; Runkel & Roosa, 1999; Scott, 2009; Seymour & Blaylock, 1999; Spinner, 1942; Stewart, Jr. & Nilsen, 1993; Wada & Uemura, 1994; Weishampel & Bedford, 2006; Wen, 1999; Wen et al., 1996; Williams, 1919; Williams & Moriarity, 2000.

Family 4.4. Tofieldiaceae [5]

Harperocallis, Isidrogalvia, Pleea, Tofieldia, and *Triantha* had long been assigned to Liliaceae (e.g., Cruden, 1991; Utech, 2002a), a classification that some continue to follow. Yet, molecular phylogenetic analyses (Azuma & Tobe, 2011; Givnish et al., 2018) clearly resolve the group as a distinct clade (false-asphodel family), which has no close relationship to *Lilium* or other genera more traditionally recognized in Liliaceae (Figure 1.29). Takhtajan (1995) circumscribed this group as the family Tolfieldiaceae. Although some molecular data have placed *Isidrogalvia* within Nartheciaceae (Tamura et al., 2004), that result was erroneous and attributable to technical errors (Azuma & Tobe, 2011). Rather, in more reliable analyses, *Isidrogalvia* resolves within Tolfieldiaceae as the sister group of *Harperocallis* (Azuma & Tobe, 2011) (Figure 1.29). Arguments to retain *Triantha* as a distinct genus or to merge it with *Tofieldia* are defensible and compatible phylogenetically in either case. Therefore, the taxonomic disposition of the groups simply becomes a matter of preference. A parallel issue involves the retention of *Harperocallis* or its merger with *Isdrogalvia* (Remizowa et al., 2011; Campbell & Dorr, 2013).

However, the precise placement of Tofieldiaceae remains unclear. Some analyses show it to be the sister group of the alismatids (e.g., Figure 1.29); however, other data resolve the family among various permutations involving Acoraceae, Araceae, Lemnaceae, and Alismatidae (Les & Tippery, 2013). Complete plastid data sets place Tofieldiaceae as the sister group to a clade comprising Araceae, Lemnaceae, and Alismatidae (Luo et al., 2016). In any case, the group usually falls among other members of Arales, in which it has been included here.

As defined by Takhtajan (2009), Tofieldiaceae are perennial herbs with creeping rhizomes, spiral, basally arranged leaves, racemes (solitary flowers in *Harperocallis*), small, bisexual, radial flowers subtended by a calyculus (i.e., a calyx-like involucre), a perianth of six petaloid tepals, six (sometimes 9+) stamens, and a gynoecium most often of three nearly apocarpous carpels with septal nectaries; the fruits are follicular or septicidal capsules. The basic chromosome number of the family is $x = 15$. The family includes many wetland plants and species associated with moist to wet habitats.

The family is of little economic importance, with some *Tofieldia* and *Triantha* species sold and grown as garden ornamentals. Tofieldiaceae are distributed in North America, South America, and Eurasia. Three genera contain OBL North American indicators:

4.4.1. ***Harperocallis*** McDaniel
4.4.2. ***Pleea*** Michx.
4.4.3. ***Triantha*** (Nutt.) Baker

4.4.1. Harperocallis
Harper's beauty

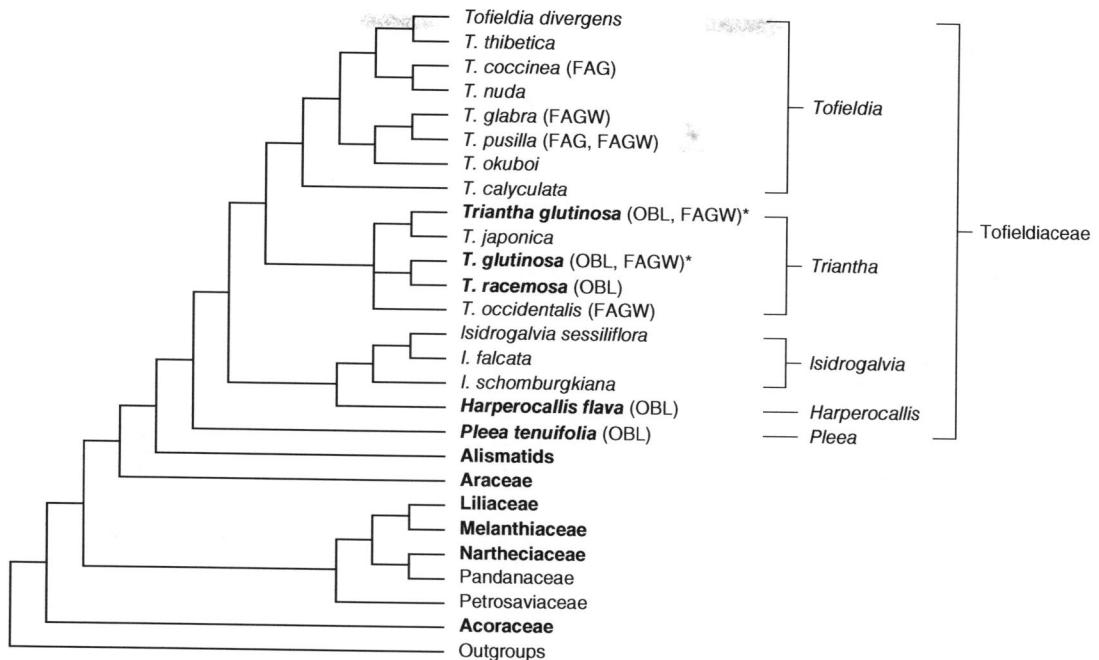

FIGURE 1.29   Relationships of genera assigned to Tofieldiaceae as indicated by analysis of *matK* sequence data (modified from Azuma & Tobe, 2011). All five genera resolve as a clade that is distinct from Liliaceae, Melanthiaceae, or Nartheciaceae (groups where they had once been placed). Although based on a chloroplast gene (*matK*), the indicated sister-group relationship of Tofieldiaceae and alismatid monocots disagrees with results from complete cpDNA sequences (Luo et al., 2016). The failure of *Triantha glutinosa* accessions (asterisked) to associate is possibly due to hybridization among the morphologically similar and closely related North American representatives of the *T. glutinosa* "complex" (*T. glutinosa*, *T. occidentalis*, *T. racemosa*). All taxa with OBL North American indicators are highlighted in bold. Wetland designations (in parentheses) are provided for the North American species.

Etymology: after Roland McMillan Harper (1878–1966) and *kalli* (Greek for "beautiful")

Synonyms: *Isidrogalvia* (in part)

Distribution: global: North America; **North America:** western Florida (panhandle)

Diversity: global: 1 species; **North America:** 1 species

Indicators (USA): obligate wetland (OBL): *Harperocallis flava*

Habitat: freshwater; palustrine; **pH:** 4.5; **depth:** <1 m; **life-form(s):** emergent herb

Key morphology: herbaceous perennials; rhizomes slender; stems (to 55 cm) erect, glabrous; leaves (21.5 cm) basal, equitant, simple, linear, sheathing, margins entire; inflorescence erect, single-flowered, with epicalyx of 2–4 small bracteoles; flowers bisexual, perianth tepaloid, tepals (to 15 mm) 6, persistent, oblanceolate, yellow above, greenish below, 10–12-veined, turning coppery brown in senescence; stamens (to 9 mm) 6, the anthers (to 3.5 mm) basifixed; gynoecium of 3 (rarely to 6) carpels, unfused distally, with interseptal (intercarpellary) nectaries, ovary superior, styles (to 1 mm) 3 (rarely to 6), recurved, stigma capitate; capsules (to 10 mm) septicidal, tuberculate, subtended by sepals; seeds (to 3 mm) pale yellow, short-appendaged at ends

Life history: duration: perennial (rhizomes); **asexual reproduction:** rhizomes; **pollination:** insect; **sexual condition:** hermaphroditic; **fruit:** septicidal capsules (occasional); **local dispersal:** rhizomes; seeds (unspecialized); **long-distance dispersal:** seeds (motor vehicles, water, wind)

Imperilment: 1. *Harperocallis flava* [G1]; S1 (FL)

Ecology: general: *Harperocallis* is monotypic (see next).

4.4.1.1. Harperocallis flava McDaniel is a rare inhabitant of bogs (pineland, titi) and ditches (flatwoods, roadside, moist) at elevations to 10 m. It has been listed as a federally endangered species since 1979. The plants thrive in full sun but also are found growing in the partial shade of shrubs. The substrates include peat and sand. Few specific site details exist, but they tend to be fairly acidic (e.g., pH: 4.5). Flowering occurs from mid-April to late May with fruits maturing in July. The reproductive rates are low, with an average of only 1.7–10.9% of ramets bearing flowers in a given year. Flower production correlates directly with the number of leaves on a plant. The flowers allegedly are self-compatible and capable of self-pollination (to an undetermined extent) but also are pollinated by sweat bees (Insecta: Hymenoptera: Halictidae: *Lasioglossum*). Genetic (allozyme) surveys have indicated a complete lack of genetic variation, with genetically uniform individuals found across all populations. The plants reproduce vegetatively by short rhizomes. The number and size of their ramets is directly proportional to the amount of annual rainfall received at a site. The recruitment rate of new ramets into a population is very low. Persistence of the plants is thought to require periodic fires and/or disturbances of the soil surfaces to restrict the growth of competitive grasses and shrubs, increase light availability, and release nutrients to growing sites. The plants suffer substantial mortality from crayfish (Crustacea: Decapoda), which bury them while building their mounds and "chimneys." **Reported associates:** *Cleistesiopsis divaricata, Cliftonia monophylla, Drosera tracyi, Gaylussacia mosieri, Lycopodium, Lyonia lucida, Macbridea alba, Myrica caroliniensis, Myrica inodora, Pinus serotina, Pleea tenuifolia, Sarracenia flava, Sarracenia psittacina, Serenoa repens.*

Use by wildlife: The flowers of *Harperocallis flava* are visited by several insects (Insecta) including bumble bees (Hymenoptera: Apidae: *Bombus*), bee flies (Diptera: Syrphidae), flower beetles (Coleoptera: Mordellidae), and katydids [nymphs] (Tettigoniidae). However, pollen appears to be transferred only by sweat bees (Halictidae: *Lasioglossum*).

Economic importance: food: not edible; **medicinal:** no uses reported; **cultivation:** not cultivated; **misc. products:** none; **weeds:** none; **nonindigenous species:** none.

Systematics: *Harperocallis* (*H. flava*) has been treated as monotypic, or among species assigned otherwise to the South American *Isidrogalvia*. Persuasive arguments have been made for either disposition, and both are compatible with phylogenetic evidence (e.g., Figure 1.29), which resolves the groups as sister taxa. The basic chromosome number of *Harperocallis* is $x = 16$. *Harperocallis flava* ($2n = 32$) is diploid genetically as indicated by allozyme data.

Distribution: *Harperocallis flava* is a rare, local endemic restricted to the Apalachicola lowlands of the western Florida panhandle.

References: Campbell & Dorr, 2013; Cook, 1979; Godt et al., 1997; Keppner & Anderson, 2008; Leonard & Baker, 1983; McDaniel, 1968; Pendergrass, 1983; Pitts-Singer et al., 2002; Remizowa et al., 2011; Utech & Anderson, 2002; Walker & Silletti, 2005.

4.4.2. Pleea

Rush featherling

Etymology: after Auguste Plée (1787–1825)

Synonyms: *Ennearina; Tofieldia* (in part)

Distribution: global: North America; **North America:** southeastern United States

Diversity: global: 1 species; **North America:** 1 species

Indicators (USA): obligate wetland (OBL): *Pleea tenuifolia*

Habitat: freshwater; palustrine; **pH:** 4.5–5.6; **depth:** <1 m; **life-form(s):** emergent herb

Key morphology: rhizomes short, thick; stems (to 85 cm) herbaceous, scape-like; leaves (to 4) mostly basal, equitant, 2-ranked, becoming smaller distally, the blade (to 50 cm) linear, short-sheathing at base; racemes (to 20 cm) terminal, with sheathing bracts (to 7 cm), flowers (3–8) subtended by an epicalyx of 3 connate bracteoles (to 8 mm) at midway of pedicel (to 3.5 cm); perianth of 6 white or yellowish, distinct tepals (to 17 mm); stamens 9, opposite outer tepals, filaments (to 5.5 mm) flattened, dilated at base, anthers (to 4 mm) versatile; ovary (to 5 mm) superior, stipitate, the 3 carpels unfused basally, septal nectaries present, styles (to 1.8 mm) 3; capsules (to 9 mm) coriaceous, ovoid to ellipsoid, septicidal; seeds (to 1.5 mm) dark reddish brown, appendaged (long or short) at ends

Life history: duration: perennial (rhizomes); **asexual reproduction:** rhizomes; **pollination:** insect [presumably]; **sexual condition:** hermaphroditic; **fruit:** capsules (common); **local dispersal:** rhizomes, seeds; **long-distance dispersal:** seeds (vector unknown)

Imperilment: 1. *Pleea tenuifolia* [G4]; SX (SC); S1 (AL); S3 (NC)

Ecology: general: *Pleea* is monotypic (see next).

4.4.2.1. *Pleea tenuifolia* Michx. occurs in bogs (pineland, seepage), depressions (seasonally inundated, seepage), ditches (swampy), fens (oligotrophic, poor), flatwoods (pine), marshes (roadside), meadows (wet), prairies (wet), roadsides, savannas (pine, seepage, wet), seepage slopes, streamheads, swales, and along the margins of pocosins and swamps at elevations to 64 m. The substrates have been described as acidic (pH: 4.5–5.6) and include clay, clayey silt, muck (saprick), Plummer (grossarenic paleaquults), Rutlege (typic humaquepts), sand (Leon series), sandy loam, spodosol, and ultisol. The plants inhabit open, sunny exposures in seasonally wet sites that are characterized by frequent fires (every 1–3 years). Flowering occurs from September to October, with fruiting extending into January. Additional information on the reproductive biology and seed ecology is unavailable and should be sought. The plants perennate by short, thick rhizomes, which results in a clump-like growth pattern. **Reported associates:** *Acer rubrum, Aletris lutea, Amphicarpum amphicarpon, Andropogon arctatus, Andropogon glomeratus, Andropogon gyrans, Andropogon liebmannii, Andropogon virginicus, Anthaenantia rufa, Aristida palustris, Aristida simpliciflora, Aristida stricta, Arnoglossum ovatum, Arnoglossum sulcatum, Aronia arbutifolia, Arundinaria tecta, Asclepias connivens, Asclepias viridula, Balduina uniflora, Bigelowia nudata, Calamagrostis canadensis, Calopogon barbatus, Carex lutea, Carex striata, Carphephorus paniculatus, Carphephorus pseudoliatris, Chaptalia tomentosa, Cladium mariscoides, Clethra alnifolia, Cliftonia monophylla, Coleataenia longifolia, Coreopsis gladiata, Ctenium aromaticum, Cyrilla racemiflora, Dichanthelium acuminatum, Dichanthelium nudicaule, Dichanthelium scabriusculum, Drosera brevifolia, Drosera capillaris, Drosera filiformis, Drosera tracyi, Eleocharis tuberculosa, Erigeron vernus, Eriocaulon compressum, Eriocaulon decangulare, Eriocaulon nigrobracteatum, Eryngium integrifolium, Eryngium yuccifolium, Eupatorium leucolepis, Eupatorium resinosum, Euphorbia inundata, Eurybia eryngiifolia, Fimbristylis puberula, Fuirena squarrosa, Gaylussacia mosieri, Harperocallis flava, Helenium vernale, Helianthus heterophyllus, Helianthus radula, Hypericum brachyphyllum, Hypericum chapmanii, Hypericum fasciculatum, Ilex coriacea, Ilex glabra, Ilex myrtifolia, Iris tridentata, Juncus trigonocarpus, Kalmia buxifolia, Lachnanthes caroliniana, Lachnocaulon anceps, Lachnocaulon digynum, Liatris spicata, Lilium catesbaei, Linum medium, Lobelia brevifolia, Lobelia glandulosa, Lophiola aurea, Lycopodiella alopecuroides, Lycopodiella appressa, Lycopodiella caroliniana, Lycopodiella prostrata, Lyonia lucida, Magnolia virginiana, Marshallia graminifolia, Muhlenbergia capillaris, Muhlenbergia expansa, Myrica cerifera, Myrica heterophylla, Nyssa biflora, Osmundastrum cinnamomeum, Oxypolis ternata, Panicum virgatum,*

Paspalum praecox, Pinguicula lutea, Pinguicula planifolia, Pinus elliottii, Pinus palustris, Pinus serotina, Pinus taeda, Pityopsis oligantha, Platanthera ciliaris, Platanthera integra, Platanthera nivea, Pogonia ophioglossoides, Polygala cruciata, Polygala cymosa, Polygala hookeri, Polygala ramosa, Quercus laurifolia, Rhexia alifanus, Rhexia lutea, Rhexia petiolata, Rhynchospora baldwinii, Rhynchospora chalarocephala, Rhynchospora chapmanii, Rhynchospora ciliaris, Rhynchospora corniculata, Rhynchospora latifolia, Rhynchospora macra, Rhynchospora oligantha, Rhynchospora pallida, Rhynchospora plumosa, Rhynchospora stenophylla, Rubus trivialis, Rudbeckia graminifolia, Sabatia decandra, Sabatia macrophylla, Sarracenia flava, Sarracenia leucophylla, Sarracenia psittacina, Scleria baldwinii, Scleria muehlenbergii, Scleria pauciflora, Scleria reticularis, Scleria triglomerata, Serenoa repens, Smilax laurifolia, Solidago patula, Solidago pulchra, Sphagnum, Sporobolus curtissii, Sporobolus pinetorum, Symphyotrichum chapmanii, Symphyotrichum dumosum, Symphyotrichum lateriflorum, Syngonanthus flavidulus, Taxodium ascendens, Tiedemannia filiformis, Tofieldia glabra, Triantha racemosa, Utricularia juncea, Utricularia subulata, Vaccinium crassifolium, Verbesina chapmanii, Viburnum nudum, Xyris ambigua, Xyris baldwiniana, Xyris caroliniana, Xyris difformis, Xyris drummondii, Xyris serotina, Zigadenus glaberrimus.

Use by wildlife: The flowers of *Pleea tenuifolia* are visited by Palamedes swallowtail butterflies (Insecta: Lepidoptera: Papilionidae: *Papilio palamedes*).

Economic importance: food: not edible; **medicinal:** no uses reported; **cultivation:** not cultivated; **misc. products:** none; **weeds:** *Pleea tenuifolia* is not weedy; **nonindigenous species:** none.

Systematics: The monotypic *Pleea tenuifolia* has been merged with *Tofieldia* but is distinct morphologically and genetically from that genus. Phylogenetically, *Pleea* resolves as the sister group to the remainder of Tofieldiaceae (Figure 1.29). The basic chromosome number of *Pleea* is $x=15$; *P. tenuifolia* ($2n=30$) is diploid. No hybrids involving *Pleea* have been reported.

Distribution: *Pleea tenuifolia* is restricted to the coastal plain of the southeastern United States (AL, FL, NC, SC [extirpated]).

References: Bridges, 2005; Carr, 2007; Carr et al., 2010; Frost, 1995; Godfrey & Wooten, 1979; Mitchell, 2011; Orzell & Bridges, 1993; Packer, 2002a; Peet, 2007; Stuart, 2018; Thornhill et al., 2014; Utech, 1978.

4.4.3. Triantha

False asphodel; tofieldie

Etymology: from the Greek *tria anthos* ("three-flowered") in reference to the clustered flowers

Synonyms: *Abama* (in part); *Amianthium* (in part); *Asphodeliris* (in part); *Melanthium* (in part); *Narthecium* (in part); *Tofieldia* (in part); *Trianthella*

Distribution: global: Asia (eastern); North America; **North America:** throughout

Diversity: global: 4 species; **North America:** 3 species

Indicators (USA): obligate wetland (OBL): *Triantha glutinosa, T. racemosa*; **facultative wetland (FACW):** *T. glutinosa*

Habitat: freshwater; palustrine; **pH:** 6.8–8.3; **depth:** <1 m; **life-form(s):** emergent herb

Key morphology: herbaceous perennials; rhizome short, slender, horizontal; leaves 2-ranked, equitant, arising basally from rhizome, sometimes 1–3 near base of flowering scape, the blades (to 35 cm) erect, linear; scapes (to 70 cm) glandular or glandular-pubescent, terminating as a dense, spike-like raceme (to 80 flowers; to 22 cm) or in cylindric spike-like heads (to 30 flowers; to 6.5 cm), the flowers in clusters of 3 (sometimes 2–7), each with an epicalyx of connate bracteoles, pedicels (to 12 mm) glandular; perianth of 6 persistent, free, white or yellow (drying to orange) tepals (to 5 mm), the innermost slightly longer; stamens (to 4.5 mm) 6, the filaments flat, dilated basally, the anthers basifixed; ovary (to 5 mm) superior, stipitate, tricarpellate (but carpels free basally), interseptal nectaries present, styles (to 1.8 mm) 3, free or connate basally; capsules (to 7.5 mm) septicidal (adaxially loculicidal), broadly ellipsoid, ovoid or subglobose, hard or chartaceous; seeds (to 1 mm) reddish brown, appendaged at ends, one appendage usually longer

Life history: duration: perennial (rhizomes); **asexual reproduction:** rhizomes; **pollination:** insect, self; **sexual condition:** hermaphroditic; **fruit:** capsules (common); **local dispersal:** rhizomes, seeds (water); **long-distance dispersal:** seeds (uncertain)

Imperilment: 1. *Triantha glutinosa* [G5]; SH (GA); S1 (NC, ND, NH, NY, <u>NS</u>, TN, VA, VT, WV); S2 (IL, IN, <u>NU</u>, OH, WI, <u>YT</u>); S3 (<u>LB</u>, ME, <u>NB</u>, <u>QC</u>); **2.** *T. racemosa* [G5]; SX (DE, MD); SH (VA); S1 (NJ, TN); S2 (LA)

Ecology: general: All *Triantha* species are wetland perennials, which inhabit exposed wet substrates (e.g., <10 cm standing water). Two of the three North American species are OBL indicators, with the excluded *T. occidentalis* ranked as FACW. Their flowers are protogynous, having a single pistillate and staminate phase. However, the shed pollen falls directly on the stigma, which is thought to result in autogamy (also in the related genus *Tofieldia*). Despite the showy perianths and carpellary (interseptal) nectaries being adapted for insect pollination, reports of insect visitation to the flowers are scarce. Yet, the limited amount of genetic data indicates at least some outcrossing. Thus, it seems likely that a mixed breeding system occurs. No reliable data on the seed ecology exists. Local dispersal of fruits and/or seeds by water seems reasonable, but vectors for long-distance dispersal have not been determined. There are no indications that a seed bank develops. All the species reproduce vegetatively from horizontal rhizomes to develop as small, clonal clumps.

4.4.3.1. ***Triantha glutinosa* (Michx.) Baker** inhabits alvars, beaches, bluffs, bog forests, bogs (calcareous, hanging, hot springs, marl, pocket, raised), canals (abandoned), crevices (rock), deltas (outwash), depressions, ditches (roadside), fens, flats (alluvial, interdunal, river, sand), floodplains (alluvial), glades, ledges (calcareous, river, seepy), marshes, meadows (boggy, marshy, shore, wet), mudflats (river), muskeg (alpine), oxbows, patterned peatlands (flarks), plains (alluvial, gravel outwash), pools (rock), prairie fens, prairies (mesic, river-scour), seeps (calcareous, gravelly, riverside), shores (calcareous, Great Lakes, lake, river), slopes (scree, seepage, talus [to 25%]), springs (boggy), streams (snowmelt), swales (dune), swamps (calcareous), tundra, and the margins of pools (bog) and streams at elevations to 2744 m. Exposures vary from open, sunny sites to partial shade, but the plants generally are shade-intolerant. The substrates include alluvium, clay, cobble, granite, gravel (calcareous), limestone, marl, muck (calcareous), peat (marly), rocky clay, sand (calcareous), sandy humus, serpentine, shale (calcareous), slate (limey), tuffa, turfy sand (i.e., impregnated with fine, tightly intertwined roots), and silt. They often are localized within cracks, crevices, or pockets in coarser material such as boulders, cobble, or rock. This calciphile is an indicator of "extreme rich" fens. Any standing water is shallow (less than 20 cm deep). Such habitats typically are characterized by the seepage of alkaline, mineral-rich waters (e.g., pH: 6.8–8.3; conductivity: 436–619 µS cm^{-1}; Ca^{2+}: 43–85 mg l^{-1}; K$^+$: 2–4 mg l^{-1}; Mg^{2+}: 21–27 mg l^{-1}; Na$^+$: 23–33 mg l^{-1}; NH$_4^+$-N: 3–77 µg l^{-1}; NO$_3^-$-N: 5–42 µg l^{-1}; P: 0–0.17 mg l^{-1}; SO$_4$: 28–60 mg l^{-1}). Flowering occurs from June to September, with fruits maturing from July to October. The flowers are protogynous and presumably are insect pollinated; however, their reproductive ecology has not been studied in any detail. Some seeds reportedly are non-dormant when shed and will germinate if incubated at 20°C. Seeds that are dormant require 2–4 weeks of cold stratification (–4°C to +4°C) to induce germination. It is not known whether a persistent seed bank can develop, and none has been reported. The seeds are thought to be dispersed by water during periods of inundation or flooding. Vegetative reproduction occurs by means of short, horizontal rhizomes. Some plants have persisted for more than a century at sites (e.g., river ledges) that are scoured periodically by floods and ice, which eliminates competing (especially woody) vegetation. Periodic fires are necessary in other sites to suppress competing woody vegetation. The roots are colonized (up to 25%) with mycorrhizal dark septate endophytic Fungi. It is thought that the glutinous stems might serve to trap potential flower or fruit predators. **Reported associates:** *Abies lasiocarpa, Aconitum uncinatum, Actaea rubra, Ageratina altissima, Agrostis hyemalis, Allium cernuum, Allium schoenoprasum, Alnus glutinosa, Alnus rugosa, Alnus serrulata, Amelanchier arborea, Amelanchier humilis, Amphicarpaea bracteata, Anaphalis margaritacea, Andromeda polifolia, Andropogon gerardii, Anemone parviflora, Aneura pinguis, Angelica lucida, Antennaria alpina, Antennaria pulcherrima, Anthoxanthum nitens, Anticlea elegans, Apocynum cannabinum, Aquilegia formosa, Arabidopsis lyrata, Arctostaphylos uva-ursi, Arctous rubra, Arethusa bulbosa, Arnica latifolia, Arnoglossum plantagineum, Asplenium trichomanes-ramosum, Astragalus eucosmus, Athyrium alpestre, Avenella flexuosa, Berula erecta, Betula glandulosa, Betula papyrifera, Betula pumila, Betula ×sandbergii, Bidens trichospermus, Bistorta vivipara, Blechnum spicant, Brachyelytrum erectum, Bromus ciliatus, Cakile edentula, Calamagrostis canadensis, Calamagrostis inexpansa, Calamagrostis lapponica, Calamagrostis stricta, Calliergon cordifolium, Calliergon stramineum, Calliergon trifarium, Calliergonella cuspidata, Caltha leptosepala, Caltha palustris, Campanula aurita, Campanula rotundifolia, Campylium stellatum, Cardamine bulbosa, Carex aquatilis, Carex aurea, Carex buxbaumii, Carex capillaris, Carex castanea, Carex conoidea, Carex cryptolepis, Carex dioica, Carex eburnea, Carex echinata, Carex flava, Carex garberi, Carex glacialis, Carex gynocrates, Carex haydeniana, Carex hoodii, Carex hystericina, Carex interior, Carex lasiocarpa, Carex lenticularis, Carex leptalea, Carex limosa, Carex livida, Carex luzulina, Carex magellanica, Carex mendocinensis, Carex microglochin, Carex muricata, Carex obnupta, Carex parryana, Carex pauciflora, Carex prairea, Carex scirpoidea, Carex scopulorum, Carex spectabilis, Carex sterilis, Carex stricta, Carex tenuiflora, Carex tetanica, Carex trisperma, Carex utriculata, Carex vaginata, Carex vesicaria, Carex viridula, Carex woodii, Cassiope tetragona, Castilleja coccinea, Catoscopium nigritum, Cetraria, Chamaedaphne calyculata, Chara, Chelone obliqua, Cicuta virosa, Cirsium palustre, Cladium mariscoides, Cladonia, Claytonia lanceolata, Clinopodium glabrum, Collema fuscovirens, Collema tenax, Comandra umbellata, Comarum palustre, Coptis aspleniifolia, Corispermum, Cornus canadensis, Cornus sericea, Cryptogramma acrostichoides, Cryptogramma stelleri, Cuscuta rostrata, Cypripedium arietinum, Cypripedium candidum, Cypripedium parviflorum, Cypripedium reginae, Cystopteris bulbifera, Danthonia spicata, Darlingtonia californica, Dasiphora floribunda, Deschampsia cespitosa, Dichanthelium acuminatum, Dichanthelium clandestinum, Doellingeria umbellata, Draba nivalis, Drosera anglica, Drosera filiformis, Drosera linearis, Drosera rotundifolia, Dryas drummondii, Dryas integrifolia, Eleocharis compressa, Eleocharis elliptica, Eleocharis palustris, Eleocharis quinqueflora, Eleocharis rostellata, Eleocharis tenuis, Empetrum nigrum, Epilobium anagallidifolium, Epilobium ciliatum, Epilobium leptophyllum, Epilobium luteum, Epilobium palustre, Epilobium strictum, Equisetum arvense, Equisetum fluviatile, Equisetum hyemale, Equisetum palustre, Equisetum scirpoides, Equisetum sylvaticum, Equisetum variegatum, Equisetum ×ferrissii, Erigeron acris, Erigeron howellii, Erigeron hyssopifolius, Eriophorum angustifolium, Eriophorum callitrix, Eriophorum chamissonis, Eriophorum crinigerum, Eriophorum virginicum, Eriophorum viridicarinatum, Eupatorium perfoliatum, Euphorbia polygonifolia, Eurybia divaricata, Eurybia radula, Euthamia graminifolia, Fallopia japonica, Festuca idahoensis, Filipendula rubra, Fragaria vesca, Fragaria virginiana, Galearis rotundifolia, Galium boreale, Galium labradoricum, Galium tinctorium, Gaultheria hispidula, Gentiana calycosa, Gentiana douglasiana, Gentianopsis crinita, Gentianopsis virgata, Geocaulon lividum, Geum calthifolium, Geum macrophyllum, Glyceria striata, Hastingsia bracteosa, Hedysarum boreale, Helenium autumnale, Houstonia caerulea, Houstonia serpyllifolia, Huperzia appalachiana, Huperzia porophila, Huperzia selago, Hydrangea arborescens, Hylocomium splendens,*

Hypericum densiflorum, Hypericum ellipticum, Hypericum kalmianum, Hypnum pratense, Iris lacustris, Iris setosa, Juncus alpinoarticulatus, Juncus alpinus, Juncus arcticus, Juncus articulatus, Juncus balticus, Juncus brachycephalus, Juncus dudleyi, Juncus ensifolius, Juncus falcatus, Juncus filiformis, Juncus nodosus, Juncus stygius, Juncus subcaudatus, Juniperus communis, Juniperus horizontalis, Kalmia angustifolia, Kalmia latifolia, Kalmia polifolia, Larix laricina, Larix lyallii, Lathyrus palustris, Liatris aspera, Liatris spicata, Ligusticum scoticum, Lilium philadelphicum, Lilium superbum, Limprichtia revolvens, Linnaea borealis, Liparis loeselii, Lobelia kalmii, Lonicera oblongifolia, Lonicera villosa, Lophiola aurea, Luetkea pectinata, Lupinus polyphyllus, Luzula spicata, Lycopodium sitchense, Lycopus americanus, Lycopus uniflorus, Lysimachia quadriflora, Lythrum salicaria, Maianthemum trifolium, Malaxis paludosa, Marshallia grandiflora, Melilotus, Menyanthes trifoliata, Menziesia pilosa, Micranthes ferruginea, Micranthes occidentalis, Micranthes rhomboidea, Mimulus glabratus, Mimulus lewisii, Mimulus moschatus, Mimulus tilingii, Mitella breweri, Mnium, Muhlenbergia glomerata, Muhlenbergia racemosa, Muhlenbergia richardsonis, Myrica gale, Myrica pensylvanica, Narthecium americanum, Nephrophyllidium crista-galli, Onoclea sensibilis, Osmunda regalis, Osmundastrum cinnamomeum, Oxypolis rigidior, Packera aurea, Packera pauciflora, Packera paupercula, Parentucellia latifolia, Parnassia caroliniana, Parnassia fimbriata, Parnassia glauca, Parnassia grandifolia, Parnassia palustris, Parnassia parviflora, Pedicularis furbishiae, Pedicularis groenlandica, Pedicularis labradorica, Phragmites australis, Phyllodoce empetriformis, Phyllodoce glanduliflora, Phyllodoce ×intermedia, Phlox maculata, Physocarpus opulifolius, Physostegia virginiana, Picea mariana, Pinguicula vulgaris, Pinus albicaulis, Pinus contorta, Piptatherum pungens, Piptatherum racemosum, Platanthera dilatata, Platanthera flava, Platanthera huronensis, Platanthera hyperborea, Platanthera leucophaea, Platanthera obtusata, Platanthera stricta, Poa alpina, Poa compressa, Polygala lutea, Polygala lutea, Populus tremuloides, Potentilla anserina, Potentilla rubricaulis, Potentilla simplex, Preissia quadrata, Prenanthes alba, Primula jeffreyi, Primula mistassinica, Proserpinaca palustris, Prunus pumila, Pteridium aquilinum, Pyrola asarifolia, Racomitrium lanuginosum, Ranunculus eschscholtzii, Ranunculus flammula, Ranunculus lapponicus, Rhamnus alnifolia, Rhinanthus minor, Rhododendron arborescens, Rhododendron canadense, Rhododendron columbianum, Rhododendron groenlandicum, Rhododendron lapponicum, Rhododendron maximum, Rhododendron tomentosum, Rhynchospora alba, Rhynchospora capillacea, Rhynchospora capitellata, Rhytidium rugosum, Ribes hudsonianum, Romanzoffia sitchensis, Rosa acicularis, Rubus arcticus, Rubus chamaemorus, Rubus pubescens, Rudbeckia californica, Rudbeckia hirta, Rudbeckia occidentalis, Salix brachycarpa, Salix candida, Salix commutata, Salix glauca, Salix myrtillifolia, Salix nivalis, Salix pedicellaris, Salix pedicellaris, Salix pseudomyrsinites, Salix pyrifolia, Salsola kali, Sanguisorba canadensis, Sanguisorba officinalis, Sarracenia purpurea, Saxifraga mertensiana, Schoenoplectus acutus, Schoenoplectus pungens, Schoenoplectus tabernaemontani, Scleria verticillata, Scorpidium scorpioides, Securigera varia, Selaginella eclipes, Selaginella selaginoides, Senecio lugens, Senecio triangularis, Shepherdia canadensis, Sibbaldiopsis tridentata, Silene acaulis, Silphium terebinthinaceum, Sinosenecio newcombei, Solidago bicolor, Solidago canadensis, Solidago multiradiata, Solidago multiradiata, Solidago nemoralis, Solidago ohioensis, Solidago ptarmicoides, Solidago rugosa, Solidago uliginosa, Sorghastrum nutans, Sphagnum austinii, Sphagnum capillifolium, Sphagnum fuscum, Sphagnum girgensohnii, Sphagnum nitidum, Sphagnum squarrosum, Sphagnum subsecundum, Sphagnum warnstorfii, Spiraea alba, Spiraea douglasii, Spiraea splendens, Spiranthes cernua, Spiranthes lucida, Spiranthes romanzoffiana, Sporobolus heterolepis, Stellaria crassifolia, Stenanthium leimanthoides, Symphyotrichum boreale, Symphyotrichum foliaceum, Symphyotrichum novi-belgii, Symphyotrichum puniceum, Symphyotrichum yukonense, Thalictrum occidentale, Thalictrum pubescens, Thuja plicata, Tofieldia pusilla, Tomenthypnum nitens, Trautvetteria caroliniensis, Trautvetteria fonticalcarea, Trichophorum alpinum, Trichophorum cespitosum, Trichophorum clintonii, Triglochin maritima, Triglochin palustris, Utricularia cornuta, Utricularia intermedia, Vaccinium deliciosum, Vaccinium myrtilloides, Vaccinium oxycoccos, Vaccinium scoparium, Vaccinium uliginosum, Vaccinium vitis-idaea, Valeriana edulis, Valeriana sitchensis, Viburnum edule, Viburnum nudum, Vicia cracca, Vincetoxicum nigrum, Viola langsdorffii, Viola sororia, Woodsia alpina, Woodsia glabella, Xyris torta.

4.4.3.2. ***Triantha racemosa*** **(Walter) Small** grows in bogs (hillside, pitcher plant, seepage), ditches (roadside), fens, flats (pineland), flatwoods, floodplains, marshes (roadside), meadows, pine barrens, roadsides (grassy, wet), savannahs (pine, pine barren), seeps (pitcher plant), swamps (hardwood, pine barren, *Sphagnum*), and along the margins of pocosins, ponds, and sloughs at elevations to 81 m. Occurrences are in open exposures that receive full sunlight in sites with no more than a sparse canopy. The substrates generally are acidic and nitrogen deficient. They have been characterized as Alapaha (arenic plinthic paleaquults), clay loam, Fuquay (plinthic paleudults), Goldsboro (aquic paleudults), loam, loamy sand, muck, Rutlege (typic humaquepts), sand (melanized), sandy clay, sandy peat, and Smithton. Flowering occurs from June to September, with fruiting extending into early December. The extent of sexual reproduction appears to be average; however, there have been no detailed studies on pollination, floral ecology, or seed ecology. The showy, nectariferous flowers are protogynous and presumably pollinated by insects; however, self-pollination also is thought to occur. Genetic data (allozymes) have indicated a potentially outcrossing breeding system with a high proportion of polymorphic loci (P_s=68.2%; P_p=47.7%) and moderate genetic diversity (H_{es}=0.134; H_{ep}=0.114). The estimated gene flow also is moderately high (N_m=2.07), which indicates that the pollen and/or seeds are dispersed widely. However, most (93%) of the total genetic diversity resides within populations. The plants inhabit fire-prone wetlands and are more abundant in sites that are burned annually than in those burned less often. Incidences of fire also have been associated with subsequent increases in the flowering frequency of adult plants. Vegetative reproduction occurs by means of short, horizontal rhizomes. Within shrub swamps, the plants occur in open habitats, which result from disturbances to the canopy. **Reported associates:** *Acer rubrum, Agalinis filicaulis, Agalinis linifolia, Aletris aurea, Aletris farinosa, Aletris lutea, Aletris obovata, Alnus rugosa, Alnus serrulata, Amelanchier canadensis, Amianthium muscitoxicum, Andropogon arctatus, Andropogon gerardii, Andropogon glomeratus, Andropogon gyrans, Andropogon virginicus, Anthaenantia rufa, Aristida palustris, Aristida stricta, Arnoglossum ovatum, Aronia arbutifolia, Aronia ×prunifolia, Arundinaria gigantea, Arundinaria tecta, Asclepias rubra, Asclepias tuberosa, Balduina uniflora, Bigelowia nudata, Burmannia capitata, Calamagrostis pickeringii, Calopogon barbatus, Calopogon multiflorus, Calopogon pallidus, Calopogon tuberosus, Carex, Carphephorus odoratissimus, Carphephorus paniculatus, Carphephorus pseudoliatris, Centella asiatica, Chamaedaphne calyculata, Chaptalia tomentosa, Chionanthus virginicus, Cirsium lecontei, Cleistesiopsis bifaria, Cleistesiopsis divaricata, Clethra alnifolia, Cliftonia monophylla, Coreopsis gladiata, Coreopsis nudata, Ctenium aromaticum, Cyperus, Cyrilla racemiflora, Dichanthelium acuminatum, Dichanthelium dichotomum, Dichanthelium ensifolium, Dichanthelium neuranthum, Dichanthelium scabriusculum, Drosera capillaris, Drosera rotundifolia, Drosera tracyi, Eleocharis microcarpa, Eleocharis tuberculosa, Erigeron vernus, Eriocaulon compressum, Eriocaulon decangulare, Eriophorum tenellum, Eryngium integrifolium, Eubotrys racemosa, Eupatorium leucolepis, Eupatorium resinosum, Eupatorium rotundifolium, Euphorbia inundata, Eurybia eryngiifolia, Fimbristylis spadicea, Gaylussacia dumosa, Gaylussacia mosieri, Gordonia lasianthus, Gratiola aurea, Harperocallis flava, Helenium vernale, Helianthus angustifolius, Helianthus heterophyllus, Helianthus radula, Hypericum brachyphyllum, Hypericum crux-andreae, Hypericum fasciculatum, Hypericum mutilum, Hypericum myrtifolium, Ilex coriacea, Ilex glabra, Ilex laevigata, Ilex myrtifolia, Iris tridentata, Iris virginica, Itea virginica, Juncus caesariensis, Juncus effusus, Juncus trigonocarpus, Kalmia angustifolia, Lachnanthes caroliniana, Lachnocaulon anceps, Leucothoe axillaris, Liatris spicata, Lilium catesbaei, Liquidambar styraciflua, Liriodendron tulipifera, Lobelia nuttallii, Lophiola aurea, Ludwigia alternifolia, Ludwigia virgata, Lycopodiella alopecuroides, Lycopodiella appressa, Lycopodiella caroliniana, Lyonia ligustrina, Lyonia lucida, Lyonia mariana, Magnolia virginiana, Mayaca fluviatilis, Mitreola sessilifolia, Mnesithea rugosa, Muhlenbergia capillaris, Muhlenbergia expansa, Myrica caroliniensis, Myrica cerifera, Myrica heterophylla, Narthecium americanum, Nyssa biflora, Nyssa sylvatica, Oclemena reticulata, Osmunda regalis, Osmundastrum cinnamomeum, Oxypolis rigidior, Packera crawfordii, Parnassia grandifolia, Penstemon digitalis, Persea borbonia, Persea palustris, Pinguicula caerulea, Pinguicula lutea, Pinguicula pumila, Pinus elliottii, Pinus palustris, Pinus serotina, Pinus taeda, Pityopsis oligantha, Platanthera blephariglottis, Platanthera ciliaris, Platanthera integra, Platanthera nivea, Pleea tenuifolia, Pluchea foetida, Pogonia ophioglossoides, Polygala brevifolia, Polygala cruciata, Polygala cymosa, Polygala hookeri, Polygala lutea, Polygala nana, Polygala ramosa, Pycnanthemum nudum, Rhexia alifanus, Rhexia lutea, Rhexia mariana, Rhexia petiolata, Rhexia virginica, Rhododendron viscosum, Rhynchospora cephalantha, Rhynchospora chapmanii, Rhynchospora ciliaris, Rhynchospora glomerata, Rhynchospora gracilenta, Rhynchospora inundata, Rhynchospora knieskernii, Rhynchospora latifolia, Rhynchospora macra, Rhynchospora microcarpa, Rhynchospora oligantha, Rhynchospora plumosa, Rhynchospora rariflora, Rudbeckia mohrii, Sabatia brevifolia, Sabatia campanulata, Sabatia decandra, Sabatia difformis, Sabatia macrophylla, Sarracenia flava, Sarracenia minor, Sarracenia psittacina, Sarracenia purpurea, Sarracenia rubra, Schizachyrium scoparium, Schwalbea americana, Scirpus longii, Scleria baldwinii, Scleria ciliata, Scleria reticularis, Scleria triglomerata, Scutellaria integrifolia, Serenoa repens, Smilax auriculata, Smilax glauca, Smilax laurifolia, Sphagnum, Spiranthes laciniata, Spiranthes longilabris, Spiranthes praecox, Stenanthium densum, Stenanthium leimanthoides, Stokesia laevis, Stylisma aquatica, Symphyotrichum adnatum, Symplocos tinctoria, Syngonanthus flavidulus, Taxodium ascendens, Taxodium distichum, Tiedemannia filiformis, Tofieldia glabra, Toxicodendron vernix, Triadenum virginicum, Utricularia juncea, Utricularia resupinata, Utricularia subulata, Vaccinium corymbosum, Vaccinium crassifolium, Vaccinium tenellum, Viburnum nudum, Viola primulifolia, Xyris ambigua, Xyris baldwiniana, Xyris caroliniana, Xyris fimbriata, Xyris jupicai, Zenobia pulverulenta, Zigadenus glaberrimus.*

Use by wildlife: The floral nectar of *Triantha glutinosa* is taken by Poweshiek skipperling butterflies (Insecta: Lepidoptera: Hesperiidae: *Oarisma poweshiek*). It is uncertain whether butterflies also function as pollinators in the genus.

Economic importance: food: no edible uses reported; **medicinal:** no medicinal uses reported; **cultivation:** not cultivated; **misc. products:** none; **weeds:** none; **nonindigenous species:** none.

Systematics: Differences of opinion persist whether to distinguish *Triantha* as a distinct genus from *Tofieldia*. Their resolution as sister clades (e.g., Figure 1.29) makes either disposition plausible phylogenetically. Furthermore, all four currently recognized species of *Triantha* have at times been treated as geographically distinct (but intergrading) subspecies of a single, widespread species (*T. glutinosa* sensu lato), a disposition that has merit considering the high degree of similarity among the taxa and the failure of molecular data to resolve them as discrete entities (e.g., Figure 1.29). Although ecologically divergent, *Triantha glutinosa* (calciphilic) and *T. racemosa* (acidic substrates) are believed to hybridize in their region of overlap (New Jersey) but that assumption has not been documented genetically. No intergeneric (i.e., *Triantha* × *Tofieldia*) hybrids have been reported. *Triantha racemosa* and *Tofieldia glabra* often are sympatric but are not known to hybridize, which has been attributed to their phenological divergence (i.e., flowering at different times of the year). Relationships within *Triantha* and its distinction from *Tofieldia* warrant further studies, which should incorporate comprehensive crossing experiments and surveys using higher resolution genetic markers. The basic chromosome number of *Triantha* is $x = 15$; *T. glutinosa* ($2n = 30$) is diploid (*T. racemosa* remains unstudied cytologically).

Distribution: *Triantha glutinosa* occurs throughout boreal North America, whereas *T. racemosa* is restricted to the southeastern United States.

References: Anderson et al., 1999; Bridges, 2005; Bridges & Orzell, 1989; Brumback, 2001; Calazza & Fairbrothers, 1980; Cázares et al., 2005; Chinnappa & Chmielewski, 1987; Choate, 1963; Cruden, 1988; Fernald, 1907; 1911; Floden & Schilling, 2018; Funston, 1896; Futyma & Miller, 2001; Gaddy, 1982; Gleason, 1917; Godt et al., 1997; Graenicher, 1935; Hall et al., 1938; Harper, 1900; 1922; Hinman & Brewer, 2007; Hitchcock, 1944; Homoya et al., 1984; Hutton, 1974; Jacobson et al., 1991; Kelly, 2001; Klinka et al., 1989; Knuth, 1906; Lahring, 2003; Lamb & Megill, 2003; Lewis & Dowding, 1926; Lewis et al., 1928; Looman, 1969; Mahmood & Strack, 2011; Mason, 1957; McMillan et al., 2002; Mitchell, 2011; Moran, 1981; Morris, 2013; Murphy & Boyd, 1999; Nelson, 1986; Nichols, 1933; Packer, 2002b; Pellatt & Mathewes, 1994; Pogue et al., 2016; Reid, 1978; Ring et al., 2013; Rochefort & Vitt, 1988; Sjörs, 1959; Slaughter & Cuthrell, 2011; Sorrie & Weakley, 2001; Standley, 1919; Swales, 1979; Swinehart, 1995; Thieret, 1964; Tobe et al., 1998; Tolman, 2006; Vitt & Chee, 1990; Walker & Peet, 1984; Weakley & Schafale, 1994; Weber & Rooney, 1994; Wells, 1928; Wells, 1996; Wright & Wright, 1932.

2 Monocotyledons II
Lilioid Monocotyledons ("Liliidae")

Historically, a prevalent North American taxonomic tradition was to recognize the lilioid monocotyledons as a subclass (Liliidae) comprising two orders: Liliales and Orchidales (Cronquist, 1981; FNA, 2003). As defined, the group excluded Cyclanthales and Pandanales, which were assigned to a different subclass (Arecidae). Subsequent molecular studies (e.g., Figure 2.1) have rendered that concept of Liliidae indefensible phylogenetically, with the former Liliid taxa resolving not within a single clade but as various disparate groups that currently are distinguished taxonomically as orders. Yet, the higher-level phylogenetic interrelationships based on molecular data are by no means compelling even when accompanied by high levels of internal support, given the all-too-familiar pattern of long external branches and short internal branches that characterize the placement of some of the more contentious groups such as Dasypogonales and Dioscoreales.

In any case, there currently is no phylogenetic evidence in support of recognizing a "Liliidae" clade that would differ materially from the order Liliales. Consequently, the subclass is regarded here as an artificial, polyphyletic group of dubious taxonomic utility; the more recent convention of grouping the former component taxa only at the rank of order (e.g., Judd et al., 2016; Soltis et al., 2018) has been followed. Three of the orders (Asparagales, Dioscoreales, and Liliales) contain OBL North American indicators.

ORDER 5: ASPARAGALES [13–25]

Modified phylogenetic concepts resulting from analyses of molecular data are exemplified well by the current circumscription of Asparagales, which represents a diverse group of families scattered previously within Liliales (e.g., Cronquist, 1981). Molecular data have provided a more reasonable partitioning of taxa among those two orders and evidence for their monophyly (Judd et al., 2016). Morphological synapomorphies that distinguish Asparagales from Liliales include seeds with an obliterated epidermis or phytomelan crust, nonspotted tepals, and interseptal nectaries (Judd et al., 2016).

However, it has been difficult to accept any particular phylogenetic scheme for the group because data from the same cpDNA genome (e.g., Rudall et al., 1997; Steele et al., 2012) or from different combined genomes (e.g., Seberg et al., 2012) have provided numerous conflicting results. It is apparent that no consensus has been reached even on the number of families that merit recognition. A pertinent example involves Alliaceae and Amaryllidaceae, which some authors recognize as distinct (e.g., Judd et al., 2016), while others treat the former as a subfamily (Allioideae) of the latter, i.e., as

Amaryllidaceae "*sensu lato*" (Meerow et al., 2007a; Chase et al., 2009). Because the two groups resolve as sister clades (e.g., Figure 2.2), either disposition would be defensible phylogenetically. The *sensu lato* concept has been accepted here in accordance with prevailing convention (e.g., Byng et al., 2016) and because of their similar morphological features (Judd et al., 2016). A parallel situation occurs with Asparagaceae, wherein Laxmanniaceae and Ruscaceae have been merged as the subfamilies Lomandroideae and Nolinoideae, respectively (Chase et al., 2009; Figure 2.2).

Yet, some consistent results have been obtained. Rudall et al. (1997) observed that a clade of the more derived families ("higher asparagoids") was characterized primarily by sulcate pollen and successive microsporogenesis, whereas trichotomosulcate pollen and simultaneous microsporogenesis dominated in a grade of less-specialized families ("lower asparagoids"). That major dichotomy has been upheld by most subsequent phylogenetic analyses despite a fair number of topological discrepancies within each group (e.g., Rudall et al., 1997; Seberg et al., 2012; Steele et al., 2012). Consequently, the phylogenetic tree included here (Figure 2.2) does not necessarily depict the most likely topology of families but is suitable for use as a reference because it retains the more consistent topological features found among the various differing analyses.

The cosmopolitan Asparagales (~36,000 species) now comprise nearly half of all monocotyledons (~74,000 species) but are dominated by the orchid family (Orchidaceae), which itself includes almost 28,000 species (Christenhusz & Byng, 2016). Four families in the order contain OBL North American indicators:

5.1. **Amaryllidaceae** J.St.-Hil.
5.2. **Asparagaceae** Juss.
5.3. **Iridaceae** Juss.
5.4. **Orchidaceae** Juss.

Family 5.1. Amaryllidaceae [75]

As currently defined, the Amaryllis family (Amaryllidaceae) consists of about 1,600 species distributed among 75 genera (Christenhusz & Byng, 2016). Taxonomically, the group is subdivided into three subfamilies (Agapanthoideae, Allioideae, and Amaryllidoideae) with the former two having been recognized previously as the families Agapanthaceae and Alliaceae (Chase et al., 2009). The monophyly of the family and the closely related subfamilies has been supported consistently by phylogenetic analyses (e.g., Figure 2.2). Morphologically,

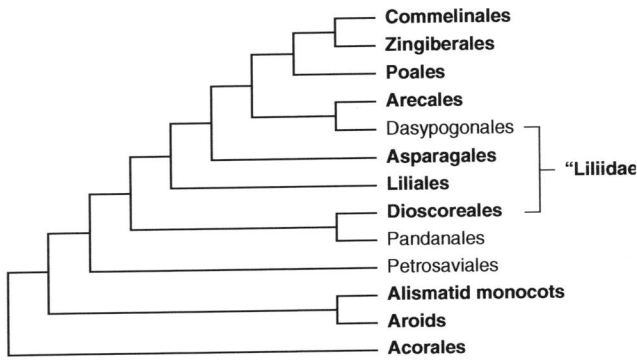

FIGURE 2.1 Hypothesized interrelationships among major monocotyledon groups based on phylogenetic analyses of complete cpDNA genome sequence data (modified from Barrett et al., 2016; Givnish et al., 2018). These results provide no support to maintain the former subclass Liliidae, which is polyphyletic by these analyses. Taxa containing OBL North American indicators are highlighted in bold.

the group is characterized by having umbels enclosed by two subtending bracts and a characteristic "amaryllis" alkaloid chemistry (Chase et al., 2009; Judd et al., 2016).

According to Judd et al. (2016) the showy, bisexual, tepaloid flowers are pollinated by assorted insects (Insecta) and birds (Aves); sometimes they self-pollinate. The seeds are dispersed abiotically (by water or wind) or occasionally by birds. All the species perennate from bulbs having contractile roots; asexual reproduction can occur by vegetative bulblets, which develop in the inflorescences of some species (Judd et al., 2016).

Many genera include cultivated ornamentals (*Acis, Allium, Amaryllis, Brunsvigia, Clivia, Crinum, Cyrtanthus, Eucharis, Galanthus, Gethyllis, Gilliesia, Habranthus, Haemanthus, Hippeasterum, Hymenocallis, Ipheion, Leucocoryne, Leucojum, Lycoris, Narcissus, Nerine, Nothoscordum, Sternbergia, Tulbaghia,* and *Zephyranthes*). *Allium* is an important source of vegetables including chives, garlic, leeks, onions, ramps, scallions, and shallots. Several species of *Allium* and *Tulbaghia* have been used medicinally (Khokar & Fenwick, 2003; Aremu & van Staden, 2013; Mnayer et al., 2014).

Amaryllidaceae are distributed broadly across temperate and tropical regions worldwide (Judd et al., 2016). OBL North American indicators occur within three genera:

5.1.1. ***Allium*** L.
5.1.2. ***Crinum*** L.
5.1.3. ***Hymenocallis*** Salisb.

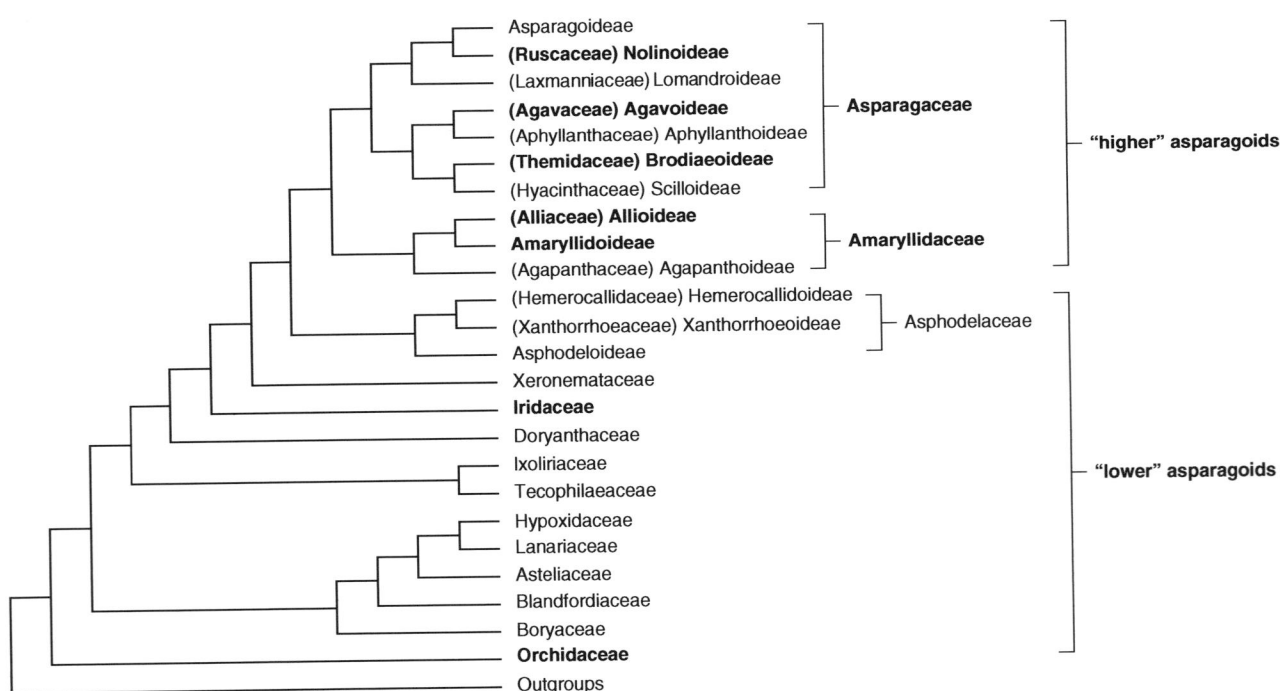

FIGURE 2.2 Phylogenetic associations of asparagoid (Asparagales) families as indicated by analysis of combined cpDNA and mtDNA sequence data (modified from Seberg et al., 2012). The clade of "higher" asparagoids consists mostly of plants having superior ovaries, sulcate pollen, and successive microsporogenesis, whereas the grade of "lower" asparagoids is dominated by species with inferior ovaries, trichotomosulcate pollen, and simultaneous microsporogenesis (Judd et al., 2016). A consensus on family circumscriptions has not yet been reached. Here Asparagaceae *sensu lato* are shown to include Agavaceae, Laxmanniaceae, Ruscaceae, and Themidaceae as subfamilies (Agavoideae, Lomandroideae, Nolinoideae, and Brodiaeoideae, respectively) and Amaryllidaceae to include Agapanthaceae and Alliaceae as subfamilies (Agapanthoideae and Allioideae) as recently proposed (Chase et al., 2009). Other *sensu stricto* family designations are indicated similarly in parentheses. Despite the large number of species in the order, OBL North American indicators occur within relatively few taxa (shown in bold), which are concentrated in the more derived groups. This topology agrees well with a previous study that sampled fewer taxa (Pires et al., 2006).

5.1.1. Allium

Swamp onion, Pacific onion; ail

Etymology: from the Latin *allium*, meaning "garlic"

Synonyms: none

Distribution: global: Northern Hemisphere; **North America:** widespread except extreme northeast arctic

Diversity: global: 850 species; **North America:** 96 species

Indicators (USA): obligate wetland (OBL): *Allium madidum, A. validum*

Habitat: freshwater; palustrine; **pH:** 5.4–7.4; **depth:** <1 m; **life-form(s):** emergent herb

Key morphology: bulbs (1–3) tunicate, solitary, or with a lateral cluster of 10–30 globose to ovoid bulbels (to 1.6 cm), or bulbs elongate (to 5 cm) and clustered (to 20) on thick rhizomes; leaves (2–6) basal, persistent, sheathed at base, the blade (to 25–80 cm) linear, channeled or flat, entire; scape (to 20–70 cm) erect, solitary, flat, terete, ridged, or narrowly winged distally; umbel compact, 10–30-flowered, hemispheric, subtended by 2 membranous bracts; flowers (to 10 mm) erect, campanulate, pedicelled (to 15 mm); tepals 6, petaloid, lanceolate, erect to spreading, pink or white, withering or papery in fruit; stamens 6, epipetalous, inserted or exserted, the filaments dilated at base, the anthers purple, white, or yellow; ovary superior, 3-lobed, 3-loculed, 2 ovules in each locule, style 1, linear, stigma capitate; capsules loculicidal; seeds black, dull, smooth, or somewhat rough

Life history: duration: perennial (bulbs, bulbils, rhizomes); **asexual reproduction:** bulbils, bulblets, rhizomes; **pollination:** bird, insect; **sexual condition:** hermaphroditic; **fruit:** capsules (common); **local dispersal:** bulblets, seeds (water); **long-distance dispersal:** seeds (water)

Imperilment: 1. *Allium madidum* [G3]; S3 (ID, OR); **2.** *A. validum* [G4]; S3 (ID)

Ecology: general: *Allium* is not considered to be a wetland genus but rather has many xerophytic species. Only 13 of the North American species (~14%) are wetland indicators, with most ranked as FACU; merely 2% of the species are ranked as OBL. All the species are perennial herbaceous geophytes, which persist from bulbs or rhizomes. The flowers are pollinated primarily by insects (Insecta). Vegetative reproduction can occur by bulbils, which develop in the inflorescence of some (but not the OBL) species, or by readily detaching bulblets, which can occur in clusters alongside the bulb or atop the thickened rhizome. The seeds and vegetative propagules of at least the OBL species are most likely dispersed by water.

5.1.1.1. ***Allium madidum* S. Watson** grows in channels (stream), depressions (vernal), draws (ephemeral), flats, meadows (drying, open, overgrazed, seasonally wet), openings (pine), prairies, seeps, slopes [to 30%] (rocky, seepy, springy), swales (ephemeral), thickets, vernal pools, and along watercourses (dry, intermittent) at elevations from 915 to 2287 m. The exposures can range from fully open to shaded sites. The substrates include alluvium, basalt, clay (baked, basaltic, stony), clay loam, gravel, lithosol, rock, sand, and sandy rock. Flowering and fruiting occur from May to July. The flowers are thought to be pollinated by honeybees (Insecta: Hymenoptera: Apidae: *Apis mellifera*) and other insects.

Presumably, the seeds are dispersed by water. Additional information on the floral and seed ecology is lacking and deserves further study. The plants persist vegetatively as a basal bulb (or bulbs), from which extends a cluster of bulblets that can function as vegetative propagules when detached. The bulblets are probably dispersed by water. The garlic-like odor of the plants is attributable to the presence of volatile allyl sulfides. **Reported associates:** *Abies grandis, Achillea millefolium, Achnatherum occidentale, Allium brandegeei, Allium geyeri, Allium tolmiei, Antennaria luzuloides, Artemisia tridentata, Balsamorhiza sagittata, Camassia quamash, Carex nebrascensis, Carex vesicaria, Claytonia lanceolata, Cryptogramma, Danthonia californica, Delphinium depauperatum, Delphinium ×burkei, Deschampsia cespitosa, Eleocharis bella, Epilobium, Erigeron, Floerkea proserpinacoides, Juncus balticus, Linanthus, Lomatium, Madia, Microsteris gracilis, Montia chamissoi, Navarretia, Oenothera flava, Orobanche, Perideridia, Pinus ponderosa, Polygonum, Populus tremuloides, Pseudoroegneria spicata, Pseudotsuga menziesii, Rorippa, Sedum, Senecio hydrophiloides, Sidalcea oregana, Trifolium eriocephalum, Trifolium longipes, Triteleia hyacinthina, Veratrum californicum, Wyethia helianthoides.*

5.1.1.2. ***Allium validum* S. Watson** inhabits alluvial fans, bogs (hanging), draws, fens (sloping, subalpine), flats (grassy), floodplains, marshes, meadows (alpine, montane, roadside, seepage, swampy), ravines (moist), seepages, slopes [to 60%] (rocky, steep), spring beds, springs (ephemeral), streams (spring-fed), swamps, thickets, vernal pools, woodlands (damp), and the margins of bogs, lakes, rivers, springs, and streams at elevations from 502 to 3657 m. The plants grow on wet substrates or in shallow standing water in open exposures with full sunlight to partial shade. The habitats are acidic to slightly alkaline (pH: 4.8–7.4) and include substrates characterized as alluvium (clayey, silty), ash, cobbles, cryofluvents, granite, granite-diorite, gravel, gravelly loam, humus (rocky), loam, mucky silt, mud (alluvial), muddy loam, peat, pumice, sand, sandy loam, serpentine, and silty loam. Flowering and fruiting extend from May to September. Flowering in high elevation alpine habitats (e.g., 3000 m) correlates with snow depth, occurring progressively later with increasing snow depth. The flowers are pollinated by bumblebees (Insecta: Hymenoptera: Apidae: *Bombus*) and rufous hummingbirds (Aves: Trochilidae: *Selasphorus rufus*). The seed germination rate is relatively high (37%) compared to other higher elevation associates. Little additional information exists on the reproductive biology of this species. Asexual reproduction occurs by means of vegetative bulbs, which develop on the thick rhizome each successive year. The onion-like odor of the plants is due to volatile *n*-propyl sulfides. **Reported associates:** *Abies concolor, Abies lasiocarpa, Abies magnifica, Acer circinatum, Achillea millefolium, Aconitum columbianum, Aconogonon phytolaccifolium, Agrostis idahoensis, Alnus tenuifolia, Alnus viridis, Amelanchier alnifolia, Antennaria alpina, Anticlea occidentalis, Aquilegia formosa, Arctostaphylos, Arnica chamissonis, Arnica mollis, Artemisia tridentata, Asarum caudatum, Athyrium filix-femina, Berberis*

aquifolium, Betula nana, Betula occidentalis, Bistorta bistortoides, Botrychium multifidum, Brachythecium frigidum, Bryum pseudotriquetrum, Calamagrostis canadensis, Calocedrus decurrens, Caltha leptosepala, Camassia quamash, Canadanthus modestus, Cardamine cordifolia, Carex alma, Carex angustata, Carex aquatilis, Carex canescens, Carex capitata, Carex echinata, Carex fissuricola, Carex hassei, Carex hendersonii, Carex illota, Carex jonesii, Carex laeviculmis, Carex lenticularis, Carex luzulina, Carex microptera, Carex nebrascensis, Carex nigricans, Carex raynoldsii, Carex scirpoidea, Carex scopulorum, Carex simulata, Carex subfusca, Carex utriculata, Carex vesicaria, Cassiope mertensiana, Castilleja chrysantha, Castilleja hispida, Castilleja miniata, Ceanothus cordulatus, Ceanothus velutinus, Chamerion angustifolium, Chrysolepis sempervirens, Cinna latifolia, Conardia compacta, Cornus nuttallii, Cornus sericea, Corydalis caseana, Danthonia californica, Danthonia intermedia, Darlingtonia californica, Delphinium ×occidentale, Deschampsia cespitosa, Descurainia pinnata, Ditrichum ambiguum, Drepanocladus, Drosera rotundifolia, Drymocallis glandulosa, Eleocharis decumbens, Eleocharis macrostachya, Eleocharis quinqueflora, Elymus elymoides, Elymus glaucus, Epilobium anagallidifolium, Epilobium ciliatum, Epilobium glaberrimum, Epilobium halleanum, Epilobium hornemannii, Epipactis gigantea, Equisetum arvense, Erigeron coulteri, Erigeron glacialis, Erigeron peregrinus, Eriophorum crinigerum, Erythronium grandiflorum, Festuca viridula, Frasera speciosa, Galium triflorum, Gaultheria humifusa, Gentiana calycosa, Gentiana newberryi, Gentianella amarella, Geranium, Glandularia pulchella, Glyceria elata, Glyceria striata, Graphephorum wolfii, Hastingsia alba, Helenium bigelovii, Heracleum sphondylium, Hordeum brachyantherum, Horkelia fusca, Hosackia pinnata, Hydrophyllum tenuipes, Hypericum anagalloides, Hypericum formosum, Ipomopsis aggregata, Iris missouriensis, Juncus arcticus, Juncus balticus, Juncus drummondii, Juncus effusus, Juncus ensifolius, Juncus howellii, Juncus macrandrus, Juncus mertensianus, Juncus nevadensis, Juncus orthophyllus, Juncus oxymeris, Juniperus occidentalis, Kalmia microphylla, Kyhosia bolanderi, Lewisia cotyledon, Lewisia pygmaea, Ligusticum grayi, Ligusticum tenuifolium, Lilium kelleyanum, Lilium pardalinum, Lilium parvum, Linnaea borealis, Lithocarpus densiflorus, Lonicera caerulea, Lonicera involucrata, Lupinus latifolius, Lupinus polyphyllus, Luzula campestris, Luzula comosa, Lysichiton americanus, Madia elegans, Maianthemum stellatum, Marchantia, Meesia triquetra, Menyanthes trifoliata, Mertensia ciliata, Micranthes odontoloma, Micranthes oregana, Mimulus cardinalis, Mimulus guttatus, Mimulus lewisii, Mimulus primuloides, Mimulus tilingii, Mitella ovalis, Mitella pentandra, Monardella odoratissima, Montia chamissoi, Muhlenbergia andina, Muhlenbergia filiformis, Muhlenbergia filiformis, Narthecium californicum, Neottia convallarioides, Neottia cordata, Oreostemma alpigenum, Orobanche fasciculata, Orthilia secunda, Oxypolis occidentalis, Packera pseudaurea, Packera subnuda, Parnassia fimbriata, Parnassia palustris, Pedicularis attollens, Pedicularis bracteosa, Pedicularis
groenlandica, Penstemon confertus, Penstemon heterodoxus, Penstemon newberryi, Penstemon procerus, Penstemon rydbergi, Perideridia parishii, Perideridia parishii, Phacelia heterophylla, Phacelia ramosissima, Philonotis fontana, Phleum alpinum, Phyllodoce empetriformis, Physocarpus capitatus, Picea breweriana, Picea engelmannii, Pinus albicaulis, Pinus contorta, Pinus flexilis, Pinus strobus, Plagiomnium, Platanthera dilatata, Platanthera sparsiflora, Platanthera sparsiflora, Poa secunda, Podagrostis humilis, Podagrostis thurberiana, Polemonium caeruleum, Polemonium occidentale, Polemonium viscosum, Populus tremuloides, Potentilla flabellifolia, Potentilla gracilis, Primula fragrans, Primula jeffreyi, Primula tetrandra, Prunella vulgaris, Prunella vulgaris, Pseudotsuga menziesii, Pteridium aquilinum, Purshia tridentata, Pyrola asarifolia, Ranunculus alismifolius, Ranunculus orthorhynchus, Ranunculus populago, Rhododendron columbianum, Ribes bracteosum, Ribes cereum, Ribes hudsonianum, Ribes inerme, Ribes nevadense, Ribes roezlii, Ribes sanguineum, Rumex paucifolius, Salix boothii, Salix commutata, Salix eastwoodiae, Salix farriae, Salix geyeriana, Salix jepsonii, Salix lasiandra, Salix lemmonii, Salix orestera, Salix scouleriana, Salix wolfii, Sanguisorba stipulata, Scapania undulata, Scirpus congdonii, Scirpus microcarpus, Scoliopus hallii, Sedum obtusatum, Senecio hydrophilus, Senecio integerrimus, Senecio triangularis, Sibbaldia procumbens, Sidalcea oregana, Sisyrinchium elmeri, Sisyrinchium idahoense, Sphagnum, Sphenosciadium capitellatum, Spiraea douglasii, Spiranthes romanzoffiana, Stellaria longipes, Streptopus amplexifolius, Swertia perennis, Symphyotrichum spathulatum, Tellima grandiflora, Thalictrum fendleri, Thalictrum sparsiflorum, Tiarella trifoliata, Toxicoscordion venenosum, Trianthema occidentalis, Trichophorum, Trifolium longipes, Trifolium monanthum, Tsuga mertensiana, Vaccinium cespitosum, Vaccinium myrtillus, Vaccinium scoparium, Vaccinium uliginosum, Veratrum californicum, Veronica wormskjoldii, Viola adunca, Viola glabella, Viola macloskeyi, Viola palustris.

Use by wildlife: *Allium validum* is host to several Fungi (Ascomycota: Davidiellaceae: *Cladosporium allii*; Didymosphaeriaceae: *Apiosporella mimuli*; Leptosphaeriaceae: *Leptosphaeria lassenensis*; Mycosphaerellaceae: *Cladosporium herbarum, Sphaerella schoenoprasi*; Thelebolaceae: *Pezizella helotioides*; Basidiomycota: Pucciniaceae: *Uromyces aemulus, U. aterrimus U. aureus, U. bicolor*). *Allium validum* provides forage for grazing animals such as sheep (Mammalia: Bovidae: *Ovis aries*) and elk (Mammalia: Cervidae: *Cervus elaphus*). However, the bulbs of *A. validum* and other wild onions contain organosulfur compounds that have caused oxidative hemolysis and severe hemolytic anemia in cattle (Mammalia: Bovidae: *Bos taurus*), sheep (Mammalia: Bovidae: *Ovis aries*), and domestic pets, which never should be allowed to feed on them. The flowers of *A. validum* are foraged by bumblebees (Insecta: Hymenoptera: Apidae: *Bombus*) and rufous hummingbirds (Aves: Trochilidae: *Selasphorus rufus*).

Economic importance: food: Wild *Allium* species never should be fed to animals that have very low erythrocytic antioxidant activity (see *Use by wildlife* above). *Allium madidum* was collected as a food by the native inhabitants of the Great Basin region. The Cahuilla people ate the raw bulbs of *A. validum* as a vegetable and used them to flavor other foods. They are said to be of good flavor; **medicinal:** no reported uses; **cultivation:** Neither *Allium madidum* nor *A. validum* is grown commercially. Attempts to cultivate *A. validum* from seed or rootstock have not been successful; **misc. products:** The annual flowering period of *Allium validum* has been used to direct studies of avian spring breeding activities; **weeds:** none; **nonindigenous species:** Both *Allium madidum* and *A. validum* are indigenous.

Systematics: Some authors place *Allium* in a separate family (Alliaceae) but most recent treatments include that clade as a subfamily (Allioideae) within a more broadly defined Amaryllidaceae (Figure 2.2). Several molecular phylogenetic studies have demonstrated that the genus is monophyletic, with a clade comprising *Dichelostemma*, *Ipheion*, *Nothoscordum*, and *Tulbaghia* as its sister group. Both OBL *Allium* species are assigned to the monophyletic subgenus *Amerallium* (~140 species), which is characterized by leaves having subepidermal laticifers and a single row of vascular bundles. The primarily western North American subgenus *Amerallium* (with subgenera *Nectaroscordum* and *Microscordum*) associates as the sister clade to the remainder of the genus. However, the two OBL species are not closely related. *Allium madidum* is most closely related to *A. fibrillum* (and possibly its polyploid derivative) in a group known as the "*Allium falcifolium* alliance," whereas *Allium validum* occupies an isolated position as the sister group of a clade that includes several eastern North American species (Figure 2.3). The basic chromosome number of subgenus *Amerallium* is *x*=7. *Allium madidum* (2*n*=28, 42) is tetraploid or hexaploid; *A. validum* (2*n*=28, 56) is tetraploid or octaploid. The genome size of the octaploid has been estimated at 148.5 pg. No hybrids have been reported that involve either *Allium madidum* or *A. validum*.

Distribution: *Allium validum* occurs at higher elevations in western North America (British Columbia, California, Idaho, Nevada, Oregon, and Washington); *A. madidum* is endemic to a small region of western Idaho and northeastern Oregon.

References: Aikens, 1986; Bottum, 2005; Chase et al., 2009; Cheng, 2004; Cope, 2005; Crowe et al., 2004; Everett, 2012; Friesen et al., 2006; Grant & Grant, 1967; Jackson & Bliss, 1982;

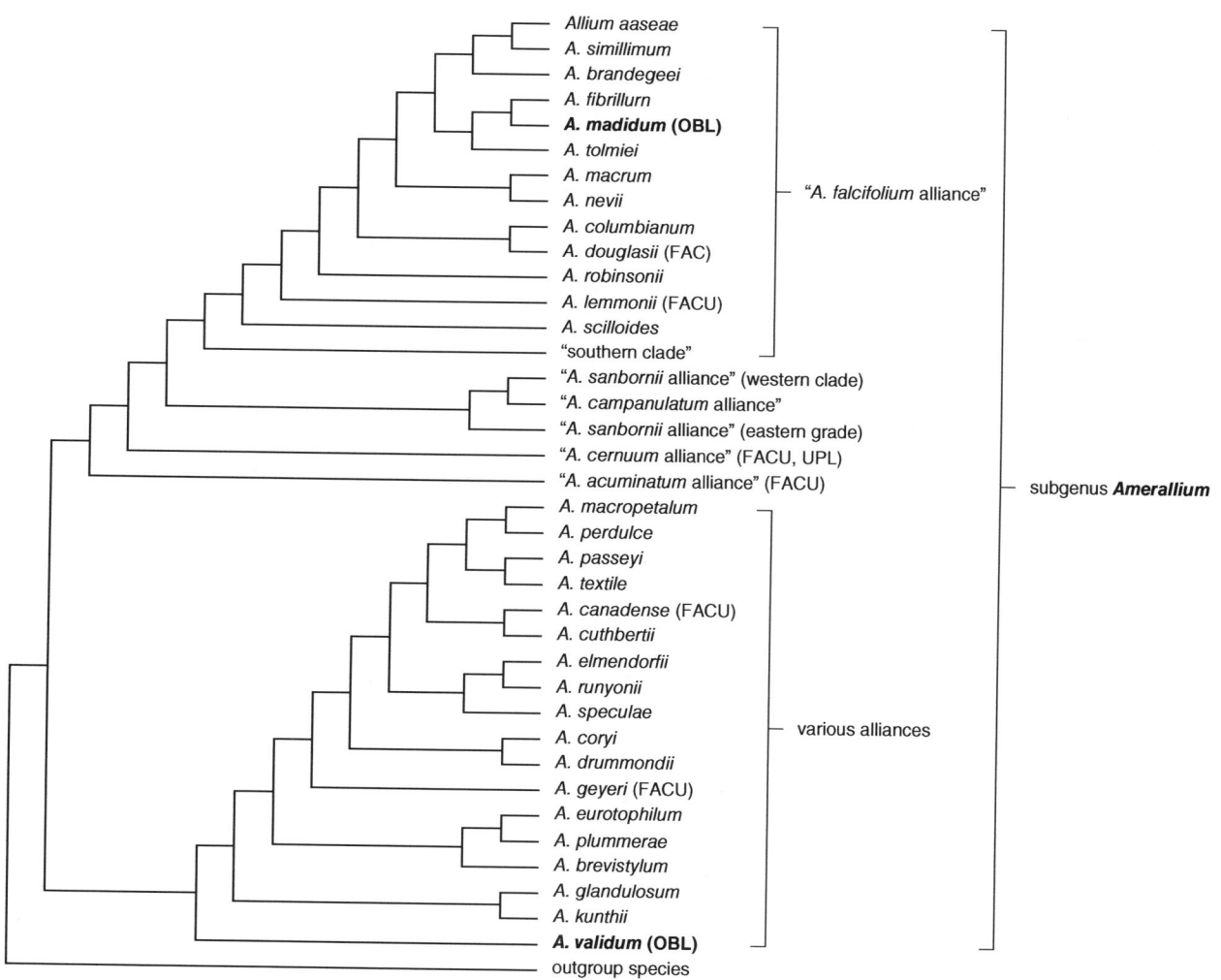

FIGURE 2.3 Interspecific relationships in *Allium* subgenus *Amerallium* as indicated by analysis of combined ETS, ITS, *rpL32-trnL*, and *trnL-F* sequence data (modified from Wheeler et al., 2013). The two OBL North American indicators (in bold) are members of quite distantly related species "alliances." Designations for other North American wetland indicators are shown in parentheses.

James & Binns, 1966; Kaltenecker, 1993; Loffland et al., 2017; Manning & Padgett, 1995; Mansfield, 1996; McNeal, 1995; 2012a; McNeal, Jr. & Jacobsen, 2002; Morton, 1994; Moseley, 1989; Munz, 1963; Murray, 2000; Nguyen et al., 2008; Ohri et al., 1998; Pereyra, 2011; Pickford & Reid, 1943; Saghir et al., 1966; Sampson, 1914; Schuller et al., 2014; Sikes et al., 2010; 2013; Spahr et al., 1991; Taylor, 1912; Todt, 1997; Van Kampen et al., 1970; Wells, 2006; Wheeler et al., 2013.

5.1.2.   Crinum

Seven sisters, stringlily, swamplily; crinole
Etymology: from the Greek *krínon* meaning "lily"
Synonyms: *Amaryllis* (in part); *Ammocharis* (in part); *Bulbine* (in part); *Scadianus*
Distribution: global: pantropical: Africa; Asia; Australia; North America; South America; **North America:** southeastern United States
Diversity: global: 102 species; **North America:** 6 species
Indicators (USA): obligate wetland (OBL); facultative (FAC): *Crinum americanum*
Habitat: brackish (coastal), freshwater; freshwater (tidal); palustrine; **pH:** 5.3–8.9; **depth:** <1 m; **life-form(s):** emergent herb
Key morphology: perennial, scapose herbs from apically elongate (to 6 cm), succulent, tunicate bulbs (to 13 cm); leaves basal, thick, the blades (to 1.5 m) lorate, flattened distally, their margins scabrous; scape (to 1.5 m) solid, terminating in an umbel of 2–7 fragrant, sessile flowers subtended by 2 scarious bracts; perianth tepaloid, connate proximally, salverform; tepals 6, the tube elongate (to 15 cm), greenish; the limb (to 14 cm) white, linear, narrowly elliptic, or oblanceolate; stamens 6, inserted on perianth tube, the filaments threadlike, purplish-pink; ovary inferior, globose, 3-locular, each locule with 1-few seeds, style slender, stigma capitate; capsules (to 4.5 cm) globose, thin-walled, beaked (to 8 cm), few-seeded; seeds large (to 2.5 cm), fleshy, corky
Life history: duration: perennial (bulbs, rhizomes); **asexual reproduction:** rhizomes; **pollination:** insect; **sexual condition:** hermaphroditic; **fruit:** capsules (common); **local dispersal:** rhizomes, seeds (water); **long-distance dispersal:** seeds (water [oceanic])
Imperilment: 1. *Crinum americanum* [G5]; SH (NC)
Ecology: general: Although many *Crinum* species have an affinity for disturbed, wet sites, only one North American species is designated as an OBL indicator. Two of the species have status as FAC, FACU, or UPL indicators. All the species are perennials, which persist from a bulb; some are rhizomatous. The flowers are large, showy, and pollinated by insects (Insecta). The capsules produce few but large, corky seeds, which are buoyant. Most dispersal is abiotic and facilitated by water. Some seeds are resistant to salinity and can be dispersed in saltwater along coastal shorelines. Their dormancy requirements (if any) have not been investigated. Due to their large size, the seeds probably do not develop a seed bank.
5.1.2.1. *Crinum americanum* **L.** grows in fresh to brackish waters (salinities to 5.3‰) in bayheads, canals (drainage, irrigation, roadside), depressions (roadside, seasonal), ditchbanks, ditches (roadside), estuaries, flats (maidencane), flatwoods (hydric), floodplains, glades (wet), hammocks (coastal, hydric), marshes (brackish, freshwater, roadside, salt, tidal), meadows (low), mudflats, prairies (marl, roadside, wet), rice fields (abandoned), sloughs, swamps (floodplain), woodlands (low, upper tidal), and along the banks and margins of alligator holes, bayous, lakes, rivers, sinkholes, and streams (blackwater, brackish, tidal) at elevations to 30 m. Exposures can range from full sunlight to shade, but the plants usually occur in open sites. The substrates have been described as muck (acid clay, black, Maurepas), mud, peat, peaty muck, sand, sandy humus, and sandy loam (fine, Harleston). The habitats typically are calcareous (pH: 5.3–8.9; $\bar{x}$ = 7.5; $\tilde{x}$ = 7.8; total alkalinity [as $CaCO_3$]: 2.0–122.5 mg l^{-1}; $\bar{x}$ = 52.3 mg l^{-1}; $\tilde{x}$ = 47.0 mg l^{-1}; conductance: 61–723 µS cm^{-1}; $\bar{x}$ = 274 µS cm^{-1}; $\tilde{x}$ = 214 µS cm^{-1}; Ca^{+2}: 5.2–100 mg l^{-1}; $\bar{x}$ = 43.3 mg l^{-1}; $\tilde{x}$ = 38.5 mg l^{-1}; Cl$^-$: 12.3–168.2 mg l^{-1}; $\bar{x}$ = 48.4 mg l^{-1}; $\tilde{x}$ = 24.2 mg l^{-1}; K$^+$: 0.7–8.4 mg l^{-1}; $\bar{x}$ = 3.5 mg l^{-1}; $\tilde{x}$ = 3.2 mg l^{-1}; Mg^{+2}: 4.4–51.8 mg l^{-1}; $\bar{x}$ = 20.0 mg l^{-1}; $\tilde{x}$ = 13.7 mg l^{-1}; Na$^+$: 6.3–89.9 mg l^{-1}; $\bar{x}$ = 26.3 mg l^{-1}; $\tilde{x}$ = 14.7 mg l^{-1}; SO$_4^{-2}$: 3.7–53.2 mg l^{-1}; $\bar{x}$ = 20.3 mg l^{-1}; $\tilde{x}$ = 11.6 mg l^{-1}; total N: 400–3230 µg l^{-1}; $\bar{x}$ = 1260 µg l^{-1}; $\tilde{x}$ = 1030 µg l^{-1}; total P: 11–152 µg l^{-1}; $\bar{x}$ = 85 µg l^{-1}; $\tilde{x}$ = 31 µg l^{-1}). The plants tolerate brackish conditions better during the spring, when a higher influx of freshwater (from rains) occurs. Flowering and fruiting occur throughout the year. The fragrant flowers are self-compatible but are adapted structurally for pollination by nocturnal hawk moths (Insecta: Lepidoptera: Sphingidae). Hand pollinations (autogamous or xenogamous) have resulted in fairly high fruit set (75–80%). Parthenocarpic fruits (seedless) have been obtained experimentally by treating the carpels with 2.0% naphthalene acetic acid. Pseudogamous seed set has been observed in some attempted interspecific hybrid crosses. As the large fruits mature, their mass causes the inflorescence to droop, eventually bringing the capsules and their seeds in contact with the ground. The large and salt-resistant seeds germinate while still within the capsule, causing it to rupture and release them. Their external seed coat is impermeable, and the internal layers are suberized, which enables them to float. They are dispersed in fresh or saltwater and have been recovered from the strand along oceanic beaches. The plants are highly flood tolerant (to at least 61 cm depths) because of their deeply set bulbs, which are connected by a network of coarse rhizomes. Small, clonal patches can develop as a consequence, with densities of up to 18.8 individuals m^{-2} reported. **Reported associates:** *Acer rubrum, Acrostichum danaeifolium, Aeschynomene pratensis, Alternanthera philoxeroides, Amaranthus cannabinus, Ambrosia artemisiifolia, Amorpha fruticosa, Ampelaster carolinianus, Ampelopsis arborea, Annona glabra, Arundo donax, Asclepias lanceolata, Asclepias perennis, Avicennia germinans, Axonopus furcatus, Baccharis angustifolia, Baccharis halimifolia, Bacopa caroliniana, Bacopa monnieri, Berchemia scandens, Bidens frondosus, Bidens laevis, Boehmeria cylindrica, Bolboschoenus maritimus, Bolboschoenus robustus, Boltonia asteroides, Campsis radicans, Carex gigantea, Carex longii, Carex lupulina, Celtis laevigata, Centella asiatica, Cephalanthus occidentalis, Chamaecrista fasciculata, Chasmanthium latifolium,*

Chasmanthium laxum, Chasmanthium nitidum, Chenopodium album, Chionanthus, Chrysobalanus icaco, Cicuta maculata, Cinna arundinacea, Cladium jamaicense, Clematis crispa, Cocculus carolinus, Coleataenia longifolia, Conoclinium coelestinum, Coreopsis leavenworthii, Crataegus viridis, Cuscuta campestris, Cyperus haspan, Cyperus odoratus, Decumaria barbara, Dichanthelium commutatum, Diodia virginiana, Diospyros virginiana, Distichlis spicata, Dyschoriste humistrata, Echinochloa crus-galli, Eichhornia crassipes, Eleocharis cellulosa, Eleocharis elongata, Eleocharis engelmannii, Eleocharis intermedia, Eleocharis interstincta, Eleocharis obtusa, Eleocharis quadrangulata, Eleocharis tuberculosa, Eriocaulon compressum, Eriocaulon decangulare, Eryngium aquaticum, Eupatorium album, Fimbristylis caroliniana, Fimbristylis spadicea, Flaveria linearis, Fraxinus caroliniana, Fraxinus profunda, Funastrum clausum, Gelsemium sempervirens, Hibiscus moscheutos, Hydrocotyle bonariensis, Hydrocotyle verticillata, Hygrophila lacustris, Hymenocallis choctawensis, Hymenocallis crassifolia, Hymenocallis duvalensis, Hymenocallis duvalensis, Hymenocallis latifolia, Hymenocallis occidentalis, Hymenocallis palmeri, Hymenocallis rotata, Hypericum galioides, Hypericum hypericoides, Hypoxis curtissii, Hyptis alata, Ilex vomitoria, Ipomoea sagittata, Iris hexagona, Iris virginica, Iva frutescens, Juncus acuminatus, Juncus biflorus, Juncus coriaceus, Juncus effusus, Juncus megacephalus, Juncus roemerianus, Juniperus virginiana, Justicia angusta, Justicia ovata, Lachnanthes caroliniana, Laguncularia racemosa, Leersia hexandra, Leersia lenticularis, Lilaeopsis chinensis, Limonium carolinianum, Liquidambar styraciflua, Ludwigia octovalvis, Ludwigia repens, Ludwigia sphaerocarpa, Lythrum lineare, Magnolia virginiana, Micranthemum glomeratum, Mikania scandens, Mitchella repens, Muhlenbergia capillaris, Myrica cerifera, Myriophyllum pinnatum, Nuphar sagittifolia, Nymphaea odorata, Nymphoides aquatica, Nyssa aquatica, Nyssa sylvatica, Oplismenus hirtellus, Orontium aquaticum, Osmunda regalis, Osmundastrum cinnamomeum, Packera glabella, Pancratium maritimum, Panicum amarum, Panicum dichotomiflorum, Panicum hemitomon, Panicum virgatum, Paspalidium geminatum, Paspalum vaginatum, Peltandra virginica, Persea borbonia, Persea palustris, Persicaria glabra, Persicaria hydropiperoides, Persicaria punctata, Persicaria punctata, Persicaria setacea, Phanopyrum gymnocarpon, Phragmites australis, Phytolacca americana, Pinus elliottii, Platanthera flava, Pluchea baccharis, Pluchea odorata, Pontederia cordata, Proserpinaca pectinata, Pteridium aquilinum, Ptilimnium capillaceum, Quercus nigra, Quercus virginiana, Rhabdadenia biflora, Rhizophora mangle, Rhynchospora caduca, Rhynchospora colorata, Rhynchospora corniculata, Rhynchospora inundata, Rhynchospora macrostachya, Rhynchospora microcarpa, Rhynchospora miliacea, Rhynchospora mixta, Rhynchospora nitens, Rhynchospora tracyi, Rubus trivialis, Rumex verticillatus, Sabal minor, Sabal palmetto, Sabatia calycina, Sabatia stellaris, Saccharum giganteum, Sacciolepis striata, Sagittaria graminea, Sagittaria lancifolia, Samolus ebracteatus, Samolus valerandi, Saururus cernuus, Schinus terebinthifolius, Schoenoplectus americanus, Schoenoplectus californicus, Schoenoplectus pungens, Schoenoplectus tabernaemontani, Schoenus nigricans, Scirpus cyperinus, Scleria triglomerata, Serenoa repens, Sesbania drummondii, Sesuvium portulacastrum, Setaria magna, Sium suave, Smilax tamnoides, Solidago leavenworthii, Solidago sempervirens, Spartina alterniflora, Spartina cynosuroides, Spartina patens, Stillingia, Symphyotrichum dumosum, Symphyotrichum subulatum, Symphyotrichum tenuifolium, Taxodium distichum, Tetrapanax papyrifer, Thelypteris kunthii, Thelypteris palustris, Tiedemannia filiformis, Toxicodendron radicans, Tradescantia hirsutiflora, Triadenum walteri, Triadica sebifera, Tripsacum dactyloides, Typha angustifolia, Typha domingensis, Typha latifolia, Ulmus americana, Utricularia, Vaccinium arboreum, Verbena scabra, Verbesina occidentalis, Viburnum obovatum, Vigna luteola, Vinca major, Vitis cinerea, Vitis rotundifolia, Woodwardia areolata, Zizania aquatica, Zizaniopsis miliacea.

Use by wildlife: White-tailed deer (Mammalia: Cervidae: *Odocoileus virginianus seminolus*) feed extensively on *Crinum americanum*, which can constitute up to 40.4% (DM) of their annual diet. The plants also are grazed by domestic cattle (Mammalia: Bovidae: *Bos taurus*) and are uprooted and eaten by whooping cranes (Aves: Gruidae: *Grus americana*). The seeds are eaten by ducks (Aves: Anatidae) and frugivorous mammals (Mammalia). The foliage is damaged or eaten by eastern lubber grasshoppers (Orthoptera: Romaleidae: *Romalea microptera*), leaf-mining flies (Insecta: Diptera: Chloropidae: *Hippelates*), plant-parasitic nematodes (Nematoda: Hoplolaimidae: *Peltamigratus christiei*), weevils (Insecta: Coleoptera: Curculionidae: Baridinae), and whiteflies (Insecta: Hemiptera: Aleyrodidae: *Aleurodicus dispersus*). It also hosts several Fungi (Ascomycota: Botryosphaeriaceae: *Phyllosticta*; Massarinaceae: *Didymella curtisii*; Mycosphaerellaceae: *Pseudocercospora pancratii*). Other larval flies (Insecta: Diptera: Chironomidae: *Polypedilum epleri*) feed on the rotting leaves. The stems provide oviposition sites for the Florida applesnail (Mollusca: Gastropoda: Ampullariidae: *Pomacea paludosa*). The plants are a habitat component of the dusky seaside sparrow (Aves: Emberizidae: *Ammodramus maritimus nigrescens*), Florida panther (Mammalia: Felidae: *Puma concolor couguar*), and Florida sandhill crane (Aves: Gruidae: *Grus canadensis pratensis*).

Economic importance: food: *Crinum* species are poisonous due to their high alkaloid content and never should be eaten; **medicinal:** The leaves and rhizomes of *Crinum americanum* contain numerous alkaloids, including Augustine, Buphamisine, Crinamine, Crinidine, Crinine, Dihydrocrinidine, Flexinine, Hamayne, Haemanthamine, Lycorine, Oxocrinine, Tazettine, and Trisphaeridine. Several *Crinum* species have been used medicinally, but the medicinal value of *C. americanum* has not been evaluated specifically. The plants contain ungeremine, which is bactericidal against some fish (Vertebrata: Teleostei) pathogens; **cultivation:** The horticultural value of *Crinum americanum* was realized early on and it was listed in American and European botanic garden catalogs by the late 18th and early 19th centuries. Luther Burbank claimed to have used *C. americanum* as the pollen donor in a wide intergeneric cross with *Amaryllis belladonna* (♀) to produce a sterile amaryllis with larger petals and misshapen leaves. Although horticulturalists were skeptical of Burbank, that cross (now known as ×*Amaricrinum*) subsequently was confirmed. However, the plants do not perform well in the garden. They are grown most successfully in an acidic clay muck, using a subirrigation system. Cultivars of *Crinum americanum* include 'Carolina Beauty' and numerous interspecific hybrids, which have resulted in more adapted garden plants. The hybrids include 'Seven Sisters' (*C. americanum* × *C. bulbispermum*); 'William Herbert' (*C. scabrum* × 'Seven Sisters'); 'Ollene' (a 'Seven Sisters' backross); *C.* ×*brownei* (*C. americanum* × *C. bracteatum*); *C.* ×*digweedii* (*C. scabrum* × *C. americanum*); *C.* ×*lajolla* ['Dwarf'] (*C. americanum* × *C. moorei*); *C.* ×*parkeri* (*C. americanum* × *C. erubescens*); and several unnamed hybrids: *C. americanum* × *C. japonicum*; *C. americanum* × *C. lugardiae*; *C. americanum* × *C. macowanii*; *C. americanum* × *C. mauritianum*; *C. americanum* × *C. subcernuum*; *C. bulbispermum* × *C. americanum*; *C. carolo-schmidtii* × *C. americanum*; *C. flaccidum* × *C. americanum*; *C. paludosum* × *C. americanum*; *C. politifolium* × *C. americanum*; *C. rautanenianum* × *C. americanum*; and *C. zeylanicum* × *C. americanum*; **misc. products:** *Crinum americanum* is able to hyperaccumulate selenium (Se) and has been evaluated for use in the phytoremediation of contaminated water; **weeds:** *Crinum americanum* is not regarded as weedy; **nonindigenous species:** *Crinum americanum* has been introduced to some Gulf coastal regions of Mexico. The other North American *Crinum* species (all non-OBL) are nonindigenous.

Systematics: *Crinum* is assigned taxonomically to subfamily Amaryllidoideae, tribe Amaryllideae, and subtribe Crininae of Amaryllidaceae (Figure 2.4). Phylogenetic studies using molecular and morphological characters resolve the group as monophyletic and sister to a clade comprising *Ammocharis* and *Cybistetes* (Figures 2.4 and 2.5). Two subgenera have been proposed based on floral symmetry (subgenus *Crinum*:

actinomorphic; subgenus *Codonocrinum*: zygomorphic), but neither group is monophyletic. *Crinum americanum* occurs with other New World species in an "American clade," which originates from two African clades. It is most closely related to *C. cruentum* and *C. erubescens*, which some authors regard as synonymous. The basic chromosome number of *Crinum* is $x = 11$. *Crinum americanum* ($2n = 22$) is a diploid. *Crinum* species are known to hybridize freely, but typically yield sterile progeny. Although no natural hybrids involving *C. americanum* have been reported (it is the only indigenous North American species), it has been crossed artificially with many other *Crinum* species (>97% sterile progeny) as well as at least one species in the distantly related genus *Amaryllis*.

Distribution: *Crinum americanum* occurs along the southeastern coastal plain of the United States and extends to coastal South America.

References: Anderson & Olsen, 2015; Austin, 2004; Becker, 2017; Belden et al., 1988; Burkhalter & Wright, 1989; Carvalho & Martin, 2001; Cleall et al., 1807; Crow & Walker, 2003; Darst et al., 2002; David, 1996; Ding et al., 2017b; Eleuterius, 1972; 1980; 1990; Evans, 2008; Fennell & Van Staden, 2001; Gabriel & de la Cruz, 1974; Gettys & Moore, 2018; Godfrey & Wooten, 1979; Grace, 1993; Grant, 1983; Gunderson & Loftus, 1993; Gustafson, 1942; Harlow, 1961; Holder et al., 1980; Holmes, 2002a; Hoyer et al., 1996; Hubert & Maton, 1939; Krausse, 1783; Kwembeya et al., 2007; Labisky et al., 2003; Lehmiller, 1992; Lewis, 1990; Meerow & Snijman, 2001; 2006; Meerow et al., 2003; Mörch, 1839; Oyewo & Jacobsen, 2007; Palmer & Mazzotti, 2004; Penfound & Hathaway, 1938; Raina, 1978; Refaat et al., 2012a; 2012b; Ribeiro et al., 2011; Roberts et al., 2008; Schrader et al., 2010; Schowalter, 2018; Smith et al., 2002; Stalter & Baden, 1994; Stegmaier, 1966; Stutzenbaker, 1999; Thomas, 2005; Thompson, 1970; Turner, 1996; Watson, 1941; White & Paine, 1992; Zemskova & Sveshnikova, 1999; Zimorski et al., 2013.

5.1.3. Hymenocallis

Alligator-lily, spider-lily; Hyménocalle, lis araignée

Etymology: from the Greek *hymên* (membrane) and *kallos* (beauty) in reference to the attractive floral corona

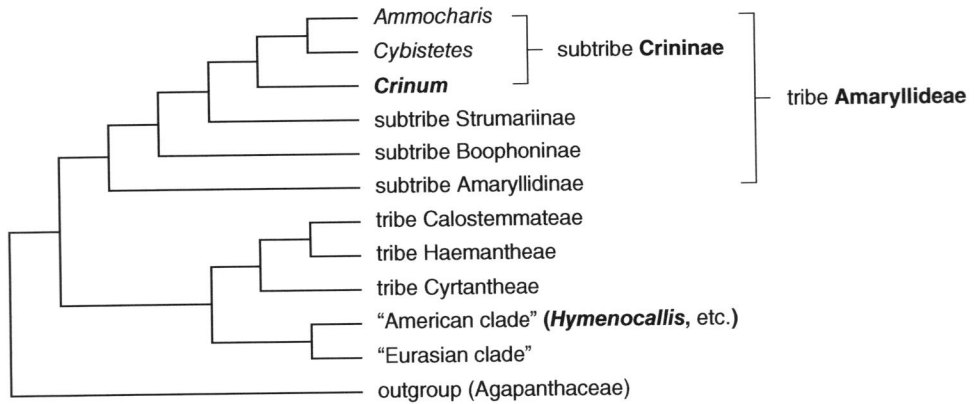

FIGURE 2.4 Tribal and subtribal relationships in Amaryllidaceae subfamily Amaryllidoideae as indicated by analyses of morphological and molecular data (modified and redrawn from Meerow & Snijman, 2006). The two OBL North American genera (*Crinum*, *Hymenocallis*) are only remotely related. Taxa containing OBL indicators are highlighted in bold.

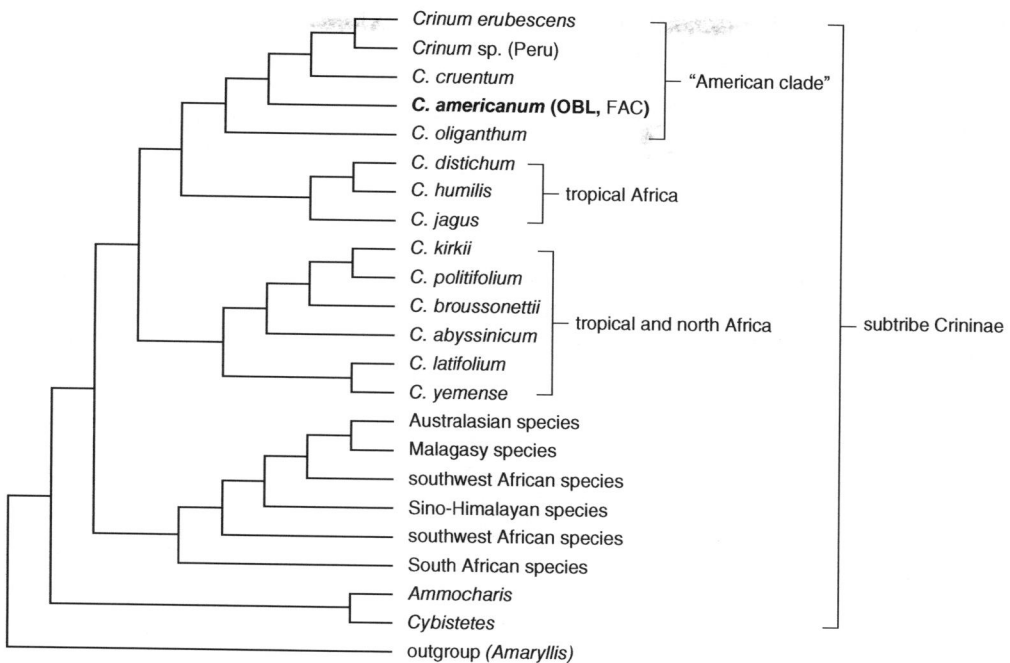

FIGURE 2.5 Phylogenetic relationships of *Crinum* species as indicated by analyses of combined nrDNA spacer and plastid *trn*L-F sequences (modified and redrawn from Meerow & Snijman, 2006). *Crinum americanum*, the only OBL North American indicator (shown in bold), is closely related to other New World species (i.e., the "American clade"), which originated from African clades.

Synonyms: *Choretis*; *Chrysiphiala* (in part); *Ismene* (in part); *Nemepiodon*; *Pancratium* (in part); *Siphotoma* (in part); *Tomodon*; *Troxistemon*

Distribution: global: New World; **North America:** southeast and south-central United States

Diversity: global: 50 species; **North America:** 18 species

Indicators (USA): obligate wetland (OBL): *Hymenocallis choctawensis, H. coronaria, H. crassifolia, H. liriosme, H. littoralis, H. occidentalis, H. palmeri, H. puntagordensis, H. rotata*; **facultative wetland (FACW):** *H. occidentalis*; **facultative (FAC):** *H. littoralis*

Habitat: brackish (coastal), freshwater; freshwater (tidal); palustrine, riverine; **pH:** 4.8–7.2; **depth:** <1 m; **life-form(s):** emergent herb

Key morphology: scapose herbs from a nonrhizomatous or rhizomatous bulb (to 8 cm), which extends into a neck (to 12 cm); leaves (2–12) sessile, deciduous or evergreen, the blade (to 12 dm) narrowly to widely liguliform or oblanceolate, coriaceous or noncoriaceous; scape (to 12.5 dm) with 2–3 bracts (to 8 cm), terminating in an umbel of 1–12 fragrant, usually sessile, showy, stellate flowers each subtended by a bract (to 6 cm); tepals (to 15 cm) green, white (sometimes green-striped or green at base) or yellow, linear, connate at base as a green tube (to 20 cm), surmounted by a funnelform or rotate corona (to 9 cm), corona white, with a small to prominent, green or yellow-green eye, the margin dentate or lacerate; stamens 6, adnate basally to corona, free (distal) filaments (to 4.5 cm) green to white, pollen golden or yellow; ovary (to 3 cm) inferior, oblong, ovoid, pyramidal, pyriform, or subglobose, ovules 1–9 per locule, style (to 26 cm) green or white, filiform, exserted beyond stamens, the stigma capitate; capsules large (to 6 cm) ellipsoid, elongate, or subglobose, leathery; seeds large (to 4.3 cm) fleshy, green, elongate or obovoid

Life history: duration: perennial (bulbs, rhizomes); **asexual reproduction:** bulb offshoots, rhizomes; **pollination:** hummingbird, insect; **sexual condition:** hermaphroditic; **fruit:** capsules (common); **local dispersal:** seeds (water); **long-distance dispersal:** seeds (fish?, water)

Imperilment: 1. *Hymenocallis choctawensis* [G3/G4]; S2 (MS); S3 (GA); **2.** *H. coronaria* [G3]; S2 (AL, GA, SC); **3.** *H. crassifolia* [GNR]; **4.** *H. liriosme* [G4]; **5.** *H. littoralis* [unranked]; **6.** *H. occidentalis* [G4]; SH (NC); S3 (IN); **7.** *H. palmeri* [G3]; **8.** *H. puntagordensis* [G1Q]; **9.** *H. rotata* [G2/G3]

Ecology: general: All the North American *Hymenocallis* species are associated with wet habitats but just over half (10/18) are designated as wetland indicators (50% as OBL). Most are palustrine species commonly found in swamps or along watercourses; one species (*H. coronaria*) is riverine. All perennate from a bulb, which can develop offshoots in some species. Several species also produce rhizomes and multiply vegetatively in that way. Surprisingly, the floral biology of these showy plants has not been documented adequately (possibly because many presume it to be the same in all the species). The flowers generally are protandrous and at least some are proven to be self-incompatible. Nearly all the species appear to be outcrossed. The long corolla tubes are filled with nectar, which attracts long-billed birds (e.g., Aves: Trochilidae) and insects with long proboscises (e.g., Insecta: Lepidoptera) as pollinators. Many of the highly fragrant flowers open nocturnally and are pollinated by hawkmoths (Insecta: Lepidoptera: Sphingidae). The seed ecology also has been poorly studied.

Existing information indicates that they lack dormancy unless they remain in cold water. The large, fleshy seeds are known to be dispersed by water in at least some species. Otherwise, they often are distributed close to the maternal plant as the inflorescence collapses, which brings the fruit and its contents to rest on the substrate. There is circumstantial evidence that some seed dispersal also occurs endozoically by frugivorous fish (Vertebrata: Teleostei).

5.1.3.1. *Hymenocallis choctawensis* **Traub** inhabits floodplains, slopes, swamps (alluvial), woodlands (low), and the margins of bayous, rivers, and streams (alluvial, blackwater, spring-fed) at elevations to 49 m. Exposures can range from full sunlight to shade. The substrates include clay, muck, mud, and silt. The plants flower from April to June. The flowers presumably are insect pollinated, but there have been no detailed studies on the floral biology or seed ecology. The bulbs are rhizomatous, which can lead to the development of dense clonal stands. **Reported associates:** *Acer rubrum, Acmella repens, Alternanthera philoxeroides, Amorpha fruticosa, Ampelopsis arborea, Arnoglossum diversifolium, Arundinaria gigantea, Asclepias perennis, Betula nigra, Brunnichia ovata, Campsis radicans, Carex crus-corvi, Carex louisianica, Carex rosea, Carex stipata, Carpinus caroliniana, Carya aquatica, Celtis laevigata, Cephalanthus occidentalis, Clematis, Cornus asperifolia, Cornus foemina, Crataegus viridis, Crinum americanum, Cyperus, Cyrilla racemiflora, Ditrysinia fruticosa, Fagus grandifolia, Fimbristylis, Fraxinus pennsylvanica, Fraxinus profunda, Gleditsia aquatica, Hygrophila lacustris, Hypericum galioides, Hypoxis curtissii, Ilex decidua, Ilex opaca, Itea virginica, Juncus, Justicia ovata, Liquidambar styraciflua, Lobelia, Lonicera japonica, Lygodium japonicum, Magnolia ashei, Magnolia grandiflora, Magnolia virginiana, Micranthemum umbrosum, Myosotis macrosperma, Myriophyllum, Nyssa aquatica, Orontium aquaticum, Osmunda regalis, Ostrya virginiana, Panicum, Persicaria, Pinus elliottii, Planera aquatica, Pontederia cordata, Quercus alba, Quercus laurifolia, Quercus lyrata, Quercus nigra, Rhapidophyllum hystrix, Rhynchospora corniculata, Sabal minor, Sagittaria graminea, Samolus valerandi, Saururus cernuus, Sideroxylon thornei, Smilax, Symplocos tinctoria, Taxodium distichum, Toxicodendron radicans, Ulmus americana, Vaccinium corymbosum, Viburnum nudum, Viburnum obovatum, Viola, Vitis, Woodwardia areolata, Zizaniopsis miliacea.*

5.1.3.2. *Hymenocallis coronaria* **(J. Le Conte) Kunth** occurs entirely within river and stream shoals (rocky) at elevations to 150 m. Exposures can range from full sunlight to shade. The substrates are crevices within hard boulder and cobble bottoms. Unlike its primarily palustrine congeners, this truly is a riverine species, which inhabits flowing water, entirely within the confines of the river channel where few associated species co-occur. Aside from flood periods, it normally grows in waters less than 1 m deep. The plants are dormant over winter and resume growth in late March. Most flowering occurs from late April to June but continues intermittently until September. The flowers open in mid to late afternoon, remain open during night, and wither within 1–2 days. During the day the flowers are pollinated primarily by bees (Insecta: Hymenoptera) and hummingbirds (Aves: Trochilidae); at night they are pollinated by hawkmoths (Insecta: Lepidoptera) (see *Use by wildlife*). The center of the corona has an ultraviolet reflective region that possibly serves as a nocturnal honey guide. The flowers are outcrossed, resulting in estimated levels of heterozygosity (determined from microsatellite markers) between 0.017 and 0.57. However, genetic differentiation occurs within some of the larger populations. Fertility is high, averaging 4–6 seeds/fruit. The seeds are photosynthetic, sink when shed, and require high water velocities to entrain them adequately for dispersal. Models indicate that subaqueous transport distances from 10.8 to 18 km are theoretically possible. The seeds germinate without pretreatment within 3–4 weeks and develop into bulblets with leaves and contractile roots, which penetrate fissures in the rocky shoal substrata. The bulbs lack rhizomes but asexual ramets (1.3–1.4 m) can develop by bulb offshoots. Seedling densities from 0.009 to 0.256 m^{-2} have been observed. Although the seeds are dispersed by water, population genetic studies have found no evidence of unidirectional (i.e., downstream) gene flow, which indicates that upstream gene flow also occurs in these habitats. This dispersal pattern is thought to result at least in part from the endozoic transport of seeds ingested by fish (e.g., Teleostei: Ictaluridae: *Ictalurus punctatus*). Analyses of cpDNA haplotypes and microsatellite markers indicate that although much of the total genetic diversity resides within populations (~73%), the variation is highly partitioned among regions, especially on either side of the former "Suwannee Strait." Proper water flow rates are critical to plant persistence. High flow rates ≥283 m^3 s^{-1} can damage plants and disrupt flowering by washing away pollen, whereas those <51.0 m^3 s^{-1} can affect seedling survival and result in inadequate seed dispersal. **Reported associates:** *Cladophora, Justicia americana, Lemna, Podostemum ceratophyllum.*

5.1.3.3. *Hymenocallis crassifolia* **Herb.** inhabits bayous, bogs, bottoms (stream), canals (drainage), ditchbanks, ditches (drainage, roadside), floodplains (river), marshes (brackish, river), meadows (low), rice fields (abandoned), seeps (low elevation), shores (tidal), sloughs, swamps (alluvial, tidal) and the margins of rivers, salt marshes, and streams (blackwater, tidewater) at elevations to 5 m. Exposures can range from full sunlight to shade. The substrates have been described as clay, clayey muck, muck, mucky sand, and mud (black); they can contain from 0.15% to 27% organic matter. The plants generally grow on saturated substrates with up to 30 cm of standing water. They can tolerate periodic exposures to brackish conditions (pH: 4.8–7.2; salinity: 1–15‰; Cl$^-$: 1.8–7.2 ppm). Flowering extends from April to June. Presumably, the flowers are pollinated by insects (Insecta); however, details on the floral or seed ecology are lacking. The bulbs are rhizomatous and can produce clonal ramets. Recorded biomass (DM) productivity varies from 203 to 248 kg ha^{-1}. **Reported associates:** *Acer rubrum, Alnus serrulata, Alternanthera philoxeroides, Amaranthus cannabinus, Ampelopsis arborea, Apios americana, Arisaema triphyllum, Arundo donax, Azolla filiculoides, Baccharis angustifolia, Baccharis halimifolia,*

Bidens laevis, Bolboschoenus robustus, Carex, Chamaecrista fasciculata, Chasmanthium latifolium, Chasmanthium laxum, Chenopodium album, Cicuta maculata, Cinna arundinacea, Cladium jamaicense, Crataegus aestivalis, Crinum americanum, Cyperus, Dichanthelium dichotomum, Echinochloa crus-galli, Eleocharis albida, Eleocharis engelmannii, Eleocharis quadrangulata, Eryngium aquaticum, Fimbristylis spadicea, Fraxinus caroliniana, Galium pilosum, Hibiscus moscheutos, Hydrocotyle umbellata, Hydrocotyle verticillata, Hypoxis curtissii, Ilex vomitoria, Impatiens capensis, Ipomoea sagittata, Iris virginica, Iva, Juncus acuminatus, Juncus biflorus, Juncus coriaceus, Juncus dichotomus, Juncus effusus, Juncus polycephalus, Juncus roemerianus, Justicia ovata, Kosteletzkya pentacarpos, Lemna minor, Lilaeopsis chinensis, Liquidambar styraciflua, Lobelia cardinalis, Lonicera japonica, Ludwigia, Lythrum lineare, Myriophyllum aquaticum, Nuphar sagittifolia, Nyssa aquatica, Nyssa biflora, Nyssa ogeche, Orontium aquaticum, Osmunda, Panicum virgatum, Peltandra virginica, Persea borbonia, Persicaria hydropiperoides, Phragmites australis, Physostegia virginiana, Planera aquatica, Pluchea, Pontederia cordata, Ptilimnium capillaceum, Quercus laurifolia, Quercus lyrata, Quercus nigra, Rhexia mariana, Rhynchosia tomentosa, Rhynchospora colorata, Rubus, Rumex verticillatus, Ruppia maritima, Sabal minor, Sabal palmetto, Saccharum giganteum, Sagittaria graminea, Sagittaria lancifolia, Salix, Salvinia minima, Saururus cernuus, Schoenoplectus pungens, Schoenoplectus tabernaemontani, Scirpus cyperinus, Serenoa repens, Sesbania punicea, Setaria, Setaria magna, Solidago sempervirens, Spartina alterniflora, Spartina cynosuroides, Symphyotrichum tenuifolium, Taxodium ascendens, Taxodium distichum, Tephrosia spicata, Triadica sebifera, Tripsacum dactyloides, Typha angustifolia, Typha latifolia, Ulmus americana, Verbesina occidentalis, Viburnum, Vigna luteola, Zizania aquatica, Zizaniopsis miliacea.

5.1.3.4. *Hymenocallis liriosme* (Raf.) Shinners

occurs in backwaters (lowland), bogs (open), bottomlands (hardwood, river), canals (roadside), channels, depressions (prairie, semipermanently flooded), ditches (drainage, irrigation, railroad, roadside), flatwoods (coastal), floodplains, marshes (open), meadows, ponds (ephemeral, shallow), prairies (wet), savannahs, sloughs, swales (marshy, roadside), swamps (cypress, river), woodlands (hardwood, low, pine, shrubby), and along the margins of ponds and streams at elevations to 110 m. Exposures from full sunlight to shade are tolerated. The substrates have been characterized as alluvium (clayey), clay (heavy, Schriever), clay loam, gumbo (heavy), humus, loam (bottomland, Nahatche), mud, sand, sandy clay, sandy loam (fine, Woodtell), sandy clay, sandy silt, silty clay, and silty loam. The plants can be found on saturated substrates or in shallow standing water. The fragrant flowers are present from March to May. There have been no detailed studies on the floral or seed ecology. The bulbs lack rhizomes but it is uncertain whether clonal offshoots are produced. **Reported associates:** *Acer negundo, Amsonia tabernaemontana, Anagallis minima, Andropogon virginicus, Azolla filiculoides, Baccharis*

halimifolia, Boltonia diffusa, Callitriche heterophylla, Carex crus-corvi, Carex joorii, Carex tetrastachya, Cephalanthus occidentalis, Chara, Crataegus opaca, Crataegus viridis, Cynodon, Cyperus, Dichanthelium dichotomum, Diospyros virginiana, Echinochloa colona, Echinochloa crus-galli, Echinodorus berteroi, Eleocharis palustris, Eleocharis quadrangulata, Eleocharis tuberculosa, Eryngium leavenworthii, Eupatorium semiserratum, Fuirena breviseta, Gratiola brevifolia, Hibiscus moscheutos, Hydrocotyle, Hydrolea ovata, Ilex decidua, Iris brevicaulis, Iris virginica, Iva annua, Juncus effusus, Juncus nodatus, Justicia ovata, Leersia hexandra, Lemna, Limnosciadium pumilum, Liquidambar styraciflua, Lonicera, Ludwigia decurrens, Ludwigia linearis, Ludwigia peploides, Ludwigia pilosa, Lycopus americanus, Marsilea, Mnesithea rugosa, Myriophyllum pinnatum, Najas guadalupensis, Nyssa, Panicum hemitomon, Panicum virgatum, Paspalidium geminatum, Paspalum praecox, Persicaria hydropiperoides, Persicaria lapathifolia, Phragmites australis, Phyla lanceolata, Physostegia intermedia, Pinus palustris, Pinus taeda, Pluchea foetida, Pontederia cordata, Proserpinaca palustris, Proserpinaca pectinata, Quercus lyrata, Quercus macrocarpa, Quercus nigra, Ranunculus laxicaulis, Ranunculus pusillus, Rhexia mariana, Rhynchospora corniculata, Rhynchospora globularis, Rhynchospora indianolensis, Rubus trivialis, Rumex altissimus, Rumex verticillatus, Sabal minor, Sagittaria graminea, Sagittaria lancifolia, Sagittaria papillosa, Salix nigra, Saururus cernuus, Schizachyrium scoparium, Schoenoplectus tabernaemontani, Scleria baldwinii, Sesbania drummondii, Smilax smallii, Symphyotrichum subulatum, Taxodium distichum, Teucrium canadense, Thalia dealbata, Triadica sebifera, Tripsacum dactyloides, Typha domingensis, Vaccinium, Vernonia gigantea, Vitis rotundifolia, Xyris laxifolia.

5.1.3.5. *Hymenocallis littoralis* (Jacq.) Salisb.

is a nonindigenous inhabitant of meadows at low elevations (to at least 370 m elsewhere). The plants can tolerate exposures from full sunlight to shade. The flowers are dichogamous (protandrous), fragrant, shed their pollen mainly in clumps of 2–10 grains, and can provide up to 220 µl of nectar. At least 26 different volatile compounds have been isolated from the flowers, including aromatic compounds, chain hydrocarbons, diterpene alcohols, monoterpenes, saturated and unsaturated fatty acids/esters, and sesquiterpenes. These features have been interpreted as adaptations to pollination by hawkmoths (Insecta: Lepidoptera: Sphingidae). Although the plants persist year-round in their indigenous tropical environments, they have displayed an altered life cycle when grown in the United States. Plants cultivated in Arizona grew from early April to late October and became dormant from early November to late March; flowering occurred from July to August. The bulbs lack rhizomes and no mechanism of asexual reproduction has been reported. The plants have been grown optimally using a combined application of 400 kg N ha^{-1} and 200 kg P$_2$O$_5$ ha^{-1}. Virtually no information exists regarding the ecological interactions of this species in North America. Even elsewhere, most of the literature has focused on the medicinal properties of the plants (see *Economic importance: medicinal* below). **Reported associates (North America):** *Paspalum urvillei.*

5.1.3.6. *Hymenocallis occidentalis* (J. Le Conte) Kunth.

inhabits bottomlands (forest, river, stream), deltas, depressions (bottomland), ditchbanks, ditches (roadside), flatwoods, floodplains (alluvial, forested, hardwood), hammocks (hardwood, mesic), hillsides, marshes (floating, freshwater), meadows (marshy, wet), openings (hardwood), roadsides (low, wet), slopes (ravine, rocky), swamps (cypress), woodlands (hardwood, low, mesic, rich, swampy), and the margins of ponds (beaver), reservoirs, and rivers (freshwater, tidal) at elevations to 519 m. Exposures can vary from full sunlight to deep shade. The reported substrates include alluvium, clay, gravel, humus, loam, muck, sand, sandy clay, sandy loam, sandy silt, and silty loam (McGary). The plants occur on saturated substrates or in shallow, fresh to brackish standing waters. The flowers have been observed from April to September. They are protandrous, self-incompatible, heavily scented, and outcrossed. Anthesis is nocturnal, with the flowers opening about an hour before sunset and remaining open for up to 48 hr. The leaves wither immediately afterward. The seeds are nondormant and are capable of germinating within 3–4 weeks of being shed. During this "after-ripening" period, the embryo undergoes continuous development. The seeds will not germinate while immersed in cold water but can achieve rates from 51% to 54% when transferred to the air (in dark or light) or up to 86% when incubated under oxygen-deficient (i.e., CO_2) but lighted conditions. The bulbs begin to develop in about 3 weeks subsequent to germination. They lack rhizomes or other apparent means of asexual reproduction, resulting in plants (genets) that often are widely spaced (50–100 m). Mechanisms of seed dispersal have not been elucidated. **Reported associates:** *Acer negundo, Acer rubrum, Acer saccharinum, Acer saccharum, Actaea pachypoda, Alternanthera philoxeroides, Amorpha fruticosa, Arisaema dracontium, Arisaema triphyllum, Aristolochia serpentaria, Arundinaria gigantea, Asarum canadense, Asclepias exaltata, Asimina triloba, Asimina triloba, Betula nigra, Boehmeria cylindrica, Boltonia asteroides, Botrychium biternatum, Carex bushii, Carex buxbaumii, Carex caroliniana, Carex cherokeensis, Carex crus-corvi, Carex flaccosperma, Carex grayi, Carex hyalinolepis, Carex intumescens, Carex joorii, Carex laxiflora, Carex lupuliformis, Carex lupulina, Carex muskingumensis, Carex socialis, Carex squarrosa, Carex tribuloides, Carpinus caroliniana, Carya cordiformis, Carya glabra, Carya illinoinensis, Carya laciniosa, Carya myristiciformis, Carya ovata, Carya texana, Celtis laevigata, Cercis canadensis, Chasmanthium latifolium, Chasmanthium laxum, Cinna arundinacea, Coleataenia anceps, Commelina diffusa, Commelina virginica, Coreopsis major, Cornus drummondii, Cornus foemina, Cynoglossum virginianum, Cyperus, Cystopteris fragilis, Desmanthus illinoensis, Desmodium pauciflorum, Dichanthelium commutatum, Diodia virginiana, Dioscorea villosa, Echinochloa walteri, Echinodorus cordifolius, Eclipta prostrata, Eleocharis tenuis, Elephantopus carolinianus, Eryngium prostratum, Fagus grandifolia, Fraxinus pennsylvanica, Galearis spectabilis, Galium circaezans, Geranium maculatum, Geum, Gillenia stipulata, Glyceria septentrionalis, Halesia carolina, Halesia diptera, Hibiscus laevis, Hydrolea uniflora, Hypericum prolificum, Ilex decidua, Iresine rhizomatos, Isoetes melanopoda, Juniperus virginiana, Justicia ovata, Leersia lenticularis, Leersia oryzoides, Leersia virginica, Lilium superbum, Liquidambar styraciflua, Liriodendron tulipifera, Lobelia cardinalis, Ludwigia glandulosa, Luzula acuminata, Lyonia ligustrina, Magnolia, Modiola caroliniana, Muhlenbergia schreberi, Myriophyllum sibiricum, Nyssa sylvatica, Obolaria virginica, Onoclea sensibilis, Oplismenus hirtellus, Oxypolis rigidior, Packera glabella, Paspalum urvillei, Persicaria hydropiperoides, Persicaria pensylvanica, Persicaria punctata, Persicaria setacea, Persicaria virginiana, Phalaris, Phaseolus polystachios, Phlox maculata, Pinus strobus, Platanthera peramoena, Pluchea camphorata, Polygonatum biflorum, Polystichum acrostichoides, Pontederia cordata, Populus heterophylla, Proserpinaca palustris, Prunus mexicana, Pteridium aquilinum, Ptilimnium capillaceum, Ptilimnium costatum, Pycnanthemum flexuosum, Quercus alba, Quercus bicolor, Quercus laurifolia, Quercus lyrata, Quercus michauxii, Quercus nigra, Quercus pagoda, Quercus palustris, Quercus phellos, Quercus shumardii, Quercus similis, Quercus stellata, Rhus copallinum, Rhynchospora corniculata, Rubus argutus, Ruellia strepens, Saccharum giganteum, Sacciolepis striata, Sagittaria lancifolia, Samolus valerandi, Sanicula canadensis, Saururus cernuus, Scirpus cyperinus, Senna obtusifolia, Sideroxylon lycioides, Sium suave, Smallanthus uvedalia, Solidago caesia, Sorghum halepense, Spermacoce glabra, Spigelia marilandica, Spiranthes ovalis, Symphoricarpos orbiculatus, Taxodium distichum, Thalia dealbata, Thelypteris noveboracensis, Toxicodendron radicans, Tragia cordata, Trepocarpus aethusae, Trillium recurvatum, Tripsacum dactyloides, Ulmus alata, Ulmus rubra, Uvularia sessilifolia, Viola pubescens, Viola sororia, Viola striata, Vitis rotundifolia, Yeatesia viridiflora.*

5.1.3.7. *Hymenocallis palmeri* S. Watson

grows in ditches (drainage, roadside), flats (sand), flatwoods (coastal, open, pine), glades, hammocks (mesic), marshes (oligohaline), prairies (acidic, disturbed, marl, wet), roadsides (grassy, low), sloughs, swamps (cypress), and along the margins of canals (drainage) and ponds (grassy) at elevations to 5 m. Unlike its congeners, this species occurs in exposures that consistently receive full sunlight. The plants are not cold tolerant and are severely damaged by frost. They inhabit seasonally wet sites having hydroperiods (water at or above surface) of 150–344 days. The substrates mostly are acidic and include muck, mud, sand, and sandy peat. The plants flower from May to September. Information on the reproductive ecology is mainly anecdotal, but most accounts indicate that the fragrant, showy flowers with their long floral tube are pollinated by the long-tongued hawk moths (Insecta: Lepidoptera: Sphingidae). The peduncles collapse as the heavy fruits mature, bringing the fruits and their seed contents to the ground, the capsules sometimes bursting open on impact. The seeds probably are dispersed locally by water. The bulbs lack rhizomes and have no apparent means of asexual reproduction. Leaf tissue analysis has indicated that the plants are likely limited entirely by nitrogen and not phosphorous. Productivity (biomass) estimates range from 0.17 to 0.61 g DM m^{-1} day^{-1}. **Reported associates:** *Acrostichum aureum, Aeschynomene pratensis, Aristida purpurascens, Bacopa caroliniana, Bolboschoenus robustus, Cassytha filiformis, Centella asiatica, Cladium jamaicense, Coleataenia tenera, Crinum americanum, Dichanthelium aciculare, Dichanthelium ensifolium, Diodia, Eleocharis cellulosa, Eleocharis elongata, Elytraria caroliniensis, Eragrostis elliotti, Fuirena scirpoidea, Hymenocallis puntagordensis, Hyptis alata, Ipomoea sagittata, Juncus roemerianus, Justicia angusta, Leersia hexandra, Ludwigia microcarpa, Muhlenbergia sericea, Panicum hemitomon, Panicum virgatum, Paspalidium geminatum, Paspalum monostachyum, Peltandra virginica, Persicaria, Phyla nodiflora, Pluchea baccharis, Pontederia cordata, Proserpinaca pectinata, Quercus laurifolia, Rhynchospora colorata, Rhynchospora divergens, Rhynchospora inundata, Rhynchospora microcarpa, Rhynchospora tracyi, Sagittaria lancifolia, Schizachyrium rhizomatum, Schoenus nigricans, Serenoa repens, Solidago stricta, Spartina bakeri, Symphyotrichum dumosum, Symphyotrichum tenuifolium, Taxodium ascendens, Typha domingensis, Utricularia purpurea.*

5.1.3.8. ***Hymenocallis puntagordensis* Traub** is known only from the type locality, which includes disturbed sites such as ditches (drainage), roadsides, and the margins of pine flatwoods at low elevations (<5 m). Virtually nothing is known about the life history of this species, which some authors do not distinguish from *H. latifolia* (see *Systematics* below). The plants have evergreen leaves and occur in exposures of full sunlight to shade. Flowering and fruiting occur from August to September. The bulbs lack rhizomes and have no apparent means of vegetative reproduction. **Reported associates:** *Aristida gyrans, Conyza canadensis, Cyperus ligularis, Fuirena, Heteropogon contortus, Hymenocallis palmeri, Ludwigia octovalvis, Myrica cerifera, Paspalum notatum, Paspalum urvillei, Pinus elliottii, Pontederia cordata, Rhynchospora latifolia, Sabal palmetto, Sagittaria lancifolia, Schinus terebinthifolius.*

5.1.3.9. ***Hymenocallis rotata* (Ker Gawl.) Herb.** occurs along backwaters (rivers), ditches, floating islands, floodplain swamps, marshes, and along the margins of rivers, sinkholes (limestone), spring runs, and streams (blackwater) at low elevations (<5 m). Exposures can vary from full sunlight to shade. The plants grow in shallow standing water (e.g., 46 cm deep) or on saturated substrates that have been described as sand, sandy clay, and sandy muck. Flowering extends from April to June. The bulbs are rhizomatous, which serves as a means of vegetative reproduction. Other life history information for this species is scarce. **Reported associates:** *Acer rubrum, Ampelaster carolinianus, Ampelopsis arborea, Asclepias perennis, Berchemia scandens, Boehmeria cylindrica, Carex comosa, Carex gigantea, Carex stipata, Carya aquatica, Celtis laevigata, Centella asiatica, Cephalanthus occidentalis, Ceratophyllum demersum, Chamaecyparis thyoides, Chara zeylanica, Chasmanthium nitidum, Cicuta maculata, Cladium jamaicense, Clematis crispa, Cocculus carolinus,* *Coleataenia longifolia, Colocasia esculenta, Cornus foemina, Crinum americanum, Decumaria barbara, Dichanthelium commutatum, Diodia virginiana, Diospyros virginiana, Eichhornia crassipes, Fraxinus caroliniana, Fraxinus profunda, Gelsemium sempervirens, Gordonia lasianthus, Hibiscus coccineus, Hydrilla verticillata, Hydrocotyle verticillata, Hymenocallis duvalensis, Hypericum hypericoides, Ilex cassine, Ilex vomitoria, Iris hexagona, Juniperus virginiana, Justicia ovata, Lactuca floridana, Lemna valdiviana, Liquidambar styraciflua, Lobelia cardinalis, Ludwigia repens, Magnolia virginiana, Mitchella repens, Myrica cerifera, Myriophyllum heterophyllum, Myriophyllum pinnatum, Najas guadalupensis, Nasturtium microphyllum, Nasturtium officinale, Nyssa aquatica, Oplismenus hirtellus, Osmunda regalis, Packera glabella, Panicum virgatum, Passiflora lutea, Peltandra virginica, Persea palustris, Persicaria hydropiperoides, Persicaria punctata, Phanopyrum gymnocarpon, Pinus serotina, Pistia stratiotes, Platanthera flava, Pluchea odorata, Polystichum acrostichoides, Ponthieva racemosa, Rhapidophyllum hystrix, Rhynchospora corniculata, Rhynchospora miliacea, Rhynchospora mixta, Rubus trivialis, Rumex verticillatus, Sabal palmetto, Sabatia calycina, Sageretia minutiflora, Sagittaria filiformis, Sagittaria graminea, Sagittaria kurziana, Sagittaria lancifolia, Sapindus saponaria, Saururus cernuus, Scleria triglomerata, Serenoa repens, Smilax laurifolia, Smilax tamnoides, Solidago sempervirens, Spirodela polyrhiza, Taxodium ascendens, Taxodium distichum, Thelypteris kunthii, Toxicodendron radicans, Ulmus americana, Vallisneria americana, Vernonia gigantea, Viburnum obovatum, Vitis cinerea, Wolffia columbiana, Woodwardia areolata, Zizania aquatica.*

Use by wildlife: *Hymenocallis coronaria* hosts numerous macroinvertebrates, provides shelter for juvenile waterfowl (Aves: Anatidae), and offers habitat for feeding fish (Vertebrata: Teleostei). The leaves are browsed by deer (Mammalia: Cervidae: *Odocoileus*) and the seeds eaten by raccoons (Mammalia: Procyonidae: *Procyon lotor*). The foliage and seeds are eaten or severely damaged by larval cutworms (Insecta: Lepidoptera: Noctuidae: *Xanthopastis timais*). Redwinged blackbirds (Aves: Icteridae: *Agelaius phoeniceus*) make their nests in stands of *H. coronaria* and are thought to transfer pollen inadvertently in the process. The flowers of *H. coronaria* are visited by hummingbirds (Aves: Trochilidae: *Archilochus colubris*) and various insects (Insecta), including bees and wasps (Hymenoptera: Apidae: *Apis mellifera*; *Bombus impatiens, Xylocopa virginica*; Crabronidae: *Microbembex*; Halictidae: *Augochlora pura, Lasioglossum*; Vespidae: *Polistes*), butterflies and moths (Lepidoptera: Crambidae: *Desmia funeralis*; Hesperiidae: *Epargyreus clarus*; Noctuidae: *Heliothis virescens, Megalographa biloba*; Nymphalidae: *Epargyreus clarus flora, Libytheana carinenta*; Papilionidae: *Battus philenor, Papilio glaucus*; Sphingidae: *Dolba hyloeus, Eumorpha pandorus, Hemaris thysbe, Manduca rustica, M. sexta, Paratrea plebeja*), and flies (Diptera: Dolichopodidae; Syrphidae: *Eristalis, Syrphus torvus, Toxomerus*; several of these have been demonstrated to be pollen carriers, with bees and hummingbirds serving that role most often. *Hymenocallis liriosme* is eaten by apple snails (Mollusca: Gastropoda: Ampullariidae: *Pomacea* *insularum*). The flowers and leaves of *Hymenocallis occidentalis* are eaten by deer (Mammalia: Cervidae: *Odocoileus*). It is a host of larval cutworms (Insecta: Lepidoptera: Noctuidae: *Xanthopastis timais*). The nocturnal flowers are visited by flies (Insecta: Diptera) but are pollinated mostly by hawkmoths and moths (Insecta: Lepidoptera: Erebidae: *Halysidota tessellaris*; Noctuidae: *Ctenoplusia oxygramma*; Sphingidae: *Dolba hyloeus, Manduca rustica*). *Hymenocallis palmeri* is eaten to a lesser extent by white-tailed deer (Mammalia: Cervidae: *Odocoileus virginianus seminolus*). *Hymenocallis rotata* is a host of larval moths (Insecta: Lepidoptera: Noctuidae: *Xanthopastis timais*). Several *Hymenocallis* species are hosts to Cercosporoid leaf-spot Fungi (Ascomycota: Mycosphaerellaceae), including *Hymenocallis coronaria, H. crassifolia*, and *H. littoralis* (*Pseudocercospora pancratii*), and *H. crassifolia* (*Zasmidium hymenocallidis*).

Economic importance: food: *Hymenocallis* species never should be eaten because of their high alkaloid content; **medicinal:** *Hymenocallis littoralis* has been used widely in folk treatments as an emetic, for wound healing, and for treating human cancer. It contains numerous potentially bioactive compounds, including lycorine alkaloids (demethylmaritidine, 5,6-dihydrobicolorine, haemanthamine, hippeastrine, homolycorine, hymenolitatine, littoraline, lycoramine, lycorenine, lycorine, macronine, *O*-methyllycorenine, pretazettine, tazettine, vittatine) and a lignan (secoisolariciresinol). Littoraline has been shown to be antineoplastic, antiviral (against the herpes simplex virus), and inhibitory to HIV reverse transcriptase activity. Haemanthamine has exhibited potent *in vitro* cytotoxic activity. The bulbs produce pancratistatin, a phenanthridone that is effectively cytotoxic against murine P-388 lymphocytic leukemia, M-5076 ovary sarcoma, melanoma, and brain, colon, liver, lung, and renal cancer lines; it also exhibits strong RNA antiviral activity. The plants additionally contain narciclasine and 7-deoxynarciclasine, which also are antineoplastic. Plant extracts (petroleum ether) are antibacterial against Gram-positive *Staphylococcus aureus* and Gram-negative *Pseudomonas aeruginosa*; ethanolic floral extracts are anti-inflammatory. *Hymenocallis rotata* contains the alkaloids 3-epimacronine, alkaloid-13, demethylmaritidine, galanthamine, haemanthamine, hipeastrine, homolycorine, ismine, lycoramine, lycorine, *N*-demethyllycoramine, *N*-demethylgalanthamine, pretazettine, tazettine, and vittatine; **cultivation:** Several *Hymenocallis* species are cultivated as ornamental garden plants. The most widely distributed species commercially is *H. occidentalis*, which is grown in gardens throughout the southern United States. *Hymenocallis liriosme* is also popular as a southern garden ornamental. *Hymenocallis rotata* has been in cultivation since the 17th century. *Hymenocallis* cultivars include 'Ephemeral Beauty' (*H. coronaria*) and 'Variegata' (*H. littoralis*); **misc. products:** *Hymenocallis littoralis* has been planted in constructed wetlands used to process seafood wastewater; **weeds:** none; **nonindigenous species:** *Hymenocallis littoralis* was first reported in the United States from Florida in 1957.

Systematics: As currently recognized, *Hymenocallis* species resolve within a group of Amaryllidaceae referred to as the "American clade" (Figure 2.4) [Note: This is a different group than that by the same name mentioned above for *Crinum*]. Although several species (e.g., *H. coronaria, H. crassifolia, H. liriosme, H. littoralis, H. occidentalis, H. rotata*) once had been included in *Pancratium*, that genus is now known to be distantly related within a sister group known as the "Eurasian clade" (Figure 2.4). The most recent evidence indicates that *Hymenocallis* is monophyletic, and a closely related sister group of *Ismene*; with *Leptochiton* and *Pamianthe*, they comprise the tribe Hymenocallidae. *Ismene* possibly is not monophyletic and further study conceivably could warrant the transfer of some species to one of the other genera in Hymenocallidae. *Hymenocallis rotata* was assigned to *Ismene* formerly, but clearly belongs within *Hymenocallis*, where it is closely related to the non-OBL *H. godfreyi* and *H. tridentata* (Figure 2.6). These three species were presumed to be related by their similar morphology and shared chromosome number ($2n = 48$). However, many of the species are similar morphologically and the taxonomic status of some has remained contentious. In particular, *H. puntagordensis* has been treated as a variety (*H. latifolia* var. *puntagordensis*) of the morphologically similar, wider ranging *H. latifolia*. Although the two taxa have distinct ISSR profiles (Figure 2.6), these genetic markers can be problematic when used to ascertain interspecific relationships, especially when only single exemplars are evaluated. A further evaluation of their distinctness seems necessary. Several interspecific relationships also remain unreconciled due to discordant results from analyses of morphological data compared to molecular data. Notably, the close relationship of *H. palmeri* and *H. henryae* indicated by their ecological and morphological similarity is not supported by ISSR data (Figure 2.7). Also, the morphologically similar and presumably closely related *H. coronaria* and *H. liriosme* share a similar pattern of coronal coloration (yellow) but do not associate as sister species by analysis of ISSR data (Figure 2.7). The placement of *H. occidentalis* also varies widely in the different analyses. The original basic chromosome number of *Hymenocallis* is postulated as $x = 11$ but has been altered secondarily given that the extant species have somatic counts of $2n = 40$ or higher and include the karyotypically distinct $x = 11$ chromosomes. The compound base number (probably $x = 23$) likely originated via allopolyploidy. Virtually all the OBL North American species are tetraploids based on the original base number and its modification. The different chromosome races observed in *Hymenocallis* are believed to have arisen by centric fusion (Robertsonian translocation), which produces telocentric chromosomes and is regarded as an important speciation mechanism in the genus. Several of the species have variable cytotypes reported: *H. choctawensis* ($2n = 44$), *H. coronaria* ($2n = 44$), *H. crassifolia* ($2n = 40$), *H. liriosme* ($2n = 40, 42$), *H. occidentalis* ($2n = 52, 54$), *H. palmeri* ($2n = 42, 46, 48$), *H. puntagordensis* ($2n = 46$), and *H. rotata* ($2n = 44, 46, 48$). Genetic analyses indicate that *H. littoralis* ($2n = 44, 46$) is a secondarily balanced segmental allotetraploid. Meiosis is irregular with multivalent formation, chromosome fragmentation, and dicentric bridge formation, which all contribute

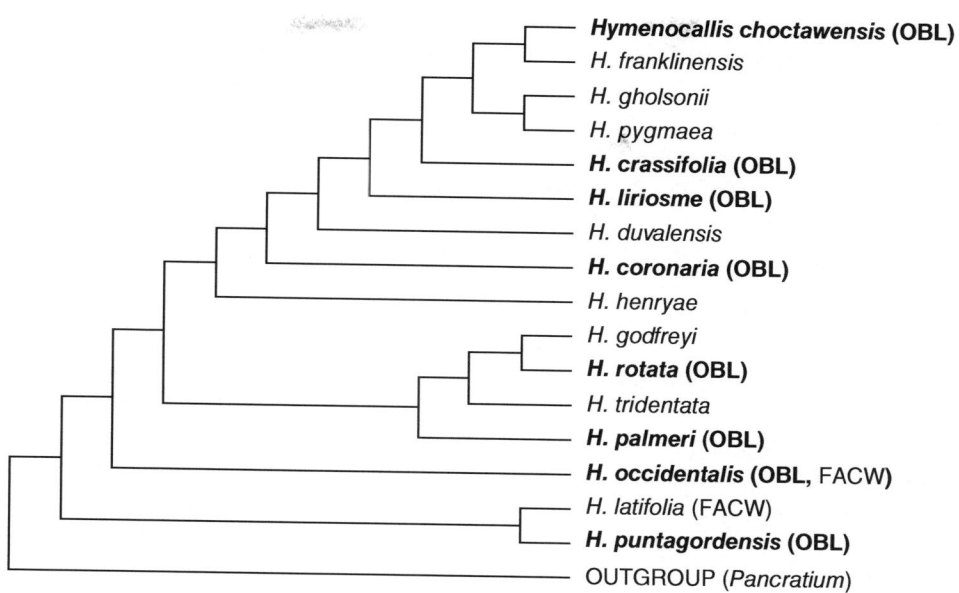

FIGURE 2.6 Interspecific relationships in *Hymenocallis* as indicated by analysis of ISSR data (redrawn and modified after Bush et al., 2010). Several of the relationships differ from those indicated by analyses of morphological data (see text). In this topology, the OBL North American indicators (in bold) are dispersed throughout the genus. The wetland indicator ranks are provided in parentheses for those North American species so designated.

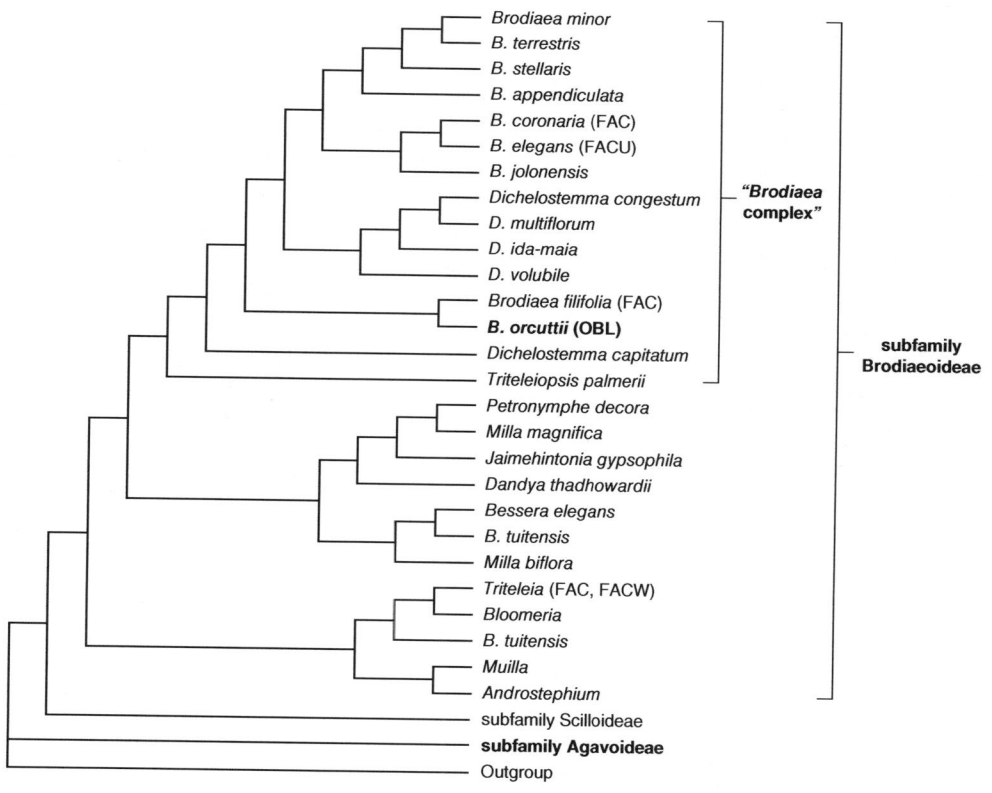

FIGURE 2.7 Phylogenetic relationships of the "*Brodiaea* complex" in Asparagaceae subfamily Brodiaeoideae as indicated by analysis of combined cpDNA sequence data (modified from Pires & Sytsma, 2002). Although *Triteleia* had been considered part of this complex, results such as these show it not to be closely related as once thought. The OBL *Brodiaea orcuttii* resolves as the sister species of *B. filifolia* (FAC), with which it is suspected to hybridize. The two species comprise a subclade that is shown here to be nested among *Dichelostemma* species; however, the internal branch support for the main *Dichelostemma* clade is poor and some analyses resolve *Brodiaea* as a single clade (see Pires & Sytsma, 2002). The wetland indicator status is given in parenthesis for ranked North American taxa; *B. orcuttii* (bold) is the only OBL taxon within the subfamily.

to its low (35%) pollen fertility. Artificial hybrids have been synthesized between *H. coronaria* and *H. occidentalis*, but the F₁ plants are sterile. Natural interspecific hybrids have not been reported in North America.

Distribution: *Hymenocallis choctawensis* grows along the Gulf coastal plain of the southern United States; *H. crassifolia* occurs along the Atlantic coastal plain of the southeastern United States; *H. occidentalis* extends broadly throughout the southern United States; *H. coronaria* is restricted to Alabama, Georgia, and South Carolina; *H. liriosme* occurs in the Mississippi embayment region of the southern United States. Three species are restricted to Florida: *H. littoralis* (introduced; also possibly to Louisiana), *H. palmeri* (mostly southern), and *H. puntagordensis* and *H. rotata* (Gulf coast).

References: Abou-Donia et al., 2008; Aldrich & Homoya, 1983; Baden, III et al., 1975; Baskin & Baskin, 1998; Beal, 1977; Bentley & Elias, 1983; Braun et al., 2014; Burlakova et al., 2009; Bush et al., 2010; Campbell et al., 2014; Caprio & Taylor, 1984; Carpenter & Chester, 1987; Carter et al., 1990; Chen et al., 2016; Crouch & Golden, 1997; Daoust & Childers, 1998; 1999; Darst et al., 2002; Drescher, 2017; East, 1940; Eleuterius & Caldwell, 1984; Eleuterius & McDaniel, 1978; Flint, 1943; Flint & Moreland, 1943; Flory, 1976; Fryxell, 1957; Garland et al., 2013; Graham, 2010; Hall & Walter, 2011; Hart, 2006; Herndon, 1987; Herring & Judd, 1995; Hubbard & Judd, 2013; Ji et al., 2014; Joye & Smith, 1993; Karthikeyan et al., 2016; Kejkar & Polara, 2015; Kihara et al., 1987; Labisky et al., 2003; Lakshmi, 1978; Lewis, 1990; Lewis, III, 1989; Lin et al., 1995; Loughmiller et al., 2018; MacRoberts et al., 2014; Markwith & Leigh, 2008; 2012; Markwith & Parker, 2007; Markwith & Scanlon, 2006; 2007; Markwith et al., 2009; Meerow & Snijman, 2006; Meerow et al., 2000; 2002; Mohlenbrock, 1959b; 1962; Nelson, 1986; Perry, 1961; Pettit et al., 1995; Rosen, 2007; Sah et al., 2010; Sasser & Gosselink, 1984; Sealy, 1954; Sundarasekar et al., 2018; Sharma & Bal, 1956; Smith & Flory, 1990; 2002; Smith & Garland, 1996; 2003; Snoad, 1955; Sohsalam et al., 2008; Stalter, 1973; Stalter & Baden, 1994; Tan & Judd, 1995; Thomas, 2017; Traub, 1957; Tveten & Tveten, 1997; Van Zandt et al., 2013; Ward, 2012; Whitehead & Brown, 1940; Wooten & Leonard, 1984; Wynn, 2012; Zaebst et al., 1995.

Family 5.2. Asparagaceae [114]

Asparagaceae have been defined by very narrow (*sensu stricto*) to very broad (*sensu lato*) circumscriptions. At one extreme, most of the order Asparagales was condensed within an extremely broad concept of the family Liliaceae (Cronquist, 1981; FNA, 2003). However, within a short time, a number of segregate families were resurrected to accommodate much smaller clades that were found consistently in phylogenetic studies (Seberg et al., 2012). The eight segregate families recognized in 2003 (APG, 2003) were condensed into a single family (Asparagaceae) within 6 years (APG, 2009), although the narrower circumscriptions (e.g., Asparagaceae as containing only two genera) are still widely adopted (e.g., Baldwin et al., 2012; Judd et al., 2016).

Chase et al. (2009) encouraged acceptance of the *sensu lato* concept (APG, 2009), in order to simplify descriptions of subfamilial characters and to facilitate teaching in angiosperm taxonomy. Defined in that way, Asparagaceae would contain about 2,900 species (Christenhusz & Byng, 2016) allocated to seven subfamilies, including six groups recognized previously as families (i.e., Agavaceae, Aphyllanthaceae, Hesperocallidaceae, Hyacinthaceae, Laxmanniaceae, Ruscaceae, Themidaceae) (see Figure 2.2). Because phylogenetic analyses have shown these groups to represent clades (e.g., Pires et al., 2006; Seberg et al., 2012), either conceptual disposition (i.e., their recognition as families or subfamilies) would be defensible. The circumscription proffered by Chase et al. (2009) has been followed here.

Although the classification of an expanded Asparagaceae has been addressed repeatedly, there is yet to be a definitive description of the group as a whole and no attempt to do so is made here.

The family is of considerable economic importance by its inclusion of comestibles like asparagus (*Asparagus officinalis*) and many ornamental species in such genera as *Agave*, *Albuca*, *Asparagus* ("asparagus ferns": *A. aethiopicus*, *A. setaceus*), *Aspidistra*, *Bellevalia*, *Brodiaea*, *Camassia*, *Chlorophytum*, *Convallaria*, *Cordyline*, *Dracaena*, *Eucomis*, *Hosta*, *Hyacinthus*, *Hyacinthoides*, *Lachenalia*, *Liriope*, *Maianthemum*, *Manfreda*, *Muscari*, *Ophiopogon*, *Ornithogalum*, *Polianthes*, *Polygonatum*, *Puschkinia*, *Ruscus*, *Sansevieria*, *Scilla*, *Triteleia*, and *Yucca*. *Agave* species are fermented in the making of the popular alcoholic beverages known as mescal and tequila. Some oral contraceptives are derived from the saponins found in *Agave* and *Yucca*; *Thuranthos* and *Urginea* contain cardenolides, which are used medicinally (Judd et al., 2016).

Members of Asparagaceae can be found worldwide from temperate to tropical regions and in wet to arid habitats. The OBL North American indicators occur within three of the seven subfamilies (Agavoideae, Brodiaeoideae, and Nolinoideae) and include:

5.2.1. *Brodiaea* Sm.
5.2.2. *Hastingsia* S. Watson
5.2.3. *Maianthemum* F.H. Wigg.
5.2.4. *Schoenolirion* Torr. ex Durand

5.2.1. Brodiaea

Cluster-lily, Orcutt's brodiaea
Etymology: after James Brodie (1744–1824)
Synonyms: *Hookera*; *Triteleia* (in part)
Distribution: global: Mexico, North America; **North America:** western
Diversity: global: 14 species; **North America:** 14 species
Indicators (USA): obligate wetland (OBL): *Brodiaea orcuttii*
Habitat: freshwater; palustrine; **pH:** alkaline; **depth:** <1 m; **life-form(s):** emergent herb
Key morphology: scapose herbs from fibrous corms; leaves (1–6) basal, the blades (to 20 cm) linear; scape (to 25 cm)

solitary, slender, terminating in an umbel of 2–8 flowers (to 20 mm) subtended by scarious bracts, pedicels (to 5 mm) erect; perianth of 6 transparent, violet tepals basally connate into a funnelform tube (to 7 mm), the free lobes (to 19 mm) widely spreading; stamens 3 (staminodes absent), the filaments (to 6 mm) adnate to tube, the anthers (to 6 mm) linear, v-notched apically; ovary (to 6 mm) superior, green, syncarpous (3 carpels), ovules several, style (to 11 mm) erect, stigma 3-lobed, spreading, recurved; capsules ovoid, loculicidal; seeds black

Life history: duration: perennial (corms); **asexual reproduction:** corm offshoots; **pollination:** insect; **sexual condition:** hermaphroditic; **fruit:** capsules (common); **local dispersal:** seeds (water); **long-distance dispersal:** seeds (animals)

Imperilment: 1. *Brodiaea orcuttii* [G2]; S2 (CA)

Ecology: general: Most *Brodiaea* species occupy terrestrial habitats such as grasslands or open woodlands. Six species (43%) are designated as wetlands indicators (OBL, FACW, FAC, FACU) but only one (7%) as OBL. All the *Brodiaea* species flower from late spring to early summer. Care should be taken when referencing the literature as nearly all pollinator and seed studies involving "*Brodiaea*" actually refer to *Dichelostemma* or *Triteleia* species (see also *Systematics* below). Those specifically evaluating *Brodiaea* species have found that the flowers are visited by many potentially pollinating insects (Insecta) including bees (Hymenoptera: Apidae; Colletidae; Halictidae; Megachilidae), beetles (Coleoptera), flies (Diptera: Acroceridae; Bombyliidae; Empididae; Rhagionidae; Syrphidae), butterflies, and moths (Lepidoptera: Adelidae; Noctuidae; Nymphalidae; Pieridae). Pollination by small-headed flies (Diptera: Acroceridae: *Eulonchus tristis*) has been well documented in one species. The plants die back completely to the corms after anthesis. There are no definitive studies on the seed ecology, leaving details on their germination requirements and dispersal largely unknown. Most of the species (but not *B. orcuttii*) reproduce vegetatively by bulblets that develop from the corm base or at the ends of delicate offshoots. Some of the species occur on serpentine.

5.2.1.1. ***Brodiaea orcuttii*** (Greene) Baker grows in depressions (clay), flats, meadows (vernal), seeps, slopes (to 15%), vernal pools, and along the margins of drying ponds and streams at elevations to 1677 m. The plants occur in exposures ranging from full sunlight to partial shade. The substrates are alkaline, metavolcanic derived, and have been described as adobe, clay (heavy), granite, gravel, gravelly loam (Redding), pebbles (gabbro), sand, sandy clay, silty loam, and stony loam (Boomer). When in pools, the plants are most prevalent during the peak of high water, extending to sites (normally edges rather than deeper bottoms) where standing waters (up to 15 cm) usually persist for 35–40 days (occasionally up to 60 days). Flowering and fruiting occur from late April to July. No specific information exists on the floral or seed ecology. Presumably, the flowers are insect pollinated (Insecta) although it is not known whether they also are capable of self-pollination. The seeds are dispersed locally by gravity and surely by water. Populations remain relatively stable over time, which indicates little annual seedling recruitment. Mechanisms for long-distance dispersal also have not been elucidated but most likely involve the movement of seeds by some animal vector. There is no means of vegetative reproduction. **Reported associates:** *Acmispon americanus, Adenostoma fasciculatum, Agnorhiza ovata, Alnus rhombifolia, Ambrosia psilostachya, Anagallis arvensis, Anagallis minima, Asclepias fascicularis, Avena, Avena barbata, Baccharis salicifolia, Bloomeria crocea, Brodiaea jolonensis, Brodiaea terrestris, Bromus hordeaceus, Bromus rubens, Callitriche marginata, Callitropsis forbsii, Callitropsis stephensonii, Calochortus splendens, Calystegia macrostegia, Carex spissa, Castilleja minor, Ceanothus cyaneus, Cotula coronopifolia, Crassula aquatica, Deinandra fasciculata, Deschampsia danthonioides, Downingia cuspidata, Elatine brachysperma, Eleocharis macrostachya, Eleocharis palustris, Elymus glaucus, Eriogonum fasciculatum, Erodium botrys, Eryngium aristulatum, Euphorbia spathulata, Gastridium ventricosum, Geranium dissectum, Grindelia hirsutula, Holocarpha virgata, Hordeum brachyantherum, Hypochaeris glabra, Isoetes howellii, Isoetes orcuttii, Iva hayesiana, Juncus acutus, Juncus bufonius, Lachnagrostis filiformis, Lythrum hyssopifolia, Madia, Malosma laurina, Microseris douglasii, Mimulus guttatus, Monardella linoides, Monardella stoneana, Muhlenbergia rigens, Muilla maritima, Myosurus minimus, Nassella pulchra, Navarretia fossalis, Navarretia intertexta, Navarretia prostrata, Orcuttia californica, Osmadenia tenella, Penstemon heterophyllus, Pilularia americana, Plagiobothrys undulatus, Plantago bigelovii, Plantago elongata, Platanus racemosa, Poa pratensis, Pogogyne abramsii, Polypogon maritimus, Polypogon monspeliensis, Psilocarphus brevissimus, Psilocarphus tenellus, Pteridium aquilinum, Quercus agrifolia, Quercus ×acutidens, Rosa californica, Rumex crispus, Salix laevigata, Salvia mellifera, Sisyrinchium bellum, Spartium junceum, Tamarix ramosissima, Toxicodendron diversilobum, Toxicoscordion venenosum, Trifolium depauperatum, Triglochin scilloides, Typha domingensis, Vulpia myuros.*

Use by wildlife: *Brodiaea orcuttii* is eaten by pocket gophers (Mammalia: Rodentia: Geomyidae: *Thomomys bottae*). It is possible that these animals also facilitate seed dispersal in some way.

Economic importance: food: The edibility of *Brodiaea* is dubious due to some taxonomic uncertainty of the specific plants reported in literature accounts; however, the native Californians reportedly gathered the corms of *Brodiaea* and related species as a food; **medicinal:** no known uses; **cultivation:** Although several *Brodiaea* species are cultivated, *B. orcuttii* is not; **misc. products:** none; **weeds:** none; **nonindigenous species:** none.

Systematics: *Brodiaea* is assigned to subfamily Brodiaeoideae of Asparagaceae, which is equivalent (in circumscription) to the segregate family Themidaceae (Figure 2.2). Although several phylogenetic studies have indicated that this subfamily is monophyletic, the monophyly of *Brodiaea* itself is unlikely. The "*Brodiaea* complex" includes the morphologically similar *Androstephium, Bloomeria, Dichelostemma, Muilla, Triteleia,* and *Triteleiopsis.* However, molecular phylogenetic analyses indicate that only *Dichelostemma* is closely related and possibly nested within *Brodiaea,* as currently defined

(Figure 2.7). The distinction between the genera is not well supported in phylogenetic studies, where critical cladogram branches have poor internal support. Thus, it is apparent that the generic limits of *Brodiaea* and *Dichelostemma* warrant reconsideration. In any case, *B. orcuttii* (OBL) associates closely and consistently with the FAC *B. filifolia* (Figure 2.7). The basic chromosome number of *Brodiaea* is uncertain and has been reported as *x*=6, 8, or 12. Accordingly, *Brodiaea orcuttii* (2*n*=24) is either diploid or tetraploid. Fertile hybrids allegedly occur between *B. orcuttii* and the closely related *B. filifolia*, but the putative hybrids have not been evaluated genetically. *Brodiaea orcuttii* also has been implicated as a possible parent of sterile hybrids involving the distantly related *B. terrestris*.

Distribution: *Brodiaea orcuttii* is restricted to a small area of military reservations in southern California.

References: Anderson, 1997; Bauder, 1994; 2000; Borkent & Schlinger, 2008; Buchmann et al., 2010; Burge, 2014; Gill & Hoppa, 2016; Hoover, 1939; Hunt, 1992; Pires, 2002; Pires & Sytsma, 2002; Purer, 1939; Solomeshch, et al., 2007; Thorne et al., 2011.

5.2.2. Hastingsia

Rush-lily

Etymology: after Serranus Clinton Hastings (1876–1880)

Synonyms: *Schoenolirion* (in part)

Distribution: global: North America; **North America:** western

Diversity: global: 4 species; **North America:** 4 species

Indicators (USA): obligate wetland (OBL): *Hastingsia alba, H. bracteosa*

Habitat: freshwater; palustrine; **pH:** 6.4–6.8; **depth:** <1 m; **life-form(s):** emergent herb

Key morphology: scapose, herbaceous perennials from solitary, fleshy bulbs (to 56 mm); leaves basal, the blades (to 53 cm) grass-like, shriveling with age; scape (to 89 cm) solitary from apex of bulb, unbranched or with 1–3 ascending branches, terminating in a raceme (to 40 cm) bearing up to 7.8 flowers per cm, the bracts small, scarious; flowers perfect, borne pedicels (to 3 mm); tepals (to 12 mm) 6, distinct, white to yellowish white, often tinged with green, pink, or purple, closed and including stamens, or partially to fully exposing stamens at anthesis, persistent but withering and shriveling to base; stamens 6, the filaments (to 7 mm) dimorphic, 3 longer with dehiscent anthers at anthesis, 3 shorter with indehiscent anthers at anthesis; ovary superior, globose, 3-lobed, with 2 ovules per locule, style single, stigma 3-lobed; capsules (to 11 mm) ellipsoid to ellipsoid–ovoid, somewhat 3-lobed; seeds (to 6 mm) fusiform, black, gray-green or yellowish brown

Life history: duration: perennial (bulbs); **asexual reproduction:** none; **pollination:** insect, self?; **sexual condition:** hermaphroditic; **fruit:** capsules (common); **local dispersal:** seeds (water); **long-distance dispersal:** seeds (unknown)

Imperilment: 1. *Hastingsia alba* [G4]; **2.** *H. bracteosa* [G2]; S2 (OR)

Ecology: general: All *Hastingsia* species are bulbous perennials of serpentine habitats. Although only two of them are categorized as wetland indicators (both OBL), the other species occupy comparable habitats (sometimes co-existing), at least for part of the year, and would seem to warrant some wetland indicator status (see also *Systematics* below). Information on the floral biology of these relatively showy plants is limited. The large racemes attract bees (Insecta: Hymenoptera) and other insects, especially at times when other nearby open flowers are scarce. Although the flowers produce nectar and are insect pollinated, the perianth of some species remains closed at anthesis (the stamens included), and the flowers of all the species are unusual in having three dehiscent anthers (on longer filaments) and three indehiscent anthers (on shorter filaments) at anthesis. The functional role of these unusual floral traits has not been elucidated, but they indicate at least the potential for self-pollination. Details on seed germination are available only for the two OBL species (see below). There is no mechanism for vegetative reproduction; therefore, all dispersal must occur by transport of the seeds. No dispersal vectors have been identified, but the seeds likely are dispersed at least locally by water.

5.2.2.1. *Hastingsia alba* **(Durand) S. Watson** occurs in bogs (seepage, serpentine), drainages (roadside), fens, flats, marshes, meadows (boggy, montane, subalpine), seepages (open, perennial, roadside), slopes (serpentine [to 45°]), meadows (rocky, streamside, wet), ravines, washes, on river bars (rocky), and along streamlet margins at elevations to 2438 m. The plants occur in full sun to shaded exposures. They grow on substrates that have been characterized as clay, clay loam (coarse), diorite, granite (decomposing), gravel, humus, mud (soft), peridotite, rock (metavolcanic, volcanic), sand, serpentine, silt, and stones. Flowering occurs from May to July, with fruiting extending from June to September. Although the nectar has a relatively dilute sugar content (16–33%; $\bar{x}$ =22%), the flowers are visited by a variety of insects (Insecta; see *Use by wildlife*) and presumably are primarily insect pollinated. The stamens are exposed only distally at anthesis, which might indicate that some self-pollination also occurs (i.e., following internal anther dehiscence); however, it is not known whether the flowers are self-compatible. The seeds germinate best (to 78%) when collected during midwinter, dry-stored until midsummer, then sown outdoors in moist towels. They continue to germinate from July to December with most germination (40%) occurring in November. **Reported associates:** *Abies magnifica, Achillea millefolium, Acmispon americanus, Aconitum columbianum, Actaea rubra, Adiantum aleuticum, Allium, Alnus rhombifolia, Alyssum murale, Aquilegia eximia, Aquilegia formosa, Arbutus menziesii, Arctostaphylos canescens, Arctostaphylos hispidula, Arctostaphylos viscida, Artemisia tridentata, Aspidotis densa, Bistorta bistortoides, Brodiaea, Bryum pseudotriquetrum, Calocedrus decurrens, Calochortus persistens, Caltha leptosepala, Camassia quamash, Campylium stellatum. Carex echinata, Carex hoodii, Carex laeviculmis, Carex mendocinensis, Carex serpenticola, Carex serratodens, Castilleja, Ceanothus cuneatus, Ceanothus pumilus, Cerastium arvense, Chamaecyparis, Chlorogalum, Circaea alpina, Cirsium douglasii, Cistanthe umbellata, Cornus nuttallii, Cypripedium californicum, Danthonia californica, Darlingtonia californica,*

Delphinium, Deschampsia cespitosa, Dichanthelium acuminatum, Drosera rotundifolia, Eleocharis quinqueflora, Epilobium ciliatum, Epilobium oregonense, Eremogone congesta, Eriodictyon, Eriophorum criniger, Erythronium citrinum, Festuca idahoensis, Festuca rubra, Galium trifidum, Garrya buxifolia, Gaultheria, Gilia, Habenaria, Hastingsia bracteosa, Helenium bigelovii, Heracleum sphondylium, Holcus, Horkelia congesta, Horkelia sericata, Hosackia pinnata, Juncus effusus, Juncus ensifolius, Leucothoe davisiae, Lewisia cotyledon, Lilium kelleyanum, Lilium maritimum, Lilium pardalinum, Lithocarpus densiflorus, Micranthes oregana, Microseris howellii, Mimulus guttatus, Mimulus moschatus, Mitella breweri, Muhlenbergia andina, Narthecium californicum, Neottia convallarioides, Oenanthe, Oreostemma alpigenum, Orobanche uniflora, Orthocarpus, Packera clevelandii, Parnassia palustris, Pedicularis attollens, Penstemon anguineus, Penstemon newberryi, Penstemon procerus, Perideridia gairdneri, Phacelia, Phlox, Phyllodoce empetriformis, Picea breweriana, Pinguicula vulgaris, Pinus attenuata, Pinus contorta, Pinus jeffreyi, Pinus monticola, Pinus ponderosa, Platanthera dilatata, Platanthera sparsiflora, Poa secunda, Primula hendersonii, Prunella vulgaris, Prunus virginiana, Pseudotsuga menziesii, Quercus vacciniifolia, Ranunculus californicus, Ranunculus occidentalis, Rhododendron occidentale, Rhynchospora alba, Rubus glaucifolius, Rudbeckia glaucescens, Salix, Sanguisorba officinalis, Schoenoplectus acutus, Sedum, Senecio triangularis, Sidalcea gigantea, Silene campanulata, Sisyrinchium bellum, Solidago elongata, Sphenosciadium capitellatum, Spiraea splendens, Spiranthes romanzoffiana, Symphyotrichum, Taraxacum officinale, Trianta occidentalis, Trifolium, Triteleia crocea, Tsuga heterophylla, Umbellularia californica, Vaccinium parvifolium, Veratrum californicum, Viola glabella, Viola lobata, Viola macloskeyi.

5.2.2.2. *Hastingsia bracteosa* S. Watson inhabits bogs, fens,
flats, ledges (shore), marshes (hillside), seepages, slopes (to 10%), springs, and the margins of streams at elevations to 536 m. The plants occur mostly under open, sunny exposures but also in densely shaded but continuously wet sites. The substrates have been characterized as peat, peridotite (Dubakella-Pearsoll series), rock, sandy clay loam, serpentine, and ultramafic. Those that have been analyzed (in fens) are slightly acidic (e.g., pH: 6.4–6.8), contain 7.9–21% organic matter, and have a relatively high cation exchange capacity (CEC: 31.5–40.7 meq 100 g^{-1}). Nitrogen levels can be low (e.g., NO$_3^-$: 0.6–1.0 mg l^{-1}); but phosphorous levels can be high (PO$_4^{3-}$: 20.5–147.5 mg l^{-1}). Flowering occurs from May to July, with fruiting extending from June to August. Although the flowers reportedly are pollinated by bees (Insecta: Hymenoptera), the stamens rarely are exposed during anthesis, raising the question of how the pollen is collected. Further study of the pollination biology of this species is advisable. The seeds have germinated well (100% within 10 weeks) after being cold-stratified (4.4°C) for 7–10 weeks and incubated at 21°C, but also (89%) after being warm-stratified (at 21°C) in the dark for 9 weeks after being transferred to colder conditions (4.4°C). Mechanisms of dispersal have not been elucidated but must involve seeds given the lack of vegetative reproduction. Local seed dispersal by water is likely. Plant densities have been observed to range from 0.5% to 11.4% cover. **Reported associates:** *Adiantum aleuticum, Agrostis, Arctostaphylos viscida, Calocedrus decurrens, Carex echinata, Carex mendocinensis, Carex utriculata, Ceanothus pumilus, Chamaecyparis lawsoniana, Cypripedium californicum, Darlingtonia californica, Deschampsia cespitosa, Epilobium oreganum, Eriophorum criniger, Festuca idahoensis, Gaillardia, Gentiana setigera, Hastingsia atropurpurea, Lilium pardalinum, Narthecium californicum, Pinus echinata, Pinus jeffreyi, Rhamnus californica, Rhododendron occidentale, Rudbeckia californica, Salix lasiolepis, Sanguisorba officinalis, Triantha glutinosa, Umbellularia californica, Viola primulifolia.*

Use by wildlife: The flowers of *Hastingsia alba* are visited by several nectar- and pollen-seeking insects (Insecta), including ants (Hymenoptera: Formicidae), bees (Hymenoptera: Apidae: *Bombus, Ceratina*; Colletidae: *Hylaeus*; Megachilidae), beetles (Coleoptera), crescent butterflies (Lepidoptera: Nymphalidae: *Phyciodes*), and wasps (Hymenoptera: Vespidae). At least some of these insects (e.g., Hymenoptera: Megachilidae) are likely to function as important pollinators. The flowers of *H. bracteosa* also reportedly are pollinated primarily by bees (Insecta: Hymenoptera).

Economic importance: food: The roots and bulbs of *Hastingsia alba* were eaten by people of the Northeastern Maidu tribe; **medicinal:** no medicinal uses have been reported; **cultivation:** *Hastingia* is not commonly cultivated; **misc. products:** none; **weeds:** none; **nonindigenous species:** none.

Systematics: *Hastingsia* is included among the "chlorogaloid" genera, a group recognized as subfamily Chlorogaloideae in treatments that maintain the family Agavaceae. This group is biphyletic (Figure 2.8) and has not been given any formal taxonomic status within Asparagaceae. Although sometimes merged with *Schoenolirion*, *Hastingsia* is distinct from the former morphologically and is unexpectedly remote phylogenetically (Figure 2.8). Molecular data consistently resolve *Hastingsia* as monophyletic and sister to *Camassia* (Figure 2.8), a close relationship that is evidenced further by microsatellite markers, which show widespread cross-reactivity between the genera. The delimitation of *Hastingsia* species is problematic. The OBL *Hastingsia alba* is presumed to be closely related to *H. serpentinicola*, which grows on dry hillsides. However, interspecific relationships depicted by DNA sequence data are perplexing with *H. alba* and *H. bracteosa* accessions associating closely, but in two different groups along with *H. serpenticola*; all three taxa occur closely with *H. atropurpurea* (Figure 2.8). Furthermore, morphological, fertility, and genetic evidence indicate that *H. atropurpurea* is indistinguishable from *H. bracteosa*. When sympatric, the plants share pollinators, are interfertile, retain high pollen viability and seed set when crossed, and exhibit no fixed allelic differences (as indicated by isozyme data). These are compelling reasons why the taxa probably should not be maintained as separate species, and some authors recognize the former as a variety of the latter (i.e., *H. bracteosa* var. *atropurpurea*). However, if *H. atropurpurea*

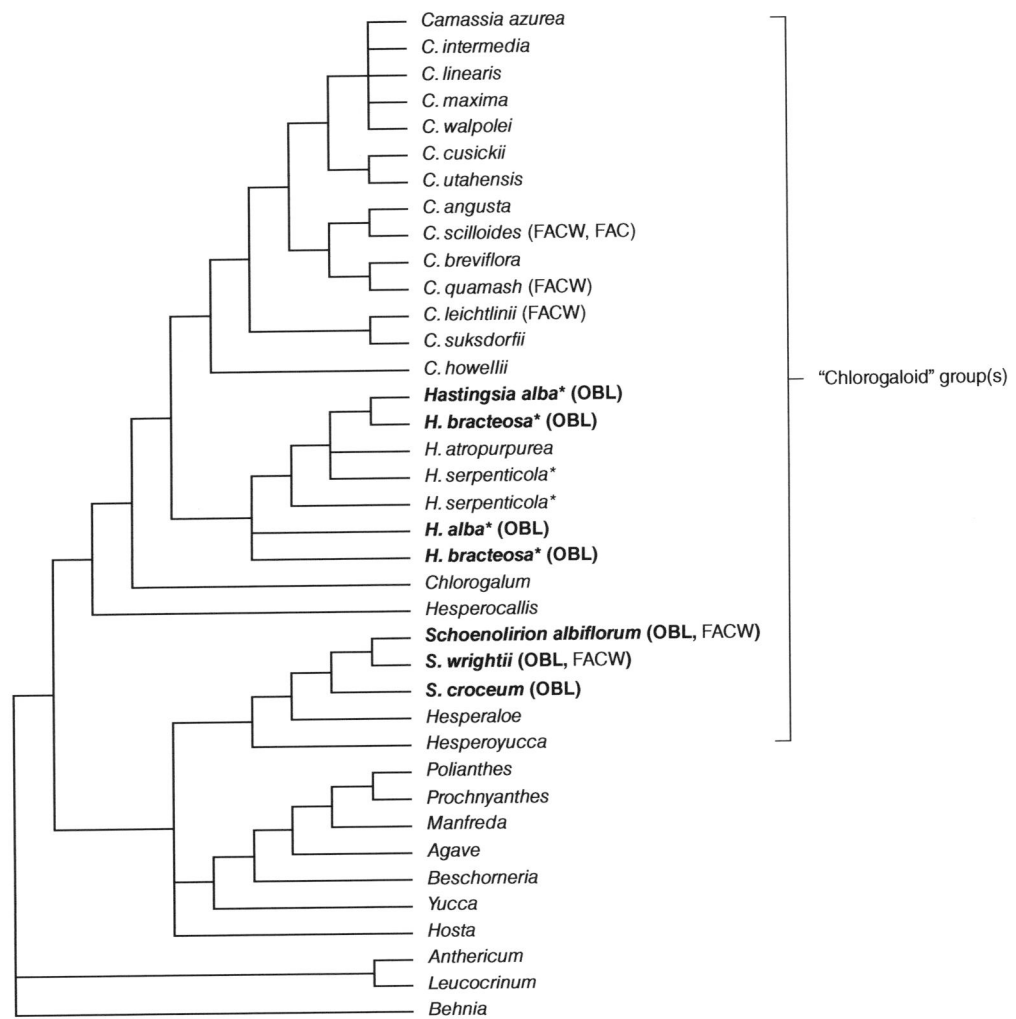

FIGURE 2.8 Phylogeny inferred for the "Chlorogaloid" group (shown here to be biphyletic) within Asparagaceae subfamily Agavoideae as reconstructed by analysis of combined nuclear and plastid DNA sequence data (modified from Archibald et al., 2015). Although once considered to be closely related (and even congeneric), *Hastingsia* and *Schoenolirion* are remote phylogenetically, with *Camassia + Hastingsia* and *Hesperaloe + Schoenolirion* comprising well-supported sister groups. Even though *Hastingsia* is monophyletic, the species are not (asterisked accessions), which could reflect a rapid, recent divergence or factors such as hybridization and polyploidy (see text). Within subfamily Agavoideae, only *Hastingsia* and *Schoenolirion* contain OBL indicators (designated in bold), with a few facultative wetland species occurring in *Camassia* (ranks are shown in parentheses for North American wetland indicators).

and *H. bracteosa* are treated as conspecific, then the results rendered by DNA sequence analysis (Figure 2.8) are not corroborative. The unusual topology indicated by DNA sequence data would suggest that all four taxa represent a single, variable species, that the species have radiated rapidly, or that the incongruencies have resulted from factors such as interspecific hybridization and/or polyploidy. The basic chromosome number of *Hastingsia* is proposed as $x=26$, which apparently represents a compound number given the base number of $x=15$ in its sister genus *Camassia*. *Hastingsia alba* ($2n=52$) arguably is a diploid. Natural interspecific hybridization likely occurs in the genus (e.g., *H. atropurpurea* × *H. bracteosa*), but a critical genetic evaluation is necessary to document its extent and to evaluate any systematic ramifications.

Distribution: *Hastingsia alba* is narrowly restricted to northern California and southwest Oregon; *H. bracteosa* is endemic to southwestern Oregon.

References: Archibald et al., 2015; Becking, 1986; 1989; 2002; Culley et al., 2013; Deno, 1993; Dixon, 1905; Gut et al., 1977; Halpin, 2011; Halpin & Fishbein, 2013; Lang & Zika, 1997; Sherman & Becking, 1991; Sikes et al., 2010; Sinkiewicz & Jules, 2003; Titus, 2017; Tolman, 2006; 2007.

5.2.3. Maianthemum

False lily-of-the-valley; mayflower; maïanthème, smilacine
Etymology: from the Greek *Máios ánthos* ("May flower") for the time of anthesis
Synonyms: *Asteranthemum; Smilacina; Convallaria* (in part); *Sigillaria* (in part); *Tovaria* (in part); *Unifolium; Vagnera*
Distribution: global: Central America; Eurasia; North America; **North America:** widespread
Diversity: global: 38 species; **North America:** 5 species
Indicators (USA): obligate wetland (OBL): *Maianthemum trifolium*

Habitat: freshwater; palustrine; **pH:** 3.0–6.8; **depth:** <1 m; **life-form(s):** emergent herb

Key morphology: perennial, rhizomatous herbs (to 15 cm); rhizomes (to 30 cm) filiform, prolific, rooting at nodes; stems (to 2.5 dm) erect, simple; leaves (2–4) cauline, distichous, sessile, the blades (to 12 cm) elliptic, the base tapering narrowly; racemes (to 9 cm) terminal, simple (to 15 flowers); flowers bisexual, 3-merous, pedicelled (to 15 mm); tepals (to 4 mm) 6, white, distinct, ovate or triangular; stamens (to 3.5 mm) fused to tepal base; ovary (to 2.5 mm) superior, globose or cylindrical, the septal wall nectariferous, styles (to 1.5 mm) 3, fused except apically, stigma 3-lobed; berries (to 9 mm) globose, baccate, shiny red at maturity; seeds (to 4 mm) 1–3, globose, with a thin, brown seed coat

Life history: duration: perennial (rhizomes); **asexual reproduction:** rhizomes; **pollination:** insect, self; **sexual condition:** hermaphroditic; **fruit:** berry (infrequent); **local dispersal:** water (?); **long-distance dispersal:** animals

Imperilment: 1. *Maianthemum trifolium* [G5]; SX (OH); S1 (NJ); S2 (CT); S3 (MA, <u>NU</u>, <u>YT</u>)

Ecology: general: Although only one *Maianthemum* species is designated as OBL, the other North American species occur often in wet sites and also have status as wetland indicators (FACW, FAC, FACU). All the North American species are perennial, rhizomatous herbs. Their flowers are small but often fragrant and generally pollinated by insects (Insecta) such as bees (Hymenoptera) and flies (Diptera). Flowers among the various species can range from being strongly self-incompatible and outcrossed to highly self-compatible and autogamous. The propensity for clonal growth enhances geitonogamous pollen transfer to adjacent ramets. The seeds of at least some species are dispersed by birds and other frugivorous animals such as brown bears (Mammalia: Ursidae: *Ursus arctos*). Those of several species are known to exhibit morphophysiological dormancy. Seed banks are not known to develop, and seedling recruitment occurs rarely. Vegetative propagation occurs by proliferation of the rhizomes.

5.3.2.1. *Maianthemum trifolium* (L.) Sloboda grows in bogs (black spruce, cold, peat, *Sphagnum*, spring), channels (drainage), depressions (sphagnous), ditches, fens (graminoid, patterned, poor, prairie), hummocks (*Sphagnum*), meadows, muskegs, pools (bog), roadsides, seeps (wooded), swamps (*Sphagnum*), tundra (sloping) woodlands (rich, wet), and along the margins of brooks, lakes (boggy), and springs at elevations to 1159 m. Although some authors have characterized this species as being highly shade intolerant, it has low light compensation and saturation points, and often is found in partially to highly shaded sites. The plants are drought intolerant and prefer hollows (flarks) and lower (wetter) sites in patterned peatlands. The substrates typically are thick layers of peat (fibric, mesic), which can be underlain by clay, fluvium, rock, sand, or silt. These plants grow primarily in mediacidic peatlands (e.g., pH: 3.0–6.8; conductivity: 66–78 µS cm^{-1}; salt: 0.2 ppt; P: 2.7–159.6 µg l^{-1}; Ca: 0.3–20.5 mg l^{-1}; Mg: 0.4–12.2 mg l^{-1}; Mn: 0.02 mg l^{-1}; Na: 0.8–3.4 mg l^{-1}; Fe: 0.3–2.9 mg l^{-1}; K: 0–0.8 mg l^{-1}; NH$_4$: 0.1–1.7 mg l^{-1}; NO$_3$: 0.01–0.14 mg l^{-1}; SO$_4$: 1.7–7.3 mg l^{-1}). They occur rarely on neutral or alkaline

substrates. The aerial shoots develop over a 3-year period before the flowers are produced. In mature plants, flowering occurs from May to August with fruits present from June to September. The fragrant flowers are pollinated by insects (e.g., Insecta: Diptera; Hymenoptera) but are self-compatible and have high potential rates of autogamy (e.g., 82% of natural pollination levels in pollinator-excluded plants); they do not appear to be pollinator limited. Fruit set in natural populations can be high (e.g., to 92%) but typically is low (e.g., <26%). No evidence of inbreeding depression has been observed in the offspring of self-pollinated plants. The mechanism of fruit dispersal has not been specified but likely involves water or their endozoic transport by birds (Aves) and other animals as in several congeners. The seed ecology essentially remains unstudied and there are no reports of a persistent seed bank; however, the seeds might be morphophysiologically dormant as in some of the congeners. These rhizomatous plants can develop into dense clonal patches. They are deeply rooted (to 40+ cm), which enables them to survive even on roadways and in motor vehicle tracks. Their individual aboveground biomass often is fairly low (e.g., 0.04–2.3 g m^{-2}) but stands have exhibited a significant increase in net primary production (by up to 266%) when fertilized with nitrogen (increasing to 14 g m^{-2} year^{-1}) or when provided with supplemental water (then increasing to 16 g m^{-2} year^{-1}). Sampled plants had on average a leaf area of 14.74 cm^2, a chlorophyll content (A+B) of 213.54 mg m$_{leaf}^{-2}$, a net photosynthetic rate of 5.10 µmol m$_{leaf}^{-2}$ s^{-1}, respiration rate of −1.15 µmol m$_{leaf}^{-2}$ s^{-1}, transpiration rate of 2.20 mmol m$_{leaf}^{-2}$ s^{-1}, and a water-use efficiency of 2.17 µmol mmol^{-1}. Net photosynthesis was found to peak during morning hours and to decline as the water table depth increases (to 80 cm below surface). Plant cover and frequency have been shown to decline as a consequence of competition by invasive shrubs (*Rhamnus frangula*). **Reported associates:** *Acer rubrum, Alnus rugosa, Andromeda polifolia, Apios americana, Arnica chamissonis, Aronia melanocarpa, Aulacomnium palustre, Betula alleghaniensis, Betula glandulosa, Betula nana, Betula papyrifera, Betula populifolia, Betula pumila, Bromus ciliatus, Bromus ciliatus, Calamagrostis canadensis, Calla palustris, Calliergon stramineum, Campylium stellatum, Carex aquatilis, Carex brunnescens, Carex canescens, Carex chordorrhiza, Carex crinita, Carex disperma, Carex exilis, Carex gynocrates, Carex haydenii, Carex heleonastes, Carex interior, Carex lasiocarpa, Carex leptalea, Carex limosa, Carex livida, Carex magellanica, Carex oligosperma, Carex pauciflora, Carex rostrata, Carex stricta, Carex tenuiflora, Carex trisperma, Carex utriculata, Chamaedaphne calyculata, Cladina rangiferina, Cladina stellaris, Cladium mariscoides, Comarum palustre, Coptis trifolia, Cornus canadensis, Cornus sericea, Cypripedium acaule, Dasiphora floribunda, Dicranum polysetum, Dicranum polysetum, Doellingeria umbellata, Drepanocladus, Drosera anglica, Drosera intermedia, Drosera rotundifolia, Elaeagnus umbellata, Eleocharis elliptica, Empetrum nigrum, Equisetum arvense, Equisetum fluviatile, Equisetum palustre, Equisetum scirpoides, Equisetum sylvaticum, Eriophorum vaginatum, Eriophorum virginicum,*

Eriophorum viridicarinatum, Eriophorum ×medium, Eutrochium maculatum, Fraxinus nigra, Galium boreale, Galium triflorum, Gaultheria hispidula, Gaultheria procumbens, Gaylussacia dumosa, Hylocomium splendens, Hypericum ellipticum, Hypericum kalmianum, Ilex mucronata, Ilex verticillata, Iris versicolor, Juniperus virginiana, Kalmia angustifolia, Kalmia microphylla, Kalmia polifolia, Larix laricina, Leersia oryzoides, Lindera benzoin, Linnaea borealis, Lonicera villosa, Lysimachia terrestris, Lythrum salicaria, Maianthemum canadense, Maianthemum stellatum, Malaxis paludosa, Menyanthes trifoliata, Mitella nuda, Myrica gale, Neottia bifolia, Nymphaea odorata, Nyssa sylvatica, Oclemena nemoralis, Onoclea sensibilis, Osmunda regalis, Osmundastrum cinnamomeum, Parnassia palustris, Petasites frigidus, Phalaris arundinacea, Physocarpus opulifolius, Picea glauca, Picea mariana, Pinus resinosa, Pinus strobus, Platanthera blephariglottis, Platanthera dilatata, Platanthera hyperborea, Pleurozium schreberi, Pogonia ophioglossoides, Pohlia nutans, Polygala paucifolia, Polytrichum commune, Polytrichum strictum, Primula nutans, Ptilium crista-castrensis, Ptilium crista-castrensis, Ranunculus lapponicus, Rhamnus alnifolia, Rhamnus frangula, Rhododendron canadense, Rhododendron groenlandicum, Rhynchospora alba, Rhytidiadelphus triquetrus, Ribes triste, Rosa acicularis, Rubus arcticus, Rubus chamaemorus, Rubus hispidus, Rubus pubescens, Salix discolor, Salix drummondiana, Salix myrtillifolia, Salix pedicellaris, Salix planifolia, Salix pyrifolia, Sarracenia purpurea, Scheuchzeria palustris, Scirpus cyperinus, Scorpidium scorpioides, Solidago uliginosa, Sphagnum angustifolium, Sphagnum balticum, Sphagnum capillifolium, Sphagnum centrale, Sphagnum compactum, Sphagnum cuspidatum, Sphagnum fallax, Sphagnum flavicomans, Sphagnum fuscum, Sphagnum girgensohnii, Sphagnum lindbergii, Sphagnum magellanicum, Sphagnum majus, Sphagnum nitidum, Sphagnum papillosum, Sphagnum pulchrum, Sphagnum recurvum, Sphagnum rubellum, Sphagnum squarrosum, Sphagnum warnstorfii, Spiraea alba, Spiraea tomentosa, Spiranthes romanzoffiana, Symplocarpus foetidus, Symplocarpus foetidus, Thelypteris palustris, Thuja occidentalis, Tomenthypnum falcifolium, Tomenthypnum nitens, Toxicodendron vernix, Triantha glutinosa, Trichophorum cespitosum, Trientalis borealis, Tsuga canadensis, Typha latifolia, Utricularia cornuta, Vaccinium angustifolium, Vaccinium corymbosum, Vaccinium macrocarpon, Vaccinium myrtilloides, Vaccinium oxycoccos, Vaccinium vitis-idaea, Viburnum nudum, Viola renifolia, Woodwardia virginica, Xyris montana.

Use by wildlife: The foliage of *Maianthemum trifolium* is eaten by several mammals (Mammalia) including hares (Leporidae: *Lepus*), mice (Rodentia: Muridae), moose (Cervidae: *Alces americanus*), white-tailed deer (Cervidae: *Odocoileus virginianus*), and woodland caribou (Mammalia: Cervidae: *Rangifer tarandus tarandus*). The berries are eaten by birds (Aves: Phasianidae; Turdidae) and martens (Mammalia: Mustelidae: *Martes americana atrata*). The plants host several Fungi (Ascomycota: Botryosphaeriaceae: *Phyllosticta smilacinae-trifoliae*; Glomerellaceae: *Colletotrichum*

liliacearum; Vibrisseaceae: *Phialocephala fortinii, P. sphaeroides*; Basidiomycota: Pucciniaceae: *Puccinia majanthemi*).

Economic importance: food: *Maianthemum trifolium* is not known to be edible and the berries allegedly cause intestinal problems if consumed; **medicinal:** Tonics and other preparations made from *M. trifolium* have been attributed with analgesic, antirheumatical, and antiseptic properties. Smoke from the burned roots was inhaled by the Ojibwa to relieve headaches and by the Menonmini to reduce throat and nasal congestion. The Potowatomi used the root smudge to treat insanity and other ailments and to revive comatose individuals; **cultivation:** Some *Maianthemum* species (but not *M. trifolium*) are cultivated for wildflower gardens; **misc. products:** none; **weeds:** not weedy; **nonindigenous species:** none.

Systematics: As currently circumscribed, *Maianthemum* now includes all members of the genus *Smilacina*, to which the OBL *M. trifolium* once had been assigned. Although preliminary molecular analyses indicated the potential distinctness of these genera, subsequent incorporation of additional genetic loci (e.g., Figure 2.9) failed to support their continued recognition. The genus is a sister group to an assemblage comprising miscellaneous genera (e.g., *Convallaria, Disporopsis*, and *Polygonatum*) within subfamily Nolinoideae of Asparagaceae (Figure 2.9). *Maianthemum trifolium* resolves as the sister species to a clade comprising the members of the formerly defined (i.e., *sensu stricto*) concept of *Maianthemum* (Figure 2.9). The four other North American species (all wetland indicators) are scattered throughout the genus (Figure 2.9). The basic chromosome number of *Maianthemum* is $x=9$; *M. trifolium* ($2n=36$) is a tetraploid. The genome size (2C) of *M. trifolium* has been estimated at 22.2 pg.

Distribution: *Maianthemum trifolium* is distributed throughout the boreal regions of North America.

References: Almquist & Calhoun, 2003; Andersen & Price, 2012; Arlen-Pouliot & Payette, 2015; Baskin & Baskin, 1998; Bérubé et al., 2017; Bharathan et al., 1994; Campbell & Bergeron, 2012; Faber-Langendoen, 2001; Garon-Labrecque et al., 2016; Gosse & Hearn, 2005; Habeck, 1992; Howe & Smallwood, 1982; Humbert et al., 2007; Humphreys et al., 2006; 2014; Jeglum, 1971; Kawano & Iltis, 1966; Kim & Lee, 2007; Kim et al., 2017; LaFrankie, 1986a; 1986b; 2002; Locky et al., 2005; Mills et al., 2009; Moore et al., 2002; Mulder, 1999; Murphy, 2009; Newmaster et al., 1997; Pavek, 1993; Pellerin et al., 2006; 2009; Pennacchio et al., 2010; Quattrocchi, 2012; Reaume, 2009; Reimer, 2001; Smith et al., 2007; Staples et al., 1999; Thompson et al., 2014; Thormann & Bayley, 1997; Vitt & Bayley, 1984; Wheelwright et al., 2006; Wherry, 1927; Willson, 1993; Wilson et al., 2005.

5.2.4. Schoenolirion

Rush-lily, sunnybell

Etymology: from the Greek *skhoínos lirion* ("rush/reed lily") after its rush-like appearance

Synonyms: *Amblostima; Anthericum* (in part); *Ornithogalum* (in part); *Oxytria; Phalangium*

Distribution: global: North America; **North America:** southeastern United States

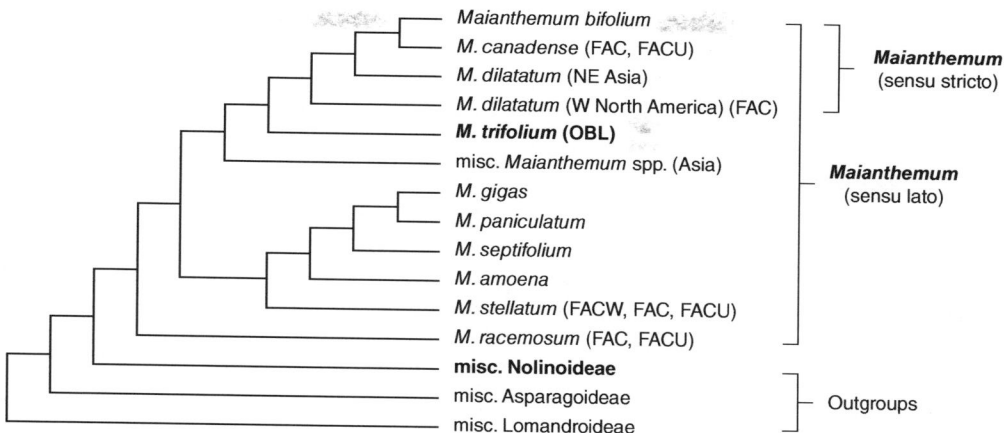

FIGURE 2.9 Interspecific relationships in *Maianthemum* as indicated by phylogenetic analysis of combined DNA sequence data from seven regions (adapted from Kim et al., 2017). Although *Mainthemum* "sensu stricto" resolves as a clade in this topology, the same is not true for former species of *Smilacina* (shown here as other members of *Maianthemum*), which resolve as a paraphyletic grade. These results have led to a revised "sensu lato" circumscription, which merges the genera. The North American species (their wetland indicator status shown in parentheses) do not resolve as a clade but occur throughout the genus. The single OBL indicator (shown in bold) is sister to the "sensu stricto" clade. Other groups containing OBL taxa also are highlighted in bold.

Diversity: global: 3 species; **North America:** 3 species
Indicators (USA): obligate wetland (OBL): *Schoenolirion albiflorum, S. croceum, S. wrightii;* **facultative wetland (FACW):** *S. albiflorum, S. wrightii*
Habitat: freshwater; saline (inland); palustrine; **pH:** 5.5–7.5; **depth:** <1 m; **life-form(s):** emergent herb
Key morphology: rootstock (to 12 cm) vertical, fleshy, sometimes with an ovoid or elongate bulb (to 17 mm) at top, roots apical, contractile; leaves (2–7) basal, sometimes coarsely fibrous, the blades (to 72 cm) linear, flat or roundish, sometimes keeled, shorter than or exceeding flowering scape, the base fleshy or not; racemes simple or branched, the flowers crowded distally, progressively becoming separated (to 6 cm), deciduous if unfertilized, pedicels (to 30 mm) conspicuous; tepals (to 6.5 mm) 6, distinct, persistent, recurved or not, ovate to ovate-lanceolate, white or yellow, with 3–7 veins, striped (green or reddish) above; stamens 6, filaments (to 2 mm) shorter than tepals, the 3 innermost nectariferous at base, anthers dehiscing inward; ovary superior, green or greenish yellow, 3-lobed, 3-locular, 2 ovules per locule; style 1; stigma unlobed or 3-lobed; capsules flattened or indented apically, prominently 3-lobed; seeds smooth, globose, shiny black, flattened
Life history: duration: perennial (bulbs, rhizomes); **asexual reproduction:** bulbs, rhizomes; **pollination:** insect; **sexual condition:** hermaphroditic; **fruit:** capsules (common); **local dispersal:** seeds; **long-distance dispersal:** seeds
Imperilment: 1. *Schoenolirion albiflorum* [G3]; S1 (GA); **2.** *S. croceum* [G4]; SH (NC); S1 (SC); S2 (AL, FL); S3 (LA, TN); **3.** *S. wrightii* [G3]; S1 (AL); S2 (AR, LA); S3 (TX)
Ecology: general: *Schoenolirion* species are ephemeral spring herbaceous perennials that persist as subterranean bulbs or from deep vertical rhizomatous rootstocks during the hot, dry summer period. All the species are obligate (OBL) indicators that grow in various wetland habitats ranging from wet prairies to moist, mossy rock outcrops. The sites typically are kept free of competing woody vegetation by periodic fires.

The flowers are nectariferous, moderately showy, and attract bees (Insecta: Hymenoptera) and other insects, which serve as potential pollinators; however, the pollination biology of this genus has not been studied in any detail. In the one species studied, the seeds exhibit an unusual type of dormancy, which is regulated primarily by light rather than by temperature. Local and long-distance dispersal must be carried out by the seeds (there are no vegetative propagules), but dispersal vectors have yet to be identified. All three species are relatively uncommon (except in local occurrences) and would benefit from additional life history studies.

5.2.4.1. *Schoenolirion albiflorum* (Raf.) R.R. Gates grows on saturated substrates or in shallow waters (e.g., 5–15 cm) of bogs (cypress), depressions (boggy), ditches, flats (low), flatwoods (mesic, pine), glades, hammocks, meadows (low, wet), prairies (cypress, marl), roadsides, savannas (pine, wetland), swales, swamps, and along the banks or margins of canals, cypress domes, ponds (cypress, flatwoods), rivers, streams, and swamps (cypress) at elevations to at least 15 m. The sites are characterized by open canopies with sparse overstory; they experience periodic fires, which reduce woody plant cover. The substrates have been characterized as Fort Drum (aeric haplaquepts), Myakka (seric haplaquods), peat, Riviera (arenic glossaqualfs), and sand (loamy, white). The habitats are calcareous, seasonally wet, and often dry out for part of the year. Flowering has been observed from April to August with fruiting occurring from June to August. The flowers are pollinated by insects (Insecta), but reliable information on specific pollinators is lacking. The principal pollinators are believed to be solitary bees (Hymenoptera). Unlike its congeners, this species lacks a bulbous base and perennates by a vertical rhizome, which is drawn into the substrate by apical contractile roots. Other life history information is lacking for this species and further studies are encouraged. **Reported associates:** *Acer rubrum, Annona glabra, Aristida stricta, Carex striata, Coreopsis nudata, Dyschoriste, Eriocaulon* compressum, *Ficus*, *Fraxinus caroliniana*, *Helenium vernale*, *Hymenocallis henryae*, *Hypericum chapmanii*, *Hypericum fasciculatum*, *Ilex myrtifolia*, *Lachnanthes caroliniana*, *Muhlenbergia sericea*, *Myrica cerifera*, *Nyssa biflora*, *Persea*, *Physostegia godfreyi*, *Pinguicula ionantha*, *Pinguicula planifolia*, *Pinus elliottii*, *Piriqueta*, *Polygala chapmanii*, *Polygala cymosa*, *Quercus*, *Rhynchospora filifolia*, *Rhynchospora harperi*, *Rhynchospora inundata*, *Sabatia decandra*, *Sarracenia flava*, *Sarracenia leucophylla*, *Sarracenia psittacina*, *Scleria baldwinii*, *Serenoa repens*, *Stillingia aquatica*, *Styrax americanus*, *Taxodium ascendens*, *Teucrium*, *Xyris elliottii*.

5.2.4.2. *Schoenolirion croceum* (Michx.) Alph. Wood

inhabits barrens (prairie), depressions, flatwoods (pine), glades (cedar), meadows, outcrops (granite), prairies, roadsides (open), savannas (railroad), seeps (glade, limestone), slopes (chalk, seepage), swales, swamps (cutover), and the margins of ponds (cypress) at elevations to at least 26 m. The habitats range from full sun to partial shade and are characterized by repeated cycles of wet winters and dry summers. The plants attain their maximum cover during the spring (April) and evade drought by becoming dormant (surviving only by corms and seeds) throughout the summer and by rooting deeply in the soil (at depths of 13–41 cm). The substrates have been described as clay (loamy), granite, limestone, loam (sandy), Meadowbrook (grossarenic ochraqualfs), muck, peat, rock, sand (black, white), and sandstone. The habitats are open sites, which experience periodic episodes of fire. Flowering extends from March to May with fruits produced through May. The flowers are pollinated by bees (Insecta: Hymenoptera). Plants at one site had a mean density of 2.1 individuals 625 cm^{-2}; they averaged 31 seeds and produced 66 seeds 625 cm^{-2}. Seed banks averaging up to at least 67 seeds m^{-2} can develop. Seeds that have been newly dispersed in late spring to early summer are dormant and their germination is regulated by light during the autumn. If buried, they germinate in autumn; however, if strewn on the surface their germination will be delayed until late in the following winter or during early spring. High germination has been obtained under dark conditions for warm-stratified seeds (12 hr:12 hr; 25°C/15°C) when exposed to 12 hr:12 hr thermoperiods from 15°C/6°C to 35°C/20°C. Seeds have not germinated in light when stratified under light. Cold-stratified seeds (5°C) germinate in the dark or in light. Dispersal mechanisms have not been elucidated. **Reported associates:** *Agalinis purpurea*, *Agrostis elliottiana*, *Agrostis hyemalis*, *Aletris*, *Allium canadense*, *Allium cernuum*, *Allium stellatum*, *Amsonia ciliata*, *Andropogon gerardii*, *Andropogon virginicus*, *Arnoglossum plantagineum*, *Asclepias viridis*, *Bignonia capreolata*, *Bouteloua curtipendula*, *Bruchia flexuosa*, *Bulbostylis capillaris*, *Callirhoe alcaeoides*, *Campylopus*, *Carex cherokeensis*, *Carex crawei*, *Carex fissa*, *Carex granularis*, *Carya*, *Chamaecrista fasciculata*, *Chaptalia tomentosa*, *Cheilanthes lanosa*, *Cladonia arbuscula*, *Cladonia caroliniana*, *Cladonia leporina*, *Cladonia rangiferina*, *Clinopodium glabellum*, *Clinopodium glabrum*, *Cnidoscolus urens*, *Commelina erecta*, *Coreopsis grandiflora*, *Croton michauxii*, *Cyperus granitophilus*, *Cypripedium candidum*, *Dalea candida*, *Dalea gattingeri*,

Dalea purpurea, *Delphinium carolinianum*, *Desmanthus illinoensis*, *Diamorpha smallii*, *Dichanthelium acuminatum*, *Diodella teres*, *Ditrichum pallidum*, *Drosera*, *Echinacea purpurea*, *Eleocharis bifida*, *Eleocharis obtusa*, *Eragrostis capillaris*, *Erigeron quercifolius*, *Erigeron strigosus*, *Eriogonum longifolium*, *Euphorbia*, *Fimbristylis puberula*, *Forestiera ligustrina*, *Gaillardia aestivalis*, *Gelsemium sempervirens*, *Glandularia canadensis*, *Gratiola quartermaniae*, *Grimmia laevigata*, *Helianthus divaricatus*, *Helianthus hirsutus*, *Helianthus porteri*, *Houstonia purpurea*, *Hypericum dolabriforme*, *Hypericum gentianoides*, *Hypericum sphaerocarpum*, *Hypoxis hirsuta*, *Isoetes butleri*, *Isoetes melanopoda*, *Isoetes piedmontana*, *Juncus filipendulus*, *Juncus georgianus*, *Juniperus virginiana*, *Krigia virginica*, *Leavenworthia alabamica*, *Leavenworthia crassa*, *Leucanthemum vulgare*, *Leucospora multifida*, *Liatris aspera*, *Liatris cylindracea*, *Liatris squarrosa*, *Lindernia monticola*, *Linum sulcatum*, *Lobelia gattingeri*, *Lobelia spicata*, *Lycopodiella alopecuroides*, *Manfreda virginica*, *Marshallia mohrii*, *Micranthes virginiensis*, *Minuartia patula*, *Minuartia uniflora*, *Monarda citriodora*, *Muhlenbergia schreberi*, *Neptunia lutea*, *Nostoc commune*, *Nothoscordum bivalve*, *Nuttallanthus canadensis*, *Oenothera biennis*, *Oenothera filipes*, *Oenothera macrocarpa*, *Ophioglossum engelmannii*, *Opuntia humifusa*, *Packera anonyma*, *Packera tomentosa*, *Panicum philadelphicum*, *Parmelia consperta*, *Parthenocissus quinquefolia*, *Pediomelum subacaule*, *Phacelia dubia*, *Phemeranthus teretifolius*, *Pilularia americana*, *Pinus elliottii*, *Pinus palustris*, *Pinus taeda*, *Pinus virginiana*, *Plantago virginica*, *Pogonia ophioglossoides*, *Polygala boykinii*, *Polygala curtissii*, *Polytrichum commune*, *Prunus serotina*, *Pycnanthemum flexuosum*, *Monarda fistulosa*, *Quercus georgiana*, *Quercus muehlenbergii*, *Ratibida pinnata*, *Rhynchospora gracilenta*, *Rudbeckia amplexicaulis*, *Rudbeckia fulgida*, *Rudbeckia hirta*, *Rudbeckia laciniata*, *Rudbeckia scabrifolia*, *Rudbeckia triloba*, *Ruellia humilis*, *Rumex hastatulus*, *Salvia azurea*, *Sarracenia alata*, *Schizachyrium scoparium*, *Scleria verticillata*, *Scutellaria integrifolia*, *Scutellaria parvula*, *Sedum pulchellum*, *Sedum pusillum*, *Selaginella rupestris*, *Silphium asteriscus*, *Silphium laciniatum*, *Silphium trifoliatum*, *Sisyrinchium albidum*, *Sisyrinchium calciphilum*, *Smilax*, *Solidago ulmifolia*, *Sorghastrum nutans*, *Spiranthes magnicamporum*, *Sporobolus junceus*, *Sporobolus neglectus*, *Sporobolus vaginiflorus*, *Symphyotrichum pratense*, *Symphyotrichum priceae*, *Taxodium ascendens*, *Tetragonotheca helianthoides*, *Tradescantia hirsuticaulis*, *Tradescantia ohiensis*, *Tragia urticifolia*, *Vaccinium arboreum*, *Verbena simplex*, *Verbena simplex*, *Vitis rotundifolia*, *Xyris tennesseensis*, *Yucca filamentosa*, *Zephyranthes atamasco*.

5.2.4.3. *Schoenolirion wrightii* Sherman

occurs in barrens, flatwoods (hardwood, pine), glades, openings (roadside, woodland), and prairies (saline) at elevations to 93 m. The plants grow under full sun or in partial shade. They occur on acidic to alkaline substrates (pH: 5.5–7.5) described as alluvium (sandy), Browndell, Kitterll, Lafe, loam, sand, sandstone (Catahoula, high sodium, tuffaceous), sandy clay loam, sandy loam (Falba), and silty loam (Bonn). Flowering occurs from March to May. The plants become dormant by mid-June. They persist by means of a cormose bulb. Although little additional life history information exists for this species, it is very similar to the previous (*S. croceum*), which some authors regard as conspecific (see *Systematics* below). **Reported associates:** *Allium drummondii, Aristida, Bigelowia nuttallii, Chaetopappa asteroides, Diamorpha smallii, Dichanthelium angustifolium, Drosera brevifolia, Fimbristylis puberula, Gratiola flava, Houstonia micrantha, Hypoxis hirsuta, Juniperus, Minuartia drummondii, Minuartia groenlandica, Nemastylis geminiflora, Neptunia lutea, Nostoc, Nothoscordum bivalve, Oenothera, Packera plattensis, Phacelia glabra, Phlox cuspidata, Plantago virginica, Quercus similis, Rhynchospora, Rumex hastatulus, Spiranthes parksii, Tetraneuris linearifolia, Valerianella.*

Use by wildlife: The flowers of *Schoenolirion albiflorum* are visited by various nectar-seeking insects (Insecta) such as bees (Hymenoptera), beetles (Coleoptera), butterflies and moths (Lepidoptera), and wasps (Hymenoptera); those of *S. croceum* also are visited by bees.

Economic importance: food: *Schoenolirion* species are not reported to be edible; **medicinal:** no uses reported; **cultivation:** not cultivated; **misc. products:** none; **weeds:** none; **non-indigenous species:** none.

Systematics: *Schoenolirion* resolves as a clade distinct from other "Chlorogaloid" members of Asparagaceae subfamily Agavoideae (Figure 2.8). It is the sister group of *Hesperaloe* and is not closely related to *Hastingsia* as once thought. The *rps19* locus is a pseudogene in *Schoenolirion* and *Hesperaloe*. Some authors have suggested that *S. wrightii* (white tepals) is simply a tepal color variant of the otherwise similar *S. croceum* (yellow tepals), but this possibility requires further evaluation. Phylogenetically, *S. wrightii* is associated more closely with *S. albiflorum* than *S. croceum* (Figure 2.8). Initially, *S. croceum* was thought to be distinct chromosomally from *S. wrightii* but more extensive surveys found that both taxa have in common a $2n=24$ cytotype. The basic chromosome number of *Schoenolirion* is $x=12$ [or $x=6$]; *S. albiflorum* ($2n=48$) is regarded as tetraploid; *S. croceum* ($2n=24, 30, 32$) and *S. wrightii* ($2n=24$) are regarded as diploid. Interspecific hybrids have not been reported or synthesized in the genus.

Distribution: *Schoenolirion* is endemic to the southeastern United States, with fairly restricted distributions for *S. albiflorum* (Florida and Georgia), *S. wrightii* (Alabama, Arkansas, Louisiana and Texas), and *S. croceum* (Alabama, Florida, Georgia, Louisiana, North Carolina, South Carolina, Tennessee, and Texas).

References: Bridges, 2005; Burbanck & Platt, 1964; Burbanck & Phillips, 1983; Carter et al., 2009; Chafin, 2007; Estes & Small, 2007; Houle & Delwaide, 1991; Houle & Phillips, 1988; 1989a; 1989b; Juras, 1997; Kral, 1973; Leslie & Burbanck, 1979; McKain et al., 2016a; Moffett, Jr., 2008; Murdy, 1966; Nixon & Ward, 1981; Peet, 2007; Sherman, 1979; 2002; Sherman & Becking, 1991; Sorrie et al., 2012; Taylor & Estes, 2012; Walck & Hidayati, 2004.

Family 5.3. Iridaceae [66]

Iridaceae occupy a rather isolated phylogenetic positions among the "lower asparagoid" groups of Asparagales (Figure 2.2), and the monophyly of this large family (approximately 2,244 species) has been confirmed in numerous studies employing both morphological and molecular data (Goldblatt et al., 2008; Christenhusz & Byng, 2016; Judd et al., 2016). More than half the species occur within only eight of the genera, the largest being *Iris* and *Gladiolus* (Cronquist, 1981). The classification of Iridaceae remains in a flux. Earlier phylogenetic studies of the group recognized six subfamilies and resolved Iridoideae (where all the OBL indicators occur) as the sister clade to Crocoideae (Meerow, 2012). However, more comprehensive phylogenetic studies later recommended the recognition of seven subfamilies and it now appears that subfamily Iridoideae is quite distant from Crocoideae (e.g., Figure 2.10). Within subfamily Iridoideae, the genera that contain OBL species are dispersed among different tribes (Irideae, Sisyrinchieae, and Tigridieae), which indicates multiple origins of the OBL habit within this primarily terrestrial family (Figure 2.10).

Iridaceae are herbaceous perennials (rarely annuals), which contain styloid calcium oxalate crystals in their vascular bundle sheaths (Judd et al., 2016). They possess equitant, 2-ranked, unifacial leaves (often gladiate), determinate inflorescences (scorpoid cymes), and flowers having three stamens and an inferior ovary (the ovary superior in the monotypic subfamily Isophysidoideae); the fruits are loculicidal capsules (Cronquist, 1981; Goldblatt, 2002; Judd et al., 2016).

The flowers are showy, colorful, usually nectariferous (bearing septal or tepal nectaries), and are pollinated by various nectar- or pollen-seeking insects (Insecta) such as bees (Hymenoptera), beetles (Coleoptera), and flies (Diptera), or sometimes by birds (Aves) (Judd et al., 2016). The seeds typically are dispersed abiotically by water or wind (Judd et al., 2016).

The family is important economically for its numerous ornamental species, which occur in many of the genera (e.g., *Crocosmia, Crocus, Dierama, Freesia, Gladiolus, Hesperantha, Iris, Moraea, Romulea, Sisyrinchium,* and *Watsonia*). It is also the source of saffron, the world's most expensive spice (Melnyk et al., 2010), which is derived from the stigmas of the saffron crocus (*Crocus sativus*).

Iridaceae are cosmopolitan, but their greatest diversity occurs in southern Africa (Cronquist, 1981). There are 16 genera in North America, with OBL wetland indicators contained within three of them:

5.3.1. ***Iris*** L.
5.3.2. ***Nemastylis*** Nutt.
5.3.3. ***Sisyrinchium*** L.

5.3.1. Iris

Flag, iris; iris, fleur-de-lis

Etymology: after *Iris*, the mythical Greek rainbow goddess, for the diverse floral colors

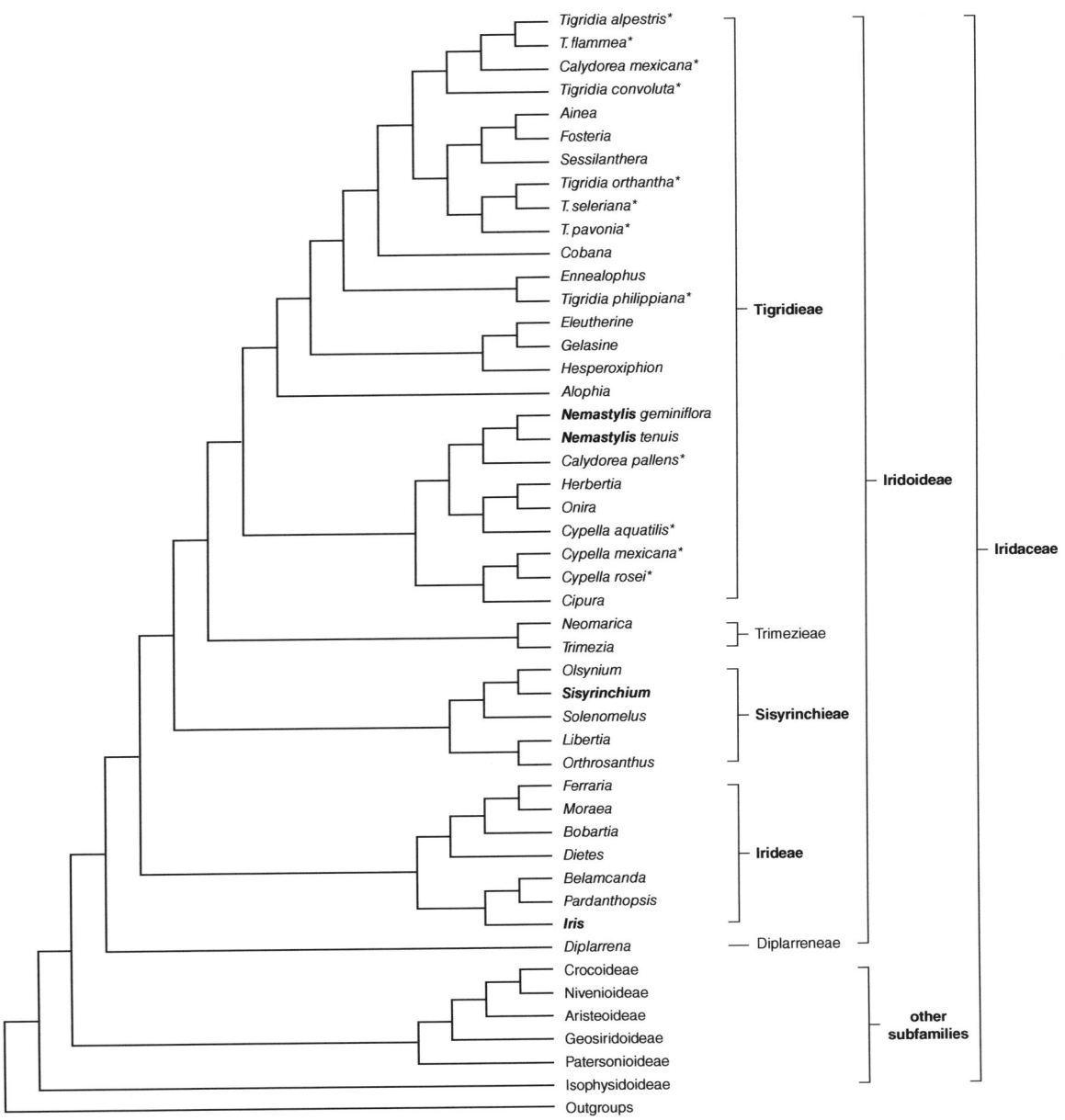

FIGURE 2.10 Subfamilial and tribal relationships within Iridaceae as evidenced by phylogenetic analysis of combined plastid DNA sequence data (modified from Goldblatt et al., 2008). *Belamcanda* and *Pardanthopsis* have since been merged with *Iris* (Wilson, 2011). Taxa containing OBL North American indicators (highlighted in bold) occur only within subfamily Iridoideae where they are dispersed within three separate tribes (Irideae, Sisyrinchieae, and Tigridieae). Although the OBL *N. floridana* was not sampled, a BLAST search of an unpublished *rbcL* sequence for *N. floridana* (GenBank: KX397869.1) shows it to be most similar to *N. tenuis*. Polyphyletic taxa are indicated by asterisks.

Synonyms: *Limnirion*; *Limniris*; *Moraea* (in part); *Neubeckia*; *Pseudo-iris*; *Vieusseuxia* (in part); *Xiphion*
Distribution: global: Northern Hemisphere; **North America:** widespread
Diversity: global: 275 species; **North America:** 42 species
Indicators (USA): obligate wetland (OBL): *Iris brevicaulis, I. fulva, I. giganticaerulea, I. hexagona, I. prismatica, I. pseudacorus, I. savannarum, I. tridentata, I. versicolor, I. virginica*
Habitat: brackish (coastal), freshwater; freshwater (tidal); palustrine; **pH:** 3.6–8.4; **depth:** <1 m; **life-form(s):** emergent herb

Key morphology: rhizomes (to 20 cm), superficial or buried, simple or branched, homogeneous, or heterogeneous with cordlike stolons (to 40 cm), producing leaves and flowering stems at apex; flowering stems (to 20+ dm) simple or 1–3-branched, erect, semi-erect, declining, or weak (falling after flowering), straight or sharply zigzag; leaves (3–10) basal, in fan-like spray, erect (sometimes recurving), prostrate, or spreading, sometimes deciduous, or dying back soon after anthesis, the blades (to 13 dm) linear-ensiform or ensiform, broad (to 3.5 cm) or narrow (to 0.7 cm), smooth or ridged, bright green, deep green, grayish-green, or yellow-green, glaucous or glaucescent, apex acute, curving asymmetrically; inflorescence 1–12-flowered, subtended by 2 spathes (to 20 cm), spathes often persistent and eventually enveloping the mature capsules; flowers (to 18 cm) epigynous, perfect, usually fragrant, pedicellate (to 8 cm), floral tube (to 2.5 cm) terete or ridged; sepals (to 11 cm) 3, prominent, spreading or reflexed, broadening from a claw to a widened limb, blue, blue-violet, coppery, reddish brown, violet, or bright yellow (rarely white), sometimes with greenish, greenish-yellow, or white to yellow "signal" spot; petals (to 9 cm) 3, erect, less-conspicuous or reduced (to 1.5 cm), blue, blue-violet, coppery, lavender, reddish brown, violet, or bright yellow (rarely white); stamens 3, opposite sepals, free but appressed to style branches; ovary (to 4.5 cm) rounded to sharply trigonal or hexagonal, the style dividing into 3 petaloid branches (to 6 cm), which arch over the stamens and their subtending sepals, each terminating apically as 2 rounded or triangular crests (to 5 cm), the stigmas 3, each an adaxial lip on style branch surface, bilobed, rounded, rounded-triangular, semicircular, sharply triangular, triangular, or unlobed; capsules (to 10 cm) dehiscent or indehiscent, cross-section hexagonal, hexagonally ribbed, obscurely to sharply 3-angled, or roundly 6-lobed; seeds (to 15 mm) 4–20, flattened, in 1 or 2 rows in each locule, D-shaped, irregular, semicircular, or pyriform, light to dark brown, the seed coat smooth or shiny and not corky, or thick and corky to very corky

Life history: duration: perennial (rhizomes); **asexual reproduction:** rhizomes (fragmentation), stolons; **pollination:** insect, self; **sexual condition:** hermaphroditic; **fruit:** capsules (common); **local dispersal:** rhizomes, seeds (water); **long-distance dispersal:** seeds (water, waterfowl)

Imperilment: 1. *Iris brevicaulis* [G4]; SH (KS); S1 (GA, KY, MS, OK, <u>ON</u>, TN); S2 (OH); S3 (IL); **2.** *I. fulva* [G5]; S1 (KY); S2 (IL, TN); S3 (MS); **3.** *I. giganticaerulea* [G3]; S3 (LA); **4.** *I. hexagona* [G4/G5]; S1 (SC); **5.** *I. prismatica* [G4/G5]; SH (GA); S1 (NC, <u>NS</u>, PA); S2 (DE, MD, ME, NH, NY, TN); S3 (VA); **6.** *I. pseudacorus*, [GNR]; **7.** *I. savannarum* [unranked]; **8.** *I. tridentata* [G3/G4]; S2 (GA, NC); **9.** *I. versicolor* [G5]; S1 (DC, <u>SK</u>); S2 (ID, <u>NU</u>); S3 (VA, <u>LB</u>, <u>MB</u>); **10.** *I. virginica* [G5]; S1 (KS, NY); S2 (OK, PA); S3 (KY, MD)

Ecology: general: Although most *Iris* species inhabit desiccated or dry, rocky, montane sites, about one-quarter of the North American species (24%) occur in hydric habitats and are designated as OBL wetland indicators. These grow on wet substrates or in shallow standing waters, which typically are seasonal and usually do not persist throughout the year. Most *Iris* flowers are large, showy, nectariferous, and well adapted for insect pollination. Many are self-incompatible, but selfing rates in the self-compatible species can exceed 74%. The flowers are adapted morphologically for outcrossing. Pollination usually is achieved by large, nectar-seeking insects such as bumblebees (e.g., Insecta: Hymenoptera: Apidae: *Bombus*), which are able to climb within the "gullet" formed by the sepal claw and style branch, thereby brushing any inborne pollen carried upon their backs onto the adhesive stigmatic surfaces that are exposed on a flap-like lip. Exiting insects force the stigmatic lip back against the style branches, which

covers the stigmas and precludes any deposition of self-pollen as it is carried away from the anthers. Some North American species also are pollinated by ruby-throated hummingbirds (Aves: Trochilidae: *Archilochus colubris*). Despite the adaptation of these features for outcrossing, much of the pollination (up to 80%) can occur between nearest neighbors. *Iris* seeds are dispersed occasionally by ants (Insecta: Hymenoptera: Formicidae), but most often are transported on the surface of still or flowing waters, facilitated by wind and their buoyant corky covering. Although the seeds of some species are known to remain afloat for more than a year, their long-range dispersal generally is believed to be relatively ineffective. Stomata, which are thought to regulate seed dormancy and water imbition, can be absent or present on *Iris* seed coats. Many *Iris* seeds are morphophysiologically dormant due primarily to their extremely hard seed coat and secondarily by a chemical inhibitor present in the endosperm. At least the northern species require a period of cold stratification ranging from 2 to 12 months in order to initiate germination. Seedling recruitment is not often observed in the field, where much of the reproduction occurs vegetatively. Strong rhizomatous growth enables clones to migrate slowly into more suitable habitats at rates of spread that can approach 1 m year^{-1}. Stolons (or "stoloniferous rhizomes") are produced by two North American species. Because *I. hexagona* formerly included both *I. giganticaerulea* and *I. savannarum* as varieties, most database and literature records attributed to the narrowly distributed *I. hexagona* clearly refer to one of the two other taxa by nature of the source of material originating from outside the restricted geographical range of *I. hexagona*. Unfortunately, it is virtually impossible to ascertain to which species these reports refer unless the varietal names have been included. Consequently, only sparse life history information could be compiled confidently for *I. hexagona* despite the existence of numerous accounts attributed [but incorrectly] to that species.

5.3.1.1. ***Iris brevicaulis* Raf.** grows in bottomlands (hardwood, rich, stream, wooded), depressions (semi-permanently flooded), ditches (roadside), floodplains (hardwood), marshes, meadows, prairies, seeps (forested), slopes, swamps (cypress, hardwood), woodlands (alluvial, open), and along the margins of bogs, lakes (marshy), ponds, rivers, swamps, and streams at elevations to 366 m. The plants occupy sunny, somewhat drier sites where they occur primarily above the mean water level on substrates that usually are waterlogged but rarely flooded for long periods of time. They can be found growing in shallow waters for short periods during the year. Reported substrates include clay (heavy), loam, sand, silty clay, and silty sand. Flowering extends from April to June with fruiting from May to July. The showy, deep blue to violet corollas attract worker bumblebees (Insecta: Hymenoptera: Apidae: *Bombus pensylvanicus*), which are the principal pollinators; however, pollination by ruby-throated hummingbirds (Aves: Trochilidae: *Archilochus colubris*) also occurs occasionally. Interestingly, the floral nectar is sucrose-dominant, which normally is associated with hummingbird rather than insect pollination. The flowers that open during the first morning are functionally male because their stigmatic lobes remain folded against the

style. The stigmatic surfaces eventually become receptive by the end of the second day. The flowers wilt after a day in the female stage as the sepals infold to block further access to the flower by floral visitors. Gene diversity in *I. brevicaulis* has been estimated (allozymically) as $H_e = 0.167$, with 1.7 mean alleles/locus. Genetic studies have demonstrated limited gene flow among populations, which is suspected to be a consequence of pollinator habitat fragmentation. The seeds are enclosed within a thick corky layer, which makes them buoyant if dispersed in water; however, the frequent occurrence of plants among more isolated, somewhat drier sites, might also restrict gene flow to some degree. The seed germination rates are low (e.g., ~20–30% after 7 months for 5-year-old seeds) but can be facilitated by physical scarification (remove seeds from their corky covering, cut the seed coat with a razor blade, remove a small piece of endosperm to expose embryo) prior to planting. The seeds remain viable for an undetermined period (but at least several years), which leads to the development of persistent seed banks. Experimentally, seed germination has been obtained only under moist, aerobic conditions, which mimic the conditions optimal for seedling survival in the wild. Germination rates are somewhat higher in shade than in the light. Seedling survivorship is also low, but those surviving seedlings are vigorous. The branching rhizomes result in the development of dense clonal clumps.
Reported associates: *Acer negundo, Acer rubrum, Acer saccharinum, Acer saccharum, Alisma, Ambrosia trifida, Amsonia tabernaemontana, Atriplex cristata, Bacopa monnieri, Bacopa rotundifolia, Bromus, Cardamine bulbosa, Carex annectens, Carex conjuncta, Carex crus-corvi, Carex davisii, Carex glaucescens, Carex grisea, Carex hyalinolepis, Carex laevivaginata, Carex lurida, Carex muskingumensis, Carex shortiana, Carex squarrosa, Carex vulpinoidea, Carya illinoinensis, Carya laciniosa, Carya ovata, Chasmanthium latifolium, Chasmanthium latifolium, Clematis pitcheri, Dichanthelium commutatum, Distichlis spicata, Eleocharis engelmannii, Eleocharis quadrangulata, Eleocharis tenuis, Euthamia graminifolia, Fimbristylis spadicea, Fraxinus nigra, Fraxinus pennsylvanica, Fraxinus profunda, Galium obtusum, Glyceria striata, Heliotropium curassavicum, Hibiscus moscheutos, Hydrolea ovata, Hymenocallis liriosme, Ipomoea hederacea, Ipomoea lacunosa, Iris virginica, Iva angustifolia, Juglans nigra, Juncus effusus, Juncus nodatus, Juncus torreyi, Justicia americana, Lathyrus, Leersia oryzoides, Liquidambar styraciflua, Ludwigia, Lythrum lineare, Myosotis macrosperma, Nyssa, Panicum hemitomon, Panicum virgatum, Paspalidium geminatum, Paspalum wrightii, Persicaria hydropiperoides, Persicaria virginiana, Phalaris arundinacea, Phlox glaberrima, Phyla nodiflora, Pinus echinata, Pluchea camphorata, Polygonum aviculare, Pontederia cordata, Proserpinaca palustris, Proserpinaca pectinata, Pycnanthemum tenuifolium, Quercus bicolor, Quercus nigra, Quercus shumardii, Ranunculus hispidus, Rhynchospora corniculata, Rosa multiflora, Rosa palustris, Rubus, Sabal minor, Sagittaria graminea, Sagittaria lancifolia, Sagittaria papillosa, Sambucus, Saururus cernuus, Schedonorus pratensis, Scirpus atrovirens, Scirpus pendulus,*

Sesbania drummondii, Setaria faberi, Smilax tamnoides, Solidago canadensis, Solidago gigantea, Solidago sempervirens Spartina pectinata, Symphyotrichum lanceolatum, Symphyotrichum subulatum, Taxodium distichum, Thalia dealbata, Toxicodendron radicans, Tradescantia occidentalis, Trepocarpus aethusae, Tridens strictus, Typha latifolia, Viola cucullata.

5.3.1.2. ***Iris fulva* Ker Gawl.** occurs in bayous, bottoms (oakwood), depressions, ditches (drainage, railroad right-of-way, roadside), flatwoods, floodplains, marshes, meadows (wet), roadsides (wet), sloughs, swamps (cypress), and along the margins of canals, lakes, levees, ponds, and streams at elevations to 161 m. The plants are fairly flood tolerant and are found across wetness gradients ranging from moist substrates to seasonal shallow standing water (e.g., 10–50 cm deep). They grow in full sun but also are shade tolerant and occur often under the shade of trees and shrubs. The substrates have been described as alluvium (clayey, loamy), clay (heavy), gumbo, humus, muck, mud, sand, sandy silt, and silty clay loam (Cancienne). Flowering and fruiting occur from April to June. The basic floral ecology is similar to that of the previous species, i.e., the flowers that open during the first morning are functionally male because their stigmatic lobes remain folded against the style. The stigmatic surfaces become receptive toward the end of the third day. The flowers wilt after 1–2 days in the female stage as the sepals infold to block further access by floral visitors. However, this species is quite anomalous in having scentless flowers with a dark red to bronze perianth (its coloration due to a combination of carotenoids, delphinidin, and malvidin), an elongate floral gullet, and by the production of large amounts of sucrose-dominant floral nectar. Consequently, they are more suitably adapted for, and are pollinated primarily by ruby-throated hummingbirds (Aves: Trochilidae: *Archilochus colubris*). Some pollination also occurs by bumblebees (Insecta: Hymenoptera: Apidae: *Bombus pensylvanicus*). The mean outcrossing rate (determined from multi-locus allozyme data) is high (0.752) but decreases as more flowers open on different stems, while increasing when more flowers are open on the same stem. A low level of biparental inbreeding also has been detected in some populations. Gene diversity in *I. fulva* has been estimated (allozymically) as $H_e = 0.122$, with 1.4 mean alleles/locus. Genetic studies have shown there to be no interpopulational restriction to gene flow, possibly because the wide-ranging pollinators (hummingbirds) are able to overcome ongoing habitat fragmentation. Within the inflorescence, the highest fruit set and seed production occurs in the first open (most apical) flower. The capsules remain green and photosynthetic as the seeds mature. The seeds are enclosed within a buoyant corky outer covering and are dispersed on the surface of water. Their germination has been achieved only under moist, aerobic conditions (which are optimal for seedling survival). Germination rates are substantially higher in the shade than in the light. **Reported associates:** *Acer rubrum, Acorus calamus, Agrimonia parviflora, Alternanthera philoxeroides, Arundinaria gigantea, Asclepias incarnata, Asclepias perennis, Bidens laevis, Bolboschoenus fluviatilis, Bromus,*

Brunnichia ovata, Carex cherokeensis, Carex comosa, Carex crus-corvi, Carex decomposita, Carex hyalina, Carex hyalinolepis, Carex lupuliformis, Carex lupulina, Carex retrorsa, Carex socialis, Carex stipata, Carya aquatica, Celtis laevigata, Cephalanthus occidentalis, Chaerophyllum tainturieri, Clematis crispa, Cornus drummondii, Cornus foemina, Crataegus viridis, Diospyros virginiana, Dulichium arundinaceum, Fraxinus pennsylvanica, Fraxinus profunda, Galium aparine, Gleditsia aquatica, Glyceria fluitans, Glyceria striata, Hibiscus moscheutos, Hydrocotyle, Hymenocallis, Iris hexagona, Iris virginica, Iris giganticaerulea, Juncus effusus, Justicia ovata, Leersia, Leitneria floridana, Lindera melissifolia, Ludwigia alternifolia, Lythrum alatum, Nyssa aquatica, Nyssa sylvatica, Penthorum sedoides, Persicaria, Phragmites australis, Planera aquatica, Populus heterophylla, Quercus lyrata, Quercus nigra, Quercus pagoda, Quercus palustris, Quercus texana, Ranunculus pusillus, Rhynchospora corniculata, Rumex verticillatus, Sagittaria graminea, Sagittaria lancifolia, Salix nigra, Saururus cernuus, Schoenoplectus lacustris, Scirpus cyperinus, Sisyrinchium angustifolium, Styrax americanus, Taxodium distichum, Tradescantia ohiensis, Typha latifolia, Ulmus crassifolia, Utricularia gibba, Utricularia macrorhiza, Viola sororia, Zizania palustris.

5.3.1.3. ***Iris giganticaerulea* Small** is a species of brackish or freshwaters that are associated with bays, bottomlands (forested), ditches, marshes (brackish, freshwater, oligohaline), meadows (wet), roadsides, slopes (to 8%), swamps, and the margins of canals, ponds, and rivers (brackish) at elevations to at least 15 m. The plants preferentially occupy sites having full sunlight exposures. The substrates have been categorized as alluvium (clayey, loamy), clay (Schriever), sandy loam (fine, Harleston), shell middens, and silty clay loam (Cancienne, Schriever). These are large plants (as their name implies), with stems reaching 2 m or more. Flowering has been observed from March to April. Specific accounts on the floral biology of this species are difficult to evaluate, given its lengthy taxonomic treatment as a variety of *I. hexagona*. Because it is known to hybridize with *I. brevicaulis* and *I. fulva* (see Systematics), it must also share their pollinators (see above). In this case, the blue perianth coloration would implicate bees (Insecta: Hymenoptera: Apidae: *Bombus*) as the most likely primary pollinators. Unlike its North American congeners, the capsules are indehiscent, which results in a distinct mechanism of seed dispersal. Following anthesis, the drooping indehiscent capsules are pushed into the water by arching of the elongating floral stem. Subsequently, the carpels deteriorate and release the seeds, which rise to surface by means of their buoyant corky covering. The seeds are dispersed abiotically by currents on the water surface. The seed coats possess stomata, which are thought to be involved in the regulation of seed germination. Because of their geographical provenance (e.g., Alabama, Georgia, Louisiana), many literature and database accounts attributed to *I. hexagona* (which is not known to occur in these states) apparently refer instead to this or another species. However, these records have not been used here because of their uncertainty. **Reported associates:** *Acer rubrum, Alternanthera philoxeroides, Ampelopsis*

cordata, Bidens laevis, Cabomba caroliniana, Campsis radicans, Carex cherokeensis, Carex longii, Centella asiatica, Cephalanthus occidentalis, Cirsium horridulum, Crinum bulbispermum, Cyperus, Eichhornia crassipes, Eleocharis cellulosa, Eleocharis fallax, Eleocharis montana, Festuca subverticillata, Fraxinus profunda, Hibiscus moscheutos, Hydrocotyle bonariensis, Hypericum hypericoides, Ilex vomitoria, Iris fulva, Iva frutescens, Juncus, Kosteletzkya pentacarpos, Leersia hexandra, Liquidambar styraciflua, Magnolia virginiana, Myrica cerifera, Najas guadalupensis, Nyssa aquatica, Operculina turpethum, Opuntia humifusa, Osmunda regalis, Pancratium maritimum, Panicum hemitomon, Persicaria punctata, Phyla lanceolata, Pinus elliottii, Pteridium aquilinum, Quercus nigra, Quercus virginiana, Sagittaria lancifolia, Schoenoplectus californicus, Schoenoplectus tabernaemontani, Serenoa repens, Smilax, Sphenopholis obtusata, Taxodium distichum, Thelypteris kunthii, Tillandsia usneoides, Toxicodendron radicans, Tradescantia hirsutiflora, Vallisneria americana, Verbena brasiliensis, Vigna luteola, Vitis rotundifolia, Zizaniopsis miliacea.

5.3.1.4. ***Iris hexagona* Walter** inhabits freshwater swamps (cypress, roadside) at low elevations. Because of taxonomic confusion (see *Ecology: general:* above) it is extraordinarily difficult to extract information on this geographically restricted species from the literature or from herbarium databases. Consequently, hardly any information ascribed to "*Iris hexagona*" can be attributed to this species reliably, and a focused study of its life history (based on the correct taxonomic circumscription) is necessary for clarification. On the other hand, the basic reproductive ecology and floral biology appear to be similar among all members of series *Hexagonae* (e.g., see details under *I. brevicaulis* and *I. fulva* above). Some plants have been observed growing on sandy substrates. Flowering occurs from April to May. The pollination biology presumably is similar to other members of series *Hexagonae*, with the blue perianths likely attracting bees (Insecta: Hymenoptera: Apidae: *Bombus*) as the primary pollinators. The buoyant, corky seeds are dispersed by water. **Reported associates:** *Crinum americanum, Hymenocallis rotata, Taxodium.*

5.3.1.5. ***Iris prismatica* Pursh ex Ker Gawl.** grows in brackish to freshwater sites within backwaters (river), bogs, depressions (sandy, swamp), ditches (marshy), floodplain terraces, marshes (freshwater, roadside), meadows (brackish, river, wet), prairies (low), roadsides, savannas (disturbed), slopes (ravine), swales (boggy, interdunal), swamps, thickets, woodlands (wet), and along the margins of ponds (coastal plain), salt marshes, and streams at elevations to 283 m. The sites typically have open, sunny exposures; shallow standing water sometimes is present. Populations are found primarily in coastal Atlantic areas but occur sporadically inland in the southeastern United States. The substrates are acidic (e.g., pH: 5.0–5.2) and have been described as chert (silicious), humus, muck, peat, and sand. Flowering occurs from April to August with fruiting extending from June to September. There have been no detailed studies on the reproductive biology, but the flowers presumably are pollinated (as in most *Iris* species) by bees (Insecta: Hymenoptera). Unlike other OBL *Iris*, the seeds are smooth rather than corky, and it is unclear if they are as buoyant or how they are principally dispersed. A germination rate of 45% (within 21–35 days) has been obtained for seeds that were cold-stratified (at ambient outdoor winter temperatures) for 83 days, and then transferred to ambient greenhouse temperatures. The seedlings are said to "come easily" in gardens from fall planted seeds. Vegetative reproduction is by rhizomes, which are buried only shallowly and lie at or near the surface. The rhizomes produce elongate, slender (40 cm×2–5 mm), branching stolons (also described as "stoloniferous rhizomes"), which give rise to more diffuse clonal plant clusters (ramets). **Reported associates:** *Acer rubrum, Acorus calamus, Agalinis maritima, Agrostis stolonifera, Alnus serrulata, Amelanchier canadensis, Andropogon glomeratus, Andropogon virginicus, Anthoxanthum nitens, Apios americana, Aronia arbutifolia, Aronia melanocarpa, Artemisia vulgaris, Asclepias incarnata, Atriplex prostrata, Baccharis halimifolia, Bidens frondosus, Bolboschoenus maritimus, Calamagrostis canadensis, Calopogon tuberosus, Carex bullata, Carex buxbaumii, Carex exilis, Carex hormathodes, Carex livida, Carex lupuliformis, Carex lurida, Carex maritima, Carex striata, Carex stricta, Cephalanthus occidentalis, Chamaecyparis thyoides, Chamaedaphne calyculata, Cicuta maculata, Cirsium horridulum, Cladium mariscoides, Clethra alnifolia, Coleataenia longifolia, Comarum palustre, Cuscuta, Cyperus dentatus, Danthonia sericea, Dichanthelium dichotomum, Dichanthelium ensifolium, Dichanthelium scabriusculum, Diospyros virginiana, Distichlis spicata, Drosera anglica, Drosera filiformis, Drosera intermedia, Drosera rotundifolia, Dulichium arundinaceum, Eleocharis acicularis, Eleocharis flavescens, Eleocharis robbinsii, Eleocharis tenuis, Eleocharis tuberculosa, Equisetum arvense, Eriocaulon aquaticum, Eriocaulon decangulare, Eubotrys racemosa, Euthamia graminifolia, Fimbristylis autumnalis, Gaultheria procumbens, Gaylussacia baccata, Gaylussacia dumosa, Glyceria obtusa, Helenium pinnatifidum, Hibiscus moscheutos, Hudsonia ericoides, Hypericum canadense, Hypericum densiflorum, Hypericum denticulatum, Hypericum mutilum, Ilex glabra, Ilex verticillata, Impatiens capensis, Iris pseudacorus, Itea virginica, Iva frutescens, Juncus canadensis, Juncus effusus, Juncus gerardii, Juncus militaris, Juncus pelocarpus, Juniperus virginiana, Kalmia angustifolia, Kalmia buxifolia, Krigia virginica, Lachnanthes caroliniana, Leersia oryzoides, Lilium superbum, Limonium carolinianum, Liquidambar styraciflua, Lobelia nuttallii, Lonicera morrowii, Lophiola aurea, Ludwigia palustris, Lycopodiella alopecuroides, Lycopodiella inundata, Lycopus americanus, Lycopus uniflorus, Lyonia mariana, Lysimachia terrestris, Lythrum salicaria, Magnolia virginiana, Muhlenbergia torreyana, Myrica gale, Myrica pensylvanica, Nymphaea odorata, Nyssa sylvatica, Oclemena nemoralis, Oenothera fruticosa, Oenothera perennis, Onoclea sensibilis, Orontium aquaticum, Osmunda regalis, Osmundastrum cinnamomeum, Panicum verrucosum, Panicum virgatum, Paspalum praecox, Peltandra virginica, Phragmites australis, Pinus rigida, Platanthera clavellata, Platanthera grandiflora, Platanthera psycodes, Pluchea camphorata, Pluchea odorata, Pogonia ophioglossoides, Polygala cruciata, Polygala lutea, Polytrichum commune, Potamogeton confervoides, Potentilla anserina, Potentilla simplex, Prunus serotina, Ptilimnium capillaceum, Puccinellia maritima, Pycnanthemum virginianum, Pyxidanthera barbulata, Quercus palustris, Rhexia virginica, Rhododendron viscosum, Rhynchospora alba, Rhamnus cathartica, Rhynchospora capitellata, Rhynchospora fusca, Rhynchospora latifolia, Rosa carolina, Rosa multiflora, Rubus hispidus, Sabatia difformis, Sagittaria engelmanniana, Salicornia maritima, Samolus valerandi, Sanguisorba canadensis, Sarracenia purpurea, Schizachyrium scoparium, Schizaea pusilla, Schoenoplectus americanus, Schoenoplectus pungens, Schoenoplectus subterminalis, Scirpus cyperinus, Sclerolepis uniflora, Scutellaria galericulata, Sisyrinchium atlanticum, Smilax glauca, Smilax herbacea, Smilax pseudochina, Smilax rotundifolia, Solanum dulcamara, Solidago sempervirens, Sparganium americanum, Spartina alterniflora, Spartina patens, Spergularia salina, Sphagnum palustre, Spiraea tomentosa, Stellaria longifolia, Suaeda linearis, Symphyotrichum novi-belgii, Symphyotrichum subulatum, Symphyotrichum tenuifolium, Symplocarpus foetidus, Taxodium ascendens, Teucrium canadense, Thelypteris palustris, Toxicodendron radicans, Triadenum virginicum, Triglochin maritima, Tripsacum dactyloides, Typha angustifolia, Utricularia cornuta, Utricularia gibba, Vaccinium angustifolium, Vaccinium corymbosum, Vaccinium macrocarpon, Vernonia noveboracensis, Viburnum dentatum, Viburnum nudum, Viburnum recognitum, Viola lanceolata, Xyris difformis, Xyris torta, Zizania aquatica.*

5.3.1.6. *Iris pseudacorus* **L.** is a nonindigenous species which grows in fresh to brackish waters (salinity to 24‰) associated with backwaters (river), bogs, bottomlands (wooded), canals (drainage, irrigation), channels (backwater), culverts, ditches (drainage, irrigation, roadside), flats, floodplains, gravel pits, marshes (freshwater, intertidal, tidal), meadows, mudflats, pools, potholes (glacial), roadsides, shorelines (coastal), slopes [to 1%] (seepage), swales (beach), swamps, woodlands (riparian), and along the margins (dried) of lagoons, lakes (artificial), ponds (artificial), reservoirs (artificial), rivers, saltmarshes, and streams at elevations to 1750 m. The plants occur in exposures of full sunlight to partial shade, but are intolerant of deep shade. They can grow on moist substrates or be inundated by up to 1 m of water, with prolonged tolerance for waters up to 25 cm deep and NaCl levels up to 10 mg g^{-1}. A wide variety of acidic or calcareous substrates (pH: 3.6–8.4; $\bar{x}=6.0$) is colonized, including alluvium (clayey), clay (heavy, sandstone, schist, Schriever), clay loam, cobble (large), gravel, muck, mud, peat, rimrock (basalt), rock, sand, sandy loam, silt, silty clay, and silty loam (Bodine-Brandon, cherty). Flowering occurs from April to July (more prolifically in open, sunny sites), with fruiting extending from June through October. The flowers are self-compatible and capable of self-pollination but are adapted morphologically for outcrossing.

Genetic analyses of some North American populations indicate virtually obligate outcrossing, with 97.7% of the sampled plants exhibiting unique (by AFLP data) genotypes. In North America, the flowers are known to be pollinated by bumblebees (Insecta: Hymenoptera: Apidae: *Bombus vagans*) and on occasion by ruby-throated hummingbirds (Aves: Trochilidae: *Archilochus colubris*); long-tongued flies (Insecta: Diptera) also have been observed as pollinators elsewhere. The nectar is neutral (pH: $\bar{x} = 7.1$) and contains from 10.9% to 27.7% sugar ($\bar{x} = 16.8\%$), comprising on average 71.4% sucrose, 10.0 % fructose, and 18.6% glucose. The unusually high sucrose content (similar to other *Iris* species evaluated) makes it suitable for hummingbird feeding. The flowers are known to open rapidly ("explosively") presumably to consistently attract the pollinators during the receptive period, provide them with a landing surface, and guide them to the nectar source. They remain open to pollinators only for about a day before wilting. During pollination, the insects crawl between a sepal and stylar lobe as they seek nectar at the base of the perianth tube, brushing any pollen adhering to their backs from previous floral visits onto the stigmatic area during the process. As they crawl deeper toward the nectar, they pass beneath the enclosed stamen, which deposits fresh pollen on their back. Because the floral parts are large and under substantial tension, only powerful, larger-bodied insects can separate them to function as effective pollinators. Individual plants develop 3–6 capsules on average, each eventually yielding about 32–60 viable seeds. The seeds lack stomata on their coats and are dispersed primarily by water. Most fresh seeds can float for at least 1 week, with up to 95% of seeds able to remain afloat after 2 months. In one study, 10% of the seeds stayed afloat after 429 days in the water. Wind-facilitated seed dispersal rates up to 0.33 m s^{-1} have been recorded from ditch systems. The large seeds also are eaten by waterfowl (Aves: Anatidae), which often regurgitate them rather than pass them digestively. The regurgitation of seeds (which can occur 10–12 hr after ingestion) is believed to facilitate long-distance dispersal, at least on occasion. The seeds germinate after they become stranded along shorelines or on other exposed, wet substrates. They are dormant, which has been attributed to an inhibitor present in the inner seed coat. About 20% of those shed in the fall (e.g., October) germinate in late winter to early spring (February–June); another 20% germinate during the following spring. Washed seeds that were left to overwinter in trays germinated well (to 85%) from early May to June. The seeds are polymorphic with respect to their relative mass (light or heavy) and shape (rounded or flat). The lighter seeds are more buoyant due to air trapped in the seed coat. Of these, the flattened ones germinate at a higher rate and more rapidly than rounded seeds. Of the heavier seeds, germination rates are equal between the flat or round variants. At 5°C, about half of fresh seeds have germinated within 2 weeks (up to 70% if scarified), with lower rates (8–40%) observed after 3–12 months. After 10 weeks of cold stratification (at 0°C), only 8.5% of seeds germinated at 25°C. Germination rates for seeds stored in distilled water for 3 weeks have varied from 5% (at 20°C) to 52% (at 30°C), with none germinating below 15°C. A germination rate of 80% has been reported using a 12/12 hr, 30°C/20°C day/night incubation regime. Fresh seeds from some North American sites are highly viable (to 99.1%) and have germinated at rates from 33% to 62% when incubated under moist or inundated conditions. In one study, the removal of the seed "caps" yielded germination rates of 97% within 30 days in contrast to a complete lack of germination in unmanipulated seeds. The extent and duration of seed banks is not well understood, with some seed bank studies finding seeds to be absent, whereas others have found germinable seeds present in up to 25% of the samples taken. Seeds that were stored in seawater have remained viable for up to 31 days. Seedlings are observed rarely in the field, with most appearing during spring. They require a moist substrate but tolerate inundation poorly. Seedling densities up to 32 m^{-2} have been recorded. Their mortality is highest within 2 months of germination and increases as the surface waters freeze. From 28% to 72% of seedlings have been observed to survive the first year, with only 6% ultimately growing into fully mature plants. The adult plants are well adapted to wetland habitats. During the growing season, the leaves can survive total anoxia in the dark for as long as 28 days (up to 60 days when overwintering). The plants are highly fire-resistant and re-sprout readily from the rhizomes when burned. Intact clones can grow to 0.66 m in diameter, but the rhizomes degrade and fragment within 6–15 years, which leads to the establishment of separate ramets. The rhizomes are highly drought tolerant and can remain viable after 3 months without water. Clonal growth is reduced under salt stress, but the plants still retain a competitive advantage over native *Iris* species under such conditions. Despite their capacity for vegetative growth, genetic studies of some North American sites have determined that the plants disperse almost exclusively by seed rather than by rhizome fragments. They can be propagated artificially using *in vitro* ovary culture methods. **Reported associates (North America):** *Acer negundo, Acer rubrum, Acer saccharinum, Achnatherum hymenoides, Acorus calamus, Adenostoma fasciculatum, Agrostis gigantea, Alnus serrulata, Alopecurus, Ammophila arenaria, Amorpha fruticosa, Ampelopsis cordata, Amphicarpa bracteata, Anemone canadensis, Angelica atropurpurea, Angelica lucida, Apocynum cannabinum, Arctium lappa, Aronia ×prunifolia, Asclepias incarnata, Betula nigra, Betula papyrifera, Bidens cernuus, Boehmeria cylindrica, Bolboschoenus fluviatilis, Brasenia schreberi, Bromus inermis, Bromus japonicus, Calamagrostis canadensis, Callitriche peploides, Caltha palustris, Cardamine pensylvanica, Carex aquatilis, Carex atherodes, Carex blanda, Carex crinita, Carex cristatella, Carex cusickii, Carex gracillima, Carex grayi, Carex gynandra, Carex laxiflora, Carex lurida, Carex molesta, Carex obnupta, Carex plantaginea, Carex projecta, Carex retrorsa, Carex sartwellii, Carex shortiana, Carex stipata, Carex torta, Carex tuckermanii, Carex utriculata, Carex vesicaria, Carex vulpinoidea, Celtis, Cephalanthus occidentalis, Chara, Cicuta bulbifera, Cinna arundinacea, Cirsium arvense, Cirsium vulgare, Comarum palustre, Cornus racemosa, Cornus sericea, Crataegus, Cryptotaenia canadensis, Cyperus squarrosus, Dactylis* *glomerata, Delairea odorata, Distichlis spicata, Dulichium arundinaceum, Eleocharis erythropoda, Eleocharis palustris, Elodea canadensis, Epipactis gigantea, Equisetum laevigatum, Erigeron annuus, Eriodictyon crassifolium, Eriogonum fasciculatum, Eutrochium maculatum, Fagus grandifolia, Festuca, Fontinalis antipyretica, Fragaria chiloensis, Fraxinus latifolia, Fraxinus pennsylvanica, Fraxinus uhdei, Galium aparine, Halesia carolina, Hemerocallis fulva, Hesperis matronalis, Holcus lanatus, Hordeum jubatum, Humulus japonicus, Ilex verticillata, Impatiens capensis, Impatiens pallida, Iris prismatica, Iris versicolor, Iris virginica, Itea virginica, Juglans nigra, Juncus effusus, Justicia americana, Kalmia, Krigia caespitosa, Laportea canadensis, Leersia oryzoides, Leersia virginica, Lemna, Leptochloa, Lilaeopsis, Liquidambar styraciflua, Liriodendron tulipifera, Ludwigia palustris, Luzula echinata, Lycopus uniflorus, Lysimachia nummularia, Lythrum alatum, Lythrum salicaria, Magnolia tripetala, Magnolia virginiana, Mazus pumilus, Melilotus officinalis, Mentha ×piperita, Mimulus ringens, Myosotis scorpioides, Myrica gale, Myriophyllum spicatum, Nasturtium, Nuphar polysepala, Nuphar variegata, Nymphaea odorata, Nyssa sylvatica, Orontium aquaticum, Onoclea sensibilis, Packera aurea, Peltandra virginica, Phalaris arundinacea, Physostegia virginiana, Pinus strobus, Plantago, Platanus occidentalis, Poa nemoralis, Poa pratensis, Pontederia cordata, Populus deltoides, Populus tremuloides, Potamogeton amplifolius, Potamogeton natans, Potamogeton praelongus, Potentilla anserina, Pseudognaphalium microcephalum, Quercus palustris, Quercus phellos, Quercus shumardii, Ranunculus hispidus, Ranunculus pensylvanicus, Ranunculus repens, Rhamnus frangula, Rhododendron groenlandicum, Robinia, Rorippa palustris, Rosa palustris, Rosa rugosa, Rubus bifrons, Rubus idaeus, Rudbeckia laciniata, Rumex altissimus, Rumex crispus, Rumex verticillatus, Sagittaria australis, Sagittaria cuneata, Sagittaria latifolia, Salix amygdaloides, Salix exigua, Salix gooddingii, Salix lasiolepis, Salix interior, Salix nigra, Salvia mellifera, Sambucus nigra, Saururus cernuus, Schedonorus arundinaceus, Schoenoplectus acutus, Schoenoplectus lacustris, Schoenoplectus subterminalis, Schoenoplectus tabernaemontani, Schoenoplectus triqueter, Scirpus atrovirens, Scutellaria galericulata, Sicyos angulatus, Silene latifolia, Sisymbrium irio, Solanum dulcamara, Solidago gigantea, Sparganium, Spartina pectinata, Spiraea douglasii, Stellaria longifolia, Symphyotrichum lanceolatum, Symphyotrichum novae-angliae, Symphyotrichum ontarionis, Taraxacum officinale, Taxodium distichum, Thalictrum dasycarpum, Thuja, Tilia americana, Toxicodendron diversilobum, Toxicodendron radicans, Triadenum fraseri, Trifolium arvense, Typha angustifolia, Typha latifolia, Ulmus americana, Urtica dioica, Utricularia macrorhiza, Vallisneria americana, Verbena hastata, Verbesina alternifolia, Viburnum dentatum, Viola cucullata, Viola sororia, Viola striata, Vitis riparia, Xanthium.*

5.3.1.7. ***Iris savannarum* Small** inhabits brackish to freshwater baygalls, canals (drainage), depressions (roadside), ditches (roadside), flatwoods (pine), floodplains, hammocks (hydric), marshes (depression), pinelands, pools (seasonal), sloughs, swamps (basin, cypress, floodplain, hardwood, strand), and the margins of lakes, ponds, rivers, and streams, and swamps at elevations to at least 5 m. The exposures can range from full sun to shade. The substrates have been described as loam (black, rich), muck, mud, and sand. This species contains ecotypes that are distinct both morphologically and genetically. The "highland" ecotypes have more slender and drooping petals and occupy drier habitats, whereas the "coastal" ecotypes have broader petals and occupy wetter habitats. Flowering occurs from February to April with fruiting extending from March to June. Because of taxonomic confusion (see *Systematics*), many literature accounts and database records attributed to "*Iris hexagona*" undoubtedly refer to this species instead; however, information has been included here only from the relatively few unambiguous sources. Consequently, it is difficult to extract information that refers specifically to this taxon, including details of its reproductive biology. Presumably, these characteristics are similar to those of closely related members of series *Hexagonae* (see details under *I. brevicaulis* and *I. fulva* above). Like other blue-flowered *Hexagonae*, the plants likely are pollinated primarily by bees (Insecta: Hymenoptera: Apidae: *Bombus*). Genetic analyses (microsatellite marker data) indicate that the plants predominantly are outcrossed with populations characterized by high mean gene diversity ($H_e = 0.457–0.749$) and from 3.3 to 10.2 alleles per locus. The loculicidal capsules release seeds with a thick, buoyant, corky covering. Seed dispersal occurs primarily by water. **Reported associates:** *Acer rubrum, Hibiscus, Juncus effusus, Pinus palustris, Polygala, Quercus, Scirpus, Teucrium, Thalia, Typha.*

5.3.1.8. ***Iris tridentata* Pursh** occurs in barrens (pine), bogs (seepage), Carolina bays, depressions (savanna), ditches (flatwoods, railroad, roadside), flatwoods (pine), meadows (depressions), savannas (wet), sloughs, swamps (floodplain), and along the margins of lakes, pocosins, ponds, pools (pineland), and streams mainly on the coastal plain at low elevations to at least 36 m. The plants can occur in sunny exposures but are found more often in shaded sites. They sometimes grow in shallow standing water (e.g., 30 cm). The substrates have been described as clay, mud, peat, peaty sand, Rains series, sand (fine, Mascotte, peaty), and sandy peat. Analyzed substrates (savanna populations) are high in silt (46.5%), with lesser proportions of sand (32.6%) and clay (20.9%); the organic matter content is fairly high (e.g., 6.6%). They often are acidic with a low cation exchange capacity (CEC), e.g., pH: 4.5; CEC: 4.42 meq 100 g^{-1}. Representative nutrient levels are: Al: 1403 ppm, B: 0.36 ppm, Ca: 196.5 ppm, Cu: 0.20 ppm, Fe: 184.5 ppm, K: 39.5 ppm, Mg: 48.5 ppm, Mn: 1.5 ppm, N: 54 ppm, Na: 30 ppm, P: 10.5 ppm, S: 29.5 ppm, and Zn: 0.7 ppm. Flowering occurs from May to June, with fruiting from August to October. There have been no detailed studies on the floral biology or seed ecology, which presumably would be similar to that of other members of subgenus *Limniris*. The seeds have a thick seed coat, but their degree of buoyancy has not yet been determined. Vegetative reproduction occurs by means of the thick rhizomes (to 12 × 2 cm), which can branch profusely to typically produce from 2 to 4 stolons, which radiate outward (to 10 cm) from the plants. This growth habit results in more scattered clusters of plants than those *Iris* species lacking stolons. The plants have been reported often from annually burned sites. **Reported associates:** *Acer rubrum, Aesculus, Agalinis linifolia, Agalinis purpurea, Aletris aurea, Aletris farinosa, Alnus serrulata, Amianthium muscitoxicum, Amphicarpum, Andropogon gerardii, Andropogon glomeratus, Andropogon virginicus, Anthaenantia rufa, Aristida palustris, Aristida purpurascens, Arundinaria gigantea, Asclepias rubra, Baccharis halimifolia, Bigelowia nudata, Buchnera americana, Calopogon barbatus, Carex glaucescens, Carex lonchocarpa, Carex striata, Carex verrucosa, Centella asiatica, Cephalanthus occidentalis, Cercis, Chamaecrista nictitans, Chaptalia tomentosa, Chrysopsis mariana, Cirsium horridulum, Cirsium virginianum, Clethra alnifolia, Clethra tomentosa, Coleataenia anceps, Coleataenia longifolia, Coreopsis gladiata, Crotalaria purshii, Ctenium aromaticum, Cuphea carthagenensis, Cyperus, Cyrilla racemiflora, Desmodium canescens, Desmodium strictum, Desmodium tenuifolium, Dichanthelium aciculare, Dichanthelium acuminatum, Dichanthelium consanguineum, Dichanthelium dichotomum, Dichanthelium ensifolium, Dichanthelium sphaerocarpon, Dichanthelium strigosum, Diodia virginiana, Diospyros virginiana, Drosera intermedia, Eleocharis, Erigeron vernus, Eriocaulon decangulare, Eryngium yuccifolium, Eupatorium linearifolium, Eupatorium mohrii, Eupatorium rotundifolium, Euthamia graminifolia, Fimbristylis puberula, Gratiola ramosa, Gymnopogon brevifolius, Helianthus angustifolius, Hibiscus moscheutos, Hypericum crux-andreae, Hypericum gymnanthum, Hypoxis hirsuta, Ilex glabra, Ilex myrtifolia, Iris hexagona, Iris prismatica, Iris virginica, Juncus biflorus, Juncus repens, Lachnanthes caroliniana, Lachnocaulon anceps, Lechea pulchella, Leersia hexandra, Lespedeza angustifolia, Lespedeza capitata, Lilium catesbaei, Linum floridanum, Liquidambar styraciflua, Liriodendron tulipifera, Lobelia elongata, Lobelia nuttallii, Ludwigia alternifolia, Ludwigia decurrens, Ludwigia octovalvis, Ludwigia pilosa, Ludwigia sphaerocarpa, Ludwigia suffruticosa, Luziola fluitans, Lycopodiella alopecuroides, Lycopus, Lyonia ligustrina, Lyonia lucida, Lyonia mariana, Lysimachia terrestris, Magnolia virginiana, Marshallia graminifolia, Mayaca fluviatilis, Muhlenbergia expansa, Myrica cerifera, Nyssa biflora, Nyssa sylvatica, Oenothera fruticosa, Osmunda, Panicum dichotomiflorum, Panicum hemitomon, Panicum verrucosum, Panicum virgatum, Paspalum floridanum, Paspalum laeve, Persea palustris, Physostegia purpurea, Pinguicula caerulea, Pinus palustris, Pinus serotina, Pinus taeda, Pityopsis graminifolia, Platanthera ciliaris, Pluchea baccharis, Pluchea camphorata, Pogonia ophioglossoides, Polygala cruciata, Proserpinaca pectinata, Pteridium aquilinum, Pterocaulon pycnostachyum, Pycnanthemum flexuosum, Quercus hemisphaerica, Quercus marilandica, Quercus pumila, Quercus stellata, Rhexia alifanus, Rhexia lutea, Rhexia mariana, Rhexia nashii, Rhexia virginica, Rhus copallinum, Rhynchospora caduca, Rhynchospora chalarocephala, Rhynchospora fascicularis, Rhynchospora filifolia, Rhynchospora globularis, Rhynchospora latifolia, Rhynchospora perplexa, Rhynchospora rariflora, Rhynchospora tracyi, Rubus argutus, Sabal, Sabatia brevifolia, Sabatia campanulata, Sabatia decandra, Sabatia difformis, Sabatia quadrangula, Saccharum brevibarbe, Salix humilis, Sarracenia flava, Sarracenia minor, Sarracenia purpurea, Sarracenia rubra, Schizachyrium scoparium, Scleria pauciflora, Scleria triglomerata, Scutellaria integrifolia, Serenoa repens, Sericocarpus tortifolius, Seymeria cassioides, Smilax glauca, Smilax rotundifolia, Solidago fistulosa, Solidago leavenworthii, Solidago puberula, Solidago rugosa, Solidago stricta, Sphagnum, Spiranthes, Sporobolus teretifolius, Symphyotrichum dumosum, Symphyotrichum walteri, Taxodium ascendens, Taxodium distichum, Tephrosia hispidula, Tiedemannia filiformis, Toxicodendron radicans, Toxicodendron vernix, Triantha racemosa, Utricularia, Vaccinium corymbosum, Vaccinium tenellum, Viola lanceolata, Viola septemloba, Woodwardia virginica, Xyris caroliniana, Xyris difformis.*

5.3.1.9. ***Iris versicolor*** **L.** inhabits bayous, bays, beaches, bogs (peat, *Sphagnum*), borrow pits (roadside), channels, depressions, ditches (drainage, roadside), fens (calcareous, minerotrophic, patterned), flats (gravelly)floating mats (sedge), floodplains, marshes (*Typha*), meadows (calcareous, open, peaty, roadside, swampy, wet), ponds (small, spring-fed), pools (permanent), prairies (wet), roadsides, rivulets, sandbars, shores (lake), sloughs, seeps (calcareous, springs (roadside), swales (meadow, prairie, seepy), swamps (beach, roadside), thickets, woodlands (bog), streambeds, and the margins of lakes, ponds (temporary), rivers, streams (muskeg), and swamps (hardwood) at elevations to 1067 m. The plants grow on wet substrates or in shallow (e.g., 15–50 cm), permanent or seasonal standing waters in exposures ranging from full sun to partial shade. Their best reproductive success occurs in shallow, open sites, with many occurrences found along the more exposed edges of various wetland communities. The reported substrates have been described as clay, cobble, gravel, Chester, DeKalb, humus (rich), loamy sand, muck (organic), mud (thick), peat, peaty sand, sand, and silt. The plants flower from May to July and fruit from June to August. The freely self-compatible flowers of this species experience three successive phases. Here, protandry limits but does not preclude self-pollination, which can occur facultatively when insect (Insecta) vectors are unavailable or ineffective. In the initial male stage, pollen is shed, but autogamy is prevented as the stigmas remain enclosed within the stigmatic flap. The female phase ensues when the flap reflexes to expose the stigmas, but usually occurs after much of the pollen already has been removed. Finally, the stigmatic flaps recurve, bringing the stigmas in physical contact with the anthers, which leads to spontaneous self-pollination. Wind also can facilitate selfing during this stage at times when insect pollinators are unavailable. The stigmas close soon after pollination, which prevents subsequent pollen deposition. Unpollinated stigmas remain receptive and reflex progressively downward, which increases their probability of contact with self-pollen. Where appropriate insect vectors occur, the plants usually are outcrossed. In one study, hand-pollinated flowers were more successful (mean seed set) when outcrossed (94%) relative to self-pollinated flowers (20%) or open-pollinated flowers (35%). Other work has shown that reproductive success remains relatively constant regardless of the pollen source (i.e., self, near-neighbor, distant-neighbor, different population). Selfing rates can approach 100% in some natural populations, where most seed set results from geitonogamous pollination facilitated by insects that forage among the different stigmatic branches when visiting a flower. Observed insect pollinators include bees (Hymenoptera: Apidae: *Anthophora terminalis*, *Bombus*; Halictidae: *Lasioglossum disparile*; Megachilidae: *Osmia distincta*) and flies (Diptera: Syrphidae: *Eristalis dimidiata*, *Parhelophilus laetus*, *Syrphus torvus*). The flowers also are pollinated by hummingbirds (Aves: Trochilidae: *Archilochus colubris*). Yet, the plants are well adapted as inbreeders and exhibit no inbreeding depression (nor heterosis) when experimentally self-pollinated. They also have been found to retain relatively high levels of genetic variation (via microsatellite marker data), even in isolated populations. The seeds are dormant when shed and lack a corky covering, which makes them less buoyant as a consequence. Over 10% of seeds tested remained afloat after 3 days, but fewer than 50% remained afloat after a week. Regardless, they still are thought to be dispersed mainly by water. They are known to germinate poorly and are not well represented in seed banks (e.g., only 4 seeds recovered from 36 sites sampled), or perhaps simply fail to germinate in seed bank assays. Cold stratification is necessary to achieve any substantial germination. In one case, the germination rate was 26% for cold-stratified seeds compared to only 5% for unstratified seeds. Stratified seeds also germinated more rapidly (within 21–42 days) than unstratified seeds (106–265 days). Germination rates from 17% to 58% have been observed after 12–34 days for seeds that were cold-stratified (4°C–5°C) for 3–4 weeks. The rhizome fragments most often grow linearly, produce few ramets, and usually produce only a single branch. When detached, they are buoyant with 50% of fragments tested remaining afloat after a week in the water. Dispersal of these fragments by water also seems likely, at least on occasion. **Reported associates:** *Abies balsamea, Acer rubrum, Acer saccharinum, Acorus calamus, Actaea, Adiantum pedatum, Agalinis purpurea, Agrostis hyemalis, Agrostis scabra, Alisma, Alnus rugosa, Ambrosia artemisiifolia, Amelanchier, Andromeda polifolia, Andropogon gerardii, Anthoxanthum nitens, Apios americana, Apocynum cannabinum, Aralia nudicaulis, Asclepias incarnata, Athyrium filix-femina, Berula erecta, Betula papyrifera, Betula pumila, Bidens aristosus, Bidens cernuus, Bidens frondosus, Bidens trichospermus, Bidens tripartitus, Bidens vulgatus, Boehmeria cylindrica, Bolboschoenus fluviatilis, Bromus ciliatus, Calamagrostis canadensis, Calamagrostis inexpansa, Calla palustris, Callitriche palustris, Caltha palustris, Campanula aparinoides, Carex appalachica, Carex arctata, Carex brunnescens, Carex buxbaumii, Carex canescens, Carex chordorrhiza, Carex comosa, Carex cristatella, Carex cryptolepis, Carex diandra, Carex disperma, Carex gracillima, Carex granularis, Carex interior, Carex intumescens, Carex lacustris, Carex lasiocarpa, Carex leptalea, Carex lurida, Carex magellanica, Carex pseudocyperus, Carex retrorsa, Carex sartwellii, Carex sterilis, Carex stipata, Carex stricta, Carex tenera, Carex trisperma, Carex tuckermanii, Carex vulpinoidea, Cephalanthus occidentalis, Chamaedaphne calyculata, Chara, Chelone glabra, Cicuta bulbifera, Cicuta maculata, Cinna arundinacea, Cinna latifolia, Circaea alpina, Cirsium muticum, Cladium mariscoides, Clintonia borealis, Comarum palustre, Coptis trifolia, Cornus amomum, Cornus canadensis, Cornus racemosa, Cornus rugosa, Cornus sericea, Corylus cornuta, Crataegus punctata, Cystopteris fragilis, Dasiphora floribunda, Dichanthelium acuminatum, Doellingeria umbellata, Drosera rotundifolia, Dryopteris carthusiana, Dryopteris cristata, Dulichium arundinaceum, Eleocharis acicularis, Eleocharis obtusa, Eleocharis ovata, Eleocharis palustris, Epilobium leptophyllum, Equisetum fluviatile, Eriophorum angustifolium, Eriophorum tenellum, Eupatorium perfoliatum, Eupatorium serotinum, Euthamia graminifolia, Eutrochium maculatum, Eutrochium purpureum, Fragaria virginiana, Fraxinus nigra, Fraxinus pennsylvanica, Galium obtusum, Galium tinctorium, Galium trifidum, Galium triflorum, Gentiana andrewsii, Gentianopsis crinita, Geranium maculatum, Glyceria canadensis Glyceria grandis, Glyceria striata, Gymnocarpium dryopteris, Helianthus giganteus, Huperzia selago, Hypericum mutilum, Hypericum majus, Hypoxis hirsuta, Ilex verticillata, Impatiens capensis, Iris pseudacorus, Juncus balticus, Juncus canadensis, Juncus effusus, Juncus torreyi, Kalmia polifolia, Larix laricina, Leersia oryzoides, Lemna minor, Linnaea borealis, Linum virginianum, Lobelia cardinalis, Lobelia siphilitica, Lonicera canadensis, Lonicera oblongifolia, Lonicera villosa, Lycopodium annotinum, Lycopus americanus, Lycopus uniflorus, Lysimachia ciliata, Lysimachia terrestris, Lysimachia thyrsiflora, Lythrum alatum, Maianthemum canadense, Maianthemum trifolium, Mentha arvensis, Mentha ×piperita, Menyanthes trifoliata, Micranthes pensylvanica, Mimulus glabratus, Mimulus ringens, Mitella nuda, Monotropa uniflora, Myrica gale, Nuphar advena, Oclemena acuminata, Oenothera perennis, Onoclea sensibilis, Osmunda claytoniana, Osmunda regalis, Osmundastrum cinnamomeum, Oxypolis rigidior, Packera pseudaurea, Parnassia palustris, Pedicularis lanceolata, Peltandra virginica, Penthorum sedoides. Persicaria amphibia, Persicaria hydropiperoides, Persicaria pensylvanica, Persicaria punctata, Persicaria sagittata, Petasites frigidus, Phalaris arundinacea, Phleum pratense, Phlox maculata, Phragmites australis, Phyla lanceolata, Picea mariana, Pilea pumila, Platanthera dilatata, Platanthera grandiflora, Platanthera hyperborea, Poa nemoralis, Poa pratensis, Polytrichum, Populus balsamifera, Populus tremuloides, Potamogeton epihydrus, Potentilla norvegica, Proserpinaca palustris, Prunella vulgaris, Prunus virginiana, Pycnanthemum virginianum, Ranunculus abortivus, Ranunculus acris, Ranunculus gmelinii, Ranunculus pensylvanicus, Rhamnus alnifolia, Rhododendron groenlandicum, Ribes aureum, Ribes cynosbati, Ribes* *glandulosum, Ribes triste, Rosa carolina, Rosa setigera, Rosa woodsii, Rotala ramosior, Rubus pubescens, Rubus sachalinensis, Rumex britannica, Sagittaria brevirostra, Sagittaria latifolia, Sagittaria rigida, Salix bebbiana, Salix exigua, Salix geyeriana, Salix lasiandra, Salix longifolia, Salix lucida, Salix monticola, Salix myrtillifolia, Salix pedicellaris, Salix petiolaris, Salix planifolia, Salix serissima, Salix wolfii, Sarracenia purpurea, Saururus cernuus, Scheuchzeria palustris, Schizachyrium scoparium, Schoenoplectus tabernaemontani, Scirpus atrocinctus, Scirpus cyperinus, Scutellaria galericulata, Scutellaria lateriflora, Sium suave, Solidago riddellii, Sparganium emersum, Sparganium eurycarpum, Sparganium glomeratum, Spartina pectinata, Sphagnum, Spiraea alba, Spiraea tomentosa, Spiranthes cernua, Stellaria longifolia, Symphyotrichum boreale, Symphyotrichum lateriflorum, Symphyotrichum novae-angliae, Thalictrum dasycarpum, Thelypteris palustris, Thuja occidentalis, Tilia americana, Toxicodendron radicans, Triadenum fraseri, Triadenum virginicum, Trichophorum alpinum, Trientalis borealis, Triglochin maritima, Typha angustifolia, Typha latifolia, Ulmus americana, Utricularia macrorhiza, Vaccinium macrocarpon, Vaccinium myrtilloides, Valeriana edulis. Verbena hastata, Vernonia fasciculata, Veronica scutellata, Veronicastrum virginicum, Viburnum opulus. Viola.*

5.3.1.10. **Iris virginica L.** grows in brackish or freshwater sites associated with bayous, beaches, bogs (raised, roadside), borrow pits, bottoms (stream), depressions (anthropogenic), ditches (roadside), fens (calcareous, prairie), flats, flatwoods, floodplains (backwater), marshes (brackish, freshwater, river, roadside, seepage, tidal), meadows (sedge, wet), oxbows, prairies (mesic, wet), ravines, right-of-ways (powerline), roadsides, savannas (flatwood, pine), seeps (acid), sloughs (cypress-tupelo), swales, swamps (buttonbush, conifer, cypress, flooded, floodplain, hardwood, tidal), thickets (alder), vernal pools, woodlands (alluvial, low, wet), and along the margins of bogs, canals, lagoons, lakes, ponds (backwater, inundated, mill), pools, rivers, springs (seepage), streams (blackwater, ephemeral, seasonal), and swamps (alluvial, roadside) at elevations to 1280 m. The plants occur on permanently wet substrates or in shallow (e.g., 3–31 cm) standing waters under fully open (sunny) to shady exposures. They can inhabit fresh, nontidal, tidal, and brackish sites (salinity: 2.1–5.3‰) but are intolerant of higher salinity and have been observed to perish in brackish estuaries after saltwater intrusion occurred. The habitats range from acidic to somewhat alkaline (pH: 4.9–7.2). The substrates have been described as alluvium (acidic), clay, clay loam, loam, muck (Dorovan, Ponzer), mucky loamy sand (Murville), mud, peat, Pickton (grossarenic paleudalfs), rock, sand, sandy loam (Coxville, fine), silt, and silty loam (Bodine-Brandon, cherty, Hartville). Analyzed substrates were acidic (pH: 5.0–5.5), brackish (chlorinity: 1.2–2.4%), had an organic matter content of 24–30%, and comprised 25% sand, 51.5% silt, and 23.5% clay. Flowering and fruiting occur from March to August. Although no study has addressed the floral biology of this species directly, it appears to be similar to other OBL taxa, i.e., primarily outcrossed by bees and hummingbirds (e.g., see *I. versicolor*, above). However, it is not clear whether the flowers

are self-compatible and the extent of self-pollination (if any) has not been reported. The seeds are very corky, thus presumably buoyant and dispersed by water. They are detected rarely in seed bank studies, even where well represented in the standing flora. Autumn-planted seeds were observed to germinate during the subsequent spring only after temperatures rose above 25°C, with the rate increasing proportionally with increasing temperature. Germination also increases proportionally with the length of the cold stratification period, with the highest rates attained for seeds kept for 10 weeks at 5°C and then subjected to a 30°C/20°C day/night temperature regime. Germination under those conditions was also highly synchronous (96% occurring within 4 weeks). Although fairly high (but unsynchronous) experimental germination rates (48%) have been obtained by exposing unstratified seeds to a 30°C/20°C thermoperiod (which normally would not occur seasonally at the time of their autumn dispersal), less than 2% of unstratified seeds germinated at lower incubation temperatures, which would be more typical of autumn conditions. These observations indicate that the seeds are conditionally dormant and that they germinate primarily on exposed mudflats, where germination would be induced as the substrates attain fairly high surface temperatures during the late spring. The resulting, highly synchronized germination pattern also provides seedlings with a full growing season during which to establish. Plants growing in freshwater marshes have attained peak biomass values ranging from 8.8 to 20.4 g m^{-2}. Vegetative reproduction occurs by the spread of the rhizomes, which branch extensively, resulting in dense clumps of plants.

Reported associates: *Acer rubrum, Acer saccharinum, Acer saccharum, Alisma subcordatum, Alnus glutinosa, Alnus rugosa, Alternanthera philoxeroides, Amaranthus cannabinus, Ammannia robusta, Amsonia, Andropogon gerardii, Andropogon glomeratus, Andropogon virginicus, Anemone canadensis, Apios americana, Apocynum cannabinum, Aronia arbutifolia, Aronia melanocarpa, Aronia ✕prunifolia, Arundo donax, Asclepias incarnata, Asclepias sullivantii, Asclepias syriaca, Asimina triloba, Bartonia virginica, Betula papyrifera, Bidens aristosus, Bidens cernuus, Bidens connatus, Bidens frondosus, Bidens laevis, Bidens trichospermus, Bidens tripartitus, Bidens vulgatus, Bignonia capreolata, Boehmeria cylindrica, Bolboschoenus fluviatilis, Bolboschoenus robustus, Boltonia asteroides, Bromus inermis, Calamagrostis canadensis, Calamagrostis inexpansa, Callitriche heterophylla, Caltha palustris, Campanula aparinoides, Cardamine bulbosa, Carex annectens, Carex atherodes, Carex aurea, Carex comosa, Carex conjuncta, Carex corrugata, Carex crinita, Carex cristatella, Carex crus-corvi, Carex davisii, Carex festucacea, Carex frankii, Carex gigantea, Carex glaucescens, Carex granularis, Carex grayi, Carex grisea, Carex hyalinolepis, Carex interior, Carex intumescens, Carex lacustris, Carex laevivaginata, Carex leavenworthii, Carex lupulina, Carex lurida, Carex meadii, Carex missouriensis, Carex muskingumensis, Carex pellita, Carex projecta, Carex scoparia, Carex seorsa, Carex shortiana, Carex squarrosa, Carex sterilis, Carex stipata, Carex striata, Carex stricta, Carex swanii, Carex tribuloides, Carex*

trichocarpa, Carex typhina, Carex viridula, Carex vulpinoi-
dea, Carya illinoinensis, Carya laciniosa, Centella asiatica,
Cephalanthus occidentalis, Chamaecrista fasciculata,
Chasmanthium latifolium, Chasmanthium latifolium,
Chasmanthium laxum, Chelone glabra, Chelone obliqua,
Chenopodium album, Cicuta maculata, Cinna arundinacea,
Cladium jamaicense, Cladium mariscoides, Clematis pitch-
eri, Comarum palustre, Convolvulus sepium, Coreopsis pal-
mata, Cornus amomum, Cornus foemina, Cornus sericea,
Crinum americanum, Cuscuta gronovii, Cyperus erythrorhizos,
Cyperus esculentus, Cyperus odoratus, Cyperus squarrosus,
Cyperus strigosus, Cypripedium calceolus, Dasiphora flori-
bunda, Decodon verticillatus, Dichanthelium scoparium,
Dioscorea villosa, Diospyros, Dipsacus fullonum, Drosera
intermedia, Drosera rotundifolia, Dryopteris cristata,
Dulichium arundinaceum, Echinochloa crus-galli,
Echinochloa walteri, Eleocharis atropurpurea, Eleocharis
compressa, Eleocharis elliptica, Eleocharis engelmannii,
Eleocharis erythropoda, Eleocharis fallax, Eleocharis flaves-
cens, Eleocharis obtusa, Eleocharis palustris, Eleocharis
quadrangulata, Elymus canadensis, Elymus virginicus,
Epilobium coloratum, Epilobium coloratum, Equisetum
arvense, Equisetum fluviatile, Equisetum variegatum,
Eragrostis hypnoides, Eragrostis pectinacea, Erechtites hier-
aciifolius, Eryngium aquaticum, Eupatorium perfoliatum,
Euthamia graminifolia, Eutrochium maculatum, Eutrochium
purpureum, Fagus grandifolia, Fimbristylis spadicea,
Fragaria virginiana, Fraxinus nigra, Fraxinus pennsylvanica,
Fraxinus profunda, Galium boreale, Galium mollugo, Galium
obtusum, Galium tinctorium, Gaylussacia baccata, Gentiana
rubricaulis, Gentianella quinquefolia, Gentianopsis crinita,
Geum laciniatum, Glyceria grandis, Glyceria septentrionalis,
Glyceria striata, Gratiola neglecta, Helianthus grosseserra-
tus, Helianthus mollis, Hibiscus moscheutos, Hydrocotyle
umbellata, Hymenocallis crassifolia, Hypericum majus,
Hypoxis hirsuta, Ilex verticillata, Impatiens capensis,
Impatiens pallida, Ipomoea cordatotriloba, Iris brevicaulis,
Iris pseudacorus, Iris verna, Itea virginica, Juglans nigra,
Juncus acuminatus, Juncus biflorus, Juncus brachycephalus,
Juncus canadensis, Juncus coriaceus, Juncus effusus, Juncus
marginatus, Juncus nodosus, Juncus roemerianus Juncus torreyi, Kosteletzkya pentacarpos, Lachnanthes caroliniana,
Larix laricina, Lathyrus palustris, Leersia oryzoides, Lemna
minor, Liatris pycnostachya, Ligustrum vulgare, Lilium michiganense, Lilium philadelphicum, Limnobium spongia,
Lindera benzoin, Lindernia dubia, Lipocarpha micrantha,
Liquidambar styraciflua, Liriodendron tulipifera, Lobelia
cardinalis, Lobelia kalmii, Lobelia siphilitica, Lonicera
japonica, Ludwigia palustris, Ludwigia peploides, Ludwigia
pilosa, Luziola fluitans, Lycopus americanus, Lycopus
amplectens, Lycopus uniflorus, Lycopus virginicus, Lyonia
lucida, Lysimachia hybrida, Lysimachia nummularia,
Lysimachia thyrsiflora, Lythrum alatum, Lythrum lineare,
Lythrum salicaria, Magnolia grandiflora, Mentha arvensis,
Mentha ×piperita, Micranthes pensylvanica, Mikania scandens, Mimulus ringens, Murdannia keisak, Myrica gale,
Myriophyllum pinnatum, Nasturtium, Nuphar sagittifolia,
Nuphar variegata, Nyssa biflora, Onoclea sensibilis,
Ophioglossum vulgatum, Orontium aquaticum, Osmunda
regalis, Osmundastrum cinnamomeum, Packera aurea,
Pancratium maritimum, Panicum virgatum, Parietaria pen-
sylvanica, Peltandra virginica, Penstemon calycosus,
Penstemon digitalis, Penthorum sedoides, Persicaria
amphibia, Persicaria arifolia. Persicaria coccinea, Persicaria
hydropiperoides, Persicaria lapathifolia, Persicaria punctata,
Persicaria sagittata, Phalaris arundinacea, Phanopyrum
gymnocarpon, Phegopteris hexagonoptera, Phlox maculata,
Phragmites australis, Phryma leptostachya, Phyla lanceolata,
Phyla nodiflora, Physostegia virginiana, Phytolacca ameri-
cana, Pilea pumila, Pinus elliottii, Platanthera peramoena,
Platanthera praeclara, Platanthera psycodes, Pluchea odo-
rata, Poa nemoralis, Polytaenia nuttallii, Pontederia cordata,
Populus tremuloides, Potamogeton natans, Potentilla anse-
rina, Primula meadia, Proserpinaca palustris, Proserpinaca
pectinata, Prunella vulgaris, Ptilimnium costatum,
Pycnanthemum virginianum, Quercus alba, Quercus bicolor,
Quercus shumardii. Quercus velutina, Ranunculus hispidus,
Ranunculus pusillus, Ranunculus sceleratus, Rhamnus fran-
gula, Rhexia alifanus, Rhexia mariana, Rhexia nashii,
Rhododendron canescens, Rhynchospora capitellata,
Rhynchospora colorata, Rhynchospora corniculata, Rorippa
islandica, Rorippa palustris, Rosa multiflora, Rosa palustris,
Rotala ramosior, Rubus hispidus, Rubus occidentalis, Rumex
maritimus, Rumex verticillatus, Saccharum giganteum,
Sagittaria lancifolia, Sagittaria latifolia, Salix amygdaloides,
Salix candida, Salix discolor, Salix exigua, Salix humilis,
Sambucus nigra, Saururus cernuus, Schedonorus arundina-
ceus, Schizachyrium scoparium, Schoenoplectus acutus,
Schoenoplectus americanus, Schoenoplectus pungens,
Schoenoplectus tabernaemontani, Scirpus atrovirens, Scirpus
cyperinus, Scirpus pendulus, Scleria triglomerata, Scolochloa
festucacea, Scutellaria galericulata, Scutellaria lateriflora,
Scutellaria parvula, Setaria magna, Silphium laciniatum,
Silphium terebinthinaceum, Sium suave, Smilax lasioneura,
Solanum dulcamara, Solidago altissima, Solidago gigantea,
Solidago patula, Solidago rigida, Solidago sempervirens,
Sparganium, Spartina alterniflora, Spartina cynosuroides,
Spartina patens, Spartina pectinata, Sphagnum, Spiraea alba,
Spiranthes cernua, Spirodela polyrhiza, Stachys
hyssopifolia, Stachys tenuifolia, Symphyotrichum ericoides,
Symphyotrichum lanceolatum, Symphyotrichum praealtum,
Symphyotrichum puniceum, Symphyotrichum racemosum,
Symphyotrichum tenuifolium, Symplocarpus foetidus,
Symplocos tinctoria, Taxodium distichum, Teucrium
canadense, Thalictrum dasycarpum, Thalictrum pubescens,
Thalictrum revolutum, Thelypteris palustris, Thuja occiden-
talis, Tilia americana, Toxicodendron vernix, Tradescantia
ohiensis, Triadenum fraseri, Triadenum virginicum,
Triadenum walteri, Triadica sebifera, Trifolium repens,
Triglochin maritima, Tripsacum dactyloides, Typha angustifo-
lia, Typha latifolia, Ulmus americana, Urtica dioica,
Utricularia radiata, Vaccinium myrtilloides, Vaccinium oxycoccos, Valeriana edulis, Valeriana edulis. Valerianella
umbilicata. Verbena urticifolia, Verbesina occidentalis,

Vernonia gigantea, Vernonia missurica, Vernonia noveboracensis, Veronica catenata, Veronicastrum virginicum, Viburnum dentatum, Viburnum opulus, Viburnum recognitum, Vigna luteola, Viola cucullata, Viola lanceolata, Viola primulifolia, Vitis riparia, Woodwardia areolata, Woodwardia virginica, Xyris difformis, Xyris fimbriata, Xyris jupicai, Zizania aquatica, Zizaniopsis miliacea, Zizia aurea.

Use by wildlife: *Iris* herbage generally is toxic to wildlife and unpalatable due to the presence of glycosides and crystal inclusions. Domestic cattle (Mammalia: Bovidae: *Bos taurus*) and sheep (Mammalia: Bovidae: *Ovis aries*) will graze the foliage when other fodder is scarce, sometimes causing gastrointestinal problems as a result. The seeds of *I. pseudacorus* are eaten (but not usually digested) by waterfowl (Aves: Anatidae: *Anas platyrhynchos*). The flowers of *Iris brevicaulis* and *I. fulva* are visited by bees (Insecta: Hymenoptera: Apidae: *Anthophora abrupta*, *Bombus pensylvanicus*, *Xylocopa*), butterflies (Insecta: Lepidoptera: Hesperiidae), and ruby-throated hummingbirds (Aves: Trochilidae: *Archilochus colubris*), which forage for the nectar. Those of *I. pseudacorus* are foraged by bumblebees (Insecta: Hymenoptera: Apidae: *Bombus vagans*) and also by ruby-throated hummingbirds. The flowers of *I. versicolor* attract nectar and pollen-seeking insects (Insecta) such as bees (Hymenoptera: Apidae: *Anthophora terminalis*, *Bombus*; Halictidae: *Lasioglossum disparile*, *L. leucozonium*; Megachilidae: *Osmia distincta*), beetles (Insecta: Coleoptera: Curculionidae: *Mononychus vulpeculus*), butterflies (Lepidoptera: Crambidae: *Evergestis stramentalis*; Hesperiidae: *Epargyreus clarus*, *Poanes hobomok*, *Polites mystic*, *Polites peckius*, *Polites themistocles*, *Thorybes pylades*; Noctuidae: *Leucania pallens*), and flies (Diptera: Chloropidae: *Chlorops proximus*, *Thaumatomyia glabra*; Sepcidae: *Sepsis vicaria*; Syrphidae: *Chalcosyrphus interruptus*, *Eristalis dimidiata*, *Orthonevra nitida*, *Parhelophilus laetus*, *Syrphus torvus*), as well as hummingbirds (Aves: Trochilidae: *Archilochus colubris*). The weevils (*Mononychus vulpeculus*) also chew through the flower tissue to the nectary, which results in copious nectar spillage to the outside of the flower. The flowing nectar attracts many other insect "nectar thieves," including beetles (Coleoptera: Cantharidae: *Podabrus basilaris*, *P. rugosulus*); Coccinellidae: *Coleomegilla maculata*, *Hippodamia tredecimpunctata*; Lampyridae: *Lucidota atra*, *Photinus carolinus*; Mordellidae: *Mordella marginata*), bugs (Hemiptera: Miridae: *Adelphocoris rapidus*, *Lygus pratensis*, *Metriorrhynchomiris dislocatus*; Pentatomidae: *Euschistus ictericus*, *E. tristigmus*, *Podisus maculiventris*), and flies (Diptera: Drosophilidae: *Drosophila*; Muscidae). The leaves and freshly opened flowers are eaten by grasshoppers and locusts (Insecta: Orthoptera: Acrididae) and larval moths (Insecta: Lepidoptera: Erebidae: *Spilosoma congrua*; Noctuidae: *Mamestra*, *Simyra*). The flowers of *I. versicolor* can be damaged severely by picture-winged flies (Insecta: Diptera: Otitidae: *Chaetopsis aenea*) whose larvae bore into the pedicels at the base of the bud and down through the peduncle, which can destroy all of the commonly originating flowers. The seeds are eaten by larval moths (Lepidoptera: Tortricidae: *Endothenia hebesana*) and weevils (Coleoptera:

Curculionidae: *Mononychus vulpeculus*), which develop within the fruits. The leaves are fed on by larval butterflies and moths (Insecta: Lepidoptera: Erebidae: *Orgyia leucostigma*; Nymphalidae: *Junonia coenia*). Leaf-miner flies (Insecta: Diptera: Agromyzidae: *Phytobia lateralis*) commonly pupate within galls that are produced on the leaves of *I. versicolor*. Female leaf-curling spiders (Arthropoda: Arachnida: Clubionidae: *Clubiona riparia*) fabricate protective capsules out of *I. versicolor* leaves to house themselves and their eggs. The foliage is low in sodium (e.g., Na: 28 ppm) but occasionally is eaten by muskrats (Mammalia: Cricetidae: *Ondatra zibethicus*). The seeds contain roughly 15% protein, 12% fat, and 12% crude fiber, whereas the leaves contain 9% crude protein and 17% crude fiber. The plants are hosts of several Fungi (Ascomycota: Botryosphaeriaceae: *Phyllosticta iridis*; Basidiomycota: Pucciniaceae: *Puccinia iridis*). The similar *I. virginica* is eaten occasionally by muskrats (*Ondatra zibethicus*) and to a slight extent by white-tailed deer (Mammalia: Cervidae: *Odocoileus virginianus*). The plants also host several Fungi (Ascomycota: Glomerellaceae: *Colletotrichum dematium*; Basidiomycota: Pucciniaceae: *Puccinia iridis*).

Economic importance: food: *Iris* plants are not edible due to their glycoside and alkaloid content. The Abnaki tribe recognized that *I. versicolor* plants were poisonous and it is known that ingestion of their leaves or rhizomes can cause severe gastric, liver, and pancreas disorders. However, the dried seeds of *I. pseudacorus* have been used as a coffee substitute; **medicinal:** In some countries, preparations made from *Iris* rhizomes have been used as an astringent, diuretic, emetic, emmenagogue, antiflatulent, and treatment for eczema. Preparations made from *I. pseudacorus* have been used for enemas and to treat colds, jaundice, kidney ailments, sore throats, toothaches, and wounds. The roots and rhizomes contain a *trans*-3-hydroxy-5,7-dimethoxyflavanone, which (at 25 μM) was found to strongly inhibit growth of HT-29 human colon carcinoma cell colonies. The plants also contain iridin, a toxic glycoside known to cause contact dermatitis in some people that have handled them. Similar reactions have occurred from skin contact with the seed endosperm. Volatile compounds extracted from the plants exhibit antibacterial activity against both Gram-positive and Gram-negative bacteria. Other *I. pseudacorus* extracts have exhibited anti-osteoporotic effects. Rhizomes of *I. versicolor* (which contain iridin) were used widely in traditional medicine. The Micmac and Penobscot people inhaled steam from the plants or chewed on their roots as a general disease preventative or cure. A macerated root poultice was applied as a dressing for burns, sores, or wounds (Algonquin, Meskwaki, Micmac, Omaha, Ponca). The Chippewa, Delaware, and Oklahoma tribes treated tuberculosis or other causes of scrofula (mycobacterial cervical lymphadenitis) by applying a root poultice. The Chippewa and Potawatomi applied a root poultice to reduce inflammation. The Cree and Hudson Bay people used the plant as a purgative. Root decoctions or infusions were used as a cathartic by the Iroquois, Ojibwa, and Creek people. Root preparations were ingested or applied externally as an antirheumatic (Delaware, Oklahoma). They also were variously used to treat blood poisoning (Iroquois Drug), cholera (Micmac, Penobscot), colds and sore throats (Malecite, Meskwaki, Micmac), earaches (Omaha, Ponca), hay fever (Iroquois), kidney disorders (Delaware, Oklahoma), liver disorders (Cree, Delaware, Hudson Bay, Oklahoma), pulmonary ailments (Meskwaki), and watery eyes (Omaha, Ponca). The Mohegan and Montagnais tribes would apply a pulverized root poultice as an analgesic. The Ojibwa administered a root decoction as an emetic. The Iroquois made an infusion from mashed roots to induce pregnancy and also to induce paralysis. Recent homeopathic reviews indicate that *I. versicolor* tinctures can affect gastrointestinal mucous membranes as well as thyroid, pancreas, salivary, and intestinal glands. Remedies made from the rhizome constituents are regarded as potentially effective treatments for thyroid dysfunction, especially hypothyroid disorders. Extracts from *I. virginica* have fairly high antioxidant activity but only slight antimicrobial activity. The Cherokee people made an infusion from *I. virginica* for treating liver maladies and prepared a compound root decoction to relieve urinary disorders and as a salve for ulcers; **cultivation:** *Iris* species have long been prized as garden ornamentals, with more than 14,000 cultivars currently listed by the Royal Horticultural Society. These include several of the OBL taxa (e.g., *I. versicolor* and *I. virginica*), whose seeds already were being offered for sale in American seed catalogues by 1804. Because some of these cultivars are of hybrid origin, they might be multiply listed under different species names. *Iris fulva* is recognized as an "Award of Garden Merit" plant by the Royal Horticultural Society. Its cultivars include 'Arkansas', 'Flash', 'Marvell Gold', and 'Terracotta Warrior.' 'Quartz' and 'Rabun White' are cultivars of *I. prismatica*. *Iris pseudacorus* is widely cultivated. The cultivars 'Chance Beauty', 'Regal Surprise,' 'Roy Davidson', and 'Variegata' are recognized as "Award of Garden Merit" plants. Its many other cultivars include 'Aketon Ivory', 'Alba,' 'Beuron,' 'Clotted Cream,' 'Come in Spinner,' 'Crème de la Crème,' 'Donau', 'Dragonfly Dance,' 'Ecru', 'Esk', 'Flore Pleno', 'Gigantea', 'Golden Daggers', 'Golden Fleece', 'Golden Queen', 'Holden Clough', 'Holden's Child', 'Ilgengold', 'Ivory', 'Kelis Choice', 'Kimboshi', 'Kinshizen', 'Krill', 'Limbo', 'Lime Sorbet', 'Mandchurica', 'Mini Mart', 'Phil Edinger', 'Roccapina', 'Roryu', 'Rowden Alchemy', 'Rowden Brimstone', 'Roy Davidson', 'Roy's Repeater', 'Seuver Fourses', 'Seuver Punch', 'Spartacus,' 'Sulphur Queen', 'Sun Cascade', 'Sun in Splendour', 'Tangarewa Cream', 'Tiger Brother', 'Tiggah', 'Turnipseed', 'Wychwood's Multifloral', 'Yarai', and 'Yasha'. The most widely cultivated OBL species is *I. versicolor*, which includes several "Award of Garden Merit" selections (i.e., 'Dark Aura', 'Regal Surprise') and numerous other cultivars including 'Algonquin', 'Anthem', 'Appointer', 'Bellerive Harmony', 'Between the Lines', 'Blue Light', 'Candystriper', 'China West Lake', 'Claret Cup', 'Dottie's Double', 'Georgia Bay', 'Gerald Darby', 'Goldbrook', 'Holden's Child', 'Kermesina', 'Limbo', 'Mainstream Spring', 'Mainstream Tempest', 'Mint Fresh', 'Mountain Brook', 'Mysterious Monique', 'Omoide', 'Oriental Touch', 'Party Line', 'Purple Fan', 'Raspberry Slurp', 'Rosea', 'Rowden Allegro', 'Rowden Anthem', 'Rowden Aria', 'Rowden Cadenza', 'Rowden Calypso', 'Rowden Cantata', 'Rowden Concerto', 'Rowden Descant', 'Rowden Electro', 'Rowden Fugue', 'Rowden Harmony', 'Rowden Jingle', 'Rowden Lullaby', 'Rowden Lyric', 'Rowden Madrigal', 'Rowden Mazurka', 'Rowden Melody', 'Rowden Minuet', 'Rowden Nocturne', 'Rowden Pastorale', 'Rowden Polka', 'Rowden Polonnaise', 'Rowden Prelude', 'Rowden Refrain', 'Rowden Rhapsody', 'Rowden Rondo', 'Rowden Scherzo', 'Rowden Sonata', 'Rowden Starlight', 'Rowden Symphony', 'Rowden Waltz', 'Scherzo', 'Seuver Fourses', 'Seuver Punch', 'Signagoniga Ridska', 'Silvington', 'Symphony', 'Thomas Variety', 'Tina', 'Version', 'Violet Minuet', and 'Whodunit'. Cultivars of *I. virginica* include 'Dark Aura' (Award of Garden Merit), 'De Luxe', 'Dottie's Double', 'Lavender Lustre', 'Mountain Brook', 'Orchid Purple', 'Pale Lavender', 'Pink Butterfly', 'Pink Perfection', 'Pond Crown Point', 'Pond Lilac Dream', 'Purple Fan', and 'Slightly Daft'; **misc. products:** *Iris pseudacorus* has been planted for erosion control and is used to rehabilitate wetland systems by reducing levels of bacteria, heavy metals, and nutrients such as nitrogen and phosphorous. The foliage was smoked as a tobacco substitute during World War II. Bioethanol yields up to 8.27 g per 100 g dry mass have been obtained from *I. pseudacorus* plants. The plants have been found to accumulate metallic nanoparticles (2 nm semispherical copper) when grown on substrates containing the metal. *Iris pseudacorus* represents the "fleur-de-lis" symbol on the coat-of-arms adopted by the French king Louis VII during his Crusade against the Saracens. *Iris virginica* and *I. pseudacorus* are regarded as practical candidates for use in wetland bioremediation systems. The "Louisiana irises" (*Iris* series *Hexagonae*) have served as a model system for studying plant hybridization, especially with respect to the genetic consequences of introgression (see *Systematics* below). Morphological data compiled by Edgar Anderson to evaluate *Iris* hybrids was published by Fisher (1936) in a paper advocating the application of multivariate statistics to resolve taxonomic problems. Subsequently, the classic "Fisher's *Iris* data" have become a standard data set evaluated in hundreds of mathematical publications that address multivariate clustering algorithms and neural networks, which are incorporated into machine learning algorithms; **weeds:** *Iris pseudacorus* is a highly invasive weed of North American wetlands. *Iris versicolor* is regarded as a weed in New Brunswick, Nova Scotia, and Prince Edward Island; **nonindigenous species:** *Iris pseudacorus* has been cultivated in North American gardens since the early 19th century and had escaped from cultivation in eastern North America by the early 20th century. The plants had reached the western United States by the 1950s, likely as a consequence of similar garden escapes.

Systematics: Classifications of the large genus *Iris* have undergone repeated revisions as phylogenetic studies continue to incorporate larger samples of taxa and genetic loci. If defined to include *Belamcanda*, *Pardanthopsis*, and several other genera (e.g., *Hemodactylus*, *Iridodictyum*, *Juno*, *Xiphium*), *Iris* resolves as the sister group to a clade containing *Bobartia*, *Dietes*, *Ferraria*, and *Moraea* (Figure 2.10), which comprise the remainder of genera assigned to tribe Irideae of subfamily Iridoideae. By this "*sensu lato*" concept (widely followed by American authors), *Iris* is monophyletic when circumscribed to include nine clades, which have been recognized taxonomically as eight formal subgenera and one informal group (i.e., *Limniris* III) whose status remains unclear (Figure 2.11). However, others have suggested subdividing *Iris* into as many as 23 (or more) different genera, which indicates that the classification of the genus remains far from being settled on a global basis. Although several studies have demonstrated the polyphyly of subgenus *Limniris* as originally defined, all the OBL *Iris* species are placed within subgenus *Limniris* as recently redefined (Figure 2.11). However, there is substantial conflict between trees generated using cpDNA vs. nuclear markers in series *Laevigatae*, leaving several relationships in that group unsettled at this time. Thus, the tree presented as Figure 2.11 should be recognized as presenting only one possible depiction of relationships. *Iris* series *Hexagonae* (the "Louisiana irises") is monophyletic and contains five species: *I. brevicaulis, I. fulva, I. giganticaerulea, I. hexagona,* and *I. savannarum.* It is clear that *I. brevicaulis* and *I. fulva* are closely related, given their morphological similarities, phylogenetic association (Figure 2.11), and propensity for hybridization (see below). Recent phylogenetic studies and genetic analyses (using microsatellite marker data) have confirmed their close relationship with *I. hexagona,* with cpDNA markers resolving *I. brevicaulis* and *I. fulva* as sister species. Despite their occasional hybridization in nature (see below), crossing studies have demonstrated that *I. fulva* and *I. hexagona* are strongly isolated genetically, as evidenced by the repeated failure of experimental interspecific crosses. The long-standing treatment of *I. giganticaerulea*

and *I. savannarum* as varieties of *I. hexagona* has resulted in befuddling literature accounts attributed to "*Iris hexagona,*" which most often refer actually to either *I. giganticaerulea* or *I. savannarum.* Although they have not yet been included in any formal phylogenetic analysis, a close relationship of both *I. giganticaerulea* and *I. savannarum* to *I. hexagona* is indicated by the similar morphology of all three taxa. Analyses using microsatellite marker data have confirmed that *I. hexagona* and *I. giganticaerulea* are distinct genetically but not necessarily the most closely related species in the series. The same analyses indicate that *I. savannarum* is also distinct from *I. hexagona* genetically; however, some populations of *I. savannarum* are more similar genetically to other species in the series than they are to other conspecific populations. Hybridization between morphologically distinct ecotypes also has been documented in *I. savannarum.* This ability of *I. giganticaerulea* and *I. savannarum* to hybridize successfully with *I. brevicaulis* and *I. fulva* (see below) attests further their genetic similarity with the other members of series *Hexagonae.* *Iris cristata* differs from other members of subgenus *Limniris* by its stolons, slender and twisted stem, and cubic-shaped seeds and is recognized as comprising the monotypic series *Prismaticae* (Figure 2.11). Molecular data have resolved series *Prismaticae* as the sister group of series *Laevigatae* (Figure 2.11), or as the sister group of a mixed clade containing members of series *Laevigatae* and *Tripetalae.* Although *I. tridentata* had been assigned to series *Tripetalae* (which contained the presumably closely related *I. hookeri* and *I. setosa*), phylogenetic analyses (e.g., Figure 2.11) have shown all three species to fall within series *Laevigatae.* Furthermore, there

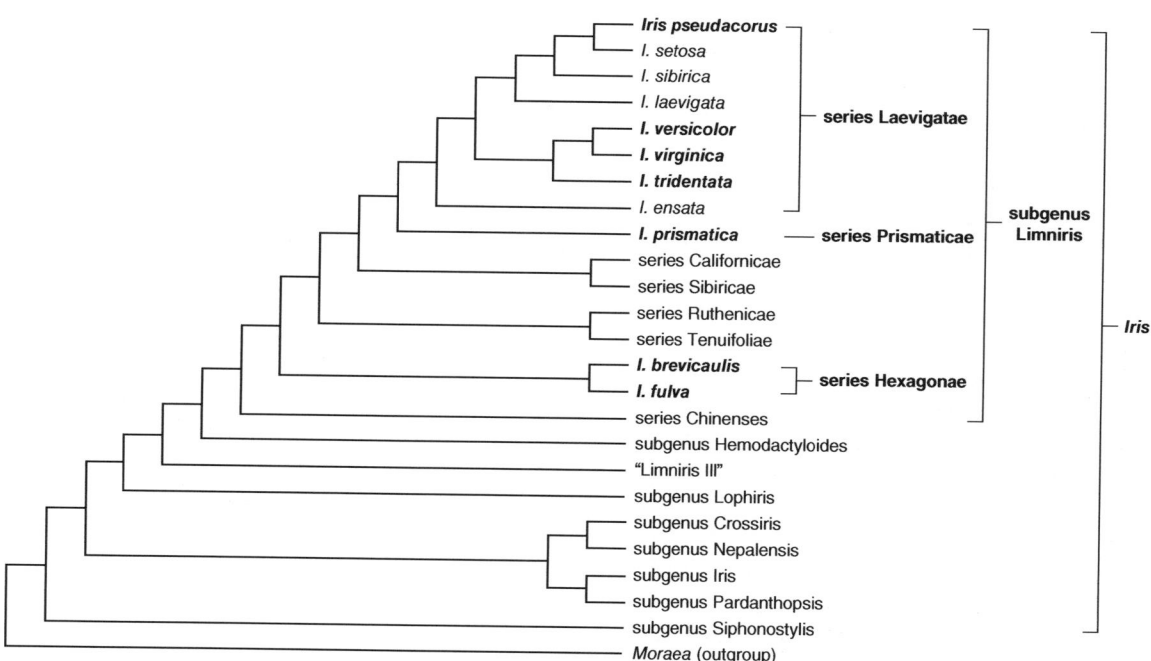

FIGURE 2.11 Interspecific relationships in *Iris* subgenus *Limniris* as indicated by phylogenetic analysis of combined cpDNA sequence data (adapted from figures in Wilson, 2009; 2011). *Limniris* is the only *Iris* subgenus containing OBL indicators (highlighted in bold), which occur here among three different series (the series limits remain contentious and require further revision – see Wheeler & Wilson, 2014). The three unsampled OBL *Iris* species (*I. giganticaerulea, I. hexagona,* and *I. savannarum*) are assigned to series Hexagonae; other studies (Hamlin et al., 2017) have confirmed the close relationship of *I. hexagona* to the *I. brevicaulis/I. fulva* clade.

has been no molecular support for a close relationship of *I. tridentata* to either *I. hookeri* or *I. setosa*. Rather, cpDNA sequence data (which are inherited maternally in the genus) place *I. tridentata* close to *I. versicolor* and *I. virginica* (Figure 2.11). Moreover, combined nuclear and plastid sequence data resolve *I. tridentata* within a grade between *I. ensata* and *I. prismatica*. Nevertheless, a lack of statistical support for the latter result precludes any definitive conclusion on the relationship of *I. tridentata* at this time. It is noteworthy that *I. tridentata* and *I. prismatica* share the rather unusual feature of stolon production, which does not occur otherwise within subgenus *Limniris*. *Iris versicolor* is an amphidiploid hybrid involving *I. virginica* and *I. setosa*, a result that was first demonstrated convincingly by crossing studies. However, it is not evident by phylogenetic analyses of combined nuclear and cpDNA sequence data, which resolve a strongly supported clade containing admixtures of multiple, highly similar *I. versicolor* and *I. virginica* accessions. That anomalous result reflects losses of nrITS regions, which have been verified using genomic *in situ* hybridization (GISH). GISH analyses have shown that *I. versicolor* retains only the 18–26S rDNA of *I. virginica* (those of *I. setosa* being absent), but possesses one copy of the 5S region from each putative parent; a second 5S locus of *I. setosa* is also lacking. The losses of these genic regions represent stages in the diploidization of the highly polyploid *I. versicolor* (see below). The basic chromosome number of *Iris* has not been determined confidently but is thought to be $x = 10$ or 12, at least ancestrally. By that interpretation, most of the OBL species would be tetraploid derivatives, e.g., *I. brevicaulis* ($2n = 42$, 44), *I. fulva* ($2n = 42$), *I. giganticaerulea*, *I. savannarum*, and *I. hexagona* (all $2n = 44$), *I. prismatica* ($2n = 42$), *I. pseudacorus* ($2n = 34$), and *I. tridentata* ($2n = 40$). However, *Iris* is suspected to represent an ancient polyploid lineage, which subsequently has become diploidized. That interpretation has been supported by the detection of only two 5S rDNA sites by FISH (fluorescence *in situ* hybridization) analysis, which clearly indicates that *I. giganticaerulea* is diploid genetically (i.e., $2n = 2x = 44$). Higher polyploids are represented by *I. virginica* ($2n = 70$, 72) and *I. versicolor* ($2n = 108$). Although differing in chromosome number, the karyotypes of *I. pseudacorus* ($2n = 34$) and *I. acutiloba* ($2n = 42$) are very similar. Protoplast culture has been used to develop artificial polyploid plants of *I. fulva* ($2n = 67$, 84). The "Louisiana irises" (i.e., series *Hexagonae*) are a well-studied group of hybridizing taxa which are involved in the parentages of *Iris* ×*cacique* (*I. fulva* × *I. savannarum*), *I.* ×*fulvala* (*I. brevicaulis* × *I. fulva*), *I.* ×*flexicaulis* (*I. brevicaulis* × *I. giganticaerulea*), and *I.* ×*vinicolor* (*I. fulva* × *I. giganticaerulea*). *Iris* ×*nelsonii* was derived from complex hybrid interactions involving *I. fulva*, *I. hexagona*, and *I. brevicaulis*. The natural (often introgressive) hybrids in series *Hexagonae* are at the root of many taxonomic and phylogenetic problems in the group. Although their effective isolating barriers result in a low frequency of interspecific hybridization, introgression can become prevalent once an F_1 hybrid is formed. Asymmetric introgression has repeatedly transferred genes of *I. fulva* to both *I. brevicaulis* and *I.*

hexagona in a predominantly unidirectional manner. The F_1 hybrids of the broadly sympatric *I. brevicaulis* × *I. fulva* have higher nectar content and concentrations, which is believed to attract pollinators preferentially over either parental species. Hybrids derived from *I. fulva* have been found to harbor the *Iris fulva* mosaic virus (IFMV), which can cause malformed leaves. No hybrids are known that involve *I. pseudacorus* as a parent. Experimental crosses between *I. pseudacorus* and *I. versicolor* are known to fail due to the premature death of the developing hybrid embryo. *Iris versicolor* ($2n = 108$) represents an allotetraploid derived from *I. virginica* ($2n = 70$) × *I. setosa* ($2n = 38$). Backcross hybrids include *I.* ×*robusta* (*I. virginica* × *I. versicolor*) and *I.* ×*sancti-cyri* (*I. versicolor* × *I. setosa*).

Distribution: Although the nonindigenous *Iris pseudacorus* is widespread throughout North America, the indigenous species have smaller ranges and are restricted to northeastern North America (*I. versicolor*), eastern North America (*I. prismatica*, and *I. virginica*), east-central United States (*I. brevicaulis*), or to the southern or southeastern United States (*I. fulva*, *I. hexagona*, *I. tridentata*); even narrower ranges characterize *I. giganticaerulea* (Alabama, Lousiana, Mississippi) and *I. savannarum* (Alabama, Florida, Georgia).

References: Anderson, 1928; 1936; Andrews, 1915; Arditti & Pray, 1969; Arnold, 1993; 1994; 2000; Arnold et al., 1990; 1991; 1993; 2010; Back et al., 1996; Baden, III et al., 1975; Baker, 1910; Barbolani et al., 1986; Barnett, 1986; Baskin & Baskin, 1998; Beal, 1977; Bennett & Grace, 1990; Bezdek et al., 1999; Blankinship, 1903; Blumenthal et al., 1986; Borchardt et al., 2008; Botkin et al., 1973; Caillet et al., 2000; Carter et al., 1990; Chung, 2015; Claassen, 1918; Cody, 1961; Coops & Van der Velde, 1995; Core et al., 1937; Corriveau & Coleman, 1988; Crespo et al., 2015; Crişan & Cantor, 2016; Crişan et al., 2018; Cruzan et al., 1994; Davidson, 1960; Denslow et al., 2011; De Steven & Harrison, 2006; Dutton & Thomas, 1991; Dyer et al., 1976; Elam et al., 2009; Eleuterius, 1972; Eleuterius & Caldwell, 1984; Ellis, 1965; Favre-Bac et al., 2017; Fisher, 1936; Flinn et al., 2008; Friedman, 2013; Galatowitsch & van der Valk, 1996; Galatowitsch & Biederman, 1998; Gaskin et al., 2016; Gates, 1911; Gleason, 1912; Godfrey & Wooten, 1979; Goldblatt, 2002a; Goldblatt & Takei, 1997; Goldblatt & Manning, 2008; Goldblatt et al., 2008; Good, 1986; Graham et al., 2015; Hamerstrom, Jr. & Blake, 1939; Hamlin & Arnold, 2014; Hamlin et al., 2017; Hazlett, 1988; He et al., 2015; Heinrich, 2015; 2016; Henderson, 2002; Hofmann et al., 1990; Homoya & Hedge, 1982; Inoue et al., 2004; Jiang et al., 2018; Johnson, 1892; Jozghasemi et al., 2016; Kauffman, 1916; Kellerman, 1905; Kelley, 1922; Kim et al., 2012; Kjellmark et al., 1998; Kleyheeg & van Leeuwen, 2015; Kron & Stewart, 1994; Kron et al., 1993; Kukula-Koch et al., 2015; Lamont, 1998; Laublin & Cappadocia, 1992; Lee et al., 2000; Lim et al., 2007; Linke, Jr., 1963; Lovell, 1905; Lynn, 1984; Mains, 1934; Mavrodiev et al., 2014; Meerow, 2012; Meerow et al., 2007b; 2011; 2017; M'Mahon, 1804; Mohlenbrock, 1959a; 1959b; Moldenke, 1936; Mopper et al., 2016; Morgan, 1990; Morton & Hogg, 1989; Needham, 1900;

Neil, 1977; Nekola & Lammers, 1989; Nelson, 1986; Nichols, 1920; 1934; Nolfo-Clements, 2006; NYNHP, 2019; Paşca et al., 2016; Penfound & Hathaway, 1938; Perry & Atkinson, 1997; Perry & Hershner, 1999; Pozo et al., 2015; Price, 1966; Ramtin et al., 2013; Raven & Thomas, 1970; Rosen, 2007; Ruch et al., 1998; Sawyer, 1925; Schlüter & Crawford, 2001; Self et al., 1975; Shaw et al., 2017; Smith, 1996; Soomers et al., 2010; Spinner & Bishop, 1950; Stang, 2018; Stansbury et al., 2012; Stearns & Goodwin, 1941; Stoltz, 1968; Stone, 2009; Stuckey & Gould, 2000; Sturtevant et al., 2016; Suter et al., 2011; Sutherland, 1990; Takos, 1947; Tarbeeva et al., 2015; Tiffany et al., 1990; Tillie et al., 2000; Tryon & Easterly, 1975; Uphof, 1922; 1941; van den Broek et al., 2005; Viosca, Jr., 1935; Walker & Peet, 1984; Walton, 1995; Wang & Hasenstein, 2016; Weerasundara et al., 2016; Wees, 2004; Werckmeister, 1969; Wesselingh & Arnold, 2000a; 2000b; 2003; Wetmore, 2001; Wheeler & Wilson, 2014; Zink & Wheelwright, 1997; Wheelwright et al., 2016; White & Simmons, 1988; Wilson, 2009; 2011; Woodcock, 1925; Yousefi & Mohseni-Bandpei, 2010; Zampella et al., 2001; Zink & Wheelwright, 1997.

5.3.2. Nemastylis

Celestial-lily, fall-flowering ixia, happyhour flower, pleatleaf, shell-flower

Etymology: from the Greek *nema stylos* ("thread stylus") for the threadlike style branches

Synonyms: *Calydorea* (in part); *Chlamydostylus*; *Colima* (in part); *Gelasine*; *Moraea* (in part)

Distribution: global: Central America to North America; **North America:** southern United States

Diversity: global: 5 species; **North America:** 4 species

Indicators (USA): obligate wetland (OBL): *Nemastylis floridana*

Habitat: freshwater; palustrine; **pH:** 6.0+; **depth:** <1 m; **life-form(s):** emergent herb

Key morphology: slender herbaceous perennials arising from a tunicate bulb (to 15 mm); stems (to 1.5 m) simple or 3–6-branched; cauline leaves 1–4, reduced and bract-like toward apex, basal leaves 2–3, the blades linear (3–10 mm wide), plicate; inflorescence 2-flowered, subtended by 2 unequal spathes (to 30 mm), the flowers (to 3.5 cm) radial, ephemeral; tepals (to 20 mm) 6, free, lanceolate, stellate, dark blue to deep violet with a central white "eyespot"; filaments (to 2 mm) connate, anthers (to 1 cm) 3, erect but collapsing spirally after dehiscence; ovary (to 2.5 mm) ovoid, the style branching above the filament tube, the 3 filiform branches (to 5 mm) each divided to base, extended horizontally between the anthers, their stigmas apical; capsule (to 9 mm) ovoid, truncate, cartilaginous, ovules 48; seeds (to 2 mm) numerous (12–38), brown, angular to prism-like

Life history: duration: perennial (bulbs); **asexual reproduction:** none; **pollination:** insect, self; **sexual condition:** hermaphroditic; **fruit:** capsules (common); **local dispersal:** seeds (ants, water); **long-distance dispersal:** seeds (ants, water)

Imperilment: 1. *Nemastylis floridana* [G2]; S2 (FL).

Ecology: general: All *Nemastylis* species are perennial herbs, which propagate from underground bulbs. Only one North American species is a wetland indicator, with the others occurring in drier open sites such as meadows, prairies, and rocky montane habitats. The flowers are self-compatible and pollinated primarily by bees (Insecta: Hymenoptera: Apidae: Halictidae). Those of most species open in the morning during spring in contrast to the later day and seasonal flowers of *N. floridana*, the only OBL species. The seeds are not dormant.

5.3.2.1. *Nemastylis floridana* **Small** inhabits ditches (roadside), flatwoods (pine), hammocks (hydric, wet), marshes, meadows (wet), swamps, and the margins of borrow pits and rivers at low elevations to at least 33 m. The habitats are wet, acidic (pH: 6.0) flatwoods to calcareous pinelands and prairies and often are in frequently burned sites where fires have reduced the extent of potentially competitive vegetation. The substrates have been characterized as Bradenton (typic ochraqualfs), Felda (arenic ochraqualfs), limestone, Malabar (grossarenic ochraqualfs), mud (organic), Riviera (arenic glossaqualfs), and sand (fine, mucky, underlain by impervious hardpan). Flowering and fruiting are prolific and extend from late July to November. A plant normally produces eight flower pairs during the growing season, which result in six mature capsules containing from 12 to 38 mature seeds. The flowers are nectarless, scentless, and short-lived, opening from 2:30 to 4:00 pm (EST) and closing from 6:00 to 7:00 pm (earlier as day length decreases). The flowers are self-compatible and often self-pollinate due to the lack of suitable pollen vectors. Because pollen grains can be deposited on the tepals, they can come in contact with the stigmatic surfaces as the tepals close, resulting in spontaneous autogamy; pollinator-mediated selfing is also likely to occur. However, the flowers are facultatively xenogamous (pollen:ovule ratio=690) and are outcrossed primarily by sweat bees (Insecta: Hymenoptera: Halictidae: *Augochlorella striata*, *Lasioglossum foxii*), which results in higher seed set and higher overall reproductive output. Maturing seeds become covered with a drying starchy substance. The capsules mature in about 4 weeks after anthesis, when they dehisce to disperse the seeds. The seeds are dispersed by water and also by ants (Insecta: Hymenoptera: Formicidae), which possibly are attracted to their starchy coating. The fresh seeds are not dormant and germinate within 2 weeks when planted on moist substrates. The bulbs occur 4.5 cm below the substrate surface on average and appear to be at least somewhat fire-resistant. The plants are not cold tolerant and are damaged as the air temperatures approach freezing point (0°C). **Reported associates:** *Acer rubrum*, *Amphicarpum muhlenbergianum*, *Aristida stricta*, *Axonopus furcatus*, *Ctenium aromaticum*, *Erigeron vernus*, *Eriocaulon decangulare*, *Eupatorium mikanioides*, *Gordonia lasianthus*, *Hypericum*, *Ilex glabra*, *Lachnocaulon anceps*, *Mikania scandens*, *Myrica cerifera*, *Panicum*, *Pinus elliottii*, *Pinus palustris*, *Pluchea baccharis*, *Quercus laurifolia*, *Sabal palmetto*, *Scutellaria integrifolia*, *Serenoa repens*, *Taxodium*, *Vernonia blodgettii*.

Use by wildlife: The flowers of *Nemastylis floridana* are visited by various bees (Insecta: Hymenoptera: Apidae:

Bombus pensylvanicus; Halictidae: *Agapostemon splendens, Augochlorella striata, Lasioglossum foxii*), which normally forage for their pollen. The capsules and seeds are eaten by grasshoppers (Insecta: Orthoptera: Acrididae).

Economic importance: food: *Nemastylis* is not edible; **medicinal:** no known uses; **cultivation:** Likely because of their rarity, *Nemastylis* species are not in cultivation; **misc. products:** none; **weeds:** not weedy; **nonindigenous species:** none.

Systematics: In phylogenetic analyses (e.g., Figure 2.10), *Nemastylis* resolves as the sister genus of *Calydorea*, which is polyphyletic as currently circumscribed. Although *N. floridana* has not yet been included in phylogenetic analyses, a BLAST search of an unpublished DNA sequence (*rbcL*) in GenBank (KX397869.1) shows it to be most similar to that of *Nemastylis tenuis*, which indicates that the genus is monophyletic. The basic chromosome number of *Nemastylis* is *x*=7; *Nemastylis floridana* (2*n*=56) is an octoploid. No hybrids involving *N. floridana* have been reported.

Distribution: *Nemastylis floridana* is endemic to southeastern Florida.

References: Foster, 1945; Goldblatt, 1975; 2002b; Goldblatt et al., 2008; Grace, 2000; Orzell & Bridges, 2006a; Prichard & Thorne, 1961; Small, 1931.

5.3.3. Sisyrinchium

Blue-eyed grass; bermudienne, sisyrinque

Etymology: after its resemblance to *Sisyrinchíon* (*sisýra* [Greek]: an animal hide garment), an ancient iris-like plant whose tunicate corms resembled a coat of shaggy animal hair

Synonyms: *Glumosia*; *Hydastylus*; *Paneguia*; *Pogadelpha*; *Souza*

Distribution: global: New World, New Zealand, United Kingdom*; **North America:** widespread throughout

Diversity: global: ~100 species; **North America:** 39 species

Indicators (USA): obligate wetland (OBL): *Sisyrinchium demissum, S. elmeri, S. idahoense, S. sarmentosum*; **facultative wetland (FACW):** *S. idahoense*

Habitat: freshwater; saline (inland); palustrine; **pH:** 5.6–9; **depth:** <1 m; **life-form(s):** emergent herb

Key morphology: rhizomes obscure; stems (to 50 cm) scapelike, simple or 2–3-branched, compressed, 2-winged; leaves (2–6) basal or basal and cauline, alternate, equitant, the blades ensiform, flat, glabrous; inflorescence single, terminal, 1–15-flowered, subtended by 2 basally connate (to 8 mm) spathes (to 65 mm); flowers radial, their pedicels erect, recurved, or spreading; tepals (to 20 mm) 6, spreading to reflexed, light or pale bluish violet to purple (the bases yellow) or yellow to yellowish orange; stamen filaments distinct to mostly connate, glabrous, or stipitate-glandular basally, tapering to apex; styles 3, erect, filiform, connate at base or higher; capsules (to 7.9 mm) globose, beige to dark brown; seeds (to 2 mm) numerous, black, globose, obconic or hemispherical, smooth, granular, or rugulose

Life history: duration: annual (fruits/seeds); perennial (rhizomes); **asexual reproduction:** rhizomes; **pollination:** insect; **sexual condition:** hermaphroditic; **fruit:** capsules

(common); **local dispersal:** seeds (unknown); **long-distance dispersal:** seeds (unknown)

Imperilment: 1. *Sisyrinchium demissum* [G5]; S2 (CO); **2.** *S. elmeri* [G4]; **3.** *S. idahoense* [G5]; S2 (BC, WY); S3 (BC, WY, UT); **4.** *S. sarmentosum* [G2]; S1 (OR); S3 (WA)

Ecology: general: *Sisyrinchium* often is found growing in wet sites, with 28 of the North American species (72%) categorized as wetland indicators (OBL, FACW, FAC, FACU). However, only four species (10%) are classified as OBL (all from western North America). Most of the species are perennials (long- or short-lived) with a few sometimes occurring as annuals. The perianth of most *Sisyrinchium* species ranges in color from light blue to deep purple (hence their common name), whereas those of a few species (including one OBL indicator) can vary from light yellow to orange-yellow. No differences in pollinator specificity have been ascribed to the different tepal colors. Interestingly, the yellow tepals will stain paper a reddish-purple color when pressed and dried. The species differ in being self-compatible (octoploids and duodecaploids) or self-incompatible (tetraploids). Although most flowers are protandrous, the temporal separation of sexual phases increases proportionally with ploidy, i.e., relatively short in the self-incompatible tetraploids, and progressively longer in the self-compatible octoploids and duodecaploids. Outcrossing is facilitated by bees (Insecta: Hymenoptera: Halictidae: *Lasioglossum*), which also can mediate selfing in the self-compatible species, especially when pollen dehiscence and stigma receptivity overlap. Most *Sisyrinchium* species are nectarless, but some have floral trichomal elaiophores that produce oil, which can function as an alternative to nectar for attracting pollinators. Neither nectar nor oil is known to be secreted by flowers of the OBL species, which offer only pollen as a reward to their floral visitors. Because vegetative propagules are lacking, all dispersal is carried out by transport of the seeds. Local dispersal potential generally is regarded as low. There is some circumstantial evidence for bird dispersal, with some species thought to have been dispersed over long distances (e.g., to islands) by seeds that occur within the mud adhering to the feet of birds (Aves). Endozoic seed transport might be possible but has not yet been demonstrated. Spontaneous appearance of plants in gardens (where unsown) has been attributed to bird dispersal, but also could result from contaminated soils brought in on other garden specimens. Some species are known to have been dispersed by humans, including nonindigenous occurrences of *Sisyrinchium montanum* in Europe, which correlate strongly with areas occupied by the American military during the First World War. The seeds of at least some species are dormant when shed or become dormant soon afterward. Many germinate within 3–4 weeks at 20°C, or after 2–4 weeks of cold stratification (−4°C–4°C). Others germinate well (within 2 weeks) after 6 weeks of cold stratification (4°C) followed by lower incubation temperatures (10°C–15°C).

5.3.3.1. *Sisyrinchium demissum* **Greene** is a perennial, which inhabits bottoms (river), channels, cienega, depressions, ditchbanks, draws, floodplains, meadows (montane, wet), marshes, oxbows, ponds (ephemeral), pools (stagnant), seeps, slopes (to 15%), springs (dripping, hot), washes, and the margins of ponds (artificial), rivers, and streams at elevations to 3109 m. The plants occur often in burned sites in exposure ranging from full sunlight to partial shade. They are more common in drier, peripheral sites, rather than in the wettest vegetation zones. The substrates are alkaline or saline and include alluvium (Quaternary, rocky, sandy, sandy/clay, silty, stony), basalt, clay (Carmel), clay loam, cobble, conglomerate (Bishop), detritus (igneous), gravel (igneous, rocky, sandy), limestone (gravelly, rocky), loam (black, rich), muck, mud, sand, sandstone (Navajo), sandy loam, silt (rocky), silty clay, silty loam, and stones. Flowering and fruiting occur from April to October. The perianth is a dark blue-violet color. Although the staminal column possesses nuptial trichomes, the flowers have been shown to lack nectar or oil production. The floral biology of this species has not been studied specifically but presumably is typical for the genus. Because both tetraploid and octoploid cytotypes occur (see *Systematics*), it is difficult to evaluate their degree of self-incompatibility or their extent of self-pollination (see *Ecology: general* above). Neither the seed ecology nor the dispersal mechanisms have been characterized. The plants are obscurely rhizomatous and grow in cespitose clumps. They are fairly tolerant of grazing, which can result in the persistence of stunted specimens. **Reported associates:** *Abies concolor, Abies lasiocarpa, Acer negundo, Achillea millefolium, Agropyron desertorum, Agrostis stolonifera, Alisma, Allium, Almutaster pauciflorus, Alnus tenuifolia, Alopecurus, Ambrosia psilostachya, Ambrosia trifida, Amorpha fruticosa, Andropogon glomeratus, Anemopsis californica, Antennaria, Apocynum cannabinum, Aquilegia chrysantha, Arctostaphylos pringlei, Arenaria, Artemisia ludoviciana, Artemisia tripartita, Asclepias subverticillata, Astragalus lentiginosus, Astragalus subcinereus, Atriplex canescens, Baccharis salicifolia, Baccharis sarothroides, Berberis haematocarpa, Betula occidentalis, Bidens aurea, Bouteloua, Bromus tectorum, Campanula parryi, Carex alma, Carex aquatilis, Carex athrostachya, Carex aurea, Carex canescens, Carex curatorum, Carex lasiocarpa, Carex nebrascensis, Carex pellita, Carex petasata, Carex praegracilis, Carex simulata, Carex stipata, Carex subfusca, Carex utriculata, Castilleja lineariiloba, Centaurium, Ceratophyllum demersum, Chenopodium album, Cirsium vulgare, Cladium californicum, Clematis ligusticifolia, Corallorhiza, Cornus sericea, Crataegus, Cryptantha setosissima, Cynodon dactylon, Cynoglossum officinale, Cyperus squarrosus, Dactylis glomerata, Danthonia intermedia, Dasiphora floribunda, Delphinium geraniifolium, Deschampsia, Distichlis spicata, Drymaria molluginea, Drymocallis arizonica, Echeandia flavescens, Elaeagnus angustifolia, Eleocharis geniculata, Eleocharis palustris, Eleocharis parishii, Eleocharis rostellata, Elymus elymoides, Elymus trachycaulus, Epilobium ciliatum, Epipactis gigantea, Equisetum ×ferrissii, Equisetum arvense, Equisetum hyemale, Equisetum laevigatum, Ericameria nauseosa, Erigeron flagellaris, Erigeron formosissimus, Eriogonum racemosum, Eriogonum umbellatum, Eryngium sparganophyllum, Erysimum capitatum, Eustoma exaltatum, Fallugia paradoxa, Festuca arizonica, Forestiera pubescens, Fraxinus pennsylvanica, Fraxinus velutina, Garrya, Gentianella amarella, Geranium lentum, Geum triflorum, Glyceria borealis, Gratiola, Gutierrezia sarothrae, Herrickia glauca, Hesperostipa, Heterotheca villosa, Hopia obtusa, Hordeum arizonicum, Hordeum brachyantherum, Hordeum jubatum, Hymenopappus filifolius, Hymenoxys hoopesii, Hymenoxys subintegra, Hymenothrix wislizeni, Hypericum formosum, Ipomopsis multiflora, Iris missouriensis, Juglans major, Juncus balticus, Juncus drummondii, Juncus dudleyi, Juncus ensifolius, Juncus interior, Juncus longistylis, Juncus saximontanus, Juncus tenuis, Juncus torreyi, Juniperus communis, Juniperus osteosperma, Juniperus scopulorum, Leersia oryzoides, Lemna minor, Ligusticum porteri, Lobelia anatina, Lobelia cardinalis, Lobelia laxiflora, Lupinus latifolius, Lysimachia maritima, Lythrum californicum, Maianthemum stellatum, Medicago lupulina, Melilotus albus, Melilotus officinalis, Mentha spicata, Mimulus guttatus, Mimulus primuloides, Muhlenbergia asperifolia, Muhlenbergia montana, Muhlenbergia rigens, Muhlenbergia tricholepis, Nolina microcarpa, Oenothera curtiflora, Oenothera elata, Oenothera rosea, Onopordum acanthium, Opuntia engelmannii, Opuntia polyacantha, Orthocarpus luteus, Orthocarpus purpureoalbus, Oxalis, Packera hartiana, Panicum virgatum, Pascopyrum smithii, Paspalum distichum, Penstemon eatonii, Penstemon procerus, Penstemon pseudoputus, Perideridia, Phacelia heterophylla, Phalaris minor, Phleum alpinum, Phragmites australis, Picea engelmannii, Picea pungens, Pinus edulis, Pinus longaeva, Pinus ponderosa, Plantago eriopoda, Plantago major, Poa compressa, Poa fendleriana, Poa nemoralis, Poa pratensis, Polygonum aviculare, Polypogon monspeliensis, Polypogon viridis, Populus angustifolia, Populus deltoides, Populus fremontii, Potamogeton diversifolius, Potamogeton foliosus, Potentilla anserina, Potentilla hippiana, Primula pauciflora, Prosopis velutina, Prunella vulgaris, Prunus, Pseudocymopterus montanus, Pseudostellaria jamesiana, Pseudotsuga menziesii, Pteridium aquilinum, Pyrrocoma lanceolata, Quercus gambelii, Ranunculus acriformis, Ranunculus cymbalaria, Ranunculus macranthus, Rhus aromatica, Ribes aureum, Ribes cereum, Ribes montigenum, Robinia neomexicana, Rorippa, Rosa woodsii, Rubus ulmifolius, Rudbeckia laciniata, Rumex acetosella, Sagina, Salix commutata, Salix exigua, Salix gooddingii, Salix lasiolepis, Salix lutea, Salix ×bebbii, Schedonorus arundinaceus, Schedonorus pratensis, Schizachyrium scoparium, Schoenoplectus acutus, Schoenoplectus americanus, Schoenoplectus pungens, Scirpus microcarpus, Senecio triangularis, Sidalcea neomexicana, Silene, Sisyrinchium montanum, Solanum ptychanthum, Solidago, Sorghum halepense, Sparganium emersum, Sphaeralcea, Sphenopholis obtusata, Spiranthes delitescens, Spiranthes romanzoffiana, Sporobolus airoides, Sporobolus wrightii, Stuckenia pectinata, Symphyotrichum ericoides, Symphyotrichum falcatum, Symphyotrichum spathulatum, Tamarix ramosissima, Taraxacum officinale, Thermopsis montana, Trifolium repens, Trifolium wormskioldii, Typha domingensis, Typha latifolia, Ulmus pumila, Valeriana,*

Veratrum californicum, Verbascum thapsus, Veronica americana, Vicia, Xanthium strumarium. Zannichellia palustris, Zeltnera arizonica.

5.3.3.2. *Sisyrinchium elmeri* Greene is an annual or a perennial, which grows in montane bogs (*Darlingtonia*), fens, marshes, meadows, seeps, slopes (marshy), and along the margins of streams at elevations from 817 to 3291 m. The plants grow in open exposures. The substrates have been described as humus, muck, and serpentine. Flowering and fruiting occur from May to August. Unlike the blue or violet hues of its OBL congeners, the tepals are yellow or orange-yellow in color. The significance of this color difference with respect to pollination has not been evaluated, nor have any other aspects of the reproductive ecology of this species. One source indicates that the seeds will germinate at 10°C after 1–6 months but mostly during spring after overwintering. It is unclear why some authors categorize the plants as annuals, while others describe them as cespitose rhizomatous perennials. Further life history details are lacking. **Reported associates:** *Aconitum columbianum, Agrostis idahoensis, Agrostis idahoensis, Allium, Bistorta bistortoides, Bryum pseudotriquetrum, Caltha leptosepala, Camassia leichtlinii, Camassia quamash, Carex echinata, Carex echinata, Carex hassei, Carex lemmonii, Carex lenticularis, Carex luzulina, Carex scopulorum, Carex utriculata, Castilleja miniata, Chamerion angustifolium, Conardia compacta, Danthonia californica, Darlingtonia californica, Deschampsia cespitosa, Ditrichum ambiguum, Drosera rotundifolia, Eleocharis decumbens, Erigeron glacialis, Eriophorum crinigerum, Gentiana newberryi, Gentianella amarella, Gentianopsis simplex, Hastingsia alba, Helenium bigelovii, Hosackia oblongifolia, Hosackia pinnata, Juncus balticus, Juncus effusus, Juncus howellii, Kyhosia bolanderi, Lilium kelleyanum, Lupinus fulcratus, Menyanthes trifoliata, Mimulus primuloides, Muhlenbergia andina, Narthecium californicum, Nuphar polysepala, Oreostemma alpigenum, Oxypolis occidentalis, Parnassia palustris, Pedicularis attollens, Perideridia parishii, Philonotis fontana, Pinus contorta, Pinus monticola, Platanthera, Podagrostis humilis, Primula, Pteridium aquilinum, Pyrola minor, Ranunculus alismifolius, Rhododendron columbianum, Rhododendron occidentale, Salix eastwoodiae, Scirpus congdonii, Senecio clarkianus, Senecio triangularis, Sisyrinchium bellum, Sisyrinchium idahoense, Sphenosciadium capitellatum, Triantha glutinosa, Triantha occidentalis, Trifolium longipes, Triteleia hyacinthina, Vaccinium uliginosum, Viola.*

5.3.3.3. *Sisyrinchium idahoense* E.P. Bicknell is a perennial, which occurs in or on beaches, bogs (marl), bottomlands (stream), depressions, ditch banks, ditches (irrigation, roadside), flats (floodplain), floodplains, grassland, ledges (moist), marshes, meadows (alluvial, calcareous, moist, roadside, wet), prairies, seeps (alkaline, ephemeral), shores, slopes (to 5%), springs (warm), swales, swamps, thickets, vernal pools, woodlands, and along the margins of channels (runoff, spring), ponds (dune), pools, rivers, and streams (intermittent) at elevations to 3277 m. Exposures consistently receive full sunlight. The substrates often are alkaline or saline but can vary substantially (e.g., pH: 5.6–9.1). They have been described as

alluvium (gravelly, sandy, silty), clay (heavy), clayey silt, clay loam, cobble, gleysols (organic, silty), granite, gravel, gravelly loam, humus (dark), limestone (Claron), loam (pumicy), muck (alluvial, grey), pumice, quartzite, rock, sand, sandy clay, sandy clay loam, sandy loam, serpentine, silty clay loam, silty loam, sinter, stony loam, and travertine. Flowering and fruiting extend from April to September. The staminal column possesses nuptial trichomes, but the flowers do not produce nectar or oil. The perianth ranges from light blue to purple in color. The flowers are self-compatible but are strongly protandrous, which is highly conducive to outcrossing. Spontaneous selfing rates (e.g., 22% fruit set) are much lower than those for artificially selfed flowers (e.g., 55% fruit set), which indicates that self-pollination in natural populations is likely to be pollinator mediated. Selfed seeds germinate well and produce fertile offspring (pollen stainability: 99–100%). Otherwise, the floral and seed ecology presumably are fairly typical of the genus; however, additional details are scarce. The plants grow vegetatively from indiscernible rhizomes, which results in a cespitose growth pattern. **Reported associates:** *Abies concolor, Achillea millefolium, Agoseris glauca, Agrostis capillaris, Agrostis gigantea, Agrostis oregonensis, Agrostis stolonifera, Aira caryophyllea, Aira elegantissima, Allium brevistylum, Allium geyeri, Allium schoenoprasum, Alnus viridis, Alopecurus pratensis, Amelanchier alnifolia, Amphiscirpus nevadensis, Anaphalis margaritacea, Anemone multifida, Anemopsis californica, Angelica arguta, Angelica lineariloba, Antennaria corymbosa, Antennaria microphylla, Antennaria rosea, Anthoxanthum nitens, Anthoxanthum odoratum, Anticlea elegans, Aquilegia formosa, Arctium minus, Arnica chamissonis, Artemisia cana, Artemisia ludoviciana, Artemisia tridentata, Astragalus agrestis, Astragalus canadensis, Astragalus lemmonii, Astragalus leptaleus, Athyrium filix-femina, Balsamorhiza deltoidea, Berberis aquifolium, Betula glandulosa, Betula occidentalis, Bistorta bistortoides, Bistorta vivipara, Botrychium virginianum, Boykinia major, Bromus carinatus, Bromus inermis, Bupleurum americanum, Calamagrostis canadensis, Calamagrostis stricta, Calocedrus decurrens, Calochortus apiculatus, Calochortus tolmiei, Camassia quamash, Campanula rotundifolia, Capsella bursa-pastoris, Cardamine cordifolia, Carex aperta, Carex athrostachya, Carex aurea, Carex canescens, Carex capitata, Carex densa, Carex deweyana, Carex douglasii, Carex flava, Carex foenea, Carex fracta, Carex geyeri, Carex inops, Carex interior, Carex lasiocarpa, Carex lenticularis, Carex livida, Carex lyngbyei, Carex mariposana, Carex microptera, Carex obnupta, Carex ovalis, Carex pachystachya, Carex parryana, Carex praegracilis, Carex raynoldsii, Carex rostrata, Carex scirpoidea, Carex scopulorum, Carex siccata, Carex simulata, Carex stipata, Carex tumulicola, Carex unilateralis, Carex utriculata, Carex vesicaria, Carex viridula, Carex xerantica, Castilleja applegatei, Castilleja miniata, Castilleja rhexiifolia, Castilleja sulphurea, Ceanothus sanguineus, Cerastium arvense, Cerastium fontanum, Cerastium glomeratum, Chamerion angustifolium, Cicuta maculata, Cirsium arvense, Cirsium remotifolium, Cirsium scariosum, Crepis* runcinata, *Cryptantha echinella, Cynosurus cristatus, Cytisus scoparius, Danthonia californica, Darlingtonia californica, Dasiphora floribunda, Daucus carota, Delphinium menziesii, Deschampsia cespitosa, Dichanthelium acuminatum, Distichlis spicata, Drosera linearis, Drosera rotundifolia, Eleocharis acicularis, Eleocharis palustris, Eleocharis rostellata, Elymus glaucus, Elymus trachycaulus, Epilobium ciliatum, Equisetum arvense, Equisetum hyemale, Equisetum laevigatum, Equisetum sylvaticum, Ericameria, Erigeron acris, Erigeron annuus, Erigeron gracilis, Erigeron leiomerus, Erigeron linearis, Erigeron lonchophyllus, Eriogonum, Eriophorum, Eriophyllum lanatum, Euthamia occidentalis, Festuca idahoensis, Festuca rubra, Fragaria vesca, Fragaria virginiana, Fragaria ×ananassa, Frasera speciosa, Fraxinus latifolia, Galium boreale, Galium oreganum, Gentiana affinis, Gentiana fremontii, Gentiana newberryi, Gentianella amarella, Gentianopsis detonsa, Geranium richardsonii, Geum macrophyllum, Glyceria striata, Glycyrrhiza lepidota, Hackelia micrantha, Hastingsia alba, Helenium autumnale, Helenium bigelovii, Heracleum sphondylium, Hieracium caespitosum, Hieracium scouleri, Holcus lanatus, Hordeum brachyantherum, Hypericum anagalloides, Hypericum perforatum, Hypochaeris radicata, Iris missouriensis, Juncus articulatus, Juncus balticus, Juncus brevicaudatus, Juncus bufonius, Juncus confusus, Juncus conglomeratus, Juncus covillei, Juncus effusus, Juncus ensifolius, Juncus filiformis, Juncus hallii, Juncus nevadensis, Juncus occidentalis, Juniperus scopulorum, Kalmia microphylla, Koeleria macrantha, Lactuca serriola, Lepidium perfoliatum, Lepidium virginicum, Leucanthemum vulgare, Leucopoa kingii, Leymus cinereus, Leymus triticoides, Lilaeopsis occidentalis, Lilium parryi, Lomatium utriculatum, Lomatium utriculatum, Lupinus bicolor, Lupinus latifolius, Lupinus lepidus, Luzula campestris, Luzula comosa, Luzula parviflora, Lycopus uniflorus. Lysimachia maritima, Maianthemum stellatum, Malus fusca, Marah, Medicago lupulina, Melica bulbosa, Melilotus officinalis, Mentha arvensis, Menyanthes trifoliata, Micranthes subapetala, Microseris laciniata, Mimulus cardinalis, Mimulus guttatus, Mimulus lewisii, Mimulus moschatus, Muhlenbergia asperifolia, Muhlenbergia richardsonis, Narthecium californicum, Nepeta cataria, Orthocarpus luteus, Oxytropis deflexa, Packera debilis, Panicum virgatum, Parnassia palustris, Pascopyrum smithii, Pedicularis attollens, Pedicularis bracteosa, Pedicularis crenulata, Pedicularis groenlandica, Penstemon rydbergii, Phleum alpinum, Phleum pratense, Phlox kelseyi, Phlox longifolia, Picea engelmannii, Pinus contorta. Pinus flexilis, Plantago eriopoda, Plantago lanceolata, Plantago major, Platanthera dilatata, Platanthera hyperborea, Poa fendleriana, Poa interior, Poa palustris Poa pratensis, Poa secunda, Polemonium occidentale, Polystichum munitum, Polytrichum juniperinum, Populus tremuloides, Populus trichocarpa, Potentilla anserina, Potentilla gracilis, Potentilla wheeleri, Potentilla ×diversifolia, Primula conjugens, Primula fragrans, Primula hendersonii, Primula incana, Primula jeffreyi, Primula pauciflora, Prunella vulgaris, Pseudoroegneria spicata, Pseudotsuga menziesii, Pteridium aquilinum,* Purshia tridentata, Pyrola asarifolia, Pyrrocoma lanceolata, Pyrrocoma linearis, Pyrrocoma uniflora, Quercus garryana, Ranunculus acris, Ranunculus cymbalaria, Ranunculus inamoenus, Ranunculus occidentalis, Ranunculus orthorhynchus, Ribes cereum, Ribes inerme, Ribes lacustre, Rosa californica, Rosa nutkana, Rosa woodsii, Rubus bifrons, Rubus ursinus, Rudbeckia occidentalis, Rumex acetosella. *Rumex crispus, Salix bebbiana, Salix boothii, Salix lasiolepis, Salix scouleriana, Salix wolfii, Sambucus nigra, Sarcobatus, Schedonorus pratensis, Schoenoplectus pungens, Senecio hydrophilus, Senecio sphaerocephalus, Senecio sphaerocephalus, Senecio triangularis, Shepherdia canadensis, Sidalcea oregana, Sisyrinchium angustifolium, Sisyrinchium bellum, Sisyrinchium elmeri, Sisyrinchium sarmentosum, Solidago canadensis, Solidago lepida, Solidago multiradiata, Solidago simplex, Solidago velutina, Sonchus arvensis, Spartina gracilis, Sphenopholis obtusata, Sphenosciadium capitellatum, Spiraea douglasii, Spiranthes romanzoffiana, Sporobolus airoides, Stachys albens, Stachys chamissonis, Stellaria longifolia, Stellaria longipes, Stipa, Symphyotrichum ascendens, Symphyotrichum ascendens, Symphyotrichum hallii, Symphyotrichum spathulatum, Synthyris reniformis, Taeniatherum caput-medusae, Tanacetum vulgare, Taraxacum officinale, Thermopsis montana, Thuja plicata, Torreyochloa pallida, Toxicodendron diversilobum, Toxicoscordion venenosum, Triantha glutinosa, Triantha occidentalis, Trientalis latifolia, Trifolium aureum, Trifolium dubium, Trifolium hybridum, Trifolium longipes, Trifolium monanthum, Trifolium pratense, Trifolium repens, Trifolium variegatum, Trifolium wormskioldii, Triglochin maritima, Triglochin palustris, Trillium chloropetalum, Triteleia hyacinthina, Typha latifolia, Urtica dioica, Vaccinium, Vahlodea atropurpurea, Valeriana edulis, Valeriana sitchensis, Vancouveria, Veratrum californicum, Verbascum thapsus, Veronica arvensis, Veronica serpyllifolia, Veronica wormskjoldii, Vicia americana, Vicia hirsuta, Vicia sativa, Vicia tetrasperma, Viola adunca.*

5.3.3.4. Sisyrinchium sarmentosum Suksd. ex Greene is a perennial, which occurs in seasonally wet flats (stream), meadows, and swales at elevations to 2140 m. The only reported substrates comprise slightly acidic (e.g., pH: 6.5) clay. Flowering and fruiting occur from April to September. The flowers have pale blue tepals and are self-compatible but somewhat protandrous, which might promote outcrossing. They open from late morning to midday and can remain closed on overcast days or during periods of poor weather conditions. The flowers are known to attract various bees (Insecta: Hymenoptera: Adrenidae; Apidae: *Bombus*; Megachilidae), which likely function as pollinators. However, the pollen is known to be shed prior to the opening of the flowers, which surely would promote self-pollination. Yet, spontaneous selfing rates are lower than those for artificially self-pollinated flowers, which indicates that selfing rates in natural populations are likely to be pollinator- mediated. Selfed seeds germinate well and produce fertile offspring (pollen stainability: 97%), with no evidence of inbreeding depression. Genetic studies (data from 16 isozyme loci) detected only one unique allele in five of six surveyed populations (fixed for the same alleles otherwise), and low levels of variation at isozyme two loci in the sixth. The low level of genetic variation has been attributed to a recent bottleneck, which may have occurred at the time of origin for this presumably recently evolved species, or as a result of repeated self-pollinations. Surveys using more sensitive genetic (RAPDs) markers detected higher levels of variation, especially between Oregon and Washington occurrences of the species. There is a dearth of information on the seed ecology, with little more than speculation on seed longevity or dispersal mechanisms. However, seed germination details have been elucidated fairly well. A 60% germination rate has been obtained after 8 weeks of cold stratification (4°C) followed by incubation under a 20°C/10°C day/night temperature regime. A reduced germination rate (40%) characterized cold-stratified seeds incubated under a constant temperature (20°C). Freshly shed seeds (unstratified) do not germinate under a constant temperature (20°C) but germinate fairly well (40%) when exposed to alternating temperatures (20°C/10°C). Thus, it seems necessary to expose seed to cooler temperatures (e.g., 10°C) in order to stimulate germination. Enhanced germination rates have been obtained by bleaching the seeds. The plants reproduce vegetatively by slender rhizomes, but it is unclear whether they are highly cespitose or capable of spreading clonally to any degree. **Reported associates:** *Abies amabilis, Abies lasiocarpa, Abies procera, Acer circinatum, Achillea millefolium, Achnatherum occidentale, Agoseris glauca, Agrostis idahoensis, Alopecurus pratensis, Antennaria microphylla, Botrychium multifidum, Camassia, Carex buxbaumii, Carex microptera, Carex rostrata, Carex vesicaria, Castilleja miniata, Cirsium arvense, Cirsium vulgare, Crataegus douglasii, Cynoglossum officinale, Dactylis glomerata, Danthonia intermedia, Delphinium trolliifolium, Deschampsia cespitosa, Dichelostemma congestum, Eleocharis quinqueflora, Elymus glaucus, Epilobium, Fragaria vesca, Fragaria virginiana, Galium triflorum, Gentiana sceptrum, Geum macrophyllum, Hackelia diffusa, Heracleum sphondylium, Hypericum anagalloides, Hypericum formosum, Hypericum perforatum, Iris tenax, Jacobaea vulgaris, Juncus confusus, Juncus ensifolius, Juncus tenuis, Linnaea borealis, Linum, Lupinus latifolius, Lycopodium, Lysichiton americanus, Mimulus guttatus, Mimulus moschatus, Mimulus primuloides, Oreostemma alpigenum, Packera pseudaurea, Phleum alpinum, Phleum pratense, Picea engelmannii, Picea sitchensis, Pinus contorta, Pinus monticola, Poa palustris, Polytrichum, Populus trichocarpa, Potentilla drummondii, Primula, Primula jeffreyi, Prunella vulgaris, Pseudotsuga menziesii, Pteridium aquilinum, Quercus garryana, Ranunculus flammula, Ribes, Rosa, Rubus lasiococcus, Salix, Senecio triangularis, Sisyrinchium idahoense, Solidago canadensis, Spiraea douglasii, Stachys chamissonis, Symphoricarpos mollis, Triantha occidentalis, Trifolium repens, Tsuga heterophylla, Urtica dioica, Vaccinium membranaceum, Vaccinium ovalifolium, Vaccinium uliginosum, Valeriana sitchensis, Vancouveria hexandra, Veratrum viride, Veronica americana, Veronica scutellata, Veronica* *wormskjoldii, Vicia americana, Viola adunca, Xerophyllum tenax.*

Use by wildlife: *Sisyrinchium demissum* is host to some Fungi (Ascomycota: Phaeosphaeriaceae: *Scolecosporiella sisyrinchii*). The flowers of *S. idahoense* reportedly are visited by "nectaring" Mardon skippers (Insecta: Lepidoptera: Hesperiidae: *Polites mardon*), despite the fact that the flowers demonstrably produce no nectar. *Sisyrinchium sarmentosum* is eaten by some mammals (Mammalia) including cattle (Bovidae: *Bos taurus*) and small rodents (Rodentia). Grazing has been identified as a serious threat to the survival of this uncommon species.

Economic importance: food: *Sisyrinchium* is not edible by humans; **medicinal:** No medicinal uses have been reported for the OBL indicators; **cultivation:** *Sisyrinchium idahoense* has been identified as a promising candidate for use in low-maintenance native landscape gardens. Bulk quantities of seed are produced for use in native habitat restoration projects; however, seedling establishment rates of only 1% have been reported; **misc. products:** *Sisyrinchium idahoense* has been evaluated for use in "green roofs," which are being developed to mitigate excessive stormwater runoff; **weeds:** none; **nonindigenous species:** all the OBL species are indigenous.

Systematics: *Sisyrinchium* is taxonomically difficult, with species estimates ranging from 80 to 100+ in different accounts. Even the incorporation of DNA "barcoding" approaches has not provided much clarification due to the low resolution obtained from those markers used routinely for species delimitation, along with numerous complicating interpretive factors such as widespread polyploidy. Although *Sisyrinchium* has been characterized as "a complex polyploid taxon," it is remarkable that polyploidy has been addressed only superficially (or not at all) in some recent molecular phylogenetic accounts of the genus, where data from nuclear loci simply have been combined *en masse* with data from mitochondrial and/or plastid loci. Although that approach has yielded some consistent results (e.g., indicating the monophyly of the genus), there are numerous inconsistencies with respect to interspecific relationships, where polyploidy would more likely arise as a complicating factor. Such problems are evident by the resolution of some species (including the OBL taxa) as being closely related by some analyses (e.g., plastid data and nrITS data) but separated in remote clades in other analyses (e.g., plastid, mtDNA, and nrITS data). The different subspecies of *S. idahoense* (which differ in chromosome number) resolve as closely related in one study but within entirely different clades in another (e.g., Figure 2.12). Similarly, *S. sarmentosum* is shown to be related closely to *S. idahoense* (subsp. *macounii*) in one study (e.g., Figure 2.12) but resolves within a distant clade in another. *Sisyrinchium elmeri* consistently is depicted as distantly related to the other OBL taxa, resolving within a clade comprising species assigned to section *Hydastylus*, which once was recognized as a separate genus (e.g., Figure 2.12); however, accessions of one species within that clade (*S. tinctorium*) resolve in three different subclades, a result that confounds any attempt to specify any precise relationships among the species. Because

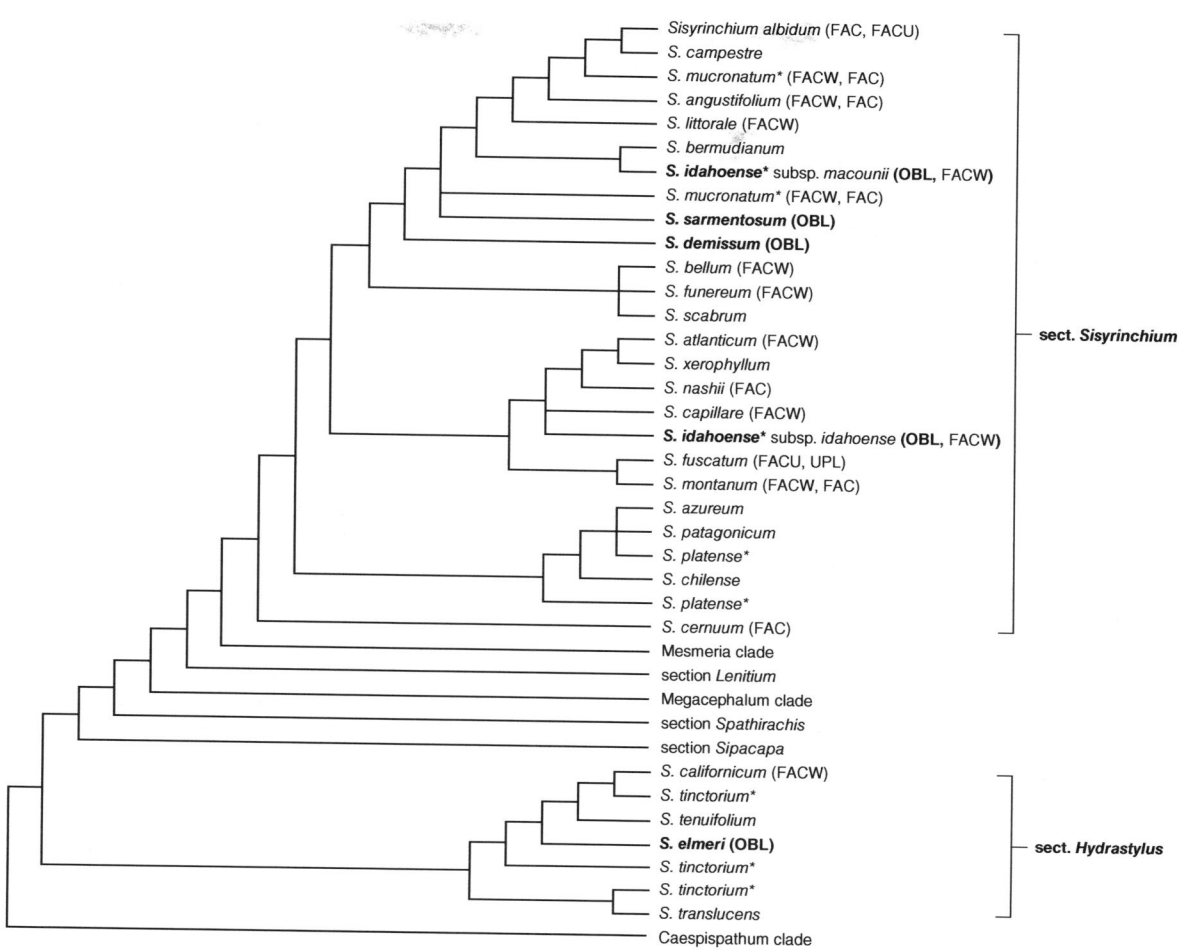

FIGURE 2.12 Relationships in *Sisyrinchium* as depicted by analysis of combined plastid and nuclear DNA sequence data (modified from Karst & Wilson, 2012). Besides the instances where conspecific accessions (possibly different paralogues) resolved in different portions of the tree (asterisked), the results depicted here also disagree substantially from those presented in a similar study by Chauveau et al. (2011), who also included mtDNA sequence data. It is likely that these anomalies are due in part to widespread polyploidy in the genus, which has led to similar interpretive problems in the related genus *Iris*, due there primarily to differential gene silencing (see text). Without fully accounting for artifacts due to polyploidy in *Sisyrinchium*, it is premature to present any tree as a reasonable estimate of relationships, or to propose any revised subgeneric taxonomic classification. The OBL North American indicators have been highlighted in bold.

all the OBL taxa are polyploid (see below), it is likely that such discrepancies are due at least in part to differential gene silencing during diploidization, such as seen in *Iris* (e.g., see *Systematics* for *Iris* above). Certainly more sophisticated genetic analyses will be necessary in order to unravel the reasons behind the anomalous phylogenetic results. Until such work has been carried out, it seems premature to use the phylogenetic results to recommend any meaningful subgeneric classification or to attempt any critical evaluation of species relationships within the group. Crossing studies indicate that *S. idahoense* retains at least partial interfertility (e.g., 5–15%) with a number of species, but exhibits the highest interfertility (44–63%) with *S. sarmentosum*, which some phylogenetic results (e.g., Figure 2.12) depict as being closely related. These two species hybridize in their zone of sympatry, where it is difficult to distinguish the parental species from the hybrid plants. Somewhat reduced interfertility (14–51%) characterizes *S. idahoense* and *S. littorale*, which phylogenetic analyses (e.g., Figure 2.12) have resolved more distantly. Fairly high

crossability also has been achieved between *S. idahoense* and *S. pallidum* (58%) or *S. radicatum* (49%). The transfer of some *Sisyrinchium* species to *Olsynium* as a segregate sister genus is consistent with phylogenetic analyses (e.g., Figures 2.10 and 2.12). The basic chromosome number of *Sisyrinchium* is $x = 8$. All the OBL species are polyploids: *S. demissum* ($2n = 32, 64$); *S. elmeri* ($2n = 34$); *S. idahoense* ($2n = 64, 96$); and *S. sarmentosum* ($2n = 96$). Genome size (mean 2C DNA) has been estimated for a duodecaploid ($2n = 96$) accession of *S. idahoense* (7.31 pg; 0.61 pg/genome). Artificial hybrids between *S. idahoense* and *S. radicatum* are highly infertile as indicated by irregular meiosis and low pollen stainability. Crosses between *S. idahoense* and *S. sarmentosum* are fertile.

Distribution: *Sisyrinchium idahoense* occurs broadly throughout the Rocky Mountains and Pacific Northwest regions of western North America. *Sisyrinchium demissum* occurs in the southwestern United States (Arizona, Colorado, Nevada, New Mexico, Texas, and Utah; extending into northern Mexico). More limited distributions characterize *S.*

sarmentosum (Oregon, Washington) and *S. elmeri* (endemic to California).

References: Alves et al., 2014; Armstrong, 1915; Beck & Peek, 2004; Bicknell, 1901; Carlquist, 1967; Carville, 1997; Chauveau et al., 2011; Cholewa & Henderson, 1994; 2002; Cross, 1991; Dawe et al., 2011; Fachinetto et al., 2018; Fairman, 1918; Fishbein et al., 1999; Froehlich, 2019; Goldblatt & Manning, 2008; Goldblatt et al., 1990; Henderson, 1976; Inácio et al., 2017; Jones, 2004; Keeler-Wolf, 1982; Kenton et al., 1986; Krock, 2016; Lotspeich et al., 1961; McLaughlin et al., 2001; Parent, 1977; Pfeifer-Meister et al., 2012; Pineda et al., 1999; Potter et al., 1999; Quattrocchi, 2012; Rocchio et al., 2001; Ruchty, 2011; Schroll et al., 2011; Shinners, 1962; Sikes et al., 2010; Silvério et al., 2012; Spence, 1996; Thomas & Schrock, 2004; Thorpe, 2008; Wilson et al., 2000.

Family 5.4. Orchidaceae [736–880+]

Numerous phylogenetic analyses have confirmed the monophyly of Orchidaceae ("orchids"), based on various combinations of morphological and molecular data (Judd et al., 2016). Molecular phylogenetic studies (e.g., Seberg et al., 2012) resolve the orchids as the sister group of the clade comprising all remaining "higher asparagoid" and "lower asparagoid" Asparagales (Figure 2.2). With an estimated 21,950 species (Judd et al., 2016), the orchids are one of the two largest angiosperm families (cf. Asteraceae); however, it is futile to attempt any precise compilation of orchid taxa given that each year approximately 500 new species and 13 new genera are described on average in this massive family (Chase et al., 2015). Orchidaceae also have presented a formidable obstacle to phylogenetic analysis given their vast size alone. Fortunately, due to the efforts of countless researchers

worldwide who have investigated many different orchid groups, a fair overview of broad-scale relationships in the family has begun to emerge. A recent synthesis of this large body of work (Chase et al., 2015) has recommended a classification that divides orchids into 5 subfamilies and 17 tribes within the 2 largest subfamilies (Figure 2.13).

Orchid species are legendary for their extremely varied and unusual floral morphology, which has made them highly sought after by plant collectors, gardeners, and as ornamental specimens. Overall, the group is distinguished by their showy, bilateral, epigynous, and typically resupinate flowers, with perianths characterized by an expanded lip-like tepal (the labellum), and 1–3 stamens that fuse with the style and stigma to form a column (or gynostemium) (Cronquist, 1981; Judd et al., 2016). The pollen usually is united into granular or waxy masses called pollinia, which are shed and dispersed as units. The capsules contain thousands of minute seeds, which are buoyant and dispersed by water or wind (Baskin & Baskin, 1998). Yet, despite their small size, the seeds often are dispersed over relatively short distances of 0.8–15.5 m (Brzosko et al., 2017). Estimates of 3.7 million seeds per capsule have been made for some of the tropical species, with more than 54,000 seeds per capsule estimated in some temperate North American species. Orchid seeds are morphologically dormant due to their undifferentiated embryos (Baskin & Baskin, 1998). The seeds of most orchid species are mycotrophic and must associate with mycorrhizal Fungi in order to germinate or for the seedlings to develop (Judd et al., 2016). Some orchids reproduce apomictically. The family includes epiphytes, saprophytes, and vines as well as "terrestrial" members, which can vary in size from millimeters to meters in length (Cronquist, 1981; Romero-González et al., 2002; Judd et al., 2016).

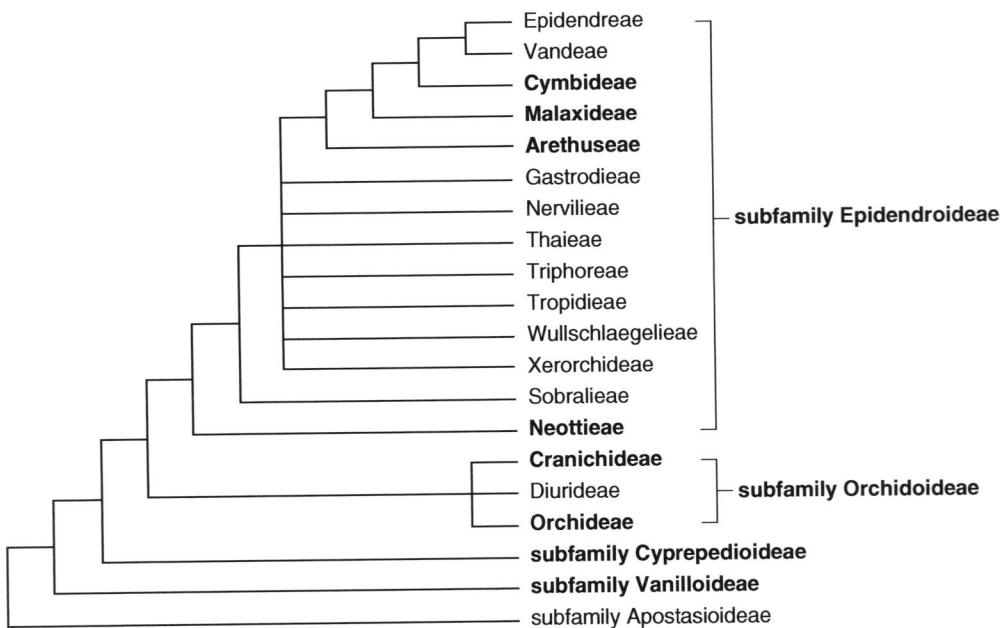

FIGURE 2.13 Subfamilies and tribes of Orchidaceae arranged in a "classification tree," which summarizes the best understanding of higher-level relationships based on numerous literature accounts (simplified from Chase et al., 2015). Taxa containing OBL North American indicators (highlighted in bold) represent a number of different tribes occurring within four of the five orchid subfamilies.

Orchids also are renowned for their elaborate pollination mechanisms. They provided compelling evidence of natural selection to Charles Darwin, who remarked that they displayed "an almost endless diversity of beautiful adaptations" (Darwin, 1877). Orchid pollinators include many types of insects (Insecta: Diptera; Hymenoptera; Lepidoptera) as well as birds (Aves) and even bats (Mammalia: Chiroptera) (Cronquist, 1981; Judd et al., 2016). Typically, the pollinators are extreme specialists whose faithful floral visitations provide an effective barrier against hybridization, which otherwise can occur readily even among different orchid genera when cross-pollinated artificially (Cronquist, 1981; Judd et al., 2016). The pollinators are attracted to the flowers by their nectar, pollen, and scent, often using the floral labellum as a "landing platform" to facilitate their foraging (Judd et al., 2016). The flowers of some species lack rewards, but are modified to resemble the structure and pheromone scent of a female insect. These features entice the male to attempt copulation with the floral mimic, thereby effecting pollination in a process known as "pseudocopulation" (Judd et al., 2016). Some orchid species are self-pollinating.

Many orchid genera are grown as garden ornamentals, conservatory specimens, corsages, or cut flowers. The most extensively cultivated genera include *Brassocattleya*, *Brassolaeliocattleya*, *Bulbophyllum*, *Calanthe*, *Cattleya*, *Coelogyne*, *Cymbidium*, *Cypripedium*, *Dactylorhiza*, *Dendrobium*, *Disa*, *Epidendrum*, *Masdevallia*, *Odontioda*, *Odontoglossum*, *Oncidium*, *Paphiopedilum*, *Phalaenopsis*, *Phragmipedium*, *Pleione*, and *Vanda*. Alcoholic extracts of *Vanilla planifolia* fruits (also from *V. pompona* and *V. tahitiensis*) provide the flavoring known universally as "vanilla." Worldwide, orchids are found in countless medicinal preparations. They also are brewed as beverages, eaten as foods, and used as flavorings, perfumes, and tobacco substitutes; some species have been used to fabricate baskets and hats (Duggal, 1971). The mucilaginous pseudobulbs of *Cyrtopodium punctatum* are the source of a glue that is used to bind books, shoe soles, and as a coating for guitar strings (Duggal, 1971).

Orchidaceae are cosmopolitan, but reach their highest diversity in tropical forests (Cronquist, 1981). Of the approximately 70 North American orchid genera, 12 (17%) contain species designated as OBL indicators:

5.4.1. *Arethusa* L.
5.4.2. *Calopogon* R. Br.
5.4.3. *Cypripedium* L.
5.4.4. *Epipactis* Zinn
5.4.5. *Galearis* Raf.
5.4.6. *Habenaria* Willd.
5.4.7. *Liparis* Rich.
5.4.8. *Malaxis* Sol. ex Sw.
5.4.9. *Platanthera* Rich.
5.4.10. *Pogonia* Juss.
5.4.11. *Ponthieva* R. Br.
5.4.12. *Spiranthes* Rich.

The distribution of these genera among four orchid subfamilies, within six tribes, i.e., Epidendroideae: Arethuseae, Cymbidieae,

Malaxideae, Neottieae; Orchidoideae: Cranichideae, Orchideae; Vanilloideae: Pogonieae (Figure 2.13), indicates widespread multiple origins of the OBL habit in the family.

5.4.1. Arethusa

Dragon's mouth, swamp-pink; aréthuse bulbeuse
Etymology: after *Arethusa*, a mythical Greek nymph of freshwater
Synonyms: none
Distribution: global: North America; **North America:** eastern and northeastern
Diversity: global: 1 species; **North America:** 1 species
Indicators (USA): obligate wetland (OBL): *Arethusa bulbosa*
Habitat: freshwater; palustrine; **pH:** 3.9–6.8; **depth:** <1 m; **life-form(s):** emergent herb
Key morphology: corms bulbous with a few fleshy roots; stems (to 40 cm) scapose; leaf solitary, basal, subtended by 2–3 sheaths, the blade (to 23 cm) linear-lanceolate, faintly plicate; flowers solitary, terminal, resupinate, erect, showy; perianth deep pink to magenta (occasionally whitish), the sepals (to 55 mm) erect, divergent, the petals (to 49 mm) arcuate, the lip (to 35 mm) obovate to oblong, arcuate to reflexed, crenulate to obscurely 3-lobed; pollinia yellow-green; capsules (to 25 mm) erect, ovoid to ellipsoid
Life history: duration: perennial (corms); **asexual reproduction:** corms, "root shoots"; **pollination:** insect; **sexual condition:** hermaphroditic; **fruit:** capsules (common); **local dispersal:** seeds (water, wind); **long-distance dispersal:** seeds (water, wind)
Imperilment: 1. *Arethusa bulbosa* [G5]; SX (IN, MD); SH (CT, DC, DE, SC, VA); S1 (LB, NC, OH, PA, SK, RI, VT); S2 (MA, MB, NH, NJ, NY, PE); S3 (MI, ME, QC)
Ecology: general: *Arethusa* is monotypic (see next).
5.4.1.1. ***Arethusa bulbosa* L.** grows in bogs (acid, *Sphagnum*, tundra), fens (patterned), meadows, swamps (coniferous), and along the margins of lakes (bog) and sloughs (peaty) at elevations to 1100 m. The plants are ecological pioneers, which establish under full light exposures but decline as shading and overstory vegetation increases. The substrates most often are thick layers (70–700 cm; $\bar{x} = 297$ cm) of peat (*Sphagnum*) but also include sand. The habitats can be calcareous but usually are acidic (pH: 3.9–6.8; $\bar{x} = 5.5$) and weakly to moderately minerotrophic (conductivity: 2–148 µS cm^{-1}; $\bar{x} = 41$ µS cm^{-1}). When occurring in alkaline sites, the plants grow atop the *Sphagnum* hummocks, in which acidic rather than alkaline conditions exist. The same is true for patterned fens, where the plants occupy the elevated sphagnous "strings" rather than the "flarks." Flowering occurs from late May to July (rarely extending into August). A fairly high proportion of plants ($\bar{x} = 27\%$) produces flowers during the season relative to other co-occurring orchids. Although morphologically adapted for outcrossing, the flowers are self-compatible and yield high seed set (e.g., 95%) when artificially self-pollinated. They are showy but deceptive, offering only a trace of nectar as a reward to the pollinators, which typically are bumblebees (Insecta: Hymenoptera: Apidae: *Bombus ternarius*, *B.*

terricola). Attracted by the floral scent and ultraviolet perianth signals, the bees seeking nectar enter the throat of the flower, remove the pollinia as they back out, and brush against the anther case, with the pollinia usually placed on the back of the thorax and the anther case snapping shut afterward. Only the larger bees (e.g., queens) are capable of jarring open the anther case. The anther contains four boat-shaped pollinia, each made up of pollen tetrads united into granular masses. Pollinia that are attached to the thorax of an arriving insect will be brushed against the stigma as the flower is entered. Because the pollinators normally will not revisit a flower once deceived, natural stands of plants often are characterized by low fruit set (e.g., 5–20%). The dust-like seeds are buoyant and dispersed abiotically by water or wind. Some studies have indicated that *A. bulbosa* tends to be more prevalent in the drier portions of its wetland habitats; however, others have described the plants as occupying the "soggiest parts" of a bog. Vegetative reproduction occurs by the production of unusual "root shoots," i.e., adventitious shoots (sometimes terminating in corms), which develop from the slender fascicular roots. Apparently, these can be quite numerous with their emerging shoots sometimes being mistaken for seedlings.

Reported associates: *Acer rubrum, Agalinis purpurea, Agrostis hyemalis, Alnus rugosa, Amelanchier canadensis, Andromeda polifolia, Andropogon glomeratus, Apios americana, Aronia arbutifolia, Aronia ×prunifolia, Aulacomnium palustre, Betula pumila, Calamagrostis canadensis, Calamagrostis coarctata, Calla palustris, Calopogon tuberosus, Caltha palustris, Calypso bulbosa, Campanula aparinoides, Carex atlantica, Carex aurea, Carex bullata, Carex buxbaumii, Carex chordorrhiza, Carex collinsii, Carex debilis, Carex disperma, Carex echinata, Carex exilis, Carex folliculata, Carex lacustris, Carex lasiocarpa, Carex leptalea, Carex livida, Carex magellanica, Carex michauxiana, Carex oligosperma, Carex pauciflora, Carex stricta, Carex tenuiflora, Carex trisperma, Carex vestita, Chamaedaphne calyculata, Chelone cuthbertii, Chionanthus virginicus, Cicuta bulbifera, Circaea alpina, Cladina rangiferina, Clintonia borealis, Comarum palustre, Coptis trifolia, Cornus canadensis, Cornus racemosa, Cornus sericea, Cypripedium acaule, Cypripedium parviflorum, Cypripedium reginae, Dichanthelium dichotomum, Drepanocladus, Drosera anglica, Drosera intermedia, Drosera rotundifolia, Dryopteris cristata, Dulichium arundinaceum, Empetrum nigrum, Equisetum fluviatile, Equisetum hyemale, Equisetum sylvaticum, Eriocaulon decangulare, Eriophorum angustifolium, Eriophorum gracile, Eriophorum vaginatum, Eriophorum virginicum, Eriophorum viridicarinatum, Eubotrys racemosa, Eupatorium perfoliatum, Eupatorium pilosum, Eurybia macrophylla, Eutrochium maculatum, Eutrochium purpureum, Fraxinus nigra, Galax urceolata, Galium labradoricum, Galium trifidum, Gaultheria hispidula, Gaultheria procumbens, Gaylussacia dumosa, Gentiana saponaria, Helianthus angustifolius, Hypericum canadense, Hypericum densiflorum, Ilex glabra, Ilex laevigata, Ilex mucronata, Iris versicolor, Isoetes engelmannii, Itea virginica, Juncus biflorus, Juncus canadensis, Juncus effusus, Juncus pelocarpus, Juncus stygius, Juncus subcaudatus, Juniperus communis, Kalmia angustifolia, Kalmia polifolia, Larix laricina, Lilium philadelphicum, Linnaea borealis, Liparis loeselii, Lonicera caerulea, Lonicera canadensis, Lonicera hirsuta, Lonicera oblongifolia, Lonicera villosa, Lycopodiella appressa, Lycopodiella caroliniana, Lycopus uniflorus, Lycopus virginicus, Lyonia ligustrina, Lysimachia quadrifolia, Lysimachia terrestris, Lysimachia thyrsiflora, Magnolia virginiana, Maianthemum trifolium, Malaxis unifolia, Mentha arvensis, Menyanthes trifoliata, Micranthes pensylvanica, Mitella nuda, Moneses uniflora, Muhlenbergia glomerata, Muhlenbergia uniflora, Myrica caroliniensis, Myrica gale, Narthecium americanum, Neottia convallarioides, Neottia cordata, Nyssa sylvatica, Oclemena nemoralis, Onoclea sensibilis, Orthilia secunda, Osmundastrum cinnamomeum, Oxypolis rigidior, Oxypolis rigidior, Packera crawfordii, Parnassia asarifolia, Pellia, Persicaria amphibia, Physocarpus opulifolius, Picea mariana, Platanthera aquilonis, Platanthera clavellata, Platanthera cristata, Platanthera dilatata, Platanthera hyperborea, Platanthera lacera, Platanthera psycodes, Pogonia ophioglossoides, Polygala cruciata, Polygala lutea, Polytrichum juniperinum, Prenanthes alba, Pycnanthemum muticum, Pyrola asarifolia, Rhamnus alnifolia, Rhexia virginica, Rhododendron canadense, Rhododendron groenlandicum, Rhododendron viscosum, Rhododendron viscosum, Rhynchospora alba, Rhynchospora capitellata, Rhynchospora fusca, Rhynchospora gracilenta, Rosa acicularis, Rosa palustris, Rosa virginiana, Rubus arcticus, Rubus chamaemorus, Rubus hispidus, Rubus pubescens, Rubus repens, Rumex britannica, Salix candida, Salix pedicellaris, Salix planifolia, Sarracenia purpurea, Scheuchzeria palustris, Schizachyrium scoparium, Schizaea pusilla, Scirpus atrovirens, Scirpus expansus, Scutellaria galericulata, Sisyrinchium atlanticum, Smilax glauca, Solidago altissima, Trientalis borealis, Solidago gigantea, Solidago patula, Solidago puberula, Solidago uliginosa, Sparganium, Sphagnum angustifolium, Sphagnum capillifolium, Sphagnum centrale, Sphagnum cuspidatum, Sphagnum flavicomans, Sphagnum fuscum, Sphagnum magellanicum, Sphagnum majus, Sphagnum palustre, Sphagnum papillosum, Sphagnum recurvum, Sphagnum rubellum, Sphagnum warnstorfii, Spiraea tomentosa, Spiranthes romanzoffiana, Stellaria longifolia, Symphyotrichum novi-belgii, Symphyotrichum puniceum, Symplocarpus foetidus, Thalictrum revolutum, Thelypteris palustris, Thuja occidentalis, Toxicodendron vernix, Triadenum fraseri, Triadenum virginicum, Triantha racemosa, Trichophorum alpinum, Trichophorum cespitosum, Usnea, Utricularia cornuta, Utricularia gibba, Utricularia intermedia, Utricularia subulata, Vaccinium angustifolium, Vaccinium corymbosum, Vaccinium macrocarpon, Vaccinium oxycoccos, Vernonia noveboracensis, Viburnum nudum, Viola macloskeyi, Viola primulifolia, Xanthorhiza simplicissima, Xyris caroliniana, Xyris torta.*

Use by wildlife: The flowers of *Arethusa bulbosa* are visited by bumblebees (Insecta: Hymenoptera: Apidae: *Bombus ternarius, B. terricola*).

Economic importance: food: *Arethusa bulbosa* is not edible; **medicinal:** no uses reported; **cultivation:** Although a remarkably beautiful plant, *Arethusa bulbosa* is rarely cultivated due to its protected status, specialized habitat requirements, and difficult growth (once described as "impatient in cultivation"). It has been grown by some botanic gardens as long ago as the early 19th century; **misc. products:** none; **weeds:** not weedy; **nonindigenous species:** none.

Systematics: *Arethusa* represents the type genus of tribe Arethuseae within the orchid subfamily Epidendroideae (Figure 2.13). Within Arethuseae, phylogenetic analyses of cpDNA sequences resolve the monotypic *Arethusa* either as the sister genus of *Calopogon* (*rbcL*) or *Eleorchis* (*matK*; *rbcL* + *matK*). The basic chromosome number of tribe Arethuseae has been interpreted as $x=20$; *A. bulbosa* ($2n=40$) then would be functionally diploid. Artificial hybrids have been obtained between *Arethusa bulbosa*, *Calopogon tuberosus*, and *Pogonia ophioglossoides* despite the absence of natural hybrids in the field. Experimental crosses between *Arethusa* and *Bletia* or *Bletilla* have been unsuccessful.

Distribution: *Arethusa bulbosa* is scattered throughout the northeastern United States and southeastern Canada.

References: Argue, 2012a; Argus, 1968; Boland & Scott, 1991; 1992; Burns, 1911; Felix & Guerra, 2010; Goldman et al., 2001; Hosack, 1811; Judd, 1968; Moisan & Pellerin, 2013; Niles, 1904; Sheviak & Catling, 2002a; Standley, 1919; Tan, 1969; Thien & Marcks, 1972; Wheeler et al., 1983; Wherry, 1918; 1920.

5.4.2. Calopogon

Grass-pink; calopogon

Etymology: from the Greek *kalo pogon* ("beautiful beard") in reference to the showy, fimbriate labellum

Synonyms: *Bletia* (in part); *Cathea*; *Cymbidium* (in part); *Helleborine* (in part); *Limodorum* (in part); *Ophrys* (in part)

Distribution: global: North America; West Indies; **North America:** eastern

Diversity: global: 5 species; **North America:** 5 species

Indicators (USA): obligate wetland (OBL): *Calopogon pallidus*, *C. tuberosus*; **facultative wetland (FACW):** *C. tuberosus*

Habitat: freshwater; palustrine; **pH:** 3.4–6.9; **depth:** <1 m; **life-form(s):** emergent herb

Key morphology: corms (to 31 mm) globose to elongate, rarely forked; stems (to 135 cm) scapose; leaves (1–3) sessile, the blade (to 50 cm) linear, or lanceolate, conduplicate, slightly to strongly curled transversely; spikes (to 135 cm) 1–3, racemose, terminal; flowers (1–25) sessile, nonresupinate, the bracts (to 30 mm) ovate to ovate-lanceolate or subulate; perianth magenta, pink, or white, bilateral, dorsal sepal (to 31 mm) lanceolate or oblanceolate, lateral sepals (to 31 mm) ovate to lanceolate, falcate, widely spreading; petals (to 28 mm) lanceolate to pandurate, falcate, the lip (to 23 mm) hinged basally, obscurely 3-lobed, the middle lobe dilated (to 21 mm) into an anvil-shaped, rounded, or triangular blade; disc with longitudinal lamellae grading distally into hairlike, yellow, or orange to magenta protuberances; column (to 25 mm) arcuate, broadly dilated (to 10 mm) distally, rostellum present or absent; anther terminal, the pollinia (4) in 2 pairs; capsules (to 30 mm) erect, ellipsoid to obconic or ovoid

Life history: duration: perennial (corms); **asexual reproduction:** corm buds; **pollination:** insect; **sexual condition:** hermaphroditic; **fruit:** capsules (common); **local dispersal:** seeds (water, wind); **long-distance dispersal:** seeds (water, wind)

Imperilment: 1. *Calopogon pallidus* [G4/G5]; S1 (VA); S2 (LA); S3 (GA, NC); **2.** *C. tuberosus* [G5]; SX (DC); SH (IA); S1 (AR, KY, DE, MD, VA, WV); S2 (IL, MO, MB, OH, OK, RI); S3 (GA, NC, PE, QC, VT)

Ecology: general: *Calopogon* often is observed in wet habitats, with four of the five species (80%) designated as wetland indicators (two as OBL). One species grows in dry pine savannas, with the others occurring in (or near) bogs, fens, or mesic to wet meadows, prairies, and savannas. The flowers are unusual for orchids in being non-resupinate, which results in the enlarged labellum being oriented at the top rather than bottom of the flower, thereby positioning the anther at the base rather than top of the flower. The plants are self-compatible but rely on insects to mediate their pollination. Although they lack nectar and a floral scent, the plants deceive pollen-foraging insects by attracting them to their showy flowers, which contain a sterile, yellow-bristled, UV-absorbing lip protuberance that resembles an anther (sometimes called "false stamens," "pseudopollen," or "pseudostamens"). Upon landing, their body mass depresses the hinged lower tepal, causing them to fall backward onto the column, where the sticky pollinia then adhere to their backs to await transport onto the next flower. Insects that are too large or too small are not able to release the pollinia from the flower properly, which limits pollination to a fairly narrow range of properly sized candidates. Because insects will avoid revisiting unrewarding flowers, fruit production typically is pollen limited. The small seeds are dispersed abiotically. Allozyme data indicate that all five species maintain high intrapopulational levels of genetic variation (e.g., $P=50.0$–94.4%; $AP=2.67$–3.32, $H_e=0.11$–0.43). A slight degree of asexual (vegetative) reproduction occurs by division of the developing corms. The root system dies back each year, which necessitates renewed inoculations of mycorrhizal fungi every new growing season.

5.4.2.1. *Calopogon pallidus* **Chapm.** grows in bogs (hillside, pitcher plant, quaking, seepage, *Sphagnum*), ditches (boggy), fens, flatwoods (coastal, mesic, pine), meadows (roadside), prairies (wet, wiregrass), roadsides (boggy), savannas (pine, seepage), seeps, slopes (seepage), swales (interdunal), swamps (pine barren, *Sphagnum*), woodlands (open), and along pocosins and rights-of-way (savanna) at elevations to 100 m. The plants can occur in exposures ranging from open sunlight to shade and are common in frequently burned sites. They typically become rooted (to a mean maximum depth of 5 cm) in acidic substrates (pH: $\bar{x}=3.4$), which remain saturated but do not become ponded. The substrates range from *Sphagnum* peat to various mineral compositions, and have been described as Basinger (spodic psammaquents), muck, Oldsmar (alfic arenic haplaquods), peat, peaty sand, Pineda, Riveria (arenic glossaqualfs), sand, sandy loam, sandy peat, and Waveland-Lawnwood complex. Several analyzed mineral substrates comprised mainly sand (56–67%) and silt (28–38%), with lesser amounts of clay (4–7%) and an organic matter content of 5.3–11.6%. They were acidic (pH: 4.3–4.5) with a low cation exchange capacity (CEC: 5.3–9.7 meq 100 g^{-1}), and mean nutrient levels (all in ppm) as follows: Al: 1035–1172; B: 0.46–0.60; Ca: 220–398; Cu: 0.20–0.34; Fe: 320–390; K: 27–44; Mg: 49–71; Mn: 1.5–3.0; N: 51–60; Na: 15–54; P: 9.0–13.0; S: 30–40; Zn: 0.8–2.3). Flowering occurs primarily from March to June but sometimes extends through August. The plants are known to flower profusely within several weeks following late winter to early spring fires. Such mass floral displays help to attract bees (Insecta: Hymenoptera: Apidae: *Bombus griseocollis*, *B. pensylvanicus*, *Xylocopa virginica*), which function as their pollinators. Otherwise, the floral biology is as described above (see *Ecology: general*) and probably quite similar to that of the next species (which has been studied in some detail). No detailed study of the seed ecology has been made. As in most orchids, the minute seeds are thought to be dispersed primarily by wind. Allozyme surveys have indicated that the populations retain substantial levels of genetic variability (P=55.6–70.0%; $\bar{x}$=64.0% [populations]; 81.0% [species level]; A=2.22–2.68; $\bar{x}$=2.22 [populations]; 3.19 [species level]; H_e=0.204–0.258; $\bar{x}$=0.233 [populations]; 0.252 [species level]) and maintain high rates of gene flow (H_T: 0.311; H_S: 0.283; G_{ST}: 0.078; F_{IS}: 0.117). **Reported associates:** *Acer rubrum, Agalinis linifolia, Agalinis purpurea, Aletris aurea, Aletris farinosa, Aletris lutea, Amphicarpum muhlenbergianum, Andropogon gerardii, Andropogon glomeratus, Andropogon gyrans, Andropogon liebmannii, Andropogon ternarius, Andropogon virginicus, Anthaenantia rufa, Aristida palustris, Aristida purpurascens, Aristida stricta, Arnica acaulis, Arnoglossum ovatum, Aronia arbutifolia, Arundinaria gigantea, Asclepias longifolia, Balduina uniflora, Berchemia scandens, Bigelowia nudata, Calopogon tuberosus, Carex glaucescens, Carphephorus paniculatus, Carphephorus pseudoliatris, Carphephorus tomentosus, Centella asiatica, Centrosema virginianum, Chamaecrista nictitans, Chaptalia tomentosa, Chasmanthium laxum, Chrysopsis mariana, Cirsium horridulum, Cirsium virginianum, Cleistesiopsis divaricata, Clematis, Clethra alnifolia, Clitoria mariana, Coleataenia anceps, Coleataenia longifolia, Coreopsis gladiata, Coreopsis major, Crotalaria purshii, Ctenium aromaticum, Cuscuta, Desmodium paniculatum, Desmodium strictum, Desmodium tenuifolium, Dichanthelium aciculare, Dichanthelium consanguineum, Dichanthelium dichotomum, Dichanthelium ensifolium, Dichanthelium latifolium, Dichanthelium scabriusculum, Dichanthelium strigosum, Dionaea muscipula, Diospyros virginiana, Drosera brevifolia, Drosera capillaris, Drosera tracyi, Elephantopus nudatus, Eragrostis refracta, Erigeron vernus, Eriocaulon compressum, Eriocaulon decangulare, Eryngium integrifolium, Eryngium yuccifolium, Eubotrys racemosa, Eupatorium album, Eupatorium capillifolium, Eupatorium leucolepis, Eupatorium mohrii, Eupatorium pilosum, Eupatorium rotundifolium, Eupatorium semiserratum, Euphorbia inundata, Eurybia eryngiifolia, Euthamia graminifolia, Fimbristylis autumnalis, Fimbristylis puberula, Fimbristylis spadicea,* *Fuirena scirpoidea, Fuirena squarrosa, Galactia volubilis, Gaylussacia dumosa, Gaylussacia frondosa, Gaylussacia mosieri, Gelsemium sempervirens, Gnaphalium, Gratiola pilosa, Gymnopogon ambiguus, Gymnopogon brevifolius, Helenium pinnatifidum, Helianthus angustifolius, Helianthus atrorubens, Helianthus heterophyllus, Hieracium gronovii, Hypericum brachyphyllum, Hypericum cistifolium, Hypericum cruxandreae, Hypericum fasciculatum, Hypericum hypericoides, Hypericum setosum, Hypoxis hirsuta, Hypoxis juncea, Ilex glabra, Ilex vomitoria, Ionactis linariifolia, Ipomoea, Juncus scirpoides, Lachnanthes caroliniana, Lachnocaulon anceps, Lactuca graminifolia, Lechea pulchella, Lespedeza angustifolia, Lespedeza capitata, Lespedeza repens, Liatris pilosa, Liatris spicata, Lilium catesbaei, Linum floridanum, Linum medium, Liquidambar styraciflua, Lobelia nuttallii, Lobelia paludosa, Lophiola aurea, Ludwigia pilosa, Ludwigia virgata, Lycopodiella alopecuroides, Lycopodiella appressa, Lycopodiella caroliniana, Lyonia ligustrina, Lyonia lucida, Lyonia mariana, Macbridea alba, Macranthera flammea, Magnolia virginiana, Marshallia graminifolia, Mitreola sessilifolia, Muhlenbergia capillaris, Muhlenbergia expansa, Myrica cerifera, Myrica heterophylla, Nyssa sylvatica, Oclemena reticulata, Osmunda regalis, Osmundastrum cinnamomeum, Panicum amarum, Panicum verrucosum, Panicum virgatum, Parthenocissus quinquefolia, Paspalum laeve, Paspalum praecox, Paspalum setaceum, Persea palustris, Physostegia purpurea, Pinguicula lutea, Pinus elliottii, Pinus palustris, Pinus serotina, Pinus taeda, Pityopsis graminifolia, Pityopsis oligantha, Platanthera blephariglottis, Platanthera ciliaris, Platanthera nivea, Pogonia ophioglossoides, Polygala cruciata, Polygala cymosa, Polygala lutea, Polygala nuttallii, Polygala ramosa, Prenanthes autumnalis, Prunus serotina, Pteridium aquilinum, Pterocaulon pycnostachyum, Pteroglossaspis ecristata, Pycnanthemum flexuosum, Quercus marilandica, Quercus minima, Quercus nigra, Quercus pagoda, Quercus phellos, Quercus pumila, Quercus stellata, Rhexia alifanus, Rhexia lutea, Rhexia mariana, Rhexia nashii, Rhexia petiolata, Rhus copallinum, Rhynchospora baldwinii, Rhynchospora caduca, Rhynchospora capitellata, Rhynchospora cephalantha, Rhynchospora chapmanii, Rhynchospora ciliaris, Rhynchospora colorata, Rhynchospora debilis, Rhynchospora divergens, Rhynchospora elliottii, Rhynchospora fascicularis, Rhynchospora filifolia, Rhynchospora globularis, Rhynchospora inexpansa, Rhynchospora latifolia, Rhynchospora oligantha, Rhynchospora plumosa, Rhynchospora rariflora, Rubus argutus, Sabal palmetto, Sabatia campanulata, Sabatia difformis, Sabatia gentianoides, Sabatia grandiflora, Sabatia quadrangula, Saccharum brevibarbe, Saccharum giganteum, Salix humilis, Sarracenia alata, Sarracenia flava, Sarracenia leucophylla, Sarracenia minor, Sarracenia psittacina, Sarracenia purpurea, Schizachyrium scoparium, Schizachyrium tenerum, Schwalbea americana, Scleria muehlenbergii, Scleria pauciflora, Scleria reticularis, Scleria triglomerata, Scutellaria integrifolia, Serenoa repens, Sericocarpus linifolius, Sericocarpus tortifolius, Seymeria cassioides, Sisyrinchium atlanticum, Smilax biltmoreana, Smilax bona-nox, Smilax glauca, Smilax* *laurifolia, Smilax rotundifolia, Solidago fistulosa, Solidago odora, Solidago puberula, Solidago stricta, Solidago tortifolia, Sphagnum, Sporobolus curtissii, Stenanthium densum, Stylosanthes biflora, Symphyotrichum dumosum, Symphyotrichum walteri, Symplocos tinctoria, Taxodium, Tephrosia hispidula, Tephrosia spicata, Tiedemannia filiformis, Toxicodendron radicans, Toxicoscordion nuttallii, Triantha racemosa, Utricularia subulata, Vaccinium corymbosum, Vaccinium myrsinites, Vaccinium stamineum, Vaccinium tenellum, Viola lanceolata, Viola primulifolia, Viola septemloba, Vitis rotundifolia, Woodwardia areolata, Xyris ambigua, Xyris baldwiniana, Xyris caroliniana, Xyris difformis, Xyris platylepis, Zigadenus glaberrimus.*

5.4.2.2. *Calopogon tuberosus* (L.) Britton, Sterns & Poggenb.

inhabits barrens (pine), bogs (floating, hillside, pitcher plant, quaking, sandy, seepage, sphagnous, spruce, tamarack), depressions (dune), ditches (grassy, peaty, roadside), fens (seepage), flats (palmetto, pine), flatwoods (burned, coastal, mesic, pine, wet), glades (sandy), marshes (floating), meadows (burned, grassy, *Sphagnum*, sandy), pinelands (annually burned), prairies (marl, wet), roadsides (boggy, damp), savannas (burned, pine, pitcher plant, seepage), seeps (calcareous, hillside), sloughs, swales (mowed, peaty, sandy), swamps (conifer, sphagnous), and the margins of canals, hammocks, lakes (cobbly, grassy) ponds (boggy, dune, gum, cypress), rivers, and streams (*Sphagnum*, swampy) at elevations to 1200 m. The habitats provide open exposures that receive full sunlight to partial shade. The plants occur on acidic substrates (pH: 3.9–6.9) to a mean maximum rooting depth of 7 cm. Reported substrates include Atmore (plinthic paleaquults), Basinger (spodic psammaquents), Bayboro, clay (red), clay loam, cobble, marl, marly gravel, limestone, loam, peat, peaty sand, Pelham (arenic paleaquults), Pottsburg (grossarenic haplaquods), sand (Felda, Myakka, Valkaria), Riviera (arenic glossaqualfs), sandy clay, sandy peat, Smithton, and Stringtown (typic hapludults). A few specimens have been found growing on floating logs. Some of the non-peat substrates analyzed comprised mainly sand (55–69%) and silt (25–40%), with lesser amounts of clay (4–7%); organic matter content ranged from 5.8% to 11.6%. They were acidic (pH: 4.3–4.7) with a low cation exchange capacity (CEC: 4.3–9.7 meq 100 g^{-1}) and mean nutrient levels (all in ppm) as follows: Al: 1084–1234; B: 0.45–0.60; Ca: 161–398; Cu: 0.20–0.34; Fe: 211–440; K: 26–44; Mg: 45–71; Mn: 1.0–3.0; N: 52–60; Na: 20–54; P: 5.0–12.7; S: 20–31; Zn: 0.6–2.3). Flowering occurs from June to August in northern localities but can extend from March to November in the south. Although adapted morphologically for outcrossing, the flowers are self-compatible and capable of geitonogamous, self-, or cross-pollination; however, insects are necessary to transfer pollen in each case. Thus, fruit set is strictly pollinator-limited. The pollen remains viable for at least 8 days. The principal pollinators are pollen-foraging bees (Insecta: Hymenoptera: Apidae: *Bombus ternarius, B. terricola, B. vagans*; Megachilidae: *Megachile melanophea*). Due to their size (which precludes them from effectively visiting the flowers), the larger *Bombus* queens usually fail in deference to the workers, which are smaller yet still able to depress the lower tepal enough to gain access to the flower. The flowers are deceptive by their lack of nectar and floral scent, but attract pollinators by their showy perianths and mass-flowering displays, which often occur after periods of fire. As a result they are thought to be "chance-pollinated," given that pollinators often avoid the flowers once they have been deceived. In one site, a low percentage of flowering (10.3–18.2%; $\bar{x} = 13.2\%$) was associated with a higher fruit set (16.7–22.2%; $\bar{x} = 18.8\%$) relative to other co-occurring orchid species. Allozyme surveys indicate that the populations retain high levels of genetic variability ($P = 31.6$–88.9%; $\bar{x} = 68.2\%$ [populations]; 90.5% [species level]; $A = 1.37$–2.38; $\bar{x} = 2.38$ [populations]; 3.10 [species level]; $H_e = 0.075$–0.250; $\bar{x} = 0.179$ [populations]; 0.215 [species level]) and maintain high rates of gene flow (H_T: 0.237; H_S: 0.197; G_{ST}: 0.085; F_{IS}: 0.090). The small seeds ($\bar{x} = 0.68$ mm) are dispersed by water or wind. They will germinate on agar or in sterile water within 2 weeks, after which they develop into green protocorms. The highest germination rates (40–45%) have been obtained for seeds cultured on Knudson C medium either under darkness or a 16-hr photoperiod. The seeds have germinated after being stored in a refrigerated desiccator for 5 years and cryopreserved seeds are known to have a higher seed germination rate (41%) than unfrozen seeds (35%). The seedlings develop through multiple growth cycles. After germination, the seedlings continue to develop for 3–4 months but then die back to small, green corms, which remain dormant at room temperature. The corms normally overwinter but their dormancy can be broken by cold stratification (4°C for 4–6 weeks or 6–8 weeks at 10°C), whereupon a new shoot develops and a second growth cycle continues through the next dormant phase. The plants grow progressively larger after each successive growth cycle. Some ecotypic differentiation has been indicated in this widespread orchid. Photoperiod has little influence on germination rates except for southern populations (e.g., Florida), which germinate better under short-day conditions. Warmer incubation temperatures also promote germination in seeds from more southern populations. Corm development (and shoot senescence) occur more rapidly in northern than in southern populations. Vegetative reproduction occurs by division of corm buds, but usually results in no more than two closely attached corms, with the older eventually disintegrating.

Reported associates: *Acer rubrum, Actaea rubra, Agalinis linifolia, Agalinis purpurea, Agalinis tenuifolia, Ageratina altissima, Aletris aurea, Aletris farinosa, Alnus rugosa, Amelanchier interior, Amianthium muscitoxicum, Amphicarpaea bracteata, Amphicarpum muhlenbergianum, Andromeda polifolia, Andropogon gerardii, Andropogon glomeratus, Andropogon gyrans, Andropogon liebmannii, Andropogon ternarius, Andropogon virginicus, Anemone quinquefolia, Annona glabra, Anthaenantia rufa, Apios americana, Aralia nudicaulis, Arisaema triphyllum, Aristida palustris, Aristida purpurascens, Aristida stricta, Arnica acaulis, Arnoglossum ovatum, Aronia arbutifolia, Aronia ×prunifolia, Arundinaria gigantea, Asclepias incarnata, Asclepias lanceolata, Asclepias longifolia, Asclepias* *sullivantii, Athyrium filix-femina, Bacopa caroliniana, Bacopa monnieri, Baptisia tinctoria, Bejaria racemosa, Berchemia scandens, Betula pumila, Bidens connatus, Bidens mitis, Bidens trichospermus, Bidens tripartitus, Bigelowia nudata, Bigelowia nuttallii, Boehmeria cylindrica, Bromus ciliatus, Bulbostylis ciliatifolia, Calamagrostis canadensis, Calla palustris, Calopogon multiflorus, Calopogon pallidus, Caltha palustris, Campanula aparinoides, Cardamine pratensis, Carex aquatilis, Carex chordorrhiza, Carex communis, Carex cristatella, Carex deweyana, Carex diandra, Carex disperma, Carex glaucescens. Carex gracillima, Carex gynocrates, Carex hystericina, Carex interior, Carex lacustris, Carex lasiocarpa, Carex leptalea, Carex leptonervia, Carex limosa, Carex livida, Carex magellanica, Carex oligosperma, Carex pedunculata, Carex prairea, Carex pseudocyperus, Carex stipata, Carex stricta, Carex tenera, Carex tenuiflora, Carex trisperma, Carex utriculata, Carex vesicaria, Carphephorus paniculatus, Carphephorus tomentosus, Castanea dentata, Centella asiatica, Centrosema virginianum, Cephalanthus occidentalis, Chamaecrista nictitans, Chamaedaphne calyculata, Chaptalia tomentosa, Chasmanthium laxum, Chelone glabra, Chrysopsis mariana, Cicuta bulbifera, Cicuta maculata, Cinna arundinacea, Circaea alpina, Cirsium horridulum, Cirsium virginianum, Cladium jamaicense, Cladium mariscoides, Cladonia leporina, Cleistesiopsis divaricata, Clematis virginiana, Clethra alnifolia, Cliftonia monophylla, Clinopodium glabrum, Clintonia borealis, Clitoria mariana, Coleataenia anceps, Coleataenia longifolia, Comarum palustre, Coptis trifolia, Coreopsis gladiata, Coreopsis major, Cornus canadensis, Cornus foemina, Cornus racemosa, Cornus sericea, Crotalaria purshii, Ctenium aromaticum, Cuscuta gronovii, Cypripedium parviflorum, Dasiphora floribunda, Decodon verticillatus, Desmodium paniculatum, Desmodium strictum, Desmodium tenuifolium, Dichanthelium aciculare, Dichanthelium acuminatum, Dichanthelium commutatum, Dichanthelium consanguineum, Dichanthelium dichotomum Dichanthelium ensifolium, Dichanthelium latifolium, Dichanthelium scabriusculum, Dichanthelium sphaerocarpon, Dichanthelium strigosum, Diodia virginiana, Dionaea muscipula, Diospyros virginiana, Doellingeria umbellata, Drosera brevifolia, Drosera capillaris, Drosera filiformis, Drosera intermedia, Drosera linearis, Drosera rotundifolia, Drosera tracyi, Dryopteris carthusiana, Dryopteris cristata, Dulichium arundinaceum, Eleocharis elliptica, Eleocharis erythropoda, Eleocharis quinqueflora, Eleocharis rostellata, Eleocharis tuberculosa, Elephantopus nudatus, Elymus virginicus, Epigaea repens, Epilobium coloratum, Epilobium leptophyllum, Equisetum arvense, Equisetum fluviatile, Eragrostis refracta, Eragrostis spectabilis, Erigeron annuus, Erigeron vernus, Eriocaulon compressum, Eriocaulon decangulare, Eriophorum vaginatum, Eriophorum virginicum, Eriophorum viridicarinatum, Eryngium integrifolium, Eryngium yuccifolium, Eubotrys racemosa, Eupatorium album, Eupatorium capillifolium, Eupatorium leucolepis, Eupatorium mohrii, Eupatorium perfoliatum, Eupatorium pilosum, Eupatorium rotundifolium, Eupatorium semiserratum, Eurybia paludosa, Euthamia graminifolia, Eutrochium maculatum, Ficus, Fimbristylis autumnalis, Fimbristylis puberula, Fimbristylis spadicea, Fragaria vesca, Fragaria virginiana, Fuirena scirpoidea, Fuirena squarrosa, Galactia elliottii, Galactia volubilis, Galium labradoricum, Galium tinctorium, Galium trifidum, Galium triflorum, Gaultheria hispidula, Gaultheria procumbens, Gaylussacia baccata, Gaylussacia dumosa, Gaylussacia frondosa, Gaylussacia ursina, Gelsemium sempervirens, Geranium maculatum, Geum canadense, Glyceria grandis, Glyceria striata, Gnaphalium, Gratiola pilosa, Gratiola ramosa, Gymnopogon ambiguus, Gymnopogon brevifolius, Helenium pinnatifidum, Helenium vernale, Helianthus angustifolius, Helianthus atrorubens, Helianthus heterophyllus, Hieracium gronovii, Hydrolea corymbosa, Hypericum cistifolium, Hypericum crux-andreae, Hypericum densiflorum, Hypericum fasciculatum, Hypericum gymnanthum, Hypericum hypericoides, Hypericum kalmianum, Hypericum myrtifolium, Hypericum setosum, Hypericum tetrapetalum, Hypoxis hirsuta, Hypoxis juncea, Ilex cassine, Ilex coriacea, Ilex glabra, Ilex mucronata, Ilex verticillata, Impatiens capensis, Ionactis linariifolia, Ipomoea, Iris prismatica, Iris versicolor, Juncus canadensis, Juncus longii, Juncus marginatus, Juncus megacephalus, Juncus scirpoides, Juncus trigonocarpus, Juniperus communis, Juniperus virginiana, Kalmia latifolia, Lachnanthes caroliniana, Lachnocaulon anceps, Lactuca graminifolia, Laportea canadensis, Larix laricina, Lechea pulchella, Leersia oryzoides, Lemna minor, Lespedeza capitata, Lespedeza repens, Liatris pilosa, Liatris pycnostachya, Liatris spicata, Lilium canadense, Lilium catesbaei, Linnaea borealis, Linum floridanum, Linum medium, Linum striatum, Liparis loeselii, Liquidambar styraciflua, Lobelia elongata, Lobelia glandulosa, Lobelia kalmii, Lobelia nuttallii, Lonicera dioica, Lonicera oblongifolia, Lonicera villosa, Lophiola aurea, Ludwigia microcarpa, Ludwigia pilosa, Ludwigia virgata, Lycopodiella alopecuroides, Lycopodiella appressa, Lycopodiella caroliniana, Lycopodiella inundata, Lycopus americanus, Lycopus uniflorus, Lyonia ligustrina, Lyonia lucida, Lyonia mariana, Lysimachia ciliata, Lysimachia terrestris, Lysimachia thyrsiflora, Macranthera flammea, Magnolia virginiana, Maianthemum canadense, Maianthemum trifolium, Malaxis unifolia, Marshallia graminifolia, Mentha arvensis, Menyanthes trifoliata, Mimosa microphylla, Mitchella repens, Mitella diphylla, Mitella nuda, Mitreola sessilifolia, Muhlenbergia capillaris, Muhlenbergia expansa, Muhlenbergia glomerata, Muhlenbergia mexicana, Myrica cerifera, Myrica gale, Myrica heterophylla, Nuphar variegata, Nyssa biflora, Nyssa sylvatica, Oclemena reticulata, Onoclea sensibilis, Osmunda regalis, Osmundastrum cinnamomeum, Oxypolis rigidior, Packera paupercula, Panicum amarum, Panicum hemitomon, Panicum verrucosum, Panicum virgatum, Parnassia glauca, Parnassia grandifolia, Parthenocissus quinquefolia, Paspalum laeve, Paspalum praecox, Paspalum setaceum, Pedicularis lanceolata, Persea palustris, Persicaria amphibia, Persicaria sagittata, Phalaris arundinacea, Phragmites australis, Physostegia purpurea,*

Picea mariana, Pieris floribunda, Pilea pumila, Pinguicula caerulea, Pinus echinata, Pinus palustris, Pinus pungens, Pinus rigida, Pinus serotina, Pinus taeda, Pinus virginiana, Pityopsis graminifolia, Platanthera blephariglottis, Platanthera ciliaris, Platanthera nivea, Pluchea baccharis, Pogonia ophioglossoides, Polygala cruciata, Polygala lutea, Polygala nana, Polygala ramosa, Polygala rugelii, Polygala sanguinea, Pontederia cordata, Prenanthes alba, Prenanthes autumnalis, Proserpinaca pectinata, Prunus serotina, Prunus virginiana, Pseudognaphalium helleri, Pteridium aquilinum, Pterocaulon pycnostachyum, Pteroglossaspis ecristata, Ptilimnium capillaceum, Pycnanthemum flexuosum, Pyrola asarifolia, Quercus hemisphaerica, Quercus marilandica, Quercus minima, Quercus nigra, Quercus pagoda, Quercus phellos, Quercus pumila, Quercus stellata, Ranunculus hispidus, Rhamnus alnifolia, Rhamnus cathartica, Rhamnus frangula, Rhexia alifanus, Rhexia lutea, Rhexia mariana, Rhexia nashii, Rhexia nuttallii, Rhexia petiolata, Rhexia virginica, Rhus copallinum, Rhynchospora alba, Rhynchospora baldwinii, Rhynchospora caduca, Rhynchospora capillacea, Rhynchospora capitellata, Rhynchospora cephalantha, Rhynchospora chapmanii, Rhynchospora ciliaris, Rhynchospora debilis, Rhynchospora divergens, Rhynchospora elliottii, Rhynchospora fascicularis, Rhynchospora filifolia, Rhynchospora globularis, Rhynchospora gracilenta, Rhynchospora inexpansa, Rhynchospora inundata, Rhynchospora latifolia, Rhynchospora oligantha, Rhynchospora plumosa, Rhynchospora rariflora, Ribes americanum, Ribes cynosbati, Ribes hirtellum, Ribes lacustre, Ribes missouriense, Ribes triste, Robinia hispida, Robinia pseudoacacia, Rosa palustris, Rubus argutus, Rubus pubescens, Rubus sachalinensis, Rudbeckia fulgida, Rudbeckia scabrifolia, Rumex crispus, Rumex orbiculatus, Sabatia decandra, Sabatia difformis, Sabatia gentianoides, Sabatia grandiflora, Sabatia quadrangula, Saccharum brevibarbe, Saccharum giganteum, Sagittaria graminea, Sagittaria latifolia, Salix bebbiana, Salix candida, Salix caroliniana, Salix discolor, Salix humilis, Salix nigra, Salix pedicellaris, Salix petiolaris, Salix sericea, Sarracenia alata, Sarracenia flava, Sarracenia leucophylla, Sarracenia minor, Sarracenia psittacina, Sarracenia purpurea, Scheuchzeria palustris, Schizachyrium scoparium, Schizaea pusilla, Schoenoplectus acutus, Schoenoplectus pungens, Schwalbea americana, Scleria baldwinii, Scleria muehlenbergii, Scleria pauciflora, Scleria reticularis, Scleria triglomerata, Scutellaria galericulata, Scutellaria integrifolia, Scutellaria lateriflora, Selaginella apoda, Serenoa repens, Sericocarpus linifolius, Sericocarpus tortifolius, Setaria corrugata, Seymeria cassioides, Sisyrinchium atlanticum, Sium suave, Smilax biltmoreana, Smilax bona-nox, Smilax glauca, Smilax laurifolia, Smilax rotundifolia, Solanum dulcamara, Solidago fistulosa, Solidago gigantea, Solidago odora, Solidago ohioensis, Solidago patula, Solidago puberula, Solidago rugosa, Solidago stricta, Solidago uliginosa, Sorghastrum nutans, Sparganium americanum, Sparganium emersum, Sphagnum palustre, Spiraea alba, Spiranthes vernalis, Sporobolus, Stellaria longifolia, Stenanthium densum, Stylosanthes biflora, Symphyotrichum boreale, Symphyotrichum dumosum, Symphyotrichum elliotii, Symphyotrichum laeve, Symphyotrichum lateriflorum, Symphyotrichum puniceum, Symphyotrichum walteri, Symplocarpus foetidus, Symplocos tinctoria, Taraxacum officinale, Taxodium, Tephrosia hispidula, Thelypteris palustris, Thuja occidentalis, Tiedemannia filiformis, Toxicodendron radicans, Toxicodendron vernix, Triadenum fraseri, Triadenum virginicum, Triantha glutinosa, Triantha racemosa, Trientalis borealis, Triglochin maritima, Typha latifolia, Typha ×glauca, Utricularia cornuta, Utricularia subulata, Vaccinium angustifolium, Vaccinium corymbosum, Vaccinium macrocarpon, Vaccinium myrsinites, Vaccinium myrtilloides, Vaccinium oxycoccos, Vaccinium stamineum, Vaccinium tenellum, Viburnum lentago, Viburnum nudum, Viburnum opulus, Viburnum prunifolium, Viola lanceolata, Viola primulifolia, Viola septemloba, Vitis riparia, Vitis rotundifolia, Woodwardia areolata, Woodwardia virginica, Xyris ambigua, Xyris baldwiniana, Xyris caroliniana, Xyris difformis, Xyris platylepis, Xyris torta, Tephrosia spicata, Zigadenus glaberrimus.

Use by wildlife: *Calopogon tuberosus* is browsed by whitetailed deer (Mammalia: Cervidae: *Odocoileus virginianus*). Its flowers are hosts of thrips (Insecta: Thysanoptera: Thripidae: *Frankliniella tritici, Pseudothrips beckhami*).

Economic importance: food: *Calopogon* species are not reported to be edible; **medicinal:** In some parts of rural America, the roots of *Arethusa bulbosa* are used to relieve toothaches; **cultivation:** *Calopogon* species are not widely cultivated; **misc. products:** none; **weeds:** none; **nonindigenous species:** none.

Systematics: *Calopogon* is assigned to the orchid subfamily Epidendroideae, tribe Arethuseae (Figure 2.13), where it is closely related to *Arethusa* and *Eleorchis*. *Arethusa* and *Calopogon* share hollow sectile pollinia and a derived *Bletia* seed type as synapomorphies (characters not evaluated for *Eleorchis*). Various analyses have demonstrated that the genus is monophyletic. From allozyme data, *C. pallidus* shared the highest mean genetic identities with *C. multiflorus* ($I = 0.888$) and *C. barbatus* ($I = 0.875$), with all three species indicated as being closely related. Various DNA data (AFLPs, RFLPs of cpDNA, nrITS sequences) have confirmed that *C. pallidus* is related most closely to *C. multiflorus*. The relationship of *C. tuberosus* to its congeners is less apparent. Allozyme data have shown that *C. tuberosus* var. *tuberosus* shares the highest genetic identity with *C. tuberosus* var. *simpsonii* ($I = 0.853$), and exhibits reduced identities ($I = 0.634–0.783$) among the other *Calopogon* species. However, there exist interpretive problems with respect to conflicting nuclear and plastid-derived data that have been used to investigate interspecific relationships in the genus. Although various explanations remain equivocal, those incongruences have been attributed to the possible involvement of *C. tuberosus* (or another species) in the suspected hybrid origin of the highly polyploid *C. oklahomensis* ($2n = 120$). Because the basic chromosome number of tribe *Arethuseae* is thought to be $x = 20$, the counts for *C. pallidus* ($2n = 40, 42$) and *C. tuberosus* ($2n = 26, 40, 42$)

would indicate that they are diploid (except for the anomalous $2n=26$ count for *C. tuberosus*). By attracting larger pollinators, the somewhat larger flowers of *C. tuberosus* isolate it from the smaller-flowered *C. pallidus* and other sympatric congeners. Isolation among the smaller-flowered species is maintained by ecological, ethological (pollinator), and phenological differences. However, hybridization is thought to occur and several naturally occurring hybrids (and their presumed parentages) have been reported, including *Calopogon* ×*floridensis* (*C. multiflorus* × *C. pallidus*), *Calopogon* ×*vulgaris* (*C. pallidus* × *C. tuberosus*), *Calopogon* ×*simulans* (*C. barbatus* × *C. tuberosus*), *Calopogon* ×*fowleri* (*C. barbatus* × *C. pallidus*), and *Calopogon* ×*goethensis* (*C. multiflorus* × *C. tuberosus*). Despite being given binomial names (intended to designate distinct species of hybrid origin), all appear simply to represent F₁ hybrids.

Distribution: *Calopogon tuberosus* occurs throughout the eastern half of North America, whereas *C. pallidus* is restricted to the southeastern United States.

References: Ackerman, 1986; Amoroso & Judd, 1995; Arditti & Ghani, 2000; Argue, 2012a; Boland & Scott, 1991; Brewer, 1998; Brewer et al., 2011; Brown, 2008; Cain, 1931; Carlson, 1943; Chapman, 1997; Emerson, 1921; Engeman et al., 2014; Felix & Guerra, 2010; Firmage & Cole, 1988; Freudenstein & Rasmussen, 1997; Funderburk et al., 2007; Gaddy, 1982; Goldman et al., 2001; 2002; 2004a; 2004b; Gregg, 1991; Hughes & Kane, 2018; Jersáková et al., 2006; Kauth & Kane, 2009; Kauth et al., 2006; 2008; 2011a; 2011b; Kjellmark et al., 1998; Lamont, 1998; Lunau, 2006; Mills et al., 2009; Munden, 2001; Ogle, 1989; Peet & Allard, 1993; Proctor, 1998; Shelingoski et al., 2005; Smith, 1993; Stewart & Richardson, 2008; Stuckey, 1967; Thien & Marcks, 1972; Trapnell et al., 2004; Urbatsch, 2013; Van der Pijl & Dodson, 1966.

5.4.3. Cypripedium

Lady's-slipper, moccasin-flower, Venus'-slipper; cypripède, sabot de la Vierge

Etymology: from the Greek *Kypris* (a former name of Venus, the mythical Cyprian goddess) and *pedilon* ("slipper") after the resemblance of the perianth to a delicately slippered foot

Synonyms: *Arietinum*; *Calceolus*; *Coelogyne* (in part); *Criosanthes*; *Fissipes*; *Sacodon*

Distribution: global: Eurasia, Central and North America; **North America:** widespread

Diversity: global: 45 species; **North America:** 12 species

Indicators (USA): obligate wetland (OBL): *Cypripedium californicum*, *C. candidum*

Habitat: freshwater; palustrine; **pH:** 7.2–8.2; **depth:** <1 m; **life-form(s):** emergent herb

Key morphology: rhizomes short to elongate, with fleshy, closely to widely spaced roots; stems (to 120 cm) erect, leafy; leaves (3–10) alternate, cauline, ascending or spreading, the bases sheathing, the blades (to 20 cm) elliptic, elliptic-lanceolate, lanceolate or oblanceolate; flowers (1–22) solitary, terminal, resupinate, showy, subtended by foliaceous bracts; perianth bilateral, green, yellow, yellow-green, or pale brownish yellow; sepals 3, often with reddish brown spots or stripes, the dorsal sepal (to 35 mm) elliptic, ovate, or ovate-lance-acuminate, lateral sepals a connate synsepal (to 35 mm); lateral petals (to 46 mm) 2, lanceolate, linear-oblong to linear-lanceolate, spreading, flat or twisted spirally, lip (to 27 mm) inflated, slipper- or sac-shaped, white or pinkish, obovoid, oblance-ovoid, or oblance-fusiform, orifice (to 15 mm) basal; lateral anthers 2, with loose, granular pollen, pollinaria absent, dorsal anther a large, subapical staminode; stigma free, 2 or 3-lobed; capsules, ellipsoid to oblong-ellipsoid

Life history: duration: perennial (rhizomes); **asexual reproduction:** rhizomes; **pollination:** insect; **sexual condition:** hermaphroditic; **fruit:** capsules (common); **local dispersal:** rhizomes; seeds (water, wind); **long-distance dispersal:** seeds (water, wind)

Imperilment: 1. *Cypripedium californicum* [G3]; S3 (OR); **2.** *C. candidum* [G4]; SX (PA); SH (SK); S1 (AL, KY, MB, MD, MO, NE, NJ, NY, OH, ON, VA, SD); S2 (IL, IN, MI, ND, WI); S3 (IA, MN)

Ecology: general: *Cypripedium* species are found often in wetlands, with eight of the North American species (67%) designated as wetland indicators (OBL, FACW, FAC, FACU). The flowers of all the species lack nectar and are pollinated deceptively by bees (Insecta: Hymenoptera: Andrenidae; Apidae; Halictidae), which are attracted to their showy flowers. Most of the flowers appear to be self-compatible but are adapted morphologically for outcrossing. Pollination typically is pollinator-limited. Several of the species are interfertile when growing sympatrically. The seed volume comprises a large percentage of air and some have remained aloft for more than 8 s when dropped into still air. Their buoyancy enables them to be dispersed over considerable distances by wind and their difficult wettability enables them to float on water for an extended duration. Like most orchids, the seeds are morphologically dormant. Some species also appear to have physiologically dormant seeds, requiring 9 weeks of cold stratification to induce germination. Most of the species are associated with mycorrhizal fungi within the family Tulasnellaceae. Vegetative reproduction occurs by the production of rhizomes.

5.4.3.1. ***Cypripedium californicum* A. Gray** grows in bogs (*Darlingtonia*), depressions (marshy), fens, forest openings (conifer, evergreen), marshes (springy), meadows (wet), ravines (deep), roadsides (mossy, rocky), seeps (*Darlingtonia*, hanging, hillside, marshy, rocky), slopes (rocky, scree, steep), springs (hillside), and along the margins of streams at elevations to 2133 m. The exposures range from full sunlight to partial shade. This species is regarded as a serpentine endemic and occurs on substrates that have been described as clay, humus, olivine gabbro, peridotite, rock, rubble (loose, rocky), schist, serpentine (red, rocky), silt (colluvial, lithosolic), and ultramafic. Flowering occurs from April to July, with fruiting extending into August. The breeding system has not been studied in detail. Although neither autogamy nor agamospermy have been evaluated adequately, the flowers are thought not to be autogamous. They are pollinated by carpenter bees (Insecta: Hymenoptera: Apidae: *Ceratina acantha*), whose frequent visits have been implicated in the high level of natural fruit set (e.g., 93% of plants; 76% of flowers yielding fruit) observed in open-pollinated plants. By their large percentage of airspace (92–95% of volume), the seeds can float and are dispersed primarily by wind (likely also by water). They are not known to be physiologically dormant and have germinated well at 23°C under light. Enhanced germination (to 20–50%) has been obtained using full- and half-strength Curtis *Cypripedium* ("CC") growth media at a pH of 7.0–7.5. The protocorms and rhizomes arise simultaneously on the seedlings in just over 13 months after being put in culture. Relative to their congeners, the plants have a relatively broader host range of mycorrhizal Fungi, which includes members of the Ascomycota (Herpotrichiellaceae), Basidiomycota (Ceratobasidiaceae, Sebacinaceae, Tulasnellaceae), and Glomeromycota (Glomeraceae). **Reported associates:** *Adiantum aleuticum, Aquilegia eximia, Artemisia tridentata, Boykinia occidentalis, Bryum pseudotriquetrum, Callitropsis sargentii, Calocedrus decurrens, Caltha leptosepala, Calycanthus occidentalis, Campylium stellatum, Carex echinata, Carex mendocinensis, Chamaecyparis lawsoniana, Cirsium, Darlingtonia californica, Deschampsia cespitosa, Drosera rotundifolia, Epilobium oreganum, Epipactis gigantea, Eriophorum criniger, Erythronium citrinum, Gaultheria shallon, Gentiana setigera, Hastingsia alba, Helenium bigelovii, Hosackia oblongifolia, Iris bracteata, Iris macrosiphon, Juncus ensifolius, Ligusticum apiifolium, Lilium bolanderi, Lilium kelleyanum, Lilium kelloggii, Lilium occidentale, Lilium pardalinum, Lithocarpus densiflorus, Lonicera hispidula, Melica torreyana, Mimulus, Muhlenbergia andina, Myrica californica, Narthecium californicum, Parnassia palustris, Pinguicula vulgaris, Pinus attenuata, Pinus contorta, Pinus jeffreyi, Pinus lambertiana, Pinus monticola, Platanthera sparsiflora, Platanthera stricta, Polygala californica, Polystichum munitum, Pseudotrillium rivale, Pseudotsuga menziesii, Quercus chrysolepis, Quercus vacciniifolia, Raillardella pringlei, Rhamnus californica, Rhododendron columbianum, Rhododendron occidentale, Rudbeckia californica, Salix breweri, Toxicodendron diversilobum, Triantha glutinosa, Triantha occidentalis, Trientalis latifolia, Vaccinium ovatum, Vaccinium parvifolium, Valeriana, Viola primulifolia, Whipplea modesta, Xerophyllum tenax.*

5.4.3.2. *Cypripedium candidum* **Muhl. ex Willd.** inhabits barrens (cedar), ditches, flats (marl), meadows (fen, peaty, sedge), fens (calcareous, hanging, sloping), prairies (mesic, wet), slopes (open, wooded), and the margins of ponds and swales at elevations to 700 m. The plants almost always occur in exposures of full sun to moderate shade, but are not shade-tolerant and usually (but not always) do poorly in areas overgrown by shrubs. The substrates primarily are of alkaline (pH: 7.2–8.2) mineral composition (less often organic) generally exceeding 3500 ppm Ca, 225 ppm Mg, and a cation exchange capacity of 20 meq 100 g^{-1}. They have been described as cherty dolomite, clay (heavy), clayey loam, limestone, marl, peat, sand, and silty loam (mucky). Flowering occurs from April to July. Anthesis begins early, as the small shoots first

emerge from the soil and surrounding vegetation has not yet developed. Growth increases substantially during and after the flowering period. The proportion of plants and flowers bearing fruits varies substantially. Several studies have found flowering to occur in 40–92% of plants with 5–54% of the flowers setting fruit. Typically, the natural level of fruit set in open-pollinated plants tends to be relatively low (e.g., 12–22%) due primarily to pollinator limitation. Higher levels of seed set (44%) have been observed in open sites relative to densely vegetated sites (18%), which is attributed to greater pollinator access to flowers in the former. The plants are self-compatible but are characterized as facultative outcrossers. Typically, they are pollinated by smaller bees (Insecta: Hymenoptera: Andrenidae: *Andrena barbilabris, A. erythrogaster, A. ziziae;* Halictidae: *Augochlorella aurata, Halictus confusus, Lasioglossum hitchensi, L. pilosum, L. versatum, L. zonulum*) and possibly also by flies (Insecta: Diptera: Syrphidae: *Paragus, Toxomerus geminatus, T. marginatus*). The flowers are able to set seed if artificially self-pollinated (which has yielded 87–100% seed set) or by pollinator-mediated selfing. Geitonogamy (but not spontaneous autogamy) also can occur. The showy and fragrant flowers are pollinated by deceit given their lack of floral nectar rewards. Nectar supplementation studies have shown rates of self-pollination and pollen discounting to increase by three-fold over non-supplemented plants, indicating a major selective advantage of deceptive pollination in this species. The perianth is designed as a "semi-trap," which prevents the bee from leaving the flower by the same entrance pathway. Instead, the bees are led through the floral lip opening so that they exit past the column, first depositing any imported pollen on the stigma. Afterward, they acquire a new pollen load as a series of stiff hairs forces them to brush against the anthers. The sticky pollen attaches to the thorax of the bee as it finally exits through one of two openings at the base of the lip. The complete pollination process can take from 5 to 15 min. The exit openings are small relative to several congeners, which discourages some of the larger-sized pollinating insects from entering the flowers. Genetic (allozyme) data have detected a high proportion of polymorphic loci=0.383 but low levels of genetic variation and heterozygosity (e.g., 1.43 mean alleles per locus; H_T=0.05–0.11) within populations; among-population differentiation is also very low (G_{ST}=0.069). The fruits are subjected to high levels of seed predation (e.g., 73%). Taller plants have higher initial seed set and less seed predation than shorter plants. By their large percentage of airspace (89% of volume), the seeds can float and are dispersed primarily by wind and also by water. There does not appear to be a persistent seed bank, but the seeds have retained the ability to germinate after cryopreservation in liquid nitrogen (at −196°C). The seeds germinate best (~45%) after 8 weeks from pollination, with rates dropping drastically (<5%) after 10 weeks. Cold stratification (4°C for 2 months) had no noticeable effect on germination rates. Treatment of 8-week-old seeds with up to 0.8 mg l^{-1} cytokinin [benzyl adenine or 6-(α,α-dimethylallylamino)-purine(2iP)] was found to enhance their germination substantially. Despite the potential for high seed production, seedlings rarely are observed in the field and apparently suffer high mortality. Only 14 seedlings were observed during one 4-year study of thousands of plants growing at two different sites. The seedlings are believed to take as long as 12 years before reaching their fully mature adult stage, with various developmental stages observed for the corms, which occur at substrate depths of 3–5 cm. A corm with root and stem primordia develops during the year of germination. The corm enlarges during the second year and produces 1–2 roots, a leaf scale, and a stem. The first aerial leaf develops during the subsequent season, followed by annual increases in the growth and size of the aerial shoot. Most populations persist primarily by vegetative reproduction, which begins to develop within 2–3 years of germination. The plants spread vegetatively by their dichotomously branching rhizome and roots (bearing numerous adventitious buds), which can lead to the development of large clonal colonies. The rhizomes occur at an average depth of 6–10 cm, with their roots extending to 16 cm. On average, each genet produces from 4.4 to 14.1 ramets, which can reach densities ranging from 0.15 to 5.37 m^{-2}. Adult plants (~40% of those surveyed) also are known to undergo dormancy lasting up to 6 years in duration. This species hosts a relatively narrow range of mycorrhizal Fungi from the Ascomycota (Herpotrichiellaceae) and Basidiomycota (Thelephoraceae, Tulasnellaceae). The plants thrive after dormant season fires or where artificial mowing has similarly removed potentially competing vegetation. The plants avoid drought stress by early drying and senescence of the leaves (e.g., during July) at sites characterized by overtopping vegetation. **Reported associates:** Agalinis purpurea, Agalinis skinneriana, Geum aleppicum, Agoseris glauca, Allium canadense, Allium cernuum, Andropogon gerardii, Anemone virginiana, Anthoxanthum hirtum, Anthoxanthum nitens, Anticlea elegans, Apocynum cannabinum, Arnoglossum plantagineum, Asclepias incarnata, Astragalus agrestis, Berula erecta, Bidens trichospermus, Bouteloua curtipendula, Brickellia eupatorioides, Calamagrostis canadensis, Calamagrostis inexpansa, Caltha palustris, Calystegia spithamaea, Camassia scilloides, Campanula aparinoides, Carex aggregata, Carex aquatilis, Carex crawei, Carex lasiocarpa, Carex leptalea, Carex lurida, Carex meadii, Carex richardsonii, Carex rosea, Carex rostrata, Carex sartwellii, Carex sterilis, Carex tetanica, Celastrus scandens, Cirsium altissimum, Cirsium muticum, Comandra umbellata, Coreopsis palmata, Coreopsis tripteris, Cornus drummondii, Cuscuta glomerata, Cypripedium parviflorum, Cypripedium reginae, Dalea candida, Dalea purpurea, Dasiphora floribunda, Desmodium illinoense, Doellingeria umbellata, Drymocallis arguta, Echinacea pallida, Echinacea simulata, Eleocharis elliptica, Elymus canadensis, Elymus canadensis, Elymus villosus, Epilobium strictum, Erigeron pulchellus, Eryngium yuccifolium, Eupatorium perfoliatum, Euphorbia corollata, Euthamia graminifolia, Eutrochium maculatum, Festuca subverticillata, Filipendula rubra, Galium boreale, Galium obtusum, Galium trifidum, Galium triflorum, Gentiana alba, Gentiana andrewsii, Gentianella quinquefolia, Gentianopsis crinita, Gentianopsis virgata, Geum laciniatum, Glyceria striata, Glycyrrhiza lepidota, Helianthus grosseserratus,

Helianthus hirsutus, Helianthus occidentalis, Heuchera richardsonii, Hieracium longipilum, Humulus lupulus, Hypericum ascyron, Hypoxis hirsuta, Impatiens, Iris virginica, Juncus brachycarpus, Juncus dudleyi, Juncus nodosus, Leersia virginica, Liatris aspera, Liatris spicata, Liatris squarrosa, Lilium philadelphicum, Lilium superbum, Liparis loeselii, Lithospermum canescens, Lobelia kalmii, Lobelia siphilitica, Lycopus americanus, Lycopus asper, Lysimachia quadriflora, Maianthemum stellatum, Manfreda virginica, Monarda fistulosa, Muhlenbergia mexicana, Muhlenbergia racemosa, Nothoscordum bivalve, Oxalis violacea, Packera aurea, Parnassia glauca, Parthenocissus quinquefolia, Pedicularis lanceolata, Eriophorum angustifolium, Phalaris arundinacea, Phlox maculata, Phlox pilosa, Phragmites australis, Platanthera hyperborea, Polygala senega, Potentilla anserina, Prenanthes racemosa, Primula, Psoralidium tenuiflorum, Pycnanthemum tenuifolium, Pycnanthemum virginianum, Ranunculus cymbalaria, Ranunculus sceleratus, Rhamnus frangula, Rhamnus lanceolata, Rhynchospora capillacea, Ribes americanum, Ribes cynosbati, Rosa arkansana, Ruellia humilis, Salix cordata, Salix interior, Sanicula canadensis, Schizachyrium scoparium, Schoenoplectus acutus, Schoenoplectus americanus, Schoenoplectus tabernaemontani, Silphium laciniatum, Silphium perfoliatum, Silphium terebinthinaceum, Silphium trifoliatum, Helenium autumnale, Sisyrinchium mucronatum, Smilax tamnoides, Solidago altissima, Solidago canadensis, Solidago ohioensis, Solidago patula, Solidago riddellii, Solidago rigida, Solidago speciosa, Sorghastrum nutans, Spartina pectinata, Spiranthes cernua, Spiranthes magnicamporum, Sporobolus compositus, Sporobolus heterolepis, Symphyotrichum cordifolium, Symphyotrichum ericoides, Symphyotrichum lanceolatum, Symphyotrichum novae-angliae, Symphyotrichum oolentangiense, Symphyotrichum puniceum, Symphyotrichum sericeum, Symplocarpus foetidus, Thalictrum dasycarpum, Thalictrum revolutum, Thaspium barbinode, Thaspium trifoliatum, Thelypteris palustris, Triglochin maritima, Typha latifolia, Utricularia macrorhiza, Valeriana edulis, Veronicastrum virginicum, Viola sororia, Zizia aptera, Zizia aurea.

Use by wildlife: The flowers of Cypripedium californicum are visited by several insects (Insecta), including bees (Hymenoptera: Apidae: Ceratina acantha; Halictidae: Lasioglossum nigrescens) and small flies (Diptera: Syrphidae: Sphegina occidentalis). Flowers of C. candidum also attract various insects, including flies (Diptera: Culicidae; Syrphidae: Paragus, Toxomerus geminatus, T. marginatus) and bees (Hymenoptera: Andrenidae: Andrena barbilabris, A. erythrogaster, A. ziziae; Apidae: Apis mellifera; Halictidae: Augochlorella aurata, Halictus confusus, Lasioglossum hitchensi, L. pilosum, L. versatum, L. zonulum). The plants are fed on by orchid weevils (Insecta: Coleoptera: Curculionidae: Stethobaris ovata). The leaves of C. candidum are parasitized by Fungi (Ascomycota: Mycosphaerellaceae: Pseudocercospora cypripedii).

Economic importance: food: Neither of the OBL Cypripedium species is edible; **medicinal:** Although Cypripedium roots and rhizome were once listed in the U.S.

Pharmacopeia, *C. californicum* and *C. candidum* have not been used medicinally; **cultivation:** *Cypripedium* species are cultivated extensively. In orchids, all offspring arising from a specific cross are regarded as belonging to the same grex (abbreviated "gx"). Various *Cypripedium* greges have been designated that involve hybrids of the OBL indicators, e.g., (seed × pollen parent): Ilse gx (*C. californicum* × *C. flavum*), Late Delight gx (*C. californicum* × *C. macranthos*), Münster gx (*C. californicum* × *C. guttatum*), Tical gx (*C. californicum* × *C. tibeticum*), Warren gx (*C. reginae* × *C. californicum*); ×*andrewsii* gx (*C. candidum* × *C. parviflorum*), ×*andrewsii* nothovar. *andrewsii* gx (*C. parviflorum* var. *makasin* × *C. candidum*), ×*andrewsii* nothovar. *favillianum* gx (*C. parviflorum* var. *pubescens* × *C. candidum*), Annette gx (*C. macranthos* × *C. candidum*), Dieter gx (*C.* ×*andrewsii* × *C. candidum*), Erika gx (*C. calceolus* × *C. candidum*), Favillianum gx (*C. parviflorum* var. *pubescens* × *C. candidum*), Geisha gx (Gisela gx × *C. candidum*), Gidget gx (*C. candidum* × *C. henryi*), GPH Charles gx (*C. candidum* × *C. montanum*), Hildegard gx (*C. candidum* × *C. cordigerum*), Lusarem gx (*C. candidum* × Sebastian gx), Mason's Birthday gx (*C. candidum* × *C. kentuckiense*), Marika gx (Aki gx × *C. candidum*), Selston High School gx (*C. candidum* × *C. fasciolatum*), and Werner gx (*C. candidum* × *C. yatabeanum*). Various wildflower nurseries distribute rootstocks of *C. candidum* to use as ornamental wildflower, or for restoration projects; **misc. products:** none; **weeds:** none; **nonindigenous species:** none.

Systematics: *Cypripedium* represents the type genus of the orchid subfamily Cypripedioideae (Figure 2.13). Phylogenetic analyses of combined DNA sequence data confirm that *Cypripedium* is monophyletic and, as long thought, indicate that it is most closely related to *Selenipedium* within the subfamily (Figure 2.14). Within the genus, *C. californicum* is assigned to section *Irapeana*, which is not monophyletic, whereas *C. candidum* is assigned to section *Cypripedium*, which is monophyletic (Figure 2.14). In any case, the two OBL indicators are not particularly closely related, with combined plastid and nuclear DNA sequence data resolving *C. californicum* as the sister group to a clade (section *Cypripedium*) comprising subclades of eastern Asian species and North American species (the latter including *C. candidum*). However, substantial differences in the topologies generated by the plastid DNA and nuclear DNA are remindful that these relationships should be regarded as hypothetical rather than factual. Allozyme data indicated a moderate genetic identity ($I=0.794$) between *C. candidum* and *C. calceolus*, which were suspected to represent a progenitor-derivative species pair; however, other studies show these species not to be particularly closely related, with the latter occurring within the "eastern Asian clade" of section *Cypripedium* (see Figure 2.14). The basic chromosome number of *Cypripedium* is $x=10$. Although counts are unavailable for *C. californicum*, all other surveyed North American species (including *C. candidum*) are uniformly $2n=20$. *Cypripedium* ×*andrewsii* is the name given to F_1 hybrids derived from *C. candidum* × *C. parviflorum*. Bidirectional introgressive hybridization occurs extensively in some natural sympatric populations of *C. candidum* and *C. parviflorum*, where genetic studies have

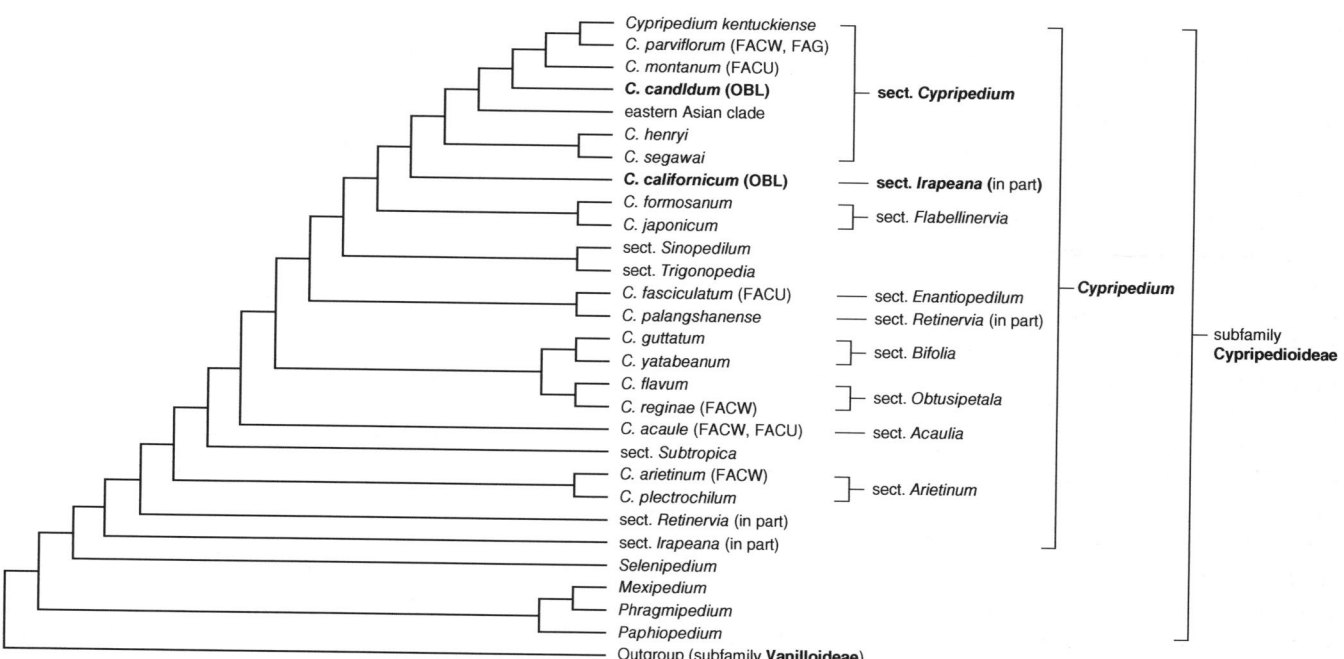

FIGURE 2.14 Interspecific relationships in *Cypripedium* as indicated by phylogenetic analysis of combined plastid and nuclear DNA sequence data (modified from Li et al., 2011). These results place *C. candidum* amidst other members of the monophyletic section *Cypripedium*, whereas *C. californicum* represents an isolated and discordant member of the polyphyletic section *Retinervia* (section *Irapeana* is also polyphyletic). The wetland indicator status for North American species is provided in parentheses; taxa containing OBL indicators are highlighted in bold.

indicated that most offspring are backcross recombinants rather than pure F_1 progeny. These recombinants are thought to be responsible for ecotypic adaptation in *C. parviflorum*. However, lower levels of introgression (~6%) also have been observed in other sympatric populations. Normally, these species are isolated to some degree phenologically (by the earlier flowering of *C. candidum*), ecologically (by the greater preference of *C. candidum* for open, sunny exposures), and ethologically (due to the larger size of some common *C. parviflorum* pollinators, which avoid *C. candidum*).

Distribution: *Cypripedium candidum* occurs throughout northeastern North America, whereas *C. californicum* is restricted to California and Oregon.

References: Anderson, 1943; Arditti & Ghani, 2000; Arditti et al., 1979; Argue, 2012b; 2012c; Atwood, Jr., 1984; Barbour et al., 2007; Baskin & Baskin, 1998; Bowles, 1983; Carroll et al., 1984; Case, 1994; Catling & Knerer, 1980; Coleman et al., 2012; Cranfill, 1991; Curtis, 1943; 1946; De Pauw & Remphrey, 1993; De Pauw et al., 1995; Faust & Harrington, 2016; From, 2007; Gleason, 1917; Homoya et al., 1984; Klier et al., 1991; Li et al., 2011; Maisch, 1872; Martin, 1978; McCormick & Jacquemyn, 2014; Niles, 1902; Oliva & Arditti, 1984; Pearn, 2012; Prena, 2017; Robertson et al., 1995; Ruch et al., 2009; Shefferson, 2006; Shefferson et al., 2005; Sheviak, 2002a; Sikes et al., 2010; Stubbs et al., 1992; Tiffany et al., 1990; Tremblay et al., 2004; Wake, 2007; Walsh & Michaels, 2017; Walsh et al., 2014; Whittaker, 1960; Worley, 2009.

5.4.4. Epipactis

Chatterbox, giant helleborine, stream orchid; Épipactis géant
Etymology: after the Greek *epipaktis*, the name of an ancient medicinal plant
Synonyms: *Amesia* (in part); *Arthrochilium*; *Cephalanthera* (in part); *Helleborine* (in part); *Limodorum* (in part); *Peramium* (in part); *Serapias* (in part)
Distribution: global: Africa (north); Eurasia; North America; Mexico; **North America:** widespread except far north
Diversity: global: 25 species; **North America:** 3 species
Indicators (USA): obligate wetland (OBL): *Epipactis gigantea*
Habitat: freshwater; palustrine, riverine; **pH:** alkaline; **depth:** <1 m; **life-form(s):** emergent herb
Key morphology: rhizome short, with fibrous roots; stems (to 1.4 m) leafy, glabrous; leaves (4–14) alternate, sheathing, plicate, the blades (to 20 cm) ovate, ovate-elliptic to narrowly lanceolate, progressively reduced as lanceolate to oblong, leaf-like bracts (to 12.7 cm); racemes terminal, lax, the flowers (2–32) showy, resupinate, pedicillate; perianth free, spreading; sepals ovate-lanceolate, concave, greenish to rose-colored, streaked with dark veins, the 2 laterals (to 24 mm) with oblique apex; petals shorter (to 17 mm), orange, pink or rose-colored with darker veins, broadly ovate to ovate-lanceolate, the lip (to 20 mm) sessile, fleshy, distinctly 3-lobed, constricted medially into 2 parts, marked with red or purple, strongly veined, with pair of wing-like calli at base; column (to 10 mm) short, stout, erect, curving over lip, broadened distally; anther terminal, sessile, green, with 4 mealy, soft,

yellow pollinia (in 2 pairs); capsule (to 25 mm) ellipsoid, pendent to spreading

Life history: duration: perennial (rhizomes); **asexual reproduction:** rhizomes; **pollination:** insect; **sexual condition:** hermaphroditic; **fruit:** capsules (common); **local dispersal:** rhizomes; seeds (water, wind); **long-distance dispersal:** seeds (water, wind)

Imperilment: 1. *Epipactis gigantea* [G4]; S1 (CO, ND, OK, WY); S2 (MT, NM); S3 (AZ, ID, TX, UT, WA)

Ecology: general: *Epipactis* species are herbaceous, rhizomatous perennials, which inhabit dry to wet sites in grasslands, meadows, prairies, swamps, and forests, often in damp microsites or along rivers or streams. Only one North American species is an OBL wetland indicator. The flowers are typical for orchids in being resupinate and bearing pollinia. They are self-compatible, but usually pollinated by flies (Insecta: Diptera: Syrphidae) or other insects. As in many orchids, the flowers attract pollinators by deception, which usually results in pollen-limited fruit production. The seeds will germinate immediately when fresh but apparently become dormant as they mature. They are buoyant and are dispersed abiotically by wind or water. Little is known regarding their potential persistence in seed banks, but the few available studies indicate that the seeds only are short-lived.

5.4.4.1. *Epipactis gigantea* **Douglas ex Hook.** grows in or on bars (gravel, sand, stream), beaches (gravel), bogs (roadside), borrow pits, canyons, channels (scour), chaparral, cienega, cliffs (wet), ditches (irrigation), fens (*Sphagnum*), floodplains, hanging gardens, hummocks (swamp), marshes, meadows (boggy), road cuts, shores, sinks (alkali), slopes [to 70%] (seepage), springs (calcareous, hot, mineral), woodlands (alluvial), and along the margins of lakes, rivers, and streams at elevations to 2743 m. The plants usually grow in open sunlight to partial shade but are not well-represented in the understory of more heavily shaded sites. They are quite versatile in their temperature tolerance and grow even in cool wet oases in the Mojave Desert, where annual rainfall is <11 cm. Although usually on or near moist ground, they also can occur in standing waters (e.g., 15 cm deep). Typical substrates are thin, calcareous, and organic, but some are slightly acidic (e.g., pH: 6.0). They are described as alluvium (silty), basalt, Camas, Chinle formation, clay (heavy, rocky), clayey silt, cobble, colluvium, gabbro (olivine), granite, gravel, limestone (cracks), loam (deep, organic), marl, mud, peat, Quartz (monzonite), rock (metamorphic), sand, sandstone (Kayenta, Navajo), sandy loam, sandy silt, serpentine, silt, silty loam, travertine, tufa, ultramafic, and Vaqueros Formation; some plants have been observed to grow on floating logs. Flowering and fruiting occur from March to September. The flowers have a mixed mating system. They are self-compatible, with hand-pollinated flowers yielding higher seed set (59%) than those of open-pollinated plants (49%), which is an indication of pollinator limitation. When no pollinators are available, selfing can occur in the aging flowers as their pollinia eventually come in contact with the stigmas. However, the flowers more often are outcrossed by small, syrphid flies (Insecta: Diptera: Syrphidae: *Copestylum satur*, *Dasysyrphus creper*,

Eupeodes americanus, Eupeodes luniger, Eupeodes volucris, Platycheirus immarginatus, Sphaerophoria philanthus) and occasionally by wasps (Insects: Hymenoptera: Vespidae: *Vespula pensylvanica*). Because syrphid flies usually lay their eggs in masses of aphids (Insecta: Hemiptera: Aphididae), and also on the lip of *E. gigantea* flowers, it is thought that these pollinators are deceived into visiting the flowers by their honeydew-scented floral nectar. The articulated floral lip is said to flutter in a mouth-like fashion with the slightest breeze (hence the common name of "chatterbox"), which is a feature of uncertain consequence with respect to the pollination biology. Isozyme surveys have indicated fairly high levels of genetic variability (e.g., $A = 2.24$; $H_e = 0.243$) along with high interpopulational differentiation (F_{ST}: 0.493; G_{ST}: 0.214). The probability of seed production by flowering plants can range from 6% to 63%. The seeds comprise 83.2–91.6% airspace, which enables them to float for 3 s when dropped in still air and for at least 3 days on water; they are dispersed by wind and water. Seed germination and protocorm formation can take from 2 to 12 months, with the aerial shoots developing in 3–18 months. Higher germination rates have been found for immature seeds than for mature seeds, which indicates the onset of dormancy in older seeds. The seeds have germinated well on Norstog medium in the light (60–70%) or after 4.3 months in the dark prior to their transfer to light (80%). When cultured, the protocorms usually develop in just under 3 months. Seedling development is facilitated by a 2–3 month period of chilling (at 5°C). In nature, seedling establishment is favored in disturbed sites. After their initial establishment, the plants often propagate either vegetatively or by self-pollination. The roots are colonized by rhizoctonia-forming mycorrhizal Fungi (Basidiomycota: Ceratobasidiaceae). Vegetative reproduction occurs by the proliferating rhizomes, which can produce up to three ramets (each potentially capable of independent growth) every season. **Reported associates:** *Abies concolor, Acer grandidentatum, Acer macrophyllum, Acer negundo, Achillea millefolium, Achnatherum hymenoides, Aconitum columbianum, Adenostoma fasciculatum, Adiantum aleuticum, Adiantum capillus-veneris, Adiantum jordanii, Agave utahensis, Ageratina adenophora, Agrimonia striata, Agropyron cristatum, Agrostis capillaris, Agrostis gigantea, Agrostis stolonifera, Ailanthus altissima, Allium validum, Alnus oblongifolia, Alnus rhombifolia, Alnus rugosa, Amelanchier utahensis, Amorpha californica, Amorpha fruticosa, Anaphalis margaritacea, Andropogon glomeratus, Angelica arguta, Anticlea elegans, Anticlea vaginata, Apium graveolens, Apocynum cannabinum, Aquilegia chrysantha, Aquilegia eximia, Aquilegia formosa, Aquilegia micrantha, Aralia racemosa, Arctostaphylos glauca, Arctostaphylos patula, Aristida purpurea, Arnica lanceolata, Artemisia bigelovii, Artemisia californica, Artemisia douglasiana, Artemisia dracunculus, Artemisia ludoviciana, Asclepias incarnata, Asclepias latifolia, Asclepias speciosa, Asparagus officinalis, Astragalus humillimus, Astragalus preussii, Atriplex canescens, Aulacomnium palustre, Baccharis pilularis, Baccharis plummerae, Baccharis salicifolia, Baccharis salicina, Baccharis sarothroides, Baccharis sergiloides, Barbarea orthoceras, Berberis fendleri, Berula erecta, Betula glandulosa, Betula occidentalis, Betula papyrifera, Betula pumila, Boechera davidsonii, Boerhavia coccinea, Bothriochloa barbinodis, Bouteloua eriopoda, Brachythecium frigidum, Brickellia brachyphylla, Brickellia californica, Brickellia longifolia, Bromus carinatus, Bryum pseudotriquetrum, Buchnera longifolia, Calamagrostis scopulorum, Calliergon giganteum, Callitropsis sargentii, Calocedrus decurrens, Calycanthus occidentalis, Calypogeia muelleriana, Campylium stellatum, Cardamine cordifolia, Carex alma, Carex amplifolia, Carex aurea, Carex buxbaumii, Carex cusickii, Carex densa, Carex flava, Carex fracta, Carex hassei, Carex hystericina, Carex interior, Carex lasiocarpa, Carex livida, Carex magellanica, Carex nudata, Carex obnupta, Carex pellita, Carex rostrata, Carex senta, Carex serratodens, Carex specuicola, Carex spissa, Carex utriculata, Carex vesicaria, Carex viridula, Castanopsis sclerophylla, Castilleja miniata, Castilleja minor, Ceanothus cuneatus, Ceanothus oliganthus, Celtis reticulata, Centaurea melitensis, Centaurea stoebe, Cephalanthus occidentalis, Cercis occidentalis, Cercocarpus ledifolius, Chamaecyparis lawsoniana, Cheilanthes clevelandii, Chilopsis linearis, Chloracantha spinosa, Chrysothamnus viscidiflorus, Chylismia walkeri, Cicuta douglasii, Cirsium arizonicum, Cirsium arvense, Cirsium douglasii, Cirsium fontinale, Cirsium occidentale, Cirsium rydbergii, Claytonia perfoliata, Clematis ligusticifolia, Clematis virginiana, Coleogyne ramosissima, Comandra umbellata, Comarostaphylis diversifolia, Conium maculatum, Conocephalum conicum, Conyza canadensis, Cornus nuttallii, Cornus sericea, Cotoneaster, Crataegus douglasii, Cratoneuron filicinum, Cynodon dactylon, Cyperus eragrostis, Cyperus erythrorhizos, Cypripedium calceolus, Cypripedium californicum, Cypripedium passerinum, Cystopteris fragilis, Dactylis glomerata, Darlingtonia californica, Dasiphora floribunda, Datisca glomerata, Deschampsia cespitosa, Desmodium metcalfei, Dichanthelium acuminatum, Dicranum polysetum, Distichlis spicata, Drosera anglica, Drosera rotundifolia, Drymocallis glandulosa, Echinacea angustifolia, Echinocereus engelmannii, Elaeagnus angustifolia, Elaeagnus commutata, Eleocharis compressa, Eleocharis macrostachya, Eleocharis montevidensis, Eleocharis palustris, Eleocharis rostellata, Eleocharis tenuis, Elymus glaucus, Elymus trachycaulus, Encelia resinifera, Ephedra torreyana, Ephedra viridis, Epilobium brachycarpum, Epilobium canum, Epilobium ciliatum, Epilobium glaberrimum, Equisetum arvense, Equisetum hyemale, Equisetum laevigatum, Equisetum telmateia, Ericameria nauseosa, Erigeron flagellaris, Erigeron kachinensis, Erigeron philadelphicus, Erigeron rhizomatus, Eriodictyon trichocalyx, Eriogonum corymbosum, Eriogonum grande, Eriogonum nudum, Eriophorum chamissonis, Eriophorum gracile, Eriophorum viridicarinatum, Eriophyllum confertiflorum, Eutrochium maculatum, Fallugia paradoxa, Fendlera rupicola, Festuca rubra, Foeniculum vulgare, Forestiera pubescens, Fraxinus anomala, Fraxinus pennsylvanica, Fraxinus velutina, Galium* bolanderi, *Gentiana calycosa, Gentianopsis detonsa, Gentianopsis holopetala, Glyceria striata, Glycyrrhiza lepidota, Gutierrezia sarothrae, Hastingsia alba, Helenium bigelovii, Helenium puberulum, Helianthella quinquenervis, Helianthus annuus, Helianthus nuttallii, Helodium blandowii, Herrickia glauca, Herrickia glauca, Hesperostipa neomexicana, Hesperoyucca whipplei, Heteromeles arbutifolia, Heterotheca villosa, Heuchera, Hoita macrostachya, Hoita orbicularis, Holcus lanatus, Holodiscus discolor, Hordeum jubatum, Humulus lupulus, Hypericum formosum, Hypericum scouleri, Hypnum pratense, Iliamna rivularis, Imperata brevifolia, Isocoma acradenia, Isocoma menziesii, Isocoma veneta, Ivesia kingii, Juglans major, Juncus balticus, Juncus bufonius, Juncus cooperi, Juncus covillei, Juncus dubius, Juncus effusus, Juncus ensifolius, Juncus nodosus, Juncus saximontanus, Juncus torreyi, Juncus xiphioides, Juniperus horizontalis, Juniperus osteosperma, Juniperus scopulorum, Keckiella antirrhinoides, Keckiella cordifolia, Kochia californica, Lactuca serriola, Larrea tridentata, Lathyrus splendens, Leptodactylon californicum, Leymus cinereus, Leymus condensatus, Leymus triticoides, Lilium humboldtii, Lilium pardalinum, Limprichtia revolvens, Liparis loeselii, Lobelia cardinalis, Lomatium californicum, Lonicera involucrata, Lotus corniculatus, Lupinus albicaulis, Lycopus americanus, Lysichiton americanus, Lysimachia ciliata, Lythrum californicum, Maianthemum stellatum, Malosma laurina, Marah fabacea, Marchantia polymorpha, Medicago lupulina, Meesia triquetra, Melilotus albus, Melilotus officinalis, Mentha arvensis, Menyanthes trifoliata, Mimulus aurantiacus, Mimulus cardinalis, Mimulus eastwoodiae, Mimulus glaucescens, Mimulus guttatus, Mimulus longiflorus, Mimulus moschatus, Monarda fistulosa, Muhlenbergia andina, Muhlenbergia asperifolia, Muhlenbergia mexicana, Muhlenbergia phleoides, Muhlenbergia rigens, Munroa squarrosa, Nasturtium officinale, Nicotiana obtusifolia, Nolina microcarpa, Oenothera elata, Oenothera longissima, Oenothera suffrutescens, Opuntia littoralis, Opuntia phaeacantha, Opuntia polyacantha, Ostrya knowltonii, Oxytenia acerosa, Packera multilobata, Palmerella debilis, Panicum capillare, Panicum virgatum, Parnassia palustris, Parryella filifolia, Parthenocissus quinquefolia, Parthenocissus vitacea, Pellaea mucronata, Penstemon ambiguus, Penstemon barbatus, Penstemon centranthifolius, Penstemon grinnellii, Pentagramma triangularis, Perityle congesta, Petasites frigidus, Petradoria pumila, Phacelia rotundifolia, Phalaris arundinacea, Phaseolus, Philadelphus lewisii, Philonotis fontana, Phleum pratense, Phragmites australis, Physocarpus capitatus, Picea engelmannii, Pinguicula vulgaris, Pinus edulis, Pinus lambertiana, Pinus monophylla, Pinus ponderosa, Pinus sabiniana, Piptatherum miliaceum, Plagiobothrys salsus, Plantago lanceolata, Plantago major, Plantago major, Plantago subnuda, Platanthera dilatata, Platanthera sparsiflora, Platanthera tescamnis, Platanthera zothecina, Platanus racemosa, Platanus wrightii, Pluchea odorata, Pluchea sericea, Poa pratensis, Polypogon viridis, Populus angustifolia, Populus deltoides, Populus fremontii, Populus tremuloides, Populus trichocarpa, Potamogeton, Primula pauciflora,* Prosopis glandulosa, *Prunella vulgaris, Prunus virginiana, Pseudognaphalium beneolens, Pseudognaphalium luteoalbum, Pseudotsuga menziesii, Psoralidium lanceolatum, Pteridium aquilinum, Pteris vittata, Purshia mexicana, Pycnanthemum californicum, Quercus agrifolia, Quercus berberidifolia, Quercus gambelii, Quercus grisea, Quercus lobata, Quercus turbinella, Quercus ×undulata, Ranunculus hydrocharoides, Ratibida columnifera, Rhamnus alnifolia, Rhamnus betulifolia, Rhamnus ilicifolia, Rhamnus pirifolia, Rhododendron, Rhododendron occidentale, Rhus aromatica, Rhus glabra, Rhus integrifolia, Rhynchospora colorata, Rhynchospora nivea, Ribes aureum, Ribes malvaceum, Ribes speciosum, Robinia neomexicana, Rosa californica, Rosa multiflora, Rosa woodsii, Rubus bifrons, Rubus leucodermis, Rubus parviflorus, Rubus ulmifolius, Rubus ursinus, Rudbeckia californica, Rumex crispus, Salix amygdaloides, Salix bebbiana, Salix breweri, Salix exigua, Salix gooddingii, Salix laevigata, Salix lasiolepis, Salix ligulifolia, Salix lucida, Salix lutea, Salix monticola, Saltugilia splendens, Salvia mellifera, Salvia spathacea, Sambucus nigra, Samolus ebracteatus, Schizachyrium scoparium, Schoenoplectus acutus, Schoenoplectus americanus, Schoenoplectus pungens, Schoenoplectus tabernaemontani, Schoenus nigricans, Scirpus microcarpus, Sedum albomarginatum, Sedum leibergii, Selaginella apoda, Selaginella bigelovii, Senecio triangularis, Senegalia greggii, Sequoia sempervirens, Shepherdia argentea, Shepherdia canadensis, Shepherdia rotundifolia, Sisyrinchium demissum, Sisyrinchium halophilum, Sisyrinchium montanum, Solanum douglasii, Solanum dulcamara, Solidago canadensis, Solidago confinis, Solidago elongata, Solidago gigantea, Solidago lepida, Solidago velutina, Sonchus arvensis, Sonchus asper, Sonchus oleraceus, Spartina pectinata, Sphagnum fuscum, Sphagnum warnstorfii, Sphenopholis obtusata, Spiraea, Spiranthes romanzoffiana, Sporobolus flexuosus, Stachys albens, Stanleya pinnata, Stephanomeria pauciflora, Strigosella africana, Suaeda nigra, Symphoricarpos, Symphyotrichum eatonii, Symphyotrichum greatae, Tamarix chinensis, Tamarix parviflora, Tamarix ramosissima, Thalictrum dasycarpum, Thalictrum fendleri, Thelypodium wrightii, Thelypteris kunthii, Thelypteris puberula, Thuidium recognitum, Thuja plicata, Thymophylla pentachaeta, Tiarella, Tomenthypnum nitens, Toxicodendron diversilobum, Toxicodendron radicans, Toxicodendron rydbergii, Trichophorum cespitosum, Triglochin, Typha domingensis, Typha latifolia, Ulmus americana, Umbellularia californica, Urtica dioica, Verbena lasiostachys, Veronica, Viola sororia, Vitis arizonica, Vitis californica, Vitis riparia, Washingtonia filifera, Woodwardia fimbriata, Yucca baccata.*

Use by wildlife: *Epipactis gigantea* is fed on by weevils (Insecta: Coleoptera: Curculionidae) and is grazed by ungulate herbivores (Mammalia: Artiodactyla). The foliage hosts leaf-damaging Fungi (Ascomycota: Glomerellaceae: *Glomerella*).

Economic importance: food: *Epipactis gigantea* is not edible; **medicinal:** Excavated in Utah, a 500-year-old Native American satchel (known as "The Patterson Bundle") included the roots of *Epipactis gigantea*, which suggests their early use as a medicinal herb. The plants reportedly were used medicinally by Native tribes in central California; **cultivation:** *Epipactis gigantea* reputedly is easy to cultivate and has been used to generate several cultivars ('Serpentine Night', 'Enchantment', and 'Trefor'), three greges (Heart of Virginia gx, Lowland Legacy gx, and Sabine gx), and the hybrid 'Serpentine Night' × *thunbergii*; **misc. products:** The Karok people of California had "no use" for *E. gigantea* (which they called "coyote shoes") other than for its attractive flowers; **weeds:** *Epipactis gigantea* is not weedy; **nonindigenous species:** *Epipactis gigantea* is the only indigenous North American species of this genus.

Systematics: *Epipactis* is assigned to tribe *Neottieae*, which resolves as the sister group to the remainder of orchid subfamily Epidendroideae (Figure 2.12). Formal molecular phylogenetic analyses of 20 *Epipactis* species (but excluding *E. gigantea*) have indicated that the genus is monophyletic. However, a BLAST search of DNA sequences in GenBank showed several unpublished sequences of *E. gigantea* to be most similar to *E. palustris* and *E. helleborine* (both included in the phylogenetic analyses), which at least confirms that it has been properly assigned to the genus. The basic chromosome number of *Epipactis* has been determined as $x=20$. Accordingly, *E. gigantea* ($2n=40, 60$) would have diploid and triploid cytotypes. No natural hybrids involving *E. gigantea* have been reported, which is understandable given that the only other North American species are nonindigenous and are concentrated in eastern North America.

Distribution: *Epipactis gigantea* occurs in western North America.

References: Arditti & Ghani, 2000; Arditti et al., 1980; 1981; 1982; Argue, 2012d; Barthlott et al., 2014; Bidartondo et al., 2004; Brown & Argus, 2002; Brunton, 1986; Chadde et al., 1998; Felix & Guerra, 2010; Forrest et al., 2004; Harrison, 2002; Hornbeck et al., 2003; Jin et al., 2014; Johansen & Rasmussen, 1992; Levine, 2000; Liggio & Liggio, 2010; Löve, 1971; Malanson, 1982; Mantas, 1993; Raven et al., 1965; Reverchon, 1886; Reznicek & Murray, 2013; Rocchio et al., 2006; Schenck & Gifford, 1952; Sheviak & Jennings, 2006; Thornhill, 1996; van der Kinderen, 1995; Wilson, 2014; Winner & Simpson, 2007.

5.4.5. Galearis

Round-leaved orchid; Orchis à feuille ronde

Etymology: from the Latin *galea* ("helmet") in reference to the hood-like envelopment of the column by the lateral petals and median sepal

Synonyms: *Amerorchis*; *Habenaria* (in part); *Orchis* (in part); *Platanthera* (in part); *Ponerorchis* (in part)

Distribution: global: Eastern Asia; North America; **North America:** throughout except the southwestern United States

Diversity: global: 9 species; **North America:** 2 species

Indicators (USA): obligate wetland (OBL); facultative wetland (FACW): *Galearis rotundifolia*

Habitat: freshwater; palustrine; **pH:** 6.9–8.4; **depth:** <1 m; **life-form(s):** emergent herb

Key morphology: rhizomes with slender, fleshy, scattered roots; stems (to 33 cm) scapose; leaf solitary, basal, conduplicate, clasping or sheathing, the blade (to 11 cm) orbiculate to broadly elliptic, ovate, obovate, or lance- or oblance-elliptic; spike terminal, solitary, lax, few to several-flowered, bracts (to 15 mm) foliaceous, lanceolate to linear, the flowers resupinate, showy; perianth white to magenta, the sepals (to 10 mm), ovate to elliptic-oblong; lateral petals (to 6 mm) ovate to lance-oblong, spreading, the lip (to 9 mm) white, magenta-spotted, ovate, deeply 3-lobed, prominently spurred at base; pollinaria 2, pollinia 2, viscidia in a bilobed bursicle; stigma flat, reniform to obcordate; capsules (to 1.5 cm) erect, ellipsoid

Life history: duration: perennial (rhizomes); **asexual reproduction:** rhizomes; **pollination:** insect; hermaphroditic; **fruit:** capsules (common); **local dispersal:** rhizomes; seeds (wind); **long-distance dispersal:** seeds (wind)

Imperilment: 1. *Galearis rotundifolia* [G5]; SX (NY); SH (NH, VT); S1 (MI, WI, WY); S2 (ME, <u>NB</u>, <u>NF</u>, <u>NU</u>); S3 (MT, <u>QC</u>)

Ecology: general: It is fruitless to summarize ecological information for *Galearis*, given widespread taxonomic disagreement in the circumscription of the genus (see *Systematics* below). The only North American wetland indicator in the genus is *Galearis rotundifolia* (= *Platanthera rotundifolia* in the 2016 wetland indicator list), whose ecological information is provided below.

5.4.5.1. ***Galearis rotundifolia* (Banks ex Pursh) R.M. Bateman** inhabits alluvial fans (gravelly), bogs (calcareous, spruce, tamarack), depressions (dune), fens (calcareous, riverside), flats (alluvial), floodplains, forests (coniferous, seepage), hollows (moist), hummocks (spruce), ledges (shaded), meadows (ravine), muskeg (spruce), seepages, swamps (cedar), taiga, thickets, tundra, and the margins of lakes, rivers, springs, and streams at elevations to 2097 m. The plants grow in exposures ranging from open sites to shaded or deeply shaded sites as long as cold temperatures are maintained in the substrate. They are adapted to the longer photoperiods of arctic climates by possessing alternative pathway respiration (where heat is generated instead of ATP), which increases during midday in a circadian-like fashion. The substrates range from acidic to calcareous conditions (e.g., pH: 6.9–8.4) and have been described as gravel, humus (rich), limestone, loam, marl, peat (*Sphagnum*), sand, silt, and silty loam. The plants often are found growing in deep moss cover, which is thought to provide favorable microsites for seed germination and seedling development. Flowering and fruiting occur from June to August. Although the flowers apparently are self-compatible, their extent of self-pollination has not been evaluated. They are pollinated (and outcrossed) primarily by bees (Insecta: Hymenoptera: Megachilidae: *Osmia proxima*) but also by syrphid flies (Insecta: Diptera: Syrphidae: *Eriozona laxus*, *Eristalis hirta*, *Eristalis rupium*, *Eupeodes lapponicus*). Most pollinator activity has been observed to occur between 1:00 and 4:00 pm, until the sunlight becomes indirect and temperatures fall. Successful floral pollination rates are estimated to range from 25% to 44%; however, the flowers yield significantly higher seed set when hand-pollinated compared to naturally open-pollinated plants, which indicates that they are pollinator limited. Pollination occurs by deception, given that the floral spur is nectarless and the flowers allegedly scentless. As the potential pollinator lands on the flower to probe the spur for nectar, its movement in and out of the flower causes the release of sticky viscidia, which attach the pollen-bearing pollinarium to the insect's head. On subsequent flower visits, the hood-like arrangement of the perianth orients the insect's head toward the stigma, which facilitates deposition of the pollinia. Geitonogamous pollinations are reduced (and outcrossing promoted) by the sequential development of flowers within an inflorescence, and by a delay of about 1 min before the pollinarium stalks (caudicles) bend into the proper orientation for pollen delivery. The minute seeds are dispersed by wind. Observed field germination rates are low (7.4–10.2%) but *in vitro*, the seeds have germinated (asymbiotically) on Fast's medium, reaching a green leaf stage in 21–27 weeks. The plants reproduce frequently by vegetative growth of the rhizomes. Root culture isolates have yielded several endophytic and mycorrhizal Fungi (Ascomycota: Vibrisseaceae: *Phialocephala fortinii*; Incertae sedis: *Leptodontidium orchidicola*; Basidiomycota: Ceratobasidiaceae: *Rhizoctonia obscura*; Tulasnellaceae: *Epulorhiza calendulina*; Zygomycota: Umbelopsidaceae: *Mortierella isabellina*). Mycorrhizal fungi are necessary for successful seed germination and seedling establishment.

Reported associates: *Acer rubrum, Actaea, Agropyron, Alnus viridis, Amelanchier alnifolia, Anemone parviflora, Angelica arguta, Antennaria pulcherrima, Antennaria racemosa, Anthoxanthum nitens, Anticlea elegans, Aquilegia flavescens, Aquilegia formosa, Aralia nudicaulis, Arctostaphylos uva-ursi, Arctous rubra, Arnica latifolia, Astragalus, Aulacomnium palustre, Aulacomnium turgidum, Berberis repens, Betula glandulosa, Betula nana, Betula occidentalis, Betula papyrifera, Betula pumila, Bromus inermis, Bromus pumpellianus, Calamagrostis canadensis, Calamagrostis purpurascens, Caltha palustris, Calypso bulbosa, Carex aurea, Carex buxbaumii, Carex capillaris, Carex concinna, Carex diandra, Carex dioica, Carex disperma, Carex eburnea, Carex gynocrates, Carex interior, Carex lasiocarpa, Carex leptalea, Carex podocarpa, Carex praegracilis, Carex rostrata, Carex scirpoidea, Carex trisperma, Carex utriculata, Carex vaginata, Carex vesicaria, Castilleja, Chamerion angustifolium, Chrysosplenium iowense, Cinclidium stygium, Circaea alpina, Cladonia, Climacium dendroides, Clintonia uniflora, Comarum palustre, Coptis trifolia, Corallorhiza trifida, Corallorhiza wisteriana, Cornus canadensis, Cornus sericea, Cypripedium calceolus, Cypripedium parviflorum, Cypripedium passerinum, Cypripedium reginae, Dasiphora floribunda, Dicranum polysetum, Drosera linearis, Drosera rotundifolia, Dryas, Dryopteris carthusiana, Elaeagnus commutata, Eleocharis, Empetrum nigrum, Equisetum arvense, Equisetum laevigatum, Equisetum palustre, Equisetum pratense, Equisetum scirpoides, Eriophorum, Eurhynchium pulchellum, Eurybia conspicua, Fragaria vesca, Fragaria virginiana, Fragaria virginiana, Galium boreale, Galium labradoricum, Galium triflorum, Gaultheria hispidula, Geocaulon lividum, Geranium richardsonii, Geum macrophyllum, Glyceria borealis, Glyceria striata, Goodyera repens, Hamatocaulis vernicosus, Harrimanella stelleriana, Helodium blandowii, Heracleum sphondylium, Hylocomium splendens, Hypnum lindbergii, Juncus balticus, Juniperus horizontalis, Lappula squarrosa, Larix laricina, Lathyrus ochroleucus, Leptarrhena pyrolifolia, Leymus innovatus, Linnaea borealis, Lonicera dioica, Lonicera involucrata, Lonicera oblongifolia, Lonicera villosa, Maianthemum canadense, Maianthemum racemosum, Maianthemum stellatum, Maianthemum trifolium, Malaxis paludosa, Marchantia, Menyanthes trifoliata, Menziesia ferruginea, Mertensia paniculata, Mitella nuda, Moneses uniflora, Neottia borealis, Neottia cordata, Oncophorus wahlenbergii, Orthilia secunda, Oryzopsis, Oxalis stricta, Parnassia fimbriata, Parnassia kotzebuei, Parnassia palustris, Pedicularis labradorica, Pedicularis racemosa, Peltigera apthosa, Petasites frigidus, Picea engelmannii, Picea glauca, Picea mariana, Pinguicula vulgaris, Pinus contorta, Plagiomnium cuspidatum, Plagiomnium ellipticum, Platanthera huronensis, Platanthera hyperborea, Platanthera obtusata, Platanthera stricta, Pleurozium schreberi, Polytrichum juniperinum, Populus balsamifera, Populus tremuloides, Primula, Prosartes trachycarpa, Pseudotsuga menziesii, Ptilium crista-castrensis, Pylaisiella polyantha, Pyrola asarifolia, Pyrola chlorantha, Pyrola grandiflora, Rhamnus alnifolia, Rhododendron columbianum, Rhododendron groenlandicum, Rhododendron tomentosum, Rhytidiadelphus triquetrus, Ribes hudsonianum, Rosa acicularis, Rubus arcticus, Rubus chamaemorus, Rubus parviflorus, Rubus pubescens, Rubus sachalinensis, Salix bebbiana, Salix candida, Salix drummondiana, Salix glauca, Salix myrtillifolia, Salix pedicellaris, Salix scouleriana, Sarracenia purpurea, Saussurea angustifolia, Saxifraga, Schizachne purpurascens, Senecio lugens, Shepherdia canadensis, Shepherdia canadensis, Sphagnum, Sphagnum capillifolium, Sphagnum girgensohnii, Sphagnum warnstorfii, Spiraea betulifolia, Stellaria longifolia, Streptopus amplexifolius, Swertia perennis, Symphoricarpos albus, Thalictrum, Thuidium recognitum, Thuja occidentalis, Tofieldia pusilla, Tomenthypnum nitens, Trichophorum alpinum, Triglochin maritima, Typha latifolia, Vaccinium oxycoccos, Vaccinium vitis-idaea, Valeriana dioica, Valeriana sitchensis, Viburnum edule, Viola adunca, Viola macloskeyi, Viola orbiculata.*

Use by wildlife: The foliage of *Galearis rotundifolia* is browsed by deer (Mammalia: Cervidae) and is eaten by weevils (Insecta: Coleoptera: Curculionidae: *Stethobaris ovata*). Though deceptive, the flowers are visited by various insects (Insecta), including bees (Hymenoptera: Megachilidae: *Osmia proxima*), butterflies (Lepidoptera: Hesperiidae: *Erynnis persius*; Lycaenidae; *Glaucopsyche lygdamus*; Papilionidae: *Papilio glaucus*), and flies (Diptera: Bombylidae; Syrphidae: *Eriozona laxus, Eristalis hirta, Eristalis rupium, Eupeodes lapponicus*).

Economic importance: food: *Galearis rotundifolia* is not edible; **medicinal:** no uses reported; **cultivation:** *Galearis rotundifolia* is not in cultivation; **misc. products:** none; **weeds:** none; **nonindigenous species:** none.

Systematics: *Galearis rotundifolia* is assigned to tribe Orchideae of the orchid subfamily Orchidoideae (Figure 2.13). This species has experienced a taxonomic "identity crisis," with various authors placing it within five other genera (*Amerorchis, Habenaria, Orchis, Platanthera, Ponerorchis*). The most recent (2016) wetland indicator list treats it as *Platanthera orbiculata*. The circumscription of *Galearis* followed here accepts the recommendation of a phylogenetic analysis of nrITS sequence data, which resolved *G. rotundifolia* within a distinct clade of species (including the monotypic *Neolindleya camtschatica*) that was sister to a larger clade comprising *Platanthera* species (Figure 2.15). However, it is evident that the inclusion of all the *Galearis* species and *N. camtschatica* in *Platanthera* (e.g., as sections) also would be compatible with the phylogenetic results. Similarly, if *Neolindleya camtschatica* is retained as distinct, then it would be equally defensible to recognize *Galearis rotundifolia* as the monotypic *Amerorchis rotundifolia*, as some authors have done. It also should be emphasized that any final circumscription of the genus should await more comprehensive sampling because only four of the nine putative *Galearis* species have been subjected to phylogenetic scrutiny. In any event, it is apparent that the currently "accepted" taxonomic status of *G. rotundifolia* basically represents a matter of opinion rather than any mandatory scheme necessary to maintain phylogenetic integrity. The basic chromosome number of subfamily Orchidoideae is estimated to be $x = 7$, with most extant species though to represent paleopolyploids. *Galearis rotundifolia* ($2n = 42$) is hexaploid. No natural hybrids involving *Galearis rotundifolia* have been reported.

Distribution: *Galearis rotundifolia* occurs across northern North America.

References: Arditti & Ghani, 2000; Bateman et al., 2003; 2009; Catling & Kostiuk, 2011a; Currah et al., 1987; Felix & Guerra, 2005; Handley & Heidel, 2005; Hilaire, 2002; Lewis & Dowding, 1926; Locky et al., 2005; McNulty et al., 1988; Moss, 1932; Nepal & Way, 2007; Prena, 2017; Proctor & Harder, 1994; Riggs, 1940; Sheviak & Catling, 2002b; Waud et al., 2014; Zelmer & Currah, 1995.

5.4.6. Habenaria

Bog orchid, creeping orchid, floating orchid, rein orchid, water orchid, water-spider orchid

Etymology: from the Latin *habena* ("rein") for the rein-like appearance of some petals

Synonyms: *Mesicera*; *Orchis* (in part); *Platanthera* (in part)

Distribution: global: pantropical; **North America:** southeastern United States

Diversity: global: 881 species; **North America:** 4 species

Indicators (USA): obligate wetland (OBL); facultative wetland (FACW): *Habenaria repens*

Habitat: freshwater, freshwater (tidal); lacustrine, palustrine, riverine; **pH:** 5.5–8.3; **depth:** <1 m; **life-form(s):** emergent herb, free-floating (in vegetation mats)

Key morphology: roots copious, long, slender, fibrous, with spherical tuberoids, producing "root shoots" in deeper water; shoots (to 90 cm) erect to decumbent, producing few to many elongate stolons, which bear terminal plantlets; leaves cauline, alternate, conduplicate, sheathing, reduced gradually to lanceolate bracts (to 9 cm), the blades (to 25 cm), ascending, linear-lanceolate to lanceolate, narrowly elliptic, or oblanceolate, 3-ribbed; racemes (to 28 cm) terminal, densely flowered, flowers small (to 12 mm), ascending, resupinate, pedicellate; sepals greenish, dorsal sepal (to 7 mm) concave, lateral sepals (to 7 mm) reflexed-spreading; petals greenish, ascending, falcate, each lateral petal divided into two, arcuate, falcate,

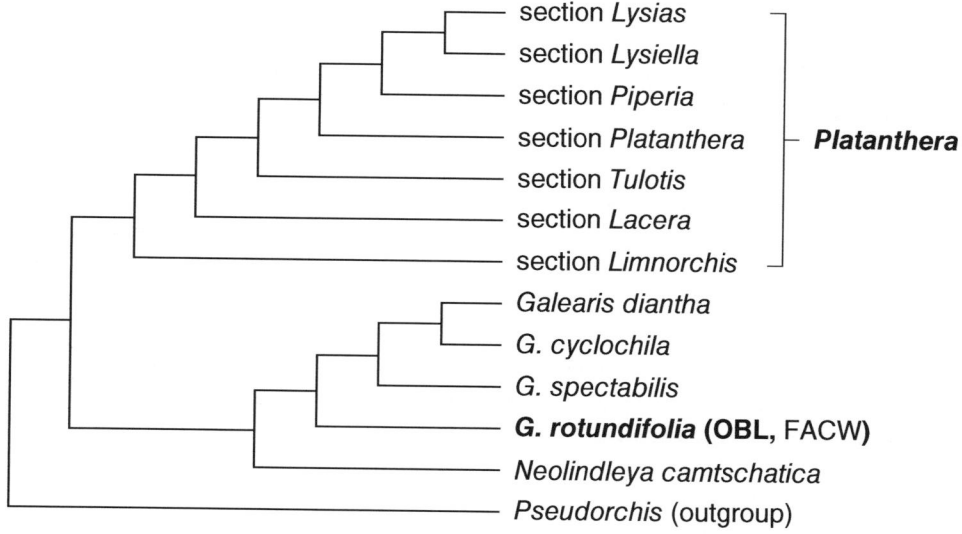

FIGURE 2.15 Phylogenetic placement of four *Galearis* species as indicated by the analysis of nrITS sequence data (simplified from Bateman et al., 2009). This result depicts *Galearis* and the monotypic *Neolindleya camtschatica* as comprising a sister clade to *Platanthera*; however, alternative taxonomic schemes that merged *Galearis* and *Neolindleya* with *Platanthera*, or that recognized *Galearis*, *Neolindleya*, and the monotypic *Amerorchis* (i.e., =*Galearis rotundifolia*) also would be compatible with these phylogenetic results. The taxa containing OBL North American indicators are highlighted in bold.

ascending filiform lobes (to 7 mm), the lip 3-lobed, greenish, spurred (to 14 mm) at base, the lateral lobes (to 11 mm) spreading, the center lobe (to 7 mm) linear, strongly deflexed; pollinaria 2, pollinia 2 with free viscidia; stigma with 2 fleshy lobes, encircling or flanking the spur opening, ovary (to 15 mm) inferior, pedicellate; capsules (to 15 mm) erect, ellipsoid, short-stalked

Life history: duration: perennial (stolons, whole plants); **asexual reproduction:** plantlets, root shoots, stolons, tuberoids; **pollination:** insect; **sexual condition:** hermaphroditic; **fruit:** capsules (common); **local dispersal:** stolons; seeds (wind); **long-distance dispersal:** seeds (wind)

Imperilment: 1. [G5]; S1 (OK); S2 (AR, NC); S3 (GA)

Ecology: general: Although technically a group of "terrestrial" (i.e., non-epiphytic) orchids, *Habenaria* species often occur in wetlands, especially in wet forests. All the North American species are wetland indicators (FACW) with one species also designated as OBL in most of its range. These are perennial herbs, which can reproduce vegetatively from small, tuber-like outgrowths from the roots, which produce shoots; some species develop stolons. The flowers range from being small and inconspicuous to larger and showy. Many species are known to be self-compatible (some being autogamous), but all are adapted for pollination by insects, either by offering nectar rewards or by deception. Many of the species are known to be pollinated by butterflies, hawkmoths, and moths (e.g., Insecta: Lepidoptera: Nymphalidae; Sphingidae; Noctuidae), with some possibly pollinated by flies (Insecta: Diptera). Some species are obligately apomictic (agamospermous). The seeds are minute and dispersed abiotically by water or by wind. They require specific mycorrhizal Fungi to germinate and for subsequent successful seedling growth.

5.4.6.1. *Habenaria repens* Nutt. grows on floating mats (mucky, peaty, vegetated) or as a shallow water emergent in "batteries" (bulge type, free-floating type), bayheads, bogs (quaking, seepage, *Sphagnum*), borrow pits, canals, depressions (dune, wet), ditches (calcareous, drainage), domes (cypress), flatwoods (calcareous), floodplains, glades, gravel pits, hammocks, impoundments, marshes (beaver, depression, floatant, freshwater, quaking, roadside, tidal), meadows (wet), pinelands, prairies (wet), rice fields, seeps, sloughs, swamps (cypress, strand), and along the margins or shores of canals, lakes, ponds (artificial, beaver, cypress, limestone sink, mill, retention, roadside), sinks, and streams at elevations to 67 m (but reaching 1700 m in South America). The plants can tolerate exposures ranging from full sun to partial shade. They are unusual in growing either on saturated substrates (in shallow standing waters from 10 to 15 cm deep) or as colonizers of the pioneer zone and open growth areas of floating mats, which comprise a diversity of vegetation and can extend over waters up to 7 m deep. In many cases the plants occur rooted along the margin of hydric sites, and then extend over the surface waters by means of stolon production. The substrates have been described as Basinger (spodic psammaquents), clayey sand, humus, loamy clay, muck, mucky sand, mud, peat, peaty mulch, Placid (typic humaquepts), sand, sandy clay, and sandy loam (Tifton).

However, the propensity of these plants to grow on floating mats has involved more unusual substrates, including floating logs, and in one case they were described as "floating on mats of *Limnobium* and alligators" (Reptilia: Alligatoridae: *Alligator mississippiensis*). Because of their ability to colonize floating vegetation, the plants also can grow in sites spanning a wide range of water clarity (e.g., Secchi depth: 0.6–3.6 m). Water chemistry data indicate a fairly broad range of circumneutral habitats (pH: 5.5–8.3; $\bar{x} = 6.7$; alkalinity [as $CaCO_3$]: 1.5–50 mg l^{-1}; $\bar{x} = 16.5$ mg l^{-1}; specific conductance [@ 25°C] 28–184 µS cm^{-1}; $\bar{x} = 101$ µS cm^{-1}) and average nutrient levels (e.g., total N: 300–1290 µg l^{-1}; $\bar{x} = 810$ µg l^{-1}; total P: 12–87 µg l^{-1}; $\bar{x} = 36$ µg l^{-1}; Ca: 2.4–45.6 mg l^{-1}; $\bar{x} = 17.2$ mg l^{-1}; K: 0.2–3.9 mg l^{-1}; $\bar{x} = 1.7$ mg l^{-1}; Mg: 2.1–12.1 mg l^{-1}; $\bar{x} = 7.2$ mg l^{-1}; Na: 2.1–16.2 mg l^{-1}; $\bar{x} = 8.6$ mg l^{-1}; Cl$^-$: 3.3–28.1 mg l^{-1}; $\bar{x} = 15.2$ mg l^{-1}; Fe: 0–0.6 mg l^{-1}; $\bar{x} = 0.2$ mg l^{-1}; SO_4^{-2}: 3.4–22 mg l^{-1}; $\bar{x} = 10.9$ mg l^{-1}). Flowering and fruiting can extend from April to December. The vanilla-like scent of the sweetly fragrant flowers attracts nocturnal moths (Insecta: Lepidoptera) as pollinators. It is thought that the perianth is structured to ensure that pollinators enter the flower from the front rather than the side. As a moth moves forward to probe the floral spur for nectar, the pollinaria are attached to its eyes by their sticky viscidia and hang in a pendant orientation. When the moth approaches the spur of different flowers, the attached pollinaria are drawn across the stigmatic lobes, which extract some of the apical pollen massulae (sometimes the entire pollinarium) by their adhesive properties. The extent of self-pollination (via autogamy or geitonogamy) has not yet been evaluated, but outcrossing appears to represent the primary breeding system. Preliminary genetic analyses of several populations (using microsatellite markers) has indicated a range of genetic variability (e.g., 4–7 alleles locus^{-1}; $\bar{x}$ number of alleles locus^{-1}: 2.8–4.0; 2.8–4.0; H_o: 0.000–0.563; H_e: 0.000–0.666). In any case, seed set typically is high. The relatively large seeds have a water content of 43%. They are not very porous, lose water at a slow rate, and are buoyant. They are thought to be dispersed by water, wind, and in mud clinging to the feet of waterfowl (Aves: Anatidae). The seeds have germinated successfully (within 21 days) *ex vitro*, after their inoculation with fungal isolates (Fungi: Basidiomycota: Tulasnellaceae: *Epulorhiza*); some seedlings have reached flowering maturity within 18 months after sowing. Seeds from southern localities have germinated successfully when sown at higher latitudes at sites where *Epulorhiza* fungi were present. Vegetative reproduction occurs by means of stolons, which terminate in small plantlets. Vegetative shoots also are known to arise directly from the roots. It is difficult to distinguish from published descriptions whether all the stolons originate from roots or if some also are derived from the shoot. Whole plants also can overwinter where climatic conditions are amenable.

Reported associates: *Acer rubrum, Aeschynomene virginica, Agalinis, Alnus serrulata, Alternanthera philoxeroides, Andropogon glomeratus, Andropogon virginicus, Annona glabra, Apios americana, Aronia arbutifolia, Arundinaria gigantea, Azolla, Bacopa caroliniana, Bacopa monnieri,*

Bidens aristosus, Bidens discoideus, Bidens laevis, Bidens mitis, Bidens pilosus, Bidens trichospermus, Blechnum serrulatum, Boehmeria cylindrica, Brasenia schreberi, Callicarpa americana, Carex decomposita, Carex glaucescens, Carya floridana, Centella asiatica, Cephalanthus occidentalis, Cicuta maculata, Cladium jamaicense, Colocasia esculenta, Conoclinium coelestinum, Cuscuta, Cyperus erythrorhizos, Cyperus flavescens, Cyperus haspan, Cyperus odoratus, Cyperus polystachyos, Cyperus sescuiflorus, Cyperus strigosus, Cyrilla racemiflora, Decodon verticillatus, Diodia virginiana, Distichlis spicata, Drosera intermedia, Dulichium arundinaceum, Echinochloa crusgalli, Eichhornia crassipes, Eleocharis acicularis, Eleocharis baldwinii, Eleocharis elongata, Eleocharis flavescens, Eleocharis microcarpa, Eleocharis obtusa, Eleocharis parvula, Elephantopus, Erechtites hieracifolius, Eriocaulon compressum, Eriocaulon decangulare, Eubotrys racemosa, Eupatorium capillifolium, Eupatorium leptophyllum, Eupatorium perfoliatum, Euphorbia, Eutrochium purpureum, Fimbristylis autumnalis, Fraxinus, Fuirena breviseta, Fuirena pumila, Gordonia lasianthus, Hibiscus grandiflorus, Hibiscus moscheutos, Hydrilla verticillata, Hydrocotyle ranunculoides, Hydrocotyle umbellata, Hydrolea quadrivalvis, Hygrophila lacustris, Hypericum mutilum, Ilex glabra, Ipomoea sagittata, Iris, Iva frutescens, Juncus effusus, Juncus trigonocarpus, Kosteletzkya pentacarpos, Kyllinga pumila, Lachnanthes caroliniana, Lachnocaulon, Leersia hexandra, Leersia oryzoides, Lemna minor, Limnobium spongia, Liquidambar styraciflua, Ludwigia alata, Ludwigia arcuata, Ludwigia grandiflora, Ludwigia lanceolata, Ludwigia leptocarpa, Ludwigia peploides, Ludwigia peruviana, Ludwigia repens, Luziola fluitans, Lycopus, Magnolia virginiana, Mayaca fluviatilis, Micranthemum umbrosum, Mikania scandens, Myrica cerifera, Myriophyllum aquaticum, Nuphar advena, Nymphaea odorata, Nymphoides aquatica, Nyssa biflora, Nyssa sylvatica, Oldenlandia uniflora, Orontium aquaticum, Panicum hemitomon, Panicum virgatum, Paspalum distichum, Peltandra virginica, Persea borbonia, Persicaria glabra, Persicaria hydropiperoides, Persicaria lapathifolia, Persicaria punctata, Phragmites australis, Pinguicula, Pinus clausa, Pinus elliottii, Pinus palustris, Pluchea foetida, Polygala rugelii, Pontederia cordata, Ptilimnium capillaceum, Quercus falcata, Quercus geminata, Quercus myrtifolia, Quercus nigra, Quercus palustris, Rhexia mariana, Rhynchospora colorata, Rhynchospora corniculata, Rhynchospora filifolia, Rhynchospora inundata, Rubus argutus, Rubus pensilvanicus, Sabal palmetto, Saccharum giganteum, Sacciolepis striata, Sagittaria australis, Sagittaria graminea, Sagittaria lancifolia, Sagittaria latifolia, Sagittaria platyphylla, Salix caroliniana, Salvinia, Saururus cernuus, Schoenoplectus tabernaemontani, Scirpus cyperinus, Scleria, Serenoa repens, Setaria parviflora, Solidago sempervirens, Spartina patens, Sphagnum, Taxodium ascendens, Taxodium distichum, Triadenum virginicum, Triadenum walteri, Typha domingensis, Utricularia gibba, Vigna luteola, Xyris jupicai, Xyris panacea, Xyris smalliana, Zizaniopsis miliacea.

Use by wildlife: The flowers of *Habenaria repens* attract moths (Insecta: Lepidoptera). The plants host adult and larval thrips (Insecta: Thysanoptera: Thripidae: *Aurantothrips orchidaceus, Frankliniella tritici, Pseudothrips beckhami*).

Economic importance: food: *Habenaria repens* is not known to be edible; **medicinal:** *Habenaria repens* is the source of habenariol (bis-*p*-hydroxybenzyl-2-isobutylmalate), an antioxidant ester that inhibits the formation of human low density lipoprotein (LDL) lipid hydroperoxides. A decoction or infusion from the tubers reportedly is used in Asia as an aphrodisiac; **cultivation:** *Habenaria repens* is distributed occasionally as an ornamental water garden or pond plant; otherwise it is of little commercial value; **misc. products:** Habenariol (see *medicinal* above), deters feeding by freshwater crayfish (Crustacea: Decapoda: Cambaridae: *Procambarus clarkii*); **weeds:** *Habenaria repens* is not weedy; **nonindigenous species:** *Habenaria repens* is indigenous.

Systematics: *Habenaria* is assigned to tribe Orchideae of the orchid subfamily Orchidoideae (Figure 2.13). Although DNA sequence data indicate that *Habenaria* (at least with respect to the New World species) is monophyletic (e.g., Figure 2.16), *H. repens* clearly illustrates the state of taxonomic confusion existing within this large genus. *Habenaria repens* is assigned to section *Clypeatae*, which is demonstrably polyphyletic in molecular phylogenetic analyses. However, molecular data ally *H. repens* most closely with *H. aranifera* and *H. warmingii*, both of which are assigned to a different section (*Pentadactylae*) that also has been shown to be broadly polyphyletic. Moreover, despite being placed in different sections, *H. repens* and *H. aranifera* are now regarded as conspecific, with the latter name being synonymous. Accordingly, the closest relative to *H. repens* then appears to be the South American *H. warmingii* (Figure 2.16), which recent revisions have now excluded from section *Pentadactylae* (Figure 2.16), leaving the sectional classification of both *H. repens* and *H. warmingii* (as well as many other species) completely unresolved. At this time, it seems reasonable to conclude only that *H. repens* is closely related to *H. warmingii*, and that both species occur within the Neotropical clade. A completely revised sectional classification of *Habenaria* likely will require years to materialize. The basic chromosome number of *Habenaria* is thought to be $x = 21$. *Habenaria repens* ($2n = 42$) is functionally diploid. There are no known hybrids involving *Habenaria repens*.

Distribution: *Habenaria repens* occurs throughout the southeastern United States and is regarded as the most widespread Neotropical species of the genus.

References: Anderson & Kral, 2008; Arditti & Ghani, 2000; Argue, 2011; Batista et al., 2013; Correll & Correll, 1975; Cypert, 1972a; do Vale et al., 2016; Felix & Guerra, 1998; 2005; Funderburk et al., 2007; Godfrey & Wooten, 1979; Gutierrez, 2010; Harper, 1916; Hawkes, 1950; Heaven et al., 2003; Holm, 1904; Hoyer et al., 1996; Hunt, 1943; Johnson et al., 1999; Keel et al., 2011; Pedron et al., 2012; 2014; Sheviak, 2002b; Singer et al., 2007; Stewart & Richardson, 2008; Stewart & Zettler, 2002; Visser et al., 1999; Wilson et al., 1999; Yoder et al., 2010; Zhang & Gao, 2018.

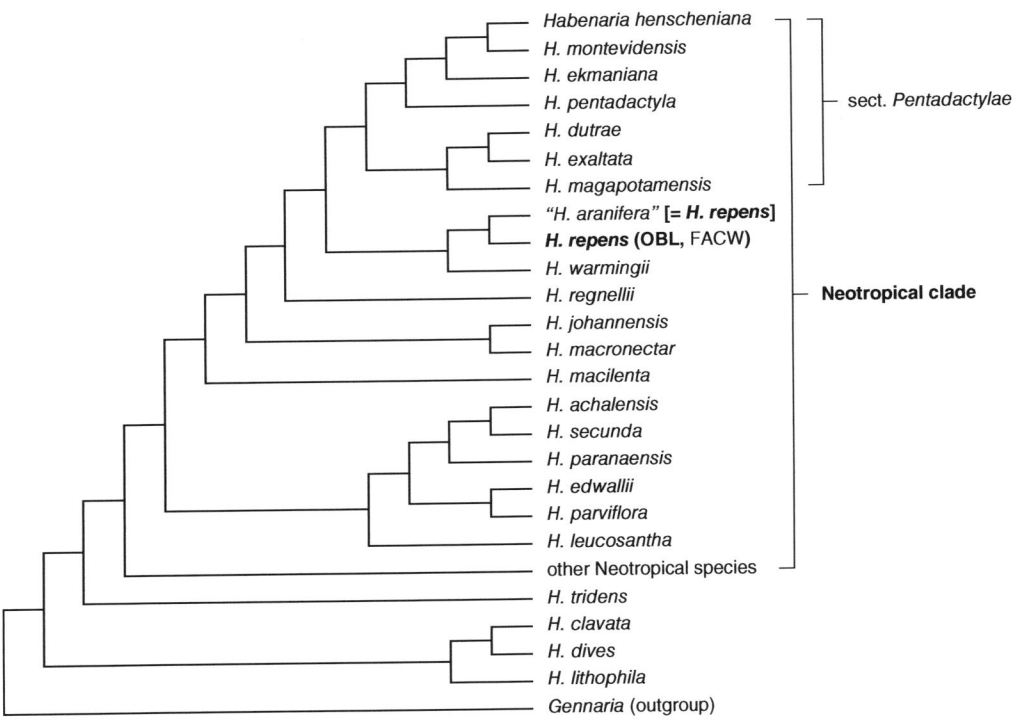

FIGURE 2.16 Interspecific relationships in *Habenaria* as indicated by phylogenetic analysis of combined nuclear and plastid DNA sequence data (simplified from Pedron et al., 2014). The problematic infrageneric classification of the group is well illustrated by *Habenaria repens* (placed traditionally in section *Clypeatae*), which resolves as the sister species of *H. aranifera* and *H. warmingii* (both placed traditionally in section *Pentadactylae*). Moreover, *H. aranifera* is now regarded as synonymous with *H. repens*. The recent revision of section *Pentadactylae* as depicted here excludes both *H. aranifera* and *H. warmingii*, whose sectional classification (as well as that of *H. repens*) remains unsettled. Taxa containing OBL indicators are highlighted in bold.

5.4.7. Liparis

Fen orchid, twayblade, wide-lip orchid; liparis

Etymology: from the Greek *liparos* ("greasy") referring to the characteristic leaf texture

Synonyms: *Anistylis*; *Cymbidium* (in part); *Diteilis*; *Epidendrum* (in part); *Iebine*; *Leptorchis* (in part); *Leptorkis*; *Malaxis* (in part); *Mesoptera* (in part); *Ophrys* (in part); *Orchis* (in part); *Paliris*; *Pseudorchis* (in part); *Serapias* (in part); *Sturmia* (in part)

Distribution: global: cosmopolitan; **North America:** eastern and central

Diversity: global: 250 species; **North America:** 3 species

Indicators (USA): obligate wetland (OBL); facultative wetland (FACW): *Liparis loeselii, L. nervosa*

Habitat: freshwater; palustrine; **pH:** 5.0–7.8; **depth:** <1 m; **life-form(s):** emergent herb

Key morphology: pseudobulbs (to 7 cm) conic or ovoid, slightly compressed, with slender, fibrous roots; stems (to 60 cm) angled, prominently or obscurely winged distally; leaves (2–7) basal, sheathing the stem and pseudobulb, reduced progressively to bracts (to 2–12 mm), the blades (to 30 cm) conduplicate or plicate, glabrous, green, glossy, ovate, elliptic, or oblong-elliptic, to elliptic- or oblong-lanceolate, membranaceous or succulent, sometimes keeled beneath; racemes (to 26 cm) terminal, lax, with 2–40 resupinate flowers, pedicels (to 5 mm) slender or stout; perianth green, greenish purple, greenish white, yellowish green, or yellowish white; dorsal sepal (to 8 mm) oblong-elliptic to linear-oblong or oblong-lanceolate to narrowly lanceolate, the margins somewhat to strongly revolute; petals (to 7.5 mm) falcate or slightly curved, filiform to narrowly spatulate, tubular, the margins revolute, lip (to 5.5 mm) translucent or green, opaque, purplish, or yellowish, arcuate-recurved, obcordate or obovate to cuneate or suborbiculate, the apex emarginate to mucronate; columns (to 5 mm) arcuate, stout, winged apically; anther apical, the cap green or yellow, with 2 pairs of waxy, yellow pollinia; capsules (to 15 mm) erect, obovate to ellipsoid, the veins sometimes winged

Life history: duration: perennial (pseudobulbs); **asexual reproduction:** pseudobulbs; **pollination:** autogamous; **sexual condition:** hermaphroditic; **fruit:** capsules (common); **local dispersal:** seeds (water, wind); **long-distance dispersal:** seeds (water, wind)

Imperilment: 1. *Liparis loeselii* [G5]; SX (DC); SH (KS); S1 (AL, AR, IL, MD, NC, NE, <u>NF</u>, RI, <u>SK</u>, SD, TN, WA); S2 (<u>AB</u>, <u>BC</u>, KY, MO, MT, ND, NH, <u>PE</u>, VA); S3 (IA, IN, <u>MB</u>, <u>NB</u>, <u>NS</u>, <u>QC</u>, VT, WV); **2.** *L. nervosa* [G3/G5]; S2 (FL)

Ecology: general: All three North American *Liparis* species are wetland indicators, with two of them ranked as OBL through part of their range. As circumscribed, the species vary widely with respect to their ecology, reproductive systems (self-incompatible or self-compatible), breeding systems (predominantly autogamous or outcrossing), pollinators, and other features. However, because *Liparis* is polyphyletic (see

Sytematics below), it is difficult to summarize any "characteristic" features for the group, which likely would represent an amalgam of information derived from several distinct genera.

5.4.7.1. *Liparis loeselii* **(L.) Rich.** grows in bogs (sandy, *Sphagnum*), depressions (roadside), ditches (roadside), dune slacks, fens (hanging, prairie, riverside, seepage), flats (sandy), flatwoods, floodplains, hollows (beach), hummocks (peaty), (*Sphagnum*), marshes, meadows (sedge), orchards (abandoned), pannes (interdunal), prairies (moist, sand, wet), quarries (abandoned, limestone), seeps (mafic, spring, wooded), shores (lake), slopes (seepage, shale), springheads, springs, swamps, thickets (alder, birch), tussocks, woodlands (rich), on berms (road), and along the margins of gravel pits, lakes, ponds, rivers, and streams at elevations to 730 m. Exposures can vary from full sunlight to shade, although the plants favor open, disturbed sites such as gravel pits, roadsides, and beaver ponds. The sites often are described as alkaline but can span a fairly broad range of acidity (e.g., pH: 5.0–7.8; specific conductivity: 161.5–400.9 μS cm^{-1}; $\bar{x} = 244.5$ μS cm^{-1}). The substrates have been described as clay, gravel, loamy sand, loess, marl, muck, peat (*Sphagnum*), peaty sand, sand, shale, and silt. The plants flower from May to August, with the fruits maturing from September to October. The nectarless and scentless flowers are not visited by potential pollinators. They are self-compatible and obligately autogamous with seed set varying from 15% to 50% (but typically only about 17%). Artificial autogamous, geitonogamous, and xenogamous pollinations have demonstrated that much higher seed set (94–100%) is possible. Spontaneous selfing occurs within 4 days of anthesis when the waxy pollinia (which lack viscidia) are released from the hinged cap as the anther degrades. Wing-like projections on the upper portion of the column guide the falling pollinia around the sides of the central stigmatic rostellum, where they adhere slightly to an adhesive ridge until eventually falling or rotating onto the stigma. However, raindrops striking the anther cap can increase the rate of autogamous seed set up to 70% by rapidly forcing the pollinia onto the stigma. As they flow across the flower (or decrease in size during evaporation), raindrops also can draw the pollinia toward the stigma by cohesion. The selfing rate is not influenced by wind currents and there is no evidence of floral apomixis. Each year, about 85% of the surviving plants in a population will flower, with 52% (33–77%) eventually bearing fruit. Ripe capsules can contain from 1,601 to 11,748 seeds ($\bar{x} = 4,270$ seeds), which are dispersed by wind or by snow-melt waters. Seed viability can be quite high (e.g., 80% by one estimate). Though perennial, the plants are short-lived, with an annual individual survivorship of about 45%. The plants overwinter by a fleshy pseudobulb, which remains attached by a short rhizome to the previous year's pseudobulb. The pseudobulb develops shallowly beneath the substrate surface and germinates during spring (May–June). It lacks any extended dormancy and rarely persists (but often remains attached in a degenerative state) after producing a new daughter pseudobulb in the fall. Consequently, multiple ramets typically do not develop but have been observed in Old World populations growing in dune slacks. The pseudobulbs are thought to provide drought resistance, which enables the plants to survive facultatively (albeit to smaller sizes) in drier sites. Studied populations have been known to experience short-term mortality rates of 45–97% while declining by only 52% in size over a longer term, which emphasizes the importance of seed bank recruitment in population maintenance. The duration of seed bank persistence has not been determined but likely is quite short. Several mycorrhizal Fungi (e.g., Basidiomycota: Tulasnellaceae: *Tulasnella*; Ceratobasidiaceae: *Ceratobasidium, Rhizoctonia repens, Thanatephorus*) have been isolated from the roots. **Reported associates:** *Acer rubrum, Acer saccharinum, Acer saccharum, Acorus americanus, Acorus calamus, Agalinis purpurea, Agalinis tenuifolia, Ageratina altissima, Agrimonia parviflora, Agrostis gigantea, Agrostis scabra, Aletris farinosa, Alisma subcordatum, Alnus viridis, Amorpha fruticosa, Andropogon gerardii, Andropogon virginicus, Antennaria plantaginifolia, Apios americana, Apocynum cannabinum, Asclepias incarnata, Asclepias sullivantii, Asplenium platyneuron, Athyrium filix-femina, Aulacomnium palustre, Betula papyrifera, Betula populifolia, Betula pumila, Bidens trichospermus, Bidens tripartitus, Boehmeria cylindrica, Botrychium dissectum, Botrychium virginianum, Bryum pseudotriquetrum, Calamagrostis canadensis, Calliergonella cuspidata, Calopogon tuberosus, Caltha palustris, Campanula aparinoides, Campylium stellatum, Cardamine bulbosa, Carex bebbii, Carex canescens, Carex conoidea, Carex crinita, Carex cristatella, Carex flava, Carex frankii, Carex haydenii, Carex hirtifolia, Carex hystericina, Carex laevivaginata, Carex lasiocarpa, Carex leptalea, Carex lurida, Carex molesta, Carex normalis, Carex prairea, Carex scoparia, Carex sterilis, Carex stricta, Carex swanii, Carex tribuloides, Carex viridula, Carex vulpinoidea, Carpinus caroliniana, Centaurium pulchellum, Cephalanthus occidentalis, Chamaecrista fasciculata, Chelone glabra, Cirsium altissimum, Cirsium muticum, Cladium mariscoides, Comarum palustre, Conoclinium coelestinum, Conyza canadensis, Corallorhiza odontorhiza, Cornus amomum, Cornus racemosa, Cornus sericea, Crataegus coccinea, Cryptotaenia canadensis, Cyperus flavescens, Cyperus strigosus, Cypripedium candidum, Dasiphora floribunda, Dichanthelium acuminatum, Dichanthelium sphaerocarpon, Doellingeria umbellata, Drepanocladus aduncus, Drosera rotundifolia, Dryopteris carthusiana, Dryopteris cristata, Elaeagnus angustifolia, Elaeagnus umbellata, Eleocharis erythropoda, Eleocharis obtusa, Eleocharis rostellata, Eleocharis tenuis, Epilobium ciliatum, Epilobium coloratum, Epilobium leptophyllum, Equisetum arvense, Equisetum fluviatile, Equisetum hyemale, Equisetum pratense, Equisetum sylvaticum, Equisetum variegatum, Erigeron philadelphicus, Erigeron strigosus, Eriophorum angustifolium, Eupatorium perfoliatum, Euthamia graminifolia, Eutrochium maculatum, Festuca subverticillata, Fimbristylis puberula, Fragaria virginiana, Fraxinus americana, Fraxinus nigra, Galearis spectabilis, Galium tinctorium, Galium triflorum, Gentianopsis crinita, Geum canadense, Glyceria striata, Goodyera pubescens, Gymnadeniopsis clavellata,*

Hamatocaulis vernicosus, Helodium blandowi, Hieracium aurantiacum, Hieracium paniculatum, Holcus lanatus, Hypericum kalmianum, Impatiens capensis, Iris virginica, Juncus acuminatus, Juncus alpinus, Juncus brachycephalus, Juncus canadensis, Juncus dudleyi, Juncus effusus, Juncus filipendulus, Juncus nodosus, Juncus scirpoides, Juncus tenuis, Juncus torreyi, Juniperus virginiana, Larix laricina, Leersia oryzoides, Leersia virginica, Leucanthemum vulgare, Lindera benzoin, Liparis liliifolia, Liriodendron tulipifera, Lobelia siphilitica, Lonicera, Lophiola aurea, Ludwigia alternifolia, Ludwigia palustris, Lycopus americanus, Lycopus asper, Lysimachia terrestris, Lythrum alatum, Lythrum salicaria, Maclura pomifera, Maianthemum stellatum, Malus ioensis, Mentha arvensis, Mentha ×piperita, Micranthes pensylvanica, Microstegium vimineum, Mimulus ringens, Muhlenbergia glomerata, Muhlenbergia mexicana, Nasturtium officinale, Onoclea sensibilis, Ophioglossum vulgatum, Osmunda regalis, Osmundastrum cinnamomeum, Oxypolis rigidior, Packera aurea, Packera paupercula, Panicum flexile, Panicum virgatum, Parnassia caroliniana, Parnassia glauca, Pedicularis lanceolata, Penstemon digitalis, Persicaria coccinea, Persicaria hydropiperoides, Persicaria maculosa, Persicaria punctata, Phalaris arundinacea, Phleum pratense, Phlox glaberrima, Phragmites australis, Physocarpus opulifolius, Picea, Pilea fontana, Pilea pumila, Pinus strobus, Platanthera flava, Platanthera huronensis, Platanthera hyperborea, Platanthera lacera, Platanthera peramoena, Poa nemoralis, Poa pratensis, Polemonium reptans, Populus deltoides, Populus tremuloides, Potentilla simplex, Prunella vulgaris, Prunus serotina, Pycnanthemum virginianum, Quercus palustris, Ranunculus recurvatus, Rhamnus cathartica, Rhamnus frangula, Rhus glabra, Rhynchospora alba, Robinia pseudoacacia, Rosa multiflora, Rosa palustris, Rubus hispidus, Rubus pubescens, Rudbeckia fulgida, Rudbeckia hirta, Rumex obtusifolius, Sagittaria australis, Salix bebbiana, Salix candida, Salix discolor, Salix eriocephala, Salix interior, Salix myricoides, Salix nigra, Salix pedicellaris, Salix sericea, Schizachyrium scoparium, Schoenoplectus acutus, Schoenoplectus pungens, Schoenoplectus tabernaemontani, Scirpus atrovirens, Scirpus cyperinus, Scirpus pendulus, Scleria verticillata, Scutellaria galericulata, Scutellaria lateriflora, Selaginella apoda, Silphium integrifolium, Sisyrinchium albidum, Sisyrinchium montanum, Solanum dulcamara, Solidago canadensis, Solidago gigantea, Solidago ohioensis, Solidago patula, Solidago riddellii, Solidago rugosa, Solidago uliginosa, Sorghastrum nutans, Sparganium americanum, Sphagnum, Sphenopholis intermedia, Spiraea alba, Spiraea tomentosa, Spiranthes cernua, Spiranthes lucida, Quercus velutina, Stachys pilosa, Stellaria longifolia, Symphyotrichum novae-angliae, Symphyotrichum puniceum, Symplocarpus foetidus, Thelypteris palustris, Toxicodendron radicans, Toxicodendron vernix, Triadenum fraseri, Triadenum virginicum, Trifolium pratense, Tsuga canadensis, Typha angustifolia, Typha latifolia, Vaccinium corymbosum, Valeriana edulis, Verbena bracteata, Verbena hastata, Vernonia missurica, Viburnum lentago, Viburnum prunifolium, Viola lanceolata, Viola macloskeyi, Viola primulifolia, Viola sagittata, Viola sororia, Xanthium strumarium.

5.4.7.2. *Liparis nervosa* (Thunb.) Lindl. occurs in hammocks (hardwood, rich, wet), prairies (cypress), seepages, swamps (cypress), and along the banks of rivers at low elevations [but to 1500 m elsewhere]. The few available North American records (i.e., from Florida) indicate that the plants grow in deep shade and have an affinity for decayed organic matter. They occur on substrates that have been characterized as decaying stumps, humus, muck, and sand. Flowering in North America has been observed from April to October with fruiting from September to January. Although this species is distributed widely throughout the tropics, and has been studied intensively for its medicinal properties (see below), details concerning its basic life history are rather vague. Florida plants have been described as autogamous; however, in Africa, the plants are pollinated by various flies (Insecta: Diptera), which results in fruit production rates up to 78.4%. Given the likelihood of self-compatibility, autogamy probably is at least facultative. It is suspected to occur in Africa during the rainy season, when pollinating flies are scarce and enhanced autogamy by raindrops (e.g., see *L. loeselii* above) would be beneficial. The seeds float and likely are dispersed by wind and water. They have a water content of about 45.5% but a fairly high rate of water loss (3.28% hr^{-1}). Other basic life-history information is lacking for this species. **Reported associates:** *Ardisia escallonioides, Roystonea regia, Sphagnum, Taxodium, Zeuxine strateumatica.*

Use by wildlife: *Liparis loeselii* occasionally is grazed by deer (Mammalia: Cervidae: *Odocoileus virginianus*), insects (Insecta), and rabbits (Mammalia: Leporidae: *Sylvilagus*).

Economic importance: food: *Liparis* species are not edible; **medicinal:** An alkaloid isolated from *L. loeselii* has been identified as auriculin (but probably intended as auricularine). The Cherokee administered a compound root infusion of *L. loeselii* to treat urinary problems. *Liparis nervosa* contains various pyrrolizidine alkaloids and has been used widely in traditional medicine. It has been studied intensively for its alleged detoxicating and hemostatic properties. Most of these alkaloids are not cytotoxic against the several human cancer cell lines tested but are effective at suppressing nitric oxide (NO) production. One compound [chloride-(*N*-chloromethyl nervosine VII] has exhibited moderate cytotoxic activity. The plants also contain biphenanthrenes, which potently are cytotoxic against human colon and stomach cancer lines. Lectins from the plants strongly inhibit growth of MCF-7 breast cancer cells. Total alkaloid extracts are highly antimicrobial and were found to inhibit all 16 bacterial and fungal strains evaluated. Whole plant extracts have been used in China as an analgesic for bruises, coughs, snakebites, sores, and toothaches, to treat dermatitis, to relieve swelling, and to stop internal and external bleeding. The alcoholic extracts are highly hemostatic and significantly shorten blood coagulation times. In Africa, the tubers are used in preparations to remedy stomach aches and to treat malignant ulcers; **cultivation:** Neither of the OBL *Liparis* species is cultivated commercially; **misc. products:** none; **weeds:** none; **nonindigenous species:** none.

Systematics: *Liparis* is assigned to tribe Malaxideae of the orchid subfamily Epidendroideae (Figure 2.13). The genus itself clearly is not monophyletic, as demonstrated by phylogenetic analyses of molecular data (e.g., Figure 2.17), where different clades of *Liparis* species are resolved along with species of *Malaxis* and *Oberonia*. All three genera bear morphologically similar flowers. Eventually, the phylogenetic results will need to be reconciled by revising the circumscriptions of these three genera, the establishment of additional genera, or by their merger. The two OBL North American indicators are only remotely related, with the temperate *L. loeselii* (OBL, FACW) being closely related to the North American *L. liliifolia* (FACU), whereas the tropical *L. nervosa* (OBL, FACW) associates with Old World species from eastern Asia and the Pacific Islands (Figure 2.17). Although *L. loeselii* is widespread geographically, the DNA sequences at representative "barcoding" regions are identical among accessions sampled from different Old and New World populations. The chloroplast genome of *L. loeselii* has been completely sequenced. Tribe Malaxideae has two commonly occurring basic chromosome numbers ($x=15$, $x=21$), which likely have been derived from an ascending polyploid lineage. The anomalous counts for *Liparis loeselii* ($2n=26$) and *L. nervosa* ($2n=42$) are consistent with their distant phylogenetic affinity (Figure 2.17). No natural hybrids involving either *L. loeselii* or *L. nervosa* have been reported.

Distribution: *Liparis loeselii* has an amphi-Atlantic distribution throughout east-central North America, which extends into central and western Europe. In North America, *L. nervosa* is restricted to the southeastern United States but extends widely into South America as well as to Africa and Asia. It has been called "the most widespread orchid in the world."

References: Arditti & Ghani, 2000; Argue, 2012e; Barksdale, 1937; Błońska et al., 2016; Cameron, 2005; Catling, 1980; Chen et al., 2018; Curtis, 1939; Ebinger & Bacone, 1980;

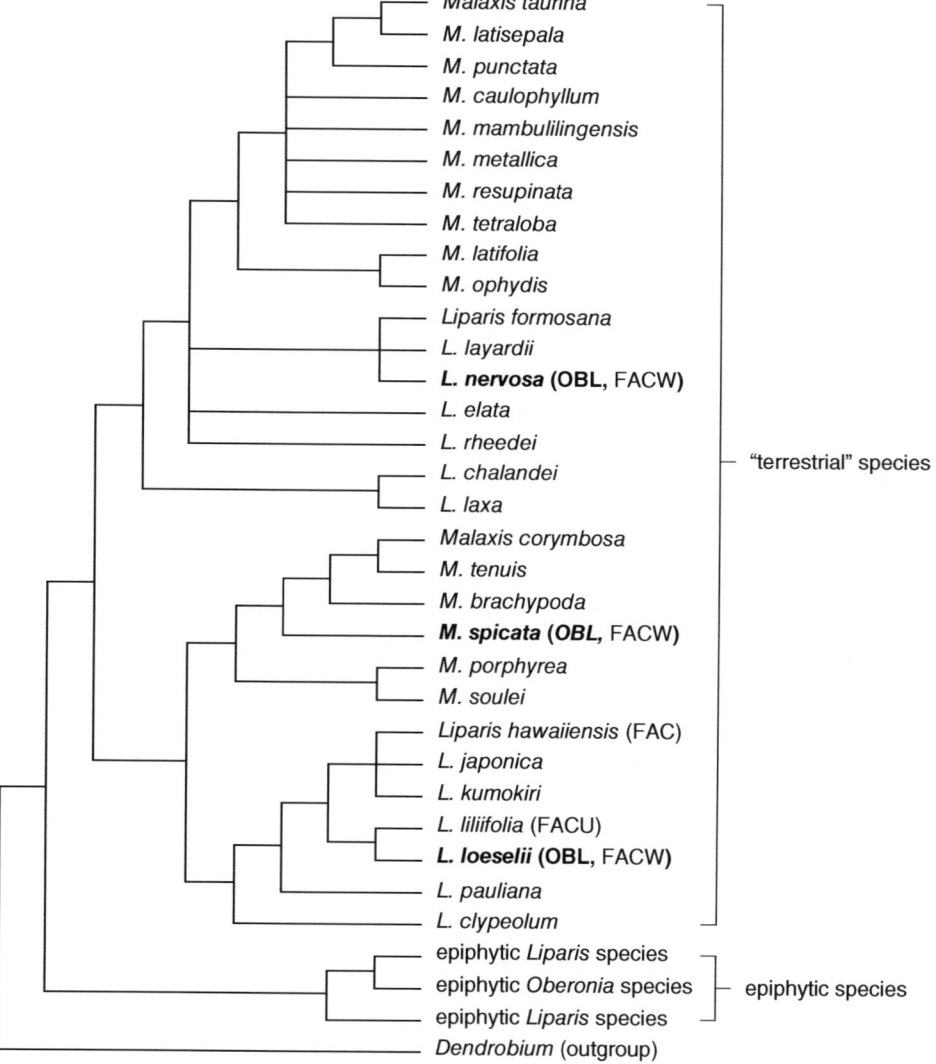

FIGURE 2.17 Phylogenetic analysis of combined nuclear and plastid DNA sequences (simplified from Cameron, 2005) indicates that *Liparis* and *Malaxis* are not monophyletic as currently circumscribed. Furthermore, some *Liparis* species resolve within a major epiphytic clade that also contains the species of *Oberonia*. It is evident that the circumscriptions of all three genera must be reconsidered and their classification revised to address these issues. In any case, the OBL indicators in both *Liparis* and *Malaxis* (highlighted in bold) do not associate together but are distributed broadly across the tree, which indicates the independent origins of the OBL habit in this group of orchids.

Essomo et al., 2016; Felix & Guerra, 2010; Godfrey & Wooten, 1979; Hamel & Chiltoskey, 1975; Huang et al., 2013; Illyés et al., 2005; Krawczyk et al., 2018; Jones, 1998; Liang et al., 2019; Lindström & Lüning, 1971; Liu et al., 2016; Löve, 1981; Magrath, 2002; Markle, 1915; McMaster, 2001; Miduno, 1940; Nekola & Lammers, 1989; Pant, 2013; Rasmussen et al., 2015; Sauleda, 2014; Thompson & McKinney, 2006; Thompson et al., 2005; Yoder et al., 2010.

5.4.8. Malaxis

Adder's mouth orchid; malaxis

Etymology: from the Greek *malakós* ("soft") in allusion to the leaf texture

Synonyms: *Achroanthes* (in part); *Epipactis* (in part); *Hammarbya*; *Microstylis* (in part); *Ophrys* (in part); *Sturmia* (in part)

Distribution: global: widespread; **North America:** widespread except north-central and northwestern United States

Diversity: global: 250 species; **North America:** 10 species

Indicators (USA): obligate wetland (OBL): *Malaxis paludosa, M. spicata*

Habitat: freshwater; palustrine; **pH:** 6.6–7.5; **depth:** <1 m; **life-form(s):** emergent herb

Key morphology: stems (to 45 cm) glabrous, expanded at base into a globose pseudobulb (to 20 mm) with few, fibrous roots; leaves 2–3, basal, the base sheathing, the blade (to 10 cm) elliptic, ovate, or suborbiculate; racemes (to 24 cm) terminal, sometimes spike-like, many-flowered (to 150), bracts (to 3.3 mm) inconspicuous, lanceolate or triangular, pedicels short (to 3 mm) or long (to 10 mm); perianth not resupinate, mostly green or greenish-yellow; sepals spreading, the dorsal (to 3.5 mm) ovate or ovate-lanceolate, the laterals (to 3 mm) reflexed, elliptic or lanceolate; petals (to 2.5 mm) recurved or strongly reflexed and crossing behind flower, the lip to (3.3 mm) ovate or ovate-lanceolate, brown, green, orange, or vermillion, the apex unlobed, the base cuneate or cordate-auriculate and nearly surrounding the column; column (to 1 mm) free, anther terminal, with 4 waxy yellow pollinia; capsules (to 8 mm) ascending or erect, ellipsoid, or ovoid

Life history: duration: perennial (corm-like pseudobulbs); **asexual reproduction:** adventitious plantlets ("brood buds," "foliar embryos"); **pollination:** insect; **sexual condition:** hermaphroditic; **fruit:** capsules (common); **local dispersal:** plantlets, seeds (wind); **long-distance dispersal:** seeds (wind)

Imperilment: 1. *Malaxis paludosa* [G3/G4]; S1 (MB, MN, ON, SK); S2 (AB, BC, YT); S3 (AK); **2.** *M. spicata* [G4]; S1 (GA, NC, SC); S3 (VA)

Ecology: general: Because *Malaxis* is known to be polyphyletic (see *Systematics* below), it would be of little value to attempt an ecological summary for the group as currently circumscribed and has not been attempted here.

5.4.8.1. ***Malaxis paludosa*** **(L.) Sw.** grows in bogs (blanket, lakeshore, roadside, *Sphagnum*), ditches, fens (poor, sloping), floating mats, heaths, meadows (wet), muskegs, swamps, terraces, on tundra, and along the margins of lakes and ponds at elevations to 766 m. The plants occur in open sites, or in semi-shaded exposures, which receive full sunlight during part of the day. The substrates include loam, peat (underlain by marl), or peaty mud. The plants often grow atop *Sphagnum* hummocks, 10–15 cm higher than the surrounding standing water (which can be up to 10 cm deep); however, they are more common on substrates made up of other mosses. Flowering and fruiting occur from June to August, with the fruits developing in September. The flowers are mildly fragrant (described as a cucumber-like odor) and are pollinated primarily by fungus gnats (Insecta: Diptera: Mycetophilidae: *Phronia digitata*). As the insects probe for nectar during floral visits, the pollinia become attached on their anterior ventral thorax, and then are deposited during the next floral visit. Unlike most orchids, the flowers are not resupinate. However, they are rotated 360° to achieve the placement of the labellum at the top of the flower, which presumably provides a more optimal configuration for the minute pollinators. Other details on the floral ecology and seed biology are unavailable, but populations often are characterized by high fruit set. The plants overwinter from a small, corm-like pseudobulb, which replicates by means of a short-lived rhizome. The pseudobulbs can attain an almost epiphytic habit when growing in mounds of mosses. The leaf apices produce vegetative, apomictic plantlets that have been referred to as "brood buds" or "foliar embryos." **Reported associates:** *Abies balsamea, Alnus rugosa, Andromeda polifolia, Aralia nudicaulis, Aulacomnium palustre, Betula nana, Betula pumila, Botrychium virginianum, Callicladium haldanianum, Calopogon tuberosus, Caltha palustris, Campylium stellatum, Campylopus atrovirens, Carex aquatilis, Carex chordorrhiza, Carex diandra, Carex disperma, Carex echinata, Carex leptalea, Carex limosa, Carex magellanica, Carex pauciflora, Carex rostrata, Circaea alpina, Coelocaulon aculeatum, Coptis trifolia, Corallorhiza trifida, Cornus canadensis, Cornus sericea, Cypripedium acaule, Cypripedium reginae, Deschampsia cespitosa, Drosera anglica, Drosera rotundifolia, Dryopteris cristata, Epilobium leptophyllum, Epilobium palustre, Equisetum fluviatile, Eriophorum angustifolium, Eriophorum chamissonis, Eriophorum tenellum, Eriophorum viridicarinatum, Fragaria vesca, Galearis rotundifolia, Galium aparine, Galium obtusum, Gaultheria hispidula, Gentiana douglasiana, Goodyera repens, Helodium blandowii, Iris versicolor, Juncus stygius, Kalmia polifolia, Larix laricina, Ligusticum calderi, Linnaea borealis, Loeskypnum badium, Lycopus uniflorus, Maianthemum trifolium, Malaxis monophyllos, Malaxis unifolia, Menyanthes trifoliata, Mitella nuda, Moneses uniflora, Neottia cordata, Orthilia secunda, Packera subnuda, Picea mariana, Pinguicula vulgaris, Plagiomnium ciliare, Plantago macrocarpa, Platanthera dilatata, Platanthera hyperborea, Platanthera obtusata, Pleurozium schreberi, Podagrostis aequivalvis, Pogonia ophioglossoides, Pohlia nutans, Primula jeffreyi, Rhamnus alnifolia, Rhododendron groenlandicum, Rhynchospora alba, Ribes lacustre, Rubus chamaemorus, Rubus pubescens, Salix pedicellaris, Sarracenia purpurea, Scheuchzeria palustris, Sphagnum angustifolium, Sphagnum magellanicum, Sphagnum papillosum, Sphagnum rubellum, Sphagnum warnstorfii, Symphyotrichum subspicatum, Thelypteris palustris, Thuidium delicatulum, Thuja* occidentalis, Tomentypnum nitens, Triadenum fraseri, Trientalis borealis, Trichophorum cespitosum, Triglochin palustris, Tsuga mertensiana, Utricularia intermedia, Vaccinium oxycoccos, Viola canadensis.

5.4.8.2. *Malaxis spicata* Sw. inhabits coppices (low), floodplains (shaded), forests (maritime, swamp), hammocks, ravines (rich), seeps (sulfur), shores, slopes, swamps, woodlands (damp, rich), and the margins of rivers and streams at elevations to 100 m. Most occurrences are reported from exposures in dense shade to filtered light. The substrates are circumneutral (pH: 6.6–7.5; Ca: 791–2144 ppm; K: 12–38 ppm; Mg: 115–404 ppm; P: 2–7 ppm) and include humus, loam (sandy), marl, muck, peat (to 46 cm deep), or sand. The plants also grow semi-epiphytically on logs (floating, rotting), tree bases, and stumps in sites up to 0.5 m above the adjacent water levels. Flowering occurs from July to February, with fruiting extending into March. The flowers can be retained by the plants for up to 3 months. The floral biology has not been studied, but pollination is thought to be by fungus gnats (Insecta: Diptera: Mycetophilidae). The plants arise from a small pseudobulb. These can become lodged in bark cavities, which results in a semi-epiphytic growth habit. Otherwise, the life history of this species is poorly understood and warrants further study. **Reported associates:** *Acer floridanum, Acer rubrum, Aletris lutea, Acer saccharum, Acmella oppositifolia, Actaea racemosa, Adiantum pedatum, Agarista populifolia, Amelanchier arborea, Amorpha fruticosa, Ampelopsis arborea, Anemone americana, Anemone virginiana, Apteria aphylla, Aquilegia canadensis, Arisaema triphyllum, Aristida stricta, Aristolochia serpentaria, Arundinaria gigantea, Asarum canadense, Asclepias perennis, Asimina triloba, Aureolaria virginica, Axonopus furcatus, Baccharis glomeruliflora, Berchemia scandens, Bidens mitis, Blechnum occidentale, Boehmeria cylindrica, Botrychium dissectum, Brachyelytrum erectum, Bumelia lycioides, Callicarpa americana, Carex albicans, Carex amphibola, Carex bromoides, Carex chapmanii, Carex decomposita, Carex digitalis, Carex godfreyi, Carex leptalea, Carpinus caroliniana, Carya tomentosa, Centella asiatica, Cephalanthus occidentalis, Cercis canadensis, Chamaecyparis thyoides, Chasmanthium laxum, Chasmanthium nitidum, Chimaphila maculata, Cirsium muticum, Cladium jamaicense, Clematis crispa, Coleataenia longifolia, Corallorhiza wisteriana, Cornus florida, Cornus foemina, Crinum americanum, Cynoglossum virginianum, Cyperus, Decumaria barbara, Desmodium glutinosum, Desmodium nudiflorum, Dichanthelium boscii, Dichanthelium commutatum, Dichanthelium dichotomum, Diodia virginiana, Dioscorea floridana, Dioscorea villosa, Diospyros virginiana, Dirca palustris, Ditrysinia fruticosa, Dryopteris ludoviciana, Dyschoriste humistrata, Elephantopus nudatus, Elytraria caroliniensis, Epidendrum magnoliae, Equisetum hyemale, Erechtites hieracifolius, Erigeron pulchellus, Erythrina herbacea, Euonymus americanus, Fagus grandifolia, Festuca subverticillata, Fimbristylis spadicea, Fraxinus americana, Fraxinus caroliniana, Fraxinus pennsylvanica, Fraxinus profunda, Galium circaezans, Galium triflorum, Gelsemium sempervirens, Geum canadense, Gonolobus suberosus, Gordonia lasianthus, Habenaria floribunda, Heuchera americana, Hexalectris spicata, Hexastylis virginica, Hieracium venosum, Houstonia caerulea, Houstonia purpurea, Hydrangea arborescens, Hydrocotyle bonariensis, Hydrocotyle verticillata, Hypericum galioides, Hypericum hypericoides, Hypolepis repens, Hypoxis curtissii, Hyptis alata, Ilex cassine, Ilex coriacea, Ilex glabra, Ilex opaca, Ilex vomitoria, Illicium parviflorum, Ipomoea cordatotriloba, Iris hexagona, Isoetes flaccida, Itea virginica, Juniperus virginiana, Justicia ovata, Leucothoe axillaris, Lindera benzoin, Liparis nervosa, Liquidambar styraciflua, Liriodendron tulipifera, Lobelia cardinalis, Lobelia georgiana, Lonicera japonica, Lonicera sempervirens, Ludwigia palustris, Luzula acuminata, Lyonia lucida, Magnolia virginiana, Maianthemum racemosum, Malaxis unifolia, Mikania cordifolia, Mikania scandens, Mimosa microphylla, Mitchella repens, Morus rubra, Myrica cerifera, Nyssa sylvatica, Nyssa biflora, Oplismenus hirtellus, Osmunda regalis, Osmundastrum cinnamomeum, Ostrya virginiana, Oxydendrum arboreum, Packera aurea, Parnassia grandifolia, Parthenocissus quinquefolia, Peltandra virginica, Persea palustris, Phanopyrum gymnocarpon, Physostegia leptophylla, Pieris phillyreifolia, Pilea fontana, Pinguicula caerulea, Pinus elliottii, Pinus taeda, Platythelys querceticola, Pleopeltis polypodioides, Pluchea odorata, Podophyllum peltatum, Polygonatum biflorum, Polystichum acrostichoides, Pontederia cordata, Ponthieva racemosa, Woodwardia virginica, Prenanthes altissima, Pteridium aquilinum, Quercus laurifolia, Quercus minima, Quercus muehlenbergii, Quercus nigra, Quercus virginiana, Quercus rubra, Rhapidophyllum hystrix, Rhododendron periclymenoides, Rhododendron viscosum, Rhynchospora caduca, Rhynchospora corniculata, Rhynchospora inundata, Rhynchospora miliacea, Rhynchospora mixta, Roystonea regia, Rubus argutus, Rubus pensilvanicus, Rubus trivialis, Ruellia caroliniensis, Rumex verticillatus, Sabal minor, Sabal palmetto, Sagittaria, Salix floridana, Sanguinaria canadensis, Sanicula canadensis, Sassafras albidum, Saururus cernuus, Scleria oligantha, Scleria triglomerata, Selaginella apoda, Serenoa repens, Sideroxylon reclinatum, Smilax glauca, Smilax laurifolia, Smilax pumila, Smilax tamnoides, Solidago bicolor, Solidago caesia, Solidago odora, Solidago sempervirens, Sphenopholis nitida, Spigelia marilandica, Symphyotrichum lateriflorum, Taxodium distichum, Teucrium canadense, Thelypteris interrupta, Thelypteris kunthii, Thelypteris palustris, Tilia americana, Tillandsia bartramii, Tillandsia usneoides, Tipularia discolor, Toxicodendron radicans, Tragia urens, Triadenum walteri, Trillium cuneatum, Triphora trianthophoros, Tripsacum dactyloides, Ulmus americana, Ulmus rubra, Uvularia perfoliata, Vaccinium corymbosum, Vaccinium elliottii, Vaccinium myrsinites, Vaccinium pallidum, Viola sororia, Vitis cinerea, Vitis rotundifolia, Vitis vulpina, Vittaria lineata, Woodwardia areolata.* **Use by wildlife:** none reported.

Economic importance: food: not edible; **medicinal:** *Malaxis paludosa* contains several alkaloids (hammarbine, paludosine); **cultivation:** *Malaxis* species are rarely cultivated; **misc.**

products: Although most orchid flowers are resupinate (by rotating180°), the 360° rotation of *M. paludosa* flowers to reposition the lip at the top of the flower was cited by Charles Darwin as a compelling example of natural selection; **weeds:** none; **nonindigenous species:** none.

Systematics: *Malaxis* is the type genus of tribe Malaxideae in the orchid subfamily Epidendroideae (Figure 2.13). As indicated in the preceding treatment of *Liparis* (see *Systematics* above), *Malaxis* is not monophyletic (e.g., Figure 2.17) and a revised circumscription is needed. Of the two *Malaxis* species formerly designated as OBL indicators, one (*M. spicata*) has been retained in *Malaxis*, while the other (*M. paludosa*) was transferred to the monotypic genus *Hammarbya* (as *H. paludosa*) in the recent 2016 indicator list. However, confusion abounds. The transfer of *M. paludosa* to *Hammarbya* was recommended because phylogenetic analyses (unpublished) allegedly resolved "*Liparis paludosa*" (i.e., *M. paludosa*) within a different clade than that containing *L. loeselii*, the type species of *Liparis*. However, the name "*Liparis paludosa*" never was published, so the preceding argument regarding inclusion of this taxon in *Liparis* is illogical. Ironically, a BLAST search of unpublished "*Hammarbya paludosa*" *matK* sequences in GenBank shows them to be highly similar to those of *Liparis japonica* and *L. kumokiri*, which actually *do* resolve within the *L. loeselii* clade (see Figure 2.17). Until the chaotic classification of tribe Malaxideae is fully resolved, it is futile to recommend how to treat the genus taxonomically. Whether compelling evidence ever will be provided to recognize *M. paludosa* as monotypic (e.g., in *Hammarbya*), to transfer the species legitimately to *Liparis*, or to retain the species within *Malaxis* cannot be anticipated. Although the eventual transfer of *Malaxis paludosa* to a different genus seems likely, it has been retained here in preference to its recognition as *Hammarbya paludosa*, which seems premature and represents a generic name that is totally unfamiliar to most North American botanists. Although the basic chromosome number of *Malaxis* is uncertain (e.g., *x* = 14, 15, 18, 20, 21, or 22), *M. paludosa*: (2*n* = 28) apparently is diploid. No natural hybrids involving *Malaxis paludosa* or *M. spicata* have been reported.

Distribution: The circumboreal *Malaxis paludosa* occurs throughout boreal North America and extends into Eurasia; *M. spicata* is limited to the southeastern United States and West Indies.

References: Argue, 2012e; Baldwin, 1961; Batygina et al., 2003; Beatty & Desjardins, 2009; Brown, 2005; Cameron, 2005; 2010; Carlson et al., 2013; Catling & Magrath, 2002; Darst et al., 2002; Darwin, 1877; Godfrey & Wooten, 1979; Lamont & Stalter, 2007; Lindström & Lüning, 1972; Massey, 1953; NAOCC, 2019; Porcher, 1981; Reeves & Reeves, 1984; Reeves & Reeves, 1985; Smith, 2012; Taylor, 1967; Vitt et al., 1990; Ward & Clewell, 1989; Ware & Ware, 1992; Wherry, 1928; Whitfeld et al., 2015.

5.4.9. Platanthera

Bog, fringed, wood, or woodland orchid; platanthère

Etymology: from the Greek *plati anthera* ("broad anther"), a feature once believed to be diagnostic for the genus

Synonyms: *Blephariglottis*; *Gymnadeniopsis*; *Habenaria* (in part); *Limnorchis*; *Peristylus* (in part); *Pseudodiphryllum*

Distribution: global: temperate Africa, Eurasia, and North America; **North America:** widespread throughout

Diversity: global: 100 species; **North America:** 45 species

Indicators (USA): obligate wetland (OBL): *Platanthera aquilonis*, *P. blephariglottis*, *P. chorisiana*, *P. clavellata*, *P. cristata*, *P. huronensis*, *P. integra*, *P. integrilabia*, *P. leucophaea*, *P. limosa*, *P. tescamnis*; **facultative wetland (FACW):** *P. aquilonis*, *P. clavellata*, *P. cristata*, *P. huronensis*, *P. leucophaea*

Habitat: freshwater; palustrine; **pH:** 3.3–8.5; **depth:** <1 m; **life-form(s):** emergent herb

Key morphology: roots fleshy, fasciculate, tuberous, lance-fusiform; stems (to 165 cm) erect, leafy; leaves 1-many, ascending or spreading, scattered on stem or subopposite at base, gradually or abruptly reduced upward to bracts, conduplicate, sheathing at base, the blades (to 35 cm) elliptic, elliptic-lanceolate, lance-oblong, lanceolate, linear-lanceolate, linear-oblong, oblanceolate, oblong, ovate-lanceolate, or suborbiculate; spikes solitary, terminal, dense to lax, few to many-flowered, the flowers partly or wholly resupinate, inconspicuous or showy, the perianth greenish, orange, white, whitish green, yellowish green, or yellow-orange; lateral sepals extending forward, reflexed, or spreading; the petals (to 6 mm) elliptic, lance-falcate, lanceolate, linear, linear-oblong, oblanceolate, oblong, oblong-elliptic, obovate, ovate, ovate-falcate, ovate-oblong, or rhombic-ovate, the margins entire, emarginate, or fringed, the lip (to 29 mm) elliptic, lanceolate, lance-spatulate, linear, linear-elliptic, linear-oblong, oblong, oblong-elliptic, orbiculate, ovate, ovate-elliptic, ovate-oblong, quadrangular-suborbiculate, rhombic-lanceolate, rhombic-ovate, or spatulate, 3-lobed, spurred (to 60 mm) at base, descending, projecting, or adhering apically to dorsal sepal; rostellum lobes curving forward, directed downward, or divergent, the pollinaria 2, the pollinia 2, viscidia free; ovary (to 31 mm) slender to stout; capsules cylindric to ellipsoid

Life history: duration: perennial (tuberous roots); **asexual reproduction:** tubers (root); **pollination:** insect, self; **sexual condition:** hermaphroditic; **fruit:** capsules (common); **local dispersal:** seeds (wind); **long-distance dispersal:** seeds (wind)

Imperilment: 1. *Platanthera aquilonis* [G5]; SX (OH); S1 (PA); S2 (AZ, PE); S3 (LB, NE, NU, YT); **2.** *P. blephariglottis* [G4/G5]; S1 (CT, DE, GA, IL, LA, OH, RI); S2 (AL, MD, MS, PA, PE, VA, VT); S3 (NB, NC, QC); **3.** *P. chorisiana* [G4]; S2 (WA); S3 (AK); **4.** *P. clavellata* [G5]; SH (ND); S1 (FL, IA, IL, OK); S2 (MO, RI); S3 (IN, MN, PE); **5.** *P. cristata* [G5]; SX (PA); S1 (AR, KY, MA, NY); S2 (DE, TN); S3 (FL, GA, MD, MS, NJ, VA); **6.** *P. huronensis* [G5]; S1 (NS, PA); S2 (IL, MA, NB); S3 (LB, YT); **7.** *P. integra* [G3/G4]; S1 (GA, NJ, SC, TN, TX); S2 (AL, NC); S3 (FL, LA, MS); **8.** *P. integrilabia* [G2/G3]; SH (NC); S1 (GA, KY, MS, SC); S2 (AL, TN); **9.** *P. leucophaea* [G2/G3]; SH (NY, OK, PA, VA); S1 (IA, IL, IN, ME, MI, MO); S2 (OH, ON, WI); **10.** *P. limosa* [G4]; **11.** *P. tescamnis* [GNR]; SH (AZ)

Ecology: general: *Platanthera* species occupy a wide variety of habitats, including forests, wetlands, and woodlands. Thirty-one of the North American species (69%) have status as wetland indicators, with 24% being categorized as OBL. All the species are perennial herbs, which overwinter by means of a tuberous, sometimes cormose rootstock. The perianth can range in color from green or white to bright orange, purple, or yellow, and sometimes has elaborately fringed or lacerate margins. The floral lip has a nectar spur, which attracts various insects (Insecta) as pollinators. Most of the species are pollinated by moths (Lepidoptera: Noctuidae; Pyralidae), while some rely on beetles (Coleoptera), bumblebees (Hymenoptera: Apidae: *Bombus*), butterflies (Lepidoptera), flies (Diptera), or hawkmoths (Lepidoptera: Sphingidae). Several species are autogamous. Rotation of pollinia (up to 90°) often is necessary to achieve a configuration enabling their insertion onto the stigma. The elasped time required to complete the rotation provides sufficient time for the insect to leave one flower and visit a different flower, thereby effecting cross-pollination. Floral features that result in the differential placement of pollinia on the eyes versus the proboscis of pollinators (e.g., divergent, separate pollinaria vs. a single pollinarium) are of importance in maintaining interspecific isolating barriers. Some specificity is also achieved by the varied length of the floral spur, which correlates inversely with the diversity of potential pollinating groups. The seeds are minute, numerous, and dispersed abiotically by water or wind. The plants persist by means of vertical, fusiform, fleshy tubers, which divide each year but do not persist to develop into clonal populations.

5.4.9.1. *Platanthera aquilonis* **Sheviak** grows in bogs, borrow pits, bottoms (ravine, stream), carrs (willow), depressions, ditches (calcareous, marshy, roadside), fens (alpine, calcareous, seepage), flats (river), floodplains, forests, gravel bars, gullies (wet), krummholz, marshes (roadside), meadows (alluvial, wet), mudflats, prairies, roadsides, seeps (open), shores, slopes [to 50%] (forest, seepage), springs, streambeds, swamps (spruce), terraces (alluvial), thickets (willow), on tundra, and along the margins of lakes (marl), ponds, rivers, sloughs, and streams at elevations to 3744 m. The plants tolerate shade but thrive in full sun and more open exposures. They often grow in disturbed habitats such as along hiking trails, where they occur primarily within 1 m of the forest edge. The substrates (pH: 5.0–8.5) have been characterized as alluvium (gravelly, sandy), clayey sand, cobble, gravel, gravelly loam, gravelly sand, loam, marl, muck (silty), mud (cobbly), organic (to 100 cm deep), peat, rego gleysol, sand (fine), sandy gravel, silt (fluvial), silty loam, and silty muck. Flowering occurs from May to August with fruiting from June to September. Although facultative outcrossing is indicated by limited interspecific hybridization, the plants primarily are autogamous, with small, scentless green flowers and loosely organized pollinia. Typically, the pollinia (sometimes large massulae) rotate out of the anther and onto the stigma by gravity, often before the flowers open. Sometimes the pollinia break apart spilling the pollen onto the stigma. Water droplets also have been observed to draw streaming massulae out of the anthers continuously as the droplets evaporate. Genetic surveys (RAPD and PCR-RFLP markers) have indicated relatively low levels of variability (polymorphic loci: 0.40; H_e: 0.137) and populations that are highly structured locally due to limited seed dispersal (which presumably occurs by wind). Little is known of the seed ecology. **Reported associates:** *Abies grandis, Abies lasiocarpa, Aconitum delphiniifolium, Acorus calamus, Agrostis capillaris, Agrostis scabra, Agrostis stolonifera, Allium schoenoprasum, Alnus tenuifolia, Andromeda polifolia, Angelica arguta, Angelica atropurpurea, Antennaria pulcherrima, Anticlea elegans, Apocynum cannabinum, Arethusa bulbosa, Arnica cordifolia, Artemisia norvegica, Artemisia tilesii, Asclepias incarnata, Athyrium filix-femina, Betula glandulosa, Bistorta vivipara, Bromus ciliatus, Calamagrostis canadensis, Calamagrostis stricta, Caltha palustris, Calypso bulbosa, Campanula parryi, Carex aquatilis, Carex aurea, Carex bebbii, Platanthera hyperborea, Carex buxbaumii, Carex crawei, Carex disperma, Carex echinata, Carex emoryi, Carex hystericina, Carex interior, Carex jonesii, Carex leptalea, Carex livida, Carex macrochaeta, Carex membranacea, Carex nebrascensis, Carex pellita, Carex pluriflora, Carex podocarpa, Carex prairea, Carex rostrata, Carex sartwellii, Carex scirpoidea, Carex simulata, Carex sterilis, Carex striata, Carex tetanica, Carex utriculata, Carex viridula, Carex vulpinoidea, Carex xerantica, Cassiope mertensiana, Castilleja linariifolia, Castilleja pallida, Chamaecyparis nootkatensis, Chamerion latifolium, Cirsium arvense, Cirsium muticum, Comarum palustre, Conyza canadensis, Cornus sericea, Cypripedium passerinum, Dasiphora floribunda, Deschampsia cespitosa, Drosera rotundifolia, Dupontia fisheri, Eleocharis erythropoda, Eleocharis palustris, Eleocharis quinqueflora, Elymus trachycaulus, Empetrum nigrum, Epilobium leptophyllum, Equisetum arvense, Equisetum scirpoides, Erigeron flagellaris, Erigeron peregrinus, Eriophorum angustifolium, Eriophorum viridicarinatum, Eupatorium perfoliatum, Euthamia graminifolia, Eutrochium maculatum, Fragaria virginiana, Fraxinus pennsylvanica, Galearis rotundifolia, Galium tinctorium, Gentianopsis detonsa, Gentianopsis virgata, Glyceria striata, Gymnocarpium dryopteris, Hedysarum boreale, Helianthus maximiliani, Helianthus nuttallii, Heracleum sphondylium, Hordeum jubatum, Hylocomium splendens, Hypoxis hirsuta, Iris missouriensis, Juncus alpinoarticulatus, Juncus arcticus, Juncus balticus, Juncus brevicaudatus, Juncus drummondii, Juncus dudleyi, Juncus mertensianus, Juncus nodosus, Juncus regelii, Kobresia simpliciuscula, Larix laricina, Lepidium densiflorum, Leptarrhena pyrolifolia, Liatris ligulistylis, Leymus, Ligusticum porteri, Lilium philadelphicum, Linnaea borealis, Liparis loeselii, Lobelia kalmii, Lycopus americanus, Lycopus asper, Lythrum salicaria, Mentha arvensis, Menziesia ferruginea, Micranthes subapetala, Mimulus guttatus, Mimulus moschatus, Moneses uniflora, Muhlenbergia asperifolia, Muhlenbergia mexicana, Muhlenbergia richardsonis, Neottia cordata, Orthilia secunda, Oxypolis fendleri, Oxytropis campestris, Packera pseudaurea, Parnassia glauca, Parnassia palustris, Parthenocissus quinquefolia,*

Pascopyrum smithii, Pedicularis groenlandica, Pedicularis lanceolata, Picea engelmannii, Picea glauca, Picea mariana, Picea pungens, Pinguicula vulgaris, Pinus albicaulis, Pinus contorta, Platanthera clavellata, Platanthera dilatata, Platanthera huronensis, Platanthera obtusata, Platanthera orbiculata, Pleurozium schreberi, Poa pratensis, Polemonium occidentale, Populus angustifolia, Populus balsamifera, Populus deltoides, Populus trichocarpa, Potentilla anserina, Primula incana, Primula pauciflora, Prunus virginiana, Pseudotsuga menziesii, Pyrola asarifolia, Ranunculus macounii, Rhododendron groenlandicum, Rhynchospora capillacea, Rhytidiadelphus squarrosus, Ribes cereum, Rosa woodsii, Rubus idaeus, Rudbeckia hirta, Rudbeckia laciniata, Salix bebbiana, Salix brachycarpa, Salix candida, Salix drummondiana, Salix eastwoodiae, Salix exigua, Salix geyeriana, Salix interior, Salix lasiandra, Salix monticola, Salix myricoides, Salix petiolaris, Salix planifolia, Salix pulchra, Salix reticulata, Salix wolfii, Sanguisorba, Schoenoplectus acutus, Schoenoplectus pungens, Scirpus pallidus, Scutellaria galericulata, Senecio lugens, Senecio triangularis, Sisyrinchium idahoense, Sisyrinchium montanum, Solidago canadensis, Solidago missouriensis, Solidago ohioensis, Solidago patula, Solidago uliginosa, Sonchus arvensis, Spartina pectinata, Sphagnum, Sphenopholis obtusata, Spiranthes romanzoffiana, Swertia perennis, Symphyotrichum boreale, Symphyotrichum eatonii, Symphyotrichum lanceolatum, Symphyotrichum novae-angliae, Symphyotrichum puniceum, Symplocarpus foetidus, Teucrium canadense, Thalictrum alpinum, Thalictrum dasycarpum, Thalictrum occidentale, Therorhodion glandulosum, Tofieldia pusilla, Trichophorum cespitosum, Triglochin maritima, Triglochin palustris, Trollius albiflorus, Trollius laxus, Tsuga heterophylla, Typha latifolia, Typha ×glauca, Vaccinium scoparium, Valeriana sitchensis, Veratrum, Veronica americana, Vicia americana, Viola sororia, Zizia aurea.

5.4.9.2. *Platanthera blephariglottis* (Willd.) Lindl. occurs

in barrens (pine, till), bogs (hillside, peat, powerline, seepage, *Sphagnum*), ditches (flatwoods, roadside), draws (pineland), easements (powerline), fens (patterned [strings], poor), flatwoods (boggy), marshes, meadows, openings, roadsides (swampy), savannas (pine, pitcher plant), seepages, slopes (seepage), swamps, woodlands (open, pine), and along the margins of pocosins and streams at elevations to 600 m. Exposures can range from open to partially shaded conditions. The plants grow in acidic sites (substrate pH: 3.3–5.7; water pH: 4.5–6.0) where standing water levels are relatively low (−5 to −22 cm). The substrates are composed primarily of peat (55–86% organic matter; Ca: 542–4251 ppm; Mg: 448–909 ppm), sandy peat, or sand, but also have been categorized as Basinger (spodic psammaquents), Escambia (plinthaquic paleudults), Lynchburg (aeric paleaquults), Meadowbrook (grossarenic ochraqualfs), Pottsburg (grossarenic haplaquods), and Surrency (arenic umbric paleaquults). Flowering and fruiting extend from late May to September. The flowers are adapted for butterfly or moth (Insecta: Lepidoptera) pollination by their fragrance, nectar production within the spur, lack of nectar guides, and protruding anther. They stay

open for up to 10 days, with the pollen remaining viable for 5 days or more. The flowers are self-compatible and produce high fruit set (98–100%) when artificially selfed or pollinated geitonogamously. However, pollen vectors (insects) are necessary for fruit set in most natural populations. Anywhere from 4% to 66% of plants have been observed to flower in a population during a given year. Natural fruit set can vary from 26% to 87% ($\bar{x} = 62.4\%$) due to pollinator limitation. The pollinators include a variety of nectar-seeking insects (see *Use by wildlife* below), which are most active during hot, humid, sunny mornings. As the longer-tongued insects probe the flowers while seeking nectar, the pollinia become attached to their eyes by adhesive viscidia (some viscidia are not sticky) and are then transported to the next visited flower where they are deposited on the stigma. On average, the pollinators spend approximately 34 s visiting 3–4 flowers within a single inflorescence. If flowers on the same plant are visited after 60 s (the time for the pollinaria to rotate into the proper orientation for insertion into the stigma), self-pollination or geitonogamy can occur. Due to their different perianth coloration, only minor pollinator competition has been observed for plants growing sympatrically with the congeneric *Platanthera ciliaris*. The seeds are dispersed (by wind) from August throughout the early winter. They are thought not to germinate immediately after dispersal, but their longevity in the seed bank has not been determined. In nature, mycorrhizal Fungi are necessary for successful seed germination and seedling development. Seeds that were cold stratified (4°C) for 11 weeks have germinated successfully (to 66%) when incubated (on *Sphagnum* moss taken from the same habitat) at 20°C in darkness for 9 weeks. The plants reproduce vegetatively by division of their vertical fusiform tuber. They can be quite rare but also have been observed at densities of 4.0–7.3 plants m⁻². **Reported associates:** *Acer rubrum, Agalinis fasciculata, Andromeda polifolia, Andropogon, Arethusa bulbosa, Aristida, Arnoglossum sulcatum, Aronia arbutifolia, Aronia ×prunifolia, Arundinaria tecta, Balduina uniflora, Bartonia paniculata, Bartonia virginica, Betula michauxii, Bigelowia nudata, Calamagrostis canadensis, Calamagrostis pickeringii, Calopogon tuberosus, Carex exilis, Carex flava, Carex glaucescens, Carex interior, Carex limosa, Carex livida, Carex oligosperma, Carex pauciflora, Carex rostrata, Carex stricta, Carex utriculata, Carphephorus odoratissimus, Carphephorus paniculatus, Centella asiatica, Chamaecrista nictitans, Chamaedaphne calyculata, Cladium mariscoides, Cleistesiopsis divaricata, Clethra alnifolia, Clintonia borealis, Coleataenia longifolia, Coptis trifolia, Coreopsis gladiata, Cornus canadensis, Ctenium, Cyperus flavescens, Desmodium marilandicum, Dichanthelium ensifolium, Dichanthelium scoparium, Doellingeria umbellata, Drosera anglica, Drosera intermedia, Drosera rotundifolia, Dulichium arundinaceum, Eleocharis flavescens, Eleocharis tenuis, Empetrum nigrum, Erigeron vernus, Eriocaulon decangulare, Eriophorum angustifolium, Eriophorum tenellum, Eriophorum virginicum, Eupatorium pilosum, Eurybia compacta, Euthamia graminifolia, Fuirena squarrosa, Galactia regularis, Gaultheria procumbens, Gaylussacia* *dumosa, Gentiana linearis, Geocaulon lividum, Helianthus floridanus, Hypericum brachyphyllum, Hypericum crux-andreae, Hyptis alata, Ilex glabra, Ilex mucronata, Iris versicolor, Juncus filiformis, Juncus pelocarpus, Juncus trigonocarpus, Juniperus communis, Kalmia angustifolia, Kalmia polifolia, Lachnanthes caroliniana, Larix laricina, Lespedeza hirta, Lilium catesbaei, Linnaea borealis, Lobelia glandulosa, Lonicera villosa, Lycopodiella alopecuroides, Lycopodium obscurum, Lygodium palmatum, Lyonia ligustrina, Lyonia lucida, Magnolia virginiana, Maianthemum canadense, Maianthemum trifolium, Marshallia graminifolia, Menyanthes trifoliata, Myrica cerifera, Myrica gale, Nyssa biflora, Nyssa sylvatica, Oclemena nemoralis, Oclemena ×blakei, Oenothera fruticosa, Orontium aquaticum, Osmunda regalis, Osmundastrum cinnamomeum, Panicum hemitomon, Picea mariana, Pinckneya bracteata, Pinus palustris, Pinus serotina, Pinus taeda, Platanthera ciliaris, Platanthera clavellata, Platanthera cristata, Platanthera dilatata, Pluchea baccharis, Pogonia ophioglossoides, Polygala chapmanii, Polygala cruciata, Polygala lutea, Polygala nana, Polytrichum commune, Pteridium aquilinum, Rhamnus frangula, Rhexia lutea, Rhexia mariana, Rhexia nashii, Rhexia petiolata, Rhexia virginica, Rhododendron canadense, Rhododendron groenlandicum, Rhynchospora alba, Rhynchospora capitellata, Rhynchospora inexpansa, Rosa nitida, Rubus hispidus, Sabatia brevifolia, Sarracenia minor, Sarracenia purpurea, Sarracenia rubra, Schizaea pusilla, Scleria pauciflora, Scleria reticularis, Scleria triglomerata, Solidago fistulosa, Solidago nemoralis, Solidago odora, Solidago uliginosa, Sphagnum capillifolium, Sphagnum fallax, Sphagnum magellanicum, Sphagnum palustre, Sphagnum papillosum, Sphagnum pulchrum, Sphagnum rubellum, Spiraea alba, Spiraea tomentosa, Stenanthium leimanthoides, Syngonanthus flavidulus, Taxodium ascendens, Thalictrum pubescens, Tiedemannia filiformis, Toxicodendron vernix, Triadenum virginicum, Trichophorum cespitosum, Trientalis borealis, Utricularia cornuta, Utricularia subulata, Vaccinium angustifolium, Vaccinium corymbosum, Vaccinium macrocarpon, Vaccinium oxycoccos, Viburnum nudum, Viola cucullata, Viola macloskeyi, Woodwardia areolata, Woodwardia virginica, Xyris ambigua.*

5.4.9.3. *Platanthera chorisiana* (Cham.) Rchb. f. is found in

bogs (hanging, minerotrophic, open, peat, perched, seepage), fens, gullies (meadow), heaths (open), hollows (*Sphagnum*), hummocks (muskeg), krummholz, meadows (alpine, bog, subalpine, wet), on muskeg (alpine), slopes [to 85°] (meadow, seepy), tundra (alpine), and along the margins of lakes and streams at elevations to 1433 m. The exposures often are indicated as indirect shade. The substrates are saturated and have been described as gravel, humus, moss, orthic numic gleysol, peat (moss, sedge), and peaty colluvium. Flowering and fruiting occur from July to August. In North American populations, the self-compatible flowers are autogamous due to their pollen massulae lying in direct contact with the stigma and germinating *in situ*. Geitonogamy and xenogamy also can occur, presumably mediated by small insects. In Japan, the

flowers were found to be pollinated by false blister beetles (Insecta: Coleoptera: Oedemeridae: *Oedemeronia lucidicollis*), which were able to reach the nectar within the short floral spur. Other life-history information is lacking. **Reported associates:** *Alnus viridis, Athyrium filix-femina, Callitropsis nootkatensis, Carex aquatilis, Carex macrochaeta, Carex obnupta, Cassiope, Coptis aspleniifolia, Cryptogramma sitchensis, Deschampsia cespitosa, Drosera rotundifolia, Gentiana douglasiana, Geranium erianthum, Geum calthifolium, Harrimanella stelleriana, Ligusticum calderi, Luetkea pectinata, Lysichiton americanus, Nephrophyllidium crista-galli, Parasenecio auriculata, Phyllodoce glanduliflora, Picea sitchensis, Pinus contorta, Primula jeffreyi, Rhynchospora, Sanguisorba menziesii, Sinosenecio newcombei, Sphagnum, Trichophorum cespitosum, Vaccinium uliginosum, Vahlodea atropurpurea.*

5.4.9.4. *Platanthera clavellata* (Michx.) Luer inhabits bogs

(pineland, slope, *Sphagnum*), depressions (wet, woodland), ditches (roadside), fens (poor, sloping), hummocks (*Sphagnum*), marshes, meadows (moist, sandy, *Sphagnum*, peaty), prairies (moist, wet), roadsides (low), seeps (acid spring, boggy, *Sphagnum*), springs, springheads, streambeds, swales, swamps (cypress, sandhill streamhead, seepage), thickets (wet), woodlands (deciduous, low, rich), and the margins of lakes, ponds (sinkhole), and streams at elevations to 1245 m. The plants are colonizers, but generally grow in shaded exposures where standing water levels are from −11 to +4 cm. The substrates are acidic (pH: 3.9–6.3; $\bar{x}$ =4.8–5.7; 30–85% organic matter; Ca: 1000–6000 ppm; Fe: 0–80000 ppm; K: 0–30000 ppm; Mg: 400–1600 ppm; N: 5000–20000 ppm; P: 150–750 ppm), and have been described as gravel, muck, peat (sand (organic), and sandy loam (red). The plants also can establish in the bases of ferns (e.g., *Osmunda* spp.) or in rock crevices. Flowering occurs from June to August. As the young plants emerge in the spring, their leaves protect the flower buds until the inflorescence expands. They produce 4–10 flowers on average. The flowers are small, not showy, and lack a floral fragrance. They only are partially resupinate (rotated about 90°) and primarily are self-pollinating. During anthesis, the friable pollen masses break apart and drop onto the receptive stigmatic lobes. Occasional hybridization (see *Systematics*) indicates that xenogamy also must occur to some extent and the pollinia are known to be deposited on the proboscises of some visiting insects (Insecta). The fruits mature from July to November. Fruit set typically is high, with each plant yielding from 3 to 10 mature capsules. The seeds are dispersed by wind. Freshly collected seeds have germinated *in vitro* (within 21 days) after inoculation with a mycorrhizal symbiont (Basidiomycota: Tulasnellaceae: *Epulorhiza inquilina*) followed by a continuous white light treatment (1 week at 21°C) and subsequent total darkness treatment (2 weeks at 21°C). Germination rates have been found to vary substantially among geographical regions, e.g., from 16.2% (Georgia) to 76.4–81.5% (Tennessee and South Carolina populations). Germination of seeds stored at 6°C was somewhat higher (62.3%) than those stored at 7°C (57.3%). The seeds have remained viable for more than 5 years when dried (to 5%

water content) immediately after collection and stored at low or sub-zero temperatures. Perennation is facilitated by a fleshy tuber, which can occur at substrate depths from 6 to 11 cm and is colonized by several mycorrhizal Fungi (Ascomycota: Hypocreaceae: *Trichoderma polysporum*; Vibrisseaceae: *Phialocephala fortinii*; Basidiomycota: Ceratobasidiaceae: *Ceratorhiza*; Tulasnellaceae: *Epulorhiza inquilina*). **Reported associates:** *Abies fraseri, Acer rubrum, Acer saccharum, Agrimonia parviflora, Aletris farinosa, Alnus rugosa, Alnus serrulata, Amelanchier canadensis, Amianthium muscitoxicum, Anemone quinquefolia, Angelica triquinata, Arethusa bulbosa, Arisaema triphyllum, Arnoglossum ovatum, Aronia arbutifolia, Aronia melanocarpa, Arundinaria gigantea, Asimina triloba, Athyrium filix-femina, Bartonia virginica, Betula alleghaniensis, Bignonia capreolata, Boehmeria cylindrica, Brachyelytrum erectum, Calamagrostis canadensis, Calamagrostis coarctata, Calamagrostis pickeringii, Calopogon tuberosus, Calypso bulbosa, Campanula, Carex aestivalis, Carex albolutescens, Carex atlantica, Carex bromoides, Carex brunnescens, Carex canescens, Carex crinita, Carex debilis, Carex echinata, Carex folliculata, Carex granularis, Carex gynandra, Carex hystericina, Carex intumescens, Carex leptalea, Carex limosa, Carex lonchocarpa, Carex lurida, Carex magellanica, Carex oligosperma, Carex pauciflora, Carex ruthii, Carex wiegandii, Carpinus caroliniana, Cephalanthus occidentalis, Chamaedaphne calyculata, Chasmanthium laxum, Chelone glabra, Chelone lyonii, Chionanthus virginicus, Cinna arundinacea, Cinna latifolia, Cladium mariscoides, Clethra alnifolia, Comandra umbellata, Cornus foemina, Cypripedium acaule, Cypripedium reginae, Dasiphora floribunda, Decumaria barbara, Dennstaedtia punctilobula, Deparia acrostichoides, Dichanthelium scabriusculum, Diervilla sessilifolia, Drosera intermedia, Drosera rotundifolia, Dryopteris campyloptera, Dryopteris carthusiana, Dryopteris cristata, Eleocharis flavescens, Eleocharis tenuis, Eleocharis tortilis, Eriocaulon aquaticum, Eriocaulon decangulare, Eriophorum tenellum, Eriophorum virginicum, Eriophorum viridicarinatum, Eryngium integrifolium, Eubotrys racemosa, Euonymus americanus, Eupatorium resinosum, Euthamia graminifolia, Eutrochium fistulosum, Fagus grandifolia, Fraxinus nigra, Fuirena squarrosa, Galium uniflorum, Gaultheria hispidula, Gaylussacia frondosa, Gentiana saponaria, Geranium maculatum, Glyceria melicaria, Glyceria melicaria, Glyceria obtusa, Glyceria striata, Goodyera pubescens, Gratiola neglecta, Gratiola virginiana, Hydrocotyle verticillata, Hypericum canadense, Hypericum ellipticum, Hypericum mutilum, Ilex laevigata, Ilex mucronata, Ilex opaca, Ilex verticillata, Itea virginica, Juncus acuminatus, Juncus caesariensis Juncus effusus, Juncus marginatus, Juncus pelocarpus, Juncus scirpoides, Juncus stygius, Juncus subcaudatus, Kalmia angustifolia, Kalmia polifolia, Larix laricina, Leersia oryzoides, Leersia virginica, Leucothoe axillaris, Lilium superbum, Lindera benzoin, Lindera subcoriacea, Liparis loeselii, Liquidambar styraciflua, Liriodendron tulipifera, Lobelia cardinalis, Lobelia nuttallii, Ludwigia palustris,* *Lycopodiella appressa, Lycopodiella inundata, Lycopodium clavatum, Lycopodium obscurum, Lycopodium digitatum, Lycopus cokeri, Lygodium palmatum, Lygodium palmatum, Lyonia ligustrina, Magnolia macrophylla, Magnolia virginiana, Maianthemum canadense, Malaxis unifolia, Melanthium virginicum, Mitchella repens, Nyssa biflora, Nyssa sylvatica, Oclemena acuminata, Oclemena nemoralis, Onoclea sensibilis, Osmunda regalis, Osmundastrum cinnamomeum, Oxalis montana, Oxydendrum arboreum, Oxypolis rigidior, Packera schweinitziana, Panicum verrucosum, Parnassia asarifolia, Pedicularis lanceolata, Persea palustris, Persicaria arifolia, Persicaria sagittata, Phlox glaberrima, Picea mariana, Picea rubens, Pinus rigida, Platanthera aquilonis, Platanthera blephariglottis, Platanthera ciliaris, Platanthera cristata, Platanthera dilatata, Platanthera hyperborea, Platanthera lacera, Poa autumnalis, Poa paludigena, Pogonia ophioglossoides, Polygala cruciata, Populus tremuloides, Pycnanthemum virginianum, Pyrola asarifolia, Quercus alba, Quercus michauxii, Quercus montana, Quercus rubra, Rhexia virginica, Rhododendron canadense, Rhododendron groenlandicum, Rhododendron maximum, Rhododendron viscosum, Rhynchospora capitellata, Rhynchospora corniculata, Rhynchospora glomerata, Rubus hispidus, Rudbeckia fulgida, Saccharum giganteum, Sagittaria latifolia, Sagittaria macrocarpa, Salix sericea, Sarracenia flava, Sarracenia purpurea, Sassafras albidum, Saururus cernuus, Schoenoplectus etuberculatus, Scirpus cyperinus, Scirpus polyphyllus, Scleria triglomerata, Solidago ohioensis, Solidago patula, Solidago rugosa, Solidago uliginosa, Sorbus americana, Sphagnum, Spiraea alba, Spiraea tomentosa, Spiranthes cernua, Symphyotrichum puniceum, Symplocarpus foetidus, Thalictrum pubescens, Thelypteris noveboracensis, Thelypteris palustris, Thuja occidentalis, Tipularia discolor, Toxicodendron radicans, Toxicodendron vernix, Tradescantia ohiensis, Triadenum virginicum, Triadenum walteri, Trichophorum alpinum, Triglochin maritima, Triglochin palustris, Trillium catesbaei, Utricularia cornuta, Utricularia gibba, Utricularia purpurea, Utricularia resupinata, Uvularia sessilifolia, Vaccinium corymbosum, Valeriana sitchensis, Veratrum viride, Viburnum acerifolium, Viburnum dentatum, Viburnum nudum, Viola cucullata, Viola lanceolata, Viola macloskeyi, Viola primulifolia, Viola sororia, Vitis riparia, Vitis rotundifolia, Waldsteinia lobata, Woodwardia areolata, Woodwardia virginica, Xanthorhiza simplicissima, Xyris jupicai, Xyris montana, Xyris scabrifolia, Xyris torta, Zizia aurea.*

5.4.9.5. *Platanthera cristata* (**Michx.**) **Lindl.** grows in bogs (hillside, quaking, seepage, *Sphagnum*), depressions (wet, woodland), ditches (roadside), fens, flatwoods (mesic, pine, wet), floodplains, hammocks, marshes, meadows (open, wet), pine barrens (moist), pinelands (annually burned, moist, wet), prairies, ravines, rights-of-way (powerline), roadsides (sandy), savannas (pine), seeps (acid, bayhead, spring), slopes (seepage), swales (roadside, *Sphagnum*), swamps (cypress, roadside), thickets (boggy), woodlands (wet), and along the margins or shores of canals, lakes, pocosins, ponds (intermittent, cypress), and streams at elevations to 503 m. Exposures vary from open to shaded sites. The substrates have been characterized as clay, Dorovan-Johnston series, Doucette (arenic plinthis paleudults), Meadowbrook (grossarenic ochraqualfs), peat, peaty muck, Plummer (grossarenic paleaquults), Pottsburg (grossarenic haplaquods), Ridgewood (aquic quartzipsamments), Rosenwall (aquic hapludults), sand, sandy clay, sandy loam, sandy peat, and Yemassee (aeric ochraquults). Flowering occurs from June to September. Pollination occurs diurnally. The showy perianths, with their fringed, orange lip, attract reward-seeking bumblebees (Insecta: Hymenoptera: Apidae: *Bombus pensylvanicus*), which pollinate them. The viscidia of the closely set pollinia adhere to the bee's head as it enters the flower to forage. The seeds are dispersed by wind. There is little additional information available on the floral biology or seed ecology. The plants perennate by means of their fleshy, tuberous roots, which produce fibrous rootlets. **Reported associates:** *Acer rubrum, Aletris aurea, Alnus serrulata, Andropogon, Anemone quinquefolia, Apteria aphylla, Arethusa bulbosa, Aristida stricta, Asimina parviflora, Bartonia paniculata, Calydorea coelestina, Carex baileyi, Carex collinsii, Carex glaucescens, Carex intumescens, Carex joorii, Carex leptalea, Carya glabra, Clethra alnifolia, Cliftonia monophylla, Ctenium aromaticum, Cyperus haspan, Cyrilla racemiflora, Deparia acrostichoides. Dichanthelium scabriusculum, Ditrysinia fruticosa, Doellingeria sericocarpoides, Drosera capillaris, Drosera intermedia, Drosera tracyi, Dulichium arundinaceum, Eleocharis tortilis, Elephantopus nudatus, Eriocaulon decangulare, Eryngium integrifolium, Eryngium yuccifolium, Eubotrys racemosa, Eupatorium rotundifolium, Eutrochium fistulosum, Fuirena bushii, Fuirena squarrosa, Gaylussacia tomentosa, Gentiana saponaria, Gratiola pilosa, Helianthus angustifolius, Hydrolea ovata, Hypericum crux-andreae, Hypericum galioides, Hypericum gymnanthum, Hypericum hypericoides, Ilex ambigua, Ilex amelanchier, Ilex myrtifolia, Ilex opaca, Ilex verticillata, Illicium floridanum, Isotria verticillata, Juncus biflorus, Juncus caesariensis, Juncus effusus, Kalmia latifolia, Lechea minor, Lespedeza capitata, Leucothoe axillaris, Liatris spicata, Liquidambar styraciflua, Liriodendron tulipifera, Lobelia cardinalis, Ludwigia hirtella, Ludwigia pilosa, Lycopodiella alopecuroides, Lycopodiella appressa, Lycopodiella caroliniana, Lycopus virginicus, Lyonia ligustrina, Magnolia grandiflora, Magnolia virginiana, Mitchella repens, Mitreola petiolata, Mitreola sessilifolia, Muhlenbergia expansa, Myrica heterophylla, Nyssa biflora, Nyssa sylvatica, Osmanthus americanus, Osmunda regalis, Osmundastrum cinnamomeum, Oxypolis rigidior, Packera crawfordii, Paspalum laeve, Paspalum praecox, Pinus elliottii, Pinus palustris, Platanthera chapmanii, Platanthera ciliaris, Platanthera clavellata, Pluchea foetida, Pogonia ophioglossoides, Polygala cruciata, Polygala lutea, Polygala mariana, Polygala nana, Polygala nuttallii, Pteridium aquilinum, Quercus geminata, Quercus laevis, Quercus virginiana, Rhexia mariana, Rhexia virginica, Rhododendron canescens, Rhododendron viscosum, Rhynchospora capitellata, Rhynchospora glomerata, Rhynchospora gracilenta, Rhynchospora inexpansa, Rhynchospora macra, Rhynchospora rariflora, Rhynchospora stenophylla, Sarracenia flava, Scleria muehlenbergii, Serenoa repens, Sisyrinchium atlanticum, Smilax laurifolia, Solidago patula, Solidago rugosa, Sparganium americanum, Sphagnum, Taxodium ascendens, Toxicodendron vernix, Tragia, Trillium sessile, Utricularia cornuta, Utricularia gibba, Vaccinium arboreum, Vaccinium stamineum, Viburnum nudum, Viola cucullata, Woodwardia areolata, Woodwardia virginica, Xanthorhiza simplicissima, Xyris ambigua, Xyris baldwiniana, Xyris torta, Zigadenus.*

5.4.9.6. ***Platanthera huronensis* (Nutt.) Lindl.** occurs in or on beaches, bogs (spruce, *Sphagnum*), bottomlands (riparian), depressions (open), ditches (roadside), fens, flats (coastal), floodplains, gravel bars, gulches, heaths, marshes, meadows (alpine, beaver, bench, hot spring, wet), mires, muskeg, roadsides, seeps (alkaline, mossy), shores, slopes [to 30°] (quartzite, seepage), springs, swamps (cedar), thickets (willow), tundra, washes, woodlands (alluvial, coniferous, wet), and along the margins or shores of gravel pits, lakes, rivers, sinkholes, sloughs, and streams at elevations to 3566 m. The exposures can range from sun to partial shade or shade. The plants usually do not occur in sites with standing water. The substrates are acidic (pH: 5.0–6.5) and include alluvium (glacial), clay, clayey silt, cobble, colluvium (Morrison), gravel, humus, loam (humic, wet), loamy cryoboralf, muck, peat, sand, sandy loam, silt, and silty loam. Flowering occurs from May to August with fruiting from June to August. The flowers are described as being strongly fragrant. The floral biology of this allotetraploid derivative of *P. aquilonis* (autogamous) and *P. dilatata* (outcrossing) has not yet been elucidated, but pollination is thought to be carried out by small insects like mosquitoes (Insecta: Diptera: Culicidae). Genetic analyses (using ISSR markers) have shown levels of genetic variability that are intermediate to those of the parental species at the population level and lower than both at the species level, i.e., gene diversity (H_E): 0.119 (population level); 0.172 (species level); percentage of polymorphic bands (PPB): 32.64% (population level); 43.0% (species level); G_{ST}: 0.36. Further details on the reproductive ecology of this species are not available. The plants persist vegetatively by means of a tuberous base with radiating fleshy roots. **Reported associates:** *Abies grandis, Abies lasiocarpa, Acer glabrum, Achillea millefolium, Aconitum maximum, Actaea rubra, Agrostis scabra, Alnus tenuifolia, Alnus viridis, Anaphalis margaritacea, Andromeda polifolia, Anemone narcissiflora, Angelica lucida, Antitrichia curtipendula, Aralia nudicaulis, Arctostaphylos uva-ursi, Arctous rubra, Arnica latifolia, Arnica longifolia, Arnica mollis, Athyrium filix-femina, Betula glandulosa, Betula pumila, Bistorta bistortoides, Bistorta vivipara, Botrychium virginianum, Calamagrostis canadensis, Campanula parryi, Cardamine pratensis, Carex aquatilis, Carex canescens, Carex capillaris, Carex chordorrhiza, Carex disperma, Carex exilis, Carex gynocrates, Carex hallii, Carex interior, Carex jonesii, Carex leptalea, Carex limosa, Carex lyngbyei, Carex macrochaeta, Carex microptera, Carex nebrascensis, Carex neurophora, Carex norvegica, Carex pluriflora, Carex rostrata, Carex stipata, Carex utriculata, Castilleja sulphurea, Castilleja* *unalaschcensis, Cerastium arvense, Chamaecyparis thyoides, Chamaedaphne calyculata, Chamerion angustifolium, Chamerion latifolium, Cicuta maculata, Circaea alpina, Cladina rangiferina, Cladina stellaris, Claytonia sibirica, Conioselinum chinense, Cornus sericea, Cornus suecica, Crataegus douglasii, Crepis runcinata, Cypripedium guttatum, Cypripedium parviflorum, Dactylis glomerata, Dactylorhiza viridis, Dasiphora floribunda, Deschampsia cespitosa, Dicranum scoparium, Drosera linearis, Drosera rotundifolia, Eleocharis palustris, Empetrum nigrum, Equisetum arvense, Equisetum fluviatile, Equisetum hyemale, Equisetum variegatum, Erigeron peregrinus, Eriophorum angustifolium, Eriophorum vaginatum, Euthamia graminifolia, Festuca rubra, Floerkea proserpinacoides, Fritillaria, Gentianopsis thermalis, Geocaulon lividum, Geranium erianthum, Geranium richardsonii, Geum rivale, Glyceria striata, Gymnocarpium dryopteris, Heracleum sphondylium, Hylocomium splendens, Hypericum scouleri, Juncus saximontanus, Kalmia polifolia, Larix laricina, Larix occidentalis, Lemna minor, Leptarrhena pyrolifolia, Leucanthemum vulgare, Ligusticum porteri, Linnaea borealis, Lobaria linita, Lonicera involucrata, Lupinus nootkatensis, Luzula multiflora, Maianthemum dilatatum, Maianthemum stellatum, Maianthemum trifolium, Menyanthes trifoliata, Mertensia ciliata, Mimulus lewisii, Mimulus moschatus, Mitella nuda, Moehringia lateriflora, Muhlenbergia filiformis, Orobanche fasciculata, Osmorhiza berteroi, Oxypolis fendleri, Packera debilis, Packera pseudaurea, Parnassia palustris, Pedicularis groenlandica, Pedicularis verticillata, Peltigera membranacea, Petasites frigidus, Phegopteris connectilis, Phleum alpinum, Picea engelmannii, Picea glauca, Picea mariana, Pinus contorta, Platanthera dilatata, Pleurozium schreberi, Poa palustris, Poa pratensis, Polemonium, Polypodium virginianum, Populus angustifolia, Populus balsamifera, Potentilla anserina, Prenanthes alata, Primula pauciflora, Pyrola, Ranunculus occidentalis, Ranunculus uncinatus, Rhododendron groenlandicum, Rhododendron tomentosum, Rhytidiadelphus squarrosus, Rhytidiadelphus triquetrus, Ribes, Rosa woodsii, Rubus arcticus, Rubus chamaemorus, Rudbeckia laciniata, Salix amygdaloides, Salix arctica, Salix bebbiana, Salix boothii, Salix candida, Salix drummondiana, Salix exigua, Salix lasiandra, Salix pedicellaris, Salix planifolia, Salix pseudomonticola, Salix wolfii, Sanguisorba stipulata, Sanicula marilandica, Sanionia orthothecioides, Sanionia uncinata, Sarracenia purpurea, Senecio triangularis, Solidago canadensis, Solidago gigantea, Solidago lepida, Solidago multiradiata, Solidago ohioensis, Solidago uliginosa, Sphagnum squarrosum, Spiranthes romanzoffiana, Streptopus amplexifolius, Symphyotrichum lanceolatum, Therorhodion camtschaticum, Thuja plicata, Triantha glutinosa, Trichophorum alpinum, Trichophorum cespitosum, Trientalis europaea, Trifolium pratense, Triglochin maritima, Trisetum spicatum, Tsuga heterophylla, Vaccinium ovalifolium, Vaccinium oxycoccos, Vaccinium vitis-idaea, Valeriana occidentalis, Veratrum album, Veronica americana, Veronica wormskjoldii, Viburnum edule, Viola langsdorffii, Xerophyllum tenax Zizia.*

5.4.9.7. ***Platanthera integra*** **(Nutt.) A. Gray ex L.C. Beck** inhabits bogs (hanging, hillside, pitcher plant, roadside, quaking, seepage, streamhead), depressions (boggy, peaty, pine savanna), fens (poor), flatwoods (damp, marshy, pine), meadows (open, swampy), pine barrens (wet), pinelands (burned, low, moist), prairies (boggy, wet), savannas (boggy, low, pine), seeps, swamps (acid), and woodlands (sandy, wet) at elevations to 128 m. The sites are susceptible to fires, which periodically remove competing overstory vegetation. The substrates typically are nitrogen-deficient and have been characterized as Alapaha (arenic plinthic paleaquults), Escambia (plinthaquic paleudults), loamy sand, Lynchburg (aeric paleaquults), muck, Plummer (grossarenic paleaquults), Pottsburg (grossarenic haplaquods), Rains (typic paleaquults), Rutledge-Pamlico (terric medisaprists), sand, sandy loam, sandy peat, Stringtown (typic hapludults), and Valkaria (spodic psammaquents). Flowering occurs from July to September, with fruiting from September to October. Pollination is diurnal, with the pollinia deposited on the proboscises of insects (Insecta) such as bees (e.g., Hymenoptera: Apidae: *Bombus pensylvanicus*) and butterflies (Lepidoptera). Because the flowers are self-compatible, selfing also presumably occurs at some level. The seeds float and are thought to be dispersed by water and wind. The seeds have germinated (symbiotically) using the following procedure: freshly collected capsules were kept for 7–10 days at 22°C–24°C and 0% relative humidity (RH); then, after capsule dehiscence, the seeds were stored in darkness at −7°C (0% RH) for 28 months or more. The seeds were then surface-scarified using bleach (1:1:1 v: deionized H_2O, absolute EtOH, 5.25% NaOCl) and then inoculated with mycorrhizal cultures (e.g., Fungi: Basidiomycota: Tulasnellaceae: *Epulorhiza inquilina*) isolated from successfully growing congeners. The seeds were then incubated in darkness (at 22°C–26°C) or after an initial white light exposure (12/12 hr light/dark) for 7 days, followed by dark incubation (at 22°C–26°C). Longer periods of stratification resulted in higher germination rates. **Reported associates:** *Acer rubrum, Aletris aurea, Andropogon, Anthaenantia rufa, Aristida stricta, Arnoglossum ovatum, Calamagrostis pickeringii, Calopogon barbatus, Calopogon multiflorus, Calopogon pallidus, Calopogon tuberosus, Carphephorus paniculatus, Carphephorus pseudoliatris, Cleistesiopsis bifaria, Cleistesiopsis divaricata, Ctenium aromaticum, Cyperus, Dichanthelium, Drosera, Eriocaulon compressum, Eriocaulon decangulare, Eriophorum tenellum, Eupatorium resinosum, Gymnopogon, Helianthus heterophyllus, Hypericum majus, Hypericum setosum, Juncus caesariensis, Lachnanthes caroliniana, Lilium catesbaei, Lophiola aurea, Ludwigia, Lycopodium, Magnolia virginiana, Marshallia grandiflora, Muhlenbergia expansa, Narthecium americanum, Nyssa biflora, Persea palustris, Pinguicula caerulea, Pinus elliottii, Pinus palustris, Pinus serotina, Pinus taeda, Platanthera blephariglottis, Platanthera ciliaris, Platanthera nivea, Pleea tenuifolia, Pogonia ophioglossoides, Polygala cymosa, Polygala ramosa, Rhexia, Rhynchospora knieskernii, Rhynchospora macra, Rhynchospora plumosa, Rhynchospora stenophylla, Sabatia, Sarracenia alata, Sarracenia flava, Sarracenia leucophylla,*

Sarracenia minor, Sarracenia minor, Sarracenia rubra, Schizachyrium, Scirpus longii, Scleria, Spiranthes laciniata, Spiranthes longilabris, Spiranthes praecox, Liquidambar styraciflua, Stenanthium densum, Stenanthium leimanthoides, Stokesia laevis, Taxodium ascendens, Taxodium distichum, Tiedemannia filiformis, Triantha racemosa, Utricularia resupinata, Verbesina chapmanii, Xyris fimbriata, Xyris stricta, Zigadenus glaberrimus.

5.4.9.8. *Platanthera integrilabia* (Correll) Luer grows in bogs (seepage, *Sphagnum*), flats (wet, wooded), marshes, roadsides, slopes (seepage), swales, woodlands (moist), and along the margins of streams at elevations to 503 m. Exposures tend to be under partial shade. The substrates include gravelly sand (acidic, organic), peat, and sandy peat. Flowering occurs from June to August. Flower production can be low (e.g., only 3.7% of plant flowering in one population studied). When flowering, the plants average 4.7 capsules per inflorescence and 3,433 seeds per capsule. The floral spurs contain from 0.8 to 19.9 μl of nectar ($\bar{x} = 4.4$ μl), which comprises 10–23% ($\bar{x} = 18.9\%$) sugar. Although the fragrant, nectariferous flowers are adapted for pollination by hawkmoths (Insecta: Lepidoptera: Sphingidae), they are pollinated diurnally (1100–1600 hr) by butterflies (Insecta: Lepidoptera: Hesperiidae: *Epargyreus clarus*; Papilionidae: *Papilio glaucus, P. troilus*), but possibly to some degree by the nocturnal hawkmoths. Pollinator activity is highest during warm (20°C–27°C) and humid (51–90% RH) conditions. However, pollination is inefficient and natural seed production typically is low (e.g., 6.9–20.3%; but up to 57% in one large population). The pollinaria do not attach readily to the compound eyes of these insects and often are not transferred until the flowers have been re-visited several successive times. Those insects with more widely spaced eyes (e.g., hawkmoths) seem to be more effective at removing the pollinaria. The flowers are self-compatible and set seed readily when artificially self-pollinated. Although the flowers usually appear to be outcrossed primarily, autogamous (or insect-mediated) self-pollination is thought to characterize at least some natural populations. Genetic analyses indicated that most of the variation (79%) was partitioned within populations rather than between populations (21%), which gives no indication of restricted gene flow. The seeds have 57.55% airspace, have a mean water content of 42.11%, and lose water at a rate of about 1.23% hr^{-1}. They are dispersed primarily by wind. High germination rates (73.1%) have been obtained for seeds that were stored at 6°C, inoculated with endophytic fungal isolates obtained from several congeners, and then incubated in darkness (at 22°C) for up to 11 months. Germination rates are higher for seed obtained from large populations compared to those gathered from small, isolated populations. Incubation with fungal isolates was found to be less important in promoting seed germination than in promoting successful seedling establishment and survival. Light also plays a significant role in seed germination. Germination was inhibited in seeds exposed to a 16-hr daily photoperiod, even if preceded by a week of storage in total darkness. The highest symbiotic germination rates have been obtained by exposing the seeds to a 16-hr daily photoperiod for 1 week after the fungal

inoculation, and then transferring them to continuous darkness. Experimental germination studies have indicated that the seeds can remain viable for at least 20 months. A study of photosynthetic light responses and water-use efficiency demonstrated that the plants can adapt to contrasting light and soil moisture conditions. The plants persist by their fleshy, tuberous root. **Reported associates:** *Acer rubrum, Acer saccharum, Alnus serrulata, Aronia arbutifolia, Athyrium filix-femina, Betula lenta, Calamagrostis canadensis, Carex baileyi, Carex intumescens, Carex joorii, Carex leptalea, Chelone lyonii, Cornus florida, Drosera, Fagus grandifolia, Gentiana saponaria, Helenium autumnale, Hydrangea arborescens, Juncus effusus, Lindera benzoin, Liquidambar styraciflua, Liriodendron tulipifera, Lobelia cardinalis, Lycopus virginicus, Magnolia virginiana, Micranthes petiolaris, Nyssa sylvatica, Oenothera fruticosa, Osmunda regalis, Osmundastrum cinnamomeum, Oxydendrum arboreum, Oxypolis rigidior, Parnassia asarifolia, Platanthera clavellata, Platanthera cristata, Prunus serotina, Rhododendron arborescens, Rhododendron maximum, Rudbeckia laciniata, Smilax laurifolia, Solidago patula, Solidago rugosa, Sphagnum, Tilia americana, Tsuga canadensis, Viola primulifolia, Woodwardia areolata, Xanthorhiza simplicissima.*

5.4.9.9. *Platanthera leucophaea* (Nutt.) Lindl. occurs in bogs (*Sphagnum*), ditches, fens (poor), marshes, meadows (acid, floodplain, sedge), pannes (lakeshore), patterned wetlands (flarks), prairies (low, mesic, sand, wet), swales (mesic), and along the shores of lakes at elevations to 235 m. The plants require open, sunny exposures and are characteristic of disturbed or successional sites, which often are disturbed routinely by fire or drought. The substrates (gravel, peat, silty loam, and sand), span a broad ecological gradient, including circumneutral base-rich organic soils, more alkaline soils with lower organic content, and somewhat acid, nutrient-poorer conditions (pH: 5.0–7.6; Ca: 1300–11311 ppm; K: 71–1235 ppm; Mg: 430–1397 ppm; Mn: 8.8–150 ppm; P: 2.5–72.5 ppm; cation exchange capacity [CEC]: 11.0–32.6; organic matter: 2.0–95.6%). Their composition can vary from 6% to 72.5% sand, 25.2–79.9% silt, and 2.3–26.2% clay. Flowering occurs from June to August with each episode lasting for about a week. Pollination is nocturnal and occurs when the pollinia become attached to the proboscises of hawkmoths (Insecta: Lepidoptera: Sphingidae: *Eumorpha achemon, E. pandorus, Lintneria eremitus*) as they probe for nectar in the long floral spur. Although the flowers are self-compatible, experimental pollinations have shown that, despite routinely high seed set (>50%) inbreeding, depression is expressed as reduced mean seed viability in self-pollinated plants (16–23%) compared to outcrossed (39–62%) or open-pollinated (59–77%) plants. Hand pollinations obtained the highest seed viability (and percent germination) from flowers that were outcrossed between populations, which indicates a facultative outcrossing breeding system. Even though the plants appear to be predominantly outcrossed, analysis of genetic data (allozyme and RAPD markers) have detected high levels of inbreeding with low levels of genetic diversity and extensive interpopulational genetic differentiation (e.g., F_{ST}: 0.754–0.889) occurring among the small, widely fragmented populations. Occasional self-pollination (e.g., by infrequent geitonogamy) is likely to be a contributing factor. The plants produce thousands of seeds, which are dispersed by the wind. The seeds are minute (0.0039 mg), contain 43.6% water, but have a low rate of water loss (0.82% hr^{-1}). Seeds that were stored at 4°C have germinated when soaked for 1.5 hr in a 5% bleach solution (at ambient room temperature) followed by a distilled water rinse and dark incubation for 24+ hr. Longer periods of stratification (e.g., 16 vs. 8 weeks) also have been found to enhance germination rates. The seeds do not survive long in the substrate and there is no persistent seed bank. Successful germination also has been obtained for seeds stratified in distilled water at 4°C–8°C under total darkness for 11 months, then surface sterilized for 1 min in an ethanol and 5.25% bleach (NaOCl) solution, followed by inoculation with a fungal endophyte culture (e.g., Basidiomycota: Ceratobasidiaceae: *Ceratorhiza*), and finally incubated in total darkness at 23°C for 95 days, with a subsequent return to 4°C–8°C for another 107 days. Mycorrhizal symbionts (Basidiomycota: Ceratobasidiaceae: *Ceratorhiza goodyerae-repentis*) are necessary for successful seedling and protocorm establishment. Dozens of *Ceratorhiza* strains have been isolated from the plants from various localities. Under optimal conditions, it will take about 5 years for a seedling to reach flowering maturity. These are long-lived perennials, which can survive up to 30 years. The plants persist by means of fusiform tubers, which regenerate each year by the development of new tubers (from perennating buds); these sometimes remain attached through the end of the season. The plants can tolerate prolonged inundation and also can endure dry-mesic conditions. Plant densities can vary widely (0.000185–1.2 plants m^{-2}), with higher densities being characteristic of the more disturbed sites. The following compilation of associated species is minimal given that most database records have been "masked" for security reasons, due to the imperiled status of this species throughout its range.

Reported associates: *Acorus americanus, Andropogon gerardii, Apios americana, Betula pumila, Calamagrostis canadensis, Carex aquatilis, Carex cristatella, Carex pellita, Carex stricta, Carex trichocarpa, Carex vulpinoidea, Chamaedaphne calyculata, Chelone glabra, Cladium mariscoides, Cornus amomum, Cornus sericea, Dasiphora floribunda, Drosera rotundifolia, Gentiana andrewsii, Gentianopsis crinita, Glyceria striata, Helianthus grosseserratus, Juncus, Larix laricina, Liatris spicata, Linum medium, Lysimachia quadrifolia, Lythrum alatum, Lythrum salicaria, Pedicularis lanceolata, Persicaria amphibia, Phalaris arundinacea, Phleum pratense, Pycnanthemum virginianum, Rumex crispus, Sarracenia purpurea, Schizachyrium scoparium, Schoenoplectus acutus, Scirpus atrovirens, Solidago riddellii, Sparganium eurycarpum, Spartina pectinata, Spartina pectinata, Sphagnum, Thelypteris palustris, Toxicodendron vernix, Trifolium pratense, Typha latifolia, Veronicastrum virginicum.*

5.4.9.10. ***Platanthera limosa* Lindl.** is found in marshes (springy), seeps, woodlands (deciduous, montane), and along the margins of streams at elevations from 1800 to 2500 m

(to 4000 m in Central America). The plants grow in open to lightly forested exposures. Flowering extends from June to August. There is a dearth of information available on the life history of this rare species, which is due in part to the redaction of information in database records for security purposes.
Reported associates: none.

5.4.9.11. ***Platanthera tescamnis* Sheviak & W.F. Jenn.** occurs in draws, floodplains, meadows (alluvial, dry, grassy), thickets (willow), and woodlands (riparian) at elevations from 1825 to 2950 m. The plants can occur in exposures ranging from full sun to dense shade but tend to occupy more mesic sites. The flowers are present from late June through early August. They are strongly scented (described as "sweet-pungent") and are thought to deposit their pollinia on the compound eyes of nectar-seeking insects (Insecta). Like the preceding species, the lack of information available on the life history of this rare species is due in part to the redaction of database record information for security purposes. Moreover, the OBL status for this species should be reconsidered (see *Systematics* below) given that it inhabits much drier sites than *P. sparsiflora* (FACW), from which it has been distinguished.

Reported associates: *Abies concolor, Aquilegia, Betula occidentalis, Cirsium, Corallorhiza maculata, Corallorhiza striata, Epipactis gigantea, Maianthemum stellatum, Picea, Pinus flexilis, Pinus longaeva, Pinus ponderosa, Platanthera aquilonis, Populus angustifolia, Pseudotsuga, Salix.*

Use by wildlife: A weevil (Insecta: Coleoptera: Curculionidae: *Stethobaris ovata*) is found on *Platanthera aquilonis*. *Platanthera blephariglottis* is eaten by rabbits (Mammalia: Leporidae) and is a host for thrips (Insecta: Thysanoptera: Thripidae: *Frankliniella tritici*). Its flowers are visited and pollinated by various insects (Insecta) including bumble bees (Hymenoptera: Apidae: *Bombus fervidus, B. vagans*), butterflies (Lepidoptera: Hesperiidae: *Epargyreus clarus, Euphyes vestris, Polites mystic, P. peckius*; Lycaenidae: *Strymon melinus*; Nymphalidae: *Danaus plexippus*; Papilionidae: *Papilio glaucus, P. troilus*; Pieridae: *Colias philodice, Pieris rapae*), and sphinx moths (Lepidoptera: Sphingidae: *Darapsa versicolor, Hemaris thysbe*). Other floral visitors (but nonpollinators) include bees (Apidae: *Apis mellifera, Bombus ternarius*), butterflies (Lepidoptera: Lycaenidae: *Lycaena epixanthe*; Nymphalidae: *Danaus plexippus, Limenitis archippus, Megisto cymela, Nymphalis antiopa, Speyeria atlantis, S. cybele, Vanessa cardui*; Papilionidae: *Papilio troilus*), flies (Diptera: Tabanidae), and wasps (Hymenoptera: Vespidae: *Vespula maculifrons*). The flowers of *P. cristata* are visited by pollinating bumblebees (Hymenoptera: Apidae: *Bombus pensylvanicus*) and nonpollinating butterflies (Lepidoptera: Papilionidae: *Papilio glaucus, P. troilus*). The seeds are eaten by larval red-banded leaf-roller moths (Insecta: Lepidoptera: Tortricidae: *Argyrotaenia velutinana*). The plants also host thrips (Insecta: Thysanoptera: Thripidae: *Frankliniella tritici*). *Platanthera integra* is another host plant of thrips (Insecta: Thysanoptera: Thripidae: *Pseudothrips beckhami*). *Platanthera integrilabia* is grazed heavily by white-tailed deer (Mammalia: Cervidae: *Odocoileus virginianus*). The flowers are visited by several insects (Insecta) including pollinating butterflies (Lepidoptera: Hesperiidae: *Epargyreus clarus*; Papilionidae: *Papilio glaucus*, *P. troilus*) and by nonpollinating bees (Hymenoptera: Apidae: *Apis mellifera*, *Bombus impatiens*), beetles (Coleoptera: Mordellidae: *Mordella*), flies (Diptera: Syrphidae: *Allograpta*, *Milesia virginiensis*), moths (Lepidoptera: Sphingidae: *Manduca*), and wasps (Hymenoptera: Vespidae: *Vespula maculifrons*). The flowers of *P. leucophaea* provide nectar for pollinating hawkmoths (Insecta: Lepidoptera: Sphingidae: *Eumorpha achemon*, *E. pandorus*, *Lintneria eremitus*). The plants are fed on by orchid weevils (Insecta: Coleoptera: Curculionidae: *Stethobaris ovata*).

Economic importance: food: *Platanthera* species are not reportedly edible; **medicinal:** Although there is no substantiating evidence, *Platanthera cristata* was believed to possess curative properties against snake bites and was often used by 19th century hunters to treat their dogs when bitten by rattlesnakes; **cultivation:** None of the OBL *Platanthera* species has any great commercial value as a cultivated specimen; **misc. products:** none; **weeds:** none; **nonindigenous species:** all the OBL *Platanthera* species are indigenous.

Systematics: As currently recognized, *Platanthera* is assigned to tribe Orchideae of orchid subfamily Orchidoideae. Although once merged with *Habenaria*, the genus has been redefined extensively as a result of several molecular phylogenetic studies (using nrITS sequence data), which clearly resolve the group as being distant from *Habenaria* and positioned instead in a more narrowly circumscribed clade that is situated closest to *Galearis* and *Neolindleya* (e.g., Figure 2.15). Although *P. clavellata* and *P. integra* had been segregated formerly in the distinct genus *Gymnadeniopsis*, phylogenetic analyses have shown that group to be included well within *Platanthera* (Figure 2.18). The genus itself has been subdivided into as many as seven sections, with some (e.g., *Blephariglottis*, *Limnorchis*) having been recognized formerly as separate genera. Currently, five sections are

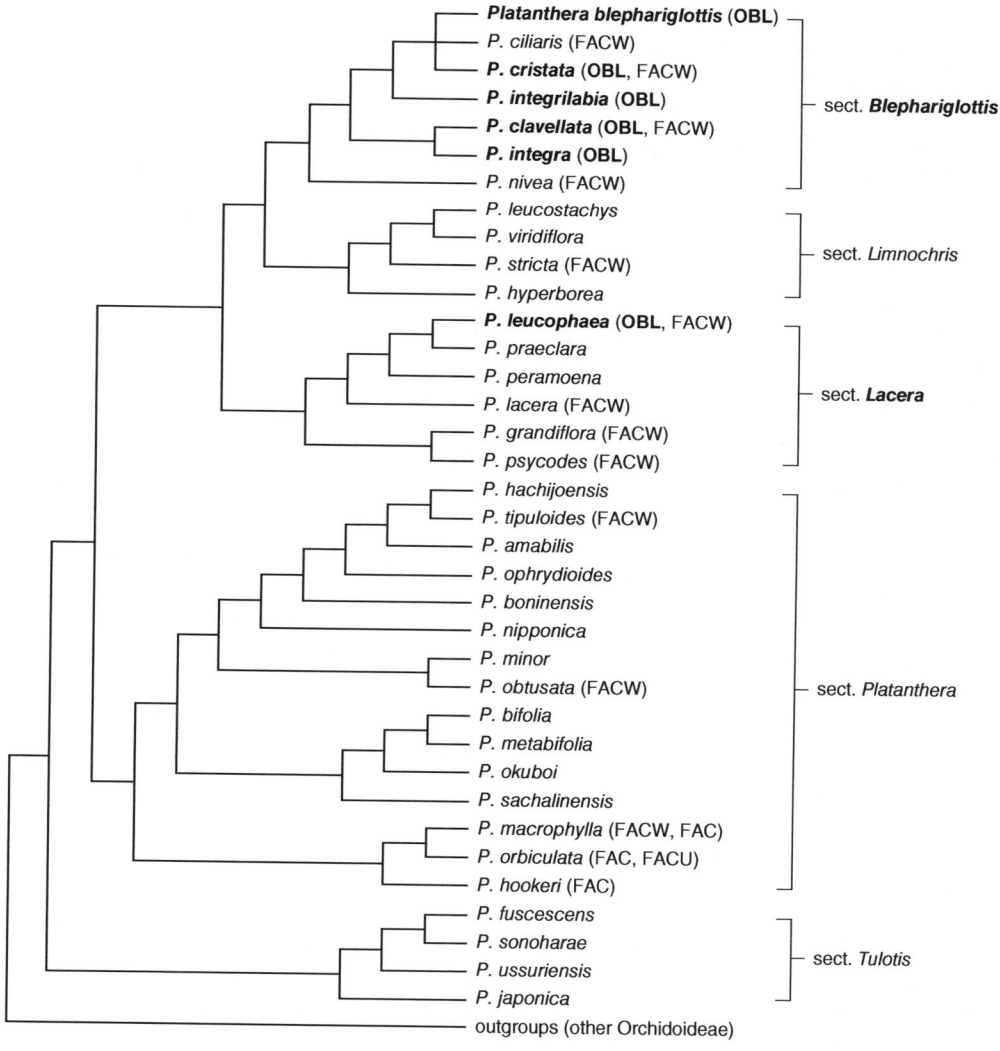

FIGURE 2.18 Interspecific relationships in *Platanthera* as indicated by phylogenetic analysis of nrITS sequence data (modified from Hapeman & Inoue, 2000). Five of the six OBL North American indicators surveyed (shown in bold) are concentrated within section *Blephariglottis*, which represents a fairly specialized group within the genus. Other wetland indicators (in parentheses) occur throughout all the sections except *Tulotis*, which lacks North American representatives.

recommended, based on the major clades recovered in phylogenetic analyses (e.g., Figure 2.18). The greatest concentration of OBL indicators occurs in section *Blephariglottis*, which represents a highly derived group in the genus (Figure 2.18). *Platanthera leucophaea* formerly included the species that has been segregated as *P. praeclara*, which occurs primarily west of the Mississippi River. *Platanthera tescamnis* (OBL) has been recognized as distinct from *P. sparsiflora* (FACW); however, it is *P. sparsiflora* that inhabits the wetter sites. The two taxa resolve as sister species (Figure 2.18). Because the ancient basic chromosome number of subfamily Orchidoideae presumably was $x=7$, most *Platanthera* species (e.g., *P. aquilonis*, *P. blephariglottis*, *P. chorisiana*, *P. clavellata*, *P. cristata*, *P. leucophaea*) would represent paleohexaploids (all are $2n=42$) but are regarded as functionally diploid. Transcriptome analysis has indicated evidence of a whole genome duplication in *Platanthera*. *Platanthera huronensis* ($2n=4x=84$) is an allotetraploid derivative of *P. aquilonis* and *P. dilatata*. Interspecific hybrids between these species are found often and there is evidence that *P. huronensis* itself is characterized by multiple independent origins. Hybrids occur commonly in *Platanthera* and putative natural interspecific F_1 hybrids involving the OBL species include *Platanthera* ×*apalachicola* (*P. chapmanii* × *P. cristata*), *P.* ×*media* (*P. aquilonis* × *P. dilatata*), *P. aquilonis* × *P. purpurascens*, *P. aquilonis* × *P. tescamnis*, *P.* ×*bicolor* (*P. blephariglottis* × *P. ciliaris*), *P.* ×*canbyi* (*P. blephariglottis* × *P. cristata*), *P.* ×*channellii* (*P. ciliaris* × *P. cristata*), *P.* ×*hollandiae* (*P. leucophaea* × *P. lacera*), *P.* ×*reznicekii* (*P. leucophaea* × *P. psycodes*), *P. sparsiflora* × *P. tescamnis*, and *P.* ×*vossii* (*P. blephariglottis* × *P. clavellata*). Successful intercrosses among *P. blephariglottis*, *P. clavellata*, *P. ciliaris*, and *P. cristata* are understandable given their close phylogenetic interrelationships (Figure 2.18).

Distribution: Most of the OBL species are dispersed widely over several geographical regions including northern North America and the Rockies (*P. huronensis*), northeastern North America (*P. leucophaea*), northern and western North America (*P. aquilonis*), northwestern North America (*P. chorisiana* – extending into eastern Asia), eastern North America (*P. blephariglottis*, *P. clavellata*), southern and eastern United States (*P. cristata*), and southeastern United States (*P. integra*, *P. integrilabia*). Two species are more restricted: *P. limosa* (Arizona and New Mexico, but extending southward into Guatemala) and *P. tescamnis* (southwestern United States).

References: Almquist & Calhoun, 2003; Altrichter et al., 2018; Arditti, 1966; Arditti & Ghani, 2000; Argue, 2011; Bateman et al., 2003; 2009; Beaven & Oosting, 1939; Birchenko, 2001; Blanton, 1939; Block & Rhoads, 2013; Bowles, 1983; 1999; Bowles et al., 1992; 2002; 2005; Boyd et al., 2016; Brant, 2018; Brown, 1995; 2005; 2006; Brown & Scott, 1997; Carlson et al., 2004; Catling, 1983a; 1984; Catling & Brownell, 1999; Catling & Catling, 1989; 1991; Catling & Kostiuk, 2011b; Cole & Firmage, 1984; Coleman, 2002; Conard, 1935; Currah et al., 1997; Darwin, 1869; Denley et al., 2002; Dunford et al., 2006; Felix & Guerra, 2005; Fleming & Van Alstine, 1999; Folsom, 1984; Fuentes, 2007; Funderburk et al., 2007;

Glenn, 2008; Grant, 1864; Gray, 1862; Hapeman & Inoue, 2000; Hill, 2007; Holsinger & Wallace, 2004; Homoya, 1982; 1983; Inoue, 1983; Johnson & Walz, 2013; Kelso et al., 2014; Laroche et al., 2012; Latham et al., 1996; Lemay et al., 2015; McCabe & Sheviak, 1980; McKenzie et al., 2012; Morris, 2013; Munden, 2001; NAOCC, 2019; Nelson, 1986; Pace & Brant, 2015; Perrette, 2010; Peter & Johnson, 2005; Prena, 2017; Rasmussen, 1995; Runkel & Roosa, 1999; Sears, 2008; Sheridan et al., 1997; Sheviak, 2001; 2002c; Sheviak & Jennings, 2006; Singhurst et al., 2014a; Sipple & Klockner, 1980; Smith, 2012; Smith & Snow, 1976; Sorrie et al., 2006; Sperduto, 2011; Standley, 1919; Stetson, 1913; Stuckey, 1967; Talbot et al., 2010; Unruh et al., 2018; USFWS, 2013; Wallace, 2002; 2003a; 2003b; 2004; 2006; Waller & Alverson, 1997; Wichman et al., 2007; Yoder et al., 2010; Zettler & Fairey, III, 1990; Zettler & Hofer, 1998; Zettler & McInnis, Jr., 1992; Zettler & Piskin, 2011; Zettler et al., 1996; 2000; 2001; 2005.

5.4.10. Pogonia

Adder's-mouth, beard-flower, rose pogonia, snake-mouth; pogonie langue-de-serpent

Etymology: from the Greek *pōgōn* ("beard") in reference to the bearded lip

Synonyms: *Arethusa* (in part); *Amesia* (in part); *Epipactis* (in part); *Helleborine* (in part)

Distribution: global: Asia (eastern); North America; **North America:** eastern

Diversity: global: 5 species; **North America:** 1 species

Indicators (USA): obligate wetland (OBL): *Pogonia ophioglossoides*

Habitat: freshwater; palustrine; **pH:** 6.3–6.7; **depth:** <1 m; **life-form(s):** emergent herb

Key morphology: roots slender, stolon-like, producing shoots at wide intervals; stems (to 70 cm) slender, solitary; leaf solitary, ascending, fleshy-leathery, inserted midway up the stem, the blade (to 12 cm) elliptic, lance-ovate, narrowly lanceolate, or oblong; inflorescence solitary or racemose (2-flowered), the floral bracts prominent (to 37 mm), leaf-like; flowers resupinate, showy, the perianth gaping, pink (rarely white or bluish); sepals (to 23 mm) oblong, oblong-elliptic, or narrowly lanceolate; petals (to 25 mm) elliptic to lance-obovate, the lip (to 25 mm) spatulate, with adaxial, fleshy crests (greenish, white or yellowish at base), and involute, lacerate to laciniate margins; pollinia 2, lacking viscidia; capsules (to 30 mm) erect, ellipsoid

Life history: duration: perennial (stoloniferous roots); **asexual reproduction:** adventitious plantlets (from stoloniferous roots); **pollination:** insect; **sexual condition:** hermaphroditic; **fruit:** capsules (common); **local dispersal:** seeds (wind); **long-distance dispersal:** seeds (wind)

Imperilment: 1. *Pogonia ophioglossoides* [G5]; SX (DC); S1 (IL, KY, LB, MB, MO, OK); S2 (AR, DE, OH, PE, TN, WV); S3 (FL, GA, MD, NC, QC, RI, VA, VT)

Ecology: general: All *Pogonia* species inhabit peatlands and other wet sites, but only one species occurs in North America. The flowers are showy, but produce minimal amounts of nectar, which has led to their categorization as being deceptively pollinated. They are self-compatible and capable of high seed set when self-pollinated artificially. Although much of the literature has reported that the flowers are outcrossed (primarily from observations of pollinators), it also seems likely that insect-mediated selfing occurs, especially given the lack of viscidia on the pollina. Seed set typically is fairly high, even when pollinator visitation is minimal. Reports of apomixis in this genus are unsubstantiated and require further evaluation. The seeds are dispersed by wind. The plants propagate vegetatively by pseudo-stoloniferous roots, which produce adventitious shoots near their apices. These frequently have been referred to mistakenly as rhizomes or stolons.

5.4.10.1. *Pogonia ophioglossoides* (L.) Ker Gawl. inhabits bogs (dune, hillside, montane, sandy, seepage, *Sphagnum*), borrow pits, depressions (floodplain, low) (dune, wet), ditches (boggy, drainage, flatwoods, roadside, wet), fens (peaty, poor), flarks (string-bog), flats (marly), flatwoods (pine, wet), floating mats, floodplains (shrubby), glades (sandstone), hummocks (*Sphagnum*), marshes (floating, marl), meadows (acidic, moist, roadside, sandy, wet), pannes (dune), pine barrens (low), pinelands (low), prairies, roadcuts, sand pits (acid), savannas (bald cypress, moist, pine), seeps (hillside, roadside, sandy), slopes (seepage), swales (boggy, interdunal, peaty, seepage, *Sphagnum*, springy), swamps (cedar, cranberry, cypress, seepage, *Sphagnum*), woodlands (open wet), and the margins (boggy) of gravel pits, lakes, pocosins, ponds (sinkhole), and streams at elevations to 1280 m. The plants are found primarily in sunny, open exposures and wetter sites (1–11 cm above the water table). Although often described as residents of calciphilic, minerotrophic habitats, they also are found quite frequently in acidic sites (e.g., pH: 6.3–6.7). The substrates have been characterized as Alapaha (arenic plintic paleaquults), Basinger (spodic psammaquents), Bibb, clay, colluvium, Dorovan (terric medisaprists), Escambia (plinthaquic paleudults), Florala (plinthaquic paleudults), gravel, muck, mud, Okeechobee (hemic medisaprists), Osier, Pactolus (aquic quartzipsamments), peat, Pelham (arenic paleaquults), Plummer (grossarenic paleaquults), Pottsburg (grossarenic haplaquods), Ridgewood (aquic quartzipsamments), Rutlege (typic humaquepts), sand, sandy loam, sandy muck, sandy peat, Tocoi (ultic haplaquods), and Valkaria (spodic psammaquents). Flowering occurs from April to July with fruiting extending from May to September. The showy flowers are self-compatible but are pollinated by bumblebees (Insecta: Hymenoptera: Apidae: *Bombus borealis*, *B. fervidus*, *B. sandersoni*, *B. ternarius*, *B. terricola*, *B. vagans*), which forage for nectar and pollen. The visiting insects land on the lip and proceed inward toward a slightly nectariferous cavity formed at its juncture with the column. As the bees exit the flower, they brush against the hinged anther to expose the pollinia, which, despite their lack of viscidia, then become attached to the top of the insect's head. Upon withdrawl from the flower, the insect is said to transfer the pollinia to the stigma of the same flower as it exits (which would represent insect-mediated selfing as the usual consequence). A selfing breeding system is understandable, given the combination of self-compatibility, minimal nectar production, weakly sectile (~granular) pollen,

lack of viscidia, and low frequency of flowering. The plants typically produce few flowers (2.3–4.4% of plants in flower) but have relatively high fruit set (28.6–33.3% of plants in flower). The association of high seed set (10–100%) in conjunction with low pollinator visitation rates has fostered speculation that some of the plants are apomictic. However, the possibility of apomixis has not been evaluated appropriately, and the high observed levels seed set also could just as well indicate autogamous or insect-mediated self-pollination. Genetic surveys (using nuclear and plastid microsatellite markers) have detected fairly high levels of intrapopulational variability but local spatial genetic structure, which has suggested limited seed dispersal. However, if the flowers actually are apomictic, then the genetic data would need to be reinterpreted accordingly. The plants reproduce vegetatively by the production of superficial, horizontal, branching, stolon-like roots, which produce adventitious shoots near their distal ends. The plants respond positively to fire, with large populations having appeared in the aftermath or prescribed burns. They are not drought tolerant. **Reported associates:** *Acer rubrum, Agalinis maritima, Agalinis purpurea, Agrostis perennans, Aletris aurea, Aletris lutea, Alnus serrulata, Alternanthera philoxeroides, Amaranthus tuberculatus, Amianthium muscitoxicum, Andromeda polifolia, Andropogon glomeratus, Andropogon virginicus, Arethusa bulbosa, Aristida palustris, Aristida purpurascens, Aristida stricta, Arnoglossum plantagineum, Arnoglossum sulcatum, Aronia arbutifolia, Arundinaria tecta, Asclepias connivens, Asclepias rubra, Bacopa, Balduina uniflora, Bartonia paniculata, Bejaria, Betula pumila, Boehmeria cylindrica, Burmannia capitata, Calla palustris, Calliergonella cuspidata, Calopogon barbatus, Calopogon pallidus, Calopogon tuberosus, Campanula aparinoides, Campylium stellatum, Carex aquatilis, Carex atlantica, Carex buxbaumii, Carex canescens, Carex chordorrhiza, Carex collinsii, Carex exilis, Carex flava, Carex glaucescens, Carex interior, Carex lasiocarpa, Carex leptalea, Carex limosa, Carex livida, Carex livida, Carex rostrata, Carex sterilis, Carex striata, Carex stricta, Carex tetanica, Carex viridula, Centella asiatica, Chamaecyparis thyoides, Chamaedaphne calyculata, Chaptalia tomentosa, Cladium mariscoides, Clethra alnifolia, Comarum palustre, Conoclinium coelestinum, Coreopsis gladiata, Coreopsis lanceolata, Cuscuta indecora, Cyperus flavescens, Cyperus polystachyos, Cypripedium acaule, Dasiphora floribunda, Decodon verticillatus, Dichanthelium ensifolium, Drosera anglica, Drosera capillaris, Drosera intermedia, Drosera linearis, Drosera rotundifolia, Drosera tracyi, Dulichium arundinaceum, Eleocharis compressa, Eleocharis flavescens, Eleocharis parvula, Eleocharis rostellata, Eleocharis tuberculosa, Equisetum fluviatile, Erigeron vernus, Eriocaulon compressum, Eriocaulon decangulare, Eriocaulon texense, Eriophorum angustifolium, Eriophorum gracile, Eriophorum tenellum, Eriophorum virginicum, Eriophorum viridicarinatum, Eryngium integrifolium, Eubotrys racemosa, Eupatorium leucolepis, Eupatorium rotundifolium, Fuirena scirpoidea, Fuirena squarrosa, Gaylussacia frondosa, Gelsemium sempervirens, Glyceria* *obtusa, Gordonia, Gratiola ramosa, Habenaria repens, Helenium brevifolium, Helenium vernale, Helianthus angustifolius, Hydrocotyle, Hypericum crux-andreae, Hypericum drummondii, Hypericum hypericoides, Hypericum mutilum, Hypoxis hirsuta, Ilex coriacea, Ilex glabra, Ipomoea sagittata, Juncus caesariensis, Juncus canadensis, Juncus debilis, Juncus gymnocarpus, Juncus marginatus, Juncus pelocarpus, Juncus stygius, Kalmia angustifolia, Kalmia hirsuta, Kalmia latifolia, Kosteletzkya pentacarpos, Pleea tenuifolia, Lachnanthes caroliniana, Lachnocaulon anceps, Lachnocaulon digynum, Larix laricina, Leersia oryzoides, Leptochloa fusca, Liatris pycnostachya, Lobelia cardinalis, Lobelia reverchonii, Lophiola aurea, Ludwigia alata, Ludwigia leptocarpa, Lycopodiella alopecuroides, Lycopodiella appressa, Lycopodiella caroliniana, Lycopodiella inundata, Lycopodiella prostrata, Lyonia ferruginea, Lyonia lucida, Lysimachia terrestris, Lythrum lineare, Magnolia virginiana, Marshallia graminifolia, Melothria pendula, Mentha arvensis, Menyanthes trifoliata, Muhlenbergia expansa, Muhlenbergia glomerata, Muhlenbergia richardsonis, Myrica caroliniensis, Myrica cerifera, Myrica gale, Myrica heterophylla, Narthecium americanum, Neottia smallii, Nyssa sylvatica, Oclemena nemoralis, Oenothera fruticosa, Orontium aquaticum, Osmanthus, Osmunda regalis, Osmundastrum cinnamomeum, Panicum hemitomon, Parnassia asarifolia, Paspalum distichum, Paspalum floridanum, Pellia epiphylla, Persea borbonia, Persea palustris, Persicaria hydropiper, Persicaria punctata, Persicaria sagittata, Phragmites australis, Phyla, Picea mariana, Pilea pumila, Eriophorum, Pinguicula primuliflora, Pinguicula pumila, Pinus elliottii, Pinus palustris, Pinus serotina, Pinus taeda, Pityopsis graminifolia, Pityopsis oligantha, Platanthera blephariglottis, Platanthera ciliaris, Platanthera clavellata, Platanthera nivea, Polygala brevifolia, Polygala lutea, Polygala mariana, Polygala ramosa, Proserpinaca palustris, Ptilimnium capillaceum, Rhamnus alnifolia, Rhamnus frangula, Rhexia lutea, Rhexia mariana, Rhynchospora alba, Rhynchospora capillacea, Rhynchospora capitellata, Rhynchospora chalarocephala, Rhynchospora ciliaris, Rhynchospora fusca, Rhynchospora gracilenta, Rhynchospora latifolia, Rhynchospora macra, Rhynchospora plumosa, Rhynchospora pusilla, Rhynchospora rariflora, Rhynchospora stenophylla, Rosa palustris, Rubus pensilvanicus, Sabatia difformis, Sabatia macrophylla, Sacciolepis striata, Sagittaria lancifolia, Sagittaria latifolia, Sarracenia alabamensis, Sarracenia alata, Sarracenia flava, Sarracenia leucophylla, Sarracenia minor, Sarracenia psittacina, Sarracenia purpurea, Sarracenia rubra, Saururus cernuus, Scheuchzeria palustris, Schizachyrium scoparium, Schoenoplectus acutus, Schoenoplectus tabernaemontani, Scleria ciliata, Scleria muhlenbergii, Scutellaria integrifolia, Serenoa repens, Smilax laurifolia, Solidago ohioensis, Solidago sempervirens, Sphagnum contortum, Sphagnum fuscum, Sphagnum magellanicum, Sphagnum papillosum, Sphagnum rubellum, Sphagnum subsecundum, Sphagnum teres, Spiranthes cernua, Spiranthes vernalis, Stenanthium densum, Stenanthium leimanthoides, Symphyotrichum dumosum, Symphyotrichum elliotii, Symphyotrichum laeve, Symphyotrichum novi-belgii, Symphyotrichum tenuifolium, Syngonanthus flavidulus, Taxodium distichum, Thelypteris palustris, Toxicodendron vernix, Triadenum fraseri, Triadenum virginicum, Trichophorum alpinum, Triglochin maritima, Typha latifolia, Utricularia cornuta, Utricularia intermedia, Utricularia minor, Utricularia subulata, Vaccinium corymbosum, Vaccinium macrocarpon, Vaccinium oxycoccos, Verbena hastata, Viburnum nudum, Vigna luteola, Viola primulifolia, Viola sororia, Woodwardia areolata, Xyris ambigua, Xyris baldwiniana, Xyris caroliniana, Xyris laxifolia, Xyris montana, Xyris scabrifolia, Xyris torta.*

Use by wildlife: The flowers of *Pogonia ophioglossoides* are visited by bumblebees (Insecta: Hymenoptera: Apidae: *Bombus borealis, B. fervidus, B. sandersoni, B. ternarius, B. terricola, B. vagans*), which forage for pollen and nectar. They also are hosts of thrips (Insecta: Thysanoptera: Thripidae: *Frankliniella tritici*).

Economic importance: food: *Pogonia ophioglossoides* is not edible; **medicinal:** no uses reported; **cultivation:** *Pogonia ophioglossoides* is not cultivated commercially; **misc. products:** none; **weeds:** none; **nonindigenous species:** none.

Systematics: *Pogonia* is the type genus of tribe Pogonieae in the orchid subfamily Vanilloideae, which represents a relatively unspecialized group in the overall orchid phylogeny (Figures 2.13 and 2.19). Morphological data have indicated that *Pogonia* is most closely related to *Isotria*; however, various molecular data have confirmed that *Pogonia* is monophyletic but related most closely to *Cleistesiopsis* and *Isotria*, either as their respective sister clade (*rbcL* + nrITS sequence data) or (combined cpDNA sequence data; rDNA sequence data; cpDNA + nrITS sequence data) as being allied most closely to *Cleistesiopsis* rather than *Isotria* (e.g., Figure 2.19). Several sources of DNA sequence data (e.g., Figure 2.19) resolve *P. ophioglossoides* as sister to the group of Asian *Pogonia* species, a result also consistent with analyses of RAPD data, which place *P. ophioglossoides* outside the cluster of Asian species. The basic chromosome number of *Pogonia* is $x = 9$; *P. ophioglossoides* ($2n = 18$) is diploid. The genome size of *P. ophioglossoides* ($1C = 55.4$ pg) is the largest among any surveyed orchid species. Synthetic reciprocal F_1 hybrids have been made between *Pogonia minor* and *P. ophioglossoides*.

Distribution: *Pogonia ophioglossoides* is distributed across eastern North America.

References: Baldwin, Jr. & Speese, 1955; Boland & Scott, 1991; 1992; Bouetard et al., 2010; Cameron, 2009; Cameron & Chase, 1999; Cameron & Molina, 2006; Carlson, 1938; Case & Case, 1974; Emerson, 1921; Freudenstein & Rasmussen, 1997; Funderburk et al., 2007; Grittinger, 1970; Ladd, 2014; Leitch et al., 2009; Namestnik et al., 2012; Nixon & Ward, 1986; Orzell & Bridges, 1991; Pansarin et al., 2008; Pandey & Sharma, 2015; Sasser et al., 1995b; Sheviak & Catling, 2002c; Singhurst et al., 2003; Sipple & Klockner, 1980; Takahashi & Kondo, 2004; Takahashi et al., 2014; Thien & Marcks, 1972; Vitt & Slack, 1975; Wheeler et al., 1983.

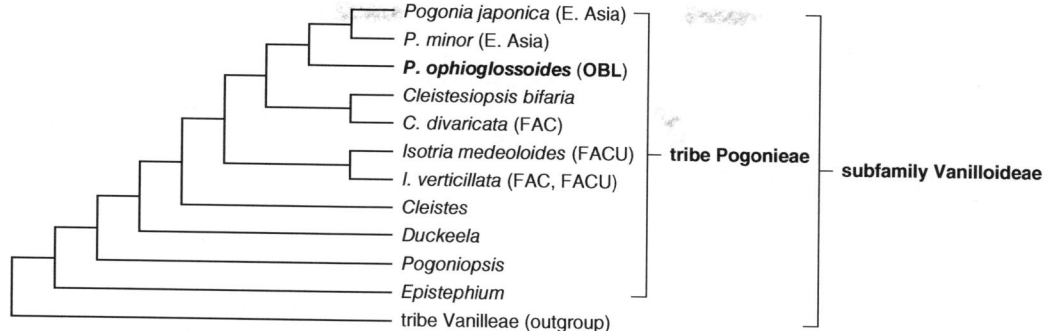

FIGURE 2.19 Phylogenetic position of *Pogonia ophioglossoides* as indicated by analysis of combined cpDNA and nrITS sequence data (modified from Pansarin et al., 2008). These and other DNA data resolve *P. ophioglossoides* as sister to the Asian *Pogonia* species. Most molecular data closely ally *Pogonia* with *Cleistesiopsis*, a group formerly recognized as part of the genus *Cleistes*. The wetland indicator status for ranked North American taxa is given in parentheses (taxa containing OBL species are highlighted in bold).

5.4.11. Ponthieva

Shadow witch

Etymology: after Henri de Ponthieu (1731–1808)

Synonyms: *Arethusa* (in part); *Cranichis* (in part); *Epipactis* (in part); *Listera* (in part); *Neottia* (in part); *Nerissa* (in part); *Ophrys* (in part); *Serapias* (in part)

Distribution: global: Neotropics and subtropics; **North America:** southeastern United States

Diversity: global: 25 species; **North America:** 2 species

Indicators (USA): obligate wetland (OBL); facultative wetland (FACW); facultative (FAC): *Ponthieva racemosa*

Habitat: freshwater; palustrine; **pH:** alkaline; **depth:** <1 m; **life-form(s):** emergent herb

Key morphology: roots in fleshy fascicles; stems short-rhizomatous; leaves evergreen, 2–8 in a basal rosette, sub-petiolate, the blade (to 17 cm) thin, dark green, elliptic to oblanceolate; raceme (to 60 cm) terminal, 20–35-flowered, the peduncle (to 25 cm) enveloped basally by leaf-like sheaths (to 10 mm); flowers not resupinate, the perianth dishlike, spreading; sepals greenish-white, pubescent above, the dorsal sepal (to 7 mm) distinct, ovate-elliptic to elliptic-lanceolate, the laterals (to 6.5 mm) broadly to obliquely ovate; petals (to 6 mm) white with green veins, glabrous, obliquely triangular, the margins entire or minutely ciliate, the lip (to 7 mm) suborbiculate, deeply concave, distinctly clawed; column (to 5 mm) dilated and slightly winged distally, the anther adaxial, erect behind rostellum, pollinia 4, joined in pairs, viscidium terminal; ovary (to 20 mm) pedicellate, pubescent; capsules (to 13 mm) suberect, ellipsoid

Life history: duration: perennial (rhizomes); **asexual reproduction:** rhizomes; **pollination:** insect; **sexual condition:** hermaphroditic; **fruit:** capsules (common); **local dispersal:** seeds (wind); **long-distance dispersal:** seeds (wind)

Imperilment: 1. *Ponthieva racemosa* [G4/G5]; S1 (TN); S2 (GA, LA, MS, NC, SC); S3 (AL, VA)

Ecology: general: *Ponthieva* is not a characteristic wetland genus but sometimes occupies moist microsites. Many of the tropical American species occur in dry habitats (often montane) and even the single OBL North American indicator (*P. racemosa*) is ranked variously as FACW or FAC in parts of its range. Little reliable information exists on the floral biology or seed ecology for which no detailed studies exist. The flowers presumably are adapted for pollination by sweat bees (Insecta: Hymenoptera: Halictidae) but more critical observations are necessary to elucidate the reproductive ecology of at least the two North American species.

5.4.11.1. *Ponthieva racemosa* (Walter) C. Mohr.
inhabits bottomlands, flats (wooded), flatwoods (mesic), glades (red cedar), gullies, hammocks (hardwood, hydric, moist), ledges (limestone), ravines (wooded), savannas (marly, wet), seeps, sink/solution holes (limestone, pinelands, shallow), slopes (calcareous, moist, wooded), springs (seepage), swamps (boggy), thickets (red cedar), woodlands (bottomland, deciduous, floodplain, hardwood, low, pine, rich, wet), and the margins ponds, sloughs, springs, and streams at elevations to 37 m [but to 2600 m in Ecuador]. The common name of "shadow witch" reflects the affinity of this species for deeply shaded exposures. Otherwise, the plants are broadly adapted and can thrive under wet or drier conditions. They occur on calcareous substrates, which have been described as Anclote, clay (prairie), clay-humus, clay loam (Greenville), coquina outcrops, limestone, loam, marl (shell), muck, mucky loam, peat, Samsula (terric medisaprists), sand (fine), sandy humus, sandy loam, and sandy silt. Flowering has been observed from September to March with fruiting from October to April. The perianth is not resupinate but forms a "pseudolip" by the two closely appressed lower petals. That structure provides a landing platform for potential pollinators, which are thought to be small insects (Insecta) such as halictid bees (Hymenoptera: Halictidae). The reproductive ecology of this species remains ambiguous and has been based much on speculation. The flowers have been described both as self-compatible (but not autogamous) and also as self-incompatible (from the lack of seed set in artificially selfed flowers). A more critical evaluation of the pollination process is needed. Natural levels of fruit set can be fairly high (e.g., 35%) but seed viability can range from 8% to 82%. The seeds are dispersed by wind. They have germinated after pretreatment (acid scarification) in 2% H_2SO_4 for 2 min, followed by sterilization in 1% sodium hypochlorite (NaOCl) for 5 min. The sterilized seeds were then

placed on modified SM medium and incubated in darkness at 20°C until they germinated. After germination, the developing protocorms have been sub-cultured on SM medium every 2–3 months in 12 hr of light at 20°C. **Reported associates:** *Acer rubrum, Acer saccharum, Actaea pachypoda, Actaea racemosa, Adiantum pedatum, Ambrosia psilostachya, Ampelopsis arborea, Andropogon glomeratus, Anemone americana, Aralia racemosa, Arisaema dracontium, Arisaema triphyllum, Aristolochia serpentaria, Arnoglossum diversifolium, Arnoglossum sulcatum, Arundinaria gigantea, Asarum canadense, Asimina triloba, Asplenium heterochroum, Asplenium platyneuron, Baccharis glomeruliflora, Berchemia scandens, Bidens mitis, Brachyelytrum erectum, Callicarpa americana, Calycocarpum lyonii, Campanula americana, Campsis radicans, Cardamine concatenata, Carex atlantica, Carex cherokeensis, Carex godfreyi, Carex leptalea, Carpinus caroliniana, Carya glabra, Carya tomentosa, Ceanothus americanus, Celtis laevigata, Cephalanthus occidentalis, Cercis canadensis, Chamaecyparis thyoides, Chasmanthium laxum, Chasmanthium ornithorhynchum, Cicuta maculata, Cirsium horridulum, Cladium jamaicense, Cocculus carolinus, Conopholis americana, Corallorhiza wisteriana, Cornus alternifolia, Cornus asperifolia, Cornus drummondii, Cornus florida, Cornus foemina, Crataegus berberifolia, Crataegus crus-galli, Crataegus viridis, Decodon verticillatus, Deparia acrostichoides, Desmodium glutinosum, Diospyros virginiana, Diplazium pycnocarpon, Dryopteris ludoviciana, Elytraria caroliniensis, Erythrina herbacea, Fagus grandifolia, Fraxinus americana, Habenaria floribunda, Helenium autumnale, Hexalectris spicata, Hybanthus concolor, Hylodesmum pauciflorum, Hymenocallis rotata, Ilex cassine, Ilex decidua, Ilex montana, Ilex opaca, Juglans nigra, Juniperus virginiana, Lactuca floridana, Lespedeza cuneata, Ligustrum sinense, Lindera benzoin, Liquidambar styraciflua, Liriodendron tulipifera, Lithospermum tuberosum, Lonicera japonica, Luzula acuminata, Lyonia lucida, Maclura pomifera, Magnolia grandiflora, Magnolia macrophylla, Magnolia tripetala, Magnolia virginiana, Malaxis spicata, Malaxis unifolia, Menispermum canadense, Mitella diphylla, Mitreola petiolata, Morus rubra, Myrica cerifera, Nasturtium microphyllum, Nothoscordum bivalve, Nyssa aquatica, Nyssa biflora, Nyssa sylvatica, Oeceoclades maculata, Osmanthus americanus, Packera aurea, Panax quinquefolius, Parthenocissus quinquefolia, Passiflora lutea, Passiflora lutea, Pedicularis canadensis, Persea palustris, Phaius tankervilleae, Pinus elliottii, Pinus glabra, Pinus taeda, Platanthera flava, Platythelys querceticola, Pleopeltis polypodioides, Podophyllum peltatum, Polystichum acrostichoides, Prenanthes trifoliolata, Prunella vulgaris, Pyrus calleryana, Quercus alba, Quercus falcata, Quercus hemisphaerica, Quercus laurifolia, Quercus marilandica, Quercus michauxii, Quercus muehlenbergii, Quercus nigra, Quercus phellos, Quercus shumardii, Quercus stellata, Quercus virginiana, Ratibida pinnata, Rhamnus caroliniana, Rhamnus lanceolata, Rhapidophyllum hystrix, Rhododendron canescens, Rhus aromatica, Rhus glabra, Rhynchospora colorata, Rhynchospora miliacea, Rosa carolina, Rosa multiflora, Rosa palustris, Ruellia caroliniensis, Sabal minor, Sabal palmetto, Sabatia angularis, Sagere21a minutiflora, Salvia lyrata, Sanguinaria canadensis, Sanicula canadensis, Sanicula marilandica, Sapindus saponaria, Sassafras albidum, Saururus cernuus, Scutellaria ovata, Sideroxylon lanuginosum, Sideroxylon lycioides, Silene catesbaei, Smilax bona-nox, Smilax lasioneura, Solidago flexicaulis, Solidago gigantea, Spigelia marilandica, Spiranthes odorata, Spiranthes ovalis, Symphyotrichum urophyllum, Symphyotrichum urophyllum, Taraxacum officinale, Taxodium distichum, Thalia geniculata, Thalictrum dioicum, Thelypteris hispidula, Thelypteris kunthii, Thelypteris palustris, Tilia americana, Tipularia discolor, Toxicodendron radicans, Trillium cuneatum, Trillium decumbens, Trillium lancifolium, Ulmus alata, Ulmus americana, Ulmus rubra, Vernonia gigantea, Viburnum obovatum, Viburnum rufidulum, Vicia minutiflora, Viola walteri, Vitis aestivalis, Xyris tennesseensis, Yeatesia viridiflora, Zephyranthes atamasco, Zizaniopsis miliacea.*

Use by wildlife: The flowers of *Ponthieva racemosa* are visited by sweat bees (Hymenoptera: Halictidae) and other small insects (Insecta).

Economic importance: food: *Ponthieva* is not edible; **medicinal:** In Costa Rica, the roots of *Ponthieva racemosa* are used as a purgative; **cultivation:** *Ponthieva racemosa* is not cultivated commercially; **misc. products:** none; **weeds:** none; **nonindigenous species:** none.

Systematics: *Ponthieva* is assigned to tribe Cranichideae (subtribe Cranichidinae) of orchid subfamily Orchidoideae (Figures 2.13 and 2.20). The genus is not monophyletic as currently circumscribed (some species associating with *Baskervilla*) but most species resolve as a derived clade within the subtribe (Figure 2.20). Other than *Baskervilla*, the group is also related closely to *Cranichis* (Figure 2.20). Analyses of combined DNA sequence data resolve *P. racemosa* as the sister species of *P. brittoniae*, the only other North American member of the genus (Figure 2.20). Chromosome counts are available for only two *Ponthieva* species: *P. mandonii* and *P. racemosa* (both $2n=46$) would indicate a possible base number of $x=23$. There are no reports of hybridization involving *P. racemosa*.

Distribution: *Ponthieva racemosa* occurs along the southeastern coastal plain of the United States and extends southward into South America.

References: Ackerman, 2002; Argue, 2012f; Brown, 2005; Bryson et al., 1994; Deutsch, 2002; Duggal, 1971; Godfrey & Wooten, 1979; Hossain, 2011; Huneycutt & Floyd, 1999; Leidolf & McDaniel, 1998; Lynch & Zomlefer, 2016; Nelson, 1986; Porcher, 1981; Robinson et al., 2011; Salazar et al., 2009; Singer & Cocucci, 1999; Szlachetko & Kolanowska, 2013; Tan & Judd, 1995; Westervelt et al., 2006; Wherry, 1928.

5.4.12.   Spiranthes

Lady's/ladies'-tresses; spiranthe

Etymology: from the Greek *speira anthos* ("coiled flower") in describing the spiral arrangement of flowers around the spike

Synonyms: *Gyrostachys* (in part); *Ibidium* (in part); *Limodorum* (in part); *Neottia* (in part); *Orchiastrum* (in part); *Triorchos* (in part)

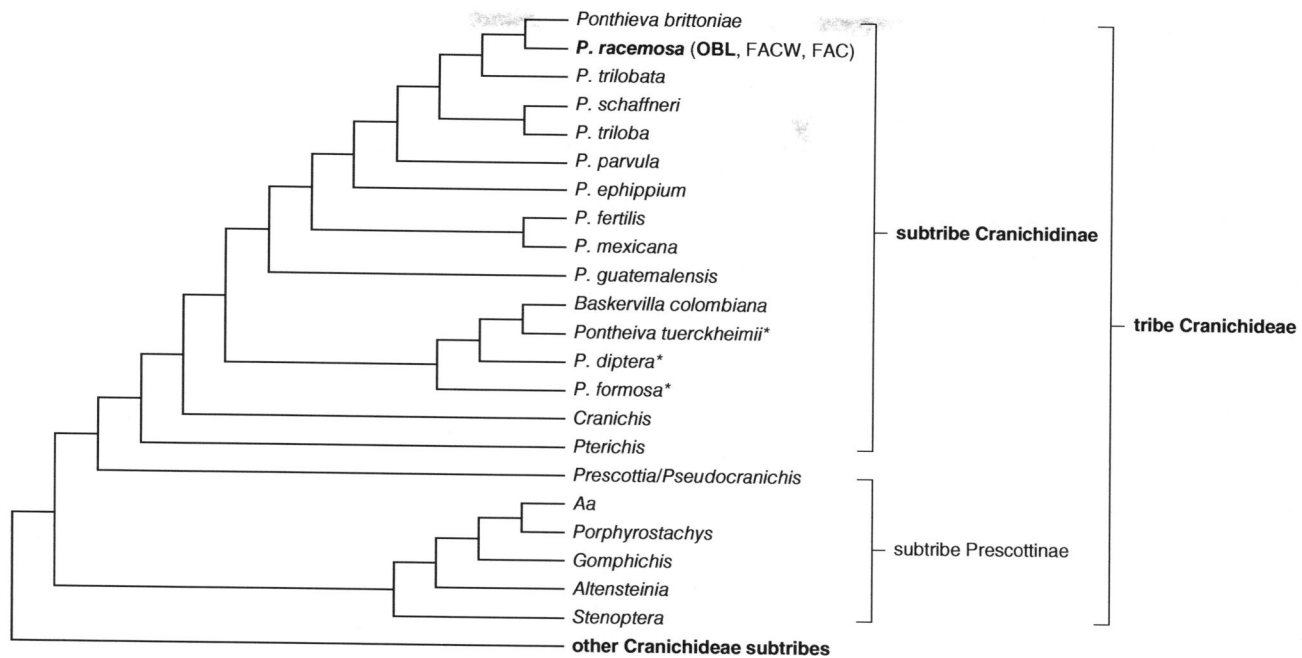

FIGURE 2.20   Relationship of *Ponthieva racemosa* as indicated by Bayesian phylogenetic analysis of combined *rbcL*, *matK-trnK*, *trnL-trnF*, and nrITS sequence data (simplified from Salazar et al., 2009). *Ponthieva* is monophyletic except for three species (asterisked) that group with *Baskervilla*; these are likely to be transferred to the latter genus pending the outcome of further taxon sampling and analyses (Salazar et al., 2009). *Ponthieva racemosa* occurs within a highly derived position in the genus. Wetland indicator designations are shown in parentheses; taxa containing OBL species are highlighted in bold. The tribe and subtribe designations refer to orchid subfamily Orchidoideae.

Distribution: global: Australia; Eurasia; New World; **North America:** widespread

Diversity: global: 36 species; **North America:** 27 species

Indicators (USA): obligate wetland (OBL): *Spiranthes delitescens*, *S. laciniata*, *S. longilabris*, *S. odorata*, *S. praecox*, *S. romanzoffiana*; **facultative wetland (FACW):** *S. laciniata*, *S. longilabris*, *S. praecox*, *S. romanzoffiana*

Habitat: freshwater; freshwater (tidal); palustrine; **pH:** acidic or alkaline; **depth:** <1 m; **life-form(s):** emergent herb

Key morphology: roots few to numerous, fleshy, fasciculate, slender to stout and tuberous (to 1 cm in diameter), sometimes stoloniferous and spreading horizontally; stems (to 1 m) with foliaceous sheaths; leaves 3–7, basal or cauline (then reduced to bracts upward), fugacious or persisting through anthesis, the blades (to 52 cm) ascending to spreading or spreading-recurved, linear to linear-lanceolate, elliptic, or oblanceolate to oblong- or linear-oblanceolate, sometimes keeled; spikes terminal, the flowers weakly to strongly gaping, recurved apically, ascending, nodding, or horizontal, arranged in a loose to dense spiral (3–9 flowers per spiral cycle) or sometimes nearly secund; perianth resupinate, colored cream, ivory, or white (rarely green or yellowish-white); sepals (to 18 mm) connivent (sepals and petals sometimes forming a hood), lanceolate, distinct or connate at base, the lateral sepals spreading, outwardly curving, or clasping lip; petals (to 18 mm) partly to strongly falcate, lanceolate to elliptic-lanceolate, linear, linear-oblanceolate or ovate, the lip (to 16 mm) lance-ovate, oblong, ovate, pandurate, or rhombic-ovate, its center creamy to yellow or green, sometimes fleshy or pleated transversely,

the apical margin crisped and crenulate, dentate, lacerate-crenulate or laciniate-dentate; column short, cylindric, pollinia clavate, the viscidia slender, linear to linear-lanceolate; ovary (to 8 mm) sessile, cylindric

Life history: duration: perennial (tuberous roots); **asexual reproduction:** stoloniferous or tuberous roots; **pollination:** insect; **sexual condition:** hermaphroditic; **fruit:** capsules (common); **local dispersal:** stoloniferous roots, seeds (wind); **long-distance dispersal:** seeds (wind)

Imperilment: 1. *Spiranthes delitescens* [G1]; S1 (AZ); **2.** *S. laciniata* [G4/G5]; S1 (NJ, SC); S2 (NC); S3 (FL, GA); **3.** *S. longilabris* [G3]; S1 (AL, GA, NC, SC, TX); S2 (MS); S3 (FL); **4.** *S. odorata* [G5]; SH (MD); S1 (AR, KY, OK, TN); S2 (NC, NJ); S3 (GA, VA); **5.** *S. praecox* [G5]; SH (DE, MD); S1 (AR, OK); S3 (GA, NC); **6.** *S. romanzoffiana* [G5]; SH (CT, IL); S1 (IA, IN, MA, ND, NE, PA); S2 (AZ, OH, NM, <u>NU</u>); S3 (<u>LB</u>, <u>PE</u>, VT, WY)

Ecology: general: Twenty North American *Spiranthes* species (74%) are wetland indicators, with nearly a quarter (22%) ranked as OBL in at least part of their range. The plants inhabit freshwaters with some species extending occasionally into tidal (but not brackish) sites. Autonomous agamospermy (adventitious embryony) has been confirmed experimentally in five North American species but so far has been detected in only one of the OBL species (*S. odorata*). The flowers are either self-compatible or self-incompatible. The flowers are nondeceptive and produce a fragrance as well as nectar. Typically, pollination occurs by long-tongued bees (Insecta: Hymenoptera: Apidae; Megachilidae) as they probe for nectar.

During that process, the pollinia attach by their rigid vis-cidia to the dorsal surface of the bee's projecting, flat-topped mouthparts (i.e., the galea). Due to the initially appressed position of the column against the lip in young flowers, it takes 2–4 days for the column to conform into an upright position, which then enables incoming pollinia to be deposited on the stigma. In this way, most of the North American species are protandrous, which facilitates outcrossing as the pollina-tors move upward on the spike, first visiting the lower, older, female-phase flowers before receiving pollen from the upper, younger, male-phase flowers. The small seeds are buoyant and are dispersed primarily by wind, or perhaps by water at times. Most seed germination information involves *in vitro* methods, which culture cold-stratified seeds either on media that lack mycorrhizal associates (asymbiotic) or in media sup-plemented with cultures of symbiotic endophytes (e.g., Fungi: Basidiomycota: Tulasnellaceae: *Epulorhiza*).

5.4.12.1. *Spiranthes delitescens* **Sheviak** occurs in cienegas (spring-fed), hummocks (stream), marshes (perennial, spring-fed), meadows (hummocky, marshy, riparian, springy, wet), seeps, slopes (seepage), and along the banks of streams at elevations from 1230 to 1530 m. The plants grow in expo-sures that receive full sunlight on substrates described as clay loam or Pima (anthropic torrifluven). Flowering commences in July and peaks during August, although a large proportion of plants (57–77%) often remains vegetative. Little is known about the floral biology or pollination ecology, except that the plants appear to be self-incompatible. The seeds have been germinated and cultured axenically following 3+ months of cold stratification (at 4°C). The seeds can be germinated *in vitro* either asymbiotically (on modified P668, O156, or W2.5 orchid media) or symbiotically using a culture known as Sbrev 266, which is an endomycorrhizal associate (Fungi: Basidiomycota). The adult plants live for an estimated 3–4 years on average, with some cultivated individuals known to have survived for 15 years or more. There is some evidence that the plants require disturbances (e.g., livestock grazing activities) to reduce competing vegetation in order to persist. However, prescribed burning (to remove accumulating lit-ter) has not been effective at stimulating the growth of pop-ulations, at least over a short term. Maximum biomass can range from 2,500 to 4,000 kg ha^{-1}. The species is listed as federally endangered in the United States. **Reported associ-ates:** *Apocynum cannabinum, Bidens aurea, Carex chihua-huensis, Eleocharis macrostachya, Eleocharis rostellata, Equisetum laevigatum, Juncus balticus, Juncus longistylis, Lythrum californicum, Mentha arvensis, Mimulus guttatus, Muhlenbergia asperifolia, Muhlenbergia utilis, Poa praten-sis, Ranunculus macranthus, Salix lasiolepis, Schoenoplectus americanus, Sisyrinchium demissum, Symphyotrichum falca-tum, Symphyotrichum praealtum.*

5.4.12.2. *Spiranthes laciniata* **(Small) Ames** inhabits bogs, depressions (seasonally inundated), ditches (roadside), flat-woods (mesic, pine, unburned), lawns, marshes (depression), meadows, pinelands, pools, prairies (wet), ravines (rich), roadsides (damp, dry), savannas (pitcher plant, pond cypress), slopes (seepage), sloughs, swamps (cypress, dome), thickets

(disturbed, wet), and the margins (shallow) of ponds (depres-sion) and streams at elevations to 50 m. The plants often occur in shallow (to 20 cm) standing water. The substrates are described as acidic and are characterized as Basinger (spodic psammaquents), Lawnwood (aeric haplaquods), Riviera (arenic glossaqualfs), sand (fine, loamy, mucky, Sellers), sandy alluvium, sandy peat, Vero (alfic haplaquods), and Winder sand (typic glossaqualfs). The plants are sexual and flower from April to September. The flowers are pro-tandrous. Additional details on the reproductive ecology are lacking. The plants sometimes have been found in recently burned habitats, but also in unburned sites. **Reported associ-ates:** *Acer rubrum, Amsonia tabernaemontana, Andropogon gyrans, Aristida palustris, Asclepias lanceolata, Bacopa caroliniana, Baptisia leucophaea, Calopogon tuberosus, Carex glaucescens, Carex verrucosa, Chaetopappa asteroi-des, Coleataenia tenera, Coreopsis nudata, Dichanthelium scabriusculum, Dichanthelium sphaerocarpon, Diospyros virginiana, Drosera capillaris, Eleocharis equisetoi-des, Eleocharis quadrangulata, Eriocaulon compressum, Eryngium yuccifolium, Eupatorium mohrii, Fimbristylis puberula, Fimbristylis puberula, Fraxinus caroliniana, Gratiola brevifolia, Helenium pinnatifidum, Hibiscus moscheutos, Hydrolea ovata, Hypericum fasciculatum, Hypericum myrtifolium, Ilex cassine, Ilex glabra, Ilex myrtifo-lia, Illicium, Itea virginica, Juncus effusus, Juncus polyceph-alus, Leersia hexandra, Liatris, Lilium catesbaei, Lobelia boykinii, Lobelia floridana, Ludwigia pilosa, Ludwigia sphaerocarpa, Lycopus rubellus, Lyonia lucida, Magnolia, Mimosa hystricina, Mnesithea rugosa, Muhlenbergia cap-illaris, Myrica cerifera, Nyssa biflora, Orontium aquati-cum, Panicum hemitomon, Panicum virgatum, Paspalum plicatulum, Persea palustris, Persicaria hydropiperoi-des, Pinguicula planifolia, Pinus elliottii, Pinus palustris, Pityopsis graminifolia, Platanthera nivea, Pleopeltis polypo-dioides, Polygala cymosa, Polygala mariana, Proserpinaca palustris, Proserpinaca pectinata, Rhexia aristosa, Rhexia virginica, Rhynchospora brachychaeta, Rhynchospora car-eyana, Rhynchospora cephalantha, Rhynchospora cor-niculata, Rhynchospora filifolia, Rhynchospora galeana, Rhynchospora latifolia, Rhynchospora tracyi, Rudbeckia texana, Sagittaria graminea, Sarracenia flava, Sarracenia minor, Saururus cernuus, Schizachyrium scoparium, Schizachyrium tenerum, Scleria baldwinii, Serenoa repens, Silphium radula, Smilax laurifolia, Taxodium ascendens, Thelypteris interrupta, Tiedemannia canbyi, Tiedemannia filiformis, Tillandsia recurvata, Tillandsia simulata, Tillandsia ×floridana, Toxicodendron radicans, Tripsacum dactyloides, Woodwardia areolata, Woodwardia virgi-nica, Xyris ambigua, Xyris fimbriata, Xyris laxifolia, Xyris smalliana.*

5.4.12.3. *Spiranthes longilabris* **Lindl.** grows in bogs (hill-side, pitcher plant, sandy, seepage), ditches (roadside), flat-woods (moist, pine), meadows (dry, moist), pine barrens (moist), pinelands (low, open, wet), prairies (wet), roadsides, savannas (pine), swamps (hardwood, pine), and along the margins of ponds at elevations to 21 m. The plants favor exposures of full sunlight and grow on substrates that have been characterized as Felda (arenic ochroqualfs), Malabar (grossarrenic ochraqualfs), Rains (typic paleaquults, and sand. Flowering occurs from October to December. A number of occurrences have been described from recently burned or fire-prone, sites. Otherwise, little life history information exists for this species. **Reported associates:** *Aletris lutea, Amphicarpum muhlenbergianum, Andropogon, Aristida palustris, Aristida stricta, Calopogon barbatus, Calopogon multiflorus, Calopogon pallidus, Calopogon tuberosus, Carphephorus pseudoliatris, Centella asiatica, Cleistesiopsis bifaria, Ctenium aromaticum, Cynodon, Cyperus, Dichanthelium, Drosera, Eriocaulon compressum, Eriocaulon decangulare, Fuirena scirpoidea, Hypericum, Ilex myrtifolia, Lachnanthes caroliniana, Liatris, Lilium catesbaei, Lophiola aurea, Ludwigia, Magnolia virginiana, Nyssa biflora, Panicum hemitomon, Paspalum, Persea palustris, Pinguicula, Pinus elliottii, Pinus palustris, Pinus taeda, Platanthera blephariglottis, Platanthera integra, Platanthera nivea, Pogonia ophioglossoides, Polygala, Ptilimnium, Rhexia, Rhynchospora inundata, Sabatia, Sarracenia, Scleria muehlenbergii, Spiranthes praecox, Sporobolus floridanus, Stenanthium densum, Stenotaphrum secundatum, Stokesia laevis, Taxodium ascendens, Triantha racemosa, Utricularia, Viola lanceolata, Xyris, Zigadenus glaberrimus.*

5.4.12.4. *Spiranthes odorata* (Nutt.) Lindl. grows in backwaters, bogs (*Sphagnum*), bottomlands (forested, hardwood), depressions, ditches (grassy, roadside), dunes, flatwoods (hardwood, pine), floodplains (river, stream), gravel pits, hammocks (swampy), marshes (brackish, oligohaline, tidal, wind-tidal), prairies (marl, wet), roadsides (grassy), sand pits (intermittent), savannas (wet), shores (tidal), slopes (seepage), sloughs (cypress), swamps (basin, bottomland, cypress, hardwood, strand), and along the margins of channels, lakes (boggy), rivers, and streams at elevations to 549 m. The plants occur mostly in shade (often deep) but also are tolerant of full sunlight. Typically they grow in seasonally inundated, subacidic, sometimes tidal (but never very brackish), sites. They tolerate shallow inundation (e.g., to 61 cm), which can cover plants completely in tidal areas during high tide. The substrates have been described as alluvium, Chowan (thapto-histic fluvaquents), clay (alluvial), humus (sandy), loam, marl, muck, mud, Okeelant-terra Ceia (medisaprists), peat (*Sphagnum*), and silt; some plants also have been found growing on cypress knees, floating logs, or other floating wood debris. Flowering and fruiting extend from September to January and can occur while the plants are emerging from shallow water. The sweetly fragrant, vanilla- or jasmine-scented flowers are self-compatible and yield numerous seeds when artificially self-pollinated. The plants are at least facultatively agamospermous, as evidenced by the documented occurrence of polyembryony (adventitious embryony) in cultivated material and from specimens in the northeastern portion of its range. Otherwise they are pollinated (outcrossed or selfed) by bumblebees (e.g., Insecta: Hymenoptera: Apidae: *Bombus pensylvanicus*). Seed production is high and can result from apomixis. The seeds

float and are dispersed abiotically by water or wind. Seed germination is impaired by light but enhanced by cold stratification. Germination has succeeded (to 29%) after drying mature capsules at 22°C followed by cold storage (6°C) of the seeds for 15 weeks. The stratified seeds are then surface sterilized, cultured along with fungal isolates (Fungi: Basidiomycota: Tulasnellaceae: *Epulorhiza*), and incubated in complete darkness at 21°C for 70 days. After germination (which commences in about 10 days), the developing protocorms are transferred to a 16/8 hr, 20°C/18°C day/night light and temperature regime. Light is necessary to induce protocorm shoot development. The plants reproduce vegetatively by producing stoloniferous roots (to 30 cm), from which vegetative adventitious shoot offshoots arise, leading to the development of large clonal colonies. They have been known to spread into brackish marshes following extended periods of low salinity. Some plants reportedly have newly appeared in sites following prescribed burns. **Reported associates:** *Acer rubrum, Acmella oppositifolia, Alnus serrulata, Alternanthera philoxeroides, Amorpha fruticosa, Apios americana, Aulacomnium palustre, Bacopa monnieri, Bidens laevis, Bignonia capreolata, Boehmeria cylindrica, Boltonia asteroides, Brunnichia ovata, Carex crus-corvi, Carex decomposita, Carex hyalinolepis, Carex tribuloides, Carpinus caroliniana, Carya ovata, Celtis laevigata, Cephalanthus occidentalis, Ceratopteris thalictroides, Cladonia, Clethra alnifolia, Commelina virginica, Crataegus viridis, Cyperus haspan, Decodon verticillatus, Dichanthelium scoparium, Dryopteris ludoviciana, Eleocharis quadrangulata, Eleocharis rostellata, Elymus virginicus, Eubotrys racemosa, Euthamia leptocephala, Fraxinus caroliniana, Fraxinus pennsylvanica, Fraxinus profunda, Gentiana catesbaei, Glyceria obtusa, Gymnopogon ambiguus, Habenaria repens, Hydrocotyle ranunculoides, Hymenocallis, Hypericum mutilum, Ilex amelanchier, Ilex glabra, Impatiens capensis, Iris, Isolepis carinata, Itea virginica, Juncus canadensis, Juncus coriaceus, Juncus megacephalus, Juncus roemerianus, Justicia ovata, Leersia lenticularis, Lemna minor, Leucobryum albidum, Lilaeopsis carolinensis, Limnobium spongia, Liquidambar styraciflua, Lobelia elongata, Ludwigia alata, Lycopodiella appressa, Lycopodiella ×copelandii, Lycopus, Magnolia virginiana, Micranthemum umbrosum, Morus rubra, Myrica cerifera, Myrica gale, Nyssa aquatica, Nyssa biflora, Nyssa ogeche, Oldenlandia uniflora, Onoclea sensibilis, Panicum virgatum, Peltandra virginica, Penthorum sedoides, Persicaria hydropiper, Persicaria hydropiperoides, Phanopyrum gymnocarpon, Phragmites australis, Phyla nodiflora, Pilea pumila, Pinus elliottii, Pinus palustris, Platanthera flava, Platanthera triloba, Pogonia ophioglossoides, Potamogeton, Proserpinaca palustris, Quercus laurifolia, Quercus lyrata, Quercus michauxii, Quercus nigra, Quercus phellos, Rhexia mariana, Rhynchospora alba, Rhynchospora colorata, Rhynchospora corniculata, Rhynchospora miliacea, Rhynchospora scirpoides, Rosa palustris, Rotala ramosior, Rudbeckia, Sabal minor, Sabal minor, Saccharum brevibarbe, Sacciolepis striata, Sagittaria, Saururus cernuus, Schoenoplectus americanus, Smilax, Solanum pseudocapsicum, Spartina cynosuroides,*

Spartina patens, Spiranthes vernalis, Symphyotrichum concolor, Symphyotrichum elliotii, Symphyotrichum lateriflorum, Symphyotrichum novi-belgii, Taxodium ascendens, Taxodium distichum, Teucrium canadense, Toxicodendron radicans, Triglochin striata, Typha angustifolia, Typha domingensis, Ulmus alata, Ulmus americana, Ulmus crassifolia, Vaccinium corymbosum, Woodwardia areolata, Xyris caroliniana, Xyris torta, Zizania.

5.4.12.5. *Spiranthes praecox* (Walter) S. Watson occurs in bogs (hillside, pineland, pitcher plant, sandy), depressions (moist, wet), ditches (wet), flats (shrubby), flatwoods (mesic, pine, roadside, scrubby), floodplains (river), glades (moist), hammocks (hardwood, hydric, sandy, wet-mesic), marshes (coastal), meadows (boggy, moist, sandy, wet), pine barrens (moist), pinelands (moist), pocosins, prairies (wet), ravines (damp, shaded, wooded), rights-of-way (powerline), roadsides (dry, low, moist), savannas (open, pine), seeps (mucky, sand), swales (dune, oak, pine), swamps (cypress), woodlands (oak, open, pine, rich), and along the margins of borrow pits (flooded), ponds, and swales (roadside) at elevations to 100 m. Exposures range from full sun to partial shade. The substrates most often are characterized as acidic and are described as Bayboro, clay, loam, loamy sand (Leefield, Pelham), muck, peat, sand (Lakeland), sandy clay, sandy loam (fine, Pooler), sandy peat, and silty loam. Flowering and fruiting extend from February to September. There have been no detailed investigations of the pollination biology or floral biology of this species, although the direction of the floral spiral up the inflorescence (clockwise vs. counterclockwise) has been found to develop randomly. No details are available on seed germination or ecology. Presumably, the seeds are dispersed by wind. The plants thrive in mowed and frequently burned sites, where competitive vegetation has been reduced. **Reported associates:** *Acer rubrum, Aesculus pavia, Aletris aurea, Aletris farinosa, Aletris lutea, Alnus serrulata, Amorpha, Ampelopsis arborea, Andropogon virginicus, Aralia spinosa, Aristida stricta, Asclepias michauxii, Baptisia lanceolata, Bejaria, Boehmeria cylindrica, Buchnera floridana, Callicarpa americana, Calopogon pallidus, Calopogon tuberosus, Camassia scilloides, Carex bullata, Carex lurida, Carex oklahomensis, Carpinus caroliniana, Carya illinoinensis, Ceanothus americanus, Centella asiatica, Cephalanthus occidentalis, Cicuta maculata, Cirsium horridulum, Cleistesiopsis divaricata, Clethra alnifolia, Clinopodium coccineum, Conradina canescens, Cornus florida, Ctenium, Cynosciadium digitatum, Cyrilla racemiflora, Desmodium rotundifolium, Dichanthelium, Dionaea muscipula, Drosera capillaris, Eleocharis tuberculosa, Eleocharis wolfii, Erigeron vernus, Eriocaulon, Eubotrys racemosa, Eupatorium mohrii, Euphorbia floridana, Galium obtusum, Gaylussacia dumosa, Gaylussacia frondosa, Gordonia, Hexastylis, Hymenocallis, Hypericum crux-andreae, Hypericum mutilum, Hypericum myrtifolium, Hypericum suffruticosum, Hypoxis juncea, Ilex coriacea, Ilex glabra, Ilex vomitoria, Isoetes melanopoda, Juncus canadensis, Juncus marginatus, Kalmia hirsuta, Lachnocaulon anceps, Lindernia monticola, Liquidambar styraciflua, Lobelia nuttallii, Lobelia paludosa, Lolium* *perenne, Lyonia ferruginea, Lyonia lucida, Magnolia grandiflora, Magnolia virginiana, Malus angustifolia, Minuartia patula, Myrica cerifera, Myrica gale, Neptunia lutea, Nolina georgiana, Nuttallanthus floridanus, Nyssa biflora, Oclemena reticulata, Osmanthus, Osmundastrum cinnamomeum, Panicum, Parthenocissus quinquefolia, Paspalum dilatatum, Persea palustris, Persicaria hydropiper, Phacelia glabra, Phlox, Pilea pumila, Pinguicula, Pinus caribaea, Pinus clausa, Pinus echinata, Pinus palustris, Pinus taeda, Pogonia ophioglossoides, Polygala lutea, Polygala nana, Polygala ramosa, Proserpinaca palustris, Prunus caroliniana, Prunus serotina, Pteridium aquilinum, Pterocaulon pycnostachyum, Quercus alba, Quercus arkansana, Quercus falcata, Quercus geminata, Quercus hemisphaerica, Quercus incana, Quercus laevis, Quercus margarettae, Quercus minima, Quercus nigra, Quercus phellos, Quercus virginiana, Rhexia alifanus, Rhus copallinum, Rhynchospora alba, Rosa palustris, Rudbeckia, Ruellia, Sabal minor, Sabal palmetto, Sabatia campestris, Sarracenia flava, Sarracenia minor, Sassafras albidum, Scleria, Serenoa repens, Smilax laurifolia, Solidago, Sphagnum, Spigelia, Spiranthes lacera, Spiranthes tuberosa, Sporobolus indicus, Sporobolus junceus, Steinchisma hians, Stylosanthes, Styrax americanus, Syngonanthus flavidulus, Taxodium ascendens, Tephrosia, Thyrsanthella difformis, Toxicodendron pubescens, Toxicodendron radicans, Tradescantia hirsutiflora, Ulmus alata, Utricularia cornuta, Vaccinium arboreum, Vaccinium corymbosum, Vaccinium stamineum, Verbena, Veronica peregrina, Vitis rotundifolia, Wisteria frutescens, Woodwardia areolata, Woodwardia virginica, Xyris caroliniana, Yucca filamentosa.*

5.4.12.6. *Spiranthes romanzoffiana* Cham. inhabits beaches, bluffs (coastal), bogs (alpine, coniferous, hanging, hot springs, *Sphagnum*), borrow pits, depressions (interdunal, seasonally wet), ditches (moist, roadside, wet), fens, flats (alluvial, muddy), floodplains, gravel bars, hummocks (bog), ledges (argyllite), marshes (freshwater), meadows (alpine, beach, boggy, marshy, moist, muskeg, peaty, seasonally wet, tundra, wet), prairies (tall grass), ravines (wet), roadsides (boggy, wet), seeps, slopes (to 40%), springs (warm), swamps, thickets (open, willow), tundra, vernal pools, woodlands, and the margins (boggy) of canals (irrigation), channels (ephemeral), lakes, muskeg, ponds (ephemeral, spring-fed), potholes, reservoirs, rivers, and streams at elevations to 3424 m. The plants occur in open to partially shaded exposures. The substrates most often are described as calcareous (e.g., pH: 7.3; total alkalinity: 48.0 ppm) and have been characterized as alluvium (cobbly, granitic, sandy, stony), clay, cobble, gravel (glacial, organic), loam (volcanic), loamy sand, marl, mica-shist (decomposed), muck, mud, peat, peaty humus, pumice, quartzite, rocky loam, sapric histosol (hydric-mesic), sand (rocky), sandy loam, sandy peat, silt (organic), silty loam, and sinter. Flowering and fruiting extend from June to October. Individual flowers can persist from 10 to 40 days, while remaining receptive for up to 30 days. The flowers are sexual and are self-compatible, with 64% seed set obtained by artificial self or geitonogamous pollination, and 100% seed set from artificially outcrossed pollinations. However, undisturbed flowers set no seed and must rely on pollination by various bees (Insecta: Hymenoptera: Apidae: *Apis mellifera, Bombus ashtoni, B. bifarius, B. borealis, B. fervidus, B. flavifrons, B. insularis, B. perplexus, B. terricola, B. vagans*; Halictidae: *Halictus confusus, Lasioglossum smilacinae, L. zonulum*; Megachilidae: *Megachile melanophaea*). The bees forage the fragrant flowers nonrandomly by preferentially seeking those plants having the tallest spikes and longest inflorescences. Within the flower, the pollinia are attached to a wedge-shaped viscidium, which is protected within a delicate membrane. Contact with the membrane causes it to split and expose the adhesive surface. The flowers are protandrous. As the bees forage for the nectar, their proboscis is directed narrowly between the column and lip, causing it to contact the rostellum, split open the viscidial membrane, and acquire the pollinia, which adhere to their upper proboscis surface. Pollinia are removed from most of the flowers (e.g., 87.2%), with little pollen remaining within the anthers (e.g., 16.7%) after anthesis. Following pollinia removal, the rostellum withers and the column moves slowly away from the lip, which only then exposes the stigmatic surface to subsequently arriving pollinators. Pollinator visitation rates can vary from 5 to 35 s. The flowers fade and dehydrate within 3 days of pollination. The seeds develop within 1–3 weeks. They are buoyant and dispersed primarily by wind. The roots are associated with mycorrhizal endophytes (Fungi: Basidiomycota: Ceratobasidiaceae: *Ceratorhiza*; Tulasnellaceae: *Epulorhiza*), which are thought to facilitate the establishment and early development of young plants. The plants persist by means of elongate, fasciculate, tuberous storage roots, which can be separated to propagate them vegetatively. Plant biomass generally is low (i.e., <1 kg ha^{-1}). **Reported associates:** *Abies grandis, Abies lasiocarpa, Abies magnifica, Abies ×shastensis, Acer glabrum, Acer rubrum, Achillea millefolium, Achnatherum occidentale, Acmispon americanus, Acmispon parviflorus, Aconitum columbianum, Agalinis purpurea, Agrostis hyemalis, Agrostis idahoensis, Agrostis pallens, Agrostis scabra, Agrostis stolonifera, Agrostis variabilis, Aira caryophyllea, Aira praecox, Alnus rhombifolia, Alnus viridis, Amelanchier alnifolia, Anaphalis margaritacea, Andromeda polifolia, Andropogon gerardii, Anemopsis californica, Angelica hendersonii, Antennaria howellii, Anthoxanthum odoratum, Apocynum androsaemifolium, Aralia nudicaulis, Arctostaphylos uva-ursi, Arctous rubra, Armeria maritima, Arnica fulgens, Arnica latifolia, Astragalus miser, Athyrium alpestre, Athyrium filix-femina, Aulacomnium acuminatum, Aulacomnium palustre, Avena barbata, Berula erecta, Betula glandulosa, Betula nana, Betula neoalaskana, Betula occidentalis, Betula pumila, Bistorta bistortoides, Blechnum spicant, Botrychium minganense, Botrychium multifidum, Briza minor, Brodiaea elegans, Bromus carinatus, Bromus inermis, Calamagrostis canadensis, Calliergon, Caltha leptosepala, Calypogeia sphagnicola, Camassia quamash, Campanula parryi, Campylium stellatum, Canadanthus modestus, Carex aquatilis, Carex aurea, Carex brunnescens, Carex buxbaumii, Carex canescens, Carex capillaris, Carex capitata, Carex chordorrhiza, Carex crawei, Carex diandra, Carex echinata, Carex flava, Carex granularis, Carex hystericina, Carex integra, Carex interior, Carex lasiocarpa, Carex lemmonii, Carex lenticularis, Carex leptalea, Carex limosa, Carex luzulina, Carex magellanica, Carex mariposana, Carex microglochin, Carex muricata, Carex nudata, Carex obnupta, Carex pauciflora, Carex paysonis, Carex pluriflora, Carex raynoldsii, Carex rostrata, Carex sartwellii, Carex saxatilis, Carex scoparia, Carex scopulorum, Carex simulata, Carex sterilis, Carex tetanica, Carex utriculata, Carex vaginata, Carex viridula, Carex xerantica, Cassiope mertensiana, Castilleja miniata, Castilleja sulphurea, Cephalozia connivens, Chamaecyparis nootkatensis, Chamaedaphne calyculata, Chamerion angustifolium, Chara, Chelone glabra, Chimaphila menziesii, Cicuta bulbifera, Cicuta douglasii, Cirsium arvense, Cirsium muticum, Cladonia cristatella, Clarkia pulchella, Clintonia borealis, Collinsia tinctoria, Comarum palustre, Comptonia peregrina, Coptis trifolia, Coreopsis tinctoria, Cornus canadensis, Cornus sericea, Corylus cornuta, Cryptogramma acrostichoides, Cynosurus echinatus, Cyperus, Cypripedium reginae, Cytisus scoparius, Danthonia spicata, Darlingtonia californica, Dasiphora floribunda, Deschampsia cespitosa, Deschampsia danthonioides, Dichanthelium acuminatum, Dichanthelium oligosanthes, Diervilla lonicera, Distichlis, Drepanocladus, Drosera anglica, Drosera rotundifolia, Dryopteris carthusiana, Dulichium arundinaceum, Elaeagnus commutata, Eleocharis elliptica, Eleocharis macrostachya, Eleocharis palustris, Eleocharis quinqueflora, Eleocharis rostellata, Eleocharis tenuis, Elymus albicans, Empetrum nigrum, Epigaea repens, Epilobium anagallidifolium, Epilobium brachycarpum, Epilobium ciliatum, Epilobium oregonense, Epipactis gigantea, Equisetum arvense, Equisetum fluviatile, Equisetum laevigatum, Equisetum palustre, Equisetum pratense, Equisetum scirpoides, Equisetum telmateia, Equisetum variegatum, Erigeron coulteri, Erigeron elmeri, Erigeron glacialis, Erigeron peregrinus, Eriogonum heracleoides, Eriophorum angustifolium, Eriophorum chamissonis, Eriophorum gracile, Eryngium petiolatum, Eurybia integrifolia, Eurybia macrophylla, Eutrema salsugineum, Festuca altaica, Festuca idahoensis, Festuca roemeri, Fragaria vesca, Fragaria virginiana, Fritillaria camschatcensis, Gaillardia, Galium trifidum, Galium triflorum, Gaultheria hispidula, Gaultheria humifusa, Gaultheria ovatifolia, Gaultheria procumbens, Gaultheria shallon, Gentiana calycosa, Gentiana douglasiana, Gentiana sceptrum, Gentianopsis crinita, Gentianopsis holopetala, Gentianopsis simplex, Gentianopsis virgata, Geocaulon lividum, Geranium oreganum, Geum macrophyllum, Glyceria elata, Glycyrrhiza lepidota, Goodyera oblongifolia, Gymnocarpium dryopteris, Hastingsia alba, Hesperostipa comata, Heterocodon rariflorus, Heuchera cylindrica, Hieracium aurantiacum, Hieracium caespitosum, Hieracium scabrum, Holodiscus discolor, Hosackia oblongifolia, Hydrocotyle ranunculoides, Hylocomium splendens, Hymenoxys hoopesii, Hypericum anagalloides, Hypericum formosum, Hypnum, Hypochaeris radicata, Hypoxis hirsuta, Iris missouriensis, Iris setosa, Ivesia campestris, Juncus alpinoarticulatus, Juncus alpinus, Juncus balticus, Juncus brachyphyllus, Juncus breweri, Juncus confusus, Juncus* *dubius, Juncus effusus, Juncus ensifolius, Juncus lesueurii, Juncus marginatus, Juncus occidentalis, Juncus orthophyllus, Juncus tenuis, Juniperus communis, Kalmia microphylla, Kalmia polifolia, Kalmia procumbens, Koeleria pyramidata, Larix laricina, Leucanthemum vulgare, Lewisia cotyledon, Ligusticum canbyi, Ligusticum grayi, Ligusticum tenuifolium, Lilium kelleyanum, Lilium pardalinum, Limprichtia revolvens, Linnaea borealis, Linum, Liparis loeselii, Lobelia kalmii, Lomatium ambiguum, Lonicera caerulea, Lonicera involucrata, Lophozia rutheana, Lupinus arcticus, Lupinus argenteus, Lupinus latifolius, Lupinus polyphyllus, Luzula campestris, Lycopodium annotinum, Lycopodium clavatum, Lycopodium obscurum, Lycopus americanus, Lycopus uniflorus, Lysichiton americanus, Lysimachia thyrsiflora, Madia glomerata, Maianthemum bifolium, Maianthemum canadense, Maianthemum stellatum, Maianthemum trifolium, Malus fusca, Marchantia polymorpha, Medicago lupulina, Meesia longiseta, Meesia triquetra, Meesia uliginosa, Melampyrum lineare, Melilotus albus, Mentha arvensis, Menyanthes trifoliata, Menziesia ferruginea, Mertensia paniculata, Micranthes nidifica, Micranthes odontoloma, Micranthes oregana, Mimulus glabratus, Mimulus guttatus, Mimulus moschatus, Mimulus primuloides, Mitella pentandra, Muhlenbergia andina, Muhlenbergia filiformis, Muhlenbergia glomerata, Mylia anomala, Myosotis scorpioides, Myrica californica, Myrica gale, Narthecium californicum, Oenanthe sarmentosa, Oreostemma alpigenum, Orthocarpus bracteosus, Osmorhiza berteroi, Oxalis montana, Oxypolis occidentalis, Packera aurea, Packera pseudaurea, Packera subnuda, Parnassia fimbriata, Parnassia glauca, Parnassia palustris, Paxistima myrsinites, Pedicularis attollens, Pedicularis groenlandica, Pedicularis labradorica, Penstemon attenuatus, Penstemon newberryi, Penstemon procerus, Perideridia gairdneri, Perideridia parishii, Petasites frigidus, Philonotis, Phleum alpinum, Phleum pratense, Phyllodoce empetriformis, Physocarpus capitatus, Picea breweriana, Picea engelmannii, Picea glauca, Picea mariana, Picea sitchensis, Pinguicula vulgaris, Pinus albicaulis, Pinus banksiana, Pinus contorta, Pinus monticola, Plantago lanceolata, Plantago major, Plantago maritima, Platanthera aquilonis, Platanthera dilatata, Platanthera elegans, Platanthera huronensis, Platanthera hyperborea, Platanthera sparsiflora, Platanthera yosemitensis, Platydictya jungermannioides, Poa nemoralis, Poa pratensis, Poa secunda, Podagrostis humilis, Podagrostis thurberiana, Polemonium occidentale, Polygala paucifolia, Polystichum munitum, Polytrichum juniperinum, Polytrichum strictum, Populus angustifolia, Populus balsamifera, Populus tremuloides, Populus trichocarpa, Potentilla anserina, Potentilla canadensis, Potentilla drummondii, Potentilla gracilis, Potentilla grayi, Primula jeffreyi, Primula pauciflora, Primula tetrandra, Prunella vulgaris, Prunus virginiana, Pseudoroegneria spicata, Pseudotsuga menziesii, Pteridium aquilinum, Pyrola elliptica, Ranunculus alismifolius, Ranunculus flammula, Ranunculus occidentalis, Rhamnus alnifolia, Rhododendron columbianum, Rhododendron groenlandicum, Rhododendron macrophyllum, Rhododendron occidentale, Rhododendron tomentosum,*

Rhynchospora capillacea, Rhytidium, Ribes, Rosa acicularis, Rosa rubiginosa, Rosa woodsii, Rubus arcticus, Rubus chamaemorus, Rubus idaeus, Rubus pedatus, Rubus pubescens, Rubus repens, Rubus ursinus, Rudbeckia occidentalis, Rumex acetosella, Salix arbusculoides, Salix boothii, Salix brachycarpa, Salix candida, Salix eastwoodiae, Salix exigua, Salix geyeriana, Salix glauca, Salix hookeriana, Salix monticola, Salix myrtillifolia, Salix orestera, Salix pedicellaris, Salix pulchra, Salix scouleriana, Salix serissima, Salix sitchensis, Salix wolfii, Salix ×bebbii, Sanguisorba officinalis, Saussurea, Scheuchzeria palustris, Schoenoplectus acutus, Schoenoplectus americanus, Scirpus atrovirens, Scirpus microcarpus, Scleria triglomerata, Scleria verticillata, Scorpidium scorpioides, Scutellaria, Sedum obtusatum, Selaginella apoda, Senecio scorzonella, Senecio triangularis, Shepherdia argentea, Sisyrinchium funereum, Sisyrinchium montanum, Solidago canadensis, Solidago gigantea, Solidago multiradiata, Solidago patula, Solidago spathulata, Solidago uliginosa, Sorbus sitchensis, Sorghastrum nutans, Sparganium natans, Spartina pectinata, Sphagnum angustifolium, Sphagnum centrale, Sphagnum fuscum, Sphagnum magellanicum, Sphagnum teres, Sphagnum warnstorfii, Sphenosciadium capitellatum, Spiraea alba, Spiraea douglasii, Spiraea spendens, Spiranthes cernua, Spiranthes lucida, Sporobolus airoides, Stellaria longifolia, Stellaria longipes, Streptopus lanceolatus, Swertia perennis, Symphoricarpos albus, Symphyotrichum foliaceum, Symphyotrichum puniceum, Symphyotrichum spathulatum, Taraxacum officinale, Thelypteris palustris, Thermopsis montana, Thuja plicata, Tomenthypnum nitens, Torreyochloa erecta, Triadenum virginicum, Triantha glutinosa, Triantha occidentalis, Trichophorum cespitosum, Trichostema oblongum, Trientalis borealis, Trientalis europaea, Trifolium dubium, Trifolium hybridum, Trifolium longipes, Trifolium repens, Trifolium wormskioldii, Triglochin maritima, Triglochin palustris, Trisetum cernuum, Trollius albiflorus, Trollius laxus, Tsuga heterophylla, Tsuga mertensiana, Typha latifolia, Utricularia minor, Vaccinium angustifolium, Vaccinium cespitosum, Vaccinium membranaceum, Vaccinium myrtilloides, Vaccinium ovalifolium, Vaccinium ovatum, Vaccinium oxycoccos, Vaccinium parvifolium, Vaccinium scoparium, Vaccinium uliginosum, Vaccinium vitis-idaea, Valeriana edulis, Veratrum californicum, Verbascum thapsus, Veronica peregrina, Veronica serpyllifolia, Veronica wormskjoldii, Viburnum, Viola adunca, Viola glabella, Viola macloskeyi, Viola orbiculata, Viola sororia, Wyethia helianthoides, Xerophyllum tenax.

Use by wildlife: The flowers of *Spiranthes odorata* are visited by bumblebees (Insecta: Hymenoptera: Apidae: *Bombus pensylvanicus*). The flowers of *S. romanzoffiana* are visited by various insects (Insecta), including pollinating bees (Hymenoptera: Apidae: *Apis mellifera, Bombus ashtoni, B. borealis, B. fervidus, B. flavifrons, B. insularis, B. perplexus, B. terricola, B. vagans*; Halictidae: *Halictus confusus, Lasioglossum smilacinae, L. zonulum*; Megachilidae: *Megachile melanophaea*) and nonpollinating beetles (Coleoptera: Curculionidae), butterflies (Lepidoptera: Hesperiidae: *Erynnis lucilius, Thymelicus* *lineola*; Noctuidae: *Syngrapha epigaea*, *S. octoscripta*), flies (Diptera: Syrphidae), and sweat bees (Hymenoptera: *Lasioglossum cinctipes*). *Spiranthes romanzoffiana* is a host to aphids (Insecta: Hemiptera: Aphididae: *Aphis fabae*). The plants are grazed by elk (Mammalia: Cervidae: *Cervus elaphus*) and livestock (Mammalia: Bovidae).

Economic importance: food: The Paiute ate the roots of some *Spiranthes* species; **medicinal:** Plants of *S. romanzoffiana* were used by the Gosiute tribe to prepare a remedy for venereal disease; **cultivation:** *Spiranthes* species are not often cultivated commercially; however, 'Chadd's Ford' (a cultivar of *Spiranthes odorata*) is an RHS "Award of Garden Merit" plant; **misc. products:** none; **weeds:** none; **nonindigenous species:** none.

Systematics: *Spiranthes* is assigned to tribe Cranichideae (subtribe Spiranthinae) of the orchid subfamily Orchidoideae (Figures 2.13 and 2.21). The genus is monophyletic (Figure 2.22) and represents a relatively specialized clade within subtribe Spiranthinae (Figure 2.21). The OBL habit appears to have evolved several times throughout the genus (Figure 2.22). DNA sequence data have shown *S. longilabris* to be very closely related to the recently described *S. igniorchis*. Care should be taken when interpreting literature accounts and database records, which often treat *S. diluvialis* as a variety of *S. romanzoffoana* (i.e., *S. romanzoffiana* var. *diluvialis*). However, *S. diluvialis* is an allopolyploid derivative of *S. romanzoffiana* (maternal) and *S. magnicamporum* (paternal) (Figure 2.22). *Spiranthes delitescens* is also presumably an allopolyploid, which was thought to have arisen from *S. porrifolia* and *S.*

vernalis; however, that hypothetical parentage is not supported by molecular data (Figure 2.22). A more critical study of the *S. odorata* accessions that were placed disparately in phylogenetic analyses has led to the recent restoration of the morphologically similar (but distinct) *S. triloba*. This factor now makes it difficult to determine which species is being described in older literature accounts for "*S. odorata*," at least in Florida, where their ranges overlap. Several floral regulatory genes (AP3/DEF-like and PI/GLO-like MADS box genes, e.g., SpodoDEF1 SpodoDEF2, SpodoDEF3, SpodoDEF4, SpodoGLO1) have been identified in *Spiranthes odorata*. The characterization of such genes has led to the development of the "orchid code," which hypothesizes that the orchid flower has evolved primarily through duplications and mutations of these genes and their subsequent functional diversification. *Spiranthes* species represent two primary chromosomal series: $x = 15$ and $x = 22$. *Spiranthes odorata* ($2n = 30$) is a diploid of the former and *S. romanzoffiana* ($2n = 44, 66, 88$) has several polyploid races based on the latter. *Spiranthes delitescens* ($2n = 74$) is thought to be of allopolyploid origin with parentages descending from both lineages (i.e., $x = 15 + 22$). Various F_1 hybrids include *Spiranthes* × *folsomii* (*S. longilabris* × *S. odorata*), *S.* ×*meridionalis* (*S. praecox* × *S. vernalis*), and *S.* ×*simpsonii* (*S. romanzoffiana* × *S. lacera*). Alleged hybrids between *S. praecox* × *S. lacera* also have been reported (as *S.* ×*intermedia*) but are recognized by some authors as *S. casei*.

Distribution: *Spiranthes romanzoffiana* is distributed widely across northern North America and at higher western montane elevations; it also extends into Great Britain and northern

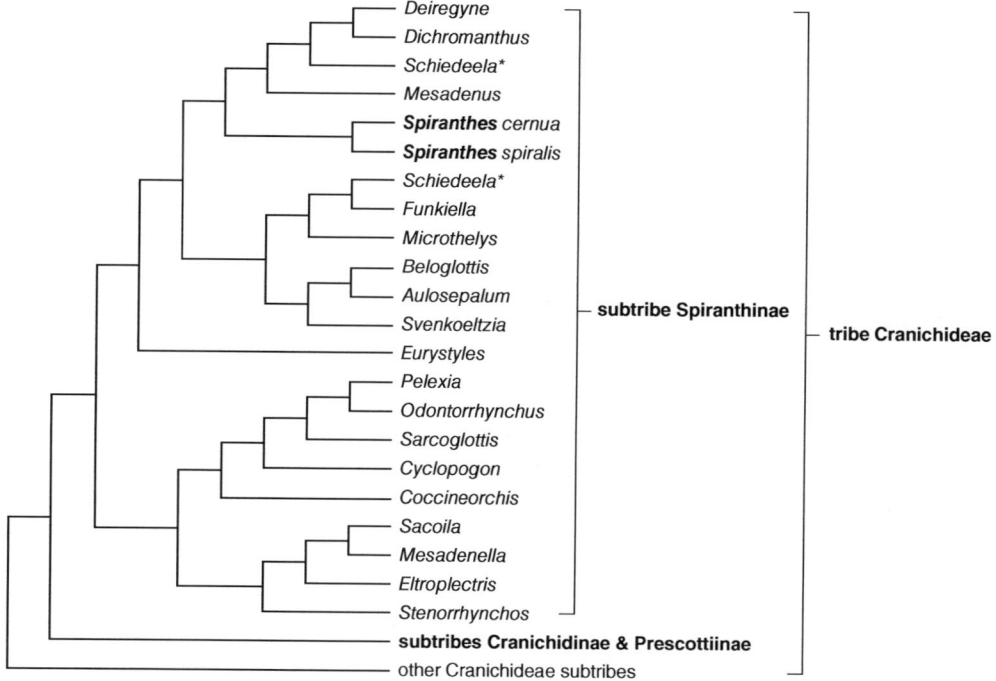

FIGURE 2.21 Placement of *Spiranthes* within subtribe Spiranthinae (tribe Cranichideae) of the orchid subfamily Orchidoideae as indicated by phylogenetic analysis of combined DNA sequence data (simplified from Salazar et al., 2009). Taxa containing OBL indicators are highlighted in bold. Although neither *Spiranthes cernua* nor *S. spiralis* are OBL indicators, they reasonably represent the *Spiranthes* clade where six OBL indicators occur (see Figure 2.22). The disparate placement of *Schiedeela* accessions (asterisked) indicates that the genus is not monophyletic as currently circumscribed.

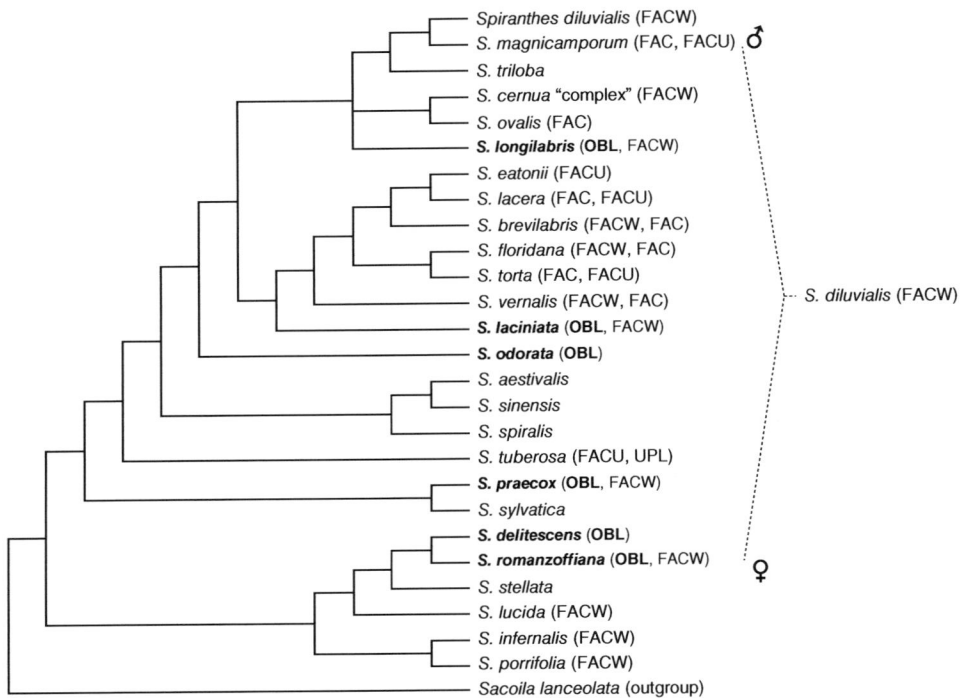

FIGURE 2.22 Interspecific relationships in *Spiranthes* as reconstructed using combined nuclear and plastid DNA sequence data (simplified from Dueck et al., 2014). The wetland indicator status is shown in parentheses for any ranked North American species. The OBL indicators (highlighted in bold) occur throughout the genus. *Spiranthes diluvialis* (FACW) is an allopolyploid whose maternal lineage traces back to *S. romanzoffiana* (OBL, FACW) and paternal lineage traces back to *S. magnicamporum* (FAC, FACU).

Ireland. Several of the species (*S. laciniata, S. longilabris, S. odorata, S. praecox*) occur along the Atlantic coastal plain of the southeastern United States. *Spiranthes delitescens* is endemic to a small region of southern Arizona.

References: Aceto & Gaudio, 2011; 2016; Allard, 1951; Ames, 1903; Arditti & Ghani, 2000; Arft & Ranker, 1998; Argue, 2012g; Armstrong & Moore, 1957; Brown, 2005; Brown & McDonald, 1995; Bursik & Moseley, 1992; Carroll et al., 1984; Catling, 1982; 1983a; 1983b; Catling & Catling, 1991; Chapman, 1997; Colwell et al., 2007; Cranfill, 1981; Dillingham, 2005; Dueck & Cameron, 2007; Dueck et al., 2014; Eilers & Roosa, 1994; Eleuterius, 1980; 1990; Eleuterius & Caldwell, 1984; Eleuterius & McDaniel, 1978; Ferguson & Wunderlin, 2006; Fishbein et al., 1999; Fowler, 2005; Harder, 2000; Hicks, 2007; 2016; Hsiao et al., 2011; Jones, 2000; Kelso et al., 2014; Lamont & Stalter, 2007; Larson & Larson, 1990; Lewis et al., 1928; Locky et al., 2005; Ludwig et al., 1991; Magrath, 2014; Massey, 1953; Maycock, 1956; McClaran & Sundt, 1992; McMillan & Porcher, 2005; McNamara et al., 1992; Mead, 1995; Mehringer, Jr. et al., 1977; Morris, 2013; Motzkin, 1994; NAOCC, 2019; Nekola, 2004; Pace & Brant, 2015; Pace & Cameron, 2016; 2017; Pace et al., 2017; Peinado et al., 2011; Persson, 1947; Picking & Veneman, 2004; Pike et al., 2000; Riggs, 1925; Salazar et al., 2009; Sheviak, 1990; Sheviak & Brown, 2002; Smith, 1996; Snyder, 1996; Stewart & Richardson, 2008; Surveswaran et al., 2018; Thompson, 2001; Tiner, 1993; Wherry, 1921; Zelmer et al., 1996; Zettler et al., 1995; Zettler & Hofer, 1997.

ORDER 6: DIOSCOREALES [3–5]

Phylogenetic analyses of entire plastid genome sequences (Barrett et al., 2016; Givnish et al., 2018) resolve Dioscoreales (Burmanniaceae, Dioscoreaceae, Nartheciaceae, Taccaceae, and Thismiaceae) as the sister clade to Pandanales (Figures 2.1 and 2.23). Some treatments (e.g., APG, 2016) retain Taccaceae, and Thismiaceae within Dioscoreaceae, which is questionably defensible phylogenetically (e.g., Figure 2.23). The order is dominated largely by vines (Dioscoreaceae) and also includes some mycoheterotrophs (Burmanniaceae, Thismiaceae) as well as autotrophs, which together account for about 800 species (Cronquist, 1981; Judd et al., 2016; Givnish et al., 2018).

There are few species of major economic importance aside from the often cultivated members of *Dioscorea* (~10 of the species), whose starchy tubers are the source of vegetables known in North America as "yams." The rhizomes and tubers of some *Dioscorea* species accumulate steroidal saponins, which have been credited as having widespread medicinal properties (Sautour et al., 2007). Several species of *Tacca* (batflowers) and *Tamus* also are cultivated.

Dioscoreales primarily are subtropical and tropical in distribution, with only two of the families containing OBL North American indicators:

6.1. **Burmanniaceae** Blume
6.2. **Nartheciaceae** Fr. ex Bjurzon

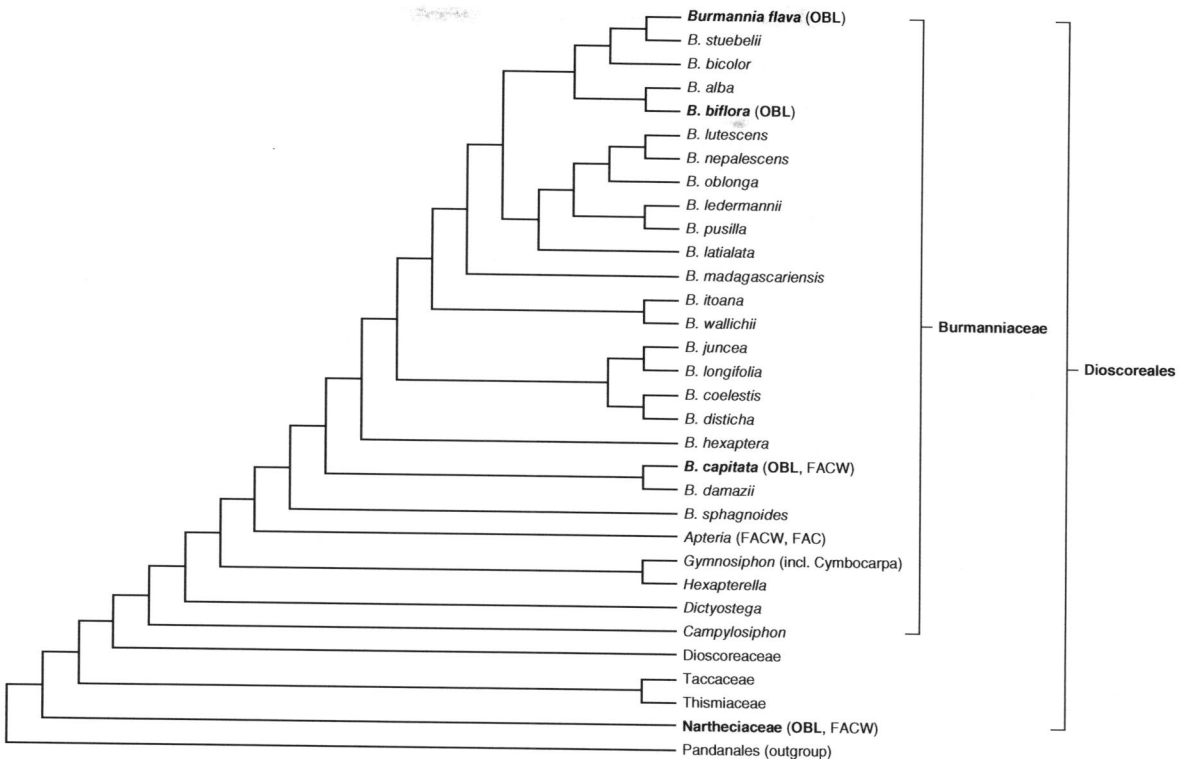

FIGURE 2.23 Phylogenetic relationships of Dioscoreales as indicated by combined analyses of nuclear and mitochondrial DNA sequence data (modified from Merckx et al., 2006, 2008 after merging the results). The wetland indicator status is shown in parentheses for the ranked North American species (taxa containing OBL indicators are highlighted in bold). OBL indicators are rare in the order and are distributed across the genus *Burmannia*. The paraphyletic association of Dioscoreaceae, Taccaceae, and Thismiaceae indicated by these (and similar) results has led recent authors to recognize the latter two families as distinct from Dioscoreaceae (as followed here).

Family 6.1. Burmanniaceae [9]

Burmanniaceae are a small family of about 70 species, of which most occur within the pantropical genera *Burmannia* and *Gymnosiphon* (Merckx et al., 2006). Earlier taxonomic treatments distinguished two major infrafamilial groups, which were recognized as tribes (Burmannieae and Thismieae); the latter most recently has been elevated to family rank as Thismiaceae (e.g., Figure 2.23; Givnish et al., 2018). The group is distinguished morphologically by having anomocytic stomata, and epigynous flowers with a persistent perianth (in fruit), three stamens that are borne opposite the petals, and seeds with a tiny, undifferentiated embryo (Cronquist, 1981; Merckx et al., 2006). The family includes green autotrophs, facultative photosynthetic mycoheterotrophs, as well as achlorophyllous species that are obligately mycoheterotrophic, i.e., they sequester carbon and nutrients indirectly from nearby autotrophic plants by usurping and digesting their arbuscular mycorrhizal Fungi (Merckx et al., 2006; Bolin et al., 2017).

Some Burmanniaceae have showy flowers but are not important commercially due to their mycotrophic habit and ensuing difficulty of cultivation.

The family is distributed throughout the subtropical and tropical regions of the New and Old Worlds. One genus contains OBL North American indicators (Figure 2.23):

6.1.1. *Burmannia* L.

6.1.1. Burmannia

Bluethread

Etymology: after Johannes Burman (1707–1780)

Synonyms: *Cryptonema*; *Cyananthus* (in part), *Cyanotis* (in part); *Gonianthes* (in part); *Gyrotheca* (in part); *Nephrocoelium*; *Tripterella*; *Vogelia* (in part)

Distribution: global: Africa, Asia, North America, South America; **North America:** southeastern United States

Diversity: global: 56 species; **North America:** 3 species

Indicators (USA): obligate wetland (OBL): *Burmannia biflora*, *B. capitata*, *B. flava*; **facultative wetland (FACW):** *B. capitata*

Habitat: freshwater; palustrine; **pH:** acidic; **depth:** <1 m; **life-form(s):** emergent herb

Key morphology: annuals with filiform roots; stems (to 33 cm) green unbranched or once-branched; basal leaves (to 11 mm) 2–7 or absent, cauline leaves (to 12 mm) subulate to lanceolate or linear; cymes solitary or with 2–25 flowers, capitate or loosely flowered, sometimes 2-cleft, floral bracts (to 5 mm) subulate to lanceolate or ovate, the flowers (to 14 mm) erect, 3-ribbed or 3-winged (to 2.5 mm), short-pedicelled (to 1 mm); perianth blue to violet, creamy to white, or greenish to yellow, persistent in fruit, perianth tube (to 0.8 mm wide)

3–6-lobed (to 1.8 mm); stamens 3, sessile; ovary 3-locular, placentation axile; capsules (to 6 mm) ellipsoid to obovoid, dehiscent transversely

Life history: duration: annual (fruits/seeds); **asexual reproduction:** none; **pollination:** self; **sexual condition:** hermaphroditic; **fruit:** capsules (common); **local dispersal:** seeds (wind); **long-distance dispersal:** seeds (wind)

Imperilment: 1. *Burmannia biflora*, [G4/G5]; SH (VA); S1 (AR, MS); S2 (NC, SC); S3 (LA); **2.** *B. capitata*, [G5]; S1 (OK); S3 (NC); **3.** *B. flava* [G5]; S1 (FL)

Ecology: general: *Burmannia* contains annual or perennial species that typically inhabit wet, shady forest understory environments. All three North American species are ranked as OBL indicators. Although the species have adapted to shaded habitats by arbuscular mycorrhizal mycoheterotrophy, most of them (including all the OBL indicators) still retain the ability to synthesize chlorophyll and are photosynthetic, at least facultatively. The reproductive biology has not been well studied for any of the species, although some clearly are self-compatible and autogamous. The seeds are minute and thought to be dispersed primarily by wind.

6.1.1.1. ***Burmannia biflora* L.** inhabits barrens (moist, pine), baygalls, bayheads, bogs (beach strand, peaty, seepage), Carolina bays, depressions (cypress dome, limesink, sandy, seasonally ponded), ditches (moist, roadside, wet), flatwoods (hydric, pine), floodplains (bayou, river), hammocks, pinelands (moist), prairies (spodic, wet-mesic), savannas (pine, wet), seeps (acid, hillside, wooded), slopes (burned, sandy, seepage), sloughs, swamps (acid, cypress, seepage, *Sphagnum*), tussocks (*Sphagnum*), woodlands (acid seep, low), and the margins (boggy, marshy, peaty, sandy, thicketed) of canals (drainage), cypress domes, lakes (interdunal), ponds (acid, depression, flatwoods), rivers, sinks, and streams at elevations to 76 m. The exposures most often are shaded. The substrates are described as acidic and have been characterized as alluvium, Basinger (spodic psammaquents), clay, Duckston (typic psammaquents), Eau Gallie (alfic haplaquods), humus (rich), loam, loamy sand, muck (organic), mud, peat, peaty sand, Placid (typic humaquepts), Rutlege-Osier (typic humaquepts), sand, sandy loam, St. Johns (typlic haplaquods), and Valkaria (spodic psammaquents). Flowering and fruiting occur from July to December. The inner tepals conceal the anthers as they curve inward above them, with the three stigmas protruding between them. The subsequent means of pollination has not been determined. The tiny seeds (0.3–0.4 mm) have been categorized as "dust seeds," which are thought to be wind-dispersed. Few other life history details are known for this species. **Reported associates:** *Acer rubrum, Alnus serrulata, Anthaenantia rufa, Apteria aphylla, Arisaema triphyllum, Aristida stricta, Bartonia paniculata, Bartonia texana, Bartonia virginica, Betula nigra, Boehmeria cylindrica, Boltonia caroliniana, Burmannia capitata, Calopogon pallidus, Carex glaucescens, Centella asiatica, Cephalanthus occidentalis, Cladonia evansii, Coleataenia longifolia, Coleataenia tenera, Commelina, Cornus florida, Ctenium aromaticum, Cyrilla racemiflora, Dichanthelium acuminatum, Dichanthelium*

sphaerocarpon, Dichanthelium strigosum, Dionaea muscipula, Diospyros virginiana, Drosera capillaris, Drosera intermedia, Eleocharis, Eriocaulon compressum, Eriocaulon decangulare, Eryngium integrifolium, Euonymus americanus, Euphorbia inundata, Eutrochium fistulosum, Fagus grandifolia, Gordonia, Habenaria repens, Hartwrightia floridana, Hydrocotyle ranunculoides, Hypericum brachyphyllum, Hypericum cistifolium, Hypericum fasciculatum, Ilex coriacea, Ilex glabra, Ilex myrtifolia, Ilex opaca, Ilex vomitoria, Iris, Isopterygium tenerum, Itea virginica, Juncus, Kalmia hirsuta, Lachnanthes caroliniana, Lachnocaulon anceps, Liquidambar styraciflua, Lobelia floridana, Lobelia glandulosa, Lophiola aurea, Ludwigia linifolia, Lycopodiella alopecuroides, Lycopodiella appressa, Lycopodiella prostrata, Lyonia ligustrina, Lyonia lucida, Lyonia mariana, Magnolia virginiana, Mayaca fluviatilis, Melanthium virginicum, Mitchella repens, Monotropa uniflora, Myrica cerifera, Myrica heterophylla, Neottia bifolia, Nyssa biflora, Nyssa sylvatica, Oldenlandia uniflora, Onoclea sensibilis, Oplismenus hirtellus, Osmunda regalis, Osmundastrum cinnamomeum, Panicum verrucosum, Parnassia asarifolia, Persea palustris, Persicaria hydropiperoides, Pinus elliottii, Pinus palustris, Pinus serotina, Pinus taeda, Platanthera ciliaris, Platanthera clavellata, Pluchea baccharis, Pogonia ophioglossoides, Polygala brevifolia, Polygala cruciata, Quercus alba, Quercus nigra, Quercus phellos, Rhexia aristosa, Rhexia mariana, Rhexia nashii, Rhexia parviflora, Rhexia petiolata, Rhexia virginica, Rhynchospora brachychaeta, Rhynchospora capitellata, Rhynchospora careyana, Rhynchospora cephalantha, Rhynchospora chalarocephala, Rhynchospora ciliaris, Rhynchospora filifolia, Rhynchospora gracilenta, Rhynchospora harperi, Rhynchospora inundata, Rhynchospora latifolia, Rhynchospora microcephala, Rhynchospora miliacea, Rhynchospora pleiantha, Rhynchospora stenophylla, Rhynchospora tracyi, Rubus, Sabal palmetto, Sabatia decandra, Sambucus nigra, Sarracenia minor, Sarracenia rubra, Saururus cernuus, Scleria muehlenbergii, Scleria reticularis, Serenoa repens, Smilax laurifolia, Smilax walteri, Solidago, Sphagnum cuspidatum, Sphagnum cyclophyllum, Sphagnum lescurii, Sphagnum magellanicum, Steinchisma hians, Taxodium ascendens, Taxodium distichum, Tiedemannia filiformis, Toxicodendron radicans, Toxicodendron vernix, Triadenum walteri, Triodanis, Utricularia juncea, Utricularia purpurea, Vaccinium corymbosum, Viburnum nudum, Viola, Vitis, Woodwardia areolata, Woodwardia virginica, Xyris fimbriata, Xyris panacea, Xyris platylepis, Xyris scabrifolia, Xyris serotina, Xyris smalliana.

6.1.1.2. ***Burmannia capitata* (J.F. Gmel.) Mart.** grows in baygalls, bogs (hillside, pitcher plant, roadside, sapric muck, seepage, *Sphagnum*), borrow pits, depressions (cypress, swampy), ditches (drainage, roadside), flatwoods (mesic, rich, well drained), gravel pits, lagoons, marshes (basin), pinelands (low, wet), pocosins, prairies (marl, wet), ravines (drainage), roadsides (moist, sandy), ruts, sandpits, savannas (drawdown, pine, wet), seeps (acid, hydric, *Sphagnum*), slopes (seepage), sloughs, swamps (dome), woodlands (low, swampy), and along the margins (acidic, sandy) of lakes, ponds (seasonal), sinks (limestone), and streams (spring-fed) at elevations to 119 m. Exposures typically are open or in full sun, with the plants found often in recently burned sites. The sites frequently are inundated at least seasonally. The substrates typically are acidic (but sometimes calcareous) and have been characterized as Basinger (spodic psammaquents), Cuthbert (typic hapludults), Dothan (plinthic paleudults), Eutaw, Felda (arenic ochraqualfs), Florala (plinthaquic paleudults), Leagueville-Henco (arenic paleaquults), Leon (aeric haplaquods), loamy sand, Meadowbrook (grossarenic ochraqualfs), muck (sapric), mucky peat, Pactolus (aquic quartzipsamments), peat, Pelham (arenic paleaquults), Pineda and Riviera (arenic glossaqualfs), Ridgewood (aquic quartzipsamments), sand, sandstone, sandy clay, sandy peat, Smithton, Stringtown (typic hapludults), and Tehran-Letney (grossarenic paleudults). Flowering and fruiting occur from August to October. The plants appear to be autogamous (self-pollinating), with the pollen being shed and deposited on the stigma prior to the opening of the flowers. No information exists on the seed ecology, but dispersal presumably occurs by wind. Although mycoheterotrophic, the plants are photosynthetic. They possess small, cauline leaves and have been reared *in vitro* (in mycorrhizae-free conditions) to flowering maturity when cultured under high light. Their ability to photosynthesize likely is related to their apparent preference for open rather than shady exposures. **Reported associates:** *Acer rubrum, Aletris, Anthaenantia, Balduina uniflora, Bigelowia nudata, Burmannia biflora, Burmannia flava, Calamagrostis coarctata, Carphephorus carnosus, Carphephorus odoratissimus, Centella asiatica, Coleataenia longifolia, Coreopsis gladiata, Ctenium aromaticum, Cuphea, Cyperus, Danthonia sericea, Dichanthelium acuminatum, Dichanthelium ensifolium, Dichanthelium scabriusculum, Diodia virginiana, Drosera brevifolia, Drosera capillaris, Drosera filiformis, Drosera tracyi, Eleocharis, Epidendrum magnoliae, Eriocaulon decangulare, Eriocaulon texense, Eryngium cuneifolium, Eryngium integrifolium, Eupatorium perfoliatum, Eupatorium rotundifolium, Eutrochium fistulosum, Fuirena squarrosa, Gordonia lasianthus, Habenaria, Hartwrightia floridana, Helianthus angustifolius, Helianthus heterophyllus, Helianthus radula, Hypericum brachyphyllum, Hypericum cistifolium, Hypericum setosum, Hypoxis rigida, Ilex coriacea, Ilex glabra, Juncus canadensis, Juncus pelocarpus, Juncus trigonocarpus, Lachnanthes caroliniana, Lachnocaulon anceps, Liatris pycnostachya, Liatris spicata, Lilium, Lobelia brevifolia, Lophiola aurea, Lycopodiella alopecuroides, Lycopodiella appressa, Lycopodiella caroliniana, Lycopodiella cernua, Lycopodiella prostrata, Lyonia lucida, Magnolia virginiana, Marshallia graminifolia, Mikania scandens, Mitreola sessilifolia, Mnesithea rugosa, Muhlenbergia capillaris, Myrica heterophylla, Nyssa biflora, Osmundastrum cinnamomeum, Panicum repens, Paspalum notatum, Paspalum urvillei, Persea palustris, Pinguicula, Pinus elliottii, Pinus palustris, Pinus taeda, Platanthera ciliaris, Platanthera integra, Pluchea baccharis, Polygala cruciata, Polygala lutea, Polygala ramosa, Polygala ramosa, Polypremum procumbens, Proserpinaca pectinata, Ptilimnium costatum, Pyrrhopappus carolinianus, Quercus laevis, Quercus nigra, Rhexia alifanus, Rhexia lutea, Rhexia mariana, Rhexia nashii, Rhexia petiolata, Rhexia virginica, Rhynchospora baldwinii, Rhynchospora chalarocephala, Rhynchospora chapmanii, Rhynchospora decurrens, Rhynchospora glomerata, Rhynchospora gracilenta, Rhynchospora inexpansa, Rhynchospora macra, Rhynchospora oligantha, Rhynchospora rariflora, Rhynchospora stenophylla, Rhynchospora tracyi, Rhynchospora wrightiana, Rudbeckia scabrifolia, Sabatia grandiflora, Saccharum giganteum, Sarracenia alata, Sarracenia flava, Sarracenia jonesii, Sarracenia leucophylla, Sarracenia psittacina, Sarracenia rubra, Schizachyrium scoparium, Scleria hirtella, Scleria reticularis, Serenoa repens, Seymeria cassioides, Smilax laurifolia, Solidago fistulosa, Solidago stricta, Sphagnum cyclophyllum, Spiranthes cernua, Symphyotrichum dumosum, Syngonanthus flavidulus, Taxodium ascendens, Tiedemannia filiformis, Toxicodendron vernix, Tridens ambiguus, Utricularia cornuta, Utricularia juncea, Utricularia subulata, Woodwardia areolata, Xyris ambigua, Xyris baldwiniana, Xyris difformis, Xyris drummondii, Xyris fimbriata, Xyris platylepis, Xyris scabrifolia, Xyris torta.*

6.1.1.3. ***Burmannia flava* Mart.** occurs in flatwoods (wet), meadows (marshy), pinelands, and swamps along the coastal plain at elevations to 50 m (to 750 m elsewhere). The plants grow in acidic, nutrient-poor sites where they appear once any standing waters recede. Flowering has been observed in October. The plants have small, chlorophyll containing leaves and are likely to be at least facultatively photosynthetic. There are only a few known sites of this species in North America (southern Florida) where life history information for this species is virtually nonexistent. **Reported associates:** *Burmannia capitata, Drosera capillaris, Pinus elliottii, Serenoa repens, Utricularia cornuta, Utricularia simulans, Utricularia subulata, Xyris.*

Use by wildlife: no uses reported.

Economic importance: food: *Burmannia* species are not edible; **medicinal:** no uses reported; **cultivation:** *Burmannia* is not cultivated; **misc. products:** none; **weeds:** none; **nonindigenous species:** none.

Systematics: Phylogenetic analyses indicate that *Burmannia* is monophyletic and is related more closely to *Apteria* than to either *Campylosiphon* or *Hexapterella* as was previously thought (Figure 2.23). The three OBL North American indicators are not closely related, but are dispersed widely across the genus (Figure 2.23). Due to a paucity of existing counts (and anomalous series among those available), the basic chromosome number of *Burmannia* has not been determined. *Burmannia capitata* ($2n = 136$) is highly polyploid. There are no reports of hybridization involving the North American species.

Distribution: All three OBL species occur in the southeastern United States with *B. biflora* extending into Cuba and the West Indies, and *B. capitata* and *B. flava* extending into South America. *Burmannia flava* is rare and occurs only within a few known sites in southwestern Florida.

References: Anderson & Kral, 2008; Austin, 1998; Bolin et al., 2017; Eriksson & Kainulainen, 2011; Fernald, 1939; Lewis, 2002; McMillan et al., 2002; Merckx et al., 2006;

2008; 2010; Nelson, 1986; Orzell & Bridges, 2006b; Popenoe, 1986; Sieren & Warr, 1992; Sorrie & Leonard, 1999; Wood, Jr., 1983.

Family 6.2. Nartheciaceae [5]

Although Nartheciaceae once were thought to be allied with Liliaceae or Melanthiaceae (Liliales) (e.g., Utech & Adanson, 2002), phylogenetic analyses of molecular data (e.g., Merckx et al., 2006; Fuse et al., 2012) have clearly established the placement of the family as a clade within Dioscoreales, where it aligns along with Dioscoreaceae, Taccaceae, and Thismiaceae (e.g., Figures 2.23 and 2.24). Nartheciaceae are regarded as relatively primitive with unspecialized flowers typical of most monocotyledons (Fuse et al., 2012). Synapomorphies for the family include airspaces in the root cortex, campylotropous ovules, hypogynous flowers, a single style, and a circular perforation on the surface of the smooth, circular orbicules, which occur on the inner anther locule wall; integument cuticles and calcium oxalate raphides are lacking (Merckx et al., 2008b). As currently circumscribed, Nartheciaceae contain 5 genera and 35 species (Merckx et al., 2008b; Fuse et al., 2012; Christenhusz et al., 2016; 2017; Tobe et al., 2018). All the species are autotrophic, rhizomatous perennials, with erect stems and terminal spikes or racemes; they are found mainly in wetland habitats throughout the Northern Hemisphere and in the highlands of northern South America (Merckx et al., 2008b; Fuse et al., 2012).

Lophiola had been treated (albeit reluctantly) as a member of Haemodoraceae (Commelinales) (Ornduff, 1979; Utech & Adanson, 2002) but eventually was assigned properly to Dioscoreales and Nartheciaceae (Figure 2.24). A close relationship is indicated for *Lophiola*, *Narthecium*, and *Nietneria* (Figure 2.24), which differ from all other Nartheciaceae by their absence of a perianth tube and septal nectaries, basal insertion of stamens on the tepals, a single stylar locule, and micro-reticulate pollen ornamentation (Merckx et al., 2008b; Tobe et al., 2018).

Narthecium contains steroidal saponins, which are thought to induce a deadly photosensitive form of jaundice in sheep (Utech & Adanson, 2002). The flowers and fruits of one species (*N. ossifragum*) have been used as a saffron substitute and to produce an orange dye (Zomlefer, 1997; Christenhusz et al., 2017). A few *Narthecium* species are cultivated as ornamental bog garden plants. Otherwise, the family is of little economic importance.

Two North American genera contain OBL indicators:

6.2.1. *Lophiola* Ker Gawl.
6.2.2. *Narthecium* Huds.

6.2.1. Lophiola
Golden crest; lophiolie dorée
Etymology: from the Greek *lophio* ("crest") for the crested tepals
Synonyms: *Argolasia* (in part); *Conostylis* (in part); *Helonias* (in part)
Distribution: global: North America; **North America:** eastern
Diversity: global: 1 species; **North America:** 1 species
Indicators (USA): obligate wetland (OBL): *Lophiola aurea*
Habitat: freshwater; palustrine; **pH:** acidic; **depth:** <1 m; **life-form(s):** emergent herb
Key morphology: rhizomes stoloniferous, densely rooted; stems (to 8.5 dm) erect, wooly distally; basal leaves linear, cauline leaves (to 40 cm) linear, erect, decreasing in size distally; corymbs open, white-wooly; tepals (to 4 mm) reflexed at anthesis, yellow, triangular, densely white-wooly below, brownish to maroon at tips, with an adaxial bright yellow crest of slender trichomes; stamens (to 3 mm) 6, shorter than tepals, erect-spreading at anthesis; ovary partly inferior to nearly superior, ovules numerous, style persistent, stigma somewhat 3-lobed; capsules (to 4 mm) globose or ovoid, enclosed by tepals before dehiscence, prominently beaked; seeds (to 1.5 mm) whitish, thin, elongate, variously curved, seed coat finely open-reticulate
Life history: duration: perennial (rhizomes [stoloniferous]); **asexual reproduction:** rhizomes; **pollination:** unknown; **sexual condition:** hermaphroditic; **fruit:** capsules (common); **local dispersal:** rhizomes; seeds (wind); **long-distance dispersal:** seeds (wind)
Imperilment: 1. *Lophiola aurea* [G4]; SX (DE); S1 (GA); S2 (LA, NC, <u>NS</u>); S3 (AL)

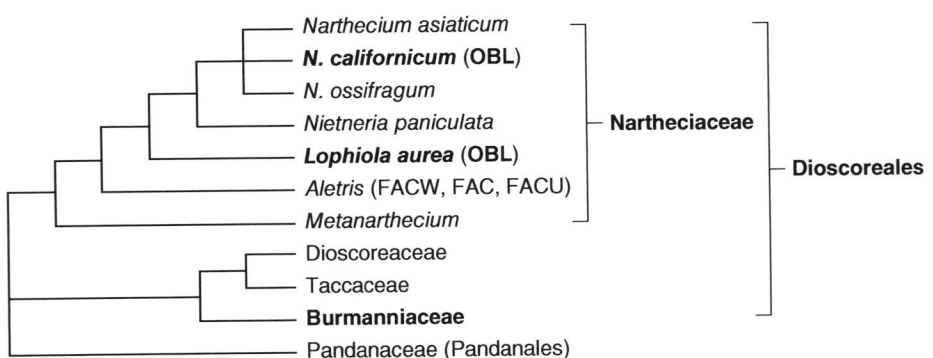

FIGURE 2.24 Parsimony analysis of combined DNA sequence data (*atpB-rbcL*, *trnL*, *trnL-F*, and 18S nrDNA) resolves a monophyletic Nartheciaceae within the order Dioscoreales (simplified from Fuse et al., 2012). The wetland indicator ranks are given in parentheses for North American representatives (taxa containing OBL indicators are highlighted in bold).

Ecology: general: *Lophiola aurea* is monotypic (see next).

6.2.1.1. *Lophiola aurea* Ker Gawl. grows in bogs (hillside, pitcher plant, seepage), depressions (moist),ditches (drainage, low, moist, roadside, wet), flats (cutover, pine), flatwoods (burned, pine), floodplains (stream), pine barrens, pocosins (roadside), roadsides (wet), savannas (boggy, cypress, dry, grassy, mesic, pitcher plant), sloughs, swales (boggy, cypress, low), swamps (*Sphagnum*), and along the margins of ponds (depression) at elevations to 128 m. Exposure are in full sunlight or under very little overstory cover. The habitats are acidic with substrates described as Atmore (plinthic paleaquults), loamy sand (Alahapa), muck (sapric), peat (over diatomite), Rains (typic paleaquults), sand, sandy loam (Bayou), sandy peat, and silty loam. Flowering occurs from April to September with fruiting extending into October. The flowers are self-compatible, but no studies have been conducted to elucidate its reproductive biology or means of pollination. Because the flowers have lost the perianth tube and nectaries associated with earlier diverging Nartheciaceae, it has been postulated that potential pollinators might include small insects such as flies (Insecta: Diptera: Syrphidae), which are known to pollinate species in the related genus *Narthecium* while foraging for pollen. The pollen grains are not starchy. Seed germination rates were found to vary little whether refrigerated (at 5°C) or unrefrigerated; however, refrigerated seeds germinated more rapidly (42–84 days) than unrefrigerated seeds (90 days). The plants occur often in periodically burned, cutover, or otherwise disturbed sites. **Reported associates:** *Aletris lutea, Andropogon virginicus, Aristida purpurascens, Aristida stricta, Arnoglossum ovatum, Asclepias connivens, Asclepias lanceolata, Asclepias longifolia, Balduina atropurpurea, Bidens, Bigelowia nudata, Burmannia capitata, Calamagrostis coarctata, Calopogon barbatus, Calopogon pallidus, Calopogon tuberosus, Carex, Carphephorus pseudoliatris, Centella asiatica, Chaptalia tomentosa, Cladium mariscoides, Cleistesiopsis bifaria, Cleistesiopsis divaricata, Cliftonia monophylla, Coreopsis gladiata, Coreopsis nudata, Ctenium aromaticum, Cyrilla racemiflora, Danthonia sericea, Dichanthelium dichotomum, Dichanthelium nudicaule, Drosera capillaris, Drosera filiformis, Drosera tracyi, Dulichium arundinaceum, Eleocharis tuberculosa, Eragrostis pilosa, Erigeron vernus, Eriocaulon compressum, Eriocaulon decangulare, Eriocaulon nigrobracteatum, Eriocaulon texense, Euphorbia inundata, Eurybia eryngiifolia, Eurybia eryngiifolia, Gaylussacia baccata, Gaylussacia dumosa, Gaylussacia frondosa, Gaylussacia mosieri, Helenium vernale, Hymenocallis henryae, Hypericum brachyphyllum, Hypericum chapmanii, Hypericum fasciculatum, Hypericum gentianoides, Hypericum microsepalum, Ilex coriacea, Ilex glabra, Ilex myrtifolia, Ilex vomitoria, Iris prismatica, Juncus acuminatus, Juncus caesariensis, Juncus gymnocarpus, Kalmia latifolia, Lachnanthes caroliniana, Lachnocaulon anceps, Liatris spicata, Lobelia brevifolia, Lobelia glandulosa, Lycopodiella alopecuroides, Lycopodiella appressa, Lycopodiella caroliniana, Lyonia lucida, Magnolia virginiana, Marshallia graminifolia, Muhlenbergia expansa,* *Muhlenbergia torreyana, Myrica caroliniensis, Myrica cerifera, Myrica heterophylla, Narthecium americanum, Nolina atopocarpa, Nyssa biflora, Orontium aquaticum, Osmunda regalis, Physostegia godfreyi, Pinguicula ionantha, Pinguicula lutea, Pinguicula planifolia, Pinus elliottii, Pinus palustris, Pityopsis oligantha, Platanthera blephariglottis, Platanthera integra, Platanthera nivea, Pleea tenuifolia, Pogonia ophioglossoides, Polygala barbeyana, Polygala brevifolia, Polygala cruciata, Polygala lutea, Polygala nana, Polygala polygama, Pycnanthemum nudum, Quercus marilandica, Rhexia alifanus, Rhexia lutea, Rhexia petiolata, Rhododendron viscosum, Rhynchospora alba, Rhynchospora baldwinii, Rhynchospora brachychaeta, Rhynchospora chalarocephala, Rhynchospora chapmanii, Rhynchospora ciliaris, Rhynchospora fascicularis, Rhynchospora glomerata, Rhynchospora gracilenta, Rhynchospora latifolia, Rhynchospora macra, Rhynchospora oligantha, Rhynchospora plumosa, Rhynchospora pusilla, Rhynchospora rariflora, Rhynchospora solitaria, Rhynchospora stenophylla, Rudbeckia graminifolia, Sabatia difformis, Sabatia macrophylla, Sarracenia alata, Sarracenia flava, Sarracenia leucophylla, Sarracenia minor, Sarracenia psittacina, Sarracenia purpurea, Schizachyrium scoparium, Schizaea pusilla, Scleria muehlenbergii, Scleria reticularis, Serenoa repens, Seymeria cassioides, Sisyrinchium atlanticum, Smilax auriculata, Smilax laurifolia, Sphagnum, Sporobolus floridanus, Stillingia aquatica, Syngonanthus flavidulus, Taxodium ascendens, Tiedemannia filiformis, Trianthia glutinosa, Triantha racemosa, Utricularia cornuta, Utricularia juncea, Utricularia resupinata, Utricularia striata, Utricularia subulata, Vaccinium corymbosum, Vaccinium macrocarpon, Xyris ambigua, Xyris baldwiniana, Xyris drummondii, Xyris scabrifolia.*

Use by wildlife: no uses reported.

Economic importance: food: *Lophiola aurea* is not edible; **medicinal:** no uses reported; **cultivation:** *Lophiola aurea* is not cultivated; **misc. products:** none; **weeds:** none; **nonindigenous species:** none.

Systematics: Phylogenetic analysis of combined DNA sequence data resolves the monotypic *Lophiola* as the sister group to a clade containing *Narthecium* and *Nietneria* (Figure 2.24). The base chromosome number of Nartheciaceae (thought to be a paleotetraploid group) is $x = 7$ or 13. *Lophiola aurea* ($2n = 42$) then would represent a polyploid derived from an $x = 7$ basic number. No natural hybrids involving *L. aurea* have been reported.

Distribution: *Lophiola aurea* is distributed along the Atlantic coastal plain from New Jersey to Louisiana but is disjunct northward in Nova Scotia.

References: Avis, 1997; Berkowitz et al., 2014; Bridges & Orzell, 1989; 1992a; Drewa et al., 2002; East, 1940; Fuse et al., 2012; Glenn, 2008; Hall et al., 1938; Keppner & Anderson, 2008; Lowry et al., 1987; Morris, 2013; Nichols, 1919; 1934; Ornduff, 1979; Robertson, 1976; 2002; Smith & Flory, 1990; White & Harley, 2016; Zhao et al., 2012; Zomlefer, 1997; Zona, 2001.

6.2.2. Narthecium

Bog-asphodel

Etymology: after *narthecium* (a Greek/Roman perfume box), which was derived from the Greek *narthex*, the name of an aromatic fennel species

Synonyms: *Abama* (in part); *Anthericum* (in part); *Phalangium* (in part)

Distribution: global: temperate Eurasia and North America; **North America:** coastal eastern and western United States

Diversity: global: 7 species; **North America:** 2 species

Indicators (USA): obligate wetland (OBL): *Narthecium californicum*

Habitat: freshwater; palustrine; **pH:** 4.3–7.0; **depth:** <1 m; **life-form(s):** emergent herb

Key morphology: rhizomes creeping, slender, with fibrous roots; stems (to 6 dm) simple; leaves basal (cauline leaves 3–6, reduced), equitant, the blades (to 3 dm); racemes (to 15 cm) simple, terminal, open, the flowers greenish-yellow to yellow, pedicellate (to 15 mm); tepals (to 10 mm) 6, distinct, persistent, spreading, erect, linear-lanceolate, the margins scarious; stamens (to 4 mm) 6, the filaments white, pubescent, the anthers brick-red; ovary superior, 3-loculed, terminating as a long-pointed stylar beak, stigma slightly 3-lobed; capsules (to 12 mm) weakly 3-lobed, oblong-lanceolate, loculicidal; seeds (to 2 mm) pale yellow, filiform-ellipsoid, narrowly bristle-tailed (to 3 mm) at ends

Life history: duration: perennial (rhizomes [creeping]); **asexual reproduction:** rhizomes; **pollination:** insect, self; **sexual condition:** hermaphroditic; **fruit:** capsules (common); **local dispersal:** creeping rhizomes; seeds (wind); **long-distance dispersal:** seeds (wind)

Imperilment: 1. *Narthecium californicum* [G4]

Ecology: general: *Narthecium* species typically are found in wetlands, with both North American species recognized as wetland indicators (OBL, FACW). The nectarless flowers attract pollen-foraging insects (Insecta) such as bees (Hymenoptera), flies (Diptera), and Lepidoptera. They are self-compatible and potentially autogamous when raindrops accumulate around the stamens, which enables the pollen to float to the stigmas. The seeds are dispersed by wind.

6.2.2.1. *Narthecium californicum* **Baker** inhabits bogs (*Sphagnum*, stream), depressions (boggy), fens (*Darlingtonia*, subalpine), forests (lowland), marshes, meadows (boggy, mountain, wet), ravines, seeps (*Darlingtonia*, roadside), slopes [to 30%] (wet), springs (marshy), streambeds (rocky), swales (moist), and the margins of lakes and streams at elevations to 2529 m. The exposures are sunny and fully open to moderately (e.g., 20%) shaded. The substrates are acidic but rich (e.g., pH: 4.3–7.0; combined $Ca^{2+} + Mg^{2+} + Na^+ = 42$ ppm) and have been described as clay (serpentine), cryumbrepts, humus (dark, wet), olivine gabbro, rocks (granite), serpentine (rocky), and stony. Flowering has been observed from April to August with fruiting from May to September. No specific details exist on the pollination or floral biology of this species; presumably it is similar to other members of the genus (see *Ecology: general* above). **Reported associates:** *Abies magnifica, Achnatherum lemmonii, Aconitum columbianum, Agrostis idahoensis, Allium validum, Alnus tenuifolia, Anemone drummondii, Arenaria, Arnica cordifolia, Arnica longifolia, Bistorta bistortoides, Brachythecium frigidum, Bryum pseudotriquetrum, Calamagrostis nutkaensis, Calocedrus decurrens, Calochortus, Caltha leptosepala, Camassia leichtlinii, Camassia quamash, Carex alma, Carex angustata, Carex aquatilis, Carex buxbaumii, Carex echinata, Carex laeviculmis, Carex lasiocarpa, Carex lemmonii, Carex limosa, Carex luzulina, Carex scabriuscula, Carex scopulorum, Carex serratodens, Carex utriculata, Castilleja miniata, Cercocarpus ledifolius, Comarum palustre, Conardia compacta, Cypripedium californicum, Danthonia californica, Darlingtonia californica, Dasiphora floribunda, Deschampsia cespitosa, Dichanthelium acuminatum, Ditrichum ambiguum, Drosera anglica, Drosera rotundifolia, Eleocharis decumbens, Eleocharis quinqueflora, Equisetum arvense, Eriophorum crinigerum, Eriophorum gracile, Gaultheria humifusa, Gentiana affinis, Gentiana newberryi, Gentiana setigera, Gentianella amarella, Hastingsia alba, Helenium bigelovii, Holodiscus, Horkelia sericata, Hosackia oblongifolia, Hypericum anagalloides, Hypericum anagalloides, Hypericum perforatum. Hypericum scouleri. Juncus balticus, Juncus drummondii, Juncus effusus, Juncus ensifolius, Juncus howellii, Juncus mertensianus, Juncus nevadensis, Juncus orthophyllus, Kalmia microphylla, Kyhosia bolanderi, Leucothoe davisiae, Ligusticum californicum, Lilium kelleyanum, Luzula comosa, Luzula subcongesta, Menyanthes trifoliata, Mimulus guttatus. Mimulus primuloides, Muhlenbergia andina, Oreostemma alpigenum, Oreostemma elatum, Oxypolis occidentalis, Panicum dichotomiflorum, Parnassia fimbriata, Parnassia palustris, Pedicularis attollens, Pedicularis bracteosa, Pedicularis groenlandica, Perideridia bolanderi, Phalacroseris bolanderi, Philonotis fontana, Pinguicula vulgaris, Pinus balfouriana, Pinus contorta, Pinus jeffreyi, Pinus monticola, Pinus ponderosa, Platanthera dilatata, Platanthera sparsiflora, Potentilla gracilis, Primula jeffreyi, Prunella vulgaris, Pseudotsuga menziesii, Pteridium aquilinum, Pyrrocoma racemosa, Quercus, Raillardella pringlei, Ranunculus alismifolius, Rhamnus californica, Rhododendron columbianum, Rhododendron occidentale, Rhynchospora alba, Rudbeckia glaucescens, Salix, Sanguisorba officinalis, Scapania undulata, Scheuchzeria palustris, Schistidium agassizii, Senecio triangularis, Sisyrinchium elmeri, Sisyrinchium idahoense, Sphagnum teres, Sphenosciadium capitellatum, Spiraea douglasii, Spiraea splendens, Spiranthes romanzoffiana, Spiranthes romanzoffiana, Symphyotrichum spathulatum, Triantha glutinosa, Triantha occidentalis, Trifolium longipes, Tsuga mertensiana, Vaccinium uliginosum, Veratrum californicum, Xerophyllum tenax.*

Use by wildlife: none reported.

Economic importance: food: *Narthecium* is not edible; **medicinal:** no uses reported; **cultivation:** *Narthecium californicum* sometimes is cultivated as an ornamental bog garden species; **misc. products:** none; **weeds:** none; **nonindigenous species:** none.

Systematics: Phylogenetic analyses indicate that *Narthecium* is monophyletic and the sister clade to *Nietneria* (Figure 2.24). The close relationship of these genera to one another, and also to *Lophiola* (Figure 2.24) is evidenced by a suite of shared morphological features (see preceding family treatment for Nartheciaceae). The basic chromosome number of *Narthecium* is *x*=13; *N. californicum* (2*n*=26) is diploid. No natural hybrids involving *N. californicum* have been reported, and are not expected due to its isolated geographical distribution.

Distribution: *Narthecium californicum* is restricted to California and Oregon.

References: Christenhusz et al., 2017; Sikes et al., 2010; 2013; Utech & Adanson, 2002; Wolf & Cooper, 2015; Zomlefer, 1997.

ORDER 7: LILIALES [10]

Over the past half a century, the circumscription of Liliales has undergone extensive revision and modification as phylogenetic studies have increasingly continued to refine hypotheses of monocotyledon relationships. The currently proposed circumscription, which is based on combined phylogenetic analyses incorporating 75 plastid genes (Figure 2.25), includes 10 families, 64 genera, and approximately 1,300–1,500 species; only the family Liliaceae is common to the other major classifications proposed during the past 50 years (Givnish et al., 2016; Judd et al., 2016). Liliaceae (600+ species) are the largest family in the order, representing nearly half of all the species (Givnish et al., 2016; Judd et al., 2016).

In the present circumscription, most (but not all) Liliales possess spotted tepals, tepal nectaries, and extrorse anthers; however, the relationships have been elucidated almost entirely on the basis of molecular characters (Givnish et al., 2016; Judd et al., 2016).

Liliales are extremely important economically as the source of countless cultivated ornamentals and medicinals. The group consists primarily of terrestrial species, but OBL North American indicators occur in three of the families:

7.1. **Liliaceae** Juss.
7.2. **Melanthiaceae** Batsch ex Borkh.
7.3. **Smilacaceae** Vent.

Family 7.1. Liliaceae [15]

Within Liliales, Liliaceae resolve closest to Smilacaceae by mtDNA sequence analysis (Petersen et al., 2013) or from analysis of plastid sequences (four genes) for many taxa (Kim & Kim, 2018), but are closer to a clade comprising Philesiaceae and Ripogonaceae by analysis of plastid genome sequences (75 genes) for fewer taxa (Figure 2.25; Givnish et al., 2018). In any case, the family is monophyletic by its current circumscription (Figure 2.26), which differs significantly from numerous previous classifications (e.g., Cronquist, 1981), where

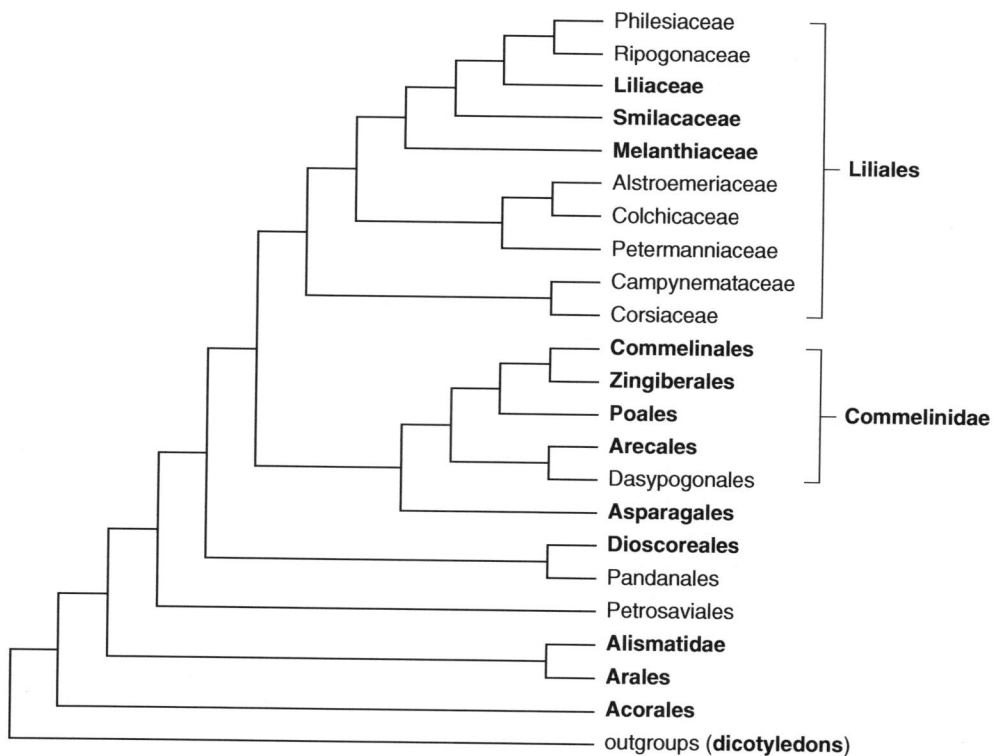

FIGURE 2.25 Maximum parsimony (MP) analysis of monocotyledon ordinal relationships based on combined DNA sequence data from 75 plastid genes (modified and simplified from Givnish et al., 2016). Taxa containing OBL North American indicators are highlighted in bold. Relationships among the ten included families of Liliales are shown in greater detail. This topology is the same as that in the comparable plastid phylogeny of Barrett et al. (2016) and is quite similar (but not identical) to that indicated by phylogenetic analysis of three mitochondrial genes (Petersen et al., 2013).

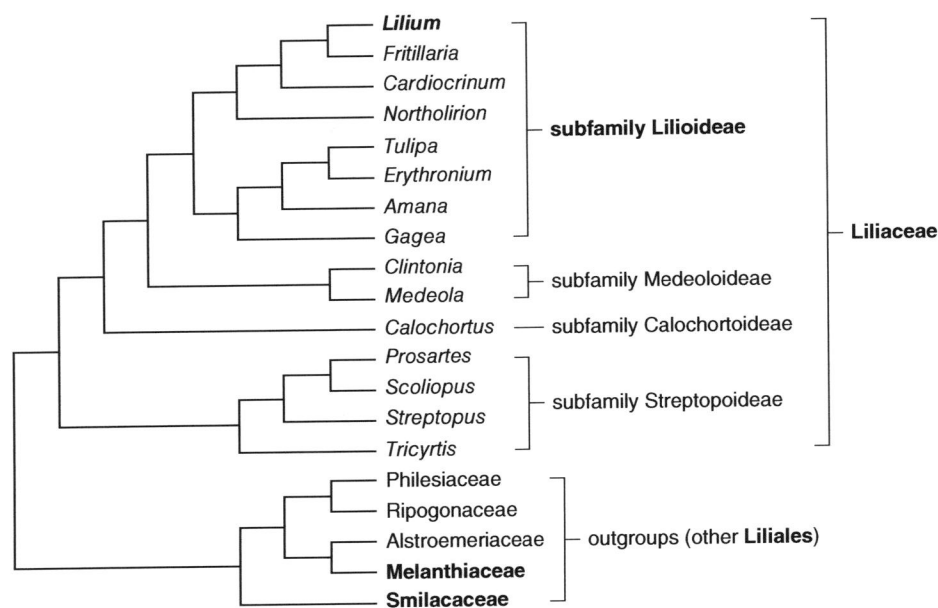

FIGURE 2.26 Classification of Liliaceae proposed on the basis of a maximum likelihood analysis of combined data from four plastid genes (simplified from Kim & Kim, 2018). This result is consistent with the recognition of 15 genera in four subfamilies. *Lilium*, the only genus of Liliaceae to include OBL North American indicators (taxa highlighted in bold), occupies a relatively derived position in the family.

the family was expanded to include many disparate groups now recognized as distinct families (e.g., Alstromeriaceae, Colchicaceae, Melanthiaceae) or placed within different orders (e.g., Amaryllidaceae, Asparagaceae, Tecophilaeaceae in Asparagales).

Even in its current circumscription of about 635 species (Judd et al., 2016), Liliaceae are diverse and difficult to characterize morphologically. The flowers are rather unspecialized for monocots in being bisexual and radially symmetric, with six free tepals (often spotted or striped), six stamens, and a superior ovary of three fused carpels (Judd et al., 2016). A recent evaluation recommended the recognition of 15 genera assigned to four subfamilies: Calochortoideae, Lilioideae, Medeoloideae, and Streptopoideae (Figure 2.26). Members of Medeoloideae and Streptopoideae are rhizomatous with reticulate leaf venation, whereas those of Calochortoideae and Lilioideae feature bulbs with contractile roots and parallel leaf venation (Kim & Kim, 2018). The subfamilies also differ in their basic chromosome number: Calochortoideae ($x=9$), Lilioideae ($x=12$), Medeoloideae ($x=7$), and Streptopoideae ($x=8$ or 13) (Kim & Kim, 2018). The flowers are showy and are pollinated by a variety of insects (Insecta) such as nectar- or pollen-seeking bees and wasps (Hymenoptera) or butterflies and moths (Lepidoptera); most of the seed dispersal is abiotic via water or wind (Judd et al., 2016).

Many genera of Liliaceae are cultivated as ornamentals, including *Calochortus*, *Erythronium*, *Fritillaria*, *Lilium*, *Tricyrtis*, and *Tulipa*. *Calochortus nuttallii* is the state flower of Utah.

Liliaceae are dispersed widely throughout the temperate Northern Hemisphere; however, obligate (OBL) wetland indicators occur within only one of the genera (Figure 2.26):

7.1.1. *Lilium* L.

7.1.1. Lilium

Lily; lis

Etymology: from *lirion*, the Greek name for the lily

Synonyms: *Fritillaria* (in part); *Martagon*; *Nomocharis*

Distribution: global: Northern Hemisphere; **North America:** widespread except far north

Diversity: global: 111 species; **North America:** 29 species

Indicators (USA): obligate wetland (OBL): *Lilium kelleyanum, L. parryi, L. parvum, L. pyrophilum*

Habitat: freshwater; palustrine; **pH:** 3.8–5.0+; **depth:** <1 m; **life-form(s):** emergent herb

Key morphology: bulbs (to 11 cm) rhizomatous, branching or unbranched, the roots contractile; stems to 2.2 m, adventitious stem roots absent or present; leaves scattered or in 1–12 whorls of 3–18, the blades (to 29 cm) elliptic, lanceolate, narrowly linear, or obovate; racemes 1–41-flowered, the flowers ascending, horizontal, or pendent, pedicelled (to 20.7 cm); perianth funnelform to strongly recurved, the tepals (to 10.7 cm) orange, red-orange, yellow, or yellow-orange, with copious to sparse purple-brown or maroon spots; stamens scarcely to strongly exserted, their filaments somewhat to widely divergent, the anthers (to 1.8 cm) dull red, magenta, magenta-brown, orange, purple, or yellow; capsules (to 5.9 cm) with 90–303 seeds

Life history: duration: perennial (bulbs); **asexual reproduction:** rhizomatous bulbs; **pollination:** insect; **sexual condition:** hermaphroditic; **fruit:** capsules (common,); **local dispersal:** seeds (gravity); **long-distance dispersal:** seeds (unknown)

Imperilment: 1. *Lilium kelleyanum* [G4]; **2.** *L. parryi* [G3]; S2 (AZ); S3 (CA); **3.** *L. parvum* [G4/G5]; **4.** *L. pyrophilum* [G2]; S1 (SC, VA); S2 (NC)

Ecology: general: Although most *Lilium* species inhabit terrestrial environments, 16 of the North American species (55%) are wetland indicators (OBL, FACW, FAC, FACU). All the species perennate by means of a basal bulb, which sometimes becomes elongate and rhizomatous. Occasionally, the branching of the bulb can result in limited asexual reproduction. The flowers of most *Lilium* species are self-incompatible and somewhat protandrous. They are showy, slightly to strongly fragrant, nectariferous, and are pollinated by hummingbirds (Aves: Trochilidae) and various insects (Insecta) including bees (Hymenoptera: Apidae), butterflies, and hawkmoths (Lepidoptera: Papilionidae; Sphingidae). The seed ecology is poorly understood for many species; however, seeds of some are characterized by deep simple epicotyle morphophysiological dormancy and require cold or warm stratification to induce germination. Likewise, little is known regarding potential vectors for dispersal of the typically large and heavy seeds. The seeds of some species (but none of the OBL indicators) are lighter, winged, and dispersed by wind.

7.1.1.1. *Lilium kelleyanum* Lemmon inhabits bogs, cienaga, forests (bottomland, coniferous, subalpine), glades, marshes, meadows (boggy, montane, streamside, wet), seeps (hillside, roadside), slopes [to 2%] (serpentine), springs, streambeds (marshy), thickets (*Salix*, shady, wet), and the margins of lakes and streams at elevations to 3261 m. Exposures range from full sun to shade. The substrates have been described as gravel, humus, limestone, loam (decomposed granite, gravelly, heavy, humus), mud, rocks, sand (gritty), serpentine, and stones. Flowering occurs from July to August. The flowers are pollinated by swallowtail butterflies (Insecta: Lepidoptera: Papilionidae: *Papilio*). Other life history information is scarce for this species. **Reported associates:** *Abies concolor, Abies magnifica, Acer macrophyllum, Aconitum columbianum, Agrostis idahoensis, Agrostis scabra, Allium validum, Alnus rhombifolia, Alnus tenuifolia, Amelanchier utahensis, Anaphalis margaritacea, Anemone quinquefolia, Angelica callii, Aquilegia formosa, Arctostaphylos patula, Arnica longifolia, Asarum hartwegii, Athyrium filix-femina, Betula occidentalis, Bistorta bistortoides, Botrychium multifidum, Boykinia major, Calamagrostis, Calocedrus decurrens, Caltha leptosepala, Cardamine breweri, Carex echinata, Carex fracta, Carex jonesii, Carex lemmonii, Carex scopulorum, Carex utriculata, Castilleja miniata, Ceanothus cordulatus, Chamaebatia foliolosa, Chamerion angustifolium, Cheilanthes gracillima, Circaea alpina, Cistanthe umbellata, Collinsia tinctoria, Cornus sericea, Cryptogramma acrostichoides, Cypripedium californicum, Cystopteris fragilis, Darmera peltata, Delphinium glaucum, Delphinium polycladon, Dicentra formosa, Draba albertina, Drosera rotundifolia, Epilobium ciliatum, Epilobium glaberrimum, Epilobium hornemannii, Equisetum arvense, Erigeron coulteri, Erigeron glacialis, Eriophorum crinigerum, Eucephalus breweri, Festuca, Frasera speciosa, Galium trifidum, Gaultheria humifusa, Geranium richardsonii, Glyceria elata, Goodyera oblongifolia, Hastingsia alba, Helenium bigelovii, Heracleum sphondylium, Horkelia fusca, Hosackia oblongifolia, Iris missouriensis, Juncus, Juncus macrandrus, Juniperus occidentalis, Leucothoe davisiae, Lewisia cotyledon, Ligusticum grayi, Lilium parvum, Linnaea borealis, Lonicera involucrata, Lupinus burkei, Lupinus latifolius, Lupinus polyphyllus, Luzula comosa, Madia, Maianthemum racemosum, Mitella pentandra, Maianthemum stellatum, Mertensia ciliata, Micranthes ferruginea, Micranthes nidifica, Micranthes oregana, Mimulus guttatus, Mimulus lewisii, Mimulus tilingii, Montia chamissoi, Muhlenbergia andina, Nasturtium officinale, Neottia convallarioides, Oreostemma alpigenum, Orobanche uniflora, Orthilia secunda, Osmorhiza berteroi, Oxypolis occidentalis, Pedicularis attollens, Physocarpus capitatus, Pinus contorta, Pinus jeffreyi, Pinus lambertiana, Pinus monticola, Pinus ponderosa, Platanthera dilatata, Platanthera sparsiflora, Podagrostis humilis, Populus tremuloides, Primula jeffreyi, Prunella vulgaris, Pseudotsuga, Pseudotsuga menziesii, Pteridium aquilinum, Quercus kelloggii, Rhododendron columbianum, Ribes nevadense, Ribes roezlii, Rosa pisocarpa, Rubus leucodermis, Salix drummondiana, Salix eastwoodiae, Salix lasiandra, Salix ligulifolia, Salix lutea, Salix melanopsis, Salix scouleriana, Sambucus nigra, Sambucus racemosa, Senecio triangularis, Sidalcea oregana, Sisyrinchium elmeri, Sphenosciadium capitellatum, Spiraea, Stachys albens, Staphylea bolanderi, Stellaria longipes, Symphoricarpos rotundifolius, Taxus brevifolia, Thalictrum fendleri, Torreya californica, Triantha occidentalis, Trifolium, Tsuga mertensiana, Umbellularia californica, Vaccinium cespitosum, Vaccinium uliginosum, Veratrum californicum, Veronica serpyllifolia, Viola adunca, Viola glabella, Viola macloskeyi, Zeltnera venusta.*

7.1.1.2. *Lilium parryi* S. Watson grows in cienaga (hillside), meadows (boggy, wet), seeps, slopes (<5%), springs (hillside), thickets (*Salix*), and along the margins of brooks and streams at elevations to 2956 m. Most occurrences are associated with shady canyon bottoms or high elevation meadows. The substrates have been described as alluvium (granitic), granite (decomposed), organic, and silty loam. Flowering occurs from May to September. The fragrant flowers open in late afternoon to early evening and are pollinated primarily by hawkmoths (Lepidoptera: Sphingidae: *Hyles lineata, Sphinx perelegans*). However, the flowers are self-compatible and self-pollination is presumed to occur at times when pollinators are scarce. The seeds are heavy with no known means of biotic dispersal. Allozyme analyses have indicated lower genetic variation in Arizona populations ($H_T=0.063$; $P=0-15.4\%$; $H_e=0.010-0.049$; $H_o=0.003-0.022$) than in California populations ($H_T=0.294$; $P=15.4-76.9\%$; $H_e=0.065-0.307$; $H_o=0.041-0.219$). Heterozygote deficiencies and fixation indices (F) have shown Arizona populations to be characterized by elevated levels of inbreeding. A subsequent survey using RAPDs markers indicated similarly low levels of polymorphism (<13%) in the Arizona populations, but discovered a small but highly variable population (47% polymorphic) in the Chiricahua Mountains. In one study, 300 seeds were produced by the artificial pollination of just two plants. When sown in appropriate (moist, shaded) habitats, the seeds exhibited 28% germination within the first year and yielded numerous new plants. The plants perennate by a rhizomatous bulb but do not reproduce vegetatively. They are known to suffer in the aftermath of severe flooding, which lowers the water table by deeply scouring the associated watercourses. **Reported associates:** *Abies concolor, Achillea millefolium, Acmispon nevadensis, Actaea rubra, Agrostis exarata, Agrostis exarata, Agrostis idahoensis, Agrostis scabra, Alnus rhombifolia, Amorpha californica, Apocynum cannabinum, Aquilegia formosa, Artemisia dracunculus, Artemisia ludoviciana, Artemisia tridentata, Athyrium filix-femina, Barbarea orthoceras, Bromus carinatus, Bromus ciliatus, Calocedrus decurrens, Calochortus invenustus, Carex alma, Carex bolanderi, Carex fracta, Carex jonesii, Carex occidentalis, Carex pellita, Carex senta, Carex spissa, Carex subfusca, Castilleja applegatei, Castilleja miniata, Castilleja minor, Castilleja montigena, Chamerion angustifolium, Chimaphila menziesii, Cicuta douglasii, Corallorhiza maculata, Erysimum capitatum, Deschampsia elongata, Descurainia californica, Dieteria canescens, Draba corrugata, Drymocallis glandulosa, Eleocharis rostellata, Elymus elymoides, Elymus glaucus, Epilobium ciliatum, Epilobium glaberrimum, Epipactis gigantea, Equisetum arvense, Equisetum hyemale, Equisetum laevigatum, Ericameria nauseosa, Erigeron breweri, Erigeron philadelphicus, Eriogonum nudum, Eriogonum parishii, Eriogonum umbellatum, Eriogonum wrightii, Euonymus occidentalis, Euphorbia lurida, Festuca rubra, Fragaria vesca, Fraxinus velutina, Fremontodendron californicum, Galium parishii, Galium triflorum, Gayophytum diffusum, Gentianella amarella, Geranium richardsonii, Geum macrophyllum, Glyceria elata, Helenium bigelovii, Helianthus nuttallii, Heracleum sphondylium, Heuchera parishii, Hoita orbicularis, Holcus lanatus, Horkelia rydbergii, Hosackia oblongifolia, Hypericum formosum, Hypericum scouleri, Juncus balticus, Juncus effusus, Juncus macrandrus, Juncus macrophyllus, Juncus xiphioides, Kelloggia galioides, Lepidium virginicum, Lilium pardalinum, Linanthus pungens, Lupinus burkei, Lupinus polyphyllus, Luzula comosa, Maianthemum stellatum, Mentha arvensis, Mentha spicata, Mimulus cardinalis, Mimulus guttatus, Mimulus moschatus, Mimulus tilingii, Monardella linoides, Muhlenbergia asperifolia, Muhlenbergia richardsonis, Nasturtium officinale, Penstemon labrosus, Pinus contorta, Pinus flexilis, Pinus jeffreyi, Pinus lambertiana, Pinus monophylla, Platanthera dilatata, Platanus racemosa, Poa palustris, Poa pratensis, Polemonium occidentale, Polygonum sawatchense, Populus tremuloides, Populus trichocarpa, Potentilla gracilis, Primula fragrans, Pteridium aquilinum, Pterospora andromedea, Pycnanthemum californicum, Pyrola asarifolia, Pyrola picta, Quercus kelloggii, Rhododendron occidentale, Ribes cereum, Ribes nevadense, Ribes roezlii, Rosa californica, Rosa woodsii, Rubus parviflorus, Sagina saginoides, Salix laevigata, Salix lasiolepis, Salix lutea, Salix scouleriana, Scirpus microcarpus, Sedum niveum, Senecio triangularis, Sidalcea malviflora, Silene menziesii, Silene verecunda, Solidago confinis, Solidago confinis, Solidago velutina, Sphenosciadium capitellatum, Stachys albens, Stellaria longipes, Symphoricarpos rotundifolius, Symphyotrichum ascendens, Taraxacum californicum, Taraxacum officinale,*

Tetradymia canescens, Thalictrum fendleri, Trifolium monanthum, Trifolium wormskioldii, Urtica dioica, Veratrum californicum, Veronica americana, Veronica serpyllifolia, Woodwardia fimbriata.

7.1.1.3. *Lilium parvum* Kellogg occurs in bogs, depressions, fens, flats (streamside), marshes, meadows (boggy, marshy, montane, wet), ravines (moist), roadsides, slopes (to 30%), thickets (alder, willow), and along the margins of streams and swamps at elevations to 3200 m. Exposures can vary from direct sunlight to dense shade. The substrates have been characterized as alluvium (pumice), cryumbrepts, granite, gravel (loose, moist), humus (rich), loam (moist), mud, rocks, sand (coarse), and volcanic. Flowering occurs from June to August, with fruiting extending into September. The flowers are pollinated primarily by hummingbirds (Aves: Trochilidae) and to a lesser extent by bumblebees (Insecta: Hymenoptera: Apidae: *Bombus*) or butterflies (Insecta: Lepidoptera: Papilionidae: *Papilio eurymedon, P. rutulus*). No information is available on the seed ecology of this species. The plants persist by a rhizomatous bulb but do not reproduce vegetatively. **Reported associates:** *Abies magnifica, Acer glabrum, Acmispon americanus, Aconitum columbianum, Allium validum, Allophyllum integrifolium, Alnus rhombifolia, Alnus tenuifolia, Anaphalis margaritacea, Aquilegia formosa, Arnica longifolia, Arnica mollis, Athyrium filix-femina, Betula occidentalis, Bistorta bistortoides, Calocedrus decurrens, Cardamine breweri, Cardamine cordifolia, Carex angustata, Carex echinata, Carex fracta, Carex illota, Carex integra, Carex jonesii, Carex pellita, Carex scopulorum, Carex spectabilis, Carex whitneyi, Castanopsis sclerophylla, Castilleja applegatei, Castilleja miniata, Ceanothus, Chamerion angustifolium, Corallorhiza maculata, Cornus sericea, Darlingtonia californica, Delphinium decorum, Drymocallis lactea, Eleocharis acicularis, Epilobium ciliatum, Equisetum, Erigeron coulteri, Erigeron glacialis, Erigeron peregrinus, Eucephalus breweri, Galium aparine, Galium trifidum, Gayophytum ramosissimum, Gentiana newberryi, Glyceria elata, Helenium bigelovii, Heracleum sphondylium, Heuchera micrantha, Hieracium albiflorum, Ipomopsis aggregata, Juncus chlorocephalus, Juncus ensifolius, Juncus mertensianus, Kelloggia galioides, Letharia vulpina, Ligusticum grayi, Lilium kelleyanum, Lonicera conjugialis, Lupinus andersonii, Lupinus lyallii, Lupinus polyphyllus, Maianthemum racemosum, Maianthemum stellatum, Mertensia ciliata, Micranthes nidifica, Micranthes oregana, Mimulus guttatus, Mimulus lewisii, Mimulus moschatus, Mimulus primuloides, Mitella breweri, Monardella odoratissima, Nasturtium officinale, Navarretia intertexta, Orobanche uniflora, Orthilia secunda, Osmorhiza depauperata, Parnassia palustris, Pedicularis groenlandica, Pedicularis semibarbata, Pinus contorta, Pinus jeffreyi, Pinus monophylla, Pinus monticola, Platanthera dilatata, Platanthera sparsiflora, Poa pratensis, Populus tremuloides, Potentilla, Primula, Pteridium aquilinum, Pterospora andromedea, Pyrola picta, Quercus vacciniifolia, Rhododendron, Ribes montigenum, Ribes viscosissimum, Rosa woodsii, Salix lasiolepis, Salix scouleriana, Senecio triangularis, Sidalcea oregana, Silene lemmonii, Sisyrinchium, Spiraea splendens,*

Stachys albens, Symphoricarpos mollis, Thalictrum, Triantha occidentalis, Trifolium kingii, Tsuga mertensiana, Vaccinium, Valeriana dioica, Veratrum californicum, Veronica serpyllifolia, Vicia americana, Viola.

7.1.1.4. *Lilium pyrophilum* M.W. Skinner & Sorrie grows in bogs (sedge, shrub), canebrakes, drainages (powerline), floodplains, pocosins (streamhead), roadsides, seeps (powerline, sandhill), slopes, and along the margins of woodlands (boggy) and streams (swampy) at elevations to 150 m. The substrates are acidic (pH: 3.8–5.0) and characterized as organic to mineral-organic with a high sulfur content ($\bar{x} = 15.4$ ppm). They include Bibb, loam (Chastain, Roanoke), loamy sand (Ailey, Blaney, Gilead, Pelion, Vaucluse), Lynn Haven, mucky loam (Johnston), sand (Candor), sandy loam (fine, Slagle), and Torhunta. Flowering occurs from July to August, with fruiting extending through October. The flowers are pollinated by ruby-throated hummingbirds (Aves: Trohilidae: *Archilochus colubris*) and swallowtail butterflies (Insecta: Papilionidae: *Papilio palamedes, P. troilus*). The seed ecology of this species has not been studied. As the specific epithet implies, this is a fire-dependent species, which is maintained by periodic burning (every 1–3 years) in order to remove potentially competitive vegetation. The plants persist by a rhizomatous bulb but are not clonal. **Reported associates:** *Acer rubrum, Alnus serrulata, Andropogon glomeratus, Aristida purpurascens, Aristida stricta, Aronia arbutifolia, Arundinaria tecta, Bignonia capreolata, Calamagrostis coarctata, Calamovilfa brevipilis, Carex glaucescens, Carex joorii, Carex turgescens, Chamaecyparis thyoides, Clethra alnifolia, Coreopsis verticillata, Ctenium aromaticum, Cyrilla racemiflora, Dichanthelium dichotomum, Dichanthelium scabriusculum, Dichanthelium scoparium, Dionaea muscipula, Dioscorea villosa, Drosera capillaris, Eleocharis tortilis, Eleocharis tuberculosa, Eriocaulon decangulare, Eriocaulon texense, Eriophorum virginicum, Eupatorium pilosum, Eupatorium resinosum, Eupatorium rotundifolium, Fothergilla gardenii, Gaylussacia frondosa, Ilex coriacea, Ilex glabra, Ilex opaca, Itea virginica, Juncus trigonocarpus, Leucothoe axillaris, Liquidambar styraciflua, Liriodendron tulipifera, Lycopodiella alopecuroides, Lycopus cokeri, Lyonia ligustrina, Lyonia lucida, Lysimachia asperulifolia, Magnolia virginiana, Muhlenbergia expansa, Myrica caroliniensis, Myrica cerifera, Myrica heterophylla, Nyssa biflora, Osmundastrum cinnamomeum, Panicum virgatum, Persea palustris, Pinus palustris, Pinus serotina, Pinus taeda, Pogonia ophioglossoides, Polygala lutea, Pteridium aquilinum, Pycnanthemum flexuosum, Quercus nigra, Quercus phellos, Rhexia mariana, Rhododendron viscosum, Rhynchospora capitellata, Rhynchospora macra, Rhynchospora oligantha, Rubus argutus, Saccharum giganteum, Sarracenia flava, Sarracenia purpurea, Sarracenia rubra, Smilax glauca, Smilax laurifolia, Smilax rotundifolia, Solidago rugosa, Sphagnum, Symphyotrichum dumosum, Taxodium ascendens, Toxicodendron radicans, Toxicodendron vernix, Vaccinium corymbosum, Vaccinium crassifolium, Vaccinium tenellum, Viburnum nudum, Vitis rotundifolia, Woodwardia areolata, Xyris platylepis, Xyris scabrifolia, Zenobia pulverulenta.*

Use by wildlife: The flowers of *Lilium kelleyanum* are visited by pollinating swallowtail butterflies (Insecta: Lepidoptera: Papilionidae: *Papilio*). *Lilium parryi* is grazed by cattle (Mammalia: Bovidae: *Bos taurus*). Its flowers are visited by hawkmoths (Insecta: Lepiodptera: Sphingidae) and hummingbirds (Aves: Trochilidae). Flowers of *L. parvum* are visited by insects (Insecta: Hymenoptera: Apidae; Lepidoptera: Papilionidae: *Papilio eurymedon, P. rutulus*) and hummingbirds (Aves: Trochilidae). Flowers of *L. pyrophilum* are visited by ruby-throated hummingbirds (Aves: Trohilidae: *Archilochus colubris*) and swallowtail butterflies (Insecta: Papilionidae: *Papilio palamedes, P. troilus*).

Economic importance: food: The Paiute tribe harvested the bulbs of *Lilium parvum* for food; **medicinal:** no uses reported for the OBL species; **cultivation:** *Lilium parryi* has been promoted as a garden plant for more than a century, with its seeds offered for sale by 1916. It includes the cultivars 'Bridesmaid' and 'Ilma Watson'; **misc. products:** none; **weeds:** none; **nonindigenous species:** none.

Systematics: Although *Lilium* has been studied quite extensively, no phylogenetic treatment currently exists that adequately addresses relationships among the OBL species. Three OBL species have been evaluated using nrITS sequence data. However, at least some results are ambiguous because of apparent contamination or paralogy; e.g., different accessions of *L. parvum* and *L. parryi* (along with several other species in the same clade) are dispersed among several different clusters. Although the accessions of the surveyed OBL species resolve overall within a North American clade, it seems pointless to propose any more specific hypotheses of relationships within the group until the phylogenetic ambiguities are accounted for satisfactorily. There is one plastid sequence (*ndhF*) available for *L. kelleyanum*, but this is also problematic, being more similar (by a GenBank BLAST search) to a species of *Fritillaria* than to other *Lilium* accessions. Putatively, the parapatric *L. kelleyanum* and *L. parvum* are closely related and are introgressive in their zone of overlap. *Lilium parryi* is thought to be closely related to *L. pardalinum*, but that hypothesis requires further evaluation. Synthetic hybrids have been produced between *L. parryi* and *L. humboldtii*, but the species are isolated in nature by habitat. Natural, mid-elevational hybrids have been reported between *L. parvum* and *L. pardalinum*. Analyses of DNA sequence data indicate that *L. pyrophilum* is a closely related peripheral isolate of *L. superbum*. Natural hybrids reportedly occur between *L. pyrophilum* and *L. michauxii*. The basic chromosome number of *Lilium* is $x = 12$. Three of the OBL indicators (*L. kelleyanum, L. parryi*, and *L. parvum*) have been counted and appear to be diploid ($2n = 24$).

Distribution: Restricted distributions characterize *Lilium kelleyanum* (California), *L. parryi* (Arizona, California), *L. parvum* (California, Nevada, Oregon), and *L. pyrophilum* (North Carolina, South Carolina, Virginia).

References: Arnett et al., 2014; Baskin & Baskin, 1998; Cheng, 2004; Douglas et al., 2011; Friar et al., 1996; Gilmour et al., 2013; Grant, 1966; Grant & Grant, 1966;

Gray et al., 2016; Gregory et al., 2010; Guo et al., 2001; Hiramatsu et al., 2001; Jekyll, 1903; Keeler-Wolf, 1982; 1991; Lee et al., 2011; Linhart & Premoli, 1994; Merrill et al., 2006; Oosting & Billings, 1943; Payne, 1916; Rundel, 2011; Skinner, 1988; 2002; Skinner & Sorrie, 2002; Wood, 1992.

Family 7.2. Melanthiaceae [17]

The circumscription of Melanthiaceae has undergone numerous revisions over the past two centuries with the current concept recognizing 17 genera and approximately 170 species (Fuse & Tamura, 2000; Kim et al., 2016). Phylogenetic analyses (e.g., Figure 2.27) have indicated that Melanthiaceae so defined are monophyletic; however, the position of the group within Liliales remains unresolved with the family having different close associates (e.g., Smilacaceae or Alstroemeriaceae) in different molecular analyses (e.g., Figures 2.25 and 2.26). Morphological synapomorphies are few but include extrorsely dehiscent anthers and three distinct styles (Kim et al., 2016).

Phylogenetic analyses also have supported the monophyly and taxonomic recognition of five tribes within the family (Zomlefer et al., 2006), with Melanthieae including the largest number of genera as well as most of the OBL indicators (Figure 2.27). Melanthieae differ from the other tribes by possessing styloid or raphide rather than cuboidal calcium oxalate crystals.

The flowers of many Melanthiaceae are small and are pollinated by various insects (Insecta: Coleoptera; Diptera: Hymnenoptera), or in some cases by wind (Judd et al., 2016). The seeds also are small and dispersed primarily by wind. Biotic seed dispersal by insects (Insecta: Hymenoptera: Formicidae) or birds (Aves) occurs in some genera (Judd et al., 2016).

Some genera (e.g., *Heloniopsis*, *Trillium*, and *Veratrum*) are cultivated as ornamentals. A number of species (notably in

Veratrum) produce toxic alkaloids (Keeler, 1975). Otherwise, the family is of little economic importance. All but one of the species (*Schoenocaulon officinale*) occur in the temperate regions of the Northern Hemisphere (Kim et al., 2016).

Six of the genera contain OBL wetland indicators:

> 7.2.1. *Helonias* L.
> 7.2.2. *Melanthium* L.
> 7.2.3. *Stenanthium* (A. Gray) Kunth
> 7.2.4. *Toxicoscordion* Rydb.
> 7.2.5. *Veratrum* L.
> 7.2.6. *Zigadenus* Michx.

7.2.1.   Helonias

Stud flower, swamp pink
Etymology: from the Greek *helos* ("marsh") in reference to the habitat
Synonyms: *Veratrum* (in part)
Distribution: global: North America; **North America:** eastern United States
Diversity: global: 1 species; **North America:** 1 species
Indicators (USA): obligate wetland (OBL): *Helonias bullata*
Habitat: freshwater; palustrine; **pH:** 4.2–7.0; **depth:** <1 m; **life-form(s):** emergent herb
Key morphology: rhizomes stout, with contractile, fibrous roots; stems (to 6 dm) erect, hollow; leaves evergreen, in basal rosettes, dark green, reduced distally to broadly triangular bracts (to 2 cm), the blades (to 35 cm) simple, oblong-spatulate to oblanceolate, the margins entire; racemes (to 17.5 cm) dense, terminal, elongating in fruit, with 30–70 fragrant, spreading flowers on pedicels (to 8 mm); perianth funnelform, the 6 tepals (to 9 mm) persistent, distinct or slightly connate at base, purplish pink to green, oblong to spatulate,

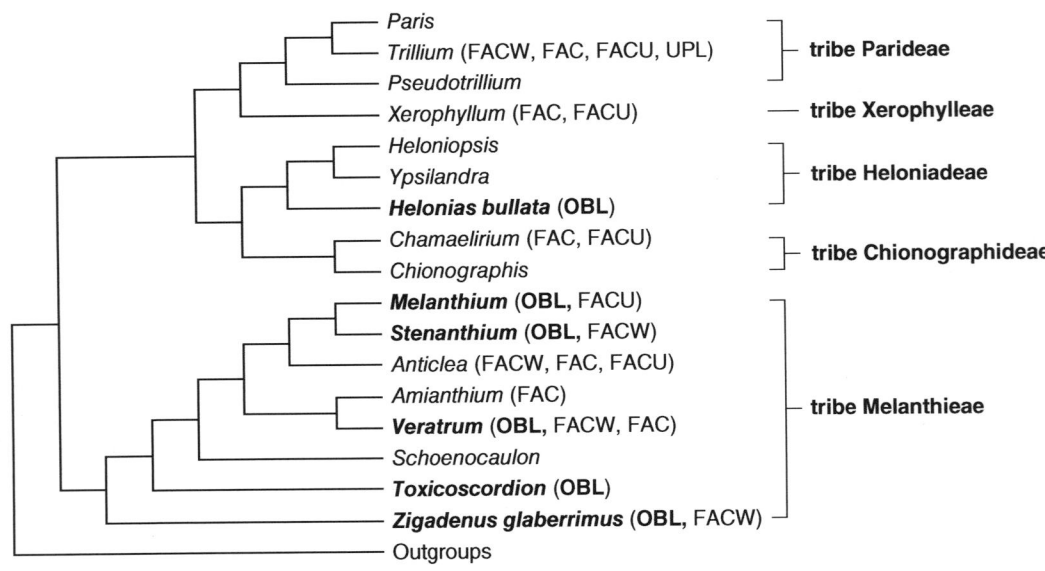

FIGURE 2.27 Relationships within Melanthiaceae as depicted from a combined phylogenetic analysis of five plastid gene sequences (simplified from Kim et al., 2016). The wetland indicator status is provided in parentheses for North American taxa (those containing OBL indicators are highlighted in bold). Note that analyses of nuclear (nrITS) sequence data (Zomlefer et al., 2001; 2003) are substantially incongruent by placing all *Melanthium* species within *Veratrum* (see discussion under *Systematics*).

with adaxial nectaries; stamens 6, the filiform filaments (to 6 mm) distinct (the innermost partially adnate to ovary), the anthers (to 1 mm) blue, unilocular, extrorse; ovary superior, styles (to 2.5 mm) 3; capsules (to 10 mm) obcordate, deeply 3-lobed, papery, loculicidal; seeds (to 6 mm) 16 per locule, linear-fusiform, whitish brown, caudate at ends

Life history: duration: perennial (rhizomes); **asexual reproduction:** rhizomes; **pollination:** insect, self; **sexual condition:** hermaphroditic; **fruit:** capsules (infrequent); **local dispersal:** rhizomes; seeds (gravity, wind); **long-distance dispersal:** seeds (ants, water)

Imperilment: 1. *Helonias bullata* [G3]; SX (NY); S1 (GA, SC); S2 (DE, MD, NC, VA); S3 (NJ)

Ecology: general: *Helonias* is monotypic (see next).

7.2.1.1. ***Helonias bullata* L.** inhabits bogs, meadows, pocosins, seeps (spring), swamps (cedar, *Sphagnum*), and the margins of streams at elevations to 1100 m. Exposures typically range from partial to dense shade (i.e., 20–100% overstory). The plants occur in continuously wet but seldom inundated sites, often growing on raised hummocks. They exhibit substantial stress after extended periods of flooding (e.g., 12–20 days). Their optimal water depth has been estimated at 5.0–9.9 cm. The substrates are acidic (4.2–4.9) to neutral, contain 5–10% organic matter, and can include gravel, loam, muck, sand, and silty loam. Flowering occurs infrequently (0–15% of plants) from March to May. The flowers are self-compatible (100% seed set obtained in artificially selfed flowers) but vary in their natural outcrossing rates ($t = 0.24$–1.36). Outcrossing occurs through pollination by a variety of insects (Insecta) such as beetles (Coleoptera) and blackflies (Diptera: Simuliidae), which are attracted to the fragrant blooms. Each flower produces 24.7 seeds on average, with about 20% estimated to arise from selfing or by biparental inbreeding. Somewhat higher seed set (89%) has been observed in naturally outcrossed flowers relative to naturally selfing flowers (77%). Allozyme data have indicated low levels of genetic variation within the species (e.g., $H_{exp} = 0.053$; $P = 6$–18%) and a high degree of population differentiation ($G_{ST} = 0.504$). The seeds are dispersed locally by gravity or over short distances by wind (to 20–160 cm depending on the velocity). They can remain afloat for at least 35 days, which facilitates their dispersal by water. Some biotic dispersal (over greater distances) potentially can occur by ants (Insecta: Hymenoptera: Formicidae), which seek the lipid-rich eliasomes as a food source (but are not common in the wet habitats). Fresh seeds can germinate readily (50–100% within 30 days), but their viability declines steadily as their age increases (e.g., 10% after 245 days). Natural germination rates are higher in wet muck than in moss or in leaf litter. If kept wet (floating or submersed), the seeds have maintained high germination rates (50–85%) after 50 days. Floating seeds begin to germinate within 3 weeks and are thought to establish readily as water levels decline. The seeds have been categorized as an "intermediate-type," which tolerate limited desiccation but not when combined with freezing temperatures. They have remained viable (35–40%) after 2 months of cold storage (5°C), but cannot endure prolonged desiccation (>21 days).

Some cryopreservation methods have been developed that retained 75% viability after 80 days. Seedlings occur rarely in natural populations. They grow slowly, have poor survival rates, and establish poorly in leaf litter (where most adult plants occur). Leaf litter is thought to provide a barrier against water loss for adult plants. Vegetative reproduction occurs by an extensive rhizome system, from which closely spaced clonal rosettes often develop (e.g., becoming 5 cm apart within 2 years). The populations are threatened by erosion/scouring and siltation, which results from increased road and residential storm drainage. The plants are more robust in areas that have experienced recent fires. **Reported associates:** *Acer rubrum, Alnus serrulata, Arethusa bulbosa, Betula populifolia, Calamovilfa brevipilis, Calopogon tuberosus, Carex atlantica, Carex austrodeflexa, Carex collinsii, Carex folliculata, Carex gynandra, Carex muricata, Carex striata, Carex styloflexa, Chamaecyparis thyoides, Chamaedaphne calyculata, Chionanthus virginicus, Clethra alnifolia, Clintonia borealis, Coptis trifolia, Drosera intermedia, Equisetum sylvaticum, Eriophorum virginicum, Eubotrys racemosa, Eurybia radula, Gaultheria procumbens, Gaylussacia baccata, Gaylussacia frondosa, Gentiana autumnalis, Ilex ambigua, Ilex glabra, Ilex laevigata, Ilex verticillata, Isotria verticillata, Kalmia angustifolia, Kalmia latifolia, Larix laricina, Lindera benzoin, Liriodendron tulipifera, Lophiola aurea, Lycopodium obscurum, Lycopus virginicus, Magnolia virginiana, Medeola virginiana, Menziesia pilosa, Mitchella repens, Muhlenbergia torreyana, Myrica pensylvanica, Nyssa sylvatica, Orontium aquaticum, Osmunda regalis, Osmundastrum cinnamomeum, Parnassia asarifolia, Picea mariana, Picea rubens, Pinus rigida, Pinus strobus, Platanthera, Pogonia ophioglossoides, Rhododendron arborescens, Rhododendron canadense, Rhododendron viscosum, Rosa palustris, Rubus hispidus, Sambucus nigra, Sarracenia purpurea, Schizaea pusilla, Smilax, Sparganium americanum, Sphagnum fallax, Sphagnum flavicomans, Sphagnum magellanicum, Sphagnum palustre, Sphagnum pulchrum, Sphagnum recurvum, Symphyotrichum puniceum, Symplocarpus foetidus, Thelypteris noveboracensis, Thelypteris palustris, Tsuga canadensis, Vaccinium corymbosum, Vaccinium macrocarpon, Viburnum dentatum, Viburnum nudum, Viola cucullata, Woodwardia areolata, Xerophyllum asphodeloides, Xyris scabrifolia.*

Use by wildlife: *Helonias* is browsed by white-tailed deer (Mammalia: Cervidae: *Odocoileus virginianus*). The flowers are visited by beetles (Insecta: Coleoptera) and blackflies (Insecta: Diptera: Simuliidae).

Economic importance: food: *Helonias bullata* is not edible; **medicinal:** Some early accounts that attribute medicinal properties to *H. bullata* are misapplied, due to confusion with *Aletris farinosa* and *Chamaelirium luteum*. However, the botanist S.C. Rafinesque remarked that a decoction made from the peeled roots of *H. bullata* was used in New Jersey to treat colic and other stomach disorders; **cultivation:** *Helonias bullata* is described as having "horticultural potential" as an ornamental water garden plant. Unfortunately, it often has been collected illegally and also has been sold commercially for decades without a permit for interstate commerce, which is required through its protection as a federally (USA) threatened species; **misc. products:** none; **weeds:** none; **nonindigenous species:** none.

Systematics: A ring of "accessory" placental vascular bundles, a compound carpellary septal bundle, and a pollen exine having an amorphous layer with spinules are synapomorphic for *Helonias*, *Heloniopsis*, and *Ypsilandra*, which resolve as a clade (tribe Heloniadeae) in molecular phylogenetic analyses (Figure 2.27). Because of their similarity, some authors have recommended the merger of these genera as a single genus (*Helonias*) of 12 species, which essentially would redefine the genera as sections. The basic chromosome number of *Helonias* (and all of tribe Heloniadeae) is *x* = 17, a number thought to be derived secondarily through ancient polyploidization and subsequent chromosome reorganization. *Helonias bullata* (2*n* =34) is diploid. The genome size of *Helonias bullata* has been estimated at 6.37–6.45 pg (2C); 3.19–3.23 pg (1C); and 3114.93–3154.05 mb (1C). No hybrids involving *H. bullata* have been reported.

Distribution: *Helonias bullata* is restricted to the eastern United States.

References: Breden et al., 2001; Fleming & Van Alstine, 1999; Floyd et al., 2018; Fuse & Tamura, 2016; Godt et al., 1995; Hale, 1867; Johnson et al., 1998; Laidig et al., 2009; McCormick, 1998; Pellicer et al., 2014; Perullo et al., 2015; Punsalan et al., 2016; Rydberg, 1926; Shirey et al., 2013; Sorrie et al., 2011; Sterling, 1980; Sutter, 1984; USFWS, 1991; Tanaka, 2019; Utech, 2002b; Zomlefer et al., 2006.

7.2.2. Melanthium

Bunchflower, bunchlily

Etymology: from the Greek *melas anthos* ("black flower") in reference to the perianth color, which blackens with age

Synonyms: *Anepsa* (in part); *Evonyxis* (in part); *Helonias* (in part); *Leimanthium* (in part); *Veratrum* (in part); *Zigadenus* (in part)

Distribution: global: 4; **North America:** 4

Diversity: global: North America; **North America:** eastern

Indicators (USA): obligate wetland (OBL); facultative wetland (FACW): *Melanthium virginicum*

Habitat: freshwater; palustrine; **pH:** acidic; **depth:** <1 m; **life-form(s):** emergent herb

Key morphology: bulbs (to 2.5 cm) erect, tunicate, producing short, vertical rhizomes (to 2 cm), the roots contractile, fleshy; stems (to 2 m) erect, hollow, thickened at base; leaves simple, basal, sheathing, becoming reduced distally, the blades (to 80 cm) linear, long-attenuate; racemes terminal, the bracts (to 6 mm) obovate to subulate, terminal racemes (to 2.6 dm) with bisexual flowers, secondary racemes (to 2.2 dm) with staminate flowers, the pedicels (to 20 mm) spreading to ascending; tepals (to 13 mm) 6, clawed (to 2.5 mm), ovate to obovate-oblong, petaloid, greenish yellow but darkening deeply with age, with 2 prominent, nectariferous glands along midvein; stamens (to 8.5 mm) 6, the filaments strongly incurving, adnate to claw, the anthers (to 0.9 mm) basifixed, fugacious; ovary (to 3.5 mm) ovoid, 3-locular, the styles (to 3.5 mm) 3,

beak-like, persistent, spreading but turning inward with age; capsules (to 18 mm) elliptic-ovoid, deeply 3-lobed, septicidal or loculicidal; seeds (to 8 mm) flat, elliptic to lanceolate, broadly winged

Life history: duration: perennial (bulbs, rhizomes); **asexual reproduction:** rhizomes; **pollination:** insect; **sexual condition:** andromonoecious; **fruit:** capsules (common); **local dispersal:** rhizomes; **long-distance dispersal:** seeds (vector unknown)

Imperilment: 1. *Melanthium virginicum* [G5]; SH (NY, WV); S1 (IN, KY, NJ, OK, PA, TN); S2 (DE, IA, IL, KS, NC, SC, OH); S3 (MD, MS)

Ecology: general: Because the present circumscription of *Melanthium* (i.e., as distinct from *Veratrum*) remains uncertain (see *Systematics*), a meaningful ecological summary for the genus is not possible. In many respects, the ecology of the two genera is similar (see treatment for *Veratrum* below). In any case, only one species assigned to *Melanthium* is designated as an OBL indicator, and for which an ecological summary has been provided below.

7.2.2.1. *Melanthium virginicum* **L.** inhabits baygalls, bogs (hillside, pitcher plant, seepage), depressions, ditches (grassy, roadside), draws (boggy), flats (grassy, low), flatwoods (burned, cleared, wet), floodplains, hammocks (wet), hummocks (*Sphagnum*), marshes, meadows (low), pinelands (low), prairies (low, railroad, wet-mesic), ravines, rights-of-way (railroad), roadsides, savannas (pine, wet), seeps (acid, boggy, spring), slopes (moist), swales (boggy, seepage, sphagnous, wooded), swamps (*Sphagnum*), thickets (moist), woodlands (boggy, sphagnous, wet), and the margins of lakes, pocosins, springs, and streams at elevations to 329 m. The exposures are open and receive full sunlight. The substrates are highly to moderately acidic and have been described as clay, Kisatchie-Rayburn (typic-vertic hapludalfs), loam, muck, peaty muck, sand, sandy loam, and silty loam (Ipava, Keomah, Tama, Virden). Flowering occurs from June to October with fruiting extending from June into late January. The extent of flowering can vary dramatically, being extensive in some years but virtually nonexistent in others. The plants are andromonoecious, with most flowers within an inflorescence being perfect, but the more uppermost being staminate. The perfect flowers are protandrous, with the three outermost anthers dehiscing first. The stigmas do not become receptive until after the anthers have dropped. Nectar is completely exposed within a depression situated between the two nectar glands at the base of each sepal limb. The flowers are pollinated by nectar-seeking insects (Insecta) such as beetles (Coleoptera) and flies (Diptera) (see *Use by wildlife*). The stamens are fused to the sepal claw, and are positioned so that the extrorse anthers come into contact with any larger-sized insect that arrives to forage for nectar. The strongly divergent styles also readily come in contact with foraging insects. There is no information available on the seed ecology of this species. **Reported associates:** *Acer rubrum, Alnus, Amianthium muscitoxicum, Andropogon gerardii, Andropogon gerardii, Andropogon glomeratus, Andropogon liebmannii, Andropogon virginicus, Arundinaria tecta, Asclepias incarnata, Asclepias purpurascens, Balduina* *atropurpurea, Baptisia, Bigelowia, Bromus inermis, Callicarpa americana, Carex atlantica, Carex collinsii, Carex debilis, Carex folliculata, Carex glaucescens, Carex intumescens, Carex stricta, Chamaecyparis thyoides, Chamaelirium luteum, Chasmanthium laxum, Cinna arundinacea, Cirsium muticum, Cladium jamaicense, Coreopsis tripteris, Corylus cornuta, Crataegus, Cyrilla racemiflora, Dichanthelium dichotomum, Dichanthelium scoparium, Drosera, Eleocharis flavescens, Eleocharis tortilis, Eriocaulon, Eryngium integrifolium, Fuirena squarrosa, Helianthus, Hypericum, Illicium floridanum, Impatiens capensis, Liatris spicata, Ligustrum, Lindera, Liquidambar styraciflua, Liriodendron tulipifera, Lophiola aurea, Ludwigia, Lycopodiella alopecuroides, Lycopodiella caroliniana, Lycopodiella prostrata, Lysimachia quadriflora, Magnolia macrophylla, Magnolia virginiana, Mikania scandens, Mitchella repens, Morus rubra, Nyssa biflora, Nyssa ogeche, Nyssa sylvatica, Oenothera fruticosa, Osmunda regalis, Osmundastrum cinnamomeum, Ostrya virginiana, Oxypolis rigidior, Panax quinquefolius, Parnassia grandifolia, Paspalum urvillei, Phlox glaberrima, Pinus echinata, Pinus palustris, Pinus taeda, Platanthera clavellata, Platanthera cristata, Primula meadia, Pycnanthemum virginianum, Quercus stellata, Rhynchospora capitellata, Rhynchospora glomerata, Rhynchospora gracilenta, Rhynchospora oligantha, Rhynchospora stenophylla, Rosa setigera, Rubus, Rudbeckia fulgida, Rudbeckia subtomentosa, Ruellia strepens, Sabatia macrophylla, Sarracenia flava, Sarracenia minor. Sarracenia psittacina, Schedonorus pratensis, Schoenoplectus americanus, Scirpus expansus, Scleria muehlenbergii, Scleria verticillata, Selaginella apoda, Serenoa repens, Sideroxylon, Silphium, Smilax walteri, Sorghastrum nutans, Spartina pectinata, Sphagnum, Stillingia, Symplocarpus foetidus, Symplocos tinctoria, Taxodium, Thalictrum dasycarpum, Toxicodendron radicans, Toxicodendron vernix, Triantha racemosa, Ulmus rubra, Veratrum viride, Veronicastrum virginicum, Viburnum nudum, Viola primulifolia, Viola rostrata, Woodwardia areolata, Xyris torta, Zizia aurea.*

Use by wildlife: The flowers of *Melanthium virginicum* are hosts to numerous nectar- and pollen-feeding insects (Insecta), including bees (Hymenoptera: Halictidae: *Halictus confusus*), beetles (Coleoptera: Chrysomelidae: *Diabrotica cristata*; Curculionidae: *Centrinites strigicollis*; Lampyridae: *Photinus pyralis*; Mordellidae: *Mordella marginata, M. melaena*; Scarabaeidae: *Trichiotinus piger*), butterflies (Lepidoptera: Lycaenidae: *Celastrina ladon*), flies (Diptera: Anthomyidae: *Anthomyia*; Muscidae: *Lucilia, Musca domestica*; Sarcophagide: *Boettcheria cimbicis*; Syrphidae: *Mesograpta marginata, Sphaerophoria contiqua, Syritta pipiens, Toxomerus marginatus*; Tachinidae: *Archytas aterrimus, Chetogena claripennis, C. edwardsii, Gymnoclytia immaculata, G. occidua, Liinnaemya comta, Paradidyma singularis, Trichopoda pennipes*), and wasps (Hymenoptera: Pteromalidae *Euperilampus triangularis*; Sphecide: *Sphex ichneumoneus*). Many of the smaller insect visitors (e.g., Diptera: Chironomidae) rob the nectar without facilitating pollination.

Economic importance: food: All *Melanthium* species are regarded as poisonous and never should be eaten; **medicinal:** *Melanthium* contains complex, steroidal alkaloids, which are toxic to humans and other mammals. Extracts are weakly antioxidative and are inhibitory to some fungal strains (e.g., Fungi: Ascomycota: Saccharomycetes: *Candida albicans*). The plants were used as an effective ("but dangerous") cure for "army itch" during the American Civil War; **cultivation:** *Melanthium virginicum* is not widely cultivated but seeds were available commercially as early as 1804; **misc. products:** *Melanthium virginicum* has been used as an insecticide; **weeds:** none; **nonindigenous species:** none.

Systematics: The taxonomic distinction of *Melanthium* and *Veratrum* has long been debated. Arguably, *Melanthium* species differ from *Veratrum* by having narrower leaves, more delicate and open inflorescences, a conspicuous pair of glands flanking the tepal midrib (indistinct or absent in *Veratrum*), and fugacious stamens, which are adnate to clawed tepals (those of *Veratrum* are not clawed). Furthermore, *Veratrum* is distributed primarily in western North America, whereas *Melanthium* is entirely eastern. The distinctness of *Melanthium* has not been clarified by phylogenetic analyses of different DNA sequence data sets, which have generated highly incongruent results. Plastid DNA sequence data resolve *Melanthium* and *Stenanthium* as sister genera within a clade that is quite distinct from *Veratrum* (e.g., Figure 2.27). The plastid phylogeny is also compatible morphologically. In Melanthiaceae, the fusion of the tepals to the ovary base occurs only in *Anticlea, Melanthium,* and *Stenanthium*; within Melanthieae, only *Melanthium* and *Stenanthium* possess beak-like styles and recurved stigmas. However, results from analyses of nuclear sequence data (nrITS-1/2) sharply contrast by resolving *Melanthium* species as a subclade embedded within a broader *Veratrum* clade. Cladogram support is not compelling for either alternative, making it difficult to decide between them. Until this inexplicable anomaly has been clarified satisfactorily, the decision was made here to retain *Melanthium* as a distinct genus, primarily because it is the disposition followed in most literature accounts. It should be noted that the 2016 wetland indicator list still recognizes *Melanthium virginicum* as a species of *Veratrum* (i.e., as *V. virginicum*). The basic chromosome number of *Melanthium* is $x = 8$; *M. virginicum* ($2n = 16$) is diploid. Its genome size has been estimated at 3.47 pg (2C); 1.74 pg (1C); and 1696.30 mb. There are no hybrids reported that involve *M. virginicum*.

Distribution: *Melanthium virginicum* occurs throughout the eastern half of the United States.

References: Aldrich et al., 1985; Bodkin & Utech, 2002; Borchardt et al., 2008; Cropley, 2006; Feinstein, 1952; Gordon & Arsenault, 2016; Graham et al., 2012; Kim et al., 2016; M'Mahon, 1804; Moorehouse et al., 2002; Pellicer et al., 2014; Robertson, 1896; 1924; Singhurst et al., 2007; Tooker & Hanks, 2000; Tooker et al., 2002; 2006; Tweedy, 1880; Wherry, 1927; Zomlefer et al., 2001; 2003.

7.2.3. Stenanthium

Crow poison, featherbells, Osceola's plume

Etymology: from the Greek *stenos anthos* (narrow flower) in reference to the narrow tepals

Synonyms: *Amianthium* (in part); *Anepsa* (in part); *Crosperma* (in part); *Helonias* (in part); *Melanthium* (in part); *Oceanoros*; *Tracyanthus*; *Veratrum* (in part); *Xerophyllum* (in part); *Zigadenus* (in part)

Distribution: global: North America; **North America:** eastern United States

Diversity: global: 2–6 species; **North America:** 2–6 species

Indicators (USA): obligate wetland (OBL); facultative wetland (FACW): *Stenanthium densum, S. leimanthoides*

Habitat: freshwater; palustrine; **pH:** acidic; **depth:** <1 m; **life-form(s):** emergent herb

Key morphology: bulbs (to 2 cm) slender, ovoid, tunicate; stems to 72 cm; blades (to 60 cm) elongate-linear, reduced distally to red-tinged, lanceolate bracts (to 12 mm); racemes (to 30 cm) conical, subcylindrical, or paniculate, with 40–100 pedicellate (to 2 cm), bisexual or staminate flowers; perianth (to 10 mm) hypogynous, the tepals (to 3.7 mm) persistent, ovate to elliptic, narrowing but not sharply clawed at base, creamy-white, greenish-white, or yellow (sometimes becoming rose or purplish in fruit), with 1 obscure gland present or absent; stamens 6; pistils present or either absent or nonfunctional in upper flowers, the ovary superior to partly inferior; capsules (to 11 mm) narrowly conic, brown or purple

Life history: duration: perennial (bulbs); **asexual reproduction:** none reported; **pollination:** probably insect; **sexual condition:** andromonoecious, hermaphroditic; **fruit:** capsules (common); **local dispersal:** seeds (unknown); **long-distance dispersal:** seeds (unknown)

Imperilment: 1. *Stenanthium densum* [G5]; 2. *S. leimanthoides* [GNR]

Ecology: general: Because of extensive taxonomic uncertainties (see *Systematics* below), the wetland status of *Stenanthium* species must be re-evaluated. All of the species persist from a subterranean bulb. Some of the species are andromonoecious; i.e., bearing perfect and staminate flowers (the latter typically in upper portions of the inflorescence) on the same individual. The flowers are protandrous, at least some are fragrant, and most possess perigonal (tepal) nectaries. The floral biology and pollination ecology is not well studied. Presumably the flowers attract insects (Insecta) as pollinators, but this assumption has not been documented beyond circumstantial evidence.

7.2.3.1. *Stenanthium densum* (Desr.) Zomlefer & Judd inhabits barrens (moist, pine), bogs (hillside, peat, pine, pitcher plant, sedge, seepage, *Sphagnum*), bottomlands, depressions (peaty), ditches (roadside), flats (pitcher plant), flatwoods (damp, dry, low, mesic, seepy, wet), marshes (burned, low, wet), meadows (boggy, wet), pinelands (burned, cleared, open wet), pocosins, prairies (burned, wet), roadsides (grassy, low, moist), savannas (low, moist, wet), slopes (seepage), swales (boggy, peaty), swamps (headwater), and the margins (boggy) of streams (baygall) at elevations to 177 m. The plants are common in wet (to dry or xeric), frequently burned sites in exposures that receive full sunlight. The substrates are acidic and have been characterized as Allanton (grossarenic haplaquods),

Basinger (spodic psammaquents), Guyton-Iuka complex, loamy sand (Betis, fine, Osier), Meadowbrook (grossarenic ochraqualfs), Myakka (aeric haplaquods), Pelham, sand (organic), sandy loam (fine, Malbis, Ruston), sandy peat, Scranton (humaqueptic psammaquents), silty loam (Eastwood), and St. Johns (typic haplaquods). Flowering occurs from March to June; otherwise, the floral biology and ecology is poorly understood. The seeds are morphophysiologically dormant and require cold stratification to induce embryo growth and germination. Embryos in seeds kept at 5°C have grown to 2.5 mm within 9 weeks, and the seeds have germinated within 12 weeks. In one study, the seeds began to germinate during winter (January–February) at daily temperatures that varied from 3.0°C to 10.9°C following their exposure to 44 days of ambient cold stratification in an unheated greenhouse. Germination occurred more quickly when the seeds were placed under a 20°C/10°C day/night temperature regime. Under natural field conditions, the seeds break dormancy during the winter and germinate by early spring. No persistent seed bank develops. Their means of dispersal remains speculative. **Reported associates:** *Acer rubrum, Aletris lutea, Aletris obovata, Aletris ×tottenii, Allium canadense, Alnus serrulata, Andropogon glomeratus, Angelica dentata, Aristida stricta, Aronia arbutifolia, Arundinaria, Asclepias cinerea, Asclepias humistrata, Asclepias michauxii, Asclepias tuberosa, Asclepias verticillata, Baptisia lanceolata, Baptisia simplicifolia, Bignonia capreolata, Calopogon barbatus, Calopogon multiflorus, Calopogon pallidus, Carex exilis, Carex glaucescens, Carex louisianica, Carphephorus paniculatus, Chaptalia tomentosa, Cirsium lecontei, Cleistesiopsis divaricata, Clethra alnifolia, Cnidoscolus urens, Ctenium aromaticum, Cyperus haspan, Cyrilla racemiflora, Dichanthelium acuminatum, Dichanthelium dichotomum, Drosera brevifolia, Drosera capillaris, Drosera rotundifolia, Drosera tracyi, Eleocharis, Erigeron vernus, Eriocaulon compressum, Eryngium integrifolium, Eupatorium album, Eupatorium rotundifolium, Euphorbia inundata, Euphorbia inundata, Eurybia eryngiifolia, Gaylussacia dumosa, Gaylussacia mosieri, Gelsemium sempervirens, Gordonia lasianthus, Helenium brevifolium, Helenium pinnatifidum, Helenium vernale, Helianthus angustifolius, Helianthus heterophyllus, Helianthus radula, Hieracium, Hypericum suffruticosum, Hypoxis hirsuta, Hypoxis juncea, Ilex coriacea, Ilex glabra, Ilex vomitoria, Ionactis linariifolia, Juncus, Lachnocaulon anceps, Liriodendron tulipifera, Lobelia paludosa, Lobelia paludosa, Lycopodiella alopecuroides, Lyonia ferruginea, Lyonia fruticosa, Lyonia lucida, Lyonia mariana, Macbridea alba, Magnolia virginiana, Mimosa microphylla, Muhlenbergia expansa, Myrica caroliniensis, Myrica cerifera, Nolina atopocarpa, Nyssa biflora, Oclemena reticulata, Osmunda regalis, Osmundastrum cinnamomeum, Oxydendrum arboreum, Persea palustris, Phoebanthus tenuifolius, Pinguicula ionantha, Pinguicula lutea, Pinguicula planifolia, Pinguicula pumila, Pinus elliottii, Pinus palustris,*

Pinus taeda, Pityopsis graminifolia, Platanthera blepha-riglottis, Platanthera integra, Platanthera nivea, Pogonia ophioglossoides, Polygala crenata, Polygala cruciata, Polygala cymosa, Polygala lutea, Polygala nana, Polygala polygama, Polygala ramosa, Prunella vulgaris, Pteridium aquilinum, Pterocaulon pycnostachyum, Quercus min-ima, Quercus pumila, Rhexia alifanus, Rhododendron viscosum, Rhynchosia, Rhynchospora chapmanii, Rubus argutus, Rudbeckia graminifolia, Sanguisorba canaden-sis, Sarracenia flava, Sarracenia leucophylla, Sarracenia minor, Sarracenia psittacina, Sarracenia purpurea, Sarracenia rubra, Schizachyrium tenerum, Scleria bald-winii, Scleria triglomerata, Serenoa repens, Smilax glauca, Smilax laurifolia, Smilax rotundifolia, Sphagnum, Spiranthes, Stillingia sylvatica, Stokesia laevis, Stylosanthes, Symphyotrichum adnatum, Taxodium ascen-dens, Tephrosia florida, Toxicodendron radicans, Trifolium, Utricularia juncea, Utricularia subulata, Vaccinium dar-rowii, Vaccinium myrsinites, Verbesina chapmanii, Viola palmata, Viola septemloba, Woodwardia virginica, Xyris ambigua, Xyris brevifolia, Xyris caroliniana.

7.2.3.2. *Stenanthium leimanthoides* (Gray) Zomlefer & Judd resides in bogs (wet), ditches, mats (*Sphagnum*), mead-ows (boggy, open, swampy, wet), pinelands (seasonally wet), plateaus (windswept), sandhills, seeps, swales (boggy), swamps (open), thickets (boggy), and along the margins of rivers and streams at elevations to 1829 m. Because of taxo-nomic issues (see *Systematics* below), much of the published information attributed to "*Stenanthium leimanthoides*" actu-ally applies to *S. densum*, leaving little information available that can be assigned confidently to *S. leimanthoides*. The plants occur in open exposures on substrates described as gravel (terraced), rock, and sand. Flowering occurs from June to August. The morphophysiologically dormant seeds require cold stratification to induce embryo growth and germination. Embryos in seeds kept at 5°C have grown to 2.5 mm within 9 weeks, and the seeds have germinated within 12 weeks. Seeds began to germinate during winter (January–February) at daily temperatures that varied from 3.0°C to 10.9°C follow-ing their exposure to 44 days of ambient cold stratification in an unheated greenhouse. Germination occurred more quickly for seeds placed under a 20°C/10°C day/night temperature regime. Under natural field conditions, the seeds break dor-mancy during the winter and germinate by early spring. No persistent seed bank develops. Once the status of this taxon has been settled satisfactorily, a thorough re-evaluation of its life history (including wetland status) will be necessary.
Reported associates: *Alnus, Sphagnum, Vaccinium.*
Use by wildlife: The floral reproductive organs (and infrequently the leaves) of *Stenanthium densum* and *S. leimanthoides* are eaten by larval sawflies (Insecta: Hymenoptera: Tenthredinidae: *Rhadinoceraea zigadenusae*); *Rhadinoceraea sodsensis* is also known to feed on *S. leiman-thoides*. *Stenanthium densum* is reported to be dangerously poisonous to livestock.
Economic importance: food: *Stenanthium* is inedible and possibly toxic; **medicinal:** Although no compelling evidence

is available, *Stenanthium* is believed to be poisonous; **culti-vation:** *Stenanthium* is not cultivated commercially; **misc. products:** none; **weeds:** none; **nonindigenous species:** none. **Systematics:** The taxonomy of *Stenanthium* is nothing short of confusing. *Stenanthium densum* and *S. leimanthoides* long had been recognized within *Zigadenus* as *Z. densus* and *Z. lei-manthoides*. However, *Zigadenus* eventually was dismantled as a consequence of phylogenetic conflicts (e.g., Figure 2.27), which ultimately led to its present monotypic circumscription (i.e., as *Zigadenus glaberrimus*). By analyses of plastid DNA sequence data, *Stenanthium* resolves within tribe Melanthieae as the sister clade to *Melanthium* in a position remote from *Zigadenus* phylogenetically (Figure 2.27). The association of *Melanthium*, and *Stenanthium* is consistent morphologi-cally with two characters (beak-like styles and recurved stig-mas) that are shared by these genera. However, analysis of nuclear DNA sequence (nrITS) nested all the *Melanthium* species within *Veratrum*, a result that would argue for the merger of the genera. The different cladogram topologies are not strongly supported in either analysis, leaving this major incongruence inexplicable at this time (see preceding discus-sion under *Systematics* for *Melanthium* above). On the other hand, analyses of plastid and nuclear sequence data similarly indicate a distant relationship of *Stenanthium* to *Zigadenus* (i.e., *Z. glaberrimus*). Although the transfer of *Z. densus* and *Z. leimanthoides* to *Stenanthium* appears to be warranted, the boundaries of these two taxa remain contentious. Some authors have regarded *Stenanthium leimanthoides* as being conspecific with *S. densum* or have recommended infraspe-cific (e.g., varietal) status for the former. In sharp contrast, a recent treatment not only retained the taxa as distinct species, but split *S. leimanthoides* into two additional segregate spe-cies (*S. tennesseense* and *S. macrum*), with the latter alleg-edly being more similar to *S. densum*. Understandably, the taxonomy of *Stenanthium* must first be clarified before any wetland status can be reasonably assigned to these species, especially given that both *S. densum* and *S. leimanthoides* were ranked differently (OBL or FACW) in various parts of their range, making it unclear which (or how many) species should actually be ranked as OBL. Whether all of these segre-gates eventually will be maintained at the species level cannot be ascertained; however, at this time they are presumed to be justified. The distribution of *S. leimanthoides* relative to the other segregates enables at least some literature accounts to be assigned confidently to that species (*sensu stricto*) in exclu-sion of the allopatric *S. tennesseense* and *S. densum*. However, the broad sympatry of *S. densum* and the recently described *S. macrum* makes it extremely difficult to distinguish these taxa in most literature accounts. As the only viable option, all accounts attributed to *S. densum* (excluding those specific to *S. macrum*) have been used to compile the information pre-sented here for *S. densum*. The basic chromosome number of *Stenanthium* appears to be $x=10$. *Stenanthium densum* and *S. leimanthoides* (both $2n=20$) are diploid. The estimated genome size for *Stenanthium densum* is 3.44 pg (2C); 1.72 pg (1C); and 1682.16 mb; and for *S. leimanthoides* it is 3.3 pg (2C); 1.65 pg (1C); and 1613.70 mb. The extent of hybridization in North American *Stenanthium* is uncertain. Although some authors have reported morphological intergradation among the taxa, the more recent demarcation of new segregate species requires that a renewed assessment be made.

Distribution: *Stenanthium densum* occurs along the coastal plain of the southeastern United States; *S. leimanthoides occurs* in the eastern montane United States.

References: Avis et al., 1997; Baskin et al., 1993; Delahoussaye et al., 2014; Knapp et al., 2011; Morris, 2012; 2013; Pellicer et al., 2014; Smith & McDearman, 1990; Smith & Barrows, 1995; Sorrie & Weakley, 2017; Wagstaff & Case, 1987; Wofford, 2006; Zomlefer et al., 2001; Zomlefer & Judd, 2002; Zomlefer & Smith, 2002.

7.2.4. Toxicoscordion

Death camus, poison camus; Zigadène

Etymology: from the Greek *toxikos scordion* ("poison onion") in reference to the toxic properties of the bulbs

Synonyms: *Amianthium* (in part); *Anticlea* (in part); *Gomphostylis* (in part); *Helonias* (in part); *Leimanthium* (in part); *Melanthium* (in part); *Zigadenus* (in part)

Distribution: global: Mexico; North America; **North America:** south-central and western

Diversity: global: 8 species; **North America:** 8 species

Indicators (USA): obligate wetland (OBL): *Toxicoscordion fontanum, T. micranthum*

Habitat: freshwater; palustrine; **pH:** unknown; **depth:** <1 m; **life-form(s):** emergent herb

Key morphology: bulbs (to 45 mm) tunicate, ovoid; stems to 9 dm; leaves basal, the blades (to 80 cm) linear, reduced upward as bracts (to 4 cm); panicles many-flowered (to 100), pyramidal to elongate; the flowers (to 15 mm) bisexual, hypogynous, pedicellate (to 45 mm); perianth campanulate, the tepals (to 12 mm) elliptic to ovate, cream colored, sometimes clawed (to 5 mm), with a basal gland; stamens 6, the filaments straight, thickened proximally; capsules (to 22 mm) septicidal, many-seeded

Life history: duration: perennial (bulbs); **asexual reproduction:** none; **pollination:** insect; **sexual condition:** andromonoecious, hermaphroditic; **fruit:** capsules (common); **local dispersal:** seeds (unknown); **long-distance dispersal:** seeds (unknown)

Imperilment: 1. *Toxicoscordion fontanum* [G3]; S3 (CA); **2.** *T. micranthum* [G4]

Ecology: general: *Toxicoscordion* is not a characteristic wetland genus. In addition to the two species designated as OBL indicators, only one other species has moderate wetland indicator status (as FAC, FACU). Moreover, the following treatments of the two OBL designees indicate that they occur quite often in very dry sites and are associated primarily with species that lack any wetland indicator status. Accordingly, it is difficult to rationalize why they have been designated as obligate wetland plants. It is likely that some confusion has resulted due to taxonomic uncertainties (see *Systematics* below), especially since one of the OBL indicators (*T. fontanum*) was recognized previously as a variety of *T. micranthum*. Confusion of these species with other genera is also

possible and likely has added anomalous accounts. Yet, even when taxonomic issues are considered, there is no compelling reason to categorize either taxon as an OBL indicator and their wetland rankings should be reconsidered. The genus is not well studied overall, and the following summary relies on information compiled from a limited number of species and studies. The flowers adapted for outcrossing. They are markedly protandrous, have high pollen:ovule ratios, and also are incompletely self-incompatible. The flowers are not autogamous but partial geitonogamous selfing (~50% seed set) has been achieved by hand pollinations. Some of the species are andromonoecious. All parts of the plants, including the pollen and nectar of at least some species, contain neurotoxic alkaloids that are poisonous to honeybees (Insecta: Hymenoptera; Apidae: *Apis mellifera*). Insect (Insecta) pollinators that apparently are not affected by the toxins have included andrenid bees (Hymenoptera: Andrenidae: *Andrena amphibola, A. astragali*), sweat bees (Halictidae: *Lasioglossum bruneri, L. nymphaearum, L. versatum*), soldier flies (Diptera: Stratiomyidae: *Stratiomys barbata*), and syrphid flies (Diptera: Syrphidae: *Eristalis hirta*). Despite their toxicity, the inflorescences of some species are eaten by moth larvae (Insecta: Lepidoptera: Geometridae: *Eupithecia zygadeniata*). Few details exist on the seed ecology of any species. Copious flower and seed production can occur in the year following a fire. However, dry heat treatments (5 days @ 120°C) have reduced germination rates in some species.

7.2.4.1. ***Toxicoscordion fontanum*** **(Eastw.) Zomlefer & Judd** grows in or on bluffs (coastal), chaparral (burned, serpentine), marshes (serpentine, vernally moist), meadows (wet), prairies (coastal), ridgetops, roadsides, seeps, slopes [to 30°] (grassy), streambeds (dried), and along the margins of rivers and streams at elevations to 1000 m. Exposures can range from full sun to partial shade. The substrates have been described as clay (red, serpentine), clay loam, granite (decomposed), gravelly loam, loam, loamy sand, rocks, sand, sandstone, serpentine (decomposing), and volcanics (Boomer series). Flowering occurs from March to August. Otherwise, the life history of this species is poorly known. This species has been characterized as a serpentine marsh or seep endemic and is reported occasionally from wet sites. However, many occurrences are in very dry habitats, almost all of the reported associates have no wetland indicator status, and only one of the nearly 70 reported associates is ranked as an OBL indicator. The OBL status of this species surely is inappropriate and its wetland classification should be re-evaluated. **Reported associates:** *Adenostoma fasciculatum, Angelica, Arbutus menziesii, Arctostaphylos glandulosa, Arctostaphylos obispoensis, Arctostaphylos viscida, Baccharis pilularis, Bromus carinatus, Callitropsis sargentii, Calochortus uniflorus, Camissonia contorta, Castilleja exserta, Castilleja rubicundula, Ceanothus cuneatus, Ceanothus leucodermis, Ceanothus papillosus, Chlorogalum pomeridianum, Clarkia, Claytonia, Cornus glabrata, Cynoglossum, Danthonia californica, Elymus glaucus, Ericameria arborescens, Eriodictyon californicum, Eriogonum elongatum, Eriogonum fasciculatum, Festuca idahoensis, Fremontodendron* *californicum, Fritillaria pluriflora, Heteromeles arbutifolia, Holozonia filipes, Horkelia yadonii, Hosackia gracilis, Juncus phaeocephalus, Juniperus californica, Lupinus subvexus, Mimulus nudatus, Muhlenbergia rigens, Nassella, Panicum, Pellaea, Pickeringia montana, Pinus coulteri, Pinus sabiniana, Plantago erecta, Plantago lanceolata, Poa secunda, Pogogyne clareana, Pseudotsuga menziesii, Quercus chrysolepis, Quercus douglasii, Quercus durata, Quercus lobata, Quercus wislizeni, Rhamnus californica, Salix breweri, Scutellaria, Sequoia sempervirens, Sisyrinchium bellum, Stachys albens, Thalictrum polycarpum, Toxicodendron, Toxicoscordion fremontii, Triphysaria eriantha, Triteleia ixioides, Umbellularia californica, Viola douglasii.*

7.2.4.2. *Toxicoscordion micranthum* (Eastw.) A. Heller
occurs on balds, bluffs (rocky), bogs, chaparral (disturbed), cliffs (granite), flats (gravelly), floodplains, grasslands (dry), gulch (damp, rocky), gullies, hillsides (rocky), meadows (dry, moist, rocky, wet), mountainsides (dry), prairies, ravines, ridges (serpentine), roadsides, seeps (dry, moist), slopes (brushy, dry, seepage), swales, uplands (dry), woodlands (burned, dry, mixed pine, oak), and along the margins of streams (vernal) at elevations to 2134 m. Exposures can range from fully open to partial shaded conditions. The substrates have been described as clay (heavy), granite, gravel, mulch, pavement (hard-packed), rhyolite/andesite, rock, rocky sand, sandy clay, schist, serpentine (clayey, gravelly, rocky). Flowering occurs from April to July, with fruiting commencing during May. Little additional life history information is available. Occurrences on dry ground have been reported about as often as those on wet ground, where the plants are observed mainly during the spring. The frequency of reports from dry upland habitats, along with the list of primarily non-OBL associates (see following), would indicate that the current OBL wetland status of this species is inappropriate and should be reconsidered. **Reported associates:** *Abies ×shastensis, Aesculus, Allium falcifolium, Arbutus menziesii, Arctostaphylos canescens, Arctostaphylos patula, Arctostaphylos stanfordiana, Arctostaphylos viscida, Aspidotis densa, Brodiaea, Callitropsis macnabiana, Callitropsis sargentii, Calocedrus decurrens, Calochortus tolmiei, Camassia quamash, Camellia, Carex praegracilis, Carex spectabilis, Castilleja, Ceanothus cuneatus, Ceanothus pumilus, Chamaecyparis lawsoniana, Chlorogalum, Chrysolepis chrysophylla, Danthonia, Eriogonum, Festuca californica, Galium, Garrya fremontii, Hastingsia alba, Horkelia tridentata, Juncus, Juniperus, Kalmia microphylla, Leucothoe davisiae, Lithocarpus densiflorus, Lomatium, Lupinus, Mimulus lewisii, Myrica californica, Pedicularis densiflora, Penstemon newberryi, Phlox, Pinguicula vulgaris, Pinus attenuata, Pinus jeffreyi, Pinus lambertiana, Pinus monticola, Pinus ponderosa, Poa, Polygala californica, Potentilla flabellifolia, Pseudotsuga menziesii, Quercus berberidifolia, Quercus chrysolepis, Quercus douglasii, Quercus durata, Quercus garryana, Quercus kelloggii, Quercus vacciniifolia, Rhamnus, Rhododendron columbianum, Rhododendron occidentale, Salix, Stellaria, Tsuga mertensiana, Umbellularia californica, Viola, Whipplea modesta.*

Use by wildlife: None reported.

Economic importance: food: Due to their alkaloid content, *Toxicoscordion* species are poisonous (to humans and livestock) and never should be eaten or fed to animals. The plants are believed to have been responsible for serious illnesses suffered by members of the Lewis and Clark expedition; **medicinal:** *Toxicoscordion fontanum* and *T. micranthum* were used by the Northern Pomo tribe as an analgesic and antirheumatic; **cultivation:** *Toxicoscordion* species are rarely cultivated; **misc. products:** none; **weeds:** none; **nonindigenous species:** none.

Systematics: Phylogenetic studies have demonstrated that *Toxicoscordion* comprises a clade that is distinct from *Zigadenus* (e.g., Figure 2.27), where most of the species had long been assigned formerly. Despite this factor, many current treatments continue to recognize *Toxicoscordion* species within *Zigadenus*, making it difficult to develop an accurate circumscription of either group. The superficial similarity of *Toxicoscordion* not only to *Zigadenus* but to other genera (e.g., *Narthecium*) undoubtedly has resulted in taxonomic confusion and misidentification. *Toxicoscordion* species are not strongly differentiated morphologically and are likely to be misidentified. There is also taxonomic disagreement regarding species limits; e.g., with *T. fontanum* sometimes treated as a variety of *T. micranthum*. A thorough taxonomic re-evaluation of the genus is recommended. The basic chromosome number of *Toxicoscordion* is $x=11$; *T. fontanum* and *T. micranthum* (both $2n=22$) are diploids. The estimated genome size of *T. fontanum* is 4.93 (2C); 2.47 (1C); and 2410.77 mb; and for *T. micranthum* it is 4.91 (2C); 2.46 (1C); and 2400.99 mb.

Distribution: *Toxicoscordion fontanum* is endemic to central California; *T. micranthum* is restricted to northwestern California and southwestern Oregon.

References: Baskin & Baskin, 1998; Emms, 1993; Irwin et al., 2014; Kim et al., 2016; McNeal & Zomlefer, 2010; Pellicer et al, 2014; Schwartz, 2002; Tepedino, 1981; Tepedino et al., 1989; Tyler & Borchert, 2003; Welch, 2013.

7.2.5. Veratrum
Corn lily, false hellebore; ellébore, varaire, vérâtre

Etymology: after *veratrum*, an ancient name for the hellebore plant used by Pliny, which is of uncertain derivation but doubtfully refers to the root color as is widely thought

Synonyms: *Acelidanthus*; *Helonias* (in part); *Zigadenus* (in part)

Distribution: global: Northern Hemisphere; **North America:** throughout except north-central

Diversity: global: 26 species; **North America:** 5 species

Indicators (USA): obligate wetland (OBL); facultative wetland (FACW): *Veratrum californicum, V. fimbriatum*

Habitat: freshwater; palustrine; **pH:** 4.2–6.5; **depth:** <1 m; **life-form(s):** emergent herb

Key morphology: rhizomes short, thick, vertical, extending from basal bulbs, the roots contractile, fleshy; stems (to 2.5 m) erect, simple, hollow, leafy, tomentose distally; leaves alternate, simple, often plicate, reduced distally and narrowing to closed, often overlapping sheaths, the blades (to 50 cm) elliptic-lanceolate or ovate (distalmost lanceolate to lance-linear), glabrous to tomentose; panicles (to 70 cm), with spreading to ascending branches, tomentose, the flowers slightly epigynous, bisexual or staminate proximally (andromonoecious), pedicellate (to 12 mm); tepals (to 17 mm) 6, white or creamy-white and greenish basally, lanceolate, rhomboid-ovate, or elliptic to oblong-ovate or ovate, the margins entire to denticulate or deeply and irregularly fimbriate, the base with 1–2 adaxial, elliptic or v-shaped glands, glabrous to tomentose beneath; stamens 6, adnate to hypanthium; ovary partially inferior (sometimes superior), 3-locular, the styles 3, distinct, short; capsules (to 3 cm) deeply 3-lobed, septicidal, narrowly ovoid or oblong-ovoid, glabrous; seeds (to 12 mm) ellipsoid to fusiform, flat or globose, winged or wingless

Life history: duration: perennial (bulbs, rhizomes); **asexual reproduction:** rhizomes; **pollination:** insect; **sexual condition:** andromonoecious or hermaphroditic; **fruit:** capsules (infrequent); **local dispersal:** rhizomes; **long-distance dispersal:** seeds (vector unknown)

Imperilment: 1. *Veratrum californicum* [G5]; S2 (MT, WY); **2.** *V. fimbriatum* [G3]; S3 (CA)

Ecology: general: All *Veratrum* species are perennials, which persist from a bulbous base, from which short, thick, vertical rhizomes develop. Three of the five North American species are wetland indicators (OBL, FACW, FAC), which generally are facultative inhabitants of wet sites and can extend into fairly dry sites where conditions permit. It is difficult to generalize the ecology of the group given that taxonomic uncertainty remains regarding its circumscription (see comments above for *Melanthium*). Flowering often is sporadic, occurring unpredictably from one year to the next. There exists little information on the specific pollination biology of any *Veratrum* species. The flowers are adapted morphologically for outcrossing by unspecialized insects (Insecta) such as flies (Diptera), which have been observed to carry pollen in some species. The flowers are fragrant, nectariferous, and protandrous, shedding pollen at the onset of anthesis before the stigmas become receptive. The seeds are morphophysiologically dormant when shed and require cold stratification to induce germination. Clonal growth can be extensive in some of the species.

7.2.5.1. *Veratrum californicum* **Durand** occurs in bottoms (stream), cirques (subalpine), depressions (roadside), draws (wet), fell fields (alpine), fens (subalpine), forest openings (coniferous), marshes, meadows (alpine, avalanche, boggy, bottomland, cienaga, dry, grassy, marshy, moist, riverine, subalpine, vernal, wet), roadsides (swampy), seeps (hillside), slopes [to 15%] (alpine, rocky, scree, springy), springs (grazed), swales (moist), swamps, and along the margins of channels (drainage, rocky), lakes (subalpine), ponds, rivers, and streams at elevations to 3689 m. The plants have been observed growing in exposures ranging from full sunlight to deep shade. The substrates typically are saturated but not usually covered by standing water. The mean water table depth has been found to range from −11.1 to −42.0 cm. The

substrates are acidic (pH: 5.8–6.3), high in calcium and iron content, and have been described as alluvium, basalt, clay, clay loam, colluvium (basaltic), decomposed granite, diorite (granite), gravel, humus, limestone, loam (granitic, gravelly, rocky, sandy, skeletal), loamy clay, muck, mucky sand, mud, muddy loam, organic, peat, pebbles, phyllite, pumice, quartz diorite, rocky, sand, sandy gravel, serpentine (Greenleaf manzanita series), shale (Mancos), silt, silty loam, and talus. Flowering and fruiting occur from May to September. Warm temperatures promote clonal growth and reduce the extent of flowering. A higher percentage of flowering plants was found to occur 2 years after the plants experienced a period of cool summer temperatures. The flowers are visited diurnally by flies (Insecta: Diptera) and nocturnally by moths (Insecta: Lepidoptera), which are thought to represent their usual pollinators. The mechanism of seed dispersal has not been elucidated. The plants occur often at sites that maintain a large winter snowpack, which ensures an adequate period of cold stratification to break seed dormancy and provides moisture for early season growth. The seeds have non-deep complex morphophysiological dormancy and require a period of cold stratification (3–4 months at 0.5°C–4.4°C) to induce germination. The seeds can germinate at 4.4°C but the seedlings develop much more rapidly at temperatures of 10°C or higher. Hundred percent germination has occurred within 13 weeks for stratified seeds that were incubated at 10°C on moist filter paper in petri dishes. Natural germination has been observed to occur beneath the snow in late February. A persistent seed bank does not develop with essentially all the seeds germinating within the first year. The seedlings develop slowly and take an estimated 7–10 years to mature from seed to adult plants. The short seedling growth periods are interrupted by episodes of bulb or crown dormancy, which require at least 90 days of cold stratification before growth resumes. A minimum of 120 days of chilling (at 5°C) is necessary to force early emergence and to promote vigorous seedling growth. The highest greenhouse survivorship has been observed for plants derived from cultures that were incubated initially at 16°C. Site productivity has been estimated at 4495 kg ha^{-1}. The plants are highly susceptible to volcanic ash and suffered high mortality following the eruption of Mount St. Helens due to overloading of the clasping leaves by ash. The plants are long-lived and undergo vigorous clonal growth from the rhizomes.

Reported associates: *Abies amabilis, Abies concolor, Abies lasiocarpa, Abies magnifica, Acer glabrum, Acer negundo, Achillea millefolium, Achnatherum lemmonii, Aconitum columbianum, Agastache breviflora, Agastache urticifolia, Agrostis idahoensis, Agrostis pallens, Allium validum, Alnus rubra, Alnus tenuifolia, Alnus viridis, Alopecurus pratensis, Amelanchier alnifolia, Amelanchier utahensis, Amorpha fruticosa, Antennaria microphylla, Anticlea elegans, Apocynum androsaemifolium, Aquilegia chrysantha, Aquilegia coerulea, Aquilegia desertorum, Aquilegia formosa, Arnica chamissonis, Arnica longifolia, Arnica mollis, Artemisia dracunculus, Artemisia tridentata, Athyrium filix-femina, Aulacomnium, Balsamorhiza deltoidea, Balsamorhiza sagittata, Barbarea orthoceras, Berberis repens, Betula, Bistorta* *bistortoides, Bromus carinatus, Bromus inermis, Calamagrostis canadensis, Calocedrus decurrens, Caltha leptosepala, Camassia quamash, Carex angustata, Carex aquatilis, Carex athrostachya, Carex aurea, Carex buxbaumii, Carex capitata, Carex densa, Carex disperma, Carex echinata, Carex feta, Carex fracta, Carex gracilior, Carex heteroneura, Carex hoodii, Carex integra, Carex interior, Carex jonesii, Carex laeviculmis, Carex lasiocarpa, Carex lenticularis, Carex leporinella, Carex luzulina, Carex mariposana, Carex microptera, Carex nebrascensis, Carex rostrata, Carex scopulorum, Carex straminiformis, Carex utriculata, Castilleja cusickii, Castilleja miniata, Ceanothus velutinus, Cerastium arvense, Chamerion angustifolium, Cirsium arizonicum, Cirsium vulgare, Claytonia sibirica, Collomia linearis, Cornus canadensis, Cornus sericea, Crataegus, Cryptantha echinella, Cryptogramma acrostichoides, Cystopteris fragilis, Dactylis glomerata, Danthonia, Darlingtonia californica, Dasiphora floribunda, Delphinium barbeyi, Delphinium glaucum, Delphinium polycladon, Delphinium ×occidentale, Deschampsia cespitosa, Descurainia californica, Draba albertina, Drepanocladus, Drosera anglica, Drosera rotundifolia, Drymocallis cuneifolia, Dryopteris patula, Eleocharis, Elymus elymoides, Elymus glaucus, Elymus trachycaulus, Epilobium anagallidifolium, Epilobium ciliatum, Equisetum arvense, Equisetum telmateia, Erigeron coulteri, Erigeron elatior, Erigeron glacialis, Erigeron oreophilus, Erigeron peregrinus, Erigeron strigosus, Eriophorum gracile, Erysimum asperum, Festuca arizonica, Festuca brachyphylla, Festuca rubra, Fragaria virginiana, Frasera fastigiata, Frasera speciosa, Galium bifolium, Galium boreale, Galium trifidum, Galium triflorum, Gentiana parryi, Gentianopsis holopetala, Gentianopsis simplex, Geranium richardsonii, Geranium viscosissimum, Geum macrophyllum, Glyceria elata, Glyceria striata, Hackelia micrantha, Helenium autumnale, Helenium bigelovii, Helianthella quinquenervis, Heracleum sphondylium, Heuchera hallii, Heuchera rubescens, Hieracium caespitosum, Holcus lanatus, Holodiscus, Hordeum brachyantherum, Horkelia fusca, Horkelia rydbergii, Hosackia pinnata, Hydrophyllum fendleri, Hydrophyllum tenuipes, Hymenoxys hoopesii, Hypericum anagalloides, Ipomopsis aggregata, Iris missouriensis, Juncus balticus, Juncus effusus, Juncus saximontanus, Juniperus communis, Kalmia polifolia, Kyhosia bolanderi, Lathyrus nevadensis, Leucanthemum vulgare, Ligusticum grayi, Ligusticum porteri, Ligusticum verticillatum, Lilium kelleyanum, Linum lewisii, Lophozia, Lorandersonia peirsonii, Lupinus arbustus, Lupinus argenteus, Lupinus burkei, Lupinus latifolius, Lupinus polyphyllus, Lupinus pratensis, Luzula orestera, Lysichiton americanus, Maianthemum stellatum, Marchantia, Meesia triquetra, Menziesia ferruginea, Mertensia franciscana, Micranthes nelsoniana, Micranthes oregana, Mimulus cardinalis, Mimulus guttatus, Mimulus moschatus, Mimulus primuloides, Mimulus tilingii, Mitella breweri, Mitella pentandra, Monarda fistulosa, Monardella, Montia chamissoi, Nasturtium, Neottia convallarioides, Oenanthe sarmentosa, Oreostemma alpigenum, Orthilia secunda, Osmorhiza berteroi, Osmorhiza depauperata, Osmorhiza occidentalis, Oxypolis fendleri, Oxypolis occidentalis, Pedicularis attollens, Pedicularis grayi, Pedicularis groenlandica, Penstemon attenuatus, Penstemon barbatus, Penstemon confertus, Penstemon heterodoxus, Penstemon rydbergii, Penstemon wilcoxii, Perideridia parishii, Phacelia heterophylla, Phleum alpinum, Phlox idahonis, Physocarpus capitatus, Physocarpus malvaceus, Picea engelmannii, Picea pungens, Pinus contorta, Pinus flexilis, Pinus lambertiana, Pinus leiophylla, Pinus monticola, Pinus ponderosa, Platanthera dilatata, Platanthera sparsiflora, Platanthera stricta, Poa palustris, Poa pratensis, Poa secunda, Podagrostis humilis, Polemonium foliosissimum, Polemonium occidentale, Populus angustifolia, Populus tremuloides, Populus trichocarpa, Potentilla anserina, Potentilla gracilis, Potentilla pulcherrima, Potentilla ×diversifolia, Primula fragrans, Primula jeffreyi, Primula latiloba, Primula pauciflora, Primula tetrandra, Prunella vulgaris, Prunus emarginata, Pseudotsuga menziesii, Pteridium aquilinum, Pyrola chlorantha, Quercus gambelii, Raillardella pringlei, Ranunculus orthorhynchus, Ranunculus uncinatus, Rhodiola integrifolia, Rhododendron columbianum, Ribes bracteosum, Ribes cereum, Ribes nevadense, Robinia neomexicana, Rorippa islandica, Rosa californica, Rosa woodsii, Rubus lasiococcus, Rubus sachalinensis, Rubus spectabilis, Rubus ursinus, Rudbeckia laciniata, Rudbeckia occidentalis, Rumex acetosella, Rumex orthoneurus, Sagina saginoides, Salix bebbiana, Salix boothii, Salix commutata, Salix drummondiana, Salix eastwoodiae, Salix exigua, Salix geyeriana, Salix irrorata, Salix lasiandra, Salix lasiolepis, Salix lemmonii, Salix ligulifolia, Salix lucida, Salix orestera, Salix wolfii, Sambucus nigra, Scirpus microcarpus, Senecio scorzonella, Senecio serra, Senecio triangularis, Sidalcea oregana, Sisyrinchium idahoense, Solidago canadensis, Sphagnum, Sphenosciadium capitellatum, Spiraea splendens, Spiranthes romanzoffiana, Stachys ajugoides, Stachys albens, Stachys chamissonis, Stellaria longipes, Streptopus amplexifolius, Streptopus lanceolatus, Symphoricarpos albus, Symphoricarpos rotundifolius, Symphyotrichum, Taraxacum officinale, Thalictrum fendleri, Thermopsis montana, Tiarella trifoliata, Torilis arvensis, Torreyochloa erecta, Trautvetteria caroliniensis, Trianthaglutinosa, Trianthaoccidentalis, Trifolium longipes, Tsuga heterophylla, Tsuga mertensiana, Turritis glabra, Urtica dioica, Vaccinium cespitosum, Vaccinium membranaceum, Vaccinium uliginosum, Valeriana capitata, Valeriana edulis, Valeriana occidentalis, Valeriana sitchensis, Veratrum viride, Veronica americana, Veronica americana, Veronica wormskjoldii, Vicia americana, Viola canadensis, Viola palustris, Wyethia amplexicaulis.*

7.2.5.2. **Veratrum fimbriatum A. Gray** grows in bogs, depressions (damp), ditches, fens, floodplains, marshes (coastal), meadows (coastal, wet), prairies (coastal), swamps (vernal), and along the margins of streams at elevations to 60 m. The exposures typically are characterized by some degree of shade. The substrates are acidic (e.g., pH: 4.2–5.1) and are associated with slightly less acidic waters (e.g., pH: 5.2–6.5). They have been described as peat or sand (humic). Flowering extends from August to October. The life history has been poorly studied and there is no specific information on the floral biology, pollination biology, or seed ecology. **Reported associates:** *Blechnum spicant, Calamagrostis nutkaensis, Calliergonella cuspidatum, Callitropsis pigmaea, Campanula californica, Carex densa, Carex obnupta, Cicuta douglasii, Comarum palustre, Drosera rotundifolia, Eleocharis acicularis, Epilobium ciliatum, Gentiana sceptrum, Glyceria ×occidentalis, Helenium bolanderi, Hydrocotyle ranunculoides, Hypericum anagalloides, Juncus bolanderi, Juncus effusus, Juncus ensifolius, Juncus lesueurii, Juncus phaeocephalus, Lysichiton americanus, Menyanthes trifoliatum, Mimulus guttatus, Myrica californica, Nasturtium officinale, Nuphar polysepala, Oenanthe sarmentosa, Pinus contorta, Platanthera dilatata, Rhododendron columbianum, Rubus, Sisyrinchium californicum, Sphagnum, Typha latifolia.*

Use by wildlife: *Veratrum californicum* provides habitat for mountain white-crowned sparrows (Aves: Emberizidae: *Zonotrichia leucophrys oriantha*). Mountain beavers (Mammalia: Rodentia: Aplodontiidae: *Aplodontia rufa*) store the plants in food caches called haystacks, where the alkaloids are thought to denature over time. *Veratrum californicum* is grazed by sheep (Mammalia: Bovidae: *Ovis aries*), but the steroidal alkaloid content (cyclopamine, cycloposine, and jervine) is teratogenic and causes cyclopean-type malformations in newly born lambs. The foliage is eaten by elk (Mammalia: Cervidae: *Cervus elaphus*) and was found to be nonteratogenic when fed to European rabbits (Mammalia: Leporidae: *Oryctolagus cuniculus*). The nectar of *V. californicum* is toxic to bees (Insecta: Hymenoptera). The plants are fed on by sawflies (Insecta: Hymenoptera: Tenthredinidae: *Rhadinoceraea aldrichi*), which sequester the toxic alkaloids as defense compounds. The foliage hosts Fungi (Ascomycota: Leotiomycetidae: *Heterosphaeria sublineolata*) and larval moths (Insecta: Lepidoptera: Noctuidae: *Xestia smithii*). The fruits (up to 6.9%) also are eaten by larval moths (Lepidoptera).

Economic importance: food: *Veratrum* species are toxic, inedible, and never should be consumed due to their numerous steroidal alkaloids (e.g., cyclopamine, germine, isorubijervine, jervine, muldamine protoveratrine A and B, protoverine, pseudojervine, rubijervine, veracevine, veratramine, veratrosine). In humans, these substances increase sodium channel permeability of nerve cells, causing continuous impulse firing, which results in what is called the Bezold-Jarisch reflex: a combination of apnea, hypotension, and reduced heart rate. The Blackfoot and Thompson people recognized that *Veratrum* species were poisonous but there are accounts of the plants being roasted in hot ashes and then eaten by members of the Miwok tribe; **medicinal:** *Veratrum californicum* has a long history of use as a medicinal plant by the indigenous North American people. The plants were used to prepare a dressing for burns (Paiute). Infusions were used to relieve fevers (Paiute). A poultice from the mashed plants was used to treat sprains and broken bones (Paiute). Decoctions, poultices, or powders prepared from the root were used as an emetic (Washo), for headaches (Blackfoot), for treating blood disorders (Paiute, Shoshoni, Thompson), as

a contraceptive (Paiute, Shoshoni, Washo), for treating colds (Paiute, Shoshoni, Thompson), to treat rheumatism (Paiute, Shoshoni, Washo), for tonsillitis (Paiute, Shoshoni), for toothaches (Paiute), to treat venereal disease (Paiute, Thompson), to promote healing of boils, bruises, cuts, sores, and swellings (Paiute, Shoshoni, Washo), as a disinfectant (Paiute, Shoshoni), and as a snakebite remedy (Paiute, Shoshoni). The therapeutic properties of *V. californicum* so widely recognized by the various indigenous North American tribes are equally valued in modern medicine. Most notably, a steroidal alkaloid (cyclopamine) found in *V. californicum* is an Hh antagonist; i.e., it inhibits an embryogenic signaling pathway known as Hedgehog (Hh), which has led to the optimistic development of treatments for psoriasis and human cancers such as nevoid basal cell carcinoma syndrome, glioblastoma, and prostate cancer. *Veratrum fimbriatum* contains several steroidal alkaloids (e.g., jervine, pseudojervine) including neogermitrine, which is hypotensive. Its alkaloid content is highest during spring, when the plants are growing rapidly; **cultivation:** *Veratrum californicum* has earned the designation an RHS *Award of Garden Merit* plant; **misc. products:** The dried root of *V. californicum* was powdered into snuff by the Blackfoot tribe. Karok girls made decorative ribbons from the white inner stem and braided them into their hair. The Paiute kept berries fresh by wrapping them in leaves of *V. californicum*; **weeds:** *Veratrum californicum* is eradicated as a weed of grazing lands due to its toxic effects on livestock. Because the plants generally are unpalatable to livestock, their density has increased substantially in many grazed localities; **nonindigenous species:** none.

Systematics: Due to irreconcilable discrepancies in phylogenetic results derived from nuclear vs. plastid DNA sequences (see *Systematics* under *Melanthium*, above), it remains uncertain whether to recognize *Melanthium* species as distinct (e.g., Figure 2.27) or as part of *Veratrum*. Without any clear resolution of this question at the present time, the groups have been treated here as separate genera, which is in accord with most traditional literature. *Veratrum californicum* and *V. fimbriatum* both have been assigned to section *Veratrum*, which is distant phylogenetically from the clade of species recognized under *Melanthium*. The basic chromosome number of *Veratrum* is $x=8$; *Veratrum californicum* and *V. fimbriatum* are tetraploid ($2n=32$). The genome size (2C) of *V. californicum* has been estimated at 5.4 pg.

Distribution: *Veratrum californicum* is distributed throughout the western United States; *V. fimbriatum* is endemic to northwestern California.

References: Adler, 2000; Baker, 1972; Barker et al., 2002; Baskin & Baskin, 1998; Beattie et al., 1973; Bharathan et al., 1994; Chandler & McDougal, 2014; Cooke, 1949; Cosgriff et al., 2004; Dillingham, 2005; Göppner & Leverkus, 2011; Halpern, 1986; Incardona et al., 1998; Johnson & Martin, 1969; Kim et al., 2016; King & Mewaldt, 1987; Korfhage et al., 1980; Mack, 1981; Maschinski, 2001; McFerren, 2006; McIlroy & Allen-Diaz, 2012; McMillan, 1956; McNeal, 2012b; McNeal & Shaw, 2002; Müller-Schwarze et al., 2001; Murray, 2000; Schaffner et al., 2001; Schep et al., 2006; Song et al., 2014; Sun et al., 2012; Taylor, 1956; Vansell & Watkins, 1933; Williams & Cronin, 1968; Williams & Cronin, 1981; Zomlefer et al., 2003.

7.2.6.   Zigadenus

Death camas, sandbog death camus; zigadène

Etymology: derived from the Greek *zygos adenos* ("yoke gland"), in reference to the paired tepal glands

Synonyms: *Helonias* (in part)

Distribution: global: North America (USA); **North America:** southeastern United States

Diversity: global: 1 species; **North America:** 1 species

Indicators (USA): obligate wetland (OBL); facultative wetland (FACW): *Zigadenus glaberrimus*

Habitat: freshwater; palustrine; **pH:** acidic; **depth:** <1 m; **life-form(s):** emergent herb

Key morphology: rhizomes thick, twisted, with contractile roots; plants (to 12 dm). subscapose; leaves (<10) basal, keeled beneath, the blades (to 42 cm) linear, glaucous beneath, alternate and reduced distally as bracts; panicles (to 60 cm) terminal, pyramidal, loosely flowered (30–75), the branches (2–6) ascending; flowers (to 30 mm) hypogynous; tepals (to 15 mm) 6, creamy or white, lanceolate to ovate, innermost clawed at base, persistent, a basal pair of nectary glands present; stamens 6, anthers versatile, introrse; styles 3, tapering apically; capsules (to 15 mm) lance- to ovate-conic; seeds (to 20) per capsule, elongate

Life history: duration: perennial (rhizomes); **asexual reproduction:** rhizomes; **pollination:** insect; **sexual condition:** hermaphroditic; **fruit:** capsules (common); **local dispersal:** rhizomes; **long-distance dispersal:** seeds (unknown)

Imperilment: 1. *Zigadenus glaberrimus* [G5]; S1 (VA)

Ecology: general: *Zigadenus* is monotypic (see ensuing treatment).

7.2.6.1. ***Zigadenus glaberrimus* Michx.** inhabits bogs (acid, hillside, pine, pitcher plant, quaking, roadside, seepage), depressions (seepage, sphagnous), ditches (roadside, shallow), flats (pine), flatwoods (pine), meadows (wet), pine barrens (low), pinelands (low), pocosins (dry), rights-of-way, roadsides (grassy, wet), savannas (acid, burned, grassy, moist, pine, seepage), seeps (hillside, roadside), slopes (seepage), swales (boggy, sphagnous), swamps, thickets (sphagnous), trenches (fire), and the margins of pocosins at elevations to 62 m. The plants occur primarily in open exposures where full sunlight is available and otherwise extend only into sites where sparse canopies exist. The habitats for this species are maintained naturally by fires, which remove competing woody vegetation. Artificially cleared sites (such as powerline easements) also have provided suitable habitat. Despite their dependency on fire, the plants decline immediately in their aftermath, then thrive during the fire-free intervals (e.g., 9 years). The substrates are acidic and have been characterized as clay, clayey silt, Florala (plinthaquic paleudults), Johns (aquic hapludults), loam, loamy sand, muck, peat, Plummer (grossarenic paleaquults), sand (Leon), sandy loam (Rains), sandy peat, silty loam, Smithton (typic paleaquults), and typic paleudults. Flowering and fruiting occur from June to November. The

flowers are visited by butterflies (Insecta: Lepidoptera), which potentially serve as pollinators. Other aspects of the floral biology or seed ecology have not been studied. The plants persist from thick, long-lived rhizomes, which occur at an average root depth of 15.25 cm. **Reported associates:** *Agalinis, Amphicarpum amphicarpon, Andropogon glomeratus, Andropogon liebmannii, Andropogon virginicus, Anthaenantia rufa, Aristida palustris, Aristida purpurascens, Aristida stricta, Arnoglossum ovatum, Bartonia virginica, Bigelowia nudata, Calamovilfa brevipilis, Calopogon barbatus, Calopogon multiflorus, Calopogon pallidus, Calopogon tuberosus, Carphephorus paniculatus, Carphephorus pseudoliatris, Carphephorus tomentosus, Chaptalia tomentosa, Cleistesiopsis divaricata, Clethra alnifolia, Coreopsis gladiata, Ctenium aromaticum, Cyperus, Cyrilla racemiflora, Dichanthelium dichotomum, Dichanthelium ensifolium, Dichanthelium scabriusculum, Dionaea muscipula, Drosera capillaris, Drosera tracyi, Eragrostis elliottii, Eragrostis refracta, Eriocaulon compressum, Eriocaulon decangulare, Eryngium integrifolium, Eupatorium leucolepis, Eupatorium mohrii. Eupatorium rotundifolium, Fimbristylis spadicea, Fothergilla gardenii, Gaylussacia, Helianthus heterophyllus, Hypericum myrtifolium, Hypoxis, Ilex glabra, Lachnanthes caroliniana, Lachnocaulon anceps, Liatris spicata, Lilium catesbaei, Lindera subcoriacea, Linum medium, Liriodendron tulipifera, Lophiola aurea, Ludwigia, Lycopodiella alopecuroides, Lycopodiella appressa, Lycopodiella caroliniana, Lycopodiella prostrata, Lycopus cokeri, Lyonia lucida, Marshallia, Muhlenbergia expansa, Myrica cerifera, Nyssa biflora, Osmunda regalis, Osmundastrum cinnamomeum, Panicum virgatum, Pinguicula lutea, Pinus elliottii, Pinus palustris, Pinus serotina, Platanthera blephariglottis, Platanthera integra, Pleea tenuifolia, Polygala cruciata, Polygala lutea, Quercus minima, Rhexia alifanus, Rhexia lutea, Rhexia petiolata, Rhexia virginica, Rhynchospora chapmanii, Rhynchospora ciliaris, Rhynchospora gracilenta, Rhynchospora latifolia, Rhynchospora oligantha, Rhynchospora plumosa, Sabatia, Saccharum giganteum, Sagittaria, Sarracenia alata, Sarracenia flava, Sarracenia leucophylla, Sarracenia minor, Sarracenia psittacina, Sarracenia purpurea, Sarracenia rubra, Schizachyrium scoparium, Schizachyrium tenerum, Schwalbea americana, Scleria reticularis, Serenoa repens, Sporobolus floridanus, Sporobolus teretifolius, Stenanthium densum, Tofieldia glabra, Trianthia racemosa, Utricularia, Vaccinium crassifolium, Vaccinium tenellum, Woodwardia virginica, Xyris ambigua, Xyris baldwiniana, Xyris caroliniana, Xyris difformis, Xyris serotina.*

Use by wildlife: *Zigadenus glaberrimus* is a host of larval webworm moths (Insecta: Lepidoptera: Urodidae; *Urodus parvula*). The flowers are visited by butterflies (Insecta: Lepidoptera).

Economic importance: food: All parts of *Zigadenus* are regarded as poisonous and never should be eaten. However, much of the information on toxicity that is ascribed to *Z. glaberrimus* seems to be based circumstantially by association with former members of the genus such as *Toxicoscordion* species (see *Systematics*); e.g., although the toxic alkaloid zygacine has been said to occur in *Z. glaberrimus*, no literature accounts could be found that document its occurrence in that species; **medicinal:** no uses reported; **cultivation:** *Zigadenus glaberrimus* is not cultivated commercially; **misc. products:** none; **weeds:** none; **nonindigenous species:** none. **Systematics:** *Zigadenus* has undergone a complex and confusing taxonomic history. Once assigned to the broadly circumscribed Liliaceae, phylogenetic studies have clarified that the group should be transferred to Melanthiaceae. At one time, the genus contained as many as 19 species, which since have been redistributed among several segregate genera (i.e., *Anticlea*, *Stenanthium*, and *Toxicoscordion*), leaving only *Z. glaberrimus* to represent the currently recognized (monotypic) genus. As such, several molecular phylogenetic studies (e.g., Figure 2.27) depict *Zigadenus* as the sister group to the remainder of Melanthiaceae tribe Melanthieae. However, other analyses have designated the genus as a "cryptaffinity" taxon, because it resolves either as the sister group to members of tribe Melanthieae or to tribe Paridae depending on the selection of outgroup and method of analysis employed. In any case, its monotypic circumscription is substantiated. The base chromosome number of tribe *Melanthieae* is $x = 9$; *Zigadenus glaberrimus* ($2n = 54$) is hexaploid (widely published reports of $2n = 52$ are erroneous). The genome size of *Z. glaberrimus* is estimated as 10.64 pg (2C); 5.32 pg (1C); and 5202.96 mb. No hybrids involving *Z. glaberrimus* have been reported.

Distribution: *Zigadenus glaberrimus* occurs along the southeastern coastal plain of the United States.

References: Brewer et al., 2011; Bridges, 2005; Chupov et al., 2007; Clark et al., 2008; Cronquist, 1981; Frost & Musselman, 1987; Godfrey & Wooten, 1979; Hinman & Brewer, 2007; Kim et al., 2016; McMillan et al., 2002; Peet & Allard, 1993; Pellicer et al., 2014; Schwartz, 2002; Sheridan et al., 1997; Zomlefer & Judd, 2002; Zomlefer et al., 2001; 2014.

Family 7.3. Smilacaceae [1]

Known as the catbrier family, Smilacaceae include only one genus (*Smilax*), which contains about 310 species (Judd et al.,

2016). Phylogenetically, the family resolves as the sister clade of Liliaceae by combined mtDNA and partial cpDNA gene sequence data (Petersen et al., 2013) or as the sister group to a clade comprising Liliaceae, Philesiaceae, and Ripogonaceae (Figure 2.25) by analysis of complete cpDNA genome sequence data (Givnish et al., 2016). In any case, its affinity lies somewhere among the latter three families. The monophyly of Smilacaceae is supported by several molecular (e.g., Figure 2.28; Cameron & Fu, 2006; Qi et al., 2013) and morphological studies (Chen et al., 2005) once the former genus *Heterosmilax* has been assimilated.

Most of the *Smilax* species are herbaceous or woody vines with fewer being erect herbs (Cronquist, 1981; Judd et al., 2016). The foliage is distinguished by leaves with converging, palmate primary veins, reticulate secondary veins, and a pair of tendrils at the petiole base; the stems (and sometimes leaf margins) often are armed with weak to strong prickles or spines; the umbellate inflorescences bear small, unisexual flowers in a dioecious arrangement (Cronquist, 1981; Judd et al., 2016). Pollination is facilitated by insects (Insecta: Diptera; Hymenoptera); the fruits are few-seeded berries, which are dispersed by birds (Aves) (Judd et al., 2016) or by other frugivorous animals.

Smilax imparts the flavor to sarsaparilla, a drink popularized by the 1957–1961 television western "Sugarfoot" where the unassuming hero would order a "frothy mug of sas'parilla" in lieu of alcohol when entering a saloon. Some medicinals known as "sarsaparilla" actually are derived from *Aralia* (Araliaceae); however, the rhizomes of about 20 *Smilax* species are rich in saponins (over 100 kinds have been reported), which exhibit antifungal, anti-inflammatory, cytotoxic, and cAMP phosphodiesterase inhibitory activities (Tian et al., 2017). The young leaves of some species are eaten as a vegetable (Tian et al., 2017). Only a few *Smilax* species have been cultivated commercially.

Smilacaceae are distributed widely throughout temperate and tropical regions. Obligate wetland indicators occur within the single genus:

7.3.1. *Smilax* L.

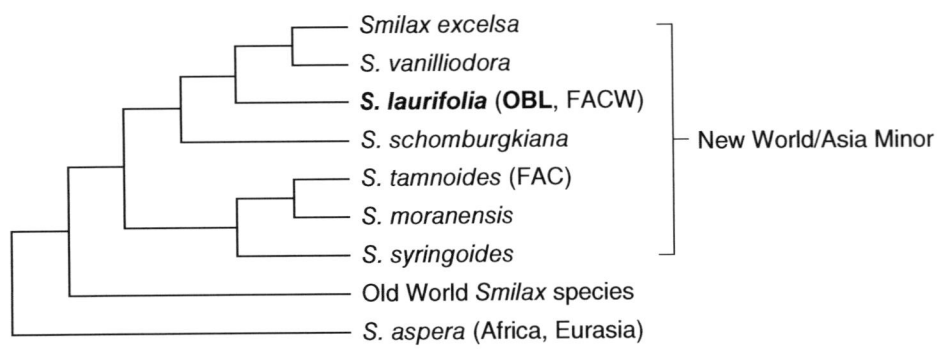

FIGURE 2.28 Interspecific relationships in *Smilax* (Smilacaceae) as indicated by phylogenetic analysis of combined nuclear and plastid DNA sequence data (simplified from Li et al., 2011). North American wetland indicator designations are provided in parentheses (OBL taxa are highlighted in bold). The relationship shown for *S. laurifolia* is only tentative given the highly incomplete sampling of North American taxa (only 2 of 21 species represented). Yet, *S. laurifolia* does resolve among other New World species (along with *S. excelsa*, which is from Asia Minor).

7.3.1. Smilax

Bamboo vine, carrionflower, catbrier, greenbrier, sarsaparilla

Etymology: derivation uncertain but thought to be after the Greek *smile* (a carving tool), in reference to the sharp prickles of some species

Synonyms: *Coprosmanthus*; *Nemexia*; *Parillax*

Distribution: global: cosmopolitan; **North America:** eastern North America and western United States

Diversity: global: 310 species; **North America:** 21 species

Indicators (USA): obligate wetland (OBL): *Smilax laurifolia*, *S. walteri*; **facultative wetland (FACW):** *S. laurifolia*

Habitat: brackish (coastal), freshwater; freshwater (tidal); palustrine; **pH:** 3.4–4.0; **depth:** <1 m; **life-form(s):** emergent vine

Key morphology: climbing vines; rhizomes tuberous and irregularly branched or slender and stoloniferous; stems (to 6 m) woody, branching, armed with prickles (to 12 mm); leaves alternate, deciduous to semi-evergreen, with paired tendrils arising from the petioles (to 1.5 cm), the blades (to 13 cm) oblong-elliptic, lance-elliptic, ovate-oblong, to ovate-lanceolate, sometimes linear or broadly ovate, 3-veined from base, coriaceous or thin, the base rounded, subcordate, or, cuneate, margins entire; umbels axillary, few to numerous, hemispherical or spherical, 5–25-flowered, peduncles (to 2 cm), usually shorter than subtending leaf petiole; flowers unisexual (plants dioecious), pedicellate (to 1 cm); tepals (to 6 mm) 6, ovate to elliptic, brownish yellow, cream, white, or yellow; stamens 6 (staminodes in ♀ flowers), the anthers basifixed, introrsely dehiscent; ovary with 1 ovule per locule, style short or absent, with 3 recurved stigmas; berries (to 9 mm) black, or bright red to orange, globose or ovoid, 5–8 mm, shining

Life history: duration: perennial (rhizomes, woody shoots with winter buds); **asexual reproduction:** rhizomes; **pollination:** insect; **sexual condition:** dioecious; **fruit:** berries (infrequent to prolific); **local dispersal:** rhizomes, berries (animals); **long-distance dispersal:** fruits (animals: Aves; Mammalia)

Imperilment: 1. *Smilax laurifolia* [G5]; S1 (TN); S3 (NJ); **2.** *Smilax walteri* [G5]; S2 (AR); S3 (DE, NJ)

Ecology: general: *Smilax* is represented by a diverse assemblage of perennial species including shrubs, vines, or herbs with erect, sprawling or, more often, climbing, herbaceous or woody stems, which are smooth or armed with prickles. Those with a viny habit grow upon trees and other tall vegetation for support. The plants occur often in wetlands with 15/21 of the North American species (71%) designated as various wetland indicators (OBL, FACW, FAC, FACU, UPL). The flowers of those species studied are pollinated by generalist insects (Insecta) including beetles (Coleoptera), bees (Hymenoptera), and flies (Diptera). The berries are eaten by birds (Aves) and mammals (Mammalia), which disperse them in their excrement. The seeds of at least some species are morphophysiologically dormant, require cold stratification to break dormancy, and germinate better in light than under darkness. Several of the species are capable of developing into dense thickets.

7.3.1.1. *Smilax laurifolia* L. occurs in tidal or nontidal freshwater to brackish baygalls, Carolina bays, bogs (cypress, hillside, pine barren, sandy, seepage, *Sphagnum*), bottoms (stream), depressions (limesink, moist, sphagnous), ditches (drainage), domes (cypress), flatwoods (mesic, pine), floodplains (forested), hammocks (high, hydric, wet), "houses" (prairie), marshes (basin), pine barrens, pocosins (high, small depression), ravines, roadsides (low, wet), seeps (boggy), slopes [to 1%] (seepage), sloughs (oxbow), swamps (bayhead, hardwood, seasonally inundated, strand), thickets (cypress, peaty, shrub), woodlands (low), and along the margins of canals, lakes, ponds, rivers, springs, and streams at elevations to 400 m. The plants thrive in open, sunny exposures but also are highly shade tolerant. The substrates are acidic (e.g., pH: 3.4–4.0), nutrient-poor, and have been characterized as Croatan (daric medisaprist), Dare (typic medisaprist), grossarenic to ultic alaquods (sandy, siliceous, thermic), grossarenic paleaquults (thermic), histosols (deep [to 2 m]), Hontoon (typic medisaprist), humus (rich), loam, loamy sand (Orangeburf, Rutlege), muck (organic), mud (loamy, organic, sandy), peat (to 1.5 m deep), peaty sand, Plummer, Pottsburg, Pungo (typic medisaprist), sandy clay loam, sandy humus, sandy loam (Bayou), sandy peat, Sapelo, silty clay loam, silty loam, silty sandy loam, and typic medisaprist series. Flowering occurs from July to October with fruiting observed until April (the fruits require 2 years to mature). The flowers are fragrant and presumably pollinated by insects (Insecta), although specific pollinators have not been reported. No details on the floral biology are available. The fresh seeds often are dormant (e.g., 83%), but germinate best (10%) when incubated at 25°C. A seed bank can develop, with one study reporting a mean emerging seedling density of 3.61 m^{-2}. The fruits are eaten and dispersed by pileated woodpeckers (Aves: Picidae: *Dryocopus pileatus*) and other birds. They also have been recovered from the fecal pellets of black bears (Mammalia: Ursidae: *Ursus americanus*) and deer (Mammalia: Cervidae: *Odocoileus virginianus*), which also implicates these mammals as dispersal agents. Their high climbing habit (to 9+ m) enables the plants to compete strongly with woody vegetation by their ability to spread over sapling trees and completely arrest their growth. Some reported densities (e.g., 0.15–0.40 plants m^{-2}) are not particularly high; however, the plants can grow so densely that they have been described as "rampant" or as presenting an "impenetrable" or "nearly impenetrable" barrier in accounts originating throughout the 18–21 centuries. Other studies have observed mean sapling/seedling densities of 620 ha^{-1} and larger stem (>1 cm diameter) mean densities of 2916.7 ha^{-1}. The plants generally are more common in close proximity of trees due to the greater seed rain occurring at those sites. The shoots are armed with sharp prickles, except at the nodes. Although a wetland species, the leaves are evergreen and xerophytic structurally. Average leaf litter production has been estimated at 5.38–17.17 g m^{-2} year^{-1}. The plants persist by means of their tuberous rhizomes. This species has been categorized as an indicator of undisturbed Carolina bay vegetation. It has been collected about as often in recently burned sites as it has from fire-suppressed sites.

Reported associates: *Acer floridanum, Acer rubrum, Agalinis purpurea, Aletris aurea, Alnus rugosa, Alnus serrulata, Amelanchier canadensis, Amorpha fruticosa, Ampelopsis arborea, Amphicarpa bracteata, Amsonia ciliata, Andropogon glomeratus, Andropogon liebmannii, Andropogon virginicus, Anthaenantia rufa, Apteria aphylla, Arisaema triphyllum, Aristida palustris, Aristida spiciformis, Aristida stricta, Aronia arbutifolia, Arundinaria gigantea, Arundinaria tecta, Asimina parviflora, Azadirachta indica, Balduina uniflora, Berchemia scandens, Betula nigra, Bigelowia nudata, Bignonia capreolata, Boehmeria cylindrica, Brasenia schreberi, Callicarpa americana, Calopogon barbatus, Calopogon pallidus, Calopogon tuberosus, Campsis radicans, Carex atlantica, Carex crinita, Carex folliculata, Carex glaucescens Carex intumescens, Carex joorii, Carex leptalea, Carex lurida, Carex striata, Carex verrucosa, Carphephorus odoratissimus, Carphephorus pseudoliatris, Carpinus caroliniana, Carya tomentosa, Cassytha filiformis, Castilleja kraliana, Centella asiatica, Cephalanthus occidentalis, Ceratiola ericoides, Chamaecyparis thyoides, Chamaedaphne calyculata, Chaptalia tomentosa, Chasmanthium laxum, Chionanthus, Chrysobalanus icaco, Cladium jamaicense, Clethra alnifolia, Cliftonia monophylla, Coreopsis gladiata, Coreopsis nudata, Croptilon divaricatum, Ctenium aromaticum, Cyperus haspan, Cyperus lecontei, Cyperus virens, Cyrilla racemiflora, Dalea pinnata, Decodon verticillatus, Decumaria barbara, Dichanthelium acuminatum, Dichanthelium clandestinum, Dichanthelium ensifolium, Dichanthelium scoparium, Dichanthelium sphaerocarpon, Diospyros virginiana, Ditrysinia fruticosa, Drosera brevifolia, Drosera capillaris, Drosera tracyi, Dulichium arundinaceum, Eleocharis tuberculosa, Elephantopus carolinianus, Erigeron vernus, Eriocaulon compressum, Eriocaulon decangulare, Eriocaulon texense, Eryngium integrifolium, Eubotrys racemosa, Eupatorium leucolepis, Eupatorium rotundifolium, Euphorbia, Fimbristylis puberula, Fraxinus caroliniana, Fraxinus pennsylvanica, Froelichia floridana, Fuirena squarrosa, Funastrum clausum, Gaylussacia dumosa, Gaylussacia frondosa, Gaylussacia mosieri, Gelsemium sempervirens, Gleditsia aquatica, Gordonia lasianthus, Harrisia eriophora, Helenium amarum, Helenium vernale, Helianthus angustifolius, Helianthus heterophyllus, Hymenocallis gholsonii, Hymenocallis henryae, Hypericum brachyphyllum, Hypericum cistifolium, Hypericum crux-andreae, Hypericum fasciculatum, Hypericum hypericoides, Hypericum tetrapetalum, Hypoxis hirsuta, Ilex cassine, Ilex coriacea, Ilex glabra, Ilex laevigata, Ilex myrtifolia, Ilex opaca, Ilex vomitoria, Iris tridentata, Itea virginica, Juncus effusus, Juncus scirpoides, Juncus trigonocarpus, Juniperus virginiana, Kalmia angustifolia, Kalmia cuneata, Kalmia hirsuta, Lachnanthes caroliniana, Lachnocaulon anceps, Leucothoe axillaris, Liatris pycnostachya, Lilium pyrophilum, Liquidambar styraciflua, Liriodendron tulipifera, Lithospermum canescens, Lobelia elongata, Lobelia reverchonii, Lonicera japonica, Lophiola aurea, Ludwigia linearis, Ludwigia linifolia, Ludwigia microcarpa, Ludwigia pilosa, Lupinus diffusus, Lycopodiella alopecuroides, Lycopodiella caroliniana, Lycopus virginicus, Lygodium japonicum, Lyonia ferruginea, Lyonia fruticosa, Lyonia ligustrina, Lyonia lucida, Lyonia mariana, Magnolia grandiflora, Magnolia virginiana, Marshallia graminifolia, Matelea floridana, Mikania scandens, Minuartia patula, Muhlenbergia capillaris, Muhlenbergia expansa, Myrica caroliniensis, Myrica cerifera, Myrica heterophylla, Myrica inodora, Nymphaea odorata, Nymphoides aquatica, Nyssa aquatica, Nyssa biflora, Nyssa ogeche, Nyssa sylvatica, Onoclea sensibilis, Onosmodium decipiens, Osmunda regalis, Osmundastrum cinnamomeum, Oxalis priceae, Oxydendrum arboreum, Panicum hemitomon, Panicum philadelphicum, Panicum verrucosum, Panicum virgatum, Parthenocissus quinquefolia, Paspalum floridanum, Paspalum plicatulum, Peltandra virginica, Persea borbonia, Persea palustris, Phoradendron serotinum, Pieris phillyreifolia, Pinckneya bracteata, Pinguicula pumila, Pinus echinata, Pinus elliottii, Pinus palustris, Pinus serotina, Pinus taeda, Pityopsis graminifolia, Planera aquatica, Platanthera clavellata, Platanthera cristata, Platanthera flava, Pluchea baccharis, Pogonia ophioglossoides, Polygala cruciata, Polygala cymosa, Polygala lutea, Polygala mariana, Polygala nuttallii, Polygala ramosa, Polystichum acrostichoides, Pontederia cordata, Proserpinaca, Prunus angustifolia, Pteridium aquilinum, Ptilimnium capillaceum, Quercus chapmanii, Quercus laevis, Quercus laurifolia, Quercus marilandica, Quercus michauxii, Quercus minima, Quercus myrtifolia, Quercus nigra, Quercus phellos, Quercus virginiana, Rhexia alifanus, Rhexia lutea, Rhexia mariana, Rhododendron viscosum, Rhus aromatica, Rhus copallinum, Rhynchospora baldwinii, Rhynchospora chalarocephala, Rhynchospora chapmanii, Rhynchospora corniculata, Rhynchospora elliottii, Rhynchospora fascicularis, Rhynchospora glomerata, Rhynchospora inundata, Rhynchospora latifolia, Rhynchospora megalocarpa, Rhynchospora miliacea, Rhynchospora oligantha, Rhynchospora plumosa, Rhynchospora rariflora, Rubus argutus, Rubus hispidus, Rubus pensilvanicus, Sabal minor, Sabal palmetto, Sabatia difformis, Sabatia macrophylla, Sagittaria, Salix caroliniana, Sarracenia alabamensis, Sarracenia alata, Sarracenia flava, Sarracenia leucophylla, Sarracenia psittacina, Sarracenia rosea, Sarracenia rubra, Saururus cernuus, Schinus terebinthifolius, Schizachyrium tenerum, Scleria pauciflora, Scleria reticularis, Scutellaria integrifolia, Serenoa repens, Silphium, Smilax glauca, Smilax pumila, Smilax rotundifolia, Smilax smallii, Smilax tamnoides, Smilax walteri, Solidago caesia, Sphagnum, Spiranthes vernalis, Styrax americanus, Symphyotrichum dumosum, Symplocos tinctoria, Taxodium ascendens, Taxodium distichum, Thelypteris palustris, Tillandsia usneoides, Toxicodendron radicans, Toxicodendron vernix, Triadenum virginicum, Triantha racemosa, Tridens, Ulmus americana, Utricularia cornuta, Utricularia subulata, Vaccinium arboreum, Vaccinium corymbosum, Vaccinium crassifolium, Vaccinium myrsinites, Vaccinium stamineum, Vaccinium stamineum, Vernonia gigantea, Viburnum nudum, Viburnum obovatum, Viola primulifolia, Vitis rotundifolia, Woodwardia* areolata, *Woodwardia virginica, Xanthorhiza simplicissima, Xyris ambigua, Xyris baldwiniana, Xyris fimbriata, Xyris laxifolia, Xyris scabrifolia, Zenobia pulverulenta, Zigadenus glaberrimus.*

7.3.1.2. Smilax walteri Pursh grows in tidal or nontidal, freshwater to brackish baygalls, bogs, bottomlands (forested), depressions (wet), ditches (roadside), flatwoods (pine), floating vegetation mats, floodplains, hammocks, marshes (brackish, freshwater, nontidal, tidal), pinelands (low), pocosins (sand-hill), prairies (saline, wet), roadsides, savannahs (burned, pine, wet), seeps (shrub), spillways, swamps (blackwater, floodplain, nonalluvial), thickets (bottomland, swampy, wet), woodlands (low, wet), and along the margins of bayous, lakes, ponds, rivers, sloughs, and streams (blackwater) at elevations to 214 m. The plants occur in exposures ranging from full sun (most often) to deep or moderate shade. They are found in very wet to inundated habitats (flood tolerance index: 3.05) where the annual maximum flooding periods sometimes average up to 119.6 days. They occur often in shallow, standing waters (0.1–1.5 m deep). The substrates have been characterized as glossic natraqualfs, muck (alluvial), mucky peat, peat, sand (organic), sandy clay, sandy humus, sandy loam, and silty loam (Bonn, Brimstone). Flowering initiates in March–April, with the fruits persisting through February. The fruits mature their first year. The flowers are devoid of fragrance. Virtually no information exists on their reproductive biology or pollination ecology. Polymorphic microsatellite markers developed for *Smilax rotundifolia* are highly cross-reactive (69.7%) with *S. walteri* and potentially could be used to study its population genetics. Fruit production can be erratic, varying sporadically from year to year. The red to orange berries are dispersed by birds (Aves). These are prickly, woody vines with deciduous to semi-evergreen foliage, which normally climb upon or sprawl over surrounding vegetation (up to 6 m in height) for support. Their biomass has been estimated at 12 g m^{-2} at an estimated productivity rate of 6 g m^{-2} year^{-1}. The plants recover well from disturbance, reaching yields of 96–861 kg ha^{-1} in the aftermath of artificial, experimental disturbance regimes. The plants persist vegetatively by means of their woody stems and slender, subterranean rhizomes. **Reported associates:** *Acer rubrum, Agrostis hyemalis, Alnus maritima, Alnus serrulata, Amelanchier canadensis, Ampelopsis arborea, Amphicarpa bracteata, Andropogon glomeratus, Andropogon virginicus, Aristida, Aronia arbutifolia, Arundinaria gigantea, Arundinaria tecta, Atriplex cristata, Baccharis halimifolia, Bacopa monnieri, Bacopa rotundifolia, Bartonia paniculata, Bartonia virginica, Berchemia scandens, Betula nigra, Bidens frondosus, Bidens mitis, Bidens trichospermus, Bignonia capreolata, Boehmeria cylindrica, Callicarpa americana, Carex albolutescens, Carex atlantica, Carex bullata, Carex crinita, Carex exilis, Carex folliculata, Carex gigantea, Carex glaucescens, Carex lupuliformis, Carex lurida, Carex striata, Carex stricta, Carex verrucosa, Carya aquatica, Carya glabra, Celtis laevigata, Centella asiatica, Cephalanthus occidentalis, Chamaecyparis thyoides, Chamaedaphne calyculata, Chasmanthium latifolium, Chasmanthium laxum, Chionanthus virginicus,* *Cicuta maculata, Cladium mariscoides, Clethra alnifolia, Coleataenia longifolia, Crataegus berberifolia, Crataegus brachyacantha, Crataegus viridis, Cuscuta, Cyrilla racemiflora, Decodon verticillatus, Decumaria barbara, Dichanthelium dichotomum, Dichanthelium scabriusculum, Distichlis spicata, Ditrysinia fruticosa, Drosera intermedia, Drosera rotundifolia, Dryopteris ludoviciana, Dulichium arundinaceum, Eleocharis flavescens, Eleocharis robbinsii, Eleocharis tenuis, Eriocaulon aquaticum, Eriocaulon compressum, Eriocaulon decangulare, Eriophorum virginicum, Eubotrys racemosa, Eupatorium mohrii, Fimbristylis spadicea, Fraxinus caroliniana, Fraxinus pennsylvanica, Fuirena squarrosa, Galium tinctorium, Gaylussacia frondosa, Gelsemium sempervirens, Glyceria obtusa, Gordonia lasianthus, Habenaria floribunda, Heliotropium curassavicum, Hibiscus moscheutos, Hypericum canadense, Hypericum densiflorum, Hypericum denticulatum, Hypericum mutilum, Hypoxis curtissii, Ilex cassine, Ilex coriacea, Ilex decidua. Ilex glabra, Ilex laevigata, Ilex myrtifolia, Ilex opaca, Ilex verticillata, Ipomoea pandurata, Iris brevicaulis, Iris tridentata, Iris versicolor, Itea virginica, Iva angustifolia, Juncus canadensis, Juncus effusus, Juncus militaris, Juncus pelocarpus, Kalmia angustifolia, Lachnanthes caroliniana, Leersia hexandra, Leersia oryzoides, Lemna valdiviana, Lilium superbum, Liquidambar styraciflua, Liriodendron tulipifera, Lobelia canbyi, Lobelia nuttallii, Lonicera japonica, Lonicera sempervirens, Ludwigia alternifolia, Ludwigia palustris, Lycopus amplectens, Lycopus angustifolius, Lycopus rubellus, Lycopus uniflorus, Lyonia fruticosa, Lyonia ligustrina, Lyonia lucida, Lyonia mariana, Lysimachia terrestris, Lythrum lineare, Magnolia virginiana, Matelea floridana, Macbridea caroliniana, Micranthemum umbrosum, Mikania scandens, Myrica cerifera, Myrica pensylvanica, Nymphaea odorata, Nyssa aquatica, Nyssa biflora, Nyssa ogeche, Nyssa sylvatica, Oclemena nemoralis, Onoclea sensibilis, Orontium aquaticum, Osmunda regalis, Osmundastrum cinnamomeum, Oxypolis rigidior, Panicum hemitomon, Panicum verrucosum, Panicum virgatum, Parthenocissus quinquefolia, Peltandra virginica, Persea borbonia, Persea palustris, Persicaria arifolia, Persicaria hydropiperoides, Phanopyrum gymnocarpon, Phyla nodiflora, Pieris phillyreifolia, Pinckneya bracteata, Pinus elliottii, Pinus glabra, Pinus rigida, Pinus serotina, Pinus taeda, Planera aquatica, Platanthera clavellata. Pluchea camphorata, Pogonia ophioglossoides, Polygala cymosa, Polygonum aviculare, Pontederia cordata, Proserpinaca, Proserpinaca pectinata, Ptilimnium capillaceum, Quercus hemisphaerica, Quercus laurifolia, Quercus lyrata, Quercus michauxii, Quercus nigra, Quercus phellos, Quercus similis, Quercus virginiana, Rhexia virginica, Rhododendron canescens, Rhododendron viscosum, Rhynchospora alba, Rhynchospora corniculata, Rhynchospora inundata, Rhynchospora miliacea, Rosa palustris, Sabal minor, Sabatia difformis, Sagittaria engelmanniana, Sagittaria lancifolia, Sagittaria latifolia, Salix caroliniana, Salix nigra, Sarracenia purpurea, Sassafras albidum, Saururus cernuus, Schoenoplectus purshianus, Schoenoplectus subterminalis, Scirpus* *cyperinus, Sideroxylon lanuginosum, Smilax glauca, Smilax laurifolia, Smilax rotundifolia, Smilax tamnoides, Solidago sempervirens, Sparganium americanum, Spartina pectinata, Sphagnum, Spiranthes cernua, Spiranthes vernalis, Stenanthium densum, Styrax americanus, Symphyotrichum novi-belgii, Symphyotrichum subulatum, Symplocos tinctoria, Taxodium ascendens, Taxodium distichum, Thelypteris palustris, Tillandsia usneoides, Toxicodendron radicans, Tradescantia occidentalis, Triadenum virginicum, Triadenum walteri, Triadica sebifera, Tridens strictus, Ulmus alata, Ulmus crassifolia, Utricularia gibba, Utricularia juncea, Utricularia striata, Utricularia subulata, Vaccinium arboreum, Vaccinium corymbosum, Vaccinium macrocarpon, Vaccinium stamineum, Vernonia noveboracensis, Viburnum dentatum, Viburnum nudum, Viola lanceolata, Viola primulifolia, Vitis labrusca, Vitis rotundifolia, Wisteria frutescens, Woodwardia areolata, Woodwardia virginica, Xyris difformis, Xyris smalliana, Zenobia pulverulenta.*

Use by wildlife: *Smilax laurifolia* provides winter roosting habitat for red-winged blackbirds (Aves: Icteridae: *Agelaius phoeniceus*). The foliage (7.8–25.7% crude protein, 1.4–6.8% crude fat, 12.5–28.1% crude fiber, 0.45–1.25% calcium, 28–111 ppm Fe, and 58–692 ppm Mn) is browsed by deer (Mammalia: Cervidae: *Odocoileus virginianus*) and hosts larval moths (Insecta: Lepidoptera: Geometridae: *Cymatophora approximaria*) and Fungi (Ascomycota: Mycosphaerellaceae: *Pseudocercospora smilacicola*). The fruits (0.47% N; 2.8–2.9% protein; 13.6% fiber; 1.3% fat) are eaten by bears (Mammalia: Ursidae: *Ursus americanus*), birds (e.g., Aves: Picidae: *Dryocopus pileatus*), and deer (Mammalia: Cervidae: *Odocoileus virginianus*). The foliage of *S. walteri* is eaten by marsh rabbits (Mammalia: Leporidae: *Sylvilagus palustris paludicola*). The fruits are said to be consumed by more than 40 species of birds (Aves). They are eaten occasionally by deer (Mammalia: Cervidae: *Odocoileus virginianus*) during early spring and to a minor degree by mallard ducks (Aves: Anatidae: *Anas platyrhynchos*).

Economic importance: food: The young shoots of *Smilax laurifolia* (sometimes known as "wild asparagus") have been eaten fresh in salads or when sautéed. The Choctaw and Houma people ground the tuberous roots into flour for bread or (Choctaw) finely mashed them in a small amount of water to produce a paste used for making small cakes, which were fried in grease; **medicinal:** The Cherokee prepared a compound from the root bark of *S. laurifolia* as an astringent wash for treating burns and sores. The Houma administered a root decoction for urinary disorders; **cultivation:** *Smilax walteri* sometimes is planted for its colorful red fruits, which provide wildlife food and fall garden color; **misc. products:** *Smilax laurifolia* was used by the Seminole as a dye plant for buckskin; **weeds:** The dense, rank growth of *S. laurifolia* and *S.*

walteri often is perceived as a nuisance; **nonindigenous species:** none.

Systematics: Morphological and molecular phylogenetic studies consistently resolve *Smilax* as a clade, but only if the segregate genus *Heterosmilax* (which is embedded within) is merged. Molecular (DNA sequence) data have been of limited use in elucidating interspecific relationships in *Smilax* because of incomplete taxon sampling and relatively low resolution, which has provided few indications of relationships. Of the OBL species, only *S. laurifolia* has been included in molecular phylogenetic analyses, where combined nuclear and plastid data resolve it among other North American (and Asian) species in a position near to *S. excelsa* and *S. vanilliodora* (Figure 2.28). Morphologically, *S. laurifolia* is related most closely to the North American *S. auriculata* and is distant from *S. walteri*, whose closest relative (by morphological data analysis) is the Asian *S. trinervula*. Both *S. laurifolia* and *S. walteri* are assigned to section *China*, which is polyphyletic; however, some authors have assigned these and other woody, North American *Smilax* species to section *Smilax*. It is evident that sectional circumscriptions in this genus require extensive re-evaluation. Chromosomal base numbers in *Smilax* range from $x = 13$–16. Chromosome counts have not been made for *Smilax laurifolia* or *S. walteri*. There have been no natural hybrids reported that involve either *S. laurifolia* or *S. walteri*.

Distribution: *Smilax laurifolia* and *S. walteri* occur throughout the southeastern United States with *S. laurifolia* extending into the West Indies.

References: Allen et al., 2002b; Allen et al., 2013; Aust et al., 1997; Baskin & Baskin, 1998; Blair, 1936; Bolin, 2007; Brewer, 1998; Brooks et al., 1993; Buell, 1946; Bullard & Allen, 2013; Cameron & Fu, 2006; Chen et al., 2005; Cypert, 1972a; DeBerry & Atkinson, 2014; De Steven & Harrison, 2006; Dressler et al., 1987; Elam et al., 2009; Eleuterius & McDaniel, 1978; Fernald & Kinsey, 1943; Freedman, 1976; Gill, 2000; Golley et al., 1965; Graves, 2001; Harlow, 1961; Harper, 1911; 1922; Harrison & Knapp, 2015; Holmes, 2002b; Hunt, 1943; Johnson & Landers, 1978; Judd, 1998; Kearney, 1901; Kevan et al., 1991; Kilham, 1976; Kirkman et al., 2000; Landers et al., 1979; Lay, 1965; Levey et al., 2000; Li et al., 2011; Light et al., 1993; Luken, 2005a; Meanley & Webb, 1961; Meehan, 1901; Mellinger, 1966; Mohlenbrock, 1976; Monk & Brown, 1965; Morong, 1894; Nash, 1895; Nixon & Ward, 1986; Perry & Uhler, 1981; Rae, 1995; Richardson, 1983; Rodgers et al., 2003; Sanchez et al., 2015; Schlesinger, 1978; Skeate, 1987; Skinner & Sorrie, 2002; Smith, 1996; Smith & Flory, 1990; Smith & Garland, 2009; Smith et al., 1956; Stalter & Dial, 1984; Sundell & Thomas, 1988; Theriot, 1993; Wang et al., 2018; Weakley, 1991; Wells, 1928; 1942; White & Stiles, 1992; Zampella et al., 2001.

3 Monocotyledons III
Commelinoid Monocots (Commelinidae)

The "Commelinoid" monocotyledons were long recognized formally within a distinct subclass known as Commelinidae. The present circumscription (e.g., APG, 2016; Judd et al., 2016) differs from earlier accounts primarily by the inclusion of Arecales and Zingiberales, which had been assigned to different subclasses (i.e., Arecidae and Zingiberidae, respectively) in some former treatments (e.g., Cronquist, 1981; FNA, 2000).

As indicated by phylogenetic analysis of full plastid coding regions (Barrett et al., 2016; Givnish et al., 2016), the Commelinidae are monophyletic and can reasonably be subdivided into four orders: Arecales, Commelinales, Poales, and Zingiberales. Arecales (allied with the family Dasypogonaceae) potentially represent the sister group to the remaining orders (Figure 2.25). Members of the subclass commonly possess an epicuticular wax known as the "*Strelitzia*-type," starchy pollen, and UV-fluorescent ferulate and *p*-coumarate (which acylates lignin) as components of their cell walls (Judd et al., 2016; Karlen et al., 2018). They also possess lower shoot concentrations of Ca and Mg relative to other angiosperms (White et al., 2015). The clade comprising Commelinales, Poales, and Zingiberales can be defined by the synapomorphy of starchy endosperm, which is lacking in the Arecales (Judd et al., 2016). Although obligate (OBL) North American indicators occur in all four orders, several families (Arecaceae, Commelinaceae, Haemodoraceae, Mayacaceae) contain only a single OBL species.

Order 8: Arecales [1]

The Arecales currently are represented only by one family, which does include OBL indicators:

8.1. **Arecaceae** Bercht. & J.Presl [=Palmae A. L. de Jussieu]

Family 8.1. Arecaceae [189]

The palms (Arecaceae or Palmae) are a distinctive group of about 2,760 species, which differ from most other monocotyledons by their shrubby to arborescent habit and compound leaves. Based on the results of extensive phylogenetic studies (reviewed in Judd et al., 2016) the family is demonstrably monophyletic and resolves as the sister group of Dasypogonaceae (e.g., Figure 3.1). Palms are divided into five subfamilies: Arecoideae, Calamoideae, Ceroxyloideae, Coryphoideae, and Nypoideae, with Calamoideae representing the sister lineage to the remainder of the family; only subfamily Coryphoideae contains OBL wetland indicators (Figure 3.1).

Some palms can become large trees (to 60 m), by which the family is most widely recognized. Their evergreen leaves have sheathing petioles and plicate blades that are divided palmately or pinnately; typically they are arranged in dense terminal crowns of unbranched trunks (Cronquist, 1981; Judd et al., 2016). The flowers (typically borne in axillary panicles) are bisexual or unisexual (then in dioecious or monoecious arrangement) with a 3-merous perianth (usually differentiated into a fleshy or leathery calyx and corolla); stamens can number anywhere from 3 to 900 or more (Cronquist, 1981; Judd et al., 2016). The gynoecium is apocarpous or syncarpous (3–10 carpels) and superior (Cronquist, 1981). The flowers often produce nectar and usually are pollinated by insects (Insecta) such as bees (Hymenoptera), beetles (Coleoptera), and flies (Diptera); pollination by bats (Mammalia: Chiroptera) or wind also occurs but less commonly (Zona, 2000). The fruits are one-seeded drupes or berries, which are dispersed biotically by animals (Aves; Mammalia) or abiotically by water (Zona, 2000).

Palms are important economically as the source of foods such as betel nuts (*Areca*), coconuts (*Cocos*), and dates (*Phoenix*); the apical buds of many species are eaten as "palm cabbage" (Judd et al., 2016). Other products derived from palms include carnauba wax (*Copernicia*), thatch (*Raffia*, etc.), and vegetable ivory (*Phytelephas*) (Judd et al., 2016). Many genera contain ornamental species, including *Brahea*, *Butia*, *Chamaedorea*, *Chamaerops*, *Cocos*, *Dypsis*, *Howea*, *Livistona*, *Phoenix*, *Rhapis*, *Roystonea*, *Trachycarpus*, and *Washingtonia*.

Arecaceae are circumtropical and extend into warmer temperate regions worldwide. Despite the large size of this family, only one genus contains a species designated as an OBL wetland indicator:

8.1.1. *Acoelorraphe* H. Wendl.

8.1.1. Acoelorraphe

Everglades palm, paurotis palm, silver saw palm (etto); palmier des Everglades
Etymology: after the Greek *akoilos raphe* ("without a hollow seam") in reference to the seeds, which lack an impressed raphe
Synonyms: *Acanthosabal*; *Brahea* (in part); *Copernicia* (in part); *Paurotis*; *Serenoa* (in part)
Distribution: global: New World; **North America:** southern Florida (USA)
Diversity: global: 1 species; **North America:** 1 species
Indicators (USA): obligate wetland (OBL): *Acoelorraphe wrightii*

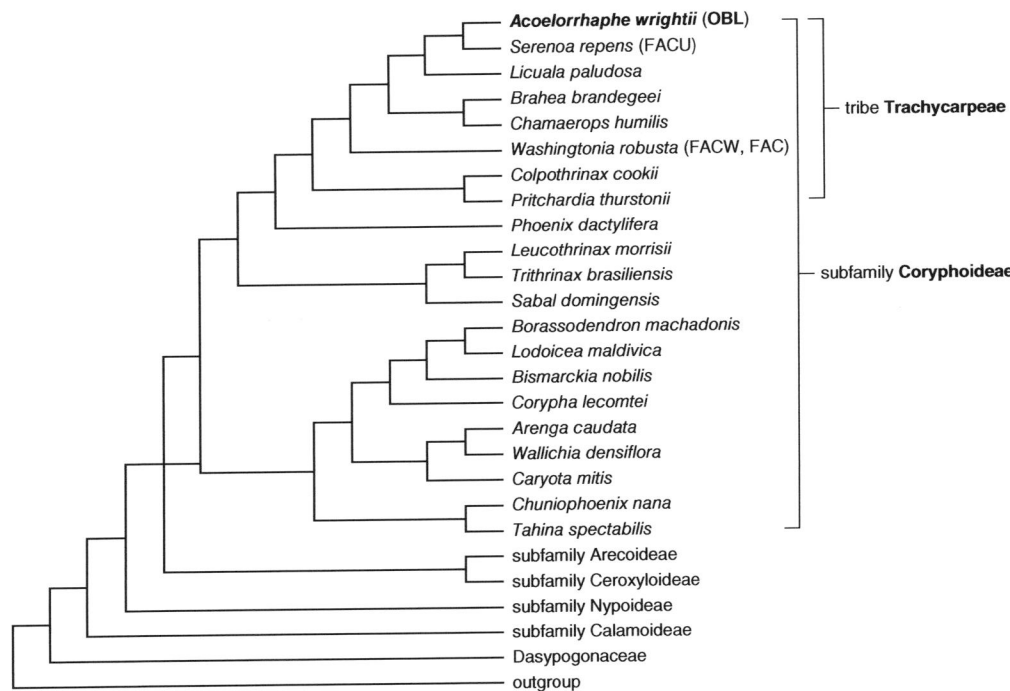

FIGURE 3.1 Phylogenetic relationships in palms (Arecaceae) as indicated by analysis of full plastid genomes (75 coding sequences) (simplified from Barrett et al., 2016). The monotypic *Acoelorraphe wrightii* (highlighted in bold), the only OBL indicator in the palm family, occurs in a highly specialized position within subfamily Coryphoideae as the sister group to the monotypic *Serenoa repens*. The wetland indicator status is given in parentheses for North American taxa.

Habitat: brackish (coastal), freshwater; freshwater (tidal); palustrine; **pH:** 5.0–6.1; **depth:** <1 m; **life-form(s):** emergent shrub

Key morphology: stems (to 9 m) multiple, caespitose, erect, clothed in persistent leaf bases; leaves with conspicuous adaxial hastula, the blades palmate, the segments lanceolate, stiff, connate basally, plication induplicate, the petioles armed with sharp spines; panicles (to 22 cm) axillary, with 3–4 orders of branching, arching within leaf crown, turning orange in fruit; flowers bisexual, sessile; sepals 3, imbricate, the margins ciliate; petals 3, connate basally; stamens 6, filaments connate basally, anthers dorsifixed, versatile; pistils 3–4, the styles connate, filiform; drupes (to 8.5 mm) globose, single-seeded, ripening through orange to black

Life history: duration: perennial (dormant apices); **asexual reproduction:** rhizomes; **pollination:** insect; **sexual condition:** hermaphroditic; **fruit:** drupes (common); **local dispersal:** rhizomes; drupes (water); **long-distance dispersal:** drupes (birds)

Imperilment: 1. *Acoelorraphe wrightii* [G3/G5]

Ecology: general: *Acoelorraphe* is monotypic (see next).

8.1.1.1. Acoelorraphe wrightii (Griseb. & H. Wendl.) H. Wendl. ex Becc. inhabits fresh or brackish waters of glades, hammocks (bayhead, mahogany), marshes (low, tidal), savannas (low), swamps (strand), and thickets at low, coastal elevations (<25 m). The plants are drought-tolerant, moderately salt-tolerant, and typically grow in shallow (0.1–1.5 m deep) standing waters for 9 or more months of the year at sites where a dry surface then becomes exposed during the low water period. Fires that occur during the dry period help to remove competing

vegetation and do not seriously harm the plants, which tolerate occasional burning. The plants can tolerate a fairly wide range of pH, but thrive under acidic (e.g., pH: 5.0–6.1) conditions. The substrates have been reported in one study as sand (mollic psammaquents). Flowering initiates in May with fruiting extending through November. The flowers are thought to be pollinated by bees (Insecta: Hymenoptera), but their reproductive and pollination biology require further study. The fruits are dispersed by white-crowned pigeons (Aves: Columbidae: *Patagioenas leucocephala*) and likely by several other birds. They also float and are dispersed locally by water. The mesocarps contain moderate levels of oxalate ($\bar{x} = 1,366\ \mu g\ g^{-1}$). Seed germination requires high temperatures, and rates up to 70% have been reported under greenhouse conditions. The highest germination has been achieved using fresh, fully ripened seeds, which were planted immediately and incubated at 33°C–39°C. Germination rates decline substantially (e.g., to 11%) at lower incubation temperatures (<30°C). Experimentally planted seeds have germinated during June, 7 months after sowing. Natural seed germination occurs only under saturated conditions but is inhibited by high water. Germination rates on saturated substrates are low (8.0%) but decline markedly if water levels are low (0.75% at 5 cm below substrate level) or high (0.25% at 10 cm above substrate level). The seeds lose viability by dehydration within 3–6 weeks; however, they can be stored successfully at 18°C–23°C if cleaned, air-dried, treated with a fungicide, and placed within a sealed container. This species appears to be highly clonal. The clumped stems arise by asexual division through basal suckering and rhizomatous branching, which allows the plants to be divided at their base for artificial propagation. Although flowers and seeds are produced sexually, no observed juveniles or evidence of seedling recruitment has been found in some studies. A period of receding water subsequent to germination is necessary in order for juveniles to attain the highest biomass, which occurs under conditions of low water and full sunlight. Peak growth occurs from June to October. Adult plants are susceptible to manganese (Mn) deficiency, which causes "frizzle top" disease; deeply planted individuals also can exhibit iron (Fe) deficiency symptoms. The roots are colonized by arbuscular mycorrhizal Fungi. This species has been designated as one being particularly vulnerable to climate change. **Reported associates:** *Bursera simaruba, Celtis laevigata, Citrus, Cladium jamaicense, Conocarpus erectus, Diospyros virginiana, Eugenia foetida, Ficus aurea, Hippomane mancinella, Lysiloma latisiliquum, Metopium toxiferum, Morus rubra, Pinus elliottii, Quercus virginiana, Rhizophora, Rhynchospora, Roystonea regia, Sabal palmetto, Serenoa repens, Sideroxylon celastrinum, Swietenia mahagoni, Taxodium.*

Use by wildlife: The foliage of *Acoelorraphe wrightii* is known to host beetles (Insecta: Coleoptera: Mycteridae: *Hemipeplus marginipennis*), false spider mites (Acari: Tenuipalpidae: *Raoiella indica*), flatid planthoppers (Insecta: Hemiptera: Flatidae: *Ormenaria rufifascia*), Florida red scale (Insecta: Hemiptera; Diaspididae: *Chrysomphalus aonidum*), and reniform nematodes (Nematoda: Pratylenchidae: *Rotylenchulus reniformis*). The dead leaves host several Fungi (Ascomycota: *incertae sedis*: *Ceratosporella basibicellularia*; *Linkosia longirostrata*; *Pleomonodictys capensis*; *Selenodriella perramosa*; Pleosporales: *Berkleasmium leonense*, Hermatomycetaceae: *Hermatomyces tucumanensis*).

Economic importance: food: *Acoelorraphe wrightii* is not known to be edible; **medicinal:** Oxalates present in the mesocarp can cause skin irritation in people who excessively handle the fruits; **cultivation:** *Acoelorraphe wrightii* is planted as an ornamental specimen; **misc. products:** The surface features of the drupes of *Acoelorraphe wrightii* have been studied as a model for developing antifouling substances for watercraft; **weeds:** none; **nonindigenous species:** *Acoelorraphe wrightii* has been introduced to Australia.

Systematics: The monotypic *Acoelorraphe* is assigned to subfamily Coryphoideae (tribe Trachycarpeae) of Arecaceae. Phylogenetic analyses of DNA sequence data resolve the genus as the sister group to the monotypic *Serenoa* in a highly derived position within the family (Figure 3.1). The basic chromosome number of *Acoelorraphe* is $x=18$; *A. wrightii* ($2n=36$) is diploid. No natural hybrids have been reported that involve *A. wrightii*.

Distribution: In North America, *Acoelorraphe wrightii* occurs naturally only in extreme southern Florida (the United States) but extends elsewhere into Mexico, Central America, South America, and the West Indies.

References: Adey et al., 1996; Armentano et al., 2002; Bezona, et al., 2009; Black, 1985; Broschat, 1994; 2011a; 2011b; Broschat & Latham, 1994; Carrillo et al., 2012; Delgado, 2013; 2014; Edelman, 2017; Edelman & Richards, 2018a; 2018b; Fisher & Jayachandran, 2005; Gunderson &

Loftus, 1993; Henderson et al., 1997; Howard & Halbert, 2005; Inserra et al., 2005; Mathis, 1947; Meerow, 2004; Meerow et al., 2001; Nachtigall & Wisser, 2015; Pollock, 1999; Potter et al., 2017; Randall, 2007; Wagner, 1982; Worden et al., 2002; Zona, 1997; 2000; Zona & Henderson, 1989.

ORDER 9: COMMELINALES [5]

Various molecular data (reviewed in Judd et al., 2016; e.g., Saarela et al., 2008) have shown Commelinales to be a well-supported clade that includes approximately 800 species assigned to five families: Commelinaceae, Haemodoraceae, Hanguanaceae, Philydraceae, and Pontederiaceae. The order is not particularly well defined morphologically, but like the closely related Zingiberales, the constituent families possess helicoid cymose inflorescences and anthers with an amoeboid tapetum (Dahlgren & Clifford, 1982; Judd et al., 2016). The relationship of Commelinales to Zingiberales and Poales is supported by molecular phylogenetic evidence (e.g., Figures 2.1 and 2.25) as well as by the shared presence of starchy endosperm in the group (Judd et al., 2016). Obligate North American wetland indicators occur in three of the families:

9.1. **Commelinaceae** Mirb.
9.2. **Haemodoraceae** R.Br.
9.3. **Pontederiaceae** Kunth

Family 9.1. Commelinaceae [41]

Commelinaceae represent about 650 species of herbaceous plants having swollen nodes and foliage containing raphides crystals and C-glycosyl flavones (Cronquist, 1981; Burns et al., 2011). The leaf blades frequently are folded or rolled lengthwise along the midrib (Judd et al., 2016). The flowers typically have bilateral, 3-merous perianths differentiated into a calyx and corolla, with ephemeral, rapidly deteriorating petals and staminal filaments that often bear slender hairs (Cronquist, 1981; Judd et al., 2016). Pollination primarily is entomophilous but the flowers lack nectaries (Cronquist, 1981). The seeds possess an opercular swelling that is known as the embryostega (Cronquist, 1981).

Phylogenetic analysis of combined plastid data (Saarela et al., 2008; Figure 3.2) shows Commelinaceae and Hanguanaceae to comprise a sister clade to the remainder of Commelinales. Although analysis of plastid *trnL-F* sequence data apparently resolves Commelinaceae and Philydraceae as sister families (Burns et al., 2011), that topology is an artifact of the outgroups used to root the tree, which focused on relationships within Commelinaceae but are inappropriate for assessing overall familial relationships within the order (i.e., the incongruent topology agrees with that of Figure 3.2 if properly rooted at Commelinaceae).

The flowers of Commelinaceae are rapidly fugacious and wither within a few hours of anthesis (hence one common name of "dayflowers"). They can be self-compatible or self-incompatible, sometimes even within a species (Owens, 1981). Although many species are autogamous (some even

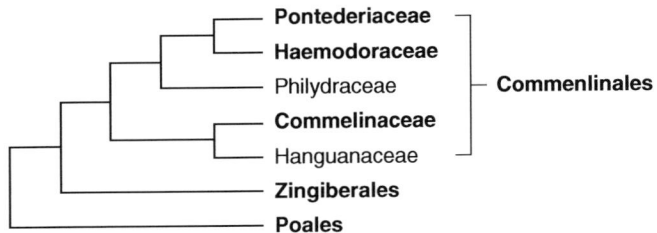

FIGURE 3.2   Interfamilial relationships in the monocot order Commelinales as inferred by phylogenetic analysis of 17 plastid genes and associated noncoding regions (simplified from Saarela et al., 2008). Taxa containing OBL North American wetland indicators have been highlighted in bold. Although a superficially different topology of the group has been indicated by analysis of sequence data from the plastid *trnL-F* region (Burns et al., 2011), that topology is an artifact of the outgroup rooting used (to focus on relationships in Commelinaceae) and is compatible with the one shown here if properly rerooted at Commelinaceae.

cleistogamous), most are pollinated by pollen-seeking insects (Insecta) such as bees (Hymenoptera) and flies (Diptera); no nectar is produced (Faden, 1998). Floral dimorphism occurs commonly with species bearing both bisexual and males flowers in an andromonoecious arrangement (Faden, 1998). The seeds are dispersed by a variety of biotic (Aves; Insecta) and abiotic (water, wind) vectors. Some species produce dimorphic seed, which potentially are dispersed by different vectors (Faden, 1998).

Commelinaceae occur worldwide but are distributed most broadly throughout the tropical and subtropical regions; none of the species is indigenous to Europe (Cronquist, 1981; Faden, 1998). Despite having ephemeral flowers, the attractive foliage of several genera has made them popular as garden ornamentals or house plants. The more notable of these are *Commelina* and *Tradescantia* (which now includes *Rhoeo* and *Zebrina*). Although diverse ecologically, Commelinaceae contain few aquatics. In North America, only a single genus includes OBL wetland indicators:

9.1.1.   *Murdannia* Royle

9.1.1.   Murdannia

Asian spiderwort, Asiatic dayflower, marsh dewflower, wart-removing herb, wart wort

Etymology: for Munshí Murdan Ali (a 19th-century plant collector at Singapore)

Synonyms: *Aneilema*; *Aphylax*; *Callisia* (in part); *Commelina* (in part); *Cyanotis* (in part); *Ditelesia*; *Phaeneilema* (in part); *Stickmannia* (in part); *Streptylis*; *Tradescantia* (in part)

Distribution: global: pantropical to warmer temperate regions; **North America:** northwestern and southeastern United States

Diversity: global: 53 species; **North America:** 3 species

Indicators (USA): obligate wetland (OBL): *Murdannia keisak*

Habitat: brackish (coastal), freshwater; freshwater (tidal); lacustrine, palustrine, riverine; **pH:** 5.3–7.7; **depth:** 0–1.5 m; **life-form(s):** emergent herb

Key morphology: annual herbs; shoots decumbent, long-trailing (to 76+ cm), stoloniferous, with fine nodal roots; leaves sessile, the blades (to 7 cm) linear-oblong to linear-lanceolate, glabrous; cymes solitary or fascicled in distal leaf

axils, 1-flowered; flowers (to 1 cm) bisexual, radial, pedicellate; sepals (to 6 mm) distinct; petals (to 8 mm) distinct, purplish lilac or purple to pink or white; stamens 3, the filaments hairy, staminodes 3; ovary 3-locular; capsules (to 9 mm) 3-valved, with 2–6 seeds per locule; seeds (to 3 mm) faintly ribbed

Life history: duration: annual (fruits/seeds); **asexual reproduction:** shoot fragments, stolons; **pollination:** unknown (self?); **sexual condition:** hermaphroditic; **fruit:** capsules (prolific); **local dispersal:** stoloniferous shoots and fragments; fruits/seeds (water); **long-distance dispersal:** seeds (waterfowl)

Imperilment: 1. *Murdannia keisak* [GNR]

Ecology: general: *Murdannia* species occur infrequently in aquatic habitats but the three that were introduced to North America all are wetland indicators (OBL, FACW, FAC, FACU). The plants are annuals or perennials. The reproductive biology is poorly understood and the following information is based on very few adequate studies. Both cleistogamous and chasmogamous can occur. The flowers of at least some of the species are self-compatible and self-pollinating, with the dehisced pollen shed directly onto the stigmas. Pollination by bees and flies (i.e., Insecta: Hymenoptera; Diptera: Syrphidae) also has been reported, but the extent of outcrossing has not been determined for any of the species. Local dispersal occurs by the transport of fruits, seeds, or stem fragments by water. Long-distance dispersal of fruits and seeds is facilitated by waterfowl (Aves: Anatidae). Several of the species have become serious weeds, especially in rice fields.

9.1.1.1. *Murdannia keisak* **(Hassk.) Hand.-Mazz.** is a nonindigenous species found in fresh to brackish backwaters (drying), bogs, bottomlands (stream, swampy), channels, deltas, depressions (wet), ditches (drainage, roadside, stormwater), floodplains, gravel bars, marshes (freshwater, intertidal, oligohaline, tidal), meadows (wet), pools (dessicated, oxbow, shallow, tidal), rice fields (abandoned), roadsides, ruts (wet),seeps (semi-wooded), slopes (damp), spillways, swales, swamps (forested), woodlands (alluvial, rich), and along the margins of canals (drainage), lakes, ponds (artificial, beaver), reservoirs, rivers, and streams at elevations to 549 m. The plants grow preferentially in full sunlight but can tolerate partial to full shade. However, plants grown in 5–10% of full sunlight required 5–10 times the stand-level biomass to support a given shoot mass than plants grown in full sunlight. The plants often occur in shallow standing waters, but have been found at depths up to 1.5 m. They can tolerate warm water temperatures (e.g., 35°C) with no adverse effects. Although sometimes reported from brackish sites (salinity: 0.1–4.4‰; $\bar{x} = 0.5$–0.9‰), the plants are intolerant of higher salinity (3–12‰). The substrates span a wide range of acidity (5.3–7.7) and have been described as alluvium (sandy-loamy), cobble, gravel, Kinston-Bibb series, loam, muck, mud, sand (fine), sandy gravel, sandy loam, sandy silt, silt (fine), and silty clay. Flowering and fruiting occur from August to October. The flowers are ephemeral and remain open only for a few hours (during morning to late afternoon) before withering. The floral biology has not been studied in detail, but Asian plants have been categorized as being insect pollinated by flies (Insecta: Diptera: Syrphidae). However, wide variability in flower size (which is correlated with autogamy) and high seed set observed in greenhouse-grown plants indicate that self-pollination is also likely to occur, especially where nonindigenous habitats might lack suitable pollinators. The plants produce numerous, small (2.62–3.94 mg) seeds, which can attain densities up to 70,000 m^{-2} and are well represented in the seed bank. Two seed size classes occur, which are characterized by different germination requirements. The larger seeds require a shorter period of cold stratification (4°C) to break dormancy than the smaller seeds. Large seeds exhibit high germination rates (93–100%) at various temperatures below 30°C, whereas the smaller, more temperature-sensitive seeds have germinated best (91–94%) under a 25°C/15°C day/night temperature regime. Dormancy can be broken artificially by treating them with acetone, which results in much higher germination rates for larger seeds (67–83%) than for smaller seeds (2–21%). The plants form thick mats with densities that can reach 12,000 individuals m^{-2} at the start of the growing season. Their annual peak (September) biomass production can be as high as 2.5 kg m^{-2} in thermally enriched localities. Although annual, the stems root adventitiously at their distal nodes, which enables any detached shoot fragment to disperse over the water surface and establish during the growing season. The stem fragment and the fruits are dispersed by water. Longer-distance dispersal occurs by the transport of fruits and seeds by waterfowl (Aves: Anatidae).

Reported associates (North America): *Acalypha rhomboidea, Acer negundo, Acer rubrum, Acmella, Aeschynomene indica, Aeschynomene virginica, Agrostis perennans, Alisma, Allium canadense, Alnus serrulata, Alternanthera philoxeroides, Amorpha fruticosa, Ampelopsis arborea, Amphicarpaea bracteata, Apios americana, Aralia spinosa, Athyrium, Betula nigra, Bidens laevis, Bignonia capreolata, Boehmeria cylindrica, Boltonia asteroides, Botrychium biternatum, Botrychium dissectum, Brasenia schreberi, Bromus pubescens, Cabomba, Calamagrostis canadensis, Campsis radicans, Cardamine hirsuta, Carex amphibola, Carex crinita, Carex intumescens, Carex lupulina, Carex lurida, Carex lyngbyei, Carex stipata, Carex stricta, Carex tribuloides, Carex typhina, Carpinus caroliniana, Carya ovata, Celtis laevigata, Centella asiatica, Cephalanthus occidentalis,* *Ceratophyllum demersum, Cicuta maculata, Cinna arundinacea, Climacium kindbergii, Commelina communis, Commelina virginica, Conoclinium, Cornus amomum, Cornus foemina, Crassula aquatica, Cyperus flavescens, Cyperus odoratus, Decodon verticillatus, Decumaria barbara, Dichanthelium dichotomum, Digitaria sanguinalis, Diodia virginiana, Echinochloa walteri, Echinodorus cordifolius, Eclipta prostrata, Egeria densa, Eleocharis geniculata, Eleocharis lanceolata, Eleocharis montevidensis, Eleocharis obtusa, Eleocharis quadrangulata, Elephantopus carolinianus, Salix nigra, Elymus virginicus, Eragrostis rufescens, Erechtites hieraciifolius, Eryngium aquaticum, Eupatorium semiserratum, Eupatorium serotinum, Fagus, Festuca subverticillata, Fontinalis sullivantii, Fraxinus pennsylvanica, Galium aparine, Galium tinctorium, Geum canadense, Glyceria striata, Gratiola brevifolia, Gratiola virginiana, Harperella nodosa, Heteranthera reniformis, Hydrocotyle bonariensis, Hydrocotyle verticillata, Hydrophyllum virginianum, Hymenocallis rotata, Hypericum mutilum, Ilex decidua, Impatiens capensis, Iris hexagona, Iris tridentata, Isoetes engelmannii, Isoetes hyemalis, Isoetes ×bruntonii, Isolepis carinata, Juncus effusus, Juncus marginatus, Juncus oxymeris, Justicia americana, Justicia ovata, Leersia oryzoides, Leersia virginica, Leptodictyum riparium, Lespedeza cuneata, Lilaeopsis occidentalis, Lindera benzoin, Lindernia dubia, Liquidambar styraciflua, Lobelia cardinalis, Lobelia glandulosa, Lonicera japonica, Ludwigia alternifolia, Ludwigia decurrens, Ludwigia grandiflora, Ludwigia leptocarpa, Ludwigia palustris, Ludwigia repens, Luziola fluitans, Lycopus rubellus, Lycopus virginicus, Lythrum salicaria, Microstegium vimineum, Mikania scandens, Mimulus ringens, Mitreola petiolata, Myriophyllum, Nymphaea odorata, Nyssa aquatica, Nyssa biflora, Nyssa sylvatica, Osmunda regalis, Packera anonyma, Panicum dichotomiflorum, Panicum virgatum, Parthenocissus quinquefolia, Paspalum urvillei, Peltandra virginica, Perilla frutescens, Persicaria arifolia, Persicaria glabra, Persicaria hydropiperoides, Persicaria lapathifolia, Persicaria longiseta, Persicaria maculosa, Persicaria meisneriana, Persicaria pensylvanica, Persicaria posumbu, Persicaria punctata, Persicaria sagittata, Persicaria virginiana, Phryma leptostachya, Phyllanthus caroliniensis, Physostegia leptophylla, Phytolacca americana, Pilea fontana, Pilea pumila, Pinus, Planera aquatica, Platanus occidentalis, Pluchea camphorata, Pluchea foetida, Podostemum ceratophyllum, Polygonatum biflorum, Pontederia cordata, Potamogeton, Proserpinaca palustris, Proserpinaca palustris, Prunus serotina, Ptilimnium capillaceum, Ptilimnium costatum, Quercus lyrata, Quercus michauxii, Quercus phellos, Ranunculus abortivus, Ranunculus pusillus, Ranunculus recurvatus, Rhynchospora miliacea, Rosa multiflora, Rubus argutus, Rumex verticillatus, Sabal minor, Sacciolepis striata, Sagittaria australis, Sagittaria fasciculata, Sagittaria latifolia, Sambucus nigra, Sarracenia, Saururus cernuus, Scapania nemorosa, Schoenoplectus americanus, Schoenoplectus purshianus, Schoenoplectus tabernaemontani, Schoenoplectus triqueter, Scirpus cyperinus, Scleria, Sisyrinchium, Sium* *suave, Smilax glauca, Smilax rotundifolia, Smilax walteri, Solidago ceasia, Solidago gigantea, Solidago sempervirens, Sparganium, Spartina cynosuroides, Sphagnum lescurii, Sphenopholis pensylvanica, Spirodela polyrrhiza, Steinchisma hians, Symphyotrichum pilosum, Taxodium distichum, Thelypteris palustris, Toxicodendron radicans, Triadenum walteri, Typha angustifolia, Typha latifolia, Ulmus alata, Ulmus americana, Ulmus rubra, Utricularia gibba, Verbena scabra, Verbesina, Viburnum rafinesqueanum, Viola palmata, Vitis aestivalis, Vitis rotundifolia, Woodwardia areolata, Xyris, Zizaniopsis miliacea.*

Use by wildlife (North America): The seeds of *Murdannia keisak* contain (by dry weight) 21.3% crude protein, 10.8% crude fiber, 0.5% fat, 0.30% Ca, and 0.50% P. They are eaten throughout the year by numerous waterfowl (Aves: Anatidae), including the American wigeon (*Anas americana*), black duck (*Anas rubripes*), blue-winged teal (*Anas discors*), green-winged teal (*Anas crecca*), lesser scaup (*Aythya affinis*), mallard (*Anas platyrhynchos*), northern pintail (*Anas acuta*), northern shoveler (*Anas clypeata*), ring-necked duck (*Aythya collaris*), and wood duck (*Aix sponsa*). The plants are host to smut Fungi (Basidiomycota: Ustilaginaceae: *Ustilago aneilemae*).

Economic importance: food: *Murdannia keisak* is not reported as being edible; **medicinal:** Extracts of *Murdannia keisak* reduce (TGF)-β-dependent signaling, which is associated with various pathologic keloid fibroblast responses associated with epidermis-proliferating diseases such as excess extracellular matrix production and overabundant collagen formation. Whole plant decoctions are used in Asia to induce diuresis and treat snake bites; **cultivation:** *Murdannia keisak* is not cultivated intentionally; **misc. products:** none; **weeds:** *Murdannia keisak* is listed as a noxious weed in the United States. Its dense growth is believed to reduce water flow and increase local sedimentation rates; **nonindigenous species:** *Murdannia keisak* (native to Asia) is nonindigenous to North America. It is thought to have first been introduced to Louisiana as a contaminant of rice seed during the 1920s and was documented from South Carolina rice fields in 1935. The plants were first reported from the Pacific Northwest in 1984–1992. This species also has been introduced to Italy and is a serious rice field weed in Asia. The abandonment of rice fields (with the concomitant cessation of weed control) has been suggested as an explanation for the spread of plants into natural habitats.

Systematics: The generic distinction between *Murdannia* and *Aneilema* has long been debated taxonomically, but both genera have resolved within distinct clades in phylogenetic analyses (of *rbcL* data). Analysis of combined nuclear (5S NTS) and plastid (*trnL-F*) sequence data (Figure 3.3) supports the monophyly of *Murdannia* as currently circumscribed, but just to the extent that only 10% of the species have been evaluated phylogenetically. *Murdannia* apparently is related closely to *Commelina* and *Pollia* (Figure 3.3); however, that result is based on taxon sampling that excluded *Aneilema* species and also the monotypic, African *Anthericopsis* (*A. sepalosa*), which is thought to be the closest relative of *Murdannia*. Analyses of *rbcL* data have resolved *Anthericopsis* and *Murdannia* as sister groups (and distinct from *Aneilema*), but included only *M. clarkeana* as a representative. Further taxon sampling in the family and more refined analyses will be necessary to better solidify the intergeneric relationships. Subtribal circumscriptions in Commelinaceae also require refinements, with Commelineae (which includes *Murdannia*) being monophyletic but Tradescantieae being paraphyletic (Figure 3.3). *Murdannia* has several basic chromosome numbers including $x=6$, 7, 9, 10, and 11. In any case, *M. keisak* ($2n=40$) is polyploid. The karyotype consists of 5 metacentric, 12 submetacentric, and 3 subtelocentric chromosome pairs. No natural hybrids have been reported that involve *M. keisak*.

Distribution: *Murdannia keisak* occurs throughout the southeastern United States and is disjunct in Oregon and Washington.

References: Aulbach-Smith & de Kozlowski, 1996; Bailey et al., 2006; Beal, 1977; Burns et al., 2011; Christy, 1994; Collins & Wein, 1995; Dee & Ahn, 2012; De Oliveira Pellegrini et al., 2016; Dunn & Sharitz, 1990; 1991; Elam et al., 2009; Ensign et al., 2014; Evans et al., 2003; Faden, 1998; 2000a; Godfrey & Wooten, 1979; Gordon et al., 1989;

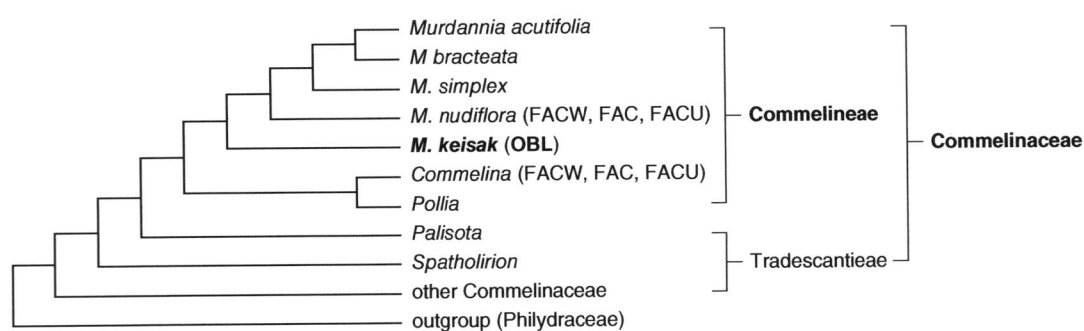

FIGURE 3.3 Intergeneric relationships in Commelinaceae as indicated by analysis of combined nuclear and plastid sequence data (adapted from Burns et al., 2011). *Murdannia keisak* (the only OBL indicator within Commelinaceae) occurs within the monophyletic subtribe Commelineae (subtribe Tradescantieae is paraphyletic in this instance). The wetland indicator status is shown in parentheses for North American taxa; those containing OBL species are highlighted in bold.

Hussner, 2012; Kato & Miura, 1996; Kim et al., 2013; Landers et al., 1976; 1977; Landman et al., 2007; Luo et al., 2018; Matthews et al., 2007; McGilvrey, 1966; Moody, 1989; Moon et al., 2000; Morris, 2003; Musselman et al., 1996; Newberry, 1991; Perry & Uhler, 1981; Peterson & Baldwin, 2004b; Saarela et al., 2008; SCEPPC, 2011; Sharpe & Baldwin, 2012; Strong & Kelloff, 1994; Tucker, 1989; Ushimaru et al., 2007; Vánky, 1994; Wetzel & Kitchens, 2007; Wetzel et al., 2004; White et al., 2012.

Family 9.2. Haemodoraceae [15]

Aside from the transfer of *Lophiola* to Nartheciaceae (Dioscoreales), which was urged by the results of molecular phylogenetic analyses (e.g., Figure 2.24), the circumscription of Haemodoraceae (bloodworts) has remained relatively unaltered from earlier, traditional concepts (e.g., Cronquist, 1981). *Lophiola* also lacks arylphenalenone pigments, which are ubiquitous in the family and uniquely so among all other angiosperms (Judd et al., 2016). Phylogenetic analyses of DNA sequence data from 17 plastid loci (Saarela et al., 2008) have indicated Pontederiaceae as the sister group of Haemodoraceae (Figure 3.2). Both families also share a basally thickened endothecium, septal nectaries, and pollen with a non-columellate exine (Simpson, 1987; 1990).

In its current circumscription, Haemodoraceae are subdivided into two subfamilies containing 15 genera (14 have been surveyed phylogenetically) and about 111 species (roughly half have been surveyed phylogenetically). Several molecular-based phylogenies (Hopper et al., 1999; 2009) have corroborated the monophyly of Haemodoraceae, its subfamilies, and its genera (Figure 3.4). Those analyses place the exclusively North American *Lachnanthes* within subfamily Haemodoroideae as the sister group of *Haemodorum* (Figure 3.4), which occurs in Australia and New Guinea. The broad geographical disjunction of these two genera was thought to reflect an ancient, pre-Gondwana origin (Hopper et al., 1999), but estimated divergence times (Hopper et al., 2009) are far too recent (<9 MYBP) to support that possibility. Similarly, broad

disjunctions in other aquatic plant groups have been explained by long-distance dispersal (Les et al., 2003).

In addition to their unique phenalones, Haemodoraceae species typically possess 2-ranked, equitant, unifacial leaves, paracytic stomata, densely pubescent inflorescences (helicoid cymes), septal nectaries, and loculicidal capsules (Cronquist, 1981; Judd et al., 2016). *Lachnanthes* is atypical of the *Dilatris/Haemodorum/Lachnanthes* subclade (Figure 3.4) by lacking sand-binding roots (Smith et al., 2011), which is understandable given the association of that feature with xeric conditions. *Lachnanthes* and the distantly related *Tribonanthes* (also helophytic) are the only genera within the family to possess leaf aerenchyma, an adaptation linked to wet habitats; *Lachnanthes* is the only member of the family to possess stem vessels (Aerne-Hains & Simpson, 2017). The flowers of Haemodoraceae are pollinated by insects (Insecta: Hymenoptera; Lepidoptera) or birds (Aves), which seek the nectar or pollen. Many of the species possess seeds that are adapted for wind dispersal.

Numerous ornamentals are derived from the "kangaroo paw" genera (*Anigozanthos* and *Macropidia*). Otherwise, the family is of minor economic importance. Haemodoraceae occur primarily in the Southern Hemisphere (Australia, southern Africa, and South America). OBL wetland indicators occur only within one North American genus:

9.2.1. *Lachnanthes* Elliott

9.2.1. Lachnanthes
Carolina redroot, redroot; lachnanthe de Caroline
Etymology: from the Greek *lachno anthos* ("wooly flower") for the pubescent inflorescence
Synonyms: *Anonymos* (in part); *Dilatris* (in part); *Gyrotheca* (in part); *Heritiera* (in part)
Distribution: global: North America; West Indies; **North America:** eastern
Diversity: global: 1 species; **North America:** 1 species
Indicators (USA): obligate wetland (OBL): *Lachnanthes caroliniana*

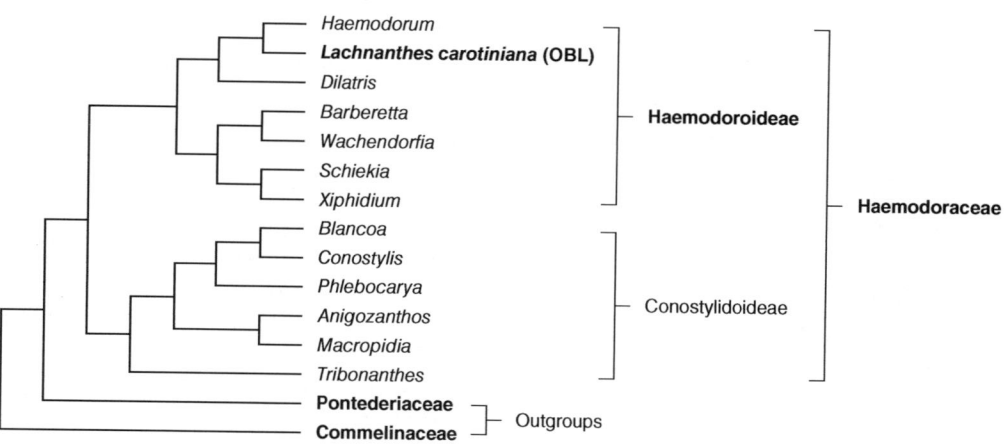

FIGURE 3.4 Intergeneric relationships in Haemodoraceae as rendered by phylogenetic analysis of combined plastid DNA sequences (simplified from Hopper et al., 1999; 2009). Within subfamily Haemodoroideae, the alliance of the OBL North American *Lachnanthes caroliniana* (highlighted in bold) with *Haemodorum* (Australia/New Guinea) raises interesting biogeographical questions (see text).

Habitat: freshwater; lacustrine, palustrine; **pH:** 4.5–7.4; **depth:** <1 m; **life-form(s):** emergent herb

Key morphology: rhizomes reddish, with fibrous roots; stems (to 10.5 dm), erect, the sap red, villous-tomentose distally when young, whitish, becoming tawny-hoary; principal leaves basal, becoming reduced as bracts upward, the blades (to 45 cm) linear-ensiform; cymes becoming open and corymbose, the branches helicoid, bracts conspicuous; flowers enantiostylous; tepals (to 9 mm) 6, erect-spreading, pale yellow, densely pubescent below, persisting as an incurving beak around fruit; stamens 3, spreading at anthesis, exceeding tepals, the filaments to 10 mm, the anthers yellow; ovary inferior, 3-locular, the style (13 mm), persistent, displaced to left or right of floral axis; capsules (to 5 mm) globose or oblate, 3-lobed; seeds (to 3 mm) reddish brown, discoid, peltate, faintly wrinkled

Life history: duration: perennial (rhizomes); **asexual reproduction:** rhizomes; **pollination:** insect; **sexual condition:** hermaphroditic (enantiostylous); **fruit:** capsules (prolific); **local dispersal:** rhizomes; seeds (water, wind); **long-distance dispersal:** seeds (water, wind)

Imperilment: 1. *Lachnanthes caroliniana* [G4]; SH (VA); S1 (CT, DE, MD, RI, NY, TN); S2 (NS); S3 (LA, MA)

Ecology: general: *Lachnanthes* is monotypic (see next).

9.2.1.1. *Lachnanthes caroliniana* **(Lam.) Dandy.** inhabits bogs (cranberry, pitcher plant, quaking, roadside, seepage, *Sphagnum*), cypress heads, depressions (flatwoods), ditches (flatwoods, pine, roadside, wet), flatwoods (cut-over, mesic, wet), glades (wet), "houses" (prairie), marshes (basin, depression, roadside), meadows (depression, wet), mudflats, pine barrens, pinelands (low), pocosins (dried), ponds (seasonal), prairies (wet), rights-of-way (powerline, wet), roadsides, savannas (damp, grassy, low, pine, wet), seeps, sinkholes (marshy), swales (roadside), swamps (peat), thickets (swampy), and along the margins or shorelines of hammocks, lakes (Carolina bay), peat islands, pocosins, ponds (depression, drying, karst, seasonal), rivers, and streams at elevations to 600 m. The plants can occur in exposures ranging from full sunlight to partial shade. They grow in clear to turbid waters (Secchi depth: 4.2–0.8 m) that are acidic (pH: 4.5–7.4; $\bar{x}$ = 5.9; $\tilde{x}$ = 5.8), soft (alkalinity [as mg l^{-1} CaCO$_3$]: 0–25.1; $\bar{x}$ = 8.1; $\tilde{x}$ = 2.1; conductance [µS cm^{-1} @ 25°C]: 40–179; $\bar{x}$ = 98; $\tilde{x}$ = 73), and nutrient-poor (total P [µg l^{-1}]: 6–45; $\bar{x}$ = 21; $\tilde{x}$ = 12; total N [µg l^{-1}]: 180–1,170; $\bar{x}$ = 660; $\tilde{x}$ = 550; Ca [mg l^{-1}]: 0.7–16.1; $\bar{x}$ = 7.8; $\tilde{x}$ = 4.6; Mg [mg l^{-1}]: 0.8–11; $\bar{x}$ = 4.3; $\tilde{x}$ = 2.7; Na [mg l^{-1}]: 3.5–16.9; $\bar{x}$ = 9.1; $\tilde{x}$ = 7.1; K [mg l^{-1}]: 0.2–3.7; $\bar{x}$ = 1.6; $\tilde{x}$ = 0.9; SO$_4^{2-}$ [mg l^{-1}]: 2.9–25.8; $\bar{x}$ = 11.4; $\tilde{x}$ = 7.8; Cl$^-$ [mg l^{-1}]: 7.1–31.9; $\bar{x}$ = 16.7; $\tilde{x}$ = 12.5; Fe [mg l^{-1}]: 0.0–0.4; $\bar{x}$ = 0.1; $\tilde{x}$ = 0.1; Si [mg l^{-1}]: 0.1–3.5; $\bar{x}$ = 1.0; $\tilde{x}$ = 0.5). Substrate conditions can vary from inundation by shallow, standing waters (~60 cm) during rainy summer periods to becoming dry and exposed during spring drawdown periods. They have been described as Allanton (grossarenic haplaquods), clay loam, gravel, loam, loamy sand (Rutledge), muck (organic), mud, Newham-Corolla complex, peat, Pottsburg (grossarenic haplaquods), Rutledge (typic humaquepts), sand (Corolla-Duckston), sandy loam (Bayou), sandy peat, and Scranton (humaqueptic psammaquents). Flowering occurs from May to November with fruiting reported from July to November. The flowers exhibit monomorphic enantiostyly, in which the styles deviate asymmetrically either to the right or left side of different flowers within the same individual plant; this feature is believed to be an adaptation to promote outcrossing. They are self-compatible, but are pollinated by insects (see *Use by wildlife* below), which forage for the nectar or pollen. Seed production is high. An average of 199.1 capsules develops in each inflorescence, with the capsules averaging 12.7 seeds. Stands have been estimated as capable of yielding potential maximum densities of 126,579 seeds m^{-2}. The marginally winged seeds are thought to be dispersed abiotically by water and wind. Most seeds germinate during the spring when any standing waters have receded. Experimental manipulations have shown that much higher seed germination rates occur on dry substrates, with only minor germination occurring under wet conditions. Germination studies have demonstrated that the seeds are dormant and require a period of cold stratification to induce germination. Seeds provided with cold stratification (77 days at ambient winter temperatures) exhibited moderate germination rates (14%). Understandably, experimentally derived germination rate for unstratified seeds is extremely low (0.35%), with any germination occurring within 51 days. Although the seeds have been characterized as very short-lived, that conclusion was based on experiments that did not control for dormancy and should be discounted. Extremely low seed bank densities have been reported, but again were based on assays without controls for dormancy (i.e., no stratification); therefore, the reports are inconclusive and should be disregarded. On the contrary, field studies have reported estimates of 600 seed germinants m^{-2}, which seems to indicate rather high seed bank densities. Viable seeds have been recovered from 10 cm deep seed bank cores (2.0 m^{-2}) and from vacuumed soil surfaces (0.2 m^{-2}). The plants grow clonally by rhizomes, which have an average maximum rooting depth of 8.5 cm. Field studies have documented higher survival rates for rhizomes that become buried at a 5 cm depth (100%) than for those exposed at the surface (62.5%). The plants are associated with both undisturbed and disturbed habitats. They are drought-resistant and have been observed to dominate in areas characterized by prolonged drought conditions. They also appear frequently in recently burned sites and are known to thrive in sites that have been disturbed by rooting feral swine (Mammalia: Suidae: *Sus scrofa*). **Reported associates:** *Acer rubrum, Agalinis, Agrostis hyemalis, Aletris, Alternanthera, Amphicarpum muhlenbergianum, Andropogon glomeratus, Andropogon liebmannii, Andropogon virginicus, Anthaenantia rufa, Aristida palustris, Aristida purpurascens, Aristida spiciformis, Aristida stricta, Aronia arbutifolia, Asclepias connivens, Baccharis halimifolia, Bacopa caroliniana, Bacopa monnieri, Balduina uniflora, Bartonia paniculata, Bejaria racemosa, Bidens mitis, Bigelowia nudata, Brasenia schreberi, Buchnera floridana, Calamagrostis canadensis, Calopogon pallidus, Calopogon tuberosus, Carex bullata, Carex hyalinolepis, Carex stricta, Carphephorus odoratissimus, Carphephorus pseudoliatris, Cenchrus, Centella asiatica, Centrosema* virginianum, Cephalanthus occidentalis, Ceratiola ericoides, Chamaedaphne calyculata, Chaptalia tomentosa, Cladium jamaicense, Cladium mariscoides, Clethra alnifolia, Coleataenia anceps, Coleataenia longifolia, Coreopsis, Cornus foemina, Ctenium aromaticum, Cuphea carthagenensis, Cyperus dentatus, Cyperus erythrorhizos, Cyperus esculentus, Cyperus lecontei, Cyperus odoratus, Cyperus surinamensis, Cyperus virens, Cyrilla racemiflora, Decodon verticillatus, Dichanthelium acuminatum, Dichanthelium commutatum, Dichanthelium dichotomum, Dichanthelium ovale, Dichanthelium sphaerocarpon, Diodia virginiana, Diodia virginiana, Drosera capillaris, Drosera intermedia, Drosera tracyi, Dulichium arundinaceum, Eleocharis baldwinii, Eleocharis cellulosa, Eleocharis elongata, Eleocharis equisetoides, Eleocharis montevidensis, Eleocharis robbinsii, Eleocharis tenuis, Eragrostis elliottii, Erigeron quercifolius, Erigeron vernus, Eriocaulon compressum, Eriocaulon decangulare, Eubotrys racemosa, Eupatorium leucolepis, Eustachys glauca, Euthamia graminifolia, Fimbristylis autumnalis, Fimbristylis spadicea, Fuirena breviseta, Fuirena scirpoidea, Fuirena scirpoidea, Galactia elliottii, Gaylussacia mosieri, Glyceria obtusa, Gordonia lasianthus, Gratiola aurea, Habenaria repens, Harperocallis flava, Helianthemum corymbosum, Helianthus heterophyllus, Hydrocotyle umbellata, Hydrocotyle verticillata, Hydrolea corymbosa, Hypericum brachyphyllum, Hypericum canadense, Hypericum cistifolium, Hypericum densiflorum, Hypericum fasciculatum, Hypericum gentianoides, Hypericum tetrapetalum, Hypoxis hirsuta, Hyptis alata, Ilex cassine, Ilex glabra, Ilex vomitoria, Ipomoea sagittata, Iris hexagona, Iris prismatica, Iris tridentata, Iris virginica, Itea virginica, Juncus canadensis, Juncus effusus, Juncus marginatus, Juncus megacephalus, Juncus militaris, Juncus pelocarpus, Juncus repens, Juncus roemerianus, Juniperus virginiana, Lachnocaulon anceps, Leersia oryzoides, Liatris laevigata, Liquidambar styraciflua, Lobelia canbyi, Lobelia nuttallii, Lophiola aurea, Ludwigia alternifolia, Ludwigia decurrens, Ludwigia lanceolata, Ludwigia maritima, Ludwigia microcarpa, Ludwigia octovalvis, Ludwigia palustris, Ludwigia pilosa, Ludwigia sphaerocarpa, Ludwigia suffruticosa, Ludwigia virgata, Lycopodiella alopecuroides, Lycopodiella appressa, Lycopodiella caroliniana, Lycopus uniflorus, Lyonia fruticosa, Lyonia lucida, Lysimachia terrestris, Magnolia virginiana, Mecardonia acuminata, Medicago minima, Mitreola sessilifolia, Muhlenbergia expansa, Muhlenbergia uniflora, Myrica cerifera, Myrica pensylvanica, Nolina atopocarpa, Nuphar advena, Nuphar variegata, Nymphaea odorata, Nymphoides aquatica, Nyssa biflora, Nyssa sylvatica, Oenothera simulans, Orontium aquaticum, Osmanthus americanus, Osmundastrum cinnamomeum, Panicum amarum, Panicum dichotomiflorum, Panicum hemitomon, Panicum repens, Panicum verrucosum, Panicum virgatum, Peltandra virginica, Persea borbonia, Persea palustris, Persicaria punctata, Pinus clausa, Pinus echinata, Pinus elliottii, Pinus palustris, Pinus taeda, Pityopsis oligantha, Platanthera cristata, Pluchea baccharis, Pogonia ophioglossoides, Polygala cruciata, Polygala cymosa, Polygala lutea, Polygala nuttallii, Polygala rugelii, Polypremum procumbens, Pontederia cordata, Potamogeton confervoides, Proserpinaca pectinata, Quercus geminata, Quercus laevis, Quercus minima, Rhexia alifanus, Rhexia lutea, Rhexia mariana, Rhexia virginica, Rhynchospora alba, Rhynchospora baldwinii, Rhynchospora chalarocephala, Rhynchospora chapmanii, Rhynchospora fascicularis, Rhynchospora glomerata, Rhynchospora inundata, Rhynchospora macrostachya, Rhynchospora microcephala, Rhynchospora nitens, Rhynchospora oligantha, Rhynchospora scirpoides, Rhynchospora tracyi, Rosa palustris, Rubus argutus, Rubus hispidus, Ruellia noctiflora, Sabal palmetto, Sabatia brevifolia, Sabatia decandra, Sabatia difformis, Sabatia stellaris, Saccharum giganteum, Sacciolepis striata, Sagittaria australis, Sagittaria engelmanniana, Sagittaria graminea, Sagittaria lancifolia, Salix caroliniana, Sarracenia alata, Sarracenia flava, Sarracenia leucophylla, Sarracenia psittacina, Sarracenia purpurea, Sarracenia rosea, Saururus cernuus, Schizachyrium maritimum, Schizachyrium tenerum, Schoenoplectus subterminalis, Scirpus cyperinus, Scleria ciliata, Scleria pauciflora, Scleria reticularis, Serenoa repens, Sesbania, Setaria corrugata, Sisyrinchium angustifolium, Sisyrinchium atlanticum, Smilax auriculata, Smilax laurifolia, Smilax rotundifolia, Sparganium americanum, Spartina cynosuroides, Spartina patens, Sphagnum cuspidatum, Spiraea tomentosa, Syngonanthus flavidulus, Taxodium ascendens, Taxodium distichum, Thelypteris palustris, Tiedemannia filiformis, Toxicodendron radicans, Triadenum virginicum, Triadica sebifera, Triantha racemosa, Uniola, Utricularia gibba, Utricularia subulata, Vaccinium corymbosum, Vaccinium macrocarpon, Vaccinium myrsinites, Vaccinium stamineum, Vicia sativa, Vigna luteola, Viola lanceolata, Vitis aestivalis, Woodwardia areolata, Woodwardia virginica, Xyris ambigua, Xyris baldwiniana, Xyris difformis, Xyris panacea, Xyris serotina, Xyris smalliana, Zigadenus glaberrimus, Zizaniopsis miliacea.

Use by wildlife: The elemental composition of *Lachnanthes caroliniana* varies within a plant, with roots containing higher levels of phosphorous and leaves containing higher levels of calcium than other plant parts. The leaves are also high in sodium. The seeds or rhizomes provide food for various birds (Aves) such as sandhill cranes (Aves: Gruidae: *Grus canadensis*) and waterfowl (Anatidae), including the American wigeon (*Anas americana*), black duck (*Anas rubripes*), gadwall (*Anas strepera*), green-winged teal (*Anas crecca*), lesser scaup (*Aythya affinis*), mallard (*Anas platyrhynchos*), northern pintail (*Anas acuta*), northern shoveler (*Anas clypeata*), and ring-necked duck (*Aythya collaris*). The plants are eaten by deer (Mammalia: Cervidae: *Odocoileus virginianus*) and by grazing domestic cattle (Mammalia: Bovidae: *Bos taurus*). The rhizomes and infructescences are eaten by round-tailed muskrats (Mammalia: Cricetidae: *Neofiber alleni*); and wild pigs (Mammalia: Suidae: *Sus scrofa*), which uproot the plants often when foraging. However, the rhizomes allegedly can be toxic when eaten by white-colored pigs. The foliage is host to phytophagous aquatic weevils (Insecta:

Coleoptera: Curculionidae: *Neobagoidus carlsoni*) and larval moths (Insecta: Lepidoptera: Hesperiidae: *Erynnis martialis*) and harbors predatory mantids (Mantodea: Mantidae: *Mantis religiosa*). The flowers are visited (and potentially pollinated by) insects (Insecta), including numerous bees (Hymenoptera: Apidae: *Bombus impatiens*, *B. pensylvanicus*, *Melissodes communis*, *Xylocopa micans*, *X. virginica*; Halictidae: *Agapostemon splendens*, *Augochloropsis metallica*, *A. sumptuosa*, *Lasioglossum coreopsis*, *L. tamiamense*; Megachilidae: *Anthidiellum perplexum*, *Anthidium maculifrons*, *Coelioxys dolichos*, *C. mexicana*, *C. octodentata*, *C. sayi*, *Megachile albitarsis*, *M. brevis*, *M. georgica*, *M. mendica*, *M. texana*, *M. xylocopoides*), butterflies (Lepidoptera: Nymphalidae: *Danaus plexippus*; Papilionidae: *Papilio glaucus*), flies (Diptera: Syrphidae: *Palpada agrorum*), and thrips (Thysanoptera: Thripidae: *Pseudothrips beckhami*).

Economic importance: food: *Lachnanthes caroliniana* potentially is poisonous and should never be eaten; **medicinal:** Extracts from the red-colored roots of *Lachnanthes* (thus its common name) contain various substances such as perinaphthenone, 9-phenylphenalenone, and phenylnaphthalene pigments, including lachnanthocarpone(IV), lachnanthofluorone, and lachnanthoside. The plants were used medicinally by the Cherokee as an antihemorrhagic and to treat bowel disorders, cancer, hemorrhoids, sore throats, and venereal disease. The Catawba administered an infusion made from the roots as a tonic to treat sick animals. *Lachnanthes caroliniana* allegedly has hallucinogenic narcotic properties and was known to the Florida Seminoles as "spirit weed." Root and fruit extracts have exhibited photodynamic antimicrobial activity against *Staphylococcus epidermidis* (Bacteria: Staphylococcaceae). Despite their potential medicinal value, the plants allegedly are poisonous "in uncontrolled doses" and were used widely as an ingredient of various ineffective "quack" medicines and treatments, including a widely purported cure for tuberculosis; **cultivation:** *Lachnanthes caroliniana* was said to have been introduced to the British horticultural trade in 1812, perhaps as an ornamental curiosity; **misc. products:** The root of *Lachnanthes caroliniana* was used in colonial America to make a red dye; **weeds:** *Lachnanthes caroliniana* is a common weed of commercial cranberry (*Vaccinium macrocarpon*) bogs; **nonindigenous species:** none.

Systematics: The monotypic North American *Lachnanthes* is assigned to subfamily Haemodoroideae, where it resolves consistently as the sister group to the Australian/New Guinean *Haemodorum* in molecular phylogenetic analyses (e.g., Figure 3.4). The biogeographical implications of this relationship are interesting and likely reflect the long-distance dispersal of Haemodoraceae to the New World (see above discussion under Haemodoraceae). The basic chromosome number of *Lachnanthes* has been cited as $x=24$, based on the counts obtained for *Lachnanthes caroliniana* ($2n=48$). Although counts for the related *Dilatris* are similarly high ($x=19$), those obtained in subfamily Conostyloideae ($x=5$–11) indicate a much lower ancestral base number for the family and a possible polyploid ancestry of subfamily Haemodoroideae.

Distribution: *Lachnanthes caroliniana* occurs along the eastern coastal plain of North America and extends into the West Indies.

References: Adams et al., 2007; Aerne-Hains & Simpson, 2017; Amoroso & Judd, 1995; Anderson & Kral, 2008; Beecher et al., 1983; Bergstrom et al., 2000; Besançon, 2019; Beshear & Howell, 1976; Bosserman, 1981; Boughton et al., 2016; Brewer, 1998; Brewer et al., 2011; Clouse et al., 1997; Cohen et al., 2004; Cooke, 1970; Cypert, 1972a; Darwin, 1872; Deyrup et al., 2002; East, 1940; Edwards & Weiss, 1970; 1974; Gerritsen & Greening, 1989; Harlow, 1961; Harms & Grodowitz, 2009; Hopper et al., 1999; 2009; Hoyer et al., 1996; Jordan et al., 1997; Kalmbacher et al., 1984a; 1984b; 2005; Keppner & Anderson, 2008; Kornfeld & Edwards, 1972; Laidig & Zampella, 1996; Landers et al., 1976; Millspaugh, 1887; Monk & Brown, 1965; Murrell, 1901; Nelson, 1986; Nichols, 1934; O'Brien, 1990; Querry & Harper, 2017; Querry et al., 2017; Robertson, 1976; 2002; Robeson et al., 2018; Scriber et al., 1998; Shelingoski et al., 2005; Simpson, 1998; Smith, 1985; Smith et al., 2011; Stuckey & Gould, 2000; Tyndall et al., 1990; Vaile & Gilbert, 2005; Valentine, Jr. & Noble, 1970.

Family 9.3. Pontederiaceae [6]

Pontederiaceae (pickerel weeds) are entirely aquatic and contain plants with life-forms that vary from emergent helophytes to free-floating and entirely submersed species (Cook, 1996). Molecular phylogenetic analyses (e.g., Figure 3.2) resolve Pontederiaceae as a well-defined sister clade to Haemodoraceae (Saarela et al., 2008). The relationship of these two families is also indicated by their shared presence of a basally thickened endothecium, septal nectaries, and pollen with a non-columellate exine (Simpson, 1987; 1990). The monophyly of Pontederiaceae never has been seriously questioned and is well supported morphologically (Eckenwalder & Barrett, 1986). The many morphological synapomorphies include involute ptyxis, early-deciduous sessile leaves, alternating xylem and phloem near the center of the leaf blades, abaxial xylem and adaxial phloem near the leaf margins, a deflexed inflorescence (post-anthesis and in fruit), sessile flowers, connate perianth with a persistent tube, an anthocarp, and bisulcate pollen grains (Pellegrini et al., 2018). The monophyly of the family also has been corroborated by several molecular phylogenetic analyses, which have incorporated plastid RFLP data, plastid *ndhF* and *rbcL* sequence data, and nuclear gene sequence data (Graham & Barret, 1995; Kohn et al., 1996; Barrett & Graham, 1997; Graham et al., 2002; Ness et al., 2011).

Although Pontederiaceae contain only an estimated 35–45 species (Judd et al., 2016; Pellegrini et al., 2018), several generic circumscriptions remain in dispute. There has been mounting evidence that *Eichhornia* is not monophyletic, with a recent evaluation based on nuclear and plastid DNA sequence data (Ness et al., 2011) continuing to indicate the disparate, polyphyletic placement of *Eichhornia* species among the other genera (Figure 3.5). There are also

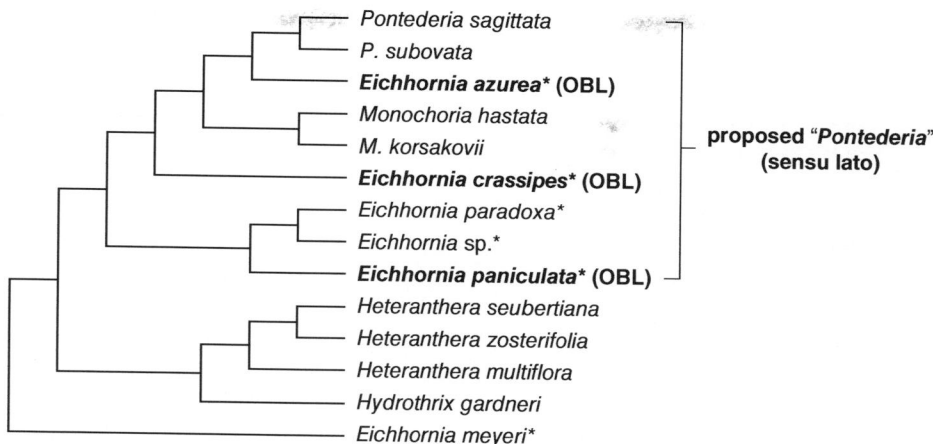

FIGURE 3.5 Phylogenetic relationships in Pontederiaceae as hypothesized by analysis of combined plastid and nuclear DNA sequence data (simplified from Ness et al., 2011). Such analyses indicate the need for revised circumscriptions of at least some of the genera like *Eichhornia*, which clearly appears to be polyphyletic (asterisks). A broad sensu lato concept of *Pontederia* recently has been proposed as indicated (Pellegrini et al., 2018) but has not been implemented here in anticipation of a more comprehensive and compelling phylogenetic assessment of the family. The North American wetland indicator status is provided in parentheses, with the OBL taxa highlighted in bold. Although all Pontederiaceae are aquatic, no wetland status is assigned to most of the remaining species, which do not occur in North America. The lack of wetland status for the North American *Heteranthera multiflora* appears to be an oversight (see text).

discrepancies for *Pontederia*, with plastid genes indicating it to be monophyletic as opposed to some (but not all) nuclear gene sequences (Ness et al., 2011). Even though some authors have now recommended the merger of all *Eichhornia*, *Monochoria*, and *Pontederia* species within a broadly circumscribed *Pontederia* (Pellegrini et al., 2018), it is highly recommended that a final reconciliation of intergeneric relationships in Pontederiaceae should await the additional sampling of taxa for further molecular analyses (Ness et al., 2011). In any case, it appears that major nomenclatural adjustments of some type eventually will be necessary, but none has yet been implemented in this treatment.

Despite their small size, the diverse Pontederiaceae contain annuals, perennials, and a variety of floral syndromes. Some species are self-compatible and self-pollinating, while others are self-incompatible and outcrossed by bees, butterflies, and flies (Insecta: Hymenoptera; Lepidoptera: Diptera); in all cases, the flowers remain open for only one day (Barrett, 1978; Judd et al., 2016). Enantiostyly and heterostyly (tristyly) occur in some species, with the latter representing the only known occurrence in monocotyledons (Barrett, 1978; Judd et al., 2016; Ness et al., 2011). The fruits and seeds often are dispersed abiotically by water. Many of the perennial taxa are highly clonal, with some becoming extremely aggressive and weedy.

Several Pontederiaceae are grown as pond or aquarium ornamentals, including members of *Eichhornia*, *Heteranthera*, and *Pontederia*. *Eichhornia crassipes* is renowned worldwide as a highly invasive aquatic weed.

Pontederiaceae are distributed nearly worldwide, especially in subtropical and tropical regions. Four North American genera contain OBL wetland indictors:

9.3.1. ***Eichhornia*** Kunth
9.3.2. ***Heteranthera*** Ruiz & Pav.

9.3.3. ***Monochoria*** C. Presl
9.3.4. ***Pontederia*** L.

9.3.1. Eichhornia
Water hyacinth; jacinthe d'eau
Etymology: after Johann Albrecht Freidrich Eichhorn (1779–1856)
Synonyms: *Cabanisia*; *Heteranthera* (in part); *Leptosomus*; *Piaropus*; *Pontederia* (in part)
Distribution: global: Africa; Neotropics; **North America:** eastern, southern, and western
Diversity: global: 7 species; **North America:** 3 species
Indicators (USA): obligate wetland (OBL): *Eichhornia azurea, E. crassipes, E. paniculata*
Habitat: freshwater; lacustrine, palustrine, riverine; **pH:** 5.2–8.7; **depth:** 0–3+ m; **life-form(s):** emergent herb, floating-leaved, free-floating
Key morphology: stems erect (to 25 cm) or rosulate, emergent and rooted in the substrate or free-floating on water surface, sometimes with proliferous stolons from which daughter rosettes develop; leaves sessile and either submersed, alternate and vittate or in a basal rosette, or petiolate and emersed or floating, the petioles (to 80 cm) not inflated or inflated with spongy airspace tissue, stipules (to 14 cm) with a truncate apex, the blades (to 16 cm) cordate, ovate, or round; inflorescence a spike (to 50-flowered) or panicle (to 35-flowered), subtended by a lanceolate, obovate or ovate spathe (to 16 cm), the flowers sessile or short-petiolate (to 3 mm), peduncle (to 20 cm) glabrous or pubescent (sometimes with orange hairs); perianth funnelform, the limbs (to 37 mm) blue, mauve-blue, or white, obovate, the margins entire or erose, the central distal lobe dark blue at base or center with a yellow spot within or distal to blue area; flowers tristylous, with adnate stamens and styles of varying lengths (1.1–35.0 mm); stigmas capitate

or 3-lobed; capsules (to 1.5 cm) many-seeded, enclosed by the marescent perianth; seeds (to 2.1 mm) with 10–14 longitudinal wings

Life history: duration: annual (fruits/seeds); perennial (rhizomes, stolons, whole plants); **asexual reproduction:** rhizomes, shoot fragments, stolons; **pollination:** insect, self; **sexual condition:** hermaphroditic, heterostylous; **fruit:** capsules (common); **local dispersal:** plant fragments; seeds (water); **long-distance dispersal:** plant fragments; seeds (fish, water)

Imperilment: 1. *Eichhornia azurea* [GNR]; **2.** *E. crassipes* [G5]; **3.** *E. paniculata* [unranked]

Ecology: general: *Eichhornia* is an entirely aquatic group of annuals and perennials, which occur as emergent wetland plants or with submersed, floating-leaved, or free-floating habits. All three North American species are designated as OBL. The plants can occupy habitats where the waters are seasonal or permanent. The flowers are showy (smaller in autogamous species), ephemeral (wilting within a day), self-compatible or self-incompatible, and pollinated by insects (Insecta: Hymenoptera; Lepidoptera). Several of the species are heterostylous (tristylous) and are primarily outcrossed. Self-pollination has evolved in the group as a consequence of the breakdown of the tristylous condition. The seeds are dispersed by water and also likely by other means. In at least some instances, they are dormant when shed in the water, maintain dormancy when dried (as a consequence of receding seasonal waters), and then germinate when wet conditions recur. Dense but short-lived seed banks can develop. Some of the species are capable of prolific vegetative reproduction (rhizomes or stolons), which enables them to rapidly colonize and spread in aquatic habitats. Vegetative reproduction occurs by the water-mediated dispersal of shoot fragments, or by the detachment of floating, clonally derived plantlets. Despite the low number of species, the literature contains tens of thousands of references on *E. crassipes* alone, making it virtually impossible to provide a truly comprehensive account of this group. Most of the information pertains either to the biology of tropical American plants (where the species are indigenous) or to methods of control or management of *E. crassipes*, a notorious aquatic weed worldwide.

9.3.1.1. *Eichhornia azurea* **(Sw.) Kunth.** is a nonindigenous perennial, which grows in permanent waters of lakes, marshes, and rivers at elevations to 100 m [to 1,200 m in its indigenous range]. North American populations of *Eichhornia azurea* presumably have been eradicated and unless noted, the information summarized here pertains primarily to native populations. Flowering in North American plants reportedly can occur from June to October. Each inflorescence contains an average of 46.2 flowers. The flowers are showy ($\bar{x} = 44$ mm), tristylous (with associated trimorphic pollen), moderately self-incompatible, and produce 169.7 ovules on average. They remain open only for one day. The mean pollen:ovule ratio is 204. Self-pollination can vary significantly with the pollen source. In one study, 94.3% fruit set in long-styled morphs was obtained using mid-stamen pollen compared to only 12.2% using short-stamen pollen. However, despite their high fruit set, seed set of the mid-style crosses was less than 50%. In another study, most legitimate (intramorph) pollinations produced more fruits than self-pollination and illegitimate pollination, but mid-styled morphs were highly self- and intermorph-compatible. Legitimate pollinations produced higher numbers of seeds per fruit, but self-pollinations and illegitimate pollinations also produced fruits and seeds. No seed production has been observed in clonal populations of tristylous plants represented by single floral morphs, which indicates the lack of any autogamy. However, self-compatible, virtually homostylous races (with dimorphic pollen) are also known, which are autogamous and represent a breakdown in the tristylous system. The flowers typically are pollinated by various bees (Insecta: Hymenoptera: Anthophoridae; Apidae: *Bombus morio*, *Ceratina sclerops*, *Florilegus fulvipes*, *Paratetrapedia*, *Tetragonisca angustula*, *Trigona spinipes*; Halictidae: *Augochlora*; Megachilidae), which forage for nectar and/or pollen. Some populations rely on an oligolectic bee (Apidae: *Ancyloscelis gigas*) to effect pollination. Seed production is high. The seeds ($\bar{x} = 1.062$ mg) have limited buoyancy (remaining afloat for ~24 hr) and are dispersed by water. Seedlings establish commonly in shallow water or on exposed mud along the periphery of water bodies. The plants are long-lived perennials (mean leaf lifespan 49–78 days; estimated shoot longevity of 28 months). They are intolerant of prolonged dry conditions due to a long prereproductive phase, which precludes their establishment in sites susceptible to seasonal desiccation. Vegetative reproduction can occur by detached, floating fragments, which are dispersed by water. In tropical sites, the apical shoot biomass production has ranged from 1.8 to 4.8 g m^{-2} day^{-1}. Individual cultivated plants have expanded to cover nearly 2 m of the water surface within a growing season. **Reported associates (North America):** none reported [presumably eradicated].

9.3.1.2. *Eichhornia crassipes* **(Mart.) Solms** is a nonindigenous perennial, which floats on the water surface, or is rooted along the margins of bayous, bottomlands (hardwood), canals (drainage, roadside), ditches (drainage, irrigation, roadside), floodplains (forested), gravel pits, lagoons, lakes (artificial), marshes (fresh), oxbows, ponds (artificial, beaver, farm, freshwater, retention), pools, reservoirs, rivers (calm), roadsides, sloughs (roadside), streams (blackwater, slow), and swamps (cypress) at elevations to 2,377 m. The plants prefer exposures of full sunlight but can tolerate partial shade. They are not cold hardy but have survived brief exposures to freezing temperatures (24 hr at 5°C). Higher temperatures (to 34°C) have been tolerated for durations of up to 4–5 weeks. They occur on various substrates, including loamy sand (Pactolus), muck (Allemands), mud, sand (granitic), sandy loam, and silty loam (Hyde). The plants are found across a broad range of habitat conditions (pH: 5.2–8.7; $\bar{x} = 7.0$; $\tilde{x} = 7.0$; alkalinity [as mg l^{-1} CaCO$_3$]: 1.5–77.6; $\bar{x} = 28.8$; $\tilde{x} = 15.7$; conductance [μS cm^{-1} @ 25°C]: 32–346; $\bar{x} = 172$; $\tilde{x} = 113$; total P [μg l^{-1}]: 10–121; $\bar{x} = 56$; $\tilde{x} = 23$; total N [μg l^{-1}]: 310–1880; $\bar{x} = 940$; $\tilde{x} = 710$; Ca [mg l^{-1}]: 2.4–67.2; $\bar{x} = 24.2$; $\tilde{x} = 13.2$; Mg [mg l^{-1}]: 1.2–31.6; $\bar{x} = 11.5$; $\tilde{x} = 7.2$; Na [mg l^{-1}]: 2.7–25.8; $\bar{x} = 16.0$; $\tilde{x} = 8.3$; K [mg l^{-1}]: 0.2–8.0; $\bar{x} = 2.9$; $\tilde{x} = 1.9$; SO$_4^{2-}$ [mg l^{-1}]: 4.1–47.0; $\bar{x} = 17.3$;

$\tilde{x} = 10.7$; Cl⁻ [mg l⁻¹]: 4.4–42.1; $\bar{x} = 28.5$; $\tilde{x} = 15.2$; Fe [mg l⁻¹]: 0.0–0.4; $\bar{x} = 0.2$; $\tilde{x} = 0.1$; Si [mg l⁻¹]: 0.1–3.0; $\bar{x} = 1.7$; $\tilde{x} = 0.8$). The water depth is irrelevant because of the ability to assume a free-floating habit, which enables the plants to tolerate conditions ranging from saturated, exposed substrates (on which they become rooted) to impoundments several meters deep where they float atop the surface with their roots suspended in the water column. Flowering (North America) occurs from March to November, but declines substantially by the end of October. A rapid onset of reproduction (flowering can occur within 10–15 weeks of seed germination) enables the plants to colonize sites with seasonal waters of shorter duration. In fertile plants, the inflorescence typically develops through the night with all the flowers ($\bar{x} = 17.2$) opening during the next day, beginning within 2 hr of sunrise. In larger inflorescences, the opening of several upper flowers may be delayed until the second day. Once open, the ephemeral flowers last for only one day and wilt by nightfall. Pollination is most effective within 2 hr of anthesis and at temperatures 20°C–25°C. The flowers are large ($\bar{x} = 58.2$ mm), self-compatible, and tristylous in their indigenous range. However, due to founder effects and clonal growth, an estimated 77% of New World populations possess only a single floral morph, with the tristylous conditions found in fewer than 5%. All North American populations lack the short-styled morph and some (western USA) also lack the long-styled morph; none contains all three floral morphs. Despite the absence of morphs, seed production has been observed in every population surveyed of these self-compatible plants. In some populations, tristyly has broken down to a semihomostylous condition (placing the anthers and stigma in close proximity), which facilitates selfing. The mean pollen:ovule ratio is 255.2 and most sexual reproduction involves outcrossing by insects (Insecta). In North America, these include honeybees (Hymenoptera: Apidae: *Apis mellifera*). Although the flowering period is long, the cooccurrence of the showy flowered *Ludwigia peploides* has been known to monopolize honeybee pollination during spring and early summer months, with foraging on *Eichhornia* flowers not commencing until August. In North America, low seed set in open-pollinated flowers (19% of hand-pollinated flowers) is indicative of their pollinator limitation. Hand pollinations have shown that flower fertility is high (94.7%). In nature, about 44.9% of the flowers produce capsules, but this figure varies regionally (tropical plants: $\bar{x} = 64.6\%$; temperate plants: $\bar{x} = 29.3\%$). Similarly, fewer capsules mature on the inflorescences of temperate plants ($\bar{x} = 2.9\%$) than on tropical plants ($\bar{x} = 5.8\%$). Each ovary produces about 150 ovules and averages 44.2 seeds. No difference in seed set was found to occur between self- and cross-pollinations in populations lacking only the short-styled morph (Florida, Louisiana); however, some outcrossed plants had significantly higher seed set than selfed plants in pollinations involving long-styled clones (California). Seed production in California was found to peak ($\bar{x} = 40.8$ seeds flower⁻¹) during September. Within a day after anthesis, the inflorescences reflex to situate the fruits beneath the water where the seeds develop. The seeds are hydrochorus and are dispersed by rain wash, downstream water flow, and

floods. They are thought to be transported over greater distances endozoically by fish (e.g., Teleostei: Cyprinidae: *Cyprinus carpio*). Seed bank densities up to 2,534 seeds m⁻² have been reported in some nonindigenous localities. The small seeds ($\bar{x} = 0.297$ mg) are physiologically dormant when shed, and require 30 days of warm stratification to induce germination. Drying the seeds for 2–4 weeks (at room temperature) can induce dormancy. Their germination can be enhanced significantly (to 98%) by artificial scarification. Continuous light is known to inhibit germination of seeds incubated at 30°C. Subsequent germination occurs primarily under flooded conditions (to 100%) with much lower rates (9%) observed under nonflooded conditions. Year-old seeds have exhibited high germination rates (87.5%) and seedling establishment under standard field conditions; however, seedlings rarely establish in most populations. If so, then establishment occurs along the habitat periphery (or where the vegetation cover has been removed by disturbance) and never within a floating mat. The seedlings fare better on firm substrates for their first 6 weeks of growth, but grow better in water when older. These plants are long-lived perennials with extensive vegetative reproduction. They often develop into large, clonal, floating mats by the prolific reproduction of stolons across the water surface. Genetic analyses have indicated that approximately 80% of all introduced populations comprise a single clone. One plant is capable of generating 43 clonal rosettes within only 53 days. Some estimates indicate that the plants are capable of producing 3,418,800 individual ramets within a 200-day period. Just 20 plants eventually are able to cover an entire hectare of water surface within a single season. Because of this tremendous capacity for clonal division, nuisance levels of plants can be reached in smaller water bodies within the growing season once a plant has been introduced, regardless of it being capable of overwintering. Relative growth rates of 1.5% day⁻¹ have been estimated. Maximum biomass accumulation in Florida has been projected to approach 20 g m⁻² day⁻¹, yielding standing crops from 2.3 to 2.5 kg m⁻². **Reported associates (North America):** *Acer rubrum, Acrostichum danaeifolium, Agalinis, Agrostis perennans, Albizia julibrissin, Alternanthera philoxeroides, Ambrosia artemisiifolia, Ammannia coccinea, Amorpha fruticosa, Ampeopsis arborea, Andropogon glomeratus, Azolla filiculoides, Baccharis halimifolia, Bacopa monnieri, Betula nigra, Bidens mitis, Brasenia schreberi, Cabomba caroliniana, Callitriche heterophylla, Callitriche stagnalis, Cephalanthus occidentalis, Ceratophyllum demersum, Cladium jamaicense, Conoclinium coelestinum, Cornus foemina, Cornus sericea, Crinum americanum, Cyperus eragrostis, Cyperus polystachyos, Cyrilla racemiflora, Digitaria ciliaris, Echinochloa crus-galli, Echinochloa walteri, Echinodorus, Eclipta prostrata, Egeria densa, Eleocharis equisetoides, Eleocharis microcarpa, Equisetum hyemale, Eryngium aquaticum, Eupatorium capillifolium, Fimbristylis autumnalis, Fimbristylis caroliniana, Fraxinus caroliniana, Fuirena pumila, Habenaria repens, Heteranthera dubia, Hibiscus moscheutos, Hydrilla verticillata, Hydrocotyle ranunculoides, Hydrocotyle umbellata, Hypericum* *hypericoides, Hypericum mutilum, Ilex cassine, Ipomoea cordatotriloba, Iris pseudacorus, Iva annua, Juncus, Lantana camara, Leersia oryzoides, Lemna minor, Lepidium latifolium, Leptochloa, Limnobium spongia, Lipocarpha maculata, Liquidambar styraciflua, Ludwigia decurrens, Ludwigia hexapetala, Ludwigia leptocarpa, Ludwigia octovalvis, Ludwigia palustris, Ludwigia peploides, Lycopus americanus, Mecardonia acuminata, Mikania scandens, Mirabilis linearis, Mitreola petiolata, Myrica cerifera, Myriophyllum aquaticum, Myriophyllum spicatum, Najas guadalupensis, Nelumbo lutea, Nymphaea mexicana, Nymphaea odorata, Nyssa aquatica, Nyssa biflora, Oenothera, Panicum repens, Panicum virgatum, Paspalum repens, Passiflora incarnata, Peltandra virginica, Pentodon pentandrus, Persea palustris, Persicaria amphibia, Persicaria maculosa, Phragmites australis, Phyla, Phyllanthus, Phytolacca americana, Pinus elliottii, Pinus palustris, Pistia stratiotes, Polypremum procumbens, Pontederia cordata, Potamogeton nodosus, Proserpinaca ×intermedia, Rhynchospora corniculata, Riccia fluitans, Ricciocarpus natans, Rubus argutus, Rubus trivialis, Rumex crispus, Sabal palmetto, Sagittaria graminea, Sagittaria lancifolia, Sagittaria latifolia, Sagittaria platyphylla, Salix exigua, Salix gooddingii, Salix interior, Salix lasiolepis, Salvinia minima, Sambucus nigra, Schoenoplectus acutus, Schoenoplectus californicus, Schoenoplectus tabernaemontani, Solidago canadensis, Sparganium, Spirodela polyrrhiza, Stuckenia pectinata, Taxodium distichum, Thelypteris palustris, Tradescantia ohiensis, Triadenum virginicum, Typha angustifolia, Typha domingensis, Typha latifolia, Utricularia gibba, Utricularia inflata, Verbena brasiliensis, Wolffiella gladiata, Woodwardia virginica, Xanthium strumarium, Zizania aquatica.*

9.3.1.3. *Eichhornia paniculata* (Spreng.) Solms is a nonindigenous annual (or short-lived perennial), which grows in depressions (roadside), ditches, meadows (low), pools (seasonal), and rice fields in waters 10–30 cm deep. Because this species presently is not known to persist in North America, the information summarized here pertains primarily to indigenous populations. The plants colonize habitats that are characterized by limited periods of inundation. Natural habitats occur in arid regions, where seasonal water is present for only a few months of the year (March–May). Flowering in Florida has been observed in August. Each inflorescence produces large numbers ($\bar{x} = 82.0$) of smaller ($\bar{x} = 24$ mm), tristylous (but self-compatible) flowers that ordinarily are pollinated by nectar-feeding bees (Insecta: Hymenoptera: Apidae: *Ancyloscelis, Florilegus*) and butterflies (Insecta: Lepidoptera), which do not discriminate among the morphs; however, pollen-foraging bees (Apidae: *Trigona*) tend to avoid the long-styled inflorescences in favor of the other morphs. Plants cultivated in North America have been pollinated by bumblebees (Insecta: Hymenoptera: Apidae: *Bombus vagans, B. fervidus*). The mean pollen:ovule ratio is 192.2. Despite the floral compatibility, genetic studies indicate that tristylous populations maintain high levels of outcrossing (e.g., $t = 0.80$–0.90), with a low frequency of self-pollination or intramorph pollination. The frequency of selfed flowers (10–17% of total seed set) increases incrementally from the lower to upper flowers of an inflorescence. The fitness (i.e., competitive ability) of outcrossed plants has been found to exceed that of inbred progeny, regardless of the combination of floral morphs involved. A low frequency of short-styled morphs in some populations has been attributed to founder events and population size fluctuations. Highly autogamous, semi-homostylous plants (mainly mid-morph) also occur commonly within monomorphic or dimorphic populations due to breakdowns in the tristylous condition thought to be caused by pollinator limitation. Genetic (allozyme) surveys indicate that populations are highly differentiated (e.g., $G_{ST} = 0.40$–0.57). Each capsule contains 109.7 ovules on average. Higher seed set occurs in upper flowers due to the characteristically upward movement of pollinators on an inflorescence. The small seeds ($\bar{x} = 0.147$ mg) are believed to be dispersed primarily by their adhesion to waterfowl (Aves: Anatidae) or other animals. They are weakly dormant, with germination rates up to 70% obtained within a month of seed harvest. A small short-term seed bank (one season) is suspected to persist in some cases. The plants have no means of vegetative reproduction but can perennate for longer times during sufficiently extended wet periods. **Reported associates (North America):** none reported [no extant populations presently documented].

Use by wildlife (North America): *Eichhornia crassipes* is eaten by West Indian manatees (Mammalia: Trichechidae: *Trichechus manatus*) and by domestic cattle (Mammalia: Bovidae: *Bos taurus*). It is a host of larval moths (Insecta: Lepidoptera: Crambidae: *Niphograpta albiguttalis, Samea multiplicalis*; Noctuidae: *Bellura densa*). The foliage is damaged by scarab beetles (Insecta: Coleoptera: Scarabaeidae: *Dyscinetus morator*) and weevils (Insecta: Coleoptera: Erirhinidae: *Neochetina bruchi, N. eichhorniae*), which have been used (or considered) as biocontrol agents. The weevils along with phytophagous mites (Arthropoda: Acari: Galumnidae: *Orthogalumna terebrantis*) are attracted to the plants by a kairomone, which is produced by actively growing, juvenile tissue. The dense roots systems provide cover for young fish (Vertebrata: Teleostei) and substrate for aquatic invertebrates.

Economic importance: food: The young leaves, petioles, and inflorescences of *Eichhornia crassipes* sometimes have been eaten when cooked or steamed; **medicinal:** Several compounds isolated from *Eichhornia crassipes* (alkaloids, phthalate derivatives, propanoid and phenyl derivatives) are promising therapeutically and allegedly have been effective as antibacterial, anticancer, antifungal, and antioxidants agents. The leaf and shoot extracts have exhibited significant anti-inflammatory activity. The plants have been used to synthesize iron oxide nanoparticles, which express antimicrobial activity; **cultivation:** *Eichhornia azurea* and *E. paniculata* are grown as ornamental water garden plants; **misc. products:** *Eichhornia paniculata* has become a useful model for studying the evolution of heterostyly in plants. *Eichhornia crassipes* has been used to generate biogas, as a fertilizer, as mulch, and for papermaking. It also has been used effectively to remove nutrients, heavy metals, and other toxic substances from domestic and industrial wastewater effluent. The plants have been investigated as a potential system for purifying liquid wastes during space travel. Pyrolysis of the dried stems produces a black pigment (carbon black) that has been used in ink and paint formulations; **weeds:** *Eichhornia crassipes* has been proclaimed as one of the world's worst weeds and is notorious globally as a pest of water bodies. In addition to its interference with navigation and recreational access, the plants can more than triple the rate of water loss through evapotranspiration; **nonindigenous species:** All *Eichhornia* species are nonindigenous to North America. *Eichhornia crassipes* is thought to have first been introduced to the United States (Louisiana) in 1884 by plants distributed as novelties at the World's Industrial and Cotton Centennial Exposition. However, seeds already were available for sale in New Jersey by 1884/1885 and could have been distributed much earlier. The plants reached California by 1920 and have since been introduced throughout the world to numerous regions where the waters remain free from winter ice cover.

Systematics: The monophyly *Eichhornia* is doubtful and the circumscription of the genus currently remains unsettled taxonomically. Despite its small number of species, the distinction of *Eichhornia* from the closely related *Pontederia* has been difficult to reconcile. Both genera are similar morphologically and tristylous, which is a condition that occurs nowhere else in Pontederiaceae or anywhere else in the monocotyledons. One proposed solution has been to simply merge the genera, which also would necessitate the inclusion of yet another genus (*Monochoria*) to conform with existing phylogenetic insights. However, that conclusion (based on a purported "total evidence" approach) excluded available nuclear molecular data and followed a phylogenetic interpretation based on the combination of highly incongruent morphological and plastid data. Results based entirely on molecular data have been no more compelling, due to uncertain, multiple rooting possibilities. Recent molecular approaches have generated yet another unusual phylogenetic topology, which situates the distinctive *Eichhornia meyeri* (not included in the "total evidence" analyses) as the sister group of the entire family (Figure 3.5). Given the large discrepancies among the various morphological and molecular-based phylogenies, it seems practical to defer any major taxonomic upheavals until further study might achieve a more compelling consensus. Yet, despite the phylogenetic attention paid to this group, no analysis has yet indicated *Eichhornia* to be monophyletic, which leads to the inescapable conclusion that some taxonomic realignments eventually will be necessary. The basic chromosome number of *Eichhornia* is $x = 8$. *Eichorrnia paniculata* ($2n = 16$) is diploid, whereas, *E. azurea* and *E. crassipes* ($2n = 32$) are tetraploid. Natural interspecific hybrids have not been reported in the genus. Attempts to hybridize *E. azurea* and *E. crassipes* artificially have been unsuccessful.

Distribution: *Eichhornia azurea* (indigenous to Central and South America) and *E. paniculata* (South America) were known historically from a few sites in Florida; presumably, they have since been eradicated or otherwise have not persisted. *Eichhornia crassipes* is widespread throughout the southern United States and has been extending its range northward into the northwestern and northeastern United States and adjacent Canada.

References: Aboul-Enein et al., 2011; Adebayo et al., 2011; Alvarado et al., 2008; Barrett, 1978; 1980a; 1980b; 1985; Barrett & Husband, 1997; Barrett & Forno, 1982; Barrett et al., 1993; 1994; Baskin & Baskin, 1998; Benton, Jr., et al., 1978; Buckingham & Bennett, 1989; Campbell & Irvine, 1977; Center & Spencer, 1981; Cook, 1996; Da Cunha & Fischer, 2009; Del Fosse & Perkins, 1977; Fernald & Kinsey, 1943; Glover & Barrett, 1986; 1987; Horn, 2002; Hoyer et al., 1996; Husband & Barrett, 1992; 1993; 1995; 1998; Ikusima & Gentil, 1993; Jagathesan & Rajiv, 2018; Jayanthi et al., 2013; Julien et al., 1999; Lomolino & Ewel, 1984; Mack, 1991; Mishra & Tripathi, 2009; Mtewa et al., 2018; Monsod, Jr., 1979; Morgan & Barrett, 1990; Ness et al., 2011; Novelo & Lot, 1994; Patel, 2012; Pellegrini et al., 2018; Penfound & Earle, 1948; Pérez et al., 2011; Perry, 1961; Pieterse & Murphy, 1993; Pott & Pott, 2000; Rapoport et al., 1995; Santos, 2002; Thyagarajan, 1984; Von Bank et al., 2018; Zhang et al., 2010c.

9.3.2. Heteranthera

Duck salad, mud plantain

Etymology: an amalgam of the Greek *héteros* and Latin *anthera* ("different anthers") in reference to the varied stamen lengths

Synonyms: *Buchozia* (in part); *Commelina* (in part); *Eichhornia* (in part); *Eurystemon*; *Heterandra*; *Hydrothrix* (in part); *Leptanthus*; *Lunania* (in part); *Phrynium* (in part); *Pontederia* (in part); *Potamogeton* (in part); *Schollera* (in part); *Triexastima*; *Zosterella*

Distribution: global: Africa; New World; **North America:** widespread except far north

Diversity: global: 12 species; **North America:** 7 species

Indicators (USA): obligate wetland (OBL): *Heteranthera dubia*, *H. limosa*, *H. mexicana*, *H. multiflora* [inferred], *H. peduncularis* [inferred], *H. reniformis*, *H. rotundifolia* [inferred]

Habitat: freshwater; freshwater (tidal); lacustrine, palustrine, riverine; **pH:** 6.0–9.5; **depth:** 0.2–5.0 [to 14] m; **life-form(s):** emergent, floating-leaved, or submersed (rosulate or vittate) herbs

Key morphology: shoots elongate (to 1+ m) when submersed, shorter or procumbent when emersed; submersed leaves vittate (alternate) or in a basal rosette, the blades (to 15 cm) linear or oblanceolate, sessile; emergent or floating leaves petiolate (to 15 cm) or sessile, the blades (to 5 cm) cordate, linear, oblong, ovate, reniform, or round, cordate, cuneate, or truncate at base, stipules (to 6 cm) adnate; inflorescence solitary or a spike of 2–30 flowers (sometimes cleistogamous), enclosed by a clasping or folded, sessile or pedunculate (to 11 cm) spathe (to 7 cm); perianth salverform or tubular, actinomorphic or zygomorphic; the tepals (to 7 cm) blue, bluish-mauve, mauve, white, or yellow (sometimes brown or yellow-spotted), connate for ½+ of length, narrowly elliptic or linear; stamens (to 10 mm) 3, equal or unequal (2 laterals shorter), the filaments glabrous to pubescent; ovary incompletely 3-locular, ovules 10–∞, style 3-lobed, glabrous or pubescent; capsules elongate; seeds (to 1.3 mm), numerous (to 200), ovoid, with 8–22 longitudinal wings

Life history: duration: annual (fruits/seeds); perennial (dormant apices, rhizomes, stolons, whole plants); **asexual reproduction:** rhizomes, shoot fragments, stolons; **pollination:** insect, self (cleistogamy); **sexual condition:** hermaphroditic; enantiostylous; **fruit:** capsules (common); **local dispersal:** seeds (water); **long-distance dispersal:** seeds (birds, fish, humans, water)

Imperilment: 1. *Heteranthera dubia* [G5]; SH (OK); S1 (CO, MT, NC, NH); S2 (CA, DE, KS, MA, MB); S3 (BC, IA, IL, KY, ME, NB, QC); **2.** *H. limosa* [G5]; SH (IA); S1 (AZ, TN); S2 (IL, KY, MN); **3.** *H. mexicana* [G2/G3]; S1 (TX); **4.** *H. multiflora* [G4]; S1 (NE, NC, PA, TN, VA); S2 (KS); **5.** *H. peduncularis* [G4/G5]; **6.** *H. reniformis* [G5]; SH (CT); S1 (IA, IL, OH, SC, WV); S2 (NC); S3 (GA, NY); **7.** *H. rotundifolia* [G4]; SH (KY)

Ecology: general: Although *Heteranthera* is entirely aquatic, just four of the seven North American species currently are designated as OBL wetland indicators. Two other North American taxa (*H. multiflora* and *H. rotundifolia*) have not been assigned an indicator status, presumably because of unresolved taxonomic questions (see *Systematics* below). *Heteranthera peduncularis* also lacks any wetland indicator status, perhaps because it is now believed to be extirpated from North America. Although not assigned a formal status at the time of this writing, two of the three unranked North American taxa (*H. rotundifolia* and *H. peduncularis*) have been included here as OBL wetland indicators, whose ranking was inferred by their similar ecology to other OBL taxa. *Heteranthera multiflora* also has been included tentatively, but probably it is not distinct from *H. reniformis* taxonomically (see *Systematics* below). *Heteranthera* is a small but diverse group of aquatic annuals and perennials, which include submersed, floating-leaved, and emergent life-forms. Their natural habitats often are characterized by alternating periods of shallow flooding and drawdown, which has resulted in their adaptation to aquaculture systems such as rice (*Oryza sativa*) production, which annually provide similar conditions. Many of the species are heterophyllous and can respond to fluctuating water conditions by altering their foliage morphology plastically (submersed or emergent) as environmental conditions require. The showy flowers are adapted for entomophily, but most are self-compatible and appear to be primarily self-pollinating (pollinators rarely are observed on the flowers). Several of the species exhibit enantiostyly, which is thought to have evolved as a means of facilitating outcrossing. The flowers sometimes develop cleistogamously when water depths interfere with the emergence of the typical aerial inflorescences. Local seed dispersal is hydrochorous. Viable seeds of some species have been recovered from the guts of some waterfowl (Aves: Anatidae), which indicates that they are dispersed over longer distances by endozoic transport. Exozoic transport of seeds embedded within the mud that adheres to external animal tissues provides yet another dispersal mechanism. Seed production typically is high with

good representation in the seed bank. The perennial species (and some annuals during the growing season) can reproduce vegetatively by the production and dispersal (by water) of shoot fragments.

9.3.2.1. *Heteranthera dubia* (Jacq.) MacMill.

is a long-lived, emergent (when stranded) or submersed perennial, which grows in still or flowing, tidal or nontidal, fresh to brackish, standing waters associated with backwaters (river), bayous, canals (irrigation), depressions (seasonal), ditches, floodplains, lakes (glacial), mudflats, ponds (farm, stock), pools, reservoirs, rivers, sandbars, sloughs, and streams at elevations to 1,493 m. Most exposures occur where the water surface receives full sunlight. Although the plants typically occur at depths from 0.03 to 5.0 m (where at least 18–23% of full sunlight is available), they are known to sustain growth at light levels as low as 2.5% of full sunlight and even have been found at unusual depths of 12–14 m where only 0.5–1.0% of the surface light was available. Plant densities can range from 3 to 400 plants m^{-2} at water depths of 1–5 m. The waters are permanent, stable, and often eutrophic. They are hard and alkaline (pH: 6.8–9.5, $\bar{x}$ =7.8; conductivity: 320.1 µS cm^{-1}) with a total alkalinity greater than 20 ppm (19.5–153.0 mg l^{-1}; $\bar{x}$ =63.5 mg l^{-1}) and a sulfate ion concentration of less than 300 ppm. The plants are tolerant of low salinities (0.2–3.5‰), but their growth declines substantially when they are exposed to elevated salinity levels (e.g., 10‰). The roots function mainly for anchorage, given that less than 1.5% of biomass is accounted for by root uptake. The substrates are described as alluvium, basalt, gravel, marl, muck, mucky clay, mucky sand, mud, rocky gravel, sand, sandy muck, rock (limestone), sandy silt, silt, silty clay, and stony loam. Flowering and fruiting occur from April to November. Cleistogamy occurs regularly on the submersed plants if the flowers develop while more than 3 cm below the water surface. Cleistogamous autogamy (self-pollination) is the prevalent means of sexual reproduction and occurs as the elongating styles contact the anthers while their pollen is shed within the closed floral buds. The anthers become coiled after anthesis, generally by the early afternoon. This reproductive system produces abundant seeds within 3 weeks of anthesis. The aerial flowers open only when the plants become stranded on the mud, but these usually already have been self-pollinated while still in bud and it is doubtful that any significant level of outcrossing ever is achieved. No potential pollinators have been reported. The seeds are not buoyant and sink to the bottom when shed. They are dispersed locally by water currents and are thought to be transported over greater distances endozoically by fish (e.g., Teleostei: Cyprinidae: *Cyprinus carpio*). The seeds are physiologically dormant when shed and require a period of cold stratification (e.g., 210 days @ 1°C–3°C) to effectively induce germination (e.g. at 30°C). A substantial seed bank (of uncertain duration) can develop, yielding mean seedling densities up to 20.4 m^{-2}. Seeds that have overwintered germinate during spring and develop into seedlings throughout the summer. The plants propagate vegetatively by their rhizomatous/stoloniferous stems, which are capable of overwintering as whole plants; however, these become dormant once temperatures fall below 8°C. Nonetheless, winter plants have maintained a biomass of 0.59–9.94 g dw m^{-2}. Clonal reproduction and dispersal (by water) occurs by vegetative shoot fragments, which subsequently root and establish under optimal conditions. The greatest flux of vegetative propagules has been observed to occur from August to October, reaching its peak (~30 mg dw m^{-2} day^{-1}) in the latter month. Plant fragments also have been found on boats and trailers, which implicates humans as another dispersal agent. Peak biomass occurs from June to July, with seasonal productivity estimated to reach 1,870 kg dw ha^{-1} year^{-1} in some sites. **Reported associates:** *Alternanthera philoxeroides, Azolla, Bidens beckii, Brasenia schreberi, Cabomba caroliniana, Ceratophyllum demersum, Chara globularis, Drepanocladus, Echinodorus cordifolius, Egeria densa, Eichhornia crassipes, Elatine minima, Eleocharis acicularis, Eleocharis cellulosa, Elodea canadensis, Elodea nuttallii, Eriocaulon aquaticum, Heteranthera limosa, Hydrilla verticillata, Hydrocotyle verticillata, Isoetes howellii, Justicia americana, Lemna minor, Lemna trisulca, Ludwigia peploides, Mimulus ringens, Myriophyllum aquaticum, Myriophyllum heterophyllum, Myriophyllum sibiricum, Myriophyllum spicatum, Najas canadensis, Najas flexilis, Najas guadalupensis, Najas marina, Najas minor, Nelumbo lutea, Nitella allenii, Nitella flexilis, Nitella tenuissima, Nitellopsis, Nuphar advena, Nuphar variegata, Nymphaea elegans, Nymphaea odorata, Podostemum ceratophyllum, Potamogeton alpinus, Potamogeton amplifolius, Potamogeton berchtoldii, Potamogeton bicupulatus, Potamogeton crispus, Potamogeton diversifolius, Potamogeton epihydrus, Potamogeton foliosus, Potamogeton friesii, Potamogeton gramineus, Potamogeton illinoensis, Potamogeton natans, Potamogeton nodosus, Potamogeton obtusifolius, Potamogeton perfoliatus, Potamogeton praelongus, Potamogeton pusillus, Potamogeton richardsonii, Potamogeton robbinsii, Potamogeton spirillus, Potamogeton strictifolius, Potamogeton vaseyi, Potamogeton zosteriformis, Ranunculus longirostris, Ranunculus trichophyllus, Rhizoclonium, Rorippa aquatica, Ruppia maritima, Sagittaria longiloba, Salvinia molesta, Schoenoplectus acutus, Schoenoplectus pungens, Sparganium natans, Stuckenia pectinata, Typha latifolia, Utricularia macrorhiza, Vallisneria americana, Veronica anagallis-aquatica, Zannichellia palustris, Zizania palustris.*

9.3.2.2. ***Heteranthera limosa* (Sw.) Willd.** is an emergent or submersed annual, which inhabits borrow pits, canals (irrigation), cattle tanks, cienega, depressions (muddy), ditches (irrigation, rice, roadside), flats (pond), floodplains, lake bottoms (playa), marshes (open), mudflats, ponds (ephemeral, pothole, seasonally flooded, sink, stock), pools (interdunal, irrigation), puddles (roadside), rice fields, roadsides, ruts (shallow), sloughs, springs, swales (ephemeral, rice field), and the margins of bayous, impoundments (beaver), lakes (oxbow, playa), ponds (drying), reservoirs, streams, and swamps (bottomland) at elevations to 1,707 m. The plants grow under exposures of full sunlight as emergents on wet ground or with floating leaves in shallow waters (to 91 cm deep). The substrates have been characterized as alkaline and are described

as alluvium, clay (Calconeus), clay loam (limestone), clayey sand, gravel, loamy clay, muck (clay loam), mucky clay, mud, peat, rocky clay, sand, sandy limestone, sandy loam, sandy muck, silt, silty clay (muddy), and silty sand. Flowering and fruiting extend from April to November, depending on the locality. The ephemeral flowers open in the morning but wilt by midafternoon. They are self-compatible and characterized by monomorphic enantiostyly; however, further details on their breeding system and floral ecology have not been elucidated. Autogamy is likely to be prevalent, given the annual habit, self-compatibility, and lack of reported pollinators. Hand-pollinated flowers (unknown whether selfed or outcrossed) have matured seeds within 14 days. The seeds disperse locally by water currents, and more broadly by adhering to animals and agricultural equipment in mud or by other means of attachment. Intact seeds also have been recovered from the guts of waterfowl (Aves: Anatidae: *Anas crecca*), which indicates their potential for endozoic, long-distance dispersal. The seeds are completely dormant when shed during fall (September) and were found to require a period of anaerobic incubation and light to induce germination. Dried seeds that were stored at 25°C–30°C have germinated best (89%) in the light while under flooded conditions, where oxygen levels are low (0.5%). Seeds that were buried under nonflooded (i.e., dry) conditions over the winter period germinated well when incubated under various day/night temperature regimes: 25°C/15°C; 30°C/15°C; 35°C/20°C (but not under 20°C/10°C or 15°C/6°C). These experiments have indicated that although winter dormancy is best broken by an extended period of drying, flooding eventually is necessary to induce seed germination. Once spring flooding occurs, germination occurs rapidly ~90% within 30 days). Under high summer temperatures, the seeds are able to germinate (under flooded conditions) throughout the season. The optimal germination conditions of winter drawdown followed by spring flooding mimic those used in commercial rice culture, which is one reason this species has become so weedy in that environment. Seed banks can be substantial, with mean densities of 213–1,137 seeds m^{-2} indicated in one study. Furthermore, the seeds retain high viability in the seed bank for at least several years (e.g., 100% for flooded or nonflooded seeds buried for 3 years; 98% for nonflooded seeds buried for over 5 years); the maximum duration of seed bank viability has not been determined. Some accessions of rice (*Oryza sativa*) allegedly are allelopathic to *H. limosa*, but the evidence presented is less than compelling. In any case, the plants easily can reach nuisance levels in rice fields where they average from 1 to 5 plants m^{-2}. The stems root along the nodes and conceivably could fragment, disperse in the water, and reestablish during the growing season; however, there is no documentation of any type of vegetative reproduction in this annual species. **Reported associates:** *Aeschynomene americana, Alisma triviale, Alternanthera philoxeroides, Ambrosia grayi, Ambrosia tomentosa, Ammannia auriculata, Ammannia coccinea, Bacopa monnieri, Bacopa rotundifolia, Bacopa repens, Bolboschoenus fluviatilis, Bromus japonicus, Calibrachoa parviflora, Carex hyalinolepis, Carex nebrascensis, Carex* *praegracilis, Cephalanthus occidentalis, Chloracantha spinosa, Coreopsis tinctoria, Cynodon dactylon, Cyperus acuminatus, Cyperus articulatus, Cyperus haspan, Cyperus iria, Cyperus ochraceus, Cyperus virens, Digitaria sanguinalis, Dopatrium junceum, Echinochloa crus-galli, Echinochloa crus-pavonis, Echinochloa muricata, Echinodorus berteroi, Echinodorus cordifolius, Eclipta prostrata, Elatine brachysperma, Eleocharis acicularis, Eleocharis macrostachya, Eleocharis minima, Eleocharis obtusa, Eleocharis ovata, Eleocharis palustris, Eleocharis parvula, Eriocaulon cinereum, Eryngium nasturtiifolium, Gratiola neglecta, Helanthium tenellum, Helianthus ciliaris, Heteranthera dubia, Heteranthera mexicana, Heteranthera reniformis, Heteranthera rotundifolia, Hibiscus laevis, Hordeum jubatum, Isoetes lithophila, Juncus balticus, Juncus brachycarpus, Juniperus scopulorum, Lemna minor, Leptochloa fusca, Leptochloa nealleyi, Leptochloa panicoides, Limnosciadium pumilum, Lindernia dubia, Ludwigia glandulosa, Ludwigia linearis, Ludwigia peploides, Lythrum alatum, Marsilea macropoda, Marsilea vestita, Mecardonia procumbens, Monochoria vaginalis, Muhlenbergia rigens, Myosurus minimus, Najas flexilis, Najas guadalupensis, Najas guadalupensis, Neptunia, Nyssa, Oryza sativa, Panicum dichotomiflorum, Pascopyrum smithii, Paspalum distichum, Paspalum hartwegianum, Persicaria amphibia, Persicaria bicornis, Persicaria coccinea, Persicaria hydropiperoides, Persicaria lapathifolia, Persicaria pensylvanica, Planera aquatica, Pluchea odorata, Poa pratensis, Polygonum aviculare, Polygonum striatulum, Potamogeton nodosus, Potamogeton nodosus, Quercus grisea, Quercus michauxii, Quercus nigra, Ranunculus pusillus, Rhynchospora indianolensis, Rotala ramosior, Rumex altissimus, Rumex chrysocarpus, Rumex crispus, Rumex stenophyllus, Sagittaria brevirostra, Sagittaria calycina, Sagittaria cuneata, Sagittaria graminea, Sagittaria guayanensis, Sagittaria latifolia, Sagittaria longiloba, Sagittaria rigida, Salix nigra, Schoenoplectus americanus, Schoenoplectus hallii, Schoenoplectus heterochaetus, Schoenoplectus pungens, Schoenoplectus saximontanus, Schoenoplectus tabernaemontani, Scirpus cyperinus, Sphenoclea zeylanica, Steinchisma hians, Symphyotrichum subulatum, Taxodium, Typha domingensis, Typha latifolia, Urochloa platyphylla, Veronica peregrina, Vitis, Xanthium strumarium.*

9.3.2.3. **Heteranthera mexicana S. Watson** is an emergent or submersed annual, which grows in depressions, ditches, mudflats, pools (ephemeral), resacas, and along the margins of channels and lakes (oxbow, playa, roadside) at elevations to 31 m. The substrates are either clay or mud. Flowering occurs from June to December. The ephemeral flowers open during morning and wilt by the afternoon. They are enantiostylous, with the stylar deflections determined randomly and occurring roughly at equal left:right frequencies within an inflorescence. The plants can remain dormant in the seed bank for extended periods of time until sufficient rainfall induces germination, which is thought to require an exposed mud surface. Additional life history information needs to be compiled for this species. **Reported associates:** *Echinochloa,*

Echinodorus berteroi, Eclipta prostrata, Heteranthera limosa, Sagittaria longiloba, Trichocoronis wrightii.

9.3.2.4. **Heteranthera multiflora (Griseb.) C.N. Horn** is an emersed or submersed annual, which occurs in canals, ponds (artificial, roadside, sink-hole), pools (roadside), rice fields, and along the margins of lakes and sloughs at elevations to 421 m. The plants are tolerant of partial shade and can occur in shallow standing water (e.g., 31 cm). The substrates have been described as alluvium, mud, sand, and silty sand. Flowering and fruiting occur from August to October. Some flowers occur in cleistogamous spikes, which remain included within the spathe. Floral enantiostyly in this taxon has been determined to be regulated by a single gene locus, with right-handed styles being dominant. This species is doubtfully distinct from *H. reniformis* (see *Systematics* below) and literature accounts often do not distinguish the two. Further taxonomic clarification will be necessary before any meaningful attempt to compile ecological data for this taxon can be made. **Reported associates:** *Acalypha rhomboidea, Alopecurus carolinianus, Amaranthus tuberculatus, Ammannia coccinea, Ampelopsis arborea, Asclepias incarnata, Bacopa rotundifolia, Bidens, Cabomba caroliniana, Cephalanthus occidentalis, Ceratophyllum echinatum, Commelina diffusa, Conyza canadensis, Cyperus acuminatus, Cyperus erythrorhizos, Cyperus esculentus, Cyperus iria, Cyperus strigosus, Dalea leporina, Diodia virginiana, Echinochloa muricata, Echinodorus berteroi, Echinodorus cordifolius, Eclipta prostrata, Eleocharis macrostachya, Eleocharis obtusa, Elymus virginicus, Eragrostis hypnoides, Fraxinus pennsylvanica, Gratiola neglecta, Heteranthera limosa, Heteranthera rotundifolia, Hibiscus laevis, Hibiscus moscheutos, Hydrilla verticillata, Ipomoea lacunosa, Iva annua, Leptochloa fusca, Leptochloa panicoides, Lindernia dubia, Ludwigia decurrens, Ludwigia palustris, Ludwigia peploides, Nymphaea odorata, Oryza sativa, Packera glabella, Panicum dichotomiflorum, Penthorum sedoides, Persicaria bicornis, Persicaria hydropiperoides, Persicaria lapathifolia, Persicaria maculosa, Persicaria pensylvanica, Persicaria punctata, Phyla lanceolata, Pluchea camphorata, Portulaca oleracea, Ranunculus sceleratus, Rorippa palustris, Rotala ramosior, Sagittaria brevirostra, Sagittaria calycina, Sagittaria latifolia, Salix nigra, Sida spinosa, Spirodela polyrrhiza, Symphyotrichum lanceolatum, Trapa natans, Typha latifolia, Veronica peregrina, Xanthium strumarium.*

9.3.2.5. **Heteranthera peduncularis Benth.** is an annual that was collected only once from stagnant waters in Arizona in the early 20th century. Subsequently, it has not been observed in North America for more than a century and presumably is extirpated from the region. No life history information is available for the North American occurrence. In other parts of its range, the plants are known to grow in shallow standing waters (10–40 cm deep), which are associated with seasonally flooded depressions, ditches, and the margins of canals, lakes, ponds, and rivers at montane elevations from 700 to 2,230 m. The flowers are autogamous or are pollinated by insects (Insecta). **Reported associates (Mexico):** *Eleocharis montana, Juncus, Ludwigia peploides.*

9.3.2.6. *Heteranthera reniformis* Ruiz & Pav. is an annual or facultative perennial, which inhabits bays (delta), bottomlands (hardwood), canal beds, channels (river), depressions, ditches, floodplains (hardwood), gravel pits, marshes (freshwater, tidal), mudflats (freshwater, intertidal, lake, tidal), mud holes, ponds (beaver, cattle, farm, roadside), puddles (roadside), rice fields, ruts (road, tire), sandbars, seepages (cold water), shoals, shores (tidal), sloughs, spillways, swamps (river), and the margins of bayous, brooks, ponds, pools (backwater), sloughs, springs, and streams at elevations to 1158 m. The plants often occur in open exposures but are shade-tolerant and exhibit the best growth under partial shade (~33% of full sunlight). With increased shading, their growth rate then declines once light levels fall below 28% of full sunlight. They prefer warm conditions and are reported to grow optimally at temperatures from 23°C to 30°C. Most habitats are characterized by intermittent cycles of flooding (shallow waters to 31 cm deep) and drying. A depth of 5 cm was found to be optimal for seedling development, with their growth decreasing substantially under shallower (0 cm depth) or deeper (10–20 cm depth) conditions. Larger plants (10-leaf stage) grow well at water depths from 5 to 15 cm, but decline in growth at 20 cm. The substrates are circumneutral (pH: 6.0–7.5) and include gravel, limestone, muck (drying), mud, sand, sandy cobble, sandy loam, silty loam (Melvin), and silty mud. Mature, fruiting plants develop within 48–62 days after seed germination. Flowering extends from June to November with fruiting occurring primarily from August through November. The flowers are ephemeral, opening from about 10:00 am until 3:00 pm and then wilting. They are self-compatible and exhibit enantiostyly (monomorphic) with skewed ratios characterized by more left-styled flowers than right-styled flowers. Although the aerial flowers are adapted morphologically for pollination by insects, cleistogamous flowers can develop in deeper waters (e.g., 15 cm) and autogamously self-pollinate. Specific pollinators have not been identified and the extent of outcrossing on this species has not been evaluated. Local seed dispersal occurs by water (the seeds both float and sink) and viable seeds also have been recovered from the guts of waterfowl (Aves: Anatidae: *Anas*), which implicates endozoic transport as a means of long-distance dispersal. The transport of seeds in mud that clings to animals has been implicated as an exozoic dispersal mechanism. The plants develop a seed bank that is thought to be long-lived, with some 15-year-old seeds (stored in water in the dark) exhibiting germination rates over 95%. The seeds germinate in mid to late spring. They require light for germination, which can be enhanced by physical scarification of the seed coat. Much higher germination (36%) has been achieved for seeds incubated in water compared to those on wet or shallowly puddled substrates (4.5–7.2%). A 30°C/20°C temperature regime (12hr/12hr photoperiod) and planting depths less than 1 cm have been most effective at inducing germination. Cold stratification (6°C for 1–8 weeks) has no noticeable effect on germination rates for dry-stored seeds, but results in reduced rates for seeds stored in water. Experimentally grown plants have reached maximum total biomass (~80 g fresh weight) in 118 days after germination.

Limited vegetative reproduction can occur by the fragmentation and dispersal of stolons. **Reported associates:** *Acer negundo, Acer rubrum, Acer saccharinum, Acorus calamus, Alisma subcordatum, Alnus serrulata, Amaranthus cannabinus, Ammannia coccinea, Amphicarpaea bracteata, Apios americana, Arthraxon hispidus, Asclepias incarnata, Azolla filiculoides, Bacopa egensis, Bacopa rotundifolia, Betula nigra, Bidens aristosus, Bidens cernuus, Bidens laevis, Bidens tripartitus, Boehmeria cylindrica, Brunnichia ovata, Cabomba caroliniana, Cardamine longii, Carex crinita, Carex frankii, Carex grayi, Carex hyalinolepis, Carex lupulina, Carex muskingumensis, Carex tribuloides, Carex typhina, Carex vulpinoidea, Carya cordiformis, Celtis laevigata, Centella asiatica, Cephalanthus occidentalis, Ceratophyllum demersum, Coleataenia longifolia, Commelina communis, Commelina virginica, Cornus foemina, Cyperus bipartitus, Cyperus difformis, Cyperus erythrorhizos, Cyperus esculentus, Cyperus filicinus, Cyperus flavescens, Cyperus iria, Cyperus polystachyos, Cyperus squarrosus, Cyperus strigosus, Dichanthelium clandestinum, Diodella teres, Diodia virginiana, Diospyros virginiana, Echinochloa crus-galli, Echinodorus cordifolius, Eclipta prostrata, Eleocharis microcarpa, Eleocharis obtusa, Eleocharis palustris, Eleocharis quadrangulata, Eragrostis hypnoides, Eupatorium serotinum, Fimbristylis autumnalis, Fimbristylis littoralis, Forestiera acuminata, Fraxinus pennsylvanica, Fraxinus profunda, Galium tinctorium, Gleditsia aquatica, Heteranthera limosa, Hibiscus moscheutos, Hydrocotyle, Hydrodictyon, Hydrolea quadrivalvis, Hypericum mutilum, Impatiens capensis, Ipomoea, Isoetes, Juncus acuminatus, Juncus debilis, Juncus diffusissimus, Juncus dudleyi, Juncus effusus, Juncus tenuis, Justicia americana, Kyllinga gracillima, Leersia oryzoides, Leersia virginica, Leptochloa fusca, Limnobium spongia, Lindernia dubia, Lonicera japonica, Ludwigia alternifolia, Ludwigia decurrens, Ludwigia grandiflora, Ludwigia leptocarpa, Ludwigia palustris, Ludwigia peploides, Ludwigia sphaerocarpa, Lysimachia ciliata, Lythrum salicaria, Microstegium vimineum, Mikania scandens, Mimosa strigillosa, Mimulus alatus, Mimulus ringens, Murdannia keisak, Myriophyllum, Najas guadalupensis, Nuphar advena, Nyssa, Orontium aquaticum, Oryza sativa, Panicum dichotomiflorum, Peltandra virginica, Persicaria arifolia, Persicaria lapathifolia, Persicaria posumbu, Persicaria punctata, Persicaria sagittata, Phalaris arundinacea, Phragmites australis, Phyla lanceolata, Pilea pumila, Planera aquatica, Plantago rugelii, Platanus occidentalis, Pluchea foetida, Poa annua, Poa autumnalis, Pontederia cordata, Populus heterophylla, Potamogeton diversifolius, Quercus palustris, Rhynchospora, Riccia, Ricciocarpus, Liquidambar styraciflua, Rubus flagellaris, Lycopus, Sagittaria graminea, Sagittaria lancifolia, Sagittaria latifolia, Sagittaria subulata, Salix nigra, Sambucus nigra, Saururus cernuus, Schoenoplectus pungens, Schoenoplectus purshianus, Schoenoplectus tabernaemontani, Scirpus cyperinus, Scirpus georgianus, Sesbania herbacea, Smilax rotundifolia, Solidago altissima, Sparganium americanum, Sphenoclea zeylanica, Spirodela polyrrhiza, Strophostyles* *helvola*, *Styrax americanus*, *Symphyotrichum lanceolatum*, *Symphyotrichum racemosum*, *Symphyotrichum subulatum*, *Taxodium distichum*, *Toxicodendron radicans*, *Triadenum walteri*, *Typha angustifolia*, *Ulmus americana*, *Utricularia gibba*, *Utricularia inflata*, *Vitis cinerea*, *Vitis labrusca*, *Vitis rotundifolia*, *Wolffia*, *Xanthium strumarium*, *Zizania aquatica*, *Zizaniopsis miliacea*.

9.3.2.7. *Heteranthera rotundifolia* (Kunth) Griseb. is a submersed, floating-leaved, or emergent annual, which inhabits depressions (roadside, swampy), ditches (roadside), fields (irrigated), impoundments (drying), lakes (playa), mudholes, ponds (artificial, depression, sinkhole, small), pools (ephemeral, interdunal), rice fields, ruts (road), sloughs, springs, and the margins of ponds and streams at elevations to 1661 m. The plants occur in seasonal, shallow waters (e.g., 25 cm) or on exposed, wet substrates, which are described as clay, clay loam, loam (Mansker, Portales), muck, mud, Potter, sandy loam, and silty loam. Flowering and fruiting extend from June to October. The ephemeral, enantiostylous flowers open within an hour of dawn and wilt by midday. They are self-compatible but are known to attract bees (Insecta: Hymenoptera: Apidae: *Apis mellifera*). Otherwise, their breeding system and pollination biology have not been studied in any detail. In any case, sexual reproduction is efficient and numerous seeds are produced as a consequence. A large seed bank develops with mean densities from 128 to 236 seeds m^{-2} reported. The plants grow rapidly, reaching maturity in about 8 weeks after germination. **Reported associates:** *Bacopa rotundifolia*, *Bolboschoenus fluviatilis*, *Bouteloua dactyloides*, *Eleocharis erythropoda*, *Grindelia squarrosa*, *Helenium*, *Heteranthera multiflora*, *Juniperus scopulorum*, *Heteranthera limosa*, *Marsilea vestita*, *Marsilea vestita*, *Najas gracillima* [Asian genotype], *Oryza sativa*, *Persicaria bicornis*, *Persicaria coccinea*, *Paspalum distichum*, *Phyla nodiflora*, *Potamogeton nodosus*, *Prosopis glandulosa*, *Quercus grisea*, *Rumex altissimus*, *Rumex crispus*, *Rumex stenophyllus*, *Sagittaria brevirostra*, *Sagittaria graminea*, *Sagittaria rigida*, *Salix*, *Schoenoplectus heterochaetus*, *Vitis*, *Vulpia octoflora*.

Use by wildlife: The spawning substrate structure provided by *Heteranthera dubia* plants has been shown to reduce predation levels on native pike (Teleostei: Esocidae: *Esox lucius*) and muskellunge (Teleostei: Esocidae: *Esox masquinongy*) eggs, which are eaten by the nonindigenous round goby (Teleostei: Gobiidae: *Neogobius melanostomus*). The foliage (13.3% crude protein) and seeds of *H. dubia* are eaten by various waterfowl (Aves: Anatidae), including the black-bellied whistling duck (*Dendrocygna autumnalis*), blue-winged teal (*Anas discors*), northern pintail *Anas acuta*), and wood duck (*Aix sponsa*). Although the alkaloid content of the plants (0.13 mg g^{-1} dry weight) is thought to deter herbivory, the foliage hosts a phytopathogenic protozoan (Cercozoa: Plasmodiophoraceae: *Membranosorus heterantherae*) and is colonized by numerous larval (and often herbivorous) insects (Insecta), including beetles (Coleoptera: Chrysomelidae; Hydrophilidae: *Berosus*), bugs (Hemiptera: Belastomatidae; Mesoveliidae: *Mesovelia*; Notonectidae: *Buenoa*), caddisflies (Trichoptera: Hydropsychidae: *Hydropsyche*;

Leptoceridae: *Leptocerus americanus*, *Nectopsyche albida*, *N. diarina*, *N. minuta*, *Oecetis cinerascens*, *O. eddlestoni*, *Triaenodes injustus*, *Ylodes*), damselflies and dragonflies (Odonata: Coenagrionidae: *Enallagma*; Libellulidae: *Libellula*, *Erythemis*), flies (Diptera: Ceratopogonidae; Ephydridae; Culicidae: *Culiseta*; Stratiomyidae: *Odontomyia*; Chironomidae: *Ablabesmyia*, *Apedilum elachistus*, *Chironomus decorus*, *Cricotopus sylvestris*, *Dicrotendipes nervosus*, *Endochironomus*, *Glyptotendipes*, *Labrundinia virescens*, *Larsia decolorata*, *Polypedilum*, *Psectrocladius vernalis*, *Pseudochironomus richardsoni*, *Tanytarsus*; Ephydridae: *Hydrellia*), mayflies (Ephemeroptera; Baetidae: *Callibaetis floridanus*; Caenidae: *Caenis latipennis*), and moths (Lepidoptera: Crambidae: *Acentria ephemerella*, *Oxyelophila callista*, *Parapoynx*, *Synclita obliteralis*, *S. occidentalis*); adult aphids (Hemiptera: Aphididae: *Rhopalosiphum nymphaeae*) and weevils (Coleoptera: Curculionidae: Curculionidae: *Bagous floridanus*) are also hosted. The aphids (*Rhopalosiphum nymphaeae*) are also herbivorous on *H. limosa* and are capable of decreasing its annual total biomass by 58–87% and fruit biomass by 82%. Seeds of *H. limosa* are eaten by the fulvous whistling-duck (Aves: Anatidae: *Dendrocygna bicolor*). The plants are also associated with larval mosquitoes (Insecta: Diptera: Culicidae: *Anopheles pseudopunctipennis*).

Economic importance: food: *Heteranthera* species are not reported as being edible; **medicinal:** The Cherokee administered a poultice made from the roots of *Heteranthera reniformis* to treat inflamed sores or wounds. In Honduras, plants of *H. reniformis* are crushed and soaked in cool water, which is taken as a tonic for coughs and sore throats; **cultivation:** *Heteranthera dubia* has been widely cultivated as an aquarium plant. *Heteranthera limosa* was recommended as a suitable carp pond plant late in the 19th century. *Heteranthera reniformis* is offered commercially as an ornamental aquarium and water garden plant; **misc. products:** *Heteranthera dubia* has been considered for use as a livestock forage plant. *Heteranthera reniformis* has been a component of some constructed wetlands used to treat agricultural sewage; **weeds:** *Heteranthera dubia* can grow densely to weedy proportions in some instances; *H. limosa*, and *H. rotundifolia* are serious rice field weeds, which can reduce rice yields by as much as 21%; **nonindigenous species:** Although indigenous to the central United States, *H. limosa* was introduced (sometime before 1948) to northern California, where it is associated almost exclusively with rice fields; *H. rotundifolia* is similarly disjunct in the California rice region where it also has been introduced. As a result of contaminated seed, several *Heteranthera* species have become rice field weeds in Europe, including *H. limosa* (France, Greece, Italy, Macedonia, Portugal, Spain), *H. reniformis* (France, Greece, Italy, Portugal), and *H. rotundifolia* (Greece, Italy, Portugal, Spain); *H. limosa* also has been introduced to Japan through rice culture.

Systematics: Various phylogenetic studies (but none comprehensive) have shown *Heteranthera* to be a monophyletic sister group to the monotypic *Hydrothrix* (e.g., Figure 3.5). Some authors have proposed the merger of *Hydrothrix gardneri* with *Heteranthera*, despite there being no credible phylogenetic conflicts with maintaining it as a morphologically and genetically distinct genus (such mergers unnecessarily create confusion in accessing information from literature accounts). *Heteranthera limosa* and *H. rotundifolia* have been merged as conspecifics by some authors, but they are quite distinct by their plastid RFLP profiles and DNA sequence data (43 differences among 3 plastid data sets). However, *Heteranthera multiflora* (= *H. reniformis* var. *multiflora*) is doubtfully a distinct species. It is poorly differentiated from *H. reniformis* with most of the putatively distinguishing characters displaying overlap. Morphologically, *H. multiflora* (spikes longer than spathes [except when cleistogamous] and purple filament hairs) essentially differs from *H. reniformis* (spikes "usually" shorter than the spathe ["sometimes" extending beyond] and white filament hairs) by color of the filament hairs. Relying on filament hair coloration to differentiate these species is ill-advised, given that geographically partitioned perianth color variation in *H. multiflora* (blue or mauve vs. white) has been concluded to be uninformative "at any taxonomic level." Moreover, Atlantic populations of *H. multiflora* are characterized by white perianths and flowers that are included within the spathe, which are states identical to those of *H. reniformis* (disjunct, western *H. multiflora* populations have blue perianths and exserted flowers). There are also several reports of *H. reniformis* (white-flowered plants) in the western portion of the alleged *H. multiflora* range. A more critical study of these taxa is needed. Even though they are thought to vary by chromosome number (see below), a comprehensive cytological survey of populations of both species never has been published. Yet, even if a consistent chromosomal distinction is found across their ranges, it is conceivable that these extremely similar taxa represent tetraploid and hexaploid races of the same species (based on an $x=8$ series) or possibly reflect a consequence of allopolyploidy (which occurs in the related *Monochoria*). In addition to cytological data, a comprehensive genetic survey of *H. reniformis* and *H. multiflora* should be conducted before a final determination of their taxonomic status is made. The basic chromosome number of *Heteranthera* is $x=7$ or 8. *Heteranthera limosa*, *H. oblongifolia*, and *H. rotundifolia* ($2n=14$) are diploid, *H. multiflora* ($2n=32$) is tetraploid, and *H. reniformis* ($2n=48$) is hexaploid. *Heteranthera dubia* ($2n=30$) exhibits a count based on an altered base number. No natural hybrids have been reported in *Heteranthera*.

Distribution: In North America *Heteranthera dubia* is widespread except for the far north; *H. limosa* and *H. rotundifolia* occur in the central and southwestern United States; *H. multiflora* and *H. reniformis* occur in central and southeastern United States; *H. mexicana* occurs only in Texas, but extends southward into northern Mexico. *Heteranthera peduncularis* extends southward into Argentina and Brasil but is historical in North America, where it has been collected only once in southern Arizona during the early 20th century. Presumably, it has become extirpated.

References: Balci & Kennedy, 2003; Barrett & Graham, 1997; Barrett & Seaman, 1980; Basinger et al., 1997; Baskin et al., 2003; Beal, 1977; Bellrose & Brown, 1941; Blackburn et al., 1961; Bolen & Beecham, 1970; Boyd, 1968; 1975; Braselton, 1983; Brim-DeForest et al., 2017a; 2017b; Brochet et al., 2010b; Bukliev, 1980; Carter & Rybicki, 1986; Champion et al., 2010; Christopher, 1983; Crocker, 1938; Delgado, 1999; Dilday et al., 1989; DiTomaso & Healy, 2003; Fassett, 1957; Forest, 1977; Gabela, 1974; Galán Demera & De Castro, 2003; Gortner, 1934; Graham et al., 1998; Harms et al., 2011; Haukos & Smith, 1994; 1997; 2001; Hellquist & Crow, 1982; Hill, 2006; Hohman et al., 1996; Holmes et al., 2005; Horn, 1983a; 1988; 1998; 2002; Horn & McClintock, 2012; Howard & Wells, 2009; Hussner, 2012; Jauzein, 1991; Jesson & Barrett, 2002; Jesson et al., 2003a; 2003b; Kadono, 2004; Kasselmann, 2003; Kohn et al., 1996; Lentz, 1993; Loczy et al., 1983; Lot et al., 1999; Manguin et al., 1996; Marcus, 1981; Marler, 1969; McFarland & Rogers, 1998; Meyer et al., 2015; Miano et al., 2019; Miller, 1987; Mora-Olivo et al., 2018; Moyle, 1945; Olofsdotter et al., 1995; Oraze & Grigarick, 1992; Ortas & Martorell, 2001; Ostrofsky & Zettler, 1986; Owens et al., 2001; Pellegrini et al., 2018; Pip & Simmons, 1986; Poirrier et al., 2010; Poole et al., 2007; Raynal & Geis, 1978; Reed, 1930; Rejmánek & Randall, 1994; Reynolds et al., 2015; Rhodes, 1978; Richardson & King, 2011; Rosatti, 1987; Rothlisberger et al., 2010; Rybicki et al., 2001; Sipaúba-Tavares & de Souza Braga, 2008; Sgattoni et al., 1989; Sheldon & Boylen, 1977; Shields & Moore, 2016; Tucker & McCaskill, 1967; Vasconcelos et al., 1999; Von Bank et al., 2018; Ward, 1883; Wilkinson, 1961; Wylie, 1919.

9.3.3.   Monochoria
False pickerelweed

Etymology: from the Greek *mono chori* ("one separate"), which refers to the single longer stamen

Synonyms: *Bootia* (in part); *Calcarunia*; *Carigola*; *Gomphima*; *Limnostachys*; *Pontederia* (in part)

Distribution: global: Africa; Asia; Australia; Europe*; North America*; **North America:** California (northern)

Diversity: global: 8 species; **North America:** 1 species

Indicators (USA): obligate wetland (OBL): *Monochoria vaginalis*

Habitat: freshwater; palustrine; **pH:** 4.0–7.0; **depth:** <1 m; **life-form(s):** emergent herb

Key morphology: plants (to 50 cm) clumped; juvenile stems submersed or floating with elongate internodes and adventitious roots; mature plants emergent, rosulate, short-stemmed (to 16.5 cm); juvenile foliage submersed, the blades (to 5 cm) linear, sessile, developing into floating or emergent, rigidly petiolate (to 28 cm) leaves, the blades (to 8 cm) linear, cordate, ovate, or ovate-orbicular, stipules (to 5.7 cm) adnate; panicles 3–8-flowered, subtended by a folded spathe (to 4 cm), the flowers pedicellate (to 40 mm), opening simultaneously; tepals (to 15 mm) connate basally, blue or white, narrowly ovate, the three innermost broadest; androecium heterandrous, having five shorter stamens (to 4 mm) with yellow anthers and toothless filaments, and 1 longer stamen (to 6 mm) with a blue anther and a conspicuous, erect tooth on the filament; ovary incompletely 3-locular, ovules numerous, style 1, filiform, glabrous; capsules (to 1 cm) ovoid, loculicidal; seeds (to 1.2 mm) numerous (to 200), brown, ovoid, with 10–11 longitudinal wings

Life history: duration: annual (fruits/seeds); perennial (whole plants); **asexual reproduction:** none; **pollination:** insect, self (cleistogamous autogamy); **sexual condition:** hermaphroditic, enantiostylous; **fruit:** capsules (prolific); **local dispersal:** seeds (water); **long-distance dispersal:** seeds (mud [vectors uncertain], water)

Imperilment: 1. *Monochoria vaginalis* [GNR]

Ecology: general: *Monochoria* is a genus of annual (or facultatively perennial), obligately aquatic herbs. They are well adapted to seasonally fluctuating waters by heterophylly, producing sessile, linear leaves when submersed, and petiolate, broad-bladed leaves when floating or emergent. The flowers are enantiostylous, nectarless, and offer only pollen as a pollinator reward. They are self-compatible but can be outcrossed by insects (Insecta) or self-pollinated (sometimes cleistogamous). The heterandrous androecium consists of five shorter stamens with yellow anthers, which function as "feeding anthers" for pollen-collecting insects (Insecta). The sixth stamen is longer, bears a larger blue anther (the "pollinating anther"), and has a conspicuous tooth on the filament. These features arguably disguise the larger anthers to deter pollen collection given that they are consistently avoided by potential pollinators. The seeds are dormant when shed and have germination requirements that reflect the seasonal nature of their habitats. The seeds are forcefully ejected from the maturing capsules and are dispersed by the water, or within mud that clings to various potential animal vectors.

9.3.3.1. *Monochoria vaginalis* **(Burm. f.) C. Presl ex Kunth** is a nonindigenous inhabitant of ditches, ponds (ephemeral, rice), and rice fields (drained, shallow) at elevations to 97 m. The plants occur typically in habitats characterized by sunny exposures and alternating wet and dry seasons. They die back as the habitat desiccates, and reproduce as annuals by seed. Some perennation can occur if waters persist, but the plants do not reproduce vegetatively or develop into clones. The substrates (California) have been described as mud. In North America, the plants flower from June to September, with fruiting extending through October. The flowers are enantiostylous but are self-compatible. In Asia, the flowers open simultaneously or over a period of several days. Both chasmogamous and cleistogamous flowers can develop, with their respective ratio decreasing as light availability is reduced. Cleistogamous autogamy is prevalent, with pollination usually already having taken place (as pollen is shed from the longer "pollinating anther") before the flowers open, which often occurs while under water. Reproductive studies have theorized that limited cross-pollination still remains possible, and although rarely observed, insect (Insecta) floral visitors have included bees (Hymenoptera: Halictidae: *Halictus*, *Lasioglossum*) and flies (Diptera: Syrphidae: *Sphaerophoria*). However, various genetic surveys (allozyme, ISSR, RAPDs data) conducted in Asia have documented imperceptible levels of outcrossing ($t=0.0$), high homozygosity, low levels of intrapopulational genetic variation, and large interpopulational differentiation in this species.

Further adaptations to selfing are evidenced by highly reduced resource allocation (16.5%) to the five "feeding anthers" that provide the pollinator rewards along with a substantially reduced pollen:ovule ratio (106:1) compared to an outcrossing congener (408:1). Seed production is prolific ($\bar{x}=110$–154 seeds capsule^{-1}). The seeds are ejected forcefully from the capsules as the valves detach explosively from the pedicel; they are dispersed subsequently by water. The freshly shed seeds are dormant, but eventually germinate after prolonged exposure to cold, wet conditions. Light and flooding (6 mm water) are required for germination to occur. The flooding requirement can be circumvented by incubation under low (0.2%) oxygen levels, although seedling development is reduced under such conditions. Even when inundated, seeds buried at depths of 4 mm exhibit reduced germination, which eventually ceases completely at burial depths of 20 mm or more. Nondormant seeds have germinated at temperatures from 14°C to 40°C, with an optimum incubation from 22°C to 32°C. Germination of some seeds has been found to cease if buried at any depth, with an absolute light requirement developing after 1 month at burial depths below 1 cm. Seed germination has been found to increase with decreasing sediment pH, if within the range of pH 4.0–7.0. **Reported associates (California):** *Alisma triviale*, *Bacopa repens*, *Heteranthera limosa*, *Oryza sativa*, *Rotala indica*, *Sagittaria montevidensis*, *Schoenoplectus mucronatus*. **Use by wildlife:** No uses reported.

Economic importance: food: *Monochoria vaginalis* is eaten as a vegetable in Asia. Its leaves (9.7 mg N 100 g^{-1}) and flowers (6.3 mg N 100 g^{-1}) contain almost all essential and nonessential amino acids as well as nutritionally competent levels of P, K, Mg, Cu, Mn, and Ca. The flowers are especially rich in protein (10.8 g 100 g^{-1}) and carbohydrates (4.6 g 100 g^{-1}); **medicinal:** Leaf and root extracts of *Monochoria vaginalis* reportedly possess significant antioxidant and anti-inflammatory properties. The juice is used medicinally in Asia, where treatments for liver or stomach disorders and toothaches also have been prepared from the roots. Caution is advised, given that extracts of some *Monochoria* species are known to be hepatotoxic; however, toxicological studies of *M. vaginalis* have indicated it to be safe; **cultivation:** *Monochoria* species are not in commercial cultivation; **misc. products:** *Monochoria vaginalis* has been used in the phytoremediation of industrial wastewaters contaminated by arsenic and other heavy metals; **weeds:** *Monochoria vaginalis* is a serious rice field weed in California as well as throughout Asia, reducing yields of rice (*Oryza sativa*) by up to 35%. Some strains have developed herbicide resistance to acetolactate synthase (ALS) inhibitors (e.g., bensulfuron-methyl; sulfonylurea); **nonindigenous species:** *Monochoria vaginalis* was introduced to northern California sometimes before 1954 as a consequence of rice (*Oryza sativa*) culture; it also has been introduced to Hawaii.

Systematics: Although phylogenetic analyses have not yet evaluated all *Monochoria* taxa simultaneously, those subsets analyzed consistently resolve as a distinct clade, which indicates that the genus is monophyletic. Some authors have proposed the merger of *Monochoria* with *Pontederia*

(see Figure 3.5); however, the two genera are distinct morphologically. Although it is evident that a revised circumscription is necessary for *Eichhornia* (which is polyphyletic), the merger of *Monochoria* and several other genera into a broadly defined *Pontederia* is not advocated, when other options are available that would result in far less nomenclatural turmoil. Although sorely needed, a revised taxonomy of Pontederiaceae at least should await more extensive taxon sampling, and hopefully, a more compelling indication of interrelationships within the family. Moreover, some confounding factors such as polyploidy (see below) have only recently been considered and should be evaluated more critically. The basic chromosome number of *Monochoria* includes $x = 6$, 7, and 13. Cytological investigations indicate that *Monochoria vaginalis* ($2n = 52$) is an allotetraploid derivative of *M. vaginalis* var. *plantaginea* ($2n = 24$) and *M. hastaefolia* ($2n = 28$).

Distribution: *Monochoria vaginalis* (indigenous to Asia and Australia) is known only from rice fields in northern California.

References: Barrett & Seaman, 1980; Barrett & Strother, 1978; Chandran & Parimelazhagan, 2012; Chandran et al., 2012; Chen & Kuo, 1995; Christopher, 1983; Cook, 1996; Graham et al., 1998; Horn, 2002; Horn & McClintock, 2012; Ileperuma et al., 2015; Imaizumi et al., 2008a; 2008b; Kim & Mercado, 1987; Kohn et al., 1996; Kuk et al., 2003; Li et al., 2005; Mason, 1957; Ness et al., 2011; Nozoe et al., 2018; Okoye et al., 2014; Oraze et al., 1988; 1989; Pellegrini et al., 2018; Pradeesh et al., 2012; Soerjani et al., 1987; Takeuchi et al., 1995; Talukdar & Talukdar, 2015; Tang & Huang, 2007; Tucker & McCaskill, 1955; Wang et al., 1998; Yanuwiadi & Polii, 2013.

9.3.4. Pontederia

Pickerelweed, wampee; pontédérie cordée
Etymology: after Giulio Pontedera (1688–1757)
Synonyms: *Eichhornia* (in part); *Hirschtia*; *Narukila*; *Reussia* (in part); *Umsema*; *Sagittaria* (in part)
Distribution: global: New World; **North America:** eastern
Diversity: global: 6 species; **North America:** 1 species
Indicators (USA): obligate wetland (OBL): *Pontederia cordata*
Habitat: brackish (coastal), freshwater; freshwater (tidal); lacustrine, palustrine, riverine; **pH:** 4.9–9.6; **depth:** <2 m; **life-form(s):** emergent, submersed (rosulate) herb
Key morphology: stems short-rhizomatous, rooted in mud; foliage rosulate, submersed leaves sessile, linear, ribbonlike, emersed leaves stalked, the petiole (to 60 cm) constricted below blade, the stipules (to 29 cm) adnate, sheathing at base, the blade (to 22 cm) cordate to lanceolate; inflorescence (to 120 cm) erect, spiciform, flowers numerous (to several hundred), spathe (to 17 cm) folded; perianth mauve, funnelform (coiling in fruit), the tube 3–9 mm, the limb (to 8 mm) oblanceolate, central lobe with a distal yellow, bilobed splotch; stamens 6, the proximal stamens (to 13 mm) longer than distal stamens (to 6.3 mm), filaments purple, glandular-pubescent; ovary 3-locular but uniloculate at maturity, ovule 1, style 3-lobed; utricle (to 6 mm) with dentate crests; seed ovoid

Life history: duration: perennial (rhizomes); **asexual reproduction:** rhizomes; **pollination:** insect; **sexual condition:** hermaphroditic, heterostylous (tristylous); **fruit:** anthocarp or utricule (common); **local dispersal:** seeds (water); **long-distance dispersal:** seeds (animals, water)
Imperilment: 1. *Pontederia cordata* [G5]; S1 (KS, KY); S3 (IA, IL)
Ecology: general: The *Systematics* section should be consulted for a taxonomic overview of this group, which explains the circumscription followed here in light of recent contrasting recommendations. *Pontederia* consists entirely of perennial aquatic plants, represented by submersed, floating-leaved, or emergent life-forms. Although North America contains only a single species, the 2016 indicator list also included *Pontederia rotundifolia* as an OBL indicator for the Atlantic and Gulf Coastal Plain region. However, that species is not known to be established there or anywhere in North America and has been excluded here from further consideration. Most of the species inhabit tropical freshwaters, sometimes extending into brackish sites. The habitats typically are characterized by intermittent standing waters of a seasonal nature. The temperate North American species is more characteristic of permanent waters. They possess showy, tristylous, strongly to weakly self-incompatible flowers, which usually are outcrossed by various insects (Insecta). They disperse and establish primarily by seed, which is transported by water or by animal vectors.

9.3.4.1. *Pontederia cordata* L. grows in tidal or nontidal, fresh to brackish waters (e.g., salinity: 1‰), of baygalls, bayheads, bayous, bays (shallow), borrow pits (roadside), bottomlands (forested), coves, culverts, depressions (roadside, seasonally flooded), ditches (drainage, dredged, roadside), estuaries, fens, flatwoods (pine, wet), floodplains, glades, hammocks (hydric, mesic, sandy), "houses" (prairie), marshes (basin, brackish, depression, flatwoods, floodplain, freshwater, interdunal, intertidal, oligohaline, salt, tidal, wind-tide), mudflats (drying), pocosins (roadside), pools (marshy, shallow), prairies (wet), puddles, roadsides (wet), sandbars, savannas (low, pine), shores (lake), sloughs (bottomland, oxbow), spillways, springs, swales (drainage), swamps (basin, blackwater, cypress, floodplain, roadside), and the margins (marshy) of bogs, canals (drainage), channels (boat, drainage), lakes, ponds (artificial, beaver, glade, intermittent, mill, roadside, sediment retention, sinkhole), rivers, and streams (blackwater) at elevations to 518 m. Exposures range from full sunlight to partial shade. The substrates have been described as clay (Schriever), grossarenic endoaqualfs, loam, marl, muck, mud (sandy), peaty muck, sand, sandy clay, sandy humus, sandy muck, sandy peat, silt, silty clay, and silty loam (cherty). The plants inhabit clear to turbid (Secchi depth: 0.4–3.6 m; $\bar{x} = 1.7$ m; $\tilde{x} = 1.4$ m), shallow to deeper (0.1–1.9 m; $\bar{x} = 0.9$ m) standing waters that span a broad range of environmental conditions (pH: 4.9–9.6; $\bar{x} = 6.8$–7.3; $\tilde{x} = 6.9$; alkalinity [as mg l^{-1} CaCO$_3$]: 1.2–280; $\bar{x} = 20.3$–45.0; $\tilde{x} = 11.1$; conductance [µS cm^{-1} @ 25°C]: 10–311; $\bar{x} = 90$–161; $\tilde{x} = 116$; total P [µg l^{-1}]: 10–92; $\bar{x} = 44$; $\tilde{x} = 20$; total N [µg l^{-1}]: 310–1,780; $\bar{x} = 870$; $\tilde{x} = 680$; Ca [mg l^{-1}]: 1.5–50.8; $\bar{x} = 19.2$; $\tilde{x} = 10.2$; Mg [mg l^{-1}]:

1.0–23.7; $\bar{x} = 9.3$; $\tilde{x} = 5.2$; Na [mg l⁻¹]: 3.6–20.9; $\bar{x} = 14.2$; $\tilde{x} = 8.5$; K [mg l⁻¹]: 0.2–8.0; $\bar{x} = 2.9$; $\tilde{x} = 2.0$; SO₄²⁻ [mg l⁻¹]: 4.2–39.7; $\bar{x} = 16.4$; $\tilde{x} = 11.0$; Cl⁻ [mg l⁻¹]: 7.0–39.4; $\bar{x} = 25.3$; $\tilde{x} = 15.8$; Fe [mg l⁻¹]: 0.0–0.4; $\bar{x} = 0.2$; $\tilde{x} = 0.1$; Si [mg l⁻¹]: 0.1–3.8; $\bar{x} = 1.5$; $\tilde{x} = 0.6$). Flowering has been observed from February to December, and fruiting from April to December. The flowers remain open for only one day. They are tristylous and are strongly to weakly self-incompatible depending on the morphs: short-styled morphs (18% self-compatibility); long-styled morphs (26% self-compatibility); mid-styled morphs (75% self-compatibility). Mean pollen production rates also vary, being highest in short stamens (20,861–20,352 grains), lower in mid-length stamens (4,589–7,995 grains), and lowest in long stamens (3,874–4,591grains). Populations typically contain all three morphs, which exhibit higher seed set from compatible pollinations (between same length stamen and style morphs), due to more rapid pollen-tube growth (compared to self-pollinated flowers). Style-length frequencies typically are anisoplethic (unequal) within populations, and are strongly influenced by the founding genotypes. Average populational style-length frequencies (long:mid:short) have been approximated as 0.25:0.35:0.40. The plants primarily are outcrossed by pollinating insects (Insecta – see *Use by wildlife*). These agents can remove a large proportion of a flower's pollen (e.g., 45%) during a single visit, with up to 95% of the total pollen load being harvested from a plant by midday. Perianth color can vary from blue or white. Genetic analyses have indicated that the polymorphism is controlled by two alleles at a single locus, with blue color completely dominant to white. The seeds are dormant when shed and require at least 6–8 weeks of cold stratification (4°C) to induce germination (~16 weeks being optimal). The best germination is achieved under flooded conditions. For adequately stratified seeds, higher germination (25–28%) has occurred under alternating day/night (12hr:12hr) incubation temperature regimes of 10°C/20°C or 15°C/20°C. Seeds also have germinated well (84–94%) at constant higher temperatures (in tap water at 20°C–30°C) after being air-dried and stored in the dark at 25°C for 6 months. Lower germination rates (43%) have been observed for seeds stored at 4°C for 6 months. Germination is extremely low in all cases for buried seeds. Those derived from the different morphs have exhibited similar germination rates (35–41%). The dormancy characteristics are thought to ensure that the seeds germinate primarily on exposed banks, where competition is minimal during their establishment phase. A seed bank develops (duration of at least several years), with densities of 4–12 emerging seedlings m⁻² reported. The seeds are dispersed by water or by their adhesion to animals. Seedlings typically establish from May to June, sometimes within floating vegetation mats. Those cultured in sand enriched with Osmocote fertilizer exhibited highly enhanced growth. The plants perennate from short, thick rhizomes. The rhizomes are not dormant (do not require cold stratification in order to continue growth), which can prolong the growing season as long as the substrates are not frozen. Average total maximum biomass (dry weight) is high, and has been estimated (after 150 days) at 1,212 g m⁻²

(aboveground: 524 g m²; belowground: 688 g m²). Corrected for carryover due to the previous season's existing root biomass, the single-season value is estimated at 1,049 g m².

Reported associates: *Acer rubrum, Acmella, Acorus calamus, Agrostis perennans, Alisma subcordatum, Alisma triviale, Alnus serrulata, Alternanthera philoxeroides, Amaranthus cannabinus, Apios americana, Asclepias incarnata, Azolla, Bacopa caroliniana, Bacopa monnieri, Betula nigra, Betula pumila, Bidens connatus, Bidens frondosus, Bidens laevis, Bidens mitis, Bidens trichospermus, Boehmeria cylindrica, Bolboschoenus fluviatilis, Bolboschoenus robustus, Brasenia schreberi, Cabomba, Calamagrostis canadensis, Calla palustris, Callitriche, Cardamine pensylvanica, Carex alata, Carex atlantica, Carex crus-corvi, Carex frankii, Carex gigantea, Carex hyalinolepis, Carex hystericina, Carex scoparia, Carex stipata, Carex stricta, Centella asiatica, Cephalanthus occidentalis, Ceratophyllum demersum, Cicuta maculata, Cladium jamaicense, Cladium mariscoides, Clethra alnifolia, Coleataenia longifolia, Commelina caroliniana, Commelina virginica, Conium maculatum, Coptis trifolia, Coreopsis nudata, Cornus amomum, Cornus foemina, Cuscuta gronovii, Cyperus esculentus, Cyperus haspan, Cyperus haspan, Cyperus pseudovegetus, Cyperus strigosus, Decodon verticillatus, Diodia virginiana, Dulichium arundinaceum, Echinochloa crus-galli, Echinochloa walteri, Egeria densa, Eichhornia crassipes, Eleocharis baldwinii, Eleocharis elongata, Eleocharis equisetoides, Eleocharis fallax, Eleocharis flavescens, Eleocharis obtusa, Eleocharis palustris, Eleocharis quadrangulata, Eleocharis robbinsii, Eleocharis rostellata, Elodea nuttallii, Equisetum arvense, Equisetum fluviatile, Eriocaulon aquaticum, Eriocaulon compressum, Eryngium aquaticum, Eubotrys racemosa, Fraxinus profunda, Galium obtusum, Glyceria acutiflora, Gratiola aurea, Gratiola ramosa, Habenaria repens, Helenium autumnale, Heteranthera reniformis, Hibiscus moscheutos, Hydrilla verticillata, Hydrocotyle umbellata, Hydrocotyle verticillata, Hydrolea ovata, Hymenocallis, Hypericum fasciculatum, Hyptis alata, Ilex cassine, Ilex cassine, Ilex myrtifolia, Impatiens capensis, Iris tridentata, Iris versicolor, Itea virginica, Juncus acuminatus, Juncus canadensis, Juncus effusus, Juncus marginatus, Juncus megacephalus, Juncus roemerianus, Justicia americana, Kosteletzkya pentacarpos, Lachnanthes caroliniana, Larix laricina, Leersia oryzoides, Lemna, Limnobium spongia, Lindernia grandiflora, Liquidambar styraciflua, Lobelia cardinalis, Ludwigia alata, Ludwigia decurrens, Ludwigia glandulosa, Ludwigia lanceolata, Ludwigia palustris, Ludwigia pilosa, Lythrum salicaria, Magnolia, Marsilea, Mayaca fluviatilis, Mikania scandens, Mimulus ringens, Mitreola petiolata, Murdannia keisak, Myrica cerifera, Myriophyllum sibiricum, Myriophyllum spicatum, Myriophyllum verticillatum, Najas flexilis, Nelumbo lutea, Nuphar advena, Nuphar variegata, Nymphaea odorata, Nymphoides aquatica, Nyssa aquatica, Nyssa biflora, Nyssa sylvatica, Onoclea sensibilis, Orontium aquaticum, Osmunda regalis, Oxalis stricta, Panicum hemitomon, Panicum verrucosum, Paspalum, Peltandra virginica, Persea borbonia,*

Persea palustris, Persicaria amphibia, Persicaria arifolia, Persicaria glabra, Persicaria hydropiper, Persicaria hydropiperoides, Persicaria punctata, Persicaria sagittata, Phalaris arundinacea, Phanopyrum gymnocarpon, Phragmites australis, Phyla lanceolata, Physostegia leptophylla, Picea mariana, Pilea pumila, Pinus elliottii, Pinus palustris, Pistia stratiotes, Plantago cordata, Pluchea baccharis, Pluchea odorata, Potamogeton berchtoldii, Potamogeton natans, Potamogeton nodosus, Potamogeton perfoliatus, Potamogeton spirillus, Potamogeton vaseyi, Proserpinaca palustris, Proserpinaca pectinata, Ptilimnium capillaceum, Quercus laevis, Quercus virginiana, Rhamnus frangula, Rhexia mariana, Rhynchospora cephalantha, Rhynchospora corniculata, Rhynchospora inundata, Rhynchospora nitens, Rorippa palustris, Rosa palustris, Sabal palmetto, Sacciolepis striata, Sagittaria calycina, Sagittaria graminea, Sagittaria lancifolia, Sagittaria latifolia, Sagittaria montevidensis, Sagittaria rigida, Sagittaria subulata, Salix exigua, Salix nigra, Salix sericea, Salvinia minima, Sarracenia flava, Saururus cernuus, Schoenoplectus acutus, Schoenoplectus americanus, Schoenoplectus californicus, Schoenoplectus pungens, Schoenoplectus subterminalis, Schoenoplectus tabernaemontani, Scirpus cyperinus, Scutellaria lateriflora, Serenoa repens, Sesbania herbacea, Sium suave, Smilax smallii, Sparganium americanum, Sparganium androcladum, Sparganium emersum, Sparganium eurycarpum, Spartina alterniflora, Spartina cynosuroides, Sphagnum, Spirodela polyrrhiza, Stillingia aquatica, Symphyotrichum firmum, Symphyotrichum novi-belgii, Symphyotrichum subulatum, Taxodium ascendens, Taxodium distichum, Thalia geniculata, Thalictrum pubescens, Thelypteris kunthii, Torreyochloa pallida, Toxicodendron vernix, Triadenum virginicum, Triadica sebifera, Typha angustifolia, Typha domingensis, Typha latifolia, Utricularia intermedia, Utricularia macrorhiza, Utricularia purpurea, Vallisneria americana, Verbena, Viola, Strophostyles umbellata, Woodwardia areolata, Woodwardia virginica, Xyris fimbriata, Xyris jupicai, Xyris smalliana, Zizania aquatica, Zizania palustris, Zizaniopsis miliacea.

Use by wildlife: The foliage of *Pontederia cordata* (8.2–10% protein) is eaten by deer (Mammalia: Cervidae: *Odocoileus virginianus*), moose (Mammalia: Cervidae: *Alces alces*), muskrats (Mammalia: Cricetidae: *Ondatra zibethicus*), nutria (Mammalia: Myocastoridae: *Myocastor coypus*), and by several herbivorous insects (Insecta), including aphids (Hemiptera: Aphididae: *Rhopalosiphum nymphaeae*), beetles (Coleoptera: Erirhinidae: *Onychylis nigrirostris*), flies (Diptera: Ephydridae: *Hydrellia pontederiae*), larval moths (Lepidoptera: Noctuidae: *Bellura densa*, *B. gortynoides*, *B. obliqua*, *Pyrrhia exprimens*), and locusts (Orthoptera: Acrididae: *Leptysma marginicollis*, *Paroxya atlantica*, *P. clavuliger*, *Stenacris vitreipennis*; Romaleidae: *Romalea microptera*). The foliage is also used by the invasive brown hoplo (Vertebrata: Teleostei: Callichthyidae: *Hoplosternum littorale*) for nest construction. Larval water beetles (Coleoptera: Chrysomelidae: *Donacia subtilis*) build their cocoons on the roots and some deer flies (Diptera: Tabanidae: *Chrysops*)

oviposit regularly on the plants. The seeds (8.94% protein; 6.98% fat) provide food for soras (Aves: Rallidae: *Porzana carolina*) and various waterfowl (Anatidae), including the American wigeon (*Anas americana*), black duck (*Anas rubripes*), gadwall (*Anas strepera*), green-winged teal (*Anas crecca*), lesser scaup (*Aythya affinis*), mallard (*Anas platyrhynchos*), northern pintail (*Anas acuta*), ring-necked duck (*Aythya collaris*), and wood duck (*Aix sponsa*). The flowers are visited by numerous insects (Insecta), including bees (Hymenoptera: Apidae: *Apis mellifera*, *Bombus affinis*, *B. bimaculatus*, *B. borealis*, *B. fervidus*, *B. griseocollis*, *B. impatiens*, *B. pensylvanicus*, *B. perplexus*, *B. ternarius*, *B. terricola*, *B. vagans*, *Euglossa viridissima*, *Melissodes apicata*, *Peponapis pruinosa*, *Xylocopa virginica*, *X. micans*; Halictidae: *Agapostemon sericeus*, *A. splendens*, *Dufourea novaeangliae* (which forages exclusively on *P. cordata* as a pollen source), *Lasioglossum*; Megachilidae: *Megachile mendica*), beetles (Coleoptera: Chrysomelidae: *Disonycha pennsylvanica*, *Trirhabda tomentosa*; Coccinellidae: *Coleomegilla maculata*; Scarabaeidae: *Strigoderma arbicola*), butterflies (Insecta: Lepidoptera: Hesperiidae: *Ancyloxypha numitor*, *Epargyreus clarus*, *Polites peckius*; Nymphalidae: *Argynnis cybele*, *Euptoieta claudia*, *Limenitis archippus*; Papilionidae: *Papilio glaucus*, *P. polyxenes*; Pieridae: *Colias philodice*, *Pieris rapae*; Sphingidae: *Hemaris diffinis*, *H. gracilis*, *H. thysbe*), and flies (Diptera: Syrphidae; *Eristalis flavipes*, *Helophilus chrysostomus*; Tabanidae: *Tabanus calens*), which forage for pollen or nectar. The nectar has a mean energy content of 0.63–0.82 J (joules). Hummingbirds (Aves: Trochilidae) are occasional floral visitors. Water in which *P. cordata* has been cultured reportedly and rhizome extracts inhibit some algae (Cyanobacteria: Chroococcaceae: *Microcystis aeruginosa*; Chlorophyta; Scenedesmaceae: *Scenedesmus acutus*) by up to 98%. Other Cyanobacteria (Oscillatoriaceae: *Lyngbya wollei*) have been inhibited by nearby growths of *P. cordata*, due to reduced light and nutrient availability.

Economic importance: food: Edible parts of *Pontederia cordata* include the young, unfurled leaves (diced for addition to salads or boiled for 10 min and served with butter) and starchy fruits (eaten raw, dried, or roasted and ground into flour); **medicinal:** The Malecite people used an infusion made from *Pontederia cordata* as a contraceptive. The Montagnais tribe brewed a drink from the plant as a curative for general illnesses; **cultivation:** *Pontederia cordata* is widely planted as an ornamental water garden specimen. It has been designated an Award of Garden Merit plant by the Royal Horticultural Society. Cultivars include 'White Pike', 'Blue Spires', 'Pink Pons', and 'Sunsplash'. The plants also have been cultured intensively for lake restoration projects; **misc. products:** *Pontederia cordata* has been used in bioremediation systems for treating polluted urban river water. Their high rate of phosphorous uptake (66 mg P m^{-2} day^{-1}) has encouraged their use in P removal from dairy wastewater; **weeds:** *Pontederia cordata* occasionally is reported as reaching "nuisance" proportions; **nonindigenous species:** Although indigenous to eastern North America, *Pontederia cordata* has escaped as a water garden ornamental in western North America.

Systematics: *Pontederia* species have consistently resolved as a clade in various phylogenetic analyses. However, problems arise at the intergeneric level, especially with respect to *Eichhornia*, which evidently is not monophyletic (see previous discussion of *Eichhornia*). Some *Eichhornia* species (e.g., *E. azurea*) resemble *Pontederia* morphologically, associate consistently with the core *Pontederia* clade in phylogenetic evaluations (e.g., Figure 3.5), and share floral characteristics (e.g., tristyly and self-incompatibility) with those species. In such cases, a taxonomic merger with *Eichhornia* seems reasonable, and in some cases it already has been proposed by past authors (i.e., *P. azurea* Sw.). On the other hand, recent recommendations to simply combine all the *Eichhornia* species (as well as those of *Monochoria*) with *Pontederia* have unnecessarily abandoned the morphologically distinct *Monochoria* and have added an additional seven new nomenclatural combinations in an attempt to establish a classification that is "more taxonomically stable." Moreover, such recommendations have been proposed even though several putative taxa (e.g., *P. parviflora*, *P. subovata*, *P. triflora*) have not yet been evaluated in phylogenetic studies. But above all, it should be emphasized that phylogenetic relationships in Pontederiaceae still are far from being settled, with numerous conflicts being evident among the trees generated for different subsets of taxa using different data combinations. For instance, if a fully comprehensive survey of the family would consistently place the distinctive *Eichhornia meyeri* as the sister group to all other species (as some analyses have indicated), would *Pontederia* then be expanded to include even *Heteranthera*? Given these issues, it is advisable to defer any taxonomic upheaval of *Pontederia* until more comprehensive phylogenetic analyses have been undertaken and other options are considered that would minimize nomenclatural confusion. Consequently, the narrower, more traditional circumscription of the genus has been followed here. The basic chromosome number of *Pontederia* is $x=8$; *P. cordata* ($2n=16$) is a diploid. The complete chloroplast genome of *P. cordata* has recently been sequenced. No natural hybrids involving *P. cordata* have been reported.

Distribution: *Pontederia cordata* is common throughout the eastern half of the United States and adjacent Canada, extending southward into South America; it is disjunct sporadically in British Columbia, California, and Oregon, where it escapes occasionally from cultivation.

References: Anderson & Barrett, 1986; Barrett et al., 1983; Beal, 1977; Bembower, 1911; Bowden, 1945; Boyd & McGinty, 1981; Capers & Les, 2005; Cook, 1996; Cottam, 1939; Cypert, 1972a; DeBusk et al., 1995; Dodds, 1960; Doyle & Smart, 1998; Eickwort et al., 1986; Fassett, 1957; Fernald & Kinsey, 1943; Fleming, 1970; Foster et al., 1973; Gettys & Wofford, 2007; Gettys & Dumroese, 2009; Graham et al., 1998; Gu et al., 2015; Harder & Barrett, 1992; 1993; Harms & Grodowitz, 2009; Hazen, 1918; Heisey & Damman, 1982; Hellquist & Crow, 1982; Hoffman, 1940; Horn, 2002; Hoyer et al., 1996; Knapton & Pauls, 1994; Kohn et al., 1996; Leck & Simpson, 1993; Lowden, 1973;

Ma & Liang, 2019; Mack, 1991; Morgan & Barrett, 1988; Ness et al., 2011; Nichols, 1999; Nico & Muench, 2004; Orndorff, 1966; Orth & Waddington, 1997; Patton & Judd, 1988; Pellegrini et al., 2018; Peterson, 1977; Price & Barrett, 1982; 1984; Qian et al., 2018; 2019; Self et al., 1975; Skov & Wiley, 2005; Spinner & Bishop, 1950; Sutton, 1991; Takos, 1947; Van der Valk et al., 2009; Webster, 1964; Wetzel et al., 2001; Whigham & Simpson, 1982; Wilsey et al., 1991; Wolfe & Barrett, 1987; 1988; 1989.

ORDER 10: POALES [14]

As currently circumscribed to contain 14 families and roughly 20,000 species, the Poales represent a clade whose monophyly has been demonstrated by various molecular (e.g., Figures 2.1, 3.2, and 3.6; McKain et al., 2016b; Hochbach et al., 2018) and morphological data (Judd et al., 2016). The group shares in common the presence of epidermal silica bodies, strongly branched styles, and the absence of raphides crystals (Judd et al., 2016). Nearly all the diversity is concentrated within three large families. Poaceae contain over half (~12,000) of the currently recognized species, with Cyperaceae (~5,500 species) and Bromeliaceae (~3,475 species) representing the other most speciose groups (Hochbach et al., 2018). Many of the species are adapted for wind pollination (anemophily) but some (Bromeliaceae) have retained floral nectaries and are pollinated by insects or other animals. Biotic pollination also occurs sporadically among several other groups (e.g., Eriocaulaceae), where nectar (and sometimes a floral odor) is produced not by nectaries, but by stigmatic appendices.

Wetland plants occur commonly in the order, with half of all the families containing OBL indicators (Figure 3.6). However, several of these families (Cyperaceae, Juncaceae, and Poaceae) have been omitted here because of various logistic reasons (see above comments under *Introduction*). Four families are given further consideration:

10.1. **Eriocaulaceae** Martinov
10.2. **Mayacaceae** Kunth
10.3. **Typhaceae** Juss.
10.4. **Xyridaceae** C. Agardh

Family 10.1. Eriocaulaceae [10]

Eriocaulaceae contain about 1,175 species of annual or perennial, herbaceous, graminoid plants (Stützel, 1998; Judd et al., 2016). The monophyly of the family has been confirmed by its unique combination of morphological features (e.g. spiaperturate pollen) and by molecular phylogenetic analyses (e.g., Figure 3.6), which usually associate the family as the sister group of Xyridaceae (Coan et al., 2010). Many of the species are emergent or submersed wetland plants with grasslike foliage arranged in a basal rosette. A few submersed species have long, vittate shoots (to 40 cm) bearing cauline, capillary to filiform leaves (Ansari & Balakrishnan, 1994; Cook, 1996; Stützel, 1998). The bases of some leaves develop into

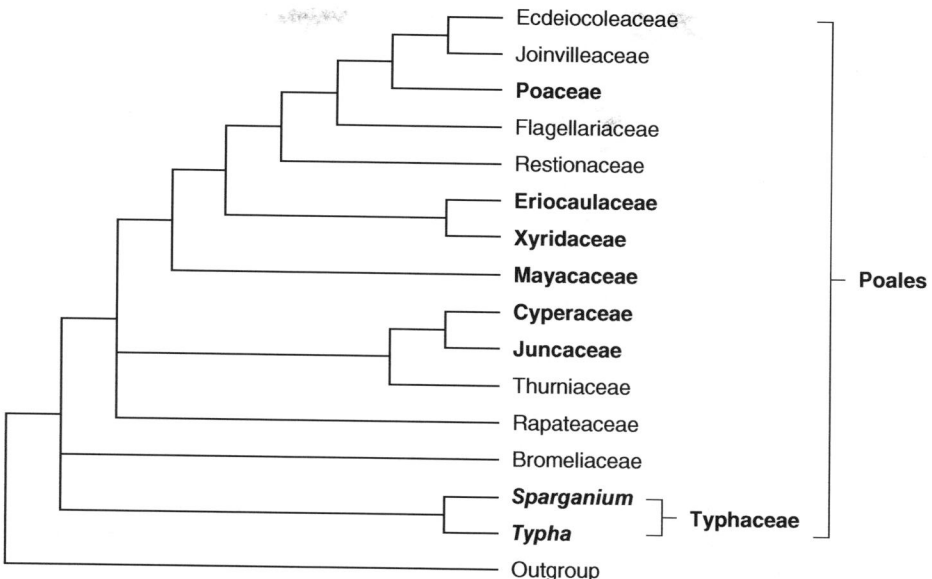

FIGURE 3.6 Analyses of combined nuclear and plastid DNA sequence data indicate the overall phylogenetic relationships within Poales (simplified from Hochbach et al., 2018). The topology represented here is largely consistent with the results of many other analyses, although the precise placements of Eriocaulaceae, Mayacaceae, and Xyridaceae remain somewhat contentious (see discussion in Hochbach et al., 2018). Taxa containing OBL wetland indicators (highlighted in bold) occur commonly throughout the order.

water-storing "cisterns," which are thought to provide protection from periodic episodes of fire (Stützel, 1998). The distinct inflorescence resembles a "hatpin," in which a dense, bracteate head of 10–1,000 minute flowers terminates a thin, elongate peduncle; depending on the species, a single plant can produce anywhere from 1 to 1,000 such inflorescences. The small flowers are hypogynous, unisexual (rarely hermaphroditic), and nearly always are borne in a monoecious configuration. They are entomophilous and pollinated mainly by flies (Insecta: Diptera) or by beetles (Coleoptera). Outcrossing occurs regularly as a consequence of temporal differences in the expression of the male and female floral phases within a head, although geitonogamous self-pollination is also possible during briefly overlapping sexual phases. The fruits are few-seeded, loculicidal capsules. Various dispersal mechanisms include the abiotic transport of the entire flower (including bract) by water or wind, the forceful ejection (up to 2 m) of female flowers from the inflorescence, or by transport of the entire head as the functional diaspore (Stützel, 1998).

Although fewer than 7% of Eriocaulaceae species have been evaluated simultaneously in allegedly "comprehensive" phylogenetic studies, several modifications to generic circumscriptions have been suggested, based primarily on the results of DNA sequence analysis. Most notably, these have included the merger of *Lachnocaulon* and several other genera with *Paepalanthus* and the splitting of *Syngonanthus* into two genera (Figure 3.7). However, given the preliminary amount of phylogenetic evidence available, it seems unwise to propose such major alterations to the taxonomy of the family pending further evaluation. Moreover, even if the topologies depicted in preliminary analyses (e.g., Figure 3.7) are later demonstrated to be accurate, there are other taxonomic alternatives to the widespread inclusion

of genera within the large genus *Paepalanthus*. One alternative would be to divide *Paepalanthus* into two genera, which would preserve the integrity of *Lachnocaulon* and *Tonina*. Because phylogenetic results from DNA analysis (Echternacht et al., 2014) support either the split of *Syngonanthus* into two genera or its retention as one genus, the inevitable nomenclatural confusion should be given adequate consideration before determining whether the former approach is truly necessary.

Most Eriocaulaceae occur in tropical or subtropical climates from humid lowland sites to alpine elevations, with relatively few extending into the temperate zone. They occur often in wetlands on poor, sandy substrates but also can colonize much drier, rocky alpine habitats (Stützel, 1998). North American plants typically inhabit wet, acidic substrates exposed to full sunlight (Kral, 2000). The plants produce essentially no specialized perennating organs (except tubers in one Asian species), but survive unfavorable conditions as whole plants (Ansari & Balakrishnan, 1994; Stützel, 1998). The family is of little economic importance with a few horticultural oddities cultivated in *Eriocaulon* and *Syngonanthus*. Some species are dried and sold commercially as ornamental flowers (Stützel, 1998).

Obligate (OBL) wetland indicators occur in two North American genera:

10.1.1. ***Eriocaulon*** L.
10.1.2. ***Lachnocaulon*** Kunth

10.1.1. Eriocaulon
Button-rods, hatpins, pipewort
Etymology: from the Greek *erion caulos* ("wool stem") for the pubescent scape base of the original species

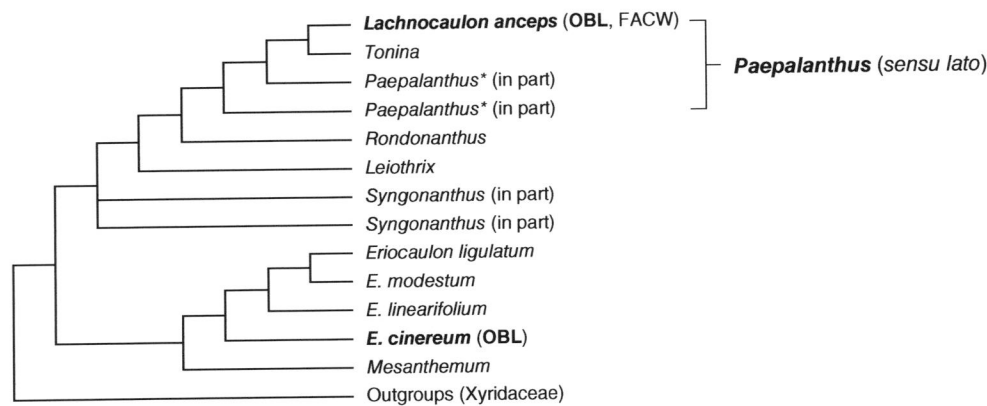

FIGURE 3.7 Intergeneric relationships in Eriocaulaceae as indicated by phylogenetic analysis of combined nuclear and plastid DNA sequence data (simplified from de Andrade et al., 2010). Some genera like the large *Paepalanthus* (~485 spp.) are not monophyletic (asterisked) unless subdivided or expanded even further (e.g., "*sensu lato*" shown) to accommodate other genera like *Lachnocaulon* and *Tonina*. Subdivision of *Syngonanthus* also has been proposed (e.g., Echternacht et al., 2014), but it also could be maintained as one genus without any phylogenetic conflict. The wetland indicator status is provided for rated North American taxa; those containing OBL indicators are highlighted in bold.

Synonyms: *Cespa*; *Leucocephala*; *Nasmythia*; *Randalia*; *Sphaerochloa*; *Symphachne*

Distribution: global: pantropical (rarely temperate); **North America:** eastern (disjunct in California)

Diversity: global: 400 species; **North America:** 12 species

Indicators (USA): obligate wetland (OBL): *Eriocaulon aquaticum, E. cinereum, E. compressum, E. decangulare, E. koernickianum, E. lineare, E. microcephalum, E. nigrobracteatum, E. parkeri, E. ravenelii, E. texense*

Habitat: brackish (coastal), freshwater; freshwater (tidal); lacustrine, palustrine, riverine; **pH:** 3.9–8.5; **depth:** to 5 m; **life-form(s):** emergent or submersed (rosulate) herb

Key morphology: stems rosulate, compressed, sometimes dividing (branching) by short, lateral offshoots; roots distinctly septate, spongy or thickened; leaves spiral in a rosette, the blades (to 50 cm) linear, linear-attenuate, or triangular-acuminate, lingulate, tapering gradually or abruptly from a conspicuously lacunate base; heads (to 20 mm) solitary, many-flowered, globose to hemispherical, subtended by chaffy to scarious involucral bracts (to 4 mm), bearing receptacular bracts (to 4 mm), heads borne terminally on 1-few filiform to linear scapes (to 110 cm), scapes with 3–8 ribs, enclosed basally by a tubular sheath; flowers unisexual (heads monoecious), 2- or 3-merous; ♂ flowers with 2–3 sepals (to 4 mm), 2–3 petals (to 1 mm), and 3–6 stamens, the filaments and petals fused into a campanulate, clavate, cylindric, or flaring androphore, the anthers 2-locular, black, exserted at anthesis; ♀ flowers with 2–3 sepals (to 3 mm) and 2–3 white to yellowish petals (to 2 mm), the calyx and corolla separated by a stipelike gynophore (to 1 mm), pistil of 2–3 carpels, style 1, with 2–3 branches; seeds (0.5–4.5 mm) ovoid or broadly ellipsoid, brown or reddish-brown, the surface faintly to conspicuously cancellate (spongelike), papillate, reticulate, or rugulose

Life history: duration: annual (fruits/seeds); biennial, perennial (dormant apices, whole plants); **asexual reproduction:** dichotomous branching of apical meristem; lateral offsets, rhizomes (short), stolons; **pollination:** insect, self; **sexual**

condition: monoecious; **fruit:** capsules (common); **local dispersal:** flowers, seeds (water); **long-distance dispersal:** seeds (animals)

Imperilment: 1. *Eriocaulon aquaticum* [G5]; SH (DC); S1 (IN, LB, MB, MD, OH, PE, VA); S2 (DE, NC); **2.** *E. cinereum* [GNR]; **3.** *E. compressum* [G5] S2 (DE, MD); S3 (NC); **4.** *E. decangulare* [G5]; SX (DC, PA); S1 (AR, DE, MD, OK, TN); S2 (VA); **5.** *E. koernickianum* [G2]; S1 (GA, TX); S2 (AR, OK); **6.** *E. lineare* [G4]; S2 (AL); **7.** *E. microcephalum* [G5]; **8.** *E. nigrobracteatum* [G1]; S1 (FL); **9.** *E. parkeri* [G3]; SX (NY, PA); SH (DC); S1 (CT, MA, NC); S2 (DE, MD, NB, NJ, VA); S3 (ME, QC); **10.** *E. ravenelii* [G3/G4]; SX (SC); S3 (FL); **11.** *E. texense* [G4]; S1 NC, SC); S2 (AL, GA, MS, TX); S3 (FL, LA)

Ecology: general: All North American *Eriocaulon* are wetland plants with most of the species (93%) designated as OBL indicators. The taxa reportedly include annuals, biennials, and perennials. However, the life history information for some species sometimes inaccurately represents their true (or at least most usual) duration, as determined in some of the more thoroughly studied species. Most of the species associate with acidic, sandy, freshwater habitats, with a few inhabiting circumneutral or brackish sites. In most cases, the habitats are characterized by fluctuating water levels, due to either seasonal desiccation or tidal influences. Flowering typically takes place on exposed substrates or during periods of lower water levels. Some of the more perennial taxa can persist in deep, permanent waters, though excessive depths make it difficult for the plants to adequately extend their inflorescence above the water surface. The period of floral anthesis usually is long, commencing in March or April and extending through November or December (until the onset of frost). During summer, the flowers begin to open by 5:00–6:00 am. Late in the season (November–December), the number of open flowers declines and their opening may be delayed until after 9:00 am. The peripheral flowers of the inflorescence (mostly ♂) open first, with their time of opening being regulated by temperature. The ♀ florets are concentrated near the center of the head and open gradually afterward. Consequently, the heads are staminate at first, then progressively exhibit a hermaphroditic phase (when ♂ and ♀ florets are open simultaneously), and eventually a pistillate phase. The flowers remain open for 3–8 hr or more, with each head producing flowers for 3–4 weeks. Seed production can range from prolific to being virtually absent, with most seeds though to be produced during the hermaphroditic floral phase. Low seed set often has been attributed to pollinator limitation, which has not been substantiated in some instances. In general, the dense heads of small flowers attract few flying insects (Insecta: Diptera; Coleoptera; Lepidoptera) except as a resting platform. Such mobile insect groups are most likely to be involved in any cross-pollination. However, pollen is scavenged more commonly by sedentary insects like thrips (e.g., Insecta: Thysanoptera: Thripidae: *Frankliniella*), which inhabit the inflorescences; however, these move only among the different flowers on a head and not between plants. In the cases studied, the flowers are self-compatible (among sexes within an individual head), have low pollen:ovule ratios, and commonly undergo geitonogamous pollination. Wind pollination has been suspected, but doubtfully occurs except under very high wind conditions. There is also evidence of agamospermous seed production in several species, and this possibility deserves further evaluation throughout the genus. Persistent seed banks develop frequently. In many cases, the seeds are dispersed while retained within the dried pistillate flowers, which detach (along with the dried bracts) as the heads deteriorate. The dried floral propagules are dispersed locally by the wind (to 2–10 m), or further with heavy winds. They also float for a time (until saturated) and can be dispersed on the water surface. Germination requirements are not commonly reported but appear to vary among the species examined. Some species can reproduce vegetatively by short rhizomes or leafy stolons (most often when submersed). Because *Eriocaulon* species can be difficult to distinguish taxonomically, it is possible that some of the life history information that follows was derived from sources where taxa had been misidentified. However, a conscious effort has been made to minimize such inconsistencies and yet provide a reasonable ecological overview for each species.

10.1.1.1. ***Eriocaulon aquaticum*** (**Hill**) **Druce** is a perennial, which occurs in bogs (*Sphagnum*), ditches, gravel pits, lakes, muskegs, ponds, rivers, sloughs, streams, or along the exposed marginal shores or swales of borrow pits, lakes, and ponds (sinkhole) at elevations to 482 m. This is a characteristic species of the rosulate "isoetid" flora, which is intolerant of turbidity and associated with acidic, clear (Secchi depth: to 18 m), soft water, oligotrophic (total P <1 μg l^{-1}) conditions. The plants occur in exposures that receive full to partial sunlight, but are susceptible to shading from floating species (e.g., *Utricularia inflata*). They take up CO_2 from the sediments, but unlike other isoetids, do not exhibit C_4 CAM photosynthesis. Growth can occur as a completely submersed life-form (in water 0.08–2.6 m deep) or under emergent conditions on exposed, wet shores. The substrates are firm, with

varying organic matter content (2.28–19.21%), and include gravel, muck, mud, peat (acidic), rock, sand, and sandy cobble. Although the plants sometimes have been associated with inorganic substrates, they have exhibited increased biomass (181.8% greater) when grown experimentally on sediments having a higher organic matter content than on less organic sediments. The waters span a wide range of acidity (pH: 4.3–8.5; $\bar{x} = 6.9$) but are characterized by low alkalinity (mg l^{-1} [as $CaCO_3$]: 2.5–55; $\bar{x} = 9.6$), conductivity (<110 μmhos cm^{-1} @ 25°C), and sulfate (<10 mg l^{-1}) levels. Flowering and fruiting occur from late June to October. Pollination and fertilization occur rapidly in the first developed flowers within the first day of their emergence from the water (the scapes can exceed 40 cm). Although the flowers produce some nectar, potential biotic pollinators are observed rarely, which has led some to implicate wind as a likely pollinating agent (but see *Ecology: general* above). Polyembryony has been observed in one instance (three embryos originating from the egg apparatus), which potentially indicates the occurrence of agamospermy (i.e., adventive polyembryony) and deserves further evaluation (see discussion under *E. parkeri* below). The seeds are transported while still within the intact flowers (and bracts), which are dispersed by the wind. An extensive seed bank develops with densities estimated up to 2,000 seeds m^{-2}. The seeds have germinated reasonably well after being subjected to 210 days of cold stratification, followed by incubation at room temperature (approximately a 19°C/15°C day/night temperature regime). The plants are evergreen perennials (overwintering as intact plants), which can develop into dense mats (500–2,500 plants m^{-2}) on the bottom substrate when conditions are favorable. However, they (at least European plants) have proven difficult to keep alive in culture for more than a half year. They can reproduce asexually (especially when submersed) by the division of the apical meristem into lateral shoot offsets, or by more elongate stoloniferous rhizomes. The roots of emergent plants have been found to contain vesicular-arbuscular mycorrhizae. **Reported associates:** *Agalinis purpurea, Agrostis scabra, Bartonia, Bidens cernuus, Bidens tripartitus, Brasenia schreberi, Cabomba caroliniana, Calamagrostis canadensis, Callitriche, Carex comosa, Carex lasiocarpa, Carex lenticularis, Cephalanthus occidentalis, Chamaedaphne calyculata, Chara, Cladium mariscoides, Coreopsis rosea, Cyperus dentatus, Cyperus flavescens, Cyperus squarrosus, Danthonia spicata, Dichanthelium acuminatum, Drosera intermedia, Drosera filiformis, Drosera intermedia, Dulichium arundinaceum, Elatine minima, Eleocharis acicularis, Eleocharis equisetoides, Eleocharis flavescens, Eleocharis melanocarpa, Eleocharis ovata, Eleocharis palustris, Eleocharis robbinsii, Eleocharis tuberculosa, Elodea canadensis, Elodea nuttallii, Equisetum fluviatile, Euthamia graminifolia, Fimbristylis autumnalis, Glossostigma cleistanthum, Glyceria canadensis, Gratiola aurea, Helonias bullata, Heteranthera dubia, Hottonia inflata, Hypericum canadense, Hypericum majus, Hypericum mutilum, Isoetes echinospora, Isoetes lacustris, Isoetes prototypus, Isoetes tuckermanii, Isoetes viridimontana, Isoetes ✕eatonii, Juncus* *canadensis, Juncus dichotomus, Juncus effusus, Juncus filiformis, Juncus greenei, Juncus militaris, Juncus pelocarpus, Leersia oryzoides, Lindernia dubia, Linum striatum, Littorella uniflora, Lobelia dortmanna, Ludwigia palustris, Lycopodiella inundata, Lycopus uniflorus, Lysimachia terrestris, Muhlenbergia uniflora, Myrica gale, Myriophyllum alterniflorum, Myriophyllum farwellii, Myriophyllum heterophyllum, Myriophyllum humile, Myriophyllum spicatum, Myriophyllum tenellum, Najas flexilis, Najas gracillima, Najas minor, Nitella, Nuphar variegata, Nymphaea odorata, Nymphoides cordata, Orontium aquaticum, Panicum verrucosum, Panicum virgatum, Persicaria careyi, Phragmites australis, Pogonia ophioglossoides, Polygala sanguinea, Pontederia cordata, Potamogeton berchtoldii, Potamogeton bicupulatus, Potamogeton confervoides, Potamogeton crispus, Potamogeton epihydrus, Potamogeton gramineus, Potamogeton natans, Potamogeton oakesianus, Potamogeton richardsonii, Potamogeton robbinsii, Potamogeton spirillus, Potamogeton vaseyi, Proserpinaca palustris, Proserpinaca pectinata, Ranunculus flammula, Rhexia virginica, Rhynchospora alba, Rhynchospora capitellata, Rhynchospora fusca, Rhynchospora scirpoides, Sagittaria graminea, Sagittaria latifolia, Sagittaria teres, Sarracenia purpurea, Schoenoplectus pungens, Schoenoplectus smithii, Schoenoplectus subterminalis, Schoenoplectus tabernaemontani, Schoenoplectus torreyi, Sium suave, Sparganium fluctuans, Sphagnum cuspidatum, Sphagnum macrophyllum, Sphagnum pulchrum, Spiraea alba, Spiraea tomentosa, Stachys hyssopifolia, Subularia aquatica, Triadenum fraseri, Triglochin maritima, Typha latifolia, Utricularia cornuta, Utricularia gibba, Utricularia inflata, Utricularia macrorhiza, Utricularia purpurea, Utricularia radiata, Utricularia resupinata, Vaccinium macrocarpon, Vallisneria americana, Viola lanceolata, Xyris difformis, Xyris montana, Xyris smalliana, Xyris torta.*

10.1.1.2. **_Eriocaulon cinereum_ R. Br.** is a nonindigenous annual or perennial (short-lived), which inhabits ditches (drainage), rice fields, and the shores (drawdown) of reservoirs at elevations to 98 m [to 2,300 m in Asia]. The plants inhabit shallow temporary waters over substrates of muck, sand, or sandy clay. In Australia, they occur in seepage areas, streams, creek beds, and along the margins of lagoons, pools, and swamps (seasonal). There they typically are found on sand and rarely are observed to grow on heavy soil. Flowering and fruiting (North America) have been documented from August to November and commence as the waters recede. Floral observations in Asia have failed to detect any potential pollinators (see also *Ecology: general* above). The seeds are dispersed by water and generally germinate under water. The plants are highly tolerant of disturbance and have been categorized as ruderal. **Reported associates (North America):** *Ammannia auriculata, Bacopa repens, Dopatrium junceum, Heteranthera limosa, Oryza sativa, Rotala ramosior, Sagittaria guayanensis, Sphenoclea zeylanica.*

10.1.1.3. **_Eriocaulon compressum_ Lam.** is a perennial, which grows in bogs (acid, cranberry, cypress, hillside, flatwoods, quaking, seepage, *Sphagnum*), borrow pits (seasonal),

Carolina bays, coves, depressions (boggy, wet), ditches (roadside), flats (pine, wet), flatwoods (low, pine, wet), "houses" (prairie), marshes (basin, depression, ephemeral, pineland), meadows (wet), mudflats (bog margin), pine barrens, ponds (flatwoods, pineland, shallow), pools, prairies (wet), rights of way (powerline), roadsides (low, wet), savannas (pine, pitcher plant, wet), scrub, seeps (flatwoods), slopes (seepage), sloughs (wet), swales (drawdown), swamps (cypress, pine barren), and along the margins (seasonal) of baygalls, canals, cypress domes, lakes, ponds, rivers, and spillways at elevations to 68 m. Most occurrences are found in exposures of full sunlight. The sites are characterized by ephemeral, standing, shallow waters from 15 to 20 cm deep; however, sterile plants can persist as a submersed life-form at depths of up to 2 m. The substrates are acidic (pH: 3.9–5.5) and have been categorized as Atmore (plinthic paleaquults), Basinger (spodic psammaquents), clay loam, Cuthbert (typic hapludults), Everdale, Floridana (arenic argiaquolls), Leefield-Stilson, loam, loamy sand, muck, mucky clay, mud, Okeechobee (hemic medisaprists), peat, peaty muck, Placid (typic humaquepts), Plummer (grossarenic paleaquults), Pomona (ultic haplaquods), Pottsburg (grossarenic haplaquods), Rutledge-Pamlico (typic humaquepts), sand (Leon, Waveland), sandy loam (Bayou, Rains), sandy peat (sphagnous), Smithton, Vero (alfic haplaquods), and Waller. The few sites analyzed indicate extremely low alkalinity (0–1.5 mg l^{-1}). The plants are described as pioneers of pure sand substrates. Flowering and fruiting occur throughout the year (January–December). Few details are available on the reproductive ecology (see *Ecology: general* above), but the flowers are visited by butterflies (Insecta: Lepidoptera), which could function as pollinators. An extensive seed bank (up to 1,200 germinants m^{-2}) can develop. The seeds germinate much better under dry or alternating dry/wet conditions relative to permanently inundated conditions, but specific germination requirements have not been detailed. Mean densities of established seedlings can range from 80 to 90 m^{-2} under dry or alternating dry/wet conditions. Seedlings will not establish in permanently wet sites. The plants can reproduce vegetatively by short rhizomes or leafy stolons. Maximum mean plant biomass (dry weight) has been estimated at 50 g m^{-2} (aboveground) and 41 g m^{-2} (belowground). Most of the belowground biomass (99%) occurs within the upper 10 cm of substrate. The aboveground biomass increases as water levels decline. The plants are intolerant of agricultural pollution and do not recover well following fires. **Reported associates:** *Aletris lutea, Amphicarpum muhlenbergianum, Andropogon virginicus, Aristida palustris, Aristida purpurascens, Aristida stricta, Arundinaria gigantea, Asclepias connivens, Bacopa caroliniana, Bartonia verna, Bejaria, Bidens mitis, Bigelowia nudata, Boltonia asteroides, Calamovilfa curtissii, Calopogon barbatus, Calopogon pallidus, Carex glaucescens, Carex striata, Carex verrucosa, Carphephorus pseudoliatris, Centella asiatica, Cephalanthus occidentalis, Chamaecyparis thyoides, Chaptalia tomentosa, Cleistesiopsis divaricata, Cliftonia monophylla, Coleataenia longifolia, Coleataenia tenera, Coreopsis nudata, Ctenium aromaticum, Cyperus ochraceus, Cyrilla racemiflora, Decodon* *verticillatus, Dichanthelium acuminatum, Dichanthelium clandestinum, Dichanthelium dichotomum, Dichanthelium hirstii, Dichanthelium nudicaule, Dichanthelium sphaerocarpon, Drosera brevifolia, Drosera capillaris, Drosera tracyi, Dulichium arundinaceum, Eleocharis baldwinii, Eleocharis elongata, Eleocharis equisetoides, Eleocharis flavescens, Eleocharis interstincta, Eleocharis quadrangulata, Eleocharis robbinsii, Eleocharis tuberculosa, Erigeron vernus, Eriocaulon decangulare, Eriocaulon nigrobracteatum, Eriocaulon texense, Eubotrys racemosa, Eupatorium leucolepis, Euphorbia inundata, Galium pilosum, Gaylussacia mosieri, Gratiola brevifolia, Harperocallis flava, Helenium vernale, Helianthus radula, Houstonia pusilla, Hydrocotyle umbellata, Hydrolea ovata, Hypericum brachyphyllum, Hypericum denticulatum, Hypericum edisonianum, Hypericum fasciculatum, Hypericum galioides, Hypericum gentianoides, Hypericum microsepalum, Ilex glabra, Ilex myrtifolia, Iris virginica, Itea virginica, Juncus dichotomus, Juncus polycephalus, Lachnanthes caroliniana, Lachnocaulon anceps, Liquidambar styraciflua, Lobelia boykinii, Lobelia brevifolia, Lophiola aurea, Ludwigia lanceolata, Ludwigia sphaerocarpa, Lycopodiella alopecuroides, Lycopodiella caroliniana, Lyonia lucida, Lythrum alatum, Magnolia grandiflora, Magnolia virginiana, Marshallia, Mayaca fluviatilis, Mnesithea rugosa, Myrica caroliniensis, Myrica cerifera, Myrica inodora, Nolina atopocarpa, Nymphaea odorata, Nymphoides aquatica, Nyssa biflora, Nyssa sylvatica, Orontium aquaticum, Osmunda regalis, Panicum hemitomon, Peltandra sagittifolia, Persea borbonia, Persicaria hydropiperoides, Pinguicula ionantha, Pinguicula planifolia, Pinus elliottii, Pinus palustris, Pinus taeda, Pityopsis oligantha, Pleea tenuifolia, Pogonia ophioglossoides, Polygala cruciata, Polygala cymosa, Polygala lutea, Pontederia cordata, Proserpinaca pectinata, Rhexia alifanus, Rhexia aristosa, Rhexia lutea, Rhexia mariana, Rhexia nuttallii, Rhexia petiolata, Rhexia virginica, Rhynchosia tomentosa, Rhynchospora baldwinii, Rhynchospora brachychaeta, Rhynchospora cephalantha, Rhynchospora chapmanii, Rhynchospora ciliaris, Rhynchospora fascicularis, Rhynchospora filifolia, Rhynchospora gracilenta, Rhynchospora harperi, Rhynchospora inundata, Rhynchospora perplexa, Rhynchospora tracyi, Sabatia decandra, Sabatia difformis, Sacciolepis striata, Sagittaria australis, Sagittaria graminea, Sagittaria lancifolia, Salix, Sarracenia alata, Sarracenia flava, Sarracenia leucophylla, Sarracenia minor, Sarracenia psittacina, Sarracenia purpurea, Sarracenia rosea, Schizachyrium rhizomatum, Schizachyrium scoparium, Schoenoplectus etuberculatus, Scleria georgiana, Scleria muehlenbergii, Sclerolepis uniflora, Serenoa repens, Seymeria cassioides, Sisyrinchium atlanticum, Smilax auriculata, Smilax glauca, Smilax laurifolia, Sphagnum macrophyllum, Sporobolus floridanus, Stenanthium leimanthoides, Stillingia aquatica, Styrax americanus, Syngonanthus flavidulus, Taxodium ascendens, Taxodium distichum, Tephrosia spicata, Tiedemannia filiformis, Triadenum virginicum, Utricularia gibba, Utricularia subulata, Viola lanceolata, Woodwardia virginica, Xyris ambigua, Xyris baldwiniana, Xyris caroliniana, Xyris drummondii, Xyris elliottii, Xyris fimbriata, Xyris isoetifolia, Xyris jupicai, Xyris laxifolia, Xyris panacea, Xyris smalliana.*

10.1.1.4. ***Eriocaulon decangulare* L.** is a perennial, which grows in fresh (sometimes brackish) baygalls, bogs (acid, cranberry, pitcher plant, roadside, seepage), depressions (wet), ditches (roadside), fens (poor, sea-level), flatwoods (pine, wet), floodplains (swampy), glades (pineland), marshes (basin, depression, ephemeral, oligohaline, salt), meadows (moist, sphagnous, wet), pine barrens, pinelands (wet), prairies (wet), puddles, ravines, roadsides, savannas (level, pine), seeps (hillside, spring), shores, sinks (roadside), sloughs (peaty), swales (boggy, pineland, seepage, sphagnous), swamps (pine barren, roadside), and along the margins of canals, cypress domes, lakes, ponds (flatwoods, pine barren), potholes, and streams at elevations to 901 m. The exposures vary from full sunlight to partial shade (96.4–50.3% full sunlight) with the plants found often along firebreaks or under clearings made for overhead powerlines. The substrates are strongly acidic (pH: 4.2–5.3) with moderate mean nutrient levels (e.g., Ca: 128–295 mg l^{-1}; K: 15–33 mg l^{-1}; Mg: 35–98 mg l^{-1}; Na: 30.11 mg l^{-1}; and P: 3–5 mg l^{-1}). They have been characterized as Basinger (apodic psammaquents), clay, clay loam, Dorovan (typic medisaprists), Dothan (plinthic paleudults), Doucette (arenic plinthic paleudults), EauGallie (alfic haplaquods), Immokalee, Leon (aeric haplaquods), loam (Bellwood), loamy sand (Pelham, Sparta), muck, mucky peat, mud, Myakka (alfic haplaquods), peat, peaty muck, Pelham (arenic paleaquults), Pomona (ultic haplaquods), Pompano, Pottsburg (grossarenic haplaquods), Riviera (arenic glossaqualfs), Rutledge (typic humaquepts), sand (Wayland), sandy gravel, sandy loam, sandy muck, sandy peat, sandy silt, silty clay loam, silty loam, silty sandy loam, Smyrna (aeric haplaquods), Stowell (typic argiabolls), Stringtown-Bonwier (typic hapludlts), and Vero (alfic haplaquods). The flowers are thought to be wind pollinated but are visited regularly by potentially pollinating insects (see *Ecology: general* above and *Use by wildlife* below). The insects either effect cross-pollination among plants or facilitate geitonogamous self-pollination among simultaneously open male and female flowers on a head. The seeds are dispersed as typical for the genus (within the dried flowers) by wind or water. A persistent seed bank develops. Seeds taken from seed bank cores have germinated when incubated at temperatures above 20°C under a 16-hr photoperiod. The plants grow vegetatively into clumps, which develop by means of lateral offsets and short, sometimes branching rhizomes.

Reported associates: *Acer rubrum, Aconitum uncinatum, Agalinis purpurea, Aldrovanda vesiculosa, Aletris aurea, Aletris farinosa, Alnus serrulata, Amphicarpum muhlenbergianum, Andropogon glomeratus, Andropogon liebmannii, Andropogon virginicus, Anthaenantia rufa, Aristida palustris, Aristida purpurascens, Arnoglossum ovatum, Aronia arbutifolia, Aronia melanocarpa, Arundinaria gigantea, Asclepias longifolia, Asclepias michauxii, Asclepias rubra, Axonopus fissifolius, Baccharis halimifolia, Balduina uniflora, Baptisia lanceolata, Baptisia leucophaea, Bartonia paniculata, Bidens mitis, Bidens trichospermus, Bigelowia* nudata, Bigelowia nuttallii, Boehmeria cylindrica, Burmannia capitata, Calamovilfa brevipilis, Calopogon tuberosus, Carex alata, Carex bullata, Carex festucacea, Carex folliculata, Carex glaucescens, Carex leptalea, Carex lurida, Carex striata, Carex stricta, Carex trichocarpa, Cassia, Centella asiatica, Cephalanthus occidentalis, Chamaecyparis thyoides, Chaptalia tomentosa, Cladium jamaicense, Cladium mariscoides, Cleistesiopsis bifaria, Cleistesiopsis divaricata, Clethra alnifolia, Cliftonia mono-phylla, Coleataenia anceps, Coleataenia longifolia, Coleataenia tenera, Commelina virginica, Coreopsis gladiata, Cortaderia selloana, Ctenium aromaticum, Cyperus haspan, Cyperus strigosus, Cyrilla racemiflora, Dicerandra odoratissima, Dichanthelium acuminatum, Dichanthelium dichotomum, Dichanthelium ensifolium, Dichanthelium ovale, Dichanthelium scabriusculum, Dichanthelium scoparium, Dichanthelium strigosum, Diospyros virginiana, Drosera brevifolia, Drosera capillaris, Drosera intermedia, Drosera rotundifolia, Drosera tracyi, Dulichium arundinaceum, Eleocharis elongata, Eleocharis equisetoides, Eleocharis fallax, Eleocharis robbinsii, Eleocharis rostellata, Eleocharis tortilis, Eleocharis tuberculosa, Eragrostis elliottii, Rhynchospora oligantha, Eremochloa ophiuroides, Erigeron vernus, Eriocaulon compressum, Eriocaulon koernickianum, Eriocaulon lineare, Eriocaulon nigrobracteatum, Eriocaulon texense, Eriophorum virginicum, Eryngium integrifolium, Eupatorium leucolepis, Eupatorium perfoliatum, Eupatorium pilosum, Eupatorium rotundifolium, Eutrochium fistulosum, Fimbristylis spadicea, Fuirena breviseta, Fuirena squarrosa, Galium asprellum, Galium obtusum, Gaylussacia frondosa, Gaylussacia mosieri, Gelsemium sempervirens, Gordonia lasianthus, Gratiola pilosa, Gratiola ramosa, Helenium drummondii, Helianthus angustifolius, Helianthus radula, Hibiscus moscheutos, Hydrocotyle umbellata, Hydrolea ovata, Hymenocallis gholsonii, Hymenocallis henryae, Hypericum brachyphyllum, Hypericum canadense, Hypericum crux-andreae, Hypericum densiflorum, Hypericum fasciculatum, Hypericum galioides, Hypericum hypericoides, Hypericum mutilum, Hypericum prolificum, Hypoxis hirsuta, Hyptis alata, Ilex cassine, Ilex coriacea, Ilex glabra, Ilex myrtifolia, Ilex vomitoria, Illicium floridanum, Iris virginica, Juncus canadensis, Juncus effusus, Juncus pelocarpus, Juncus polycephalus, Juncus repens, Juncus scirpoides, Juncus subcaudatus, Juncus tenuis, Juncus trigonocarpus, Juncus validus, Justicia, Lachnanthes caroliniana, Lachnocaulon anceps, Lachnocaulon digynum, Lachnocaulon minus, Liatris acidota, Liatris pycnostachya, Lilium catesbaei, Linum medium, Liquidambar styraciflua, Liriodendron tulipifera, Lobelia brevifolia, Lobelia puberula, Lobelia reverchonii, Lophiola aurea, Ludwigia hirtella, Ludwigia linearis, Ludwigia pilosa, Lycopodiella alopecuroides, Lycopodiella appressa, Lycopodiella caroliniana, Lycopodiella prostrata, Lycopus amplectens, Lycopus cokeri, Macranthera flammea, Magnolia virginiana, Marshallia graminifolia, Mikania, Mimulus alatus, Mitreola petiolata, Mitreola sessilifolia, Muhlenbergia capillaris, Muhlenbergia expansa, Myrica cerifera, Myrica heterophylla, Myriophyllum,

Nuphar advena, Nymphaea odorata, Nyssa sylvatica, Oenothera fruticosa, Oenothera lindheimeri, Oldenlandia uniflora, Orontium aquaticum, Osmunda regalis, Osmundastrum cinnamomeum, Oxypolis rigidior, Packera anonyma, Packera crawfordii, Panicum hemitomon, Panicum verrucosum, Panicum virgatum, Paspalum floridanum, Persea borbonia, Persea palustris, Persicaria sagittata, Phyla, Physostegia godfreyi, Pinguicula caerulea, Pinguicula lutea, Pinguicula pumila, Pinus echinata, Pinus elliottii, Pinus palustris, Pinus rigida, Pinus taeda, Pityopsis graminifolia, Platanthera ciliaris, Platanthera cristata, Platanthera integra, Pleea tenuifolia, Pluchea foetida, Pogonia ophioglossoides, Polygala cruciata, Polygala lutea, Polygala mariana, Polygala ramosa, Polytrichum commune, Pontederia cordata, Potamogeton diversifolius, Prunus serotina, Pterocaulon, Ptilimnium capillaceum, Ptilimnium costatum, Pycnanthemum flexuosum, Quercus laurifolia, Quercus nigra, Quercus virginiana, Rhexia alifanus, Rhexia cubensis, Rhexia lutea, Rhexia mariana, Rhexia virginica, Rhododendron canescens, Rhododendron maximum, Rhododendron viscosum, Rhus copallinum, Rhynchospora alba, Rhynchospora baldwinii, Rhynchospora caduca, Rhynchospora capitellata, Rhynchospora cephalantha, Rhynchospora chalarocephala, Rhynchospora chapmanii, Rhynchospora corniculata, Rhynchospora curtissii, Rhynchospora divergens, Rhynchospora elliottii, Rhynchospora filifolia, Rhynchospora filiformis, Rhynchospora globularis, Rhynchospora gracilenta, Rhynchospora inundata, Rhynchospora latifolia, Rhynchospora macra, Rhynchospora microcarpa, Rhynchospora nitens, Rhynchospora perplexa, Rhynchospora plumosa, Rhynchospora pusilla, Rhynchospora rariflora, Rhynchospora tracyi, Rosa palustris, Rubus argutus, Rubus hispidus, Rubus pensilvanicus, Rudbeckia scabrifolia, Sabatia difformis, Sabatia gentianoides, Sagittaria engelmanniana, Sagittaria lancifolia, Sagittaria latifolia, Sanguisorba canadensis, Sarracenia alata, Sarracenia flava, Sarracenia leucophylla, Sarracenia minor, Sarracenia oreophila, Sarracenia psittacina, Sarracenia purpurea, Sarracenia rubra, Saururus cernuus, Schizachyrium scoparium, Schizachyrium tenerum, Scirpus cyperinus, Scleria ciliata, Scleria georgiana, Scleria lithosperma, Scleria muehlenbergii, Scleria pauciflora, Scleria reticularis, Scutellaria integrifolia, Setaria parviflora, Smilax laurifolia, Smilax walteri, Solidago, Sparganium americanum, Spartina bakeri, Sphagnum, Spiranthes cernua, Spiranthes vernalis, Steinchisma hians, Stenanthium leimanthoides, Stenanthium texanum, Symphyotrichum dumosum, Symplocos tinctoria, Syngonanthus flavidulus, Taxodium ascendens, Taxodium distichum, Thelypteris palustris, Tiedemannia canbyi, Toxicodendron radicans, Toxicodendron vernix, Triadenum virginicum, Triadica sebifera, Triantha racemosa, Trichostema dichotomum, Typha latifolia, Utricularia cornuta, Utricularia geminiscapa, Utricularia gibba, Utricularia juncea, Utricularia juncea, Utricularia macrorhiza, Utricularia striata, Utricularia subulata, Vaccinium arboreum, Vaccinium corymbosum, Veratrum virginicum,

Vernonia blodgettii, Viburnum nudum, Viola cucullata, Viola lanceolata, Viola primulifolia, Woodwardia areolata, Xyris ambigua, Xyris baldwiniana, Xyris caroliniana, Xyris difformis, Xyris drummondii, Xyris isoetifolia, Xyris jupicai, Xyris laxifolia, Xyris scabrifolia, Xyris stricta, Xyris torta, Zigadenus glaberrimus.

10.1.1.5. *Eriocaulon koernickianum* Van Heurck & Müll. Arg. is an annual or weak perennial, which inhabits permanent waters of bogs (hillside, seepage), depressions (seasonal), and glades at elevations to 441 m. The substrates are acidic (pH: 4.5–5.0) and include granite outcrops, sand (arenic paleaquults, Carrizo, Pelham), sandstone, sandy loam (fine, Pandina), and sandy silt. Flowering and fruiting occur from May to June. Each head has on average 17.8 flowers, with 2–4 times as many pistillate than staminate flowers. The staminate flowers occur primarily around the periphery of the head, with a solitary staminate flower occasionally occurring in the center. On average the anthers produce 105.3 pollen grains and the heads 16.8 ovules (~40% develop into seeds). Mean seed set varies across heads and can range from 0% (45% of heads) to 100% (37% of heads); intermediate seed set occurs in roughly 18% of heads. Plants excluded from abiotic and biotic pollination vectors have produced some seeds, which indicates either autogamy (geitonogamous) or apomixis. Outcrossing is unlikely given that mobile insects have not been observed visiting the flowers and wind pollination is highly unlikely. However, geitonogamy probably is facilitated by insects such as ants (Insecta: Hymenoptera: Formicidae), which have been observed to crawl across the heads. Strict inbreeding is indicated by genetic (allozyme) analyses, which found nearly all individuals sampled to be genetically identical at the 20–21 loci surveyed. All the loci were homozygous except for one (with 4 alleles), but no individuals were heterozygous aside from several loci, which exhibited fixed heterozygosity. Even though some plants set no seed, the genetic and erratic seed set data could indicate that facultative apomixis occurs, a possibility that deserves further evaluation, especially given its suspected occurrence in other congeners (see discussions under *E. aquaticum* and *E. parkeri*). A seed bank with densities averaging up to 4,934 seeds m^{-2} is produced. Successful seed germination (63%) has been obtained only after exposure to a 10-month overripening period, wherein seeds that were stored at room temperature received a 48-hr wash in warm (37°C tap water), followed by 7 weeks of cold stratification (3.5°C) in the dark, and 3 weeks of incubation at 29°C under a 14-hr photoperiod. Although the seeds are highly viable, natural seedling establishment was found to be poor at the sites studied. The plants have been observed to occur at densities from 1.6 to 116 individuals m^{-2}. Their outlook for prolonged survival of this rare species is poor given its low seed set, genetic homogeneity, low recruitment, weak competitive ability, and a lack of any means for vegetative reproduction (although reproduction by short lateral offsets has been reported). Plants in the western portion of the range are generally larger than those in the disjunct eastern populations, but are not distinct genetically by allozyme surveys

(21 loci). **Reported associates:** *Allium, Andropogon gerardii, Asclepias rubra, Boltonia diffusa, Cladium mariscoides, Coreopsis basilis, Coreopsis tinctoria, Cyperus haspan, Drosera capillaris, Eleocharis compressa, Eleocharis equisetoides, Eleocharis lanceolata, Eleocharis microcarpa, Eleocharis obtusa, Eleocharis tenuis, Eleocharis wolfii, Eriocaulon decangulare, Eriocaulon texense, Eupatorium perfoliatum, Fuirena simplex, Helianthus angustifolius, Hydrocotyle verticillata, Hypericum mutilum, Isoetes melanopoda, Isolepis pseudosetacea, Juncus biflorus, Juncus coriaceus, Juncus debilis, Juncus effusus, Juncus georgianus, Juncus scirpoides, Juncus secundus, Juncus validus, Liatris pycnostachya, Lindernia monticola, Linum medium, Ludwigia sphaerocarpa, Minuartia uniflora, Panicum hemitomon, Phemeranthus mengesii, Rhexia, Rhynchospora globularis, Rhynchospora glomerata, Rhynchospora recognita, Rotala ramosior, Saccharum giganteum, Sarracenia alata, Scleria verticillata, Sphagnum, Utricularia cornuta, Utricularia subulata, Viburnum nudum, Xyris ambigua, Xyris jupicai.*

10.1.1.6. *Eriocaulon lineare* Small is a perennial, which grows in bogs (pitcher plant, seepage), borrow pits, flatwoods (pine), marshes, ponds (borrow pit, depression, *Hypericum*), roadsides, savannas (wet), shores, springs, terraces, and along the margins (desiccated, exposed, fluctuating, marshy) of lakes (interdunal, small) and ponds (artificial, depression, limesink, sandhill) at elevations to 50 m. The plants generally occupy open sites in shallow, standing waters from 5 to 60 cm deep, but submerged plants have been reported at depths up to 5.0 m. The substrates have been characterized as Chipley (aquic quartzipsamments), Duckston (typic psammaquents), muck, mud, Ortega (typic quartzipsamments), peat, peaty muck, Plummer (grossarenic paleaquults), Rutledge (typic humaquepts), sand, sandy peat, and sandy peaty muck. Flowering and fruiting occur from April to October. Vegetative reproduction can occur (especially under wetter conditions) by lateral shoot offsets, or by more elongate stoloniferous rhizomes. Little additional life history information is available for this species. **Reported associates:** *Acer rubrum, Amphicarpum muhlenbergianum, Aristida, Bacopa caroliniana, Bidens mitis, Calopogon, Carphephorus odoratissimus, Ctenium aromaticum, Cyperus, Dichanthelium dichotomum, Drosera capillaris, Eleocharis elongata, Eleocharis melanocarpa, Eleocharis minima, Eriocaulon decangulare, Fuirena pumila, Fuirena scirpoidea, Habenaria repens, Hypericum fasciculatum, Ilex glabra, Juncus repens, Lachnanthes caroliniana, Lachnocaulon minus, Lechea, Liatris spicata, Liquidambar styraciflua, Lophiola aurea, Ludwigia linifolia, Ludwigia suffruticosa, Lycopodiella, Mayaca fluviatilis, Nuphar, Nymphaea, Panicum hemitomon, Paronychia chartacea, Pinus taeda, Polygala lutea, Polygala nana, Rhexia cubensis, Rhexia lutea, Rhexia petiolata, Rhexia salicifolia, Rhynchospora nitens, Rhynchospora tracyi, Sabal palmetto, Sacciolepis striata, Sagittaria graminea, Sagittaria isoetiformis, Sarracenia alata, Sarracenia flava, Sarracenia leucophylla, Sarracenia psittacina, Scleria reticularis, Serenoa repens, Sphagnum, Taxodium, Utricularia* *resupinata, Verbena, Xyris jupicai, Xyris laxifolia. Xyris longisepala.*

10.1.1.7. *Eriocaulon microcephalum* Kunth is known only historically from a single 19th-century collection made in California, where it occurred in upland meadows (boggy, moist) at montane elevations above 1,000 m. No other life history information relative to North America is available. **Reported associates:** none.

10.1.1.8. *Eriocaulon nigrobracteatum* E.L. Bridges & Orzell is a perennial, which occurs in bogs (quaking, seepage, sphagnous, streamhead) or poor fens at elevations to 49 m. The habitats have been characterized as "poor fens"; i.e., herbaceous, oligotrophic to minimally minerotrophic, acidic (pH: 5.4–5.6), nutrient deficient, lower seepage slope mires. The substrates are permanently saturated and comprise deep deposits (0.3–2.2 m) of sapric muck, which originate from autochthonous sedge decomposition and usually are underlain by coarse sand. Some are underlain (at a depth of 30–45 cm) by a layer of sandy clay. They have been categorized as Dorovan (typic medisaprists), Pamlico (terric medisaprists), Pantego (umbric paleaquults), Pelham (arenic paleaquults), Rains (typic paleaquults), or Rutledge (typic humaquepts). The plants are in flower from March to April and produce fruits from April to May. They persist vegetatively by the development of dense, long-lived clumps from basal offshoots. **Reported associates:** *Aristida purpurascens, Aristida stricta, Balduina uniflora, Bigelowia nudata, Carphephorus pseudoliatris, Chaptalia tomentosa, Cliftonia monophylla, Coreopsis gladiata, Ctenium aromaticum, Cyrilla racemiflora, Dichanthelium nudicaule, Dichanthelium scabriusculum, Drosera capillaris, Drosera tracyi, Eleocharis tuberculosa, Eriocaulon compressum, Eriocaulon decangulare, Gaylussacia mosieri, Helenium vernale, Hypericum brachyphyllum, Hypericum chapmanii, Hypericum fasciculatum, Ilex glabra, Pinguicula ionantha, Pinguicula planifolia, Pinguicula planifolia, Pleea tenuifolia, Pogonia ophioglossoides, Polygala cruciata, Polygala lutea, Rhynchospora baldwinii, Rhynchospora chalarocephala, Rhynchospora chapmanii, Rhynchospora corniculata, Rhynchospora gracilenta, Rhynchospora macra, Rhynchospora macra, Rhynchospora oligantha, Rhynchospora stenophylla, Rhynchospora stenophylla, Rudbeckia graminifolia, Sabatia macrophylla, Sarracenia flava, Sarracenia leucophylla, Sarracenia psittacina, Sarracenia psittacina, Schoenolirion albiflorum, Scleria baldwinii, Smilax laurifolia, Syngonanthus flavidulus, Syngonanthus flavidulus, Taxodium ascendens, Tiedemannia greenmanii, Triantha racemosa, Utricularia subulata, Utricularia subulata, Xyris ambigua, Xyris baldwiniana, Xyris drummondii, Xyris isoetifolia, Xyris serotina.*

10.1.1.9. *Eriocaulon parkeri* B.L. Rob. is an annual or short-lived perennial, which inhabits tidally influenced, brackish to freshwater (salinity: 0–5‰) estuaries, flats, marshes (freshwater, intertidal), mud flats, pools, rivers, and streams at low, coastal (Atlantic) elevations to 100 m. The habitats are characterized by highly dynamic, fluctuating water levels, which can strand plants on wet ("mudflat") sediments during low tides but completely submerse them (e.g., to 2 m depths)

during high tides. The sediments are characterized as circumneutral to brackish and include gravel, mud, peat, sand (alluvial), and sandy mud. The flowers have been observed from July to October. Sexual reproduction is difficult to maintain due to the daily immersion of floral structures during high tidal periods, which deters or precludes pollinator access. Coverage of emerging inflorescences by algal "slime caps" further reduces pollinator access to the reproductive organs. Although the flowers are unisexual (monoecious), protandrous within an inflorescence, and produce echinate/spinulate pollen (associated with insect syndromes), they are also self-compatible, exhibit overlap between the ♀ and ♂ phases (which are in close spatial proximity), and have low pollen:ovule ratios (14:1–396:1; $\bar{x} = 196$:1) within the range associated with facultative autogamy. They are visited by flies (Insecta: Dolichopodidae: *Condylostylus*; Syrphidae: *Somula decora*), which have been verified as carrying *E. parkeri* pollen; however, insect visitation rates are low, occurring in only 8.3% of inflorescences monitored. Interestingly, the pollen remains viable and germinates even when exposed to water; however, no evidence of water-pollination has been found. Pollinator limitation does not account for the low seed set, with hand-pollinated flowers showing no significant increase in seed set over unmanipulated flowers. Moreover, pollinator-deprived (i.e., "bagged") greenhouse plants have exhibited the highest seed set (89%), which indicates that effective spontaneous, within-inflorescence autogamy (geitonogamy) and/or apomixis occurs. The latter has been demonstrated on emasculated inflorescences, where some female flowers (5.0%) successfully set seed along with showing no evidence of pollen tube growth. The low apparent level of agamospermy recalls the infrequent incidence of polyembryony in the related (if not conspecific) *E. aquaticum* (see discussion above). Although local dispersal probably is typical for the genus (see *Ecology: general* above), the tiny seeds (7.7 million kg⁻¹) are thought to be dispersed over longer distances by adhering to waterfowl (Aves: Anatidae). Nothing is known of their germination requirements. Although described in many sources as a perennial (from short, lateral shoot offsets), the plants apparently display an annual life history most often. In any case, prolific seed production and high seed viability (34–100%) enable populations to persist primarily through annual recruitment from the seed bank. The plants are susceptible to siltation and loss of habitat due to dredging and other hydrologic alterations. They are also threatened with habitat usurpation by the invasive *Glossostigma cleistanthum*. **Reported associates:** *Amaranthus cannabinus, Bidens bidentoides, Bidens connatus, Bidens eatonii, Bidens hyperborea, Bidens trichospermus, Bolboschoenus fluviatilis, Cardamine longii, Carex stricta, Cephalanthus occidentalis, Cicuta maculata, Crassula aquatica, Cyperus bipartitus, Dulichium arundinaceum, Echinochloa walteri, Elatine americana, Elatine minima, Eleocharis aestuum, Eleocharis erythropoda, Eleocharis flavescens, Eryngium aquaticum, Glossostigma cleistanthum, Gratiola aurea, Gratiola virginiana, Hypericum mutilum, Isoetes mattaponica, Isoetes riparia, Juncus pelocarpus, Juncus stygius, Justicia americana, Kalmia polifolia,*

Kyllinga gracillima, Lilaeopsis chinensis, Limosella australis, Lindernia dubia, Lobelia cardinalis, Ludwigia palustris, Micranthemum micranthemoides, Mikania scandens, Myrica gale, Nuphar, Orontium aquaticum, Panicum virgatum, Peltandra virginica, Persicaria punctata, Pohlia, Pontederia cordata, Ptilimnium capillaceum, Rhynchospora macrostachya, Riccardia, Sagittaria calycina, Sagittaria latifolia, Sagittaria subulata, Schoenoplectus pungens, Schoenoplectus smithii, Schoenoplectus subterminalis, Schoenoplectus tabernaemontani, Smilax walteri, Spartina pectinata, Utricularia intermedia, Viola primulifolia, Xyris difformis, Zizania aquatica.

10.1.1.10. *Eriocaulon ravenelii* Chapm. is a biennial or perennial, which occurs in depressions (flatwoods), ditches, flatwoods (mesic), pools (shallow), prairies (calcareous, marl), puddles, savannas (calcareous, flatwoods, pine), swamps (pineland), and along the margins or shores of ponds (coastal, cypress, shallow) at elevations to 21 m. The plants occur on wet, fluctuating shores in standing water up to 15 cm deep, or in drier mesic sites. Their substrates have been described as mildly acid, while others have characterized the plants as calciphiles (no actual pH data are available). The substrates have been characterized as Basinger, Eau Gallie (alfic haplaquods), Felda (arenic ochraqualfs), Fort Drum, marl, Myakka (aeric haplaquods), Malabar (grossarenic ochraqualfs), Meggett (typic albaqualfs), peaty sand, Pompano, sand, sandy peat, sandy peaty muck, and Winder (typic glossaqualfs). Other life history information is scarce for this species. **Reported associates:** *Amphicarpum muhlenbergianum, Aristida stricta, Centella asiatica, Cirsium nuttallii, Cladium jamaicense, Ctenium aromaticum, Dichanthelium aciculare, Eleocharis geniculata, Elionurus tripsacoides, Eragrostis elliottii, Eriocaulon decangulare, Eryngium yuccifolium, Fimbristylis autumnalis, Fimbristylis caroliniana, Fimbristylis schoenoides, Fuirena breviseta, Fuirena scirpoidea, Gratiola ramosa, Helenium pinnatifidum, Hymenocallis palmeri, Hypericum fasciculatum, Hyptis alata, Ilex glabra, Iva microcephala, Iva microcephala, Hyptis alata, Lachnocaulon anceps, Liatris chapmanii, Liatris garberi, Lipocarpha maculata, Ludwigia alata, Ludwigia microcarpa, Muhlenbergia capillaris, Muhlenbergia sericea, Myrica cerifera, Pinus elliottii, Piriqueta cistoides, Pluchea baccharis, Quercus geminata, Quercus minima, Rhexia nuttallii, Rhynchospora colorata, Rhynchospora divergens, Rhynchospora fascicularis, Rhynchospora filifolia, Sabal palmetto, Sacciolepis indica, Schizachyrium rhizomatum, Scleria hirtella, Scleria reticularis, Scleria verticillata, Solidago stricta, Taxodium, Tiedemannia filiformis, Woodwardia areolata, Xyris calcicola, Xyris platylepis.*

10.1.1.22. *Eriocaulon texense* Körn. is a perennial, which inhabits bogs (hillside, quaking, *Sarracenia*, seepage, *Sphagnum*), depressions (peaty), ditches, flatwoods (mesic), pocosins (open, pine), pools, potholes (bog), puddles, savannas (cypress, low, pine), seeps, slopes (seepage), and the margins (boggy) of streams at elevations to 116 m. The plants are found in open (well-burned) sites under exposures that receive full sunlight. The substrates are acidic and have been

categorized as Alapaha (arenic plinthic paleaquults), clay, Escambia (plinthaquic paleudults), Leefield-Stilson complex, muck, peat, Pelham (arenic paleaquults), Pickton (grossarenic paleudults), Plummer (grossarenic paleaquults), Rains (typic paleaquults), sand, and sandy peat. Flowering occurs during the spring from April to June. The inflorescences as well as the foliage die back and virtually disappear by early summer. The reproductive ecology has not been studied in any detail, but is likely typical for the genus (see remarks under *Ecology: general* above). The plants perennate by short, lateral off-shoots and develop into clumps. **Reported associates:** *Acer rubrum, Aletris aurea, Andropogon gyrans, Andropogon liebmannii, Aristida stricta, Asclepias connivens, Balduina uniflora, Bigelowia nudata, Carex turgescens, Centella asiatica, Chaptalia tomentosa, Cirsium muticum, Cladium mariscoides, Cliftonia monophylla, Ctenium aromaticum, Danthonia sericea, Dichanthelium acuminatum, Dichanthelium ensifolium, Drosera brevifolia, Drosera capillaris, Drosera tracyi, Dulichium arundinaceum, Eriocaulon compressum, Eriocaulon decangulare, Eriocaulon koernickianum, Eryngium integrifolium, Eupatorium rotundifolium, Fuirena squarrosa, Gaylussacia mosieri, Hypericum brachyphyllum, Hypericum tenuifolium, Ilex glabra, Juncus, Lachnocaulon anceps, Liatris spicata, Linum macrocarpum, Lycopodiella appressa, Lycopodiella caroliniana, Lycopodiella prostrata, Magnolia virginiana, Muhlenbergia expansa, Myrica cerifera, Myrica heterophylla, Persea palustris, Pinus elliottii, Pinus palustris, Pleea tenuifolia, Polygala hookeri, Polygala lutea, Polygala ramosa, Rhexia alifanus, Rhynchospora capitellata, Rhynchospora cephalantha, Rhynchospora chapmanii, Rhynchospora galeana, Rhynchospora latifolia, Rhynchospora macra, Rhynchospora oligantha, Rhynchospora stenophylla, Sabatia macrophylla, Sarracenia alata, Sarracenia flava, Sarracenia leucophylla, Sarracenia minor, Sarracenia psittacina, Sarracenia purpurea, Scleria muhlenbergii, Smilax glauca, Solidago patula, Solidago pulchra, Sphagnum, Sporobolus teretifolius, Syngonanthus flavidulus, Utricularia subulata, Xyris ambigua, Xyris baldwiniana, Xyris drummondii, Xyris elliottii, Xyris scabrifolia, Xyris torta.*

Use by wildlife: *Eriocaulon* leaves are eaten by several duck species (Aves: Anatidae), including American wigeon (*Anas americana*), black duck (*Anas rubripes*), and wood duck (*Aix sponsa*). The inflorescences often are used as pupation sites by larval flies (Insecta: Diptera). The inflorescences and seeds of some species (*E. aquaticum, E. compressum, E. texense*) can become infected with smut Fungi (e.g., Basidiomycota: Anthracoideaceae: *Moesziomyces eriocauli*; Ustilaginaceae: *Eriocaulago eriocauli*). The dead leaves are colonized by saprophytic chytrids (Fungi: Chytridiomycota: Endochytriaceae: *Entophlyctis luteolus, E. texana*). *Eriocaulon aquaticum* and *E. compressum* are also weakly parasitized by chytrids (Fungi: Chytridiomycota: Cladochytriaceae: *Cladochytrium replicatum*). *Eriocaulon aquaticum* is known to harbor dense epiphytic invertebrate communities. Fifteen genera of nematodes (Nematoda) have been found associated with the roots of *E. compressum*, especially species of *Tripyla* (Tripylidae).

The flowers of *E. compressum* allegedly provide a nectar source for some butterflies (Insecta: Lepidoptera). Flowers of *E. decangulare* are visited fairly regularly by several insects (Insecta), including leafcutting bees (Hymenoptera: Megachilidae: *Megachile albitarsis*), plasterer bees (Hymenoptera: Colletidae: *Hylaeus confluens*), sweat bees (Hymenoptera: Halictidae: *Halictus ligatus, Lasioglossum coreopsis, L. creberrimum, L. longifrons, L. tamiamense*), and flies (Diptera: Ephydridae: *Typopsilopa atra*), which likely facilitate pollination in some way.

Economic importance: food: *Eriocaulon* species are not reported as edible; **medicinal:** Ethanolic extracts of *E. cinereum* reportedly are inhibitory to the growth of HeLa cervical cancer cell lines. In India, a paste made from the plants is used as a hair tonic and a remedy for ringworm; **cultivation:** *Eriocaulon compressum* was cultivated as an aquarium plant during the early part of the 20th century; **misc. products:** *Eriocaulon aquaticum* has been used as a biomonitor to detect trace elements (Cd, Fe, Hg, Pb); **weeds:** *Eriocaulon* species are not regarded as weedy; **nonindigenous species:** *Eriocaulon cinereum* was introduced to California and Louisiana (and parts of Europe) through rice culture. It has not been reported in California since 1947.

Systematics: The monophyly of the morphologically distinctive *Eriocaulon* has been confirmed by phylogenetic analyses of molecular data, which associate *Eriocaulon* and *Mesanthemum* as sister genera within a clade diverging early in the evolution of Eriocaulaceae (Figure 3.7). However, since few *Eriocaulon* species have been evaluated phylogenetically, many questions remain with respect to species boundaries and interrelationships in the genus. Because these plants can inhabit such a wide range of environments, taxonomic uncertainty often is due to an inability to distinguish phenotypic variability from genetically based variation. The *E. aquaticum* complex provides a good overview of some of the taxonomic problems encountered. In North America, this taxon occurs throughout the northeast, but is also amphi-Atlantic with populations in Ireland and the Scottish Hebrides. Some minor morphological variation has been reported between plants from the different regions, but has not been evaluated adequately to exclude the possibility that the differences simply reflect broad, plastic variation within a single wide-ranging species. Also within North America occurs a third taxon (*E. parkeri*), which is extremely similar to *E. aquaticum*, but is smaller in stature, has hemispherical rather than globular inflorescence heads, and occurs under intermittent tidal conditions rather than in more permanent nontidal waters. Some authors have pointed out that any distinction between *E. parkeri* and *E. aquaticum* becomes blurred as one moves from the tidal portions of a habitat to the plants in the upper, nontidal reaches. The seeds of the "doubtfully distinct" *E. aquaticum* and *E. parkeri* are also extremely similar. Furthermore, extensive phenotypic studies of Scottish plants (which comprise but one taxon) have shown that the stature of the plants and shape of the inflorescence can differ markedly between those growing under intermittently exposed conditions (shorter, with hemispherical heads) and

those in still, permanent waters (larger with globular heads). In this case, the main phenotypic differences used to differentiate the species also correlate with phenotypically plastic, habitat-induced variability. Yet, preliminary cytological studies indicated that the North American ($2n=64$), European ($2n=32$), and *E. parkeri* ($2n=48$) plants all differed from one another chromosomally, which suggested their status as distinct species. If that logic is followed, then the North American plants would adopt the name *E. pellucidum*, with *E. aquaticum* necessarily retained for the European specimens. However, a subsequent cytological investigation also found the $2n=32$ "European" cytotype in Canadian material, which renders the proposed cytological distinction ineffective in distinguishing the Old and New World material as different species. Moreover, the intermediate chromosome number of "*E. parkeri*" could represent a hybrid cytotype derived from the two North American cytotypes (see discussion below). Yet, further complicating matters is a highly anomalous count ($2n=20$) reported for other Canadian plants of *E. parkeri*. It is evident that a more complete cytological survey (and comprehensive genetic analysis) of all three taxa will be necessary before any final assessment of this taxon can be made. Adding to the problem are several other species (i.e., *E. nigrobracteatum* and *E. lineare*), which also have been regarded as questionably distinct from *E. aquaticum* or as closely related sister species. Thus, the species boundaries and interspecific relationships of North American *Eriocaulon* remain dubious and deserve further study, which eventually must include extensive genetic analyses and a more thorough population sampling of the putative taxa. Chromosomal data are quite incomplete for *Eriocaulon* and are also unreliable; no single species has been surveyed adequately. The basic chromosome number of *Eriocaulon* is uncertain. The Old World *E. cinereum* ($2n=18$) indicates a base number of $x=9$, whereas *E. compressum* ($2n=40$) indicates a base number of 10. However, $x=8$ seems to pertain at least to several other North American taxa. In this interpretation, *E. aquaticum* is either tetraploid ($2n=32$ reported from Québec, Canada, and Ireland) or octoploid ($2n=64$ reported from Nova Scotia, Canada). An approximate count reported for *E. parkeri* from the estuaries of the St. Lawrence River (~$2n=48$) indicates it to be hexaploid. The provenance of some cytological material is dubious. A tetraploid count ($2n=32$) allegedly has been reported for *E. aquaticum* based on material originating from "Tropical and warm areas"; however, because *E. aquaticum* is entirely temperate, it is uncertain where the material originated or even if it belongs to the correct taxon (the DNA content of that tetraploid was estimated at 4C: 16.74 pg; 1C=4,101 Mbp). A complete chloroplast genome sequence is available for *E. sexangulare*.

Distribution: *Eriocaulon aquaticum* occurs in northeastern North America and is disjunct in Ireland and the Scottish Hebrides (but see *Systematics* above). *Eriocaulon cinereum* is a nonindigenous species introduced to California and Louisiana; it occurs elsewhere in Africa, Asia, Australia, and South America. *Eriocaulon compressum, E. decangulare, E. lineare, E. ravenelii,* and *E. texense* occur along the Atlantic

Coastal Plain of the southeastern United States. *Eriocaulon koernickianum* occurs in more inland portions of the southern United States. *Eriocaulon parkeri* is restricted to eastern North America. Even more limited distributions characterize *E. microcephalum*, which extends southward into South America, but in North America it is known only historically (and presumably is extirpated) from one locality in southern California; and *E. nigrobracteatum*, which occurs only in northwestern Florida.

References: Anderson & Kalff, 1986; Arsenault, 1984; Bartholomew et al., 2006; Baskin & Baskin, 1998; Blankinship, 1903; Breden et al., 2001; Bridges & Orzell, 1989; 1992b; Brunton & Britton, 1993; Carter et al., 2004; Clark et al., 2008; Clinton, 1901; Clough & Best, 1991; Cohen et al., 2004; Cook, 1996a; Cox & Smart, Jr., 1994; Cross et al., 2015; Cusick, 1970; Cypert, 1972a; Day et al., 1988; de Andrade et al., 2010; Dennis, 1980; Deyrup et al., 2002; Deyrup & Deyrup, 2008; Drewa et al., 2002; Eggeling & Ehrenberg, 1908; Eleuterius & McDaniel, 1978; Ewel & Atmosoedirdjo, 1987; Fassett, 1957; Fernald, 1941; Ferren, Jr. & Schuyler, 1980; Ferren, Jr. et al., 1981; France & Stokes, 1988; Freeman & Urban, 2012; Gerritsen & Greening, 1989; Glück, 1924; Godfrey & Wooten, 1979; Haines, 2000b; 2001; Halbritter et al., 2015; Han et al., 2019; Hanson et al., 2003; Harper, 1905a; Harrison & Knapp, 2015; Hellquist & Crow, 1982; Hinman & Brewer, 2007; Hitchin et al., 1984; Kapoor & Ramcharitar, 1982; Karling, 1931; 1941; Kato et al., 2008; Keddy & Reznicek, 1982; Keppner & Anderson, 2008; Kral, 1966a; 1966b; 1989; 2000; Lacoul et al., 2011; Leach, 2017; LeBlond & Sorrie, 2001; Leck et al., 1988; Les et al., 2006b; Longcore, 1995; Löve & Löve, 1958; 1982; Maass, 1966; 1967; McAvoy & Wilson, 2014; McKenzie et al., 2009; McLaughlin, 1932; McMillan et al., 2002; Moldenke, 1957; 1969; Morgan & Philipp, 1986; Morgan & Adams, 2017; Morris, 2012; Moyle, 1945; Murphy & Boyd, 1999; Musselman et al., 2001; Nichols, 1999; Nicholson & Keddy, 1983; Nixon & Ward, 1986; Nugraha et al., 2017; Orzell & Bridges, 1993; 2006a; 2006b; Owens et al., 2018; Padgett et al., 2004; Poole et al, 2007; Prasad, 2007; Preston & Croft, 1997; Quattrocchi, 2012; Raven et al., 1988; Rhoads, 1984; Rosenthal et al., 2014; Rossell & Losure, 2005; Sawyer et al., 2005; Sheldon & Boylen, 1977; Singer et al., 1983; Singhurst et al., 2010; 2012; 2014b; Smith, 1910; Smith & Flory, 1990; Smith & Garland, 2009; Soerjani et al., 1987; Sorrie et al., 1997; Srivastava et al., 1995; Stace, 1999; Streng & Harcombe, 1982; Stützel, 1984; 1998; Thieret, 1970; 1971a; Uphof, 1927; Urban et al., 2009; Watson et al., 1994; 2002; Wells, 1928; 1942; Wilson & Keddy, 1988; Zebryk, 2004; Zona et al., 2012.

10.1.2. Lachnocaulon

Bogbuttons, hat-pins

Etymology: from the Greek *lachnos chaulos* ("wooly stem") for the long hairs on some scapes
Synonyms: *Eriocaulon* (in part)
Distribution: global: Cuba; North America; **North America:** southeastern United States
Diversity: global: 7 species; **North America:** 5 species

Indicators (USA): obligate wetland (OBL): *Lachnocaulon anceps, L. beyrichianum, L. minus*; **facultative wetland (FACW):** *L. anceps*
Habitat: freshwater; palustrine; **pH:** 5.8–7.3; **depth:** <1 m; **life-form(s):** emergent herb
Key morphology: stems cespitose or solitary, short, few-branched or elongate, forming rosettes (to 4–40 cm); roots branched, not septate; leaves in crowded spirals, the blade (to 12 cm) narrowly to broadly linear or linear-triangular; scapes (to 40 cm) 1-several per stem, filiform, obscurely to distinctly ribbed or ridged, glabrous or pubescent, with a tubular sheath (to 12 cm); heads (to 9 mm) monoecious, globose or short-cylindric, dull gray-brown, pale gray, or whitish, the receptacle densely pilose to copiously hairy, outer involucral bracts (to 2 mm) reflexed, hairy or pilose beneath, inner bracts (to 2 mm) with white or translucent, clavate hairs; flowers 3-merous, ♂ and ♀ mixed in heads; ♂: sepals (to 2 mm) 3, connivent, with translucent or white hairs, forming club-shaped flower; petals absent or reduced to minute scales or hairs; stamens 3, the filaments adnate to rim of a tubular, clavate or narrowly obconic androphore (to 1.5 mm), anthers bisporangiate, unilocular; ♀: sepals (to 3 mm) 3; petals absent or reduced to minute scales or hairs; gynophore short, pistil 3-carpellate, styles 3, bifid, appendaged between branches; capsules loculicidal; seeds (to 0.6 mm) ellipsoid, pale to dark brown, dark red-brown or deep brown, not lustrous to very lustrous, with conspicuous, faint, or low longitudinal ribs and fine transverse striae
Life history: duration: perennial (lateral offshoots, rhizomes); **asexual reproduction:** lateral offshoots, rhizomes; **pollination:** insect, self; **sexual condition:** monoecious; **fruit:** capsules (common); **local dispersal:** rhizomes; seeds (water, wind); **long-distance dispersal:** seeds (water, wind)
Imperilment: 1. *Lachnocaulon anceps* [G5]; SH (TN); S1 (VA); **2.** *L. beyrichianum* [G4]; S1 (GA); S2 (FL, SC); **3.** *L. minus* [G3/G4]; S1 (AL, SC); S2 (NC); S3 (GA)
Ecology: general: All North American *Lachnocaulon* species are wetland indicators designated either as OBL or FACW. The assigned designations should be reevaluated, given that one species (*L. digynum*) is ranked only as FACW, but has been described as tolerating the "wettest" conditions relative to its OBL congeners (*L. anceps, L. beyrichianum*), which extend into the "least-moist" sites. Some ecological characterizations probably are jumbled by the fact that the species are difficult to differentiate (even by experts), making it difficult to apply ecological information accurately from literature accounts. Fortunately, the species are quite similar ecologically, with most growing sympatrically and often cooccurring at a site. In all cases, the plants optimally colonize open, acidic, sandy substrates, which can range from virtually pure sand to combinations of sand, loam, and peat (*Sphagnum*). The substrates are saturated and gleyed, but consistently are associated with intermittent rather than permanent standing waters. These conditions often characterize ecotones between wetter and drier communities, to which all the species are well adapted. In general, *Lachnocaulon* species are found more toward the drier end of the hydrologic spectrum

than *Eriocaulon* species. Their root systems are shallow but adapted to withstand desiccating conditions during low water periods. The plants are monoecious and possess heads of inconspicuous, minute, unisexual, flowers, which lack petals and nectar. Their reproductive and pollination biology have not been well studied, but most available evidence indicates that the flowers are self-compatible and are pollinated (geitonogamously selfed or outcrossed) by small insects such as flies (Insecta: Diptera). Wind pollination has been attributed to some species, but it is not likely to occur other than on rare occasions. The fruits and seeds are produced regularly and persist in seed banks. They are thought to be dispersed primarily by water or wind. The North American species are perennial, but some are short-lived.

10.1.2.1. *Lachnocaulon anceps* **(Walter) Morong** is a perennial, which occurs in bogs (flatwoods, hillside, seepage, *Sphagnum*), borrow pits, Carolina bays, depressions (flatwoods), ditches (roadside), ecotones (savanna), flats (palmetto, pine), flatwoods (disturbed, pine), hammocks (hardwood), ledges (wet), marshes (roadside), meadows (moist), pine barrens (low), pinelands (moist), pocosins (streamhead), prairies (alfic, calcareous, mesic, wet, wet-mesic, spodic), ravines, rights-of-way (bog, powerline, seepage), savannas (open, pine, seepage, wet), seeps (sphagnous), swales (boggy, moist, sphagnous), swamps (floodplain, headwater, pinewood), and along the margins or shores of glades, lakes, ponds, and streams at elevations to 335 m. The plants are quite facultative and have been found growing under a variety of hydrological conditions from saturated to very dry soil. Exposures usually are open with the plants reported often from frequently burned or mowed sites. The substrates (Cl: 0–10 ppm) primarily are acidic (pH: 5.8–7.3) and have been characterized as Alapaha (arenic plinthic-paleaquults), Basinger (spodic psammaquents), Cuthbert (typic hapludults), Dothan (plinthic paleudults), Doucette (arenic plinthic paleudults), Floridana (arenic argiaquolls), Immokalee (arenic haplaquods), Leon (aeric haplaquods), loamy sand, Myakka (aeric haplaquods), peat, Paisley (typic albaqualfs), Pomello (arenic haplohumods), Pomona (ultic haplaquods), Pottsburg (grossarenic haplaquods), Rutledge (typic humaquepts), sand (Basinger, Kureb, Leon, Tarboro), sandy loam, sandy muck, sandy peat, sandy peaty loam, Satellite (aquic quartzipsamments), silty loam, Smyrna (aeric haplaquods), spodic psammaquents (hyperthermic), St. Johns (typic haplaquods), Tehran-Letney (grossarenic and arenic paleudults), and Tocoi (ultic haplaquods). Flowering and fruiting occur throughout most of the year (March–December). The inconspicuous heads (the flowers completely lack petals) are small (4–7 mm) and, despite having spinulose pollen, attract few insects (Insecta), which rarely are observed visiting the plants. The staminate and pistillate flowers usually are "indiscriminately mixed" in the heads compared to *Eriocaulon*. Pollination occasionally is facilitated by flies (Diptera), which are known to feed on the pollen. Mites (Arthropoda: Acari) also have been reported as effective pollinators. The flowers open from 5:00 am to 6:00 am during summer and later (9:00–11:00 am) as cooler weather approaches; anthesis ceases during extremely cold

periods. The seeds or fruits are dispersed by water or wind. A persistent seed bank develops. Seed bank seeds that were stratified (at 6°C) for 4 months have germinated under ambient greenhouse conditions. The plants are clump-forming perennials with short rhizomes bearing roots that extend down to a mean maximum depth of 10 cm. **Reported associates:** *Agalinis aphylla, Agalinis flexicaulis, Agalinis obtusifolia, Agalinis purpurea, Aletris aurea, Aletris farinosa, Alnus rugosa, Alnus serrulata, Amorpha herbacea, Amphicarpum muhlenbergianum, Andropogon brachystachyus, Andropogon glomeratus, Andropogon liebmannii, Andropogon ternarius, Andropogon virginicus, Aristida purpurascens, Aristida spiciformis, Aristida stricta, Aristida virgata, Asclepias connivens, Asimina reticulata, Baccharis halimifolia, Balduina uniflora, Baptisia tinctoria, Bartonia paniculata, Bejaria racemosa, Bigelowia nudata, Bigelowia nuttallii, Calamagrostis coarctata, Calamovilfa brevipilis, Calamovilfa curtissii, Callisia graminea, Calopogon barbatus, Calopogon multiflorus, Calopogon pallidus, Calopogon tuberosus, Carduus virginiana, Carex glaucescens, Carex turgescens, Carex verrucosa, Carphephorus odoratissimus, Carphephorus paniculatus, Carphephorus pseudoliatris, Carphephorus tomentosus, Centella asiatica, Chamaecrista fasciculata, Chaptalia tomentosa, Cirsium nuttallii, Cladium jamaicense, Cleistesiopsis divaricata, Clethra alnifolia, Conoclinium coelestinum, Conyza canadensis, Coreopsis gladiata, Crotalaria rotundifolia, Ctenium aromaticum, Cyperus bipartitus, Cyperus haspan, Cyperus retrorsus, Desmodium lineatum, Desmodium tenuifolium, Dichanthelium aciculare, Dichanthelium acuminatum, Dichanthelium ensifolium, Dichanthelium portoricense, Dichanthelium scabriusculum, Dichanthelium sphaerocarpon, Dichanthelium strigosum, Dichondra carolinensis, Dionaea muscipula, Diospyros virginiana, Drosera brevifolia, Drosera capillaris, Drosera intermedia, Drosera rotundifolia, Drosera tracyi, Eleocharis tortilis, Eleocharis tuberculosa, Eragrostis elliottii, Eremochloa ophiuroides, Erigeron quercifolius, Erigeron vernus, Eriocaulon compressum, Eriocaulon decangulare, Eriocaulon nigrobracteatum, Eriocaulon ravenelii, Eriocaulon texense, Eriophorum virginicum, Eryngium baldwinii, Eryngium integrifolium, Eupatorium compositifolium, Eupatorium leucolepis, Eupatorium mohrii, Eupatorium pilosum, Eupatorium rotundifolium, Euphorbia curtisii, Euphorbia inundata, Euphorbia telephioides, Euthamia graminifolia, Fimbristylis autumnalis, Fimbristylis cymosa, Fimbristylis dichotoma, Fimbristylis spadicea, Fothergilla gardenii, Fuirena scirpoidea, Galactia elliottii, Gaylussacia frondosa, Gratiola hispida, Gratiola pilosa, Gymnopogon breviseta, Harperocallis flava, Helenium pinnatifidum, Helenium vernale, Helianthemum corymbosum, Helianthus angustifolius, Helianthus heterophyllus, Helianthus radula, Hydrocotyle umbellata, Hypericum brachyphyllum, Hypericum cistifolium, Hypericum cruxandreae, Hypericum fasciculatum, Hypericum hypericoides, Hypericum mutilum, Hypericum tetrapetalum, Hypoxis juncea, Hypoxis micrantha, Hyptis alata, Ilex coriacea, Ilex glabra, Ilex myrtifolia, Ilex vomitoria, Juncus brachycarpus,*

Juncus caesariensis, Juncus marginatus, Juncus trigonocarpus, Juncus validus, Kalmia hirsuta, Kyllinga pumila, Lachnanthes caroliniana, Lachnocaulon beyrichianum, Lechea mucronata, Liatris graminifolia, Liatris garberi, Liatris gracilis, Liatris laevigata, Liatris tenuifolia, Licania michauxii, Lilium catesbaei, Lilium pyrophilum, Linum medium, Linum virginianum, Lipocarpha maculata, Lobelia brevifolia, Lobelia glandulosa, Lobelia nuttallii, Lobelia paludosa, Lobelia reverchonii, Lophiola aurea, Ludwigia alternifolia, Ludwigia decurrens, Ludwigia hirtella, Ludwigia microcarpa, Ludwigia virgata, Lycopodiella alopecuroides, Lycopodiella appressa, Lycopodiella caroliniana, Lycopodiella cernua, Lycopodiella prostrata, Lyonia fruticosa, Lyonia ligustrina, Lyonia lucida, Macbridea alba, Magnolia virginiana, Marshallia graminifolia, Micranthemum umbrosum, Mimosa microphylla, Mitreola petiolata, Mitreola sessilifolia, Muhlenbergia expansa, Muhlenbergia sericea, Myrica cerifera, Myrica heterophylla, Nolina atopocarpa, Oldenlandia uniflora, Oclemena reticulata, Ophioglossum petiolatum, Osmunda regalis, Osmundastrum cinnamomeum, Oxypolis rigidior, Panicum ciliatum, Panicum hemitomon, Panicum verrucosum, Panicum virgatum, Paspalum floridanum, Paspalum setaceum, Paspalum urvillei, Penstemon multiflorus, Persea palustris, Persicaria hydropiperoides, Phyla nodiflora, Physostegia, Pinguicula ionantha, Pinguicula lutea, Pinguicula primuliflora, Pinguicula pumila, Pinus clausa, Pinus elliottii, Pinus palustris, Pinus serotina, Piriqueta cistoides, Pityopsis graminifolia, Pityopsis oligantha, Platanthera, Pluchea baccharis. Pogonia ophioglossoides, Polygala cruciata, Polygala lutea, Polygala nana, Polygala polygama, Polygala ramosa, Polygala rugelii, Polypremum procumbens, Prenanthes autumnalis, Pteridium aquilinum, Pterocaulon pycnostachyum, Ptilimnium capillaceum, Pycnanthemum flexuosum, Quercus laevis, Quercus minima, Quercus virginiana, Rhexia alifanus, Rhexia lutea, Rhexia mariana, Rhexia nuttallii, Rhexia petiolata Rhododendron viscosum, Rhynchospora alba, Rhynchospora capitellata, Rhynchospora chapmanii, Rhynchospora ciliaris, Rhynchospora colorata, Rhynchospora divergens, Rhynchospora fascicularis, Rhynchospora globularis, Rhynchospora glomerata, Rhynchospora gracilenta, Rhynchospora grayi, Rhynchospora inexpansa, Rhynchospora latifolia, Rhynchospora macra, Rhynchospora oligantha, Rhynchospora plumosa, Rhynchospora pusilla, Rhynchospora rariflora, Rhynchospora stenophylla, Rhynchospora wrightiana, Rubus, Rudbeckia mohrii, Rudbeckia scabrifolia, Sabatia brevifolia, Sabatia campanulata, Sabatia difformis, Sabatia gentianoides, Sagittaria lancifolia, Sagittaria latifolia, Salix nigra, Sarracenia alabamensis, Sarracenia alata, Sarracenia flava, Sarracenia leucophylla, Sarracenia minor, Sarracenia psittacina, Sarracenia purpurea, Saururus cernuus, Schizachyrium rhizomatum, Schizachyrium scoparium, Schoenolirion croceum, Schwalbea americana, Scleria ciliata, Scleria georgiana, Scleria muhlenbergii, Scleria pauciflora, Scleria reticularis, Scleria triglomerata, Scleria verticillata,

Scutellaria integrifolia, Serenoa repens, Seymeria cassioides, Sisyrinchium arenicola, Smilax auriculata, Smilax glauca, Smilax laurifolia, Smilax pumila, Solidago stricta, Solidago villosicarpa, Sorghastrum secundum, Spermacoce prostrata, Sphagnum, Spiranthes praecox, Sporobolus floridanus, Stenanthium densum, Stipulicida setacea, Symphyotrichum dumosum, Symphyotrichum walteri, Syngonanthus flavidulus, Taxodium ascendens, Tiedemannia filiformis, Toxicodendron vernix, Triantha racemosa, Trilisa paniculata, Typha latifolia, Utricularia cornuta, Utricularia inflata, Utricularia subulata, Vaccinium crassifolium, Vaccinium darrowii, Vaccinium elliottii, Vaccinium myrsinites, Vaccinium stamineum, Vaccinium tenellum, Verbesina chapmanii, Viburnum nudum, Viola lanceolata, Viola primulaefolia, Viola septemloba, Vitis rotundifolia, Woodwardia virginica, Xyris ambigua, Xyris baldwiniana, Xyris brevifolia, Xyris caroliniana, Xyris difformis, Xyris drummondii, Xyris elliottii, Xyris isoetifolia, Xyris platylepis, Xyris smalliana, Zenobia pulverulenta, Zigadenus glaberrimus.

10.1.2.2. ***Lachnocaulon beyrichianum* Sporl. ex Körn.** is a long-lived perennial, which inhabits baygalls (streamside), depressions (limesink, wet), ditches, ecotones (Carolina bay, pocosin, sandhills), flats (open), flatwoods (moist, pine, scrubby, wet), prairies (dry, mesic), savannas (pine, wet), scrub (pine), swales, and the shores (dry) of ponds at elevations to 34 m. The plants often occupy the ecotone habitats that occur between pocosins and dry sand ridges, especially in fire-cleared sites. The substrates have been characterized as entisols, Immokalee (arenic haplaquods), Leon, Paola-Basinger (spodic quartzipsamments), Pomona (ultic haplaquods), Pompano (typic psammaquents), sand (EauGallie, Myakka, Pomello, sterile white), sandy peat, Satellite (aquic quartzipsamments), spodosols, St. Johns (typic haplaquods), and Tavares. Flowering has been observed from May to August. There is no additional information available on the reproductive ecology or pollination biology. The plants are mat-forming and perennate by means of long, persistent, subterranean rhizomes. **Reported associates:** *Acer rubrum, Aletris lutea, Andropogon brachystachyus, Andropogon ternarius, Andropogon virginicus, Aristida stricta, Asclepias pedicellata, Asemeia grandiflora, Asimina obovata, Bejaria racemosa, Bigelowia nudata, Carphephorus carnosus, Carphephorus odoratissimus, Carphephorus paniculatus, Centella asiatica, Ceratiola ericoides, Chamaecyparis thyoides, Chaptalia tomentosa, Clethra alnifolia, Cliftonia monophylla, Coleataenia tenera, Conradina grandiflora, Coreopsis gladiata, Ctenium aromaticum, Cyrilla racemiflora, Dichanthelium acuminatum, Dichanthelium ensifolium, Dichanthelium portoricense, Drosera capillaris, Eleocharis baldwinii, Eragrostis, Erigeron vernus, Eriocaulon decangulare, Eryngium aquaticum, Eupatorium, Euthamia graminifolia, Galactia elliottii, Gaylussacia frondosa, Gordonia lasianthus, Gratiola hispida, Hypericum edisonianum, Hypericum fasciculatum, Hypericum tenuifolium, Hypericum tetrapetalum, Hypoxis juncea, Ilex glabra, Juncus scirpoides, Kalmia angustifolia, Kalmia buxifolia, Lachnocaulon anceps, Lachnocaulon minus, Lechea deckertii, Liatris* garberi, *Licania michauxii, Lobelia glandulosa, Lyonia ferruginea, Lyonia fruticosa, Lyonia lucida, Lyonia mariana, Magnolia virginiana, Muhlenbergia capillaris, Myrica cerifera, Osmanthus americanus, Persea borbonia, Persea humilis, Pinus clausa, Pinus elliottii, Pinus palustris, Pinus taeda, Pityopsis graminifolia, Polygala lutea, Polygala setacea, Pteridium aquilinum, Quercus geminata, Quercus laevis, Quercus laurifolia, Quercus minima, Quercus myrtifolia, Quercus nigra, Quercus virginiana, Rhexia cubensis, Rhexia mariana, Rhexia petiolata, Rhynchospora fascicularis, Rhynchospora megalocarpa, Rhynchospora plumosa, Sabal palmetto, Schizachyrium scoparium, Scleria reticularis, Selaginella arenicola, Serenoa repens, Seymeria cassioides, Sideroxylon reclinatum, Smilax auriculata, Solidago stricta, Spiranthes longilabris, Stillingia aquatica, Stipulicida setacea, Symphyotrichum dumosum, Syngonanthus flavidulus, Tiedemannia filiformis, Utricularia juncea, Utricularia subulata, Vaccinium corymbosum, Vaccinium crassifolium, Vaccinium tenellum, Viola lanceolata, Ximenia americana, Xyris brevifolia, Xyris caroliniana, Xyris flabelliformis, Xyris platylepis.*

10.1.2.3. ***Lachnocaulon minus* (Chapm.) Small** is a short-lived perennial, which grows in bogs (flatwoods, pine), borrow pits (sandy), depressions (mesic, pineland, sandy), ditches (moist, roadside, sandy), flatwoods (low, pine), marshes (depression), meadows (low), pinelands (moist), roadsides (sandy, wet), savannas, scrub, seeps (exposed, moist, wet), and along the margins (drawdown) or shores (fluctuating, marshy) of canals, lakes, and ponds (depression, karst, limesink) at elevations to 50 m. Although found in a variety of habitats, the plants thrive along the intermittent shorelines of karst ponds. The substrates (Cl: 0–10 ppm) are acidic (pH: 3.6–5.8) and have been described as Basinger (spodic psammaquents), Chipley (aquic quartzipsamments), Fuquay (plinthic paleudults), muck (damp), Myakka (aeric haplaquods), peat, peaty sand, Placid (typic humaquepts), Plummer (grossarenic palequults), sand (moist), sandy peat, Sellers (cumulic humaquepts), and St. Johns (spodic psammaquents). Flowering and fruiting extend from April to November. There is no information on the floral or pollination biology; however, the pollen ($\bar{x} = 22$ μm) is among the smallest observed in Eriocaulaceae. **Reported associates:** *Acer rubrum, Amphicarpum muhlenbergianum, Andropogon virginicus, Aristida stricta, Bigelowia nudata, Bulbostylis, Burmannia capitata, Calamovilfa curtissii, Carphephorus pseudoliatris, Centella asiatica, Cephalanthus occidentalis, Clethra alnifolia, Cliftonia monophylla, Clinopodium coccineum, Coleataenia longifolia, Ctenium aromaticum, Cyperus lecontei, Cyrilla racemiflora, Dichanthelium acuminatum, Dichanthelium dichotomum, Dichanthelium ensifolium, Drosera capillaris, Drosera filiformis, Drosera intermedia, Drosera tracyi, Dulichium arundinaceum, Eleocharis microcarpa, Erechtites hieracifolius, Erigeron vernus, Eriocaulon compressum, Eriocaulon decangulare, Eriocaulon lineare, Eriocaulon nigrobracteatum, Eubotrys racemosa, Eupatorium capillifolium. Eupatorium leptophyllum, Eupatorium mohrii, Fimbristylis autumnalis, Fimbristylis schoenoides, Fuirena breviseta,* *Fuirena pumila, Fuirena scirpoidea, Gaylussacia mosieri, Gordonia lasianthus, Gratiola ramosa, Hydrocotyle umbellata, Hypericum brachyphyllum, Hypericum fasciculatum, Hypericum lissophloeus, Hypericum tenuifolium, Ilex cassine, Ilex glabra, Ilex myrtifolia, Itea virginica, Juncus debilis, Juncus scirpoides, Lachnanthes caroliniana, Lachnocaulon anceps, Lachnocaulon beyrichianum, Lachnocaulon digynum, Lachnocaulon engleri, Liatris spicata, Linum medium, Lophiola aurea, Ludwigia suffruticosa, Lycopodiella alopecuroides, Lycopodiella prostrata, Lyonia lucida, Lyonia mariana, Muhlenbergia capillaris, Myrica cerifera, Nymphoides aquatica, Nyssa biflora, Panicum hemitomon, Persicaria hirsuta, Physostegia godfreyi, Pieris phillyreifolia, Pinguicula lutea, Pinus elliottii, Pinus palustris, Pleea tenuifolia, Polygala cruciata, Polygala lutea, Polygala nuttallii, Quercus geminata, Rhexia alifanus, Rhexia cubensis, Rhexia lutea, Rhexia mariana, Rhexia parviflora, Rhexia salicifolia, Rhynchospora baldwinii, Rhynchospora chapmanii, Rhynchospora corniculata, Rhynchospora curtissii, Rhynchospora filifolia, Rhynchospora inexpansa, Rhynchospora oligantha, Rhynchospora pleiantha, Rhynchospora plumosa, Rhynchospora scirpoides, Sabatia grandiflora, Sacciolepis indica, Sagittaria isoetiformis, Sarracenia flava, Sarracenia leucophylla, Sarracenia psittacina, Schizachyrium, Scleria muhlenbergii, Scleria reticularis, Serenoa repens, Smilax laurifolia, Syngonanthus flavidulus, Taxodium ascendens, Tiedemannia, Triantha racemosa, Utricularia cornuta, Utricularia gibba, Utricularia subulata, Woodwardia areolata, Woodwardia virginica, Xyris ambigua, Xyris baldwiniana, Xyris caroliniana, Xyris drummondii, Xyris elliottii, Xyris isoetifolia, Xyris longisepala.*

Use by wildlife: *Lachnocaulon anceps* is a host for leafhoppers (Insecta: Hemiptera: Cicadellidae: *Lonatura bicolor*).

Economic importance: food: *Lachnocaulon* species are not edible; **medicinal:** no uses reported; **cultivation:** *Lachnocaulon* species are not cultivated; **misc. products:** none; **weeds:** none; **nonindigenous species:** none.

Systematics: *Lachnocaulon* faces taxonomic issues at many levels. Although the group resolves as a clade in phylogenetic analyses, it is nested within *Paepalanthus* if the traditional circumscription of that genus is maintained (e.g., Figure 3.7). Consequently, some authors have advocated merging *Lachnocaulon* (also *Tonina*) within a broad (*sensu lato*) generic concept of *Paepalanthus*. However, other nomenclatural options are possible; e.g., *Lachnocaulon* and *Tonina* could be retained if *Paepalanthus* was split into different genera. In order to minimize nomenclatural confusion, *Lachnocaulon* has been retained here as a distinct genus until this issue has been settled satisfactorily. As such, it appears to be related most closely to the monotypic *Tonina* (i.e., *T. fluviatilis*; Figure 3.7), which differs from all other Eriocaulaceae (including *Lachnocaulon*) by its spiny floral heads and bilocular stamens. Although comprising a relatively small number of species, *Lachnocaulon* is a highly technical genus taxonomically, which is known to create "considerable confusion" when attempting to distinguish some of the species from one another (e.g., *L. beyrichianum* and *L. minus*). A comprehensive phylogenetic study of the genus has not yet been published but is sorely needed in order to adequately assess species limits and interrelationships. Distinctive seed-coat features (appendages) have indicated a possible close relationship among *L. anceps*, *L. digynum*, and *L. minus*; however, results based on preliminary (but unpublished) molecular evidence have provided conflicting indications. No chromosome counts are available for any species of *Lachnocaulon*. No hybrids have been reported in the genus.

Distribution: *Lachnocaulon anceps*, *L. beyrichianum*, and *L. minus* all occur along the southeastern coastal plain of the United States.

References: Amoroso & Judd, 1995; Avis et al., 1997; Beal, 1977; Brewer et al., 2011; Bridges, 2005; Bridges & Orzell, 1992b; 2002; Buchmann et al., 2010; Chafin, 2007; Cook, 1996; de Borges et al., 2009; Drewa et al., 2002; Elam et al., 2009; Godfrey & Wooten, 1979; Harper, 1906; Hays, 2010; Huffman & Judd, 1998; Iltis, 1950; Kral, 2000; Kramer, 1967; McMillan et al., 2002; Murphy & Boyd, 1999; Orzell & Bridges, 2006a; 2006b; Peet & Allard, 1993; Poiani & Dixon, 1995; Sheridan et al., 1997; Skinner & Sorrie, 2002; Stützel, 1998; Stützel & Gansser, 1996; Thornhill et al., 2014; Unwin, 2004; Uphof, 1927; Walker & Peet, 1984; Wooten & Leonard, 1984; Zona et al., 2012.

Family 10.2. Mayacaceae [1]

Mayacaceae are a small family comprising a single genus and six species, which all inhabit aquatic or wetland environments where they experience at least seasonal submersion in freshwater. Phylogenetic studies have failed to provide any evidence of a close relationship between Mayacaceae and Commelinaceae, which had once been assumed (Cronquist, 1981; Stevenson, 1998; Thieret, 1975). Mayacaceae and Xyridaceae uniquely share several embryological features such as an exothecium, secretory tapetum, and parietal placentation (Stevenson, 1998). An operculate seed apparently does not develop as often is attributed to the family (Venturelli & Bouman, 1986). Moreover, Mayacaceae differ from Commelinaceae by their unsheathed leaves, presence of scalariform vessels, lateral roots that arise opposite the protophloem poles, and the absence of raphide crystals and silica (Stevenson, 1998). Plastid DNA sequence data (e.g., Figure 3.6) have indicated a fairly isolated position of Mayacaceae in proximity of Eriocaulaceae and Xyridaceae; however, contrasting indications of relationships (but always within Poales) have been depicted in other analyses of various sequence data (Barrett et al., 2016; Judd et al., 2016; Givnish et al., 2018; Hochbach et al., 2018). At this time, it seems best to regard Mayacaceae simply as an isolated lineage of Poales.

Little information has been published on the life history attributes of the family. The flowers allegedly are entomophilous, but probably are primarily autogamous; their seeds are buoyant and dispersed by water (Venturelli & Bouman, 1986; Cook, 1996). Some species (e.g., *Mayaca fluviatilis*) are cultivated on occasion as aquarium plants (Kasselmann, 2003); otherwise, the group is of little commercial value.

Mayacaceae are distributed in tropical or semitropical regions of western Africa and the New World. One genus contains OBL North American indicators:

10.2.1. ***Mayaca*** Aubl.

10.2.1. Mayaca

Bog-moss

Etymology: derived from *maiacá*, the common name of the plants in the Guianas

Synonyms: *Biaslia*; *Coletia*; *Syena*

Distribution: global: Africa; New World; **North America:** southeastern United States

Diversity: global: 6 species; **North America:** 1 species

Indicators (USA): obligate wetland (OBL): *Mayaca fluviatilis*

Habitat: freshwater; saline (inland); lacustrine, palustrine, riverine; **pH:** 4.6–7.8; **depth:** to 2 m; **life-form(s):** emergent or submersed (vittate) herb

Key morphology: stems (to 60 cm) herbaceous, submerged or terrestrial, decumbent to erect, sometimes branched and matted; leaves (to 30 mm), cauline, alternate, spiral, the blade linear to narrowly lanceolate, single-veined, apex entire or bifid, margins entire; inflorescence solitary, terminal but appearing lateral; flowers (to 1 cm) chasmogamous or cleistogamous, bisexual, radial, pedicellate (to 12 mm but elongating in fruit to 20 mm); sepals (to 4.5 mm) 3, green, lanceolate-elliptic to ovate; petals (to 5 mm) 3, imbricate, broadly ovate, short-clawed, lilac, maroon, pink, or rose, sometimes whitish at base; stamens (to 3 mm) 3, antisepalous, the filaments (to 2 mm) slender, the anthers (to 1 mm) yellow, dehiscing apically by pore-like slits; pistil (to 2.3 mm) tricarpellate, ovary superior, unilocular, style 1, terminal, stigma simple or 3-fid; capsules (to 4 mm) globose to ellipsoid, loculicidal, 3-valved; seeds (to 1.3 mm) numerous (to 25), nearly globose, seed coat pitted or ridged

Life history: duration: perennial (dormant apices, whole plants); **asexual reproduction:** shoot fragments; **pollination:** self; **sexual condition:** hermaphroditic; **fruit:** capsules (common); **local dispersal:** seeds (water); **long-distance dispersal:** seeds (water)

Imperilment: 1. *Mayaca fluviatilis* [G5]; S2 (LA); S3 (NC)

Ecology: general: All *Mayaca* species are perennial hydrophytes, which inhabit freshwater communities as submersed or emergent life-forms. Their flowers are self-compatible, lack nectaries, have fairly showy perianths, and exhibit varied pollination systems among the species. Some of the flowers are "buzz-pollinated" by bees (Insecta: Hymenoptera: Halictidae), which vibrate the pollen out from the distinctly poricidal anthers. Other flowers are cleistogamous and autogamous, with precociously germinating pollen and weakly defined anther pores that are directed toward the stigma. The seeds float and are dispersed locally by water currents. Other means of longer-distance dispersal have not been elucidated. Germination requirements have been poorly characterized for the seeds, but those of some species appear to require a period of desiccation in order to induce germination.

10.2.1.1. *Mayaca fluviatilis* **Aubl.** inhabits exposed, saturated substrates or deeper waters (to 2 m) in baygalls (seep, wooded), bogs (acid, beach strand, hillside, peat, pitcher plant), borrow pits (roadside), canals (drainage), depressions (boggy, roadside, wet), ditches (drainage, roadside, wet), flatwoods (pine), marshes (roadside), mudflats, pools (swamp), rights-of-way (gas pipeline, powerline), savannas (pine, wet), seepages (spring-fed), slopes (seepage), sloughs, springheads (miry), spring runs, streams, swamps (bayhead, nonalluvial), tire ruts, and along the margins or shores of lagoons, lakes (coastal, dune), ponds (ephemeral, farm, fish, karst), pools, streams, and swamps at elevations to 140 m. The plants are amphibious and able to grow well either as a submersed lifeform in permanent, still or flowing, standing waters (to 2 m deep) or as an emergent on saturated substrates where no standing water is present. They can grow in full sunlight but are shade-tolerant and have been recorded in deeper waters where only 8.7% of the surface illumination was available. The substrates have been described as clay, loamy sand (Plummer), Melhomes (humaqueptic psammaquents), muck, mud, Pamlico (terric medisaprists), peat, sand (Sparta), sandy gravel, sandy loam, sandy muck, and silt. The plants have a general preference for waters (often darkly colored) that are warm (23°C–30°C), acidic (pH: 4.6–7.8; $\bar{x}=5.9$; $\tilde{x}=5.8$), soft (alkalinity [as mg l^{-1} CaCO$_3$]: 0.2–26.2; $\bar{x}=7.5$; $\tilde{x}=2.1$; conductance [µS cm^{-1} @ 25°C]: 10–189; $\bar{x}=102$; $\tilde{x}=70$), and nutrient-poor (total P [µg l^{-1}]: 6–27; $\bar{x}=17$; $\tilde{x}=11$; total N [µg l^{-1}]: 160–1,090; $\bar{x}=550$; $\tilde{x}=410$; Ca [mg l^{-1}]: 0.9–18.3; $\bar{x}=8.8$; $\tilde{x}=3.7$; Mg [mg l^{-1}]: 0.8–12.4; $\bar{x}=6.2$; $\tilde{x}=2.4$; Na [mg l^{-1}]: 2.6–16.7; $\bar{x}=10.4$; $\tilde{x}=7.0$; K [mg l^{-1}]: 0.2–5.3; $\bar{x}=1.8$; $\tilde{x}=0.9$; SO$_4^{2-}$ [mg l^{-1}]: 4.0–27.7; $\bar{x}=13.3$; $\tilde{x}=7.8$; Cl$^-$ [mg l^{-1}]: 4.2–27.5; $\bar{x}=18.6$; $\tilde{x}=12.1$; Fe [mg l^{-1}]: 0.0–0.4; $\bar{x}=0.1$; $\tilde{x}=0.1$; Si [mg l^{-1}]: 0.1–3.4; $\bar{x}=0.9$; $\tilde{x}=0.3$). Flowering and fruiting occur from March to November. Several misconceptions exist regarding the pollination biology of this species. Some accounts indicate that the plants are sterile when submersed and flower only when emersed, whereas others indicate that flowering and fruiting occurs beneath the surface in deeper waters. In actuality, the plants produce chasmogamous and cleistogamous flowers, with the latter present on both submersed and emergent plants. The cleistogamous flowers are self-compatible and autogamous while in bud, wherein the poricidal anthers are directed at the stigma. The pollen grains germinate within the anthers, with the pollen tubes then extending to the stigma to effect self-pollination. Chasmogamous flowers do not develop on plants growing in deep water, which produce only cleistogamous flowers. On stranded or emergent plants, the chasmogamous flowers open primarily during bright, sunny days and remain closed when overcast or rainy. Although the chasmogamous flowers were presumed to be pollinated by insects (Insecta), no insect floral visitors have been observed or reported. There also has been no evidence of wind pollination, though that mechanism also has been suggested. Thus, the principal (if not only) mode of sexual reproduction in this species appears to be by autogamous self-pollination. The fruits can mature and release their seeds while above or below the water surface. The buoyant seeds float by means of airspaces in their exotesta and are dispersed (at least locally) by water. Germinating seeds occur often near the parental plants. Germination has occurred in seeds that were dried for 6 weeks, while those kept in water remained dormant. A seed bank of uncertain duration develops. Seeds taken from seed bank cores have germinated under saturated (but not inundated) conditions. The plants apparently perennate as intact shoots, perhaps with dormant apices during colder periods. Vegetative reproduction occurs readily by fragmentation of the shoots, from which adventitious roots can arise. Dislodged plants often develop into dense floating mats. The plants do not compete well with *Hydrilla verticillata* in experimental manipulations. **Reported associates:** *Acer rubrum, Agalinis maritima, Alnus serrulata, Alternanthera philoxeroides, Amphicarpum muhlenbergianum, Arnoglossum, Azolla, Baccharis, Bacopa caroliniana, Bacopa monnieri, Brasenia schreberi, Burmannia biflora, Cabomba, Carex alata, Centella asiatica, Cephalanthus occidentalis, Ceratophyllum demersum, Chamaecyparis, Cladium jamaicense, Clethra alnifolia, Cliftonia monophylla, Cyperus lecontei, Cyperus surinamensis, Cyrilla racemiflora, Decodon verticillatus, Dichanthelium ensifolium, Didiplis diandra, Diodia virginiana, Drosera capillaris, Drosera intermedia, Dulichium arundinaceum, Egeria densa, Eichhornia crassipes, Eleocharis acicularis, Eleocharis baldwinii, Eleocharis cellulosa, Eleocharis equisetoides, Eleocharis obtusa, Eleocharis quadrangulata, Eleocharis tuberculosa, Eriocaulon compressum, Eriocaulon decangulare, Eriocaulon lineare, Fimbristylis spadicea, Fuirena scirpoidea, Galium tinctorium, Gordonia lasianthus, Habenaria repens, Helanthium tenellum, Hibiscus grandiflorus, Humbertacalia, Hydrocotyle umbellata, Hypericum canadense, Hypericum crux-andreae, Ilex coriacea, Iris tridentata, Iris virginica, Itea virginica, Juncus effusus, Juncus megacephalus, Juncus repens, Juncus roemerianus, Lemna, Limnobium spongia, Liriodendron tulipifera, Lobelia boykinii, Ludwigia leptocarpa, Ludwigia repens, Luziola fluitans, Lyonia lucida, Magnolia grandiflora, Magnolia virginiana, Mikania cordifolia, Mimosa strigillosa, Mitchella repens, Myrica cerifera, Myriophyllum heterophyllum, Myriophyllum laxum, Najas filifolia, Najas guadalupensis, Nelumbo lutea, Nuphar advena, Nymphaea odorata, Nymphoides aquatica, Nyssa biflora, Nyssa sylvatica, Orontium aquaticum, Osmanthus americanus, Osmundastrum cinnamomeum, Panicum hemitomon, Panicum repens, Panicum virgatum, Persicaria hirsuta, Persicaria hydropiperoides, Persicaria punctata, Pinus glabra, Pinus palustris, Platanthera, Polygala ramosa, Pontederia cordata, Potamogeton floridanus, Potamogeton pulcher, Quercus, Reimarochloa oligostachya, Rhexia mariana, Rhododendron viscosum, Rhynchospora capitellata, Rhynchospora chalarocephala, Rhynchospora inundata, Rhynchospora stenophylla, Ricciocarpus natans, Ruppia maritima, Sagittaria graminea, Sagittaria lancifolia, Sagittaria latifolia, Sagittaria macrocarpa, Sarracenia flava, Sarracenia leucophylla, Sphagnum, Taxodium ascendens, Taxodium distichum, Thelypteris, Toxicodendron vernix, Triantha racemosa, Typha latifolia, Uniola paniculata,*

Utricularia cornuta, Utricularia gibba, Utricularia inflata, Vaccinium corymbosum, Vallisneria americana, Viburnum nudum, Vitis rotundifolia, Xyris baldwiniana, Xyris correlliorum, Xyris drummondii, Xyris elliottii, Xyris scabrifolia.

Use by wildlife: *Mayaca fluviatilis* plants afford cover for invertebrates (e.g., Crustacea: Amphipoda: Hyalellidae: *Hyalella azteca*) and small fish (Vertebrata: Teleostei). The foliage provides substrate for epiphytic diatoms (Bacillariophyceae). The plants are consumed by Amazonian manatees (Mammalia: Trichechidae: *Trichechus inunguis*). In Brazil, they are eaten by female slider turtles (Reptilia: Emydidae: *Trachemys dorbigni*). In northern South America, they are hosts for larval moths (Insecta: Lepidoptera: Erebidae: *Paracles laboulbeni*), which feed on the foliage.

Economic importance: food: *Mayaca fluviatilis* is not reported as edible; **medicinal:** Extracts of *Mayaca fluviatilis* have effectively reduced the growth (by >65%) of *Trypanosoma brucei*, the causative agent of African Trypanosomiasis in humans. *Mayaca* extracts also have been considered as an antifungal treatment for *Candida albicans* (Fungi: Ascomycota), which causes oral thrush in humans; **cultivation:** *Mayaca fluviatilis* is cultivated as an ornamental aquarium plant; **misc. products:** Extracts of *Mayaca fluviatilis* have been used as the reducing agent in the synthesis of silver nanoparticles; **weeds:** Thriving Florida populations of *Mayaca fluviatilis* have interfered with boat traffic and have become a nuisance in some lakes. The plants are regarded as potentially invasive in Sri Lanka where they are cultivated as aquarium specimens; **nonindigenous species:** *Mayaca fluviatilis* has been introduced to Sri Lanka as an escape from cultivation.

Systematics: *Mayaca* has been represented by only one or two taxa in phylogenetic studies. However, by its morphological distinctness and cohesiveness, there is little doubt that the group is monophyletic. Yet, additional sampling of *Mayaca* taxa needs to be carried out in phylogenetic studies in order to corroborate this assumption. In various phylogenetic analyses, the genus resolves consistently within Poales, either within a clade containing Cyperaceae, Eriocaulaceae, Juncaceae, and Xyridaceae (the "cyperid clade") or in an isolated position in proximity of those groups (e.g., Figure 3.6). Its more precise phylogenetic placement remains uncertain. In the past, two *Mayaca* species were thought to occur in North America. However, only *M. fluviatilis* currently is recognized, with material assigned formerly to *M. aubletii* arguably representing phenotypically plastic variability associated with its terrestrial form. The basic chromosome number of Mayacaceae presumably is $x=8$; however, no counts are known to have been published for any *Mayaca* species. Nevertheless, the genome size (1C) of *M. fluviatilis* has been estimated at 0.49 pg. The plants lack two introns (4, 9) in *ABCB1*, a gene that encodes a P-glycoprotein involved in basipetal shoot auxin transport. No hybrids involving any *Mayaca* species have been reported.

Distribution: *Mayaca fluviatilis* occurs along the Atlantic coastal plain of the southeastern United States and extends southward into South America.

References: Adis, 1983; Avis et al., 1997; Beal, 1977; Birkenholz, 1963; Bridges & Orzell, 2003; Bunghez et al.,

2010; Canfield et al., 1985; Chathurangani et al., 2016; Cherry & Gough, 2006; Christenhusz et al., 2017; Crema et al., 2019; Crocker, 1907; Faden, 2000b; Godfrey & Wooten, 1979; Hahn et al., 2014; Hanlon et al., 2000; Hoyer et al., 1996; Jain et al., 2016; Kasselmann, 2003; Leitch et al., 2010; Oriani & Scatena, 2012; Palis, 1998; Parvathaneni et al., 2017; Penn, 1950; Pott & Pott, 2000; Reena et al., 2013; Spence, 1981; Tarver et al., 1978; Thieret, 1975; Uphof, 1933; 1938; VanTassel & Janosik, 2019; Venturelli & Bouman, 1986; Yakandawala & Dissanayake, 2010; Zimba & Bates, 1996.

Family 10.3. Typhaceae [2]

In earlier literature (e.g., Cronquist, 1981; FNA, 2000), the cattail family (Typhaceae) included only the genus *Typha*, with *Sparganium* assigned to its own family (Sparganiaceae). However, morphological studies (Müller-Doblies, 1969; 1970) had emphasized the highly similar floral and inflorescence structure of both genera, which indicated that they should be combined within the same family. As subsequent phylogenetic studies (summarized by Judd et al., 2016) demonstrated that these two, small, morphologically similar genera resolved as sister clades (e.g., Figures 3.6 and 3.8), they eventually were combined into one family as Typhaceae (APG, 2016), a circumscription that still includes fewer than 55 species. As such, the family can be defined as comprising monoecious, aquatic, and wetland herbs, with broadly linear, 2-ranked, sheathing leaves and reduced, unisexual flowers, which are aggregated into dense heads or spikes. Although tricarpellate, only one carpel remains functional and produces a single, apical ovule. The flowers lack nectaries and have a perianth modified as hairlike or scalelike tepals (Cronquist, 1981; Judd et al., 2016). All the species are wind pollinated; the seeds are dispersed primarily by water or wind, but also by animals (Cronquist, 1981; Cook, 1996).

To quell any doubt whether these genera should be placed within the same family, the author's own laboratory obtained interesting results when genetically screening sterile, cattail-like material from several sites in Connecticut and New Hampshire (USA). We found all the unusual material to be polymorphic for biparentally inherited (nrDNA) sequences, which when cloned and resequenced, yielded one copy matching *Typha* and another matching *Sparganium* (Les et al., unpublished). Thus, the ability to form intergeneric hybrids provides compelling evidence that the genera are quite similar genetically.

Both *Sparganium* and *Typha* have some commercial value as ornamental pond or water garden specimens. Indigenous North Americans used the plants extensively as food (inflorescences, pollen, rhizomes, seeds, young shoots), medicine, for weaving (baskets and roofing), and various other purposes.

Typhaceae are iconic inhabitants of wetland communities and occur widely throughout hydric habitats worldwide. Two genera contain OBL North American indicators:

10.3.1. ***Sparganium*** L.
10.3.2. ***Typha*** L.

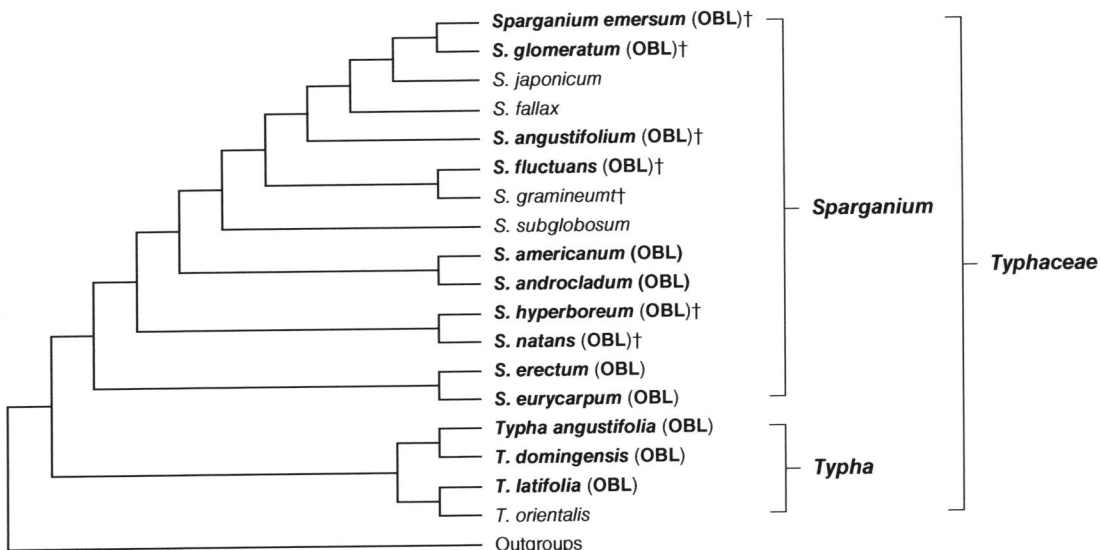

FIGURE 3.8 Phylogenetic relationships in Typhaceae as indicated by analysis of combined nuclear and plastid DNA sequence data (modified from Sulman et al., 2013). *Sparganium* and *Typha* resolve as sister genera and are classified appropriately within the same family. The entire Typhaceae are aquatic with the North American taxa (highlighted in bold) consistently designated as OBL wetland indicators (shown in parentheses). Floating-leaved species (marked by †) have evolved several times within *Sparganium*.

10.3.1. Sparganium

Bur-reed; ruban d'eau, rubanier

Etymology: an ancient name, allegedly derived from the Greek *sparganon* ("band" or "strip") for the strap-like leaves

Synonyms: *Platanaria*

Distribution: global: cosmopolitan; **North America:** widespread

Diversity: global: 14–22 species; **North America:** 9 species

Indicators (USA): obligate wetland (OBL): *Sparganium americanum, S. androcladum, S. angustifolium, S. emersum, S. eurycarpum, S. fluctuans, S. glomeratum, S. hyperboreum, S. natans*

Habitat: brackish, freshwater; freshwater (tidal); lacustrine, palustrine, riverine; **pH:** 2.0–9.8; **depth:** to 2.5 m; **life-form(s):** emergent or floating-leaved herbs

Key morphology: plants (to 2.5 m) herbaceous, slender and grasslike or robust; rhizomes slender or stout; leaves grasslike or ribbonlike, flat or keeled, spongy, emergent and erect (to 2.5 m) or limp and floating (to 2.0 m); inflorescences emergent or floating, monoecious, rachis 0–3-branched, erect or flexuous, the flowers sessile and unisexual, ♀ heads (to 5.0 cm) 1–6, lowermost, glomerate, axillary or supraaxillary, sessile or peduncled; ♂ heads 1–40+, uppermost, glomerate; tepals persistent, with or without a prominent subapical dark spot, attached basally or partly adnate to stipe; stigmas 1 or 2, lanceolate, lance-ovate, linear, linear-lanceolate, or ovate; fruits (to 10 mm) sessile or stipitate, ellipsoid, elliptic, fusiform, obovoid or obpyramidal, the body unfaceted or 3–7-faceted, constricted medially or not, tapering abruptly or gradually to the beak, beak (to 7 mm) curved or straight, or essentially absent (<0.5 mm); seeds 1–3, slender-ovoid, seed coat thin, appressed to endocarp

Life history: duration: perennial (rhizomes); **asexual reproduction:** rhizomes, vegetative fragments; **pollination:** wind;

sexual condition: monoecious; **fruit:** multiple nutlets (prolific); **local dispersal:** fragments, fruits (water); **long-distance dispersal:** fragments, fruits (animals, water)

Imperilment: 1. *Sparganium americanum* [G5]; S1 (IL, KS); S2 (PE); S3 (NF, QC); **2.** *S. androcladum* [G4/G5]; SH (NH, ON); S1 (IA, PA, TN, VT); S2 (IN, OH, QC, WV); S3 (IL, MA, NY); **3.** *S. angustifolium* [G5]; SH (NJ); S1 (WV); S2 (MA, NU, PA); S3 (NY, WY); **4.** *S. emersum* [G5]; S1 (IL, OH, SD, NC, VA); S2 (NE); S3 (LB, MB, MD, WY); **5.** *S. eurycarpum* [G5]; S1 (DC, KY, WY); S2 (CO, NH, UT); S3 (MD, NJ, VA); **6.** *S. fluctuans* [G5]; S1 (AB, CT, PE, WA); S2 (LB, MA, NF); S3 (BC, NY, VT); **7.** *S. glomeratum* [G4]; SH (ON, QC); S1 (AB, MB, SK); S2 (LB, WI); S3 (MN); **8.** *S. hyperboreum* [G5]; S1 (NS, SK); S2 (ON); S3 (AB, MB, NF); **9.** *S. natans* [G5]; SX (IN, PA); S1 (CT, IL, MA, NJ, UT); S2 (NH, NY, NU, PE, VT); S3 (CA, LB, NF, NS, WI, WY, QC, YT)

Ecology: general: *Sparganium* species are entirely aquatic with all the North American representatives designated as OBL wetland indicators. The species vary in being strictly emergent, strictly floating-leaved, or capable of either growth form in response to varying water depths. The emergent species possess structural foliar adaptations (e.g., fiber masses, large bundle sheaths, extra photosynthetic tissue) compared to floating leaves of the natant species, which lack fiber masses and possess only moderately developed bundle sheaths. The flowers of all the species are unisexual and are produced in glomerate heads on emergent inflorescences. The plants are monoecious. At least some of the species are self-compatible. However, individual plants are protogynous (temporal dioecy), with the pistillate flowers being receptive before the staminate heads mature. There is only a single pistillate phase followed by a single staminate phase. Flowers of all the species are odorless and wind pollinated. The fruits are achene-like, but can contain from 1-several seeds. They are buoyant and dispersed locally on the water surface, or are dispersed over greater distances by birds (Aves) and other animals, which eat them, transport them endozoically, and then release them upon defecation. The fresh seeds germinate rarely but can remain viable for decades in persistent seed banks. The thickly sclerified pericarps resist water imbibition, which results in an extended seed dormancy that can last up to several years. However, removal or disruption of the "micropyle cap" from a seed will result in rapid germination. For several species, the optimal treatment for promoting germination of intact fruits consists of aqueous storage at 20°C for 6–8 months followed by 1 month of cold storage at 5°C. Storage at temperatures below 20°C for a 1-year period or longer usually will promote germination. The highest germination rates also have been observed for seeds incubated under an alternating day/night (8/16 hr) temperature regime (30°C/5°C). The plants develop from a bulblike base, which produces rhizomes that facilitate clonal, vegetative reproduction. As often is the case with taxonomically difficult groups, literature reports can be misleading. In particular, any reference to "*S. emersum*" (without indication of subspecies) is particularly problematic given that *S. emersum* var. *multipedunculatum* actually is synonymous with *S. angustifolium* and *S. emersum* subsp. *acaule* recently has been distinguished as *S. acaule*, a potentially distinct species (see *Systematics* below). An appropriate awareness should be implemented when referencing the following treatments for *S. angustifolium* and *S. emersum*, for which at least some of the information has been based on the prevailing (but not necessarily correct) taxonomy.

10.3.1.1. *Sparganium americanum* **Nutt.** is an emergent plant, which grows in canals (drainage), channels, depressions (low, wet), ditches (roadside, spring-fed), fens, floodplains, hammocks (hydric), marshes (spring-fed), mats (floating), meadows (low), oxbows, pools (beaver), roadsides, seeps, shoals, sloughs (spring-fed), swales (saturated), swamps (bottomland, cypress, hardwood, nonalluvial, roadside), and along the margins and shores of bogs (*Sphagnum*), lakes (artificial), ponds (beaver, farm, roadside, sinkhole), reservoirs, rivers, and streams (seepage) at elevations to 631 m. Exposures can range from full sun to partial shade. The plants are adapted to a broad range of environmental conditions. They can grow on saturated substrates, or in up to 1 m of still or flowing (0–38 cm s^{-1}), standing water. The substrates can include gravel, muck, mucky peat, mucky sand, mud, muddy gravel, peat, peaty muck, Pelham (arenic paleaquults), rocks, sand (Sparta), sandy cobble, sandy gravel, sandy silt, silt, or silty sand (pH: 2.0–9.8; $\bar{x} = 5.7$–6.9; alkalinity [mg l^{-1}]: 0.5–123.5; $\bar{x} = 17.7$–27.6; Al [μeq l^{-1}]: 2–16; Ca [μeq l^{-1}]: 20–60; Cl [μeq l^{-1}]: 104–138; Fe [μeq l^{-1}]: 1–14; HCO$_3$ [μeq l^{-1}]: 0–44; K [μeq l^{-1}] 3–5; Mg [μeq l^{-1}]: 25–41; Na [μeq l^{-1}]: 104–139; NH$_4$ [μeq l^{-1}]: 0.7–2.1; SO$_4$ [μeq l^{-1}]: 52–79). Flowering and fruiting occur from May to October. The frequency of flowering is highest in shallow water (12–45 cm depth) and decreases with increasing water depth. The production of staminate heads can vary with greater numbers produced at higher levels of pH and alkalinity. The floral biology and seed dispersal are typical for

the genus. A germination rate of 25% has been reported for seeds buried in damp sand within mesh bags for 2–12 months at 5°C and then incubated at 22°C under a 14-hr photoperiod. Seeds stored in water at 1°C–3°C have retained high viability (80–90%) after 5 years. Total biomass estimates of 3.5 kg m^{-2} have been reported. The rhizome:shoot biomass ratio has been found to average about 2:1 (1.92) but can vary from 0.64 to 4.23. The geographical "absence" of this and other aquatic plant species from the "prairie peninsula" has been attributed to a post-glacial warming period and the unavailability of acidic habitats; however, at least in this species, that pattern is due more likely to agricultural development given its broad range of tolerance to acidity and temperature. **Reported associates:** *Acorus calamus, Aldrovanda vesiculosa, Alisma subcordatum, Alnus serrulata, Alternanthera philoxeroides, Amaranthus tuberculatus, Ambrosia artemisiifolia, Ammannia coccinea, Amorpha fruticosa, Apios americana, Asclepias incarnata, Azolla cristata, Azolla filiculoides, Bacopa rotundifolia, Betula pumila, Bidens beckii, Bidens cernuus, Bidens frondosus, Bidens laevis, Bidens trichospermus, Bidens tripartitus, Boehmeria cylindrica, Bolboschoenus fluviatilis, Boltonia asteroides, Brasenia schreberi, Calamagrostis canadensis, Calla palustris, Callitriche heterophylla, Callitriche palustris, Carex alata, Carex albolutescens, Carex aquatilis, Carex atherodes, Carex comosa, Carex crinita, Carex debilis, Carex gigantea, Carex gynandra, Carex hystericina, Carex intumescens, Carex lacustris, Carex lasiocarpa, Carex lonchocarpa, Carex lupuliformis, Carex lupulina, Carex lurida, Carex projecta, Carex pseudocyperus, Carex retrorsa, Carex rostrata, Carex scoparia, Carex stipata, Carex stricta, Carex suberecta, Carex tribuloides, Carex typhina, Carex vulpinoidea, Cephalanthus occidentalis, Ceratophyllum demersum, Chamaecyparis thyoides, Chasmanthium laxum, Cicuta bulbifera, Cicuta maculata, Clethra alnifolia, Coleataenia anceps, Coleataenia longifolia, Commelina virginica, Cuphea carthagenensis, Cuscuta gronovii, Cyperus bipartitus, Cyperus erythrorhizos, Cyperus haspan, Cyperus odoratus, Cyperus strigosus, Cyrilla racemiflora, Dichanthelium ravenelii, Dichanthelium sphaerocarpon, Didiplis diandra, Diodia virginiana, Drosera rotundifolia, Dulichium arundinaceum, Echinochloa crusgalli, Echinochloa walteri, Echinodorus cordifolius, Eclipta prostrata, Elatine minima, Eleocharis acicularis, Eleocharis erythropoda, Eleocharis obtusa, Eleocharis palustris, Eleocharis quadrangulata, Eleocharis robbinsii, Eleocharis tortilis, Elodea canadensis, Elodea nuttallii, Epilobium coloratum, Equisetum fluviatile, Eragrostis hypnoides, Erechtites hieracifolius, Eriocaulon aquaticum, Eubotrys racemosa, Eupatorium perfoliatum, Eutrochium maculatum, Fimbristylis autumnalis, Galium obtusum, Galium tinctorium, Geum laciniatum, Glyceria borealis, Glyceria canadensis, Glyceria grandis, Glyceria septentrionalis, Glyceria striata, Gratiola neglecta, Gratiola virginiana, Helonias bullata, Heteranthera dubia, Heteranthera limosa, Hibiscus laevis, Hibiscus moscheutos, Hippuris vulgaris, Hydrocotyle ranunculoides, Hydrocotyle verticillata, Hydrolea ovata, Hydrolea quadrivalvis, Hygrophila lacustris, Hypericum* fasciculatum, *Hypericum mutilum, Hypoxis curtissii, Ilex coriacea, Ilex glabra, Ilex laevigata, Ilex verticillata, Impatiens capensis, Iris hexagona, Iris virginica, Isoetes acadiensis, Isoetes hyemalis, Isoetes lacustris, Isoetes tuckermanii, Itea virginica, Juncus canadensis, Juncus coriaceus, Juncus diffusissimus, Juncus effusus, Juncus militaris, Justicia ovata, Kalmia latifolia, Leersia oryzoides, Lemna minor, Lemna trisulca, Ligustrum sinense, Lindernia dubia, Lobelia cardinalis, Lobelia dortmanna, Lobelia siphilitica, Lonicera japonica, Ludwigia alternifolia, Ludwigia decurrens, Ludwigia palustris, Ludwigia polycarpa, Ludwigia sphaerocarpa, Lycopus americanus, Lycopus rubellus, Lycopus uniflorus, Lycopus virginicus, Lyonia lucida, Lysimachia ciliata, Lysimachia hybrida, Lysimachia quadriflora, Lysimachia terrestris, Lysimachia thyrsiflora, Lythrum alatum, Lythrum salicaria, Mentha arvensis, Menyanthes trifoliata, Microstegium vimineum, Mimulus ringens, Myosotis laxa, Myrica cerifera, Myrica gale, Myriophyllum aquaticum, Myriophyllum farwellii, Myriophyllum pinnatum, Myriophyllum verticillatum, Najas flexilis, Najas gracillima, Najas guadalupensis, Nasturtium officinale, Nelumbo lutea, Nuphar advena, Nuphar variegata, Nymphaea odorata, Nymphoides aquatica, Nymphoides cordata, Nyssa biflora, Nyssa sylvatica, Oldenlandia uniflora, Onoclea sensibilis, Orontium aquaticum, Osmunda regalis, Osmundastrum cinnamomeum, Panicum hemitomon, Panicum verrucosum, Peltandra virginica, Penthorum sedoides, Persicaria amphibia, Persicaria arifolia, Persicaria coccinea, Persicaria glabra, Persicaria hydropiper, Persicaria hydropiperoides, Persicaria lapathifolia, Persicaria pensylvanica, Persicaria punctata, Persicaria sagittata, Phalaris arundinacea, Phoradendron serotinum, Phragmites australis, Phyla lanceolata, Physostegia parviflora, Picea mariana, Pilea pumila, Planera aquatica, Plantago lanceolata, Pluchea camphorata, Poa palustris, Pontederia cordata, Potamogeton confervoides, Potamogeton epihydrus, Potamogeton foliosus, Potamogeton natans, Potamogeton nodosus, Potamogeton oakesianus, Potamogeton pulcher, Potamogeton pusillus, Potamogeton spirillus, Potamogeton vaseyi, Potamogeton zosteriformis, Proserpinaca palustris, Prunella vulgaris, Ptilimnium, Quercus phellos, Ranunculus flabellaris, Ranunculus longirostris, Ranunculus sceleratus, Rhexia virginica, Rhynchospora careyana, Rhynchospora corniculata, Rhynchospora scirpoides, Riccia fluitans, Rorippa islandica, Rorippa sessiliflora, Rosa multiflora, Rotala ramosior, Rumex orbiculatus, Rumex verticillatus, Saccharum giganteum, Sacciolepis striata, Sagittaria australis, Sagittaria calycina, Sagittaria cuneata, Sagittaria engelmanniana, Sagittaria graminea, Sagittaria latifolia, Sagittaria montevidensis, Sagittaria rigida, Salix eriocephala, Salix nigra, Saururus cernuus, Schoenoplectus acutus, Schoenoplectus californicus, Schoenoplectus subterminalis, Schoenoplectus tabernaemontani, Scirpus atrovirens, Scirpus cyperinus, Scirpus pendulus, Scutellaria galericulata, Scutellaria lateriflora, Sium suave, Smilax laurifolia, Smilax walteri, v Sparganium angustifolium, Sparganium emersum, Sparganium erectum, Sparganium eurycarpum, Sparganium fluctuans, Spartina* pectinata, *Sphagnum, Spiraea tomentosa, Spirodela polyrrhiza, Stachys palustris, Stachys tenuifolia, Stuckenia pectinata, Styrax, Symphyotrichum puniceum, Taxodium ascendens, Thelypteris palustris, Torreyochloa pallida, Toxicodendron radicans, Toxicodendron vernix, Triadenum virginicum, Typha angustifolia, Typha latifolia, Utricularia cornuta, Utricularia geminiscapa, Utricularia gibba, Utricularia intermedia, Utricularia macrorhiza, Utricularia purpurea, Utricularia radiata, Verbena hastata, Verbena urticifolia, Vernonia fasciculata, Veronica anagallis-aquatica, Viburnum nudum, Wolffia brasiliensis, Wolffia columbiana, Wolffiella gladiata, Woodwardia areolata, Xyris difformis, Xyris jupicai, Zizania aquatica, Zizania palustris.*

10.3.1.2. ***Sparganium androcladum* (Engelm.) Morong** is an emergent plant, which occurs in brooks, ditches, fens, flatwoods (seasonal), marshes (roadside), meadows (wet), sloughs, swales (dune), and along the margins or shores of impoundments, lakes, ponds (sand, sinkhole), rivers, streams, and swamps (shrub) at elevations to 424 m. Full sunlight exposures are typical. The waters are shallow (to 61 cm), quiet, and circumneutral (e.g., pH: 6.8; alkalinity: 15 mg l^{-1}). Substrates include muck, mud, and sand. Flowering and fruiting occur from June to September. Sexual reproduction and seed dispersal are typical for the genus. A long-lived seed bank can develop. The seeds have remained viable after more than 4 years of dark storage in cool water. The plants perennate by means of a rhizome. Their lateral and axile roots completely die by the end of October, which is thought to represent an adaptation to the disturbance regime characteristic of their usual habitats. **Reported associates:** *Acer rubrum, Achillea millefolium, Acorus, Agrostis perennans, Alisma subcordatum, Alnus rugosa, Alnus serrulata, Beckmannia syzigachne, Bidens frondosus, Boehmeria cylindrica, Bolboschoenus fluviatilis, Butomus umbellatus, Calamagrostis canadensis, Calamagrostis stricta, Callitriche palustris, Carex atherodes, Carex comosa, Carex decomposita, Carex lasiocarpa, Carex sartwellii, Carex scoparia, Carex stricta, Carex utriculata, Carex vesicaria, Cephalanthus occidentalis, Ceratophyllum demersum, Cornus amomum, Cuscuta, Decodon verticillatus, Dulichium arundinaceum, Echinochloa muricata, Elatine minima, Eleocharis obtusa, Eleocharis ovata, Eleocharis palustris, Eleocharis robbinsii, Equisetum fluviatile, Eriocaulon aquaticum, Eupatorium perfoliatum, Galium tinctorium, Glyceria canadensis, Glyceria grandis, Glyceria septentrionalis, Glyceria striata, Heteranthera dubia, Heteranthera reniformis, Hypericum densiflorum, Hypericum ellipticum, Hypericum mutilum, Ilex verticillata, Impatiens capensis, Iris versicolor, Juncus acuminatus, Juncus alpinus, Juncus balticus, Juncus brevicaudatus, Juncus canadensis, Juncus dudleyi, Juncus effusus, Juncus pelocarpus, Juncus subcaudatus, Juncus tenuis, Kalmia angustifolia, Leersia oryzoides, Lemna minor, Lobelia cardinalis, Lycopus uniflorus, Lysimachia terrestris, Lythrum salicaria, Microstegium vimineum, Myriophyllum heterophyllum, Myriophyllum tenellum, Najas gracillima, Nasturtium officinale, Nuphar advena, Nymphaea odorata, Osmunda regalis, Peltandra virginica, Persicaria amphibia,*

Persicaria coccinea, Persicaria hydropiper, Persicaria hydropiperoides, Persicaria lapathifolia, Persicaria sagittata, Phalaris arundinacea, Phragmites australis, Pontederia cordata, Potamogeton diversifolius, Potamogeton epihydrus, Potamogeton foliosus, Potamogeton natans, Potamogeton praelongus, Potamogeton pusillus, Proserpinaca palustris, Rhododendron viscosum, Riccia fluitans, Rorippa aquatica, Rosa palustris, Rumex verticillatus, Sagittaria latifolia, Sagittaria rigida, Salix sericea, Saururus cernuus, Schoenoplectus tabernaemontani, Scirpus ancistrochaetus, Scirpus atrovirens, Scirpus cyperinus, Sparganium americanum, Sparganium eurycarpum, Spartina pectinata, Spirodela polyrrhiza, Stachys tenuifolia, Thelypteris palustris, Triadenum fraseri, Triadenum walteri, Typha latifolia, Utricularia gibba, Utricularia macrorhiza, Vaccinium corymbosum, Vallisneria americana, Zizania palustris.

10.3.1.3. *Sparganium angustifolium* Michx.

is a floating-leaved plant, which grows in backwaters (river), bogs (*Sphagnum*), channels (irrigation), depressions (roadside), ditches (drainage, roadside), fens (subalpine), flats, floodplains, lagoons, lakes (muskeg), marshes, meadows (boggy, wet), mudflats, oxbows, ponds (alpine, beaver, mill, montane, stock), pools (muskeg, roadside), potholes (glacial), reservoirs, rivers, seeps, sloughs, springs (hot, muddy), streams (muskeg), swales (river bottom), and swamps (riverine) at elevations to 3,628 m. Exposures are in full sunlight. The substrates include alluvium (lakeshore), clay, clay loam, cobble, gravel, gravelly loam, humus loam, loam, loamy clay, muck (organic), mucky clay, mucky peat, mud, muddy sand, sand, sandy muck, sandy mud, sandy ooze, silt, silty clay, silty loam, and silty muck. The plants achieve greater cover as the sediment organic content increases. The waters typically are shallow ($\tilde{x}$: ~50 cm; but to 2.5 m deep), clear (Secchi depth: 1.7–4.1 m), and oligotrophic (total N [μg^{-1}]: 657.3–862.8; $\bar{x} = 744.5$; total P [μg^{-1}]: 13.3–49.5; $\bar{x} = 25.2$). They span a broad range of acidity (pH: 4.6–9.5; $\bar{x} = 7.2$; $\tilde{x}$: 6.8) but are characterized by low alkalinity (as mg l^{-1} CaCO$_3$: 2.5–103.5; $\bar{x} = 26.1$; $\tilde{x} = 15$) and low conductivity (<125 μmhos cm^{-1}@ 25°C). Flowering and fruiting can occur from June to November. Sexual reproduction and seed dispersal are typical for the genus. Details on the seed ecology are unavailable. **Reported associates:** *Acorus americanus, Alisma triviale, Alnus rubra, Alopecurus aequalis, Alopecurus saccatus, Athyrium filix-femina, Beckmannia syzigachne, Betula glandulosa, Bidens beckii, Bidens cernuus, Brasenia schreberi, Calamagrostis canadensis, Calla palustris, Callitriche hermaphroditica, Callitriche heterophylla, Callitriche palustris, Callitriche stagnalis, Caltha palustris, Carex aquatilis, Carex athrostachya, Carex canescens, Carex cusickii, Carex echinata, Carex lacustris, Carex lenticularis, Carex limosa, Carex lyngbyei, Carex macrochaeta, Carex microptera, Carex nebrascensis, Carex obnupta, Carex pauciflora, Carex pluriflora, Carex rostrata, Carex scopulorum, Carex simulata, Carex spectabilis, Carex stipata, Carex utriculata, Carex vesicaria, Ceratophyllum demersum, Chara, Cicuta douglasii, Comarum palustre, Cylindrospermum, Dasiphora floribunda, Drosera anglica, Dulichium arundinaceum, Egeria densa, Elatine minima,*

Eleocharis acicularis, Eleocharis macrostachya, Eleocharis palustris, Eleocharis quinqueflora, Elodea canadensis, Epilobium ciliatum, Equisetum arvense, Equisetum fluviatile, Eriocaulon aquaticum, Eriophorum crinigerum, Eriophorum gracile, Fontinalis, Galium trifidum, Glyceria borealis, Glyceria canadensis, Glyceria grandis, Glyceria septentrionalis, Hippuris vulgaris, Howellia aquatilis, Hymenoxys hoopesii, Hypericum anagalloides, Isoetes bolanderi, Isoetes echinospora, Isoetes occidentalis, Juncus articulatus, Juncus balticus, Juncus bufonius, Juncus effusus, Juncus ensifolius, Juncus filiformis, Juncus militaris, Lemna minor, Lemna trisulca, Leptodictyum riparium, Limosella aquatica, Lobelia dortmanna, Lotus corniculatus, Ludwigia palustris, Lysichiton americanus, Meesia triquetra, Menyanthes trifoliata, Mimulus lewisii, Mimulus moschatus, Mimulus tilingii, Mougeotia, Myosotis laxa, Myrica gale, Myriophyllum alterniflorum, Myriophyllum aquaticum, Myriophyllum farwellii, Myriophyllum quitense, Myriophyllum sibiricum, Myriophyllum tenellum, Myriophyllum verticillatum, Najas flexilis, Najas gracillima, Nardia scalaris, Nephrophyllidium crista-galli, Nitella, Nuphar microphylla, Nuphar polysepala, Nuphar variegata, Nymphaea odorata, Nymphaea tetragona, Nymphoides cordata, Oenanthe sarmentosa, Pedicularis groenlandica, Persicaria amphibia, Persicaria coccinea, Persicaria maculosa, Phalaris arundinacea, Philonotis fontana, Plantago major, Platanthera dilatata, Pontederia cordata, Potamogeton alpinus, Potamogeton amplifolius, Potamogeton berchtoldii, Potamogeton confervoides, Potamogeton epihydrus, Potamogeton friesii, Potamogeton gramineus, Potamogeton illinoensis, Potamogeton natans, Potamogeton nodosus, Potamogeton oakesianus, Potamogeton obtusifolius, Potamogeton perfoliatus, Potamogeton praelongus, Potamogeton pusillus, Potamogeton richardsonii, Potamogeton robbinsii, Potamogeton spirillus, Potamogeton vaseyi, Potamogeton zosteriformis, Prunella vulgaris, Ptychostomum pseudotriquetrum, Ranunculus flammula, Ranunculus gmelinii, Ranunculus longirostris, Ranunculus pallasii, Ranunculus sceleratus, Ranunculus trichophyllus, Rorippa palustris, Rorippa sphaerocarpa, Rumex maritimus, Sagittaria cuneata, Sagittaria graminea, Sagittaria latifolia, Sagittaria teres, Salix bebbiana, Salix boothii, Salix hookeriana, Sarmenthypnum sarmentosum, Schoenoplectus acutus, Schoenoplectus subterminalis, Schoenoplectus tabernaemontani, Scirpus microcarpus, Sium suave, Sparganium americanum, Sparganium emersum, Sparganium eurycarpum, Sparganium fluctuans, Sparganium glomeratum, Sparganium hyperboreum, Sparganium natans, Spiraea douglasii, Spirodela polyrrhiza, Stuckenia filiformis, Stuckenia pectinata, Stuckenia vaginata, Subularia aquatica, Taraxacum officinale, Tephroseris palustris, Torreyochloa erecta, Torreyochloa pallida, Trichophorum cespitosum, Trientalis europaea, Typha latifolia, Utricularia gibba, Utricularia intermedia, Utricularia macrorhiza, Utricularia minor, Utricularia ochroleuca, Utricularia purpurea, Utricularia resupinata, Vallisneria americana, Veronica americana, Veronica anagallis-aquatica, Veronica scutellata, Zannichellia palustris, Zizania palustris.

10.3.1.4. *Sparganium emersum* **Rehmann** is an emergent or floating-leaved plant, which inhabits backwaters (river), bogs, canals, depressions (peat), ditches (drainage, roadside), fens, floodplains, gravel pits, lakes, marshes, meadows, mudflats, oxbows, ponds (beaver, kettle, mill, roadside, spring-fed), pools (brook, roadside, seasonal), potholes, reservoirs, rivers, sloughs (river, tidal), streams, swales, and swamps at elevations to 2,774 m. Exposures can range from full to partial sunlight. The substrates often are soft but can include clay loam, clayey peat, gravel, muck, mucky peat, mud, peat (*Sphagnum*), sand, sandy mud, silt, silty clay, silty loam, and silty muck. The waters are shallow to moderately deep (0.1–2.1 m deep; $\bar{x}=0.9$ m), still to flowing (0–0.58 m s^{-1}), eutrophic and mesotrophic (conductivity: 20–310 µmhos cm^{-1} @ 25°C), and circumneutral to somewhat alkaline (pH: 5.4–8.6; $\bar{x}=6.9$; $\tilde{x}=7.2$; alkalinity [mg l^{-1} CaCO$_3$]: 3.0–170; $\bar{x}=17.2$; $\tilde{x}=45$). European plants have been found in waters as acidic at pH 3.1. The highest plant densities have been observed in underflow currents where streambed temperatures were between 13°C and 19°C. Flowering and fruiting occur from May to October. Seed dispersal occurs primarily by water. Two seed types are produced: short-floating seeds (~71% of total) that sink within 4 weeks and long-floating seeds (~28% of total) that remain afloat for at least 6 months. The short-floating seeds are larger ($\bar{x}$ seed mass = 15.17 mg), germinate faster (8.7 days), and at a higher rate (89.9%) than the long-floating seeds (11.25 mg; 9.3 days; 32.6%, respectively). The seeds are also capable of endozoic transport by fish (Vertebrata: Teleostei) at potential distances of up to 27 km and by waterfowl (Aves: Anatidae), which can retain them in their gut for over 60 hr. Once ingested, the seeds no longer are buoyant and sink soon after being egested. Seeds stored for 5 years in cold water (1°C–3°C) have retained high viability (80–90%). Dispersal also occurs by means of waterborne vegetative fragments. The seeds are completely dormant when shed. They have germinated most effectively following 25 months of cold (3°C), moist stratification. Individual leaves live 31–39 days on average. The plants reportedly are associated with arbuscular mycorrhizae. **Reported associates:** *Acorus calamus, Agrostis stolonifera, Alisma subcordatum, Alopecurus aequalis, Beckmannia syzigachne, Bidens cernuus, Bolboschoenus fluviatilis, Brasenia schreberi, Calamagrostis canadensis, Calla palustris, Callitriche palustris, Caltha palustris, Carex aquatilis, Carex lasiocarpa, Carex leporinella, Carex lurida, Carex nebrascensis, Carex pseudocyperus, Carex retrorsa, Carex rostrata, Carex scoparia, Carex simulata, Carex stipata, Carex utriculata, Carex vesicaria, Carex vulpinoidea, Cephalanthus occidentalis, Ceratophyllum demersum, Ceratophyllum echinatum, Chara vulgaris, Cicuta, Comarum palustre, Downingia, Drosera rotundifolia, Dulichium arundinaceum, Elatine, Eleocharis acicularis, Eleocharis palustris, Elodea canadensis, Equisetum fluviatile, Equisetum hyemale, Equisetum palustre, Eriophorum gracile, Eupatorium perfoliatum, Euthamia graminifolia, Glyceria borealis, Glyceria septentrionalis, Glyceria striata, Glyceria ×occidentalis, Helenium bigelovii, Hippuris vulgaris, Howellia aquatilis, Hypericum mutilum, Iris pseudacorus, Isoetes echinospora, Juncus acuminatus, Juncus balticus, Juncus brevicaudatus, Juncus effusus, Juncus nevadensis, Juncus nodosus, Leersia oryzoides, Lemna minor, Lemna trisulca, Lilaeopsis, Ludwigia palustris, Lycopodiella inundata, Lycopus uniflorus, Menyanthes trifoliata, Mimulus moschatus, Mimulus ringens, Myosotis laxa, Myriophyllum sibiricum, Myriophyllum spicatum, Myriophyllum verticillatum, Nasturtium officinale, Nitella, Nuphar advena, Nuphar polysepala, Nymphaea odorata, Oenanthe sarmentosa, Peltandra virginica, Penthorum sedoides, Persicaria amphibia, Persicaria coccinea, Persicaria hydropiperoides, Phalaris arundinacea, Physostegia virginiana, Pontederia cordata, Potamogeton alpinus, Potamogeton amplifolius, Potamogeton berchtoldii, Potamogeton crispus, Potamogeton epihydrus, Potamogeton foliosus, Potamogeton gramineus, Potamogeton natans, Potamogeton pusillus, Potamogeton richardsonii, Potamogeton zosteriformis, Proserpinaca palustris, Ranunculus flabellaris, Ranunculus hispidus, Ranunculus sceleratus, Ranunculus subrigidus, Riccia fluitans, Ricciocarpus natans, Rorippa curvisiliqua, Rubus bifrons, Rubus hispidus, Sagittaria brevirostra, Sagittaria cuneata, Sagittaria latifolia, Sagittaria rigida, Salix exigua, Schoenoplectus acutus, Schoenoplectus tabernaemontani, Scirpus cyperinus, Scirpus hattorianus, Scirpus microcarpus, Scutellaria lateriflora, Sium suave, Solidago canadensis, Sparganium americanum, Sparganium angustifolium, Sparganium eurycarpum, Sparganium fluctuans, Sparganium glomeratum, Sparganium natans, Sphagnum, Spiraea alba, Spiraea tomentosa, Spirodela polyrrhiza, Sporobolus compositus, Stuckenia filiformis, Stuckenia pectinata, Symphyotrichum puniceum, Torreyochloa pallida, Triglochin maritima, Triglochin scilloides, Typha latifolia, Utricularia macrorhiza, Verbena hastata, Veronica americana, Veronica anagallis-aquatica, Veronica catenata, Wolffia columbiana, Zizania aquatica, Zizania palustris.*

10.3.1.5. *Sparganium eurycarpum* **Engelm.** is an emergent plant, which inhabits beaches (wet), bogs (prairie, *Sphagnum*), borrow pits, bottoms (swampy), canals (drainage), depressions (beach, low, wet), ditches (flood control, roadside), fens, flats (peaty), floodplains, gravel pits, hummocks (peaty), marshes (freshwater, lakeside, saline, streamside, tidal), meadows (floodplain, swampy), muskrat mounds, oxbows, ponds (artificial, carp, ephemeral, lagoon, stock), pools (stagnant), seeps (marshy, riverside), sloughs (interdunal, meandering, river), swales (back dune), swamps, and the shores (damp, wave-washed) or margins of brooks, lagoons, lakes, ponds (roadside), reservoirs, rivers, and streams at elevations to 2,743 m. The site exposures range from full sunlight to light shade. The substrates have been described as boulders, clay, gravel, gravelly sand, loam (alluvial), marl, muck (marly, organic), mud (oozy), muddy peat, peat, peaty mud, sand, sandy clay, sandy gravel, sandy loam, sandy muck, sandy silt, silt (alluvial), silty clay loam, silty mud, and silty sand. The plants tolerate some desiccation and can persist in formerly wet sites that have since dried out. They usually grow in shallow (0.10–1.65 m; $\tilde{x}=0.35$ m), nontidal, or tidal waters; however, their maximum biomass is produced at depths <0.4 m. The waters are hard (alkalinity [mg l^{-1}]: 20–320; $\bar{x}=48$; $\tilde{x}=95$; conductivity [μmhos cm^{-1} @ 25°C]: 20–640; $\tilde{x}=230$), neutral to alkaline (pH: 6.6–9.3; $\bar{x}=7.4$; $\tilde{x}=7.8$) or brackish (e.g., salinity: to 0.44‰), and nutrient-enriched (total P >0.5 mg l^{-1}), where sulfate ion concentrations typically exceed 50 ppm. Flowering and fruiting extend from April to November. The pistillate phase lasts 5–7 days and is followed by a longer staminate phase, which averages 8.4 days. Small, young plants sometimes produce only male inflorescences. Individual plants can produce more than 450 fruits, with each containing 1–3 seeds (usually 2). Average seed volumes range from 158 to 288 cm^3 m^{-2}. The fruits are dispersed locally by water but over greater distances (to 1,400 km) by waterfowl (Aves: Anatidae) via endozoic transport. The seeds are dormant when shed with their germinability increasing over at least the first 5 years as they age. A persistent seed bank develops, with numerous seedlings (to 2,175 m^{-2}) capable of emerging from seeds buried at depths up to 2 cm. The seeds are known to remain viable in the seed bank for more than 20 years. Seeds stored in water at 1°C–3°C have retained high viability (80–90%) after 5 years. Vegetative reproduction can be extensive. Within 2–3 weeks, freshly planted corms have produced an average of 36 new shoots, which can reach 135 cm in length. Rhizomes develop within 4–6 weeks, spreading in all directions and producing 250 buds plant^{-1}, while reaching a maximum cumulative length of 2,500 cm plant^{-1}. Production rates can vary considerably. The average annual belowground productivity (rhizomes and roots) has been estimated at 225 g m^{-2} year^{-1}. A maximum total biomass (dry weight) up to 1.2 kg m^{-2} has been recorded from Canada. Total biomass in an Ohio site ranged from 200 to 300 g m^{-2} year^{-1}. The following standing crop biomass and nutrient levels (g m^{-2}) were recorded at sites in New York: total biomass (378); N (7.0); P (1.6); K (5.1); Ca (1.8); Mg (0.7). An average of 99 shoots m^{-2} (@ 0.2 g shoot^{-1}) was estimated for the same sites. Lesser biomass values (e.g., 18.6–48.2 g m^{-2}) have been recorded elsewhere. The roots are associated with various methanotrophic bacteria and are weakly colonized by arbuscular mycorrhizae. **Reported associates:** Acalypha rhomboidea, Acer negundo, Acer rubrum, Acorus americanus, Acorus calamus, Aesculus glabra, Agrostis gigantea, Agrostis stolonifera, Alisma subcordatum, Alisma triviale, Alnus serrulata, Alopecurus aequalis, Alopecurus pratensis, Amaranthus cannabinus, Amaranthus tuberculatus, Ambrosia trifida, Amorpha fruticosa, Anemone canadensis, Apios americana, Apocynum cannabinum, Arctium minus, Arenaria paludicola, Asclepias incarnata, Asclepias syriaca, Athyrium filix-femina, Baccharis douglasii, Baccharis pilularis, Bacopa rotundifolia, Beckmannia syzigachne, Berula erecta, Bidens beckii, Bidens cernuus, Bidens frondosus, Bidens laevis, Bidens trichospermus, Bidens tripartitus, Boehmeria cylindrica, Bolboschoenus fluviatilis, Bolboschoenus maritimus, Bromus ciliatus, Butomus umbellatus, Calamagrostis canadensis, Calamagrostis stricta, Callitriche hermaphroditica, Callitriche heterophylla, Callitriche palustris, Caltha palustris, Campanula aparinoides, Cardamine bulbosa, Carex alata, Carex albolutescens, Carex annectens, Carex aquatilis, Carex atherodes, Carex buxbaumii, Carex comosa, Carex crinita, Carex cusickii, Carex diandra, Carex emoryi, Carex frankii, Carex hyalinolepis, Carex hystericina, Carex interior, Carex lacustris, Carex lasiocarpa, Carex lupuliformis, Carex lupulina, Carex lurida, Carex muskingumensis, Carex nebrascensis, Carex obnupta, Carex pellita, Carex pseudocyperus, Carex retrorsa, Carex rostrata, Carex sartwellii, Carex sheldonii, Carex sterilis, Carex stipata, Carex stricta, Carex sychnocephala, Carex tribuloides, Carex utriculata, Carex vesicaria, Carex viridula, Cephalanthus occidentalis, Ceratophyllum demersum, Chaiturus marrubiastrum, Chara, Chelone glabra, Chenopodium rubrum, Chenopodium simplex, Cicuta bulbifera, Cicuta douglasii, Cicuta maculata, Cirsium arvense, Cirsium muticum, Cirsium palustre, Comarum palustre, Convolvulus sepium, Conyza canadensis, Coreopsis tinctoria, Cornus amomum, Cornus sericea, Cyperus bipartitus, Cyperus erythrorhizos, Cyperus odoratus, Cyperus refractus, Cyperus squarrosus, Cyperus strigosus, Cypripedium parviflorum, Dasiphora floribunda, Decodon verticillatus, Distichlis spicata, Drosera rotundifolia, Dulichium arundinaceum, Echinochloa crus-galli, Echinochloa muricata, Echinochloa walteri, Elaeagnus angustifolia, Eleocharis acicularis, Eleocharis erythropoda, Eleocharis flavescens, Eleocharis macrostachya, Eleocharis obtusa, Eleocharis palustris, Eleocharis parvula, Elodea canadensis, Epilobium ciliatum, Epilobium coloratum, Epilobium leptophyllum, Equisetum fluviatile, Erechtites hieracifolius, Eriophorum angustifolium, Eupatorium perfoliatum, Euthamia occidentalis, Eutrochium maculatum, Galium labradoricum, Galium obtusum, Galium tinctorium, Galium trifidum, Galium triflorum, Geum canadense, Glyceria borealis, Glyceria grandis, Glyceria septentrionalis, Glyceria striata, Glycyrrhiza lepidota, Gratiola virginiana, Helianthus nuttallii, Heteranthera dubia, Hibiscus laevis, Hibiscus moscheutos, Hordeum jubatum, Hydrocotyle, Hypericum canadense, Hypericum ×dissimulatum, Impatiens capensis, Iris versicolor, Isoetes echinospora, Juncus acuminatus, Juncus alpinoarticulatus, Juncus balticus, Juncus canadensis, Juncus effusus, Juncus nodosus, Juncus torreyi, Lactuca canadensis, Leersia oryzoides, Lemna minor, Lemna trisulca, Lepidium latifolium, Lindernia dubia, Lipocarpha micrantha, Lobelia cardinalis, Lobelia siphilitica, Ludwigia alternifolia, Ludwigia palustris, Ludwigia peploides, Lycopus americanus, Lycopus asper, Lycopus uniflorus, Lycopus virginicus, Lysimachia ciliata, Lysimachia terrestris, Lysimachia thyrsiflora, Lysimachia vulgaris, Lythrum salicaria, Mentha arvensis, Mikania scandens, Mimulus glabratus, Mimulus guttatus, Mimulus ringens, Myosotis laxa, Myosoton aquaticum, Myriophyllum heterophyllum, Myriophyllum sibiricum, Myriophyllum spicatum, Myriophyllum tenellum, Najas flexilis, Napaea dioica, Nasturtium officinale, Nuphar advena, Nuphar variegata, Oenanthe sarmentosa, Onoclea sensibilis, Orontium aquaticum, Oxalis stricta, Panicum virgatum, Pedicularis lanceolata, Peltandra virginica, Penthorum sedoides, Persicaria amphibia, Persicaria arifolia, Persicaria bicornis, Persicaria coccinea, Persicaria hydropiper, Persicaria hydropiperoides, Persicaria lapathifolia, Persicaria maculosa, Persicaria pensylvanica, Persicaria punctata, Persicaria sagittata, Phalaris arundinacea, Phleum pratense,

Phragmites australis, Picea mariana, Pilea pumila, Platanthera hyperborea, Pluchea odorata, Poa nemoralis, Poa palustris, Pontederia cordata, Populus deltoides, Potamogeton amplifolius, Potamogeton berchtoldii, Potamogeton crispus, Potamogeton epihydrus, Potamogeton foliosus, Potamogeton gramineus, Potamogeton natans, Potamogeton richardsonii, Potamogeton zosteriformis, Potentilla anserina, Potentilla anserina, Potentilla norvegica, Prunus virginiana, Ptilimnium capillaceum, Ranunculus gmelinii, Ranunculus sceleratus, Ranunculus subrigidus, Ribes americanum, Rorippa aquatica, Rorippa islandica, Rorippa palustris, Rorippa sylvestris, Rosa palustris, Rosa setigera, Rubus ulmifolius, Rumex orbiculatus, Rumex verticillatus, Sagittaria cuneata, Sagittaria latifolia, Sagittaria subulata, Salix amygdaloides, Salix bebbiana, Salix candida, Salix discolor, Salix eriocephala, Salix exigua, Salix lasiolepis, Salix nigra, Salix petiolaris, Sarracenia purpurea, Schizachyrium scoparium, Schoenoplectus acutus, Schoenoplectus americanus, Schoenoplectus californicus, Schoenoplectus heterochaetus, Schoenoplectus lacustris, Schoenoplectus pungens, Schoenoplectus tabernaemontani, Scirpus atrovirens, Scirpus cyperinus, Scirpus microcarpus, Scutellaria galericulata, Scutellaria lateriflora, Sium suave, Solanum dulcamara, Solidago altissima, Solidago canadensis, Solidago ohioensis, Solidago rugosa, Solidago uliginosa, Sonchus, Sorghastrum nutans, Sparganium americanum, Sparganium androcladum, Sparganium emersum, Spartina pectinata, Sphagnum, Spiraea alba, Stachys albens, Stachys palustris, Stachys tenuifolia, Stuckenia pectinata, Symphyotrichum praealtum, Symphyotrichum puniceum, Symphyotrichum subulatum, Symplocarpus foetidus, Taraxacum officinale, Thalictrum revolutum, Thelypteris palustris, Toxicodendron diversilobum, Toxicodendron radicans, Toxicodendron vernix, Triadenum fraseri, Typha angustifolia, Typha domingensis, Typha latifolia, Typha ×glauca, Ulmus americana, Urtica dioica, Utricularia macrorhiza, Utricularia minor, Valeriana sitchensis, Vallisneria americana, Verbena hastata, Vernonia fasciculata, Vernonia gigantea, Veronica anagallis-aquatica, Veronica catenata, Vitis riparia, Wolffia columbiana, Xanthium strumarium, Zannichellia palustris, Zizania aquatica, Zizania palustris, Zizaniopsis miliacea.

10.3.1.6. *Sparganium fluctuans* (Engelm. ex Morong) B.L. Rob.
is a floating-leaved plant, which is found in backwaters (stream), bays (lake), channels, coves, lagoons, lakes, ponds (beaver), pools, rivers (marshy), shores (boggy, lake), sloughs, and streams at elevations to 1,173 m. The plants are uncommon but can be abundant locally. Although floating-leaved, they can survive temporarily when stranded on mud during periods of low water. They grow up to 2 m deep but usually occur in shallower conditions (0.5–1.3 m). The substrates include clay, cobble, gravel, gravelly sand, muck, ooze (dense), sand (organic), sandy gravel, sandy silt, silt, and silty rock. The habitats are characterized by quiet, cold, acidic to alkaline (pH: 4.9–8.5; $\bar{x} = 6.9$; $\tilde{x} = 6.7$), oligotrophic waters (conductivity [μmhos cm^{-1} @ 25°C]: 10–180; $\tilde{x} = 40$) of low alkalinity (mg l^{-1} CaCO$_3$: 3.0–90; $\bar{x} = 13.5$; $\tilde{x} = 20$). Chloride, magnesium, and sulfate levels (<5 ppm) are also very low.

Flowering and fruiting occur from July to September. Plants growing in lakes having high mercury levels were found to harbor mercury-resistant bacteria, which are thought to ameliorate their tolerance to the toxins. **Reported associates:** *Acorus calamus, Asclepias incarnata, Bidens beckii, Brasenia schreberi, Calla palustris, Campanula aparinoides, Carex canescens, Carex lacustris, Ceratophyllum demersum, Ceratophyllum echinatum, Chara, Cladophora glomerata, Dulichium arundinaceum, Elatine minima, Eleocharis acicularis, Eleocharis palustris, Eleocharis robbinsii, Elodea canadensis, Equisetum fluviatile, Eriocaulon aquaticum, Eutrochium maculatum, Glyceria borealis, Heteranthera dubia, Hygroamblystegium tenax, Impatiens capensis, Isoetes echinospora, Isoetes lacustris, Juncus canadensis, Juncus pelocarpus, Lemna minor, Lemna trisulca, Lobelia dortmanna, Lycopus uniflorus, Lythrum salicaria, Myriophyllum alterniflorum, Myriophyllum farwellii, Myriophyllum sibiricum, Myriophyllum tenellum, Myriophyllum verticillatum, Najas flexilis, Najas gracillima, Nitella, Nuphar polysepala, Nuphar variegata, Nymphaea odorata, Persicaria amphibia, Pontederia cordata, Potamogeton amplifolius, Potamogeton epihydrus, Potamogeton foliosus, Potamogeton friesii, Potamogeton gramineus, Potamogeton illinoensis, Potamogeton natans, Potamogeton oakesianus, Potamogeton obtusifolius, Potamogeton perfoliatus, Potamogeton praelongus, Potamogeton richardsonii, Potamogeton robbinsii, Potamogeton spirillus, Potamogeton vaseyi, Potamogeton zosteriformis, Potentilla simplex, Ranunculus flammula, Riccia fluitans, Sagittaria graminea, Sagittaria latifolia, Sagittaria rigida, Schoenoplectus acutus, Schoenoplectus subterminalis, Schoenoplectus tabernaemontani, Scirpus cyperinus, Scirpus microcarpus, Sium suave, Sparganium americanum, Sparganium angustifolium, Sparganium emersum, Sparganium natans, Spiraea alba, Spirodela polyrrhiza, Stuckenia pectinata, Utricularia geminiscapa, Utricularia gibba, Utricularia intermedia, Utricularia macrorhiza, Utricularia minor, Utricularia purpurea, Utricularia resupinata, Vallisneria americana, Zizania aquatica.*

10.3.1.7. *Sparganium glomeratum* (Beurl. ex Laest.) Neuman
is an emergent or floating-leaved plant, which grows in bogs (*Sphagnum*), depressions (open), ditches, fens (poor), marshes (desiccated, lakeshore), meadows, pools (ephemeral, marshy, seasonal, shallow, woodland), ponds (seasonal), streams (intermittent), swales, and swamps (black ash) at elevations to 1,000 m. The plants occur under exposures of full sun to shade. The substrates have been characterized as clay (red), muck, mucky peat, sand, and silt. The habitats typically are characterized by shallow (1–46 cm), quiet, neutral, mesotrophic waters. Flowering occurs from June to July with fruiting from July to September. Vegetative reproduction occurs by means of slender rhizomes. **Reported associates:** *Abies balsamea, Acer rubrum, Acorus calamus, Agalinis, Agrostis stolonifera, Alisma subcordatum, Alisma triviale, Alnus rugosa, Alopecurus aequalis, Athyrium filix-femina, Beckmannia syzigachne, Betula pumila, Bidens cernuus, Bidens frondosus, Bidens trichospermus, Brasenia schreberi, Calamagrostis canadensis, Calla palustris, Callitriche*

palustris, Caltha natans, Caltha palustris, Campanula aparinoides, Carex aquatilis, Carex aurea, Carex brunnescens, Carex buxbaumii, Carex canescens, Carex comosa, Carex crawfordii, Carex diandra, Carex disperma, Carex hystericina, Carex interior, Carex intumescens, Carex lacustris, Carex lasiocarpa, Carex leptalea, Carex lupulina, Carex magellanica, Carex praticola, Carex pseudocyperus, Carex retrorsa, Carex rostrata, Carex stipata, Carex tenera, Carex tuckermanii, Carex utriculata, Carex vaginata, Chamaedaphne calyculata, Chelone glabra, Chrysosplenium americanum, Cicuta bulbifera, Cicuta maculata, Clintonia borealis, Comarum palustre, Coptis trifolia, Cornus canadensis, Cornus sericea, Doellingeria umbellata, Dryopteris carthusiana, Dryopteris cristata, Dulichium arundinaceum, Eleocharis acicularis, Eleocharis erythropoda, Eleocharis nitida, Eleocharis obtusa, Eleocharis ovata, Eleocharis palustris, Epilobium coloratum, Epilobium leptophyllum, Equisetum fluviatile, Equisetum sylvaticum, Eriophorum angustifolium, Euthamia, Eutrochium maculatum, Fraxinus nigra, Galium tinctorium, Galium trifidum, Glyceria borealis, Glyceria canadensis, Glyceria grandis, Glyceria striata, Gnaphalium uliginosum, Hypericum canadense, Hypericum majus, Hypericum mutilum, Impatiens capensis, Iris versicolor, Isoetes echinospora, Juncus brevicaudatus, Juncus dudleyi, Juncus effusus, Juncus vaseyi, Larix laricina, Lemna minor, Lycopus asper, Lycopus uniflorus, Lysimachia terrestris, Lysimachia thyrsiflora, Maianthemum trifolium, Mentha arvensis, Menyanthes trifoliata, Nuphar, Nymphaea odorata, Onoclea sensibilis, Persicaria amphibia, Persicaria sagittata, Petasites frigidus, Phalaris arundinacea, Picea mariana, Poa alsodes, Poa nemoralis, Pogonia ophioglossoides, Potamogeton epihydrus, Potamogeton natans, Potamogeton obtusifolius, Potentilla norvegica, Ranunculus flabellaris, Ranunculus gmelinii, Ranunculus pensylvanicus, Rhamnus alnifolia, Rubus pubescens, Rubus sachalinensis, Rumex orbiculatus, Rumex verticillatus, Sagittaria latifolia, Salix bebbiana, Salix discolor, Salix lucida, Salix pedicellaris, Salix petiolaris, Salix planifolia, Salix purpurea, Salix pyrifolia, Salix serissima, Scheuchzeria palustris, Schoenoplectus acutus, Schoenoplectus tabernaemontani, Scirpus atrovirens, Scirpus cyperinus, Scutellaria galericulata, Scutellaria lateriflora, Sium suave, Solidago gigantea, Solidago uliginosa, Sparganium angustifolium, Sparganium emersum, Sparganium fluctuans, Sparganium natans, Spergularia rubra, Sphagnum, Spiraea alba, Spirodela polyrrhiza, Symphyotrichum lanceolatum, Symphyotrichum lateriflorum, Symphyotrichum puniceum, Thelypteris palustris, Torreyochloa pallida, Triadenum fraseri, Trientalis borealis, Typha latifolia, Typha ×glauca, Urtica dioica, Utricularia intermedia, Utricularia macrorhiza, Utricularia minor, Veronica americana, Veronica scutellata, Viola.

10.3.1.8. *Sparganium hyperboreum* Beurl. ex Laest. is a floating-leaved plant, which occurs in bogs (marshy, *Sphagnum*), depressions, ditches (roadside), fens, flats, marshes (drawdown), mires, mudflats, pools (bog, fen, ice wedge, roadside, rock, spring, subalpine), potholes, streambeds, and along the shallow margins of bogs, lakes (kettle, moraine, oxbow,

thermokarst), and ponds (alpine, beaded stream, beaver, borrow pit, drying, floodplain, muskeg, roadside, subalpine, thermokarst, tundra) at elevations to 1,900 m. The substrates have been described as cobbles (basalt), gravel, mucky peat, mud, peat, pergelic cryofibrists, sand, sandy gravel, sandy silt, silt, and silty clay. The usual habitats (e.g., pH: 5.9) are characterized by cold, fresh (<800 µS cm^{-1}), quiet, shallow (<60 cm; rarely to 150 cm), oligotrophic to mesotrophic, arctic, subarctic, or alpine waters. Flowering and fruiting occur from June to September. Vegetative reproduction occurs by means of slender, creeping rhizomes. **Reported associates:** *Alnus viridis, Alopecurus aequalis, Arctophila fulva, Betula nana, Calamagrostis canadensis, Calla palustris, Calliergon giganteum, Callitriche hermaphroditica, Callitriche palustris, Caltha palustris, Carex aquatilis, Carex chordorrhiza, Carex diandra, Carex lacustris, Carex lenticularis, Carex limosa, Carex lyngbyei, Carex pauciflora, Carex pluriflora, Carex rostrata, Carex rotundata, Carex saxatilis, Carex utriculata, Ceratophyllum demersum, Comarum palustre, Drosera rotundifolia, Eleocharis palustris, Empetrum nigrum, Equisetum arvense, Equisetum fluviatile, Eriophorum angustifolium, Eriophorum russeolum, Eriophorum vaginatum, Gentiana douglasiana, Geum calthifolium, Glyceria borealis, Hippuris lanceolata, Hippuris tetraphylla, Hippuris vulgaris, Isoetes, Juncus bulbosus, Juncus mertensianus, Myrica gale, Myriophyllum sibiricum, Nephrophyllidium cristagalli, Nuphar polysepala, Nymphaea tetragona, Paludella squarrosa, Polytrichum swartzii, Potamogeton amplifolius, Potamogeton epihydrus, Potamogeton gramineus, Potamogeton natans, Potamogeton perfoliatus, Potamogeton praelongus, Potamogeton richardsonii, Potamogeton subsibiricus, Potamogeton zosteriformis, Ranunculus flammula, Ranunculus gmelinii, Ranunculus hyperboreus, Ranunculus pallasii, Ranunculus trichophyllus, Rhododendron tomentosum, Rubus chamaemorus, Salix barclayi, Sarmenthypnum sarmentosum, Scirpus cyperinus, Scirpus microcarpus, Sparganium angustifolium, Sparganium glomeratum, Sparganium natans, Sphagnum lindbergii, Sphagnum obtusum, Stuckenia filiformis, Stuckenia pectinata, Torreyochloa pallida, Trichophorum cespitosum, Utricularia macrorhiza, Vaccinium vitis-idaea, Warnstorfia exannulata.*

10.3.1.9. *Sparganium natans* L. is a floating-leaved plant, which inhabits backwaters (marly), bays, bogs (peat), depressions, ditches (slough), fens (graminoid), gravel pits, kettle holes, marshes, meadows (open, wet), mudflats, muskeg, oxbows, ponds (beaver, bog, interdunal, kettle, roadside, subalpine), pools (drying, stagnant, swamp), potholes (glacial), shores, sloughs (roadside), streams, swales (dune), swamps (cedar), and the margins of reservoirs at elevations to 3,475 m. Exposures can vary from full sun to shade. The substrates include alluvium, gravel, gravelly loam, humus, marl, muck, mucky peat, mud, peat, peaty muck, sand, sandy mud, silt, and silty clay loam. The plants grow in waters that are clear (Secchi depth: 4.0–4.1 m), cool (or geothermal), quiet, shallow (to 70 cm depth [to 1.5 m when growing on floating mats]), and somewhat acidic to basic (pH: 6.2–8.5; $\bar{x}$ =6.2–7.3; mean conductivity 229.08 µS cm^{-1}; total alkalinity: 3.5–69.5 mg l^{-1};

$\bar{x} = 35.4$ mg l^{-1}). Flowering and fruiting extend from June to October. Vegetatively reproduction occurs by means of slender, creeping rhizomes. **Reported associates:** *Agrostis idahoensis, Agrostis pallens, Alnus, Bidens cernuus, Bistorta bistortoides, Brasenia schreberi, Calamagrostis canadensis, Calla palustris, Callitriche hermaphroditica, Callitriche palustris, Carex aquatilis, Carex aurea, Carex buxbaumii, Carex canescens, Carex capitata, Carex comosa, Carex diandra, Carex dioica, Carex exilis, Carex integra, Carex lacustris, Carex lasiocarpa, Carex leptalea, Carex limosa, Carex mariposana, Carex membranacea, Carex pseudocyperus, Carex rostrata, Carex scopulorum, Carex utriculata, Carex utriculata, Carex vesicaria, Carex viridula, Chamaedaphne calyculata, Chara, Cicuta virosa, Comarum palustre, Crataegus, Dasiphora floribunda, Deschampsia cespitosa, Drepanocladus aduncus, Drosera anglica, Drosera intermedia, Drosera rotundifolia, Dulichium arundinaceum, Eleocharis acicularis, Eleocharis compressa, Eleocharis palustris, Eleocharis quinqueflora, Equisetum fluviatile, Erigeron coulteri, Erigeron glacialis, Eriophorum angustifolium, Eriophorum viridicarinatum, Festuca filiformis, Galium trifidum, Gentianopsis holopetala, Glyceria borealis, Glyceria canadensis, Glyceria grandis, Hippuris vulgaris, Hymenoxys hoopesii, Isoetes bolanderi, Isoetes echinospora, Juncus brevicaudatus, Juncus bulbosus, Juncus stygius, Kalmia polifolia, Lemna minor, Lemna trisulca, Ligusticum grayi, Lupinus burkei, Lysimachia terrestris, Lysimachia thyrsiflora, Meesia triquetra, Menyanthes trifoliata, Mimulus primuloides, Myrica gale, Myriophyllum sibiricum, Myriophyllum verticillatum, Nostoc, Nuphar polysepala, Nuphar variegata, Nymphaea tetragona, Oreostemma alpigenum, Pedicularis attollens, Pedicularis groenlandica, Perideridia parishii, Persicaria amphibia, Phleum alpinum, Podagrostis humilis, Potamogeton alpinus, Potamogeton amplifolius, Potamogeton berchtoldii, Potamogeton epihydrus, Potamogeton foliosus, Potamogeton friesii, Potamogeton gramineus, Potamogeton natans, Potamogeton obtusifolius, Potamogeton praelongus, Potamogeton pusillus, Potamogeton richardsonii, Potamogeton spirillus, Potamogeton zosteriformis, Potentilla gracilis, Ranunculus flammula, Ranunculus gmelinii, Rhododendron columbianum, Sagittaria cuneata, Sagittaria latifolia, Sagittaria rigida, Salix drummondiana, Salix eastwoodiae, Salix lucida, Schoenoplectus acutus, Schoenoplectus subterminalis, Scirpus cyperinus, Scutellaria galericulata, Senecio scorzonella, Sium suave, Sparganium angustifolium, Sparganium emersum, Sparganium fluctuans, Sparganium glomeratum, Sparganium hyperboreum, Sphagnum, Spiranthes romanzoffiana, Stellaria longifolia, Stuckenia filiformis, Stuckenia pectinata, Swertia perennis, Thelypteris palustris, Thuja occidentalis, Torreyochloa erecta, Torreyochloa pallida, Triadenum fraseri, Trichophorum alpinum, Triglochin maritima, Typha latifolia, Typha ×glauca, Utricularia gibba, Utricularia intermedia, Utricularia macrorhiza, Utricularia minor, Veratrum californicum, Veronica americana.*

Use by wildlife: *Sparganium americanum* (crude protein: 23.7%; soluble protein: 2.71 mg ml^{-1}; crude fat: 8.11%; 4.17 kcal g^{-1}) is eaten (up to 30% of the biomass) by beavers (Mammalia: Castoridae: *Castor canadensis*), Canada geese (Aves: Anatidae: *Branta canadensis*), crayfish (Crustacea: Decapoda: Cambaridae: *Procambarus spiculifer*), muskrats (Mammalia: Cricetidae: *Ondatra zibethicus*), and trumpeter swans (Aves: Anatidae: *Cygnus buccinator*). The plants provide habitat for aquatic beetles (Insecta: Coleoptera: Chrysomelidae: *Donacia fulgens, D. subtilis*), black flies (Insecta: Diptera: Simuliidae: *Simulium krebsorum*), and are hosts to larval moths (Lepidoptera: Crambidae: *Langessa nomophilalis, Munroessa icciusalis, Paraponyx badiusalis, P. obscuralis, P. seminealis, Synclita obliteralis*). The stems are mined by adult midges (Insecta: Diptera: Chironomidae: *Demeijerea atrimana*). The leaves are used (sometimes selectively) as oviposition sites by several damselflies (Insecta: Odonata: Calopterygidae: *Calopteryx aequabilis, C. maculata*; Lestidae: *Lestes eurinus*; Coenagrionidae: *Chromagrion conditum*). The achenes are consumed by various waterfowl (Aves: Anatidae), including the black duck (*Anas rubripes*), blue-winged teal (*Anas discors*), canvasback (*Aythya valisineria*), cinnamon teal (*Anas cyanoptera*), greater scaup (*Aythya marila*), lesser scaup (*Aythya affinis*), mallard (*Anas platyrhynchos*), pintail (*Anas acuta*), and wood duck (*Aix sponsa*). Numerous actinomycetes (Bacteria: Actinobacteria: *Actinoplanes, Microbiospora, Micromonospora, Nocardia, Nocardiodes, Pseudonocardia, Streptoalloteichus, Streptomyces*) have been isolated from the foliage. *Sparganium androcladum* is used as a nesting site for least bitterns (Aves: Ardeidae: *Ixobrychus exilis*). The fruits are eaten by lesser scaup (Aves: Anatidae: *Aythya affinis*) and ring-necked ducks (Aves: Anatidae: *Aythya collaris*). *Sparganium angustifolium* (2.22% crude fat; 19.0% crude protein; Ca: 1.01%; Fe: 0.25%; K: 2.42%; Mg: 0.47%; Na: 0.60%; SO$_4$: 1.06% [as dry matter]) is fed upon by moose (Mammalia: Cervidae: *Alces alces*) and muskrats (Mammalia: Cricetidae: *Ondatra zibethicus*). The achenes are eaten by the lesser scaup (Aves: Anatidae: *Aythya affinis*). The leaves and roots host adult and pupal water beetles (Insecta: Coleoptera: Chrysomelidae: *Donacia fulgens, D. hirticollis; D. subtilis*). The seeds of *S. emersum* are eaten by many waterfowl (Aves: Anatidae) such as black ducks (*Anas rubripes*), blue-winged teal (*Anas discors*), green-winged teal (*Anas crecca*), mallard ducks (*Anas platyrhynchos*), and wood ducks (*Aix sponsa*). The leaves host several Fungi (Ascomycota: Botryosphaeriaceae: *Phyllosticta*; Basidiomycota: Doassansiaceae: *Nannfeldtiomyces anomalus*). The foliar detritus (249 mg protein g^{-1}) is consumed by larval caddisflies (Insecta: Trichoptera: Limnephilidae: *Nemotaulius hostilis*) and is used by them to build cases. *Sparganium eurycarpum* is an important food of muskrats (Mammalia: Cricetidae: *Ondatra zibethicus*), which also use the plants in their house construction. The plants provide habitat for marsh flies (Insecta: Diptera: Sciomyzidae: *Sepedon fuscipennis*). The foliage is fed on by adult beetles (Insecta: Coleoptera: Chrysomelidae: *Donacia subtilis*) and provides cover for larval damselflies (Insecta: Odonata: Lestidae: *Lestes disjunctus*) and larval northern pike (Vertebrata: Teleostei: Esocidae: *Esox lucius*). The fruits (6.13% protein, 1.88% fat) are eaten by waterfowl (Aves: Anatidae) such as black duck (*Anas rubripes*), Canada goose (*Branta canadensis*), canvasback (*Aythya valisineria*), common goldeneye (*Bucephala clangula*), mallard (*Anas platyrhynchos*), greater scaup (*Aythya marila*), lesser scaup (*Aythya affinis*), and wood duck (*Aix sponsa*). Additionally, the fruits are eaten by sandhill cranes (Aves: Gruidae: *Grus canadensis*), which use the plants as nesting sites. Other birds (Aves) that use the plants for nest construction include the American bittern (Ardeidae: *Botaurus lentiginosus*), American coot (Rallidae: *Fulica americana*), blackbirds (Icteridae: *Agelaius phoeniceus, Xanthocephalus xanthocephalus*), gallinules (Rallidae: *Gallinula chloropus*), pied-billed grebes (Podicipedidae: *Podilymbus podiceps*), and marsh wrens (Troglodytidae: *Cistothorus palustris*). The plants host several Fungi (Ascomycota: Botryosphaeriaceae: *Phyllosticta*; Massarinaceae: *Stagonospora sparganii*; Basidiomycota: Pucciniaceae: *Uromyces sparganii*). *Sparganium fluctuans* provides cover for larval northern pike (Vertebrata: Teleostei: Esocidae: *Esox lucius*). The foliage (by dry wt) contains 13.19% crude protein, 0.77% Ca, 2.11% K, 0.264% Mg, 0.397% Na, and 0.27% P as well as high amounts of Cu (3.6 ppm), Fe (342 ppm), Mn (484 ppm), Mo (13.6), and Zn (166.8 ppm). It is eaten by beavers (Mammalia: Castoridae: *Castor canadensis*) and eastern moose (Mammalia: Cervidae: *Alces americanus americanus*). Its flowers are host to silken fungus beetles (Insecta: Coleoptera: Cryptophagidae: *Telmatophilus americanus*) and the fruits are consumed by waterfowl (Aves: Anatidae). *Sparganium glomeratum* is fed upon by muskrats (Mammalia: Cricetidae: *Ondatra zibethicus*). *Sparganium hyperboreum* provides habitat for the northern pike (Vertebrata: Teleostei: Esocidae: *Esox lucius*). Its fruits are eaten by waterfowl (Aves: Anatidae).

Economic importance: food: *Sparganium* has not been used as a human food; **medicinal:** *Sparganium fluctuans* contains condensed tannins, saponins, and steroids (beta-sitosterol). Benzene extracts have exhibited antifungal activity against human yeast pathogens (Fungi: Ascomycota: Saccharomycetidae: *Candida albicans*). The flowers and roots (but not stems or leaves) of *S. glomeratum* have exhibited antimicrobial activity against *Staphylococcus aureus* (Bacteria: Staphylococcaceae); **cultivation:** *Sparganium* species generally are not cultivated, except occasionally as ornamental water garden oddities; **misc. products:** *Sparganium americanum* and *S. eurycarpum* have been investigated for the bioremediation of agriculturally enriched nutrients (N, P), pesticides, and various other toxins. *Sparganium angustifolium* has been evaluated for aluminum (Al) removal from wastewater; **weeds:** *Sparganium* species are not perceived as weedy in North America; **nonindigenous species:** None. The Eurasian *Sparganium erectum* has been reported mistakenly from central North America based on records of *S. eurycarpum* (see *Systematics* below).

Systematics: Phylogenetic analyses (e.g., Figure 3.8) have indicated consistently that *Sparganium* is monophyletic and the sister clade to *Typha*. As mentioned above under Typhaceae, the close relationship of these genera is evidenced further by their ability to form intergeneric hybrids. *Sparganium*

americanum exhibits latitudinal variation in stigma length and robustness, which has been attributed to phenotypic plasticity, hybridization, or racial differentiation, but has not been investigated experimentally. *Sparganium androcladum* has been treated as a variety of *S. americanum*, but has larger pollen with different surface features and resolves as a genetically distinct sister species (e.g., Figure 3.8). A close relationship of *S. americanum* and *S. emersum* (assumed by some) is not supported by phylogenetic analysis (Figure 3.8), which indicates that the species actually are quite distantly related. Natural, fertile hybrids reportedly occur between *Sparganium angustifolium* and *S. emersum*, which are extremely similar morphologically, and have been treated as conspecific by some authors. Although the two species are distinct in phylogenetic analyses (e.g., Figure 3.8), the subspecies of *S. emersum* are polyphyletic, with *S. emersum* subsp. *emersum* (Eurasia; North America) allied with *S. angustifolium*, and *S. emersum* subsp. *acaule* (E. North America) allied with *S. glomeratum*. Accordingly, the latter subspecies has been recognized taxonomically as *S. acaule*, which if maintained will require a reassessment of eastern North American material of this taxon. Seven polymorphic microsatellite loci have been developed for *S. emersum*, which should facilitate further genetic evaluations. *Sparganium erectum* (*S. erectum* subsp. *stoloniferum*) has been reported from North America, but these accessions have been assigned instead to *Sparganium eurycarpum* (as *S. eurycarpum* var. *greenei*). *Sparganium erectum* (Eurasian) and *S. eurycarpum* (North American) are extremely similar, varying primarily by their usual stigma number (one on the former, two in the latter). *Sparganium natans* was long known in North America by the synonym *S. minimum*, to which much information on North American populations has been attributed. The basic chromosome number of *Sparganium* is $x = 15$. All published counts for the genus (including *S. angustifolium*, *S. emersum*, *S. erectum*, *S. erectum*, *S. eurycarpum*, *S. glomeratum*, *S. hyperboreum*, and *S. natans*) are diploid [$2n = 30$]. Putative hybrids involving *Sparganium hyperboreum* and *S. natans* have been reported, but require further documentation beyond morphological criteria. Both species also allegedly hybridize with *S. angustifolium*. The genome size of *S. angustifolium* has been estimated at 1.18 pg (1C = 577 mbp). The chloroplast genome of *S. eurycarpum* has been sequenced completely.

Distribution: *Sparganium americanum* is widespread in eastern North America and extends southward into Mexico; *S. androcladum* occurs in mid-eastern North America. *Sparganium angustifolium* and *S. emersum* are circumboreal and are found across northern and western North America. *Sparganium eurycarpum* extends broadly across the central portion of North America, becoming rarer in the northern or southern regions. *Sparganium fluctuans*, *S. glomeratum*, *S. hyperboreum*, and *S. natans* occur across northern North America; *S. glomeratum*, *S. hyperboreum*, and *S. natans* are circumboreal.

References: Albanese, 2000; Allen et al., 2002a. Anderson et al., 1968; Barnes, 1976; Barrat-Segretain & Bornette, 2000; Beal, 1960; 1977; Beas et al., 2013; Belyakov & Lapirov, 2015;

Bernard & Bernard, 1973; 1977; Bick et al., 1976; Bontrager et al., 2014; Boyd, 1968; Brackney & Bookhout, 1982; Buckley & Hicks, 1962; Calhoun & King, 1998; Caslake et al., 2006; Catling et al., 1986; Cellot et al., 1998; Ceska et al., 1986; Christy, 2004; 2013; Cook & Nicholls, 1986; 1987; Cottam, 1939; Coulter, 1955; Craven & Hunt, 1984; Crow & Hellquist, 1981; Cruden, 1988; Cruden & Lloyd, 1995; Cruden & Lyon, 1985; Davidson, 1960; Dechant et al., 2004; DeVelice et al., 1999; DeVere, 1970; Duffy, 1994; Ewing, 1924; Farmer et al., 2017; Fassett, 1957; Forbes et al., 1989; Fortner & White, 1988; Fraser et al., 1980; 1982; 2014; Fyson, 2000; Gallon et al., 2004; Garon-Labrecque et al., 2016; Gil et al., 2019; Gleason et al., 2003; Green et al., 2016; Griffin et al., 2009; Hall & Penfound, 1943; Harms, 1973; Harms & Grodowitz, 2009; Harvey & Haines, 2003; Havera, 1986; Hellquist et al., 2014; Heyn, 1992; Hicks et al., 2005; Hidalgo et al., 2015; Hoffman, 1940; Homoya & Hedge, 1982; Hopkins, 1969; Hudon, 1997; Ito et al., 2016b; Jackson & Charles, 1988; Jelinski, 1989; Jorgenson & Ely, 2001; Kangas & Hannan, 1985; Kao et al., 2003; Kaul, 1972; 2000; Keddy & Reznicek, 1986; Knapton & Petrie, 1999; Konkel, 2006; Kullberg, 1974; Lakela, 1941; Lichthardt et al., 2000; Linn et al., 1973; Littlefield, 2001; Low & Bellrose, 1944; Lutz & Pittman, 1968; Majka & Langor, 2010; Martin, 1939; Martin & Uhler, 1951; Marx, 1957; Matteson et al., 1986; McAtee, 1916; Meinshausen, 1895; Mendall & Gashwiler, 1940; Mitsch et al., 2004; Moldenke, 1936; Moore et al., 2013; Moulton & Adler, 1995; Moyle, 1945; Mueller & Van der Valk, 2002; Mullins & Bizeau, 1978; Murray, 2000; Nichols, 1988; 1999; Nichols & Buchan, 1997; Parker, 2003; Parker et al., 2007a; 2007b; Payette & Delwaide, 2000; Perleberg, 2006; Perry & Uhler, 1981; Pfauth & Sytsma, 2005; Pierre & Kovalenko, 2014; Pollux, 2011; Pollux & Ouborg, 2006; Pollux et al., 2007; 2009; Pritchard, 1987; Provost, 1947; Puijalon et al., 2008; Rober et al., 2014; Ryser & Kamminga, 2009; Seyer, 1979; Shacklette, 1961; Sharpe & Baldwin, 2012; Sherff, 1912; Shull, 1914; Singhurst et al., 2007; Smith, 2017; Spinner & Bishop, 1950; Srivastava et al., 1995; Steinbauer & Neil, 1948; Stollberg, 1950; Stoops et al., 1998; Stoudt, 1944; Strong & Kelloff, 1994; Stuckey, 1983; Su & Staba, 1972; Su et al., 1972; Sulman et al., 2013; Sundue, 2006; Takos, 1947; Thieret, 1982; Timm & Pierce, 2015; Townsend, 1953; Truman, 1931; Tsuchiya, 1991; Van der Valk & Davis, 1978; Viereck et al., 1992; Vincent, 1958; Walker & Medve, 1976; Walker & Everett, 1991; Walker et al., 1987; 1989; 1994; Walton, 1995; Wang & Qiu, 2006; Ward, 1882; Weinhold & Van der Valk, 1988; Werblan, 1978; Whigham & Simpson, 1978; Wilcox, 2012; Wilcox & Meeker, 1992; Wilson, 1935; Witmer, 1964; Wohl & McArthur, 1998; Yasukawa, 1981; Yocom & Hansen, 1960.

10.3.2. Typha

cattail, reed-mace; quenouille

Etymology: from *týphē*, the Greek name for the plants, which were used for stuffing

Synonyms: *Massula* (in part); *Rohrbachia*

Distribution: global: cosmopolitan; **North America:** widespread

Diversity: global: 10–15 species; **North America:** 3 species

Indicators (USA): obligate wetland (OBL): *Typha angustifolia, T. domingensis, T. latifolia*

Habitat: brackish (coastal), freshwater; freshwater (tidal); saline (inland); palustrine; **pH:** 5.1–10.3; **depth:** <1.5 m; **life-form(s):** emergent herb

Key morphology: rhizomes (to 70 cm) horizontal, unbranched; shoots (to 4 m) erect, unbranched; leaves (to 3.5 m) few to 15, linear, loosely twisted helically, the margins entire, the apex tapering, rounded, the sheaths tapering to the blade or auriculate, the sides papery or membranous, the margins narrowly to broadly clear; inflorescence spikelike, monoecious, the ♂ (upper) and ♀ (lower) portions contiguous or separated by a gap (to 12 cm), ♂ portion (to 50 cm) yellow (during anthesis), ♀ portion (to 35 cm) brown, cinnamon-brown, or greenish (in fruit); ♂ flowers (to 12 mm) numerous, the anthers (to 3 mm) 4-sporangiate, dehiscing longitudinally, the pollen in monads or tetrads; ♀ flowers (to 3 mm in flower; to 15 mm in fruit), numerous, the pedicels (to 1.2 mm) compound, the pistil hairs (to 0.6 mm) brown or colorless, filiform, or enlarged apically (by 10–20×), the stigmas (to 1.4 mm) deciduous or persistent, the carpodia (sterile ♀ flowers) green or straw-colored

Life history: duration: perennial (rhizomes); **asexual reproduction:** rhizomes; **pollination:** wind; **sexual condition:** monoecious; **fruit:** achenes (prolific); **local dispersal:** rhizomes, fruits (water, wind); **long-distance dispersal:** fruits (wind)

Imperilment: 1. *Typha angustifolia* [G5]; S2 (WY); S3 (MB, NC); **2.** *T. domingensis* [G4/G5]; S1 (MO); S2 (NC); S3 (VA); **3.** *T. latifolia* [G5]; S2 (NU, YT); S3 (WY)

Ecology: general: All *Typha* plants inhabit wetlands and all three North American species are ranked as OBL indicators throughout their ranges. Because of their ubiquitous occurrence in hydric habitats, cattails represent some of the most reliable single-species indicators of wetland sites. However, the ability of cattail species to persist in intermittently wet areas that can dry out completely (especially in arid regions) often puts them in association with upland (and even xeric) species. They occur across a remarkably broad spectrum of wetland habitats ranging from early to late successional stages and can produce considerable standing crop biomass yields up to 22.4 Mg ha^{-1} (foliage) and 30.9 Mg ha^{-1} (rhizomes). The characteristic spiraling of cattail leaves represents a biomechanical adaptation, which enables them to achieve a greater length than is possible for uniformly flat leaves. The leaves also contain specialized "cables" of long, nonlignified fibers enclosed by crystal-containing cells, which add not only strength but flexibility to the foliage. The monoecious inflorescences comprise staminate (upper) and pistillate (lower) flowers, which are "self" (within inflorescence)-compatible, allowing for geitonogamous self-pollination to occur. The inflorescences are protogynous, with xenogamous pollen transfer occurring exclusively by wind. Fruit production is profuse, often reaching numbers from 70,000 to 250,000 fruits per inflorescence. The fruits are wind dispersed, facilitated by their hairlike perianth, which provides buoyancy. Seed banks can be extensive with as many as 1,800 *Typha* seedlings m^{-2} emerging from sediment samples. If the fruits of some species land on water, their hygroscopic pericarp opens rapidly to release the nonbuoyant seeds, which then sink beneath the water. In other species, the seeds are fused to the pericarp. Germination characteristics apparently are differentiated ecotypically. Seeds from lowland plants have germinated (100%) within a week, whereas those from higher elevations have exhibited dormancy, with a low percent germination even after several months. The seeds can remain viable in the seed bank for extended periods of time. There have been reports that cattails produce autotoxic compounds as they grow, which inhibit seed germination, thereby preventing invasion of potentially competitive congener genotypes. However, experimental evidence has not supported that contention. In any case, seed germination and seedlings are observed rarely in natural, established populations, with sexual reproduction functioning primarily for dispersal and clonal growth for establishment and persistence. Vegetative reproduction occurs by an effective, extensive rhizome system, which also enables sterile (or partially sterile) hybrid clones to persist. Cattail plants can be killed by prolonged oxygen deprivation to the rhizomes, which can occur if water levels are maintained above the shoots throughout the winter for one or more seasons. Aeration of the rhizomes normally is facilitated by convective and humidity-induced pressurized gas flow from the living shoots or from the dead, standing shoots that remain above the water surface. Diazotrophic microbes associated with *Typha* plants can fix nitrogen at rates (estimated via acetylene reduction) from 2.9 to 7.1 μmol ramet^{-1} day^{-1}.

10.3.2.1. ***Typha angustifolia*** L. inhabits brackish to freshwater beaches, bogs (dune, *Sphagnum*), borrow pits, bottoms (river), canal beds (dry), channels (ephemeral, stream), coves, depressions (wet), ditches (boggy, damp, drainage, irrigation, marshy, open, roadside, spring-fed, tidal, wet), draws (hardwood), estuaries (coastal), fens (calcareous, open, prairie), flats (marl), floodplains (alluvial), flowages, lakebeds (dry), loam, marshes (brackish, estuarine, freshwater, high, tidal), mats (bog), meadows (beaver, bottomland, floodplain, moist, wet), mudflats, oxbows, pools (ephemeral, quarry, strip mine), potholes (prairie), prairies (mesic), rice fields, rights of way (railroad), roadsides, sandbars, seeps (calcareous, roadside), shores (marshy), sloughs (dry), sludge pits, spillways, swales, swamps (alluvial, roadside), thickets, and the margins of gravel pits, lakes, ponds (beaver, quarry, sand pit, sinkhole), reservoirs, rivers, and streams (perennial) at elevations to 2,715 m. The plants occur in exposures that receive full sunlight and are more commonly associated with disturbed or saline sites than are other cattail species. Salinities as high as 16‰ can be tolerated, but the plants tend to occur in shallower waters as salinity increases. Although the greatest densities occur at high salinity sites, maximum plant height and biomass occur at low salinity sites. Plants have thrived (increased biomass) in greenhouse experiments where salinities were as high as 32‰, when nutrients were added. The long leaves (2.5–2.7 m) and thick rhizomes enable greater water colonization depth than other North American

congeners. Most occurrences are in waters of intermediate depth. Although reported at depths from 0.2 to 1.2 m, the plants grow most often in waters deeper than 0.5 m and optimally at 0.8 m. The substrates have been characterized as alluvium (wet), clay (alkali, hardpan, heavy), clay loam, clay muck, gravel, loam (wet), loamy clay, loamy muck, muck (marly), mucky silt, mud (hard-packed), peat (saturated), peaty clay, peaty muck, rocky silt, sand (alkali), sandy clay, sandy gravel, sandy loam, sandy marl, sandy mud, sandy silt, serpentine, shale, silt, and silty loam. Besides being quite salinity-tolerant (chloride: 3–1,380 ppm), the plants occur across a broad ecological spectrum of habitat conditions (pH: 5.8–9.9; $\bar{x} = 6.7$–7.2; $\tilde{x} = 8.1$; conductivity [μmhos cm^{-1} @ 25°C]: 10–640; $\tilde{x} = 290$; total alkalinity [mg l^{-1} CaCO$_3$]: 5–275; $\bar{x} = 56.0$; $\tilde{x} = 135$). Flowering and fruiting have been observed from April to October. Foliar resource allocation (40–57%) is low relative to broader-leaved species but sexual allocation is relatively higher, varying from 5% to 39% over various depths. Pollen is dispersed xenogamously by the wind, facilitated by its lightweight and relatively higher physical position on the inflorescence due to an intervening sterile region that separates the male from female flowers. Yet, geitonogamous self-pollination also occurs by pollen that simply falls from the upper staminate to the lower pistillate flowers. The average seed set is relatively high (63%). The plumose seeds are dispersed over vast distances (estimated up to 83 ha) by the wind. In nature, much higher germination (88%) occurs under flooded than under nonflooded conditions (8%), but good germination has been observed at water levels from 5 to +10 cm. The seeds will germinate in salt concentrations up to 20‰, but the seedlings cannot tolerate salt levels above 10‰. Seeds have retained viability following 4 months of submergence in a 2% salt solution, when placed subsequently in tap water for germination. Seeds have germinated successfully under ambient greenhouse temperatures after being air-dried at 22°C for 5 days, stored in the dark at 50°C for 2 months, and then transferred to dark, moist, or wet storage at 2°C for 5 months. Vegetative reproduction is prolific and occurs by expansion of rhizomes, which vary in average length from 13 to 21 cm. Ramet densities can vary from 2 to 43 ramets m^{-2}, with the biomass of a single ramet averaging from 28.4 to 34.5 g ash-free dw. Maximum biomass recorded from sites in Texas peaked at 2,895 g ash-free dw m^{-2} (aboveground) and 2,506 g ash-free dw m^{-2} (belowground). A standing crop biomass of 2.9 kg m^{-2} has been reported from some brackish sites. The roots can be colonized by arbuscular mycorrhizal Fungi, except under high (>100 μM) phosphorous conditions.

Reported associates: *Acalypha rhomboidea, Acer negundo, Acer saccharinum, Achillea millefolium, Acorus calamus, Agave utahensis, Agrostis gigantea, Agrostis stolonifera, Alisma subcordatum, Alisma triviale, Alnus oblongifolia, Alnus rubra, Ambrosia psilostachya, Ammannia robusta, Amorpha fruticosa, Andropogon glomeratus, Anemopsis californica, Angelica atropurpurea, Apios americana, Apocynum cannabinum, Aristida purpurea, Aristida ternipes, Artemisia californica, Artemisia dracunculus, Asclepias fascicularis, Asclepias incarnata, Astragalus, Atriplex canescens, Atriplex* *phyllostegia, Atriplex torreyi, Baccharis salicifolia, Baccharis salicina, Bassia hyssopifolia, Berula erecta, Betula pumila, Bidens cernuus, Bidens frondosus, Bidens laevis, Bidens polylepis, Boehmeria cylindrica, Bolboschoenus fluviatilis, Bolboschoenus maritimus, Bothriochloa laguroides, Bouteloua curtipendula, Brasenia schreberi, Brickellia californica, Brickellia longifolia, Bromus ciliatus, Bromus inermis, Bromus japonicus, Bromus pubescens, Bromus rubens, Butomus umbellatus, Calamagrostis canadensis, Calamagrostis stricta, Calibrachoa parviflora, Calla palustris, Caltha palustris, Campanula aparinoides, Campsis radicans, Capsella bursa-pastoris, Carex annectens, Carex aquatilis, Carex atherodes, Carex brevior, Carex buxbaumii, Carex canescens, Carex comosa, Carex cristatella, Carex disperma, Carex douglasii, Carex hystericina, Carex interior, Carex lacustris, Carex lasiocarpa, Carex leptalea, Carex lurida, Carex lyngbyei, Carex molesta, Carex nebrascensis, Carex prairea, Carex pseudocyperus, Carex sartwellii, Carex serratodens, Carex sterilis, Carex stipata, Carex stricta, Carex tribuloides, Carex utriculata, Carex vulpinoidea, Castilleja minor, Cephalanthus occidentalis, Ceratophyllum demersum, Chamaedaphne calyculata, Chara, Chelone glabra, Chenopodium rubrum, Chloropyron maritimum, Cicuta bulbifera, Cicuta maculata, Cirsium arvense, Cirsium fontinale, Cladium jamaicense, Cladium mariscoides, Cleomella obtusifolia, Comarum palustre, Conyza canadensis, Corispermum, Cornus foemina, Cornus sericea, Cyperus acuminatus, Cyperus diandrus, Cyperus eragrostis, Cyperus squarrosus, Cyperus strigosus, Cyperus virens, Cypripedium parviflorum, Dasiphora floribunda, Datisca glomerata, Daucus carota, Decodon verticillatus, Deschampsia cespitosa, Deschampsia danthonioides, Descurainia sophia, Distichlis spicata, Doellingeria umbellata, Drosera rotundifolia, Dulichium arundinaceum, Echinochloa crus-galli, Echinochloa walteri, Eclipta prostrata, Elaeagnus angustifolia, Elaeagnus commutata, Elatine minima, Eleocharis acicularis, Eleocharis erythropoda, Eleocharis macrostachya, Eleocharis montevidensis, Eleocharis obtusa, Eleocharis palustris, Eleocharis parishii, Eleocharis rostellata, Elymus canadensis, Elymus repens, Epilobium ciliatum, Epilobium leptophyllum, Epipactis gigantea, Equisetum fluviatile, Equisetum laevigatum, Equisetum palustre, Equisetum variegatum, Equisetum ×ferrissii, Eragrostis curvula, Ericameria nauseosa, Eriogonum polycladon, Eriophorum angustifolium, Eupatorium perfoliatum, Euphorbia esula, Euthamia graminifolia, Euthamia occidentalis, Eutrochium maculatum, Fallugia paradoxa, Fraxinus velutina, Galium asprellum, Galium trifidum, Garrya flavescens, Gentiana andrewsii, Glyceria striata, Gnaphalium, Gutierrezia sarothrae, Hedeoma drummondii, Hedeoma oblongifolia, Helenium autumnale, Helianthus annuus, Helianthus annuus, Helianthus grosseserratus, Herrickia glauca, Heterotheca subaxillaris, Heterotheca villosa, Hibiscus moscheutos, Holodiscus discolor, Hordeum jubatum, Hornungia procumbens, Hydrocotyle americana, Hymenothrix loomisii, Impatiens capensis, Imperata brevifolia, Iris pseudacorus,*

Iris virginica, Iva annua, Iva axillaris, Juglans major, Juncus alpinoarticulatus, Juncus articulatus, Juncus balticus, Juncus bufonius, Juncus effusus, Juncus nevadensis, Juncus nodosus, Juncus tenuis, Juncus torreyi, Juniperus monosperma, Lactuca serriola, Larix laricina, Lathyrus palustris, Leersia oryzoides, Lemna minor, Lemna trisulca, Lepidium appelianum, Lepidium perfoliatum, Leptochloa fusca, Leymus cinereus, Lindernia dubia, Lobelia cardinalis, Lobelia kalmii, Lobelia siphilitica, Lolium perenne, Lorandersonia linifolia, Ludwigia palustris, Ludwigia peploides, Lycopus americanus, Lycopus asper, Lycopus uniflorus, Lysimachia nummularia, Lysimachia quadriflora, Lysimachia thyrsiflora, Lythrum alatum, Lythrum salicaria, Melilotus albus, Melilotus officinalis, Mentha arvensis, Micranthes pensylvanica, Mimulus cardinalis, Mimulus guttatus, Mimulus ringens, Mirabilis albida, Mirabilis linearis, Monarda fistulosa, Monardella odoratissima, Muhlenbergia asperifolia, Muhlenbergia glomerata, Muhlenbergia porteri, Muhlenbergia rigens, Muhlenbergia thurberi, Myriophyllum sibiricum, Myriophyllum tenellum, Najas flexilis, Najas gracillima, Nasturtium officinale, Nitrophila occidentalis, Nuphar advena, Nymphaea odorata, Oenanthe, Oenothera curtiflora, Oenothera longissima, Onoclea sensibilis, Oxypolis rigidior, Panicum capillare, Parnassia glauca, Paspalum distichum, Paspalum urvillei, Peltandra virginica, Penthorum sedoides, Persicaria amphibia, Persicaria coccinea, Persicaria hydropiperoides, Persicaria lapathifolia, Persicaria maculosa, Persicaria pensylvanica, Persicaria sagittata, Petasites frigidus, Phacelia egena, Phalaris arundinacea, Phlox maculata, Phragmites australis, Physostegia virginiana, Picea glauca, Picea mariana, Pilea fontana, Pinus edulis, Plagiobothrys salsus, Plantago major, Platanthera hyperborea, Platanus wrightii, Pluchea odorata, Poa nemoralis, Poa palustris, Poa pratensis, Polypogon monspeliensis, Polypogon viridis, Pontederia cordata, Populus deltoides, Populus fremontii, Populus tremuloides, Potamogeton crispus, Potamogeton foliosus, Potamogeton nodosus, Potamogeton zosteriformis, Prosopis velutina, Pseudognaphalium luteoalbum, Ptilimnium, Puccinellia distans, Purshia mexicana, Pycnanthemum virginianum, Quercus agrifolia, Quercus lobata, Quercus turbinella, Ranunculus hispidus, Ranunculus pensylvanicus, Ranunculus recurvatus, Ranunculus sceleratus, Rhamnus alnifolia, Rhamnus frangula, Rhus aromatica, Rhynchospora capillacea, Ribes aureum, Rorippa palustris, Rosa californica, Rosa palustris, Rubus pubescens, Rubus sachalinensis, Rudbeckia fulgida, Rudbeckia hirta, Rudbeckia laciniata, Rumex altissimus, Rumex crispus, Rumex mexicanus, Rumex salicifolius, Rumex triangulivalvis, Sagittaria cuneata, Sagittaria latifolia, Salix amygdaloides, Salix candida, Salix discolor, Salix exigua, Salix gooddingii, Salix hookeriana, Salix interior, Salix lasiolepis, Salix pedicellaris, Sambucus nigra, Sarcobatus vermiculatus, Sarracenia purpurea, Schedonorus arundinaceus, Schizachyrium scoparium, Schoenoplectus acutus, Schoenoplectus americanus, Schoenoplectus pungens, Schoenoplectus tabernaemontani, Scirpus atrovirens, Scirpus cyperinus, Scirpus pendulus, Scutellaria galericulata, Scutellaria lateriflora, Sesbania drummondii, Setaria leucopila, Shepherdia rotundifolia, Silphium compositum, Silphium laciniatum, Silphium perfoliatum, Sisymbrium altissimum, Sium suave, Smilax lasioneura, Smilax walteri, Solanum carolinense, Solanum dulcamara, Solidago altissima, Solidago canadensis, Solidago gigantea, Solidago ohioensis, Solidago patula, Solidago rugosa, Solidago uliginosa, Solidago velutina, Sonchus oleraceus, Sorghastrum nutans, Sorghum halepense, Spartina cynosuroides, Spartina patens, Spartina pectinata, Spergularia salina, Sphagnum, Spiraea alba, Sporobolus airoides, Stachys palustris, Stuckenia filiformis, Stuckenia pectinata, Suaeda nigra, Symphoricarpos, Symphyotrichum firmum, Symphyotrichum frondosum, Symphyotrichum lanceolatum, Symphyotrichum novae-angliae, Symphyotrichum praealtum, Symphyotrichum puniceum, Symplocarpus foetidus, Tamarix chinensis, Tamarix ramosissima, Thalictrum dasycarpum, Thelypodium integrifolium, Thelypteris palustris, Toxicodendron diversilobum, Toxicodendron radicans, Toxicodendron vernix, Triadica sebifera, Trifolium pinetorum, Trifolium repens, Triglochin maritima, Triosteum perfoliatum, Typha domingensis, Typha latifolia, Typha ×glauca, Ulmus americana, Urtica dioica, Utricularia intermedia, Utricularia macrorhiza, Valeriana edulis, Valeriana sitchensis, Vallisneria americana, Verbascum thapsus, Verbena lasiostachys, Vernonia missurica, Veronica americana, Veronica anagallis-aquatica, Viola cucullata, Viola palustris, Vitis riparia, Wolffia, Xanthium strumarium, Zizania aquatica, Zizaniopsis miliacea.

10.3.2.2. *Typha domingensis* Pers. grows in or on fresh to brackish backwaters (river), beaches, bog mats, bottoms (canyon, meadow, river), channels (brackish, flood-control, low-flow, stream, tidal), depressions (artificial, seasonally wet, wet), ditches (desiccated, drainage, flooded, irrigation, roadside, swampy), estuaries, flats (alkaline, tidal), floodplains (hardwood), gravel bars, gravel pits, gullies, hammocks (hardwood, hydric), inlets (river), marshes (brackish, depression, floodplain, freshwater, oligohaline, riparian, roadside, saline, spring-fed, tidal), meadows (bottomland), mudflats, oxbows, ponds (artificial, beaver, ephemeral, freshwater, holding, irrigation, oasis, retention, sag), pools (brackish, dune, permanent, stagnant, streambed, swampy, vernal), prairies (marlaceous, wet), rice fields, riverbeds, roadsides, salt marshes, seepages (alkaline, marshy, roadside, sulfur-spring), sinks (alkaline, desiccating, wet), slopes [to 10°] (dry), sloughs, springs (desert, dry), stock tanks, streambeds (dried), swales (drainage, hillside, moist), swamps (ephemeral, mangrove), thickets (willow), washes (alluvial, desert, intermittent, rocky, saline, sandy, wet), and along the banks, margins, or shores of bayous, borrow pits, canals (drainage), lagoons, lakes (artificial, brackish, playa), ponds, reservoirs, rivers, and streams (intermittent, perennial, permanent) at elevations from 70 to 2,341 m. Most stands occur in exposures of full sunlight to partial shade. Although the plants are adapted to brackish conditions (salinities to 8‰; chloride: 20–6,500 ppm), their growth is optimal at 0–1.1‰, decreases by half at 3.5‰, and is negligible above 6‰. Substantial mortality (75%) has been

reported at salinities of 15‰ and the complete absence of growth at 25‰. However, in one experiment, a small number of rhizomatous plants (5% of those evaluated) survived for 9 months at a salinity of 45‰. The plants can grow in shallow standing waters (e.g., 60–115 cm deep), but their flowering and shoot density decrease while their total ramet size increases (as much as fourfold) as they extend into deeper water; however, the percentage of biomass allocated to leaves remains fixed across different water depths. Maximum densities approaching 200 plants m^{-2} have been observed at water depths of 22 cm. Plants have survived experimental flooding to depths of 91 cm, when followed by a recovery period in shallower (40 cm) water, whereas those stressed by exposure to deeper flooding (137 cm) did not recover. This species occurs commonly along arid riparian zones, where it establishes during ephemeral, inundated conditions and persists as numerous facultative and upland species (e.g., *Opuntia*) colonize the sites after the waters subside and the substrates desiccate. Many of these species have been excluded from the list of associates below, despite their being listed as such in various database records. The substrates mostly are alkaline (pH: 6.7–8.6) and have been described as adobe, alluvium, basalt, clay (Houston black), clay loam (black), cobble (river), granite (decomposed), gravel, gravelly loam, loam (dried), muck (Allemands), mud (organic), Newhan Corolla complex, peat, rocky cobble, rocky gravel, sand (alkali, brackish, Duckston, granite, Leon), sandy alluvium, sandy clay, sandy clay loam, sandy gravel, sandy loam (Clovis, Ramona), sandy silt, sandy silty loam, shale (Lodo, Monterey), silt (alkaline), silty clay (Austin), silty clay loam, and silty loam (Claiborne-Peridge). Flowering and fruiting extend from March to November. Flowering frequency is highest at water depths from 5 to 5 cm. Fruiting is prolific, with a single inflorescence capable of producing upward of 250,000 seeds. Light is required for seed germination, which will not occur under buried conditions. A long-term seed bank can develop, with emerging seedling densities as high as 33 m^{-2} reported. The seeds can germinate under moist, saturated, and even flooded conditions. Field studies have shown that seed germination can reach 100% in freshwater but is reduced to only 2% at salinities of 20‰, which restricts invasive potential in highly saline sites. High germination has been obtained for freshly collected seeds that have been stored in the dark at room temperature (20°C) for several months, then incubated (averaging from 1.1 to 9.5 days) under constant (15°C–30°C) or fluctuating (25°C/10°C or 30°C/20°C) day/night temperatures. Germination on peat occurs more rapidly than in water but is not influenced by phosphate concentrations. Natural seed germination typically occurs within 3 days and can reach 40% after 1 week. The growth rate (to 89 mg g^{-1} day^{-1}) and above/belowground and shoot/root biomass ratios increase proportionally with phosphorous concentrations, which can be substantially enhanced in muck-burned sites. High transpiration rates (>11 mmol $m^2 s^1$) can be achieved during periods of low ambient temperature and vapor pressure (i.e., winter and spring). The encroachment of *T. domingensis* into sites dominated formerly by sawgrass (*Cladium jamaicense*) is viewed as indicative of long-term habitat degradation and has been attributed to a higher tolerance of the former for enriched phosphorous and sulfide levels (to 0.69 mM sulfide). The plant density in burn sites has been observed to increase rapidly during the first 1–2 years following fire before stabilizing at preburn levels.

Reported associates: Acacia cyclops, Acer macrophyllum, Acer negundo, Achillea millefolium, Achnatherum hymenoides, Acmella oppositifolia, Acmispon glabrus, Acmispon parviflorus, Acrostichum aureum, Adenostoma sparsifolium, Adiantum capillus-veneris, Agave palmeri, Ageratina adenophora, ×Agropogon lutosus, Agrostis stolonifera, Agrostis stolonifera, Allenrolfea occidentalis, Alnus oblongifolia, Alnus rhombifolia, Alternanthera philoxeroides, Amaranthus albus, Amaranthus australis, Amaranthus australis, Ambrosia psilostachya, Ambrosia trifida, Ammannia auriculata, Amorpha californica, Amorpha fruticosa, Anemopsis californica, Andropogon glomeratus, Anisacanthus thurberi, Apios americana, Apium graveolens, Apocynum, Arctostaphylos glandulosa, Arctostaphylos patula, Arctostaphylos pringlei, Aristida purpurea, Aristida ternipes, Artemisia californica, Artemisia douglasiana, Artemisia ludoviciana, Artemisia tridentata, Arundo donax, Asclepias fascicularis, Asclepias incarnata, Astragalus trichopodus, Atriplex canescens, Atriplex elegans, Atriplex lentiformis, Atriplex patula, Atriplex phyllostegia, Avicennia germinans, Baccharis halimifolia, Baccharis neglecta, Baccharis pilularis, Baccharis salicifolia, Baccharis salicina, Baccharis sarothroides, Baccharis sergiloides, Bacopa caroliniana, Bacopa monnieri, Bahiopsis parishii, Bassia hyssopifolia, Batis maritima, Bebbia juncea, Berberis haematocarpa, Berula erecta, Bidens laevis, Boehmeria cylindrica, Boerhavia coccinea, Bolboschoenus maritimus, Bolboschoenus robustus, Bothriochloa laguroides, Brassica nigra, Brassica tournefortii, Brickellia californica, Brickellia floribunda, Brickellia longifolia, Brodiaea orcuttii, Bromus hordeaceus, Bromus japonicus, Bromus rubens, Bromus tectorum, Buchnera americana, Butomus umbellatus, Cakile maritima, Calamagrostis stricta, Canotia holacantha, Carduus pycnocephalus, Carex alata, Carex canescens, Carex leptalea, Carex praegracilis, Carex senta, Carex spissa, Castilleja lanata, Castilleja linariifolia, Ceanothus leucodermis, Celtis pallida, Celtis reticulata, Centaurea melitensis, Centella asiatica, Cephalanthus occidentalis, Chaetopappa asteroides, Chara, Cheilanthes lindheimeri, Chenopodium album, Chenopodium fremontii, Chenopodium rubrum, Chilopsis linearis, Chloracantha spinosa, Chloropyron maritimum, Chylismia walkeri, Cicuta maculata, Cirsium arvense, Cirsium undulatum, Cirsium vulgare, Cladium californicum, Cladium jamaicense, Clematis ligusticifolia, Cleomella obtusifolia, Cleomella obtusifolia, Cnidoscolus angustidens, Colocasia esculenta, Comarum palustre, Conium maculatum, Conoclinium betonicifolium, Convolvulus arvensis, Convolvulus sepium, Conyza canadensis, Cordylanthus rigidus, Corethrogyne filaginifolia, Cornus foemina, Cortaderia selloana, Cotoneaster lacteus, Cotula coronopifolia, Cressa truxillensis, Croton capitatus, Crypsis schoenoides, Crypsis vaginiflora, Cryptantha barbigera, Cucurbita foetidissima, Cuscuta pentagona, Cynodon dactylon, Cyperus elegans, Cyperus entrerianus, Cyperus eragrostis, Cyperus esculentus, Cyperus haspan, Cyperus involucratus, Cyperus ochraceus, Cyperus odoratus, Cyperus strigosus, Dactylis glomerata, Dasylirion wheeleri, Datisca glomerata, Datura wrightii, Daucus pusillus, Deinandra mohavensis, Deschampsia elongata, Desmanthus illinoensis, Dieteria canescens, Distichlis spicata, Draba, Drosera rotundifolia, Dudleya lanceolata, Echinochloa colona, Echinochloa crusgalli, Echinochloa walteri, Echinodorus berteroi, Eclipta prostrata, Egeria densa, Eichhornia crassipes, Elaeagnus angustifolia, Eleocharis acicularis, Eleocharis erythropoda, Eleocharis macrostachya, Eleocharis montevidensis, Eleocharis occulta, Eleocharis palustris, Eleocharis parishii, Eleocharis parvula, Eleocharis quadrangulata, Eleocharis rostellata, Elymus elymoides, Encelia farinosa, Engelmannia peristenia, Ephedra viridis, Epilobium canum, Epilobium ciliatum, Epipactis gigantea, Equisetum arvense, Equisetum hyemale, Equisetum laevigatum, Equisetum telmateia, Eragrostis cilianensis, Eragrostis reptans, Ericameria albida, Ericameria laricifolia, Ericameria nauseosa, Erigeron divergens, Erigeron strigosus, Eriogonum corymbosum, Eriogonum fasciculatum, Eriogonum inflatum, Eriogonum leptophyllum, Eriogonum polycladon, Eriophyllum confertiflorum, Erodium cicutarium, Erythrina flabelliformis, Eupatorium serotinum, Euphorbia, Euthamia occidentalis, Fallugia paradoxa, Fimbristylis autumnalis, Fimbristylis spadicea, Foeniculum vulgare, Forestiera, Fraxinus velutina, Fuirena simplex, Funastrum cynanchoides, Glinus radiatus, Gossypium thurberi, Gutierrezia sarothrae, Hazardia squarrosa, Hedeoma drummondii, Hedeoma oblongifolia, Helenium thurberi, Helianthus annuus, Helianthus annuus, Helianthus californicus, Helianthus gracilentus, Heliotropium curassavicum, Helminthotheca echioides, Heteranthera dubia, Heteromeles arbutifolia, Heteropogon contortus, Heterotheca grandiflora, Heterotheca subaxillaris, Heterotheca villosa, Hibiscus grandiflorus, Hilaria jamesii, Hirschfeldia incana, Holcus lanatus, Hordeum jubatum, Horkelia rydbergii, Hosackia oblongifolia, Hydrocotyle umbellata, Hydrocotyle verticillata, Hymenocallis palmeri, Hyptis emoryi, Ilex cassine, Ipomoea cordatotriloba, Ipomopsis aggregata, Isocoma acradenia, Isocoma menziesii, Isolepis cernua, Iva annua, Iva frutescens, Iva hayesiana, Iva imbricata, Juglans major, Juncus acutus, Juncus arcticus, Juncus articulatus, Juncus balticus, Juncus bufonius, Juncus diffusissimus, Juncus dubius, Juncus effusus, Juncus ensifolius, Juncus interior, Juncus nevadensis, Juncus phaeocephalus, Juncus roemerianus, Juncus saximontanus, Juncus torreyi, Juncus validus, Juncus xiphioides, Juniperus osteosperma, Justicia americana, Kochia scoparia, Lactuca saligna, Lactuca serriola, Larrea tridentata, Leersia oryzoides, Leersia virginica, Lemna minor, Lepidospartum squamatum, Leptochloa fusca, Leymus triticoides, Lilaeopsis chinensis, Limnobium spongia, Lindernia dubia, Linum alatum, Lobelia cardinalis, Logfia filaginoides, Lolium perenne, Lotus corniculatus, Ludwigia hexapetala, Ludwigia peploides, Ludwigia repens, Lycium, Lythrum californicum,

Lythrum lineare, Machaeranthera, Magnolia virginiana, Malacothamnus fasciculatus, Malosma laurina, Marsilea vestita, Melilotus albus, Melilotus indicus, Melilotus officinalis, Mentha arvensis, Mentha spicata, Mikania scandens, Mimosa aculeaticarpa, Mimulus cardinalis, Mimulus guttatus, Mimulus parishii, Monarda citriodora, Monarda fistulosa, Monardella odoratissima, Morus alba, Muhlenbergia asperifolia, Muhlenbergia capillaris, Muhlenbergia richardsonis, Muhlenbergia rigens, Myoporum laetum, Myrica cerifera, Myriophyllum sibiricum, Nassella leucotricha, Nasturtium officinale, Nelumbo lutea, Nicotiana glauca, Nicotiana obtusifolia, Nitrophila occidentalis, Oenothera curtiflora, Oenothera curtiflora, Oenothera elata, Oenothera humifusa, Oenothera pallida, Oenothera suffrutescens, Olneya tesota, Onopordum acanthium, Oonopsis foliosa, Opuntia engelmannii, Opuntia phaeacantha, Opuntia santarita, Oxytenia acerosa, Panicum amarum, Panicum hemitomon, Panicum virgatum, Parkinsonia aculeata, Parkinsonia florida, Parryella filifolia, Paspalum dilatatum, Paspalum pubiflorum, Paspalum urvillei, Persicaria amphibia, Persicaria arifolia, Persicaria glabra, Persicaria hydropiperoides, Persicaria lapathifolia, Persicaria maculosa, Persicaria pensylvanica, Persicaria punctata, Petunia, Phacelia crenulata, Phacelia egena, Phacelia grandiflora, Phalaris angusta, Phalaris aquatica, Phalaris minor, Phragmites australis, Phyla nodiflora, Physalis longifolia, Pinus elliottii, Pinus monophylla, Pinus taeda, Piptatherum miliaceum, Plantago coronopus, Plantago major Plantago ovata, Plantago subnuda, Plantago wrightiana, Platanus racemosa, Platanus wrightii, Pluchea camphorata, Pluchea odorata, Pluchea sericea, Poa annua, Poa compressa, Poa pratensis, Polanisia dodecandra, Polygala, Polygonum argyrocoleon, Polygonum aviculare, Polypogon interruptus, Polypogon monspeliensis, Polypogon viridis, Pontederia cordata, Populus deltoides, Populus fremontii, Potamogeton crispus, Potamogeton foliosus, Potamogeton nodosus, Prosopis glandulosa, Prosopis pubescens, Prosopis velutina, Pseudognaphalium luteoalbum, Pseudognaphalium microcephalum, Ptilimnium capillaceum, Pulicaria paludosa, Pyracantha, Quercus agrifolia, Quercus berberidifolia, Quercus engelmannii, Quercus gambelii, Quercus pacifica, Quercus palmeri, Quercus turbinella, Ranunculus cymbalaria, Rhamnus californica, Rhaponticum repens, Rhus aromatica, Rhus integrifolia, Rhus ovata, Rhynchospora colorata, Rhynchospora corniculata, Rhynchospora miliacea, Ribes aureum, Ribes malvaceum, Robinia neomexicana, Robinia pseudoacacia, Rorippa, Rosa californica, Rotala ramosior, Rubus ursinus, Rudbeckia hirta, Rumex chrysocarpus, Rumex conglomeratus, Rumex crispus, Rumex dentatus, Rumex obovatus, Rumex pulcher, Rumex verticillatus, Sabal palmetto, Sacciolepis striata, Sagittaria lancifolia, Sagittaria latifolia, Sagittaria papillosa, Salazaria mexicana, Salicornia depressa, Salix amygdaloides, Salix exigua, Salix gooddingii, Salix laevigata, Salix lasiolepis, Salix lucida, Salix lutea, Salix nigra, Salsola kali, Salsola tragus, Salvia columbariae, Salvia mellifera, Sambucus nigra, Samolus valerandi, Sarcobatus vermiculatus, Saururus cernuus, Schedonorus arundinaceus, Schedonorus pratensis, Schoenoplectus acutus, Schoenoplectus americanus, Schoenoplectus californicus, Schoenoplectus mucronatus, Schoenoplectus pungens, Schoenoplectus tabernaemontani, Scirpus cyperinus, Scutellaria lateriflora, Scutellaria lateriflora, Senegalia greggii, Sesbania vesicaria, Sesuvium portulacastrum, Sesuvium verrucosum, Setaria leucopila, Setaria parviflora, Solanum dimidiatum, Solanum elaeagnifolium, Solidago altissima, Solidago confinis, Solidago spectabilis, Solidago velutina, Sonchus asper, Sorghastrum nutans, Sorghum halepense, Sparganium eurycarpum, Spartina bakeri, Spartina patens, Spartium junceum, Spergularia salina, Sphaeralcea grossulariifolia, Sphaeralcea laxa, Sporobolus airoides, Stachys albens, Stachys bullata, Stanleya pinnata, Stenotaphrum secundatum, Stephanomeria exigua, Stephanomeria pauciflora, Stuckenia filiformis, Stuckenia pectinata, Suaeda nigra, Symphyotrichum subulatum, Tamarix chinensis, Tamarix gallica, Tamarix parviflora, Tamarix ramosissima, Taxodium, Tetracoccus hallii, Teucrium canadense, Thalia geniculata, Thelypteris palustris, Toxicodendron diversilobum, Triadica sebifera, Tridens muticus, Trifolium pinetorum, Typha angustifolia, Typha latifolia, Ulmus parvifolia, Umbellularia californica, Urtica dioica, Utricularia, Veronica anagallis-aquatica, Vicia acutifolia, Vigna luteola, Viola palustris, Vitis arizonica, Vitis girdiana, Woodwardia, Xanthium spinosum, Xanthium strumarium, Zeltnera exaltata, Zizaniopsis miliacea.

10.3.2.3. **Typha latifolia** L. occurs in or on alluvial plains, bays (shallow), bogs (peat), borrow pits, bottoms (canyon, drying, ravine, reservoir, stream), canals (feeder), channels (river), depressions (desiccated, freshwater, railroad, roadside, wet), ditches (agricultural, desiccated, disturbed, drainage, flatwoods, flooded, irrigation, moist, railroad, roadside, swampy, wet), draws (wet), dunes (beach), embayments (boggy), fens (calcareous, patterned [flarks], sloping), flats, flatwoods (pine), floodplains (bottomland, river), gravel bars, impoundments, inlets (river), lagoons (desiccating), marshes (basin, coastal, degraded, depression, desiccated, dune, freshwater, open, prairie, riparian, roadside), mats (floating, sedge), meadows (beaver, boggy, dune, mesic, seepage, wet), muskeg (spruce), oxbows, pools (beach, muskeg, roadside), potholes (moist), prairies (lowland, mesic, sand, short-grass, wet, wet-mesic), quarries (abandoned, limestone), rice fields, riverbeds, roadsides (disturbed, low), seeps (alkaline, calcareous, freshwater, grassy, open, roadside, sandy), shores (lake, shallow), slopes [to 2%] (dry, rocky), sloughs (dried), springs, streambeds (seasonal), swales (beach, dune, moist), swamps (cypress, floodplain, hardwood, roadside, *Sphagnum*), thickets (poplar, willow), vernal pools, washes (rocky, sandy, shallow), woodlands (alluvial, deciduous, low, riparian), and along the margins of bogs, brooks, canals, gravel pits, impoundments, lagoons, lakes, ponds (backwater, beaver, cattle, containment, dune, freshwater, holding, logging, prairie, silt, stock, stormwater, thermokarst), pools (artificial, drainage), potholes (seasonal), reservoirs, rivers, sloughs, streams (artificial, intermittent, thermal), and salt marshes at elevations to 2,880 m. The plants grow in exposures ranging from full sunlight to shade (to 9% light), and are more shade-tolerant than their North American congeners. The substrates have been characterized as alluvium (Kaiparowits, Quarternary, sandy), Capay-Clear Lake soil, clay (acid, slipping, sticky), clay loam (calcareous, moist, wet), cobbles, granite (decomposed), gravel (alluvial), gravel loam, humus, limestone (Claron), loam, loamy sand (Burbank), muck (Barbar, organic), mucky clay, mud, muddy clay loam, Newhan-Corolla complex, peat (saturated), peaty muck (saturated), rock, rocky silt, sand, sandstone, sandy clay, sandy gravel, sandy loam (fine, Harleston, red), sandy muck, sandy silt, silt (dolomitic), silty clay, silty clay loam, silty loam (Bodine-Brandon, cherty, Esquatzel, granitic), Travertine, and volcanics. As in other cattail species, the plants grow taller as water depth increases. However, their relatively shorter leaves (1.5–1.9 m) and thinner rhizomes preclude any long-term persistence in deeper waters. Although reported at depths from 0.15 to 1.2 m, the plants usually occur much shallower at depths <80 cm ($\tilde{x}$ ~25 cm) and often can be found in drying sites or where soils are wet but not overlain by standing water. They do not persist when water levels rise above 95 cm. Competitive superiority over other cattail species occurs only in shallow water (<15 cm). Individuals are highly drought-tolerant and able to withstand soil moisture levels as low as 5% before complete root mortality occurs. Habitats include fresh to brackish, often eutrophic sites (pH: 5.1–10.3; $\bar{x}$ =6.7–7.4; $\tilde{x}$ =7.4; total alkalinity [mg l^{-1} CaCO$_3$]: 1–200; $\bar{x}$ =39.2; $\tilde{x}$ =45; Ca: ~2.72 mg/l, Mg: ~0.78 mg/l; Cl: 0.1–1,380 ppm; conductivity [μmhos cm^{-1}@ 25°C]: 10–700; $\tilde{x}$ =105). The plants generally occur at salinities below 3‰, but have shown no ill effects when grown in sodium (Na) concentrations up to 300 mg l^{-1}. Flowering and fruiting have been observed from May to October. Flowering and shoot density decrease and more biomass is allocated to leaves as water depth increases. Foliar resource allocation (47–67%) is high relative to narrower-leaved species but sexual allocation is lower, varying across depths from 1% to 17%. Flowering does not occur in deeper waters and is most common when water levels are low (i.e., 5 cm). Unlike other North American cattails, the pollen is produced in tetrads (cf. monads), which number about 90 × 10^6 per inflorescence. A highly active form of phytase has been isolated from the pollen. Dead tetrad grains are thought to result from chromosomal non-disjunction during meiotic divisions. The pollen is dispersed entirely by strong winds and can remain viable for 4 or more weeks. The female flowers remain pollen receptive for 4 weeks, with the highest proportion of receptive flowers occurring during the first 2 weeks of anthesis. However, the pollen is heavy (due to tetrad formation) and falls readily from the upper staminate to the contiguous, lower region of pistillate flowers, which can result in a high degree of geitonogamous self-pollination that leads to higher levels of seed set (to 54%) in the upper portion of the pistillate region. Simulations and field data have estimated that 99% of pollen is dispersed within a 2 m radius, but that inter-shoot pollination is more likely to occur, albeit among closely neighboring shoots. Genetic analyses (allozymes and VNTR locus diversity) have shown the plants to be characterized by extremely low levels of detectable variation (genetic similarity: 0.89–0.95), low heterozygosity (0.05–0.13), high clonality (~61%), and extensive among population subdivision (F_{st}: 0.32–0.41). A substantial seed bank can develop, with studies reporting up to 5,610 seedlings m^{-2} emerging from sediment samples. The seeds have germinated well in flats of saturated topsoil (pH: 6.5). Germination is optimal when seeds are exposed to white light in low-oxygen environments. Under low-light conditions, germination is enhanced by widely fluctuating temperatures. Germination rates increase proportionally with temperature, with the highest germination observed between 20°C and 35°C and little (<10%) to no germination evident at 10°C–15°C. Faster germination also has been observed under higher nutrient (i.e., phosphate) levels. Complete germination (100%) has been obtained for seeds after the seed coat ends were ruptured. The seeds can germinate in salt concentrations up to 20‰, but the seedlings cannot tolerate salt levels above 10‰. The seeds have retained viability after 4 months of submergence in a 2% salt solution, when placed subsequently in tap water for germination. Most of the seedlings develop while they are submersed in water. Perennation and clonal development occur by means of the rhizome and root system, which can store large amounts of starch (45.03% and 22.80% dw, respectively). Individual rhizome length averages from 27 to 28 cm. The plants produce from 11 to 44 ramets m^{-2} annually. They can achieve a high clonal spread (to 39 m in extent) that is relatively greater than other North American taxa, as they expand radially by 3–3.5 m each year. Ramet biomass has been found to remain fairly constant across the habitable depth gradient. Single ramet biomass averages from 40.5 to 57.6 g ash-free dw but can reach nearly 150 g ash-free dw. Although productivity has been correlated with levels of soluble hydrosoil phosphorous, controlled experiments have implicated sediment nitrogen as the most limiting nutrient. These C$_3$ plants can achieve photosynthetic rates comparable to those of tropical C$_4$ species. A range of standing crop biomass from 428 to 2,252 g dw m^{-2} and annual production rates of 15,300 kg ha^{-1} have been reported. Although some early sources have reported the absence of mycorrhizal associations, it has been demonstrated that vesicular arbuscular mycorrhizal colonization does occur (to 34%; averaging 13%) under flooded and unflooded conditions, but is much reduced under drawdown states. Greenhouse experiments have indicated less growth in mycorrhizal inoculated plants, despite their higher photosynthetic rate and levels of mineralization. Although an aggressive species, *T. latifolia* does not compete well with *Phalaris arundinacea*, which can reduce light levels by growing quickly to overtop the plants. **Reported associates:** *Abies balsamea, Abronia maritima, Acalypha gracilens, Acer macrophyllum, Acer rubrum, Acer saccharinum, Achillea millefolium, Acmispon glabrus, Acorus americanus, Adenostoma fasciculatum, Adiantum capillus-veneris, Agrostis exarata, Agrostis gigantea, Agrostis idahoensis, Agrostis scabra, Agrostis stolonifera, Alisma subcordatum, Alisma triviale, Alnus oblongifolia, Alnus rhombifolia, Alnus rubra, Alnus rugosa, Alnus serrulata, Alnus tenuifolia, Alnus viridis, Alopecurus aequalis, Alopecurus pratensis,*

Alternanthera philoxeroides, Ambrosia chamissonis, Ambrosia psilostachya, Ambrosia trifida, Ammannia coccinea, Amorpha fruticosa, Anaphalis margaritacea, Andropogon gerardii, Anemopsis californica, Angelica atropurpurea, Apios americana, Apocynum cannabinum, Aquilegia chrysantha, Arctium minus, Arctous rubra, Arisaema triphyllum, Artemisia californica, Artemisia douglasiana, Artemisia tridentata, Arundinaria tecta, Asclepias incarnata, Asclepias syriaca, Asclepias verticillata, Atriplex canescens, Atriplex patula, Atriplex prostrata, Avenella flexuosa, Baccharis douglasii, Baccharis halimifolia, Baccharis pilularis, Baccharis salicifolia, Baccharis sarothroides, Baccharis sergiloides, Bacopa monnieri, Beckmannia syzigachne, Berteroa incana, Berula erecta, Betula nana, Betula neoalaskana, Betula occidentalis, Betula pumila, Bidens cernuus, Bidens frondosus, Bidens trichospermus, Bidens vulgatus, Boehmeria cylindrica, Bolboschoenus fluviatilis, Bolboschoenus maritimus, Bolboschoenus robustus, Bouteloua curtipendula, Brasenia schreberi, Brickellia desertorum, Brickellia longifolia, Bromus ciliatus, Bromus diandrus, Bromus japonicus, Bromus tectorum, Cakile maritima, Calamagrostis canadensis, Calamagrostis stricta, Calla palustris, Caltha palustris, Camissoniopsis cheiranthifolia, Campanula aparinoides, Campsis radicans, Carduus nutans, Carex aquatilis, Carex atherodes, Carex athrostachya, Carex bebbii, Carex bigelowii, Carex buxbaumii, Carex canescens, Carex chordorrhiza, Carex comosa, Carex cristatella, Carex diandra, Carex disperma, Carex echinata, Carex flava, Carex frankii, Carex hystericina, Carex interior, Carex lacustris, Carex lasiocarpa, Carex leptalea, Carex limosa, Carex lyngbyei, Carex nebrascensis, Carex occidentalis, Carex pellita, Carex praegracilis, Carex prairea, Carex retrorsa, Carex rostrata, Carex sartwellii, Carex schottii, Carex simulata, Carex sterilis, Carex stipata, Carex stricta, Carex utriculata, Carex vesicaria, Carpobrotus chilensis, Castilleja minor, Ceanothus integerrimus, Ceanothus leucodermis, Celtis reticulata, Centaurea stoebe, Centella asiatica, Cephalanthus occidentalis, Cercis occidentalis, Chamaedaphne calyculata, Chamerion angustifolium, Chara, Chasmanthium, Chelone glabra, Cicuta bulbifera, Cicuta douglasii, Cicuta maculata, Cicuta virosa, Cirsium arvense, Cirsium vulgare, Cladium mariscoides, Clematis virginiana, Coleataenia anceps, Comarum palustre, Conium maculatum, Constancea nevinii, Convolvulus arvensis, Convolvulus sepium, Conyza canadensis, Coreopsis gigantea, Cornus amomum, Cornus foemina, Cornus sericea, Coronilla, Cotula coronopifolia, Crataegus, Crypsis schoenoides, Cucurbita foetidissima, Cynodon dactylon, Cyperus difformis, Cyperus eragrostis, Cyperus erythrorhizos, Cyperus involucratus, Cyperus iria, Cyperus odoratus, Cyperus strigosus, Cyperus virens, Cypripedium parviflorum, Cyrilla racemiflora, Dasiphora floribunda, Datura wrightii, Decodon verticillatus, Deschampsia elongata, Dichanthelium acuminatum, Dichanthelium sphaerocarpon, Dieteria canescens, Diodella teres, Distichlis spicata, Doellingeria umbellata, Drosera, Dulichium arundinaceum, Dysphania ambrosioides, Dysphania botrys, Echinochloa crus-galli, Echinodorus berteroi, Eclipta prostrata, Elaeagnus angustifolia, Elaeagnus commutata, Eleocharis acicularis, Eleocharis elliptica, Eleocharis erythropoda, Eleocharis interstincta, Eleocharis macrostachya, Eleocharis montevidensis, Eleocharis obtusa, Eleocharis ovata, Eleocharis palustris, Eleocharis parishii, Eleocharis robbinsii, Eleocharis rostellata, Eleusine indica, Elodea canadensis, Elymus canadensis, Elymus elymoides, Elymus repens, Elymus virginicus, Ephedra viridis, Epilobium canum, Epilobium ciliatum, Epilobium coloratum, Epilobium hirsutum, Epilobium leptophyllum, Epipactis, Equisetum arvense, Equisetum fluviatile, Equisetum hyemale, Equisetum laevigatum, Equisetum telmateia, Eragrostis spectabilis, Erechtites hieracifolius, Ericameria nauseosa, Ericameria palmeri, Erigeron lonchophyllus, Eriocaulon aquaticum, Eriogonum fasciculatum, Eriophorum angustifolium, Eriophorum chamissonis, Eriophorum gracile, Eriophorum tenellum, Eriophorum vaginatum, Eryngium yuccifolium, Eupatorium perfoliatum, Eupatorium serotinum, Euphorbia corollata, Euthamia graminifolia, Euthamia leptocephala, Euthamia occidentalis, Eutrochium maculatum, Eutrochium purpureum, Extriplex californica, Fagus grandifolia, Fimbristylis autumnalis, Fraxinus americana, Fraxinus pennsylvanica, Fraxinus velutina, Galeopsis tetrahit, Galium asprellum, Galium boreale, Galium labradoricum, Galium tinctorium, Gentianopsis virgata, Glebionis coronarium, Gleditsia triacanthos, Glyceria borealis, Glyceria canadensis, Glyceria elata, Glyceria grandis, Glyceria septentrionalis, Glyceria striata, Grindelia squarrosa, Habenaria repens, Helenium autumnale, Helenium puberulum, Helianthus annuus, Helianthus nuttallii, Heliopsis helianthoides, Heliotropium curassavicum, Hemerocallis fulva, Heteranthera dubia, Heterotheca sessiliflora, Hibiscus grandiflorus, Hippuris montana, Hippuris vulgaris Hirschfeldia incana, Holcus lanatus, Hopia obtusa, Hordeum jubatum, Hordeum murinum, Hordeum pusillum, Hosackia oblongifolia, Hydrocotyle americana, Hydrocotyle ranunculoides, Hypericum mutilum, Impatiens capensis, Iris pseudacorus, Iris tridentata, Iris versicolor, Iris virginica, Isocoma veneta, Isoetes echinospora, Iva axillaris, Iva frutescens, Ivesia kingii, Juglans nigra, Juncus acuminatus, Juncus alpinoarticulatus, Juncus balticus, Juncus breweri, Juncus bufonius, Juncus diffusissimus, Juncus dudleyi, Juncus effusus, Juncus ensifolius, Juncus nevadensis, Juncus nodosus, Juncus roemerianus, Juncus tenuis, Juncus torreyi, Juncus xiphioides, Juniperus deppeana, Juniperus monosperma, Juniperus osteosperma, Kickxia, Kosteletzkya pentacarpos, Kyllinga brevifolia, Lactuca serriola, Landoltia punctata, Laportea canadensis, Larix laricina, Lathyrus palustris, Leersia oryzoides, Lemna gibba, Lemna minor, Lemna trisulca, Lepidium densiflorum, Lepidospartum squamatum, Leptochloa fusca, Lespedeza capitata, Leymus condensatus, Leymus triticoides, Liatris spicata, Lindernia dubia, Liparis loeselii, Liquidambar styraciflua, Lobelia dortmanna, Lobelia kalmii, Lobelia siphilitica, Lonicera involucrata, Lonicera villosa, Lorandersonia salicina, Ludwigia decurrens, Ludwigia erecta, Ludwigia palustris, Ludwigia peploides, Ludwigia polycarpa, Lycium cooperi, Lycopus asper, Lycopus uniflorus, Lyonia ligustrina, Lysimachia nummularia, Lysimachia quadriflora, Lysimachia thyrsiflora, Lythrum alatum, Lythrum californicum, Lythrum salicaria, Machaeranthera, Maclura pomifera, Maianthemum stellatum, Malacothamnus fasciculatus, Malacothrix saxatilis, Malus coronaria, Malva moschata, Malva neglecta, Marchantia polymorpha, Marrubium vulgare, Marrubium vulgare, Medicago lupulina, Melilotus albus, Melilotus officinalis, Melochia pyramidata, Mentha arvensis, Mentha spicata, Mentha ×piperita, Populus ×acuminata, Menyanthes trifoliata, Mesembryanthemum nodiflorum, Micranthes pensylvanica, Microstegium vimineum, Mikania scandens, Mimulus cardinalis, Mimulus floribundus, Mimulus guttatus, Mimulus ringens, Morus alba, Muhlenbergia asperifolia, Muhlenbergia glomerata, Muhlenbergia richardsonis, Muhlenbergia rigens, Myosotis laxa, Myrica cerifera, Myrica gale, Myriophyllum sibiricum, Nasturtium officinale, Nelumbo lutea, Neptunia pubescens, Nitrophila occidentalis, Nuphar polysepala, Nyssa biflora, Oenanthe, Oenothera curtiflora, Oenothera elata, Onoclea sensibilis, Osmorhiza, Osmundastrum cinnamomeum, Oxypolis rigidior, Panicum capillare, Panicum repens, Panicum urvilleanum, Panicum virgatum, Parkinsonia aculeata, Parnassia glauca, Parnassia palustris, Paspalum dilatatum, Paspalum distichum, Paspalum vaginatum, Pedicularis canadensis, Pedicularis lanceolata, Peltandra virginica, Penthorum sedoides, Persicaria amphibia, Persicaria coccinea, Persicaria hydropiperoides, Persicaria lapathifolia, Persicaria maculosa, Persicaria pensylvanica, Persicaria punctata, Persicaria sagittata, Petasites frigidus, Phalaris angusta, Phalaris arundinacea, Phragmites australis, Phyla lanceolata, Phyla nodiflora, Physostegia virginiana, Picea engelmannii, Picea glauca, Picea mariana, Picea rubens, Pilea pumila, Pinus banksiana, Pinus echinata, Pinus edulis, Pinus elliottii, Pinus monophylla, Pinus resinosa, Pinus strobus, Plantago lanceolata, Plantago major, Platanthera psycodes, Platanus racemosa, Platanus wrightii, Pluchea baccharis, Pluchea odorata, Poa palustris, Poa pratensis, Polygala cymosa, Polygala lutea, Polygonum aviculare, Polygonum douglasii, Polypogon australis, Polypogon monspeliensis, Polypogon viridis, Pontederia cordata, Populus angustifolia, Populus balsamifera, Populus fremontii, Populus tremuloides, Portulaca oleracea, Potamogeton crispus, Potamogeton foliosus, Potamogeton natans, Potentilla anserina, Potentilla norvegica, Prenanthes alba, Proserpinaca palustris, Prosopis glandulosa, Prunella vulgaris, Prunus fasciculata, Prunus serotina, Pseudognaphalium luteoalbum, Pteridium aquilinum, Ptilimnium capillaceum, Pulicaria, Pycnanthemum virginianum, Pyrrocoma racemosa, Quercus agrifolia, Quercus cornelius-mulleri, Quercus emoryi, Quercus palustris, Quercus velutina, Quercus ×acutidens, Quercus ×macdonaldii, Ranunculus gmelinii, Ratibida, Rhamnus alnifolia, Rhamnus californica, Rhamnus frangula, Rhaponticum repens, Rhexia alifanus, Rhexia nashii, Rhexia virginica, Rhododendron, Rhus integrifolia, Rhynchospora alba, Rhynchospora capillacea, Rhynchospora chapmanii, Rhynchospora corniculata, Ribes aureum, Ribes glandulosum, Ribes hirtellum, Rorippa palustris, Rosa californica, Rosa palustris, Rosa pisocarpa, Rosa woodsii, Rotala ramosior, Rubus allegheniensis, Rubus pubescens, Rubus trivialis, Rubus ulmifolius, Rubus ursinus, Rudbeckia fulgida, Rudbeckia laciniata, Rumex altissimus, Rumex crispus, Rumex obtusifolius, Rumex orbiculatus, Rumex salicifolius, Rumex verticillatus, Sabal palmetto, Sagittaria brevirostra, Sagittaria latifolia, Salicornia depressa, Salix amygdaloides, Salix arbusculoides, Salix bebbiana, Salix candida, Salix caroliniana, Salix exigua, Salix geyeriana, Salix gooddingii, Salix hookeriana, Salix interior, Salix laevigata, Salix lasiandra, Salix lasiolepis, Salix lucida, Salix myricoides, Salix nigra, Salix niphoclada, Salix pedicellaris, Salix pulchra, Salsola tragus, Salvia mellifera, Sambucus nigra, Samolus valerandi, Sarcobatus vermiculatus, Sarracenia flava, Sarracenia purpurea, Schedonorus arundinaceus, Schedonorus pratensis, Scheuchzeria palustris, Schizachyrium scoparium, Schoenoplectus acutus, Schoenoplectus americanus, Schoenoplectus californicus, Schoenoplectus lacustris, Schoenoplectus pungens, Schoenoplectus tabernaemontani, Scirpus atrovirens, Scirpus cyperinus, Scirpus microcarpus, Scleria verticillata, Scutellaria galericulata, Scutellaria lateriflora, Senecio flaccidus, Senegalia greggii, Sesuvium verrucosum, Setaria magna, Seutera angustifolia, Sidalcea neomexicana, Sidalcea oregana, Sisymbrium irio, Sium suave, Smilax, Solanum americanum, Solanum dulcamara, Solanum ptychanthum, Solanum rostratum, Solidago altissima, Solidago canadensis, Solidago gigantea, Solidago ohioensis, Solidago patula, Solidago riddellii, Solidago uliginosa, Sonchus arvensis, Sonchus asper, Sonchus oleraceus, Sorbus, Sorghastrum nutans, Sorghum halepense, Sparganium americanum, Sparganium angustifolium, Sparganium emersum, Sparganium eurycarpum, Spartina bakeri, Spartina patens, Spartina pectinata, Sphagnum angustifolium, Sphagnum centrale, Sphagnum teres, Spiraea alba, Spiraea douglasii, Spiraea tomentosa, Spirodela polyrrhiza, Sporobolus heterolepis, Stachys ajugoides, Stachys palustris, Stuckenia vaginata, Symphyotrichum boreale, Symphyotrichum laeve, Symphyotrichum lanceolatum, Symphyotrichum novae-angliae, Symphyotrichum praealtum, Symphyotrichum puniceum, Symphyotrichum subulatum, Symplocarpus foetidus, Symplocos tinctoria, Tamarix chinensis, Tamarix ramosissima, Tanacetum vulgare, Taraxacum officinale, Taxodium distichum, Tetradymia glabrata, Tetragonia tetragonioides, Teucrium canadense, Thalictrum dasycarpum, Thelypteris palustris, Thinopyrum elongatum, Thlaspi arvense, Thuja occidentalis, Thuja plicata, Torilis japonica, Toxicodendron diversilobum, Toxicodendron vernix, Tragopogon dubius, Triadenum virginicum, Triadica sebifera, Triantha glutinosa, Trichophorum alpinum, Trichophorum pumilum, Tridens flavus, Trifolium aureum, Triglochin maritima, Triglochin palustris, Tsuga canadensis, Typha angustifolia, Typha domingensis, Typha ×glauca, Ulmus americana, Ulmus pumila, Ulmus rubra, Urtica dioica, Utricularia intermedia, Utricularia macrorhiza, Vaccinium membranaceum, Vaccinium oxycoccos, Valeriana edulis, Valeriana sitchensis,

Verbena bracteata, Verbena hastata, Verbena urticifolia, Verbena ×illicita, Verbesina alternifolia, Veronica anagallis-aquatica, Veronica peregrina, Veronica scutellata, Viburnum nudum, Vigna luteola, Viola, Vitis girdiana, Vitis labrusca, Washingtonia filifera, Xanthium strumarium, Xylococcus bicolor, Yucca brevifolia, Zizania aquatica, Zizania palustris, Zizaniopsis miliacea, Zizia aurea, Ziziphus obtusifolia.

Use by wildlife: *Typha* seeds are consumed by bitterns and herons (Aves: Ardeidae), ducks and geese (Aves: Anatidae), king rails (Aves: Rallidae: *Rallus elegans*), and Virginia rails (Aves: Rallidae: *Rallus limicola*). *Typha angustifolia* is eaten by muskrats (Mammalia: Cricetidae: *Ondatra zibethicus*), which also use the plants for house and mound construction. The rhizomes are fed on by nutria (Mammalia: Rodentia: Myocastoridae: *Myocastor coypus*) and geese (Aves: Anatidae: *Anser albifrons, Chen caerulescens*). The plants are used as nesting sites by red-winged blackbirds (Aves: Icteridae: *Agelaius phoeniceus*) and as forage sites for marsh wrens (Aves: Troglodytidae: *Cistothorus palustris*). They host larval moths (Insecta: Lepidoptera: Noctuidae: *Bellura brehmei, B. obliqua, Capsula oblonga*) and several Fungi (Ascomycota: *Cryptomela typhae*; Leptosphaeriaceae: *Leptosphaeria hydrophila*; Phaeosphaeriaceae: *Phaeosphaeria culmorum, Scolecosporiella typhae*; Basidiomycota: Psathyrellaceae: *Psathyrella typhae*). In brackish sites the plants provide favorable nursery conditions for topminnows (Vertebrata: Teleostei: Fundulidae: *Fundulus*). *Typha domingensis* is eaten by desert mule deer (Mammalia: Cervidae: *Odocoileus hemionus eremicus*). In some sites, the rhizomes have contained up to 320 µg g^{-1} dry weight selenium, which has been associated with various types of toxicity in waterfowl (Aves: Anatidae). *Typha latifolia* is eaten by beavers (Mammalia: Castoridae: *Castor canadensis*), boreal caribou (Mammalia: Cervidae: *Rangifer tarandus*), eastern lubber grasshoppers (Insecta: Orthoptera: Romaleidae: *Romalea microptera*), grass carp (Vertebrata: Teleostei: Cyprinidae: *Ctenopharyngodon idella*), moose (Mammalia: Cervidae: *Alces alces*), muskrats (Mammalia: Cricetidae: *Ondatra zibethicus*), nutria (Mammalia: Myocastoridae: *Myocastor coypus*), and western chicken turtles (Reptilia: Emydidae: *Deirochelys reticularia miaria*). Muskrats also use the plants preferentially for house building and mound construction. American alligators (Reptilia: Alligatoridae: *Alligator mississippiensis*) use the plants to construct their nests. The plants are also used frequently as nesting sites by Franklin's gull (Aves: Laridae: *Leucophaeus pipixcan*), king rails (Aves: Rallidae: *Rallus elegans*), red-winged blackbirds (Aves: Icteridae: *Agelaius phoeniceus*), and ruddy ducks (Aves: Anatidae: *Oxyura jamaicensis*). They are hosts for larval moths (Insecta: Lepidoptera: Cosmopterigidae: *Limnaecia phragmitella*; Crambidae: *Dicymolomia julianalis*; Gelechiidae: *Phthorimaea operculella*; Noctuidae: *Acronicta oblinita, Capsula oblonga, Bellura densa, B. obliqua, Leucania pseudargyria, Simyra insularis*; Tortricidae: *Bactra maiorina, Choristoneura rosaceana*). The foliage also hosts numerous Chromista (Oomycota: Pythiaceae: *Phytopythium helicoides*) and Fungi (Ascomycota: *Cryptomela typhae; Hymenopsis hydrophila,*

H. typhae; Neomassariosphaeria typhicola; Phoma orthosticha, P. typharum, P. typhicola, P. typhina; Scolicotrichum typhae; Apiosporaceae: *Arthrinium phaeospermum;* Botryosphaeriaceae: *Phyllosticta renouana;* Cladosporiaceae/Davidiellaceae: *Cladosporium astroideum, C. herbarum, C. heleophilum; C. macrocarpum;* Dictyosporiaceae: *Dictyosporium elegans;* Clavicipitaceae: *Epichloe typhina;* Didymellaceae: *Didymella viridimontana, Epicoccum nigrum;* Didymosphaeriaceae: *Didymosphaeria typhae;* Glomerellaceae: *Colletotrichum typhae;* Leptosphaeriaceae: *Leptosphaeria punctillum;* Lophiostomataceae: *Lophiostoma heterosporum;* Massarinaceae: *Lindgomyces ingoldianus, Stagonospora typhoidearum;* Melanconidaceae: *Hymenopsis typhae;* Mycosphaerellaceae: *Mycosphaerella typhae;* Peniophoraceae: *Hyphoderma typhicola;* Phaeosphaeriaceae: *Ophiobolus; Phaeosphaeria barriae, P. caricis, P. culmorum, P. eustoma, P. fuckelii, P. herpotrichoides, P. licatensis, P. luctuosa, Scolecosporiella typhae, Typhicola typharum;* Phomatosporaceae: *Phomatospora berkeleyi, P. muskellungensis;* Pleosporaceae: *Comoclathris typhicola, Lewia infectoria, Pleospora herbarum, P. typhae;* Rhytismataceae: *Lophodermium typhinum;* Basidiomycota: Psathyrellaceae: *Psathyrella typhae;* Tricholomataceae: *Cellypha goldbachii;* Typhulaceae: *Sclerotium hydrophilum, Typhula latissima*). The pollen (14.5% starch; 17.0% protein) is believed to be toxic to bees (Insecta: Hymenoptera), which avoid collecting it.

Economic importance: food: All the North American *Typha* species are quite edible and have long been used as cooked vegetables, flour (sifted pollen mixed 50% with wheat flour), in salads (young sprouts), and as a substitute for asparagus (boiled young shoots), corn (cooked green inflorescences), pickles (pickled young sprouts), and potato (starchy base of shoot). The seeds contain up to 20% of an edible oil comprising 69% linolenic acid. The foliage (e.g., *T. domingensis, T. latifolia*) can contain from 6.6% to 8.8% crude protein and yield up to 4,552 calories g^{-1}. Flour can be obtained from the rhizomes after washing, removal of the coarse outer tissue, and maceration in cold water for removal of the fibers. Cattail plants have provided food for many indigenous North American tribes. The ripe, fruiting heads of *T. angustifolia* were eaten "like corn" by the Havasupai. The pollen of *T. angustifolia* was baked into brownish biscuits or mixed with ground wheat, which was added to boiling water to make porridge (Pima). The green seeds, tender shoot bases, or flowers of *T. angustifolia* and *T. domingensis* were eaten raw as vegetables (Kawaiisu, Northern, Paiute, Pima). The peeled rootstalks of *T. domingensis* were eaten fresh (by swallowing the juice and expectorating the pulp) or were dried and ground into flour, which was made into a sweet porridge or into cakes (Paiute, Pima, Gila River). The seeds were eaten fresh and raw, ground and powdered, cooked, or roasted (Paiute, Northern). Roasted seeds of *T. domingensis* were ground into flour, boiled, and sun dried into cakes, or were ground into meal, which was eaten with water without boiling (Paiute); they were also boiled into a mush or soup. The pollen of *T. domingensis* was mixed with water, kneaded, and baked into cakes (Paiute). *Typha latifolia* was used widely as a comestible. The rootstocks (baked in ashes, cooked with meat, dipped in boiling water, peeled, pit-cooked, raw, or salted) were eaten by the Acoma, Apache, Chehalis, Chiricahua, Costanoan, Cree, Keres, Klamath, Laguna, Mendocino Indians, Montana Indians, Mescalero Blackfoot, Navajo, Okanagan-Colville, Paiute, Pomo, Ramah, Sioux, Thompson, Tubatulabal, Western, Woodlands, and Yuma tribes. Peeled rootstocks were also dried, pounded, and boiled with fish (Yuma) or ground into a flour or meal for making cakes, bread, and porridge (Cahuilla, Cree, Hualapai, Paiute, Woodlands). The tender shoots (baked in ashes, boiled, ground, and mixed with corn meal, peeled, pit cooked, raw, roasted, or salted) were a food of the Acoma, Alaska Native, Apache, Carrier, Chehalis, Clallam, Costanoan, Cree, Keres, Laguna, Mescalero Blackfoot, Mendocino, Montana, Navajo, Okanagan-Colville, Paiute, Pomo, Ramah, San Felipe, Sioux, Tanana, Western, Woodlands, and Yuma people. The young (green) inflorescences were eaten when dried, boiled in salt water, or roasted (Alaska Native, Ojibwa, Okanagan-Colville, Paiute). The pollen was used as flavoring or made into dough or flour for making cakes and porridge (Cahuilla, Costanoan, Lakota, Ojibwa, Paiute, Yuma); it was also sifted and eaten raw (Yuma). The seeds (boiled, cooked, fresh, powdered, raw, or roasted) were eaten alone, with a little water without boiling, and were made into a dough, fine flour, or meal for preparing cakes, porridge, and soups (Gosiute, Paiute); **medicinal:** The Malecite and Micmac tribes made a root infusion from *T. angustifolia* to use as a medicine for treating kidney stones. The rhizomes of *T. latifolia* have hemostatic properties and were crushed into a poultice to treat boils, carbuncles, infections, inflammations, sores, sprains, and bleeding wounds in humans (Algonquin, Cahuilla, Iroquois, Mahuna, Ojibwa, Potawatomi, Quebec) and horses (Iroquois). An infusion made from dried, pulverized roots and leaves was taken for abdominal cramps (Cheyenne). Rhizomes were also prepared as a medicine to dissolve kidney stones (Delaware, Oklahoma) and were chewed by women as a treatment for gonorrhea (Iroquois). The Houma administered a decoction from the stalks for whooping cough. The leaves were slept upon as a mattress to treat breast cysts (Iroquois) or (when greased) were applied to sores (Malecite, Micmac). Young inflorescences were eaten to alleviate diarrhea (Washo). The pollen was used as a medicine by the Apache and Mescalero tribes. The downy fruits were used as a poultice to dress burns, pustules (smallpox), scalds, sores and wounds (Dakota, Meskwaki, Montana, Okanagan-Colville, Omaha, Pawnee, Ponca, Sioux, Winnebago), and to prevent chafing in infants (Dakota, Omaha, Pawnee, Plains, Ponca, Winnebago). Methanol extracts of *T. angustifolia* have strongly inhibited the growth of several bacterial strains (*Salmonella typhimurium*, *Pseudomonas aeruginosa*, and *Escherichia coli*) in experimental trials. Dietary supplements containing *T. angustifolia* rhizome flour (10%) have been shown to be as effective as prednisolone (a steroid), which is used in treating human inflammatory bowel disease (IBD). Pollen from the plants contains antioxidant flavonoids, and has been evaluated as a potential therapeutic strategy to treat LPS-induced

inflammation in humans. Ointments comprising 5% of pistillate flowers (or their alcoholic extracts) from *T. domingensis* have demonstrated exceptional wound healing properties. The antioxidant fruit extracts of *T. domingensis* exhibit highly potent glucosidase inhibitory and free-radical scavenging activity. They have been regarded as having potential to develop natural antihyperglycemic agents. Extracts of *T. latifolia* exhibit similarly high levels of antioxidant activity; **cultivation:** Several *Typha* species are cultivated as ornamental water garden marginals. Cultivars include 'Zebratails' (*T. angustifolia*) and 'Variegata' (*T. latifolia*). Most of the species are planted commonly for wildlife habitat and wetland restoration projects; **misc. products:** The Hopi chewed the mature heads of *T. angustifolia* with tallow as a type of gum. The Pima tribe used the dried, split flower stalks as a fiber for basket weaving and the leaves as a fiber to weave into mats or as a roofing material. They stuffed pillows with the downlike fruits. The Pima also decorated their bodies using the dry, yellow pollen of *T. angustifolia* and *T. domingensis*. The leaves and stalks of *T. domingensis* were used in house construction, for shingles, and as thatch (Havasupai, Kawaiisu, Paiute, Northern). The Havasupai used the pollen as a face paint and the stalks to make toy arrows. The leaves were used for basketry, in mat and sandal making, and in various types of clothing (Paiute, Northern). The plant fibers were also used in the construction of duck decoys and canoes (Paiute, Northern). The Hopi tribe used *T. angustifolia* in ceremonies associated with water. *Typha latifolia* was an important source of fiber for Native Americans, who wove the leaves and shoots into bedding, building materials (canoes, houses), clothing (capes and raincoats), duck decoys, headdresses (for doctors), house mats (for canoes and tipi flooring), packsacks, raincoats, roof thatching, sandals, storage bags, and trays (Apache, Cahuilla, Chippewa, Clallam, Coast, Cowlitz, Hesquiat, Hopi, Isleta, Klallam, Kwakiutl, Makah, Menominee, Mescalero, Meskwaki, Montana, Southern, Navajo, Nimpkish, Nitinaht, Ojibwa, Okanagon, Okanagan-Colville, Paiute, Pomo, Potawatomi, Quinault, Ramah, Shuswap, Snohomish, Squaxin, Thompson, Tolowa, White Mountain, Yurok). The leaves were also used widely in basket weaving (Chehalis, Cheyenne, Chippewa, Clallam, Coast, Cowlitz, Hesquiat, Klallam, Makah, Meskwaki, Navajo, Nitinaht, Quinault, Ramah, Salish, Snohomish, Squaxin) and to insulate sweathouse entrance frames (Okanagan-Colville). The rootstalks yielded a natural oakum used for caulking canoes (Menominee, Meskwaki). The downlike fruits were used to line moccasins, and to pad or stuff baby wraps, cradleboards, diapers, mattresses, pillows, and quilts (Algonquin, Blackfoot, Chehalis, Dakota, Iroquois, Lakota, Mendocino, Montana, Ojibwa, Okanagan-Colville, Potawatomi, Shuswap, Tete-de-Boule, Thompson). Various tribes (Apache, Cahuilla, Cheyenne, Chiricahua, Keresan, Mescalero, Navajo, Omaha, Ramah, White Mountain) used the leaves, pollen, and shoots of *Typha latifolia* ceremoniously as bundles, medicines, necklaces, tipi floor coverings, and wristbands. The Navajo and Ramah hung mats made from the leaves to bring rain and to protect their hogans, people, and sheep from lightning. Ripe inflorescences were shaken in rain dances to summon clouds (Keres, Western). The downy fruits were used as a war medicine by the Ojibwa, who threw them into the eyes of their enemies to impair them during combat. *Typha angustifolia* has been used to prepare a partially biodegradable natural fiber–reinforced polyester composite. Fibers from *T. domingensis* have been used as a source of cellulose nanocrystals, which are used to reinforce polymeric materials. All three of the North American *Typha* species (*T. angustifolia*, *T. domingensis*, and *T. latifolia*) exhibit similarly high tolerances to toxic pollutants (e.g., Al, As, Cd, Cr, Cu, Hg, Mn, Ni, Pb, and Zn) and are regarded as prime candidates for heavy metal phytoremediation of industrial wastewaters and for purification of polluted natural waters. All the *Typha* species are used widely in wetland restoration projects. Ethanolic extracts of *T. angustifolia* have exhibited insecticidal activity against invasive red fire ants (Insecta: Hymenoptera: Formicidae: *Solenopsis invicta*). Lignocellulose derived from *T. angustifolia* has been used to produce ethanol as a biofuel. Fiber derived from *T. angustifolia* and *T. latifolia* have been used to manufacture a low-density particleboard. Those of *T. latifolia* have been evaluated as an additive for strengthening gypsum board. *Typha domingensis* has been used to build sediment plant microbial fuel cells for generating electrical current. *Typha domingensis* and *T. latifolia* also have been evaluated as a promising source for renewable bioethanol production; **weeds:** *Typha* species often are regarded as weeds due to their prolific growth, especially in drainage canals, ditches, or other shallow sites. Some authors categorize them as "invasive" but there is sufficient evidence to indicate that all three taxa are indigenous (but see next); **nonindigenous species:** Molecular data indicate that North America comprises both indigenous and nonindigenous populations of *Typha angustifolia* and *T. latifolia* (also possibly *T. domingensis*) as a consequence of introductions from water garden nursery stock.

Systematics: The morphologically distinctive *Typha* traditionally has been regarded as monophyletic, which is firmly supported by phylogenetic evidence (e.g., Figures 3.8 and 3.9). Phylogenetic analyses also consistently resolve the genus as the sister clade of *Sparganium*, which in early serological studies exhibited high similarity to *Typha*. As mentioned previously (see Typhaceae above), the close relationship of these genera is also evidenced by their ability to form intergeneric hybrids. Phylogenetic analyses of DNA sequence data (e.g., Figure 3.9) resolve *T. angustifolia* within a clade that includes *T. domingensis* and *T. elephantina*. Phytochemical studies have characterized *T. domingensis* as the most distinctive North American cattail species serologically. *Typha latifolia* is related more distantly as the sister species to the morphologically similar *T. shuttleworthii* (Figure 3.9). The plastid genome of *T. latifolia* has been sequenced. *Typha minima* (not North American) is the sister species to the remainder of the genus. The basic chromosome number of *Typha* is $x = 15$. *Typha angustifolia*, *T. domingensis*, and *T. latifolia* all are diploid ($2n = 30$). Hybridization occurs throughout *Typha* and artificial pollinations of the North American species have confirmed

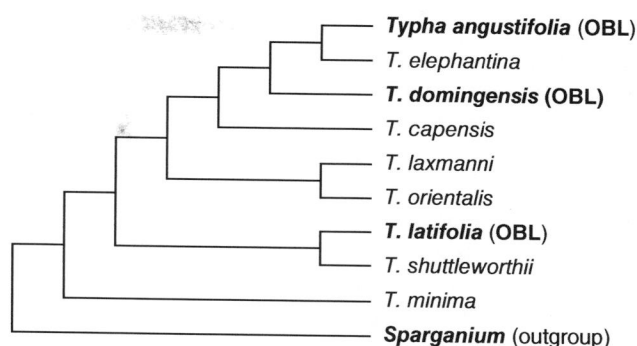

FIGURE 3.9 Phylogenetic relationships in *Typha* analyzed using combined nuclear and plastid DNA sequence data (modified from Kim & Choi, 2011) have confirmed the monophyly of *Typha* and its sister-group relationship with *Sparganium*. These results indicate that the obligate (OBL) North American species (highlighted in bold) are interspersed among several non-North American species across the genus.

that all hybrid combinations and backcross progeny can be synthesized, at least experimentally. *Typha latifolia* and *T. angustifolia* hybridize to generate the F_1 known widely as *T. ×glauca*, which exhibits heterosis by achieving a greater height. However, different floral phenologies and other factors (e.g., univalent formation) often restrict hybridization, even where the two species grow intermixed. In one study, this hybrid was synthesized only after five separate attempts involving hundreds of experimental pollinations. Considerable overlap in floral phenology in the potentially parental species also has been found to occur in sites where hybrids are rare, which indicates that other, more effective isolating barriers (i.e., genetic) must exist. Although once regarded as a prime example of introgressive hybridization, some molecular data have indicated little gene flow between these species and consistent recovery of F_1 hybrids exhibiting the maternal (chloroplast) haplotype of *T. angustifolia*. However, studies using microsatellite (SSR) markers have provided evidence of backcrossing, albeit at a much lower frequency than F_1 hybridization. Crossing studies have verified that the mating asymmetry extends to backcrosses, with *T. latifolia* producing no seeds when pollinated by hybrid plants. This factor has helped preserve the integrity of *T. latifolia* despite repeated hybridizations. The *T. ×glauca* hybrids can coexist with either of the parental associates and no niche segregation among the parental and hybrid taxa has been detected. Hybrids between *T. angustifolia* and *T. domingensis* are fertile, exhibit 15 bivalents, and allegedly can develop into hybrid swarms. Interspecific hybrids between *Typha domingensis* and *T. latifolia* are sterile, with univalent formation, bridges, and micronuclei occurring. Trihybrids involving all three of the North American species also reportedly occur. Serological studies have detected novel proteins that are expressed only by some hybrid plants. **Distribution:** *Typha angustifolia* occurs in most of North America but the far north. *Typha domingensis* is found throughout the southern half of the United States. *Typha latifolia* exists throughout nearly all parts of North America.

References: Abideen et al., 2014; Ahee et al., 2015; Akkol et al., 2011; Asamoah & Bork, 2010; Bajwa et al., 2015; Ball & Freeland, 2013; Barnard, 1882; Baskin & Baskin, 1998; Beal, 1977; Beare & Zedler, 1987; Bendix et al., 1994; Berdnikov et al., 2002; Bonanno & Cirelli, 2017; Bonnewell et al., 1983; Boyd, 1970a; 1970b; Boyd & Hess, 1970; Boyd & McGinty, 1981; Cervantes-Alcalá et al., 2012; César et al., 2015; Chai et al., 2015; Chen et al., 2010; 2017; Ciotir & Freeland, 2016; Ciotir et al., 2013; Claassen, 1921; Clarke & Dalrymple, 2003; Crow & Hellquist, 1981; Culling & Cichowski, 2017; Dickerman & Wetzel, 1985; Dozier, 1945; Dubbe et al., 1988; Dunham et al., 2003; Eckardt & Biesboer, 1988; Ekstam & Forseby, 1999; Farrell et al., 2010; Fassett, 1957; Fassett & Calhoun, 1952; Fell et al., 2003; Fiala, 1978; Fraser et al., 1982; Freeland et al., 2013; Fruet et al., 2012; Glenn et al., 1995; Grace, 1983; 1984; 1988; 1989; Grace & Wetzel, 1981a; 1981b; 1982; 1998; Guisinger et al., 2010; Hara et al., 1985; Harris, 1927; Hegazy et al., 2011; Hill, 1987; Joanen & McNease, 1975; Kausch et al., 1981; Keane et al., 1999; Keddy & Constabel, 1986; Keddy & Ellis, 1985; Kim & Choi, 2011; Kirk et al., 2011; Koch & Rawlik, 1993; Koropchak & Vitt, 2013; Krattinger, 1975; Krattinger et al., 1979; Kuehn et al., 1999; Leck & Graveline, 1979; Lee & Fairbrothers, 1969; 1972; Les & Philbrick, 1993; Li et al., 2009; Lombardi et al., 1997; Londonkar et al., 2013; Lorenzen et al., 2000; 2001; Marshal et al., 2004; Mashburn et al., 1978; McColl & Burger, 1976; McKenzie-Gopsill et al., 2012; McKnight et al., 2015; McMillan, 1959; McNaughton, 1966; 1968; McNaughton & Fullem, 1970; Meanley, 1953; Mitchell et al., 2011; Morinaga, 1926; Morton, 1975; Nichols, 1999; Olson et al., 2009; Özesmi & Mitsch, 1997; Parker & Leck, 1985; Penfound, 1956; Penko & Pratt, 1986; Peterson, 1977; Pieper et al., 2017; Ponzio et al., 2004; Ray & Inouye, 2006; Rea, 1997; Rebaque et al., 2017; Sale & Wetzel, 1983; Santos, 2004; Sasser et al., 2018; Schmidt et al., 1989; Schowalter, 2018; Schuler et al., 1990; Schulgasser & Witztum, 2004; Selbo & Snow, 2004; Shih & Finkelstein, 2008; Shukla et al., 2012; Siegfried, 1976; Sifton, 1959; Smith, 1967; 1986; 1987; 2000; Smith & Newman, 2001; Smith et al., 2002; 2015; Snow et al., 2010; Soons, 2006; Sopajarn & Sangwichien, 2015; Stenlund & Charvat, 1994; Stewart et al., 1997; Stutzenbaker, 1999; Tang et al., 2001; Tornberg et al., 1994; Toth & Galloway, 2009; Ungar et al., 1969; Weberg et al., 2015; Wetzel & Van Der Valk, 1998; Wetzel et al., 2001; Whigham et al., 1989; Witztum & Wayne, 2014; Yeo, 1964; Zapfe & Freeland, 2015; Zhang et al., 2013b.

Family 10.4. Xyridaceae [5]

The "yellow-eyed grass" family (Xyridaceae) contains roughly 300+ species, which occur mostly within *Xyris* (Kral, 1998). Recent phylogenetic analyses (e.g., Coan et al., 2010; Hochbach et al., 2018) have resolved Xyridaceae within Poales as the sister family of Eriocaulaceae (Figure 3.6). Some genera (*Abolboda, Orectanthe*) have been segregated as a distinct family (i.e., Abolbodaceae), which most contemporary authors regard instead as a subfamily of Xyridaceae (i.e., Abolbodoideae). Nearly all members of the family are aquatic or helophytic, and occur frequently in acidic, wet habitats (Kral, 1998).

The North American species are distinctive in having laterally or dorsiventrally compressed, distichous and equitant leaves, and somewhat bilateral flowers (due to their asymmetric calyx), with yellow (less often white) corollas, which emerge from elongate, scapose inflorescences that resemble small woody cones (Kral, 1998). The sessile, nectarless flowers remain open for only a few hours (Judd et al., 2016). They are pollinated mainly by bees (Insecta: Hymenoptera) or occasionally by flies (Insecta: Diptera), which are attracted to the pollen (Kral, 1998; Ramirez & Brito, 1990; 1992); one South American genus (*Orectanthe*) contains bird-pollinated species (Kral, 1998). The fruits are many-seeded, loculicidal capsules (Kral, 1998). The seeds are dispersed abiotically by gravity, water, or wind (Kral, 1998; Judd et al., 2016). They can germinate rapidly when conditions are favorable, or maintain a prolonged period of dormancy (Kral, 1998).

Xyris species are cultivated occasionally as water garden ornamentals or as aquarium plants (Kral, 1966c; Oakes, 1990). The flowers of some species are dried and sold commercially (mainly in Europe) as "everlasting" flowers (Giulietti et al., 1988; 2000); otherwise, the family is of little economic importance.

Xyridaceae primarily are pantropical (Kral, 1998), yet are fairly well represented in the warmer parts of North America. Four genera (*Achlyphila, Abolboda, Aratitiyopea,* and *Orectanthe*) are restricted to South America, but only one more widespread genus occurs in North America and contains OBL wetland indicators:

10.4.1. *Xyris* L.

10.4.1. Xyris
Hardhead, vare-goldies, yellow-eyed grass; herbe aux yeux jaunes
Etymology: from the Greek *xyron* (razor), in reference to the flat, double-edged leaves
Synonyms: *Jupica* (in part); *Kotsjilettia*; *Ramotha*; *Schizmaxon*; *Synoliga*; *Xuris*; *Xyroides*
Distribution: global: Africa, Australasia, South America; **North America:** eastern
Diversity: global: 361 species; **North America:** 21 species
Indicators (USA): obligate wetland (OBL): *Xyris ambigua, X. baldwiniana, X. brevifolia, X. difformis, X. drummondii, X. elliottii, X. fimbriata, X. flabelliformis, X. isoetifolia, X. jupicai, X. laxifolia, X. longisepala, X. montana, X. platylepis, X. scabrifolia, X. serotina, X. smalliana, X. stricta, X. tennesseensis, X. torta;* **facultative wetland (FACW):** *X. elliottii, X. platylepis*
Habitat: brackish (coastal), freshwater; freshwater (tidal); lacustrine, palustrine; **pH:** 3.7–7.0; **depth:** <1 m; **life-form(s):** emergent herb
Key morphology: plants (to 150 cm), cespitose or solitary; the stems rosulate, simple, erect, short to branching, the base bulbous or compact; leaves alternate, equitant, 2-ranked, green [deep, lustrous, olive, pale] or pale red-brown, sometimes

red-tinged, ascending or in fanlike arrays, the blade (to 70 cm) linear to filiform, flattened to nearly terete, sometimes twisted, the margins smooth to scabrous, sometimes maroon or reddish-tinged, the sheath dull green, brown [glossy, reddish], greenish, maroon, red [pale], reddish, pink, straw-colored, or tan, with or without a chestnut brown patch; inflorescence scapose, variously elongate, linear to flexuous, terete or ribbed, sometimes twisted, the sheaths shorter than, equaling, or exceeding leaves, terminating in a cylindric, ellipsoid, globose, lanceoloid, oblate, or ovoid, conelike spike (to 35 mm) of imbricate bracts (to 9 mm); flowers solitary in bract axils; sepals 3, brown [dark, reddish, yellow], slightly to strongly curved, asymmetric, keeled, winged, the laterals (to 9 mm) chaffy, free or connate, included or partially exserted, the medial sepal membranous; petals 3, free or connate, strongly clawed, the limb (to 10 mm) obovate [broadly], obtri-angular, or nearly orbiculate, ephemeral, spreading, yellow to white; staminodes 3, distinct, strongly clawed, apically yoked, bearded or beardless, stamens 3, epipetalous; ovary thin-walled, style elongate, 3-branched, stigmas 3; fruit a 3-valved, loculicidal capsule; the seeds (to 90+; to 1 mm) cylindric, cylindro-fusiform, ellipsoid [broadly, narrowly], ellipsoid-cylindric, fusiform, oblong-fusiform, or ovoid, amber, mealy, opaque, or translucent, faintly, finely, or prominently lined, ribbed [flat, pale], or ridged [irregularly] longitudinally

Life history: duration: annual (seeds); perennial (dormant shoot buds, rhizomes); **asexual reproduction:** lateral off-shoots, rhizomes; **pollination:** insect; **sexual condition:** hermaphroditic; **fruit:** capsules (common); **local dispersal:** seeds (water, wind); **long-distance dispersal:** seeds (animals)

Imperilment: 1. *Xyris ambigua* [G5]; S1 (TN); **2.** *X. baldwiniana* [G5]; S1 (AR); S2 (NC); **3.** *X. brevifolia* [G4/G5]; S1 (SC); S3 (NC); **4.** *X. difformis* [G5]; SH (VT); S1 (AR, KY, <u>NB</u>, NC, OH, VA); S2 (AR, IN, NC, RI, SC); S3 (NC); **5.** *X. drummondii* [G3/G4]; S1 (GA); S2 (MS, TX); S3 (AL, FL, LA); **6.** *X. elliottii* [G4]; S2 (SC); **7.** *X. fimbriata* [G5]; S1 (DE, NJ, MD, TN, VA); S2 (NC); **8.** *X. flabelliformis* [G4]; S1 (SC); S3 (GA, NC); **9.** *X. isoetifolia* [G1]; SH (AL); S1 (FL); **10.** *X. jupicai* [G5]; S1 (IL, MO); S3 (NC, NJ); **11.** *X. laxifolia* [G4/G5]; S1 (VA); S2 (NC, TN); **12.** *X. longisepala* [G2/G3]; S1 (AL); S2 (FL); **13.** *X. montana* [G5]; S1 (NJ, RI, VT); S2 (CT, MA); S3 (MN, <u>NB</u>, <u>NF</u>, <u>QC</u>); **14.** *X. platylepis* [G5]; S1 (TX); S2 (LA, VA); S3 (NC); **15.** *X. scabrifolia* [G3]; S1 (AL, FL, GA, NJ, SC); S2 (MS, NC, TX); S3 (FL, MS); **16.** *X. serotina* [G3/G4]; S1 (AL, LA, MS, NC, SC); S3 (GA); **17.** *X. smalliana* [G5]; S1 (CT, LA, MD, ME); S2 (DE, NY, RI); S3 (NC); **18.** *X. stricta* [G2/G3/G4]; S1 (FL, MS, TX); S3 (LA); **19.** *X. tennesseensis* [G2]; S1 (AL, GA, TN); **20.** *Xyris torta* [G5]; S1 (IA, MN, MO. PA, SC); S2 (NE, RI, WV); S3 (AR, DE, IL, KY, OH, NC, NY)

Ecology: general: *Xyris* comprises perennial or annual/bien-nial wetland plants with shallow, unbranched root systems. All but one of the North American species (95%) are ranked as OBL indicators in some portion of their range. The species occur primarily in warm climates where the habitats are at least partially acidic (except *X. tennesseensis*), and often are domi-nated by sandy substrates. Many species favor ecotones, and occur frequently in open, disturbed conditions. Although the

reproductive biology of *Xyris* has not been studied thoroughly, some species are known to be self-incompatible, whereas others are self-compatible and are highly autogamous. The flowers are short-lived with trifid styles that disarticulate and are shed soon after anthesis. They are pollinated by various bees (Insecta: Hymenoptera: Apidae; Andrenidae; Halictidae; Megachilidae) and flies (Insecta: Diptera: Syrphidae). The seeds of some species (e.g., from more northerly localities) become dormant, whereas many of those from southern localities have no specialized dormancy requirements and can germinate year-round in the presence of adequate light, mois-ture, and temperature. Though detailed studies on dispersal are lacking for *Xyris*, the seeds are thought to be transported abiotically by water or wind, facilitated by their external mor-phology. In some cases they have remained viable within the digestive tract of mammalian herbivores, to be dispersed upon their defecation. Persistent seed banks can develop but tend to be fairly short-lived (~4 years). In most cases the seedlings mature rapidly and many will flower within a year of germi-nation. Overwintering occurs as seeds, seedlings, or by means of dormant, paired lateral buds, which develop in the upper leaf axils of the short, contracted stem. Vegetative reproduc-tion by lateral offsets can occur in most species, including annuals (during the growing season). Several species that are categorized as OBL indicators in the 2016 list have been excluded for taxonomic reasons. These are: *X. chapmanii* (included in *X. scabrifolia*), *X. floridana* (included in *X. dif-formis*), and *X. louisianica* (included in *X. stricta*). The group is complex taxonomically (see *Systematics*), which makes it difficult to evaluate ecological information reliably for several of the species.

10.4.1.1. *Xyris ambigua* Bey. ex Kunth is a perennial, which grows in baygalls, bogs (burned, hillside, *Hypericum*, pitcher-plant, quaking, savanna, seepage, *Sphagnum*), bottomlands (hardwood), depressions (limesink, roadside, seasonal, wet), ditches (damp, flatwoods, hydric, moist, roadside), flats, flat-woods (disturbed, mesic, palmetto, pine, wet), glades, ham-mocks (low), marshes (ephemeral, overgrown), meadows (moist, seasonally wet, successional), pannes (dune), pine bar-rens (low), pinelands (low, seasonally wet, wet), prairies (burned, seasonally wet, spodic, wet, wet-mesic), roadsides (moist, wet), savannas (moist, oak, pine, seepage, wet), seeps (hillside, sphagnous, springhead, wet), slopes [to 2%] (seep-age), swales (boggy, peaty, sandy, sphagnous, springy), and along the margins or shores of canals, lakes, pocosins (over-grown), ponds (cypress-gum, sinkhole), and streams at eleva-tions to 134 m. The plants occur in wet sites but usually not in standing water. They can grow under exposures of full sun-light to light shade. The substrates are acidic (e.g., pH: $\bar{x} = 4.2–4.7$) and have been characterized as alfisols (Waller series), Basinger (spodic psammaquents), Blanton (grossare-nic paludults), clay silt, Compass (plinthic paleudults), Dorovan-Johnston series, Doucette (arenic plinthic paleudults), Floridana (arenic argiaquolls), Goldsboro (aquic paleudults), Immokalee, Leagueville-Henco, marl, Myakka (aeric haplaquods), muck (black, drying), Pelham (arenic paleaquults), Plummer (grossarenic paleaquults), Pompano,

Pottsburg (grossarenic haplaquods), Rains (typic paleaquults), Riviera (arenic glossaqualfs), Rutledge (typic humaquepts), sand (Leon, Mascotte, Queen City, Sparta), sandy clay, sandy loam (Atmore, Bayou), sandy peat, sandy silt (seepy), silty clay loam, silty loam (Amy, Bonn), silty sand (organic), silty sandy loam, Stringtown (typic hapludults), Surrency (arenic umbric paleaquults), and Tehran-Letney (grossarenic paleudults). Flowering and fruiting can occur all year in the southernmost populations. The flowers are short-lived, opening only from mid-morning (e.g., 9–11 am) until early afternoon of the same day (e.g., 1–3 pm). The seeds have germinated well under greenhouse conditions (minimum temperature: 20°C; 16-h photoperiod; misted daily). A seed bank of undetermined duration can develop. The plants perennate by lateral offsets, which arise from a hard, fibrous, thickened base. They are fire-tolerant and dependent on periodic fires to reduce competition. Several studies have shown their frequency to increase in the aftermath of fires. **Reported associates:** *Acer rubrum, Agalinis fasciculata, Agalinis linifolia, Aletris aurea, Aletris lutea, Amphicarpum amphicarpon, Amphicarpum muhlenbergianum, Andropogon arctatus, Andropogon glomeratus, Andropogon gyrans, Andropogon liebmannii, Andropogon virginicus, Anthaenantia rufa, Aristida palustris, Aristida purpurascens, Aristida stricta, Arnoglossum ovatum, Aronia arbutifolia, Arundinaria tecta, Asclepias hirtella, Asclepias lanceolata, Asclepias rubra, Baccharis halimifolia, Balduina uniflora, Bartonia verna, Bidens polylepis, Bigelowia nudata, Buchnera floridana, Calamovilfa brevipilis, Calamovilfa curtissii, Callicarpa americana, Calopogon barbatus, Calopogon pallidus, Calopogon tuberosus, Carex glaucescens, Carex striata, Carphephorus odoratissimus, Carphephorus paniculatus, Carphephorus pseudoliatris, Carphephorus tomentosus, Centella asiatica, Cephalanthus occidentalis, Chamaedaphne calyculata, Chaptalia tomentosa, Cleistesiopsis bifaria, Cleistesiopsis divaricata, Clethra alnifolia, Cliftonia monophylla, Coleataenia tenera, Coreopsis gladiata, Cornus florida, Ctenium aromaticum, Cyperus haspan, Cyperus polystachyos, Cyrilla racemiflora, Decodon verticillatus, Dichanthelium aciculare, Dichanthelium acuminatum, Dichanthelium dichotomum, Dichanthelium ensifolium, Dichanthelium nudicaule, Dichanthelium ovale, Dichanthelium scabriusculum, Dichanthelium scoparium, Dichanthelium sphaerocarpon, Dichanthelium strigosum, Diodella teres, Dionaea muscipula, Diospyros virginiana, Ditrysinia fruticosa, Doellingeria sericocarpoides, Drosera brevifolia, Drosera capillaris, Drosera intermedia, Drosera rotundifolia, Drosera tracyi, Eleocharis equisetoides, Eleocharis microcarpa, Eleocharis vivipara, Elephantopus elatus, Elephantopus nudatus, Eragrostis refracta, Erechtites, Erigeron vernus, Eriocaulon compressum, Eriocaulon decangulare, Eriocaulon lineare, Eriocaulon texense, Eryngium integrifolium, Eupatorium capillifolium, Eupatorium leucolepis, Eupatorium mohrii, Eupatorium pilosum, Eupatorium rotundifolium, Euphorbia inundata, Eurybia eryngiifolia, Euthamia graminifolia, Fagus grandifolia, Fimbristylis littoralis, Fimbristylis spadicea, Fraxinus pennsylvanica, Fuirena breviseta, Fuirena bushii, Fuirena pumila, Fuirena squarrosa, Gaylussacia dumosa, Gaylussacia frondosa, Gaylussacia mosieri, Gratiola brevifolia, Gratiola hispida, Gratiola pilosa, Habenaria repens, Helenium brevifolium, Helenium vernale, Helianthus angustifolius, Helianthus carnosus, Helianthus floridanus, Helianthus heterophyllus, Helianthus radula, Hibiscus aculeatus, Hydrolea corymbosa, Hypericum brachyphyllum, Hypericum chapmanii, Hypericum cistifolium, Hypericum crux-andreae, Hypericum fasciculatum, Hypericum microsepalum, Hypoxis juncea, Hyptis alata, Ilex coriacea, Ilex glabra, Ilex myrtifolia, Ilex vomitoria, Juncus acuminatus, Juncus caesariensis, Juncus canadensis, Juncus gymnocarpus, Juncus marginatus, Juncus pelocarpus, Juncus repens, Juncus scirpoides, Juncus trigonocarpus, Juncus validus, Justicia crassifolia, Justicia ovata, Lachnanthes caroliniana, Lachnocaulon anceps, Lechea minor, Liatris pycnostachya, Liatris spicata, Lilium catesbaei, Linum medium, Liquidambar styraciflua, Lobelia boykinii, Lobelia brevifolia, Lobelia glandulosa, Lobelia paludosa, Lobelia puberula, Lophiola aurea, Ludwigia hirtella, Ludwigia linearis, Ludwigia microcarpa, Ludwigia palustris, Ludwigia pilosa, Ludwigia virgata, Lycopodiella alopecuroides, Lycopodiella appressa, Lycopodiella caroliniana, Lycopodiella caroliniana, Lycopus rubellus, Lygodium japonicum, Lyonia ferruginea, Lyonia lucida, Magnolia virginiana, Marshallia graminifolia, Mimosa quadrivalvis, Mitreola sessilifolia, Mnesithea rugosa, Muhlenbergia capillaris, Muhlenbergia expansa, Myrica caroliniensis, Myrica cerifera, Myrica heterophylla, Myrica inodora, Nymphaea odorata, Nyssa biflora, Nyssa sylvatica, Oldenlandia uniflora, Osmunda regalis, Osmundastrum cinnamomeum, Packera tomentosa, Panicum verrucosum, Parnassia caroliniana, Paspalum laeve, Paspalum praecox, Paspalum setaceum, Paspalum urvillei, Persea palustris, Pinguicula caerulea, Pinguicula ionantha, Pinguicula lutea, Pinguicula planifolia, Pinus echinata, Pinus elliottii, Pinus palustris, Pinus serotina, Pinus taeda, Pityopsis, Plantago sparsiflora, Platanthera blephariglottis, Platanthera cristata, Platanthera integra, Pleea tenuifolia, Pluchea bacchari, Polygala cruciata, Polygala incarnata, Polygala lutea, Polygala polygama, Polygala ramosa, Polygala rugelii, Polypremum procumbens, Pteridium aquilinum, Quercus geminata, Quercus laevis, Quercus laurifolia, Quercus minima, Quercus nigra, Quercus virginiana, Rhexia alifanus, Rhexia aristosa, Rhexia lutea, Rhexia mariana, Rhexia nashii, Rhexia nuttallii, Rhexia petiolata, Rhexia virginica, Rhododendron viscosum, Rhus copallinum, Rhynchospora alba, Rhynchospora baldwinii, Rhynchospora brachychaeta, Rhynchospora careyana, Rhynchospora chalarocephala, Rhynchospora chapmanii, Rhynchospora ciliaris, Rhynchospora compressa, Rhynchospora corniculata, Rhynchospora debilis, Rhynchospora fascicularis, Rhynchospora filifolia, Rhynchospora galeana, Rhynchospora globularis, Rhynchospora glomerata, Rhynchospora gracilenta, Rhynchospora inexpansa, Rhynchospora inundata, Rhynchospora latifolia, Rhynchospora macra, Rhynchospora* macrostachya, Rhynchospora oligantha, Rhynchospora perplexa, Rhynchospora plumosa, Rhynchospora pusilla, Rhynchospora rariflora, Rhynchospora recognita, Rhynchospora solitaria, Rhynchospora stenophylla, Rhynchospora tracyi, Rubus, Rudbeckia hirta. Rudbeckia scabrifolia, Sabal minor, Sabatia brevifolia, Sabatia campanulata, Saccharum giganteum, Sagittaria graminea, Salix nigra, Sarracenia alata, Sarracenia flava, Sarracenia leucophylla, Sarracenia minor, Sarracenia psittacina, Sarracenia rosea, Sarracenia rubra, Schizachyrium scoparium, Schizachyrium tenerum, Scleria georgiana, Scleria muhlenbergii, Scleria pauciflora, Scleria reticularis, Scutellaria integrifolia, Serenoa repens, Seymeria cassioides, Smilax auriculata, Smilax glauca, Smilax laurifolia, Smilax pumila, Solidago fistulosa, Solidago odora, Solidago patula, Spiranthes laciniata, Sporobolus floridanus, Sporobolus teretifolius, Stenanthium leimanthoides, Stenaria nigricans, Stylisma aquatica, Stylosanthes biflora, Symphyotrichum dumosum, Syngonanthus flavidulus, Taxodium ascendens, Taxodium distichum, Tiedemannia canbyi, Tiedemannia filiformis, Toxicodendron radicans, Toxicodendron vernix, Triadenum virginicum, Triadica sebifera, Triantha racemosa, Utricularia cornuta, Utricularia gibba, Utricularia purpurea, Utricularia subulata, Utricularia subulata, Vaccinium corymbosum, Vaccinium crassifolium, Vaccinium macrocarpon, Viburnum nudum, Vitis rotundifolia, Woodwardia areolata, Woodwardia virginica, Xyris baldwiniana, Xyris brevifolia, Xyris caroliniana, Xyris difformis, Xyris drummondii, Xyris elliottii, Xyris fimbriata, Xyris isoetifolia, Xyris jupicai, Xyris laxifolia, Xyris platylepis, Xyris scabrifolia, Xyris serotina, Xyris smalliana, Xyris stricta, Xyris torta, Zenobia pulverulenta, Zigadenus glaberrimus.

10.4.1.2. **Xyris baldwiniana Schult.** is a long-lived perennial, which occurs in bogs (acid, cypress, hillside, pitcher-plant, quaking, roadside, seepage), ditches (boggy, drainage, roadside, seepy, wet), flats (pitcher-plant), flatwoods (pine), marshes (depression), meadows (moist), pine barrens, pinelands (moist), pools (desiccated), rights of way (powerline), roadsides (swampy), savannas (open, pine, pitcher-plant, wet), seepage slopes (hillside), swales (moist), and along the margins of borrow pits (roadside), ponds, and lakes (marshy) at elevations to 380 m. The plants avoid standing waters and occur typically in exposures that receive full sunlight. The substrates are acidic (e.g., pH: 5.9), heavier, finer-textured soils categorized as Atmore (plinthic paleaquults), Bayboro, Dorovan-Johnstons series, Florala (plinthic paleudults), Leagueville-Henco, Leon (aeric haplaquods), Meadowbrook (grossarenic ochraqualfs), muck (sapric), Pelham (arenic paleaquults), Rutledge (typic humaquepts), peaty muck, sand (moist, Queen City, Sparta), sandy clay, sandy loam, sandy peat, sandy rock, and Wahee (aeric ochraquults). Flowering commences in early spring (April) with fruiting extending through July. The flowers open during the morning (e.g., 11 am) and are known to be visited (and likely pollinated) by sweat bees (Insecta: Hymenoptera: Halictidae). A seed bank of uncertain duration develops. The plants develop into extremely large clumps (up to hundreds of flowering culms), which overwinter by paired lateral buds, which develop on the contracted stem in the upper axils of their hardened, thickened aggregates of leaf bases. They are fire-resistant and average flowering stalk density was found to increase significantly in the first flowering season following fire (but declined in the second post-fire year). **Reported associates:** Acer rubrum, Aletris aurea, Alnus serrulata, Andropogon arctatus, Andropogon glomeratus, Andropogon gyrans, Andropogon liebmannii, Andropogon virginicus, Aristida palustris, Aristida purpurascens, Aristida stricta, Arnoglossum ovatum, Arundinaria tecta, Asclepias connivens, Axonopus, Balduina atropurpurea, Bidens laevis, Bigelowia nudata, Bigelowia nuttallii, Calamovilfa brevipilis, Calopogon tuberosus, Carex atlantica, Carex canescens, Carphephorus pseudoliatris, Centella asiatica, Chamaecyparis thyoides, Chaptalia tomentosa, Clethra alnifolia, Cliftonia monophylla, Coreopsis gladiata, Ctenium aromaticum, Cyperus haspan, Dichanthelium ensifolium, Dichanthelium scabriusculum, Doellingeria sericocarpoides, Drosera capillaris, Drosera filiformis, Drosera intermedia, Eleocharis tortilis, Eriocaulon compressum, Eriocaulon decangulare, Eriocaulon texense, Eryngium integrifolium, Eupatorium leucolepis, Eupatorium mohrii, Eupatorium perfoliatum, Eupatorium rotundifolium, Fuirena scirpoidea, Gaylussacia mosieri, Gordonia lasianthus, Helianthus angustifolius, Helianthus heterophyllus, Hydrocotyle verticillata, Hypericum fasciculatum, Hypericum mutilum, Ilex coriacea, Ilex glabra, Ilex myrtifolia, Juncus effusus, Juncus polycephalus, Juncus scirpoides, Juncus trigonocarpus, Juncus validus, Kalmia hirsuta, Lachnocaulon anceps, Liatris pycnostachya, Liatris spicata, Lobelia glandulosa, Lobelia puberula, Lophiola aurea, Ludwigia, Lycopodiella appressa, Lycopodiella caroliniana, Lycopus cokeri, Lyonia ferruginea, Magnolia virginiana, Marshallia graminifolia, Mikania scandens, Mitreola sessilifolia, Mnesithea tuberculosa, Muhlenbergia expansa, Myrica cerifera, Myrica heterophylla, Nyssa biflora, Onoclea sensibilis, Osmunda regalis, Osmundastrum cinnamomeum, Oxypolis rigidior, Panicum virgatum, Paspalum praecox, Persea palustris, Physostegia, Pinguicula pumila, Pinus palustris, Pinus serotina, Pityopsis oligantha, Platanthera ciliaris, Platanthera cristata, Platanthera integra, Platanthera nivea, Pleea tenuifolia, Polygala cruciata, Polygala lutea, Polygala mariana, Polygala ramosa, Proserpinaca palustris, Ptilimnium capillaceum, Ptilimnium costatum, Pycnanthemum nudum, Rhexia alifanus, Rhexia lutea, Rhexia petiolata, Rhynchospora alba, Rhynchospora baldwinii, Rhynchospora capitellata, Rhynchospora chalarocephala, Rhynchospora ciliaris, Rhynchospora compressa, Rhynchospora globularis, Rhynchospora gracilenta, Rhynchospora latifolia, Rhynchospora macra, Rhynchospora oligantha, Rhynchospora plumosa, Rhynchospora pusilla, Rhynchospora rariflora, Rhynchospora solitaria, Rhynchospora stenophylla, Rubus, Rudbeckia scabrifolia, Sabatia difformis, Sabatia gentianoides, Sabatia macrophylla, Saccharum giganteum, Sagittaria graminea, Sarracenia alata, Sarracenia flava, Sarracenia minor, Sarracenia psittacina, Sarracenia rubra, Schizachyrium scoparium,

Scleria muhlenbergii, Scleria pauciflora, Scleria reticularis, Scutellaria integrifolia, Smilax laurifolia, Solidago patula, Sphagnum, Symphyotrichum chapmanii, Syngonanthus flavidulus, Tephrosia hispidula, Tiedemannia filiformis, Toxicodendron vernix, Triantha racemosa, Utricularia cornuta, Utricularia juncea, Utricularia subulata, Vaccinium stamineum, Viburnum nudum, Xyris ambigua, Xyris caroliniana, Xyris difformis, Xyris drummondii, Xyris fimbriata, Xyris flabelliformis, Xyris isoetifolia, Xyris jupicai, Xyris scabrifolia, Xyris stricta, Xyris torta, Zenobia pulverulenta.

10.4.1.3. *Xyris brevifolia* **Michx.** is an annual, which inhabits bogs (grassy), depressions (interdunal, moist, sandy), ditches (roadside, swampy, wet), flatwoods (burned, damp, disturbed, mesic, palmetto, pine, scrubby, upland, wet), hammocks (mesic, xeric, wet), marshes (depression, flatwoods), meadows (wet), prairies (dry, ungrazed, wet), rights of way (hydric, powerline), roadsides (wet), savannas (burned, spodosol), seeps (acid, roadside, wet), tire ruts, washes, and the margins (ephemeral) of baygalls (disturbed), canals (sandy), hammocks, lakes (interdunal), ponds (borrow-pit, pineland, seasonal), and swamps (cypress, hardwood) at elevations to 70 m. The plants occur under open exposures in wet but rarely flooded sites. The substrates are coarse, acidic, nutrient-poor, and have been characterized as Basinger (spodic psammaquents), Chipley (aquic quartzipsamments), Duckston (typic psammaquents), Floridana (arenic argiaquolls), Immokalee (arenic haplaquods), Mascotte (ultic haplaquods), Myakka (aeric haplaquods), Pomello (arenic haplohumods), Rutledge (typic humaquepts), sand, sandy clay, sandy peat, Sapelo (ultic haplaquods), Satellite (aquic quartzipsamments), St. Johns (typic haplaquods), Surrency (arenic umbric paleaquults), and Wabasso (alfic haplaquods). The plants can progress from seed to flower within 4 months. Flowering and fruiting commence early as January–February but can occur throughout the year in warmer sites. Consequently, 2–3 flowering generations of plants can develop during the season under ideal soil moisture conditions. The flowers open in the morning and close by early afternoon. They are known to be pollinated by leaf-cutting bees (Insecta: Hymenoptera: Megachilidae: *Megachile brevis*) and sweat bees (Insecta: Hymenoptera: Halictidae: *Lasioglossum coreopsis, L. nymphale, L. tamiamense*). A seed bank of uncertain duration develops. Seed sampling studies have found up to an average of 6.5 seeds m^{-2} on the substrate surface and 18 seeds m^{-2} in sediment cores (10 cm depth). Seeds collected during January have germinated readily (under greenhouse conditions) by March with the seedlings reaching flowering maturity by June of the same year. The plants are fire-tolerant and often are associated with periodically burned sites. They have been described both as perennials and as cespitose, hard-based annuals. However, the former designation is inaccurate, given that the plants overwinter as seeds; however, they can reproduce vegetatively to form dense tufts during the growing season by means of lateral offshoots. **Reported associates:** *Aletris, Andropogon glomeratus, Andropogon ternarius, Aristida purpurascens, Aristida spiciformis, Aristida stricta, Asclepias cinerea, Asimina reticulata, Axonopus, Bartonia verna, Bejaria racemosa, Burmannia capitata, Calopogon, Campanula floridana, Carphephorus odoratissimus, Carya floridana, Centella asiatica, Chaptalia tomentosa, Cladonia, Clethra alnifolia, Cliftonia monophylla, Conradina canescens, Crotalaria rotundifolia, Cyrilla racemiflora, Dichanthelium, Dichanthelium aciculare, Dichanthelium ensifolium, Dichanthelium strigosum, Drosera brevifolia, Drosera capillaris, Drosera intermedia, Drosera tracyi, Eleocharis vivipara, Elephantopus elatus, Erigeron quercifolius, Erigeron vernus, Eriocaulon decangulare, Eupatorium mohrii, Eupatorium rotundifolium, Euphorbia telephioides, Euthamia graminifolia, Fuirena scirpoidea, Galactia elliottii, Gaylussacia, Gratiola hispida, Helenium pinnatifidum, Hypericum brachyphyllum, Hypericum cistifolium, Hypericum microsepalum, Hypericum tenuifolium, Hypericum tetrapetalum, Ilex cassine, Ilex coriacea, Ilex glabra, Itea virginica, Juncus effusus, Juncus validus, Kalmia buxifolia, Kalmia hirsuta, Lachnanthes caroliniana, Lachnocaulon anceps, Liatris tenuifolia, Lilium catesbaei, Lobelia feayana, Ludwigia maritima, Lycopodiella alopecuroides, Lycopodiella appressa, Lycopodiella caroliniana, Lycopodiella cernua, Lycopodiella inundata, Lyonia ferruginea, Lyonia fruticosa, Lyonia lucida, Muhlenbergia, Myrica cerifera, Nyssa, Oclemena reticulata, Oenothera biennis, Oldenlandia uniflora, Osmanthus americanus, Osmundastrum cinnamomeum, Panicum, Paspalum setaceum, Persea humilis, Piloblephis rigida, Pinguicula lutea, Pinus elliottii, Pinus palustris, Pinus serotina, Pinus taeda, Pityopsis graminifolia, Platanthera, Polygala lutea, Polygala nana, Polygala ramosa, Polygala setacea, Polygonella gracilis, Pterocaulon, Pterocaulon pycnostachyum, Quercus chapmanii, Quercus minima, Quercus myrtifolia, Quercus virginiana, Rhexia nuttallii, Rhexia petiolata, Rhus copallinum, Rhynchospora megalocarpa, Rhynchospora plumosa, Sabal palmetto, Sabatia brevifolia, Sarracenia flava, Sarracenia leucophylla, Sarracenia psittacina, Schizachyrium scoparium, Scleria, Selaginella arenicola, Serenoa repens, Sisyrinchium, Smilax auriculata, Smilax laurifolia, Smilax pumila, Solidago fistulosa, Solidago petiolaris, Sorghastrum secundum, Spartina patens, Spiranthes, Sporobolus floridanus, Stillingia sylvatica, Symphyotrichum elliotii, Syngonanthus flavidulus, Taxodium ascendens, Toxicodendron radicans, Utricularia cornuta, Utricularia juncea, Utricularia subulata, Vaccinium darrowii, Vaccinium myrsinites, Vitis rotundifolia, Woodwardia virginica, Xyris ambigua, Xyris elliottii, Xyris flabelliformis, Xyris isoetifolia, Xyris jupicai, Xyris platylepis.*

10.4.1.4. *Xyris difformis* **Chapm.** is a perennial, which grows in or on bars (cobble, sand), baygalls, bogs (argillaceous, cedar, hillside, pitcher plant, *Sphagnum*, sandy, seepage, spring-fed), depressions (damp, intermittent, peaty), ditches (brackish, peaty, roadside, seepage-fed, wet), fens (poor), flats (disturbed, low, pine), flatwoods (burned, pine, wet, wet-mesic), floodplains, gullies (drainage, moist), hummocks, marshes (beaver, floating, tidal), meadows (sedge, wet), mudflats, pine barrens (damp), pinelands (low), pools (depressions), prairies (burned, coastal, marl, mesic, wet), (depression, wet), puddles (sandy), riverbeds, roadsides (boggy, moist, wet), sand pits (moist), savannas (disturbed, level, mesic, pineland, seepage, wet), seeps (acid, hillside, low, streamhead), slopes (seepage), sloughs (peaty), swales (boggy, dune, inundated, sphagnous), woodlands (alluvial, sandy), and along the margins (marshy, peaty) or shores of canals (drainage), lakes, pocosins (cutover, streamhead), ponds (beaver, brackish, borrow pit, depression, flatwoods, freshwater, mill, seasonal, swampy), rivers, savannas, streams, and swamps (boggy, dome, sandy) at elevations to 494 m. The plants occur often in open exposures but also can tolerate fairly dense shade. They are characteristic of shores with intermediate wave exposure, although their maximum biomass is produced in sites ranging from intermediate to high wave exposure. Although characteristic of poor conditions, the plants have grown larger when transplanted to more fertile sites; however, they cannot compete with superiorly adapted rich-site species. The typical habitats are represented by shallow (to 46 cm), acidic (e.g., pH: 4.0–7.0; $\tilde{x}=4.0$–4.5; $\bar{x}=6.4$; alkalinity: 1.0–8.5 mg l^{-1}); $\bar{x}=4.3$ mg l^{-1}), and nutrient-poor (e.g., specific conductance: $\tilde{x}=28$–48 µS cm^{-1}) waters. The substrates are acidic, nutrient-poor, and have been characterized as Alapaha (arenic plinthic paleaquults), Allanton (grossarenic haplaquods), alluvium, Basinger (spodic psammaquents), Bibb (typic flubaquents), Bonwier (typic hapludults), Escambia (plinthaquic paleudults), Florala (plinthaquic paleudults), Fuquay (plinthic paleudults), gravel, Hallandale (lithic psammaquents), Immokalee, Kinston (typic fluvaquents), Leon (aeric haplaquods), Letney-Tehran (arenic paleudults), loam (Pantego), loamy sand (Pelham), Mandarin (typic haplohumods), Meadowbrook (grossarenic ochraqualfs), Melhomes (humaqueptic psammaquents), muck, mucky sand, mud, Ona (typic haplaquods), Pamlico-Donovan (terric medisaprists), peaty sand, Pineda (arenic glossaqualfs), Plummer (grossarenic paleaquults), Pompano, Pottsburg (grossarenic haplaquods), Rains (typic paleaquults), Riviera (arenic glossaqualfs), Rutledge (typic humaquepts), Samsula (terric medisaprists), sand (Carrizo, Leon), sandy loam, sandy peat, Smyrna (aeric haplaquods), Stringtown (typic hapludults), and Valkaria (spodic psammaquents). Flowering and fruiting occur from July through September. The flowers open during the morning and fade by 7:30 pm. The seeds have fairly high viability (66–85%). Those that were cold stratified for 270 days at 4°C have germinated well (to 66%) in light under a 30°C/20°C day/night temperature regime. Higher germination (69%) has been obtained for seeds (flooded to 2 cm) sown on sand than for those sown on an organic substrate (27%). Lower germination also has been observed for seeds sown below the waterline of substrates from sheltered shores than for those on substrates from exposed shores. However, a seed bank is maintained at water depths to 90 cm. The plants perennate by pairs of low, lateral shoot buds. **Reported associates:** *Acer rubrum, Agalinis aphylla, Agalinis purpurea, Agrostis scabra, Aletris farinosa, Amelanchier canadensis, Amphicarpum amphicarpon, Amphicarpum muhlenbergianum, Andropogon arctatus, Andropogon glomeratus, Andropogon gyrans, Andropogon liebmannii, Andropogon virginicus, Aristida longespica, Aristida palustris, Aristida stricta, Arnoglossum ovatum,*

Aronia arbutifolia, Aronia ×prunifolia, Arundinaria tecta, Asclepias incarnata, Asclepias rubra, Aulacomnium palustre, Avenella flexuosa, Bacopa caroliniana, Bacopa monnieri, Balduina uniflora, Bartonia virginica, Betula populifolia, Bidens cernuus, Bidens connatus, Bidens discoideus, Bidens tripartitus, Bigelowia nudata, Boltonia, Brasenia schreberi, Bulbostylis capillaris, Burmannia capitata, Calamagrostis canadensis, Carex atlantica, Carex barrattii, Carex debilis, Carex elliottii, Carex glaucescens, Carex joorii, Carex lasiocarpa, Carex limosa, Carex livida, Carex lonchocarpa, Carex louisianica, Carex striata, Carphephorus pseudoliatris, Centella asiatica, Cephalanthus occidentalis, Cephalozia connivens, Chamaecrista fasciculata, Chamaecyparis thyoides, Chamaedaphne calyculata, Chasmanthium laxum, Cladium mariscoides, Cladopodiella fluitans, Clethra alnifolia, Coleataenia longifolia, Coleataenia tenera, Comptonia peregrina, Coreopsis gladiata, Coreopsis rosea, Croton glandulosus, Ctenium aromaticum, Cyperus dentatus, Cyperus virens, Danthonia sericea, Danthonia spicata, Decodon verticillatus, Desmodium glabellum, Dichanthelium acuminatum, Dichanthelium dichotomum, Dichanthelium sphaerocarpon, Diodella teres, Drosera brevifolia, Drosera capillaris, Drosera filiformis, Drosera intermedia, Drosera rotundifolia, Drosera tracyi, Dulichium arundinaceum, Echinochloa, Elatine minima, Eleocharis acicularis, Eleocharis equisetoides, Eleocharis flavescens, Eleocharis melanocarpa, Eleocharis microcarpa, Eleocharis obtusa, Eleocharis palustris, Eleocharis robbinsii, Eleocharis tortilis, Eleocharis tricostata, Eleocharis tuberculosa, Elephantopus nudatus, Epilobium coloratum, Eragrostis elliottii, Eragrostis hypnoides, Eragrostis lugens, Eriocaulon aquaticum, Eriocaulon compressum, Eriocaulon decangulare, Eriocaulon texense, Eryngium integrifolium, Eubotrys racemosa, Eupatorium leptophyllum, Eupatorium leucolepis, Eupatorium mohrii, Eupatorium perfoliatum, Eupatorium pilosum, Eupatorium rotundifolium, Euthamia graminifolia, Eutrochium dubium, Fimbristylis autumnalis, Fuirena breviseta, Fuirena pumila, Fuirena scirpoidea, Fuirena squarrosa, Galium tinctorium, Gaultheria procumbens, Gaylussacia baccata, Gaylussacia frondosa, Glyceria canadensis, Gratiola aurea, Gratiola ramosa, Helenium autumnale, Helenium brevifolium, Helianthus heterophyllus, Helianthus radula, Hibiscus moscheutos, Hydrocotyle umbellata, Hydrolea corymbosa, Hypericum adpressum, Hypericum brachyphyllum, Hypericum canadense, Hypericum cistifolium, Hypericum denticulatum, Hypericum ellipticum, Hypericum fasciculatum, Hypericum hypericoides, Hypericum majus Hypericum mutilum, Hypericum myrtifolium, Ilex coriacea, Ilex glabra, Ilex myrtifolia, Ilex opaca, Iris tridentata, Isoetes lacustris, Isoetes tuckermanii, Juncus balticus, Juncus biflorus, Juncus caesariensis, Juncus canadensis, Juncus debilis, Juncus diffusissimus, Juncus effusus, Juncus greenei, Juncus longii, Juncus marginatus, Juncus militaris, Juncus pelocarpus, Juncus pylaei, Juncus scirpoides, Juncus trigonocarpus, Kalmia angustifolia, Kalmia buxifolia, Kalmia latifolia, Lachnanthes caroliniana, Lactuca canadensis, Larix laricina, Leersia hexandra, Leersia oryzoides, Leucobryum glaucum, Liatris pycnostachya, Lilium catesbaei, Lindernia dubia, Linum striatum, Liquidambar styraciflua, Lobelia canbyi, Lobelia cardinalis, Lobelia dortmanna, Lobelia nuttallii, Ludwigia alternifolia, Ludwigia linearis, Ludwigia linifolia, Ludwigia palustris, Ludwigia suffruticosa, Ludwigia virgata, Lycopodiella appressa, Lycopodiella caroliniana, Lycopodiella inundata, Lycopus amplectens, Lycopus uniflorus, Lyonia mariana, Lysimachia terrestris, Lythrum salicaria, Magnolia virginiana, Marshallia graminifolia, Melampyrum lineare, Mikania scandens, Mimulus ringens, Mnesithea rugosa, Muhlenbergia capillaris, Muhlenbergia expansa, Muhlenbergia torreyana, Muhlenbergia uniflora, Mylia anomala, Myrica cerifera, Myrica gale, Myrica heterophylla, Myrica pensylvanica, Myriophyllum farwellii, Myriophyllum tenellum, Narthecium americanum, Nelumbo lutea, Nuphar variegata, Nymphaea odorata, Nyssa biflora, Nyssa sylvatica, Oclemena nemoralis, Orontium aquaticum, Osmunda regalis, Osmundastrum cinnamomeum, Panicum amarum, Panicum hemitomon, Panicum philadelphicum, Panicum verrucosum, Panicum virgatum, Peltandra virginica, Persea palustris, Persicaria careyi, Persicaria pensylvanica, Persicaria punctata, Pinus elliottii, Pinus rigida, Pinus taeda, Platanthera ciliaris, Platanthera clavellata, Platanthera flava, Platanthera integra, Pleea tenuifolia, Pluchea baccharis, Pluchea foetida, Pogonia ophioglossoides, Polygala brevifolia, Polygala cruciata, Polygala incarnata, Polygala lutea, Polygala nuttallii, Polygala ramosa, Polygala rugelii, Polygala sanguinea, Polypremum procumbens, Pontederia cordata, Potamogeton bicupulatus, Proserpinaca palustris, Proserpinaca pectinata, Ptilimnium costatum, Quercus ilicifolia, Quercus laurifolia, Quercus marilandica, Quercus minima, Quercus nigra, Quercus virginiana, Ranunculus flammula, Rhexia aristosa, Rhexia lutea, Rhexia mariana, Rhexia nashii, Rhexia petiolata, Rhexia salicifolia, Rhexia virginica, Rhododendron canadense, Rhododendron viscosum, Rhynchospora alba, Rhynchospora caduca, Rhynchospora capitellata, Rhynchospora cephalantha, Rhynchospora chalarocephala, Rhynchospora chapmanii, Rhynchospora ciliaris, Rhynchospora debilis, Rhynchospora decurrens, Rhynchospora divergens, Rhynchospora fascicularis, Rhynchospora filifolia, Rhynchospora fusca, Rhynchospora galeana, Rhynchospora globularis, Rhynchospora glomerata, Rhynchospora gracilenta, Rhynchospora inexpansa, Rhynchospora latifolia, Rhynchospora macra, Rhynchospora macrostachya, Rhynchospora microcephala, Rhynchospora mixta, Rhynchospora nitens, Rhynchospora oligantha, Rhynchospora perplexa, Rhynchospora rariflora, Rhynchospora recognita, Rhynchospora scirpoides, Rhynchospora torreyana, Rhynchospora tracyi, Rubus hispidus, Rudbeckia scabrifolia, Sabal minor, Sabatia difformis, Sabatia kennedyana, Sabatia macrophylla, Saccharum giganteum, Sagittaria engelmanniana, Sagittaria graminea, Sagittaria lancifolia, Sagittaria latifolia, Sagittaria platyphylla, Salix nigra, Sarracenia alata, Sarracenia flava, Sarracenia jonesii, Sarracenia leucophylla, Sarracenia psittacina, Sarracenia purpurea, Sarracenia rosea, Sassafras albidum, Saururus cernuus, Schizachyrium rhizomatum, Schizachyrium scoparium, Schizachyrium tenerum, Schizaea pusilla, Schoenoplectus pungens, Schoenoplectus subterminalis, Schoenoplectus torreyi, Scirpus cyperinus, Scleria baldwinii, Scleria georgiana, Scleria minor, Scleria muehlenbergii, Scleria reticularis, Scutellaria lateriflora, Serenoa repens, Smilax glauca, Smilax laurifolia, Smilax rotundifolia, Smilax walteri, Solidago odora, Solidago pulchra, Sphagnum bartlettianum, Sphagnum compactum, Sphagnum cuspidatum, Sphagnum fimbriatum, Sphagnum magellanicum, Sphagnum papillosum, Sphagnum pulchrum, Sphagnum recurvum, Sphagnum rubellum, Spiraea alba, Spiraea tomentosa, Symphyotrichum lateriflorum, Taxodium ascendens, Taxodium distichum, Tetraphis pellucida, Thelypteris palustris, Tiedemannia filiformis, Triadenum fraseri, Triadenum virginicum, Triantha racemosa, Trientalis borealis, Utricularia cornuta, Utricularia geminiscapa, Utricularia gibba, Utricularia macrorhiza, Utricularia purpurea, Utricularia resupinata, Utricularia striata, Utricularia subulata, Vaccinium corymbosum, Vaccinium macrocarpon, Vaccinium myrsinites, Vaccinium oxycoccos, Verbena hastata, Viburnum nudum, Viola lanceolata, Viola sagittata, Woodwardia virginica, Xyris ambigua, Xyris baldwiniana, Xyris caroliniana, Xyris drummondii, Xyris elliottii, Xyris fimbriata, Xyris flabelliformis, Xyris jupicai, Xyris laxifolia, Xyris montana, Xyris platylepis, Xyris scabrifolia, Xyris serotina, Xyris smalliana, Xyris stricta, Xyris torta, Zigadenus glaberrimus.

10.4.1.5. **Xyris drummondii** Malme is a perennial, which occurs in bogs (flatwoods, hillside, peaty, pitcher plant, roadside, savanna, seepage, streamhead), ditches (roadside, wet), flatwoods (pine), lowlands (disturbed), poor fens (quaking), powerline easements, prairies (wet), roadsides (baygall), savannas (cutover, pine), and the shores (exposed) of ponds (karst) at elevations to 116 m. The plants occur only in full sunlight exposures. The substrates are acidic, finer-textured, heavier types described as Dorovan (typic medisaprists), Escambia (plinthaquic paleudults), Leefield (arenic plinthaquic paleudults), Leon (aeric haplaquods), Lynchburg (aeric paleaquults), Meadowbrook (grossarenic ochraqualfs), muck, Pamlico-Donovan (terric medisaprists), peat (sphagnous), Pelham (arenic paleaquults), Plummer (grossarenic paleaquults), Rains (typic paleaquults), Rutledge (typic humaquepts), sand, sandy clay, sandy peat, sandy peaty clay, and Stringtown (typic hapludults). Flowering extends from midsummer (June) to fall (August). The flowers open in early morning and typically wither by midday. The compressed, hardened shoots perennate by means of low-set, lateral buds. The number of plants initially increases substantially following a fire, but then declines in subsequent years. **Reported associates:** Aletris aurea, Aristida purpurascens, Balduina atropurpurea, Bigelowia nudata, Burmannia capitata, Cliftonia monophylla, Coreopsis gladiata, Ctenium aromaticum, Drosera capillaris, Drosera tracyi, Eriocaulon decangulare, Eriocaulon nigrobracteatum, Eriocaulon texense, Eryngium integrifolium, Gratiola pilosa, Helianthus heterophyllus, Hypericum brachyphyllum, Ilex glabra, Juncus trigonocarpus, Lachnocaulon digynum, Liatris pycnostachya, Liatris *spicata, Lobelia glandulosa, Lophiola aurea, Lycopodiella appressa, Lycopodiella caroliniana, Lycopodiella prostrata, Marshallia graminifolia, Mitreola sessilifolia, Muhlenbergia expansa, Orontium aquaticum, Oxypolis ternata, Pinguicula pumila, Pinus palustris, Platanthera ciliaris, Pleea tenuifolia, Pogonia ophioglossoides, Polygala cruciata, Polygala ramosa, Polypremum procumbens, Pycnanthemum nudum, Rhexia alifanus, Rhexia lutea, Rhexia petiolata, Rhynchospora chalarocephala, Rhynchospora gracilenta, Rhynchospora macra, Rhynchospora oligantha, Rhynchospora solitaria, Rhynchospora stenophylla, Rudbeckia scabrifolia, Sabatia macrophylla, Sarracenia alata, Sarracenia flava, Sarracenia leucophylla, Sarracenia minor, Sarracenia psittacina, Scleria reticularis, Syngonanthus flavidulus, Tiedemannia filiformis, Triantha racemosa, Utricularia juncea, Xyris ambigua, Xyris baldwiniana, Xyris difformis, Xyris isoetifolia, Xyris scabrifolia, Xyris serotina.*

10.4.1.6. *Xyris elliottii* **Chapm.** is a long-lived perennial, which inhabits baygalls, bogs (flatwoods, seepage), borrow pits, clay pits, depressions (boggy, desiccated, flatwoods, Karst, shallow), ditches (disturbed, exposed, flatwoods, low, roadside, seepage), firebreaks (mowed), flats (boggy, interdunal, pine, *Serenoa repens*, wet), flatwoods (burned, clearcut, disturbed, infrequently burned, low, mesic, palmetto, pine, scrubby, wet), marshes (depression, sandhill), meadows (burned, depression, grassy, mesic, wet), pinelands (moist), pine barrens (low, moist), prairies (dry, moist, palmetto, wet, wet-mesic), savannas (cypress, depression, open, pine, wet), seeps (flatwood, wet), slopes [to 2%] (roadside, seepage), sloughs, swales (dune, intradunal, moist, pinelands, sand ridge), swamps (sandy), and the margins or shores of bayheads, bogs, canals, hammocks (hardwood, oak, *Sabal palmetto*), lakes (interdunal, marshy), pocosins, ponds (artificial, depression, desiccated, dry, oxidation, pineland, seasonal), and rivers at elevations to 58 m. Exposures can range from full sunlight to partial shade. The substrates are acidic and have been described as Basinger (spodic psammaquents), Duckston (typic psammaquents), Eau Gallie (alfic haplaquods), Floridana (arenic argiaquolls), Immokalee, Lawnwood (aeric haplaquods), loamy sand (Pelham), Myakka (aeric haplaquods), peat, peaty muck, Pineda (arenic glossaqualfs), Pomona (ultic haplaquods), Pompano, Pottsburg (grossarenic haplaquods), sand (Corolla-Duckston, Leon), sandy muck, sandy peat, sandy peaty loam, Sellers (cumulic humaquepts), St. Johns (typic haplaquods), and Surrency (arenic umbric paleaquults). Flowering and fruiting occur primarily from early spring through summer, but can take place throughout the year in more southern sites. The flowers open in the morning (e.g., 10:00–11:00 am) and usually persist only until midday (but have been observed at 7:30 pm). They are pollinated by various bees (Insecta: Hymenoptera), including bumblebees (Apidae: *Bombus impatiens*), honeybees (Apidae: *Apis mellifera*), leaf-cutting bees (Megachilidae: *Megachile brevis*), plasterer bees (Colletidae: *Colletes distinctus*), and sweat bees (Halictidae: *Augochloropsis sumptuosa, Lasioglossum coreopsis, L. nymphale*). The hardened, compacted stems perennate by elongate, fleshy, lateral buds and develop into large tufts (frequently of 100+ plants to 10+ cm). **Reported associates:** *Acer rubrum, Amphicarpum muhlenbergianum, Andropogon glomeratus, Andropogon gyrans, Andropogon virginicus, Anthaenantia rufa, Aristida palustris, Aristida purpurascens, Aristida spiciformis, Aristida stricta, Asclepias longifolia, Asimina pygmea, Axonopus fissifolius, Axonopus furcatus, Baptisia calycosa, Bigelowia nudata, Burmannia capitata, Calamovilfa curtissii, Calydorea coelestina, Carphephorus pseudoliatris, Centella asiatica, Chaptalia tomentosa, Cladium jamaicense, Clethra alnifolia, Coleataenia longifolia, Coleataenia tenera, Coreopsis gladiata, Ctenium aromaticum, Cyperus lecontei, Cyperus polystachyos, Dichanthelium acuminatum, Dichanthelium ensifolium, Dichanthelium pinetorum, Dichanthelium scabriusculum, Dichanthelium sphaerocarpon, Dichanthelium strigosum, Drosera brevifolia, Drosera capillaris, Drosera tracyi, Eragrostis elliottii, Erigeron vernus, Erigeron vernus, Eriocaulon compressum, Eriocaulon decangulare, Eriocaulon texense, Eryngium yuccifolium, Eupatorium album, Eupatorium hyssopifolium, Eupatorium leptophyllum, Eupatorium mohrii, Eupatorium pilosum, Fuirena scirpoidea, Gaylussacia dumosa, Gaylussacia frondosa, Gaylussacia mosieri, Gratiola ramosa, Hypericum brachyphyllum, Hypericum cistifolium, Hypericum edisonianum, Hypericum fasciculatum, Hypericum gentianoides, Hypericum myrtifolium, Hypericum tenuifolium, Hypericum tetrapetalum, Ilex glabra, Juncus marginatus, Juncus scirpoides, Kalmia, Lachnanthes caroliniana, Lachnocaulon anceps, Lachnocaulon engleri, Lachnocaulon minus, Lindera melissifolia, Linum medium, Litsea aestivalis, Lobelia brevifolia, Lobelia puberula, Lophiola aurea, Ludwigia linearis, Ludwigia linifolia, Lycopodiella appressa, Lycopodiella caroliniana, Lyonia lucida, Magnolia grandiflora, Magnolia virginiana, Mitreola sessilifolia, Mnesithea rugosa, Muhlenbergia capillaris, Muhlenbergia expansa, Myrica cerifera, Panicum hemitomon, Paspalum setaceum, Persea borbonia, Persea palustris, Phoebanthus grandiflorus, Phyla nodiflora, Pinus elliottii, Pinus palustris, Piriqueta cistoides, Pityopsis graminifolia, Pleea tenuifolia, Pluchea baccharis, Pluchea foetida, Pluchea foetida, Polygala cruciata, Polygala lutea, Polygala ramosa, Polygala rugelii, Polygala setacea, Pterocaulon pycnostachyum, Quercus minima, Quercus pumila, Rhexia alifanus, Rhexia lutea, Rhexia mariana, Rhododendron viscosum, Rhynchospora chapmanii, Rhynchospora divergens, Rhynchospora fascicularis, Rhynchospora filifolia, Rhynchospora inundata, Rhynchospora oligantha, Rhynchospora plumosa, Rhynchospora tracyi, Sabal palmetto, Sabatia brevifolia, Sabatia grandiflora, Sagittaria lancifolia, Sarracenia flava, Sarracenia minor, Schizachyrium rhizomatum, Schizachyrium scoparium, Scleria baldwinii, Scleria ciliata, Scleria georgiana, Scleria hirtella, Scleria muhlenbergii, Scleria reticularis, Serenoa repens, Setaria parviflora, Sisyrinchium atlanticum, Smilax auriculata, Smilax glauca, Smilax laurifolia, Smilax rotundifolia, Spartina bakeri, Spartina patens, Spiranthes igniorchis, Sporobolus curtissii, Sporobolus floridanus, Sporobolus junceus, Steinchisma hians, Syngonanthus* *flavidulus, Taxodium ascendens, Tiedemannia filiformis, Tragia, Utricularia cornuta, Utricularia juncea, Vaccinium, Viola lanceolata, Vitis rotundifolia, Woodwardia, Xyris ambigua, Xyris brevifolia, Xyris caroliniana, Xyris difformis, Xyris flabelliformis, Xyris isoetifolia, Xyris jupicai, Xyris platylepis, Xyris stricta, Zigadenus glaberrimus.*

10.4.1.7. ***Xyris fimbriata*** **Elliott** is a biennial or short-lived perennial, which grows in bogs (hillside, pineland, springhead seepage), bottomlands (forested), clearings (peaty), ditches (boggy, drainage, roadside), depressions (cypress, fire-maintained, flatwood, shallow, sphagnous, wooded), flats (clear-cut, pine), flatwoods, floating logs, floating mats (peat), floodplains, marshes (depression), meadows (wet), mudflats (open), pine barrens (wet), pools (pineland), ponds (*Hypericum*, seepage), prairies (peaty, wet), rights of way (depression, powerline), roadside (weedy, wet), savannas (cutover, cypress, disturbed, low, pine, seasonally flooded, wet), seeps, slopes (seepage), sloughs (backwater, pond cypress), streams (shallow, sluggish), stumps (waterlogged), swamps (cypress, dome, flatwoods, shallow, *Sphagnum*), and along the margins (boggy, peaty, springy) of baygalls, bayheads (springy), borrow pits, canals, lakes (depression marsh, seepage-fed), pocosins (overgrown), ponds (beaver, cypress, flatwoods), rivers, and streams (spring-fed) at elevations to 283 m. The plants have been observed in exposures of full sun to partial (e.g., 5%) shade. Their tall stature enables survival in conditions ranging from wet ground to deeper standing waters (e.g., 31 cm). However, the plants do not tolerate desiccation and extensive die-offs have been reported in areas where the substrates have dried. The substrates are acidic (e.g., pH: 5.6–6.8), often coarse, and have been described as Basinger (spodic psammaquents), Dothan (plinthic paleudults), Escambia (plinthaquic paleudults), Florala (plinthaquic paleudults), Kaliga (terric medisaprists), Myakka (aeric haplaquods), muck (Ponzer), mucky loamy sand (Murville), mud, peaty muck, Placid (typic humaquepts), Plummer (grossarenic paleaquults), Pottsburg (grossarenic haplaquods), Rutledge-Osier (typic humaquepts), sand, sandy muck, sandy peat, St. Johns (typic haplaquods), Scranton (humaqueptic psammaquents), Surrency (umbric paleaquults), and Waveland (arenic haplaquods). Flowering and fruiting occur mainly during spring and summer, but extend throughout the year in favorable localities. The flowers open by late morning (e.g., >10:00 am) and deteriorate by the afternoon. The soft, compacted stems perennate by low-set lateral buds. As in several other *Xyris* species, the plants expand initially in the year following a fire, but decline in frequency during subsequent years. **Reported associates:** *Acer rubrum, Agalinis fasciculata, Alnus maritima, Alnus serrulata, Amsonia tabernaemontana, Andropogon glomeratus, Andropogon gyrans, Andropogon liebmannii, Andropogon virginicus, Anthaenantia rufa, Apios americana, Aristida palustris, Aristida purpurascens, Arnoglossum ovatum, Arundinaria gigantea, Bacopa caroliniana, Bartonia paniculata, Bidens mitis, Burmannia biflora, Calamagrostis coarctata, Calopogon tuberosus, Carex glaucescens, Carex hyalinolepis, Carex joorii, Carex striata, Carex verrucosa, Centella asiatica, Cephalanthus*

occidentalis, Chamaecyparis thyoides, Chaptalia tomentosa, Cladium mariscoides, Clethra alnifolia, Coleataenia longifolia, Coleataenia tenera, Coreopsis gladiata, Ctenium aromaticum, Cyperus lecontei, Cyrilla racemiflora, Decodon verticillatus, Dichanthelium acuminatum, Dichanthelium roanokense, Dichanthelium scabriusculum, Dichanthelium sphaerocarpon, Doellingeria sericocarpoides, Drosera capillaris, Drosera filiformis, Drosera intermedia, Drosera rotundifolia, Dulichium arundinaceum, Eleocharis baldwinii, Eleocharis elongata, Eleocharis equisetoides, Eleocharis flavescens, Eleocharis microcarpa, Eleocharis quadrangulata, Eleocharis robbinsii, Eleocharis tuberculosa, Eleocharis vivipara, Eragrostis refracta, Erigeron vernus, Eriocaulon compressum, Eriocaulon decangulare, Eriophorum virginicum, Eubotrys racemosa, Eupatorium capillifolium, Eupatorium leucolepis, Euthamia graminifolia, Fuirena breviseta, Fuirena squarrosa, Gaylussacia dumosa, Gelsemium rankinii, Gordonia lasianthus, Gratiola brevifolia, Habenaria repens, Heterotheca, Hibiscus moscheutos, Hydrolea corymbosa, Hydrolea ovata, Hypericum cistifolium, Hypericum crux-andreae, Hypericum fasciculatum, Hypericum galioides, Hypericum mutilum, Ilex cassine, Ilex glabra, Iris prismatica, Iris tridentata, Iris verna, Iris virginica, Itea virginica, Iva frutescens, Juncus canadensis, Juncus effusus, Juncus marginatus, Juncus polycephalus, Juncus scirpoides, Lachnanthes caroliniana, Lachnocaulon anceps, Leersia hexandra, Lilium catesbaei, Litsea aestivalis, Lobelia cardinalis, Ludwigia lanceolata, Ludwigia linearis, Ludwigia palustris, Ludwigia pilosa, Ludwigia sphaerocarpa, Lycopodiella alopecuroides, Lycopodiella appressa, Lycopus amplectens, Lycopus rubellus, Lyonia lucida, Magnolia virginiana, Mayaca fluviatilis, Mitreola petiolata, Mitreola sessilifolia, Mnesithea rugosa, Muhlenbergia expansa, Myrica cerifera, Myrica heterophylla, Myriophyllum pinnatum, Narthecium americanum, Nymphaea odorata, Nymphoides cordata, Nyssa biflora, Nyssa sylvatica, Oldenlandia uniflora, Orontium aquaticum, Osmunda regalis, Osmundastrum cinnamomeum, Oxypolis rigidior, Panicum hemitomon, Panicum verrucosum, Panicum virgatum, Paspalum, Persea borbonia, Persicaria hydropiperoides, Pieris phillyreifolia, Pinus elliottii, Pinus serotina, Platanthera ciliaris, Platanthera integra, Pluchea baccharis, Pogonia ophioglossoides, Polygala rugelii, Pontederia cordata, Proserpinaca palustris, Proserpinaca pectinata, Rhexia alifanus, Rhexia aristosa, Rhexia mariana, Rhexia nashii, Rhexia petiolata, Rhexia virginica, Rhynchospora alba, Rhynchospora baldwinii, Rhynchospora capitellata, Rhynchospora careyana, Rhynchospora cephalantha, Rhynchospora chalarocephala, Rhynchospora ciliaris, Rhynchospora corniculata, Rhynchospora elliottii, Rhynchospora fascicularis, Rhynchospora filifolia, Rhynchospora fusca, Rhynchospora gracilenta, Rhynchospora harperi, Rhynchospora inexpansa, Rhynchospora inundata, Rhynchospora macrostachya, Rhynchospora oligantha, Rhynchospora perplexa, Rhynchospora pleiantha, Rhynchospora pusilla, Rhynchospora solitaria, Rhynchospora stenophylla, Rhynchospora tracyi, Rudbeckia texana, Sabatia dodecandra, Saccharum baldwinii, Saccharum giganteum, Sacciolepis striata, Sagittaria graminea, Sagittaria isoetiformis, Sarracenia alata, Sarracenia flava, Sarracenia minor, Sarracenia psittacina, Sarracenia purpurea, Sarracenia rubra, Scirpus cyperinus, Scleria baldwinii, Scleria georgiana, Scleria reticularis, Sclerolepis uniflora, Silphium, Smilax laurifolia, Solidago patula, Sphagnum lescurii, Sphagnum macrophyllum, Sphagnum portoricense, Spiranthes laciniata, Styrax americanus, Symphyotrichum novi-belgii, Syngonanthus flavidulus, Taxodium ascendens, Taxodium distichum, Tephrosia, Tiedemannia canbyi, Tiedemannia filiformis, Toxicodendron radicans, Triadenum virginicum, Tripsacum dactyloides, Typha latifolia, Utricularia gibba, Utricularia purpurea, Utricularia radiata, Vaccinium corymbosum, Viola primulifolia, Vitis rotundifolia, Woodwardia areolata, Woodwardia virginica, Xyris ambigua, Xyris baldwiniana, Xyris difformis, Xyris jupicai, Xyris laxifolia, Xyris platylepis, Xyris scabrifolia, Xyris smalliana, Xyris stricta.

10.4.1.8. *Xyris flabelliformis* Chapm. is an annual, which occurs in bogs (savanna, seepage, streamside), clearings, depressions (flatwoods, roadside), ditches (roadside), flats (interdunal), flatwoods (burned, disturbed, mesic, peaty, pine, wet), hammocks (hardwood), marshes (tidal), rights of way (powerline), roadbeds, roadsides (sandy, wet), savannas (pine, seasonally inundated, wet), seepage slopes (overgrown), swales (dune, roadside), and along the margins (peaty) or shores of cypress domes, lakes, ponds (flatwoods, pineland, seasonal), and pools (pineland) at elevations to 32 m. The plants occur mostly toward the drier end of the soil moisture continuum and not in inundated sites or those characterized by heavier substrates. The substrates are acidic, usually coarser-textured, and have been described as Basinger (spodic psammaquents), Compass (plinthic paleudults), Estero (typic haplaquods), Leon, loamy sand (wet), Myakka (aeric haplaquods), Olustee (ultic haplaquods), Pomona (ultic haplaquods), Rutledge (typic humaquepts), sand (disturbed, moist, sterile), sandstone (seepy), sandy peat (moist), spodosols, and Wabasso (alfic haplaquods). Flowering normally initiates early in the year (February–March) and proceeds through the summer with fruiting extending into late fall (November); however, several flowering episodes can occur year-long in some favorable sites. The flowers open during the morning and close around midday. The seeds (excavated from 5 cm of sediment) have germinated under greenhouse conditions following a 1-week period of cold stratification (at 8°C). A seed bank (e.g., 9 seeds m^{-2}) of unspecified duration develops. The plants have gone from seed to flower within 4 months. **Reported associates:** *Amphicarpum muhlenbergianum, Andropogon brachystachyus, Andropogon virginicus, Aristida purpurascens, Aristida spiciformis, Aristida spiciformis, Aristida stricta, Asclepias pedicellata, Asimina reticulata, Bacopa monnieri, Boehmeria cylindrica, Carex tenuiflora, Centella asiatica, Cephalanthus occidentalis, Coreopsis gladiata, Cyperus croceus, Cyperus lecontei, Dichanthelium acuminatum, Dichanthelium ensifolium, Dichanthelium portoricense, Drosera brevifolia, Drosera capillaris, Drosera intermedia, Drosera tracyi, Eclipta*

prostrata, Eleocharis baldwinii, Eleocharis montevidensis, Eragrostis elliottii, Eragrostis pectinacea, Erigeron vernus, Eupatorium resinosum, Euthamia graminifolia, Fimbristylis puberula, Fuirena scirpoidea, Gratiola hispida, Gymnopogon chapmanianus, Hydrocotyle bonariensis, Hypericum fasciculatum, Hypericum myrtifolium, Hypericum tenuifolium, Hypericum tetrapetalum, Hypoxis hirsuta, Ilex cassine, Ilex glabra, Ilex myrtifolia, Ilex vomitoria, Isoetes riparia, Juncus brachycarpus, Juncus scirpoides, Juncus trigonocarpus, Krigia virginica, Lachnocaulon beyrichianum, Lechea torreyi, Ludwigia maritima, Lycopodiella alopecuroides, Lygodesmia aphylla, Lyonia fruticosa, Lyonia mariana, Macbridea caroliniana, Nyssa biflora, Oldenlandia uniflora, Panicum hemitomon, Phyla nodiflora, Pinus caribaea, Pinus elliottii, Pinus palustris, Polygala lutea, Polygala ramosa, Polygala rugelii, Polygala setacea, Proserpinaca, Pterocaulon, Quercus geminata, Quercus laevis, Rhexia nuttallii, Rhynchospora chapmanii, Rhynchospora fernaldii, Rhynchospora macra, Rhynchospora plumosa, Rubus trivialis, Sarracenia leucophylla, Schizachyrium maritimum, Schizachyrium scoparium, Schoenoplectus americanus, Scleria ciliata, Scleria muehlenbergii, Scleria reticularis, Serenoa repens, Spartina patens, Sphagnum, Symphyotrichum dumosum, Syngonanthus flavidulus, Utricularia subulata, Vaccinium, Xyris ambigua, Xyris baldwiniana, Xyris brevifolia, Xyris caroliniana, Xyris difformis, Xyris platylepis, Xyris serotina.

10.4.1.9. *Xyris isoetifolia* Kral is a long-lived perennial, which grows in bogs (savanna, seepage, sphagnous), clearings (boggy), depressions (roadside), ditches (roadside, seepage-fed), flatwoods (coastal, scrubby), savannas (burned, cypress, disturbed, flatwoods (wet), low, pine, seepage), seeps (sandy), and along the margins (sandy) or shores (fluctuating, sandy) of dolines (sinkholes), lakes (flatwoods), and ponds (flatwoods, limesink) at elevations to 67 m. The plants occur in open exposures to partial shade and thrive in areas of artificial or natural disturbance. They can grow in shallow water but are atypical of inundated sites. The substrates have been described as Leon (aeric haplaquods), muck, muddy sand, peat, peaty sand, Plummer (grossarenic paleaquults), sand, sandy peat, silty sand, and Troup (grossarenic paleudults). Flowering and fruiting occur from early summer (June) through fall (September). The flowers open during morning and wither by early afternoon. Reported pollinators include bees (Insecta: Hymenoptera: Apidae: *Bombus*; Halictidae: *Agapostemon, Lasioglossum zephyrum*) and flies (Insecta: Diptera: Syrphidae: *Ocyptamus fuscipennis, Toxomerus boscii, T. geminatus*). The densely cespitose plants overwinter by dormant, paired lateral buds on their compacted, hardened, thickened bases. **Reported associates:** *Amphicarpum muhlenbergianum, Anthaenantia rufa, Asclepias cinerea, Bigelowia nudata, Burmannia capitata, Carphephorus odoratissimus, Carphephorus pseudoliatris, Centella asiatica, Clethra alnifolia, Cliftonia monophylla, Coleataenia longifolia, Conradina canescens, Cyrilla racemiflora, Dichanthelium dichotomum, Drosera capillaris, Drosera filiformis, Drosera tracyi, Erigeron* *vernus, Eriocaulon compressum, Eriocaulon decangulare, Eriocaulon lineare, Eupatorium leptophyllum, Euphorbia telephioides, Fuirena breviseta, Fuirena scirpoidea, Gaylussacia mosieri, Gratiola hispida, Hypericum brachyphyllum, Hypericum fasciculatum, Hypericum lissophloeus, Hypericum tenuifolium, Hypericum tetrapetalum, Ilex glabra, Juncus scirpoides, Kalmia hirsuta, Lachnanthes caroliniana, Lachnocaulon anceps, Lachnocaulon digynum, Lachnocaulon minus, Liatris tenuifolia, Linum medium, Lobelia puberula, Lophiola aurea, Lycopodiella alopecuroides, Lycopodiella caroliniana, Lycopodiella prostrata, Lyonia ferruginea, Lyonia lucida, Physostegia godfreyi, Pinguicula lutea, Pinus palustris, Pleea tenuifolia, Polygala cruciata, Polygala lutea, Polygonella gracilis, Quercus minima, Rhexia alifanus, Rhexia lutea, Rhexia mariana, Rhexia salicifolia, Rhynchospora baldwinii, Rhynchospora chapmanii, Rhynchospora curtissii, Rhynchospora filifolia, Rhynchospora oligantha, Rhynchospora pleiantha, Rhynchospora plumosa, Sagittaria isoetiformis, Sarracenia flava, Sarracenia psittacina, Scleria reticularis, Smilax auriculata, Smilax laurifolia, Syngonanthus flavidulus, Triantha racemosa, Utricularia cornuta, Utricularia juncea, Utricularia subulata, Woodwardia areolata, Xyris ambigua, Xyris baldwiniana, Xyris brevifolia, Xyris elliottii, Xyris drummondii, Xyris jupicai, Xyris longisepala.*

10.4.1.10. **Xyris jupicai Rich.** is an annual (rarely biennial), which inhabits bogs (hillside, pineland, pitcher plant, roadside, seepage, *Sphagnum*), borrow pits (roadside), bottomlands, canals (drainage, roadside), channels (river), clay pits, clearings (pocosin), depressions (cedar, cypress, desiccated, ephemeral, limesink, moist, open), ditches (moist, pine barren, roadside, wet), flats (cleared, cutover, pine), flatwoods (cutover, disturbed, hickory, mesic, oak, pine, scrubby, wet), floating mats, glades (marl), lagoons, marshes (basin, brackish, disturbed, flatwoods, freshwater, slough), meadows (depression, seepage, streamside, wet), mudflats, pocosins (seepage, streamside), potholes (moist, seasonal), prairies (marl, sand, wet), rights of way (powerline, seasonally wet), riverbeds (dry, sandy), roadbeds (drying), roadsides (disturbed), sand bars, sand pits (abandoned), savannas (pine, wet), seepages (acidic, hillside, roadside, sandy, *Sphagnum*, spring-fed), slopes [to 5%] (seepage), swales (interdunal, intradunal, sandy, sphagnous, springy), and the margins (desiccated, ephemeral) or shores of canals, Carolina bays (disturbed), lakes (drawn down, interdunal, marshy), ponds (artificial, beaver, burned, depression, farm, flatwoods, grassy, limesink, mill, pineland, sand pit), pools (sandy, shallow, stream channel), reservoirs (unshaded), rivers, sloughs, streams, and swamps (boggy, cleared, cypress, floodplain) at elevations to 183 m. This species occurs almost exclusively in exposures of full sunlight but otherwise has a broad ecological amplitude and can persist on wet substrates to shallow standing water (e.g., 15 cm), particularly in artificially (e.g., mechanically) disturbed situations. It also tolerates a much wider range of substrates (especially heavier types) relative to its congeners; these can include acidic (e.g., pH: 3.8–5.9) to slightly acidic/calcareous alluvium, Basinger (spodic

psammaquents), Blanton (grossarenic paleudults), clay, cobbles, Doucette (arenic plinthic paleudults), Duckston (typic psammaquents), Felda (arenic ochraqualfs), Galveston (typic udipsamments), Gourdin (typic ochraquults), gravel, Hornsville (aquic hapladults), loam, loamy sand, muck, mucky sand, mud, Ortega (typic quartzipsamments), peat, peaty muck, Pelham (arenic paleaquults), Pineda/Riviera (arenic glossaqualfs), Plummer (grossarenic paleaquults), Pompano (typic psammaquents), Pottsburg (grossarenic haplaquods), rocks, Rutledge (typic humaquepts), sand (chalky, Duckston, Leon, moist, Queen City, Resota, Sparta, Waveland, wet), sandy clay, sandy loam, sandy muck, sandy mud, sandy peat, silt, silty clay loam, silty loam (Amy, Bonn), Tehran-Letney (grossarenic paleudults), and Waveland (arenic haplaquods). The plants can develop from seed to flower in as short as 4 months time. Flowering and fruiting occur primarily from March through November, but can extend year-round in more southern localities. The flowers open during morning (e.g., 9:30–10:30 am) and close by early afternoon. A seed bank of uncertain duration develops. In South America, the seeds have germinated in fecal samples taken from wild tapirs (Mammalia: Tapiridae: *Tapirus terrestris*), which demonstrates the potential of these small seeds ($\bar{x} = 0.010$ mg) to be dispersed endozoically by mammalian herbivores. **Reported associates:** *Acer rubrum, Agalinis fasciculata, Agalinis linifolia, Alnus serrulata, Andropogon glomeratus, Andropogon virginicus, Anthaenantia, Aristida patula, Aristida stricta, Aronia arbutifolia, Asclepias hirtella, Asclepias lanceolata, Axonopus furcatus, Baccharis halimifolia, Bacopa caroliniana, Bacopa monnieri, Balduina uniflora, Bartonia paniculata, Berchemia scandens, Betula nigra, Bidens mitis, Bigelowia nudata, Boehmeria cylindrica, Boltonia diffusa, Brasenia schreberi, Buchnera americana, Bulbostylis ciliatifolia, Calamovilfa arcuata, Calamovilfa curtissii, Carex atlantica, Carex canescens Carex glaucescens, Carex joorii, Carex lupulina, Carphephorus odoratissimus, Carphephorus pseudoliatris, Centella asiatica, Cephalanthus occidentalis, Chamaecyparis thyoides, Chamaedaphne calyculata, Cirsium nuttallii, Cladium jamaicense, Clethra alnifolia, Coleataenia longifolia, Coleataenia tenera, Commelina diffusa, Commelina virginica, Conoclinium coelestinum, Coreopsis tripteris, Cornus amomum, Crataegus, Croton glandulosus, Ctenium aromaticum, Cyperus erythrorhizos, Cyperus haspan, Cyperus ochraceus, Cyperus oxylepis, Cyperus polystachyos, Cyperus pseudovegetus, Cyperus strigosus, Cyrilla racemiflora, Decodon verticillatus, Dichanthelium acuminatum, Dichanthelium dichotomum, Dichanthelium scoparium, Dichanthelium sphaerocarpon, Dichanthelium strigosum, Diodella teres, Diodia virginiana, Diospyros, Drosera brevifolia, Drosera capillaris, Drosera filiformis, Drosera intermedia, Dulichium arundinaceum, Echinochloa walteri, Eleocharis baldwinii, Eleocharis equisetoides, Eleocharis flavescens, Eleocharis geniculata, Eleocharis melanocarpa, Eleocharis microcarpa, Eleocharis montevidensis, Eleocharis obtusa, Eleocharis ovata, Eleocharis parvula, Eleocharis quadrangulata, Eleocharis tortilis, Eleocharis tricostata, Eleocharis tuberculosa,*

Elephantopus carolinianus, Eragrostis lugens, Eriocaulon compressum, Eriocaulon decangulare, Eryngium prostratum, Eryngium yuccifolium, Eupatorium capillifolium, Eupatorium compositifolium, Eupatorium hyssopifolium, Eupatorium leptophyllum, Eupatorium linearifolium, Eupatorium mohrii, Eupatorium perfoliatum, Eupatorium rotundifolium, Eupatorium serotinum, Eustachys glauca, Euthamia graminifolia, Fagus, Fimbristylis autumnalis, Fimbristylis caroliniana, Fimbristylis littoralis, Flaveria linearis, Fraxinus caroliniana, Fuirena breviseta, Fuirena bushii, Fuirena longa, Fuirena scirpoidea, Fuirena squarrosa, Gaylussacia dumosa, Glyceria obtusa, Gratiola brevifolia, Gratiola pilosa, Habenaria repens, Harperella nodosa, Helanthium tenellum, Helenium flexuosum, Helenium pinnatifidum, Helianthus angustifolius, Hydrocotyle ranunculoides, Hydrocotyle umbellata, Hydrocotyle verticillata, Hydrolea ovata, Hypericum cistifolium, Hypericum crux-andreae, Hypericum fasciculatum, Hypericum gentianoides, Hypericum gymnanthum, Hypericum hypericoides, Hypericum mutilum, Hypericum setosum, Hyptis alata, Ilex decidua, Ilex glabra, Ionactis linariifolia, Iris tridentata, Isoetes, Itea virginica, Iva microcephala, Juncus acuminatus, Juncus brachycarpus, Juncus bufonius, Juncus canadensis, Juncus coriaceus, Juncus dichotomus, Juncus diffusissimus, Juncus effusus, Juncus interior, Juncus pelocarpus, Juncus repens, Juncus scirpoides, Juncus validus, Justicia americana, Kalmia hirsuta, Kyllinga odorata, Lachnanthes caroliniana, Lachnocaulon anceps, Lechea, Leersia virginica, Liatris spicata, Lindernia dubia, Linum striatum, Lipocarpha micrantha, Liquidambar styraciflua, Liriodendron tulipifera, Lobelia brevifolia, Lobelia cardinalis, Lobelia glandulosa, Lobelia puberula, Lophiola aurea, Ludwigia alternifolia, Ludwigia arcuata, Ludwigia decurrens, Ludwigia linearis, Ludwigia microcarpa, Ludwigia octovalvis, Ludwigia palustris, Ludwigia pilosa, Ludwigia sphaerocarpa, Luziola fluitans, Lycopodiella appressa, Lycopus rubellus, Lyonia lucida, Lythrum alatum, Magnolia virginiana, Mayaca fluviatilis, Mikania scandens, Mitchella repens, Mitreola petiolata, Mitreola sessilifolia, Mnesithea rugosa, Muhlenbergia capillaris, Murdannia keisak, Myrica cerifera, Myrica inodora, Nelumbo lutea, Nuphar advena, Nymphaea odorata, Nymphoides cordata, Nyssa biflora, Nyssa sylvatica, Oldenlandia boscii, Oldenlandia uniflora, Onoclea sensibilis, Osmunda regalis, Osmundastrum cinnamomeum, Oxydendrum arboreum, Panicum amarum, Panicum hemitomon, Panicum verrucosum, Panicum virgatum, Paspalum plicatulum, Paspalum praecox, Paspalum urvillei, Persicaria hydropiperoides, Persicaria punctata, Phyla nodiflora, Pinus elliottii, Pinus palustris, Pinus serotina, Pinus taeda, Pityopsis graminifolia, Platanthera blephariglottis, Pluchea baccharis, Pluchea camphorata, Pluchea odorata, Polygala cruciata, Polygala cymosa, Polygala lutea, Polygala rugelii, Polypremum procumbens, Pontederia cordata, Populus deltoides, Proserpinaca pectinata, Quercus geminata, Quercus laurifolia, Quercus minima, Quercus nigra, Quercus velutina, Rhexia alifanus, Rhexia cubensis, Rhexia mariana, Rhexia nashii, Rhexia salicifolia, Rhexia virginica,

Rhynchospora alba, Rhynchospora caduca, Rhynchospora capitellata, Rhynchospora chalarocephala, Rhynchospora chapmanii, Rhynchospora colorata, Rhynchospora corniculata, Rhynchospora debilis, Rhynchospora divergens, Rhynchospora elliottii, Rhynchospora eximia, Rhynchospora fascicularis, Rhynchospora filifolia, Rhynchospora globularis, Rhynchospora glomerata, Rhynchospora gracilenta, Rhynchospora inexpansa, Rhynchospora inundata, Rhynchospora macrostachya, Rhynchospora microcarpa, Rhynchospora microcephala, Rhynchospora nitens, Rhynchospora oligantha, Rhynchospora perplexa, Rhynchospora pleiantha, Rhynchospora rariflora, Rhynchospora scirpoides, Rhynchospora tracyi, Rotala ramosior, Rubus trivialis, Rudbeckia mohrii, Rudbeckia subtomentosa, Sabal palmetto, Sabal palmetto, Sabatia brevifolia, Sabatia difformis, Sabatia grandiflora, Saccharum giganteum, Sacciolepis striata, Sagittaria graminea, Sagittaria lancifolia, Sagittaria papillosa, Salix caroliniana, Salix nigra, Sarracenia alata, Sarracenia flava, Sarracenia purpurea, Schizachyrium scoparium, Schizachyrium tenerum, Schoenoplectus hallii, Schoenoplectus pungens, Scirpus cyperinus, Scleria georgiana, Scleria muehlenbergii, Scleria pauciflora, Scleria reticularis, Serenoa repens, Sesbania drummondii, Sesbania herbacea, Sesbania punicea, Setaria magna, Sideroxylon reclinatum, Smilax auriculata, Smilax glauca, Smilax laurifolia, Solidago pulchra, Sparganium americanum, Spartina patens, Spartina bakeri, Sphagnum recurvum, Sphagnum rubellum, Spiranthes vernalis, Steinchisma hians, Stillingia aquatica, Syngonanthus flavidulus, Taxodium ascendens, Taxodium distichum, Thyrsanthella difformis, Tiedemannia filiformis, Toxicodendron radicans, Triadenum virginicum, Triadenum walteri, Typha, Utricularia cornuta, Utricularia gibba, Utricularia juncea, Utricularia subulata, Vaccinium arboreum, Vaccinium macrocarpon, Vaccinium myrsinites, Verbena, Vernonia lettermannii, Vigna luteola, Viola lanceolata, Vitis rotundifolia, Woodwardia areolata, Woodwardia virginica, Xyris ambigua, Xyris baldwiniana, Xyris brevifolia, Xyris caroliniana, Xyris difformis, Xyris elliottii, Xyris fimbriata, Xyris isoetifolia, Xyris laxifolia, Xyris longisepala, Xyris platylepis, Xyris serotina, Xyris smalliana, Xyris stricta, Xyris torta.

10.4.1.11. ***Xyris laxifolia* Mart.** is a perennial, which occurs in bogs (open, roadside, sandy, seepage, *Sphagnum*), bottomlands (boggy, deciduous), depressions (muddy, wet), ditch banks, ditches (drainage, roadside, wet), drains (gravel pit), flatwoods (pine), lakes (shallow), marshes (beaver, depression, floating, freshwater, open), meadows (seepage slope), mudflats (drying, stream bottom), pinelands (marshy), ponds (coastal plain, ephemeral, flatwoods, shallow), pools (pineland, river bottom, roadside), rights of way (powerline, wet), roadsides (wet), savannas (cutover, cypress, mesic, pine, wet), seeps (acid, bayhead, forested, gravelly, sandy), shores, slopes [to 5%], spring runs, swales, swamps (cutover, low, pineland), and along the margins (peaty) of baygalls (impounded, spring-fed), borrow pits (ponded), canals, drains, gravel pits, lakes, ponds (beaver, shallow), pools (railroad), rivers, streams (blackwater, seepage, spring-fed), and thickets at elevations to 432 m. The plants occur in exposures that receive full sunlight. They can grow in shallow water (e.g., 25–31 cm) and are especially common in wet, disturbed areas. The substrates are acidic (e.g., pH: 4.3–4.5; Ca: 62–113 ppm; Cu: 0.4–0.7 ppm; Mg: 23–36 ppm; P: 2.9–6.1 ppm; K: 12–17 ppm; Na: 6.5–9.9 ppm; S: 9.1–11.9 ppm; Zn: 0.5–0.9 ppm) and include alluvium (clayey), Bibb (typic fluvaquents), clay loam, cobbles, gravel, Kinston (typic fluvaquents), Melhomes (humqueptic psammaquents), muck (Kenner, organic), mud, peat, peaty sand, Rutledge (typic humapuepts), sand (fine, moist, Queen City, Sparta), sandy clay loam, sandy loam (fine, Harleston, Lynchburg, wet), sandy peat, silt (fine), and silty loam (Atmore). Flowering and fruiting occur from summer (June) through fall (November). The flowers of the South American variety (*X. laxifolia* var. *delta*) are self-compatible and are highly autogamous (up to 84% fruit set in self-pollinated flowers). The flowers open during morning hours and are pollinated by bees (Insecta: Hymenoptera: Halictidae) and flies (Insecta: Diptera: Syrphidae). The cespitose plants perennate by means of paired lateral buds at the base of their compacted stem. Annual biomass production can reach 5.4 g m^{-2} year^{-1} ($\bar{x}$ = 0.5 g m^{-2} year^{-1}). **Reported associates:** *Acer rubrum, Agalinis maritima, Ageratina altissima, Alnus serrulata, Alternanthera philoxeroides, Amaranthus australis, Amsonia tabernaemontana, Andropogon glomeratus, Andropogon virginicus, Apios americana, Aristida palustris, Asemeia grandiflora, Bacopa caroliniana, Carex atlantica, Carex glaucescens, Carex joorii, Carex leptalea, Carex lupulina, Carex striata, Carex stricta, Carex verrucosa, Cephalanthus occidentalis, Cladium jamaicense, Coleataenia tenera, Conoclinium coelestinum, Crataegus opaca, Cyperus flavescens, Cyperus haspan, Cyperus polystachyos, Cyrilla racemiflora, Decodon verticillatus, Dichanthelium dichotomum, Dichanthelium scabriusculum, Dichanthelium scoparium, Diodia virginiana, Diospyros virginiana, Drosera, Dulichium arundinaceum, Eleocharis cellulosa, Eleocharis equisetoides, Eleocharis microcarpa, Eleocharis quadrangulata, Eleocharis rostellata, Eleocharis tortilis, Eleocharis tuberculosa, Eriocaulon compressum, Eupatorium leucolepis, Eupatorium perfoliatum, Eupatorium semiserratum, Fuirena breviseta, Gratiola brevifolia, Gratiola pilosa, Harperella nodosa, Hibiscus dasycalyx, Hibiscus moscheutos, Hydrocotyle, Hydrolea ovata, Hymenocallis liriosme, Hypericum crux-andreae, Hypericum fasciculatum, Hypericum galioides, Hypericum hypericoides, Hypericum mutilum, Hyptis alata, Ilex verticillata, Ipomoea sagittata, Juncus acuminatus, Juncus marginatus, Juncus nodatus, Juncus validus, Justicia americana, Justicia ovata, Kosteletzkya pentacarpos, Kyllinga odorata, Lachnanthes caroliniana, Leersia hexandra, Leersia oryzoides, Lemna minor, Leptochloa fusca, Liquidambar styraciflua, Lobelia canbyi, Ludwigia glandulosa, Ludwigia leptocarpa, Ludwigia pilosa, Ludwigia sphaerocarpa, Lycopus americanus, Lyonia ligustrina, Lythrum lineare, Magnolia virginiana, Mnesithea rugosa, Mayaca fluviatilis, Mikania scandens, Mnesithea rugosa, Mnesithea tuberculosa, Myrica cerifera, Myrica heterophylla, Myriophyllum pinnatum, Nymphaea odorata, Nyssa aquatica, Nyssa biflora, Orontium aquaticum, Osmunda regalis, Osmundastrum cinnamomeum, Panicum hemitomon, Panicum virgatum, Paspalum floridanum, Paspalum modestum, Paspalum praecox, Paspalum vaginatum, Persicaria hydropiperoides, Persicaria punctata, Persicaria sagittata, Pinus palustris, Pinus taeda, Platanthera clavellata, Pluchea baccharis, Pluchea foetida, Polypremum procumbens, Pontederia cordata, Proserpinaca palustris, Proserpinaca pectinata, Quercus, Rhexia mariana, Rhynchospora caduca, Rhynchospora capitellata, Rhynchospora cephalantha, Rhynchospora corniculata, Rhynchospora filifolia, Rhynchospora globularis, Rhynchospora glomerata, Rhynchospora inexpansa, Rhynchospora macrostachya, Rhynchospora microcephala, Rhynchospora perplexa, Rhynchospora rariflora, Rhynchospora tracyi, Rubus, Sabal minor, Sacciolepis striata, Sagittaria graminea. Sagittaria lancifolia, Sagittaria latifolia, Sarracenia, Scirpus cyperinus, Saururus cernuus, Scleria baldwinii, Smilax laurifolia, Solidago fistulosa, Solidago sempervirens, Sphagnum macrophyllum, Spiranthes cernua, Spirodela polyrrhiza, Styrax americanus, Taxodium ascendens, Taxodium distichum, Thelypteris kunthii, Thelypteris palustris, Toxicodendron vernix, Triadenum virginicum, Triadenum walteri, Triadica sebifera, Utricularia gibba, Utricularia juncea, Utricularia purpurea, Viburnum nudum, Vigna luteola, Woodwardia areolata, Woodwardia virginica, Xyris ambigua, Xyris difformis, Xyris fimbriata, Xyris jupicai, Xyris smalliana, Xyris stricta, Xyris torta.*

10.4.1.12. ***Xyris longisepala* Kral** is a rare perennial, which occurs in flatwoods (moist, wet), swales (pine hill, sandy), and along the margins (fluctuating) or shores (moist, sandy, wet) of borrow pits, dolines/sinkholes, lakes (sandhill), and ponds (depression, karst, limesink, sandhill) at elevations to 46 m. The site exposures are open. The substrates have been characterized as Chipley (aquic quartzipsamments), Lakeland, loamy sand, muck (loosely consolidated), Ortega (typic quartzipsamments), peaty sand, sand (coarse, moist, wet), sandy peat, and Troup (grossarenic paleudults). Flowering and fruiting extend from summer (July) through fall (October). The flowers do not open until noon or later in the afternoon. Documented pollinators include bees (Insecta: Hymenoptera: Apidae: *Bombus*; Halictidae: *Agapostemon, Lasioglossum zephyrum*) and flies (Insecta: Diptera: Syrphidae: *Ocyptamus fuscipennis, Toxomerus boscii, T. geminatus*). The seeds can germinate profusely along shorelines where the waters have receded. These cespitose plants perennate by the paired lateral buds on their compacted stem base. They can develop into extensive stands where conditions are optimal. **Reported associates:** *Amphicarpum muhlenbergianum, Bulbostylis ciliatifolia, Centella asiatica, Dichanthelium dichotomum, Drosera filiformis, Eleocharis melanocarpa, Eriocaulon lineare, Eupatorium leptophyllum, Fuirena pumila, Hypericum fasciculatum, Hypericum lissophloeus, Hypericum tenuifolium, Juncus pelocarpus, Juncus repens, Lachnocaulon minus, Ludwigia suffruticosa, Mayaca fluviatilis, Panicum hemitomon, Panicum virgatum, Pinus palustris, Rhexia* salicifolia, Rhynchospora nitens, Rhynchospora pleiantha, Sacciolepis striata, Sagittaria isoetiformis, Scleria reticularis, Syngonanthus flavidulus, Utricularia cornuta, Utricularia floridana, Utricularia juncea, Xyris isoetifolia, Xyris jupicai, Xyris smalliana.

10.4.1.13. **Xyris montana Ries** is a perennial, which inhabits bogs (kettle-hole, peat, powerline right-of-way, *Sphagnum*), bog mats (floating), borrow pits (sandy), depressions (moist, eat), patterned fens [in flarks and pools], poor fens, seeps (acid), swamps (tamarack), and the shores (gently sloping, gneiss, sphagnous) of lakes (glacial, oligotrophic, sphagnous), muskeg, rivers, sloughs (peaty), and streams at elevations to 396 m. The plants occur most commonly in open, coldwater, *Sphagnum*-dominated communities. The substrates are acidic (e.g., pH: 3.7–6.7; Ca^{2+}: <1.6 mg l^{-1}; 1.2 µeq l^{-1}; total hardness: <4 mg l^{-1} $CaCO_3$; total alkalinity: 1 ppm; specific conductance: 66–78 µs cm^{-1}) and include gravel, muck, mucky peat, mud (wet), sand, and peat (*Sphagnum*). Flowering has been observed in winter (February), but flowering and fruiting typically occur from summer (June) through fall (October). The flowers open during the morning; observed seed set reportedly is quite low. The cespitose shoots perennate by the production of short rhizomes. **Reported associates:** *Acer rubrum, Agalinis purpurea, Alnus rugosa, Andromeda polifolia, Arethusa bulbosa, Aronia melanocarpa, Aronia ×prunifolia, Bartonia paniculata, Betula populifolia, Calamagrostis stricta, Calla palustris, Calopogon tuberosus, Carex atlantica, Carex buxbaumii, Carex canescens, Carex chordorrhiza, Carex echinata, Carex exilis, Carex interior, Carex intumescens, Carex lasiocarpa, Carex lenticularis, Carex leptalea, Carex limosa, Carex livida, Carex lurida, Carex pauciflora, Carex rostrata, Carex stricta, Carex tenuiflora, Carex trisperma, Carex utriculata, Cephalanthus occidentalis, Chamaecyparis thyoides, Chamaedaphne calyculata, Cicuta bulbifera, Cladium mariscoides, Cladopodiella fluitans, Comarum palustre, Decodon verticillatus, Drosera anglica, Drosera intermedia, Drosera linearis, Drosera rotundifolia, Dulichium arundinaceum, Eleocharis compressa, Eleocharis flavescens, Equisetum fluviatile, Eriocaulon aquaticum, Eriophorum angustifolium, Eriophorum gracile, Eriophorum tenellum, Eriophorum vaginatum, Eriophorum virginicum, Eriophorum viridicarinatum, Galium labradoricum, Gaylussacia baccata, Gaylussacia frondosa, Ilex verticillata, Iris versicolor, Juncus canadensis, Juncus filiformis, Juncus pelocarpus, Juncus stygius, Kalmia angustifolia, Kalmia polifolia, Larix laricina, Lycopodiella inundata, Lycopus uniflorus, Lysimachia terrestris, Maianthemum trifolium, Menyanthes trifoliata, Muhlenbergia uniflora, Myrica gale, Nymphaea odorata, Oclemena nemoralis, Onoclea sensibilis, Peltandra virginica, Phragmites australis, Picea mariana, Pinus rigida, Pinus strobus, Platanthera clavellata, Pogonia ophioglossoides, Potamogeton gramineus, Rhododendron viscosum, Rhynchospora alba, Rhynchospora fusca, Sarracenia purpurea, Scheuchzeria palustris, Schizaea pusilla, Schoenoplectus subterminalis, Schoenoplectus tabernaemontani, Scutellaria lateriflora, Sphagnum cuspidatum, Spiraea alba, Spiraea tomentosa,* Thelypteris palustris, Toxicodendron vernix, Triadenum virginicum, Trichophorum alpinum, Trichophorum cespitosum, Triglochin maritima, Tsuga canadensis, Typha latifolia, Utricularia cornuta, Utricularia intermedia, Utricularia minor, Utricularia radiata, Utricularia subulata, Vaccinium corymbosum, Vaccinium macrocarpon, Vaccinium oxycoccos, Viola lanceolata, Woodwardia virginica, Xyris difformis, Xyris torta.

10.4.1.14. **Xyris platylepis Chapm.** is a perennial, which grows in bayheads (cypress), bogs (blanket, cutover, granitic, hillside, recently burned, seepage), borrow pits (roadside), bottomlands (desiccated lake), boggy, hardwood), depressions (flatwoods), ditches (drainage, flatwood, roadside, wet), ditch banks, flatwoods (disturbed, low, mesic, moist, pine), hammocks (hydric, leached, low, xeric), marshes (depression), meadows (low, wet), pinelands (low, moist, flat, wet), pocosins (remnant, sandhill), pools (pineland, temporary), prairies (dry, mesic, recently burned, wet), roadsides (flatwoods, seepy, wet), savannas (mesic, pine), seeps (roadside), seepage slopes (annually burned, open), swales (pineland, sphagnous, springy, wet), swamps (alluvial, cutover, cypress-bay, hardwood, pine), and along the margins (disturbed, sandy, seepy) of marshes, ponds (ephemeral, flatwoods, pineland, sink), pools and swamps (cypress) at elevations to 193 m. The plants can occur in open to shaded exposures but seldom are found at inundated sites; occasionally they grow in shallow waters of temporary pools. The substrates are moist to wet, acidic, coarse-textured, and include Alapaha (arenic plinthic paleaquults), Basinger (spodic psammaquents), Bibb-Osier, clayey sand, Fuquay (plinthic paleudults), Immokalee (arenic haplaquods), Kitsatchie-Rayburn (typic/vertic hapludalfs), loam, Lynchburg (aeric paleaquults), Malabar (grossarenic ochraqualfs), Meadowbrook (grossarenic ochraqualfs), Myakka (aeric haplaquods), mud, Ona (typic haplaquods), peaty sand, Pamlico (terric medisaprists), Placid (typic humaquepts), Plummer (grossarenic paleaquults), Pomona (ultic haplaquods), Pompano (typic psammaquents), Pottsburg (grossarenic haplaquods), Rosenwall (aquic hapludults), sand, sandy clay (organic, wet), Samsula (terric medisaprists), sandy peat, silty loam (Atmore), Smyrna (aeric haplaquods), St. Johns (typic haplaquods), Surrency (arenic umbric paleaquults), Wahee (aeric ochraquults), and Waveland (arenic haplaquods). Flowering and fruiting occur from spring (April) to fall (December), but can continue all year in the south. The flowers open during the afternoon (e.g., 1:30 pm) and can possess either white or yellow petals. Seeds planted in the fall are capable of developing into flowering plants by the end of the ensuing growing season. Perennation occurs by paired, lateral buds arising from their bulbous, thickened, and mucilaginous base. The cespitose plants can develop into a dense turf where suitable conditions prevail. **Reported associates:** *Acer rubrum, Aletris lutea, Alnus serrulata, Amelanchier, Amorpha paniculata, Amphicarpum muhlenbergianum, Andropogon arctatus, Andropogon glomeratus, Andropogon gyrans, Andropogon liebmannii, Andropogon virginicus, Apios americana, Aristida stricta, Arnoglossum ovatum, Aronia arbutifolia, Arundinaria* tecta, *Balduina uniflora*, *Bignonia capreolata*, *Burmannia capitata*, *Calamagrostis canadensis*, *Carex austrodeflexa*, *Carex glaucescens*, *Carya floridana*, *Castanea pumila*, *Centella asiatica*, *Chamaecyparis thyoides*, *Chaptalia tomentosa*, *Cirsium nuttallii*, *Clethra alnifolia*, *Coleataenia longifolia*, *Coreopsis gladiata*, *Coreopsis major*, *Coreopsis tripteris*, *Croton elliottii*, *Croton michauxii*, *Ctenium aromaticum*, *Cyperus*, *Danthonia sericea*, *Dennstaedtia punctilobula*, *Dichanthelium aciculare*, *Dichanthelium acuminatum*, *Dichanthelium dichotomum*, *Dichanthelium ensifolium*, *Dichanthelium sphaerocarpon*, *Dichanthelium strigosum*, *Doellingeria umbellata*, *Drosera capillaris*, *Drosera filiformis*, *Drosera tracyi*, *Eleocharis tuberculosa*, *Elephantopus elatus*, *Elionurus tripsacoides*, *Erigeron vernus*, *Eriocaulon compressum*, *Eriocaulon decangulare*, *Eriocaulon ravenelii*, *Eriocaulon ravenelii*, *Eriocaulon texense*, *Eryngium integrifolium*, *Eryngium yuccifolium*, *Eupatorium capillifolium*, *Eupatorium mohrii*, *Eupatorium pilosum*, *Eupatorium rotundifolium*, *Euphorbia inundata*, *Euthamia graminifolia*, *Eutrochium*, *Fimbristylis autumnalis*, *Fuirena breviseta*, *Fuirena scirpoidea*, *Gaylussacia nana*, *Gordonia lasianthus*, *Gratiola pilosa*, *Gratiola ramosa*, *Gymnopogon brevifolius*, *Helenium brevifolium*, *Helianthus angustifolius*, *Hymenocallis palmeri*, *Hypericum brachyphyllum*, *Hypericum brachyphyllum*, *Hypericum cistifolium*, *Hypericum densiflorum*, *Hypericum fasciculatum*, *Hypericum hypericoides*, *Hypericum myrtifolium*, *Hyptis alata*, *Ilex coriacea*, *Ilex glabra*, *Ilex vomitoria*, *Iva microcephala*, *Juncus scirpoides*, *Kalmia latifolia*, *Lachnanthes caroliniana*, *Lachnocaulon beyrichianum*, *Lespedeza capitata*, *Liatris spicata*, *Liatris spicata*, *Lilium catesbaei*, *Linum floridanum*, *Linum medium*, *Liquidambar styraciflua*, *Liriodendron tulipifera*, *Lobelia elongata*, *Lobelia glandulosa*, *Lobelia nuttallii*, *Lobelia paludosa*, *Lobelia reverchonii*, *Ludwigia alternifolia*, *Ludwigia maritima*, *Ludwigia microcarpa*, *Ludwigia spathulata*, *Lyonia fruticosa*, *Lyonia ligustrina*, *Lyonia lucida*, *Magnolia virginiana*, *Muhlenbergia capillaris*, *Myrica heterophylla*, *Nyssa sylvatica*, *Oldenlandia boscii*, *Osmunda regalis*, *Osmundastrum cinnamomeum*, *Oxydendrum arboreum*, *Oxypolis rigidior*, *Panicum virgatum*, *Paspalum praecox*, *Persea palustris*, *Physostegia purpurea*, *Pinguicula planifolia*, *Pinus clausa*, *Pinus echinata*, *Pinus elliottii*, *Pinus palustris*, *Pinus serotina*, *Pinus taeda*, *Pinus virginiana*, *Pleea tenuifolia*, *Pluchea baccharis*, *Polygala cruciata*, *Polygala lutea*, *Polygala nana*, *Polygala rugelii*, *Pontederia cordata*, *Potentilla canadensis*, *Prunus serotina*, *Ptilimnium costatum*, *Quercus geminata*, *Quercus minima*, *Quercus virginiana*, *Rhexia mariana*, *Rhexia petiolata*, *Rhexia virginica*, *Rhododendron canescens*, *Rhododendron viscosum*, *Rhus copallinum*, *Rhynchospora capitellata*, *Rhynchospora cephalantha*, *Rhynchospora chalarocephala*, *Rhynchospora chapmanii*, *Rhynchospora filifolia*, *Rhynchospora gracilenta*, *Rhynchospora latifolia*, *Rhynchospora oligantha*, *Rhynchospora pallida*, *Rhynchospora plumosa*, *Rhynchospora rariflora*, *Rhynchospora recognita*, *Rhynchospora stenophylla*, *Rubus*, *Rudbeckia scabrifolia*, *Sabal palmetto*, *Sabatia brevifolia*,

Sabatia brevifolia, *Sabatia grandiflora*, *Sacciolepis indica*, *Salix nigra*, *Sarracenia jonesii*, *Sarracenia psittacina*, *Sarracenia purpurea*, *Sarracenia rubra*, *Schizachyrium rhizomatum*, *Schizachyrium scoparium*, *Scleria hirtella*, *Scleria oligantha*, *Scleria reticularis*, *Scleria triglomerata*, *Selaginella apoda*, *Serenoa repens*, *Sericocarpus linifolius*, *Smilax glauca*, *Smilax laurifolia*, *Solidago fistulosa*, *Solidago fistulosa*, *Solidago patula*, *Sphagnum*, *Sporobolus curtissii*, *Symphyotrichum dumosum*, *Syngonanthus flavidulus*, *Tephrosia onobrychoides*, *Tiedemannia filiformis*, *Tofieldia glabra*, *Utricularia subulata*, *Vaccinium corymbosum*, *Vaccinium darrowii*, *Viburnum nudum*, *Viburnum obovatum*, *Viola lanceolata*, *Viola primulifolia*, *Vitis rotundifolia*, *Woodwardia areolata*, *Woodwardia virginica*, *Xanthorhiza simplicissima*, *Xyris ambigua*, *Xyris brevifolia*, *Xyris difformis*, *Xyris fimbriata*, *Xyris flabelliformis*, *Xyris jupicai*, *Xyris scabrifolia*, *Xyris torta*.

10.4.1.15. *Xyris scabrifolia* **R.M. Harper** is a perennial, which grows in baygalls, bogs (hillside, open, pineland, seepage, *Sphagnum*, streamhead), ditches (roadside), ecotones (pocosin, sphagnous), ponds (cypress-gum, karst), prairies (wet), savannas (wet), seeps (sandhill), and slopes (boggy, pineland, sandy, seepage) at elevations to 236 m. The plants occur in open exposures to light shade. They grow on acidic, heavier, finer-textured substrates, which have been characterized as Doucette (arenic plinthic paleudults), Escambia (plinthaquic paleudults), Florala (plinthaquic paleudults), Fuquay (plinthic paleudults), Johns (aquic hapludults), Leefield-Stilson complex, Leon (aeric haplaquods), Lynchburg (aeric paleaquults), Melholmes (humaqueptic psammaquents), muck (seepy), Pamlico (terric hedisaprists), Pantego (umbric paleaquults), Plummer (grossarenic paleaquults), Rains (typic paleaquults), Rutledge (typic humaquepts), sand, sandy peat (moist, wet), Scranton (humaqueptic psammaquents), Stringtown (typic hapludults), and Tehran (arenic paleudults). Flowering and fruiting extend from summer (July) through the fall (October). The flowers begin to open in the afternoon. Seeds collected in January and planted in March reached floral maturity by midsummer of the same year. The plants are solitary or in small tufts, perennating from elongate, bulblike lateral buds, which arise under thickened leaf bases from a compact, bulbous stem. They respond positively to fires, with increased abundance being observed in the aftermath years. **Reported associates:** *Acer rubrum*, *Agalinis aphylla*, *Agalinis linifolia*, *Agalinis purpurea*, *Aletris aurea*, *Andropogon arctatus*, *Andropogon glomeratus*, *Andropogon gyrans*, *Anthaenantia rufa*, *Apteria aphylla*, *Aristida palustris*, *Aristida purpurascens*, *Aristida simpliciflora*, *Aristida stricta*, *Arnoglossum ovatum*, *Aronia arbutifolia*, *Arundinaria tecta*, *Asclepias lanceolata*, *Asclepias rubra*, *Balduina uniflora*, *Bartonia paniculata*, *Bigelowia nudata*, *Burmannia capitata*, *Calopogon barbatus*, *Calopogon oklahomensis*, *Calopogon tuberosus*, *Carex exilis*, *Carex glaucescens*, *Carex lutea*, *Carphephorus pseudoliatris*, *Centella asiatica*, *Chaptalia tomentosa*, *Chasmanthium laxum*, *Cleistesiopsis bifaria*, *Coleataenia longifolia*, *Coleataenia tenera*, *Coreopsis gladiata*, *Croton michauxii*, *Ctenium aromaticum*, *Dichanthelium* acuminatum, Dichanthelium dichotomum, Dichanthelium ensifolium, Dichanthelium nudicaule, Dichanthelium scabriusculum, Drosera capillaris, Drosera tracyi, Eleocharis tuberculosa, Eragrostis refracta, Eriocaulon compressum, Eriocaulon decangulare, Eriocaulon texense, Eryngium integrifolium, Eupatorium leucolepis, Eupatorium pilosum, Eupatorium rotundifolium, Fuirena squarrosa, Gelsemium sempervirens, Gratiola pilosa, Helianthus angustifolius, Helianthus heterophyllus, Hypericum brachyphyllum, Hypericum crux-andreae, Hypericum galioides, Hypericum hypericoides, Hypoxis rigida, Ilex coriacea, Ilex glabra, Ilex myrtifolia, Juncus scirpoides, Juncus trigonocarpus, Lachnocaulon anceps, Lachnocaulon digynum, Liatris pycnostachya, Liatris spicata, Liquidambar styraciflua, Lobelia reverchonii, Lophiola aurea, Ludwigia hirtella, Lycopodiella appressa, Lycopodiella caroliniana, Lycopodiella prostrata, Lyonia ligustrina, Macranthera flammea, Magnolia virginiana, Marshallia graminifolia, Melanthium virginicum, Mitchella repens, Mitreola sessilifolia, Muhlenbergia capillaris, Myrica cerifera, Myrica heterophylla, Nyssa sylvatica, Osmunda regalis, Osmundastrum cinnamomeum, Oxypolis rigidior, Panicum verrucosum, Panicum virgatum, Parnassia caroliniana, Parnassia grandifolia, Paspalum floridanum, Persea borbonia, Persea palustris, Pinus elliottii, Pinus palustris, Pinus taeda, Pityopsis graminifolia, Platanthera blephariglottis, Platanthera ciliaris, Platanthera integra, Pleea tenuifolia, Pogonia ophioglossoides, Polygala cruciata, Polygala nana, Polygala ramosa, Pteridium aquilinum, Ptilimnium capillaceum, Ptilimnium costatum, Rhexia alifanus, Rhexia lutea, Rhexia mariana, Rhexia petiolata, Rhododendron canescens, Rhynchospora capitellata, Rhynchospora cephalantha, Rhynchospora chalarocephala, Rhynchospora gracilenta, Rhynchospora macra, Rhynchospora oligantha, Rhynchospora plumosa, Rhynchospora rariflora, Rhynchospora stenophylla, Rudbeckia scabrifolia, Sabatia gentianoides, Sabatia macrophylla, Sarracenia alata, Sarracenia flava, Sarracenia leucophylla, Sarracenia psittacina, Sarracenia ✕catesbaei, Schizachyrium scoparium, Schizachyrium tenerum, Scleria baldwinii, Scleria reticularis, Scleria triglomerata, Scutellaria integrifolia, Smilax laurifolia, Smilax walteri, Solidago pulchra, Solidago rugosa, Sphagnum, Spiranthes cernua, Sporobolus teretifolius, Symphyotrichum dumosum, Symphyotrichum lateriflorum, Taxodium ascendens, Thalictrum cooleyi, Tiedemannia filiformis, Toxicodendron vernix, Triantha racemosa, Tridens ambiguus, Utricularia cornuta, Utricularia gibba, Utricularia juncea, Utricularia subulata, Vaccinium corymbosum, Viburnum nudum, Viola primulifolia, Woodwardia virginica, Xyris ambigua, Xyris baldwiniana, Xyris difformis, Xyris drummondii, Xyris fimbriata, Xyris platylepis, Xyris serotina, Xyris torta.

10.4.1.16. *Xyris serotina* Chapm. is a short-lived perennial, which inhabits bogs (seepage), cypress domes, depressions (flatwoods), ditches (roadside), flatwoods (cypress, mesic, wet), lakebeds (receded), marshes (burned, depression, fresh, roadside), prairies (burned, wet), savannas (cutover, depression, mucky, pine, roadside, wet), seepage slopes (hillside),

swales, swamps (pine barren), and the margins (desiccated) of ponds (beaver, cypress, depression, marshy, pineland) and potholes at elevations to 70 m. The plants have been reported from open to deeply shaded exposures and occur often in shallow standing water. The substrates are uniformly wet and have been described as alluvium, Basinger (spodic psammaquents), Covxille (typic paleaquults), Cuthbert (typic hapludults), Evergreen-Leon (typic haplaquods), Lynchburg (aeric paleaquults), muck, Myakka (aeric haplaquods), Pamlico-Donovan (terric medisaprists), Pantego (umbric paleaquults), peat, peaty muck, Placid (typic humaquepts), Plummer (grossarenic paleaquults), Pomona (ultil haplaquods), Rains (typic paleaquults), Rutledge (typic humapuepts), sand, sandy clay, sandy clayey peat, sandy loam (fine), sandy muck, sandy peat, sandy peaty muck, Smyrna (aeric haplaquods), St. Johns (typic haplaquods), and Valkaria (spodic psammaquents). Flowering and fruiting are prevalent from summer (June) through the fall (October), but can occur throughout the year in the warmer south. The flowers are open only from late morning through early afternoon. Unlike many other congeners, the seedlings take more than a year before they will mature to flowering stage. The densely cespitose plants perennate by dormant lateral buds on their soft, compact shoots. They are found commonly in sites that are subjected frequently to fires. **Reported associates:** *Agalinis, Amphicarpum muhlenbergianum, Aristida stricta, Carex striata, Carphephorus carnosus, Clethra alnifolia, Ctenium aromaticum, Cyrilla racemiflora, Dichanthelium, Drosera capillaris, Eriocaulon decangulare, Eriocaulon nigrobracteatum, Hypericum brachyphyllum, Hypericum chapmanii, Hypericum fasciculatum, Juncus polycephalus, Lachnanthes caroliniana, Lycopodiella alopecuroides, Nyssa biflora, Nyssa ogeche, Pinguicula planifolia, Pinus serotina, Pleea tenuifolia, Proserpinaca pectinata, Rhynchospora baldwinii, Rhynchospora careyana, Rhynchospora cephalantha, Rhynchospora elliottii, Rhynchospora fascicularis, Rhynchospora latifolia, Rhynchospora macra, Rhynchospora oligantha, Rhynchospora perplexa, Rhynchospora plumosa, Rhynchospora stenophylla, Saccharum giganteum, Sarracenia flava, Sarracenia psittacina, Serenoa repens, Sporobolus teretifolius, Taxodium ascendens, Tiedemannia filiformis, Utricularia purpurea, Woodwardia virginica, Xyris ambigua, Xyris difformis, Xyris drummondii, Xyris fimbriata, Xyris flabelliformis, Xyris jupicai, Xyris scabrifolia, Xyris stricta.*

10.4.1.17. *Xyris smalliana* Nash is a short-lived perennial, which occurs in "batteries," bogs (hillside, ombrotrophic, peat, seepage, streamhead), borrow pits, Carolina bays, ditches (roadside), flatwoods (cypress-pine, hydric, mesic, pine, wet), hammocks (moist), lake bottoms (desiccated), marshes (basin, flatwoods, impounded, pineland), meadows (burned, cutthroat, depression, grassy, limesink, low, moist, salt), pinelands (dry, wet), pococins (streamside), pools (pineland), prairies (moist, wet), roadsides (sandy), ruts (ATV), quagmires (peaty), rights of way (powerline), savannas (cutover, mesic, pine), seepages (acidic, roadside), swales (coastal, interdunal, seepage, wet), swamps (branch, cypress, drying, floodplain, gum), and in the shallows or along the margins (peaty) or shores of cypress domes, lakes (interdunal), ponds (cypress, depression, flat-woods, freshwater, interdunal, intermittent, limesink, mucky, open, sandhill, upland), rivers, and streams (acidic, sluggish) at elevations to 65 m. The plants grow characteristically in deeper standing water (e.g., 30–91 cm). The substrates are acidic (e.g., pH: $\tilde{x}=4.6$; specific conductance: $\tilde{x}=23$ μS cm^{-1}; alkalinity: 0.5 mg l^{-1}), usually coarser-textured, and include Basinger (spodic psammaquents), clay, Duckston (typic psammaquents), Floridana (arenic argiaquolls), Floridana variant (typic umbraqualfs), Holopaw (grossarenic ochraqualfs), Isles (arenic ochraqualfs), Kinston (typic fluvaquents), Lawnwood (aeric haplaquods), muck, Ona (typic haplaquods), peat (floating), peaty muck, Pineda (arenic glossaqualfs), Plummer (grossarenic paleaquults), Pocomoke series, Rains (typic paleaquults), Rutledge (typic humaquepts), sand (exposed), sandy peat, Scranton (humaqueptic psammaquents), Sellers (cumulic humaquepts), Smyrna (aeric haplaquods), Wabasso (alfic haplaquods), Waveland (arenic haplaquods), and Zephyr (typic ochraquults). Flowering and fruiting occur from January to October but can extend throughout the year where suitable conditions exist. The flowers do not open until the afternoon or near evening. The best seed germination (spring or summer) from deep sediment sources has been observed for seeds exposed to 1 month of dry conditions, followed by 2 months of inundation; however, the highest seedling biomass (5–10 mg) develops from seeds that have germinated under constantly inundated conditions. Seeds recovered from shallow sediments also germinate best under comparable dry-wet treatments, with comparable germination rates for seeds kept under dry conditions in summer and fall; fall seedling weight (6–12+ mg) was also highest for seeds that germinated under strictly inundated conditions. Greater seedling emergence has been reported under elevated sediment phosphorous conditions. A large seed bank (>1,500 seeds m^{-2}) of undetermined duration can develop. The plants perennate by paired lateral buds on their compacted stems. The reduced growth rates observed for plants reared under elevated nutrient levels have been used to explain their absence in habitats characterized by higher specific conductivity. The plants are also known to increase in density with decreasing organic matter content. **Reported associates:** *Acer rubrum, Amphicarpum muhlenbergianum, Amsonia tabernaemontana, Andropogon glomeratus, Andropogon gyrans, Andropogon virginicus, Aristida palustris, Aristida stricta, Aronia ×prunifolia, Bacopa caroliniana, Bartonia paniculata, Bidens mitis, Bidens trichospermus, Boltonia asteroides, Brasenia schreberi, Burmannia biflora, Calamovilfa brevipilis, Carex barrattii, Carex bullata, Carex livida, Carex striata, Carex verrucosa, Centella asiatica, Cephalanthus occidentalis, Chamaecyparis thyoides, Chamaedaphne calyculata, Cladium mariscoides, Coleataenia longifolia, Coleataenia tenera, Coreopsis rosea, Ctenium aromaticum, Cyperus dentatus, Cyperus polystachyos, Cyrilla racemiflora, Decodon verticillatus, Dichanthelium acuminatum, Dichanthelium hirstii, Dichanthelium scabriusculum, Dichanthelium sphaerocarpon, Diodia virginiana, Dionaea muscipula, Drosera intermedia, Dulichium arundinaceum, Eleocharis baldwinii, Eleocharis elongata, Eleocharis equisetoides,* *Eleocharis flavescens, Eleocharis microcarpa, Eleocharis obtusa, Eleocharis robbinsi, Eleocharis tenuis, Eleocharis tricostata, Eleocharis tuberculosa, Eleocharis vivipara, Eriocaulon aquaticum, Eriocaulon compressum, Eriocaulon decangulare, Eubotrys racemosa, Eupatorium leucolepis, Eupatorium mohrii, Fimbristylis autumnalis, Fuirena pumila, Fuirena scirpoidea, Fuirena squarrosa, Gaylussacia frondosa, Gratiola aurea, Gratiola brevifolia, Habenaria repens, Hydrocotyle umbellata, Hydrolea ovata, Hypericum adpressum, Hypericum canadense, Hypericum denticulatum, Hypericum fasciculatum, Hypericum galioides, Ilex glabra, Iris virginica, Itea virginica, Juncus biflorus, Juncus canadensis, Juncus militaris, Juncus pelocarpus, Lachnanthes caroliniana, Lachnocaulon anceps, Lilium catesbaei, Lindernia dubia, Linum striatum, Lobelia boykinii, Lobelia canbyi, Lobelia nuttallii, Lophiola aurea, Ludwigia brevipes, Ludwigia lanceolata, Ludwigia linearis, Ludwigia palustris, Ludwigia pilosa, Ludwigia sphaerocarpa, Lycopodiella appressa, Lycopus uniflorus, Lysimachia terrestris, Mayaca fluviatilis, Mnesithea rugosa, Muhlenbergia torreyana, Myrica gale, Myriophyllum, Nuphar advena, Nymphaea odorata, Nymphoides aquatica, Nymphoides cordata, Nyssa biflora, Orontium aquaticum, Panicum hemitomon, Panicum verrucosum, Panicum virgatum, Peltandra virginica, Persicaria hydropiperoides, Pinguicula, Pinus elliottii, Pinus palustris, Platanthera blephariglottis, Polygala cruciata, Polygala cymosa, Polygala lutea, Pontederia cordata, Proserpinaca pectinata, Rhexia aristosa, Rhexia mariana, Rhexia nashii, Rhexia virginica, Rhododendron viscosum, Rhynchospora alba, Rhynchospora brachychaeta, Rhynchospora capitellata, Rhynchospora careyana, Rhynchospora cephalantha, Rhynchospora chapmanii, Rhynchospora corniculata, Rhynchospora fascicularis, Rhynchospora filifolia, Rhynchospora fusca, Rhynchospora galeana, Rhynchospora globularis, Rhynchospora gracilenta, Rhynchospora harperi, Rhynchospora inexpansa, Rhynchospora inundata, Rhynchospora macrostachya, Rhynchospora nitens, Rhynchospora pallida, Rhynchospora perplexa, Rhynchospora pleiantha, Rhynchospora scirpoides, Rhynchospora tracyi, Rotala ramosior, Sabatia difformis, Saccharum giganteum, Sacciolepis striata, Sagittaria engelmanniana, Sagittaria graminea, Sagittaria isoetiformis, Sagittaria teres, Sarracenia flava, Schoenoplectus pungens, Schoenoplectus subterminalis, Scirpus cyperinus, Scleria minor, Scleria muhlenbergii, Scleria reticularis, Sclerolepis uniflora, Serenoa repens, Smilax rotundifolia, Sparganium americanum, Sphagnum cuspidatum, Sphagnum cyclophyllum, Sphagnum macrophyllum, Spiranthes laciniata, Styrax americanus, Syngonanthus flavidulus, Taxodium ascendens, Thelypteris palustris, Tiedemannia canbyi, Tillandsia usneoides, Triadenum virginicum, Utricularia cornuta, Utricularia geminiscapa, Utricularia juncea, Utricularia striata, Utricularia subulata, Vaccinium corymbosum, Vaccinium macrocarpon, Viola lanceolata, Woodwardia virginica, Xyris ambigua, Xyris difformis, Xyris fimbriata, Xyris jupicai, Xyris laxifolia, Xyris longisepala, Xyris stricta, Xyris torta.*

10.4.1.18. **Xyris stricta Chapm.** is a perennial, which grows in the shallow waters of bayheads, bogs (deep, hillside, pine barren, seepage, wet), borrow pits, Carolina bays, clearings, depressions (cutover, flatwoods), ditchbanks, ditches (flatwoods, mesic, roadside, wet), domes, flats (cypress, *Nyssa*, pine-cypress), flatwoods (burned, hydric, mesic-hydric, wet), hammocks (baygall), ponds (acidic, pineland), pools (pineland), potholes (flatwoods), prairies (sandy, wet), rights of way (powerline, wet), savannas (cutover, drying, flatwoods, pineland, pine-gum, pine-oak, pitcher plant, recently burned, swale, wet, wet-mesic), seepage slopes, swales (seasonally wet), and swamps (acid) at elevations to 67 m. The plants occur commonly in shallow, standing waters (e.g., 5 cm) under open to shaded exposures. The sediments are finer-textured, wet, and include alluvium, Chipley (aquic quartzipsamments), clay, Covxille (typic paleaquults), Leon (typic haplaquods), Lynchburg (aeric paleaquults; fine sandy loam), Malbis-Kirbyville (plinthaquic paleudults), muck, Olustee (fine sand; ultic haplaquods), Pansey (plinthic paleaquults), Pantego (umbric paleaquults), peat, peaty muck, Plummer (grossarenic paleaquults), Pomona (ultic haplaquods), Rains (typic paleaquults), Rutledge (typic humaquepts), sand (wet), sandy clay (fine, organic), sandy clay peat, sandy peat (fine), silty loam, silty sandy loam, and Stowell (fine sandy loam). Flowering and fruiting extend from June to October. The flowers are open from late morning through early afternoon (by 4 pm). Although there has been some conjecture that the plants might be apomictic, there exists no compelling evidence to support that supposition. The cespitose plants perennate by paired lateral buds, which are protected beneath thickened leaf bases on the compacted stems. **Reported associates:** *Amphicarpum muhlenbergianum, Andropogon gyrans, Andropogon virginicus, Anthaenantia rufa, Aristida palustris, Aristida stricta, Arnoglossum ovatum, Arundinaria tecta, Bigelowia nuttallii, Boltonia diffusa, Borrichia frutescens, Chamaecyparis thyoides, Chasmanthium ornithorhynchum, Clethra alnifolia, Coleataenia tenera, Dichanthelium acuminatum, Ctenium aromaticum, Cyrilla racemiflora, Dichanthelium acuminatum, Dichanthelium scoparium, Dichanthelium sphaerocarpon, Diospyros virginiana, Drosera brevifolia, Eleocharis tuberculosa, Eriocaulon compressum, Eriocaulon decangulare, Eupatorium linearifolium, Eupatorium rotundifolium, Euthamia graminifolia, Fimbristylis spadicea, Fuirena breviseta, Gratiola pilosa, Gratiola ramosa, Helianthus angustifolius, Hypericum crux-andreae, Hypericum drummondii, Hypericum fasciculatum, Hyptis alata, Ilex coriacea, Ilex myrtifolia, Iva angustifolia, Juncus marginatus, Lachnocaulon anceps, Liatris acidota, Lilium catesbaei, Ludwigia hirtella, Ludwigia linearis, Ludwigia linifolia, Ludwigia microcarpa, Ludwigia pilosa, Ludwigia sphaerocarpa, Lycopodiella alopecuroides Lythrum alatum, Magnolia virginiana, Muhlenbergia capillaris, Myrica cerifera, Myrica heterophylla, Nyssa biflora, Panicum hemitomon, Paspalum floridanum, Paspalum plicatulum, Persea palustris, Pinguicula pumila, Pinus elliottii, Pinus palustris, Pinus serotina, Pinus taeda, Platanthera integra, Pluchea baccharis, Polygala cruciata, Polygala incarnata, Polygala mariana, Proserpinaca pectinata, Quercus laurifolia, Quercus virginiana, Rhexia alifanus, Rhexia aristosa, Rhexia lutea,* *Rhexia mariana, Rhynchospora baldwinii, Rhynchospora careyana, Rhynchospora cephalantha, Rhynchospora corniculata, Rhynchospora elliottii, Rhynchospora fascicularis, Rhynchospora gracilenta, Rhynchospora harperi, Rhynchospora inundata, Rhynchospora latifolia, Rhynchospora perplexa, Rhynchospora pleiantha, Rhynchospora plumosa, Rhynchospora rariflora, Rhynchospora tracyi, Sabatia difformis, Sabatia gentianoides, Sarracenia alata, Schizachyrium scoparium, Schizachyrium tenerum, Schoenolirion croceum, Scleria ciliata, Scleria georgiana, Scleria muhlenbergii, Scleria pauciflora, Scleria reticularis, Serenoa repens, Smilax laurifolia, Solidago tortifolia, Taxodium ascendens Tiedemannia filiformis; Triadica sebifera, Vaccinium, Viola lanceolata, Xyris ambigua, Xyris baldwiniana, Xyris difformis, Xyris elliottii, Xyris fimbriata, Xyris jupicai, Xyris laxifolia, Xyris serotina, Xyris smalliana.*

10.4.1.19. ***Xyris tennesseensis* Kral** is a perennial, which inhabits fens, meadows, seeps, swales, and the margins or banks of ponds (spring-fed) and streams at elevations to 250 m. The plants occur in sunny exposures and compete poorly in the presence of overtopping (e.g., shrubby) vegetation. The substrates are calcareous and have been described as mud, rock, sandstone, sandy peat, and shale. Flowering occurs for about 8 weeks from July to September, with the flowers opening from late morning for up to 3 hr ($\bar{x} = 130$ min). Most floral visitation occurs during 9:00–11:00 am. The flowers are self-compatible and often produce seed by selfing due to the frequent release of pollen prior to the opening of the flowers. They are also pollinated by bees (Insecta: Hymenoptera: Apidae; Halictidae) and to a much lesser degree by flies (Insecta: Diptera: Syrphidae) (see *Use by wildlife* below). Apomictic seed production has been postulated, but has not yet been evaluated adequately. One bee species (Halictidae: *Lasioglossum zephyrum*) has been observed to manipulate the closed flowers in order to expedite their opening for priority pollen collection. A small, persistent seed bank (up to 2,933 m^{-2}) can (but does not always) develop. Natural germination occurs from mid-May through early August. Germination will occur under water regardless of depth. The seeds are conditionally dormant but do not require a period of cold stratification. Fresh seeds have germinated best (100%) in the light after being incubated for 4 weeks under a 25°C/15°C day/night temperature regime. Germination rates were found to be lower for outcrossed seeds than for self-pollinated seeds. Light (e.g., 6 days under a 14-h photoperiod) is essential for promoting germination. Buried seeds kept in an unheated greenhouse have retained high germination rates (80–100%) in light (incubated at 25°C/15°C or 30°C/15°C) over a 39.5-month period. High spring germination (79–86% at 20°C/10°C) has been observed for seeds after their first year of burial, with much reduced rates (0–1%) occurring in the autumn. High germination rates (e.g., 97%) also have been obtained for cryopreserved seeds. Seedling survival is poor when they are transplanted to nonnative soils. Several micropropagation techniques have been developed successfully for this species. The solitary or cespitose plants perennate by means of dormant lateral buds. The artificial clearing of competitive shrubs has been found to increase floral visitation rates, ramet numbers (by up to 16.5%), and seedling densities (by up to 3,300%). **Reported associates:** *Acer rubrum, Andropogon glomeratus, Andropogon virginicus, Arthraxon hispidus, Boehmeria cylindrica, Cardamine hirsuta, Carex granularis, Carex hystericina, Carex lurida, Carex vulpinoidea, Chamaecrista fasciculata, Cirsium, Clematis virginiana, Conyza canadensis, Cyperus flavescens, Cyperus strigosus, Dichanthelium acuminatum, Eclipta prostrata, Eleocharis obtusa, Erechtites hieracifolius, Erigeron quercifolius, Eupatorium capillifolium, Eupatorium mohrii, Eupatorium perfoliatum, Eupatorium serotinum, Festuca, Fimbristylis autumnalis, Fimbristylis littoralis, Galium bermudense, Galium tinctorium, Gamochaeta purpurea, Glyceria striata, Helianthus angustifolius, Hypericum densiflorum, Hypericum densiflorum, Hypericum mutilum, Juncus brachycephalus, Juncus coriaceus, Juncus marginatus, Leersia virginica, Lemna, Lespedeza cuneata, Leucanthemum vulgare, Lobelia amoena, Lobelia puberula, Lonicera japonica, Ludwigia alternifolia, Ludwigia microcarpa, Ludwigia palustris, Lysimachia quadriflora, Mecardonia acuminata, Microstegium vimineum, Mikania scandens, Mitreola petiolata, Mollugo verticillata, Myriophyllum spicatum, Nasturtium officinale, Packera anonyma, Parthenocissus quinquefolia, Persicaria hydropiper, Persicaria pensylvanica, Persicaria setacea, Pinus taeda, Platanus occidentalis, Prunus serotina, Pseudognaphalium obtusifolium, Rhynchospora caduca, Rhynchospora capillacea, Rhynchospora glomerata, Rhynchospora thornei, Rotala ramosior, Rubus, Rudbeckia fulgida, Rumex, Sabatia angularis, Sabatia quadrangula, Salix nigra, Sambucus nigra, Samolus valerandi, Scleria verticillata, Solidago altissima, Solidago patula, Sonchus, Symphyotrichum dumosum, Toxicodendron radicans, Triodanis perfoliata, Vernonia angustifolia, Vitis cinerea, Vitis rotundifolia.*

10.4.1.20. ***Xyris torta*** **Sm.** is a perennial, which grows in or on barrens (sandy), beaches (sand), bogs (drained, graminoid, gravel, hillside, mountain, open, pitcher plant, sandy, seepage, sphagnous), channels (stream), clearings (boggy, open, sphagnous, wet), depressions (open), ditches (acid, boggy, drainage, roadside, seasonally moist, wet), fen (peaty, seepage), flats (moist, sand), flatwoods (pine), floodplain (low), hummocks (moist), marshes (lowland, roadside), meadows (grazed, interdunal, open, sandy, *Sphagnum*, swampy, wet), prairies (open, wet, wet-mesic), quarries (rock), ravines (boggy, low, swampy), rights of way (low, powerline, railroad, roadside), roadsides (acid, exposed, low, sandy, wet), savannas (pine, sand), seepages (graminoid, gravelly, hillside, open, ravine head, sandy, sphagnous, treeless, wet), slopes [to 40%] (sandy, sandy clay, seepage), swales (boggy, burned, interdunal, open, sandy, seepage, sphagnous, wet), swamps (acid, open, peaty, roadside, sandy, seepage, wet), woodlands (low, seasonally wet, upland), and along the banks or shores (drying, moist, wet) of borrow pits, gravel pits, lakes, ponds (desiccated, ephemeral, farm, sand, seasonal), pools (freshwater), sloughs, springs, and streams (seasonal) at elevations to 1,158 m. The

plants are common in disturbed, open to lightly shaded sites, but do not persist in standing water. The reported substrates include clay, Colita (typic glossaqualfs), gravel, Leagueville-Henco (arenic paleaquults), Letney (arenic paleudults), mucky peat, mucky sand, mud, peat, Pelham (arenic paleaquults), sand (moist, Queen City, Sparta), sandstone (shaley), sandy clay, sandy gravel, sandy gravelly clay, sandy loam, sandy peat, sandy silt, silty clay loam, and Stringtown-Bonwier (typic hapludults). Flowering and fruiting extend from June to October. The short-lived flowers are open during the morning. The cespitose or solitary, compacted shoots perennate from paired, bulblike, lateral buds that occur beneath the thickened leaf bases. **Reported associates:** *Agalinis purpurea, Agalinis skinneriana, Agrostis hyemalis, Alnus serrulata, Andropogon gerardii, Andropogon glomeratus, Andropogon virginicus, Anthoxanthum nitens, Aristida longespica, Aronia arbutifolia, Aronia ×prunifolia, Asclepias hirtella, Athyrium filix-femina, Axonopus furcatus, Bartonia paniculata, Bartonia virginica, Betula nigra, Bidens aristosus, Bidens polylepis, Boltonia diffusa, Bulbostylis capillaris, Calopogon tuberosus, Campanula aparinoides, Carex albolutescens, Carex atlantica, Carex bullata, Carex buxbaumii, Carex crinita, Carex folliculata, Carex longii, Carex striata, Carex torta, Centella asiatica, Cephalanthus occidentalis, Chamaedaphne calyculata, Chasmanthium laxum, Chelone glabra, Chrysopsis mariana, Cladium mariscoides, Cleistesiopsis bifaria, Coleataenia longifolia, Cyperus dentatus, Cyperus haspan, Decumaria barbara, Dichanthelium acuminatum, Doellingeria sericocarpoides, Drosera filiformis, Drosera intermedia, Drosera rotundifolia, Drymocallis arguta, Dulichium arundinaceum, Elatine minima, Eleocharis flavescens, Eleocharis microcarpa, Eleocharis quadrangulata, Eleocharis tenuis, Eleocharis tuberculosa, Eriocaulon aquaticum, Eriocaulon decangulare, Eriocaulon kornickianum, Eriophorum virginicum, Eupatorium leucolepis, Eupatorium perfoliatum, Eupatorium pilosum, Eupatorium rotundifolium, Eurybia compacta, Euthamia gymnospermoides, Eutrochium fistulosum, Fimbristylis autumnalis, Fuirena squarrosa, Gaylussacia, Gentianopsis crinita, Gentianopsis virgata, Glyceria obtusa, Gratiola aurea, Gratiola pilosa, Harperella nodosa, Helianthemum, Helianthus angustifolius, Hypericum adpressum, Hypericum canadense, Hypericum crux-andreae, Hypericum denticulatum, Hypericum frondosum, Hypericum gentianoides, Hypericum gymnanthum, Hypericum hypericoides, Hypericum kalmianum, Hypericum majus, Hypericum mutilum, Ilex verticillata, Iris prismatica, Juncus acuminatus, Juncus balticus, Juncus biflorus, Juncus brachycephalus, Juncus brevicaudatus, Juncus canadensis, Juncus coriaceus, Juncus diffusissimus, Juncus effusus, Juncus greenei, Juncus marginatus, Juncus pelocarpus, Juncus scirpoides, Juncus validus, Justicia americana, Lachnanthes caroliniana, Lachnocaulon anceps, Lechea, Liatris microcephala, Linum medium, Linum striatum, Lobelia canbyi, Lobelia nuttallii, Lophiola aurea, Ludwigia alternifolia, Luzula multiflora, Lycopodiella appressa, Lycopodiella inundata, Lycopus, Lycopus americanus, Magnolia virginiana, Marshallia grandiflora, Mitreola* *sessilifolia, Muhlenbergia glomerata, Muhlenbergia racemosa, Muhlenbergia torreyana, Muhlenbergia uniflora, Myrica gale, Myrica heterophylla, Myrica pensylvanica, Oclemena nemoralis, Osmunda claytoniana, Osmunda regalis, Osmundastrum cinnamomeum, Oxypolis rigidior, Panicum hemitomon, Panicum verrucosum, Panicum virgatum, Parnassia glauca, Parnassia grandifolia, Paspalum, Persicaria careyi, Phragmites australis, Pinguicula, Pinus palustris, Pinus taeda, Pityopsis ruthii, Platanthera ciliaris, Platanthera clavellata, Platanthera cristata, Platanthera lacera, Pogonia ophioglossoides, Polygala brevifolia, Polygala cruciata, Polygala nuttallii, Polygala polygama, Polygala sanguinea, Polygonella articulata, Polygonum tenue, Polypremum procumbens.Polytrichum, Pontederia cordata, Populus tremuloides, Pycnanthemum virginianum, Quercus lyrata Quercus palustris, Ranunculus flammula, Rhexia mariana, Rhexia virginica, Rhododendron canescens, Rhus copallinum, Rhynchospora alba, Rhynchospora caduca, Rhynchospora capitellata, Rhynchospora cephalantha, Rhynchospora chalarocephala, Rhynchospora fusca, Rhynchospora globularis, Rhynchospora glomerata, Rhynchospora gracilenta, Rhynchospora inexpansa, Rhynchospora knieskernii, Rhynchospora latifolia, Rhynchospora rariflora, Rhynchospora recognita, Rosa carolina, Rubus hispidus, Sabatia difformis, Sabatia gentianoides, Saccharum giganteum, Sagittaria engelmanniana, Salix humilis, Salix nigra, Sarracenia oreophila, Sarracenia purpurea, Schoenoplectus pungens, Schoenoplectus subterminalis, Scirpus cyperinus, Scirpus georgianus, Scleria muehlenbergii, Scleria reticularis, Scleria triglomerata, Scutellaria integrifolia, Sericocarpus linifolius, Solidago patula, Solidago uliginosa, Spartina pectinata, Sphagnum magellanicum, Sphagnum papillosum, Sphagnum pulchrum, Spiraea alba, Spiraea tomentosa, Spiranthes cernua, Spiranthes lacera, Spiranthes lucida, Spiranthes praecox, Symphyotrichum lateriflorum, Thelypteris palustris, Toxicodendron vernix, Triadenum virginicum, Triadenum walteri, Triantha glutinosa, Triantha racemosa, Utricularia cornuta, Utricularia gibba, Utricularia purpurea, Utricularia striata, Utricularia subulata, Vaccinium corymbosum, Vaccinium macrocarpon, Valeriana edulis, Vernonia missurica, Viburnum nudum, Viola lanceolata, Viola pedata, Viola pedatifida, Viola primulifolia, Woodwardia areolata, Woodwardia virginica, Xyris ambigua, Xyris baldwiniana, Xyris caroliniana, Xyris difformis, Xyris jupicai, Xyris laxifolia, Xyris montana, Xyris platylepis, Xyris scabrifolia, Xyris smalliana.*

Use by wildlife: *Xyris* seeds are eaten by mallard ducks (Aves: Anatidae: *Anas platyrhynchos*) and wild turkeys (Aves: Phasianidae: *Meleagris gallopavo*). The foliage ($\bar{x}$ =Ca: 0.46%; K: 0.33%; Mg: 0.22%; P: 0.09%) is grazed by cattle (Mammalia: Bovidae: *Bos taurus*) and white-tailed deer (Mammalia: Cervidae: *Odocoileus virginianus*). *Xris ambigua* flowers are a nectar source for various butterflies (Insecta: Lepidoptera), and the plants produce a mucous-like substance, which is thought to protect the eggs of reticulated flatwoods salamanders (Vertebrata: Amphibia: Ambystomatidae:

Ambystoma bishopi) from desiccation. *Xyris brevifolia* and *X. elliottii* are used for nest construction by the Florida grasshopper sparrow (Aves: Emberizidae: *Ammodramus savannarum floridanus*). *Xyris elliottii* is an important habitat indicator for Florida grasshoppers (Insecta: Orthoptera: Acrididae: *Paroxya clavuliger*). The roots of *Xyris jupicai* are associated with many nematodes (Nematoda: Actinolaimidae: *Actinolaimus*; Bastianiidae: *Bastiania*; Cephalobidae: *Zeldia*; Cryptonchidae: *Cryptonchus*; Dorylaimidae: *Dorylaimus, Labronema, Laimydorus, Mesodorylaimus*; Leptolaimidae: *Aphanolaimus*; Prismatolaimidae: *Prismatolaimus*; Nygolaimidae: *Nygolaimus*; Prismatolaimidae: *Onchulus*; Rhabditidae: *Rhabditis*; Tripylidae: *Tobrilus, Tripyla*; Tylenchidae: *Malenchus*). The pollen of *X. tennesseensis* is collected by bees (Insecta: Hymenoptera: Apidae: *Bombus*; Halictidae: *Augochlorella striata, Lasioglossum zephyrum*) and is consumed by syrphid flies (Insecta: Diptera: Syrphidae: *Chalcosyrphus metallicus, Toxomerus boscii, Toxomerus geminatus, Ocyptamus fuscipennis*). The flowers are also used as hunting platforms by robber flies (Diptera: Asilidae: *Holcocephala abdominalis*). Other incidental insect visitors include weevils (Coleoptera: Curculionidae) and skipper butterflies (Lepidoptera: Hesperiidae). *Xyris torta* is a host plant for larval red satyr butterflies (Insecta: Lepidoptera: Nymphalidae: *Megisto rubricata*). The flowers are visited by little wood satyr butterflies (*Megisto cymela*).

Economic importance: **food:** *Xyris* plants are not eaten by humans; **medicinal:** The juice of macerated *Xyris jupicai* roots has been used by indigenous Paraguayan people as a contraceptive (sterilizer). Extracts from *X. jupicai* are a source of acetylcholinesterase inhibitors and exhibit antioxidant activity. Roots of *X. laxiflora* have been used in Brazil in treatments for various gynecological problems; **cultivation:** *Xyris* species are cultivated occasionally as ornamental water garden plants; **misc. products:** none; **weeds:** *Xyris ambigua, X. brevifolia, X. flabelliformis, X. laxifolia, X. platylepis*, and especially *X. jupicai* have been reported as weeds of disturbed sites; **nonindigenous species:** There is some disagreement on the native status of *Xyris jupicai*. Its penchant for artificially disturbed sites, recent spread, and dense population development have raised the possibility that it might represent an adventive introduction from South America. *Xyris platylepis* was introduced accidentally to Hawaii sometime before 1951.

Systematics: Despite the limited inclusion of *Xyris* taxa in phylogenetic analyses, it appears that the genus is monophyletic (as also indicated by its distinctive morphology) and arguably is related most closely to Eriocaulaceae (e.g., Figure 3.6). However, the existing taxonomy of *Xyris* is difficult to embrace given its reliance entirely on morphology, numerous discrepancies noted among species descriptions, common misidentifications, confusing keys, and a dearth of supportive empirical evidence. Although extensive variability has been observed among progeny reared from the seeds of a single individual, no attempt has been made to elucidate the extent of morphological variation that can be attributed to genetic rather than phenotypically plastic expression. A thorough, comprehensive, genetically based study of the genus is necessary in order to better circumscribe taxa, to evaluate possible instances of hybridization, and to adequately estimate interspecific relationships; no comprehensive phylogenetic evaluation of the genus has yet been carried out. It would not be surprising if such efforts might reduce the number of alleged species considerably. The basic chromosome number of *Xyris* is $x=9$. The genus is uniform chromosomally, with *Xyris ambigua*, *X. baldwiniana*, *X. brevifolia*, *X. difformis*, *X. drummondii*, *X. elliottii*, *X. fimbriata*, *X. flabelliformis*, *X. isoetifolia*, *X. jupicai*, *X. laxifolia*, *X. platylepis*, *X. scabrifolia*, *X. serotina*, *X. smalliana*, *X. stricta*, *X. tennesseensis*, and *X. torta* all being diploid ($2n=18$). Chromosomes of the short-lived species (*X. brevifolia*, *X. flabelliformis*, *X. jupicai*) are relatively smaller in size than those of the perennials. There are few obvious barriers to hybridization, with most species being pollinated by generalist insects; differences in floral phenology do exist in some cases. *Xyris jupicai* is suspected to hybridize with *X. difformis* and *X. laxifolia*; *X. laxifolia* is thought to hybridize with *Xyris platylepis* (neither case has been investigated experimentally). Consistently low seed set in *X. montana* has raised the possibility that it might be of hybrid origin, but this possibility also has not been investigated empirically. *Xyris stricta* is thought to have arisen as a hybrid between *X. ambigua* and *X. laxifolia*.

Distribution: *Xyris ambigua*, *X. baldwiniana*, and *X. elliottii* occur in the southeastern United States, and extend through Mexico into Central America; the distribution of *X. laxifolia* is similar but expands only into Mexico. *Xyris brevifolia* and *X. jupicai* are distributed in the southeastern United States but extend southward into South America. *Xyris difformis* occurs sporadically throughout eastern North America and extends into Central America. *Xyris drummondii*, *X. fimbriata*, *X. flabelliformis*, *X. platylepis*, *X. scabrifolia*, *X. serotina*, and *X. stricta* are restricted to the southeastern coastal plain of the United States (*X. fimbriata* occasionally is found more inland). *Xyris smalliana* occurs along the eastern and southeastern coastal plain of the United States and extends into Central America. *Xyris montana* is restricted to northeastern North America, whereas *X. torta* is common throughout the eastern half of the United States. The distribution of *Xyris tennesseensis* is restricted to a small adjacent region of Alabama, Georgia, and Tennessee. *Xyris isoetifolia* and *X. longisepala* occur only within a small portion of northwestern Florida and southern Alabama.

References: Adams et al., 2010; Aldrich et al., 1985; Almquist & Calhoun, 2003; Anderson, 1991; Andreas & Bryan, 1990; Arenas & Azorero, 1977; Avis et al., 1997; Barcelos et al., 2013; Baskin & Baskin, 2003; Berkowitz et al., 2014; Birkenholz, 1963; Blomquist, 1955; Bowers, 1972; Boyd & Moffett, Jr., 2003; Boyd et al., 2011; Bridges & Orzell, 1987; 1989; 1992a; 1992b; 2002; 2003; Bush, 1988; Buthod & Hoagland, 2013a; 2013b; Calderón et al., 2010; Carr, 1940; Carr, 2007; Carr et al., 2010; Carter et al., 2004; Cohen et al., 2004; Cox & Smart, Jr., 1994; Cypert, 1972; Delany & Linda, 1998; Deyrup et al., 2002; Dighton et al., 2013; Drewa et al., 2002; Elam et al., 2009; Fassett, 1957; Fell, 1957; Ferren, Jr. & Schuyler, 1980; Fleming & Van Alstine, 1999; Gerritsen & Greening, 1989; Glaser, 1992; Glenn, 2008; Gordon, 2002; 2004; 2011; Gordon & Demitroff, 2009; Gorman et al., 2014; Gregg & Klotz, 2015; Haag et al., 2005; Halbritter et al., 2015; Harlow, 1961; Harper, 1903; 1905b; Harper, 1920; Harper et al., 1998; Harrison & Knapp, 2010; Hartley, 1959; Hellquist & Crow, 1982; Hill, 2003; Hinman & Brewer, 2007; Howell et al., 2016; Hubbard & Judd, 2013; Huffman & Judd, 1998; Hutton & Clarkson, 1961; Johnson et al., 2012; Kalmbacher et al., 1984a; 1984b; 2005; Karlin & Lynn, 1988; Keddy & Reznicek, 1982; Keppner & Anderson, 2008; Keith & Carrie, 2002; Kneitel & Miller, 2002; Kral, 1960; 1966b; 1973; 1978; 1998; 1999; Laidig, 2012; Laidig et al., 2001; Lemon, 1949; Lewis, 1970; Luken, 2005b; Lynn & Karlin, 1985; MacRoberts & MacRoberts, 1995; 2005; MacRoberts et al., 2014; Maliakal et al., 2000; McCormick et al., 2011; McLaughlin, 1932; McMillan & Porcher, 2005; McMillan et al., 2002; Minno & Slaughter, 2003; Moffett, Jr. & Boyd, 2013; Moore, 2004; Morris, 1988; 2012; Moyer & Bridges, 2015; Nardi et al., 2016; Nolfo-Clements, 2006; Oriani & Scatena, 2017; Orzell & Bridges, 1992; 2006b; Pace et al., 2017; Peet & Allard, 1993; Penfound & O'Neill, 1934; Pitts-Singer et al., 2002; Poole et al., 2007; Pyne, 2005; Ramirez & Brito, 1990; 1992; Reid & Faulkner, 2010; Reid & Urbatsch, 2012; Reznicek, 1994; Ruth et al., 2008; Schilling & Grubbs, 2016; Schneider, 1994; Scudder, 1889; Shelingoski et al., 2005; Shipley & Parent, 1991; Singhurst et al., 2007; 2009; 2011; 2014c; Smith, 1996; Smith & Capinera, 2005; Sommers et al., 2011; Sorrie & Dunwiddie, 1990; Sorrie et al., 1997; Strong & Sheridan, 1991; Sundue, 2006; 2007; Tyndall, 2000; Visser & Sasser, 2009; Walker & Peet, 1984; Wall et al., 2002; Weiher & Keddy, 1995; Wester, 1992; Wheeler et al., 1983; Whigham & Richardson, 1988; Wichman et al., 2007; Wilson & Keddy, 1985a; 1985b; Wilson et al., 1985; Wisheu & Keddy, 1989; Yazbek et al., 2016; Zaremba & Lamont, 1993; Zaremba & Lamont, 1993; Zebryk, 2004.

ORDER 11: ZINGIBERALES [8]

The order Zingiberales (Zingiberid monocotyledons) represents a monophyletic assemblage of 8 families, roughly 110 genera, and 2,600+ species distributed primarily in tropical and subtropical regions (Carlsen et al., 2018). Although this taxon had been recognized formerly as a distinct subclass (Zingiberidae) by Cronquist (1981) and others, various phylogenetic analyses have shown it to be embedded within the clade of "commelinid monocots" (i.e., subclass Commelinidae), where it resolves as the sister group to the order Commelinales (e.g., Figures 2.25 and 3.2). In addition to support by molecular data from nuclear and plastid genomes (e.g., Deng et al., 2016; Sass et al., 2016; Carlsen et al., 2018), the monophyly of the order is also evidenced by fossil data (Smith et al., 2018) and morphologically, with potential synapomorphies that include vessels that essentially are restricted to the roots, the presence of silica cells in the bundle sheath, tubular leaves in bud that differentiate into a blade and petiole (with enlarged air chambers), pinnate veins that often tear between the secondary veins, epigynous and bilaterally symmetrical flowers, arillate seeds with perisperm, and exineless pollen (Judd et al., 2016).

The order contains only a few OBL wetland species, which are distributed among three of the eight families (Figure 3.10):

11.1. **Cannaceae** Juss.
11.2. **Marantaceae** R.Br.
11.3. **Zingiberaceae** Martinov

Family 11.1. Cannaceae [1]

The Canna family (Cannaceae) comprises a single genus (*Canna*), which contains an estimated 22 species (Prince, 2010). The monophyly of Cannaceae and their sister group relationship to Marantaceae (e.g., Figure 3.10) have been supported by both morphological (Kress, 1990) and molecular (Kress & Prince, 2000; Kress et al., 2001; Prince, 2010; Carlsen et al., 2018) data. The species are characterized by mucilage canals in their rhizome and aerial shoots, an externally papillate and inferior ovary, flattened, petaloid styles, a single stigma that extends along the stylar margin, warty capsules, and also by the seeds, which arise from a tuft of funicular hairs (Judd et al., 2016).

Most of the species appear to be self-pollinating due to the precocious deposition of pollen on or near the stigma prior to anthesis (Kubitzki, 1998). However, putative pollinating agents include birds (Aves), bats (Mammalia: Chiroptera), and various insects (Insecta) such as bees (Hymenoptera), butterflies, and moths (Lepidoptera) (Kubitzki, 1998; Judd et al., 2016). The long-lived seeds (some still viable after 600 years) initially are impermeable and dispersed primarily by water (Kubitzki, 1998; Kress & Prince, 2000; Judd et al., 2016; Lerman & Cigliano, 1971).

Cannaceae contain more than 1,000 cultivars, which are grown commercially as ornamentals (Kubitzki, 1998). The rhizomes of some species (e.g. *C. edulis*, *C. indica*) are the source of one type of "arrowroot" starch; the flowers of

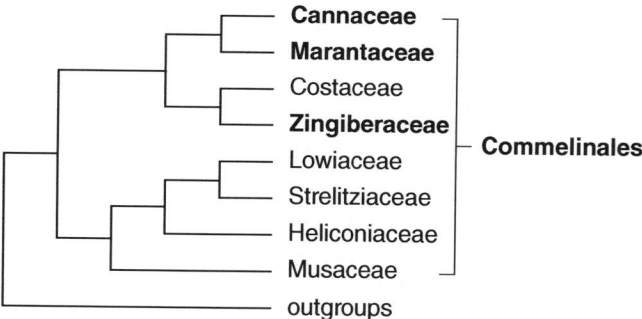

FIGURE 3.10 Interfamilial relationships in Commelinales as evidenced by analysis of sequence data from 378 nuclear genes (adapted from Carlsen et al., 2018). The placement of families is fairly consistent with other analyses of nuclear and plastid sequence data, except for the position of Musaceae, which resolves as the sister group of the entire order in some analyses (e.g., Deng et al., 2016; Sass et al., 2016). Families that include OBL North American wetland indicators are highlighted in bold.

C. indica have been used to produce green or pink to purple dyes for staining cotton textiles (Kubitzki, 1998; Kress & Prince, 2000; Vankar & Srivastava, 2018; Algar et al., 2019). Some of the plants have been used as medicinals; their seeds have been used as rosary beads or, when placed in gourds, as rattles (Kress & Prince, 2000).

Cannas grow in wetlands, along rivers, and in moister forested sites throughout the New World tropics and subtropics (Kress & Prince, 2000; Judd et al., 2016). In North America, OBL wetland indicators occur within one genus:

11.1.1. *Canna* L.

11.1.1. Canna
Bandana-of-the-Everglades, golden canna, maraca amarilla
Etymology: from the Greek *kanna*, a name that referred to a reedlike plant
Synonyms: *Eurystylus*; *Xyphostylis*
Distribution: global: Africa*; Asia*; New World; **North America:** southeastern United States
Diversity: global: 22 species; **North America:** 3 species
Indicators (USA): obligate wetland (OBL): *Canna flaccida*, *C. glauca*
Habitat: brackish (coastal), freshwater; palustrine; **pH:** acidic; **depth:** <1 m; **life-form(s):** emergent herb
Key morphology: rhizomes (to 5 m) fleshy or creeping; leaves green, sheathed, often glaucous, the blade (to 70 cm) narrowly ovate to narrowly elliptic, the base cuneate or gradually tapered to sheath; racemes simple (occasionally branched), bearing few (<5) to 10+ flowers in one or two-flowered circinni, peduncle green, often glaucous, 1° bracts (to 30 cm) green, often glaucous, 2° bracts absent or present (to 20 cm), floral bracts (to 2.5 cm) persistent, ovate-triangular, glaucous; flowers (to 15 cm) short-pedicelled (to 0.5 cm), epigynous; sepals (to 3.5 cm) green, narrowly elliptic-triangular or oblong-elliptic; corolla bilateral, pure yellow, the petals (to 12 cm) erect or strongly reflexed, differentiated into a tube (to 4 cm) and narrowly oblong-elliptic or narrowly ovate lobes (to 8 cm); functional stamen one, staminodes (to 11 cm) 3–4, pale yellow to deep crimson red, broadly ovate, narrowly elliptic, or narrowly ovate, the outermost ("labellum"; to 10 cm) not reflexed to strongly reflexed, broadly ovate or linear; ovary green; capsules (to 6 cm) brown, globose to ellipsoid, warty, becoming papery; seeds (to 8 mm) 5–75, brown, globose or ovoid
Life history: duration: perennial (rhizomes); **asexual reproduction:** rhizomes; **pollination:** bird (Aves), insect (Insecta), self; **sexual condition:** hermaphroditic; **fruit:** capsules (common); **local dispersal:** seeds (gravity, water), rhizomes; **long-distance dispersal:** seeds (gravity, water)
Imperilment: 1. *Canna flaccida* [G4]; S1 (AL, MS); S2 (SC); **2.** *C. glauca* [G5]; S2 (LA)
Ecology: general: *Canna* species are herbaceous, rhizomatous perennials, which have an affinity for marshes and other moist sites. Two of the three North American species are designated as OBL indicators, with the third species categorized as a facultative wetland indicator (FACW, FAC, FACU). The

flowers produce nectar (from septal nectaries), but are self-compatible and often self-pollinating (their pollen is shed prior to anthesis). Otherwise, the nectar attracts pollinators such as hummingbirds (Aves: Trochilidae) and various insects (Insecta) like hawkmoths (Lepidoptera: Sphingidae). The seeds are large with no obvious adaptations for dispersal other than an impervious coat. They are thought to be dispersed abiotically by gravity or water. The seeds lack dormancy and can germinate and grow into reproductively mature plants within one growing season.

11.1.1.1. ***Canna flaccida*** **Salisb.** is an indigenous species, which inhabits baygalls, cypress domes, depressions (marshy, wet), ditches (drainage, irrigation, moist, roadside, shaded, wet), dune bogs, flatwoods (inundated, mesic, pine), floodplains, hammocks (coastal, hydric, maritime), marshes (depression, floodplain, freshwater, roadside, salt, seasonally flooded/wet), rights of way (powerline, roadside), roadsides, sloughs, swamps (burned over, deciduous, drying, gum, hardwood, overgrown, strand), woodlands (alluvial, bottomland, rich, swampy, wet), and the margins or shores of canals, draws (wooded), lakes, and rivers at elevations to 231 m. The plants occur in open exposures to partial shade, often in shallow standing water (to 10 cm). The substrates are acidic (e.g., pH: 4.5–5.8) and have been described as muck, mud, peaty muck, sand, and sandy loam. Flowering and fruiting can occur nearly throughout the year (February–December). The yellow flowers are self-pollinating, but open at dusk and are also thought to be pollinated by hawkmoths (Insecta: Lepidoptera: Sphingidae). Biomass production rates as high as 39.3 (g m^{-2} day^{-1}) have been reported for plants grown on nutrient enriched sediments. **Reported associates:** *Acer rubrum, Alternanthera philoxeroides, Amorpha fruticosa, Ampelopsis arborea, Annona glabra, Baccharis glomeruliflora, Carex striata, Carya glabra, Cephalanthus occidentalis, Chionanthus virginicus, Chrysobalanus icaco, Cicuta maculata, Cladium jamaicense, Colocasia esculenta, Crinum americanum, Dendrophylax lindenii, Fraxinus caroliniana, Helenium flexuosum, Hydrocotyle, Hymenocallis franklinensis, Hypericum, Ilex vomitoria, Iris hexagona, Itea virginica, Juncus, Leitneria floridana, Liriodendron tulipifera, Ludwigia octovalvis, Lysimachia hybrida, Magnolia grandiflora, Magnolia virginiana, Myrica cerifera, Nyssa biflora, Nyssa ogeche, Orontium aquaticum, Osmunda regalis, Packera glabella, Panicum hemitomon, Peltandra virginica, Persea palustris, Persicaria glabra, Phyla, Physostegia leptophylla, Pleopeltis polypodioides, Pontederia cordata, Psilotum nudum, Quercus nigra, Rhododendron viscosum, Rhynchospora colorata, Rumex verticillatus, Sagittaria lancifolia, Salix caroliniana, Salix nigra, Sambucus, Samolus valerandi, Saururus cernuus, Sesbania drummondii, Sesbania herbacea, Spartina bakeri, Taxodium ascendens, Taxodium distichum, Thalia geniculata, Tillandsia paucifolia, Toxicodendron radicans, Triadenum virginicum, Tripsacum dactyloides, Vicia acutifolia, Vigna luteola, Zizaniopsis miliacea.*

11.1.1.2. ***Canna glauca*** **L.** is an indigenous (and nonindigenous) plant, which grows in ditches (drainage, wet), on spoilbanks, and along the margins of bayous, canals (freshwater), marshes (fresh), ponds (beaver), and swamps (low, wet) at elevations to 100 m. The exposures are in full sunlight. The substrates have been described as muck (Allemands, Kenner). Flowering occurs from April to July, with fruiting observed from June to September. The flowers open at dusk and are thought to be pollinated by butterflies (Insecta: Lepidoptera), hawkmoths (Insecta: Lepidoptera: Sphingidae), and hummingbirds (Aves: Trochilidae). An orange-flowered form is partially self-compatible, whereas the more common yellow-flowered form is fully self-compatible. No inbreeding depression has been observed in the offspring derived from self-pollinated yellow flowers. **Reported associates:** *Amaranthus australis, Convolvulus sepium, Hibiscus moscheutos, Ipomoea sagittata, Panicum hemitomon, Sagittaria lancifolia, Sesbania drummondii, Teucrium canadense, Verbascum thapsus.*

Use by wildlife: *Canna flaccida* and *C. glauca* are hosts to larval Brazilian skippers (Insecta: Lepidoptera: Hersperiidae: *Calpodes ethlius*), which protect themselves within cases made from the rolled leaves. *Canna flaccida* is also host to larval leafroller moths (Lepidoptera: Crambidae: *Geshna cannalis*) and to rugose spiraling whitefly (Insecta: Hemiptera: Aleyrodidae: *Aleurodicus rugioperculatus*). *Canna glauca* is eaten by apple snails (Mollusca: Gastropoda: Ampullariidae: *Pomacea insularum*) and is known to be grazed by cattle (Mammalia: Bovidae: *Bos taurus*).

Economic importance: food: The rhizomes of *Canna glauca* are higher in Mg content (350 mg 100 g^{-1}) than cassava (*Manihot esculenta*). In Paraguay (where they are known as *bacaó*) they are baked in ashes for 20 min before eating. There, the rhizomes allegedly taste better when harvested in autumn and winter (but reputedly lose their flavor if a woman sings when gathering them!); **medicinal:** The roots of *Canna glauca* have been used in South America (Argentina) as a diuretic, diaphoretic, and as a treatment for earaches; **cultivation:** *Canna flaccida* and *C. glauca* are implicated as parents in the origins of the ornamental hybrids known as *C.* ×*generalis* and *C.* ×*orchioides* (see *Systematics* below). Cultivars of *C. glauca* include 'Bolivia', 'Erebus', 'Java', 'Montevideo', 'Pure Yellow', and 'Ra'; **misc. products:** *Canna flaccida* has been evaluated for the bioremediation of nitrogen and phosphorous from agriculturally enriched wastewaters and stormwaters. The seeds of *C. flaccida* were used by the Florida Seminoles as pellets for making turtle shell rattles; they were also used at one time in primitive guns as bullet substitutes. Aqueous whole-plant extracts of *C. flaccida* reportedly are toxic to cockroaches (Insecta: Blattodea: Blattidae: *Periplaneta americana*; Blattellidae: *Blattella germanica*). *Canna glauca* has been investigated for the phytoremediation of metal-contaminated sites, and can remove arsenic (As) at the rate of 61 mg m^{-2} day^{-1}. The plants also have a high uptake rate for nitrogen and phosphorous; **weeds:** The OBL *Canna* species are not regarded as weedy; **nonindigenous species:** *Canna flaccida* is native to the United States but reportedly is nonindigenous to Virginia. *Canna glauca* presumably is native to Louisiana and Texas, but is thought to have been introduced to Florida as an escape from cultivation. The earliest record for Florida is 1967; however, the first record for Louisiana is only a few years earlier (1964); it was first collected in Alabama in 2003. *Canna glauca* has become naturalized in the Iberian Peninsula.

Systematics: *Canna* is the only genus of Cannaceae, which resolve as the sister group to Marantaceae (Figure 3.10). Cladograms derived from plastid and nuclear DNA sequence data situate *Canna flaccida* as the sister species to the remainder of the genus, whereas *C. glauca* resolves in a more derived portion of the genus, where it is allied with the facultative *C. indica* and several other species (Figure 3.11). The interspecific relationships in *Canna* remain unresolved at least in part due to many discrepancies that exist between phylogenetic trees generated from plastid vs. nuclear (nrITS) markers, and also between different analyses (i.e., Bayesian vs. MP) of the same data. The basic chromosome number of *Canna* is *x* = 9. *Canna flaccida* and *C. glauca* (2*n* = 18) are diploid. The parentage of the ornamental hybrid cultivar known as *Canna* × *generalis* (ranked as FACW, OBL) allegedly involves *C. indica* and *C. glauca* (or *C. iridifolia*); the hybrids known as Année involve crosses between *C. indica* and *C. glauca*; *C.* × *orchioides* represents a cross between *C.* ×*generalis* (♀) and *C. flaccida* (♂).

Distribution: *Canna flaccida* and *C. glauca* both occur along the southeastern coastal plain of the United States, with the latter extending into South America.

References: Austin, 2004; Burlakova et al., 2009; DeBusk et al., 1995; Francis et al., 2016; Goleniowski et al., 2006; Grootjen & Bouman, 1988; Gupta et al., 2003; Jacobson, 1958; Johnson, 2018; Jomjun et al., 2010; Kanwal et al., 2018; Kinupp & de Barros, 2008; Kress & Prince, 2000; Kunze, 1985; Laguna & Gallego, 2009; Loveless, 1959; McMillan et al., 2002; Mújica et al., 2018; Offerijns, 1936; Prince, 2010; Schmeda-Hirschmann, 1994; Smith et al., 2001; Tanaka et al., 2009; Thieret, 1971b; White & Cousins, 2013; Woods et al., 2006; Wunderlin et al., 2002; Yu et al., 2013.

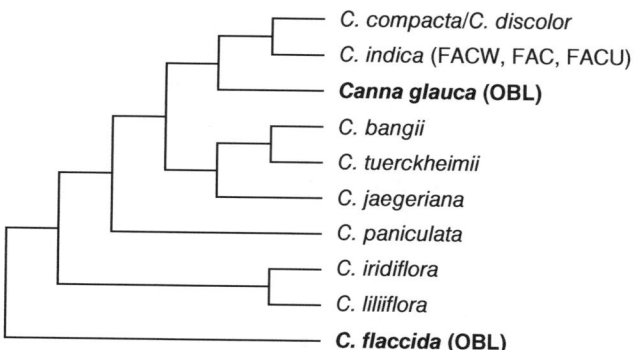

FIGURE 3.11 Phylogenetic relationships in *Canna* estimated using nrITS sequence data (modified from Prince, 2010). Although there are numerous discrepancies between the topology shown, and those generated from plastid sequence data or by using different analytical methods (Prince, 2010), consistent results show the sister relationship of *C. flaccida* to the remainder of the genus, and the distant relationship between the two OBL North American indicators (highlighted in bold); the wetland indicator status is provided in parentheses.

Family 11.2. Marantaceae [31]

The prayer plant family (Marantaceae) is represented by ~550 species, which are distributed throughout warm temperate and tropical regions (Cronquist, 1981; Andersson, 1998; Kennedy, 2000; Prince & Kress, 2006). About 450 species (82%) occur in the New World, with most (~300) found within the genus *Calathea* (Prince & Kress, 2006). Various phylogenetic studies (e.g., Kress, 1990; 1995; Prince & Kress, 2006; Figure 3.12) have demonstrated the monophyly of Marantaceae and consistently resolve the family as the sister group of Cannaceae (e.g., Figure 3.10). The family possesses several distinguishing features, most notably leaves (often 2-ranked) with an upper pulvinus that facilitates a nocturnal, upward, prayer-like folding of the blades (Cronquist, 1981; Judd et al., 2016). Other characteristics include closed sheaths, flowers arranged in mirror-image pairs, a hooded, inner androecial staminode that exerts tension on the style prior to floral visitation (but is released elastically during pollination), a callose-thickened staminode that provides a landing platform form pollinators, a single ovule per functional carpel, and a curved embryo (Judd et al., 2016).

Although most of the species are self-compatible, outcrossing is facilitated by bees (Insecta: Hymenoptera: Apidae), which effect pollination as they forage for the floral nectar (Andersson, 1998; Kennedy, 2000; Judd et al., 2016). Approximately 8% of the species are self-pollinating, including both OBL indicators, which shed their pollen onto the stigma within the bud prior to anthesis (Kennedy, 2000). Seed set typically is low throughout the family except for the autogamous taxa (Andersson, 1998). Many of the species produce colorful, arillate seeds, which attract ants (Insecta: Hymenoptera: Formicidae) or birds (Aves) as dispersal agents; a few species are water dispersed (Andersson, 1998; Judd et al., 2016).

Cultivated ornamental species are found in *Calathea*, *Ctenanthe*, *Goeppertia*, *Maranta*, *Saranthe*, and *Stromanthe*. *Maranta arundinacea* is grown throughout the tropics as a source of arrowroot starch (Andersson, 1998; Madineni et al., 2012). The stems of *Ischnosiphon* are split and used in basket weaving (Andersson, 1998).

Most Marantaceae grow along tropical rainforest margins or in light gaps, with some inhabiting riparian forests or swamps and other wetlands (Andersson, 1998; Judd et al., 2016). Two genera (*Maranta* and *Thalia*) grow in North America, with OBL indicators occurring only in the latter:

11.2.1. ***Thalia* L.**

11.2.1. Thalia

Alligator-flag, arrowroot, fire-flag, thalia

Etymology: after Johannes Thal (1542–1583)

Synonyms: *Malacarya*; *Maranta* (in part); *Peronia*; *Renealmia* (in part); *Spirostalis*

Distribution: global: Africa; New World; **North America:** southern United States

Diversity: global: 6 species; **North America:** 2 species

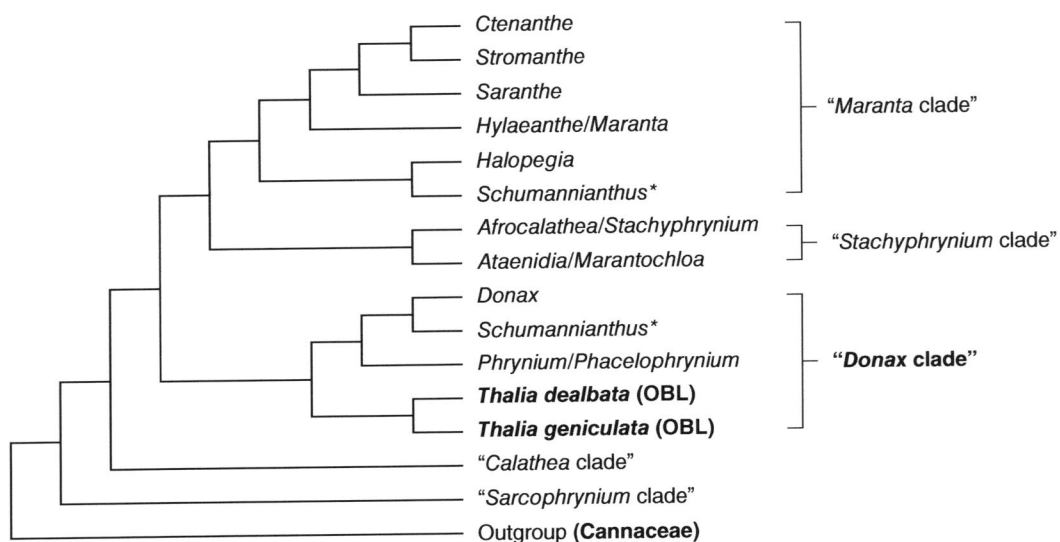

FIGURE 3.12 Phylogenetic relationships in Marantaceae reconstructed using combined plastid sequence data depict *Thalia* as the sister genus to the "*Donax* clade" (modified from Prince & Kress, 2006). The genus *Schumannianthus* was polyphyletic (marked by *) in the analysis. Although only *T. geniculata* was evaluated in the combined analysis tree, both *T. dealbata* and *T. geniculata* resolved as a clade (as shown) in a separate, more thoroughly sampled *trnL-F* intergenic spacer analysis (Prince & Kress, 2006), a result also obtained by the analysis of combined *rps16* intron, ITS1, and 5S-NTS DNA sequence data (Suksathan et al., 2009). Taxa with OBL North American indicators are highlighted in bold.

Indicators (USA): obligate wetland (OBL): *Thalia dealbata, T. geniculata*

Habitat: freshwater; palustrine; **pH:** 5.4–8.8; **depth:** <2 m; **life-form(s):** emergent herb

Key morphology: plants (to 3.5 m) upright, the stems unbranched beneath the inflorescence; leaves (2–6) basal, occasionally also cauline (1–2) above an elongate internode (to 2.5 m), sheath spongy, with prominent airspaces, petiole (to 2.5 cm) glabrous, pulvinus brown (caramel, reddish, yellowish, purplish), green (olive), or red-purple, blade (to 60 cm) green, ovate to elliptic; inflorescences (to 31 cm) erect and tightly clustered, or (to 1 m) lax, broadly spreading to pendant, and panicle-like, the scapes to 2.5 m, the branches short to elongate and arching, the rachis internodes conspicuously zigzagged, the flowers in pairs and subtended by a bract (to 2.8 cm); sepals (to 2.5 mm) membranous, persistent; corolla pale to dark purple, the tube to 6 mm, the lobes subequal to strongly unequal; outer staminode (to 20 mm) petaloid, showy, lavender to dark purple, callose staminode fleshy, the base white, yellow, or purple, cucullate staminode with 2 subterminal, finger-like appendages; style elongate, straplike, with 1 appendage; fruits (to 12 mm) utricular, 1-seeded, ellipsoid, globose, or broadly obovoid, the pericarp thin; seeds to 10 mm, dark brown to black, nearly globose or ellipsoid, with a reduced aril

Life history: duration: perennial (rhizomes); **asexual reproduction:** rhizomes; **pollination:** self; **sexual condition:** hermaphroditic; **fruit:** caryopsis-like utricles (common); **local dispersal:** seeds (water), rhizomes; **long-distance dispersal:** seeds (water)

Imperilment: 1. *Thalia dealbata* [G4]; SH (IL); S1 (AL, GA, MS); S2 (LA, MO, SC); S3 (AR, OK); **2.** *T. geniculata* [G4]; S1 (GA)

Ecology: general: *Thalia* species are rhizomatous perennials, which typically occur in wetlands; both North American species are designated as OBL indicators. The plants typically grow in shallow standing waters under open exposures. At least the North American species have an affinity for more alkaline conditions. *Thalia* species have complex inflorescences and exhibit "false resupination" (a rotation of <180°) in order to orient their flower pairs in the proper position for pollination; e.g., those of *T. geniculata* (pendulous inflorescence) rotate laterally by 90° before anthesis, whereas those of *T. dealbata* (erect inflorescence) curve backward by 90° as they develop. The flowers are self-compatible and shed their pollen onto the style prior to anthesis, which readily facilitates self-pollination. However, floral mechanisms also can present pollen secondarily by an explosive system that involves a tensioned style, which is released elastically (along with the shed pollen) when triggered by an appropriate pollinator. The most frequent, successful pollinators are bees (Insecta: Hymenoptera: Apidae) and hummingbirds (Aves: Trochilidae). Observed seed set is low in many natural populations. The fruits contain a gas-filled space, which makes them buoyant; they are dispersed by water. Both of the North American species perennate and reproduce vegetatively by rhizomes, which lack apparent modifications for starch storage. The plants typically die back to the rhizome in winter or during dry periods.

11.2.1.1. ***Thalia dealbata* Fraser ex Roscoe** inhabits depressions (wet prairie), ditches (agricultural, cotton field, drainage, mowed, ricefield, roadside, seasonally wet, wet), marshes (streamside), mudflats, ponds, pools (roadside), ricefields, rights of way (powerline), river bottoms, swamps (cypress-gum, open), and the margins of borrow pits, canals, lakes (artificial), and streams at elevations to 112 m. The plants occur commonly in exposures of full sunlight and in standing, shallow waters (e.g., 15–46 cm). The substrates are alkaline and have been described as clay (gray), Edna-Aris complex, muck, mud, and silty clay loam. Flowering occurs from May to September with fruiting from June to October. The plants are autogamous. The large seeds (~3,184 kg^{-1}) have internal dormancy, which requires a cold, moist treatment to induce germination. They typically ripen from late August through September. Their natural germination rate is poor. Good germination (35%) has been obtained for dry seeds that were stratified for 3–4 months at 13°C and 45% relative humidity, for moist-stored seeds, and for dry seeds that were scarified mechanically (40%). The seedlings grow best under moist but noninundated conditions. The plants have survived cultivation in northern latitudes (to 49°N) when their rhizomes remained submersed during the winter. **Reported associates:** *Ambrosia trifida, Bacopa rotundifolia, Boehmeria cylindrica, Carex glaucescens, Centella asiatica, Cephalanthus occidentalis, Cyperus acuminatus, Cyperus haspan, Echinochloa, Echinodorus cordifolius, Eleocharis quadrangulata, Eleocharis tuberculosa, Fimbristylis autumnalis, Glyceria, Hibiscus laevis, Hydrolea ovata, Hymenocallis liriosme, Hymenocallis occidentalis, Hyptis alata, Iris brevicaulis, Juncus debilis, Juncus effusus, Juncus nodatus, Leersia oryzoides, Leitneria floridana, Lemna, Leptochloa panicoides, Ludwigia, Luziola fluitans, Mikania scandens, Myriophyllum aquaticum, Nelumbo lutea, Nyssa biflora, Panicum hemitomon, Paspalidium geminatum, Paspalum wrightii, Persicaria amphibia, Persicaria hydropiperoides, Persicaria pensylvanica, Phanopyrum gymnocarpon, Phragmites australis, Physostegia, Pluchea, Pontederia cordata, Populus, Proserpinaca palustris, Rhynchospora corniculata, Rumex, Saccharum giganteum, Sagittaria graminea, Sagittaria lancifolia, Sagittaria montevidensis, Sagittaria papillosa, Salix nigra, Saururus cernuus, Sesbania drummondii, Taxodium ascendens, Typha, Woodwardia virginica.*

11.2.1.2. ***Thalia geniculata*** L. grows in depressions (roadside, wet), ditches (roadside, swampy, wet), floodplain forests, hammocks (coastal, disturbed, floodplain, hydric), marshes (basin, created, freshwater, slough), meadows (lowland), ponds (cypress, lowland, marshy), pools, roadsides, seeps (acid), sloughs (cypress), streambeds (drying), swamps (cypress, dome, strand), and along the margins or shores (swampy) of canals, lakes, rivers, and streams at elevations to 350 m. Exposures range from full sun to shade. The plants grow most often in shallow standing water (e.g., 21–200 cm). The substrates have been described as sandy loam. The plants occur primarily in alkaline, hard water, nutrient-rich sites (pH: 5.4–8.8; $\bar{x} = 7.7$; $\tilde{x} = 7.9$; total alkalinity [mg l^{-1} as CaCO$_3$]: 2.0–114.5; $\bar{x} = 50.0$; $\tilde{x} = 49.3$; conductance [μS cm^{-1} @ 25°C]: 74–818; $\bar{x} = 334$; $\tilde{x} = 204$; total N [μg l^{-1}]: 480–2,092; $\bar{x} = 116$; $\tilde{x}. = 1,110$; total P [μg l^{-1}]: 16–139; $\bar{x} = 64$; $\tilde{x} = 42$; Ca [mg l^{-1}]: 8.6–99.4; $\bar{x} = 51.6$; $\tilde{x} = 44.4$; Cl$^-$ [mg l^{-1}]: 13.9–223.2; $\bar{x} = 69.2$; $\tilde{x} = 26.5$; Fe [mg l^{-1}]: 0–0.7; $\bar{x} = 0.3$; $\tilde{x} = 0.2$; K [mg l^{-1}]: 1.0–7.4; $\bar{x} = 3.6$; $\tilde{x} = 2.6$; Mg [mg l^{-1}]: 2.8–55.0; $\bar{x} = 23.4$; $\tilde{x} = 11.3$; Na [mg l^{-1}]: 6.7–122.9; $\bar{x} = 38.4$; $\tilde{x} = 15.0$; Si [mg l^{-1}]: 0.3–5.2; $\bar{x} = 2.7$; $\tilde{x} = 3.1$; SO$_4^{2-}$ [mg l^{-1}]: 7.0–69.8; $\bar{x} = 26.9$; $\tilde{x} = 13.7$). Flowering occurs from June to December with fruiting extending from August to January. The flowers are self-compatible and are perfectly capable of successful self-pollination (autogamy) when pollinators are scarce. However, they also exhibit "explosive secondary pollen presentation." By that mechanism, the pollen is shed into a depression behind the stigma prior to anthesis. The style is hidden behind a staminode, which is reflexed under tension by a pressure-sensitive spur as the flower opens. If a potential pollinator exerts sufficient force on the spur, the style springs toward the front of the flower (in as short as 0.03 s) and deposits its pollen on the floral visitor. These can include butterflies (Insecta: Lepidoptera), carpenter bees (Insecta: Hymenoptera: Apidae: *Xylocopa*), and hummingbirds (Aves: Trochilidae), with the bees releasing the pollen more effectively than hummingbirds and butterflies essentially being ineffective. The fruits are buoyant utricles that are dispersed by water; the seeds remain afloat for less than a day. **Reported associates:** *Acer rubrum, Acrostichum, Aeschynomene pratensis, Ampelaster carolinianus, Annona glabra, Arundo donax, Asclepias lanceolata, Azolla filiculoides, Bacopa caroliniana, Bacopa monnieri, Bulbostylis ciliatifolia, Carex comosa, Centella asiatica, Cephalanthus occidentalis, Ceratophyllum, Chrysobalanus icaco, Cladium jamaicense, Coreopsis integrifolia, Crinum americanum, Cyperus odoratus, Cyperus polystachyos, Dalbergia ecastaphyllum, Dichanthelium dichotomum, Eleocharis cellulosa, Eleocharis interstincta, Fuirena breviseta, Helenium pinnatifidum, Hyptis alata, Ilex cassine, Ipomoea sagittata, Iris hexagona, Iris pseudacorus, Juncus effusus, Juncus megacephalus, Kosteletzkya pentacarpos, Leitneria floridana, Limnobium spongia, Liquidambar styraciflua, Ludwigia microcarpa, Ludwigia peruviana, Ludwigia repens, Mikania scandens, Myrica cerifera, Nuphar, Nymphaea odorata, Nymphoides aquatica, Panicum hemitomon, Paspalidium geminatum, Peltandra virginica, Persea borbonia, Persicaria hydropiperoides, Persicaria punctata, Phragmites australis, Pinus elliottii, Pistia stratiotes, Pluchea baccharis, Pontederia cordata, Proserpinaca palustris, Rhynchospora colorata, Rhynchospora inundata, Rhynchospora microcarpa, Rhynchospora tracyi, Riccia, Ruellia simplex, Sagittaria lancifolia, Sagittaria latifolia, Salix caroliniana, Salvinia minima, Schoenoplectus californicus, Schoenoplectus tabernaemontani, Spartina bakeri, Taxodium ascendens, Taxodium distichum, Typha angustifolia, Typha domingensis, Typha latifolia, Utricularia foliosa, Utricularia gibba, Utricularia purpurea, Utricularia resupinata.*

Use by wildlife: The rootstocks of *Thalia dealbata* occasionally are eaten by nutria (Mammalia: Rodentia: Myocastoridae: *Myocastor coypus*). The seeds are eaten by ducks (Aves: Anatidae). Its leaves are host to several Fungi including a leaf spot (Ascomycota: Mycosphaerellaceae: *Cercospora thaliae*) and a rust (Basidiomycota: Pucciniaceae: *Puccinia thaliae*). *Thalia geniculata* is a spring food of the Florida black bear (Mammalia: Ursidae: *Ursus americanus floridanus*). The fruits and seeds are eaten by ducks (Aves: Anatidae), purple gallinules (Aves: Rallidae: *Porphyrio martinica*), and yellow-knobbed curassows (Aves: Cracidae: *Crax daubentoni*). The plants host various insects (Insecta), including "crazy" ants

(Hymenoptera: Formicidae: *Paratrechina longicornis*), heteropterans (Insecta: Hemiptera: Blissidae: *Ischnodemus fulvipes, I. variegatus*), plume moths (Lepidoptera: Pterophoridae: *Sphenarches anisodactylus*), as well as rust Fungi (Basidiomycota: Pucciniaceae: *Puccinia cannae, P. thaliae*). They provide important nesting habitat for red-winged blackbirds (Aves: Icteridae: *Agelaius phoeniceus*) and yellow-hooded blackbirds (Icteridae: *Chrysomus icterocephalus*).

Economic importance: food: *Thalia geniculata* is of minor use as a food but was eaten by the Seminoles of Florida; **medicinal:** *Thalia geniculata* is used in Africa to treat malaria and other parasitic diseases. The plants contain geranylfarnesol, which exhibits significant antiprotozoal activity against *Plasmodium falciparum* (Chromalveolata: Plasmodiidae), and *Leishmania donovani* (Excavates: Trypanosomatidae); **cultivation:** *Thalia dealbata* is grown as a fish pond or water garden ornamental; **misc. products:** Root extracts of *Thalia dealbata* have inhibited the growth of some Cyanobacteria (Cyanophyceae: Chroococcaceae: *Microcystis aeruginosa*; Nostocaceae: *Anabaena flosaquae*). Microspheres derived from *T. dealbata* have been used as a sorbent for cadmium, phosphate, and sulfamethoxazole removal from contaminated waters. *Thalia geniculata* has been evaluated as potentially effective for the phytoremediation of waters contaminated by heavy metals. The plants have been used to produce biogas and as a reinforcement (known as "babadua bamboo") for concrete beams; **weeds:** *Thalia dealbata* sometimes has been regarded as weedy due to its ability to form dense stands that crowd out other species; **nonindigenous species:** none.

Systematics: Although only two *Thalia* species (*T. dealbata, T. geniculata*) have been analyzed simultaneously in a phylogenetic analysis, the morphologically distinct genus appears to be monophyletic. As currently evaluated, *Thalia* shows no particularly close affinity to any particular genus and resolves as the sister group to the "*Donax* clade," an assemblage that also includes the genera *Ataenidia, Donax, Marantochloa*, and *Schumannianthus* (Figure 3.12). Further study of *Thalia* species is necessary. Previous morphometric investigations have indicated that some taxa recognized formerly as distinct species (i.e., *T. trichocalyx* and *T. welwitschii*) merely represented characteristics better ascribed to the variational continuum associated with *T. geniculata*. The basic chromosome number of *Thalia* is thought to be $x=6$. However, in this interpretation, counts reported for cultivated material of *Thalia dealbata* ($2n=12$) would indicate it to be diploid, whereas those for *T. geniculata* [from Africa] ($2n=18$) would indicate it to be triploid, an interpretation that hardly is tenable. Anomalous counts (e.g., $2n=26$ for cultivated material of *T. geniculata*) further indicate the need for a more refined karyological study of the genus. No hybrids involving either of the OBL North American species have been reported.

Distribution: *Thalia dealbata* is endemic to the southeastern coastal plain and Mississippi embayment of the United States; *T. geniculata* also grows in the southeastern United States but extends into Africa, Mexico, and South America.

References: Andersson, 1981; 1998; Anning et al., 2013; Baranowski, 1979; Bertsch & Barreto, 2008; Buthod &

Hoagland, 2011; Carter et al., 2009; Cassani et al., 1990; Cui et al., 2016; Davis, 1987; De Gruchy, 1938; Dworaczek & Claßen-Bockhoff, 2016; Ellis & Langlois, 1890; Grabowski, 2001; 2002; Hoagland & Buthod, 2017; Hoyer et al., 1996; Jiang et al., 2014; Kankam & Odum-Ewuakye, 2000; Kaur et al., 2011; Kennedy, 2000; Lagnika et al., 2008; Ley & Claßen-Bockhoff, 2012; Maehr & Brady, 1984; Mali & Datta, 2018; Martin & Uhler, 1951; Mossman, 2009; Orians, 1973; Pischtschan & Claßen-Bockhoff, 2008; Prince & Kress, 2006; Rosen, 2007; Smith, 2001; Stutzenbaker, 1999; Sykes, Jr. et al., 1999; Tárano et al., 1995; Tao et al., 2019; Tipping et al., 2012; Uphof, 1922; West, 1941; Wetterer et al., 1999; Wiley & Wiley, 1980; Zhang et al., 2011.

Family 11.3. Zingiberaceae [53]

As currently circumscribed, the ginger family (Zingiberaceae) includes 50–53 genera and approximately 1,000–1,300 species (Larsen et al., 1998; Kress et al., 2001; Judd et al., 2016). Molecular and morphological phylogenetic evidence (Kress, 1990; 1995) have indicated the monophyly of the group and situate it as the sister family to Costaceae (e.g., Figure 3.10), which some authors have treated as a subfamily within Zingiberaceae (Judd et al., 2016). However, relationships within the family are not as well understood, with numerous conflicts, poor internal support, and low resolution characterizing the results of different molecular analyses. Although the proposed subdivision of Zingiberaceae into two subfamilies (Alpinioideae and Zingiberoideae) appears to be justified phylogenetically on the basis of molecular data (e.g., Kress et al., 2002), the same data provide no compelling support for any of the proposed tribal subdivisions, which require further evaluation.

The group is notable by its characteristically aromatic, spicy herbage, 2-ranked leaves, and ephemeral flowers with two larger staminodes fused into a liplike structure known as the labellum (Larsen et al., 1998; Judd et al., 2016). The nectariferous flowers are pollinated by birds (Aves: Nectariniidae; Trochilidae), bees (Insecta: Apidae: Hymenoptera: Halictidae), and hawk moths (Insecta: Lepidoptera: Sphingidae) (Larsen et al., 1998). Many of the species are self-pollinating. The seeds are dispersed primarily by birds, which are attracted to the colorful fruits or arillate seeds (Judd et al., 2016).

Commercial ginger is obtained from the rhizomes of *Zingiber officinale*, cardamon from *Amomum* and *Elattaria*, and turmeric from various species of *Curcuma* (Rogers, 1984; Larsen et al., 1998; Judd et al., 2016). Ornamental species are cultivated in the genera *Alpinia, Cautleya, Curcuma, Etlingera, Globba, Hedychium, Kaempferia, Roscoea*, and *Zingiber*.

Gingers primarily are a pantropical family, which is represented most prominently throughout the shaded forest understory; a few species occupy temporarily inundated wetlands (Larsen et al., 1998; Judd et al., 2016). Although most of the species occur in humid tropical lowlands, true aquatics are absent and some species persist at altitudes as great as 4,800 m (Larsen et al., 1998). In North America, only one species has been designated as an OBL indicator, which occurs in the following genus:

11.3.1. *Hedychium* J. Koenig

11.3.1. Hedychium

Butterfly ginger, butterfly ginger lily, butterfly lily, garland flower, garland lily, ginger lily, white ginger

Etymology: from the Greek *hedy chion* ("sweet snow") with respect to its fragrant flowers

Synonyms: *Amomum* (in part); *Gandasulium*; *Kaempferia* (in part)

Distribution: global: Asia; New World*; **North America:** southern United States

Diversity: global: 50 species; **North America:** 1 species

Indicators (USA): obligate wetland (OBL); facultative (FAC): *Hedychium coronarium*

Habitat: freshwater; palustrine; **pH:** unknown; **depth:** <1 m; **life-form(s):** emergent herb

Key morphology: pseudostems (to 3 m) well developed; leaves sessile, the blades (to 48 cm) oblanceolate or narrowly elliptic; inflorescences (to 19 cm) an erect, dense, conelike thyrse, the cincinni ($\bar{x} = 9.3$ per thyrse) sessile, 1–6-flowered ($\bar{x} = 3.4$ per circinnum), bracts (to 5 cm) imbricate, ovate; perianth white; calyx (to 4 cm) cylindric, 3-toothed or 3-lobed; corolla tube (to 8 cm) slender, the lateral lobes (to 5 cm) lanceolate, central lobe spatulate; lateral staminodes (to 5 cm) white, petal-like, oblong-lanceolate, labellum (to 6 cm) white, obcordate, pale at base; filament (to 3 cm) linear, tubular-incurved, enclosing style, anther (to 1.5 cm) long-exserted; capsules globose, sericeous; seeds numerous ($\bar{x} = 133.5$ ovules per flower), the arils lacerate

Life history: duration: perennial (rhizomes); **asexual reproduction:** rhizomes; **pollination:** bird, insect; **sexual condition:** hermaphroditic; **fruit:** capsules (common); **local dispersal:** rhizomes; **long-distance dispersal:** seeds (birds)

Imperilment: 1. *Hedychium coronarium* [GNR]

Ecology: general: *Hedychium* primarily comprises terrestrial or epiphytic herbs, with even the single OBL North American indicator (*H. coronarium*) being ranked alternatively as FAC in part of its range. The flowers of at least some species are self-compatible and self-pollinating (autogamous). Other species are at least facultatively outcrossed by insects (Insecta). All of the species reproduce vegetatively by means of rhizomes.

11.3.1.1. *Hedychium coronarium* **J. Koenig** is a nonindigenous perennial, which inhabits depressions (swampy), hammocks (disturbed, hydric), roadsides (damp), seepages, swamps (floodplain), and the margins or shores of impoundments, lakes, and streams at elevations to 260 m. Exposures can range from partial shade to full sunlight. North American populations have occurred primarily on sand. Flowering (North America) occurs from June to November. The sweetly fragrant flowers are pollinated by hawk moths (Insecta: Lepidoptera: Sphingidae). Each flower produces 14,210 pollen grain on average; the mean pollen:ovule ratio ranges from 106.5 to 138.5:1, which indicates a facultatively

xenogamous breeding system. The red, arillate seeds probably are dispersed by birds. They allegedly have poor germination rates. The plants have been micropropagated successfully from axillary buds, rhizome buds, and by somatic embryogenesis. *Hedychium coronarium* reduces the biodiversity of invaded habitats by negatively influencing the recruitment of native species. Where soil moisture and competition are high, the plants grow taller instead of increasing their rhizome production. Ramets are more likely to expand when soil conditions are dry and competitive pressures are low. **Reported associates (North America):** *Acer rubrum*, *Amaranthus spinosus*, *Amorpha fruticosa*, *Begonia cucullata*, *Campsis*, *Celtis laevigata*, *Cephalanthus occidentalis*, *Cocculus carolinus*, *Crinum americanum*, *Eriogonum tomentosum*, *Fraxinus*, *Osmundastrum cinnamomeum*, *Phytolacca americana*, *Pinus serotina*, *Sabal palmetto*, *Taxodium*, *Tetrapanax papyrifer*, *Thelypteris palustris*, *Ulmus americana*, *Vinca major*.

Use by wildlife: *Hedychium coronarium* is a host for burrowing nematodes (Nematoda: Hoplolaimidae: *Radopholus similis*). In Brazil, the plants are eaten by capybaras (Mammalia: Caviidae: *Hydrochoerus hydrochaeris*).

Economic importance: food: The spicy rhizomes and flowers of *H. coronarium* have been eaten as vegetables or used to flavor soups and other foods; **medicinal:** *Hedychium coronarium* has many traditional medicinal uses and the literature contains hundreds of studies addressing its potential medicinal applications. Various plant extracts have been attributed with analgesic, antiallergic, antiangiogenic, antibacterial, antidiabetic, antifungal, anti-inflammatory, antioxidant, antiurolithiatic, chemopreventive, coagulant, cytotoxic, fibrinogenolytic, hepatoprotective, and neuropharmacological properties. The efficacy of the plant constituents for developing anticancer treatments is also under investigation. Aqueous acetone extracts of the flowers have exhibited hepatoprotective activity in laboratory animals. The rhizomes contain numerous (53+) labdane-type diterpenes, which exhibit cytotoxic properties. The diterpenes have inhibited increased vascular permeability and nitric oxide production in laboratory animals. The rhizomes also contain several essential oils (i.e., 1,8-cineole, α-terpineol, and β-pinene), which have antibacterial and antifungal properties; **cultivation:** *Hedychium coronarium* is an RHS "Award of Garden Merit" plant. Its cultivars include 'Andromeda', 'Corelli', 'F.W. Moore', 'Gold Spot', and 'Orange Spot'; **misc. products:** Foliar and rhizome essential oils (α-pinene β-pinene, and 1,8-cineol) from *H. coronarium* are anthelminthic and have exhibited and larvicidal activity against mosquitoes (Insecta: Diptera: Culicidae). The oils are also used in the manufacture of perfume. The plants have been used to synthesize silver nanoparticles for use in controlling disease-carrying mosquitoes (Insecta: Diptera: Culicidae: *Aedes aegypti*). The plants have been evaluated as a potential source of pulp for paper-making; **weeds:** Although *H. coronarium* is becoming rare in parts of its indigenous range, it is regarded as an invasive weed in Brazil and Peru; **nonindigenous species:** *Hedychium coronarium* (native to Asia) is

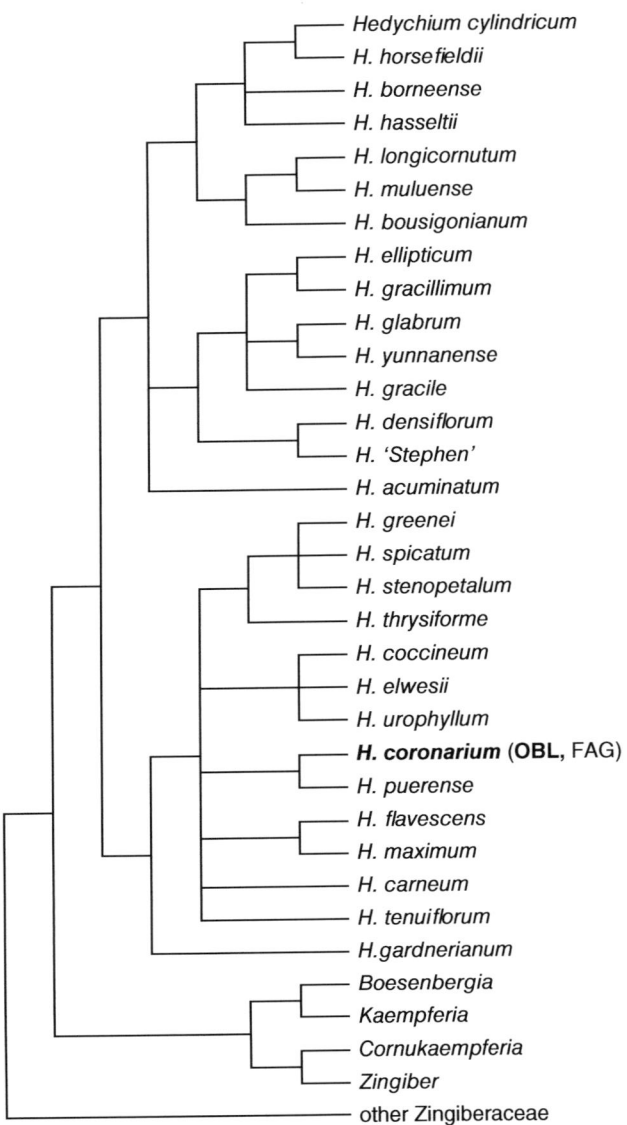

FIGURE 3.13 Interspecific relationships in *Hedychium* as indicated by a phylogenetic analysis of nrITS sequence data (modified from Wood et al., 2000). Although these results resolve the OBL North American *H. coronarium* (highlighted in bold) as related closely to *H. puerense*, an analysis based on SRAP markers (Gao et al., 2008) places it with *H. neocarneum* in a cluster that is fairly distant from *H. puerense*.

nonindigenous to the New World and likely was introduced as an escape from cultivation.

Systematics: Phylogenetic studies of 28 *Hedychium* species based on nrITS sequence data have indicated a close relationship between *H. coronarium* and *H. puerense* (Figure 3.13); however, a study using SRAP markers (which included additional species from China) indicated a closer relationship to *H. neocarneum* and a more distant relationship to *H. puerense*. Both analyses indicate that the genus is monophyletic; however, the relationship of *Hedychium* to other genera of Zingiberaceae remains uncertain. The base chromosome number of *Hedychium* is $x = 17$; *Hedychium coronarium* ($2n = 34$) is diploid. Some plants are autotriploids ($2n = 3x = 51$), which exhibit larger flowers and leaves. Tetraploid plants ($2n = 4x = 68$) also have been synthesized; these grow taller and have larger rhizomes than the diploids.

Hedychium species readily hybridize. Hybrids have been synthesized between *H. coronarium* (♀) and *H. forrestii* (♂).

Distribution: In North America, *Hedychium coronarium* (Asian) currently occurs along the southern Gulf coastal plain of the United States, but extends through Mexico and into South America.

References: Behera et al., 2018; Brooks, 1954; Chan & Wong, 2015; Dash, 2016; De Castro et al., 2016; Gao et al., 2008; Ho, 2011; Itokawa et al., 1988; Joy et al., 2007; Kalimuthu et al., 2017; Knudsen & Tollsten, 1993; Kress et al., 2002; Larsen et al., 1998; Matsuda et al., 2002; Mohanty et al., 2013; Nakamura et al., 2008; Ochoa & Andrade, 2003; Panigrahy et al., 2018; Parida et al., 2013; Ramachandran, 1969; Rogers, 1984; Sakhanokho & Rajasekaran, 2010; Shu, 2000; Suresh et al., 2010; Tu et al., 2018; Verma & Bansal, 2010; 2012; Whittemore, 2000; Wood et al., 2000.

References

Aalto, M. 1970. Potamogetonaceae fruits. I. Recent and subfossil endocarps of the Fennoscandian species. *Acta Bot. Fenn.* **88**: 1–85.

Aalto, M. 1974. Potamogetonaceae fruits. II. *Potamogeton robbinsii*, a seldom fruiting North American pondweed species. *Ann. Bot. Fenn.* **11**: 29–33.

Abbitt, R. J. & J. M. Scott. 2001. Examining differences between recovered and declining endangered species. *Conserv. Biol.* **15**: 1274–1284.

Abbott, J. M., K. DuBois, R. K. Grosberg, S. L. Williams & J. J. Stachowicz. 2018. Genetic distance predicts trait differentiation at the subpopulation but not the individual level in eelgrass, *Zostera marina*. *Ecol. Evol.* **8**: 7476–7489.

Abdel-Hamid, A. A., M. A. Al-Ghobashy, M. Fawzy, M. B. Mohamed & M. M. S. A. Abdel-Mottaleb. 2013. Phytosynthesis of Au, Ag, and Au–Ag bimetallic nanoparticles using aqueous extract of sago pondweed (*Potamogeton pectinatus* L.). *ACS Sustain. Chem. Eng.* **1**: 1520–1529.

Abdul-Alghaffar, H. N. & A. S. Al-Dhamin. 2016. Phytoremediation of chromium and copper from aqueous solutions using *Hydrilla verticillata*. *Iraqi J. Sci.* **57**: 78–86.

Abernethy, V. J., M. R. Sabbatini & K. J. Murphy. 1996. Response of *Elodea canadensis* Michx. and *Myriophyllum spicatum* L. to shade, cutting and competition in experimental culture. *Hydrobiologia* **340**: 219–224.

Abideen, Z., A. Hameed, H.-W. Koyro, B. Gul, R. Ansari & M. A. Khan. 2014. Sustainable biofuel production from non-food sources - an overview. *Emir. J. Food Agric.* **26**: 1057–1066.

Abou-Donia, A. H., S. M. Toaima, H. M. Hammoda, E. Shawky, E. Kinoshita & H. Takayama. 2008. Phytochemical and biological investigation of *Hymenocallis littoralis* Salisb. *Chem. Biodivers.* **5**: 332–340.

Aboul-Enein, A. M., A. M. Al-Abd, E. Shalaby, F. Abul-Ela, A. A. Nasr-Allah, A. M. Mahmoud & H. A. El-Shemy. 2011. *Eichhornia crassipes* (Mart) solms: from water parasite to potential medicinal remedy. *Plant Signal. Behav.* **6**: 834–836.

Aceto, S. & L. Gaudio. 2011. The MADS and the beauty: genes involved in the development of orchid flowers. *Curr. Genomics* **12**: 342–356.

Aceto, S. & L. Gaudio. 2016. The MADS-box genes involved in orchid flower development. Pp. 113–142 *In*: Atta-ur-Rahman (ed.), *Advances in Genome Science. Volume 4: Genes in Health and Disease*. Bentham Science Publishers, Sharjah, United Arab Emirates.

Ackerman, J. D. 1986. Mechanisms and evolution of food-deceptive pollination systems in orchids. *Lindleyana* **1**: 108–113.

Ackerman, J. D. 1995. Convergence of filiform pollen morphologies in seagrasses: functional mechanisms. *Evol. Ecol.* **9**: 139–153.

Ackerman, J. D. 1997a. Submarine pollination in the marine angiosperm *Zostera marina* (Zosteraceae). I. The influence of floral morphology on fluid flow. *Am. J. Bot.* **84**: 1099–1109.

Ackerman, J. D. 1997b. Submarine pollination in the marine angiosperm *Zostera marina* (Zosteraceae). II. Pollen transport in flow fields and capture by stigmas. *Am. J. Bot.* **84**: 1110–1119.

Ackerman, J. D. 2002. *Ponthieva* R. Brown. Pp. 547–548 *In*: Flora North America Editorial Committee (eds.), *Flora of North America North of Mexico, Vol. 26: Magnoliophyta: Liliidae: Liliales and Orchidales*. Oxford University Press, New York, NY.

Ackerman, J. D. 2006. Sexual reproduction of seagrasses: pollination in the marine context. Pp. 89–109 *In*: A. W. D. Larkum, R. J. Orth & C. M. Duarte (eds.), *Seagrasses: Biology, Ecology and Conservation*. Springer-Verlag, Dordrecht, The Netherlands.

Adair, S. E., J. L. Moore & C. P. Onuf. 1994. Distribution and status of submerged vegetation in estuaries of the upper Texas coast. *Wetlands* **14**: 110–121.

Adair, R. J., B. R. Keener, R. M. Kwong, J. L. Sagliocco & G. E. Flower. 2012. The biology of Australian weeds 60. *Sagittaria platyphylla* (Engelmann) J. G. Smith and *Sagittaria calycina* Engelmann. *Plant Prot. Q.* **27**: 47–58.

Adamec, L. 2018. Ecophysiological characteristics of turions of aquatic plants: a review. *Aquat. Bot.* **148**: 64–77.

Adams, J. 1927. The germination of the seeds of some plants with fleshy fruits. *Am. J. Bot.* **14**: 415–428.

Adams, P. & R. K. Godfrey. 1961. Observations on the *Sagittaria subulata* complex. *Rhodora* **63**: 247–266.

Adams, J. B. & G. C. Bate. 1994a. The tolerance to desiccation of the submerged macrophytes *Ruppia cirrhosa* (Petagna) Grande and *Zostera capensis* Setchell. *J. Exp. Mar. Biol. Ecol.* **183**: 53–62.

Adams, J. B. & G. C. Bate. 1994b. The ecological implications of tolerance to salinity by *Ruppia cirrhosa* (Petagna) Grande and *Zostera capensis* Setchell. *Bot. Mar.* **37**: 449–456.

Adams, M., F. Gmünder & M. Hamburger. 2007. Plants traditionally used in age related brain disorders—a survey of ethnobotanical literature. *J. Ethnopharmacol.* **113**: 363–381.

Adams, L. D., S. Buchmann, A. D. Howell & J. Tsang. 2010. A study of insect pollinators associated with DoD TER-S flowering plants, including identification of habitat types where they co-occur by military installation in the southeastern United States. No. LRMP-09-391. U.S. Department of Defense, Legacy Resource Management Program, Arlington, VA. 83 pp.

Adams, M., S. Gschwind, S. Zimmermann, M. Kaiser & M. Hamburger. 2011. Renaissance remedies: antiplasmodial protostane triterpenoids from *Alisma plantago-aquatica* L. (Alismataceae). *J. Ethnopharmacol.* **135**: 43–47.

Adebayo, A. A., E. Briski, O. Kalaci, M. Hernandez, S. Ghabooli, B. Beric, F. T. Chan, A. Zhan, E. Fifield, T. Leadley & H. J. MacIsaac. 2011. Water hyacinth (*Eichhornia crassipes*) and water lettuce (*Pistia stratiotes*) in the Great Lakes: playing with fire? *Aquat. Invasions* **6**: 91–96.

Adey, W. H., M. Finn, P. Kangas, L. Lange, C. Luckett & D. M. Spoon. 1996. A Florida Everglades mesocosm—model veracity after four years of self-organization. *Ecol. Eng.* **6**: 171–224.

Adis, J. 1983. Eco-entomological observations from the Amazon. IV. Occurrence and feeding habits of the aquatic caterpillar *Palustra laboulbeni* Bar, 1873 (Arctiidae: Lepidoptera) in the vicinity of Manaus. *Acta Amazon.* **13**: 31–36.

Adler, L. S. 2000. The ecological significance of toxic nectar. *Oikos* **91**: 409–420.

Adler, J. M., S. C. Barry, G. R. Johnston, C. A. Jacoby & T. K. Frazer. 2018. An aggregation of turtles in a Florida spring yields insights into effects of grazing on vegetation. *Freshw. Sci.* **37**: 397–403.

Aerne-Hains, L. & M. G. Simpson. 2017. Vegetative anatomy of the Haemodoraceae and its phylogenetic significance. *Int. J. Plant. Sci.* **178**: 117–156.

Afton, A. D., R. H. Hier & L. Paulus. 1991. Lesser scaup diets during migration and winter in the Mississippi flyway. *Can. J. Zool.* **69**: 328–333.

Agami, M. & Y. Waisel. 1984. Germination of *Najas marina* L. *Aquat. Bot.* **19**: 37–44.

Agami, M. & Y. Waisel. 1986. The role of mallard ducks (*Anas platyrhynchos*) in distribution and germination of seeds of the submerged hydrophyte *Najas marina* L. *Oecologia* **68**: 473–475.

Agami, M. & Y. Waisel. 1988. The role of fish in distribution and germination of seeds of the submerged macrophytes *Najas marina* L. and *Ruppia maritima* L. *Oecologia* **76**: 83–88.

Agami, M., S. Beer & Y. Waisel. 1980. Growth and photosynthesis of *Najas marina* L. as affected by light intensity. *Aquat. Bot.* **9**: 285–289.

Agami, M., S. Beer & Y. Waisel. 1984. Seasonal variations in the growth capacity of *Najas marina* L. as a function of various water depths at the Yarkon Springs, Israel. *Aquat. Bot.* **19**: 45–51.

Agami, M., A. Eshel & Y. Waisel. 1984a. *Najas marina* in Israel: it is a halophyte or a glycophyte? *Physiol. Plant.* **61**: 634–636.

Agami, M., S. Beer & Y. Waisel. 1986. The morphology and physiology of turions in *Najas marina* L. in Israel. *Aquat. Bot.* **26**: 371–376.

Ahee, J. E., W. E. Van Drunen & M. E. Dorken. 2015. Analysis of pollination neighbourhood size using spatial analysis of pollen and seed production in broadleaf cattail (*Typha latifolia*). *Botany* **93**: 91–100.

Ahmad, Z., N. S. Hossain, S. G. Hussain & A. H. Khan. 1990. Effect of duckweed (*Lemna minor*) as complement to fertilizer nitrogen on the growth and yield of rice. *Int. J. Trop. Agric.* **8**: 72–79.

Ahmadi, M., H. Saki, A. Takdastan, M. Dinarvand, S. Jorfi & B. Ramavandi. 2017. Advanced treatment of saline municipal wastewater by *Ruppia maritima*: a data set. *Data Brief* **13**: 545–549.

Aikens, C. M. 1986. *Archaeology of Oregon*, 2nd ed. U.S. Department of the Interior, Bureau of Land Management, Oregon State Office, Portland, OR. 133 pp.

Ailstock, S. & D. Shafer. 2004. Restoration potential of *Ruppia maritima* and *Potamogeton perfoliatus* by seed in the mid-Chesapeake Bay. Technical Report No. ERDC/EL-TN-04-02. Engineer Research and Development Center, Vicksburg, MS. 8 pp.

Ailstock, S. & D. Shafer. 2006. Protocol for large-scale collection, processing, and storage of seeds of two mesohaline submerged aquatic plant species. Technical Report No. ERDC/TN-SAV-06-3. Engineer Research and Development Center, Vicksburg, MS. 8 pp.

Ailstock, M. S., D. J. Shafer & A. D. Magoun. 2010a. Effects of planting depth, sediment grain size, and nutrients on *Ruppia maritima* and *Potamogeton perfoliatus* seedling emergence and growth. *Restor. Ecol.* **18**: 574–583.

Ailstock, M. S., D. J. Shafer & A. D. Magoun. 2010b. Protocols for use of *Potamogeton perfoliatus* and *Ruppia maritima* seeds in large-scale restoration. *Restor. Ecol.* **18**: 560–573.

Ainsworth, C. (ed.). 2006. *Flowering and Its Manipulation. Annual Plant Reviews, Volume 20*. Blackwell Publishing, Oxford, United Kingdom. 320 pp.

Airaksinen, M. M., P. Peura, L. Ala-Fossi-Salokangas, S. Antere, J. Lukkarinen, M. Saikkonen & F. Stenbäck. 1986. Toxicity of plant material used as emergency food during famines in Finland. *J. Ethnopharmacol.* **18**: 273–296.

Akkol, E. K., I. Süntar, H. Keles & E. Yesilada. 2011. The potential role of female flowers inflorescence of *Typha domingensis* Pers. in wound management. *J. Ethnopharmacol.* **133**: 1027–1032.

Akridge, R. E. & P. J. Fonteyn. 1981. Naturalization of *Colocasia esculenta* (Araceae) in the San Marcos River, Texas. *Southwest. Nat.* **26**: 210–211.

Alam, Z., M. M. Huq, A. Jabbar & C. M. Hasan. 1999. *Potamogeton nodosus*: an aquatic source for bioactive compounds. *Fitoterapia* **70**: 523–525.

Albanese, B. 2000. Reproductive behavior and spawning microhabitat of the flagfin shiner *Pteronotropis signipinnis*. *Am. Midl. Nat.* **143**: 84–94.

Alderson, H. E. & A. G. Rawlins. 1925. Rice workers' dermatitis. *Cal. West. Med.* **23**: 42–45.

Alderton, E., C. D. Sayer, R. Davies, S. J. Lambert & J. C. Axmacher. 2017. Buried alive: aquatic plants survive in 'ghost ponds' under agricultural fields. *Biol. Conserv.* **212**: 105–110.

Aldrich, J. R. & M. A. Homoya. 1983. Natural barrens and post oak flatwoods in Posey and Spencer counties. *Proc. Indiana Acad. Sci.* **93**: 291–302.

Aldrich, J. R., L. A. Casebere, M. A. Homoya & H. Starcs. 1985. The discovery of native rare vascular plants in northern Indiana. *Proc. Indiana Acad. Sci.* **95**: 421–428.

Alexander, M. L. & C. T. Phillips. 2012. Habitats used by the endangered fountain darter (*Etheostoma fonticola*) in the San Marcos River, Hays County, Texas. *Southwest. Nat.* **57**: 449–452.

Alexander, S. A., K. A. Hobson, C. L. Gratto-Trevor & A. W. Diamond. 1996. Conventional and isotopic determinations of shorebird diets at an inland stopover: the importance of invertebrates and *Potamogeton pectinatus* tubers. *Can. J. Zool.* **74**: 1057–1068.

Algar, A. F. C., A. B. Umali & R. R. P. Tayobong. 2019. Physicochemical and functional properties of starch from Philippine edible canna (*Canna indica* L.) rhizomes. *J. Microbiol. Biotechnol. Food Sci.* **9**: 34–37.

Aliferis, K. A., S. Materzok, G. N. Paziotou & M. Chrysayi-Tokousbalides. 2009. *Lemna minor* L. as a model organism for ecotoxicological studies performing 1H NMR fingerprinting. *Chemosphere* **76**: 967–973.

Alix, M. S. & R. W. Scribailo. 2006. First report of *Potamogeton × undulatus* (*P. crispus × P. praelongus*, Potamogetonaceae) in North America, with notes on morphology and stem anatomy. *Rhodora* **108**: 329–346.

Allard, H. A. 1951. The ratios of clockwise and counterclockwise spirality observed in the phyllotaxy of some wild plants. *Castanea* **16**: 1–6.

Allen, L., J. D. Johnson & K. Vujnovic. 2002a. *Small Patch Communities of La Butte Creek Wildland Provincial Alberta*. Natural Heritage Information Centre, Alberta Community Development, Edmonton, AB. 38 pp.

Allen, G. M., M. D. Bond & M. B. Main. 2002b. 50 common native plants important in Florida's ethnobotanical history. Circular 1439. Florida Cooperative Extension Service, Institute of Food and Agricultural Sciences, University of Florida, Gainseville, FL. 21 pp.

Allen, C., R. Erwin, J. McMillian & J. McMillian. 2013. A quantitative study of the vegetation surrounding a *Xanthorhiza simplicissima* (Ranunculaceae) population at Fort Polk in West Central Louisiana. *J. Bot. Res. Inst. Texas* **7**: 519–528.

Allen, A. C., C. A. Beck, R. K. Bonde, J. A. Powell & N. A. Gomez. 2017. Diet of the Antillean manatee (*Trichechus manatus manatus*) in Belize, Central America. *J. Mar. Biol. Assoc. U. K.* doi: 10.1017/S0025315417000182.

Almquist, H. & A. J. K. Calhoun. 2003. A coastal, southern-outlier patterned fen: Lily Fen, Swans Island, Maine. *Northeast. Nat.* **10**: 119–130.

Altrichter, K. M., E. S. DeKeyser, B. Kobiela & C. L. M. Hargiss. 2018. A comparison of five wetland communities in a North Dakota fen complex. *Nat. Areas J.* **38**: 275–285.

Alvarado, S., M. Guédez, M. P. Lué-Merú, G. Nelson, A. Alvaro, A. C. Jesús & Z. Gyula. 2008. Arsenic removal from waters by bioremediation with the aquatic plants water hyacinth (*Eichhornia crassipes*) and lesser duckweed (*Lemna minor*). *Bioresour. Technol.* **99**: 8436–8440.

Alves, T. L. S., O. Chauveau, L. Eggers & T. T. de Souza-Chies. 2014. Species discrimination in *Sisyrinchium* (Iridaceae): assessment of DNA barcodes in a taxonomically challenging genus. *Mol. Ecol. Resour.Molec. Ecol. Resources* **14**: 324–335.

Alves Pagotto, M., R. D. M. L. Silveira, C. N. D. Cunha & I. Fantin-Cruz. 2011. Distribution of herbaceous species in the soil seed bank of a flood seasonality area, northern Pantanal, Brazil. *Int. Rev. Hydrobiol.* **96**: 149–163.

Amano, M., S. Iida & K. Kosuge. 2011. Comparative studies of thermotolerance: different modes of heat acclimation between tolerant and intolerant aquatic plants of the genus *Potamogeton*. *Ann. Bot.* **109**: 443–452.

Ames, O. 1903. Natural hybrids in *Spiranthes* and *Habenaria*. *Rhodora* **5**: 261–264.

Amoroso, J. L. & W. S. Judd. 1995. A floristic study of the Cedar Key Scrub State Reserve, Levy County, Florida. *Castanea* **60**: 210–232.

Andersen, R. & J. S. Price. 2012. Microbial communities in boreal peatlands of the Athabasca region, Canada: building a reference for fen creation. *Extended abstract No. 251.* 14th International Peat Congress, Stockholm, Sweden. 6 pp.

Anderson, E. 1928. The problem of species in the northern blue flags, *Iris versicolor* L. and *Iris virginica* L. *Ann. Missouri Bot. Gard.* **15**: 241–332.

Anderson, E. 1936. The species problem in *Iris*. *Ann. Missouri Bot. Gard.* **23**: 457–509.

Anderson, W. A. 1943. A fen in northwestern Iowa. *Am. Midl. Nat.* **29**: 787–791.

Anderson, H. G. 1959. Food habits of migratory ducks in Illinois. *Illinois Nat. Hist. Surv. Bull.* **27**: 288–344.

Anderson, R. R. 1969. Temperature and rooted aquatic plants. *Chesapeake Sci.* **10**: 157–164.

Anderson, L. W. J. 1978a. Abscisic acid induces formation of floating leaves in the heterophyllous aquatic angiosperm *Potamogeton nodosus*. *Science* **201**: 1135–1138.

Anderson, M. G. 1978b. Distribution and production of sago pondweed (*Potamogeton pectinatus* L.) on a northern prairie marsh. *Ecology* **59**: 154–160.

Anderson, L. W. J. 1982. Effects of abscisic acid on growth and leaf development in American pondweed (*Potamogeton nodosus* Poir.). *Aquat. Bot.* **13**: 29–44.

Anderson, L. C. 1991. Noteworthy plants from north Florida. V. *Sida* **14**: 467–474.

Anderson, M. K. 1997. From tillage to table: the indigenous cultivation of geophytes for food in California. *J. Ethnobiol.* **17**: 149–169.

Anderson, L. W. J. 2003. A review of aquatic weed biology and management research conducted by the United States Department of Agriculture—Agricultural Research Service. *Pest Manag. Sci.* **59**: 801–813.

Anderson, L. C. 2007. Noteworthy plants from north Florida. VIII. *J. Bot. Res. Inst. Texas* **1**: 741–751.

Anderson, J. M. & S. C. H. Barrett. 1986. Pollen tube growth in tristylous *Pontederia cordata* (Pontederiaceae). *Can. J. Bot.* **64**: 2602–2607.

Anderson, M. R. & J. Kalff. 1986. Regulation of submerged aquatic plant distribution in a uniform area of a weedbed. *J. Ecol.* **74**: 953–961.

Anderson, L. C. & R. Kral. 2008. *Xyris panacea* (Xyridaceae)—a new yellow-eyed grass from the Florida Panhandle. *J. Bot. Res. Inst. Texas* **2**: 1–5.

Anderson, N. O. & R. T. Olsen. 2015. A vast array of beauty: the accomplishments of the father of American ornamental breeding, Luther Burbank. *HortScience* **50**: 161–188.

Anderson, R. R., R. G. Brown & R. D. Rappleye. 1968. Water quality and plant distribution along the upper Patuxent River, Maryland. *Chesapeake Sci.* **9**: 145–156.

Anderson, L. C., C. D. Zeis & S. F. Alam. 1974. Phytogeography and possible origins of *Butomus* in North America. *Bull. Torrey Bot. Club* **101**: 292–296.

Anderson, R. C., J. S. Fralish & J. M. Baskin (eds.). 1999. *Savannas, Barrens, and Rock Outcrop Plant Communities of North America*. Cambridge University Press, New York, NY. 470 pp.

Andersson, L. 1981. Revision of the *Thalia geniculata* complex (Marantaceae). *Nordic J. Bot.* **1**: 48–56.

Andersson, L. 1998. Marantaceae. Pp. 278–293 *In*: K. Kubitzki, H. Huber, P. J. Rudall, P. S. Stevens & T. Stützel (eds.), *The Families and Genera of Vascular Plants, Vol. IV. Flowering Plants, Monocotyledons: Alismatanae and Commelinanae (Except Gramineae)*. Springer-Verlag, Berlin, Germany.

Andersson, E., C. Nilsson & M. E. Johansson. 2000. Plant dispersal in boreal rivers and its relation to the diversity of riparian flora. *J. Biogeogr.* **27**: 1095–1106.

Andreas, B. K. & G. R. Bryan. 1990. The vegetation of three *Sphagnum*-dominated basin-type bogs in northeastern Ohio. *Ohio J. Sci.* **90**: 54–66.

Andrews, O. V. 1915. An ecological survey of Lake Butte des Morts bog, Oshkosh, Wisconsin. Bull. Wisconsin Nat. Hist. Soc. **13**: 196–211.

Anning, A. K., P. E. Korsah & P. Addo-Fordjour. 2013. Phytoremediation of wastewater with *Limnocharis flava*, *Thalia geniculata* and *Typha latifolia* in constructed wetlands. *Int. J. Phytoremed.* **15**: 452–464.

Ansari, R. & N. P. Balakrishnan. 1994. *The Family Eriocaulaceae in India*. Bishen Singh Mahendra Pal Singh, Dehra Dun. 195 pp.

Anuradha, J., T. Abbasi & S. A. Abbasi. 2015. An eco-friendly method of synthesizing gold nanoparticles using an otherwise worthless weed pistia (*Pistia stratiotes* L.). *J. Adv. Res.* **6**: 711–720.

Angier, B. 1974. *Field Guide to Edible Wild Plants*. Stackpole Books, Mechanicsburg, PA. 255 pp.

Ankutowicz, E. J. & R. A. Laird. 2018. Offspring of older parents are smaller—but no less bilaterally symmetrical—than offspring of younger parents in the aquatic plant *Lemna turionifera*. *Ecol. Evol.* **8**: 679–687.

APG. 2003. An update of the Angiosperm Phylogeny Group classification for the orders and families of flowering plants: APG II. *Bot. J. Linn. Soc.* **141**: 399–436.

APG. 2009. An update of the Angiosperm Phylogeny Group classification for the orders and families of flowering plants: APG III. *Bot. J. Linn. Soc.* **161**: 105–121.

APG. 2016. An update of the Angiosperm Phylogeny Group classification for the orders and families of flowering plants: APG IV. *Bot. J. Linn. Soc.* **181**: 1–20.

Appenroth, K.-J. 2002. Co-action of temperature and phosphate in inducing turion formation in *Spirodela polyrhiza* (great duckweed). *Plant Cell Environ.* **25**: 1079–1085.

Appenroth, K.-J. & G. Nickel. 2010. Turion formation in *Spirodela polyrhiza*: the environmental signals that induce the developmental process in nature. *Physiol. Plant.* **138**: 312–320.

Appenroth, K.-J., J. Opfermann, W. Hertel & H. Augsten. 1989. Photophysiology of turion germination in *Spirodela polyrhiza* (L.) Schleiden. II. Influence of after-ripening on germination kinetics. *J. Plant Physiol.* **135**: 274–279.

Appenroth, K.-J., W. Hertel & H. Augsten. 1990a. Phytochrome control of turion formation in *Spirodela polyrhiza* L. Schleiden. *Ann. Bot.* **66**: 163–168.

Appenroth, K.-J., W. Hertel & H. Augsten. 1990b. Photophysiology of turion germination in *Spirodela polyrhiza* (L.) Schleiden. The cause of germination inhibition by overcrowding. *Biol. Plant.* **32**: 420–428.

Appenroth, K.-J., H. Augsten & H. Mohr. 1992. Photophysiology of turion germination in *Spirodela polyrhiza* (L.) Schleiden. X. Role of nitrate in the phytochrome-mediated response. *Plant Cell Environ.* **15**: 743–748.

Appenroth, K.-J., N. Borisjuk & E. Lam. 2013a. Telling duckweed apart: genotyping technologies for the *Lemnaceae. Chin. J. Appl. Environ. Biol.* **19**: 1–10.

Appenroth, K.-J., L. Palharini & P. Ziegler. 2013b. Low-molecular weight carbohydrates modulate dormancy and are required for post-germination growth in turions of *Spirodela polyrhiza. Plant Biol.* **15**: 284–291.

Appenroth, K.-J., D. J. Crawford & D. H. Les. 2015. After the genome sequencing of duckweed–how to proceed with research on the fastest growing angiosperm? *Plant Biol.* **17**: 1–4.

Appenroth, K.-J., K. S. Sree, V. Böhm, S. Hammann, W. Vetter, M. Leiterer & G. Jahreis. 2017. Nutritional value of duckweeds (*Lemnaceae*) as human food. *Food Chem.* **217**: 266–273.

Arata, A. A. 1959. Ecology of muskrats in strip-mine ponds in southern Illinois. *J. Wildl. Manag.* **23**: 177–186.

Archer, S. K., E. W. Stoner & C. A. Layman. 2015. A complex interaction between a sponge (Halichondria *melanadocia*) and a seagrass (*Thalassia testudinum*) in a subtropical coastal ecosystem. *J. Exp. Mar. Biol. Ecol.* **465**: 33–40.

Archibald, J. K., S. R. Kephart, K. E. Theiss, A. L. Petrosky & T. M. Culley. 2015. Multilocus phylogenetic inference in subfamily Chlorogaloideae and related genera of *Agavaceae*–Informing questions in taxonomy at multiple ranks. *Mol. Phylogenet. Evol.* **84**: 266–283.

Ardenghi, N. M. G., W. P. Armstrong & D. Paganelli. 2017. *Wolffia columbiana* (*Araceae, Lemnoideae*): first record of the smallest alien flowering plant in southern Europe and Italy. *Bot. Lett.* **164**: 121–127.

Arditti, J. 1966. Orchids. *Sci. Am.* **214**(1): 70–81.

Arditti, J. & T. R. Pray. 1969. Dormancy factors in iris (*Iridaceae*) seeds. *Am. J. Bot.* **56**: 254–259.

Arditti, J. & A. K. A. Ghani. 2000. Tansley review no. 110. Numerical and physical properties of orchid seeds and their biological implications. *New Phytol.* **145**: 367–421.

Arditti, J., J. D. Michaud & P. L. Healey. 1979. Morphometry of orchid seeds. I. *Paphiopedilum* and native California and related species of Cypripedium. *Am. J. Bot.* **66**: 1128–1137.

Arditti, J., J. D. Michaud & P. L. Healey. 1980. Morphometry of orchid seeds. II. Native California and related species of *Calypso, Cephalanthera, Corallorhiza* and *Epipactis. Am. J. Bot.* **67**: 347–360.

Arditti, J., J. D. Michaud & A. P. Oliva. 1981. Seed germination of North American orchids. I. Native California and related species of Calypso, *Epipactis, Goodyera, Piperia*, and *Platanthera. Bot. Gaz.* **142**: 442–453.

Arditti, J., J. D. Michaud & A. P. Oliva. 1982. Practical germination of North American and related orchids. I. *Epipactis atrorubens, Epipactis gigantea* and *Epipactis helleborine. Am. Orchid Soc. Bull.* **51**: 162–171.

Aremu, A. & J. van Staden. 2013. The genus *Tulbaghia* (*Alliaceae*) - a review of its ethnobotany, pharmacology, phytochemistry and conservation needs. *J. Ethnopharmacol.* **149**: 387–400.

Arenas, P. & R. M. Azorero. 1977. Plants used as means of abortion, contraception, sterilization and fecundation by Paraguayan indigenous people. *Econ. Bot.* **31**: 302–306.

Aresco, M. J. 2010. Competitive interactions of two species of freshwater turtles, a generalist omnivore and an herbivore, under low resource conditions. *Herpetologica* **66**: 259–268.

Arft, A. M. & T. A. Ranker. 1998. Allopolyploid origin and population genetics of the rare orchid *Spiranthes diluvialis. Am. J. Bot.* **85**: 110–122.

Argue, C. L. 2011. *The Pollination Biology of North American Orchids: Volume 1: North of Florida and Mexico.* Springer, New York, NY. 228 pp.

Argue, C. L. 2012a. Tribe *Arethuseae* (*Calopogon* R. Brown and Arethusa L.) and subfamily Vanilloideae (part one) (Pogonia Jussieu). Pp. 147–164 *In: The Pollination Biology of North American Orchids*: Volume 2. Springer, New York, NY.

Argue, C. L. 2012b. Sections Irapeana and Bifolia. Pp. 75–78 *In: The Pollination Biology of North American Orchids*: Volume 1. Springer, New York, NY.

Argue, C. L. 2012c. Section Cypripedium. Pp. 47–62 *In: The Pollination Biology of North American Orchids*: Volume 1. Springer, New York, NY.

Argue, C. L. 2012d. Tribe *Neottieae*. Pp. 55–83 *In: The Pollination Biology of North American Orchids*: Volume 2. Springer, New York, NY.

Argue, C. L. 2012e. Tribe *Malaxideae*. Pp. 91–104 *In: The Pollination Biology of North American Orchids*: Volume 2. Springer, New York, NY.

Argue, C. L. 2012f. Subtribes Goodyerinae and Cranichidinae. Pp. 3–18 *In: The Pollination Biology of North American Orchids*: Volume 2. Springer, New York, NY.

Argue, C. L. 2012g. Subtribe Spiranthinae. Pp. 19–52 *In: The Pollination Biology of North American Orchids*: Volume 2. Springer, New York, NY.

Argus, G. W. 1968. Contributions to the flora of boreal Saskatchewan. *Rhodora* **70**: 200–214.

Arlen-Pouliot, Y. & S. Payette. 2015. The influence of climate on pool inception in boreal fens. *Botany* **93**: 637–649.

Armellina, A. D., C. R. Bezic & O. A. Gajardo. 1996. Propagation and mechanical control of *Potamogeton illinoensis* Morong in irrigation canals in Argentina. *J. Aquat. Plant Manag.* **34**: 12–14.

Armentano, T. V., D. T. Jones, M. S. Ross & B. W. Gamble. 2002. Vegetation pattern and process in tree islands of the southern Everglades and adjacent areas. Pp. 225–281 *In:* F. H. Sklar & A. van der Valk (eds.), *Tree Islands of the Everglades.* Springer, Dordrecht, The Netherlands.

Armitage, J. & B. Phillips. 2011. A hybrid swamp lantern. *Plantsman* **10**: 155–157.

Armstrong, M. 1915. *Field Book of Western Wild Flowers.* The Knickerbocker Press, New York, NY. 598 pp.

Armstrong, W. P. 2012. Wolffia. Watermeal. Pp. 1301–1302 *In:* B. G. Baldwin, D. H. Goldman, D. J. Keil, R. Patterson, T. J. Rosatti & D. H. Wilken (eds.), *The Jepson Manual: Vascular Plants of California*, 2nd ed. University of California Press, Berkeley, CA.

Armstrong, R. & D. M. Moore. 1957. Botanical aspects of Massard Prairie Arkansas. *J. Arkansas Acad. Sci.* **10**: 44–57.

Armstrong, W. P. & R. F. Thorne. 1984. The genus *Wolffia* (*Lemnaceae*) in California. *Madroño* **31**: 171–179.

Armstrong, W. P., W. Kelley & D. H. Wilken. 1989. Noteworthy collections. *Madroño* **36**: 283–286.

Armstrong, N., D. Planas & E. Prepas. 2003. Potential for estimating macrophyte surface area from biomass. *Aquat. Bot.* **75**: 173–179.

Arnett, M., A. M. Huber, K. M. Stevenson & S. Haultain. 2014. Vascular flora of Devils Postpile National Monument, Madera County, California. *Madroño* **61**: 367–388.

Arnold, M. L. 1993. *Iris nelsonii* (*Iridaceae*): origin and genetic composition of a homoploid hybrid species. *Am. J. Bot.* **80**: 577–583.

Arnold, M. L. 1994. Natural hybridization and Louisiana irises. *BioScience* **44**: 141–147.

Arnold, M. L. 2000. Anderson's paradigm: Louisiana irises and the study of evolutionary phenomena. *Mol. Ecol.* **9**: 1687–1698.

Arnold, M. L., J. L. Hamrick & B. D. Bennett. 1990. Allozyme variation in Louisiana irises: a test for introgression and hybrid speciation. *Heredity* **65**: 297–306.

Arnold, M. L., C. M. Buckner & J. J. Robinson. 1991. Pollen-mediated introgression and hybrid speciation in Louisiana irises. *Proc. Natl. Acad. Sci. U.S.A.* **88**: 1398–1402.

Arnold, M. L., J. L. Hamrick & B. D. Bennett. 1993. Interspecific pollen competition and reproductive isolation in Iris. *J. Hered.* **84**: 13–16.

Arnold, M. L., S. Tang, S. J. Knapp & N. H. Martin. 2010. Asymmetric introgressive hybridization among Louisiana *Iris* species. *Genes* **1**: 9–22.

Arnold, T., C. Mealey, H. Leahey, A. W. Miller, J. M. Hall-Spencer, M. Milazzo & K. Maers. 2012. Ocean acidification and the loss of phenolic substances in marine plants. *PLoS One* **7**(4): e35107.

Aronson, J. A. 1989. *HALOPH: A Data Base of Salt Tolerant Plants of the World*. Office of Arid Lands Studies, The University of Arizona, Tuscon, AZ. 77 pp.

Arsenault, J. 1984. Field trip reports. *Bartonia* **50**: 71.

Arts, G. H. P., J. G. M. Roelofs & M. J. H. De Lyon. 1990. Differential tolerances among soft-water macrophyte species to acidification. *Can. J. Bot.* **68**: 2127–2134.

Asamoah, S. A. & E. W. Bork. 2010. Drought tolerance thresholds in cattail (*Typha latifolia*): a test using controlled hydrologic treatments. *Wetlands* **30**: 99–110.

Ashton, M. J. & J. B. Layzer. 2010. Summer microhabitat use by adult and young-of-year snail darters (*Percina tanasi*) in two rivers. *Ecol. Freshw. Fish* **19**: 609–617.

Aston, H. I. 1967. Aquatic angiosperms: records of four introduced species new to Victoria. *Muelleria* **1**: 169–174.

Aston, H. I. & S. W. L. Jacobs. 1980. *Hydrocleys nymphoides* (*Butomaceae*) in Australia. *Muelleria* **4**: 285–293.

Atapaththu, K. S. S. & T. Asaeda. 2015. Growth and stress responses of Nuttall's waterweed *Elodea nuttallii* (Planch) St. John to water movements. *Hydrobiologia* **747**: 217–233.

Atwood, E. L. 1950. Life history studies of nutria, or coypu, in coastal Louisiana. *J. Wildl. Manag.* **14**: 249–265.

Atwood, Jr., J. T. 1984. The relationships of the slipper orchids (subfamily Cypripedioideae, Orchidaceae). *Selbyana* **7**: 129–247.

Auclair, A. N., A. Bouchard & J. Pajaczkowski. 1973. Plant composition and species relations on the Huntingdon Marsh, Quebec. *Can. J. Bot.* **51**: 1231–1247.

Aulbach-Smith, C. A. & S. J. de Kozlowski. 1996. *Aquatic and Wetland Plants of South Carolina*, 2nd ed. South Carolina Department of Natural Resources, Columbia, SC. 128 pp.

Aust, W. M., S. H. Schoenholtz, T. W. Zaebst & B. A. Szabo. 1997. Recovery status of a tupelo-cypress wetland seven years after disturbance: silvicultural implications. *For. Ecol. Manag.* **90**: 161–169.

Austin, D. F. 1998. Plant adaptations in the Fakahatchee. *Palmetto* **18**(3): 7–8.

Austin, D. F. 2004. *Florida Ethnobotany*. CRC Press, Boca Raton, FL. 952 pp.

Austin, G. & D. J. Cooper. 2016. Persistence of high elevation fens in the Southern Rocky Mountains, on Grand Mesa, Colorado, U.S.A. *Wetl. Ecol. Manag.* **24**: 317–334.

Avis, P., B. Brown, K. Buscher, F. Coskun, P. Coulling, S. Hanford, J. Harrod, J. Horn, C. Mankoff, B. Peet, L. Prince, A. Stanford, S. Seiberling, R. White & K. Wurdack. 1997. Proceedings. UNC sometimes annual phytogeographical excusion to the Florida panhandle. March 8–14, 1997. Available online: http://labs.bio.unc.edu/Peet/PEL/PGE/PGE1997_report.pdf [accessed 16 March, 2019].

Ayyad, S.-E. N., A. S. Judd, W. T. Shier & T. R. Hoye. 1998. Otteliones A and B: potently cytotoxic 4-methylene-2-cyclohexenones from *Ottelia alismoides*. *J. Organic Chem.* **63**: 8102–8106.

Azuma, H. & H. Tobe. 2011. Molecular phylogenetic analyses of Tofieldiaceae (Alismatales): family circumscription and intergeneric relationships. *J. Plant Res.* **124**: 349–357.

Azuma, H. & M. Toyota. 2012. Floral scent emission and new scent volatiles from *Acorus* (Acoraceae). *Biochem. Syst. Ecol.* **41**: 55–61.

Babayemi, O. J., M. A. Bamikole & A. B. Omojola. 2006. Evaluation of the nutritive value and free choice intake of two aquatic weeds (*Nephrolepis biserrata* and *Spirodela polyrhiza*) by West African dwarf goats. *Trop. Subtrop. Agroecosyst.* **6**: 15–21.

Back, A. J., P. Kron & S. C. Stewart. 1996. Phenological regulation of opportunities for within-inflorescence geitonogamy in the clonal species, *Iris versicolor* (Iridaceae). *Am. J. Bot.* **83**: 1033–1040.

Backman, T. W. H. 1991. Genotypic and phenotypic variability of *Zostera marina* on the west coast of North America. *Can. J. Bot.* **69**: 1361–1371.

Baden, III, J., W. T. Batson & R. Stalter. 1975. Factors affecting the distribution of vegetation of abandoned rice fields, Georgetown Co., South Carolina. *Castanea* **40**: 171–184.

Badzinski, S. S., C. D. Ankney & S. A. Petrie. 2006. Influence of migrant tundra swans (*Cygnus columbianus*) and Canada geese (*Branta canadensis*) on aquatic vegetation at Long Point, Lake Erie, Ontario. *Hydrobiologia* **567**: 195–211.

Bailey, D. E., J. E. Perry & D. A. DeBerry. 2006. *Aeschynomene virginica* (Fabaceae) habitat in a tidal marsh, James City County, Virginia. *Banisteria* **27**: 3–9.

Bailey, M., S. A. Petrie & S. S. Badzinski. 2008. Diet of mute swans in lower Great Lakes coastal marshes. *J. Wildl. Manag.* **72**: 726–732.

Bains, J. S., V. Dhuna, J. Singh, S. S. Kamboj, K. K. Nijjar & J. N. Agrewala. 2005. Novel lectins from rhizomes of two *Acorus* species with mitogenic activity and inhibitory potential towards murine cancer cell lines. *Int. Immunopharmacol.* **5**: 1470–1478.

Bajwa, D. S., E. D. Sitz, S. G. Bajwa & A. R. Barnick. 2015. Evaluation of cattail (*Typha* spp.) for manufacturing composite panels. *Ind. Crops Prod.* **75**: 195–199.

Bakenhaster, M. D. & J. S. Knight-Gray. 2016. New diet data for *Mola mola* and *Masturus lanceolatus* (Tetraodontiformes: Molidae) off Florida's Atlantic coast with discussion of historical context. *Bull. Mar. Sci.* **92**: 497–511.

Baker, F. C. 1910. The ecology of the Skokie Marsh area, with special reference to the Mollusca. *Illinois Nat. Hist. Surv. Bull.* **8**: 441–499.

Baker, H. G. 1972. A fen on the northern California coast. *Madroño* **21**: 405–416.

Baker, F. C. 1975. The littoral macrophyte vegetation of southeastern Devils Lake. *Trans. Wisconsin Acad. Sci.* **63**: 66–71.

Baker, P., F. Zimmanck & S. M. Baker. 2010. Feeding rates of an introduced freshwater gastropod *Pomacea insularum* on native and nonindigenous aquatic plants in Florida. *J. Molluscan Stud.* **76**: 138–143.

Balci, P. & J. H. Kennedy. 2003. Comparison of chironomids and other macroinvertebrates associated with *Myriophyllum spicatum* and *Heteranthera dubia*. *J. Freshw. Ecol.* **18**: 235–247.

Balciunas, J. K. & M. C. Minno. 1985. Insects damaging hydrilla in the USA. *J. Aquat. Plant Manag.* **23**: 77–83.

Baldridge, A. K. & D. M. Lodge. 2014. Long-term studies of crayfish-invaded lakes reveal limited potential for macrophyte recovery from the seed bank. *Freshw. Sci.* **33**: 788–797.

Baldwin, W. K. W. 1961. *Malaxis paludosa* (L.) Sw. in the Hudson Bay lowlands. *Can. Field-Nat.* **75**: 74–77.

Baldwin, Jr., J. T. & B. M. Speese. 1955. Chromosomes of taxa of the Alismataceae in the range of Gray's Manual. *Am. J. Bot.* **42**: 406–411.

Baldwin, J. R. & J. R. Lovvorn. 1994. Expansion of seagrass habitat by the exotic *Zostera japonica*, and its use by dabbling ducks and brant in Boundary Bay, British Columbia. *Mar. Ecol. Prog. Ser.* **103**: 119–127.

Baldwin, A. H. & I. A. Mendelssohn. 1998. Effects of salinity and water level on coastal marshes: an experimental test of disturbance as a catalyst for vegetation change. *Aquat. Bot.* **61**: 255–268.

Baldwin, B. G., D. H. Goldman, D. J. Keil, R. Patterson, T. J. Rosatti & D. H. Wilken (eds.). 2012. *The Jepson Manual: Vascular Plants of California*, 2nd ed. University of California Press, Berkeley, CA. 1,568 pp.

Ball, D. & J. R. Freeland. 2013. Synchronous flowering times and asymmetrical hybridization in *Typha latifolia* and *T. angustifolia* in northeastern North America. *Aquat. Bot.* **104**: 224–227.

Ballantine, D. & H. J. Humm. 1975. Benthic algae of the Anclote Estuary I. Epiphytes of seagrass leaves. *Florida Sci.* **38**: 150–162.

Baltisberger, M., A. Kocyan, V. V. Chepinoga, A. A. Gnutikov & I. V. Enushchenko. 2010. IAPT/IOPB chromosome data 9. *Taxon* **59**: 1298–1302.

Bance, H. M. 1946. A comparative account of the structure of *Potamogeton filiformis* Pers. and *P. pectinatus* L. in relation to the identity of a supposed hybrid of these species. *Trans. Bot. Soc. Edinburgh* **34**: 361–367.

Bando, K. J. 2006. The roles of competition and disturbance in a marine invasion. *Biol. Invasions* **8**: 755–763.

Baniszewski, J., J. P. Cuda, S. A. Gezan, S. Sharma & E. N. I. Weeks. 2016. Stem fragment regrowth of *Hydrilla verticillata* following desiccation. *J. Aquat. Plant Manag.* **54**: 53–60.

Barabe, D. & M. Labrecque. 1985. Vascularisation de la fleur d'*Orontium aquaticum* L. (Aracées). *Bull. Soc. Bot. France* **132**: 133–145.

Baranowski, R. M. 1979. Notes on the biology of *Ischnodemus oblongus* and *I. fulvipes* with descriptions of the immature stages (Hemiptera: Lygaeidae). *Ann. Entomol. Soc. Am.* **72**: 655–658.

Barber, B. J. & P. J. Behrens. 1985. Effects of elevated temperature on seasonal *in situ* leaf productivity of *Thalassia testudinum* Banks ex König and *Syringodium filiforme* Kützing. *Aquat. Bot.* **22**: 61–69.

Barbolani, E., M. Clauser, F. Pantani & R. Gellini. 1986. Residual heavy metal (Cu and Cd) removal by *Iris Pseudacorus*. *Water Air Soil Pollut.* **28**: 277–282.

Barbour, M. G. & S. R. Radosevich. 1979. ^{14}C Uptake by the marine angiosperm *Phyllospadix scouleri*. *Am. J. Bot.* **66**: 301–306.

Barbour, M. G., A. I. Solomeshch, R. F. Holland, C. W. Witham, R. L. Macdonald, S. S. Cilliers, J. A. Molina, J. J. Buck & J. M. Hillman. 2005. Vernal pool vegetation of California: communities of long-inundated deep habitats. *Phytocoenologia* **35**: 177–200.

Barbour, M. G., T. Keeler-Wolf & A. A. Schoenherr. 2007. *Terrestrial Vegetation of California*, 3rd ed. University of California Press, Berkeley, CA. 712 pp.

Barcelos, A. R., P. E. D. Bobrowiec, T. M. Sanaiotti & R. Gribel. 2013. Seed germination from lowland tapir (*Tapirus terrestris*) fecal samples collected during the dry season in the northern Brazilian Amazon. *Integr. Zool.* **8**: 63–73.

Barker, A., U. Schaffner & J.-L. Boevé. 2002. Host specificity and host recognition in a chemically-defended herbivore, the tenthredinid sawfly *Rhadinoceraea nodicornis*. *Entomol. Exp. Appl.* **104**: 61–68.

Barko, J. W. & R. M. Smart. 1983. Effects of organic matter additions to sediment on the growth of aquatic plants. *J. Ecol.* **71**: 161–175.

Barko, J. W., D. G. Hardin & M. S. Matthews. 1982. Growth and morphology of submersed freshwater macrophytes in relation to light and temperature. *Can. J. Bot.* **60**: 877–887.

Barks, P. M., Z. W. Dempsey, T. M. Burg & R. A. Laird. 2018. Among-strain consistency in the pace and shape of senescence in duckweed. *J. Ecol.* **106**: 2132–2145.

Barksdale, L. 1937. The occurrence of *Liparis loeselii* and *Habenaria bracteata* in North Carolina. *J. Elisha Mitchell Sci. Soc.* **53**: 137–138.

Barnard, J. 1882. Economic value of the aquatic plant *Typha latifolia*. *Pap. Proc. Royal Soc. Tasmania* **17**: 163–167.

Barnes, J. K. 1976. Effect of temperature on development, survival, oviposition, and diapause in laboratory populations of *Sepedon fuscipennis* (Diptera: Sciomyzidae). *Environ. Entomol.* **5**: 1089–1098.

Barnes, M. A., C. L. Jerde, D. Keller, W. L. Chadderton, J. G. Howeth & D. M. Lodge. 2013. Viability of aquatic plant fragments following desiccation. *Invasive Plant Sci. Manag.* **6**: 320–325.

Barnett, O. W. 1986. *Iris fulva* mosaic virus. *In*: J. Antoniw & M. Adams (eds.), *Descriptions of Plant Viruses*. Association of Applied Plant Biologists. Available online: http://www.dpvweb.net/dpv/showdpv.php?dpvno=310 [accessed 28 January, 2019].

Baron, J. L. & M. L. Ostrofsky. 2010. The effects of macrophyte tannins on the epiphytic macroinvertebrate assemblages in Sandy Lake, Pennsylvania. *J. Freshw. Ecol.* **25**: 457–465.

Barrat-Segretain, M.-H. 1996. Strategies of reproduction, dispersion, and competition in river plants: a review. *Vegetatio* **123**: 13–37.

Barrat-Segretain, M.-H. 2004. Growth of *Elodea canadensis* and *Elodea nuttallii* in monocultures and mixture under different light and nutrient conditions. *Archiv Hydrobiol.* **161**: 133–144.

Barrat-Segretain, M.-H. & G. Bornette. 2000. Regeneration and colonization abilities of aquatic plant fragments: effect of disturbance seasonality. *Hydrobiologia* **421**: 31–39.

Barrat-Segretain, M.-H. & B. Cellot. 2007. Response of invasive macrophyte species to drawdown: the case of *Elodea* sp. *Aquat. Bot.* **87**: 255–261.

Barrat-Segretain, M.-H., G. Bornette & A. Hering-Vilas-Bôas. 1998. Comparative abilities of vegetative regeneration among aquatic plants growing in disturbed habitats. *Aquat. Bot.* **60**: 201–211.

Barrat-Segretain, M.-H., A. Elger, P. Sagnes & S. Puijalon. 2002. Comparison of three life-history traits of invasive *Elodea canadensis* Michx. and *Elodea nuttallii* (Planch.) H. St. John. *Aquat. Bot.* **74**: 299–313.

Barrett, S. C. H. 1978. Floral biology of *Eichhornia azurea* (Swartz) Kunth (Pontederiaceae). *Aquat. Bot.* **5**: 217–228.

Barrett, S. C. H. 1980a. Sexual reproduction in *Eichhornia crassipes* (water hyacinth). I. Fertility of clones from diverse regions. *J. Appl. Ecol.* **17**: 101–112.

Barrett, S. C. H. 1980b. Sexual reproduction in *Eichhornia crassipes* (water hyacinth). II. Seed production in natural populations. *J. Appl. Ecol.* **17**: 113–124.

Barrett, S. C. H. 1985. Floral trimorphism and monomorphism in continental and island populations of *Eichhornia paniculata* (Spreng.) Solms (Pontederiaceae). *Biol. J. Linn. Soc.* **25**: 41–60.

Barrett, S. C. H. 1988. Evolution of breeding systems in *Eichhornia* (Pontederiaceae): a review. *Ann. Missouri Bot. Gard.* **75**: 741–760.

Barrett, S. C. H. & J. L. Strother. 1978. Taxonomy and natural history of *Bacopa* (Scrophulariaceae) in California. *Syst. Bot.* **3**: 408–419.

Barrett, S. C. H. & D. E. Seaman. 1980. The weed flora of Californian rice fields. *Aquat. Bot.* **9**: 351–376.

Barrett, S. C. H. & I. W. Forno. 1982. Style morph distribution in new world populations of *Eichhornia crassipes* (Mart.) Solms-Laubach (water hyacinth). *Aquat. Bot.* **13**: 299–306.

Barrett, S. C. H. & S. W. Graham. 1997. Adaptive radiation in the aquatic plant family Pontederiaceae: insights from phylogenetic analysis. Pp. 225–258 *In*: T. J. Givnish & K. J. Sytsma (eds.), *Molecular Evolution and Adaptive Radiation*. Cambridge University Press, New York, NY.

Barrett, S. C. H. & B. C. Husband. 1997. Ecology and genetics of ephemeral plant populations: *Eichhornia paniculata* (Pontederiaceae) in northeast Brazil. *J. Hered.* **88**: 277–284.

Barrett, S. C. H., S. D. Price & J. S. Shore. 1983. Male fertility and anisoplethic population structure in tristylous *Pontederia cordata* (Pontederiaceae). *Evolution* **37**: 745–759.

Barrett, S. C. H., B. C. Husband & W. W. Cole. 1993. Variation in outcrossing rates in *Eichhornia paniculata*: temporal changes in populations of contrasting style morph structure. *Plant Species Biol.* **8**: 141–148.

Barrett, S. C. H., L. D. Harder & W. W. Cole. 1994. Effects of flower number and position on self-fertilization in experimental populations of *Eichhornia paniculata* (Pontederiaceae). *Funct. Ecol.* **8**: 526–535.

Barrett, S. C. H., R. I. Colautti & C. G. Eckert. 2008. Plant reproductive systems and evolution during biological invasion. *Mol. Ecol.* **17**: 373–383.

Barrett, C. F., W. J. Baker, J. R. Comer, J. G. Conran, S. C. Lahmeyer, J. H. Leebens-Mack, J. Li, G. S. Lim, D. R. Mayfield-Jones, L. Perez, J. Medina, J. C. Pires, C. Santos, D. W. Stevenson, W. B. Zomlefer & J. I. Davis. 2016. Plastid genomes reveal support for deep phylogenetic relationships and extensive rate variation among palms and other commelinid monocots. *New Phytol.* **209**: 855–870.

Barriault, I., M. Gibernau & D. Barabe. 2009. Flowering period, thermogenesis, and pattern of visiting insects in *Arisaema triphyllum* (Araceae) in Quebec. *Botany* **87**: 324–329.

Barry, S. C., T. K. Frazer & C. A. Jacoby. 2013. Production and carbonate dynamics of *Halimeda incrassata* (Ellis) Lamouroux altered by *Thalassia testudinum* Banks and Soland ex König. *J. Exp. Mar. Biol. Ecol.* **444**: 73–80.

Bart, H. L. 1989. Fish habitat association in an Ozark stream. *Environ. Biol. Fishes* **24**: 173–186.

Bartgis, R. L. 1992. The endangered sedge *Scirpus ancistrochaetus* and the flora of sinkhole ponds in Maryland and West Virginia. *Castanea* **57**: 46–51.

Bartgis, R. L. 1997. The distribution of the endangered plant *Ptilimnium nodosum* (Rose) Mathias (Apiaceae) in the Potomac River drainage. *Castanea* **62**: 55–59.

Barthlott, W., B. Große-Veldmann & N. Korotkova. 2014. Orchid seed diversity. *Englera* **32**: 1–245.

Bartholomew, C. S., D. Prowell & T. Griswold. 2006. An annotated checklist of bees (Hymenoptera: Apoidea) in longleaf pine savannas of southern Louisiana and Mississippi. *J. Kansas Entomol. Soc.* **79**: 184–198.

Bartling, F. T. 1830. *Ordines Naturales Plantarum eorumque Characteres et Affinitates adjecta Generum Enumeratione*. Sumtibus Dieterichianis, Gottingae. 498 pp.

Bartodziej, W. I. & G. Weymouth. 1995. Waterbird abundance and activity on waterhyacinth and egeria in the St. Marks River, Florida. *J. Aquat. Plant Manag.* **33**: 19–22.

Bartonek, J. C. & J. J. Hickey. 1969. Food habits of canvasbacks, redheads, and lesser scaup in Manitoba. *Condor* **71**: 280–290.

Basinger, M. A., J. S. Huston, R. J. Gates & P. A. Robertson. 1997. Vascular flora of Horseshoe Lake Conservation Area, Alexander County, Illinois. *Castanea* **62**: 82–99.

Basiouny, F. M., W. T. Haller & L. A. Garrard. 1978. Survival of hydrilla (*Hydrilla verticillata*) plants and propagules after removal from the aquatic habitat. *Weed Sci.* **26**: 502–504.

Baskin, C. C. & J. M. Baskin. 1998. *Seeds: Ecology, Biogeography, and Evolution of Dormancy and Germination*. Academic Press, San Diego, CA. 666 pp.

Baskin, C. C. & J. M. Baskin. 2003. Seed germination and propagation of *Xyris tennesseensis*, a federal endangered wetland species. *Wetlands* **23**: 116–124.

Baskin, J. M. & C. C. Baskin. 2015. Inbreeding depression and the cost of inbreeding on seed germination. *Seed Sci. Res.* **25**: 355–385.

Baskin, C. C., J. M. Baskin & W. W. McDearman. 1993. Seed germination ecophysiology of two *Zigadenus* (Liliaceae) species. *Castanea* **58**: 45–53.

Baskin, C. C., J. M. Baskin & E. W. Chester. 2003. Ecological aspects of seed dormancy-break and germination in *Heteranthera limosa* (Pontederiaceae), a summer annual weed of rice fields. *Weed Res.* **43**: 103–107.

Bateman, R. M., P. M. Hollingsworth, J. Preston, L. Yi-Bo, A. M. Pridgeon & M. W. Chase. 2003. Molecular phylogenetics and evolution of Orchidinae and selected Habenariinae (Orchidaceae). *Bot. J. Linn. Soc.* **142**: 1–40.

Bateman, R. M., K. E. James, Y.-B. Luo, R. K. Lauri, T. Fulcher, P. J. Cribb & M. W. Chase. 2009. Molecular phylogenetics and morphological reappraisal of the *Platanthera* clade (Orchidaceae: Orchidinae) prompts expansion of the generic limits of *Galearis* and *Platanthera*. *Ann. Bot.* **104**: 431–445.

Batista, J. A. N., K. S. Borges, M. W. F. de Faria, K. Proite, A. J. Ramalho, G. A. Salazar & C. van den Berg. 2013. Molecular phylogenetics of the species-rich genus *Habenaria* (Orchidaceae) in the New World based on nuclear and plastid DNA sequences. *Mol. Phylogenet. Evol.* **67**: 95–109.

Battauz, Y. S., S. B. José de Paggi & J. C. Paggi. 2017. Macrophytes as dispersal vectors of zooplankton resting stages in a subtropical riverine floodplain. *Aquat. Ecol.* **51**: 191–201.

Batygina, T. B., E. A. Bragina & V. E. Vasilyeva. 2003. The reproductive system and germination in orchids. *Acta Biol. Cracov.* Ser. Bot. **45**: 21–34.

Bauder, E. T. 1994. *Vernal Pool Habitat Restoration: Eastgate Mall, NAS Miramar, San Diego, California*. California Department of Fish and Game, Sacramento, CA. 33 pp.

Bauder, E. T. 2000. Inundation effects on small-scale plant distributions in San Diego, California vernal pools. *Aquat. Ecol.* **34**: 43–61.

Baxter, R. 2007. Hydrogeologic characterization of the bunched arrowhead, *Sagittaria fasciculata. Geological Society of America, Southeastern Section, 56th Annual Meeting (29–30 March, 2007).* [Abstract].

Beal, E. O. 1960. *Sparganium* (Sparganiaceae) in the southeastern United States. *Brittonia* **12**: 176–181.

Beal, E. O. 1977. A manual of marsh and aquatic vascular plants of North Carolina with habitat data. Technical Bulletin No. 247. North Carolina Agricultural Experiment Station, Raleigh, NC. 298 pp.

Beal, E. O., J. W. Wooten & R. B. Kaul. 1982. Review of the *Sagittaria engelmanniana* complex (Alismataceae) with environmental correlations. *Syst. Bot.* **7**: 417–432.

Beare, P. A. & J. B. Zedler. 1987. Cattail invasion and persistence in a coastal salt marsh: the role of salinity reduction. *Estuaries* **10**: 165–170.

Beas, B. J., L. M. Smith, K. R. Hickman, T. G. LaGrange & R. Stutheit. 2013. Seed bank responses to wetland restoration: do restored wetlands resemble reference conditions following sediment removal? *Aquat. Bot.* **108**: 7–15.

Beasley, V. 1999. Plants of the Araceae family (plants containing oxalate crystals and histamine releasers). *In*: V. Beasley (ed.), *Veterinary Toxicology*. International Veterinary Information Service, Ithaca, NY.

Beattie, A. J., D. E. Breedlove & P. R. Ehrlich. 1973. The ecology of the pollinators and predators of *Frasera speciosa*. *Ecology* **54**: 81–91.

Beatty, J. & E. C. Desjardins. 2009. Natural selection and history. *Biol. Philos.* **24**: 231–246.

Beaven, G. F. & H. J. Oosting. 1939. Pocomoke swamp: a study of a cypress swamp on the eastern shore of Maryland. *Bull. Torrey Bot. Club* **66**: 367–389.

Beck, J. L. & J. M. Peek. 2004. Herbage productivity and ungulate use of northeastern Nevada mountain meadows. *J. Range Manag.* **57**: 376–383.

Beck, M. W. & J. Alahuhta. 2017. Ecological determinants of *Potamogeton* taxa in glacial lakes: assemblage composition, species richness, and species-level approach. *Aquat. Sci.* **79**: 427–441.

Becker, R. 2017. Biologia reprodutiva de *Crinum americanum* L. – Amaryllidaceae. B.S. thesis. Universidade Federal do Rio Grande do Sul, Porto Alegre, Brazil. 16 pp.

Beckett, D. C., T. P. Aartila & A. C. Miller. 1992a. Invertebrate abundance on *Potamogeton nodosus*: effects of plant surface area and condition. *Can. J. Zool.* **70**: 300–306.

Beckett, D. C., T. P. Aartila & A. C. Miller. 1992. Contrasts in density of benthic invertebrates between macrophyte beds and open littoral patches in Eau Galle Lake, Wisconsin. *Am. Midl. Nat.* **127**: 77–90.

Becking, R. W. 1986. *Hastingsia atropurpurea* (Liliaceae: Asphodeleae). A new species from southwestern Oregon. *Madroño* **33**: 175–181.

Becking, R. W. 1989. Segregation of *Hastingsia serpentinicola*, sp. nov. from *Hastingsia alba* (Liliaceae: Asphodeleae). *Madroño* **36**: 208–216.

Becking, R. W. 2002. *Hastingsia* S. Watson. Pp. 310–312 *In*: Flora North America Editorial Committee (eds.), *Flora of North America North of Mexico, Vol. 26: Magnoliophyta: Liliidae: Liliales and Orchidales*. Oxford University Press, New York, NY.

Beck-Nielsen, D. & T. V. Madsen. 2001. Occurrence of vesicular–arbuscular mycorrhiza in aquatic macrophytes from lakes and streams. *Aquat. Bot.* **71**: 141–148.

Beecher, C. W. W., T. M. Sarg & J. M. Edwards. 1983. Occurrence and biosynthesis of 9-phenylphenalenones in callus tissue of *Lachnanthes tinctoria*. *J. Nat. Prod.* **46**: 932–933.

Beer, S. & R. G. Wetzel. 1982a. Photosynthetic carbon fixation pathways in *Zostera marina* and three Florida seagrasses. *Aquat. Bot.* **13**: 141–146.

Beer, S. & R. G. Wetzel. 1982b. Photosynthesis in submersed macrophytes of a temperate lake. *Plant Physiol.* **70**: 488–492.

Beer, S. & J. Rehnberg. 1997. The acquisition of inorganic carbon by the seagrass *Zostera marina*. *Aquat. Bot.* **56**: 277–283.

Beer, S., M. Bjork, F. Hellblom & L. Axelsson. 2002. Inorganic carbon utilization in marine angiosperms (seagrasses). *Funct. Plant Biol.* **29**: 349–354.

Begum, M., A. S. Juraimi, S. O. O. B. S. Rastan, R. Amartalingam & A. B. Man. 2006. Seedbank and seedling emergence characteristics of weeds in ricefield soils of the muda granary area in north-west peninsular Malaysia. *Biotropia* **13**: 11–21.

Behera, S., S. Rath, M. S. Akhtar & S. K. Naik. 2018. Biotechnological intervention through tissue culture in *Hedychium coronarium*: a potential anticancer plant. Pp. 551–564 *In*: M. Akhtar & M. Swamy (eds.), *Anticancer Plants: Natural Products and Biotechnological Implements*. Springer, Singapore.

Belden, R. C., W. B. Frankenberger, R. T. McBride & S. T. Schwikert. 1988. Panther habitat use in southern Florida. *J. Wildl. Manag.* **52**: 660–663.

Bell, S. S., M. S. Fonseca & W. J. Kenworthy. 2008. Dynamics of a subtropical seagrass landscape: links between disturbance and mobile seed banks. *Landscape Ecol.* **23**: 67–74.

Bellis, E. D. 1962. Cover value and escape habits of the wood frog in a Minnesota bog. *Herpetologica* **17**: 228–231.

Bellrose, F. C. 1941. Duck food plants of the Illinois River Valley. *Bull. Illinois. Nat. Hist. Surv.* **21**: 237–280.

Bellrose, F. C. & L. G. Brown. 1941. The effect of fluctuating water levels on the muskrat population of the Illinois River Valley. *J. Wildl. Manag.* **5**: 206–212.

Belyakov, E. A. & A. G. Lapirov. 2015. Fruit germination of some representatives of the family Sparganiaceae Rudolphi under laboratory conditions. *Inland Water Biol.* **8**: 33–37.

Bembower, W. 1911. Pollination notes from the Cedar Point region. *Ohio Nat.* **11**: 378–383.

Bendix, M., T. Tornbjerg & H. Brix. 1994. Internal gas transport in *Typha latifolia* L. and *Typha angustifolia* L. 1. Humidity-induced pressurization and convective throughflow. *Aquat. Bot.* **49**: 75–89.

Benedict, N. B. 1983. Plant associations of subalpine meadows, Sequoia National Park, California. *Arct. Alp. Res.* **15**: 383–396.

Bengtson, J. L. 1983. Estimating food consumption of free-ranging manatees in Florida. *J. Wildl. Manag.* **47**: 1186–1192.

Bennett A. 1901. *Potamogeton polygonifolius* in Newfoundland. *Bot. Gaz.* **32**: 58–59.

Bennett, B. D. & J. B. Grace. 1990. Shade tolerance and its effect on the segregation of two species of Louisiana iris and their hybrids. *Am. J. Bot.* **77**: 100–107.

Bennett, D. H. & J. M. DuPont. 1993. Fish habitat associations of the Pend Oreille River, Idaho. Project Report: Project F-73-R-15, Subproject VI, Study VII. University of Idaho, Moscow, ID. 124 pp.

Benoit, L. K. 2011. Cryptic speciation, genetic diversity and herbicide resistance in the invasive aquatic plant *Hydrilla verticillata* (L.f.) Royle (Hydrocharitaceae). Ph.D. dissertation. University of Connecticut, Storrs, CT. 117 pp.

Benoit, L. K. & D. H. Les. 2013. Rapid identification and molecular characterization of phytoene desaturase mutations in fluridone-resistant hydrilla (*Hydrilla verticillata*). *Weed Sci.* **61**: 32–40.

Benoit, L. K., D. H. Les, U. M. King, H. R. Na, L. Chen & N. P. Tippery. 2019. Extensive interlineage hybridization in the predominantly clonal *Hydrilla verticillata* (Hydrocharitaceae). *Am. J. Bot.* **106**: 1622–1637.

Benson, E. R., J. M. O'Neil & W. C. Dennison. 2008. Using the aquatic macrophyte *Vallisneria americana* (wild celery) as a nutrient bioindicator. *Hydrobiologia* **596**: 187–196.

Bentley, B. & T. S. Elias (eds.). 1983. *The Biology of Nectaries*. Columbia University Press, New York, NY. 259 pp.

Benton, Jr., A. R., W. P. James & J. W. Rouse, Jr. 1978. Evaoptranspiration from water hyacinth (*Eichhomia crassipes* (Mart.) Solms) in Texas reservoirs. *J. Am. Water Resour. Assoc.* **14**: 919–930.

Berdnikov, V. A., O. E. Kosterin & V. S. Bogdanova. 2002. Mortality of pollen grains may result from errors of meiosis: study of pollen tetrads in *Typha latifolia* L. *Heredity* **89**: 358–362.

Berkowitz, J. F., S. Page & C. V. Noble. 2014. Potential disconnect between observations of hydrophytic vegetation, wetland hydrology indicators, and hydric soils in unique pitcher plant bog habitats of the southern Gulf Coast. *Southeast. Nat.* **13**: 721–735.

Beshear, R. J. & J. O. Howell. 1976. A new species of *Pseudothrips*, with a key to the North American species. *Ann. Entomol. Soc. Am.* **69**: 1082–1084.

Bezona, N., D. Hensley, J. Yogi, J. Tavares, F. Rauch, R. Iwata, M. Kellison, M. Won & P. Clifford. 2009. Salt and wind tolerance of landscape plants for Hawai'i. Document L13. Cooperative Extension Service, University of Hawai'i at Mānoa, Honolulu, HI. 9 pp.

Bercu, R. 2015. Histoanatomical features of the aquatic plant *Helanthium tenellum* (Mart.) Britt. (Alismataceae). *Ann. West Univ. Timişoara, Ser. Biol.* **18**: 67–72.

Berg, C. O. 1949. Limnological relations of insects to plants of the genus *Potamogeton*. *Trans. Am. Microsc. Soc.* **68**: 279–291.

Berg, C. O. 1950a. Biology of certain Chironomidae reared from *Potamogeton*. *Ecol. Monogr.* **20**: 83–101.

Berg, C. O. 1950b. *Hydrellia* (Ephydridae) and some other acalyptrate Diptera reared from *Potamogeton*. *Ann. Entomol. Soc. Am.* **43**: 374–398.

Berger, M., R. O. Stephenson, P. Karczmarczyk & C. C. Gates. 1995. *Habitat Inventory of the Yukon Flats as Potential Wood Bison Range*. Alaska Department of Fish and Game, Fairbanks, AK. 38 pp.

Bergeron, A. L. & S. T. Pellerin. 2014. *Carex×cayouettei* (Cyperaceae), a new intersectional sedge hybrid from southern Québec, Canada. *Phytoneuron* **52**: 1–11.

Bergersen, E. P. 1969. Some factors affecting fish forage production in four Arizona lakes. M.S. thesis. University of Arizona, Tuscon, AZ. 89 pp.

Bergerud, A. T. 1972. Food habits of Newfoundland caribou. *J. Wildl. Manag.* **36**: 913–923.

Bergey, E. A., S. F. Balling, J. N. Collins, G. A. Lamberti & V. H. Resh. 1992. Bionomics of invertebrates within an extensive *Potamogeton pectinatus* bed of a California marsh. *Hydrobiologia* **234**: 15–24.

Bergman, R. D. 1973. Use of southern boreal lakes by postbreeding canvasbacks and redheads. *J. Wildl. Manag.* **37**: 160–170.

Bergman, B. G. & J. K. Bump. 2014. Mercury in aquatic forage of large herbivores: impact of environmental conditions, assessment of health threats, and implications for transfer across ecosystem compartments. *Sci. Total Environ.* **479**: 66–76.

Bergmann, B. A., J. Cheng, J. Classen & A.-M. Stomp. 2000. In vitro selection of duckweed geographical isolates for potential use in swine lagoon effluent renovation. *Bioresour. Technol.* **73**: 13–20.

Bergstrom, B. J., T. Farley, H. L. Hill, Jr. & T. Hon. 2000. Ecology and conservation of a frontier population of the round-tailed muskrat (*Neofiber alleni*). Pp. 74–82 *In*: B. R. Chapman &

J. Laerm (eds.), *Fourth Colloquium on the Conservation of Mammals in the Southeastern United States*. Occasional Paper Number 12. North Carolina Museum of Natural Sciences and the North Carolina Biological Survey, Raleigh, NC.

Berkowitz, J. F., S. Page & C. V. Noble. 2014. Potential disconnect between observations of hydrophytic vegetation, wetland hydrology indicators, and hydric soils in unique pitcher plant bog habitats of the southern Gulf Coast. *Southeast. Nat.* **13**: 721–735.

Bernard, J. M. & F. A. Bernard. 1973. Winter biomass in *Typha glauca* Godr. and *Sparganium eurycarpum* Engelm. *Bull. Torrey Bot. Club* **100**: 125–127.

Bernard, J. M. & F. A. Bernard. 1977. Winter standing crop and nutrient contents in five central New York wetlands. *Bull. Torrey Bot. Club* **104**: 57–59.

Bernardello, L. M. & E. A. Moscone. 1986. The karyotype of *Limnobium spongia* (Hydrocharitaceae). *Plant Syst. Evol.* **153**: 31–36.

Bernhardt, E. A. & J. M. Duniway. 1985. A new *Papulaspora* species from the pondweeds *Potamogeton nodosus* and *P. crispus* in California. *Can. J. Bot.* **63**: 429–431.

Bertsch, C. & G. R. Barreto. 2008. Diet of the yellow-knobbed curassow in the central Venezuelan llanos. *Wilson J. Ornithol.* **120**: 767–778.

Bérubé, V., L. Rochefort & C. Lavoie. 2017. Fen restoration: defining a reference ecosystem using paleoecological stratigraphy and present-day inventories. *Botany* **95**: 731–750.

Besançon, T. E. 2019. Carolina Redroot (*Lachnanthes caroliniana*) in cranberry: assessment of shoot and rhizome control with POST herbicides. *Weed Technol.* **33**: 210–216.

Bezdek, J. C., J. M. Keller, R. Krishnapuram, L. I. Kuncheva & N. R. Pal. 1999. Will the real iris data please stand up? *IEEE Trans. Fuzzy Syst.* **7**: 368–369.

Bhalla, P. R. & P. S. Sabharwal. 1972. Induction of flowering in *Lemna minor* by EDDHA. *Acta Bot. Neerl.* **21**: 200–202.

Bhanthumnavin, K. & M. G. McGarry. 1971. *Wolffia arrhiza* as a possible source of inexpensive protein. *Nature* **232**: 495.

Bharathan, G., G. Lambert & D. W. Galbraith. 1994. Nuclear DNA content of monocotyledons and related taxa. *Am. J. Bot.* **81**: 381–386.

Bhardwaj, M. & C. G. Eckert. 2001. Functional analysis of synchronous dichogamy in flowering rush, *Butomus umbellatus* (Butomaceae). *Am. J. Bot.* **88**: 2204–2213.

Bhatia, H. L. 1970. Grass carps can control aquatic weeds. *Indian Farming* **20**: 36–37.

Bhunia, D. & A. K. Mondal. 2014. The free amino acids (FAA) composition of seeds among some members of *Blyxa* Thou. (Hydrocharitaceae): a systematic approach. *Int. J. Curr. Res.* **6**: 10437–10445.

Bianchini, I., M. B. Cunha-Santino, J. A. M. Milan, C. J. Rodrigues & J. H. P. Dias. 2010. Growth of *Hydrilla verticillata* (L.f.) Royle under controlled conditions. *Hydrobiologia* **644**: 301–312.

Biber, P. D. & H. J. Cho. 2017. Shoalgrass in the Gulf of Mexico: a Mississippi perspective. *Southeast Geogr.* **57**: 203–206.

Bick, G. H., J. C. Bick & L. E. Hornuff. 1976. Behavior of *Chromagrion conditum* (Hagen) adults (Zygoptera: Coenagrionidae). *Odonatologica* **5**: 129–141.

Bicknell, E. P. 1901. Studies in *Sisyrinchium*-IX: the species of Texas and the southwest. *Bull. Torrey Bot. Club* **28**: 570–592.

Bidartondo, M. I., B. Burghardt, G. Gebauer, T. D. Bruns & D. J. Read. 2004. Changing partners in the dark: isotopic and molecular evidence of ectomycorrhizal liaisons between forest orchids and trees. *Proc. Roy. Soc. London Ser. B Biol. Sci.* **271**: 1799–1806.

Bielefeld, R. R. & A. D. Afton. 1992. Canvasback food density in the Mississippi River Delta, Louisiana: habitat and temporal differences. *Proc. Ann. Conf. S. E. Assoc. Game Fish Agencies* **46**: 97–103.

Bigley, R. E. 1981. The population biology of two intertidal seagrasses, *Zostera japonica* and *Ruppia maritima*, at Roberts Bank, British Columbia. M.S. thesis. University of British Columbia, Vancouver, BC. 205 pp.

Bigley, R. E. & J. L. Barreca. 1982. Evidence for synonymizing *Zostera americana* den Hartog with *Zostera japonica* Aschers. & Graebn. *Aquat. Bot.* **14**: 349–356.

Bigley, R. E. & P. G. Harrison. 1986. Shoot demography and morphology of *Zostera japonica* and *Ruppia maritima* from British Columbia, Canada. *Aquat. Bot.* **24**: 69–82.

Bini, L. M. & S. M. Thomaz. 2005. Prediction of *Egeria najas* and *Egeria densa* occurrence in a large subtropical reservoir (Itaipu Reservoir, Brazil-Paraguay). *Aquat. Bot.* **83**: 227–238.

Bintz, J. C. & S. W. Nixon. 2001. Responses of eelgrass *Zostera marina* seedlings to reduced light. *Mar. Ecol. Prog. Ser.* **223**: 133–141.

Birchenko, I. V. 2001. Genetic diversity and microsite characterization of the rare monkeyface orchid, *Platanthera integrilabia* (Orchidaceae), in the southeastern United States. M.S. thesis. Ohio University, Athens, OH. 61 pp.

Bird, K. T. & J. Jewett-Smith. 1994. Development of a medium and culture system for in vitro propagation of the seagrass *Halophila engelmannii*. *Can. J. Bot.* **72**: 1503–1510.

Bird, K. T., J. Jewett-Smith & M. S. Fonseca. 1994. Use of *in vitro* propagated *Ruppia maritima* for seagrass meadow restoration. *J. Coast. Res.* **10**: 732–737.

Bird, K. T., J. R. Johnson & J. Jewett-Smith. 1998. In vitro culture of the seagrass *Halophila decipiens*. *Aquat. Bot.* **60**: 377–387.

Birkenholz, D. E. 1963. A study of the life history and ecology of the round-tailed muskrat (*Neofiber alleni* True) in north-central Florida. *Ecol. Monogr.* **33**: 255–280.

Birks, H. H. 2002. Plant macrofossils. Pp. 49–74 *In*: J. P. Smol, H. J. B. Birks & W. M. Last (eds.), *Tracking Environmental Change Using Lake Sediments, Vol. 3: Terrestrial, Algal, and Siliceous Indicators*. Kluwer Academic Publishers, Dordrecht, The Netherlands.

Bishop, P. L. & T. T. Eighmy. 1989. Aquatic wastewater treatment using *Elodea nuttallii*. *J. Water Pollut. Control Fed.* **61**: 641–648.

Bitton, G., M. Ward & R. Dagan. 2005. Determination of the heavy metal binding capacity (HMBC) of environmental samples. Pp. 215–231 *In*: C. Blaise & J.-F. Férard (eds.), *Small-Scale Freshwater Toxicity Investigations*. Springer, Dordrecht, The Netherlands.

Björk, C. R. & P. W. Dunwiddie. 2004. Floristics and distribution of vernal pools on the Columbia Plateau of eastern Washington. *Rhodora* **106**: 327–347.

Björkqvist, I. 1967. Studies in *Alisma* L. I. Distribution, variation and germination. *Opera Bot.* **17**: 1–128.

Björkqvist, I. 1968. Studies in *Alisma* L. II. Chromosome studies, crossing experiments and taxonomy. *Opera Bot.* **19**: 1–138.

Bjorndal, K. A. 1996. Foraging ecology and nutrition of sea turtles. Pp. 199–231 *In*: P. L. Lutz & J. A. Musick (eds.), *The Biology of Sea Turtles, Volume 1*. CRC Press, Boca Raton, FL.

Bjorndal, K. A. & A. B. Bolten. 1992. Body size and digestive efficiency in a herbivorous freshwater turtle: advantages of small bite size. *Physiol. Zool.* **65**: 1028–1039.

Black, R. J. 1985. Salt tolerant plants for Florida. Document ENH26. University of Florida Cooperative Extension Service, Institute of Food and Agriculture Sciences, Gainesville, FL. 10 pp.

Black, M., K. J. Bradford & J. Vázquez-Ramos. 2000. *Seed Biology: Advances and Applications: Proceedings of the Sixth International Workshop on Seeds, Mérida, México, 1999*. CABI Publishing, Cambridge, MA. 508 pp.

Blackburn, R. D., J. M. Lawrence & D. E. Davis. 1961. Effects of light intensity and quality on the growth of *Elodea densa* and *Heteranthera dubia*. *Weeds* **9**: 251–257.

Blackwell, Jr., W. H. & K. P. Blackwell. 1974. The taxonomy of *Peltandra* (Araceae). *J. Elisha Mitchell Sci. Soc.* **90**: 137–140.

Blair, W. F. 1936. The Florida marsh rabbit. *J. Mammal.* **17**: 197–207.

Blair, A. 1975. Karyotypes of five plant species with disjunct distributions in Virginia and the Carolinas. *Am. J. Bot.* **62**: 833–837.

Blake, S. F. 1912. The forms of *Peltandra virginica*. *Rhodora* **14**: 102–106.

Blanchette, C. A., S. E. Worcester, D. Reed & S. J. Holbrook. 1999. Algal morphology, flow, and spatially variable recruitment of surfgrass *Phyllospadix torreyi*. *Mar. Ecol. Prog. Ser.* **184**: 119–128.

Blanchette, C. A., C. M. Miner, P. T. Raimondi, D. Lohse, K. E. K. Heady & B. R. Broitman. 2008. Biogeographical patterns of rocky intertidal communities along the Pacific coast of North America. *J. Biogeogr.* **35**: 1593–1607.

Blaney, S., D. Mazerolle, S. Robinson & A. Belliveau. 2014. *Ecological Surveys and Vascular Plant Inventory of the Bay of Islands Conservation Area*. Atlantic Canada Conservation Data Centre, Sackville, NB. 83 pp.

Blankinship, J. W. 1903. The plant-formations of eastern Massachusetts. *Rhodora* **5**: 124–137.

Blanton, F. S. 1939. Notes on some thrips collected in the vicinity of Babylon, Long Island, NY. *J. New York Entomol. Soc.* **47**: 83–94.

Blazey, E. B. & J. W. McClure. 1968. The distribution and taxonomic significance of lignin in the Lemnaceae. *Am. J. Bot.* **55**: 1240–1245.

Bleuel, M. T. 1923. Some unusual food plants. *School Sci. Math.* **23**: 369–376.

Bliss, S. A. & P. H. Zedler. 1998. The germination process in vernal pools: sensitivity to environmental conditions and effects on community structure. *Oecologia* **113**: 67–73.

Bliss, B. J. & J. Y. Suzuki. 2012. Genome size in *Anthurium* evaluated in the context of karyotypes and phenotypes. *AoB Plants* **2012**. doi: 10.1093/aobpla/pls006.

Block, T. A. & A. F. Rhoads. 2011. *Aquatic Plants of Pennsylvania. A Complete Reference Guide*. University of Pennsylvania Press, Philadelphia, PA. 308 pp.

Block, T. A. & A. F. Rhoads. 2013. *Critical Resources of Bald Mountain Section Lehigh Gorge State Park. Research Works (Botany) 7*. Western Pennsylvania Conservancy and the Pennsylvania Department of Conservation and Natural Resources, Bureau of State Parks, Harrisburg, PA. 24 pp.

Block, T. A., A. F. Rhoads & C. Loeffler. 2013. Resource recovery plan for glade spurge *Euphorbia purpurea* (Raf.) Fernald in Pennsylvania. *Res. Works (Botany)* **17**. Available online: https://repository.upenn.edu/morrisarboretum_botany-works/17. [accessed 4 February, 2020].

Blomquist, H. L. 1955. The genus *Xyris* L. in North Carolina. *J. Elisha Mitchell Sci. Soc.* **71**: 35–46.

Błońska, A., D. Halabowski & A. Sowa. 2016. Population structure of *Liparis loeselii* (L.) Rich. in relation to habitat conditions in the Warta River valley (Poland). *Biodivers. Res. Conserv.* **43**: 41–52.

Blüm [Bluem], V. & F. Paris. 2001. Aquatic food production modules in bioregenerative life support systems based on higher plants. *Adv. Space Res.* **27**: 1513–1522.

Blüm, V., M. Andriske, M. Eichhorn, K. Kreuzberg & M. P. Schreibman. 1995. A controlled aquatic ecological life support system (CAELSS) for combined production of fish and higher plant biomass suitable for integration into a lunar or planetary base. *Acta Astronautica* **37**: 361–371.

Blumenthal, A., H. R. Lerner, E. Werker & A. Poljakoff-Mayber. 1986. Germination preventing mechanisms in *Iris* seeds. *Ann. Bot.* **58**: 551–561.

Board, V. V., C. Allen & A. Reeves. 1993. Botanical survey of a cypress-tupelo swamp. *J. Arkansas Acad. Sci.* **47**: 20–22.

Bodkin, N. L. & F. H. Utech. 2002. *Melanthium* Linnaeus. Pp. 77–79 *In*: Flora North America Editorial Committee (eds.), *Flora of North America North of Mexico, Vol. 26: Magnoliophyta: Liliidae: Liliales and Orchidales*. Oxford University Press, New York, NY.

Bodner, M. 1994. Inorganic carbon source for photosynthesis in the aquatic macrophytes *Potamogeton natans* and *Ranunculus fluitans. Aquat. Bot.* **48**: 109–120.

Boedeltje, G. 2005. The role of dispersal, propagule banks and abiotic conditions in the establishment of aquatic vegetation. Doctoral dissertation. Radboud University, Nijmegen, The Netherlands. 224 pp.

Boedeltje, G., G. N. J. Ter Heerdt & J. P. Bakker. 2002. Applying the seedling-emergence method under waterlogged conditions to detect the seed bank of aquatic plants in submerged sediments. *Aquat. Bot.* **72**: 121–128.

Boedeltje, G., J. P. Bakker & G. N. J. ter Heerdt. 2003a. Potential role of propagule banks in the development of aquatic vegetation in backwaters along navigation canals. *Aquat. Bot.* **77**: 53–69.

Boedeltje, G., J. P. Bakker, R. M. Bekker, J. M. Van Groenendael & M. Soesbergen. 2003b. Plant dispersal in a lowland stream in relation to occurrence and three specific life-history traits of the species in the species pool. *J. Ecol.* **91**: 855–866.

Boedeltje, G. E. R., J. P. Bakker, A. T. Brinke, J. M. Van Groenendael & M. Soesbergen. 2004. Dispersal phenology of hydrochorous plants in relation to discharge, seed release time and buoyancy of seeds: the flood pulse concept supported. *J. Ecol.* **92**: 786–796.

Boedeltje, G., T. Spanings, G. Flik, B. J. A. Pollux, F. A. Sibbing & W. C. E. P. Verberk. 2015. Effects of seed traits on the potential for seed dispersal by fish with contrasting modes of feeding. *Freshw. Biol.* **60**: 944–959.

Boedeltje, G., E. Jongejans, T. Spanings & W. C. E. P. Verberk. 2016. Effect of gut passage in fish on the germination speed of aquatic and riparian plants. *Aquat. Bot.* **132**: 12–16.

Boehm, R., C. Kruse, D. Voeste, S. Barth & H. Schnabl. 2001. A transient transformation system for duckweed (*Wolffia columbiana*) using *Agrobacterium*-mediated gene transfer. *J. Appl. Bot.* **75**: 107–111.

Boese, B. L., K. E. Alayan, E. F. Gooch & B. D. Robbins. 2003. Desiccation index: a measure of damage caused by adverse aerial exposure on intertidal eelgrass (*Zostera marina*) in an Oregon (USA) estuary. *Aquat. Bot.* **76**: 329–337.

Boese, B. L., B. D. Robbins & G. Thursby. 2005. Desiccation is a limiting factor for eelgrass (*Zostera marina* L.) distribution in the intertidal zone of a northeastern Pacific (USA) estuary. *Bot. Mar.* **48**: 274–283.

Boestfleisch, C. & J. Papenbrock. 2017. Changes in secondary metabolites in the halophytic putative crop species *Crithmum maritimum* L., *Triglochin maritima* L. and *Halimione portulacoides* (L.) Aellen as reaction to mild salinity. *PLoS One* **12**(4): e0176303.

Bog, M., H. Baumbach, U. Schween, F. Hellwig, E. Landolt & K.-J. Appenroth. 2010. Genetic structure of the genus *Lemna* L. (Lemnaceae) as revealed by amplified fragment length polymorphism. *Planta* **232**: 609–619.

Bog, M., P. Schneider, F. Hellwig, S. Sachse, E. Z. Kochieva, E. Martyrosian, E. Landolt & K.-J. Appenroth. 2013. Genetic characterization and barcoding of taxa in the genus *Wolffia* Horkel ex Schleid. (Lemnaceae) as revealed by two plastidic markers and amplified fragment length polymorphism (AFLP). *Planta* **237**: 1–13.

Bog, M., U. Lautenschlager, M. F. Landrock, E. Landolt, J. Fuchs, K. S. Sree, C. Oberprieler & K.-J. Appenroth. 2015. Genetic characterization and barcoding of taxa in the genera *Landoltia* and *Spirodela* (Lemnaceae) by three plastidic markers and amplified fragment length polymorphism (AFLP). *Hydrobiologia* **749**: 169–182.

Bog, M., M. F. Landrock, D. Drefahl, K. S. Sree & K.-J. Appenroth. 2018. Fingerprinting by amplified fragment length polymorphism (AFLP) and barcoding by three plastidic markers in the genus *Wolffiella* Hegelm. *Plant Syst. Evol.* **304**: 373–386.

Bogner, J. 2009. The free-floating Aroids (Araceae) – living and fossil. *Zitteliana* **48/49**: 113–128.

Bogner, J. & S. J. Mayo. 1998. Acoraceae. Pp. 7–11 *In*: K. Kubitzki, H. Huber, P. J. Rudall, P. S. Stevens & T. Stützel (eds.), *The Families and Genera of Vascular Plants, Vol. IV. Flowering Plants, Monocotyledons: Alismatanae and Commelinanae (Except Gramineae)*. Springer-Verlag, Berlin, Germany.

Bohlmann, H., J. Apple, N. Burnett & S. Shull. 2018. Warm water temperature regimes in eelgrass beds (*Z. marina* and *Z. japonica*) of Padilla Bay, WA. *Salish Sea Ecosystem Conference*, Western Washington University, Bellingham, WA. [Abstract].

Bokhari, S. H., I. Ahmad, M. Mahmood-Ul-Hassan & A. Mohammad. 2016. Phytoremediation potential of *Lemna minor* L. for heavy metals. *Int. J. Phytoremed.* **18**: 25–32.

Boland, J. T. & P. J. Scott. 1991. Ecological aspects of *Arethusa bulbosa, Calopogon tuberosus*, and *Pogonia ophioglossoides* (Orchidaceae) in eastern Newfoundland. I. Flowering and fruiting patterns. *Rhodora* **93**: 248–255.

Boland, W. & C. J. Burk. 1992. Some effects of acidic growing conditions on three emergent macrophytes: *Zizania aquatica, Leersia oryzoides* and *Peltandra virginica. Environ. Pollut.* **76**: 211–217.

Boland, J. T. & P. J. Scott. 1992. Ecological aspects of *Arethusa bulbosa, Calopogon tuberosus* and *Pogonia ophioglossoides* (Orchidaceae) in eastern Newfoundland. II. Partitioning of the microhabitat. *Rhodora* **94**: 374–380.

Bolda, E. & R. C. Anderson. 2011. Southeast Wisconsin's Pewaukee Lake aquatic plant survey 2010. Technical Bulletin 013. Wisconsin Lutheran College, Milwaukee, WI. 22 pp.

Bolduan, B. R., G. C. Van Eeckhout, H. W. Quade & J. E. Gannon. 1994. *Potamogeton crispus* – The other invader. *Lake Reserv. Manag.* **10**: 113–125.

Bolen, E. G. & J. J. Beecham. 1970. Notes on the foods of juvenile black-bellied tree ducks. *Wilson Bull.* **82**: 325–326.

Bolin, J. F. 2007. Seed bank response to wet heat and the vegetation structure of a Virginia pocosin1. *J. Torrey Bot. Soc.* **134**: 80–89.

Bolin, J. F., K. U. Tennakoon, M. B. A. Majid & D. D. Cameron. 2017. Isotopic evidence of partial mycoheterotrophy in *Burmannia coelestis* (Burmanniaceae). *Plant Species Biol.* **32**: 74–80.

Bonanno, G. & G. L. Cirelli. 2017. Comparative analysis of element concentrations and translocation in three wetland congener plants: *Typha domingensis, Typha latifolia* and *Typha angustifolia. Ecotoxicol. Environ. Saf.* **143**: 92–101.

Bonnewell, V., W. L. Koukkari & D. C. Pratt. 1983. Light, oxygen, and temperature requirements for *Typha latifolia* seed germination. *Can. J. Bot.* **61**: 1330–1336.

Bontrager, M., K. Webster, M. Elvin & I. M. Parker. 2014. The effects of habitat and competitive/facilitative interactions on reintroduction success of the endangered wetland herb, *Arenaria paludicola*. *Plant Ecol.* **215**: 467–478.

Boonyapookana, B., E. S. Upatham, M. Kruatrachue, P. Pokethitiyook & S. Singhakaew. 2002. Phytoaccumulation and phytotoxicity of cadmium and chromium in duckweed *Wolffia globosa*. *Int. J. Phytoremed.* **4**: 87–100.

Borchardt, J. R., D. L. Wyse, C. C. Sheaffer, K. L. Kauppi, R. G. Fulcher, N. J. Ehlke, D. D. Biesboer & R. F. Bey. 2008. Antioxidant and antimicrobial activity of seed from plants of the Mississippi river basin. *J. Med. Plant Res.* **2**: 81–93.

Borisjuk, N., P. Chu, R. Gutierrez, H. Zhang, K. Acosta, N. Friesen, K. S. Sree, C. Garcia, K. J. Appenroth & E. Lam. 2015. Assessment, validation and deployment strategy of a two-barcode protocol for facile genotyping of duckweed species. *Plant Biol.* **17**: 42–49.

Borkent, C. J. & E. I. Schlinger. 2008. Pollen loads and pollen diversity on bodies of *Eulonchus tristis* (Diptera: Acroceridae): implications for pollination and flower visitation. *Can. Entomol.* **140**: 257–264.

Borman, S., R. Korth & J. Temte. 1997. *Through the Looking Glass: A Field Guide to Aquatic Plants. FH-207-97*. Department of Natural Resources. University of Wisconsin, Stevens Point, WI. 248 pp.

Bornette, G. & S. Puijalon. 2011. Response of aquatic plants to abiotic factors: a review. *Aquat. Sci.* **73**: 1–14.

Bortone, S. A. (ed.). 2000. *Seagrasses: Monitoring, Ecology, Physiology, and Management*. CRC Press, Boca Raton, FL. 336 pp.

Borum, J., O. Pedersen, T. M. Greve, T. A. Frankovich, J. C. Zieman, J. W. Fourqurean & C. J. Madden. 2005. The potential role of plant oxygen and sulphide dynamics in die-off events of the tropical seagrass, *Thalassia testudinum*. *J. Ecol.* **93**: 148–158.

Boschilia, S. M., E. F. de Oliveira & A. Schwarzbold. 2012. The immediate and long-term effects of water drawdown on macrophyte assemblages in a large subtropical reservoir. *Freshw. Biol.* **57**: 2641–2651.

Bosserman, R. W. 1981. Elemental composition of aquatic plants from Okefenokee Swamp. *J. Freshw. Ecol.* **1**: 307–320.

Botero, J. E. & D. H. Rusch. 1994. Foods of blue-winged teal in two neotropical wetlands. *J. Wildl. Manag.* **58**: 561–565.

Botkin, D. B., P. A. Jordan, A. S. Dominski, H. S. Lowendorf & G. E. Hutchinson. 1973. Sodium dynamics in a northern ecosystem. *Proc. Natl. Acad. Sci. U.S.A.* **70**: 2745–2748.

Bottum, E. 2005. *Wetland Conservation Strategy for the Weiser River Basin, Idaho*. Idaho Conservation Data Center, Idaho Department of Fish and Game, Boise, ID. 156 pp.

Bouchard, S. S. & K. A. Bjorndal. 2005. Microbial fermentation in juvenile and adult pond slider turtles, *Trachemys scripta*. *J. Herpetol.* **39**: 321–324.

Bouchard, A., S. Hay, Y. Bergeron & A. Leduc. 1991. The vascular flora of Gros Morne National Park, Newfoundland: a habitat classification approach based on floristic, biogeographical and life-form data. Pp. 123–157 *In*: P. L. Nimis & T. J. Crovello (eds.), *Quantitative Approaches to Phytogeography*. Springer, Dordrecht, The Netherlands.

Bouetard, A., P. Lefeuvre, R. Gigant, S. Bory, M. Pignal, P. Besse & M. Grisoni. 2010. Evidence of transoceanic dispersion of the genus *Vanilla* based on plastid DNA phylogenetic analysis. *Mol. Phylogenet. Evol.* **55**: 621–630.

Boughton, E. H. & R. K. Boughton. 2014. Modification by an invasive ecosystem engineer shifts a wet prairie to a monotypic stand. *Biol. Invasions* **16**: 2105–2114.

Boughton, E. H., R. K. Boughton, C. Griffith & J. Bernath-Plaisted. 2016. Reproductive traits of *Lachnanthes caroliniana* (Lam.) Dandy related to patch formation following feral swine rooting disturbance. *J. Torrey Bot. Soc.* **143**: 265–274.

Boutin, C., B. Jobin, L. Bélanger & L. Choinière. 2002. Plant diversity in three types of hedgerows adjacent to cropfields. *Biodivers. Conserv.* **11**: 1–25.

Bowden, W. M. 1945. A list of chromosome numbers in higher plants. II. Menispermaceae to Verbenaceae. *Am. J. Bot.* **32**: 191–201.

Bowers, F. D. 1972. The existence of *Heterotheca ruthii* (Compositae). *Castanea* **37**: 130–132.

Bowes, G., T. K. Van, L. A. Garrard & W. T. Haller. 1977. Adaptation to low light levels by hydrilla. *J. Aquat. Plant Manag.* **15**: 32–35.

Bowes, G., A. S. Holaday & W. T. Haller. 1979. Seasonal variation in the biomass, tuber density, and photosynthetic metabolism of hydrilla in three Florida lakes. *J. Aquat. Plant Manag.* **17**: 61–65.

Bowes, G., S. K. Rao, G. M. Estavillo & J. B. Reiskind. 2002. C_4 mechanisms in aquatic angiosperms: comparisons with terrestrial C_4 systems. *Funct. Plant Biol.* **29**: 379–392.

Bowker, R. G. 1991. *Kral's Water-Plantain (Sagittaria secundifolia) Recovery Plan*. United States Fish and Wildlife Service, Jackson, MS. 15 pp.

Bowles, M. L. 1983. The tallgrass prairie orchids *Platanthera leucophaea* (Nutt.) Lindl. and *Cypripedium candidum* Muhl. ex Willd.: some aspects of their status, biology, and ecology, and implications toward management. *Nat. Areas J.* **3**: 14–37.

Bowles, M. L. 1991. Some aspects of the status and ecology of seven rare wetland plant species in the Chicago region of northeastern Illinois. *Erigenia* **11**: 52–66.

Bowles, M. L. 1999. *Eastern Prairie Fringed Orchid Platanthera leucophaea (Nuttall) Lindley Recovery Plan*. U.S. Fish & Wildlife Service, Region 3, Fort Snelling, MN. 58 pp.

Bowles, D. E. 2013. Missouri: first record of *Limnobium spongia* from the Ozarks physiographic region. *Castanea* **78**: 137.

Bowles, D. E. & B. D. Bowles. 2015. Non-native species of the major spring systems of Texas, USA. *Texas J. Sci.* **67**: 51–78.

Bowles, D. E. & H. R. Dodd. 2015. Floristics and community ecology of aquatic vegetation occurring in seven large springs at Ozark National Scenic Riverways, Missouri (USA), 2007–2012. *J. Bot. Res. Inst. Texas* **9**: 235–249.

Bowles, M. L., R. Flakne & R. Dombeck. 1992. Status and population fluctuations of the eastern prairie fringed orchid [*Platanthera leucophaea* (Nutt.) Lindl] in Illinois. *Erigenia* **12**: 26–40.

Bowles, M. L., K. A. Jacobs, L. W. Zettler & T. W. Delaney. 2002. Crossing effects on seed viability and experimental germination of the federal threatened *Platanthera leucophaea* (Orchidaceae). *Rhodora* **104**: 14–30.

Bowles, M., L. Zettler, T. Bell & P. Kelsey. 2005. Relationships between soil characteristics, distribution and restoration potential of the federal threatened eastern prairie fringed orchid, *Platanthera leucophaea* (Nutt.) Lindl. *Am. Midl. Nat.* **154**: 273–286.

Bown, D. 2000. *Aroids. Plants of the Arum Family*, 2nd ed. Timber Press, Portland, OR. 392 pp.

Bowyer, M. W., J. D. Stafford, A. P. Yetter, C. S. Hine, M. M. Horath & S. P. Havera. 2005. Moist-soil plant seed production for waterfowl at Chautauqua National Wildlife Refuge, Illinois. *Am. Midl. Nat.* **154**: 331–341.

Boxell, V. 2014. *Edible and Useful Plants for the Swan Coastal Plain*. Lulu.com. Morrisville, NC. 168 pp.

Boyd, C. E. 1968. Fresh-water plants: a potential source of protein. *Econ. Bot.* **22**: 359–368.

Boyd, C. E. 1969. The nutritive value of three species of water weeds. *Econ. Bot.* **23**: 123–127.

Boyd, C. E. 1970a. Vascular aquatic plants for mineral nutrient removal from polluted waters. *Econ. Bot.* **24**: 95–103.

Boyd, C. E. 1970b. Amino acid, protein, and caloric content of vascular aquatic macrophytes. *Ecology* **51**: 902–906.

Boyd, C. E. 1975. Competition for light by aquatic plants in fish ponds. Circular 215. Agricultural Experiment Station, Auburn University, Auburn, AL. 19 pp.

Boyd, C. E. 1987. Evapotranspiration/evaporation (E/Eo) ratios for aquatic plants. *J. Aquat. Plant Manag.* **25**: 1–3.

Boyd, C. E. & R. D. Blackburn. 1970. Seasonal changes in the proximate composition of some common aquatic weeds. *Hyacinth Control J.* **8**: 42–44.

Boyd, C. E. & L. W. Hess. 1970. Factors influencing shoot production and mineral nutrient levels in *Typha latifolia*. *Ecology* **51**: 296–300.

Boyd, C. E. & C. P. Goodyear. 1971. Nutritive quality of food in ecological systems. *Arch. Hydrobiol.* **69**: 256–270.

Boyd, C. E. & W. W. Walley. 1972. Studies of the biogeochemistry of boron. I. Concentrations in surface waters, rainfall and aquatic plants. *Am. Midl. Nat.* **88**: 1–14.

Boyd, C. E. & P. S. McGinty. 1981. Percentage digestible dry matter and crude protein in dried aquatic weeds. *Econ. Bot.* **35**: 296–299.

Boyd, R. S. & J. M. Moffett, Jr. 2003. Management of *Xyris tennesseensis*, a federally-endangered plant species. GDOT Research Project No. 2003; Final Report. Georgia Department of Transportation, Atlanta, GA. 24 pp.

Boyd, R. S., A. Teem & M. A. Wall. 2011. Floral biology of an Alabama population of the federally endangered plant, *Xyris tennesseensis* Kral (Xyridaceae). *Castanea* **76**: 255–265.

Boyd, J. N., G. A. Raymond, G. P. Call & M. J. Pistrang. 2016. Ecophysiological performance of the rare terrestrial orchid *Platanthera integrilabia* across contrasting habitats. *Plant Ecol.* **217**: 1259–1272.

Boylen, C. W. & R. B. Sheldon. 1976. Submergent macrophytes: growth under winter ice cover. *Science* **194**: 841–842.

Brackney, A. & T. A. Bookhout. 1982. Population ecology of common gallinules in southwestern Lake Erie marshes. *Ohio J. Sci.* **82**: 229–237.

Bradfield, G. E. & G. L. Porter. 1982. Vegetation structure and diversity components of a Fraser estuary tidal marsh. *Can. J. Bot.* **60**: 440–451.

Bragg, L. H. & C. McMillan. 1986. SEM comparison of fruits of a seagrass, *Halodule* (Cymodoceaceae), from Australia and Texas. *Am. J. Bot.* **73**: 815–821.

Brahma, S., H. Narzary & S. Basumatary. 2013. Wild edible fruits of Kokrajhar district of Assam, North-East India. *Asian J. Plant Sci. Res.* **3**: 95–100.

Brain, R. A., D. J. Johnson, S. M. Richards, H. Sanderson, P. K. Sibley & K. R. Solomon. 2004. Effects of 25 pharmaceutical compounds to *Lemna gibba* using a seven-day static-renewal test. *Environ. Toxicol. Chem.* **23**: 371–382.

Brant, A. E. 2018. *Thelypteris noveboracensis* (New York fern) new to Missouri from the Southeastern Missouri Ozarks. *Missouriensis* **36**: 18–20.

Braselton, J. P. 1983. The plasmodiophoromycete parasitic on *Heteranthera dubia*. *Can. J. Bot.* **61**: 45–52.

Bratton, S. P. 1974. The effect of the European wild boar (*Sus scrofa*) on the high-elevation vernal flora in Great Smoky Mountains National Park. *Bull. Torrey Bot. Club* **101**: 198–206.

Bräuchler, C. 2015. Towards a better understanding of the *Najas marina* complex: notes on the correct application and typification of the names *N. intermedia*, *N. major*, and *N. marina*. *Taxon* **64**: 1028–1030.

Braun, U., P. W. Crous & C. Nakashima. 2014. Cercosporoid fungi (Mycosphaerellaceae) 2. Species on monocots (Acoraceae to Xyridaceae, excluding Poaceae). *IMA Fungus* **5**: 203–390.

Breden, T. F., Y. R. Alger, K. S. Walz & A. G. Windisch. 2001. *Classification of Vegetation Communities of New Jersey: Second Iteration*. Association for Biodiversity Information and New Jersey Natural Heritage Program, Office of Natural Lands Management, Division of Parks and Forestry, New Jersey Department of Environmental Protection, Trenton, NJ. 230 pp.

Brewer, C. A. 1995. The submersed aquatic plant community in Jackson Lake, Grand Teton National Park, Wyoming. *Univ. Wyoming Natl. Park Serv. Res. Center Ann. Rep.* **19**: 14–18.

Brewer, J. S. 1998. Patterns of plant species richness in a wet slash-pine (*Pinus elliottii*) savanna. *J. Torrey Bot. Soc.* **125**: 216–224.

Brewer, C. A. & M. Parker. 1990. Adaptations of macrophytes to life in moving water: upslope limits and mechanical properties of stems. *Hydrobiologia* **194**: 133–142.

Brewer, J. S., D. J. Baker, A. S. Nero, A. L. Patterson, R. S. Roberts & L. M. Turner. 2011. Carnivory in plants as a beneficial trait in wetlands. *Aquat. Bot.* **94**: 62–70.

Brewis, A. 1975. *Sagittaria subulata* (L.) Buch. in the British Isles. *Watsonia* **10**: 411.

Bridges, E. L. 2005. *Assessment of Biodiversity and Conservation Status Priorities for Pitcher Plant Bogs and Wetland Savannas of the Apalachicola National Forest*. E. L. Bridges, Bremerton, WA. 93 pp.

Bridges, E. L. & S. L. Orzell. 1987. A new species of *Xyris* (sect. *Xyris*) from the Gulf Coastal Plain. *Phytologia* **64**: 56–61.

Bridges, E. L. & S. L. Orzell. 1989. *Syngonanthus flavidulus* (Eriocaulaceae) new to Mississippi. *Sida* **13**: 512–515.

Bridges, E. L. & S. L. Orzell. 1992a. The rediscovery of *Rhynchospora solitaria* Harper (Cyperaceae) in Georgia. *Phytologia* **72**: 369–372.

Bridges, E. L. & S. L. Orzell. 1992b. *Xyris isoetifolia* Kral (Xyridaceae) new to Alabama and its range and habitats in Florida. *Phytologia* **72**: 152–156.

Bridges, E. L. & S. L. Orzell. 2002. *Euphorbia* (Euphorbiaceae) section *Tithymalus* subsection *Inundatae* in the southeastern United States. *Lundellia* **5**: 59–79.

Bridges, E. L. & S. L. Orzell. 2003. Two new species and a new combination in southeastern United States *Xyris* (Xyridaceae) from Florida. *Novon* **13**: 16–25.

Brim-DeForest, W. B., K. Al-Khatib, B. A. Linquist & A. J. Fischer. 2017a. Weed community dynamics and system productivity in alternative irrigation systems in California rice. *Weed Sci.* **65**: 177–188.

Brim-DeForest, W. B., K. Al-Khatib & A. J. Fischer. 2017b. Predicting yield losses in rice mixed-weed species infestations in California. *Weed Sci.* **65**: 61–72.

Brinckmann-Voss, A. 1996. Seasonality of hydroids (Hydrozoa, Cnidaria) from an inter-tidal pool and adjacent subtidal habitats at Race Rocks, off Vancouver Island, Canada. *Sci. Mar.* **60**: 89–97.

Brinckmann-Voss, A., D. M. Lickey & C. E. Mills. 1993. *Rhysia fletcheri* (Cnidaria, Hydrozoa, Rhysiidae), a new species of colonial hydroid from Vancouver Island (British Columbia, Canada) and the San Juan Archipelago (Washington, U.S.A.). *Can. J. Zool.* **71**: 401–406.

Broadhurst, L. & C. Chong. 2011. Examining clonal propagation of the aquatic weed *Sagittaria platyphylla*. RIRDC Publication No. 11/020. CSIRO Plant Industry, Canberra, ACT. 17 pp.

Brochet, A.-L., M. Guillemain, H. Fritz, M. Gauthier-Clerc & A. J. Green. 2009. The role of migratory ducks in the long-distance dispersal of native plants and the spread of exotic plants in Europe. *Ecography* **32**: 919–928.

Brochet, A.-L., M. Guillemain, M. Gauthier-Clerc, H. Fritz & A. J. Green. 2010a. Endozoochory of Mediterranean aquatic plant seeds by teal after a period of desiccation: determinants of seed survival and influence of retention time on germinability and viability. *Aquat. Bot.* **93**: 99–106.

Brochet, A. L., M. Guillemain, H. Fritz, M. Gauthier-Clerc & A. J. Green. 2010b. Plant dispersal by teal (*Anas crecca*) in the Camargue: duck guts are more important than their feet. *Freshw. Biol.* **55**: 1262–1273.

Brochet, A.-L., J.-B. Mouronval, P. Aubry, M. Gauthier-Clerc, A. J. Green, H. Fritz & M. Guillemain. 2012. Diet and feeding habitats of Camargue dabbling ducks: what has changed since the 1960s? *Waterbirds* **35**: 555–576.

Brodie, B. S., A. Renyard, R. Gries, H. Zhai, S. Ogilvie, J. Avery & G. Gries. 2018. Identification and field testing of floral odorants that attract the rove beetle *Pelecomalium testaceum* (Mannerheim) to skunk cabbage, *Lysichiton americanus* (L.). *Arthropod-Plant Interact.* **12**: 591–599.

Brook, I. M. 1978. Comparative macrofaunal abundance in turtle-grass (*Thalassia testudinum*) communities in south Florida characterized by high blade density. *Bull. Mar. Sci.* **28**: 212–217.

Brooks, T. L. 1954. Host range of the burrowing nematode internationally and in Florida. *Citrus Industry* **35**(7–8): 14–15.

Brooks, A. R., E. S. Nixon & J. A. Neal. 1993. Woody vegetation of wet creek bottom communities in eastern Texas. *Castanea* **58**: 185–196.

Brooks, R. P. & D. H. Wardrop (eds.). 2013. *Mid-Atlantic Freshwater Wetlands: Advances in Wetlands Science, Management, Policy, and Practice*. Springer, New York, NY. 491 pp.

Brophy, T. E. 1980. Food habits of sympatric larval *Ambystoma tigrinum* and *Notophthalmus viridescens*. *J. Herpetol.* **14**: 1–6.

Broschat, T. K. 1994. Palm seed propagation. *Acta Hort.* **360**: 141–147.

Broschat, T. K. 2011a. Uptake and distribution of boron in coconut and paurotis palms. *HortScience* **46**: 1683–1686.

Broschat, T. K. 2011b. Iron deficiency in palms. Document ENH 1013. IFAS Extension, University of Florida, Gainesville, FL. 3 pp.

Broschat, T. K. & W. G. Latham. 1994. Oxalate content of palm fruit mesocarp. *Biochem. Syst. Ecol.* **22**: 389–392.

Brotherson, J. D. 1981. Aquatic and semiaquatic vegetation of Utah Lake and its bays. *Great Basin Nat. Mem.* **5**: 68–84.

Brousil, M., C. Humann & D. Fischer. 2015. Plant-pollinator interactions in a northwest *Arum* related to plant traits but not riparian forest management. *Northwest Sci.* **89**: 297–307.

Brouwer, E., R. Bobbink & J. G. M. Roelofs. 2002. Restoration of aquatic macrophyte vegetation in acidified and eutrophied softwater lakes: an overview. *Aquat. Bot.* **73**: 405–431.

Brown, W. V. 1942. A note on *Sagittaria kurziana*. *Rhodora* **44**: 211–213.

Brown, C. J. 1995. An investigation of environmental parameters influencing the distribution of *Platanthera blephariglottis* (Willdenow) Lindley and *P. clavellata* (Michaux) Luer (Orchidaceae) in peatlands, and some aspects of their population dynamics. Ph.D. dissertation. Memorial University of Newfoundland, St. John's, NF. 114 pp.

Brown, P. M. 2005. *Wild Orchids of Florida*, expanded ed. University Press of Florida, Gainesville, FL. 409 pp.

Brown, P. M. 2006. Resurrection of the genus *Gymnadeniopsis* Rydberg. *North Am. Native Orchid J.* **12**: 33–40.

Brown, P. M. 2008. Hybrids in the genus *Calopogon*. *North Am. Native Orchid J.* **14**: 177–182.

Brown, J. H. & W. McDonald. 1995. Livestock grazing and conservation on southwestern rangelands. *Conserv. Biol.* **9**: 1644–1647.

Brown, C. J. & P. J. Scott. 1997. Environmental parameters influencing the distribution of *Platanthera blephariglottis* and *Platanthera clavellata* (Orchidaceae) in peatlands on the Avalon Peninsula, Newfoundland. *Can. J. Bot.* **75**: 974–980.

Brown, P. M. & G. W. Argus. 2002. *Epipactis* Zinn. Pp. 584–586 *In*: Flora North America Editorial Committee (eds.), *Flora of North America North of Mexico, Vol. 26: Magnoliophyta: Liliidae: Liliales and Orchidales*. Oxford University Press, New York, NY.

Brown, J. S. & C. G. Eckert. 2005. Evolutionary increase in sexual and clonal reproductive capacity during biological invasion in an aquatic plant *Butomus umbellatus* (Butomaceae). *Am. J. Bot.* **92**: 495–502.

Brown, W. Y., M. Choct & J. R. Pluske. 2013. Duckweed (*Landoltia punctata*) in dog diets decreases digestibility but improves stool consistency. *Anim. Prod. Sci.* **53**: 1188–1194.

Brown, M. T., T. Boyer, R. J. Sindelar, S. Arden, A. Persaud & S. Brandt-Williams. 2018. A floating island treatment system for the removal of phosphorus from surface waters. *Engineering*. doi: 10.1016/j.eng.2018.08.002.

Brugiolo, A. S. S., C. C. de Souza Alves, A. C. C. Gouveia, A. T. Dias, M. F. Rodrigues, L. G. G. Pacífico, B. J. V. Aarestrup, M. A. Machado, R. Domingues, H. C. Teixeira, J. Gameiro & A. P. Ferreira. 2011. Effects of aqueous extract of *Echinodorus grandiflorus* on the immune response in ovalbumin-induced pulmonary allergy. *Ann. Allergy Asthma Immunol.* **106**: 481–488.

Brumback, W. E. 2001. *Carex garberi Fern. (Garber's Sedge) and Triantha glutinosa (Michx.) Baker (Sticky False Asphodel) Conservation Plan*. New England Wild Flower Society, Framingham, MA. 43 pp.

Brunel, S. 2009. Pathway analysis: aquatic plants imported in 10 EPPO countries. *EPPO Bull.* **39**: 201–213.

Brunton, D. F. 1986. Status of giant helleborine, *Epipactis gigantea* (Orchidaceae) in Canada. *Can. Field-Nat.* **100**: 414–417.

Brunton, D. F. & D. M. Britton. 1993. *Isoetes prototypus* (Isoetaceae) in the United States. *Rhodora* **95**: 122–128.

Brunton, D. F., D. M. Britton & W. C. Taylor. 1994. *Isoetes hyemalis*, sp. nov. (Isoetaceae): a new quillwort from the southeastern United States. *Castanea* **59**: 12–21.

Brux, H., D. Todeskino & G. Wiegleb. 1987. Growth and reproduction of *Potamogeton alpinus* Balbis growing in disturbed habitats. *Arch. Hydrobiol.* **27**: 115–127.

Bryson, C. T., J. R. MacDonald & R. Warren. 1994. Notes on *Carex* (Cyperaceae), with *C. godfreyi* new to Alabama and *C. scoparia* new to Mississippi (USA). *Sida* **16**: 355–360.

Brzosko, E., B. Ostrowiecka, J. Kotowicz, M. Bolesta, A. Gromotowicz, M. Gromotowicz, A. Orzechowska, J. Orzolek & M. Wojdalska. 2017. Seed dispersal in six species of terrestrial orchids in Biebrza National Park (NE Poland). *Acta Soc. Bot. Poloniae* **86**: 3557.

Buchmann, S., L. D. Adams, A. D. Howell & M. Weiss. 2010. A study of insect pollinators associated with DoD TER-S flowering plants, including identification of habitat types where they co-occur by military installation in the Western United States. No. LRMP-PN-08-391. U.S. Department of Defense, Legacy Resource Management Program, Arlington, VA. 67 pp.

Buchsbaum, R., I. Valiela & T. Swain. 1984. The role of phenolic compounds and other plant constituents in feeding by Canada geese in a coastal marsh. *Oecologia* **63**: 343–349.

Buckel, C. A., C. A. Blanchette, R. R. Warner & S. D. Gaines. 2012. Where a male is hard to find: consequences of male rarity in the surfgrass *Phyllospadix torreyi. Mar. Ecol. Prog. Ser.* **449**: 121–132.

Buckingham, G. R. 1989. *Lemnaphila scotlandae* (Diptera: Ephydridae) and three of its parasites discovered in Florida. *Florida Entomol.* **72**: 219–221.

Buckingham, G. R. & C. A. Bennett. 1989. *Dyscinetus morator* (Fab.)(Coleoptera: Scarabaeidae) adults attack waterhyacinth, *Eichhornia crassipes* (Pontederiaceae). *Coleopt. Bull.* **43**: 27–33.

Buckley, R. A. & E. A. Hicks. 1962. An analysis of mite populations in muskrat houses. *Proc. Iowa Acad. Sci.* **69**: 541–556.

Buddington, R. K. 1979. Digestion of an aquatic macrophyte by *Tilapia zillii* (Gervais). *J. Fish Biol.* **15**: 449–455.

Buell, M. F. 1946. Jerome Bog, a peat-filled "Carolina bay." *Bull. Torrey Bot. Club* **73**: 24–33.

Buffington, K. J., B. D. Dugger & K. M. Thorne. 2018. Climate-related variation in plant peak biomass and growth phenology across Pacific Northwest tidal marshes. *Estuar. Coast. Shelf Sci.* **202**: 212–221.

Bukliev, R. 1980. *Heteranthera limosa* Vahl – new adventive species for the flora of Yugoslavia and Europe. *Fragmenta Balcanica* **11**: 11–17.

Bull, J. S., D. C. Reed & S. J. Holbrook. 2004. An experimental evaluation of different methods of restoring *Phyllospadix torreyi* (surfgrass). *Restor. Ecol.* **12**: 70–79.

Bullard, A. J. & C. M. Allen. 2013. Synopsis of the woody species of *Smilax* in the Eastern United States north of peninsular Florida. *J. North Carolina Acad. Sci.* **129**: 37–43.

Bunghez, I. R., R. M. Ion, S. Pop, M. Ghiurea, I. Dumitriu & R. C. Fierascu. 2010. Silver nanoparticles fabrication using marine plant (*Mayaca fluviatilis*) resources. *Analele Sti. Univ. "Al. I. Cuza" Iasi, Sect. 2.a, Genet. Biol. Mol.* **11**: 89–94.

Burbanck, M. P. & R. B. Platt. 1964. Granite outcrop communities of the Piedmont Plateau in Georgia. *Ecology* **45**: 292–306.

Burbanck, M. P. & D. L. Phillips. 1983. Evidence of plant succession on granite outcrops of the Georgia Piedmont. *Am. Midl. Nat.* **109**: 94–104.

Burchill, C. A. & N. C. Kenkel. 1991. Vegetation–environment relationships of an inland boreal salt pan. *Can. J. Bot.* **69**: 722–732.

Burge, D. O. 2014. The role of soil chemistry in the geographic distribution of *Ceanothus otayensis* (Rhamnaceae). *Madroño* **61**: 276–289.

Burgess, T. E. 1970. Foods and habitat of four anatinids wintering on the Fraser Delta tidal marshes. M.S. thesis. University of British Columbia, Vancouver, BC. 124 pp.

Burk, C. J. 1962. The North Carolina Outer Banks: a floristic interpretation. *J. Elisha Mitchell Sci. Soc.* **78**: 21–28.

Burkhalter, J. R. & J. C. Wright. 1989. Fourteen additions to the known stranded seeds and fruits of Northwest Florida beaches. *Sida* **13**: 345–349.

Burkholder, J. M., H. B. Glasgow, Jr. & J. E. Cooke. 1994. Comparative effects of water-column nitrate enrichment on eelgrass *Zostera marina*, shoalgrass *Halodule wrightii*, and widgeongrass *Ruppia maritima. Mar. Ecol. Prog. Ser.* **105**: 121–138.

Burks, K. C. 1995. Noteworthy collections: Florida. *Castanea* **60**: 169–170.

Burlakova, L. E., A. Y. Karatayev, D. K. Padilla, L. D. Cartwright & D. N. Hollas. 2009. Wetland restoration and invasive species: apple snail (*Pomacea insularum*) feeding on native and invasive aquatic plants. *Restor. Ecol.* **17**: 433–440.

Burnham, S. H. 1917. The Naiadales of the flora of the Lake George region. *Torreya* **17**: 80–84.

Burns, G. P. 1911. A botanical survey of the Huron river valley. VIII. Edaphic conditions in peat bogs of southern Michigan. *Bot. Gaz.* **52**: 105–125.

Burns, Jr., J. W., M. A. Poirrier & K. P. Preston. 1995. The status of *Potamogeton perfoliatus* (Potamogetonaceae) in Lake Pontchartrain, Louisiana. *Sida* **16**: 757–763.

Burns, J. H., R. B. Faden & S. J. Steppan. 2011. Phylogenetic studies in the Commelinaceae subfamily Commelinoideae inferred from nuclear ribosomal and chloroplast DNA sequences. *Syst. Bot.* **36**: 268–276.

Bursik, R. J. & R. K. Moseley. 1992. *Vegetation and Water Chemistry Monitoring and Twenty-Year Floristic Changes at Huff Lake fen, Kaniksu National Forest*. Conservation Data Center, Idaho Department of Fish and Game, Boise, ID. 17 pp.

Bush, E. M. 1988. A floristic study of a wet meadow in Barbour County, West Virginia. *Castanea* **53**: 132–139.

Bush, C. M., D. Rollins & G. L. Smith. 2010. The phylogeny of the southeastern United States *Hymenocallis* (Amaryllidaceae) based on ISSR fingerprinting and morphological data. *Castanea* **75**: 368–380.

Bushmann, P. J. & M. S. Ailstock. 2006. Antibacterial compounds in estuarine submersed aquatic plants. *J. Exp. Mar. Biol. Ecol.* **331**: 41–50.

Bushnell, Jr., J. H. 1966. Environmental relations of Michigan Ectoprocta, and dynamics of natural populations of *Plumatella repens. Ecol. Monogr.* **36**: 95–123.

Bushnell, Jr., J. H. & T. W. Porter. 1967. The occurrence, habitat, and prey of *Craspedacusta sowerbyi* (particularly polyp stage) in Michigan. *Trans. Am. Microsc. Soc.* **86**: 22–27.

Buthod, A. K. & B. W. Hoagland. 2011. New to Oklahoma: *Leptochloa panicoides* (Poaceae). *Phytoneuron* 2011–55: 1–2.

Buthod, A. K. & B. W. Hoagland. 2013a. Noteworthy collections: Oklahoma. *Castanea* **78**: 213–215.

Buthod, A. K. & B. W. Hoagland. 2013b. New to Oklahoma: *Murdannia keisak* (Commelinaceae). *Phytoneuron* **2013–93**: 1–3.

Buzgo, M., D. E. Soltis, P. S. Soltis, S. Kim, H. Ma, B. A. Hauser, J. Lebens-Mack & B. Johansen. 2006. Perianth development in the basal monocot *Triglochin maritima* (Juncaginaceae) *Aliso* **22**: 107–125.

Buznego, M. T. & H. Pérez-Saad. 2006. Behavioral and antiepileptic effect of acute administration of the extract of the aquatic plant *Echinodorus berteroi* (Sprengel) Fassett (upright burhead). *Epilepsy Behav.* **9**: 40–45.

Buzzelli, C., R. Robbins, P. Doering, Z. Chen, D. Sun, Y. Wan, B. Welch & A. Schwarzschild. 2012. Monitoring and modeling of *Syringodium filiforme* (manatee grass) in southern Indian River lagoon. *Estuaries Coast.* **35**: 1401–1415.

Byng, J. W., M. W. Chase, M. J. M. Christenhusz, M. F. Fay, W. S. Judd, D. J. Mabberley, A. N. Sennikov, D. E. Soltis, P. S. Soltis & P. F. Stevens. 2016. An update of the Angiosperm Phylogeny Group classification for the orders and families of flowering plants: APG IV. *Bot. J. Linn. Soc.* **181**: 1–20.

Bytnerowicz, T. A. & R. I. Carruthers. 2014. Germination characteristics of Zannichellia palustris from a northern California spring-fed river. *Aquat. Bot.* **119**: 44–50.

Cabaço, S., R. Santos & C. M. Duarte. 2008. The impact of sediment burial and erosion on seagrasses: a review. *Estuar. Coast. Shelf Sci.* **79**: 354–366.

Cabrera, L. I., G. A. Salazar, M. W. Chase, S. J. Mayo, J. Bogner & P. Dávila. 2008. Phylogenetic relationships of aroids and duckweeds (Araceae) inferred from coding and noncoding plastid DNA. *Am. J. Bot.* **95**: 1153–1165.

Caffrey, J. M. & W. M. Kemp. 1990. Nitrogen cycling in sediments with estuarine populations of *Potamogeton perfoliatus* and *Zostera marina*. *Mar. Ecol. Prog. Ser.* **66**: 147–160.

Caffrey, J. M. & W. M. Kemp. 1991. Seasonal and spatial patterns of oxygen production, respiration and root-rhizome release in *Potamogeton perfoliatus* L. *Zostera marina* L. *Aquat. Bot.* **40**: 109–128.

Caffrey, J. M. & W. M. Kemp. 1992. Influence of the submersed plant, *Potamogeton perfoliatus*, on nitrogen cycling in estuarine sediments. *Limnol. Oceanogr.* **37**: 1483–1495.

Caicedo, J. R., N. P. Van der Steen, O. Arce & H. J. Gijzen. 2000. Effect of total ammonia nitrogen concentration and pH on growth rates of duckweed (*Spirodela polyrrhiza*). *Water Res.* **34**: 3829–3835.

Caillet, M., J. F. Campbell, K. C. Vaughn & D. Vercher. 2000. *The Louisiana Iris: The Taming of a Native American Wildflower*, 2nd ed. Timber Press, Portland, OR. 211 pp.

Cain, S. A. 1931. Ecological studies of the vegetation of the Great Smoky Mountains of North Carolina and Tennessee. I. Soil reaction and plant distribution. *Bot. Gaz.* **91**: 22–41.

Calazza, N. & D. E. Fairbrothers. 1980. *Threatened and Endangered Vascular Plant Species of the New Jersey Pinelands and Their Habitats*. New Jersey Pinelands Commission, Pemberton, NJ. 38 pp.

Calderón, A. I., M. Cubilla, A. Espinosa & M. P. Gupta. 2010. Screening of plants of Amaryllidaceae and related families from Panama as sources of acetylcholinesterase inhibitors. *Pharm. Biol.* **48**: 988–993.

Calha, I. M., C. Machado & F. Rocha. 1998. New *Alisma* spp. biotypes resistant to sulfonylurea herbicides. Pp. 245–248 *In*: A. Monteiro, T. Vasconcelos & L. Catarino (eds.), *Management and Ecology of Aquatic Plants. Proceedings of the 10th EWRS International Symposium on Aquatic Weeds, Lisbon, Portugal, 21-25 September 1998*. European Weed Research Society, Poznañ, Poland.

Calhoun, A. & G. M. King. 1998. Characterization of root-associated methanotrophs from three freshwater macrophytes: *Pontederia cordata*, *Sparganium eurycarpum*, and *Sagittaria latifolia*. *Appl. Environ. Microbiol.* **64**: 1099–1105.

Calicioglu, O. & R. A. Brennan. 2018. Sequential ethanol fermentation and anaerobic digestion increases bioenergy yields from duckweed. *Bioresour. Technol.* **257**: 344–348.

Calvo-Polanco, M., M. A. Equiza, J. Señorans & J. J. Zwiazek. 2014. Responses of rat root (*Acorus americanus* Raf.) plants to salinity and pH conditions. *J. Environ. Qual.* **43**: 578–586.

Camazine, S. & K. J. Niklas. 1984. Aerobiology of *Symplocarpus foetidus*: interactions between the spathe and spadix. *Am. J. Bot.* **71**: 843–850.

Cameron, K. M. 2005. Leave it to the leaves: a molecular phylogenetic study of Malaxideae (Epidendroideae, Orchidaceae). *Am. J. Bot.* **92**: 1025–1032.

Cameron, K. M. 2009. On the value of nuclear and mitochondrial gene sequences for reconstructing the phylogeny of vanilloid orchids (Vanilloideae, Orchidaceae). *Ann. Bot.* **104**: 377–385.

Cameron, K. M. 2010. On the value of taxonomy, phylogeny, and systematics to orchid conservation: implications for China's Yachang Orchid Reserve. *Bot. Rev.* **76**: 165–173.

Cameron, K. M. & M. W. Chase. 1999. Phylogenetic relationships of Pogoniinae (Vanilloideae, Orchidaceae): an herbaceous example of the eastern North America-eastern Asia phytogeographic disjunction. *J. Plant Res.* **112**: 317–329.

Cameron, K. M. & C. Fu. 2006. A nuclear rDNA phylogeny of *Smilax* (Smilacaceae). *Aliso* **22**: 598–605.

Cameron, K. M. & M. C. Molina. 2006. Photosystem II gene sequences of *psbB* and *psbC* clarify the phylogenetic position of *Vanilla* (Vanilloideae, Orchidaceae). *Cladistics* **22**: 239–248.

Camill, P. 1999. Patterns of boreal permafrost peatland vegetation across environmental gradients sensitive to climate warming. *Can. J. Bot.* **77**: 721–733.

Camp, W. H. 1933. Distribution and flowering in *Wolffia papulifera*. *Ohio J. Sci.* **33**: 163.

Campanella, J. J., P. A. X. Bologna, M. Carvalho, J. V. Smalley, M. Elakhrass, R. W. Meredith & N. Zaben. 2015. Clonal diversity and connectedness of turtle grass (*Thalassia testudinum*) populations in a UNESCO Biosphere Reserve. *Aquat. Bot.* **123**: 76–82.

Campbell, D. H. 1897. A morphological study of *Naias* and *Zannichellia*. *Proc. Calif. Acad. Sci. Ser. 3* **1**: 1–61.

Campbell, R. W. & D. Stirling. 1968. Notes on the vertebrate fauna associated with a Brandt's Cormorant colony in British Columbia. *Murrelet* **49**: 7–9.

Campbell, H. W. & A. B. Irvine. 1977. Feeding ecology of the West Indian manatee *Trichechus manatus* Linnaeus. *Aquaculture* **12**: 249–251.

Campbell, D. & J. Bergeron. 2012. Natural revegetation of winter roads on peatlands in the Hudson Bay lowland, Canada. *Arct. Antarct. Alp. Res.* **44**: 155–163.

Campbell, L. M. & L. J. Dorr. 2013. A synopsis of *Harperocallis* (Tofieldiaceae, Alismatales) with ten new combinations. *PhytoKeys* **21**: 37–52.

Campbell, J. E. & J. W. Fourqurean. 2013. Mechanisms of bicarbonate use influence the photosynthetic carbon dioxide sensitivity of tropical seagrasses. *Limnol. Oceanogr.* **58**: 839–848.

Campbell, J. W., A. M. Starring & G. L. Smith. 2014. Flower visitors of *Hymenocallis coronaria* (rocky shoals spider-lily) of Landsford Canal State Park — South Carolina, USA. *Nat. Areas J.* **34**: 332–337.

Canfield, D. E., K. A. Langeland, S. B. Linda & W. T. Haller. 1985. Relations between water transparency and maximum depth of macrophyte colonization in lakes. *J. Aquat. Plant Manag.* **23**: 25–28.

Cangiano, T., M. DellaGreca, A. Fiorentino, M. Isidori, P. Monaco & A. Zarrelli. 2001. Lactone diterpenes from the aquatic plant *Potamogeton natans*. *Phytochemistry* **56**: 469–473.

Cannings, S. G. 2003. Status of river jewelwing (*Calopteryx aequabilis* Say) in British Columbia. Wildlife Bulletin No. B-110. British Columbia Ministry of Water, Land and Air Protection, Biodiversity Branch, Victoria, BC. 10 pp.

Cannings, R. A. & J. P. Simaika. 2005. *Lestes disjunctus* and *L. forcipatus* (Odonata: Lestidae): an evaluation of status and distribution in British Columbia. *J. Entomol. Soc. British Columbia* **102**: 57–64.

Cao, H. X., G. T. H. Vu, W. Wang, K. J. Appenroth, J. Messing & I. Schubert. 2016. The map-based genome sequence of *Spirodela polyrhiza* aligned with its chromosomes, a reference for karyotype evolution. *New Phytol.* **209**: 354–363.

Capers, R. S. 2003a. Macrophyte colonization in a freshwater tidal wetland (Lyme, CT, USA). *Aquat. Bot.* **77**: 325–338.

Capers, R. S. 2003b. Six years of submerged plant community dynamics in a freshwater tidal wetland. *Freshw. Biol.* **48**: 1640–1651.

Capers, R. S. & D. H. Les. 2005. Plant community structure in a freshwater tidal wetland. *Rhodora* **107**: 386–407.

Capers, R. S., R. Selsky & G. J. Bugbee. 2010. The relative importance of local conditions and regional processes in structuring aquatic plant communities. *Freshw. Biol.* **55**: 952–966.

Cappers, R. T. J. 1993. Seed dispersal by water: a contribution to the interpretation of seed assemblages. *Veg. Hist. Archaeobot.* **2**: 173–186.

Caprio, A. C. & D. L. Taylor. 1984. Effect of frost on a subtropical *Muhlenbergia* prairie in south Florida. *Florida Sci.* **47**: 27–32.

Carefoot, T. H. 1973. Feeding, food preference, and the uptake of food energy by the supralittoral isopod *Ligia pallasii*. *Mar. Biol.* **18**: 228–236.

Carlquist, S. 1967. The biota of long-distance dispersal. V. Plant dispersal to Pacific Islands. *Bull. Torrey Bot. Club* **94**: 129–162.

Carlquist, S. & E. L. Schneider. 2014. Origins and nature of vessels in Monocotyledons. 14. Vessellessness in Orontioideae (Araceae): adaptation or relicticulism? *Nordic J. Bot.* **32**: 493–502.

Carlsen, M. M., T. Fér, R. Schmickl, J. Leong-Škorničková, M. Newman & W. J. Kress. 2018. Resolving the rapid plant radiation of early diverging lineages in the tropical Zingiberales: pushing the limits of genomic data. *Mol. Phylogenet. Evol.* **128**: 55–68.

Carlson, M. C. 1938. Origin and development of shoots from the tips of roots of *Pogonia ophioglossoides*. *Bot. Gaz.* **100**: 215–225.

Carlson, M. C. 1943. The morphology and anatomy of *Calopogon pulchellus*. *Bull. Torrey Bot. Club* **70**: 349–368.

Carlson, M. L., R. Lipkin, M. Sturdy & J. A. Michaelson. 2004. Kenai Fjords National Park vascular plant inventory. Final technical report. NPS Report: NPS/AKR/SWAN/NRTR-2004/02. National Park Service, Southwest Alaska Network, Anchorage, AK. 29 [+40 Appendix] pp.

Carlson, M. L., R. Lipkin, C. Roland & A. E. Miller. 2013. New and important vascular plant collections from south-central and southwestern Alaska: a region of floristic convergence. *Rhodora* **115**: 61–95.

Carpenter, S. R. & J. E. Titus. 1984. Composition and spatial heterogeneity of submersed vegetation in a softwater lake in Wisconsin. *Vegetatio* **57**: 153–165.

Carpenter, J. S. & E. W. Chester. 1987. Vascular flora of the Bear Creek Natural Area, Stewart County, Tennessee. *Castanea* **52**: 112–128.

Carr, L. G. 1940. Further notes on coastal floral elements in the bogs of Augusta County, Virginia. *Rhodora* **42**: 86–93.

Carr, S. C. 2007. Floristic and environmental variation of pyrogenic pinelands in the Southeastern Coastal Plain: description, classification, and restoration. Ph.D. dissertation. University of Florida, Gainesville, FL. 184 pp.

Carr, S. C., K. M. Robertson & R. K. Peet. 2010. A vegetation classification of fire-dependent pinelands of Florida. *Castanea* **75**: 153–189.

Carrillo, D., D. Amalin, F. Hosein, A. Roda, R. E. Duncan & J. E. Peña. 2012. Host plant range of *Raoiella indica* (Acari: Tenuipalpidae) in areas of invasion of the New World. *Exp. Appl. Acarol.* **57**: 271–289.

Carroll, S., R. L. Miller & P. D. Whitson. 1984. Status of four orchid species at Silver Lake Fen complex. *Proc. Iowa Acad. Sci.* **91**: 132–139.

Carter, J. W. 1982. Natural history observations on the gastropod shell-using amphipod *Photis conchicola* Alderman, 1936. *J. Crustacean Biol.* **2**: 328–341.

Carter, Jr., J. W. & S. B. Jones, Jr. 1968. The vascular flora of Johnson State Park, Mississippi. *Castanea* **33**: 194–205.

Carter, V. & N. Rybicki. 1986. Resurgence of submersed aquatic macrophytes in the tidal Potomac River, Maryland, Virginia, and the District of Columbia. *Estuaries Coast.* **9**: 368–375.

Carter, M. C. & M. D. Sytsma. 2001. Comparison of the genetic structure of North and South American populations of a clonal aquatic plant. *Biol. Invasions* **3**: 113–118.

Carter, R., M. W. Morris & C. T. Bryson. 1990. Some rare or otherwise interesting vascular plants from the Delta Region of Mississippi. *Castanea* **55**: 40–55.

Carter, R. E., M. D. MacKenzie, D. H. Gjerstad & D. Jones. 2004. Species composition of fire disturbed ecological land units in the southern loam hills of south Alabama. *Southeast. Nat.* **3**: 297–309.

Carter, R., W. W. Baker & M. W. Morris. 2009. Contributions to the flora of Georgia, USA. *Vulpia* **8**: 1–54.

Carvalho, K. M. & D. F. Martin. 2001. Removal of aqueous selenium by four aquatic plants. *J. Aquat. Plant Manag.* **39**: 33–36.

Carvalho, A. T., S. Dötterl & C. Schlindwein. 2014. An aromatic volatile attracts oligolectic bee pollinators in an interdependent bee-plant relationship. *J. Chem. Ecol.* **40**: 1126–1134.

Carville, J. S. 1997. *Hiking Tahoe's Wildflower Trails*. Lone Pine Publishing, Vancouver, BC. 352 pp.

Casadoro, G., N. Rascio, M. Pagiusco & N. Ravagnan. 1982. Flowers of *Orontium aquaticum* L.: membrane rearrangement in chloroplast—chromoplast interconversions. *J. Ultrastruct. Res.* **81**: 202–208.

Case, M. A. 1994. Extensive variation in the levels of genetic diversity and degree of relatedness among five species of *Cypripedium* (Orchidaceae). *Am. J. Bot.* **81**: 175–184.

Case, F. W. & R. B. Case. 1974. *Sarracenia alabamensis*, a newly recognized species from central Alabama. *Rhodora* **76**: 650–665.

Casierra-Posada, F., M. M. Blanke & J. C. Guerrero-Guío. 2014. Iron tolerance in calla lilies (*Zantedeschia aethiopica*). *Gesunde Pflanzen* **66**: 63–68.

Caslake, L. F., S. S. Harris, C. Williams & N. M. Waters. 2006. Mercury-resistant bacteria associated with macrophytes from a polluted lake. *Water Air Soil Pollut.* **174**: 93–105.

Casper, S. J. & H.-D. Krausch. 1980. *Süsswasserflora von Mitteleuropa. Band 23: Pteridophyta und Anthophyta. 1. Teil: Lycopodiaceae bis Orchidaceae*. Gustav Fischer Verlag, Stuttgart, Germany. 403 pp.

Cassani, J. R. 1981. Feeding behaviour of underyearling hybrids of the grass carp, Ctenopharyngodon *idella*♀ and the bighead, *Hypophthalmicthys nobilis*♂ on selected species of aquatic plants. *J. Fish Biol.* **18**: 127–133.

Cassani, J. R., W. E. Caton & T. H. Hansen, Jr. 1982. Culture and diet of hybrids grass carp fingerlings. *J. Aquat. Plant Manag.* **20**: 30–32.

Cassani, J. R., D. H. Habeck & D. L. Matthews. 1990. Life history and immature stages of a plume moth *Sphenarches anisodactylus* (Lepidoptera: Pterophoridae) in Florida. *Florida Entomol.* **73**: 257–266.

Castellanos, D. L. & L. P. Rozas. 2001. Nekton use of submerged aquatic vegetation, marsh, and shallow unvegetated bottom in the Atchafalaya River Delta, a Louisiana tidal freshwater ecosystem. *Estuaries* **24**: 184–197.

Catling, P. M. 1980. Rain-assisted autogamy in *Liparis loeselii* (L.) L. C. Rich. (Orchidaceae). *Bull. Torrey Bot. Club* **107**: 525–529.

Catling, P. M. 1982. Breeding systems of northeastern North American *Spiranthes* (Orchidaceae). *Can. J. Bot.* **60**: 3017–3039.

Catling, P. M. 1983a. Autogamy in eastern Canadian Orchidaceae: a review of current knowledge and some new observations. *Nat. Can.* **110**: 37–54.

Catling, P. M. 1983b. Pollination of northeastern North American *Spiranthes* (Orchidaceae). *Can. J. Bot.* **61**: 1080–1093.

Catling, P. M. 1984. Self-pollination and probable autogamy in chamisso's orchid *Platanthera chorisiana* (Cham.) Reichb. f. *Naturaliste Can.* **111**: 451–453.

Catling, P. M. & G. Knerer. 1980. Pollination of the small white lady's-slipper (*Cypripedium candidum*) in Lambton County, Southern Ontario. *Can. Field-Nat.* **94**: 435–438.

Catling, P. M. & W. G. Dore. 1982. Status and identification of *Hydrocharis morsus-ranae* and *Limnbobium spongia* (Hydrocharitaceae) in Northeastern North America. *Rhodora* **84**: 523–545.

Catling, P. M. & I. Dobson. 1985. The biology of Canadian weeds. 69. *Potamogeton crispus* L. *Can. J. Plant Sci.* **65**: 655–668.

Catling, P. M. & W. Wojtas. 1986. The waterweeds (*Elodea* and *Egeria*, Hydrocharitaceae) in Canada. *Can. J. Bot.* **64**: 1525–1541.

Catling, P. M. & V. R. Catling. 1989. Observations on the pollination of *Platanthera huronensis* in southwest Colorado. *Lindleyana* **4**: 78–84.

Catling, P. M. & V. R. Catling. 1991. A synopsis of breeding systems and pollination in North American orchids. *Lindleyana* **6**: 187–210.

Catling, P. M. & Z. S. Porebski. 1995. The spread and current distribution of European Frogbit, *Hydrocharis morsus-ranae* L., in North America. *Can. Field-Nat.* **109**: 236–241.

Catling, P. M. & V. R. Brownell. 1999. *Platanthera lacera × leucophaea*, a new cryptic natural hybrid, and a key to northeastern North American fringed-orchids. *Can. J. Bot.* **77**: 1144–1149.

Catling, P. M. & G. Mitrow. 2001. *Egeria najas* at the Canadian border and its separation from the related aquatic weeds *Egeria densa* and *Hydrilla verticillata* (Hydrocharitaceae). *Bot. Electronic Newslett.* **278**: 1–3.

Catling, P. M. & L. K. Magrath. 2002. *Malaxis* Solander ex Swartz. Pp. 627–631 *In*: Flora North America Editorial Committee (eds.), *Flora of North America North of Mexico, Vol. 26: Magnoliophyta: Liliidae: Liliales and Orchidales*. Oxford University Press, New York, NY.

Catling, P. M. & B. Kostiuk. 2011a. Some observations on the pollination of round-leaf orchid, *Galearis* (*Amerorchis*) *rotundifolia*, near Jasper, Alberta. *Can. Field-Nat.* **125**: 47–54.

Catling, P. M. & B. Kostiuk. 2011b. Some wild Canadian orchids benefit from woodland hiking trails-and the implications. *Can. Field-Nat.* **125**: 105–115.

Catling, P. M., B. Freedman & Z. Lucas. 1984. The vegetation and phytogeography of Sable Island, Nova Scotia. *Proc. Nova Scotian Inst. Sci.* **34**: 181–247.

Catling, P. M., B. Freedman, C. Stewart, J. J. Kerekes & L. P. Lefkovitch. 1986. Aquatic plants of acid lakes in Kejimkujik National Park, Nova Scotia; floristic composition and relation to water chemistry. *Can. J. Bot.* **64**: 724–729.

Catling, P. M., K. W. Spicer & L. P. Lefkovitch. 1988. Effects of the floating *Hydrocharis* morsus-ranae (Hydrocharitaceae), on some North American aquatic macrophytes. *Naturaliste Can.* **115**: 131–137.

Catling, P. M., K. W. Spicer, M. Biernacki & J. Lovett Doust. 1994. The biology of Canadian weeds. 103. *Vallisneria americana* Michx. *Can. J. Plant Sci.* **74**: 883–897.

Catling, P. M., G. Mitrow, E. Haber, U. Posluszny & W. A. Charlton. 2003. The biology of Canadian weeds. 124. *Hydrocharis morsus-ranae* L. *Can. J. Plant Sci.* **83**: 1001–1016.

Cattaneo, A. & J. Kalff. 1978. Seasonal changes in the epiphyte community of natural and artificial macrophytes in Lake Memphremagog (Que. & Vt.). *Hydrobiologia* **60**: 135–144.

Cázares, E., J. M. Trappe & A. Jumpponen. 2005. Mycorrhiza-plant colonization patterns on a subalpine glacier forefront as a model system of primary succession. *Mycorrhiza* **15**: 405–416.

Cedergreen, N. & T. V. Madsen. 2002. Nitrogen uptake by the floating macrophyte *Lemna minor*. *New Phytol.* **155**: 285–292.

Celdran, D. 2017. Photosynthetic activity detected in the seed epidermis of *Thalassia testudinum*. *Aquat. Bot.* **136**: 39–42.

Cellot, B., F. Mouillot & C. P. Henry. 1998. Flood drift and propagule bank of aquatic macrophytes in a riverine wetland. *J. Veg. Sci.* **9**: 631–640.

Center, T. D. & N. R. Spencer. 1981. The phenology and growth of water hyacinth (*Eichhornia crassipes* (Mart.) Solms) in a eutrophic north-central Florida lake. *Aquat. Bot.* **10**: 1–32.

Cervantes-Alcalá, R., A. A. Arrocha-Arcos, L. A. Peralta-Peláez & L. A. Ortega-Clemente. 2012. Electricity generation in sediment plant microbial fuel cells (SPMFC) in warm climates using *Typha domingensis* Pers. *Int. Res. J. Biotechnol.* **3**: 166–173.

César, N. R., M. A. Pereira-da-Silva, V. R. Botaro & A. J. de Menezes. 2015. Cellulose nanocrystals from natural fiber of the macrophyte *Typha domingensis*: extraction and characterization. *Cellulose* **22**: 449–460.

Ceschin, S., I. Leacche, S. Pascucci & S. Abati. 2016. Morphological study of *Lemna minuta* Kunth, an alien species often mistaken for the native *L. minor* L. (Araceae). *Aquat. Bot.* **131**: 51–56.

Ceschin, S., S. Abati, N. T. W. Ellwood & V. Zuccarello. 2018. Riding invasion waves: spatial and temporal patterns of the invasive *Lemna minuta* from its arrival to its spread across Europe. *Aquat. Bot.* **150**: 1–8.

Ceska, A. & P. D. Warrington. 1976. *Myriophyllum farwellii* (Haloragaceae) in British Columbia. *Rhodora* **78**: 75–78.

Ceska, O., A. Ceska & P. D. Warrington. 1986. *Myriophyllum quitense* and *Myriophyllum ussuriense* (Haloragaceae) in British Columbia, Canada. *Brittonia* **38**: 73–81.

Chabreck, R. H. 1958. Beaver-forest relationships in St. Tammany Parish, Louisiana. *J. Wildl. Manag.* **22**: 179–183.

Chabreck, R. H. & A. W. Palmisano. 1973. The effects of Hurricane Camille on the marshes of the Mississippi River delta. *Ecology* **54**: 1118–1123.

Chabreck, R. H., R. K. Yancey & L. McNease. 1974. Duck usage of management units in the Louisiana coastal marsh. *Proc. S. E. Assoc. Game Fish Commiss. Conf.* **28**: 507–516.

Chadde, S. W., J. S. Shelly, R. J. Bursik, R. K. Moseley, A. G. Evenden, M. Mantas, F. Rabe & B. Heidel. 1998. Peatlands on national forests of the northern Rocky Mountains: ecology and conservation. General Technical Report RMRS-GTR-11. USDA Forest Service, Rocky Mountain Research Station, Ogden, UT. 75 pp.

Chadin, I., V. Volodin, P. Whiting, T. Shirshova, N. Kolegova & L. Dinan. 2003. Ecdysteroid content and distribution in plants of genus *Potamogeton*. *Biochem. Syst. Ecol.* **31**: 407–415.

Chafin, L. G. 2007. *Field Guide to the Rare Plants of Georgia*. University of Georgia Press, Athens, GA. 526 pp.

Chai, T.-T., M.-J. Chiam, C.-H. Lau, N. I. M. Ismail, H.-C. Ong, F. A. Manan & F.-C. Wong. 2015. Alpha-glucosidase inhibitory and antioxidant activity of solvent extracts and fractions of *Typha domingensis* (Typhaceae) fruit. *Trop. J. Pharmaceut. Res.* **14**: 1983–1990.

Chaïr, H., R. E. Traore, M.-F. Duval, R. Rivallan, A. Mukherjee, L. M. Aboagye, W. J. Van Rensburg, V. Andrianavalona, M. A. A. Pinheiro de Carvalho, F. Saborio, M. S. Prana, B. Komolong, F. Lawac & V. Lebot. 2016. Genetic diversification and dispersal of taro (*Colocasia esculenta* (L.) Schott). *PLoS One* **11**(6): e0157712.

Chamberlain, J. L. 1959. Gulf coast marsh vegetation as food of wintering waterfowl. *J. Wildl. Manag.* **23**: 97–102.

Chambers, P. A. & J. Kalff. 1987. Light and nutrients in the control of aquatic plant community structure. I. In situ experiments. *J. Ecol.* **75**: 611–619.

Chambers, P. A., E. E. Prepas, H. R. Hamilton & M. L. Bothwell. 1991a. Current velocity and its effect on aquatic macrophytes in flowing waters. *Ecol. Appl.* **1**: 249–257.

Chambers, P. A., J. M. Hanson & E. E. Prepas. 1991b. The effect of aquatic plant chemistry and morphology on feeding selectivity by the crayfish, *Orconectes virilis*. *Freshw. Biol.* **25**: 339–348.

Champion, P. D., J. S. Clayton & D. E. Hofstra. 2010. Nipping aquatic plant invasions in the bud: weed risk assessment and the trade. *Hydrobiologia* **656**: 167–172.

Chan, K. L. & J. R. Linley. 1989. A new Florida species of *Forcipomyia* (*Euprojoannisia*) (Diptera: Ceratopogonidae) from leaves of the water lettuce, *Pistia stratiotes*. *Florida Entomol.* **72**: 252–262.

Chan, E. W. C. & S. K. Wong. 2015. Phytochemistry and pharmacology of ornamental gingers, *Hedychium coronarium* and *Alpinia purpurata*: a review. *J. Integrat. Med.* **13**: 368–379.

Chandler, D. C. 1937. Fate of typical lake plankton in streams. *Ecol. Monogr.* **7**: 445–479.

Chandler, C. M. & O. M. McDougal. 2014. Medicinal history of North American *Veratrum*. *Phytochem. Rev.* **13**: 671–694.

Chandran, R. & T. Parimelazhagan. 2012. Nutritional assessment of *Monochoria vaginalis*, a wild edible vegetable supplement to the human diet. *Int. J. Veg. Sci.* **182**: 199–207.

Chandran, R., P. Thangaraj, S. Shanmugam, S. Thankarajan & A. Karuppusamy. 2012. Antioxidant and anti-inflammatory potential of *Monochoria vaginalis* (Burm. f.) C. Presl.: a wild edible plant. *J. Food Biochem.* **36**: 421–431.

Chang, E. R., R. L. Jefferies & T. J. Carleton. 2001. Relationship between vegetation and soil seed banks in an arctic coastal marsh. *J. Ecol.* **89**: 367–384.

Chang, I.-H., K.-T. Cheng, P.-C. Huang, Y.-Y. Lin, L.-J. Cheng & T.-S. Cheng. 2012. Oxidative stress in greater duckweed (*Spirodela polyrhiza*) caused by long-term NaCl exposure. *Acta Physiol. Plant.* **34**: 1165–1176.

Chantiratikul, A., O. Chinrasri, P. Chantiratikul, A. Sangdee, U. Maneechote & C. Bunchasak. 2010. Effect of replacement of protein from soybean meal with protein from wolffia meal [*Wolffia globosa* (L). Wimm.] on performance and egg production in laying hens. *Int. J. Poult. Sci.* **9**: 283–287.

Chanton, J. P., G. J. Whiting, W. J. Showers & P. M. Crill. 1992. Methane flux from *Peltandra virginica*: stable isotope tracing and chamber effects. *Global Biogeochem. Cycles* **6**: 15–31.

Chapman, W. K. 1997. *Orchids of the Northeast: A Field Guide*. Syracuse University Press, Syracuse, NY. 200 pp.

Charalambidou, I. & L. Santamaría. 2002. Waterbirds as endozoochorous dispersers of aquatic organisms: a review of experimental evidence. *Acta Oecol.* **23**: 165–176.

Charalambidou, I., L. Santamaria & O. Langevoord. 2003. Effect of ingestion by five avian dispersers on the retention time, retrieval and germination of *Ruppia maritima* seeds. *Funct. Ecol.* **17**: 747–753.

Chareontesprasit, N. & W. Jiwyam. 2001. An evaluation of *Wolffia* meal (*Wolffia arrhiza*) in replacing soybean meal in some formulated rations of Nile tilapia (*Oreochromis niloticus* L.). *Pak. J. Biol. Sci.* **4**: 618–620.

Charlton, W. A. 1979. Studies in the Alismataceae. VII. Disruption of phyllotactic and organogenetic patterns in pseudostolons of *Echinodorus tenellus* by means of growth-active substances. *Can. J. Bot.* **57**: 215–222.

Charlton, W. A. & U. Posluszny. 1991. Meristic variation in *Potamogeton* flowers. *Bot. J. Linn. Soc.* **106**: 265–293.

Chartier, M., M. Gibernau & S. S. Renner. 2014. The evolution of pollinator–plant interaction types in the Araceae. *Evolution* **68**: 1533–1543.

Chase, S. S. 1947. Preliminary studies in the genus *Najas* in the United States. Ph.D. dissertation. Cornell University, Ithaca, NY. 118 pp.

Chase, J. M. 2003. Strong and weak trophic cascades along a productivity gradient. *Oikos* **101**: 187–195.

Chase, J. M. 2007. Drought mediates the importance of stochastic community assembly. *Proc. Natl. Acad. Sci. U.S.A.* **104**: 17430–17434.

Chase, M. W., K. M. Cameron, J. V. Freudenstein, A. M. Pridgeon, G. Salazar, C. Van den Berg & A. Schuiteman. 2015. An updated classification of Orchidaceae. *Bot. J. Linn. Soc.* **177**: 151–174.

Chase, M. W., D. E. Soltis, R. G. Olmstead, D. Morgan, D. H. Les, B. D. Mishler, M. R. Duvall, R. A. Price, H. G. Hills, Y.-L. Qiu, K. A. Kron, J. H. Rettig, E. Conti, J. D. Palmer, J. R. Manhart, K. J. Sytsma, H. J. Michaels, W. J. Kress, K. G. Karol, W. D. Clark, M. Hedren, B. S. Gaut, R. K. Jansen, K.-J. Kim, C. F. Wimpee, J. F. Smith, G. R. Furnier, S. H. Strauss, Q.-Y. Xiang, G. M. Plunkett, P. S. Soltis, S. M. Swensen, S. E. Williams, P. A. Gadek, C. J. Quinn, L. E. Eguiarte, E. Golenberg, G. H. Learn, Jr., S. W. Graham, S. C. H. Barrett, S. Dayanandan & V. A. Albert. 1993. Phylogenetics of seed plants: an analysis of nucleotide sequences from the plastid gene *rbcL*. *Ann. Missouri Bot. Gard.* **80**: 528–580.

Chase, M. W., M. R. Duvall, H. G. Hills, J. G. Conran, A. V. Cox, L. E. Eguiarte, J. Hartwell, M. F. Fay, L. R. Caddick, K. M. Cameron & S. Hoot. 1995. Molecular phylogenetics of Lilianae. Pp. 109–137 *In*: P. J. Rudall, P. J. Cribb, D. F. Cutler & C. J. Humphries (eds.), *Monocotyledons: Systematics and Evolution*. Royal Botanic Gardens, Kew, United Kingdom.

Chase, M. W., J. L. Reveal & M. F. Fay. 2009. A subfamilial classification for the expanded asparagalean families Amaryllidaceae, Asparagaceae and Xanthorrhoeaceae. *Bot. J. Linn. Soc.* **161**: 132–136.

Chathurangani, D., K. Yakandawala & D. Yakandawala. 2016. A study on competition between *Hydrilla verticillata* and *Mayaca fluviatilis*. *J. Environ. Profess. Sri Lanka* **5**: 11–22.

Chauveau, O., L. Eggers, C. Raquin, A. Silvério, S. Brown, A. Couloux, C. Cruaud, E. Kaltchuk-Santos, R. Yockteng, T. T. Souza-Chies & S. Nadot. 2011. Evolution of oil-producing trichomes in Sisyrinchium (Iridaceae): insights from the first comprehensive phylogenetic analysis of the genus. *Ann. Bot.* **107**: 1287–1312.

Chee, W.-L. & D. H. Vitt. 1989. The vegatation, surface water chemistry and peat chemistry of moderate-rich fens in central Alberta, Canada. *Wetlands* **9**: 227–261.

Chen, P.-H. & W. H. J. Kuo. 1995. Germination conditions for the non-dormant seeds of *Monochoria vaginalis*. *Taiwania* **40**: 419–432.

Chen, S.-C., Y.-X. Qiu, A.-L. Wang, K. M. Cameron & C.-X. Fu. 2005. A phylogenetic analysis of the Smilacaceae based on morphological data. *J. Syst. Evol.* **44**: 113–125.

Chen, Y.-Y., X.-L. Li, L.-Y. Yin & W. Li. 2008. Genetic diversity of the threatened aquatic plant *Ottelia alismoides* in the Yangtze River. *Aquat. Bot.* **88**: 10–16.

Chen, H., M. F. Zamorano & D. Ivanoff. 2010. Effect of flooding depth on growth, biomass, photosynthesis, and chlorophyll fluorescence of *Typha domingensis*. *Wetlands* **30**: 957–965.

Chen, Q., Y. Jin, G. Zhang, Y. Fang, Y. Xiao & H. Zhao. 2012. Improving production of bioethanol from duckweed (*Landoltia punctata*) by pectinase pretreatment. *Energies* **5**: 3019–3032.

Chen, L.-Y., G. W. Grimm, Q.-F. Wang & S. S. Renner. 2015a. A phylogeny and biogeographic analysis for the Cape-Pondweed family Aponogetonaceae (Alismatales). *Mol. Phylogenet. Evol.* **82**: 111–117.

Chen, L., Y. Fang, Y. Jin, Q. Chen, Y. Zhao, Y. Xiao & H. Zhao. 2015b. Biosorption of Cd^{2+} by untreated dried powder of duckweed *Lemna aequinoctialis*. *Desalin. Water Treat.* **53**: 183–194.

Chen, N., Y. Ji, W. Zhang, Y. Xu, X. Yan, X. Sun, H. Song. 2016. Chemical constituents from *Hymenocallis littoralis*. *Lett. Org. Chem.* **13**: 536–539.

Chen, P., Y. Cao, B. Bao, L. Zhang & A. Ding. 2017. Antioxidant capacity of *Typha angustifolia* extracts and two active flavonoids. *Pharm. Biol.* **55**: 1283–1288.

Chen, L., S. Huang, C. Y. Li, F. Gao & X. L. Zhou. 2018. Pyrrolizidine alkaloids from *Liparis nervosa* with antitumor activity by modulation of autophagy and apoptosis. *Phytochemistry* **153**: 147–155.

Cheng, S. 2004. Forest Service research natural areas in California. General Technical Paper PSW-GTR-188. U.S. Department of Agriculture, Forest Service, Pacific Southwest Research Station, Albany, CA. 338 pp.

Cheng, T.-S. 2011. NaCl-induced responses in giant duckweed (*Spirodela polyrhiza*). *Can. J. Bot.* **59**: 104–105.

Cheng, J. J. & A.-M. Stomp. 2009. Growing duckweed to recover nutrients from wastewaters and for production of fuel ethanol and animal feed. *Clean–Soil Air Water* **37**: 17–26.

Cheplick, G. P. 2006. A modular approach to biomass allocation in an invasive annual (*Microstegium vimineum*; Poaceae). *Am. J. Bot.* **93**: 539–545.

Cherry, J. A. & L. Gough. 2006. Temporary floating island formation maintains wetland plant species richness: the role of the seed bank. *Aquat. Bot.* **85**: 29–36.

Cheruvelil, K. S., P. A. Soranno & J. D. Madsen. 2001. Epiphytic macroinvertebrates along a gradient of Eurasian watermilfoil cover. *J. Aquat. Plant Manag.* **39**: 67–72.

Chester, K. S. 1930. The *Phytophthora* disease of the calla in America. *J. Arnold Arbor.* **11**: 169–171.

Chester, E. W. & K. Souza. 1984. *Echinodorus tenellus* var. *parvulus* (Alismataceae) in Kentucky. *Sida* **10**: 262–263.

Chester, E. W. & B. L. Palmer-Ball. 2011. Second county records for two Kentucky endangered species, *Echinodorus tenellus* (Alismataceae) and *Schoenoplectus hallii* (Cyperaceae). *Phytoneuron* 2011–43: 1–4.

Chinnappa, C. C. & J. G. Chmielewski. 1987. Documented plant chromosome numbers 1987: 1. Miscellaneous counts from western North America. *Sida* **12**: 409–417.

Cho, H. R. & H.-S. Choi. 2003. Effects of anticoagulant from *Spirodela polyrhiza* in rats. *Biosci. Biotechnol. Biochem.* **67**: 881–883.

Cho, H. J. & M. A. Poirrier. 2005. Seasonal growth and reproduction of *Ruppia maritima* L. s.l. in Lake Pontchartrain, Louisiana, USA. *Aquat. Bot.* **81**: 37–49.

Cho, H. J. & C. A. May. 2008. Short-term spatial variations in the beds of *Ruppia maritima* (Ruppiaceae) and *Halodule wrightii* (Cymodoceaceae) at Grand Bay National Estuarine Research Reserve, Mississippi, USA. *J. Mississippi Acad. Sci.* **53**: 133–145.

Cho, H. J. & Y. L. Sanders. 2009. Note on organic dormancy of estuarine *Ruppia maritima* L. seeds. *Hydrobiologia* **617**: 197–201.

Cho, T. O., S. Fredericq & K. K. Yates. 2002. Characterization of macroalgal epiphytes on *Thalassia testudinum* in Tampa Bay, Florida. *J. Phycol.* **38**: 4.

Cho, H. J., A. Lu, P. Biber & J. D. Caldwell. 2012. Aquatic plants of the Mississippi Coast. *J. Mississippi Acad. Sci.* **57**: 240–249.

Choate, C. M. 1963. Ordination of the alpine plant communities at Logan Pass Glacier National Park Montana. M.A. thesis. Montana State University, Bozeman, MT. 106 pp.

Choi, K. S., K. T. Park & S. J. Park. 2017. The chloroplast genome of *Symplocarpus renifolius*: a comparison of chloroplast genome structure in Araceae. *Genes* **8**(11): 324.

Cholewa, A. F. & D. M. Henderson. 1994. Iridaceae *Iris* family: part one *Sisyrinchium* L. *J. Arizona-Nevada Acad. Sci.* **27**: 215–218.

Cholewa, A. F. & D. M. Henderson. 2002. *Sisyrinchium* Linnaeus. Pp. 351–371 *In*: Flora North America Editorial Committee (eds.), *Flora of North America North of Mexico, Vol. 26: Magnoliophyta: Liliidae: Liliales and Orchidales*. Oxford University Press, New York, NY.

Chollett, I., D. Bone & D. Pérez. 2007. Effects of heavy rainfall on *Thalassia testudinum* beds. *Aquat. Bot.* **87**: 189–195.

Chouhan, A. P. S. & A. K. Sarma. 2013. Biodiesel production from *Jatropha curcas* L. oil using *Lemna perpusilla* Torrey ash as heterogeneous catalyst. *Biomass Bioenergy* **55**: 386–389.

Christenhusz, M. J. M. & J. W. Byng. 2016. The number of known plants species in the world and its annual increase. *Phytotaxa* **261**: 201–217.

Christenhusz, M. J. M., M. F. Fay & M. W. Chase. 2017. *Plants of the World: An Illustrated Encyclopedia of Vascular Plants*. University of Chicago Press, Chicago, IL. 816 pp.

Christopher, J. 1983. Cytology of *Monochoria vaginalis* complex Presl. *Cytologia* **48**: 627–631.

Christy, J. A. 1994. Noteworthy collections - Washington. *Madroño* **41**: 332.

Christy, J. A. 2004. *Native Freshwater Wetland Plant Associations of Northwestern Oregon*. Oregon Natural Heritage Information Center, Oregon State University, Corvallis, OR. 246 pp.

Christy, J. A. 2013. *Wet Meadow Plant Associations, Malheur National Wildlife Refuge, Harney County, Oregon*. Oregon Biodiversity Information Center, Institute for Natural Resources, Portland State University, Portland, OR. 73 pp.

Christy, J. A. & M. Garvey. 2015. *Existing Vegetation and Site Observations at Killin Wetland, Washington County, Oregon*. The Wetlands Conservancy and Oregon Biodiversity Information Center. Institute for Natural Resources, Portland State University, Portland, OR. 23 pp.

Chung, M.-C. 2015. Chromosome techniques and FISH. Pp. 287–309 *In*: E. C. T. Yeung, C. Stasolla, M. J. Sumner & B. Q. Huang (eds.), *Plant Microtechniques and Protocols*. Springer International Publishing, Basel, Switzerland.

Chupov, V. S., E. O. Punina, E. M. Machs & A. V. Rodionov. 2007. Nucleotide composition and CpG and CpNpG content of ITS1, ITS2, and the 5.8 S rRNA in representatives of the phylogenetic branches Melanthiales-Liliales and Melanthiales-Asparagales (Angiospermae, Monocotyledones) reflect the specifics of their evolution. *Molec. Biol.* **41**: 737–755.

Churchill, A. C. 1983. Field studies on seed germination and seedling development in *Zostera marina* L. *Aquat. Bot.* **16**: 21–29.

Churchill, A. C. & M. I. Riner. 1978. Anthesis and seed production in *Zostera marina* L. from Great South Bay, New York, USA. *Aquat. Bot.* **4**: 83–93.

Cicero, P., J. Kiser, J. Dunlavy & J. Daly. 2011. *Aquatic Plant Management Plan for Lower Spring Lake, 2011*. Jefferson County Land and Water Conservation Department, Jefferson, WI. 48 pp.

Ciotir, C. & J. Freeland. 2016. Cryptic intercontinental dispersal, commercial retailers, and the genetic diversity of native and non-native cattails (*Typha* spp.) in North America. *Hydrobiologia* **768**: 137–150.

Ciotir, C., H. Kirk, J. R. Row & J. R. Freeland. 2013. Intercontinental dispersal of *Typha angustifolia* and *T. latifolia* between Europe and North America has implications for *Typha* invasions. *Biol. Invasions* **15**: 1377–1390.

Ciutti, F., M. E. Beltrami, I. Confortini, S. Cianfanelli & C. Cappelletti. 2011. Non-indigenous invertebrates, fish and macrophytes in Lake Garda (Italy). *J. Limnol.* **70**: 315–320.

Claassen, P. W. 1918. Observations on the life history and biology of *Agromyza laterella* Zetterstedt. (Diptera). *Ann. Entomol. Soc. Am.* **11**: 9–16.

Claassen, P. W. 1921. *Typha Insects: Their Ecological Relationships*. Memoir 47. Agricultural Experiment Station, Cornell University, Ithaca, NY. 531 pp.

Clark, W. A. 1956. Plant distribution in the Western Isles. *Proc. Linn. Soc. London* **167**: 96–103.

Clark, C. F. 1990. Movements of northern pike tagged in waters tributary to Lake Erie. *Ohio J. Sci.* **90**: 41–45.

Clark, M. A., J. Siegrist & P. A. Keddy. 2008. Patterns of frequency in species-rich vegetation in pine savannas: effects of soil moisture and scale. *Ecoscience* **15**: 529–535.

Clarke, A. L. & G. H. Dalrymple. 2003. $7.8 billion for Everglades restoration: why do environmentalists look so worried? *Popul. Environ.* **24**: 541–569.

Clausen, R. T. 1936. Studies in the genus *Najas* in the northern United States. *Rhodora* **38**: 333–345.

Clawson, A. B. & E. A. Moran. 1937. Toxicity of arrowgrass for sheep and remedial treatment. Technical Bulletin No. 580. U.S. Department of Agriculture, Washington, DC. 16 pp.

Clayton, J. S. 1996. Aquatic weeds and their control in New Zealand lakes. *Lake Reserv. Manag.* **12**: 477–486.

Cleall, E., J. Wadman, J. Wadman, J. Baker, J. Pinney, J. Chaffey, W. Bond, W. Baine & A. Anderson. 1807. Papers in colonies and trade. *Trans. Soc. London Encour. Arts* **25**: 143–212.

Cleland, C. F. & W. R. Briggs. 1967. Flowering responses of the long-day plant *Lemna gibba* G3. *Plant Physiol.* **42**: 1553–1561.

Cleland, C. F., O. Tanaka & L. J. Feldman. 1982. Influence of plant growth substances and salicylic acid on flowering and growth in the Lemnaceae (duckweeds). *Aquat. Bot.* **13**: 3–20.

Clinton, G. P. 1901. Two new smuts on *Eriocaulon septangulare.* *Rhodora* **28**: 79–82.

Clough, K. S. & G. R. Best. 1991. *Wetland Macrophyte Production and Hydrodynamics in Hopkins Prairie, Ocala National Forest, Florida March 1989–December-1990.* Center for Wetlands, University of Florida, Gainesville, FL. 87 pp.

Clouse, R. M., B. Ferster & M. A. Deyrup. 1997. Observations of insects associated with an infestation of sand pine (*Pinus clausa*) by the aphid *Cinara pinivora.* *Florida Sci.* **60**: 89–93.

Coan, A. I., T. Stützel & V. L. Scatena. 2010. Comparative embryology and taxonomic considerations in Eriocaulaceae (Poales). *Feddes Repert.* **121**: 268–284.

Cody, W. J. 1961. *Iris pseudacorus* L. escaped from cultivation in Canada. *Can. Field-Nat.* **75**: 139–142.

Cody, W. J., K. L. Reading & J. M. Line. 2003. Additions and range extensions to the vascular plant flora of the continental Northwest Territories and Nunavut, Canada, II. *Can. Field-Nat.* **117**: 448–465.

Cohen, S., R. Braham & F. Sanchez. 2004. Seed bank viability in disturbed longleaf pine sites. *Restor. Ecol.* **12**: 503–515.

Cole, F. R. & D. H. Firmage. 1984. The floral ecology of *Platanthera blephariglottis.* *Am. J. Bot.* **71**: 700–710.

Cole, C. T. & M. I. Voskuil. 1996. Population genetic structure in duckweed (*Lemna minor*, Lemnaceae). *Can. J. Bot.* **74**: 222–230.

Coleman, R. A. 2002. *The Wild Orchids of Arizona and New Mexico.* Cornell University Press, Ithaca, NY. 248 pp.

Coleman, T. S. & D. A. Boag. 1987. Canada goose foods: their significance to weight gain. *Wildfowl* **38**: 82–88.

Coleman, R. A., D. H. Wilkin & W. F. Jennings. 2012. Orchidaceae. Orchid family. Pp. 1398–1405 *In*: B. G. Baldwin, D. H. Goldman, D. J. Keil, R. Patterson, T. J. Rosatti & D. H. Wilken (eds.), *The Jepson Manual*, 2nd ed. University of California Press, Berkeley, CA.

Coler, R. A. & H. B. Gunner. 1969. The rhizosphere of an aquatic plant (*Lemna minor*). *Can. J. Microbiol.* **15**: 964–966.

Colle, D. E., J. V. Shireman & R. W. Rottmann. 1978. Food selection by grass carp fingerlings in a vegetated pond. *Trans. Am. Fish. Soc.* **107**: 149–152.

Collectanea, M. 1976. Aquatic crops vs. organic soil subsidence. *Proc. Florida State Hort. Soc.* **89**: 125–129.

Collins, F. S. 1905. *Chlorochytrium lemnae* in America. *Rhodora* **7**: 97–99.

Collins, B. & G. Wein. 1995. Seed bank and vegetation of a constructed reservoir. *Wetlands* **15**: 374–385.

Collins, C. D., R. B. Sheldon & C. W. Boylen. 1987. Littoral zone macrophyte community structure: distribution and association of species along physical gradients in Lake George, New York, USA. *Aquat. Bot.* **29**: 177–194.

Collon, E. G. & J. Velasquez. 1989. Dispersion, germination and growth of seedlings of *Sagittaria lancifolia* L. *Folia Geobot.* **24**: 37–49.

Colt, L. C. & C. B. Hellquist. 1974. The role of some Haloragaceae in algal ecology. *Rhodora* **76**: 446–459.

Colt, L. C., C. B. Hellquist & W. J. L. Zubrin. 1971. An interesting association of rare aquatic plants from New Hampshire. *Rhodora* **73**: 296–299.

Colwell, M. A. & L. W. Oring. 1988. Habitat use by breeding and migrating shorebirds in southcentral Saskatchewan. *Wilson Bull.* **100**: 554–566.

Colwell, A. E. L., C. J. Sheviak & P. E. Moore. 2007. A new *Platanthera* (Orchidaceae) from Yosemite National Park, California. *Madroño* **54**: 86–94.

Combroux, I., G. Bornette, N. J. Willby & C. Amoros. 2001. Regenerative strategies of aquatic plants in disturbed habitats: the role of the propagule bank. *Archiv. Hydrobiol.* **152**: 215–235.

Comes, R. D., V. F. Bruns & A. D. Kelley. 1978. Longevity of certain weed and crop seeds in fresh water. *Weed Sci.* **26**: 336–344.

Conard, H. S. 1935. The plant associations of central Long Island. A study in descriptive plant sociology. *Am. Midl. Nat.* **16**: 433–516.

Conover, D. & S. Pelikan. 2010. Earlier flowering in a restored wetland–prairie correlated with warmer temperatures (Ohio). *Ecol. Restor.* **28**: 428–430.

Conrad, J. L., A. J. Bibian, K. L. Weinersmith, D. De Carion, M. J. Young, P. Crain, E. L. Hestir, M. J. Santos & A. Sih. 2016. Novel species interactions in a highly modified estuary: association of largemouth bass with Brazilian Waterweed *Egeria densa.* *Trans. Am. Fish. Soc.* **145**: 249–263.

Cook, R. S. 1979. Determination that *Harperocallis flava* is an endangered species. *Fed. Reg.* **44**: 56862–56863.

Cook, C. D. K. 1982. Pollination mechanisms in the Hydrocharitaceae. Pp. 1–15 *In*: J. J. Symoens, S. S. Hooper & P. Compère (eds.), *Studies on Aquatic Vascular Plants. Proceedings of the International Colloquium on Aquatic Vascular Plants (Brussels, 23–25 January, 1981).* Royal Botanical Society of Belgium, Brussels.

Cook, C. D. K. 1985. Range extensions of aquatic vascular plant species. *J. Aquat. Plant Manag.* **23**: 1–6.

Cook, C. D. K. 1988. Wind pollination in aquatic angiosperms. *Ann. Missouri Bot. Gard.* **75**: 768–777.

Cook, C. D. K. 1996a. *Aquatic Plant Book.* SPB Academic Publishing bv, Amsterdam, The Netherlands. 228 pp.

Cook, C. D. K. 1996b. *Aquatic and Wetland Plants of India: A Reference Book and Identification Manual for the Vascular Plants Found in Permanent or Seasonal Fresh Water in the Subcontinent of India South of the Himalayas.* Oxford University Press, New York, NY. 385 pp.

Cook, C. D. K. 2004. *Aquatic and Wetland Plants of Southern Africa.* Backhuys Publishers BV, Leiden, The Netherlands. 281 pp.

Cook, C. D. K. & R. Lüönd. 1982a. A revision of the genus *Hydrilla* (Hydrocharitaceae). *Aquat. Bot.* **13**: 485–504.

Cook, C. D. K. & R. Lüönd. 1982b. A revision of the genus *Hydrocharis* (Hydrocharitaceae). *Aquat. Bot.* **14**: 177–204.

Cook, C. D. K. & R. Lüönd. 1983. A revision of the genus *Blyxa* (Hydrocharitaceae). *Aquat. Bot.* **15**: 1–52.

Cook, C. D. K. & K. Urmi-König. 1983. A revision of the genus *Limnobium* (Hydrocharitaceae). *Aquat. Bot.* **17**: 1–27.

Cook, C. D. K. & K. Urmi-König. 1984a. A revision of the genus *Egeria* (Hydrocharitaceae). *Aquat. Bot.* **19**: 73–96.

Cook, C. D. K. & K. Urmi-König. 1984b. A revision of the genus *Ottelia* (Hydrocharitaceae). 2. The species of Eurasia, Australasia, and America. *Aquat. Bot.* **20**: 131–177.

Cook, C. D. K. & K. Urmi-König. 1985. A revision of the genus *Elodea* (Hydrocharitaceae). *Aquat. Bot.* **21**: 111–156.

Cook, C. D. K. & M. S. Nicholls. 1986. A monographic study of the genus *Sparganium*. Part 1: subgenus Xanthosparganium. *Bot. Helvet.* **96**: 213–267.

Cook, C. D. K. & M. S. Nicholls. 1987. A monographic study of the genus *Sparganium*. Part 2: subgenus Sparganium. *Bot. Helvet.* **97**: 1–44.

Cook, C. D. K., J.-J. Symoens & K. Urmi-König. 1984. A revision of the genus *Ottelia* (Hydrocharitaceae) I. Generic considerations. *Aquat. Bot.* **18**: 263–274.

Cook, R. T. A., A. J. Inman & C. Billings. 1997. Identification and classification of powdery mildew anamorphs using light and scanning electron microscopy and host range data. *Mycol. Res.* **101**: 975–1002.

Cooke, W. B. 1949. Western Fungi–I. *Mycologia* **41**: 601–622.

Cooke, R. G. 1970. Phenylnaphthalene pigments of *Lachnanthes tinctoria*. *Phytochemistry* **9**: 1103–1106.

Cooke, S. S. (ed.). 1997. *A Field Guide to the Common Wetland Plants of Western Washington and Northwestern Oregon*. Trailside Series. Seattle Audubon Society, Seattle, WA. 417 pp.

Cooper, D. J. 1996. Water and soil chemistry, floristics, and phytosociology of the extreme rich High Creek fen, in South Park, Colorado, USA. *Can. J. Bot.* **74**: 1801–1811.

Cooper, L. W. & C. P. McRoy. 1988a. Anatomical adaptations to rocky substrates and surf exposure by the seagrass genus *Phyllospadix*. *Aquat. Bot.* **32**: 365–381.

Cooper, L. W. & C. P. McRoy. 1988b. Stable carbon isotope ratio variations in marine macrophytes along intertidal gradients. *Oecologia* **77**: 238–241.

Cooper, D. J. & R. E. Andrus. 1994. Patterns of vegetation and water chemistry in peatlands of the west-central Wind River Range, Wyoming, U.S.A. *Can. J. Bot.* **72**: 1586–1597.

Cooper, D. J. & L. H. MacDonald. 2000. Restoring the vegetation of mined peatlands in the southern Rocky Mountains of Colorado, USA. *Restor. Ecol.* **8**: 103–111.

Cooperrider, T. S. 1955. The Ophioglossaceae of Iowa. *Am. Fern J.* **45**: 156–159.

Coops, H. & G. Van der Velde. 1995. Seed dispersal, germination and seedling growth of six helophyte species in relation to water-level zonation. *Freshw. Biol.* **34**: 13–20.

Coops, H., M. A. A. de la Haye & F. W. B. van der Brink. 1994. Studies on germination and growth of a river macrophyte (*Potamogeton nodosus* Poir.) in relation to substrate type and water quality. *Verh. Int. Vereinigung Limnol.* **25**: 2247–2250.

Cope, R. B. 2005. *Allium* species poisoning in dogs and cats. *Veterin. Med.* **100**: 562–566.

Cordero, G. A. & C. W. Swarth. 2010. Notes on the movement and aquatic behavior of some kinosternid turtles. *Acta Zool. Mexicana* **26**: 233–235.

Core, E. L. 1941. *Butomus umbellatus* in America. *Ohio J. Sci.* **41**: 79–85.

Core, E. L. 1967. Ethnobotany of the southern Appalachian aborigines. *Econ. Bot.* **21**: 199–214.

Core, E. L., J. M. Fogg & R. H. Torrey. 1937. Field trips of the club. *Torreya* **37**: 130–137.

Corogin, P. T. & W. S. Judd. 2009. Floristic inventory of Tiger Creek preserve and Saddle Blanket Scrub preserve, Polk County, Florida. *Rhodora* **111**: 449–503.

Correll, D. S. & H. B. Correll. 1975. *Aquatic and Wetland Plants of Southwestern United States*. 2 volumes. Stanford University Press, Stanford, CA. 1777 pp.

Corriveau, J. L. & A. W. Coleman. 1988. Rapid screening method to detect potential biparental inheritance of plastid DNA and results for over 200 angiosperm species. *Am. J. Bot.* **75**: 1443–1458.

Cortinovis, C. & F. Caloni. 2013. Epidemiology of intoxication of domestic animals by plants in Europe. *Veterin. J.* **197**: 163–168.

COSEWIC. 2005. *COSEWIC Assessment and Update Status Report on the Hill's Pondweed Potamogeton hillii in Canada*. Committee on the Status of Endangered Wildlife in Canada. Ottawa, ON. vi + 19 pp.

Cosgriff, R., V. J. Anderson & S. Monson. 2004. Restoration of communities dominated by false hellebore. *J. Range Manag.* **57**: 365–370.

Costa, J. Y. 2004. A triploid cytotype of *Echinodorus tenellus*. *Aquat. Bot.* **79**: 325–332.

Cottam, C. 1939. Food habits of North American diving ducks. *USDA Tech. Bull.* **643**: 1–129.

Cottam, C., J. J. Lynch & A. L. Nelson. 1944. Food habits and management of American sea brant. *J. Wildl. Manag.* **8**: 36–56.

Coughlan, N. E., T. C. Kelly, J. Davenport & M. A. K. Jansen. 2015a. Humid microclimates within the plumage of mallard ducks (*Anas platyrhynchos*) can potentially facilitate long distance dispersal of propagules. *Acta Oecol.* **65**: 17–23.

Coughlan, N. E., T. C. Kelly & M. A. K. Jansen. 2015b. Mallard duck (*Anas platyrhynchos*)-mediated dispersal of Lemnaceae: a contributing factor in the spread of invasive *Lemna minuta*? *Plant Biol.* **17**: 108–114.

Coughlan, N. E., T. C. Kelly, J. Davenport & M. A. K. Jansen. 2017a. Up, up and away: bird-mediated ectozoochorous dispersal between aquatic environments. *Freshw. Biol.* **62**: 631–648.

Coughlan, N. E., T. C. Kelly & M. A. K. Jansen. 2017b. "Step by step": high frequency short-distance epizoochorous dispersal of aquatic macrophytes. *Biol. Invasions* **19**: 625–634.

Coughlan, N. E., R. N. Cuthbert, T. C. Kelly & M. A. K. Jansen. 2018. Parched plants: survival and viability of invasive aquatic macrophytes following exposure to various desiccation regimes. *Aquat. Bot.* **150**: 9–15.

Coulter, M. W. 1955. Spring food habits of surface-feeding ducks in Maine. *J. Wildl. Manag.* **19**: 263–267.

Countryman, W. D. 1968. *Alisma gramineum* in Vermont. *Rhodora* **70**: 577–579.

Countryman, W. D. 1970. The history, spread and present distribution of some immigrant aquatic weeds in New England. *Hyacinth Control J.* **8**: 50–52.

Cowell, B. C. & C. H. Resico, Jr. 1975. Life history patterns in the coastal shiner, *Notropis petersoni*, Fowler. *Florida Sci.* **38**: 113–121.

Cox, R. J. & G. C. Smart, Jr. 1994. Nematodes associated with plants from naturally acidic wetlands soil. *J. Nematol.* **26**: 535–537.

Cox, T. E. & S. N. Murray. 2006. Feeding preferences and the relationships between food choice and assimilation efficiency in the herbivorous marine snail *Lithopoma undosum* (Turbinidae). *Mar. Biol.* **148**: 1295–1306.

Cox, P. A., T. Elmqvist & P. B. Tomlinson. 1990. Submarine pollination and reproductive morphology in *Syringodium filiforme* (Cymodoceaceae). *Biotropica* **22**: 259–265.

Cox, P. A., P. B. Tomlinson & K. Nieznanski. 1992a. Hydrophilous pollination and reproductive morphology in the seagrass *Phyllospadix scouleri* (Zosteraceae). *Plant Syst. Evol.* **180**: 65–75.

Cox, P. A., R. H. Laushman & M. H. Ruckelshaus. 1992b. Surface and submarine pollination in the seagrass *Zostera marina* L. *Bot. J. Linn. Soc.* **109**: 281–291.

Coyer, J. A., K. A. Miller, J. M. Engle, J. Veldsink, A. Cabello-Pasini, W. T. Stam & J. L. Olsen. 2008. Eelgrass meadows in the California Channel Islands and adjacent coast reveal a mosaic of two species, evidence for introgression and variable clonality. *Ann. Bot.* **101**: 73–87.

Coyer, J. A., G. Hoarau, J. Kuo, A. Tronholm, J. Veldsink & J. L. Olsen. 2013. Phylogeny and temporal divergence of the seagrass family Zosteraceae using one nuclear and three chloroplast loci. *Syst. Biodivers.* **11**: 271–284.

Craig, C., S. Wyllie-Echeverria, E. Carrington & D. Shafer. 2008. Short-term sediment burial effects on the seagrass *Phyllospadix scouleri*. EMRRP Technical Notes Collection (ERDC TN-EMRRP-EI-03). U.S. Army Engineer Research and Development Center, Vicksburg, MS. 10 pp.

Cranfill, R. 1981. Bog clubmosses (*Lycopodiella*) in Kentucky. *Am. Fern J.* **71**: 97–100.

Cranfill, R. 1991. Flora of Hardin County, Kentucky. *Castanea* **56**: 228–267.

Craven, S. R. & R. A. Hunt. 1984. Food habits of Canada Geese on the coast of Hudson Bay. *J. Wildl. Manag.* **48**: 567–569.

Crawford, D. J. 2010. Progenitor-derivative species pairs and plant speciation. *Taxon* **59**: 1413–1423.

Crawford, D. J. & E. Landolt. 1993. Allozyme studies in *Spirodela* (Lemnaceae): variation among conspecific clones and divergence among the species. *Syst. Bot.* **18**: 389–394.

Crawford, D. J. & E. Landolt. 1995. Allozyme divergence among species of *Wolffia* (Lemnaceae). *Plant Syst. Evol.* **197**: 59–69.

Crawford, D. J., E. Landolt & D. H. Les. 1996. An allozyme study of two sibling species of *Lemna* (Lemnaceae) with comments on their morphology, ecology and distribution. *Bull. Torrey Bot. Club* **123**: 1–6.

Crawford, D. J., E. Landolt, D. H. Les & E. Tepe. 1997. Allozyme variation and the taxonomy of *Wolffiella* (Lemnaceae). *Aquat. Bot.* **58**: 43–54.

Crawford, D. J., E. Landolt, D. H. Les & R. T. Kimball. 2001. Allozyme studies in Lemnaceae: variation and relationships in *Lemna* sections *Alatae* and *Biformes*. *Taxon* **50**: 987–999.

Crawford, D. J., E. Landolt, D. H. Les, R. T. Kimball & J. K. Archibald. 2005. Allozyme variation within and divergence between *Lemna gibba* and L. *disperma* (Lemnaceae): systematic and biogeographic implications. *Aquat. Bot.* **83**: 119–128.

Crawford, D. J., E. Landolt, D. H. Les & R. T. Kimball. 2006. Speciation in duckweeds (Lemnaceae): phylogenetic and ecological inferences. *Aliso* **22**: 231–242.

Crema, L. C., V. M. F. da Silva & M. T. F. Piedade. 2019. Riverine people's knowledge of the Vulnerable Amazonian manatee *Trichechus inunguis* in contrasting protected areas. *Oryx* **831**: 1–10.

Crespo, M. B., M. Martínez-Azorín & E. V. Mavrodiev. 2015. Can a rainbow consist of a single colour? A new comprehensive generic arrangement of the 'Iris sensu latissimo' clade (Iridaceae), congruent with morphology and molecular data. *Phytotaxa* **232**: 1–78.

Crişan, I. & M. Cantor. 2016. New perspectives on medicinal properties and uses of *Iris* sp. *Hop Med. Plants* 24: 24–36.

Crişan, I., R. Vidican, I. Oltean, A. Stoie & V. Stoian. 2018. *Iris* spp. flower visitors: pollinators or nectar thieves? *Romanian J. Grassland Forage Crops* 17: 11–19.

Crocker, W. 1907. Germination of seeds of water plants. *Bot. Gaz.* **44**: 375–380.

Crocker, W. 1938. Life-span of seeds. *Bot. Rev.* **4**: 235–274.

Crombie, L. 1999. Natural product chemistry and its part in the defence against insects and fungi in agriculture. *Pestic. Sci.* **55**: 761–774.

Cronan, Jr., J. M. 1957. Food and feeding habits of the scaups in Connecticut waters. *Auk* **74**: 459–468.

Cronin, G. & D. M. Lodge. 2003. Effects of light and nutrient availability on the growth, allocation, carbon/nitrogen balance, phenolic chemistry, and resistance to herbivory of two freshwater macrophytes. *Oecologia* **137**: 32–41.

Cronin, G., D. M. Lodge, M. E. Hay, M. Miller, A. M. Hill, T. Horvath, R. C. Bolser, N. Lindquist & M. Wahl. 2002. Crayfish feeding preferences for freshwater macrophytes: the influence of plant structure and chemistry. *J. Crustacean Biol.* **22**: 708–718.

Cronk, K. L., D. T. L. Myers & M. L. Claucherty. 2015. *Mullett Lake Partial Aquatic Plant Survey 2015*. Tip of the Mitt Watershed Council, Petoskey, MI. 49 pp.

Cronquist, A. 1981. *An Integrated System of Classification of Flowering Plants*. Columbia University Press, New York, NY. 1262 pp.

Cropley, T. G. 2006. The "army itch:" A dermatological mystery of the American civil War. *J. Am. Acad. Dermatol.* **55**: 302–308.

Cross, A. F. 1991. Vegetation of two southeastern Arizona desert marshes. *Madroño* **38**: 185–194.

Cross, A. T., L. M. Skates, L. Adamec, C. M. Hammond, P. M. Sheridan & K. W. Dixon. 2015. Population ecology of the endangered aquatic carnivorous macrophyte *Aldrovanda vesiculosa* at a naturalised site in North America. *Freshw. Biol.* **60**: 1772–1783.

Crothers, G. M. 2012. Early woodland ritual use of caves in eastern North America. *Am. Antiquity* **77**: 524–541.

Crouch, C. A. 1991. Infaunal polychaetes of a rocky intertidal surfgrass bed in southern California. *Bull. Mar. Sci.* **48**: 386–394.

Crouch, V. E. & M. S. Golden. 1997. Floristics of a bottomland forest and adjacent uplands near the Tombigbee River, Choctaw County, Alabama. *Castanea* **62**: 219–238.

Crow, G. E. 1969. An ecological analysis of a southern Michigan bog. *Michigan Bot.* **8**: 11–27.

Crow, J. H. 1979. Distribution and ecological characteristics of *Zannichellia* palustris L. along the Alaska Pacific coast. *Bull. Torrey Bot. Club* **106**: 346–349.

Crow, G. E. & C. B. Hellquist. 1981. Aquatic vascular plants of New England: Part 2. Typhaceae and Sparganiaceae. Station Bulletin 517. New Hampshire Agricultural Experiment Station, University of New Hampshire, Durham, NH. 21 pp.

Crow, G. E. & C. B. Hellquist. 1982. Aquatic vascular plants of New England: part 4. Juncaginaceae, Scheuchzeriaceae, Butomaceae, Hydrocharitaceae. Station Bulletin 520. New Hampshire Agricultural Experiment Station, University of New Hampshire, Durham, NH. 20 pp.

Crow, G. E. & C. B. Hellquist. 2000. *Aquatic and Wetland Plants of Northeastern North America, Volume Two: Angiosperms: Monocotyledons*. The University of Wisconsin Press, Madison, WI. 400 pp.

Crow, W. T. & N. R. Walker. 2003. Diagnosis of *Peltamigratus christiei*, a plant-parasitic nematode associated with warm-season turfgrasses in the southern United States. *Plant Health Progr.* **4**: 1–8.

Crowder, A. A., J. M. Bristow, M. R. King & S. Vanderkloet. 1977a. Distribution, seasonality, and biomass of aquatic macrophytes in Lake Opinicon (Eastern Ontario). *Naturaliste Can.* **104**: 451–456.

Crowder, A. A., J. M. Bristow, M. R. King & S. Vanderkloet. 1977b. The aquatic macrophytes of some lakes in southeastern Ontario. *Naturaliste Can.* **104**: 457–464.

Crowe, E. A., B. L. Kovalchik & M. J. J. Kerr. 2004. *Riparian and Wetland Vegetation of Central and Eastern Oregon.* Oregon Natural Heritage Information Center, Institute for Natural Resources, Oregon State University, Corvallis, OR. 483 pp.

Cruden, R. W. 1988. Temporal dioecism: systematic breadth, associated traits, and temporal patterns. *Bot. Gaz.* **149**: 1–15.

Cruden, R. W. 1991. A revision of *Isidrogalvia* (Liliaceae): recognition for Ruiz and Pavon's genus. *Syst. Bot.* **16**: 270–282.

Cruden, R. W. & D. L. Lyon. 1985. Patterns of biomass allocation to male and female functions in plants with different mating systems. *Oecologia* **66**: 299–306.

Cruden, R. W. & R. M. Lloyd. 1995. Embryophytes have equivalent sexual phenotypes and breeding systems: why not a common terminology to describe them? *Am. J. Bot.* **82**: 816–825.

Crutchfield, Jr., J. U., D. H. Schiller, D. D. Herlong & M. A. Mallin. 1992. Establishment and impact of redbelly tilapia in a vegetated cooling reservoir. *J. Aquat. Plant Manag.* **30**: 28–35.

Cruz, A. J. U., E. V. Hernández, R. N. Carbó & M. L. López. 1998. The Najadaceae of Cuba. *Anales Jard. Bot. Madrid* **56**: 85–94.

Cruzan, M. B., J. L. Hamrick, M. L. Arnold & B. D. Bennett. 1994. Mating system variation in hybridizing irises: effects of phenology and floral densities on family outcrossing rates. *Heredity* **72**: 95–105.

Cuenca, A., G. Petersen & O. Seberg. 2013. The complete sequence of the mitochondrial genome of *Butomus umbellatus*–a member of an early branching lineage of Monocotyledons. *PLoS One* **8**(4): e61552.

Cui, W., J. Xu, J. J. Cheng & A. M. Stomp. 2011. Starch accumulation in duckweed for bioethanol production. *Biol. Eng. Trans.* **3**: 187–197.

Cui, X., X. Dai, K. Y. Khan, T. Li, X. Yang & Z. He. 2016. Removal of phosphate from aqueous solution using magnesium-alginate/chitosan modified biochar microspheres derived from *Thalia dealbata. Bioresour. Technol.* **218**: 1123–1132.

Cullain, N., R. McIver, A. L. Schmidt & H. K. Lotze. 2018. Spatial variation of macroinfaunal communities associated with *Zostera marina* beds across three biogeographic regions in Atlantic Canada. *Estuaries Coast.* **41**: 1381–1396.

Culley, T. M., J.-F. Leng, S. R. Kephart, F. J. Cartieri & K. E. Theiss. 2013. Development of 16 microsatellite markers within the *Camassia* (Agavaceae) species complex and amplification in related taxa. *Appl. Plant Sci.* **1**: 1300001.

Culling, D. E. & D. B. Cichowski. 2017. *Boreal Caribou (Rangifer tarandus) in British Columbia: 2017 Science Review.* British Columbia Oil and Gas Research and Innovation Society, Victoria, BC. 141 pp.

Cunningham, D. D. 1887. On the phenomenon of gaseous evolution from the flowers of *Ottelia alismoides. Sci. Mem. Med. Officers Army India* 2: 39–46.

Currah, R. S., L. Sigler & S. Hambleton. 1987. New records and new taxa of fungi from the mycorrhizae of terrestrial orchids of Alberta. *Can. J. Bot.* **65**: 2473–2482.

Currah, R. S., L. W. Zettler & T. M. McInnis. 1997. *Epulorhiza inquilina* sp. nov. from *Platanthera* (Orchidaceae) and a key to *Epulorhiza* species. *Mycotaxon* **61**: 335–342.

Curtis, J. T. 1939. The relation of specificity of orchid mycorrhizal fungi to the problem of symbiosis. *Am. J. Bot.* **26**: 390–399.

Curtis, J. T. 1943. Germination and seedling development in five species of *Cypripedium* L. *Am. J. Bot.* **30**: 199–206.

Curtis, J. T. 1946. Use of mowing in management of white ladyslipper. *J. Wildl. Manag.* **10**: 303–308.

Cusick, A. W. 1970. Noteworthy Mississippi plant records. *Castanea* **35**: 322–324.

Cusimano, N., J. Bogner, S. J. Mayo, P. C. Boyce, S. Y. Wong, M. Hesse, W. L. A. Hetterscheid, R. C. Keating & J. C. French. 2011. Relationships within the Araceae: comparison of morphological patterns with molecular phylogenies. *Am. J. Bot.* **98**: 654–668.

Custer, C. M. & T. W. Custer. 1986. Food habits of diving ducks in the Great Lakes after the zebra mussel invasion. *J. Field Ornithol.* **67**: 86–99.

Cutler, A. J., W. Hösel, M. Sternberg & E. E. Conn. 1981. The in vitro biosynthesis of taxiphyllin and the channeling of intermediates in *Triglochin maritima. J. Biol. Chem.* **256**: 4253–4258.

Cutter, V. M. 1943. An undescribed *Lagenidium* parasitic upon *Potamogeton. Mycologia* **35**: 2–12.

Cypert, E. 1961. The effects of fires in the Okefenokee Swamp in 1954 and 1955. *Am. Midl. Nat.* **66**: 485–503.

Cypert, E. 1972a. The origin of houses in the Okefenokee prairies. *Am. Midl. Nat.* **87**: 448–458.

Cypert, E. 1972b. Plant succession on burned areas in Okefenokee swamp following the fires of 1954 and 1955. *Proc. Annual Tall Timbers Fire Ecol. Conf.* **12**: 199–217.

Da Cunha, N. L. & E. Fischer. 2009. Breeding system of tristylous *Eichhornia azurea* (Pontederiaceae) in the southern Pantanal, Brazil. *Plant Syst. Evol.* **280**: 53–58.

Da Silva, S. L., K. M. Magalhães & R. de Carvalho. 2017. Karyotype variations in seagrass (*Halodule wrightii* Ascherson—Cymodoceaceae). *Aquat. Bot.* **136**: 52–55.

Dahlgren, R. M. T. & H. T. Clifford. 1982. *The Monocotyledons: A Comparative Study.* Academic Press, London, UK. 378 pp.

Dalby, R. 2000. A trio of spring bee plants: prairie crocus, skunk cabbage, and manzanita. *Am. Bee J.* **140**: 147–149.

Dale, H. M. & T. J. Gillespie. 1977. The influence of submersed aquatic plants on temperature gradients in shallow water bodies. *Can. J. Bot.* **55**: 2216–2225.

Dammer, U. 1888. *Beiträge zur Kenntnis der vegetativen Organe von Limnobium stoloniferum Grisebach nebst einigen betrachtungen über die phylogenetische Dignität von Diclinie und Hermaphroditismus.* Becker & Hornberg, Berlin, Germany. 20 pp.

Daniëls, F. J. A., S. S. Talbot, S. L. Talbot & W. B. Schofield. 1998. Geobotanical aspects of Simeonof Island, Shumagin Islands, southwestern Alaska. *Ber. Reinh. Tüxen Ges.* **10**: 125–138.

Daoust, R. J. & D. L. Childers. 1998. Quantifying aboveground biomass and estimating net aboveground primary production for wetland macrophytes using a non-destructive phenometric technique. *Aquat. Bot.* **62**: 115–133.

Daoust, R. J. & D. L. Childers. 1999. Controls on emergent macrophyte composition, abundance, and productivity in freshwater Everglades wetland communities. *Wetlands* **19**: 262–275.

Darnell, R. M. 1958. Food habits of fishes and large invertebrate of Lake Pontchartrain, Louisiana, an estuarine community. *Publ. Inst. Mar. Sci. Univ. Texas* **5**: 353–416.

Darnell, K. M. & K. H. Dunton. 2015. Consumption of turtle grass seeds and seedlings by crabs in the western Gulf of Mexico. *Mar. Ecol. Prog. Ser.* **520**: 153–163.

Darnell, K. M. & K. H. Dunton. 2016. Reproductive phenology of the subtropical seagrasses *Thalassia testudinum* (turtle grass) and *Halodule wrightii* (shoal grass) in the northwest Gulf of Mexico. *Bot. Mar.* **59**: 473–483.

Darnell, K. M. & K. H. Dunton. 2017. Plasticity in turtle grass (*Thalassia testudinum*) flower production as a response to porewater nitrogen availability. *Aquat. Bot.* **138**: 100–106.

Darst, M. R., H. M. Light & L. J. Lewis. 2002. Ground-cover vegetation in wetland forests of the lower Suwannee River floodplain, Florida, and potential impacts of flow reductions. Water Resources Investigations Report 02-4027. U.S. Geological Survey, Tallahassee, FL. 46 pp.

Daru, B. H. & K. Yessoufou. 2016. A search for a single DNA barcode for seagrasses of the world. Pp. 313–330 *In*: S. Trivedi, A. Ansari, S. Ghosh & H. Rehman (eds.), *DNA Barcoding in Marine Perspectives*. Springer International Publishing, Basel, Switzerland.

Darwin, C. 1869. Notes on the fertilization of orchids. *Ann. Mag. Nat. Hist.* **4**: 143–159.

Darwin, C. 1872. *The Origin of Species by Means of Natural Selection or the Preservation of Favored Races in the Struggle for Life and the Descent of Man and Selection in Relation to Sex*. Modern Library, New York, NY. 1000 pp.

Darwin, C. 1877. *On the Various Contrivances by Which British and Foreign Orchids Are Fertilised by Insects, and on the Good Effects of Intercrossing*, 2nd ed. John Murray, London. 300 pp.

Das, R. R. & B. Gopal. 1969. Vegetative propagation in *Spirodela polyrhiza*. *Trop. Ecol.* **10**: 270–277.

Das, N. J., S. P. Saikia, S. Sarkar & K. Devi. 2006. Medicinal plants of North-Kamrup district of Assam used in primary healthcare system. *Indian J. Tradit. Knowl.* **5**: 489–493.

das Neves, H. J. C. & M. S. S. Pais. 1980. Identification of a spathe regreening factor in *Zantedeschia aethiopica*. *Biochem. Biophys. Res. Commun.* **95**: 1387–1392.

Dash, P. R. 2016. *Phytochemical Screening and Pharmacological Investigations on Hedychium coronarium*. Anchor Academic Publishing, Hamburg, Germany. 72 pp.

Daubs, E. H. 1962. The occurrence of *Spirodela oligorrhiza* in the United States. *Rhodora* **64**: 83–85.

Daubs, E. H. 1965. *A Monograph of Lemnaceae*. Illinois Biological Monographs 34. The University of Illinois Press, Urbana, IL. 118 pp.

David, P. G. 1996. Changes in plant communities relative to hydrologic conditions in the Florida Everglades. *Wetlands* **16**: 15–23.

Davidson, R. A. 1960. Plant communities of southeastern Iowa. *Proc. Iowa Acad. Sci.* **67**: 162–173.

Davidson, D. & J.-P. Simon. 1981. Thermal adaptation and acclimation of ecotypic populations of *Spirodela polyrhiza* (L.) Schleid. (Lemnaceae): morphology and growth rates. *J. Therm. Biol.* **6**: 121–128.

Davidson, D. W., R. D. Gitar & D. S. Anderson. 2008. Floristic and ecological studies on the Nelson Mitigation Site, near Grantsburg, Burnett County, Wisconsin. *Wulfenia* **15**: 25–34.

Davies, P., C. Morvan, O. Sire & C. Baley. 2007. Structure and properties of fibres from sea-grass (*Zostera marina*). *J. Mater. Sci.* **42**: 4850–4857.

Davis, J. J. 1903. Third supplementary list of parasitic fungi of Wisconsin. *Trans. Wisconsin Acad. Sci.* 14: 83–106.

Davis, F. W. 1985. Historical changes in submerged macrophyte communities of upper Chesapeake Bay. *Ecology* **66**: 981–993.

Davis, M. A. 1987. The role of flower visitors in the explosive pollination of *Thalia geniculata* (Marantaceae), a Costa Rican marsh plant. *Bull. Torrey Bot. Club* **114**: 134–138.

Davis, A. F. 1993. Rare wetland plants and their habitats in Pennsylvania. *Proc. Acad. Nat. Sci. Philadelphia* **144**: 254–262.

Davis, J. L., D. L. Childers & D. N. Kuhn. 1999. Clonal variation in a Florida Bay *Thalassia testudinum* meadow: molecular genetic assessment of population structure. *Mar. Ecol. Prog. Ser.* **186**: 127–136.

Davis, J. I., D. W. Stevenson, G. Petersen, O. Seberg, L. M. Campbell, J. V. Freudenstein, D. H. Goldman, C. R. Hardy, F. A. Michelangeli, M. P. Simmons & C. D. Specht. 2004. A phylogeny of the monocots, as inferred from *rbcL* and *atpA* sequence variation, and a comparison of methods for calculating jackknife and bootstrap values. *Syst. Bot.* **29**: 467–510.

Davy, A. J. & G. F. Bishop. 1991. *Triglochin maritima* L. *J. Ecol.* **79**: 531–555.

Dawe, N. K., W. S. Boyd, R. Buechert & A. C. Stewart. 2011. Recent, significant changes to the native marsh vegetation of the Little Qualicum River estuary, British Columbia; a case of too many Canada Geese (*Branta canadensis*)? *Br. Columbia Birds* **21**: 11–31.

Dawes, C. J. 1986. Seasonal proximate constituents and caloric values in seagrasses and algae on the west coast of Florida. *J. Coast. Res.* **2**: 25–32.

Dawes, C. J. & J. M. Lawrence. 1980. Seasonal changes in the proximate constituents of the seagrasses *Thalassia testudinum*, *Halodule wrightii*, and *Syringodium filiforme*. *Aquat. Bot.* **8**: 371–380.

Dawes, C. J., C. S. Lobban & D. A. Tomasko. 1989. A comparison of the physiological ecology of the seagrasses *Halophila decipiens* Ostenfeld and *H. johnsonii* Eiseman from Florida. *Aquat. Bot.* **33**: 149–154.

Dawes, C. J., D. Hanisak & J. W. Kenworthy. 1995. Seagrass biodiversity in the Indian River lagoon. *Bull. Mar. Sci.* **57**: 59–66.

Day, R. T., P. A. Keddy, J. M. T. J. McNeill & T. Carleton. 1988. Fertility and disturbance gradients: a summary model for riverine marsh vegetation. *Ecology* **69**: 1044–1054.

Dayton, P. K. 1975. Experimental evaluation of ecological dominance in a rocky intertidal algal community. *Ecol. Monogr.* **45**: 137–159.

De Abreu Pietrobelli, J. M. T., A. N. Módenes, F. R. Espinoza-Quiñones, M. R. Fagundes-Klen & A. Kroumov. 2009. Removal of copper ions by non-living aquatic macrophytes *Egeria densa*. *Bioautomation* **12**: 21–32.

De Almeida, J. D. & H. Freitas. 2006. Exotic naturalized flora of continental Portugal - a reassessment. *Bot. Complut.* **30**: 117–130.

De Andrade, M. J. G., A. M. Giulietti, A. Rapini, L. P. de Queiroz, A. de Souza Conceição, P. R. M. de Almeida & C. van den Berg. 2010. A comprehensive phylogenetic analysis of Eriocaulaceae: evidence from nuclear (ITS) and plastid (*psbA-trnH* and *trnL-F*) DNA sequences. *Taxon* **59**: 379–388.

De Borges, R. L. B., F. D. A. R. dos Santos & A. M. Giulietti. 2009. Comparative pollen morphology and taxonomic considerations in Eriocaulaceae. *Rev. Palaeobot. Palynol.* **154**: 91–105.

De Castro, W. A. C., M. L. Moitas, G. M. Lobato, M. B. da Cunha-Santino & D. M. da Silva Matos. 2013. First record of herbivory of the invasive macrophyte *Hedychium coronarium* J. König (Zingiberaceae). *Biota Neotrop.* **13**: 368–370.

De Cock, A. W. A. M. 1980. Flowering, pollination and fruiting in *Zostera marina* L. *Aquat. Bot.* **9**: 201–220.

De Castro, W. A. C., R. V. Almeida, M. B. Leite, R. H. Marrs & D. M. S. Matos. 2016. Invasion strategies of the white ginger lily *Hedychium coronarium* J. König (Zingiberaceae) under different competitive and environmental conditions. *Environ. Exp. Bot.* **127**: 55–62.

Dechant, J. A., M. L. Sondreal, D. H. Johnson, L. D. Igl, C. M. Goldade, A. L. Zimmerman & B. R. Euliss. 2004. Effects of management practices on grassland birds: American Bittern. USGS Northern Prairie Wildlife Research Center, Jamestown, ND. 14 pp.

De Faria Garcia, E., M. A. de Oliveira, A. M. Godin, W. C. Ferreira, L. F. S. Bastos, M. de Matos Coelho & F. C. Braga. 2010. Antiedematogenic activity and phytochemical composition of preparations from *Echinodorus grandiflorus* leaves. *Phytomedicine* **18**: 80–86.

De Gruchy, J. H. B. 1938. Preliminary study of the larger aquatic plants of Oklahoma with special reference to their value in fish culture. Technical Bulletin No. 4. Oklahoma Agricultural and Mechanical College, Agricultural Experiment Station, Stillwater, OK. 31 pp.

De la Vega, E. L., J. R. Cassani & H. Allaire. 1993. Seasonal relationship between southern Naiad and associated periphyton. *J. Aquat. Plant Manag.* **31**: 84–88.

de Lange, L. & E. Westinga. 1979. The distinction between *Lemna gibba* L. and *Lemna minor* L. on the basis of vegetative characters. *Acta Bot. Neerl.* **28**: 169–176.

De Oliveira, P. C., J. M. D. Torezan & C. N. da Cunha. 2015. Effects of flooding on the spatial distribution of soil seed and spore banks of native grasslands of the Pantanal wetland. *Acta Bot. Brasilica* **29**: 400–407.

De Oliveira Pellegrini, M. O., R. B. Faden & R. F. de Almeida. 2016. Taxonomic revision of Neotropical *Murdannia* Royle (Commelinaceae). *PhytoKeys* **74**: 35–78.

De Pauw, M. A. & W. R. Remphrey. 1993. In vitro germination of three *Cypripedium* species in relation to time of seed collection, media, and cold treatment. *Can. J. Bot.* **71**: 879–885.

De Pauw, M. A., W. R. Remphrey & C. E. Palmer. 1995. The cytokinin preference for *in vitro* germination and protocorm growth of *Cypripedium candidum*. *Ann. Bot.* **75**: 267–275.

De Steven, D. & C. A. Harrison. 2006. Hydrology, vegetation, and landscape distribution of depression wetlands on the Francis Marion National Forest. Final Report 11/16/2005. United States Forest Service, Southern Research Station, Stoneville, MS. 24 pp.

De Vos, A. 1958. Summer observations on moose behavior in Ontario. *J. Mammal.* **39**: 128–139.

De Vynck, J. C., B.-E. Van Wyk & R. M. Cowling. 2016. Indigenous edible plant use by contemporary Khoe-San descendants of South Africa's Cape South Coast. *S. Afr. J. Bot.* **102**: 60–69.

De Wit, H. C. D. 1964. *Aquarium Plants*. Blandford Press, London. 255 pp.

Dean, R. J. & M. J. Durako. 2007. Carbon sharing through physiological integration in the threatened seagrass *Halophila johnsonii*. *Bull. Mar. Sci.* **81**: 21–35.

DeBerry, J. W. & R. B. Atkinson. 2014. Aboveground forest biomass and litter production patterns in Atlantic white cedar swamps of differing hydroperiods. *Southeast. Nat.* **13**: 673–691.

DeBusk, T. A., J. E. Peterson & K. R. Reddy. 1995. Use of aquatic and terrestrial plants for removing phosphorus from dairy wastewaters. *Ecol. Eng.* **5**: 371–390.

Dechant, J. A., M. L. Sondreal, D. H. Johnson, L. D. Igl, C. M. Goldade, A. L. Zimmerman & B. R. Euliss. 1999 (revised 2002). *Effects of Management Practices on Grassland Birds: American Bittern*. Northern Prairie Wildlife Research Center, Jamestown, ND. 14 pp.

Dee, S. M. & C. Ahn. 2012. Soil properties predict plant community development of mitigation wetlands created in the Virginia Piedmont, USA. *Environ. Manage.* **49**: 1022–1036.

Degani, N., E. Ben-Hur & E. Riklis. 1980. DNA damage and repair: induction and removal of thymine dimers in ultraviolet light irradiated intact water plants. *Photochem. Photobiol.* **31**: 31–36.

Del Fosse, E. S. & B. D. Perkins. 1977. Discovery and bioassay of a kairomone from waterhyacinth, *Eichhornia crassipes*. *Florida Entomol.* **60**: 217–222.

del Moral, R. & A. F. Watson. 1978. Vegetation on the Stikine Flats, southeast Alaska. *Northwest Sci.* **52**: 137–150.

Delahoussaye, J., C. Allen, S. Huskins & A. Dauzart. 2014. A quantitative study of the vegetation surrounding populations of *Zigadenus densus* (Melianthiaceae) at Fort Polk in west Central Louisiana, USA. *J. Bot. Res. Inst. Texas* **8**: 253–259.

Delany, M. F. & S. B. Linda. 1998. Characteristics of Florida grasshopper sparrow nests. *Wilson Bull.* **110**: 136–139.

Delesalle, V. A. & S. Blum. 1994. Variation in germination and survival among families of *Sagittaria latifolia* in response to salinity and temperature. *Int. J. Plant Sci.* **155**: 187–195.

Delgado, E. S. 1999. *Flora Acuática y Subacuática de Aguascalientes*, 2nd ed. Universidad Autonoma De Aguascalientes, Aguascalientes, Mexico. 75 pp.

Delgado, G. 2013. South Florida microfungi: a new species of *Ellisembia* (Hyphomycetes) with new records from the USA. *Mycotaxon* **123**: 445–450.

Delgado, G. 2014. South Florida microfungi: *Linkosia longirostrata*, a new hyphomycete on paurotis palm. *Mycotaxon* **129**: 41–46.

Delisle, F., C. Lavoie, M. Jean & D. Lachance. 2003. Reconstructing the spread of invasive plants: taking into account biases associated with herbarium specimens. *J. Biogeogr.* **30**: 1033–1042.

DellaGreca, M., A. Fiorentino, M. Isidori, P. Monaco & A. Zarrellia. 2000. Antialgal *ent*-labdane diterpenes from *Ruppia maritima*. *Phytochemistry* **55**: 909–913.

DellaGreca, M., A. Fiorentino, P. Monaco & A. Zarrelli. 2001a. Two new polyhydroxylated sterols from *Ruppia maritima*. *Nat. Prod. Lett.* **15**: 111–118.

DellaGreca, M., A. Fiorentino, M. Isidori, P. Monaco, F. Temussi & A. Zarrelli. 2001b. Antialgal furano-diterpenes from *Potamogeton natans* L. *Phytochemistry* **58**: 299–304.

DeMarco, K. E., E. R. Hillmann & M. G. Brasher. 2016. Brackish marsh zones as a waterfowl habitat resource in submerged aquatic vegetation beds in the northern Gulf of Mexico. *J. Southeast. Assoc. Fish Wildl. Agen.* **3**: 261–269.

Deng, J., G. Gao, Y. Zhang, F. He, X. Luo, F. Zhang, X. Liao, K. S. Ahmad & R. Yang. 2016. Phylogenetic and ancestral area reconstruction of Zingiberales from plastid genomes. *Biochem. Syst. Ecol.* **66**: 123–128.

Denley, K. K., C. T. Bryson & R. A. Stewart. 2002. Vascular flora of Yalobusha County, Mississippi. *Castanea* **67**: 402–415.

Dennis, W. M. 1980. *Sarracenia oreophila* (Kearny) Wherry in the Blue Ridge Province of Northeastern Georgia. *Castanea* **45**: 101–103.

Dennis, W. M., A. M. Evans & B. E. Wofford. 1979. Disjunct populations of *Isoëtes macrospora* in southeastern Tennessee. *Am. Fern J.* **69**: 97–99.

Dennison, W. C. & R. S. Alberte. 1985. Role of daily light period in the depth distribution of *Zostera marina* (eelgrass). *Mar. Ecol. Prog. Ser.* **25**: 51–61.

Dennison, W. C., G. J. Marshall & C. Wigand. 1989. Effect of "brown tide" shading on eelgrass (*Zostera marina* L.) distributions. Pp. 675–692 *In*: E. M. Cosper, V. M. Bricelj & E. J. Carpenter (eds.), *Novel Phytoplankton Blooms*. Springer, Berlin, Germany.

Deno, N. C. 1993. *Seed Germination, Theory and Practice*, 2nd ed. 139 Lenor Drive, State College, PA. 242 pp.

Denslow, M. W., G. L. Katz & W. F. Jennings. 2011. First record of *Iris pseudoacorus* (Iridaceae) from Colorado. *J. Bot. Res. Inst. Texas* **5**: 327–329.

Dethier, M. N. 1984. Disturbance and recovery in intertidal pools: maintenance of mosaic patterns. *Ecol. Monogr.* **54**: 99–118.

Dethier, M. N. & D. O. Duggins. 1988. Variation in strong interactions in the intertidal zone along a geographical gradient: a Washington-Alaska comparison. *Mar. Ecol. Prog. Ser.* **50**: 97–105.

Deutsch, G. 2002. *Ponthieva orchioides* Schlechter (Orchidaceae-Spiranthoideae): in vitro-propagation, culture and chromosome number. *Stapfia* **80**: 327–331.

DeVere, E. N. 1970. Habitat selection and species interactions of some marsh passerines. M.S. thesis. Iowa State University, Ames, IA. 75 pp.

DeVlaming, V. & V. W. Proctor. 1968. Dispersal of aquatic organisms: viability of seeds recovered from the droppings of captive killdeer and mallard ducks. *Am. J. Bot.* **55**: 20–26.

DeVelice, C. J. Hubbard, K. Boggs, S. Boudreau, M. Potkin, T. Boucher & C. Wertheim. 1999. Plant community types of the Chugach National Forest: southcentral Alaska. Technical Publication R10-TP-76. United States Department of Agriculture Forest Service, Chugach National Forest, Alaska Region, Anchorage, AK.

Dewanji, A., S. Chanda, L. Si, S. Barik & S. Matai. 1997. Extractability and nutritional value of leaf protein from tropical aquatic plants. *Plant Foods Hum. Nutr.* **50**: 349–357.

Deyrup, M. & L. Deyrup. 2008. Flower visitation by adult shore flies at an inland site in Florida (Diptera: Ephydridae). *Florida Entomol.* **91**: 504–507.

Deyrup, M., J. Edirisinghe & B. Norden. 2002. The diversity and floral hosts of bees at the Archbold Biological Station, Florida (Hymenoptera: Apoidea). *Insecta Mundi* **16**: 87–120.

Diamond, A. R. 2016. Range extension and first record of *Fimbristylis perpusilla* (Cyperaceae) for Alabama. *Phytoneuron* 2016–10: 1–4.

Dias, T. L., I. L. Rosa & J. K. Baum. 2002. Threatened fishes of the world: *Hippocampus erectus* Perry, 1810 (Syngnathidae). *Environ. Biol. Fishes* **65**: 326.

Dibble, E. D. & S. L. Harrel. 1997. Largemouth bass diets in two aquatic plant communities. *J. Aquat. Plant Manag.* **35**: 74–78.

Dickerman, J. A. & R. G. Wetzel. 1985. Clonal growth in *Typha latifolia*: population dynamics and demography of the ramets. *J. Ecol.* **73**: 535–552.

Dickinson, M. B. & T. E. Miller. 1998. Competition among small, free-floating, aquatic plants. *Am. Midl. Nat.* **140**: 55–67.

Dieffenbacher-Krall, A. & W. Halteman. 2000. The relationship of modern plant remains to water depth in alkaline lakes in New England, USA. *J. Paleolimnol.* **24**: 213–229.

Dighton, J., T. Gordon, R. Mejia & M. Sobel. 2013. Mycorrhizal status of Knieskern's beaked sedge (*Rhynchospora knieskernii*) in the New Jersey Pine Barrens. *Bartonia* **66**: 24–27.

Dilday, R. H., P. Nastasi & R. J. Smith, Jr. 1989. Allelopathic observations in rice (*Oryza sativa* L.) to ducksalad (*Heteranthera limosa*). *J. Arkansas Acad. Sci.* **43**: 21–22.

Dillingham, C. 2005. Conservation assessment for Meesia triquetra (L.) Aongstr. (three-anked hump-moss) and Meesia uliginosa Hedwig (broad-nerved hump-moss) in California with a focus on the Sierra Nevada bioregion. Internal Document, USDA Forest Service, VMS Enterprise Team, Quincy, CA. 29 pp.

Dindorf, C. & C. Kjelland. 2006. *Lake George Aquatic Plant Survey, Anoka County, MN. August 2 – 3, 2006.* Fortin Consulting, Inc., Hamel, MN. 12 pp.

Ding, Y., Y. Fang, L. Guo, Z. Li, K. He, Y. Zhao & H. Zhao. 2017a. Phylogenic [sic!] study of Lemnoideae (duckweeds) through complete chloroplast genomes for eight accessions. *PeerJ* **5**: e4186.

Ding, Y., D. Qu, K.-M. Zhang, X.-X. Cang, Z.-N. Kou, W. Xiao & J.-B. Zhu. 2017b. Phytochemical and biological investigations of Amaryllidaceae alkaloids: a review. *J. Asian Nat. Prod. Res.* **19**: 53–100.

Dirrigl, Jr., F. J. & R. H. Mohlenbrock. 2012. Land snails in ephemeral pools at Ottine Swamp, Gonzales County, Texas. *Southwest. Nat.* **57**: 353–355.

Dissanayake, C., M. Hettiarachchi & M. C. M. Iqbal. 2007. Sustainable use of *Cryptocoryne wendtii* and *Echinodorus cordifolius* in the aquaculture industry of Sri Lanka by micropropagation. *Sri Lanka J. Aquat. Sci.* **12**: 89–101.

Dister, D. C. 2017. The vascular flora of Ludington State Park, Mason County, Michigan. *Great Lakes Bot.* **56**: 52–90.

DiTomaso, J. M. & E. A. Healy. 2003. Aquatic and riparian weeds of the west. Publication 3421. Agriculture and Natural Resources, University of California, Oakland, CA. 442 pp.

Dixon, R. B. 1905. The Northern Maidu. *Bull. Am. Mus. Nat. Hist.* **17**: 119–346.

Dixon, L. K. & J. R. Leverone. 1995. Light requirements of *Thalassia testudinum* in Tampa Bay, Florida. Final report. Mote Technical Report No. 425. Mote Marine Laboratory, Sarasota, FL. 77 pp.

do Vale, A. A., B. L. Lau, B. S. S. Leal, J. A. N. Batista, E. L. Borba & K. Proite. 2016. Development and transferability of microsatellite markers in *Habenaria nuda* and *H. repens* (Orchidaceae). *Brazilian J. Bot.* **39**: 387–392.

Dodd, J. D. & R. T. Coupland. 1966. Vegetation of saline areas in Saskatchewan. *Ecology* **47**: 958–968.

Dodd, S. R., R. S. Haynie, S. M. Williams & S. B. Wilde. 2016. Alternate food-chain transfer of the toxin linked to Avian Vacuolar Myelinopathy and implications for the endangered Florida snail kite (*Rostrhamus sociabilis*). *J. Wildl. Dis.* **52**: 335–344.

Dodds, D. G. 1960. Food competition and range relationships of moose and snowshoe hare in Newfoundland. *J. Wildl. Manag.* **24**: 52–60.

Doering, P. H. & R. H. Chamberlain. 2000. Experimental studies on the salinity tolerance of turtle grass, *Thalassia testudinum*. Pp. 81–98 *In*: S. A. Bortone (ed.), *Seagrasses: Monitoring, Ecology, Physiology, and Management*. CRC Press, Boca Raton, FL.

Doering, M. & D. Coxson. 2010. Riparian alder ecosystems as epiphytic lichen refugia in sub-boreal spruce forests of British Columbia. *Botany* **88**: 144–157.

Doherty, J. M., J. C. Callaway & J. B. Zedler. 2011. Diversity–function relationships changed in a long-term restoration experiment. *Ecol. Appl.* **21**: 2143–2155.

Dolan, R. W. 2014. Bacon's Swamp - ghost of a central Indiana natural area past. *Proc. Indiana Acad. Sci.* **123**: 138–160.

Donnermeyer, G. N. & M. M. Smart. 1985. The biomass and nutritive potential of *Vallisneria americana* Michx in navigation pool 9 of the upper Mississippi River. *Aquat. Bot.* **22**: 33–44.

Dooley, F. D., S. Wyllie-Echeverria, E. Gupta & P. D. Ward. 2015. Tolerance of *Phyllospadix scouleri* seedlings to hydrogen sulfide. *Aquat. Bot.* **123**: 72–75.

Dore, W. G. 1954. European frogbit (*Hydrocharis morsus-ranae* L.) in the Ottawa River. *Can. Field-Nat.* **68**: 180–181.

Dore, W. G. 1968. Progress of the European frogbit in Canada. *Can. Field-Nat.* **82**: 76–84.

Dorken, M. E. & S. C. Barrett. 2003. Life-history differentiation and the maintenance of monoecy and dioecy in *Sagittaria latifolia* (Alismataceae). *Evolution* **57**: 1973–1988.

Dorken, M. E. & S. C. Barrett. 2004a. Phenotypic plasticity of vegetative and reproductive traits in monoecious and dioecious populations of *Sagittaria latifolia* (Alismataceae): a clonal aquatic plant. *J. Ecol.* **92**: 32–44.

Dorken, M. E. & S. C. Barrett. 2004b. Sex determination and the evolution of dioecy from monoecy in *Sagittaria latifolia* (Alismataceae). *Proc. Biol. Sci.* **271**: 213–219.

Dorken, M. E. & S. C. Barrett. 2004c. Chloroplast haplotype variation among monoecious and dioecious populations of *Sagittaria latifolia* (Alismataceae) in eastern North America. *Mol. Ecol.* **13**: 2699–2707.

Dorken, M. E., J. Friedman & S. C. H. Barrett. 2002. The evolution and maintenance of monoecy and dioecy in *Sagittaria latifolia* (Alismataceae). *Evolution* **56**: 31–41.

Dostine, P. L. & S. R. Morton. 1989. Food of the black-winged Stilt *Himantopus himantopus* in the Alligator Rivers Region, Northern Territory. *Emu-Austral Ornith.* **89**: 250–253.

Douglas, N. A., W. A. Wall, Q.-Y. Xiang, W. A. Hoffmann, T. R. Wentworth, J. B. Gray & M. G. Hohmann. 2011. Recent vicariance and the origin of the rare, edaphically specialized Sandhills lily, *Lilium pyrophilum* (Liliaceae): evidence from phylogenetic and coalescent analyses. *Mol. Ecol.* **20**: 2901–2915.

Doumlele, D. G. 1981. Primary production and seasonal aspects of emergent plants in a tidal freshwater marsh. *Estuaries* **4**: 139–142.

Downing, J. A. 1981. In situ foraging responses of three species of littoral cladocerans. *Ecol. Monogr.* **51**: 85–104.

Doyle, R. D. 2000. Expansion of the exotic aquatic plant *Cryptocoryne beckettii* (Araceae) in the San Marcos River, Texas. *Sida* **19**: 1027–1038.

Doyle, R. D. 2001. Effects of waves on the early growth of *Vallisneria americana. Freshw. Biol.* **46**: 389–397.

Doyle, G. J. & J. G. Duckett. 1985. The occurrence of *Lysichiton americanus* Hultén & St. John on Woodfield Bog, County Offaly (H18). *Ir. Nat. J.* **21**: 536–538.

Doyle, R. D. & R. M. Smart. 1998. Competitive reduction of noxious *Lyngbya wollei* mats by rooted aquatic plants. *Aquat. Bot.* **61**: 17–32.

Doyle, R. D. & R. M. Smart. 2001. Impacts of water column turbidity on the survival and growth of *Vallisneria americana* winterbuds and seedlings. *Lake Reserv. Manag.* **17**: 17–28.

Doyle, R., S. Hester & C. Williams. 2014. Edwards Aquifer Authority: 2014 ecomodeling: vegetation percent cover to biomass. Report of Research Activities. Center For Reservoir and Aquatic Systems Research, Baylor University, Waco, TX. 13 pp.

Dozier, H. L. 1945. Sex ratio and weights of muskrats from the Montezuma National Wildlife Refuge. *J. Wildl. Manag.* **9**: 232–237.

Dray, Jr., F. A. & T. D. Center. 1989. Seed production by *Pistia stratiotes* L. (water lettuce) in the United States. *Aquat. Bot.* **33**: 155–160.

Dray, Jr., F. A., C. R. Thompson, D. H. Habeck, J. K. Balciunas & T. D. Center. 1988. A survey of the fauna associated with *Pistia stratiotes* L. (water lettuce) in Florida. Technical Report A-88-6. US Army Engineer Waterways Experiment Station, Vicksburg, MS. 36 pp.

Dray, Jr., F. A., F. Allen, T. D. Center & D. H. Habeck. 1993. Phytophagous insects associated with *Pistia stratiotes* in Florida. *Environ. Entomol.* **22**: 1146–1155.

Drescher, B. 2017. An ecological examination of Johnson Bayou (Pass Christian, MS) with a reproductive histological analysis of *Rangia cuneata*, and a comparative morphological study of the foot and shell of *Rangia cuneata* and *Polymesoda caroliniana*. Ph.D. dissertation. University of Southern Mississippi, Hattiesburg, MS. 176 pp.

Dressler, R. L., D. W. Hall, K. D. Perkins & N. H. Williams. 1987. *Identification Manual for Wetland Plant Species of Florida*. Publication SP-35.Institute of Food and Agricultural Sciences, University of Florida, Gainesville, FL. 297 pp.

Drewa, P. B., W. J. Platt & E. B. Moser. 2002. Community structure along elevation gradients in headwater regions of longleaf pine savannas. *Plant Ecol.* **160**: 61–78.

Driever, S. M., E. H. van Nes & R. M. M. Roijackers. 2005. Growth limitation of *Lemna minor* due to high plant density. *Aquat. Bot.* **81**: 245–251.

Drobney, R. D. & L. H. Fredrickson. 1979. Food selection by wood ducks in relation to breeding status. *J. Wildl. Manag.* **43**: 109–120.

Drouet, F. 1933. Algal vegetation of the large Ozark springs. *Trans. Am. Microsc. Soc.* **52**: 83–100.

Drummond, D. C. 1960. The food of *Rattus norvegicus* Berk. in an area of sea wall, saltmarsh and mudflat. *J. Anim. Ecol.* **29**: 341–347.

Drysdale, F. R. & M. G. Barbour. 1975. Response of the marine angiosperm *Phyllospadix torreyi* to certain environmental variables: a preliminary study. *Aquat. Bot.* **1**: 97–106.

Du, Y., J. Feng, R. Wang, H. Zhang & J. Liu. 2015. Effects of flavonoids from *Potamogeton crispus* L. on proliferation, migration, and invasion of human ovarian cancer cells. *PLoS One* **10**: e0130685.

Duarte, C. M. 1991. Seagrass depth limits. *Aquat. Bot.* **40**: 363–377.

Duarte, C. M. & C. L. Chiscano. 1999. Seagrass biomass and production: a reassessment. *Aquat. Bot.* **65**: 159–174.

Dubbe, D. R., E. G. Garver & D. C. Pratt. 1988. Production of cattail (*Typha* spp.) biomass in Minnesota, USA. *Biomass* **17**: 79–104.

DuBois, R. B., J. M. Pleski, W. A. Smith & E. J. Epstein. 2009. Odonata of coastal peatland habitats adjacent to Lake Superior in Wisconsin. *Great Lakes Entomol.* **42**: 52–66.

Dudley, W. R. 1893. The genus *Phyllospadix*. Pp. 403–418 *In: Wilder Quarter-Century Book*. Comstock, Ithaca, NY.

Dudley, W. R. 1894. *Phyllospadix*, its systematic characters and distribution. *Zoe* **4**: 381–385.

Dudley, M. G. 1937. Morphological and cytological studies of *Calla palustris. Bot. Gaz.* **98**: 556–571.

Dudley, J. L. 1987. Turion formation in strains of *Lemna minor* (6591) and *Lemna turionifera* (6573, A). *Aquat. Bot.* **27**: 207–215.

Dueck, L. A. & K. Cameron. 2007. Sequencing re-defines *Spiranthes* relationships, with implications for rare and endangered taxa. *Lankesteriana* **7**: 190–195.

Dueck, L. A., D. Aygoren & K. M. Cameron. 2014. A molecular framework for understanding the phylogeny of *Spiranthes* (Orchidaceae), a cosmopolitan genus with a North American center of diversity. *Am. J. Bot.* **101**: 1551–1571.

Duffy, W. G. 1994. Demographics of *Lestes disjunctus disjunctus* (Odonata: Zygoptera) in a riverine wetland. *Can. J. Zool.* **72**: 910–917.

Dugdale, T. M., D. Clements & T. D. Hunt. 2012. Survival of a submerged aquatic weed (*Egeria densa*) during lake drawdown within mounds of stranded vegetation. *Lake Reserv. Manag.* **28**: 153–157.

Duggal, S. C. 1971. Orchids in human affairs (a review). *Quart. J. Crude Drug Res.* **11**: 1727–1734.

Duke, J. A. 2000. *Handbook of Edible Weeds*. CRC Press, Boca Raton, FL. 256 pp.

Duncan, B. L. & D. L. DeGiusti. 1976. Three new lissorchiid Cercariaea of the Mutabile group from *Laevapex fuscus* (Adams, 1841) and *Ferrisia rivularis* (Say, 1917). *Proc. Helminthol. Soc. Washington* **43**: 1–9.

Dunford, J. C., D. K. Young & S. J. Krauth. 2006. *Stethobaris ovata* (LeConte) (Curculionidae) on eastern prairie fringed orchid [*Platanthera leucophaea* (Nuttall) Lindley] in Wisconsin. *Coleopterists Bull.* **60**: 51–53.

Dunham, R. M., A. M. Ray & R. S. Inouye. 2003. Growth, physiology, and chemistry of mycorrhizal and nonmycorrhizal *Typha latifolia* seedlings. *Wetlands* **23**: 890–896.

Dunn, C. P. & R. R. Sharitz. 1990. The history of *Murdannia keisak* (Commelinaceae) in the southeastern United States. *Castanea* **55**: 122–129.

Dunn, C. P. & R. R. Sharitz. 1991. Population structure, biomass allocation, and phenotypic plasticity in *Murdannia keisak* (Commelinaceae). *Am. J. Bot.* **78**: 1712–1723.

Dunton, K. H. 1990. Production ecology of *Ruppia maritima* L. s.l. and *Halodule wrightii* Aschers, in two subtropical estuaries. *J. Exp. Mar. Biol. Ecol.* **143**: 147–164.

Dunton, K. H. 1994. Seasonal growth and biomass of the subtropical seagrass *Halodule wrightii* in relation to continuous measurements of underwater irradiance. *Mar. Biol.* **120**: 479–489.

Dunton, K. H. 1996. Photosynthetic production and biomass of the subtropical seagrass *Halodule wrightii* along an estuarine gradient. *Estuaries Coast.* **19**: 436–447.

Duong, T. P. & J. M. Tiedje. 1985. Nitrogen fixation by naturally occurring duckweed–cyanobacterial associations. *Can. J. Microbiol.* 31: 327–330.

Durako, M. J. & M. D. Moffler. 1985. Observations on the reproductive ecology of *Thalassia testudinum* (Hydrocharitaceae). III. Spatial and temporal variations in reproductive patterns within a seagrass bed. *Aquat. Bot.* **22**: 265–276.

Durako, M. J. & M. D. Moffler. 1987. Factors affecting the reproductive ecology of *Thalassia testudinum* (Hydrocharitaceae). *Aquat. Bot.* **27**: 79–95.

Durako, M. J. & J. F. Howarth. 2017. Leaf spectral reflectance shows *Thalassia testudinum* seedlings more sensitive to hypersalinity than hyposalinity. *Front. Plant Sci.* **28**. doi: 10.3389/fpls.2017.01127.

Durako, M. J., J. I. Kunzelman, W. J. Kenworthy & K. K. Hammerstrom. 2003. Depth-related variability in the photobiology of two populations of *Halophila johnsonii* and *Halophila decipiens*. *Mar. Biol.* **142**: 1219–1228.

Dutton, B. E. & R. D. Thomas. 1991. The vascular flora of Cameron Parish, Louisiana. *Castanea* **56**: 1–37.

Duvall, M. R., G. H. Learn, L. E. Eguiarte & M. T. Clegg. 1993a. Phylogenetic analysis of *rbcL* sequences identifies *Acorus calamus* as the primal extant monocotyledon. *Proc. Natl. Acad. Sci. U.S.A.* **90**: 4641–4644.

Duvall, M. R., M. T. Clegg, M. W. Chase, W. D. Clark, W. J. Kress, H. G. Hills, L. E. Eguiarte, J. F. Smith, B. S. Gaut, E. A. Zimmer & G. H. Learn, Jr. 1993b. Phylogenetic hypotheses for the monocotyledons constructed from *rbcL* sequence data. *Ann. Missouri Bot. Gard.* **80**: 607–619.

Dworaczek, E. & R. Claßen-Bockhoff. 2016. 'False resupination' in the flower-pairs of *Thalia* (Marantaceae). *Flora* **221**: 65–74.

Dyer, A. F., T. H. N. Ellis, E. Lithgow, S. Lowther, I. Mason & D. Williams. 1976. The karyotype of *Iris pseudacorus* L. *Trans. Bot. Soc. Edinburgh* **42**: 421–429.

Earnst, S. L & T. C. Rothe. 2004. Habitat selection by Tundra Swans on northern Alaska breeding grounds. *Waterbirds* **27**: 224–233.

Easley, J. F. & R. L. Shirley. 1974. Nutrient elements for livestock in aquatic plants. *Hyacinth Control J.* **12**: 82–85.

Easley, M. C. & W. S. Judd. 1993. Vascular flora of Little Talbot Island, Duval County, Florida. *Castanea* **58**: 162–177.

East, E. M. 1940. The distribution of self-sterility in the flowering plants. *Proc. Am. Philos. Soc.* **82**: 449–518.

Ebinger, J. E. & J. A. Bacone. 1980. Vegetation survey of hillside seeps at Turkey Run State Park. *Proc. Indiana Acad. Sci.* **90**: 390–394.

Echternacht, L., P. T. Sano, C. Bonillo, C. Cruaud, A. Couloux & J.-Y. Dubuisson. 2014. Phylogeny and taxonomy of *Syngonanthus* and *Comanthera* (Eriocaulaceae): evidence from expanded sampling. *Taxon* **63**: 47–63.

Eckardt, N. A. & D. D. Biesboer. 1988. A survey of nitrogen fixation (acetylene reduction) associated with *Typha* in Minnesota. *Can. J. Bot.* **66**: 2419–2423.

Eckenwalder, J. E. & S. C. H. Barrett. 1986. Phylogenetic systematics of Pontederiaceae. *Syst. Bot.* **11**: 373–391.

Eckert, C. G., B. Massonnet & J. J. Thomas. 2000. Variation in sexual and clonal reproduction among introduced populations of flowering rush, *Butomus umbellatus* (Butomaceae). *Can. J. Bot.* **78**: 437–446.

Edelman, S. M. 2017. Morphology, architecture and growth of a clonal palm, *Acoelorrhaphe wrightii*. Ph.D. dissertation. Florida International University, Miama, FL. 187 pp.

Edelman, S. & J. H. Richards. 2018a. Morphology and architecture of the threatened Florida palm *Acoelorrhaphe wrightii* (Arecaceae: Coryphoideae). *Candollea* **73**: 49–59.

Edelman, S. & J. H. Richards. 2018b. Germination and juvenile growth of the clonal palm, *Acoelorrhaphe wrightii*, under different water and light treatments: a mesocosm study. *Feddes Repert.* **129**: 92–104.

Edgerton, E. A. 2014. Prevention and management of aquatic invasive plants in Texas. Ph.D. dissertation. Texas A&M University, College Station, TX. 254 pp.

Edwards, T. I. 1933. The germination and growth of *Peltandra virginica* in the absence of oxygen. *Bull. Torrey Bot. Club* **60**: 573–581.

Edwards, T. I. 1934. Seed frequencies in *Cytisus* and *Peltandra*. *Am. Nat.* **68**: 283–286.

Edwards, P. 1980. *Food Potential of Aquatic Macrophytes*. ICLARM Studies and Reviews No 5. International Center for Living Aquatic Resources Management, Manila, Philippines. 51 pp.

Edwards, A. L. & R. R. Sharitz. 2000. Population genetics of two rare perennials in isolated wetlands: *Sagittaria isoetiformis* and *S. teres* (Alismataceae). *Am. J. Bot.* **87**: 1147–1158.

Edwards, A. L. & R. R. Sharitz. 2003. Clonal diversity in two rare perennial plants: *Sagittaria isoetiformis* and *Sagittaria teres* (Alismataceae). *Int. J. Plant Sci.* **164**: 181–188.

Edwards, J. M. & U. Weiss. 1970. Perinaphthenone pigments from the fruit capsules of *Lachnanthes tinctoria*. *Phytochemistry* **9**: 1653–1657.

Edwards, J. M. & U. Weiss. 1974. Phenalenone pigments of the root system of *Lachnanthes tinctoria*. *Phytochemistry* **13**: 1597–1602.

Egertson, C. J., J. A. Kopaska & J. A. Downing. 2004. A century of change in macrophyte abundance and composition in response to agricultural eutrophication. *Hydrobiologia* **524**: 145–156.

Eggeling, O. & F. Ehrenberg. 1908. *The Freshwater Aquarium and Its Inhabitants: A Guide for the Amateur Aquarist*. Henry Holt & Company, New York, NY. 352 pp.

Eichler, L. 2009. *Aquatic Vegetation of Lake Luzerne, NY*. Lake Luzerne Association, Lake Luzerne, NY. 18 + 45 pp.

Eichler, L. 2011. *An Assessment of Aquatic Plant Growth in Galway Lake, Saratoga County, New York*. Darrin Fresh Water Institute, Bolton Landing, NY. 14 + 21 pp.

Eichler, L. & C. Boylen. 2013. Aquatic vegetation of Canada Lake, West Lake and Green Lake, Town of Caroga, New York. DFWI Technical Report 2013-6. Darrin Fresh Water Institute, Bolton Landing, NY. 27 pp.

Eichler, L. W., L. E. Ahrens-Franklin & C. W. Boylen. 2006. A survey of tributaries to Lake George, New York for the presence of Eurasian watermilfoil. DFWI Technical Report 2006-7. Darrin Fresh Water Institute, Bolton Landing, NY. 14 + 29 pp.

Eickwort, G. C., P. F. Kukuk & F. R. Wesley. 1986. The nesting biology of *Dufourea novaeangliae* (Hymenoptera: Halictidae) and the systematic position of the Dufoureinae based on behavior and development. *J. Kansas Entomol. Soc.* **59**: 103–120.

Eid, E. M., T. M. Galal, M. A. Dakhil & L. M. Hassan. 2016. Modeling the growth dynamics of *Pistia stratiotes* L. populations along the water courses of south Nile Delta, Egypt. *Rend. Fis. Accad. Lincei* **27**: 375–382.

Eighmy, T. T., L. S. Jahnke & W. R. Fagerberg. 1991. Studies of *Elodea nuttallii* grown under photorespiratory conditions. II. Evidence for bicarbonate active transport. *Plant Cell Environ.* **14**: 157–165.

Eilers, L. J. & D. M. Roosa. 1994. *The Vascular Plants of Iowa: An Annotated Checklist and Natural History.* University of Iowa Press, Iowa City, IA. 304 pp.

Eiseman, N. J. & C. McMillan. 1980. A new species of seagrass, *Halophila johnsonii*, from the Atlantic coast of Florida. *Aquat. Bot.* **9**: 15–19.

Ekstam, B. & Å. Forseby. 1999. Germination response of *Phragmites australis* and *Typha latifolia* to diurnal fluctuations in temperature. *Seed Sci. Res.* **9**: 157–163.

Elam, C. E., J. M. Stucky, T. R. Wentworth & J. D. Gregory. 2009. Vascular flora, plant communities, and soils of a significant natural area in the middle Atlantic coastal plain (Craven County, North Carolina). *Castanea* **74**: 53–77.

Elder, C. L. 1997. Reproductive biology of *Darlingtonia californica*. M.A. thesis. Humboldt State University, Arcata, CA. 67 pp.

Eleuterius, L. N. 1972. The marshes of Mississippi. *Castanea* **37**: 153–168.

Eleuterius, L. N. 1980. An illustrated guide to tidal marsh plants of Mississippi and adjacent states. Publication No. MASGP-77-039. Mississippi-Alabama Sea Grant Consortium, Ocean Springs, MS. 131 pp.

Eleuterius, L. N. 1990. *Tidal Marsh Plants.* Pelican Publishing Company, Inc., Gretna, LA. 168 pp.

Eleuterius, L. N. & S. McDaniel. 1978. The salt marsh flora of Mississippi. *Castanea* **43**: 86–95.

Eleuterius, L. N. & J. D. Caldwell. 1984. Flowering phenology of tidal marsh plants in Mississippi. *Castanea* **49**: 172–179.

Elger, A. & N. J. Willby. 2003. Leaf dry matter content as an integrative expression of plant palatability: the case of freshwater macrophytes. *Funct. Ecol.* **17**: 58–65.

El-Hady, H. H., S. M. Daboor & A. E. Ghoniemy. 2007. Nutritive and antimicrobial profiles of some seagrasses from Bardawil Lake, Egypt. *Egypt. J. Aquat. Res.* **33**: 103–110.

Ellis, E. A. 1965. *The Broads.* Collins, London. 401 pp.

Ellis, J. B. & B. M. Everhart. 1885. New fungi. *J. Mycol.* **1**: 148–154.

Ellis, J. B. & A. B. Langlois. 1890. New species of Louisiana fungi. *J. Mycol.* **6**: 35–37.

Elmore, A. J., K. A. Engelhardt, D. Cadol & C. M. Palinkas. 2016. Spatial patterns of plant litter in a tidal freshwater marsh and implications for marsh persistence. *Ecol. Appl.* **26**: 846–860.

Elser, J. J., C. Junge & C. R. Goldman. 1994. Population structure and ecological effects of the crayfish *Pacifastacus leniusculus* in Castle Lake, California. *Great Basin Nat.* **54**: 162–169.

El-Shafai, S. A., F. A. El-Gohary, J. A. J. Verreth, J. W. Schrama & H. J. Gijzen. 2004. Apparent digestibility coefficient of duckweed (*Lemna minor*), fresh and dry for Nile tilapia (*Oreochromis niloticus* L.). *Aquac. Res.* **35**: 574–586.

El-Shamy, M. A., A. H. M. El-Feky & N. Y. L. Eliwa. 2009. Propagation of calla lily (*Zantedeschia aethiopica* Spreng) plants by tissue culture technique. *Bull. Fac. Agric. Cairo Univ.* **60**: 99–105.

Elzenga, J. T. M., L. De Lange & A. H. Pieterse. 1980. Further indications that ethylene is the gibbosity regulator of the *Lemna gibba*/*Lemna minor* complex in natural waters. *Acta Bot. Neerl.* **29**: 225–229.

Emerson, F. W. 1921. Subterranean organs of bog plants. *Bot. Gaz.* **72**: 359–374.

Emms, S. K. 1993. Andromonoecy in *Zigadenus paniculatus* (Liliaceae): spatial and temporal patterns of sex allocation. *Am. J. Bot.* **80**: 914–923.

Enders, R. K. 1932. Food of the muskrat in summer. *Ohio J. Sci.* **32**: 21–30.

Enerstvedt, K. H., A. Lundberg & M. Jordheim. 2017. Characterization of polyphenolic content in the aquatic plants *Ruppia cirrhosa* and *Ruppia maritima*—a source of nutritional natural products. *Molecules* **23**(1): 16. doi: 10.3390/molecules23010016.

Engel, S. & S. A. Nichols. 1994. Aquatic macrophyte growth in a turbid windswept lake. *J. Freshw. Ecol.* **9**: 97–109.

Engelhardt, K. A. M. & M. E. Ritchie. 2001. Effects of macrophyte species richness on wetland ecosystem functioning and services. *Nature* **411**: 687–689.

Engelhardt, K. A. M. & M. E. Ritchie. 2002. The effect of aquatic plant species richness on wetland ecosystem processes. *Ecology* **83**: 2911–2924.

Engeman, R. M., T. Guerrant, G. Dunn, S. F. Beckerman & C. Anchor. 2014. Benefits to rare plants and highway safety from annual population reductions of a "native invader," white-tailed deer, in a Chicago-area woodland. *Environ. Sci. Pollut. Res.* **21**: 1592–1597.

Enser, R. W. & C. A. Caljouw. 1989. Plant conservation concerns in Rhode Island—a reappraisal. *Rhodora* **91**: 121–130.

Ensign, S. H., C. R. Hupp, G. B. Noe, K. W. Krauss & C. L. Stagg. 2014. Sediment accretion in tidal freshwater forests and oligohaline marshes of the Waccamaw and Savannah Rivers, USA. *Estuaries Coast.* **37**: 1107–1119.

Epler, J. H. 2016. A new species of *Dicrotendipes* (Diptera: Chironomidae) from Florida. *Zootaxa* **4208**: 77–83.

Erfanzadeh, R., A. Garbutt, J. Pétillon, J.-P. Maelfait & M. Hoffmann. 2010. Factors affecting the success of early salt-marsh colonizers: seed availability rather than site suitability and dispersal traits. *Plant Ecol.* **206**: 335–347.

Eriksson, O. & K. Kainulainen. 2011. The evolutionary ecology of dust seeds. *Perspect. Plant Ecol. Evol. Syst.* **13**: 73–87.

Ernst-Schwarzenbach, M. 1945. Kreuzungsversuche an Hydrocharitaceen. *Arch. Julius Klaus Stiftung (Zürich)* **20**: 22–41.

Ernst-Schwarzenbach, M. 1953. Zur Kompatibilitiit von Art- und Gattungs-Bastardierungen bei Hydrocharitaceae. *Oester. Bot. Z.* **100**: 403–423.

Esau, K. 1977. *Anatomy of Seed Plants*, 2nd ed. John Wiley & Sons, New York, NY. 550 pp.

Escher, R. L. & L. P. Lounibos. 1993. Insect associates of *Pistia stratiotes* (Arales: Araceae) in southeastern Florida. *Florida Entomol.* **76**: 473–500.

Eshel, A. & S. Beer. 1986. Inorganic carbon assimilation by *Spirodela polyrrhiza*. *Hydrobiologia* **131**: 149–153.

Esler, D. 1990. Avian community responses to hydrilla invasion. *Wilson Bull.* **102**: 427–440.

Espinosa-García, F. J., J. L. Villaseñor & H. Vibrans. 2006. Mexico: biodiversity, distribution, and possible economic impact of exotic weeds. Pp. 43–52 *In*: T. R. Van Devender, F. J. Espinosa-García, B. L. Harper-Lore & T. Hubbard (eds.), *Invasive Plants on the Move: Controlling Them in North America.* Arizona-Sonora Desert Museum Press, Tucson, AZ.

Essomo, S. E., B. A. Fonge, E. E. Bechem, P. T. Tabot & B. D. Arrey. 2016. Flowering phenology and reproductive success of the orchids of Mt. Cameroon in relation to a changing environment. *Int. J. Curr. Res. Biosci. Plant Biol.* **3**: 21–35.

Estes, D. & R. L. Small. 2007. Two new species of *Gratiola* (Plantaginaceae) from eastern North America and an updated circumscription for *Gratiola neglecta*. *J. Bot. Res. Inst. Texas* **1**: 149–170.

Eugelink, A. H. 1998. Phosphorus uptake and active growth of *Elodea canadensis* Michx. and *Elodea nuttallii* (Planch.) St. John. *Water Sci. Technol.* **37**: 59–65.

Evans, G. A. 2008. *The Whiteflies (Hemiptera: Aleyrodidae) of the World and Their Host Plants and Natural Enemies.* Version 2008-09–23. USDA Animal Plant Health Inspection Service (APHIS), Riverdale, MD. 703 pp.

Evans, J. M. 2013. *Pistia stratiotes* L. in the Florida Peninsula: biogeographic evidence and conservation implications of native tenure for an 'invasive' aquatic plant. *Conserv. Soc.* **11**: 233–246.

Evans, J. M. & A. C. Wilkie. 2010. Life cycle assessment of nutrient remediation and bioenergy production potential from the harvest of hydrilla (*Hydrilla verticillata*). *J. Environ. Manag.* **91**: 2626–2631.

Evans, A. S., K. L. Webb & P. A. Penhale. 1986. Photosynthetic temperature acclimation in two coexisting seagrasses, *Zostera marina* L. and *Ruppia maritima* L. *Aquat. Bot.* **24**: 185–197.

Evans, T. M., K. J. Sytsma, R. B. Faden & T. J. Givnish. 2003. Phylogenetic relationships in the Commelinaceae: II. A cladistic analysis of *rbcL* sequences and morphology. *Syst. Bot.* **28**: 270–292.

Everett, P. C. 2012. *A Second Summary of the Horticulture and Propagation of California Native Plants at the Rancho Santa Ana Botanic Garden, 1950–1970.* Version 1.1. Rancho Santa Ana Botanic Garden, Claremont, CA. 514 pp.

Everitt, J. H., D. L. Drawe & R. I. Lonard. 1999. *Field Guide to the Broad-Leaved Herbaceous Plants of South Texas: Used by Livestock and Wildlife.* Texas Tech University Press, Lubbock, TX. 277 pp.

Ewanchuk, P. J. & S. L. Williams. 1996. Survival and re-establishment of vegetative fragments of eelgrass (*Zostera marina*). *Can. J. Bot.* **74**: 1584–1590.

Ewel, K. C. & S. Atmosoedirdjo. 1987. Flower and fruit production in three north Florida ecosystems. *Florida Sci.* **50**: 216–222.

Ewing, J. 1924. Plant successions of the brush-prairie in north-western Minnesota. *J. Ecol.* **12**: 238–266.

Expósito, A. B., A. Siverio, L. A. Bermejo & E. Sobrino-Vesperinas. 2018. Checklist of alien plant species in a natural protected area: Anaga Rural Park (Tenerife, Canary Islands); effect of human infrastructures on their abundance. *Plant Ecol. Evol.* **151**: 142–152.

Eyjólfsson, R. 1970. Isolation and structure determination of triglochinin, a new cyanogenic glucoside from *Triglochin maritimum*. *Phytochemistry* **9**: 845–851.

Faber-Langendoen, D. (ed.). 2001. *Plant Communities of the Midwest: Classification in an Ecological Context.* Association for Biodiversity Information, Arlington, VA. 61 pp + appendix (705 pp).

Fachinetto, J., E. Kaltchuk-Santos, C. D. Inácio, L. Eggers & T. T. de Souza-Chies. 2018. Multidisciplinary approaches for species delimitation in *Sisyrinchium* (Iridaceae). *Plant Species Biol.* **33**: 3–15.

Faden, R. B. 1998. Commelinaceae. Pp. 109–128 *In*: K. Kubitzki, H. Huber, P. J. Rudall, P. S. Stevens & T. Stützel (eds.), *The Families and Genera of Vascular Plants, Vol. IV. Flowering Plants, Monocotyledons: Alismatanae and Commelinanae (Except Gramineae).* Springer-Verlag, Berlin, Germany.

Faden, R. B. 2000a. Commelinaceae R. Brown. Spiderwort family. Pp. 170–197 *In*: Flora North America Editorial Committee (eds.), *Flora of North America North of Mexico, Vol. 22: Magnoliophyta: Alismatidae, Arecidae, Commelinidae (in Part), and Zingiberidae.* Oxford University Press, New York, NY.

Faden, R. B. 2000b. Mayacaceae Kunth. Mayaca or bog-moss family. Pp. 168–169 *In*: Flora North America Editorial Committee (eds.), *Flora of North America North of Mexico, Vol. 22: Magnoliophyta: Alismatidae, Arecidae, Commelinidae (in Part), and Zingiberidae.* Oxford University Press, New York, NY.

Fagúndez, G. A. & M. A. Caccavari. 2006. Pollen analysis of honeys from the central zone of the Argentine province of Entre Ríos. *Grana* **45**: 305–320.

Fahn, A. 1982. *Plant Anatomy*, 3rd ed. Pergamon Press, New York, NY. 544 pp.

Fairman, C. E. 1918. New or noteworthy ascomycetes and lower fungi from New Mexico. *Mycologia* **10**: 239–264.

Fang, Y. Y., O. Babourina, Z. Rengel, X. E. Yang & P. M. Pu. 2007. Ammonium and nitrate uptake by the floating plant *Landoltia punctata*. *Ann. Bot.* **99**: 365–370.

Farmer, J. A., E. B. Webb, R. A. Pierce & K. W. Bradley. 2017. Evaluating the potential for weed seed dispersal based on waterfowl consumption and seed viability. *Pest Manag. Sci.* **73**: 2592–2603.

Farney, R. A. & T. A. Bookhout. 1982. Vegetation changes in a Lake Erie marsh (Winous Point, Ottawa County, Ohio) during high water years. *Ohio J. Sci.* **82**: 103–107.

Farrell, J. M., B. A. Murry, D. J. Leopold, A. Halpern, M. B. Rippke, K. S. Godwin & S. D. Hafner. 2010. Water-level regulation and coastal wetland vegetation in the upper St. Lawrence River: inferences from historical aerial imagery, seed banks, and *Typha* dynamics. *Hydrobiologia* **647**: 127–144.

Fasakin, E. A., A. M. Balogun & B. E. Fasuru. 1999. Use of duckweed, *Spirodela polyrrhiza* L. Schleiden, as a protein feedstuff in practical diets for tilapia, *Oreochromis niloticus* L. *Aquacult. Res.* **30**: 313–318.

Fassett, N. C. 1955. *Echinodorus* in the American tropics. *Rhodora* **57**: 133–156.

Fassett, N. C. 1957. *A Manual of Aquatic Plants. With Revision Appendix by E. C. Ogden.* The University of Wisconsin Press, Madison, WI. 405 pp.

Fassett, N. C. & B. Calhoun. 1952. Introgression between *Typha latifolia* and *T. angustifolia*. *Evolution* **6**: 367–379.

Faust, M. & J. A. Harrington. 2016. Pilot study: limitations to pollination and ovary development in the small white lady's-slipper (*Cypripedium candidum*). *N. Am. Prairie Conf. Proc.* **11**: 96–103.

Favre-Bac, L., B. Lamberti-Raverot, S. Puijalon, A. Ernoult, F. Burel, L. Guillard & C. Mony. 2017. Plant dispersal traits determine hydrochorous species tolerance to connectivity loss at the landscape scale. *J. Veg. Sci.* **28**: 605–615.

Fehr, A. W. 1989. Environmental factors limiting success of revegetation on man-made waterfowl nesting islands in east-central Alberta. Ph.D. dissertation. University of British Columbia, Vancouver, BC. 192 pp.

Feijoó, C. S. & R. J. Lombardo. 2007. Baseline water quality and macrophyte assemblages in Pampean streams: a regional approach. *Water Res.* **41**: 1399–1410.

Feinstein, L. 1952. Insecticides from plants. Pp. 222–229 *In*: F. C. Bishopp, G. J. Haeussler, H. L. Haller, W. L. Popham, B. A. Porter, E. R. Sasscer and J. S. Wade (eds.), *Insects. The Yearbook of Agriculture 1952.* Superintendent of Documents, Washington, DC.

Feitoza, L. L., L. P. Felix, A. A. J. F. Castro & R. Carvalho. 2009. Cytogenetics of Alismatales ss: chromosomal evolution and C-banding. *Plant Syst. Evol.* **280**: 119–131.

Felger, R. & M. B. Moser. 1973. Eelgrass (*Zostera marina* L.) in the Gulf of California. *Science* **181**: 355–356.

Felix, L. P. & M. Guerra. 1998. Cytogenetic studies on species of *Habenaria* (Orchidoideae: Orchidaceae) occurring in the northeast of Brazil. *Lindleyana* **13**: 224–230.

Felix, L. P. & M. Guerra. 2005. Basic chromosome numbers of terrestrial orchids. *Plant Syst. Evol.* **254**: 131–148.

Felix, L. P. & M. Guerra. 2010. Variation in chromosome number and the basic number of subfamily Epidendroideae (Orchidaceae). *Bot. J. Linn. Soc.* **163**: 234–278.

Fell, E. W. 1957. Plants of a northern Illinois sand deposit. *Am. Midl. Nat.* **58**: 441–451.

Fell, P. E., R. S. Warren, J. K. Light, R. L. Rawson & S. M. Fairley. 2003. Comparison of fish and macroinvertebrate use of *Typha angustifolia*, *Phragmites australis*, and treated *Phragmites* marshes along the lower Connecticut River. *Estuaries* **26**: 534–551.

Fennell, C. W. & J. Van Staden. 2001. *Crinum* species in traditional and modern medicine. *J. Ethnopharmacol.* **78**: 15–26.

Ferasol, J., L. Lovett Doust, J. Lovett Doust & M. Biernacki. 1995. Seed germination in *Vallisneria americana*: effects of cold stratification, scarification, seed coat morphology and PCB concentration. *Ecoscience* **2**: 368–376.

Ferguson, E. & R. P. Wunderlin. 2006. A vascular plant inventory of Starkey Wilderness Preserve, Pasco County, Florida. *Sida* **22**: 635–659.

Ferguson, R. L., B. T. Pawlak & L. L. Wood. 1993. Flowering of the seagrass *Halodule wrightii* in North Carolina, USA. *Aquat. Bot.* **46**: 91–98.

Fernald, M. L. 1907. The soil preferences of certain alpine and subalpine plants. *Rhodora* **9**: 149–193.

Fernald, M. L. 1911. A botanical expedition to Newfoundland and southern Labrador. *Contr. Gray Herb. Harvard Univ.* **40**: 109–162.

Fernald, M. L. 1923a. Notes on the distribution of *Najas* in northeastern America. *Rhodora* **25**: 105–109.

Fernald, M. L. 1923b. The American variety of *Scheuchzeria palustris*. *Rhodora* **25**: 177–179.

Fernald, M. L. 1932. The linear-leaved North American species of *Potamogeton* section *Axillares*. Memoirs of the Gray Herbarium of Harvard University III. *Mem. Am. Acad. Arts Sci.* **17**: 1–183.

Fernald, M. L. 1939. Last survivors in the flora of tidewater Virginia (continued). *Rhodora* **41**: 529–559.

Fernald, M. L. 1941. *Elatine americana* and *E. triandra*. *Rhodora* **43**: 208–211.

Fernald, M. L. 1946. The North American representatives of *Alisma plantago-aquatica*. *Rhodora* **48**: 86–88.

Fernald, M. L. & K. M. Wiegand. 1914. The genus *Ruppia* in eastern North America. *Rhodora* **16**: 119–127.

Fernald, M. L. & A. C. Kinsey. 1943. *Edible Wild Plants of Eastern North America*. Idlewild Press, Cornwall-On-Hudson, New York, NY. 452 pp.

Fernández-Pascual, E., B. Jiménez-Alfaro & T. E. Díaz. 2013. The temperature dimension of the seed germination niche in fen wetlands. *Plant Ecol.* **214**: 489–499.

Fernando, D. D. & D. D. Cass. 1996. Genotypic differentiation in *Butomus umbellatus* (Butomaceae) using isozymes and random amplified polymorphic DNAs. *Can. J. Bot.* **74**: 647–652.

Ferren, Jr., W. R. 1973. Range extensions of *Sagittaria montevidensis* in the Delaware River system. *Bartonia* **42**: 1–4.

Ferren, Jr., W. R. & A. E. Schuyler. 1980. Intertidal vascular plants of river systems near Philadelphia. *Proc. Acad. Nat. Sci. Philadelphia* **132**: 86–120.

Ferren, Jr., W. R., R. E. Good, R. Walker & J. Arsenault. 1981. Vegetation and flora of Hog Island, a brackish wetland in the Mullica River, New Jersey. *Bartonia* **48**: 1–10.

Feuchtmayr, H., R. Moran, K. Hatton, L. Connor, T. Heyes, B. Moss, I. Harvey & D. Atkinson. 2009. Global warming and eutrophication: effects on water chemistry and autotrophic communities in experimental hypertrophic shallow lake mesocosms. *J. Appl. Ecol.* **46**: 713–723.

Fiala, K. 1978. Underground organs of *Typha angustifolia* and *Typha latifolia*, their growth, propagation and production. *Acta Sci. Nat. Acad. Sci. Bohem. Brno* **12**: 1–43.

Fields, J. R., T. R. Simpson, R. W. Manning & F. L. Rose. 2003. Food habits and selective foraging by the Texas river cooter (*Pseudemys texana*) in Spring Lake, Hays County, Texas. *J. Herpetol.* **37**: 726–729.

Figuerola, J. & A. J. Green. 2002. Dispersal of aquatic organisms by waterbirds: a review of past research and priorities for future studies. *Freshw. Biol.* **47**: 483–494.

Figuerola, J. & A. J. Green. 2004. Effects of seed ingestion and herbivory by waterfowl on seedling establishment: a field experiment with wigeongrass *Ruppia maritima* in Doñana, south-west Spain. *Plant Ecol.* **173**: 33–38.

Firmage, D. H. & F. R. Cole. 1988. Reproductive success and inflorescence size of *Calopogon tuberosus* (Orchidaceae). *Am. J. Bot.* **75**: 1371–1377.

Fishbein, M., D. Gori & D. Meggs. 1999. Prescribed burning as a management tool for Sky Island bioregion wetlands with reference to the management of the endangered orchid *Spiranthes delitescens*. Pp. 468–477 *In*: L. F. DeBano (ed.), *Biodiversity and the Management of the Madrean Archipelago: The Sky Islands of Southwestern United States and Northwestern Mexico*. Gen. Tech. Rep. RM-GTR-264. U.S. Department of Agriculture, Forest Service, Rocky Mountain Forest and Range Experiment Station, Fort Collins, CO.

Fisher, R. A. 1936. The use of multiple measurements in taxonomic problems. *Ann. Eugen.* **7**: 179–188.

Fisher, J. B. & K. Jayachandran. 2005. Presence of arbuscular mycorrhizal fungi in South Florida native plants. *Mycorrhiza* **15**: 580–588.

Fishman, J. R. & R. J. Orth. 1996. Effects of predation on *Zostera marina* L. seed abundance. *J. Exp. Mar. Biol. Ecol.* **198**: 11–26.

Fleming, R. C. 1970. Food plants of some adult sphinx moths (Lepidoptera: Sphingidae). *Great Lakes Entomol.* **3**: 17–23.

Fleming, G. P. & N. E. Van Alstine. 1999. Plant communities and floristic features of sinkhole ponds and seepage wetlands in southeastern Augusta County, Virginia. *Banisteria* **13**: 67–94.

Fleming, J. P., J. D. Madsen & E. D. Dibble. 2011. Macrophyte re-establishment for fish habitat in Little Bear Creek Reservoir, Alabama, USA. *J. Freshw. Ecol.* **26**: 105–114.

Flensburg, T. & J. H. Sparling. 1973. The algal microflora of a string mire in relation to the chemical composition of water. *Can. J. Bot.* **51**: 743–749.

Flinn, K. M., M. J. Lechowicz & M. J. Waterway. 2008. Plant species diversity and composition of wetlands within an upland forest. *Am. J. Bot.* **95**: 1216–1224.

Flint, L. H. 1943. Note on the germination of seeds of the spider lily (*Hymenocallis occidentalis*). *Louisiana Acad. Sci.* **7**: 20–23.

Flint, L. H. & C. F. Moreland. 1943. Note on photosynthetic activity in seeds of the spider lily. *Am. J. Bot.* **30**: 315–317.

Flint, N. A. & J. D. Madsen. 1995. The effect of temperature and daylength on the germination of *Potamogeton nodosus* tubers. *J. Freshw. Ecol.* **10**: 125–128.

Floden, A. J. & E. E. Schilling. 2018. *Trautvetteria fonticalcarea* (Ranunculaceae: Ranunculeae), a new tassel rue species endemic to calcareous seepage habitats in Tennessee, USA. *Nordic J. Bot.* **36**: e01738.

Flory, W. S. 1976. Distribution, chromosome numbers and types of various species and taxa of *Hymenocallis*. *Nucleus* **19**: 204–227.

Floyd, R. H., S. Ferrazzano, B. W. Josey, A. L. Garey & J. R. Applegate. 2018. *Helonias bullata* (swamp pink) habitat characteristics under different landscape settings at Fort AP Hill, Virginia. *Southeast. Nat.* **17**: 484–512.

FNA. 2000. Flora of North America Editorial Committee (eds.), *Flora of North America North of Mexico, Vol. 22: Magnoliophyta: Alismatidae Arecidae, Commelinidae (in Part), and Zingiberidae*. Oxford University Press, New York, NY. 352 pp.

FNA. 2003. Flora of North America Editorial Committee (eds.), *Flora of North America North of Mexico, Vol. 26: Magnoliophyta: Liliidae: Liliales and Orchidales*. Oxford University Press, New York, NY. 723 pp.

Foerste, A. F. 1883. Plants of Belle Isle, Michigan. *Bot. Gaz.* **8**: 202–203.

Folsom, J. P. 1984. A reinterpretation of the status of relationships of taxa of the yellow-fringed orchid complex. *Orquidea* **9**: 320–345.

Fonseca, M. S. & J. A. Cahalan. 1992. A preliminary evaluation of wave attenuation by four species of seagrass. *Estuar. Coast. Shelf Sci.* **35**: 565–576.

Fonseca, M. S., J. C. Zieman, G. W. Thayer & J. S. Fisher. 1983. The role of current velocity in structuring eelgrass (*Zostera marina* L.) meadows. *Estuar. Coast. Shelf Sci.* **17**: 367–380.

Fonseca, M. S., W. J. Kenworthy, K. M. Cheap, C. A. Currin & G. W. Thayer. 1984. A low-cost transplanting technique for shoalgrass (*Halodule wrightii*) and manatee grass (*Syringodium filiforme*). Final report. Instruction Report EL-84-1. U.S. Army Engineer Waterways Experiment Station, Vicksburg, MS. 16 pp.

Forbes, H. C., W. R. Ferren, Jr. & J. R. Haller. 1988. The vegetation and flora of Fish Slough and vicinity, Inyo and Mono Counties, California. Pp. 99–138 *In*: C. Hall and V. Doyle-Jones (eds.), *Plant Biology of Eastern California*. University of California, Los Angeles, CA.

Forbes, M. R. L., H. P. Barkhouse & P. C. Smith. 1989. Nest-site selection by Pied-billed Grebes *Podilymbus podiceps*. *Ornis Scand.* **20**: 211–218.

Ford, B. A. & P. W. Ball. 1988. A reevaluation of the *Triglochin maritimum* complex (Juncaginaceae) in eastern and central North America and Europe. *Rhodora* **90**: 313–337.

Forest, H. S. 1977. Study of submerged aquatic vascular plants in northern glacial lakes (New York State, USA). *Folia Geobot. Phytotax.* **12**: 329–341.

Forrest, A. D., M. L. Hollingsworth, P. M. Hollingsworth, C. Sydes & R. M. Bateman. 2004. Population genetic structure in European populations of *Spiranthes romanzoffiana* set in the context of other genetic studies on orchids. *Heredity* **92**: 218–227.

Fournier, M. A. & J. E. Hines. 2001. Breeding ecology of sympatric greater and lesser scaup (*Aythya marila* and *Aythya affinis*) in the subarctic Northwest Territories. *Arctic* **54**: 444–456.

Forni-Martins, E. R. & K. P. Calligaris. 2002. Chromosomal studies on Neotropical Limnocharitaceae (Alismatales). *Aquat. Bot.* **74**: 33–41.

Fortner, S. L. & D. S. White. 1988. Interstitial water patterns: a factor influencing the distributions of some lotic aquatic vascular macrophytes. *Aquat. Bot.* **31**: 1–12.

Foster, R. C. 1945. Studies in the Iridaceae, III. *Contr. Gray Herb.* **155**: 3–54.

Foster, C. H., G. D. Renaud & K. L. Hays. 1973. Some effects of the environment on oviposition by *Chrysops* (Diptera: Tabanidae). *Environ. Entomol.* **2**: 1048–1050.

Fourqurean, J. W., G. V. N. Powell, W. J. Kenworthy & J. C. Zieman. 1995. The effects of long-term manipulation of nutrient supply on competition between the seagrasses *Thalassia testudinum* and *Halodule wrightii* in Florida Bay. *Oikos* **72**: 349–358.

Fourqurean, J. W., A. Willsie, C. D. Rose & L. M. Rutten. 2001. Spatial and temporal pattern in seagrass community composition and productivity in south Florida. *Mar. Biol.* **138**: 341–354.

Fowler, J. A. 2005. *Wild Orchids of South Carolina: A Popular Natural History*. University of South Carolina Press, Columbia, SC. 242 pp.

Fox, C. H., R. El-Sabaawi, P. C. Paquet & T. E. Reimchen. 2014. Pacific herring *Clupea pallasii* and wrack macrophytes subsidize semi-terrestrial detritivores. *Mar. Ecol. Prog. Ser.* **495**: 49–64.

Fox, C. H., P. C. Paquet & T. E. Reimchen. 2018. Pacific herring spawn events influence nearshore subtidal and intertidal species. *Mar. Ecol. Prog. Ser.* **595**: 157–169.

França, G. M. de O., F. Melo, C. M. Pereira, G. A. Faria, F. V. S. T. de Melo & J. G. dos Santos. 2009. Nutritional value of *Lemna valdiviana* Phil (Araceae) submitted to different concentrations of fertilization with excrement of birds. *Biotemas* **22**: 19–26.

France, R. L. & P. M. Stokes. 1988. Isoetid-zoobenthos associations in acid-sensitive lakes in Ontario, Canada. *Aquat. Bot.* **32**: 99–114.

Francis, A. W., I. C. Stocks, T. R. Smith, A. J. Boughton, C. M. Mannion & L. S. Osborne. 2016. Host plants and natural enemies of rugose spiraling whitefly (Hemiptera: Aleyrodidae) in Florida. *Florida Entomol.* **99**: 150–153.

Francois, R., F. T. Short & J. H. Weber. 1989. Accumulation and persistence of tributyltin in eelgrass (*Zostera marina* L.) tissue. *Environ. Sci. Technol.* **23**: 191–196.

Frank, P. A. 1966. Dormancy in winter buds of American pondweed, *Potamogeton nodosus* Poir. *J. Exp. Bot.* **17**: 546–555.

Frankovich, T. A., E. E. Gaiser, J. C. Zieman & A. H. Wachnicka. 2006. Spatial and temporal distributions of epiphytic diatoms growing on *Thalassia testudinum* Banks ex König: relationships to water quality. *Hydrobiologia* **569**: 259–271.

Fraser, D. & H. Hristienko. 1983. Effects of moose, *Alces alces*, on aquatic vegetation in Sibley Provincial Park, Ontario. *Can. Field-Nat.* **97**: 57–61.

Fraser, D., D. Arthur, J. K. Morton & B. K. Thompson. 1980. Aquatic feeding by moose *Alces alces* in a Canadian lake. *Ecography* **3**: 218–223.

Fraser, D., B. K. Thompson & D. Arthur. 1982. Aquatic feeding by moose: seasonal variation in relation to plant chemical composition and use of mineral licks. *Can. J. Zool.* **60**: 3121–3126.

Fraser, D., E. R. Chavez & J. E. Palohelmo. 1984. Aquatic feeding by moose: selection of plant species and feeding areas in relation to plant chemical composition and characteristics of lakes. *Can. J. Zool.* **62**: 80–87.

Fraser, L. H., K. Mulac & F. B.-G. Moore. 2014. Germination of 14 freshwater wetland plants as affected by oxygen and light. *Aquat. Bot.* **114**: 29–34.

Freedman, R. L. 1976. Native North American food preparation techniques. *Bol. Bibliogr. Antropol. Am.* **38**: 101–159.

Freeland, J., C. Ciotir & H. Kirk. 2013. Regional differences in the abundance of native, introduced, and hybrid *Typha* spp. in northeastern North America influence wetland invasions. *Biol. Invasions* **15**: 2651–2665.

Freeman, C. W. & R. A. Urban. 2012. Sediment oxidation capabilities of four submersed aquatic macrophytes. *J. Freshw. Ecol.* **27**: 259–271.

French, J. C., M. G. Chung & Y. K. Hur. 1995. Chloroplast DNA phylogeny of the Ariflorae. Pp. 255–275 *In*: P. J. Rudall, P. J. Cribb, D. F. Cutler & C. J. Humphries (eds.), *Monocotyledons: Systematics and Evolution*. Royal Botanic Gardens, Kew, United Kingdom.

Freudenstein, J. V. & F. N. Rasmussen. 1997. Sectile pollinia and relationships in the Orchidaceae. *Plant Syst. Evol.* **205**: 125–146.

Friar, E., M. Falk & D. W. Mount. 1996. Use of random amplified polymorphic DNA (RAPD) markers for genetic analysis of *Lilium parryi*, a rare Arizona plant. Pp. 86–91 *In*: J. Maschinski,

J., H. D. Hammond & L. Holter (eds.), *Southwestern Rare and Endangered Plants: Proceedings of the Second Conference.* General Technical Report RM-GTR-283. USDA Forest Service, Rocky Mountain Forest and Range Experiment Station, Fort Collins, CO.

Frick, H. 1994. Heterotrophy in the Lemnaceae. *J. Plant Physiol.* **144**: 189–193.

Friedman, M. 2013. Thyroid autoimmune disease. *J. Restorative Med.* **2**: 70–81.

Friesen, N., R. M. Fritsch & F. R. Blattner. 2006. Phylogeny and new intrageneric classification of *Allium* (Alliaceae) based on nuclear ribosomal DNA ITS sequences. *Aliso* **22**: 372–395.

Frodge, J. D., G. L. Thomas & G. B. Pauley. 1990. Effects of canopy formation by floating and submergent aquatic macrophytes on the water quality of two shallow Pacific Northwest lakes. *Aquat. Bot.* **38**: 231–248.

Froehlich, S. 2019. Blue-eyed grass. *NatureNorth.com.* Available online: http://www.naturenorth.com/spring/flora/begrass/Blue-eyed_Grass.html [accessed 9 February, 2019].

From, M. M. 2007. Drought, peril, and survival in the Great Plains: *Cypripedium candidum. North Am. Native Orchid J.* **13**: 66–74.

Frost, C. C. 1980. State record for *Potamogeton confervoides* in South Carolina. *Castanea* **45**: 146–147.

Frost, C. C. 1995. Presettlement fire regimes in southeastern marshes, peatlands, and swamps. Pp. 39–60 *In*: S. I. Cerulean & R. T. Engstrom (eds.), *Fire in Wetlands: A Management Perspective. Proceedings of the Tall Timbers Fire Ecology Conference, No. 19.* Tall Timbers Research Station, Tallahassee, FL.

Frost, C. C. & L. J. Musselman. 1987. History and vegetation of the Blackwater Ecologic Preserve. *Castanea* **52**: 16–46.

Frost, P. C. & A. L. Hicks. 2012. Human shoreline development and the nutrient stoichiometry of aquatic plant communities in Canadian Shield lakes. *Can. J. Fish. Aquat. Sci.* **69**: 1642–1650.

Frost-Christensen, H. & K. Sand-Jensen. 1995. Comparative kinetics of photosynthesis in floating and submerged *Potamogeton* leaves. *Aquat. Bot.* **51**: 121–134.

Fruet, A. C., L. N. Seito, V. L. M. Rall & L. C. Di Stasi. 2012. Dietary intervention with narrow-leaved cattail rhizome flour (*Typha angustifolia* L.) prevents intestinal inflammation in the trinitrobenzenesulphonic acid model of rat colitis. *BMC Complement. Altern. Med.* **12**: 62.

Fry, B. 1983. Leaf growth in the seagrass *Syringodium filiforme* Kütz. *Aquat. Bot.* **16**: 361–368.

Fry, B. & R. W. Virnstein. 1988. Leaf production and export of the seagrass *Syringodium filiforme* Kütz. In Indian river Lagoon, Florida. *Aquat. Bot.* **30**: 261–266.

Fryer, A., A. Bennett & A. H. Evans. 1915. *The Potamogetons (Pond Weeds) of the British Isles, with Descriptions of all the Species, Varieties and Hybrids.* L. Reeve & Co., Ltd., Ashford, Kent. 94 pp [+ 60 plates].

Fryxell, P. A. 1957. Mode of reproduction of higher plants. *Bot. Rev.* **23**: 135–233.

Fu, L., M. Huang, B. Han, X. Sun, K. S. Sree, K.-J. Appenroth & J. Zhang. 2017. Flower induction, microscope-aided cross-pollination, and seed production in the duckweed *Lemna gibba* with discovery of a male-sterile clone. *Sci. Rep.* **7**: 3047.

Fuentes, T. L. 2007. Collection history of *Platanthera chorisiana* (Orchiddaceae) in Washington State. *Madroño* **54**: 164–167.

Fujita, K. & P. Hallock. 1999. A comparison of phytal substrate preferences of *Archaias angulatus* and *Sorites orbiculus* in mixed macroalgal-seagrass beds in Florida Bay. *J. Foraminiferal Res.* **29**: 143–151.

Fujita, M., K. Mori & T. Kodera. 1999. Nutrient removal and starch production through cultivation of *Wolffia arrhiza. J. Biosci. Bioeng.* **87**: 194–198.

Fujiwara, A., S. Matsuhashi, H. Doi, S. Yamamoto & T. Minamoto. 2016. Use of environmental DNA to survey the distribution of an invasive submerged plant in ponds. *Freshw. Sci.* **35**: 748–754.

Fulmer, J. E. & A. T. Robinson. 2008. Aquatic plant species distributions and associations in Arizona's reservoirs. *J. Aquat. Plant Manag.* **46**: 100–106.

Funderburk, J., L. Mound & J. Sharma. 2007. Thysanoptera inhabiting native terrestrial orchids in northern Florida and southern Georgia. *J. Entomol. Sci.* **42**: 573–581.

Funston, F. 1896. Botany of Yakutat Bay, Alaska. I. Field report. *Contr. U.S. Natl. Herb.* **3**: 325–333.

Furlow, B. M. & K. L. Hays. 1972. Some influences of aquatic vegetation on the species and number of Culicidae (Diptera) in small pools of water. *Mosquito News* **32**: 595–599.

Fuse, S. & M. N. Tamura. 2000. A phylogenetic analysis of the plastid *matK* gene with emphasis on Melanthiaceae sensu lato. *Plant Biol.* **2**: 415–427.

Fuse, S. & M. N. Tamura. 2016. Biosystematic studies on the genus *Heloniopsis* (Melanthiaceae) I. Phylogeny inferred from plastid DNA sequences and taxonomic implications. *Nordic J. Bot.* **34**: 584–595.

Fuse, S., N. S. Lee & M. N. Tamura. 2012. Biosystematic studies on the family Nartheciaceae (Dioscoreales) I. Phylogenetic relationships, character evolution and taxonomic re-examination. *Plant Syst. Evol.* **298**: 1575–1584.

Futyma, R. P. & N. G. Miller. 2001. Postglacial history of a marl fen: vegetational stability at Byron-Bergen Swamp, New York. *Can. J. Bot.* **79**: 1425–1438.

Fyson, A. 2000. Angiosperms in acidic waters at pH 3 and below. *Hydrobiologia* **433**: 129–135.

Gabela, F. J. A. 1974. Biology of mudplantain (*Heteranthera reniformis* Ruiz et Pavon) and its control in flooded rice. M.S. thesis. Oregon State University, Corvallis, OR. 212 pp.

Gabriel, B. C. & A. A. de la Cruz. 1974. Species composition, standing stock, and net primary production of a salt marsh community in Mississippi. *Chesapeake Sci.* **15**: 72–77.

Gabriel, C., D. W. Kerstetter & A. C. Hirons. 2015. Trophic linkages of Intracoastal Waterway seagrass beds in Broward County, Florida. *Florida Sci.* **78**: 156–166.

Gaddy, L. L. 1982. The floristics of three South Carolina pine savannahs. *Castanea* **47**: 393–402.

Galán Demera, A. & E. De Castro. 2003. *Heteranthera* Ruiz & Pav. (Pontederiaceae) en la Península Ibérica. *An. Jard. Bot. Madr.* **60**: 241–242.

Galatowitsch, S. M. & A. G. van der Valk. 1996. The vegetation of restored and natural prairie wetlands. *Ecol. Appl.* **6**: 102–112.

Galatowitsch, S. M. & L. A. Biederman. 1998. Vegetation and seedbank composition of temporarily flooded *Carex* meadows and implications for restoration. *Int. J. Ecol. Environ. Sci.* **24**: 253–270.

Gallardo-Williams, M. T., V. A. Whalen, R. F. Benson & D. F. Martin. 2002. Accumulation and retention of lead by cattail (*Typha domingensis*), hydrilla (*Hydrilla verticillata*), and duckweed (*Lemna obscura*). *J. Environ. Sci. Health* **37**: 1399–1408.

Gallegos, M., M. Menno, A. Rodriguez, N. Marba & C. M. Duarte. 1994. Growth patterns and demography of pioneer Caribbean seagrasses (*Halodule wrightii* and *Syringodium filiforme*). *Mar. Ecol. Prog. Ser.* **109**: 99–104.

Gallon, C., C. Munger, S. Prémont & P. G. C. Campbell. 2004. Hydroponic study of aluminum accumulation by aquatic plants: effects of fluoride and pH. *Water Air Soil Pollut.* **153**: 135–155.

Gama, L. R., C. Domit, M. K. Broadhurst, M. M. P. B. Fuentes & R. B. Millar. 2016. Green turtle *Chelonia mydas* foraging ecology at 25°S in the western Atlantic: evidence to support a feeding model driven by intrinsic and extrinsic variability. *Mar. Ecol. Prog. Ser.* **542**: 209–219.

Gandhi, K. N., J. H. Wiersema & L. Brouillet. 2012. Validation of the name *Wolffia borealis* (Lemnaceae). *Harvard Pap. Bot.* **17**: 47–50.

Ganie, A. H., Z. A. Reshi & B. A. Wafai. 2016. Reproductive ecology of *Potamogeton pectinatus* L. (= *Stuckenia pectinata* (L.) Börner) in relation to its spread and abundance in freshwater ecosystems of the Kashmir Valley, India. *Trop. Ecol.* **57**: 787–803.

Ganter, B. 2000. Seagrass (*Zostera* spp.) as food for brent geese (*Branta bernicla*): an overview. *Helgoland Mar. Res.* **54**: 63–70.

Gao, L., N. Liu, B. Huang & X. Hu. 2008. Phylogenetic analysis and genetic mapping of Chinese *Hedychium* using SRAP markers. *Sci. Hort.* **117**: 369–377.

García, K. L. G., O. V. Iglesias, A. Laguna, M. D. Martínez & J. A. G. Lavaut. 2011. Efecto antioxidante y contenido polifenólico de *Syringodium filiforme* (Cymodoceaceae). *Revista Biol. Trop.* **59**: 465–472.

García, K. L. G., M. Rodríguez, Á. Concepción, O. Valdés, J. G. Marrero, M. Macías-Alonso, O. Valdés-Iglesias, Y. H. Rivera, A. Fagundo, I. Rodeiro & R. G. Cuesta. 2017. Phytochemical profile and evaluation of photoprotective potential of *Syringodium filiforme* Kützing. *Biotecnia* **19**: 18–22.

García, T. E., R. Menéndez, F. Rivera, A. Garateix, R. A. Morales, E. Regalado, J. C. Rodríguez & F. Dajas. 2017. Neuroprotective effects of *Thalassia testudinum* leaf extract BM-21 on focal ischemia in rats. *J. Pharm. Pharmacogn. Res.* **5**: 174–186.

Garcias-Bonet, N., T. D. Sherman, C. M. Duarte & N. Marbà. 2011. Distribution and pathogenicity of the protist *Labyrinthula* sp. in western Mediterranean seagrass meadows. *Estuaries Coast.* **34**: 1161. doi: 10.1007/s12237-011-9416-4.

Garland, M. A., G. L. Smith & P. J. Anderson. 2013. The spider lilies (*Hymenocallis*) native to Florida. Botany Circular No. 39. FDACS-P-01843. Florida Department of Agriculture and Consumer Services, Division of Plant Industry, Gainesville, FL. 5 pp.

Garner, K. M. 1963. Some new nutria foods. *J. Mammal.* **44**: 261.

Garon-Labrecque, M.-È., É. Léveillé-Bourret, K. Higgins & O. Sonnentag. 2016. Additions to the boreal flora of the Northwest Territories with a preliminary vascular flora of Scotty Creek. *Can. Field-Nat.* **129**: 349–367.

Garrote-Moreno, A., A. McDonald, T. D. Sherman, J. L. Sánchez-Lizaso, K. L. Heck & J. Cebrian. 2015. Short-term impacts of salinity pulses on ionic ratios of the seagrasses *Thalassia testudinum* and *Halodule wrightii*. *Aquat. Bot.* **120**: 315–321.

Garvin, E. M., C. F. Bridge & M. S. Garvin. 2018. Edible wild plants growing in contaminated floodplains: implications for the issuance of tribal consumption advisories within the Grand Lake watershed of northeastern Oklahoma, USA. *Environ. Geochem. Health* **40**: 999–1025.

Gaskin, J. F., M. L. Pokorny & J. M. Mangold. 2016. An unusual case of seed dispersal in an invasive aquatic; yellow flag iris (*Iris pseudacorus*). *Biol. Invasions* **18**: 2067–2075.

Gates, F. C. 1910. The plant associations of the recent and fossil beaches of Lake Michigan, between Kenosha, Wisconsin and Waukegan, Illinois. B.A. thesis. University of Illinois, Urbana-Champaign, IL. 97 pp.

Gates, F. C. 1911. A bog in central Illinois. *Torreya* **11**: 205–211.

Gavin, N. M. & M. J. Durako. 2011. Localization and antioxidant capacity of flavonoids from intertidal and subtidal *Halophila johnsonii* and *Halophila decipiens*. *Aquat. Bot.* **95**: 242–247.

Gavin, N. M. & M. J. Durako. 2012. Localization and antioxidant capacity of flavonoids in *Halophila johnsonii* in response to experimental light and salinity variation. *J. Exp. Mar. Biol. Ecol.* **416**: 32–40.

Gavin, N. M. & M. J. Durako. 2014. Population-based variation in resilience to hyposalinity stress in *Halophila johnsonii*. *Bul. Mar. Sci.* **90**: 781–794.

Ge, X., J. Liu & R. Wang. 2013. Effects of flooding on the germination of seed banks in the Nansi Lake wetlands, China. *J. Freshw. Ecol.* **28**: 225–237.

Gehrs, C. W. 1974. Horizontal distribution and abundance of *Diaptomus clavipes* Schacht in relation to *Potamogeton foliosus* in a pond and under experimental conditions. *Limnol. Oceanogr.* **19**: 100–104.

Geiger, D. R. & M. G. Banker. 2012. Resolution of sweet flag species identity provides basis for managing contrasting impacts of native and introduced sweet flag in North American ecosystems. *Ecol. Restor.* **30**: 160–162.

Genaust, H. 1999. *Calla*--an enigmatic aroid taxon and its etymological solution. *Aroideana* **22**: 7–9.

George, C. J., C. W. Boylen & R. B. Sheldon. 1977. The presence of the red-spotted newt, *Notophthalmus viridescens* Rafinesque (Amphibia, Urodela, Salamandridae), in waters exceeding 12 meters in Lake George, New York. *J. Herpetol.* **11**: 87–90.

Gerloff, G. C. & P. H. Krombholz. 1966. Tissue analysis as a measure of nutrient availability for the growth of angiosperm aquatic plants. *Limnol. Oceanogr.* **11**: 529–537.

Gerritsen, J. & H. S. Greening. 1989. Marsh seed banks of the Okefenokee swamp: effects of hydrologic regime and nutrients. *Ecology* **70**: 750–763.

Gesti, J., A. Badosa & X. D. Quintana. 2005. Reproductive potential in *Ruppia cirrhosa* (Petagna) Grande in response to water permanence. *Aquat. Bot.* **81**: 191–198.

Gettys, L. A. & D. S. Wofford. 2007. Inheritance of flower color in pickerelweed (*Pontederia cordata* L.). *J. Hered.* **98**: 629–632.

Gettys, L. A. & R. K. Dumroese. 2009. Optimum storage and germination conditions for seeds of pickerelweed (*Pontederia cordata* L.) from Florida. *Native Plant J.* **10**: 4–12.

Gettys, L. A. & K. A. Moore. 2018. Greenhouse culture and production of four ornamental native wetland plants. *HortTechnology* **28**: 332–336.

Getz, L. L. 1959. Notes on the ecology of slugs: *Arion circumscriptus*, *Deroceras reticulatum*, and *D. laeve*. *Am. Midl. Nat.* **61**: 485–498.

Getz, L. L. 1961. Factors influencing the local distribution of shrews. *Am. Midl. Nat.* **65**: 67–88.

Ghimire, B. K., C. Y. Yu, H. J. Kim & I. M. Chung. 2012. Karyotype and nucleic acid content in *Zantedeschia aethiopica* Spr. and *Zantedeschia elliottiana* Engl. *Afr. J. Biotechnol.* **11**: 11604–11609.

Giardelli, M. L. 1935. Las flores de *Wolffiella oblonga*. *Revista Argent. Agron.* **2**: 17–20.

Gibemau, M. 2011. Pollinators and visitors of aroid inflorescences: an addendum. *Aroideana* **34**: 70–83.

Gibernau, M. 2003. Pollinators and visitors of aroid inflorescences. *Aroideana* **26**: 73–91.

Gibernau, M. 2016. Pollinators and visitors of Aroid inflorescences III–phylogenetic & chemical insights. *Aroideana* **39**: 4–22.

Gibson, G. D. & F.-S. Chia. 1989. Description of a new species of *Haminoea*, *Haminoea callidegenita* (Mollusca: Opisthobranchia), with a comparison with two other *Haminoea* species found in the northeast Pacific. *Can. J. Zool.* **67**: 914–922.

Gibson, C. M., L. E. Chasmer, D. K. Thompson, W. L. Quinton, M. D. Flannigan & D. Olefeldt. 2018. Wildfire as a major driver of recent permafrost thaw in boreal peatlands. *Nat. Commun.* **9**: 3041.

Gifford, J. 1893. Indian relics in south Jersey. *Science* **552**: 113–114.

Gil, H.-Y., Y.-H. Ha, K. S. Choi, J. S. Lee, K. S. Chang & K. Choi. 2019. The chloroplast genome sequence of an aquatic plant, *Sparganium eurycarpum* subsp. *coreanum* (Typhaceae). *Mitochondrial DNA Part B* **4**: 684–685.

Giles, Ba. E. 1977. A preliminary study of the MDH variability in *Lemna minor*/*Lemna turionifera*. M.S. thesis. Brock University, St. Catharines, ON. 186 pp.

Gill, D. A. 2000. Landscaping with native plants. D. A. Gill, Charlotte, NC. 40 pp. Unpublished compilation. Available online: http://citeseerx.ist.psu.edu/viewdoc/download?doi=10.1.1.387.9223&rep=rep1&type=pdf [accessed 8 April, 2019].

Gill, K. M. & K. M. Hoppa. 2016. Evidence for an Island Chumash geophyte-based subsistence economy on the Northern Channel Islands. *J. Calif. Gt. Basin Anthropol.* **36**: 51–71.

Gilman, B., J. Foust & B. Zhu. 2008. Composition, seasonal standing crop biomass and estimated annual productivity of macrophyte communities in Owasco Lake. Pp. 1–17 *In*: J. D. Halfman, M. E. Balyszak & S. A. Meyer (eds.), *A 2007 Water Quality Study of Owasco Lake, New York*. Finger Lakes Institute, Hobart and William Smith Colleges, Geneva, NY.

Gilmour, C. N., J. R. Starr & R. F. C. Naczi. 2013. *Calliscirpus*, a new genus for two narrow endemics of the California Floristic Province, *C. criniger* and *C. brachythrix* sp. nov. (Cyperaceae). *Kew Bull.* **68**: 85–105.

Giulietti, N., A. M. Giulietti, J. R. Pirani & N. L. Menezes. 1988. Estudos em sempre-vivas: importância econômica do extrativismo em Minas Gerais, Brasil. *Acta Bot. Brasilica* **1**: 179–193.

Giulietti, N., V. L. Scatena, P. T. Sano, L. Parra, L. P. Queiroz, R. M. Harley, N. L. Menezes, A. M. B. Ysepon, A. Salatino, M. L. Salatino, W. Vilegas, L. C. Santos, C. V. Ricci, M. C. P. Bonfim & E. B. Miranda. 2000. Multidisciplinary studies on neotropical Eriocaulaceae. Pp. 580–589 *In*: K. L. Wilson & D. A. Morrison (eds.), *Monocots: Systematics and Evolution*. CSIRO, Melbourne, VIC.

Givnish, T. J., A. Zuluaga, I. Marques, V. K. Y. Lam, M. S. Gomez, W. J. D. Iles, M. Ames, D. Spalink, J. R. Moeller, B. G. Briggs, S. P. Lyon, D. W. Stevenson, W. Zomlefer & S. W. Graham. 2016. Phylogenomics and historical biogeography of the monocot order Liliales: out of Australia and through Antarctica. *Cladistics* **32**: 581–605.

Givnish, T. J., A. Zuluaga, D. Spalink, M. S. Gomez, V. K. Y. Lam, J. M. Saarela, C. Sass, W. J. D. Iles, D. J. L. de Sousa, J. Leebens-Mack, J. C. Pires, W. B. Zomlefer, M. A. Gandolfo, J. I. Davis, D. W. Stevenson, C. dePamphilis, C. D. Specht, S. W. Graham, C. F. Barrett & C. Ané. 2018. Monocot plastid phylogenomics, timeline, net rates of species diversification, the power of multi-gene analyses, and a functional model for the origin of monocots. *Am. J. Bot.* **105**: 1–23.

Glaettli, M. & S. C. H. Barrett. 2008. Pollinator responses to variation in floral display and flower size in dioecious *Sagittaria latifolia* (Alismataceae). *New Phytol.* **179**: 1193–1201.

Glaser, P. H. 1992. Raised bogs in eastern North America–regional controls for species richness and floristic assemblages. *J. Ecol.* **80**: 535–554.

Glaser, P. H., G. A. Wheeler, E. Gorham & H. E. Wright, Jr. 1981. The patterned mires of the Red Lake peatland, northern Minnesota: vegetation, water chemistry and landforms. *J. Ecol.* **69**: 575–599.

Glaser, P. H., J. A. Janssens & D. I. Siegel. 1990. The response of vegetation to chemical and hydrological gradients in the Lost River peatland, northern Minnesota. *J. Ecol.* **78**: 1021–1048.

Glawe, D. A., F. M. Dugan, Y. Liu & J. D. Rogers. 2005. First record and characterization of a powdery mildew on a member of the Juncaginaceae: *Leveillula taurica* on *Triglochin maritima*. *Mycol. Progr.* **4**: 291–298.

Gleason, H. A. 1912. An isolated prairie grove and its phytogeographical significance. *Bot. Gaz.* **53**: 38–49.

Gleason, H. A. 1917. A prairie near Ann Arbor, Michigan. *Rhodora* **19**: 163–165.

Gleason, R. A., N. H. Euliss, Jr., D. E. Hubbard & W. G. Duffy. 2003. Effects of sediment load on emergence of aquatic invertebrates and plants from wetland soil egg and seed banks. *Wetlands* **23**: 26–34.

Gledhill, D. 2002. *The Names of Plants*, 3rd ed. Cambridge University Press, Cambridge, United Kingdom. 326 pp.

Glenn, S. 2008. Field trip reports. *J. Torrey Bot. Soc.* **135**: 149–153.

Glenn, E., T. L. Thompson, R. Frye, J. Riley & D. Baumgartner. 1995. Effects of salinity on growth and evapotranspiration of *Typha domingensis* Pers. *Aquat. Bot.* **52**: 75–91.

Glover, D. E. & S. C. H. Barrett. 1986. Variation in the mating system of *Eichhornia paniculata* (Spreng.) Solms (Pontederiaceae). *Evolution* **40**: 1122–1131.

Glover, D. E. & S. C. H. Barrett. 1987. Genetic variation in continental and island populations of *Eichhornia paniculata* (Pontederiaceae). *Heredity* **59**: 7–17.

Glück, H. 1924. *Biologische und morphologische Untersuchungen über Wasser- und Sumpfgewächse. Teil IV. Untergetauchte und Schwimmblattflora*. Verlag von Gustav Fischer, Jena. 746 pp.

Gluck, H. 1927. A new *Sagittaria* from Florida: *Sagittaria kurziana*. *Bull. Torrey Bot. Club* **54**: 257–261.

Godfrey, R. K. & P. Adams. 1964. The identity of *Sagittaria isoetiformis* (Alismataceae). *Sida* **1**: 269–273.

Godfrey, R. K. & J. W. Wooten. 1979. *Aquatic and Wetland Plants of Southeastern United States. Vol. I: Monocotyledons*. University of Georgia Press, Athens, GA. 712 pp.

Godt, M. J., J. L. Hamrick & S. Bratton. 1995. Genetic diversity in a threatened wetland species, *Helonias bullata* (Liliaceae). *Conserv. Biol.* **9**: 596–604.

Godt, M. J. W., J. Walker & J. L. Hamrick. 1997. Genetic diversity in the endangered lily *Harperocallis flava* and a close relative, *Tofieldia racemosa*. *Conserv. Biol.* **11**: 361–366.

Godziemba-Czyż, J. 1970. Characteristic of vegetative and resting forms in *Wolffia arrhiza* (L.) Wimm. II. Anatomy, physical and physiological properties. *Acta Soc. Bot. Polon.* **39**: 421–443.

Goecker, M. E., K. L. Heck, Jr. & J. F. Valentine. 2005. Effects of nitrogen concentrations in turtlegrass *Thalassia testudinum* on consumption by the bucktooth parrotfish *Sparisoma radians*. *Mar. Ecol. Prog. Ser.* **286**: 239–248.

Goldberg, B. 1941. Life history of *Peltandra virginica*. *Bot. Gaz.* **102**: 641–662.

Goldblatt, P. 1975. Revision of the bulbous Iridaceae of North America. *Brittonia* **27**: 373–385.

Goldblatt, P. 2002a. Iridaceae Jussieu. Pp. 348–349 *In*: Flora North America Editorial Committee (eds.), *Flora of North America North of Mexico, Vol. 26: Magnoliophyta: Liliidae: Liliales and Orchidales*. Oxford University Press, New York, NY.

Goldblatt, P. 2002b. *Nemastylis* Nuttall. Pp. 398–400. *In*: Flora North America Editorial Committee (eds.), *Flora of North America North of Mexico, Vol. 26: Magnoliophyta: Liliidae: Liliales and Orchidales*. Oxford University Press, New York, NY.

Goldblatt, P. & M. Takei. 1997. Chromosome cytology of Iridaceae-patterns of variation, determination of ancestral base numbers, and modes of karyotype change. *Ann. Missouri Bot. Gard.* **84**: 285–304.

Goldblatt, P. & J. C. Manning. 2008. *The Iris Family: Natural History and Classification*. Timber Press, Portland, OR. 290 pp.

Goldblatt, P., P. Rudall & J. E. Henrich. 1990. The genera of the *Sisyrinchium* alliance (Iridaceae: Iridoideae): phylogeny and relationships. *Syst. Bot.* **15**: 497–510.

Goldblatt, P., A. Rodriguez, M. P. Powell, J. T. Davies, J. C. Manning, M. Van der Bank & V. Savolainen. 2008. Iridaceae 'out of Australasia'? Phylogeny, biogeography, and divergence time based on plastid DNA sequences. *Syst. Bot.* **33**: 495–508.

Goldman, D. H., J. V. Freudenstein, P. J. Kores, M. Molvray, D. C. Jarrell, W. M. Whitten, K. M. Cameron, R. K. Jansen & M. W. Chase. 2001. Phylogenetics of Arethuseae (Orchidaceae) based on plastid *matK* and *rbcL* sequences. *Syst. Bot.* **26**: 670–695.

Goldman, D. H., L. K. Magrath & P. M. Catling. 2002. *Calopogon* R. Brown. Pp. 597–601. *In*: Flora North America Editorial Committee (eds.), *Flora of North America North of Mexico, Vol. 26: Magnoliophyta: Liliidae: Liliales and Orchidales.* Oxford University Press, New York, NY.

Goldman, D. H., R. K. Jansen, C. van den Berg, I. J. Leitch, M. F. Fay & M. W. Chase. 2004a. Molecular and cytological examination of *Calopogon* (Orchidaceae, Epidendroideae): circumscription, phylogeny, polyploidy, and possible hybrid speciation. *Am. J. Bot.* **91**: 707–723.

Goldman, D. H., C. Van den Berg & M. P. Griffith. 2004b. Morphometric circumscription of species and infraspecific taxa in *Calopogon* R. Br. (Orchidaceae). *Plant Syst. Evol.* **247**: 37–60.

Goldsborough, W. J. & W. M. Kemp. 1988. Light responses of a submersed macrophyte: implications for survival in turbid tidal waters. *Ecology* **69**: 1775–1786.

Goleniowski, M. E., G. A. Bongiovanni, L. Palacio, C. O. Nuñez & J. J. Cantero. 2006. Medicinal plants from the "Sierra de Comechingones", Argentina. *J. Ethnopharmacol.* **107**: 324–341.

Golley, F. B., G. A. Petrides & J. F. McCormick. 1965. A survey of the vegetation of the Boiling Springs Natural Area, South Carolina. *Bull. Torrey Bot. Club* **92**: 355–363.

Gonzalez, J., E. Garcia & M. Perdomo. 1983. Important rice weeds in Latin America. Pp. 119–132 *In*: International Rice Research Institute and International Weed Science Society (eds.), *Proceedings of the Conference on Weed Control in Rice.* Los Baños, Laguna, Philippines.

González-Soriano, E., O. Delgado-Hernández & G. L. Harp. 2004. Biological notes on *Neoerythromma gladiolatum* Williamson & Williamson, 1930 with description of its female (Zygoptera: Coenagrionidae). *Odonatologica* **33**: 327–331.

Good, J. A. 1986. Insect visitors to *Iris pseudacorus* (Iridaceae) in Ireland. *Ir. Nat. J.* **22**: 71–74.

Göppner, D. & M. Leverkus. 2011. Basal cell carcinoma: from the molecular understanding of the pathogenesis to targeted therapy of progressive disease. *J. Skin Cancer* **2011**: 650258.

Gordón, E. 1997. Notas sobre la ecología de *Echinodorus grandiflorus* (Cham. et Shl.) Mich. (Alismataceae). *Ecotropicos* **10**: 33–39.

Gordon, T. 2002. 1997–1999 field trips. *Bartonia* **61**: 155–173.

Gordon, T. 2004. 2000–2002 field trips. *Bartonia* **62**: 113–128.

Gordon, T. 2011. 2007–2008 field trips. *Bartonia* **65**: 126–147.

Gordon, T. & M. Demitroff. 2009. Sprungs, cripples, blue holes, and savannahs (savannas) of the pine barrens of Atlantic and Gloucester counties, New Jersey. *Bartonia* **64**: 59–62.

Gordon, T. & J. Arsenault. 2016. Flora of Burden Hill Forest: a checklist for a Salem County, New Jersey landscape. *Bartonia* **69**: 20–46.

Gordon, D. H., B. T. Gray, R. D. Perry, M. B. Prevost, T. H. Strange & R. K. Williams. 1989. South Atlantic coastal wetlands. Pp. 57–92 *In*: L. M. Smith, R. L. Pederson & R. M. Kaminski (eds.), *Habitat Management for Migrating and Wintering Waterfowl in North America.* Texas Tech University Press, Lubbock, TX. 560 pp.

Goremykin, V. V., B. Holland, K. I. Hirsch-Ernst & F. H. Hellwig. 2005. Analysis of *Acorus calamus* chloroplast genome and its phylogenetic implications. *Mol. Biol. Evol.* **22**: 1813–1822.

Gorham, E. & W. H. Pearsall. 1956. Acidity, specific conductivity and calcium content of some bog and fen waters in northern Britain. *J. Ecol.* **44**: 129–141.

Gorham, E., J. A. Janssens, G. A. Wheeler & P. H. Glaser. 1987. The natural and anthropogenic acidification of peatlands. Pp. 493–512 *In*: T. C. Hutchinson & K. M. Meema (eds.), *Effects of Atmospheric Pollutants on Forests, Wetlands and Agricultural Ecosystems.* NATO ASI Series (Series G: Ecological Sciences), vol 16. Springer, Berlin, Germany.

Gorman, T. A., S. D. Powell, K. C. Jones & C. A. Haas. 2014. Microhabitat characteristics of egg deposition sites used by reticulated flatwoods salamanders. *Herpetol. Conserv. Biol.* **9**: 543–550.

Gortner, R. A. 1934. Lake vegetation as a possible source of forage. *Science* **80**: 531–533.

Goslee, S. C., R. P. Brooks & C. A. Cole. 1997. Plants as indicators of wetland water source. *Plant Ecol.* **131**: 199–206.

Gosse, J. W. & B. J. Hearn. 2005. Seasonal diets of Newfoundland martens, *Martes americana atrata. Can. Field-Nat.* **119**: 43–47.

Gotceitas, V., S. Fraser & J. A. Brown. 1997. Use of eelgrass beds (*Zostera marina*) by juvenile Atlantic cod (*Gadus morhua*). *Can. J. Fish. Aquat. Sci.* **54**: 1306–1319.

Goulet, R. R., J. D. Lalonde, C. Munger, S. Dupuis, G. Dumont-Frenette, S. Prémont & P. G. C. Campbell. 2005. Phytoremediation of effluents from aluminum smelters: a study of Al retention in mesocosms containing aquatic plants. *Water Res.* **39**: 2291–2300.

Grabowski, A. 1973. The biomass, organic matter contents and calorific values of macrophytes in the lakes of the Szeszupa drainage area. *Polsk. Arch. Hydrobiol.* **20**: 269–282.

Grabowski, J. M. 2001. Observations on seed propagation of 5 Mississippi wetland species. *Native Plant J.* **2**: 67–68.

Grabowski, J. M. 2002. Three Mississippi ecotypes of wetland plants. Pp. 94–97 *In*: M. M. Holland, M. L. Warren & J. A. Stanturf (eds.), *Proceedings of a Conference on Sustainability of Wetlands and Water Resources: How Well Can Riverine Wetlands Continue to Support Society into the 21st Century?* Gen. Tech. Rep. SRS-50. U.S. Department of Agriculture, Forest Service, Southern Research Station, Asheville, NC.

Grace, J. B. 1983. Autotoxic inhibition of seed germination by *Typha latifolia*: an evaluation. *Oecologia* **59**: 366–369.

Grace, J. B. 1984. Effects of tubificid worms on the germination and establishment of *Typha. Ecology* **65**: 1689–1693.

Grace, J. B. 1988. The effects of nutrient additions on mixtures of *Typha latifolia* L. and *Typha domingensis* Pers. along a water-depth gradient. *Aquat. Bot.* **31**: 83–92.

Grace, J. B. 1989. Effects of water depth on *Typha latifolia* and *Typha domingensis. Am. J. Bot.* **76**: 762–768.

Grace, J. B. 1993. The adaptive significance of clonal reproduction in angiosperms: an aquatic perspective. *Aquat. Bot.* **44**: 159–180.

Grace, S. L. 2000. Short-term response of plant species of special concern and exotics to the 1998 Florida wildfires. Final Report. U.S. Geological Survey, National Wetlands Research Center, Lafayette, LA. 13 pp.

Grace, J. B. & R. G. Wetzel. 1981a. Habitat partitioning and competitive displacement in cattails (*Typha*): experimental field studies. *Am. Nat.* **118**: 463–474.

Grace, J. B. & R. G. Wetzel. 1981b. Effects of size and growth rate on vegetative reproduction in *Typha. Oecologia* **50**: 158–161.

Grace, J. B. & R. G. Wetzel. 1982. Niche differentiation between two rhizomatous plant species: *Typha latifolia* and *Typha angustifolia. Can. J. Bot.* **60**: 46–57.

Grace, J. B. & J. S. Harrison. 1986. The biology of Canadian weeds: 73. *Typha latifolia* L., *Typha angustifolia* L. and *Typha* ×*glauca* Godr. *Can. J. Plant Sci.* **66**: 361–379.

Grace, J. B. & M. A. Ford. 1996. The potential impact of herbivores on the susceptibility of the marsh plant *Sagittaria lancifolia* to saltwater intrusion in coastal wetlands. *Estuaries Coast.* **19**: 13–20.

Grace, J. B. & R. G. Wetzel. 1998. Long-term dynamics of *Typha* populations. *Aquat. Bot.* **61**: 137–146.

Graenicher, S. 1935. Bee-fauna and vegetation of Wisconsin. *Ann. Entomol. Soc. Am.* **28**: 285–310.

Graham, S. P. 2010. Visitors to southeastern hawkmoth flowers. *Southeast. Nat.* **9**: 413–426.

Graham, S. W. & S. C. H. Barrett. 1995. Phylogenetic systematics of Pontederiales: implications for breeding-system evolution. Pp. 415–441 *In*: P. J. Rudall, P. J. Cribb, D. F. Cutler & C. J. Humphries (eds.), *Monocotyledons: Systematics and Evolution*. Royal Botanic Gardens, Kew, United Kingdom.

Graham, S. W., J. R. Kohn, B. R. Morton, J. E. Eckenwalder & S. C. H. Barrett. 1998. Phylogenetic congruence and discordance among one morphological and three molecular data sets from Pontederiaceae. *Syst. Biol.* **47**: 545–567.

Graham, S. W., R. G. Olmstead & S. C. H. Barrett. 2002. Rooting phylogenetic trees with distant outgroups: a case study from the commelinoid monocots. *Molec. Biol. Evol.* **19**: 1769–1781.

Graham, E. E., J. F. Tooker & L. M. Hanks. 2012. Floral host plants of adult beetles in central Illinois: an historical perspective. *Ann. Entomol. Soc. Am.* **105**: 287–297.

Graham, J. R., E. Willcox & J. D. Ellis. 2015. The potential management of a ground-nesting, solitary bee: *Anthophora abrupta* (Hymenoptera: Apidae). *Florida Entomol.* **98**: 528–535.

Grainger, J. 1947. Nutrition and flowering of water plants. *J. Ecol.* **35**: 49–64.

Grajczyk, A. M., W. A. Overholt, J. P. Cuda, S. D. Brown & D. A. Williams. 2009. Characterization of microsatellite loci in *Hydrilla verticillata*. *Mol. Ecol. Resour.* **9**: 1460–1559.

Grant, W. T. 1864. Indigenous medicinal plants. *Confederate States Med. Surg. J.* **1**(6): 84.

Grant, K. A. 1966. A hypothesis concerning the prevalence of red coloration in California hummingbird flowers. *Am. Nat.* **100**: 85–97.

Grant, V. 1983. The systematic and geographical distribution of hawkmoth flowers in the temperate North American flora. *Bot. Gaz.* **144**: 439–449.

Grant, V. & K. A. Grant. 1966. Records of hummingbird pollination in the western American flora: I. Some California plant species. *Aliso* **6**: 51–66.

Grant, V. & K. A. Grant. 1967. Records of hummingbird pollination in the western American flora: II. Additional California records. *Aliso* **6**: 103–105.

Grant, T. A., P. Henson & J. A. Cooper. 1994. Feeding ecology of trumpeter swans breeding in south central Alaska. *J. Wildl. Manag.* **58**: 774–780.

Grau, C. R. & K. J. Leonard. 1978. *Zea mays*, a new host for *Ligniera junci* (Plasmodiophorales). *Mycologia* **70**: 41–46.

Graves, A. H. 1908. The morphology of *Ruppia maritima*. *Trans. Connecticut Acad. Arts Sci.* **14**: 59–170.

Graves, H. B. 1984. Behavior and ecology of wild and feral swine (*Sus scrofa*). *J. Anim. Sci.* **58**: 482–492.

Graves, G R. 2001. Factors governing the distribution of Swainson's Warbler along a hydrological gradient in Great Dismal Swamp. *Auk* **118**: 650–664.

Gray, A. 1862. Fertilization of orchids through the agency of insects. *Am. J. Sci.* **34**: 420–429.

Gray, J. B., B. A. Sorrie & W. Wall. 2016. Canebrakes of the Sandhills region of the Carolinas and Georgia: fire history, canebrake area, and species frequency. *Castanea* **81**: 280–291.

Grayum, M. H. 1987. A summary of evidence and arguments supporting the removal of *Acorus* from the Araceae. *Taxon* **36**: 723–729.

Grear, J. W. 1966. Cytogeography of *Orontium aquaticum* (Araceae). *Rhodora* **68**: 25–34.

Grear, Jr., J. W. 1973. Observations on the stomatal apparatus of *Orontium aquaticum* (Araceae). *Bot. Gaz.* **134**: 151–153.

Green, A. J. 2016. The importance of waterbirds as an overlooked pathway of invasion for alien species. *Divers. Distrib.* **22**: 239–247.

Green, E. P. & F. T. Short. 2003. *World Atlas of Seagrasses*. University of California Press, Berkeley, CA. 298 pp.

Green, A. J., M. Soons, A.-L. Brochet & E. Kleyheeg. 2016. Dispersal of plants by waterbirds. Pp. 147–195 *In*: Ç. H. Sekercioglu, D. G. Wenny & C. J. Whelan (eds.), *Why Birds Matter: Avian Ecological Function and Ecosystem Services*. University of Chicago Press, Chicago, IL.

Greenwood, M. E. & P. J. DuBowy. 2005. Germination characteristics of *Zannichellia palustris* from New South Wales, Australia. *Aquat. Bot.* **82**: 1–11.

Gregg, K. B. 1991. Defrauding the deceitful orchid: pollen collection by pollinators of *Cleistes divaricata* and *C. bifaria*. *Lindleyana* **6**: 214–220.

Gregg, K. B. & L. H. Klotz. 2015. The flora of Beavers' Meadow, Barbour County, West Virginia, revisited after a quarter century. *Castanea* **80**: 130–143.

Gregory, C., R. Braham, G. Blank & J. Stucky. 2010. Habitat and search criteria of the rare sandhills lily, *Lilium pyrophilum* M. W. Skinner and Sorrie. *Castanea* **75**: 198–204.

Greimler, J., T. F. Stuessy, U. Swenson, P. López-Sepúlveda & C. M. Baeza. 2017. Invasive species. Pp. 134–148 *In*: T. F. Stuessy, D. J. Crawford, P. López-Sepúlveda, C. M. Baeza & E. A. Ruiz (eds.), *Plants of Oceanic Islands: Evolution, Biogeography, and Conservation of the Flora of the Juan Fernández (Robinson Crusoe) Archipelago*. Cambridge University Press, New York, NY.

Greulich, S. & M. Tremolieres. 2006. Present distribution of the genus *Elodea* in the Alsatian Upper Rhine floodplain (France) with a special focus on the expansion of *Elodea nuttallii* St. John during recent decades. *Hydrobiologia* **570**: 249–255.

Grewell, B. J. 2008. Parasite facilitates plant species coexistence in a coastal wetland. *Ecology* **89**: 1481–1488.

Grewell, B. J., J. C. Callaway, W. R. Ferren & R. Wayne. 2007. Estuarine wetlands. Pp. 124–154 *In*: M. Barbour, T. Keeler-Wolf, A. A. Schoenherr (eds.), *Terrestrial Vegetation of California*, 3rd ed. University of California Press, Berkeley, CA.

Grier, N. M. 1916. A new species of *Opercularia*. *Trans. Am. Microsc. Soc.* **35**: 138–139.

Griese, H. J., R. A. Ryder & C. E. Braun. 1980. Spatial and temporal distribution of rails in Colorado. *Wilson Bull.* **92**: 96–102.

Griffen, K. O. 1975. Vegetation studies and modern pollen spectra from the Red Lake Peatland, northern Minnesota. *Ecology* **56**: 531–546.

Griffin, N. E. & M. J. Durako. 2012. The effect of pulsed versus gradual salinity reduction on the physiology and survival of *Halophila johnsonii* Eiseman. *Mar. Biol.* **159**: 1439–1447.

Griffin, A. D., F. E. Durbian, D. A. Easterla & R. L. Bell. 2009. Spatial ecology of breeding Least Bitterns in northwest Missouri. *Wilson J. Ornithol.* **121**: 521–528.

Grimaldi, D. & J. Jaenike. 1983. The Diptera breeding on skunk cabbage, *Symplocarpus foetidus* (Araceae). *J. New York Entomol. Soc.* **91**: 83–89.

Grippo, M. A., I. Hlohowskyj, L. Fox, B. Herman, J. Pothoff, C. Yoe & J. Hayse. 2017. Aquatic nuisance species in the great lakes and Mississippi river basin – a risk assessment in support of GLMRIS. *Environ. Manage.* **59**: 154–173.

Grisé, D., J. E. Titus & D. J. Wagner. 1986. Environmental pH influences growth and tissue chemistry of the submersed macrophyte *Vallisneria americana. Can. J. Bot.* **64**: 306–310.

Grittinger, T. F. 1970. String bog in southern Wisconsin. *Ecology* **51**: 928–930.

Grootjen, C. J. & F. Bouman. 1988. Seed structure in Cannaceae: taxonomic and ecological implications. *Ann. Bot.* **61**: 363–371.

Gross, A. O. 1944. Food of the snowy owl. *Auk* **61**: 1–18.

Gross, E. M., R. L. Johnson & N. G. Hairston, Jr. 2001. Experimental evidence for changes in submersed macrophyte species composition caused by the herbivore *Acentria ephemerella* (Lepidoptera). *Oecologia* **127**: 105–114.

Gu, D., H. Xu, Y. He, F. Zhao & M. Huang. 2015. Remediation of urban river water by *Pontederia cordata* combined with artificial aeration: organic matter and nutrients removal and root-adhered bacterial communities. *Int. J. Phytoremed.* **17**: 1105–1114.

Guard, B. J. 1995. *Wetland Plants of Oregon and Washington.* Lone Pine Publishing, Redmond, WA. 239 pp.

Guerra, I. R., S. L. Hernandez, I. Hernandez, S. A. Padron, J. A. Herrera, S. Olguin, R. Camacho, M. D. Ronquillo, M. D. Fernandez, R. Menendez & J. Espinosa-Aguirre. 2014. *Thalassia testudinum* extract modulates hepatic cytochome P450 system in Wistar rats. *J. Fed. Am. Soc. Exp. Biol.* **28**: 657.16.

Guisinger, M. M., T. W. Chumley, J. V. Kuehl, J. L. Boore & R. K. Jansen. 2010. Implications of the plastid genome sequence of *Typha* (Typhaceae, Poales) for understanding genome evolution in Poaceae. *J. Mol. Evol.* **70**: 149–166.

Gulnaz, O., A. Sahmurova & S. Kama. 2011. Removal of Reactive Red 198 from aqueous solution by *Potamogeton crispus. Chem. Eng. J.* **174**: 579–585.

Gunderson, L. H. & W. F. Loftus. 1993. The Everglades. Pp. 199–255 *In*: W. E. Martin, S. G. Boyce & A. C. Echternacht (eds.), *Biodiversity of the Southeastern United States. Volume 2: Lowland Terrestrial Communities.* John Wiley & Sons, Inc., New York, NY.

Gunderson, M. D., K. L. Kapuscinski, D. P. Crane & J. M. Farrell. 2016. Habitats colonized by non-native flowering rush *Butomus umbellatus* (Linnaeus, 1753) in the Niagara River, USA. *Aquat. Invasions* **11**: 369–380.

Guo, Y.-H. & C. D. K. Cook. 1989. Pollination efficiency of *Potamogeton pectinatus* L. *Aquat. Bot.* **34**: 381–384.

Guo, Y.-H., R. Sperry, C. D. K. Cook & P. A. Cox. 1990. The pollination ecology of *Zannichellia palustris* L. (Zannichelliaceae). *Aquat. Bot.* **38**: 341–356.

Guo, L. D., K. D. Hyde & E. C. Y. Liew. 2001. Detection and taxonomic placement of endophytic fungi within frond tissues of *Livistona chinensis* based on rDNA sequences. *Mol. Phylogenet. Evol.* **20**: 1–13.

Guppy, H. B. 1906. *Observations of a Naturalist in the Pacific Between 1896 and 1899. Volume II. Plant-Dispersal.* Macmillan and Company, Limited, London, UK. 627 pp.

Gupta, A. L. K. A. & V. N. Pandey. 2014. Herbal remedies of aquatic macrophytes of Gorakhpur district, Uttar Pradesh (India). *Int. J. Pharm. Bio. Sci.* **5**: 300–308.

Gupta, M. P., S. S. Handa, G. Longo & D. D. Rakesh (eds.). 2011. *Compendium of Medicinal and Aromatic Plants: The Americas.* The International Centre for Science and High Technology, Trieste, Italy. 411 pp.

Gupta, A., R. Maurya, R. K. Roy, S. V. Sawant & H. K. Yadav. 2013. AFLP based genetic relationship and population structure analysis of *Canna*—An ornamental plant. *Sci. Hort.* **154**: 1–7.

Gusain, R. & S. Suthar. 2017. Potential of aquatic weeds (*Lemna gibba*, *Lemna minor*, *Pistia stratiotes* and *Eichhornia* sp.) in biofuel production. *Process Saf. Environ. Prot.* **109**: 233–241.

Gustafson, F. G. 1942. Parthenocarpy: natural and artificial. *Bot. Rev.* **8**: 599–654.

Gut, L. J., R. A. Schlising & C. E. Stopher. 1977. Nectar-sugar concentrations and flower visitors in the western Great Basin. *Great Basin Nat.* **37**: 523–529.

Guterres-Pazin, M. G., M. Marmontel, F. C. W. Rosas, V. F. V. Pazin & E. M. Venticinque. 2014. Feeding ecology of the Amazonian manatee (*Trichechus inunguis*) in the Mamiraua and Amana sustainable development reserves, Brazil. *Aquat. Mamm.* **40**: 139–149.

Gutierrez, R. M. P. 2010. Orchids: a review of uses in traditional medicine, its phytochemistry and pharmacology. *J. Med. Plant Res.* **4**: 592–638.

Gutierrez, M. A., A. A. Cardona & D. L. Smee. 2010. Growth patterns of shoal grass *Halodule wrightii* and manatee grass *Syringodium filiforme* in the western Gulf of Mexico. *Gulf Caribbean Res.* **22**: 71–75.

Guy, C. J. 1988. A seasonal investigation of nonstructural carbohydrates in submerged macrophytes of Shoal Lake in relation to water depth. M.S thesis. University of Manitoba, Winnipeg, MB. 311 pp.

Guzmán del Próo, S. A., E. Serviere-Zaragoza & D. A. S. Beltrones. 2003. Natural diet of juvenile abalone *Haliotis fulgens* and *H. corrugata* (Mollusca: Gastropoda) in Bahia Tortugas, Mexico. *Pac. Sci.* **57**: 319–324.

Haag, R. W. 1983. Emergence of seedlings of aquatic macrophytes from lake sediments. *Can. J. Bot.* **61**: 148–156.

Haag, R. W. & P. R. Gorham. 1977. Effects of thermal effluent on standing crop and net production of *Elodea canadensis* and other submerged macrophytes in Lake Wabamun, Alberta. *J. Appl. Ecol.* **14**: 835–851.

Haag, K. H., T. M. Lee, D. C. Herndon, P. County & T. B. Water. 2005. Bathymetry and vegetation in isolated marsh and cypress wetlands in the Northern Tampa Bay Area, 2000–2004. Scientific Investigations Report 2005–5109. U.S. Department of the Interior, Geological Survey, Tampa, FL. 49 pp.

Habeck, R. J. 1992. *Maianthemum stellatum. In: Fire Effects Information System.* U.S. Department of Agriculture, Forest Service, Rocky Mountain Research Station, Fire Sciences Laboratory (Producer). Available online: https://www.fs.fed.us/database/feis/plants/forb/maiste/all.html [accessed 23 January, 2019].

Hafez, M. B., N. Hafez & Y. S. Ramadan. 1992. Uptake of cerium, cobalt and cesium by *Potamogeton crispus. J. Chem. Technol. Biotechnol.* **54**: 337–340.

Hagley, C. A., D. Wright, C. J. Owen, P. Eiler & M. Banks. 1996. Changes in aquatic macrophytes after liming Thrush Lake, Minnesota. *Restor. Ecol.* **4**: 307–312.

Hahn, A. T., C. A. Rosa, A. Bager & L. Krause. 2014. Dietary variation and overlap in D'Orbigny's slider turtles *Trachemys dorbigni* (Duméril and Bibron 1835) (Testudines: Emydidae). *J. Nat. Hist.* **48**: 721–728.

Haines, A. 2000a. Taxonomy and distribution of Acorus in Maine. *Bot. Notes* **2**: 4–6.

Haines, A. 2000b. *Eriocaulon parkeri Robinson Parker's Pipewort. New England Plant Conservation Program Conservation and Research Plan.* New England Wild Flower Society, Framingham, MA. 15 pp.

Haines, A. 2001. *Eleocharis aestuum* (Cyperaceae), a new tidal river shore spikesedge of the Eastern United States. *Novon* **11**: 45–49.

Hakim, L. & H. Miyakawa. 2015. Exotic plant species in the restoration project area in Ranupani recreation forest, Bromo Tengger Semeru National Park (Indonesia). *Biodiv. J.* **6**: 831–836.

Halbritter, D. A., J. C. Daniels, D. C. Whitaker & L. Huang. 2015. Reducing mowing frequency increases floral resource and butterfly (Lepidoptera: Hesperioidea and Papilionoidea) abundance in managed roadside margins. *Florida Entomol.* **98**: 1081–1092.

Halder, S. & P. Venu. 2012. The taxonomy and report of flowering in *Lemna* L. (Lemnaceae) in India. *Curr. Sci.* **102**: 1629–1632.

Hale, E. M. 1867. *Homoeopathic Materia Medica of the New Remedies: Their Botanical Description, Medical History, Pathogenetic Effects and Therapeutical Application in Homoeopathic Practice.* Dr. E. A. Lodge, Detroit, MI. 1142 pp.

Hall, T. F. & W. T. Penfound. 1943. Cypress-gum communities in the Blue Girth Swamp near Selma, Alabama. *Ecology* **24**: 208–217.

Hall, J. B. & D. U. U. Okali. 1974. Phenology and productivity of *Pistia stratiotes* L. on the Volta Lake, Ghana. *J. Appl. Ecol.* **11**: 709–725.

Hall, D. J. & E. E. Werner. 1977. Seasonal distribution and abundance of fishes in the littoral zone of a Michigan lake. *Trans. Am. Fish. Soc.* **106**: 545–555.

Hall, M. O. & S. S. Bell. 1993. Meiofauna on the seagrass *Thalassia testudinum*: population characteristics of harpacticoid copepods and associations with algal epiphytes. *Mar. Biol.* **116**: 137–146.

Hall, J. A. & G. H. Walter. 2011. Does pollen aerodynamics correlate with pollination vector? Pollen settling velocity as a test for wind versus insect pollination among cycads (Gymnospermae: Cycadaceae: Zamiaceae). *Biol. J. Linn. Soc.* **104**: 75–92.

Hall, E. C., J. L. Rodda & J. A. Small. 1938. Field trips of the club. *Torreya* **38**: 129–133.

Hall, L. M., M. D. Hanisak & R. W. Virnstein. 2006. Fragments of the seagrasses *Halodule wrightii* and *Halophila johnsonii* as potential recruits in Indian River Lagoon, Florida. *Mar. Ecol. Prog. Ser.* **310**: 109–117.

Haller, W. T., D. L. Sutton & W. C. Barlowe. 1974. Effects of salinity on growth of several aquatic macrophytes. *Ecology* **55**: 891–894.

Haller, W. T., J. L. Miller & L. A. Garrard. 1976. Seasonal production and germination of hydrilla vegetative propagules. *J. Aquat. Plant Manag.* **14**: 26–29.

Halloran, A. F. 1943. Management of deer and cattle on the Aransas National Wildlife Refuge, Texas. *J. Wildl. Manag.* **7**: 203–216.

Halpern, C. B. 1986. Montane meadow plant associations of Sequoia National Park, California. *Madroño* **33**: 1–23.

Halpin, K. M. 2011. A chloroplast phylogeny of Agavaceae subfamily Chlorogaloideae with a focus on species relationships in Hastingsia. M.S. thesis. Oklahoma State University, Stillwater, OK. 95 pp.

Halpin, K. M. & M. Fishbein. 2013. A chloroplast phylogeny of Agavaceae subfamily Chlorogaloideae: implications for the tempo of evolution on serpentine soils. *Syst. Bot.* **38**: 996–1011.

Hamel, P. B. & M. U. Chiltoskey. 1975. *Cherokee Plants and Their Uses – A 400 Year History.* Herald Publishing, Sylva, NC. 59 pp.

Hamel, K. S. & J. K. Parsons. 2001. Washington's aquatic plant quarantine. *J. Aquat. Plant Manag.* **39**: 72–75.

Hamerstrom, Jr., F. N. & J. Blake. 1939. Central Wisconsin muskrat study. *Am. Midl. Nat.* **21**: 514–520.

Hamlin, J. A. P. & M. L. Arnold. 2014. Determining population structure and hybridization for two iris species. *Ecol. Evol.* **4**: 743–755.

Hamlin, J. A. P., T. J. Simmonds & M. L. Arnold. 2017. Niche conservatism for ecological preference in the Louisiana iris species complex. *Biol. J. Linn. Soc.* **120**: 144–154.

Hammer, U. T. & J. M. Heseltine. 1988. Aquatic macrophytes in saline lakes of the Canadian prairies. Pp. 101–116 *In*: J. M. Melack (ed.), *Saline Lakes*. Dr. W. Junk Publishers, Dordrecht, The Netherlands.

Hammerstrom, K. K., W. J. Kenworthy, M. S. Fonseca & P. E. Whitfield. 2006. Seed bank, biomass, and productivity of *Halophila decipiens*, a deep water seagrass on the west Florida continental shelf. *Aquat. Bot.* **84**: 110–120.

Han, B., G. Tan, Z. Hu, Y. Wang, Y. Liu, R. Zhou & Q. Zhou. 2019. The complete chloroplast genome of *Eriocaulon sexangulare* (Eriocaulaceae). *Mitochondrial DNA Part B* **4**: 666–667.

Handa, I. T., R. Harmsen & R. L. Jefferies. 2002. Patterns of vegetation change and the recovery potential of degraded areas in a coastal marsh system of the Hudson Bay lowlands. *J. Ecol.* **90**: 86–99.

Handley, R. J. & A. J. Davy. 2000. Discovery of male plants of *Najas marina* L. (Hydrocharitaceae) in Britain. *Watsonia* **23**: 331–334.

Handley, R. J. & A. J. Davy. 2002. Seedling root establishment may limit *Najas marina* L. to sediments of low cohesive strength. *Aquat. Bot.* **73**: 129–136.

Handley, J. & B. Heidel. 2005. *Amerorchis rotundifolia* (Banks ex Pursh) Hultén (roundleaf orchid): a technical conservation assessment. USDA Forest Service, Rocky Mountain Region. Available online: http://www.fs.fed.us/r2/projects/scp/assessments/amerorchisrotundifolia.pdf [accessed 19 February, 2019].

Handley, R. J. & A. J. Davy. 2005. Temperature effects on seed maturity and dormancy cycles in an aquatic annual, *Najas marina*, at the edge of its range. *J. Ecol.* **93**: 1185–1193.

Hangelbroek, H. H., N. J. Ouborg, L. Santamaría & K. Schwenk. 2002. Clonal diversity and structure within a population of the pondweed *Potamogeton pectinatus* foraged by Bewick's swans. *Mol. Ecol.* **11**: 2137–2150.

Hangelbroek, H. H., L. Santamaría & T. De Boer. 2003. Local adaptation of the pondweed *Potamogeton pectinatus* to contrasting substrate types mediated by changes in propagule provisioning. *J. Ecol.* **91**: 1081–1092.

Hanley, T. A., M. P. Gillingham & K. L. Parker. 2014. Composition of diets selected by Sitka black-tailed deer on Channel Island, central southeast Alaska. Research Note PNW-RN-570. U.S. Department of Agriculture, Forest Service, Pacific Northwest Research Station, Portland, OR. 21 pp.

Hanlon, S. G., M. V. Hoyer, C. E. Cichra & D. E. Canfield. 2000. Evaluation of macrophyte control in 38 Florida lakes using triploid grass carp. *J. Aquat. Plant Manag.* **38**: 48–54.

Hann, J. H. 1986. The use and processing of plants by Indians of Spanish Florida. *Southeast. Archaeol.* **5**: 91–102.

Hann, B. J. 1995. Invertebrate associations with submersed aquatic plants in a prairie wetland. *UFS (Delta Marsh) Ann. Rep.* **30**: 78–84.

Hanson, H. C. 1951. Characteristics of some grassland, marsh, and other plant communities in western Alaska. *Ecol. Monogr.* **21**: 317–378.

Hanson, B. A., N. H. Euliss, Jr. & D. M. Mushet. 2002. First records of loosely coiled valve snail in North Dakota. *Prairie Nat.* **34**: 63–65.

Hanson, L., R. L. Brown, A. Boyd, M. A. T. Johnson & M. D. Bennett. 2003. First nuclear DNA C-values for 28 angiosperm genera. *Ann. Bot.* **91**: 31–38.

Hapeman, J. R. & K. Inoue. 2000. Plant-pollinator interactions and floral radiation in *Platanthera* (Orchidaceae). Pp. 433–454 *In*: T. J. Givnish & K. J. Sytsma (eds.), *Molecular Evolution and Adaptive Radiation*. Cambridge University Press, Cambridge, United Kingdom.

Hara, A., S. Ebina, A. Kondo & T. Funaguma. 1985. A new type of phytase from pollen of *Typha latifolia* L. *Agric. Biol. Chem.* **49**: 3539–3544.

Haramoto, T. & I. Ikusima. 1988. Life cycle of *Egeria densa* Planch., an aquatic plant naturalized in Japan. *Aquat. Bot.* **30**: 389–403.

Harder, L. D. 2000. Pollen dispersal and the floral diversity of monocotyledons. Pp. 243–257 *In*: K. L. Wilson & D. A. Morrison (eds.), *Monocots: Systematics and Evolution*. CSIRO Publishing, Melbourne, VIC.

Harder, L. D. & S. C. H. Barrett. 1992. The energy cost of bee pollination for *Pontederia cordata* (Pontederiaceae). *Funct. Ecol.* **6**: 226–233.

Harder, L. D. & S. C. H. Barrett. 1993. Pollen removal from tristylous *Pontederia cordata*: effects of anther position and pollinator specialization. *Ecology* **74**: 1059–1072.

Hardy, L. M. & L. R. Raymond. 1991. Observations on the activity of the pickerel frog, *Rana palustris* (Anura: Ranidae), in northern Louisiana. *J. Herpetol.* **25**: 220–222.

Harley, K. L. S. 1990. Production of viable seeds by water lettuce, *Pistia stratiotes* L., in Australia. *Aquat. Bot.* **36**: 277–279.

Harley, M. T. & S. Findlay. 1994. Photosynthesis-irradiance relationships for three species of submersed macrophytes in the tidal freshwater Hudson River. *Estuaries* **17**: 200–205.

Harlin, M. M. 1973. "Obligate" algal epiphyte: *Smithora naidaum* grows on a synthetic substrate. *J. Phycol.* **9**: 230–232.

Harlow, R. F. 1961. Fall and winter foods of Florida white-tailed deer. *Quart. J. Florida Acad. Sci.* 24: 19–38.

Harman, W. N. 1974. Phenology and physiognomy of the hydrophyte community in Otsego Lake, NY. *Rhodora* **76**: 497–508.

Harms, V. L. 1973. Taxonomic studies of North American *Sparganium*. I. *S. hyperboreum* and *S. minimum*. *Can. J. Bot.* **51**: 1629–1641.

Harms, N. E. & M. J. Grodowitz. 2009. Insect herbivores of aquatic and wetland plants in the United States: a checklist from literature. *J. Aquat. Plant Manag.* **47**: 73–96.

Harms, N. E. & J. F. Shearer. 2015. Apparent herbivory and indigenous pathogens of invasive flowering rush (*Butomus umbellatus* L.) in the Pacific Northwest. ERDC/TN APCRP-BC-35. U.S. Army Corps of Engineers, Vicksburg, MS. 11 pp.

Harms, N., M. Grodowitz & J. Kennedy. 2011. Insect herbivores of water stargrass (*Heteranthera dubia*) in the US. *J. Freshw. Ecol.* **26**: 185–194.

Harper, R. M. 1900. Notes on the flora of south Georgia. *Bull. Torrey Bot. Club* **27**: 413–436.

Harper, R. M. 1903. Botanical explorations in Georgia during the summer of 1901.-II. Noteworthy species. *Bull. Torrey Bot. Club* **30**: 319–342.

Harper, R. M. 1905a. Phytogeographical explorations in the coastal plain of Georgia in 1904. *Bull. Torrey Bot. Club* **32**: 451–467.

Harper, R. M. 1905b. Two misinterpreted species of *Xyris*. *Torreya* **5**: 128–130.

Harper, R. M. 1906. Some more coastal plain plants in the Palaeozoic Region of Alabama. *Torreya* **6**: 111–117.

Harper, R. M. 1911. Early spring aspects of the coastal plain vegetation of South Carolina, Georgia, and northeastern Florida. *Bull. Torrey Bot. Club* **38**: 223–236.

Harper, R. M. 1916. *Habenaria repens* and *Piaropus crassipes* in Leon County, Florida. *Torreya* **16**: 267–270.

Harper, F. 1920. The Florida water-rat (*Neofiber alleni*) in Okefinokee Swamp, Georgia. *J. Mammal.* **1**: 65–66.

Harper, R. M. 1922. Some pine-barren bogs in central Alabama. *Torreya* **22**: 57–60.

Harper, H. J. & H. A. Daniel. 1934. Chemical composition of certain aquatic plants. *Bot. Gaz.* **96**: 186–189.

Harper, M. G., A.-M. Trame & M. G. Hohman. 1998. Management of herbaceous seeps and wet savannas for threatened and endangered species. Technical Report 98/70. U.S. Army Corps of Engineers, Construction Engineering Research Lab. Champaign, IL, 83 pp.

Harrel, S. L. & E. D. Dibble. 2001. Foraging efficiency of juvenile bluegill, *Lepomis macrochirus*, among different vegetated habitats. *Environ. Biol. Fishes* **62**: 441–453.

Harris, J. A. 1927. The cat tail, *Typha angustifolia*, in Utah. *Torreya* **27**: 9–11.

Harris, S. A. 1990. Dynamics and origin of saline soils on the Slims River delta, Kluane National Park, Yukon Territory. *Arctic* **43**: 159–175.

Harris, D. D. & M. P. Gutzmer. 1996. Macrophyte production, fish herbivory, and water quality in a tailwater reservoir-Lake Ogallala, Nebraska. *Trans. Nebraska Acad. Sci.* **23**: 29–35.

Harrison, P. G. 1979. Reproductive strategies in intertidal populations of two co-occurring seagrasses (*Zostera* spp.). *Can. J. Bot.* **57**: 2635–2638.

Harrison, P. G. 1982a. Comparative growth of *Zostera japonica* Aschers. & Graebn. and *Z. marina* L. under simulated intertidal and subtidal conditions. *Aquat. Bot.* **14**: 373–379.

Harrison, P. G. 1982b. Seasonal and year-to-year variations in mixed intertidal populations of *Zostera japonica* Aschers. & Graebn. and *Ruppia maritima* L. s.l. *Aquat. Bot.* **14**: 357–371.

Harrison, M. L. 2002. The Patterson bundle. *HerbalGram* **55**: 35–41.

Harrison, P. G. & R. E. Bigley. 1982. The recent introduction of the seagrass *Zostera japonica* Aschers. and Graebn. to the Pacific coast of North America. *Can. J. Fish. Aquat. Sci.* **39**: 1642–1648.

Harrison, J. W. & W. M. Knapp. 2010 [2015]. *Ecological Classification of Groundwater-Fed Seepage Wetlands of the Maryland Coastal Plain*. Maryland Department of Natural Resources, Wildlife and Heritage Service, Natural Heritage Program, Annapolis, MD. 98 pp.

Hart, K. A. 2006. Evaluation of the nutrient removal efficiency of a constructed wetland system. M.S. thesis. Texas A&M University, College Station, TX. 103 pp.

Hartke, K. M., K. H. Kriegel, G. M. Nelson & M. T. Merendino. 2009. Abundance of wigeongrass during winter and use by herbivorous waterbirds in a Texas coastal marsh. *Wetlands* **29**: 288–293.

Hartley, T. G. 1959. Notes on some rare plants of Wisconsin. *Trans. Wisconsin Acad. Sci.* **8**: 57–64.

Hartley, T. G. 1960. Plant communities of the LaCrosse area in western Wisconsin. *Proc. Iowa Acad. Sci.* **67**: 174–188.

Hartman, G. 1985. Foods of male Mallard, before and during moult, as determined by faecal analysis. *Wildfowl* **36**: 65–71.

Hartman, R. T. & S. M. English. 1959. *Wolffiella floridana* in western Pennsylvania. *Castanea* **24**: 45–47.

Hartog, C. D. 1970. *The Sea-Grasses of the World*. Verhandelingen der Koninklijke Nederlandse Akademie van Wetenschappen, Afd. Natuurkunde Tweede Reeks, Deel 59, No 1. North-Holland Publishing Co., Amsterdam, The Netherlands. 294 pp.

Hartog, C. D. 1989. Distribution of *Plasmodiophora bicaudata*, a parasitic fungus on small *Zostera* species. *Dis. Aquat. Organ.* **6**: 227–229.

Hartog, C. D. & F. V. D. Plas. 1970. A synopsis of the Lemnaceae. *Blumea* **18**: 355–368.

Hartog, C. D., P. J. Van Loenhoud, J. G. M. Roelofs & J. C. P. M. Van De Sande. 1979. Chromosome numbers of three seagrasses from The Netherlands Antilles. *Aquat. Bot.* **7**: 267–271.

Harvey, R. M. & J. L. Fox. 1973. Nutrient removal using *Lemna minor*. *J. Water Pollut. Control Fed.* **45**: 1928–1938.

Harvey, L. & A. Haines. 2003. *Sagittaria teres* (Alismataceae) in New Hampshire. *Rhodora* **105**: 282–285.

Harwell, M. C. & R. J. Orth. 2002. Long-distance dispersal potential in a marine macrophyte. *Ecology* **83**: 3319–3330.

Hassan, M. S. & P. Edwards. 1992. Evaluation of duckweed (*Lemna perpusilla* and *Spirodela polyrrhiza*) as feed for Nile tilapia (*Oreochromis niloticus*). *Aquaculture* **104**: 315–326.

Hauber, D. P. & L. Lege. 1999. A survey of allozymic variation among three members of the *Sagittaria graminea* complex (Alismataceae) from the southeastern United States. *J. Torrey Bot. Soc.* **126**: 181–187.

Haug, E. J., J. T. Harris & R. J. Richardson. 2019. Monoecious *Hydrilla verticillata* development in complete darkness. *Aquat. Bot.* **154**: 28–34.

Haukos, D. A. & L. M. Smith. 1994. Composition of seed banks along an elevational gradient in playa wetlands. *Wetlands* **14**: 301–307.

Haukos, D. A. & L. M. Smith. 1997. *Common Flora of the Playa Lakes*. Texas Tech University Press, Lubbock, TX. 196 pp.

Haukos, D. A. & L. M. Smith. 2001. Temporal emergence patterns of seedlings from playa wetlands. *Wetlands* **21**: 274–280.

Haustein, A. T., R. H. Gilman, P. W. Skillicorn, H. Hannan, F. Diaz, V. Guevara, V. Vergara, A. Gastanaduy & J. B. Gilman. 1994. Performance of broiler chickens fed diets containing duckweed (*Lemna gibba*). *J. Agric. Sci.* **122**: 285–289.

Havera, S. P. 1986. *Completion of Illinois Waterfowl Studies*. Illinois Department of Conservation, Division of Wildlife Resources, Springfield, IL. 103 pp.

Hawkes, A. D. 1950. Studies in Florida Botany 8. The genus *Habenaria* in Florida. *Am. Midl. Nat.* **44**: 622–629.

Hay, F. R. & J. S. Muir. 2000. Low temperature survival of slender naiad (*Najas flexilis*) seeds. *Cryo Lett.* **21**: 271–278.

Haynes, R. R. 1974. A revision of North American *Potamogeton* subsection *Pusilli* (Potamogetonaceae). *Rhodora* **76**: 564–649.

Haynes, R. R. 1979. Revision of North and Central American *Najas* (Najadaceae). *Sida* **8**: 34–56.

Haynes, R. R. 1985a. A new species of *Najas* (Najadaceae) from the southeastern U.S.A. *Brittonia* **37**: 392–393.

Haynes, R. R. 1985b. A revision of the clasping-leaved *Potamogeton* (Potamogetonaceae). *Sida* **11**: 173–188.

Haynes, R. R. 1988. Reproductive biology of selected aquatic plants. *Ann. Missouri Bot. Gard.* **75**: 805–810.

Haynes, R. R. 2000a. Butomaceae Linnaeus. Pp. 3–4 *In*: Flora North America Editorial Committee (eds.), *Flora of North America North of Mexico, Vol. 22: Magnoliophyta: Alismatidae, Arecidae, Commelinidae (in Part), and Zingiberidae*. Oxford University Press, New York, NY.

Haynes, R. R. 2000b. Hydrocharitaceae Jussieu. Pp. 26–38 *In*: Flora North America Editorial Committee (eds.), *Flora of North America North of Mexico, Vol. 22: Magnoliophyta: Alismatidae, Arecidae, Commelinidae (in Part), and Zingiberidae*. Oxford University Press, New York, NY.

Haynes, R. R. 2000c. Najadaceae Jussieu. Pp. 77–83 *In*: Flora North America Editorial Committee (eds.), *Flora of North America North of Mexico, Vol. 22: Magnoliophyta: Alismatidae, Arecidae, Commelinidae (in Part), and Zingiberidae*. Oxford University Press, New York, NY.

Haynes, R. R. 2000d. Ruppiaceae Hutchinson. Pp. 75–76 *In*: Flora North America Editorial Committee (eds.), *Flora of North America North of Mexico, Vol. 22: Magnoliophyta: Alismatidae, Arecidae, Commelinidae (in Part), and Zingiberidae*. Oxford University Press, New York, NY.

Haynes, R. R. 2000e. The aquatic vascular flora of the southeastern United States: endemism and origins. *Sida* **18**: 23–28.

Haynes, R. R. 2000f. Cymodoceaceae N. Taylor. Pp. 86–89 *In*: Flora North America Editorial Committee (eds.), *Flora of North America North of Mexico, Vol. 22: Magnoliophyta: Alismatidae, Arecidae, Commelinidae (in Part), and Zingiberidae*. Oxford University Press, New York, NY.

Haynes, R. R. 2000g. Zosteraceae Dumortier. Pp. 90–94 *In*: Flora North America Editorial Committee (eds.), *Flora of North America North of Mexico, Vol. 22: Magnoliophyta: Alismatidae, Arecidae, Commelinidae (in Part), and Zingiberidae*. Oxford University Press, New York, NY.

Haynes, R. R. 2004. Limnocharitaceae. Pp. 456–457 *In*: N. Smith, S. A. Mori, A. Henderson, D. W. Stevenson & S. V. Heald (eds.), *Flowering Plants of the Neotropics*. Princeton University Press, Princeton, NJ.

Haynes, R. R. & L. B. Holm-Nielsen. 1992. *The Limnocharitaceae. Monograph 56, Flora Neotropica*. The New York Botanical Garden, Bronx, NY. 34 pp.

Haynes, R. R. & L. B. Holm-Nielsen. 1994. *The Alismataceae. Monograph 64, Flora Neotropica*. The New York Botanical Garden, Bronx, NY. 112 pp.

Haynes, R. R. & J. R. Burkhalter. 1998. A new species of *Echinodorus* (Alismataceae) from the United States of America. *Castanea* **63**: 180–182.

Haynes, R. R. & C. B. Hellquist. 2000a. Alismataceae Ventenat. Pp. 7–25 *In*: Flora North America Editorial Committee (eds.), *Flora of North America North of Mexico, Vol. 22: Magnoliophyta: Alismatidae, Arecidae, Commelinidae (in Part), and Zingiberidae*. Oxford University Press, New York, NY.

Haynes, R. R. & C. B. Hellquist. 2000b. Juncaginaceae Richard. Pp. 43–46 *In*: Flora North America Editorial Committee (eds.), *Flora of North America North of Mexico, Vol. 22: Magnoliophyta: Alismatidae, Arecidae, Commelinidae (in Part), and Zingiberidae*. Oxford University Press, New York, NY.

Haynes, R. R. & C. B. Hellquist. 2000c. Potamogetonaceae Dumortier. Pp. 47–74 *In*: Flora North America Editorial Committee (eds.), *Flora of North America North of Mexico, Vol. 22: Magnoliophyta: Alismatidae, Arecidae, Commelinidae (in Part), and Zingiberidae*. Oxford University Press, New York, NY.

Haynes, R. R. & D. H. Les. 2004. Alismatales. *In*: *Nature Encyclopedia of Life Sciences*. Nature Publishing Group, London. doi: 10.1038/npg.els.0003702.

Haynes, R. R., D. H. Les & L. B. Holm-Nielsen. 1998a. Alismataceae. Pp. 11–18 *In*: K. Kubitzki (ed.), *The Families and Genera of Vascular Plants, Vol. IV, Flowering Plants: Monocotyledons, Alismatanae and Commelinanae (Except Gramineae)*. Springer-Verlag, Berlin, Germany.

Haynes, R. R., D. H. Les & L. B. Holm-Nielsen. 1998b. Limnocharitaceae. Pp. 271–275 *In*: K. Kubitzki (ed.), *The Families and Genera of Vascular Plants, Vol. IV, Flowering Plants: Monocotyledons, Alismatanae and Commelinanae (Except Gramineae)*. Springer-Verlag, Berlin, Germany.

Haynes, R. R., L. B. Holm-Nielsen & D. H. Les. 1998c. Najadaceae. Pp. 301–306 *In*: K. Kubitzki (ed.), *The Families and Genera of Vascular Plants, Vol. IV, Flowering Plants: Monocotyledons, Alismatanae and Commelinanae (Except Gramineae)*. Springer-Verlag, Berlin, Germany.

Haynes, R. R., L. B. Holm-Nielsen & D. H. Les. 1998d. Potamogetonaceae. Pp. 408–415 *In*: K. Kubitzki (ed.), *The Families and Genera of Vascular Plants, Vol. IV, Flowering Plants: Monocotyledons, Alismatanae and Commelinanae (Except Gramineae)*. Springer-Verlag, Berlin, Germany.

Haynes, R. R., D. H. Les & M. Král. 1998e. Two new combinations in *Stuckenia*, the correct name for *Coleogeton* (Potamogetonaceae). *Novon* **8**: 241.

Haynes, R. R., D. H. Les & L. B. Holm-Nielsen. 1998f. Scheuchzeriaceae. Pp. 449–451 In: K. Kubitzki (ed.), *The Families and Genera of Vascular Plants, Vol. IV, Flowering Plants: Monocotyledons, Alismatanae and Commelinanae (Except Gramineae)*. Springer-Verlag, Berlin, Germany.

Hays, J. F. 2010. *Agalinis flexicaulis* sp. nov. (Orobanchaceae: Lamiales), a new species from northeast Florida. *J. Bot. Res. Inst. Texas* **4**: 1–6.

Hazen, T. 1918. The trimorphism and insect visitors of *Pontederia*. *Mem. Torrey Bot. Club* **17**: 459–484.

Hazlett, B. T. 1988. Aquatic vegetation and flora of Sleeping Bear Dunes National Lakeshore, Benzie and Leelanau Counties, Michigan. Technical Report No. 15. Douglas Lake, Pellston, MI. 66 pp.

He, M.-X., Q. Hu, Q. Zhu, K. Pan & Q. Li. 2015. The feasibility of using constructed wetlands plants to produce bioethanol. *Environ. Prog. Sustain. Energy* **34**: 276–281.

Heal, R. E., E. F. Rogers, R. T. Wallace & O. Starnes. 1950. A survey of plants for insecticidal activity. *Lloydia* **13**: 89–162.

Heaven, J. B., F. E. Gross & A. T. Gannon. 2003. Vegetation comparison of a natural and a created emergent marsh wetland. *Southeast. Nat.* **2**: 195–207.

Heck, Jr., K. L. & G. S. Wetstone. 1977. Habitat complexity and invertebrate species richness and abundance in tropical seagrass meadows. *J. Biogeog.* **4**: 135–142.

Heck, Jr., K. L. & R. J. Orth. 1980. Structural components of eelgrass (*Zostera marina*) meadows in the lower Chesapeake Bay—decapod Crustacea. *Estuaries* **3**: 289–295.

Heck, Jr., K. L., F. J. Fodrie, S. Madsen, C. J. Baillie & D. A. Byron. 2015. Seagrass consumption by native and a tropically associated fish species: potential impacts of the tropicalization of the northern Gulf of Mexico. *Mar. Ecol. Prog. Ser.* **520**: 165–173.

Heckscher, S. 1984. *Potamogeton confervoides* in Cumberland County, New Jersey. *Bartonia* **50**: 63–64.

Hedeen, Stanley E. 1972. Food and feeding behavior of the mink frog, *Rana septentrionalis* Baird, in Minnesota. *Am. Midl. Nat.* **88**: 291–300.

Heenatigala, P. P. M., J. Yang, Z. Sun, G. Li, S. Kumar, S. Hu, Z. Wu, W. Lin, L. Yao, P. Duan & H. Hou. 2018. Development of efficient protocols for stable and transient gene transformation for *Wolffia globosa* using *Agrobacterium*. *Front. Chem.* **6**: 227.

Heffernan, J. J. & R. A. Gibson. 1983. A comparison of primary production rates in Indian River, Florida seagrass systems. *Florida Sci.* **46**: 295–306.

Hegazy, A. K., N. T. Abdel-Ghani & G. A. El-Chaghaby. 2011. Phytoremediation of industrial wastewater potentiality by *Typha domingensis*. *Int. J. Environ. Sci. Technol.* **8**: 639–648.

Hegelmaier, F. 1868. *Die Lemnaceen. Eine monographische Untersuchung*. Verlag von Wilhelm Engelmann, Leipzig. 169 pp.

Hegnauer, R. & H. W. L. Ruijgrok. 1971. *Lilaea scilloides* und *Juncus bulbosus* zwei neue cyanogene pflanzen. *Phytochemistry* **10**: 2121–2124.

Heidbüchel, P., K. Kuntz & A. Hussner. 2016. Alien aquatic plants do not have higher fragmentation rates than native species: a field study from the River Erft. *Aquat. Sci.* **78**: 767–777.

Heilman, M. A. & R. G. Carlton. 2001. Ebullitive release of lacunar gases from floral spikes of *Potamogeton angustifolius* and *Potamogeton amplifolius*: effects on plant aeration and sediment CH_4 flux. *Aquat. Bot.* **71**: 19–33.

Heimbinder, E. 2001. Revegetation of a San Francisco coastal salt marsh. *Native Plant J.* **2**(1): 54–59.

Heinrich, B. 2015. Rapid flower-opening in *Iris pseudacorus*. *Northeast. Nat.* **22**: N11–N14.

Heinrich, B. 2016. A note on Iris flower anthesis: mechanism and meaning of sudden flower opening. *Northeast. Nat.* **23**: N12–N17.

Heisey, R. M. & A. W. H. Damman. 1982. Biomass and production of *Pontederia cordata* and *Potamogeton epihydrus* in three Connecticut rivers. *Am. J. Bot.* **69**: 855–864.

Hellblom, F. & L. Axelsson. 2003. External HCO_3^- dehydration maintained by acid zones in the plasma membrane is an important component of the photosynthetic carbon uptake in *Ruppia cirrhosa*. *Photosynth. Res.* **77**: 173–181.

Hellquist, C. B. 1971. Vascular flora of Ossipee Lake, New Hampshire and its shoreline. *Rhodora* **73**: 249–261.

Hellquist, C. B. 1972. Range extensions of vascular aquatic plants in New England. *Rhodora* **74**: 131–141.

Hellquist, C. B. 1975. Correlation of selected dissolved substances and the distribution of Potamogeton in New England. Ph.D. dissertation. Univeristy of New Hampshire, Durham, NH. 269 pp.

Hellquist, C. B. 1977. Observations on some uncommon vascular aquatic plants in New England. *Rhodora* **79**: 445–452.

Hellquist, C. B. 1980. Correlation of alkalinity and the distribution of *Potamogeton* in New England. *Rhodora* **82**: 331–344.

Hellquist, C. B. 1984. Observations of *Potamogeton hillii* Morong in North America. *Rhodora* **86**: 101–111.

Hellquist, C. B. 1997. *A Guide to Invasive Non-Native Aquatic Plants in Massachusetts*. Massachusetts Department of Environmental Management, Lakes & Ponds Program, Boston, MA. 15 pp.

Hellquist, C. B. & G. E. Crow. 1980. Aquatic vascular plants of New England: part 1. Zosteraceae, Potamogetonaceae, Zannichelliaceae, Najadaceae. Station Bulletin 515. New Hampshire Agricultural Experiment Station, University of New Hampshire, Durham, NH. 68 pp.

Hellquist, C. B. & G. E. Crow. 1981. Aquatic vascular plants of New England: part 3. Alismataceae. Station Bulletin 518. New Hampshire Agricultural Experiment Station, University of New Hampshire, Durham, NH. 32 pp.

Hellquist, C. B. & G. E. Crow. 1982. Aquatic vascular plants of New England: part 5. Araceae, Lemnaceae, Xyridaceae, Eriocaulaceae, and Pontederiaceae. Station Bulletin 523. New Hampshire Agricultural Experiment Station, University of New Hampshire, Durham, NH. 46 pp.

Hellquist, C. B. & R. L. Hilton. 1983. A new species of *Potamogeton* (Potamogetonaceae) from northeastern United States. *Syst. Bot.* **8**: 86–92.

Hellquist, C. B. & R. R. Haynes. 2000. Aponogetonaceae J. Agardh. Pp. 39–40 In: Flora North America Editorial Committee (eds.), *Flora of North America North of Mexico, Vol. 22: Magnoliophyta: Alismatidae, Arecidae, Commelinidae (in Part), and Zingiberidae*. Oxford University Press, New York, NY.

Hellquist, C. B. & A. R. Pike. 2004. *Potamogeton strictifolius A. Bennett (Straight-Leaf Pondweed) Conservation and Research Plan for New England*. New England Plant Conservation Program, Framingham, MA. 24 pp.

Hellquist, C. B., C. T. Philbrick & R. L. Hilton. 1988. The taxonomic status of *Potamogeton lateralis* Morong (Potamogetonaceae). *Rhodora* **90**: 15–20.

Hellquist, C. B., R. F. Thorne & R. R. Haynes. 2012. Potamogetonaceae. Pondweed family. Pp. 1500–1503 In: B. G. Baldwin, D. H. Goldman, D. J. Keil, R. Patterson, T. J. Rosatti & D. H. Wilken (eds.), *The Jepson Manual: Vascular Plants of California*, 2nd ed. University of California Press, Berkeley, CA.

Hellquist, C. E., C. B. Hellquist & J. J. Whipple. 2014. New records for rare and under-collected aquatic vascular plants of Yellowstone National Park. *Madroño* **61**: 159–176.

Henderson, D. M. 1976. A biosystematic study of Pacific Northwestern blue-eyed grasses (*Sisyrinchium*, Iridaceae). *Brittonia* **28**: 149–176.

Henderson, N. 2002. *Iris* Linnaeus. Pp. 371–395 *In*: Flora North America Editorial Committee (eds.), *Flora of North America North of Mexico, Vol. 26: Magnoliophyta: Liliidae: Liliales and Orchidales*. Oxford University Press, New York, NY.

Henderson, A., G. Galeano-Garces & R. Bernal. 1997. *Field Guide to the Palms of the Americas*. Princeton University Press, Princeton, NJ. 352 pp.

Hendricks, A. J. 1957. A revision of the genus *Alisma* (Dill.) L. *Am. Midl. Nat.* **58**: 470–493.

Hendricks, E. L. & M. H. Goodwin, Jr. 1952. Observations on surface-water temperatures in limesink ponds and evaporation pans in southwestern Georgia. *Ecology* **33**: 385–397.

Henriquez, C. L., T. Arias, J. C. Pires, T. B. Croat & B. A. Schaal. 2014. Phylogenomics of the plant family Araceae. *Mol. Phylogenet. Evol.* **75**: 91–102.

Henry-Silva, G. G., A. F. M. Camargo & M. M. Pezzato. 2008. Growth of free-floating aquatic macrophytes in different concentrations of nutrients. *Hydrobiologia* **610**: 153–160.

Herb, W. R. & H. G. Stefan. 2003. Integral growth of submersed macrophytes in varying light regimes. *Ecol. Modell.* **168**: 77–100.

Herbert, D. A. 1986. The growth dynamics of *Halophila hawaiiana*. *Aquat. Bot.* **23**: 351–360.

Herndon, A. 1987. A morphometric comparison of *Hymenocallis palmeri* and *Hymenocallis floridana* (Amaryllidaceae) in southern Florida. *Sida* **12**: 295–305.

Herrera-Silveira, J. A. 1994. Phytoplankton productivity and submerged macrophyte biomass variation in a tropical coastal lagoon with groundwater discharge. *Vie Milieu* **44**: 257–266.

Herring, J. L. 1950. The aquatic and semiaquatic Hemiptera of northern Florida. Part 1: Gerridae. *Florida Entomol.* **33**: 23–32.

Herring, B. J. & W. S. Judd. 1995. A floristic study of Ichetucknee Springs State Park, Suwannee and Columbia Counties, Florida. *Castanea* **60**: 318–369.

Hertweck, K. L., M. S. Kinney, S. A. Stuart, O. Maurin, S. Mathews, M. W. Chase, M. A. Gandolfo & J. C. Pires. 2015. Phylogenetics, divergence times and diversification from three genomic partitions in monocots. *Bot. J. Linn. Soc.* **178**: 375–393.

Herzka, S. Z. & K. H. Dunton. 1998. Light and carbon balance in the seagrass *Thalassia testudinum*: evaluation of current production models. *Mar. Biol.* **132**: 711–721.

Herzog, M. P. & J. S. Sedinger. 2004. Dynamics of foraging behavior associated with variation in habitat and forage availability in captive black Brant (*Branta bernicla nigricans*) goslings in Alaska. *Auk* **121**: 210–223.

Heslop-Harrison, J. W. 1952. Occurrence of the American pondweed, *Potamogeton epihydrus* Raf., in the Hebrides. *Nature* **169**: 548.

Hesse, M. 2006. Pollen wall ultrastructure of Araceae and Lemnaceae in relation to molecular classifications. *Aliso* **22**: 204–208.

Hestand, R. S. & C. C. Carter. 1974. The effects of a winter drawdown on aquatic vegetation in a shallow water reservoir. *Hyacinth Control J.* **12**: 9–12.

Heuschele, D. J. & F. K. Gleason. 2014. Two stages of dormancy in turions of *Potamogeton crispus* L. *Aquat. Bot.* **119**: 100–104.

Heyn, M. W. 1992. A review of the systematic position of the North American species of the genus *Glyptotendipes*. *Netherlands J. Aquat. Ecol.* **26**: 129–137.

Hicks, L. E. 1932a. Flower production in the Lemnaceae. *Ohio J. Sci.* **32**: 115–131.

Hicks, L. E. 1932b. Ranges of pH-tolerance of the Lemnaceae. *Ohio J. Sci.* **32**: 237–244.

Hicks, L. E. 1937. The Lemnaceae of Indiana. *Am. Midl. Nat.* **18**: 774–789.

Hicks, A. 2007. On the germination and subsequent culture of *Spiranthes delitescens* Sheviak in sterile culture. *Orch. Dig.* **71**(3): 158–160.

Hicks, A. 2016. Growing the elusive Canelo Hills lady's tresses of Arizona's cienegas (*Spiranthes delitescens* Sheviak). *Plant Press* **39**(2): 14–15.

Hicks, B. J., M. S. Wipfli, D. W. Lang & M. E. Lang. 2005. Marine-derived nitrogen and carbon in freshwater-riparian food webs of the Copper River Delta, southcentral Alaska. *Oecologia* **144**: 558–569.

Hidalgo, O., S. Garcia, T. Garnatje, M. Mumbrú, A. Patterson, J. Vigo & J. Vallès. 2015. Genome size in aquatic and wetland plants: fitting with the large genome constraint hypothesis with a few relevant exceptions. *Plant Syst. Evol.* **301**: 1927–1936.

Hidding, B., B. A. Nolet, M. R. van Eerden, M. Guillemain & M. Klaassen. 2009. Burial depth distribution of fennel pondweed tubers (*Potamogeton pectinatus*) in relation to foraging by Bewick's swans. *Aquat. Bot.* **90**: 321–327.

Hiebert, R. D., D. A. Wilcox & N. B. Pavlovic. 1986. Vegetation patterns in and among pannes (calcareous intradunal ponds) at the Indiana Dunes National Lakeshore, Indiana. *Am. Midl. Nat.* **116**: 276–281.

Hilaire, L. S. 2002. *Amerorchis rotundifolia (Banks ex Pursh) Hultén. Small Round-Leaved Orchis*. Conservation and Research Plan for New England, New England Conservation Program, Framingham, MA. 48 pp.

Hildebrand, T. 2012. Correlation between wetland vegetative and microbial community diversity of Bryce Canyon National Park's southern regions–Phase I. Colorado Plateau Cooperative Ecosystems Studies Unit; Agreement No.: H1200-09–0005; 30 December 2012. Bryce Canyon National Park & Southern Utah University, Bryce/Cedar City, UT. 48 pp.

Hill, E. J. 1898. *Potamogeton robbinsii*. *Bot. Gaz.* **25**: 195–196.

Hill, B. H. 1987. *Typha* productivity in a Texas pond: implications for energy and nutrient dynamics in freshwater wetlands. *Aquat. Bot.* **27**: 385–394.

Hill, S. R. 2003. Conservation Assessment for twining screwstem (Bartonia paniculata (Michx.) Muhl.). Technical Report 2003 (7). Illinois Natural History Survey, Center for Biodiversity, Champaign, IL. 32 pp.

Hill, S. R. 2006. Conservation assessment for the Kidneyleaf Mudplantain (Heteranthera reniformis Ruiz & Pavon). Technical Report 2006 (5). Illinois Natural History Survey, Center for Wildlife and Plant Ecology, Champaign, IL. 34 pp.

Hill, S. R. 2007. *Conservation Assessment for the Green Wood Orchid Platanthera clavellata (Michx.) Luer*. USDA Forest Service, Eastern Region (Region 9), Shawnee and Hoosier National Forests, Milwaukee, WI. 48 pp.

Hillman, W. S. 1961. The Lemnaceae, or duckweeds. *Bot. Rev.* **27**: 221–287.

Hinman, S. E. & J. S. Brewer. 2007. Responses of two frequently-burned wet pine savannas to an extended period without fire. *J. Torrey Bot. Soc.* **134**: 512–526.

Hiramatsu, M., K. Ii, H. Okubo, K. L. Huang & C. W. Huang. 2001. Biogeography and origin of *Lilium longiflorum* and *L. formosanum* (Liliaceae) endemic to the Ryukyu Archipelago and Taiwan as determined by allozyme diversity. *Am. J. Bot.* **88**: 1230–1239.

Hitchcock, C. L. 1944. The *Tofieldia glutinosa* complex of western North America. *Am. Midl. Nat.* **31**: 487–498.

Hitchin, G. G., I. Wile, G. E. Miller & N. D. Yan. 1984. Macrophyte data from 46 southern Ontario soft-water lakes of varying pH. Data Report DR 84/2. Water Resources Branch, Ontario Ministry of the Environment, Dorset, ON. 132 pp.

Ho, J.-C. 2011. Antimicrobial, mosquito larvicidal and antioxidant properties of the leaf and rhizome of *Hedychium coronarium*. *J. Chin. Chem. Soc.* **58**: 563–567.

Hoagland, B. W. & A. Buthod. 2007. Updated Oklahoma Ozark flora. *Oklahoma Nat. Plant Rec.* **7**: 54–66.

Hoagland, B. W. & A. K. Buthod. 2017. Vascular flora of the Deep Fork National Wildlife Refuge, Okmulgee County, Oklahoma. *Castanea* **82**: 32–45.

Hoagland, B. W., I. H. Butler & N. A. McCarty. 2001. *Identification and Assessment of Ecologically Significant Wetland Communities in North Central, Northwestern, and the Panhandle of Oklahoma: Final Report*. Oklahoma Biological Survey, University of Oklahoma, Norman, OK. 42 pp.

Hoang, P. T. N. & I. Schubert. 2017. Reconstruction of chromosome rearrangements between the two most ancestral duckweed species *Spirodela polyrhiza* and *S. intermedia*. *Chromosoma* **126**: 729–739.

Hochbach, A., H. P. Linder & M. Röser. 2018. Nuclear genes, *matK* and the phylogeny of the Poales. *Taxon* **67**: 521–536.

Hodgdon, A. R., P. Giguere, S. B. Krochmal & A. Riel. 1952. New *Potamogeton* records in New Hampshire. *Rhodora* **54**: 237–246.

Hoffman, C. E. 1940. Limnological relationships of some northern Michigan Donaciini (Chrysomelidae; Coleoptera). *Trans. Am. Microsc. Soc.* **59**: 259–274.

Hoffmann, M. A., U. Raeder & A. Melzer. 2014a. Influence of environmental conditions on the regenerative capacity and the survivability of *Elodea nuttallii* fragments. *J. Limnol.* **74**: 12–20.

Hoffmann, M. A., U. Raeder & A. Melzer. 2014b. Influence of the gender on growth and phenology of the dioecious macrophyte *Najas marina* ssp. *intermedia*. *Hydrobiologia* **727**: 167–176.

Hofmann, J. E., E. Gardner & M. J. Morris. 1990. Distribution, abundance, and habitat of the marsh rice rat (*Oryzomys palustris*) in southern Illinois. *Trans. Illinois State Acad. Sci.* **83**: 162–180.

Hofstra, D. E., K. D. Adam & J. S. Clayton. 1995. Isozyme variation in New Zealand populations of *Myriophyllum* and *Potamogeton* species. *Aquat. Bot.* **52**: 121–131.

Hogsden, K. L., E. P. S. Sager & T. C. Hutchinson. 2007. The impacts of the non-native macrophyte *Cabomba caroliniana* on littoral biota of Kasshabog Lake, Ontario. *J. Great Lakes Res.* **33**: 497–504.

Hohman, W. L. & C. D. Ankney. 1994. Body size and condition, age, plumage quality, and foods of prenesting male cinnamon teal in relation to pair status. *Can. J. Zool.* **72**: 2172–2176.

Hohman, W. L., D. W. Woolington & J. H. Devries. 1990. Food habits of wintering canvasbacks in Louisiana. *Can. J. Zool.* **68**: 2605–2609.

Hohman, W. L., T. M. Stark & J. L. Moore. 1996. Food availability and feeding preferences of breeding fulvous whistling-ducks in Louisiana ricefields. *Wilson Bull.* **108**: 137–150.

Holaday, A. S. & G. Bowes. 1980. C_4 acid metabolism and dark CO_2 fixation in a submersed aquatic macrophyte (*Hydrilla verticillata*). *Plant Physiol.* **65**: 331–335.

Holbrook, S. J., D. C. Reed, K. Hansen & C. A. Blanchette. 2000. Spatial and temporal patterns of predation on seeds of the surfgrass *Phyllospadix torreyi*. *Mar. Biol.* **136**: 739–747.

Holbrook, S. J., D. C. Reed & J. S. Bull. 2002. Survival experiments with outplanted seedlings of surfgrass (*Phyllospadix torreyi*) to enhance establishment on artificial structures. *ICES J. Mar. Sci.* **59**(suppl.): S350–S355.

Holder, G. L., M. K. Johnson & J. L. Baker. 1980. Cattle grazing and management of dusky seaside sparrow habitat. *Wildl. Soc. Bull.* **8**: 105–109.

Hollingsworth, P. M., C. D. Preston & R. J. Gornall. 1996a. Isozyme evidence for the parentage and multiple origins of *Potamogeton* × *suecicus* (*P. pectinatus* × *P. filiformis*, Potamogetonaceae). *Plant Syst. Evol.* **202**: 219–232.

Hollingsworth, P. M., C. D. Preston & R. J. Gornall. 1996b. Genetic variability in two hydrophilous species of *Potamogeton*, *P. pectinatus* and *P. filiformis* (Potamogetonaceae). *Plant Syst. Evol.* **202**: 233–254.

Hollingsworth, P. M., C. D. Preston & R. J. Gornall. 1998a. Euploid and aneuploid evolution in *Potamogeton* (Potamogetonaceae): a factual basis for interpretation. *Aquat. Bot.* **60**: 337–358.

Hollingsworth, P. M., C. D. Preston & R. J. Gornall. 1998b. Lack of detectable isozyme variability in British populations of *Potamogeton epihydrus* (Potamogetonaceae). *Aquat. Bot.* **60**: 433–437.

Holm, T. 1891. Contributions to the knowledge of the germination of some North American plants. *Mem. Torrey Bot. Club* **2**: 57–108.

Holm, T. 1904. The root-structure of North American terrestrial Orchideae. *Am. J. Sci.* **18**: 197–212.

Holmes, W. C. 1978. Range extension for *Ottelia alismoides* (L.) Pers. (Hydrocharitaceae). *Castanea* **43**: 193–194.

Holmes, W. C. 2002a. *Crinum* Linnaeus. Pp. 278–279 *In*: Flora North America Editorial Committee (eds.), *Flora of North America North of Mexico, Vol. 26: Magnoliophyta: Liliidae: Liliales and Orchidales*. Oxford University Press, New York, NY.

Holmes, W. C. 2002b. Smilacaceae Ventenat. Catbrier family. Pp. 468–478 *In*: Flora North America Editorial Committee (eds.), *Flora of North America North of Mexico, Vol. 26: Magnoliophyta: Liliidae: Liliales and Orchidales*. Oxford University Press, New York, NY.

Holmes, W. C., A. E. Rushing & J. R. Singhurst. 2005. Taxonomy and identification of *Isoetes* (Isoetaceae) in Texas based on megaspore features. *Lundellia* **2005**: 1–7.

Holsinger, K. E. & L. E. Wallace. 2004. Bayesian approaches for the analysis of population genetic structure: an example from *Platanthera leucophaea* (Orchidaceae). *Mol. Ecol.* **13**: 887–894.

Holt, C. R., G. W. Folkerts & D. R. Folkerts. 2011. A floristic study of a steephead stream in northwestern Florida. *Southeast. Nat.* **10**: 289–302.

Homoya, M. A. 1982. Additions to the flora of southern Indiana. *Proc. Indiana Acad. Sci.* **92**: 379–382.

Homoya, M. A. 1983. A floristic survey of acid seep springs in Martin and Dubois countries, Indiana. *Proc. Indiana Acad. Sci.* **93**: 323–332.

Homoya, M. A. & C. L. Hedge. 1982. The upland sinkhole swamps and ponds of Harrison County, Indiana. *Proc. Indiana Acad. Sci.* **92**: 383–388.

Homoya, M. A., D. B. Abrell, J. A. Aldrich & T. W. Post. 1984. The natural regions of Indiana. *Proc. Indiana Acad. Sci.* **94**: 245–268.

Hooper, F. F. 1951. Limnological features of a Minnesota seepage lake. *Am. Midl. Nat.* **46**: 462–481.

Hootsmans, M. J. M., J. E. Vermaat & W. Van Vierssen. 1987. Seed-bank development, germination and early seedling survival of two seagrass species from The Netherlands: *Zostera marina* L. and *Zostera noltii* Hornem. *Aquat. Bot.* **28**: 275–285.

Hoover, R. F. 1939. A revision of the genus *Brodiaea Am. Midl. Nat.* **22**: 551–574.

Hopfensperger, K. N. & A. H. Baldwin. 2009. Spatial and temporal dynamics of floating and drift-line seeds at a tidal freshwater marsh on the Potomac River, USA. *Plant Ecol.* **201**: 677–686.

Hopkins, C. E. O. 1969. Vegetation of fresh-water springs of Southern Illinois. *Castanea* **34**: 121–145.

Hopper, S. D., M. F. Fay, M. Rossetto & M. W. Chase. 1999. A molecular phylogenetic analysis of the bloodroot and kangaroo paw family, Haemodoraceae: taxonomic, biogeographic and conservation implications. *Bot. J. Linn. Soc.* **131**: 285–299.

Hopper, S. D., R. J. Smith, M. F. Fay, J. C. Manning & M. W. Chase. 2009. Molecular phylogenetics of Haemodoraceae in the Greater Cape and southwest Australian floristic regions. *Mol. Phylogenet. Evol.* **51**: 19–30.

Horn, C. N. 1983a. The annual growth cycle of *Heteranthera dubia* in Ohio. *Michigan Bot.* **23**: 29–34.

Horn, M. H. 1983b. Optimal diets in complex environments: feeding strategies of two herbivorous fishes from a temperate rocky intertidal zone. *Oecologia* **58**: 345–350.

Horn, C. N. 1988. Developmental heterophylly in the genus *Heteranthera* (Pontederiaceae). *Aquat. Bot.* **31**: 197–209.

Horn, C. N. 1998. Pontederiaceae. Pickerelweed family. *J. Arizona-Nevada Acad. Sci.* **30**: 133–136.

Horn, C. N. 2002. Pontederiaceae Kunth. Pp. 37–46 *In*: Flora North America Editorial Committee (eds.), *Flora of North America North of Mexico, Vol. 26: Magnoliophyta: Liliidae: Liliales and Orchidales*. Oxford University Press, New York, NY.

Horn, C. N. & E. McClintock. 2012. Pontederiaceae. Pp. 1498–1500 *In*: B. G. Baldwin, D. H. Goldman, D. J. Keil, R. Patterson, T. J. Rosatti & D. H. Wilken (eds.), *The Jepson Manual: Vascular Plants of California*, 2nd ed. University of California Press, Berkeley, CA.

Hornbeck, J. H., D. Reyher, C. H. Sieg & R. W. Crook. 2003. *Conservation Assessment for Southern Maidenhair Fern and Stream Orchid in the Black Hills National Forest South Dakota and Wyoming*. USDA Forest Service, Rocky Mountain Region, Custer, SD. 41 pp.

Hosack, D. 1811. *Hortus Elginensis: Or Catalogue of Plants Indigenous and Exotic, Cultivated in the Elgin Botanic Garden in the Vicinity of the City of New York, Established in 1801*. T. & J. Swords, New York, NY. 65 pp.

Hossain, M. M. 2011. Therapeutic orchids: traditional uses and recent advances—an overview. *Fitoterapia* **82**: 102–140.

Hossain, M. L., S. Sarker & N. J. Sarker. 2010. Food habits and feeding behaviour of Bengal eyed turtle *Morenia petersi* (Anderson: 1879) in Bangladesh. *Bangladesh J. Zool.* **38**: 213–222.

Hossell, J. C. & J. H. Baker. 1979. Estimation of the growth rates of epiphytic bacteria and *Lemna minor* in a river. *Freshw. Biol.* **9**: 319–327.

Hotchkiss, N. & R. E. Stewart. 1947. Vegetation of the patuxent research refuge, Maryland. *Am. Midl. Nat.* **38**: 1–75.

Hothem, R. L. & H. M. Ohlendorf. 1989. Contaminants in foods of aquatic birds at Kesterson Reservoir, California, 1985. *Arch. Environ. Contam. Toxicol.* **18**: 773–786.

Houle, G. & D. L. Phillips. 1988. The soil seed bank of granite outcrop plant communities. *Oikos* **52**: 87–93.

Houle, G. & D. L. Phillips. 1989a. Seasonal variation and annual fluctuation in granite outcrop plant communities. *Vegetatio* **80**: 25–35.

Houle, G. & D. L. Phillips. 1989b. Seed availability and biotic interactions in granite outcrop plant communities. *Ecology* **70**: 1307–1316.

Houle, G. & A. Delwaide. 1991. Population structure and growth-stress relationship of *Pinus taeda* in rock outcrop habitats. *J. Veg. Sci.* **2**: 47–58.

Howard, R. J. & I. A. Mendelssohn. 1995. Effect of increased water depth on growth of a common perennial freshwater-intermediate marsh species in coastal Louisiana. *Wetlands* **15**: 82–91.

Howard, F. W. & S. Halbert. 2005. Flatid planthopper, *Oormenaria rufifascia* (Walker) (Insecta: Hemiptera: Auchenorrhyncha: Flatidae). Document EENY351. IFAS Extension, University of Florida, Gainesville, FL. 5 pp.

Howard, R. J. & C. J. Wells. 2009. Plant community establishment following drawdown of a reservoir in southern Arkansas, USA. *Wetl. Ecol. Manag.* **17**: 565–583.

Howe, H. F. & J. Smallwood. 1982. Ecology of seed dispersal. *Ann. Rev. Ecol. Syst.* **13**: 201–228.

Howell, L. N. 2012. Ontogenetic shifts in diet and habitat by juvenile green sea turtles (*Chelonia mydas*) along the middle and lower Texas coast. M.S. thesis. Texas A&M University, Galveston, TX. 87 pp.

Howell, N., A. Krings & R. R. Braham. 2016. Guide to the littoral zone vascular flora of Carolina bay lakes (U.S.A.). *Biodivers. Data J.* **2016**(4). doi: 10.3897/BDJ.4.e7964.

Howells, R. G., L. E. Burlakova, A. Y. Karatayev, R. K. Marfurt & R. L. Burks. 2006. Native and introduced Ampullariidae in North America: history, status, and ecology. Pp. 73–112 *In*: R. C. Joshi & L. S. Sebastian (eds.), *Global Advances in the Ecology and Management of Golden Apple Snails*. Philippine Rice Research Institute, Muñoz, Nueva Ecija, Philippines. 588 pp.

Howie, S. A. & I. van Meerveld. 2016. Classification of vegetative lagg types and hydrogeomorphic lagg forms in bogs of coastal British Columbia, Canada. *Can. Geogr.* **60**: 123–134.

Hoyer, M. V., D. E. Canfield, Jr., C. A. Horsburgh & K. Brown. 1996. *Florida Freshwater Plants: A Handbook of Common Aquatic Plants in Florida Lakes*. University of Florida, Institute of Food & Agricultural Sciences, Gainesville, FL. 264 pp.

Hroudová, Z. & P. Zákravský. 1993. Ecology of two cytotypes of *Butomus umbellatus* II. Reproduction, growth and biomass production. *Folia Geobot.* **28**: 413–424.

Hroudová, Z. & P. Zákravský. 2003. Germination responses of diploid *Butomus umbellatus* to light, temperature and flooding. *Funct. Ecol. Plants* **198**: 37–44.

Hroudová, Z., P. Zákravský & O. Čechurová. 2004. Germination of seed of *Alisma gramineum* and its distribution in the Czech Republic. *Preslia* **76**: 97–118.

Hsiao, Y.-Y., Z.-J. Pan, C.-C. Hsu, Y.-P. Yang, Y.-C. Hsu, Y.-C. Chuang, H.-H. Shih, W.-H. Chen, W.-C. Tsai & H.-H. Chen. 2011. Research on orchid biology and biotechnology. *Plant Cell Physiol.* **52**: 1467–1486.

Hua, K.-F., H.-Y. Hsu, Y.-C. Su, I.-F. Lin, S.-S. Yang, Y.-M. Chen & L. K. Chao. 2006. Study on the antiinflammatory activity of methanol extract from seagrass *Zostera japonica*. *J. Agric. Food Chem.* **54**: 306–311.

Huang, S.-Q. 2003. Flower dimorphism and the maintenance of andromonoecy in *Sagittaria guyanensis* ssp. *lappula* (Alismataceae). *New Phytol.* **15**: 357–364.

Huang, S.-Q. & X.-X. Tang. 2008. Discovery of gynoecium color polymorphism in an aquatic plant. *J. Integr. Plant Biol.* **50**: 1178–1182.

Huang, S.-Q., N. Song, Q. Wang, L.-L. Tang & X.-F. Wang. 2000. Sex expression and the evolutionary advantages of male flowers in an andromonoecious species, *Sagittaria guyanensis* subsp. *lappula* (Alismataceae). *Acta Bot. Sin.* **36**: 310–316.

Huang, S.-Q., Y.-H. Guo, G. W. Robert, Y.-H. Shi & K. Sun. 2001. Mechanism of underwater pollination in *Najas marina* (Najadaceae). *Aquat. Bot.* **70**: 67–78.

Huang, S., X.-L. Zhou, C.-J. Wang, Y.-S. Wang, F. Xiao, L.-H. Shan, Z.-Y. Guo & J. Weng. 2013. Pyrrolizidine alkaloids from *Liparis nervosa* with inhibitory activities against LPS-induced NO production in RAW264. 7 macrophages. *Phytochemistry* **93**: 154–161.

Hubbard, J. R. & W. S. Judd. 2013. Floristics of Silver River State Park, Marion County, Florida. *Rhodora* **115**: 250–280.

Hubert, B. & J. Maton. 1939. Parthenocarpie en Groeistof. *Natuurw. Tijdschr. Antwerp.* **21**: 339–348.

Hudon, C. 1997. Impact of water level fluctuations on St. Lawrence River aquatic vegetation. *Can. J. Fish. Aquat. Sci.* **54**: 2853–2865.

Hudon, C. 2004. Shift in wetland plant composition and biomass following low-level episodes in the St. Lawrence River: looking into the future. *Can. J. Fish. Aquat. Sci.* **61**: 603–617.

Huffman, R. T. & R. I. Lonard. 1983. Successional patterns on floating vegetation mats in a southwestern Arkansas bald cypress swamp. *Castanea* **48**: 73–78.

Huffman, R. T. & G. E. Tucker. 1984. Preliminary guide to the onsite identification and delineation of the wetlands of Alaska. Technical Report Y-78-9. U.S. Army Engineer Waterways Experiment Station, Vicksburg, MS. 88 pp.

Huffman, J. M. & W. S. Judd. 1998. Vascular flora of Myakka River State Park, Sarasota and Manatee Counties, Florida. *Castanea* **63**: 25–50.

Hughes, J. H. & E. L. Young. 1982. Autumn foods of dabbling ducks in southeastern Alaska. *J. Wildl. Manag.* **46**: 259–263.

Hughes, A. R. & J. J. Stachowicz. 2009. Ecological impacts of genotypic diversity in the clonal seagrass *Zostera marina*. *Ecology* **90**: 1412–1419.

Hughes, B. A. & M. E. Kane. 2018. Seed cryopreservation of selected Florida native orchid species. *Seed Sci. Technol.* **46**: 431–446.

Huiskes, A. H. L., B. P. Koutstaal, P. M. J. Herman, W. G. Beeftink, M. M. Markusse & W. De Munck. 1995. Seed dispersal of halophytes in tidal salt marshes. *J. Ecol.* **83**: 559–567.

Humaida, N., K. Krisdianto & S. B. Peran. 2016. Estimation of carbon storage in water lettuce (*Pistia stratiotes*) at freshwater swamps. *Trop. Wetland J.* **2**: 38–46.

Humbert, L., D. Gagnon, D. Kneeshaw & C. Messier. 2007. A shade tolerance index for common understory species of northeastern North America. *Ecol. Indicators* **7**: 195–207.

Humm, H. J. 1956. Sea grasses of the northern Gulf Coast. *Bull. Mar. Sci.* **6**: 305–308.

Humphreys, E. R., P. M. Lafleur, L. B. Flanagan, N. Hedstrom, K. H. Syed, A. J. Glenn & R. Granger. 2006. Summer carbon dioxide and water vapor fluxes across a range of northern peatlands. *J. Geophys. Res. Biogeosci.* **111**: G04011. doi: 10.1029/2005JG000111.

Humphreys, E. R., C. Charron, M. Brown & R. Jones. 2014. Two bogs in the Canadian Hudson Bay lowlands and a temperate bog reveal similar annual net ecosystem exchange of CO. *Arct. Antarct. Alp. Res.* **46**: 103–113.

Huneycutt, M. B. & M. D. Floyd. 1999. *Ponthieva racemosa* (Orchidaceae) in Trace State Park, Pontotoc County, Mississippi. *J. Mississippi Acad. Sci.* **44**: 229–230.

Hunt, K. W. 1943. Floating mats on a southeastern coastal plain reservoir. *Bull. Torrey Bot. Club* **70**: 481–488.

Hunt, J. 1992. Feeding ecology of valley pocket gophers (*Thomomys bottae sanctidiegi*) on a California coastal grassland. *Am. Midl. Nat.* **127**: 41–51.

Hunt, D. M., K. B. Searcy, R. E. Zaremba & C. R. Lombardi. 1995. The vascular plants of Fort Devens, Massachusetts. *Rhodora* **97**: 208–244.

Hunter, Jr., M. L., J. J. Jones & J. W. Witham. 1986. Biomass and species richness of aquatic macrophytes in four Maine (U.S.A.) lakes of different acidity. *Aquat. Bot.* **24**: 91–95.

Huotari, T. & H. Korpelainen. 2012. Complete chloroplast genome sequence of *Elodea canadensis* and comparative analyses with other monocot plastid genomes. *Gene* **508**: 96–105.

Huotari, T., H. Korpelainen, E. Leskinen & K. Kostamo. 2011. Population genetics of the invasive water weed *Elodea canadensis* in Finnish waterways. *Plant Syst. Evol.* **294**: 27–37.

Husband, B. C. & M. Hickman. 1989. The frequency and local abundance of *Ruppia occidentalis* in relation to sediment texture and lake salinity. *Can. J. Bot.* **67**: 2444–2449.

Husband, B. C. & S. C. H. Barrett. 1992. Pollinator visitation in populations of tristylous *Eichhornia paniculata* in northeastern Brazil. *Oecologia* **89**: 365–371.

Husband, B. C. & S. C. H. Barrett. 1993. Multiple origins of self-fertilization in tristylous *Eichhornia paniculata* (Pontederiaceae): inferences from style morph and isozyme variation. *J. Evol. Biol.* **6**: 591–608.

Husband, B. C. & S. C. H. Barrett. 1995. Estimates of gene flow in *Eichhornia paniculata* (Pontederiaceae): effects of range substructure. *Heredity* **75**: 549–560.

Husband, B. C. & S. C. H. Barrett. 1998. Population genetic structure in the tropical plant *Eichhornia paniculata*: implications for gene flow in ephemeral habitats. *Am. J. Bot.* **85**: 60–61.

Hussner, A. 2012. Alien aquatic plant species in European countries. *Weed Res.* **52**: 297–306.

Hussner, A., H. P. Hoelken & P. Jahns. 2010a. Low light acclimated submerged freshwater plants show a pronounced sensitivity to increasing irradiances. *Aquat. Bot.* **93**: 17–24.

Hussner, A., K. van de Weyer, E. M. Gross & S. Hilt. 2010b. Comments on increasing number and abundance of non-indigenous aquatic macrophyte species in Germany. *Weed Res.* **50**: 519–526.

Hussner, A., P. Heidbuechel & S. Heiligtag. 2014. Vegetative overwintering and viable seed production explain the establishment of invasive *Pistia stratiotes* in the thermally abnormal Erft River (North Rhine-Westphalia, Germany). *Aquat. Bot.* **119**: 28–32.

Hussner, A., T. Mettler-Altmann, A. P. M. Weber & K. Sand-Jensen. 2016. Acclimation of photosynthesis to supersaturated CO_2 in aquatic plant bicarbonate users. *Freshw. Biol.* **61**: 1720–1732.

Hussey, J. S. 1974. Some useful plants of early New England. *Econ. Bot.* **28**: 311–337.

Hutchings, M. J. & P. J. Russell. 1989. The seed regeneration dynamics of an emergent salt marsh. *J. Ecol.* **77**: 615–637.

Hutton, E. E. 1974. A large vegetational formation of Cretaceous and Tertiary origin in West Virginia. *Castanea* **39**: 71–76.

Hutton, E. E. & R. B. Clarkson. 1961. Two plants new for North America and some new or otherwise interesting plants in West Virginia. *Castanea* **26**: 84–88.

Iida, S., K. Kosuge & Y. Kadono. 2004. Molecular phylogeny of Japanese *Potamogeton* species in light of noncoding chloroplast sequences. *Aquat. Bot.* **80**: 115–127.

Iida, S., A. Miyagi, S. Aoki, M. Ito, Y. Kadono & K. Kosuge. 2009. Molecular adaptation of *rbcL* in the heterophyllous aquatic plant *Potamogeton*. *PLoS One*, 4(2): e4633.

Ikusima, I. & J. G. Gentil. 1993. Vegetative growth and productivity of *Eichhornia azurea* with special emphasis on leaf dynamics. *Ecol. Res.* **8**: 287–295.

Ileperuma, V., S. Udage, D. Yakandawala, L. Jayasinghe, S. Kumar & A. Ratnatilleke. 2015. Does ingestion of plants from a phenetic group of *Monochoria* ('Diyahabarala') cause hepatotoxicity? *Ceylon Med. J.* **60**: 28–30.

Iler, A. M. & D. W. Inouye. 2013. Effects of climate change on mast-flowering cues in a clonal montane herb, *Veratrum tenuipetalum* (Melanthiaceae). *Am. J. Bot.* **100**: 519–525.

Iles, W. J. D., S. Y. Smith & S. W. Graham. 2013. A well-supported phylogenetic framework for the monocot order Alismatales reveals multiple losses of the plastid NADH dehydrogenase complex and a strong long-branch effect. Pp. 1–28 *In*: P. Wilkin & S. J. Mayo (eds.), *Early Events in Monocot Evolution*. Cambridge University Press, Cambridge, United Kingdom.

Illyés, Z., S. Rudnóy & Z. Bratek. 2005. Aspects of *in situ*, *in vitro* germination and mycorrhizal partners of *Liparis loeselii*. *Acta Biol. Szegediensis* **49**: 137–139.

Iltis, H. H. 1950. Studies in Virginia plants I. List of bryophytes from the vicinity of Fredericksburg, Virginia. *Castanea* **15**: 38–50.

Imaizumi, T., G. X. Wang, T. Ohsako & T. Tominaga. 2008a. Genetic diversity of sulfonylurea-resistant and-susceptible *Monochoria vaginalis* populations in Japan. *Weed Res.* **48**: 187–196.

Imaizumi, T., G.-X. Wang & T. Tominaga. 2008b. Pollination of chasmogamous flowers and the effects of light and emergence time on chasmogamy and cleistogamy in *Monochoria vaginalis*. *Weed Biol. Manag.* **8**: 260–266.

Inácio, C. D., O. Chauveau, T. T. Souza-Chies, H. Sauquet & L. Eggers. 2017. An updated phylogeny and infrageneric classification of the genus *Sisyrinchium* (Iridaceae): challenges of molecular and morphological evidence. *Taxon* **66**: 1317–1348.

Incardona, J. P., W. Gaffield, R. P. Kapur & H. Roelink. 1998. The teratogenic *Veratrum* alkaloid cyclopamine inhibits sonic hedgehog signal transduction. *Development* **125**: 3553–3562.

Inglis, G. J. 1999. Variation in the recruitment behaviour of seagrass seeds: implications for population dynamics and resource management. *Pac. Conserv. Biol.* **5**: 251–259.

Inoue, K. I. 1983. Systematics of the genus *Platanthera* (Orchidaceae) in Japan and adjacent regions with special reference to pollination. *J. Fac. Sci. Univ. Tokyo, Sect. 3, Bot.* **13**: 285–374.

Inoue, K., T. Kato, H. Kunitake & T. Yabuya. 2004. Efficient production of polyploid plants via protoplast culture of *Iris fulva*. *Cytologia* **69**: 327–333.

Inserra, R. N., J. D. Stanley, J. H. O'Bannon & R. P. Esser. 2005. Nematode quarantine and certification programmes implemented in Florida. *Nematol. Medit.* **33**: 113–123.

In't Veen, J. 1981. *Cryptocorynen. Wissenswertes über die Pflege asiatischer Wasserkelche im Aquarium*. Lehrmeister Bücherei Nr. 68. Albrecht Philler Verlag, Minden. 136 pp.

Irwin, R. E., D. Cook, L. L. Richardson, J. S. Manson & D. R. Gardner. 2014. Secondary compounds in floral rewards of toxic rangeland plants: impacts on pollinators. *J. Agric. Food Chem.* **62**: 7335–7344.

Ismail, B. S. & K. Phaik-Hong. 2004. A study of weed populations and their buried seeds in the soil of MARDI research station and at farmers' rice fields in Sungai Burung, Tanjung Karang, Selangor, Malaysia. *Pertanika J. Trop. Agric. Sci.* **27**: 113–120.

Ismail Sahid, Z. & N. K. Ho. 1995. Weed populations and their buried seeds in rice fields of the Muda area, Kedah, Malaysia. *Pertanika J. Trop. Agric. Sci.* **18**: 21–28.

Ito, K. 1999. Isolation of two distinct cold-inducible cDNAs encoding plant uncoupling proteins from the spadix of skunk cabbage (*Symplocarpus foetidus*). *Plant Sci.* **149**: 167–173.

Ito, Y. & N. Tanaka. 2011. Hybridisation in a tropical seagrass genus, *Halodule* (Cymodoceaceae), inferred from plastid and nuclear DNA phylogenies. *Telopea* **13**: 219–231.

Ito, Y., T. Ohi-Toma, J. Murata & N. Tanaka. 2010. Hybridization and polyploidy of an aquatic plant, *Ruppia* (Ruppiaceae), inferred from plastid and nuclear DNA phylogenies. *Am. J. Bot.* **97**: 1156–1167.

Ito, Y., T. Ohi-Toma, J. Murata & N. Tanaka. 2013. Comprehensive phylogenetic analyses of the *Ruppia maritima* complex focusing on taxa from the Mediterranean. *J. Plant Res.* **126**: 753–762.

Ito, Y., T. Ohi-Toma, N. Tanaka, J. Murata & A. M. Muasya. 2015. Phylogeny of *Ruppia* (Ruppiaceae) revisited: molecular and morphological evidence for a new species from Western Cape, South Africa. *Syst. Bot.* **40**: 942–949.

Ito, Y., N. Tanaka, P. García-Murillo & A. M. Muasya. 2016a. A new delimitation of the Afro-Eurasian plant genus *Althenia* to include its Australasian relative, *Lepilaena* (Potamogetonaceae)–evidence from DNA and morphological data. *Mol. Phylogenet. Evol.* **98**: 261–270.

Ito, Y., N. Tanaka, C. Kim, R. B. Kaul & D. C. Albach. 2016b. Phylogeny of *Sparganium* (Typhaceae) revisited: nonmonophyletic nature of S. *emersum* sensu lato and resurrection of S. *acaule*. *Plant Syst. Evol.* **302**: 129–135.

Ito, Y., N. Tanaka, S. W. Gale, O. Yano & J. Li. 2017a. Phylogeny of *Najas* (Hydrocharitaceae) revisited: implications for systematics and evolution. *Taxon* **66**: 309–323.

Ito, Y., T. Ohi-Toma, C. Nepi, A. Santangelo, A. Stinca, N. Tanaka & J. Murata. 2017b. Towards a better understanding of the *Ruppia maritima* complex (Ruppiaceae): notes on the correct application and typification of the names R. *cirrhosa* and R. *Spiralis*. *Taxon* **66**: 167–171.

Itokawa, H., H. Morita, I. Katou, K. Takeya, A. J. Cavalheiro, R. C. B. de Oliveira, M. Ishige & M. Motidome. 1988. Cytotoxic diterpenes from the rhizomes of *Hedychium coronarium*. *Plant Med.* **54**: 311–315.

Ivancic, A. & V. Lebot. 2000. *The Genetics and Breeding of Taro*. CIRAD, Montpellier, France. 194 pp.

Ivancic, A., V. Lebot, O. Roupsard, J. Q. Garcia & T. Okpul. 2004. Thermogenic flowering of taro (*Colocasia esculenta*, Araceae). *Can. J. Bot.* **82**: 1557–1565.

Iverson, S. J., J. E. McDonald, Jr. & L. K. Smith. 2001. Changes in the diet of free-ranging black bears in years of contrasting food availability revealed through milk fatty acids. *Can. J. Zool.* **79**: 2268–2279.

Izzati, M. 2016. Salt tolerance of several aquatic plants. *Proc. Int. Conf. Glob. Resour. Conserv.* **6**: 154–157.

Jackson, L. E. & L. C. Bliss. 1982. Distribution of ephemeral herbaceous plants near treeline in the Sierra Nevada, California, USA. *Arct. Alp. Res.* **14**: 33–43.

Jackson, S. T. & D. F. Charles. 1988. Aquatic macrophytes in Adirondack (New York) lakes: patterns of species composition in relation to environment. *Can. J. Bot.* **66**: 1449–1460.

Jacobson, A. & M. Hedrén. 2007. Phylogenetic relationships in *Alisma* (Alismataceae) based on RAPDs, and sequence data from ITS and *trnL*. *Pl. Syst. Evol.* **265**: 27–44.

Jacobs, D. L. 1947. An ecological life-history of *Spirodela polyrhiza* (Greater Duckweed) with emphasis on the turion phase. *Ecol. Monogr.* **17**: 437–469.

Jacobs, S. W. L. & D. H. Les. 2009. New combinations in *Zostera* (Zosteraceae). *Telopea* **12**: 419–423.

Jacobsen, N. 1982. *Cryptocorynen*. Alfred Kernen Verlag, Stuttgart. 112 pp.

Jacobson, M. 1958. *Insecticides from Plants*. Agriculture Handbook No. 154. U.S. Government Printing Office, Washington, DC. 299 pp.

Jacobson, G. L., H. Almquist-Jacobson & J. C. Winne. 1991. Conservation of rare plant habitat: insights from the recent history of vegetation and fire at Crystal Fen, northern Maine, USA. *Biol. Conserv.* **57**: 287–314.

Jacono, C. C. 2001. *Scleria lacustris* (Cyperaceae), an aquatic and wetland sedge introduced to Florida. *Sida* **19**: 1163–1170.

Jacono, C. C. 2002. *Cryptocoryne beckettii* complex (Araceae) introduced at a Florida spring. *Sida* **20**: 819–832.

Jagathesan, G. & P. Rajiv. 2018. Biosynthesis and characterization of iron oxide nanoparticles using *Eichhornia crassipes* leaf extract and assessing their antibacterial activity. *Biocatal. Agric. Biotechnol.* **13**: 90–94.

Jäger, E. 1964. Zur Deutung des Arealbildes von *Wolffia arrhiza* (L.) Wimm. *Ber. Deutsch. Bot. Ges.* **77**: 101–111.

Jain, S. K., M. R. Jacob, L. A. Walker & B. L. Tekwani. 2016. Screening North American plant extracts in vitro against *Trypanosoma brucei*, the causative agent for Human African Trypanosomiasis. *BMC Complement. Altern. Med.* **16**: 131.

James, L. F. & W. Binns. 1966. Effects of feeding wild onions (*Allium validum*) to bred ewes. *J. Am. Veterin. Med. Assoc.* **149**: 512–514.

Janousek, C. N. & C. L. Folger. 2013. Inter-specific variation in salinity effects on germination in Pacific Northwest tidal wetland plants. *Aquat. Bot.* **111**: 104–111.

Jansen, M. A. K., R. E. van den Noort, S. J. Boeke, S. A. M. Huggers & J. H. de Haan. 1999. Differences in UV-B tolerance among *Spirodela punctata* ecotypes. *J. Photochem. Photobiol. B* **48**: 194–199.

Jarvis, J. C. & K. A. Moore. 2008. Influence of environmental factors on *Vallisneria americana* seed germination. *Aquat. Bot.* **88**: 283–294.

Jarvis, J. C. & K. A. Moore. 2010. The role of seedlings and seed bank viability in the recovery of Chesapeake Bay, USA, *Zostera marina* populations following a large-scale decline. *Hydrobiologia* **649**: 55–68.

Jauzein, P. 1991. *Eclipta prostrata* (L.) L. a weed of rice fields in the Camargue. *Monde Plantes* **86**: 15–16.

Jayanthi, P., P. Lalitha, R. Sujitha & A. Thamaraiselvi. 2013. Anti-inflammatory activity of the various solvent extracts of *Eichhornia crassipes* (Mart.) Solms. *Int. J. Pharmtech Res.* **5**: 641–644.

Jeanmonod, D., A. Schlüssel & J. Gamisans. 2011. Status and trends in the alien flora of Corsica. *EPPO Bull.* **41**: 85–99.

Jeglum, J. K. 1971. Plant indicators of pH and water level in peatlands at Candle Lake, Saskatchewan. *Can. J. Bot.* **49**: 1661–1676.

Jekyll, G. 1903. *Lilies for English Gardens: A Guide for Amateurs*, 2nd ed. Country Life, Covent Garden, England. 72 pp.

Jelinski, D. E. 1989. Seasonal differences in habitat use and fat reserves in an arctic muskrat population. *Can. J. Zool.* **67**: 305–313.

Jennings, O. E. 1919. *Potamogeton vaseyi* in northeastern Ohio. *Ohio J. Sci.* **19**: 343.

Jensen, A. S. 2000. Eight new species of *Macrosiphum* Passerini (Hemiptera: Aphididae) from western North America, with notes on four other poorly known species. *Proc. Entomol. Soc. Washington* **102**: 427–472.

Jensen, K. 2004. Dormancy patterns, germination ecology, and seed-bank types of twenty temperate fen grassland species. *Wetlands* **24**: 152–166.

Jersáková, J., S. D. Johnson & P. Kindlmann. 2006. Mechanisms and evolution of deceptive pollination in orchids. *Biol. Rev.* **81**: 219–235.

Jervis, R. A. 1969. Primary production in the freshwater marsh ecosystem of Troy Meadows, New Jersey. *Bull. Torrey Bot. Club* **96**: 209–231.

Jesson, L. K. & S. C. H. Barrett. 2002. Enantiostyly: solving the puzzle of mirror-image flowers. *Nature* **417**: 707.

Jesson, L. K., J. Kang, S. L. Wagner, S. C. H. Barrett & N. G. Dengler. 2003a. The development of enantiostyly. *Am. J. Bot.* **90**: 183–195.

Jesson, L. K., S. C. H. Barrett & T. Day. 2003b. A theoretical investigation of the evolution and maintenance of mirror-image flowers. *Am. Nat.* **161**: 916–930.

Jewett-Smith, J. & C. McMillan. 1990. Germination and seedling development of *Halophila engelmannii* Aschers. (Hydrocharitaceae) under axenic conditions. *Aquat. Bot.* **36**: 167–177.

Jewett-Smith, J., C. McMillan, W. J. Kenworthy & K. Bird. 1997. Flowering and genetic banding patterns of *Halophila johnsonii* and conspecifics. *Aquat. Bot.* **59**: 323–331.

Jha, K. K. & S. Chaudhary. 2011. Resource production and consumption system: focus on wetland biodiversity of Uttar Pradesh. Pp. 9–23 *In*: *National Conference on Forest Biodiversity: Earth's Living Treasure, 22 May, 2011*. Uttar Pradesh Biodiversity Board, Lucknow, India.

Ji, Y.-B., N. Chen, H.-W. Zhu, N. Ling, W.-L. Li, D.-X. Song, S.-Y. Gao, W.-C. Zhang & N.-N. Ma. 2014. Alkaloids from beach spider lily (*Hymenocallis littoralis*) induce apoptosis of HepG-2 cells by the fas-signaling pathway. *Asian Pac. J. Cancer Prev.* **15**: 9319–9325.

Jiang, M. & Y. Kadono. 2001a. Seasonal growth and reproductive ecology of two threatened aquatic macrophytes, *Blyxa aubertii* and *B. echinosperma* (Hydrocharitaceae), in irrigation ponds of south-western Japan. *Ecol. Res.* **16**: 249–256.

Jiang, M. & Y. Kadono. 2001b. Growth and reproductive characteristics of an aquatic macrophyte *Ottelia alismoides* (L.) Pers. (Hydrocharitaceae). *Ecol. Res.* **16**: 687–695.

Jiang, K., H. Gao, N.-N. Xu, E. P. K. Tsang & X.-Y. Chen. 2011. A set of microsatellite primers for *Zostera japonica* (Zosteraceae). *Am. J. Bot.* **98**: e236–e238.

Jiang, X., X. Song, Y. Chen & W. Zhang. 2014. Research on biogas production potential of aquatic plants. *Renew. Energy* **69**: 97–102.

Jiang, Y.-L., Z. Huang, J.-Q. Liao, H.-X. Song, X.-M. Luo, S.-P. Gao, T. Lei, M.-Y. Jiang, Y. Jia, Q.-B. Chen, X.-F. Yu & Y.-H. Zhou. 2018. Phylogenetic analysis of *Iris* L. from China on [sic!] chloroplast *trnL-F* sequences. *Biologia* **73**: 459–466.

Jin, H.-J., J.-H. Kim, C. H. Sohn, R. E. DeWreede, T.-J. Choi, G. H. N. Towers, J. B. Hudson & Y.-K. Hong. 1997. Inhibition of Taq DNA polymerase by seaweed extracts from British Columbia, Canada and Korea. *J. Appl. Phycol.* **9**: 383–388.

Jin, X.-H., Z.-X. Ren, S.-Z. Xu, H. Wang, D.-Z. Li & Z.-Y. Li. 2014. The evolution of floral deception in *Epipactis veratrifolia* (Orchidaceae): from indirect defense to pollination. *BMC Plant Biol.* **14**(1): 63.

Joanen, T. & L. McNease. 1975. Notes on the reproductive biology and captive propagation of the American alligator. *Proc. Southeast Assoc. Game Fish Comm.* **29**: 407–414.

Johansen, B. & H. Rasmussen. 1992. *Ex situ* conservation of orchids. *Opera Bot.* **113**: 43–48.

Johansson, R., W. Avery, K. B. Hennenfent & J. J. Pacowta. 2009. Restoration of seagrass habitat in Tampa Bay using large manatee grass (*Syringodium filiforme*) sod units and a discussion of planting site sediment elevation dynamics. *Reports* **52**. University of South Florida, Tampa, FL. accessed 4 February, 2020. Available online: http://scholarcommons.usf.edu/basgp_report/52.

Johnson, L. N. 1892. Notes on the flora of southwestern Connecticut. *Bull. Torrey Bot. Club* **19**: 88–91.

Johnson, K. R. 1941. Vegetation of some mountain lakes and shores in northwestern Colorado. *Ecology* **22**: 306–316.

Johnson, T. 1999. *CRC Ethnobotany Desk Reference*. CRC Press, Boca Raton, FL. 1211 pp.

Johnson, S. D. 2018. Reproductive biology of *Canna* species naturalized in southern Africa. M.S. thesis. University of KwaZulu-Natal, Pietermaritzburg, South Africa. 84 pp.

Johnson, W. E. & A. R. Martin. 1969. Nonteratogenicity of *Veratrum californicum* in rabbits. *J. Pharm. Sci.* **58**: 1165–1166.

Johnson, A. S. & J. L. Landers. 1978. Fruit production in slash pine plantations in Georgia. *J. Wildl. Manag.* **42**: 606–613.

Johnson, W. P. & F. C. Rohwer. 2000. Foraging behavior of green-winged teal and mallards on tidal mudflats in Louisiana. *Wetlands* **20**: 184–188.

Johnson, J. B. & D. A. Steingraeber. 2003. The vegetation and ecological gradients of calcareous mires in the South Park valley, Colorado. *Can. J. Bot.* **81**: 201–219.

Johnson, R. K. & M. L. Ostrofsky. 2004. Effects of sediment nutrients and depth on small-scale spatial heterogeneity of submersed macrophyte communities in Lake Pleasant, Pennsylvania. *Can. J. Fish. Aquat. Sci.* **61**: 1493–1502.

Johnson, A. & S. Johnson. 2006. *Primefact 359: Garden Plants Poisonous to People.* New South Wales Department of Primary Industries, Orange, NSW, Australia. 12 pp.

Johnson, E. A. & K. S. Walz. 2013. *Integrated Management Guidelines for Four Habitats and Associated State Endangered Plants and Wildlife Species of Greatest Conservation Need in the Skylands and Pinelands Landscape Conservation Zones of the New Jersey State Wildlife Action Plan.* Center for Biodiversity and Conservation and New Jersey Department of Environmental Protection, Natural Heritage Program, for NatureServe, Arlington, VA. 140 pp.

Johnson, E. A., T. F. Breden & R. J. Cartica. 1998. *Monitoring of Helonias bullata Populations in Camden and Gloucester Counties, New Jersey.* New Jersey Department of Environmental Protection, Division of Parks and Forestry, Office of Natural Lands Management, Trenton, NJ. 71 pp.

Johnson, M. K., K. E. Alexander, N. Lindquist & G. Loo. 1999. A phenolic antioxidant from the freshwater orchid, *Habenaria repens. Comp. Biochem. Physiol., C, Toxicol. Pharmacol. Endocrinol.* **122**: 211–214.

Johnson, T., J. M. Cruse-Sanders & G. S. Pullman. 2012. Micropropagation and seed cryopreservation of the critically endangered species Tennessee yellow-eye grass, *Xyris tennesseensis* Kral. *In Vitro Cell. Dev. Biol. Plant.* **48**: 369–376.

Johnson-Green, P. C., N. C. Kenkel & T. Booth. 1995. The distribution and phenology of arbuscular mycorrhizae along an inland salinity gradient. *Can. J. Bot.* **73**: 1318–1327.

Johnston, C. A., B. L. Bedford, M. Bourdaghs, T. Brown, C. Frieswyk, M. Tulbure, L. Vaccaro & J. B. Zedler. 2007. Plant species indicators of physical environment in Great Lakes coastal wetlands. *J. Great Lakes Res.* **33**: 106–124.

Johnston, R. F. 1956a. Predation by short-eared owls on a *Salicornia* salt marsh. *Wilson Bull.* **68**: 91–102.

Johnston, R. F. 1956b. Population structure in salt marsh song sparrows: part I. Environment and annual cycle. *Condor* **58**: 24–44.

Johri, B. M. (ed.). 1984. *Embryology of Angiosperms.* Springer-Verlag, NY. 830 pp.

Jomjun, N., T. Siripen, S. Maliwan, N. Jintapat, T. Prasak, C. Somporn & P. Petch. 2010. Phytoremediation of arsenic in submerged soil by wetland plants. *Int. J. Phytoremed.* **13**: 35–46.

Jones, P. S. 1998. Aspects of the population biology of *Liparis loeselii* (L.) Rich. var. *ovata* Ridd. ex Godfery (Orchidaceae) in the dune slacks of South Wales, UK. *Bot. J. Linn. Soc.* **126**: 123–139.

Jones, G. P. 2000. 1999 survey of BLM-managed lands along the Snake River in Jackson Hole, Wyoming for Ute ladies tresses (*Spiranthes diluvialis*). Bureau of Land Management, Wyoming State Office, Cheyenne, WY. 18 pp.

Jones, G. P. 2004. *Inventory and Mapping of Plant Communities in the Sweetwater Canyon Wilderness Study Area, Fremont County, Wyoming.* Wyoming Natural Diversity Database, University of Wyoming, Laramie, WY. 103 pp.

Jones, J. J. & R. D. Drobney. 1986. Winter feeding ecology of scaup and common goldeneye in Michigan. *J. Wildl. Manag.* **50**: 446–452.

Jones, R. L., C. T. Witsell & G. L. Nesom. 2008. Distribution and taxonomy of *Symphyotrichum sericeum* and *S. pratense* (Asteraceae: Astereae). *J. Bot. Res. Inst. Texas* **2**: 731–739.

Jones, J., H. B. Massicotte & A. L. Fredeen. 2016. Calcium and pH co-restrict abundance of *Drosera rotundifolia* (Droseraceae) in a *Sphagnum* bog in central British Columbia. *Botany* **94**: 139–146.

Jordan, W. C., M. W. Courtney & J. E. Neigel. 1996. Low levels of intraspecific genetic variation at a rapidly evolving chloroplast DNA locus in North American duckweeds (Lemnaceae). *Am. J. Bot.* **83**: 430–439.

Jordan, F., H. L. Jelks & W. M. Kitchens. 1997. Habitat structure and plant community composition in a northern Everglades wetland landscape. *Wetlands* **17**: 275–283.

Jorgenson, T. & C. Ely. 2001. Topography and flooding of coastal ecosystems on the Yukon-Kuskokwim Delta, Alaska: implications for sea-level rise. *J. Coast. Res.* **17**: 124–136.

Josselyn, M. 1986. Biomass, production and decomposition of a deep water seagrass, *Halophila decipiens* Ostenf. *Aquat. Bot.* **25**: 47–61.

Jovet, P. & S. Jovet-Ast. 1967. Floraison, fructification, germination du *Lemna valdiviana* au lac Marion (B. P.). *Bull. Cent. Etud. Rech. Sci., Biarritz* **6**: 729–734.

Joy, B., A. Rajan & E. Abraham. 2007. Antimicrobial activity and chemical composition of essential oil from *Hedychium coronarium. Phytother. Res.* **21**: 439–443.

Joye, D. B. & G. L. Smith. 1993. Biosystematic investigations of a hybrid between *Hymenocallis occidentalis* and *Hymenocallis coronaria. Cancas* **39**: 95–103.

Jozghasemi, S., V. Rabiei, A. Soleymani & A. Khalighi. 2016. Karyotype analysis of seven *Iris* species native to Iran. *Caryologia* **69**: 351–361.

Juday, G. P. 1992. Alaska research natural areas: 3. Serpentine slide. Gen. Tech. Rep. PNW-GTR-271. U.S. Department of Agriculture, Forest Service, Pacific Northwest Research Station, Portland, OR. 66 pp.

Judd, W. W. 1953. A study of the population of insects emerging as adults from the Dundas Marsh, Hamilton, Ontario, during 1948. *Am. Midl. Nat.* **49**: 801–824.

Judd, W. W. 1961. Insects and other invertebrates associated with flowering skunk cabbage, *Symplocarpus foetidus* (L.) Nutt., at Fanshawe Lake, Ontario. *Can. Entomol.* **93**: 241–249.

Judd, W. W. 1964. A study of the population of insects emerging as adults from Saunders Pond at London, Ontario. *Am. Midl. Nat.* **71**: 402–414.

Judd, W. W. 1968. Studies of the Byron bog in southwestern Ontario. XXX. Distribution of orchids in the bog. *Rhodora* **70**: 193–199.

Judd, W. S. 1998. The Smilacaceae in the southeastern United States. *Harvard Pap. Bot.* **3**: 147–169.

Judd, W. S., C. S. Campbell, E. A. Kellogg, P. F. Stevens & M. J. Donoghue. 2016. *Plant Systematics: A Phylogenetic Approach*, 4th ed. Sinauer Associates, Inc., Sunderland, MA. 677 pp.

Julien, M. H., M. W. Griffiths & A. D. Wright. 1999. *Biological Control of Water Hyacinth.* Australian Centre for International Agricultural Research (ACIAR), Canberra, ACT. 87 pp.

June-Wells, M., F. Gallagher, J. Gibbons & G. Bugbee. 2013. Water chemistry preferences of five nonnative aquatic macrophyte species in Connecticut: a preliminary risk assessment tool. *Lake Reserv. Manag.* **29**: 303–316.

Jung, M. E., J. W. Hong, J. I. Lee, C.-S. Kong, J.-S. Chang & Y. Seo. 2012. Inhibitory effect of *Zostera japonica* on growth of human cancer cells. *Ocean Polar Res.* **34**: 385–394.

Jupp, B. J. & D. H. N. Spence. 1977. Limitations of macrophytes in a eutrophic lake, Loch Leven: II. Wave action, sediments and waterfowl grazing. *J. Ecol.* **65**: 431–446.

Juraimi, A. S., M. S. Ahmad-Hamdani, A. R. Anuar, M. Azmi, M. P. Anwar & M. K. Uddin. 2012. Effect of water regimes on germination of weed seeds in a Malaysian rice field. *Aust. J. Crop Sci.* **6**: 598–605.

Juras, P. 1997. The presettlement piedmont savanna: a model for landscape design and management. M.S. thesis. University of Georgia, Athens, GA. 98 pp.

Jutila, H. M. 2002. Seed banks of river delta meadows on the west coast of Finland. *Ann. Bot. Fenn.* **39**: 49–61.

Kadlec, J. A. & L. M. Smith. 1984. Marsh plant establishment on newly flooded salt flats. *Wildl. Soc. Bull.* **12**: 388–394.

Kadono, Y. 1980. Photosynthetic carbon sources in some *Potamogeton* species. *Bot. Mag. (Tokyo)* **93**: 185–194.

Kadono, Y. 1982. Occurrence of aquatic macrophytes in relation to pH, alkalinity, Ca++, Cl- and conductivity. *Jap. J. Ecol.* **32**: 39–44.

Kadono, Y. 1984. Comparative ecology of Japanese *Potamogeton*: an extensive survey with special reference to growth form and life cycle. *Jap. J. Bot.* **34**: 161–172.

Kadono, Y. 2004. Alien aquatic plants naturalized in Japan: history and present status. *Glob. Environ. Res.* **8**: 163–169.

Kadono, Y., T. Nakamura & T. Suzuki. 1997. Genetic uniformity of two aquatic plants, *Egeria densa* Planch. and *Elodea nuttallii* (PLANCH.) ST. JOHN, introduced in Japan. *Jap. J. Limnol.* **58**: 197–203.

Kahn, A. E. & M. J. Durako. 2005. The effect of salinity and ammonium on seed germination in *Ruppia maritima* from Florida Bay. *Bull. Mar. Sci.* **77**: 453–458.

Kahn, A. E. & M. J. Durako. 2006. *Thalassia testudinum* seedling responses to changes in salinity and nitrogen levels. *J. Exp. Mar. Biol. Ecol.* **335**: 1–12.

Kahn, A. E. & M. J. Durako. 2008. Photophysiological responses of *Halophila johnsonii* to experimental hyposaline and hyper-CDOM conditions. *J. Exp. Mar. Biol. Ecol.* **367**: 230–235.

Kahn, A. E. & M. J. Durako. 2009. Wavelength-specific photosynthetic responses of *Halophila johnsonii* from marine-influenced versus river-influenced habitats. *Aquat. Bot.* **91**: 245–249.

Kahn, A. E., J. L. Beal & M. J. Durako. 2013. Diurnal and tidal variability in the photobiology of the seagrass *Halophila johnsonii* in a riverine versus marine habitat. *Estuaries Coast.* **36**: 430–443.

Kaisar, M. I., R. K. Adhikary, M. Dutta & S. Bhowmik. 2016. Diversity of aquatic weeds at Noakhali Sadar in Bangladesh. *Am. J. Sci. Ind. Res.* **7**: 117–128.

Kaldy, J. E. 2006. Production ecology of the non-indigenous seagrass, dwarf eelgrass (*Zostera japonica* Ascher. & Graeb.), in a Pacific Northwest estuary, USA. *Hydrobiologia* **553**: 201–217.

Kaldy, J. E. & K. H. Dunton. 1999. Ontogenetic photosynthetic changes, dispersal and survival of *Thalassia testudinum* (turtle grass) seedlings in a sub-tropical lagoon. *J. Exp. Mar. Biol. Ecol.* **240**: 193–212.

Kaldy, J. E. & D. J. Shafer. 2013. Effects of salinity on survival of the exotic seagrass *Zostera japonica* subjected to extreme high temperature stress. *Bot. Mar.* **56**: 75–82.

Kaldy, J. E., D. J. Shafer, M. S. Ailstock & A. D. Magoun. 2015a. Effects of temperature, salinity and seed age on induction of *Zostera japonica* germination in North America, USA. *Aquat. Bot.* **126**: 73–79.

Kaldy, J. E., D. J. Shafer & A. D. Magoun. 2015b. Duration of temperature exposure controls growth of *Zostera japonica*: implications for zonation and colonization. *J. Exp. Mar. Biol. Ecol.* **464**: 68–74.

Kalimuthu, K., C. Panneerselvam, C. Chou, L.-C. Tseng, K. Murugan, K.-H. Tsai, A. A. Alarfaj, A. Higuchi, A. Canale, J.-S. Hwang & G. Benelli. 2017. Control of dengue and Zika virus vector *Aedes aegypti* using the predatory copepod *Megacyclops formosanus*: synergy with *Hedychium coronarium*-synthesized silver nanoparticles and related histological changes in targeted mosquitoes. *Process Saf. Environ. Prot.* **109**: 82–96.

Kalkman, L. & R. J. Van Wijk. 1984. On the variation in chromosome number in *Potamogeton pectinatus* L. *Aquat. Bot.* **20**: 343–349.

Kalle, R. & R. Sõukand. 2013. Wild plants eaten in childhood: a retrospective of Estonia in the 1970s–1990s. *Bot. J. Linn. Soc.* **172**: 239–253.

Kalmbacher, R. S., K. R. Long & F. G. Martin. 1984a. Seasonal mineral concentration in diets of esophageally fistulated steers on three range areas. *J. Range Manag.* **37**: 36–39.

Kalmbacher, R. S., K. R. Long, M. K. Johnson & F. G. Martin. 1984b. Botanical composition of diets of cattle grazing south Florida rangeland. *J. Range Manag.* **37**: 334–340.

Kalmbacher, R., N. Cellinese & F. Martin. 2005. Seeds obtained by vacuuming the soil surface after fire compared with soil seedbank in a flatwoods plant community. *Native Plant J.* **6**: 233–240.

Kaltenecker, J. H. 1993. *Field Investigation of Allium madidum (Swamp Onion), a USFS Region 4 Sensitive Species, on the Payette National Forest.* Idaho Conservation Data Center, Idaho Department of Fish and Game, Boise, ID. 11 pp.

Kaminski, A., B. Bober, E. Chrapusta & J. Bialczyk. 2014. Phytoremediation of anatoxin-a by aquatic macrophyte *Lemna trisulca* L. *Chemosphere* **112**: 305–310.

Kan, J. & S. Song. 2008. Effects of dehydration, chilling, light, phytohormones and nitric oxide on germination of *Pistia stratiotes* seeds. *Seed Sci. Technol.* **36**: 38–45.

Kanabkaew, T. & P. Udomphon. 2004. Aquatic plants for domestic wastewater treatment: lotus (*Nelumbo nucifera*) and hydrilla (*Hydrilla verticillata*) systems. *Songklanakarin J. Sci. Technol.* **26**: 749–756.

Kandalepas, D., K. J. Stevens, G. P. Shaffer & W. J. Platt. 2010. How abundant are root-colonizing fungi in Southeastern Louisiana's degraded marshes? *Wetlands* **30**: 189–199.

Kandeler, R. 1984. Flowering in the *Lemna* system. *Phyton* **24**: 113–124.

Kangas, P. C. & G. L. Hannan. 1985. Vegetation on muskrat mounds in a Michigan marsh. *Am. Midl. Nat.* **113**: 392–396.

Kankam, C. K. & B. Odum-Ewuakye. 2000. Flexural strength and behavior of babadua-reinforced concrete beams. *J. Mater. Civil Eng.* **12**: 39–45.

Kantrud, H. A. 1990. Sago pondweed (*Potamogeton pectinatus* L.): a literature review. Resource Publication 176. United States Department of the Interior, Fish and Wildlife Service. Washington, DC. 89 pp.

Kantrud, H. A. 1991. Wigeongrass (*Ruppia maritima* L.): a literature review. *Fish Wildl. Res.* **10**: 1–58.

Kanwal, H., M. Hameed, N. Akhter, A. Ilyas, A. Mahmood & N. Noreen. 2018. Ecological and taxonomic significance of root anatomy in some species and cultivars of *Canna* L. *Int. J. Agric. Environ. Res.* **5**: 128–137.

Kao, J. T., J. E. Titus & W.-X. Zhu. 2003. Differential nitrogen and phosphorus retention by five wetland plant species. *Wetlands* **23**: 979–987.

Kaplan, Z. 2002. Phenotypic plasticity in *Potamogeton* (Potamogetonaceae). *Folia Geobot.* **37**: 141–170.

Kaplan, Z. 2008. A taxonomic revision of *Stuckenia* (Potamogetonaceae) in Asia, with notes on the diversity and variation of the genus on a worldwide scale. *Folia Geobot.* **43**: 159–234.

Kaplan, Z. & J. Štěpánek. 2003. Genetic variation within and between populations of *Potamogeton pusillus* agg. *Plant Syst. Evol.* **239**: 95–112.

Kaplan, Z. & J. L. Reveal. 2013. Taxonomic identity and typification of selected names of North American Potamogetonaceae. *Brittonia* **65**: 452–468.

Kaplan, Z., J. Fehrer & C. B. Hellquist. 2009. New hybrid combinations revealed by molecular analysis: the unknown side of North American pondweed diversity (*Potamogeton*). *Syst. Bot.* **34**: 625–642.

Kaplan, Z., V. Jarolímová & J. Fehrer. 2013. Revision of chromosome numbers of Potamogetonaceae: a new basis for taxonomic and evolutionary implications. *Preslia* **85**: 421–482.

Kaplan, Z., J. Fehrer, V. Bambasová & C. B. Hellquist. 2018. The endangered Florida pondweed (*Potamogeton floridanus*) is a hybrid: why we need to understand biodiversity thoroughly. *PLoS One* **13**: e0195241.

Kapoor, B. M. & S. Ramcharitar. 1982. IOPB chromosome number reports LXXVI. *Taxon* **31**: 596–597.

Kapoor, T. & M. R. Vijayaraghavan. 1991. Seed biology of *Potamogeton nodosus* Poir. *Aquat. Bot.* **40**: 261–273.

Kapuscinski, K. L., J. M. Farrell, S. V. Stehman, G. L. Boyer, D. D. Fernando, M. A. Teece & T. J. Tschaplinski. 2014. Selective herbivory by an invasive cyprinid, the rudd *Scardinius erythrophthalmus*. *Freshw. Biol.* **59**: 2315–2327.

Karadeniz, A., G. Alexie, H. J. Greten, K. Andersch & T. Efferth. 2015. Cytotoxicity of medicinal plants of the West-Canadian Gwich' in Native Americans towards sensitive and multidrug-resistant cancer cells. *J. Ethnopharmacol.* **168**: 191–200.

Karlen, S. D., H. C. A. Free, D. Padmakshan, B. G. Smith, J. Ralph & P. J. Harris. 2018. Commelinid monocotyledon lignins are acylated by p-coumarate. *Plant Physiol.* **177**: 513–521.

Karlin, E. F. & L. M. Lynn. 1988. Dwarf-shrub bogs of the southern Catskill Mountain region of New York State: geographic changes in the flora of peatlands in northern New Jersey and southern New York. *Bull. Torrey Bot. Club* **115**: 209–217.

Karling, J. S. 1931. Studies in the Chytridiales VI. The occurrence and life history of a new species of *Cladochytrium* in cells of *Eriocaulon septangulare*. *Am. J. Bot.* **18**: 526–557.

Karling, J. S. 1941. Texas chytrids. *Torreya* **41**: 105–108.

Karol, K. G., P. M. Skawinski, R. M. McCourt, M. E. Nault, R. Evans, M. E. Barton, M. S. Berg, D. J. Perleberg & J. D. Hall. 2017. First discovery of the charophycean green alga *Lychnothamnus barbatus* (Charophyceae) extant in the New World. *Am. J. Bot.* **104**: 1108–1116.

Karthikeyan, R., O. S. Koushik, P. S. Babu & J. Chunduru. 2016. Anti-inflammatory activity of ethanolic extract of flowers [sic!] *Hymenocallis littoralis* (Jacq.) Salisb. by HRBC membrane stabilization method. *Transl. Biomed.* **7**: 2.

Kaskey, J. B. & D. R. Tindall. 1979. Physiological aspects of growth and heteroblastic development of *Nasturtium officinale* under natural conditions. *Aquat. Bot.* **7**: 209–229.

Kasselmann, C. 2001. *Echinodorus. Die beliebtesten Aquarienpflanzen*. Dähne Verlag, Ettlingen, Germany. 168 pp.

Kasselmann, C. 2003. *Aquarium Plants*. Krieger Publishing Co., Malabar, FL. 518 pp.

Karst, L. & C. A. Wilson. 2012. Phylogeny of the New World genus *Sisyrinchium* (Iridaceae) based on analyses of plastid and nuclear DNA sequence data. *Syst. Bot.* **37**: 87–95.

Kato, M. & R. Miura. 1996. Flowering phenology and anthophilous insect community at a threatened natural lowland marsh at Nakaikemi in Tsuruga, Japan. *Contr. Biol. Lab. Kyoto Univ.* **29**: 1–48.

Kato, M., Y. Kosaka, A. Kawakita, Y. Okuyama, C. Kobayashi, T. Phimminith & D. Thongphan. 2008. Plant–pollinator interactions in tropical monsoon forests in Southeast Asia. *Am. J. Bot.* **95**: 1375–1394.

Kauffman, C. H. 1916. *Unreported Michigan fungi* 1911, 1912, 1913 and 1914. *Ann. Rep. Michigan Acad. Sci.* 17: 194–216.

Kaul, R. B. 1969. Morphology and development of the flowers of *Boottia cordata*, *Ottelia alismoides*, and their synthetic hybrid (Hydrocharitaceae). *Am. J. Bot.* **56**: 951–959.

Kaul, R. B. 1970. Evolution and adaptation of inflorescences in the Hydrocharitaceae. *Am. J. Bot.* **57**: 708–715.

Kaul, R. B. 1972. Adaptive leaf architecture in emergent and floating *Sparganium*. *Am. J. Bot.* **59**: 270–278.

Kaul, R. B. 1976. Conduplicate and specialized carpels in the Alismatales. *Am. J. Bot.* **63**: 175–182.

Kaul, R. B. 1978. Morphology of germination and establishment of aquatic seedlings in Alismataceae and Hydrocharitaceae. *Aquat. Bot.* **5**: 139–147.

Kaul, R. B. 1979. Inflorescence architecture and flower sex ratios in *Sagittaria brevirostra* (Alismataceae). *Am. J. Bot.* **66**: 1062–1066.

Kaul, R. B. 1985. Reproductive phenology and biology in annual and perennial Alismataceae. *Aquat. Bot.* **22**: 153–164.

Kaul, R. B. 1991. Foliar and reproductive responses of *Sagittaria calycina* and *Sagittaria brevirostra* (Alismataceae) to varying natural conditions. *Aquat. Bot.* **40**: 47–59.

Kaul, R. B. 1993. Meristic and organogenetic variation in *Ruppia occidentalis* and *R. maritima*. *Int. J. Plant Sci.* **154**: 416–424.

Kaul, R. B. 2000. Sparganiaceae F. Rudolphi. Pp. 270–277 *In*: Flora of North America Editorial Committee (eds.), *Flora of North America North of Mexico, Vol. 22: Magnoliophyta: Alismatidae Arecidae, Commelinidae (in Part), and Zingiberidae*. Oxford University Press, New York, NY.

Kaur, R., T. A. Rush, D. M. Ferrin & M. C. Aime. 2011. First report of *Puccinia thaliae* rust on canna lily in Louisiana. *Plant Dis.* **95**: 353–353.

Kausch, A. P., J. L. Seago, Jr. & L. C. Marsh. 1981. Changes in starch distribution in the overwintering organs of *Typha latifolia* (Typhaceae). *Am. J. Bot.* **68**: 877–880.

Kauth, P. J. & M. E. Kane. 2009. In vitro ecology of *Calopogon tuberosus* var. *tuberosus* (Orchidaceae) seedlings from distant populations: implications for assessing ecotypic differentiation. *J. Torrey Bot. Soc.* **136**: 433–444.

Kauth, P. J. & P. D. Biber. 2015. Moisture content, temperature, and relative humidity influence seed storage and subsequent survival and germination of *Vallisneria americana* seeds. *Aquat. Bot.* **120**: 297–303.

Kauth, P. J., W. A. Vendrame & M. E. Kane. 2006. In vitro seed culture and seedling development of *Calopogon tuberosus*. *Plant Cell Tissue Organ Cult.* **85**: 91–102.

Kauth, P. J., M. E. Kane, W. A. Vendrame & C. Reinhardt-Adams. 2008. Asymbiotic germination response to photoperiod and nutritional media in six populations of *Calopogon tuberosus* var. *tuberosus* (Orchidaceae): evidence for ecotypic differentiation. *Ann. Bot.* **102**: 783–793.

Kauth, P. J., M. E. Kane & W. A. Vendrame. 2011a. Comparative in vitro germination ecology of *Calopogon tuberosus* var. tuberosus (Orchidaceae) across its geographic range. *In Vitro Cell. Dev. Biol.-Plant* **47**: 148–156.

Kauth, P. J., M. E. Kane & W. A. Vendrame. 2011b. Chilling relieves corm dormancy in *Calopogon tuberosus* (Orchidaceae) from geographically distant populations. *Environ. Exp. Bot.* **70**: 283–288.

Kautsky, L. 1987. Life-cycles of three populations of *Potamogeton pectinatus* L. at different degrees of wave exposure in the Askö area, northern Baltic proper. *Aquat. Bot.* **27**: 177–186.

Kautsky, L. 1988. Life strategies of aquatic soft bottom macrophytes. *Oikos* **53**: 126–135.

Kautsky, L. 1990. Seed and tuber banks of aquatic macrophytes in the Askö area, northern Baltic proper. *Ecography* **13**: 143–148.

Kautsky, L. 1991. In situ experiments on interrelationships between six brackish macrophyte species. *Aquat. Bot.* **39**: 159–172.

Kawano, S. & H. H. Iltis. 1966. Cytotaxonomy of the genus *Smilacina* (Liliaceae) II. *Cytologia* **31**: 12–28.

Keane, B., S. Pelikan, G. P. Toth, M. K. Smith & S. H. Rogstad. 1999. Genetic diversity of *Typha latifolia* (Typhaceae) and the impact of pollutants examined with tandem-repetitive DNA probes. *Am. J. Bot.* **86**: 1226–1238.

Kearney, T. H. 1901. *Report on a Botanical Survey of the Dismal Swamp Region.* U.S. Department of Agriculture, Washington, DC. 585 pp.

Keddy, P. A. 1983. Shoreline vegetation in Axe Lake, Ontario: effects of exposure on zonation patterns. *Ecology* **64**: 331–344.

Keddy, C. J. 1987. Reproduction of annual eelgrass: variation among habitats and comparison with perennial eelgrass (*Zostera marina* L.). *Aquat. Bot.* **27**: 243–256.

Keddy, P. A. &. A. A. Reznicek. 1982. The role of seed banks in the persistence of Ontario's coastal plain flora. *Am. J. Bot.* **69**: 13–22.

Keddy, P. A. & T. H. Ellis. 1985. Seedling recruitment of 11 wetland plant species along a water level gradient: shared or distinct responses? *Can. J. Bot.* **63**: 1876–1879.

Keddy, P. A. & P. Constabel. 1986. Germination of ten shoreline plants in relation to seed size, soil particle size and water level: an experimental study. *J. Ecol.* **74**: 133–141.

Keddy, P. A. &. A. A. Reznicek. 1986. Great Lakes vegetation dynamics: the role of fluctuating water levels and buried seeds. *J. Great Lakes Res.* **12**: 25–36.

Kee, N. L. A., N. Mnonopi, H. Davids, R. J. Naudé & C. L. Frost. 2008. Antithrombotic/anticoagulant and anticancer activities of selected medicinal plants from South Africa. *Afr. J. Biotechnol.* **7**: 217–223.

Keel, B. G., L. W. Zettler & B. A. Kaplin. 2011. Seed germination of *Habenaria repens* (Orchidaceae) in situ beyond its range, and its potential for assisted migration imposed by climate change. *Castanea* **76**: 43–54.

Keeler, R. F. 1975. Teratogenic effects of cyclopamine and jervine in rats, mice and hamsters. *Proc. Soc. Exp. Biol. Med.* **149**: 302–306.

Keeler-Wolf, T. 1982. An ecological survey of the proposed Cedar Basin research natural area, Shasta-Trinity National Forest, California. Unpublished Report. Available online: https://www.fs.fed.us/psw/rna/publications_pdf/psw:1982_rna_es_cedar_basin.pdf [accessed 9 February, 2019].

Keeler-Wolf, T. 1991. Ecological survey of the proposed Mountaineer Creek research natueral area, Sequoia National Forest, Tulare, County, California. Unpublished Report. Available online: https://www.fs.fed.us/psw/rna/publications_pdf/22south_mountaineer_creek_es.pdf [accessed 19 March, 2019].

Keener, B. R. 2005. Molecular systematics and revision of the aquatic monocot genus *Sagittaria* (Alismataceae). Ph.D. dissertation. The University of Alabama, Tuscaloosa, AL. 336 pp.

Keer, G. H. & J. B. Zedler. 2002. Salt marsh canopy architecture differs with the number and composition of species. *Ecol. Appl.* **12**: 456–473.

Keil, D. J. 2012. Juncaginaceae. Arrow-grass family. Pp. 1375–1377 *In*: B. G. Baldwin, D. H. Goldman, D. J. Keil, R. Patterson, T. J. Rosatti & D. H. Wilken (eds.), *The Jepson Manual: Vascular Plants of California*, 2nd ed. University of California Press, Berkeley, CA.

Keith, L. B. 1961. A study of waterfowl ecology on small impoundments in southeastern Alberta. *Wildl. Monogr.* **6**: 3–88.

Keith, L. B. & R. P. Stanislawski. 1960. Stomach contents and weights of some flightless adult pintails. *J. Wildl. Manag.* **24**: 95–96.

Kejkar, P. K. & N. D. Polara. 2015. Effect of N, P and K on growth, bulb yield and nutrient content in ratoon spider lily (*Hymenocallis littoralis* L.) cv. local. *HortFlora Res. Spectr.* **4**: 22–27.

Kellerman, W. A. 1905. Ohio Fungi. Fascicle X. *J. Mycol.* **11**: 38–45.

Kelley, A. P. 1922. Plant indicators of soil types. *Soil Sci.* **13**: 411–423.

Kelley, A. D. & V. F. Bruns. 1975. Dissemination of weed seeds by irrigation water. *Weed Sci.* **23**: 486–493.

Kellogg, C. H., S. D. Bridgham & S. A. Leicht. 2003. Effects of water level, shade and time on germination and growth of freshwater marsh plants along a simulated successional gradient. *J. Ecol.* **91**: 274–282.

Kelly, C. M. 2001. Prehistoric land-use patterns in the North Santiam subbasin on the western slopes of the Oregon Cascade Range. M.A. thesis. Oregon State University, Corvallis, OR. 231 pp.

Kelly, J. P. & G. Fletcher. 1994. Habitat correlates and distribution of *Cordylanthus maritimus* (Scrophulariaceae) on Tomales Bay, California. *Madroño* **41**: 316–327.

Kelso, S., L. Fugere, M. Kummel & S. Tsocanos. 2014. Vegetation and vascular flora of tallgrass prairie and wetlands, Black Squirrel Creek drainage, south-central Colorado: perspectives from the 1940s and 2011. *J. Bot. Res. Inst. Texas* **8**: 203–225.

Kemp, A. C., S. E. Engelhart, S. J. Culver, A. Nelson, R. W. Briggs & P. J. Haeussler. 2013. Modern salt-marsh and tidal-flat foraminifera from Sitkinak and Simeonof Islands, southwestern Alaska. *J. Foraminif. Res.* **43**: 88–98.

Keddy, P. A. 1976. Lakes as islands: the distributional ecology of two aquatic plants, *Lemna minor* L. and *L. trisulca* L. *Ecology* **57**: 353–359.

Kendall, M. S., T. Battista & Z. Hillis-Starr. 2004. Long term expansion of a deep *Syringodium filiforme* meadow in St. Croix, US Virgin Islands: the potential role of hurricanes in the dispersal of seeds. *Aquat. Bot.* **78**: 15–25.

Kendrick, G. A., M. Waycott, T. J. B. Carruthers, M. L. Cambridge, R. Hovey, S. L. Krauss, P. S. Lavery, D. H. Les, R. J. Lowe, O. Mascaró I Vidal, J. L. S. Ooi, R. J. Orth, D. O. Rivers, L. Ruiz-Montoya, E. A. Sinclair, J. Statton, J. K. van Dijk & J. J. Verduin. 2012. The central role of dispersal in the maintenance and persistence of seagrass populations. *BioScience* **62**: 56–65.

Kendrick, G. A., R. J. Orth, J. Statton, R. Hovey, L. Ruiz Montoya, R. J. Lowe, S. L. Krauss & E. A. Sinclair. 2017. Demographic and genetic connectivity: the role and consequences of reproduction, dispersal and recruitment in seagrasses. *Biol. Rev.* **92**: 921–938.

Kennedy, K. 1994a. *Little Aguja Pondweed (Potamogeton clystocarpus) Recovery Plan.* U.S. Fish and Wildlife Service, Albuquerque, NM. 78 pp.

Kennedy, K. 1994b. Endangered and threatened wildlife and plants; 90-day finding for a petition to delist the plant *Potamogeton clystocarpus* (Little Aguja Pondweed). *Fed. Reg.* **59**: 67267–67268.

Kennedy, H. 2000. Marantaceae Petersen in Engler & Prantl, Arrowroot or Prayer-plant Family, maranta. Pp. 315-319 *In*: Flora North America Editorial Committee (eds.), *Flora of North America North of Mexico, Vol. 22: Magnoliophyta: Alismatidae, Arecidae, Commelinidae (in part), and Zingiberidae.* Oxford University Press, New York, NY.

Kennedy, M. A. 2016. Impacts of wintering Redhead ducks (*Aythya americana*) on shoal grass (*Halodule wrightii*) and widgeon grass (*Ruppia maritima*) in the northern Gulf of Mexico. M.S. thesis. University of South Alabama, Mobile, AL. 64 pp.

Kenow, K. P. & J. E. Lyon. 2009. Composition of the seed bank in drawdown areas of Navigation Pool 8 of the Upper Mississippi River. *River Res. Appl.* **25**: 194–207.

Kenton, A. 1981. A Robertsonian relationship in the chromosomes of two species of *Hydrocleys* (Butomaceae sens. lat.). *Kew Bull.* **36**: 487–492.

Kenton, A. Y., P. J. Rudall & A. R. Johnson. 1986. Genome size variation in *Sisyrinchium* L. (Iridaceae) and its relationship to phenotype and habitat. *Bot. Gaz.* **147**: 342–354.

Kentula, M. E. & C. D. McIntire. 1986. The autecology and production dynamics of eelgrass (*Zostera marina* L.) in Netarts Bay, Oregon. *Estuaries* **9**: 188–199.

Kentula, M. E. & T. H. DeWitt. 2003. Abundance of seagrass (*Zostera marina* L.) and macroalgae in relation to the salinity-temperature gradient in Yaquina Bay, Oregon, USA. *Estuaries* **26**: 1130–1141.

Kenworthy, W. J. 1999. The role of sexual reproduction in maintaining populations of *Halophila decipiens*: implications for the biodiversity and conservation of tropical seagrass ecosystems. *Pac. Conserv. Biol.* **5**: 260–268.

Kenworthy, W. J. & M. S. Fonseca. 1996. Light requirements of seagrasses *Halodule wrightii* and *Syringodium filiforme* derived from the relationship between diffuse light attenuation and maximum depth distribution. *Estuaries* **19**: 740–750.

Kenworthy, W. J. & A. C. Schwarzschild. 1998. Vertical growth and short-shoot demography of *Syringodium filiforme* in outer Florida Bay, USA. *Mar. Ecol. Prog. Ser.* **173**: 25–37.

Keppner, L. A. & L. C. Anderson. 2008. Notes on Harper's beauty, *Harperocallis flava* (Tofieldiaceae), in Bay County, Florida. *Southeast. Nat.* **7**: 180–184.

Kevan, P. G. 1989. How honey bees forage for pollen at skunk cabbage, *Symplocarpus foetidus* (Araceae). *Apidologie* **20**: 485–490.

Kevan, P. G., J. D. Ambrose & J. R. Kemp. 1991. Pollination in an understorey vine, *Smilax rotundifolia*, a threatened plant of the Carolinian forests in Canada. *Can. J. Bot.* **69**: 2555–2559.

Khan, M. A. & I. A. Ungar. 1997. Effects of light, salinity, and thermoperiod on the seed germination of halophytes. *Can. J. Bot.* **75**: 835–841.

Khan, M. A. & I. A. Ungar. 1999. Effect of salinity on seed germination of *Triglochin maritima* under various temperature regimes. *Great Basin Nat.* **59**: 144–150.

Khan, M. A. & I. A. Ungar. 2001. Seed germination of *Triglochin maritima* as influenced by salinity and dormancy relieving compounds. *Biol. Plant.* **44**: 301–303.

Khan, M. A., B. Gul & D. J. Weber. 2001. Germination of dimorphic seeds of *Suaeda moquinii* under high salinity stress. *Aust. J. Bot.* **49**: 185–192.

Khan, M. A., K. B. Marwat, B. Gul, F. Wahid, H. Khan & S. Hashim. 2014. *Pistia stratiotes* L. (Araceae): phytochemistry, use in medicines, phytoremediation, biogas and management options. *Pak. J. Bot.* **46**: 851–860.

Khokar, S. & G. R. Fenwick. 2003. Onions and related crops. Pp. 4267–4272 *In*: L. Trugo & P. M. Finglas (eds.), *Encyclopedia of Food Sciences and Nutrition*, 2nd ed. Academic Press, New York, NY.

Khoshoo, T. N. & I. Mukherjee. 1970. Genetic-evolutionary studies on cultivated cannas. *Theoret. Appl. Genet.* **40**: 204–217.

Khurana, J. P. & S. C. Maheshwari. 1980. Some effects of salicylic acid on growth and flowering in *Spirodela polyrrhiza* SP20. *Plant Cell Physiol.* **21**: 923–927.

Khuroo, A. A., I. Rashid, Z. Reshi, G. H. Dar & B. A. Wafai. 2007. The alien flora of Kashmir Himalaya. *Biol. Invasions* **9**: 269–292.

Khvatkov, P., M. Chernobrovkina, A. Okuneva, A. Pushin & S. Dolgov. 2015. Transformation of *Wolffia arrhiza* (L.) Horkel ex Wimm. *Plant Cell Tissue Organ Cult.* **123**: 299–307.

Khvatkov, P., A. Firsov, A. Shvedova, L. Shaloiko, O. Kozlov, M. Chernobrovkina, A. Pushin, I. Tarasenko, I. Chaban & S. Dolgov. 2018. Development of *Wolffia arrhiza* as a producer for recombinant human granulocyte colony-stimulating factor. *Front. Chem.* **6**: 304.

Keith, E. L. & N. R. Carrie. 2002. Effects of fire on two pitcher plant bogs with comments on several rare and interesting plants. *Sida* **20**: 387–395.

Kihara, M., T. Koike, Y. Imakura, K. Kida, T. Shingu & S. Kobayashi. 1987. Alkaloidal constituents of *Hymenocallis rotata* Herb. (Amaryllidaceae). *Chem. Pharmaceut. Bull.* **35**: 1070–1075.

Kikuta, K., L. D. Whitney & G. K. Parris. 1938. Seeds and seedlings of the taro, *Colocasia esculenta*. *Am. J. Bot.* **25**: 186–188.

Kilgen, R. H. & R. O. Smitherman. 1971. Food habits of the white amur stocked in ponds alone and in combination with other species. *Progr. Fish-Cult.* **33**: 123–127.

Kilgo, J. C. & R. F. Labisky. 1995. Nutritional quality of three major deer forages in pine flatwoods of northern Florida. *Florida Sci.* **58**: 327–334.

Kilham, L. 1976. Winter foraging and associated behavior of pileated woodpeckers in Georgia and Florida. *Auk* **93**: 15–24.

Kim, J. S. & B. L. Mercado. 1987. Viability and emergence of buried seeds of *Echinochloa glabrescens*, *Monochoria vaginalis* and *Cyperus difformis*. Pp. 469–476 *In*: *Proceedings, 11th Asian Pacific Weed Science Society Conference*, Volume 2. Asian Pacific Weed Science Society, Taipei, China.

Kim, S.-C. & N. S. Lee. 2007. Generic delimitation and biogeography of *Maianthemum* and *Smilacina* (Ruscaceae sensu lato): preliminary results based on partial 3′ *matK* gene and *trnK* 3′ intron sequences of cpDNA. *Plant Syst. Evol.* **265**: 1–12.

Kim, C. & H.-K. Choi. 2011. Molecular systematics and character evolution of *Typha* (Typhaceae) inferred from nuclear and plastid DNA sequence data. *Taxon* **60**: 1417–1428.

Kim, J. S. & J.-H. Kim. 2018. Updated molecular phylogenetic analysis, dating and biogeographical history of the lily family (Liliaceae: Liliales). *Bot. J. Linn. Soc.* **187**: 579–593.

Kim, J.-P., I.-S. Lee, J.-J. Seo, M.-Y. Jung, Y.-H. Kim, N.-H. Yim & K.-H. Bae. 2010. Vitexin, orientin and other flavonoids from *Spirodela polyrhiza* inhibit adipogenesis in 3T3-L1 cells. *Phytother. Res.* **24**: 1543–1548.

Kim, J.-L., H. M. Li, Y.-H. Kim, Y.-J. Lee, J.-H. Shim, S. S. Lim & Y.-H. Kang. 2012. Osteogenic activity of yellow flag *Iris* (*Iris pseudacorus*) extract modulating differentiation of osteoblasts and osteoclasts. *Am. J. Chinese Med.* **40**: 1289–1305.

Kim, W.-S., J.-S. Lee, G.-Y. Bae, J.-J. Kim, Y.-W. Chin, Y. Y. Bahk, H. G. Min & H.-J. Cha. 2013. Extract of *Aneilema keisak* inhibits transforming growth factor-β-dependent signalling by inducing Smad2 downregulation in keloid fibroblasts. *Exp. Dermatol.* **22**: 69–71.

Kim, M.-J., N.-Y. Bae, K.-B.-W.-R. Kim, J.-H. Park, S.-H. Park, Y.-J. Cho & D.-H. Ahn. 2015. Anti-inflammatory effect of *Zostera marina* ethanolic extract on LPS-induced RAW264. 7 cells and mouse model. *Korean Soc. Biotechnol. Bioeng. J.* **30**: 182–190.

Kim, S.-C., J. S. Kim, M. W. Chase, M. F. Fay & J.-H. Kim. 2016. Molecular phylogenetic relationships of Melanthiaceae (Liliales) based on plastid DNA sequences. *Bot. J. Linn. Soc.* **181**: 567–584.

Kim, C., K. M. Cameron & J.-H. Kim. 2017. Molecular systematics and historical biogeography of *Maianthemum* s.s. *Am. J. Bot.* **104**: 939–952.

Kim, S.-H., M.-S. Cho, P. Li & S.-C. Kim. 2018. Phylogeography and ecological niche modeling reveal reduced genetic diversity and colonization patterns of skunk cabbage (*Symplocarpus foetidus*; Araceae) from glacial refugia in eastern North America. *Front. Plant Sci.* **9**: 648.

Kimball, R. T., D. J. Crawford, D. H. Les & E. Landolt. 2003. Out of Africa: molecular phylogenetics and biogeography of *Wolffiella* (Lemnaceae). *Biol. J. Linn. Soc.* **79**: 565–576.

Kimber, A., A. G. van der Valk & C. E. Korschgen. 1995. The distribution of *Vallisneria americana* seeds and seedling light requirements in the Upper Mississippi River. *Can. J. Bot.* **73**: 1966–1973.

Kimble, R. B. & A. Ensminger. 1959. Duck food habits in southwestern Louisiana marshes following a hurricane. *J. Wildl. Manag.* **23**: 453–455.

Kindscher, K. & D. P. Hurlburt. 1998. Huron Smith's ethnobotany of the Hocąk (Winnebago). *Econ. Bot.* **52**: 352–372.

King, J. R. & L. R. Mewaldt. 1987. The summer biology of an unstable insular population of White-crowned Sparrows in Oregon. *Condor* **89**: 549–565.

King, U. M. & D. H. Les. 2016. A significant new record for *Hydrilla verticillata* (Hydrocharitaceae) in central Connecticut. *Rhodora* **118**: 306–309.

King, U. M., D. H. Les, E. Peredo & L. K. Benoit. 2017. Adaptive evolution of the chloroplast genome in the submersed monocotyledon *Najas* (Hydrocharitaceae). Pp. 52–68 *In*: L. M. Campbell, J. I. Davis, A. W. Meerow, R. F. C. Naczi, D. W. Stevenson & W. W. Thomas (eds.), *Diversity and Phylogeny of the Monocotyledons*. New York Botanical Garden Press, Bronx, NY.

Kinupp, V. F. & I. B. I. de Barros. 2008. Protein and mineral contents of native species, potential vegetables, and fruits. *Food Sci. Technol.* **28**: 846–857.

Kipriyanova, L. M., L. A. Dolmatova, B. B. Bazarova, B. B. Naydanov, R. E. Romanov, G. T. Tsybekmitova & A. V. Dyachenko. 2017. On the ecology of some species of genus *Stuckenia* (Potamogetonaceae) in lakes of Zabaykalsky krai and the Republic of Buryatia. *Inland Water Biol.* **10**: 73–82.

Király, G., D. Steták & Á. Bányász. 2007. Spread of invasive macrophytes in Hungary. *Neobiota* **7**: 123–131.

Kirby, L. J. & N. H. Ringler. 2015. Associations of epiphytic macroinvertebrates within four assemblages of submerged aquatic vegetation in a recovering urban lake. *Northeast. Nat.* **22**: 672–689.

Kirk, H., C. Connolly & J. R. Freeland. 2011. Molecular genetic data reveal hybridization between *Typha angustifolia* and *Typha latifolia* across a broad spatial scale in eastern North America. *Aquat. Bot.* **95**: 189–193.

Kirkman, L. K., P. C. Goebel, L. West, M. B. Drew & B. J. Palik. 2000. Depressional wetland vegetation types: a question of plant community development. *Wetlands* **20**: 373–385.

Kirschner, J. & Z. Kaplan. 2002. Taxonomic monographs in relation to global Red Lists. *Taxon* **51**: 155–158.

Kirsten, J. H., C. J. Dawes & B. J. Cochrane. 1998. Randomly amplified polymorphism detection (RAPD) reveals high genetic diversity in *Thalassia testudinum* banks ex König (Turtlegrass). *Aquat. Bot.* **61**: 269–287.

Kishima, J., S. Harada & R. Sakurai. 2011. Suitable water temperature for seed storage of *Zostera japonica* for subtropical seagrass bed restoration. *Ecol. Eng.* **37**: 1416–1419.

Kissoon, L. T. T., D. L. Jacob, M. A. Hanson, B. R. Herwig, S. E. Bowe & M. L. Otte. 2013. Macrophytes in shallow lakes: relationships with water, sediment and watershed characteristics. *Aquat. Bot.* **109**: 39–48.

Kissoon, L. T. T., D. L. Jacob, M. A. Hanson, B. R. Herwig, S. E. Bowe & M. L. Otte. 2015. Multi-elements in waters and sediments of shallow lakes: relationships with water, sediment, and watershed characteristics. *Wetlands* **35**: 443–457.

Kitamura, H., M. Mori, Y. Nishibori & H. Ohtani. 2008. A study on the mass propagation of skunk cabbages (*Symplocarpus foetidus*). *Bull. Shiga Prefect. Agric. Technol. Promotion Center* 46. Available online: http://agris.fao.org/agris-search/search.do?recordID=JP2008003781.

Kitner, M., R. Prausova & L. Adamec. 2013. Present status of genetic diversity of *Potamogeton praelongus* populations in the Czech Republic. *Phyton* **53**: 73–86.

Kittiwongwattana, C. & C. Thawai. 2014. *Rhizobium lemnae* sp. nov., a bacterial endophyte of *Lemna aequinoctialis*. *Int. J. Syst. Evol. Microbiol.* **64**: 2455–2460.

Kittiwongwattana, C. & C. Thawai. 2015. *Paenibacillus lemnae* sp. nov., an endophytic bacterium of duckweed (*Lemna aequinoctialis*). *Int. J. Syst. Evol. Microbiol.* **65**: 107–112.

Kiviniemi, K. 1996. A study of adhesive seed dispersal of three species under natural conditions. *Acta Bot. Neerl.* **45**: 73–83.

Kiviniemi, K. & A. Telenius. 1998. Experiments on adhesive dispersal by wood mouse: seed shadows and dispersal distances of 13 plant species from cultivated areas in southern Sweden. *Ecography* **21**: 108–116.

Kjellmark, E. W., P. D. McMillan & R. K. Peet. 1998. Longleaf pine vegetation of the South Carolina and Georgia Coastal Plain: A preliminary classification. Final Report, USDA Forest Service, Southern Forest Experiment Station, Asheville, NC. 84 pp.

Kleyheeg, E. & C. H. A. van Leeuwen. 2015. Regurgitation by waterfowl: an overlooked mechanism for long-distance dispersal of wetland plant seeds. *Aquat. Bot.* **127**: 1–5.

Klier, K., M. J. Leoschke & J. F. Wendel. 1991. Hybridization and introgression in white and yellow ladyslipper orchids (*Cypripedium candidum* and *C. pubescens*). *J. Hered.* **82**: 305–318.

Kline, L. & B. McCune. 1987. Factors influencing the distribution of *Wolffia columbiana* and *Wolffia punctata* (Lemnaceae). *Northwest Sci.* **61**: 41–43.

Kling, G. W. 1986. The physicochemistry of some dune ponds on the Outer Banks, North Carolina. *Hydrobiologia* **134**: 3–10.

Klinka, K., V. J. Krajina, A. Ceska & A. M. Scagel. 1989. *Indicator Plants of Coastal British Columbia*. UBC Press, Vancouver, BC. 296 pp.

Klotz, L. H. 1992. On the biology of *Orontium aquaticum* L. (Araceae), golden club or floating arum. *Aroideana* **15**: 25–33.

Klotzsch, J. F. 1852. Über *Pistia*. *Abh. Königl. Akad. Wiss. Berlin* **1852**: 329–359.

Knapp, W. M., R. F. C. Naczi, W. D. Longbottom, C. A. Davis, W. A. McAvoy, C. T. Frye, J. W. Harrison & P. Stango, III. 2011. Floristic discoveries in Delaware, Maryland, and Virginia. *Phytoneuron* 2011–64: 1–26.

Knapton, R. W. & K. Pauls. 1994. Fall food habits of American wigeon at Long Point, Lake Erie, Ontario. *J. Great Lakes Res.* **20**: 271–276.

Knapton, R. W. & S. A. Petrie. 1999. Changes in distribution and abundance of submerged macrophytes in the inner bay at Long Point, Lake Erie: implications for foraging waterfowl. *J. Great Lakes Res.* **25**: 783–798.

Kneitel, J. M. & T. E. Miller. 2002. Resource and top-predator regulation in the pitcher plant (*Sarracenia purpurea*) inquiline community. *Ecology* **83**: 680–688.

Knight, R. R. 1965. Vegetative characteristics and waterfowl usage of a Montana water area. *J. Wildl. Manag.* **29**: 782–788.

Knight, N. S., C. Prentice, M. Tseng & M. I. O'Connor. 2015. A comparison of epifaunal invertebrate communities in native eelgrass *Zostera marina* and nonnative *Zostera japonica* at Tsawwassen, BC. *Mar. Biol. Res.* **11**: 564–571.

Knobloch, I. W. 1972. Intergeneric hybridization in flowering plants. *Taxon* **21**: 97–103.

Knollenberg, W. G., R. W. Merritt & D. L. Lawson. 1985. Consumption of leaf litter by *Lumbricus terrestris* (Oligochaeta) on a Michigan woodland floodplain. *Am. Midl. Naturalist* **113**: 1–6.

Knudsen, J. T. & L. Tollsten. 1993. Trends in floral scent chemistry in pollination syndromes: floral scent composition in moth-pollinated taxa. *Bot. J. Linn. Soc.* **113**: 263–284.

Knuth, P. 1906. *Handbook of Flower Pollination: Vol. I*. Clarendon Press, Oxford, United Kingdom. 382 pp.

Knuth, P. 1909. *Handbook of Flower Pollination. Vol. III*. Translated by J. R. Ainsworth Davis. Clarendon Press, Oxford, United Kingdom. 644 pp.

Knutson, R. M. 1972. Temperature measurements of the spadix of *Symplocarpus foetidus* (L.) Nutt. *Am. Midl. Nat.* **88**: 251–254.

Koca Çalişkan, U., C. Aka & E. Bor. 2017. Melatonin in edible and non-edible plants. *Turkish J. Pharmaceut. Sci.* **14**: 75–83.

Kočić, A., J. Horvatić & S. D. Jelaska. 2014. Distribution and morphological variations of invasive macrophytes *Elodea nuttallii* (Planch.) H. St. John and *Elodea canadensis* Michx in Croatia. *Acta Bot. Croatica* **73**: 437–446.

Koch, R. G. 1970. The vascular flora of Cowley County, Kansas. *Trans. Kansas Acad. Sci.* **73**: 135–168.

Koch, E. W. & C. J. Dawes. 1991. Influence of salinity and temperature on the germination of *Ruppia maritima* L. from the North Atlantic and Gulf of Mexico. *Aquat. Bot.* **40**: 387–391.

Koch, M. S. & P. S. Rawlik. 1993. Transpiration and stomatal conductance of two wetland macrophytes (*Cladium jamaicense* and *Typha domingensis*) in the subtropical Everglades. *Am. J. Bot.* **80**: 1146–1154.

Koch, M. S., S. A. Schopmeyer, C. Kyhn-Hansen, C. J. Madden & J. S. Peters. 2007. Tropical seagrass species tolerance to hypersalinity stress. *Aquat. Bot.* **86**: 14–24.

Koch, E. W., M. S. Ailstock, D. M. Booth, D. J. Shafer & A. D. Magoun. 2010. The role of currents and waves in the dispersal of submersed angiosperm seeds and seedlings. *Restor. Ecol.* **18**: 584–595.

Koch, M., G. Bowes, C. Ross & X.-H. Zhang. 2013. Climate change and ocean acidification effects on seagrasses and marine macroalgae. *Glob. Chang. Biol.* **19**: 103–132.

Koecke, V., S. von Mering, L. Mucina & J. W. Kadereit. 2010. Revision of the Mediterranean and southern African *Triglochin bulbosa* complex (Juncaginaceae). *Edinburgh J. Bot.* **67**: 353–398.

Koehler, S. & C. P. Bove. 2001. Hydrocharitaceae from central Brazil: a new species of *Egeria* and a note on *Apalanthe granatensis*. *Novon* **11**: 63–66.

Kohn, J. R., S. W. Graham, B. R. Morton, J. J. Doyle & S. C. H. Barrett. 1996. Reconstruction of the evolution of reproductive characters in Pontederiaceae using phylogenetic evidence from chloroplast DNA restriction-site variation. *Evolution* **50**: 1454–1469.

Konkel, D. 2006. *Changes in the Aquatic Plant Community of Round Lake, Chippewa County, Wisconsin: 1988–2004*. Wisconsin Department of Natural Resources, West Central Region, Eau Claire, WI. 53 pp.

Kopec, D. 2015. Cazenovia Lake: a comprehensive management plan. Occasional Paper No. 50. State University of New York College at Oneonta, Cooperstown, NY. 102 pp.

Kopecko, K. J. P. & E. W. Lathrop. 1975. Vegetation zonation in a vernal marsh on the Santa Rosa Plateau of Riverside County, California. *Aliso* **8**: 281–288.

Korfhage, R. C., J. R. Nelson & J. M. Skovlin. 1980. Summer diets of Rocky Mountain elk in northeastern Oregon. *J. Wildl. Manag.* **44**: 746–750.

Körner, S. & J. E. Vermaat. 1998. The relative importance of *Lemna gibba* L., bacteria and algae for the nitrogen and phosphorus removal in duckweed-covered domestic wastewater. *Water Res.* **32**: 3651–3661.

Körner, S., S. K. Das, S. Veenstra & J. E. Vermaat. 2001. The effect of pH variation at the ammonium/ammonia equilibrium in wastewater and its toxicity to *Lemna gibba*. *Aquat. Bot.* **71**: 71–78.

Kornfeld, J. M. & J. M. Edwards. 1972. An investigation of the photodynamic pigments in extracts of *Lachnanthes tinctoria*. *Biochim. Biophys. Acta* **286**: 88–90.

Koropchak, S. & D. Vitt. 2013. Survivorship and growth of *Typha latifolia* L. across a NaCl gradient: a greenhouse study. *Int. J. Min. Reclam. Environ.* **27**: 143–150.

Korschgen, L. J. & D. L. Moyle. 1955. Food habits of the bullfrog in central Missouri farm ponds. *Am. Midl. Nat.* **54**: 332–341.

Korschgen, C. E., L. S. George & W. L. Green. 1988. Feeding ecology of canvasbacks staging on Pool 7 of the Upper Mississippi River. Pp. 237–249 *In*: M. W. Weller (ed.), *Waterfowl in Winter: Selected Papers from Symposium and Workshop Held in Galvestion, Texas, 7–10 January 1985*. University of Minnesota Press, Minneapolis, MN.

Korschgen, C. E., W. L. Green & K. P. Kenow. 1997. Effects of irradiance on growth and winter bud production by *Vallisneria americana* and consequences to its abundance and distribution. *Aquat. Bot.* **58**: 1–9.

Koryak, M. 1978. Emergent aquatic plants in the upper Ohio River and major navigable tributaries, West Virginia and Pennsylvania. *Castanea* **43**: 228–237.

Koschnick, T. J., W. T. Haller & L. Glasgow. 2006. Documentation of landoltia (*Landoltia punctata*) resistance to diquat. *Weed Sci.* **54**: 615–619.

Kowalski, J. L. & H. R. DeYoe. 2016. Flowering and seed production in the subtropical seagrass, *Halodule wrightii* (shoal grass). *Bot. Mar.* **59**: 193–199.

Kowarik, I. 2003. Human agency in biological invasions: secondary releases foster naturalisation and population expansion of alien plant species. *Biol. Invasions* **5**: 293–312.

Kozen, E. N. 2013. The scent of eastern skunk cabbage, *Symplocarpus foetidus* (Araceae): qualification of floral volatiles and sex differences in floral scent composition. M.S. thesis. Indiana University of Pennsylvania, Indiana, PA. 59 pp.

Krajnčić, B. & Z. Devidé. 1982. Photoperiodic responses in Lemnaceae from North Croatia. *Acta Bot. Croatica* **41**: 57–63.

Kral, R. 1960. The genus *Xyris* in Florida. *Rhodora* **62**: 295–319.

Kral, R. 1966a. Eriocaulaceae of continental North America north of Mexico. *Sida* **2**: 285–332.

Kral, R. 1966b. Observations on the flora of the southeastern United States with special reference to northern Louisiana. *Sida* **2**: 395–408.

Kral, R. 1966c. Xyris (Xyridaceae) of the continental United States and Canada. *Sida* **2**: 177–260.

Kral, R. 1973. Some notes on the flora of the southern states, particularly Alabama and middle Tennessee. *Rhodora* **75**: 366–410.

Kral, R. 1978. A new species of *Xyris* (sect. *Xyris*) from Tennessee and northwestern Georgia. *Rhodora* **80**: 444–447.

Kral, R. 1982. A new phyllodial-leaved *Sagittaria* (Alismaceae) from Alabama. *Brittonia* **34**: 12–17.

Kral, R. 1989. The genera of Eriocaulaceae in the southeastern United States. *J. Arnold Arbor.* **70**: 131–142.

Kral, R. 1998. Xyridaceae. Pp. 461–469 *In*: K. Kubitzki, H. Huber, P. J. Rudall, P. S. Stevens & T. Stützel (eds.), *The Families and Genera of Vascular Plants, Vol. IV. Flowering Plants, Monocotyledons: Alismatanae and Commelinanae (Except Gramineae)*. Springer-Verlag, Berlin, Germany.

Kral, R. 1999. A revised taxonomy for two north american *Rhynchospora* (Cyperaceae) and for two north american *Xyris* (Xyridaceae). *Novon* **9**: 205–219.

Kral, R. 2000. Eriocaulaceae Palisot de Beauvois ex Desvaux. Pp. 198–210 *In*: Flora North America Editorial Committee (eds.), *Flora of North America North of Mexico, Vol. 22: Magnoliophyta: Alismatidae, Arecidae, Commelinidae (in Part), and Zingiberidae*. Oxford University Press, New York, NY.

Kramer, J. P. 1967. A taxonomic study of the brachypterous North American leafhoppers of the genus *Lonatura* (Homoptera: Cicadellidae: Deltocephalinae). *Trans. Am. Entomol. Soc.* **93**: 433–462.

Krapu, G. L. 1974. Foods of breeding pintails in North Dakota. *J. Wildl. Manag.* **38**: 408–417.

Krattinger, K. 1975. Genetic mobility in *Typha*. *Aquat. Bot.* **1**: 57–70.

Krattinger, K., D. Rast & H. Karesch. 1979. Analysis of pollen proteins of *Typha* species in relation to identification of hybrids. *Biochem. Syst. Ecol.* **7**: 125–128.

Kratzer, L. A. 2014. Effect of herbivory on the growth and competitive ability of reed canary grass (*Phalaris arundinacea*). M.S. thesis. Rochester Institute of Technology, Rochester, NY. 30 pp.

Krausse, A. F. W. 1783. *Catalogus von perennirenden Pflanzen, welche um billige preise zu haben*. Botanischer Garten und Botanisches Museum Berlin, Berlin, Germany. 71 pp.

Kravtsova, L. S., L. A. Izhboldina, I. V. Mekhanikova, G. V. Pomazkina & O. I. Belykh. 2010. Naturalization of *Elodea canadensis* Mich. in Lake Baikal. *Russ. J. Biol. Invasions* **1**: 162–171.

Krawczyk, K., J. Wiland-Szymańska, K. Buczkowska-Chmielewska, M. Drapikowska, M. Maślak, K. Myszczyński, M. Szczecińska, M. Ślipiko & J. Sawicki. 2018. The complete chloroplast genome of a rare orchid species *Liparis loeselii* (L.). *Conserv. Genet. Resour.* **10**: 305–308.

Krecker, F. H. 1939. A comparative study of the animal population of certain submerged aquatic plants. *Ecology* **20**: 553–562.

Kreiling, R. M., Y. Yin & D. T. Gerber. 2007. Abiotic influences on the biomass of *Vallisneria americana* Michx. in the Upper Mississippi River. *River Res. Appl.* **23**: 343–349.

Kress, W. J. 1990. The phylogeny and classification of the Zingiberales. *Ann. Missouri Bot. Gard.* **77**: 698–721.

Kress, W. J. 1995. Phylogeny of the Zingiberanae: morphology and molecules. Pp. 443–460 *In*: P. J. Rudall, P. J. Cribb, D. F. Cutler & C. J. Humphries (eds.), *Monocotyledons: Systematics and Evolution*. Royal Botanic Gardens, Kew, United Kingdom.

Kress, W. J. & L. M. Prince. 2000. Cannaceae Jussieu. Canna family. Pp. 310–314 *In*: Flora North America Editorial Committee (eds.), *Flora of North America North of Mexico, Vol. 22: Magnoliophyta: Alismatidae, Arecidae, Commelinidae (in Part), and Zingiberidae*. Oxford University Press, New York, NY.

Kress, W. J., L. M. Prince, W. J. Hahn & E. A. Zimmer. 2001. Unraveling the evolutionary radiation of the families of the Zingiberales using morphological and molecular evidence. *Syst. Biol.* **50**: 926–944.

Krock, S. 2016. Effects of sowing time and relative prairie quality on first year establishment of 23 native prairie species. M.E.S. thesis. Evergreen State College, Olympia, WA. 88 pp.

Kron, P. & S. C. Stewart. 1994. Variability in the expression of a rhizome architecture model in a natural population of *Iris versicolor* (Iridaceae). *Am. J. Bot.* **81**: 1128–1138.

Kron, P., S. C. Stewart & A. Back. 1993. Self-compatibility, autonomous self-pollination, and insect-mediated pollination in the clonal species *Iris versicolor*. *Can. J. Bot.* **71**: 1503–1509.

Krull, J. N. 1969. Factors affecting plant die-offs in shallow water areas. *Am. Midl. Nat.* **82**: 293–295.

Krull, J. N. 1970. Aquatic plant-macroinvertebrate associations and waterfowl. *J. Wildl. Manag.* **34**: 707–718.

Krull, J. N. 1976. Abundance and diversity of benthos during the spring waterfowl migration. *Am. Midl. Nat.* **95**: 459–462.

Kruse, C., R. Boehm, D. Voeste, S. Barth & H. Schnabl. 2002. Transient transformation of *Wolffia columbiana* by particle bombardment. *Aquat. Bot.* **72**: 175–181.

Ktita, S. R., A. Chermiti & M. Mahouachi. 2010. The use of seaweeds (*Ruppia maritima* and *Chaetomorpha linum*) for lamb fattening during drought periods. *Small Rumin. Res.* **91**: 116–119.

Ktita, S. R., A. Chermiti & M. Mahouachi. 2014. The use of marine plants for growing rabbits. *Livest. Res. Rural Dev.* **26**: 209. Retrieved February 12, 2018 from: http://www.lrrd.org/lrrd26/11/rjib26209.html.

Kubitzki, K. 1998. Cannaceae. Pp. 103–106 *In*: K. Kubitzki, H. Huber, P. J. Rudall, P. S. Stevens & T. Stützel (eds.), *The Families and Genera of Vascular Plants, Vol. IV. Flowering Plants, Monocotyledons: Alismatanae and Commelinanae (Except Gramineae)*. Springer-Verlag, Berlin, Germany.

Kuehdorf, K. & K.-J. Appenroth. 2012. Influence of salinity and high temperature on turion formation in the duckweed *Spirodela polyrhiza*. *Aquat. Bot.* **97**: 69–72.

Kuehdorf, K., G. Jetschke, L. Ballani & K.-J. Appenroth. 2014. The clonal dependence of turion formation in the duckweed *Spirodela polyrhiza*—an ecogeographical approach. *Physiol. Plant.* **150**: 46–54.

Kuehn, M. M., J. E. Minor & B. N. White. 1999. An examination of hybridization between the cattail species *Typha latifolia* and *Typha angustifolia* using random amplified polymorphic DNA and chloroplast DNA markers. *Mol. Ecol.* **8**: 1981–1990.

Kuehne, L. M., J. D. Olden & E. S. Rubenson. 2016. Multi-trophic impacts of an invasive aquatic plant. *Freshw. Biol.* **61**: 1846–1861.

Kujawski, J. & R. Thompson. 2000. Propagation of redhead grass (*Potamogeton perfoliatus* L.) transplants for restoration projects. *Native Plant J.* **1**: 124–127.

Kuk, Y. I., H. I. Jung, O. D. Kwon, N. R. Burgos & J. O. Guh. 2003. Rapid diagnosis of resistance to sulfonylurea herbicides in monochoria (*Monochoria vaginalis*). *Weed Sci.* **51**: 305–311.

Kukula-Koch, W., E. Sieniawska, J. Widelski, O. Urjin, P. Głowniak & K. Skalicka-Woźniak. 2015. Major secondary metabolites of *Iris* spp. *Phytochem. Rev.* **14**: 51–80.

Kula, R. R. & G. Zolnerowich. 2008. Revision of New World *Chaenusa* Haliday sensu lato (Hymenoptera: Braconidae: Alysiinae), with new species, synonymies, hosts, and distribution records. *Proc. Entomol. Soc. Washington* **110**: 1–60.

Kullberg, R. G. 1974. Distribution of aquatic macrophytes related to paper mill effluents in a southern Michigan stream. *Am. Midl. Nat.* **91**: 271–281.

Kumar, K. & M. Singh. 2015. Chromosomal diversity among different ecotypes of *Acorus calamus* L. reported from Ranchi Jharkhand, India. *Int. J. Bioassays* **4**: 3656–3658.

Kunii, H. 1981. Characteristics of the winter growth of detached *Elodea nuttallii* (Planch.) St. John in Japan. *Aquat. Bot.* **11**: 57–66.

Kuntz, K., P. Heidbüchel & A. Hussner. 2014. Effects of water nutrients on regeneration capacity of submerged aquatic plant fragments. *Int. J. Limnol.* **50**: 155–162.

Kunze, H. 1985. Comparative studies of the flowers in Cannaceae and Marantaceae. *Flora (Jena)* **175**: 301–318.

Kunzelman, J. I., M. J. Durako, W. J. Kenworthy, A. Stapleton & J. L. C. Wright. 2005. Irradiance-induced changes in the photobiology of *Halophila johnsonii*. *Mar. Biol.* **148**: 241–250.

Kuo, J. 2013. Chromosome numbers of the Australian Cymodoceaceae. *Plant Syst. Evol.* **299**: 1443–1448.

Kuoh, C.-S., M.-J. Yang & G.-I. Liao. 2000. The flower structure and anther dehiscence of *Wolffia arrhiza* (Lemnaceae). *Taiwania* **45**: 30–37.

Kurugundla, C. N. 2014. Seed dynamics and control of *Pistia stratiotes* in two aquatic systems in Botswana. *Afr. J. Aquat. Sci.* **39**: 209–214.

Kurz, H. & D. Crowson. 1948. The flowers of *Wolffiella floridana* (JD Sm.) Thompson. *Quart. J. Florida Acad. Sci.* **11**: 87–98.

Kuzmina, M. L., T. W. A. Braukmann & E. V. Zakharov. 2018. Finding the pond through the weeds: eDNA reveals underestimated diversity of pondweeds. *Appl. Plant Sci.* **6**(5): e1155.

Kwembeya, E. G., C. S. Bjorå, B. Stedje & I. Nordal. 2007. Phylogenetic relationships in the genus *Crinum* (Amaryllidaceae) with emphasis on tropical African species: evidence from *trnL-F* and nuclear ITS DNA sequence data. *Taxon* **56**: 801–810.

Laberge, V., S. Hugron, L. Rochefort & M. Poulin. 2015. Influence of different bryophyte carpets on vascular plant establishment around pools in restored peatlands. *Land Degrad. Dev.* **26**: 813–818.

Labisky, R. F., C. C. Hurd, M. K. Oli & R. S. Barwick. 2003. Foods of white-tailed deer in the Florida Everglades: the significance of *Crinum*. *Southeast. Nat.* **2**: 261–270.

Lacor, M. A. M. 1968. Flowering of *Spirodela polyrhiza* (L.) Schleiden. *Acta Bot. Neerl.* **17**: 357–359.

Lacor, M. A. M. 1969. On the influence of gibberellic acid and kinetin on the germination of turions of *Spirodela polyrhiza* (L.) Schleiden. *Acta Bot. Neerl.* **18**: 550–557.

Lacor, M. A. M. 1970. Some physiological and morphogenetic aspects of flowering of *Spirodela polyrhiza* (L.) Schleiden. *Acta Bot. Neerl.* **19**: 53–60.

Lacoul, P. & B. Freedman. 2006. Relationships between aquatic plants and environmental factors along a steep Himalayan altitudinal gradient. *Aquat. Bot.* **84**: 3–16.

Lacoul, P., B. Freedman & T. Clair. 2011. Effects of acidification on aquatic biota in Atlantic Canada. *Environ. Rev.* **19**: 429–460.

Lacroix, C. R. & J. R. Kemp. 1997. Developmental morphology of the androecium and gynoecium in *Ruppia maritima* L.: considerations for pollination. *Aquat. Bot.* **59**: 253–262.

Ladd, D. 2014. Ecologically appropriate fire in the Missouri landscape: a 35 year reflection. *Missouri Nat. Areas Newslett.* **14**(1): 31–35.

Lafabrie, C., K. M. Major, C. S. Major, M. M. Miller & J. Cebrián. 2011. Comparison of morphology and photo-physiology with metal/metalloid contamination in *Vallisneria neotropicalis*. *J. Hazard. Mater.* **191**: 356–365.

Lafferty, K. D., R. F. Hechinger, J. C. Shaw, K. L. Whitney & A. M. Kuris. 2006. Food webs and parasites in a salt marsh ecosystem. Pp. 119–134 *In*: S. Collinge & C. Ray (eds.), *Disease Ecology: Community Structure and Pathogen Dynamics*. Oxford University Press, Oxford, United Kingdom.

LaFrankie, J. V. 1985. A note on seedling morphology and establishment growth in the genus *Smilacina* (Liliaceae). *Bull. Torrey Bot. Club* **112**: 313–317.

LaFrankie, J. V. 1986a. Transfer of the species of *Smilacina* to *Maianthemum* (Liliaceae). *Taxon* **35**: 584–589.

LaFrankie, J. V. 1986b. Morphology and taxonomy of the New World species of *Maianthemum* (Liliaceae). *J. Arnold Arbor.* **67**: 371–439.

LaFrankie, J. V. 2002. *Maianthemum* F. H. Wiggers. Pp. 206–207 *In*: Flora North America Editorial Committee (eds.), *Flora of North America North of Mexico, Vol. 26: Magnoliophyta: Liliidae: Liliales and Orchidales*. Oxford University Press, New York, NY.

Lagler, K. F. 1943. Food habits and economic relations of the turtles of Michigan with special reference to fish management. *Am. Midl. Nat.* **29**: 257–312.

Lagnika, L., B. Attioua, B. Weniger, M. Kaiser, A. Sanni & C. Vonthron-Senecheau. 2008. Phytochemical study and antiprotozoal activity of compounds isolated from *Thalia geniculata*. *Pharm. Biol.* **46**: 162–165.

Lagueux, C. J., K. A. Bjorndal, A. B. Bolten & C. L. Campbell. 1995. Food habits of *Pseudemys concinna suwanniensis* in a Florida spring. *J. Herpetol.* **29**: 122–126.

Laguna, E. & P. F. Gallego. 2009. *Canna glauca* L. (Cannaceae), un nuevo taxón naturalizado para la Flora Peninsular Ibérica. *Lagascalia* **29**: 292–295.

Lahring, H. 2003. *Water and Wetland Plants of the Prairie Provinces*. Canadian Plains Research Center, University of Regina, Regina, Saskatchewan. 327 pp.

Laidig, K. J. 2012. Simulating the effect of groundwater withdrawals on intermittent-pond vegetation communities. *Ecohydrology* **5**: 841–852.

Laidig, K. J. & R. A. Zampella. 1996. *Stream Vegetation Data for Twenty Long-Term Study Sites in the New Jersey Pinelands*. Pinelands Commission, New Lisbon, NJ. 78 pp.

Laidig, K. J., R. A. Zampella, J. F. Bunnell, C. L. Dow & T. M. Sulikowski. 2001. *Characteristics of Selected Pine Barrens Treefrog Ponds in the New Jersey Pinelands*. Pinelands Commission, New Lisbon, NJ. 43 pp.

Laidig, K. J., R. A. Zampella & C. Popolizio. 2009. Hydrologic regimes associated with *Helonias bullata* L. (swamp pink) and the potential impact of simulated water-level reductions. *J. Torrey Bot. Soc.* **136**: 221–233.

Laing, K. K. & D. G. Raveling. 1993. Habitat and food selection by emperor goose goslings. *Condor* **95**: 879–888.

Lakela, O. 1941. *Sparganium glomeratum* in Minnesota. *Rhodora* **43**: 83–85.

Lakshmi, N. 1978. Cytological studies in two allopolyploid species of the genus *Hymenocallis*. *Cytologia* **43**: 555–563.

Lal, C. & B. Gopal. 1993. Production and germination of seeds in *Hydrilla verticillata*. *Aquat. Bot.* **45**: 257–261.

Lamb, E. G. & W. Megill. 2003. The shoreline fringe forest and adjacent peatlands of the southern central British Columbia coast. *Can. Field-Nat.* **117**: 209–217.

Lamont, E. E. 1998. Status of *Schizaea pusilla* in New York, with notes on some early collections. *Am. Fern J.* **88**: 158–164.

Lamont, E. E. & R. Stalter. 2007. Orchids of Atlantic coast barrier islands from North Carolina to New York. *J. Torrey Bot. Soc.* **134**: 540–552.

Lamont, E. E., R. Sivertsen, C. Doyle & L. Adamec. 2013. Extant populations of *Aldrovanda vesiculosa* (Droseraceae) in the New World. *J. Torrey Bot. Soc.* **140**: 517–522.

LaMontagne, J. M., L. J. Jackson & R. M. R. Barclay. 2003. Compensatory growth responses of *Potamogeton pectinatus* to foraging by migrating trumpeter swans in spring stop over areas. *Aquat. Bot.* **76**: 235–244.

Lamoureux, G. (ed.). 1987. *Plantes sauvages des lacs, rivières et tourbières. Guide d'identification Fleurbec*. Fleurbec, Saint-Augustin (Portneuf), QC. 399 pp.

Lampe, K. F. 1986. Dermatitis-producing plants of south Florida and Hawaii. *Clin. Dermatol.* **4**: 83–93.

Landers, J. L., A. S. Johnson, P. H. Morgan & W. P. Baldwin. 1976. Duck foods in managed tidal impoundments in South Carolina. *J. Wildl. Manag.* **40**: 721–728.

Landers, J. L., T. T. Fendley & A. S. Johnson. 1977. Feeding ecology of wood ducks in South Carolina. *J. Wildl. Manag.* **41**: 118–127.

Landers, J. L., R. J. Hamilton, A. S. Johnson & R. L. Marchinton. 1979. Foods and habitat of black bears in southeastern North Carolina. *J. Wildl. Manag.* **43**: 143–153.

Landesman, L., N. C. Parker, C. B. Fedler & M. Konikoff. 2005. Modeling duckweed growth in wastewater treatment systems. *Livest. Res. Rural Dev.* **17**: 1–8.

Landman, G. B. & E. S. Menges. 1999. Dynamics of woody bayhead invasion into seasonal ponds in south central Florida. *Castanea* **64**: 130–137.

Landman, G. B., R. K. Kolka & R. R. Sharitz. 2007. Soil seed bank analysis of planted and naturally revegetating thermally-disturbed riparian wetland forests. *Wetlands* **27**: 211–223.

Landolt, E. 1975. Morphological differentiation and geographical distribution of the *Lemna gibba-Lemna minor* group. *Aquat. Bot.* **1**: 345–363.

Landolt, E. 1981. Distribution pattern of the family Lemnaceae in North Carolina. *Veröff. Geobot. Inst. E. T. H. Stiftung Rübel Zürich* **77**: 112–148.

Landolt, E. 1986. Biosystematic investigations in the family of duckweeds (Lemnaceae), vol. 2. The family of Lemnaceae – a monographic study, vol. 1. *Veröff. Geobot. Inst. E. T. H. Stiftung Rübel Zürich* **71**: 1–566.

Landolt, E. 1994. Taxonomy and ecology of the section *Wolffia* of the genus *Wolffia* (Lemnaceae). *Ber. Geobot. Inst. E. T. H., Stiftung Rübel* **60**: 137–151.

Landolt, E. 1997. How do Lemnaceae (duckweed family) survive dry conditions? *Bull. Geobot. Inst. E. T. H. Zürich* **63**: 25–31.

Landolt, E. 1998. *Lemna yungensis*, a new duckweed species from rocks of the Andean Yungas in Bolivia. *Bull. Geobot. Inst. E. T. H. Zürich* **64**: 15–21.

Landolt, E. 1998. Lemnaceae. Pp. 264–270 *In*: K. Kubitzki (ed.), *The Families and Genera of Vascular Plants, Vol. IV, Flowering Plants: Monocotyledons, Alismatanae and Commelinanae (Except Gramineae)*. Springer-Verlag, Berlin, Germany.

Landolt, E. 2000. Lemnaceae Gray. Duckweed family. Pp. 143–153 *In*: Flora North America Editorial Committee (eds.), *Flora of North America North of Mexico, Vol. 22: Magnoliophyta: Alismatidae, Arecidae, Commelinidae (in Part), and Zingiberidae*. Oxford University Press, New York, NY.

Landolt, E. & O. Wildi. 1977. Ökologische Felduntersuchungen bei Wasserlinsen (Lemnaceae) in den südwestlichen Staaten der USA. *Ber. Geobot. Inst. E. T. H. Stiftung Rübel Zürich* **44**: 104–146.

Landolt, E. & R. Kandeler. 1987. Biosystematic investigations in the family of duckweeds (Lemnaceae), vol. 4. The family of Lemnaceae – a monographic study, vol. 2. *Veröff. Geobot. Inst. E. T. H. Stiftung Rübel Zürich* **95**: 1–638.

Landolt, E., I. Jäger-Zürn & R. A. A. Schnell. 1998. *Extreme Adaptations in Angiospermous Hydrophytes*. Gebrüder Borntraeger, Berlin, Germany. 290 pp.

Lang, F. A. & P. F. Zika. 1997. A nomenclatural note on *Hastingsia bracteosa* and *Hastingsia atropurpurea*. *Madroño* **44**: 189–192.

Langeland, K. A. 1989. Karyotypes of hydrilla (Hydrocharitaceae) populations in the United States. *J. Aquat. Plant Manag.* **27**: 111–115.

Langeland, K. A. & D. L. Sutton. 1980. Regrowth of hydrilla from axillary buds. *J. Aquat. Plant Manag.* **18**: 27–29.

Langeland, K. A., D. G. Shilling, J. L. Carter, F. B. Laroche, K. K. Steward & P. T. Madiera. 1992. Chromosome morphology and number in various populations of *Hydrilla verticillata* (L.f.) Royle. *Aquat. Bot.* **42**: 253–263.

Lansdown, R. V. 2008. Red duckweed (*Lemna turionifera* Landolt) new to Britain. *Watsonia* **27**: 127–130.

Lansdown, R. V., P. Anastasiu, Z. Barina, I. Bazos, H. Çakan, D. Caković, P. Delipetrou, V. Matevski, B. Mitić, E. Ruprecht, G. Tomović, A. Tosheva & G. Király. 2016. Review of alien freshwater vascular plants in south-east Europe. Pp. 137–154 *In*: M. Rat, T. Trichkova, R. Scalera, R. Tomov & A. Uludag (eds.), *State of the Art of Invasive Alien Species in South-Eastern Europe*. ESENIAS, IBER BAS, Sofia, Bulgaria.

Lantz, D. E. 1910. The muskrat. Farmers' Bulletin 396. U.S. Department of Agriculture, Government Printing Office, Washington, DC. 38 pp.

La Peyre, M. K. & S. Rowe. 2003. Effects of salinity changes on growth of *Ruppia maritima* L. *Aquat. Bot.* **77**: 235–241.

Lapolli, F. R., M. A. Lobo-Recio, P. B. Filho, J. B. R. Rodrigues & F. A. Tavares. 2008. Performance of the macrophyte *Lemna valdiviana* in tertiary pig waste treatment and its contribution to the sustainability of swine production. *Biotemas* **21**: 17–27.

Lara, M. V., P. Casati & C. S. Andreo. 2002. CO_2-concentrating mechanisms in *Egeria densa*, a submersed aquatic plant. *Physiol. Plant.* **115**: 487–495.

Larkin, P., T. Schonacher, M. Barrett & M. Paturzzio. 2012. Development and characterization of microsatellite markers for the seagrass *Halodule wrightii*. *Conserv. Genet. Res.* **4**: 511–513.

Larkin, P. D., T. J. Maloney, S. Rubiano-Rincon & M. M. Barrett. 2017. A map-based approach to assessing genetic diversity, structure, and connectivity in the seagrass *Halodule wrightii*. *Mar. Ecol. Prog. Ser.* **567**: 95–107.

Larkin, D. J., A. K. Monfils, A. Boissezon, R. S. Sleith, P. M. Skawinski, C. H. Welling, B. C. Cahill & K. G. Karol. 2018. Biology, ecology, and management of starry stonewort (*Nitellopsis obtusa*; Characeae): a red-listed Eurasian green alga invasive in North America. *Aquat. Bot.* **148**: 15–24.

Larkum, A. W. D. 1995. *Halophila capricorni* (Hydrocharitaceae): a new species of seagrass from the Coral Sea. *Aquat. Bot.* **51**: 319–328.

Larned, S. T. 2003. Effects of the invasive, nonindigenous seagrass *Zostera japonica* on nutrient fluxes between the water column and benthos in a NE Pacific estuary. *Mar. Ecol. Prog. Ser.* **254**: 69–80.

Laroche, V., S. Pellerin & L. Brouillet. 2012. White fringed orchid as indicator of *Sphagnum* bog integrity. *Ecol. Indicators* **14**: 50–55.

Larsen, K., J. M. Lock, H. Maas & P. J. M. Maas. 1998. Zingiberaceae. Pp. 474–495 *In*: K. Kubitzki (ed.), *The Families and Genera of Vascular Plants, Vol. IV, Flowering Plants: Monocotyledons, Alismatanae and Commelinanae (Except Gramineae)*. Springer-Verlag, Berlin, Germany.

Larson, K. S. & R. J. Larson. 1990. Lure of the locks: showiest ladies-tresses orchids, *Spiranthes romanzoffiana*, affect bumblebee, *Bombus* spp., foraging behavior. *Can. Field-Nat.* **104**: 519–525.

Larson, J. S. & K. B. Searcy. 2007. *Lemna minuta* (Lemnaceae) discovered in Massachusetts. *Rhodora* **109**: 456–458.

Lastra, M., H. M. Page, J. E. Dugan, D. M. Hubbard & I. F. Rodil. 2008. Processing of allochthonous macrophyte subsidies by sandy beach consumers: estimates of feeding rates and impacts on food resources. *Mar. Biol.* **154**: 163–174.

Lastrucci, L., R. Calamassi, G. Ferretti, G. Galasso & B. Foggi. 2012. Contributo alla conoscenza della flora esotica dell'Isola di Capraia (Arcipelago Toscano, Italia). *Nat. Hist. Sci.* **153**: 127–134.

Latham, R. E., J. E. Thompson, S. A. Riley & A. W. Wibiralske. 1996. The Pocono till barrens: shrub savanna persisting on soils favoring forest. *Bull. Torrey Bot. Club* **123**: 330–349.

Laublin, G. & M. Cappadocia. 1992. In vitro ovary culture of some apogon garden irises (*Iris pseudacorus* L., *I. setosa* Pall., *I. versicolor* L.). *Bot. Acta* **105**: 319–322.

Lauer, N., M. Yeager, A. E. Kahn, D. R. Dobberfuhl & C. Ross. 2011. The effects of short term salinity exposure on the sublethal stress response of *Vallisneria americana* Michx. (Hydrocharitaceae). *Aquat. Bot.* **95**: 207–213.

Lavoie, C., M. Jean, F. Delisle & G. Létourneau. 2003. Exotic plant species of the St Lawrence River wetlands: a spatial and historical analysis. *J. Biogeogr.* **30**: 537–549.

Lay, D. W. 1965. Fruit utilization by deer in southern forests. *J. Wildl. Manag.* **29**: 370–375.

Layne, J. N. & B. S. Johns. 1965. Present status of the beaver in Florida. *Quart. J. Florida Acad. Sci.* 28: 212–220.

Leach, G. J. 2017. A revision of Australian *Eriocaulon* (Eriocaulaceae). *Telopea* **20**: 205–259.

LeBlond, R. J. & B. A. Sorrie. 2001. Additions to and noteworthy records for the flora of the coastal plain of North Carolina. *Castanea* **66**: 288–302.

Leck, M. A. & K. J. Graveline. 1979. The seed bank of a freshwater tidal marsh. *Am. J. Bot.* **66**: 1006–1015.

Leck, M. A. & R. L. Simpson. 1987. Seed bank of a freshwater tidal wetland: turnover and relationship to vegetation change. *Am. J. Bot.* **74**: 360–370.

Leck, M. A. & R. L. Simpson. 1993. Seeds and seedlings of the Hamilton Marshes, a Delaware River tidal freshwater wetland. *Proc. Acad. Nat. Sci. Philadelphia* **144**: 267–281.

Leck, M. A. & R. L. Simpson. 1995. Ten-year seed bank and vegetation dynamics of a tidal freshwater marsh. *Am. J. Bot.* **82**: 1547–1557.

Leck, M. A., R. L. Simpson, D. F. Whigham & C. F. Leck. 1988. Plants of the Hamilton Marshes: a Delaware River freshwater tidal wetland. *Bartonia* **54**: 1–17.

Lee, W. L. 1966. Color change and the ecology of the marine isopod *Idothea* (Pentidothea) *montereyensis* Maloney, 1933. *Ecology* **47**: 930–941.

Lee, P. F. 1987. Ecological relationships of wild rice, *Zizania aquatica*. 5. Enhancement of wild rice production by *Potamogeton robbinsii*. *Can. J. Bot.* **65**: 1433–1438.

Lee, D. W. & D. E. Fairbrothers. 1969. A serological and disc electrophoretic study of North American *Typha*. *Brittonia* **21**: 227–243.

Lee, D. W. & D. E. Fairbrothers. 1972. Taxonomic placement of the Typhales within the monocotyledons: preliminary serological investigation. *Taxon* **21**: 39–44.

Lee, K.-S. & K. H. Dunton. 1999. Influence of sediment nitrogen-availability on carbon and nitrogen dynamics in the seagrass *Thalassia testudinum*. *Mar. Biol.* **134**: 217–226.

Lee, C.-L., T. C. Wang, C.-K. Lin & H.-K. Mok. 1999. Heavy metals removal by a promising locally available aquatic plant, *Najas graminea* Del., in Taiwan. *Water Sci. Technol.* **39**: 177–181.

Lee, A., G. Gordon, R. T. Saucier, B. D. Maygarden & M. Godzinski. 2000. *Cultural Resource Survey for the West Bank Vicinity of New Orleans, Louisiana, Hurricane Protection Project*. Earth Search Inc., New Orleans, LA. 57 pp.

Lee, C. S., S.-C. Kim, S. H. Yeau & N. S. Lee. 2011. Major lineages of the genus *Lilium* (Liliaceae) based on nrDNA ITS sequences, with special emphasis on the Korean species. *J. Plant Biol.* **54**: 159–171.

Lee, H. J., M. H. Kim, Y. Y. Choi, E. H. Kim, J. Hong, K. Kim & W. M. Yang. 2016. Improvement of atopic dermatitis with topical application of *Spirodela polyrhiza*. *J. Ethnopharmacol.* **180**: 12–17.

Lefebvre, L. W., J. A. Provancha, D. H. Slone & W. J. Kenworthy. 2017. Manatee grazing impacts on a mixed species seagrass bed. *Mar. Ecol. Prog. Ser.* **564**: 29–45.

Legner, E. F. & T. W. Fisher. 1980. Impact of *Tilapia zillii* (Gervais) on *Potamogeton pectinatus* L., *Myriophyllum spicatum* var. *exalbescens* Jepson, and mosquito reproduction in lower Colorado Desert irrigation canals. *Acta Oecol.* **1**: 3–14.

Lehman, R. L., R. O'Brien & T. White. 2009. *Plants of the Texas Coastal Bend*. Texas A&M University Press, College Station, TX. 368 pp.

Lehmann, N. L. & R. Sattler. 1992. Irregular floral development in *Calla palustris* (Araceae) and the concept of homeosis. *Am. J. Bot.* **79**: 1145–1157.

Lehmiller, D. J. 1992. Interspecific hybrids of *Crinum americanum* L. (Amaryllidaceae). *Acta Hort.* **325**: 591–596.

Lehtonen, S. 2007. *Natural History of Echinodorus (Alismataceae)*. Turun Yliopisto, Turku, Finland. 120 pp.

Lehtonen, S. 2009. Systematics of the Alismataceae—a morphological evaluation. *Aquat. Bot.* **91**: 279–290.

Lehtonen, S. 2009a. On the origin of *Echinodorus grandiflorus* (Alismataceae) in Florida ("*E. floridanus*"), and its estimated potential as an invasive species. *Hydrobiologia* **635**: 107–112.

Lehtonen, S. & L. Myllys. 2008. Cladistic analysis of *Echinodorus* (Alismataceae): simultaneous analysis of molecular and morphological data. *Cladistics* **24**: 218–239.

Leidolf, A. & S. McDaniel. 1998. A floristic study of black prairie plant communities at sixteen section prairie, Oktibbeha County, Mississippi. *Castanea* **63**: 51–62.

Leif, J. W. & E. A. Oelke. 1990. Growth and development of giant burreed (*Sparganium eurycarpum*). *Weed Technol.* **4**: 849–854.

Leighton, D. & R. A. Boolootian. 1963. Diet and growth in the black abalone, *Haliotis cracerodii*. *Ecology* **44**: 227–238.

Leitch, I. J., I. Kahandawala, J. Suda, L. Hanson, M. J. Ingrouille, M. W. Chase & M. F. Fay. 2009. Genome size diversity in orchids: consequences and evolution. *Ann. Bot.* **104**: 469–481.

Leitch, I. J., J. M. Beaulieu, M. W. Chase, A. R. Leitch & M. F. Fay. 2010. Genome size dynamics and evolution in monocots. *J. Bot.* **2010**: 862516.

Lemay, M.-A., L. De Vriendt, S. Pellerin & M. Poulin. 2015. Ex situ germination as a method for seed viability assessment in a peatland orchid, *Platanthera blephariglottis*. *Am. J. Bot.* **102**: 390–395.

Lemke, D. E. 1989. Aquatic macrophytes of the upper San Marcos River, Hays Co., Texas. *Southwest. Nat.* **34**: 289–291.

Lemon, P. C. 1949. Successional responses of herbs in the longleaf-slash pine forest after fire. *Ecology* **30**: 135–145.

Lemon, G. D., U. Posluszny & B. C. Husband. 2001. Potential and realized rates of vegetative reproduction in *Spirodela polyrhiza*, *Lemna minor*, and *Wolffia borealis*. *Aquat. Bot.* **70**: 79–87.

Lentz, D. L. 1993. Medicinal and other economic plants of the Paya of Honduras. *Econ. Bot.* **47**: 358–370.

Leonard, S. W. & W. W. Baker. 1983. Additional populations of *Harperocallis flava* McDaniel (Liliaceae). *Castanea* **48**: 151–152.

LeResche, R. E. & J. L. Davis. 1973. Importance of nonbrowse foods to moose on the Kenai Peninsula, Alaska. *J. Wildl. Manag.* **37**: 279–287.

Lerman, J. C. & E. M. Cigliano. 1971. New carbon-14 evidence for six hundred years old *Canna compacta* seed. *Nature* **232**: 568–570.

Les, D. H. 1983. Taxonomic implications of aneuploidy and polyploidy in *Potamogeton* (Potamogetonaceae). *Rhodora* **85**: 301–323.

Les, D. H. 1988. Breeding systems, population structure, and evolution in hydrophilous angiosperms. *Ann. Missouri Bot. Gard.* **75**: 819–835.

Les, D. H. 2017. *Aquatic Dicotyledons of North America: Ecology, Life History, and Systematics.* CRC Press, Boca Raton, FL. 1334 pp.

Les, D. H. & D. J. Sheridan. 1990. Biochemical heterophylly and flavonoid evolution in North American *Potamogeton* (Potamogetonaceae). *Am. J. Bot.* **77**: 453–465.

Les, D. H. & C. T. Philbrick. 1993. Studies of hybridization and chromosome number variation in aquatic plants: evolutionary implications. *Aquat. Bot.* **44**: 181–228.

Les, D. H. & R. R. Haynes. 1995. Systematics of subclass Alismatidae: a synthesis of approaches. Pp. 353–377 *In*: P. J. Rudall, P. J. Cribb, D. F. Cutler & C. J. Humphries (eds.), *Monocotyledons: Systematics and Evolution.* Royal Botanic Gardens, Kew, United Kingdom.

Les, D. H. & R. R. Haynes. 1996. *Coleogeton* (Potamogetonaceae), a new genus of pondweeds. *Novon* **6**: 389–391.

Les, D. H. & R. S. Capers. 1999. *Limnobium spongia* (Hydrocharitaceae) discovered in New England. *Rhodora* **101**: 419–423.

Les, D. H. & D. J. Crawford. 1999. *Landoltia* (Lemnaceae), a new genus of duckweeds. *Novon* **9**: 530–533.

Les, D. H. & L. J. Mehrhoff. 1999. Introduction of nonindigenous aquatic vascular plants in southern New England: a historical perspective. *Biol. Invasions* **1**: 281–300.

Les, D. H. & N. Tippery. 2013. In time and with water … the systematics of alismatid monocotyledons. Pp. 118–164 *In*: P. Wilkin & S. J. Mayo (eds.), *Early Events in Monocot Evolution.* Cambridge University Press, Cambridge, United Kingdom.

Les, D. H., D. K. Garvin & C. F. Wimpee. 1993. Phylogenetic studies in the monocot subclass Alismatidae: evidence for a reappraisal of the aquatic order Najadales. *Mol. Phylogenet. Evol.* **2**: 304–314.

Les, D. H., M. A. Cleland & M. Waycott. 1997a. Phylogenetic studies in Alismatidae, II: evolution of marine angiosperms ('seagrasses') and hydrophily. *Syst. Bot.* **22**: 443–463.

Les, D. H., E. Landolt & D. J. Crawford. 1997b. Systematics of Lemnaceae: inferences from micromolecular and morphological data. *Plant Syst. Evol.* **204**: 161–177.

Les, D. H., L. J. Mehrhoff, M. A. Cleland & J. D. Gabel. 1997c. *Hydrilla verticillata* (Hydrocharitaceae) in Connecticut. *J. Aquat. Plant Manag.* **35**: 10–14.

Les, D. H., D. J. Crawford, E. Landolt, J. D. Gabel & R. T. Kimball. 2002. Phylogeny and systematics of Lemnaceae, the duckweed family. *Syst. Bot.* **27**: 221–240.

Les, D. H., D. J. Crawford, R. T. Kimball, M. L. Moody & E. Landolt. 2003. Biogeography of discontinuously distributed hydrophytes: a molecular appraisal of intercontinental disjunctions. *Int. J. Plant Sci.* **164**: 917–932.

Les, D. H., M. L. Moody & S. W. L. Jacobs. 2005. Phylogeny and systematics of *Aponogeton* (Aponogetonaceae): the Australian species. *Syst. Bot.* **30**: 503–519.

Les, D. H., M. L. Moody & C. Soros. 2006a. A reappraisal of phylogenetic relationships in the monocotyledon family Hydrocharitaceae. *Aliso* **22**: 211–230.

Les, D. H., R. S. Capers & N. P. Tippery. 2006b. Introduction of *Glossostigma* (Phrymaceae) to North America: a taxonomic and ecological overview. *Am. J. Bot.* **93**: 927–939.

Les, D. H., S. W. L. Jacobs, N. Tippery, L. Chen, M. L. Moody & M. Wilstermann. 2008. Systematics of *Vallisneria* L. (Hydrocharitaceae Juss.). *Syst. Bot.* **33**: 49–65.

Les, D. H., N. M. Murray & N. P. Tippery. 2009. Systematics of two imperiled pondweeds (*Potamogeton vaseyi, P. gemmiparus*) and taxonomic ramifications for subsection *Pusilli* (Potamogetonaceae). *Syst. Bot.* **34**: 643–651.

Les, D. H., S. P. Sheldon & N. P. Tippery. 2010. Hybridization in hydrophiles: natural interspecific hybrids in *Najas* L. (Hydrocharitaceae). *Syst. Bot.* **35**: 736–744.

Les, D. H., N. P. Tippery & H. Razifard. 2012a. Noteworthy collections. California. *Najas minor. Madroño* **59**: 232.

Les, D. H., N. P. Tippery & H. Razifard. 2012b. Noteworthy collections. Texas. *Najas minor. Madroño* **59**: 232–233.

Les, D. H., E. L. Peredo, L. K. Benoit, N. P. Tippery, U. M. King & S. P. Sheldon. 2013. Phytogeography of *Najas gracillima* (Hydrocharitaceae) in North America and its cryptic introduction to California. *Am. J. Bot.* **100**: 1905–1915.

Les, D. H., E. L. Peredo, N. P. Tippery, L. K. Benoit, H. Razifard, U. M. King, H. R. Na, H.-K. Choi, L. Chen, R. K. Shannon & S. P. Sheldon. 2015a. *Najas minor* (Hydrocharitaceae) in North America: a reappraisal. *Aquat. Bot.* **126**: 60–72.

Les, D. H., E. L. Peredo, U. M. King, L. K. Benoit, N. P. Tippery, C. J. Ball & R. K. Shannon. 2015b. Through thick and thin: cryptic sympatric speciation in the submersed genus *Najas* (Hydrocharitaceae). *Mol. Phylogenet. Evol.* **82**: 15–30.

Les, D. H., A. M. Les, U. M. King & E. L. Peredo. 2015c. *Najas flexilis* (Hydrocharitaceae) in Alaska: a reassessment. *Rhodora* **117**: 354–370.

Lesica, P. 1986. Vegetation and flora of Pine Butte Fen, Teton County, Montana. *Great Basin Nat.* **46**: 22–32.

Leslie, K. A. & M. P. Burbanck. 1979. Vegetation of granitic outcroppings at Kennesaw Mountain, Cobb County, Georgia. *Castanea* **44**: 80–87.

Lessa, M. A., C. V. Araújo, M. A. Kaplan, D. Pimenta, M. R. Figueiredo & E. Tibirica. 2008. Antihypertensive effects of crude extracts from leaves of *Echinodorus grandiflorus*. *Fundam. Clin. Pharmacol.* **22**: 161–168.

Letty, C. 1973. The genus *Zantedeschia*. *Bothalia* **11**: 5–26.

Léveillé-Bourret, É., M.-È. Garon-Labrecque & E. Thomson. 2017. Le statut de la naïade grêle (*Najas gracillima*, Najadaceae) au Québec. *Naturaliste Can.* **141**: 6–14.

Levey, D. J., H. A. Bissell & S. F. O'Keefe. 2000. Conversion of nitrogen to protein and amino acids in wild fruits. *J. Chem. Ecol.* **26**: 1749–1763.

Levine, J. M. 2000. Complex interactions in a streamside plant community. *Ecology* **81**: 3431–3444.

Lewis, C. E. 1970. Responses to chopping and rock phosphate on south Florida ranges. *J. Range Manag.* **23**: 276–282.

Lewis III, R. R. 1989. Creation and restoration of coastal plain wetlands in Florida. Pp. 73–94 *In*: J. A. Kusler & M. E. Kentula (eds.), *Wetland Creation and Restoration: The Status of the Science. Volume 1.* United States Environmental Protection Agency, Environmental Research Laboratory, Corvallis, OR.

Lewis, J. R. 1990. Amaryllidaceae alkaloids. *Nat. Prod. Rep.* **7**: 549–556.

Lewis, D. Q. 2002. Burmanniaceae Blume. *Burmannia* Family. Pp. 486–488 *In*: Flora North America Editorial Committee (eds.), *Flora of North America North of Mexico, Vol. 26: Magnoliophyta: Liliidae: Liliales and Orchidales.* Oxford University Press, New York, NY.

Lewis, F. J. & E. S. Dowding. 1926. The vegetation and retrogressive changes of peat areas ("muskegs") in central Alberta. *J. Ecol.* **14**: 317–341.

Lewis, R. R. & R. C. Phillips. 1980. Occurrence of seeds and seedlings of *Thalassia testudinum* Banks ex König in the Florida Keys (USA). *Aquat. Bot.* **9**: 377–380.

Lewis, J. B. & C. E. Hollingworth. 1982. Leaf epifauna of the sea-grass *Thalassia testudinum*. *Mar. Biol.* **71**: 41–49.

Lewis, F. J., E. S. Dowding & E. H. Moss. 1928. The vegetation of Alberta: II. The swamp, moor and bog forest vegetation of central Alberta. *J. Ecol.* **16**: 19–70.

Ley, A. C. & R. Claßen-Bockhoff. 2012. Floral synorganization and its influence on mechanical isolation and autogamy in Marantaceae. *Bot. J. Linn. Soc.* **168**: 300–322.

Lichvar, R. W., D. L. Banks, W. N. Kirchner & N. C. Melvin. 2016. The national wetland plant list: 2016 wetland ratings. *Phytoneuron* **2016–30**: 1–15.

Li, H. & P. C. Boyce. 2010. *Colocasia* Schott. Pp. 73–76 *In*: Flora of China Editorial Committee (eds.), *Flora of China. Vol. 23 (Acoraceae Through Cyperaceae)*. Science Press, Beijing/Missouri Botanical Garden Press, St. Louis, MO.

Li, W. G., J. J. Shen & J. B. Wang. 2005. Genetic diversity of the annual weed *Monochoria vaginalis* in southern China detected by random amplified polymorphic DNA and inter-simple sequence repeat analyses. *Weed Res.* **45**: 424–430.

Li, S., I. A. Mendelssohn, H. Chen & W. H. Orem. 2009. Does sulphate enrichment promote the expansion of *Typha domingensis* (cattail) in the Florida Everglades? *Freshw. Biol.* **54**: 1909–1923.

Li, J.-H., Z.-J. Liu, G. A. Salazar, P. Bernhardt, H. Perner, Y. Tomohisa, X.-H. Jin, S.-W. Chung & Y.-B. Luo. 2011. Molecular phylogeny of *Cypripedium* (Orchidaceae: Cypripedioideae) inferred from multiple nuclear and chloroplast regions. *Mol. Phylogenet. Evol.* **61**: 308–320.

Li, J., Q.-F. Wang, R. W. Gituru, C.-F. Yang & Y.-H. Guo. 2012. Reversible anther opening enhances male fitness in a dichogamous aquatic plant *Butomus umbellatus* L., the flowering rush. *Aquat. Bot.* **99**: 27–33.

Li, Z., W. Lu, L. Yang, X. Kong & X. Deng. 2015. Seed weight and germination behavior of the submerged plant *Potamogeton pectinatus* in the arid zone of northwest China. *Ecol. Evol.* **5**: 1504–1512.

Liang, W., X. Guo, D. G. Nagle, W.-D. Zhang & X.-H. Tian. 2019. Genus *Liparis*: a review of its traditional uses in China, phytochemistry and pharmacology. *J. Ethnopharmacol.* **234**: 154–171.

Lichthardt, J., R. K. Moseley & J. Mallet. 2000. *Ecological Assessment of Howellia aquatilis Habitat at the Harvard–Palouse River Flood Plain Site, Idaho*. Idaho Department of Fish and Game, Idaho Conservation Data Center, Boise, ID. 32 pp.

Lieffers, V. J. 1984. Emergent plant communities of oxbow lakes in northeastern Alberta: salinity, water-level fluctuation, and succession. *Can. J. Bot.* **62**: 310–316.

Liehrmann, O., F. Jégoux, M.-A. Guilbert, F. Isselin-Nondedeu, S. Saïd, Y. Locatelli & C. Baltzinger. 2018. Epizoochorous dispersal by ungulates depends on fur, grooming and social interactions. *Ecol. Evol.* **8**: 1582–1594.

Lien, E. J.-C., L. L.-M. Lien, R. Wang & J. Wang. 2012. Phytochemical analysis of medicinal plants with kidney protective activities. *Chin. J. Integ. Med.* **18**: 790–800.

Lieneman, C. 1929. A host index to the North American species of the genus *Cercospora*. *Ann. Missouri Bot. Gard.* **16**: 1–52.

Liggio, J. & A. O. Liggio. 2010. *Wild Orchids of Texas*. University of Texas Press, Austin, TX. 240 pp.

Light, H. M., M. R. Darst, M. T. MacLaughlin & S. W. Sprecher. 1993. Hydrology, vegetation, and soils of four north Florida river flood plains with an evaluation of state and federal wetland determinations. Water Resources Investigation Report 93-4033. U.S. Geological Survey, Tallahassee, FL. 94 pp.

Lillie, R. A. 1986. The spread of Eurasian watermilfoil *Myriophyllum spicatum* in Devils Lake, Sauk County, Wisconsin. *Lake Reserv. Manage.* **2**: 64–68.

Lim, K. Y., R. Matyasek, A. Kovarik & A. Leitch. 2007. Parental origin and genome evolution in the allopolyploid *Iris versicolor*. *Ann. Bot.* **100**: 219–224.

Lin, Y.-L. & B.-K. Li. 2016. Removal of pharmaceuticals and personal care products by *Eichhornia crassipes* and *Pistia stratiotes*. *J. Taiwan Inst. Chem. Eng.* **58**: 318–323.

Lin, L.-Z., S.-F. Hu, H.-B. Chai, T. Pengsuparp, J. M. Pezzuto, G. A. Cordell & N. Ruangrungsi. 1995. Lycorine alkaloids from *Hymenocallis littoralis*. *Phytochemistry* **40**: 1295–1298.

Lind, C. T. & G. Cottam. 1969. The submerged aquatics of University Bay: a study in eutrophication. *Am. Midl. Nat.* **81**: 353–369.

Lindig-Cisneros, R. & J. B. Zedler. 2002. Halophyte recruitment in a salt marsh restoration site. *Estuaries* **25**: 1174–1183.

Lindqvist, C., J. De Laet, R. R. Haynes, L. Aagesen, B. R. Keener & V. A. Albert. 2006. Molecular phylogenetics of an aquatic plant lineage, Potamogetonaceae. *Cladistics* **22**: 568–588.

Lindström, B. & B. Lüning. 1971. Studies on orchidaceae alkaloids. 23. Alkaloids from *Liparis loeselii* (L.) LC Rich. and *Hammarbya paludosa* (L.) OK. *Acta Chem. Scand.* **25**: 895–897.

Lindström, B. & B. Lüning. 1972. Studies on Orchidaceae alkaloids XXXV. Alkaloids from *Hammarbya paludosa* (L.) OK and *Liparis keitaoensis* Hay. *Acta Chem. Scand.* **25**: 895–897.

Linhart, Y. B. & A. C. Premoli. 1994. Genetic variation in central and disjunct populations of *Lilium parryi*. *Can. J. Bot.* **72**: 79–85.

Linke, Jr., W. R. 1963. *Drosera filiformis* in Connecticut. *Rhodora* **65**: 273.

Linn, J. G., E. J. Staba, R. D. Goodrich, J. C. Meiske & D. E. Otterby. 1975. Nutritive value of dried or ensiled aquatic plants. I. Chemical composition. *J. Animal Sci.* **41**: 601–609.

Linn, J. G., R. D. Goodrich, J. C. Meiske & E. J. Staba. 1973. Aquatic plants from Minnesota. Part 4 - nutrient composition. Bulletin 73. Water Resources Research Center, University of Minnesota, Minneapolis, MN. 22 pp.

Little, Jr., E. L. 1938. The vegetation of Muskogee County, Oklahoma. *Am. Midl. Nat.* **19**: 559–572.

Littlefield, C. D. 2001. Sandhill crane nest and egg characteristics at Malheur National Wildlife Refuge, Oregon. *Proc. N. Am. Crane Workshop* **8**: 40–44.

Littler, M. M. & D. S. Littler. 1981. Intertidal macrophyte communities from Pacific Baja California and the upper Gulf of California: relatively constant vs. environmentally fluctuating systems. *Mar. Ecol. Prog. Ser.* **4**: 145–158.

Littler, M. M., D. R. Martz & D. S. Littler. 1983. Effects of recurrent sand deposition on rocky intertidal organisms: importance of substrate heterogeneity in a fluctuating environment. *Mar. Ecol. Prog. Ser.* **11**: 129–139.

Liu, G.-H., W. Li, J. Zhou, W.-Z. Liu, D. Yang & A. J. Davy. 2006. How does the propagule bank contribute to cyclic vegetation change in a lakeshore marsh with seasonal drawdown? *Aquat. Bot.* **84**: 137–143.

Liu, L., Q.-M. Yin, X.-W. Zhang, W. Wang, X.-Y. Dong, X. Yan & R. Hu. 2016. Bioactivity-guided isolation of biphenanthrenes from *Liparis nervosa*. *Fitoterapia* **115**: 15–18.

Liu, Y., T. Sanguanphun, W. Yuan, J. J. Cheng & M. Meetam. 2017. The biological responses and metal phytoaccumulation of duckweed *Spirodela polyrhiza* to manganese and chromium. *Environ. Sci. Pollut. Res.* **24**: 19104–19113.

Lloyd, M. W., K. A. M. Engelhardt & M. C. Neel. 2009. Development of 11 polymorphic microsatellite markers in a macrophyte of conservation concern, *Vallisneria americana* Michaux (Hydrocharitaceae). *Mol. Ecol. Resour.* **9**: 1427–1429.

Lloyd, M. W., R. K. Burnett, K. A. M. Engelhardt & M. C. Neel. 2012. Does genetic diversity of restored sites differ from natural sites? A comparison of *Vallisneria americana* (Hydrocharitaceae) populations within the Chesapeake Bay. *Conserv. Genet.* **13**: 753–765.

Lobel, P. S. & J. C. Ogden. 1981. Foraging by the herbivorous parrotfish *Sparisoma radians*. *Mar. Biol.* **64**: 173–183.

Lock, I. E., L. D. Ashurkova, O. A. Belova, I. G. Kvasha, N. B. Chashkina, M. V. Remizowa & D. D. Sokoloff. 2009. A continuum between open and closed inflorescences? Inflorescence tip variation in *Potamogeton* (Potamogetonaceae: Alismatales). *Wulfenia* **16**: 33–50.

Locky, D. A., S. E. Bayley & D. H. Vitt. 2005. The vegetational ecology of black spruce swamps, fens, and bogs in southern boreal Manitoba, Canada. *Wetlands* **25**: 564–582.

Loczy, S., R. Carignan & D. Planas. 1983. The role of roots in carbon uptake by the submersed macrophytes *Myriophyllum spicatum*, *Vallisneria americana*, and *Heteranthera dubia*. *Hydrobiologia* **98**: 3–7.

Lodge, D. M., M. W. Kershner, J. E. Aloi & A. P. Covich. 1994. Effects of an omnivorous crayfish (*Orconectes rusticus*) on a freshwater littoral food web. *Ecology* **75**: 1265–1281.

Loeffler, C. C. & B. C. Wegner. 2000. Demographics and deer browsing in three Pennsylvania populations of the globally rare glade spurge, *Euphorbia purpurea* (Raf.) Fern. *Castanea* **65**: 273–290.

Loffland, H. L., J. S. Polasik, M. W. Tingley, E. A. Elsey, C. Loffland, G. Lebuhn & R. B. Siegel. 2017. Bumble bee use of post-fire chaparral in the central Sierra Nevada. *J. Wildl. Manag.* **81**: 1084–1097.

Lohammar, G. 1954. Bulbils in the inflorescences of *Butomus umbellatus*. *Svensk Bot. Tidskr.* **48**: 485–488.

Lokker, C., D. Susko, L. Lovett-Doust & J. Lovett-Doust. 1994. Population genetic structure of *Vallisneria americana*, a dioecious clonal macrophyte. *Am. J. Bot.* **81**: 1004–1012.

Lokker, C., L. Lovett-Doust & J. Lovett-Doust. 1997. Seed output and the seed bank in *Vallisneria americana* (Hydrocharitaceae). *Am. J. Bot.* **84**: 1420–1420.

Lombardi, T., T. Fochetti, A. Bertacchi & A. Onnis. 1997. Germination requirements in a population of *Typha latifolia*. *Aquat. Bot.* **56**: 1–10.

Lombardo, P. 1997. Predation by *Enallagma* nymphs (Odonata, Zygoptera) under different conditions of spatial heterogeneity. *Hydrobiologia* **356**: 1–9.

Lomolino, M. V. & K. C. Ewel. 1984. Digestive efficiencies of the West Indian manatee (*Trichechus manatus*). *Florida Sci.* **47**: 176–179.

Londonkar, R. L., U. M. Kattegouga, K. Shivsharanappa & J. V. Hanchinalmath. 2013. Phytochemical screening and in vitro antimicrobial activity of *Typha angustifolia* Linn leaves extract against pathogenic gram negative microorganisms. *J. Pharm. Res.* **6**: 280–283.

Longcore, J. E. 1995. Morphology and zoospore ultrastructure of *Entophlyctis luteolus* sp. nov. (Chytridiales): implications for chytrid taxonomy. *Mycologia* **87**: 25–33.

Looman, J. 1969. The fescue grasslands of western Canada. *Vegetatio* **19**: 128–145.

Looman, J. 1976. Biological flora of the Canadian Prairie Provinces: IV. *Triglochin* L., the genus. *Can. J. Plant Sci.* **56**: 725–732.

Looman, J. 1983. Water meal, *Wolffia arrhiza* (Lemnaceae) in Saskatchewan. [actually refers to *Wolffia columbiana*]. *Can. Field-Nat.* **97**: 220–222.

López-Mendilaharsu, M., S. C. Gardner, J. A. Seminoff & R. Riosmena-Rodriguez. 2005. Identifying critical foraging habitats of the green turtle (*Chelonia mydas*) along the Pacific coast of the Baja California peninsula, Mexico. *Aquat. Conserv. Mar. Freshw. Ecosyst.* **15**: 259–269.

Lorenzen, B., H. Brix, K. L. McKee, I. A. Mendelssohn & S. L. Miao. 2000. Seed germination of two Everglades species, *Cladium jamaicense* and *Typha domingensis*. *Aquat. Bot.* **66**: 169–180.

Lorenzen, B., H. Brix, I. A. Mendelssohn, K. L. McKee & S. L. Miao. 2001. Growth, biomass allocation and nutrient use efficiency in *Cladium jamaicense* and *Typha domingensis* as affected by phosphorus and oxygen availability. *Aquat. Bot.* **70**: 117–133.

Lot, A., A. Novelo-Retana, M. Olvera García & P. Ramírez-García. 1999. *Catálogo de Angiospermas Acuáticas de México. Hidrófitas Estrictas Emergentes, Sumergidas y flotantes.* Cuardernos del Instituto de Biología 33. Instituto de Biología, Universidad Nacional Autónoma de México, México. 161 pp.

Lot, A., F. Ramos & P. Ramírez-García. 2002. *Sagittaria demersa* (Alismataceae) en la Sierra Tarahumara, México. *Anales Inst. Biol. Univ. Nac. Autón. México, Bot.* **73**: 95–97.

Lot-Helgueras, A. 2004. Flora and vegetation of freshwater wetlands in the coastal zone of the Gulf of Mexico. Pp. 314–339 *In*: M. Caso, I. Pisanty & E. Ezcurra (eds.), *Environmental Analysis of the Gulf of Mexico.* Harte Research Institute for Gulf of Mexico Studies, Texas A&M University, Corpus Christi, TX. 710 pp.

Lotspeich, F. B., J. B. Secor, R. Okazaki & H. W. Smith. 1961. Vegetation as a soil-forming factor on the Quillayute physiographic unit in western Clallam County, Washington. *Ecology* **42**: 53–68.

Lougheed, V. L., B. Crosbie & P. Chow-Fraser. 2001. Primary determinants of macrophyte community structure in 62 marshes across the Great Lakes basin: latitude, land use, and water quality effects. *Can. J. Fish. Aquat. Sci.* **58**: 1603–1612.

Loughmiller, C., L. Loughmiller & J. Marcus. 2018. *Texas Wildflowers: A Field Guide.* University of Texas Press, Austin, TX. 488 pp.

Löve, Á. 1971. IOPB chromosome number reports XXXII. *Taxon* **20**: 349–356.

Löve, Á. 1981. Chromosome number reports LXXII. *Taxon* **30**: 694–708.

Löve, Á. 1982. IOPB chromosome number reports LXXV. *Taxon* **31**: 363–364.

Löve, A. & D. Löve. 1954. Vegetation of a prairie marsh. *Bull. Torrey Bot. Club* **81**: 16–34.

Löve, A. & D. Löve. 1958. The American element in the flora of the British Isles. *Bot. Not.* **111**: 376–388.

Löve, D. & H. Lieth. 1961. *Triglochin gaspense*, a new species of arrow grass. *Can. J. Bot.* **39**: 1261–1272.

Löve, A. & D. Löve. 1982. IOPB chromosome number reports LXXVII. *Taxon* **31**: 766–768.

Loveless, C. M. 1959. A study of the vegetation in the Florida Everglades. *Ecology* **40**: 1–9.

Love, R. 1975. The primary production of submersed macrophytes in West Blue Lake, Manitoba. M.S. thesis. University of Manitoba, Winnipeg, MB. 111 pp.

Lovell, J. H. 1899. The colors of northern monocotyledonous flowers. *Am. Nat.* **33**: 493–504.

Lovell, J. H. 1905. Some Maine species of *Halictus*. *Can. Entomol.* **37**: 299–300.

Lovett Doust, J. & G. Laporte. 1991. Population sex ratios, population mixtures and fecundity in a clonal dioecious macrophyte, *Vallisneria americana*. *J. Ecol.* **79**: 477–489.

Lovett Doust, L., J. Lovett Doust & M. Biernacki. 1994. American wildcelery, *Vallisneria americana*, as a biomonitor of organic contaminants in aquatic ecosystems. *J. Great Lakes Res.* **20**: 333–354.

Lovvorn, J. R. & J. R. Baldwin. 1996. Intertidal and farmland habitats of ducks in the Puget Sound region: a landscape perspective. *Biol. Conserv.* **77**: 97–114.

Low, J. B. & F. C. Bellrose, Jr. 1944. The seed and vegetative yield of waterfowl food plants in the Illinois River valley. *J. Wildl. Manag.* **8**: 7–22.

Low, K. S., C. K. Lee & L. L. Heng. 1994. Sorption of basic dyes by *Hydrilla verticillata*. *Environ. Technol.* **15**: 115–124.

Low-Décarie, E., C. Chivers & M. Granados. 2014. Rapidly spreading seagrass invades the Caribbean with unknown ecological consequences. *Ecology* **69**: 974–83.

Lowden, R. M. 1973. Revision of the genus *Pontederia* L. *Rhodora* **75**: 426–487.

Lowden, R. M. 1982. An approach to the taxonomy of *Vallisneria* L. (Hydrocharitaceae). *Aquat. Bot.* **13**: 269–298.

Lowden, R. M. 1986. Taxonomy of the genus *Najas* L. (Najadaceae) in the Neotropics. *Aquat. Bot.* **24**: 147–184.

Lowden, R. M. 1992. Floral variation and taxonomy of *Limnobium* LC Richard (Hydrocharitaceae). *Rhodora* **94**: 111–134.

Lowry, P. P., P. Goldblatt & H. Tobe. 1987. Notes on the floral biology, cytology, and embryology of *Campynemanthe* (Liliales: Campynemataceae). *Ann. Missouri Bot. Gard.* **74**: 573–576.

Lucas, C., T. Thangaradjou & J. Papenbrock. 2012. Development of a DNA barcoding system for seagrasses: successful but not simple. *PLoS One* **7**: e29987.

Luckenbach, M. W. & R. J. Orth. 1999. Effects of a deposit-feeding invertebrate on the entrapment of *Zostera marina* L. seeds. *Aquat. Bot.* **62**: 235–247.

Ludwig, J. C., J. B. Wright & N. E. Van Alstine. 1991. The rare plants of False Cape State Park, Virginia Beach City, Virginia. Pp. 249–256 *In*: H. G. Marshall & M. D. Norman (eds.), *Proceedings of the Back Bay Ecological Symposium*. Old Dominion University, Norfolk, VA.

Lui, K., F. L. Thompson & C. G. Eckert. 2005. Causes and consequences of extreme variation in reproductive strategy and vegetative growth among invasive populations of a clonal aquatic plant, *Butomus umbellatus* L. (Butomaceae). *Biol. Invasions* **7**: 427–444.

Luken, J. O. 2005a. *Dionaea muscipula* (Venus flytrap) establishment, release, and response of associated species in mowed patches on the rims of Carolina bays. *Restor. Ecol.* **13**: 678–684.

Luken, J. O. 2005b. Habitats of *Dionaea muscipula* (Venus' fly trap), Droseraceae, associated with Carolina bays. *Southeast. Nat.* **4**: 573–585.

Luken, J. O. & W. Thieret. 2001. Floristic relationships of mud flats and shorelines at Cave Run Lake, Kentucky. *Castanea* **66**: 336–351.

Lunau, K. 2006. Stamens and mimic stamens as components of floral colour patterns. *Bot. Jahrb.* **127**: 13–41.

Luo, Y., P.-F. Ma, H.-T. Li, J.-B. Yang, H. Wang & D.-Z. Li. 2016. Plastid phylogenomic analyses resolve Tofieldiaceae as the root of the early diverging monocot order Alismatales. *Genome Biol. Evol.* **8**: 932–945.

Luo, B., Y. Liu, B. Liu, S. Liu, B. Zhang, L. Zhang, C. Lin, Y. Liu, E. J. Kennelly, Z. Guo & C. Long. 2018. Yao herbal medicinal market during the Dragon Boat Festival in Jianghua County, China. *J. Ethnobiol. Ethnomed.* **14**: 61.

Lüning, K. & W. Freshwater. 1988. Temperature tolerance of northeast Pacific marine algae. *J. Phycol.* **24**: 310–315.

Lupoae, P., V. Cristea, D. Borda, M. Lupoae, G. Gurau & R. M. Dinica. 2015. Phytochemical screening: antioxidant and antibacterial properties of *Potamogeton* species in order to obtain valuable feed additives. *J. Oleo Sci.* **64**: 1111–1123.

Lutz, P. E. & A. R. Pittman. 1968. Oviposition and early developmental stages of *Lestes eurinus* (Odonata: Lestidae). *Am. Midl. Nat.* **80**: 43–51.

Lynch, P. S. & W. B. Zomlefer. 2016. Vascular plant flora of the south Atlantic coastal plain limestone forest: a globally imperiled association endemic to central Georgia. *Southeast. Nat.* **15**: 331–346.

Lynch, J. J., T. O'Neil & D. W. Lay. 1947. Management significance of damage by geese and muskrats to Gulf Coast marshes. *J. Wildl. Manag.* **11**: 50–76.

Lynn, L. M. 1984. The vegetation of Little Cedar Bog, southeastern New York. *Bull. Torrey Bot. Club* **111**: 90–95.

Lynn, L. M. & E. F. Karlin. 1985. The vegetation of the low-shrub bogs of northern New Jersey and adjacent New York: ecosystems at their southern limit. *Bull. Torrey Bot. Club* **112**: 436–444.

Lyon, J. & T. Eastman. 2006. Macrophyte species assemblages and distribution in a shallow, eutrophic lake. *Northeast. Nat.* **13**: 443–453.

Ma, Q. & J. Liang. 2019. The first complete chloroplast genome of *Pontederia cordata* (Pontederiaceae). *Mitochondrial DNA Part B* **4**: 555–557.

Maass, W. S. G. 1966. Studies on the taxonomy and distribution of *Sphagnum* I. *Sphagnum pylaesii* and *Sphagnum angermanicum* in Quebec and some phytogeographic considerations. *Bryologist* **69**: 95–100.

Maass, W. S. G. 1967. Studies on the taxonomy and distribution of *Sphagnum* III. Observations on *Sphagnum macrophyllum* in the northern part of its range. *Bryologist* **70**: 177–192.

Mabbott, D. C. 1920. Food habits of seven species of American shoalwater ducks. Bulletin No. 862. U.S. Department of Agriculture. Government Printing Office, Washington, DC. 68 pp.

Maberly, S. C. 1993. Morphological and photosynthetic characteristics of *Potamogeton obtusifolius* from different depths. *J. Aquat. Plant Manag.* **31**: 34–34.

Maberly, S. C. & D. H. N. Spence. 1983. Photosynthetic inorganic carbon use by freshwater plants. *J. Ecol.* **71**: 705–724.

Mabona, U., A. Viljoen, E. Shikanga, A. Marston & S. Van Vuuren. 2013. Antimicrobial activity of southern African medicinal plants with dermatological relevance: from an ethnopharmacological screening approach, to combination studies and the isolation of a bioactive compound. *J. Ethnopharmacol.* **148**: 45–55.

MacGillivray, A. D. 1903. Aquatic Chrysomelidae and a table of the families of Coleopterous larvae. New York State Museum Bulletin 68. State Museum of New York, Albany, NY. 44 pp.

Mach, M. E., S. Wyllie-Echeverria & K. M. A. Chan. 2014. Ecological effect of a nonnative seagrass spreading in the Northeast Pacific: a review of *Zostera japonica*. *Ocean Coast. Manag.* **102**: 375–382.

Mack, R. N. 1981. Initial effects of ashfall from Mount St. Helens on vegetation in eastern Washington and adjacent Idaho. *Science* **213**: 537–539.

Mack, R. N. 1991. The commercial seed trade: an early disperser of weeds in the United States. *Econ. Bot.* **45**: 257–273.

MacRae, I. V., N. N. Winchester & R. A. Ring. 1990. Feeding activity and host preference of the milfoil midge, *Cricotopus myriophylli* Oliver (Diptera: Chironomidae). *J. Aquat. Plant Manag.* **28**: 89–92.

MacRoberts, M. H. & B. R. MacRoberts. 1995. Noteworthy vascular plant collections on the Kisatchie National Forest, Louisiana. *Phytologia* **78**: 291–313.

MacRoberts, M. H. & B. R. MacRoberts. 2005. The ecology of *Trillium texanum* (Trilliaceae) on the Angelina National Forest, Texas. *Sida* **21**: 1893–1903.

MacRoberts, M. H. & B. R. MacRoberts. 2010. *Hydrocleys nymphoides* (Limnocharitaceae): new to Louisiana. *Phytoneuron* **29**: 1–2.

MacRoberts, B. R., M. H. MacRoberts, D. C. Rudolph & D. W. Peterson. 2014. Floristics of ephemeral ponds in east-central Texas. *Southeast. Nat.* **13**: 15–25.

Mader, E., W. Van Vierssen & K. Schwenk. 1998. Clonal diversity in the submerged macrophyte *Potamogeton pectinatus* L. inferred from nuclear and cytoplasmic variation. *Aquat. Bot.* **62**: 147–160.

Madeira, P. T., C. C. Jacono & T. K. Van. 2000. Monitoring hydrilla using two RAPD procedures and the Nonindigenous Aquatic Species database. *J. Aquat. Plant Manag.* **38**: 33–40.

Madineni, M. N., S. Faiza, R. S. Surekha, R. Ravi & M. Guha. 2012. Morphological, structural, and functional properties of maranta (*Maranta arundinacea* L) starch. *Food Sci. Biotechnol.* **21**: 747–752.

Madsen, J. D. 1991. Resource allocation at the individual plant level. *Aquat. Bot.* **41**: 67–86.

Madsen, J. D. & M. S. Adams. 1988. The germination of *Potamogeton pectinatus* tubers: environmental control by temperature and light. *Can. J. Bot.* **66**: 2523–2526.

Madsen, J. D. & M. S. Adams. 1989. The light and temperature dependence of photosynthesis and respiration in *Potamogeton pectinatus* L. *Aquat. Bot.* **36**: 23–31.

Madsen, J. D. & R. M. Wersal. 2009. Aquatic plant community and Eurasian watermilfoil (*Myriophyllum spicatum* L.) management assessment in Lake Pend Oreille, Idaho for 2008. Geosystems Research Institute Report 5032. Geosystems Research Institute, Mississippi State, MS. 65 pp.

Madsen, J. D., C. F. Hartleb & C. W. Boylen. 1991. Photosynthetic characteristics of *Myriophyllum spicatum* and six submersed aquatic macrophyte species native to Lake George, New York. *Freshw. Biol.* **26**: 233–240.

Madsen, J. D., M. S. Adams & W. Kleindl. 1992. The aquatic macrophyte community of Black Earth Creek, Wisconsin: 1981 to 1986. *Trans. Wisconsin Acad. Sci.* 80: 101–114.

Madsen, J. D., R. M. Wersal, M. Tyler & P. D. Gerard. 2006. The distribution and abundance of aquatic macrophytes in Swan Lake and Middle Lake, Minnesota. *J. Freshw. Ecol.* **21**: 421–429.

Madsen, J. D., R. M. Wersal, M. D. Marko & J. G. Skogerboe. 2012. Ecology and management of flowering rush (*Butomus umbellatus*) in the Detroit Lakes, Minnesota. Geosystems Research Institute Report 5054. Mississippi State University, Mississippi State, MS. 43 pp.

Madsen, J. D., R. M. Wersal & M. D. Marko. 2016. Distribution and biomass allocation in relation to depth of flowering rush (*Butomus umbellatus*) in the Detroit Lakes, Minnesota. *Invasive Plant Sci. Manag.* **9**: 161–170.

Maehr, D. S. & J. R. Brady. 1984. Food habits of Florida black bears. *J. Wildl. Manag.* **48**: 230–235.

Magallán, F., M. Martínez, L. Hernández-Sandoval, A. González-Rodríguez & K. Oyama. 2013. Diversidad genética de *Lilaea scilloides* (Juncaginaceae) en el centro de México. *Revista Mexicana Biodiversidad* **84**: 240–248.

Magrath, L. K. 2002. *Liparis* Richard. Pp. 624–626 *In*: Flora North America Editorial Committee (eds.), *Flora of North America North of Mexico, Vol. 26: Magnoliophyta: Liliidae: Liliales and Orchidales*. Oxford University Press, New York, NY.

Magrath, L. K. 2014. Native orchids of Oklahoma. *Oklahoma Nat. Plant Rec.* **1**: 39–66.

Maheshwari, S. C. 1958. *Spirodela polyrrhiza*: the link between the aroids and the duckweeds. *Nature* **181**: 1745–1746.

Mahmood, M. S. & M. Strack. 2011. Methane dynamics of recolonized cutover minerotrophic peatland: implications for restoration. *Ecol. Eng.* **37**: 1859–1868.

Mains, E. B. 1934. Host specialization in the rust of *Iris, Puccinia iridis*. *Am. J. Bot.* **21**: 23–33.

Maisch, J. M. 1872. Pharmacognostical notes. *Am. J. Pharm.* (May, 1872). Available online: https://ezproxy.lib.uconn.edu/login?url=https://search.proquest.com/docview/89684270?accountid=14518 [accessed 16 February, 2019].

Maia, V. H., M. A. Gitzendanner, P. S. Soltis, G. K.-S. Wongand & D. E. Soltis. 2014. Angiosperm phylogeny based on 18S/26S rDNA sequence data: constructing a large dataset using next-generation sequence data. *Int. J. Plant Sci.* **175**: 613–650.

Majak, W., R. E. McDiarmid, A. Van Ryswyk & J. W. Hall. 1980. Seasonal variation in the cyanide potential of arrowgrass (*Triglochin maritima*). *Can. J. Plant Sci.* **60**: 1235–1241.

Majka, C. & D. Langor. 2010. Contributions towards an understanding of the Cryptophaginae (Coleoptera, Cryptophagidae) of Atlantic Canada. *ZooKeys* **35**: 13–35.

Majumdar, K. & B. K. Datta. 2009. Folklore herbal formulations by the tribes of Tripura. Pp. 155–162 *In*: S. D. Ramashankar & B. K. Sharma (eds.), *Proceeding on Traditional Healing Practices in North East India*. North Eastern Institute of Folk Medicine (NEIFM), Pasighat, Arunachal Pradesh, India.

Majure, L. C., J. Hill, C. Doffitt & T. C. Majure. 2011. The vascular flora of Lauderdale County, Mississippi, USA. *Rhodora* **113**: 365–418.

Maki, K. & S. Galatowitsch. 2004. Movement of invasive aquatic plants into Minnesota (USA) through horticultural trade. *Biol. Conserv.* **118**: 389–396.

Maki, K. C. & S. M. Galatowitsch. 2008. Cold tolerance of the axillary turions of two biotypes of *Hydrilla* and northern watermilfoil. *J. Aquat. Plant Manag.* **46**: 42–50.

Makkay, K., F. R. Pick & L. Gillespie. 2008. Predicting diversity versus community composition of aquatic plants at the river scale. *Aquat. Bot.* **88**: 338–346.

Malek, L. 1981. The effect of drying on *Spirodela polyrhiza* turion germination. *Can. J. Bot.* **59**: 104–105.

Mali, P. R. & D. Datta. 2018. Experimental evaluation of bamboo reinforced concrete slab panels. *Constr. Build. Mater.* **188**: 1092–1100.

Maliakal, S. K., E. S. Menges & J. S. Denslow. 2000. Community composition and regeneration of Lake Wales Ridge wiregrass flatwoods in relation to time-since-fire. *J. Torrey Bot. Soc.* **127**: 125–138.

Mallik, S., M. Nayak, B. B. Sahu, A. K. Panigrahi & B. P. Shaw. 2011. Response of antioxidant enzymes to high NaCl concentration in different salt-tolerant plants. *Biol. Plant.* **55**: 191–195.

Mallin, M. A., S. H. Ensign, T. L. Wheeler & D. B. Mayes. 2002. Pollutant removal efficacy of three wet detention ponds. *J. Environ. Qual.* **31**: 654–660.

Malanson, G. P. 1982. The assembly of hanging gardens: effects of age, area, and location. *Am. Nat.* **119**: 145–150.

Manandhar, N. P. 1985. Ethnobotanical notes on certain medicinal plants used by Tharus of Dang-Deokhuri District, Nepal. *Int. J. Crude Drug Res.* **23**: 153–159.

Mancera, J. E., G. C. Meche, P. P. Cardona-Olarte, E. Castañeda-Moya, R. L. Chiasson, N. A. Geddes, L. M. Schile, H. G. Wang, G. R. Guntenspergen & J. B. Grace. 2005. Fine-scale spatial variation in plant species richness and its relationship to environmental conditions in coastal marshlands. *Plant Ecol.* **178**: 39–50.

Mandal, R. N., A. K. Datta, N. Sarangi & P. K. Mukhopadhyay. 2010. Diversity of aquatic macrophytes as food and feed components to herbivorous fish - a review. *Indian J. Fish.* **57**: 65–73.

Mandossian, A. & R. P. McIntosh. 1960. Vegetation zonation on the shore of a small lake. *Am. Midl. Nat.* **64**: 301–308.

Manguin, S., D. R. Roberts, E. L. Peyton, E. Rejmankova & J. Pecor. 1996. Characterization of *Anopheles pseudopunctipennis* larval habitats. *J. Am. Mosq. Control Assoc.* **12**: 619–626.

Manning, M. E. & W. G. Padgett. 1995. Riparian community type classification for Humboldt and Toiyabe National Forests, Nevada and eastern California. R4-Ecol-95-01. U.S. Department of Agriculture, Forest Service, Intermountain Region, Ogden, UT. 306 pp.

Manolis, T. 2016. Odonate exuviae used for roosts and nests by *Sassacus vitis* and other jumping spiders (Araneae: Salticidae). *Peckhamia* 142(1): 1–17.

Mansfield, D. 1996. The unique botany of Steens Mountain: the rare and endemic plants. *Kalmiopsis* 5: 10–17.

Mansor, M. 1996. Noxious floating weeds of Malaysia. *Hydrobiologia* 340: 121–125.

Mantas, M. 1993. Ecology and reproductive biology of *Epipactis gigantea* Doug. (Orchidaceae) in northwestern Montana. M.S. thesis. University of Idaho, Moscow, ID. 73 pp.

Marburger, J. E. 1993. Biology and management of *Sagittaria latifolia* Willd (broad-leaf arrow-head) for wetland restoration and creation. *Restor. Ecol.* 1: 248–255.

Marchant, C. J. 1970. Chromosome variation in Araceae: I: Pothoeae to Stylochitoneae. *Kew Bull.* 24: 315–322.

Marco-Méndez, C., P. Prado, L. M. Ferrero-Vicente, C. Ibáñez & J. L. Sánchez-Lizaso. 2015. Seasonal effects of waterfowl grazing on submerged macrophytes: the role of flowers. *Aquat. Bot.* 120: 275–282.

Marcus, B. A. 1981. *Hydropsyche* larvae (Trichoptera: Hydropsychidae) from a lake weedbed. *J. New York Entomol. Soc.* 89: 56–58.

Mardanov, A. V., N. V. Ravin, B. B. Kuznetsov, T. H. Samigullin, A. S. Antonov, T. V. Kolganova & K. G. Skyabin. 2008. Complete sequence of the duckweed (*Lemna minor*) chloroplast genome: structural organization and phylogenetic relationships to other angiosperms. *J. Mol. Evol.* 66: 555–564.

Marie-Victorin, F. 1943. Les Vallisnéries américaines. *Contr. Inst. Bot. Univ. Montréal* 46: 1–38.

Marion, C., J.-L. Fernandez, S.-M. d'Hères, F. D. Prehsler, J. Schönenberger & M. Gibernau. 2017. Note on the pollination of *Calla palustris* L. (Araceae). *Aroideana* 40: 71–83.

Markle, M. S. 1915. The phytecology of peat bogs near Richmond, Indiana. *Proc. Indiana Acad. Sci.* 25: 359–376.

Markwith, S. H. & M. J. Scanlon. 2006. Characterization of six polymorphic microsatellite loci isolated from *Hymenocallis coronaria* (J. LeConte) Kunth (Amaryllidaceae). *Mol. Ecol. Notes* 6: 72–74.

Markwith, S. H. & K. C. Parker. 2007. Conservation of *Hymenocallis coronaria* genetic diversity in the presence of disturbance and a disjunct distribution. *Conserv. Genet.* 8: 949–963.

Markwith, S. H. & M. J. Scanlon. 2007. Multiscale analysis of *Hymenocallis coronaria* (Amaryllidaceae) genetic diversity, genetic structure, and gene movement under the influence of unidirectional stream flow. *Am. J. Bot.* 94: 151–160.

Markwith, S. H. & D. S. Leigh. 2008. Subaqueous hydrochory: open-channel hydraulic modelling of non-buoyant seed movement. *Freshw. Biol.* 53: 2274–2286.

Markwith, S. H. & D. S. Leigh. 2012. Comparison of estimated and experimental subaqueous seed transport. *Ecohydrology* 5: 346–350.

Markwith, S. H., L. J. Davenport, J. Shelton, K. C. Parker & M. J. Scanlon. 2009. Ichthyochory, closure of the Suwannee Strait, and population divergence in *Hymenocallis coronaria*. *Florida Sci.* 72: 28–36.

Marler, J. E. 1969. A study of the germination process of seeds of *Heteranthera limosa*. Ph.D. dissertation. Louisiana State University, Baton Rouge, LA. 92 pp.

Marles, R. J. 2001. Non-timber forest products and Aboriginal traditional knowledge. Pp. 53–65 *In*: I. Davidson-Hunt, L. C. Duchesne & J. C. Zasada (eds.), *Forest Communities in the Third Millennium: Linking Research, Business, and Policy Toward a Sustainable Non-Timber Forest Product Sector*. Gen. Tech. Rep. NC-217. U.S. Department of Agriculture, Forest Service, North Central Research Station, St. Paul, MN.

Marsden, J. E. & B. J. Ladago. 2017. The Champlain Canal as a non-indigenous species corridor. *J. Great Lakes Res.* 43: 1173–1180.

Marsden, B. W., K. A. M. Engelhardt & M. C. Neel. 2013. Genetic rescue versus outbreeding depression in *Vallisneria americana*: implications for mixing seed sources for restoration. *Biol. Conserv.* 167: 203–214.

Marsh, Jr., J. A., W. C. Dennison & R. S. Alberte. 1986. Effects of temperature on photosynthesis and respiration in eelgrass (*Zostera marina* L.). *J. Exp. Mar. Biol. Ecol.* 101: 257–267.

Marshal, J. P., V. C. Bleich, N. G. Andrew & P. R. Krausman. 2004. Seasonal forage use by desert mule deer in southeastern California. *Southwest. Nat.* 49: 501–506.

Martin, G. 1886. The Phyllostictas of North America. *J. Mycol.* 2: 25–27.

Martin, R. D. C. 1939. Life histories of *Agrion aequabile* and *Agrion maculatum*. *Ann. Entomol. Soc. Am.* 32: 601–619.

Martin, M. A. 1978. A unique natural area in Dane county, Wisconsin. Pp. 187–189 *In*: R. L. Stuckey & K. J. Reese (eds.), *The Prairie Peninsula--In the 'Shadow' of Transeau: Proceedings of the Sixth North American Prairie Conference*. Ohio Biological Survey Biological Notes No. 15. Columbus, OH.

Martin, A. C. & F. M. Uhler. 1951. Food of game ducks in the United States and Canada. Research Report 30. Fish & Wildlife Service, U.S. Department of the Interior, Washington, DC. 308 pp.

Martin, P. G. & J. M. Dowd. 1986. A phylogenetic tree for some monocotyledons and gymnosperms derived from protein sequences. *Taxon* 35: 469–475.

Martin, K. & J. Sauerborn. 2000. An aquatic wild plant as a keystone species in a traditional Philippine rice growing system: its agroecological implications. *Ann. Trop. Res.* 22: 1–15.

Martin, S. B. & G. P. Shaffer. 2005. *Sagittaria* biomass partitioning relative to salinity, hydrologic regime, and substrate type: implications for plant distribution patterns in coastal Louisiana, United States. *J. Coast. Res.* 21: 167–174.

Martin, T. H., L. B. Crowder, C. F. Dumas & J. M. Burkholder. 1992. Indirect effects of fish on macrophytes in Bays Mountain Lake: evidence for a littoral trophic cascade. *Oecologia* 89: 476–481.

Martine, C. T., S. F. Langdon, T. M. Shearman, C. Binggeli & T. B. Mihuc. 2015. European frogbit (*Hydrocharis morsus-ranae*) in the Champlain/Adirondack region: recent inferences. *Rhodora* 117: 499–504.

Martínez-Garrido, J., E. A. Serrão, A. H. Engelen, C. J. Cox, P. García-Murillo & M. González-Wangüemert. 2016. Multilocus genetic analyses provide insight into speciation and hybridization in aquatic grasses, genus *Ruppia*. *Biol. J. Linn. Soc.* 117: 177–191.

Martínez-Garrido, J., R. Bermejo, E. A. Serrão, J. Sánchez-Lizaso & M. González-Wangüemert. 2017a. Regional genetic structure in the aquatic macrophyte *Ruppia cirrhosa* suggests dispersal by waterbirds. *Estuaries Coast.* 40: 1705–1716.

Martínez-Garrido, J., J. C. Creed, S. Martins, C. H. Almada & E. A. Serrão. 2017b. First record of *Ruppia maritima* in West Africa supported by morphological description and phylogenetic classification. *Bot. Mar.* 60: 583–589.

Martirosyan, E. V., N. N. Ryzhova, E. Z. Kochieva & K. G. Skryabin. 2009. Analysis of chloroplast *rps16* intron sequences in Lemnaceae. *Mol. Biol.* 43: 32–38.

Marx, E. J. F. 1957. A review of the subgenus *Donacia* in the Western Hemisphere (Coleoptera, Donaciidae). *Bull. Am. Mus. Nat. Hist.* 112: 191–278.

Maschinski, J. 2001. Impacts of ungulate herbivores on a rare willow at the southern edge of its range. *Biol. Conserv.* **101**: 119–130.

Mashburn, S. J., R. R. Sharitz & M. H. Smith. 1978. Genetic variation among *Typha* populations of the southeastern United States. *Evolution* **32**: 681–685.

Mason, H. L. 1938. The flowering of *Wolffiella lingulata* (Hegelm.) Hegelm. *Madroño* **4**: 241–251.

Mason, H. L. 1957. *A Flora of the Marshes of California*. University of California Press, Berkeley, CA. 878 pp.

Massey, A. B. 1953. Orchids in Virginia. *Castanea* **18**: 107–115.

Massoud, M. S. 2012. Mycoflora associated with aquatic plants in ponds and lakes in central west of Florida, USA. *Sci. Res. Rep.* **2**: 1–6.

Mata, J. L. & J. Cebrián. 2013. Fungal endophytes of the seagrasses *Halodule wrightii* and *Thalassia testudinum* in the north-central Gulf of Mexico. *Bot. Mar.* **56**: 541–545.

Matsuda, H., T. Morikawa, Y. Sakamoto, I. Toguchida & M. Yoshikawa. 2002. Labdane-type diterpenes with inhibitory effects on increase in vascular permeability and nitric oxide production from *Hedychium coronarium*. *Bioorg. Med. Chem.* **10**: 2527–2534.

Matsuhashi, S., T. Minamoto & H. Doi. 2019. Seasonal change in environmental DNA concentration of a submerged aquatic plant species. *Freshw. Sci.* **38**: 654–660.

Matteson, S. W., T. A. Andryk & J. Wetzel. 1986. Wisconsin trumpeter swan recovery plan. *Passeng. Pigeon* **50**: 119–130.

Matthews, E. R., R. K. Peet & A. S. Weakley. 2007. Natural vegetation of the Carolinas: classification and description of Piedmont alluvial plant communities of the Cape Fear River Basin. Unpublished Report. Carolina Vegetation Survey, University of North Carolina, Chapel Hill, NC. 29 pp.

Mathis, W. 1947. Biology of the Florida red scale in Florida. *Florida Entomol.* **29**: 13–35.

Mathur, A. C. & B. P. Saxena. 1975. Induction of sterility in male houseflies by vapors of *Acorus calamus* L. oil. *Naturwissenschaften* **62**: 576–577.

Mattila, J., G. Chaplin, M. R. Eilers, K. L. Heck, Jr., J. P. O'Neal & J. F. Valentine. 1999. Spatial and diurnal distribution of invertebrate and fish fauna of a *Zostera* marina bed and nearby unvegetated sediments in Damariscotta River, Maine (USA). *J. Sea Res.* **41**: 321–332.

Mattson, R. A., J. H. Epler & M. K. Hein. 1995. Description of benthic communities in karst, spring-fed streams of north central Florida. *J. Kansas Entomol. Soc.* **68**: 18–41.

Mattson, D. J., S. R. Podruzny & M. A. Haroldson. 2005. Consumption of pondweed rhizomes by Yellowstone grizzly bears. *Ursus* **16**: 41–46.

Matulewich, V. A. & M. S. Finstein. 1978. Distribution of autotrophic nitrifying bacteria in a polluted river (the Passaic). *Appl. Environ. Microbiol.* **35**: 67–71.

Mauermann, K. J. 1995. Feeding and habitat preferences of the red-eared slider, *Trachemys scripta elegans* Wied. M.S. thesis. Texas Woman's University, Denton, TX. 106 pp.

Maurice, D. V., J. E. Jones, C. R. Dillon & J. M. Weber. 1984. Chemical composition and nutritional value of Brazilian elodea (*Egeria densa*) for the chick. *Poult. Sci.* **63**: 317–323.

Mavrodiev, E. V., M. Martínez-Azorín, P. Dranishnikov & M. B. Crespo. 2014. At least 23 genera instead of one: the case of *Iris* L. s.l. (Iridaceae). *PLoS One* **9**(8): e106459.

Maycock, P. F. 1956. Composition of an upland conifer community in Ontario. *Ecology* **37**: 846–848.

Mayes, R. A., A. W. MacIntosh & V. L. Anderson. 1977. Uptake of cadmium and lead by a rooted aquatic macrophyte (*Elodea canadensis*). *Ecology* **58**: 1176–1180.

Mayo, S. J., J. Bogner & P. C. Boyce. 1998a. Araceae. Pp. 26–74 *In*: K. Kubitzki (ed.), *The Families and Genera of Vascular Plants, Vol. IV, Flowering Plants: Monocotyledons, Alismatanae and Commelinanae (Except Gramineae)*. Springer-Verlag, Berlin, Germany.

Mayo, S. J., J. Bogner & P. C. Boyce. 1998b. The genera of Araceae project. *Acta Bot. Yunnan.* **Suppl. X**: 4–11.

Mazerolle, D., S. Blaney & A. Belliveau. 2014. *Rare Vascular Plant Surveys in the Polletts Cove and LaHave River Areas of Nova Scotia*. Atlantic Canada Conservation Data Centre, Sackville, NB, Canada. 71 pp.

McAllister, C. T., M. B. Leite & R. Tumlison. 2017. The fishes of Chadron Creek, Dawes County, Nebraska. *J. Arkansas Acad. Sci.* **71**: 62–68.

McAtee, W. L. 1913. Some local names of plants. *Torreya* **13**: 225–236.

McAtee, W. L. 1916. Plants collected on Matinicus Island, Maine, in late fall, 1915. *Rhodora* **18**: 29–45.

McAtee, W. L. 1918. Food habits of the mallard ducks of the United States. U.S. Department of Agriculture Bulletin No. 720. Government Printing Office, Washington, DC. 36 pp.

McAtee, W. L. 1922. Notes on food habits of the shoveller or spoonbill duck (*Spatula clypeata*). *Auk* **39**: 380–386.

McAtee, W. L. 1925. Notes on drift, vegetable balls, and aquatic insects as a food product of inland waters. *Ecology* **6**: 288–302.

McAtee, W. L. 1939. *Wildfowl Food Plants*. Collegiate Press, Inc., Ames, IA. 141 pp.

McAuley, D. G. & J. R. Longcore. 1988. Foods of juvenile ring-necked ducks: relationship to wetland pH. *J. Wildl. Manag.* **52**: 177–185.

McAvoy, W. A. & R. M. Wilson. 2014. Rediscovery of *Lobelia boykinii* (Campanulaceae) in Delaware. *Phytoneuron* **23**: 1–4.

McCabe, T. L. & C. J. Sheviak. 1980. *Platanthera cristata* (Michx.) Lindl., a new host for the red-banded leaf roller. *J. New York Entomol. Soc.* **88**: 197–198.

McCall, D. D. & C. F. Rakocinski. 2007. Grass shrimp (*Palaemonetes* spp.) play a pivotal trophic role in enhancing *Ruppia maritima*. *Ecology* **88**: 618–624.

McCann, M. J. 2016. Response diversity of free-floating plants to nutrient stoichiometry and temperature: growth and resting body formation. *PeerJ* **4**: e1781.

McClaran, M. P. & P. C. Sundt. 1992. Population dynamics of the rare orchid, *Spiranthes delitescens*. *Southwest. Nat.* **37**: 299–303.

McClure, J. W. & R. E. Alston. 1966. A chemotaxonomic study of Lemnaceae. *Am. J. Bot.* **53**: 849–860.

McColl, J. G. & J. Burger. 1976. Chemical inputs by a colony of Franklin's gulls nesting in cattails. *Am. Midl. Nat.* **96**: 270–280.

McCormick, J. 1998. The vegetation of the New Jersey Pine Barrens. Pp. 229–243 *In*: R. T. T. Forman (ed.), *Pine Barrens: Ecosystem and Landscape*. Rutgers University Press, New Brunswick, NJ. 601 pp.

McCormick, M. K. & H. Jacquemyn. 2014. What constrains the distribution of orchid populations? *New Phytol.* **202**: 392–400.

McCormick, P. V., J. W. Harvey & E. S. Crawford. 2011. Influence of changing water sources and mineral chemistry on the Everglades ecosystem. *Crit. Rev. Environ. Sci. Technol.* **41**(S1): 28–63.

McDaniel, S. 1968. *Harperocallis*, a new genus of the Liliaceae from Florida. *J. Arnold Arbor.* **49**: 35–40.

McDermid, K. J., B. Stuercke & G. H. Balazs. 2007. Nutritional composition of marine plants in the diet of the green sea turtle (*Chelonia mydas*) in the Hawaiian Islands. *Bull. Mar. Sci.* **81**: 55–71.

McDermott, J. J. 1988. The role of hoplonemerteans in the ecology of seagrass communities. Pp. 1–11 *In*: P. Sundberg, R. Gibson & G. Berg (eds.), *Recent Advances in Nemertean Biology*. Dr. W. Junk, Publishers, Dordrecht, The Netherlands.

McDonald, C. 1991. Endangered and threatened wildlife and plants; final rule to list *Potamogeton clystocarpus* (Little Aguja pondweed) as endangered. *Fed. Reg.* **56**: 57844–57849.

McDowell, L. R., L. C. Lizama, J. E. Marion & C. J. Wilcox. 1990. Utilization of aquatic plants *Elodea canadensis* and *Hydrilla verticillata* in diets for laying hens. 1. Performance and egg-yolk pigmentation. *Poult. Sci.* **69**: 673–678.

McFarland, D. 2006. Reproductive ecology of *Vallisneria americana* Michaux. SAV Technical Notes Collection (ERDC/TN SAV-06-4). U.S. Army Engineer Research and Development Center, Vicksburg, MS. 27 pp.

McFarland, D. G. & S. J. Rogers. 1998. The aquatic macrophyte seed bank in Lake Onalaska, Wisconsin. *J. Aquat. Plant Manag.* **36**: 33–39.

McFarland, D. G., J. W. Barko & N. J. McCreary. 1992. Effects of sediment fertility and initial plant density on growth of *Hydrilla verticillata* (LF) Royle and *Potamogeton nodosus* Poiret. *J. Freshw. Ecol.* **7**: 191–200.

McFerren, M. A. 2006. Useful plants of dermatology. VIII. The false hellebore (*Veratrum californicum*). *J. Am. Acad. Dermatol.* **54**: 718–720.

McGaha, Y. J. 1952. The limnological relations of insects to certain aquatic flowering plants. *Trans. Am. Microsc. Soc.* **71**: 355–381.

McGilvrey, F. B. 1966. Fall food habits of ducks near Santee Refuge, South Carolina. *J. Wildl. Manag.* **30**: 577–580.

McGovern, T. M. & K. Blankenhorn. 2007. Observation of fruit production by the seagrass *Halodule wrightii* in the northeastern Gulf of Mexico. *Aquat. Bot.* **87**: 247–250.

McIlraith, A. L., G. G. C. Robinson & J. M. Shay. 1989. A field study of competition and interaction between *Lemna minor* and *Lemna trisulca*. *Can. J. Bot.* **67**: 2904–2911.

McIlroy, S. K. & B. H. Allen-Diaz. 2012. Plant community distribution along water table and grazing gradients in montane meadows of the Sierra Nevada Range (California, USA). *Wetl. Ecol. Manag.* **20**: 287–296.

McIntyre, S. & M. R. Newnham. 1988. Distribution and spread of the Alismataceae in the rice-growing region of New South Wales. *Cunninghamia* **2**: 25–38.

McKain, M. R., J. R. McNeal, P. R. Kellar, L. E. Eguiarte, J. C. Pires & J. Leebens-Mack. 2016a. Timing of rapid diversification and convergent origins of active pollination within Agavoideae (Asparagaceae). *Am. J. Bot.* **103**: 1717–1729.

McKain, M. R., H. Tang, J. R. McNeal, S. Ayyampalayam, J. I. Davis, C. W. dePamphilis, T. J. Givnish, J. C. Pires, D. W. Stevenson & J. H. Leebens-Mack. 2016b. A phylogenomic assessment of ancient polyploidy and genome evolution across the Poales. *Genome Biol. Evol.* **8**: 1150–1164.

McKee, K. L. & I. A. Mendelssohn. 1989. Response of a freshwater marsh plant community to increased salinity and increased water level. *Aquat. Bot.* **34**: 301–316.

McKenzie, R. J. & P. H. Lovell. 1992. Flower senescence in monocotyledons: a taxonomic survey. *New Zealand J. Crop Hort. Sci.* **20**: 67–71.

McKenzie, P. M., C. T. Witsell, L. R. Phillippe, C. S. Reid, M. A. Homoya, S. B. Rolfsmeier & C. A. Morse. 2009. Status assessment of *Eleocharis wolfii* (Cyperaceae) in the United States. *J. Bot. Res. Inst. Texas* **3**: 831–854.

McKenzie, P. M., T. Nagel, D. Ashley & N. Paothong. 2012. A second recent record of eastern prairie-fringed orchid (*Platanthera leucophaea*) for Missouri. *Missouriensis* **31**: 5–8.

McKenzie-Gopsill, A., H. Kirk, W. Van Drunen, J. R. Freeland & M. E. Dorken. 2012. No evidence for niche segregation in a North American cattail (*Typha*) species complex. *Ecol. Evol.* **2**: 952–961.

McKnight, D. T., A. C. Jones & D. B. Ligon. 2015. The omnivorous diet of the western chicken turtle (*Deirochelys reticularia miaria*). *Copeia* **103**: 322–328.

McLaughlin, W. T. 1932. Atlantic coastal plain plants in the sand barrens of northwestern Wisconsin. *Ecol. Monogr.* **2**: 335–383.

McLaughlin, S. P., E. L. Geiger & J. E. Bowers. 2001. Flora of the Appleton-Whittell Research Ranch, northeastern Santa Cruz County, Arizona. *J. Arizona-Nevada Acad. Sci.* **33**: 113–131.

McLay, C. L. 1974. The distribution of duckweed *Lemna perpusilla* in a small southern California lake: an experimental approach. *Ecology* **55**: 262–276.

McLay, C. L. 1976. The effect of pH on the population growth of three species of duckweed: *Spirodela oligorrhiza*, *Lemna minor* and *Wolffia arrhiza*. *Freshw. Biol.* **6**: 125–136.

McMahan, C. A. 1970. Food habits of ducks wintering on Laguna Madre, Texas. *J. Wildl. Manag.* **34**: 946–949.

McMaster, R. T. 2001. The population biology of *Liparis loeselii*, Loesel's twayblade, in a Massachusetts wetland. *Northeast. Nat.* **8**: 163–179.

McMillan, J. F. 1953. Some feeding habits of moose in Yellowstone Park. *Ecology* **34**: 102–110.

McMillan, C. 1956. The edaphic restriction of *Cupressus* and *Pinus* in the coast ranges of central California. *Ecol. Monogr.* **26**: 177–212.

McMillan, C. 1959. Salt tolerance within a *Typha* population. *Am. J. Bot.* **46**: 521–526.

McMillan, C. 1976. Experimental studies on flowering and reproduction in seagrasses. *Aquat. Bot.* **2**: 87–92.

McMillan, C. 1979. Differentiation in response to chilling temperatures among populations of three marine spermatophytes, *Thalassia testudinum*, *Syringodium filiforme* and *Halodule wrightii*. *Am. J. Bot.* **66**: 810–819.

McMillan, C. 1980. Reproductive physiology in the seagrass, *Syringodium filiforme*, from the Gulf of Mexico and the Caribbean. *Am. J. Bot.* **67**: 104–110.

McMillan, C. 1981. Seed reserves and seed germination for two seagrasses, *Halodule wrightii* and *Syringodium filiforme*, from the western Atlantic. *Aquat. Bot.* **11**: 279–296.

McMillan, C. 1982. Reproductive physiology of tropical seagrasses. *Aquat. Bot.* **14**: 245–258.

McMillan, C. 1983. Seed germination in *Halodule wrightii* and *Syringodium filiforme* from Texas and the U.S. Virgin Islands. *Aquat. Bot.* **15**: 217–220.

McMillan, C. 1984. The distribution of tropical seagrasses with relation to their tolerance of high temperatures. *Aquat. Bot.* **19**: 369–379.

McMillan, C. 1985. The seed reserve for *Halodule wrightii*, *Syringodium filiforme* and *Ruppia maritima* in Laguna Madre, Texas. *Contrib. Mar. Sci.* **28**: 141–149.

McMillan, C. 1987. Seed germination and seedling morphology of the seagrass, *Halophila engelmannii* (Hydrocharitaceae). *Aquat. Bot.* **28**: 179–188.

McMillan, C. 1988. The seed reserve of *Halophila engelmannii* (Hydrocharitaceae) in Redfish Bay, Texas. *Aquat. Bot.* **30**: 253–259.

McMillan, C. 1988a. The seed reserve of *Halophila decipiens* Ostenfeld (Hydrocharitaceae) in Panama. *Aquat. Bot.* **31**: 177–182.

McMillan, C. 1988b. Seed germination and seedling development of *Halophila decipiens* Ostenfeld (Hydrocharitaceae) from Panama. *Aquat. Bot.* **31**: 169–176.

McMillan, C. 1989. Timing of anthesis for staminate flowers on *Halophila* engelmannii Aschers. from Texas and *Halophila decipiens* Ostenfeld from Panama. *Aquat. Bot.* **33**: 141–147.

McMillan, C. 1990. Testing the influence of night length on the anthesis of staminate flowers of *Halophila engelmannii* Aschers. (Hydrocharitaceae). *Aquat. Bot.* **37**: 383–385.

McMillan, C. 1991. The longevity of seagrass seeds. *Aquat. Bot.* **40**: 195–198.

McMillan, C. & R. C. Phillips. 1981. Morphological variation and isozymes of North American *Phyllospadix* (Potamogetonaceae). *Can. J. Bot.* **59**: 1494–1500.

McMillan, C. & J. Jewett-Smith. 1988. The sex ratio and fruit production of laboratory-germinated seedlings of *Halophila engelmannii* Aschers. (Hydrocharitaceae) from Redfish Bay, Texas. *Aquat. Bot.* **32**: 329–339.

McMillan, C., S. C. Williams, L. Escobar & O. Zapata. 1981. Isozymes, secondary compounds and experimental cultures of Australian seagrasses in *Halophila, Halodule, Zostera, Amphibolis* and *Posidonia. Aust. J. Bot.* **29**: 247–260.

McMillan, P. D. & R. D. Porcher. 2005. Noteworthy collections: South Carolina. *Castanea* **70**: 237–240.

McMillan, P. D., R. K. Peet, R. D. Porcher & B. A. Sorrie. 2002. Noteworthy botanical collections from the fire-maintained pineland and wetland communities of the coastal plain of the Carolinas and Georgia. *Castanea* **67**: 61–83.

McMullan, J. J., R. J. Gornall & C. D. Preston. 2011. ITS rDNA polymorphism among species and hybrids of *Potamogeton* subgenus *Coleogeton* (Potamogetonaceae) in north-western Europe. *New J. Bot.* **1**: 111–115.

McNair, D. B. & C. Cramer-Burke. 2006. Breeding ecology of American and Caribbean coots at Southgate Pond, St. Croix: use of woody vegetation. *Wilson J. Ornithol.* **118**: 208–217.

McNair, D. M. & M. H. Alford. 2014. *Blyxa aubertii* (Hydrocharitaceae) new to Mississippi, USA. *J. Bot. Res. Inst. Texas* **8**: 267–270.

McNamara, J. P., D. I. Siegel, P. H. Glaser & R. M. Beck. 1992. Hydrogeologic controls on peatland development in the Malloryville Wetland, New York (USA). *J. Hydrol.* **140**: 279–296.

McNaughton, S. J. 1966. Ecotype function in the *Typha* community-type. *Ecol. Monogr.* **36**: 297–325.

McNaughton, S. J. 1968. Autotoxic feedback in relatin to germination and seedling growth in *Typha latifolia. Ecology* **49**: 367–369.

McNaughton, S. J. & L. W. Fullem. 1970. Photosynthesis and photorespiration in *Typha latifolia. Plant Physiol.* **45**: 703–707.

McNeal, D. W. 1995. Report on *Allium.* Columbia Basin scientific assessment project. Unpublished Report. Available online: https://www.google.com/url?sa=t&rct=j&q=&esrc=s&source=web&cd=2&ved=2ahUKEwjG_ZWY8_rgAhXFmeAKHQFBA6IQFjABegQICRAC&url=https%3A%2F%2Fwww.fs.fed.us%2Fr6%2Ficbemp%2Fscience%2Fmcneal.pdf&usg=AOvVaw0f_MCBdjQSCzdgMlEUZOHi [accessed 11 February, 2019], 21 pp.

McNeal, D. W. 2012a. Alliaceae. Onion or garlic family. Pp. 1395–1398 *In*: B. G. Baldwin, D. H. Goldman, D. J. Keil, R. Patterson, T. J. Rosatti & D. H. Wilken (eds.), *The Jepson Manual*, 2nd ed. University of California Press, Berkeley, CA.

McNeal, D. W. 2012b. Melanthiaceae. False hellebore family. Pp. 1289–1297 *In*: B. G. Baldwin, D. H. Goldman, D. J. Keil, R. Patterson, T. J. Rosatti & D. H. Wilken (eds.), *The Jepson Manual*, 2nd ed. University of California Press, Berkeley, CA.

McNeal, Jr., D. W. & T. D. Jacobsen. 2002. *Allium* Linnaeus. Pp. 224–276 *In*: Flora North America Editorial Committee (eds.), *Flora of North America North of Mexico, Vol. 26: Magnoliophyta: Liliidae: Liliales and Orchidales*. Oxford University Press, New York, NY.

McNeal, Jr., D. W. & A. D. Shaw. 2002. *Veratrum* Linnaeus. Pp. 72–76 *In*: Flora North America Editorial Committee (eds.), *Flora of North America North of Mexico, Vol. 26: Magnoliophyta: Liliidae: Liliales and Orchidales*. Oxford University Press, New York, NY.

McNeal, D. W. & W. B. Zomlefer. 2010. Documentation of the chromosome number for the California endemic, *Toxicoscordion exaltatum* (Liliales: Melanthiaceae). *Madroño* **57**: 180–184.

McNeal, D. W. & W. B. Zomlefer. 2012. *Toxicoscordion.* Death camus. Pp. 1395–1396 *In*: B. G. Baldwin, D. H. Goldman, D. J. Keil, R. Patterson, T. J. Rosatti & D. H. Wilken (eds.), *The Jepson Manual*, 2nd ed. University of California Press, Berkeley, CA.

McNulty, A. K., W. R. Cummins & A. Pellizzari. 1988. A field survey of respiration rates in leaves of arctic plants. *Arctic* **41**: 1–5.

McPherson, J. E. & S. M. Paskewitz. 1984. Laboratory rearing of *Amaurochrous cinctipes* (Hemiptera: Pentatomidae: Podopinae) with descriptions of immature stages. *J. New York Entomol. Soc.* **92**: 61–68.

McShane, D. & K. Mehigan. 2012. *Aquatic Macrophyte Survey of Otsego Lake.* SUNY Oneonta Biological Field Station, Cooperstown, NY. 23 pp.

Meanley, B. 1953. Nesting of the King Rail in the Arkansas rice fields. *Auk* **70**: 261–269.

Meanley, B. & J. S. Webb. 1961. Distribution of winter redwinged blackbird populations on the Atlantic Coast. *Bird-Banding* **32**: 94–97.

Meehan, T. 1901. *Smilax walteri. Meehan's Monthly* **12**(12): 181–182.

Meeker, J. E., D. A. Wilcox & A. G. Harris. 2018. Changes in wetland vegetation in regulated lakes in northern Minnesota, USA ten years after a new regulation plan was implemented. *Wetlands*. doi: 10.1007/s13157-017-0986-1.

Mead, B. R. 1995. Plant biomass in the Tanana River basin, Alaska. Research Paper PNW-RP-477. U.S. Department of Agriculture, Forest Service, Pacific Northwest Research Station, Portland, OR. 78 pp.

Meerow, A. W. 2004. Palm seed germination. Document BUL274. IFAS Extension, University of Florida, Gainesville, FL. 11 pp.

Meerow, A. W. 2012. Taxonomy and phylogeny. Pp. 17–56 *In*: R. Kamenetsky & H. Okubo (eds.), *Ornamental Geophytes: Basic Science to Sustainable Production*. CRC Press, Boca Raton, FL.

Meerow, A. W. & D. A. Snijman. 2001. Phylogeny of Amaryllidaceae tribe Amaryllideae based on nrDNA ITS sequences and morphology. *Am. J. Bot.* **88**: 2321–2330.

Meerow, A. W. & D. A. Snijman. 2006. The never-ending story: multigene approaches to the phylogeny of Amaryllidaceae. *Aliso* **22**: 355–366.

Meerow, A. W., C. L. Guy, Q.-B. Li & S.-L. Yang. 2000. Phylogeny of the American Amaryllidaceae based on nrDNA ITS sequences. *Syst. Bot.* **25**: 708–726.

Meerow, A. W., H. M. Donselman & T. K. Broschat. 2001. *Native Trees for South Florida*. Florida Cooperative Extension Service, Institute of Food and Agricultural Sciences, University of Florida, Gainesville, FL. 17 pp.

Meerow, A. W., C. L. Guy, Q.-B. Li & J. R. Clayton. 2002. Phylogeny of the tribe Hymenocallideae (Amaryllidaceae) based on morphology and molecular characters. *Ann. Missouri Bot. Gard.* **89**: 400–413.

Meerow, A. W., D. J. Lehmiller & J. R. Clayton. 2003. Phylogeny and biogeography of *Crinum* L. (Amaryllidaceae) inferred from nuclear and limited plastid non-coding DNA sequences. *Bot. J. Linn. Soc.* **141**: 349–363.

Meerow, A. W., J. L. Reveal, D. A. Snijman & J. H. Dutilh. 2007a. Proposal to conserve the name Amaryllidaceae against Alliaceae, a "superconservation" proposal. *Taxon* **56**: 1299–1300.

Meerow, A. W., M. Gideon, D. N. Kuhn, J. C. Motamayor & K. Nakamura. 2007b. Genetic structure and gene flow among south Florida populations of *Iris hexagona* Walt. (Iridaceae) assessed with 19 microsatellite DNA loci. *Int. J. Plant Sci.* **168**: 1291–1309.

Meerow, A. W., M. Gideon, D. N. Kuhn, S. Mopper & K. Nakamura. 2011. The genetic mosaic of *Iris* series *Hexagonae* in Florida: inferences on the Holocene history of the Louisiana irises and anthropogenic effects on their distribution. *Int. J. Plant Sci.* **172**: 1026–1052.

Meerow, A. W., M. Gideon & K. Nakamura. 2017. Hybridization between ecotypes in a phenotypically and ecologically heterogeneous population of *Iris savannarum* (Iridaceae) in Florida. *Plant Species Biol.* **32**: 309–322.

Megateli, S., S. Semsari & M. Couderchet. 2009. Toxicity and removal of heavy metals (cadmium, copper, and zinc) by *Lemna gibba. Ecotoxicol. Environ. Saf.* **72**: 1774–1780.

Megonigal, J. P. & W. H. Schlesinger. 1997. Enhanced CH_4 emission from a wetland soil exposed to elevated CO_2. *Biogeochemistry* **37**: 77–88.

Megonigal, J. P., C. D. Vann & A. A. Wolf. 2005. Flooding constraints on tree (*Taxodium distichum*) and herb growth responses to elevated CO_2. *Wetlands* **25**: 430–438.

Mehringer, Jr., P. J., S. F. Arno & K. L. Petersen. 1977. Postglacial history of lost trail pass bog, Bitterroot Mountains, Montana. *Arct. Alp. Res.* **9**: 345–368.

Meinshausen, K. F. 1895. Das genus *Sparganium* L. Systematische Beschreibung der Arten nebst Darstellung ihrer Verbreitung auf Grundlage ihres Vorkommens im Gouv. St. Petersburg. *Bull. Acad. Imper. Sci. St. Petersburg* **36**: 21–41.

Mejbel, H. S. & A. M. Simons. 2018. Aberrant clones: birth order generates life history diversity in Greater Duckweed, *Spirodela polyrhiza. Ecol. Evol.* **8**: 2021–2031.

Mellinger, M. B. 1966. Some plant associations of *Pinckneya pubens. Castanea* **31**: 310–313.

Melnyk, J. P., S. Wang & M. F. Marcone. 2010. Chemical and biological properties of the world's most expensive spice: Saffron. *Food Res. Int.* **43**: 1981–1989.

Mendall, H. L. 1949. Food habits in relation to black duck management in Maine. *J. Wildl. Manag.* **13**: 64–101.

Mendall, H. L. & J. S. Gashwiler. 1940. Water bulrush as a food of waterfowl. *Auk* **57**: 245–246.

Menéndez, M. 2002. Net production of *Ruppia cirrhosa* in the Ebro Delta. *Aquat. Bot.* **73**: 107–113.

Menéndez, M. & F. A. Comín. 1989. Seasonal patterns of biomass variation of *Ruppia cirrhosa* (Petagna) Grande and *Potamogeton pectinatus* L. in a coastal lagoon. *Sci. Mar.* **53**: 633–638.

Meng, Y., A. J. Krzysiak, M. J. Durako, J. I. Kunzelman & J. L. C. Wright. 2008. Flavones and flavone glycosides from *Halophila johnsonii. Phytochemistry* **69**: 2603–2608.

Merckx, V., P. Schols, H. Maas-van de Kamer, P. Maas, S. Huysmans & E. Smets. 2006. Phylogeny and evolution of Burmanniaceae (Dioscoreales) based on nuclear and mitochondrial data. *Am. J. Bot.* **93**: 1684–1698.

Merckx, V., L. W. Chatrou, B. Lemaire, M. N. Sainge, S. Huysmans & E. F. Smets. 2008a. Diversification of myco-heterotrophic angiosperms: evidence from Burmanniaceae. *BMC Evol. Biol.* **8**: 178.

Merckx, V., P. Schols, K. Geuten, S. Huysmans & E. Smets. 2008b. Phylogenetic relationships in Nartheciaceae (Dioscoreales), with focus on pollen and orbicule morphology. *Belg. J. Bot.* **141**: 64–77.

Merckx, V., M. Stöckel, A. Fleischmann, T. D. Bruns & G. Gebauer. 2010. ^{15}N and ^{13}C natural abundance of two mycoheterotrophic and a putative partially mycoheterotrophic species associated with arbuscular mycorrhizal fungi. *New Phytol.* **188**: 590–596.

Meriläinen, J. 1968. *Najas minor* All. in North America. *Rhodora* **70**: 161–175.

Merrill, A. G., T. L. Benning & J. A. Fites. 2006. Factors controlling structural and floristic variation of riparian zones in a mountainous landscape of the western United States. *West. N. Am. Nat.* **66**: 137–155.

Merz, J. E., J. R. Smith, M. L. Workman, J. D. Setka & B. Mulchaey. 2008. Aquatic macrophyte encroachment in Chinook salmon spawning beds: lessons learned from gravel enhancement monitoring in the Lower Mokelumne River, California. *North Am. J. Fish. Manag.* **28**: 1568–1577.

Messner, T. C. & B. Schindler. 2010. Plant processing strategies and their affect [sic!] upon starch grain survival when rendering *Peltandra virginica* (L.) Kunth, Araceae edible. *J. Archaeol. Sci.* **37**: 328–336.

Meyer, M. D., C. A. Davis & D. Dvorett. 2015. Response of wetland invertebrate communities to local and landscape factors in north central Oklahoma. *Wetlands* **35**: 533–546.

Miano, A. J., J. P. Leblanc & J. M. Farrell. 2019. Laboratory evaluation of spawning substrate type on potential egg predation by round goby (*Neogobius melanostomus*). *J. Great Lakes Res.* **45**: 390–393.

Michael, T. P., D. Bryant, R. Gutierrez, N. Borisjuk, P. Chu, H. Zhang, J. Xia, J. Zhou, H. Peng, M. E. Baidouri, B. ten Hallers, A. R. Hastie, T. Liang, K. Acosta, S. Gilbert, C. McEntee, S. A. Jackson, T. C. Mockler, W. Zhang & E. Lam. 2017. Comprehensive definition of genome features in *Spirodela polyrhiza* by high-depth physical mapping and short-read DNA sequencing strategies. *Plant J.* **89**: 617–635.

Michel, L. U. C., J. G. Baldwin & H. Arnold. 1986. *Pratylenchus morettoi* n. sp. (Nemata: Pratylenchidae). *Rev. Nematol.* **9**: 119–123.

Michel, A., R. S. Arias, B. E. Scheffler, S. O. Duke, M. Netherland & F. E. Dayan. 2004. Somatic mutation-mediated evolution of herbicide resistance in the nonindigenous invasive plant hydrilla (*Hydrilla verticillata*). *Mol. Ecol.* **13**: 3229–3237.

Middleton, B. A. 1989. Succession and goose herbivory in monsoonal wetlands of the Keoladeo National Park, Bharatpur, India. Ph.D. dissertation. Iowa State University, Ames, IA. 144 pp.

Middleton, B. A. 1995. Seed banks and species richness potential of coal slurry ponds reclaimed as wetlands. *Restor. Ecol.* **3**: 311–318.

Middleton, B. A. 2009. Regeneration potential of *Taxodium distichum* swamps and climate change. *Plant Ecol.* **202**: 257–274.

Miduno, T. 1940. Chromosomenstudien an Orchidazeen. *Cytologia* **11**: 179–185.

Mielecki, M. & E. Pieczynska. 2005. The influence of fragmentation on the growth of *Elodea canadensis* Michx. in different light conditions. *Polish J. Ecol.* **53**: 155–164.

Mifsud, S. 2010. First occurences of *Lemna minuta* Kunth (fam. Lemnaceae) in the Maltese islands. *Cent. Mediterr. Nat.* **5**: 1–4.

Miguel, V., J. A. Otero, B. Barrera, I. Rodeiro, J. G. Prieto, G. Merino & A. I. Álvarez. 2015. ABCG2/BCRP interaction with the sea grass *Thalassia testudinum. Drug Metab. Pers. Ther.* **30**: 251–256.

Miki, S. 1934. On fresh water plants new to Japan. *Jap. J. Pharmacogn.* **48**: 326–337.

Miller, M. R. 1987. Fall and winter foods of northern pintails in the Sacramento Valley, California. *J. Wildl. Manag.* **51**: 405–414.

Miller, T. G. & H. M. Hoven. 2007. *Ecological and Beneficial Use Assessment of Farmington Bay Wetlands: Assessment and Site-Specific Nutrient Criteria.* U.S. Environmental Protection Agency (Region VIII), Denver, CO. 32 pp (plus appendices).

Miller, J. D., W. T. Haller & M. S. Glenn. 1993. Turion production by dioecious hydrilla in north Florida. *J. Aquat. Plant Manag.* **31**: 101–105.

Miller, M. M., S. W. Phipps, C. S. Major & K. M. Major. 2011. Effects of environmental variation and non-point source (NPS) nutrient pollution on aquatic plant communities in Weeks Bay National Estuarine Research Reserve (WBNERR), AL. *Estuaries Coast.* **34**: 1182–1193.

Mills, J. E., J. A. Reinartz, G. A. Meyer & E. B. Young. 2009. Exotic shrub invasion in an undisturbed wetland has little community-level effect over a 15-year period. *Biol. Invasions* **11**: 1803–1820.

Millsap, B. A. 2009. Endangered and threatened wildlife and plants; 5-year reviews of 23 southwestern species. *Fed. Reg.* **74**: 6917–6919.

Millspaugh, C. F. 1887. *American Medicinal Plants: An Illustrated and Descriptive Guide to the American Plants Used as Homopathic Remedies: Their History, Preparation, Chemistry and Physiological Effects.* Boericke & Tafel, New York, NY. [irregular pagination].

Minayeva, T. Y. 2010. The peculiarities of seed reproduction biology of some monocotyledonous peatland plant species. *Bot. Zhurn.* **95**: 482–495.

Minciullo, P. L., E. Fazio, M. Patafi & S. Gangemi. 2007. Allergic contact dermatitis due to *Zantedeschia aethiopica*. *Cont. Dermat.* **56**: 46–46.

Minckley, W. L. & D. R. Tindall. 1963. Ecology of *Batrachospermum* sp. (Rhodophyta) in Doe Run, Meade County, Kentucky. *Bull. Torrey Bot. Club* **90**: 391–400.

Ming, W., S. Zhe & L. Qinzhong. 1994. Study of *Hydrilla verticillata* (L.F.) Royle as protein resource. 1. Analysis of the biological characters and nutrition elements of *Hydrilla verticilata* (L. F.) Royle. *J. Hunan Agric. College* **20**: 457–463.

Minno, M. C. & C. Slaughter. 2003. New record of the endangered lakeside sunflower, *Helianthus carnosus* (Asteraceae), from Putnam County, Florida. *Florida Sci.* **66**: 291–293.

Miretzky, P., A. Saralegui & A. F. Cirelli. 2004. Aquatic macrophytes potential for the simultaneous removal of heavy metals (Buenos Aires, Argentina). *Chemosphere* **57**: 997–1005.

Mishra, V. K. & B. D. Tripathi. 2009. Accumulation of chromium and zinc from aqueous solutions using water hyacinth (*Eichhornia crassipes*). *J. Hazard. Mater.* **164**: 1059–1063.

Misnikov, O. S. 2006. Physicochemical principles of hydrophobization of mineral binders by additives produced from peat raw material. *Theor. Found. Chem. Eng.* **40**: 423–430.

Misnikov, O. 2018. The hydrophobic modification of gypsum binder by peat products: physico-chemical and technological basis. *Mires Peat* **21**: 1–14.

Misra, M. P. 1974. Cytological studies in *Ottelia alismoides* Pers. *Cytologia* **39**: 419–427.

Mitchell, E. 1926. Germination of seeds of plants native to Dutchess County, New York. *Bot. Gaz.* **81**: 108–112.

Mitchell, D. K. 2011. Wet flatwoods restoration after decades of fire suppresion [sic!]. M.S. thesis. University of Florida, Gainesville, FL. 77 pp.

Mitchell, C. C. & W. A. Niering. 1993. Vegetation change in a topogenic bog following beaver flooding. *Bull. Torrey Bot. Club* **120**: 136–147.

Mitchell, C. A., T. W. Custer & P. J. Zwank. 1994. Herbivory on shoalgrass by wintering redheads in Texas. *J. Wildl. Manag.* **58**: 131–141.

Mitchell, M. E., S. C. Lishawa, P. Geddes, D. J. Larkin, D. Treering & N. C. Tuchman. 2011. Time-dependent impacts of cattail invasion in a Great Lakes coastal wetland complex. *Wetlands* **31**: 1143–1149.

Mitsch, W. J., D. F. Fink & L. Zhang. 2004. Net primary productivity of macrophyte communities after eight growing seasons in experimental planted and unplanted marshes. Pp. 43–48 *In*: W. J. Mitsch, L. Zhang & C. J. Anderson (eds.), *Macrophyte Production.* Annual Report 2004. The Olentangy River Wetland Research Park, The Ohio State University, Columbus, OH.

Miyazawa, M., H. Nakahashi, Y. Kashima, R. Motooka, N. Hara, H. Nakagawa, T. Yoshii, A. Usami & S. Marumoto. 2015. Chemical composition and aroma evaluation of essential oils from skunk cabbage (*Symplocarpus foetidus*). *J. Oleo Sci.* **64**: 1329–1336.

Mkandawire, M., J. A. Teixeira da Silva & E. G. Dudel. 2014. The *Lemna* bioassay: contemporary issues as the most standardized plant bioassay for aquatic ecotoxicology. *Crit. Rev. Environ. Sci. Technol.* **44**: 154–197.

M'Mahon, B. 1804. *A Catalogue of American Seeds &c. Sold by Bernard M'Mahon, Seedsman, Philadelphia. 1804.* Bartholomew Graves, Philadelphia, PA. 30 pp.

Mnayer, D., A.-S. Fabiano-Tixier, E. Petitcolas, T. Hamieh, N. Nehme, C. Ferrant, X. Fernandez & F. Chemat. 2014. Chemical composition, antibacterial and antioxidant activities of six essentials oils from the Alliaceae family. *Molecules* **19**: 20034–20053.

MNFS (National Marine Fisheries Service). 2002. Recovery plan for Johnson's seagrass (*Halophila johnsonii*). Prepared by the Johnson's Seagrass Recovery Team for the National Marine Fisheries Service, Silver Spring, MD. 134 pp.

Moeller, R. W. 1984. The Ivory Pond mastodon project. *N. Am. Archaeol.* **5**: 1–12.

Moeller, R. E., J. M. Burkholder & R. G. Wetzel. 1988. Significance of sedimentary phosphorus to a rooted submersed macrophyte (*Najas flexilis* (Willd.) Rostk. and Schmidt) and its algal epiphytes. *Aquat. Bot.* **32**: 261–281.

Moen, T. 1953. Food habits of the carp in northwest Iowa lakes. *Proc. Iowa Acad. Sci.* **60**: 665–686.

Moffett, Jr., J. M. 2008. *Xyris tennesseensis*: status survey, habitat restoration/management concerns, and relation to a new xyrid, *Xyris spathifolia*. Ph.D. dissertation. Auburn University, Auburn, AL. 178 pp.

Moffett, Jr., J. M. & R. S. Boyd. 2013. Management of a population of the federally endangered *Xyris tennesseensis* (Tennessee yellow-eyed grass). *Castanea* **78**: 198–212.

Mohagheghzadeh, A., P. Faridi, M. Shams-Ardakani & Y. Ghasemi. 2006. Medicinal smokes. *J. Ethnopharmacol.* **108**: 161–184.

Mohanty, P., S. Behera, S. S. Swain, D. P. Barik & S. K. Naik. 2013. Micropropagation of *Hedychium coronarium* J. Koenig through rhizome bud. *Physiol. Mol. Biol. Plant.* **19**: 605–610.

Mohlenbrock, R. H. 1959a. A floristic study of a southern Illinois swampy area. *Ohio J. Sci.* **59**: 89–100.

Mohlenbrock, R. H. 1959b. Plant communities in Jackson County, Illinois. *Bull. Torrey Bot. Club* **86**: 109–119.

Mohlenbrock, R. H. 1962. On the occurrence of *Lilium superbum* L. in Illinois. *Castanea* **27**: 173–176.

Mohlenbrock, R. H. 1976. Woody plants of the Ocala National Forest, Florida. *Castanea* **41**: 309–319.

Mohlenbrock, R. H., G. E. Dillard & T. S. Abney. 1961. A survey of southern Illinois aquatic vascular plants. *Ohio J. Sci.* **61**: 262–273.

Mohr, C. T. 1901. Plant life of Alabama. *Contrib. U.S. Natl. Herb.* **6**: 1–921.

Moisan, C. & S. Pellerin. 2013. Factors associated with the presence of flowering individuals of *Arethusa bulbosa* (Orchidaceae) in peatlands of southern Quebec. *Ecoscience* **20**: 1–8.

Mokni, R. E. & M. H. E. Aouni. 2012. *Zantedeschia aethiopica* (Araceae) a new species naturalized in the Northwest of Tunisia. *Fl. Medit.* **22**: 191–196.

Moldenke, H. N. 1936. The flora of the Watchung Mountains. Part II—the Flora. *Torreya* **36**: 88–93.

Moldenke, H. N. 1957. Additional notes on the Eriocaulaceae. XII. *Bull. Jard. Bot. 'Etat. Bruxelles* **27**: 115–141.

Moldenke, H. N. 1969. Additional notes on the Eriocaulaceae. XXII. *Phytologia* **18**: 344–396.

Moldenke, H. N., G. F. Dillman, F. D. Kern & F. Place. 1939. Field trips of the club. *Torreya* **39**: 143–151.

Mollik, M. A. H., M. S. Hossan, A. K. Paul, M. Taufiq-Ur-Rahman, R. Jahan & M. Rahmatullah. 2010. A comparative analysis of medicinal plants used by folk medicinal healers in three districts of Bangladesh and inquiry as to mode of selection of medicinal plants. *Ethnobot. Res. Appl.* **8**: 195–218.

Monda, M. J., J. T. Ratti & T. R. McCabe. 1994. Reproductive ecology of tundra swans on the Arctic National Wildlife Refuge, Alaska. *J. Wildl. Manag.* **58**: 757–773.

Monk, C. D. & T. W. Brown. 1965. Ecological consideration of cypress heads in northcentral Florida. *Am. Midl. Nat.* **74**: 126–140.

Monsod, Jr., G. G. 1979. *Man and the Water Hyacinth*. Vantage Press, New York, NY. 48 pp.

Montz, G. N. 1978. The submerged vegetation of Lake Pontchartrain, Louisiana. *Castanea* **43**: 115–128.

Mony, C., T. J. Koschnick, W. T. Haller & S. Muller. 2007. Competition between two invasive Hydrocharitaceae (*Hydrilla verticillata* (L.f.) (Royle) and *Egeria densa* (Planch)) as influenced by sediment fertility and season. *Aquat. Bot.* **86**: 236–242.

Moodie, G. E. E. 1976. Heat production and pollination in Araceae. *Can. J. Bot.* **54**: 545–546.

Moody, K. 1989. *Weeds Reported in Rice in South and Southeast Asia*. International Rice Research Institute, Los Baños, Laguna, Philippines. 442 pp.

Moon, B.-C., C.-S. Kim, T.-S. Park, J.-R. Jo, I.-Y. Lee & J.-E. Park. 2000. Germination and dormancy by different seed types of marsh dayflower (*Aneilema keisak* Hassk.). *Korean J. Weed Sci.* **20**: 191–196.

Moore, E. 1913. The Potamogetons in relation to pond culture. *Bull. U.S. Bur. Fish.* **33**: 251–291.

Moore, D. L. 2000. The aquatic macrophyte community at Put-in-Bay, Ohio. *Great Lakes Res. Rev.* **5**: 37–42.

Moore, G. 2004. Field trip reports. *J. Torrey Bot. Soc.* **131**: 403–419.

Moore, K. A., R. J. Orth & J. F. Nowak. 1993. Environmental regulation of seed germination in *Zostera marina* L. (eelgrass) in Chesapeake Bay: effects of light, oxygen and sediment burial. *Aquat. Bot.* **45**: 79–91.

Moore, T. R., J. L. Bubier, S. E. Frolking, P. M. Lafleur & N. T. Roulet. 2002. Plant biomass and production and CO$_2$ exchange in an ombrotrophic bog. *J. Ecol.* **90**: 25–36.

Moore, M. T., H. L. Tyler & M. A. Locke. 2013. Aqueous pesticide mitigation efficiency of *Typha latifolia* (L.), *Leersia oryzoides* (L.) Sw., and *Sparganium americanum* Nutt. *Chemosphere* **92**: 1307–1313.

Moorehouse, A., A. Mankowski, W. E. McClain & J. E. Ebinger. 2002. Status of the known populations of the Virginia bunchflower (*Melanthium virginicum*) in Illinois. *Castanea* **67**: 188–192.

Mopper, S., K. C. Wiens & G. A. Goranova. 2016. Competition, salinity, and clonal growth in native and introduced irises. *Am. J. Bot.* **103**: 1575–1581.

Moran, R. C. 1981. Prairie fens in northeastern Illinois: floristic composition and disturbance. Pp. 164–168 *In*: R. L. Stuckey & K. J. Reese (eds.), *The Prairie Peninsula--In the "Shadow" of Transeau: Proceedings of the Sixth North American Prairie Conference*. Ohio Biological Survey Biological Notes No. 15. Columbus, OH.

Moran, K. L. & K. A. Bjorndal. 2005. Simulated green turtle grazing affects structure and productivity of seagrass pastures. *Mar. Ecol. Prog. Ser.* **305**: 235–247.

Moran, P. J. & C. Yang. 2012. Distribution of wild taro (*Colocasia esculenta*) in subtropical Texas, growth of young colonies, and tolerance to simulated herbivory. *Subtrop. Plant Sci.* **64**: 18–28.

Moran, E. A., R. R. Briese & J. F. Couch. 1940. Some new cyanogenetic plants. *J. Washington Acad. Sci.* **30**: 237–239.

Mora-Olivo, A., J. G. Martínez-Ávalos & E. de la Rosa-Manzano. 2018. *Heteranthera peduncularis* (Pontederiaceae) en Tamaulipas, México. *Phytoneuron* **2018–7**: 1–4.

Moravcová, L., P. Zákravský & Z. Hroudová. 2001. Germination and seedling establishment in *Alisma gramineum*, *A. plantago-aquatica* and *A. lanceolatum* under different environmental conditions. *Folia Geobot.* **36**: 131–146.

Mörch, O. J. N. 1839. *Catalogus plantarum horti botanici Hafniensis*. Copenhagen Botanical Garden. Hauniæ [Copenhagen], Denmark. 102 pp.

Morgan, G. R. 1980. The ethnobotany of sweet flag among North American Indians. *Bot. Mus. Leafl.* **28**: 235–246.

Morgan, M. D. 1990. Seed germination characteristics of *Iris virginica*. *Am. Midl. Nat.* **124**: 209–213.

Morgan, M. D. & K. R. Philipp. 1986. The effect of agricultural and residential development on aquatic macrophytes in the New Jersey Pine Barrens. *Biol. Conserv.* **35**: 143–158.

Morgan, M. T. & S. C. H. Barrett. 1988. Historical factors and anisoplethic population structure in tristylous *Pontederia cordata*: a reassessment. *Evolution* **42**: 496–504.

Morgan, M. T. & S. C. H. Barrett. 1990. Outcrossing rates and correlated mating within a population of *Eichhornia paniculata* (Pontederiaceae). *Heredity* **64**: 271–280.

Morgan, E. C. & W. A. Overholt. 2005. New records of invasive exotic plant species in St. Lucie County, Florida. *Castanea* **70**: 59–62.

Morgan, V. H. & M. Sytsma. 2009. *Introduction to Common Native and Potential Invasive Freshwater Plants in Alaska*. Center for Lakes and Reservoirs Publications and Presentations. Paper 26. Portland State University, Portland, OR. 195 pp.

Morgan, P. A. & M. D. O. Adams. 2017. Tidal marshes in the Saco River Estuary, Maine: a study of plant diversity and possible effects of shoreline development. *Rhodora* **119**: 304–331.

Mori, E. S., C. F. Gouvea, S. M. M. Leite, C. L. Marino, D. Martins & E. D. Velini. 1999. Genetic characterization of *Egeria najas* presented in the Jupia lake and its tributaries. *Planta Daninha* **17**: 217–225.

Mori, E. S., D. Martins, E. D. Velini, C. L. Marino, C. F. Gouvêa, S. M. M. Leite, E. Camacho & R. P. Guries. 2012. Genetic diversity in *Egeria densa* and *E. najas* in Jupiá Reservoir, Brazil. *Ci. Invest. Agraria* **39**: 321–330.

Morinaga, T. 1926. The favorable effect of reduced oxygen supply upon the germination of certain seeds. *Am. J. Bot.* **13**: 159–166.

Morita, T., A. Miyamatsu, H. Fujii, H. Kokubu, M. Abe, A. Kurashima & M. Maegawa. 2011. Germination in *Zostera japonica* is determined by cold stratification, tidal elevation and sediment type. *Aquat. Bot.* **95**: 234–241.

Morong, T. 1893. The Naiadaceae of North America. *Mem. Torrey Bot. Club* **3**: 1–65.

Morong, T. 1894. The Smilaceae of North and Central America. *Bull. Torrey Bot. Club* 21: 419–443.

Morris, M. W. 1988. Noteworthy vascular plants from Grenada County, Mississippi. *Sida* 13: 177–186.

Morris, E. C. 2003. How does fertility of the substrate affect intraspecific competition? Evidence and synthesis from self-thinning. *Ecol. Res.* 18: 287–305.

Morris, M. W. 2012. *Stenanthium leimanthoides* (Melanthiaceae) and *Nestronia umbellula* (Santalaceae) in Alabama. *Castanea* 77: 375–380.

Morris, M. W. 2013. The genus *Platanthera* (Orchidaceae) in Mississippi. *J. Bot. Res. Inst. Texas* 7: 323–339.

Morris, D. M. 2014. Aquatic habitat use by North American moose (*Alces alces*) and associated richness and biomass of submersed and floating-leaved aquatic vegetation in north-central Minnesota. M.S. thesis. Lakehead University, Thunder Bay, ON. 130 pp.

Morris, P. F. & W. G. Barker. 1977. Oxygen transport rates through mats of *Lemna minor* and *Wolffia* sp. and oxygen tension within and below the mat. *Can. J. Bot.* 55: 1926–1932.

Morris, C. D., J. L. Callahan & R. H. Lewis. 1990. Distribution and abundance of larval *Coquillettidia perturbans* in a Florida freshwater marsh. *J. Am. Mosq. Control Assoc.* 6: 452–460.

Morrone, J. J. 2013. The subtribes and genera of the tribe Listroderini (Coleoptera, Curculionidae, Cyclominae): phylogenetic analysis with systematic and biogeographical accounts. *Zookeys* 273: 15–71.

Morton, J. F. 1963. Principal wild food plants of the United States excluding Alaska and Hawaii. *Econ. Bot.* 17: 319–330.

Morton, J. F. 1975. Cattails (*Typha* spp.) – weed problem or potential crop? *Econ. Bot.* 29: 7–29.

Morton, M. L. 1994. Comparison of reproductive timing to snow conditions in wild onions and White-crowned Sparrows at high altitude. *Great Basin Nat.* 54: 371–375.

Morton, J. K. & E. H. Hogg. 1989. Biogeography of island floras in the Great Lakes. II. Plant dispersal. *Can. J. Bot.* 67: 1803–1820.

Morzaria-Luna, H. N. & J. B. Zedler. 2007. Does seed availability limit plant establishment during salt marsh restoration? *Estuaries Coast.* 30: 12–25.

Morzaria-Luna, H. N. & J. B. Zedler. 2014. Competitive interactions between two salt marsh halophytes across stress gradients. *Wetlands* 34: 31–42.

Moseley, R. K. 1989. Field investigations of *Allium validum* (tall swamp onion) and *Douglasia idahoensis* (Idaho Douglasia), region 1 sensitive species, on the Nez Pierce National Forest. Unpublished Report. Idaho Department of Fish and Game, Conservation Data Center, Boise, ID. 18 pp.

Moscone, E. A. & L. M. Bernardello. 1985. Chromosome Studies on *Hydromystria laevigata* (Hydrocharitaceae). *Ann. Missouri Bot. Gard.* 72: 480–484.

Moss, E. H. 1932. The vegetation of Alberta: IV. The poplar association and related vegetation of central Alberta. *J. Ecol.* 20: 380–415.

Mossman, R. E. 2009. Seed dispersal and reproduction patterns among Everglades plants. Ph.D. dissertation. Florida International University, Miami, FL. 125 pp.

Motley, T. J. 1994. The ethnobotany of sweet flag, *Acorus calamus* (Araceae). *Econ. Bot.* 48: 397–412.

Motzkin, G. 1994. Calcareous fens of western New England and adjacent New York state. *Rhodora* 96: 44–68.

Moulton, J. K. & P. H. Adler. 1995. Revision of the *Simulium jenningsi* species-group (Diptera: Simuliidae). *Trans. Am. Entomol. Soc.* 121: 1–57.

Mouissie, A. M., W. Lengkeek & R. Van Diggelen. 2005. Estimating adhesive seed-dispersal distances: field experiments and correlated random walks. *Funct. Ecol.* 19: 478–486.

Moulton, O. M. & S. D. Hacker. 2011. Congeneric variation in surfgrasses and ocean conditions influence macroinvertebrate community structure. *Mar. Ecol. Prog. Ser.* 433: 53–63.

Mowat, G., P. J. Curtis & D. J. R. Lafferty. 2017. The influence of sulfur and hair growth on stable isotope diet estimates for grizzly bears. *PloS One* 12: e0172194.

Moyer, R. D. & E. L. Bridges. 2015. *Xyris chapmanii*, an overlooked *Xyris* species of the New Jersey Pine Barrens. *Bartonia* 67: 58–74.

Moyer, R. D. & R. F. C. Naczi. 2017. *Rhynchospora leptocarpa*, an overlooked species of the New Jersey Pine Barrens. *Brittonia* 69: 127–132.

Moyle, J. B. 1944. Wild rice in Minnesota. *J. Wildl. Manag.* 8: 177–184.

Moyle, J. B. 1945. Some chemical factors influencing the distribution of aquatic plants in Minnesota. *Am. Midl. Nat.* 34: 402–420.

Moyle, J. B. & N. Hotchkiss. 1945. The aquatic and marsh vegetation of Minnesota and its value to waterfowl. Technical Bulletin No. 3. Minnesota Dept. of Conservation, St. Paul, MN. 122 pp.

Mrachek, R. J. 1966. Macroscopic invertebrates on the higher aquatic plants at Clear Lake, Iowa. *Proc. Iowa Acad. Sci.* 73: 168–177.

Mtewa, A. G., S. Deyno, K. Ngwira, F. Lampiao, E. L. Peter, L. Y. Ahovegbe, P. E. Ogwang & D. C. Sesaazi. 2018. Drug-like properties of anticancer molecules elucidated from *Eichhornia crassipes*. *J. Pharmacogn. Phytochem.* 7: 2075–2079.

Muehlstein, L. K., D. T. S. F. Porter & F. T. Short. 1988. *Labyrinthula* sp., a marine slime mold producing the symptoms of wasting disease in eelgrass, *Zostera marina*. *Mar. Biol.* 99: 465–472.

Muehlstein, L. K., D. Porter & F. T. Short. 1991. *Labyrinthula zosterae* sp. nov., the causative agent of wasting disease of eelgrass, *Zostera marina*. *Mycologia* 83: 180–191.

Mueller, B. 2004. Quality of *Halodule wrightii* growing near marinas. *Bios* 75: 53–57.

Mueller, M. H. & A. G. van der Valk. 2002. The potential role of ducks in wetland seed dispersal. *Wetlands* 22: 170–178.

Muenchow, G. 1998. Subandrodioecy and male fitness in *Sagittaria lancifolia* subsp. *lancifolia* (Alismataceae). *Am. J. Bot.* 85: 513–513.

Muenchow, G. & V. A. Delesalle. 1992. Patterns of weevil herbivory on male, monoecious and female inflorescences of *Sagittaria latifolia*. *Am. Midl. Nat.* 127: 355–367.

Muenchow, G. & V. Delesalle. 1994. Pollinator response to male floral display size in two *Sagittaria* species (Alismataceae). *Am. J. Bot.* 81: 568–573.

Muenscher, W. C. 1936a. *Storage and Germination of Seeds of Aquatic Plants*. Agricultural Experiment Station, Cornell University, Ithaca, NY. 17 pp.

Muenscher, W. C. 1936b. The germination of seeds of *Potamogeton*. *Ann. Bot. (Oxford)* 50: 805–821.

Muenscher, W. C. 1943. *Potamogeton spirillus* may grow as an annual. *Rhodora* 45: 329–330.

Muenscher, W. C. 1948. *Potamogeton latifolius* in Texas. *Madroño* 9: 220–223.

Mühlberg, H. 1982. *The Complete Guide to Water Plants*. EP Publishing Limited, Leipzig, Germany. 392 pp.

Muhonen, M., J. Showman & R. Couch. 1983. Nutrient absorption by *Spirodela polyrrhiza*. *J. Aquat. Plant Manag.* 21: 107–109.

Mújica, E. B., J. J. Mably, S. M. Skarha, L. L. Corey, L. W. Richardson, M. W. Danaher, E. H. González & L. W. Zettler. 2018. A comparison of ghost orchid (*Dendrophylax lindenii*) habitats in Florida and Cuba, with particular reference to seedling recruitment and mycorrhizal fungi. *Bot. J. Linn. Soc.* 186: 572–586.

Mulder, R. 1999. Wolf Lake area. Wolf Lake Report #2. A preliminary report on the findings of 3 biological surveys at Nisutlin Lake, Wolf River and Morris Lake. Canadian Parks & Wilderness Society (Yukon Chapter), Whitehorse, Yukon Territory, CA. 103 pp.

Mulder, C. P. H. & R. W. Ruess. 1998a. Relationships between size, biomass allocation, reproduction, and survival in *Triglochin palustris*: implications for the effects of goose herbivory. *Can. J. Bot.* **76**: 2164–2176.

Mulder, C. P. H. & R. W. Ruess. 1998b. Effects of herbivory on arrowgrass: interactions between geese, neighboring plants, and abiotic factors. *Ecol. Monogr.* **68**: 275–293.

Mulder, C. P. H., R. W. Ruess & J. S. Sedinger. 1996. Effects of environmental manipulations on *Triglochin palustris*: implications for the role of goose herbivory in controlling its distribution. *J. Ecol.* **84**: 267–278.

Mulhouse, J. M. 2004. Vegetation change in herbaceous Carolina bays of the upper coastal plain: dynamics during drought. M.S. thesis. University of Georgia, Athens, GA. 108 pp.

Müller-Doblies, D. 1969. Über die Blütenstände und Blüten sowie zue embryologie von *Sparganium. Bot. Jahrb. Syst.* **89**: 359–450.

Müller-Doblies, D. 1970. Über die Verwandtschaft von *Typha* und *Sparganium* im Infloreszenz- und Blütenbau. *Bot. Jahrb. Syst.* **89**: 451–462.

Müller-Schwarze, D., H. Brashear, R. Kinnel, K. A. Hintz, A. Lioubomirov & C. Skibo. 2001. Food processing by animals: do beavers leach tree bark to improve palatability? *J. Chem. Ecol.* **27**: 1011–1028.

Mullins, W. H. & E. G. Bizeau. 1978. Summer foods of sandhill cranes in Idaho. *Auk* **95**: 175–178.

Munden, C. 2001. *Native Orchids of Nova Scotia: A Field Guide.* University College of Cape Breton Press, Sydney, NS. 96 pp.

Muñoz Escobar, M. M., M. Voyevoda, C. Fühner & A. Zehnsdorf. 2011. Potential uses of *Elodea nuttallii*-harvested biomass. *Energy Sustain. Soc.* **1**: 4. doi: 10.1186/2192-0567-1-4.

Munro, J. A. 1936. Food of the common Mallard in the lower Fraser Valley, British Columbia. *Condor* **38**: 109–111.

Munro, M. C., R. E. Newell & N. M. Hill. 2014. 4-18 Potamogetonaceae, pondweed family. Pp. 1421–1442 *In*: M. C. Munro, R. E. Newell & N. M. Hill (eds.), *Nova Scotia Plants: Part 4: Monocots.* Nova Scotia Museum, Halifax, NS.

Munz, P. A. 1963. *California Mountain Wildflowers.* University of California Press, Berkeley, CA. 122 pp.

Munz, P. A. & I. M. Johnston. 1922. Miscellaneous notes on plants of southern California - I. *Bull. Torrey Bot. Club* **49**: 31–44.

Murdy, W. H. 1966. The systematics of *Phacelia maculata* and *P. dubia* var. *georgiana*, both endemic to granite outcrop communities. *Am. J. Bot.* **53**: 1028–1036.

Murphy, M. T. 2009. Getting to the root of the matter: variations in vascular root biomass and production in peatlands and responses to global change. Ph.D. dissertation. McGill University, Montreal, CA. 173 pp.

Murphy, P. B. & R. S. Boyd. 1999. Population status and habitat characterization of the endangered plant, *Sarracenia rubra* subspecies *alabamensis. Castanea* **64**: 101–113.

Murphy, L. R., S. T. Kinsey & M. J. Durako. 2003. Physiological effects of short-term salinity changes on *Ruppia maritima. Aquat. Bot.* **75**: 293–309.

Murray, M. 2000. Wetland plant associations of the western hemlock zone in the central coastal and Cascade Mountains. Interim Report. Oregon Natural Heritage Program, Portland, OR. 83 pp.

Murrell, W. 1901. *Lachnanthes. Brit. Medical J.* **2**(2124): 747.

Musselman, L. J. 1972. Root parasitism of *Macranthera flammea* and *Tomanthera auriculata* (Scrophulariaceae). *J. Elisha Mitchell Sci. Soc.* **88**: 58–60.

Musselman, L. J., D. A. Knepper, R. D. Bray, C. A. Caplen & C. Ballou. 1995. A new *Isoetes* hybrid from Virginia. *Castanea* **60**: 245–254.

Musselman, L. J., R. D. Bray & D. A. Knepper. 1996. *Isoetes* ×*bruntonii* (*Isoetes engelmannii* × *I. hyemalis*), a new hybrid quillwort from Virginia. *Am. Fern J.* **86**: 8–15.

Musselman, L. J., W. C. Taylor & R. D. Bray. 2001. *Isoetes mattaponica* (Isoetaceae), a new diploid quillwort from freshwater tidal marshes of Virginia. *Novon* **11**: 200–204.

Myers, F. J. 1942. The rotatorian fauna of the Pocono Plateau and environs. *Proc. Acad. Nat. Sci. Philadelphia* **94**: 251–285.

Nachtigall, W. & A. Wisser. 2015. Procedures and processes. Pp. 185–219 *In*: *Bionics by Examples: 250 Scenarios from Classical to Modern Times.* Springer Academic Publishing, Dordrecht, Switzerland.

Nachtrieb, J. G., M. J. Grodowitz & R. M. Smart. 2011. Impact of invertebrates on three aquatic macrophytes: American pondweed, Illinois pondweed, and Mexican water lily. *Nordic J. Bot.* **11**: 179–203.

Nahrsted, A., W. Hösel & A. Walther. 1979. Characterization of cyanogenic glucosides and β-glucosidases in *Triglochin maritima* seedlings. *Phytochemistry* **18**: 1137–1141.

Naidoo, G. 1994. Growth, water and ion relationships in the coastal halophytes *Triglochin bulbosa* and *T. striata. Environ. Exp. Bot.* **34**: 419–426.

Naidoo, G. & K. Naicker. 1992. Seed germination in the coastal halophytes *Triglochin bulbosa* and *Triglochin striata. Aquat. Bot.* **42**: 217–229.

Nakamura, G. & J. K. Nelson. 2001. *Illustrated Field Guide to Selected Rare Plants of Northern California.* Publication 3395. University of California Agriculture and Natural Resources, Oakland, CA. 370 pp.

Nakamura, S., Y. Okazaki, K. Ninomiya, T. Morikawa, H. Matsuda & M. Yoshikawa. 2008. Medicinal flowers. XXIV. Chemical structures and hepatoprotective effects of constituents from flowers of *Hedychium coronarium. Chem. Pharmaceut. Bull.* **56**: 1704–1709.

Namestnik, S. A., J. R. Thomas & B. S. Slaughter. 2012. Two recent plant discoveries in Missouri: *Cladium mariscus* subsp. *jamaicense* (Cyperaceae) and *Utricularia minor* (Lentibulariaceae). *Phytoneuron* **92**: 1–6.

NAOCC. 2019. North American Orchid Conservation Center. Go orchids. Available online: https://goorchids.northamericanorchidcenter.org/ [accessed 26 February, 2019].

Nardi, K. de O., A. Oriani & V. L. Scatena. 2016. Seed micromorphology and its taxonomic significance to *Xyris* (Xyridaceae, Poales). *Brazilian J. Bot.* **39**: 721–727.

Nash, G. V. 1895. Notes on some Florida plants. *Bull. Torrey Bot. Club* **22**: 141–161.

Nauheimer, L., D. Metzler & S. S. Renner. 2012. Global history of the ancient monocot family Araceae inferred with models accounting for past continental positions and previous ranges based on fossils. *New Phytol.* **195**: 938–950.

Nault, M. E. & A. Mikulyuk. 2009. European Frog-bit (*Hydrocharis morsus-ranae*): a technical review of distribution, ecology, impacts, and management. Bureau of Science Services, PUB-SS-1048 2009. Wisconsin Department of Natural Resources, Madison, WI. 13 pp.

Nava, H. S., M. Á. F. Casado & F. J. S. Pérez. 2000. *Lilaea scilloides* (Poir.) Hauman (Lilaeaceae), en Asturias. *Ann. Jard. Bot. Madrid* **58**: 191.

Near, K. A. & R. O. Belcher. 1974. New localities for *Najas minor* and *N. marina* in southeastern Michigan. *Michigan Bot.* **13**: 181–185.

Ņečajeva, J. & G. Ievinsh. 2007. Interacting influence of cold stratification treatment and osmotic potential on seed germination of *Triglochin maritima* L. *Biology* **723**: 115–122.

Neck, R. W. & H. L. Schramm. 1992. Freshwater molluscs of selected playa lakes of the Southern High Plains of Texas. *Southwest. Nat.* **37**: 205–209.

Nedeau, E. J., R. W. Merritt & M. G. Kaufman. 2003. The effect of an industrial effluent on an urban stream benthic community: water quality vs. habitat quality. *Environ. Pollut.* **123**: 1–13.

Needham, J. G. 1900. The fruiting of the blue flag (*Iris versicolor* L.). *Am. Nat.* **34**: 361–386.

Neil, K. 1977. An annotated list of the macrolepidoptera of Sable Island. *Proc. Nova Scotia Inst. Sci.* **28**: 41–46.

Neinhuis, C. & W. Barthlott. 1997. Characterization and distribution of water-repellent, self-cleaning plant surfaces. *Ann. Bot.* **79**: 667–677.

Nejrup, L. B. & M. F. Pedersen. 2008. Effects of salinity and water temperature on the ecological performance of *Zostera marina*. *Aquat. Bot.* **88**: 239–246.

Nekola, J. C. 2004. Vascular plant compositional gradients within and between Iowa fens. *J. Veg. Sci.* **15**: 771–780.

Nekola, J. C. & T. G. Lammers. 1989. Vascular flora of Brayton-Horsley Prairie: a remnant prairie and spring fen complex in eastern Iowa. *Castanea* **54**: 238–254.

Nekrasova, G. F., D. A. Ronzhina, M. G. Maleva & V. I. P'yankov. 2003. Photosynthetic metabolism and activity of carboxylating enzymes in emergent, floating, and submersed leaves of hydrophytes. *Russ. J. Plant Physiol.* **50**: 57–67.

Nelsen, Jr., J. E. & R. N. Ginsburg. 1986. Calcium carbonate production by epibionts on *Thalassia* in Florida Bay. *J. Sediment. Petrol.* **56**: 622–628.

Nelson, J. B. 1986. *The Natural Communities of South Carolina. Initial Classification and Description*. South Carolina Wildlife & Marine Resources Department, Charleston, SC. 64 pp.

Nepal, S. K. & P. Way. 2007. Comparison of vegetation conditions along two backcountry trails in Mount Robson Provincial Park, British Columbia (Canada). *J. Environ. Manage.* **82**: 240–249.

Nesom, G. L. 2009. Assessment of invasiveness and ecological impact in non-native plants of Texas. *J. Bot. Res. Inst. Texas* **3**: 971–991.

Ness, R. W., S. W. Graham & S. C. H. Barrett. 2011. Reconciling gene and genome duplication events: using multiple nuclear gene families to infer the phylogeny of the aquatic plant family Pontederiaceae. *Mol. Biol. Evol.* **28**: 3009–3018.

Netherland, M. D. 1997. Turion ecology of hydrilla. *J. Aquat. Plant Manag.* **35**: 1–10.

Neuenschwander, P., M. H. Julien, T. D. Center & M. P. Hill. 2009. *Pistia stratiotes* L. (Araceae). Pp. 332–352 *In*: R. Muniappan, G. V. Reddy & A. Raman (eds.), *Biological Control of Tropical Weeds Using Arthropods*. Cambridge University Press, NY.

Newberry, G. 1991. Factors affecting the survival of the rare plant, *Sagittaria fasciculata* EO Beal (Alismataceae). *Castanea* **56**: 59–64.

Newmaster, S. G., A. G. Harris & L. J. Kershaw. 1997. *Wetland Plants of Ontario*. Lone Pine Publishing, Redmond, WA. 240 pp.

Newton, R. E. 2008. A floristic inventory of selected Bureau of Land Management wetlands in Wyoming. M.S. thesis. University of Wyoming, Laramie, WY. 93 pp.

Newton, R. J., D. R. Shelton, S. Disharoon & J. E. Duffey. 1978. Turion formation and germination in *Spirodela polyrhiza*. *Am. J. Bot.* **65**: 421–428.

Ngai, J. T. & R. L. Jefferies. 2004. Nutrient limitation of plant growth and forage quality in Arctic coastal marshes. *J. Ecol.* **92**: 1001–1010.

Ngamau, K. 2001. Development of an in vitro culture procedure using seeds from *Zantedeschia aethiopica* 'Green Goddess' as explants. *Gartenbauwissenschaft* **66**: 133–139.

Ngamau, K. 2006. Selection for early flowering, temperature and salt tolerance of *Zantedeschia aethiopica* 'Green Goddess'. *Acta Hort.* **766**: 155–162.

Nguyen, N. H., H. E. Driscoll & C. D. Specht. 2008. A molecular phylogeny of the wild onions (Allium; Alliaceae) with a focus on the western North American center of diversity. *Mol. Phylogenet. Evol.* **47**: 1157–1172.

Nguyen, X.-V., S. Höfler, Y. Glasenapp, T. Thangaradjou, C. Lucas & J. Papenbrock. 2015. New insights into DNA barcoding of seagrasses. *Syst. Biodivers.* **13**: 496–508.

Nichols, G. E. 1919. *Lophiola aurea* in Nova Scotia. *Rhodora* **21**: 68.

Nichols, G. E. 1920. The vegetation of Connecticut. VII. The associations of depositing areas along the seacoast. *Bull. Torrey Bot. Club* **47**: 511–548.

Nichols, G. E. 1933. Notes on Michigan bryophytes. II. *Bryologist* **36**: 69–78.

Nichols, G. E. 1934. The influence of exposure to winter temperatures upon seed germination in various native American plants. *Ecology* **15**: 364–373.

Nichols, S. A. 1975. The impact of overwinter drawdown on the aquatic vegetation of the Chippewa Flowage, Wisconsin. *Trans. Wisconsin Acad. Sci.* **63**: 176–186.

Nichols, S. A. 1984. Macrophyte community dynamics in a dredged Wisconsin lake. *J. Am. Water Res. Assoc.* **20**: 573–576.

Nichols, S. A. 1988. Vegetation of Wisconsin's benchmark lakes. *Trans. Wisconsin Acad. Sci.* **76**: 1–15.

Nichols, S. A. 1990. Interspecific association of some Wisconsin lake plants. *Trans. Wisconsin Acad. Sci.* **78**: 111–128.

Nichols, S. A. 1999. Distribution and habitat descriptions of Wisconsin lake plants. Bulletin 96. Wisconsin Geological & Natural History Survey, Madison, WI. 268 pp.

Nichols, S. A. & B. H. Shaw. 1986. Ecological life histories of the three aquatic nuisance plants, *Myriophyllum spicatum*, *Potamogeton crispus* and *Elodea canadensis*. *Hydrobiologia* **131**: 3–21.

Nichols, S. A. & L. A. J. Buchan. 1997. Use of native macrophytes as indicators of suitable Eurasian watermilfoil habitat in Wisconsin lakes. *J. Aquat. Plant Manag.* **35**: 21–24.

Nichols, S., S. Weber & B. Shaw. 2000. A proposed aquatic plant community biotic index for Wisconsin lakes. *Environ. Manage.* **26**: 491–502.

Nicolson, D. H. 1984. Suprageneric names attributable to Araceae. *Taxon* **33**: 680–690.

Nicholson, A. & P. A. Keddy. 1983. The depth profile of a shoreline seed bank in Matchedash Lake, Ontario. *Can. J. Bot.* **61**: 3293–3296.

Nicholson, B. J. & D. H. Vitt. 1994. Wetland development at Elk Island National Park, Alberta, Canada. *J. Paleolimnol.* **12**: 19–34.

Nico, L. G. & A. M. Muench. 2004. Nests and nest habitats of the invasive catfish *Hoplosternum littorale* in lake Tohopekaliga, Florida: a novel association with non-native *Hydrilla verticillata*. *Southeast. Nat.* **3**: 451–467.

Nie, Z.-L., H. Sun, H. Li & J. Wen. 2006. Intercontinental biogeography of subfamily Orontioideae (*Symplocarpus*, *Lysichiton*, and *Orontium*) of Araceae in eastern Asia and North America. *Mol. Phylogenet. Evol.* **40**: 155–165.

Nielsen, J. S. & B. L. Møller. 1999. Biosynthesis of cyanogenic glucosides in *Triglochin maritima* and the involvement of cytochrome P450 enzymes. *Arch. Biochem. Biophys.* **368**: 121–130.

Nielsen, J. S. & B. L. Møller. 2000. Cloning and expression of cytochrome P450 enzymes catalyzing the conversion of tyrosine to p-hydroxyphenylacetaldoxime in the biosynthesis of cyanogenic glucosides in *Triglochin maritima*. *Plant Physiol.* **122**: 1311–1322.

Nielson, P. 2016. Presence and habitat requirements of secretive marshbirds in the Washington, DC metropolitan area. Ph.D. dissertation. University of Maryland, College Park, MD. 147 pp.

Nienaber, M. A. 2000. Scheuchzeriaceae Rudolfi. Pp. 41–42 *In*: Flora North America Editorial Committee (eds.), *Flora of North America North of Mexico, Vol. 22: Magnoliophyta: Alismatidae, Arecidae, Commelinidae (in Part), and Zingiberidae*. Oxford University Press, New York, NY.

Nies, G. & T. B. H. Reusch. 2004. Nine polymorphic microsatellite loci for the fennel pondweed *Potamogeton pectinatus* L. *Mol. Ecol. Notes* **4**: 563–565.

Niles, G. S. 1902. Origin of plant names. H. The lady's slippers and moccasin flowers. *Plant World* **5**: 201–204.

Niles, G. G. 1904. *Bog-trotting for Orchids*. G. P. Putnam's Sons, Knickerbocker Press, New York, NY. 310 pp.

Nishihara, G. N. & J. D. Ackerman. 2006. The effect of hydrodynamics on the mass transfer of dissolved inorganic carbon to the freshwater macrophyte *Vallisneria americana*. *Limnol. Oceanogr.* **51**: 2734–2745.

Nixon, E. S. & J. R. Ward. 1981. Distribution of *Schoenolirion wrightii* (Liliaceae) and *Bartonia texana* (Gentianaceae). *Sida* **9**: 64–69.

Nixon, E. S. & J. R. Ward. 1986. Floristic composition and management of East Texas pitcher plant bogs. Pp. 283–287. *In*: D. L. Kulhavy & R. H. Conner (eds.), *Wilderness and Natural Areas in the Eastern United States: A Management Challenge*. Center for Applied Studies, School of Forestry, Stephen F. Austin State University, Nacogdoches, TX.

Njambuya, J., I. Stiers & L. Triest. 2011. Competition between *Lemna minuta* and *Lemna minor* at different nutrient concentrations. *Aquat. Bot.* **94**: 158–164.

Nolfo-Clements, L. E. 2006. Vegetative survey of wetland habitats at Jean Lafitte National Historical Park and Preserve in southeastern Louisiana. *Southeast. Nat.* **5**: 499–514.

Noonpui, S. & P. Thiravetyan. 2011. Treatment of reactive azo dye from textile wastewater by burhead (*Echinodorus cordifolius* L.) in constructed wetland: effect of molecular size. *J. Environ. Sci. Health* **46**: 709–714.

Norquist, C. 1990. Threatened status for *Sagittaria secundifolia* (Kral's water-plantain). *Fed. Reg.* **55**: 13907–13910.

Novelo, A. & A. Lot. 1994. Pontederiaceae. Pp. 65–71 *In*: G. Davidse, M. Sousa & A. Chater (eds.), *Flora Mesoamericana. Vol. 6. Alismataceae a Cyperaceae*. Missouri Botanical Garden Press, St. Louis, MO.

Noyce, K. V., P. B. Kannowski & M. R. Riggs. 1997. Black bears as ant-eaters: seasonal associations between bear myrmecophagy and ant ecology in north-central Minnesota. *Can. J. Zool.* **75**: 1671–1686.

Nozoe, T., J. Tazawa, A. Uchino & S. Miura. 2018. Promotive effect of soil solution on germination of *Monochoria vaginalis* under paddy conditions. *Soil Sci. Plant Nutr.* **64**: 396–405.

Nugraha, A. T., V. Ramadhan, H. Pandapotan & F. Romadhonsyah. 2017. A study of proliferative activity of herbs *Eriocaulon cinereum* R. Br on cervical cancer cells (HeLa) with MTT Assay Method. *Int. J. Pharm. Med. Biol. Sci.* **6**: 73–76.

Nuissier, G., B. Rezzonico & M. Grignon-Dubois. 2010. Chicoric acid from *Syringodium filiforme*. *Food Chem.* **120**: 783–788.

NYNHP. 2019. New York Natural Heritage Program. 2019. Online Conservation Guide for *Iris prismatica*. Available online: https://guides.nynhp.org/slender-blue-flag/ [accessed 30 January, 2019].

Oakes, A. J. 1990. *Ornamental Grasses and Grasslike Plants*. Van Nostrand Reinhold, New York, NY. 618 pp.

Oberholser, H. C. & W. L. McAtee. 1920. Waterfowl and their food plants in the sandhill region of Nebraska. Bulletin No. 794. United States Department of Agriculture, Washington, DC. 77 pp.

O'Brien, C. W. 1990. *Neobagoidus carlsoni*, new genus, new species of aquatic weevil from Florida. *Southwest. Entomol.* **15**: 71–76.

O'Brien, C. W. 1997. *A Catalogue of the Coleoptera of America North of Mexico*. Agriculture Handbook Number 529-143g. United States Department of Agriculture, Washington, DC. 48 pp.

Ochoa, J. G. & G. I. Andrade. 2003. The introduced flora to Machu Picchu Sanctuary: an inventory and management priorities for biodiversity conservation. *Ecol. Bolivia* **38**: 141–160.

Oda, Y. 1962. Effect of light quality on flowering of *Lemna perpusilla* 6746. *Plant Cell Physiol.* **3**: 415–417.

Odum, W. E. & M. A. Heywood. 1978. Decomposition of intertidal freshwater marsh plants. Pp. 89–97 *In*: R. E. Good, D. F. Whigham & R. L. Simpson (eds.), *Freshwater Wetlands: Ecological Processes and Management Potential*. Academic Press, New York, NY.

Odum, W. E., P. W. Kirk & J. C. Zieman. 1979. Non-protein nitrogen compounds associated with particles of vascular plant detritus. *Oikos* **32**: 363–367.

Offerijns, F. J. M. 1936. Meiosis in the pollen mother cells of some Cannas. *Genetica* **18**: 1–60.

Ogle, D. W. 1989. Barns Chapel Swamp: an unusual arbor-vitae (*Thuja occidentalis* L.) site in Washington County, Virginia. *Castanea* **54**: 200–202.

Ogle, B. M., P. H. Hung & H. T. Tuyet. 2001. Significance of wild vegetables in micronutrient intakes of women in Vietnam: an analysis of food variety. *Asia Pac. J. Clin. Nutr.* **10**: 21–30.

Ohanjanian, P. & C. Carli. 2010. Northern leopard frog. An assessment of potential reintroduction sites in the Columbia marshes. Report to the Local Conservation Fund. The Columbia Basin Environmental Initiatives Fund and The Columbia Wetlands Stewardship Partners. Nelson, BC. 26 pp.

Ohri, D., R. Fritsch & P. Hanelt. 1998. Evolution of genome size in *Allium* (Alliaceae). *Plant Syst. Evol.* **210**: 57–86.

O'Kennon, R. J. & C. McLemore. 2004. *Schoenoplectus hallii* (Cyperaceae), a globally threatened species new for Texas. *Sida* **21**: 1201–1204.

Okoye, T. C., P. F. Uzor, C. A. Onyeto & E. K. Okereke. 2014. Safe African medicinal plants for clinical studies. Pp. 535–555 *In*: V. Kuete (ed.), *Toxicological Survey of African Medicinal Plants*. Elsevier, London, United Kingdom.

Oláh, V., A. Hepp & I. Mészáros. 2016. Assessment of giant duckweed (*Spirodela polyrhiza* L. Schleiden) turions as model objects in ecotoxicological applications. *Bull. Environ. Contam. Toxicol.* **96**: 596–601.

Oliva, A. P. & J. Arditti. 1984. Seed germination of North American orchids. II. Native California and related species of *Aplectrum*, *Cypripedium*, and *Spiranthes*. *Bot. Gaz.* **145**: 495–501.

Olney, P. J. S. 1960. Lead poisoning in wildfowl. *Wildfowl* **11**: 123–134.

Olofsdotter, M., D. Navarez & K. Moody. 1995. Allelopathic potential in rice (*Oryza sativa* L.) germplasm. *Ann. Appl. Biol.* **127**: 543–560.

Olsen, J. L., J. A. Coyer & B. Chesney. 2014. Numerous mitigation transplants of the eelgrass *Zostera marina* in southern California shuffle genetic diversity and may promote hybridization with *Zostera pacifica*. *Biol. Conserv.* **176**: 133–143.

Olsen, J. L., P. Rouzé, B. Verhelst, Y.-C. Lin, T. Bayer, J. Collen, E. Dattolo, E. De Paoli, S. Dittami, F. Maumus, G. Michel, A. Kersting, C. Lauritano, R. Lohaus, M. Töpel, T. Tonon, K. Vanneste, M. Amirebrahimi, J. Brakel, C. Boström, M. Chovatia, J. Grimwood, J. W. Jenkins, A. Jueterbock, A. Mraz, W. T. Stam, H. Tice, E. Bornberg-Bauer, P. J. Green, G. A. Pearson, G. Procaccini, C. M. Duarte, J. Schmutz, T. B. H. Reusch & Y. Van de Peer. 2016. The genome of the seagrass *Zostera marina* reveals angiosperm adaptation to the sea. *Nature* **530**: 331–335.

Olson, E. R. & J. M. Doherty. 2014. Macrophyte diversity–abundance relationship with respect to invasive and native dominants. *Aquat. Bot.* **119**: 111–119.

Olson, A., J. Paul & J. R. Freeland. 2009. Habitat preferences of cattail species and hybrids (*Typha* spp.) in eastern Canada. *Aquat. Bot.* **91**: 67–70.

Onuf, C. P. 1996. Biomass patterns in seagrass meadows of the Laguna Madre, Texas. *Bull. Mar. Sci.* **58**: 404–420.

Oosting, H. J. & W. D. Billings. 1943. The red fir forest of the Sierra Nevada: Abietum magnificae. *Ecol. Monogr.* **13**: 259–274.

Ophel, I. L. & C. D. Fraser. 1970. Calcium and strontium discrimination by aquatic plants. *Ecology* **51**: 324–327.

Oraze, M. J. & A. A. Grigarick. 1992. Biological control of ducksalad (*Heteranthera limosa*) by the waterlily aphid (*Rhopalosiphum nymphaeae*) in rice (*Oryza sativa*). *Weed Sci.* **40**: 333–336.

Oraze, M. J., A. A. Grigarick, J. H. Lynch & K. A. Smith. 1988. Spider fauna of flooded rice fields in northern California. *J. Arachnol.* **16**: 331–337.

Oraze, M. J., A. A. Grigarick & K. A. Smith. 1989. Population ecology of *Pardosa ramulosa* (Araneae, Lycosidae) in flooded rice fields of northern California. *J. Arachnol.* **17**: 163–170.

Oriani, A. & V. L. Scatena. 2012. Floral anatomy of xyrids (Poales): contributions to their reproductive biology, taxonomy, and phylogeny. *Int. J. Plant Sci.* **173**: 767–779.

Oriani, A. & V. L. Scatena. 2017. Ovule, fruit, and seed development of *Orectanthe sceptrum* and its systematic relevance to Xyridaceae (Poales). *Int. J. Plant Sci.* **178**: 104–116.

Orians, G. H. 1973. The red-winged blackbird in tropical marshes. *Condor* **75**: 28–42.

Orndruff, R. 1966. The breeding system of *Pontederia cordata* L. *Bull. Torrey Bot. Club* **93**: 407–416.

Orndruff, R. 1979. Chromosome numbers and relationships of certain African and American genera of Haemodoraceae. *Ann. Missouri Bot. Gard.* **66**: 577–580.

Oron, G., D. Porath & L. R. Wildschut. 1986. Wastewater treatment and renovation by different duckweed species. *J. Environ. Eng.* **112**: 247–263.

Orpurt, P. A., S. P. Meyers, L. L. Boral & J. Sims. 1964. Thalassiomycetes V. A new species of *Lindra* from turtle grass, *Thalassia testudinum* König. *Bull. Mar. Sci.* **14**: 405–417.

Ortas, L. & J. A. Martorell. 2001. *Heteranthera rotundifolia* Griseb., nueva planta adventicia de los arrozales en España. *Anales Jard. Bot. Madrid* **59**: 161.

Orth, R. J. 1973. Benthic infauna of eelgrass, *Zostera marina*, beds. *Chesapeake Sci.* **14**: 258–269.

Orth, R. J. 1975. Destruction of eelgrass, *Zostera marina*, by the cownose ray, *Rhinoptera bonasus*, in the Chesapeake Bay. *Chesapeake Sci.* **16**: 205–208.

Orth, R. J. & K. L. Heck, Jr. 1980. Structural components of eelgrass (*Zostera marina*) meadows in the lower Chesapeake Bay—fishes. *Estuaries* **3**: 278–288.

Orth, R. J. & K. A. Moore. 1986. Seasonal and year-to-year variations in the growth of *Zostera marina* L. (eelgrass) in the lower Chesapeake Bay. *Aquat. Bot.* **24**: 335–341.

Orth, R. J. & K. A. Moore. 1988. Distribution of *Zostera marina* L. and *Ruppia maritima* L. sensu lato along depth gradients in the lower Chesapeake Bay, USA. *Aquat. Bot.* **32**: 291–305.

Orth, A. I. & K. D. Waddington. 1997. The movement patterns of carpenter bees *Xylocopa micans* and bumblebees *Bombus pennsylvanicus* on *Pontederia cordata* inflorescences. *J. Insect Behav.* **10**: 79–86.

Orth, R. J., K. A. Moore & J. F. Nowak. 1990. Monitoring seagrass distribution and abundance patterns: a case study from the Chesapeake Bay. Federal coastal wetland mapping programs. *Biol. Rep.* **90**(18): 111–123.

Orth, R. J., M. C. Harwell, E. M. Bailey, A. Bartholomew, J. T. Jawad, A. V. Lombana, K. A. Moore, J. M. Rhode & H. E. Woods. 2000. A review of issues in seagrass seed dormancy and germination: implications for conservation and restoration. *Mar. Ecol. Prog. Ser.* **200**: 277–288.

Orth, R. J., J. R. Fishman, M. C. Harwell & S. R. Marion. 2003. Seed-density effects on germination and initial seedling establishment in eelgrass *Zostera marina* in the Chesapeake Bay region. *Mar. Ecol. Prog. Ser.* **250**: 71–79.

Orth, R. J., M. C. Harwell & G. J. Inglis. 2006. Ecology of seagrass seeds and seagrass dispersal strategies. Pp. 111–133 *In*: A. W. D. Larkum, R. J. Orth & C. M. Duarte (eds.), *Seagrasses: Biology, Ecology and Conservation*. Springer-Verlag, Dordrecht, The Netherlands.

Orzell, S. L. & E. L. Bridges. 1991. *Carex exilis* Dewey (Cyperaceae) new to Alabama. *Phytologia* **70**: 400–403.

Orzell, S. L. & E. L. Bridges. 1992. *Xyris isoetifolia* Kral (Xyridaceae) new to Alabama and its range and habitats in Florida. *Phytologia* **72**: 152–156.

Orzell, S. L. & E. L. Bridges. 1993. *Eriocaulon nigrobracteatum* (Eriocaulaceae), a new species from the Florida panhandle, with a characterization of its poor fen habitat. *Phytologia* **74**: 104–124.

Orzell, S. L. & E. Bridges. 2006a. Floristic composition and species richness of subtropical seasonally wet *Muhlenbergia sericea* prairies in portions of central and south Florida. Pp. 136–175 *In*: R. F. Noss (ed.), *Land of Fire and Water: The Florida Dry Prairie Ecosystem. Proceedings of the Florida Dry Prairie Conference*. E. O. Painter Printing Co., DeLeon Springs, FL.

Orzell, S. L. & E. Bridges. 2006b. Species composition and environmental characteristics of Florida dry prairies from the Kissimmee River region of south-central Florida. Pp. 100–135 *In*: R. F. Noss (ed.), *Land of Fire and Water: The Florida Dry Prairie Ecosystem. Proceedings of the Florida Dry Prairie Conference*. E. O. Painter Printing Co., DeLeon Springs, FL.

Ostrofsky, M. L. & E. R. Zettler. 1986. Chemical defences in aquatic plants. *J. Ecol.* **74**: 279–287.

Osborne, J. A. & N. M. Sassic. 1981. The size of grass carp as a factor in the control of *Hydrilla*. *Aquat. Bot.* **11**: 129–136.

Ostrofsky, M. L. & C. Miller. 2017. Photosynthetically-mediated calcite and phosphorus precipitation by submersed aquatic vascular plants in Lake Pleasant, Pennsylvania. *Aquat. Bot.* **143**: 36–40.

Othman, A. S. 1997. Molecular systematics of the tropical aquatic plant genus, *Cryptocoryne* Fischer ex Wydler (Araceae). Ph.D. dissertation. University of St. Andrews, St. Andrews, Scotland. 223 pp.

Ott-Conn, C. N., D. Clifford, T. Branston, R. Klinger & J. Foley. 2014. Pathogen infection and exposure, and ectoparasites of the federally endangered Amargosa vole (*Microtus californicus scirpensis*), California, USA. *J. Wildl. Dis.* **50**: 767–776.

Owens, S. J. 1981. Self-incompatibility in the Commelinaceae. *Ann. Bot.* **47**: 567–581.

Owens, C. S., J. D. Madsen, R. M. Smart & R. M. Stewart. 2001. Dispersal of native and nonnative aquatic plant species in the San Marcos River, Texas. *J. Aquat. Plant Manag.* **39**: 75–79.

Owens, C. S., A. Shad, J. Tripe, L. Dodd & L. Glomski. 2010. *San Marcos River Survey-2009 Feasibility.* U.S. Army Corps of Engineers, Lewisville Aquatic Ecosystem Research Facility, Lewisville, TX. 41 pp.

Owens, B. E., L. Allain, E. C. Van Gorder, J. L. Bossart & C. E. Carlton. 2018. The bees (Hymenoptera: Apoidea) of Louisiana: an updated, annotated checklist. *Proc. Entomol. Soc. Washington* **120**: 272–308.

Oyewo, E. A. & R. E. Jacobsen. 2007. *Polypedilum* (*Pentapedilum*) *epleri*, a new species from the eastern USA (Diptera: Chironomidae). Pp. 225–234 *In*: T. Andersen (ed.), *Contributions to the Systematics and Ecology of Aquatic Diptera–A Tribute to Ole A. Sæther.* The Caddis Press, Columbus, OH.

Özbay, H. & A. Alim. 2009. Antimicrobial activity of some water plants from the northeastern Anatolian region of Turkey. *Molecules* **14**: 321–328.

Özesmi, U. & W. J. Mitsch. 1997. A spatial habitat model for the marsh-breeding red-winged blackbird (*Agelaius phoeniceus* L.) in coastal Lake Erie wetlands. *Ecol. Modell.* **101**: 139–152.

Pace, M. C. & A. E. Brant. 2015. Missouri: *Spiranthes praecox* (Orchidaceae) and *Carex atlantica* subsp. *capillacea* (Cyperaceae): new to the Flora of Missouri. *Castanea* **80**: 144–146.

Pace, M. C. & K. M. Cameron. 2016. Reinstatement, redescription, and emending of *Spiranthes triloba* (Orchidaceae): solving a 118 year old cryptic puzzle. *Syst. Bot.* **41**: 924–939.

Pace, M. C. & K. M. Cameron. 2017. The systematics of the *Spiranthes cernua* species complex (Orchidaceae): untangling the Gordian knot. *Syst. Bot.* **42**: 640–669.

Pace, M. C., S. L. Orzell, E. L. Bridges & K. M. Cameron. 2017. *Spiranthes igniorchis* (Orchidaceae), a new and rare cryptic species from the south-central Florida subtropical grasslands. *Brittonia* **69**: 323–339.

Packer, J. G. 2002a. *Pleea* Michaux. Pp. 59–60 *In*: Flora North America Editorial Committee (eds.), *Flora of North America North of Mexico, Vol. 26: Magnoliophyta: Liliidae: Liliales and Orchidales.* Oxford University Press, New York, NY.

Packer, J. G. 2002b. *Triantha* (Nuttall) Baker. Pp. 61–64 *In*: Flora North America Editorial Committee (eds.), *Flora of North America North of Mexico, Vol. 26: Magnoliophyta: Liliidae: Liliales and Orchidales.* Oxford University Press, New York, NY.

Packer, J. G. & G. S. Ringius. 1984. The distribution and status of *Acorus* (Araceae) in Canada. *Can. J. Bot.* **62**: 2248–2252.

Padgett, D. J. & G. E. Crow. 1994a. A vegetation and floristic analysis of a created wetland in southeastern New Hampshire. *Rhodora* **96**: 1–29.

Padgett, D. J. & G. E. Crow. 1994b. Foreign plant stock: concerns for wetland mitigation. *Restor. Manag. Notes* **12**: 168–171.

Padgett, D. J., L. Cook, L. Horky, J. Noris & K. Vale. 2004. Seed production and germination in long's bittercress (*Cardamine longii*) of Massachusetts. *Northeast. Nat.* **11**: 49–56.

Pagano, A. M. & J. E. Titus. 2007. Submersed macrophyte growth at low pH: carbon source influences response to dissolved inorganic carbon enrichment. *Freshw. Biol.* **52**: 2412–2420.

Pagotto, M. A., R. de M. L. Silveira, C. N. da Cunha & I. Fantin-Cruz. 2011. Distribution of herbaceous species in the soil seed bank of a flood seasonality area, northern Pantanal, Brazil. *Int. Rev. Hydrobiol.* **96**: 149–163.

Pai, A. & B. C. McCarthy. 2005. Variation in shoot density and rhizome biomass of *Acorus calamus* L. with respect to environment. *Castanea* **70**: 263–275.

Pai, A. & B. C. McCarthy. 2010. Suitability of the medicinal plant, *Acorus calamus* L., for wetland restoration. *Nat. Areas J.* **30**: 380–386.

Palis, J. G. 1998. Breeding biology of the gopher frog, *Rana capito*, in western Florida. *J. Herpetol.* **32**: 217–223.

Palma, Á. T., A. Schwarz, L. A. Henríquez, X. Alvarez, J. M. Fariña & Q. Lu. 2013. Do subtoxic levels of chlorate influence the desiccation tolerance of *Egeria densa*? *Environ. Toxicol. Chem.* **32**: 417–422.

Palmer, M. L. & F. J. Mazzotti. 2004. Structure of Everglades alligator holes. *Wetlands* **24**: 115–122.

Pandey, M. & J. Sharma. 2015. Disjunct populations of a locally common North American orchid exhibit high genetic variation and restricted gene flow. *Open J. Genet.* **5**: 159–175.

Pandi, P. S. & J. Rajkumar. 2015. Phytochemical screening and bioactive potential of *Hydrilla verticillata*. *J. Chem. Pharmaceut. Res.* **7**: 1809–1815.

Panetta, F. D. 1988. Studies on the seed biology of arum lily (*Zantedeschia aethiopica* (L.) Spreng.). *Plant Protect. Quart.* **3**: 169–171.

Panigrahy, S. K., A. Kumar & R. Bhatt. 2018. *Hedychium coronarium* rhizomes: promising antidiabetic and natural inhibitor of α-amylase and α-glucosidase. *J. Dietary Suppl.* doi: 10.1080/19390211.2018.1483462.

Pankey, R. D., H. N. Draudt & N. W. Desrosier. 1965. Characterization of the starch of *Spirodela polyrrhiza*. *J. Food Sci.* **30**: 627–631.

Pansarin, E. R., A. Salatino & M. L. F. Salatino. 2008. Phylogeny of South American Pogonieae (Orchidaceae, Vanilloideae) based on sequences of nuclear ribosomal (ITS) and chloroplast (*psaB, rbcL, rps16,* and *trnL-F*) DNA, with emphasis on *Cleistes* and discussion of biogeographic implications. *Org. Divers. Evol.* **8**: 171–181.

Pant, B. 2013. Medicinal orchids and their uses: tissue culture a potential alternative for conservation. *Afr. J. Plant Sci.* **7**: 448–467.

Paolacci, S., S. Harrison & M. A. K. Jansen. 2018. The invasive duckweed *Lemna minuta* Kunth displays a different light utilisation strategy than native *Lemna minor* Linnaeus. *Aquat. Bot.* **146**: 8–14.

Pardee, K. I., P. Ellis, M. Bouthillier, G. H. N. Towers & C. J. French. 2004. Plant virus inhibitors from marine algae. *Can. J. Bot.* **82**: 304–309.

Parent, G. H. 1977. L'écologie de *Sisyrinchium montanum* Greene (Iridaceae) en forêt d'Argonne et en Lorraine. *Bull. Soc. Roy. Bot. Belgique* **110**: 77–84.

Parida, R., S. Mohanty & S. Nayak. 2013. In vitro propagation of *Hedychium coronarium* Koen. through axillary bud proliferation. *Plant Biosyst.* **147**: 905–912.

Parker, G. 2003. *Status Report on the Eastern Moose (Alces alces americana Clinton) in Mainland Nova Scotia.* Nova Scotia Department of Natural Resources, Halifax, NS, Canada. 77 pp.

Parker, V. T. & M. A. Leck. 1985. Relationships of seed banks to plant distribution patterns in a freshwater tidal wetland. *Am. J. Bot.* **72**: 161–174.

Parker, J. D. & M. E. Hay. 2005. Biotic resistance to plant invasions? Native herbivores prefer non-native plants. *Ecol. Lett.* **8**: 959–967.

Parker, B. C., N. Schanen & R. Renner. 1969. Viable soil algae from the herbarium of the Missouri Botanical Garden. *Ann. Missouri Bot. Gard.* **56**: 113–119.

Parker, J. D., C. C. Caudill & M. E. Hay. 2007a. Beaver herbivory on aquatic plants. *Oecologia* **151**: 616–625.

Parker, J. D., D. E. Burkepile, D. O. Collins, J. Kubanek & M. E. Hay. 2007b. Stream mosses as chemically-defended refugia for freshwater macroinvertebrates. *Oikos* **116**: 302–312.

Parkes, K. C., D. P. Kibbe & E. L. Roth. 1978. First records of the Spotted Rail (*Pardirallus maculatus*) for the United States, Chile, Bolivia and western Mexico breeding range. *Am. Birds* **32**: 295–299.

Parkinson, H., J. Mangold & C. McLane. 2016. Biology, ecology and management of curlyleaf pondweed (*Potamogeton crispus*). EBO223. Montana State University Extension, Bozeman, MT. 9 pp.

Partridge, C., A. Boettcher & A. G. Jones. 2012. Population structure of the gulf pipefish in and around Mobile Bay and the northern Gulf of Mexico. *J. Hered.* **103**: 821–830.

Parvathaneni, R. K., V. L. DeLeo, J. J. Spiekerman, D. Chakraborty & K. M. Devos. 2017. Parallel loss of introns in the *ABCB1* gene in angiosperms. *BMC Evol. Biol.* **17**: 238.

Paşca, M. B., D. Gîtea, A. Pallag, S. Nemeth & I. Ileş. 2016. Research regarding the capitalization of *Iris pseudacorus* L. species. *Analele Univ. Oradea Fascic. Protecţ. Med.* **27**: 135–140.

Patel, S. 2012. Threats, management and envisaged utilizations of aquatic weed *Eichhornia crassipes*: an overview. *Rev. Environ. Sci. Bio/Technol.* **11**: 249–259.

Paterson, M. 1993. The distribution of microcrustacea in the littoral zone of a freshwater lake. *Hydrobiologia* **263**: 173–183.

Pates, A. L. & G. C. Madsen. 1955. Occurrence of antimicrobial substances in chlorophyllose plants growing in Florida. II. *Bot. Gaz.* **116**: 250–261.

Patrock, R. J. W. 2007. The use by red imported fire ants, *Solenopsis invicta* (Hymenoptera: Formicidae), of *Potamogeton nodosus* (Potamogetonaceae) leaves as platforms into the littoral zone in Texas, USA. *Entomol. News* **118**: 527–529.

Patt, J. M., T. G. Hartman, R. W. Creekmore, J. J. Elliott, C. Schal, J. Lech & R. T. Rosen. 1992. The floral odour of *Peltandra virginica* contains novel trimethyl-2, 5-dioxabicyclo [3.2. 1] nonanes. *Phytochemistry* **31**: 487–491.

Patt, J. M., J. C. French, C. Schal, J. Lech & T. G. Hartman. 1995. The pollination biology of Tuckahoe, *Peltandra virginica* (Araceae). *Am. J. Bot.* **82**: 1230–1240.

Patterson, C. T. 1982. Foods of migrating coots (*Fulica americana*) and sympatric ducks during fall and spring in northeastern Oklahoma. Ph.D. dissertation. Oklahoma State University, Norman, OK. 31 pp.

Patten, M. V. 2016. Phenotypic plasticity and morphological variation in a native submerged aquatic plant. M.S. thesis. San Francisco State University, San Francisco, CA. 59 pp.

Patton, J. E. & W. S. Judd. 1986. Vascular flora of Paynes Prairie Basin and Alachua Sink Hammock, Alachua County, Florida. *Castanea* **51**: 88–110.

Patton, J. E. & W. S. Judd. 1988. A phenological study of 20 vascular plant species occurring on the Paynes Prairie Basin, Alachua County, Florida. *Castanea* **53**: 149–163.

Paulus, S. L. 1982a. Feeding ecology of gadwalls in Louisiana in winter. *J. Wildl. Manag.* **46**: 71–79.

Paulus, S. L. 1982b. Gut morphology of gadwalls in Louisiana in winter. *J. Wildl. Manag.* **46**: 483–489.

Paulus, S. L. 1984. Activity budgets of nonbreeding gadwalls in Louisiana. *J. Wildl. Manag.* **48**: 371–380.

Pavek, D. S. 1993. *Maianthemum canadense. In: Fire Effects Information System*. U.S. Department of Agriculture, Forest Service, Rocky Mountain Research Station, Fire Sciences Laboratory (Producer). Available online: https://www.fs.fed.us/database/feis/plants/forb/maican/all.html [accessed 23 January, 2019].

Payette, S. & A. Delwaide. 2000. Recent permafrost dynamics in a subarctic floodplain associated with changing water levels, Québec, Canada. *Arct. Antarct. Alp. Res.* **32**: 316–323.

Payne, T. 1916. *California Wild Flowers, Their Culture and Care: A Treatise Describing Over a Hundred Beautiful Species, with a Few Notes on Their Habits and Characteristics*, 3rd ed. Theodore Payne, Los Angeles, CA. 24 pp.

PCA (Parks Canada Agency). 2012. Management plan for Hill's pondweed (*Potamogeton hillii*) in Canada [Proposed]. Species at Risk Act Management Plan Series. Parks Canada Agency, Ottawa, ON, Canada. v + 23 pp.

Pearn, M. 2012. Pollination and comparative reproductive success of lady's slipper orchids *Cypripedium candidum, C. parviflorum*, and their hybrids in southern Manitoba. M.S. thesis. University of Maitoba, Winnipeg, MB. 193 pp.

Peck, M. E. 1925. A preliminary sketch of the plant regions of Oregon. I. Western Oregon. *Am. J. Bot.* **12**: 33–49.

Pedron, M., C. R. Buzatto, R. B. Singer, J. A. N. Batista & A. Moser. 2012. Pollination biology of four sympatric species of *Habenaria* (Orchidaceae: Orchidinae) from southern Brazil. *Bot. J. Linn. Soc.* **170**: 141–156.

Pedron, M., C. R. Buzatto, A. J. Ramalho, B. M. Carvalho, J. A. Radins, R. B. Singer & J. A. N. Batista. 2014. Molecular phylogenetics and taxonomic revision of *Habenaria* section *Pentadactylae* (Orchidaceae, Orchidinae). *Bot. J. Linn. Soc.* **175**: 47–73.

Peet, R. K. 2007. Ecological classification of longleaf pine woodlands. Pp. 51–93 *In*: S. Jose, E. J. Jokela & D. L. Miller (eds.), *The Longleaf Pine Ecosystem*. Springer, New York, NY.

Peet, R. K. & D. J. Allard. 1993. Longleaf pine vegetation of the southern Atlantic and eastern Gulf Coast regions: a preliminary classification. Pp. 45–81 *In*: S. M. Hermann (ed.), *Proceedings of the Tall Timbers Fire Ecology Conference, Vol. 18: The Longleaf Pine Ecosystem: Ecology, Restoration and Management*. Tall Timbers Research Station, Tallahassee, FL.

Peeters-Van der Meijden, K. & T. Rotteveel. 2006. *Lysichiton americanus* Hultén & H. St. John, een imposante indringer. *Gorteria* **32**: 37–44.

Peinado, M., J. L. Aguirre, M. Á. Macías & J. Delgadillo. 2011. A phytosociological survey of the dune forests of the Pacific Northwest. *Plant Biosyst.* **145**: 105–117.

Pellatt, M. G. & R. W. Mathewes. 1994. Paleoecology of postglacial tree line fluctuations on the Queen Charlotte Islands, Canada. *Ecoscience* **1**: 71–81.

Pellegrini, M. O. O., C. N. Horn & R. F. Almeida. 2018. Total evidence phylogeny of Pontederiaceae (Commelinales) sheds light on the necessity of its recircumscription and synopsis of *Pontederia* L. *PhytoKeys* **108**: 25–83.

Pellerin, S., J. Huot & S. D. Côté. 2006. Long-term effects of deer browsing and trampling on the vegetation of peatlands. *Biol. Conserv.* **128**: 316–326.

Pellerin, S., L.-A. Lagneau, M. Lavoie & M. Larocque. 2009. Environmental factors explaining the vegetation patterns in a temperate peatland. *Comp. Rend. Biol.* **332**: 720–731.

Pellicer, J., L. J. Kelly, I. J. Leitch, W. B. Zomlefer & M. F. Fay. 2014. A universe of dwarfs and giants: genome size and chromosome evolution in the monocot family Melanthiaceae. *New Phytol.* **201**: 1484–1497.

Pellmyr, O. & J. M. Patt. 1986. Function of olfactory and visual stimuli in pollination of *Lysichiton americanum* (Araceae) by a staphylinid beetle. *Madrono* **33**: 47–54.

Pemberton, R. W. 1998. Waterblommetjie, an unusual aquatic food plant, new crop and cultural symbol in the Cape of South Africa. Pp. 223–228 *In*: H. Walker (ed.), *Fish: Food from the Waters*. Prospect Books, Devon, England.

Pemberton, R. W. 2000. Waterblommetjie (*Aponogeton distachyos*, Aponogetonaceae), a recently domesticated aquatic food crop in Cape South Africa with unusual origins. *Econ. Bot.* **54**: 144–149.

Pendergrass, L. 1983. *Harper's Beauty (Harperocallis flava) Recovery Plan*. U.S. Fish & Wildlife Service, Atlanta, GA. 32 pp.

Penfound, W. T. 1953. Plant communities of Oklahoma lakes. *Ecology* **34**: 561–583.

Penfound, W. T. 1956. Primary production of vascular aquatic plants. *Limnol. Oceanogr.* **1**: 92–101.

Penfound, W. T. & M. E. O'Neill. 1934. The vegetation of Cat Island, Mississippi. *Ecology* **15**: 1–16.

Penfound, W. T. & E. S. Hathaway. 1938. Plant communities in the marshlands of southeastern Louisiana. *Ecol. Monogr.* **8**: 1–56.

Penfound, W. T. & T. T. Earle. 1948. The biology of the water hyacinth. *Ecol. Monogr.* **18**: 447–472.

Penfound, W. T., T. F. Hall & D. Hess. 1945. The spring phenology of plants in and around the reservoirs in North Alabama with particular reference to malaria control. *Ecology* **26**: 332–352.

Peng, K., C. Luo, L. Lou, X. Li & Z. Shen. 2008. Bioaccumulation of heavy metals by the aquatic plants *Potamogeton pectinatus* L. and *Potamogeton malaianus* Miq. and their potential use for contamination indicators and in wastewater treatment. *Sci. Total Environ.* **392**: 22–29.

Penhale, P. A. 1977. Macrophyte-epiphyte biomass and productivity in an eelgrass (*Zostera marina* L.) community. *J. Exp. Mar. Biol. Ecol.* **26**: 211–224.

Penko, J. M. & D. C. Pratt. 1986. The growth and survival of early instars of *Bellura obliqua* (Lepidoptera: Noctuidae) on *Typha latifolia* and *Typha angustifolia*. *Great Lakes Entomol.* **19**: 35–42.

Penn, G. H. 1950. The genus *Cambarellus* in Louisiana (Decapoda, Astacidae). *Am. Midl. Nat.* **44**: 421–426.

Pennacchio, M., L. Jefferson & K. Havens. 2010. *Uses and Abuses of Plant-Derived Smoke: Its Ethnobotany as Hallucinogen, Perfume, Incense, and Medicine*. Oxford University Press, New York, NY. 264 pp.

Pennington, T. G. & M. D. Sytsma. 2009. Seasonal changes in carbohydrate and nitrogen concentrations in Oregon and California populations of Brazilian Egeria (*Egeria densa*). *Invasive Plant Sci. Manag.* **2**: 120–129.

Perdomo, P., P. Nitzsche & D. Drake. 2004. Landscape plants rated by deer resistance. Bulletin E271. Rutgers Cooperative Research & Extension, NJAES, Rutgers, The State University of New Jersey, New Brunswick, NJ. 6 pp.

Peredo, E. L., D. H. Les, U. M. King & L. K. Benoit. 2012. Extreme conservation of the *psaA/psaB* intercistronic spacer reveals a translational motif coincident with the evolution of land plants. *J. Mol. Evol.* **75**: 184–197.

Peredo, E. L., U. M. King & D. H. Les. 2013. The plastid genome of *Najas flexilis*: adaptation to submersed environments is accompanied by the complete loss of the NDH complex in an aquatic angiosperm. *PLoS One* **8**: e68591.

Pereyra, M. E. 2011. Effects of snow-related environmental variation on breeding schedules and productivity of a high-altitude population of Dusky Flycatchers (*Empidonax oberholseri*). *Auk* **128**: 746–758.

Pérez, E. A., J. A. Coetzee, T. Ruiz Téllez & M. P. Hill. 2011. A first report of water hyacinth (*Eichhornia crassipes*) soil seed banks in South Africa. *S. Afr. J. Bot.* **77**: 795–800.

Perkin, J. S., Z. R. Shattuck & T. H. Bonner. 2012. Life history aspects of a relict ironcolor shiner *Notropis chalybaeus* population in a novel spring environment. *Am. Midl. Nat.* **167**: 111–126.

Perleberg, D. 2006. *Aquatic Vegetation of Birch Lake (DOW 11-0412-00) Cass County, Minnesota*. Minnesota Department of Natural Resources, Ecological Resources Division, Brainerd, MN. 20 pp.

Perleberg, D. 2008. *Aquatic Vegetation of Long Lake (DOW 11-0142-00), Cass County, Minnesota, 2007*. Minnesota Department of Natural Resources, Ecological Resources Division, Brainerd, MN. 24 pp.

Perleberg, D. & S. Loso. 2009a. *Aquatic Vegetation of Thunder Lake (DOW 11-0062-00) Cass County, Minnesota, 2008*. Minnesota Department of Natural Resources, Ecological Resources Division, Brainerd, MN. 21 pp.

Perleberg, D. & S. Loso. 2009b. *Aquatic Vegetation Survey of Trout Lake (DOW #31-0216-00) Itasca County, Minnesota, 2005*. Minnesota Department of Natural Resources, Ecological Resources Division, Brainerd, MN. 18 pp.

Perrette, G. 2010. Comparative dendrochronology study of black spruce (*Picea mariana*) on lowlands & uplands in the Hudson Bay lowland. B.S. thesis. Laurentian University, Sudbury, ON. 43 pp.

Perry, F. 1961. *Water Gardening*, revised ed. Country Life Limited, London, UK. 338 pp.

Perry, T. O. 1968. Dormancy, turion formation, and germination by different clones of *Spirodela polyrrhiza*. *Plant Physiol.* **43**: 1866–1869.

Perry, M. C. & F. M. Uhler. 1981. Asiatic clam (*Corbicula manilensis*) and other foods used by waterfowl in the James River, Virginia. *Estuaries* **4**: 229–233.

Perry, M. C. & F. M. Uhler. 1988. Food habits and distribution of wintering canvasbacks, *Aythya valisineria*, on Chesapeake Bay. *Estuaries* **11**: 57–67.

Perry, J. E. & R. B. Atkinson. 1997. Plant diversity along a salinity gradient of four marshes on the York and Pamunkey Rivers in Virginia. *Castanea* **62**: 112–118.

Perry, J. E. & C. H. Hershner. 1999. Temporal changes in the vegetation pattern in a tidal freshwater marsh. *Wetlands* **19**: 90–99.

Perry, L. E. & M. E. Dorken. 2011. The evolution of males: support for predictions from sex allocation theory using mating arrays of *Sagittaria latifolia* (Alismataceae). *Evolution* **65**: 2782–2791.

Persson, H. 1947. Further notes on Alaskan-Yukon bryophytes. *Bryologist* **50**: 279–310.

Persson, H. & H. T. Shacklette. 1959. *Drepanocladus trichophyllus* found in North America. *Bryologist* **62**: 251–254.

Perullo, N., R. O. Determann, J. M. Cruse-Sanders & G. S. Pullman. 2015. Seed cryopreservation and micropropagation of the critically endangered species swamp pink (*Helonias bullata* L.). *In Vitro Cell. Dev. Biol.-Plant* **51**: 284–293.

Pervin, F., M. M. Hossain, S. Khatun, S. P. Siddique, K. A. Salam, M. R. Karim & N. Absar. 2006. Comparative citotoxicity study of six bioactive lectins purified from pondweed (*Potamogeton nodosus* Poir) rootstock on Brine Shrimp. *J. Med. Sci.* **6**: 999–1002.

Petch, T. 1928. Notes on *Cryptocoryne*. *Ann. Roy. Bot. Gard. Peradeniya* **11**: 11–26.

Peter, C. I. & S. D. Johnson. 2005. Doing the twist: a test of Darwin's cross-pollination hypothesis for pollinarium reconfiguration. *Biol. Lett.* **2**: 65–68.

Peters, J. A., T. Kreps & D. M. Lodge. 2008. Assessing the impacts of rusty crayfish (*Orconectes rusticus*) on submergent macrophytes in a north-temperate US lake using electric fences. *Am. Midl. Nat.* **159**: 287–297.

Peters, D., A. Morales, S. Morales & R. Hernández. 2009. Feeding quality evaluation of *Lemna obscura* meal as ingredients in the elaboration of food for red tilapia (*Orechromis* spp.). *Revista Ci. Fac. Ci. Veterin. Univ. Zulia* **19**: 303–310.

Petersen, G., O. Seberg, J. I. Davis & D. W. Stevenson. 2006. RNA editing and phylogenetic reconstruction in two monocot mitochondrial genes. *Taxon* **55**: 871–886.

Petersen, G., O. Seberg & J. I. Davis. 2013. Phylogeny of the Liliales (Monocotyledons) with special emphasis on data partition congruence and RNA editing. *Cladistics* **29**: 274–295.

Petersen, G., O. Seberg, F. T. Short & M. D. Fortes. 2014. Complete genomic congruence but non-monophyly of *Cymodocea* (Cymodoceaceae), a small group of seagrasses. *Taxon* **63**: 3–8.

Petersen, G., O. Seberg, A. Cuenca, D. W. Stevenson, M. Thadeo, J. I. Davis, S. Graham & T. G. Ross. 2016. Phylogeny of the Alismatales (Monocotyledons) and the relationship of *Acorus* (Acorales?). *Cladistics* **32**: 141–159.

Petersen, G., A. Cuenca, A. Zervas, G. T. Ross, S. W. Graham, C. F. Barrett, J. I. Davis & O. Seberg. 2017. Mitochondrial genome evolution in Alismatales: size reduction and extensive loss of ribosomal protein genes. *PLoS One* **12**: e0177606.

Peterson, L. A. 1977. *A Field Guide to Edible Wild Plants of Eastern and Central North America*. Houghton Mifflin, Company, Boston, MA. 330 pp.

Peterson, J. E. & A. H. Baldwin. 2004a. Seedling emergence from seed banks of tidal freshwater wetlands: response to inundation and sedimentation. *Aquat. Bot.* **78**: 243–254.

Peterson, J. E. & A. H. Baldwin. 2004b. Variation in wetland seed banks across a tidal freshwater landscape. *Am. J. Bot.* **91**: 1251–1259.

Peterson, P. M., R. J. Soreng, D. Styer, D. Neubauer, R. Morgan & V. Yadon. 2011. *Agrostis lacuna-vernalis* (Pooideae: Poeae: Agrostidinae), a new species from California. *J. Bot. Res. Inst. Texas* **5**: 421–426.

Pettit, G. R., G. R. Pettit III, R. A. Backhaus & F. E. Boettner. 1995. Antineoplastic agents, 294. Variations in the formation of pancratistatin and related isocarbostyrils in *Hymenocallis littoralis*. *J. Nat. Prod.* **58**: 37–43.

Pettitt, J. M. & A. C. Jermy. 1974. Pollen in hydrophilous angiosperms. *Micron* **5**: 377–405.

Petty, R. O. & A. M. Petty. 2005. *Wild Plants in Flower–Wetlands and Quiet Waters of the Midwest*. Quarry Books, Bloomington, IN. 100 pp.

Pezeshki, S. R., P. H. Anderson & R. D. DeLaune. 2000. Effects of nursery pre-conditioning on *Panicum hemitomon* and *Sagittaria lancifolia* used for wetland restoration. *Restor. Ecol.* **8**: 57–64.

Pfauth, M. & M. Sytsma. 2004. Coastal lakes aquatic plant survey report. Paper 25. Center for Lakes and Reservoirs, Portland State University, Portland, OR. 41 pp.

Pfauth, M. & M. Sytsma. 2005. Alaska aquatic plant survey report 2005. Paper 12. Center for Lakes and Reservoirs, Portland State University, Portland, OR. 20 pp.

Pfeifer-Meister, L., B. R. Johnson, B. A. Roy, S. Carreño, J. L. Stewart & S. D. Bridgham. 2012. Restoring wetland prairies: tradeoffs among native plant cover, community composition, and ecosystem functioning. *Ecosphere* **3**(12): 121. doi: 10.1890/ES12-00261.1.

Philbrick, C. T. 1984. Aspects of floral biology in three species of *Potamogeton* (pondweeds). *Michigan Bot.* **23**: 35–38.

Philbrick, C. T. 1988. Evolution of underwater outcrossing from aerial pollination systems: a hypothesis. *Ann. Missouri Bot. Gard.* **75**: 836–841.

Philbrick, C. T. & G. J. Anderson. 1987. Implications of pollen/ovule ratios and pollen size for the reproductive biology of *Potamogeton* and autogamy in aquatic angiosperms. *Syst. Bot.* **12**: 98–105.

Philbrick, C. T. & D. H. Les. 1996. Evolution of aquatic angiosperm reproductive systems. *BioScience* **46**: 813–826.

Phillips, J. C. 1911. Ten years of observation on the migration of Anatidae at Wenham Lake, Massachusetts. *Auk* **28**: 188–200.

Phillips, R. C. 1960. Observations on the ecology and distribution of the Florida seagrasses. Professional Papers Series No. 2. Florida State Board of Conservation, Marine Laboratory, St. Petersburg, FL. 72 pp.

Phillips, E. C. 2008. Invertebrate colonization of native and invasive aquatic macrophytes in Presque Isle Bay, Lake Erie. *J. Freshw. Ecol.* **23**: 451–457.

Phillips, R. C. & E. G. Meñez. 1988. *Seagrasses*. Smithsonian Contributions to the Marine Sciences, No. 34. Smithsonian Institution Press, Washington, DC. 104 pp.

Phillips, R. C. & S. W. Echeverria. 1990. *Zostera asiatica* Miki on the Pacific coast of North America. *Pac. Sci.* **44**: 130–134.

Phillips, R. C., C. McMillan & K. W. Bridges. 1983a. Phenology of eelgrass, *Zostera marina* L., along latitudinal gradients in North America. *Aquat. Bot.* **15**: 145–156.

Phillips, R. C., W. S. Grant & C. P. McRoy. 1983b. Reproductive strategies of eelgrass (*Zostera marina* L.). *Aquat. Bot.* **16**: 1–20.

Phillips, C. T., M. L. Alexander & A. M. Gonzales. 2011. Use of macrophytes for egg deposition by the endangered fountain darter. *Trans. Am. Fish. Soc.* **140**: 1392–1397.

Pholhiamhan, R., S. Saensouk & P. Saensouk. 2017. Ethnobotany of Phu Thai ethnic group in Nakhon Phanom Province, Thailand. *Walailak J. Sci. Technol.* **15**: 679–699.

Pickford, G. D. & E. H. Reid. 1943. Competition of elk and domestic livestock for summer range forage. *J. Wildl. Manag.* **7**: 328–332.

Picking, D. J. & P. L. M. Veneman. 2004. Vegetation patterns in a calcareous sloping fen of southwestern Massachusetts, USA. *Wetlands* **24**: 514–528.

Pieper, S. J., A. A. Nicholls, J. R. Freeland & M. E. Dorken. 2017. Asymmetric hybridization in cattails (*Typha* spp.) and its implications for the evolutionary maintenance of native *Typha latifolia*. *J. Hered.* **108**: 479–487.

Pierce, J. R. & M. E. Jensen. 2002. A classification of aquatic plant communities within the Nothern Rocky Mountains. *West. N. Am. Nat.* **62**: 257–265.

Pierinia, S. A. & S. M. Thomaz. 2004. Effects of inorganic carbon source on photosynthetic rates of *Egeria najas* Planchon and *Egeria densa* Planchon (Hydrocharitaceae). *Aquat. Bot.* **78**: 135–146.

Pierre, J. I. St. & K. E. Kovalenko. 2014. Effect of habitat complexity attributes on species richness. *Ecosphere* **5**: 1–10.

Pieterse, A. H. 1981. *Hydrilla verticillata* - a review. *Abstr. Trop. Agric.* **7**: 9–34.

Pieterse, A. H. 2013. Is flowering in Lemnaceae stress-induced? A review. *Aquat. Bot.* **104**: 1–4.

Pieterse, A. H. & K. J. Murphy (eds.). 1993. *Aquatic Weeds: The Ecology and Management of Nuisance Aquatic Vegetation*. Oxford University Press, Oxford, United Kingdom. 593 pp.

Pieterse, A. H., P. R. Bhalla & P. S. Sabharwal. 1970. Chemical induction of turions in *Wolffiella floridana* (JD Smith) Thompson. *Acta Bot. Neerl.* **19**: 901–905.

Pieterse, A. H., P. R. Bhalla & P. S. Sabharwal. 1971. Endogenous gibberellins in floating plants and turions of *Wolffiella floridana*. *Physiol. Plant.* **24**: 512–516.

Pieterse, A. H., L. de Lange & L. Verhagen. 1981. A study on certain aspects of seed germination and growth of *Pistia stratiotes* L. *Acta Bot. Neerl.* **30**: 47–57.

Pike, K. S., P. Starý, T. Miller, G. Graf, D. Allison, L. Boydston & R. Miller. 2000. Aphid parasitoids (Hymenoptera: Braconidae: Aphidiinae) of northwest USA. *Proc. Entomol. Soc. Washington* **102**: 688–740.

Pilon, J. & L. Santamaría. 2002. Clonal variation in morphological and physiological responses to irradiance and photoperiod for the aquatic angiosperm *Potamogeton pectinatus*. *J. Ecol.* **90**: 859–870.

Pilon, J., L. Santamaría, M. Hootsmans & W. van Vierssen. 2003. Latitudinal variation in life-cycle characteristics of *Potamogeton pectinatus* L.: vegetative growth and asexual reproduction. *Plant Ecol.* **165**: 247–262.

Pindel, Z. 2001. Influence of stratification time and colour of light on the germination of *Calla palustris* seeds. *Zesz. Nauk. Akad. Roln. Kollataja Krakowie* **24**: 49–56.

Pine, R. T. & L. W. J. Anderson. 1991. Plant preferences of triploid grass carp. *J. Aquat. Plant Manag.* **29**: 80–82.

Pineda, P. M., R. J. Rondeau & A. Ochs. 1999. *A Biological Inventory and Conservation Recommendations for the Great Sand Dunes and San Luis Lakes, Colorado.* Colorado Natural Heritage Program, Colorado State University, Ft. Collins, CO. 67 pp.

Pinkava, D. J. 1963. Vascular flora of the Miller blue hole and stream, Sandusky County, Ohio. *Ohio J. Sci.* **63**: 113–127.

Pip, E. 1979. Survey of the ecology of submerged aquatic macrophytes in central Canada. *Aquat. Bot.* **7**: 339–357.

Pip, E. 1987. The ecology of *Potamogeton* species in central North America. *Hydrobiologia* **153**: 203–216.

Pip, E. 1988. Niche congruency of aquatic macrophytes in central North America with respect to 5 water chemistry parameters. *Hydrobiologia* **162**: 173–182.

Pip, E. & J. M. Stewart. 1976. The dynamics of two aquatic plant–snail associations. *Can. J. Zool.* **54**: 1192–1205.

Pip, E. & K. Simmons. 1986. Aquatic angiosperms at unusual depths in Shoal Lake, Manitoba-Ontario. *Can. Field-Nat.* **100**: 354–358.

Pípalová, I. 2002. Initial impact of low stocking density of grass carp on aquatic macrophytes. *Aquat. Bot.* **73**: 9–18.

Pípalová, I. 2003. Grass carp (*Ctenopharyngodon idella*) grazing on duckweed (*Spirodela polyrhiza*). *Aquac. Int.* **11**: 325–336.

Pires, J. C. 2002. *Brodiaea* Smith. Pp. 321–327 *In*: Flora North America Editorial Committee (eds.), *Flora of North America North of Mexico, Vol. 26: Magnoliophyta: Liliidae: Liliales and Orchidales.* Oxford University Press, New York, NY.

Pires, J. C. & K. J. Sytsma. 2002. A phylogenetic evaluation of a biosystematic framework: *Brodiaea* and related petaloid monocots (Themidaceae). *Am. J. Bot.* **89**: 1342–1359.

Pires, J. C., I. J. Maureira, T. J. Givnish, K. J. Systma, O. Seberg, G. Peterson, J. I. Davis, D. W. Stevenson, P. J. Rudall, M. F. Fay & M. W. Chase. 2006. Phylogeny, genome size, and chromosome evolution of Asparagales. *Aliso* **22**: 287–304.

Pischtschan, E. & R. Claßen-Bockhoff. 2008. Setting-up tension in the style of Marantaceae. *Plant Biol.* **10**: 441–450.

Pitts-Singer, T. L., J. L. Hanula & J. L. Walker. 2002. Insect pollinators of three rare plants in a Florida longleaf pine forest. *Florida Entomol.* **85**: 308–316.

Platt, S. G., R. M. Elsey, H. Liu, T. R. Rainwater, J. C. Nifong, A. E. Rosenblatt, M. R. Heithaus & F. J. Mazzotti. 2013. Frugivory and seed dispersal by crocodilians: an overlooked form of saurochory? *J. Zool.* **291**: 87–99.

Plowman, T. 1969. Folk uses of new world aroids. *Econ. Bot.* **23**: 97–122.

Pogan, E. 1963. Taxonomical value of *Alisma triviale* Pursh and *Alisma subcordatum* Rafin. *Can. J. Bot.* **41**: 1011–1013.

Pogue, C. D., M. J. Monfils, D. L. Cuthrell, B. W. Heumann & A. K. Monfils. 2016. Habitat suitability modeling of the federally endangered poweshiek skipperling in Michigan. *J. Fish Wildl. Manag.* **7**: 359–368.

Poiani, K. A. & W. C. Johnson. 1989. Effect of hydroperiod on seedbank composition in semipermanent prairie wetlands. *Can. J. Bot.* **67**: 856–864.

Poiani, K. A. & P. M. Dixon. 1995. Seed banks of Carolina bays: potential contributions from surrounding landscape vegetation. *Am. Midl. Nat.* **134**: 140–154.

Poirrier, M. A., K. Burt-Utley, J. F. Utley & E. A. Spalding. 2010. Submersed aquatic vegetation of the Jean Lafitte National Historical Park and Preserve. *Southeast. Nat.* **9**: 477–486.

Pollock, D. A. 1999. Review of the New World Hemipeplinae (Coleoptera: Mycteridae) with descriptions of ten new species. *Insect Syst. Evol.* **30**: 47–73.

Pollux, B. J. A. 2011. The experimental study of seed dispersal by fish (ichthyochory). *Freshw. Biol.* **56**: 197–212.

Pollux, B. J. A. & N. J. Ouborg. 2006. Isolation and characterization of microsatellites in *Sparganium emersum* and cross-species amplification in the related species *S. erectum*. *Mol. Ecol. Notes* **6**: 530–532.

Pollux, B. J. A., N. J. Ouborg, J. M. Van Groenendael & M. Klaassen. 2007. Consequences of intraspecific seed-size variation in *Sparganium emersum* for dispersal by fish. *Funct. Ecol.* **21**: 1084–1091.

Pollux, B. J. A., E. Verbruggen, J. M. Van Groenendael & N. J. Ouborg. 2009. Intraspecific variation of seed floating ability in *Sparganium emersum* suggests a bimodal dispersal strategy. *Aquat. Bot.* **90**: 199–203.

Ponzio, K. J., S. J. Miller & M. A. Lee. 2004. Long-term effects of prescribed fire on *Cladium jamaicense* crantz and *Typha domingensis* Pers. densities. *Wetl. Ecol. Manag.* **12**: 123–133.

Poole, J. M., W. R. Carr, D. M. Price & J. R. Singhurst. 2007. *Rare Plants of Texas: A Field Guide.* Texas A&M University Press, College Station, TX. 640 pp.

Popenoe, J. 1986. *Burmannia flava* Mart. in Florida. *Florida Sci.* **49**: 126.

Popov, S. V., G. Y. Popova, N. M. Paderin, O. A. Koval, R. G. Ovodova & Y. S. Ovodov. 2007. Preventative antiinflammatory effect of potamogetonan, a pectin from the common pondweed *Potamogeton natans* L. *Phytother. Res.* **21**: 609–614.

Porath, D., B. Hepher & A. Koton. 1979. Duckweed as an aquatic crop: evaluation of clones for aquaculture. *Aquat. Bot.* **7**: 273–278.

Porcher, R. D. 1981. The vascular flora of the Francis Beidler Forest in Four Holes Swamp, Berkeley and Dorchester Counties, South Carolina. *Castanea* **46**: 248–280.

Posey, M. H. 1988. Community changes associated with the spread of an introduced seagrass, *Zostera japonica*. *Ecology* **69**: 974–983.

Posey, M. H., C. Wigand & J. C. Stevenson. 1993. Effects of an introduced aquatic plant, *Hydrilla verticillata* on benthic communities in the upper Chesapeake Bay. *Estuar. Coast. Shelf Sci.* **37**: 539–555.

Posluszny, U. 1981. Unicarpellate floral development in *Potamogeton zosteriformis*. *Can. J. Bot.* **59**: 495–504.

Posluszny, U. 1983. Re-evaluation of certain key relationships in the Alismatidae: floral organogenesis of *Scheuchzeria palustris* (Scheuchzeriaceae). *Am. J. Bot.* **70**: 925–933.

Posluszny, U. & R. Sattler. 1974. Floral development of *Potamogeton richardsonii*. *Am. J. Bot.* **61**: 209–216.

Posluszny, U. & R. Sattler. 1976. Floral development of *Zannichellia palustris*. *Can. J. Bot.* **54**: 651–662.

Posluszny, U. & A. Charlton. 1993. Evolution of the helobial flower. *Aquat. Bot.* **44**: 303–324.

Posluszny, U., W. A. Charlton & D. K. Jain. 1986. Morphology and development of the reproductive shoots of *Lilaea scilloides* (Poir.) Hauman (Alismatidae). *Bot. J. Linn. Soc.* **92**: 323–342.

Posluszny, U., W. A. Charlton & D. H. Les. 2000. Modularity in Helobial flowers. Pp. 63–74 *In*: K. L. Wilson & D. Morrison (eds.), *Systematics and Evolution of Monocots*. CSIRO Publishing, Collingwood, VIC, Australia.

Poster, L. S., A. F. Rhoads & T. A. Block. 2013. Vascular flora and community assemblages of Delhaas Woods, a coastal plain forest in Bucks County, Pennsylvania. *J. Torrey Bot. Soc.* **140**: 101–124.

Potgieter, M. 1940. Taro (*Colocasia esculenta*) as a food. *J. Am. Diet. Assoc.* **16**: 536–540.

Pott, V. J. & A. Pott. 2000. *Plantas Aquáticas do Pantanal*. EMBRAPA. Comunicação para Transferência de Tecnologia, Brasília, Brazil. 404 pp.

Potter, A., J. Fleckenstein, S. Richardson & D. Hays. 1999. *Washington State Status Report for the Mardon Skipper*. Washington Department of Fish and Wildlife, Olympia, WA. 39 pp.

Potter, K. M., B. S. Crane & W. W. Hargrove. 2017. A United States national prioritization framework for tree species vulnerability to climate change. *New Forests* **48**: 275–300.

Poulin, M., T. Landry, V. Laberge & L. Rochefort. 2012. Establishing vascular plants from seeds around pool margins in restored peatlands. Extended abstract No. 264. *Proceedings of the 14th International Peat Congress*, Stockholm, Sweden. 5 pp.

Power, P. 1996. Direct and indirect effects of floating vegetation mats on Texas wildrice (*Zizania texana*). *Southwest. Nat.* **41**: 462–464.

Pozo, M. I., C. M. Herrera, W. Van den Ende, K. Verstrepen, B. Lievens & H. Jacquemyn. 2015. The impact of nectar chemical features on phenotypic variation in two related nectar yeasts. *FEMS Microbiol. Ecol.* **91**: fiv055.

Pradeesh, S., G. N. Archana, M. D. Chinmayee, C. S. Sarika, I. Mini & T. S. Swapna. 2012. Biochemical and nutritional evaluation of *Monochoria vaginalis* Presl. Pp. 93–101 *In*: A. Sabu & A. Augustine (eds.), *Prospects in Bioscience: Addressing the Issues*. Springer, New Delhi, India.

Prado, P. & K. L. Heck, Jr. 2011. Seagrass selection by omnivorous and herbivorous consumers: determining factors. *Mar. Ecol. Prog. Ser.* **429**: 45–55.

Prakash, P. S. & N. M. Nayar. 2000. Flowering and pollination in taro. *J. Root Crops* **26**: 15–19.

Prameela, R. & M. Venkaiah. 2016. Recent study on aquatic monocots of Vizianagaram district, Andhra Pradesh, India. *J. Med. Plant Stud.* **4**: 1–4.

Prasad, M. N. V. 2007. Aquatic plants for phytotechnology. Pp. 259–274 *In*: S. N. Singh & R. D. Tripathi (eds.), *Environmental Bioremediation Technologies*. Springer, Berlin, Germany.

Prasad, V. 2017. Some tetranychoid mites of Michigan. *Great Lakes Entomol.* **3**: 24–31.

Prausová, R., J. Janová & L. Šafářová. 2013. Testing achene germination of *Potamogeton praelongus* Wulfen. *Centr. Eur. J. Biol.* **8**: 78–86.

Prausová, R., P. Sikorová & L. Šafářová. 2014. Generative reproduction of long stalked pondweed (*Potamogeton praelongus* Wulfen) in the laboratory. *Aquat. Bot.* **120**: 268–274.

Prausová, R., Z. Kozelková & L. Šafářová. 2015. Protocol for acclimatization of in vitro cultured *Potamogeton praelongus*–aspect of plantlet size and type of substrate. *Acta Soc. Bot. Polon.* **84**: 35–41.

Prena, J. 2017. Orchid weevils (Coleoptera: Curculionidae) in Canada. *Can. Entomol.* **149**: 38–47.

Prepas, E. E., J. Babin, T. P. Murphy, P. A. Chambers, G. J. Sandland, A. Ghadouani & M. Serediak. 2001. Long-term effects of successive Ca (OH)$_2$ and CaCO$_3$ treatments on the water quality of two eutrophic hardwater lakes. *Freshw. Biol.* **46**: 1089–1103.

Prescott-Allen, R. & C. Prescott-Allen. 1990. How many plants feed the world? *Conserv. Biol.* **4**: 365–374.

Preston, C. D. 1995. *Pondweeds of Great Britain and Ireland*. B.S.B.I. Handbook No. 8. Botanical Society of the British Isles, London. UK. 352 pp.

Preston, C. D. & J. M. Croft. 1997. *Aquatic Plants in Britain and Ireland*. Harley Books, Colchester, Essex, United Kingdom. 365 pp.

Prevett, J. P., I. F. Marshall & V. G. Thomas. 1979. Fall foods of lesser snow geese in the James Bay region. *J. Wildl. Manag.* **43**: 736–742.

Prevett, J. P., I. F. Marshall & V. G. Thomas. 1985. Spring foods of snow and Canada geese at James Bay. *J. Wildl. Manag.* **49**: 558–563.

Price, M. 1966. *The Iris Book*. D. Van Nostrand Co., Inc., Princeton, NJ. 204 pp.

Price, S. D. & S. C. H. Barrett. 1982. Tristyly in *Pontederia cordata* (Pontederiaceae). *Can. J. Bot.* **60**: 897–905.

Price, S. D. & S. C. H. Barrett. 1984. The function and adaptive significance of tristyly in *Pontederia cordata* L. (Pontederiaceae). *Biol. J. Linn. Soc.* **21**: 315–329.

Prichard, E. C. & F. T. Thorne. 1961. A new locality for *Nemastylis floridana* Small. *Castanea* **26**: 97–97.

Prince, L. M. 2010. Phylogenetic relationships and species delimitation in *Canna* (Cannaceae). Pp. 301–331 *In*: O. Seberg, G. Petersen, A. S. Barfod & J. I. Davis (eds.), *Diversity, Phylogeny, and Evolution in the Monocotyledons*. Aarhus University Press, Aarhus, Denmark.

Prince, L. M. & W. J. Kress. 2006. Phylogenetic relationships and classification in Marantaceae: insights from plastid DNA sequence data. *Taxon* **55**: 281–296.

Pritchard, A. L. 1935. The higher aquatic plants of Lake Abitibi, Ontario. *Univ. Toronto Stud., Biol. Ser.* **39**: 79–85.

Pritchard, G. 1987. Growth and food choice by two species of limnephilid caddis larvae given natural and artificial foods. *Freshw. Biol.* **18**: 529–535.

Proctor, H. C. 1998. Effect of pollen age on fruit set, fruit weight, and seed set in three orchid species. *Can. J. Bot.* **76**: 420–427.

Proctor, H. C. & L. D. Harder. 1994. Pollen load, capsule weight, and seed production in three orchid species. *Can. J. Bot.* **72**: 249–255.

Profous, G. V. & R. E. Loeb. 1984. Vegetation and plant communities of Van Cortlandt Park, Bronx, New York. *Bull. Torrey Bot. Club* **111**: 80–89.

Promdee, K., D. Phihusut, A. Monthienvichienchai, Y. Tongaram & P. Khongsuk. 2018. Conversion of *Hydrilla verticillata* to bio-oil and charcoal using a continuous pyrolysis reactor. *Biofuels* **2018**. doi.org/10.1080/17597269.2018.1448633.

Provancha, J. A. & C. R. Hall. 1991. Observations of associations between seagrass beds and manatees in east central Florida. *Florida Sci.* **54**: 87–98.

Provost, M. W. 1947. Nesting of birds in the marshes of northwest Iowa. *Am. Midl. Nat.* **38**: 485–503.

Puijalon, S., F. Piola & G. Bornette. 2008. Abiotic stresses increase plant regeneration ability. *Evol. Ecol.* **22**: 493–506.

Pulgar, Í. & J. Izco. 2005. *Egeria densa* Planchon (Hydrocharitaceae) in Pontevedra province (Spain). *Acta Bot. Malacitana* **30**: 173–175.

Pulich, Jr., W. M. 1982. Culture studies of *Halodule wrightii* Aschers. Edaphic requirements. *Bot. Mar.* **25**: 477–482.

Pulich, Jr., W. M. 1985. Seasonal growth dynamics of *Ruppia maritima* L. s.l. and *Halodule wrightii* Aschers. in southern Texas and evaluation of sediment fertility status. *Aquat. Bot.* **23**: 53–66.

Punsalan, A. P., B. Collins & L. E. DeWald. 2016. The germination ecology of *Helonias bullata* L. (Swamp Pink) with respect to dry, saturated, and flooded conditions. *Aquat. Bot.* **133**: 17–23.

Purer, E. A. 1939. Ecological study of vernal pools, San Diego County. *Ecology* **20**: 217–229.

Putman, W. L. 1953. Notes on the bionomics of some Ontario cercopids (Homoptera). *Can. Entomol.* **85**: 244–248.

Pyne, M. 2005. West Gulf Coastal Plain flatwoods pond ecological system. Ecological integrity assessment. NatureServe, Ecology Department (Southeast), Durham, NC. 32 pp.

Qais, N., M. R. Mandal, M. A. Rashid, A. Jabbar, H. Koshino, K. Nagasawa & T. Nakata. 1998. A furanoid labdane diterpene from *Potamogeton nodosus*. *J. Nat. Prod.* **61**: 156–157.

Qi, Z., K. M. Cameron, P. Li, Y. Zhao, S. Chen, G. Chen & C. Fu. 2013. Phylogenetics, character evolution, and distribution patterns of the greenbriers, Smilacaceae (Liliales), a near-cosmopolitan family of monocots. *Bot. J. Linn. Soc.* **173**: 535–548.

Qian, Y., N. Xu, J. Liu & R. Tian. 2018. Inhibitory effects of *Pontederia cordata* on the growth of *Microcystis aeruginosa*. *Water Sci. Technol.* **1**: 99–107.

Qian, Y.-P., X.-T. Li & R.-N. Tian. 2019. Effects of aqueous extracts from the rhizome of *Pontederia cordata* on the growth and interspecific competition of two algal species. *Ecotoxicol. Environ. Saf.* **168**: 401–407.

Quade, H. W. 1969. Cladoceran faunas associated with aquatic macrophytes in some lakes in northwestern Minnesota. *Ecology* **50**: 170–179.

Quattrocchi, U. 2012. *CRC World Dictionary of Medicinal and Poisonous Plants: Common Names, Scientific Names, Eponyms, Synonyms, and Etymology.* 5 vols. CRC Press, Boca Raton, FL. 3960 pp.

Quayyum, H. A., A. U. Mallik & P. F. Lee. 1999. Allelopathic potential of aquatic plants associated with wild rice (*Zizania palustris*): I. Bioassay with plant and lake sediment samples. *J. Chem. Ecol.* **25**: 209–220.

Querry, N. D. & K. A. Harper. 2017. Structural diversity as a habitat indicator for endangered lakeshore flora using an assemblage of common plant species in Atlantic Canada. *Plant Ecol.* **218**: 1339–1353.

Querry, N. D., X. Bordeleau, K. A. Harper & S. P. Basquill. 2017. Multiscale habitat characterization of herbaceous Atlantic coastal plain flora on lakeshores in Nova Scotia. *Botany* **95**: 587–598.

Quinlan, E. L., E. J. Phlips, K. A. Donnelly, C. H. Jett, P. Sleszynski & S. Keller. 2008. Primary producers and nutrient loading in Silver Springs, FL, USA. *Aquat. Bot.* **88**: 247–255.

Racine, C. H., M. T. Jorgenson & J. C. Walters. 1998. Thermokarst vegetation in lowland birch forests on the Tanana Flats, interior Alaska, USA. Pp. 927–933 *In*: A. G. Lewkowicz & M. Allard (eds.), *Proceedings of the Seventh International Conference on Permafrost, Université Laval, Québec, Collection Nordicana, Vol. 57.* Université Laval, Quebec City, QC, Canada.

Radomski, P. & D. Perleberg. 2012. Application of a versatile aquatic macrophyte integrity index for Minnesota lakes. *Ecol. Indicators* **20**: 252–268.

Rae, J. G. 1995. Aspects of the population and reproductive ecology of the endangered fragrant prickly-apple cactus [*Cereus eriophorus* var. *fragrans* (Small) L. Benson]. *Castanea* **60**: 255–269.

Raina, S. N. 1978. Genetic mechanisms underlying evolution in *Crinum*. *Cytologia* **43**: 575–580.

Ramachandran, K. 1969. Chromosome numbers in Zingiberaceae. *Cytologia* **34**: 213–221.

Ramachandran, R. & M. Nosonovsky. 2014. Surface micro/nanotopography, wetting properties and the potential for biomimetic icephobicity of skunk cabbage *Symplocarpus foetidus*. *Soft Matter* **10**: 7797–7803.

Ramaley, F. 1919. Vegetation of undrained depressions on the Sacramento plains. *Bot. Gaz.* **68**: 380–387.

Ramanaiah, K., A. V. R. Prasad & K. H. C. Reddy. 2011. Mechanical properties and thermal conductivity of *Typha angustifolia* natural fiber–reinforced polyester composites. *Int. J. Polym. Anal. Charact.* **16**: 496–503.

Ramirez, N. & Y. Brito. 1990. Reproductive biology of a tropical palm swamp community in the Venezuelan llanos. *Am. J. Bot.* **77**: 1260–1271.

Ramirez, N. & Y. Brito. 1992. Pollination biology in a palm swamp community in the Venezuelan central plains. *Bot. J. Linn. Soc.* **110**: 277–302.

Ramirez, Jr., P. & K. Dickerson. 1997. Follow-up investigation of selenium and other trace elements in biota from the Riverton reclamation project, Fremont County, Wyoming. U.S. Fish & Wildlife Publication 213. U.S. Fish & Wildlife Service, Ecological Services, Wyoming Field Office, Cheyenne, WY. 20 pp.

Ramirez, Jr., P. & K. Dickerson. 1999. Monitoring of selenium concentrations in biota from the Kendrick reclamation project, Natrona County, Wyoming 1992–1996. Contaminant Report No.: R6/714C/99. U.S. Fish & Wildlife Service, Ecological Services, Wyoming Field Office, Cheyenne, WY. 16 pp.

Ramírez-Babativa, D. F. & A. J. E. Ramírez. 2018. Comparison of the populational characteristics of *Lemna minuta* (Araceae: Lemnoideae) in three culture media. *Revista Colombiana Biotecnol.* **20**: 84–96.

Ramírez-García, P., A. Lot, C. M. Duarte, J. Terrados & N. S. R. Agawin. 1998. Bathymetric distribution, biomass and growth dynamics of intertidal *Phyllospadix scouleri* and *Phyllospadix torreyi* in Baja California (Mexico). *Mar. Ecol. Prog. Ser.* **173**: 13–23.

Ramırez-Garcıa, P., J. Terrados, F. Ramos, A. Lot, D. Ocaña & C. M. Duarte. 2002. Distribution and nutrient limitation of surfgrass, *Phyllospadix scouleri* and *Phyllospadix torreyi*, along the Pacific coast of Baja California (Mexico). *Aquat. Bot.* **74**: 121–131.

Ramtin, M., M. R. M. K. Pahlaviani, A. Massiha, K. Issazadeh & S. Heidari. 2013. Comparative evaluation of the antibacterial activities of essential oils of *Iris pseudacorus* and *Urtica dioica* native north of Iran. *J. Pure Appl. Microbiol.* **7**: 1065–1070.

Randall, R. P. 2007. *The Introduced Flora of Australia and Its Weed Status.* CRC for Australian Weed Management, University of Adelaide, Adelaide, SA, Australia. 524 pp.

Randall, R. P. & S. G. Lloyd. 2002. Weed warning from down under – the weed potential of selected South African plants in cultivation in California. Pp. 192–195 *In*: H. Spafford-Jacob, J. Dodd & J. Moore (eds.), *Proceedings of the 13th Australian Weeds Conference, Perth, Western Australia September 8–13th 2002.* Plant Protection Society of Western Australia, Perth, WA, Australia.

Ransom, J. K. & E. A. Oelke. 1983. Cultural control of common waterplantain (*Alisma triviale*) in wild rice (*Zizania palustris*). *Weed Sci.* **31**: 562–566.

Rapoport, E. H., E. Raffaele, L. Ghermandi & L. Margutti. 1995. Edible weeds: a scarcely used resource. *Bull. Ecol. Soc. Am.* **76**: 163–166.

Rasmussen, H. N. 1995. *Terrestrial Orchids: From Seed to Mycotrophic Plant*. Cambridge University Press, Oxford, United Kingdom. 444 pp.

Rasmussen, H. N., K. W. Dixon, J. Jersáková & T. Těšitelová. 2015. Germination and seedling establishment in orchids: a complex of requirements. *Ann. Bot.* **116**: 391–402.

Rataj, K. 1975. *Revizion [sic!] of the genus Echinodorus Rich*. Studie Československá Akademie Věd. Československá akademie věd, Praha, Czech Republic. 156 pp.

Rataj, K. 2004. *A New Revision of the Swordplant Genus Echinodorus Richard, 1848 (Alismataceae)*. Aquapress, Miradolo Terme (Pavia), Italy. 142 pp.

Raulings, E., K. A. Y. Morris, R. Thompson & R. Mac Nally. 2011. Do birds of a feather disperse plants together? *Freshw. Biol.* **56**: 1390–1402.

Raven, P. H. & J. H. Thomas. 1970. *Iris pseudacorus* in western North America. *Madroño* **20**: 390–391.

Raven, P. H., D. W. Kyhos & A. J. Hill. 1965. Chromosome numbers of spermatophytes, mostly Californian. *Aliso* **6**: 105–113.

Raven, J. A., L. L. Handley, J. J. MacFarlane, S. McInroy, L. McKenzie, J. H. Richards & G. Samuelsson. 1988. The role of CO_2 uptake by roots and CAM in acquisition of inorganic C by plants of the isoetid life-form: a review, with new data on *Eriocaulon decangulare* L. *New Phytol.* **108**: 125–148.

Ray, A. M. & R. S. Inouye. 2006. Effects of water-level fluctuations on the arbuscular mycorrhizal colonization of *Typha latifolia* L. *Aquat. Bot.* **84**: 210–216.

Raynal, D. J. & J. W. Geis. 1978. Winter studies of littoral vegetation. Technical Report G. State University College of Environmental Science & Forestry, Institute of Environmental Program Affairs, Syracuse, NY. 21 pp.

Rea, A. M. 1997. *At the Desert's Green Edge: An Ethnobotany of the Gila River Pima*. University of Arizona Press, Tuscon, AZ. 430 pp.

Reaume, T. 2009. *620 Wild Plants of North America: Fully Illustrated*. Canadian Plains Research Center, Regina, Saskatchewan, CA. 784 pp.

Rebaque, D., R. Martínez-Rubio, S. Fornalé, P. García-Angulo, A. Alonso-Simón, J. M. Álvarez, D. Caparros-Ruiz, J. L. Acebes & A. Encina. 2017. Characterization of structural cell wall polysaccharides in cattail (*Typha latifolia*): evaluation as potential biofuel feedstock. *Carbohydr. Polym.* **175**: 679–688.

Redekop, P., D. Hofstra & A. Hussner. 2016. *Elodea canadensis* shows a higher dispersal capacity via fragmentation than *Egeria densa* and *Lagarosiphon major*. *Aquat. Bot.* **130**: 45–49.

Redmond, K., J. A. Reinartz & S. Critchley. 1993. Flowering phenology along the UWM Field Station boardwalk in the Cedarburg Bog. *Field Sta. Bull.* **26**(2): 1–23.

Reed, E. L. 1930. Vegetation of the playa lakes in the staked plains of western Texas. *Ecology* **11**: 597–600.

Reed, D. C., S. J. Holbrook, E. Solomon & M. Anghera. 1998. Studies on germination and root development in the surfgrass *Phyllospadix torreyi*: implications for habitat restoration. *Aquat. Bot.* **62**: 71–80.

Reed, D. C., S. J. Holbrook, C. A. Blanchette & S. Worcester. 2009. Patterns and sources of variation in flowering, seed supply and seedling recruitment in surfgrass *Phyllospadix torreyi*. *Mar. Ecol. Prog. Ser.* **384**: 97–106.

Reena, T., R. Prem, M. S. Deepthi, R. B. Ramachanran & S. Sujatha. 2013. Comparative effect of natural commodities and commercial medicines against oral thrush causing fungal organism of *Candida albicans*. *Sci. J. Clin. Med.* **2**: 75–80.

Reese, M. C. & K. S. Lubinski. 1983. A survey and annotated checklist of late summer aquatic and floodplain vascular flora, middle and lower Pool 26, Mississippi and Illinois rivers. *Castanea* **48**: 305–316.

Reeves, L. M. & T. Reeves. 1984. Life history and reproduction of *Malaxis paludosa* in Minnesota. *Am. Orchid Soc. Bull.* **53**: 1280–1291.

Reeves, T. & L. Reeves. 1985. Rediscovery of *Malaxis paludosa* (L.) Sw. (Orchidaceae) in Minnesota. *Rhodora* **87**: 133–136.

Refaat, J., M. S. Kamel, M. A. Ramadan & A. A. Ali. 2012a. *Crinum*; an endless source of bioactive principles: a review, Part II. *Crinum* alkaloids: crinine-type alkaloids. *Int. J. Pharmaceut. Sci. Res.* **3**: 3091–3100.

Refaat, J., M. S. Kamel, M. A. Ramadan & A. A. Ali. 2012b. *Crinum*; an endless source of bioactive principles: a review. Part III; crinum alkaloids: Belladine-, galanthamine-, lycorenine-, tazettine-type alkaloids and other minor types. *Int. J. Pharmaceut. Sci. Res.* **3**: 3630–3638.

Reid, T. C. 1978. Vegetation and environment patterns of Liard River Hot Springs Provincial Park, British Columbia. M.S. thesis. Simon Fraser University, Burnaby, BC. 206 pp.

Reid, C. S. & P. L. Faulkner. 2010. Noteworthy collections: Louisiana. *Castanea* **75**: 138–140.

Reid, C. S. & L. Urbatsch. 2012. Noteworthy plant records from Louisiana. *J. Bot. Res. Inst. Texas* **6**: 273–278.

Reid, C. S., P. L. Faulkner, M. H. MacRoberts, B. R. MacRoberts & M. Bordelon. 2010. Vascular flora and edaphic characteristics of saline prairies in Louisiana. *J. Bot. Res. Inst. Texas* **4**: 357–379.

Reimer, A. 2001. The role of bog plants in the exchange of carbon dioxide and water between the atmosphere and the Mer Bleue peatland. M.S. thesis. McGill University, Montreal, CA. 83 pp.

Reiskind, J. B., T. V. Madsen, L. C. Van Ginkel & G. Bowes. 1997. Evidence that inducible C_4-type photosynthesis is a chloroplastic CO_2-concentrating mechanism in *Hydrilla*, a submersed monocot. *Plant Cell Environ.* **20**: 211–220.

Rejmánek, M. & J. M. Randall. 1994. Invasive alien plants in California: 1993 summary and comparison with other areas in North America. *Madroño* **41**: 161–177.

Rejmánková, E. 1976. Germination of seeds of *Lemna gibba*. *Folia Geobot. Phytotax.* **11**: 261–267.

Remizowa, M. V., D. D. Sokoloff, L. M. Campbell, D. W. Stevenson & P. J. Rudall. 2011. *Harperocallis* is congeneric with *Isidrogalvia* (Tofieldiaceae, Alismatales): evidence from comparative floral morphology. *Taxon* **60**: 1076–1094.

Ren, D. & S. Zhang. 2008. Separation and identification of the yellow carotenoids in *Potamogeton crispus* L. *Food Chem.* **106**: 410–414.

Renn, C. E. 1935. A mycetozoan parasite of *Zostera marina*. *Nature* **135**: 544–545.

Renzi, M., A. Giovani & S. E. Focardi. 2013. Biofuel production from the Orbetello Lagoon macrophytes: efficiency of lipid extraction using accelerate solvent extraction technique. *J. Environ. Protect.* **4**: 1224–1229.

Reusch, T. B. H. 2001. Fitness-consequences of geitonogamous selfing in a clonal marine angiosperm (*Zostera marina*). *J. Evol. Biol.* **14**: 129–138.

Reveal, J. L. 1990. The neotypification of *Lemna minuta* Humb., Bonpl. & Kunth, an earlier name for *Lemna minuscula* Herter (Lemnaceae). *Taxon* **39**: 328–330.

Reveal, J. L. 2012. An outline of a classification scheme for extant flowering plants. *Phytoneuron* **2012–37**: 1–221.

Reverchon, J. 1886. Botanizing in Texas. II. *Bot. Gaz.* **11**: 211–216.

Reynolds, C., N. A. F. Miranda & G. S. Cumming. 2015. The role of waterbirds in the dispersal of aquatic alien and invasive species. *Divers. Distrib.* **21**: 744–754.

Reznicek, A. A. 1994. The disjunct coastal plain flora in the Great Lakes region. *Biol. Conserv.* **68**: 203–215.

Reznicek, A. A. & R. S. W. Bobbette. 1976. The taxonomy of *Potamogeton* subsection *hybridi* in North America. *Rhodora* **78**: 650–673.

Reznicek, A. A. & D. F. Murray. 2013. A re-evaluation of *Carex specuicola* and the *Carex parryana* complex (Cyperaceae). *J. Bot. Res. Inst. Texas* **7**: 37–51.

Rhoades, R. W. 1962. The aquatic form of *Alisma subcordatum* Raf. *Rhodora* **64**: 227–229.

Rhoads, A. F. 1984. Rare Pennsylvania plants. *Bartonia* **50**: 61–63.

Rhodes, M. J. 1978. Habitat preferences of breeding waterfowl on the Texas high plains. M.S. thesis. Texas Tech University, Lubbock, TX. 48 pp.

Ribeiro, J. P. N., R. S. Matsumoto, L. K. Takao, A. C. Peret & M. I. S. Lima. 2011. Spatial distribution of *Crinum americanum* L. in tropical blind estuary: hydrologic, edaphic and biotic drivers. *Environ. Exp. Bot.* **71**: 287–291.

Rich, P. H., R. G. Wetzel & N. Thuy. 1971. Distribution, production and role of aquatic macrophytes in a southern Michigan marl lake. *Freshw. Biol.* **1**: 3–21.

Richards, J. H. & C. T. Ivey. 2004. Morphological plasticity of *Sagittaria lancifolia* in response to phosphorus. *Aquat. Bot.* **80**: 53–67.

Richardson, F. D. 1980. Ecology of *Ruppia maritima* L. in New Hampshire (U.S.A.) tidal marshes. *Rhodora* **82**: 403–439.

Richardson, C. J. 1983. Pocosins: vanishing wastelands or valuable wetlands? *Bioscience* **33**: 626–633.

Richardson, J. S. & H. F. Clifford. 1983. Life history and microdistribution of *Neureclipsis bimaculata* (Trichoptera: Polycentropodidae) in a lake outflow stream of Alberta, Canada. *Can. J. Zool.* **61**: 2434–2445.

Richardson, A. & K. King. 2011. *Plants of Deep South Texas: A Field Guide to the Woody and Flowering Species.* Texas A&M University Press, College Station, TX. 448 pp.

Richardson, S. M., J. M. Hanson & A. Locke. 2002. Effects of impoundment and water-level fluctuations on macrophyte and macroinvertebrate communities of a dammed tidal river. *Aquat. Ecol.* **36**: 493–510.

Riddin, T. & J. B. Adams. 2009. The seed banks of two temporarily open/closed estuaries in South Africa. *Aquat. Bot.* **90**: 328–332.

Ridley, H. N. 1930. *The Dispersal of Plants Throughout the World.* L. Reeve & Co. Ltd., Ashford, Kent, United Kingdom. 744 pp.

Riemer, D. N. 1975. Seed dormancy and longevity in *Potamogeton pulcher* Tuckerm. *Proc. Northeast. Weed Control Conf.* **29**: 114–117.

Riemer, D. N. & S. J. Toth. 1969. A survey of the chemical composition of *Potamogeton* and *Myriophyllum* in New Jersey. *Weed Sci.* **17**: 219–223.

Riggs, G. B. 1925. Some *Sphagnum* bogs of the north Pacific coast of America. *Ecology* **6**: 260–278.

Riggs, G. B. 1940. Comparisons of the development of some sphagnum bogs of the Atlantic coast, the interior, and the Pacific coast. *Am. J. Bot.* **27**: 1–14.

Riis, T., T. V. Madsen & R. S. H. Sennels. 2009. Regeneration, colonisation and growth rates of allofragments in four common stream plants. *Aquat. Bot.* **90**: 209–212.

Riis, T., C. Lambertini, B. Olesen, J. S. Clayton, H. Brix & B. K. Sorrell. 2010. Invasion strategies in clonal aquatic plants: are phenotypic differences caused by phenotypic plasticity or local adaptation? *Ann. Bot.* **106**: 813–822.

Riis, T., B. Olesen, J. S. Clayton, C. Lambertini, H. Brix & B. K. Sorrell. 2012. Growth and morphology in relation to temperature and light availability during the establishment of three invasive aquatic plant species. *Aquat. Bot.* **102**: 56–64.

Ring, R. M., E. A. Spencer & K. S. Walz. 2013. *Vulnerability of 70 Plant Species of Greatest Conservation Need to Climate Change in New Jersey.* New York Natural Heritage Program, Albany, NY and New Jersey Natural Heritage Program, Department of Environmental Protection, Office of Natural Lands Management, Trenton, NJ. 38 pp.

Ritchie, J. C. 1957. The vegetation of northern Manitoba: II. A prisere on the Hudson Bay lowlands. *Ecology* **38**: 429–435.

Rixon, C. A. M., I. C. Duggan, N. M. N. Bergeron, A. Ricciardi & H. J. Macisaac. 2005. Invasion risks posed by the aquarium trade and live fish markets on the Laurentian Great Lakes. *Biodivers. Conserv.* **14**: 1365–1381.

Rizwana, M., M. Darshan & D. Nilesh. 2014. Phytoremediation of textile waste water using potential wetland plant [sic!]: eco sustainable approach. *Int. J. Interdiscipl. Multidiscipl. Stud.* **1**: 130–138.

Rober, A. R., K. H. Wyatt, R. J. Stevenson & M. R. Turetsky. 2014. Spatial and temporal variability of algal community dynamics and productivity in floodplain wetlands along the Tanana River, Alaska. *Freshw. Sci.* **33**: 765–777.

Roberts, T. H. & D. H. Arner. 1984. Food habits of beaver in east-central Mississippi. *J. Wildl. Manag.* **48**: 1414–1419.

Roberts, M. L. & R. R. Haynes. 1986. Flavonoid systematics of *Potamogeton* subsections *Perfoliati* and *Praelongi* (Potamogetonaceae). *Nordic J. Bot.* **6**: 291–294.

Roberts, B. A. & A. Robertson. 1986. Salt marshes of Atlantic Canada: their ecology and distribution. *Can. J. Bot.* **64**: 455–467.

Roberts, D. A., R. Singer & C. W. Boylen. 1985. The submersed macrophyte communities of Adirondack lakes (New York, U.S.A.) of varying degrees of acidity. *Aquat. Bot.* **21**: 219–235.

Roberts, R. E., M. Y. Hedgepeth & T. R. Alexander. 2008. Vegetational responses to saltwater intrusion along the Northwest Fork of the Loxahatchee River within Jonathan Dickinson State Park. *Florida Sci.* **71**: 383–397.

Robertson, C. 1896. Flowers and insects. XVI. *Bot. Gaz.* **21**: 266–274.

Robertson, C. 1924. Flowers and insects. XXIII. *Bot. Gaz.* **78**: 68–84.

Robertson, K. R. 1976. The genera of Haemodoraceae in the southeastern United States. *J. Arnold Arb.* **57**: 205–216.

Robertson, K. R. 2002. Haemodoraceae R. Brown. Pp. 47–48 *In*: Flora North America Editorial Committee (eds.), *Flora of North America North of Mexico, Vol. 26: Magnoliophyta: Liliidae: Liliales and Orchidales.* Oxford University Press, New York, NY.

Robertson, A. I. & K. H. Mann. 1982. Population dynamics and life history adaptations of *Littorina neglecta* Bean in an eelgrass meadow (*Zostera marina* L.) in Nova Scotia. *J. Exp. Mar. Biol. Ecol.* **63**: 151–171.

Robertson, K., G. A. Levin & L. R. Phillippe. 1995. Vascular plants and natural areas of Site M, Cass County, Illinois. Technical Report 1995 (17). Illinois Natural History Survey, Center for Biodiversity, Champaign, IL. 74 pp.

Robeson, M. S., K. Khanipov, G. Golovko, S. M. Wisely, M. D. White, M. Bodenchuck, T. J. Smyser, Y. Fofanov, N. Fierer & A. J. Piaggio. 2018. Assessing the utility of metabarcoding for diet analyses of the omnivorous wild pig (*Sus scrofa*). *Ecol. Evol.* **8**: 185–196.

Robles, W., J. D. Madsen & V. L. Maddox. 2008. *Reservoir Survey for Invasive and Native Aquatic Plants Species Within the Pat Harrison Waterway District.* GeoResources Institute, Mississippi State University, Mississippi State, MS. 28 pp.

Robinson, F. D. 1956. An ecological survey of the vascular aquatic vegetation of the Cumberland Plateau in Tennessee. M.S. thesis. University of Tennessee, Knoxville, TN. 44 pp.

Robinson, J. V. & B. L. Frye. 1986. Survivorship, mating and activity pattern of adult *Telebasis salva* (Hagen) (Zygoptera: Coenagrionidae). *Odonatologica* **15**: 211–217.

Robinson, A. T., J. E. Fulmer, L. D. Avenetti. 2007. Aquatic plant surveys and evaluation of aquatic plant harvesting in Arizona reservoirs. Federal Aid in Sport Fish Restoration Project F-14-R. Arizona Game and Fish Department, Phoenix, AZ. 39 pp.

Robinson, D. J., E. Gandy, C. VanHoek & R. W. Pemberton. 2011. Naturalization of the nun's hood orchid (*Phaius tankervilleae*: Orchidaceae) in central Florida. *J. Bot. Res. Inst. Texas* **5**: 337–339.

Robson, D. B., J. H. Wiersema, C. B. Hellquist & T. Borsch. 2016. Distribution and ecology of a new species of water-lily, *Nymphaea loriana* (Nymphaeaceae), in western Canada. *Can. Field-Nat.* **130**: 25–31.

Rocchio, J., J. Sovell & P. Lyon. 2001. *Survey of Seeps and Springs Within the Bureau of Land Management's Grand Junction Field Office management Area (Garfield County, CO)*. Colorado Natural Heritage Program, Colorado State University, Fort Collins, CO. 19 pp.

Rocchio, J., M. March & D. G. Anderson. 2006. *Epipactis gigantea Dougl. ex Hook. (Stream Orchid): A Technical Conservation Assessment*. USDA Forest Service, Rocky Mountain Region. Available online: http://www.fs.fed.us/r2/projects/scp/assessments/epipactisgigantea.pdf [accessed 17 February, 2019].

Rochefort, L. & D. H. Vitt. 1988. Effects of simulated acid rain on *Tomenthypnum nitens* and *Scorpidium scorpioides* in a rich fen. *Bryologist* **91**: 121–129.

Rodger, E. A. 1933. Wound healing in submerged plants. *Am. Midl. Nat.* **14**: 704–713.

Rodgers, H. L., F. P. Day & R. B. Atkinson. 2003. Fine root dynamics in two Atlantic white-cedar wetlands with contrasting hydroperiods. *Wetlands* **23**: 941–949.

Rogers, G. K. 1983. The genera of Alismataceae in the southeastern United States. *J. Arnold Arbor.* **64**: 383–420.

Rogers, G. K. 1984. The Zingiberales (Cannaceae, Marantaceae, and Zingiberaceae) in the southeastern United States. *J. Arnold Arbor.* **65**: 5–55.

Rogers, J. G. 1993. Availability of a draft recovery plan for the Little Aguja pondweed for review and comment. *Fed. Reg.* **58**: 67808.

Rogers, J. P. & L. J. Korschgen. 1966. Foods of lesser scaups on breeding, migration, and wintering areas. *J. Wildl. Manag.* **30**: 258–264.

Rolim, R. G., P. M. A. de Ferreira, A. A. Schneider & G. E. Overbeck. 2015. How much do we know about distribution and ecology of naturalized and invasive alien plant species? A case study from subtropical southern Brazil. *Biol. Invasions* **17**: 1497–1518.

Romanello, G. A., K. L. Chuchra-Zbytniuk, J. L. Vandermer & B. W. Touchette. 2008. Morphological adjustments promote drought avoidance in the wetland plant *Acorus americanus. Aquat. Bot.* **89**: 390–396.

Romero-González, G. A., G. C. Fernández-Concha, R. L. Dressler, L. K. Magrath & G. W. Argus. 2002. Orchidaceae Jussieu. Orchid family. Pp. 490–499 *In*: Flora North America Editorial Committee (eds.), *Flora of North America North of Mexico, Vol. 26: Magnoliophyta: Liliidae: Liliales and Orchidales.* Oxford University Press, New York, NY.

Ronse, A. 2011. Botanic garden escapes' [sic!] from the living collections at the Botanic Garden. *Scripta Bot. Belg.* **47**: 89–111.

Rooney, R. C., C. Carli & S. E. Bayley. 2013. River connectivity affects submerged and floating aquatic vegetation in floodplain wetlands. *Wetlands* **33**: 1165–1177.

Rosatti, T. J. 1987. The genera of Pontederiaceae in the southeastern United States. *J. Arnold Arb.* **68**: 35–71.

Rose, C. D., W. C. Sharp, W. J. Kenworthy, J. H. Hunt, W. G. Lyons, E. J. Prager, J. F. Valentine, M. O. Hall, P. E. Whitfield & J. W. Fourqurean. 1999. Overgrazing of a large seagrass bed by the sea urchin *Lytechinus variegatus* in Outer Florida Bay. *Mar. Ecol. Prog. Ser.* **190**: 211–222.

Rosen, D. J. 2002. *Cryptocoryne beckettii* (Araceae), a new aquatic plant in Texas. *Sida* **19**: 399–401.

Rosen, D. J. 2007. The vascular flora of Nash Prairie: a coastal prairie remnant in Brazoria County, Texas. *J. Bot. Res. Inst. Texas* **1**: 679–692.

Rosendahl, C. O. & F. K. Butters. 1935. The genus *Najas* in Minnesota. *Rhodora* **37**: 345–348.

Rosenthal, S. K., S. S. Stevens & D. M. Lodge. 2006. Whole-lake effects of invasive crayfish (*Orconectes* spp.) and the potential for restoration. *Can. J. Fish. Aquat. Sci.* **63**: 1276–1285.

Rosenthal, M. A., S. R. Rosenthal, G. Johnson, W. C. Taylor & E. A. Zimmer. 2014. *Isoetes viridimontana*: a previously unrecognized quillwort from Vermont, USA. *Am. Fern J.* **104**: 7–16.

Rosine, W. N. 1955. The distribution of invertebrates on submerged aquatic plant surfaces in Muskee Lake, Colorado. *Ecology* **36**: 308–314.

Ross, C., M. P. Puglisi & V. J. Paul. 2008. Antifungal defenses of seagrasses from the Indian River Lagoon, Florida. *Aquat. Bot.* **88**: 134–141.

Ross, T. G., C. F. Barrett, M. S. Gomez, V. K.-Y. Lam, C. L. Henriquez, D. H. Les, J. I. Davis, A. Cuenca, G. Petersen, O. Seberg, M. Thadeo, T. J. Givnish, J. Conran, D. W. Stevenson & S. W. Graham. 2016. Plastid phylogenomics and molecular evolution of Alismatales. *Cladistics* **32**: 160–178.

Rossell, I. M. & D. A. Losure. 2005. The habitat and plant associates of *Eriocaulon decangulare* L. in three southern Appalachian wetlands. *Castanea* **70**: 129–136.

Rossi, C. C., A. P. Aguilar, M. A. N. Diaz & A. D. O. B. Ribon. 2011. Aquatic plants as potential sources of antimicrobial compounds active against bovine mastitis pathogens. *Afr. J. Biotechnol.* **10**: 8023–8030.

Rosso, W. A. 1977. Acid lake renovation. Pp. 61–70 *In: Seventh Symposium on Coal Mine Drainage Research*. National Coal Association, Washington, DC.

Rothe, S. P. 2011. Exotic medicinal plants from West Vidarbha region - V. *Int. Multidiscip. Res. J.* **2011**: 14–16.

Rothlisberger, J. D., W. L. Chadderton, J. McNulty & D. M. Lodge. 2010. Aquatic invasive species transport via trailered boats: what is being moved, who is moving it, and what can be done. *Fisheries* **35**: 121–132.

Rouger, R. & A. S. Jump. 2013. Isolation and characterization of 20 microsatellite loci for the saltmarsh plant *Triglochin maritima* L. *Conserv. Genet. Resour.* **5**: 1157–1158.

Rouger, R. & A. S. Jump. 2014. A seascape genetic analysis reveals strong biogeographical structuring driven by contrasting processes in the polyploid saltmarsh species *Puccinellia maritima* and *Triglochin maritima. Mol. Ecol.* **23**: 3158–3170.

Rout, G. K., H.-S. Shin, S. Gouda, S. Sahoo, G. Das, L. F. Fraceto & J. K. Patra. 2018. Current advances in nanocarriers for biomedical research and their applications. *Artif. Cells Nanomed. Biotechnol.* doi: 10.1080/21691401.2018.14788431-10.

Rowell, Jr., C. M. 1983. *Status Report, Potamogeton clystocarpus Fern.* U.S. Fish and Wildlife Service. Albuquerque, NM. 9 pp.

Roze, E. 1887. Le mode de fecondation du *Zannichellia palustris* L. *J. Bot. (Paris)* **1**: 296–299.

Rozema, J., W. Arp, J. Van Diggelen, M. Van Esbroek, R. Broekman & H. Punte. 1986. Occurrence and ecological significance of vesicular arbuscular mycorrhiza in the salt marsh environment. *Acta Bot. Neerl.* **35**: 457–467.

Rozentsvet, O. A., N. M. Dembitsky & V. S. Zhuicova. 1995a. Lipids from macrophytes of the middle Volga. *Phytochemistry* **38**: 1209–1213.

Rozentsvet, O. A., E. R. Ponomareva, Y. N. Mazepova & N. V. Koneva. 1995b. Lipids of some aquatic plants of the central Volga region. *Chem. Nat. Compd.* **31**: 169–171.

Rozentsvet, O. A., S. V. Saksonov & V. M. Dembitsky. 2002. Hydrocarbons, fatty acids, and lipids of freshwater grasses of the Potamogetonaceae family. *Biochemistry (Moscow)* **67**: 351–356.

Rubtsov, N. I. 1975. *Sagittaria platyphylla*-novyi adventivnyi vid flory Evropeiskoi chasti SSSR. *Bot. Zhurn.* **60**: 387–388.

Ruch, D. G., A. Schoultz & K. S. Badger. 1998. The flora and vegetation of Ginn Woods, Ball State University, Delaware County, Indiana. *Proc. Indiana Acad. Sci.* **107**: 17–60.

Ruch, D. G., B. G. Torke, B. R. Hess, K. S. Badger & P. E. Rothrock. 2008. The vascular flora and vegetational communities of the wetland complex on the IMI property in Henry County, near Luray, Indiana. *Proc. Indiana Acad. Sci.* **117**: 142–158.

Ruch, D. G., B. G. Torke, B. R. Hess & K. S. Badger. 2009. The vascular flora and plant communities of the Bennett wetland complex in Henry County, Indiana. *Proc. Indiana Acad. Sci.* **118**: 39–54.

Ruchty, A. 2011. *Conservation Strategy for Sisyrinchium sarmentosum Suks. ex. Greene*. USDA Forest Service Region 6, Oregon and Washington, Pacific Northwest Region, Portland, OR. 68 pp.

Ruckelshaus, M. H. 1995. Estimates of outcrossing rates and of inbreeding depression in a population of the marine angiosperm *Zostera marina*. *Mar. Biol.* **123**: 583–593.

Ruckelshaus, M. H. 1996. Estimation of genetic neighborhood parameters from pollen and seed dispersal in the marine angiosperm *Zostera marina* L. *Evolution* **50**: 856–864.

Rudall, P. J., C. A. Furness, M. W. Chase & M. F. Fay. 1997. Microsporogenesis and pollen sulcus type in Asparagales (Lilianae). *Can. J. Bot.* **75**: 408–430.

Rudgley, R. 1999. *The Encyclopaedia of Psychoactive Substances*. St. Martin's Press, New York, NY. 302 pp.

Rüegg, S., U. Raeder, A. Melzer, G. Heubl & C. Bräuchler. 2017. Hybridisation and cryptic invasion in *Najas marina* L. (Hydrocharitaceae)? *Hydrobiologia* **784**: 381–395.

Ruekaewma, N., S. Piyatiratitivorakul & S. Powtongsook. 2015. Culture system for *Wolffia globosa* L. (Lemnaceae) for hygiene human food. *Songklanakarin J. Sci. Technol.* **37**: 575–580.

Ruenglertpanyakul, W., S. Attasat & P. Wanichpongpan. 2004. Nutrient removal from shrimp farm effluent by aquatic plants. *Water Sci. Technol.* **50**: 321–330.

Ruesink, J. L., J.-S. Hong, L. Wisehart, S. D. Hacker, B. R. Dumbauld, M. Hessing-Lewis & A. C. Trimble. 2010. Congener comparison of native (*Zostera marina*) and introduced (*Z. japonica*) eelgrass at multiple scales within a Pacific Northwest estuary. *Biol. Invasions* **12**: 1773–1789.

Ruijgrok, H. W. L. 1974. Cyanogenese bei *Scheuchzeria palustris*. *Phytochemistry* **13**: 161–162.

Rundel, P. W. 2011. The diversity and biogeography of the alpine flora of the Sierra Nevada, California. *Madroño* **58**: 153–185.

Runkel, S. T. & D. M. Roosa. 1999. *Wildflowers and Other Plants of Iowa Wetlands*. Iowa State University Press, Ames, IA. 372 pp.

Rusoff, L. L., E. W. Blakeney, Jr. & D. D. Culley, Jr. 1980. Duckweeds (Lemnaceae family): a potential source of protein and amino acids. *J. Agric. Food Chem.* **28**: 848–850.

Ruth, A. D., S. Jose & D. L. Miller. 2008. Seed bank dynamics of sand pine scrub and longleaf pine flatwoods of the Gulf Coastal Plain (Florida). *Ecol. Restor.* **26**: 19–21.

Ryan, J. B., D. N. Riemer & S. J. Toth. 1972. Effects of fertilization on aquatic plants, water, and bottom sediments. *Weed Sci.* **20**: 482–486.

Rybicki, N. B. & V. Carter. 1986. Effect of sediment depth and sediment type on the survival of *Vallisneria americana* Michx grown from tubers. *Aquat. Bot.* **24**: 233–240.

Rybicki, N. B., D. G. McFarland, H. A. Ruhl, J. T. Reel & J. W. Barko. 2001. Investigations of the availability and survival of submersed aquatic vegetation propagules in the tidal Potomac River. *Estuaries Coast.* **24**: 407–424.

Rydberg, P. A. 1914. Phytogeographical notes on the Rocky Mountain region. III. Formations in the alpine zone. *Bull. Torrey Bot. Club* **41**: 459–474.

Rydberg, P. A. 1926. Two new species from the mountains of West Virginia. *Torreya* **26**: 29–33.

Ryman, S. & A. Anderberg. 1999. Five species of introduced duckweeds. *Svensk Bot. Tidskr.* **93**: 129–138.

Ryser, P. & A. T. Kamminga. 2009. Root survival of six cool-temperate wetland graminoids in autumn and early winter. *Plant Ecol. Divers.* **2**: 27–35.

Ryser, P., H. K. Gill & C. J. Byrne. 2011. Constraints of root response to waterlogging in *Alisma triviale*. *Plant Soil* **343**: 247–260.

Ryuk, J. A., Y. S. Kim, H. W. Lee & B. S. Ko. 2014. Identification of *Acorus gramineus*, *A. calamus*, and *A. tatarinowii* using sequence characterized amplified regions (SCAR) primers for monitoring of *Acori graminei rhizoma* in Korean markets. *Int. J. Clin. Exp. Med.* **7**: 2488–2496.

Saarela, J. M., P. J. Prentis, H. S. Rai & S. W. Graham. 2008. Phylogenetic relationships in the monocot order Commelinales, with a focus on Philydraceae. *Botany* **86**: 719–731.

Saarela, J. M., P. C. Sokoloff, L. J. Gillespie, L. L. Consaul & R. D. Bull. 2013. DNA barcoding the Canadian Arctic flora: core plastid barcodes (*rbcL+ matK*) for 490 vascular plant species. *PLoS One* **8**(10): e77982.

Saarela, J. M., P. C. Sokoloff & R. D. Bull. 2017. Vascular plant biodiversity of the lower Coppermine River valley and vicinity (Nunavut, Canada): an annotated checklist of an Arctic flora. *PeerJ* **5**: e2835.

Sabbatini, M. R. & K. J. Murphy. 1996. Submerged plant survival strategies in relation to management and environmental pressures in drainage channel habitats. *Hydrobiologia* **340**: 191–195.

Sable, N., S. Gaikwad, S. Bonde, A. Gade & M. Rai. 2012. Phytofabrication of silver nanoparticles by using aquatic plant *Hydrilla verticillata*. *Nusantara Biosci.* **4**: 45–49.

Saeger, A. 1933. Manganese and the growth of Lemnaceae. *Am. J. Bot.* **20**: 234–245.

Saeger, A. 1934. *Spirodela oligorrhiza* collected in Missouri. *Bull. Torrey Bot. Club* **61**: 233–236.

Saghir, A. R. B., L. K. Mann, M. Ownbey & R. Y. Berg. 1966. Composition of volatiles in relation to taxonomy of American Alliums. *Am. J. Bot.* **53**: 477–484.

Sah, J. P., M. S. Ross & S. Stofella. 2010. Developing a data-driven classification of South Florida plant communities. SERC Research Reports 93. Available online: http://digitalcommons. fiu.edu/sercrp/93. [accessed 4 February, 2020].

Saini, N. K., M. Singhal, B. Srivastava & O. P. Sharma. 2011. Antitubercular plants: a review. *Inventi* **2011**: 1–6.

Šajna, N., M. Haler, S. Škornik & M. Kaligarič. 2007. Survival and expansion of *Pistia stratiotes* L. in a thermal stream in Slovenia. *Aquat. Bot.* **87**: 75–79.

Sakhanokho, H. F. & K. Rajasekaran. 2010. Pollen biology of ornamental ginger (*Hedychium* spp. J. Koenig). *Sci. Hort.* **125**: 129–135.

Salazar, G. A., L. I. Cabrera, S. Madrinán & M. W. Chase. 2009. Phylogenetic relationships of Cranichidinae and Prescottiinae (Orchidaceae, Cranichideae) inferred from plastid and nuclear DNA sequences. *Ann. Bot.* **104**: 403–416.

Sale, P. J. M. & R. G. Wetzel. 1983. Growth and metabolism of *Typha* species in relation to cutting treatments. *Aquat. Bot.* **15**: 321–334.

Salisbury, E. 1976. Seed output and the efficacy of dispersal by wind. *Proc. Royal Soc. London, Ser. B Biol. Sci.* **192**: 323–329.

Sampson, A. W. 1914. Natural revegetation of range lands based upon growth requirements and life history of the vegetation. *J. Agric. Res.* **3**: 93–148.

Sánchez, J. A., L. Montejo, A. Gamboa, D. Albert-Puentes & F. Hernández. 2015. Germinación y dormancia de arbustos y trepadoras del bosque siempreverde de la Sierra del Rosario, Cuba. *Pastos y Forrajes* **38**: 11–28.

Sánchez-Galván, G., F. J. Mercado & E. J. Olguín. 2013. Leaves and roots of *Pistia stratiotes* as sorbent materials for the removal of crude oil from saline solutions. *Water Air Soil Pollut.* **224**: 1421.

Sandberg, D. C., L. J. Battista & A. E. Arnold. 2014. Fungal endophytes of aquatic macrophytes: diverse host-generalists characterized by tissue preferences and geographic structure. *Microbial Ecol.* **67**: 735–747.

Sand-Jensen, K. & D. M. Gordon. 1986. Variable HCO_3^- affinity of *Elodea canadensis* Michaux in response to different HCO_3^- and CO_2 concentrations during growth. *Oecologia* **70**: 426–432.

Santamaría, L. & A. I. Llano García. 2004. Latitudinal variation in tuber production in an aquatic pseudo-annual plant, *Potamogeton pectinatus. Aquat. Bot.* **79**: 51–64.

Santamaría, L., I. Charalambidou, J. Figuerola & A. J. Green. 2002. Effect of passage through duck gut on germination of fennel pondweed seeds. *Arch. Hydrobiol.* **156**: 11–22.

Santos, I. A. D. 2002. Flower-visiting bees and the breakdown of the tristylous breeding system of *Eichhornia azurea* (Swartz) Kunth (Pontederiaceae). *Biol. J. Linn. Soc.* **77**: 499–507.

Santos, A. G. 2004. Constructive applications of composite gypsum reinforced with *Typha latifolia* fibres. *Material. Construcc.* **54**: 73–77.

Santos, M. J., L. W. Anderson & S. L. Ustin. 2011. Effects of invasive species on plant communities: an example using submersed aquatic plants at the regional scale. *Biol. Invasions* **13**: 443–457.

Sarkar, S., D. Deka & N. Devi. 2008. Studies on some medicinally important wetland angiosperms used by the Bodo tribe of Kamrup District in Assam, India. *Pleione* **2**: 20–26.

Sarkissian, T. S., S. C. Barrett & L. D. Harder. 2001. Gender variation in *Sagittaria latifolia* (Alismataceae): is size all that matters? *Ecology* **82**: 360–373.

Sarma, K., R. Tandon, K. R. Shivanna & H. Y. M. Ram. 2007. Snail-pollination in *Volvulopsis nummularium. Curr. Sci.* **93**: 826–831.

Sarneel, J. M. 2013. The dispersal capacity of vegetative propagules of riparian fen species. *Hydrobiologia* **710**: 219–225.

Sarneel, J. M. & M. B. Soons. 2012. Post-dispersal probability of germination and establishment on the shorelines of slow-flowing or stagnant water bodies. *J. Veg. Sci.* **23**: 517–525.

Sass, L. L., M. A. Bozek, J. A. Hauxwell, K. Wagner & S. Knight. 2010. Response of aquatic macrophytes to human land use perturbations in the watersheds of Wisconsin lakes, USA. *Aquat. Bot.* **93**: 1–8.

Sass, C., W. J. D. Iles, C. F. Barrett, S. Y. Smith & C. D. Specht. 2016. Revisiting the Zingiberales: using multiplexed exon capture to resolve ancient and recent phylogenetic splits in a charismatic plant lineage. *PeerJ* **4**: e1584.

Sasser, C. E. & J. G. Gosselink. 1984. Vegetation and primary production in a floating freshwater marsh in Louisiana. *Aquat. Bot.* **20**: 245–255.

Sasser, C. E., J. G. Gosselink, E. M. Swenson & D. E. Evers. 1995a. Hydrologic, vegetation, and substrate characteristics of floating marshes in sediment-rich wetlands of the Mississippi river delta plain, Louisiana, USA. *Wetl. Ecol. Manag.* **3**: 171–187.

Sasser, C. E., J. M. Visser, D. E. Evers & J. G. Gosselink. 1995b. The role of environmental variables on interannual variation in species composition and biomass in a subtropical minerotrophic floating marsh. *Can. J. Bot.* **73**: 413–424.

Sasser, C. E., G. O. Holm, E. Evers-Hebert & G. P. Shaffer. 2018. The nutria in Louisiana: a current and historical perspective. Pp. 39–60 *In*: J. W. Day & J. A. Erdman (eds.), *Mississippi Delta Restoration*. Springer, Cham, Switzerland.

Sastroutomo, S. S. 1980. Environmental control of turion formation in curly pondweed (*Potamogeton crispus*). *Physiol. Plant.* **49**: 261–264.

Sastroutomo, S. S. 1981a. Germination of turions in *Potamogeton berchtoldii. Bot. Gaz.* **142**: 454–460.

Sastroutomo, S. S. 1981b. Turion formation, dormancy, and germination of curlyleaf pondweed, *Potamogeton crispus* L. *Aquat. Bot.* **10**: 161–173.

Sastroutomo, S. S., I. Ikusima, M. Numata & S. Iizumi. 1979. Importance of turions in the propagation of pondweed (*Potamogeton crispus* L.). *Ecol. Rev.* **19**: 75–88.

Satake, K. & S. Shimura. 1983. Carbon dioxide assimilation from air and water by duckweed *Spirodela polyrrhiza* (L.) Schleid. *Hydrobiologia* **107**: 51–55.

Sauleda, R. P. 2014. Vertical stratification of epiphytic orchids in a south Florida swamp forest. *New World Orchidaceae – Nomencl. Notes* **11**: 1–8.

Saunders, K. 2005. First record of *Nymphoides indica* (Menyanthaceae) in Texas. *Sida* **21**: 2441–2443.

Sautour, M., A.-C. Mitaine-Offer & M.-A. Lacaille-Dubois. 2007. The *Dioscorea* genus: a review of bioactive steroid saponins. *J. Nat. Med.* **61**: 91–101.

Sawyer, M. L. 1925. Crossing *Iris pseudacorus* and *I. versicolor. Bot. Gaz.* **79**: 60–72.

Sawyer, N. W., D. S. Mertins & L. A. Schuster. 2005. Pollination biology of *Eriocaulon parkeri* in Connecticut. *Aquat. Bot.* **82**: 113–120.

SCEPPC. 2011. *South Carolina Exotic Pest Plant Council. Invasive Plant Pest Species of South Carolina*. Forestry Leaflet 28, Clemson Cooperative Extension Service, Clemson, SC. 4 pp.

Schaffner, J. H. 1924. Expression of the sexual state in *Sagittaria latifolia. Bull. Torrey Bot. Club* **51**: 103–112.

Schaffner, U., D. Kleijn, V. Brown & H. Muller-Scharer. 2001. *Veratrum album* in montane grasslands: a model system for implementing biological control in land management practices of high biodiversity habitats. *Biocontrol News Inform.* **22**: 19N–27N.

Scharfetter, E., C. Lesemann & R. Kandeler. 1987. Ethylene as a flower-promoting agent in *Lemna. Phyton* **27**: 31–37.

Schenck, S. M. & E. W. Gifford. 1952. Karok ethnobotany. *Anthropol. Rec.* **13**: 377–392.

Schep, L. J., D. M. Schmierer & J. S. Fountain. 2006. *Veratrum* poisoning. *Toxicol. Rev.* **25**: 73–78.

Schilling, E. E. & K. C. Grubbs. 2016. Systematics of the *Eupatorium mohrii* complex (Asteraceae). *Syst. Bot.* **41**: 787–795.

Schincariol, R. A., M. A. Maun, J. N. Steinbachs, J. A. Wiklund & A. C. Crowe. 2004. Response of an aquatic ecosystem to human activity: hydro-ecology of a river channel in a dune watershed. *J. Freshw. Ecol.* **19**: 123–139.

Schlesinger, W. H. 1978. On the relative dominance of shrubs in Okefenokee Swamp. *Am. Nat.* **112**: 949–954.

Schlickeisen, E., T. E. Tietjen, T. L. Arsuffi & A. W. Groeger. 2003. Detritus processing and microbial dynamics of an aquatic macrophyte and terrestrial leaf in a thermally constant, spring-fed stream. *Microbial Ecol.* **45**: 411–418.

Schloesser, D. W. & B. A. Manny. 1986. Distribution of submersed macrophytes in the St. Clair-Detroit River system, 1978. *J. Freshw. Ecol.* **3**: 537–544.

Schloesser, D. W. & B. A. Manny. 2007. Restoration of wildcelery, *Vallisneria americana* Michx., in the lower Detroit River of the Lake Huron-Lake Erie corridor. *J. Great Lakes Res.* **33**: 8–19.

Schloesser, D. W., P. L. Hudson & S. J. Nichols. 1986. Distribution and habitat of *Nitellopsis obtusa* (Characeae) in the Laurentian Great Lakes. *Hydrobiologia* **133**: 91–96.

Schlüter, U. & R. M. M. Crawford. 2001. Long-term anoxia tolerance in leaves of *Acorus calamus* L. and *Iris pseudacorus* L. *J. Exp. Bot.* **52**: 2213–2225.

Schmalzer, P. A., T. S. Patrick & H. R. DeSelm. 1985. Vascular flora of the Obed Wild and Scenic River, Tennessee. *Castanea* **50**: 71–88.

Schmeda-Hirschmann, G. 1994. Plant resources used by the Ayoreo of the Paraguayan Chaco. *Econ. Bot.* **48**: 252–258.

Schmid, W. D. 1965. Distribution of aquatic vegetation as measured by line intercept with SCUBA. *Ecology* **46**: 816–823.

Schmidt, J. O., S. L. Buchmann & M. Glaum. 1989. The nutritional value of *Typha latifolia* pollen for bees. *J. Apicult. Res.* **28**: 155–165.

Schmitt, R. & S. Holbrook. 1984. Ontogeny of prey selection by black surfperch *Embiotoca jacksoni* (Pisces: Embiotocidae): the roles of fish morphology, foraging behavior, and patch selection. *Mar. Ecol. Prog. Ser.* **18**: 225–239.

Schmitz, U., S. Köhler & A. Hussner. 2014. First records of American *Wolffia columbiana* in Europe–Clandestine replacement of native *Wolffia arrhiza*? *BioInvas. Rec.* **3**: 213–216.

Schneider, R. 1994. The role of hydrologic regime in maintaining rare plant communities of New York's coastal plain pondshores. *Biol. Conserv.* **68**: 253–260.

Scholtens, B. 1996. Moths of the Douglas Lake region (Emmet and Cheboygan Counties), Michigan: V. Crambidae and Pyralidae (Lepidoptera). *Great Lakes Entomol.* **29**: 141–160.

Schöpfel, H. 1975. *Keine Probleme bei Cryptocorynen*. AT-Ratgeber Reihe 3. Urania-Verlag, Leipzig. 51 pp.

Schowalter, T. D. 2018. Biology and management of the eastern lubber grasshopper (Orthoptera: Acrididae). *J. Integr. Pest Manag.* **9**(10): 1–7.

Schrader, K. K., A. Andolfi, C. L. Cantrell, A. Cimmino, S. O. Duke, W. Osbrink, D. E. Wedge & A. Evidente. 2010. A survey of phytotoxic microbial and plant metabolites as potential natural products for pest management. *Chem. Biodivers.* **7**: 2261–2280.

Schreinemakers, W. A. C. 1986. The interaction between Cd-absorption and Cd compartmentation in *Wolffiella gladiata*. *Acta Bot. Neerl.* **35**: 23–34.

Schroll, E., J. Lambrinos, T. Righetti & D. Sandrock. 2011. The role of vegetation in regulating stormwater runoff from green roofs in a winter rainfall climate. *Ecol. Eng.* **37**: 595–600.

Schuler, C. A., R. G. Anthony & H. M. Ohlendorf. 1990. Selenium in wetlands and waterfowl foods at Kesterson Reservoir, California, 1984. *Arch. Environ. Contam. Toxicol.* **19**: 845–853.

Schulgasser, K. & A. Witztum. 2004. Spiralling upward. *J. Theoret. Biol.* **230**: 275–280.

Schuller, R., R. Showalter, T. Kaye & B. Lawrence. 2014. North Fork Silver Creek Research Natural Area: Guidebook Supplement 47. Gen. Tech. Rep. PNW-GTR-894. U.S. Department of Agriculture, Forest Service, Pacific Northwest Research Station, Portland, OR. 25 pp.

Schults, D. W., K. W. Malueg & P. D. Smith. 1976. Limnological comparison of culturally eutrophic Shagawa Lake and adjacent oligotrophic Burntside Lake, Minnesota. *Am. Midl. Nat.* **96**: 160–178.

Schutten, J. & A. J. Davy. 2000. Predicting the hydraulic forces on submerged macrophytes from current velocity, biomass and morphology. *Oecologia* **123**: 445–452.

Schutten, J., J. Dainty & A. J. Davy. 2005. Root anchorage and its significance for submerged plants in shallow lakes. *J. Ecol.* **93**: 556–571.

Schuyler, A. E. & T. Gordon. 2002. Rare plants in the middle branch of the Forked River watershed, Lacey Township, Ocean County, New Jersey. *Bartonia* **61**: 117–121.

Schuyler, A. E., S. B. Andersen & V. J. Kolaga. 1993. Plant zonation changes in the tidal portion of the Delaware River. *Proc. Acad. Nat. Sci. Philadelphia* **144**: 263–266.

Schwartz, F. C. 2002. *Zigadenus* Michaux. Pp. 81–88 *In*: Flora North America Editorial Committee (eds.), *Flora of North America North of Mexico, Vol. 26: Magnoliophyta: Liliidae: Liliales and Orchidales*. Oxford University Press, New York, NY.

Schwartz, F. J. & B. W. Dutcher. 1963. Age, growth, and food of the oyster toadfish near Solomons, Maryland. *Trans. Am. Fish. Soc.* **92**: 170–173.

Schwarzschild, A. C. & J. C. Zieman. 2008. Effects of physiological integration on the survival and growth of ramets and clonal fragments in the seagrass *Syringodium filiforme*. *Mar. Ecol. Prog. Ser.* **372**: 97–104.

Scotland, M. B. 1934. The animals of the *Lemna* association. *Ecology* **15**: 290–294.

Scotland, M. B. 1940. Review and summary of studies of insects associated with *Lemna minor*. *J. New York Entomol. Soc.* **48**: 319–333.

Scott, R. K. 2009. The vascular flora of Turkey Run State Park, Parke County, Indiana. *Proc. Indiana Acad. Sci.* **118**: 55–75.

Scott, J. K. 2012. *Zantedeschia aethiopica* (L.) Spreng. – arum lily. Pp. 609–613 *In*: J. K. Scott, M. Julien, R. McFadyen & J. Cullen (eds.), *Biological Control of Weeds in Australia*. CSIRO Publishing, Collingwood, VIC, Australia.

Scott, J. A. 2014. *Lepidoptera of North America. 13. Flower Visitation by Colorado Butterflies (40,615 Records) with a Review of the Literature on Pollination of Colorado Plants and Butterfly Attraction (Lepidoptera: Hesperioidea and Papilionoidea)*. C. P. Gillette Museum of Arthropod Diversity, Colorado State University, Fort Collins, CO. 190 pp.

Scott, S. L. & J. A. Osborne. 1981. Benthic macroinvertebrates of a hydrilla infested central Florida lake. *J. Freshw. Ecol.* **1**: 41–49.

Scott, J. K. & S. Neser. 1996. Prospects for the biological control of the environmental weed, *Zantedeschia aethiopica* (arum lily). Pp. 413–416 *In*: R. C. H. Shepherd (ed.), *Proceedings of the 11th Australian Weeds Conference, Melbourne, Australia, 30 September-3 October 1996*. Weed Science Society of Victoria Inc., Latrobe University, Bundoora, Victoria, VIC, Australia.

Scribailo, R. W. & U. Posluszny. 1983. Morphology and establishment of seedlings of *Hydrocharis morsus-ranae*. *Am. J. Bot.* **70**(5, part 2): 31.

Scribailo, R. W. & U. Posluszny. 1984. The reproductive biology of *Hydrocharis morsus-ranae*. I. Floral biology. *Can. J. Bot.* **62**: 2779–2787.

Scribailo, R. W. & U. Posluszny. 1985. The reproductive biology of *Hydrocharis morsus-ranae*. II. Seed and seedling morphology. *Can. J. Bot.* **63**: 492–496.

Scribailo, R. W. & M. S. Alix. 2002. First reports of *Ceratophyllum echinatum* A. Gray from Indiana with notes on the distribution, ecology and phytosociology of the species. *J. Torrey Bot. Soc.* **129**: 164–171.

Scribailo, R. W., K. Carey & U. Posluszny. 1984. Isozyme variation and the reproductive biology of *Hydrocharis morsus-ranae* L. (Hydrocharitaceae). *Bot. J. Linn. Soc.* **89**: 305–312.

Scriber, J. M., M. D. Deering, L. N. Francke, W. F. Wehling & R. C. Lederhouse. 1998. Notes on swallowtail population dynamics of three *Papilio* species in south-central Florida (Lepidoptera: Papilionidae). *Holarctic Lepidoptera* **5**: 53–62.

Scudder, S. H. 1889. Classified list of food plants of American butterflies, drawn from Scudder's "*Butterflies of the Eastern United States.*" *Psyche* **5**: 274–278.

Sculthorpe, C. D. 1967. *The Biology of Aquatic Vascular Plants.* Edward Arnold (Publishers) Ltd., London, United Kingdom. 610 pp.

Seabloom, E. W., A. G. van der Valk & K. A. Moloney. 1998. The role of water depth and soil temperature in determining initial composition of prairie wetland coenoclines. *Plant Ecol.* **138**: 203–216.

Sealy, J. R. 1954. Review of the genus *Hymenocallis*. *Kew Bull.* **9**: 201–240.

Seaman, D. E. & W. A. Porterfield. 1964. Control of aquatic weeds by the snail *Marisa cornuarietis*. *Weeds* **12**: 87–92.

Seaman, D., M. Morse, M. Miller, W. Harvey, L. Buschmann, C. Wick & B. Fischer. 1968. Controlling submersed weeds in rice. *Calif. Agric.* **22**: 11.

Sears, C. J. 2008. Morphological discrimination of *Platanthera aquilonis*, *P. huronensis*, and *P. dilatata* (Orchidaceae) herbarium specimens. *Rhodora* **110**: 389–406.

Seberg, O., G. Petersen, J. I. Davis, J. C. Pires, D. W. Stevenson, M. W. Chase, M. F. Fay, D. S. Devey, T. Jøgensen, K. J. Sytsma & Y. Pillon. 2012. Phylogeny of the Asparagales based on three plastid and two mitochondrial genes. *Am. J. Bot.* **99**: 875–889.

Sedinger, J. S. & D. G. Raveling. 1984. Dietary selectivity in relation to availability and quality of food for goslings of cackling geese. *Auk* **101**: 295–306.

Sedki, M., M. B. Mohamed, M. Fawzy, D. A. Abdelrehim & M. M. S. A. Abdel-Mottaleb. 2015. Phytosynthesis of silver–reduced graphene oxide (Ag–RGO) nanocomposite with an enhanced antibacterial effect using *Potamogeton pectinatus* extract. *RSC Adv.* **5**: 17358–17365.

See, R. B., D. L. Naftz, D. A. Peterson, J. G. Crock, J. A. Erdman, R. C. Severson, P. Ramirez, Jr. & J. A. Armstrong. 1992. Detailed study of selenium in soil, representative plants, water, bottom sediment, and biota in the Kendrick reclamation project area, Wyoming, 1988–90. Resources Investigations Report 91–4131. U.S. Geological Survey, Cheyenne, WY. 142 pp.

Seeliger, U., C. Cordazzo & E. W. Koch. 1984. Germination and algal-free laboratory culture of widgeon grass, *Ruppia maritima*. *Estuaries* **7**: 176–178.

Seigler, D. S. 1976. Plants of the northeastern United States that produce cyanogenic compounds. *Econ. Bot.* **30**: 395–407.

Sekercioglu, Ç. H., D. G. Wenny & C. J. Whelan. 2016. *Why Birds Matter: Avian Ecological Function and Ecosystem Services.* University of Chicago Press, Chigago, IL. 368 pp.

Selbo, S. M. & A. A. Snow. 2004. The potential for hybridization between *Typha angustifolia* and *Typha latifolia* in a constructed wetland. *Aquat. Bot.* **78**: 361–369.

Self, C. A., R. H. Chabreck & T. Joanen. 1975. Food preferences of deer in Louisiana coastal marshes. Pp. 548–556 *In*: W. A. Rogers (ed.), *Proceedings of the Twenty-eighth Annual Conference, November 17-20, 1974.* Southeastern Association of Game and Fish Commissioners, White Sulphur Springs, WV.

Seo, C.-S., M.-Y. Lee, I.-S. Shin, J.-A. Lee, H. Ha & H.-K. Shin. 2012. *Spirodela polyrhiza* (L.) Sch. ethanolic extract inhibits LPS-induced inflammation in RAW264. 7 cells. *Immunopharmacol. Immunotoxicol.* **34**: 794–802.

Serie, J. R. & G. A. Swanson. 1976. Feeding ecology of breeding gadwalls on saline wetlands. *J. Wildl. Manag.* **40**: 69–81.

Serviss, B. E., S. T. McDaniel & C. T. Bryson. 2000. Occurrence, distribution, and ecology of *Alocasia, Caladium, Colocasia,* and *Xanthosoma* (Araceae) in the southeastern United States. *Sida* **19**: 149–174.

Setchell, W. A. 1924. *Ruppia* and its environmental factors. *Proc. Natl. Acad. Sci. U.S.A.* **10**: 286–288.

Seyer, S. C. 1979. Vegetative ecology of a montane mire, Crater Lake National Park, Oregon. M.S. thesis. Oregon State University, Corvallis, OR. 187 pp.

Seymour, R. S. & A. J. Blaylock. 1999. Switching off the heater: influence of ambient temperature on thermoregulation by eastern skunk cabbage *Symplocarpus foetidus*. *J. Exp. Bot.* **50**: 1525–1532.

Sexton, O. J. 1959. Spatial and temporal movements of a population of the painted turtle, *Chrysemys picta marginata* (Agassiz). *Ecol. Monogr.* **29**: 113–140.

Sgattoni, P., V. Ticchiati, F. Arosio, P. Villani & C. Mallegni. 1989. Distribution and importance of the main rice weeds in Italy: results from a 1988 technical survey. Pp. 301–311 *In*: *Problems of Weed Control in Fruit, Horticultural Crops and Rice. Proceedings of the 4th EWRS Symposium on Weed Problems in Mediterranean Climates (Vol. 2).* Valencia, Spain.

Shabnam, N. & P. Pardha-Saradhi. 2013. Photosynthetic electron transport system promotes synthesis of Au-nanoparticles. *PLoS One* **8**(8): e71123.

Shabnam, N. & P. Pardha-Saradhi. 2016. Floating and submerged leaves of *Potamogeton nodosus* exhibit distinct variation in the antioxidant system as an ecophysiological adaptive strategy. *Funct. Plant Biol.* **43**: 346–355.

Shabnam, N., P. Sharmila, A. Sharma, R. J. Strasser & P. Pardha-Saradhi. 2015. Mitochondrial electron transport protects floating leaves of long leaf pondweed (*Potamogeton nodosus* Poir) against photoinhibition: comparison with submerged leaves. *Photosyn. Res.* **125**: 305–319.

Shacklette, H. T. 1961. Substrate relationships of some bryophyte communities on Latouche Island, Alaska. *Bryologist* **64**: 1–16.

Shafer, D. J., T. D. Sherman & S. Wyllie-Echeverria. 2007. Do desiccation tolerances control the vertical distribution of intertidal seagrasses? *Aquat. Bot.* **87**: 161–166.

Shafer, D. J., S. Wyllie-Echeverria & T. D. Sherman. 2008. The potential role of climate in the distribution and zonation of the introduced seagrass *Zostera japonica* in North America. *Aquat. Bot.* **89**: 297–302.

Shafer, D. J., J. E. Kaldy, T. D. Sherman & K. M. Marko. 2011. Effects of salinity on photosynthesis and respiration of the seagrass *Zostera japonica*: a comparison of two established populations in North America. *Aquat. Bot.* **95**: 214–220.

Shafer, D. J., J. E. Kaldy & J. L. Gaeckle. 2014. Science and management of the introduced seagrass *Zostera japonica* in North America. *Environ. Manage.* **53**: 147–162.

Shaffer, G. P., C. E. Sasser, J. G. Gosselink & M. Rejmanek. 1992. Vegetation dynamics in the emerging Atchafalaya Delta, Louisiana, USA. *J. Ecol.* **80**: 677–687.

Shaffer-Fehre, M. 1991a. The endotegmen tuberculae: an account of little-known structures from the seed coat of the Hydrocharitoideae (Hydrocharitaceae) and *Najas* (Najadaceae). *Bot. J. Linn. Soc.* **107**: 169–188.

Shaffer-Fehre, M. 1991b. The position of *Najas* within the subclass Alismatidae (Monocotyledones) in the light of new evidence from seed coat structures in the Hydrocharitoideae (Hydrocharitales). *Bot. J. Linn. Soc.* **107**: 189–209.

Shanks, A. L., B. A. Grantham & M. H. Carr. 2003. Propagule dispersal distance and the size and spacing of marine reserves. *Ecol. Appl.* **13**: 159–169.

Shannon, R. D., J. R. White, J. E. Lawson & B. S. Gilmour. 1996. Methane efflux from emergent vegetation in peatlands. *J. Ecol.* **84**: 239–246.

Shao, H., B. Gontero, S. C. Maberly, H. Sheng, J. Yu, C. Wei, L. Wen & M. Huang. 2017. Responses of *Ottelia alismoides*, an aquatic plant with three CCMs, to variable CO_2 and light. *J. Exp. Bot.* **68**: 3985–3995.

Sharitz, R. R., P. Stankus & L. Lee. 2010. Vegetation community of the H-02 wetlands: importance to amphibians. Pp. 31–41 *In*: D. Scott & T. Tuberville (eds.), *FY-2009 Annual Report*. Savannah River Ecology Laboratory, Aiken, SC.

Sharma, B. M. 1984. Ecophysiological studies on water lettuce in a polluted lake. *J. Aquat. Plant Manag.* **22**: 17–21.

Sharma, A. K. & A. K. Bal. 1956. A cytological study of a few genera of Amaryllidaceae with a view to find out the basis of their phylogeny. *Cytologia* **21**: 329–352.

Sharpe, P. J. & A. H. Baldwin. 2012. Tidal marsh plant community response to sea-level rise: a mesocosm study. *Aquat. Bot.* **101**: 34–40.

Shaw, J. P., S. J. Taylor, M. C. Dobson & N. H. Martin. 2017. Pollinator isolation in Louisiana iris: legitimacy and pollen transfer. *Evol. Ecol. Res.* **18**: 429–441.

Shay, J. M. 1999. Annotated vascular plant species list for the Delta Marsh, Manitoba and surrounding area. Occasional Publication No. 2. University of Manitoba Field Station (Delta Marsh), Winnipeg, MB, Canada. 52 pp.

Shefferson, R. P. 2006. Survival costs of adult dormancy and the confounding influence of size in lady's slipper orchids, genus *Cypripedium*. *Oikos* **115**: 253–262.

Shefferson, R. P., M. Weiss, T. I. I. U. Kull & D. L. Taylor. 2005. High specificity generally characterizes mycorrhizal association in rare lady's slipper orchids, genus *Cypripedium*. *Mol. Ecol.* **14**: 613–626.

Sheldon, S. P. 1986. Factors influencing the numbers of branches and inflorescences of *Potamogeton richardsonii* (A. Benn.) Rydb. *Aquat. Bot.* **24**: 27–34.

Sheldon, S. P. 1987. The effects of herbivorous snails on submerged macrophyte communities in Minnesota lakes. *Ecology* **68**: 1920–1931.

Sheldon, R. B. & C. W. Boylen. 1975. Factors affecting the contribution by epiphytic algae to the primary productivity of an oligotrophic freshwater lake. *Appl. Microbiol.* **30**: 657–667.

Sheldon, R. B. & C. W. Boylen. 1977. Maximum depth inhabited by aquatic vascular plants. *Am. Midl. Nat.* **97**: 248–254.

Shelingoski, S., R. J. LeBlond, J. M. Stucky & T. R. Wentworth. 2005. Flora and soils of Wells Savannah, an example of a unique savanna type. *Castanea* **70**: 101–114.

Shelton, A. O. 2008. Skewed sex ratios, pollen limitation, and reproductive failure in the dioecious seagrass *Phyllospadix*. *Ecology* **89**: 3020–3029.

Shelton, A. O. 2010a. The origin of female-biased sex ratios in intertidal seagrasses (*Phyllospadix* spp.). *Ecology* **91**: 1380–1390.

Shelton, A. O. 2010b. The ecological and evolutionary drivers of female-biased sex ratios: two-sex models of perennial seagrasses. *Am. Nat.* **175**: 302–315.

Sherff, E. E. 1912. The vegetation of Skokie Marsh, with special reference to subterranean organs and their interrelationships. *Bot. Gaz.* **53**: 415–435.

Sheridan, P. F. & R. J. Livingston. 1983. Abundance and seasonality of infauna and epifauna inhabiting a *Halodule wrightii* meadow in Apalachicola Bay, Florida. *Estuaries Coast.* **6**: 407–419.

Sheridan, P. M., S. L. Orzell & E. L. Bridges. 1997. Powerline easements as refugia for state rare seepage and pineland plant taxa. Pp. 451–460 *In*: J. R. Williams, J. W. Goodrich-Mahoney, J. R. Wisniewski & J. Wisniewski (eds.), *The Sixth International Symposium on Environmental Concerns in Rights-of-Way Management*. Elsevier, Oxford, England.

Sher-Kaul, S., B. Oertli, E. Castella & J.-B. Lachavanne. 1995. Relationship between biomass and surface area of six submerged aquatic plant species. *Aquat. Bot.* **51**: 147–154.

Sherman, H. L. 1979. Evidence of misapplication of the name *Schoenolirion texanum* (Scheele) Gray (Liliaceae). *Southwest. Nat.* **24**: 123–126.

Sherman, H. L. 2002. *Schoenolirion* Torrey ex Durand. Pp. 312–314 *In*: Flora North America Editorial Committee (eds.), *Flora of North America North of Mexico, Vol. 26: Magnoliophyta: Liliidae: Liliales and Orchidales*. Oxford University Press, New York, NY.

Sherman, H. L. & R. W. Becking. 1991. The generic distinctness of *Schoenolirion* and *Hastingsia*. *Madroño* **38**: 130–138.

Sherman, D. E., R. W. Kroll & T. L. Engle. 1996. Flora of a diked and an undiked southwestern Lake Erie wetland. *Ohio J. Sci.* **96**: 4–8.

Sheviak, C. J. 1990. A new *Spiranthes* (Orchidaceae) from the cienegas of southernmost Arizona. *Rhodora* **92**: 213–231.

Sheviak, C. J. 2001. A role for water droplets in the pollination of *Platanthera aquilonis* (Orchidaceae). *Rhodora* **103**: 380–386.

Sheviak, C. J. 2002a. *Cypripedium* Linnaeus. Pp. 499–506 *In*: Flora North America Editorial Committee (eds.), *Flora of North America North of Mexico, Vol. 26: Magnoliophyta: Liliidae: Liliales and Orchidales*. Oxford University Press, New York, NY.

Sheviak, C. J. 2002b. *Habenaria* Willdenow. Pp. 581–582 *In*: Flora North America Editorial Committee (eds.), *Flora of North America North of Mexico, Vol. 26: Magnoliophyta: Liliidae: Liliales and Orchidales*. Oxford University Press, New York, NY.

Sheviak, C. J. 2002c. *Platanthera* Richard. Pp. 551–571 *In*: Flora North America Editorial Committee (eds.), *Flora of North America North of Mexico, Vol. 26: Magnoliophyta: Liliidae: Liliales and Orchidales*. Oxford University Press, New York, NY.

Sheviak, C. J. & P. M. Brown. 2002. *Spiranthes* Richard. Pp. 530–544 *In*: Flora North America Editorial Committee (eds.), *Flora of North America North of Mexico, Vol. 26: Magnoliophyta: Liliidae: Liliales and Orchidales*. Oxford University Press, New York, NY.

Sheviak, C. J. & P. M. Catling. 2002a. *Arethusa* Linnaeus. Pp. 596–597 *In*: Flora North America Editorial Committee (eds.), *Flora of North America North of Mexico, Vol. 26: Magnoliophyta: Liliidae: Liliales and Orchidales*. Oxford University Press, New York, NY.

Sheviak, C. J. & P. M. Catling. 2002b. *Amerorchis* Hultén. Pp. 550–551 *In*: Flora North America Editorial Committee (eds.), *Flora of North America North of Mexico, Vol. 26: Magnoliophyta: Liliidae: Liliales and Orchidales*. Oxford University Press, New York, NY.

Sheviak, C. J. & P. M. Catling. 2002c. *Pogonia* Jussieu. Pp. 513–514 *In*: Flora North America Editorial Committee (eds.), *Flora of North America North of Mexico, Vol. 26: Magnoliophyta: Liliidae: Liliales and Orchidales*. Oxford University Press, New York, NY.

Sheviak, C. J. & W. F. Jennings. 2006. A new *Platanthera* (Orchidaceae) from the intermountain west. *Rhodora* **108**: 19–32.

Shields, E. C. & K. A. Moore. 2016. Effects of sediment and salinity on the growth and competitive abilities of three submersed macrophytes. *Aquat. Bot.* **132**: 24–29.

Shields, E. C., K. A. Moore & D. B. Parrish. 2012. Influences of salinity and light availability on abundance and distribution of tidal freshwater and oligohaline submersed aquatic vegetation. *Estuaries Coast.* **35**: 515–526.

Shih, J. G. & S. A. Finkelstein. 2008. Range dynamics and invasive tendencies in *Typha latifolia* and *Typha angustifolia* in eastern North America derived from herbarium and pollen records. *Wetlands* **28**: 1–16.

Shin, H. T., M. H. Yi, J. W. Yoon, J. W. Sun & G. S. Kim. 2012. Status of alien plant species in the Seongeup folk village in Jeju Island. *J. Korean Nat.* **5**: 299–304.

Shinners, L. H. 1962. Annual sisyrinchiums (Iridaceae) in the United States. *Sida* **1**: 32–42.

Shipley, B. & M. Parent. 1991. Germination responses of 64 wetland species in relation to seed size, minimum time to reproduction and seedling relative growth rate. *Funct. Ecol.* **5**: 111–118.

Shirey, P. D., B. N. Kunycky, D. T. Chaloner, M. A. Brueseke & G. A. Lamberti. 2013. Commercial trade of federally listed threatened and endangered plants in the United States. *Conserv. Lett.* **6**: 300–316.

Shoemaker, G. & S. Wyllie-Echeverria. 2013. Occurrence of rhizomal endophytes in three temperate northeast pacific seagrasses. *Aquat. Bot.* **111**: 71–73.

Short, F. T. 1983. The seagrass, *Zostera marina* L.: plant morphology and bed structure in relation to sediment ammonium in Izembek Lagoon, Alaska. *Aquat. Bot.* **16**: 149–161.

Short, F. T. 1985. A method for the culture of tropical seagrasses. *Aquat. Bot.* **22**: 187–193.

Short, F. T. & C. P. McRoy. 1984. Nitrogen uptake by leaves and roots of the seagrass *Zostera marina* L. *Bot. Mar.* **27**: 547–556.

Short, F. T. & S. Wyllie-Echeverria. 1996. Natural and human-induced disturbance of seagrasses. *Environ. Conserv.* **23**: 17–27.

Short, F. T., W. C. Dennison & D. G. Capone. 1990. Phosphorus-limited growth of the tropical seagrass *Syringodium filiforme* in carbonate sediments. *Mar. Ecol. Prog. Ser.* **62**: 169–174.

Short, F. T., J. Montgomery, C. F. Zimmermann & C. A. Short. 1993. Production and nutrient dynamics of a *Syringodium filiforme* Kütz. seagrass bed in Indian River Lagoon, Florida. *Estuaries* **16**: 323–334.

Short, F. T., G. E. Moore & K. A. Peyton. 2010. *Halophila ovalis* in the tropical Atlantic Ocean. *Aquat. Bot.* **93**: 141–146.

Shu, J. H. 2000. *Hedychium*. Pp. 370–377 *In*: Z. Wu & P. Raven (eds.), *Flora of China. Vol. 24 (Flagellariaceae Through Marantaceae)*. Science Press, Beijing/Missouri Botanical Garden Press, St. Louis, MO.

Shukla, R., S. Srivastava, P. K. Dwivedi, S. Sarkar, S. Gupta & A. Mishra. 2012. Evaluation of free radical scavenging activity of the different fractions of *Typha latifolia* (Typhaceae). *J. Harmonized Res. Pharm.* **1**: 33–43.

Shull, G. H. 1914. The longevity of submerged seeds. *Plant World* **17**: 329–337.

Siegfried, W. R. 1976. Breeding biology and parasitism in the Ruddy Duck. *Wilson Bull.* **88**: 566–574.

Sieren, D. J. & K. R. Warr. 1992. The flora of limesink depressions in Carolina Beach State Park (North Carolina). *Rhodora* **94**: 156–166.

Sifton, H. B. 1959. The germination of light-sensitive seeds of *Typha latifolia* L. *Can. J. Bot.* **37**: 719–739.

Sikes, K., D. Roach, J. Buck, J. Nelson & S. Erwin. 2010. Classification and mapping of vegetation from three fen sites of the Shasta-Trinity National Forest, California. Unpublished Report. U.S. Forest Service, Pacific Southwest Region, Vallejo, CA. 36 pp.

Sikes, K., D. J. Cooper, S. Weis, T. Keeler-Wolf, M. Barbour, D. Ikeda, D. Stout & J. Evens. 2013. Fen conservation and vegetation assessment in the National Forests of the Sierra Nevada and adjacent mountains, California. Revised public version 2. Unpublished Document. University of California-Davis, Davis, CA. 20 pp.

Silberhorn, G. M., R. J. Orth & K. A. Moore. 1983. Anthesis and seed production in *Zostera marina* L. (eelgrass) from the Chesepeake Bay. *Aquat. Bot.* **15**: 133–144.

Siegal-Willott, J. L., K. Harr, L. C. Hayek, K. C. Scott, T. Gerlach, P. Sirois, M. Reuter, D. W. Crewz & R. C. Hill. 2010. Proximate nutrient analyses of four species of submerged aquatic vegetation consumed by Florida manatee (*Trichechus manatus latirostris*) compared to romaine lettuce (*Lactuca sativa* var. *longifolia*). *J. Zoo Wildl. Med.* **41**: 594–602.

Siegfried, W. R. 1973. Summer food and feeding of the ruddy duck in Manitoba. *Can. J. Zool.* **51**: 1293–1297.

Sikes, K., D. Roach, J. Buck, J. Nelson & S. Erwin. 2010. Classification and mapping of vegetation from three fen sites of the Shasta-Trinity National Forest, California. Unpublished Report to the United States Forest Service, Region 5. Available online: http://www.academia.edu/download/41910128/Classification_and_Mapping_of_Vegetation20160202-15687-1dz4wxa.pdf [accessed 9 February, 2019].

Silva, J., N. Dantas-Santos, D. L. Gomes, L. S. Costa, S. L. Cordeiro, M. S. S. P. Costa, N. B. Silva, M. L. Freitas, K. C. Scortecci, E. L. Leite & H. A. O. Rocha. 2012. Biological activities of the sulfated polysaccharide from the vascular plant *Halodule wrightii*. *Rev. Bras. Farmacogn.* **22**: 94–101.

Silveira, M. J., S. M. Thomaz, R. P. Mormul & F. P. Camacho. 2009. Effects of desiccation and sediment type on early regeneration of plant fragments of three species of aquatic macrophytes. *Int. Rev. Hydrobiol.* **94**: 169–178.

Silvério, A., S. Nadot, T. T. Souza-Chies & O. Chauveau. 2012. Floral rewards in the tribe Sisyrinchieae (Iridaceae): oil as an alternative to pollen and nectar? *Sex. Plant Reprod.* **25**: 267–279.

Simberloff, D. & C. Leppanen. 2019. Plant somatic mutations in nature conferring insect and herbicide resistance. *Pest Manag. Sci.* **75**: 14–17.

Simpson, M. G. 1987. Pollen ultrastructure of the Pontederiaceae. *Grana* **26**: 113–126.

Simpson, M. G. 1990. Phylogeny and classification of the Haemodoraceae. *Ann. Missouri Bot. Gard.* **77**: 722–784.

Simpson, M. G. 1998. Haemodoraceae. Pp. 212–222 *In*: K. Kubitzki, H. Huber, P. J. Rudall, P. S. Stevens & T. Stützel (eds.), *The Families and Genera of Vascular Plants, Vol. IV. Flowering Plants, Monocotyledons: Alismatanae and Commelinanae (Except Gramineae)*. Springer-Verlag, Berlin, Germany.

Simpson, R. L., D. F. Whigham & K. Brannigan. 1979. Mid-summer insect communities of freshwater tidal wetland macrophytes. *Bull. New Jersey Acad. Sci.* **24**: 22–28.

Singer, R. B. & A. A. Cocucci. 1999. Pollination mechanism in southern Brazilian orchids which are exclusively or mainly pollinated by halictid bees. *Plant Syst. Evol.* **217**: 101–117.

Singer, R., D. A. Roberts & C. W. Boylen. 1983. The macrophytic community of an acidic lake in Adirondack (New York, USA): a new depth record for aquatic angiosperms. *Aquat. Bot.* **16**: 49–57.

Singer, R. B., T. B. Breier, A. Flach & R. Farias-Singer. 2007. The pollination mechanism of *Habenaria pleiophylla* Hoehne & Schlechter (Orchidaceae: Orchidinae). *Funct. Ecosyst. Communities* **1**: 10–14.

Singh, Y., A. E. Van Wyk & H. Baijnath. 1996. Floral biology of *Zantedeschia aethiopica* (L.) Spreng. (Araceae). *S. Afr. J. Bot.* **62**: 146–150.

Singh, D., R. Gupta & S. A. Saraf. 2012. Herbs–are they safe enough? An overview. *Crit. Rev. Food Sci. Nutr.* **52**: 876–898.

Singh, M., U. N. Rai, U. Nadeem & A. A. David. 2014. Role of *Potamogeton pectinatus* in phytoremediation of metals. *Chem. Sci. Rev. Lett.* **3**: 123–129.

Singhurst, J. R., J. C. Cathy, D. Prochaska, H. Haucke, G. C. Kroh & W. C. Holmes. 2003. The vascular flora of Gus Engeling Wildlife Management Area, Anderson County, Texas. *Southeast. Nat.* **2**: 347–369.

Singhurst, J. R., E. L. Bridges & W. C. Holmes. 2007. Two additions to the flora of Oklahoma and notes on *Xyris jupicai* (Xyridaceae) in Oklahoma. *Phytologia* **89**: 211–218.

Singhurst, J. R., D. J. Rosen & W. C. Holmes. 2009. Two additions to the vascular flora of Texas. *Phytologia* **91**: 69–72.

Singhurst, J. R., J. N. Mink & W. C. Holmes. 2010. New and noteworthy plants of Texas. *Phytologia* **92**: 249–255.

Singhurst, J. R., A. E. Rushing, C. K. Hanks & W. C. Holmes. 2011. *Isoetes texana* (Isoetaceae): a new species from the Texas Coastal Bend. *Phytoneuron* **2011–22**: 1–6.

Singhurst, J. R., B. A. Sorrie & W. C. Holmes. 2012. *Andropogon glaucopsis* (Poaceae) in Texas. *Phytoneuron* **2012–16**: 1–3.

Singhurst, J. R., C. T. Witsell, E. Sundell & W. C. Holmes. 2014a. *Utricularia cornuta* (Lentibulariaceae) new to the Arkansas and Oklahoma flora. *Phytoneuron* **2014–109**: 1–5.

Singhurst, J. R., A. K. Buthod & W. C. Holmes. 2014b. *Pluchea foetida* (Asteraceae) confirmed in the Oklahoma flora. *Phytoneuron* **2014–97**: 1–4.

Singhurst, J. R., A. Cooper, D. J. Rosen & W. C. Holmes. 2014c. The vascular flora and plant communities of Candy Abshier Wildlife Managament Area, Chamber County, Texas, USA. *J. Bot. Res. Inst. Texas* **8**: 665–675.

Singleton, J. R. 1951. Production and utilization of waterfowl food plants on the east Texas Gulf coast. *J. Wildl. Manag.* **15**: 46–56.

Sinkiewicz, C. & E. Jules. 2003. Port orford cedar and the non-native pathogen, *Phytophthora lateralis*. *Fremontia* **31**: 14–20.

Sipaúba-Tavares, L. H. & F. M. de Souza Braga. 2008. Constructed wetland[s] in wastewater treatment. *Acta Sci. Biol. Sci.* **30**: 261–265.

Sipple, W. S. & W. A. Klockner. 1980. A unique wetland in Maryland. *Castanea* **45**: 60–69.

Siripahanakul, T., S. Thongsila, T. Tanuthong & S. Chockchaisawasdee. 2013. Product development of *Wolffia*-pork ball. *Int. Food Res. J.* **20**: 213–217.

Sivakumar, D., R. Anand, J. Rajaganapathy & M. Balasubramanian. 2008. Textile industry wastewater color removal using *Lemna minuta* Lin. *J. Chem. Pharmaceut. Sci.* **8**: 563–570.

Siver, P. A. 1978. Development of diatom communities on *Potamogeton robbinsii* Oakes. *Rhodora* **80**: 417–430.

Siver, P. A. 1980. Microattachment patterns of diatoms on leaves of *Potamogeton robbinsii* Oakes. *Trans. Am. Microsc. Soc.* **99**: 217–220.

Siver, P. A., A. M. Coleman, G. A. Benson & J. T. Simpson. 1986. The effects of winter drawdown on macrophytes in Candlewood Lake, Connecticut. *Lake Reserv. Manag.* **2**: 69–73.

Sjöberg, K. & K. Danell. 1982. Feeding activity of ducks in relation to diel emergence of chironomids. *Can. J. Zool.* **60**: 1383–1387.

Sjörs, H. 1959. Bogs and fens in the Hudson Bay lowlands. *Arctic* **12**: 2–19.

Skeate, S. T. 1987. Interactions between birds and fruits in a northern Florida hammock community. *Ecology* **68**: 297–309.

Skillicorn, P., W. Spira & W. Journey. 1993. *Duckweed Aquaculture: A New Aquatic Farming System for Developing Countries*. The World Bank, Washington, DC. 77 pp.

Skinner, M. W. 1988. Comparative pollination ecology and floral evolution in Pacific Coast *Lilium*. Ph.D. dissertation. Harvard University, Cambridge, MA.

Skinner, M. W. 2002. *Lilium* Linnaeus. Pp. 172–197 *In*: Flora North America Editorial Committee (eds.), *Flora of North America North of Mexico, Vol. 26: Magnoliophyta: Liliidae: Liliales and Orchidales*. Oxford University Press, New York, NY.

Skinner, W. R. & E. S. Telfer. 1974. Spring, summer, and fall foods of deer in New Brunswick. *J. Wildl. Manag.* **38**: 210–214.

Skinner, M. W. & B. A. Sorrie. 2002. Conservation and ecology of *Lilium pyrophilum*, a new species of Liliaceae from the Sandhills Region of the Carolinas and Virginia, USA. *Novon* **12**: 94–105.

Skougard, M. G. & J. D. Brotherson. 1979. Vegetational response to three environmental gradients in the salt playa near Goshen, Utah County, Utah. *Great Basin Nat.* **39**: 44–58.

Skov, C. & J. Wiley. 2005. Establishment of the Neotropical orchid bee *Euglossa viridissima* (Hymenoptera: Apidae) in Florida. *Florida Entomol.* **88**: 225–228.

Slaughter, B. S. & D. L. Cuthrell. 2011. A survey and characterization of Michigan's coastal fen communities. Report No. 2011-12. Michigan Natural Features Inventory, Lansing, MI. 151 pp.

Slaughter, R. J., D. M. G. Beasley, B. S. Lambie, G. T. Wilkins & L. J. Schep. 2012. Poisonous plants in New Zealand: a review of those that are most commonly enquired about to the National Poisons Centre. *New Zealand Med. J.* **125**: 87–118.

Sledge, W. A. 1948. The distribution and ecology of *Scheuchzeria palustris* L. *Watsonia* **1**: 24–35.

Sletten, K. K. & G. E. Larson. 1984. Possible relationships between surface water chemistry and aquatic plants in the Northern Great Plains. *Proc. South Dakota Acad. Sci.* **63**: 70–76.

Small, J. K. 1931. Celestial lilies. *J. New York Bot. Gard.* **32**: 260–269.

Small, J. A., D. Smiley, Jr., P. W. Zimmerman, E. B. Matzke, J. W. Thomson, Jr., O. P. Medsger, H. C. Bold, E. Ashwell & J. J. Copeland. 1941. Field trips of the club. *Torreya* **41**: 135–142.

Smith, J. G. 1895. A revision of the North American species of *Sagittaria* and *Lophotocarpus*. *Missouri Bot. Gard. Ann. Rep.* **1895**: 1–64.

Smith, J. G. 1900. Revision of the species of *Lophotocarpus* of the United States: and description of a new species of *Sagittaria*. *Missouri Bot. Gard. Ann. Rep.* **1900**: 145–151.

Smith, R. W. 1910. The floral development and embryogeny of *Eriocaulon septangulare*. *Bot. Gaz.* **49**: 281–289.

Smith, R. H. 1953. A study of waterfowl production on artificial reservoirs in eastern Montana. *J. Wildl. Manag.* **17**: 276–291.

Smith, S. G. 1967. Experimental and natural hybrids in north American *Typha* (Typhaceae). *Am. Midl. Nat.* **78**: 257–287.

Smith, Jr., R. J. 1983. Weeds of major economic importance in rice and yield losses due to weed competition. Pp. 19–36 *In*: *Weed Control in Rice*. International Rice Research Institute, Los Baños, Laguna, Philippines.

Smith, B. 1985. Gullible's travails: tuberculosis and quackery 1890–1930. *J. Contemp. Hist.* **20**: 733–756.

Smith, S. G. 1986. The cattails (*Typha*): interspecific ecological differences and problems of identification. *Lake Reserv. Manag.* **2**: 357–362.

Smith, S. G. 1987. *Typha*: its taxonomy and the ecological significance of hybrids. *Arch. Hydrobiol., Beih. Ergebn. Limnol.* **27**: 129–138.

Smith, W. R. 1993. *Orchids of Minnesota*. University of Minnesota Press, Minneapolis, MN. 172 pp.

Smith, L. M. 1996. *The Rare and Sensitive Natural Wetland Plant Communities of Interior Louisiana*. Louisiana Natural Heritage Program, Louisiana Department of Wildlife and Fisheries, Baton Rouge, LA. 40 pp.

Smith, S. G. 2000. Typhaceae A. L. Jussieu. Pp. 278–285 *In*: Flora North America Editorial Committee (eds.), *Flora of North America North of Mexico, Vol. 22: Magnoliophyta: Alismatidae, Arecidae, Commelinidae (in Part), and Zingiberidae*. Oxford University Press, New York, NY.

Smith, C. W. G. 2001. *Easy-Care Water Garden Plants*. Storey Publishing, North Adams, MA. 32 pp.

Smith, W. R. 2012. *Native Orchids of Minnesota*. University of Minnesota Press, Minneapolis, MN. 254 pp.

Smith, S. D. P. 2014. The roles of nitrogen and phosphorus in regulating the dominance of floating and submerged aquatic plants in a field mesocosm experiment. *Aquat. Bot.* **112**: 1–9.

Smith, R. 2017. Plant species richness and diversity of northern white-cedar (*Thuja occidentalis*) swamps in northern New York: effects and interactions of multiple variables. M.S. thesis. State University of New York, Syracuse, NY. 87 pp.

Smith, G. R. & G. E. Snow. 1976. Pollination ecology of *Platanthera* (*Habenaria*) *ciliaris* and *P. blephariglottis* (Orchidaceae). *Bot. Gaz.* **137**: 133–140.

Smith, L. M. & J. A. Kadlec. 1983. Seed banks and their role during drawdown of a North American marsh. *J. Appl. Ecol.* **20**: 673–684.

Smith, L. M. & J. A. Kadlec. 1985. The effects of disturbance on marsh seed banks. *Can. J. Bot.* **63**: 2133–2137.

Smith, G. L. & W. S. Flory. 1990. Studies on *Hymenocallis henryae* (Amaryllidaceae). *Brittonia* **42**: 212–220.

Smith, D. R. & W. McDearman. 1990. A new *Rhadinoceraea* (Hymenoptera: Tenthredinidae) feeding on *Zigadenus* (Liliaceae) from southeastern United States. *Entomol. News* **101**: 13–19.

Smith, D. R. & E. M. Barrows. 1995. *Rhadinoceraea* n. sp. (Hymenoptera: Tenthredinidae) from West Virginia, a second species on *Zigadenus* (Liliaceae). *Entomol. News* **106**: 237–240.

Smith, G. L. & M. A. Garland. 1996. Taxonomic status of *Hymenocallis choctawensis* and *Hymenocallis puntagordensis* (Amaryllidaceae). *Sida* **17**: 305–319.

Smith, S. M. & S. Newman. 2001. Growth of southern cattail (*Typha domingensis* Pers.) seedlings in response to fire-related soil transformations in the northern Florida Everglades. *Wetlands* **21**: 363–369.

Smith, G. L. & W. S. Flory. 2002. *Hymenocallis* Salisbury. Pp. 283–292 *In*: Flora North America Editorial Committee (eds.), *Flora of North America North of Mexico, Vol. 26: Magnoliophyta: Liliidae: Liliales and Orchidales*. Oxford University Press, New York, NY.

Smith, G. L. & M. A. Garland. 2003. Nomenclature of *Hymenocallis* taxa (Amaryllidaceae) in southeastern United States. *Taxon* **52**: 805–817.

Smith, T. R. & J. L. Capinera. 2005. Host preferences and habitat associations of some Florida grasshoppers (Orthoptera: Acrididae). *Environ. Entomol.* **34**: 210–224.

Smith, G. L. & M. A. Garland. 2009. A new species of *Hymenocallis* (Amaryllidaceae) from the Apalachicola Forest of the Florida panhandle, USA. *Novon* **19**: 234–239.

Smith, F. H., K. C. Beeson & W. E. Price. 1956. Chemical composition of herbage browsed by deer in two wildlife management areas. *J. Wildl. Manag.* **20**: 359–367.

Smith, G. L., L. C. Anderson & W. S. Flory. 2001. A new species of *Hymenocallis* (Amaryllidaceae) in the lower central Florida panhandle. *Novon* **11**: 233–240.

Smith, S. M., P. V. McCormick, J. A. Leeds & P. B. Garrett. 2002. Constraints of seed bank species composition and water depth for restoring vegetation in the Florida Everglades, USA. *Restor. Ecol.* **10**: 138–145.

Smith, K. B., C. E. Smith, S. F. Forest & A. J. Richard. 2007. *A Field Guide to the Wetlands of the Boreal Plains Ecozone of Canada*. Ducks Unlimited Canada, Western Boreal Office, Edmonton, AB. 98 pp.

Smith, R. J., S. D. Hopper & M. W. Shane. 2011. Sand-binding roots in Haemodoraceae: global survey and morphology in a phylogenetic context. *Plant Soil* **348**: 453–470.

Smith, S. M., A. R. Thime, B. Zilla & K. Lee. 2015. Responses of narrowleaf cattail (*Typha angustifolia*) to combinations of salinity and nutrient additions: implications for coastal marsh restoration. *Ecol. Restor.* **33**: 297–302.

Smith, S. Y., W. J. D. Iles, J. C. Benedict & C. D. Specht. 2018. Building the monocot tree of death: progress and challenges emerging from the macrofossil-rich Zingiberales. *Am. J. Bot.* **105**: 1389–1400.

Smits, A. J. M., M. J. H. De Lyon, G. Van der Velde, P. L. M. Steentjes & J. G. M. Roelofs. 1988. Distribution of three nymphaeid macrophytes (*Nymphaea alba* L., *Nuphar lutea* (L.) Sm. and *Nymphoides peltata* (Gmel.) O. Kuntze) in relation to alkalinity and uptake of inorganic carbon. *Aquat. Bot.* **32**: 45–62.

Smits, A. J. M., R. Van Ruremonde & G. Van der Velde. 1989. Seed dispersal of three nymphaeid macrophytes. *Aquat. Bot.* **35**: 167–180.

Smock, L. A. & K. L. Harlowe. 1983. Utlization and processing of freshwater wetland macrophytes by the detritivore *Asellus forbesi*. *Ecology* **64**: 1556–1565.

Smock, L. A., D. L. Stoneburner & D. R. Lenat. 1981. Littoral and profundal macroinvertebrate communities of a coastal brown-water lake. *Arch. Hydrobiol.* **92**: 306–320.

Smreciu, A., S. Wood, K. Gould & B. Wood. 2014. Propagation protocol for ratroot (*Acorus americanus*). *Native Plant J.* **15**: 219–222.

Smreciu, A., K. Gould & S. Wood. 2015. Stratification and light promote germination of ratroot (*Acorus americanus* (Raf.) Raf. [Acoraceae]) seeds harvested in northeastern Alberta. *Native Plant J.* **16**: 19–22.

Sneddon, L. & E. E. Lamont. 2010. Diversity and classification of tidal wetlands on Long Island, New York. *Mem. Torrey Bot. Soc.* **26**: 14–33.

Snoad, B. 1955. Somatic instability of chromosome number in *Hymenocallis calathinum*. *Heredity* **9**: 129–134.

Snow, A. A., S. E. Travis, R. Wildová, T. Fér, P. M. Sweeney, J. E. Marburger, S. Windels, B. Kubátová, D. E. Goldberg & E. Mutegi. 2010. Species-specific SSR alleles for studies of hybrid cattails (*Typha latifolia* × *T. angustifolia*; Typhaceae) in North America. *Am. J. Bot.* **97**: 2061–2067.

Snyder, D. B. 1988. *Heteranthera multiflora* in New Jersey: a first look. *Bartonia* **54**: 21–23.

Snyder, D. 1996. The genus *Rhexia* in New Jersey. *Bartonia* **59**: 55–70.

Soda, S., T. Ohchi, J. Piradee, Y. Takai & M. Ike. 2015. Duckweed biomass as a renewable biorefinery feedstock: ethanol and succinate production from *Wolffia globosa*. *Biomass Bioenergy* **81**: 364–368.

Soerjani, M., A. J. G. H. Kostermans & G. Tjitrosoepomo (eds.). 1987. *Weeds of Rice in Indonesia*. Balai Pustaka, Jakarta. 716 pp.

Sohsalam, P., A. J. Englande & S. Sirianuntapiboon. 2008. Seafood wastewater treatment in constructed wetland: Tropical case. *Bioresour. Technol.* **99**: 1218–1224.

Solomeshch, A. I., M. G. Barbour & R. F. Holland. 2007. Vernal pools. Pp. 394–424 *In*: M. Barbour, T. Keeler-Wolf & A. A. Schoenherr (eds.), *Terrestrial Vegetation of California*, 3rd ed. University of California Press, Berkeley, CA.

Soltis, D., P. Soltis, P. Endress, M. W. Chase, S. Manchester, W. Judd, L. Majure & E. Mavrodiev. 2018. *Phylogeny and Evolution of the Angiosperms*. University of Chicago Press, Chicago, IL. 560 pp.

Sommers, K. P., M. Elswick, G. I. Herrick & G. A. Fox. 2011. Inferring microhabitat preferences of *Lilium catesbaei* (Liliaceae). *Am. J. Bot.* **98**: 819–828.

Søndergaard, M., L. S. Johansson, T. L. Lauridsen, T. B. Jørgensen, L. Liboriussen & E. Jeppesen. 2010. Submerged macrophytes as indicators of the ecological quality of lakes. *Freshw. Biol.* **55**: 893–908.

Song, J. Y., J. Naylor-Adelberg, S. A. White, D. A. Mann & J. Adelberg. 2014. Establishing clones of *Veratrum californicum*, a native medicinal species, for micropropagation. *In Vitro Cell. Dev. Biol.-Plant* **50**: 337–344.

Sonnenholzner, J. I., G. Montaño-Moctezuma, R. Searcy-Bernal & A. Salas-Garza. 2011. Effect of macrophyte diet and initial size on the survival and somatic growth of sub-adult *Strongylocentrotus purpuratus*: a laboratory experimental approach. *J. Appl. Phycol.* **23**: 505–513.

Soomers, H., D. N. Winkel, Y. U. N. Du & M. J. Wassen. 2010. The dispersal and deposition of hydrochorous plant seeds in drainage ditches. *Freshw. Biol.* **55**: 2032–2046.

Soons, M. B. 2006. Wind dispersal in freshwater wetlands: knowledge for conservation and restoration. *Appl. Veg. Sci.* **9**: 271–278.

Sopajarn, A. & C. Sangwichien. 2015. Optimization of enzymatic saccharification of alkali pretreated *Typha angustifolia* for glucose production. *Int. J. Chem. Eng. Applic.* **6**: 232–236.

Soreng, R. J., W. P. Armstrong, A. Tiehm & T. K. Todsen. 1984. Noteworthy collections. *Madroño* **31**: 123–127.

Soros-Pottruff, C. L. & U. Posluszny. 1994. Developmental morphology of reproductive structures of *Phyllospadix* (Zosteraceae). *Int. J. Plant Sci.* **155**: 405–420.

Soros-Pottruff, C. L. & U. Posluszny. 1995. Developmental morphology of reproductive structures of *Zostera* and a reconsideration of *Heterozostera* (Zosteraceae). *Int. J. Plant Sci.* **156**: 143–158.

Sorrie, B. A. 2000. *Rhynchospora leptocarpa* (Cyperaceae), an overlooked species of the southeastern United States. *Sida* **19**: 139–147.

Sorrie, B. A. & P. W. Dunwiddie. 1990. *Amphicarpum purshii* (Poaceae), a genus and species new to New England. *Rhodora* **92**: 105–107.

Sorrie, B. A. & S. W. Leonard. 1999. Noteworthy records of Mississippi vascular plants. *Sida* **18**: 889–908.

Sorrie, B. A. & A. S. Weakley. 2001. Coastal plain vascular plant endemics: phytogeographic patterns. *Castanea* **66**: 50–82.

Sorrie, B. A. & A. S. Weakley. 2017. *Stenanthium leimanthoides* and *S. densum* (Melanthiaceae) revisited, with the description of two new species. *J. Bot. Res. Inst. Texas* **11**: 275–286.

Sorrie, B. A., B. Van Eerden & M. J. Russo. 1997. Noteworthy plants from Fort Bragg and Camp MacKall, North Carolina. *Castanea* **62**: 239–259.

Sorrie, B. A., J. B. Gray & P. J. Crutchfield. 2006. The vascular flora of the longleaf pine ecosystem of Fort Bragg and Weymouth Woods, North Carolina. *Castanea* **71**: 129–161.

Sorrie, B. A., B. R. Keener & A. L. Edwards. 2007. Reinstatement of *Sagittaria macrocarpa* (Alismataceae). *J. Bot. Res. Inst. Texas* **1**: 345–350.

Sorrie, B. A., P. D. McMillan, B. van Eerden, R. J. LeBlond, P. E. Hyatt & L. C. Anderson. 2011. *Carex austrodeflexa* (Cyperaceae) a new species of *Carex* sect. *Acrocystis* from the Atlantic coastal plain of the southeastern United States. *J. Bot. Res. Inst. Texas* **5**: 45–51.

Sorrie, B. A., W. M. Knapp, D. Estes & D. D. Spaulding. 2012. A new *Sisyrinchium* (Iridaceae) from cedar glades in northern Alabama. *J. Bot. Res. Inst. Texas* **6**: 323–329.

Sosiak, A. 2002. Long-term response of periphyton and macrophytes to reduced municipal nutrient loading to the Bow River (Alberta, Canada). *Can. J. Fish. Aquat. Sci.* **59**: 987–1001.

Sousa, W. T. Z. 2011. *Hydrilla verticillata* (Hydrocharitaceae), a recent invader threatening Brazil's freshwater environments: a review of the extent of the problem. *Hydrobiologia* **669**: 1–20.

Southwick, C. H. & F. W. Pine. 1975. Abundance of submerged vascular vegetation in the Rhode River from 1966 to 1973. *Chesapeake Sci.* **16**: 147–151.

Spahr, R., L. Armstrong, D. Atwood & M. Rath. 1991. *Threatened, Endangered, and Sensitive Species of the Intermountain Region*. U.S. Department of Agriculture, Forest Service, Intermountain Region, Ogden, UT. 560 pp.

Sparrow, F. K. 1974. Observations on chytridiaceous parasites of phanerogams. XX. Resting spore germination and epibiotic stage of *Physoderma butomi* Schroeter. *Am. J. Bot.* **61**: 203–208.

Spence, J. F. 1981. The diversity and abundance of the benthic macroinvertebrates in an oligo-mesotrophic central Florida lake. M.S. thesis. University of Central Florida, Orlando, FL. 50 pp.

Spence, J. R. 1996. Demography and monitoring of the autumn buttercup, *Ranunculus aestivalis* (Benson) Van Buren & Harper, southcentral Utah. Pp. 19–27 *In*: J. Maschinski, H. D. Hammond & L. Holter (tech. eds.), *Southwestern Rare and Endangered Plants: Proceeding of the Second Conference; 1995 September 11–14; Flagstaff, AZ*. Gen. Tech. Rep. RM-GTR-283. U.S. Department of Agriculture, Forest Service, Rocky Mountain Forest and Range Experiment Station, Fort Collins, CO.

Spence, D. H. N. & J. Chrystal. 1970. Photosynthesis and zonation of freshwater macrophytes I. Depth distribution and shade tolerance. *New Phytol.* **69**: 205–215.

Spence, D. H. N. & H. M. Dale. 1978. Variations in the shallow water form of *Potamogeton richardsonii* induced by some environmental factors. *Freshw. Biol.* **8**: 251–268.

Spencer, D. F. 1986. Early growth of *Potamogeton pectinatus* L. in response to temperature and irradiance: morphology and pigment composition. *Aquat. Bot.* **26**: 1–8.

Spencer, D. F. 1987. Tuber size and planting depth influence growth of *Potamogeton pectinatus* L. *Am. Midl. Nat.* **118**: 77–84.

Spencer, D. F. & L. W. J. Anderson. 1987. Influence of photoperiod on growth, pigment composition and vegetative propagule formation for *Potamogeton nodosus* Poir. and *Potamogeton pectinatus* L. *Aquat. Bot.* **28**: 103–112.

Spencer, D. F. & G. G. Ksander. 1992. Influence of temperature and moisture on vegetative propagule germination of *Potamogeton* species: implications for aquatic plant management. *Aquat. Bot.* **43**: 351–364.

Spencer, D. F. & G. G. Ksander. 1995. Influence of propagule size, soil fertility, and photoperiod on growth and propagule production by three species of submersed macrophytes. *Wetlands* **15**: 134–140.

Spencer, D. F. & G. G. Ksander. 1996. Growth and carbon utilization by sprouted propagules of two species of submersed rooted aquatic plants grown in darkness. *Hydrobiologia* **317**: 69–78.

Spencer, D. F. & G. G. Ksander. 1997. Influence of anoxia on sprouting of vegetative propagules of three species of aquatic plant propagules. *Wetlands* **17**: 55–64.

Spencer, D. F. & G. G. Ksander. 1999. Phenolic acids and nutrient content for aquatic macrophytes from Fall River, California. *J. Freshw. Ecol.* **14**: 197–209.

Spencer, D. F. & G. G. Ksander. 2001. Comparison of light compensation points for two submersed macrophytes. *J. Freshw. Ecol.* **16**: 509–515.

Spencer, D. F. & G. G. Ksander. 2002. Sedimentation disrupts natural regeneration of *Zannichellia palustris* in Fall River, California. *Aquat. Bot.* **73**: 137–147.

Spencer, D. & G. Ksander. 2003. Nutrient limitation of *Zannichellia palustris* and *Elodea canadensis* growing in sediments from Fall River, California. *J. Freshw. Ecol.* **18**: 207–213.

Spencer, D. F. & M. Rejmánek. 2010. Competition between two submersed aquatic macrophytes, *Potamogeton pectinatus* and *Potamogeton gramineus*, across a light gradient. *Aquat. Bot.* **92**: 239–244.

Spencer, D. & G. Ksander. 2011. Spatial pattern analysis for underground propagules of *Potamogeton gramineus* L. in two northern California irrigation canals. *J. Freshw. Ecol.* **8**: 297–303.

Spencer, D. F., L. W. J. Anderson & G. G. Ksander. 1994. Field and greenhouse investigations on winter bud production by *Potamogeton gramineus* L. *Aquat. Bot.* **48**: 285–295.

Spencer, D. F., F. J. Ryan & G. G. Ksander. 1997. Construction costs for some aquatic plants. *Aquat. Bot.* **56**: 203–214.

Spencer, D. F., G. G. Ksander, J. D. Madsen & C. S. Owens. 2000. Emergence of vegetative propagules of *Potamogeton nodosus*, *Potamogeton pectinatus*, *Vallisneria americana*, and *Hydrilla verticillata* based on accumulated degree-days. *Aquat. Bot.* **67**: 237–249.

Spencer, D. F., C. L. Elmore, G. G. Ksander & J. A. Roncoroni. 2003. Influence of dilute acetic acid treatments on American pondweed winter buds in the Nevada irrigation district, California. *J. Aquat. Plant Manag.* **41**: 65–68.

Sperduto, D. D. 2011. *Natural Community Systems of New Hampshire*. New Hampshire Natural Heritage Bureau, Concord, NH. 119 pp.

Spindler, M. A. & K. F. Hall. 1991. Local movements and habitat use of Tundra or Whistling Swans *Cygnus columbianus* in Kobuk-Selawik Lowlands of northwest Alaska. *Wildfowl* **42**: 17–32.

Spinner, G. P. 1942. A studey of herbaceous plants of value to wildlife in Connecticut. M.S. thesis. University of Michigan, Ann Arbor, MI. 63 pp.

Spinner, G. P. & J. S. Bishop. 1950. Chemical analysis of some wildlife foods in Connecticut. *J. Wildl. Manag.* **14**: 175–180.

Sprenkle, E. S., L. A. Smock & J. E. Anderson. 2004. Distribution and growth of submerged aquatic vegetation in the Piedmont section of the James River, Virginia. *Southeast. Nat.* **3**: 517–530.

Squires, J. R. & S. H. Anderson. 1995. Trumpeter swan (*Cygnus buccinator*) food habits in the Greater Yellowstone ecosystem. *Am. Midl. Nat.* **133**: 274–282.

Squires, M. M. & L. F. W. Lesack. 2003. The relation between sediment nutrient content and macrophyte biomass and community structure along a water transparency gradient among lakes of the Mackenzie Delta. *Can. J. Fish. Aquat. Sci.* **60**: 333–343.

Srivastava, D. S. & R. L. Jefferies. 2002. Intertidal plant communities of an Arctic salt marsh: the influence of isostatic uplift and herbivory. *Écoscience* **9**: 112–118.

Srivastava, D. S., C. A. Staicer & B. Freedman. 1995. Aquatic vegetation of Nova Scotian lakes differing in acidity and trophic status. *Aquat. Bot.* **51**: 181–196.

Stace, C. A. (ed.). 1999. IOPB chromosome data 14. *IOPB Newslett.* **30**: 10–15.

Stafford, N. B. & S. S. Bell. 2006. Space competition between seagrass and *Caulerpa prolifera* (Forsskaal) Lamouroux following simulated disturbances in Lassing Park, FL. *J. Exp. Mar. Biol. Ecol.* **333**: 49–57.

Stalter, R. 1973. Factors influencing the distribution of vegetation of the Cooper River estuary. *Castanea* **38**: 18–24.

Stalter, R. 1985. The flora of Hunting Island, Beaufort County, South Carolina. *Bartonia* **51**: 99–104.

Stalter, R. & S. C. Dial. 1984. Hammock vegetation of Little Talbot Island State Park, Florida. *Bull. Torrey Bot. Club* **111**: 494–497.

Stalter, R. & J. Baden. 1994. A twenty year comparison of vegetation of three abandoned rice fields, Georgetown County, South Carolina. *Castanea* **59**: 69–77.

Standley, P. C. 1919. A new locality for *Senecio crawfordii*. *Rhodora* **21**: 117–120.

Stang, M. 2018. Flowers in lace designs. *Datatèxtil* **38**: 2–11.

Stansbury, J., P. Saunders & D. Winston. 2012. Promoting healthy thyroid function with iodine, bladderwrack, guggul and iris. *J. Restor. Med.* **1**: 83–90.

Staples, T. E., K. C. J. Van Rees & C. van Kessel. 1999. Nitrogen competition using ^{15}N between early successional plants and planted white spruce seedlings. *Can. J. For. Res.* **29**: 1282–1289.

Stearns, L. A. & M. W. Goodwin. 1941. Notes on the winter feeding of the muskrat in Delaware. *J. Wildl. Manag.* **5**: 1–12.

Steele, P. R., K. L. Hertweck, D. Mayfield, M. R. McKain, J. Leebens-Mack & J. C. Pires. 2012. Quality and quantity of data recovered from massively parallel sequencing: examples in Asparagales and Poaceae. *Am. J. Bot.* **99**: 330–348.

Steenkamp, V. 2003. Traditional herbal remedies used by South African women for gynaecological complaints. *J. Ethnopharmacol.* **86**: 97–108.

Stegmaier, C. E. 1966. A leaf-mining *Hippelates* in south Florida (Diptera, Chloropidae). *Florida Entomol.* **49**: 19–21.

Stehle, M. E. 1920. A preliminary survey of the protozoa of Mirror Lake, on the Ohio State University campus. *Ohio J. Sci.* **20**: 89–127.

Steinbauer, G. P. & D. Neil. 1948. Dormancy and germination of seeds of the burreeds. *Pap. Michigan Acad. Sci. Arts Lett.* **34**: 33–37.

Stenlund, D. L. & I. D. Charvat. 1994. Vesicular arbuscular mycorrhizae in floating wetland mat communities dominated by *Typha*. *Mycorrhiza* **4**: 131–137.

Sterling, C. 1980. Comparative morphology of the carpel in the Liliaceae: Helonieae. *Bot. J. Linn. Soc.* **80**: 341–356.

Stetson, S. 1913. The flora of Copake Falls, NY. *Torreya* **13**: 121–133.

Stevens, O. A. 1957. Weights of seeds and numbers per plant. *Weeds* **5**: 46–55.

Stevenson, D. W. 1998. Mayacaceae. Pp. 294–296 *In*: K. Kubitzki (ed.), *The Families and Genera of Vascular Plants, Vol. IV, Flowering Plants: Monocotyledons, Alismatanae and Commelinanae (Except Gramineae)*. Springer-Verlag, Berlin, Germany.

Stevenson, J. C., L. W. Staver & K. W. Staver. 1993. Water quality associated with survival of submersed aquatic vegetation along an estuarine gradient. *Estuaries* **16**: 346–361.

Stevenson, D. W., J. I. Davis, J. V. Freudenstein, C. R. Hardy, M. P. Simmons & C. D. Specht. 2000. A phylogenetic analysis of the monocotyledons based on morphological and molecular character sets, with comments on the placement of *Acorus* and Hydatellaceae. Pp. 17–24 *In*: K. L. Wilson and D. Morrison (eds.), *Systematics and Evolution of Monocots*. CSIRO Publishing, Collingwood, VIC, Australia.

Steward, K. K. 1991. Growth of various hydrilla races in waters of differing pH. *Florida Sci.* **54**: 117–125.

Steward, K. K. 1993. Seed production in monoecious and dioecious populations of *Hydrilla*. *Aquat. Bot.* **46**: 169–183.

Steward, K. K. & T. K. Van. 1987. Comparative studies of monoecious and dioecious hydrilla (*Hydrilla verticillata*) biotypes. *Weed Sci.* **35**: 204–210.

Stewart, R. E. & J. H. Manning. 1958. Distribution and ecology of Whistling Swans in the Chesapeake Bay region. *Auk* **75**: 203–212.

Stewart, R. E. & H. A. Kantrud. 1972. Vegetation of prairie potholes, North Dakota, in relation to quality of water and other environmental factors. Geological Survey Professional Paper 585-D. U.S. Government Printing Office, Washington, DC. 36 pp.

Stewart, G. R. & J. A. Lee. 1974. The role of proline accumulation in halophytes. *Planta* **120**: 279–289.

Stewart, J. G. & L. Rüdenberg. 1980. Microsporocyte growth and meiosis in *Phyllospadix torreyi*, a marine monocotyledon. *Am. J. Bot.* **67**: 949–954.

Stewart, Jr., C. N. & E. T. Nilsen. 1993. Association of edaphic factors and vegetation in several isolated Appalachian peat bogs. *Bull. Torrey Bot. Club* **120**: 128–135.

Stewart, S. L. & L. W. Zettler. 2002. Symbiotic germination of three semi-aquatic rein orchids (*Habenaria repens*, *H. quinquiseta*, *H. macroceratitis*) from Florida. *Aquat. Bot.* **72**: 25–35.

Stewart, S. L. & L. W. Richardson. 2008. Orchid flora of the Florida Panther National Wildlife Refuge. *North Am. Native Orchid J.* **14**: 70–104.

Stewart, H., S. L. Miao, M. Colbert & C. E. Carraher. 1997. Seed germination of two cattail (*Typha*) species as a function of Everglandes nutrient levels. *Wetlands* **17**: 116–122.

Steyermark, J. A. & J. A. Moore. 1933. Report of a botanical expedition into the mountains of western Texas. *Ann. Missouri Bot. Gard.* **20**: 791–806.

Stieglitz, W. O. 1972. Food habits of the Florida duck. *J. Wildl. Manag.* **36**: 422–428.

Stollberg, B. P. 1949. Competition of American coots and shoal-water ducks for food. *J. Wildl. Manag.* **13**: 423–424.

Stollberg, B. P. 1950. Food habits of shoal-water ducks on Horicon Marsh, Wisconsin. *J. Wildl. Manag.* **14**: 214–217.

Stoltz, L. P. 1968. Iris seed dormancy. *Physiol. Plant.* **21**: 1328–1331.

Stoops, C. A., P. H. Adler & J. W. McCreadie. 1998. Ecology of aquatic Lepidoptera (Crambidae: Nymphulinae) in South Carolina, USA. *Hydrobiologia* **379**: 33–40.

Storch, T. A., J. D. Winter & C. Neff. 1986. The employment of macrophyte transplanting techniques to establish *Potamogeton amplifolius* beds in Chautauqua Lake, New York. *Lake Reserv. Manag.* **2**: 263–266.

Stoudt, J. H. 1944. Food preferences of mallards on the Chippewa National Forest, Minnesota. *J. Wildl. Manag.* **8**: 100–112.

Strayer, D. L., C. Lutz, H. M. Malcom, K. Munger & W. H. Shaw. 2003. Invertebrate communities associated with a native (*Vallisneria americana*) and an alien (*Trapa natans*) macrophyte in a large river. *Freshw. Biol.* **48**: 1938–1949.

Strazisar, T., M. S. Koch, E. Dutra & C. J. Madden. 2013a. *Ruppia maritima* L. seed bank viability at the Everglades-Florida Bay ecotone. *Aquat. Bot.* **111**: 26–34.

Strazisar, T., M. S. Koch, C. J. Madden, J. Filina, P. U. Lara & A. Mattair. 2013b. Salinity effects on *Ruppia maritima* L. seed germination and seedling survival at the Everglades-Florida Bay ecotone. *J. Exp. Mar. Biol. Ecol.* **445**: 129–139.

Strazisar, T., M. S. Koch & C. J. Madden. 2015. Seagrass (*Ruppia maritima* L.) life history transitions in response to salinity dynamics along the Everglades-Florida Bay ecotone. *Estuaries Coast.* **38**: 337–352.

Streng, D. R. & P. A. Harcombe. 1982. Why don't east Texas savannas grow up to forest? *Am. Midl. Nat.* **108**: 278–294.

Stricker, S. A. 1985. A new species of *Tetrastemma* (Nemertea, Monostilifera) from San Juan Island, Washington, USA. *Can. J. Zool.* **63**: 682–690.

Strong, W. L. 2000. Vegetation development on reclaimed lands in the Coal Valley Mine of western Alberta, Canada. *Can. J. Bot.* **78**: 110–118.

Strong, M. T. & P. M. Sheridan. 1991. *Juncus caesariensis* Coville (Juncaceae) in Virginia peat bogs. *Castanea* **56**: 65–69.

Strong, M. T. & C. L. Kelloff. 1994. Intertidal vascular plants of Brent Marsh, Potomac River, Stafford County, Virginia. *Castanea* **59**: 354–366.

Sree, K. S., S. Sudakaran & K.-J. Appenroth. 2015. How fast can angiosperms grow? Species and clonal diversity of growth rates in the genus *Wolffia* (Lemnaceae). *Acta Physiol. Plant.* **37**: 204.

Sree, K. S., M. Bog & K. J. Appenroth. 2016. Taxonomy of duckweeds (Lemnaceae), potential new crop plants. *Emir. J. Food Agric.* **28**: 291–302.

Stone, K. R. 2009. *Iris pseudacorus*. *In*: *Fire Effects Information System*. U.S. Department of Agriculture, Forest Service, Rocky Mountain Research Station, Fire Sciences Laboratory. Available online: https://www.fs.fed.us/database/feis/plants/forb/iripse/all.html [accessed 31 January, 2019].

Stratman, K., W. A. Overholt, J. P. Cuda, M. D. Netherland & P. C. Wilson. 2013. The diversity of Chironomidae associated with *Hydrilla* in Florida, with special reference to *Cricotopus lebetis* (Diptera: Chironomidae). *Florida Entomol.* **96**: 654–657.

Stuart, W. 2018. Cape Fear native flora focus. *Native Plant News* **15**(4): 1, 5.

Stubbs, C. S., H. A. Jacobson, E. A. Osgood & F. A. Drummond. 1992. Alternative forage plants for native (wild) bees associated with lowbush blueberry, *Vaccinium* spp., in Maine. Technical Bulletin 148. Maine Agricultural Experiment Station, Orono, ME. 54 pp.

Stubler, A. D., L. J. Jackson, B. T. Furman & B. J. Peterson. 2017. Seed production patterns in *Zostera marina*: effects of patch size and landscape configuration. *Estuaries Coast.* **40**: 564–572.

Stuckey, I. H. 1967. Environmental factors and the growth of native orchids. *Am. J. Bot.* **54**: 232–241.

Stuckey, R. L. 1968. Distributional history of *Butomus umbellatus* (flowering-rush) in the western Lake Erie and Lake St. Clair region. *Michigan Bot.* **7**: 134–142.

Stuckey, R. L. 1971. Changes of vascular aquatic flowering plants during 70 years in Put-in-Bay Harbor, Lake Erie, Ohio. *Ohio J. Sci.* **71**: 321–342.

Stuckey, R. L. 1979. Distributional history of *Potamogeton crispus* (curly pondweed) in North America. *Bartonia* **46**: 22–42.

Stuckey, R. L. 1983. Absence of certain aquatic vascular plants from the prairie peninsula. Pp. 97–103 *In*: R. Brewer (ed.), *Proceedings of the Eighth North American Prairie Conference*. Western Michigan University, Kalamazoo, MI.

Stuckey, R. L. 1985. Distributional history of *Najas marina* (spiny naiad) in North America. *Bartonia* **51**: 2–16.

Stuckey, R. L. & D. H. Les. 1984. *Pistia stratiotes* (Water Lettuce) recorded from Florida in Bartram's Travels, 1765–74. *Aquaphyte* **4**(2): 1, 6.

Stuckey, R. L. & D. L. Moore. 1995. Return and increase in abundance of aquatic flowering plants in Put-In-Bay Harbor, Lake Erie, Ohio. *Ohio J. Sci.* **95**: 261–266.

Stuckey, I. H. & L. L. Gould. 2000. *Coastal Plants from Cape Cod to Cape Canaveral*. The University of North Carolina Press, Chapel Hill, NC. 305 pp.

Stuckey, R. L., J. R. Wehrmeister & R. J. Bartolotta. 1978. Submersed aquatic vascular plants in ice-covered ponds of central Ohio. *Rhodora* **80**: 575–580.

Sturtevant, R., L. Berent, T. Makled, W. Conard, A. Fusaro & E. Rutherford. 2016. An overview of the management of established nonindigenous species in the Great Lakes. NOAA Technical Memorandum GLERL-168. GLERL Information Services, Ann Arbor, MI. 273 pp.

Stützel, T. 1984. Blüten- und infloreszenmorphologische Untersuchungen zur Systematik der Eriocaulaceen. *Diss. Bot.* **71**: 1–108.

Stützel, T. 1998. Eriocaulaceae. Pp. 197–207 *In*: K. Kubitzki, H. Huber, P. J. Rudall, P. S. Stevens & T. Stützel (eds.), *The Families and Genera of Vascular Plants, Vol. IV. Flowering Plants, Monocotyledons: Alismatanae and Commelinanae* (*Except Gramineae*). Springer-Verlag, Berlin, Germany.

Stützel, T. & N. Gansser. 1996. Floral morphology of North American Eriocaulaceae and its taxonomic implications. *Feddes Repert.* **106**: 495–502.

Stutzenbaker, C. D. 1999. *Aquatic and Wetland Plants of the Western Gulf Coast.* Texas Parks & Wildlife Press/University of Texas Press, Austin, TX. 465 pp.

Su, K. L. & E. J. Staba. 1972. Aquatic plants from Minnesota. Part 1 - chemical survey. Bulletin 46. Water Resources Research Center, University of Minnesota, Minneapolis, MN. 50 pp.

Su, K. L., E. J. Staba & Y. Abul-Hajj. 1972. Aquatic plants from Minnesota. Part 3 - antimicrobial effects. Bulletin 48. Water Resources Research Center, University of Minnesota, Minneapolis, MN. 36 pp.

Su, K. L., Y. Abul-Hajj & E. J. Staba. 1973. Antimicrobial effects of aquatic plants from Minnesota. *Lloydia* **36**: 80–87.

Su, H., Y. Zhao, J. Jiang, Q. Lu, Q. Li, Y. Luo, H. Zhao & M. Wang. 2014. Use of duckweed (*Landoltia punctata*) as a fermentation substrate for the production of higher alcohols as biofuels. *Energy Fuels* **28**: 3206–3216.

Subudhi, H. N., S. P. Panda, P. K. Behera & C. Patnaik. 2015. A check list of weeds in rice fields of coastal Orissa, India. Pp. 1–10 *In*: H. Wang (ed.), *Soil Ecology and Land-Use Management.* White Word Publications, New York, NY.

Sugden, L. G. & E. A. Driver. 1980. Natural foods of mallards in Saskatchewan parklands during late summer and fall. *J. Wildl. Manag.* **44**: 705–709.

Suksathan, P., M. H. Gustafsson & F. Borchsenius. 2009. Phylogeny and generic delimitation of Asian Marantaceae. *Bot. J. Linn. Soc.* **159**: 381–395.

Sullivan, V. I. 1981. *Najas minor* (Najadaceae) in Louisiana. *Sida* **9**: 88–90.

Sullivan, M. J. 1977. Structural characteristics of a diatom community epiphytic on *Ruppia maritima*. *Hydrobiologia* **53**: 81–86.

Sullivan, G. & J. E. Titus. 1996. Physical site characteristics limit pollination and fruit set in the dioecious hydrophilous species, *Vallisneria americana*. *Oecologia* **108**: 285–292.

Sulman, J. D., B. T. Drew, C. Drummond, E. Hayasaka & K. J. Sytsma. 2013. Systematics, biogeography, and character evolution of *Sparganium* (Typhaceae): diversification of a widespread, aquatic lineage. *Am. J. Bot.* **100**: 2023–2039.

Sultana, N., S. S. Sultana & S. S. Alam. 2013. Karyotype and RAPD analysis to elucidate taxonomic status in two morphological forms of *Egeria densa* Planch. and *Hydrilla verticillata* (L.f.) Royle. *Cytologia* **78**: 277–284.

Sumoski, S. E. & R. J. Orth. 2012. Biotic dispersal in eelgrass *Zostera marina*. *Mar. Ecol. Prog. Ser.* **471**: 1–10.

Sun, Y., S. A. White, D. Mann & J. Adelberg. 2012. Chilling requirements to break dormancy of *Veratrum californicum*. *HortScience* **47**: 1710–1713.

Sundarasekar, J., G. Sahgal, V. Murugaiyah, L. K. Lay, O. M. Thong & S. Subramaniam. 2018. Wound healing activity of *Hymenocallis littoralis*-moving beyond ornamental plant. *Pak. J. Pharm. Sci.* **31**: 2537–2543.

Sundell, E. & R. D. Thomas. 1988. Four new records of *Cyperus* (Cyperaceae) in Arkansas. *Sida* **13**: 259–261.

Sundue, M. 2006. Addendum to field trip reports. *J. Torrey Bot. Soc.* **133**: 210–211.

Sundue, M. 2007. Field trip reports. *J. Torrey Bot. Soc.* **134**: 144–152.

Surdick, J. A. & A. M. Jenkins. 2010. *Population Surveys of Rare Lauraceae Species to Assess the Effect of Laurel wilt Disease in Florida.* Florida Natural Areas Inventory, Tallahassee, FL. 15 pp.

Suresh, G., P. P. Reddy, K. S. Babu, T. B. Shaik & S. V. Kalivendi. 2010. Two new cytotoxic labdane diterpenes from the rhizomes of *Hedychium coronarium*. *Bioorg. Med. Chem. Lett.* **20**: 7544–7548.

Surveswaran, S., V. Gowda & M. Sun. 2018. Using an integrated approach to identify cryptic species, divergence patterns and hybrid species in Asian ladies' tresses orchids (Spiranthes, Orchidaceae). *Mol. Phylogenet. Evol.* **124**: 106–121.

Suter, R. B., P. R. Miller & G. E. Stratton. 2011. Egg capsule architecture and siting in a leaf-curling sac spider, *Clubiona riparia* (Araneae: Clubionidae). *J. Arachnol.* **39**: 76–83.

Sutherland, S. 1986. Floral sex ratios, fruit-set, and resource allocation in plants. *Ecology* **67**: 991–1001.

Sutherland, W. J. 1990. *Iris pseudacorus* L. *J. Ecol.* **78**: 833–848.

Sutter, R. 1984. The status of *Helonias bullata* L. (Liliaceae) in the southern Appalachians. *Castanea* **49**: 9–16.

Sutter, L. A., J. E. Perry & R. M. Chambers. 2014. Tidal freshwater marsh plant responses to low level salinity increases. *Wetlands* **34**: 167–175.

Sutton, D. L. 1991. Culture and growth of pickerelweed from seedlings. *J. Aquat. Plant Manag.* **29**: 39–42.

Sutton, D. L., T. K. Van & K. M. Portier. 1992. Growth of dioecious and monoecious hydrilla from single tubers. *J. Aquat. Plant Manag.* **30**: 15–20.

Svedelius, N. 1932. On the different types of pollination in *Vallisneria spiralis* L. and *Vallisneria americana* Michx. *Svensk Bot. Tidskr.* **26**: 1–12.

Swales, D. E. 1979. Nectaries of certain arctic and sub-arctic plants with notes on pollination. *Rhodora* **81**: 363–407.

Swanson, G. A. & J. C. Bartonek. 1970. Bias associated with food analysis in gizzards of blue-winged teal. *J. Wildl. Manag.* **34**: 739–746.

Swindale, D. N. & J. T. Curtis. 1957. Phytosociology of the larger submerged plants in Wisconsin lakes. *Ecology* **38**: 397–407.

Swinehart, A. L. 1995. Paleoecology of an alkaline peatland in Elkhart County, Indiana. *Proc. Indiana Acad. Sci.* **104**: 43–46.

Sykes, Jr., P. W., C. B. Kepler, K. L. Litzenberger, H. R. Sansing, E. T. R. Lewis & J. S. Hatfield. 1999. Density and habitat of breeding swallow-tailed kites in the Lower Suwannee Ecosystem, Florida. *J. Field Ornithol.* **70**: 321–336.

Sytsma, M. D. & M. Pfauth. 2006. 2005 Diamond Lake submersed aquatic vegetation survey. Center for Lakes and Reservoirs Publications and Presentations Paper 43. Portland State University, Portland, OR. 17 pp.

Szlachetko, D. L. & M. Kolanowska. 2013. Three new species of *Ponthieva* (Orchidaceae, Spiranthoideae) from Colombia and Venezuela. *Plant Syst. Evol.* **299**: 1671–1678.

Takahashi, M. & S. Kawano. 1989. Pollen morphology of the Melanthiaceae and its systematic implications. *Ann. Missouri Bot. Gard.* **76**: 863–876.

Takahashi, C. & K. Kondo. 2004. A comparison of karyotypes in two artificial reciprocal hybrids between *Pogonia minor* and *Pogonia ophioglossoides* (Orchidaceae). *Chromosome Sci.* **8**: 11–16.

Takahashi, K. & T. Asaeda. 2014. The effect of spring water on the growth of a submerged macrophyte *Egeria densa*. *Landscape Ecol. Eng.* **10**: 99–107.

Takahashi, C. T., M. Itsuji, T. Yagame & K. Kondo. 2014. Newly naturally appeared *Pogonia* isolated from *Pogonia japonica* and *P. minor* in Japan analyzed by RAPD (randomly amplified polymorphic DNA). *Chromosom. Bot.* **9**: 77–82.

Takayanagi, S., Y. Takagi, A. Shimizu & H. Hasegawa. 2012. The shoot is important for high-affinity nitrate uptake in *Egeria densa*, a submerged vascular plant. *J. Plant Res.* **125**: 669–678.

Takemoto, B. K. & R. D. Noble. 1986. Differential sensitivity of duckweeds (Lemnaceae) to sulphite. *New Phytol.* **103**: 525–539.

Takeuchi, Y., T. Sassa, S. Kawaguchi, M. Ogasawara, K. Yoneyama & M. Konnai. 1995. Stimulation of germination of *Monochoria vaginalis* seeds by seed coat puncture and cotylenins. *Weed Res. Japan* **40**: 221–224.

Takhtajan, A. 1969. *Flowering Plants: Origin and Dispersal.* Smithsonian Institution Press, Washington, DC. 310 pp.

Takhtajan, A. 1995. New families of the monocotyledons [in Russian]. *Bot. Zhur.* **79**: 65–66.

Takhtajan, A. 2009. *Flowering Plants.* Springer Science & Business Media B. V., Berlin, Germany. 871 pp.

Takimoto, A. & O. Tanaka. 1973. Effects of some SH-inhibitors and EDTA on flowering in *Lemna perpusilla* 6746. *Plant Cell Physiol.* **14**: 1133–1141.

Takos, M. J. 1947. A semi-quantitative study of muskrat food habits. *J. Wildl. Manag.* **11**: 331–339.

Talavera, S., P. Garcia-Murillo & J. Herrera. 1993. Chromosome numbers and a new model for karyotype evolution in *Ruppia* L. (Ruppiaceae). *Aquat. Bot.* **45**: 1–13.

Talbot, S. L., S. Wyllie-Echeverria, D. H. Ward, J. R. Rearick, G. K. Sage, B. Chesney & R. C. Phillips. 2006. Genetic characterization of *Zostera asiatica* on the Pacific Coast of North America. *Aquat. Bot.* **85**: 169–176.

Talbot, S. S., S. L. Talbot & L. R. Walker. 2010. Post-eruption legacy effects and their implications for long-term recovery of the vegetation on Kasatochi Island, Alaska. *Arct. Antarct. Alp. Res.* **42**: 285–296.

Talbot, S. L., G. K. Sage, J. R. Rearick, M. C. Fowler, R. Muñiz-Salazar, B. Baibak, S. Wyllie-Echeverria, A. Cabello-Pasini & D. H. Ward. 2016. The structure of genetic diversity in eelgrass (*Zostera marina* L.) along the north Pacific and Bering Sea coasts of Alaska. *PLoS One* **11**(4): e0152701.

Talukdar, T. & D. Talukdar. 2015. Heavy metal accumulation as phytoremediation potential of aquatic macrophyte, *Monochoria vaginalis* (Burm. F.) K. Presl ex Kunth. *Int. J. Appl. Sci. Biotechnol.* **3**: 9–15.

Tamada, K., K. Itoh, Y. Uchida, S. Higuchi, D. Sasayama & T. Azuma. 2015. Relationship between the temperature and the overwintering of water lettuce (*Pistia stratiotes*) at Kowataike, a branch of Yodogawa River, Japan. *Weed Biol. Manag.* **15**: 20–26.

Tamura, M. N., S. Fuse, H. Azuma & M. Hasebe. 2004. Biosystematic studies on the family Tofieldiaceae I. Phylogeny and circumscription of the family inferred from DNA sequences of *matK* and *rbcL*. *Plant Biol.* **6**: 562–567.

Tan, K. W. 1969. The systematic status of the genus *Bletilla* (Orchidaceae). *Brittonia* **21**: 202–214.

Tan, B. H. & W. S. Judd. 1995. A floristic inventory of O'Leno State Park and Northeast River Rise State Preserve, Alachua and Columbia Counties, Florida. *Castanea* **60**: 141–165.

Tanaka, H. 2004. Reproductive biology of *Lysichiton camtschatcense* (Araceae) in Japan. *Aroideana* **27**: 167–171.

Tanaka, N. 2019. Taxonomy, evolution and phylogeography of the genus *Helonias* (Melanthiaceae) revisited. *Phytotaxa* **390**: 1–84.

Tanaka, N., K. Uehara & J. Murata. 2004. Correlation between pollen morphology and pollination mechanisms in the Hydrocharitaceae. *J. Plant Res.* **117**: 265–276.

Tanaka, N., J. Kuo, Y. Omori, M. Nakaoka & K. Aioi. 2003. Phylogenetic relationships in the genera *Zostera* and *Heterozostera* (Zosteraceae) based on *matK* sequence data. *J. Pl. Res.* **116**: 273–279.

Tanaka, N., H. Uchiyama, H. Matoba & T. Koyama. 2009. Karyological analysis of the genus *Canna* (Cannaceae). *Plant Syst. Evol.* **280**: 45–51.

Tang, L.-L. & S.-Q. Huang. 2007. Evidence for reductions in floral attractants with increased selfing rates in two heterandrous species. *New Phytol.* **175**: 588–595.

Tang, F., J. A. White & I. Charvat. 2001. The effect of phosphorus availability on arbuscular mycorrhizal colonization of *Typha angustifolia*. *Mycologia* **93**: 1042–1047.

Tang, Y., L. Chen, X. Wei, Q. Yao & T. Li. 2013. Removal of lead ions from aqueous solution by the dried aquatic plant, *Lemna perpusilla* Torr. *J. Hazard. Mater.* **244**: 603–612.

Tang, J., F. Zhang, W. Cui & J. Ma. 2014. Genetic structure of duckweed population of *Spirodela*, *Landoltia* and *Lemna* from Lake Tai, China. *Planta* **239**: 1299–1307.

Tang, J., Y. Li, X. Wang & M. Daroch. 2017. Effective adsorption of aqueous Pb^{2+} by dried biomass of *Landoltia punctata* and *Spirodela polyrhiza*. *J. Clean. Prod.* **145**: 25–34.

Tao, Q., B. Li, Q. Li, X. Han, Y. Jiang, R. Jupa, C. Wang & T. Li. 2019. Simultaneous remediation of sediments contaminated with sulfamethoxazole and cadmium using magnesium-modified biochar derived from *Thalia dealbata*. *Sci. Total Environ.* **659**: 1448–1456.

Tárano, Z., S. Strahl & J. Ojasti. 1995. Feeding ecology of the purple gallinule *Porphyrula martinica* in the Central Llanos of Venezuela. *Ecotropicos* **8**: 53–61.

Tarbeeva, D. V., S. A. Fedoreev, M. V. Veselova, A. I. Kalinovskii, P. G. Gorovoi, O. S. Vishchuk, S. P. Ermakova & P. A. Zadorozhnyi. 2015. Polyphenolic metabolites from *Iris pseudacorus* roots. *Chem. Nat. Compd.* **51**: 451–455.

Tarver, D. P., J. A. Rodgers, M. J. Mahler & R. L. Lazor. 1978. *Aquatic and Wetland Plants of Florida.* Bureau of Aquatic Plant Research and Control, Florida Department of Natural Resources, Tallahassee, FL. 127 pp.

Tatkowska, E. & J. Buczek. 1983. Effect of ammonium nutrition on the nitrate utilization, nitrate reductase actvity and growth of *Spirodela polyrrhiza*. *Acta Soc. Bot. Polon.* **52**: 241–252.

Tavares, F., F. R. Lapolli, R. Roubach, M. K. Jungles, D. M. Fracalossi & A. M. Moraes. 2010. Use of domestic effluent through duckweeds and red tilapia farming in integrated system. *Pan-Am. J. Aquat. Sci.* **5**: 1–10.

Tavechio, W. L. G. & S. M. Thomaz. 2003. Effects of light on the growth and photosynthesis of *Egeria najas* Planchon. *Brazilian Arch. Biol. Technol.* **46**: 203–209.

Taylor, W. P. 1912. Field notes on amphibians, reptiles and birds of northern Humboldt County, Nevada. *Univ. Calif. Publ. Zool.* **7**: 319–436.

Taylor, C. A. 1956. Alkaloid yields of *Veratrum fimbriatum* as influenced by site, season and other factors. *Econ. Bot.* **10**: 166–173.

Taylor, R. L. 1967. The foliar embryos of *Malaxis paludosa*. *Can. J. Bot.* **45**: 1553–1556.

Taylor, B. R. & J. Helwig. 1995. Submergent macrophytes in a cooling pond in Alberta, Canada. *Aquat. Bot.* **51**: 243–257.

Taylor, K. N. & D. Estes. 2012. The floristic and community ecology of seasonally wet limestone glade seeps of Tennessee and Kentucky. *J. Bot. Res. Inst. Texas* **6**: 711–724.

Taylor, B. R., J. Ferrier, R. Lauff & D. J. Garbary. 2008. New distribution records for flowering plants in Antigonish County, Nova Scotia. *Proc. Trans. Nova Scotian Inst. Nat. Sci.* **44**: 109–123.

Taylor, M. L., K. M. Altrichter & L. B. Aeilts. 2018. Pollen ontogeny in *Ruppia* (Alismatidae). *Int. J. Plant Sci.* **179**: 217–230.

Teamkao, P. & P. Thiravetyan. 2010. Phytoremediation of ethylene glycol and its derivatives by the burhead plant (*Echinodorus cordifolius* (L.)): effect of molecular size. *Chemosphere* **81**: 1069–1074.

Teltscherová, L. & S. Hejný. 1973. The germination of some *Potamogeton* species from south-Bohemian fishponds. *Folia Geobot. Phytotax.* **8**: 231–239.

Tennessen, K. J. 1993. New distribution records for *Ophiogomphus howei* (Odonata: Gomphidae). *Great Lakes Entomol.* **26**: 245–249.

Tepedino, V. J. 1981. Notes on the reproductive biology of *Zigadenus paniculatus*, a toxic range plant. *Great Basin Nat.* **41**: 427–430.

Tepedino, V. J., A. K. Knapp, G. C. Eickwort & D. C. Ferguson. 1989. Death camas (*Zigadenus nuttallii*) in Kansas: pollen collectors and a florivore. *J. Kansas Entomol. Soc.* **62**: 411–412.

Terrados, J. & S. L. Williams. 1997. Leaf versus root nitrogen uptake by the surfgrass *Phyllospadix torreyi*. *Mar. Ecol. Prog. Ser.* **149**: 267–277.

Terrell, E. E., W. H. Emery & H. E. Beaty. 1978. Observations on *Zizania texana* (Texas wildrice), an endangered species. *Bull. Torrey Bot. Club* **105**: 50–57.

Teryokhin, E. S. 1996. The mechanism of hyphydrophilous pollination in *Potamogeton filiformis* Pers. (Potamogetonaceae). *Repr. Biol.* **96**: 11.

Teryokhin, E. S., S. I. Chubarov & V. O. Romanova. 2002. Pollination, mating systems and self-incompatibility in some species of the genus *Potamogeton* L. *Phytomorphology* **52**: 249–261.

Thacker, R. W., B. A. Hazlett, L. A. Esman, C. P. Stafford & T. Keller. 1993. Color morphs of the crayfish *Orconectes virilis*. *Am. Midl. Nat.* **129**: 182–199.

Tharaldson, T. M. 2018. The ability of *Phyllospadix* spp., a pair of intertidal foundation species, to maintain biodiversity and ameliorate CO_2 stress in rocky shore tidepools. M.S. thesis. Humboldt State University, Arcata, CA. 104 pp.

Thiébaut, G., Y. Gross, P. Gierlinski & A. Boiché. 2010. Accumulation of metals in *Elodea canadensis* and *Elodea nuttallii*: implications for plant–macroinvertebrate interactions. *Sci. Total Environ.* **408**: 5499–5505.

Thien, L. B. & B. G. Marcks. 1972. The floral biology of *Arethusa bulbosa*, *Calopogon tuberosus*, and *Pogonia ophioglossoides* (Orchidaceae). *Can. J. Bot.* **50**: 2319–2325.

Thieret, J. W. 1963. Botanical survey along the Yellowknife Highway, Northwest Territories, Canada. I. Catalogue of the flora. *Sida* **1**: 117–170.

Thieret, J. W. 1964. Botanical survey along the Yellowknife Highway, Northwest Territories, Canada: II. Vegetation. *Sida* **1**: 187–239.

Thieret, J. W. 1969. *Sagittaria guayanensis* (Alismataceae) in Louisiana: new to the United States. *Sida* **3**: 445.

Thieret, J. W. 1970. *Bacopa repens* (Scrophulariaceae) in the conterminous United States. *Castanea* **35**: 132–136.

Thieret, J. W. 1971a. *Eriocaulon cinereum* in Louisiana. *Southwest. Nat.* **15**: 391.

Thieret, J. W. 1971b. Additions to the Louisiana flora. *Castanea* **36**: 219–222.

Thieret, J. W. 1975. The Mayacaceae in the southeastern United States. *J. Arnold Arb.* **56**: 248–255.

Thieret, J. W. 1982. The Sparganiaceae in the southeastern United States. *J. Arnold Arb.* **63**: 341–355.

Thieret, J. W., R. R. Haynes & D. H. Dike. 1969. *Blyxa aubertii* (Hydrocharitaceae) in Louisiana: new to North America. *Sida* **3**: 343–344.

Theriot, R. F. 1993. Flood tolerance of plant species in bottomland forests of the southeastern United States. Wetlands Research Program Technical Report WRP-DE-6. U.S. Army Corps of Engineers, Vicksburg, MS. 50 pp.

Thom, R., B. Miller & M. Kennedy. 1995. Temporal patterns of grazers and vegetation in a temperate seagrass system. *Aquat. Bot.* **50**: 201–205.

Thomas, M. C. 2005. An exotic baridine weevil pest (Coleoptera: Curculionidae) of Amaryllidaceae in Florida. Florida Department of Agriculture and Consumer Services, Division of Plant Industry. Available online: www.doacs.state.fl.us/pi/enpp/ento/amaryllisweevil.html [accessed 25 November, 2018].

Thomas, J. R. 2017. New additions, vouchers of old additions, and a new combination (*Dichanthelium inflatum*) for the Missouri flora. *Missouriensis* **34**: 4–19.

Thomas, V. G. & J. P. Prevett. 1980. The nutritional value of arrowgrasses to geese at James Bay. *J. Wildl. Manag.* **44**: 830–836.

Thomas, A. L. & D. Schrock. 2004. Performance of 67 native Midwestern U.S. perennials in a low-maintenance landscape. *HortTechnology* **14**: 381–388.

Thomaz, S. M., P. A. Chambers, S. A. Pierini & G. Pereira. 2007. Effects of phosphorus and nitrogen amendments on the growth of *Egeria najas*. *Aquat. Bot.* **86**: 191–196.

Thompson, C. H. 1898. A revision of the American Lemnaceae occurring north of Mexico. *Rep. Missouri Bot. Gard.* **9**: 21–42.

Thompson, R. L. 1970. Florida sandhill crane nesting on the Loxahatchee National Wildlife Refuge. *Auk* **87**: 492–502.

Thompson, S. A. 2000a. Acoraceae Martinov. Pp. 124–127 *In*: Flora North America Editorial Committee (eds.), *Flora of North America North of Mexico, Vol. 22: Magnoliophyta: Alismatidae, Arecidae, Commelinidae (in Part), and Zingiberidae*. Oxford University Press, New York, NY.

Thompson, S. A. 2000b. Araceae Jussieu. Pp. 128–142 *In*: Flora North America Editorial Committee (eds.), *Flora of North America North of Mexico, Vol. 22: Magnoliophyta: Alismatidae, Arecidae, Commelinidae (in Part), and Zingiberidae*. Oxford University Press, New York, NY.

Thompson, R. L. 2001. Botanical survey of Myrtle Island Research Natural Area, Oregon. General Technical Report PNW-GTR-507. U.S. Department of Agriculture, Forest Service, Pacific Northwest Research Station, Portland, OR. 27 pp.

Thompson, J. S. 2010. *Sarracenia* minor var. *okefenokeensis* (Sarraceniaceae) discovered outside of the Okefenokee Swamp area. *J. Bot. Res. Inst. Texas* **4**: 771–773.

Thompson, R. L. & L. E. McKinney. 2006. Vascular flora and plant habitats of an abandoned limestone quarry at Center Hill Dam, DeKalb County, Tennessee. *Castanea* **71**: 54–64.

Thompson, R. L., J. R. Abbott & A. E. Shupe. 2005. Vascular flora from five plant habitats of an abandoned limestone quarry in Clark County, Kentucky. *J. Kentucky Acad. Sci.* **66**: 24–35.

Thompson, K., D. Perleberg & S. Loso. 2009. *Final Report on the Sensitive Lakeshore Surveys for Little Boy Lake (11-0167-00), Wabedo Lake (11-0171-00), and Louise Lake (11-0573-00), Cass County, MN*. Division of Ecological Resources, Minnesota Department of Natural Resources, St. Paul, MN. 85 pp.

Thompson, I. D., P. A. Wiebe, E. Mallon, A. R. Rodgers, J. M. Fryxell, J. A. Baker & D. Reid. 2014. Factors influencing the seasonal diet selection by woodland caribou (*Rangifer tarandus tarandus*) in boreal forests in Ontario. *Can. J. Zool.* **93**: 87–98.

Thomson, E. R. 2005. Papillate watermeal, *Wolffia brasiliensis*, in eastern Ontario: an addition to the flora of Canada. *Can. Field-Nat.* **119**: 137–138.

Thorhaug, A., A. D. Richardson & G. P. Berlyn. 2006. Spectral reflectance of *Thalassia testudinum* (Hydrocharitaceae) seagrass: low salinity effects. *Am. J. Bot.* **93**: 110–117.

Thormann, M. N. & S. E. Bayley. 1997. Response of aboveground net primary plant production to nitrogen and phosphorus fertilization in peatlands in southern boreal Alberta, Canada. *Wetlands* **17**: 502–512.

Thorne, R. F. 1982. The desert and other transmontane plant communities of southern California. *Aliso* **10**: 219–257.

Thorne, R. F. & E. W. Lathrop. 1969. A vernal marsh on the Santa Rosa Plateau of Riverside County, California. *Aliso* **7**: 85–95.

Thorne, R. F. & E. W. Lathrop. 1970. *Pilularia americana* on the Santa Rosa Plateau, Riverside County, California. *Aliso* **7**: 149–155.

Thorne, J. H., P. R. Huber & S. Harrison. 2011. Systematic conservation planning: protecting rarity, representation, and connectivity in regional landscapes. Pp. 309–328 *In*: S. Harrison & N. Rajakaruna (eds.), *Serpentine: The Evolution and Ecology of a Model System*. University of California Press, Berkeley, CA.

Thornhill, A. D. 1996. Species and population-level patterns of genetic variation in *Epipactis gigantea* (Orchidaceae), with examination of local genetic and clonal structure in riparian and bog populations inferred from allozyme analysis. Ph.D. dissertation. University of California-Irvine, Irvine, CA. 73 pp.

Thornhill, R., A. Krings, D. Lindbo & J. Stucky. 2014. Guide to the vascular flora of the savannas and flatwoods of Shaken Creek Preserve and vicinity (Pender & Onslow counties, North Carolina, USA. *Biodivers. Data J.* **2**: e1099.

Thorp, A. G., R. C. Jones & D. P. Kelso. 1997. A comparison of water-column macroinvertebrate communities in beds of differing submersed aquatic vegetation in the tidal freshwater Potomac River. *Estuaries* **20**: 86–95.

Thorpe, A. S. 2008. *Habitat Sampling at Hanson, Long Tom, North Taylor, Speedway, and Turtle Swale*. Institute for Applied Ecology, Corvallis, OR. 19 pp.

Thursby, G. B. 1984. Root-exuded oxygen in the aquatic angiosperm *Ruppia maritima*. *Mar. Ecol. Prog. Ser.* **16**: 303–305.

Thursby, G. B. & M. M. Harlin. 1984. Interaction of leaves and roots of *Ruppia maritima* in the uptake of phosphate, ammonia and nitrate. *Mar. Biol.* **83**: 61–67.

Thyagarajan, G. (ed.). 1984. *Water Hyacinth. Proceedings of the International Conference on Water Hyacinth*. United Nations Environment Programme, Nairobi, Kenya. 1005 pp.

Tian, L.-W., Z. Zhang, H.-L. Long & Y.-J. Zhang. 2017. Steroidal saponins from the genus *Smilax* and their biological activities. *Nat. Prod. Bioprospect.* **7**: 283–298.

Tibiriçá, E., A. Almeida, S. Cailleaux, D. Pimenta, M. A. Kaplan, M. A. Lessa & M. R. Figueiredo. 2007. Pharmacological mechanisms involved in the vasodilator effects of extracts from *Echinodorus grandiflorus*. *J. Ethnopharmacol.* **111**: 50–55.

Tietje, W. D. & J. G. Teer. 1996. Winter feeding ecology of northern shovelers on freshwater and saline wetlands in south Texas. *J. Wildl. Manag.* **60**: 843–855.

Tiffany, L. H., J. F. Shearer & G. Knaphus. 1990. Plant parasitic fungi of four tallgrass prairies of northern Iowa: distribution and prevalence. *J. Iowa Acad. Sci.* **97**: 157–166.

Tjia, B. 1985. Hybrid calla lilies: a potential new crop for Florida. *Proc. Florida State Hort. Soc.* **98**: 127–130.

Tiling, K. & C. E. Proffitt. 2017. Effects of *Lyngbya majuscula* blooms on the seagrass *Halodule wrightii* and resident invertebrates. *Harmful Algae* **62**: 104–112.

Tillie, N., M. W. Chase & T. Hall. 2000. Molecular studies in the genus *Iris* L.: a preliminary study. *Ann. Bot. (Rome)* **58**: 105–112.

Timbrook, J. 1990. Ethnobotany of Chumash indians, California, based on collections by John P. Harrington. *Econ. Bot.* **44**: 236–253.

Timm, A. L. & R. B. Pierce. 2015. Vegetative substrates used by larval northern pike in Rainy and Kabetogama Lakes, Minnesota. *Ecol. Freshw. Fish* **24**: 225–233.

Timoney, K. P. 2001. String and net-patterned salt marshes: rare landscape elements of Boreal Canada. *Can. Field-Nat.* **115**: 406–412.

Tiner, R. W. 1987. *A Field Guide to Coastal Wetland Plants of the Northeastern United States*. The University of Massachusetts Press, Amherst, MA. 286 pp.

Tiner, R. W. 1993. *Field Guide to Coastal Wetland Plants of the Southeastern United States*. The University of Massachusetts Press, Amherst, MA. 328 pp.

Tippery, N. P. & D. H. Les. 2020. Tiny plants with enormous potential: phylogeny and evolution of duckweeds. Pp. 19–38 *In*: X. H. Cao, P. Fourounjian & W. Wang (eds.), *The Duckweed Genomes. Compendium of Plant Genomes*. Springer-Verlag, New York, NY.

Tippery, N. P., D. H. Les & D. J. Crawford. 2015. Evaluation of phylogenetic relationships in Lemnaceae using nuclear ribosomal data. *Plant Biol.* **17**(suppl. 1): 50–58.

Tippery, N. P., G. J. Bugbee & S. E. Stebbins. 2020. Evidence for a genetically distinct strain of introduced *Hydrilla verticillata* (Hydrocharitaceae) in North America. *J. Aquat. Plant Manag.* **58**: 1–6.

Tipping, P. W., L. Bauer, M. R. Martin & T. D. Center. 2009. Competition between *Salvinia minima* and *Spirodela polyrhiza* mediated by nutrient levels and herbivory. *Aquat. Bot.* **90**: 231–234.

Tipping, P. W., M. R. Martin, L. Bauer, R. M. Pierce & T. D. Center. 2012. Ecology of common salvinia, *Salvinia minima* Baker, in southern Florida. *Aquat. Bot.* **102**: 23–27.

Titcomb, J. W. 1923. Aquatic plants in pond culture, second ed. Bureau of Fisheries Document No. 948. Appendix II to the Report of the U.S. Commissioner of Fisheries for 1923. Department of Commerce, Government Printing Office, Washington, DC. 24 pp.

Titus, J. E. 1983. Submersed macrophyte vegetation and depth distribution in Chenango Lake, New York. *Bull. Torrey Bot. Club* **110**: 176–183.

Titus, J. H. 2017. *Unpublished Wetland Plot Data Recorded in Southwestern Oregon*. Oregon Natural Heritage Program, Corvallis, OR. 6 pp.

Titus, J. E. & M. S. Adams. 1979. Coexistence and the comparative light relations of the submersed macrophytes *Myriophyllum spicatum* L. and *Vallisneria americana* Michx. *Oecologia* **40**: 273–286.

Titus, J. E. & W. H. Stone. 1982. Photosynthetic response of 2 submersed macrophytes to dissolved inorganic carbon concentration and pH. *Limnol. Oceanogr.* **27**: 151–160.

Titus, J. E. & M. D. Stephens. 1983. Neighbor influences and seasonal growth patterns for *Vallisneria americana* in a mesotrophic lake. *Oecologia* **56**: 23–29.

Titus, J. E. & D. J. Grisé. 2009. The invasive freshwater macrophyte *Utricularia inflata* (inflated bladderwort) dominates Adirondack Mountain lake sites. *J. Torrey Bot. Soc.* **136**: 479–486.

Tobe, J. D., K. C. Burks, R. W. Cantrell, M. A. Garland, M. E. Sweeley, D. W. Hall, P. Wallace, G. Anglin, G. Nelson, J. R. Cooper, D. Bickner, K. Gilbert, N. Aymond, K. Greenwood & N. Raymond. 1998. *Florida Wetland Plants: An Identification Manual*. University of Florida, Institute of Food & Agricultural Sciences, Gainesville, FL. 598 pp.

Tobe, H., Y.-L. Huang, T. Kadokawa & M. N. Tamura. 2018. Floral structure and development in Nartheciaceae (Dioscoreales), with special reference to ovary position and septal nectaries. *J. Plant Res.* **131**: 411–428.

Tobiessen, P. & P. D. Snow. 1984. Temperature and light effects on the growth of *Potamogeton crispus* in Collins Lake, New York State. *Can. J. Bot.* **62**: 2822–2826.

Todorova, M., N. Grozeva, M. Gospodinova & J. Ivanova. 2013. Relationships between soil salinity and vascular plants in inland salt meadow near the town of Radnevo. *Sci. Pap. Ser. A Agron.* **56**: 119–125.

Todt, D. L. 1997. Cross-cultural folk classifications of ethnobotanically important geophytes in southern Oregon and northern California. *J. Calif. Gt. Basin Anthropol.* **19**: 250–259.

Tol, S. J., J. C. Jarvis, P. H. York, A. Grech, B. C. Congdon & R. G. Coles. 2017. Long distance biotic dispersal of tropical seagrass seeds by marine mega-herbivores. *Sci. Rep.* **7**: 4458.

Tolman, D. A. 2006. Characterization of the ecotone between Jeffrey pine savannas and *Darlingtonia* fens in southwestern Oregon. *Madroño* **53**: 199–210.

Tolman, D. A. 2007. Soil patterns in three *Darlingtonia* fens of southwestern Oregon. *Nat. Areas J.* **27**: 374–384.

Tom, B. & S. K. Karr. 2013. Management considerations for the restoration of bunched arrowhead *Sagittaria fasciculata*. *Nat. Areas J.* **33**: 105–108.

Tomas, W. M. & S. M. Salis. 2000. Diet of the marsh deer (*Blastocerus dichotomus*) in the Pantanal wetland, Brazil. *Stud. Neotrop. Fauna Environ.* **35**: 165–172.

Tomlinson, P. B. 1969. On the morphology and anatomy of turtle grass, *Thalassia testudinum* (Hydrocharitaceae). III. Floral morphology and anatomy. *Bull. Mar. Sci.* **19**: 286–305.

Tooker, J. F. & L. M. Hanks. 2000. Flowering plant hosts of adult hymenopteran parasitoids of central Illinois. *Ann. Entomol. Soc. Am.* **93**: 580–588.

Tooker, J. F., P. F. Reagel & L. M. Hanks. 2002. Nectar sources of day-flying Lepidoptera of central Illinois. *Ann. Entomol. Soc. Am.* **95**: 84–96.

Tooker, J. F., M. Hauser & L. M. Hanks. 2006. Floral host plants of Syrphidae and Tachinidae (Diptera) of central Illinois. *Ann. Entomol. Soc. Am.* **99**: 96–112.

Topuzovic, M. D., I. D. Radojevic, M. S. Dekic, N. S. Radulovic, S. M. Vasic, L. R. Comic & B. Z. Licina. 2015. Phytomedical investigation of *Najas minor* All. in the view of the chemical constituents. *Excli J.* **14**: 496–503.

Torit, J., W. Siangdung & P. Thiravetyan. 2012. Phosphorus removal from domestic wastewater by *Echinodorus cordifolius* L. *J. Environ. Sci. Health* **47**: 794–800.

Toriyama, K. (ed.). 2005. *Rice Is Life: Scientific Perspectives for the 21st Century*. International Rice Research Institute, Manila, Philippines. 133 pp.

Tornberg, T., M. Bendix & H. Brix. 1994. Internal gas transport in *Typha latifolia* L. and *Typha angustifolia* L. 2. Convective throughflow pathways and ecological significance. *Aquat. Bot.* **49**: 91–105.

Tomlinson, P. B. & U. Posluzny. 2001. Generic limits in the seagrass family Zosteraceae. *Taxon* **50**: 429–437.

Toner, M., N. Stow & C. J. Keddy. 1995. Arrow arum, *Peltandra virginica*: a nationally rare plant in the Ottawa Valley region of Ontario. *Can. Field-Nat.* **109**: 441–442.

Torquemada, Y. F., M. J. Durako & J. L. S. Lizaso. 2005. Effects of salinity and possible interactions with temperature and pH on growth and photosynthesis of *Halophila johnsonii* Eiseman. *Mar. Biol.* **148**: 251–260.

Toth, L. A. & J. P. Galloway. 2009. Clonal expansion of cattail (*Typha domingensis*) in Everglades stormwater treatment areas: implications for alternative management strategies. *J. Aquat. Plant Manag.* **47**: 151–155.

Touchette, B. W. 2007. Seagrass-salinity interactions: physiological mechanisms used by submersed marine angiosperms for a life at sea. *J. Exp. Mar. Biol. Ecol.* **350**: 194–215.

Townsend, J. E. 1953. Beaver ecology in western Montana with special reference to movements. *J. Mammal.* **34**: 459–479.

Tracy, M., J. M. Montante, T. E. Allenson & R. A. Hough. 2003. Long-term responses of aquatic macrophyte diversity and community structure to variation in nitrogen loading. *Aquat. Bot.* **77**: 43–52.

Transeau, E. N. 1903. On the geographic distribution and ecological relations of the bog plant societies of northern North America. *Bot. Gaz.* **36**: 401–420.

Trapnell, D. W., J. L. Hamrick & D. E. Giannasi. 2004. Genetic variation and species boundaries in *Calopogon* (Orchidaceae). *Syst. Bot.* **29**: 308–315.

Traub, H. P. 1957. *Hymenocallis littoralis* from Florida. *Rhodora* **59**: 99.

Travis, S. E. & P. Sheridan. 2006. Genetic structure of natural and restored shoalgrass *Halodule wrightii* populations in the NW Gulf of Mexico. *Mar. Ecol. Prog. Ser.* **322**: 117–127.

Tremblay, R. L., J. D. Ackerman, J. K. Zimmerman & R. N. Calvo. 2004. Variation in sexual reproduction in orchids and its evolutionary consequences: a spasmodic journey to diversification. *Biol. J. Linn. Soc.* **84**: 1–54.

Trevathan-Tackett, S. M., A. L. Lane, N. Bishop & C. Ross. 2015. Metabolites derived from the tropical seagrass *Thalassia testudinum* are bioactive against pathogenic *Labyrinthula* sp. *Aquat. Bot.* **122**: 1–8.

Triest, L. 1988. A revision of the genus *Najas* L. (Najadaceae) in the Old World. *Mém. Acad. Roy. Sci. Outre-Mer, Cl. Sci. Nat. Méd., Collect. 8vo* **22**: 1–172.

Triest, L. 1989. Electrophoretic polymorphism and divergence in *Najas marina* L. (Najadaceae): molecular markers for individuals, hybrids, cytodemes, lower taxa, ecodemes and conservation of genetic diversity. *Aquat. Bot.* **33**: 301–380.

Triest, L. & L. Vanhecke. 1991. Isozymes in European and Mediterranean *Zannichellia* (Zannichelliaceae) populations: a situation of predominant inbreeders. *Opera Bot. Belgica* **4**: 133–166.

Triest, L. & S. Fénart. 2014. Clonal diversity and spatial genetic structure of *Potamogeton pectinatus* in managed pond and river populations. *Hydrobiologia* **737**: 145–161.

Triest, L. & T. Sierens. 2014. Seagrass radiation after messinian salinity crisis reflected by strong genetic structuring and out-of-Africa scenario (Ruppiaceae). *PLoS One* **9**(8): e104264.

Triest, L. & T. Sierens. 2015. Strong bottlenecks, inbreeding and multiple hybridization of threatened European *Ruppia maritima* populations. *Aquat. Bot.* **125**: 31–43.

Triest, L., Y. Viinikka & M. Agami. 1989. Isozymes as molecular markers for diploid and tetraploid individuals of *Najas marina* (Najadaceae). *Plant Syst. Evol.* **166**: 131–139.

Triest, L., V. T. Thi & T. Sierens. 2007. Chloroplast microsatellite markers reveal *Zannichellia* haplotypes across Europe using herbarium DNA. *Belgian J. Bot.* **140**: 109–120.

Triest, L., D. L. Thi, T. Sierens & A. Van Geert. 2010. Genetic differentiation of submerged plant populations and taxa between habitats. *Hydrobiologia* **656**: 15–27.

Trocine, R. P., J. D. Rice & G. N. Wells. 1981. Inhibition of seagrass photosynthesis by ultraviolet-B radiation. *Plant Physiol.* **68**: 74–81.

Trocine, R. P., J. D. Rice & G. N. Wells. 1982. Photosynthetic response of seagrasses to ultraviolet-A radiation and the influence of visible light intensity. *Plant Physiol.* **69**: 341–344.

True-Meadows, S., E. J. Haug & R. J. Richardson. 2016. Monoecious hydrilla—a review of the literature. *J. Aquat. Plant Manag.* **54**: 1–11.

Truman, H. V. 1931. Pollen of *Sparganium americanum* and *S. androcladum*. *Rhodora* **33**: 141–142.

Tryon, C. A. & N. W. Easterly. 1975. Plant communities of the Irwin Prairie and adjacent wooded areas. *Castanea* **40**: 201–213.

Tsao, T. H., H. W. Zhong, S. P. Jiao & Z. Y. Tan. 1986. Changes in endogenous ABA and GA contents during floral induction of *Lemna aequinoctialis*. *Acta Bot. Neerl.* **35**(4): 443–448.

Tsuchiya, T. 1991. Leaf life span of floating-leaved plants. *Vegetatio* **97**: 149–160.

Tu, H.-Y., A.-L. Zhang, W. Xiao, Y.-R. Lin, J.-H. Shi, Y.-W. Wu, S.-T. Wu, C.-H. Zhong & S.-X. Mo. 2018. Induction and identification of tetraploid *Hedychium coronarium* through thin cell layer culture. *Plant Cell Tissue Organ Cult.* **135**: 395–406.

Tucker, G. C. 1989. The genera of Commelinaceae in the southeastern United States. *J. Arnold Arbor.* **70**: 97–130.

Tucker, J. M. & B. J. McCaskill. 1955. *Monochoria vaginalis* in California. *Madroño* **13**: 112.

Tucker, J. M. & B. J. McCaskill. 1967. *Heteranthera limosa* in California. *Madroño* **19**: 64.

Tucker, G. C., J. E. Ebinger, W. E. McClain, E. L. Smith, P. B. Marcum & R. Jansen. 2015. Marsh vegetation of the Margaret Guzy pothole wetlands land and water reserve, Shelby County, Illinois. *Trans. Illinois State Acad. Sci.* **108**: 7–12.

Turnage, G. & C. Shoemaker. 2018. 2017 survey of aquatic plant species in MS waterbodies. GRI Report 5077. Geosystems Research Institute, Mississippi State University, Starkville, MS. 69 pp.

Turnage, L. G., C. Duncan & J. D. Madsen. 2012. Aquatic invasive plant survey of selected Montana waters for 2012. GRI Report 5055. Geosystems Research Institute, Mississippi State University, Starkville, MS. 84 pp.

Turner, C. E. 1980. Noteworthy collections: *Ottelia alismoides* (L.) Pers. (Hydrocharitaceae). *Madroño* **27**: 177.

Turner, T. 1983. Facilitation as a successional mechanism in a rocky intertidal community. *Am. Nat.* **121**: 729–738.

Turner, T. 1985. Stability of rocky intertidal surfgrass beds: persistence, preemption, and recovery. *Ecology* **66**: 83–92.

Turner, N. J. 1995. *Food Plants of Coastal First Peoples*. UBC Press, Vancouver, BC. 164 pp.

Turner, R. L. 1996. Use of stems of emergent plants for oviposition by the Florida applesnail, *Pomacea paludosa*, and the implications for marsh management. *Florida Sci.* **59**: 34–49.

Turner, N. J. 2003. The ethnobotany of edible seaweed (*Porphyra abbottae* and related species; Rhodophyta: Bangiales) and its use by First Nations on the Pacific Coast of Canada. *Can. J. Bot.* **81**: 283–293.

Turner, T. & J. Lucas. 1985. Differences and similarities in the community roles of three rocky intertidal surfgrasses. *J. Exp. Mar. Biol. Ecol.* **89**: 175–189.

Turner, W. M. & D. H. Nelson. 2001. Composition of the diet of the Alabama redbelly turtle (*Psuedemys* [sic!] *alabamensis*). *J. Alabama Acad. Sci.* **72**: 97–97.

Turner, M. A., D. B. Huebert, D. L. Findlay, L. L. Hendzel, W. A. Jansen, R. A. Bodaly, L. M. Armstrong & S. E. M. Kasian. 2005. Divergent impacts of experimental lake-level drawdown on planktonic and benthic plant communities in a boreal forest lake. *Can. J. Fish. Aquat. Sci.* **62**: 991–1003.

Turner, C. E., R. R. Haynes & C. B. Hellquist. 2012. Alismataceae. Water plantain family. Pp. 1288–1289 *In*: B. G. Baldwin, D. H. Goldman, D. J. Keil, R. Patterson, T. J. Rosatti & D. H. Wilken (eds.), *The Jepson Manual*, 2nd ed. University of California Press, Berkeley, CA.

Turesson, G. 1916. *Lysichiton camtschatcense* (L) Schott, and its behavior in sphagnum bogs. *Am. J. Bot.* **3**: 189–209.

Tveten, J. & G. Tveten. 1997. *Wildflowers of Houston and Southeast Texas*. University of Texas Press, Austin, TX. 319 pp.

Tweedy, F. 1880. Notes on the flora of Plainfield, N. J. *Bull. Torrey Bot. Club* **7**: 26–28.

Twilley, R. R. & J. W. Barko. 1990. The growth of submersed macrophytes under experimental salinity and light conditions. *Estuaries* **13**: 311–321.

Tyler, C. & M. Borchert. 2003. Reproduction and growth of the chaparral geophyte, *Zigadenus fremontii* (Liliaceae), in relation to fire. *Plant Ecol.* **165**: 11–20.

Tyndall, R. W. 2000. Vegetation change in a Carolina bay on the Delmarva Peninsula of Maryland during an eleven-year period (1987–1997). *Castanea* **65**: 155–164.

Tyndall, R. W., K. A. McCarthy, J. C. Ludwig & A. Rome. 1990. Vegetation of six Carolina bays in Maryland. *Castanea* **55**: 1–21.

Uchimura, M., E. J. Faye, S. Shimada, T. Inoue & Y. Nakamura. 2008. A reassessment of *Halophila* species (Hydrocharitaceae) diversity with special reference to Japanese representatives. *Bot. Mar.* **51**: 258–268.

Uhler, F. M. 1982. Food habits of diving ducks in the Carolinas. *Proc. Ann. Conf. Southeast. Assoc. Fish Wildl. Agencies* **36**: 492–504.

Ulrich, S., M. Hesse, D. Bröderbauer, J. Bogner, M. Weber & H. Halbritter. 2013. *Calla palustris* (Araceae): new palynological insights with special regard to its controversial systematic position and to closely related genera. *Taxon* **62**: 701–712.

Umetsu, C. A., H. B. A. Evangelista & S. M. Thomaz. 2012. The colonization, regeneration, and growth rates of macrophytes from fragments: a comparison between exotic and native submerged aquatic species. *Aquat. Ecol.* **46**: 443–449.

Ungar, I. A. 1964. A phytosociological analysis of the Big Salt Marsh, Stafford County, Kansas. *Trans. Kansas Acad. Sci.* **67**: 50–64.

Ungar, I. A. 1970. Species-soil relationships on sulfate dominated soils of South Dakota. *Am. Midl. Nat.* **83**: 343–357.

Ungar, I. A. 1974. Halophyte communities of Park county, Colorado. *Bull. Torrey Bot. Club* **101**: 145–152.

Ungar, I. A., W. Hogan & M. McClelland. 1969. Plant communities of saline soils at Lincoln, Nebraska. *Am. Midl. Nat.* **82**: 564–577.

Unruh, S. A., M. R. McKain, Y.-I. Lee, T. Yukawa, M. K. McCormick, R. P. Shefferson, A. Smithson, J. H. Leebens-Mack & J. C. Pires. 2018. Phylotranscriptomic analysis and genome evolution of the Cypripedioideae (Orchidaceae). *Am. J. Bot.* **105**: 631–640.

Unwin, M. M. 2004. Molecular systematics of the Eriocaulaceae Martinov. Ph.D. dissertation. Miami University, Oxford, OH. 127 pp.

Uphof, J. C. T. 1922. Ecological relations of plants in southeastern Missouri. *Am. J. Bot.* **9**: 1–17.

Uphof, J. C. T. 1927. The floral behavior of some Eriocaulaceae. *Am. J. Bot.* **14**: 44–48.

Uphof, J. C. T. 1933. Die Blütenbiologie von *Mayaca fluviatilis* Aub. *Ber. Deutsch. Bot. Ges.* **51**: 78–85.

Uphof, J. C. T. 1938. Cleistogamic flowers. *Bot. Rev.* **4**: 21–49.

Uphof, J. C. T. 1941. Halophytes. *Bot. Rev.* **7**: 1–58.

Urban, L., S. G. M. Bridgewater & D. J. Harris. 2006. The Macal River: a floristic and phytosociological study of a threatened riverine vegetation community in Belize. *Edinburgh J. Bot.* **63**: 95–118.

Urban, R. A., J. E. Titus & W.-X. Zhu. 2009. Shading by an invasive macrophyte has cascading effects on sediment chemistry. *Biol. Invasions* **11**: 265–273.

Urbanska-Worytkiewicz, K. 1975. Cytological variation within *Lemna* L. *Aquat. Bot.* **1**: 377–394.

Urbanska-Worytkiewicz, K. 1980. Cytological variation within the family of Lemnaceae. *Veröff. Geobot. Inst. E. T. H. Stiftung Rübel Zürich* **70**: 30–101.

Urbatsch, L. E. 2013. Plants new and noteworthy for Louisiana and Mississippi. *Phytoneuron* **14**: 1–7.

USACE. 2016. U.S. Army Corps of Engineers. National Wetland Plant List, version 3.3. Available online: http://wetland-plants.usace.army.mil/. [accessed 4 February, 2020].

USFWS. 1983. *Bunched Arrowhead Recovery Plan.* U.S. Fish and Wildlife Service, Atlanta, GA. 37 pp.

USFWS. 1991. *Swamp Pink (Helonias bullata) Recovery Plan.* U.S. Fish and Wildlife Service, Region V. Newton Corner, MA. 56 pp.

USFWS. 2013. *Platanthera integrilabia. Species Assessment and Listing Priority Assignment Form.* U.S. Fish and Wildlife Service, Region IV (Southeast Region), Atlanta, GA. 15 pp.

USFWS. 2014. *Kral's Water-Plantain (Sagittaria secundifolia) 5-Year Review: Summary and Evaluation.* U.S. Fish and Wildlife Service, Alabama Ecological Services Field Office, Daphne, AL. 14 pp.

USFWS. 2019. *Swamp Pink (Helonias bullata). 5-Year Review.* U.S. Fish & Wildlife Service. New Jersey Field Office, Pleasantview, NJ. 50 pp.

Ushimaru, A., S. Kikuchi, R. Yonekura, A. Maruyama, N. Yanagisawa, M. Kagami, M. Nakagawa, S. Mahoro, Y. Kohmatsu, A. Hatada, S. Kitamura & K. Nakata. 2007. The influence of floral symmetry and pollination systems on flower size variation. *Nordic J. Bot.* **24**: 593–598.

Usinger, R. L. 1956. *Aquatic Insects of California: With Keys to North American Genera and California Species.* University of California Press, Berkeley, CA. 508 pp.

Utech, F. H. 1978. Floral vascular anatomy of *Pleea tenuifolia* Michx. (Liliaceae-Tofieldieae) and its reassignment to *Tofieldia. Ann. Carnegie Mus. Pittsburgh* **47**: 423–454.

Utech, F. H. 2002a. Liliaceae Jussieu. Lily Family. Pp. 49–347 *In*: Flora North America Editorial Committee (eds.), *Flora of North America North of Mexico, Vol. 26: Magnoliophyta: Liliidae: Liliales and Orchidales.* Oxford University Press, New York, NY.

Utech, F. H. 2002b. *Helonias* Linnaeus. Pp. 69–70 *In*: Flora North America Editorial Committee (eds.), *Flora of North America North of Mexico, Vol. 26: Magnoliophyta: Liliidae: Liliales and Orchidales.* Oxford University Press, New York, NY.

Utech, F. H. & A. Adanson. 2002. *Narthecium* Hudson. Pp. 66–67 *In*: Flora North America Editorial Committee (eds.), *Flora of North America North of Mexico, Vol. 26: Magnoliophyta: Liliidae: Liliales and Orchidales.* Oxford University Press, New York, NY.

Utech, F. H. & L. C. Anderson. 2002. *Harperocallis* McDaniel. Pp. 58–59 *In*: Flora North America Editorial Committee (eds.), *Flora of North America North of Mexico, Vol. 26: Magnoliophyta: Liliidae: Liliales and Orchidales.* Oxford University Press, New York, NY.

Vaile, M. & S. Gilbert. 2005. The curious case of Dr Alabone—heterodoxy in 19th century medicine. *J. Roy. Soc. Med.* **98**: 281–286.

Vaithiyanathan, P. & C. J. Richardson. 1999. Macrophyte species changes in the Everglades: examination along a eutrophication gradient. *J. Environ. Qual.* **28**: 1347–1358.

Valentine, D. H. 1976. Patterns of variation in north temperate taxa with a wide distribution. *Taxon* **25**: 225–231.

Valentine, Jr., J. M. & R. E. Noble. 1970. A colony of sandhill cranes in Mississippi. *J. Wildl. Manag.* **34**: 761–768.

Vamosi, J. C., S. M. Vamosi & S. C. H. Barrett. 2006. Sex in advertising: dioecy alters the net benefits of attractiveness in *Sagittaria latifolia* (Alismataceae). *Proc. Royal Soc. London B: Biol. Sci.* **273**: 2401–2407.

Van, T. K. & K. K. Steward. 1990. Longevity of monoecious hydrilla propagules. *J. Aquat. Plant Manag.* **28**: 74–76.

Van, T. K., W. T. Haller, G. Bowes & L. A. Garrard. 1977. Effects of light quality on growth and chlorophyll composition in *Hydrilla. J. Aquat. Plant Manag.* **15**: 29–31.

Van Bruggen, H. W. E. 1973. Revision of the genus *Aponogeton* (Aponogetonaceae) VI. The species of Africa. *Bull. Jard. Bot. Natl. Belg.* **43**: 193–233.

Van Bruggen, H. W. E. 1985. Monograph of the genus *Aponogeton* (Aponogetonaceae). *Biblioth. Bot.* **137**: 1–97.

Van Bruggen, H. W. E. 1990. Die Gattung *Aponogeton* (Aponogetonaceae). *Aqua-Planta (Sonderheft)* **2**: 1–84.

Van den Berg, M. S., H. Coops, J. Simons & A. de Keizer. 1998. Competition between *Chara aspera* and *Potamogeton pectinatus* as a function of temperature and light. *Aquat. Bot.* **60**: 241–250.

Van den Broek, T., R. van Diggelen & R. Bobbink. 2005. Variation in seed buoyancy of species in wetland ecosystems with different flooding dynamics. *J. Veg. Sci.* **16**: 579–586.

Van der Bijl, L., K. Sand-Jensen & A.-L. Hjermind. 1989. Photosynthesis and canopy structure of a submerged plant, *Potamogeton pectinatus*, in a Danish lowland stream. *J. Ecol.* **77**: 947–962.

Van der Kinderen, G. 1995. A method for the study of field germinated seeds of terrestrial orchids. *Lindleyana* **10**: 68–73.

Van der Pijl, L. & C. H. Dodson. 1966. *Orchid Flowers: Their Pollination and Evolution.* University of Miami Press, Coral Gables, FL. 214 pp.

Van der Valk, A. G. 1975. Floristic composition and structure of fen communities in northwest Iowa. *Proc. Iowa Acad. Sci.* **82**: 113–118.

Van der Valk, A. G. & C. B. Davis. 1978. The role of seed banks in the vegetation dynamics of prairie glacial marshes. *Ecology* **59**: 322–335.

Van der Valk, A. G. & T. R. Rosburg. 1997. Seed bank composition along a phosphorus gradient in the northern Florida Everglades. *Wetlands* **17**: 228–236.

Van der Valk, A. G., S. D. Swanson & R. F. Nuss. 1983. The response of plant species to burial in three types of Alaskan wetlands. *Can. J. Bot.* **61**: 1150–1164.

Van der Valk, A. G., L. A. Toth, E. B. Gibney, D. H. Mason & P. R. Wetzel. 2009. Potential propagule sources for reestablishing vegetation on the floodplain of the Kissimmee River, Florida, USA. *Wetlands* **29**: 976–987.

Van Dijk, G. M. & W. van Vierssen. 1991. Survival of a *Potamogeton pectinatus* L. population under various light conditions in a shallow eutrophic lake (Lake Veluwe) in The Netherlands. *Aquat. Bot.* **39**: 121–129.

Van Dijk, J. K. & B. I. Van Tussenbroek. 2010. Clonal diversity and structure related to habitat of the marine angiosperm *Thalassia testudinum* along the Atlantic coast of Mexico. *Aquat. Bot.* **92**: 63–69.

Van Dine, D. L. 1922. Impounding water in a bayou to control breeding of malaria mosquitoes. Bulletin 1098. U.S. Department of Agriculture, Washington, DC. 22 pp.

Van Drunen, W. E. & M. E. Dorken. 2012. Trade-offs between clonal and sexual reproduction in *Sagittaria latifolia* (Alismataceae) scale up to affect the fitness of entire clones. *New Phytol.* **196**: 606–616.

Van Kampen, K. R., L. F. James & A. E. Johnson. 1970. Hemolytic anemia in sheep fed wild onion (*Allium validum*). *J. Am. Veterin. Med. Assoc.* **156**: 328–332.

Van Montfrans, J., R. J. Orth & S. A. Vay. 1982. Preliminary studies of grazing by *Bittium varium* on eelgrass periphyton. *Aquat. Bot.* **14**: 75–89.

Van Montfrans, J., R. L. Wetzel & R. J. Orth. 1984. Epiphyte-grazer relationships in seagrass meadows: consequences for seagrass growth and production. *Estuaries Coast.* **7**: 289–309.

Van Onsem, S., J. Rops & L. Triest. 2018. Submerged seed, turion and oospore rain: a trap quantifying propagule deposition under aquatic vegetation. *Aquat. Bot.* **145**: 21–28.

Van Tussenbroek, B. I. 1994. Aspects of the reproductive ecology of *Thalassia testudinum* in Puerto Morelos Reef Lagoon, Mexico. *Bot. Marina* **37**: 413–420.

Van Tussenbroek, B. I. & M. Muhlia-Montero. 2013. Can floral consumption by fish shape traits of seagrass flowers? *Evol. Ecol.* **27**: 269–284.

Van Tussenbroek, B. I., N. Villamil, J. Marquezguzman, R. Wong, L. V. Monroy-Velázquez & V. Solis-Weiss. 2016a. Experimental evidence of pollination in marine flowers by invertebrate fauna. *Nat. Commun.* **7**. doi: 10.1038/ncomms12980.

Van Tussenbroek, B. I., T. Valdivia-Carrillo, I. T. Rodríguez-Virgen, S. N. M. Sanabria-Alcaraz, K. Jiménez-Durán, K. J. Van Dijk & G. J. Marquez-Guzmán. 2016b. Coping with potential bi-parental inbreeding: limited pollen and seed dispersal and large genets in the dioecious marine angiosperm *Thalassia testudinum*. *Ecol. Evol.* **6**: 5542–5556.

Van Valkenburg, J. L. C. H. & R. Pot. 2008. *Landoltia punctata* (G. Mey.) DH Les & DJ Crawford (Smal kroos), nieuw voor Nederland. *Gorteria* **33**: 41–49.

Van Vierssen, W. 1982a. Some notes on the germination of seeds of *Najas marina* L. *Aquat. Bot.* **12**: 201–203.

Van Vierssen, W. 1982b. The ecology of communities dominated by *Zannichellia* taxa in western Europe. I. Characterization and autecology of the *Zannichellia* taxa. *Aquat. Bot.* **12**: 103–155.

Van Vierssen, W., R. J. Van Wijk & J. R. Van der Zee. 1981. Some additional notes on the cytotaxonomy of *Ruppia* taxa in Western Europe. *Aquat. Bot.* **11**: 297–301.

Van Vierssen, W., R. J. Van Wijk & J. R. Van der Zee. 1982. On the pollination mechanism of some eurysaline Potamogetonaceae. *Aquat. Bot.* **14**: 339–347.

Van Wijck, C., C.-J. de Groot & P. Grillas. 1992. The effect of anaerobic sediment on the growth of *Potamogeton pectinatus* L.: the role of organic matter, sulphide and ferrous iron. *Aquat. Bot.* **44**: 31–49.

Van Wijk, R. J. 1988. Ecological studies on *Potamogeton pectinatus* L. I. General characteristics, biomass production and life cycles under field conditions. *Aquat. Bot.* **31**: 211–258.

Van Wijk, R. J. 1989a. Ecological studies on *Potamogeton pectinatus* L. III. Reproductive strategies and germination ecology. *Aquat. Bot.* **33**: 271–299.

Van Wijk, R. J. 1989b. Ecological studies on *Potamogeton pectinatus* L. IV. Nutritional ecology, field observations. *Aquat. Bot.* **35**: 301–318.

Van Wijk, R. J. 1989c. Ecological studies on *Potamogeton pectinatus* L. V. Nutritional ecology, in vitro uptake of nutrients and growth limitation. *Aquat. Bot.* **35**: 319–335.

Van Wijk, R. J., E. M. J. Van Goor & J. A. C. Verkley. 1988. Ecological studies on *Potamogeton pectinatus* L. II. Autecological characteristics, with emphasis on salt tolerance, intraspecific variation and isoenzyme patterns. *Aquat. Bot.* **32**: 239–260.

Van Wyk, B.-E. 2011. The potential of South African plants in the development of new food and beverage products. *S. Afr. J. Bot.* **77**: 857–868.

Van Zandt, P. A., P. L. Freeman & L. J. Davenport. 2013. Sporadic destructive occurrence of convict caterpillars (*Xanthopastis timais*) on Cahaba lilies (*Hymenocallis coronaria*). *S. Lepid. News* **35**: 5–14.

Van Zuidam, J. P., E. P. Raaphorst & E. T. H. M. Peeters. 2012. The role of propagule banks from drainage ditches dominated by free-floating or submerged plants in vegetation restoration. *Restor. Ecol.* **20**: 416–425.

Vanitha, K., P. Subhashini & T. Thangaradjou. 2016. Karyo-morphometric analysis of somatic chromosomes of selected seagrasses of families Hydrocharitaceae and Cymodoceaceae. *Aquat. Bot.* **133**: 45–49.

Vankar, P. S. & J. Srivastava. 2018. A review - canna the wonder plant. *J. Textile Eng. Fashion Technol.* **4**: 158–162.

Vánky, K. 1994. Ustilaginales of Commelinaceae. *Mycoscience* **35**: 353–360.

Vannatta, J. 2016. The giant liver fluke: a review, intermediate host habitat, and infection in a white-tailed deer population in Minnesota. M.S. thesis. University of Minnesota, St. Paul, MN. 137 pp.

Vansell, G. H. & W. G. Watkins. 1933. A plant poisonous to adult bees. *J. Econ. Entomol.* **26**: 168–170.

VanTassel, N. M. & A. M. Janosik. 2019. A compendium of coastal dune lakes in northwest Florida. *J. Coast. Conserv.* **23**: 385–416.

Vári, Á. 2013. Colonisation by fragments in six common aquatic macrophyte species. *Fundam. Appl. Limnol.* **183**: 15–26.

Vasconcelos, T., M. Tavares & N. Gaspar. 1999. Aquatic plants in the rice fields of the Tagus Valley, Portugal. Pp. 59–65 *In:* J. Caffrey, P. R. F. Barrett, M. T. Ferreira, I. S. Moreira, K. J. Murphy & P. M. Wade (eds.), *Biology, Ecology and Management of Aquatic Plants.* Springer, Dordrecht, The Netherlands.

Vasey, M. C., V. T. Parker, J. C. Callaway, E. R. Herbert & L. M. Schile. 2012. Tidal wetland vegetation in the San Francisco Bay-Delta estuary. *San Francisco Estuary Watershed Sci.* **10**: 1–16.

Vasseur, L., L. W. Aarssen & T. Bennett. 1993. Allozymic variation in local apomictic populations of *Lemna minor* (Lemnaceae). *Am. J. Bot.* **80**: 974–979.

Velichkova, K., I. Sirakov, E. Valkova, S. Stoyanova & G. Kostadinova. 2017. Bioaccumulation and protein content of *Lemna minuta* Kunth and *Lemna valdiviana* Phil. in Bulgarian water reservoirs. *Land Reclam. Earth Observ. Surv. Environ. Eng.* **6**: 104–107.

Venturelli, M. & F. Bouman. 1986. Embryology and seed development in *Mayaca fluviatilis* (Mayacaceae). *Acta Bot. Neerl.* **35**: 497–516.

Verhoeven, J. T. A. 1979. The ecology of *Ruppia*-dominated communities in Western Europe. I. Distribution of *Ruppia* representatives in relation to their autecology. *Aquat. Bot.* **6**: 197–268.

Verkleij, J. A. C., A. H. Pieterse, G. J. T. Horneman & M. Torenbeek. 1983. A comparative study of the morphology and isoenzyme patterns of *Hydrilla verticillata* (Lf) Royle. *Aquat. Bot.* **17**: 43–59.

Verma, M. & Y. K. Bansal. 2010. Butterfly lilly [sic!] (*Hedychium coronarium* Koenig): an endangered medicinal plant. *Plant Arch.* **10**: 841–843.

Verma, M. & Y. K. Bansal. 2012. Induction of somatic embryogenesis in endangered butterfly ginger *Hedychium coronarium* J. Koenig. *Indian J. Exp. Biol.* **50**: 904–909.

Vermaat, J. E. & M. J. M. Hootsmans. 1994. Growth of *Potamogeton pectinatus* L. in a temperature-light gradient. Pp. 40–61 *In*: W. van Vierssen, M. Hootsmans & J. Vermaat (eds.), *Lake Veluwe, A Macrophyte-Dominated System Under Eutrophication Stress*. Springer, Dordrecht, The Netherlands.

Vermaat, J. E., L. Santamaria & P. J. Roos. 2000. Water flow across and sediment trapping in submerged macrophyte beds of contrasting growth form. *Arch. Hydrobiol.* **148**: 549–562.

Vest, J. L. & M. R. Conover. 2011. Food habits of wintering waterfowl on the Great Salt Lake, Utah. *Waterbirds* **34**: 40–50.

Vickery, R. 1983. *Lemna minor* and Jenny Greenteeth. *Folklore* **94**: 247–250.

Vidal, J. 1960. Les Plantes utiles du Laos. *J. Agric. Trop.* **7**: 560–587.

Vieira, M. F. & N. A. de Souza Lima. 1997. Pollination of *Echinodorus grandiflorus* (Alismataceae). *Aquat. Bot.* **58**: 89–98.

Viereck, L. A., C. T. Dyrness, A. R. Batten & K. J. Wenzlick. 1992. The Alaska vegetation classification. General Technical Report PNW-GTR-286. U.S. Department of Agriculture, Forest Service, Pacific Northwest Research Station, Portland, OR. 278 pp.

Viereck, L. A., C. T. Dyrness & M. J. Foote. 1993. An overview of the vegetation and soils of the floodplain ecosystems of the Tanana River, interior Alaska. *Can. J. For. Res.* **23**: 889–898.

Viinikka, Y. 1973. The occurrence of B chromosomes and their effect on meiosis in *Najas marina*. *Hereditas* **75**: 207–212.

Viinikka, Y. 1976. *Najas marina* L. (Najadaceae). Karyotypes, cultivation and morphological variation. *Ann. Bot. Fenn.* **13**: 119–131.

Viinikka, Y. 1977. The role of U-type exchanges in the differentiation of karyotypes in *Najas marina*. *Hereditas* **86**: 91–101.

Viinikka, Y. & M. Kotimäki. 1979. The meiotic and postmeiotic consequences of spontaneous chromosome breakage in *Najas marina*. *Hereditas* **91**: 273–277.

Viinikka, Y., M. Kotimäki & K. Litmanen. 1978. Spontaneous chromosome breakage in natural populations of *Najas marina*. *Hereditas* **88**: 279–283.

Viinikka, Y., M. Agami & L. Triest. 1987. A tetraploid cytotype of *Najas marina* L. *Hereditas* **106**: 289–291.

Vincent, R. E. 1958. The larger plants of Little Kitoi lake. *Am. Midl. Nat.* **60**: 212–218.

Vincent, W. F. & C. Bertola. 2012. François Alphonse FOREL. *Archiv. Sci.* **65**: 51–64.

Vinson, M. & B. Bushman. 2005. *An Inventory of Aquatic Invertebrate Assemblages in Wetlands in Utah*. National Aquatic Monitoring Center, Utah State University, Logan, UT. 224 pp.

Viosca, Jr., P. 1935. The irises of southeastern Louisiana, a taxonomic and ecological interpretation. *Bull. Am. Iris Soc.* **57**: 3–56.

Visser, J. M. & C. E. Sasser. 2009. The effect of environmental factors on floating fresh marsh end-of-season biomass. *Aquat. Bot.* **91**: 205–212.

Visser, J. M., C. E. Sasser, R. H. Chabreck & R. G. Linscombe. 1999. Long-term vegetation change in Louisiana tidal marshes, 1968–1992. *Wetlands* **19**: 168–175.

Vitt, D. H. & N. G. Slack. 1975. An analysis of the vegetation of *Sphagnum*-dominated kettle-hole bogs in relation to environmental gradients. *Can. J. Bot.* **53**: 332–359.

Vitt, D. H. & S. Bayley. 1984. The vegetation and water chemistry of four oligotrophic basin mires in northwestern Ontario. *Can. J. Bot.* **62**: 1485–1500.

Vitt, D. H. & W.-L. Chee. 1990. The relationships of vegetation to surface water chemistry and peat chemistry in fens of Alberta, Canada. *Vegetatio* **89**: 87–106.

Vitt, D. H., D. G. Horton, N. G. Slack & N. Malmer. 1990. Sphagnum-dominated peatlands of the hyper-oceanic British Columbia coast: patterns in surface water chemistry and vegetation. *Can. J. For. Res.* **20**: 696–711.

Vogel, R. L. 1977. Aquatic plant communities of East-Central Illinois. M.S. thesis. Eastern Illinois University, Charleston, IL. 28 pp.

Volker, R. & S. G. Smith. 1965. Changes in the aquatic vascular flora of Lake East Okoboji in historic times. *Proc. Iowa Acad. Sci.* **72**: 65–72.

Volkova, O. A., M. V. Remizowa, D. D. Sokoloff & E. E. Severova. 2016. A developmental study of pollen dyads and notes on floral development in *Scheuchzeria* (Alismatales: Scheuchzeriaceae). *Bot. J. Linn. Soc.* **182**: 791–810.

Von Bank, J. A., J. A. DeBoer, A. F. Casper & H. M. Hagy. 2018. Ichthyochory in a temperate river system by common carp (*Cyprinus carpio*). *J. Freshw. Ecol.* **33**: 83–96.

Von Mering, S. & J. W. Kadereit. 2010. Phylogeny, systematics, and recircumscription of Juncaginaceae – a cosmopolitan wetland family. Pp. 55–79 *In*: O. Seberg, G. Petersen, A. S. Barfod & J. I. Davis (eds.), *Diversity, Phylogeny, and Evolution in the Monocotyledons*. Aarhus University Press, Aarhus, Denmark.

Von Mering, S. & J. W. Kadereit. 2015. Phylogeny, biogeography and evolution of *Triglochin* L. (Juncaginaceae)--morphological diversification is linked to habitat shifts rather than to genetic diversification. *Mol. Phylogenet. Evol.* **83**: 200–212.

Vossler, F. G., M. C. Telleria & M. Cunningham. 2010. Floral resources foraged by *Geotrigona argentina* (Apidae, Meliponini) in the Argentine Dry Chaco forest. *Grana* **49**: 142–153.

Vuille, F. L. 1987. Reproductive biology of the genus *Damasonium* (Alismataceae). *Plant Syst. Evol.* **157**: 63–71.

Vymaza, J. 2013. Emergent plants used in free water surface constructed wetlands: a review. *Ecol. Eng.* **61P**: 582–592.

Wada, N. & S. Uemura. 1994. Seed dispersal and predation by small rodents on the herbaceous understory plant *Symplocarpus renifolius*. *Am. Midl. Nat.* **132**: 320–327.

Wadkar, S. S., C. C. Shete, F. R. Inamdar, S. S. Wadkar & R. V. Gurav. 2017. Phytochemical screening and antibacterial activity of *Cryptocoryne spiralis* var. *spiralis* and *Cryptocoryne retrospiralis* (Roxb) Kunth. *Med. Aromat. Plants* **6**: 2167–0412.

Wagner, R. 1982. Raising ornamental palms. *Principes* **26**: 86–101.

Wagner, E. J. & R. W. Oplinger. 2017. Effect of overwinter hydration, seed storage time, temperature, photoperiod, water depth, and scarification on seed germination of some *Schoenoplectus*, *Polygonum*, *Eleocharis* and *Alisma* species. *Aquat. Bot.* **136**: 164–174.

Wagstaff, D. J. & A. A. Case. 1987. Human poisoning by *Zigadenus*. *J. Toxicol.* **25**: 361–367.

Wake, C. M. F. 2007. Micro-environment conditions, mycorrhizal symbiosis, and seed germination in *Cypripedium candidum*: strategies for conservation. *Lankesteriana* **7**: 423–426.

Wakeman, R. W. & D. H. Les. 1994a. Interspecific competition between *Potamogeton amplifolius* and *Myriophyllum spicatum*. *Lake Reserv. Manag.* **9**: 125–129.

Wakeman, R. W. & D. H. Les. 1994b. Optimum growth conditions for *Potamogeton amplifolius*, *Myriophyllum spicatum* and *Potamogeton richardsonii*. *Lake Reserv. Manag.* **9**: 129–133.

Walck, J. L. & S. N. Hidayati. 2004. Differences in light and temperature responses determine autumn versus spring germination for seeds of *Schoenolirion croceum*. *Can. J. Bot.* **82**: 1429–1437.

Walker, B. H. & C. F. Wehrhahn. 1971. Relationships between derived vegetation gradients and measured environmental variables in Saskatchewan wetlands. *Ecology* **52**: 85–95.

Walker, B. N. & R. J. Medve. 1976. The effects of acid mine drainage on *Sparganium americanum* Nutt. *Proc. Pennsylvania Acad. Sci.* **50**: 170–172.

Walker, J. & R. K. Peet. 1984. Composition and species diversity of pine-wiregrass savannas of the Green Swamp, North Carolina. *Vegetatio* **55**: 163–179.

Walker, D. A. & K. R. Everett. 1991. Loess ecosystems of northern Alaska: regional gradient and toposequence at Prudhoe Bay. *Ecol. Monogr.* **61**: 437–464.

Walker, J. L. & A. M. Silletti. 2005. A three-year demographic study of Harper's beauty (*Harperocallis flava* McDaniel), an endangered Florida endemic. *J. Torrey Bot. Soc.* **132**: 551–560.

Walker, D. A., N. D. Lederer & M. D. Walker. 1987. Permanent vegetation plots. Site factors, soil physical and chemical properties, and plant species cover. Data Report, Department of Energy R4D Project. Plant Ecology Laboratory, Institute of Arctic and Alpine Research, University of Colorado, Boulder, CO. 89 pp.

Walker, D. A., E. Binnian, B. M. Evans, N. D. Lederer, E. Nordstrand & P. J. Webber. 1989. Terrain, vegetation and landscape evolution of the R4D research site, Brooks Range Foothills, Alaska. *Ecography* **12**: 238–261.

Walker, M. D., D. A. Walker & N. A. Auerbach. 1994. Plant communities of a tussock tundra landscape in the Brooks Range Foothills, Alaska. *J. Veg. Sci.* **5**: 843–866.

Walker, D. I., B. Olesen & R. C. Phillips. 2001. Reproduction and phenology in seagrasses. Pp. 59–78 *In*: F. T. Short & R. G. Coles (eds.), *Global Seagrass Research Methods*. Elsevier Science B.V., Amsterdam, The Netherlands.

Wall, M. A., A. P. Teem & R. S. Boyd. 2002. Floral manipulation by *Lasioglossum zephyrum* (Hymenoptera: Halictidae) ensures first access to floral rewards by initiating premature anthesis of *Xyris tennesseensis* (Xyridaceae) flowers. *Florida Entomol.* **85**: 290–292.

Wallace, G. J. 1942. More Berkshire plants. *Rhodora* **44**: 332–334.

Wallace, L. E. 2002. Examining the effects of fragmentation on genetic variation in *Platanthera leucophaea* (Orchidaceae): inferences from allozyme and random amplified polymorphic DNA markers. *Plant Species Biol.* **17**: 37–49.

Wallace, L. E. 2003a. Molecular evidence for allopolyploid speciation and recurrent origins in *Platanthera huronensis* (Orchidaceae). *Int. J. Plant Sci.* **164**: 907–916.

Wallace, L. E. 2003b. The cost of inbreeding in *Platanthera leucophaea* (Orchidaceae). *Am. J. Bot.* **90**: 235–242.

Wallace, L. E. 2004. A comparison of genetic variation and structure in the allopolyploid *Platanthera huronensis* and its diploid progenitors, *Platanthera aquilonis* and *Platanthera dilatata* (Orchidaceae). *Can. J. Bot.* **82**: 244–252.

Wallace, L. E. 2006. Spatial genetic structure and frequency of interspecific hybridization in *Platanthera aquilonis* and *P. dilatata* (Orchidaceae) occurring in sympatry. *Am. J. Bot.* **93**: 1001–1009.

Waller, D. M. & D. A. Steingraeber. 1995. Opportunities and constraints in the placement of flowers and fruits. Pp. 51–73 *In*: B. L. Gartner (ed.), *Plant Stems: Physiology and Functional Morphology*. Academic Press, New York, NY.

Waller, D. M. & W. S. Alverson. 1997. The white-tailed deer: a keystone herbivore. *Wildl. Soc. Bull.* **25**: 217–226.

Wallis, C. S. 2007. Vascular plants of the Oklahoma Ozarks. *Oklahoma Nat. Plant Rec.* **7**: 4–20.

Walsh, G. E. & T. E. Grow. 1972. Composition of *Thalassia testudinum* and *Ruppia maritima*. *Quart. J. Florida Acad. Sci.* **35**: 97–108.

Walsh, R. P. & H. J. Michaels. 2017. When it pays to cheat: examining how generalized food deception increases male and female fitness in a terrestrial orchid. *PLoS One* **12**(1): e0171286.

Walsh, R. P., P. M. Arnold & H. J. Michaels. 2014. Effects of pollination limitation and seed predation on female reproductive success of a deceptive orchid. *AoB Plants* **6**(1): plu031.

Walton, G. B. 1995. *Report for Field Season 1994 Status Survey for Caltha natans and Sparganium glomeratum in Minnesota*. Natural Heritage Program, Minnesota Department of Natural Resources, Minneapolis, MN. 43 pp.

Wan, T., X.-L. Zhang, J. Gregan, Y. Zhang, P. Guo & Y.-H. Guo. 2012. A dynamic evolution of chromosome in subgenus *Potamogeton* revealed by physical mapping of rDNA loci detection. *Plant Syst. Evol.* **298**: 1195–1210.

Wang, W. 1990. Literature review on duckweed toxicity testing. *Environ. Res.* **52**: 7–22.

Wang, B. & Y.-L. Qiu. 2006. Phylogenetic distribution and evolution of mycorrhizas in land plants. *Mycorrhiza* **16**: 299–363.

Wang, W. H. & J. Min. 2007. Comparison on salt tolerance of nine submerged macrophytes. *J. Agro-Environ. Sci.* **26**: 1259–1263.

Wang, W. & J. Messing. 2011. High-throughput sequencing of three Lemnoideae (duckweeds) chloroplast genomes from total DNA. *PLoS One* **6**: e24670.

Wang, W. & J. Messing. 2015. Status of duckweed genomics and transcriptomics. *Plant Biol.* **17**: 10–15.

Wang, L. & K. H. Hasenstein. 2016. Seed coat stomata of several *Iris* species. *Flora* **224**: 24–29.

Wang, G., Y. Yamasue, K. Itoh & T. Kusanagi. 1998. Outcrossing rates as affected by pollinators and the heterozygote advantage of *Monochoria korsakowii*. *Aquat. Bot.* **62**: 135–143.

Wang, H., W. Li, Z. Gu & Y. Chen. 2001. Cytological study on *Acorus* L. in southwestern China, with some cytogeographical notes on *A. calamus*. *Acta Bot. Sin.* **43**: 354–358.

Wang, X.-F., Y.-Y. Tan, J.-H. Chen & Y.-T. Lu. 2006. Pollen tube reallocation in two preanthesis cleistogamous species, *Ranalisma rostratum* and *Sagittaria guyanensis* ssp. *lappula* (Alismataceae). *Aquat. Bot.* **85**: 233–240.

Wang, B., L. Peng, L. Zhu & P. Ren. 2007. Protective effect of total flavonoids from *Spirodela polyrrhiza* (L.) Schleid on human umbilical vein endothelial cell damage induced by hydrogen peroxide. *Colloids Surf. B Biointerfaces* **60**: 36–40.

Wang, H., H. Zhu, K. Zhang, L. Zhang & Z. Wu. 2010. Chemical composition in aqueous extracts of *Najas marina* and *Najas minor* and their algae inhibition activity. *Sci. Res.* **2010**: 806–809.

Wang, W., R. A. Kerstetter & T. P. Michael. 2011a. Evolution of genome size in duckweeds (Lemnaceae). *J. Bot.* **2011**. doi: 10.1155/2011/570319.

Wang, G., L. Chen, Z. Hao, X. Li & Y. Liu. 2011b. Effects of salinity stress on the photosynthesis of *Wolffia arrhiza* as probed by the OJIP test. *Fresenius Environ. Bull.* **20**: 432–438.

Wang, X., W. Zhou, J. Lu, H. Wang, C. Xiao, J. Xia & G. Liu. 2012a. Effects of population size on synchronous display of female and male flowers and reproductive output in two monoecious *Sagittaria* species. *PLoS One* **7**(10): e48731.

Wang, W., Y. Wu & J. Messing. 2012b. The mitochondrial genome of an aquatic plant, *Spirodela polyrhiza*. *PLoS One* **7**(10): e46747.

Wang, W., Y. Wu & J. Messing. 2014a. RNA-Seq transcriptome analysis of *Spirodela* dormancy without reproduction. *BMC Genomics* **15**(1): 60.

Wang, H. Q., H. J. Zhu, L. Y. Zhang, W. J. Xue & B. Yuan. 2014b. Identification of antialgal compounds from the aquatic plant *Elodea nuttallii*. *Allelopathy J.* **34**: 207–214.

Wang, W., G. Haberer, H. Gundlach, C. Gläßer, T. C. L. M. Nussbaumer, M. C. Luo, A. Lomsadze, M. Borodovsky, R. A. Kerstetter, J. Shanklin, D. W. Byrant, T. C. Mockler, K. J. Appenroth, J. Grimwood, J. Jenkins, J. Chow, C. Choi,

C. Adam, X.-H. Cao, J. Fuchs, I. Schubert, D. Rokhsar, J. Schmutz, T. P. Michael, K. F. X. Mayer & J. Messing. 2014c. The *Spirodela polyrhiza* genome reveals insights into its neotenous reduction fast growth and aquatic lifestyle. *Nat. Commun.* **5**: ncomms4311.

Wang, W., W. Zhang, Y. Wu, P. Maliga & J. Messing. 2015. RNA editing in chloroplasts of *Spirodela polyrhiza*, an aquatic monocotelydonous species. *PLoS One* **10**(10): e0140285.

Wang, P., P. Alpert & F.-H. Yu. 2016. Clonal integration increases relative competitive ability in an invasive aquatic plant. *Am. J. Bot.* **103**: 2079–2086.

Wang, C.-J., J.-Z. Wan, H. Qu & Z.-X. Zhang. 2017. Climatic niche shift of aquatic plant invaders between native and invasive ranges: a test using 10 species across different biomes on a global scale. *Knowl. Manag. Aquat. Ecosyst.* **418**: 1–27.

Wang, W., A. Dong, C. Li, Y. Zhou & Y. Wu. 2018a. Genomes and transcriptomes of duckweeds. *Front. Chem.* **6**: 230. doi: 10.3389/fchem.2018.00230.

Wang, R., M. Li, X. Wu, C. Shen, W. Yu, J. Liu, Z. Qi & P. Li. 2018b. Characterization of polymorphic microsatellite loci for North American common greenbrier, *Smilax rotundifolia* (Smilacaceae). *Appl. Plant Sci.* **6**: e01163.

Ward, L. F. 1882. Proterogyn in *Sparganium eurycarpum*. *Bot. Gaz.* **7**: 100.

Ward, L. F. 1883. Marsh and aquatic plants of the Northern United States, many of which are suitable for carp ponds. *Bull. U.S. Fish Commiss.* **3**: 257–265.

Ward, D. B. 2011. *Spirodela oligorrhiza* (Lemnaceae) is the correct name for the lesser greater duckweed. *J. Bot. Res. Inst. Texas* **5**: 197–203.

Ward, D. B. 2012. New combinations in the Florida flora III. *Phytologia* **94**: 459–485.

Ward, C. H. & S. S. Wilks. 1963. Use of algae and other plants in the development of life support systems. *Am. Biol. Teach.* **25**: 512–521.

Ward, D. B. & A. F. Clewell. 1989. Atlantic white cedar (*Chamaecyparis thyoides*) in the southern states. *Florida Sci.* **52**: 8–47.

Ware, D. M. E. & S. Ware. 1992. An *Acer barbatum*-rich ravine forest community in the Virginia coastal plain. *Castanea* **57**: 110–122.

Waridel, P., J.-L. Wolfender, J.-B. Lachavanne & K. Hostettmann. 2003. *ent*-Labdane diterpenes from the aquatic plant *Potamogeton pectinatus*. *Phytochemistry* **64**: 1309–1317.

Warrington, P. D. 1986. *The pH Tolerance of the Aquatic Plants of British Columbia. Part I. Literature Survey of the pH Limits of Aquatic Plants of the World*. Water Management Branch, British Columbia Ministry of Environment, Victoria, BC. 157 pp.

Wasowicz, P., E. M. Przedpelska-Wasowicz, L. Guðmundsdóttir & M. Tamayo. 2014. *Vallisneria spiralis* and *Egeria densa* (Hydrocharitaceae) in arctic and subarctic Iceland. *New J. Bot.* **4**: 85–89.

Watson, J. R. 1941. Migrations and food preferences of the lubberly locust. *Florida Entomol.* **24**: 40–42.

Watson, L. E., G. E. Uno, N. A. McCarty & A. B. Kornkven. 1994. Conservation biology of a rare plant species, *Eriocaulon kornickianum* (Eriocaulaceae). *Am. J. Bot.* **81**: 980–986.

Watson, L. E., A. B. Kornkven, C. R. Miller, J. R. Allison, N. B. McCarty & M. M. Unwin. 2002. Morphometric and genetic variation in *Eriocaulon koernickianum* Van Heurck & Muller-Argoviensis (Eriocaulaceae): a disjunct plant species of the southeastern United States. *Castanea* **67**: 416–426.

Waud, M., P. Busschaert, S. Ruyters, H. Jacquemyn & B. Lievens. 2014. Impact of primer choice on characterization of orchid mycorrhizal communities using 454 pyrosequencing. *Mol. Ecol. Resour.* **14**: 679–699.

Way, M. O. 1982. The aster leafhopper, *Macrosteles fascifrons* Stål: conditions affecting abundance, tactics for control and effects on yield of rice in California. Ph.D. dissertation. University of California, Davis, CA.

Waycott, M. & D. H. Les. 1996. An integrated approach to the evolutionary study of seagrasses. Pp. 71–78 *In*: R. C. Philipps, D. I. Walker & H. Kirkman (eds.), *Seagrass Biology: Proceedings of an International Workshop, Rottnest Island, Western Australia, 25-29 January, 1996*. Faculty of Sciences, University of Western Australia, Nedlands, WA, Australia.

Waycott, M., D. W. Freshwater, R. A. York, A. Calladine & W. J. Kenworthy. 2002. Evolutionary trends in the seagrass genus *Halophila* (Thouars): insights from molecular phylogeny. *Bull. Mar. Sci.* **71**: 1299–1308.

Waycott, M., G. Procaccini, D. H. Les & T. B. H. Reusch. 2006. Seagrass evolution, ecology and conservation: a genetic perspective. Pp. 25–50 *In*: A. W. D. Larkum, R. J. Orth & C. M. Duarte (eds.), *Seagrasses: Biology, Ecology and Conservation*. Springer-Verlag, Dordrecht, The Netherlands.

Weakley, A. S. 1991. Classification of pocosins of the Carolina coastal plain. *Wetlands* **11**: 355–375.

Weakley, A. S. & M. P. Schafale. 1994. Non-alluvial wetlands of the southern Blue Ridge—diversity in a threatened ecosystem. *Water Air Soil Pollut.* **77**: 359–383.

Weatherbee, P. B. & G. E. Crow. 1992. Natural plant communities of Berkshire County, Massachusetts. *Rhodora* **94**: 171–209.

Weber, W. A. 1940. *Potamogeton hillii* in Berkshire County, Massachusetts. *Rhodora* **42**: 95.

Weber, J. & S. Rooney. 1994. Josselyn Botanical Society's Annual Meeting, Unity College, July 1993. *Maine Nat.* **2**: 50–52.

Weberg, M. A., B. R. Murphy, A. L. Rypel & J. R. Copeland. 2015. A survey of the New River aquatic plant community in response to recent triploid grass carp introductions into Claytor Lake, Virginia. *Southeast. Nat.* **14**: 308–318.

Webster, C. G. 1964. Fall foods of soras from two habitats in Connecticut. *J. Wildl. Manag.* **28**: 163–165.

Weerasundara, L., C. N. Nupearachchi, P. Kumarathilaka, B. Seshadri, N. Bolan & M. Vithanage. 2016. Bio-retention systems for storm water treatment and management in urban systems. Pp. 175–200 *In*: A. A. Ansari, S. S. Gill, R. Gill, G. R. Lanza & L. Newman (eds.), *Phytoremediation. Management of Environmental Contaminants, Volume 4*. Springer International Publishing, Basel, Switzerland.

Wees, D. 2004. Stratification and priming may improve seed germination of purple coneflowers, blue-flag iris, and evening primrose. *Acta Hort.* **629**: 391–395.

Wehrmeister, J. R. 1978. An ecological life history of the pondweed *Potamogeton crispus* L. in North America. CLEAR Technical Report No. 99. The Ohio State University Center for Lake Erie Area Research, Columbus, OH. 157 pp.

Wei, Z.-Z., L.-B. Luo, H.-L. Zhang, M. Xiong, X. Wang & D. Zhou. 2012. Identification and characterization of 43 novel polymorphic EST-SSR markers for arum lily, *Zantedeschia aethiopica* (Araceae). *Am. J. Bot.* **99**: e493–e497.

Weiher, E. & P. A. Keddy. 1995. The assembly of experimental wetland plant communities. *Oikos* **73**: 323–335.

Weiher, E., S. Peot & K. Voss. 2003. Experimental restoration of lake shoreland in western Wisconsin. *Ecol. Restor.* **21**: 186–191.

Weiher, E. R., C. W. Boylen & P. A. Bukaveckas. 2011. Alterations in aquatic plant community structure following liming of an acidic Adirondack lake. *Can. J. Fish. Aquat. Sci.* **51**: 20–24.

Wein, K. 1939. Die älteste Einführungs- und Ausbreitungsgeschichte von *Acorus calamus*. Erster Teil. *Hercynia (Special Edition)* **1**: 367–450.

Wein, K. 1940a. Die älteste Einführungs- und Ausbreitungsgeschichte von *Acorus calamus*. Zweiter Teil. *Hercynia (Special Edition)* **3**: 72–128.

Wein, K. 1940b. Die älteste Einführungs- und Ausbreitungsgeschichte von *Acorus calamus*. Dritter Teil. *Hercynia (Special Edition)* **3**: 241–291.

Weinhold, C. E. & A. G. Van der Valk. 1988. The impact of duration of drainage on the seed banks of northern prairie wetlands. *Can. J. Bot.* **67**: 1878–1884.

Weishampel, P. A. & B. L. Bedford. 2006. Wetland dicots and monocots differ in colonization by arbuscular mycorrhizal fungi and dark septate endophytes. *Mycorrhiza* **16**: 495–502.

Welch, J. R. 2013. *Sprouting Valley: Historical Ethnobotany of the Northern Pomo from Potter Valley, California*. Society of Ethnobiology, Department of Geography, University of North Texas, Denton, TX. 207 pp.

Wellborn, G. A. & J. V. Robinson. 1987. Microhabitat selection as an antipredator strategy in the aquatic insect *Pachydiplax longipennis* Burmeister (Odonata: Libellulidae). *Oecologia* **71**: 185–189.

Weller, M. W. (ed.). 1988. *Waterfowl in Winter: Selected Papers from Symposium and Workshop Held in Galveston, Texas, 7–10 January 1985*. University of Minnesota Press, Minneapolis, MN. 624 pp.

Wells, B. W. 1928. Plant communities of the coastal plain of North Carolina and their successional relations. *Ecology* **9**: 230–242.

Wells, B. W. 1942. Ecological problems of the southeastern United States coastal plain. *Bot. Rev.* **8**: 533–561.

Wells, E. D. 1996. Classification of peatland vegetation in Atlantic Canada. *J. Veg. Sci.* **7**: 847–878.

Wells, A. F. 2006. Deep canyon and subalpine riparian and wetland plant associations of the Malheur, Umatilla, and Wallowa-Whitman National Forests. Gen. Tech. Rep. PNW-GTR-682. U.S. Department of Agriculture, Forest Service, Pacific Northwest Research Station, Portland, OR. 277 pp.

Wells, C. & M. Alexander. 2014. *Bunched Arrowhead (Sagittaria fasciculata) 5-Year Review: Summary and Evaluation*. U.S. Fish and Wildlife Service, Southeast Region, Asheville Ecological Services Field Office, Asheville, NC. 22 pp.

Weltzin, J. F., J. Pastor, C. Harth, S. D. Bridgham, K. Updegraff & C. T. Chapin. 2000. Response of bog and fen plant communities to warming and water-table manipulations. *Ecology* **81**: 3464–3478.

Wen, J. 1999. Evolution of eastern Asian and eastern North American disjunct distributions in flowering plants. *Ann. Rev. Ecol. Syst.* **30**: 421–455.

Wen, J., R. K. Jansen & K. Kilgore. 1996. Evolution of the eastern Asian and eastern North American disjunct genus *Symplocarpus* (Araceae): insights from chloroplast DNA restriction site data. *Biochem. Syst. Ecol.* **24**: 735–747.

Wentz, W. A. & R. L. Stuckey. 1971. The changing distribution of the genus *Najas* (Najadaceae) in Ohio. *Ohio J. Sci.* **71**: 292–302.

Werblan, D., R. J. Smith, A. G. van der Valk & C. B. Davis. 1978. Treatment of waste from a confined hog feeding unit by using artificial marshes. Pp. 149–155 *In*: H. L. McKim (ed.), *State of Knowledge in Land Treatment of Wastewater, Vol. 2*. U.S. Army Corps of Engineers, Cold Regions Research and Engineering Laboratory, Hanover, NH.

Werckmeister, P. 1969. Red irises and cyanidin. *Bull. Am. Iris Soc.* **194**: 7–13.

Wersal, R. M., B. R. McMillan & J. D. Madsen. 2005. Food habits of dabbling ducks during fall migration in a prairie pothole system, Heron Lake, Minnesota. *Can. Field-Nat.* **119**: 546–550.

Wersal, R. M., J. D. Madsen & J. C. Cheshier. 2010. Aquatic plant monitoring in Noxon Rapids Reservoir and Cabinet Gorge Reservoir for 2010. Geosystem Research Institute Report #5042. Geosystems Research Institute, Mississippi State University, Mississippi State, MS. 18 pp.

Wesselingh, R. A. & M. L. Arnold. 2000a. Pollinator behaviour and the evolution of Louisiana iris hybrid zones. *J. Evol. Biol.* **13**: 171–180.

Wesselingh, R. A. & M. L. Arnold. 2000b. Nectar production in Louisiana *Iris* hybrids. *Int. J. Plant Sci.* **161**: 245–251.

Wesselingh, R. A. & M. L. Arnold. 2003. A top-down hierarchy in fruit set on inflorescences in *Iris fulva* (Iridaceae). *Plant Biol.* **5**: 651–660.

West, E. 1941. Notes on Florida fungi. II. *Mycologia* **33**: 38–49.

West, E. 1945. Notes on Florida fungi. III. *Mycologia* **37**: 65–79.

West, D. & D. F. Whigham. 1975. Seed germination of arrow arum (*Peltandra virginica* L.). *Bartonia* **44**: 44–49.

Westcott, K., T. H. Whillans & M. G. Fox. 1997. Viability and abundance of seeds of submerged macrophytes in the sediment of disturbed and reference shoreline marshes in Lake Ontario. *Can. J. Bot.* **75**: 451–456.

Wester, L. 1992. Origin and distribution of adventive alien flowering plants in Hawai'i. Pp. 99–154 *In*: C. P. Stone, C. W. Smith & J. T. Tunison (eds.), *Alien Plant Invasions in Native Ecosystems of Hawaii: Management and Research*. University of Hawaii Press, Honolulu, HI.

Westervelt, K., E. Largay, R. Coxe, W. McAvoy, S. Perles, G. Podniesinski, L. Sneddon & K. Walz. 2006. *A Guide to the Natural Communities of the Delaware Estuary: Version 1*. NatureServe, Arlington, VA. 336 pp.

Wetmore, C. M. 2001. *Rare Lichens Habitats in Superior National Forest*. USDA Forest Service, Duluth, MN. 20 pp.

Wetterer, J. K., S. E. Miller, D. E. Wheeler, C. A. Olson, D. A. Polhemus, M. Pitts, I. W. Ashton, A. G. Himler, M. M. Yospin, K. R. Helms, E. L. Harken, J. Gallaher, C. E. Dunning, M. Nelson, J. Litsinger, A. Southern & T. L. Burgess. 1999. Ecological dominance by *Paratrechina longicornis* (Hymenoptera: Formicidae), an invasive tramp ant, in biosphere 2. *Florida Entomol.* **82**: 381–388.

Wetzel, R. G. 1960. Marl encrustation on hydrophytes in several Michigan lakes. *Oikos* **11**: 223–236.

Wetzel, R. G. & D. L. McGregor. 1968. Axenic culture and nutritional studies of aquatic macrophytes. *Am. Midl. Nat.* **80**: 52–64.

Wetzel, P. R. & A. G. Van Der Valk. 1998. Effects of nutrient and soil moisture on competition between *Carex stricta*, *Phalaris arundinacea*, and *Typha latifolia*. *Plant Ecol.* **138**: 179–190.

Wetzel, P. R. & W. M. Kitchens. 2007. Vegetation change from chronic stress events: detection of the effects of tide gate removal and long-term drought on a tidal marsh. *J. Veg. Sci.* **18**: 431–442.

Wetzel, P. R., A. G. van der Valk & L. A. Toth. 2001. Restoration of wetland vegetation on the Kissimmee River floodplain: potential role of seed banks. *Wetlands* **21**: 189–198.

Wetzel, P. R., W. M. Kitchens, J. M. Brush & M. L. Dusek. 2004. Use of a reciprocal transplant study to measure the rate of plant community change in a tidal marsh along a salinity gradient. *Wetlands* **24**: 879–890.

Wheeler, A. S. & C. A. Wilson. 2014. Exploring phylogenetic relationships within a broadly distributed northern hemisphere group of semi-aquatic *Iris* species (Iridaceae). *Syst. Bot.* **39**: 759–766.

Wheeler, G. A., P. H. Glaser, E. Gorham, C. M. Wetmore, F. D. Bowers & J. A. Janssens. 1983. Contributions to the flora of the Red Lake Peatland, northern Minnesota, with special attention to Carex. *Am. Midl. Nat.* **110**: 62–96.

Wheeler, E. J., S. Mashayekhi, D. W. McNeal, J. T. Columbus & J. C. Pires. 2013. Molecular systematics of *Allium* subgenus *Amerallium* (Amaryllidaceae) in North America. *Am. J. Bot.* **100**: 701–711.

Wheelwright, N. T., E. E. Dukeshire, J. B. Fontaine, S. H. Gutow, D. A. Moeller, J. G. Schuetz, T. M. Smith, S. L. Rodgers & A. G. Zink. 2006. Pollinator limitation, autogamy and minimal inbreeding depression in insect-pollinated plants on a boreal island. *Am. Midl. Nat.* **155**: 19–38.

Wheelwright, N. T., E. Begin, C. Ellwanger, S. H. Taylor & J. L. Stone. 2016. Minimal loss of genetic diversity and no inbreeding depression in blueflag iris (*Iris versicolor*) on islands in the Bay of Fundy. *Botany* **94**: 543–554.

Wherry, E. T. 1918. The reactions of the soils supporting the growth of certain native orchids. *J. Washington Acad. Sci.* **8**: 589–598.

Wherry, E. T. 1920. Soil tests of Ericaceae and other reaction-sensitive families in northern Vermont and New Hampshire. *Rhodora* **22**: 33–49.

Wherry, E. T. 1921. The soil reactions of *Spiranthes cernua* and its relatives. *Rhodora* **23**: 127–129.

Wherry, E. T. 1927. Divergent soil reaction preferences of related plants. *Ecology* **8**: 197–206.

Wherry, E. T. 1928. Northward range-extensions of some southern orchids in relation to soil reaction. *J. Washington Acad. Sci.* **18**: 212–216.

Whetstone, R. D., C. L. Lawler, L. H. Hopkins, A. L. Martin & C. C. Dickson. 1987. Kral's water-plantain, *Sagittaria secundifolia* Kral (Alismataceae), new to Georgia. *Castanea* **52**: 313–314.

Whigham, D. F. & R. L. Simpson. 1978. The relationship between aboveground and belowground biomass of freshwater tidal wetland macrophytes. *Aquat. Bot.* **5**: 355–364.

Whigham, D. F. & R. L. Simpson. 1982. Germination and dormancy studies of *Pontederia cordata* L. *Bull. Torrey Bot. Club* **109**: 524–528.

Whigham, D. F. & C. J. Richardson. 1988. Soil and plant chemistry of an Atlantic white cedar wetland on the inner coastal plain of Maryland. *Can. J. Bot.* **66**: 568–576.

Whigham, D. F., R. L. Simpson & M. A. Leck. 1979. The distribution of seeds, seedlings, and established plants of arrow arum (*Peltandra virginica* (L.) Kunth) in a freshwater tidal wetland. *Bull. Torrey Bot. Club* **106**: 193–199.

Whigham, D. F., T. E. Jordan & J. Miklas. 1989. Biomass and resource allocation of *Typha angustifolia* L. (Typhaceae): the effect of within and between year variations in salinity. *Bull. Torrey Bot. Club* **116**: 364–370.

White, D. J. 2010. *Plants of Lanark County Ontario*. Published by the author, Lanark, ON. 100 pp.

White, D. H. & D. James. 1978. Differential use of fresh water environments by wintering waterfowl of coastal Texas. *Wilson Bull.* **90**: 99–111.

White, D. A. & M. J. Simmons. 1988. Productivity of the marshes at the mouth of the Pearl River, Louisiana. *Castanea* **53**: 215–224.

White, W. A. & J. G. Paine. 1992. Wetland plant communities, Galveston Bay system. GBNEP-16. Galveston Bay National Estuary Program, Galveston, TX. 124 pp.

White, D. W. & E. W. Stiles. 1992. Bird dispersal of fruits of species introduced into eastern North America. *Can. J. Bot.* **70**: 1689–1696.

White, S. L. & R. R. Wise. 1998. Anatomy and ultrastructure of *Wolffia columbiana* and *Wolffia borealis*, two nonvascular aquatic angiosperms. *Int. J. Plant Sci.* **159**: 297–304.

White, S. A. & M. M. Cousins. 2013. Floating treatment wetland aided remediation of nitrogen and phosphorus from simulated stormwater runoff. *Ecol. Eng.* **61**: 207–215.

White, C. R. & G. L. Harley. 2016. Historical fire in longleaf pine (*Pinus palustris*) forests of south Mississippi and its relation to land use and climate. *Ecosphere* **7**: e01458.

White, S. A., M. D. Taylor, S. L. Chandler, T. Whitwell & S. J. Klaine. 2010. Remediation of nitrogen and phosphorus from nursery runoff during the spring via free water surface constructed wetlands. *J. Environ. Hort.* **28**: 209–217.

White, S. A., M. D. Taylor & D. Z. Damrel. 2012. Floral colonization of a free-water surface constructed wetland system in Grady County, Georgia. *Castanea* **77**: 159–171.

White, P. J., H. C. Bowen, E. Farley, E. K. Shaw, J. A. Thompson, G. Wright & M. R. Broadley. 2015. Phylogenetic effects on shoot magnesium concentration. *Crop Pasture Sci.* **66**: 1241–1248.

Whitehead, M. R. & C. A. Brown. 1940. The seed of the spider lily, *Hymenocallis occidentalis*. *Am. J. Bot.* **27**: 199–203.

Whitfeld, T. J. S., E. R. Rowe, M. D. Lee & W. R. Smith. 2015. New occurrences of the elusive *Malaxis paludosa* (Orchidaceae) in Minnesota. *Rhodora* **117**: 98–105.

Whittaker, R. H. 1960. Vegetation of the Siskiyou mountains, Oregon and California. *Ecol. Monogr.* **30**: 279–338.

Whittall, J. B., C. B. Hellquist, E. L. Schneider & S. A. Hodges. 2004. Cryptic species in an endangered pondweed community (*Potamogeton*, Potamogetonaceae) revealed by AFLP markers. *Am. J. Bot.* **91**: 2022–2029.

Whittemore, A. T. 2000. Zingiberaceae Lindley. Ginger family. Pp. 305–309 *In*: Flora North America Editorial Committee (eds.), *Flora of North America North of Mexico, Vol. 22: Magnoliophyta: Alismatidae Arecidae, Commelinidae (in Part), and Zingiberidae*. Oxford University Press, New York, NY.

Whyte, R. J. & B. W. Cain. 1981. Wildlife habitat on grazed or ungrazed small pond shorelines in South Texas. *J. Range Manag.* **34**: 64–68.

Wichman, B., R. K. Peet & T. R. Wentworth. 2007. *Natural Vegetation of the Carolinas: Classification and Description of Montane Non-Alluvial Wetlands of the Southern Appalachian Region*. Carolina Vegetation Survey, University of North Carolina, Chapel Hill, NC. 15 pp.

Wicksten, M. K. 1979. Decorating behavior in *Loxorhynchus crispatus* Stimpson and *Loxorhynchus grandis* Stimpson (Brachyura, Majidae). *Crustaceana (Suppl.)* **5**: 37–46.

Wieffering, J. H. 1972. Some notes on the diploid chromosome number of the genus *Acorus* L. (Araceae). *Acta Bot. Neerl.* **21**: 555–559.

Wiegleb, G. & H. Brux. 1991. Comparison of life history characters of broad-leaved species of the genus *Potamogeton* L. I. General characterization of morphology and reproductive strategies. *Aquat. Bot.* **39**: 131–146.

Wiegleb, G. & Z. Kaplan. 1998. An account of the species of *Potamogeton* L. (Potamogetonaceae). *Folia Geobot.* **33**: 241–316.

Wiersema, J. H. 2015. Application of the name *Lemna punctata* G. Mey., the type of *Landoltia* Les & DJ Crawford. *Plant Biol.* **17**: 5–9.

Wigand, C. 1997. Facilitation of phosphate assimilation by aquatic mycorrhizae of *Vallisneria americana* Michx. *Hydrobiologia* **342**: 35–41.

Wiland-Szymańska, J., K. Buczkowska, M. Drapikowska, M. Maślak, A. Bączkiewicz & A. Czylok. 2016. Genetic structure and barcode identification of an endangered orchid species, Liparis loeselii, in Poland. *Syst. Biodivers.* **14**: 345–354.

Wilcox, D. A. 2012. Response of wetland vegetation to the post-1986 decrease in Lake St. Clair water levels: seed-bank emergence and beginnings of the *Phragmites australis* invasion. *J. Great Lakes Res.* **38**: 270–277.

Wilcox, D. A. & J. E. Meeker. 1991. Disturbance effects on aquatic vegetation in regulated and unregulated lakes in northern Minnesota. *Can. J. Bot.* **69**: 1542–1551.

Wilcox, D. A. & J. E. Meeker. 1992. Implications for faunal habitat related to altered macrophyte structure in regulated lakes innorthern Minnesota. *Wetlands* **12**: 192–203.

Wilder, G. J. 1974. Symmetry and development of *Limnobium spongia* (Hydrocharitaceae). *Am. J. Bot.* **61**: 624–642.

Wiley, R. H. & M. S. Wiley. 1980. Territorial behavior of a blackbird: mechanisms of site-dependent dominance. *Behaviour* **73**: 130–154.

Wiley, M. J., S. M. Pescitelli & L. D. Wike. 1986. The relationship between feeding preferences and consumption rates in grass carp and grass carp × bighead carp hybrids. *J. Fish Biol.* **29**: 507–514.

Wilhelm, G. S. & R. H. Mohlenbrock. 1986. Rediscovery of *Potamogeton floridanus* Small (Potamogetonaceae). *Sida* **11**: 340–346.

Wilkinson, R. E. 1961. Effects of reduced sunlight on water stargrass (*Heteranthera dubia*). *Weeds* **9**: 457–462.

Wilkinson, F. & P. J. Beckett. 2002. Aquatic plant establishment on nickel tailings five years after flooding. *Proc. Am. Soc. Min. Reclam.* **2002**: 178–193.

Willcocks, P. A. 1982. Colonization and distribution of the red algal epiphytes *Melobesia mediocris* and *Smithora naiadum* on the seagrass *Phyllospadix torreyi*. *Aquat. Bot.* **12**: 365–373.

Willette, D. A. & R. F. Ambrose. 2012. Effects of the invasive seagrass *Halophila stipulacea* on the native seagrass, *Syringodium filiforme*, and associated fish and epibiota communities in the Eastern Caribbean. *Aquat. Bot.* **103**: 74–82.

Williams, K. A. 1919. A botanical study of skunk cabbage, *Symplocarpus foetidus*. *Torreya* **19**: 21–29.

Williams, L. 1981. Foods for early man. *Revista Ceiba* **24**: 4–39.

Williams, S. L. 1987. Competition between the seagrasses *Thalassia testudinum* and *Syringodium filiforme* in a Caribbean lagoon. *Mar. Ecol. Prog. Ser.* **35**: 91–98.

Williams, S. L. 1995. Surfgrass (*Phyllospadix torreyi*) reproduction: reproductive phenology, resource allocation, and male rarity. *Ecology* **76**: 1953–1970.

Williams, S. L. 2007. Introduced species in seagrass ecosystems: status and concerns. *J. Exp. Mar. Biol. Ecol.* **350**: 89–110.

Williams, M. C. & E. H. Cronin. 1968. Dormancy, longevity, and germination of seed of three larkspurs and western false hellebore. *Weed Sci.* **16**: 381–384.

Williams, M. C. & E. H. Cronin. 1981. Ten-year control of western false hellebore (*Veratrum californicum*). *Weed Sci.* **29**: 22–23.

Williams, C. E. & W. J. Moriarity. 2000. Composition and structure of hemlock-dominated riparian forests of the northern Allegheny Plateau: a baseline assessment. Pp. 216–224 *In*: K. A. McManus, K. S. Shields & D. R. Souto (eds.), *GTR-NE-267. Proceedings: Symposium on Sustainable Management of Hemlock Ecosystems in Eastern North America*. U.S. Department of Agriculture, Forest Service, Northeastern Research Station, Newtown Square, PA.

Williams, S. L. & E. D. Grosholz. 2002. Preliminary reports from the *Caulerpa taxifolia* invasion in southern California. *Mar. Ecol. Prog. Ser.* **233**: 307–310.

Williams, D. A., N. E. Harms, L. Dodd, M. J. Grodowitz & G. O. Dick. 2017. Do the US dioecious and monoecious biotypes of *Hydrilla verticillata* hybridize? *J. Aquat. Plant Manag.* **55**: 35–38.

Willner, G. R., J. A. Chapman & D. Pursley. 1979. Reproduction, physiological responses, food habits, and abundance of nutria on Maryland marshes. *Wildl. Monogr.* **65**: 3–43.

Willson, M. F. 1993. Mammals as seed-dispersal mutualists in North America. *Oikos* **67**: 159–176.

Willson, M. F. & P. E. Hennon. 1997. The natural history of western skunk cabbage (*Lysichiton americanum*) in southeast Alaska. *Can. J. Bot.* **75**: 1022–1025.

Wilsey, B. J., R. H. Chabreck & R. G. Linscombe. 1991. Variation in nutria diets in selected freshwater forested wetlands of Louisiana. *Wetlands* **11**: 263–278.

Wilson, L. R. 1935. Lake development and plant succession in Vilas County, Wisconsin. *Ecol. Monogr.* **5**: 207–247.

Wilson, L. R. 1937. A quantitative and ecological study of the larger aquatic plants of Sweeney Lake, Oneida County, Wisconsin. *Bull. Torrey Bot. Club* **64**: 199–208.

Wilson, B. 1998. Epiphytal foraminiferal assemblages on the leaves of the seagrasses *Thalasia testudinum* and *Syringodium filiforme*. *Caribbean J. Sci.* **34**: 131–131.

Wilson, C. A. 2009. Phylogenetic relationships among the recognized series in *Iris* section *Limniris*. *Syst. Bot.* **34**: 277–284.

Wilson, C. A. 2011. Subgeneric classification in *Iris* re-examined using chloroplast sequence data. *Taxon* **60**: 27–35.

Wilson, D. C. 2014. Pollination biology of the stream orchid, *Epipactis gigantea*. *Aquilegia* **38**: 9–11.

Wilson, S. D. & P. A. Keddy. 1985a. The distribution of *Xyris difformis* along a gradient of exposure to waves: an experimental study. *Can. J. Bot.* **63**: 1226–1230.

Wilson, S. D. & P. A. Keddy. 1985b. Plant zonation on a shoreline gradient: physiological response curves of component species. *J. Ecol.* **73**: 851–860.

Wilson, S. D. & P. A. Keddy. 1988. Species richness, survivorship, and biomass accumulation along an environmental gradient. *Oikos* **53**: 375–380.

Wilson, C. J. & K. H. Dunton. 2012. *Assessment of Seagrass Habitat Quality and Plant Physiological Condition in Texas Coastal Waters*. Marine Science Institute, University of Texas at Austin, Port Aransas, TX. 46 pp.

Wilson, S. S. & K. H. Dunton. 2017. Hypersalinity during regional drought drives mass mortality of the seagrass *Syringodium filiforme* in a subtropical lagoon. *Estuaries Coast.* doi: 10.1007/s12237-017-0319-x.

Wilson, S. D., P. A. Keddy & D. L. Randall. 1985. The distribution of *Xyris difformis* along a gradient of exposure to waves: an experimental study. *Can. J. Bot.* **63**: 1226–1230.

Wilson, J. B., W. M. King, M. T. Sykes & T. R. Partridge. 1996. Vegetation zonation as related to the salt tolerance of species of brackish riverbanks. *Can. J. Bot.* **74**: 1079–1085.

Wilson, D. M., W. Fenical, M. Hay, N. Lindquist & R. Bolser. 1999. Habenariol, a freshwater feeding deterrent from the aquatic orchid *Habenaria repens* (Orchidaceae). *Phytochemistry* **50**: 1333–1336.

Wilson, B. L., D. L. Doede & V. D. Hipkins. 2000. Isozyme variation in *Sisyrinchium sarmentosum* (Iridaceae). *Northwest Sci.* **74**: 346–354.

Wilson, A. S. G., B. J. van der Kamp & C. Ritland. 2005. Opportunities for geitonogamy in the clonal herb *Maianthemum dilatatum*. *Can. J. Bot.* **83**: 1082–1087.

Winge, Ö. 1927. Chromosome behaviour in male and female individuals of *Vallisneria spiralis* and *Najas marina*. *J. Genet.* **18**: 99–107.

Winkel, A. & J. Borum. 2009. Use of sediment CO_2 by submersed rooted plants. *Ann. Bot.* **103**: 1015–1023.

Winner, A. L. & M. G. Simpson. 2007. A new subspecies of *Pentagramma triangularis* (Pteridaceae). *Madroño* **54**: 345–354.

Winter, K. 1978. Short-term fixation of ^{14}carbon by the submerged aquatic angiosperm *Potamogeton pectinatus*. *J. Exp. Bot.* **29**: 1169–1172.

Wirth, F. F., K. J. Davis & S. B. Wilson. 2004. Florida nursery sales and economic impacts of 14 potentially invasive landscape plant species. *J. Environ. Hort.* **22**: 12–16.

Wisheu, I. C. 1994. Disjunct Atlantic coastal plain species in Nova Scotia: distribution, habitat and conservation priorities. *Biol. Conserv.* **68**: 217–224.

Wisheu, I. C. & P. A. Keddy. 1989. The conservation and management of a threatened coastal plain plant community in eastern North America (Nova Scotia, Canada). *Biol. Conserv.* **48**: 229–238.

Witmer, S. W. 1964. *Butomus umbellatus* L. in Indiana. *Castanea* **29**: 117–119.

Witty, M. 2009. *Wolffia columbiana* can switch between two anatomically and physiologically separate states: buoyant for invasion and starch rich for colonization. *Int. J. Bot.* **5**: 307–313.

Witty, M. 2011. Suppression of the invasive plant watermeal (*Wolffia columbiana*) by interfering with floatation. *Int. J. Plant Pathol.* **2**: 96–100.

Witz, M. J. A. & C. J. Dawes. 1995. Flowering and short shoot age in three *Thalassia testudinum* meadows off west-central Florida. *Bot. Marina* **38**: 431–436.

Witztum, A. 1977. An ecological niche for *Lemna gibba* L. that depends on seed formation. *Israel J. Bot.* **26**: 36–38.

Witztum, A. 1986. Seed viability in *Lemna gibba* L. *Israel J. Bot.* **35**: 279–279.

Witztum, A. & M. Chaouat. 1991. Contributions to the flora of Israel and Sinai. VI. *Najas guadalupensis* (Sprengel) Magnus in Israel. *Israel J. Bot.* **40**: 65–66.

Witztum, A. & R. Wayne. 2014. Fibre cables in the lacunae of *Typha* leaves contribute to a tensegrity structure. *Ann. Bot.* **113**: 789–797.

Wofford, B. E. 2006. A new species of Stenanthium (Melanthiaceae) from Tennessee, USA. *Sida* **22**: 447–459.

Wohl, D. L. & J. V. McArthur. 1998. Actinomycete-flora associated with submersed freshwater macrophytes. *FEMS Microbiol. Ecol.* **26**: 135–140.

Wohler, J. R., I. M. Wohler & R. T. Hartman. 1965. The occurrence of *Spirodela oligorrhiza* in western Pennsylvania. *Castanea* **30**: 230–231.

Wohler, J. R., D. B. Robertson & H. R. Laube. 1975. Studies on the decomposition of *Potamogeton diversifolius. Bull. Torrey Bot. Club* **102**: 76–78.

Wolf, E. C. & D. J. Cooper. 2015. Fens of the Sierra Nevada, California, USA: patterns of distribution and vegetation. *Mires Peat* **15**: 1–22.

Wolfe, L. M. & S. C. H. Barrett. 1987. Pollinator foraging behavior and pollen collection on the floral morphs of tristylous *Pontederia cordata* L. *Oecologia* **74**: 347–351.

Wolfe, L. M. & S. C. H. Barrett. 1988. Temporal changes in the pollinator fauna of tristylous *Pontederia cordata*, an aquatic plant. *Can. J. Zool.* **66**: 1421–1424.

Wolfe, L. M. & S. C. H. Barrett. 1989. Patterns of pollen removal and deposition in tristylous *Pontederia cordata* L. (Pontederiaceae). *Biol. J. Linn. Soc.* **36**: 317–329.

Wolfer, S. R. & D. Straile. 2004. Spatio-temporal dynamics and plasticity of clonal architecture in *Potamogeton perfoliatus. Aquat. Bot.* **78**: 307–318.

Wolfer, S. R. & D. Straile. 2012. To share or not to share: clonal integration in a submerged macrophyte in response to light stress. *Hydrobiologia* **684**: 261–269.

Wolters, M., A. Garbutt & J. P. Bakker. 2005. Plant colonization after managed realignment: the relative importance of diaspore dispersal. *J. Appl. Ecol.* **42**: 770–777.

Won, B. Y., K. K. Yates, S. Fredericq & T. O. Cho. 2010. Characterization of macroalgal epiphytes on *Thalassia testudinum* and *Syringodium filiforme* seagrass in Tampa Bay, Florida. *Algae* **25**: 141–153.

Wong, S. Y., P. C. Boyce, A. S. bin Othman & L. C. Pin. 2010. Molecular phylogeny of tribe Schismatoglottideae (Araceae) based on two plastid markers and recognition of a new tribe, Philonotieae, from the Neotropics. *Taxon* **59**: 117–124.

Wood, R. D. 1950. Stability and zonation of Characeae. *Ecology* **31**: 642–647.

Wood, Jr., C. E. 1983. The genera of Burmanniaceae in the southeastern United States. *J. Arnold Arbor.* **64**: 293–307.

Wood, T. 1992. Management of a rare lily, *Lilium parryi*, at Ramsey Canyon Preserve. Pp. 50–52 *In*: A. M. Barton & S. A. Sloane (eds.), *Chiricahua Mountains Research Symposium. Proceedings.* Southwest Parks and Monuments Association, Tucson, AZ.

Wood, J. D. & M. D. Netherland. 2017. How long do shoot fragments of hydrilla (*Hydrilla verticillata*) and Eurasian watermilfoil (*Myriophyllum spicatum*) remain buoyant? *J. Aquat. Plant Manag.* **55**: 76–82.

Wood, J. R., V. H. Resh & E. M. McEwan. 1982. Egg masses of nearctic sericostomatid caddisfly genera (Trichoptera). *Ann. Entomol. Soc. Am.* **75**: 430–434.

Wood, T. H., W. M. Whitten & N. H. Williams. 2000. Phylogeny of *Hedychium* and related genera (Zingiberaceae) based on ITS sequence data. *Edinburgh J. Bot.* **57**: 261–270.

Woodcock, E. F. 1925. Observations on the poisonous plants of Michigan. *Am. J. Bot.* **12**: 116–131.

Woodruffe-Peacock, E. A. 1917. The means of plant dispersal. *Selborne Mag. Nat. Notes* **28/29**: 80–83; 97–101; 114–116.

Woods, M., A. R. Diamond, Jr. & C. N. Horn. 2006. Noteworthy collections. Alabama. *Castanea* **71**: 251–252.

Woolf, T. E. & J. D. Madsen. 2003. Seasonal biomass and carbohydrate allocation patterns in southern Minnesota curlyleaf pondweed populations. *J. Aquat. Plant Manag.* **41**: 113–118.

Wooten, J. W. 1970. Experimental investigations of the *Sagittaria graminea* complex: transplant studies and genecology. *J. Ecol.* **58**: 233–242.

Wooten, J. W. 1971. The monoecious and dioecious conditions in *Sagittaria latifolia* L. (Alismataceae). *Evolution* **25**: 549–553.

Wooten, J. W. 1973a. Taxonomy of seven species of *Sagittaria* from eastern North America. *Brittonia* **25**: 64–74.

Wooten, J. W. 1973b. Edaphic factors in species and ecotype differentiation in *Sagittaria. J. Ecol.* **61**: 151–156.

Wooten, J. W. 1978. Effects of photoperiod, light intensity, and stage of development on flower initiation in *Sagittaria graminea* Michx. (Alismataceae). *Aquat. Bot.* **4**: 245–255.

Wooten, J. W. 1986. Edaphic factors associated with eleven species of *Sagittaria* (Alismataceae). *Aquat. Bot.* **24**: 35–41.

Wooten, J. W. & S. A. Brown. 1983. Observations of inflorescence patterns and flower sex in a population of *Sagittaria papillosa* Buch. (Alismataceae). *Aquat. Bot.* **15**: 409–418.

Wooten, J. W. & R. L. Leonard. 1984. Community organization based on medians of various parameters amog marsh and aquatic plants in North Carolina. *J. Elisha Mitchell Sci. Soc.* **100**: 12–22.

Worden, E. C., T. K. Broschat & C. Yurgalevitch. 2002. Care and maintenance of landscape palms in South Florida. Document ENH 866. University of Florida Extension, Institute of Food and Agricultural Sciences, Gainseville, FL. 8 pp.

Worley, A. C., L. Sawich, H. Ghazvini & B. A. Ford. 2009. Hybridization and introgression between a rare and a common lady's slipper orchid, *Cypripedium candidum* and *C. parviflorum* (Orchidaceae). *Botany* **87**: 1054–1065.

Wright, A. H. & A. A. Wright. 1932. The habitats and composition of the vegetation of Okefinokee Swamp, Georgia. *Ecol. Monogr.* **2**: 109–232.

Wu, H., X. Zhang, S. Zheng, I. J. Shi & Y. Bi. 2008. Polyploid induction of coloured *Zantedeschia aethiopica. Acta Hort. Sin.* **35**: 443–446.

Wu, S.-H., T. Y. A. Yang, Y.-C. Teng, C.-Y. Chang, K.-C. Yang & C.-F. Hsieh. 2010. Insights of the latest naturalized flora of Taiwan: change in the past eight years. *Taiwania* **55**: 139–159.

Wu, X., H. Wu, J. Ye & B. Zhong. 2015. Study on the release routes of allelochemicals from *Pistia stratiotes* Linn., and its anti-cyanobacteria mechanisms on *Microcystis aeruginosa. Environ. Sci. Pollut. Res.* **22**: 18994–19001.

Wu, K., C.-N. N. Chen & K. Soong. 2016. Long distance dispersal potential of two seagrasses *Thalassia hemprichii* and *Halophila ovalis. PLoS One* **11**: e0156585.

Wunderlin, T. F. & R. P. Wunderlin. 1968. A preliminary survey of the algal flora of Horseshoe Lake, Alexander County, Illinois. *Am. Midl. Nat.* **79**: 534–539.

Wunderlin, R. P., B. F. Hansen & L. C. Anderson. 2002. Plants new to the United States and Florida. *Sida* **20**: 813–817.

Wylie, R. B. 1917. The pollination of *Vallisneria spiralis. Bot. Gaz.* **63**: 135–145.

Wylie, R. B. 1919. Cleistogamy in *Heteranthera dubia. Bull. Lab. Nat. Hist. State Univ. Iowa* **7**: 48–58.

Wynn, T. D. 2012. Habitat-specific production of a fall line river shoal macroinvertebrate assemblage. Ph.D. dissertation. University of Alabama, Tuscaloosa, AL. 123 pp.

Xiao, C. & S. S. Desser. 1998. The oligochaetes and their actinosporean parasites in Lake Sasajewun, Algonquin Park, Ontario. *J. Parasitol.* **84**: 1020–1026.

Xiao, C., W.-F. Dou & G.-H. Liu. 2010. Variation in vegetation and seed banks of freshwater lakes with contrasting intensity of aquaculture along the Yangtze River, China. *Aquat. Bot.* **92**: 195–199.

Xie, W.-Y., Q. Huang, G. Li, C. Rensing & Y.-G. Zhu. 2013. Cadmium accumulation in the rootless macrophyte *Wolffia globosa* and its potential for phytoremediation. *Int. J. Phytoremed.* **15**: 385–397.

Xie, D., D. Yu, C. Xia & W. You. 2014. Stay dormant or escape sprouting? Turion buoyancy and sprouting abilities of the submerged macrophyte *Potamogeton crispus* L. *Hydrobiologia* **726**: 43–51.

Xu, N.-N., X. Tong, P.-K. E. Tsang, H. Deng & X.-Y. Chen. 2011. Effects of water depth on clonal characteristics and biomass allocation of *Halophila ovalis* (Hydrocharitaceae). *J. Plant Ecol.* **4**: 283–291.

Xu, J., H. Zhao, A.-M. Stomp & J. J. Cheng. 2012. The production of duckweed as a source of biofuels. *Biofuels* **3**: 589–601.

Xu, Y., S. Ma, M. Huang, M. Peng, M. Bog, K. S. Sree, K.-J. Appenroth & J. Zhang. 2015. Species distribution, genetic diversity and barcoding in the duckweed family (Lemnaceae). *Hydrobiologia* **743**: 75–87.

Xu, H., C. Yu, X. Xia, M. Li, H. Li, Y. Wang, S. Wang, C. Wang, Y. Ma & G. Zhou. 2018a. Comparative transcriptome analysis of duckweed (*Landoltia punctata*) in response to cadmium provides insights into molecular mechanisms underlying hyperaccumulation. *Chemosphere* **190**: 154–165.

Xu, N., F. Hu, J. Wu, W. Zhang, M. Wang, M. Zhu & J. Ke. 2018b. Characterization of 19 polymorphic SSR markers in *Spirodela polyrhiza* (Lemnaceae) and cross-amplification in *Lemna perpusilla. Appl. Plant Sci.* **6**(5): e01153.

Xue, H., Y. Xiao, Y. Jin, X. Li, Y. Fang, H. Zhao, Y. Zhao & J. Guan. 2012. Genetic diversity and geographic differentiation analysis of duckweed using inter-simple sequence repeat markers. *Mol. Biol. Rep.* **39**: 547–554.

Yacoub, H. 2009. *Najas* spp. growth in relation to environmental factors in Wadi Allaqi (Nasser Lake, Egypt). *Transylv. Rev. Syst. Ecol. Res.* **8**: 1–40.

Yakandawala, K. & D. M. G. S. Dissanayake. 2010. *Mayaca fluviatilis* Aubl.: an ornamental aquatic with invasive potential in Sri Lanka. *Hydrobiologia* **656**: 199–204.

Yan, N. D., G. E. Miller, I. Wile & G. G. Hitchin. 1985. Richness of aquatic macrophyte floras of soft water lakes of differing pH and trace metal content in Ontario, Canada. *Aquat. Bot.* **23**: 27–40.

Yan, Y., J. Candreva, H. Shi, E. Ernst, R. Martienssen, J. Schwender & J. Shanklin. 2013. Survey of the total fatty acid and triacylglycerol composition and content of 30 duckweed species and cloning of a Δ6-desaturase responsible for the production of γ-linolenic and stearidonic acids in *Lemna gibba. BMC Plant Biol.* **13**: 201.

Yang, L., Y. Han, D. Wu, W. Yong, M. Liu, S. Wang, W. Liu, M. Lu, Y. Wei & J. Sun. 2017. Salt and cadmium stress tolerance caused by overexpression of the *Glycine max* Na+/H+ Antiporter (*GmNHX1*) gene in duckweed (*Lemna turionifera* 5511). *Aquat. Toxicol.* **192**: 127–135.

Yannitsaros, A. 1998. Additions to the flora of Kithira (Greece) I. *Willdenowia* **28**: 77–94.

Yanovsky, E. 1936. *Food Plants of the North American Indians*. U.S. Department of Agriculture, Washington, DC. 84 pp.

Yanuwiadi, B. & B. Polii. 2013. Phytoremediation of arsenic from geothermal power plant waste water using *Monochoria vaginalis*, *Salvinia molesta* and *Colocasia esculenta. Int. J. Biosci.* **3**: 104–111.

Yao, J. l., R. E. Rowland & D. Cohen. 1994. Karyotype studies in the genus *Zantedeschia* (Araceae). *S. Afr. J. Bot.* **60**: 4–7.

Yarrow, M., V. H. Marin, M. Finlayson, A. Tironi, L. E. Delgado & F. Fischer. 2009. The ecology of *Egeria densa* Planchon (Liliopsida: Alismatales): a wetland ecosystem engineer? *Revista Chilena de Historia Natural* **82**: 299–313.

Yasukawa, K. 1981. Male quality and female choice of mate in the red-winged blackbird (*Agelaius phoeniceus*). *Ecology* **62**: 922–929.

Yates, K. K., C. Dufore, N. Smiley, C. Jackson & R. B. Halley. 2007. Diurnal variation of oxygen and carbonate system parameters in Tampa Bay and Florida Bay. *Mar. Chem.* **104**: 110–124.

Yatskievych, G. & C. E. Jenkins. 1981. Fall vegetation and zonation of Hooker Cienega, Graham County, Arizona. *J. Arizona-Nevada Acad. Sci.* **16**: 7–11.

Yatskievych, G. & J. A. Raveill. 2001. Notes on the increasing proportion of non-native angiosperms in the Missouri flora, with reports of three new genera for the state. *Sida* **23**: 701–709.

Yazbek, P. B., J. Tezoto, F. Cassas & E. Rodrigues. 2016. Plants used during maternity, menstrual cycle and other women's health conditions among Brazilian cultures. *J. Ethnopharmacol.* **179**: 310–331.

Yeo, R. R. 1964. Life history of common cattail. *Weeds* **12**: 284–288.

Yeo, R. R. 1966. Yields of propagules of certain aquatic plants I. *Weeds* **14**: 110–113.

Yeo, R. R. 1967. Silver dollar fish for biological control of submersed aquatic weeds. *Weeds* **15**: 27–31.

Yilmaz, D. D. 2007. Effects of salinity on growth and nickel accumulation capacity of *Lemna gibba* (Lemnaceae). *J. Hazard. Mater.* **147**: 74–77.

Yılmaz, E., İ. Akyurt & G. Günal. 2004. Use of duckweed, *Lemna minor*, as a protein feedstuff in practical diets for common carp, *Cyprinus carpio*, fry. *Turk. J. Fish. Aquat. Sci.* **4**: 105–109.

Yin, L., C. Wang, Y. Chen, C. Yu, Y. Cheng & W. Li. 2009. Cold stratification, light and high seed density enhance the germination of *Ottelia alismoides. Aquat. Bot.* **90**: 85–88.

Yin, L., R. Zhang, Z. Xie, C. Wang & W. Li. 2013. The effect of temperature, substrate, light, oxygen availability and burial depth on *Ottelia alismoides* seed germination. *Aquat. Bot.* **111**: 50–53.

Yin, L., W. Li, T. V. Madsen, S. C. Maberly & G. Bowes. 2017. Photosynthetic inorganic carbon acquisition in 30 freshwater macrophytes. *Aquat. Bot.* **140**: 48–54.

Yocom, C. F. & H. A. Hansen. 1960. Population studies of waterfowl in eastern Washington. *J. Wildl. Manag.* **24**: 237–250.

Yoder, J. A., S. M. Imfeld, D. J. Heydinger, C. E. Hart, M. H. Collier, K. M. Gribbins & L. W. Zettler. 2010. Comparative water balance profiles of Orchidaceae seeds for epiphytic and terrestrial taxa endemic to North America. *Plant Ecol.* **211**: 7–17.

York, H. H. 1905. The hibernacula of Ohio water plants. *Ohio Nat.* **5**: 291–293.

York, R. A., M. J. Durako, W. J. Kenworthy & D. W. Freshwater. 2008. Megagametogenesis in *Halophila johnsonii*, a threatened seagrass with no known seeds, and the seed-producing *Halophila decipiens* (Hydrocharitaceae). *Aquat. Bot.* **88**: 277–282.

You, J., X. Z. Sun & H. Q. Wang. 1985. Taxonomy of *Najas*: a synthetical analysis with evidences on cytology, isozymes and SEM examination. *J. Wuhan Univ. Nat. Sci.* **4**: 111–118. [in Chinese].

Youngken, H. W. 1919. Notes on the Dasheen and Chayote. *Am. J. Bot.* **6**: 380–386.

Yousefi, Z. & A. Mohseni-Bandpei. 2010. Nitrogen and phosphorus removal from wastewater by subsurface wetlands planted with *Iris pseudacorus*. *Ecol. Eng.* **36**: 777–782.

Yu, L.-F. & D. Yu. 2009. Responses of the threatened aquatic plant *Ottelia alismoides* to water level fluctuations. *Fundam. Appl. Limnol.* **174**: 295–300.

Yu, S., M.-Y. Cui, B. Liu, X.-Y. Wang & X.-Y. Chen. 2009. Development and characterization of microsatellite loci in *Ruppia maritima* L. (Ruppiaceae). *Conserv. Genet. Resour.* **1**: 241–243.

Yu, G., H. Liu, K. Venkateshan, S. Yan, J. Cheng, X. S. Sun & D. Wang. 2011. Functional, physiochemical, and rheological properties of duckweed (*Spirodela polyrhiza*) protein. *Trans. Am. Soc. Agric. Biol. Eng.* **54**: 555–561.

Yu, H.-B., Z.-J. Yang, R.-L. Xiao, S.-N. Zhang, F. Liu & Z.-X. Xiang. 2013. Absorption capacity of nitrogen and phosphorus of aquatic plants and harvest management research. *Acta Pratacult. Sinica* **22**: 294–299.

Yuan, J. & K. Xu. 2017. Effects of simulated microgravity on the performance of the duckweeds *Lemna aequinoctialis* and *Wolffia globosa*. *Aquat. Bot.* **137**: 65–71.

Yuan, J.-X., J. Pan, B.-S. Wang & D.-M. Zhang. 2011. Genetic differentiation of *Wolffia globosa* in China. *J. Syst. Evol.* **49**: 509–517.

Zaebst, T. W., W. M. Aust, S. H. Schoenholtz & C. Fristoe. 1995. Recovery status of a tupelo-cypress wetland seven years after disturbance: silvicultural implications. Pp. 229–235 *In*: M. B. Edwards (ed.), *Proceedings of the Eighth Biennial Southern Silvicultural Research Conference*. General Technical Report SRS-1. USDA Forest Service, Southern Research Station, Asheville, NC.

Zákravsky, P. & Z. Hroudová. 1998. Reproductive ecology of *Alisma gramineum* Gmel. Pp. 155–158 *In*: A. Monteiro, T. Vasconcelos & L. Catarino (eds.), *Management and Ecology of Aquatic Plants. Proceedings of the 10th EWRS International Symposium on Aquatic Weeds, Lisbon, Portugal, 21-25 September 1998*. European Weed Research Society, Poznañ, Poland.

Zalewska-Gałosz, J. & M. Ronikier. 2012. Molecular evidence for two rare *Potamogeton natans* hybrids with reassessment of *Potamogeton* hybrid diversity in Poland. *Aquat. Bot.* **103**: 15–22.

Zaman, T. & T. Asaeda. 2013. Effects of NH_4–N concentrations and gradient redox level on growth and allied biochemical parameters of *Elodea nuttallii* (Planch.). *Flora, Morphol. Distrib. Funct. Ecol. Plant* **208**: 211–219.

Zampella, R. A., J. F. Bunnell, K. J. Laidig & C. L. Dow. 2001. The Mullica River basin. A report to the Pinelands Commission on the status of the landscape and selected aquatic and wetland resources. Pinelands Commission, New Lisbon, NJ. 371 pp.

Zanetti, A. 2014. Taxonomic revision of North American *Eusphalerum* Kraatz, 1857 (Coleoptera, Staphylinidae, Omaliinae). *Insecta Mundi* **0379**: 1–80.

Zapfe, L. & J. R. Freeland. 2015. Heterosis in invasive F_1 cattail hybrids (*Typha* × *glauca*). *Aquat. Bot.* **125**: 44–47.

Zaremba, R. E. & E. E. Lamont. 1993. The status of the coastal plain pondshore community in New York. *Bull. Torrey Bot. Club* **120**: 180–187.

Zebryk, T. M. 2004. Inland sandy acid pondshores in the lower Connecticut River Valley, Hampden County, Massachusetts. *Rhodora* **106**: 66–76.

Zedler, J. B., J. C. Callaway, J. S. Desmond, G. Vivian-Smith, G. D. Williams, G. Sullivan, A. E. Brewster & B. K. Bradshaw. 1999. Californian salt-marsh vegetation: an improved model of spatial pattern. *Ecosystems* **2**: 19–35.

Zedler, J. B., H. Morzaria-Luna & K. Ward. 2003. The challenge of restoring vegetation on tidal, hypersaline substrates. *Plant Soil* **253**: 259–273.

Zefferman, E. 2014. Increasing canopy shading reduces growth but not establishment of *Elodea nuttallii* and *Myriophyllum spicatum* in stream channels. *Hydrobiologia* **734**: 159–170.

Zelmer, C. D. & R. S. Currah. 1995. *Ceratorhiza pernacatena* and *Epulorhiza calendulina* spp. nov.: mycorrhizal fungi of terrestrial orchids. *Can. J. Bot.* **73**: 1981–1985.

Zelmer, C. D., L. Cuthbertson & R. S. Currah. 1996. Fungi associated with terrestrial orchid mycorrhizas, seeds and protocorms. *Mycoscience* **37**: 439–448.

Zemskova, E. A. & L. I. Sveshnikova. 1999. Karyological study of some representatives of the family Amaryllidaceae. *Bot. Zhurn. S.S.S.R.* **84**: 86–98.

Zepeda, C., A. Lot, X. A. Nemiga & J. Manjarrez. 2014. Seed bank and established vegetation in the last remnants of the Mexican Central Plateau wetlands: the Lerma marshes. *Revista Biol. Tropic.* **62**: 455–472.

Zettler, L. W. & J. E. Fairey III. 1990. The status of *Platanthera integrilabia*, an endangered terrestrial orchid. *Lindleyana* **5**: 212–217.

Zettler, L. W. & T. M. McInnis, Jr. 1992. Propagation of the endangered terrestrial orchid *Platanthera integrilabia* (Correll) Luer through symbiotic seed germination. *Lindleyana* **7**: 154–161.

Zettler, L. W. & T. M. McInnis, Jr. 1994. Light enhancement of symbiotic seed germination and development of an endangered terrestrial orchid (*Platanthera integrilabia*). *Plant Sci.* **102**: 133–138.

Zettler, L. W. & C. J. Hofer. 1997. Sensitivity of *Spiranthes odorata* seeds to light during in vitro symbiotic seed germination. *Lindleyana* **12**: 26–29.

Zettler, L. W. & C. J. Hofer. 1998. Propagation of the little club-spur orchid (*Platanthera clavellata*) by symbiotic seed germination and its ecological implications. *Environ. Exp. Bot.* **39**: 189–195.

Zettler, L. W. & K. A. Piskin. 2011. Mycorrhizal fungi from protocorms, seedlings and mature plants of the eastern prairie fringed orchid, *Platanthera leucophaea* (Nutt.) Lindley: a comprehensive list to augment conservation. *Am. Midl. Nat.* **166**: 29–40.

Zettler, L. W., F. V. Barrington & T. M. McInnis. 1995. Developmental morphology of *Spiranthes odorata* seedlings in symbiotic culture. *Lindleyana* **10**: 211–216.

Zettler, L. W., N. S. Ahuja & T. M. McInnis, Jr. 1996. Insect pollination of the endangered monkey-face orchid (*Platanthera integrilabia*) in McMinn County, Tennessee: one last glimpse of a once common spectacle. *Castanea* **61**: 14–24.

Zettler, L. W., J. A. Sunley & T. W. Delaney. 2000. Symbiotic seed germination of an orchid in decline (*Platanthera integra*) from the Green Swamp, North Carolina. *Castanea* **65**: 207–212.

Zettler, L. W., S. L. Stewart, M. L. Bowles & K. A. Jacobs. 2001. Mycorrhizal fungi and cold-assisted symbiotic germination of the federally threatened eastern prairie fringed orchid, *Platanthera leucophaea* (Nuttall) Lindley. *Am. Midl. Nat.* **145**: 168–176.

Zettler, L. W., K. A. Piskin, S. L. Stewart, J. J. Hartsock, M. L. Bowles & T. J. Bell. 2005. Protocorm mycobionts of the federally threatened eastern prairie fringed orchid, *Platanthera leucophaea* (Nutt.) Lindley, and a technique to prompt leaf elongation in seedlings. *Stud. Mycol.* **53**: 163–171.

Zhang, W. & J. Gao. 2018. High fruit sets in a rewardless orchid: a case study of obligate agamospermy in *Habenaria*. *Aust. J. Bot.* **66**: 144–151.

Zhang, Y. W., S. J. Huang, X. N. Zhao, F. Liu & J. M. Zhao. 2010a. New record of an invasive species, *Sagittaria graminea*, in Yalu River Estuary wetland. *J. Wuhan Bot. Res.* **28**: 631–633.

Zhang, Y., Y. Hu, B. Yang, F. Ma, P. Lu, L. Li, C. Wan, S. Rayner & S. Chen. 2010b. Duckweed (*Lemna minor*) as a model plant system for the study of human microbial pathogenesis. *PLoS One* **5**(10): e13527.

Zhang, Y.-Y., D.-Y. Zhang & S. C. H. Barrett. 2010c. Genetic uniformity characterizes the invasive spread of water hyacinth (*Eichhornia crassipes*), a clonal aquatic plant. *Mol. Ecol.* **19**: 1774–1786.

Zhang, T.-T., L.-L. Wang, Z.-X. He & D. Zhang. 2011. Growth inhibition and biochemical changes of cyanobacteria induced by emergent macrophyte *Thalia dealbata* roots. *Biochem. Syst. Ecol.* **39**: 88–94.

Zhang, X., M. K. Uroic, W.-Y. Xie, Y.-G. Zhu, B.-D. Chen, S. P. McGrath, J. Feldmann & F.-J. Zhao. 2012. Phytochelatins play a key role in arsenic accumulation and tolerance in the aquatic macrophyte *Wolffia globosa*. *Environ. Pollut.* **165**: 18–24.

Zhang, Y., L. Zhang, X. Zhao, S. Huang & J. Zhao. 2013a. Effects of tidal action on pollination and reproductive allocation in an estuarine emergent wetland plant–*Sagittaria graminea* (Alismataceae). *PLoS One* **8**(11): e78956.

Zhang, Z., Y. Zhou, X. Song, H. Xu & D. Cheng. 2013b. Insecticidal activity of the whole grass extract of *Typha angustifolia* and its active component against *Solenopsis invicta*. *Sociobiology* **60**: 362–366.

Zhang, L.-H., Y.-W. Zhang, X.-N. Zhao, S.-J. Huang, J.-M. Zhao & Y.-F. Yang. 2014a. Effects of different nutrient sources on plasticity of reproductive strategies in a monoecious species, *Sagittaria graminea* (Alismataceae). *J. Syst. Evol.* **52**: 84–91.

Zhang, Y., L. Yin, H.-S. Jiang, W. Li, B. Gontero & S. C. Maberly. 2014b. Biochemical and biophysical CO$_2$ concentrating mechanisms in two species of freshwater macrophyte within the genus *Ottelia* (Hydrocharitaceae). *Photosyn. Res.* **121**: 285–297.

Zhao, Y.-M., W. Wang & S.-R. Zhang. 2012. Delimitation and phylogeny of *Aletris* (Nartheciaceae) with implications for perianth evolution. *J. Syst. Evol.* **50**: 135–145.

Zhen, G., T. Kobayashi, X. Lu, G. Kumar & K. Xu. 2016. Biomethane recovery from *Egeria densa* in a microbial electrolysis cell-assisted anaerobic system: performance and stability assessment. *Chemosphere* **149**: 121–129.

Zhongqiang, L., D. Yu & T. Manghui. 2005. Seed germination of three species of *Vallisneria* (Hydrocharitaceae), and the effects of freshwater microalgae. *Hydrobiologia* **544**: 11–18.

Zhou, X., Z. He, F. Ding, L. Li & P. J. Stoffella. 2018. Biomass decaying and elemental release of aquatic macrophyte detritus in waterways of the Indian River Lagoon basin, South Florida, USA. *Sci. Total Environ.* **635**: 878–891.

Zhu, B. 2014. Investigating snails as potential biological control agents for invasive European frogbit (Hydrocharis morsus-ranae). *J. Aquat. Plant Manag.* **52**: 102–105.

Zhu, B., D. G. Fitzgerald, C. M. Mayer, L. G. Rudstam & E. L. Mills. 2006. Alteration of ecosystem function by zebra mussels in Oneida Lake: impacts on submerged macrophytes. *Ecosystems* **9**: 1017–1028.

Zhu, B., M. S. Ellis, K. L. Fancher & L. G. Rudstam. 2014. Shading as a control method for invasive European frogbit (*Hydrocharis morsus-ranae* L.) *PLoS One* **9**(6): e98488.

Zhu, J., X. Xu, Q. Tao, P. Yi, D. Yu & X. Xu. 2017. High invasion potential of *Hydrilla verticillata* in the Americas predicted using ecological niche modeling combined with genetic data. *Ecol. Evol.* **7**: 4982–4990.

Zidorn, C. 2016. Secondary metabolites of seagrasses (Alismatales and Potamogetonales; Alismatidae): chemical diversity, bioactivity, and ecological function. *Phytochemistry* **124**: 5–28.

Ziegler, P., K. Adelmann, S. Zimmer, C. Schmidt & K.-J. Appenroth. 2015. Relative *in vitro* growth rates of duckweeds (Lemnaceae)–the most rapidly growing higher plants. *Plant Biol.* **17**: 33–41.

Zieman, J. C., R. L. Iverson & J. C. Ogden. 1984. Herbivory effects on *Thalassia testudinum* leaf growth and nitrogen content. *Mar. Ecol. Prog. Ser.* **15**: 151–158.

Zika, P. F. 1996. *Pilularia americana* A. Braun in Klamath County, Oregon. *Am. Fern J.* **86**: 26–26.

Zika, P. F., E. R. Alverson & L. Wilson. 2000. Noteworthy collections: Oregon; Washington. *Madroño* **47**: 213–216.

Zimba, P. V. & S. R. Bates. 1996. Mineralogical and microscopic analyses of material deposited on submersed macrophytes in Florida lakes. *Hydrobiologia* **340**: 37–41.

Zimba, P. V., M. S. Hopson & D. E. Colle. 1993. Elemental composition of five submersed aquatic plants collected from Lake Okeechobee, Florida. *J. Aquat. Plant Manag.* **31**: 137–137.

Zimorski, S. E., T. L. Perkins & W. Selman. 2013. Chelonian species in the diet of reintroduced whooping cranes (*Grus americana*) in Louisiana. *Wilson J. Ornithol.* **125**: 420–423.

Zink, R. A. & N. T. Wheelwright. 1997. Facultative self-pollination in island irises. *Am. Midl. Nat.* **137**: 72–78.

Zomlefer, W. B. 1997. The genera of Nartheciaceae in the southeastern United States. *Harvard Pap. Bot.* **2**: 195–211.

Zomlefer, W. B. & W. S. Judd. 2002. Resurrection of segregates of the polyphyletic genus *Zigadenus* s.l. (Liliales: Melanthiaceae) and resulting new combinations. *Novon* **12**: 299–308.

Zomlefer, W. B. & G. L. Smith. 2002. Documented chromosome numbers 2002: 1. Chromosome number of *Stenanthium* (Liliales: Melanthiaceae) and its significance in the taxonomy of tribe Melanthieae. *Sida* **20**: 221–226.

Zomlefer, W. B., N. H. Williams, W. M. Whitten & W. S. Judd. 2001. Generic circumscription and relationships in the tribe Melanthieae (Liliales, Melanthiaceae), with emphasis on *Zigadenus*: evidence from ITS and *trnL-F* sequence data. *Am. J. Bot.* **88**: 1657–1669.

Zomlefer, W. B., W. M. Whitten, N. H. Williams & W. S. Judd. 2003. An overview of *Veratrum* s.l. (Liliales: Melanthiaceae) and an infrageneric phylogeny based on ITS sequence data. *Syst. Bot.* **28**: 250–269.

Zomlefer, W. B., W. S. Judd, W. M. Whitten & N. H. Williams. 2006. A synopsis of Melanthiaceae (Liliales) with focus on character evolution in tribe Melanthieae. *Aliso*: **22**: 566–578.

Zomlefer, W. B., M. McKain & J. Rentsch. 2014. Documentation of the chromosome number for *Zigadenus glaberrimus* (Liliales: Melanthiaceae) and its significance in the taxonomy of tribe Melanthieae. *Syst. Bot.* **39**: 411–414.

Zona, S. 1997. The genera of Palmae (Arecaceae) in the southeastern United States. *Harvard Pap. Bot.* **2**: 77–107.

Zona, S. 2000. Arecaceae Schultz Schultzenstein. Palm family. Pp. 95–123 *In*: Flora North America Editorial Committee (eds.), *Flora of North America North of Mexico, Vol. 22: Magnoliophyta: Alismatidae Arecidae, Commelinidae (in Part), and Zingiberidae.* Oxford University Press, New York, NY.

Zona, S. 2001. Starchy pollen in commelinoid monocots. *Ann. Bot.* **87**: 109–116.

Zona, S. & A. Henderson. 1989. A review of animal-mediated seed dispersal of palms. *Selbyana* **11**: 6–21.

Zona, S., P. Davis, L. A. A. H. Gunathilake, J. Prince & J. W. Horn. 2012. Seeds of Eriocaulaceae of the United States and Canada. *Castanea* **77**: 37–45.

Zundel, G. L. 1920. Some Ustilagineae of the state of Washington. *Mycologia* **12**: 275–281.

Zupo, V. & W. G. Nelson. 1999. Factors influencing the association patterns of *Hippolyte zostericola* and *Palaemonetes interme-dius* (Decapoda: Natantia) with seagrasses of the Indian River Lagoon, Florida. *Mar. Biol.* **134**: 181–190.

Zurita, F., M. A. Belmont, J. De Anda & J. Cervantes-Martinez. 2008. Stress detection by laser-induced fluorescence in *Zantedeschia aethiopica* planted in subsurface-flow treatment wetlands. *Ecol. Eng.* **33**: 110–118.

Zurzycki, J. 2015. The energy of chloroplast movements in *Lemna trisulca* L. *Acta Soc. Bot. Poloniae* **34**: 637–666.

Index

Index of systematically relevant taxa (excluding associated species, cultivated groups, etc.). Primary treatments for OBL taxa [indicated by bracketed numbers] begin on pages highlighted in bold.